NEUROTOXICOLOGY
Approaches
and
Methods

NEUROTOXICOLOGY
Approaches
and
Methods

LOUIS W. CHANG, Editor
Departments of Pathology, Pharmacology, and Toxicology
University of Arkansas for Medical Sciences
Little Rock, Arkansas

WILLIAM SLIKKER, JR., Coeditor
Neurotoxicology Division
National Center for Toxicological Research
Jefferson, Arkansas

Academic Press

San Diego   New York   Boston   London   Sydney   Tokyo   Toronto

Cover photograph: Neurological networks of the rat cerebellum as demonstrated by special techniques. See Chapter 1, Figure 12. Photograph reprinted with permission of Charles C. Thomas, Publisher.

This book is printed on acid-free paper. ∞

Academic Press, Inc.
A Division of Harcourt Brace & Company
525 B Street, Suite 1900, San Diego, California 92101-4495

United Kingdom Edition published by
Academic Press Limited
24-28 Oval Road, London NW1 7DX

Library of Congress Cataloging-in-Publication Data

Neurotoxicology : approaches and methods / edited by Louis W. Chang,
    William Slikker, Jr.
        p.   cm.
    Includes index.
    ISBN 0-12-168055-X
    1. Neurotoxicology.   I. Chang, Louis W.   II. Slikker, William,
    [DNLM: 1. Nervous System Diseases—chemically induced.   2. Nervous
System—drug effects.   3. Nervous System—physiology.   WL 100
N49675   1995]
    RC347.5.N485   1995
    616.8--dc20
    DNLM/DLC
    for Library of Congress   94-40509
                                                               CIP

Printed and bound in the United Kingdom
Transferred to Digital Printing, 2011

This book is dedicated to my daughters
Jennifer Michelle Chang
and
Stephanie Monee Chang
May their world be a healthier and safer world.

L.W.C.

CONTENTS

CONTRIBUTORS

Numbers in parentheses indicate the pages on which authors' contributions begin.

Elizabeth M. Abdulla (495, 595) Wellcome Research Laboratories, Beckenham, Kent BR3 3BS, United Kingdom

Daniel Acosta, Jr. (493) Department of Pharmacology and Toxicology, College of Pharmacy, The University of Texas, Austin, Texas 78712

Syed F. Ali (385) Division of Neurotoxicology, National Center for Toxicological Research Food and Drug Administration, Jefferson, Arkansas 72079

Michael Aschner (439, 549) Department of Physiology and Pharmacology, Bowman Gray School of Medicine, Winston-Salem, North Carolina 27106

William D. Atchison (157) Department of Pharmacology and Toxicology, Neurosciences Program and Institute for Environmental Toxicology, Michigan State University, East Lansing, Michigan 48824

Gerald Audesirk (137) Biology Department, University of Colorado at Denver, Denver, Colorado 80217

R. Bagnell (81) Department of Pathology, School of Medicine, University of North Carolina at Chapel Hill, Chapel Hill, North Carolina 27599

Melvin L. Billingsley (423) Department of Pharmacology, Macromolecular Core Lab., Milton S. Hershey Medical Center, Penn State College of Medicine, Hershey, Pennsylvania 17033

Marc Bonnefoi (67) Drug Safety Division, Rhône-Poulenc Rorer Central Research, Horsham, Pennsylvania 19044

Murray B. Bornstein (573) Department of Neurology and Neuroscience, Albert Einstein College of Medicine, Bronx, New York 10467

John F. Bowyer (433) Division of Neurotoxicology, National Center for Toxicological Research, Jefferson, Arkansas 72079

William K. Boyes (133, 205) Neurotoxicology Division, U.S. Environmental Protection Agency, Research Triangle Park, North Carolina 27711

H. W. Broening (371) Division of Neurotoxicology, National Center for Toxicological Research, Food and Drug Administration, Jefferson, Arkansas 72079

L. J. Burdette (183) Department of Neurology, Graduate Hospital, Philadelphia, Pennsylvania 19131

Iain C. Campbell (495, 595) Institute of Psychiatry, London 5E5 8AF, United Kingdom

Terrence Cascino (657) Department of Neurology, The Mayo Clinic, Rochester, Minnesota 55905

Louis W. Chang (3, 5) Departments of Pathology, Pharmacology, and Toxicology, College of Medicine, University of Arkansas for Medical Sciences, Little Rock, Arkansas 72205

Yung-Chi Cheng (483) Department of Pharmacology, Yale University School of Medicine, New Haven, Connecticut 06510

Chang-Ming Chern (695) Neurology Institute, Veterans General Hospital, Taipai, Taiwan Republic of China

Deborah A. Cory-Slechta (225, 333) Department of Environmental Medicine, University of Rochester School of Medicine and Dentistry, Rochester, New York 14642

Kevin M. Crofton (789) Division of Neurotoxicology, Health Effects Research Laboratory, U.S. Environmental Protection Agency, Research Triangle Park, North Carolina 27711

George H. DeVries (563) Department of Biochemistry and Molecular Biophysics, Medical College of Virginia, Richmond, Virginia 23298

David C. Dorman (67) Chemical Industry Institute of Toxicology, Research Triangle Park, North Carolina 27709

Robert G. Feldman (687, 689, 695) Department of Neurology, Boston University School of Medicine, Boston, Massachusetts 02118

Kevin T. Finnegan (641) Departments of Psychiatry, Pharmacology, and Toxicology, University of Utah School of Medicine, Salt Lake City, Utah 84148

Stephen B. Fountain (517) Department of Psychology, Kent State University, Kent, Ohio 44242

David W. Gaylor (771) Division of Biometry and Risk Assessment, National Center for Toxicological Research, Food and Drug Administration, Jefferson, Arkansas 72079

M. E. Gilbert (183) ManTech Environmental Technology, Inc., Research Triangle Park, North Carolina 27709

John R. Glowa (777) Behavioral Pharmacology Unit, Laboratory of Medicinal Chemistry, National Institute of Diabetes and Digestive and Kidney Diseases, National Institutes of Health, Bethesda, Maryland 20892

Alan M. Goldberg (493) Johns Hopkins School of Public Health, Baltimore, Maryland 21205

Joseph S. Handler (747) Department of Neurology, Boston University School of Medicine, Boston, Massachusetts 02118

Dale Hattis (789) Center for Toxicology, Environment, and Development Hazard Assessment Group, Clark University, Worcester, Massachusetts 01610

David W. Herr (205) Neurotoxicology Division, U.S. Environmental Protection Agency, Research Triangle Park, North Carolina 27711

Paul Honegger (507) Institute of Physiology, University of Lausanne, CH-1005 Lausanne, Switzerland

Joseph F. Jabre (737) Department of Neurology, Boston University School of Medicine, Boston, Massachusetts 02118

Karl F. Jensen (27) Neurotoxicology Division, Health Effect Research Laboratory, U.S. Environmental Protection Agency, Research Triangle Park, North Carolina 27711

Jan N. Johannessen (399) Neurobehavioral Toxicology Team, Division of Toxicological Research, Center for Food Safety and Applied Nutrition, U.S. Food and Drug Administration, Laurel, Maryland 20708

H. K. Kimelberg (439) Department of Pharmacology and Toxicology, and Division of Neurosurgery, Albany Medical College, Albany, New York 12208

Ken Kulig (629) Division of Emergency Medicine and Trauma, University of Colorado Health Sciences Center, Morrison, Colorado 80465

C. Langaman (81) Department of Pathology, School of Medicine, University of North Carolina at Chapel Hill, Chapel Hill, North Carolina 27599

Trese Leinders-Zufall (603) Research Institute of Toxicology, University of Utrecht, NL 3508 TD Utrecht, The Netherlands

Richard M. LoPachin (445) Department of Anesthesiology, Montefiore Medical Center, Albert Einstein Medical School, Bronx, New York 10467

A. C. Ludolph (671) Department of Neurology, Universitätsklinikum Charité, Medizinische Fakultät der Humboldt-Universität zu Berlin, Neurologische Klinik und Poliklinik 10098 Berlin, Germany

Robert C. MacPhail (225, 231, 777) Neurotoxicology Division, Health Effects Research Laboratories, U.S. Environmental Protection Agency, Research Triangle Park, North Carolina 27711

V. Madden (81) Department of Pathology, School of Medicine, University of North Carolina at Chapel Hill, Chapel Hill, North Carolina 27599

Jacques P. J. Maurissen (239) The Dow Chemical Company, The Toxicology Research Laboratory, Midland, Michigan 48674

Donald E. McMillan (323) Department of Pharmacology and Toxicology, University of Arkansas for Medical Sciences, Little Rock, Arkansas 72205

Donna Mergler (727) Centre pour létude des Interactions Biologiques entre la Santé et lenvironnement (CINBIOSE), Université du Québec Montréal, Quebec Montreal, Canada H3C 3P8

Kevin T. Morgan (67) Chemical Industry Institute of Toxicology, Research Triangle Park, North Carolina 27709

William R. Mundy (359) Neurotoxicology Division, Health Effects Research Laboratory, U.S.

Environmental Protection Agency, Research Triangle Park, North Carolina 27711

M. Christopher Newland (265) Department of Psychology, Auburn University, Auburn, Alabama 36849

John W. Olney (455) Departments of Psychiatry and Pathology, Washington University School of Medicine, St. Louis, Missouri 63110

Marga Oortgiesen (603) Research Institute of Toxicology, University of Utrecht, NL 3508 TD Utrecht, The Netherlands

James A. D. Otis (747) EEG and EP Laboratory, Boston University School of Medicine, Boston, Massachusetts 02118

S. Michael Owens (323) Department of Pharmacology and Toxicology, University of Arkansas for Medical Sciences, Little Rock, Arkansas 72205

William B. Parker (483) Southern Research Institute, Birmingham, Alabama 35205

Merle G. Paule (301, 371) Division of Neurotoxicology, National Center for Toxicological Research, Food and Drug Administration, Jefferson, Arkansas 72079

Leon D. Prockop (753) Department of Neurology, University of South Florida School of Medicine, Tampa, Florida 33612

Susan P. Proctor (695, 711) Environmental and Occupational Neurology Program, Department of Neurology, Boston University School of Medicine, Boston, Massachusetts 02118

Anthony L. Riley (309) Psychopharmacology Laboratory, Department of Psychology, The American University, Washington D.C. 20016

Gabriele V. Ronnett (581) Departments of Neuroscience and Neurology, Johns Hopkins University School of Medicine, Baltimore, Maryland 21205

Neil L. Rosenberg (615, 617) Department of Medicine, Clinical Pharmacology and Medical Toxicology, University of Colorado School of Medicine, Englewood, Colorado 80110

Mohammad I. Sabri (465) Center for Research on Occupational and Environmental Toxicology, and Department of Neurology, Oregon Health Sciences University, Portland, Oregon 97201

Andrew C. Scallet (99) Division of Neurotoxicology, National Center for Toxicological Research, Jefferson, Arkansas 72079

Benoît Schilter (507) Institute of Physiology, University of Lausanne, CH-1005 Lausanne, Switzerland

Richard F. Seegal (347) New York State Department of Health, Wadsworth Center for Laboratories and Research, Albany, New York 12201

Timothy J. Shafer (157) Neurotoxicology Division, Health Effects Research Laboratories, U.S. Environmental Protection Agency, Research Triangle Park, North Carolina 27711

William Slikker, Jr. (371, 383, 385, 771) Division of Neurotoxicology, National Center for Toxicological Research, Food and Drug Administration, Jefferson, Arkansas 72079

K. Suzuki (81) Department of Pathology, School of Medicine, University of North Carolina at Chapel Hill, Chapel Hill, North Carolina 27599

Timothy J. Teyler (517) Department of Neurobiology, Northeastern Ohio Universities College of Medicine, Rootstown, Ohio 44272

Hugh A. Tilson (359, 767, 805) Neurotoxicology Division, Health Effects Research Laboratory, U.S. Environmental Protection Agency, Research Triangle Park, North Carolina 27711

Stephanie M. Toggas (423) Department of Neuropharmacology, Division of Virology, The Scripps Research Institute, La Jolla, California 92037

Joep van den Bercken (603) Research Institute of Toxicology, University of Utrecht, NL 3508 TD Utrecht, The Netherlands

M. Anthony Verity (537) Department of Pathology (Neuropathology) and Brain Research Institute, University of California Los Angeles Medical Center, Los Angeles, California 90024

Henk P. M. Vijverberg (603) Research Institute of Toxicology, University of Utrecht, NL 3508 TD Utrecht, The Netherlands

Domenico Vitarella (439, 549) Department of Pharmacology and Toxicology, and Division of Neurosurgery, Albany Medical College, Albany, New York 12208

Bernard Weiss (815) Department of Environmental Medicine, University of Rochester, School of Medicine and Dentistry, Rochester, New York 14642

Roberta F. White (711) Department of Neurology, Boston University School of Medicine, Boston, Massachusetts 02118

FOREWORD

The techniques involved in toxicological studies range from molecular analytical assays to behavioral methods in primates to epidemiological studies in human populations, a range almost as wide as all of science. Even in a single field of toxicology, such as neurotoxicology, chemical, *in vitro,* morphological, physiological, behavioral, and clinical techniques are involved, as is evident in the contents of this volume. All make valuable contributions.

It is impossible to conceive of the science of toxicology in the absence of analytical chemistry. Without such assays, the amounts of toxicants to which people might be exposed and the consequent body burden could not be determined, so there could be no abscissae for dose–effect curves. Similarly, the science of toxicology needs bioassays in whole animals.

There have been great advances in understanding the so-called mechanisms of toxicological action, the chemistry and physics of how toxicants produce their biological effects. The advances will continue. One consequence is that less blind screening of agents in whole animals is necessary. Bioassays in whole animals can be conducted more intelligently and can be more efficiently targeted when a great deal is known about the biological properties of the agent from *in vitro* studies on cells and isolated organs and tissues. Fewer animals are then needed for bioassay in whole animals. Sometimes agents may be discarded on the basis of *in vitro* studies before bioassay, at the risk of losing valuable agents to *in vitro* "false" positions. *In vitro* studies have always had a strong attraction for toxicologists: these studies are faster and cheaper and often yield simpler and clearer results than *in vivo* studies. Antivivisectionists do not seem to realize that scientists have a strong attraction to "alternative methods" and will use them to the limits of their usefulness.

Yet bioassay of potential toxicants in whole animals will continue for the foreseeable future to be a crucial part of the assessment of hazard. Just as astronomers, despite their tremendous body of knowledge and powerful theory, still need to keep pointing their instruments out into the universe, toxicologists must continue to assess effects of agents in intact, integrating, self-regulating whole mammals before predicting effects in humans. It is important, therefore, to conduct the bioassay in whole animals in an efficient and informative manner.

History is helpful in assessing current practices of bioassay. Biological effects of agents have been measured, more or less crudely, for centuries, but the formal bioassay to provide information on, for example, the quantitative toxicity of an agent is relatively recent. An early example, developed around the beginning of the 20th century, was the assay of toxins (such as diphtheria toxin) to determine the "minimum lethal dose" (MLD). MLD assays are clearly unsatisfactory; for example, if two guinea pigs are given the same dose of toxin and one dies and the other does not, what can be said about the MLD? In spite of the unsatisfactory nature of the MLD it was used for a quarter of a century, because there was no accepted alternative. In practice, people have continued to try to determine "minimum" or "lowest observable effective" or "no observable effect level" (NOEL) doses up to the present time. Some reasons for the unsatisfactory nature of the MLD were clearly enunciated in 1927 (J. W. Trevan, *Proc. R. Soc. London Series B,* **101,** 483, 1927). Dose–effect (or dose–lethality) curves are typically sigmoid. Vast experience since has confirmed that (log-)dose–effect curves are rarely distinguishable from integrated normal or logistic curves. Sigmoid curves have asymptotes, for example, at 0% and 100% mortality for a dose–mortality curve. Theoretically, as the dose is progressively lowered the mortality becomes less and less but never reaches 0, which may be a biological absurdity; however, the slope of the dose–mortality curve does become very low as mortality approaches the lower detectable limit. In practice, there is "noise" in the system: if the number of animals is large enough, there is certain to be mortality due to natural causes unrelated to the agent, to accidents, and to errors. There are also errors in dosing and mistakes in recording. In consequence, we cannot measure very low mortality rates (or any other slight effect) caused by a specific agent. How low a rate we can measure will obviously depend on circumstances and care; I have speculated that it can rarely be lower than 10% of maximum. Whatever the numerical value, the combination of a curve approaching the MLD as an asymptote and the noise factor makes MLD indeterminate.

The reason that NOELs are unsound is not because there are levels of agent that have no effect, but because there is no determinable unique level for a NOEL on the uncertain, low-slope curve. Trevan said to express toxicity, not in terms of the MLD, where the slope of the dose–mortality curve is approaching 0 and precision infinitesimal, but rather in terms of LD50, the estimate of the dose killing 50% of the population, where the curve is at its steepest.

The strictures and suggestions of Trevan apply to all sigmoid curves, not just to lethality curves, both quantal (or digital as we would probably say today) and quantitative (or continuous). An example of a quantal assay is one based on lethality. Subjects are given various doses of an agent and each either survives or dies. The increased frequency of dying as dose is increased generates the sigmoid curve. An example of a quantitative assay is the old way of measuring histamine in body fluids by how much a strip of guinea pig ileum contracts. The sigmoid curve is generated by exposing the gut to increasing known concentrations of histamine until further increase in dose produces no increment in contraction, defining 100% response. The solution containing the unknown amount of histamine is then added, the contraction measured, and the concentration of the unknown read from the dose–effect curve. For all these sigmoid curves, ED50s, the dose effective in producing 50% of maximum effect, can be defined. Use of ED50s was accepted rather quickly and generally in pharmacology; certainly by the mid 1940s and probably a decade before that, the fundamental unsoundness of attempting to specify minimum or lowest effective or highest noneffective doses was recognized and the practice eschewed.

Unfortunately, the pursuit of estimates of NOELs persisted in toxicology, and, worse, was canonized in regulations. The reasons are not hard to divine. Pharmacologists are interested in effects, including maximum effects, so it is natural to seek to explore the entire dose–effect curve. To use the midpoint of the curve to characterize the location of the curve is also natural. Toxicologists, in contrast, have very little interest in the top part of the curve but great interest in the lower part of the curve all the way down to safe levels. Hence they weight their studies in terms of numbers of subjects heavily toward the low end of the dose–effect curve. They are driven by regulations to conduct many of their studies with this bias. But a value such as a NOEL that is inherently indeterminate cannot serve as a satisfactory starting point for anything: extrapolation, regulation, litigation, or metaphysics. The fact that we earnestly desire to specify safe levels, and that the EPA and the FDA require us to do so, does not mean that we can do so in a rational scientific manner, at least at present. Congress could mandate that the National Cancer Institute specify within a decade how to prevent new cases of cancer, but that would not guarantee the NCI could do so. It is likely that progress will be made in our ability to obtain better estimates of safe levels, but it is certain that the progress will not come by pursuing fundamentally unsound science, such as trying to estimate NOELs or other values close to the asymptote. Alternative approaches are necessary and have been suggested, for example, by Glowa *et al.* (J. R. Glowa, J. DeWeese, M. E. Natale, J. J. Holland, and P. B. Dews *J. Environ. Pathol. Toxicol. Oncol.* **6,** 153, 1986.)

Another consequence of the responsibility of toxicologists to advise on what levels of agents are "safe" is seen in the experimental design of studies. In the pursuit of NOEL, dose levels with unmeasurable effects must be included in the design. It is arguable that the highest dose level that does not have a measurable effect contributes some information: namely, that this dose level does not have clearly measurable effects, within the range of error. However, it contributes essentially no information about the slope of the dose–effect curve and imprecise information about the location of the dose–effect curve. Moreover, what if in a four-dose-level study the three lower doses have effects too small to be measured? Certainly the results from the two lowest doses yield essentially no information, and, as above, the next higher dose yields limited information; thus, the effort is wasted.

I conjecture that studies designed to give good information about the slope as well as the position of the dose–effect curve would provide a better basis for making a rational decision on what to promulgate as a safe level. Much work will be necessary to validate such an approach, but, of course, decisions based on putative NOELs have not been validated. It would not be appropriate to replace NOELs with ED50s. A level affecting 50% of the population, for example, is too far from the levels of prime interest to the toxicologist, the safe levels. The level chosen must be sufficiently high that a reliable estimate can be made: reliable to within, say, a hemibel when error is estimated from replication. As before, such a level is unlikely to be less than 10% in most circumstances. The final specification of the safe level could be made by probabilistic inference rather than by extrapolation into completely unknown extremes of the dose–effect curve.

PETER B. DEWS
New England Regional Primate Research Center
Harvard Medical School
Southborough, Massachusetts

PREFACE

When I was asked to organize a book on neurotoxicology, I was confronted with the problem of what to include in the text. With much deliberation, I decided that for a comprehensive coverage of neurotoxicology, at least three major areas must be addressed: basic principles of neurotoxicology, current concepts on the effects and mechanisms of various classes of neurotoxicants, and state-of-the-art approaches and methods in toxicological research and assessment. This thought led to the birth of the "Trilogy of Neurotoxicology" with which I am so proudly involved.

The volumes *Principles of Neurotoxicology* and *Handbook of Neurotoxicology* (Marcel Dekker, Inc.) represent the first two installments of this trilogy. The current title, *Neurotoxicology: Approaches and Methods,* is the final epic of this series. This ambitious project was accomplished by the vision of one and the efforts of many. I cannot thank all the contributors enough for making this vision a reality.

Neurotoxicity has been defined as any adverse effect on the structure and function of the nervous system. It becomes obvious that the structural aspects (neuromorphology and neuropathology) and the functional aspects (neurophysiology and neurobehavior) of neural alterations are critical for the assessment of neurotoxicity. Furthermore, biochemical and molecular events underlying the toxic effects can provide clues to the mechanisms of action of the toxicants involved and may serve as markers for early neurotoxicity detection. It is hoped that methods developed for neurotoxicology will aid in accomplishing one or more of the following objectives: identifying toxic substances, detecting for toxic effects and lesion development, elucidating the mechanism of action of toxicity, and providing means for health risk assessment.

The nervous system is perhaps the most complex biological system in mammalian species. Therefore the approaches and methods available for its assessment are also broad and multiple. The first four sections of the book are devoted to the four major disciplines in neuroscience or neurotoxicology: neuroanatomy/neuropathology, neurophysiology, neurobehavioral science, and neurobiochemistry. Each of these approaches provides an "index" (morphological index, functional index, biochemical index, etc.) for the assessment of the integrity of the nervous system. Both qualitative and quantitative methods for such assessments are presented and discussed. It may be surprising to some readers that quantitative evaluations of morphological alterations are actually possible and that biochemical bases for neurobehavioral and neuropathological alterations are also being established. Some of these approaches are included in this volume.

The first four sections of the book provide the reader with the opportunity to examine the animal as a whole: its structures, physiology, chemistry, and function. Indeed, it would be a grave error to view neurotoxicology, or any other medical science, with tunnel vision. It is my hope that the reader will appreciate the "world of science" and the "facet of truth" that each of the disciplines (anatomy, physiology, etc.) can offer. We frequently encounter debates on the relative sensitivity of each disciplinary approach in the detection of neurotoxicity. While it is important to understand the strengths and limitations of each approach, it is foolish to argue, as some may, the relative "importance" of each discipline and its contributions to neurotoxicology. It is my hope that the reader will come to the conclusion that each disciplinary approach has much to offer and a combination of these approaches is needed for comprehensive assessment of the nervous system.

It is important and obvious, for authoritative reasons, that each section be organized by experts in that specialization of neurotoxicology. I am most fortunate and grateful that Drs. William Boyes and William Atchison (neurophysiology), Drs. Deborah Cory-Slechta and Robert MacPhail (neurobehavioral toxicology), and Dr. William Slikker, Jr. (neurobiochemistry/molecular biology) agreed to serve as section heads for their disciplines. I thank them for their gallant efforts in helping me select both the chapter topics and the best contributors for these chapters.

Aside from the four major areas of neurotoxicology, three issues in neurotoxicology have emerged and have gained increasing attention and importance in the past decade. These issues are animal alternative models (*in vitro* neurotoxicology), toxiconeurology (clinical neurotoxicology), and neurotoxicity risk assessment. The last three sections of this volume provide comprehensive coverage on these issues of importance.

In contrast to the *in vivo* model (whole animal study), *in vitro* neurotoxicology focuses on the exploration of neurotoxicological phenomena via cell or tissue cultures. The promotion of *in vitro* neurotoxicology is not so much for the politics of animal rights, but because of the undeniable scientific merits and advantages of this approach. A comprehensive section in this volume is devoted to the various methods and approaches via cell or tissue culture for the investigation of neurotoxicology. It becomes apparent that aside from neurobehavioral testings, many methods from neuromorphology/neuropathology, neurophysiology, and neurobiochemistry can also be performed on the *in vitro* system. I am greatly indebted to Professor Alan Goldberg and Professor Daniel Acosta for their willingness to serve as co-section heads of the *In Vitro* Neurotoxicology section. Their devoted assistance in the planning and organization of this section provided the needed strength and expertise for this important subject. I am certain that the excellent collection of information in this section will provide the reader with a comprehensive view on and solid foundation for *in vitro* neurotoxicology.

All methods and techniques have their own inherent strengths and weaknesses. *In vitro* approaches are no exception. Readers and practioners of these methods are encouraged to recognize the limitations of the methods in which they are interested. Studies correlating different approaches (e.g., *in vitro*/*in vivo* correlation) will reveal more of the scientific issue than studies in which each approach is performed alone.

Similarly, a correlation between experimental models (basic research) and clinical practice (patient care) is very much needed. The final frontier of all medical research is applicable health care in clinical situations. The increasing interest in and attention to bridging experimental neurotoxicology and toxiconeurology

therefore become logical. Such bridging led to the development of clinical neurotoxicology. A special section in this volume is devoted to this important area of neurotoxicology. I am thankful to Dr. Neil Rosenberg and Dr. Robert Feldman for their enthusiastic support and assistance in the development and organization of this section, in which the basic principles and clinical approaches in clinical neurotoxicology are thoroughly presented and discussed. I am sure that this information will be extremely important and useful to clinical practitioners in toxiconeurology as well as to basic researchers in neurotoxicology. It is also my hope that the reader will be stimulated and motivated to play a part in the future development of this vital area of neurotoxicology. Extrapolation of experimental data to clinical practice and the transfer of basic research knowledge to health care applications are the ultimate goals and responsibilities of all medical (both basic and clinical) scientists.

The control and prevention of toxic episodes can be best accomplished via early detection and prediction of toxic events. This concept leads to one of the most important and rapidly developing fields in neurotoxicology: neurotoxicity risk assessment. One may view risk assessment as "applied toxicology" in which information generated from both basic and clinical researches is used to develop methods for early detection and prediction of toxicity. Despite the effort devoted to this area of science, neurotoxicity risk assessment is still in its infancy. In the Neurotoxicity Risk Assessment section of this volume, Dr. Hugh Tilson, the section head, has put together chapters surveying the current concepts and approaches in this important area of neurotoxicology. The concepts of quantitative risk assessment, the development of biomarkers for neurotoxic risks, and the issues of risk assessment with developmental neurotoxicity are all covered and presented by active researchers in this field. It is my hope that readers will be stimulated by the critical importance of this area of neurotoxicology. All the methodologies that we present in this volume can become useful tools for early detection or prediction of neurotoxicity. Biological marker (biomarker) research has become one of the most important research areas in cancer research. Such "markers" are used as early warning signs or confirmational indices of cancer development. Biomarker research for neurotoxicity has also developed during the past decade. Encouraging findings have been obtained and are presented in some of the chapters in this section. Future advances in this area will include the development of "peripheral markers" for central (central nervous system) effects (e.g., a marker in the peripheral blood or tissue to reflect changes in the CNS) and the correlation of animal markers (biomarkers found in animal models) with clinical applications. Biomarker research and risk assessment development are "new kids on the block" for neurotoxicology. I hope that the reader will find this section exciting and stimulating.

The objective of this volume is to present the general principles and concepts of the various approaches to the assessment of neurotoxicity. It is my hope that through the information provided in this volume, readers will be in a better position to select the approaches and methods for their individual needs. Readers may also become aware of the many state-of-the-art methods available to them as investigators of neurotoxicology.

This volume is not intended to serve as a technical manual. For the convenience of the readers, some contributors (particularly those contributing to the Neuromorphology/Neuropathology section) were encouraged to include step-by-step procedures for some of the techniques described in their chapters. The precise procedures and technical performance of many of the methods are difficult to describe. Many methods are subject to individual laboratory variations and modifications. Readers interested in those methods are encouraged to consult the references provided in each chapter as well as to observe the performance of such techniques in laboratories that have experience with them. For technical mastery, there is no substitute for repeated practice.

Readers must also be reminded that all methods and techniques have their advantages (strengths) and limitations (weaknesses). No single approach or method can provide all of the answers. A combination of methods and correlated approaches is needed to probe and to explore the complexity of the nervous system and the secrets of its diseases. This volume certainly provides many state-of-the-art options for these combinations and correlations. It is my hope that with the aid of this volume readers will make these choices wisely.

I am grateful to all of my distinguished colleagues for their enthusiastic assistance with and support of this project. The number of distinguished contributors involved in this project is overwhelming indeed. (The list of contributors in this series of the trilogy reads like an international Who's Who in Neurotoxicology.) This strong endorsement by my colleagues of this project is especially comforting because it reflects the importance and need for these volumes to the community of neurotoxicology. Without the genuine support and assistance from my fellow scientists, this project could never have been accomplished.

While we are at the twilight of the 20th century, we may look back and be proud of all the exciting developments in neurotoxicology. The trilogy that I have put together serves as a summation and document of most, if not all, the major developments and accomplishments in neurotoxicology in the past decades. It also provides an overview of our present standing on many issues in neurotoxicology. As we look ahead to the new century, we can sense the exciting and formidable challenges that we will be facing. It is my hope that this trilogy in neurotoxicology will serve as the stepping stone for some and the inspiration for many to carry this torch forth and bring neurotoxicology to a new era of excellence.

LOUIS W. CHANG

PART

I

Neuromorphological and Neuropathological Approaches

Neuromorphological and Neuropathological Approaches
An Introductory Overview

LOUIS W. CHANG
Departments of Pathology, Pharmacology, and Toxicology
University of Arkansas for Medical Sciences
Little Rock, Arkansas 72205

Neurotoxicity has been defined as any adverse effects on the structures and functions of the nervous system. It becomes obvious that structural (morphological) changes in the nerve cells and tissues can be used as good indices or markers for neurotoxicity assessment.

Pathology, including neuropathology, is basically a "visual" science, where cells and tissues are examined visually with the aid of optical instruments (microscopes) for the determination of lesion (pathological changes) development. To enhance such optical images, various chemicals (e.g., dyes and stains) are used to bind to or react with tissue components so that the final reaction product may be visually detectable. This specialty of science has been referred to as histochemistry.

The nervous system is perhaps the most complex of all biological systems. The nerve cell (neuron), likewise, may be the most complex cell among all parikymal units. Aside from the main cell body, most neurons are associated with structural entities such as axons, dendrites, neurofilaments, Nissl substances, synaptic terminals, and associated myelin sheaths. All these structural entities are subjected to alterations under toxic conditions giving rise to various neurotoxicological conditions: neuronopathy, axonopathy, dendropathy, myelinopathy, and neuropathy (Chang, 1994). These structures, unfortunately, are not readily visible or detectable with the general hematoxylin and eosin (H&E) stain that was used as a routine pathological screening method by the general pathologist. In the first chapter, Chang presents selected histochemical methods for identifying these specialized structures of the nerve cell and tissue. These methods allow neurotoxicologists or neuropathologists to detect structural changes which may have gone unnoticed by the H & E method. A general screening scheme is also introduced in this chapter to assist those who wish to use neuropathology as a screening tool for neurotoxicity.

Because pathology is basically a "visual" science, many people bear the misconception that this science is largely "descriptive" and that quantitations on the tissue or cellular changes are not possible. The concepts and approaches of "quantitative histochemistry" are included in Chang's chapter to show that many of the histochemical reactions (stainings) can actually be quantitatively measured by special instrumentations. With quantitative histochemistry, pathological changes in cells and tissues can therefore be quantitatively evaluated.

In the chapter by Scallet, the approaches and methods in quantitatively measuring structural changes in the nerve cells and their components are presented. Dr. Scallet discusses the various methods in quantitative morphometry through which the "physical properties" and their changes (cell number, population density, cell size, etc.) can be measured and quantitatively evaluated. Examples of such approaches are also provided.

Although only the more basic histochemical methods are presented in Chang's chapter, the chapters by Drs. Jensen and Dorman *et al.* introduce more specialized and specific methods. Dr. Jensen discusses the various "labeling" techniques for the nerve cells and their components. The use of fluorescent stains and radioactive labels for specific tissue or cellular components is presented. Other approaches, such as immunoantibody labeling, are also discussed. Dorman and his colleagues, on the other hand, focus on the most important "functional proteins," the enzymes, in the cells and tissues. Changes in the enzymatic levels in the nerve cells not only alter the functional capacity of the nervous system, but also serve as sensitive markers for neurotoxicity. The various enzymes that are of special importance and interest to neurotoxicologists are introduced. General techniques and tissue preparations for enzyme histochemistry are also discussed.

Although the chapters by Chang, Jensen, and Dorman *et al.* present a broad spectrum of approaches and methods to evaluate tissue and cellular changes, both qualitatively and quantitatively, at the light microscopic level, Bagnell and colleagues discuss cytological changes at the ultrastructural level. In addition to the basic transmission electron microscopy, concepts on more specialized microscopies such as scanning electron microscopy (SEM), immunocytochemistry, enzyme cytochemistry, EM autoradiography, and high voltage electron microscopy (HVEM) are also introduced so that the readers can be aware of the availability and applications of these various specialized electron microscopic approaches. The basic protocols on tissue preparations for electron microscopy are also presented.

This section presents a series of chapters demonstrating the state-of-the-art approaches and methods avail-

able for the evaluation of the nervous system "morphologically" in terms of both the structure (anatomical features) and chemistry (associated chemical components) of the nerve cells and supporting tissues. As the scope of this volume is to present an overview concept and methods available for the evaluation of the nervous system, the precise step-by-step procedures for some of the methods presented are not included in these chapters. These procedures, however, are readily available in standard technological manuals for these techniques. Appropriate references are provided in each chapter for those readers who are interested in practicing some of these methods.

Like biochemistry, physiology, or behavioral science, pathology has advanced from simple screening methods to much more sophisticated cellular and molecular evaluations. As evidenced in the chapters in this section, methods for well-correlated anatomical and biochemical alterations *in situ*, as well as quantitative evaluations of cellular and tissue changes via optical analyses, are indeed available. These approaches, together with data generated from other approaches (biochemical, physiological, and behavioral), help in evaluating the nervous sytem *in toto*. Quantitative pathology has become a rapidly developing field in pathology. The various methods in pathology used in detecting lesion development and the increasing capabilities in quantitation have made pathological approaches important tools in the risk assessment of toxic substances. Since it cannot be denied that all approaches or methods have their strengths and limitations, it is hoped that this section provides the readers with enough information and appropriate guidance for the wise selection of methods that satisfy their individual needs. It must also be emphasized that evaluations of the integrity of the nervous sytem are complicated tasks. A single approach usually only leads to limited information. A coordinated and correlated study in multidisciplinary fashion is needed. Approaches and methods by other disciplines (neurophysiology, biochemistry/molecular biology, and behavioral toxicology) are presented in the following sections in this volume.

Reference

Chang, L. W. (1944). Introduction to basic principles of neurocytology and general concepts on neurotoxicopathology. *In* Principles of Neurotoxicology (L. W. Chang, Ed.), pp. 3–34. Dekker, New York.

CHAPTER

1

Selected Histopathological and Histochemical Methods for Neurotoxicity Assessment

LOUIS W. CHANG

Departments of Pathology, Pharmacology, and Toxicology
University of Arkansas for Medical Sciences
Little Rock, Arkansas 72205

I. Introduction

As defined by the U.S. Office of Technology Assessment, neurotoxicity is "an adverse change in the structure or function of the nervous system following exposure to a chemical agent" (US-OTA, 1990). Although the "preciseness" of this definition may be debated, it is an acceptable notion that structural or morphological changes in the nervous system can serve as reliable markers for neurotoxicity.

Neuropathology has a long-rooted tradition in medicine. Its basic objectives are:

1. To determine the *existence* of pathological lesions in the nerve cells and tissues.
2. If lesions exist, to define the *topographical loci* of the lesion(s) within the nervous system (e.g., granule cells of the cerebellum, myelin sheath of the axons in the sciatic nerve).
3. To characterize the *nature* of the alterations or damages involved (e.g., neuronal swelling, neuronal necrosis, myelin loss).

4. To provide quantitative information on the *extent* of pathological involvement (see Chapter 5).
5. To provide a *time course* progression and *dose/response* correlation on lesion development.
6. To provide anatomicophysiological, anatomicobiochemical, and anatomicofunctional (behavioral) *correlates* on the disease or toxicological situation.
7. To provide pathogenetic information on the *biomechanism* of the disease or toxicological agent(s) involved.

The basic neurocytological and neuropathological aspects of the nerve cells and their supporting elements have been fully presented and discussed already (Chang, 1994) and will, therefore, not be repeated here.

This chapter reviews and presents some of the most frequently employed histopathological and histochemical methods that are helpful for pathological screening and toxicological assessment. It is hoped that this information will be helpful to those wishing to establish a basic histopathological battery for neuropathological study in his/her own laboratory.

5

II. General Tissue Preparations

The general steps or procedures for tissue preparation for histopathology are:

Tissue fixation → dehydration → clearing →
embedding → sectioning → staining.

A. Tissue Fixation and Animal Perfusion

Neural tissues, particularly those of the central nervous system (CNS), are extremely fragile and are easily subjected to rapid anoxic and postmortem changes. Immediate fixation of the nervous tissues is therefore required to avoid artifacts from occurring.

The primary goals for fixation are:

1. to preserve cells in their native state in regard to their morphology and localization of chemical constituents;
2. to prevent cellular autolysis and breakdown;
3. to precipitate and to coagulate tissue components so that loss of diffusible substance from the cell can be minimized;
4. to fortify the tissue against the deleterious effects of various stages of tissue processing; and
5. to enhance tissue condition for optimal staining results.

Perfusion fixation is important for CNS. Nonperfused tissues may risk anoxic artifacts and "dark cell" formation (Figs. 1 and 2). Animals should be anesthetized and perfused intracardially with physiologic saline solution followed by an appropriate fixative such as 4% buffered paraformaldehyde, 2.5% buffered glutaraldehyde, or 10% buffered formalin. Perfusion with fixative solution without the initial saline "flushing" may risk intravascular coagulation of the red blood cells and poor perfusion results. It must also be cautioned that perfusion should be performed under physiological pressure (approximately 120 mm of mercury; perfusion pumps are also commercially available). Hypopressured perfusion may lead to incomplete perfusion and hyperpressured perfusion may result in "perfusion artifacts" (Fig. 3).

Tissue removal and sampling, especially from the brain and cord, should be performed with care. Traumatizing pressure is one of the most frequently seen artifact-producing causes in neuropathology. After removal or sampling, the tissues should be further subjected to immersion fixation for at least 24 hr prior to other tissue processing procedures.

Neutral-buffered 10% formalin is an excellent general fixative and can be used for all types of tissues for light microscopic examination. For the rodent central nervous system, a brief immersion fixation (4–6 hr) in Bouin's fixative prior to neutral-buffered formalin

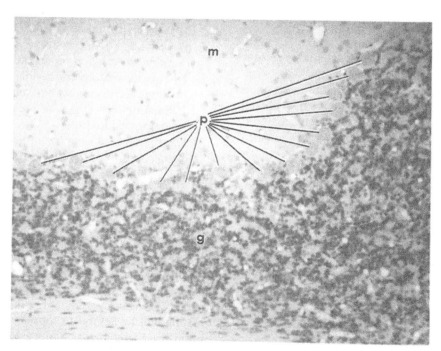

FIGURE 1 Rat cerebellum: perfused (saline followed by 10% buffered formalin) and immersion fixed (10% buffered formalin). g, granule cell layer; p, Purkinje cells; m, molecular layer. H & E, ×200.

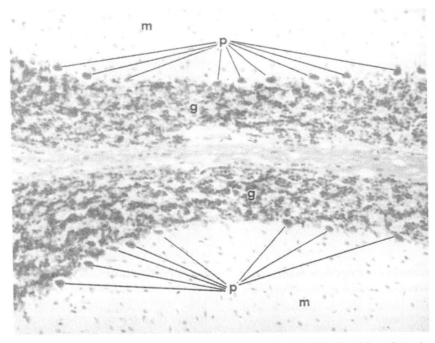

FIGURE 2   Rat cerebellum: nonperfused but immersion fixed (10% buffered formalin) only. Note the shrunken, densely stained Purkinje cells (p), a "dark cell" phenomenon that represents a common artifact seen in nonperfused, poorly fixed, or malhandled CNS tissues. g, granule cell layer; m, molecular layer. H & E, ×200.

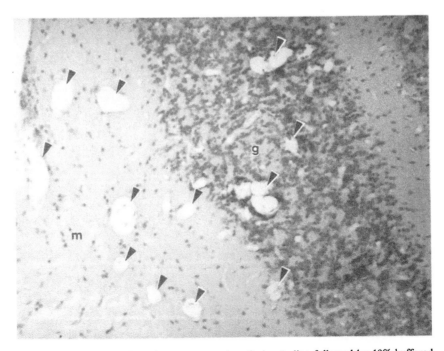

FIGURE 3   Rat cerebellum: hyper-pressured perfusion (saline followed by 10% buffered formalin). Some of the overly distended vasculatures ruptured to give the tissue an artifactual vacuolated appearance (→). g, granule layer; m, molecular layer. H & E, ×200.

fixation has proven to produce excellent tissue sections and staining results.

B. Tissue Embedding and Sectioning

The purpose for tissue embedding is to provide the tissues with enough hardness and support for sectioning. A good embedding media should:

1. infiltrate the tissue samples rapidly;
2. not disrupt the cellular morphology and chemical constituents;
3. provide appropriate hardness and support of the tissue without interfering with sectioning; and
4. not interfere with the subsequent staining characteristics.

There are various choices of embedding media, such as agar, carbowax, celloidin, ester wax, paraffin (Paraplast), and methacrylate (epoxy resin). The most commonly used embedding media are paraffin (Paraplast) or methacrylate. Because many embedding media are not water miscible, water within the cells and tissues must be removed first via the alcohol dehydration process so that proper infiltration of the clearing agent (xylene) and the water-inmiscible embedding materials (e.g., paraffin or paraplast) can be achieved.

Although the most basic and commonly used embedding medium is still paraffin or Paraplast, various attempts have been made for the development of a new embedding medium. 2-Hydroxy ethyl methacrylate or glycol methacrylate probably represent the most satisfactory development. These media are aqueous-soluble monomers and will harden to epoxy resin on polymerization. The advantage of this technique is that very thin sections (1–2 nm) can be cut from these tissue blocks, thus increasing the resolution of the microscopic image tremendously as compared to much thicker (8–10 nm) paraffin sections (Figs. 4 and 5). The disadvantages are, however, that (1) infiltration is slow, only very small tissue size (both surface area and thickness) can be used; (2) the time required for infiltration and processing is longer than the routine paraffin method; (3) only a limited number of special histological stains can be applied; and (4) it is much more costly than the routine paraffin method. For paraffin embedding, not only can larger tissue sizes be used, but multiple tissue samples (e.g., brain, cord, ganglia) can also be embedded together in one single tissue block for sectioning and staining. Furthermore, since most, if not all, staining procedures were originally developed with paraffin-embedded tissues in mind, one probably will have more success on the staining results with paraffin sections. Therefore, for general purposes, paraffin embedding still remains the method of choice for routine evaluations.

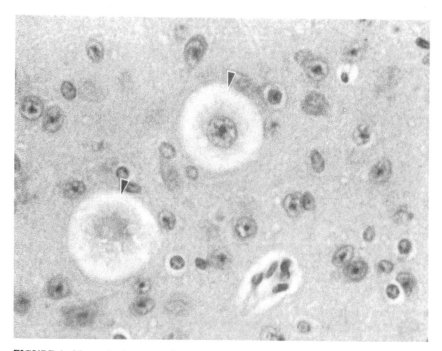

FIGURE 4 Mouse brain stem: trimethyl tin intoxication, paraffin-embedded tissue sectioned at 8 nm thick. Two neurons (→) appear to be swollen (edematous) with floccular cytoplasm. H & E, ×450. (From Chang, 1994.)

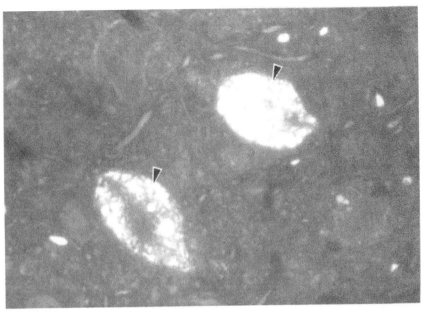

FIGURE 5 Mouse spinal cord: trimethyl tin intoxication, epoxy-embedded tissue sectioned at 1 nm thick. Two anterior horn motoneurons (→) appear edematous and swollen. Microvacuoles (distended endoplasmic reticula) within the neuronal cytoplasms are clearly demonstrated. Toluidine blue O staining, H & E, ×450. (Reprinted from Chang *et al.*, 1984.)

III. Selected Histochemical Staining Methods for the Nervous System

Pathology is basically a "visual" science. The lesion(s) must be morphologically "visible" or detectable by the viewer. For all general purposes, the hematoxylin–eosin (H&E) stain serves well in demonstrating the basic structures and cytoarchitectures of the tissues. Therefore, H&E has been recognized as the time-honored general stain in pathology. Although this may be generally true for many organ systems (liver, kidney, lung), it does not work as well for the nervous system.

The nervous system is a highly complex system and the nerve cells, likewise, are highly complex with many associated components such as axons, dendrites, Nissl substances, and myelin sheaths (Chang, 1994). Many of these components, such as axons, dendrites, and myelin sheaths, are not demonstrable with the common H & E stain (compare Fig. 1 and Fig. 12). For a detailed surveillance on the morphological integrity of the nervous tissues, all these components, therefore, must be made "visible" to the examiner. "Special histochemical stains" are therefore needed for these purposes.

This chapter reviews and presents some of the most reliable staining methods for the various components in the nervous system. Detailed step-by-step proce-

dures of these methods are available in a histochemical and histopathological manual previously published by this author (Chang, 1979) or from many other standard histotechnology manuals (AFIP, 1968; Sheehan and Hrapchak, 1973; Culling, 1974; Bancroft, 1975; Luna, 1992). Selected procedures are also provided in the Appendix of this chapter.

A. H & E Method

H & E is a good general screening staining method. All the nuclei of the cells in the nervous tissue stain blue (hematoxylin) against a pink (eosin) background.

This method is effective in screening for neuronal necrosis (with pyknotic nuclear changes) (Fig. 6), neuronal losses (Fig. 7), and neuronal edema (cytoplasmic swelling and vacuolation; Fig. 4). This method is therefore useful in screening for acute or extensive neurotoxicity. More subtle pathological changes involving the axons, dendrites, or myelin sheaths are not demonstrable with this technique.

B. Einarson's Gallocyanin Method (Nissl Stain)

Many large neurons, such as the motoneurons in the anterior horn of the spinal cord or in the brain stem

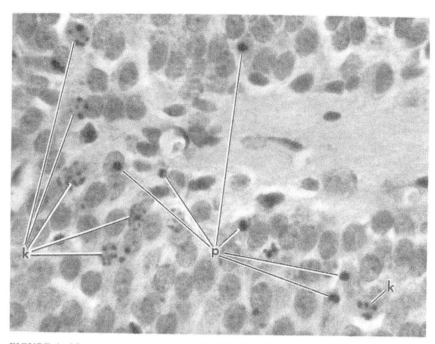

FIGURE 6 Mouse dentate fascia: trimethyl tin intoxication. Nuclear changes, such as pyknosis (p) and karyorrhexis (k), are observed among the granule cells denoting neuronal necrosis in this cell layer. H & E, ×450. (From Chang and Dyer, 1983.)

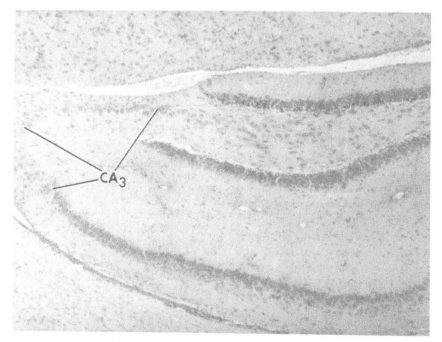

FIGURE 7 Hippocampal formation, rat exposed to trimethyl tin on postnatal day 7 of life. Destruction with significant cell loss selectively at the CA_3 sector ($\rightarrow$) of the Ammon's horn is shown. H & E, ×250. (From Chang, 1984.)

nuclei, are rich in Nissl substances (Chang, 1994). Electron microscopy reveals that the Nissl substance consists of aggregations of rough endoplasmic reticula and polyribosomes, the protein synthesis machinery of the nerve cell. The presence or alterations of Nissl substance, therefore, may be used as an indicator or marker for the general condition of these nerve cells. Disintegration of the Nissl substance (chromatolysis) (Chang, 1994) usually signifies early degenerative change or injury of a neuron.

Although the H & E stain may reveal the chromatolytic phenomenon as a bleaching of the neuronal cytoplasm (Fig. 8), Nissl stain, such as gallocyanine, should be used to confirm if this phenomenon is truly related to the disintegration of the Nissl substance. Chromatolytic changes in nerve cells can be demonstrated in the large brain stem neurons as well as in the anterior horn motoneurons in trimethyl tin and alkyllead poisoning (Fig. 9).

The Gallocyanin method also provides a clear and well-defined nuclear staining. Thus it can also be used as a stain to estimate cellular density (neuronal population) changes in toxic conditions.

Cresyl violet staining is an alternate method for Nissl substance. The choice between the Gallocyanin and cresyl violet methods is purely a personal one as either method serves the purpose well.

C. Bielschowsky's Method for Neurofilaments

This is a staining method selectively for neurofilaments in the nerve cells and their axons. Many disease states, such as aluminum poisoning, Alzheimer's disease, and cadmium intoxication (Lukiw and McLachlan, 1985; Hastings, 1995), show an increase of neurofilamentous accumulation in the neuronal bodies (Fig. 10). Similar neurofilamentous changes are also described in various forms of toxic axonopathies (Gupta and Abou-Donia, 1994; Abou-Donia and Gupta, 1994). These changes, although "invisible" or difficult to detect via the routine H & E method, can be clearly demonstrated with the Bielschowsky's method (Fig. 11).

D. Bodian Method for Axonal Process

Axonopathy is a distinct neuropathological phenomenon that results from various neurotoxic (Gupta and Abou-Donia, 1994; Abou-Donia and Gupta, 1994; Abou-Donia, 1995) and nutritional conditions (Albee *et al.*, 1987). Axonal swellings or ballooings are usually seen. Although the specific site of the staining reaction is still speculative, the Bodian method is selective for the demonstration of neuronal processes (Fig. 12), especially axons. Axonal swellings, atrophy, or other

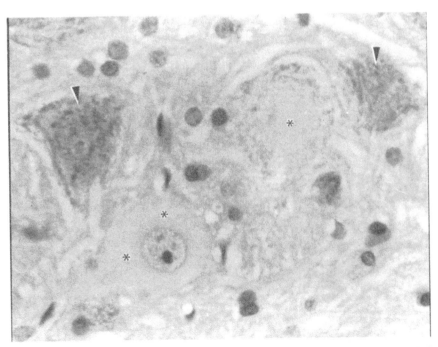

FIGURE 8 Mouse brain stem: trimethyl lead intoxication. The depletion of the Nissl substance (*) (chromatolysis) is observed in two large neurons. The Nissl substance is still demonstrated as dense particles in two other neurons (→). H & E, ×450. (From Chang, 1992.)

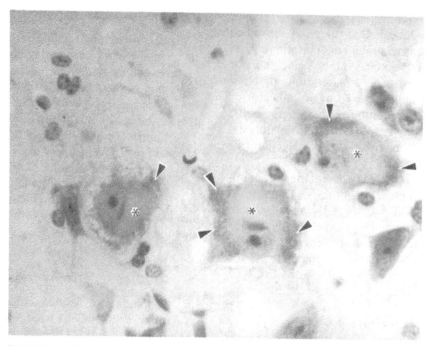

FIGURE 9 Mouse brain stem: trimethyl tin intoxication. Central chromatolytic changes (*) can be seen in the neurons. Remnants of the Nissl substance can still be detected at the periphery of these cells (→). Gallocyanin stains, ×400. (From Chang, 1987.)

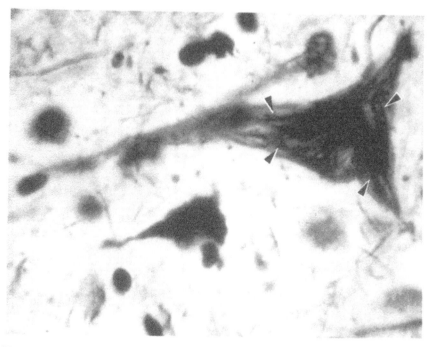

FIGURE 10 Rat hippocampus: aluminum intoxication. Accumulations of neurofilamentous bundles (→) can be detected in the cytoplasm of a large neuron. Bielschowsky's stain, ×650. (From Chang, 1994.)

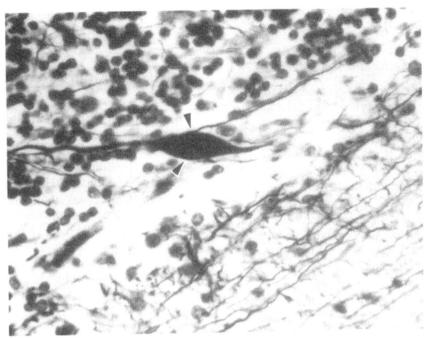

FIGURE 11   Rat cerebellar cortex: IDPN intoxication. Bielschowsky's stain. The axonal balloon (→) is well demonstrated. ×400. (From Chang, 1994.)

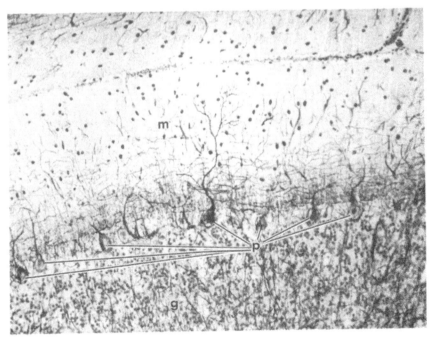

FIGURE 12   Rat cerebellum: normal. Complex fiber networks representing the neuronal processes at the granular–Purkinje–molecular layer junctions. Such complex fiber networks are otherwise "invisible" with the routine H & E staining methods (see Fig. 1). g, Granule layer; p, Purkinje cells; m, molecular layer. Bodian stain, ×250. (From Chang, 1979.)

alterations can be easily detected with this staining method (Fig. 13).

E. Rapid Golgi Technique for Dendrites

Many neurotoxicants, e.g., methyl mercury or inorganic lead, injure the dendritic process of the neuron (Chang and Verity, 1995; Cory-Slechta and Pounds, 1995). Since the dendritic processes of neurons are "invisible" with routine H & E stains, perhaps the best method to use is the rapid Golgi technique. The "dendritic tree," together with the neuronal body, are well stained by this technique against a light-gold background (Fig. 14). The well-demarcated outline of the dendritic structure can be subjected to quantitative morphometry for the estimation of reduction in the various morphology (branching, dendritic spines, etc.) of the dendrites (see Chapter 5 in this section).

It is of interest to note that not all the neurons (even the same type of neurons, e.g., cerebellar Purkinje cells) are stained simultaneously by this technique. Only isolated neurons are stained. It is this author's belief that the "selective stainability" of individual neurons by this technique may be related to the state of neuronal activity of the neuron at the time the animal was sacrificed. Since the "staining" of the neurons in this technique depends on the ability of the neuron to attract and reduce the silver from the staining solution, the various states of the neuronal activity (e.g., depolarization) may alter the "stainability" of the nerve cells. The "unstained" neurons may simply represent those nerve cells at the resting state.

F. Kluver's Luxol Fast Blue (LFB) Method for Myelin

Myelinopathy represents another major neuropathological entity in neurotoxicology. Numerous chemicals are known to induce degenerative changes in the myelin sheath (Bouldin and Goodrum, 1994; Morell, 1994; Chang, 1995). Although H & E stains can stain the entire white matter pink, they cannot reveal the myelin sheaths associated with the fiber tracts. Kluver's LFB method is perhaps the most commonly used technique for the demonstration of myelin sheaths in both central and peripheral nervous systems (Figs. 15 and 16). This method stains the myelin sheaths a vivid blue color. Any "fading" or "pallor" in the staining at sites of normal myelination (e.g., white matter) denotes a loss of myelin (demyelination or dysmyelination) in those areas.

G. Nerve Fiber Teasing Technique for Peripheral Nerve

While all the staining techniques (e.g., Bielschowsky's, Bodian, LFB) work well for the CNS and PNS, the "weaviness" of the peripheral nerve fibers always poses a problem in embedding as well as in sectioning.

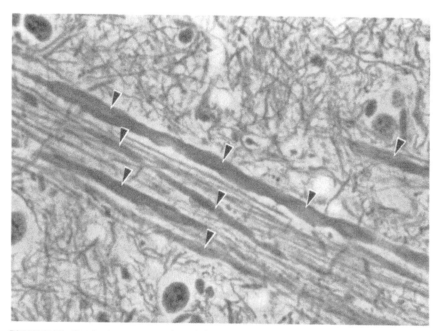

FIGURE 13 Rat brain stem: protein-deficient diet. Axonal swellings (→) in the fiber tract are shown. Bodian stain, ×400. (From Chang, 1992.)

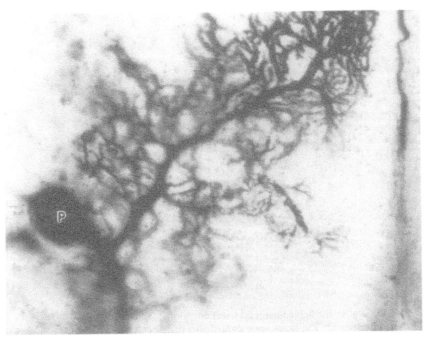

FIGURE 14   Human cerebral cortex: rapid Golgi stain. The elaborate dendritic aborization of a Purkinje neuron (P) is demonstrated. ×400. (From Chang. 1994.)

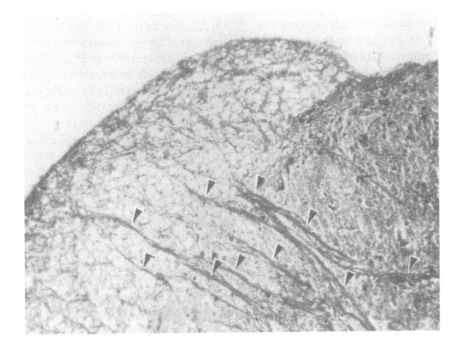

FIGURE 15   Rat spinal cord: normal. Kluver's Luxol fast blue with H & E counterstaining. The myelinated fiber tracts (→) were heavily stained blue and are clearly demonstrated. ×450. (From Chang, 1979.)

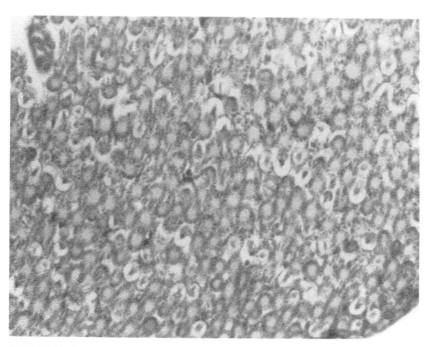

FIGURE 16 Rat sciatic nerve fiber, cross-section: normal. Kluver's Luxol fast blue with H & E counterstaining. The axons were stained pink (light center) and the myelin sheaths surrounding the axons were stained blue (dark rim). The fine cartwheeled pattern of the myelin sheaths can also be recognized. ×650. (From Chang, 1979.)

When embedded and sectioned, the nerve fibers (axons) always appear to be irregular in thickness and have multiple discontinuities ("breaks" or "gaps") along the fibers (Fig. 17). It is therefore disturbing when attempting to determine axonal changes from these sections.

The fiber teasing technique is a simple method by which small segments of peripheral nerves are fixed and stained with osmium tetroxide, teased with a fine pin, and then mounted whole on a slide for examination without sectioning (Gessford, 1991). Under the microscope, the fibers can be viewed in three-dimensional planes. It is a useful method for the study of axonal and myelin changes in the peripheral nerves (Fig. 18).

H. Guillery's Method for Degenerating Nerve Fibers

This method is useful for the screening of degenerating nerve fibers in the central nervous system. The degenerating fibers appear to have the potential in reducing silver more than intact fibers. Thus degenerating fibers appear much darker than normal fibers (Fig. 19).

I. Holzer's Method for Astrocytic Fiber

Although fibroblasts react to injuries and scarring in tissues of peripheral organs, astrocytes enlarge and proliferate in the CNS as a result of tissue injury and cellular degeneration in the central nervous system (Chang, 1994; Aschner *et al.*, 1994). The identification of proliferation or hypertrophy of the astrocytic fibers in the CNS, therefore, becomes good evidence of tissue injury (e.g., proliferation of Bergmann's glial fibers in methyl mercury poisoning).

Holzer's method represents a reliable technique in the demonstration of such astrocytic changes (Figs. 20 and 21). The enlargement and proliferation of astrocytes may be correlated with the increased level of glial fibrillary acid protein (GFAP) that has been proposed as a sensitive biomarker for neurotoxicity (O'Callaghan, 1991).

J. Controls: False Positive and False Negative

As in any chemical reaction, histopathological and histochemical staining techniques are subjected to both false positive and false negative results. To guard against these occurrences, proper "controls" should be used. These controls include: (1) positive control, a tissue sample known to lead to a positive staining result; (2) negative control, a tissue sample known to lack the tissue component in question, which therefore will certainly give a negative staining result; and (3) normal control, tissue samples obtained from a normal animal (same species, sex, and age as the test animals)

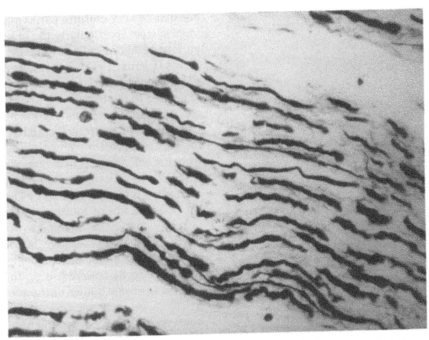

FIGURE 17 Rat sciatic nerve: normal, paraffin-embedded, longitudinal section, Bielschowsky's stain. All the nerve fibers (axons) are demonstrated as darkly stained argentophilic structures. Because of the "waviness" of the fibers, artifactual variations in fiber thickness and "breaks" (or "gaps") along the fibers are expected. ×250. (From Chang, 1979.)

that has not been exposed to the chemical (neurotoxicant) as the test animals have. These three "control tissues" should be stained *together* simultaneously with the test tissues in the *same* staining solution.

K. Neuropathological Screenings for Known and Potential Neurotoxicants

It has been pointed out that while H & E is an excellent general stain, it is a very nonspecific dye. Therefore, unless the lesions involved are of obvious or extensive nature, the H & E method may fail to demonstrate changes in the nervous tissues. Special techniques are useful to selectively demonstrate the various cellular and tissue components in the nervous system: axons, dendrites, Nissl substance, and myelin sheath. These neurohistochemical methods are important for identifying or confirming specific structural changes that may go undetected by the H & E staining method. The concern in neuropathological screening is therefore not so much the lesions that are detectable by the H & E method, but those that are not readily demonstrable by routine examination methodology (H & E staining).

Neuropathological screening schemes may be established for either "known" neurotoxicants (i.e., the general sites and nature of lesions produced by the neurotoxicant are known; e.g., in methyl mercury intoxication, lesions occur in cerebellar cortex, calcarine cortex, and dorsal root ganglia) or "potential" neurotoxicants (the sites and nature of lesions to be produced are not known and have to be determined by the investigator). These schemes are presented in Diagrams 1 and 2 (modified from Chang, 1992) and are self-explanatory. These screening schemes are also helpful approaches for conducting risk assessment on the neurotoxicity of chemicals.

L. Quantitative Histochemistry

Morphological methods frequently give the false impression that histological analyses are purely "descriptive." Some investigators may also be misled to believe that lesion developments and the changes in the staining characteristics of cellular components cannot be quantitatively analyzed.

The quantitation of morphological or structural alterations can actually be achieved by quantitative morphometry. Changes in "physical properties" of the cells and their components (number, size, volume, ratios, etc.) can be measured and analyzed via morphometric analytical instrumentations. Detailed approaches in quantitative morphometry will be further presented and discussed in a chapter by Dr. A. C. Scalett.

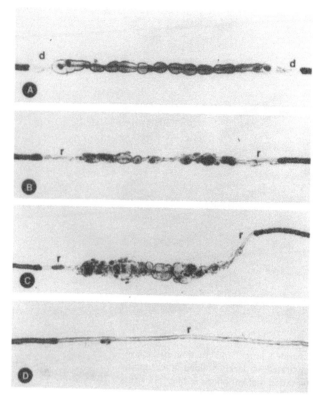

FIGURE 18 Teased tibial nerve, AETT intoxication. The nerve fiber teasing technique shows various morphological changes in individual nerve fibers: (A) Myelin bubbling and paranodal demyelination (d); (B) reduced myelin bubbling with paranodal remyelination (r); (C) phagocytic removal of myelin debris and paranodal remyelination (r); and (D) advanced remyelination (r). (From Spencer *et al.*, 1980.)

"Staining" of cells is the result of binding of a visually identifiable marker (dye, reduced silver product, etc.) to a target ligand of the cellular component. This type of chemical interaction or reaction is referred to as histochemistry. "Quantitative histochemistry" has become a very specialized area in the field of histochemistry. Simply defined, quantitative histochemistry is the quantitative chemical analysis of cells and their components *in situ*. Although it is beyond the basic scope of this chapter and book to present quantitative histochemistry in detail, a synoptic introduction may affirm the concept that methods are indeed available to quantify the chemical constituents in the biological system *in situ* without mechanical disruption of cells and tissues (e.g., tissue homogenation or cellular fractionation).

In general biochemical techniques, cellular or tissue components have to be "isolated" or "extracted" first before analysis via specific chemical reactions can be performed. This approach is generally good and informative, particularly for an organ, such as liver, that

consists of very uniform parikymal cell types. The nervous system, particularly the brain, consists of a great variety of nerve cells, synapses, and supporting elements (Chang, 1994). Analyses generated from homogenation of brain tissues can seldom be pinpointed or restricted to the target cell type of interest. (For example, one cannot *just* analyze the mitochondrial enzyme changes in the neurons located at the third striatum of the calcarine cortex. When mitochondria are isolated from the calcarine cortical homogenates, mitochondria from *all* nerve cells and supporting glial cells in the calcarine cortex are included.) The substances of interest obtained are of multiorigin and certainly compromise the analytical outcome to some extent.

In contrast, quantitative histochemistry is carried out *in situ* in intact tissues using a microscope. The substance to be assayed, e.g., the Nissl substance in specific neuronal bodies or the acetylcholinesterase activity related to a specific synaptic terminal in the nervous system, can be demonstrated by an appropriate histochemical reaction leading to a deposition of a final reaction product (FRP) at the specific site of the intended substance. (Examples of such techniques, such as immunofluorescence labeling, autoradiographic tagging, and enzyme histochemistry, are presented in the following chapters by Dr. K. Jensen and Dr. D. Dorman and colleagues.) The specimen is then examined under a microscope that is linked to special computer-assisted analyzers (e.g., microfluorometers or microspectrophotometer). The amount of FRP can then be quantitatively measured by optical means.

The direct quantitation of some substance, e.g., nucleic acids, via ultraviolet absorption, without staining is also possible. It must be pointed out that the "light source" from the microscope need not be limited to visible light. Infrared, ultraviolet, X-rays, and electron beam have been used for specific purposes. In addition to the basic light microscope, other specialized microscopes, such as fluorescent, phase-contrast, dark field, interference, infrared, polarized, and various types of electron microscope, are also important tools for the enhancement of cellular analyses. (The various forms and techniques for electron microscopy will be presented and discussed in the chapter by Dr. Bagnell and colleagues).

Although the various quantitative histochemical approaches are not discussed in detail here, the reader may benefit from a brief outline of these methods. The methods for optical quantitation (cytophotometry) may be summarized as:

1. Absorptiometry: for quantitation of those FRP that absorb light (ultraviolet to infrared) or X-rays. A basic light microscope and an infrared

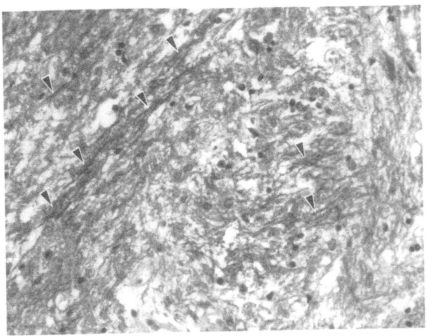

FIGURE 19   Guillery's method for degenerating fibers. Degenerating fibers (→) are stained heavily (dark) by this method. ×250. (From Chang, 1979.)

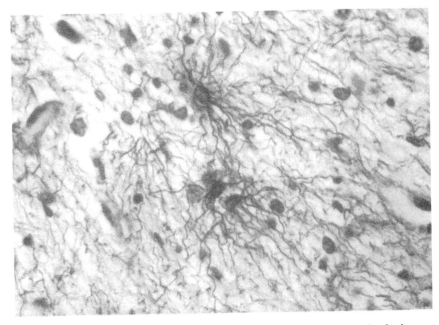

FIGURE 20   Human cerebral cortex. Astrocytic fibers and processes are clearly demonstrated by the Holzer method. ×400. (From Chang, 1994.)

microscope are important for these types of studies.

2. Interferometry: for measurement of the dry mass of a translucent FRP or whole cell. The thickness or refractive index of cells and tissues can also be measured. An interference microscope is needed.

3. Microfluorimetry: for measurement of fluorescent FRPs. Obviously a fluorescent microscope is required. This type of analysis has a high optical specificity because one can select the excitation and emission wavelengths for analysis.

4. Reflectiometry: for quantitative measurements of

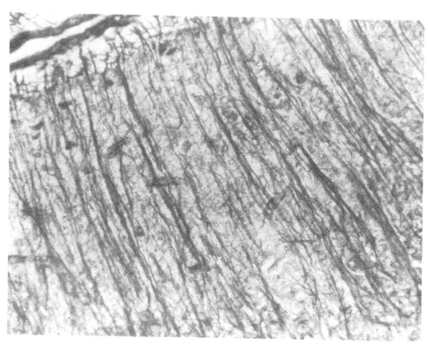

FIGURE 21 Rat cerebellum: methyl mercury intoxicated. The proliferation of Bergmann's glial fibers, projecting from the Purkinje layer through the molecular layer of the cerebellar cortex, is well demonstrated by the Holzer method. ×450. (From Chang, 1979.)

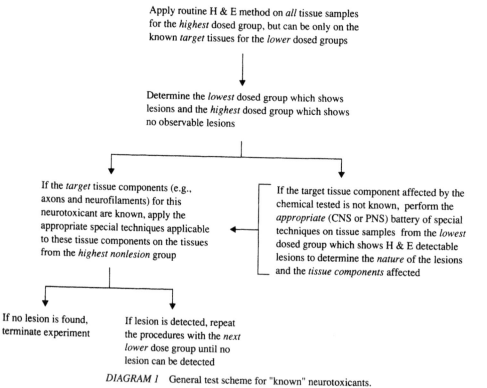

DIAGRAM 1   General test scheme for "known" neurotoxicants.

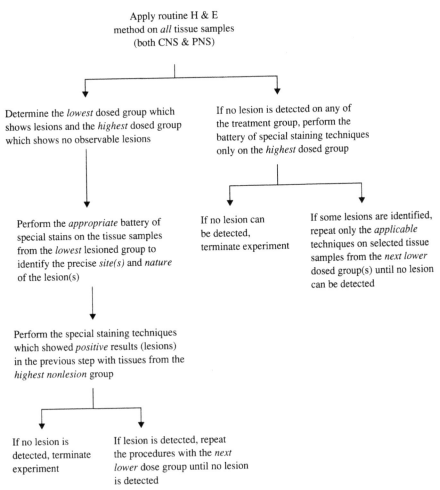

Apply routine H & E
method on *all* tissue samples
(both CNS & PNS)

Determine the *lowest* dosed group which
shows lesions and the *highest* dosed group
which shows no observable lesions

If no lesion is detected on any of
the treatment group, perform the
battery of special staining techniques
only on the *highest* dosed group

Perform the *appropriate* battery of
special stains on the tissue samples
from the *lowest* lesioned group to
identify the precise *site(s)* and *nature*
of the lesion(s)

If no lesion can
be detected,
terminate experiment

If some lesions are identified,
repeat only the *applicable*
techniques on selected tissue
samples from the *next lower*
dosed group(s) until no lesion
can be detected

Perform the special staining techniques
which showed *positive* results (lesions)
in the previous step with tissues from the
highest nonlesion group

If no lesion is
detected, terminate
experiment

If lesion is detected, repeat
the procedures with the *next
lower* dose group until no lesion
is detected

DIAGRAM 2   General test scheme for "potential" neurotoxicants.

particulate deposits such as silver grains (e.g., autoradiography). Phase-contrast and polarized microscopes would be the most helpful.

Since quantitative histochemical analyses are performed on intact tissue sections, the preparation, staining, and sectioning of the tissue influence the outcome of the FRPs and, therefore, the analytical results. Consistency in tissue preparation/staining procedures and thickness of tissue section are extremely important. Proper standarization and control system should be established. The analytical results obtained may be compared with standards or control tissues and expressed as percentage of control values.

It is hoped that the just-mentioned brief outline on the various approaches in quantitative histochemistry and microscopy helps to affirm the notion that sophisticated quantitative methods for morphological studies are indeed available. The morphological analysis of

tissue and its constituent changes does not have to be purely "descriptive," "static," or "nonquantitative." Quantitative histochemistry can be an exciting and rewarding experience.

Indeed, quantitative histochemistry probably offers the best of the two worlds of anatomy and biochemistry. It allows biochemical analyses in *intact* cells and tissues. With the modern developments in computer-assisted analyses, quantitative histochemistry can be performed with even greater accuracy and speed. Autoscanning devices are also available with some instruments for rapid analysis of tissue sites (e.g., synapses, cells, etc.) containing the FRPs. Quantitative histochemistry offers the direct anatomicochemical correlation on tissue and cellular alterations. Such information is extremely helpful in elucidating the biomolecular mechanism of toxicants.

Many, if not all, of the basic histochemical staining techniques introduced in this chapter, together with those presented in the following chapters, are suitable for quantitative histochemical analysis. Because quantitative histochemistry is an extremely diversified and rapidly advancing subject, literatures involved are too many to be cited for the purpose in this chapter. Interested readers are encouraged to consult the classic volume on histochemistry by A. G. Everson Pearse (1972) for more basic principles on this subject and to seek recent publications from the literature on the most advanced techniques and procedures, as well as instruments available, for the various aspects of quantitative histochemistry. The outline and brief presentation in this chapter should offer the reader a good starting point on this important subject.

IV. Concluding Remarks

This chapter only serves as an introduction to the basic techniques and approaches in neuropathology that are important for the investigation of neurotoxicology. Other special methods and techniques certainly exist. Many of these approaches are presented in the following chapters in this section.

Morphological approaches provide a "visual" demonstration on the structural (e.g., via light and electron microscopy) and chemical (e.g., via enzyme histochemistry) changes within the cells and tissues. Although basic morphological approaches offer only "descriptive" and "qualitative" observations on tissue and cellular changes, more advanced instrumentations allow quantitative analyses on these changes via quantitative histochemistry and morphometry. When coupled with data generated via other approaches (e.g., biochemical and neurobehavioral), neuromorphology can be a powerful and important tool for the elucidation of toxic mechanisms.

The techniques described in this chapter have all been "time tested" for their reliability and are simple and inexpensive to perform. Precise step-by-step procedures for these techniques are readily available in most standard histotechnological manuals. It is hoped that this chapter will be useful for those investigators wishing to establish neuropathological screening potentials in their laboratories.

A P P E N D I X

Procedures for Special Neurohistochemical Techniques[1]

Einarson's Gallocyanin Method

Fixation:

    Any general fixative.

Solution required:

    Gallocyanin solution

Gallocyanine	0.3 g
Chromalum	10.0 g
Distilled water	200.0 ml

Dissolve the chromalum in the distilled water by heating, add gallocyanine, boil gently for 10 min. Allow it to simmer for 30 min and cool. Add distilled water to restore to original volume and filter. Solution can be kept for only about 1 week.

Procedures:

1. Bring deparaffinized sections to water.
2. Stain in gallocyanin solution at 36°C for 1 hr or at room temperature for 12 hr.
3. Rinse with three changes of distilled water.
4. Dehydrate, clear, and mount.

Results (see Fig. 9):

Nissl substance	bluish purple
Nuclei	dark blue
Background	pale gray

Bielschowsky's Method

Fixation:

    Buffered 10% formalin is adequate.

Solutions required:

1. *Ammoniacal silver.* Add 6 drops of 40% sodium hydroxide to 5 ml of 20% silver nitrate. Add con-

---

[1] All the procedures presented here have been previously published in Chang (1979). They are reprinted here with permission from the original publisher, Charles C. Thomas, Springfield, Illinois.

centrated ammonia drop by drop until the resultant precipitate is just dissolved. Add distilled water to a total volume of 25 ml and filter.

2. *20% formaldehyde solution*

Formaldehyde (37–40% formalin)	20.0 ml
Tap water	80.0 ml

3. *Gold chloride solution*

Gold chloride, 1%	3 drops
Distilled water	10.0 ml

General procedures:

1. Bring deparaffinized section to water.
2. Put in 2% silver nitrate solution in the dark for 48 hr.
3. Rinse with double distilled water.
4. Transfer to ammoniacal silver solution for 10–20 min until section becomes deep brown.
5. Rinse in distilled water.
6. Reduce in 20% formaldehyde solution for 5 min.
7. Rinse in distilled water.
8. Tone in gold chloride solution for 1 hr.
9. Rinse in distilled water.
10. Dip in sodium thiosulfate solution for 1 min.
11. Wash in running water for 5 min.
12. Dehydrate, clear, and mount.

Results (see Figs. 10, 11, and 17):

Axons and intracellular neurofibrils	black
Background	golden yellow

Bodian Method

Fixation:

Formalin (40.0 ml) mixed with glacial acetic acid (10.0 ml), 80% ethanol (100.0 ml), and picric acid (2.0 g) gives best results.

Fixatives containing chromate, chromic acid, osmic acid, or mercuric chloride should be avoided. Ten percent formalin may cause excessive nonspecific staining.

Solutions required:

1. *Protargol solution*

Sprinkle 1.0 g of Protargol on the surface of water (100.0 ml) in a beaker. Do not stir. Keep solution on 37°C hot plate until Protargol is dissolved.

2. *Reducing solution*

Hydroquinone	1.0 g
Formalin, 27–40%	5.0 ml
Distilled water	100.0 ml

3. *Gold chloride, 1%*

Gold chloride	1.0 g
Distilled water	100.0 ml

4. *Acidified water*

Glacial acetic acid	0.5 g
Distilled water	100.0 ml

5. *Oxalic acid, 2%*

Oxalic acid	2.0 g
Distilled water	100.0 ml

General procedures:

1. Bring deparaffinized section to water.
2. Impregnate in Protargol solution and add 5–6 g of clean copper (wire or shot) per 100 ml of solution. Let stand at 37°C for 12–24 hr.
3. Wash in distilled water, several changes.
4. Reduce in reducing solution.
5. Rinse in distilled water, three changes.
6. Tone in gold chloride for 5 min.
7. Rinse in distilled water, three changes.
8. Develop in 2% aqueous oxalic acid for about 3 min until fibers are sharply defined.
9. Wash in distilled water.
10. Fix in 5% sodium thiosulfate solution for 5 min.
11. Wash in running water for 10 min.
12. Counterstain if desired.
13. Treat with acidified water for 5 min.
14. Dehydrate, clear, and mount.

Results (see Figs. 12 and 13):

Neuronal processes (esp. axons)	black
Background	depending on counterstain

Rapid Golgi Method

The Golgi technique is invaluable for the study of the nervous system. However, it is difficult to obtain the same result in different preparations, even of the same tissue. The basis of the original method is a treatment of the central nervous tissue with potassium dichromate followed by impregnation with silver nitrate. The nature of the impregnation is still not clear. By the Golgi method, selected nerve cells and their processes are impregnated with a dense black color while myelinated fibers are usually unstained.

Fixation:

Fix small tissue sample in the following for 1–7 days. Change fixative if it becomes cloudy.

Potassium dichromate, 2.5%	40.0 ml
Osmic acid, 1%	10.0 ml

Staining:

1. Blot dry tissue with filter paper and place tissue in 0.75% aqueous silver nitrate for 24–48 hr. Renew solution when it becomes yellowish.

2. Transfer to several changes of 40% ethyl alcohol for 1–2 hr.
3. Transfer to 80% and 90% ethanol, 1 hr each.

Embedding and sectioning:

1. Dehydrate with absolute ethanol for 12 hr.
2. Transfer to absolute ethanol–ether (1 : 1) for 2–4 hr.
3. Infiltrate with 4% celloidin or nitrocellulose for 1–2 days.
4. Embed and cut section at 20 μm or more.

Mounting:

1. Wash sections with 80% alcohol to remove any excess silver.
2. Dehydrate with absolute ethanol for a few minutes. (A few milliliters of chloroform may be added to the alcohol to prevent loss of nitrocellulose.)
3. Clear in clove oil or terpinol.
4. Mount on slide.

Results (see Fig. 14):

Nerve cells and processes	golden black
Background	yellow

Comments:

1. If tissue was fixed in 10% formalin, it can be treated with 5% potassium dichromate (100 ml) in glacial acetic acid (7 ml) for 1–2 days.
2. Prolonged staining may result in impregnation of glial fibers and blood vessels as well.

Klüver's (Luxol Fast Blue) Method

This method is widely used because it is simple and does not require mordantization of the tissue.

Solution required:

1. *Luxol fast blue (LFB) solution, 0.1%*
Luxol fast blue, MBS	0.1 g
Ethanol, 95%	100.0 ml
Acetic acid, 10%	0.5 ml

   Dissolve the dye in the alcohol and add the glacial acetic acid slowly.
   Solution should be stable for several months.
2. *Cresyl echt violet solution, 0.1%*
Cresyl echt violet	0.1 g
Distilled water	100.0 ml

   Just before using, add 15 drops of 10% glacial acetic acid. Filter.
3. *Lithium carbonate solution, 0.05%*
Lithium carbonate	0.05 g
Distilled water	100.0 ml

General procedures:

1. Deparaffinize and hydrate to 95% ethanol.
2. Stain in LFB at 60°C for 8–11 hr.
3. Rinse in 95% ethanol to remove excess stain.
4. Rinse in distilled water, three changes, for 10 min.
5. Differentiate quickly in lithium carbonate solution for about 10 sec.
6. Continue to differentiate with 70% ethanol until gray–white matters can be clearly distinguished.
7. Wash thoroughly and examine under microscope. Repeat Steps 5 and 6 if necessary.
8. Stain in 0.1% cresyl echt violet solution for 10 min at 37°C or with hematoxylin and eosin.
9. Rinse in distilled water.
10. Differentiate in 95% ethanol until only nuclei and Nissl substances are purple.
11. Dehydrate in absolute alcohol, clear, and mount.

Results (see Figs. 15 and 16):

Myelin	blue
Cells	purple/violet

Comments:

1. The purity of the ethanol used is essential. If poor grades of ethanol are used, difficulties in subsequent differentiation may result.
2. Other counterstains such as hematoxylin–eosin, nuclear fast red, or PAS can also be used.

Guillery's Method for Degenerating Nerve Fibers

Fixation and embedding:

Tissue samples may be fixed with 10% formal saline and embedded in paraffin.

Solutions required:

1. *Alcoholic ammonium hydroxide solution*
Ammonium hydroxide, 28%	1.0 ml
Alcohol, 50%	99.0 ml
2. *Silver nitrate–white pyridine solution*
Pyridine	5.0 ml
Silver nitrate, 1.5%	95.0 ml
3. *Ammoniacal silver solution*
Silver nitrate solution, 4.5%	20.0 ml
Ethanol, 100%	10.0 ml
Ammonium hydroxide, 28%	1.8 ml
Sodium hydroxide, 2.5%	1.5 ml

   Prewarm the solution to 40–45°C before use.
4. *Reducing solution*
Citric acid, 1%	17.5 ml
Unbuffered 10% formalin	17.5 ml
Ethanol, 10%	450.0 ml

5. *Sodium thiosulfate (hypo) solution, 1%*
   Sodium thiosulfate .... 1.0 g
   Distilled water .... 100.0 ml

General procedures:

1. Bring deparaffinized sections to water.
2. Place in alcoholic ammonium hydroxide solution for 6–8 hr.
3. Rinse in distilled water, three changes, for 15 min.
4. Place in silver nitrate–white pyridine solution for 24 hr in the dark.
5. Without washing, place the slides into the prewarmed ammoniacal silver solution for 3 min. Use no more than three slides at a time and use fresh solution for each batch.
6. Without washing, place the slides directly into the reducing solution for 1 min until the section appears to be golden brown.
7. Rinse in distilled water.
8. Fix in the sodium thiosulfate solution for 2–5 min.
9. Wash in running water.
10. Dehydrate, clear, and mount.

Results (see Fig. 19):
   Degenerating fibers .... black
   Normal fibers .... light brown

Comments:

1. After formalin fixation, tissues should be washed for several hours in distilled water before dehydration. Clear the tissues in cedarwood oil, benzene, and paraffin (melting point of 45°C) at 37°C for 2–3 hr before embedding in paraffin (melting point of 54°C).
2. Pyridine is used as the suppressor of normal fiber impregnation.
3. This method should not be used on frozen sections.

Holzer's Method for Astrocytic Fibers

This is an excellent method in the demonstration of fibrillar gliosis. Normal glial fibers are not stained.

Fixation and embedding:
   Formalin-fixed and paraffin-embedded tissues are adequate for this technique.

Solutions required:

1. *Phosphomolybdic–alcohol solution*
   Phosphomolybdic acid, 0.5% .... 50.0 ml
   Ethanol, 95% .... 100.0 ml
2. *Alcohol–chloroform solution*
   Ethanol, 100% .... 40.0 ml
   Chloroform .... 160.0 ml
3. *Crystal violet solution*
   Crystal violet .... 5.0 g
   Ethanol, 100% .... 20.0 ml
   Chloroform .... 80.0 ml
4. *Potassium bromide solution*
   Potassium bromide .... 100.0 g
   Distilled water .... 1000.0 ml
5. *Differentiating solution*
   Aniline oil .... 120.0 ml
   Chloroform .... 180.0 ml
   Ammonium hydroxide, 28% .... 20 drops

General procedures:

1. Bring deparaffinized slides to water.
2. Place in phosphomolybdic–alcohol solution for 3 min.
3. Drain off the mordant solution and flood with absolute alcohol.
4. Place in absolute alcohol–chloroform solution until section becomes translucent.
5. Place slides in staining rack and flood section with crystal violet solution for 30 sec, then blot dry.
6. Flood section with potassium bromide solution for 1 min. Blot dry.
7. Differentiate in differentiating solution for approximately 30 sec; restain if overdifferentiated.
8. Rinse in xylene, several changes.
9. Mount in Permount.

Results (see Figs. 20 and 21):

   Glial fibers .... deep violet purple
   Background .... pale violet

Comments:

1. Crystal violet precipitates may be removed with straight aniline oil.
2. If precipitates remain on the section, after differentiation, in the form of glistening gold crystals, the section should be dried completely and redifferentiated.
3. Since aniline fumes are extremely toxic, the differentiation step should be done under a well-ventilated hood.

References

Abou-Donia, M. B. (1985). Organophosphorous compounds and cholinergic insecticides. In *Handbook of Neurotoxicology* (L. W. Chang, Ed.), pp. 419–474, Dekker, New York.

Abou-Donia, M. B., and Gupta, R. P. (1994). Involvement of cytoskeletal proteins in chemically induced neuropathies. In *Princi-*

ples of Neurotoxicology (L. W. Chang, Ed.), pp. 153–210, Dekker, New York.

Albee, R. R., Matsson, J. L., Yano, B. L., and Chang, L. W. (1987). Neurobehavioral effects of dietary restriction. *Neurotoxicol. Teratol.* **9:**203–212.

Armed Forces Institute of Pathology (AFIP) (1968). *Manual of Histologic Staining Methods.* McGraw Hill, New York.

Aschner, M., Aschner, J. L., and Kimelberg, H. K. (1994). The role of glia in central nervous system induced injuries. In *Principles of Neurotoxicology* (L. W. Chang, Ed.), pp. 93–110, Dekker, New York.

Bancroft, J. D. (1975). *Histochemical Techniques.* Butterworth, London.

Bouldin, T. W., and Goodrum, J. F. (1994). Toxicant-induced demyelinating neuropathy. In *Principles of Neurotoxicology* (L. W. Chang, Ed.), pp. 221–236, Dekker, New York.

Chang, L. W. (1979). *A Color Atlas and Manual for Applied Histochemistry.* Thomas, Springfield, Illinois.

Chang, L. W. (1984). Trimethyltin induced hippocampal lesions in various neonatal ages. *Bull. Environ. Contam. Toxicol.* **33,** 295–301.

Chang, L. W. (1987). Neuropathological changes associated with accidental or experimental exposure to organometallic compounds: CNS effects. In *Neurotoxicants and Neurobiological Function: Effects of Organoheavy Metals* (H. A. Tilson and S. B. Sparber, Eds.), pp. 82–116, Wiley & Sons, New York.

Chang, L. W. (1992). Basic histopathological alternatives in the central and peripheral nervous systems: classification, identification, approaches, and techniques. In *Neurotoxicology* (M. B. Abou-Donia, Ed.), pp. 223–252, CRC Press, Boca Raton, Florida.

Chang, L. W. (1994). Introduction to basic principles of neurocytology and general concepts on neurotoxicopathology. In *Principles of Neurotoxicology* (L. W. Chang, Ed.), pp. 3–34, Dekker, New York.

Chang, L. W. (1995). Neurotoxicology of organotins and organoleads. In *Handbook of Neurotoxicology* (L. W. Chang, Ed.), pp. 143–170, Dekker, New York.

Chang, L. W., and Dyer, R. S. (1983). A time-course study of trimethyltin induced neuropathology in rats. *Neurobeh. Toxicol. Teratol.* **5,** 443–459.

Chang, L. W., and Verity, M. A. (1995). Mercury neurotoxicity: effects and mechanisms. In *Handbook of Neurotoxicology* (L. W. Chang, Ed.), pp. 31–60, Dekker, New York.

Chang, L. W., Wenger, G. R., and McMillan, D. E. (1984). Neuropathology of trimethyltin intoxication. IV. Changes in the spinal cord. *Environ. Res.* **34,** 123–134.

Cory-Slechta, D., and Pounds, J. (1995). Neurotoxicology of lead. In *Handbook of Neurotoxicology* (L. W. Chang, Ed.), pp. 61–90, Dekker, New York.

Culling, C. F. A. (1974). *Handbook of Histopathological and Histochemical Techniques.* Butterworth, London.

Gessford, M. K. (1991). Preparation of teased nerves. *J. Histotech.* **14**(2), 105–108.

Gupta, R. P., and Abou-Donia, M. B. (1994). Axonopathy. In *Principles of Neurotoxicology* (L. W. Chang, Ed.), pp. 135–152, Dekker, New York.

Hastings, L. (1995). Neurotoxicity of cadmium. In *Handbook of Neurotoxicology* (L. W. Chang and R. S. Dyer, Eds.), pp. 171–212, Dekker, New York.

Lukiw, W. J., and McLachlan, D. R. (1995). Aluminum neurotoxicity. In *Handbook of Neurotoxicology* (L. W. Chang and R. S. Dyer, Eds.), pp. 105–142, Dekker, New York.

Luna, L. G. (1992). *Histopathologic Methods and Color Atlas of Special Stains and Tissue Artifacts.* American Histolabs, Publication Division, Gaithersburg, MD.

Morell, P. (1994). Biochemical and molecular bases of myelinopathy. In *Principles of Neurotoxicology* (L. W. Chang, Ed.), pp. 583–608, Dekker, New York.

O'Callaghan, J. P. (1991). Assessment of neurotoxicity: use of glial fibrillary acidic protein as a biomarker. *Biomed. Environ. Sci.* **4,** 197–206.

Pearse, A. G. E. (1972). *Histochemistry: Theoretical and Applied,* Vol. 2, Williams & Wilkins, Baltimore.

Sheehan, D. C., and Hrapchak, B. B. (1973). *Theory and Practice of Histotechnology.* C. V. Mosby, St. Louis, MO.

Spencer, P. S., Foster, G. V., Sterman, A. B., and Horoupian, D. (1980). Acetyl ethyl tetramethyl tetralin. In *Experimental and Clinical Neurotoxicology* (P. S. Spencer and H. H. Schaumburg, Eds.), pp. 296–308, Williams & Wilkins, Baltimore/London.

United States Office of Technology Assessment (1990). Neurotoxicity: identifying and controlling poisons of the nervous system. U.S. Government Printing Office, Washington, D.C.

Neuroanatomical Techniques for Labeling Neurons and Their Utility in Neurotoxicology

KARL F. JENSEN
Neurotoxicology Division
Health Effects Research Laboratory
U.S. Environmental Protection Agency
Research Triangle Park, North Carolina 27711

I. Introduction: Characterizing Alterations in Neural Circuitry Is Critical to Understanding the Action of Neurotoxicants

Toxin-induced neuropathology is generally taken as incontrovertible evidence of neurotoxicity and is therefore essential in efforts to identify neurotoxins. The value of characterizing toxin-induced structural damage is also gaining appreciation as neurotoxicity is increasingly being considered critical in the etiology of a variety of neurologic diseases. In this context, the role of structural assessment has grown significantly beyond identifying neurotoxins per se and is being applied more broadly to investigations of mechanistic hypotheses of neurologic disease. In addition to advances in neuropathology, recent developments in molecular biology, biochemistry, anatomy, physiology, and behavior have also resulted in a variety of hypotheses concerning the role of toxins in the pathogenesis of neurologic disease. Inherent in many of these hypotheses are assumptions concerning structural al-

terations that can only be assessed with morphological approaches. This chapter describes neuroanatomical methods that are particularly well suited for the characterization of such structural alterations.

This chapter also addresses the value of neuroanatomical approaches in ascertaining the significance of structural alterations. The ample 1500 pages of the most recent edition of *Greenfield's Neuropathology* (Adams and Duchen, 1992) attests to the tremendous variety of morphological alterations that are currently considered pathological. The significance of such pathological alterations, however, is not always apparent in the broader context of public health. Indeed, it is rare that a specific pathological alteration has a unique one-to-one relationship with a specific neurologic disease. This may be partly due to limitations in our knowledge of structure–function relationships of the nervous system. But to a greater extent, the adaptability of the nervous system itself contributes to the difficulty in establishing a precise correspondence between deleterious functional effects and pathological alterations. Consequently, one of the most important challenges facing neurotoxicology is indistinguishable from the fundamental challenge facing the neurosciences in general: understanding how the nervous system is organized and how this organization is disrupted in disease. Neuroanatomical approaches, by evaluating structural alterations in the context of neural circuitry, provide a strategic link between pathological and neurobehavioral assessments. They also provide a context for examining the relationship between neuropathology and neuronal structure at the cellular level. For these reasons, this chapter focuses on the value of neuroanatomical studies of neural circuitry as a means to identify, characterize, and ascertain the significance of toxin-induced structural alterations to the nervous system.

This chapter discusses the assessment of neural circuitry at three levels of organization. The first level corresponds to that of major brain regions and their interconnecting pathways. The integrity of this level of organization is typically assessed with classical stains that employ dyes and metals as well as more contemporary methods that employ neuroanatomical tracers. Neurochemical properties of circuits such as neurotransmitters, their synthetic enzymes, and their receptors can be visualized with immunohistochemistry and radioactivity labeled ligands. The second level is the individual neuron. Methods used to assess the morphology of neurons include the Golgi method, injections of tracers or dyes, and immunohistochemistry. The third level is the subcellular organization of molecules critical to neuronal integrity. Alterations in the expression, modification, and location of cellular constituents, such as the components of the cytoskeleton or elements of signal transduction pathways, can be assessed with *in situ* hybridization and immunohistochemistry. For the methods applied to each of these three levels, this chapter briefly describes the methods, their advantages and disadvantages with regard to making inferences concerning neural structure and provides examples where such methods have been applied to the assessment of neurotoxicity or related injury.

The range of approaches discussed is not exhaustive nor are the discussions of the methods technically detailed since numerous excellent and comprehensive tomes already exist whose length attests to need of space far beyond that of a chapter. Instead, the goal of this chapter is to describe a "neuroanatomical tool box" of sorts, providing the reader with an overview of how selected tools can be used to develop creative approaches to challenging questions. By having a variety of tools at one's disposal it may be easier to overcome limitations alluded to in the adage, "When the only tool one has is a hammer everything begins to look like a nail."

II. Assessment of Major Brain Regions and Their Associated Pathways

The first step in characterizing toxicant-induced changes in neural circuitry is the assessment of major brain regions and their associated pathways. This structural assessment is regularly undertaken in many laboratories. The wide application of the more routine of these methods attests to their robust tolerance of variation and modification. The most widely used are those that label nucleic acids with Nissl stains or hematoxylin dyes. Histochemical methods, silver staining for normal and degenerating axons, neuroanatomical tracers, and immunohistochemistry are also being used with increasing frequency in regional assessments of the effects of toxicants on the nervous system.

A. Nissl and Chromatin Stains Label Nucleic Acids and Can Be Used to Assess the Integrity of Neurons

Nissl stains (e.g., methylene blue, cresyl echt violet, cresyl violet acetate, thionin, toluidine blue) were used to assess the integrity of neurons long before it was known that the Nissl substance corresponds to clustered polysomes associated with the highly structured endoplasmic reticulum characteristic of neurons (Peters *et al.*, 1991). This appreciation of Nissl staining as a reflection of the metabolic integrity of a neuron was largely based on the changes in the appearance of the Nissl substance after axonal injury. This retrograde response, called "chromatolysis," is highly stereotypical for a given cell type and was one of the earliest experimental approaches to assessing neural connectivity (Lieberman, 1971; Grafstein, 1975; Brodal, 1969). The response is characterized by several changes first apparent several days after the injury: the cell body swells, the Nissl substance fragments, and the nucleus moves to an eccentric position. The fragmentation and diffusion of the Nissl substance give the cell body a characteristic pallor. If the cell survives, basophilic staining begins to appear around the nucleus and the Nissl substance gradually becomes apparent again as the nucleus returns to a normal position. The ultrastructural correlates of these alterations are consistent with the view that the appearance of the Nissl substance can reveal much about the physiological state of a neuron (Peters *et al.*, 1991). It is important to note, however, that the time course and characteristic appearance of these changes can vary dramatically between different types of neurons. The primary retrograde response can be observed as early as several hours after axon transection in certain neurons whereas the primary response may not appear for months in other neuronal types. Similarly, later aspects of the reaction to axon damage may occur in some cases within hours or days whereas in other cases they may not be apparent for weeks or months. The reason for this variance in the response is not completely understood but one consideration shown to be significant is the extent of injury to the entire axonal projection, as intact collateral projections can "sustain" the cell body (Brodal, 1969). Furthermore, an apparent loss of Nissl staining should not necessarily be taken as an indication that neurons have died and been phagocytized as alterations in Nissl staining can result from neurons entering a quiescent state in response to injury. Neurons can persist in such a state for relatively long periods of time (Naumann *et al.*, 1992; Peterson *et al.*, 1990, 1992). The relationship of Nissl staining to the metabolic status of a neuron, while advantageous in detecting subtle injury, is disadvantageous when attempting to determine the loss of neurons. In this regard, hematoxylin stains are more reliable.

Hematoxylin is a "powerful nuclear stain and chromatin stain par excellence" (Lillie, 1965), but its utility extends beyond this ubiquitous application. Coupled with an eosin counterstain, "H&E" provides a foundation for neuropathological assessment (Duchen, 1992; Chang, 1992a; Adams and Lee, 1982). Staining with hematoxylin is based on the oxidation of hematoxylin to hematein and the binding of hematein to a metal complexed with particular tissue components (Lillie, 1965). Numerous variations exists regarding the particular metals (e.g., iron, aluminum, copper, tungsten, molybdenum) and the manner in which the dye and the metal are applied to the tissue section (Kierman and Berry, 1975; Kierman, 1981). The choice of metal and pH influences which particular tissue elements are stained as well as their chromatic appearance. Procedures range from the Mayer's or Erlich's (aluminum) and Wiegert's (iron), which are useful in characterizing inflammation, tumors, intracellular inclusions, ischemia, and neuronal loss, to the more specialized Mallory's PTAH (tungsten), used to differentiate neurons and glia, or Wieger-Pal and Loyez (chromium), used to stain myelinated tracts.

The nuclear staining typically associated with Mayer's or Wiegert's methods is most likely due to staining of the nucleoproteins rather than the DNA itself (Kierman, 1981). The robust nature and the striking contrast of this staining make it ideal for characterizing the presence and pattern of distribution of specific neuronal populations. In addition to demonstrating the presence of the characteristic distribution of neurons, H&E also reveals a variety of neuropathological alterations, including cytoplasmic inclusions as well as stages of in-

jury associated with hypoxia–ischemia or granulovesicular degeneration (Duchen, 1992; Chang, 1992a; Adams and Lee, 1982). The wealth of information H&E provides the experienced eye is, no doubt, the likely reason it retains its dominance as a tool for the initial assessment of injury.

Some of the limitations of H&E and Nissl stains are derived more from the context in which they are used rather than in their staining properties per se. For example, when selected sections of the brain are examined for evidence of neuropathological alterations in the context of a neurotoxicological screen, the underlying presumption is that neurotoxic compounds will produce sufficiently dramatic alterations as to be readily detected. There are at least two important considerations that should be kept in mind with regard to this presumption (Jensen, 1994). One consideration is the extent of tissue that needs to be altered in order for detection to occur. For example, it has been estimated that as much as a 30% reduction in the number of neurons must occur before such a neuronal loss is recognizable in a typical neuropathological screen. A second consideration is the intrinsic structural heterogeneity of the nervous system, such that similarly appearing neurons can have dramatically different vulnerabilities to a toxicant. Given these considerations, extensive sampling of brain regions is critical and needs to be considered a priori in the design of the neuropathological analysis. When sampling is adequate, H&E and Nissl can be important aids in guiding more sophisticated and targeted approaches to characterizing structural damage to the nervous system. Furthermore, the vital role of these stains is also related to what may be considered a fundamental dilemma in characterizing neurotoxicant-induced structural alterations: no one specialized technique has been demonstrated to be universally applicable in detecting the wide variety of toxin-induced neuropathological alterations. Consequently, the fact that H&E and Nissl stain neurons nonselectively makes them essential in characterizing toxin-induced structural damage to neural circuitry.

B. Neural Tissue's Affinity for Silver Can Be Used to Label Intact and Degenerating Axons

Argyrophilia, the affinity for silver, is a property of components of nervous tissue that is responsible for the phenomenon of silver staining. Two broad categories of silver stains are considered here: those designed to stain normal axons and those designed to selectively stain degenerating axons. The former procedures result in the staining of fibrillar structures within axons and historically have been called "neurofibrillary stains." The latter have developed from efforts to selectively enhance the staining of degenerating axons while suppressing or eliminating the staining of normal axons and are referred to in this chapter as "silver degeneration stains."

The mechanism by which particular tissue components become stained with silver is only partially understood and may be different for various procedures. Consequently, most procedures have been developed empirically by trial and error in an attempt to optimize results concerning detection, resolution, contrast, and reproducibility. Different variations represent optimal procedures for specific conditions of fixation, embedding, and sectioning, as well as particular tissues and their cellular constituents.

The neurofibrillar methods provide a crisp image of axons within central tracts and peripheral nerves and can highlight the complex pattern of connections within the neuropil that appears vacant with Nissl stains. However, alterations in such complex staining patterns are not easy to detect, particularly since such patterns vary dramatically in different regions of the nervous system. It is for the most part impossible to precisely follow a particular set of axons from their origin to their termination using neurofibrillary stains. To the experienced eye, however, alterations in the normal pattern of silver staining can indicate subtle but important alterations in neural connectivity. In a neuropathological context, neurofibrillary stains can reveal a variety of classical neuropathological hallmarks such as the plaques and tangles diagnostic for various degenerative neurological disorders (Duchen, 1992).

Since their initial development these neurofibrillary stains have also been used to demonstrate degenerative events following neuronal injury. Their ability to reveal the detail of the response to injury is central to their historical importance in tract tracing as well as their usefulness in neurotoxicology. Ramon ý Cajal (1928) described silver staining procedures, based on en bloc staining following fixation with chloral hydrate, that he considered ". . . favorable for the analysis of the phenomena occurring in wounded nerves . . . ." Bielschowsky (1935) described a method of silver staining of axons in frozen sections of formalin-fixed tissue. Numerous investigators have attempted to improve the reliability of these methods, most notably by including proteinates (Peters *et al.*, 1991; Bodian 1936) and more recently with the replacement of chemical developers with physical developers (Gallyas, 1971b; 1979a; Gallyas and Wolff, 1986; Gallyas *et al.*, 1990). The commonly used versions of Bodian's and Bielschowsky's methods are provided in histological reference texts (e.g., Kierman, 1981; Armed Forces Institute of Pathology, 1968; Chang, 1979; Ralis *et al.*, 1973).

Gray and Guillery (1966; Guillery, 1965) reviewed studies of neural degeneration employing the neurofibrillar methods. They concluded the increased staining of "neurofibrillar hypertrophy" associated with the degeneration of axon terminals was the result of an increase in neurofilaments within the degenerating terminals. Substantial differences in the intensity of staining of degenerating terminals in different brain regions correspond to the number of neurofibrils present in the terminals at various stages of degeneration. In line with Ramon ý Cajal's (1928) interpretation that neurofibrillary hypertrophy in the proximal stump of a cut axon is an "abortive attempt at regeneration," Gray and Guillery (1966) suggest that the extent of neurofibrillary hypertrophy observed in degenerating terminals was indicative of their regenerative capacity. However, the basis for such differences in such regenerative capacity is not clear. In addition, the "neurofibrillar hypertrophy" also varies in its time course for different brain regions, peaking from 7 to 21 days after acute injury. Despite these limitations, neurofibrillary stains have found wide application in characterizing the effects of toxicants, including some of the classic studies of organophosphates and hexacarbons (Cavanagh, 1954, 1964, 1982).

Neurofibrillar stains can be used to visualize Wallerian degeneration, the anterograde degeneration of the portion of the axon distal to injury (Brodal, 1982). Used in conjunction with lesions targeted to specific brain regions, this approach has provided a wealth of information about the organization of neural circuitry. This endeavor led several laboratories to develop variations of neurofibrillary methods with an enhanced capacity to stain degenerating fibers (for a historical perspective, see Beltramino *et al.*, 1993; Brodal, 1969; Nauta and Ebbesson, 1970; Voogd and Feiradbend, 1981). The first substantial advance over the low signal to noise ratio characteristic of classical neurofibrillary stains in revealing degeneration was achieved by Nauta and Gygax (1954), who developed a method to suppress the staining of normal fibers while retaining the staining of degenerating fibers. But the suppression of normal fiber staining reduced the staining of degenerating elements, particularly terminals. To address this problem, Fink and Heimer (1967) developed methods that avoid the suppression of staining of axon terminals, providing for the clear visualization of axons and their terminal fields.

Additional methods have been derived from attempts to improve the reliability of silver staining in revealing injured neurons. Gallyas and co-workers (Gallyas *et al.*, 1980a,b,c, 1992a,b; Gallyas and Zoltay, 1992) have developed methods based on physical instead of chemical development, and de Olmos and co-workers (Beltramino *et al.*, 1993; de Olmos *et al.*, 1991, 1994; Carlsen and de Olmos, 1981; de Olmos, 1969; de Olmos and Ingram, 1971) have developed methods based on the inclusion of copper. Recent versions of both the Gallyas and de Olmos methods demonstrate injury neurons within minutes to hours after injury, can be optimized for the staining of particular elements of injured neurons (i.e., perikarya, axons, terminals), and are noted for providing exquisitely detailed images of neuronal morphology. Several excellent reviews discuss the advantages and disadvantages of the various methods, considerations essential to their interpretation, and important technical details (Beltramino *et al.*, 1993; Kiernan and Berry, 1975; Balaban, 1992; de Olmos *et al.*, 1991; Heimer, 1967; Nauta and Ebbesson, 1970; Voogd and Feiradbend, 1981).

Gallyas and co-workers (Gallyas, 1971a, 1979b, 1980a,b) have presented evidence that silver staining by physical development is similar to a photographic process in that it involves enhancement of a latent image by the selective enlargement of very fine silver deposits. Selectivity of the method is related to both the state of degeneration and the preservation of degeneration products in the tissue. In a neurotoxicological context, these findings indicate that the earliest evidence of staining may not always indicate the primary site of action of a toxin. Furthermore, early staining can represent reversible trauma unrelated to later degenerative events and the extent to which injury is reversible or irreversible can only be inferred.

These considerations are also important with regard to survival time after injury. The survival time for the optimal demonstration of neuronal injury varies with the silver method employed and the nature of the insult as well as the particular kind of neurons injured. Consequently, not all the injured neural elements will necessarily be demonstrated at the same time. The optimal time course for the staining of particular degenerating elements after a given kind of insult must be empirically determined. Thus, when using silver degeneration stains to characterize neurotoxic injury, one needs to be cautious in attempting to determine the sequence of events when the pattern of degeneration is temporally and spatially complex. It is not always possible to distinguish a "primary" versus a "secondary" site of damage based on the sequence of appearance of staining. A complex temporal and spatial pattern may reflect differences in inherent vulnerability, difference in the time course of development of argyrophilia, and transneuronal degenerative effects. These considerations are a small but significant part of the much broader issue of "direct" and "indirect" neurotoxicity, concepts that are critical in relating the biochemical action

of a toxin to the nature of the resulting lesions as well as neurobehavioral effects (Chang, 1992b).

Also of central importance in applying silver degeneration stains to characterizing neurotoxic injury is distinguishing between various forms of artifact and the staining of degenerating neural elements. The occurrence of particular kinds of artifacts also depends on the species being studied, the particular method employed, the age of the animal, and the quality of fixation. Such considerations are therefore particularly important when trying to assess the differential vulnerability of sensitive subpopulations such as the developing organism (Leonard, 1975; Janssen *et al.*, 1991; Schweitzer *et al.*, 1991). Several varieties of artifact are described in detail by de Olmos and colleagues (1991). These include deposits that can be confused for evidence of terminal degeneration as well as incomplete staining of normal fibers and myelin artifacts that can be confused for evidence of degenerating axons. In addition, the extent to which a method can visibly label the full complement of degenerating neural elements is dependent on a variety of experimental variables such as fixation, water and reagent quality, and even relatively minor contaminants on glassware. Although a number of methods specify aldehyde fixation by vascular perfusion, there are also methods adapted for use with immersion-fixed tissue from human autopsy (Grafe and Leonard, 1980; Albrecht and Fernstrom, 1959).

One of the drawbacks that silver degeneration stains share with other specialized stains is that the absence of staining cannot be construed as the absence of damage. In addition, as previously noted, the converse is also true. The fact that a neuron is unambiguously stained may not necessarily indicate that the neuron would have been irreversibly committed to death. Despite these caveats, consistent dose- and time-dependent patterns of silver degeneration staining constitute compelling evidence of toxicant-induced injury. Silver degeneration stains have been used to demonstrate the patterns of injury induced by a variety of toxins, including organometals (Beltramino *et al.*, 1993; Desclin and Escubi, 1974; Balaban, 1985; Balaban *et al.*, 1988; O'Callaghan and Jensen, 1992), 6-hydroxydopamine (Maler *et al.*, 1973), pyridine compounds (Desclin and Escubi, 1974; Balaban, 1985; O'Callaghan and Jensen, 1992), organophosphates (Tanaka *et al.*, 1991, 1992; Tanaka and Bursian, 1989; Inui *et al.*, 1994; Glees and Jansik, 1965), cocaine (Ellison and Switzer, 1993), amphetamine analogs (Frith *et al.*, 1987; Ricaurte *et al.*, 1982, 1984, 1985, 1988; Scallet *et al.*, 1988; Slikker *et al.*, 1988; Jensen *et al.*, 1993; Harvey and McMaster, 1976; Ellison and Switzer, 1993), phencyclidine (Ellison and Switzer, 1993), capsacian (Ritter and Dinh, 1988), and excitotoxicants (Beltramino *et al.*, 1993; Janssen *et al.*, 1991; Schweitzer *et al.*, 1991; Scallet *et al.*, 1993).

Silver degeneration stains are particularly valuable in defining the location and extent of injury. This is important where toxicant-induced damage to the brain is localized to several regions, such as with methylene-deoxymethamphetamine (MDMA) (Fig. 1). It is interesting in this regard that the same dose of a MDMA can in one animal produce evidence of damage in the striatum (Fig. 1B) whereas in another animal that same brain region can be free of evidence of injury (Fig. 1D). The consistency of degeneration staining in the neocortex in homologous sections from these two animals suggests that this difference is not due to variation in toxicant administration or in histological processing. Quantitative estimates of the dose dependence of toxin-

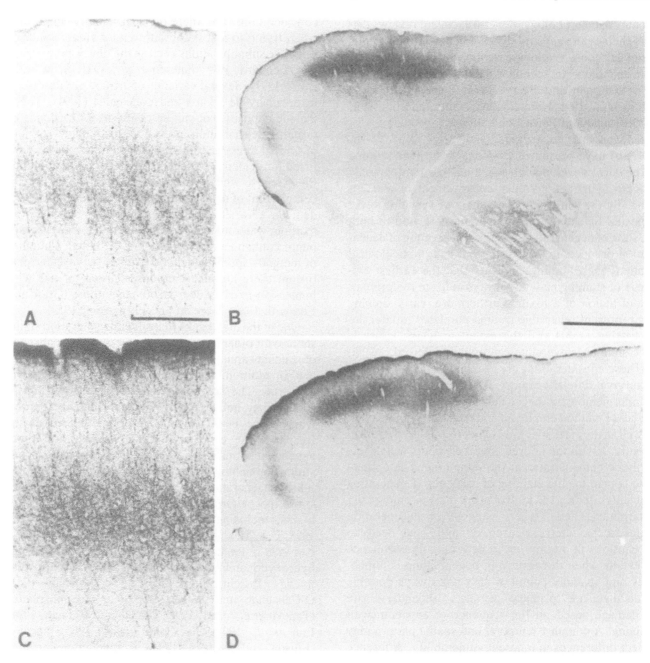

FIGURE 1   Silver degeneration staining in the rat forebrain following MDMA administration. Animals were sacrificed 48 hr after four twice daily injections of MDMA. (A and B) Sagittal section from a rat in which silver staining was evident in the striatum. (C and D) Section from a rat in which silver staining was absent from the striatum. (A and C) High power of neocortex in B and D, respectively. (de Olmos method, from Jensen *et al.*, 1993.)

induced neural injury have also been demonstrated with silver staining (Fig. 2A). In a similar fashion, time-dependent changes in the extent of degeneration following toxicant administration have also been quantified (Fig. 2B). But perhaps most importantly, silver stains have been used to identify populations of neurons that may be vulnerable to particular toxins that were not revealed by other methods (Commins *et al.*, 1987; Jensen *et al.*, 1993).

As a final note with regard to methods that visualize neural degeneration, the *p*-phenylenediamine (PPD) method deserves mention (Sadun *et al.*, 1983). Although this method does not utilize silver, it does label myelin-encapsulated remnants of degenerated axons in a manner compatible with light microscopy. The basis of this method, the rendering of osmophilic tissue elements light opaque, is compatible with many routine histological methods. The myelin-encoated aggregates of degenerated axons can be easily distinguished from normal axons and remain detectable for substantially longer periods of time than degenerating axons visualized with silver stains (Sadun and Schaechter, 1985; Johnson and Sadun, 1988). This feature is particularly valuable when histological samples from studies utilizing long-term exposure are only available at the termination of the study. The persistence of these myelin-encapsulated aggregates increases the likelihood of detecting axonal degeneration that may have occurred early during exposure. It should be noted that the number of these aggregates is qualitatively rather than quantitatively related to the extent of axonal degenera-tion. When there is evidence of such remnants of degeneration, however, subsequent quantitative analysis of the remaining intact axons, which are also visible in PPD-stained material, can provide reliable estimates of the number and size of axons (Tenhula *et al.*, 1992; Sadun, 1989; Hinton *et al.*, 1986). Such data are valuable when subpopulations of axons, such as those with large diameters, are small in number but preferentially vulnerable to injury and their loss would go undetected with more routine methods.

C. Neuroanatomical Tracers Can Be Used to Assess the Integrity of Specific Pathways Interconnecting Major Brain Regions

Neuroanatomical tracers are exogenous substances that when injected into the brain can be transported along axons in the anterograde direction to reveal patterns of termination or in the retrograde direction to reveal the origin of a projection. In addition to being taken up by intact perikarya, dendrite, and terminals, tracers can also be taken up and transported by axons damaged during injection. Some tracers are also taken up intact axons coursing through the injection site. Since neuroanatomical tracers do not require that the processes of neurons be degenerating for pathways to be visualized, they can render a more reliable and detailed image of neurons and their projections than silver degeneration methods. Neuroanatomical tracers have the potential to demonstrate alterations in the organization of major pathways resulting from toxic

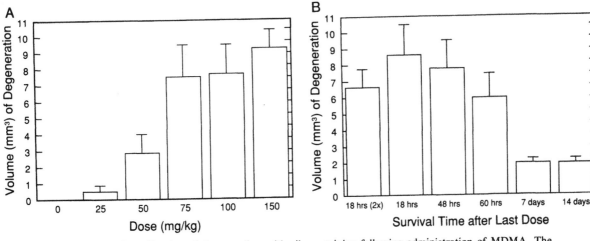

FIGURE 2 Quantification of degeneration with silver staining following administration of MDMA. The volume of tissue in which evidence of silver was present was determined by visually outlining staining areas from homologous sections from a 0.72-mm-thick sagittal block. (A) Dose–response. Animals received each indicated dose of MDMA twice a day for 2 days and were sacrificed 48 hr after the last dose. Vertical bars indicate standard errors. (B) Time course. Animals received 4 × 100 mg/kg MDMA and were sacrificed 18 hr later, with the exception of the group indicated with 2×, which only received 2 × 100 mg/kg. (From Jensen *et al.* 1993.)

insult. Such alterations in connectivity can result from direct neural damage from toxin, from secondary degenerative events, or from adaptive responses to injury. Regardless of how such changes relate to the direct actions of a toxin, characterizing the changes in connections between major brain regions can be of strategic value in relating neuropathological findings to behavioral deficits. Broad categories of tracers include amino acids, enzymes, lectins, fluorescent dyes, and dextran-amines (for reviews see Kierman and Berry, 1975; Heimer and Robards, 1981; Heimer and Zaborszky, 1989; Mesulam, 1982; Bolam, 1992; Cowan and Cuenod, 1975).

Axonal connections can be demonstrated by injecting radioactivity labeled amino acids into particular brain regions where they are taken up and transported anterogradely to axonal terminals where they can be visualized with autoradiography (Edwards and Hendrickson, 1991; Cowan *et al.*, 1972; Groenewegen *et al.*, 1994; Rogers, 1975). Since only neurons at the site of injection transport the injected amino acids, this procedure affords greater reliability in determining the origin of projections than silver degeneration methods. Autoradiography does not, however, provide a clear image of the morphology of axons or terminals since the silver grains are exposed to radiation emanating primarily from the uppermost portion of the section next to the emulsion. This limitation has been overcome with tracers such as biocytin and neurobiotin composed of amino acids conjugated with biotin (Jacquin *et al.*, 1992; Izzo, 1991; King *et al.*, 1989; Lapper and Bolam, 1991). These tracers are also transported for long distances and can be detected in the finest ramifications of neuronal processes. The biotin moiety retains its affinity for avidin, and consequently can be linked with avidin complexes containing peroxidase which can be visualized histochemically to provide detailed images of axonal morphology (see Section III.A).

Horseradish peroxidase (HRP) is retrogradely transported down an axon to the cell body as well as anterogradely transported to an axon's terminal fields (Mesulam, 1982; Kristensson, 1975; Cowan and Cuenod, 1975; LaVail, 1975; Warr *et al.*, 1981). The conjugation of HRP to a lectin (e.g., agglutinin-coupled wheat germ lectin, WGA-HRP) improves uptake and transport, particularly in the anterograde direction, and has been widely used to trace connections over long distances (Mesulam, 1982; Zaborszky and Heimer, 1989). Another lectin, *Phaseolus vulgaris*–leucoagglutinin (PHA-L) has been demonstrated to be very effective in providing a highly detailed image of the morphology of anterogradely labeled axons (Gerfen and Sawchenko, 1984, 1985; Gerfen *et al.*, 1989; Wouterloud and Groenewegen, 1985; Zaborszky and Heimer,

1989). PHA-L is injected iontophoretically; consequently the injection and uptake sites are more restricted. PHA-L can be visualized immunohistochemically, with either immunoperoxidase or avidin–biotin techniques (see Section III.A).

HRP, or tracers with peroxidase labels, can be visualized histochemically using a variety of chromogens (Mesulam, 1982; Warr *et al.*, 1981; Zaborszky and Heimer, 1989). Most commonly, peroxidase catalyzes the oxidation of a benzidine derivative as a chromogen, which polymerizes to an insoluble light-opaque precipitate. Detection, color, and stability can be enhanced with supplementary methods using silver, cobalt, or nickel. Since benzidine and some of its derivatives are considered carcinogenic, alternative chromogen systems have been and continue to be developed. The relative sensitivity of different chromogens, their suitability for light and electron microscopy, and advantages and disadvantage of different peroxidase reactions, as well as sources of artifact, have been reviewed (Kierman and Berry, 1975; Kierman, 1981; Mesulam, 1982; Heimer and Robards, 1981; Heimer and Zaborszky, 1989; Bolam, 1992; Cowan and Cuenod, 1975; LaVail, 1975; Winer, 1977; Fox and Powley, 1989; Robertson, 1975).

Another class of neuroanatomical tracers are dyes that can be directly visualized with fluorescence microscopy. Examples of such fluorescent dyes include Lucifer yellow, true blue, Evans blue, fast blue, nuclear yellow, propidium iodide, flouro-gold, and SITS (Kuypers and Huisman, 1984; Schmued and Fallon, 1985, 1986; Schmued, 1990; Blum and Reed, 1994; Skirboll *et al.*, 1984; Wessendork, 1990). Fluorescent molecules have also been conjugated to dextran-amines and latex microspheres resulting in improved stability, emission characteristics, or transport (Bolam, 1992; Schmued *et al.*, 1990; Katz and Iarovici, 1984; Nance and Burns, 1990). Since various dyes fluoresce in response to different wavelengths of light, they can be discriminated when simultaneously present in individual neurons. Such multiple labeling can be used to characterize several kinds of phenomena, including the multiple projections of neurons to several brain regions, changes in neuronal projections during development, loss of projections resulting from injury, and the reorganization of connections that can occur secondary to insult (Ivy and Killackey, 1982; Kaas *et al.*, 1983; O'Leary and Stanfield, 1989; O'Leary *et al.*, 1981).

Neuroanatomical tracers have also been used in conjunction with immunohistochemistry or *in situ* hybridization to identify the neurotransmitter characteristics of neurons projecting to particular brain regions (Schmidt, 1987; Chronwall *et al.*, 1989; Zaborszky and Heimer, 1989). The reaction products of some of the benzidine derivatives are osmophilic, thus compatible with the ultrastructural analysis of synaptic relationships with electron microscopy (Bolam, 1992; Schmidt, 1987; Björklund and Hokfelt, 1983). Many of the fluorescent tracers can be photoconverted to a stable, insoluble, osmophilic benzidine precipitate also suitable for electron microscopy (Schmued and Snavely, 1993; Maranto, 1982; Sandell and Masland, 1988; Balercia *et al.*, 1992; Buhl *et al.*, 1990). Immunohistochemical and avidin–biotin procedures can also be used with immungold labels for superior ultrastructural localization (Polyak and Varndell, 1984; Slater, 1993; Chan *et al.*, 1990).

Several considerations regarding tracer methodology are particularly important in a neurotoxicological context. Interaction between variations in the placement of injections can interact with variation in the extent of toxicant-induced injury to complicate interpretation. Perhaps more importantly, neuroanatomical tracers are dependent on the axonal transport mechanism (LaVail, 1975; Kristensson, 1975; Mesulam, 1982; Cowan and Cuenod, 1975; Robertson, 1975; Bolam, 1992) that can be disrupted by neurotoxicants (Blum and Reed, 1994; Gupta and Abou-Donia, 1994). Consequently, significant alterations in the labeling of pathways in animals treated with such toxicants might be erroneously interpreted as evidence of significant structural damage, even when axons within the pathway are intact. Furthermore, optimal survival time depends on characteristics of the circuit being investigated. Variability in transport time can make interpretation a challenge if the tracer is injected at a time when degenerative or regenerative events are taking place. Despite these limitations, neuroanatomical tracers have been proven to be valuable tools in assessing the normal patterns of connectivity as well as characterizing changes in connections that follow injury.

Neuroanatomical tracers have been particularly useful in characterizing alterations in connections following developmental insult. One example is the demonstration of persistent changes in the projection from the thalamus to neocortex that occurs consequent to neonatal peripheral nerve damage (Fig. 3). The normal somatotopic organization of thalamocortical projections can be demonstrated by injecting WGA-HRP into the ventrobasal thalamus and visualizing the tracer in tangential sections of the somatosensory cortex (Fig. 3A). Patches of label are arranged in a pattern corresponding to the representation of the body surface. Most notable are large patches corresponding to the representation of the mystacial vibrissae on the face of the rat. When an identical procedure is performed in an adult rat in which the trigeminal nerve (which innervates the mystacial vibrissae) was cut at birth,

the pattern in the corresponding neocortical region is disrupted (Fig. 3B). This altered pattern indicates that neonatal damage to the sensory periphery can result in a persistent reorganization in a central thalamocortical pathway (Jensen and Killackey, 1987a,b).

In a toxicological context, the use of tracers is probably most appropriate in characterizing alterations in circuitry produced by toxicants whose sites of action are known. Some examples include the characterization of effects of 2,5-hexanedione on the visual system (Pasternak *et al.*, 1985), the organization of forebrain pathways in 1-methyl-4-phenyl-1,2,3,6 tetrahydropyridine (MPTP)-treated animals (Page *et al.*, 1993), and alterations in the corticospinal or callosal projections following developmental exposure to ionizing radiation or ethanol (Jensen and Killackey, 1984; Miller *et al.*, 1990; Miller, 1986, 1987).

D. Toxicant-Induced Alterations Can Be Detected with Methods That Reveal Transmitter-Related Characteristics of Neural Circuits

The identification and cloning of genes for neurotransmitters, their synthetic enzymes, and their receptors have dramatically advanced the characterization of neural circuitry. Most notably, screening for nucleotide sequences homologous to cloned receptors has revealed a regulatory mechanism responsible for the diversity of neurotransmitter receptors, including alternate splicing and mRNA editing (Hollmann and Heinemann, 1994; Gringich and Caron, 1993). Neuroanatomical mapping of the expression of neurotransmitter-related genes and their products with *in situ* hybridization and immunohistochemistry has played a central role in determining the importance of these regulatory mechanisms to the functional integrity of specific neural pathways.

From a toxicological perspective, the multiplicity of regulatory mechanisms for gene expression and translocation and the assembly of gene products represent a diversity in potential sites at which toxicants can act as well as a variety of mechanisms potentially involved in responses to toxicant-induced injury. Although this knowledge of regulatory mechanisms enhances our ability to identify and characterize toxicant-induced neurochemical alterations, it has not yet provided easy solutions to more difficult challenges in interpreting such alterations. These challenges include determining the degree of adversity attributable to toxicant-induced neurochemical alterations, determining the extent to which such alterations are indicative of the disruption of the structural integrity of a circuit, and establishing causal relationships between the neurochemical alter-

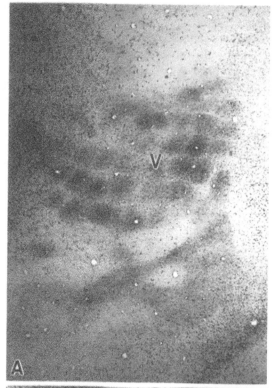

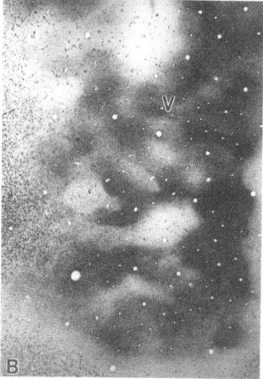

ations and behavioral deficits. In the absence of clearcut approaches to these questions, several considerations can assist in determining the extent of adversity ascribed to a toxicant-induced neurochemical alteration. These include persistence and reversibility, correlative evidence of structural damage, indications of altered development, concordance with behavioral deficits, and a compromised response to a physiological challenge (National Research Council, 1992, 1993). Neuroanatomical localization contributes to addressing these considerations by providing an integrated assessment of neurochemical and anatomical properties of circuitry and thus specifies the structural context for relating neurochemical alterations to behavioral deficits.

Advances in our understanding of gene regulation have also highlighted the numerous ways the neurochemical and morphological properties of a circuit can be independently regulated and altered. Thus, reliable inferences regarding structural and neurochemical properties of a circuit depend on the specific properties of probes, antisera, or ligands employed; methods of visualization; and supportive evidence from correlative assessments.

Histochemistry, immunohistochemistry, receptor–ligand autoradiography, and *in situ* hybridization with nucleic acid probes are all used to characterize neurochemical properties of circuitry. The use of histochemistry to characterize effects of neurotoxicants is addressed by Chang in Chapter 1 of this volume. Experimental approaches that address neurochemical properties of circuits that are not related to specific neurotransmitters are discussed in Section II.E of this chapter. This section focuses on neuroanatomical methods that localize neurotransmitters, their synthetic enzymes, and their associated receptors.

1. Neurotransmitters and Their Synthetic Enzymes

Immunohistological labeling of neurotransmitters and their synthetic enzymes can reveal the organization of specific pathways in the central nervous system. The distinctive appearance of these pathways in histo-

FIGURE 3 Pattern of thalamocortical projection in rat somatosensory cortex following neonatal peripheral nerve damage. (A) Tangential section of adult control rat showing pattern of anterograde HRP labeling. (B). Adult rat in which the trigeminal nerve was cut at birth. Thalamocortical projections were anterogradely labeled with injections of WGA-HRP into the region of the ventrobasal thalamus. Survival time was 24 hr and TMB was used as the chromogen. The region of the vibrissae representation (V) exhibits "patches" that in the control correspond to individual vibrissae. In the neonatal nerve cut animal this pattern of patches is disrupted. (From Jensen and Killackey, 1987b.)

logical sections can be particularly valuable in localizing toxicant-induced effects. Procedures used to make polyclonal and monoclonal antisera against neurotransmitters and their synthetic enzymes can involve the use of a variety of sources of antigen such as fixed tissue, homogenates, purified holoenzymes, subunits, cloned peptide sequences, or post-translationally modified proteins. In addition, the choice of a procedure for generating antisera for immunohistochemistry may be influenced by the nature of antigenic sites on the molecule and its sensitivity to fixation and histological processing (Sternberger, 1986; Larsson, 1988; Hockfield *et al.*, 1993).

In practice, primary antisera are usually selected for use in immunohistochemistry on the basis of their specificity demonstrated *in vitro* and the ability to stain sections of fixed tissue in a manner based on the expected distribution of the enzyme or transmitter. A variety of factors can influence differences in the performance of an antibody between *in vitro* and histological applications, including penetration, accessibility, conformation of epitopes, novel cross-reactivity induced by fixation, and differential sensitivity of post-translationally modified proteins. Several, but not all, of these factors can be ascertained by comparing staining in fixed and unfixed tissue. There are extensive discussions of the many challenges in demonstrating the specificity of antisera in a histological context that should be carefully considered before undertaking experiments using immunohistochemistry as well as when interpreting the results (Sternberger, 1986; Larsson, 1988; Polak and Van Noorden, 1984; Van Leeuwen, 1982; Fuxe *et al.*, 1985; Pool *et al.*, 1983; Petruz *et al.*, 1980; Swaab, 1982).

Two of the most widely used methods to label primary antisera bound to the section are the peroxidase–anti-peroxidase method (PAP) and the avidin–biotin method (ABC) (Hsu *et al.*, 1981; Sternberger, 1986; Larsson, 1988). Both methods provide reliable and robust labeling. In brief, the PAP method uses a secondary antisera that recognizes the IgG of the host in which the primary antisera was made. The secondary antisera is then linked to a complex composed of peroxidase molecules bound to anti-peroxidase antibodies made in the same host as the primary antisera. In contrast, the ABC method is based on the high affinity between avidin and biotin. This method uses a biotinylated secondary antisera that recognizes IgG of the host in which the primary antisera was made, which is then linked to the avidin–peroxidase complex. Both methods typically employ peroxidase histochemistry to visualize the label that was previously described for HRP (see Section II.C), but they can also employ colloidal gold for ultrastructural analysis or fluorescent conjugates for studies using multiple labeling (Sternberger, 1986; Larsson, 1988).

The immunohistochemical staining for specific neurotransmitters or their synthetic enzymes can provide a striking image of major pathways within histological sections of the brain. For example, Fig. 4 demonstrates the immunohistochemical labeling of tyrosine hydroxylase in sections of brain from control mice and mice treated with MPTP, a toxicant known to damage the nigrostriatal system. The nigrostriatal projection is one of the major dopaminergic projection systems in the rodent forebrain. Tyrosine hydroxylase is the rate-limiting enzyme for the synthesis of dopamine. In sagittal sections of the brains of control mice, the nigrostriatal projection is delineated by the tyrosine hydroxylase immunoreactivity (Fig. 4A). At low power, the cell

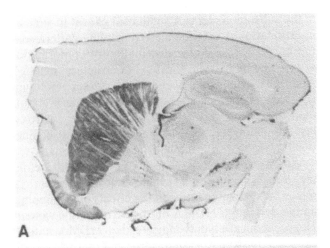

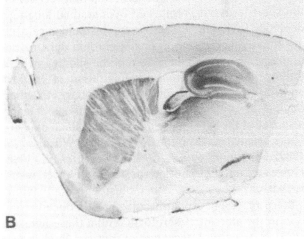

FIGURE 4 Altered immunohistochemical labeling of tyrosine hydroxylase in MPTP-treated mice. The striatum is intensity stained in the sagittal section from control mice (A) and staining is reduced in the section from the MPTP mouse (B). There is a reduction in staining in the nigrostriatal pathway as well. Polyclonal antisera against tyrosine hydroxylase. ABC method with metal-intensified DAB as chromogen. (From O'Callaghan and Jensen, 1992.)

bodies within the substantia nigra stain darkly whereas the terminal fields of their axons in the stiatum appear as a more uniform and diffuse pattern. Some staining of axons can also be detected as they course from the nigra to the striatum. In animals treated with MPTP, this pattern of staining for tyrosine hydroxylase is altered (Fig. 4B). There is an apparent reduction in the staining intensity of the striatum, whereas there is still a large number of neurons stained in the substantia nigra. When material from an extensive time course is examined, alterations in staining within the striatum are prominent before those in substantia nigra (K. F. Jensen *et al.*, unpublished observations). Such time-dependent localized changes in the appearance of tyrosine hydroxylase staining are consistent with the hypothesis that MPTP's initial insult is to axon terminals within the striatum with subsequent retrograde effects in the substantia nigra. A significant caveat to such an interpretation is that the altered staining pattern may represent changes in the distribution or conformation of tyrosine hydroxylase instead of structural damage. Additional evidence, however, from studies employing silver degenerations stains support the interpretation that the changes observed with tyrosine hydroxylase immunohistochemistry reflect structural damage (K. F. Jensen *et al.*, unpublished observations).

Another example of the use of immunohistochemistry to assess injury to a specific neurotransmitter system is the characterization of the effects of the amphetamine derivative MDMA on the serotonergic system. Neurochemical studies indicate that MDMA initially causes a massive release of serotonin that recovers in the short term but is followed by a gradual long-term depletion (Schmidt, 1987). Immunohistochemical studies employing antisera against serotonin demonstrate dramatic and complex changes in the staining of serotonergic axons following administration of MDMA, and Molliver and co-workers have interpreted these findings as evidence that MDMA selectively damages a subset of the serotonergic axons, the small diameter fibers originating from the dorsal raphe (Wilson *et al.*, 1989; Mamounas *et al.*, 1991; O'Hearn *et al.*, 1988). However, an alternate interpretation of these results is that changes in the immunohistochemical staining pattern represent a pharmacologically induced reduction in the amount of serotonin within these fibers instead of structural degeneration per se. Studies with silver degeneration stains following administration of MDMA reveal a distinctly different pattern of damage that does not correspond to the pattern of the fine serotonergic axons (Commins *et al.*, 1987; Jensen *et al.*, 1993). A comparison of the result obtained with these two different methods highlights a number of considerations that are important when interpret-

ing neurotransmitter-specific immunohistochemistry. First, neurotransmitter-specific immunohistochemistry can reflect pharmacological as well as structural alterations. In this case, the protracted time course of changes in serotonergic staining are also consistent with neurochemical data that indicate MDMA induces a substantial and prolonged disruption in the availability of serotonin. Nonetheless, a dramatic and persistent neurochemical change, even in the absence of overt structural damage, may in some cases be considered evidence of neurotoxicity. Second, the pattern of silver degeneration staining differs significantly from the pattern of serotonergic innervation, indicating that nonserotonergic elements are vulnerable to MDMA. Such damage to nonserotonergic elements would not be revealed by neurochemical or immunohistochemical methods selective for serotonin. This observation demonstrates that the exclusive dependence on neurotransmitter-specific methods can preclude detection of injury to diverse targets. The identification of the full complement of a neurotoxicant's targets is of prime importance to neurotoxicity assessment and a critical foundation for investigating pathogenesis. More generally, this example demonstrates that results obtained with different methods do not always provide correlative evidence of a neurotoxin's specificity but instead can reveal that the action of a toxicant is more complex than indicated by the results of either method alone.

The injury-related changes in the regulation of synthetic enzymes for neurotransmitters can also contribute to difficulties in the interpretation of immunohistochemical staining patterns. For example, the elimination of immunohistochemical staining for choline acetyltransferase (ChAt) in perikarya of septal neurons following fimbria-fornix lesions has been interpreted as evidence that these neurons die, but more careful analysis revealed that they persist but no longer express sufficient ChAt to be labeled, presumably because of the elimination of retrogradely transported nerve growth factor as a result of the axotomy (Peterson *et al.*, 1990). It has also been hypothesized that toxicant-induced damage may stimulate the production of growth factors which may in turn modulate the expression of ChAt (Barone *et al.*, 1992).

Toxicant-induced alterations in the transcription of neurotransmitter synthetic enzymes can be localized at the cellular level with *in situ* hybridization. The advantages and disadvantages of cDNA probes, riboprobes, and oligoprobes; procedures for demonstrating their specificity; conditions of stringency; the significance of melting temperature; the importance of appropriate tissue preparation; and the relative merits of radioactive and avidin–biotin labeling as applied to the nervous system have been reviewed (Trask, 1991; Emson, 1993; Viale and Dell'Orto, 1992; Gerfen *et al.*, 1992; Chronwall *et al.*, 1989; Uhl, 1986; Valentino *et al.*, 1987; Chesselet, 1990; Emson, 1989; Wilson and Higgens, 1990; Hockfield *et al.*, 1993). The mRNA encoding for a variety of neurotransmitter synthetic enzymes has been localized with *in situ* hybridization (Kawata *et al.*, 1991) and, when coupled with immunohistochemistry, can assist in determining whether toxicant-induced alterations in neurotransmitters are the result alterations in synthesis, modification, or degradation.

2. Receptors

The classical approach to mapping the distribution of neurotransmitter receptors is the autoradiographic labeling of ligands. The utility of a ligand is dependent on the pharmacological criteria of specificity, affinity, and saturability as demonstrated both *in vitro* and *in vivo*. Additional methodological variables that influence the resolution of the mapping as well as the reliability of the data include tissue preparation, the isotope, incubation conditions, film or emulsion characteristics, exposure conditions, preparation of standards, and the application of densitometry. The major advantage of receptor mapping with ligands is that it reveals the regional localization of binding sites that have close correspondence to pharmacologically characterized receptors. The major disadvantage is that autoradiographic resolution is generally insufficient for cellular or subcellular localization. These issues are discussed in detail in several monographs and reviews (Kuhar *et al.*, 1986, 1991; Rogers, 1973, 1975; Hockfield *et al.*, 1993; Baker, 1989; Yamamura *et al.*, 1985, 1990; Stewart, 1992; Baskin and Dorsa, 1986; Leslie and Altar, 1988; Frost and Magner, 1990). An alternative to autoradiographic localization is the use of ligands derivatized using fluorescent compounds. These ligands can be viewed directly with fluorescence microscopy immediately after incubation and provide greater detail with regard to cellular localization (Ariano *et al.*, 1989).

The cloning of genes for receptor subunits provides an approach independent of the need for pharmacologically defined ligands. Recent reviews have described various families of receptor subunits for glutamatergic (Hollmann and Heinemann, 1994), dopaminergic (Gringich and Caron, 1993), and cholinergic (Bonner, 1989; Brann *et al.*, 1993; Patrick *et al.*, 1993) receptors. Oligonucleotide probes for *in situ* hybridization and antipeptide antisera have been used to localize the expression and accumulation of specific receptor subunits (Fisher *et al.*, 1994; Aoki *et al.*, 1994; Ariano and Sibley, 1994; Wullner *et al.*, 1994). These methods have the potential to map mRNA and receptor subunits at the cellular and subcellular levels. They do, however,

have limitations. Localization of mRNA is primarily cytoplasmic so that mapping is more practically viewed in terms of the cellular, instead of subcellular, resolution. On the other hand, the receptor subunit can be effectively localized at the subcellular level with ultrastructural immunohistochemistry. Immunohistochemical labeling can encompass the range from newly synthesized protein, its translocation, and assembly into receptors, as well as its association with postsynaptic or presynaptic membranes. In contrast to ligand binding, immunohistochemical labeling does not typically distinguish whether a receptor can bind pharmacologically active compounds. In addition, a variety of factors may regulate the post-transcriptional assembly of subunits into the receptor subtypes that exhibit differential pharmacological sensitivities. Post-translational modifications may also influence the translocation of receptor subunits between different subcellular compartments. The disparity in labeling patterns between receptor ligand binding and immunohistochemical localization may reflect in part the relative importance of these regulatory processes. Alternatively, differences between the labeling patterns produced by ligands and antisera could reflect a differential vulnerability of particular molecular species to modifications during histological processing and the subsequent accessibility of epitopes to antisera or binding sites to ligands. Consequently, the characterization of toxicant-induced changes in the distribution of receptors may be far more compelling when more than one approach is used to map alterations in receptors.

Examples of toxicant-induced alterations in the localization of receptors include changes in the dopaminergic and opiate receptors in the striatum following MPTP (Przedborski *et al.*, 1991; Graybiel *et al.*, 1993), alterations in dopaminergic receptors following 6-hydroxydopamine (Breese *et al.*, 1992, 1994), and alterations in cholinergic receptors following developmental exposure to parathion (Dvergston and Meeker, 1994).

E. Responses of Neural Tissue to Activation and Insult Can Be Useful in Detecting the Vulnerability of Particular Circuits to Specific Toxicants

In contrast to approaches that focus on neurotransmitter-related characteristics of circuits, there are several methods that focus on their more general biochemical properties. Such methods include 2-deoxyglucose labeling, cytochrome oxidase histochemistry, the expression of early intermediate genes, and the immunohistochemical labeling of reactive astrocytes. Although each of these methods uses a different approach to labeling cells that are activated or injured as a result of toxic exposure or insult, the general nature of the phenomena on which they are based is suggestive of a value in assessing chemicals for whose neurotoxic potential is unknown. There are, however, two important limitations with regard to such general applicability. First, these methods address phenomena that play a role in normal as well as pathological processes and, consequently, care must be taken when using them as a basis to infer neurotoxicity. Second, these methods may be optimal for demonstrating particular forms of neurotoxicity, such as excitotoxicity, while refractory to other forms. These limitations may be overcome as their application in a toxicological context is refined. In particular, criteria may be developed to distinguish responses related to pathological alterations from those associated with normal physiological processes. Such limitations notwithstanding, these methods can provide unique information not readily available from other approaches.

The [$^{2-14}$C]deoxyglucose (2DG) method reveals patterns of local glucose metabolism *in vivo* (Sokoloff *et al.*, 1977). Its utility is based on the fact that the brain derives its energy almost exclusively from aerobic catabolism of glucose continuously supplied by cerebral blood flow and that functional activity is directly related to energy consumption. 2DG is injected intravenously, and its uptake by cells and subsequent phosphorylation by hexokinase parallels that of endogenous glucose. The altered structure of the phosphorylated 2DG precludes subsequent modification and its polarity prevents diffusion from the cell, resulting in a half-life of about 8 hr (Sokoloff *et al.*, 1977). The 2DG method has been used to study the organization and development of sensory pathways (for a review see Hand, 1981). Due to the low signal-to-noise ratio resulting from the high basal level of glucose utilization of the brain, sensory stimuli may need to be presented for an extended period of time (e.g., 45 min). The brain is subsequently fixed by perfusion, and cryostat sections are dried onto emulsion-coated slides or onto plain slides upon which film is placed. Exposure periods range from 1 week to 4 months. The intensity of the silver grains reveals the brain regions that actively process the sensory stimulation. Similar approaches have been used to demonstrate alterations in patterns of activity following deafferentation, central lesions, drugs, and toxicants (Hand, 1981; Miller and Dow-Edwards, 1993; Juliano *et al.*, 1990). This method may be particularly valuable in characterizing alteration patterns of brain activity associated with exposure to the wide variety of neurotoxicants known to alter glucose metabolism (Damstra and Bondy, 1982). The extent that the pattern of neuropathological alterations corre-

late with alterations in glucose metabolism may provide clues as to pathogenesis. Alternatively, the 2DG method can be used to demonstrate that functional activity can modify toxicant-induced damage, such as in the case of auditory stimulation influencing the size of dintrobenzene-induced lesions in auditory pathways (Ray *et al.*, 1992).

Histochemistry for mitochondrial enzymes such as cytochrome oxidase and succinate dehydrogenase also deserve mention here because of their value in demonstrating the particular aspects of the organization of neural circuitry. The histochemical demonstrations of these mitochondrial enzymes, and cytochrome oxidase in particular, have been used to reveal unique patterns of "patches" or modules" in numerous brain regions (Wong-Riley, 1989). Perhaps the most striking example of such segmented organization is in the somatosensory system of the rodent where this histochemical staining pattern reveals the somatotopic organization of connections at each level of the synaptic pathway from the periphery to neocortex (Belford and Killackey, 1979, 1980; Killackey *et al.*, 1976). In tangential sections of the neocortex of the rat, a segmented pattern of histochemical staining reveals the region in which there is a somatotopic representation of the body surface (Fig. 5). The most dramatic aspects of this pattern are the rows of patches whose arrangement corresponds to the

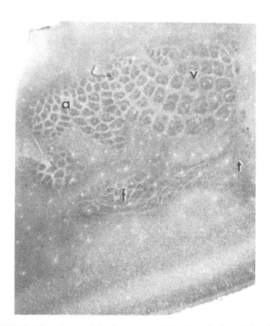

FIGURE 5 Succinate dehydrogenase histochemical staining in a tangential section of rat neocortex. The pattern of staining reveals regional specializations associated with representation of the body surface. Patches in the upper right hand region (v) that are arranged in rows correspond to the arrangement of vibrissae on the face of the rat. Additional patches of staining correspond to the anterior sinus hairs (a), trunk (t), and forelimb (f).

arrangement of the vibrissae on the face and is altered following injury to the sensory periphery during development (Killackey *et al.,* 1976). These alterations correspond to changes in the organization of thalamocortical projections (Jensen and Killackey, 1987a,b). The basis for such staining patterns is likely due to higher sustained levels of functional activity in intensely stained cells. In contrast to 2DG labeling, which labels acute phases of activity, cytochrome oxidase staining appears to reflect long-term changes in activation (Wong-Riley, 1989). This is consistent with the observation that the intensity of histochemical staining closely parallels the enzyme content as determined with immunohistochemistry (Hevner and Wong-Riley, 1989; Karmy *et al.,* 1991). The extent to which histochemistry of mitochondrial enzymes may prove of value in neurotoxicity assessment depends on the extent to which toxicants disrupt functional organization of major pathways. This may be most likely to occur when toxic exposure during development alters ontogenetic events, resulting in aberrantly organized circuitry (Jensen, 1987; O'Kusky, 1992). Noteworthy is that the normal appearance of this pattern has been offered as evidence for the degree to which developmental mechanisms responsible for the patterning of neural circuitry operate normally in a mutant mouse strain that lacks the regulatory subunit of a major postsynaptic protein (Silva *et al.,* 1992).

Early intermediate genes, such as c-*fos* and c-*jun,* can be induced by stimulation and thus have been proposed to be of utility in assessing the functional status of neural circuits (Morgan and Curran, 1991; Sheng and Greenburg, 1990). The products of these genes form heterologous dimers that bind to AP1 sites of late response genes. Many pharmacological agents are capable of inducing c-*fos* expression, and it has been proposed that it be used in the assessment of stimulus–transcription coupling. In a toxicological context, the advantage of such an assay is that no specific information is required regarding the source of activation. Expression of c-*fos* has been proposed as an approach to identify the neuronal populations vulnerable to the actions of a toxicant and has been used to localize injury from ischemia and a variety of toxicants (Sharp *et al.,* 1993; Sharp and Sagar, 1994). Immunohistochemical detection of c-*fos* has also been used to demonstrate a persistent supersensitivity to dopamine agonists in a subpopulation of neurons in the striatum of adult rats neonatally lesioned with 6-hydroxydopamine (Johnson *et al.,* 1992). Thus, it may also be useful in characterizing persistent effects resulting from developmental insult.

Toxicant-induced injury can also be localized with immunohistochemical staining for glial fibrillary acidic protein (GFAP). Astrocytic hypertrophy occurs in response to neural tissue injury and is characterized by the accumulation of glial filaments, the major component of which is GFAP (Eng, 1985, 1988; Norton *et al.,* 1993). Immunohistochemical labeling with antisera against GFAP reveals astrocytes that have complex and intricate processes. The density and morphological characteristics of astrocytes vary dramatically in different brain regions, being high in regions such as the hippocampus and low in regions such as the striatum. Toxicant-induced astrocytic hypertrophy can appear dramatic in immunohistochemically stained sections. For example, in coronal sections of control mouse striatum, very few astrocyes are visible (Figs. 6A and 6C). In a corresponding section from a MPTP-treated mouse, the striatum appears filled with astrocytes that have complex processes (Figs. 6B and 6D). The short survival times in which such dramatic changes can occur suggest that the apparent increase in astrocyte number may not be due to proliferation. Furthermore, the rapidity with which the complexity of the astrocytic processes can become apparent suggests that such a change may not necessarily represent the outgrowth of processes. Instead, there may be rapid transcription and transport of newly synthesized GFAP within astrocytes. This possibility is supported by the observation that mRNA encoding GFAP is spatially dispersed in glial processes (Sarthy *et al.,* 1989). Another possibility is that hypertrophy entails conformational changes in GFAP that dramatically alter the availability of epitopes recognized by the antisera (Norton *et al.,* 1993). The number of visible astrocytes and the complexity of their processes may be related to the extent of neural injury (O'Callaghan and Jensen, 1993). The quantitative assessment of morphological characteristics of hypertrophic astrocytes has not yet been extensively analyzed in a quantitative fashion and correlated with the extent of toxicant-induced injury. But as has been demonstrated with radiometric and ELISA immunoassays for GFAP (O'Callaghan, 1991a,b), quantification of immunohistochemical labeling for GFAP may be useful for both localizing toxicant-induced injury and evaluating the dose dependence of such effects.

III. The Alteration in the Morphology of Individual Neurons Is an Important Aspect of Toxicant-Induced Changes in Neuron Circuitry

Toxicant-induced alterations in the morphology of individual neurons can occur as a consequence of direct insult to the neurons themselves or as the result of

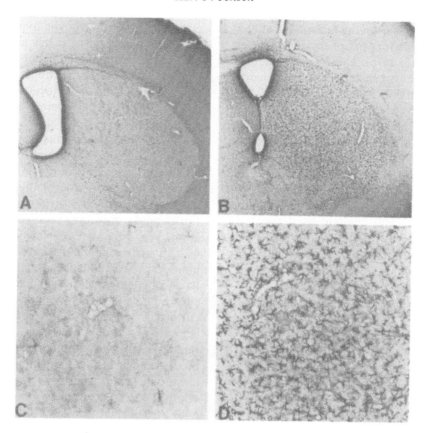

FIGURE 6   Altered immunohistochemical labeling of GFAP in MPTP-treated mice.
(A) In the coronal sections from control mice very few astrocytes are labeled in the
striatum. (C) The few labeled astrocytes visible in control at higher power have relatively
simple processes. (B) In sections from MPTP-treated mice, the striatum is densely
filled with astrocytes. (D) At higher power these astrocytes appear to have a complex
array of processes. ABC method with metal-intensified DAB as the chromogen. (From
O'Callaghan and Jensen 1992.)

transynaptic influences. There are many approaches to revealing the morphological details of individual neurons. Those considered here fall into three general categories: the Golgi method, intracellular injections of tracers, and immunohistochemical labeling of endogenous antigens.

A. The Golgi Method Is a Classic Approach to Studying Dendritic Architecture

The Golgi method visualizes the delicate features of the processes and cell body of neurons. Because only a few neurons stain and very thick sections can be used, the processes of neurons can be accurately traced for long distances. The Golgi method has been most successfully applied to the study of the architecture of dendrites (e.g., Ramón-Moliner and Nauta, 1966; Ramon ý Cajal, 1928; Nauta and Ebbeson, 1970; Sholl, 1956; Conel, 1939; Lund, 1988), but significant contributions with regard to axon morphology have also been made (e.g., Lorente de No, 1949; Valverde, 1986; Marin-Padilla, 1985). The relative merits of specialized variances of the Golgi methods have been discussed in detail in several reviews (Kierman and Berry, 1975; Millhouse, 1991; Heimer and Robards, 1981; Nauta and Ebbesson, 1970; Santini, 1975; Scheibel and Scheibel, 1978; Morest, 1981).

There are several aspects of the results of the Golgi method that are important in the context of detecting toxicant-induced alterations. Perhaps most important is that it is difficult to predict the extent to which specific neurons will consistently stain. Since the mechanism underlying the selectivity of the method remains unknown, there are invariably two possible effects that could account for a toxicant-induced change in the appearance of Golgi-impregnated neurons. The most commonly invoked explanation is that the structure of the neuron under examination is altered. A less widely discussed possibility is that the alteration is not in the structure of the neuron per se, but in the receptivity of that structure to the staining process (Nauta and Ebbesson, 1970; Powell, 1967). The results of a study by Spacek (1992) suggest that close control of the chromanation step may afford greater reliability. Currently the only confirmation available for the former interpretation is to demonstrate the same morphological alterations with alternate techniques. Another consideration is the selectivity of Golgi staining for particular classes of neurons. Less than 10% of total number of neurons in a section are visualized with the Golgi method (Sholl, 1956; Smit and Colon, 1969; Pasternak and Woolsey, 1975; Shimono and Tsuji, 1987). Consequently, if the staining of cells is nonrandom (Ramön-Moliner, 1961), then interpretation of differences between control and experimentally treated animals may be difficult. Several investigations have demonstrated, however, that for at least particular brain regions and variants of the Golgi method, there does not appear to be any selectivity for particular sizes of neurons (Sholl, 1956; Smit and Colon, 1969; Pasternak and Woolsey, 1975; Shimono and Tsuji, 1987).

Despite these considerations, the Golgi method has been supremely important in studies of dendritic architecture. A variety of quantitative approaches have been developed for the analysis of dendritic structure (Capowski, 1989; Shipley *et al.*, 1989; Sholl *et al.*, 1956), and changes in dendritic morphology have been characterized after various kinds of developmental manipulations and insults (for a review see Jacobson, 1991). In this regard, studies of dendritic morphology are probably of greatest value in the area of developmental neurotoxicity. Developmental exposure to toxicants could produce alterations in dendritic morphology in a variety of ways. As in the adult, they could result from a direct action of toxicant on a developing neuron or as the result of transsynaptic effects. Toxic exposure during development could also alter hormonal influences on neural growth and maturation. Toxicants that act at particular neurotransmitter receptors could modify the way extrinsic influences, such as sensory experience, participate in the patterning of dendritic architecture.

B. Neuroanatomical Tracers Can Provide Detailed Images of the Morphology of Axons and Dendrites of Selected Neurons

Another approach to characterizing the morphology of individual neurons is to inject neuroanatomical tracers directly into the neuron either *in vivo* or *in vitro*. *In vivo* "intracellular fills" can be performed in conjunction with intracellular recording by injecting a tracer directly into the neuron with the recording pipette (Kitai and Bishop, 1991; Kater and Nicholson,

1973). Alternatively, larger pipettes can be used to make extracellular injections into pathways. The tracer can be taken up by axons severed in this process and travel to both the cell bodies and terminal axonal fields that are in close proximity to the injection site. Tracers used for these purposes are in most cases the same ones that are used trace pathways (e.g., HRP, WGA-HRP, biotin conjugates, fluorescent dyes) and the methods of visualization are similar (see Section II.C).

An alternative approach to *in vivo* injection of tracers is the use of slices of fresh or lightly fixed brain and injection of the tracer under visual guidance. This approach has the advantage of having a high success rate of labeling neurons within visually selected regions of the slice (Horikawa and Armstrong, 1988; Jacquin *et al.*, 1992; Buhl *et al.*, 1990). Another approach is the injection of carbocyanine dyes (Honig and Hume, 1986) into blocks of fixed brain tissue. This approach appears optimal for studies of the prenatal or early postnatal brain since it allows accurate placement of injections and provides a detailed image of the morphology of developing neurons (Catalano *et al.*, 1991).

The two major caveats that need to be considered in interpreting results obtained from these kind of approaches are sampling bias and extent of labeling. Both these considerations are inherently addressed in the experimental design. The criteria by which neurons are selected as well as the criteria by which a neuron of axon will be considered adequately labeled will determine the generality of inferences that can be made from the data obtained.

An example of anterograde labeling of the local terminal arbors of axons injected *in vivo* with WGA-HRP is shown in Fig. 7. In this coronal section of the somatosensory cortex of the rat, the area of the diffuse label above the white matter corresponds to the anterior aspect of the injection site. Many of the axons extending from the injection site have branches that reach layer IV. The short survival time resulted in a preferential labeling of axons. Intensely labeled neuronal perikarya and dendrites would be visible with longer survival times. These labeled axons have sufficient definition that they can be reconstructed across serial sections. Examples of such reconstructed axons from the region of cortex in which the vibrissae are represented is shown in Fig. 8. The terminal arbor of an axon from an adult control animal exhibits a compact projection area densely filled with branches (Fig. 8A). In contrast, a terminal arbor from an axon from an adult animal in which the trigeminal nerve was cut at birth has a larger projection area and a much lower density of branches (Fig. 8B). Quantitative analysis of

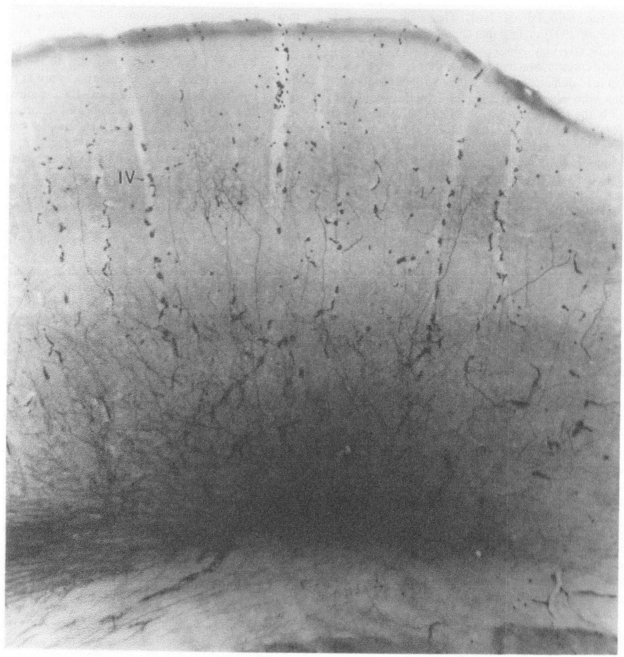

FIGURE 7 Terminal arbors of axons in the somatosensory cortex of the rat anterogradely labeled with HRP. HRP-WGA was injected into internal capsule-labeling individual terminal arbors of axons visible in layer IV. The anterior aspect of the injection site is visible as dark staining at the bottom of the micrograph. Heavy metal-intensified DAB was the chromogen. (From Jensen and Killackey 1987b.)

the characteristics of populations of axons confirmed that terminal arbors from animals with neonatal nerve cuts have a lower branch density that those of corresponding size in control animals (Fig. 9).

The key contribution of this kind of morphological approach is the identification of alterations at the cellu-

lar level that contribute to changes in more complex circuits. In the context of developmental neurotoxicology, this kind of evidence can link biochemical evidence of toxicant-induced changes in mechanisms of neural growth with alterations in the organization of circuitry.

A

B

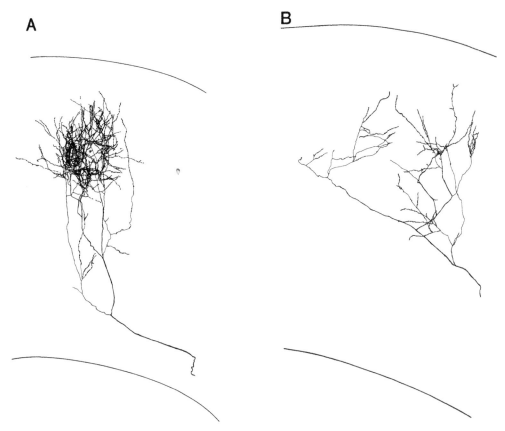

FIGURE 8 Reconstruction of terminal arbors of HRP-labeled axons. (A) Terminal arbor of a thalamocortical axon from the region of the vibrissae representation of the somatosensory cortex of a control adult rat. (B) Terminal arbor of a thalamocortical axon from a rat whose vibrissae were deafferented at birth. The apparent differences in size and density of these terminal arbors can be measured, and the results of such a quantitative analysis are provided in Fig. 10. (Adapted from Jensen and Killackey 1987a,b.)

C. Immunohistochemistry Can Reveal Morphological Details of Neurons That Express Significant Amounts of Specific Antigens

Immunohistochemical approaches can take advantage of endogenous antigens, such as neurotransmitters, to obtain information with regard to morphology of neurons. For example, the detailed morphology of serotonergic axons in the spinal cord of the rat can be clearly delineated with immunohistochemical labeling (Fig. 10). In this case the amount of antigen in the terminals was optimized by treatment of the animals with tryptophan and pargyline prior to perfusion. Since immunohistochemical labeling is related to the amount of antigen present, it is reasonable to expect enhancement of staining for neurotransmitters by pharmacological treatments that increase that amount of transmitter in the structures of interest (Wallace *et al.*, 1982). However, caution should be used in assuming the morphological image reflects

details that are present in the normal state. This is particularly true when investigating the action of toxicants. Alterations in labeling may reflect the influence of the toxicant on the ability of the pretreatments to enhance the morphological image rather than effects on the neural structure itself.

Although it is reasonable to expect pharmacological manipulations known to alter neurotransmitter levels in particular brain regions to have a corresponding influence on immunohistochemical staining, they can also alter the immunohistochemical staining for synthetic enzymes. For example, the immunohistochemical labeling of tyrosine hydroxylase positive elements can be altered by pharmacological agents such as pargyline or γ butryl lactone (Fig. 11). The enhanced images of nigro-striatal terminals provided by these two pharmacological agents are qualitatively different and when the agents are used together their effects appear to be additive. Since these treatments did not increase in amount of tyrosine hydroxylase in the

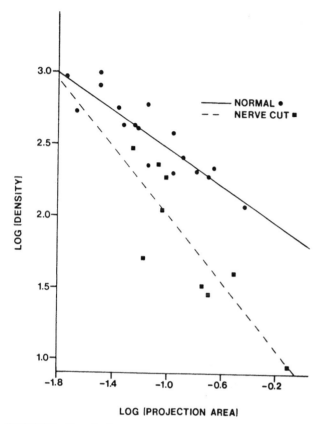

FIGURE 9 Quantitative analysis of reconstructed terminal arbors in the somatosensory cortex of control and neonatally deafferented adult rats. The quantitative analysis revealed a relationship between the size of the projection area and the density of axonal branches in both groups of animals. The density of branches of terminal arbors from neonatally different animals was less than that of control animals for arbors having comparably sized projection areas. (From Jensen and Killackey, 1987b.)

IV. Identifying Toxicant-Induced Alterations in Cellular and Subcellular Localization of Neuronal Components Can Provide Vital Clues about Pathogenesis

One advantage of immunohistochemistry and *in situ* hybridization in mechanistic studies of neurotoxicity is the precision with which the expression and location of specific cellular constituents can be determined. At the light microscopic level, immunohistochemistry can provide a regional assessment of altered cellular components. Such regional assessments are indispensible in characterizing the spatial–temporal pattern of toxicant-induced alterations that are pivotal to unraveling the process of pathogenesis. Ultrastructural immunohistochemistry can provide persuasive confirmation of subcellular location and colocalization inferred from light microscopic or biochemical data.

A. Immunohistochemistry Provides a Means for Regional Assessments of the Integrity of Cellular and Subcellular Components of Neurons Revealing the Sequence of Pathogenesis

Although immunohistochemistry can be used to reveal a wide variety of toxicant alterations in the nervous system, one area that appears to be a particularly productive inroad is the study of the cytoskeleton. Some of the most consistent pathological hallmarks of a variety of neurotoxicants are distinguishing and dramatic cytoskeleton alterations (Spencer and Schaumburg, 1980; Graham *et al.*, 1985). Cytoskeletal regulation is critical to the morphological integrity of a neuron (for a revew see Burgoyne, 1991). Thus, understanding the pathogenesis of cytoskeletal alterations is a key element to revealing how toxicants disrupt neural circuitry.

A number of possible hypotheses address ways in which toxicants disrupt the neuronal cytoskeleton. One possibility is that a toxicant or metabolite disrupts transcription or translation altering the quantity or quality of cytoskeletal elements. A toxicant might also disrupt the post-translational regulation of cytoskeletal elements influencing the transport or conformation. A third possibility is that a toxicant or metabolite directly interacts with a cytoskeletal element and interferes with its stability or interactions with other cytoskeletal elements. These potential mechanisms would presumably result in different types and distributions of cytoskeletal alterations. A toxicant may also simultaneously affect more than one of these processes, particularly at exposure levels that are associated with overt pathology. In such cases, the interaction between

striatum, it is likely that these agents induce conformational changes in the enzyme resulting in increased availability of antigenic epitopes to the antisera (Haykal-Coates *et al.*, 1991). These results indicate that changes in the morphological appearance of neurons can reflect functional as well as structural alterations.

When it is desirable to characterize both the neurotransmitter identity and detailed morphological characteristics of selected neurons, "combined" methods can be used [for a collection of "combined" methods see Heimer and Zaborszky (1989)]. One particular method that provides excellent morphological details of individual neurons is the "Golgi slice" method. This method of obtaining Golgi impregnations within already cut sections is compatible with a variety of tracer and immunohistochemical procedures (Freund and Somogyi, 1989).

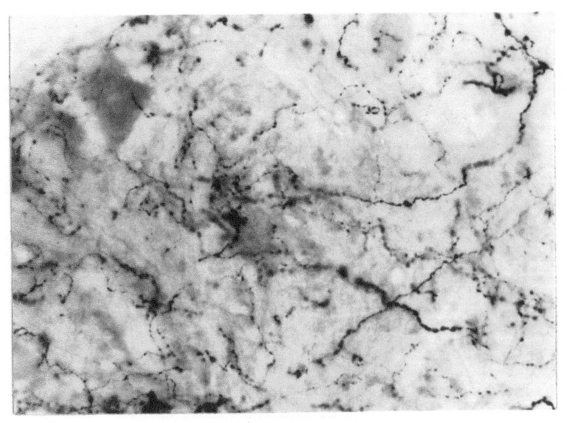

FIGURE 10   Immunohistochemical labeling of the serotonergic axons in the spinal cord of the rat. Animals were pretreated with tryptophan and pargyline. Polyclonal antisera against BSA-conjugated serotonin. The PAP method was used with metal-intensified DAB as the chromogen.

these multiple effects may be responsible for the emergence of unique types of alterations. The temporal–spatial pattern of alterations observable with immunohistochemistry can provide vital evidence to distinguish between various hypotheses of cytoskeletal pathogenesis.

One example in which immunohistochemistry has been valuable in this context is in studies of organophosphorus-induced delayed neuropathy (OPIDN). Abou-Donia and co-workers observed biochemical evidence of elevated CaM II kinase activity after organophosphate exposure and an associated increase in the phosphorylation of cytoskeletal proteins (Lapadula *et al.*, 1991; Abou-Donia and Lapadula, 1990; Abou-Donia *et al.*, 1988). They suggest that excessive phosphorylation of cytoskeletal proteins may be antecedent to the axonal degeneration characteristic of OPIDN. This hypothesis is consistent with observations that excessive phosphorylation of cytoskeletal proteins is common in a number of neuropathic conditions (Saitoh *et al.*, 1991). However, axonal degeneration in OPIDN primarily occurs in the large long axons of peripheral nerves and the spinal cord. Homogenates typically

used in biochemical studies contain both vulnerable and nonvulnerable neural elements. An immunohistochemical approach is uniquely suited to determine whether phosphorylated cytoskeletal proteins are selectively associated with damaged axons (Jensen *et al.*, 1992). As has been demonstrated previously in other species, axons in the hen spinal cord contain an abundance of phosphorylated neurofilaments (Figs. 12C and 12D). In animals sacrificed at 7 days after the administration of the organophosphate TOCP, aberrant aggregations of phosphorylated neurofilaments could be observed in enlarged axons of the spinal cord (Figs. 12A and 12B). Whether such aggregations are an antecedent to degeneration or are characteristic of late degenerative events remains to be determined. Furthermore, biochemical evidence indicates that increases in the phosphorylation of the other cytoskeletal proteins also occur in OPIDN, indicating that a more global dysregulation of phosphorylation may be a determinant of degeneration (Lapadula *et al.*, 1991; Abou-Donia and Lapadula, 1990; Abou-Donia *et al.*, 1988). Nonetheless, this example illustrates how immunohistochemical observations can serve as a unique link be-

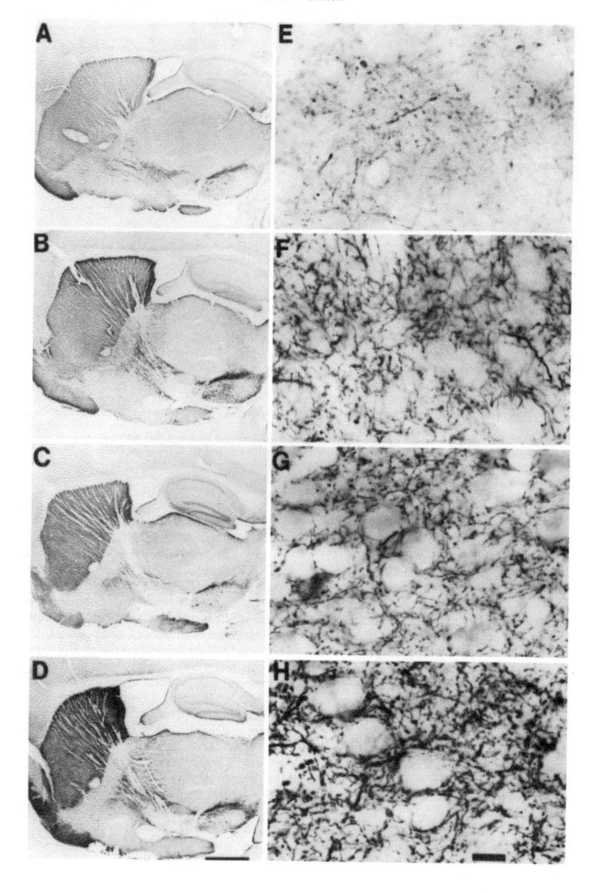

tween a classical neuropathological observation and biochemical data in support of a mechanistic hypothesis.

A more complex situation exists with toxicants such as methyl mercury that affect a variety of cellular functions with convergent deleterious consequences on the nervous system, particularly during development. The pathogenesis of methyl mercury-induced structural alterations is therefore extremely complex and determining the relative contribution of different mechanisms remains a challenge (Chang *et al.*, 1980; Reuhl and Chang, 1979; Chang and Verity, 1995). Even methyl mercury's action on a single cytoskeletal component, such as microtubules, can alter a variety of cellular functions. The binding of methyl mercury to the sulfhydryl groups on a tubulin monomer can inhibit their assembly and, at higher concentrations, result in the disassembly of polymerized microtubules (Reuhl *et al.*, 1994). The extent of a cell's dependence on the assembly of microtubules appears to contribute to its vulnerability. Dividing neuronal precursors, for which tubule polymerization is critical to the regulation of the mitotic spindle, appears the most sensitive. Postmitotic neurons, for which migration is to a lesser degree dependent on tubule polymerization, appear correspondingly less sensitive. Methyl mercury exposure, particularly in *in vitro*, could also disrupt microtubule assembly by effects such as enhanced proteolysis. Reuhl and colleagues (1994) have attempted to address this question using double labeling for α-tubulin and F-actin positive filaments in embryonal carcinoma (EC) cells (Figs. 13 and 14). When EC cells are labeled with a fluorescein-conjugated antisera against α-tubulin, brightly fluorescent microtubules are apparent. When EC are incubate for 2 hr in 3.3×10^{-6} M of methyl mercury, brightly fluorescent microtubules are no longer visible. However, when F-actin is labeled with rhodamine-conjugated phalloidin in the same cells, there is no obvious difference in the appearance of the F-actin-positive filaments between control and methyl mercury-exposed cells. Thus, methyl mercury appears to disrupt microtubules but not F-actin-positive filaments. This observation is consistent with the hypothesis that the loss of the distinct microtubular staining is the result of a selective effect of methyl mercury on microtubule stability instead of an extensive breakdown of the cytoskeleton. This vulnerability of microtubules may be a significant factor in the greater sensitivity of the developing nervous system than that of the adult to methyl mercury.

One fundamental goal of mechanistic hypotheses of pathogenesis is to provide a foundation for relating findings in animals studies to human disease. Perhaps one of the most controversial areas in which cytoskeletal alterations have been in the spotlight is in regards to the suggestion that aluminum may play a role in Alzheimer's dementia (Perl and Brody, 1980). Cytoskeletal pathology can occur when aluminum is introduced into the brains of experimental animals and in dialysis encephalopathy (Yokel *et al.*, 1988; McLachlan and Massiah, 1992). But the cytoskeletal alterations observed in these cases are different from the neurofibrillary tangles associated with Alzheimer's dementia (Yokel *et al.*, 1988). However, hyperphosphorylation of the cytoskeletal protein, τ, can be detected in neurofibrillary tangles and in aluminum-induced cytoskeletal alterations as well as in a number of pathological conditions (Saitoh *et al.*, 1991). Thus, the possibility exists that neurofibrillary tangles are end-stage pathology resulting from a protracted pathogenic process to which aluminum may contribute. Alternatively, aluminum may preferentially accumulate in compromised neurons as a secondary event. More extensive biochemical studies in which the effect of aluminum on the transcription, translation, and post-translational regulation of the cytoskeletal proteins may shed light on the extent to which aluminum accumulation in the brain may contribute to neurological disease. Immunohistochemical approaches may, however, provide the data crucial to sorting out whether earlier events associated with aluminum-induced cytoskeletal alterations are related to the risk of developing Alzheimer's dementia.

In all three of the examples just given of toxicant-induced cytoskeletal pathology, the potential involve-

FIGURE 11 Pharmacological treatments alter immunohistochemical labeling of tyrosine hydroxylase. Photomicrographs are from corrresponding sagittal sections from mice receiving no pretreatments (A,C), pargyline 80 min before perfusion (B,F), γ butryl lactone approximately 20 min before perfusion (C,G), or pargyline and gamma butryl lactone (D,H). The high power photomicrographs (E,F,G,H) reveal that pargyline and γ butryl lactone resulted in qualitatively different images of tyrosine hydroxylase positive axons in the striatum. Pargyline increased the apparent number and intensity of staining of the axons. With γ butryl lactone, axonal staining was also increased but the apparent size of fibers and varicosities was not as great as with pargyline. When these two pretreatments were combined, the intensity of staining was further increased, indicating a possible additivity or synergism between the actions of these two pharmacological pretreatments with respect to increasing the availability of epitopes. The ABC method was used with metal-intensified DAB as the chromogen. (From Haykal-Coates *et al.*, 1991.)

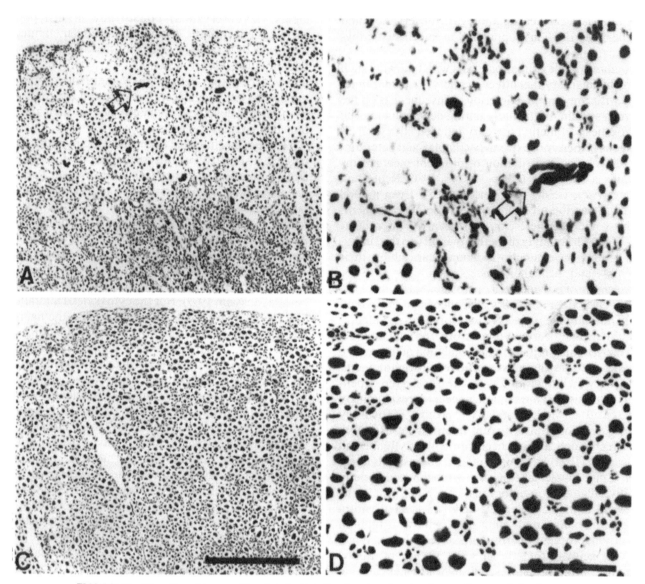

FIGURE 12 Alteration of immunohistochemical labeling for phosphorylated neurofilaments following organophosphate (TOCP) treatment. (A and C) Low power photomicrographs of cross section of hen spinal cord. (A) Enlarged axons with aggregations of phosphorylated neurofilaments including a spiral configuration (arrow) from TOCP treated animal. (C) Corresponding region from control animal. Scale = 200 μm. (B and D) Higher power photomicrographs. (B) Enlargement showing details of aggregations. (D) Photomicrograph of corresponding sections from control animal. Scale = 50 μm. (From Jensen *et al.* 1992.)

ment of the disruption of post-translational regulation of cytoskeletal proteins was a significant consideration. In particular, increased phosphorylation of cytoskeletal proteins is associated with a variety of pathological considerations (Saitoh *et al.,* 1991). The investigations of the potential involvement of kinases and phosphatases in neurotoxicity, however, are still relatively uncharted territory (O'Callaghan, 1994). Consequently, assessing the effect of toxicants on the status of various signal transduction pathways is also an area in which immunohistochemistry can make a substantial contribution. An example of the role of immunohisto-

chemistry in the localization of components of a signal transduction pathway is the characterization of CaM kinase-Gr. This kinase was discovered by screening libraries of rat DNA for a calmodulin-sensitive epitope, and immunoblot analysis of homogenates of various brain regions revealed the highest concentrations in the cerebellum (Ohmstede *et al.,* 1989). Immunohistochemical analysis was able to demonstrate that within the adult cerebellum, the greatest intensity of staining was in the granule cell and the molecular layer (Fig. 15A). Greater magnification reveals that granule cell perikarya but not Purkinje cells are labeled. The molecular layer, but not the medullary layer, is labeled. This is consistent with the presence of granule cell axons in the molecular layer (Fig. 15B). When the antisera used for immunohistochemistry was preadsorbed with the fusion protein, all staining was abrogated, indicating that the pattern of labeling represents the pattern of enzyme-specific antigenic sites (Fig. 15C). The localization of CaM kinase-Gr to the nuclei, axons, dendrites, and perikarya of cerebellar granule cells was confirmed by ultrastructural immunohistochemistry (Jensen *et al.*, 1991a,b).

The correlative biochemical and histological observations present convincing evidence of the regional and cellular distribution of CaM kinase-Gr (Ohmstede *et al.*, 1989). With regard to subcellular localization, however, biochemical evidence suggested that there were higher concentrations of this enzyme in the cytosol than in the nucleus. In contrast, histological and ultrastructural observations suggested a higher concentration in the nucleus than in the cytosol. One possible explanation for this apparent discrepancy is that the cytosolic kinase was in a conformation that may be more readily denatured by fixative than was the nuclear kinase. To address this issue, staining of fixed and unfixed tissue was compared using fluorescein-conjugated secondary antisera as a label (Fig. 16). In unfixed tissue, however, the most robust staining was in the molecular layer, presumably in the axons of granule cells (Fig. 16B). The lack of staining in the granule cell layer was accounted for by the short incubation time (2 hr) used with fresh tissue sections as compared with the longer incubation time (72 hr) required to label fixed sections. In neither case, however, were Purkinje cells observed to be stained. These observations are consistent with the hypothesis that the cytosolic kinase may be more abundant yet more sensitive to denaturation by aldehyde fixatives. The difference between the results obtained with fresh and fixed tissue points to the importance of careful consideration of methodological details in interpreting immunohistochemical results, particularly with regard to subcellular localization. Such considerations may be crucial to localizing elements of signal transduction pathways that may undergo subcellular translocation as part of a pathogenic process (Bronstein *et al.*, 1993).

B. In Situ *Hybridization and Immunohistochemistry Can Be Used to Localize Patterns of Gene Expression That May Underlie the Selective Vulnerability of Particular Neural Populations to Specific Toxicants*

The brain has a large number of diverse transcripts, a large proportion of which are of relatively low abundance, presumably related to a restricted neuroanatomical distribution (Bantle and Hahn, 1976; Milner and Sutcliffe, 1983; Chandi and Hahn, 1983; Sutcliffe, 1988). The extent to which regionally restricted gene expression is related to the selective vulnerability of particular neuronal populations to specific toxicants is not known. However, the cloning of differentially expressed mRNAs (Usui *et al.*, 1994) holds promise for elucidating such potential relationships. For example, Billingsley and co-workers (Krady *et al.*, 1990; Toggas *et al.*, 1992, 1993; also see the chapter by Drs. Billingsley and Toggas in this volume) have used subtractive hybridization to isolate and clone cDNA to mRNAs that were significantly decreased in the brains of rats treated with trimethyltin. *In situ* hybridization with two of the identified clones produced a regional pattern of labeling strikingly similar to the regional pattern of injury revealed by silver degeneration staining in animals treated with trimethyltin. One of the clones encoded for a novel 88 amino acid peptide. Immunohistochemistry with an antisera recognizing this peptide also produced a pattern of staining corresponding to the distribution of neurons sensitive to trimethyltin. Furthermore, transfection of this gene into bacteria conferred a sensitivity to trimethyltin toxicity. It is important to note that these studies have neither defined a specific role for this protein in trimethyltin neurotoxicity nor have they determined what other cellular constituents are necessary for the expression of this toxicity. Nonetheless, these studies do demonstrate an approach that can be used to identify circuits predisposed to injury to a specific toxicant. Subtractive hybridization can also be used to identify genes induced by toxic exposure. Demonstrating the time course of novel gene expression following toxicant exposure may be particularly valuable in characterizing early transcriptional events that occur prior to detectable pathology in cases of delayed neurotoxicity such as that observed with carbon monoxide (Penny and White, 1994) and certain organophosphates (Abou-Donia and Lapadula, 1990; Gupta and Abou-Donia, 1994).

V. Methodological Considerations Are Vital in Interpreting Neuroanatomical Results

Consistent treatment-related morphological alterations are often viewed as hallmarks of neurotoxicity. The interpretation of toxicant-induced morphological alterations, however, involves two challenges. One challenge results from the practical limits of histological sampling and involves the difficulty of detecting

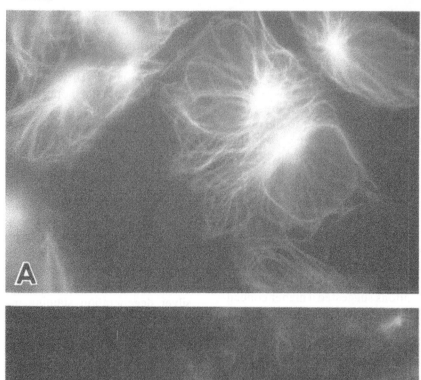

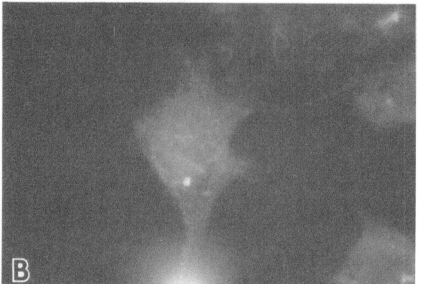

FIGURE 13 The disruption of α-tubulin-labeled elements in cultured embryonal carcinoma cells at interphase by methyl mercury. (A) Control cells in which bright green (white
in the black and white figure) fluorescent microtubules have been labeled with fluorescein-
conjugated antisera against α-tubulin. (B) Cells treated with $3.3 \times 10^{-6}\,M$ of methyl mercury
for 2 hr and stained as in A. Bright green labeling is no longer visible, indicating destruction
of α-tubulin-positive elements. (Courtesy of Prof. Ken Reuhl via Prof. L. W. Chang.)

subtle but important structural alterations. A second
challenge results from the inherent complexity of the
nervous system and involves understanding the significance of observed morphological alterations. The
purpose of this chapter has been to outline how neuroanatomical methods, in general, and a focus on neural
circuitry, in particular, can address these challenges.
The strengths and limitations of neuroanatomical methods in meeting these challenges can be summarized
with regards to three major issues: consistency, sensitivity, and interpretability.

A. The Consistency of Labeling Depends on Identifying Critical Methodical Variables

The fact that the physiochemical basis is not fully
understood for many of the neuroanatomical methods
contributes to their mystique. But repetition and subsequent familiarity result in an appreciation of the variables critical to consistently obtaining reproducible results. The critical interrelationship between procedure
and results was elegantly addressed for Nissl's stain
by Lee (1950).

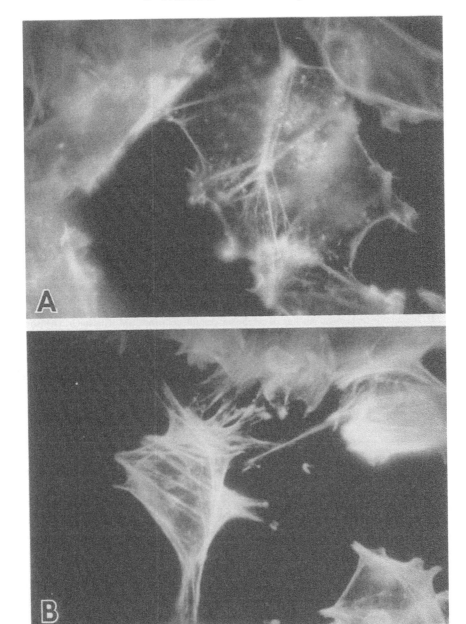

FIGURE 14 The persistence of F-actin-labeled elements in cultured embryonal carcinoma cells in the presence of methyl mercury. (A) Control cells in which bright red (white in the black and white figure) fluorescent F-actin-containing stress fibers have been labeled with rhodamine-conjugated phalloidin. (B) Cells treated with 3.3×10^{-6} M of methyl mercury for 2 hr and stained as in A. Bright red-staining F-actin-containing stress fibers are preserved. These results, together with those presented in Fig. 13, indicate that cytoskeletal alterations induced by methyl mercury may be mediated by a selective effect on microtubules. (Note: Figures 13 and 14 resulted from the "double staining" technique in the same cells.) (Courtesy of Prof. Ken Reuhl via Prof. L. W. Chang.)

"It is now universally admitted that this substance exists in the living cells as a fluid or semi-fluid "plasm rich in nutritive value," and that the blocks, granules, or patches are appearances chiefly due to the coagulation of this plasm, as brought about by the fixing agent employed for their demonstration. As, however, these bodies or granules appear always the same under constant optical condition in healthy cells fixed and stained in a constant manner, they are said to be the equivalent of such healthy cells during life. It follows that if the cells prepared by the same method and examined under the same conditions show a difference from the equivalent or symbol of healthy

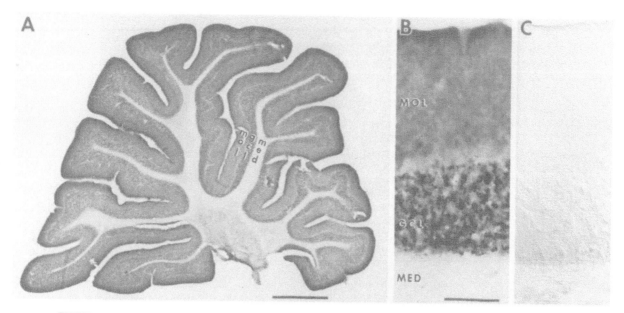

FIGURE 15   Immunohistochemical labeling of cerebellar granule cells for CaM Kinase-Gr. (A) Low power photomicrograph of sagittal section of cerebellum showing intense labeling of molecular and granule cell layers. (B) Higher power photomicrograph showing the intense labeling of granule cell perikarya. The lighter labeling in the molecular layer presumably reflects the staining of parallel fibers, the axons of the granule cells. Staining was minimal in the medullary layer. (C) Section stained with antisera preabsorbed with CaM kinase-Gr protein, indicating that the staining by the antisera was a result of its affinity for the enzyme. (From Jensen *et al.*, 1991b.)

cells the difference is the measure of something that has occurred during life."

This passage articulates the "uncertainty principle" in neuroanatomy—we can only study the brain microscopically after we have altered the very structure that is of interest. The most obvious example is the need for fixation. Fixation renders the tissue resistant to autolysis, providing stability to its structure. Such stability is inherently artifactual and consequently interpretation and conjecture are essential in extrapolating to the living state. The acceptability of interpretation rests on the consistency of the images obtained, which in turn, is dependent on the reliability of method and technique. This issue is not unique to the Nissl stains. It is universally applicable and fundamental to interpreting the results obtained with any of the techniques discussed in this chapter.

Advances in molecular biology have brought powerful new capabilities to neuroanatomical investigations of neurotoxicity. The specificity of a chemical reaction, an antibody, a ligand, or a nucleic acid probe for a particular cellular constituent can seldom be ascertained with complete certainty in histochemical applications. The specificity of such reagents determined *in vitro* cannot be directly extrapolated to the complex biochemical milieu of the histological section. The

demonstration of specificity is a prime concern when such reagents are first used, when they are applied to new species or brain regions, or when they are used with novel procedures. Under these conditions, the most convincing evidence of specificity is when the same pattern of labeling can be demonstrated with different reagents known to recognize the same element *in vitro*. When investigating toxicant-induced alterations with well-characterized reagents in a system in which the pattern of normal labeling is established, such specificity is rarely a primary concern. Instead, the consistency of the pattern of labeling is often the prime determinant in detecting meaningful changes associated with toxicant exposure. Such consistency is ultimately dependent on the identification and control of critical methodological variables. In such cases, routine "methods controls" are essential in eliminating the possibility that interactions between treatment and method could be responsible for an observed alteration in a pattern of labeling.

B. Sensitivity Is a Function of Experimental Design

One might reasonably consider a sequential approach to detecting structural damage would involve

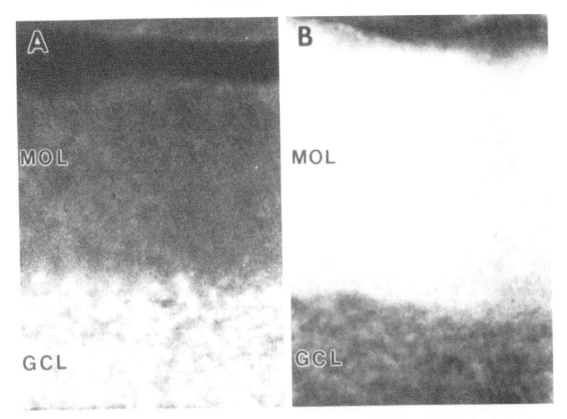

FIGURE 16 Differential patterns of immunohistochemical staining for CaM-kinase Gr in fixed and unfixed sections with a fluorescein label. (A) A fixed sagittal section of cerebellum preserved by perfusion with aldehyde fixative and vibratome sectioned. (B) An unfixed sagittal section of cerebellum, and cryostat sectioned. A dramatically different relative intensity of staining is observed in the molecular and granule cell layers depending on the fixation and processing procedures employed. This difference may reflect the relative amount of the enzyme that exists in different conformational states in different subcellular locations. Such different conformational states may influence the enzyme's susceptibility to denaturation by fixation, which in turn influences the availability of epitopes. (From Jensen *et al.*, 1991b.)

(1) mapping out alterations in major brain regions, (2) identifying alterations in specific neuronal types within those regions, and (3) identifying subcellular alterations related to pathogenesis within these specific neuronal types. Such a tiered approach has an important limitation. In a crude sense, the "sensitivity" of a method tends to be inversely proportional to the level of analysis. Techniques used to map out alterations in major brain regions are less capable of detecting subtle but significant changes than approaches used to detect alterations at the cellular and subcellular levels. Consequently, the process must be considered iterative, with repetitive cycles involving the evaluation of more subtle effects following the evaluation of more gross alterations. How then can one achieve reliable detection of toxicant effects? Basic toxicological principles are essential to this iterative process. Dose–response assessment is critical in terms of the observed location, extent, and type of morphological alterations. Establishing dose-dependence requires precision and accuracy in both qualitative and quantitative characterizations.

Characterizing the time course of the emergence of the morphological alterations is also critical to detection. The inherently static nature of anatomical methods invariably influences how we think about the structure of the nervous system. Little is known about the dynamic nature of the structural changes that occur in response to toxic injury. Many controversies continue over the extent of this response, the variables that influence it, how rapidly such alterations develop, and the manner and extent of repair. As one investigator has suggested, detecting toxicant-induced damage can be like trying to hit a moving target—blindfolded.

Taken together, these considerations suggest that sensitivity should not be thought of as a property of a method or a test, but rather a consequence of experimental design.

C. Interpretability: Neuroanatomical Alterations Are Not Inherently Indicative of Adverse Effects

Neuroanatomical methods can reveal dynamic aspects of nervous system structure that may not necessarily correspond to pathological processes. Furthermore, no one technique can be promoted on its ability to demonstrate the most proximate or earliest alterations induced by a toxicant. Alterations in neural circuitry may be secondary to the toxin-induced neuronal loss or they can occur when a loss of neurons is not detectable at all. Alterations can consist of a massive degeneration of axonal terminals in a major projection system or they can be subtle changes in the subcellular distribution of a single element of a signal transduction pathway. As a consequence of these uncertainties, no one method should be relied upon, to the exclusion of others, as demonstrative of adverse structural alterations. Wherever feasible, attempts should be made to confirm results indicative of an adverse change with alternate techniques. For example, if evidence of toxicant-induced alterations is observed with immunohistochemistry, it should be possible to confirm these results with another method, such a silver degeneration stains. Although confirmation by multiple techniques is not essential to the interpretation of a given experiment, it reduces the ambiguity that is inherent in the use of methods whose basis is not fully understood.

VI. Neuroanatomical Approaches Serve as Important Adjuncts to Neuropathology in Risk Assessment

Overviews of neurotoxicity risk assessment have been developed by the National Academy of Sciences (National Research Council, 1992) and by federal interagency panel (U.S. EPA, 1993). The process involves four steps: hazard identification, dose–response assessment, exposure assessment, and risk characterization. A ''reference dose'' (RfD) approach to risk assessment is widely employed when only minimal information is known concerning the effect of toxicants, as is the case for many potentially neurotoxic agents (Cote *et al.*, 1994; Sette and MacPhail, 1992). The RfD is an estimate that corresponds to an exposure level expected not to be associated with any adverse effect. An important objective of the RfD approach is to make estimates of risk that are sufficiently conservative to balance the degree of uncertainty resulting from the limitations of the data on which they are based.

In the context of risk assessment, neurotoxicity can be considered the disruption of ''. . . neurophysiologi-

cal, neurochemical, or structural integrity of the nervous system or the integration of the nervous system function expressed as modified behavior. . . .'' (U.S. EPA, 1993). Such disruptions are usually assessed by a battery of tests that includes standardized observations resembling a clinical neurological exam, the assessment of spontaneous motor activity, and a neuropathological evaluation. The measure of the intrinsic toxicity of a pollutant used in the RfD approach is commonly the lowest dose at which no adverse affect is observed (NOAEL). The neuropathological evaluation is one of the essential elements in establishing this NOAEL (WHO, 1986; U.S. EPA, 1991). Important technical aspects of the neuropathological evaluation have been discussed in detail elsewhere (Chang 1992a; Jensen, 1994; Spencer and Schaumburg, 1980a; World Health Organization, 1986; Koestner and Norton, 1991; O'Donoghue, 1989; Krinke, 1989; Broxup, 1989).

The process of risk assessment ultimately involves extrapolating from the NOAEL to a level of exposure that minimizes risk to the human population and the environment. Understanding how adverse effects relate to the pathogenesis of human disease is critical to the credibility of the extrapolation process. Structural assessments are unique in their ability to provide information about the precise location and morphological characteristics of toxicant-induced structural alterations. In this regard, neuroanatomical methods can contribute to risk assessment in several ways. They can contribute to hazard identification by expanding and refining categories of pathological alterations associated with toxic insult. In addition, many neuroanatomical approaches are amenable to quantitative analysis and can be useful in establishing dose–response relationships. Furthermore, information on cellular and subcellular localization is valuable in testing hypotheses concerning the cascade of events in pathogenesis, thus contributing to several aspects of the extrapolation process. Perhaps most importantly, assessing the integrity of neural circuitry is of strategic value in developing concordance between structural damage and functional deficits. Such concordance provides a foundation for relating functional deficits observed in animals to human neurological disease.

A. Neuroanatomical Approaches Assist in Developing a More Complete Catalog of Toxicant-Induced Structural Alterations

The description and reporting of toxicant-induced structural alterations observed in the neuropathological evaluation require a reliable and systematic classification of neuropathological changes. Perhaps the most widely adapted classification scheme of toxicant-induced neuropathological alterations was introduced by Spencer and Schaumburg (1980b). This system defines types of neuropathy based on the cellular and subcellular locus of initial injury and has three major categories: neuronopathy, axonopathy, and myelinopathy where primary damage is first detected in the neuronal cell body, the axon, and the myelin sheath, respectively. This classification system is particularly valuable in relating the axonopathies to the emergence of functional deficits. For example, a distal axonopathy may be characterized by degeneration of the long motor and sensory pathways and is associated with a slowly evolving distal motor impairment observable initially in the hind limbs. Such concordant functional and structural alterations provide a convincing picture of the hazard posed by a chemical and contribute substantially to the weight of evidence given results derived from neurotoxicity testing.

One advantage of this classification system is related to the significant number of neurotoxicants whose initial site of injury appears to be in the axon. Such toxicants typically result in a characteristic pattern of neuropathology in peripheral nerves and spinal cord tracts. However, other toxicants can selectively damage neuronal subpopulations of the brain, producing a pattern of damage that does not readily implicate a specific cellular or subcellular site of injury. In such cases, techniques such as immunohistochemistry may reveal alterations in neurotransmitter-related antigens or components of the cytoskeleton that can serve as pathological "hallmarks" of the action of a toxicant.

B. Neuroanatomical Methods Amenable to Quantitative Analysis Can Be Used to Evaluate Dose–Response Relationships

To maximize the probability of detecting toxic-induced adverse effects, toxicity testing can involve a wide range of doses and can include very high exposure levels. A fundamental challenge in risk assessment is determining whether effects observed at elevated exposure levels are of significance for the lower levels of exposure experienced by the public. A number of sophisticated approaches to risk assessment are being developed to address this problem, most notably quantitative biologically based dose–response modeling (Rees and Glowa, 1993; Cote *et al.*, 1994). Another approach is to develop tests that can detect toxicant-induced alterations at lower levels more closely approximating human exposure. The capacity of neuroanatomical approaches to detect subtle structural alterations holds promise for reducing the high level of exposure necessary for detection and thus reducing the need for extrapolation across a wide range of exposure levels.

Many neuroanatomical approaches are amenable to quantitative analysis and can therefore be used in the evaluation of dose–response relationships. Specific approaches can be designed for the kind of alterations that are of interest. Some examples include gross measurements to ascertain growth reductions, counting of neurons to establish patterns of cell loss, measuring length and branch angles of neuronal processes, volume determinations of tissue exhibiting evidence of silver degeneration staining, and areal measurements of regions of sections exhibiting immunohistochemical labeling.

C. Neuroanatomical Approaches Can Be Used to Test Specific Hypotheses Concerning the Pathogenesis of Neurotoxicity

Neuroanatomical approaches are particularly valuable in characterizing regional, cellular, and subcellular alterations that can provide critical information in formulating and testing hypotheses concerning pathogenesis. Several of the approaches discussed are uniquely suited to characterizing the time course and evolution of toxicant-induced structural damage. For example, silver degeneration stains can reveal signs of acute damage within hours of exposure and thus can be used to characterize the time course of development of pathological abnormalities. Likewise, immunohistochemical approaches, such as those used to label cytoskeletal proteins, can reveal novel alterations that, while not typical of end stage pathology, may be important to understanding pathogenesis (Jensen *et al.*, 1992).

The use of neuroanatomical approaches to characterize the integrity of neural circuitry can also be used in investigating extrinsic factors that influence pathogenesis. For example, toxicant-induced damage can be influenced by the state of activation of a circuit, as in the case of where the size of dintrobenzene-induced lesions in auditory pathways are modulated by afferent stimulation (Ray *et al.*, 1992). This observation demonstrates the insight that can be gained when the pattern of damage induced by a toxicant is interpreted in the context of the selective vulnerability of a circuit instead of in the selective vulnerability of a specific type of neuron.

D. Neuroanatomical Information Can Have Strategic Value in Developing Concordance between Toxicant-Induced Structural and Functional Alterations

Basic neuroscience has yet to establish universal principles by which inference to humans can be drawn from pathological changes in animals. Integrative assessments, however, can relate damage to specific brain regions to particular functional deficits. Such concordance between structural and functional observations is important for weight of evidence considerations in extrapolating observations of animals to the human situation. The adaptability of the nervous system can mask the functional consequences of structural damage. Presumably this adaptability arises from the complex interwoven patterns of neural connections. The distributed nature of this connectivity may preserve function even as connections are altered or lost in toxicant-induced injury. Unfortunately, neuroanatomical characterization of the counterparts of neurobehavioral deficits affecting learning and memory remains sufficiently experimental as to preclude their undertaking in routine testing. Experimental investigations using sophisticated structural and functional approaches can, however, characterize the actions of neurotoxicants in sufficient detail to develop animal models of human neurological disease. Such models can then be used to identify potential pathological "hallmarks" that may be related to identified functional impairments. These pathological "hallmarks" can then be of use in more standard neuropathological evaluations.

An example of one such animal model is MPTP neurotoxicity. MPTP produces Parkinson-like effects in humans and nonhuman primates, with damage to the nigrostriatal pathway and a substantial loss of neurons in substantia nigra (Irwin *et al.*, 1990; Kopin and Markey, 1988; Langston, 1987; Kopin, 1987; Burns *et al.*, 1983). MPTP produces parallel but less dramatic effects in the nigrostriatal pathway of the mouse (Heikkila and Sonsalla, 1987; Reinhard and Nichol, 1986; Heikkila *et al.*, 1984; Hess *et al.*, 1990. A proposed mechanism by which MPTP damages the nigrostriatal pathway is by the selective uptake of the MPTP's metabolite, MPP+. MPP+ inhibits NADH-linked oxidation at the level of complex I, disrupting mitochondrial respiration. Since complex I is also vulnerable to disruption by certain pesticides, these findings are consistent with hypotheses by which environmental influences may be a factor in the development of Parkinson's disease (Semchuk *et al.*, 1993, 1994; Koller *et al.*, 1990). Even though the mouse does not exhibit the motor impairments characteristic of Parkinson's disease, damage to the nigrostriatal pathway appears to occur by the same mechanism that operates in primates. Thus, by focusing on parallels in neural circuitry that may exist between species, it may be possible to develop a more realistic approach to cross-species extrapolation for risk assessment.

VII. Conclusion

This chapter's goal has been to demonstrate how neuroanatomical approaches, when applied in conjunction with basic toxicological considerations, can help reveal how toxicants alter the structure of the nervous system. Morphological approaches are unique in their capacity to provide detailed information about the location and extent of toxicant-induced damage. This kind of information is critical for recognizing the significance of structural alterations in pathogenesis. Furthermore, by interpreting toxicant-induced alterations in terms of their impact on neural circuitry, an important link can be established between neuropathology and functional deficits. Such concordance between structural and functional alterations is critical to improving the credibility of the extrapolation process that is the foundation of neurotoxicity risk assessment.

This chapter has assumed that neural circuits can be delineated by their morphological characteristics and exist as concrete entities. This assumption has been questioned. For example, Isaacson (1992) has asserted that the boundaries of the limbic system are not distinct and the involvement of particular structures is dynamic instead of concrete. Isaacson suggests that by applying "fuzzy logic," a mathematical approach that can take into account the inherent uncertainties of biological systems (Klir and Folger, 1988), we can more accurately describe the limbic system. From a different perspective, Bullock (1993) proposes that the integrative capacity of neurons and their interactions exceed what is encompassed by the currently accepted concept of a neural circuit. The questions raised by these two eminent neuroscientists may help to heighten our awareness of the dynamic nature of brain structure and, in so doing, emphasize the formidable nature of the challenge that exists in trying to relate the functional and structural alterations induced by neurotoxicants. Bullock concludes his essay, "I, for one, believe that the brain still conceals a host of unrecognized qualitative and quantitative traits, including basic principles, still awaiting description and, only then, analysis." If Bullock's belief is correct, then descriptive morphology will continue to play a pivotal role in basic neuroscience and in neurotoxicology, even as we expand and refine our concept of a neural circuit.

Acknowledgments

The author extends special thanks to Dr. Linda Ide and to Dr. John March. This manuscript has been reviewed by the Health Effects Research Laboratory in the Office of Research and Development of the U.S. Environmental Protection Agency and approved for publication. Approval does not signify that the contents necessarily reflect the views and policies of the Agency nor does mention of trade names or commercial products constitute endorsement or recommendation for use.

References

Abou-Donia, M. B., Lapadula, D. M., and Suwita, E. (1988). Cytoskeletal proteins as targets for organophosphorus compound and aliphatic hexacarbon-induced neurotoxicity. *Toxicology* **49**, 469–477.

Abou-Donia, M. B., and Lapadula, D. M. (1990). Mechanisms of organophosphorus ester-induced delayed neurotoxicity: type I and type II. *Annu. Rev. Pharmacol. Toxicol.* **30**, 405–440.

Adams, J. H., and Duchen, L. W. (1992). *Greenfield's Neuropathology*, 5th ed., Oxford University Press, New York.

Adams, R. D., and Lee, J. C. (1982). Neurons and neuronal reactions to disease states. In *Histology and Histopathology of the Nervous System* (W. Haymaker and R. D. Adams, Eds.), pp. 174–275, Thomas, Springfield, Il.

Albrecht, M. H., and Fernstrom, R. C. (1959). A modified Nauta-Gygax method for human brain and spinal cord. *Stain Technol.* **34**, 91–94.

Aoki, C., Go, C.-G., Venkatesan, C., and Kurose, H. (1994). Perikaryal and synaptic localization of alpha-2a adrenergic receptor-like immunoreactivity. *Brain Res.* **650**, 181–204.

Ariano, M. A., Monsma, F. J., Jr., Barton, A. C., Kang, H. C., Haugland, R. P., and Sibley, D. R. (1989). Direct visualization and cellular localization of D₁ and D₂ dopamine receptors in rat forebrain by use of fluorescent ligands. *Proc. Natl. Acad. Sci. USA* **86**, 8570–8574.

Ariano, M. A., and Sibley, D. R. (1994). Dopamine receptor distribution in the rat CNS: elucidation using anti-peptide antisera directed against D1A and D3 subtypes. *Brain Res.* **649**, 95–110.

Armed Forces Institute of Pathology, (1968). *Manual of Histologic Staining Methods*. McGraw Hill, New York.

Baker, J. R. (1989). *Autoradiography: A Comprehensive Overview.* Oxford University Press, New York.

Balaban, C. D. (1985). Central neurotoxic effects of intraperitoneally administered 3-acetylpyridine, harmaline and niacinamide in Sprague-Dawley and Long-Evans rats: A critical review of central 3-acetylpyridine neurotoxicity. *Brain Res.* **9**, 21–42.

Balaban, C. D., O'Callaghan, J. P., and Billingsley, M. L. (1988). Trimethyltin-induced neuronal damage in the rat brain: comparative studies using silver degeneration stains, immunocytochemistry and immunoassay for neuronotypic and gliotypic proteins. *Neuroscience* **26**, 337–361.

Balaban, C. D. (1992). The use of selective silver degeneration stains in neurotoxicology: Lessons from studies of selective neurotoxicants. In *Vulnerable Brain and Environmental Risks*, Vol. 1, *Malnutrition and Hazard Assessment* (R. L. Isaacson and K. F. Jensen, Eds.), pp. 223–238. Plenum, New York.

Balercia, G., Chen, S., and Bentivoglio, M. (1992). Electron microscopic analysis of fluorescent neuronal labeling after photoconversion. *J. Neurosci. Methods* **45**, 87–89.

Bantle, J. A., and Hahn, W. E. (1976). Complexity and characterization of polyadenylated RNA in the mouse brain. *Cell* **8**, 139–150.

Barone, S. J., Bonner, M., Tandon, P., McGinty, J. F., and Tilson, H. A. (1992). The neurobiological effects of colchicine: modulation by nerve growth factor. *Brain Res. Bull.* **28**, 265–274.

Baskin, D. G., and Dorsa, D. M. (1986). Quantitative autoradiography and in vitro radioligand binding. *Exp. Biol. Med.* **11**, 204–234.

Beatty, R. M., Sadun, A. A., Smith, L., Vonsattel, J. P., and Richardson, E. P. (1982). Direct demonstration of transsynaptic degeneration in the human visual system: a comparison of retrograde and anterograde changes. *J. Neurol. Neurosurg. Psychiatry* **45**, 143–146.

Belford, G., and Killackey, H. P. (1979). The development of vibrissae representation in subcortical trigeminal centers of the neonatal rat. *J. Comp. Neurol.* **188**(1), 63–73.

Belford, G., and Killackey, H. P. (1980). The sensitive period in the development of the trigeminal system of the neonatal rat. *J. Comp. Neurol.* **193**, 335–350.

Beltramino, C. A., de Olmos, J. S., Gallyas, F., Heimer, L., and Zaborszky, L. (1993). Silver staining as a tool for neurotoxic assessment. *NIDA Res. Monogr.* **136**, 101–132.

Bielschowsky, M. (1935). Allgemeine Histologie und Histopathologie des Nervensystem. In *Handbuch der Neurologie*, (O. Bumke and O. Foerster, Eds.), Vol. I, pp. 35–226, Springer-Verlag, Berlin.

Björklund, A., and Hokfelt, T. (1983). *Handbook of Chemical Neuroanatomy* Vol. 1, *Methods in Chemical Neuroanatomy*, Elsevier, New York.

Blanks, J. C., Hinton, D. R., Sadun, A. A., and Miller, C. A. (1989). Retinal ganglion cell degeneration in Alzheimer's disease. *Brain Res.* **501**, 364–372.

Blum, J. J., and Reed, M. C. (1994). Models of axonal transport: application to understanding certain neuropathies. In *Principles of Neurotoxicology* (L. W. Chang, Ed.), pp. 113–134. Dekker, New York.

Bodian, D. (1936). A new method of staining nerve fibers and nerve endings mounted in paraffin sections. *Anat. Rec.* **65**, 89–97.

Bolam, J. P. (1992). *Experimental Neuroanatomy. A Practical Approach*. Oxford University Press, New York.

Bonner, T. I. (1989). The molecular basis of muscarinic receptor diversity. *Trends Neurosci.* **12**, 148–151.

Brann, M. R., Ellis, J., Jorgensen, H., Hill-Eubanks, D., and Jones, S. V. P. (1993). Muscarinic acetycholine receptor subtypes: localization and structure/function. *Prog. Brain Res.* **98**, 121–126.

Breese, G. R., Criswell, H. E., Duncan, G. E., Johnson, K. B., Simson, P. E., Mueller, R. A., Jensen, K. F., and O'Callaghan, J. P. (1992). Investigations of adaptive changes associated with lesioning dopaminergic neurons with 6-hydroxydopamine. In *Progress in Parkinson's Disease Research* (F. Hefti and W. J. Weiner, Eds.), pp. 259–274, Futura Publishing, Mount Kisco, New York.

Breese, G. R., Criswell, H. E., Johnson, K. B., O'Callaghan, J. P., Duncan, G. E., Jensen, K. F., Simson, P. E., and Mueller, R. A. (1994). Neonatal destruction of dopaminergic neurons. *Neurotoxicology* **15**, 149–160.

Brodal, A. (1969). *Neurological Anatomy In Relation to Clinical Medicine*, 2nd Ed. Oxford University Press, New York.

Brodal, A. (1982). Anterograde and retrograde degeneration of the nerve cells in the central nervous system. In *Histology and Histopathology of the Nervous System* (W. Haymaker and R. D. Adams, Eds.), pp. 276–362, Thomas, Springfield, IL.

Bronstein, J. M., Farber, D. B., and Wasterlain, C. G. (1993). Regulation of type II calmodulin kinase: functional implications. *Brain Res. Rev.* **18**, 135–147.

Broxup, B. (1989). Neuropathology as a screen for neurotoxicity assessment. *J. Am. Coll. Toxicol.* **8**, 689–695.

Buhl, E. H., Schwerdtfeger, W. K., and Germroth, P. (1990). Intracellular injection of neurons in fixed brain tissue combined with other neuroanatomical techniques at the light and electron microscopic level. In *Handbook of Chemical Neuroanatomy* Vol. 8, *Analysis of Neuronal Microcircuits and Synaptic Connections* (A. Borklund, T. Höckfelt, F. G. Wouterland, and A. N. van der Pol, Eds.), pp. 273–304, Elsevier, New York.

Bullock, T. H. (1993). Integrative systems research on the brain: resurgence and new opportunities. *Annu. Rev. Neurosci.* **16**, 1–15.

Burgoyne, R. D. (1991). *The Neuronal Cytoskeleton.* Wiley-Liss, New York.

Burns, R. S., Chiuch, C. C., Markey, S. P., Ebert, M. H., Jacobowitz, D. M., and Kopin, I. J. (1983). The primate model of parkinsonism: selective destruction of dopaminergic neurons in the pars compacta of the substantia nigra by N-methyl-4-phenyl-1,2,3,6-tetrahydropyridine. *Proc. Natl. Acad. Sci. U.S.A.* **80,** 4546–4550.

Capowski, J. (1989). *Computer Techniques in Neuroanatomy.* Plenum, New York.

Carlsen, J., and de Olmos, J. S. (1981). A modified cupric silver technique for the impregnation of degenerating neurons and their processes. *Brain Res.* **208,** 426–431.

Catalano, S. M., Robertson, R. T., and Killackey, H. P. (1991). Early ingrowth of thalamocortical afferents to the neocortex of the prenatal rat. *Proc. Natl. Acad. Sci. U.S.A.* **88,** 2999–3003.

Cavanagh, J. B. (1954). The toxic effects of tri-ortho-cresyl phosphate on the nervous system. An experimental study in hens. *J. Neurol. Neurosurg. Psychiatry* **17,** 163–172.

Cavanagh, J. B. (1964). Peripheral nerve changes in orthro-cresyl phosphate poisoning in the cat. *J. Path. Bact.* **87,** 365–383.

Cavanagh, J. B. (1982). The pattern of recovery of axons in the nervous system of rats following 2,5-hexanediol intoxication: a question of rheology? *Neuropathol. Appl. Neurobiol.* **8,** 19–34.

Chan, J., and Aoki, C., and Pickel, V. M. (1990). Optimization of differential immunogold-silver and peroxidase labeling with maintenance of ultrasound in brain sections before plastic embedding. *J. Neurosci. Methods* **33,** 113–127.

Chandi, N., and Hahn, W. E. (1983). Genetic expression in the developing mouse brain. *Science* **220,** 924–928.

Chang, L. W. (1979). *A Color Atlas and Manual for Applied Histochemistry.* Thomas, Springfield, IL.

Chang, L. W., Wade, P. R., Pounds, J. G., and Reuhl, K. R. (1980). Prenatal and neonatal toxicology and pathology of heavy metals. *Adv. Pharmacol. Chemother.* **17,** 195–231.

Chang, L. W. (1992a). Basic histopathological alterations in the central and peripheral nervous system: classification, identification, approaches, and techniques. In *Neurotoxicology* (M. B. Abou-Donia, Ed.), pp. 223–252, CRC Press, Boca Raton, Florida.

Chang, L. W. (1992b). The concept of direct and indirect neurotoxicity and the concept of toxic metal/essential element interactions as a common biomechanism underlying metal toxicity. In *The Vulnerable Brain and Environmental Risks,* Vol. 2, *Toxins in Food* (R. L. Isaacson and K. F. Jensen, Eds.), pp. 61–82, Plenum, New York.

Chang, L. W., and Verity, M. A. (1995). Mercury neurotoxicity: effects and mechanisms. In *Handbook of Neurotoxicology* (L. W. Chang, and R. S. Dyer Eds.), pp. 31–59, Dekker, New York.

Chesselet, M. (1990). *In Situ Hybridization Histochemistry,* CRC Press, Boca Raton, Florida.

Chronwall, B. M., Lewis, M. E., Schwaber, J. S., and O'Donohue, T. L. (1989). In situ hybridization combined with retrograde fluorescent tract tracing. In *Neuroanatomical Tract-Tracing Methods* (L. Heimer and L. Zaborszky, Eds.), pp. 265–310, Plenum, New York.

Commins, D. L., Vosmer, G., Virus, R. M., Woolverton, W. L., Schuster, C. R., and Seiden, L. S. (1987). Biochemical and histological evidence that methylenedioxymethylamphetamine (MDMA) is toxic to neurons in the rat brain. *J. Pharmacol. Exp. Ther.* **241,** 338–345.

Conel, J. L. (1939). *The Postnatal Developmental of the Human Cerebral Cortex. Six Volumes,* Harvard University Press, Cambridge.

Cote, I. L., Vandenberg, J. J., and Hassett-Sipple, B. M. (1994). An introduction to the principles and methods of risk assessment.

In *The Vulnerable Brain and Environmental Risks,* Vol. 3, *Toxins in Air and Water* (R. L. Isaacson and K. F. Jensen, Eds.), pp. 231–246, Plenum, New York.

Cowan, W. M., Gottlieb, D. I., Hendrickson, A., Price, J. L., and Woolsey, T. A. (1972). The autoradiographic demonstration of axonal connections in the central nervous system. *Brain Res.* **37,** 21–51.

Cowan, W. M., and Cuenod, M. (1975). *The Use of Axonal Transport for Studies of Neuronal Connectivity.* Elsevier, New York.

Damstra, T., and Bondy, S. C. (1982). Neurochemical approaches to the detection of neurotoxicity. In *Nervous System Toxicology* (C. L. Mitchell, Ed.), pp. 349–374, Raven Press, New York.

de Olmos, J. S. (1969). A cupric-silver method for impregnation of terminal axon degeneration and its further use in staining granular argyrophilic neurons. *Brain Behav. Evol.* **2,** 210–237.

de Olmos, J. S., Ebbesson, S. O. E., and Heimer, L. (1991). Silver methods for the impregnation of degenerating axoplasm. In *Neuroanatomical Tract-Tracing Methods* (L. Heimer and M. J. RoBards, Eds.), pp. 117–170, Plenum Press, New York.

de Olmos, J. S., Beltramino, C. A., and de Olmos de Lorenzo, S. (1994). The use of an amino-cupric-silver technique for the detection of early and semiacute neuronal degeneration caused by neurotoxicants and hypoxia. *Neurotoxicol. Teratol.,* **16,** 545–561.

de Olmos, J. S., and Ingram, W. R. (1971). An improved cupric-silver method for impregnation of axonal and terminal degeneration. *Brain Res.* **33,** 523–529.

Desclin, J. C., and Escubi, J. (1974). Effects of 3-acetylpyridine on the central nervous system of the rat, as demonstrated by silver methods. *Brain Res.* **77,** 349–364.

Duchen, L. W. (1992). General pathology of neurons and neuroglia. In *Greenfield's Neuropathology* (L. W. Duchen and J. H. Adams, Eds.), 5th ed., pp. 1–68, Oxford University Press, New York.

Dvergston, C. L., and Meeker, R. B. (1994). Muscarinic cholinergic receptor regulation and acetylcholinesterase inhibition in response to chronic insecticide exposure during development. *Int. J. Dev. Neurosci.* **12,** 63–75.

Edwards, S. B., and Hendrickson, A. (1991). The autoradiographic tracing of axonal connection in the central nervous system. In *Neuroanatomical Tract-Tracing Methods* (L. Heimer and M. J. RoBards, Eds.), pp. 171–206, Plenum Press, New York.

Ellison, G., and Switzer III, R. C. (1993). Dissimilar patterns of degeneration in brain following four different addictive stimulants. *Neuroreport* **5,** 17–20.

Emson, P. C. (1989). Hybridization histochemistry or in situ hybridization in the study of neuronal gene expression. *Comp. Biochem. Physiol.* **93,** 233–239.

Emson, P. C. (1993). In-situ hybridization as a methodological tool for the neuroscientist. *Trends Neurosci.* **16,** 9–16.

Eng, L. F. (1985). Glial fibrillary acidic protein (GFAP): the major protein of glial intermediate filaments in differentiated astrocytes. *J. Neuroimmunol.* **8,** 203–214.

Eng, L. F. (1988). Regulation of glial intermediate filaments in astrogliosis. In *Biochemical Pathology of Astrocytes* (M. D. Norenberg, L. Hertz, and A. Schousboe, Eds.), pp. 79–90, Liss, New York.

Fink, R. P., and Heimer, L. (1967). Two methods for selective silver impregnation of degenerating axons and their synaptic endings in the central nervous system. *Brain Res.* **4,** 369–374.

Fisher, R. S., Levine, M. S., Sibley, D. R., and Ariano, M. A. (1994). D2 dopamine receptor protein location: Golgi impregnation-gold toned and ultrastructural analysis of the rat neostriatum. *J. Neurosci. Res.* **38,** 551–564.

Fox, E. A., and Powley, T. L. (1989). False-positive artifacts of tracer strategies distort autonomic connectivity maps. *Brain Res. Rev.* **14,** 53–77.

Freund, T. F., and Somogyi, P. (1989). Synaptic relationships of Golgi-impregnated neurons as identified by electrophysiological or immunocytochemical techniques. In *Neuroanatomical Tract-Tracing Methods 2* (L. Heimer and L. Zaborszky, Eds.), pp. 201–264, Plenum, New York.

Frith, C. H., Chang, L. W., Lattin, D. L., Walls, R. C., Hamm, J., and Doblin, R. (1987). Toxicity of methylenedioxymethamphetamine (MDMA) in the dog and the rat. *Fundam. Appl. Toxicol.* **9**, 110–119.

Frost, J. J., and Magner Jr., H. N. (1990). *Quantitative Imaging.* Raven Press, New York.

Fuxe, K., Aganti, L. F., Hokfelt, T., Calza, L., Benfenati, F., Mascagni, F., and Goldstein, M. (1985). Immunocytochemistry of central neurons. In *Central Nervous System Plasticity and Repair* (A. Bigname, F. E. Bloom, C. L. Bolis, and A. Adeloye, Eds.), pp. 47–56, Raven Press, New York.

Gallyas, F. (1971a). A principle for silver staining of tissue elements by physical development. *Acta Morphol. Acad. Sci. Hung.* **19**, 57–71.

Gallyas, F. (1971b). Silver staining of Alzheimer's neurofibrillary changes by means of physical development. *Acta Morphol. Acad. Sci. Hung.* **19**, 1–8.

Gallyas, F. (1979a). Light insensitive physical developers. *Stain Technol.* **54**, 173–176.

Gallyas, F. (1979b). Kinetics of formation of metallic silver and binding of silver ions by tissue components. *Histochemistry* **64**, 87–96.

Gallyas, F. (1980a). Chemical nature of the first products (nuclei) of the argyrophil staining. *Acta Histochem.* **67**, 145–158.

Gallyas, F. (1980b). Determination of the development time for the characterization of the nucleus formation in the argyrophil stainings. *Acta Histochem.* **67**, 1–5.

Gallyas, F., Wolff, J. R., Böttcher, H., and Zaborszky, L. (1980a). A reliable and sensitive method to localize terminal degeneration and lysosomes in the central nervous system. *Stain Technol.* **55**, 299–306.

Gallyas, F., Wolff, J. R., Böttcher, H., and Zaborszky, L. (1980b). A reliable method for demonstrating axonal degeneration shortly after axotomy. *Stain Technol.* **55**, 291–297.

Gallyas, F., Zaborszky, L., and Wolff, J. R. (1980c). Experimental studies of mechanisms involved in methods demonstrating axonal and terminal degeneration. *Stain Technol.* **55**, 281–290.

Gallyas, F., Güldner, F. H., Zoltay, G., and Wolff, J. R. (1990). Golgi-like demonstration of "dark" neurons with an argyrophil iii method for experimental neuropathology. *Acta Neuropathol. (Berl.)* **79**, 620–628.

Gallyas, F., Zoltay, G., and Dames, W. (1992a). Formation of "dark" (argyrophilic) neurons of various origin proceeds with a common mechanism of biophysical nature (a novel hypothesis). *Acta Neuropathol. (Berl.)* **83**, 504–509.

Gallyas, F., Zoltay, G., and Horvath, Z. (1992b). Light microscopic response of neuronal somata, dendrites and axons to post-mortem concussive head injury. *Acta Neuropathol. (Berl.)* **83**, 499–503.

Gallyas, F., and Wolff, J. R. (1986). Metal-catalyzed oxidation renders silver intensification selective. Applications for the histochemistry of diaminobenzidine and neurofibrillary changes. *J. Histochem. Cytochem.* **34**, 1667–1672.

Gallyas, F., and Zoltay, G. (1992). An immediate light microscopic response of neuronal somata, dendrites and axons to non-contusing concussive head injury in the rat. *Acta Neuropathol. (Berl.)* **83**, 386–393.

Gerfen, C. R., Sawchenko, P. E., and Carlsen, J. (1989). The PHA-L anterograde axonal tracing method. In *Neuroanatomical Tract-Tracing Methods 2* (L. Heimer and L. Zaborszky, Eds.), pp. 19–48, Plenum, New York.

Gerfen, C. R., Young III, W. S., and Emson, P. C. (1992). In situ hybridization histochemistry with oligonucleotides for localization of messenger RNA in neurons. In *Experimental Neuroanatomy* (J. P. Bolam, Ed.), pp. 173–196, Oxford University Press, New York.

Gerfen, C. R., and Sawchenko, P. E. (1984). An anterograde neuroanatomical tracing method that shows the detailed morphology of neurons, their axons and terminals: immunohistochemical localization of an axonally transported plant lectio, *Phaseolus vulgaris*-leucoagglutinin (PHA-L). *Brain Res.* **290**, 219–238.

Gerfen, C. R., and Sawchenko, P. E. (1985). A method for anterograde axonal tracing chemically specified circuits in the central nervous system: combined *Phaseolus vulgaris*-leucoagglutinin (PHA-L) tract-tracing with immunohistochemistry. *Brain Res.* **343**, 144–150.

Glees, P., and Jansik, H. (1965). Chemically (TCP) induced fiber degeneration in the central nervous system, with reference to clinical and neuropharmacological aspects. *Prog. Brain Res.* **14**, 97–121.

Grafe, M. R., and Leonard, C. M. (1980). Successful silver impregnation of degenerating axons after long survivals in the human brain. *J. Neuropathol. Exp. Neurol.* **39**, 555–574.

Grafstein, B. (1975). The nerve cell body response to axotomy. *Exp. Neurol.* **48**, 32–51.

Graham, D. G., Lowndes, H. E., and Cranmer, J. M. (1985). *Neurofilamentous Axonopathies*, Intox Press, Little Rock, Arkansas.

Gray, E. G., and Guillery, R. W. (1966). Synaptic morphology in the normal and degenerating nervous system. *Int. Rev. Cytol.* **19**, 111–182.

Graybiel, A. M., Moratalla, R., Quinn, B., DeLanney, L. E., Irwin, I., and Langston, J. W. (1993). Early-stage loss of dopamine uptake-site binding in MPTP-treated monkeys. *Adv. Neurol.* **60**, 34–39.

Gringich, J. A., and Caron, M. G. (1993). Recent advances in the molecular biology of dopamine receptors. *Annu. Rev. Neurosci.* **16**, 299–321.

Groenewegen, H. J., Gerrits, N. M., and Vrensen, G. (1994). Autoradiography in the nervous system. In *Methods in Neurobiology* (R. Lahue, Ed.), Vol. 2, pp. 543–597, Plenum, New York.

Guillery, R. W. (1965). Some electron microscopical observations of degenerative changes in central nervous system synapses. *Prog. Brain Res.* **14**, 57–76.

Gupta, R. P., and Abou-Donia, M. B. (1994). Axonopathy. In *Principles of Neurotoxicology* (L. W. Chang, Ed.), pp. 135–152, Dekker, New York.

Hand, P. J. (1981). The 2-deoxyglucose method. In *Neuroanatomical Tract-Tracing Methods* (L. Heimer and M. J. RoBards, Eds.), pp. 511–538, Plenum Press, New York.

Harvey, J. A., and McMaster, S. E. (1976). Neurotoxic action of para-chloroamphetamine in the rat as revealed by Nissl and silver stains. *Psychopharmacol. Bull.* **12**, 62–64.

Haykal-Coates, N., O'Callaghan, J. P., Reinhard, J. F., Jr., and Jensen, K. F. (1991). Pargyline and gamma-butyrolactone enhance tyrosine hydroxylase immunostaining of nigrostriatal axons. *Brain Res.* **556**, 353–357.

Heikkila, R. E., Hess, A., and Duvosin, R. C. (1984). Dopaminergic neurotoxicity of 1-methyl-4-phenyl-1,2,3,6-tetrahydropyridine in mice. *Science* **224**, 1451–1453.

Heikkila, R. E., and Sonsalla, P. K. (1987). The use of the MPTP-treated mouse as an animal model of parkinsonism. *Can. J. Neurol. Sci.* **14**, 436–440.

Heimer, L. (1967). Silver impregnation of terminal degeneration in some forebrain fibers systems: a comparative evaluation of current methods. *Brain Res.* **5**, 86–108.

Heimer, L., and Robards, M. J. (1981). *Neuroanatomical Tract-Tracing Methods*. Plenum Press, New York.

Heimer, L., and Zaborszky, L. (1989). *Neuroanatomical Tract-Tracing Methods 2*. Plenum Press, New York.

Hess, A., Desiderio, C., and McAuliffe, W. G. (1990). Acute neuropathological changes in the caudate nucleus caused by MPTP and methamphetamine: immunohistochemical studies. *J. Neurocytol.* **19**, 338–342.

Hevner, R. F., and Wong-Riley, M. T. T. (1989). Brain cytochrome oxidase: purification, antibody production, and immunohistochemical/histochemical correlations in the CNS. *J. Neurosci.* **9**, 3884–3898.

Hinton, D. R., Sadun, A. A., Blanks, J. C., and Miller, C. A. (1986). Optic-nerve degeneration in Alzheimer's disease. *N. Engl. J. Med.* **315**, 485–487.

Hockfield, S., Carlson, S., Evans, C., Levitt, P., Pintar, J., and Silberstein, L. (1993). *Molecular Probes of the Nervous System*, Cold Spring Harbor Laboratory Press, Cold Spring Harbor, New York.

Hollmann, M., and Heinemann, S. (1994). Cloned glutamate receptors. *Annu. Rev. Neurosci.* **17**, 31–108.

Honig, M. G., and Hume, R. L. (1986). Fluorescent carbocyanine dyes allow living neurons of identified origin to be studied in long term cultures. *J. Cell Biol.* **103**, 171–187.

Horikawa, K., and Armstrong, W. E. (1988). A verstile mean of intracellular labeling: Injection of biocytin and its detection with avidin conjugates. *J. Neurosci. Methods* **25**, 1–11.

Hsu, S.-M., Raine, L., and Fanger, H. (1981). Use of avidin-biotin-peroxidase (ABC) in immunoperoxidase techniques: a comparsion between ABC and unlabeled antibody (PAP) procedures. *J. Histochem. Cytochem.* **29**, 577.

Inui, K., Mitsumori, K., Harada, T., and Maita, K. (1994). Quantitative analysis of neuronal damage induced by tri-ortho-cresyl phosphate in wistar rats. *Fundam. Appl. Toxicol.* **20**, 111–119.

Irwin, I., DeLanney, L. E., Forno, L. S., Finnegan, K. T., Di Monte, D. A., and Langston, J. W. (1990). The evolution of nigrostriatal neurochemical changes in the MPTP-treated squirrel monkey. *Brain Res.* **531**, 242–252.

Isaacson, R. L. (1992). A fuzzy limbic system. *Behav. Brain Res.* **52**, 129–131.

Ivy, G., and Killackey, H. P. (1982). Ontogenetic changes in the projections of neurocortical neurons. *J. Neurosci.* **2**(6), 735–743.

Izzo, P. N. (1991). A note on the use of biocytin in anterograde tracing studies in the central nervous system: application at both light and electron microscopic level. *J. Neurosci. Methods* **36**, 155–166.

Jacobson, M. (1991). *Developmental Neurobiology*, 3rd ed., Plenum Press, New York.

Jacquin, M. F., Hu, J. W., Sessle, B. J., Renehan, W. E., and Waite, P. M. E. (1992). Intra-axonal neurobiotin injection rapidly stain the long-ranges projection of identified trigeminal primary afferents in vivo: comparisons with HRP and PHAL. *J. Neurosci. Methods* **45**, 71–86.

Janssen, R., Schweitzer, L., and Jensen, K. F. (1991). Glutamate neurotoxicity in the developing rat cochlea: physiological and morphological approaches. *Brain Res.* **552**, 255–264.

Jensen, K. F. (1987). Altered thalamocortical axon morphology following neonatal peripheral nerve damage: implications for the study of neurotoxicology. *Neurotoxicology* **7**, 169–182.

Jensen, K. F., Ohmstede, C. A., Fisher, R. S., Olin, J. K., and Sahyoun, N. (1991a). Acquisition and loss of a neuronal Ca2+/calmodulin-dependent protein kinase during neuronal differentiation. *Proc. Natl. Acad. Sci. U.S.A.* **88**, 4050–4053.

Jensen, K. F., Ohmstede, C. A., Fisher, R. S., and Sahyoun, N. (1991b). Nuclear and axonal localization of Ca2+/calmodulin-dependent protein kinase type Gr in rat cerebellar cortex. *Proc. Natl. Acad. Sci. U.S.A.* **88**, 2850–2853.

Jensen, K. F., Lapadula, D. M., Anderson, J. K., Haykal-Coates, N., and Abou-Donia, M. B. (1992). Anomalous phosphorylated neurofilament aggregations in central and peripheral axons of hens treated with tri-ortho-cresyl phosphate (TOCP). *J. Neurosci. Res.* **33**, 455–460.

Jensen, K. F., Olin, J., Haykal-Coates, N., O'Callaghan, J. P., Miller, D. B., and de Olmos, J. S. (1993). Mapping toxicant-induced nervous system damage with a cupric silver stain: a quantitative analysis of neural degeneration induced by 3,4-methylenedioxymethamphetamine. *NIDA Res. Monogr.* **136**, 133–149; discussion 150–154.

Jensen, K. F. (1994). Evaluating the structural integrity of the nervous system for risk assessment. In *Neurobehavioral Plasticity: Learning Development, Response to Brain Insult* (L. Spear, N. Spear, and J. Woodruff, Eds.), Lawrence Erbaum Associates, Hillsdale, New Jersey, in press.

Jensen, K. F., and Killackey, H. P. (1984). Subcortical projections from ectopic neocortical neurons. *Proc. Natl. Acad. Sci. U.S.A.* **81**, 964–968.

Jensen, K. F., and Killackey, H. P. (1987a). Terminal arbors of axons projecting to the somatosensory cortex of the adult rat. I. The normal morphology of specific thalamocortical afferents. *J. Neurosci.* **7**, 3529–3543.

Jensen, K. F., and Killackey, H. P. (1987b). Terminal arbors of axons projecting to the somatosensory cortex of the adult rat. II. The altered morphology of thalamocortical afferents following neonatal infraorbital nerve cut. *J. Neurosci.* **7**, 3544–3553.

Johnson, B. M., and Sadun, A. A. (1988). Ultrastructural and paraphenylene studies of degeneration in the primate visual system: degenerative remnants persist for much longer than expected. *J. Electron Microsc. Tech.* **8**, 179–183.

Johnson, K. B., Criswell, H. E., Jensen, K. F., Simson, P. E., Mueller, R. A., and Breese, G. R. (1992). Comparison of the D1-dopamine agonists SKF-38393 and A-68930 in neonatal 6-hydroxydopamine-lesioned rats: behavioral effects and induction of c-fos-like immunoreactivity. *J. Pharmacol. Exp. Ther.* **262**, 855–865.

Juliano, S. L., Ma, W., Bear, M. F., and Eslin, D. (1990). Cholinergic manipulation alters stimulus-evoked metabolic activity in cat somatosensory cortex. *J. Comp. Neurol.* **297**, 106–120.

Kaas, J. H., Merzenich, M. M., and Killackey, H. P. (1983). The reorganization of somatosensory cortex following peripheral nerve damage in adult and developing animals. *Annu. Rev. Neurosci.* **6**, 325–356.

Karmy, G., Carr, P. A., Yamamoto, T., Chan, S. H. P., and Nagy, J. I. (1991). Cytochrome oxidase immunohistochemistry in rat brain and dorsal root ganglia: visualization of enzyme in neuronal perikarya and in parvalbumin-positive neurons. *Neuroscience* **40**, 825–839.

Kater, S. D., and Nicholson, C. (1973). *Intracellular Staining in Neurobiology*, Springer-Verlag, New York.

Katz, L. C., and Iarovici, D. M. (1984). Fluorescent latex microspheres as a retrograde neuronal marker for *in vivo* and *in vitro* studies of visual cortex. *Nature* **310**, 498–500.

Kawata, M., Yuri, K., and Sano, Y. (1991). Localization and regulation of mRNAs in the nervous tissue as revealed by in situ hybridization. *Comp. Biochem. Physiol.* **98**, 41–50.

Kierman, J. A. (1981). *Histological and Histochemical Methods: Theory and Practice*. Pergamon Press, New York.

Kierman, J. A., and Berry, M. (1975). Neuroanatomical Methods. In *Methods in Brain Research* (P. B. Bradley, Ed.), pp. 2–77, Wiley and Sons, New York.

Killackey, H. P., Belford, G., Ryugo, R., and Ryugo, D. (1976). Anomalous organization of thalamocortical projections consequent to vibrissae removal in the newborn rat and mouse. *Brain Res.* **104**, 309–315.

King, M. A., Louis, P. M., Hunter, B. E., and Walker, O. W. (1989). Biocytin:a versatile anterograde neuroanatomical tract-tracing alternative. *Brain Res.* **497**, 361–367.

Kitai, S. T., and Bishop, G. A. (1991). Horseradish peroxidase: intracellular staining of neurons. In *Neuroanatomical Tract-Tracing Methods* (L. Heimer and M. J. RoBards, Eds.), pp. 263–278, Plenum Press, New York.

Klir, G. J., and Folger, T. A. (1988). *Fuzzy Sets, Uncertainty and Information.* Prentice-Hall, Englewood Cliffs, New Jersey.

Koestner, A., and Norton, S. (1991). Nervous system. In *Handbook of Toxicologic Pathology* (W. M. Haschek and C. G. Rousseaux, Eds.), pp. 625–674, Academic Press, New York.

Koller, W., Vetere-Overfield, B., Gray, C., Alexander, C., Chin, T., Dolezal, J., Hassanein, R., and Tanner, C. (1990). Environmental risk factors in Parkinson's disease. *Neurology* **40**, 1218–1221.

Kopin, I. J., and Markey, S. P. (1988). MPTP toxicity implications for research in Parkinson's disease. *Annu. Rev. Neurosci.* **11**, 81–96.

Kopin, I. J. (1987). MPTP: an industrial chemical and contaminant of illicit narcotics stimulates a new era in research on Parkinson's Disease. *Environ. Health Perspect.* **75**, 45–51.

Krady, J. K., Oyler, G. A., Balaban, C. D., and Billingsley, M. L. (1990). Use of avidin-biotin subtractive hybridization to characterize mRNA common to neurons destroyed by the selective neurotoxicant trimethyltin. *Mol. Brain Res.* **7**, 287–297.

Krinke, G. J. (1989). Neuropathologic screening in rodent and other species. *J. Am. Coll. Toxicol.* **8**, 141–145.

Kristensson, K. (1975). Retrograde axonal transport of protein tracers. In *The Use of Axonal Transport for Studies of Neuronal Connectivity* (W. M. Cowan and M. Cuenod, Eds.), pp. 69–82. Elsevier, New York.

Kuhar, M. J., DeSourza, E. B., and Ungerstall, J. R. (1986). Neurotransmitter receptor mapping by autoradiography and other methods. *Annu. Rev. Neurosci.* **9**, 27–59.

Kuhar, M. J., Lloyd, D. G., Appel, N., and Loats, H. L. (1991). Imaging receptors by autoradiography: computer-assisted approaches. *J. Chem. Neuroanat.* **4**, 319–327.

Kuypers, H. G. J. M., and Huisman, A. M. (1984). Fluorescent neuronal tracers. *Adv. Cell. Neurobiol.* **5**, 307–340.

Langston, J. (1987). The Discovery of MPTP: how far will it take us? In *Neurotoxins and their pharmalogical implications* (P. Jenner, Ed.), pp. 153–161, Raven Press, New York.

Lapadula, E. S., Lapadula, D. M., and Abou-Donia, M. B. (1991). Persistent alterations of calmodulin kinase II activity in chickens after an oral dose of tri-o-cresyl phosphate. *Biochem. Pharmacol.* **42**, 171–180.

Lapper, S. R., and Bolam, J. P. (1991). The anterograde and retrograde transport of neurobiotin in the central nervous system of the rat: comparison with biocytin. *J. Neurosci. Methods* **39**, 163–174.

Larsson, L.-I. (1988). *Immunocytochemistry: Theory and Practice.* CRC Press, Boca Raton, Florida.

LaVail, J. H. (1975). A review of the retrograde transport technique. In *Neuroanatomical Research Techniques* (R. T. Robertson, Ed.), pp. 355–384, Academic Press, New York.

Lee, B. (1950). *The Microtomist's Vade-Mecum.* 11th ed., Blakiston, Philadelphia.

Leonard, C. M. (1975). Developmental changes in olfactory bulb projections revealed by degeneration argyrophilia. *J. Comp. Neurol.* **162**, 467–486.

Leslie, F. M., and Altar, C. A. (1988). *Receptor Localization: Ligand Autoradiography.* Liss, New York.

Lieberman, A. R. (1971). The axon reaction: a review of the principal features of perikaryl response to axonal injury. *Int. Rev. Neurobiol.* **14**, 49–124.

Lillie, R. D. (1965). *Histopathologic Technique and Practical Histochemistry,* 3rd ed., McGraw Hill, New York.

Lorente de No, R. (1949). Cerebral cortex: architectonics, intracortical connections. In *Physiologicay of the Nervous System* (J. Fulton, Ed.), pp. 274–301, Oxford University Press, New York.

Lund, J. S. (1988). Anatomical organization of macaque monkey striate cortex. *Annu. Rev. Neurosci.* **11**, 253–288.

Maler, L., Fibiger, H. C., and McGeer, P. L. (1973). Demonstration of the nigrostriatal projection by silver staining after nigral injection of 6-hydroxydopamine. *Exp. Neurol.* **40**, 505–515.

Mamounas, L. A., Mullen, C. A., O'Hearn, E., and Molliver, M. E. (1991). Dual serotoninergic projections to forebrain in the rat: morphologically distinct 5-HT axon terminals exhibit differential vulnerbility to neurotoxic amphetamine derivatives. *J. Comp. Neurol.* **314**, 558–586.

Maranto, A. R. (1982). Neuronal mapping: a photooxidation reaction make Lucifer useful for electron microscopy. *Science* **217**, 953–955.

Marin-Padilla, M. (1985). Neurogenesis of the climbing fibers in the human cerebellum: a Golgi study. *J. Comp. Neurol.* **235**, 82–96.

McLachlan, D. R., and Massiah, J. (1992). Aluminum ingestion: a risk factor for alzheimer's disease. In *The Vulnerable Brain and Environmental Risks,* Vol. 2, *Toxins in Food* (R. L. Isaacson and K. F. Jensen, Eds.), pp. 49–50, Plenum, New York.

Mesulam, M. (1982). *Tracing neural connections with horseradish peroxidase.* Wiley, New York.

Miller, M. W. (1986). Effects of alcohol on the generation and migration of cerebral cortical neurons. *Science* **233**, 1308–1311.

Miller, M. W. (1987). Effect of prenatal exposure to alcohol on the distribution and time of origin of corticospinal neurons in the rat. *J. Comp. Neurol.* **257**, 372–382.

Miller, M. W., Chiaia, N. L., and Rhoades, R. W. (1990). Intracellular recording and injection study of corticospinal neurons in the rat somatosensory cortex: effect of prenatal exposure to ethanol. *J. Comp. Neurol.* **297**, 91–105.

Miller, M. W., and Dow-Edwards, D. L. (1993). Vibrissal stimulation affects glucose utilization in the trigeminal/somatosensory system of normal rats and rats prenatally exposed to ethanol. *J. Comp. Neurol.* **335**, 283–284.

Millhouse, O. E. (1991). The golgi methods. In *Neuroanatomical Tract-Tracing Methods* (L. Heimer and M. J. RoBards, Eds.), pp. 311–344, Plenum Press, New York.

Milner, R. J., and Sutcliffe, J. G. (1983). Gene expression in rat brain. *Nucleic Acids Res.* **11**, 5497–5520.

Morest, D. K. (1981). The golgi methods. In *Techniques in Neuroanatomical Research* (C. Heym and W.-G. Forssmann, Eds.), pp. 124–142, Springer-Verlag, New York.

Morgan, J. I., and Curran, T. (1991). Stimulus-transcription coupling in the nervous system: involvement of the inducible protooncogenes *fos* and *jun. Annu. Rev. Neurosci.* **14**, 421–451.

Nance, D. M., and Burns, J. (1990). Fluorescent dextrans as sensitive anterograde neuroanatomical tracers: applications and pitfalls. *Brain Res. Bull.* **25**, 139–145.

National Research Council, (1992). *Environmental Neurotoxicology.* National Academy Press. Washington D.C.

National Research Council, (1993). *Pesticides in the Diets of Infants and Children.* National Academy Press, Washington D.C.

National Research Council, and Committee on Neurotoxicology and Models for Assessing Risks, (1992). *Environmental Neurotoxicology.* National Academy Press, Washington D.C.

Naumann, T., Peterson, G. M., and Frotscher, M. (1992). Fine structure of rat septohippocampal neurons: II. A time course analysis following axotomy. *J. Comp. Neurol.* **325**, 219–242.

Nauta, W. J. H., and Gygax, P. A. (1954). Silver impregnation of degeneration axons in the central nervous system: a modified technique. *Stain Technol.* **29**, 91–93.

Nauta, W. J. H., and Ebbesson, S. O. E. (1970). *Contemporary Research Methods in Neuroanatomy.* Springer-Verlag, New York.

Norton, W. T., Aquino, D. A., Hozumi, I., Chiu, F.-C., and Brosnan, C. F. (1993). Quantitative aspects of reactive gliosis: a review. *Neurochem. Res.* **17**, 877–885.

O'Callaghan, J. P. (1991a). Quantification of glial fibrillary acidic protein: comparison of slot-immunobinding assays with a novel sandwich ELISA. *Neurotoxicol. Teratol.* **13**, 275–281.

O'Callaghan, J. P. (1991b). Assessment of neurotoxicity: use of glial fibrillary acidic protein as a biomarker. *Biomed. Environ. Sci.* **4**, 197–206.

O'Callaghan, J. P. (1994). A potential role for altered protein phosphorylation in the mediation of developmental neurotoxicity. *Neurotoxicology* **15**, 29–40.

O'Callaghan, J. P., and Jensen, K. F. (1992). Enhanced expression of glial fibrillary acidic protein and the cupric silver degeneration reaction can be used as sensitive and early indicators of neurotoxicity. *Neurotoxicology* **13**, 113–112.

O'Callaghan, J. P., and Jensen, K. F. (1993). Enhanced expression of glial fibrillary acidic protein and the cupric silver degeneration reaction can be used as sensitive and early indicators of neurotoxicity. *Neurotoxicology* **13**, 113–122.

O'Donoghue, J. L. (1989). Screening for neurotoxicity using a neurologically based examination and neuropathology. *J. Am. Coll. Toxicol.* **8**, 97–115.

O'Hearn, E., Battaglia, G., De Souza, E. B., Kuhar, M. J., and Milliver, M. E. (1988). Methylenedioxyamphetamine (MDA) and methylenedioxymethamphetamine (MDMA) cause selective ablation of serotonergic axon terminals in forebrain: immunocytochemical evidence for neurotoxicity. *J. Neurosci.* **8**, 2788–2803.

O'Kusky, J. R. (1992). The neurotoxicity of methylmercury in the developing nervous system. In *The Vulnerable Brain and Environmental Risks*, Vol. 2, *Toxins in Food* (R. L. Isaacson and K. F. Jensen, Eds.), pp. 19–48, Plenum, New York.

O'Leary, D. D. M., Stanfield, B. B., and Cowan, W. M. (1981). Evidence that the early postnatal restriction of the cells of origin of the callosal projection is due to the elimination of axonal collaterals rather than to the death of the neurons. *Dev. Brain Res.* **1**, 607–617.

O'Leary, D. D. M., and Stanfield, B. B. (1989). Selective elimination of axons extended by developing cortical neurons is dependent on regional locale: experiments using fetal cortical transplants. *J. Neurosci.* **7**, 2230–2246.

Ohmstede, C. A., Jensen, K. F., and Sahyoun, N. E. (1989). Ca^{2+}/calmodulin-dependent protein kinase enriched in cerebellar granule cells. Identification of a novel neuronal calmodulin-dependent protein kinase. *J. Biol. Chem.* **264**, 5866–5875.

Pagge, R. D., Sambrook, M. A., and Crossman, A. R. (1993). Thalamotomy for the alleviation of levodopa-induced dyskinesia: experimental studies in t he 1-methyl-4-phenyl-1,2,3,6-tetrahydropyridine-treated parkinsonian monkey. *Neuroscience* **55**, 147–165.

Pasternak, J. F., and Woolsey, T. A. (1975). On the "selectivity" of the Golgi-Cox method. *J. Comp. Neurol.* **160**, 307–312.

Pasternak, T., Flood, D. G., Eskin, T. A., and Merigan, W. H. (1985). Selective drainage to large cells in the cat retinogeniculate pathways with 2,5 hexadione. *J. Neurosci.* **5**, 1641–1652.

Patrick, J., Sequela, P., Vernino, S., Amador, M., Luetje, C., and Dani, J. A. (1993). Functional diversity of neuronal nicotinic acetylcholine receptors. *Prog. Brain Res.* **98**, 113–119.

Penny, D. G., and White, S. R. (1994). The neural and behavioral effects of carbon monoxide. In *The Vulnerable Brain and Environmental Risks*, Vol. 3, *Toxins in Air and Water* (R. L. Isaacson and K. F. Jensen, Eds.), pp. 123–143, Plenum Press, New York.

Perl, D. P., and Brody, A. R. (1980). Alzheimer's disease: X-ray spectrometric evidence of aluminum accumulation in neurofibrillary tangle-bearing neurons. *Science* **208**, 297–299.

Peters, A., Palay, S. L., and Webster, H. D. (1991). *The Fine Structure of the Nervous System*, 3rd ed., Oxford University Press, New York.

Peterson, G. M., Lanford, G. W., and Powell, E. W. (1980). Fate of septohippocampal neurons following fimbria-fornix transection: a time course analysis. *Brain Res. Bull.* **25**, 129–137.

Peterson, G. M., Naumann, T., and Frotscher, M. (1992). Identified septohippocampal neurons survive axotomy: a fine-structural analysis in the rat. *Neurosci. Lett* **138**, 81–85.

Petruz, P., Ordronneau, P., and Finley, J. C. W. (1980). Criteria of reliability for light microscopic immunocytochemical staining. *Histochem. J.* **12**, 333–348.

Polak, J. M., and Van Noorden, S. (1984). *An Introduction to Immunocytochemistry: Current Techniques and Problems.* Oxford University Press, New York.

Polyak, J. M., and Varndell, I. M. (1984). *Immunolabeling for Electron Microscopy.* Elsevier, New York.

Pool, C. W., Bujis, R. M., Swaab, D. F., Boer, G. J., and Van Leeuwen, F. (1983). On the way to a specific immunocytochemical localization. In *Immunocytochemistry* (A. C. Cuello, Ed.), pp. 2–46, IBRO, New York.

Powell, T. P. S. (1967). Transneuronal cell degeneration in the olfactory bulb shown by that Golgi method. *Nature* **215**, 425–426.

Przedborski, S., Jackson-Lewis, V., Popilskis, S., Kostic, V., Levivier, M., Fahn, S., and Cadet, J. L. (1991). Unilateral MPTP-induced parkinsonism in monkeys. A quantitative autoradiographic study of dopamine D1 and D2 receptors and re-uptake sites. *Neurochirurgie* **37**, 377–382.

Ralis, H. M., Beasley, R. A., and Ralis, Z. A. (1973). *Techniques in Neurohistology*, Butterworths & Co. Ltd., London.

Ramon y Cajal, S. (1928). *Degeneration and Regeneration of the Nervous System.* Oxford University Press, New York.

Ramön-Moliner, E. (1961). The histology of the postcruciate gyrus in the cat. II A statistical analysis of dendritic distribution. *J. Comp. Neurol.* **117**, 63–76.

Ramön-Moliner, E., and Nauta, W. J. H. (1966). The isodendritic core of the brain stem. *J. Comp. Neurol.* **126**, 311–335.

Ray, D. E., Brown, A. W., Cavanagh, J. B., Nolan, C. C., Richards, H. K., and Wylie, S. P. (1992). Functional/metabolic modulation of the brain stem lesions caused by 1,3-dinitrobenzene in the rat. *Neurotoxicology* **13**, 379–388.

Rees, D. C., and Glowa, J. R. (1993). Extrapolation to humans for neurotoxicants: issues and challenges. In *The Vulnerable Brain and Environmental Risks*, Vol. 3, *Toxins in Air and Water* (R. L. Isaacson and K. F. Jensen, Eds.), Plenum, New York, pp. 207–230.

Reinhard, J. F., Jr., and Nichol, C. A. (1986). The neurotoxin MPTP decreases striatal dopamine and tetrahydrobiopterin levels and increases striatal dopamine turnover in mice. In *MPTP: A neurotoxin producing a Parkinsonian syndrome* (S. P. Markey, N. Castagnoli, A. J. Trevor, and I. J. Kopin, Eds.), pp. 523–528, Academic Press, Orlando, Florida.

Reuhl, K. R., Lagunowich, L. A., and Brown, D. L. (1994). Cytoskeleton and cell adhesion molecules: critical targets of toxic agents. *Neurotoxicology* **15**, 133–146.

Reuhl, K. R., and Chang, L. W. (1979). Effects of methyl mercury on the development of the nervous system: a review. *Neurotoxicology* **1**, 21–55.

Ricaurte, G. A., Guillery, R. W., Seiden, L. S., Schuster, C. R., and Moore, R. Y. (1982). Dopamine nerve terminal degeneration produced by high doses of methylamphetamine in the rat brain. *Brain Res.* **235**, 93–103.

Ricaurte, G. A., Guillery, R. W., Seiden, L. S., and Schuster, C. R. (1984). Nerve terminal degeneration after a single injection of da-amphetamine in iprindole-treated rats: relation of selective long-lasting dopamine depletion. *Brain Res.* **291**, 378–382.

Ricaurte, G. A., Bryan, G., Strauss, L., Seiden, L. S., and Schuster, C. R. (1985). Hallucinogenic amphetamine selectively destroys brain serotonin nerve terminals. *Science* **229**, 986–988.

Ricaurte, G. A., Forno, L. S., Wilson, M. A., DeLanney, L. E., Irwin, I., Molliver, M. E., and Langston, J. W. (1988). (+/−)3,4-Methylenedioxymethamphetamine selectively damages central serotonergic neurons in nonhuman primates. *JAMA* **260**, 51–55.

Ritter, S., and Dinh, T. T. (1988). Capsaicin-induced neuronal degeneration: silver impregnation of cell bodies, axons, and terminals in the central nervous system of the adult rat. *J. Comp. Neurol.* **271**, 79–90.

Robertson, R. T. (1975). *Neuroanatomical Research Techniques.* Academic Press, New York.

Rogers, A. W. (1973). *Techniques in Autoradiography.* Elsevier, New York.

Rogers, A. W. (1975). Autoradiography and the study of the central nervous system. In *Methods in Brain Research* (P. B. Bradely, Ed.), pp. 79–112, Wiley, New York.

Sadun, A. A., Smith, L. E., and Kenyon, K. R. (1983). Paraphenylenediamine: a new method for tracing human visual pathways. *J. Neuropathol. Exp. Neurol.* **42**, 200–206.

Sadun, A. A. (1989). The optic neuropathy of Alzheimer's disease. *Metab. Pediatr. Syst. Ophthalmol.* **12**, 64–68.

Sadun, A. A., and Schaechter, J. D. (1985). Tracing axons in the human brain: a method utilizing light and TEM techniques. *J. Electron Microsc. Tech.* **2**, 175–186.

Saitoh, T., Masliah, E., Jin, L., Cole, G. M., Wieloch, T., and Shapiro, I. P. (1991). Biology of disease. Protein kinases and phosphorylation in neurological disorders and cell death. *Lab. Invest.* **64**, 596–616.

Sandell, J. H., and Masland, R. H. (1988). Photoconversion of some fluorescent markers to a diaminobenzidine product. *J. Histochem. Cytochem.* **5**, 555–559.

Santini, M. (1975). *Golgi Centennial Symposium: Perspectives in Neurobiology.* Raven Press, New York.

Sarthy, P. V., Fu, M., and Huang, J. (1989). Subcellular localization of an intermediate filament protein and its mRNA in glial cells. *Mol. Cell. Biol.* **9**, 4556–4559.

Scallet, A. C., Lipe, G. W., Ali, S. F., Holson, R. R., Frith, C. H., and Slikker W., Jr., (1988). Neuropathological evaluation by combined immunohistochemistry and degeneration-specific methods: application to methylenedioxymethamphetamine. *Neurotoxicology* **9**, 529–538.

Scallet, A. C., Binienda, Z., Caputo, F. A., Hall, S., Paule, M. G., Routree, R. L., Schmued, L. C., Sobotka, T., and Slikker, W. Jr., (1993). Domoic acid-treated cynomolgus monkeys (M. fascicularis): effects of dose on hippocampal neuronal and terminal degeneration. *Brain Res.* **627**, 307–313.

Schiebel, M. E., and Scheibel, A. B. (1978). The methods of Golgi. In *Neuroanatomical Research Techniques* (R. T. Robertson, Ed.), pp. 89–114, Academic Press, New York.

Schimdt, C. J. (1987). Neurotoxicity of the psychedelic amphetamine, methylenedioxymethamphetamine. *J. Pharmacol. Exp. Ther.* **240**, 1–7.

Schmued, L. C. (1990). Fluoro-gold and SITS: use of substituted stilbenes in neuroanatomical studies. *Meth. Neurosci.* **3**, 317–330.

Schmued, L. C., Kyriakidis, K., and Heimer, L. (1990). In vivo anterograde and retrograde axonal transport of the fluorescent rhodamine-dextran-amine, Fluoro-Ruby, within the CNS. *Brain Res.* **526**, 127–134.

Schmued, L. C., and Fallon, J. H. (1985). Selective neuronal uptake of 4-acetamido-4′-isothiocyanostilbene-2,2′-disulfonic acid (SITS) and other related substituted stilbenes in vivo: a fluorescent whole cell staining technique. *Brain Res.* **346**, 124–129.

Schmued, L. C., and Fallon, J. H. (1986). Fluoro-Gold: a new fluorescent retrograde axonal tracer with numerous unique properties. *Brain Res.* **377**, 147–154.

Schmued, L. C., and Snavely, L. F. (1993). Photoconversion and electron microscopic localization of the fluorescent axon tracer fluoro-ruby. *J. Histochem. Cytochem.* **41**, 777–782.

Schweitzer, L., Jensen, K. F., and Janssen, R. (1991). Glutamate neurotoxicity in rat auditory system: cochlear nuclear complex. *Neurotoxicol. Teratol.* **13**, 189–193.

Semchuk, K. M., Love, E. J., and Lee, R. G. (1993). Parkinson's disease: a test of the multifactorial etiologic hypothesis. *Neurology* **43**, 1173–1180.

Semchuk, K. M., Love, E. J., and Lee, R. G. (1994). Parkinson's disease and exposure to agricultural work and pesticide chemicals. *Neurology* **42**, 1328–1335.

Sette, W. F., and MacPhail, R. C. (1992). Qualitative and quantitative issues in assessment of neurotoxic effects. In *Neurobehavioral Toxicology* (H. Tilson and C. Mitchell, Eds.), pp. 345–361, Raven Press, New York.

Sharp, F. R., Sagar, S. M., and Swanson, R. A. (1993). Metabolic mapping with cellular resolution: c-fos vs. 2-deoxyglucose. *Crit. Rev. Neurobiol.* **7**, 205–228.

Sharp, F. R., and Sagar, S. M. (1994). Alterations in gene expression as an index of neuronal injury: heat shock and the immediate early gene response. *Neurotoxicology* **15**, 51–60.

Sheng, M., and Greenberg, M. E. (1990). The regulation and function of c-fos and other immediate early genes in the nervous system. *Neuron* **4**, 477–485.

Shimono, M., and Tsuji, N. (1987). Study of the selectivity of the impregnation of neurons by the Golgi method. *J. Comp. Neurol.* **259**, 122–130.

Shipley, M. T., Luna, J., and McLean, J. H. (1989). Processing and analysis of neuroanatomical images. In *Neuroanatomical Tract-Tracing Methods 2* (L. Heimer and L. Zaborszky, Eds.), pp. 331–389, Plenum Press, New York.

Sholl, D. A. (1956). *The Organization of the Cerebral Cortex,* Methuen, London.

Silva, A. J., Paylor, R., Wehner, J. M., and Tonegawa, S. (1992). Impaired spatial learning in alpha-calcium calmdoulin kinase II mutant mouse. *Science* **257**, 206–211.

Skirboll, L. R., Hokfelt, T., Norell, G., Phillipson, O., Kuypers, H. G. J. M., Bentivoglio, M., Catmans-Berrevoets, C. E., Visser, T. J., Steinbush, H., Verhofstad, A., Cuello, A. C., and Goldstein, M., and Brownstein, M. (1984). A method for specific transmitter identification of retrogradely labeled neurons: immunofluorescence combined with fluoresence tracing. *Brain Res. Rev.* **8**, 99–127.

Slater, M. (1993). Ultrastructural double labelling using colloidal gold. *Micron* **24**, 661–675.

Slikker, W., Jr., Ali, S. F., Scallet, A. C., Frith, C. H., Newport, G. D., and Bailey, J. R. (1988). Neurochemical and neurohistological alterations in the rat and monkey produced by orally administered methylenedioxymethamphetamine (MDMA). *Toxicol. Appl. Pharmacol.* **94**, 448–457.

Smit, G. J., and Colon, E. J. (1969). Quantitative analysis of the cerebral cortex. I. A selectivity of the Golgi Cox staining technique. *Brain Res.* **13**, 485–510.

Sokoloff, L., Reivich, M., Kennedy, C., Des Rosiers, M. H., Patlak, C. S., Pettigrew, K. D., Sakurda, O., and Shinohara, M. (1977). The [14C] deoxyglucose method for measurement of local cerebral glucose utilization: theory, procedure, normal values in the conscious and anesthetized albino rat. *J. Neurochem.* **28**, 897–916.

Spacek, J. (1992). Dynamics of Goldi impregnation of neurons. *Microsc. Res. Tech.* **23**, 264–274.

Spencer, P. S., and Schaumburg, H. H. (1980a). *Experimental and Clinical Neurotoxicology.* Williams & Wilkins, Baltimore.

Spencer, P. S., and Schaumburg, H. H. (1980b). Classification of neurotoxic disease: a morphological approach. In *Experimental and Clinical Neurotoxicology* (Spencer, P.S. and Schaumburg, H. H. Eds.), pp. 92–99, Williams & Wilkins, Baltimore.

Sternberger, L. A. (1986). *Immunocytochemistry.* 3rd ed., Wiley, New York.

Stewart, M. (1992). *Quantitative Methods in Neuroanatomy.* Wiley & Sons, New York.

Sutcliffe, J. G. (1988). mRNA in the mammalian central nervous system. *Annu. Rev. Neurosci.* **11**, 157–198.

Sutcliffe, J. G., Travis, G. H., Danielson, P. E., Wong, K. K., Ottiger, H. P., Burton, F. H., Hasel, K. W., Bloom, F. E., and Forss-Petter, S. (1991). Molecular approaches to genes of the CNS. *Epilepsy Res. Suppl.* **4**, 213–223.

Swaab, D. F. (1982). Comments on the validity of immunocytochemical methods. In *Cytochemical Methods in Neuroanatomy* (V. Chan-Palay and S. L. Palay, Eds.), pp. 423–440, Liss, New York.

Tanaka, D., Bursian, S. J., and Lehning, E. J. (1992). Silver impregnation of organphosphorus-induced delayed neuropathy in the central nervous system. In *The Vulnerable Brain and Environmental Risks,* Vol. 2, *Toxins in Food* (R. L. Isaacson and K. F. Jensen, Eds.), pp. 215–251, Plenum Press, New York.

Tanaka, D., Jr., Bursian, S. J., Lehning, E. J., and Aulerich, R. J. (1991). Delayed neurotoxic effects of bis (1-methylethyl) phosphorofluoridate (DFP) in the European ferret: a possible mammalian model for organophosphorus-induced delayed neurotoxicity. *Neurotoxicology* **12**, 209–224.

Tanaka, D., Jr., and Bursian, S. J. (1989). Degeneration patterns in the chicken central nervous system induced by ingestion of the organophosphorus delayed neurotoxin tri-*ortho*-tolyl phosphate. A silver impregnation study. *Brain Res.* **484**, 240–256.

Tenhula, W. N., Xu, S. Z., Madigan, M. C., Heller, K., Freeman, W. R., and Sadun, A. A. (1992). Morphometric comparisons of optic nerve axon loss in acquired immunodeficiency syndrome. *Am. J. Ophthalmol.* **113**, 14–20.

Toggas, S. M., Kady, J. K., and Billingsley, M. L. (1992). Molecular neurotoxicology of trimethyltin: identification of stannin, a novel protein expressed in trimethyltin-sensitive cells. *Mol. Pharmacol.* **42**, 44–56.

Toggas, S. M., Kady, J. K., Thompson, A. T., and Billingsley, M. L. (1993). Molecular mechanisms of selective neurotoxicants: studies on organotin compounds. *Ann. N.Y. Acad. Sci.* **679**, 157–177.

Trask, B. J. (1991). Gene mapping by in situ hybridization. *Curr. Opin. Genet. Dev.* **1**, 82–87.

U.S.EPA, (1991). Office of Pesticide Programs, Subdivision F. Hazard Evaluation: Human and Domestic Animals Addendum 10 Neurotoxicity Series 81, 82, 83. In *Pesticide Assessment Guidelines* (*EPA Pub Np 540/9-82-025*), National Technical Information Service, Springfield, Virginia.

U.S.EPA (1993). Draft report: principles of neurotoxicity assessment. *Fed. Reg.* **58**, 41556–41599.

Uhl, G. R. (1986). *In Situ Hybridization in Brain.* Plenum, New York.

Usui, H., Falk, J. D., Dopazo, A., de Lecea, L., Erlander, M. G., and Sutcliffe, J. G. (1994). Isolation of clones of rat striatum-specific mRNAs by directional Tag PCR substraction. *J. Neurosci.* **14**, 4915–4926.

Valentino, K. L., Eberwine, J. H., and Barchas, J. D. (1987). *In Situ Hybridization: Applications to Neurobiology.* Oxford University Press, New York.

Valverde, F. (1986). Intrinsic neocortical organization: some comparative aspects. *Neuroscience* **18**, 1–23.

Van Leeuwen, F. (1982). Specific immunocytochemical localization of neuropeptides: a utopian goal? In *Techniques in Immunocytochemistry* (A. C. Cuello, Ed.), pp. 238–299, Academic Press, New York.

Viale, G., and Dell'Orto, P. (1992). Non-radioactive nucleic acid probes: labelling and detection procedures. *Liver* **12**, 243–251.

Voogd, J., and Feiradbend, H. K. P. (1981). Classical methods in neuroanatomy. In *Methods in Neurobiology* (R. Lahue, Ed.), Vol. 2, pp. 301–364, Plenum Press, New York.

Wallace, J. A., Petrusz, P., and Lauder, J. M. (1982). Serotonin immunocytochemistry in the adult and developing rat brain: methodological and pharmacological considerations. *Brain Res. Bull.* **9**, 117–129.

Warr, W. B., de Olmos, J. S., and Heimer, L. (1981). Horseradish peroxidase. The basic procedure. In *Neuroanatomical Tract-Tracing Methods* (L. Heimer and M. J. RoBards, Eds.), pp. 207–277, Plenum, New York.

Wessendork, M. W. (1990). Characterization and use of multi-color fluorescence microscopic techniques. In *Handbook of Chemical Neuroanatomy* Vol. 8, *Analysis of Neuronal Microcircuits and Synaptic Interconnections* (A. Bjorklund, T. Hokfelt, F. G. Wouterlood, and A. N. Van den Pol, Eds.), pp. 1–46, Elsevier, New York.

Wilson, M. A., Ricaurte, G. A., and Milliver, M. E. (1989). Distinct morphologic classes of serotonergic axons in primates exhibit differential vulnerability to the psychotropic drug 3,4-methylene-dioxymethamphetamine. *Neuroscience* **28**, 121–137.

Wilson, M. C., and Higgens, G. A. (1990). In situ hybridization. *Neuromethods* **16**, 239–283.

Winer, J. A. (1977). A review of the status of the horseradish peroxidase method in neuroanatomy. *Biobehav. Rev.* **1**, 45–54.

Wong-Riley, M. T. T. (1989). Cytochrome oxidase: an endogenous metabolic marker for neuronal activity. *Trends Neurosci.* **12**, 94–101.

World Health Organization (WHO), (1986). *Principles and methods for the assessment of neurotoxicity associated with exposure to chemicals. environmental health criteria document* 60, World Health Organization, Geneva.

Wouterloud, F. G., and Groenewegen, H. J. (1985). Neuroanatomical tracing by use of *Phaseolus vulgaris*-leucoagglutinin (PHA-L): electron microscopy of PHA-L fill neuronal somata, dendrites, and axon terminals. *Brain Res.* **326**, 188–191.

Wullner, U., Standaert, D. G., Testa, C. M., Landwehrmeyer, G. B., Catania, M. V., Penney, J. B., Jr., and Young, A. B. (1994). Glutamate receptor expression in rat striatum: effects of deafferentation. *Brain Res.* **647**, 209–219.

Yamamura, H. I., Enna, S. J., and Kuhar, M. J. (1985). *Neurotransmitter receptor binding,* Raven Press, New York.

Yamamura, H. I., Enna, S. J., and Kuhar, M. J. (1990). *Methods in Neurotransmitter Receptor Analysis,* Raven Press, New York.

Yokel, R. A., Provan, S. D., Meyer, J. J., and Campbell, S. R. (1988). Aluminum intoxication and the victim of Alzheimer's disease: similarities and differences. *Neurotoxicology* **9**, 429–442.

Zaborszky, L., and Heimer, L. (1989). Combinations of tracer techniques, especially HRP and PHA-L, with transmitter identification for correlated light and electron microscopic studies. In *Neuroanatomical Tract-Tracing Methods 2* (L. Heimer and L. Zaborszky, Eds.), pp. 49–96, Plenum, New York.

Enzyme Histochemical Methods and Techniques

DAVID C. DORMAN
Chemical Industry Institute
of Toxicology
Research Triangle Park
North Carolina 27709

MARC BONNEFOI
Rhône-Poulenc Rorer
Central Research
Horsham, Pennsylvania 19044

KEVIN T. MORGAN
Chemical Industry Institute
of Toxicology
Research Triangle Park
North Carolina 27709

Histochemistry can be defined as the identification, localization, and quantitation, in cells and tissues and by chemical or physical tests, of specific substances, reactive groups, and enzyme-catalyzed activities (Pearse, 1980). The first part of this chapter deals with general principles of enzyme histochemical techniques, whereas the second part discusses particular applications relevant to neurotoxicology.

I. Enzyme Histochemical Techniques

Enzyme histochemistry provides information about an enzymes identity, location, and quantity. Histochemistry offers a sensitive way to detect and monitor localized and initial insults caused by toxic compounds in the nervous system (Horobin, 1991). The primary advantage of enzyme histochemistry over conventional biochemical enzyme assays is that enzyme histochemistry allows for the visualization of location of enzyme activity at the level of the tissue, cell, or organelle. A neurotoxicant may affect one cell type which represents only a small percentage of the total volume of

neural tissue. Such effects may be missed when biochemical analysis is performed on a tissue homogenate. When enzyme localization is less critical, then biochemical assays of disrupted whole tissues generally provide a more sensitive measure of an enzyme's activity. The goal of enzyme histochemistry, however, is to achieve an understanding of the localization and the quantification of enzyme constituents and metabolic functions of cells and tissues under various conditions.

Enzyme histochemistry allows for the easy and economical evaluation of small samples for enzyme activity. Enzyme histochemistry offers the potential to contribute to the understanding of the biochemical basis of neurotoxicant-induced tissue injury. The investigation of enzyme and metabolic changes in the development of a lesion can give insight into the mechanism of neurotoxicity, as well as the extent and progression of the injury and the degree of repair. Changes in enzyme function may also be related to the structure of the tissue or the cell type(s) involved. In every qualitative and quantitative histochemical investigation of enzymes in the brain, appropriate conditions should be considered and fulfilled with respect to the preparation of brain tissue, the detection methods, and the incubation (Kugler, 1990a).

II. Preparation of Tissue

The microscopic observation of an enzyme's activity in a neural tissue requires the stabilization of the sample of interest, preparation of thin sections, and the application of an appropriate stain. Neural tissue must be processed in such a way that normal morphology is maintained and the enzymes are neither inactivated nor displaced from their normal site. Improperly stabilized samples may lose enzyme activity through autolysis and by leaching into the aqueous and organic solvents used for specimen processing. A significant loss of tissue architecture may also occur in the storage or processing of improperly stabilized samples.

Theoretically, the best method for the demonstration of enzyme activity in the absence of *in vivo* approaches (e.g., PET scan) would be by the use of fresh frozen tissues sections. For some nervous tissues, it is difficult to obtain sections that maintain their architectural integrity. In addition, when prolonged incubation of a sample with an enzyme substrate occurs, then loss of enzymes and other proteins to the incubation media may occur. These diffusion artifacts may lead to the erroneous localization of enzyme as well as diminished quality to the sample. Stabilization of cryostat sections or tissues for enzyme histochemistry is sometimes carried out by adding colloid protecting agents (e.g., polyvinylalcohol) to the staining solution or by coating the preparation surface with semipermeable membranes (e.g., glycol methacrylate).

These techniques prevent the diffusion of enzymes away from the sample while allowing for the movement of low molecular weight stain components to the sample surface (Van Noorden and Vogels, 1989). A similar problem may be encountered in the application of substrate to unfixed cell culture preparations. The inclusion of gelatin to the incubation media has been used to decrease the diffusion of acetylcholinesterase activity from its normal site (Hefti *et al.*, 1989).

One traditional approach to stabilizing tissue specimens is fixation (Hopwood, 1991). Although fixation often maintains the architectural integrity of the tissue, enzyme histochemistry may be compromised by fixation-induced denaturation of enzymes and other proteins. Formaldehyde and glutaraldehyde, for example, crosslink proteins and may alter enzyme substrate specificity or activity. Fixation may be carried out by the immersion of a block of tissue in an excess of fixative. Fixation by perfusion, the removal of blood by the infusion of the first isotonic saline followed by

fixative, is a viable alternative in studies conducted on rodents and other small laboratory animals. Perfusion is generally superior to immersion fixation for the preservation of the central nervous system and is associated with fewer artifacts, such as astroglial swelling resulting from postmortem cellular hypoxia. Fixation of tissues or cell cultures with ice-cold ethanol or acetone has been used for the study of esterases and phosphatases in the olfactory sensory epithelium of rodents with good preservation of the neural component of this tissue (Randall *et al.*, 1987). Ice-cold acetone fixation combined with cold glycol methacrylate processing and embedding also stabilize formaldehyde dehydrogenase activity in the central nervous system (Keller *et al.*, 1990). Acetone and ethanol denaturation is compatible with paraffin-embedding techniques and may have improved maintenance of enzyme activity when compared to fixation with aldehydes. Microwave irradiation has also been successfully applied to brain tissues immersed in saline followed by cryostat sectioning (Marani *et al.*, 1988) or paraffin embedding (Marani *et al.*, 1987). Trial and error are often required, however, for the identification of the "ideal" fixative or stabilization technique for many tissues and enzyme combinations.

Histochemical staining may be performed on whole cells (e.g., neural cell cultures) or tissue blocks. In most cases, however, samples are formed into thin layers either before or after stabilization. Enzyme histochemistry may incorporate the application of confocal microscopic techniques to visualize thick-stained specimens, or ultrastructural studies of normal tissue enzymes may be performed (Rutenberg *et al.*, 1969; Ogawa and Barka, 1993). Molecular biological techniques have been developed to incorporate specific enzymes as markers into tissues, and the location of these markers may be subsequently demonstrated at the ultrastructural level (Ferrante *et al.*, 1993). However, the routine microscopic evaluation of 5-μm-thick tissue sections is used more commonly in neurotoxicology. Thicker sections (>10 μm) may be required to visualize neural synaptic nets. Common methods to obtain thin tissue sections include using a cryostat for frozen sections or the use of a suitable microtome on embedded samples. Fixed frozen sections may also provide a good approach, using a cold support to maintain freezing of the tissue combined with a knife at room temperature. Two widely used embedding media are paraffin wax and plastic resins. Either medium is infiltrated into the tissue as a liquid or monomer and then solidified by either cooling (e.g., paraffin) or polymerization (plastics). Resinous media permit cutting thinner sections and allow harder specimens to be sectioned than does paraffin. Thin sections or cryostat-prepared frozen sections are then commonly transferred to a microscope slide for subsequent processing.

Once suitable specimens have been obtained, they are stained with the objective of producing a stable-colored reaction product at or near the active site of the enzyme. Wax-embedding media are generally dissolved away from the specimen prior to staining. In contrast, most cryostat sections and many resin-embedded sections are stained directly. Any embedding method has associated artifactual changes that must be accounted for when examining a specimen, and appropriate controls are essential. The stained samples are generally mounted with a resin or oil film and cover-slipped. These mounted specimens may be subject to storage-related damage or loss of staining intensity. Knowledge of such artifacts should be incorporated into protocols for tissue storage and examination.

The most common precipitation reactions that are used in quantitative enzyme determinations in brain sections are those with metallic ions, as for the determination of acetylcholinesterase (Kugler, 1988). The diaminobenzidine method has proven useful for the quantitation of cytochrome *c* oxidase (Kugler *et al.*, 1988). Numerous dehydrogenases and reductases may be demonstrated using tetrazolium methods in which a water-soluble tetrazolium salt is reduced to a water-insoluble formazan chromophore by hydrogen ions that originate in dehydrogenase reactions (Kugler, 1990b).

III. Enzyme Substrates and Sample Staining

Once a suitable specimen has been obtained, then they are stained with an appropriate agent. Most commonly, a substrate known to be metabolized by the enzyme in question is applied to the tissue sample in an appropriate incubation medium for a suitable duration of time. As with most enzyme-catalyzed reactions the amount of product formed is dependent on both the substrate concentration and enzyme activity and amount. For an enzyme that demonstrates first-order kinetics there is a maximal rate of product formation, and the addition of excessive amounts of substrate will not lead to increased product formation. Awareness of underlying biochemical principles is mandatory during the development of enzyme histochemical staining procedures. Knowledge of metabolic pathways influencing biotransformation of xenobiotics during the application of enzyme histochemistry is essential for successful application of these procedures to neurotoxicology.

Similar to biochemical assays, enzyme histochemical techniques are influenced by many factors during incubation. For example, acetylcholinesterase activity in neural tissues is pH sensitive with maximal activity

occurring at different media pH depending on the tissue and species examined. Axonal acetylcholinesterase activity as demonstrated using enzyme histochemistry was more intense at pH 8.0 in human brain tissue (Mesulam and Moran, 1987) whereas rat cerebral cortical tissues do not display this differential staining intensity at either pH 6.8 or 8.0 (Emre *et al.*, 1992). Careful consideration should be given to selection of substrates, especially for enzymes or enzyme families capable of metabolizing a wide range of related chemicals. For instance, a number of esterases are actively under investigation using both biochemical and histochemical methods for assessment of toxic responses in the neural olfactory mucosa of laboratory rodents (Bogdanffy, 1990) and humans (Mattes and Mattes, 1992). These studies have revealed marked differences in substrate specificity for carboxylesterases in the olfactory mucosa, making interspecies comparisons of toxic responses a complex issue. It is recommended that wherever possible the substrate used in enzyme histochemistry be identical to that of interest to the toxicologist (e.g., substrates used in conventional biochemical assays). It is important to keep in mind the fact that marked differences in enzyme kinetics, and thus staining intensity, may be associated with different alternative substrates for a single enzyme or group of isozymes present in the tissue under investigation.

IV. Qualitative Evaluation of Staining Intensity

Applying an enzyme histochemical staining method may or may not result in the production of a demonstrable coloration of the tissue specimen. If specimen staining does occur, then its specificity and localization must be determined. In addition the relationship between the staining intensity and amount of enzyme present should be established. Specificity of an enzyme stain can be established by application of the enzyme stain to a sample pretreated with a blocking agent (e.g., a specific enzyme inhibitor or denaturing agents). If such blockage fails to inhibit the staining characteristics of a sample, then nonspecific (i.e., nonenzymatic) staining may be occurring.

Along with nonspecific staining, enzyme staining may also occur in "inappropriate" or unexpected locations. Inaccurate enzyme localization may occur if the enzyme in question has diffused from its normal *in vivo* location following either sample processing or staining. In addition, following staining, the chromagen may also diffuse from its original site of generation. For instance, the red reaction product in cold GMA sections stained for γ-glutamyl transpeptidase (Rutenberg *et al.*, 1969) is stable at $-20°C$ but rapidly diffuses from its original location at room temperature. The amount of stain present may also be influenced by the sample. Inter-sample variation in staining intensity may be the result of variations in sample thickness (commonly observed in cryostat-sectioned frozen samples) in addition to true sample variation. The stain distribution patterns produced by different fixatives, processing regimens, staining techniques, and so forth should be determined for each tissue and stain combination.

The absence of staining may be interpreted as a lack of enzyme activity. Negative artifacts may also arise, however, due to loss of enzyme activity prior to staining or to loss of stain intensity from successfully stained specimens. Enzymes vary widely in their stability, and a thorough understanding of the necessary storage conditions is essential. Failure to stain may also represent technical failure or a lack of sensitivity in the staining method. Biochemical assays of representative tissue samples may be needed to demonstrate the presence of the anticipated enzyme activity in the sample in question. Staining procedures may be compared to an enzyme-rich test specimen to validate technical methods. It is strongly recommended that histochemical studies be carried out in combination with appropriate biochemical studies whenever possible.

V. Quantitative Evaluation of Staining Intensity

The determination of the amount of enzyme activity present is known as quantitative enzyme histochemistry (Van Noorden and Gossrau, 1991). Quantitative enzyme histochemistry is most commonly performed by using microdensitometers or microscope photometers (Kugler, 1988). Ultramicrochemistry may be used to analyze the enzyme activity in single cells or organelles (Ogawa and Barka, 1993) and will not be considered further in this chapter. In cytospectrophotometry, cell preparations or tissue sections are treated histochemically for the demonstration of a parameter of interest (e.g., enzyme activity). After the staining procedure, the tissue or cell preparation is examined cytophotometrically. The amount of the chromophore present within an area of a tissue section or cell is measured most commonly by a photometer fitted to a microscope. Microphotometric measurements can be performed as kinetic (i.e., continuous monitoring of enzyme reaction) or end point measurements (i.e., single measure after predetermined incubation interval; Kugler, 1991). Staining intensity may be converted into gray values by means of a video camera attached to a microscope which can produce a digital picture via an image analysis system. It is critical in all cases that the amount of chromophore present correlate with the enzymatic function to be studied. In general, quantitative enzyme histochemistry is more time consuming and requires greater expertise and instrumentation than biochemical assays.

In most quantitative enzyme histochemical studies the data are presented in arbitrary gray scale units. Therefore, comparison of data from different tissues or preparations or from different laboratories is difficult. Direct comparisons between biochemical assays and enzyme histochemical data are also problematic. Enzyme activity can be expressed in absolute units when the stoichiometry of the enzyme stain reaction and tissue section thickness and area, as well as the chromophore molecular extinction coefficient, are known. However, relative values for treatment groups within a study, with respect to these end points, may be very informative.

Quantitation requires the assessment of staining intensity. The major approaches used to assess staining intensity include the extraction of stain from a histologic preparation with comparison to photomicrographs of the stained specimen. This method assumes that the staining is specific to the tissue of interest (i.e., little background staining). More commonly, analysis systems (e.g., localized microdensitometry or computerized image analysis) are used to perform *in situ* determination of the staining intensity of a selected area of a specimen.

Results from these image analyses may be influenced by neurotoxicant treatment and histological sample preparation and handling, as well as the staining technique used. The staining intensity of tissues from toxicant-exposed animals or neural cultures may then be compared to untreated control specimens. Staining intensity could also be calibrated using samples of known enzyme activity. Alternatively, biochemical assays of enzyme activity and quantitative enzyme histochemistry can be performed in parallel on the same tissue sample.

Several reports on quantitative enzyme histochemistry in the mature and developing brain are available. Quantification has been reported for numerous enzymes, including cytochrome oxidase (Kugler *et al.*, 1988), succinate dehydrogenase (Kugler *et al.*, 1988), hexokinase (Kugler, 1990c), GABA transaminase (Kugler and Baier, 1990; Kugler, 1993), NAD-linked glutamate dehydrogenase (Kugler, 1990d), NAD-linked isocitrate dehydrogenase, NAD-linked malate dehydrogenase, glycerol-3-phosphate dehydrogenase, and acetylcholinesterase (Kugler, 1988). Chieco and co-workers (1988) used microphotometry to examine the distribution of aldehyde, glucose-6-phosphate, lactate, malate, NAD-linked glycerophosphate, and succinate dehydrogenases, as well as α-glycerophosphate-menadione oxidoreductase and cytochrome c oxidase in the brain and trigeminal ganglion of the developing rat fetus.

VI. Enzyme Histochemistry Stains of Interest to Neurotoxicologists

Several histochemical enzyme reactions have been used as probes for the functional integrity of cell membranes and metabolic compartments in toxicologic studies. For instance, damage to the cerebral cortex of sheep suffering from a thiaminase disease, polioencephalomalacia, results in almost total tissue destruction in affected regions of the brain. However, staining of cerebrocortical capillaries for alkaline phosphatase using the Gomori method (Thompson and Hunt, 1966) clearly demonstrates that the capillary bed of these regions is intact, supporting the proposal that this condition, which can be mimicked by poisoning with thiamine antimetabolites, is the result of a cytopathic rather than a vasogenic edema (Morgan, 1972). The histochemical analysis of cellular enzymes may also prove useful in the assessment of neurotoxicity resulting from exposure to xenobiotics (Chieco *et al.*, 1988).

A. Xenobiotic Metabolism

Enzyme histochemistry can be used to examine the nervous system for the presence or absence of enzymes involved in the metabolism of xenobiotics. Many neurotoxicants may be activated or detoxified by metabolic enzyme systems (e.g., cytochromes P450, esterases,

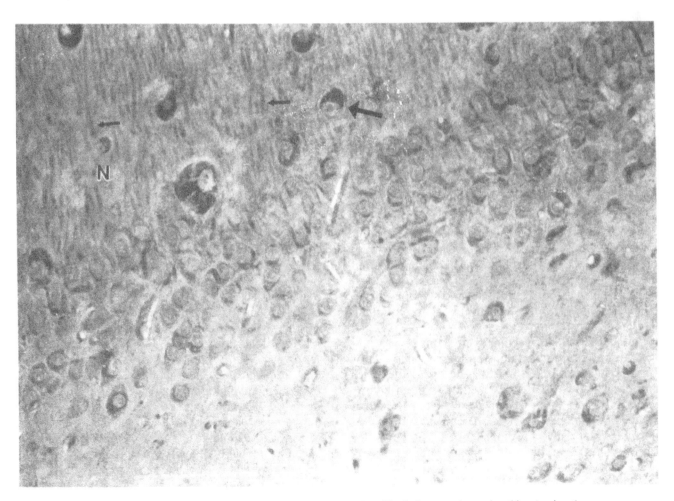

FIGURE 1 Rat brain sections were stained for nonprotein sulfhydryl groups (e.g., glutathione) using the methods of Larrauri *et al.* (1987) as described by Keller and co-workers (1990). Brain neuropil stains strongly, whereas neuronal perikarya (large arrows) and their processes (small arrows) in the gray matter are generally negative.

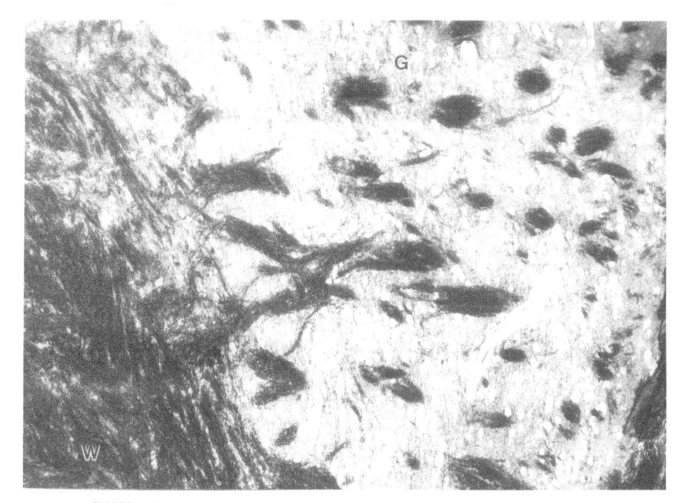

FIGURE 2 Rat brain sections stained for formaldehyde dehydrogenase were obtained using the techniques of Keller and co-workers (1990). Note the intense staining of white matter containing myelinated axons.

etc.). The presence or absence of these metabolic pathways may contribute to the differential cellular toxicity observed with many toxicants. Thus, the identification of intracellular sites at which xenobiotics are activated and detoxified is critical to our understanding of the biochemical basis of the cell-selective nature of many neurotoxicants.

In Figure 1 rat brain sections were stained for sulfhydryl groups (e.g., glutathione) using the methods of Larrauri and co-workers (1987) as described by Keller and co-workers (1990). This method is described as specific for glutathione when staining times of less than 5 min are used. Glycol methacrylate monomer (GMA)-imbedded rat brains were dipped in 9 : 1 acetone : water for 30 sec to 2 min, then rinsed in acetone : water, air-dried, and coverslipped. Sulfhydryls were visualized by fluorescence microscopy, with excitation wavelengths of 450–490 nm and emission being monitored at >515 nm.

Rat brain sections stained for formaldehyde dehydrogenase were obtained using the techniques of Keller

and co-workers (1990), (Figure 2). Oxidation of S-hydroxymethylglutathione by FDH results in the reduction of NAD^+ to NADH, and NADH in turn reduces nitroblue tetrazolium to a blue formazan precipitate. Tissue sections were incubated in Coplin jars for 24 to 120 hr at room temperature in an incubation medium containing 70 mM phosphate buffer (pH 7.5), 4.8 mM formaldehyde, 1 mM glutathione, 1.2 mM NAD^+, 1.5 mM pyrazole, 20 μM disulfiram, and 0.24 mM nitroblue tetrazolium. The substrate, S-hydroxymethylglutathione, concentration was approximately 0.73 mM. Pyrazole was added to inhibit alcohol dehydrogenase-catalyzed reduction of formaldehyde to methanol, while disulfiram was used to inhibit nonspecific aldehyde dehydrogenases. Stain specificity was determined by incubating tissue sections in solutions lacking glutathione which prevented the formation of S-hydroxymethylglutathione.

Glutathione (γ-glutamylcysteinylglycine, GSH) is a ubiquitous tripeptide that protects tissues from the toxic effects of endogenous substances and xenobiotics. Although glutathione is not an enzyme, it may function as a co-substrate in certain enzyme-catalyzed reactions (Keller *et al.*, 1990) and it plays an important role in the phase II reactions (generally conjugations) of xenobiotic biotransformation (Sipes and Gandolfi, 1991). A brief description of GSH is in order. The cysteinyl thiol group of GSH may conjugate electrophilic toxicants either nonenzymatically or via glutathione *S*-transferase to form a premercapturic compound (Reed and Beatty, 1980). Glutathione also provides reducing power to prevent oxidative damage to tissues. The neurotoxic effects of some chemicals may be partly modulated by brain GSH levels. For example, manganese exposure to rats results in a depletion of brain GSH and glutathione peroxidase (Liccione and Maines, 1988). Rapid tissue reactivity (3- to 4-min incubation) with the sulfhydryl reagent mercury orange [1-(4-chloromercuriphenyazo)-2-naphthol] may be used to visualize small soluble tissue sulfhdryl groups including those found in GSH. Histochemical localization of GSH indicates a primarily glial distribution that can be depleted by exposure to the toxic metabolite of styrene, styrene oxide (Trenga *et al.*, 1991), or the glutathione depletor diethyl maleate (Slivka *et al.*, 1987).

B. Enzymes as Targets of Neurotoxicant Action

Many toxic responses result from the inhibition of enzymes through allosteric, competitive, or noncompetitive mechanisms. Some toxic agents inhibit certain classes of enzymes while others may inhibit a specific enzyme. Studies of enzyme changes using histochemical techniques have provided important information on the effect of toxic agents on the nervous system. For instance, acetylcholinesterase is intimately involved in the termination of acetylcholine action at cholinergic synapses. Organophosphate and carbamate insecticides are commonly encountered neurotoxicants that inhibit acetylcholinesterase leading to increased acetylcholine concentrations at cholinergic synapses. Another related cholinesterase is butyryl or pseudocholinesterase. Decreased brain acetylcholinesterase activity has been observed in humans (Finkelstein *et al.*, 1988) and animals (Hahn *et al.*, 1991) lethally poisoned with these insecticides.

Among the diverse metabolic activities of the neuron, ion pumping requires the most ATP to maintain the ionic concentration gradients required to generate neuronal membrane potentials. The most important ion pump in terms of energy requirements is the sodium–potassium ATPase with this sole enzyme consuming 40 to 60% of all brain ATP (Ericinska and Silver, 1989). The supply of ATP is primarily dependent on oxidative glucose metabolism which is influenced by the activity of cytochrome oxidase. Cytochrome oxidase is the terminal component of the mitochondrial electron transport chain and is an integral protein complex of the inner mitochondrial membrane. This enzyme participates in the reduction of oxygen to water with the consequent synthesis of ATP. Mitochondrial cytochrome oxidase and membrane-bound Na^+/K^+ ATPase are common targets for a variety of neurotoxicants that disrupt neural energy metabolism. Both enzyme complexes may be visualized using enzyme histochemical techniques (Hevner and Wong-Riley, 1989; Mayahara and Ogawa, 1988). For example, Hevner and co-workers (1992) demonstrated tetrodotoxin-induced inhibition of Na^+/K^+ ATPase in the monkey retina following intraocular injection. Likewise, Kim and co-workers (1987) demonstrated a decrease in both cytochrome oxidase and Na^+/K^+ ATPase in the choroid plexus of rabbits given the organic acid herbicide 2,4,5-trichlorophenoxyacetic acid. Decreased cytochrome oxidase activity has also been observed in cerebellar Purkinje cells from copper-deficient suckling mice given D-penicillamine (Yamamoto *et al.*, 1990).

An example of acetylcholinesterase histochemistry applied to neural cultures is presented in Figure 3. Mature primary dissociated cerebrocortical cultures containing both neurons and glia (primarily astrocytes) were prepared from gestational day -15 fetal CD-1 mice as described previously by Bolon and co-workers (1993). Acetylcholinesterase staining of formaldehyde-fixed cultures was prepared using the method described by Gahwiler and co-workers (1989). Fixed cultures were rinsed once with 0.1 *M* phosphate buffer (pH 7.3) followed by a rinse with an aqueous incubation stock solution containing 2 m*M* cupric sulfate (pentahydrate), 10 m*M* glycine, 15 m*M* acetic acid, and 35 m*M* sodium acetate. The pH of this incubation mixture was adjusted to 5.0 using either 0.2 *M* acetic acid or 0.2 *M* sodium acetate. Cultures were then exposed to an incubation stock solution containing the substrate, acetythiocholine iodide (4 m*M*), and a blocking agent, ethopropazine (0.2 m*M*), for 24 hr at 37°C. Cultures were then rinsed with distilled water (5×). The reaction product was visualized by the addition of a 1.25% aqueous sodium sulfide solution for 15 min. Cultures were then rinsed again with distilled water (5×) and exposed to a 1% aqueous solution of silver nitrate for 5 min (to intensify stain production). Cultures were then rinsed with water and photographed.

Dorman and co-workers (1993) have also demonstrated formate-induced neurotoxicity to large pyramidal neurons in a primary neural cell culture system through the use of mitochondrial 3-[4,5-dimethylthiazol-2-yl]-2,5-diphenyltetrazolium bromide (MTT) cyto-

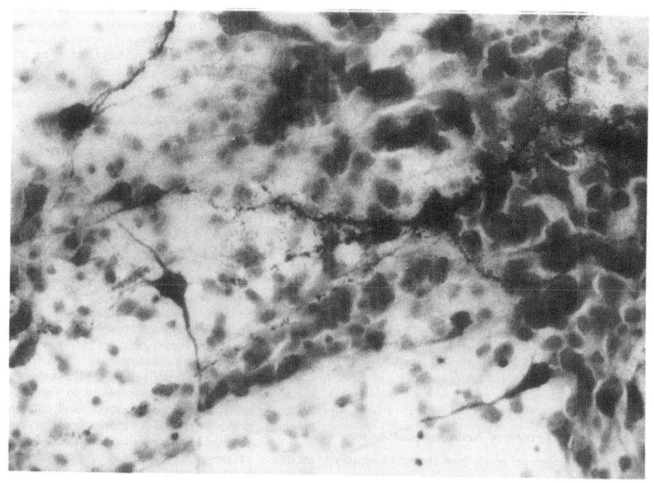

FIGURE 3 Acetylcholinesterase-positive neurons in a mature mouse primary neural cell culture. Acetylcholinesterase staining of formaldehyde-fixed cultures was prepared using the method described by Gahwiler and co-workers (1989). Note granular staining around neuronal processes.

chemical staining. Formic acid weakly inhibits cytochrome oxidase with an apparent inhibition constant between 5 and 30 mM (Nicholls, 1976). Inhibition occurs secondarily to the binding of formic acid with the ferric heme iron of cytochrome oxidase (Keyhani and Heyhani, 1980). Succinate-dependent MTT reduction occurs between cytochrome c and cytochrome oxidase and also with succinate-ubiquinone oxidoreductase (Berridge and Tan, 1992).

In Figure 4 similarly prepared primary neural cell cultures have been stained for MTT reactivity. The procedures used in this study were based on the methods of Dorman and co-workers (1993), with minor modifications. Another group of cultures were treated with 5 mM iodoacetamide, an inhibitor of MTT metabolism, to act as a positive control (no staining). The medium was removed by suction, and cells were rinsed once with 2.0 ml of 0.1 M phosphate buffer saline (pH 7.6). One milliliter of Eagle's MEM supplemented with glu-

tamine (2 mM), glucose (total 21 mM), and bicarbonate (26 mM) and containing 0.5 mg MTT/ml was then added to each well, and plates returned to the incubator for another hour. The MTT treatment was terminated by removal of the media, followed by two gentle rinsings of the cells with 0.15 M Dulbecco's phosphate buffer (pH 7.2).

C. Enzymes as Cellular Markers for Neurotoxicity

Although rarely considered as an enzyme histochemical technique, the visualization of exogenous horseradish peroxidase has been widely used by neurotoxicologists to study chemical-induced damage to vascular endothelium as well as a neuroanatomical tracing technique. For instance, the earliest evidence of damage to the brains of mice injected with *Clostridium perfringens (welchii)* toxin is leakage of intravascularly administered horseradish peroxidase, which occurs within 20 min of toxin treatment (Morgan *et al.*, 1975). Horseradish peroxidase is also used extensively for the study of neuroanatomy. Following injection into discrete cerebral areas, this enzyme is taken up by synaptic termini and nerve cell bodies (Damstra and Bondy, 1982), permitting detailed mapping of arborizations of neuronal processes.

A commonly used enzyme marker of endothelial integrity is alkaline phosphatase. Figure 5 is a photomicrograph of the cerebral cortex of a sheep killed *in extremis* while dying of the thiaminase type II disease, polioencephalomalacia. This formaldehyde-fixed frozen section was stained with the Gomori's method for alkaline phosphatase. An aqueous incubation solution containing substrate, sodium β-glycerophosphate (20 mM), calcium chloride (70 mM), magnesium chloride (5 mM), and sodium barbital (100 mM) was adjusted to pH 9.2 with 0.1 N NaOH. This substrate incubation solution was added to tissue sections mounted on glass slides or free floating using a glass "hockey stick" for transfer between solutions for 1 to 6 hr at 37°C. After incubation, the samples were rinsed twice with distilled water and an aqueous cobalt sulfate solution (2%) was added for 3 min. The samples were rinsed three times with distilled water and an aqueous 1% ammonium sulfide solution was added for 2 min. Samples were then water-rinsed twice and an aqueous nuclear red stain containing 3 mM nuclear red and 150 mM aluminum sulfate was added for 5 min. Following staining, samples were dehydrated and mounted using alcohol and xylenes.

Along with its potential as a target site for certain neurotoxicants, acetylcholinesterase activity has also been used as a marker for cholinergic neurons. For example, acetylcholinesterase activity in the hippocampus and striatum is significantly reduced following the administration of a saporin-conjugated monoclonal antibody (to nerve growth factor) to rats (Nilsson *et al.*, 1992). A similar histochemically demonstrated loss of cholinergic neurons has been reported in rodents following exposure to the excitotoxin quisqualic acid (Unger and Schmidt, 1993), colchicine (Tandon *et al.*, 1991), and lead (Alfano *et al.*, 1983). Disulfiram administration to rats has also been associated with decreased ileal acetylcholinesterase activity in animals that developed peripheral neuropathy (Savolainen *et al.*, 1984). Immunohistochemical techniques that stain for choline acetyltransferase can also be used to complement enzyme histochemistry in the visualization of neurotoxicant-induced damage to cholinergic neurons.

5'-Nucleotidase hydrolyzes 5-AMP to adenosine and is associated with the modulation of purinergic neurotransmission (Kreutzberg *et al.*, 1986). In addition, 5'-nucleotidase has been ascribed an adhesive function in nervous tissue because of its *in vitro* binding affinity for laminin and fibronectin (Schoen and Graybiel, 1992). Enzyme histochemical staining for 5'-nucleotidase is also a useful marker for striosomes (opiate patches) in rat caudoputamen. Destruction of intrastriatal neurons and gliosis with increased 5'-nucleotidase activity occurs in rats following caudoputamenal injection of an excitatory neurotoxicant, ibotenic acid. Other enzyme histochemical alterations associated with animal exposure to excitatory neurotransmitters include changes on acetylcholinesterase (Boegman and Parent, 1988), γ-aminobutyrate, and NADPH-diaphorase-reactive neurons (Emerich *et al.*, 1991; Sagar, 1990).

Methyl mercury exposure in young rats results in clinical signs including psychomotor retardation, microcephaly, hyperreflexia, ataxia, paralysis, blindness, and seizures. Reduced NADPH-diaphorase activity indicative of neuron loss has been demonstrated in methyl mercury-exposed juvenile rats through the use of histochemical techniques (O'Kusky *et al.*, 1988). NADPH-diaphorase histochemistry selectively stains striatal neurons that contain both somatostatin and neuropeptide Y (Kowall *et al.*, 1987; Vincent *et al.*, 1988; Scherer-Singler *et al.*, 1983). Interestingly, the messenger RNA for both NADPH-diaphorase and nitric oxide synthase colocalize, and nitric oxide synthase itself can give a positive NADPH-diaphorase stain (Bredt *et al.*, 1991). This observation illustrates the need for caution in the interpretation of results obtained by enzyme histochemistry.

VII. Future Directions of Enzyme Histochemistry in Neurotoxicology

It is evident that enzyme histochemistry will continue to be a valuable tool for the neurotoxicologist. However, in recent years many of the classic enzyme histochemical stains are being superceded by immunocytochemical procedures, to locate enzyme proteins, or molecular biological techniques, such as *in situ* hybridization for the localization of mRNA that translates these proteins. The latter techniques provide exquisite details on the location of enzyme proteins, the genes from which they are transcribed, and the ribosomal machinery that permits their translation. However, it is important to remember that in many cases these procedures do not confirm the normal tertiary structure of the active site or that catalytic activity is intact in the location demonstrated. Furthermore, such issues as substrate specificity cannot be studied directly. If

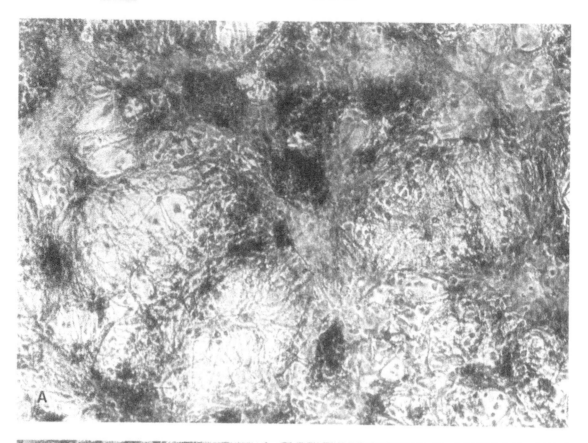

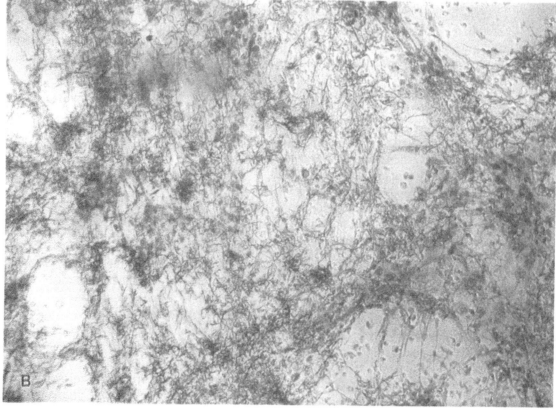

FIGURE 5 Cerebral cortex of a sheep killed *in extremis* while dying of the thiaminase type II disease, polioencephalomalacia. This formaldehyde-fixed frozen section was stained with the Gomori's method for alkaline phosphatase. The lower right of the figure indicates intense staining of the neuropil at the margins of the necrotic cortex. Cerebrocortical capillaries are evidently intact, however, as indicated by their staining for alkaline phosphatase; this observation was confirmed by electron microscopy.

investigators are aware of these limitations appropriate interpretations will be made.

Although a valuable tool, enzyme histochemistry does have important limitations. For instance, a number of enzymes that play a crucial role in xenobiotic transformation are not readily localized by enzyme histochemistry. Notable among such enzymes are the cytochromes P450 of the mixed function oxidase (MFO) system. It has been known for many years that the latter enzymes play a central role in xenobiotic metabolism in the liver (Sipes and Gandolfi, 1991). More recently the MFO system has been found to be very active in the biotransformation of xenobiotics in the

olfactory mucosa of the nose (Dahl and Hadley, 1991); in this location these enzymes probably play a crucial role in the sense of smell (Dahl, 1988). Less is known about the role of the MFO system in the central nervous system. However, studies of the cellular and subcellular localization of specific isoenzymes of cytochrome P450 require demonstration of enzyme activity using specific and relevant substrates. The development of such approaches has considerable potential for work on interactions between xenobiotics, systemically (or hepatically) generated metabolites of these compounds, and toxic responses in the central nervous system.

FIGURE 4 Control neural cell culture showing intense neuronal MTT staining (A). Similarly prepared neural cell cultures treated with 5 m*M* iodoacetamide treatment (B) demonstrate reduced mitochondrial activity manifested by decreased neuronal MTT reactivity. Staining procedures used in this study were based on the methods of Dorman and co-workers (1993).

Occasionally, enzyme histochemistry may reveal enzyme activity in the nervous system as an incidental finding. This was true of recent studies of formaldehyde dehydrogenase (FDH) in this laboratory (Keller *et al.,* 1990). Extensive FDH activity was found in the white matter of the central nervous system and in myelinated peripheral nerves in the nose of untreated control rats; the function of this enzyme(s) in these sites is unknown. Perhaps it plays a role in scavenging formaldehyde, thus reducing the risk of crosslinks and protecting the fluidity of the myelin sheaths. Observations such as these, encountered during the application of enzyme histochemistry to toxicology studies, should ideally be exploited in investigations of the normal biology of the nervous system.

References

Alfano, D. P., Petit, T. L., and LeBoutillier, J. C. (1983). Development and plasticity of the hippocampal-cholinergic system in normal and early lead exposed rats. *Brain Res.* **312,** 117–124.

Berridge, M. V., and Tan, A. S. (1992). The protein kinase C inhibitor, calphostin C, inhibits succinate-dependent mitochondrial reduction of MTT by a mechanism that does not involve protein kinase C. *Biochem. Biophys. Res. Commun.* **185,** 806–811.

Boegman, R. J., and Parent, A. (1988). The response of striatal neuropeptide Y and cholinergic neurons to excitatory amino acid agonists. *Brain Res.* **452,** 219–226.

Bolon, B., Dorman, D. C., Bonnefoi, M. S., Randall, H., and Morgan, K. T. (1993). Neuropathologic endpoints of chemical toxicity in primary dissociated neural cell cultures: application of routine microscopic techniques. *Toxicol. Pathol.* **21,** 465–479.

Bogdanffy, M. (1990). Biotransformation enzymes in the rodent nasal mucosa: the value of a histochemical approach. *Environ. Health Perspect.* **85,** 177–186.

Bredt, D. S., Hwang, P. M., Glatt, C. E., Lowenstein, C., Reed, R. R., and Snyder, S. H. (1991). Cloned and expressed nitric oxide synthase structurally resembles cytochrome P-450 reductase. *Nature* **351,** 714–718.

Chieco, P., Hrellia, P., Lisignoli, G., and Cantelli-Forti, G. (1988). Quantitative enzyme histochemistry of rat foetal brain and trigeminal ganglion. *Histochem. J.* **20,** 455–463.

Dahl, A. R. (1988). The effect of chtochrome P-450-dependent metabolism and other enzyme activities on olfaction. In *Molecular Neurobiology of the Olfactory System, Molecular, Membranous, and Cytological Studies* (F. L. Margolis, and T. V. Getchell, Eds.), pp. 51–70. Plenum Press, New York.

Dahl, A. R., and Hadley, W. H. (1991). Nasal cavity enzymes involved in xenobiotic metabolism: effects on the toxicity of inhalants. *Crit. Rev. Toxicol.* **21,** 345–372.

Damstra, T., and Bondy, S. C. (1982). Neurochemical approaches to the detection of neurotoxicity. In *Nervous System Toxicology* (C. L. Mitchell, Ed.), pp. 349–373.

Dorman, D. C., Bolon, B., and Morgan, K. T. (1993). The toxic effects of formate in dissociated primary mouse neural cell cultures. *Toxicol. Appl. Pharmacol.* **122,** 265–272.

Emerich, D. F., Zubricki, E. M., Shipley, M. T., Norman, A. B., and Sanberg, P. R. (1991). Female rats are more sensitive to the locomotor alterations following quiolinic acid-induced striatal lesions: effects of striatal transplants. *Exp. Neurol.* **111,** 369–378.

Emre, M., Geula, C., Ransil, B. J., and Mesulam, M. M. (1992). The acute neurotoxicity and effects upon cholinergic axons of intracerebrally injected β-amyloid in the rat brain. *Neurobiol. Aging* **13,** 553–559.

Ericinska, M., and Silver, I. A. (1989). ATP and brain function. *J. Cereb. Blood Flow Metab.* **9,** 2–19.

Ferrante, R. J., Kowall, N. W., Cipolloni, P. B., Storey, E., and Beal, M. F. (1993). Excitotoxin lesions in primates as a model for Huntington's disease: histopathologic and neurochemical characterization. *Exp. Neurol.* **119,** 46–71.

Finkelstein, Y., Wolff, M., and Biegon (1988). Brain acetylcholinesterase after parathion poisoning: a comparative quantitative histochemical analysis post-mortem. *Toxicology* **49,** 165–169.

Gahwiler, B. H., Zimmer, J., and Robertson, R. T. (1989). Staining for acetylcholinesterase. In *A Dissection and Tissue Culture Manual of the Nervous System,* pp. 306–307. Liss, New York.

Gale, K., Sarvey, C., Stork, J., Childs, J. A., Yalisove, B. L., and Dayhoff, R. E. (1984). Quantitative histochemical measurement of GABA-transaminase: method for evaluation of intracerebral lesions produced by excitotoxic agents. *Brain Res.* **307,** 255–262.

Hahn, T., Ruhnke, M., and Luppa, H. (1991). Inhibition of acetylcholinesterase and butyrylcholinesterase by the organophosphorus insecticide methylparathion in the central nervous system of the golden hamster (*Mesocricetus auratus*). *Acta Histochem.* **91,** 13–19.

Hefti, F., Hartikka, J., and Sanchez-Ramos, J. (1989). Dissociated cholinergic neurons of the basal forebrain in culture. In *A Dissection and Tissue Culture Manual of the Nervous System,* pp. 172–182, Liss, New York.

Hevner, R. F., Duff, R. S., and Wong-Riley, M. T. T. (1992). Coordination of ATP production and consumption in brain: parallel regulation of cytochrome oxidase and Na^+, K^+-ATPase. *Neurosci. Lett.* **138,** 188–192.

Hevner, R. F., and Wong-Riley, M. T. T. (1989). Brain cytochrome oxidase: purification, antibody production, and immunohistochemical/histochemical correlations in the CNS. *J. Neurosci.* **9,** 3884–3898.

Hopwood, D. (1991). Fixation of tissue of histochemistry. In *Histochemical and Immunohistochemical Techniques* (P. H. Bach and J. R. J. Baker, Eds.), pp. 147–167. Chapman & Hall, London.

Horobin, R. W. (1991). Why use histochemistry for a better understanding of pharmacology and toxicology. In *Histochemical and Immunohistochemical Techniques* (P. H. Bach and J. R. J. Baker, Eds.), pp. 1–11. Chapman & Hall, London.

Keller, D. A., Heck, H. D'A, Randall, H. W., and Morgan, K. T. (1990). Histochemical localization of formaldehyde dehydrogenase in the rat. *Toxicol. Appl. Pharmacol.* **106,** 311–326.

Keyhani, J., and Keyhani, E. (1980). EPR study of the effect of formate on cytochrome c oxidase. *Biochem. Biophys. Res. Commun.* **92,** 327–333.

Kim, C. S., Keizer, R. F., Ambrose, W. W., and Breese, G. R. (1987). Effects of 2,4,5-trichlorophenoxyacetic acid and quinolinic acid on 5-hydroxy-3-indoleacetic acid transport by the rabbit choroid plexus: pharmacology and electron microscopy cytochemistry. *Toxicol. Appl. Pharmacol.* **90,** 436–444.

Kowall, N. W., Ferrante, R. J., Beal, M. F., Richardson, E. P., Jr., Sofroniew, M. V., Cuello, A. C., and Martin, J. B. (1987). Neuropeptide Y, somatostatin, and reduced nicotinamide adenine dinucleotide phosphate diaphorase in the human striatum a combined immunocytochemical and enzyme histochemical study. *Neuroscience* **20,** 817–828.

Kreutzberg, G. W., Heymann, D., and Reddington, M. (1986). 5-Nucleotidase in the nervous system. In *Cellular Biology of Ectoenzymes* (G. W. Kreutzberg and M. Reddington, and H. Zimmermann, Eds.), pp. 147–164. Springer, New York.

Kugler, P. (1988). Quantitative enzyme histochemistry in the brain. *Histochemistry* **90**, 295–298.

Kugler, P. (1990a). Enzyme histochemical methods applied to the brain. *Eur. J. Morphol.* **28**, 109–120.

Kugler, P. (1990b). Quantification of enzyme activities in brain sections by microphotometry. *Eur. J. Morphol.* **23**, 657–661.

Kugler, P. (1990c). Microphotometric determination of enzymes in brain sections. I. Hexokinase. *Histochemistry* **90**, 295–298.

Kugler, P. (1990d). Microphotometric determination of enzymes in brain sections. III. Glutamate dehydrogenase. *Histochemistry* **93**, 537–540.

Kugler, P. (1993). *In situ* measurements of enzyme activities in the brain. *Histochem. J.* **25**, 329–338.

Kugler, P., and Baier, G. (1990). Microphotometric determination of enzymes in brain sections. II. GABA transaminase. *Histochemistry* **93**, 501–505.

Kugler, P., Vogel, S., Volk, H., and Schiebler, T. H. (1988). Cytochrome oxidase histochemistry in the rat hippocampus. A quantitative methodological study. *Histochemistry* **89**, 269–275.

Larrauri, A., Lopez, P., Gomez-Lechon, M. J., and Castell, J. V. (1987). A cytochemical stain for glutathione in rat hepatocytes cultured on plastic. *J. Histochem. Cytochem.* **35**, 271–274.

Liccione, J. J., and Maines, M. D. (1988). Selective vulnerability of glutathione metabolism and cellular defense mechanisms in rat striatum to manganese. *J. Pharmacol. Exp. Ther.* **247**, 156–161.

Marani, E., Bolhuis, P., and Boon, M. E. (1988). Brain enzyme histochemistry following stabilization by microwave irradiation. *Histochem. J.* **20**, 397–404.

Marani, E., Boon, M. E., and Kok, L. P. (1987). Neuropathological techniques. In *Microwave Cookbook of Pathology* (M. E. Boon and L. P. Kok, Eds.), 3rd ed., Vol. 2, Churchill Livingstone, Edinburgh/London.

Mattes, P. M., and Mattes, W. B. (1992). α-Naphthyl butyrate carboxylesterase activity in human and rat nasal tissue. *Toxicol. Appl. Pharmacol.* **114**, 71–76.

Mayahara, H., and Ogawa, K. (1988). Histochemical localization of Na^+, K^+-ATPase. *Methods Enzymol.* **156**, 417–430.

Mesulam, M. M., and Moran, A. (1987). Cholinesterases within neurofibrillary tangles related to age and Alzheimer's disease. *Ann. Neurol.* **22**, 223–228.

Morgan, K. T. (1972). An ultrastructural study of ovine polioencephalomalacia. *J. Pathol.* **110**, 123–130.

Morgan, K. T., Kelly, B. G., and Buxton, D. (1975). Vascular leakage produced in the brains of mice by Clostridium welchii type D toxin. *J. Comp. Pathol.* **85**, 461–466.

Nicholls, P. (1975). Formate as an inhibitor of cytochrome c oxidase. *Biochem. Biophys. Res. Commun.* **67**, 610–616.

Nilsson, O. G., Leanza, G., Rosenblad, C., Lappi, D. A., Wiley, R. G., and Bjorklund, A. (1992). Spatial learning impairments in rats with selective immunolesion of the forebrain cholinergic system. *Neuroreport* **3**, 1005–1008.

Ogawa, K., and Barka, T. (Eds.) (1993). *Electron Microscopic Cytochemistry and Immunocytochemistry in Biomedicine.* CRC Press, Boca Raton, Florida.

Pearse, A. G. E. (1980). In *Histochemistry Theoretical and Applied*, 4th ed., Churchill Livingstone, Edinburgh.

Randall, H. W., Bogdanffy, M. S., and Morgan, K. T. (1987). Enzyme histochemistry of the rat nasal mucosa embedded in cold glycol methacrylate. *Am. J. Anat.* **179**, 10–17.

Reed, D. J., and Beatty, P. W. (1980). Biosynthesis and regulation of glutathione: toxicological implications. In *Reviews in Biochemical Toxicology 2* (E. Hodgson *et al.*, Eds.), pp. 213–241. Elsevier/North Holland, New York.

Rutenburg, A. M., Kim, H., Fischbein, J. W., Hanker, J. S., Wassenburg, H. L., and Seligman, A. M. (1969). Histochemical and ultrastructural demonstration of gamma-glutamyl transpeptidase activity. *J. Histochem. Cytochem.* **17**, 517–526.

Sagar, S. M. (1990). NADPH-diaphorase reactive neurons of the rabbit retina: differential sensitivity to excitotoxins and unusual morphologic features. *J. Comp. Neurol.* **300**, 309–319.

Savolainen, K., Hervonen, H., Lehto, V. P., and Mattila, M. J. (1984). Neurotoxic effects of disulfiram on autonomic nervous system in rat. *Acta Pharmacol. Toxicol.* **55**, 339–344.

Scherer-Singler, U., Vincent, S. R., Kimura, H., and McGeer, E. G. (1983). Demonstration of a unique population of neurons with NADPH-diaphorase histochemistry. *J. Neurosci. Methods* **9**, 229–234.

Schoen, S. W., and Graybiel, A. M. (1992). 5′Nucleotidase: a new marker for striosomal organization in the rat caudoputamen. *J. Comp. Neurol.* **322**, 566–576.

Sipes, I. G., and Gandolfi, A. J. (1991). Biotransformation of toxicants. In *Casarett and Doull's Toxicology, The Basic Science of Poisons* (M. O. Amdur, J. Doull and K. Klaassen, Eds.), 4th ed., pp. 88–126. Pergamon Press, New York.

Slivka, A., Mytillineou, C., and Cohen, G. (1987). Histochemical evaluation of glutathione in brain. *Brain Res.* **409**, 275–284.

Tandon, P., Barone, S. Jr., Drust, E. G., and Tilson, H. A. (1991). Long-term behavioral and neurochemical effects of intradentate administration of colchicine in rats. *Neurotoxicology* **12**, 67–78.

Thompson, S. W., and Hunt, R. D. (1966). *Selected Histochemical and Histopathological Methods*, pp. 629–633. Thomas, Springfield.

Trenga, C. A., Kunkel, D. D., Eaton, D. L., and Costa, L. G. (1991). Effect of styrene oxide on rat brain glutathione. *Neurotoxicology* **12**, 165–178.

Unger, J. W., and Schmidt, Y. (1993). Galanin-immunoreactivity in the nucleus basalis of Meynert in the rat: age-related changes and differential response to lesion-induced chloinergic cell loss. *Neurosci. Lett.* **153**, 140–143.

Van Noorden, C. J. F., and Vogels, I. M. C. (1989). Polyvinyl alcohol and other tissue protectants in enzyme histochemistry: a consumer's guide. *Histochem. J.* **21**, 373–379.

Van Noorden, C. J. F., and Gossrau, R. (1991). Quantitative histochemical and cytochemical assays. In *Histochemical and Immunohistochemical Techniques* (P. H. Bach and J. R. J. Baker, Eds.), pp. 119–147. Chapman & Hall, London.

Vincent, S. R., McIntosh, C. H. S., Buchan, A. M. J., and Brown, J. C. (1983). NADPH-diaphorase: a selective histochemical marker for striatal neurons containing both somatostatin- and avian pancreatic polypeptide (APP)-like immunoreactivities. *J. Comp. Neurol.* **238**, 169–186.

Yamamoto, M., Akiyama, C., and Aikawa, H. (1990). d-Penicillamine-induced copper deficiency in suckling mice: neurological abnormalities and brain mitochondrial enzyme activities. *Dev. Brain Res.* **55**, 51–55.

4

Ultrastructural Methods for Neurotoxicology and Neuropathology

R. BAGNELL
Department of Pathology, School of Medicine
University of North Carolina at Chapel Hill
Chapel Hill, North Carolina 27599-7525

V. MADDEN
Department of Pathology, School of Medicine
University of North Carolina at Chapel Hill
Chapel Hill, North Carolina 27599-7525

C. LANGAMAN
Department of Pathology, School of Medicine
University of North Carolina at Chapel Hill
Chapel Hill, North Carolina 27599-7525

K. SUZUKI
Department of Pathology, School of Medicine
University of North Carolina at Chapel Hill
Chapel Hill, North Carolina 27599-7525

I. Introduction

The morphological investigation is an essential procedure in evaluating the effects of various neurotoxic agents. Not uncommonly, morphological changes can be found in the tissues in apparently normal or clinically unremarkable individuals or experimental animals exposed to neurotoxic agents. Subtle early structural changes of certain cellular organelles or structures may give an important clue as to the pathogenetic mechanism(s) of the effect of toxicants. For detailed morphological investigations, use of the electron microscopic technique is highly recommended. The purpose of this chapter is to present an outline of methods that are universal in the study of ultrastructure using examples that are specific to the central (CNS) and peripheral nervous systems (PNS). For those readers whose interest is more focused on ultrastructural features of the nervous systems (CNS and PNS) caused by various neurotoxic agents, detailed descriptions and references can be found elsewhere (Spencer and Shaumberg, 1980; Vrensen *et al.*, 1981; Chang, 1992).

Because of page limitation, the procedural description is largely limited to the application of transmission electron microscopy (TEM) and scanning electron microscopy (SEM) and a brief description of special methods. Artifacts of these methods are indicated. Special methods included are immunocytochemistry, enzyme cytochemistry, autoradiography, tracers, freeze-fracture, Golgi EM, and high voltage electron microscopy.

II. Transmission Electron Microscopy

The methods for tissue preparation for TEM, including fixation, dehydration, embedding, ultramicrotomy, staining, and microscopy, are covered.

A. Fixation

1. Fixatives

Fixatives are required to prevent deterioration of ultrastructure, to add electron contrast to the tissue, and to harden the tissue for ultramicrotomy. Fixatives are composed of fixing agents, a buffer, and sometimes additives. The role of each of these is described here. For a full discussion of fixation for electron microscopy see Hayat (1981).

a. Fixing Agents These chemicals react with various tissue components to prevent their physical alteration or loss during tissue processing.

i. Glutaraldehyde Glutaraldehyde is the best general fixative for electron microscopy. Used at concentrations between 2 and 4% in buffer, its dialdehyde nature stabilizes proteins, carbohydrates, and nucleic acids, but not lipids. Its method of action is by binding of the aldehyde ends of the molecule with tissue components, thus crosslinking and stabilizing them. It penetrates tissue at a rate of about 0.5 mm/hr.

ii. Paraformaldehyde Formaldehyde is a monoaldehyde that preserves proteins and nucleic acids but not carbohydrates or lipids. Its method of action is by combining with tissue components, denaturing them, and making them insoluble. It is a much smaller molecule than glutaraldehyde and thus penetrates tissue more quickly (~2.5 mm/hr). For this reason it is usually included as part of the primary fixative. Formaldehyde in solution as 37% formalin is not suitable for preserving ultrastructure because it contains methanol as a stabilizing agent. Alcohols, such as methanol, coagulate proteins, destroying their ultrastructural conformation and also dissolve lipids. This makes alcohols very undesirable in ultrastructural fixatives. Formaldehyde produced from the dissociation of paraformaldehyde is suitable for ultrastructural preservation (Robertson et al., 1963). It is normally used at concentrations of from 2 to 4% with glutaraldehyde.

iii. Osmium Tetroxide Osmium tetroxide is a strong heavy metal oxidant that preserves unsaturated lipids, proteins, and nucleic acids but not carbohydrates. Its method of action is by forming an insoluble, additive, oxidation product at double bond sites. Palade (1952) originally described osmium tetroxide as a primary fixative for electron microscopy. However, because of the slow rate of osmium penetration into tissues (~0.4 mm/hr), it is now more widely used as a secondary fixative following initial fixation with glutaraldehyde. Normally used as a 1–2% buffered solution,

not only is osmium tetroxide a good fixative, it is the best substance known for providing electron contrast. For optimal preservation of the myelin sheath, Dalton's chrome–osmium fixative (Dalton, 1955) or a combined osmium tetroxide–potassium ferrocyanide/ferricyanide fixative is recommended (Karnovsky, 1971; Langford and Coggeshall, 1980; Nagara and Suzuki, 1982).

2. Buffers

These solutions control the pH and osmotic pressure of the fixative.

a. pH Fixative pH should be maintained in the range of 7.2 to 7.4. Proteins are mainly responsible for tissue ultrastructure and a change in pH can alter the tertiary structure of protein molecules by changing their isoelectric point (Hayat, 1981).

b. Osmolarity Osmotic pressure of the buffer should be isotonic to that of the tissues. Buffer osmolarity is the main contributor to fixative osmolarity, but fixing agents, especially formaldehyde, also contribute to the total. If the buffer is hypotonic the tissue will swell whereas if it is hypertonic the tissue will shrink (Webster et al., 1969). Osmolarity can vary within a tissue and between cell organelles, thus the osmolarity of a fixative cannot exactly match that of tissues. The average osmolarity of mammalian tissues is 400 mOsm. For CNS and PNS, a buffer osmolarity close to that of the cerebrospinal fluid is the best. For the rat, this is 320 mOsm (Karlsson and Schultz, 1965; Glauert, 1975).

Osmolarity is measured by freezing point depression using an osmometer. However, the appropriate buffer osmolarity for tissue fixation may be determined by examining the mitochondria, which are particularly sensitive to osmotic changes. If the mitochondria appear swollen, then the buffer is hypotonic whereas if the mitochondria appear shrunken, the buffer is hypertonic. Tonicity of the buffer may be altered by changing the buffer concentration (molarity) or with buffer additivies. Hayat (1981, 1986) has many useful tables of osmolarities for buffers, fixatives, and additives.

3. Additives

Nonelectrolytes such as sucrose, polyvinylpyrrolidone, or dextran may be added to the buffer to adjust osmolarity. Hayat (1981) recommends the use of dextran to preserve myelin and node of Ranvier and to prevent brain slices from swelling during incubation and fixation. Electrolytes such as $CaCl_2$, $MgCl_2$, and NaCl may be added to adjust the ionic constitution of the buffer which may reduce extraction of cellular material and aid in membrane preservation. For electron microscopy of rat central nervous system, the addition of Na^+ or K^+ to the fixative buffer may not be desirable as it produces structures on the internal and external sides of the plasma membrane which may be artifactual (Karlsson *et al.*, 1975).

4. Types of Buffers

Sodium phosphate, sodium cacodylate, and PIPES are the most commonly used buffers. Phosphate buffers are the universal favorite because they are more physiological, are less hazardous and less expensive, and are easy to prepare and store. PIPES buffer is becoming more widely used and may be preferred over arsenic-containing cacodylate because it is nontoxic and is less extracting. Cacodylate or PIPES buffer should be used when the electrolytes Ca^+ and Mg^{2+} are added to the fixative or for certain specialized techniques such as enzyme cytochemistry, immunocytochemistry, or cell culture preparations.

5. Primary Fixation Methods

a. Perfusion Vascular perfusion, in which the fixative is perfused through the vascular system, usually through the left cardiac ventricle of an anesthetized animal, is the method of choice whenever possible. Hayat (1981) lists several advantages of perfusion: (1) fixation begins immediately after systemic circulation stops; (2) rate, depth, and uniformity of fixation are enhanced by rapid penetration of fixative through the vascular bed; (3) artifacts due to handling tissue prior to fixation are eliminated; (4) a variety of tissue types are fixed simultaneously; (5) *in vivo* morphology and spatial relationship are retained; (6) cellular substances are stabilized against dissolution or translocation; and (7) mild but thorough fixation can be employed to preserve enzyme activity and cellular antigenicity. Perfusion is particularly essential for the ultrastructural study of the nervous system since the brain or peripheral nerves removed without prior fixation almost invariably show many artifacts which make correct interpretation of the result difficult.

b. Immersion Immersion fixation is used for cells and tissues grown *in vitro* and for cell fractions and other specimens for which perfusion is inappropriate (see Table 1). It is the least desirable method of fixation for *in vivo* tissues and organs. If immersion is the only way of fixing tissue, for example, brain or nerve biopsies, fix the tissue *in toto* for 10–20 min before cutting into 1-mm³ pieces to avoid artifacts caused by the mechanical handling of unfixed tissue. An excellent fixative for immersion fixation is a modification of Karnovsky's original: 2% paraformaldehyde/2.5%

glutaraldehyde in 0.15 *M* Karlsson and Schultz (1965) sodium phosphate buffer.

For *in vitro* specimens, the vehicle should be a serum-free version of the growth medium adjusted for pH and applied at the incubation temperature. Cell cultures may be fixed *in situ* in polystyrene culture plates or flasks. There are several excellent plastic substrates for cell culture that resist the chemicals used in EM processing and add to the ease of handling cell culture preparations: Permanox petri dishes and slide wells; Thermanox coverslips (Lux-Miles Scientific); Aclar embedding film (Ted Pella, Inc.); and polycarbonate filters and filter wells. Most major electron microscopy catalogs carry these products. For cells in suspension or cell fractions, centrifuge gently into a loose pellet, discard the supernatant, and resuspend the specimen in the fixative. After fixation, cells or cell fractions are encapsulated in 1% buffered agarose for ease of handling without further subjection to centrifugation.

6. Secondary Fixation Methods

Secondary fixation is carried out by immersing the specimen in a solution of osmium tetroxide or ruthenium tetroxide. This step serves two purposes: (1) tissue constituents that are not well fixed by aldehydes, such as unsaturated lipids, are stabilized; and (2) heavy metal atoms are added to ultrastructural elements, greatly enhancing contrast in the electron microscope.

Care must be taken to remove any unreacted aldehyde from the tissue prior to secondary fixation because precipitates will form from the reaction of osmium with aldehyde at physiological pH. This is especially true when phosphate buffers are used. Thus a thorough washing with buffer is done prior to secondary fixation.

Generally, immersion with agitation in a 1% buffered osmium tetroxide solution for 30 min to 1 hr at room temperature followed by a thorough wash in buffer or deionized water is all that is required. Membranes lose their semipermeable nature after osmium treatment, thus the use of deionized water at this step is allowable (if maintenance of pH is not required) since the tissue is no longer subject to osmotic pressures. Washing is very important after osmium treatment as osmium will react with the dehydration agent, causing a dense precipitate that may remain in the tissue. Overexposure to either osmium or ruthenium tetroxide causes tissue to become brittle and hard to section.

En Bloc Treatment with Uranyl Acetate Following secondary fixation, further stabilization and contrast enhancement may be achieved by immersing the specimen in a solution of 0.5–2.0% aqueous uranyl acetate. Not only does uranyl acetate bind to nucleic acids, but

TABLE 1　Sample Protocols for Fixation of Cells and Cell Fractions

	Cell monolayers *in situ*	Free cells and cell fractions
Washes before fixation	If cells are grown in serum-supplemented media, rinse gently two times with serum-free growth medium warmed to 37°C. This is *very* necessary; glutaraldehyde will strongly crosslink the serum in solution which could result in insufficient fixation or artifact. If for some reason the growth medium is incompatible with glutaraldehyde, 0.1 M PIPES or sodium cacodylate buffer warmed to 37°C should be used.	Gently centrifuge cells or fractions into a loose pellet, discard supernatant, resuspend in warmed serum-free medium, 0.1 M PIPES or sodium cacodylate buffer.
Fixation	Decant the majority of the wash solution (always keep a small amount of solution over cells to prevent drying). Fix cells with 3% glutaraldehyde in serum-free growth medium, made freshly and warmed to 37°C directly before use. Note: if the fixative solution turns cloudy, turns yellow in the absence of phenol red indicator, or does not maintain pH, use 0.1 M PIPES or sodium cacodylate buffer instead of culture medium for the fixative vehicle. Cell monolayers only need to be fixed for 5–10 min if left at room temperature. Tightly sealed dishes may be kept at 4°C overnight until further processing. If cultures need to be held after fixation for longer than 24 hr, replace the serum-free medium fixative with one that contains one of the aforementioned alternative buffers.	Centrifuge gently into a loose pellet, decant most of the wash solution, and resuspend cells or cell fractions in several milliliters of warmed 3% glutaraldehyde in serum-free growth medium or 3% glutaraldehyde in one of the alternative buffers mentioned for cell monolayers. Following 5–10 min in the fixative, 1–2 ml of the fixative/cell suspension may be gently spun down onto a small amount of gelled buffered 1% agarose in the bottom of a 1.5-ml centrifuge tube. The pellet should not be larger than 0.5 mm thick. Suction off most of the supernatant, and add a thin layer of liquid 1% buffered agarose. Let gel, then gently scoop the agarose-encapsulated pellet from the tube and process further in the same manner as a piece of tissue.

it will also bind to various proteins and react with the phosphate groups present in lecithins (Hayat, 1993). In studies of rat brain, de Silva *et al.* (1968) noted that the amount of protein extracted by ethanol dehydration was much less in tissues that received the *en bloc* treatment in uranyl acetate. For cell culture and cell fractions, treatment with uranyl acetate will greatly improve ultrastructural preservation and contrast. A hindrance to *en bloc* treatment with uranyl acetate is the occasional extraction or destaining effect on glycogen which is largely pH dependent. This problem may be overcome using an uranyl acetate–oxalate mixture at neutral pH (Mumaw and Munger, 1971). Since uranyl acetate will precipitate in the presence of phosphate and cacodylate buffers, care must be taken to wash specimens well with water or veronal acetate buffer before treatment with uranyl acetate.

B. Dehydration and Infiltration

Gradual replacement of specimen fluids (mostly water) with an embedding medium capable of hardening into a sectionable block is the process of dehydration and infiltration. The most commonly used embedding media are not miscible with water. In this case, specimen water is first replaced with a fluid miscible with the embedding medium and water, then this fluid is gradually replaced with the medium. Some embedding media, such as methacrylates, are miscible with water, and in this case the medium itself can be used to gradually replace specimen water.

Ethanol and acetone are the most common dehydrating fluids. Ethanol is easier to keep dry and is less hazardous than acetone. Phosphate buffer will precipitate in 70% ethanol. Therefore, dehydration of phosphate-buffered specimens should start with ethanol concentrations less than 70% in order to avoid the deposition of phosphate granules within the tissue.

When epoxy resins are used as the embedding medium, propylene oxide is generally used after ethanol dehydration to replace the ethanol and to act as the epoxy solvent. Epoxy resins are soluble in ethanol, but propylene oxide has the advantage that if solvent is left after infiltration with 100% resin, it will chemically combine with the resin to become part of the cured block whereas ethanol will not.

Ethanol, acetone, and propylene oxide are powerful lipid solvents. Specimens that have not been fixed with osmium tetroxide will lose most of their lipid content during dehydration and infiltration.

C. Embedding

The variety of epoxy embedding media has been nicely summarized by Ellis (1986), and their hazards by Ellis (1989).

Properties of an embedding medium that are important are: (1) solubility in transition fluids, (2) viscosity of the pure medium, (3) chemical effects on tissue components, (4) factors affecting polymerization, (5) shrinkage on curing, (6) sectioning qualities, (7) staining qualities, and (8) stability in the electron beam. Three types of embedding media are commonly used: epoxy resins, polyester resins, and methacrylates. These materials can be polymerized to a block that is hard enough to produce ultrathin sections.

1. Solubility

Epoxy resins and methacrylates are soluble in ethanol and acetone. Methacrylates will not cure if residual acetone is present. Some methacrylates and epoxies are soluble in water. Polyester resins are not soluble in ethanol; acetone must be used for them.

2. Viscosity

Time required to infiltrate the tissue block with pure embedding medium depends on the viscosity of the medium (which is measured in centipoise, cps). A low viscosity medium infiltrates faster than a high viscosity medium but has the disadvantage of greater shrinkage on polymerization.

3. Chemical Effects

Extraction of unfixed lipids by methacrylates is the only significant chemical effect of an embedding medium that is currently recognized. Chemical components or contaminants in the medium may affect studies using X-ray microanalysis. For example, Epon 812 contains a chlorine contaminant whereas Araldite 6002 does not.

4. Polymerization

Each of the three types of media can be cured by heat, ultraviolet light, or with a catalyst at room temperature. Ultraviolet light is used when heat would adversely affect some tissue property, such as antigenicity. Polyester resins will polymerize at room temperature in the presence of light and oxygen, so they must be stored in the dark. Methacrylates require the absence of oxygen to cure and will not cure if residual acetone is present. Catalysts or accelerators are generally used in combination with heat; however, the catalyzed resin will polymerize eventually at room temperature. Uniformity of polymerization affects the sectioning qualities of the final block. Epoxies and polyesters cure very uniformly; methacrylates can cure unevenly. Epoxies and polyesters also form crosslinkages during curing which affect the block-sectioning qualities. Methacrylates produce a straight chain polymer without crosslinkages.

5. Shrinkage

All three media types shrink some on polymerization. The extent of this for epoxies and polyesters is slight (2 to 5%) and depends on the viscosity of the medium with less viscous media shrinking the most. Methacrylates are much worse (15 to 20% shrinkage) and this cannot be ignored in interpreting results.

6. Sectioning Qualities

Two physical properties of the polymerized embedding material affect its sectionability: hardness and elasticity. Hard materials can be sectioned thinner than soft materials. Fixation is used to stabilize and harden tissue. Ideally the hardness of the embedding material would closely match that of the tissue, and indeed the hardness of embedding media can be adjusted within certain limits. If the medium is much harder or softer than the tissue, the tissue will tend to separate from the block during sectioning. Elasticity is important because the block is deformed by the knife during sectioning. The block must recover from this prior to the next section in order to maintain uniform section thickness.

The components of epoxy resins can be varied to produce different block qualities. Epoxy resins always consist of an epoxide (such as Epon or Araldite) and an anhydride (such as DDSA, NMA, and NSA) which is the hardener. The epoxide to anhydride ratio determines the hardness of the block (0.7 : 1 is normal), the more anhydride the harder the block. Epoxides vary from batch to batch and the manufacturer will indicate the weight per epoxide on the epoxide container label. Other components that may be added are plasticizers (such as dibutyl phthalate), flexibilizers (such as DER 736, Cardolite NC-513, and Araldite RD-2), and accelerators or catalysts (such as BDMA and DMP-30). Flexibilizers and plasticizers are added to reduce viscosity and to improve cutting qualities. Flexibilizers react with the epoxide–anhydride mixture to become part of the crosslinked structure, plasticizers do not. Accelerators catalyze the curing reaction causing it to complete quickly. They can adversely affect sectioning qualities by making the block brittle if used in excessive amounts. In general, fully cured methacrylates form a more brittle block than epoxy resins and may be more difficult to section.

7. Staining Qualities

Poorly fixed tissues do not stain well. Prolonged dehydration and infiltration degrade staining due to oxidation of membrane proteins. Ellis (1986) states that a loss of differential contrast may result from better infiltration and more crosslinkage that are possible with low viscosity epoxies. Methacrylates require less staining time than epoxy or polyester resins.

8. Beam Stability

The electron beam damages an ultrathin section by heating, evaporating (thinning), and eroding it. Epoxy and polyester resins are very stable in the electron beam whereas methacrylates tend to be very unstable. The beam's thinning effect on methacrylates often contributes to an increase in contrast.

9. Routine Processing Protocols

Table 2 lists protocols commonly used for the fixation and embedment of neurological tissue for electron microscopy.

For each procedure, fixative concentration as well as fixation duration should be evaluated and adjusted to provide the optimal ultrastructural preservation. In immunological and enzyme localization techniques, this becomes particularly important as the goal is to provide both the best localization and ultrastructural preservation. Hayat (1993) is an excellent reference that includes, from fixation to staining, many of the protocols used in neuropathology at the electron microscopic level. Another recommended text dealing with immunolabeling processing protocols is Polak and Varndell (1989).

D. Ultramicrotomy

Ultrathin Sections

Two properties of the electron microscope make ultrathin sections necessary. First, chromatic aberra-

TABLE 2   Routine Processing Protocols for TEM

	Routine tissues	Cell culture	Immunocytochemistry
Primary fixative	2% paraformaldehyde, 2.5% glutaraldehyde in 0.15 *M* Karlsson and Schultz sodium phosphate buffer.	3% glutaraldehyde in serum-free tissue culture or 0.1 *M* sodium cacodylate buffer.	2% paraformaldehyde, 0.1–0.5% glutaraldehyde, 0.2% picric acid (Somogyi and Takagi, 1982) in 0.15 *M* Karlsson and Schultz sodium phosphate buffer (0.1 *M* sodium cacodylate buffer is recommended for enzyme cytochemistry).
Postfixation	1.0% osmium tetroxide and 1.25% potassium ferrocyanide (Russell, 1978) in 0.15 *M* Karlsson and Schultz sodium phosphate buffer for 1 hr. Rinse tissue thoroughly in deionized water, three times for 10 min.	1.0% osmium tetroxide and 1.25% potassium ferrocyanide in 0.15 *M* Karlsson and Schultz sodium phosphate buffer or 0.1 *M* sodium cacodylate buffer for 30 min. Rinse cultures three times for 5 min each.	Osmium postfixation not recommended. If tissues or cultures are fixed with a fixative containing sodium phosphate or cacodylate, rinse three times for 5 min in 50% ethanol; picrates will precipitate in the tissue and sodium phosphate and cacodylate will precipitate with uranyl acetate if not well rinsed.
En bloc stain	2% aqueous uranyl acetate for 1 hr.	2% aqueous uranyl acetate for 20 min.	2% uranyl acetate in 50% ethanol for 20 min.
Dehydration	Ethanol: 30, 50, 75, 95% in water for 10 min each, followed by two 10-min steps in absolute ethanol.	Ethanol: 30, 50, 75, 95% in water for 5 min each, followed by two 5-min steps in absolute ethanol.	Ethanol: 70% in water twice, for 10 min each, followed by an additional step for 10 min in either 95% or absolute ethanol. Note: when using LR White resin, the sample need not be fully dehydrated and may have greater antigenicity if dehydrated only through 70% ethanol.
Intermediate solvent	Two changes for 10 min each in propylene oxide.	Propylene oxide not recommended unless cells are in pellet form or grown on a solvent-resistant substrate such as glass, Thermanox, or Aclar film.	None
Infiltration	1 part propylene oxide : 1 part epoxy resin (PolyBed 812 or Spurr's) for 2 hr, followed by 100% epoxy resin for 3–8 hr.	1 part absolute ethanol : 1 part PolyBed 812 epoxy resin for 1 hr, followed by two changes in 100% epoxy resin for 1–2 hr each.	1 part 70% ethanol or absolute ethanol : 2 parts LR White acrylic resin for 1–3 hr, followed by two changes in 100% LR White for 2 hr each.
Embedment	PolyBed 812 or Spurr's epoxy resin for 24–48 hr at 60°C.	PolyBed 812 epoxy resin for 24–48 hr at 60°C.	LR White acrylic resin for 12–24 hr at 50°C or polymerize by long wave UV light (365 nm) at 4°C for 24 hr. Please refer to Newman and Hobot (1993) for full details.

tion, which is caused by the slowing down of primary electrons by the specimen, greatly degrades resolution. This is summarized in "Cosslett's Rule" which states that for biological specimens the expected resolution is about one-tenth the section thickness. Indeed, chromatic aberration is the limiting factor in resolving biological ultrastructure. Second, the electron microscope's objective lens has a large depth of field (about 4000 nm) compared to the section thickness (about 60 nm), and a very large depth of focus (about 5000 m). The entire section thickness is imaged in focus, resulting in superimposition of structures with an accompanying loss of resolution. Sections for standard transmission electron microscopy must be sectioned using an ultramicrotome and glass or diamond knives. For a detailed account of the art of ultramicrotomy see Reid (1975).

E. Staining

There are two approaches to staining in electron microscopy: (1) staining to enhance general morphology, which is covered in this section, and (2) staining to reveal specific tissue components, which is covered in Section IV. The most common methods for staining specific tissue components can be grouped as: (1) general cytochemical methods in which the target component consists of a broad range of structures such as acidic or basic groups, nucleic acids, DNA, proteins, carbohydrates, and lipids; (2) enzymatic reactions in which the site of action of a specific enzyme is demonstrated, such as acid phosphatase or ATPase; and (3) immunological reactions in which the site of attachment of an antibody is demonstrated. Lewis and Knight (1977) cover specific staining methods in great detail with extensive references. Recipes and procedures for a variety of staining methods are given in Hayat (1986).

1. Semithin (1 μm Thick) Sections

Sections 0.5 to 1 μm thick for light microscopy are often taken prior to ultrathin sectioning to judge fixation quality and to check specimen orientation. Normal histological stains do not work well on plastic-embedded material. Methacrylate is the best choice if standard histological stains are necessary. A few general purpose stains have become standard for plastic-embedded sections. Of these, 1% toluidine blue in 1% borax is the simplest (Richardson *et al.,* 1960). This stain gives a blue to purple coloration to cellular membranes and dark blue to myelin. A brown stain, good for black and white photomicrography, is also simple to achieve with 1% aqueous *p*-phenylenediamine (Estable-Puig *et al.,* 1965). This stain is an especially good myelin stain. Lewis and

Knight (1977) describe a number of other stains for semithin plastic sections.

2. Ultrathin Sections

Staining for electron microscopy is quite different from staining for light microscopy. In light microscopy, alteration in the color and intensity of white light by absorption and reemission is the principal mechanism for achieving contrast. In electron microscopy, the object of a stain is to remove electrons from stained areas by diverting them out of the field of view of the objective lens. Electrons are not absorbed by the stain but instead are scattered by it. Fast moving electrons are more easily scattered by atoms of high atomic number than low atomic number. Thus, electron microscopy stains are all heavy metals. Osmium, uranium, and lead are the three most common general stains used in electron microscopy. Osmium has already been discussed under Section II.A.

Uranium can be introduced into a specimen by two techniques. The first technique involves soaking tissue in a solution of acidic uranyl acetate prior to dehydration, a process called *en bloc* staining. Uranyl acetate used in this way also serves as a tertiary fixative. Tissue must be thoroughly washed with water prior to *en bloc* staining to remove buffer since uranyl acetate will precipitate in phosphate and cacodylate buffers. The second technique soaks ultrathin sections in a solution of uranyl acetate. Uranyl acetate is a general stain for membranes and nucleic acids.

Lead can be applied to tissue that has had osmium treatment. Tissue that has not been osmium treated does not stain as intensely with lead. Sections are soaked in an alkaline lead solution, usually lead citrate. Lead stains lipids and any other tissue components that have been osmium stained. Lead stains are extremely sensitive to carbon dioxide. A lead carbonate precipitate will form on sections if care is not taken to exclude carbon dioxide from the staining environment.

The most often used staining method, and the one that provides the most contrast, is that of double staining: a uranyl acetate stain followed by a lead citrate stain (Venable and Coggeshall, 1965) (Table 3).

III. Scanning Electron Microscopy

Methods for tissue preparation, including fixation, dehydration, critical point drying, mounting, coating, and microscopy, are covered in the following sections (Table 4).

The scanning electron microscope is used primarily to examine the surface of objects. It produces images with a large depth of field, giving objects a three-

TABLE 3 Uranyl Acetate/Lead Citrate
Staining Procedure

Prepare 4% aqueous uranyl acetate by adding 2 g of uranyl
acetate to 50 ml of deionized water. Add 5 drops of
concentrated glacial acetic acid and mix thoroughly until the
stain is clear. The pH of the stain should be 3.5 for optimal
results. Store uranyl acetate in a brown glass bottle. To prepare
0.4% lead citrate, add one carbonate-free pellet of sodium
hydroxide (0.2 g) and 0.2 g of commercially prepared lead
citrate powder to 50 ml of boiled deionized water. Shake until
dissolved. Do not let the boiled water cool before adding the
chemicals and store in a tightly sealed plastic bottle. Optimal
pH for staining is 11.9–12.1. Microfilter both stains just before
use with a 0.22-μm syringe filter.

Invert grids section side down onto drops of 4% microfiltered
uranyl acetate placed on a clean sheet of Parafilm inside a
closed petri dish. Stain for 5 min. Remove grids from drops and
rinse well with microfiltered deionized water. Transfer rinsed
grids, still wet, to microfiltered drops of 0.4% lead citrate
placed on clean Parafilm. To prevent lead carbonate formation
on the surface of the stain drops, place 5–10 wet sodium
hydroxide pellets around the sheet of Parafilm in a closed petri
dish during the staining. Stain for 5–8 min, remove grids
promptly from the stain, and rinse well in microfiltered
deionized water. Let grids air dry.

dimensional appearance and larger specimens, compared with the specimens for TEM, can be examined.
SEM is widely used in developmental neurobiology
(Scot *et al.*, 1976) (Fig. 1) and in experimental teratology (Smith *et al.*, 1982). Its top resolution is in the
3- to 5-nm range; about 100 times better than a light
microscope but about 10 times worse than a transmission electron microscope. Goldstein and colleagues

(1981) have dealt with the techniques of scanning electron microscopy and X-ray microanalysis in detail.

In SEM the integrity of surface structure is the main
concern. Some specimens, such as rocks and minerals,
metals, and ceramics, require very little preparation.
Biological materials usually require fixation, dehydration, and drying. A specimen must be electrically conductive to be viewed by SEM. Those that are not inherently conductive must be coated with a conductive
film.

A. Fixation

Before fixation of a sample of SEM, it may be necessary to wash or otherwise clean away particles, cells,
and other debris by gently rinsing the specimen in an
appropriate buffer.

The fixatives and buffers used in TEM can be used
as a starting point for SEM. In any fixation procedure,
it is important to keep in mind that the fixative itself
may cause changes in surface ultrastructure. Correlative techniques should always be used to add support
to any interpretation of structure.

B. Dehydration

In the ordinary SEM, specimens must be dry; otherwise, their structure will change as water sublimes in
the vacuum. Dehydration is done by the same methods
used for TEM. A special type of SEM called the environmental SEM is constructed in such a way as to keep
the area immediately around the specimen at nearly

TABLE 4 Routine Processing Protocols for SEM

	Tissues	Cell culture
Fixation	3% glutaraldehyde in 0.15 *M* Karlsson and Schultz phosphate buffer, pH 7.4, for 2 hr or overnight.	3% glutaraldehyde in serum-free growth media or 0.1 *M* sodium cacodylate buffer, pH 7.4, for 1 hr.
Dehydration	Ethanol: 30, 50, 70, 80, 95% in water for 10 min each, 100% twice for 10 min each.	Ethanol: 10, 30, 50, 70, 85, 95% in water for 5 min each, 100% twice for 5 min each.
Intermediate/transfer solvent	Freon 113 solvent for 15 min; two changes.	Freon 113 solvent for 5 min; two changes. If cells are grown on a substrate that is affected adversely by Freon 113 (i.e., silastic, other plastics sensitive to Freon 113), then use an additional change in 100% ethanol and use 100% ethanol as the transferring fluid.
Critical point drying	Transfer specimens to the critical point dryer covered in Freon 113, use CO_2 as the transitional solvent. Exchange fluid for three to five cycles depending on size of tissue.	Transfer to the critical point dryer covered in Freon 113 (or 100% ethanol) and in an appropriate holder to minimize surface damage, use CO_2 as the transitional solvent and exchange fluid for three cycles.
Mounting and coating	Mount specimens onto SEM stubs with colloidal silver paste, sputter, or evaporate 10–15 nm of gold/palladium alloy 60 : 40 onto specimens.	Mount cells with substrate onto SEM stubs with either double-sided plastic tape or carbon adhesive tabs, coat by sputter coating or vacuum evaporation with 10 nm of gold/palladium allow 60 : 40.

atmospheric pressure, thus allowing hydrated specimens to be viewed.

C. Drying

Once water is removed the specimen still must be dried. This process of drying is very important in order to preserve the delicate surface features of the specimen.

The surface of a liquid is maintained by an attraction of the liquid molecules toward one another. This surface tension is greater in some liquids than in others. For example, water at 20°C has a surface tension of 72.8 dynes/cm while ethyl alcohol at 20°C has 22.3 dynes/cm. When a liquid dries from around a specimen the specimen is subjected to a force equal to the surface tension of the liquid. As the volume of a structure becomes smaller, the relative surface area becomes larger and thus very small structures are subjected to enormous surface forces as a liquid dries off them. This force is often sufficient to distort or destroy delicate structures. This fact cannot be overlooked when drying specimens for SEM. Drying can be achieved by sublimation or by the use of low surface tension fluids such as ethyl alcohol or acetone.

1. Critical Point Drying

This is the best drying method because it eliminates surface tension altogether. For an explanation of the physical mechanism of critical point see Bozzola and Russell (1992).

A critical point apparatus is used to carry a liquid through its critical point. The most common liquid used today is carbon dioxide, whose critical temperature is 31°C and whose critical pressure is 73 atmospheres (1022 lb/sq in.). After dehydration in ethanol and replacement of 100% ethanol with 100% Freon 113, the specimen is placed into the critical point bomb in Freon 113. (The Freon 113 may be omitted at the expense of longer exchange times with carbon dioxide and some additional shrinkage of the tissue.) The bomb is closed and the Freon is replaced with dry liquid carbon dioxide. The carbon dioxide is taken through its critical point by heating the closed bomb to a temperature slightly above the critical temperature. Pressure in the bomb is then gradually released while the temperature is kept high. Once atmospheric pressure is reached the bomb is opened and the dry specimens are removed.

2. Frozen Hydrated Specimens

Using special low temperature specimen stages, it is possible to place frozen specimens into the SEM and to keep them frozen while examining them. This requires that the specimen be kept close to liquid nitrogen temperature to prevent materials from subliming from the specimen in the vacuum of the SEM. This technique is used primarily for the localization of soluble substances within the specimen (Goldstein *et al.*, 1981).

D. Mounting

The dry specimen must be placed on a conductive holder for insertion into the SEM. These holders are called stubs. Some of the methods for mounting specimens to stubs are double sticky tapes, conductive pasts, carbon dags (for use in X-ray analysis studies), and mechanical clamps. Regardless of the mounting method, it must be remembered that an electrical path from the specimen to ground must be maintained. Sticky tapes must not completely cover the surface of the metal stub, and if the specimen is fixed to a nonconducting substrate, such as a glass coverslip, the substrate must not cover the entire surface of the stub.

E. Coating

Many specimens, especially biological ones, are not inherently conducting. To make them so, one of two methods may be used. Specimens may be chemically treated with conducting materials prior to dehydration and drying or they may be coated with a layer of conducting material. In the first method, osmium is added to the specimen through a number of steps that use a mordant, such as thiocarbohydrazide or tannic acid, to build up the osmium content. In the second method a very thin layer of metal is deposited onto the specimen's surface. This is usually a 10- to 30-nm-thick layer of a gold–paladium alloy. When these metals condense on the specimen they form grains of a certain size. Gold grains are in the 5- to 7-nm range, whereas palladium is in the 5-nm range. These grains must be smaller than the expected resolution. Experience has shown that an alloy of 60% gold and 40% palladium gives the best grain size for most SEM work. For X-ray microanalytical work, metals may interfere with the analysis. In this case, carbon, which conducts poorly, is used.

Two common methods for depositing metals onto the specimen are used: high vacuum evaporation and sputter coating.

1. High Vacuum Evaporation

Metals and carbon may be evaporated under high vacuum. Since the atoms of the metal travel in straight lines away from the evaporating source, the specimen must be tilted and rotated continuously in order not to miss some spots. This is a beautiful and somewhat amusing procedure. Despite all this motion, a specimen

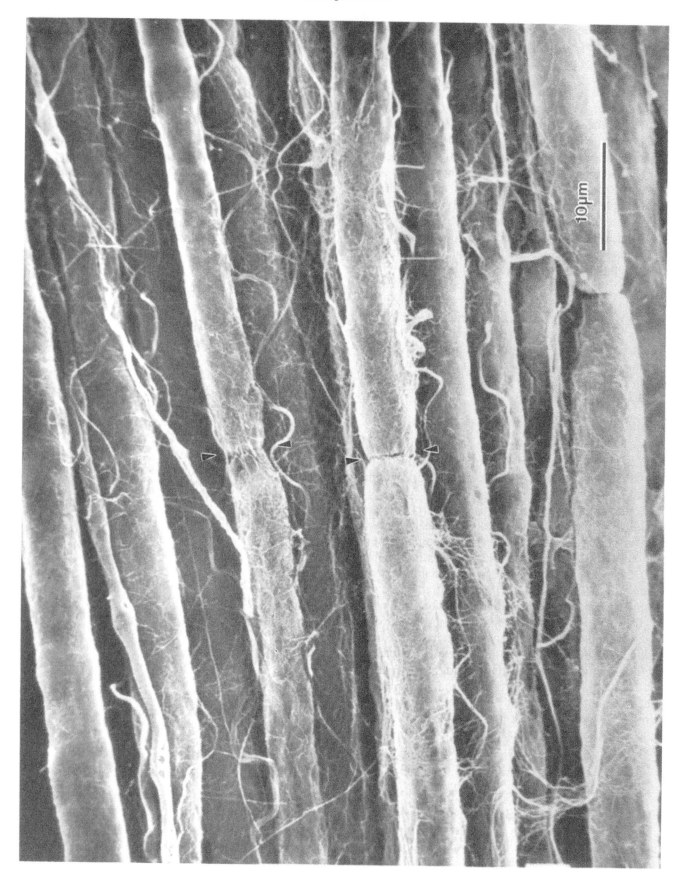

with a very convoluted or pocketed surface will not be uniformly coated.

2. Sputter Coating

When atoms of a metal are freed by knocking them loose with other particles the process is called sputtering. Gold and palladium atoms are knocked out of a target by atoms of ionized argon. The argon atoms then interact with the metal atoms to produce a random distribution of metal atoms in space. If a metal atom runs into the surface of the specimen (or any other surface), it will stick and become part of the metal coating. Because the paths of the metal atoms are random, even very convoluted surfaces of the specimen stand a good chance of being coated. This process is carried out at a vacuum of 0.5 mm Hg and thus a mechanical rough pump is all that is needed to produce the vacuum.

F. Other Specimen Preparation Methods

Many methods have been devised to examine special materials in the SEM. For example, there is a method that reveals the surface of blood vessels and preserves the three-dimensional arrangement of the vessels themselves. This is called the corrosion cast method. It requires that the vessels be filled with a quick curing plastic after which the tissue is dissolved away and the "cast" is examined in the SEM (Akima et al., 1987).

IV. Specialized Methods Using Electron Microscopy

There are many special methods for the elucidation of ultrastructure and the localization of biological molecules using electron microscopy. The following is a brief review of the most common ones.

A. Immunocytochemistry

1. Background

Polak and Priestly (1992) give a good survey of this technique and Ironside (1992) covers electron microscopic immunocytochemistry in neuropathology. Reviews of immunocytochemistry in neuroscience research have been published by Pickel (1981), Priestley and Cuello (1983), Priestley (1984), and van den Pol (1984).

2. Outline of the Technique

Any biological molecule to which an antibody can be formed can be localized at the ultrastructural level by this technique. An electron-dense tag is attached to the antibody and the tag is detected by electron microscopy. This method was first demonstrated by Singer (1959) with the conjugation of ferritin to antibodies.

a. Methods

i. Direct Method When the electron-dense tag is associated with the primary antibody and no secondary antibodies are used, the immunostaining method is referred to as direct. This method requires that the tag be conjugated to the primary antibody, which may be in short supply, and there is no method of amplifying the staining. For these two reasons, this method is used infrequently.

ii. Indirect Method When the electron-dense tag is associated with a secondary antibody that has been raised against the primary antibody, the immunostaining method is referred to as indirect. Immunostaining can be greatly amplified by this method and there is usually an abundant supply of the secondary antibody.

iii. Preembedding/Postembedding Immunocytochemical staining can be carried out prior to infiltration of the tissue with resin (preembedding technique) or after infiltration and sectioning (postembedding). Preembedding methods allow better access of the antibody to its antigen, but these methods are difficult to carry out. Postembedding methods are easier to carry out but may limit access of the antibody to its antigen due to the presence of the plastic resin.

b. Tags

i. Structured Tags These tags exhibit a characteristic shape when viewed in the electron microscope. Ferritin was the first such tag used. Its small size allows it to cross cell membranes. Hemocyanin and viruses are also used as structured tags. Their large size makes them useful for labeling the exterior of cells as seen by scanning electron microscopy.

ii. Enzymes The conjugation of an enzyme to an antibody (Nakane and Pierce, 1966; Avrameas and Ur-

---

FIGURE 1 A scanning electron micrograph of the spinal roots of the mouse. Arrowheads indicate node of Ranvier.

iel, 1966) opened the possibility for greatly enhancing the detectability of an antigen. The enzyme peroxidase can be used to catalyze the oxidation of diaminobenzidine to form an insoluble precipitate. This precipitate can be stained for electron microscopy by the addition of osmium. The detection of peroxidase can be further enhanced by use of the peroxidase–antiperoxidase method (Sternberger, 1970) or by the use of the avidin–biotin–peroxidase complex method (Hsu, 1981) (Fig. 2).

iii. Colloidal Gold   The adsorption of colloidal gold particles to an antibody and its visualization directly in the electron microscope was demonstrated by Faulk and Taylor (1971). This method is becoming the most commonly used technique in immunocytochemistry for

the following reasons: (1) the ease of adsorbing colloidal gold to an antibody, (2) the large size ranges in which gold particles can be produced (enabling multiple labeling methods), and (3) the fact that the gold particles can be seen by transmission and scanning electron microscopy without further enhancement (Foster *et al.*, 1985; Wiley *et al.*, 1987; Gross *et al.*, 1990).

B. Enzyme Cytochemistry

1. Background

Enzyme cytochemistry is a method for localizing the site of action of endogenous enzymes at the subcellular level. It is also a method for determining which

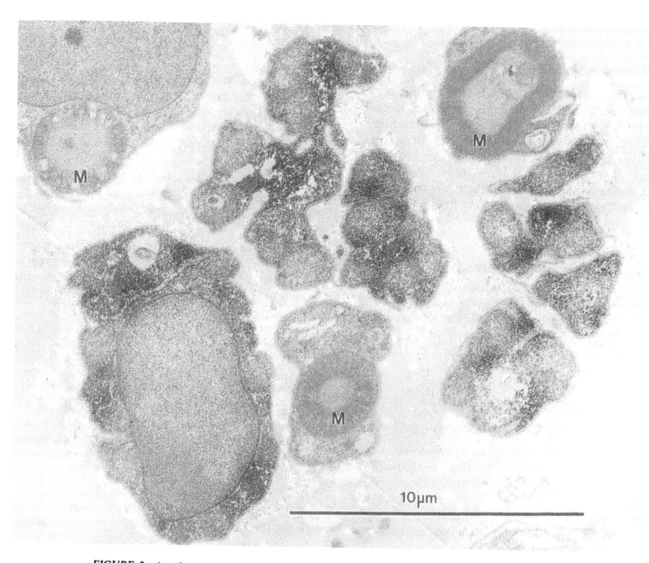

FIGURE 2 An electron micrograph of the peripheral nerve of the mouse, which was processed for immunocytochemical demonstration of glial fibrillary acidic protein (GFAP) with avidin–biotin–peroxidase complex methods. The electron-dense reaction product indicating GFAP is localized on the Schwann cells of unmyelinated fibers. Myelinating Schwann cells (M) do not express GFAP immunoreactivity.

specific enzymes a cell contains. Enzyme cytochemistry is an outgrowth of enzyme histochemistry which has been reviewed by Pearse (1985). Marani (1981) has reviewed enzyme cytochemistry as a method in neurobiology.

2. Outline of the Technique

Bozzola and Russell (1992) have given a concise description of the enzyme cytochemistry technique. Lewis (1977) and Hayat (1993) give an extensive practical guide to the use of this method. In general, the method consists of the addition of an excess of suitable exogenous substrate to the cells or tissue. The enzymatic reaction converts this substrate into an insoluble reaction product that, in the presence of a trapping agent, produces an electron-dense precipitate.

C. Autoradiography

1. Background

Subcellular sites of metabolism or synthesis of compounds can be localized by use of autoradiography. Any compound that can be made radioactive can be traced intracellularly by this technique. The method has its roots in light microscopic autoradiography which has been reviewed by Rogers (1973). Liquier-Milward (1956) first demonstrated the feasibility of autoradiography at the electron microscope level. Salpeter and McHenry (1973) have described the technique in detail.

2. Outline of the Technique

Bozzola and Russell (1992) give an excellent brief description of the autoradiography technique. The compound to be studied must be made radioactive. This is usually done by replacing one of its stable isotopes with a radioactive one. Carbon-14 and tritium are the most commonly used isotopes. Many such compounds are available commercially. The animal or physiological system is exposed to the radioactive compound. Tissue is prepared for electron microscopy in the usual way except sections are placed on glass slides. The slides are coated with a photographic emulsion and set aside allowing the radioactive decay process to expose the emulsion. The emulsion is developed, stripped from the slide along with the section, placed on a grid, and examined in the electron microscope. Silver grains in the emulsion will indicate the location of the radioactive compound within the tissue.

D. Tracers

1. Background

Certain materials which, because of their size or electric charge, cannot penetrate the cell membrane

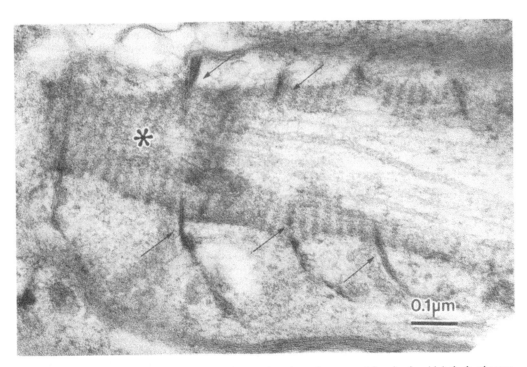

FIGURE 3 A longitudinal section of the nerve fiber of rat through a paranodal region in which the lanthanum is visible between transverse bands (*) and between adjacent lateral loops (arrows). (From Hirano and Dembitzer, 1969.)

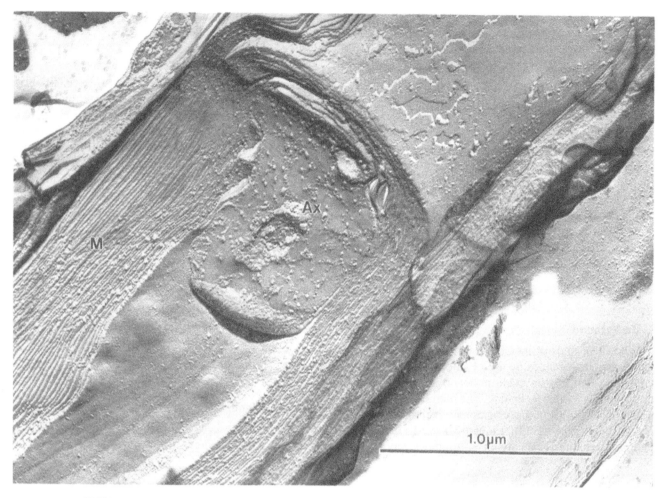

FIGURE 4   An electron micrograph of a peripheral nerve fiber of the mouse, prepared with the freeze-fracture technique. Fractured face demonstrates cross section of the axon with neurofilaments (Ax) and cross and longitudinal section of myelin sheaths (M).

have been used to study extracellular structures such as junctional complexes. Ferritin, lanthanum, and horseradish peroxidase, among others, have been used for this purpose. This method is important in studies of the neuronal extracellular matrix, blood–brain, and blood–nerve barriers (Reese and Karnovsky, 1967; Krishnan and Singer, 1973) (Fig. 3).

2. Outline of the Technique

Depending on the study, tracers may be applied in the same manner as an *en bloc* stain just prior to embed-ding or they may be added to all the solutions used in the specimen preparation technique, including the fixatives (Lewis and Knight, 1977). Care must be taken with the buffering systems since some tracers, such as lanthanum, will precipitate in phosphate buffers. Tracers that are not bound to the tissue either by electric charge or chemical bonds will tend to leach out. This makes interpretation difficult. Leaching can be minimized by using thicker than normal tissue samples and shortened dehydration times. Embedding and sectioning are done as usual.

FIGURE 5   Typical findings of combined Golgi-EM studies. (Inset) Light micrograph of gold-toned neurite-bearing (arrow) pyramidal neurons as seen in a 3-μm plastic section prior to sectioning for EM. Electron micrograph reveals typical findings for neurons of this type. Gold stippling readily identifies cell body (not shown), axon hillock (AH), and axonal initial segment (ax). Numerous gold-stippled profiles, which are cross sections of neurites, are evident in the adjacent neuropil (straight arrows). These profiles commonly display asymmetrical synapses and often contain one prominent multivesicular body. Thie presynaptic axon contacting the profile shown in the lower right corner also synapses on an unlabeled dendrite (curved arrow). Calibration bars = 20 μm (inset) and 0.5 μm. (From Walkley *et al.*, 1990.)

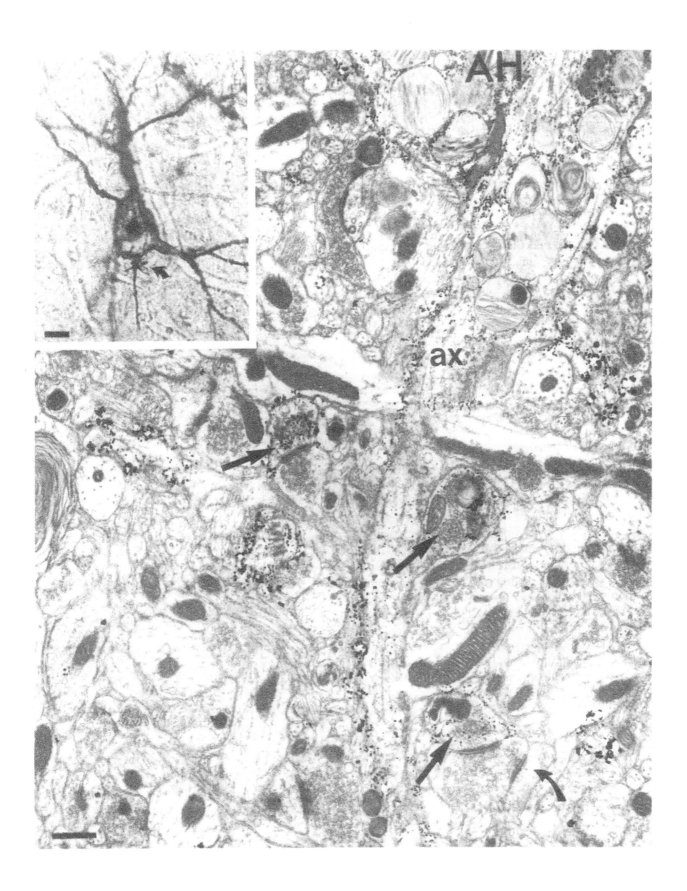

E. Freeze-Fracture

1. Background

This technique involves the separation of frozen biological membranes along the interior of their phospholipid bilayer followed by the deposition of a metal film onto the fractured membranes. The metal film forms a replica of the membrane's surface and the film is examined in the electron microscope. Originally described by Moor and Mühlethaler (1963), this seemingly complex method has produced most of our detailed information about the ultrastructure of membranes beginning with the original work of Singer and Nicolson (1972).

Neurobiology has benefited greatly from the use of this technique: Major studies included are on the synaptic membrane structure (Landis *et al.*, 1974; Gulley *et al.*, 1977); myelin membrane (Schnapp and Mugnaini, 1976; de Silva and Miller, 1975), myelin tight junctions (Tetzlaff, 1978; Dermietzel and Kroczek, 1980), axon–glial membranes at the node of Ranvier (Rosenbluth, 1976, 1978; Ellisman, 1979; Wiley and Ellisman, 1979); and growth cones (Pfenninger and Bunge, 1974). For more detailed applications in neurobiology, the readers are referred to the review by Pfenninger (1981) (Fig. 4).

2. Outline of the Technique

Bozzola and Russell (1992) give a very good description of this method. The steps involved are as follows: Tissue is fixed, usually with glutaraldehyde; this is followed by cryoprotection of the tissue by submersion in glycerol or DMSO; next, freezing is carried out using liquid freon or liquid propane. After being frozen, the tissue is placed in a freeze-fracture apparatus where it is fractured, then coated, and shadowed with palladium and carbon. The replica is removed from the apparatus, cleaned of tissue by soaking in sodium hypochloride, and thoroughly washed. Finally the replica is picked up on an EM grid and examined in an electron microscope. Interpretation of the image and detection of artifacts are acquired skills which assume a reasonable foreknowledge of the ultrastructure of the tissue.

F. Golgi EM

1. Background

The Golgi silver impregnation technique has been used for more than a century to visualize neuronal soma as well as their processes. At the light microscopic level, many classes of neurons have been identified. Abnormal configurations of the soma and neuritic processes have been well demonstrated in pathological conditions. Since successful application of this technique for the ultrastructural study (Fairen *et al.*, 1977), the Golgi EM technique has been used not only for anatomical study but for the study of neuronal structural abnormalities in various pathological conditions (Braak *et al.*, 1982, 1985) (Fig. 5).

2. Outline of the Technique

The fixation of the tissue is carried out either by perfusion or immersion in the buffered aldehyde solution. Then the tissue is processed for Golgi impregnation. The impregnated tissue is cut and the sections are examined at the light microscopic level and the neuron of interest is selected. The sections are gold-toned (Fairen *et al.*, 1977) or treated with ammonia solution (Braak *et al.*, 1982), and the depigmented neurons of interest are identified and trimmed for further electron microscopic study.

G. High Voltage Electron Microscopy

1. Background

In order to view thick sections or whole cell preparations by electron microscopy, intermediate and high voltage electron microscopes (HVEM) are used. These instruments operate in the 300 kV to 1 mV range. At these voltages, the electron beam can penetrate a thick specimen imaging the entire thickness in focus on film. Three-dimensional information may be obtained by recording images at different angles of tilt of the specimen. One disadvantage of HVEM is a reduction in image contrast. Hama and Kosaka (1981) have reviewed the use of HVEM in neurobiology.

2. Outline of the Technique

Specimens to be viewed by HVEM may be thick (2 to 3 mm) plastic sections or whole cells that have been critical point dried. Conventional electron contrast as well as immunological, enzymatic, and cytological staining procedures can be used to highlight specific areas of interest.

V. Artifacts

Artifacts may result from any of the specimen preparation processes and from exposure to the electron beam. An understanding of artifacts is vital in interpreting electron micrographs. A meaningful discussion of artifacts is beyond the space limitations of this chapter. The reader is strongly encouraged to see Crang and Klomparens (1988) who have thoroughly reviewed artifacts in biological electron microscopy.

Acknowledgments

This work was supported in part by U.S. Public Health Service Grants NS-24454, HD-03110, and ES-01104.

References

Akima, M., Nonaka, H., Kagesawa, M., and Tanaka, K. (1987). A study on the microvasculature of the cerebellar cortex. *Acta Neuropathol. (Berl)* **75**, 69–76.

Avrameas, S., and Uriel, J. (1966). Méthode de marquage d'antigès et d'anticopts avec des enzymes et son application en immunodiffusion. *C.R. Acam. Sci. Paris,* Series D, **262**, 2543.

Bozzola, J. J., and Russell, L. D. (1992). Enzyme Cytochemistry. In *Electron Microscopy,* ch. 10, pp. 252–261, Jones and Bartlett, Boston.

Braak, H., and Braak, E. (1982). A simple procedure for electron microscopy of Golgi-impregnated nerve cells. *Neurosci Lett.* **32**, 1–4.

Braak, H., and Braak, E. (1985). Golgi preparations as a tool in neuropathology with particular reference to investigations of the human telencephalic cortex. *Progr. Neurobiol.* **25**, 93–139.

Chang, L. W. (1992). Pathological studies. In *Neurotoxicology* (M. Abou-Dania, Ed.), Part III, Section C, CRC Press, Boca Raton, Florida.

Crang, R. F. E., and Klomparens, K. L. (1988). *Artifacts in Biological Electron Microscopy.* Plenum Press, New York.

Dalton, A. J. (1955). A chrome-osmium fixative for electron microscopy. *Anat. Rec.* **121**, 281.

de Silva, P. P., and Miller, R. G. (1975). Membrane particles on fracture faces of frozen myelin. *Proc. Nat. Acad. Sci. USA* **72**, 4046–4050.

Dermietzel, R., and Kroczek, H. (1980). Interlamellar tight junction of central myelin. I. Developmental mechanisms during myelinogenesis. *Cell Tissue Res.* **213**, 81–94.

Ellis, E. A. (1986). Araldites, low viscosity epoxy resins and mixed resin embedding: formulations and uses. *EMSA Bull.* **16**, 53.

Ellis, E. A. (1989). Embedding media: an overview of hazards and safe handling. *EMSA Bull.* **19**, 83.

Ellisman, M. H. (1979). Molecular specializations of the axon membrane at nodes of Ranvier are not dependent upon myelination. *J. Neurocytol.* **8**, 719–735.

Estable-Puig, J. F., et al. (1965). Technical note: paraphenylenediamine staining of osmium-fixed plastic-embedded tissue for light and phase microscopy. *J. Neuropath. Exp. Neurol.* **24**, 531–535.

Fairen, A., Peters, A., and Saldanha, J. (1977). A new procedure for examining Golgi impregnated neurons by light and electron microscopy. *J. Neurocytol.* **5**, 311–337.

Faulk, W. P., and Taylor, G. M. (1971). An immunocolloid method for the electron microscope. *Immunocytochemistry* **8**, 1081–1083.

Foster, G. A., and Johansson, O. (1985). Ultrastructural morphometric analysis of somatostatin-like immunoreactive neurons in the rat central nervous system after labeling with colloidal gold. *Brain Res.* **342**, 117–127.

Glauert, A. M. (1975). Fixation, dehydration and embedding of biological specimens. In *Practical Methods in Electron Microscopy* (A. M. Glauert, Ed.), North-Holland, New York.

Goldstein, J. I., Newbury, E. I., Echlin, P., Joy, D. C., Fiori, C., and Lifshin, E. (1981). *Scanning Electron Microscopy and X-Ray Microanalysis.* Plenum Press, New York.

Gross, D. K., and de Boni, U. (1990). Colloidal gold labeling of intracellular ligands in dorsal root sensory neurons, visualized by scanning electron microscopy. *J. Histochem. Cytochem.* **38**, 775–784.

Gulley, R., and Reese, T. S. (1977). Freeze-fracture studies on the synapses in the organ of corti. *J. Comp. Neurol.* **171**, 517–544.

Hama, K., and Kosaka, T. (1981). Neurobiological applications of high voltage electron microscopy. *Trends Neurosci.* **4**, 193–196.

Hayat, M. A. (1981). *Fixation for Electron Microscopy.* Academic Press, New York.

Hayat, M. A. (1986). *Basic Techniques for Transmission Electron Microscopy.* Academic Press, New York.

Hayat, M. A. (1993). *Stains and Cytochemical Methods.* Plenum Press, New York.

Hirano, A., and Dembitzer, H. M. (1969). The transverse bands as a means of access to the periaxonal space of the central myelinated nerve fiber. *J. Ultrastruct. Res.* **28**, 141–149.

Hsu, S. M., Raine, L., and Fanger, H. (1981). Use of avidin-biotin-peroxidase complex (ABC) in immunoperoxidase techniques: a comparison between ABC and unlabelled antibody (PAP) procedures. *J. Histochem. Cytochem.* **29**, 577–580.

Ironside, J. W. (1992). Electron microscopic immunocytochemistry in neuropathology. In *Electron Microscopic Immunocytochemistry, Principles and Practice* (J. M. Polak and J. V. Priestley, Eds.), pp. 1185–2000, Oxford University Press, New York.

Karlsson, U. L., and Schultz, R. L. (1965). Fixation of the central nervous system for electron microscopy by aldehyde perfusion. I. Preservation with aldehyde perfusates versus direct perfusion with osmium tetroxide with special reference to membranes and the extracellular space. *J. Ultrastruct. Res.* **12**, 160–186.

Karlsson, U. L., Schultz, R. L., and Hooker, W. M. (1975). Cation-dependent structures associated with membranes in the rat central nervous system. *J. Neurocytol.* **4**, 537–542.

Karnovsky, M. J. (1971). Use of ferrocyanide reduced osmium tetroxide in electron microscopy. Proceedings of the 11th Annual Meeting of American Society of Cell Biology, p. 146.

Krishnan, N., and Singer, M. (1973). Penetration of peroxidase into peripheral nerve fibers. *J. Anat.* **136**, 1–14.

Landis, D. M. D., and Reese, T. S. (1974). Differences in membrane structure between excitatory and inhibitory synapses in the cerebellar cortex. *J. Comp. Neurol.* **155**, 93–126.

Langford, L., and Coggeshall, R. E. (1980). The use of potassium ferricyanate in neural fixative. *Anat. Rec.* **197**, 297–303.

Lewis, P. R. (1977). Metal precipitation methods for hydrolytic enzymes, Ch. 4, and Other cytochemical methods for enzymes, Ch. 5. In *Staining Methods for Sectioned Material,* (P. R. Lewis and D. P. Knight, Eds.), pp. 137–278, North-Holland, Amsterdam.

Lewis, P. R., and Knight, D. P. (1977). Staining methods for sectioned material. In *Practical Methods in Electron Microscopy* (A. M. Glauert, Ed.), North-Holland, New York.

Liquier-Milward, J. (1956). Electron microscopy and radioautography as coupled techniques. In *Tracer Experiments,* pp. 177–617. Nature, London.

Marani, E. (1981). In *Methods in Neurobiology* (Robert Lehue, Ed.), Vol. 1, Ch. 9, Plenum Press, New York.

Moor, H., and Mühlethaler, K. (1963). Fine structure in frozen-etched yeast cells. *J. Cell Biol.* **17**, 609–628.

Mumaw, V. R., and Munger, B. L. (1971). Uranyl acetate as a fixative from pH 2.0 to 8.0. Proceedings of the 29th Annual EMSA Meeting, p. 490. Claitors Publications Division, Baton Rouge, Louisiana.

Nagara, H., and Suzuki, K. (1982). Radial component of the central myelin in neurologic mutant mice. *Lab. Invest.* **47**, 51–59.

Nakane, P. K., and Pierce, G. N. (1966). Enzyme-labelled antibodies: preparation and application for the localization of antigens. *J. Histochem. Cytochem.* **14**, 929–931.

Newman, G. R., and Hobot, J. A. (1993). *Resin Microscopy and On-Section Immunocytochemistry.* Springer Verlag, Heidelberg.

Palade, G. E. (1952). A study of fixation for electron microscopy. *J. Exper. Med.* **95**, 285–297.

Pearse, A. G. E. (1985). *Histochemistry: Theoretical and Applied.* Vols. 1–3, Churchill Livingstone, Edinburgh.

Pfenninger, K. H. (1981). *Methods in Neurobiology* (R. Lahue, Ed.), Vol. 1, Ch. 8, Plenum Press, New York.

Pfenninger, K. H., and Bunge, R. P. (1974). Freeze-fracturing of nerve growth cones and young fibers: a study of developing plasma membrane. *J. Cell Biol.* **63**, 180–196.

Pickel, V. M. (1981). Immunocytochemical methods. In *Neuroanatomical Tract-Tracing Methods* (L. Heimer and M. J. Robards, Eds.), pp. 483–509, Plenum, New York.

Polak, J. M., and Priestley, J. V. (1992). *Electron Microscopic Immunocytochemistry, Principles and Practice.* Oxford University Press, New York.

Polak, J. M., and Varndell, I. M. (1984). *Immunolabelling for Electron Microscopy.* Elsevier Science Publishers B. V., Amsterdam.

Priestley, J. V. (1984). Pre-embedding ultrastructural immunocytochemistry; immunoenzyme techniques. In *Immunolabelling for Electron Microscopy* (J. M. Polak and I. M. Varndell, Eds.), pp. 37–52, Elsevier Biomedical, Amsterdam.

Priestly, J. V., and Cuello, A. C. (1983). Electron microscopic immunocytochemistry for CNS transmitters and transmitter markers. In *Immunohistochemistry* (A. C. Cuello, Ed.), pp. 273–322, John Wiley, Chichester.

Reese, T. S., and Karnovsky, M. J. (1967). Fine structural localization of a blood-brain barrier to exogenous peroxidase. *J. Cell Biol.* **34**, 207–217.

Reid, N. (1975). Ultramicrotomy. In *Practical Methods in Electron Microscopy* (A. M. Glauert, Ed.), North-Holland, New York.

Richardson, K. C., Jarett, I., and Finke, F. H. (1960). Embedding in epoxy resins for ultrathin sectioning in electron microscopy. *Stain Technol.* **35**, 313–323.

Robertson, J. D., Bodenheimer, T., and State, D. E. (1963). The ultrastructure of Maunther cell synapses and nodes in goldfish brain. *J. Cell Biol.* **19**, 159–199.

Rogers, A. W. (1973). *Technique of autoradiography,* 2nd Ed., Elsevier, Amsterdam.

Rosenbluth, J. (1976). Intramembranous particle distribution at the node of Ranvier and adjacent axolemma in myelinated axons of the frog brain. *J. Neurocytol.* **5**, 731–745.

Rosenbluth, J. (1978). Glial membrane specializations in extraparanodal regions. *J. Neurocytol.* **7**, 709–718.

Russell, L. D., and Burguet, S. (1978). Ultrastructure of Leydig cells as revealed by secondary tissue treatment with a ferrocyanide: osmium mixture. *Tissue Cell* **9**, 99–112.

Salpeter, M. M., and McHenry, F. A. (1973). Electron microscope autoradiography. In *Advanced Techniques in Biological Electron Microscopy* (J. K. Koehler, Ed.), pp. 113–152, Springer-Verlag, New York.

Schnapp, B., and Mugnaini, E. (1976). Freeze-fracture properties of central myelin in the bullfrog. *Neuroscience* **1**, 459–467.

Scott, D. E., Paull, W. K., and Kozlowski, G. P. (1976). Scanning electron microscopy. Application and implications in developmental neurobiology. *Am. J. Dis. Child.* **130**, 555–561.

Singer, S. J. (1959). Preparation of an electron-dense antibody conjugate. *Nature* **183**, 1523–1524.

Singer, S. J., and Nicolson, G. L. (1972). The fluid mosaic model of the structure of cell membranes. *Science* **175**, 720–731.

Smith, M. T., Wood, L. B., and Honig, S. R. (1982). Scanning electron microscopy of experimental anencephaly development. *Neurology* **32**, 992–999.

Somogyi, P., and Takagi, H. (1982). A note on the use of picric acid-paraformaldehyde-glutaraldehyde fixative for correlated light and electron microscopic immunocytochemistry. *Neuroscience* **7**, 1779–1783.

Spencer, P., and Shaumberg, H. (1980). *Experimental and Clinical Neurotoxicology.* Williams & Wilkins, Baltimore.

Sternberger, L. A., Hardy, P. H., Cuculis, J. J., and Meyer, H. G. (1970). The unlabeled antibody-enzyme method of immunocytochemistry. Preparation and properties of soluble antigen-antibody complex (horseradish peroxidase-antihorseradish peroxidase) and its use in identification of spirocheters. *J. Histochem. Cytochem.* **18**, 315–333.

Tani, E., and Ametant, T. (1970). Substructure of microtubules in brain nerve cells as revealed by ruthenium red. *J. Cell Biol.* **46**, 159–165.

Tetzlaff, W. (1978). The development of a zonula occludens in peripheral myelin of the chick embryo. A freeze-fracture study. *Cell Tissue Res.* **189**, 187–201.

van den Pol, A. N. (1984). Colloidal gold and biotin-avidin conjugates as ultrastructural markers for neural antigens. *J. Exp. Physiol.* **69**, 1–33.

Venable, J. H., and Coggeshall, R. (1965). A simplified lead citrate stain for use in electron microscopy. *J. Cell. Biol.* **25**, 407.

Vrensen, De Grout, B. (1981). An electron microscopy in neurobiology. In *Methods in Neurobiology* (R. Lahue, Ed.), Vol. 2, Ch. 7, pp. 433–500, Plenum Press, New York.

Walkley, S. U., Wurzelmann, S., Rattazzi, M. C., and Baker, H. J. (1990). Distribution of ectopic growth and other geometrical distortions of CNS neurons in feline GM2 gangliosidosis. *Brain Res.* **510**, 63–73.

Webster, H. deF., Ames, A. III, and Nesbett, F. B. (1969). A quantitative morphological study of osmotically induced swelling and shrinkage in nervous tissue. *Tissue Cell* **1**, 201–216.

Wiley, C. A., Burrola, P. G., Buchmeier, M. J., Wooddell, M. K., Barry, R. A., Prusiner, S. B., and Lampert, P. W. (1987). Immuno Gold localization of prion filaments in scrapie infected hamster brain. *Lab. Invest.* **57**, 646–656.

Wiley, C. A., and Ellisman, M. H. (1979). Development of axonal membrane specializations defines nodes of Ranvier and precedes Schwann cell myelin formation. *J. Cell Biol.* **83**, 83a.

Quantitative Morphometry for Neurotoxicity Assessment

ANDREW C. SCALLET

Division of Neurotoxicology
National Center for Toxicological Research
Jefferson, Arkansas 72079

I. Introduction

Measurement is an essential part of the experimental procedure in most areas of science, and histological studies of neurotoxic chemicals are no exception. Historically, measurement of the features of images has seemed more difficult than measuring the rate of an enzyme reaction or the amount of a neurochemical, but since the advent of stereological methods some years ago (e.g., "Weibel's Bible," Weibel, 1979), tremendous progress has been made. At present, the availability of numerous commercial image analysis systems designed to support quantitative neurohistological studies represents both an opportunity and a challenge for the neurotoxicologist. In order to meet this challenge, studies must be reliable, conclusive, and replicable.

This chapter reviews various neurohistological tissue preparation approaches, describes their application to neurotoxicology, and compares their ease in lending themselves to quantitation and statistical analysis. It also discusses the conditions under which the various procedures might be expected to be most useful, their limitations, and describes in a general way the types of instrumentation available to measure the results. A great deal of attention is directed to describing specific procedures since often different methods provide complementary information about neurotoxic agents and must be combined. The most economic way to accomplish this requires a set of methods that are compatible with each other so that all can be applied to identically prepared tissue samples. In most cases, a general approach to fixation will suffice for several different procedures, but there are notable instances of incompatibilities that must be considered as experiments are designed. This chapter includes some example applications drawn from neurohistological research on excitotoxins, trimethlytin, and marijuana.

II. Methods: General Procedures

For optimal results in brain tissue with most of the methods, it is desirable to perfuse the subject through the aorta with an appropriate fixative to minimize postmortem deterioration or artifacts. Appendix 1 outlines perfusion procedures suitable for mice, neonatal to adult rats, and monkeys. Although numerous choices of fixatives are available and may be needed to optimize a particular procedure, the routine use of 3.75% formaldehyde suffices for most procedures. It is available ready-to-use or may be prepared as a 10% v/v dilution with 0.1 *M* phosphate buffer from stock, commercially available formalin, which is 37.5% formaldehyde. The use of formalin simplifies fixative preparation and is compatible with virtually all the light microscopic techniques described in this chapter. For electron microscopic studies, 1–3% paraformaldehyde mixed with 0.1–4% gluteraldehyde (also in 0.1 *M* phosphate or cacodylate buffer) can be substituted to improve membrane preservation, depending on whether immunohistochemical labeling is needed, and the characteristics of the primary antisera selected as well as the particular antigen to be labeled.

For histochemical light microscopic work (such as immunohistochemistry, Golgi methods, and degeneration-specific stains), sections can be readily prepared using a vibrating microtome to cut the cooled (but unfrozen) brain tissue. This avoids the creation of freezing artifacts, especially troublesome for ultrastructural studies, due to formation of inter- and intracellular ice crystals. It also limits the potential loss of antigens (particularly ethanol-soluble ones) during the preparation of paraffin or plastic-embedded tissue blocks. The vibratome sections can be immediately sampled, after staining, with a punch to obtain blocks for postembedding immunoelectron microscopy and still maintain excellent ultrastructural appearance. The brain tissue can be sectioned within a day of perfusion, with no additional embedding steps, thus increasing the throughput of samples. The major disadvantage of vibratome sections is a superficial "chatter" due to blade vibration, which is sometimes unavoidable. Their relative thickness (about 20 up to 200 μm) compared to paraffin (5–7 μm) or plastic sections 1-2 μm) is a disadvantage for viewing subcellular detail, but is an advantage when measuring large structures such as dendritic arborizations. Another disadvantage is the relative impermanence of leftover sections or blocks, which must be stored either in fixative or frozen for subsequent use. Plastic and paraffin blocks can be stored indefinitely as prepared. If ultrastructural preservation is unimportant, use of a cryostat (with chilled blade and sectioning chamber, capable of 5- to 25-μm sections) or a sliding, freezing microtome (with chilled specimen stage, but not blade or chamber, capable of 20- to 200-μm sections) may be preferable. The freezing microtome rapidly and conveniently produces good-quality sections that can be float-stained and handled exactly like vibra-

tome sections. Cryostat sections are preferable for preservation of highly water-soluble chemicals since they can be directly mounted on glass slides and stored dry and frozen.

For temporary storage of vibratome or freezing-microtome sections of rat or mouse brain, it is convenient to use 24-well culture dishes containing buffered 1% formaldehyde (silver degeneration stain) or simply 0.1 *M* buffer (all other stains) which are readily stacked in a refrigerator at 4°C. After sectioning, a sequence of experiments can be undertaken with the aim of minimizing the required storage period prior to staining. Staining baskets can be made by drilling 24 holes in a solid block of plastic and fastening nylon window-screening material to one side with cyanoacrylate adhesive. If made to fit snugly into the plastic tops of the tissue culture wells, they displace fluid up into each hole and allow incubation of tissues with a minimum amount of (sometimes expensive) reagents, about 25–30 ml per basket.

All experiments are run as complete balanced sets including an equal number of sections from each animal and an equal number of animals from each experimental condition. For example, experiments are often conducted with four dose levels and 3 subjects per dose, which allows convenient staining of 2 sections from each of the 12 subjects in each staining experiment. These could be two nearly adjacent duplicate sections focused on a single area such as hippocampus or two sections from different regions of the brain (e.g., brain stem and corpus striatum). To cover even more brain areas, multiple sets of baskets can be used. With these manual methods, 48 sections of tissue can be immunohistochemically processed in a day, and perhaps twice as many sections can be processed for silver degeneration staining. If it is necessary to increase the sample size, replications of the experiment can be scheduled for subsequent times, allowing perhaps 2–6 weeks between experiments. It is extremely important for quantitative purposes to maintain equal balance between the animals of all the groups in order to avoid biasing the results in case of run-to-run variability, which can be difficult to avoid with some histochemical methods.

III. Methods: Acute Neurotoxicity

A. Paraffin and Plastic Sections

The procedures available for neurohistological toxicity studies can be roughly divided into those especially useful for chemicals that produce relatively rapid, acute damage to neurons and their processes versus procedures capable of detecting the accumulation of slower, more subtle changes in structure that may be produced by the continual, chronic presence of a chemical.

The simple use of 5- to 7-μm-thick hematoxylin and eosin (H&E)-stained paraffin sections, through the area of interest, is the starting point for many investigations. These sections are thin enough to facilitate the recognition of acutely necrotic, pyknotic neuronal cell bodies as "bull's eye" profiles with dense, dark, compacted nuclei surrounded by clear cytoplasm. The neuroglial cells can also be recognized and evaluated. Thin (about 1 μm) plastic sections stained with toluidine blue, Richardson's stain (toluidine blue and Azure II), or methylene blue and fuschin provide somewhat better resolution for the evaluation of acutely necrotic cells (see Fig. 1). In H&E paraffin sections, neuronal dendrites cannot be distinguished, whereas they are readily visible in plastic sections (see Fig. 1). Neither technique

provides sufficient resolution at the light microscopic level to visualize axons, although the plastic sections lend themselves to subsequent electron microscopic evaluation. The main disadvantage of these simple methods is that acutely necrotic cells rapidly degenerate and are phagocytized: the evidence of the damage can be removed within a period of hours following the neurotoxic insult. Therefore, there is a real need for the investigator to very carefully and completely characterize the post-treatment time course in order not to draw the "false negative" conclusion that no damage occurred.

B. Silver Impregnation of Degenerating Processes

Vibratome, frozen, or paraffin sections can be utilized in silver staining procedures designed to impregnate only axon terminals or neuronal cell bodies that

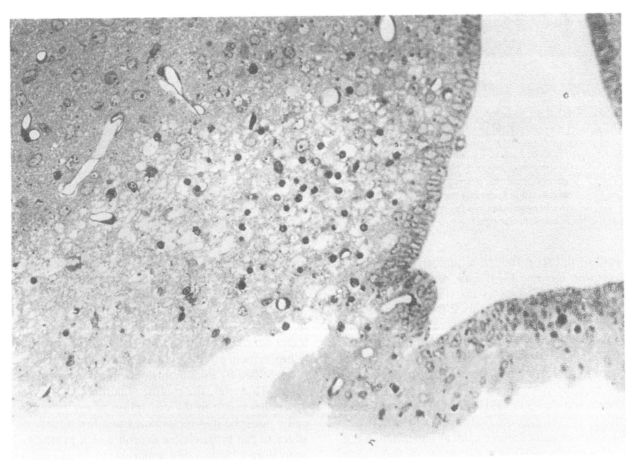

FIGURE 1 A 1-μm plastic section (Richardson stain) illustrating the arcuate nucleus adjacent to the third ventricle and median eminence of a 10-day-old rat pup 6 hr after exposure to a single subcutaneous dose of 4 mg/g of the food additive monosodium glutamate. Note the dark, condensed nuclear material in the centers of many pyknotic cells and the absence of Nissl substance producing the clear appearance of the surrounding cytoplasm. Note also the clusters of small, clear, circular structures (identifiable as dendrites by electron microscopy) visible in the rarefied tissue between perikarya.

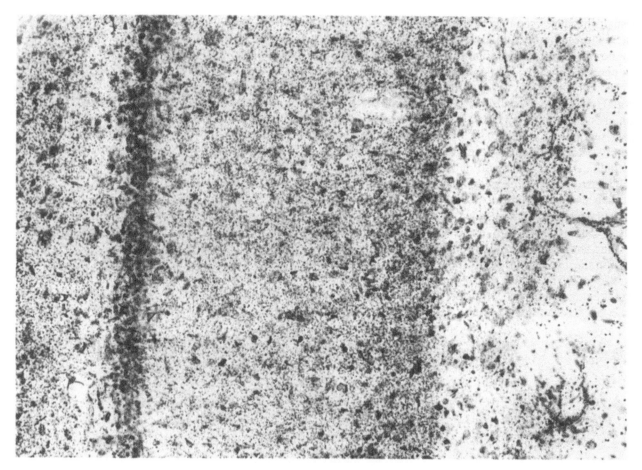

FIGURE 2   A 40-μm vibratome section (silver degeneration stain) illustrating the laminar pattern of damage to the hippocampus 1 month following a 4.5-mg/kg intraperitoneal dose of trimethyltin. At this survival interval, note that despite the considerable thinning of the CA1 pyramidal neurons (dark band at left), the degenerating axons and terminals of CA1 stratum oriens (narrow band at far left) and CA1 stratum radiatum (wider band in the middle) remain easily demonstrated. Also note the relative absence of silver in the stratum lacunosum (far right).

are dead or dying. Although a number of techniques are available (Nauta, 1954; Fink and Heimer, 1967; de Olmos, 1969; Gallyas *et al.*, 1980a,b, 1990), the procedure outlined in Appendix 2 (Nadler and Evenson, 1983) is simple and provides excellent results. The neurotoxicity of trimethyltin (Chang *et al.*, 1982; Miller and O'Callaghan, 1984) has been studied with silver degeneration-selective methods (Fig. 2) applied up to several months postexposure (Balaban *et al.*, 1988; Matthews and Scallet, 1991). This method has been used to study the neurotoxicity of the "excitotoxic" seafood contaminant domoic acid, as illustrated in Fig. 3 (Scallet *et al.*, 1993). An advantage of these tech-

niques stems from the apparent mechanism by which the procedures distinguish damaged from intact neurons.

It has previously been demonstrated (Gambetti *et al.*, 1981) that the Bodian silver stain procedure for intact neurons results in the deposition of silver on the neurofilament protein components of the cytoskeleton, as identified by immunohistochemical electron microscopy. However, the routine conditions of the Bodian procedure result in the disruption of even the normal cell's phospholipid membranes to allow access of the silver to the intracellular neurofilament proteins. The procedure may thus be comparable to many commonly

FIGURE 3   (a) Silver degeneration stains (40-μm vibratome section) distinguish necrotic (dark black) from viable (light-gold colored, with prominent nuclei and nucleoli) neurons between CA1 and subiculum subfields of the hippocampus of a cynomolgus monkey treated with 4 mg/kg i.v. of the convulsant neurotoxicant domoic acid, a seafood contaminant. (b) Higher magnification of one necrotic and several viable neurons.

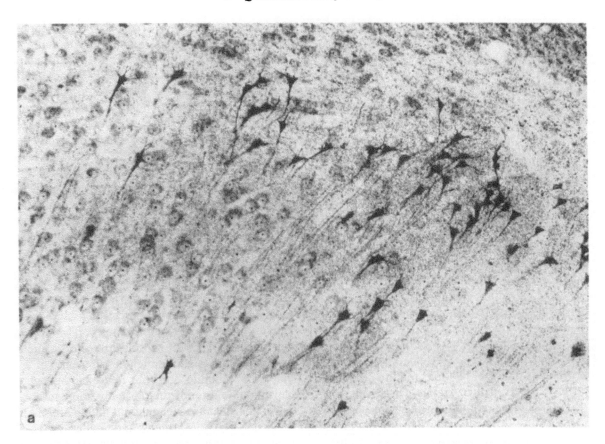

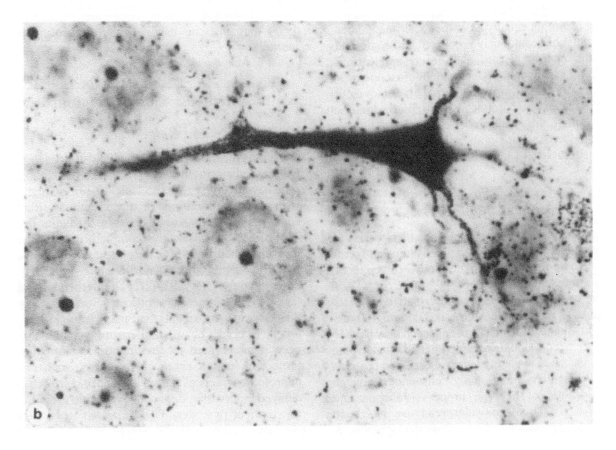

used biochemical methods of using silver to stain proteins separated by gel electrophoresis. The nature of this staining is uncertain, but may involve the reduction of the divalent silver in $AgNO_3$ by free sulfhydryl groups on cysteines in the denatured intracellular neurofilament proteins to form Cys–S–Ag (monovalent) complexes.

Silver staining procedures selective for neurodegeneration have the ability to differentiate healthy neurons with intact phospholipid membranes from impaired neurons with damaged membranes. A possible explanation may depend on the ability of the silver to penetrate into the cytoplasm (only when the membrane is compromised as it is in a damaged cell) to bind to the proteinaceous cytoskeleton. Since the neurofilament proteins have an extended life compared to the diffusible cytoplasmic Nissl substance stained in conventional plastic and paraffin sections, the presence of prior damage can be inferred for at least several weeks after the time of dosing by using the silver methods. Another advantage is that the presence of degenerating axon terminals can be readily ascertained by light microscopy, whereas the other methods require complex interpretation of electron micrographs. Disadvantages are that it is difficult to reproduce silver stain procedures exactly from run to run, and that the thickness of the frozen or vibratome sections may make it challenging to quantitate the results. Repeated sets of sections, representing equally all of the experimental conditions, must be assiduously processed and sometimes failed runs must be discarded. In future applications, the use of a confocal laser-scanning microscope or an image array processor to computationally remove out-of-focus haze may help "optically section" the material so that a small, discrete slab in near-perfect focus can be provided to the frame-grabber board of a computerized densitometer for measurement.

C. Immunohistochemistry of Neurotoxicity Biomarkers

Immunohistochemical staining is a highly flexible and useful technique by which nearly any antigenic chemical component of the tissue section can be localized. The same method (e.g., Sternberger *et al.*, 1970; Scallet *et al.*, 1988) can be used to visualize a variety of antigens, as indicated in Appendix 3, although sometimes adjustments of the fixation approach may have to be made to accommodate an individual antigen. Some progress has been made in identifying specific proteins that may selectively mark only cells undergoing an acute apoptotic or necrotic process (Johnson and Deckwerth, 1993). More widespread use has been made of markers such as c-fos/c-jun protein immunore-

activities (Morgan *et al.*, 1987; Zawia and Harry, 1993; Dragunow *et al.*, 1993), which indicate acute cellular metabolic activity in some but not all neuronal populations, and heat shock protein (HSP-72 and others) antibodies (Shimosaka *et al.*, 1992; Li *et al.*, 1992), which may label neurons responding acutely to a neurotoxic exposure. Although these markers cannot be interpreted as indicating dead or dying neurons, they do serve the potentially valuable purpose of marking neuroanatomical regions of acute cellular activation. The neuroanatomy may then aid in understanding the mechanisms by which certain types of neurotoxins may propagate damage throughout the brain. For example, these markers have been used to indicate regions of cellular activation in the brains of gestational day 21 rat pups exposed to hypoxia–ischemia (Fig. 4; Binienda *et al.*, 1995). Immunohistochemical markers such as glial fibrillary acidic protein (GFAP), found only in nonneuronal astrocytic cells, can nevertheless be considered biomarkers of acute neurotoxicity. The presence of increased GFAP with cytoplasmic and/or nuclear enlargement may signal either the presence of a gliosis that was "reactive" to prior neuronal damage or perhaps an alteration of the normally cooperative role of astrocytes in neuronal ammonia/glutamate/GABA metabolism (Fig. 5; Matthews and Scallet, 1991). Macrophage cell surface markers for microglia (for their application to ibogaine neurotoxicity see O'Hearn *et al.*, 1993) may also be used as neurotoxicological biomarkers.

IV. Methods: Chronic Neurotoxicity

A. Golgi Methods

Among the most useful and widely utilized histological methods in the neurosciences are the Golgi methods, developed by Camillo Golgi in 1873. Golgi, along with a gifted practioner of these methods (Ramon ý Cajal), shared the Nobel Prize in Medicine in 1906. These methods allow the complete impregnation of individual neurons, revealing in great detail the cell body, dendritic arbor, dendritic spines, and, in some cases, axons. They also offer the tremendous advantage of reflecting even subtle structural changes produced by neurotoxicants that may be too slight to kill a cell. Even developmental rearing experiences have been demonstrated to alter dendritic branching patterns (Greenough *et al.*, 1976). Also, such possible neurotoxic effects as failure of dendritic growth due to delayed innervation or absence of a vital growth factor at a critical time would be expected to be detected by these methods. One example of the application of

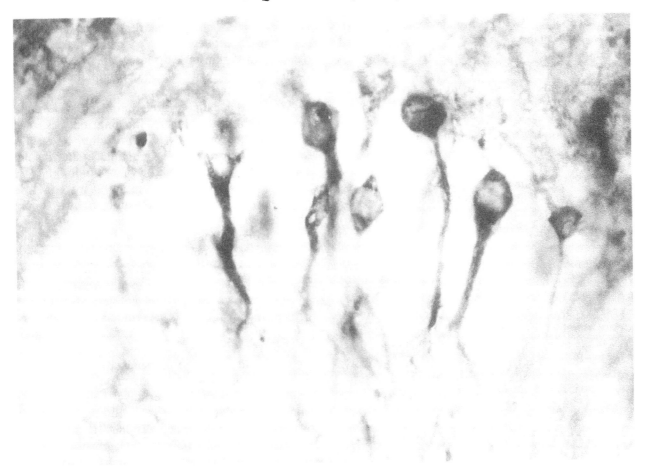

FIGURE 4 A 40-µm vibratome section showing heat shock protein immunopositive pyramidal neurons in the hippocampal CA3 zone of a gestational day 21 rat pup 85 min after it was rendered hypoxic/ischemic by clamping shut the maternal vasculature to its uterine horn.

quantitative Golgi procedures to neurotoxicology is provided by Uemura and co-workers (1985a): chronic gestational inhalation exposure to low amounts of halothane produced alterations in dendritic arborization. There are several different methods of Golgi staining (e.g., Vaisamraut and Hess, 1953; Riley, 1978; D'Amelio, 1983) in common use, generally divided into procedures for previously formaldehyde-fixed tissue and those for nonfixed tissue. Most of these methods are *en bloc* stains, and the few single section methods published (e.g., Gabbott and Somogyi, 1984) appear difficult and unrewarding. Opinions are divided as to whether the methods for formaldehyde-fixed or unfixed tissues are superior. In this author's experience, the methods for unfixed tissue impregnate a greater depth of tissue in a more uniform fashion than fixed-tissue methods. Appendix 4 describes a simple *en bloc* Golgi procedure for unperfused brain tissue that yields excellent results (Uemura *et al.*, 1985a; Scallet *et al.*, 1987; Fig. 6). A potential limitation of the Golgi methods is the still unresolved uncertainty about which particular neurons will be impregnated and why, suggesting cautious interpretation as to the universe of cells to which any data extrapolations may apply. A possible way around these limitations has been offered by procedures involving dye injections to label individual cells that thus can be randomly sampled or chosen on the basis of neurophysiological characteristics. Nevertheless, the practicality of such methods for large, quantitative morphometric studies has not yet been demonstrated.

B. Synaptic Density

Another useful procedure sensitive to chronic, persistent, and/or gradual neurotoxic effects of chemicals is the estimation of synaptic density by electron microscopic counting methods (Fig. 7). These methods have been sensitive to subtle neurotoxic effects (e.g., Uemura *et al.*, 1985b), even to developmental social experiences (Greenough *et al.*, 1976), and can complement the quantitative Golgi technique. For example, Golgi analysis was applied to groups of monkeys' brains and it was found that high levels of operant testing experience were correlated with well-arborized CA3 hippocampal pyramidal cell dendrites, consistent with (and extending the anatomical specificity of) Greenough's observations on environmental enrichment (Scallet *et al.*, 1990). This observation was complemented by a completely separate study using identically treated monkeys that found decreased synaptic density in CA3 stratum radiatum of the more experienced animals, as also reported by Greenough and colleagues (1976). Although it may seem that the complexity of the requirement for large-scale electron microscopy work would favor the use of light microscopic methods such as the silver impregnation of degenerating axon terminals, it is important to note that the degeneration methods can only reveal recently damaged axons. Gradual changes such as slow loss of small numbers of axons or a failure to support synapses due to absence of a needed growth factor might not be revealed by the silver degeneration stains. On the other hand, evaluations of synaptic density are sensitive to any change in synaptic number, regardless of cause or time course, if the effect is sufficiently large.

C. Immunohistochemistry of Neurotransmitters and Receptors

Immunohistochemical methods have great potential as biomarkers of chronic neurotoxicity, especially where there is some available neurochemical information to help choose an appropriate marker. For example, neonatal glutamate (GLU) neurotoxicity can be readily seen in plastic sections or silver degeneration stains obtained 4–6 hr after exposure (Fig. 1), but it is very difficult to detect histologically when the adult animal's hypothalamus is evaluated long after the initial dosing. This is because it is quite hard to detect the resultant diminished neuronal density in the GLU-treated arcuate nucleus visually. However, the immunohistochemical staining of proopiomelanocortin neurons, which are primarily located in the arcuate nucleus, provides a highly sensitive method to evaluate the adult lesion (Scallet and Olney, 1986; Scallet, 1987; Alessi *et al.*, 1988; Fig. 8). Similarly, counts of tyrosine hydroxylase-positive dopamine cells of the substantia

nigra can be a useful biomarker of the development of chronic neurodegenerative or neurotoxic changes resulting in Parkinson's disease.

However, there are some important interpretational caveats regarding immunohistochemical observations of this type that cannot be overemphasized. For example, immunohistochemical labeling only succeeds when enough antigen is present within or on a cell for it to acquire sufficient amounts of the insoluble diaminobenzidine reaction product or the fluorescent second antibody to visualize in the microscope. Cells may be present that contain the antigen, but in concentrations too low to visualize them. This implies that there may be at least two reasons why fewer immunohistochemically labeled cells or cell processes might be seen following "neurotoxic" doses of a compound. First, it is possible that cells synthesizing the antigen or processes containing the antigen have been destroyed and are no longer present to synthesize antigen or serve as a vessel for antigen. Second, the amount of antigen synthesized may have been reduced below the threshold required for visualization. For example, in neurohistological studies of methylenedioxymethamphetamine (MDMA, "ecstasy") and the related compound fenfluramine, it has been suggested that serotonergic neurons are actually lost (Commins *et al.*, 1987; Molliver and Molliver, 1990; Appel *et al.*, 1989). However, it is also possible that such chemicals disrupt the capability of uptake sites to load viable neurons with extracellular serotonin (5HT), especially when 5HT levels are enhanced by pretreatment with MAO inhibitors (such as tranylcypromine) often used to boost immunohistochemical sensitivity (Slikker *et al.*, 1988; Scallet *et al.*, 1988, Kalia, 1991). In either case, there is a "neurotoxicity" of the drugs, but the interpretation of the specificity, site of action, and extent of damage produced by the compounds vary. Careful consideration of these factors is essential to rationally evaluate the therapeutic benefit vs risk of such drugs.

As specific antisera for receptors of known amino acid sequences become increasingly available, the use of receptor immunohistochemistry (Henry *et al.*, 1991; Good *et al.*, 1993) should be of increasing importance to neurotoxicology. Methods using radiolabeled receptor ligands for autoradiography (Bondy and Ali, 1992) will also remain important for some years to come. These

---

FIGURE 5 (a) A 40-μm vibratome section showing normal astrocytes (immunohistochemically labeled for glial fibrillary acidic protein) of the CA1 region of a control rat hippocampus. Note the flattened appearance of the perikarya and the prominent fibrous processes. (b) The astrocytes of the CA1 region of a TMT-treated rat, even 1 month after dosing, remain swollen with greatly enlarged nuclei, an appearance which corresponds to Alzheimer's type II gliosis in conventionally stained H&E material.

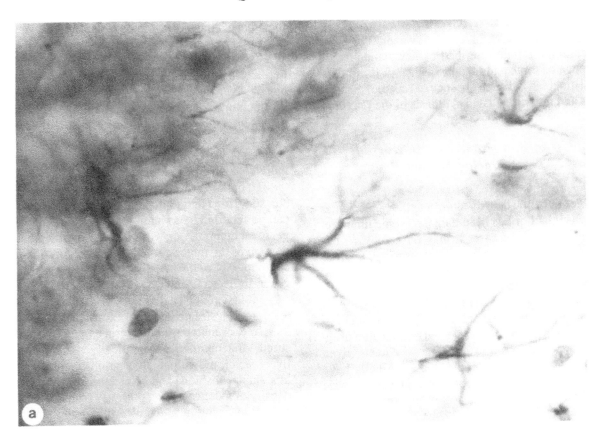

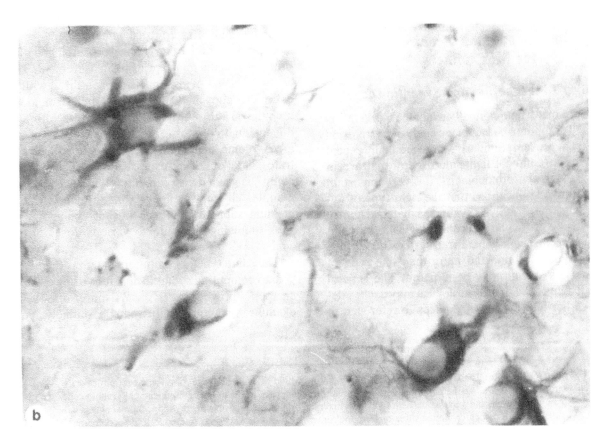

methods can indicate the selective loss of cells expressing the particular receptor type; however, the same interpretational pitfall described earlier maintains. In the absence of some additional evidence, it is impossible to distinguish between absence of receptor-bearing neurons and down-regulated synthesis of the receptor.

D. Use of Retrograde and Anterograde Tract-Tracing Methods

It may be uncertain based on silver degeneration methods or synaptic density alone whether a particular pathway is intact or has been damaged some time previously. If this is the case, it may be helpful to use an anterogradely or retrogradely transported marker compound such as wheat germ agglutinin-coupled horseradish peroxidase (WGA-HRP), Fluoro-Gold, or some similar marker (for a review see Schmued, 1994). Observing whether transport of these compounds following direct injection into a selected brain region progresses normally can aid in inferring whether a given pathway is intact. For example, these methods have been used to support the concept that serotonergic axons innervating the frontal cortex may remain intact despite the large losses of HPLC-measurable and immunohistochemically visible serotonin (Kalia, 1991).

V. Instrumentation Requirements for Morphometry

A. Silver Impregnation and Synaptic Density

There are many commercially available workstation or personal computer-based image analysis systems on the market at present. Similarly, there are many recent innovations in confocal light microscopy that are of relevance to the neurotoxicologist. It is very important for quantitative applications that the right type of microscope and image analysis system be matched appropriately to the particular neurohistological approach. In some cases, no single generalized system may suffice for every type of stain, and more than one type of image analysis system may be required. The general characteristics of the neurohistological approaches described earlier and in the Appendices in reference to

the image analysis systems required for successful measurements will be discussed.

The silver impregnation procedures described earlier and in Appendix 2 result in deposits of black, reduced silver that are selectively retained in degenerating (membrane-damaged) neurons. If the cell bodies are damaged, a high-contrast image of distinctly separated neurons can usually be obtained (see Fig. 3). If axonal damage is present, its appearance is like a dusting of individual silver grains or small groups of grains across the region innervated by the damaged axons (Figs. 2, 3a, and 3b). High-contrast images of the cell bodies can be acquired by a frame-grabber board installed in a personal computer or workstation, and then the digital version of the image can be measured by software that computes its optical density ("darkness" in terms of generally 256 gray-level quanta). The number of objects present that are at least as dark or darker than a desired level, meet a size criteria or a "form factor" describing its shape, or match other criteria can be selected, measured, and counted automatically. Analysis of relatively small, individual silver grains can be conducted in a similar fashion, and offers the advantage that the density of a degenerating pathway from disparate neurons converging in a given terminal zone may reflect an integration of the toxicity to a large group of cells otherwise difficult to count. These systems generally deal with images of a single microscopic plane through a sample and offer contrast enhancement, edge erosion, and several other image-processing options to precondition the image prior to measurement. The current generation of systems is being utilized with confocal microscopy and much larger image memories to recompose the multiple confocal slices into a three-dimensional composite image for analysis of surface area, volume, and other such parameters. Figure 9 illustrates data obtained using a single-plane system to count the grain density in the projection of the dentate granule cells' mossy fiber axons to their CA2 stratum lucidum subfield terminal zone of cynomolgus monkeys following exposure to various doses of domoic acid.

The same type of computer system is readily adaptable to computing synaptic densities using electron microscopic procedures (e.g., Wilson *et al.*, 1990). For example, synaptic densities can be routinely computed

FIGURE 6 (a) The Golgi procedure for impregnating whole neurons (150-μm vibratome section) reveals the soma, as well as the dendrites with their dendritic spines, of (right to left) a hippocampal granule cell, an interneuron, and a pyramidal neuron in a normal rat. (b) A normal rhesus monkey hippocampal pyramidal neuron with computer-plotted three-dimensional drawings (head-on view and rotated 90°) of a comparable neuron shown as insets.

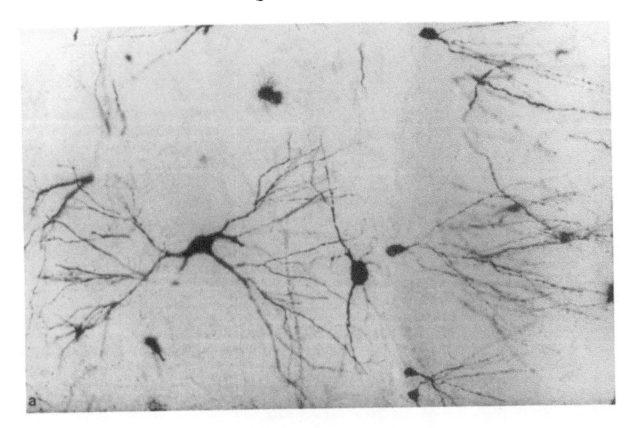

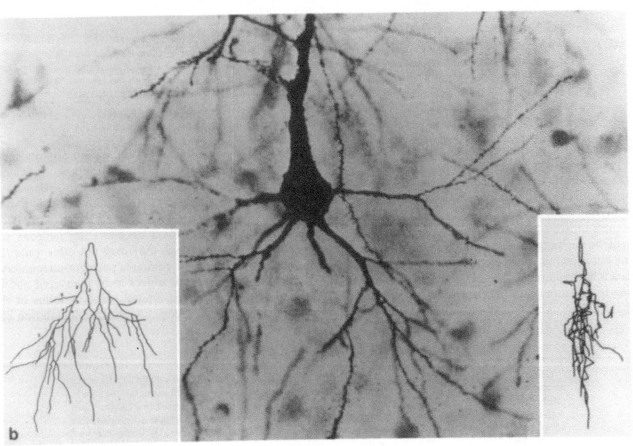

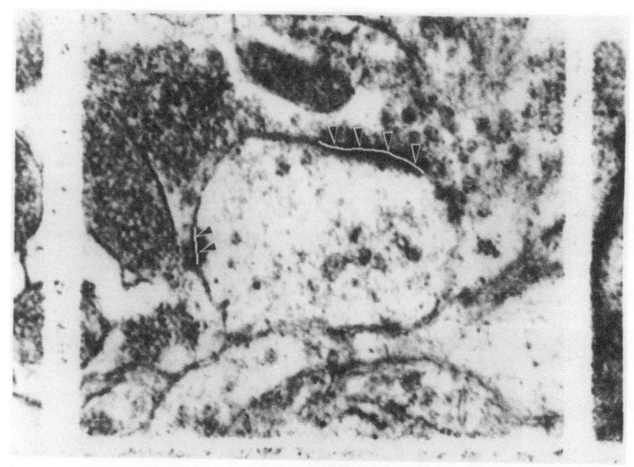

FIGURE 7 The photograph was taken with a 35-mm camera from a computer monitor that displayed captured images of sectors (demarked by the white bars from a ruled glass overlay) of an electron micrograph one at a time, until the entire micrograph was sampled. The white lines ($\rightarrow$) visible between the pre- and postsynaptic membranes of the two synapses shown illustrate how the length and number of synaptic contacts were measured.

based on the method of Collonier and Beaulieu (1985) by placing electron microscopic negatives on a light box underneath a video camera relaying images to the frame grabber board and scanning small squares of the negative image (converted to positive by the computer) one at a time by means of a ruled glass plate laid on top of the micrograph (Fig. 7). The length of the contact zone between the opposing membranes of each synaptic cleft encountered in the electron micrographic negatives is measured and the computer stores a tabular record of the number and measurements of the syn-

apses as they are obtained. Then a simple spreadsheet program applying the computational approach of Collonier and Beaulieu (1985) yields the synaptic density in terms of number of synapses per unit volume of tissue. Immunostained tissues can be analyzed very similarly to the silver-impregnated degenerating neurons described earlier, although they may not offer quite the degree of contrast. A common problem encountered is that the edges of the object to be measured and/or counted are out of the limited plane of focus of the standard light microscope. A reasonable solution can

FIGURE 8 (a) An immunohistochemical stain reveals numerous proopiomelanocortin (POMC) cells of the arcuate nucleus, adjacent to the third ventricle and median eminence, of a 1-year old control rat. (b) Significantly fewer cells are visible in this micrograph of a 1-year old rat that had been treated as a neonatal pup with monosodium glutamate. Note that the lateral portion of the POMC neurons (appearing as a distinguishable subgroup in the control animal) remains visible while very few cells located adjacent to the ventricular wall can be labeled.

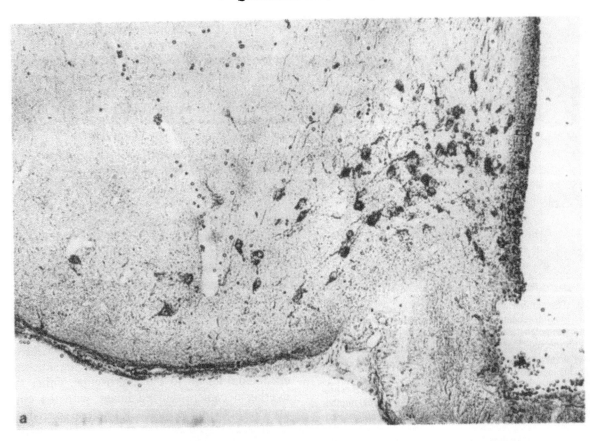

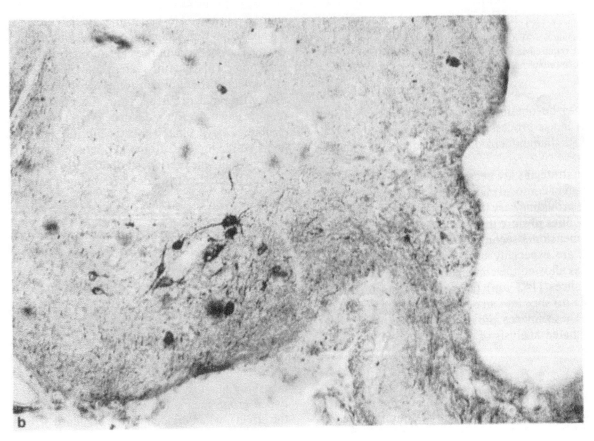

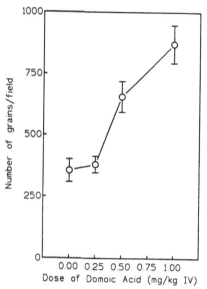

FIGURE 9 The number of grains per visual field (means ± SEM, averaged across four repeated sets of slides) counted in CA3 stratum lucidum in cynomolgus monkeys given i.v. doses of domoic acid. There was an inequality of variances among the four sets of slides prepared for silver degeneration and measured by computerized densitometry. This inequality results from run-to-run "capriciousness" inherent to the silver degeneration technique. Presumably, it represents a different ratio of silver grain "markers" to the actual number of underlying degenerating terminals, depending on unstable experimental conditions during the grain development. To appropriately combine all the data for analysis, a nonparametric Friedman's analysis of variance was performed and indicated a significant effect of domoic acid: $\chi^2(3\,df) = 8.1\,(P < 0.05)$. Alternately, parametric analyses (assuming ANOVA is "robust" to departures from the equality of variance requirements) were performed and agreed with the nonparametric approach.

sometimes be obtained from the edge enhancement or other image-processing algorithms or using metal-intensified diaminobenzidine procedures to enhance contrast.

Other strategies are becoming available with the development of confocal microscopy. More recently, systems are providing very large RAM and hard disk memory capacities to store the large image files and provide three-dimensional reconstructions of the data. These features are especially useful where confocal microscopy has allowed the combination of many individual optical slices (1–2 μm) taken from a single thicker (30–150 μm) slice into an optimally focused composite image. These images provide more suitable contrast for automated analysis of digitized micrographs.

B. Golgi Methods

Measurement of the features of Golgi-impregnated neurons has often involved special attention to measurable features of the branching patterns of the neurons'

dendrites. The current generation of commercial instruments suitable for this purpose generally function quite similarly to one another and provide measurement of soma diameter, soma area, number of apical or basilar dendrites forming from the soma, length of each dendrite from the soma to the first branch point, mean length of such "first order" segments, number of branch points per dendrite, distance between first and second order branch points, total number of such segments, and mean length of such "second order" (and higher order!) branches per neuron. Surface area of the dendrite, including the dendritic spines, is another important parameter, considering that this accounts for most of the contact surface for synaptic input to the cell. The available instruments utilize a variation of computer-aided-drawing (CAD) methods and are semiautomatic, requiring the operator to use a mouse or stylus to outline the neuron. In some systems, the cell being drawn is observed through the microscope while the developing drawing is viewed as an overlapping image on a computer CRT screen viewed through a beam-splitting drawing sidearm (camera lucida) added to the microscope. Other systems overlap a camera-relayed image with the computer drawing both on the same CRT screen so an image can be directly traced without the operator spending long periods of time looking into the microscope. More recently, systems are becoming available that are designed to analyze a stack of confocal Golgi-type images by measuring the dendrites in each slab considered one at a time. This type of system may be best suited for dye-injected three-dimensional neuron reconstruction so as to avoid problems from the presence of more than a single neuron's dendrites in a given slab. Such problems have limited most dendritic analytical systems so far to semiautomatic operation, requiring the oversight of an experienced operator to distinguish the neuronal source of the dendrite being measured.

VI. Experimental Design and Statistical Analysis of Morphometric Studies

A. Numbers of Animals

The number of animals (n) required per experimental group to detect an effect, as in any study, depends on the minimum difference one would like to detect (precision) as well as the underlying variability (as represented by the standard deviation "sigma") of the measured quantity. More precisely, the difference between group means (an estimate of the magnitude of the effect), when divided by the standard error of such differences (computed as σ/square root of n), provides a Student's *t* statistic which determines the probability that the means were that far apart by chance alone. The smaller the standard error, the larger the *t* statistic and the less likely the effect is due to chance alone. The magnitude of the effect is a *priori* unknown (or else, why do the experiment?), but it *is* known that the standard error (in the denominator) decreases as the square root of the sample size, *n,* increases. For this reason, as the sample size increases, the value of the *t* statistic for a given magnitude of effect increases, but only as the square root of the sample size. For this reason, increasing "*n*" from 5 to 20 per group is required to double "*t*". This law of diminishing returns with increased sample size is mitigated partially by the increase in degrees of freedom (reducing the *t* value required for a given level of significance) with sample size. As a practical matter, the cost of morphometric sampling and analysis must be divided in some ratio between all the samples chosen from the same anatomical site or fewer samples divided among a broader range of anatomical sites sampled. For this reason, an *n* of 5–10 per region is suggested as a minimum for an initial experiment for most applications, where the precise region of greatest anatomical sensitivity to neurotoxicity may be unknown. Then, within lab replication of any important (but statistically marginal) effects should be undertaken. Also, the desired sample size (and anatomical location) of subsequent experiments may be better predicted on the basis of power computations, once the variability has been estimated from the initial experiment. Ultimately, interlab replication is important to establish the veracity of experimental evidence in any case.

B. Experimental Design

Although anatomical measurements usually require a more complex design than a simple *t* test, the basic principles of experimental design and statistical analysis, beyond the scope of the discussion here, are the same for morphometric studies as for any other scientific investigations (Winer, 1971).

Selection of treatment levels (independent variables) may be dictated by available information on minimally effective or lethal doses, and if the goal of the research is to develop data for quantitative risk analysis, enough dose levels (five or more is preferable) should be included to characterize the shape of the dose–response relationship. Subjects should be randomly assigned to treatment groups, and if the experiment cannot be conducted on all subjects treated as a single set, it should be arranged as identical, replicated blocks. This means that each treatment (dose level) should be repeated several times on blocks of subjects assigned randomly

to treatment groups and replicates. This randomized assignment protects against systematic bias that may occur with variation in such things as accuracy of dose preparation, intensity of staining between repeated batches of tissue sections, and seasonal or other periodic hormonal fluctuations. The randomization is also necessary to meet the assumption that all observations are independent that underlies the theory of analysis of variance procedures.

Balanced designs with equal numbers of animals per group are preferable to unbalanced designs with missing observations. Sometimes more than one treatment (or "independent") variable is employed such as in the examples of (1) marijuana treatment and differential operant performance and (2) two types of hypothalamic neurotoxicants that are described next. In such cases it is preferable, if possible, to use a complete factorial design; subgroups of each group of subjects assigned to each level of variable 1 should be present and randomly assigned to each level of variable 2. These complete factorial designs are easier to interpret in the presence of interactions between the variables, which cannot be evaluated in partial factorial designs.

C. Stereological Considerations

Attention must also be paid to the selection of the dependent variables to be measured (as described earlier and in the Appendices) so that they represent unbiased estimates of means that are biologically meaningful. The theoretical underpinnings of the choice of appropriate parameters to measure are within the field of stereology, as discussed by Weibel (1979), Russ (1986), and Reith and Mayhew (1988). The basic principles of stereology address the measurement of structural parameters such as volume density (V_v) and surface density (S_v). These parameters are independent of any assumptions as to the shape of the objects being measured (the stereologist, like the topologist, is someone who cannot tell the difference between a doughnut and a coffee cup!). As a practical matter, the implication is that the fractional area (A_a) of an irregularly shaped structure of interest (such as mitochondria) can be directly measured. After outlining the "mitochondrial phase" of the section manually (or as the computer defines it from relative optical density measurements), the ratio of mitochondrial area to section area (measured as number of pixels in each phase) is an unbiased estimate of the value of A_a. Mathematically, this ratio is also an unbiased estimate of V_v, the ratio of mitochondrial volume to tissue volume (or perikaryal volume, if extracellular space is excluded) (Weibel, 1979, p. 26). No assumptions have to be made about the average size or shape of mitochondria. Previously, most studies of this type used square grid overlays to compute the volume density from P_p, the ratio of the number of line intersections (points) over mitochondria compared to the total points over the section. However, the ease of directly measuring area (either semi-automatically or automatically) offered by modern image analysis systems has largely obviated the older "grid" methods. In "point" of fact, though, the computer image measurement approach is actually the same as the earlier "point counting" methods, with the computer providing the 512×480 pixel grid (depending on its resolution) over the image in lieu of the manual approach! This stereological model underlies the measurement of relative area of hippocampal subregions in example application "c".

As pointed out by Weibel, stereological "structure parameters" such as the volume (V_v) density or the surface density (S_v) sometime conceal a large amount of information: they reflect the average proportion over the entire volume of interest. For example, a decrease in volume density may occur when the same number of objects is present, if the size of each object is reduced. Alternately, a decrease in volume density may also reflect a reduction in number of objects with no change in their size. Stereological "structure parameters" such as V_v cannot distinguish between these alternatives. In some cases, an alternate stereological approach ("particle parameters") that does require assumptions about particle shape and size can be used, for example, to compute N_v, the numerical density of particles in a given volume. This is the approach often used in the computation of synaptic density as number of synapses per unit volume of tissue. This stereological approach underlies the measurements of neuron density in example "a," as well as the examples of synaptic density ("b"), number of MSH-immunopositive cells ("e"), and numeric density of degenerating axon terminals ("f"). The measurement of whole neurons via the Golgi approach (example "d") is a special case: stereological principles are not evoked explicitly because an attempt is made to directly measure the whole cell instead of a section through it. Sterio (1984) introduced a method, the "dissector", which is now receiving widespread application (e.g., Coggeshall, 1992) because it does not require assumptions about particle shape and size in order to estimate "particle parameters" such as N_v.

D. Statistical Approaches

It is extremely important to designate effects appropriately since most computer-based statistical analysis programs will allow the data to be entered a number of ways which look correct, but will have the wrong

associated F statistic and probabilities. For example, many of the early quantitative Golgi studies of the effects of environmental complexity on dendritic branching incorrectly used the total number of cells measured (instead of the number of animals treated) as the "n" for statistical comparisons. Often the most appropriate statistical analyses for neurohistology studies will be mixed model repeated measures analyses of variance. Thus, with anatomical morphometric measurements, multiple levels of the same variable are often measured on the same subject. For example, the area of a conical object or brain structure may be measured on each of several two-dimensional cross sections through it. If an experimental treatment damaged neurons selectively in the narrow tip of the cone, it could be shown as a treatment by level of cross section interaction in a mixed model repeated measures analysis of variance design. The level of cross section variable is a "within," repeated, or correlated measure since each section's area depends on the area of the preceding one. The values of the cross-sectional area are all measured on the same subject. The experimental treatment is an independent variable since each estimate of the effects of the different treatment levels is made on a separate, independent animal.

Continual attention must also be directed at the correlations between different variables so that the investigator recognizes how to deal with strongly interrelated variables, such as the total length of dendrites and the total number of branch points per cell in Golgi analyses. Separate analyses in such a case are of limited value because of their redundancy, and a multivariate analytical approach may be more appropriate (if less easily publishable!). The best practical course in such cases may be to report the data as multiple and separate univariate analyses, but consider the intercorrelations between the separate analyses in the discussion and interpretation of the findings. Refer to a standard *Statistics and Design* text for more detailed discussions of these points, such as Winer (1971).

VII. Applications to Neurotoxicology

A. Example: Neuron Density in Thin, Plastic Sections of Hippocampus Following Marijuana Exposure and Operant Conditioning

A group of peripubertal male rhesus monkeys (*M. mulatta*) smoked one marijuana cigarette daily for 7 days per week for a year (HI dose group). A second group smoked marijuana 2 days per week (LO dose group). A "sham" (SH) control group and a group that smoked extracted (EX) cannabis cigarettes (with no tetrahydrocannabinol) 7 days a week were also included. Further details of the four exposure conditions are reported elsewhere (Slikker *et al.,* 1991, 1992).

Prior to initiation of smoke exposure, all the monkeys had been trained for approximately a year to perform a battery of five operant behavioral tasks (Paule *et al.,* 1992). Within each dosing group, half continued to perform their daily (55 min) learning sessions (TEST monkeys) whereas the other half was deprived of continued testing during the 1 year of smoke exposure (REST monkeys).

Six monkeys per drug condition (HI, LO, SH, or EX) were identified by a sequence number. Their experimental history was unknown to the neurohistology staff. Monkeys were anesthetized and perfused as described in Appendix 1. Tissue was held in 4% gluteraldehyde in phosphate buffer until a series of 3-mm slices were made within 2 weeks using a slot box to obtain uniform slices. Samples from the anterior, middle, and posterior hippocampus of each animal, among other brain regions, were then further dissected, trimmed, osmicated, dehydrated, and embedded in Epon/Araldite (Fig. 10). Richardson-stained sections (1 μm) were digitized from two separate locations within each of the anterior, middle, and posterior hippocampal blocks. As in Landfield and colleagues (1988), neuronal density

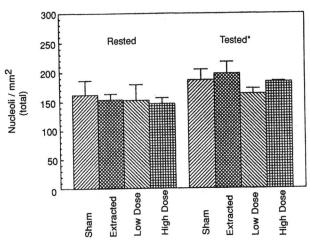

FIGURE 11 Monkeys tested in operant training boxes daily throughout the drug exposure period had significantly (*, $F_{(3,11)} = 6.31$, $P<0.05$) greater CA3 neuronal density than cage-rested cohorts (as shown). Data shown are for total number of neurons with visible nucleoli counted per square millimeter of CA3 s. pyramidale, whether single or multiple, anywhere within the hippocampus. If data are restricted to only neurons with single nucleoli in the middle block [$F_{(3,11)} = 18.7$, $P<0.01$, data not illustrated]. There were no significant differences in neuronal density in CA4, CA1, or in the dentate gyrus granule cell layer.

was then measured as the number of nucleoli per unit area of the hippocampal dentate gyrus, CA1, CA3, and CA4 subregions. The rationale for counting nucleoli instead of nuclei or whole parikarya is that nucleoli are only 2–3 μm in diameter, whereas nuclei are on the order of 7–20 μm and whole neurons are even larger. Theoretically, counting the relatively tiny nucleoli decreases the counting error that occurs when larger structures actually centered outside the section are falsely counted (Weibel, 1979, p. 316). Unfortunately, numerous hippocampal neurons have two or even three nucleoli per nucleus, so counts have to be corrected by either subtracting the extra nucleoli or by separately counting them as multiple nucleoli (as was done in this experiment). Although there were no effects of marijuana exposure on number of neurons per area in any hippocampal regions, the animals that continued to perform operant testing daily throughout exposure had higher neuronal density than rested animals, but only in the CA3 subfield (see Fig. 11).

B. Example: Synaptic Density and Width Following Marijuana Exposure and Operant Conditioning

Ultrathin sections from the region of mossy fiber axons near the proximal apical dendrites of the hippocampal pyramidal cells were obtained. Sections from two separate locations within each of the three (ante-

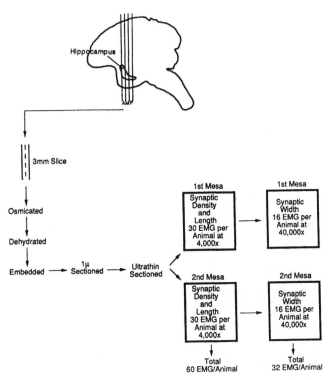

FIGURE 10 The sampling procedure used for obtaining thin plastic sections and grids for electron microscopic analyses of marijuana-treated monkeys.

rior, middle, and posterior) hippocampal regions (total of 6 grids) were obtained and mounted on 300 mesh copper grids (Fig. 10). After staining with lead citrate/uranyl acetate, 10 electron microscopic negatives (one each from the four corners plus the middle of two adjacent grid squares) were taken from each of the 6 grids at a magnification of 4000×. The resulting 60 negatives (as digitized positive images displayed on an MCID image analysis system, Imaging Research, Inc., Ontario, Canada) were examined a sector at a time by means of an overlay ruled with 9-mm squares (see Fig. 7). Axodendritic synapses as identified by distinct pre- and postsynaptic membranes, presynaptic vesicles, and postsynaptic density were counted and their lengths measured. Data transformations as described by Collonier *et al.* (1985) were performed by means of a standard spreadsheet program (Lotus Development Corp., Cambridge, MA) prior to statistical analysis. The transformations estimated the number of synapses per unit volume of tissue, a quantity known as N_V (Weibl, 1979, pp. 42–45). The number per unit area (N_A) is directly obtained from the total number of synapses observed divided by the area of tissue examined in the 20 micrographs per region. Collonier and Beaulieu (1985) compared predicted to actual volume densities of several formulae applied to an artificial sectioning example: various shaped pieces of fruit were suspended in gelatin and then sectioned. To estimate N_V, two formulae were relatively accurate at predicting the actual volume density of the particles: one approach involved simply dividing N_A by the average

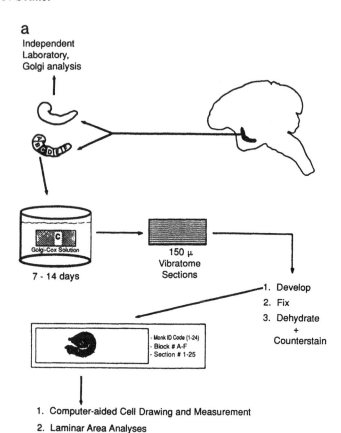

FIGURE 13 (a) The sampling procedure used for obtaining the vibratome sections used for measurements of the areas of hippocampal subfields and for Golgi dendritic measurements. (b) Computer-generated outlines were used to measure hippocampal subareas.

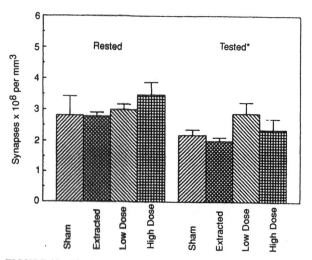

FIGURE 12 Monkeys tested daily throughout drug exposure had significantly (*, $F(1,11) = 8.80$, $P<0.05$) less synaptic density than cage-rested cohorts. Data shown were transformed by the DeHoff and Rhines procedure, as described in the text. Analysis of the raw data (simply total number of synapses counted) leads to the identical conclusion [$F(1,11) = 5.31$, $P<0.05$]. The slightly weaker F value represents the absence of any shape correction for total synaptic density based on the length of synaptic profiles encountered.

length of the synapses. The more complicated approach (attributed to DeHoff and Rhines) that was also accurate involved multiplying N_A by eight times the average of the reciprocals of the synaptic lengths, and then dividing by π^2. These are typical transformations used to report synaptic density measurements. The "tested" group of monkeys given extensive operant behavioral experience during the year of dosing had reduced synaptic density compared to "rested" animals, although marijuana was without effect (see Fig. 12). However, it is also important to note that both "accurate transformations" are linear functions of the original data combining the number and length of synapses observed per unit area of tissue that was examined. For this reason, it may be preferable to statistically examine the raw parameters of number of synapse per area and average length of synapse separately. The neurotoxicologist is more interested in evidence that a treatment caused a shift in synaptic density (however expressed) than in knowledge of the actual volume density (N_V) of synapses. It is also helpful to keep in mind the limitations of the relatively tiny sample of tissue that can be examined by electron microscopy

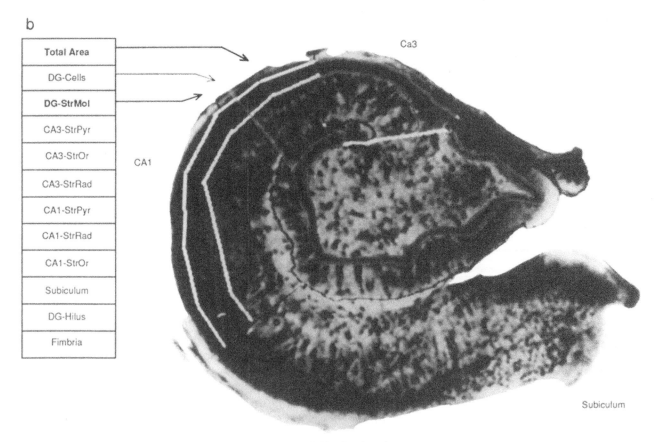

FIGURE 13 Continued.

even with 60 micrographs per brain. These limitations tend to be obscured when the few hundreds of synapses actually measured per animal are converted to hundred millions of synapses per cubic millimeter.

C. Example: Regional Measurement of Areas of Hippocampal Layers

For regional area measurements and Golgi dendritic measurements, monkeys were intravenously administered a lethal overdose of pentobarbital (Nembutal) as described in Appendix 1, and their unperfused brains were rapidly dissected. The whole hippocampus from each hemisphere was removed and further divided into six approximately equal (5 mm thick) blocks which were processed for Golgi impregnations as described in Appendix 4. Blocks A–E were then serially sectioned on a vibratome (150 or 200 μm) for computerized morphometry (Fig. 13). Block F was the narrow tail of the hippocampus and could not be readily sectioned.

Measurements were performed at the macroscopic level by outlining the total as well as regional subareas of hippocampus on one section chosen from the middle of each block from each monkey using a mouse-driven cursor and a microcomputer imaging device (MCID, Imaging Research, Inc., Ontario, Canada). In order to preserve statistical power to detect potential treatment-related changes in the proportion of the total hippocampal area represented by each different subfield, data for each animal's subfield were expressed as a percentage of its total area. This was intended to eliminate the increase in variance that differences between animals in total area would cause on estimates of the absolute mean values of the subfield areas. Table 1 indicates that while there were no effects of marijuana treatment on total hippocampal areas in any of the blocks, the continually tested or "environmentally enriched" group had significantly greater total areas and a proportionally greater area of CA3 s. oriens (the basilar dendrite layer).

D. Example: Three-Dimensional Measurement of Neurons and Dendrites Following Marijuana Exposure and Operant Conditioning

Measurements at the microscopic level were performed by a Tree Analysis program on computer-aided three-dimensional drawings of the soma, dendrites, and

TABLE 1   F Values for the Analyses of Variance of the Hippocampal Size Data and for the
Subfields as a Proportion of the Total Area

	Drug df (3,16)	Behav. (1,16)	Drug × behav. (3,16)	Blocks (3,48)	Blocks × behav. (3,48)	Blocks × drug (9,48)	Blocks × drug × behav. (9,48)
Total area	0.2	4.8[a]	4.2[a]	56.9[b]	1.1	0.6	0.9
CA1 (s. oriens)	0.1	3.5	2.3	1.3	0.5	0.4	0.6
CA1 (s. pyramidale)	0.8	0.1	0.1	34.7[b]	0.4	0.4	0.6
CA1 (s. radiatum)	0.2	0.8	1.5	2.3	1.3	1.6	0.7
CA3 (s. oriens)	0.6	6.8[a]	0.2	19.2[b]	0.1	0.4	1.1
CA3 (s. pyramidale)	1.0	0.9	0.7	14.5[b]	0.2	0.4	1.3
CA3 (s. radiatum)	0.2	1.3	0.8	27.4[b]	0.2	0.5	1.0
Dentate (s. moleculare)	0.7	2.6	0.7	5.8[b]	0.6	1.1	0.5
Dentate (granule cells)	1.2	0.1	3.0	1.8	0.3	0.3	1.9
Dentate (hilus)	3.1	0.1	4.4[a]	32.4[b]	0.3	1.1	0.9
Myelinated axons	2.2	1.7	6.6[b]	36.3[b]	1.3	1.1	1.9

[a] $P < 0.05$.
[b] $P < 0.01$.

branch points (see Fig. 6) made with a LEP motorized microscope stage and a Matrox CAD board installed in a personal computer (R and M Biometrics, Nashville, TN). A minimum of six neurons per regional subarea per block per monkey were averaged to derive a "prototypical neuron" reflecting that brain region and block for further analysis. There were no differences in soma size, number of dendrites per neuron, number of branches per neuron, length of dendrites of any given branch order, or total length with marijuana treatment. There were also no effects of continual operant testing or "enrichment" during the exposure period on any of the dendritic parameters, except as indicated in Fig. 14; sham-treated "enriched" monkeys

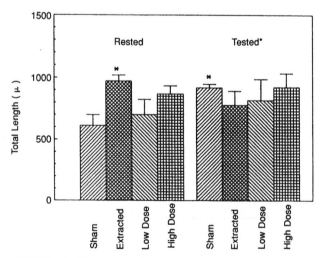

FIGURE 14 Sham-exposed monkeys tested in operant training boxes daily throughout the drug exposure period had significantly (*, $t(4df) = 6.4$, $P < 0.01$) greater CA3 basilar dendritic length than cage-rested cohorts.

had significantly longer basilar dendrites in CA3 than cage-rested sham animals.

The statistical approach taken in examples a, b, c, and d (e.g., Table 1) recognized that each experimental animal contributed an *n* of 1 for comparisons based on drug or test/rest conditions and treated multiple parameters obtained from the same animal as "within" or "repeated" measures. Thus most analyses were mixed-model factorial ANOVAs with two "between" factors (Drug with four levels, Test/Rest with two levels) and several "repeated" factors chosen to evaluate potential effects of anatomical location on the measurements (SAS, GLM, SAS Institute, Cary, NC).

It is instructive to note how these separate analyses considered together indicate that there are consistent and statistically significant differences between the continuously tested "enriched" monkeys and the cage-rested, relatively "impoverished" group. The selective increase of neuron density observed only in CA3 of the continuously tested monkeys may be related to the reduced synaptic density also measured in CA3 s. radiatum near the perikarya of these same "enriched" monkeys. Although it is uncertain why the synaptic density was lower in "enriched" monkeys with a greater density of target neurons available, the same decrease in synaptic density with "enrichment" has been previously reported (Greenough, 1976).

The measurement of hippocampal areas was done on a completely different set of animals, but again CA3, this time s. oriens, was the only hippocampal subregion that measured larger in the continuously tested ("enriched") than cage-rested ("impoverished") groups. The measurement under higher magnification of the basilar CA3 pyramidal cell dendrites forming the

s. oriens layer additionally revealed that their length was greater in "enriched" than in "impoverished" monkeys. This observation, together with the higher CA3 neuronal density in the separate set of monkeys providing the 1-μm plastic sections, could account for the greater area of CA3 s. oriens in these "enriched" monkeys. The results consistently indicate that the greatest region of plasticity in relation to the monkeys behavioral experience was the CA3 subregion of hippocampus.

E. Example: Loss of Melanocyte-Stimulating Hormone (MSH)-Immunoreactive Cells Following Hypothalamic Neurotoxins

A study using a chemical lesioning approach evaluated the acute histology in thin plastic sections (e.g., Fig. 1) and the functional effects of neurotoxic hypothalamic lesions in rats treated neonatally with glutamic acid (GLU), treated as weanlings with bipiperidyl mustard (BPM), or treated with both chemical agents (Scallet and Olney, 1986). The location of hypothalamic neurons immunoreactive for MSH (Fig. 8) was mapped and counted in year-old animals from this study ($n = 5-8$ per group) in order to evaluate the potential role of the brain hypothalamic proopiomelanocortin system in the functional effects. The major bed nucleus of the proopiomelanocortin system lies within the hypothalamic region (arcuate nucleus and the ventral part of the ventromedial hypothalamic nucleus) acutely damaged by the GLU/BPM combination as based on evaluation of necrotic neurons in thin plastic sections. It is a region shaped somewhat like a sunken rowboat: the fluid-filled third ventricle separates two distinct sides of the arcuate nucleus (joined by the median eminence) from each other as the inside of the boat separates the sides from the bottom. The anterior part of the arcuate nucleus joins together into a single group of cells like the prow of the boat joins the gunnels together, and the posterior part of the arcuate nucleus again becomes a single group like the stern of the boat. Mathematically, the distribution of MSH-immunoreactive cells from anterior to posterior within the arcuate nucleus (Fig. 15) represents this shape in control rats: there are about twice as many cells per section in the middle portion of the nucleus, where it is present on both sides of the third ventricle, than in the anterior or posterior portions, where it is a single grouping of cells. The dimensions of the arcuate nucleus containing the MSH neurons were also altered by treatment, but the cell-free zone separating arcuate from ventromedial nuclei was difficult to discern reliably. For this reason, data are expressed as number per section instead of as N_a or N_v as an estimate of the concentration of MSH

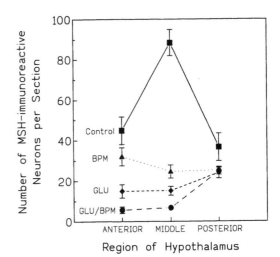

FIGURE 15 Both bipiperidyl mustard (BPM, main effect of drug on total cell counts: $F(1,22) = 22.8$, $P < 0.01$) and glutamic acid (GLU, main effect of drug on total cell counts: $F(1,22) = 151.8$, $P < 0.01$) greatly reduced the total number of hypothalamic melanocyte-stimulating hormone (MSH) immunoreactive neurons per section. Also, both agents were considerably more toxic to the anterior and middle regions of the hypothalamus than the posterior part (which has a different blood supply, perhaps not outside the blood–brain barrier). This is indicated by the significant drug by regions interactions for both MSG [$F(2,44) = 47.1$, $P < 0.01$] and for BPM [$F(2,44) = 31.2$, $P < 0.01$]. The combination of the two toxins was more effective than either alone, reflecting the more lateral distribution of BPM damage compared to GLU damage, which is periventricular.

cells per hypothalamus, which is probably the more biologically relevant quantity. There is an interaction between the neurotoxins (two independent treatment factors) and the region of the arcuate nucleus evaluated (a repeated measure) readily appreciated as a much greater effect of both neurotoxicants in the anterior two-thirds of the nucleus compared to the posterior one-third (Fig. 15). The blood supply of these regions is different, suggesting the possibility that differences in blood/brain barrier permeability may underlie the relative differences in sensitivity between regions. In any event (as in the marijuana studies), this type of analysis indicates the kind of information about anatomical regions that may be obtained by including them as variables (albeit the dependent, repeated measures kind of variable) in the analyses. This study also serves as an example of how immunohistochemical analysis, even a year after the administration of the neurotoxins, reveals reduced numbers of arcuate MSH-positive neurons. In the present case, because of the observation of necrotic arcuate neurons in animals studied 6–48 hr after dosing, it can be inferred that the absence of MSH-immunoreactive neurons in the adult arcuate hypothalamus probably represents a persistent, permanent loss of these neurons instead of a diminished synthesis of MSH. Some such demonstration of acute neurotoxicity is necessary for a definitive interpretation of reduced counts of immunolabeled cells, axons, or dendrites as "cell loss" instead of a simply reduced synthesis of the antigen.

F. Example: A Dose–Response for CA2 Silver-Impregnated Degenerating Axon Terminals Following Domoic Acid Exposure

An incident of neurotoxic shellfish poisoning of humans with a chemical subsequently extracted from the suspect mussels and identified as domoic acid has been reported (see discussion in Scallet *et al.*, 1993). Structurally, domoic acid is a tricarboxylic amino acid similar to glutamate, kainic acid, and related excitatory amino acids (EAAs) also termed "excitotoxins" by Olney (Olney *et al.*, 1983; Olney 1980). Domoic acid is a convulsive agent that produces necrosis of hippocampal pyramidal neurons. Scallet and co-workers (1993) used the silver-staining method described in Appendix 2 to reveal degenerating axon terminals in the hippocampal CA2 s. lucidum region of the cynomolgus monkey ("Type A" lesion) at apparently subconvulsant doses, as well as more extensive damage ("Type B" lesion, Fig. 3) at higher doses. Two other biomarkers, c-fos protein and glial fibrillary acidic protein immunohistochemistry, were further identified as useful biomarkers of domoic acid exposure. This work indicated that the primary brain target tissues of domoic acid were presynaptic axons terminating in CA2, i.e., elsewhere than the sites usually damaged by seizures, the CA3 and CA1 regions. It also determined a threshold dose of around 1 mg/kg in adult cynomolgus monkeys, and around 4 mg/kg in juvenile monkeys for the occurrence of a "Type B" lesion (Scallet *et al.*, 1993). The threshold dose for extensive hippocampal damage was very close to a lethal dose. Because of the potential importance of the CA2 s. lucidum in the low-dose effects of domoic acid and perhaps in initiation of the more extensive "Type B" lesions, the number of silver grains present in this region in animals given 0, 0.25, 0.50, or 1.0 mg/kg of domoic acid was quantitated. The grain counts were obtained from digitized images of the CA2 s. lucidum (X40) evaluated by the MCID image analysis system (Imaging Research, Inc.). The system was set by measuring the average optical density (gray-scale value over a 1 to 256 range) of 16 silver grain "targets" per image compared to backround gray-scale readings. The system was then set to count all grain-sized particles darker than half-way between "target" and "backround" gray-scale values for that image. Composite

grains or clumps larger than a single grain were excluded based on a size criteria. Then grains per image were counted and averaged for the monkeys of each experimental group. Figure 9 shows the dose–response obtained from this lower range of doses of domoic acid. A major difficulty associated with conducting quantitative studies of this type is the need to perform many sets of silver stains (each including equal numbers of sections from all the subjects) in order to obtain several successfully impregnated batches to use for quantitation since this silver procedure (like many others) is often "capricious." A second problem is that it is absolutely essential to define a systematic and reproducible way of setting the threshold for computerized grain counting of the images, as described earlier. Otherwise, it is too easy to bias the results by setting the threshold arbitrarily high or low.

VIII. Summary

The application of quantitative morphometric methods to neurotoxicology is a relatively recent endeavour, and appropriate techniques are still evolving. However, such methods are essential for subsequent use in mathematical representations of the risk of exposure to neurotoxicants (McMillan, 1987; Slikker and Gaylor, 1990; Slikker, 1991; Scallet and Slikker, 1992) whenever histological biomarkers are the most sensitive. Initially, decisions about the most appropriate histological procedure for a given neurotoxicological problem must be made. The rationale for such decisions with regard to a number of common histochemical techniques was discussed. Random sampling of a section or sections from defined anatomical loci must first be performed. Then, consistent and reproducible staining methods must be simultaneously and uniformly applied to a set made up of sections gathered from comparable anatomical loci of animals from each of the different experimental treatment groups. The importance of assembling sections in a balanced manner, so that at least one section from each animal of each experimental condition is represented in a procedure, was emphasized. A random, balanced selection of sections in this manner is vital to avoid a systematic bias that might differentially affect the morphometric measurements of one of the experimental groups compared to the others. Since run-to-run variations are possible with all histochemical procedures, it is also necessary to strive for consistency so that morphometric measurements have the minimum possible uncontrolled error variance. The equipment, materials, and standard operating procedures needed to prepare the neurotoxicology laboratory embarking on quantitative histochemical morphometry were considered, and representative photomicrographs and quantitative examples were presented. The appropriate use of the resulting quantitative data in mathematical risk analysis models is described in detail elsewhere in this volume in the chapter by Slikker and Gaylor.

APPENDICES

Appendix 1: Perfusion Fixation

Materials:

Suitable chemical fume hood or vented necropsy table.
Rodent (5–1200 g); monkey (1–15 kg).
Exsanguinating solution, usually saline [0.9% NaCl in distilled, deionized water (ddw)].
Perfusion solution, usually 10% buffered formalin (3.7% formaldehyde (aq) in phosphate buffer, 0.1 *M* final concentration).
Peristaltic pump, calibrated and set for
  35–40 ml/min flow rate (adult rat).
  5–10 ml/min (adult mouse, fetal rat).
  330–350 ml/min (adult monkey).
Chemically resistant tubing (Norprene or equivalent) leading from the bottoms of the exsanguinating and perfusion bottles through a three-way stopcock valve on the inlet side of the pump and from the outlet to the experimental animal.
Surgical scissors, iris scissors, hemostats, forceps, rongeurs, etc. as appropriate for the experimental animal
Blunt perfusion needle
  18–22 gauge rat
  26–27 gauge, mouse, neonatal rat
  14–16 gauge, monkey

Anesthesia (rodents):

Administer 1.5 ml (cc) of 60 mg/ml sodium pentabarbital per kilogram of body weight by intraperitoneal (i.p.) injection. Verify that the rodent is in a deep plane of anesthesia (does not flinch if tail is pinched) before proceeding with the surgical preparation.

Anesthesia (monkeys):

Administer 3 ml of sodium pentobarbital (60 mg/ml) i.v. and supplement if necessary until animal loses consciousness.

Surgical preparation:

Rodents:
1. Place the anesthesized subject on its back on a chemically resistant work tray. The tray should be tilted slightly with a drain hole over a collection vessel for disposal of perfusates as hazardous waste.
2. Dampen the thoracic and abdominal regions with water or alcohol to reduce the inconvenience of loose hair.
3. Lift the skin overlying the xyphoid process (the cartilagenous posterior extension of the sternum) with the forceps and cut through the skin, underlying muscle, and peritoneum with the large scissors.
4. Extend the cut laterally along both posterior margins of the ribcage so that the intact diaphragm muscle is completely visable.
5. Retract the sternum upward to stretch the diaphragm and cut it free along the ventral line of attachment to the ribs. Be careful to avoid cutting the heart or major vessels.
6. Quickly use the large scissors to cut upwards along the two sides of the ribcage.
7. Carefully snip the fatty attachment of the heart to the ribcage, removing the pericardium.
8. Use the large straight hemostat to clamp the xyphoid process to hold the ribcage out of the way.
9. Use the iris scissors to make a small incision into the right atrium (the animal's right atrium, which will be on the left when looking down at the supine subject). This will serve as a point for fluids to exit.
10. Use the iris scissors to make a small incision through the apex of the left ventricle, with the direction of the incision parallel to the

longitudinal axis of the heart and not through the interventricular septum.

11. Blotting or washing away any blood that obstructs your vision, insert the blunt perfusion needle through the previously created ventricular opening and into the proximal portion of the aorta (about 5 mm) and clamp into place (about 3 mm from the needle opening) with a hemostat. You should be able to see the tip of the needle just as it enters the aorta. Do not clamp the needle too far into the aorta or else it may turn to one side so that the wall of the aorta occludes the tip. Optionally (especially for mice or perinatal rats), a sharp perfusion needle may be inserted and held or clamped into the left ventricle instead of the aorta.

Monkeys:

1. Following anesthesia, make a longitudinal scalpel incision from the top of the ribcage to well past the xyphoid process, ending above the stomach.
2. Retract the skin laterally by stretching it with hemostats and using a scalpel to trim it away from the underlying muscle.
3. Use the scalpel to make an incision along the lower margin of the ribcage through the muscle and peritoneum, exposing the intact diaphragm.
4. Cut through the diaphragm with scissors and use bone cutters to cut through the lateral portion of the ribcage on both sides. Reflect the ribcage upwards, exposing the heart, and quickly snip through the pericardium.
5. When the heart is completely exposed, use scissors or the scalpel tip to make a small hole in the left ventricle.

6. Insert the perfusion needle through the left ventricle into the aorta and clamp it in place with a hemostat. Alternately, the needle can be clamped in place in the left ventricle.
7. To open an exit for the circulating fixative, either snip a hole in the right atrium or remove about half of one of the lobes of the liver.

Perfusion:

1. The apparatus must be prepared prior to surgery by first running perfusate through the tubing to get rid of any bubbles, then switching the valve to saline until the perfusate is completely flushed out of the tubing leading to the perfusion needle. Bubbles, or formaldehyde, in the initial flush will obstruct the vasculature and ruin the perfusion.
2. When the perfusion needle is secured in the heart, run the pump for up to 1 min to exsanguinate the animal (usually about 45 ml of saline for rat; 10 ml for mice or neonatal rats; 330 ml for monkeys). Long flushes risk loss of water-soluble antigens.
3. Turn the valve so that the perfusate, usually a buffered formalin solution, flows through the line.
4. Perfuse for 10–15 min with 450–700 ml (rats); 50–150 ml (mice, neonatal rats); or 3–5 liters (monkeys) of buffered formalin.
5. Turn the valve back to the setting for saline and flush thoroughly.
6. Remove the brain using appropriate rongeurs and/ or heavy-duty scissors or a bone saw and postfix in the perfusate or other solution as desired. Some investigators, especially for electron microscopy, recommend postfixing the brain in the skull overnight before removing it.

Appendix 2: Silver Stain for Degenerating Terminals and Cell Bodies

Preparations (based on the method of Nadler and Evenson, 1983):

A. Cut *perfused* brains on a vibratome into 30- to 50-μm sections.

   Alternately, cryoprotect in 0.1 *M* phosphate or cacodylate-buffered 20% sucrose (until brains no longer float). Freeze onto metal chucks in embedding compound by immersing the chuck (but not the tissue itself) in a dry ice/alcohol slurry or in liquid nitrogen. Cut sections with a sliding, freezing microtome (25–50 μm) or cryostat (5–50 μm).

B. Store unstained, free-floating sections up to 10 days.

   Store at 4°C in 0.1 *M* phosphate or cacodylate-buffered 1% formalin in 24-well disposable plastic culture plates (Costar Co.).

C. Place sections (one per well) in dd water in staining baskets.

   Suitable baskets can be made by drilling 24 round, 1.6-cm-diameter holes (4 rows of 6) into an 8 × 12.5 × 1.1-cm polycarbonate block. A small hole is drilled in the center of the block and then tapped for a screw-in Teflon or plastic handle. A nylon window screen is secured with cyanoacrylate glue to the bottom of the block, forming individual wells. The size of the block is chosen to be the same as the culture plates used to store unstained sections. Rinsed solutions and reagents may be placed in the top covers of *thoroughly rinsed* culture plates in a series and staining baskets sequentially immersed in them for the appropriate time. The tight-fitting blocks force solutions up into the individual wells, allowing a minimum of expensive reagents (e.g., for immunohistochemistry, about 30 ml/basket) to be required. No metal or organic contaminants can be allowed at any of the handling steps, as this would interfere with the silver deposition procedure.

Stock Solutions:

These quantities are sufficient to process 8 sets of 24 slides:

Solution "A"—9% (w/v) sodium hydroxide (NaOH).
   Dissolve 63 g in 700 ml final volume with dd water.

Solution "B"—16% (w/v) ammonium nitrate (NH_4NO_3).
   Dissolve 32 g in 200 ml final volume with dd water.

Solution "C"—50% (w/v) silver nitrate ($AgNO_3$).
   Dissolve 1.25 g in 2.5 ml final volume with dd water.
   Store in the dark in a well-sealed amber bottle.

Solution "D"—1.2% (w/v) ammonium nitrate (NH_4NO_3).
   Dissolve 6.0 g in 500 ml final volume with dd water.

Solution "E"—0.5% (w/v) anhydrous sodium carbonate (Na_2CO_3).
   Mix 360 ml of 95% ethanol with 700 ml of dd water. Dissolve 6.0 g (Na_2CO_3); dilute to 1200 ml with dd water.

Solution "F"—0.05% (w/v) anhydrous citric acid ($C_6H_8O_7$)
   Mix 6 ml of 37% formaldehyde with 40 ml of 95% ethanol, and 280 ml of dd water.
   Dissolve 0.2 g anhydrous citric acid ($C_6H_8O_7$).
   Add Solution "A", while stirring, until the pH reaches 5.8–6.1, then dilute to 400 ml final volume with dd water.

Solution "G"—0.5% (v/v) acetic acid (CH_3COOH).
   2.5 ml to 500 ml final volume with dd water.

These solutions are stable for at least 8 weeks at room temperature when tightly sealed and protected from light.

Procedures:

Working solutions:

Prepare no more than 1 hr ahead; volumes indicated are sufficient for one basket of tissue (24 sections):

1. Pretreating solution
   Mix 50 ml of Solution "A" with 50 ml of Solution "D".

2. Impregnating solution
   Mix 30 ml of Solution "A" with 20 ml of Solution "B".
   Add 0.25–0.30 ml of Solution "C" to the combined solution.

3. Washing solution
   Mix 1.5 ml of Solution "D" with 150 ml of Solution "E".

4. Developing solution

Mix 0.5 ml of Solution "D" with 50 ml of Solution "F".

Staining procedure:

1. Rinse thoroughly in three changes of dd water at 5 min/change. Residual phosphate buffer may react with silver nitrate; cacodylate or glycine may be substituted if rinsing is insufficient.

Notes: For Steps 2–5, keep baskets covered to minimize ammonia loss. Drain thoroughly between changes. Do not let sections float and agitate for at least 1 min into each change (except developer).

2. Two changes of pretreating solution at 5 min/change.
3. One change of impregnating solution at 10 min/change.
4. Three changes of washing solution at 100 sec/change.
5. One change of developing solution for at least 1 min.
6. Wet mount sections, from saline, onto gelatin-coated slides.
7. Three changes of fixing solution "G" at 10 min/change.
8. Counterstain (optional), dehydrate, and coverslip sections.

Dark-field optics may help distinguish signal from backround for quantitation.

Appendix 3: Immunohistochemistry (IHC: Unlabeled Antibody or PAP Method)

Methods: (based on Sternberger *et al.*, 1970)

1. Vibratome or frozen sections are obtained from animals perfused and sectioned as described in Appendices 1 and 2 and are placed in a staining basket. The use of vibratome sections (30–50 μm) avoids the freezing artifacts caused by subcellular ice crystal formation and only partially prevented by cryoprotection with sucrose. Vibratome sections prepared for IHC (or silver stains) may also be observed by light microscopy while wet mounted but not yet dehydrated, cleared, or coverslipped. A needle punch can then readily be removed for gluteraldehyde postfixation, osmification, and plastic embedding for preembedding immunohistochemical electron microscopy. Gluteraldehyde concentrations may have to be reduced to 0.1–0.2%, depending on the antigen, to preserve immunoreactivity for this method. Alternately, frozen sections are useful for light microscopic applications and may be necessary for optimal preservation of weakly fixed water-soluble antigens that may be unstable at 4°C in solution. Sections may be stored for at least a week in 24-well culture dishes containing 0.1 M phosphate or cacodylate buffer prior to immunostaining. Transfer sections to staining baskets or individual wells of a sterile culture dish for immunohistology.

2. If backround staining is a problem, an *optional* 30-min preincubation in 10% normal goat serum (or the normal serum of the species providing the second antibody) diluted in 0.1 M phosphate buffer should be performed. The normal goat serum floods any nonspecific surface-binding sites on the tissue that may have some affinity for goat IgG, competing out any backround that could be caused by the second antibody nonspecifically binding to tissue.

3. If undesirable endogenous peroxidase staining occurs from retention of red blood cells in the vasculature, an *optional* preincubation in 0.3% hydrogen peroxide (H_2O_2)/0.1 M phosphate buffer for 30 min should be performed. With good perfusions and quality immunoreagents, these steps are usually not necessary.

4. Rinse three times for 5 min each in 0.1 M phosphate buffer.

5. Incubate sections in the basket or tissue culture wells in primary antisera diluted at least 1 : 500 in antibody diluent and keep at 4°C for 16–72 hr, depending on the primary antisera. Usually, overnight incubation is sufficient, but results may be improved by longer staining in some circumstances. Antisera maintained at 4°C may be stored and reused for up to 4–5 years, depending on the frequency of use, if not contaminated with formaldehyde.

6. Remove from primary antisera and wash three times for 5 min each in 0.1 M phosphate buffer.

7. Incubate 1 hr in GAR (goat anti-rabbit IgG, diluted 1:100 with antibody diluent).

8. Remove from GAR and wash three times, for 5 min each in 0.1 *M* phosphate buffer.

9. Incubate 1 hr in rabbit PAP (peroxidase–antiperoxidase complex diluted 1:200 in 0.1 *M* phosphate buffer). Avoid using the antibody diluent here since detergents such as the Triton it contains disrupt the PAP complex).

10. During PAP incubation (Step 9), prepare the DAB solution for Step 12. Mix 0.05% DAB (3,3'-diaminobenzidine tetrahydrochloride) in 0.1 *M* phosphate buffer (50 mg/100 ml buffer) under the hood (wear gloves and filter mask when weighing DAB, a potential carcinogen). The solution may have to be heated slightly on a stirring hotplate for proper dissolution. Filter the solution when dissolution is complete. Alternately, use two 10-mg DAB tablets (Sigma Chemical Co.) into 40 ml of 0.1 *M* phosphate buffer (enough for one staining basket) and filter. Add 33 μl of 50% H_2O_2 (hydrogen peroxide) per 100 ml (13 μl H_2O_2 per 40 ml) of filtered DAB solution immediately prior to use. Mix DAB with a bleach solution and dispose in a liquid hazardous waste container when finished. Wash glassware before DAB has time to adhere.

11. Remove sections from PAP and wash three times for 5 min each in 0.1 *M* phosphate buffer.

12. Immerse sections for 5–12 min in DAB–H_2O_2. Remove when tissue backround appears a light to moderate brown color.

13. Remove sections from DAB and wash three times for 5 min each in 0.1 *M* phosphate buffer.

14. Wet mount sections on subbed, labeled slides. Remove excess moisture with laboratory tissues or *light* suction and dry on slide warmer. Rehydrate sections for 60 sec in Millipore water, counterstain 3 min in 0.02% cresyl violet if desired, and dehydrate through a progressive series of 70, 95, and 100% ethyl alcohol baths for two 60-sec dips in each concentration. Place the slides in xylene or a suitable substitute for two 60-sec changes and then coverslip with Permount or a suitable alternate such as DPX.

Appendix 4: Simple Golgi Whole Neuron "En Bloc" Impregnation Method

1. Remove the fresh, unperfused brain tissue with rongeurs or scissors following euthanasia of the subject.

2. Using a brain-blocking mold or slicing aid, prepare slices of uniform thickness (3–5 mm) and immerse in Golgi solution (prepared as described below) in a glass bottle or scintillation vial. Slices should be suspended off the bottom with pieces of nylon mesh (nonmetallic) to allow free circulation of the solution through both sides of the block.

3. Store in the dark at room temperature for 7–14 days.

4. Serial section the blocks at 150–250 μm on a vibratome and collect tissue in 0.1 *M* phosphate buffer.

5. Rinse three times for 1 min each in 0.1 *M* phosphate buffer.

6. Develop a staining basket full of sections for 20 min in 15% ammonium hydroxide (50% v/v of 30% stock NH_4OH in 0.1 *M* phosphate buffer).

7. Rinse three times for 1 min each in 0.1 *M* phosphate buffer.

8. Fix for 10 min in sodium thiosulfate ($Na_2S_2O_3$), 10% w/v in 0.1 *M* phosphate buffer.

9. Rinse three times for 1 min each in 0.1 *M* phosphate buffer.

10. Wet mount sections on subbed, labeled slides. Remove excess moisture with laboratory tissues or *light* suction and dry gently. *Do not overdry or cracking of sections may occur!* Rehydrate sections for 60 sec in Millipore water, counterstain 3 min in 0.02% cresyl violet if desired, and dehydrate through a progressive series of 70, 95, and 100% ethyl alcohol baths for two 60-sec dips in each concentration. Place the slides in xylene or a suitable substitute for two 60-sec changes and then coverslip with Permount.

Golgi solution:

Stock A = 5% (w/v) potassium dichromate.
Stock B = 5% (w/v) mercuric chloride.
Stock C = 5% (w/v) potassium chromate.

For 400 ml of working solution (can be stored at room temperature in the dark for up to 2 months):

1. Mix 100 ml of A with 100 ml of B.

2. Dilute 80 ml of C to 200 ml with deionized, distilled water.
3. Add solution 1 slowly into 2, stirring constantly.
4. Allow a precipitate to form for about 5 days and then filter for use.

Appendix 5: General Recipes and Procedures

A. Buffers
1. Potassium phosphate buffer
   Stock solution (1.0 M PPB)

Monobasic potassium phosphate (KH_2PO_4)	264 g
Dibasic potassium phosphate (K_2HPO_4)	1402 g
Dissolve and dilute with dd water to	10 liters

   For 0.1 M working solution at pH 7.4

1.0 M stock solution	400 ml
Dilute with dd water to	4 liters

2. Cacodylate buffer
   Stock solution (0.2 M cacodylate)

Sodium cacodylate ($C_2H_6As\text{-}NaO_2 \cdot 3H_2O$; F.W. = 214)	42.8 g
Dissolve in 975 ml dd water. Titrate to pH desired (7.2–7.6) with 1 N HCl	
Dilute with dd water to	1000 ml

   For 0.1 M working solution
   Dilute 0.2 M stock solution 1:1 (v:v) with dd water

3. Sorensen's sodium phosphate buffer
   Stock solution A

Monobasic sodium phosphate ($NaH_2PO_4 \cdot H_2O$)	27.8 g
Dilute with distilled water to	1000 ml

   Stock Solution B

Dibasic sodium phosphate ($Na_2HPO_4 \cdot 7H_2O$)	53.6 g
(or $Na_2HPO_4 \cdot 12H_2O$	71.7 g
Dilute with distilled water to	1000 ml

   For working solution at pH 7.4
   Add 19 ml of A to 81 ml of B

Dilute with distilled water to	200 ml

4. Sodium phosphate buffer (0.5 M stock)

Monobasic sodium phosphate ($NaH_2PO_4 \cdot H_2O$)	132 g
Dibasic sodium phosphate ($Na_2HPO_4 \cdot 7H_2O$)	1086 g
Dilute with dd water to	10 liters

   For 0.1 M working solution at pH 7.4

0.5 M stock solution	800 ml
Dilute with distilled water to	4 liters

B. Exsanguinates/perfusates
1. 0.9% saline

Sodium chloride (NaCl)	9 g
Dilute with dd water to	1000 ml

2. 3.7% fomaldehyde in 0.1 M buffer

1.0 M potassium phosphate stock	100 ml
or 0.2 M cacodylate stock	500 ml
or 0.5 M sodium phosphate stock	200 ml
37% formaldehyde (100% Formalin)	100 ml
Dilute with dd water to	1000 ml

3. 10% stock paraformaldehyde

Add paraformaldehyde powder to a Pyrex beaker	100 g
Add dd water and heat to about 60°C	700 ml
Add 10 N NaOH dropwise while stirring until clear	
Dilute with dd water to	1000 ml

4. Routine EM fixative (1% PF/1.5% gluteraldehyde)

1.0 M potassium phosphate stock	100 ml
or 0.2 M cacodylate stock	500 ml
or 0.5 M sodium phosphate stock	200 ml
10% stock paraformaldehyde	100 ml
50% gluteraldehyde	30 ml
Dilute with distilled water to	1000 ml

C. Subbing slides

Warm dd water to 80°C	1000 ml

Remove heat; add gelatin while stirring.	5–10 g
Add sodium azide (NaN$_3$, bacterial growth inhibitor or, optionally, thimerasol) and filter	1 g
Load slides in glass or metal staining racks	
Preclean by agitation in water, 1 min; and in 95% ethanol, 1 min	
Dip each rack: air-dry 24 hr or for 2–3 hr at 40–50°C	

D. Antibody diluent

1.0 *M* potassium phosphate stock	100 ml
or 0.5 *M* sodium phosphate stock	200 ml
or 0.2 *M* cacodylate stock	500 ml
Sodium chloride (NaCl)	9 g
Normal goat serum (or serum from the species providing the secondary antibody)	20 ml
Sodium azide (NaN$_3$)	0.2 g
Triton X-100 detergent	3 ml
Dilute with dd water to	1000 ml

Acknowledgments

The author thanks Ms. Annette Andrews and Mr. Robert Rountree for help in the preparation of the standard operating procedures, and Ms. Shelly Lensing and Dr. Kettie Terry for very helpful comments on the manuscript. He also thanks his collaborators in the Division of Neurotoxicology (Drs. William Slikker, Jr., Syed Ali, Merle Paule, Larry Schmued, Zbigniew Binienda, Florence Caputo, and John Bowyer) as well as Dr. Etsuro Uemura from Iowa State University for assisting in many ways with this research. Finally, he especially thanks his postdoctoral mentor, Dr. John W. Olney, in whose laboratory some of the work reported here was done, and much of it inspired.

References

Alessi, N. E., Quinlan, P., and Khachaturian, H. (1988). MSG effects on beta-endorphin and alpha-MSH in the hypothalamus and caudal medulla. *Peptides* **9,** 689–695.

Appel, N. M., Contrera, J. F., and De Souza, E. B. (1989). Fenfluramine selectively and differentially decreases the density of serotonergic nerve terminals in rat brain: evidence from immunocytochemical studies. *J. Pharmacol. Exp. Ther.* **249,** 928–943.

Balaban, C. D., O'Callaghan, J. P., and Billingsley, M. L. (1988). Trimethyltin-induced neuronal damage in the rat brain: comparative studies using silver degeneration stains, immunocytochemistry and immunoassay for neuronotypic and gliotypic proteins. *Neuroscience* **26,** 337–361.

Binienda, Z., and Scallet, A. C. (1995). The effects of reduced perfusion and reperfusion on c-fos and HSP-72 protein immunohistochemistry in gestational day 21 rat brains. *Int. J. Dev. Neurosci.,* in press.

Bondy, S. C., and Ali, S. F. (1992). Neurotransmitter receptors, In *Neurotoxicology* (M. B. Abou-Donia, Ed.), pp. 121–151. CRC Press, Boca Raton, Florida.

Chang, L. W., Tiemeyer, T. M., Wenger, G. R., and McMillan, D. E. (1982). Neuropathology of mouse hippocampus in acute trimethyltin intoxication. *Neurobehav. Toxicol. Teratol.* **4,** 149–156.

Coggeshall, R. E. (1992). A consideration of neural counting methods. *T.I.N.S.* **15,** 9–13.

Collonier, M., and Beaulieu, C. (1985). An empirical assessment of stereological formulae applied to the counting of synaptic disks in the cerebral cortex. *J. Comp. Neurol.* **231,** 175–179.

Commins, D. L., Vosmer, G., Virus, R. M., Woolverton, W. L., Schuster, C. R., and Seiden, L. S. (1987). Biochemical and histological evidence that methylenedioxymethamphetamine (MDMA) is toxic to neurons in the rat brain. *J. Pharmacol. Exp. Ther.* **241,** 338–345.

D'Amelio, F. E. (1983). The Golgi-Hortega-Lavilla technique, with a useful additional step for application to brain tissue after prolonged fixation. *Stain Technol.* **58,** 79–84.

de Olmos, J. S. (1969). A cupric-silver method for impregnation of terminal axon degeneration and its further use in staining granular argyrophilic neurons. *Brain Behav. Evol.* **2,** 210–237.

Dragunow, M., Young, D., Hughes, P., MacGibbon, G., Lawlor, P., Singleton, K., Sirimanne, E., Beilharz, E., and Gluckman, P. (1993). Is c-jun involved in nerve cell death following status epilepticus and hypoxic-ischemic brain injury? *Mol. Brain Res.* **18,** 347–352.

Fink, R. P., and Heimer, L. (1967). Two methods for selective impregnation of degenerating axons and their synaptic endings in the central nervous system. *Brain Res.* **4,** 369–374.

Gabbott, P. L., and Somogyi, J. (1984). The 'single' section Golgi-impregnation procedure: methodological description. *J. Neurosci. Methods* **11,** 221–230.

Gallyas, F., Wolff, J. R., Bottcher, H., and Zaborszky, L. (1980a). A reliable method for demonstrating axonal degeneration shortly after axotomy. *Stain Technol.* **55,** 291–297.

Gallyas, F., Wolff, J. R., Bottcher, H., and Zaborszky, L. (1980b). A reliable and sensitive method to localize terminal degeneration and lysosomes in the central nervous system. *Stain Technol.* **55,** 299–306.

Gallyas, F., Guldner, F. H., Zoltay, G., and Wolff, J. R. (1990). Golgi-like demonstration of "dark" neurons with an argyrophil III method for experimental neuropathology. *Acta Neuropathol.* **79,** 620–628.

Gambetti, P., Gambetti, L. A., and Papasozomenos, S. C. (1981). Bodian's silver method stains neurofilament polypeptides. *Science* **213,** 1521–1522.

Gaylor, D. W., and Slikker, W., Jr. (1990). Risk assessment for neurotoxic effects. *Neurotoxicology* **11**, 211–218.

Good, P. F., Huntley, G. W., Rogers, S. W., Heinemann, S. F., and Morrison, J. H. (1993). Organization and quantitative analysis of kainate receptor subunit GluR5-7 immunoreactivity in monkey hippocampus. *Brain Res.* **624**, 347–353.

Greenough, W. T. (1976). Enduring brain effects of differential experience and training. In *Neural Mechanisms of Learning and Memory* (M. R. Rosenzweig and E. L. Bennett, Eds.), pp. 255–278, MIT Press, Cambridge.

Henry, W. W., Jr., Medlock, K. L., Sheehan, D. M., and Scallet, A. C. (1991). Detection of estrogen receptor (ER) in the rat hypothalamus using rat anti-ER monoclonal IgG with the unlabeled antibody method. *Histochemistry* **96**, 157–162.

Johnson, E. M., Jr., and Deckwerth, T. L. (1993). Molecular mechanisms of developmental neuronal death. *Annu. Rev. Neurosci.* **16**, 31–46.

Kalia, M. (1991). Reversible, short-lasting, and dose-dependent effect of fenfluramine on neocortical serotonergic axons. *Brain Res.* **548**, 111–125.

Landfield, P. W., Cadwallader, L. B., and Vinsant, S. (1988). Quantitative changes in hippocampal structure following long-term exposure to delta-9-tetrahydrocannabinol: possible mediation by the glucocorticoid system. *Brain Res.* **443**, 47–62.

Li, Y., Chopp, M., Garcia, J. H., Yoshida, Y., Zhang, Z. G., and Levine, S. R. (1992). Distribution of the 72-kd heat-shock protein as a function of transient focal ischemia in rats. *Stroke* **23**, 1292–1298.

Matthews, J. C., and Scallet, A. C. (1991). Nutrition, neurotoxicants, and age-related neurodegeneration. *Neurotoxicology* **12**, 547–558.

McMillan, D. E. (1987). Risk assessment for neurobehavioral toxicity. *Environ. Health Perspect.* **76**, 155–161.

Miller, D. B., and O'Callaghan, J. P. (1984). Biochemical, functional, and morphological indicators of neurotoxicity: effects of acute administration of trimethyltin to the developing rat. *J. Pharm. Exp. Ther.* **231**, 744–751.

Molliver, D. C., and Molliver, M. E. (1990). Anatomic evidence for a neurotoxic effect of (±)-fenfluramine upon serotonergic projections in the rat. *Brain Res.* **511**, 165–168.

Morgan, J. I., Cohen, D. R., Hempstead, J. L., and Curran, T. (1987). Mapping patterns of c-fos expression in the central nervous system after seizure. *Science* **237**, 192–197.

Nauta, J. H. (1954). Silver impregnation of degenerating axons in the central nervous system: a modified technique. *Stain Technol.* **27**, 91–93.

Nadler, J. V., and Evenson, D. A. (1983). Use of excitatory amino acids to make axon sparing lesions of the hypothalamus. *Methods Enzymol.* **103**, 393–400.

O'Hearn, E., Long, D. B., and Molliver, M. E. (1993). Ibogaine induces glial activation in parasaggital zones of the cerebellum. *Neuroreport* **4**, 299–302.

Olney, J. W. (1980). Excitotoxic mechanisms of neurotoxicity. In *Experimental and Clinical Neurotoxicity* (P. S. Spencer and H. H. Schaumberg, Eds.), pp. 272–294. Williams and Wilkins, Baltimore.

Olney, J. W., de Gubareff, T., and Sloviter, R. S. (1983). "Epileptic" brain damage in rats induced by sustained electrical stimulation of the perforant path. II. Ultrastructural analysis of acute hippocampal pathology. *Brain Res. Bull.* **10**, 699–712.

Paule, M. G., Allen, R. R., Bailey, J. R., Scallet, A. C., Ali, S. F., Brown, R. M., and Slikker, W., Jr., (1992). Chronic marijuana smoke exposure in the rhesus monkey II: effects on progressive ratio and conditioned position responding. *J. Pharmacol. Exp. Ther.* **260**(1), 210–222.

Reith, A., and Mayhew, T. M. (Eds.) (1988). *Stereology and morphometry in electron microscopy: some problems and their solutions.* Hemisphere Publications, New York.

Riley, J. N. (1978). A reliable Golgi-Kopsch modification. *Brain Res. Bull.* **4**, 127–129.

Russ, J. C. (1986). *Practical stereology.* Plenum Press, New York.

Scallet, A. C., and Olney, J. W. (1986). Components of hypothalamic obesity: bipiperidyl mustard lesions add hyperphagia to monosodium glutamate-induced hyperinsulinemia. *Brain Res.* **374**, 380–384.

Scallet, A. C. (1987). The hypothalamic-pituitary-adrenal axis of rhesus monkeys and rats: a role in energy balance. *Ann. N.Y. Acad. Sci.* **512**, 491–494.

Scallet, A. C., Lipe, G. W., Ali, S. F., Holson, R. R., Frith, C. H., and Slikker, W., Jr. (1988). Neuropathological evaluation by combined immunohistochemistry and degeneration-specific methods: application to methylenedioxymethamphetamine. *Neurotoxicology* **9**, 529–538.

Scallet, A. C., Uemura, E., Andrews, A. M., Ali, S. F., McMillan, D. E., Paule, M. G., Brown, R. M., and Slikker, W., Jr. (1987). Morphometric studies of the rat hippocampus following chronic delta-9-tetrahydrocannabinol (THC). *Brain Res.* **436**, 193–198.

Scallet, A. C., Uemura, E., Andrews, A. M., Craven, J. M., Rountree, R. L., Wilson, S. W., Ali, S. F., Bailey, J. R., Paule, M. G., and Slikker, W., Jr., (1990). Morphometric neurohistological studies of rhesus monkeys after chronic marijuana smoke exposure. *Soc. Neurosci. Abstr.* **16**, 1116.

Scallet, A. C., and Slikker, W., Jr. (1992). Biomarkers of developmental neurotoxicity. In *Risk Assessment of Prenatally-Induced Adverse Health Effects* (D. Neubert, R. J. Kavlock, H-J. Merker, and J. Klein, Eds.), pp. 63–78. Springer-Verlag, Berlin.

Scallet, A. C., Binienda, Z., Caputo, F. A., Hall, S., Paule, M. G., Rountree, R. L., Schmued, L., Sobotka, T., and Slikker, W., Jr. (1993). Domoic Acid-treated Cynomolgus monkeys (M. fascicularis): effects of dose on hippocampal neuronal and terminal degeneration. *Brain Res.* **627**, 307–313.

Schmued, L. C. (1994). Anterograde and retrograde neuroanatomical tract-tracing with fluorescent compounds. In *Neuroscience Protocols,* Elsevier Press, Amsterdam, 1–15.

Shimosaka, S., So, Y. T., and Simon, R. P. (1992). Distribution of HSP-72 induction and neuronal death following limbic seizures. *Neurosci. Lett.* **138**, 202–206.

Slikker, W., Jr., Ali, S. F., Scallet, A. C., Frith, C. H., and Newport, G. D. (1988). Neurochemical and neurohistological alterations produced by orally administered methylenedioxymethamphetamine (MDMA). *Toxicol. Appl. Pharmacol.* **94**, 448–457.

Slikker, W., Jr., and Gaylor, D. W. (1990). Biologically-based dose-response model for neurotoxicity risk assessment. *Korean J. Toxicol.* **6**, 205–213.

Slikker, W., Jr. (1991). Biomarkers of neurotoxicity: an interview. *Biomed. Environ. Sci.* **4**, 192–196.

Slikker, W., Jr., Paule, M. G., Ali, S. F., Scallet, A. C., and Bailey, J. R. (1991). Chronic marijuana smoke exposure in the rhesus monkey I: Plasma cannabinoid and blood carboxyhemoglobin concentrations and clinical chemistry parameters. *Fundam. Appl. Toxicol.* **17**, 321–334.

Slikker, W., Jr., Paule, M. G., Ali, S. F., Scallet, A. C., and Bailey, J. R. (1992). Behavioral, neurochemical, and neurohistological effects of chronic marijuana smoke exposure in the non-human primate. In *Marijuana/Cannabinoids: Neurobiology and Neurophysiology* (L. L. Murphy and A. Bartke, Eds.), pp. 219–273. CRC Press, Boca Raton, Florida.

Sterio, D. C. (1984). The unbiased estimation of number and sizes of arbitrary particles using the disector. *J. Microscopy* **134**, 127–136.

Sternberger, L. A., Hardy, P.H., Jr., Cuculis, J. J., and Meyer, H. G. (1970). The unlabeled antibody enzyme method of immunohistochemistry. *J. Histochem. Cytochem.* **13,** 315–333.

Uemura, E., Ireland, W. P., Levin, E. D., and Bowman, R. E. (1985a). Effects of halothane on the development of rat brain: a Golgi study of the dendritic growth. *Exp. Neurol.* **89,** 503–519.

Uemura, E., Levin, E. D., and Gowman, R. E. (1985b). Effects of halothane on synaptogenesis and learning behavior in rats. *Exp. Neurol.* **89,** 520–529.

Vaisamraut, V., and Hess, A. (1953). Golgi impregnation after formalin fixation. *Stain Technol.* **28,** 303–304.

Weibel, E. R. (1979). *Stereological Methods: Practical Methods for Biological Morphometry.* Academic Press, London.

Wilson, S. W., Andrews, A. M., Scallet, A. C., Ali, S. F., Bailey, J. R., Paule, M. G., and Slikker, W., Jr. (1990). Evaluation of hippocampus in rhesus monkeys after chronic (1-year) inhalation exposure to marijuana: II. Morphometric methods. *Proc. Int. Conf. Electron Micros.* **12,** 400–401.

Winer, B. J. (1971). *Statistical Principles in Experimental Design.* 2nd ed., pp. 127–201, McGraw-Hill, New York.

Zawia, N. H., and Harry, G. J. (1993). Trimethyltin-induced c-fos expression: Adolescent vs. neonatal rat hippocampus. *Toxicol. Appl. Pharmacol.* **121,** 99–102.

PART
II

Neurophysiological Approaches and Methods

PART

II

Neurophysiological Approaches and Methods

An Introductory Overview

WILLIAM K. BOYES

Neurophysiological Toxicology Branch
Neurotoxicology Division
U.S. Environmental Protection Agency
Research Triangle Park, North Carolina 27711

The scale of neurophysiological techniques that can be used to investigate the function of components of the nervous system ranges from the very small, such as single membrane ion channel methods, to measuring the response of large anatomical pathways (e.g., somatosensory afferents extending from the tip of the toe to the somatosensory cortex). The operation of discrete parts of nerve cells, single neurons, simple neural circuits, or complex brain systems can be studied. A great number of techniques and preparations have evolved to study specific functions along this continuum, and many of these procedures have been applied to neurotoxicological problems.

A full understanding of the actions of neurotoxic compounds requires knowledge of their impact on neural function across levels of analysis. The mechanisms of action of neurotoxicants can be studied best at the smallest level. Thus, studies of ion channel function are at the heart of understanding the actions of many neurotoxicants. The nervous system operates, however, not only as many individual units, but also as a highly organized communication and information processing network. Therefore, the actions of neurotoxic compounds on synaptic communication between cells are vitally important. Neurons form many, often reciprocal, synaptic contacts with nearby cells, and local neural circuits are crucial functional units. In turn, the action of single neurons and local neuronal circuits build into larger functional organizations, which culminate in the activity of vast neuronal systems. The latter are the likely determinants of such apical processes as sensory perception, cognition, and behavior.

The goal of this section is to provide a sample of the range of neurophysiological techniques representing diverse levels of analysis, particularly regarding the application and interpretation of the results to neurotoxicological problems. Beginning at the finest level of neurophysiological analysis, the chapter by Audesirk describes electrophysiological procedures used in studying the function of membrane ion channels. Both

This manuscript has been reviewed by the Health Effects Research Laboratory, U.S. Environmental Protection Agency, and approved for publication. Mention of trade names and commercial products does not constitute endorsement or recommendation for use.

133

single channel and whole cell techniques are described, including several voltage clamp and patch clamp procedures, and the use of these procedures in studying neurotoxicant effects. Newer techniques and their applications are introduced, including optical imaging using voltage- and ion-sensitive dyes. Finally, Audesirk foresees an interface between electrophysiology and molecular biology. Combining the study of ion channel function with techniques to control the expression of ion channel proteins could be a productive approach in investigating the actions of neurotoxic compounds on ion channels. In addition, such a combination could be used to study mechanisms through which toxicants alter gene expression. Such powerful combinations of technologies will surely contribute to advancing the understanding of neurotoxicants.

The communication of information between neurons via the synapse is an essential component of the function of the nervous system. Processes involved in synaptic transmission are also primary targets for the action of many neurotoxicants. In Chapter 7, Shafer and Atchison review neurophysiological methods for analysis of the effects of neurotoxicants on synaptic transmission. The authors provide an overview of recent developments in understanding the process of synaptic transmission and discuss several procedures for studying transmission processes in peripheral and central nervous system preparations. Both intracellular and extracellular recording procedures are discussed, as well as the use of patch and voltage clamp techniques. Analyses of presynaptic processes, including evoked and spontaneous neurotransmitter release, are covered. In addition, the assessment of postsynaptic function, including receptor-operated ion channel function, is discussed. The actions of neurotoxic compounds are discussed with regard to all of these functions. This chapter provides a bridge between Chapter 6, in the relationship between ion channel and synaptic function, and Chapter 8, in providing a basis for understanding the function of simple neural circuits.

The chapter by Gilbert and Burdette presents a model system to characterize neurotoxicity: hippocampal field potentials. The hippocampus contains a trisynaptic circuit involving the dentate gyrus, pyramidal cells of CA3, and pyramidal cells of CA1. This circuit has been intensively studied in basic neurobiology and has contributed much to our knowledge of the physiology of information processing in the central nervous system. Of particular interest are neuroplastic mechanisms, which are studied in this preparation, and possible relationships to phenomena such as learning and memory. Chapter 8 focuses primarily on extracellular field potentials recorded *in vivo*. This discussion is supplemented by data recorded from *in vitro* preparations

such as the brain slice, which allows more readily both intra- and extracellular recordings to be made. Thus, this chapter builds on the foundation provided in the previous chapters. It also represents an intermediate stage of analysis between neurophysiological investigation at the level of the ion channel or synapse and the analysis of complex systems such as that of some of the sensory systems covered in Chapter 9.

Neurophysiological analyses of increasingly complex systems are evident in Chapter 9. In this chapter, Herr and Boyes describe a set of procedures known as sensory-evoked potentials, which have been used to assess the functional status of sensory systems in a variety of situations, including neurotoxicological applications. Like the preparation described in the previous chapter, sensory-evoked potentials are capable of measuring the responses of relatively simple neural pathways. Examples of this include specific peaks in flash-elicited electroretinograms generated by known retinal cells, and early peaks in the brainstem auditory and somatosensory-evoked potentials reflecting the initial stages of processing in the auditory and somatosensory pathways, respectively. As the latency of sensory-evoked peaks increases, however, the neural generators of each component grow progressively more complex. The use of sensory-evoked potentials in toxicology has been reviewed extensively in recent years and, accordingly, Chapter 9 concentrates on the topic of the neurogenerators of sensory-evoked potentials. Potentials elicited by visual, auditory, and somatosensory stimulation are discussed. A description of the current understanding of the neural circuits contributing to sensory-evoked potentials completes the presentation of the continuum of neurophysiological procedures. Thus, neurophysiological analysis stretches from the study of single membrane ion channels, to the study of complex systems involving vast networks of interconnecting cells in both the central and peripheral nervous systems.

This set of chapters dealing with neurophysiological procedures is accompanied by fitting companions in other sections of this book. As a rule, the results of neurophysiological investigations should not be interpreted in isolation of the results of other experimental procedures. Understanding the effects of neurotoxic compounds on biochemical, neuroanatomical, and behavioral measures can help tremendously in understanding the results of neurophysiological experiments. The converse is also true. Neurophysiological results can be invaluable in interpreting other neurotoxicity data. Together, a compendium of neurotoxicity information from a variety of disciplinary approaches provides the best understanding of the nature and consequences of exposure to neurotoxic compounds.

Finally, another section of this volume deals with risk assessment. What is the role of neurophysiological procedures in risk assessment? Most risk assessments require a determination of whether the measured consequence of exposure to a compound can be considered "adverse." Adverse neurotoxic effects have been defined as including ". . . both unwanted effects and any alteration from baseline that diminishes the ability to survive, reproduce, or adapt to the environment" (U.S. Environmental Protection Agency, 1993). Certainly, changes in neurophysiological endpoints can fulfill this definition. Few would desire to experience changes in the function of their nervous system as a result of inadvertent exposure to environmental pollutants. Such changes are clearly distinct, from a risk assessment point of view, from changes such as the beneficial and desired consequences of therapeutic drugs, which also might be measured neurophysiologically. A harder distinction, however, comes in regard to the second part of the definition of adversity. To what extent can neurophysiological measures reflect on the ability to survive, reproduce, or adapt to the environment? Understanding the context of neurophysiological changes both in terms of their implications across levels of analysis, as discussed earlier, and in relationship to other disciplinary endpoints will certainly help in this regard.

Reference

U.S. Environmental Protection Agency. (1993). Draft report: Principles of neurotoxicity risk assessment. *Fed. Reg.* **58**, 41556–41599.

CHAPTER

6

Electrophysiological Analysis of Ion Channel Function

GERALD AUDESIRK
Biology Department
University of Colorado at Denver
Denver, Colorado 80217-3364

I. Introduction

Ion movement across the plasma membrane regulates virtually every aspect of neuronal function, including not only electrical activity and neurotransmitter release, usually on a millisecond scale, but also long-lasting changes in intracellular metabolism and structure. For example, an influx of Ca^{2+} modulates neuronal development (Kater et al., 1988), activates some intracellular messenger molecules [e.g., calmodulin (Means et al., 1982; Stoclet et al., 1987), calcineurin (Klee et al., 1988)], and activates transcription of certain genes [e.g., c-fos (Morgan and Curran, 1988; Sheng et al., 1990)]. Therefore, any neurotoxicant that alters the function of ion channels, by enhancing or inhibiting ion flux or by changing the conditions under which the channels open to permit ion flux, is likely to affect neuronal structure and function. Some of these neurotoxicant effects are described in other chapters in this series (van den Berken et al.).

Until about 1980, most techniques for studying ion fluxes and/or their associated cellular sequelae were often relatively slow and, in many instances, required the use of large populations of cells or subcellular particles (e.g., measurement of calcium influx by ^{45}Ca uptake into synaptosomes depolarized with elevated K^+ solutions) or were practical only in selected types of neurons (e.g., voltage clamping of squid giant axon or large gastropod neurons with microelectrodes). With the development of the patch clamp technique, however, it became a relatively straightforward procedure to measure ion fluxes through both voltage-sensitive and ligand-regulated channels, in either single intact cells or in isolated patches of membrane (Auerbach and Sachs, 1985; Hamill et al., 1981; Sachs and Auerbach, 1983; Sakmann and Neher, 1983). The small size of most mammalian neurons is no longer a major obstacle, but instead is an advantage because small neurons allow greater speed and accuracy of voltage clamping. Nevertheless, it is still relatively uncommon to see these powerful techniques applied to neurotoxicology. This chapter begins with a brief overview of voltage-sensitive and ligand-regulated ion channels as targets for neurotoxicant action (for more detail, see the chapters by van den Berken et al., in this series) and the

137

methodologies appropriate to their study in individual neurons. Most of the chapter is devoted to descriptions of voltage and patch clamping, including some of the advantages and disadvantages of the various configurations and potential disparities in the resulting data. The chapter concludes with a description of more recent techniques, including combinations of electrophysiology and molecular biology, that promise to provide new insights into molecular mechanisms of neurotoxicity.

A. Role of Ion Channels in Neuronal Function

The electrical activity and metabolism of neurons is regulated by voltage-sensitive mechanisms and ligand-sensitive mechanisms (although there is some overlap between these two). Voltage-sensitive mechanisms involve ion-selective channels in the plasma membrane whose permeability is regulated by the transmembrane potential. The familiar voltage-sensitive sodium, calcium, and potassium channels, each of which exists in several molecular forms, are the most thoroughly studied of these channels, from the perspectives both of basic structure and function and of neurotoxicity. Ligand-sensitive mechanisms fall into two broad categories. Many plasma membrane receptors are associated with or are an integral part of transmembrane ion channels. Binding of ligand (e.g., neurotransmitter, hormone, toxicant) to a receptor opens (or, more rarely, closes) ion-selective channels. The GABA$_A$ receptor, nicotinic acetylcholine receptors, and the NMDA and kainate subtypes of glutamate receptors are important and quite well-understood members of ligand-regulated ion channels. Other receptors, often associated with G-proteins, modulate intracellular metabolism rather than ion fluxes across the plasma membrane. Dopamine receptors and the metabotropic glutamate receptors fall into this category.

Not surprisingly, these regulatory mechanisms interact extensively. Calcium influx through voltage-sensitive calcium channels, for example, may affect intracellular metabolism in several ways, such as activating transcription of immediate early genes (Morgan and Curran, 1988; Sheng *et al.*, 1990). Metabotropic receptors may trigger activation of protein kinase C or cyclic AMP-dependent protein kinase which may in turn phosphorylate voltage-sensitive calcium channels and modulate their permeability and/or response to depolarization (Armstrong and Eckert, 1987; Chad *et al.*, 1987; Hartzell *et al.*, 1991; Nunoki *et al.*, 1989). Nevertheless, distinctive techniques are often employed in studying each mechanism and, in principle, neurotoxic actions may be pinpointed to individual aspects of individual mechanisms. This chapter focuses

on techniques for studying ion-selective channels, with particular emphasis on voltage-sensitive channels. Receptor-operated ion channels are discussed in the chapter by Shafer and Atchison in this volume. Where appropriate, I will point out where intracellular metabolism affects such channels.

B. Techniques of Study

The commonly used techniques for studying the function of ion channels and the effects of neurotoxicants on those channels can be grouped into two broad categories: whole cell techniques and single channel techniques. Despite these names, there are some circumstances in which single channel techniques may provide more information about channel function in intact cells than most whole cell techniques do, as discussed in Section III.

1. Whole Cell Techniques

Whole cell techniques can be subdivided into recording from neurons or groups of neurons, either intracellularly or extracellularly, and voltage clamping individual neurons. Generally speaking, recording techniques measure the voltage response of a neuron to external or internal stimuli and may provide a reasonable approximation of the reponse that a neuron would produce *in vivo* to the same stimulus. However, with recording techniques alone, it is often difficult to determine the mechanism(s) whereby a stimulus (a neurotoxicant, for example) elicits the observed response. Voltage clamping measures the transmembrane current elicited in a neuron in response to an imposed voltage. With suitable pharmacological manipulations and the proper choice of voltage control, it is often possible to limit the observed currents to those that flow through one specific type of ion channel. Voltage clamping allows the investigator to measure voltages and kinetics of channel activation and inactivation and, in some cases, to assess the effects of alterations of intracellular metabolism on ion channel function.

2. Single Channel Techniques

Single channel techniques involve sealing a micropipette to a patch of plasma membrane and recording the currents that flow through individual ion channels. If the pipette tip (and hence the membrane patch) is small and/or the channels are sparsely distributed, the pipette may unambiguously record currents flowing through single channels. If the pipette tip is relatively large, the pipette may record several channels simultaneously; if there are only a few channels, it is often possible to determine current amplitudes and durations for individual channels. Single channel techniques can be subdivided into cell-attached and cell-free patch clamping (Hamill *et al.,* 1981; Rae and Levis, 1984). In a cell-attached patch, the membrane patch sealed onto the pipette tip remains attached to the intact cell. In principle, the normal cellular metabolism remains intact, and this configuration allows one to observe effects of intracellular metabolism (e.g., phosphorylation) on ion channel function. Usually, cell-attached patches are not used to assess the direct effects of toxicants on ion channels because the patch beneath the pipette tip is sealed off from the bath. In a cell-free patch, a piece of plasma membrane is sealed onto a pipette tip and pulled free from the rest of the cell. With appropriate manipulations, either the external surface of the membrane patch can be exposed to the bath (the "outside-out" configuration) or the intracellular surface can be exposed to the bath (the "inside-out" configuration). Precise control of the chemical composition of the solutions bathing both inside and outside of the membrane patch can be achieved with isolated patches. However, normal intracellular modulating influences are lost, and any yet to be discovered influences, by definition, cannot be duplicated.

3. Other Techniques

The relatively recent, widespread availability of voltage-sensitive and ion-sensitive dyes, with appropriate microscopic digital imaging capability, offers additional, and often complementary, methods for evaluating the function of ion channels. Voltage-sensitive dyes of several varieties are available, including relatively "fast" dyes such as the styryls, which can be used to monitor action potentials, and "slow" dyes such as the oxonols, which provide a time-averaged integration of changes in membrane potential over seconds to minutes. Ion-sensitive dyes change their fluorescence upon binding hydrogen, sodium, potassium, calcium, or magnesium ions. The best of these dyes is quite selective for individual ions in the presence of physiological concentrations of potentially competing ions. Some of these dyes can be used simultaneously, and many can be used in conjunction with electrophysiological techniques. Finally, recent developments allow electrophysiological measurements and determination of mRNAs to be made in the same individual neuron. These techniques are briefly described at the end of this chapter.

II. Whole Cell Electrophysiological Techniques

A. Voltage Recording with Microelectrodes

A variety of microelectrode techniques have been developed for recording electrical activity from single neurons or groups of neurons, in culture, slices of brain tissues, or *in situ*. Microelectrode recording is essential in determining many aspects of neuronal function, particularly the dynamics of neuronal interactions in reflexes, long-term potentiation, etc. Many of these applications of microelectrode recording are described in the chapters by Shafer and Atchison, Gilbert and Burdette, and Herr and Boyes in this volume. It is usually difficult to determine mechanisms of effect of neurotoxicants at the level of the ion channel with voltage recordings, and these techniques will not be described here.

B. Voltage Clamping

In essence, voltage clamping involves five steps. (1) The experimenter determines the voltage at which the cell is to be maintained. (2) A microelectrode records the transmembrane potential. (3) A "clamp amplifier" measures the difference between the desired and measured (presumably actual) potentials. (4) The clamp amplifier causes current to be injected into the cell through a microelectrode (which may be the same as, or separate from, the voltage-sensing electrode). Ideally, the injected current should precisely and instantaneously counterbalance the transmembrane current that flows through ion channels, so that it is an exact replica of the transmembrane current but of opposite polarity. (5) Steps one through four are repeated for the duration of the clamp experiment.

The similarity between desired and actual potentials, and the speed and faithfulness with which the actual transmembrane potential mimics the desired potential during changes, or "steps" of voltage, determines how "good" the clamp is. Many factors affect the precision of voltage clamping. With the fast clamp amplifiers commercially available, such as those from Axon Instruments, Dagan Instruments, or World Precision Instruments, the first limiting factor is the accuracy of voltage measurement. Obviously, if the voltage-sensing microelectrode does not faithfully record the true transmembrane potential, then the experimenter does not in fact know to what voltage the cell is clamped. For example, the determination of activation and inactivation voltages requires accurate voltage measurement. The second limiting factor is the maximum possible amplitude of current injection. The larger the amplitude of current that can be injected, the faster the cell's own transmembrane currents can be neutralized, and the more accurately the experimenter will be able to determine the amplitudes and especially kinetics of current flow. Accuracy of voltage measurement and amplitude of current injection often interact in the various voltage clamp configurations, as described next.

1. Voltage Clamp Configurations

There are four basic configurations for voltage clamping whole cells: a two-microelectrode, a single-microelectrode, a whole cell patch, and a perforated patch. Axon Instruments has published *The Axon Guide,* which should prove invaluable to any investigator, new or established, who wishes to pursue any of these techniques (Sherman-Gold, 1993). *The Axon Guide* discusses theoretical considerations and provides practical tips for implementing the full range of voltage clamp techniques.

a. A Two-Microelectrode Voltage Clamp

In a two-electrode clamp, a single neuron is simultaneously impaled with two micropipettes. One pipette is used to record the voltage of the neuron, whereas the second pipette is used to inject current to maintain the transmembrane potential at the desired voltage (i.e., to "clamp" the membrane potential). In principle, the two electrodes are completely independent of one another and do not influence each other's performance (although in most practical applications there is some cross-talk between electrodes, this is often negligible). Since the two electrodes are independent, the voltage-sensing electrode, when connected to a suitable amplifier headstage, provides an extremely accurate measurement of the true transmembrane potential. If the current micropipette has a large enough tip diameter, large amounts of current can be injected into the cell. Therefore, the ideal two-electrode clamp can be both very accurate and very fast. Principles and applications of two-electrode clamps are described in articles by Finkel (1985), Sachs (1985), Finkel and Gage (1985), and Lecar and Smith (1985), all of which may be found in a very useful text by Smith and co-workers (1985).

The necessity of impaling a single neuron with two micropipettes, one of which (the current-passing electrode) should preferably be fairly large, has restricted two-electrode clamps to large invertebrate neurons and a few vertebrate neurons (in the hands of dedicated and extremely skilled electrophysiologists). Double-barreled micropipettes ease the impalement problem, but suffer from the disadvantages of increased cross-talk between electrodes and (usually) relatively small diameter tip openings.

b. A Single-Microelectrode Voltage Clamp

In a single-microelectrode clamp, a neuron is impaled with only one micropipette (Finkel and Redman, 1984, 1985; Merickel, 1980; Wilson and Goldner, 1975). Because the resistance of the micropipette is usually a substan-

tial fraction of the resistance of the neuron, current injected into the cell simultaneously causes significant potential changes both across the membrane of the neuron and across the resistance of the pipette. The resistance of the neuron changes, usually by an unknown amount, as channels open and close; the resistance of the pipette often changes as well. If simultaneous voltage measurement and current injection were attempted, there would be very large and unquantified errors in both voltage control and current measurement. To reduce or eliminate these difficulties, the single micropipette is "time shared" between voltage measurement and current injection. A fast-switching circuit in the clamp amplifier first samples the voltage, then passes current, waits until the current-induced potential in the micropipette has declined virtually to zero, and samples the voltage again, usually repeating this sequence at a rate of several kilohertz.

The decay of current-induced potential in the pipette limits both the magnitude of current that can be passed into the cell and the switching frequency: in general, large currents induce large potentials in the pipette, which take a long time to decay to zero. The magnitude of current that can be passed depends on the resistance and capacitance of the micropipette: the higher the micropipette resistance, the larger the potential induced in the pipette during current passing; the higher the capacitance, the more slowly the potential decays. A compromise is usually reached between switching frequency and current magnitude to allow the best possible clamp. Even in the best of circumstances, pulses of current must be relatively small and of brief duration so that the potential induced in the micropipette can decay completely before the voltage is sampled. Therefore, single-electrode clamps suffer from two inherent defects when attempting to voltage clamp cells with fast and/or large transmembrane currents. (1) The amount of current that can be injected during a given switching cycle may not be large enough to completely clamp the transmembrane voltage to the desired potential, so that the clamp lags behind the actual voltage changes in the neuron. (2) The switching frequency may be so slow that rapidly changing voltages, such as those that occur during opening of sodium channels, are not accurately detected by the clamp system; i.e., in any given cycle, the clamp amplifier injects current to clamp the previous voltage sample, but meanwhile the actual transmembrane voltage may have changed appreciably.

To optimize single-electrode voltage clamps, it is important to reduce the electrode resistance and capacitance as much as possible; for example, by beveling the electrode tip (reduces resistance while maintaining sharpness; Brown and Flaming, 1986) and insulating the tip with wax or Sylgard (reduces capacitance; Sachs, 1985). The smaller the time constant of the micropipette, the larger the currents that can be injected and the higher the switching frequency that can be used.

As these considerations suggest, the speed and accuracy of single-electrode clamps depend on the relative resistances and capacitances of micropipette and neuron. Generally, small neurons have large resistances and generate small transmembrane currents. Therefore, relatively small currents are needed to clamp a small neuron. For a constant pipette resistance, the larger resistance of small neurons also means that the voltage drop across the pipette is reduced. However, to avoid damaging the neurons during impalement, smaller neurons usually require sharper pipettes with smaller tip diameters (and consequently higher resistance), which reduces the advantage of using small neurons. Nevertheless, if care is taken to reduce pipette capacitance, and if electrode resistance can be kept reasonably low (e.g., by beveling pipettes), the single-electrode configuration may be able to clamp even fast sodium currents in small neurons fairly well.

c. Whole Cell Patch Clamping Although the whole cell patch clamp was not designed as a successor to the single-microelectrode switching clamp, it is, in fact, a logical solution to the problems inherent in the single-electrode configuration. As outlined earlier, the single-electrode configuration suffers from an inherent contradiction: it would perform best if one could employ pipettes of large tip diameter in very small neurons. In fact, with the combination of a very large pipette tip (on the order of 1 μm in diameter), a typical neuron from mammalian CNS (5 to 15 μm in diameter), and electronic resistance compensation techniques, the voltage drop across the pipette would become negligible, even while passing very large currents, and one would no longer need to time share the pipette between measuring voltage and passing current. Such a large micropipette tip would usually seriously damage the neuron during impalement, and therefore cannot normally be used.

The whole cell version of patch clamping bypasses the impalement problem. As Hamill and colleagues (1981) described, when a clean, smooth, fire-polished pipette tip is brought into contact with the plasma membrane of a neuron, it sometimes forms a "gigohm seal" with the membrane patch, in which the resistance across the glass–membrane junction between the inside of the pipette and the bathing solution may be 1 to 10 GΩ. Apparently, the electrode glass interacts with the membrane lipids at distances on the same order of magnitude as chemical bonds. Applying gentle

suction improves the frequency of attaining gigohm seals. This configuration is termed the "cell-attached patch" and is discussed in Section III.

To produce the whole cell clamp configuration, stronger suction and/or brief pulses of relatively high voltage are applied, which disrupt the patch of membrane beneath the pipette tip opening, providing direct, low-resistance access between the saline in the pipette and the cytoplasm of the neuron (Hamill *et al.,* 1981; Marty and Neher, 1983; Rae and Levis, 1984). Naturally, either suction or voltage sometimes disrupts not only the patch beneath the pipette, but also the seal between the pipette glass and the rest of the neuron's membrane, and the investigator must try again on another neuron. When patch disruption goes well, however, it is not uncommon for the access resistance to be only a few megohms while the neuron's transmembrane resistance may exceed 100 megohms. Under these conditions, particularly when investigating channel types that produce relatively low whole cell currents, many investigators dispense with electronic series resistance compensation.

Some investigators use the whole cell patch configuration with switching electronics similar to those normally used for single-electrode clamping. Because of the extremely low pipette resistance, large currents can be combined with fast-switching frequencies to produce fast, accurate clamps.

Failure to achieve a gigohm seal (probably because of debris on the pipette tip or membrane surface) and loss of the seal during patch disruption are the two most common sources of failure and frustration in the whole cell configuration. Eventually, the seal always fails, but with suitable electrode glass, neuron type, and physical stability (particularly a high-quality vibration isolation table and a micromanipulator devoid of drift), the whole cell clamp configuration may be maintained for many minutes, even an hour or more.

For an introduction to the methodology of whole cell patch clamping, including electrode manufacture and shielding, electronic circuitry, techniques, and data analysis and interpretation, see Hamill *et al.* (1981), Rae and Levis (1984), *The Axon Guide,* and chapters by Corey and Stevens, Marty and Neher, Sakmann and Neher, and Sigworth in the "patch clamp bible" by Sakmann and Neher (1983). Even for those who already own the Sakmann and Neher text, *The Axon Guide* and the article by Rae and Levis (1984) are extremely worthwhile, for they contain a wealth of useful tips and practical information that is very difficult to find elsewhere.

d. Perforated Patch, Whole Cell Voltage Clamping Whole cell patch clamping has one unique feature that can be both an advantage and a disadvantage. Because the pipette tip opening is so large, there is a rapid diffusional exchange of solutes between the pipette solution and the neuronal cytoplasm. The volume of the pipette is, of course, vastly larger than the volume of the neuron, so that quite rapidly the composition of the neuronal "cytoplasm" becomes identical to that of the pipette solution, at least as far as small, easily diffusible molecules are concerned. This can be an advantage, for example, in blocking potassium currents: the pipette is filled with a solution devoid of potassium but containing cesium, which is not permeable through potassium channels, and tetraethylammonium, which at least partially blocks several types of potassium channels even in the presence of potassium. In this way, one can often virtually eliminate "contamination" by potassium currents during studies of sodium and especially calcium currents.

However, diffusional exchange can also be a disadvantage. It was discovered quite early on that certain types of calcium currents (principally the L-type current) diminished very rapidly during whole cell clamps, presumably because one or more diffusible constituents of the cytoplasm, which were essential to the functioning of the channels, "washed out" of the neuron (e.g., Fenwick *et al.*, 1982; Kostyuk *et al.*, 1981). In many cases, it has proved possible to compensate, at least partially, for diffusion exchange by including appropriate substances in the pipette solution. [Indeed, this has even been turned to an advantage by providing a means to study which intracellular molecular events help to regulate channel function (see Armstrong, 1988; Chad *et al.*, 1988; Levitan, 1985).] In general, however, neuronal responses during whole cell clamps may differ, in unknown ways, from responses that would be observed in truly intact neurons that did not suffer from "wash out."

The perforated patch technique (Falke *et al.*, 1989; Horn and Marty, 1988; Korn and Horn, 1989) limits washout. In this configuration, pore-forming substances (usually nystatin or amphotericin B) are included in the pipette solution. A gigohm seal is attained, as for a normal whole cell clamp, but the membrane patch beneath the pipette tip is not disrupted. Instead, the cell-attached patch is maintained, and the pore-forming molecules insert themselves in the membrane patch. Generally, molecules are used that form pores with high but relatively indiscriminant permeability to monovalent cations and chloride (and perhaps some other very small, charged molecules, but this has not been extensively studied). The pores have low permeability to most organic molecules, even quite small ones such as glucose. Over perhaps 20 to 30 min, as more and more pore-forming molecules insert them-

selves into the membrane patch, the access resistance between pipette and cytoplasm falls, sometimes to levels comparable to the normal whole cell configuration. The diffusional exchange of monovalent cations and chloride between the pipette solution and the neuronal cytoplasm occurs during pore formation. Therefore, the ionic composition of the cytoplasm is partially under experimental control (e.g., for blocking potassium channels by replacing intracellular potassium with cesium).

Ideally, the perforated patch technique reduces or eliminates two difficulties of normal whole cell clamping: First, the membrane patch beneath the pipette is not disrupted, which eliminates one common source of experimental failure. The perforated patch can be stable for several hours under ideal conditions (e.g., an extremely stable, drift-free micromanipulator). Second, although monovalent cations and chloride are exchanged between pipette and cytoplasm, both multivalent cations such as Ca^{2+} and most organic molecules remain in the cytoplasm, and intracellular regulation of channel function presumably proceeds relatively normally. This second point is difficult to prove, and indeed some investigators have found different results when comparing perforated patch clamp experiments with otherwise identical cell-attached patch clamp experiments (O'Dell and Alger, 1991).

The principal disadvantages of the perforated patch technique appear to be the rather tedious preparation of the pipette solutions, the fact that both nystatin and amphotericin appear to reduce the likelihood of achieving a gigohm seal, and the fairly long delay while enough pores insert into the patch to reduce access resistance to a low level.

For extensive descriptions of the perforated patch technique, the reader should consult Korn and colleagues (1991) and Rae and co-workers (1991). Axon Instruments' *The Axon Guide* and their *Axobits* newsletter also contain useful pratical hints on the perforated patch technique.

2. Applications of Voltage Clamp Configurations

As alluded to in the previous section, each configuration for voltage clamping has advantages and disadvantages, as described here and summarized in Table 1.

a. Neuron Size As the name implies, two-microelectrode voltage clamps require the simultaneous impalement of a single neuron with two micropipettes. This is possible but difficult with small (<100 μm) neurons, and therefore this clamp configuration has seldom been used with mammalian neurons. Although the necessity for large neurons is a serious disadvantage of

TABLE 1   Whole Cell Voltage Clamp Configurations

Consideration	Two electrode	Single electrode	Whole cell patch	Perforated patch
Useful with small neurons	No	Sometimes	Yes	Yes
Useful in slices or *in situ*	No	Yes	Sometimes	Sometimes
Provides very fast clamp	Yes	Sometimes	Yes	Yes
Avoids "washout" of intracellular messengers and metabolites	Yes	Yes	No	Probably in most uses
Allows control of intracellular composition	No	No	Yes	For small ions only

the two-electrode clamp, such a clamp, or some hybrid of one (Brown *et al.,* 1985; Byerly and Hagiwara, 1982), is also the only clamp configuration really suited for large neurons. The inherent limitations of the other three configurations in current-passing capability and/or accurate voltage measurement when passing large currents make them less powerful and/or less accurate when used with large neurons.

The standard whole cell patch clamp is ideally suited for small vertebrate neurons, as long as the cell surface is visible and free of any overlying materials. Whole cell clamps are seldom used with very large neurons, in which neuron resistance and pipette resistance may be fairly similar. However, a hybrid system, impaling the neuron with a microelectrode, for voltage measurement, and simultaneously accessing the cytoplasm with a patch pipette, the better to pass extremely large currents, has been used in large neurons (Brown *et al.,* 1985; Byerly and Hagiwara, 1982). The perforated patch technique has the same cell size requirements and limitations as the standard whole cell configuration.

Single-microelectrode clamping is not as commonly used as whole cell clamping because of its speed limitations. However, single-electrode clamps are applicable to small vertebrate neurons, particularly those *in situ* or in brain slices (see the following discussion). Single-electrode clamps are not generally used with large neurons.

b. *Usefulness in Slices or in Situ*   Generally, the two-microelectrode clamp cannot be used in brain slices or *in situ,* except in gastropod nervous systems in which extremely large neurons are found at or near the surface of small ganglia. Single-microelectrode clamps are particularly useful in situations in which neurons cannot be visualized through a microscope and/or are overlain by other cells or materials. For example, single-electrode clamps, especially with beveled pipettes, can be used for "blind" impalements of cells in intact brains or brain sections (e.g., Gahwiler and Brown, 1987), which would be impossible with any of the other configurations.

As mentioned earlier, the cell surface must be exposed and free of debris in order to make the gigohm seal required for whole cell and perforated patch configurations. Until recently, these limitations have largely limited whole cell clamping to small vertebrate neurons, either acutely isolated from intact brains or cultured for days to weeks. However, in slice preparations, with sufficient care, one can expose individual neurons by blowing the overlying cells away with gentle streams of saline (Blanton *et al.,* 1989; Edwards *et al.,* 1989; Sakmann *et al.,* 1989; see also *The Axon Guide*). The exposed neurons can then be visualized through a microscope and clamped in the standard way.

c. *Speed of Clamp*   The two-microelectrode clamp and its hybrids probably provide the fastest clamping speed, particularly with larger neurons, and are applicable for all current types in any neuron large enough to accommodate the two electrodes. Single-microelectrode clamps are generally the slowest clamps and, in most cell types, are not really suitable for accurate clamping of fast, large currents, such as sodium currents. In their normal application with small neurons, whole cell and perforated patch clamps are fast enough for good accuracy even with sodium currents.

d. *Washout*   The microelectrode clamps do not usually wash out cytoplasmic constituents to any significant extent. The standard whole cell configuration essentially completely washes out all soluble molecules within a few minutes. The perforated patch configuration washes out monovalent cations and perhaps some small organic molecules, but any moderately large organic molecule (glucose or larger) will not wash out. It should be noted that calcium currents still run down in perforated patch clamps, although much more slowly than in whole cell clamps, suggesting that perhaps some essential small molecule is capable of diffusing out of the neuron through the patch pores.

e. *Control of Cytoplasmic Composition*   The other side of the "washout" coin is the control of cytoplasmic composition. Microelectrode clamps do not provide any significant control over cytoplasmic composition, with the principal exception that substances, such as ion-selective dyes or cesium ions (to block potassium channels), may be injected into the neuron either by pressure or electrical potential. The perforated patch configuration exchanges monovalent cations and chloride (therefore cesium can be used in the pipette solution to block potassium channels). Most other molecules are probably not controlled to any significant extent. The neuron cannot be loaded with large molecules (e.g., proteins) through the perforated patch. The standard whole cell configuration allows almost complete control over the soluble constituents of the cytoplasm. Components may be added to the pipette solution to investigate such diverse phenomena as the role of phosphorylation in channel function (Armstrong, 1988; Armstrong and Eckert, 1987; Chad *et al.,* 1987), the role of intracellular calcium buffering in calcium-dependent channel inactivation (e.g., Kalman *et al.,* 1988; Kohr and Mody, 1991), or changes in intracellular ionic composition in response to channel activation (e.g., Thayer and Miller, 1990).

C. Applications of Voltage Clamping to Neurotoxicology

In this section, two related topics are considered: What types of information relevant to neurotoxicology may be obtained with voltage clamp techniques? What types of information have already appeared in the literature?

1. Neurotoxicological Interpretation of Voltage Clamp Studies of Whole Cells

Depending on the configuration employed, voltage clamps of whole cells may be used to investigate several phenomena.

a. Change in Current Amplitude by Neurotoxicants By far the simplest and most widely used application of voltage clamping in neurotoxicology has been to determine if acute exposure to a toxicant reduces the amplitude of the current flowing through a particular type of channel. In the most straightforward experiment, identical voltage clamp steps are used to elicit current flow during perfusion with control medium or with control medium plus added toxicant (Fig. 1). If the current amplitude is reduced during toxicant exposure, this is usually taken to indicate channel block; if the current amplitude is increased, this may indicate channel activation by the toxicant. It should be noted, however, that such experiments cannot readily distinguish among actions such as plugging the ion-selective pore

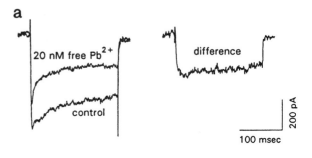

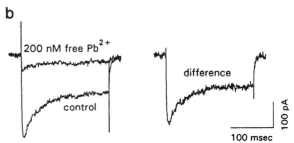

FIGURE 1 Whole cell voltage clamp examples of the effects of inorganic Pb^{2+} on voltage-sensitive calcium channels in cultured embryonic rat hippocampal neurons. In each part of the figure, raw current traces are on the left. The difference trace (right) is a digital subtraction of the current recorded during acute Pb^{2+} exposure from the current recorded in Pb^{2+}-free control medium. The shape of the difference current indicates the type of current that has been inhibited by each concentration of Pb^{2+}. (a) The difference current caused by exposure to 20 nM free Pb^{2+} shows almost no inactivation; therefore, this concentration of Pb^{2+} probably inhibits mostly L-type calcium channels. (b) The difference current caused by exposure to 200 nM free Pb^{2+} has both transient and sustained components; therefore, this concentration of Pb^{2+} probably inhibits both N-type and L-type channels. Note that whole cell voltage clamp recordings such as these cannot determine mechanisms of block at the single channel level. (From Audesirk and Audesirk, 1993.)

of the channel, binding to other sites that alter the likelihood of channel opening or the physical configuration of the pore, or even altering intracellular metabolism in ways that affect channel function, such as stimulating or inhibiting channel phosphorylation by any one of several protein kinases.

These limitations of current amplitude measurements must be taken into account when interpreting experimental results. In particular, if a toxicant appears to have no effect in the standard whole cell patch clamp configuration, this cannot be taken as unambiguous evidence that the toxicant has no effect on channel function. Because of cytoplasmic washout, a toxicant with intracellular actions on a channel may never achieve intracellular concentrations needed to produce an effect. Perhaps even more likely, the whole cell configuration may wash out intracellular components which, when acted upon by a toxicant, might alter channel function. Although in this latter case the toxicant would have no direct effect on the channel, its indirect effects on channel function may be profound.

b. Voltage Dependence of Activation and Inactivation

The likelihood of channel opening is a function both of the present voltage (e.g., the phase of an action potential or postsynaptic potential, or the voltage step during clamping) and of recent voltage history (e.g., the resting potential, or the holding potential during clamping). Voltage dependence of channel activation is determined by measuring current amplitudes in response to a series of voltage steps from a constant holding potential (Fig. 2, right-hand curves in each graph). It provides information about the absolute voltage (usually depolarization) required to open the channel under study. Voltage dependence of inactivation is determined by measuring current amplitudes in response to voltage steps of a constant absolute potential (usually at or near the voltage required for maximum current) from a series of holding potentials (Fig. 2, left-hand curves in each graph). It provides information about the likelihood that a significant current can be elicited through that channel by proper stimulation, depending on the "resting potential" of the neuron. (In this context, "resting potential" is placed in quotes to denote that many neurons do not have true resting

potentials, but have varying potentials, including spontaneous spiking, over time. In the case of a tonically firing neuron, for example, at times the transmembrane potential is very negative, during the undershoot or afterhyperpolarization of an action potential, and at times it is relatively depolarized, when the voltage approaches threshold for the next action potential.)

Different channels, even those that may be permeable to the same ion, such as potassium or calcium, may play very different roles in neuronal physiology, depending on their voltage dependence of activation and inactivation. For example, one type of potassium channel (the "fast transient" or A channel; Connor and Stevens, 1971) can be activated only from very negative potentials (near the undershoot of an action potential) but then requires relatively small depolarizations to open the channel. This channel helps to regulate firing frequency in some spontaneously active neurons: the channel opens during the depolarization leading to an action potential, and the resulting outflow of potassium partially counteracts the ongoing depolarization, thus slowing the approach to threshold for the action potential.

In principle, a toxicant could exert significant effects on neuronal function by altering the voltage dependence of activation or inactivation, even if it exerted no effect on current amplitude at all. In the case of the fast transient potassium channel, for example, if a

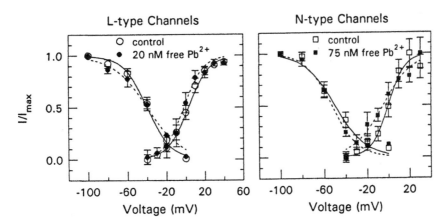

FIGURE 2 Activation and inactivation curves for L-type (left) and N-type (right) channels, during perfusion with control medium (open symbols) or the indicated concentrations of free Pb^{2+} (filled symbols). Inactivation of both channel types was determined by measuring currents elicited by a 250-msec step to +20 mV from the indicated holding potentials. Activation of L-type currents was determined by plotting the amplitude of tail currents measured at the end of a 250-msec step from −80 mV to the indicated potentials. Activation of N-type channels was determined by plotting the amplitude of currents elicited by 250-msec voltage steps from a holding potential of −80 mV to the indicated potentials. All currents are normalized to the maximum current. Activation curves are best fits to the equation $I/I_{max} = [1 + \exp(V - V_{1/2})/k)]^{-1}$, where $V_{1/2}$ is the voltage that elicits currents equal to $(I_{max})/2$ and k is a constant determined during the curve-fitting routine. Inactivation curves are best fits to the equation $I/I_{max} = [1 + \exp(-(V - V_{1/2})/k)]^{-1}$. (From Audesirk and Audesirk, 1993.)

toxicant reduced the voltage needed for activation, the channel would open sooner and probably reduce the spiking frequency of the cell. If the toxicant increased the activation voltage beyond threshold, the fast transient potassium channel opening might be overwhelmed by the simultaneous sodium and calcium channels opening at threshold. In this situation, the fast transient channel might completely lose its function of slowing spiking, and the neuron would increase its spiking frequency.

c. Time Dependence of Activation and Inactivation
All channels open and close with a particular time course, even in the face of constant depolarization. Most channels open fairly rapidly, whereas some, such as delayed rectifier potassium channels, open more slowly. For some channels, such as delayed rectifier potassium channels and L-type calcium channels, time-dependent channel closing is so slow as to be physiologically irrelevant under most circumstances, and it is normally the end of depolarization that closes the channels. For other channels, such as sodium channels and T-type calcium channels, channel closing is fairly rapid even during sustained depolarization, and time-dependent channel closing is physiologically relevant. It is obvious that changes in time dependence of either activation or inactivation might have major impact on neuronal function. For example, many insecticides slow inactivation of sodium channels and cause rapid, repetitive firing (see the chapter by van den Brec *et al.* in this series, and Section C.2).

2. Examples of Voltage Clamp Studies in Neurotoxicology

Considering the importance of voltage-sensitive ion channels in neuronal physiology and the likelihood that their functioning may be affected by toxicants, either directly or through effects on intracellular metabolism, surprisingly few neurotoxicology studies have employed voltage clamping. The majority of studies have probably been conducted by Narahashi and colleagues and Vijverberg and colleagues; many of these are discussed in a chapter by van den Brec *et al.* in another volume of this series and therefore will not be described here in detail.

Many heavy metals appear to enter and block the ion-selective pore of calcium channels (for reviews, see Audesirk, 1989, 1993; for mechanisms, see Lansman *et al.*, 1986), usually (at least at low concentration) with little or no effect on voltage or time dependence of activation or inactivation (see Fig. 2). In voltage clamps of whole cells, this block appears as a reduction in current amplitude with no change in the overall shape of the current–voltage relation (Fig. 1). In contrast,

DDT and the pyrethroid insecticides have no effect on the permeability of sodium channels, but greatly slow the time-dependent inactivation of these channels. These phenomena are described in detail by Lund and Narahashi (1981, 1982) and Vijverberg and de Weille (1985) and in the chapter by van den Brec in this series.

III. Single Channel Techniques

The currents recorded while voltage clamping intact cells are composed of currents flowing through many individual channels. The characteristics of individual channels cannot always be determined from the characteristics of the whole cell currents. For example, although a whole cell record of calcium channel current seems to indicate a relatively smooth, continuous process, it is composed of the sum of rapid openings and closings of many channels, whose individual behavior at any given time is only statistically predictable. Even a single channel varies in its behavior to repeated identical voltage steps.

To a first approximation, the time and voltage dependence of activation and inactivation of currents measured under a whole cell voltage clamp are reasonably good predictors of the much more probabilistic single channel parameters. However, current amplitude is more complicated. Whole cell current amplitude is a function of three single channel parameters: single channel conductance, the likelihood of channel opening, and the mean channel open time (these parameters are discussed in more detail later). Whole cell currents will be larger with larger single channel conductance, a higher likelihood of channel opening at a given voltage, and with longer mean open times. A toxicant-induced change in current amplitude, therefore, could be due to effects on one or several parameters, even to opposite but unequal changes in two or more parameters. Whole cell voltage clamp configurations cannot reveal these details, but configurations that examine single channels can.

Most experiments involving single channel recording rely on the fact that channels are relatively sparsely distributed over the membrane of a cell, so that a reasonably small patch of membrane will contain only one, or at most a few, channels of any given type. Using pipettes similar to those described earlier for whole cell patch clamping, and with appropriate pharmacological and electrophysiological manipulations, one can study the currents flowing through a single channel of a known type. Single channel recording techniques are applicable to any cell type in which gigohm seals can be achieved, i.e., any cell with a plasma membrane not covered by debris, basement membranes, or other cells. What one can find out about the channel greatly depends on the recording configuration.

There are two basic configurations for single channel recordings (Hamill *et al.*, 1981; Rae and Levis, 1984; Sakmann and Neher, 1983): the cell-attached path and the isolated patch. In all cases, a gigohm seal is made over a small patch of plasma membrane, exactly as is done in the initial step of whole cell patch clamping. For cell-attached patch clamping, the membrane is left intact and only the small patch of membrane beneath the pipette is voltage clamped. For isolated patch clamping, a patch of membrane is physically removed from the neuron. Depending on the method of removal, the patch may be inside-out (the formerly extracellular side of the membrane faces the pipette solution, whereas the cytoplasmic side faces the bathing solution) or outside-out [the formerly extracellular side of the membrane faces the bathing solution, whereas the cytoplasmic side faces the pipette solution; see Hamill *et al.* (1981), Rae and Levis (1984), and several chapters in Sakmann and Neher (1983) for descriptions of the techniques needed to produce these configurations].

A. Data Analysis in Single Channel Voltage Clamping

Ideally, a single voltage-sensitive channel exists in one of two, all-or-none states: open or closed [transition configurations, which are not directly visible in current measurements, and subconductance states, which have been reported for some types of channels (e.g., dihydropyridine-sensitive calcium channel in GH$_3$ pituitary cells: Kunze and Ritchie, 1990), will not be considered here]. In the closed state, the channel conducts no current at all. In the open state, every channel of a given type in a given neuronal type has the same conductance when recorded under the same conditions of voltage and charge carriers. Channel opening is more or less instantaneous on the time scales normally employed in patch clamping. Therefore, what one sees in single channel records is a series of essentially square-wave current pulses as individual channels open and close (Fig. 3). If the patch contains only one channel, then all current pulses are identical in amplitude. However, current pulse durations vary greatly because the channel may open for very different periods of time, even to a prolonged constant stimulus or to repetitions of an identical stimulus.

In single channel voltage clamps, one normally repeatedly steps the patch (regardless of the patch configuration) from the holding potential to the desired depolarizing step. Because the behavior of single channels is stochastic, a single depolarizing step is relatively

uninformative. Even if only a single copy of the desired channel type is present in the patch, any given sweep may result in no openings, a few openings, or many openings, and the openings will vary in duration. By analyzing the current responses to many identical voltage steps, one develops a picture of overall channel characteristics.

A wide variety of parameters may be measured in, or derived from, single channel experiments; for thorough discussions, the reader should consult chapters by Aldrich, Colquhoun and Hawkes, Lauger, Colquhoun and Sigworth, Sachs, Sigworth, and DeFelice and Clay in Sakmann and Neher (1983). Most commonly, however, single channel data are used to measure three parameters: single channel conductance, mean open time, and open probability.

Single channel conductance is obtained by measuring the single-channel current during steps to a number of voltages. The slope of the current–voltage relation gives the single channel conductance, although there are a number of potential pitfalls, particularly nonlinearity. Single channel conductance varies with certain experimental conditions, particularly the concentration of permeant ions. Generally, conductance is low when recorded in low concentrations of permeant ions and increases asymptotically as ion concentration increases. Therefore, one must always specify the ionic conditions when reporting single channel conductance. Although there are circumstances in which single channel conductance, recorded with a constant concentration of permeant ions, may appear to change (see the following discussion), the single channel conductance is not usually altered by exposure to toxicants or drugs.

The mean open time is a measure of how long a channel, once opened, tends to remain open. The mean open time is usually distributed exponentially, with a high frequency of short openings and progressively lower frequencies of long openings (Sigworth, 1983). Many substances that block ion channels do so by reducing the mean open time. Heavy metals, for example, reduce the mean open time of L-type calcium channels in heart ventricle cells (Fig. 3; Lansman *et al.*, 1986). Some drugs increase mean open time; for example, the dihydropyridine BAY K8644 greatly prolongs the mean open time of L-type calcium channels (Nowycky *et al.*, 1985).

The open probability measures the likelihood, over relatively long time periods (usually hundreds of milliseconds or more), of a channel being in the open state. For a patch with a single channel, the open probability is derived by dividing the time spent in the open state (the sum of all individual channel openings) by the total time of recording. A high open probability may be achieved by relatively infrequent channel opening

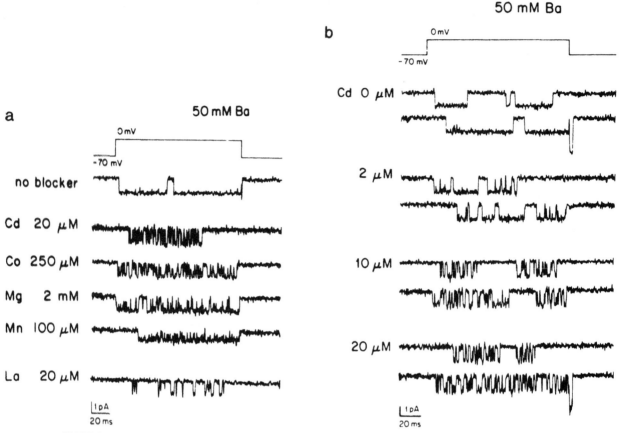

FIGURE 3 Effects of heavy metals on L-type voltage-sensitive calcium channels in guinea pig ventricular cells, using the cell-attached patch clamp configuration. Both 50 m*M* barium and the indicated heavy metals were included in the patch pipette. The different traces were recorded from different patches; therefore, the "control" and "0 μM" (b) traces do not represent "same patch" controls for the heavy metal traces. Note the reduction in mean open time by all of the heavy metals and the "flickery block" of current. The apparent reductions in single channel current amplitude is an artifact of inadequate resolution of extremely fast channel openings and closings during channel block, as explained in the text. (Reproduced from *The Journal of General Physiology*, 1986, **88**, 321–347, by copyright permission of the Rockefeller University Press.)

events combined with long mean open times or by shorter mean open times combined with a high frequency of channel opening events. Some substances decrease open probability without altering mean open time. Tetrodotoxin, for example, blocks sodium channels but has no effect on mean open time (Quandt *et al.*, 1985). Tetrodotoxin probably acts, not directly on channel opening, but by preventing channel conductance even though the channel does open; in single channel recordings, these two mechanisms are indistinguishable. At least superficially, inorganic lead blocks the NMDA-type of glutamate-sensitive ion channel in a similarmanner, by reducing the frequency of channel opening without altering the mean open time (Alkondon *et al.*, 1990).

Probably the most intractible difficulty in analyzing single channel recordings is the problem of "flicker." The simplest assumption of channel behavior, as indi-

cated earlier, is that a single channel opens instantaneously, remains open at a constant conductance for a period of time, and closes instantaneously once again. [Whether the channel is simply "closed," that is, can be immediately reopened, or is "inactivated," and can be reopened only after removal of inactivation, will not be discussed here; see Colquhoun and Hawkes (1983) for details of these states and analyses.] Because of the electrical characteristics of the patch pipettes and clamp amplifiers, the recorded current reaches its true "open channel" value more slowly than the channel actually opens. If the channel opens and closes very rapidly, the recorded current may not have reached its true open channel value before the channel closes again. This "flickery" behavior of the channel causes the single channel conductance to appear to have been reduced. Heavy metals, for example, cause a flickery block of L-type calcium channels (greatly reducing mean open time), resulting in single channel openings that may be difficult to resolve (Fig. 3; Lansman *et al.,* 1986).

B. Relationship between Single Channel Data and Whole Cell Data

Aside from the fact that some data, such as single channel conductance or mean open time, can be reliably obtained only by single channel recording, what is the relationship between data obtained by single channel vs whole cell methods? Obviously, whole cell currents are the composite of currents flowing through all of the individual channels of a neuron. Therefore, if all channels of a single type are essentially identical, and if pharmacological and/or electrophysiological isolation of currents is sufficiently precise, then single channel currents should be accurate predictors of whole cell currents. Indeed, many studies have found that summing many single channel records yields ''pseudo whole cell'' currents that are virtually identical to those obtained in traditional whole cell voltage clamps (e.g., Kostyuk *et al.,* 1988; Pietrobon and Hess, 1990; Plummer *et al.,* 1989). However, this is not always the case, particularly if intracellular metabolism (e.g., soluble second messengers) affects channel function (e.g., O'Dell and Alger, 1991).

C. Cell-Attached Patch Clamping

In cell-attached patch clamping, the tip of a pipette is sealed to a small patch of neuronal membrane. Although the membrane patch usually deforms somewhat, especially if suction is used to assist in making the gigohm seal, the patch remains attached to the rest of the neuronal membrane, and the cytoplasm of the neuron is essentially unaffected by the presence of the pipette. The experimenter controls the voltage and measures the resulting currents only in channels within the patch; because the patch is small relative to the total surface area of the neuron, voltage clamping the patch has little effect on potentials or currents in the rest of the cell (Hamill *et al.,* 1981; Rae and Levis, 1984).

Cell-attached patch clamping has both advantages and disadvantages over whole cell or isolated patch clamping. Cell-attached patches place the least stress on the membrane and therefore are most easily achieved and are usually quite stable (Rae and Levis, 1984). The desired channel type(s) may not be present within the patch, but this is equally a problem with isolated patches. If one wishes to be certain that intracellular metabolism remains intact, and particularly to investigate the effects of intracellular metabolism on ion channel function, the cell-attached patch may be the configuration of choice. Further, the gigohm seal of pipette to

membrane effectively prevents any bath-applied substances from reaching the channels beneath the pipette, which are, of course, the only ones recorded. Therefore, any substances added to the bath can only affect channel function through intracellular mechanisms.

For most applications, however, there are significant disadvantages to cell-attached patches, and they are relatively seldom used. First, as mentioned earlier, the gigohm seal prevents bath-applied substances from reaching the recorded channels. Second, although methods have been described for changing solutions within the pipette (Cull-Candy *et al.,* 1980), these have not been widely applied. Therefore, in most circumstances the solution bathing the outside of the channels beneath the pipette remains constant for the duration of the experiment, and a channel cannot serve as its own control in studies of substances that act on the extracellular side of the channel. Third, it is impossible to control the solution bathing the inside of the channels because this remains normal cytoplasm. Membrane-permeable substances (e.g., diacylglycerol, phorbol esters) may be introduced into the bath and diffuse through the plasma membrane into the cytoplasm, but their concentrations are unknown. Fourth, the resting potential of the neuron is usually unknown (unless a second micropipette is inserted into the neuron, which would be a formidable task in most vertebrate neurons). Since the voltage to which the patch is clamped is the sum of the resting potential and the voltage applied through the pipette, absolute clamp voltages cannot be determined. One may know, from other experiments, the ''normal'' resting potential of the cell type in question, and use that in calculating voltage steps, but inaccuracies are inevitably introduced. Alternatively, one may ''zero'' the resting potential by bathing the neuron in a solution containing a concentration of potassium ions comparable to that of the cytoplasm (e.g., Lipscombe *et al.,* 1988; Fisher *et al.,* 1990). Even in this situation, it is seldom known if the resting potential is truly zero. Further, prolonged depolarization to zero may significantly alter intracellular metabolism and consequently give misleading data about the influences of intracellular events on channel function.

D. Isolated Patch Clamping

Isolating membrane patches takes advantage of the great mechanical stability of the gigohm seal between pipette and membrane. A piece of membrane is literally torn away from the neuron, which is possible if the strength of the gigohm seal exceeds the strength of the membrane. Unlike the cell-attached patch, voltage control of an isolated patch is simple. Depending on the methodology used, the patch may be outside-out or inside-out. At the risk of some oversimplification, outside-out patches are most suited to study the effects of extracellular substances on channel function, whereas inside-out patches are most suited to study the effects of intracellular substances, because the bath solution can be easily changed, while the pipette solution cannot.

1. Outside-Out Patch Clamping

This is by far the most commonly used patch clamp technique. In most respects, it is the single channel equivalent of whole cell patch clamping, in which the inside face of the membrane contacts the pipette solution (and therefore the "cytoplasm" of the patch is the pipette solution) whereas the outside face of the membrane contacts the bath. For some channel types, outside-out patches suffer from washout, as do whole cell clamps. However, unlike cell-attached patches, the "cytoplasm" is completely controlled and the solution contacting the outside of the patch can be changed in less than a millisecond (e.g., Maconochie and Knight, 1989), which is especially useful for studying ligand-operated channels.

2. Inside-Out Patch Clamping

Forming inside-out patches involves some technical difficulties (see Rae and Levis, 1984), but, once achieved, precise control and rapid change of solutions contacting the cytoplasmic face of the patch is straightforward. Therefore, inside-out patches are ideal for defining mechanisms of action of intracellular molecules on channel function.

E. The Choice of Single-Channel Methods

As the previously discussions indicate, each single channel methodology has advantages and disadvantages. Which configuration to use depends on the question to be addressed, prior knowledge of the channel under study, and the hypotheses one makes about the control of channel function by intracellular processes. The following general rules apply to most experimental situations. (1) If it is essential that cytoplasmic integrity be preserved (e.g., a toxicant is thought to alter channel function by actions on intracellular metabolism, but at least some of the molecules involved are unknown or are not readily available to add to pipette or bath solutions), then cell-attached patches are the only choice. (2) If a toxicant is thought to act on the extracellular side of a channel, then outside-out patches are indicated. (3) To determine what specific intracellular molecules are involved in the control of channel function (i.e., where the hypothesis predicts effects of known, available molecules), inside-out patches are indicated.

F. Single Channel Studies in Neurotoxicology

Unfortunately, single channel methods have thus far seen little application to neurotoxicological studies. Some of these scarce studies are discussed in the chapter by van den Brec *et al.* in this series. A few studies that either directly focused on neurotoxicological questions or that produced results useful to neurotoxicology have been mentioned earlier and illustrate the types of information about toxicological mechanisms that can be obtained through these powerful techniques. For example, Alkondon and colleagues (1990) showed that inorganic lead reduces current flow through NMDA-type glutamate-sensitive ion channels by reducing the frequency of channel opening without reducing the mean open time or channel conductance. In contrast, heavy metals (e.g., Cd^{2+}:inorganic Pb^{2+} was not tested) reduce current flow through voltage-sensitive L-type calcium channels by reducing the mean open time with no effect on channel conductance or (probably) frequency of channel opening (Lansman *et al.*, 1986).

IV. Optical Techniques for Studying Ion Channel Function

The voltage clamp techniques described earlier, both whole cell and single channel, suffer from two major deficiencies. First, of necessity they allow the investigator to study one cell or a patch of one cell at a time and usually to study only one channel type at a time. In some studies, it may be desirable to investigate the electrical activity of many cells simultaneously. Second, it is not always obvious, from voltage clamp records, how the activity of a particular channel type might affect other aspects of neuronal physiology, e.g., spike frequency or intracellular ion concentrations. Optical recording methods, using dyes that are sensitive to voltage or to the presence of specific ions or other small molecules, may be used to address some facets of these two questions. Introductions to various optical recording methods and equipment may be found in Aikens and colleagues (1989), Mason and colleagues (1990), Takamatsu and Wier, (1990), Tsien and Harootunian (1990), and Wampler and co-workers (1989).

A major drawback to optical recordings is the normal requirement for thin, optically clear specimens. In most experimental setups, even a section only a few cells thick is either too opaque for efficient imaging or yields a confusing signal composed of simultaneous responses from several cells. Therefore, optical recording is usually performed on cultured or acutely isolated cells. There are two principal approaches that may be used in larger tissues, such as brain slices, but they both have their own disadvantages. First, the dye that is observed during recording may be injected into one or a few cells with micropipettes, so that these are the only cells capable of generating an optical response (e.g., Muller and Connor, 1991; Guthrie *et al.*, 1991). This has the obvious disadvantage of requiring impalement of small, often unseen neurons with micropipettes, without damaging the neurons. Second, "optical sectioning" may be used to examine only a limited plane of section of a thicker specimen, usually using confocal scanning microscopes (e.g., Hernandez-Cruz *et al.*, 1990; Williams, 1990). However, these are usually relatively slow procedures, with poor time resolution, and the required equipment is quite expensive.

A. Imaging of Voltage-Sensitive Dyes

A considerable number of molecules are now commercially available that in some way report the electrical potential of cells (for a survey of molecules and methods, see Freedman and Novak, 1989; Gross and Loew, 1989; Tsien, 1989a,b). These "voltage-sensitive" dyes, although chemically diverse, can be classified into two general categories: fast and slow. The fast dyes, such as the styryls (e.g., di-4-ANEPPS), usually incorporate themselves directly into the neuronal membrane and change absorbance or fluorescence when the electric field across the molecule changes. The best of these dyes change their optical properties on a millisecond time scale and can be used to detect individual action potentials or even, in particularly favorable situations, synaptic potentials. Suitably dense photosensor arrays, or individually positionable photosensors, can record electrical events in several, potentially dozens, of neurons simultaneously [for an elegant application of this approach in cultured vertebrate neurons, see Chien and Pine (1991)]. Fast voltage-sensitive dyes may prove suitable for studying changes in ongoing electrical activity, particularly in neuronal circuits established in culture, during toxicant exposure. However, to this author's knowledge they have not yet been used in this way.

The slow dyes, such as the oxonols, usually distribute themselves across the plasma membrane in a ratio that is a function of the transmembrane potential (Freedman and Novak, 1989; Tsien, 1989a,b). Their absorbance or fluorescence usually changes in the intracellular environment, although the cause is not always understood. Therefore, it is the distribution of the dye, and not its the optical properties, that changes with potential. The result is an optical signal that reports a time-weighted "average" transmembrane potential. Slow voltage-sensitive dyes (in conjunction with calcium-sensitive dyes) have been used by Oyama

and colleagues (1991) to show that triphenyl tin hyperpolarizes thymocytes, presumably by promoting Ca^{2+} influx and the subsequent opening of Ca^{2+}-dependent potassium channels.

B. Imaging of Ion-Sensitive Dyes

Since the early 1980s there has been a veritable explosion of information about the regulation of intracellular ion concentrations and modulation of cellular physiology and morphology by changes in intracellular ion concentrations (for brief reviews see Negulescu and Machen, 1990; Tsien, 1989a,b). Useful, reasonably selective ion-sensitive dyes have been developed for Ca^{2+}, Mg^{2+}, Na^+, K^+, and pH. When used with care, the dual emission and dual excitation dyes such as fura-2 (calcium sensitive), SNARF, SNAFL (both pH sensitive), and SBFI (sodium sensitive) allow reasonably accurate estimations of intracellular ion concentrations independent of dye concentration, illumination intensity, and cell path-length thickness. Most of these dyes can be easily loaded into cells as the membrane-permeable acetoxymethyl ester, which is subsequently deesterified by intracellular esterases into an ion-sensitive, membrane-impermeable form that remains trapped within the cells. Because intracellular ion concentrations are often strongly influenced by fluxes through plasma membrane ion channels, ion-sensitive dyes have been used to assess the intracellular effects of alterations of ion channel function, particularly when the dyes are used in conjunction with electrophysiological techniques. Methodologies, advantages, and pitfalls of calcium-sensitive dyes (probably the most commonly used ion-sensitive dyes) are discussed in a series of articles in *Cell Calcium* (Goldman *et al.*, 1990; Milani *et al.*, 1990; Moore *et al.*, 1990; Roe *et al.*, 1990; Ryan *et al.*, 1990; Williams and Fay, 1990) and in many other sources (e.g., Bright *et al.*, 1989; O'Rourke *et al.*, 1990). Anyone contemplating use of ion-sensitive dyes for the first time would be well advised to obtain literature from Molecular Probes (Eugene, OR), which supplies application information and a bibliography of hundreds of references, periodically updated. Turnkey systems for imaging fluorescent ion-sensitive dyes are available from several commercial sources, often coupled with standard image-analysis capabilities.

Ion-sensitive dyes have an enormous range of applications in neurobiology. Within the context of this review, ion-sensitive dyes also may be used to supplement electrophysiological methods in addressing questions of ion channel function and the role of ion channels in neuronal physiology. Some of these applications may be illustrated by studies concerned with the roles of voltage-sensitive calcium channels and the regulation of intracellular calcium. Most, if not all, neurons possess several distinctly different types of voltage-sensitive calcium channels, which differ in single channel conductances, voltages of activation and inactivation, kinetics, and pharmacology. In almost all cases, the overall numbers and subcellular distribution of the different channel types in any given type of neuron are unknown. Therefore, it is not obvious from voltage clamp records which channel types might admit physiologically significant numbers of calcium ions into which parts of the neuron. Further, various physiological stimuli cause the release of calcium from intracellular stores. Finally, it is not necessarily true that large calcium influxes and/or large changes in intracellular calcium are required to trigger important physiological processes. Calcium-sensitive dyes can be valuable in addressing these issues. For example, Suzuki and coworkers (1990) found that T-type, but not L-type, calcium channels were most important in producing a sustained increase in intracellular calcium in GH_3 pituitary cells in response to thyrotropin-releasing hormone stimulation. Silver and colleagues (1990) showed that L-type, but not T-type, calcium channels admit significant amounts of calcium into growth cones of N1E-115 neuroblastoma cells and trigger morphological changes of the growth cones. As a third example, Thayer and Miller (1990) combined voltage clamping with calcium-sensitive dyes to investigate the regulation of intracellular calcium concentrations in rat dorsal root ganglion cells by organelles and plasma membrane transport systems in response to calcium influx through voltage-sensitive calcium channels.

Ion-sensitive dyes have found a niche in neurotoxicological studies. For example, intracellular calcium concentrations increase in synaptosomes in response to acute triethyl lead exposure (Komulainen and Bondy, 1987), but thus far ion-sensitive dyes have not been used in combination with electrophysiological techniques in neurotoxicology. However, in the future, these combined techniques may prove very valuable in the precise definition of mechanisms of neurotoxicant action.

V. Future Directions

New techniques, or new applications of existing techniques, continue to be developed, but these are difficult to predict. However, one area that is receiving increasing attention in the basic neuroscience community, and that should prove important in neurotoxicological studies in the future, is the interface between molecular biology and electrophysiology. Two avenues

of investigation that should prove fruitful are briefly summarized here.

A. Amino Acid Sequence and Detailed Structure of Ion Channels

A number of ion channels, both voltage-sensitive and ligand-regulated, have been cloned and their amino acid sequence deduced, including a number of voltage-sensitive sodium, potassium, and calcium channels, and receptor/channels for acetylcholine, glutamate, and GABA (e.g., Mori *et al.*, 1991; Noda *et al.*, 1983, 1984; Tanabe *et al.*, 1987). In many cases, there are families of ion channels of a given type, with different combinations of subunits joining to produce channels that are superficially similar (e.g., that all conduct sodium ions) but that differ in important details (e.g., kinetics of activation, sensitivity to blocking agents, permeability). Many of these cloned channels have been expressed in *Xenopus* oocytes or other cell types and have been analyzed electrophysiologically (e.g., Lacerda *et al.*, 1991; Mori *et al.*, 1991; Varadi *et al.*, 1991). Finally, altered forms of channel proteins or altered combinations of channel subunits have been used to analyze the roles of specific parts of the channel proteins in channel function, for example, in regulating kinetics of channel activation and inactivation (Lacerda *et al.*, 1991; Varadi *et al.*, 1991) or in regulating phosphorylation by protein kinase C (West *et al.*, 1991).

Many ion channels show extensive sequence homology with one another (Catterall, 1988). As more information is gathered about channel structure and function, it is becoming possible to recognize motifs in amino acid sequences that generate specific aspects of channel function and to predict the physiology of newly cloned channel proteins. Therefore, it may become possible in the future to predict the effects of some neurotoxicants on ion channel function by examining the amino acid sequences of channels and comparing them to channels that have been well characterized.

In the more immediate future, cloned channels expressed in oocytes, lipid bilayers, or other systems may prove valuable in studying mechanisms of action of neurotoxicants on ion channels. Although normal neurons will always remain the key to investigations of neurotoxicant actions, the multiplicity of channels in normal neurons, including multiple types of the "same" channel, may complicate electrophysiological analyses. Cloned channels can be studied more nearly in isolation from physiologically similar channels, which should assist in defining targets and mechanisms of neurotoxicant action more precisely. With the information from cloned channels in hand, investigators may then be able to return to specific neuronal types in the brain and predict toxicant effects, and perhaps design experiments that would yield less ambiguous results than may now be possible. Further, if molecular biological techniques can determine which specific molecular forms of ion channels exist in different brain regions, then results from cloned channels may allow prediction of target area sensitivity to specific neurotoxicants.

B. Analysis of Electrophysiology and Gene Expression in Single Neurons

Many physiological stimuli, and presumably many neurotoxicants as well, alter gene expression in the nervous system. Until recently, regulation of gene expression could be studied only in cultured neurons or in pieces of the brain. However, neither cultures nor even very small pieces of the brain consist of truly homogeneous populations of neurons. If a stimulus or toxicant alters gene expression in a small percentage of the neuronal population, this may be physiologically significant but difficult to analyze. Even *in situ* hybridization, which in principle allows one to examine gene expression in single neurons, may not yield the full spectrum of RNA species and, in any case, it is usually impossible to determine what the physiological response of the cell was before it was fixed and analyzed. Recently, these obstacles have been overcome by simultaneous application of whole cell patch clamping and molecular biological techniques (Eberwine *et al.*, 1992).

This procedure takes advantage of the exchange of material between the cytoplasm and pipette during whole cell clamping. Oligo(dT) primer, nucleotides, and reverse transcriptase are included in a patch pipette. After establishing the whole cell voltage clamp configuration, any mRNA present in the cytoplasm will encounter the reverse transcription reagents diffusing into the cell from the pipette, and cDNA will be synthesized using the mRNA as a template. The cell may be used for standard voltage clamp procedures, if desired, to analyze the electrophysiological responses of the cell to the experimental procedure (e.g., toxicant exposure). After up to 30 min of recording or simple incubation, the cell contents are sucked up into the pipette. The pipette solution is then subjected to fairly standard molecular biological techniques for second-strand DNA synthesis, amplification, and preparation of probes or clones.

This combination of electrophysiology and molecular biology would allow an investigator to assess the electrophysiological effects of a neurotoxicant in a given neuron and then determine the effects of that

toxicant on gene expression in the same cell. This opens the door for highly integrated studies: How do cells in a population differ in their responses to a neurotoxicant, at the level of gene expression? At least for some substances, the electrophysiological responses to a toxicant or other stimuli may cause changes in gene expression (e.g., stimulation of c-*fos* transcription by depolarization; Murphy *et al.*, 1991; Sheng *et al.*, 1990). Do pharmacological or other manipulations that block the electrophysiological effects of the toxicant (as actually measured in the specific cell under study) alter the toxicant's effects on gene expression in that cell? One can include substances in the pipette solution that alter intracellular metabolism in known ways. Do such manipulations alter the genetic response to the toxicant? If so, this would assist in elucidating pathways from toxicant exposure to effects on gene expression. Although these procedures have not yet been used in neurotoxicological experiments, they promise a rich harvest of mechanistic explanations of neurotoxicant action in the future.

At this time, detailed methodological procedures are not available; however, Axon Instruments markets an antisense RNA kit, and should be consulted by anyone wishing to begin using this technique.

Acknowledgments

Preparation of this manuscript was supported in part by a grant from the National Institute of Environmental Health Sciences. I thank Dr. Teresa Audesirk for critical reading of the manuscript.

References

Aikens, R. S., Agard, D. A., and Sedat, J. W. (1989). Solid-state imagers for microscopy. In *Fluorescent Microscopy of Living Cells in Culture* (Y.-L. Wang and D. L. Taylor, Eds.), Part A, pp. 292–314, Academic Press, New York.

Aldrich, R. W., and Yellen, G. (1983). Analysis of nonstationary channel kinetics. In *Single-channel Recording* (B. Sakmann and E. Neher, Eds.), pp. 287–300, Plenum Press, New York.

Alkondon, M., Costa, A. C. S., Radhakrishnan, V., Aronstam, R. S., and Albuquerque, E. X. (1990). Selective blockade of NMDA-activated channel currents may be implicated in learning deficits caused by lead. *FEBS Lett.* **261**, 124–130.

Armstrong, D. L. (1988). Calcium channel regulation by protein phosphorylation in a mammalian tumor cell line. *Biomed. Res.* 9(2), 11–15.

Armstrong, D., and Eckert, R. (1987). Voltage-activated calcium channels that must be phosphorylated to respond to membrane depolarization. *Proc. Natl. Acad. Sci. U.S.A.* **84**, 2518–2522.

Audesirk, G. (1989). Effects of heavy metals on neuronal calcium channels. In *Biological Effects of Heavy Metals* E. C. Foulkes, Ed. Vol. 1, pp. 1–17, CRC Press, Boca Raton, Florida.

Audesirk, G. (1993). Electrophysiology of lead intoxication: effects on voltage-sensitive ion channels. *Neurotoxicology* **14**, 137–147.

Audesirk, G., and Audesirk, T. (1993). The effects of inorganic lead on voltage-sensitive calcium channels differ among cell types and among channel subtypes. *Neurotoxicology* **14**, 259–265.

Auerbach, A., and Sachs, F. (1985). High-resolution patch-clamp techniques. In *Voltage and Patch Clamping with Microelectrodes* (T. G. Smith, Jr., H. Lecar, S. J. Redman, and P. W. Gage, Eds.), pp. 121–149, American Physiological Society, Bethesda, Maryland.

Blanton, M. G., Loturco, J. J., and Kriegstein, A. R. (1989). Whole cell recording from neurons in slices of reptilian and mammalian cerebral cortex. *J. Neurosci. Methods* **30,** 203–210.

Bright, G. R., Fisher, G. W., Rogowska, J., and Taylor, D. L. (1989). Fluorescence ratio imaging microscopy. In *Fluorescent Microscopy of Living Cells in Culture* (Y.-L. Wang, and D. L. Taylor, Eds.) Part B, pp. 157–192, Academic Press, New York.

Brown, A. M., Wilson, D. L., and Tsuda, Y. (1985). Voltage clamp and internal perfusion with suction-pipette method. In *Voltage and Patch Clamping with Microelectrodes* (T. G. Smith, Jr., H. Lecar, S. J. Redman, and P. W. Gage, Eds.), pp. 151–169, American Physiological Society, Bethesda, Maryland.

Brown, K. T., and Flaming, D. G. (1986). *Advanced Micropipette Techniques for Cell Physiology,* Wiley, New York.

Byerly, L., and Hagiwara, S. (1982). Calcium currents in internally perfused nerve cell bodies of *Limnaea stagnalis. J. Physiol.* **322,** 503–528.

Catterall, W. A. (1988). Structure and function of voltage-sensitive ion channels. *Science* **242,** 50–61.

Chad, J., Kalman, D., and Armstrong, D. (1987). The role of cyclic AMP-dependent phosphorylation in the maintenance and modulation of voltage-activated calcium channels. In *Cell Calcium and the Control of Membrane Transport* (D. C. Eaton, and L. J. Mandel, Eds.), pp. 167–186, Rockefeller University Press, New York.

Chien, C.-B., and Pine, J. (1991). Voltage-sensitive dye recording of action potentials and synaptic potentials from sympathetic microcultures. *Biophys. J.* **60,** 697–711.

Colquhoun, D., and Hawkes, A. G. (1983). The principles of the stochastic interpretation of ion-channel mechanisms. In *Single-channel Recording* (B. Sakmann and E. Neher, Eds.), pp. 135–176, Plenum Press, New York.

Colquhoun, D., and Sigworth, F. J. (1983). Fitting and statistical analysis of single-channel records. In *Single-channel Recording* (B. Sakmann, and E. Neher, Eds.), pp. 191–264, Plenum Press, New York.

Connor, J. A., and Stevens, C. F. (1971). Voltage clamp studies of a transient outward membrane current in gastropod neural somata. *J. Physiol.* **213,** 21–30.

Corey, D. P., and Stevens, C. F. (1983). Science and technology of patch-recording electrodes. In *Single-channel Recording* (B. Sakmann and E. Neher, Eds.), pp. 53–68. Plenum Press, New York.

Cull-Candy, S. G., Miledi, R., and Parker, I. (1980). Single glutamate-activated channels recorded from locust muscle fibers with perfused patch-clamp electrodes. *J. Physiol.* **321,** 195–210.

DeFelice, L. J., and Clay, J. R. (1983). Membrane current and membrane potential from single-channel kinetics. In *Single-channel Recording* (B. Sakmann, and E. Neher, Eds.), pp. 323–344, Plenum Press, New York.

Eberwine, H., Yeh, H., Miyashiro, K., Cao, Y., Nair, S., Finnell, R., Zettel, M., and Coleman, P. (1992). Analysis of gene expression in single live neurons. *Proc. Natl. Acad. Sci. U.S.A.* **89,** 3010–3014.

Edwards, F. A., Konnerth, A., Sakmann, B., and Takahashi, T. (1989). A thin slice preparation for patch clamp recordings from neurones of the mammalian central nervous system. *Pfluegers Arch.* **414,** 600–612.

Falke, L. C., Gillis, K. D., Pressel, D. M., and Misler, S. (1989). Perforated patch recording allows long-term monitoring of metabolite-induced electrical activity and voltage-dependent Ca^{2+} currents in pancreatic islet B-cells. *FEBS Lett.* **251,** 167–172.

Fenwick, E. M., Marty, A., and Neher, E. (1982). Sodium and calcium channels in bovine chromaffin cells. *J. Physiol.* **331,** 599–635.

Finkel, A. S. (1985). Useful circuits for voltage clamping with microelectrodes. In *Voltage and Patch Clamping with Microelectrodes* (T. G. Smith, Jr., H. Lecar, S. J. Redman, and P. W. Gage, Eds.) pp. 9–24, American Physiological Society, Bethesda, Maryland.

Finkel, A. S., and Gage, P. W. (1985). Conventional voltage clamping with two intracellular microelectrodes. In *Voltage and Patch Clamping with Microelectrodes* (T. G. Smith, Jr., H. Lecar, S. J. Redman, and P. W. Gage, Eds.) pp. 47–94, American Physiological Society, Bethesda, Maryland.

Finkel, A. S., and Redman, S. J. (1984). Theory and operation of a single microelectrode voltage clamp. *J. Neurosci. Methods* **11,** 101–127.

Finkel, A. S., and Redman, S. J. (1985). Optimal voltage clamping with single microelectrode. In *Voltage and Patch Clamping with Microelectrodes* (T. G. Smith, Jr., H. Lecar, S. J. Redman, and P. W. Gage, Eds.) pp. 95–120, American Physiological Society, Bethesda, Maryland.

Fisher, R. E., Gray, R., and Johnston, D. (1990). Properties and distribution of single voltage-gated calcium channels in adult hippocampal neurons. *J. Neurophysiol.* **64,** 91–104.

Freedman, J. C., and Novak, T. S. (1989). Optical measurement of membrane potential in cells, organelles, and vesicles. *Methods Enzymol.* **172,** 102–122.

Gahwiler, B. H., and Brown, D. A. (1987). Effects of dihydropyridines on calcium currents in CA3 pyramidal cells in slice cultures of rat hippocampus. *Neuroscience* **20,** 731–738.

Goldman, W. F., Bova, S., and Blaustein, M. P. (1990). Measurement of intracellular Ca^{2+} in cultured arterial smooth muscle cells using fura-2 and digital imaging microscopy. *Cell Calcium* **11,** 221–231.

Gross, D., and Loew, L. M. (1989). Fluorescent indicators of membrane potential: microspectrofluorometry and imaging. In *Fluorescent Microscopy of Living Cells in Culture* (Y.-L. Wang, and D. L. Taylor, Eds.) pp. 193–219, Part B, Academic Press, New York.

Guthrie, P. B., Segal, M., and Kater, S. B. (1991). Independent regulation of calcium revealed by imaging dendritic spines. *Nature* **354,** 76–80.

Hamill, O. P., Marty, A., Neher, E., Sakmann, B., and Sigworth, F. J. (1981). Improved patch-clamp techniques for high-resolution current recording from cells and cell-free membrane patches. *Pfluegers Arch.* **391,** 85–100.

Hartzell, H. C., Mery, P.-F., Fischmeister, R., and Szabo, G. (1991). Sympathetic regulation of cardiac calcium current is due exclusively to cAMP-dependent phosphorylation. *Nature* **351,** 573–576.

Hernandez-Cruz, A., Sala, F., and Adams, P. R. (1990). Subcellular calcium transients visualized by confocal microscopy in a voltage-clamped vertebrate neuron. *Science* **247,** 858–862.

Horn, R., and Marty, A. (1988). Muscarinic activation of ionic currents measured by a new whole-cell recording method. *J. Gen. Physiol.* **92,** 145–159.

Kalman, D., O'Lague, P. H., Erxleben, C., and Armstrong, D. L. (1988). Calcium-dependent inactivation of the dihydropyridine-sensitive calcium channels in GH_3 cells. *J. Gen. Physiol.* **92,** 531–548.

Kater, S. B., Mattson, M. P., Cohan, C., and Connor, J. (1988). Calcium regulation of the neuronal growth cone. *Trends Neurosci.* **11,** 315–321.

Klee, C. B., Draetta, G. F., and Hubbard, M. J. (1988). Calcineurin. *Adv. Enzymol.* **61**, 149–200.

Kohr, G., and Mody, I. (1991). Endogenous intracellular calcium buffering and the activation/inactivation of HVA calcium currents in rat dentate gyrus granule cells. *J. Gen. Physiol.* **98**, 941–967.

Komulainen, H., and Bondy, S. C. (1987). Increased free intrasynaptosomal Ca^{2+} by neurotoxic organometals: distinctive mechanisms. *Toxicol. Appl. Pharmacol.* **88**, 77–86.

Korn, S. J., and Horn, R. (1989). Influence of sodium-calcium exchange on calcium current rundown and the duration of calcium-dependent chloride currents in pituitary cells, studied with whole cell and perforated patch recording. *J. Gen. Physiol.* **94**, 789–812.

Korn, S. J., Marty, A., Connor, J. A., and Horn, R. (1991). Perforated patch recording. *Methods Neurosci.* **4**, 264–373.

Kostyuk, P. G., Veselovsky, N. S., Fedulova, S. A. (1981). Ionic currents in the somatic membrane of rat dorsal root ganglion neurons. II. Calcium currents. *Neuroscience* **7**, 2431–2437.

Kostyuk, P. G., Shuba, Y. M., and Savchenko, A. N. (1988). Three types of calcium channels in the membrane of mouse sensory neurons. *Pfluegers Arch.* **411**, 661–669.

Kunze, D. L., and Ritchie, A. K. (1990). Multiple conductance levels of the dihydropyridine-sensitive calcium channel in GH_3 cells. *J. Membr. Biol.* **118**, 171–178.

Lacerda, A. E., Kim, H. S., Ruth, P., Perez-Reyes, E., Flockerzi, V., Hofmann, F., Birnbaumer, L., and Brown, A. M. (1991). Normalization of current kinetics by interaction between the α_1 and B subunits of the skeletal muscle dihydropyridine-sensitive Ca^{2+} channel. *Nature* **352**, 527–530.

Lansman, J. B., Hess, P., and Tsien, R. W. (1986). Blockade of current through single calcium channels by Cd^{2+}, Mg^{2+}, and Ca^{2+}. Voltage and concentration dependence of calcium entry into the pore. *J. Gen. Physiol.* **88**, 321–347.

Lauger, P. (1983). Conformational transitions of ionic channels. In *Single-channel Recording* (B. Sakmann, and E. Neher, Eds.) pp. 177–190, Plenum Press, New York.

Levitan, I. B. (1985). Phosphorylation of ion channels. *J. Membr. Biol.* **87**, 177–190.

Lipscombe, D., Madison, D. V, Poenie, M., Reuter, H., Tsien, R. Y., and Tsien, R. W. (1988). Spatial distribution of calcium channels and cytosolic calcium transients in growth cones and cell bodies of sympathetic neurons. *Proc. Natl. Acad. Sci. U.S.A.* **85**, 2398–2402.

Lund, A. E., and Narahashi, T. (1981). Modification of sodium channel kinetics by the insecticide tetramethrin in crayfish giant axons. *Neurotoxicology* **2**, 213–229.

Lund, A. E., and Narahashi, T. (1982). Dose-dependent interaction of the pyrethroid isomers with sodium channels of squid axon membranes. *Neurotoxicology* **3**, 11–24.

Maconochie, D. J., and Knight, D. E. (1989). A method for making solution changes in the sub-millisecond range at the tip of a patch pipette. *Pfluegers Arch.* **414**, 589–596.

Marty, A., and Neher, E. (1983). Tight-seal whole-cell recording. In *Single-channel Recording* (B. Sakmann, and E. Neher, Eds.) pp. 107–122, Plenum Press, New York.

Mason, W. T., Hoyland, J., Rawlings, S. R., and Relf, G. T. (1990). Techniques and technology for dynamic video imaging of cellular fluorescence. In *Quantitative and Qualitative Microscopy* (P. M. Conn, Ed.) pp. 109–135, Academic Press, New York.

Means, A. R., Tash, J. S., and Chafouleas, J. G. (1982). Physiological implications of the presence, distribution, and regulation of calmodulin in eukaryotic cells. *Physiol. Rev.* **62**, 1–39.

Merickel, M. (1980). Design of a single electrode voltage clamp. *J. Neurosci. Methods* **2**, 87–96.

Milani, D., Malgaroli, A., Guidolin, D., Fasolato, C., Skaper, S. D., Meldolesi, J., and Pozzan, T. (1990). Ca^{2+} channels and intracellular Ca^{2+} stores in neuronal and neuroendocrine cells. *Cell Calcium* **11**, 191–199.

Moore, E. D. W., Becker, P. L., Fogarty, K. E., Williams, D. A., and Fay, F. S. (1990). Ca^{2+} imaging in single living cells: theoretical and practical issues. *Cell Calcium* **11**, 157–179.

Morgan, J. I., and Curran, T. (1988). Calcium as a modulator of the immediate-early gene cascade in neurons. *Cell Calcium* **9**, 303–311.

Mori, Y., Friedrich, T., Kim, M.-S., Mikami, A., Nakai, J., Ruth, P., Bosse, E., Hofmann, F., Flockerzi, V., Furuichi, T., Mikoshiba, K., Imoto, K., Tanabe, T., and Numa, S. (1991). Primary structure and functional expression from complementary DNA of a brain calcium channel. *Nature* **350**, 398–402.

Muller, W., and Connor, J. A. (1991). Dendritic spines as individual neuronal compartments for synaptic Ca^{2+} responses. *Nature* **354**, 73–76.

Murphy, T. H., Worley, P. F., Nakabeppu, Y., Christy, B., Gastel, J., and Baraban, J. M. (1991). Synaptic regulation of immediate early gene expression in primary cultures of cortical neurons. *J. Neurochem.* **57**, 1862–1872.

Negulescu, P. A., and Machen, T. E. (1990). Intracellular ion activities and membrane transport in parietal cells measured with fluorescent dyes. *Methods Enzymol.* **192**, 38–81.

Noda, M., Furutani, Y., Takahashi, H., Toyosata, M., Tanabe, T., Shimizu, S., Kikyotani, S., Kayano, T., Hirose, T., Inayama, S., and Numa, S. (1983). Cloning and sequence analysis of calf cDNA and human genomic DNA encoding α-subunit precursor of muscle acetylcholine receptor. *Nature* **305**, 818–823.

Noda, M., Shimizu, S., Tanabe, T., Takai, T., Kanano, T., Ikeda, T., Takahashi, H., Nakayama, H., Kanaoka, Y., Minamino, N., Kangawa, K., Matsuo, H., Raftery, M. A., Hirose, T., Inayama, S., Hayashida, H., Miyata, T., and Numa, S. (1984). Primary structure of *Electrophorus electricus* sodium channel deduced from cDNA sequence. *Nature* **312**, 121–127.

Nowycky, M. C., Fox, A. P., and Tsien, R. W. (1985). Long-opening mode of gating of neuronal calcium channels and its promotion by the dihydropyridine calcium agonist Bay K 8644. *Proc. Natl. Acad. Sci. U.S.A.* **82**, 2178–2182.

Nonoki, K., Florio, V., and Catterall, W. A. (1989). Activation of purified calcium channels by stoichiometric protein phosphorylation. *Proc. Natl. Acad. Sci. U.S.A.* **86**, 6816–6820.

O'Dell, T. J., and Alger, B. E. (1991). Single calcium channels in rat and guinea-pig hippocampal neurons. *J. Physiol.* **436**, 739–767.

O'Rourke, B., Reibel, D. K., and Thomas, A. P. (1990). High-speed digital imaging of cytosolic Ca^{2+} and contraction in single cardiomyocytes. *Am. J. Physiol.* **259**, H230–H242.

Oyama, Y., Chikahisa, L., Tomiyoshi, F., and Hayaashi, H. (1991). Cytotoxic action of triphenyltin on mouse thymocytes: a flow-cytometric study using fluorescent dyes for membrane potential and intracellular Ca^{2+}. *Jpn J. Pharmacol.* **57**, 419–424.

Pietrobon, D., and Hess, P. (1990). Novel methanism of voltage-dependent gating in L-type calcium channels. *Nature* **346**, 651–655.

Plummer, M. R., Logothetis, D. E., and Hess, P. (1989). Elementary properties and pharmacological sensitivities of calcium channels in mammalian peripheral neurons. *Neuron* **2**, 1453–1463.

Quandt, F. N., Yeh, J. Z., and Narahashi, T. (1985). All or none block of single Na^+ channels by tetrodotoxin. *Neurosci. Lett.* **54**, 77–83.

Rae, J., Cooper, K., Gates, G., and Watsky, M. (1991). Low access resistance perforated patch recordings using amphotericin B. *J. Neurosci. Methods* **37**, 15–26.

Rae, J. L., and Levis, R. A. (1984). Patch voltage clamp of lens epithelial cells: theory and practice. *Mol. Physiol.* **6**, 115–161.

Roe, M. W., Lemasters, J. J., and Herman, B. (1990). Assessment of fura-2 for measurements of cytosolic free calcium. *Cell Calcium* **11**, 63–73.

Ryan, T. A., Millard, P. J., and Webb, W. W. (1990). Imaging $[Ca^{2+}]_i$ dynamics during signal transduction. *Cell Calcium* **11**, 145–155.

Sachs, F. (1983). Automated analysis of single-channel records. In *Single-channel Recording* (B. Sakmann, and E. Neher, Eds.) pp. 265–286, Plenum Press, New York.

Sachs, F. (1985). Microelectrode shielding. In *Voltage and Patch Clamping with Microelectrodes* (T. G. Smith, Jr., H. Lecar, S. J. Redman, and P. W. Gage, Eds.) pp. 25–46, American Physiological Society, Bethesda, Maryland.

Sachs, F., and Auerbach, A. (1983). Single channel electrophysiology: use of the patch clamp. *Methods Enzymol.* **103**, 147–176.

Sakmann, B., Edwards, F., Konnerth, A., and Takahashi, T. (1989). Patch clamp techniques used for studying synaptic transmission in slices of mammalian brain. *Q. J. Exp. Physiol.* **74**, 1107–1118.

Sakmann, B., and Neher, E. (1983). *Single-channel Recording.* Plenum Press, New York.

Sakmann, B., and Neher, E. (1983). Geometric parameters of pipettes and membrane patches. In *Single-channel Recording* (B. Sakmann, and E. Neher, Eds.) pp. 37–52, Plenum Press, New York.

Sigworth, F. (1983). An example of analysis. In *Single-channel Recording* (B. Sakmann, and E. Neher, Eds.) pp. 301–322, Plenum Press, New York.

Sheng, M., McFadden, G., and Greenberg, M. E. (1990). Membrane depolarization and calcium induce c-*fos* transcription via phosphorylation of transcription factor CREB. *Neuron* **4**, 571–582.

Sherman-Gold, R. (1993). *The Axon Guide.* Axon Instruments, Foster City, California.

Sigworth, F. J. (1983). Electronic design of the patch clamp. In *Single-channel Recording* (B. Sakmann, and E. Neher, Eds.) pp. 3–36, Plenum Press, New York.

Silver, R. A., Lamb, A. G., and Bolsover, S. R. (1990). Calcium hotspots caused by L-channel clustering promote morphological changes in neuronal growth cones. *Nature* **343**, 751–754.

Smith, T. G., Jr., and Lecar, H. (1985). Voltage clamping small cells. In *Voltage and Patch Clamping with Microelectrodes* (T. G. Smith, Jr., H. Lecar, S. J. Redman, and P. W. Gage, Eds.) pp. 221–256. American Physiological Society, Bethesda, Maryland.

Smith, T. G., Jr., Lecar, H., Redman, S. J., and Gage, P. W. (Eds.) (1985). *Voltage and Patch Clamping with Microelectrodes,* American Physiological Society, Bethesda, Maryland.

Stoclet, J.-C., Gerard, D., Kilhoffer, M.-C., Lugnier, C., Miller, R., and Schaeffer, P. (1987). Calmodulin and its role in intracellular calcium regulation. *Prog. Neurobiol.* **29**, 321–364.

Suzuki, N., Kudo, Y., Takagi, H., Yoshioka, T., Tanakadate, A., and Kano, M. (1990). Participation of transient-type Ca^{2+} channels in the sustained increase of Ca^{2+} level in GH_3 cells. *J. Cell. Physiol.* **144**, 62–68.

Takamatsu, T., and Wier, W. G. (1990). High temporal resolution video imaging of intracellular calcium. *Cell Calcium* **11**, 111–120.

Tanabe, T., Takeshima, H., Mikami, A., Flockerzi, V., Takahashi, H., Kangawa, K., Kojima, M., Matsuo, H., Hirose, T., and Numa, S. (1987). Primary structure of the receptor for calcium channel blockers from skeletal muscle. *Nature* **328**, 313–318.

Thayer, S. A., and Miller, R. J. (1990). Regulation of the intracellular free calcium concentration in single rat dorsal root ganglion neurones *in vitro. J. Physiol.* **425**, 85–115.

Tsien, R. Y. (1989a). Fluorescent probes of cell signalling. *Ann. Rev. Neurosci.* **12**, 227–253.

Tsien, R. Y. (1989b). Fluorescent indicators of ion concentrations. In *Fluorescent Microscopy of Living Cells in Culture* (Y.-L. Wang, and D. L. Taylor, Eds.) Part B. pp. 127–156, Academic Press, New York.

Tsien, R. Y., and Harootunian, A. T. (1990). Practical design criteria for a dynamic ratio imaging system. *Cell Calcium* **11**, 93–109.

Varadi, G., Lory, P., Schultz, D., Varadi, M., and Schwartz, A. (1991). Acceleration of activation and inactivation by the B subunit of the skeletal muscle calcium channel. *Nature* **352**, 159–162.

Vijverberg, H. P. M., and de Weille, J. R. (1985). The interaction of pyrethroids with voltage-dependent Na channels. *Neurotoxicology* **6**, 23–34.

Wampler, J. E., and Klutz, K. (1989). Quantitative fluorescence microscopy using photomultiplier tubes and imaging detectors. In *Fluorescent Microscopy of Living Cells in Culture* (Y.-L. Wang, and D. L. Taylor, Eds.) Part A, pp. 239–269, Academic Press, New York.

West, J. W., Numann, R., Murphy, B. J., Scheuer, T., and Catterall, W. A. (1991). A phosphorylation site in the Na^+ channel required for modulation by protein kinase C. *Science* **254**, 866–868.

Williams, D. A. (1990). Quantitative intracellular calcium imaging with laser-scanning confocal microscopy. *Cell Calcium* **11**, 589–597.

Williams, D. A., and Fay, F. S. (1990). Intracellular calibration of the fluorescent calcium indicator fura-2. *Cell Calcium* **11**, 75–83.

Wilson, W. A., and Goldner, M. M. (1975). Voltage clamping with a single microelectrode. *J. Neurobiol.* **6**, 411–432.

Electrophysiological Methods for Analysis of Effects of Neurotoxicants on Synaptic Transmission

TIMOTHY J. SHAFER
Neurotoxicology Division
Health Effects Research Laboratories
U.S. Environmental Protection Agency
Research Triangle Park, North Carolina 27711

WILLIAM D. ATCHISON
Department of Pharmacology and Toxicology
Neurosciences Program and Institute for Environmental Toxicology
Michigan State University
East Lansing, Michigan 48824-1317

I. Introduction

Transmission of information from one cell to another in the nervous system is of primary importance for this system to receive, process, and respond to the external environment, control or modulate homeostatic functions, and engage in cognitive function including learning and memory. Chemical synaptic transmission involves the conversion of an electrical signal into a chemical signal in the presynaptic neuron, release of the chemical signal into the synapse, and generation of a response in the postsynaptic cell following action of the chemical signal on receptors on the postsynaptic cell. Because there are a number of critical steps in this process, disruption of any step by chemicals or toxins can result in drastic alterations of synaptic transmission which range from overaction (for example, cholinesterase inhibitors) to total failure (block by heavy metals).

The mammalian nervous system has evolved to respond rapidly to its environment. As such, at so-called "fast synapses," such as those at the neuromuscular junction, synaptic transmission occurs within 4–5 msec, including termination of the signal. In addition, the presynaptic terminal of most vertebrates is extremely small in size. The rapidity of synaptic transmission and the small size of the nerve terminal at most synapses present special problems for those who desire to study this process; the range of techniques that respond with sufficient rapidity to measure changes in synaptic transmission on a real-time basis *and* can quantitatively measure changes in function at the level of a single synapse are limited. Although biochemical, anatomical, and behavioral measures all provide useful

This manuscript has been reviewed by the Health Effects Research Laboratory, U.S. Environmental Protection Agency, and is approved for publication. Mention of commercial products or trade names does not constitute endorsement or recommendation for use.

157

information about neurophysiology to the toxicologist, electrophysiological measurements have proven especially useful for quantitative, real-time measurement of synaptic function. Although the small size of the presynaptic terminal still presents difficulties in making direct measurements at this site, improvements in recording techniques during the last 15 years have increased the reliability and sensitivity of measurements made even at the nerve terminal.

The electrical excitability of neurons and muscle cells was recognized by pioneers in the field of neurophysiology as a property that could be exploited to measure neuronal function accurately. In the late 1930s, Curtis and Cole used electrophysiological measurements to examine the nature of the action potential in squid giant axon, and suggested that the change in potential was driven by changes in the conductance of the membrane (Cole and Curtis, 1938, 1939). The development of the voltage clamp technique by Marmont (1949) allowed major leaps in understanding of action potential propagation and synaptic transmission. The ionic nature of the action potential was described by Hodgkin and colleagues (1949, 1952) using voltage clamp techniques. Many of the current techniques in electrophysiology make use of voltage clamp in some form. Voltage clamp techniques and theory have been the subject of several excellent reviews and book chapters (Smith *et al.*, 1985; Hille, 1993), including the previous chapter in this volume (Audesirk, 1995). Interested readers should refer to those sources for more detailed discussions of this technique. In addition to traditional methods of voltage clamp recording, intracellular microelectrode recording, iontophoresis, and patch clamp recording have all been utilized to study synaptic transmission. Most of these techniques provide quantitative information, allow real-time analysis, and, in some cases, serve as a continuous bioassay of nerve function.

As noted earlier, disruption of any step in the process of synaptic transmission can have drastic consequences. For the neurotoxicologist to determine the mechanism by which a compound disrupts synaptic transmission, it is crucial to have a thorough understanding of the sequence of events that occur at the synaptic cleft during normal transmission. Figure 1 presents the sequence of events that occur at an idealized synapse. The process begins with the invasion of an action potential into the presynaptic terminal, resulting in activation of clusters of voltage-sensitive Ca^{2+} channels located very near the release sites or "active zone" of the nerve terminal (Pumplin *et al.*, 1981; Robitaille *et al.*, 1990; Cohen *et al.*, 1991). Opening of the Ca^{2+} channels allows Ca^{2+} to move down its electrochemical gradient into the nerve terminal,

resulting in dramatic increases in the $[Ca^{2+}]_i$ in the immediate vicinity of the Ca^{2+} channel clusters (Augustine *et al.*, 1987, 1991; Smith and Augustine, 1988). Ca^{2+} binds to a protein or proteins (see the following discussion) which triggers prepackaged neurotransmitter vesicles positioned close to the terminal membrane to fuse with the membrane and release their contents into the synaptic cleft. The transmitter diffuses across the cleft and interacts with receptors located on the postsynaptic cell to initiate an electrical or biochemical response. Depending on the receptors present, the electrical excitability of the postsynaptic cell may be increased or decreased by this action. Changes in excitability may be an immediate response if the neurotransmitter activates a so-called "receptor-gated" ion channel (for example, nicotinic acetylcholine, glutamate, GABA receptors), resulting in current flow across the membrane. Alternatively, the response may develop more slowly if the receptor triggers the release of intracellular second messengers (for example, muscarinic acetylcholine, noradrenergic receptors) which modulate the activity of voltage-sensitive ion channels in the membrane. Persistent action of the neurotransmitter is prevented by rapid elimination of the transmitter from the synaptic cleft by enzymatic inactivation or by reuptake into the presynaptic terminal or surrounding glial cells.

Advances in the understanding of the proteins involved in triggering neurotransmitter release and the Ca^{2+} channels which mediate Ca^{2+} entry into the nerve terminal deserve discussion. The type(s) of Ca^{2+} channel responsible for Ca^{2+} entry required for neurotransmitter release is a subject of intense research (see later). In addition, the Ca^{2+}-binding protein(s) that serves as the "trigger" for vesicle fusion has not been characterized fully to date. As with Ca^{2+} channels, several potential candidates are currently under rigorous investigation, and progress in this area has been summarized in recent reviews (De Bello *et al.*, 1993; Jahn and Südhof, 1993, 1994; Söllner *et al.*, 1993; Südhof *et al.*, 1993; O'Connor *et al.*, 1994). Present evidence indicates that the machinery involved in synaptic vesicle fusion with the nerve terminal membrane is similar to that used in nonneuronal exocytotic processes. Synaptic vesicles may be directed to the active zone by Rab3A, a neuron-specific protein that associates with vesicles after their dissociation from the Golgi complex and dissociates from the vesicle membrane following exocytosis (Fischer von Mollard *et al.*, 1990, 1991; Matteoli *et al.*, 1991). Rabphilin 3A is a putative Rab3A target protein that is highly concentrated at neurite tips near active release sites (Wada *et al.*, 1994). Once at the active site, a second vesicular membrane protein, synaptobrevin (or VAMP), forms a complex with proteins in-

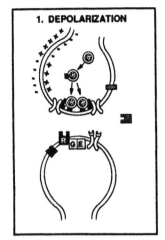
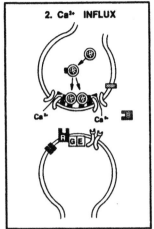
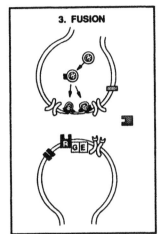
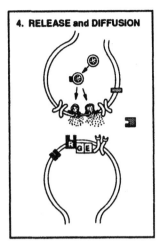

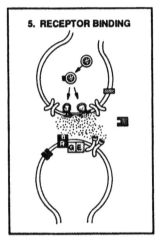
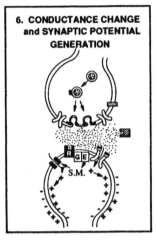
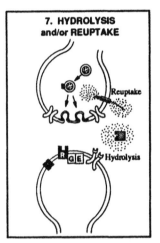

FIGURE 1 Sequence of events during synaptic transmission. (1) Depolarization of the terminal region by electrotonic spread of the action potential into the terminal region from preterminal nodes of Ranvier. (2) Activation of voltage-sensitive Ca^{2+} channels and influx of Ca^{2+} into the nerve terminal. (3) Ca^{2+}-triggered fusion of synaptic vesicles with the terminal membrane. (4) Release of neurotransmitter into the synaptic cleft and diffusion to the postsynaptic cell where (5) binding to receptors occurs. (6) Conductance changes and synaptic potentials are generated when receptor binding results in activation of a ligand-gated ion channel or when receptor activation stimulates formation of a second messenger (SM) which modulates ion channel activity. (7) Transmitter action is stopped by rapid hydrolysis to inactive products and/or carrier-mediated reuptake of neurotransmitter into the presynaptic terminal.

volved in fusion processes, α-soluble NSF attachment protein (α-SNAP) and N-ethylmaleimide sensitive factor (NSF). Two proteins found on the nerve terminal plasma membrane, syntaxin and SNAP-25, also bind NSF and α-SNAP and are thought to be responsible for docking of the vesicle–fusion protein complex to the plasma membrane. NSF binds and hydrolyzes ATP, which may initiate the fusion process. However, ATP hydrolysis is not required for synaptic transmission (Alnaes and Rahamimoff, 1975) and the just-mentioned mechanism does not account for the requirement of Ca^{2+} for neurotransmitter release. Thus, it has been proposed that hydrolysis of ATP occurs prior to Ca^{2+} influx, resulting in an energized "prefusion" complex and the dissociation of α-SNAP and NSF.

Synaptotagmin, a Ca^{2+}-binding protein, replaces α-SNAP and NSF, stabilizing the prefusion vesicle. Upon influx, Ca^{2+} binds to synaptotagmin, resulting in a conformational change that facilitates rapid fusion of the vesicular and plasma membranes (O'Connor *et al.*, 1994).

The hypothesis outlined here is strengthened by several observations. Syntaxin and synaptotagmin are also associated with N-type Ca^{2+} channels (see later) (Bennett *et al.*, 1992, 1993; Inoue *et al.*, 1992; Morita *et al.*, 1992; Leveque *et al.*, 1992), suggesting close localization of the putative release mechanisms to points of Ca^{2+} entry. Augustine and colleagues (1991) have suggested that release mechanisms be localized within no more than 10 or 20 nm from the point of Ca^{2+} entry.

Furthermore, it has been demonstrated that botulinum toxins and tetanus toxin, potent inhibitors of neurotransmitter release, act specifically on synaptobrevin, SNAP-25, or syntaxin (Link *et al.*, 1992; Schiavo *et al.*, 1992; Blasi *et al.*, 1993a,b). However, it should be noted that other proteins or mechanisms may be involved in the release process, and ongoing research is likely to change or modify this hypothesis.

Whether at central or peripheral synapses, the role of Ca^{2+} in evoking neurotransmitter release is essential. Katz and Miledi (1967a,b) demonstrated that neurotransmitter release had an absolute requirement for extracellular Ca^{2+} at the critical moment when the action potential depolarized the nerve terminal. Since the demonstration of the importance of Ca^{2+} action in neurotransmitter release, considerable effort has been made to characterize the channel(s) responsible for Ca^{2+} entry at the release site. Based on their electrophysiological and/or pharmacological characteristics, up to six types of Ca^{2+} channels have been proposed to be present in neuronal preparations (Llinàs and Yarom, 1981; Carbone and Lux, 1984; Nowycky *et al.*, 1985; Fox *et al.*, 1987a,b; Kostyuk *et al.*, 1988; Llinàs *et al.*, 1989; Adams *et al.*, 1993; Randall *et al.*, 1993). Each of the channel types is reportedly distinct with respect to the membrane potential at which it opens or "activates," whether or not it closes or "inactivates" during a depolarization and the rate at which it does so, the amount of current carried through a single channel, and the ability of different pharmacological agents to block the channel. The existence of a heterogeneous family of Ca^{2+} channel genes (Snutch *et al.*, 1990) which undergo alternative splicing patterns (Snutch *et al.*, 1991) supports the existence of multiple types of Ca^{2+} channels with differing pharmacological, biophysical, and functional characteristics (for review see

Tsien *et al.*, 1991; Hofmann *et al.*, 1994). The calcium channel nomenclature system of Nowycky and colleagues (1985), which assigned the letters T, N, and L to Ca^{2+} channels with different characteristics, will be used in this discussion. More recent descriptions of novel Ca^{2+} channels (P, Q, R) have continued the use of letter names, although the letter name may not necessarily correspond to biophysical properties of the channel. For example, T-type Ca^{2+} channels exhibit transient currents and L-type Ca^{2+} channels exhibit long lasting currents, whereas P-type Ca^{2+} channels were first described in cerebellar Purkinje cells. The biophysical and pharmacological profiles of the different Ca^{2+} channel types are summarized in Table 1. It should be noted that the P, Q, and R channel types are less well characterized and are not yet as widely accepted as the L, N, and T channel types.

Indirect information on the types of Ca^{2+} channels which mediate Ca^{2+} influx associated with neurotransmitter release can be obtained by examining the pharmacological sensitivity of neurotransmitter release. Generally, a hallmark of L-type channels is their sensitivity to members of the dihydropyridine (DHP) class of Ca^{2+} channel agents (Fox *et al.*, 1987a,b), whereas N-type channels are sensitive to inhibition by ω-conotoxin-GVIA (ω-CTx-GVIA) (Plummer *et al.*, 1989) and P-type channels are blocked by Funnel-web spider toxin (FTX) (Lin *et al.*, 1990) and ω-agatoxin-IVA (ω-Aga-IVA) Mintz *et al.*, 1992a,b). Q-type channels are reportedly sensitive to ω-conotoxin (MVIIC (ω-CTx-MVIIC) and R-type channels are resistant to block by all of these antagonists (Randall *et al.*, 1993). However, it should be emphasized that none of these antagonists is absolutely specific for a given channel type, as at higher concentrations they will also block other Ca^{2+} channel subtypes. There is as yet no specific antagonist

TABLE 1   Biophysical and Pharmacological Characteristics of Ca^{2+} Channel Types

	T[a]	N[a,b]	L[a]	P[c]	Q[d]	R[e]
Activation range (mV)	−70	−10	−10	−20 to −30	−40	−40
Inactivation kinetics	Fast	Moderate	Slow	Slow	Moderate	Fast
	$\tau = 20–50$ msec	$\tau = 95$ msec	$\tau = 0.5$ sec		$\tau = 116$ msec	$\tau_{1/2} = 12$ msec
Divalent cation sensitivity	$Ni^{2+} > Cd^{2+}$	$Cd^{2+} > Ni^{2+}$	$Cd^{2+} > Ni^{2+}$	$Cd^{2+}, <100 \mu M$	Cd^{2+}	$Ni^{2+}, 66 \mu M$
						$Cd^{2+}, 1 \mu M$
Pharmacological sensitivity	None	ω-CTx-GVIA	DHPs	ω-Aga-IVA FTX	ω-CTx-MVIIC	None
Single channel conductance (PS in 110 mM Ba^{2+})	8	13	25	10–12 (80 mM Ba^{2+})	15	14

[a] Modified from Tsien *et al.* (1988).
[b] From Plummer *et al.* (1989).
[c] P channel data compiled from Llinàs *et al.* (1989) and Lin *et al.* (1990).
[d] Q channel data from Sather *et al.* (1992, 1993) and Randall *et al.* (1993).
[e] Zhang *et al.* (1993), Randall *et al.* (1993), and Ellinor *et al.* (1993).

for the T-type channel; however, it opens at relatively negative membrane potentials (-70 mV) and is said to be "low voltage activated" (Llinàs and Yarom, 1981). In comparison, all of the other Ca^{2+} channel subtypes open (are activated) at depolarized membrane potentials when compared to T-type channels. Thus they are said to be "high voltage activated" (Llinàs and Yarom, 1981). It is unlikely that T-type channels remain open long enough or carry a large enough Ca^{2+} current to support transmitter release.

There is no reason to suspect *a priori* that all synapses utilize the same Ca^{2+} channel type to mediate neurotransmitter release. In fact, at least four of the channel types have been demonstrated to be involved in neurotransmitter release from various central and peripheral preparations (Table 2) and, in some cases, more than one channel type may be involved at the same synapse (Turner *et al.*, 1993). The range of results from the studies presented in Table 2 suggests that release of neurotransmitters may be controlled by different Ca^{2+} channels depending on the location of and/ or transmitter present in the nerve terminal. These differences should be kept in mind when examining neurotoxicant effects on Ca^{2+}-dependent synaptic transmission.

Two approaches can be taken to study the effects of neurotoxicants on synaptic transmission. In one, potential or known synaptic effects can be investigated at the sites of known or suspected intoxication (e.g.,

a paralytic toxin such as α-bungarotoxin at nicotinic receptors of the neuromuscular junction). Alternatively, one can simply study a model synapse to examine potential actions of a chemical on synaptic transmission. In the latter case, the model synapse chosen must be well characterized with respect to its physiology, biochemistry, and anatomy. For this reason, most investigators have employed simple preparations such as neuromuscular junction preparations, cell culture, or invertebrate preparations. Information collected at peripheral synapses has been applied to central synapses by arguing that the process of synaptic transmission is essentially the same at both sites. In most cases, this is a reasonable presumption, although distinct differences do exist. The last decade has seen advances in the popularity and ability to use central nervous system preparations to study synaptic transmission. Thus, it is now possible to examine synaptic transmission in central synapses with the same degree of rigor as has been applied to peripheral synapses.

This chapter presents an overview of methods that can be used to study the effects of neurotoxic compounds on these processes in both peripheral and central nervous system preparations. The methodology and the nature of the data collected using each technique are described briefly, as well as the advantages and disadvantages of different techniques. Whenever possible, examples from the literature will be used to demonstrate how neurotoxic compounds affect neuro-

TABLE 2 Selected Examples of Ca^{2+} Channel Types Involved in Neurotransmitter Release

Type	Preparation (transmitter)	Reference
L	Rat NMJ (modulate ACh release)	Atchison (1989)
	Chick dorsal root ganglion (substance P)	Rane *et al.* (1987); Holz *et al.* (1988)
	Undifferentiated PC12 cells (NE)	Kongsamut and Miller (1986)
N	Rat sympathetic neurons (NE)	Hirning *et al.* (1988)
	Peripheral neurons	Perney *et al.* (1986)
	NGF-differentiated PC12 cells (NE)	Kongsamut and Miller (1986)
	Cerebellar Purkinje cells (GABA)	Takahashi and Momiyama (1993)
	Spinal interneurons (glycine)	
	Hippocampal cholinergic EPSPs	Dutar *et al.* (1989)
	Striatal synaptosomes (DA, not NE)	Turner *et al.* (1993)
	Hippocampal CA_3/CA_1 synapse (glutamate)	Dutar *et al.* (1989); Takahashi and Momiyama (1993); Wheeler *et al.* (1994)
P	Squid giant synapse	Charlton and Augustine (1990)
	Mouse NMJ (ACh)	Uchitel *et al.* (1992); Protti and Uchitel (1993)
	Striatal synaptosomes (DA, NE)	
	Cerebellar Purkinje cells (GABA)	Turner *et al.* (1993)
	Spinal interneurons (glycine)	Takahashi and Momiyama (1993)
	Hippocampal CA_3/CA_1 synapse (glutamate)	
Q	Hippocampal CA_3/CA_1 synapse (glutamate)	Wheeler *et al.* (1994)
	Mouse NMJ (ACh)	Bowersox *et al.* (1993)

transmission and how these effects are manifested. More comprehensive reviews of the actions of anticholinesterases (Albuquerque *et al.*, 1984, 1985), metals (Cooper and Manalis, 1983; Cooper *et al.*, 1984; Atchison *et al.*, 1984; Atchison, 1987), environmental agents (Bierkamper, 1981, 1987), and other compounds (Atchison and Spitsbergen, 1994) on synaptic transmission are available.

II. Analysis of Presynaptic Function in the Peripheral Nervous System

A. Intracellular Recording at the Motor End Plate

One of the most widely employed techniques in investigating synaptic transmission is intracellular recording at the neuromuscular junction (NMJ) of vertebrates. The NMJ has been the exclusive subject of several book chapters and reviews (Martin, 1966; Hubbard, 1970, 1973; Silinsky, 1985; Atchison, 1988; Prior *et al.*, 1993; Atchison and Spitsbergen, 1994) which will provide additional information not presented here. Examples of such preparations are the rat phrenic nerve-hemidiaphragm (Bülbring, 1946) or the frog cutaneous pectoris (Blioch *et al.*, 1968) preparation, both of which are thin sheets of muscle. Thin preparations are easily transilluminated, equilibrate rapidly with solutions containing drugs or toxins, and allow an adequate oxygen supply throughout the tissue. The preparation is trimmed of excess tissue, anchored into a Sylgard-coated chamber, and perfused with an oxygenated biological buffer solution. The nerve is usually drawn up into a suction electrode or placed across two wire stimulating electrodes in a separate chamber.

Electrophysiological assessment of synaptic transmission by intracellular recording at the end plate relies on measuring the changes in end plate potential produced by acetylcholine (ACh) action on nicotinic receptors. Stimulation of the nerve in a viable preparation results in a muscle action potential (MAP) and contraction of the muscle fibers. When recording with intracellular microelectrodes at the motor end plate, the MAP obscures the end plate potential, making analysis of synaptic events impossible. In addition, contraction of the muscle dislodges the electrode from the recording site. Four principal methods are used to eliminate these problems. Inclusion of *d*-tubocurarine (dTC, $1-3 \times 10^{-6} M$) or magnesium (6–15 mM) prevents the MAP (del Castillo and Engbaek, 1954) by decreasing the sensitivity of the end plate to ACh or by decreasing the release of ACh from the motor nerve terminal, respectively. Both of these techniques decrease the

magnitude of the end plate potential to below the threshold necessary to trigger the MAP. The third method of preventing muscle twitch is to disrupt the transverse tubular system of the muscle fibers osmotically using glycerol (Howell and Jenden, 1967; Miyamoto, 1975), formamide (del Castillo and De Motta, 1978), or ethylene glycol (Sevcik and Narahashi, 1972). This method is most successful in extremely thin and amphibian preparations. The method used most often for mammalian preparations is to cut the muscle fiber close to the motor nerve (Hubbard and Schmidt, 1963; Glavinovic, 1979). This depolarizes the fiber and prevents generation of a MAP. A fifth and more recently developed technique in preventing MAP is to block muscle Na$^+$ channels with μ-conotoxin (Hong and Chang, 1989). This prevents the MAP but allows for normal nerve conduction due to the relative specificity of μ-conotoxin for muscle over neuronal Na$^+$ channels (Cruz *et al.*, 1985).

The large diameter of the muscle cell allows recording of electrical events with low resistance (5–25 MΩ), microelectrodes with tip diameters less than 1 μm. For intracellular recording the electrode is filled with 3 M KCl or K$^+$ acetate solution (if changes in [Cl$^-$]$_i$ are a concern) and a micromanipulator is used to position the electrode within the end plate region. The potential at the tip of the microelectrode is compared to a Ag/AgCl ground pellet in the bath, and is zero when the electrode is in the bath solution. Upon impalement of a muscle fiber, the voltage difference due to the membrane potential (negative with respect to ground) is sensed by the microelectrode. If the microelectrode is placed within an end plate region, spontaneous miniature end plate potentials (MEPPs) can be observed using a high gain setting and AC coupling of the ocilloscope. Evoked end plate potentials (EPPs) are recorded in response to stimulation of the motor nerve.

When recording at the motor end plate, qualitative information concerning the ability of a compound to affect synaptic transmission can be obtained simply by examining its ability to reduce or block the EPP. However, such qualitative analysis is insufficient to determine whether a toxicant alters synaptic transmission by interfering with processes at presynaptic, postsynaptic, or both pre- and postsynaptic sites. By making use of mathematical descriptions of the process of neurotransmitter release, neurotoxicologists can gain valuable information concerning the pre- and/or postsynaptic site of action of a compound as well as insight into its mechanism of action.

In early work at motor nerve terminals, Fatt and Katz (1952) and del Castillo and Katz (1954) provided evidence that evoked release of ACh from the nerve terminal occurs in multiples of a discrete unit, or quantum. Fatt and Katz (1952) proposed that the MEPP, which occurs randomly in the absence of nerve stimulation, corresponds to the action of a single quantum on the end plate region and is a packet (or vesicle) containing hundreds of molecules of ACh. In 1954, del Castillo and Katz demonstrated that the EPP results from the simultaneous action of many quanta of ACh on the nerve terminal and presented a statistical model for the process of quantal transmission. Release of packets of neurotransmitter (vesicular release) in response to nerve terminal depolarization is a process described by binomial statistics. The mean number of vesicles discharged in response to a single nerve impulse is the product of the probability of discharge for an individual vesicle times the total population of vesicles, or in mathematical terms: $m = np$. In this equation, m, the mean quantal content, is the mean number of vesicles released by an impulse; p is the probability of release of any individual vesicle; and n is the population immediately available for release. It should be noted that m, n, and p all define presynaptic processes, thus, estimation of these release parameters is a valuable tool in investigating possible presynaptic mechanisms of drugs and toxins on evoked transmitter release. There are several different methods for determination of m, n, and p in neuromuscular preparations, each with advantages and disadvantages. Because methods for determining these values have not changed significantly since they were last reviewed (Atchison, 1988; Prior *et al.*, 1993), the effects of neurotoxicants on neurotransmitter release at the neuromuscular junction will be focused on here.

Neurotoxicants that affect Ca^{2+} entry alter mean quantal content, m. Divalent and trivalent ions, including Pb^{2+} (Manalis and Cooper, 1973; Cooper and Manalis, 1983; Atchison and Narahashi, 1984; Manalis *et al.*, 1984; Pickett and Bornstein, 1984), Cd^{2+} (Forshaw, 1977; Satoh *et al.*, 1982; Cooper and Manalis, 1984a,b), Hg^{2+} (Juang, 1976; Binah *et al.*, 1978; Cooper and Manalis, 1983) La^{3+} (DeBassio *et al.*, 1971; Heuser and Miledi, 1971), Co^{2+} (Weakly, 1973), and Sn^{2+} (Allen *et al.*, 1980), all cause clear decreases in quantal content at the NMJ. The organic metal methyl mercury (MeHg), which produces a distinct neurologic syndrome, also decreases m (Barrett *et al.*, 1974; Atchison and Narahashi, 1982; Atchison *et al.*, 1984; Traxinger and Atchison, 1987a).

Changes in n, the immediately available store of neurotransmitter, are more difficult to interpret with respect to the biological substrate that may be affected. However, it is hypothesized that changes in this statistic may reflect changes in active release sites or the number of vesicles prepositioned for release (McLachlan, 1978). It is evident from the nature of this variable that com-

pounds that alter n act within the nerve terminal. Quantal analysis indicates that MeHg also decreases n at the NMJ of rats (Atchison and Narahashi, 1982).

Alterations in the probability of release, p, are most often associated with effects of neurotoxicants on the free $[Ca^{2+}]$ in the nerve terminal (Bennett *et al.*, 1975; McLachlan, 1978). The source of Ca^{2+} may be either intracellular or extracellular. Agents that cause release of Ca^{2+} from intracellular stores, prevent sequestration or extrusion of Ca^{2+}, or are capable of mimicking Ca^{2+} at the binding sites of release proteins increase p. The neurotoxic agents Pb^{2+} and MeHg increase p (Atchison and Narahashi, 1982, 1984) after *in vitro* exposure. Spontaneous release of ACh is increased by Pb^{2+} and MeHg even in the absence of extracellular Ca^{2+} (Kolton and Yaari, 1982; Atchison and Narahashi, 1984; Atchison, 1986), therefore it is unlikely that the increase in p is due to an influx of extracellular Ca^{2+}. Evidence exists that MeHg increases intraterminal Ca^{2+} in part by interfering with the mitochondrial sequestration of Ca^{2+} (Levesque and Atchison, 1987; 1988) and depolarizes the mitochondrial membrane (Hare and Atchison, 1992; Hare *et al.*, 1993). Silbergeld and Adler (1978) have shown that Pb^{2+} interferes with Ca^{2+} sequestration by mitochondria, which may underlie its effect on p. However, there is also evidence that suggests that Pb^{2+} may act on release proteins as a surrogate for Ca^{2+} (Shao and Suszkiw, 1991; Tomsig and Suszkiw, 1993).

MeHg and Pb^{2+} both cause similar alterations in the statistical analysis of neurotransmitter release. These metals decrease m and n but increase p. This is reflected most notably as decreased EPP amplitude and increased MEPP frequency (see later). The effect of MeHg and Pb^{2+} on m are due, at least in part, to their ability to block Ca^{2+} channels, decreasing depolarization-dependent Ca^{2+} influx (Nachshen, 1984; Atchison *et al.*, 1986; Shafer and Atchison, 1989). Despite the effect of both of these compounds to increase the probability of release, they also decrease the immediate store available for release. Thus, in addition to increasing intraterminal Ca^{2+}, these metals must also interfere with Ca^{2+}-dependent mechanisms responsible for initiating vesicle fusion at release sites.

As discussed in the introduction, Ca^{2+} entry into the nerve terminal is critical to the process of synaptic transmission. If the reductions in m are because of inhibition of Ca^{2+} entry into the terminal, then increases in the extracellular $[Ca^{2+}]$ may cause partial or full reversal of the effects of the neurotoxicant. Such effects have been demonstrated for divalent and trivalent heavy metals. It is possible to produce a quantitative analysis of the effects of a compound on Ca^{2+}-dependent transmission by determining the cooperativity factor for Ca^{2+}, i.e., the number of Ca^{2+} ions required to release one quanta of ACh (Dodge and Rahamimoff, 1967). This is established by determining the slope of the line described by plotting the log EPP amplitude vs. log $[Ca^{2+}]$. By determining this relationship in the presence and absence of toxicant, the K_D for antagonism of the neurotoxicant on the release process can be determined from the following equation (Gaddum, 1957):

$$K_D = [\text{toxicant}]/([Ca]_2/[Ca]_1]) - 1],$$

where $[Ca]_1$ and $[Ca]_2$ correspond to equal amplitude EPPs in the absence and presence of toxicant, respectively.

MeHg differs from Pb^{2+}, Cd^{2+}, and other inorganic divalent cations with respect to the ability of Ca^{2+} to antagonize its block of EPPs. The block of evoked release by inorganic divalent cations is reduced or reversed by increasing the $[Ca^{2+}]$ in the bath solution or by washing in metal-free buffer solution (Atchison and Narahashi, 1984). Inorganic divalent cations, because of their charge and ionic radius similar to Ca^{2+}, compete with Ca^{2+} for entry into the pore of Ca^{2+} channels. However, these metals bind with high affinity to a site in the pore of the Ca^{2+} channel, impeding the flow of Ca^{2+} through the channel (Hagiwara *et al.*, 1974; Hess and Tsien, 1984; Lansman *et al.*, 1986; Hess *et al.*, 1986). Increasing $[Ca^{2+}]_e$ favors Ca^{2+} entry by mass action, resulting in the restoration of the EPP, whereas washing with a metal-free buffer restores the EPP due to the reversible nature of interaction of inorganic divalent heavy metals with Ca^{2+} channels (Nachshen, 1984; Atchison *et al.*, 1986). In comparison, the block of EPPs by MeHg at the NMJ initially is not reversed by increasing $[Ca^{2+}]$ or washing with MeHg-free solution (Traxinger and Atchison, 1987a), but is reversed by increasing the stimulus intensity in MeHg-free solutions. If MeHg is reintroduced under conditions of increased stimulus intensity, a subsequent block of EPP occurs that is partially antagonized by increasing $[Ca^{2+}]$ (Traxinger and Atchison, 1987a). The actions of MeHg on neuronal Ca^{2+} channels are qualitatively different from inorganic divalent heavy metals in that the block of Ca^{2+} channels is not reversible by washing (Shafer and Atchison, 1991) or by increasing $[Ca^{2+}]$ (Atchison *et al.*, 1986; Shafer and Atchison, 1989, 1991; Hewett and Atchison, 1992). Thus, MeHg apparently inhibits evoked ACh release by an irreversible, noncompetitive block of Ca^{2+} entry into the nerve terminal. However, the requirement that stimulus intensity be increased to initially relieve the block of EPP suggests that MeHg also decreased axonal excitability and may prevent invasion of the action potential into the nerve terminal.

B. Perineurial Recording

The extremely small size of the vertebrate nerve terminal precludes impalement with even the finest microelectrode, thus it is impossible to measure directly the flow of ionic current at this structure. The lack of ability to perform intracellular or voltage clamp recording at the nerve terminal has prevented a highly sophisticated analysis of channel function in this region. In some measure, this problem has contributed to the difficulty in determining the type(s) of Ca^{2+} channel that is involved in neurotransmitter release. When it is suspected that a compound may block invasion of the action potential into the terminal (Hatt and Smith, 1976) or block Ca^{2+} or K^+ channel function in the nerve terminal, methods other than intracellular recording must be employed to study ionic currents. Isolated nerve terminal preparations from the central nervous system have been widely used to study channel function and neurotransmitter release. However, these preparations have four major disadvantages: (1) they are not purified nerve terminal preparations, (2) the time course over which measurements are made is sometimes several orders of magnitude longer than the process of synaptic transmission, (3) these structures are generally also too small to impale or patch onto with a microelectrode, and (4) synaptosomes are heterologous with respect to neurotransmitter released.

It is possible to study indirectly the flow of ionic current at the nerve terminal using perineurial recording techniques. Information gained from perineurial recordings, though still not direct, can substantially strengthen inferences made concerning effects of compounds on the critical processes of action potential invasion, Ca^{2+} channel activation, and K^+ channel function at the nerve terminal.

The perineurium is a layer of flattened cells that surrounds mammalian nerve trunks, forming a relatively impermeant barrier to small molecules and ions. Because the perineurium functions as an electrical insulator around the nerve trunk, axial currents generated in distal nodes of Ranvier and/or the nerve terminal form local circuits in the perineurium by returning to their origin via the perineurial space (Gundersen *et al.*, 1982; Mallart, 1985). By inserting a microelectrode (4–10 MΩ) filled with physiological buffer or 1 mM NaCl inside the perineurial sheath near motor nerve terminal regions, voltage changes resulting from inward Na^+ currents in preterminal regions and outward K^+ currents in the nerve terminal can be recorded simultaneously (Mallart, 1985; Penner and Dreyer, 1986). Using pharmacological interventions, inward Ca^{2+} currents in nerve terminals can also be recorded. Furthermore, the signal-to-noise ratio is increased compared to that of "loose patch" recordings (Brigant and Mallart, 1982; Mallart and Brigant, 1982) due to the presence of multiple nerve fibers contained in the perineurium.

Perineurial techniques require an extremely thin muscle and higher magnification than traditional neuro-muscular junction preparations. It is important to position the electrode within the finest branches of the motor nerve bundle so that it will be close to the nerve terminals. If the distance is too great, the Ca^{2+}-dependent component will be too small to be measured reliably. A thin muscle preparation is necessary to be able to see the finest branches and accurately place the microelectrode. The triangularis sterni muscle of the mouse is a commonly used mammalian preparation. This muscle, located on the inner surface of the ribcage, is only two or three fiber layers thick and inserts into the sternum and the intercalations of the ribs. Successful dissection (McArdle et al., 1981) of this muscle is a tedious process, but is not difficult if performed patiently. Branches of the second, third, and fourth intercostal nerves innervate the triangularis and are used for stimulation. As with other nerve-muscle preparations, it is necessary to prevent muscle contractions. Most investigators use d-tubocurarine to suppress muscle twitch.

Using this technique, it is possible to record the Na^+ channel component of the action potential at the last two or three nodes of Ranvier. This potential passively depolarizes the nerve terminal. In addition K^+ channel-mediated repolarization of the nerve terminal can also be observed. In the typical response shown in Fig. 2A, the negative deflection in voltage consists of two components: a small fast component (arrow) due to the Na^+ current in the nodal and heminodal regions slightly precede the larger K^+-dependent component due to outward K^+ current at the nerve terminal (Brigant and Mallart, 1982; Mallart, 1985; Penner and Dreyer, 1986; Anderson and Harvey, 1988). Two types of K^+ waveforms can be distinguished by pharmacological manipulation: a fast K^+ component which is blocked by 3,4-diaminopyridine (3,4-DAP) and a Ca^{2+}-dependent component which is insensitive to 3,4-DAP but is blocked by tetraethylammonium (TEA) and Ba^{2+} (Dreyer and Penner, 1987). If K^+ channel-blocking agents such as TEA, Ba^{2+}, or 3,4-DAP are included in the bath solution, the depolarization of the nerve terminal is prolonged and two Ca^{2+}-dependent components of the response are unmasked (Figs. 2B and 2C). In addition to prolonging the action potential and unmasking the Ca^{2+}-dependent components of the response, blocking K^+ channels also increases the excitability of the motor nerve. Therefore most inves-

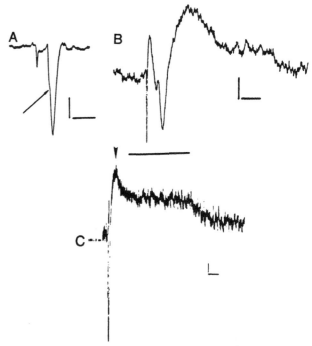

FIGURE 2 Voltage responses recorded from perineurial sheaths. (A) Na^+/K^+ voltage potential recorded from perineurial sheaths near nerve terminal regions. The arrow indicates the shoulder arising from Na^+ current in the nodal and heminodal regions. (B) Na^+/Ca^{2+} voltage potential recorded in the presence of TEA and 3,4-diaminopyridine in 5 mM Ba^{2+} solutions. (C) Na^+/Ca^{2+} voltage potential demonstrating fast and slow Ca^{2+}-dependent potentials. The arrowhead indicates the peak of the fast Ca^{2+} voltage potential, whereas the bar indicates the slow Ca^{2+} potential. Responses are not averaged or summed and were recorded in response to intercostal nerve stimulation at a rate of 0.1 Hz. Scale bars are 0.35 mV and 2 msec for A, 0.5 mV and 4 msec for B, and 0.5 mV and 40 msec for C.

tigators also include low concentrations of a local anesthetic in the bath to prevent repetitive firing of the motor nerve in response to a single stimulus. The fast component of the Ca^{2+} potential, which lasts approximately 60 msec, is presumed to be associated with neurotransmitter release (Mallart, 1985; Penner and Dreyer, 1986). This component is not affected by organic Ca^{2+} channel blockers (Penner and Dreyer, 1986) or ω-CTx-GVIA (Anderson and Harvey, 1987), but is blocked by millimolar concentrations of divalent cations (Penner and Dreyer, 1986), FTX (Uchitel et al., 1992), and ω-Aga-IVA (Protti and Uchitel, 1993). The ability of FTX and ω-Aga-IVA to block this component in mammalian preparations suggests that P-type channels may mediate this component of Ca^{2+} entry. However, as noted earlier, the type of Ca^{2+} channel which mediates neurotransmitter release may depend on the preparation. In amphibian preparations, the fast Ca^{2+} component is sensitive to ω-CTx-GVIA, suggesting that N-type

Ca^{2+} channels may mediate transmitter release (Sano *et al.*, 1987). At the NMJ of the rat extensor digitorum longus muscle, Hamilton and Smith (1993) have shown that a component of the Ca^{2+} current is sensitive to ω-CgTx-GVIA. Although the relationship of the ω-CgTx-GVIA-sensitive component to neurotransmitter release is not yet clear, it suggests the presence of N-type Ca^{2+} channels at the motor nerve terminals of this preparation. The long-lasting component of the potential is not thought to be coupled directly to neurotransmitter release (Penner and Dreyer, 1986). However, no evidence exists to preclude this mechanism from contributing to nerve-evoked release. This component is sensitive to FTX (Uchitel *et al.*, 1992), verapamil, and diltiazem but is insensitive to DHPs (Penner and Dreyer, 1986) and ω-CTx-GVIA (Anderson and Harvey, 1987). Thus, the Ca^{2+} channel type(s) mediating this component is unclear.

Perineurial recording techniques have been used to investigate the effects of several different classes of pharmacological compounds as well as toxicants on ionic currents at the nerve terminal. Guanidine and catechol decrease the K^+ component of the perineurial response whereas phencyclidine and norepinephrine produce small and transient increases in both the Na^+ and K^+ components (Anderson and Harvey, 1988). Bourret and Mallart (1989) used this technique to demonstrate that polycationic antibiotics decrease the Ca^{2+} current at motor nerve terminals. Dendrotoxin, β-bungarotoxin, crotoxin, and taipoxin inhibit a slowly activating K^+ response at the nerve terminal (Dreyer and Penner, 1987). As previously discussed, evidence from the rat neuromuscular junction suggested that MeHg decreases axonal excitability and interferes with Ca^{2+}-dependent neurotransmitter release (Traxinger and Atchison, 1987a). Using synaptosomal preparations (Atchison *et al.*, 1986; Shafer and Atchison, 1989; Shafer *et al.*, 1990) and whole cell patch clamp techniques (Hamill *et al.*, 1981) in pheochromocytoma cells (Shafer and Atchison, 1991), it has been demonstrated that MeHg blocks $^{45}Ca^{2+}$ influx into nerve terminals and blocks somatic current carried by N- and L-type Ca^{2+} channels, which have been associated with neurotransmitter release. However, recordings of synaptic potentials at the rat NMJ also indicate that MeHg may affect action potential conduction (Traxinger and Atchison, 1987a) whereas others have demonstrated previously that MeHg blocks Na^+ channels in neuroblastoma cells (Quandt *et al.*, 1982) and squid giant axon (Shrivastav *et al.*, 1976). Thus, to determine if MeHg blocks Ca^{2+} channels at motor nerve terminals and to examine its effect on excitability of motor neurons, the effects of MeHg on Na^+- and Ca^{2+}-dependent

current were examined using perineurial recording techniques (Shafer and Atchison, 1992). Because of the potential effects of MeHg on excitability of the motor nerve, local anesthetics were not included in the bath solution, and the resulting repetitive discharge of the nerve is evident in Fig. 3A. Application of MeHg suppressed repetitive firing, suggesting that it decreases neuronal excitability (Figs. 3B–3F). In addition, MeHg blocked both the Na^+- and Ca^{2+}-dependent components of the perineurial waveforms. Blocking the Ca^{2+}-dependent component often precedes blocking the Na^+-dependent component, suggesting that MeHg directly affects nerve terminal Ca^{2+} channels. These observations provide convincing evidence that MeHg disrupts the evoked release of ACh at the NMJ by two mechanisms: blocking Ca^{2+} entry into the nerve terminal and decreasing axonal excitability, resulting in failure of the action potential to invade the nerve terminal.

C. Focal Patch Recording

An alternative method to perineurial recording for examining ionic current at nerve terminals is to record extracellular currents with focal electrodes placed directly over the nerve terminal (Hubbard and Schmidt, 1963; del Castillo and Katz, 1956; Brigant and Mallart, 1982; Mallart and Brigant, 1982; Konishi, 1985). Focal patch recordings offer two advantages over perineurial recordings: electrical events at a single nerve terminal are measured and both presynaptic and postsynaptic responses can be monitored if the electrode is placed properly. Electrodes (0.3–1.0 MΩ) for focal patch recording have tip diameters of 2–4 μm, which is slightly larger than the size of the mammalian nerve terminal. When the tip of the electrode comes into contact with the connective tissue over the nerve terminal, a low resistance (1–3 MΩ) seal is formed which allows the measurement of local circuit currents in the terminal area. If the electrode is properly placed, both the presynaptic Na^+/K^+ or Na^+/Ca^{2+} and postsynaptic end plate current can be recorded. However, because of the low resistance of the seal and the small size of the currents, the signal-to-noise ratio of this type of recording is low, and most investigators average several responses.

In order to use this technique, one must be able to visualize the neuromuscular junction itself, which requires high magnification (400–500×) with a microscope with Normarski optics (Brigant and Mallart, 1982; Mallart and Brigant, 1982; Konishi, 1985) or a combination of fluorescent dyes to stain the nerve terminal (Kelly *et al.*, 1985) and an epifluorescene microscope (Hamilton and Smith, 1991). Because of the need

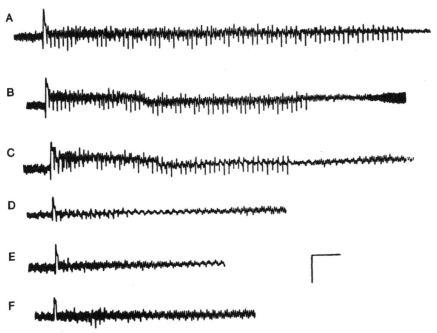

FIGURE 3 Effects of MeHg on repetitive firing of motor nerves. An example of repetitive firing in a motor nerve recorded prior to (A,B) and 20 (C), 70 (D), 80 (E), and 90 (F) sec after the addition of 50 μM MeHg. Note also the loss of the long-lasting portion of the Ca^{2+} response in this example. Scale bar is 200 msec. (From Shafer and Atchison, 1992.)

for high magnification, very thin muscle preparations such as the triangularis sterni are also often used for this technique, although thicker preparations such as the rat extensor digitalis longus muscle have also been used (Hamilton and Smith, 1991).

D. Analysis of Spontaneous Release

The techniques just discussed are useful for analysis of effects of neurotoxicants on nerve-evoked synaptic transmission. However, analysis of spontaneous release can also provide valuable information concerning the effects of a neurotoxicant on pre- and postsynaptic function. It is presumed that the spontaneous release of ACh at the neuromuscular junction utilizes the same cellular machinery as evoked release. However, spontaneous release differs from evoked release in that it is much less dependent on extracellular Ca^{2+}, does not require precise timing of Ca^{2+} presence, and does not require invasion of the action potential into the nerve terminal. Because of these properties, analysis of spontaneous release can provide information on additional effects of a neurotoxicant on the release process when evoked release has been blocked. For example, if invasion of the action potential is blocked by depolarization of the membrane or Ca^{2+} entry is prevented by channel block, analysis of spontaneous release can demonstrate additional effects on neurotransmitter synthesis and/

or storage, Ca^{2+} sequestration and homeostasis, or release mechanisms. In addition, changes in spontaneous release can be used as a bioassay for effects of neurotoxicants on the postsynaptic membrane. Neurotoxicants that alter ACh receptor channel function may produce alterations in the amplitude or rate of rise and decay of MEPPs.

As discussed earlier, spontaneous release is measured simultaneously with evoked release when recording at motor nerve end plates. Analysis of MEPP frequency ($MEPP_f$) and MEPP amplitude can be done in conjunction with analysis of evoked release in many instances. However, in many cases, only $MEPP_f$ or MEPP amplitude is examined and the effects of pharmacological agents on these measures are examined to gain further information about the effects of a neurotoxicant on the release process.

The frequency of MEPPs is dependent entirely on presynaptic function of the motor nerve (Hubbard *et al.*, 1968; Katz, 1962). Increases in $MEPP_f$ often result from increased intraterminal $[Ca^{2+}]$. A number of mechanisms exist by which $[Ca^{2+}]$ may be increased, including depolarization (Baker *et al.*, 1971), anoxia/inhibition of Na^+/K^+ ATPase (Baker and Crawford, 1975; Elmqvist and Feldman, 1965; Hubbard and Løyning, 1966), and disruption of mitochondrial or endoplasmic reticulum Ca^{2+} buffering (Glagoleva *et al.*, 1970; Rahamimoff and Alnaes, 1973). However, $MEPP_f$ can also be increased by physical alterations, underscoring the need for careful experimental procedure. Increased osmotic pressure, increased temperature, and stretch or mechanical distention result in increased $MEPP_f$. Conversely, decreases in $MEPP_f$ are often related to decreases in $[Ca^{2+}]$ in the nerve terminal. Agents that hyperpolarize the terminal decrease $MEPP_f$. Agents that deplete ACh stores or transmitter vesicles may also decrease $MEPP_f$ as well as agents that interfere directly with processes involved with fusion of the vesicle with the terminal membrane. Decreases in temperature also decrease $MEPP_f$. Black widow spider venom and La^{3+} cause a massive release of neurotransmitter (Heuser and Miledi, 1971; Clark *et al.*, 1970, 1972), therefore these compounds can be used to examine whether decreases in $MEPP_f$ or block of MEPPs by a neurotoxicant are due to depletion of ACh stores.

In the case in which evoked release has been blocked by conduction block, analysis of $MEPP_f$ can provide further information on effects of the compound on release processes. Since $MEPP_f$ is affected by changes in $[Ca^{2+}]_e$ and membrane potential, these parameters can be altered and their effects on $MEPP_f$ examined. Depolarization of the nerve terminal with elevated K^+ solutions causes Ca^{2+} influx and increases $MEPP_f$, circumventing the need for action potential invasion. In addition, since $MEPP_f$ is somewhat influenced by $[Ca^{2+}]_e$, the effects of supraphysiological $[Ca^{2+}]$ (4–8 mM) on $MEPP_f$ can also be examined and compared in control and treated preparations. Other methods for increasing $MEPP_f$ include use of Ca^{2+} ionophores and preparation of Ca^{2+}-containing liposomes.

A large number of neurotoxicants alter $MEPP_f$, including heavy metals such as Pb^{2+} and MeHg. Increased $MEPP_f$ by heavy metals often occurs following a delay (Juang and Yonemura, 1975; Miyamoto, 1975; Juang, 1976; Binah *et al.*, 1978; Atchison and Narahashi, 1982; Cooper and Manalis, 1983; Atchison *et al.*, 1984; Manalis *et al.*, 1984; Weigand, 1984; Atchison, 1986; 1987) which may reflect the time required for the neurotoxicant to enter the terminal. Effects of MeHg on MEPPs are biphasic. The $MEPP_f$ is dramatically increased by MeHg within a short time following exposure in the bath solution. However, $MEPP_f$ decreases and MEPPs are eventually no longer observed following prolonged perfusion with MeHg (Barrett *et al.*, 1974; Juang, 1976; Atchison and Narahashi, 1982). When MEPPs are blocked by MeHg, La^{3+} causes restoration of MEPPs, indicating that MeHg blocks release instead of depleting neurotransmitter stores. Mechanisms underlying MeHg-induced increases in $MEPP_f$ were examined by physiological and pharmacological manipulations. In Ca^{2+}-free solutions, MeHg increased $MEPP_f$ (Atchison, 1986), suggesting that

nerve terminal depolarization and Ca^{2+} entry are not responsible for the increased frequency, but rather that effects of MeHg on intracellular Ca^{2+} buffering may underlie increased $MEPP_f$. Furthermore, $MEPP_f$ is increased following tetanic stimulation and decays in a time-dependent manner which is thought to reflect extrusion and buffering of Ca (or Sr^{2+} or Ba^{2+}) in the nerve terminal (Magleby, 1979; Silinsky, 1978). Following tetanic stimulation in Ca^{2+}, Sr^{2+}, or Ba^{2+} solutions, the rate of $MEPP_f$ decay was not significantly altered by MeHg (Traxinger and Atchison, 1987b). These data suggest that MeHg may act to cause release of Ca^{2+} from intraterminal stores instead of altering extrusion/sequestration mechanisms. To determine if MeHg acts on mitochondrial Ca^{2+} stores, mitochondrial stores of Ca^{2+} were depleted with ruthenium red (Moore, 1971) or inhibited from releasing Ca^{2+} by N,N-bis(3,4-dimethoxy phenylethyl)-N-methylamine prior to MeHg exposure. These compounds blocked the effect of MeHg to increase $MEPP_f$, suggesting that release of Ca^{2+} from mitochondria underlies the effect of MeHg on $MEPP_f$ (Levesque and Atchison, 1987, 1988).

In contrast to $MEPP_f$, alterations in MEPP amplitude are usually indicative of postsynaptic effects of a neurotoxicant, such as alterations in the sensitivity or number of postsynaptic receptors or channels or changes in the input resistance of the postsynaptic fiber (Katz and Thesleff, 1957). Having made this statement, it should be cautioned that there are examples of neurotoxicants that alter MEPP amplitude via presynaptic mechanisms. For example, botulinum toxin decreases MEPP amplitude presynaptically, possibly by decreasing the amount of ACh contained in each vesicle (Harris and Miledi, 1971; Spitzer, 1972). Pre- vs postsynaptic effects of neurotoxicants on MEPP amplitude can be separated by examining the effects of neurotoxicants on the response of the postsynaptic fiber to iontophoretic application of ACh. Lack of effects of neurotoxicants on iontophoretic responses demonstrate that effects on MEPP amplitude are presynaptic in origin. However, presynaptic contributions due to altered MEPP amplitude cannot be eliminated from consideration when neurotoxicants affect iontophoretic responses.

III. Analysis of Postsynaptic Function in the Peripheral Nervous System

The previous section focused largely on methods for examining presynaptic function in peripheral nerve preparations. However, as discussed in the introduction, successful neurotransmission requires that neurotransmitter diffuses across the synaptic cleft and activates receptors on the postsynaptic cell, generating a response in this cell. Many pharmacological agents and neurotoxicants disrupt function at postsynaptic sites. Therefore, determining their mechanism of action requires the ability to examine postsynaptic function in a precise manner. Methods range from iontophoretic application of transmitter to sophisticated analysis of end plate currents measured by voltage clamp techniques.

Iontophoretic application of ACh to the end plate region of muscle fibers allows aspects of postsynaptic function to be examined when it is suspected or demonstrated that a neurotoxicant causes alterations in presynaptic function. Using this method, changes in the postsynaptic sensitivity to ACh are determined by examining electrical responses in the muscle end plate following direct application of ACh or other nicotinic agonists to the end plate region (Nastuk, 1951; del Castillo and Katz, 1955). In addition to the recording electrode placed in the end plate region as described earlier, iontophoretic application requires that a second (iontophoretic) electrode (50–100 MΩ)-containing drug be placed (extracellularly) as close as possible to the end plate region being recorded from. The proximity of the iontophoresis electrode affects the magnitude and rate of rise of the end plate response, as concentrations of drug are higher nearer the tip of the iontophoretic electrode because of less diffusion. Thus, more molecules of ACh or other agonist will act on receptors at any given moment, generating a faster and a larger response. Because ACh and most other nicotinic agonists are charged compounds, this physical characteristic can be taken advantage of to provide a pulsed release of ACh and prevent leakage of ACh from the tip of the electrode by passing current through the iontophoretic electrode. To prevent leakage of the drug, a small constant current is applied to the iontophoretic electrode which is opposite in polarity to the charge of the drug. Ejection of a constant amount of drug is accomplished by application of a current across a series resistance of 10^9 Ω. Dedicated iontophoretic units are available from a variety of vendors but a functional unit can be assembled by utilizing a two channel stimulus isolation unit with one channel used to provide a braking current to prevent leakage of the drug and the other channel for delivery of a pulse to eject the contents of the electrode.

If the iontophoretic electrode is placed close enough to the end plate region, responses to the inotophoretic application of the agonist will have a rapid rate of rise and fall and will resemble EPPs (Kuffler and Yoshikami, 1975). The magnitude of the iontophoretic response (in mV) should be expressed as a function of the iontophoretic pulse (in coulombs). The frequency

of application of iontophoretic pulses should be kept low to avoid causing desensitization of end plate ACh receptors. Because iontophoretic potentials are extremely dependent on placement of the electrode, comparisons must be limited to responses recorded at the same end plate region. For this same reason it is also required that an extremely stable recording setup (micromanipulators, etc.) be used for iontophoretic experiments.

Iontophoretic techniques also have been widely employed to study function of postsynaptic receptors found in the CNS. Because the basic technique of iontophoresis is similar to that used to study ACh receptor function at the NMJ, the application of this technique to study receptor function in the CNS will be briefly discussed at this juncture. The study of postsynaptic receptor function in CNS neurons has been limited largely to recording single channel or macroscopic current from individual cells grown in tissue culture because of the difficulties associated with recording from individual neurons in more intact preparations such as slices (see later). However, in some cases, patches have been obtained from specific areas (e.g., CA$_1$ or CA$_3$ neurons or dendrites) of slice preparations (Jonas and Sakmann, 1992; Colquhoun *et al.*, 1992; Stuart *et al.*, 1993). Postsynaptic receptor function is examined following the iontophoretic application of agonists to neurons grown in culture and recording the whole cell current response or by iontophoretic application of agonist to outside-out patches obtained from neurons in slices or grown in culture. Patch clamp techniques allow current amplitude, ionic specificity, and gating characteristics, as well as the effects of neurotoxicants on these properties, to be examined.

At the NMJ, heavy metals including Pb^{2+} (Manalis and Cooper, 1973; Manalis *et al.*, 1984) and La^{3+} (Colton, 1976) alter iontophoretic currents, although only at concentrations much higher than those that exert effects on presynaptic processes. However, heavy metals, especially Pb^{2+}, may alter postsynaptic receptor function in the CNS. Potent effects of Pb^{2+} have been reported on NMDA (Alkondon *et al.*, 1990; Ujihara and Albuquerque, 1992; Uteshev *et al.*, 1993) and ACh receptors (Oortgiesen *et al.*, 1990). Inorganic Hg augments GABA-induced Cl$^-$ current in dorsal root ganglion cells at concentrations between 1 and 10 μM, whereas MeHg did not augment GABA-induced Cl$^-$ current at any concentration (Arakawa *et al.*, 1991). In fact, MeHg decreased the GABA-induced current at a concentration of 100 μM (Arakawa *et al.*, 1991). MeHg did not alter iontophoretic responses at the NMJ at a time when nerve-evoked transmission was completely blocked (Atchison and Narahashi, 1982). However, MeHg decreased the response of muscarinic receptors to iontophoretically applied ACh in neuro-blastoma cells (Quandt *et al.*, 1982), suggesting that nicotinic and muscarinic ACh receptors may differ in their sensitivity to MeHg. Thus, effects on postsynaptic function in the central nervous system may be an important difference between the peripheral and central effects of metals.

Iontophoretic application of ACh or other agonist to the end plate region provides information concerning the ability of a neurotoxicant to interact with the postsynaptic membrane but provides little mechanistic information about the effect of the compound. Voltage clamp analysis of postsynaptic currents can provide more detailed information about how a neurotoxicant disrupts postsynaptic function because there is direct control of the membrane voltage and precise measurement of ionic currents generated. This allows changes in the current voltage relationship, time course of current activation and decay, as well as the absolute amplitudes of evoked and spontaneous end plate currents to be measured with a high degree of accuracy. Using information from voltage-clamped preparations, one can differentiate between blockage of receptor-operated channels in the open configuration and either receptor or closed channel block.

A complete discussion of voltage clamping techniques and their application can be found in a previous chapter (Audesirk, 1995) and in Atchison and Spitsbergen (1994) and Smith *et al.* (1985). Because muscle fibers are substantially larger than the average neuron, the two-electrode voltage clamp technique can be used readily to examine postsynaptic function. The microelectrodes (2–5 MΩ) are filled with 3 M KCl solution and are inserted into the end plate region. To achieve adequate space clamp, the minimum distance between the electrodes should be at least 50 μm. MAP generation and muscle twitch must be eliminated by inclusion of dTC, magnesium, or cutting of the muscle fibers in order to record end plate currents. A disadvantage to using dTC is that spontaneous miniature end plate currents (MEPCs) may be obscured. Stimulation of the motor nerve will result in generation of an end plate current (EPC) whose magnitude will depend on the voltage at which the postsynaptic membrane is clamped. By measuring EPCs at a series of different voltages in the presence and absence of a neurotoxicant, the current–voltage relationship can be determined by plotting the peak current vs the holding potential. Furthermore, by digitizing the EPC, the decay time constant, τ_{EPC}, can be determined by least-squares regression analysis of the decay phase of the EPC. Neurotoxicants that block agonist binding (competitively or noncompetitively) or block the ACh receptor channel in the closed state produce decreases in EPC amplitude. On the other hand, neurotoxicants that block the ACh receptor channel only after it has opened cause the EPC to decay in a biphasic manner. In this case, the normal number of receptors/channels are activated, but are blocked rapidly, resulting in a fast component of current decay. A second, longer component of decay results from current flowing through channels as they become unblocked, then close (Adams and Sakmann, 1978; Lambert *et al.*, 1980).

IV. Analysis of Synaptic Transmission in the Central Nervous System

While the neuromuscular junction is an extremely valuable preparation for studying synaptic transmission, it cannot serve as a model for central nervous system transmission in all respects. The involvement of dendrites and dendritic spines in CNS neurons in the integration of information, the interaction of excitatory and inhibitory transmission, and use-dependent plasticity are examples of processes that may be influenced by neurotoxicants, but (typically) do not occur at the NMJ. Methods for analysis of synaptic transmission in the central nervous system have been developed only recently. This is due in part to the much smaller size of postsynaptic neurons in comparison to muscle cells, making impalement and intracellular recording more difficult. As with peripheral synapses, in order to characterize effects of a neurotoxicant on synaptic transmission it is important to use a preparation that is well understood in terms of its physiology and anatomy. Synaptic transmission has been characterized in slices made from cortex, hypothalamus, inferior olive, lateral olfactory tract, and cerebellum, but the process of synaptic function has been characterized most thoroughly in transverse slices from the hippocampus. Briefly, the synaptic pathways, anatomy of the pre- and postsynaptic structures, and pharmacology of receptors present in this region of the brain are maintained following slicing and have been well characterized. The major transmitters in the hippocampus are glutamate, which typically produces excitatory responses, and GABA, which typically produces inhibitory responses. Synapses in the hippocampus demonstrate activity-dependent plasticity including long-term potentiation (LTP) and depression (LTD), making this tissue a popular model in which to examine biochemical and physiological mechanisms of synaptic plasticity. This discussion focuses on the use of hippocampal slices to examine synaptic function in the central nervous system but it should be understood that the techniques discussed could be applied to other regions as well.

The preparation of hippocampal slices for electrophysiological recording has been described in detail by Teyler (1980). Briefly, transverse slices are prepared by manually slicing the hippocampus into sections 300–400 μm thick. However, slice thickness may vary between 100 and 600 μm, depending on the intended use. For example, thinner slices (100–300 μm) are typically used for patch clamp recordings. Transverse slices preserve the synaptic circuitry in the hippocampus and are thin enough to allow adequate oxygenation and equilibration of drugs and/or chemicals. Slices are placed into a perfusion chamber that can be built (Teyler, 1980) or purchased from one of several vendors. Warmed, oxygenated physiological saline solution is perfused over the slices which are usually illuminated from below to aid in the identification of structures for the placement of stimulus and recording electrodes. Hippocampal slices are extremely sensitive to oxygen, pH, temperature, and chemical contaminants from glues and plastics which may be used in construction of recording chambers (Teyler, 1980; see also Reid *et al.*, 1988). Therefore, care must be taken to control environmental conditions as closely as possible and nontoxic materials should be used in chamber construction. If conditions are properly maintained, slices may be viable for 10–14 hr or longer, allowing for relatively long exposure durations.

Four basic electrophysiological techniques have been used to examine synaptic function of slices *in vitro*: extracellular recording of field potentials, intracellular recording of potential changes, intracellular single electrode voltage clamp recording, and patch clamp recording. Each of these techniques has unique advantages. Extracellular recording of field potentials is technically the least difficult method to use and can be performed in awake animals as well as *in vitro*. Recordings from awake animals and slices *in vitro* have many similarities, allowing greater ease of comparison between data obtained in awake animals and isolated slices. In addition, *in vitro* slice preparations allow electrodes to be placed under direct visual control in discrete pathways, without the relative uncertainty imposed by stereotaxic manipulations. Also, since the electrode does not have to pass through a mass of tissue, blockage and/or breakage of the electrode is avoided. However, the limitations of this technique are that field potential responses are compromised of the activity of populations of neurons instead of a single or several synapses. In addition, this method does not allow for voltage clamp techniques to be applied or administration of pharmacological agents intracellularly. Intracellular recording from central neurons allows quantal analysis of synaptic events in many cases. This is an obvious advantage over extracellular

techniques in slices or awake animals. In addition, use of single electrode voltage clamp (SEC) techniques allows precise measurement of synaptic events. Patch clamp recording techniques offer many of the same advantages as intracellular techniques plus the ability to introduce intracellularly pharmacological agents to study the involvement of second messengers in modulation of synaptic transmission. In addition, single and macroscopic channel events can be examined using patch techniques.

A. Extracellular Recording

Extracellular recordings in hippocampal slices can provide information on both neurotransmitter release and neuronal excitability. Excitatory postsynaptic potentials (EPSPs) provide a measure of the viability of synaptic transmission, whereas measurement of orthodromic and antidromic population spikes (PS) allows membrane excitability to be examined. Recordings made in the CA_1 region of the hippocampus will be discussed as an example. However, similar methods have been applied to other regions of the hippocampus and other tissues. Extracellular recordings in region CA_1 are made using microelectrodes with impedances of 5–10 MΩ when filled with artificial cerebrospinal fluid (ACSF). These electrodes are placed in the pyramidal cell layer of the CA_1 region which receives synaptic input via the Schaffer collaterals and the commissural pathway. By placing a bipolar or monopolar tungsten electrode into the stratum radiatum, the Schaffer collaterals can be stimulated and the response recorded in the CA_1 pyramidal cells. Low intensity stimulation of the Schaffer collaterals results in the generation of a positive-going EPSP in the pyramidal cell layer, whereas higher intensity stimulation triggers a PS which will be a negative deflection on the positive EPSP (Fig. 4, left side). By placing the recording electrode in the dendritic region of CA_1 cells, negative-going EPSPs are recorded (Fig. 4, right side). The am-

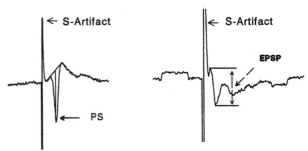

FIGURE 4 Measurement of amplitude of population spike (PS) (left) or EPSP (right). S-artifact indicates the stimulus artifact. (From Yuan and Atchison, 1993.)

plitude of the population spike depends on a number of factors, including synchrony of discharge, the distribution of the activated cell population, and tissue resistance. However, all other things being equal, the amplitude is related to the number of cells firing. Thus, changes in the amplitude of the extracellularly recorded PS indicate grossly that the ability of CA_1 neurons to generate action potentials in response to stimulation of the Schaffer collaterals is altered. An inhibitory postsynaptic potential (IPSP) can also be recorded from hippocampal cells. Subthreshold stimulation of the alveus evokes a negative potential; this deflection is associated with an increased conductance and inhibition of spike generation. The IPSP is produced in hippocampal pyramidal cells by a recurrent pathway through basket cells synapsing on the somata of pyramidal cells (Ramon ý Cajal, 1911; Anderson *et al.*, 1964a,b). Antidromically generated population spikes can be recorded by placing the recording electrode in the cell body layer of CA_1 and stimulating electrodes in the alveus. Effects of neurotoxicants are determined by comparing the amplitudes of the EPSP and PS prior to and following exposure. Input/output functions can be determined for the EPSP slope and PS amplitude which can be generated at several time points following exposure to neurotoxicant (Fountain *et al.*, 1992). Normally, orthodromic stimulation should elicit only a single spike. The presence of repetitive firing following orthodromic stimulation is indicative of cell deterioration.

Calcium appears to play an important role in impulse conduction in hippocampal neurons in certain regions. When pyramidal cells from the CA_1 region of the hippocampus are treated with tetrodotoxin (TTX, 1 μg/ml), normal, sodium-dependent action potentials are abolished. Electrical activation of the cells now produces slower spikes (20- to 50-msec duration), with amplitudes ranging from 20 to 30 mV in the soma. These slow spikes are presumed to be driven by Ca^{2+}, as they can be blocked by Mn^{2+} and enhanced by application of Ca^{2+} or Ba^{2+} (Schwartzkroin and Slawsky, 1977). The stimulation threshold for these putative Ca^{2+} spikes is higher than for the faster, TTX-sensitive spikes. Calcium action potentials can also be recorded in the apical dendrites of CA_3 cells; calcium spikes in this region are much larger and often attain amplitudes of 60 mV (Wong and Prince, 1979). Using histological techniques to verify recording electrodes placement, Wong and colleagues (1979) showed that TTX-resistant action potentials could be recorded from both the somal and dendritic regions of the CA_3 cells. The amplitude of TTX-sensitive spikes are always larger in the somal region whereas amplitudes of TTX-resistant spikes are larger in the dendritic region. The hypothesis proposed

to account for these findings is that Na^+ spikes predominate in the soma, whereas Ca^{2+} spikes predominate in the dendrites. An additional slow, voltage-dependent inward current has been recorded in CA_2/CA_3 cells using voltage clamp (Johnston *et al.*, 1980). This current is also suppressed by Mn^{2+} and enhanced by Ba^{2+}. It is thought to underlie the repetitive Ca^{2+} potentials which are observed in CA_2/CA_3 cells. Thus, Ca^{2+}-dependent action potentials appear to play a major role in synaptic transmission in the hippocampus. They may contribute significantly to burst potentials which are thought to underlie epileptiform activity (Prince, 1978; Schwartzkroin and Wyler, 1980). Thus, agents that affect Ca^{2+} channels may affect impulse conduction in the hippocampus, although none have been tested to date.

Effects of neurotoxicants on synaptic transmission in central nervous system preparations have not been investigated as widely as in peripheral preparations due to the relatively more recent development of techniques for use in the former. Organic heavy metals MeHg (Yuan and Atchison, 1993), trimethyl tin (TMT), and triethyl tin (TET) (Fountain *et al.*, 1988) have been examined for their effects on field potentials in hippocampal slices. TMT and TET at concentrations of 1 and 10 μM, respectively, decrease the slope of the input/output relationship for EPSP slope and PS amplitude. The decrease in excitability may be due to effects of these compounds directly on ion channels, receptors, energy metabolism or synthesis, storage, or release of neurotransmitter (Fountain *et al.*, 1988). MeHg decreased the amplitude of the PS as well as the EPSP following *in vitro* exposure. In addition, MeHg blocked both orthodromically and antidromically stimulated PSs and altered the induction of LTP in hippocampal slices (Yuan and Atchison, 1993). As in peripheral preparations, the effects of MeHg on population spikes indicate that it interferes with axonal excitability, whereas the ability of MeHg to block the EPSP suggests that it also interferes with the process of neurotransmitter release. The effects of MeHg on excitability and neurotransmitter release in the central nervous system preparations are consistent with its effects in peripheral preparations (Traxinger and Atchison, 1987a; Shafer and Atchison, 1992). Effects of lead on LTP have been examined in hippocampal slices. Perfusion of lead acetate (10–20 μM) blocked LTP formation in 50% of slices tested, whereas potentiation occurred in the remainder of the slices following removal of lead (Altmann *et al.*, 1991). In other experiments, 10 μM lead did not affect normal synaptic transmission or postsynaptic potentiation, but completely blocked induction of LTP (Hori *et al.*, 1993). Although the concentrations of lead used in the above studies are relatively high, electrophysiological alterations have been reported in developmentally exposed rats having blood lead concentrations below 20 $\mu g/dl$ (Altmann *et al.,* 1994). Extracellular recordings made *in vitro* from cortical and hippocampal slices of control and lead-exposed animals showed differential sensitivity of the two regions to lead effects. In cortical slices from 12–20 day old animals, lead treatment was associated with decreased ability to exhibit LTP, development of LTD following high frequency stimulation and weaker paired pulse inhibition. By contrast, only hippocampal slices from 16–20 day old animals showed decreased ability to exhibit LTP (Altmann *et al.,* 1994).

B. Intracellular Recording

Intracellular recording in hippocampal slices offers two advantages over field potential recording. First, the response of a single neuron to a stimulus is measured instead of the population response. This allows one to make quantal analysis of the process of synaptic transmission in addition to measuring changes in membrane potential and input resistance. Second, voltage clamp techniques can be utilized to facilitate measurements of synaptic function. When recording in the voltage clamp configuration, the small size of CNS neurons precludes use of the two-electrode voltage clamp technique. This problem is overcome by use of a time-sharing electrode that cycles between current injection and recording (see Audesirk, 1995). Because current flow is being measured with the voltage clamp technique, excitatory and inhibitory postsynaptic currents (EPSCs and IPSCs) are measured. For intracellular recordings, glass microelectrodes have impedances of 40 to 100 MΩ and are filled with $2\,M\,K^+$ acetate solution (Sayer *et al.,* 1990; Brown *et al.,* 1981). Intracellular recordings have been made from both CA$_1$ and CA$_3$ pyramidal cell bodies. Because of the relatively "slow" speed of response of the SEC, "fast" currents such as the Na$^+$ currents responsible for the action potential cannot be measured accurately. However, slower membrane currents such as those carried through some K$^+$ channels or most Ca^{2+} channels can be measured. When interpreting data from SEC experiments of effects of neurotoxicants on membrane currents in hippocampal slices, it is important to keep in mind that the possibility exists that the cells may be coupled electrotonically (Taylor and Dudek, 1982). The input resistance is an important parameter for estimating the viability of the cell membrane during clamp. Thus, the recording preamplifier should have a bridge circuit that permits current passage through the intracellular recording electrode. Significant decreases in input resistance indicate cell deterioration.

In contrast to the neuromuscular junction, where only one synapse/muscle fiber typically exists, the dendritic trees of hippocampal neurons contain hundreds of synapses. Because the activity measured on impalement of a hippocampal neuron results from that of numerous synapses, quantal analysis is more difficult, but not impossible, to perform in central preparations. In addition to evoked responses, spontaneous events can also be recorded in hippocampal slices. In central neurons synaptic activity persists even after impulse propagation and synaptic transmission have been blocked. Brown and colleagues (1979) reported the presence of spontaneous miniature excitatory postsynaptic potentials (MEPSPs) from neurons in the CA$_3$ region of the stratum pyramidale in the guinea pig hippocampal slice. MEPSPs were observed under conditions in which regenerative spiking activity and evoked release of transmitter were blocked by use of TTX (1 $\mu g/ml$). The timing of occurrence of these MEPSPs appeared to be random. The mean amplitude of the spontaneous potentials was approximately 2 mV; however, the mode of the distribution was buried in the background estimate. These spontaneous potentials do not represent so-called "d spike" or "fast prepotentials" which are thought to be nonpropagating dendritic spikes (Schwartzkroin, 1975; Spencer and Kandel, 1961; Schwartzkroin and Slawsky, 1977). These MEPSPs are thought to be functional correlates to MEPPs recorded from spinal neurons (Colomo and Erulkar, 1968) and from vertebrate (Fatt and Katz, 1952) and invertebrate neuromuscular junctions (Dudel and Kuffler, 1961; Usherwood, 1963).

MEPSPs were examined further under current and voltage clamp using SEC (Brown and Johnston, 1983). Under current clamp conditions, adjustment of the membrane potential to values between −125 and −50 mV led to miniature currents that were always depolarizing. However, when cells were depolarized further to −20 mV, hyperpolarizing miniature currents were observed. To distinguish between the possibility that these potentials were simply reversed MEPSCs (at membrane potentials past the reversal potential for the transmitter) or actually inhibitory synaptic potentials (MIPSCs), agents known to block synaptic inhibition in the hippocampus were used (bicuculline, 10 μM; picrotoxin, 10 μM; and penicillin, 3.3 mM). In the presence of the blockers of inhibition, the usual depolarizing responses were still obtained at membrane potentials between −125 and −50 mV, but now, as the cells were depolarized to values close to 0 mV, no trace of hyperpolarizing miniature current was found. Thus, MIPSCs can also be measured from isolated hippocampal slices.

C. Patch Clamp Recording

Patch recording in slice preparations offers the benefits of voltage clamp techniques available from intracellular recording. In addition, in the whole cell configuration the background noise is much lower than in intracellular recordings and pharmacological agents can be introduced into the cell via the patch pipette. This has been used to the advantage of those who study the biochemical events which underlie the generation of LTP in hippocampal neurons.

One of the main obstacles in slice preparations to obtaining the high-resistance seals necessary for patch clamp recording is that access to neurons is often obstructed by glial processes and the neuropil. Three basic methodologies have been used to overcome this problem. To obtain high resistance seals, Edwards and co-workers (1989) used thinner slices (150 μm, maximum thickness) than those used for conventional intra- or extracellular recording and improved access to neurons in the slice by microdissecting away the surrounding tissue with a gentle stream of fluid from a "cleaning pipette". A detailed description of this technique, including technical and methodological tips, can be found in Edwards and Konnerth, 1992. The second, or so-called "blind poke," method involves applying positive pressure through the tip of the patch pipette itself, resulting in a pressure wave which pushes obstructing tissue out of the path of the pipette. Current amplitude in response to a small voltage step is monitored as the pipette is inserted into the tissue, and will decrease as a cell is approached due to increased resistance. Suction is then applied to the pipette in an attempt to form a high resistance seal with the membrane of the neuron (Fig. 5) (Blanton *et al.*, 1989; Malinow and Tsien, 1990). Finally, in perhaps the most sophisticated method employed to date, infrared videomicroscopy (Dodt, 1993) has been utilized to improve visualization of slice structures and facilitate extremely accurate placement of the patch electrode (Stuart *et al.*, 1993). In this method, positive pressure is also applied through the patch pipette to remove obstructing tissue. Because of the excellent visual control of electrode placement using infrared videomicroscopy, recordings can be made in the whole cell, cell attached, or outside-out configuration from dendrites as well as cell bodies, and infrequently occurring neuronal types can be more readily identified and recorded from (Stuart *et al.*, 1993).

To record in the whole cell mode from the cell soma, patch pipettes have resistances between 2 and 7 MΩ (Malinow and Tsien, 1990; Stuart *et al.*, 1993). For recording from smaller structures such as dendrites, pipettes with higher resistances (12 MΩ) are required (Stuart *et al.*, 1993). As with conventional patch clamp recording, patch pipettes are fire polished to improve seal formation and coated with Sylgard to decrease pipette noise. However, these precautions may not always be necessary (Malinow and Tsien, 1990). The extracellular solution is typically ACSF, whereas the pipette solution will vary depending on experimental conditions (i.e., whole cell or cell-attached patch). For recording from CA$_1$ cells in the hippocampus, bipolar tungsten electrodes placed in the stratum radiatum are used for stimulating the Schaffer collaterals which synapse with CA$_1$ dendrites. However, it is also possible to record from paired CA$_3$/CA$_1$ neurons by obtaining a patch on a CA$_1$ neuron, then stimulating CA$_3$ neurons with an intracellular electrode until a response is observed in the CA$_1$ neuron (Malinow, 1991). Using patch clamp techniques, recordings in excess of 10 hr have been obtained from a single neuron (Malinow and Tsien, 1990). Thus, the effects of neurotoxicants can be examined on the function of a single neuron over a relatively long period of exposure. In some experiments, 20–100 μM picrotoxin is included in the bath solution to block inhibitory responses (Malinow and Tsien, 1990; Hessler *et al.*, 1993). In this case, the CA$_3$ input into CA$_1$ is separated by cutting between CA$_3$ and CA$_1$ or removing CA$_3$ to prevent epileptiform activity (Malinow and Tsien, 1990).

Quantal analysis of synaptic transmission in the CNS is more complex than at the NMJ; however, it is possible to estimate quantal content and quantal amplitude (Malinow and Tsien, 1990; Malinow, 1991; Lisman and Harris, 1993) as well as the probability of release (Hessler *et al.*, 1993). In order to limit the number of synapses activated, the intensity of the stimulus is usually adjusted to just above the level that produces failures consistently. Alternatively, one can record from paired CA$_3$/CA$_1$ cells. A simplified measure of presynaptic function in the hippocampus is determined by measuring M, the mean synaptic current, and σ^2, the variance about M, and then computing M^2/σ^2 (see Malinow and Tsien, 1990 for further details). In central neurons, M^2/σ^2 is affected by the same manipulations that act presynaptically at the NMJ to alter quantal content. Thus, M^2/σ^2 is increased by increasing $[Ca^{2+}]_e$, lowering $[Mg^{2+}]_e$, or applying 4-aminopyridine (Malinow and Tsien, 1990). In contrast, treatments that alter postsynaptic function, such as blockage of glutamate receptors, do not alter M^2/σ^2, despite having large effects on the current amplitude (Malinow and Tsien, 1990).

Because techniques for patch clamp recording from slices have been developed so recently, they have not yet been used in neurotoxicological investigations.

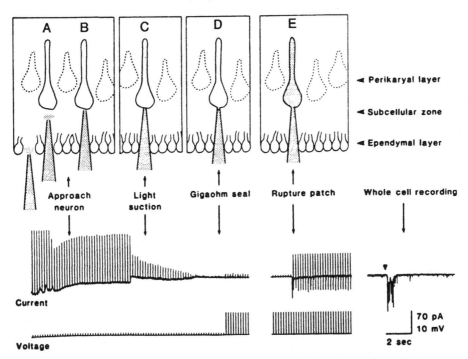

FIGURE 5 Events in the formation of a whole cell recording in intact turtle cerebral cortex. The top portion of the figure depicts the electrode (stippled) bypassing the ependymal glial layer, then approaching and sealing onto a neuron; these events are monitored (bottom portion) by observing the amplitude of current produced by small voltage steps applied to the pipette. A decrease in the current amplitude (A,B) reveals a resistance increase as a neuron is approached. Application of suction (C) results in a further decrease in current amplitude as a gigaohm seal forms (D), more easily measured if the voltage step amplitude is increased. The patch is ruptured by additional suction, and electrical continuity between pipette and cell interior is obtained (E), yielding a whole cell recording. Stimulation of the optic tract in a turtle hemisphere preparation (triangle) produced a barrage of synaptic currents (Vh = −70 mV). (From Blanton *et al.*, 1989.)

Patch clamp recordings have been used to investigate mechanisms underlying synaptic plasticity in the hippocampus. Induction of LTP in CA_1 neurons results in an increase in quantal content, decreased failures, and increased quantal amplitude (Malinow and Tsien, 1990; Malinow, 1991; Kullmann and Nicholl, 1992). Presynaptic mechanisms may therefore be important in expression of LTP in CA_1, although postsynaptic mechanisms may still contribute to this phenomenon. The ability to apply quantal analysis to mechanisms of plasticity in the CNS has great potential for the study of neurotoxic compounds which produce effects on learning and memory.

V. Conclusion

Electrophysiological analysis of synaptic transmission can provide detailed information on the mechanisms by which a neurotoxic compound may disrupt nerve function. The ability to examine nerve function in the central or peripheral nervous system using quantitative techniques on a real-time basis provides information about the actions of a compound which, generally, cannot be provided by biochemical or histological techniques. Thus, these techniques compliment each other, providing a better overall understanding of the mechanism of action of a neurotoxic compound. Information collected using peripheral preparations such as the NMJ has already provided an understanding of the mechanisms of action of a number of metals and other compounds on synaptic transmission at this synapse. With the recent advances in recording techniques in the central nervous system, it is expected that, in the near future, much additional information will be collected on the mechanism of action of neurotoxicants at central synapses as well.

Acknowledgments

The authors thank Mr. Douglas Whitehouse of the Environmental Protection Agency for making the drawings presented in Fig. 1 and Ms. Rachel Barton for her excellent clerical assistance. In addition, we thank Dr. John M. Spitsbergen of the University of Virginia and Dr. William K. Boyes of the Environmental Protection Agency for their helpful comments and suggestions on an earlier version of this chapter. This work was supported in part by NIH Grants ES03299 and NS 20683 to WDA.

References

Adams, M. E., Myers, R. A., Imperial, J. S., and Olivera, B. M. (1993). Toxityping rat brain calcium channels with ω-toxins from spider and cone snail venoms. *Biochemistry* **32,** 12566–12570.

Adams, P. R., and Sakmann, B. (1978). Decamethonium both opens and blocks end-plate channels. *Proc. Natl. Acad. Sci. U.S.A.* **75,** 2994–2998.

Albuquerque, E. X., Akaike, A., Shaw, K.-P., and Rickett, D. L. (1984). The interaction of anticholinesterase agents with the acetylcholine receptor-ionic channel complex. *Fundam. Appl. Toxicol.* **4,** S27–S33.

Albuquerque, E. X., Deshpande, S. S., Kawabuchi, M., Aracava, Y., Idriss, M., Rickett, D. L., and Boyne, A. F. (1985). Multiple actions of anticholinesterase agents on chemosensitive synapses: molecular basis for prophylaxis and treatment of organophosphate poisoning. *Fundam. Appl. Toxicol.* **5,** S182–S203.

Alkondon, M., Costa, A. C. S., Radhakrishnan, V., Aronstam, K. S., and Albuquerque, E. X. (1990). Selective blockade of NMDA-activated channel currents may be implicated in learning deficits caused by lead. *FEBS Lett.* **261,** 124–130.

Allen, J. E., Gage, P. W., Leaver, D. D., and Leow, A. T. (1980). Triethyltin depresses evoked transmitter release at the mouse neuromuscular junction. *Chem. Biol. Interact.* **31,** 227–231.

Alnaes, A., and Rahamimoff, R. (1975). On the role of mitochondria in transmitter release from motor nerve terminals. *J. Physiol. (Lond.)* **248,** 285–306.

Altmann, L., Sveinsson, K., and Wiegand, H. (1991). Long-term potentiation in rat hippocampal slices is impaired following acute lead perfusion. *Neurosci. Lett.* **128,** 109–112.

Altmann, L., Gutowski, M., and Wiegand, H. (1994). Effects of maternal lead exposure on functional plasticity in the visual cortex and hippocampus of immature rats. *Dev. Brain Res.* **81,** 50–56.

Andersen, P., Løyning, A., and Eccles, J. C. (1964a). Location of post-synaptic inhibitory synapses of hippocampal pyramids. *J. Neurophysiol.* **27,** 592–607.

Andersen, P., Løyning, A., and Eccles, J. C. (1964b). Pathway of postsynaptic inhibition in the hippocampus. *J. Neurophysiol.* **27,** 608–619.

Anderson, A. J., and Harvey, A. M. (1987). ω-Conotoxin does not block the verapamil-sensitive calcium channels at mouse motor nerve terminals. *Neurosci. Lett.* **82,** 177–180.

Anderson, A. J., and Harvey, A. M. (1988). Effects of the facilitatory compounds catechol, guanidine, noradrenaline and phencyclidine on presynaptic currents of mouse motor nerve terminals. *Naünyn-Schmied. Arch. Pharm.* **338,** 133–137.

Arakawa, O., Nakahiro, M., and Narahashi, T. (1991). Mercury modulation of GABA-induced chloride channels and non-specific cation channels in dorsal root ganglion neurons. *Brain Res.* **551,** 58–63.

Atchison, W. D. (1986). Extracellular calcium-dependent and -independent effects of methylmercury on spontaneous and potassium-evoked release of acetylcholine from the neuromuscular junction. *J. Pharmacol. Exp. Ther.* **237,** 672–678.

Atchison, W. D. (1987). Neurophysiological effects of mercurials. In *Neurotoxicity of Methylmercury* (Z. Annau and C. Eccles, Eds.), pp. 189–219, Johns Hopkins Press, New York.

Atchison, W. D. (1988). Effects of neurotoxicants on synaptic transmission: lessons learned from electrophysiological studies. *Neurotoxicol. Teratol.* **10,** 393–416.

Atchison, W. D. (1989). Dihydropyridine-sensitive and -insensitive components of acetylcholine release from rat motor nerve terminals. *J. Pharmacol. Exp. Ther.* **251,** 672–678.

Atchison, W. D., Clark, A. W., and Narahashi, T. (1984). Presynaptic effects of methylmercury at the mammalian neuromuscular junction. In *Cellular and Molecular Neurotoxicology* (T. Narahashi, Ed.), pp. 23–43, Raven Press, New York.

Atchison, W. D., Joshi, U., and Thornburg, J. E. (1986). Irreversible suppression of calcium entry into nerve terminals by methylmercury. *J. Pharmacol. Exp. Ther.* **238,** 618–624.

Atchison, W. D., and Narahashi, T. (1982). Methylmercury-induced depression of neuromuscular transmission in the rat. *Neurotoxicology* **3,** 37–50.

Atchison, W. D., and Narahashi, T. (1984). Mechanism of action of lead. *Neurotoxicology* **5,** 267–284.

Atchison, W. D., and Spitsbergen, J. M. (1994). The neuromuscular junction as a target for toxicity. In *Principles of Neurotoxicology* (L. W. Chang, Ed.), pp. 265–307, Dekker, New York.

Audesirk, (1995). Electrophysiological analysis of ion channel function. In *Neurotoxicology: Approaches and Methods* (L. W. Chang and W. Slikker, Eds.), pp. 137–156, Academic Press, New York.

Augustine, G. J., Charlton, M. P., and Smith, S. J. (1987). Calcium action in synaptic transmitter release. *Ann. Rev. Neurosci.* **10,** 633–693.

Augustine, G. J., Adler, E. M., and Charlton, M. P. (1991). The calcium signal for transmitter secretion from presynaptic nerve terminals. *Ann. N. Y. Acad. Sci.* **635,** 365–381.

Baker, P. F., and Crawford, A. C. (1975). A note on the mechanism by which inhibitors of the sodium pump accelerate spontaneous release of transmitter from motor nerve terminals. *J. Physiol. (Lond.)* **247,** 209–226.

Baker, P. F., Hodgkin, A. L., and Ridgway, E. B. (1971). Depolarization and calcium entry in squid giant axons. *J. Physiol. (Lond.)* **218,** 709–755.

Barrett, J., Botz, D., and Chang, D. B. (1974). Block of neuromuscular transmission by methylmercury. Behavioral toxicology. In *Early Detection of Occupational Hazards* (C. Zintaras, B. L. Johnson, and I. de Groot, Eds.), pp. 277–287. U.S. Dept. of Health, Education, and Welfare.

Bennett, M. R., Florin, T., and Hall, R. (1975). The effect of calcium ions on the binomial statistic parameters which control acetylcholine release at synapses in striated muscle. *J. Physiol. (Lond.)* **247,** 429–446.

Bennett, M. K., Calakos, N., and Scheller, R. H. (1992). Syntaxin: a synaptic protein implicated in the docking of synaptic vesicles at presynaptic active zones. *Science* **257,** 255–259.

Bennett, M. K., Garcia-Arràs, J. E., Elferink, L. A., Peterson, K., Fleming, A., Hazuka, C. D., and Scheller, R. H. (1993). The syntaxin family of vesicular transport receptors. *Cell* **74,** 863–873.

Bierkamper, G. G. (1981). Electrophysiological effects of diisopropylfluorophosphate on neuromuscular transmission. *Eur. J. Pharmacol.* **73,** 343–348.

Bierkamper, G. G. (1987). Synaptic toxicology of environmental agents. In *Electrophysiology in Neurotoxicology* (H. E. Lowndes, Ed.), Vol. 2, pp. 99–133, CRC Press, Boca Raton, Florida.

Binah, O., Meiri, U., and Rahamimoff, H. (1978). The effects of HgCl$_2$ and mersalyl on mechanisms regulating intracellular calcium and transmitter release. *Eur. J. Pharmacol.* **51,** 453–457.

Blanton, M. G., Lo Turco, J. J., and Kreigstein, A. R. (1989). Whole cell recording from neurons in slices of reptilian and mammalian cortex. *J. Neurosci. Methods* **30,** 203–210.

Blasi, J., Chapman, E. R., Link, E., Binz, T., Yamasaki, S., De-Camilli, P., Südhof, T. C., Niemann, H., and Jahn, R. (1993a). Botulinum neurotoxin A selectively cleaves the synaptic protein SNAP-25. *Nature (Lond.)* **365**, 160–163.

Blasi, J., Chapman, E. R., Yamasaki, S., and Binz, T. (1993b). Botulinum neurotoxin C1 blocks neurotransmitter release by means of cleaving HPC-1/syntaxin. *EMBO J.* **12**, 4821–4828.

Blioch, Z. L., Glagoleva, I. M., Liberman, E. A., and Nenashev, V. A. (1968). A study of the mechanism of quantal transmitter release at a chemical synapse. *J. Physiol. (Lond.)* **199**, 11–35.

Bourret, C., and Mallart, A. (1989). Depression of calcium current at mouse motor nerve endings by polycationic antibiotics. *Brain Res.* **478**, 403–406.

Bowersox, C., Ko, C-P., Sugiura, Y., Li, C. Z., Fox, J., Hoffman, B. B., and Miljanich, G. (1993). Omega-conopeptide SNX-230 (MVIIC) blocks calcium current in mouse neuromuscular junction nerve terminals. *Neurosci. Abstr.* **19**, 1478.

Brigant, J. L., and Mallart, A. (1982). Presynaptic currents in mouse motor nerve endings. *J. Physiol. (Lond.)* **333**, 619–636.

Brown, T. H., Wong, R. K. S., and Prince, D. A. (1979). Spontaneous miniature synaptic potentials in hippocampal neurons. *Brain Res.* **177**, 194–199.

Brown, T. H., Fricke, R. A., and Perkel, D. H. (1981). Passive electrical constants in three classes of hippocampal neurons. *J. Neurophysiol.* **46**, 812–827.

Brown, T. H., and Johnston, D. (1983). Voltage-clamp analysis of mossy fiber synaptic input to hippocampal neurons. *J. Neurophysiol.* **50**, 487–507.

Bülbring, E. (1946). Observations on the isolated phrenic nerve diaphragm preparation of the rat. *Br. J. Pharmacol.* **1**, 38–61.

Carbone, E., and Lux, L. D. (1984). A low voltage-activated, fully inactivating Ca channel in vertebrate sensory neurons. *Nature (Lond.)* **310**, 501–503.

Charlton, M. P., and Augustine, G. J. (1990). Classification of presynaptic calcium channels at the squid giant synapse: neither T-, L-nor N-type. *Brain Res.* **525**, 133–139.

Clark, A. W., Harlbut, W. F., and Mauro, A. (1972). Changes in fine structure of the neuromuscular junction caused by black widow spider venom. *J. Cell. Biol.* **52**, 1–14.

Clark, A. W., Mauro, A., Longenecker, H. E., and Hurlbut, W. P. (1970). Effects of black widow spider venom on the frog neuromuscular junction. *Nature (Lond.)* **225** 703–705.

Cohen, M. W., Jones, O. T., and Angelides, K. J. (1991). Distribution of Ca^{2+} channels in frog motor nerve terminals revealed by fluorescent ω-conotoxin. *J. Neurosci.* **11**, 1032–1039.

Cole, K. S., and Curtis, H. J. (1938). Electric impedance of *Nitella* during activity. *J. Gen. Physiol.* **22**, 37–64.

Cole, K. S., and Curtis, H. J. (1939). Electric impedance of the squid giant axon during activity. *J. Gen. Physiol.* **22**, 649–670.

Colomo, F., and Erulkar, S. D. (1968). Miniature synaptic potentials at frog spinal neurons in the presence of tetrodotoxin. *J. Physiol. (Lond.)* **199**, 205–221.

Colquhoun, D., Jonas, P., and Sakmann, B. (1992). Action of brief pulses of glutamate on AMPA/kainate receptors in patches from different neurons of rat hippocampal slices. *J. Physiol. (Lond.)* **458**, 261–287.

Colton, C. A. (1976). Postsynaptic effect of La^{3+} at the frog neuromuscular junction. *J. Neurobiol.* **7**, 87–91.

Cooper, G. P., and Manalis, R. S. (1983). Influence of heavy metals on synaptic transmission: a review. *Neurotoxicology* **4**, 69–84.

Cooper, G. P., and Manalis, R. S. (1984a). Interactions of lead and cadmium on acetylcholine release at the frog neuromuscular junction. *Toxicol. Appl. Pharm.* **74**, 411–416.

Cooper, G. P., and Manalis, R. S. (1984b). Cadmium: effects on transmitter release at the frog neuromuscular junction. *Eur. J. Pharmacol.* **99**, 251–256.

Cooper, G. P., Suszkiw, J. B., and Manalis, R. S. (1984). Presynaptic effects of heavy metals. In *Cellular and Molecular Neurotoxicology* (T. Narahashi, Ed.), pp. 1–21, Raven Press, New York.

Cruz, L. J., Gray, W. R., Oliviera, B. M., Zeikus, R. D., Kerr, L., Yoshikami, D., and Modzdlowski, E. (1985). *Conus geographicus* toxins that discriminate between neuronal and muscle sodium channels. *J. Biol. Chem.* **260**, 9280–9288.

deBassio, W. A., Schnitzler, R. M., and Parsons, R. L. (1971). Influence of lanthanum on transmitter release at the neuromuscular junction. *J. Neurobiol.* **2**, 263–278.

De Bello, W. M., Betz, H., and Augustine, G. J. (1993). Synaptotagmin and neurotransmitter release. *Cell* **74**, 947–950.

del Castillo, J., and De Motta, G. E. (1978). A new method for excitation-contraction uncoupling in frog skeletal muscle. *J. Cell Biol.* **78**, 782–784.

del Castillo, J., and Engbaek, L. (1954). The nature of the neuromuscular block produced by magnesium. *J. Physiol. (Lond.)* **124**, 370–384.

del Castillo, J., and Katz, B. (1954). Quantal components of the endplate potential. *J. Physiol. (Lond.)* **124**, 560–573.

del Castillo, J., and Katz, B. (1955). On the localization of acetylcholine receptors. *J. Physiol. (Lond.)* **128**, 157–181.

del Castillo, J., and Katz, B. (1956). Localization of active spots within the neuromuscular junction of the frog. *J. Physiol. (Lond.)* **132**, 630.

Dodge, F., Jr., and Rahamimoff, R. (1967). Co-operative action of calcium ions in the transmitter release at the neuromuscular junction. *J. Physiol. (Lond.)* **193**, 419–432.

Dodt, H.-U. (1993). Infrared-interference videomicroscopy of living brain slices. *Adv. Exp. Biol. Med.* **333**, 245–249.

Dreyer, F., and Penner, R. (1987). The actions of presynaptic snake toxins on membrane currents of mouse motor nerve terminals. *J. Physiol. (Lond.)* **386**, 455–463.

Dudel, J., and Kuffler, S. W. (1961). Presynaptic inhibition at the crayfish neuromuscular junction. *J. Physiol. (Lond.)* **155**, 543–562.

Dutar, P., Rascol, O., and Lamour, Y. (1989). ω-Conotoxin GVIA blocks synaptic transmission in the CA1 field of the hippocampus. *Eur. J. Pharmacol.* **174**, 261–266.

Edwards, F. A., and Konnerth, A. (1992). Patch-clamping cells in sliced tissue preparations. In *Methods in Enzymology* **207**, 208–222.

Edwards, F. A., Konnerth, A., Sakmann, B., and Takahashi, T. (1989). A thin slice preparation for patch clamp recordings from neurons of the mammalian central nervous system. *Pflügers Arch.* **414**, 600–612.

Ellinor, P. T., Zhang, J.-F., Randall, A. D., Zhou, M., Schwarz, T. L., Tsien, R. W., and Horne, W. A. (1993). Functional expression of a rapidly inactivating neuronal calcium channel. *Nature (Lond.)* **363**, 455–458.

Elmqvist, D., and Feldman, D. S. (1965). Calcium dependence of spontaneous acetylcholine release at mammalian motor nerve terminals. *J. Physiol. (Lond.)* **181**, 487–497.

Fatt, P., and Katz, B. (1952). Spontaneous subthreshold activity at motor nerve endings. *J. Physiol. (Lond.)* **117**, 109–128.

Fischer von Mollard, G., Mignery, G. A., Baumert, M., Burger, P. R., Perin, M., Jahn, R., and Südhof, T. C. (1990). Rab3 is a small GTP-binding protein exclusively localized to synaptic vesicles. *Proc. Natl. Acad. Sci. U.S.A.* **87**, 1988–1992.

Fischer von Mollard, G., Südhof, T. C., and Jahn, R. (1991). A small GTP-binding protein (*rab3A*) dissociates from synaptic vesicles during exocytosis. *Nature (Lond.)* **349**, 79–81.

Forshaw, P. J. (1977). The inhibitory effect of cadmium on neuromuscular transmission in the rat. *Eur. J. Pharmacol.* **42**, 371–377.

Fountain, S. B., Ting, Y. L. T., Hennes, S. K., and Teyler, T. J. (1988). Triethyltin exposure suppresses synaptic transmission in area CA1 of the rat hippocampal slice. *Neurotoxicol. Teratol.* **10,** 539–548.

Fountain, S. B., Ting, Y. L. T., and Teyler, T. J. (1992). The *in vitro* hippocampal slice preparation as a screen for neurotoxicity. *Toxicol. in Vitro* **6,** 77–87.

Fox, A. P., Nowycky, M. C., and Tsien, R. W. (1987a). Kinetic and pharmacological properties distinguishing three types of calcium currents in chick sensory neurones. *J. Physiol. (Lond.)* **394,** 149–172.

Fox, A. P., Nowycky, M. C., and Tsien, R. W. (1987b). Single channel recordings of three types of calcium channels in chick sensory neurons. *J. Physiol. (Lond.)* **394,** 173–200.

Gaddum, J. H. (1957). Theories of drug antagonism. *Pharmacol. Rev.* **9,** 211–218.

Glagoleva, I. M., Liberman, Y. A., and Khashayev, Z. K. M. (1970). Effect of uncoupling agents of oxidative phosphorylation on the release of acetylcholine from nerve ending. *Biofizika* **15,** 76–83.

Glavinovic, M. I. (1979). Voltage clamping of unparalyzed cut rat diaphragm for study of transmitter release. *J. Physiol. (Lond.)* **290,** 467–480.

Gundersen, C. B., Katz, B., and Miledi, R. (1982). The antagonism between botulinum toxin and calcium in motor nerve terminals. *Proc. R. Soc. London B.* **216,** 369–376.

Hagiwara, S., Fukuda, J., and Eaton, D. (1974). Membrane currents carried by Ca, Sr and Ba in barnacle muscle during voltage clamp. *J. Gen. Physiol.* **63,** 564–578.

Hamill, O. P., Marty, A., Neher, E., Sakmann, B., and Sigworth, F. J. (1981). Improved patch-clamp techniques for high resolution current recording from cells and cell-free membrane patches. *Pfluegers Arch.* **391,** 85–100.

Hamilton, B. R., and Smith, D. O. (1991). Autoreceptor-mediated purinergic and cholinergic inhibition of motor nerve terminal calcium currents in the rat. *J. Physiol. (Lond.)* **432,** 327–341.

Hamilton, B. R., and Smith, D. O. (1993). Calcium currents in rat motor nerve terminals. *Brain Res.* **584,** 123–131.

Hare, M. F., and Atchison, W. D. (1992). Comparative action of methylmercury and divalent inorganic mercury on nerve terminal and intraterminal mitochondrial membrane potentials. *J. Pharmacol. Exp. Ther.* **261,** 166–172.

Hare, M. F., McGinnis, K. M., and Atchison, W. D. (1993). Methylmercury increases intracellular concentrations of Ca^{2+} and heavy metals in NG108-15 cells. *J. Pharmacol. Exp. Ther.* **266,** 1626–1635.

Harris, A. J., and Miledi, R. (1971). The effect of Type D botulinum toxin on frog neuromuscular functions. *J. Physiol. (Lond.)* **217,** 497–515.

Hatt, H., and Smith, D. O. (1976). Synaptic depression related to presynaptic axon conduction block. *J. Physiol. (Lond.)* **259,** 367–393.

Hess, P., and Tsien, R. W. (1984). Mechanism of ion permeation through Ca^{2+} channels. *Nature (Lond.)* **309,** 453–456.

Hess, P., Lansman, J. B., and Tsien, R. W. (1986). Calcium channel selectivity for divalent and monovalent cations. Voltage and concentration dependence of single channel current in ventricular heart cells. *J. Gen. Physiol.* **88,** 293–319.

Hessler, N. A., Shirke, A. M., and Malinow, R. (1993). The probability of transmitter release at a mammalian central synapse. *Nature (Lond.)* **366,** 569–572.

Heuser, J., and Miledi, R. (1971). Effects of lanthanum ions on function and structure of frog neuromuscular junctions. *Proc. R. Soc. Biol.* **179,** 247–260.

Hewett, S. J., and Atchison, W. D. (1992). Effects of charge and lipophilicity on mercurial-induced reduction of $^{45}Ca^{2+}$ uptake in isolated nerve terminals of the rat. *Toxicol. Appl. Pharmacol.* **113,** 267–273.

Hille, B. (1993). *Ionic Channels of Excitable Membranes.* 2nd ed., Sinauer Associates Sunderland, Massachusetts.

Hirning, L. D., Fox, A. P., McCleskey, E. W., Olivera, B. M., Thayer, S. A., Miller, R. J., and Tsien, R. W. (1988). Dominant role of N-type Ca^{2+} channels in evoked release of norepinephrine from sympathetic neurons. *Science* **239,** 57–61.

Hodgkin, A. L., and Huxley, A. F. (1952). A quantative description of membrane current and its application to conduction and excitation in the nerve. *J. Physiol. (Lond.)* **117,** 500–544.

Hodgkin, A. L., Huxley, A. F., and Katz, B. (1949). Ionic currents underlying activity in the giant axon of the squid. *Arch. Sci. Physiol.* **3,** 129–150.

Hodgkin, A. L., Huxley, A. F., and Katz, B. (1952). Measurements of current-voltage relations in the membrane of the giant axon of *Loligo*. *J. Physiol. (London)* **116,** 428–448.

Hofmann, F., Biel, M., and Flockerzi, V. (1994). Molecular basis for Ca^{2+} channel diversity. *Annu. Rev. Neurosci.* **17,** 399–418.

Holz, G. G., Dunlap, K., and Kream, R. M. (1988). Characterization of the electrically-evoked release of substance P from dorsal root ganglion neurones: Methods and dihydropyridine sensitivity. *J. Neurosci.* **8,** 463–471.

Hong, S. J., and Chang, C. C. (1989). Use of geographutoxin II (μ-conotoxin) for the study of neuromuscular transmission in the mouse. *Br. J. Pharmacol.* **97,** 934–940.

Hori, N., Büsselberg, D., Matthews, M. R., Parsons, P. J., and Carpenter, D. O. (1993). Lead blocks LTP by an action not at NMDA receptors. *Exp. Neurol.* **119,** 192–197.

Howell, J. N., and Jenden, D. J. (1967). Tubules of skeletal muscle; morphological alterations which interrupt excitation-contraction coupling. *Fed. Proc.* **26,** 553.

Hubbard, J. I. (1970). Mechanisms of transmitter release. *Prog. Biophys. Mol. Biol.* **21,** 33–124.

Hubbard, J. I. (1973). Microphysiology of the vertebrate neuromuscular junction. *Physiol. Rev.* **53,** 674–723.

Hubbard, J. I., and Løyning, Y. (1966). The effects of hypoxia on neuromuscular transmission in a mammalian preparation. *J. Physiol. (Lond.)* **185,** 205–223.

Hubbard, J. I., and Schmidt, R. F. (1963). An electrophysiological investigation of mammalian motor nerve terminals. *J. Physiol. (Lond.)* **166,** 145–167.

Hubbard, J. I., Jones, S. F., and Landau, E. M. (1968). On the mechanism by which calcium and magnesium affect the spontaneous release of transmitter from mammalian motor nerve terminals. *J. Physiol. (Lond.)* **194,** 353–380.

Inoue, A., Obata, K., and Akagawa, K. (1992). Cloning and sequence analysis of cDNA for a neuronal membrane antigen, HPC-1. *J. Biol. Chem.* **267,** 10613–10619.

Jahn, R., and Südhof, T. C. (1993). Synaptic vesicle traffic: rush hour in the nerve terminal. *J. Neurochem.* **61,** 12–21.

Jahn, R., and Südhof, T. C. (1994). Synaptic vesicles and exocytosis. *Ann. Rev. Neurosci.* **17,** 219–246.

Johnston, D., Hablitz, J. J., and Wilson, W. A. (1980). Voltage clamp discloses slow inward current in hippocampal burst-firing neurons. *Nature (Lond.)* **286,** 391–393.

Jonas, P., and Sakmann, B. (1992). Glutamate receptor channels in isolated patches from CA1 and CA3 pyramidal cells of rat hippocampal slices. *J. Physiol. (Lond.)* **455,** 143–171.

Juang, M. S. (1976). An electrophysiological study of the action of methylmercuric chloride and mercuric chloride on the sciatic nerve-sartorius muscle preparation of the frog. *Toxicol. Appl. Pharmacol.* **35,** 339–348.

Juang, M. S., and Yonemura, K. (1975). Increased spontaneous transmitter release from presynaptic nerve terminals by methylmercuric chloride. *Nature (Lond.)* **256,** 211–213.

Katz, B. (1962). The transmission of impulses from nerve to muscle, and the subcellular unit of synaptic action. *Proc. R. Soc. London B.* **155**, 455–477.

Katz, B., and Miledi, R. (1967a). The timing of calcium action during neuromuscular transmission. *J. Physiol. (Lond.)* **189**, 535–544.

Katz, B., and Miledi, R. (1967b). A study of synaptic transmission in the absence of nerve impulses. *J. Physiol. (Lond.)* **192**, 407–436.

Katz, B., and Thesleff, S. (1957). On the factors which determine the amplitude of the 'miniature end-plate potential'. *J. Physiol. (Lond.)* **137**, 267–278.

Kelly, S. S., Anis, N., and Robbins, N. (1985). Fluorescent staining of living mouse neuromuscular junctions. *Pfluegers Arch.* **404**, 97–99.

Kolton, L., and Yaari, Y. (1982). Sites of action of lead on spontaneous transmitter release from motor nerve terminals. *Isr. J. Med. Sci.* **18**, 165–170.

Konishi, T. (1985). Electrical excitability of motor nerve terminals in the mouse. *J. Physiol. (Lond.)* **366**, 411–421.

Kongsamut, S., and Miller, R. J. (1986). Nerve growth factor modulates the drug sensitivity of neurotransmitter release from PC12 cells. *Proc. Natl. Acad. Sci. U.S.A.* **83**, 2243–2247.

Kostyuk, P. G., Shuba, Y. M., and Savchenko, A. N. (1988). Three types of calcium channels in the membrane of mouse sensory neurones. *Pfluegers Arch.* **411**, 661–669.

Kuffler, S. W., and Yoshikami, D. (1975). The number of ACh molecules in a quantum: an estimate from iontophoretic application of acetylcholine at the neuromuscular synapse. *J. Physiol. (Lond.)* **251**, 465–482.

Kullmann, D. M., and Nicholl, R. A. (1992). Long-term potentiation is associated with increases in quantal content and quantal amplitude. *Nature (Lond.)* **357**, 240–244.

Lambert, J. J., Durant, N. N., Reynolds, L. S., Volle, R. L., and Henderson, E. G. (1980). Characterization of end-plate conductance in transected frog muscle: modification by drugs. *J. Pharmacol. Exp. Ther.* **216**, 62–69.

Lansmann, J. B., Hess, P., and Tsien, R. W. (1986). Blockade of current through single calcium channels by Cd^{2+}, Mg^{2+} and Ca^{2+}. Voltage and concentration dependence of calcium entry into the pore. *J. Gen. Physiol.* **88**, 321–347.

Leveque, C., Hoshino, T., David, P., Shoji-Kasai, Y., Leys, K., Omori, A., Lang, B., El Far, O., Sato, K., Martin-Moutot, N., Newsom-Davis, J., Takahashi, M., and Seagar, M. J. (1992). The synaptic vesicle protein synaptotagmin associates with calcium channels and is a putative Lambert-Eaton myasthenic syndrome antigen. *Proc. Natl. Acad. Sci. U.S.A.* **89**, 3625–3629.

Levesque, P. C., and Atchison, W. D. (1987). Interactions of mitochondrial inhibitors with methylmercury on spontaneous quantal release of acetylcholine. *Toxicol. Appl. Pharmacol.* **8**, 315–324.

Levesque, P. C., and Atchison, W. D. (1988). Effect of alteration of nerve terminal Ca^{2+} regulation on increased spontaneous quantal release of acetylcholine by methylmercury. *Toxicol. Appl. Pharmacol.* **94**, 55–65.

Lin, Y. W., Rudy, B., and Llinàs, R. R. (1990). Funnel-web spider venom and a toxin fraction block calcium current expressed from rat brain mRNA in *Xenopus* oocytes. *Proc. Natl. Acad. Sci. U.S.A.* **87**, 4538–4532.

Link, E., McMahon, H., Fischer von Mollard, G., Yamasaki, S., Niemann, H., Südhof, T. C., and Jahn, R. (1992). Cleavage of cellubrevin by tetanus toxin does not affect fusion in early endosomes. *J. Biol. Chem.* **268**, 18423–18426.

Llinàs, R., and Yarom, Y. (1981). Properties and distribution of ionic conductances generating electroresponsiveness of mammalian inferior olivary neurones in vitro. *J. Physiol. (Lond.)* **315**, 569–584.

Llinàs, R., Sugimori, M., Lin, J.-W., and Cherksey, B. (1989). Blocking and isolation of a calcium channel from neurons in mammals and cephalopods utilizing a toxin fraction (FTX) from funnel-web spider poison. *Proc. Natl. Acad. Sci. U.S.A.* **86**, 1689–1693.

Lisman, J. E., and Harris, K. M. (1993). Quantal analysis and synaptic anatomy: integrating two views of hippocampal plasticity. *Trends Neurosci.* **16**, 141–147.

Magleby, K. L. (1979). Facilitation, augmentation, and potentiation of transmitter release. *Prog. Brain Res.* **49**, 175–182.

Malinow, R. (1991). Transmission between pairs of hippocampal slice neurons: quantal levels, oscillations, and LTP. *Science* **252**, 722–724.

Malinow, R., and Tsien, R. W. (1990). Presynaptic enhancement shown by whole-cell recordings of long-term potentiation in hippocampal slices. *Nature (Lond.)* **346**, 177–180.

Mallart, A. (1985). Electric current flow inside perineurial sheaths of mouse motor nerves. *J. Physiol. (Lond.)* **368**, 565–575.

Mallart, A., and Brigant, J. L. (1982). Presynaptic currents in mouse motor endings. *J. Physiol. (Lond.)* **333**, 619–636.

Manalis, R. S., Cooper, G. P., and Pomeroy, S. L. (1984). Effects of lead on neuromuscular transmission in the frog. *Brain Res.* **294**, 95–109.

Manalis, R. S., and Cooper, G. P. (1973). Presynaptic and postsynaptic effects of lead at the neuromuscular junction. *Nature (Lond.)* **243**, 354–356.

Marmont, G. (1949). Studies on the axon membrane. I. A new method. *J. Cell. Comp. Physiol.* **34**, 351–382.

Martin, A. R. (1966). Quantal nature of synaptic transmission. *Physiol. Rev.* **46**, 51–66.

Matteoli, M., Takei, K., Cameron, R., Hurlbut, P., Johnson, P. A., Südhof, T. C., Jahn, R., and De Camilli, P. (1991). Association of Rab3A with synaptic vesicles at late stages of the secretory pathway. *J. Cell Biol.* **115**, 625–632.

McArdle, J., Angaut-Petit, D., Mallart, A., Faille, L., and Brigant, J. L. (1981). Advantage of the *triangularis sterni* muscle of the mouse for investigations of synaptic phenomena. *J. Neurosci. Methods* **4**, 109–113.

McLachlan, E. M. (1978). The statistics of transmitter release at a chemical synapse. In *International Review of Neurophysiology, III.* (R. Porter, Ed.), pp. 49–117. University Park Press, Baltimore.

Mintz, I. M., Venema, V. J., Swiderek, K. M., Lee, T. D., Bean, B. P., and Adams, M. E. (1992a). P-type calcium channels blocked by the spider toxin ω-Aga-IVA. *Nature (Lond.)* **355**, 827–829.

Mintz, I. M., Adams, M. E., and Bean, B. P. (1992b). P-type calcium channels in rat central and peripheral neurons. *Neuron* **9**, 85–95.

Miyamoto, M. D. (1975). Binomial analysis of quantal transmitter release at glycerol treated frog neuromuscular junctions. *J. Physiol. (Lond.)* **250**, 121–142.

Moore, C. L. (1971). Specific inhibition of mitochondrial Ca^{2+} transport by ruthenium red. *Biochem. Biophys. Res. Commun.* **42**, 298–305.

Morita, T., Mori, H., Sakimura, K., Mishina, M., Sekine, Y., Tsugita, A., Odani, S., Horikawa, H. P. M., Saisu, H., and Abe, T. (1992). Synaptocanalin I, a protein associated with brain ω-conotoxin-sensitive calcium channels. *Biomed. Res.* **13**, 357–364.

Nachshen, D. A. (1984). Selectivity of the Ca binding site in synaptosome Ca channels. Inhibition by multivalent metal cations. *J. Gen. Physiol.* **83**, 941–967.

Nastuk, W. L. (1951). Membrane potential changes at a single muscle end-plate produced by acetylcholine. *Fed. Proc.* **10**, 96.

Nowycky, M. C., Fox, A. P., and Tsien, R. W. (1985). Three types of neuronal calcium channels with different calcium agonist sensitivity. *Nature (Lond.)* **316**, 440–443.

O'Connor, V., Augustine, G. J., and Betz, H. (1994). Synaptic vesicle exocytosis: molecules and models. *Cell* **76**, 785–787.

Oortgiesen, M., VanKleef, R. G. D. M., Bajnath, R. B., and Vijverberg, H. P. M. (1990). Nanomolar concentrations of lead selectively block neuronal nicotinic acetylcholine responses in mouse neuroblastoma cells. *Toxicol. Appl. Pharmacol.* **103**, 165–174.

Penner, R., and Dreyer, F. (1986). Two different presynaptic calcium currents in mouse motor nerve terminals. *Pfluegers Arch.* **406**, 190–197.

Perney, T. M., Hirning, L. D., Leeman, S. E., and Miller, R. J. (1986). Multiple calcium channels mediate neurotransmitter release from peripheral neurons. *Proc. Natl. Acad. Sci. U.S.A.* **83**, 6656–6659.

Pickett, J. B., and Bornstein, J. C. (1978). Lead competitively inhibits calcium-dependent evoked release at rat neuromuscular junctions. *Neurology* **28**, 335.

Pickett, J. B., and Bornstein, J. C. (1984). Some effects of lead at mammalian neuromuscular junction. *Am. J. Physiol.* **246**, C271–C276.

Plummer, M. R., Logothetis, D. E., and Hess, P. (1989). Elementary properties and pharmacological sensitivities of calium channels in mammalian peripheral neurons. *Neuron* **2**, 1453–1463.

Prince, D. A. (1978). Neurophysiology of epilepsy. *Ann. Rev. Neurosci.* **1**, 395–415.

Prior, C., Dempster, J., and Marshall, I. G. (1993). Electrophysiological analysis of transmission at the skeletal neuromuscular junction. *J. Pharmacol. Toxicol. Methods* **30**, 1–17.

Protti, D. A., and Uchitel, O. D. (1993). Transmitter release and presynaptic Ca^{2+} currents blocked by the spider toxin ω-Aga-IVA. *Neuroreport* **5**, 333–336.

Pumplin, D. W., Reese, T. S., and Llinás, R. (1981). Are the presynaptic membrane particles the calcium channels? *Proc. Natl. Acad. Sci. U.S.A.* **78**, 7210–7213.

Quandt, F. N., Kato, E., and Narahashi, T. (1982). Effects of methylmercury on electrical responses of neuroblastoma cells. *Neurotoxicology* **3**, 205–220.

Rahamimoff, R., and Alnaes, A. (1973). Inhibitory action of ruthenium red on neuromuscular transmission. *Proc. Natl. Acad. Sci. U.S.A.* **70**, 3413–3416.

Ramon ý Cajal, S. (1911). Histologia du systeme nerveux de l'homme et des vertebres. Vol. II, Maloine, Paris.

Randall, A. D., Wendland, B., Schweizer, F., Miljanich, G., Adams, M. E., and Tsien, R. W. (1993). Five pharmacologically distinct high voltage-activated Ca^{2+} channels in cerebellar granule cells. *Neurosci. Abstr.* **19**, 1478.

Rane, S. G., Holtz, G. G., and Dunlap, K. (1987). Dihydropyridine inhibition of neuronal calcium current and substance P release. *Pfluegers Arch.* **409**, 361–366.

Reid, K. H., Edmonds, H. L., Schurr, A., Tseng, M. T., and West, C. A. (1988). Pitfalls in the use of brain slices. *Prog. Neurobiol.* **31**, 1–18.

Robitaille, R., Adler, E. M., and Charlton, M. P. (1990). Strategic location of calcium channels at transmitter release sites of frog neuromuscular synapses. *Neuron* **5**, 733–739.

Sano, K., Enomoto, K., and Maeno, T. (1987). Effects of synthetic ω-conotoxin, a new type of Ca^{2+} channel antagonist, on frog and mouse neuromuscular transmission. *Eur. J. Pharmacol.* **141**, 235–241.

Sather, W. A., Tanabe, T., Mori, Y., Adams, M. E., Miljanich, G., Numa, S., and Tsien, R. W. (1992). Contrasts between cloned BI and cardiac L-type Ca^{2+} channels expressed in *Xenopus* oocytes. *Neurosci. Abstr.* **18**, 10.

Sather, W. A., Tanabe, T., Zhang, J. F., Mori, Y., Adams, M. E., and Tsien, R. W. (1993). Distinctive biophysical and pharmacological properties of class A (BI) calcium channel α_1 subunits. *Neuron* **11**, 291–303.

Satoh, E., Asai, F., Itoh, K., Nishimura, M., and Urakawa, N. (1982). Mechanism of cadmium-induced blockade of neuromuscular transmission. *Eur. J. Pharmacol.* **77**, 251–257.

Sayer, R. J., Friedlander, M. J., and Redman, S. J. (1990). The time course and amplitude of EPSPs evoked at synapses between pairs of CA3/CA1 neurons in the hippocampal slice. *J. Neurosci.* **10**, 826–836.

Schiavo, G., Benfenati, F., Poulain, B., Rossetto, O., Polverino de Laureto, P., DasGupta, B. R., and Montecucco, C. (1992). Tetanus and botulinum-B neurotoxins block neurotransmitter release by protelytic cleavage of synaptobrevin. *Nature (Lond.)* **359**, 832–835.

Schwartzkroin, P. A. (1975). Characteristics of CA1 neurons recorded intracellularly in the hippocampal slice. *Brain Res.* **85**, 423–435.

Schwartzkroin, P. A., and Slawsky, M. (1977). Probable calcium spikes in hippocampal systems. *Brain Res.* **135**, 157–161.

Schwartzkroin, P. A., and Weyler, M. D. (1980). Mechanisms underlying epileptiform burst discharge. *Ann. Rev. Neurol.* **7**, 95–107.

Sevcik, C., and Narahashi, T. (1972). Electrical properties and excitation-contraction coupling in skeletal muscle treated with ethylene glycol. *J. Gen. Physiol.* **60**, 221–236.

Shafer, T. J., and Atchison, W. D. (1989). Block of ^{45}Ca uptake into synaptosomes by methylmercury: Ca^{2+}- and Na^+-dependence. *J. Pharmacol. Exp. Ther.* **248**, 696–702.

Shafer, T. J., and Atchison, W. D. (1991). Methylmercury blocks N- and L-type Ca^{2+} channels in nerve growth factor-differentiated pheochromocytoma cells. *J. Pharmacol. Exp. Ther.* **258**, 697–702.

Shafer, T. J., and Atchison, W. D. (1992). Effects of methylmercury on perineurial Na^+ and Ca^{2+}-dependent potentials at neuromuscular junctions of the mouse. *Brain Res.* **595**, 215–219.

Shafer, T. J., Contreras, M. L., and Atchison, W. D. (1990). Characterization of interactions of methylmercury with Ca^{2+} channels in synaptosomes and pheochromocytoma cells: radiotracer flux and binding studies. *Mol. Pharmacol.* **38**, 102–113.

Shao, Z., and Suszkiw, J. B. (1991). Ca^{2+} surrogate action of Pb^{2+} on acetylcholine release from rat brain synaptosomes. *J. Neurochem.* **56**, 568–574.

Shrivastav, B. B., Brodwick, M. S., and Narahashi, T. (1976). Methylmercury: effects on electrical properties of squid axon membranes. *Life Sci.* **18**, 1077–1082.

Sibergeld, E. K., and Adler, H. S. (1978). Subcellular mechanisms of lead neurotoxicity. *Brain Res.* **148**, 451–467.

Silinsky, E. M. (1978). On the role of barium in supporting the asynchronous release of acetylcholine quanta by motor nerve impulses. *J. Physiol. (Lond.)* **274**, 157–171.

Silinsky, E. M. (1985). The biophysical pharmacology of calcium-dependent acetylcholine secretion. *Pharmacol. Rev.* **37**, 81–132.

Smith, S. J., and Augustine, G. J. (1988). Calcium ions, active zones and synaptic transmitter release. *Trends Neurosci.* **11**, 458–464.

Smith, T. G., Lecar, H., Redman, S. J., and Gage, P. W. (1985). Voltage and patch clamping with microelectrodes. *American Physiological Society,* Bethesda, Maryland.

Snutch, T. P., Leonard, J. P., Gilbert, M. M., Lester, H. A., and Davidson, N. (1990). Rat brain expresses a heterogeneous family of calcium channels. *Proc. Natl. Acad. Sci. U.S.A.* **87**, 3391–3395.

Snutch, T. P., Tomlinson, W. J., Leonard, J. P., and Gilbert, M. M. (1991). Distinct calcium channels are generated by alternative splicing and are differentially expressed in mammalian CNS. *Neuron* **7**, 45–57.

Söllner, T., Bennett, M. K., Whiteheart, S. W., Scheller, R. H., and Rothman, J. E. (1993). A protein assembly-disassembly pathway *in vitro* that may correspond to sequential steps of synaptic vesicle docking, activation and fusion. *Cell* **75**, 409–418.

Spencer, W. A., and Kandel, E. R. (1961). Electrophysiology of hippocampal neurons. IV. Fast prepotentials. *J. Neurophysiol.* **24**, 272–285.

Spitzer, N. (1972). Miniature end-plate potentials at mammalian neuromuscular junctions poisoned by botulinum toxin. *Nature New Biol.* **237**, 26–27.

Stuart, G. J., Dodt, H.-U., and Sakmann, B. (1993). Patch-clamp recordings from the soma and dendrites of neurons in brain slices using infrared video microscopy. *Pfluegers Arch.* **423**, 511–518.

Südhof, T. C., De Camilli, P., Niemann, H., and Jahn, R. (1993). Membrane fusion machinery: Insights from synaptic proteins. *Cell* **75**, 1–4.

Takahashi, T., and Momiyama, A. (1993). Different types of calcium channels mediate central synaptic transmission. *Nature (Lond.)* **366**, 156–158.

Taylor, C. P., and Dudek, F. E. (1982). Synchronous neuronal afterdischarges in rat hippocampal slices without active chemical synapses. *Science* **218**, 810–812.

Teyler, T. J. (1980). Brain slice preparation: hippocampus. *Brain Res. Bull.* **5**, 391–403.

Tomsig, J. L., and Suszkiw, J. B. (1993). Intracellular mechanism of Pb^{2+}-induced norepinephrine release from bovine chromaffin cells. *Am. J. Physiol.* **265**, C1630–C1636.

Traxinger, D. L., and Atchison, W. D. (1987a). Reversal of methylmercury-induced block of nerve-evoked release of acetylcholine at the neuromuscular junction. *Toxicol. Appl. Pharmacol.* **90**, 23–33.

Traxinger, D. L., and Atchison, W. D. (1987b). Comparative effects of divalent cations on the methylmercury-induced alterations of acetylcholine release. *J. Pharmacol. Exp. Ther.* **240**, 451–459.

Tsien, R. W., Lipscombe, D., Madison, D. V., Bley, K. R., and Fox, A. P. (1988). Multiple types of neuronal calcium channels and their selective modulation. *Trends Neurosci.* **11**, 431–438.

Tsien, R. W., Ellinor, P. T., and Horne, W. A. (1991). Molecular diversity of voltage-dependent Ca channels. *Trends Pharmacol. Sci.* **12**, 349–354.

Turner, T. J., Adams, M. E., and Dunlap, K. (1993). Multiple Ca^{2+} channel types coexist to regulate synaptosomal neurotransmitter release. *Proc. Natl. Acad. Sci. U.S.A.* **90**, 9518–9522.

Uchitel, O. D., Protti, D. A., Sanchez, V., Cherksey, B. D., Sugimori, M., and Llinás, R. (1992). P-type voltage-dependent calcium channel mediates presynaptic calcium influx and transmitter release in mammalian synapses. *Proc. Natl. Acad. Sci. U.S.A.* **89**, 3330–3333.

Ujihara, H., and Albuquerque, E. X. (1992). Developmental change of the inhibition by lead of NMDA-activated currents in cultured hippocampal neurons. *J. Pharmacol. Exp. Ther.* **263**, 868–875.

Usherwood, P. R. N. (1963). Spontaneous miniature potentials from insect muscle fibers. *J. Physiol. (Lond.)* **169**, 149–160.

Uteshev, V., Büsselberg, D., and Haas, H. L. (1993). Pb modulates the NMDA-receptor-channel complex. *Naunyn-Schmiedeberg Arch. Pharmacol.* **347**, 209–213.

Wada, K., Mizoguchi, A., Kaibuchi, K., Shirataki, H., Ide, C., and Takai, Y. (1994). Localization of rabophilin 3A, a putative target protein for Rab3A, at the sites of Ca^{2+}-dependent exocytosis in PC12 cells. *Biochem. Biophys. Res. Comm.* **198**, 158–165.

Weakly, J. N. (1973). The action of cobalt ions on neuromuscular transmission in the frog. *J. Physiol. (Lond.)* **234**, 597–612.

Wheeler, D. B., Randall, A., and Tsien, R. W. (1994). Roles of N-type and Q-type Ca^{2+} channels in supporting hippocampal synaptic transmission. *Science* **264**, 107–111.

Wiegand, H. (1984). The action of thallium acetate on spontaneous transmitter release in the rat neuromuscular junction. *Toxicology* **55**, 253–257.

Wong, R. K. S., and Prince, D. A. (1979). Dendritic mechanisms underlying penicillin-induced epileptiform activity. *Science* **204**, 1228–1231.

Wong, R. K. S., Prince, D. A., and Basbaum, A. I. (1979). Intradendritic recordings from hippocampal neurons. *Proc. Natl. Acad. Sci. U.S.A.* **76**, 986–990.

Yuan, Y., and Atchison, W. D. (1993). Disruption by methylmercury of membrane excitability and synaptic transmission of CA1 neurons of hippocampal slices of the rat. *Toxicol. Appl. Pharmacol.* **120**, 203–215.

Zhang, J. F., Ellinor, P. T., Randall, A. D., Zhou, M., Schwarz, T. L., Tsien, R. W., and Horne, W. A. (1993). Functional expression of a novel neuronal voltage-dependent calcium channel, Doe-1. *Neurosci. Abstr.* **19**, 11.

8

Hippocampal Field Potentials: A Model System to Characterize Neurotoxicity

M. E. GILBERT
ManTech Environmental Technology, Inc.
Research Triangle Park, North Carolina 27709

L. J. BURDETTE
Department of Neurology
Graduate Hospital
Philadelphia, Pennsylvania 19131

Research on the neuroanatomy of the hippocampus dates back to over a century ago. The classic trisynaptic circuit of the dentate gyrus, pyramidal cell subfield CA3, and pyramidal cell subfield CA1 has contributed to the hippocampus being one of the most extensively studied regions in the central nervous system (CNS). The deceptively simple architecture, in which afferents are received from neocortical and subcortical regions and terminate in specific lamellae, is suggestive of a highly integrative function with the hippocampus serving as a communication interface between higher and lower order brain centers. This role assumes even greater significance within the context of the large body of evidence documenting the critical involvement of the hippocampus in some forms of learning and mem-

ory (for review see Eichenbaum *et al.,* 1992). The neurological processes subserving memory are encompassed in the term plasticity, one form of which, long-term potentiation (LTP), has been studied extensively in the hippocampus (see reviews by Bliss and Collingridge, 1993; Massicotte and Baudry, 1991). Although evidence obtained within the past decade now challenges the concept of the hippocampus as a simple serial processor (for reviews see Schwartzkroin *et al.,* 1990; Yeckel and Berger, 1990), the complexity of the emerging picture only strengthens the hypothesis of the hippocampus as an important center for the integration of information.

The cytoarchitecture of the hippocampus lends itself to neurophysiological studies of excitatory and inhibitory synaptic inputs, and therefore an analysis of the functional implications of neural insult. This chapter summarizes information provided by *in vivo* field potential recordings, supplemented by findings with field and intracellular recordings obtained from the *in vitro* hippocampal slice preparation. An emphasis is placed on field recordings from the dentate gyrus as this is the site where the majority of the *in vivo* work

Research described in this chapter was supported by the U.S. Environmental Protection Agency (Contract #68-02-4450 to ManTech Environmental Technology Services Inc.) and NIH grant MH-45961. The manuscript has been reviewed by the Health Effects Research Laboratory, U.S. EPA, and approved for publication. Approval does not signify that the contents necessarily reflect the views and policies of the Agency nor does mention of trade names or commercial products constitute endorsement or recommendation for use.

183

has been performed. It describes how changes in the relative contribution of excitatory and inhibitory components of field potentials are determined as a function of the afferent pathways that are stimulated and the types of stimulus protocols that are applied. Both the advantage and disadvantage of field potential recordings is that they reflect the summated output of excitatory and inhibitory influences on an entire population of cells. Although results obtained with field potentials cannot be extrapolated to the cellular level, this chapter demonstrates how these measures have confirmed or directed studies of a mechanistic nature. Finally, the utility of field potentials in identifying processes by which toxicant exposure results in more broadly defined changes in CNS function, such as cell death, learning/memory deficits, and seizure activity, is considered.

I. The Trisynaptic Hippocampal Circuit

The hippocampus is composed of two principal cell types, pyramidal and granule, which form interlocking C and V configurations, respectively (see Fig. 1). Pyramidal cells defining the hippocampus proper have been subdivided by differences in cell size and density into three subfields: CA1, CA2, and CA3. The pyramidal cells are oriented with their apical dendrites extended toward the center of the C. Granule cells comprise the dentate gyrus area and are arranged in a V-shaped band with their dendrites oriented away from the center of the V.

The hippocampus has been traditionally viewed as a trisynaptic serial relay circuit. Axons arising from neurons in entorhinal cortex join to form the perforant path and run medial to the longitudinal axis of the hippocampus. Upon entering the dentate gyrus, the perforant path fibers synapse on the outer two-thirds of the dentate granule cell dendrites (Fig. 1), forming the first link in the trisynaptic circuit. Excitation of dentate granule cells is conveyed by the mossy fibers that course through the internal portion of the V, the hilus, to terminate on the apical dendrites of CA3 pyramidal cells. The third component of the trisynaptic circuit, CA1 pyramidal cells, is richly innervated by CA3 pyramidal cell efferents, the Schaffer collaterals. The relay is completed by CA1 axons that send information back to the entorhinal cortex. Within each subregion, several local inhibitory circuits exist that can be activated by principal cell excitation of interneurons

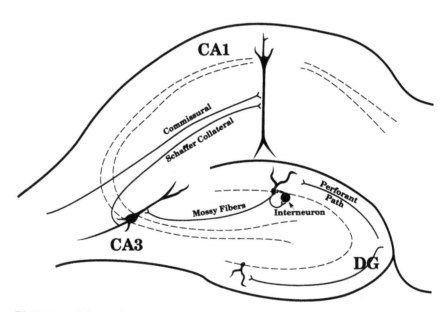

FIGURE 1 Trisynaptic circuit of the hippocampus. Axons of cells in entorhinal cortex course through the perforant path to synapse on the apical dendrites of granule cells in the dentate gyrus. The axons of the granule cells form the mossy fibers that travel through the hilus of the dentate gyrus to synapse on CA3 pyramidal cells. The Schaffer collaterals are the axons of the CA3 pyramidal cells and synapse on the apical dendrites of the CA1 pyramidal cells. The commissural input from the contralateral hippocampus synapses on CA1 pyramidal cell dendrites, dentate granule cells, and cells within the hilus (not shown). Local GABAergic interneurons within each subregion (shown here in the dentate gyrus only) provide powerful inhibitory control over principal cell activity.

which feed back onto the principal cells, by direct feedforward excitation of interneurons by the principal cell afferents, or by subcortical inputs.

A. Excitatory Neurotransmission

The primary excitatory neurotransmitter to dentate granule cells and CA1 pyramidal cells is glutamate, whereas CA3 pyramidal cells respond preferentially to aspartate (Fagg and Foster, 1983; Fonnum, 1984; Wieraszko, 1983). Glutamate agonists such as quisqualate, α-amino-3-hydroxy-5-methylisoxazole-4-proprionic acid (AMPA), and kainate discriminate among glutamate receptor subtypes that are directly linked to ion channels. Quisqualate and AMPA initiate a large inward current that underlies fast excitatory synaptic transmission (Collingridge and Lester, 1989; Tang *et al.*, 1989) whereas the role that kainate receptors play in the hippocampus has not been clearly established. A distinct family of metabotropic glutamate receptors are coupled to various signal transduction processes through guanine triphosphate (GTP)-binding proteins (i.e., G-proteins) (Conn and Desai, 1991). Metabotropic receptors may also contribute to synaptic transmission and have been implicated in synaptic plasticity. For the purpose of subsequent discussion, all of these receptor subtypes (quisqualate, AMPA, metabotropic) will be collectively considered as one component contributing to field potentials.

Ion channels coupled to the other class of ionotropic glutamate receptors, the N-methyl-D-aspartate (NMDA) subtype, are clearly differentiated from AMPA and metabotropic potentials by their delayed onset and long-lasting currents (Dingledine, 1986; Jahr and Stevens, 1990). In addition, the NMDA receptor ionophore is blocked by Mg^{2+} in a voltage-dependent fashion such that these currents are preferentially activated under unique circumstances involving large or prolonged cell depolarizations that relieve the Mg^{2+} block (Coan and Collingridge, 1985; Nowak *et al.*, 1989). With the arrival of more selective antagonists for glutamate receptor subtypes, it has become apparent that NMDA currents also contribute to normal synaptic transmission (Blanpied and Berger, 1992; Lambert and Jones, 1990; Mayer and Westbrook, 1987). Because the NMDA receptor ion channel is highly permeable to Ca^{2+} (Dingledine, 1983), its activation is believed to be critical for regulating Ca^{2+}-dependent processes involved in synaptic growth and plasticity (Collingridge and Bliss, 1987) as well as cell death (Choi, 1988, 1990).

B. Inhibitory Neurotransmission

Hippocampal inhibitory interneurons are predominantly GABAergic. There are two general subclasses of GABA receptors that contribute to field potentials: $GABA_A$ and $GABA_B$. $GABA_A$ receptors are directly linked to chloride (Cl^-) ionophores which permit entry of hyperpolarizing Cl^- ions upon activation. The $GABA_B$ receptor is a metabotropic receptor linked to a G-protein whose activation hyperpolarizes cells by initiating an outward K^+ conductance with a delayed onset (for reviews see Alger, 1991; Dutar and Nicoll, 1988; Schwartzkroin *et al.*, 1990). Changes in field potentials that have been attributed to either ionic $GABA_A$ or metabotropic $GABA_B$ receptor activation generally have been differentiated by fast and slow time courses, respectively. Peptides such as somatostatin, vasopressin, calretinin, neuropeptide Y, and cholecystokinin also are localized within a variety of interneurons, some of which also contain GABA (see for example Ascady *et al.*, 1993; Gahwiler, 1983; Miettinen and Freund, 1992; Rich-Bennett *et al.*, 1993). Others, such as enkephalin and dynorphin, are found in the mossy fiber terminals of dentate granule cells and modulate excitatory transmission, feedforward, and feedback inhibition (Lupica and Dunwiddie, 1991). Apart from their association predominantly with interneurons, the functional significance of these peptides is poorly understood.

C. Chronic Preparations in Vivo

Selection of the preparation to be used in field potential studies depends on the experimental question under investigation. For chronic recordings, the electrode assembly is permanently attached to the skull during surgery. Following a period of postoperative recovery, stable responses may be monitored for periods of 1–12 months. Chronic recordings are reported more frequently using the perforant path–dentate gyrus circuit, although the intact animal also has been used in studies of the Schaffer collateral–CA1 pathway. In the latter case, stimulation leads typically are placed in the contralateral CA3 area and Schaffer collateral activation is achieved via the commissural pathway that connects the two hippocampi.

In addition to the obvious advantage of evaluating changes that may be time dependent, the chronic preparation permits the assessment of physiological activity in the awake, unrestrained animal. Given the broad spectrum of nervous system effects exerted by anesthetic agents, interpretation of waveform components in the chronic preparation is more straightforward than that associated with the acute, anesthetized preparation. This advantage must be qualified, however, by the influence of hippocampal EEG frequencies on field potentials which, in turn, are strongly linked to the behavior of the animal at the time the stimulus is delivered (e.g., Cao and Leung, 1991; Green *et al.*, 1993; Winson and Azburg, 1978). These confounding factors can be minimized by adapting the animal to the recording chamber before testing is initiated, by administering stimuli only during specific behavior states, and by waveform averaging procedures.

D. Acute Preparations in Vivo

Acute preparations offer the advantage of evaluating the integrity of the entire trisynaptic circuit in the same animal. In an anesthetized animal, the distribution of current flow can be mapped to identify the origin of voltages contributing to the field potential. This is accomplished by sequential lowering of a single or multipolar electrode to the optimal position for recording CA1 or dentate granule cell field potentials while stimulating the afferent input fibers (see Fig. 2B for a simplified current–source–density analysis in the dentate gyrus with perforant path stimulation). The acute preparation also simplifies intraventricular or direct intracranial administration of compounds that do not readily cross the blood–brain barrier and requires significantly less time investment than chronic preparations. The major limitations of this technique are the temporal constraints of a single testing session and the difficulty of isolating anesthetic effects from the experimental measures utilized to test the hypothesis. Interestingly, urethane is the anesthetic of choice in most acute hippocampal experiments, yet the mechanism whereby it produces its anesthetic properties is unknown.

E. In Vitro Preparation

Finally, stable field potentials also may be recorded from the *in vitro* hippocampal slice preparation. Because of the technical difficulty of recording from single neurons *in vivo*, most intracellular recordings have utilized the *in vitro* preparation (but see Andersen *et al.*, 1966; Lomo, 1971). The technique frequently is used to address experimental questions of a mechanistic nature that cannot easily be assessed *in vivo*. Examples include the study of the mossy fiber–CA3 circuit, inhibitory synaptic transmission using paired antidromic–orthodromic stimulation of specific cell populations, and comparisons of field and intracellular recordings. Drug effects also are readily assessed in hippocampal slices by identifying changes in responses before and during drug treatment and following drug washout. Although there is generally good agreement between results obtained *in vivo* and *in vitro*, the major limitation of the hippocampal slice preparation is the inability to interpret findings within the context of coincident modula-

tory influences present in the intact brain. Transection of afferent pathways and inhibitory collaterals due to the orientation of fiber tracts with respect to plane of section also restricts evaluation of field potentials in the hippocampal slice preparation to a highly localized area with limited inhibitory input.

II. Field Potential Recordings

As the name implies, field potentials reflect the summated output of an entire population of neurons which includes both excitatory and inhibitory influences. The presence or contribution of each of these components can be isolated only in intracellular recordings, experiments typically performed *in vitro* with the aid of pharmacological antagonists to eliminate competing currents. However, the orientation of the dendritic fields to the cell body layers in the hippocampus presents a unique situation in which field potentials recorded *in vitro* and *in vivo* faithfully reflect intracellular activity (Andersen *et al.*, 1966; Lomo, 1971). As depicted in Fig. 2B, stimulation of perforant path fibers generates a dipole of current flowing from the synaptic region (negative-going sink) to the cell body layer (positive-going source). An extracellular recording electrode reflects the pattern of current flow as a negative potential at and above the region of incoming perforant path fibers, which reverses in polarity below the region of synaptic contact. Intracellular recordings reveal an initial excitatory postsynaptic potential (EPSP) that is followed by a fast inhibitory postsynaptic potential (IPSP) and a slower, late IPSP or afterhyperpolarization. In the field potential, these inhibitory influences may be masked by the EPSP, but effectively operate to limit the amplitude and time course of the population EPSP. A current sink is created at the soma layer when the applied stimulation is strong enough to bring granule cells to fire action potentials. A negative potential (the population spike) is then superimposed on the positive population EPSP (Andersen *et al.*, 1971) and produces the typical dentate gyrus field potential depicted in Fig. 2A. The higher the stimulus intensity, the more synchronous the firing of the cells and the larger the amplitude and narrower the peak of the negative potential reflecting the population spike (Lomo, 1971).

Accurate placement of stimulating and recording electrodes in the *in vivo* preparation is best accomplished by monitoring the evoked field potential for changes in morphology, polarity, amplitude, and latency in the EPSP and population spike during surgery as the recording electrode is lowered through the dendritic fields to the dentate hilus (Fig. 2B). An estimate of the postsynaptic potential (PSP) is derived from the

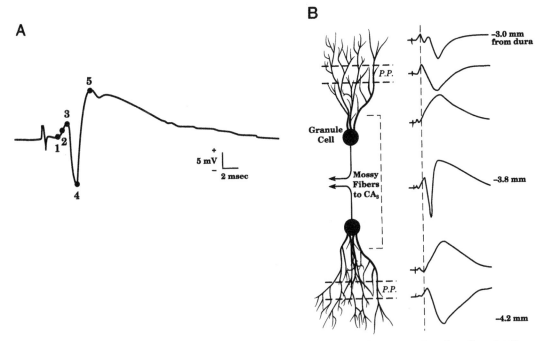

FIGURE 2 (A) Field potential recorded from dentate gyrus in response to a single pulse of moderate stimulus intensity delivered to the perforant path. The amplitude of the postsynaptic potential (PSP), although contaminated by inhibitory currents, is approximated by the amplitude of the most positive peak on the waveform (point 5). A better estimate of the excitatory PSP (EPSP) is derived by taking the slope of the line between points 1 and 2 on the rising phase of the potential. The amplitude of the population spike represents the number of granule cells reaching threshold for action potential firing and is reflected in the negative potential that is superimposed on the PSP. It is quantified by taking the average of the difference between points 3 and 4, and between 4 and 5. (B) A current source density analysis provides a profile of field potentials evoked by perforant path stimulation as an electrode is lowered through the dendritic, somatic, and hilar regions of the dentate gyrus. The initially negative EPSP (current sink) reverses polarity as the electrode passes cell dendrites (current source). Deeper penetration beyond the cell body layer and sufficient stimulation intensities into the hilar region produce a maximal EPSP with a negative population spike superimposed on the positive waveform. If the trajectory of the electrode is continued, the EPSP again reverses polarity at the dipole created by perforant path synaptic contact on the granule cells of the ventral blade of the dentate gyrus. (Adapted from McNaughton, 1983.)

amplitude of the most positive peak in the potential (point 5 of Fig. 2A). This measure, however, is contaminated by activation of inhibitory circuits and is not typically used in field potential studies *in vivo*. Instead, the slope of the EPSP is used as an index of the strength of the synaptic response (Lomo, 1971) and is optimally measured as the rate of amplitude change for the initial waveform segment (approximately 1 msec after the stimulus) before the onset of the population spike (i.e., slope of the line between points 1 and 2 in Fig. 2A). The amplitude of the population spike is measured by averaging the two peaks comprising the negative potential (mean of amplitude between points 3 and 4 and between points 4 and 5 in Fig. 2A). Alternatively, the tangent measure of the population spike is computed as the voltage difference between the most negative point of the population spike (point 4, Fig. 2A) and the midpoint of the tangent connecting the onset of the population spike (point 3, Fig. 2A) and the peak of the

PSP (point 5, Fig. 2A). Waveform area or response duration measures have proven useful in differentiating NMDA from non-NMDA contributions to field potentials. A significant increase in the area and response duration has been attributed to the increasing recruitment of NMDA receptor activation with moderate to high stimulus intensities (Muller and Lynch, 1988; Pacelli *et al.*, 1991; Racine *et al.*, 1991).

III. Measures of Excitatory Synaptic Transmission

Input-Output Functions

An input–output (I/O) function refers to the relationship between the amount of current applied to the afferent fiber bundle (input-current) and the resulting synaptic response (output-voltage). Figure 3A illustrates

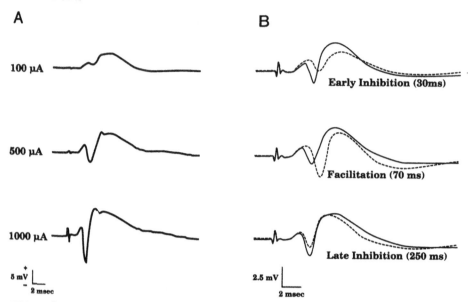

FIGURE 3 (A) Input/output (I/O) function. Evoked field potentials recorded in the dentate gyrus following perforant path stimulation at increasing stimulus intensities in an unanesthetized animal. (B) Paired pulse function. Sample waveforms taken from the dentate gyrus following pairs of stimulus pulses delivered to the perforant path of a conscious rat. Interpulse intervals (IPIs) of 30, 70, and 250 msec produce periods of early inhibition, facilitation, and late inhibition, respectively. The solid line is the conditioning pulse and the broken line is the test pulse. Note the reduction of the amplitude of the PSP (point 5 of Fig. 2A) with early inhibition at brief IPIs, not evident at longer interpulse intervals (see text).

responses evoked in a typical I/O test of the dentate granule cells in response to perforant path stimulation in an unanesthetized rat. Each response is an average of 10 stimuli delivered at the same intensity (0.1-msec pulse duration, administered every 10–20 sec).

The absolute voltage of the field potential for a given stimulus intensity varies from animal to animal (or from slice to slice *in vitro*). To obtain a complete profile of the excitability of the cell population, a range of stimulus intensities is administered, extending from just above threshold for evoking an EPSP to that which evokes a maximal population spike. Normalizing the data obtained from each preparation by expressing the EPSP slope and population spike amplitude measurements as a percentage of the maximum response recorded under control conditions facilitates the description and analysis of group data. Pre- and post-treatment curves are typically presented with each animal (or hippocampal slice) serving as its own control.

A selective effect on the EPSP slope suggests a primarily dendritic locus of action, whereas changes in population spike amplitude in the absence of EPSP changes indicate a primarily somatic focus. Similar changes are often observed in the I/O function of the EPSP slope and population spike amplitude which has led to the common practice of expressing the I/O function as a relationship between these two variables. Al-

though the intercept of this function may fluctuate according to experimental treatment, the slope of the I/O relationship between the two waveform parameters remains relatively constant, suggesting that the cellular coupling mechanism of EPSP to action potential generation does not change. A leftward shift in the I/O relationship of EPSP slope/population spike amplitude is interpreted as reflecting greater synchrony in cell firing for the same level of synaptic activation, whereas a rightward shift in this function reflects a decrease in synchronous cell firing. Synaptic events that could mediate a directional shift in the I/O function include cell loss, a change in neurotransmitter release, a change in postsynaptic sensitivity or number of non-NMDA glutamate receptors, or a change in inhibition. Local inhibitory influences that contribute to the observed change may be elucidated using homosynaptic and heterosynaptic paired pulse stimulation procedures.

IV. Measures of Inhibitory Synaptic Transmission

A. Paired Pulse Depression

Excitability of the hippocampal network is heavily influenced by powerful inhibitory circuits. The functional integrity of hippocampal inhibition can be assessed in field potentials using paired pulse techniques. Paired pulse depression is measured by delivering two pulses to the same pathway (homosynaptic inhibition), separated by an interval (interpulse interval or IPI) ranging from 10 to 4000 msec. The degree of depression exerted by the first (conditioning) pulse is reflected in the smaller amplitude response to the second (test) pulse and typically is expressed as the ratio of the second response amplitude to that of the first response amplitude. A triphasic pattern of depression, facilitation, and depression of the test response is recorded from dentate granule cells as a function of increasing IPIs (Fig. 3B and Fig. 4). A second type of modulatory influence of hippocampal activity can be assessed by pairing activation of commissural, septal, and brain stem circuits with perforant path stimulation, a discussion of which is included under extrinsic modulation of hippocampal output.

B. Early Inhibition in the Dentate Gyrus

At brief IPIs (10–40 msec), a dramatic reduction is observed in the amplitude of the test population spike responses in the chronic recording preparation (Fig. 3B). Early paired pulse depression reflects the influence of the fast IPSP, a Cl^--mediated current, resulting from activation of $GABA_A$ receptors. A prominent role of $GABA_A$-mediated feedback inhibition was suggested in early recordings of hippocampal field potentials by the dependence of paired pulse depression upon stimulus intensities that were sufficient to evoke a discharge from the principal cell population (Andersen *et al.*, 1964, 1969; Kandel *et al.*, 1961; Lomo, 1971). Based on this historical precedent, measures of paired pulse depression in hippocampal field potentials *in vivo* typically have been limited to the population spike amplitude. Anatomical evidence of a feedback circuit (see Fig. 1) consists of the presence of GABA-containing interneurons, innervated by axon collaterals of principal cells, that terminate back on the principal cell soma (Amaral, 1978; Ribak and Seress, 1983).

C. Late Inhibition in Dentate Gyrus

A relatively modest depression in the test population spike amplitude is recorded from dentate granule cells at longer IPIs (150–1000 msec in Figs. 3B and 4). Although less is known about late paired pulse depression, especially in the dentate gyrus, the emerging picture is more complex than that described for early paired pulse depression. Intracellular recordings indicate the presence of a late hyperpolarizing current in dentate granule cells that peaks at 200–300 msec and

reflects an increase in outward K^+ conductance (Rausche *et al.*, 1990; Thalmann and Ayala, 1982). Some K^+ conductances rely on activation of postsynaptic $GABA_B$ receptors (Rausche *et al.*, 1990; Heinemann *et al.*, 1992), whereas others are Ca^{2+} dependent (Rausche *et al.*, 1990; Thalmann and Ayala, 1982). A comparison of intracellular IPSPs evoked by antidromic and orthodromic stimulation of hippocampal pyramidal cells suggests the possibility that the late component of paired pulse depression in this hippocampal subfield may be entirely feedforward in origin. Antidromic stimulation results only in a simple IPSP that exhibits the early latency and Cl^- sensitivity characteristic of $GABA_A$ receptor activation. Orthodromic stimulation, on the other hand, results in a biphasic IPSP with both early and late components, the latter of which is K^+ dependent and corresponds to the time course of late paired pulse depression (see Alger, 1991).

In the dentate gyrus, anatomical evidence of feedforward inhibition derives from the presence of GABA synapses on dentate granule cell dendrites in the outer molecular layer. The source of these afferents has been attributed to hilar interneurons that receive excitatory input from perforant path fibers (Leranth *et al.*, 1990; Scharfman, 1991; Schwartzkroin *et al.*, 1990; Zipp *et al.*, 1989). A GABAergic projection originating in the entorhinal cortex that runs in the perforant path to synapse directly on granule cells has also been observed (Fifkova *et al.*, 1992; Germroth *et al.*, 1989). The threshold for late paired pulse depression in dentate granule cell recordings is observed at stimulus intensities that are well below the levels believed necessary to activate feedback inhibitory circuits (Burdette and Gilbert, in press). Although these findings are consistent with the low threshold exhibited by feedforward interneurons (Buzsaki, 1984; Buzsaki and Eidelberg, 1982; Buzsaki and Czech, 1981; Scharfman, 1991), further work is necessary to determine the contribution of feedforward inhibition to late paired pulse depression in the dentate gyrus.

D. Paired Pulse Depression in Area CA1

Paired pulse depression in CA1 pyramidal cells exhibits some of the same cellular properties described earlier for dentate granule cells. However, two important distinctions can be readily seen in Fig. 4. First, the time course of early paired pulse depression is prolonged in CA1 compared to that of the dentate gyrus. Second, the triphasic function of inhibition/potentiation/inhibition seen in dentate gyrus is replaced by a biphasic function of inhibition/potentiation with no evidence of late paired pulse depression in CA1 field recordings. These differences in paired pulse inhibition

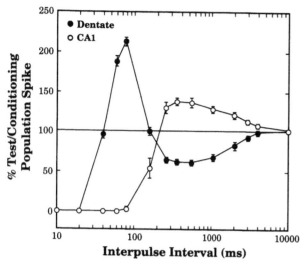

FIGURE 4 Paired pulse stimulation of the perforant path produces a triphasic function of inhibition/potentiation/inhibition in the dentate gyrus. Stimulation of the commissural input with pairs of stimulus pulses produces a biphasic function of inhibition/potentiation in area CA1. Differences in distribution and density of presynaptic GABAB receptors on interneurons associated with each of these hippocampal subregions may account for the differences in paired pulse functions in CA1 and dentate gyrus (see text). Data were acquired in halothane-anesthetized rats and are taken with permission from Steffensen and Henriksen (1991).

may reflect characteristics of local inhibitory circuits, as well as the density and distribution of GABA receptor subtypes. For example, potentiation in CA1 field recordings at long IPIs has been attributed to a paired pulse depression in the fast GABA$_A$-mediated IPSP (Nathan *et al.*, 1990; Nathan and Lambert, 1991) by GABA$_B$ receptors located on the presynaptic terminals of interneurons (Nathan *et al.*, 1990). Activation of these receptors results in decreased GABA release and, consequently, produces a disinhibitory effect and paired pulse potentiation in area CA1 (Davies *et al.*, 1990). Although the presence of a similar GABA disinhibitory process has yet to be demonstrated in the dentate gyrus, the presence of late paired pulse depression in the dentate granule cell field recordings suggests that a comparable disinhibitory mechanism must exhibit a significantly higher threshold than that observed in CA1 recordings. Paired pulse facilitation is discussed further in the section on plasticity in the hippocampus.

E. Extrinsic Modulation of Hippocampal Output

Modulation of hippocampal excitability is further influenced by intrahippocampal associational fibers, commissural fibers arising from the contralateral hippocampus, and by fibers originating in a variety of subcor-

tical structures that terminate on both the principal cells and local interneurons. Heterosynaptic paired pulse stimulation has provided insight into the summated excitatory and inhibitory influences exerted by commissural, septal, locus coeruleus, and median raphé afferents on CA1 and dentate granule cell populations.

The influence of commissural stimulation has been studied predominantly in the dentate gyrus where terminals are localized in the inner one-third of the dendritic region (Frotscher and Zimmer, 1983; Gottlieb and Cowan, 1973). Although single pulse stimulation of commissural afferents evokes a small excitatory postsynaptic potential in dentate granule cells, the most frequently observed response is a long lasting inhibition of spontaneous activity (Deadwyler *et al.*, 1975). When commissural stimulation precedes perforant path stimulation (heterosynaptic paired pulses), a GABA$_A$-mediated suppression of the dentate grnaule cell field potential is observed at IPIs less than 40 msec; paired pulse facilitation occurs at longer IPIs (Buzsaki and Eidelberg, 1982; Buzsaki and Czeh, 1981; Douglas *et al.*, 1983). The facilitation phase may be related to the inhibition of spontaneous activity that also occurs in interneurons following paired pulse stimulation (Buzsaki and Eidelberg, 1982). Anatomical evidence demonstrating commissural terminals synapsing on interneurons located in the dentate molecular layer and hilar regions (Frotscher and Zimmer, 1983; Seress and Ribak, 1984), as well as physiological evidence showing lower firing thresholds of interneurons relative to granule cells (Buzsaki and Eidelberg, 1982; Scharfman *et al.*, 1991), is suggestive of a feedforward inhibitory circuit (for reviews see Buzsaki, 1984; Schwartzkroin *et al.*, 1990).

The dentate gyrus is also innervated by cholinergic and GABAergic fibers arising from the medial septum and diagonal band. Septohippocampal cholinergic axons terminate on both dentate granule cells and dentate interneurons (Frotscher and Leranth, 1985), whereas GABAergic axons arising in the septum terminate preferentially on interneurons (Freund and Antal, 1988). In dentate granule cell field recordings, a conditioning stimulus applied to the medial septum enhances the test response to perforant path stimulation at brief IPIs (Bilkey and Goddard, 1987; Fantie and Goddard, 1982; McNaughton and Miller, 1984; Robinson and Racine, 1986). The disinhibitory effect of septal stimulation on the principal cell population response has been attributed to inhibition of interneurons by the septohippocampal GABAergic input. Enhanced excitation in dentate gyrus field potentials may also be augmented by direct septocholinergic excitation of granule cells.

Anatomical and physiological evidence supports a similar disinhibitory modulating influence of monoamine-containing afferents from the locus coeruleus and the raphé nucleus (Freund *et al.,* 1990; Halasy *et al.,* 1992; Harley and Sara, 1992; Klancnik and Phillips, 1991). Although less extensively studied than the commissural and septal inputs, stimulation of these subcortical structures may enhance excitation in the dentate gyrus by reducing activation of local inhibitory interneurons. However, excitation of principal cells and inhibitory interneurons by noradrenalin and serotonin has also been reported (Andrade and Chaput, 1991; Andreasen and Lambert, 1991; Harley and Sara, 1992; Ropert and Guy, 1991; Sara and Bergis, 1991).

To summarize, the paired pulse stimulation paradigm provides a method to assess moderating influences on hippocampal excitability arising from local inhibitory circuits, as well as commissural and subcortical afferents. Interpretation of these findings, however, is subject to the same qualifications as those described earlier for field potentials in general, and is compounded by the excitatory/inhibitory influences present in both responses of the pair. An additional caveat placed upon paired pulse depression studies is that the degree of feedback inhibition is dependent on the amplitude of the conditioning population spike (Andersen *et al.,* 1969; Lomo, 1971). This has led to the common practice of adjusting stimulus intensity to match the amplitude of the conditioning population spike between control and experimental conditions. Despite these limitations, results obtained from field potential studies using paired pulse stimulation have significantly advanced our understanding of the complexity of hippocampal circuitry and have identified areas for further exploration.

V. Plasticity in the Hippocampus

The ability of the central nervous system to modify its behavior based on experience is one of the most fundamental concepts of neurobiology. Among brain structures, the hippocampus has demonstrated a remarkable propensity for synaptic plasticity following activation. Activity-dependent synaptic potentiation occurs within milliseconds and persists for hours in anesthetized animals and in *in vitro* hippocampal slices. In awake animals, potentiation is observed for days to weeks. A number of temporally and mechanistically distinct components are incorporated into this time span, and inconsistency in the nomenclature can be cause for confusion. The terminology of McNaughton (1983) and Bliss and Collingridge (1993) that is outlined in Table 1 has been adopted here.

A. Short-Lasting Potentiation

Three types of relatively short-lasting facilitation have been identified in the neuromuscular junction (Zucker, 1989) and in mammalian forebrain pathways (Andersen *et al.,* 1969; Creager *et al.,* 1980; Hess *et al.,* 1987; Hotson and Prince, 1980; Konnerth and Heinemann, 1983; McNaughton, 1982; Racine and Milgram, 1983; White *et al.,* 1979). Paired pulse facilitation, augmentation, and post-tetanic potentiation have been differentiated by their time course and by the stimulus parameters used to induce them. Paired pulse facilitation reflects a short duration (decay constant of approximately 100 msec) increase in synaptic transmission. In this form of plasticity, application of a single electrical pulse followed by a pulse of equal intensity within a restricted time range results in a synaptic response of increased amplitude (e.g., 70 ms IPI of Fig. 4B). Augmentation and post-tetanic potentiation are longer lasting and are typically produced with a train (tetanus) instead of pairs of stimulus pulses. The time constant for augmentation is on the order of 5 sec, and post-tetanic potentiation has a decay time constant of 20–240 sec (McNaughton, 1982; Racine and Milgram, 1983).

At the neuromuscular junction and other peripheral synapses, these short-term facilitation processes affect the number of quanta of transmitter released, and decay in an exponential fashion (Zucker, 1989). Although the cellular processes responsible for short-lasting potentiation phenomena in the hippocampus have not been determined, it is widely believed that an increase in glutamate release as a consequence of calcium accumulation in the presynaptic terminal is a contributing factor (Zalutsky and Nicoll, 1990; Zucker, 1989). It is also clear that a decrease in GABA release contributes to paired pulse facilitation in hippocampal field potentials. The prolonged phase of paired pulse facilitation in CA1 pyramidal cell recordings has been attributed to a decrease in the fast IPSP resulting from a presynaptic GABA$_B$ receptor-mediated inhibition of GABA release from interneurons (Davies *et al.,* 1990; Morrisett *et al.,* 1991; Nathan and Lambert, 1991). On the other hand, Joy and Albertson (1993) have suggested that a postsynaptic mechanism accounts for paired pulse facilitation of the population spike in the dentate gyrus. Under urethane anesthesia, a reduction of paired pulse facilitation was induced by the NMDA antagonists MK-801, ketamine, and dextromethorphan. However, these observations differ from those obtained in the unanesthetized preparation where MK-801 failed to exert an effect on paired pulse facilitation, but substantially increased late paired pulse inhibition (200–1000 msec IPIs; Gilbert *et al.,* 1992). Thus, the relative contribu-

TABLE 1   Duration, Stimulus Requirements, and Possible Mechanisms Underlying
Postactivation Potentiation in the Hippocampus[a]

Process	Duration	Stimulus parameters	Type of potentiation	Proposed mechanism
Facilitation	Milliseconds	Pulse induced	Paired pulse potentiation	Presynaptic/GABA$_B$?
Augmentation	Seconds	Train induced	Frequency potentiation	Presynaptic
Short-term potentiation	Seconds to minutes	Train induced	STP	Combination of augmentation and LTP-1 (?)
Long-term potentiation 1	30–60 min	Train induced	Induction (LTP-1)	Ca^{2+}/NMDA dependent
Long-term potentiation 2	Hours	Train induced	Induction expression (LTP-2)	PKC? CaMKII?
Long-term potentiation 3	Days–weeks	Train induced	Maintenance (LTP-3)	Retrograde messenger/structural alterations

Note: See text and Fig. 5 for further description. STP, short-term potentiation; LTP, long-term potentiation; GABA, γ-aminobutyric acid; NMDA, N-methyl-D-aspartate; Ca^{2+}, calcium; PKC, protein kinase C; CaMKII, Ca^{2+} – calmodulin kinase II.
   [a] Adapted from Bliss and Collingridge (1993).

tions of pre- and postsynaptic mechanisms to short-lasting potentiation phenomena in hippocampus await further research.

B. Long-Term Potentiation

Long-term potentiation is defined as a persistent, activity-dependent increase in the strength of synaptic transmission and is one of the most intensively studied forms of synaptic plasticity in the mammalian brain. The hippocampus exhibits the most dramatic and longest duration LTP of any structure examined to date (Racine *et al.*, 1983). Many theorists assume that the mechanisms underlying LTP are similar to those supporting learning and memory processes (Massicotte and Baudry, 1991). LTP investigations conducted in the hippocampus have focused on different cell populations depending on the preparation used. Characteristics of LTP exhibited by the perforant path-dentate granule cell synapse have been explored primarily in the intact animal, whereas LTP in CA1 pyramidal cells has been studied almost exclusively in the hippocampal slice preparation. Despite some differences that are discussed later, the basic mechanisms supporting LTP in these two regions appear to be qualitatively the same.

Traditionally, LTP has been induced by the delivery of a high frequency tetanus to a monosynaptic pathway. In the intact animal, a series (5–10) of 50-msec trains (0.1-msec pulse duration; 10–20 pulses/train) are typically administered at a frequency of 400 Hz. In the hippocampal slice preparation, a slower frequency (100 Hz), longer duration (1 sec) train typically is used. Several laboratories have recently reported that the magnitude and persistence of LTP are enhanced when train delivery mimics the natural 4- to 7-Hz θ rhythm

that dominates hippocampal EEG recordings (Arai and Lynch, 1992; Diamond *et al.*, 1988; Larson *et al.*, 1986; Racine *et al.*, 1991). For example, trains limited to as few as 4-10 stimulus pulses (100–400 Hz) delivered at an intertrain interval of 200 msec (5 Hz) appear to be the most efficacious (Diamond *et al.*, 1988; Larson *et al.*, 1986; Leung *et al.*, 1992; Pacelli *et al.*, 1989, 1991). The robust LTP produced by these stimulation patterns has been linked to the optimal activation of NMDA currents in principal cells that follows from the disinhibitory action of GABA$_B$ receptor activation (Mott and Lewis, 1991). As discussed earlier, GABA$_B$ autoreceptors activated at IPIs of 200–400 msec provide a physiological means of suppressing synaptic inhibition during high frequency transmission, facilitating the activation of voltage-dependent NMDA receptors, and thereby promoting the initiation of LTP.

LTP is also characterized by a threshold of minimal stimulation required for its induction. Regardless of the specific train parameters used, train intensity functions can be performed to establish a threshold for LTP induction. Trains of increasing stimulus intensity are delivered and responses evoked by single pulse stimulation at the midrange of the I/O function are monitored for 10–15 min to verify the persistence of the potentiated response. Following the final train, responses are monitored for at least 1 hr and are compared to pre-train responses. The presence of LTP is defined by an increase (>25–200%) in the EPSP slope or the population spike amplitude. The feature that distinguishes LTP from other forms of synaptic potentiation is the longevity of the observed synaptic enhancement. In chronic preparations, responses are typically examined over a number of days and LTP has been observed in the dentate gyrus for periods lasting from 2 to 4 weeks.

C. Mechanisms of LTP

Many excellent reviews of the proposed mechanisms of LTP are available, so only a simplified overview will be presented here (see Bliss and Collingridge, 1993; Colley and Routtenberg, 1993; Collingridge, 1992; Lynch *et al.*, 1990; Malenka, 1992; Malenka *et al.*, 1990; Massicotte and Baudry, 1991; McNaughton, 1993). LTP is composed of three distinct phases, induction, expression, and maintenance, each displaying a different time course and relying upon different cellular mechanisms (see Table 1). As the name implies, LTP induction (LTP-1 in Table 1) refers to the initiation of cellular events upon which the rest of the cascade depends. A large body of evidence indicates that LTP induction in the dentate granule cells and CA1 pyramidal cells is a calcium-dependent process (Collingridge, 1992; Dunwiddie and Lynch, 1979; Malenka, 1992; Malenka *et al.*, 1988) and requires postsynaptic depolarization sufficient to recruit NMDA receptor activation and subsequent Ca^{2+} influx into the cell through receptor-activated voltage-dependent channels (Collingridge and Bliss, 1987).

The processes activated following induction of LTP are less well understood; an attempt to represent them pictorially in the schematic of a presynaptic terminal and postsynaptic spine can be seen in Fig. 5. Second messenger systems such as protein kinase C (PKC) (Akers *et al.*, 1986; Colley and Routtenberg, 1993; Malinow *et al.*, 1989) and calcium–calmodulin-dependent protein kinase II (CaMKII) (Malinow *et al.*, 1989; Malenka, 1992; Malenka *et al.*, 1989) are activated by metabotropic receptor stimulation and the initial NMDA-mediated influx of extracellular calcium. Activation of these kinases, together with an increase of intracellular calcium by release from internal stores, may in turn initiate the release of retrograde messengers such as nitrous oxide (O'Dell *et al.*, 1992; Zhou *et al.*, 1993) or arachidonic acid (Lynch *et al.*, 1991). These messengers initiate presynaptic changes that support LTP. Expression of LTP (LTP-2) is mediated through the non-NMDA receptors, the class of glutamate receptors predominantly active during low frequency synaptic transmission (Bliss and Collingridge, 1993). Alterations in the number or properties of ion channels that mediate synaptic transmission in the hippocampus may contribute to the expression phase of LTP (Bliss and Collingridge, 1993; Colley and Routenberg, 1993; Lynch *et al.*, 1990).

Maintenance of LTP (LTP-3) is achieved when the retrograde signal emanating from the postsynaptic cell triggers events in the presynaptic terminal that promote enhanced neurotransmitter release (Bliss *et al.*, 1986; Lynch *et al.*, 1991). These may include translocation of PKC from the cytosol to the membrane, direct increases in cyclic nucleotide levels (i.e., cyclic GMP), and phosphorylation of voltage-sensitive calcium channels (see Fig. 5). Increased expression of immediate early genes such as *zif*-268 (Cole *et al.*, 1989) and tPA (Qian *et al.*, 1993, 1994) may initiate gene transcription contributing to the structural changes that are believed to contribute to the longevity of LTP (Fazeli *et al.*, 1993; Geinisman *et al.*, 1991; Lynch *et al.*, 1990; Wallace *et al.*, 1991).

D. Interpretation of Effects on LTP

As with other physiological measures discussed in this chapter, interpretation of potentiation-induced changes in field potentials is subject to several qualifications. The magnitude of potentiation varies both within hippocampal slices obtained from the same animal and from animal to animal in the chronic preparation. Also, a certain percentage of hippocampal slices or animals fail to demonstrate LTP. In chronic studies, this can occur despite large field potentials of characteristic morphology and histological verification of accurate electrode placement (Bliss and Lomo, 1973; Diamond *et al.*, 1988; Gilbert and Mack, 1990; Racine *et al.*, 1983; Leung *et al.*, 1992). Although the reasons for failure to induce LTP are unclear, the use of a tetanus delivered at high stimulus intensities may also potentiate inhibitory interneurons (Korn *et al.*, 1992; Leung *et al.*, 1992), thereby masking the excitatory effects of LTP. Potentiation of inhibition in field potentials recorded in the dentate gyrus is evidenced by enhanced paired pulse inhibition that persists for days (Fig. 6) following a tetanus of sufficient strength to induce focal seizure activity (deJonge and Racine; 1986; Gilbert, 1991; Robinson *et al.*, 1991). It also is possible that the robust LTP elicited at more biological stimulation frequencies may be due to a preferential potentiation of excitatory synapses. This hypothesis awaits a parametric investigation directly comparing the duration and magnitude of LTP using different train parameters and the influence of LTP elicited by different stimulation frequencies on paired pulse depression.

Generalization of the cellular processes underlying LTP also must be qualified by the differences in waveform components exhibiting potentiation depending on the preparation that is used and the cell population that is studied. For example, the increased amplitude of the EPSP is the most common measure of potentiation in CA1 pyramidal cell recordings obtained from hippocampal slices. In contrast, the population spike recorded from the dentate gyrus exhibits the most reliable incidence and magnitude of LTP in the intact animal (Racine *et al.*, 1983). It is not clear if the waveform

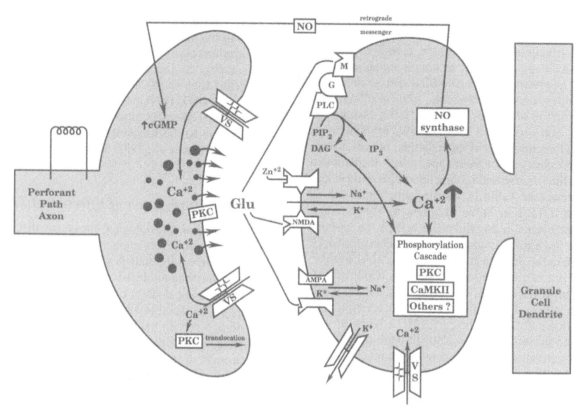

FIGURE 5 Presynaptic terminal (left) and postsynaptic spine (right) illustrating the proposed processes underlying long-term potentiation (LTP) (see text). Briefly, induction of LTP is initiated by depolarization and activation of voltage-sensitive NMDA receptors by glutamate leading to an increase in intracellular calcium. A series of secondary changes occurs in response to elevated calcium levels, including phosphorylation of proteins by calcium-dependent kinases. Presynaptic alterations important for the maintenance of enhanced synaptic efficacy are brought about by retrograde messengers such as nitrous oxide (NO) and arachidonic acid that travel from the postsynaptic spine to the presynaptic terminal to enhance transmitter release. Phosphorylation of pre- and postsynaptic substrates in addition to structural alterations may contribute to the expression and persistence of LTP. cGMP, cyclic-guanosine monophosphate; VS, voltage-sensitive calcium channel; PKC, protein kinase C; Glu, glutamate; AMPA, α-amino-3-hydroxy-5-methyl-4-isoxazole propionic acid; NMDA, N-methyl-D-aspartate; Zn, zinc; NO, nitrous oxide; M, metabotropic glutamate receptor; G, guanine triphosphate-binding protein; DAG, diacylglycerol; PLC, phospholipase C; CaMKII, calcium–calmodulin-dependent protein kinase II; PIP_2, phosphatidylinositol bisphosphate.

components most often reported as exhibiting robust LTP are due to different cell populations studied between these two preparations (CA1 vs dentate gyrus), the placement of the recording electrode (dendritic vs somatic regions), the placement of the stimulating electrode (see the following discussion), or the transection of inhibitory pathways in the hippocampal slice that participate in LTP in the intact brain.

E. LTP in Vivo and in Vitro

Teyler and colleagues (1977) conducted a comparison of LTP recorded from CA1 pyramidal cells *in vivo* (under urethane anesthesia) and *in vitro,* and concluded that the two preparations exhibited similar responses to LTP-inducing stimulation. However, Leung *et al.*

(1992) demonstrated in unanesthetized animals that the site of stimulation, basilar or apical dendrites, is critical to the magnitude and longevity of LTP evoked in CA1 pyramidal cells. Stimulation of CA1 basilar dendrites produces a high threshold, low magnitude, but persistent potentiation. In contrast, stimulation of the contralateral hippocampus to activate CA1 apical dendrites produces a low threshold, high amplitude LTP with rapid decay. LTP is induced in the hippocampal slice preparation by stimulating apical dendrites in the stratum radiatum. *In vitro* LTP is also low threshold and high amplitude, but exhibits little decay over several hours. The sectioning of longitudinal inhibitory pathways or differences in extracellular ionic concentrations artificially maintaining the hippocampal slice relative to the extracellular milieu of the intact brain may

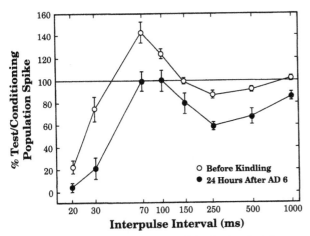

FIGURE 6 Electrographic seizure activity in the dentate gyrus augments paired pulse inhibition. Paired pulse inhibition in the dentate gyrus following perforant path stimulation in awake rats before kindling and 24 hr after six evoked epileptiform afterdischarges (ADs). ADs were induced with a once daily stimulation of the perforant path (800 μA, 1-sec train, 1.0-msec pulse duration, 60 Hz). A long-lasting increase in inhibition does not require the elicitation of overt convulsive behavioral seizure patterns.

account for these differences and may limit the ability to generalize findings between different recording preparations.

Unlike the short duration LTP recorded in CA1 *in vivo*, LTP induced in chronic perforant path–dentate gyrus preparations has been monitored for several weeks in numerous laboratories (see Racine *et al.*, 1983; Robinson and Reed, 1992). A comparison of *in vivo* and *in vitro* LTP effects in the dendate gyrus has been precluded by the difficulty of obtaining dentate granule cell LTP in hippocampal slice recordings (Wigstrom and Gustafsson, 1983). Mott and Lewis (1991), however, have reported robust LTP in the dentate gyrus of hippocampal slices elicited by θ stimulation frequencies. To understand the properties of LTP and its role in memory functions, studies on conscious animals are imperative. These caveats aside, LTP provides a potent tool for investigating cellular changes involved in synaptic plasticity in the adult mammalian brain.

VI. Application of Hippocampal Field Potentials in Neurotoxicology

Perhaps related to its plasticity, the hippocampus exhibits a unique sensitivity to cell injury resulting from a spectrum of causes, including hypoxia, seizures, and exposure to neurotoxic compounds. Hippocampal vulnerability has been exploited in anatomical, behavioral, biochemical, and pharmacological studies of the mechanisms of neurotoxicity. To a lesser extent, neurophys-

iological examination of the hippocampus has been utilized to elucidate the functional effects of neurotoxicants. The unique homogeneity of the three principal cell populations in the hippocampus, each with well-defined afferents and efferents, however, presents an ideal system for studying the effects of chemical exposure on neuronal function.

The application of hippocampal field potentials to the study of neurotoxicology offers several advantages. In the chronic preparation, administration of a compound of unknown action to an awake, unrestrained animal permits an assessment of chemical effects on both the behavioral and neurophysiological integrity of the subject. There are no time limitations that are a prevailing concern in the acutely anesthetized animal or in the hippocampal slice preparation. The effects of slowly acting metabolites and repeated low level exposure to the compound of interest can only be assessed in the chronic preparation. An examination of the time course of toxicant action also is possible, characterizing both the onset and the decay or permanence of observed changes in the functional output of a vulnerable cell population.

Despite the widespread use of *in vivo* hippocampal field potentials in pharmacology and physiology, it is surprising how underutilized these measures have been in neurotoxicological studies. The systemic administration of chemicals with unknown neurotoxic potential, however, may indirectly affect hippocampal field recordings in the unanesthetized animal, thereby confounding interpretation of the results. Appropriate controls are needed when the following conditions prevail. First, many neurotoxicants disrupt motor activity patterns (Crofton *et al.*, 1991) which can profoundly influence the amplitude and variability of hippocampal field potentials (Cao and Leung, 1991; Green *et al.*, 1992; Hargreaves *et al.*, 1992; Winson and Azburg, 1978). This problem can be minimized by delivering stimulation during comparable behavioral states (i.e., quiet alert state) or by using an anesthetized preparation. Second, in studies where gestational, perinatal, or long-term dosing regimens are utilized, the effects of the treatment on the nutritional state of the animal must be considered. Austin and colleagues (1992) have reported persistent detrimental effects of prenatal protein malnutrition on inhibition and facilitation in the dentate gyrus. Third, neurotoxicants have been repeatedly shown to alter thermoregulation (Gordon, 1988) which can effect both motor activity patterns and hippocampal field potentials (Eichenbaum and Otto, 1993; Moser *et al.*, 1993). Within limits, this problem may be circumvented by adjusting environmental settings.

Finally, a significant proportion of neurotoxic compounds are convulsants when administered at high doses (see Gilbert, 1992a). A transient decrement in hippocampal excitatory transmission immediately follows seizure activity (Burdette *et al.*, 1991; Milgram *et al.*, 1991), but in the absence of significant cell loss, this suppression is replaced by a persistent augmentation of excitatory and inhibitory synaptic transmission at later time points (deJonge and Racine, 1986; Gilbert and Mack, 1990; Gilbert, 1991; Milgram *et al.*, 1991). Figure 6 demonstrates the augmentation of paired pulse depression 24 hr following a brief electrographic seizure evoked by electrical kindling stimulation applied to the perforant path. Monitoring the behavior of the animal and the electroencephalogram, and restricting exposure to subconvulsant levels are critical when convulsant activity is suspected.

The following section summarizes the findings of neurotoxicant actions using *in vivo* hippocampal field potentials. This discussion is not intended to be an exhaustive review of the literature, but rather an illustration of the contribution these measures have made to our understanding of the action of select compounds on the function of the nervous system.

VII. Pesticides and Insecticides

A. Chlorinated Hydrocarbons

Lindane is a member of the chlorinated hydrocarbon group of insecticides that binds to the picrotoxin site on the $GABA_A$ receptor ionophore complex (Abalis *et al.*, 1985, 1986; Lawrence and Casida, 1984). Joy and Albertson (1985, 1987) have shown that lindane reduces the EPSP threshold and increases the EPSP slope and population spike amplitude in the dentate gyrus. An increase in excitatory drive, as indicated by the leftward shift in the I/O function, may result from reduction in $GABA_A$-mediated inhibition. More direct evidence of the functional impact of lindane on GABA-mediated inhibition was provided by a selective reduction of paired pulse depression at brief IPIs known to reflect $GABA_A$ activity (Joy and Albertson, 1985). Consistent with its effects on $GABA_A$ receptor function and enhanced excitation, lindane exhibits convulsant properties at high dosages (Tusell *et al.*, 1987, 1988), acts as a proconvulsant in other seizure models (i.e., electrical kindling) with repeated, subchronic exposure (Joy *et al.*, 1982; 1983), and can promote a persistent increase in seizure predisposition when administered at low dosages for prolonged periods (chemical kindling) (Gilbert, 1992d). Endosulfan, a member of the cyclodiene class of chlorinated hydrocarbons, shows effects similar to lindane on $GABA_A$ receptor binding *in vitro* (Abalis *et al.*, 1985, 1986; Lawrence

and Casida, 1984), is convulsant when administered at high dosages *in vivo*, facilitates electrical kindling, and produces chemical kindling when administered repeatedly at low dosages (Gilbert, 1992a,b,c). The effect of endosulfan on inhibition in the hippocampus has not been examined.

B. Formamidines

The formamidine class of pesticides (e.g., chlordimeform and amitraz) also alters both excitatory and inhibitory function in the hippocampus. Amitraz exerts an excitatory influence that is depicted in recordings from dentate granule cells as an increase in the EPSP slope and population spike amplitude, effects that may be related to an observed reduction in paired pulse depression (Gilbert and Dyer, 1988). Like lindane, the functional consequences of enhanced hippocampal excitation are reflected in the potent proconvulsant properties exhibited by amitraz and another formamidine pesticide, chlordimeform (Gilbert, 1988; Gilbert and Mack, 1989). It is unlikely that this class of pesticides exerts its action on hippocampal function by a direct interference with the $GABA_A$ receptor, but rather via its action on modulating noradrenergic input to the hippocampus (Boyes and Moser, 1988; Costa *et al.*, 1988; Gilbert and Dyer, 1988) or by its local anesthetic properties (Mack and Gilbert, 1989).

C. Pyrethroids

The synthetic pyrethroids constitute a diverse group of chemicals that are widely used for their insecticidal action. These compounds generally are recognized as excitatory neurotoxicants, with a primary site of action identified as a prolongation of the sodium current evoked by normal membrane depolarization (Narahashi, 1987). Pyrethroids have been subdivided into two categories, differentiated by the nomenclature Type I and Type II. Findings obtained in receptor-binding studies have suggested that Type II pyrethroids, in addition to their sodium channel effects, may also enhance excitability by blocking $GABA_A$-mediated inhibition (Casida *et al.*, 1983). However, in the intact animal, administration of the Type II pyrethroid, deltamethrin, produces a dramatic augmentation (see Fig. 7A) rather than a reduction in paired pulse depression recorded from the dentate gyrus (Gilbert *et al.*, 1989; Joy and Albertson, 1991; Joy *et al.*, 1989). These studies serve as an example of the pitfalls of relying exclusively on *in vitro* preparations to assess neurotoxic effects, without consideration of the functional implications.

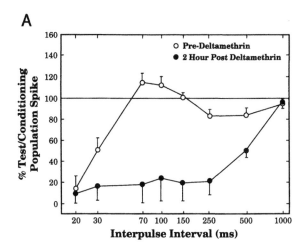

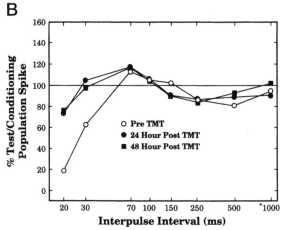

FIGURE 7 (A) Deltamethrin (10 mg/kg p.o.) produces a profound augmentation of paired pulse depression 2 hr after dosing. (B) Trimethyl tin (8 mg/kg i.v.) produces a persistent reduction in paired pulse depression in the dentate gyrus in an unanesthetized rat. Effects are limited to inhibition at brief interpulse intervals.

VIII. Heavy Metals

A. Trimethyl Tin

Although trimethyl tin (TMT) numbers among one of the most widely studied neurotoxic compounds, its primary mechanism of action is yet to be identified. The hallmark characteristic of TMT exposure is cell loss in limbic brain structures that is especially prominent in the hippocampal CA3 pyramidal cell region (Balaban *et al.*, 1988; Chang and Dyer, 1985). Some animals experience seizures following administration of high doses of TMT (Dyer *et al.*, 1982; Sloviter *et al.*, 1986), but cell loss does not appear to depend on the presence of seizure activity (Hassan *et al.*, 1984; O'Callaghan *et al.*, 1988; Zimmer *et al.*, 1985). Consistent with the report of Hassan and colleagues (1984),

continuous electrographic recordings for 12–18 hr each day for 5–7 days postdosing failed to reveal evidence of seizure activity following high dosages of TMT (M. E. Gilbert, unpublished observations).

Perhaps related to its proconvulsant potential, exposure to TMT results in a decrease in early paired pulse depression in dentate granule cell field potentials recorded within 2 hr of dosing (Dyer and Boyes, 1984). As can be seen in Fig. 7B, this effect persists for at least 48 hr and is restricted to brief IPIs (O'Callaghan *et al.*, 1988), suggesting a selective interference of TMT with $GABA_A$-mediated inhibition. In contrast to the expected increase in hippocampal excitability that frequently accompanies a decrease in $GABA_A$-mediated inhibition, the I/O function is shifted to the right following TMT administration, demonstrating an increase in threshold and a decrease in population spike amplitudes in both the dentate gyrus (M. E. Gilbert, unpublished observations) and CA1 pyramidal cells (Allen and Fonnum, 1984; Armstrong *et al.*, 1987; Harkins and Armstrong, 1992).

B. Lead

It has been documented extensively that developmental lead exposure leads to cognitive deficits in young children (for a review see Riess and Needleman, 1992). A significant literature exists demonstrating the ubiquitous actions of lead on cell biochemistry and physiology *in vitro*, yet few *in vivo* studies have been conducted at lead concentrations typically encountered in the environment. An interaction between lead and calcium function, and the calcium dependence of LTP, suggested that lead could produce cognitive deficits by disrupting LTP processes. In hippocampal slices from normal animals, Altmann and co-workers (1991) have shown interference with LTP induction in hippocampal slices bathed in lead-containing media. Their findings were inconclusive, however, with some exposed slices failing to exhibit lasting LTP, whereas others exhibited LTP of comparable magnitude to that observed under control conditions. Administration of lead acetate to the bathing medium may produce quite different effects relative to prolonged low level lead exposure during brain development. Recently Altman and colleagues (1993) have reported a reduction in the magnitude of LTP in area CA1 of hippocampal slices taken from developmentally lead-exposed animals.

Lasley and colleagues (1993) were the first to apply *in vivo* LTP recording procedures to characterize the effects of developmental lead exposure in anesthetized animals. Their findings suggested that continual exposure to lead in the drinking water from birth impaired LTP of the EPSP in the dentate gyrus. A more detailed analysis revealed that the train intensity required to produce LTP of population spike was increased in lead-exposed animals. The shift in LTP threshold was detected in the absence of any change in the baseline I/O function of dentate gyrus field potentials, suggesting a selective interference of developmental lead exposure with LTP induction processes rather than a general decrease in cell excitability. LTP of equal magnitude was induced in all animals when saturating trains of high stimulus intensity were applied (Gilbert and Lasley, 1993). No differential rate of decay was observed over the course of the 1-hr recording period in acute preparations or in animals prepared with chronically implanted electrodes and monitored for 1 month.

C. Methyl Mercury

Animals exposed to methyl mercury during prenatal or early postnatal development exhibit disorders ranging from gross congenital malformations to subtle behavioral abnormalities in the absence of obvious neurological impairment (see reviews by Chang and Annau, 1984; Eccles and Annau, 1987). Disrupting effects on synaptic transmission in peripheral synapses (peripheral ganglia, neuromuscular junction) have been attributed to the interference of sodium and calcium currents by methyl mercury interference. Yuan and Atchison (1993) have demonstrated a pronounced decrement in membrane excitability and synaptic transmission in CA1 of hippocampal slices bathed in media containing methyl mercury. LTP was induced in these slices, but deteriorated over the course of the 2-hr post-train period. However, a similar pattern and time course of deterioration in the population spike was observed in slices not administered LTP-inducing stimulation. It is difficult to interpret these toxicant-induced effects on LTP in the presence of compromised synaptic function and increased thresholds for evoked responses. Preliminary data fail to reveal any persistent effects of neonatal methyl mercury exposure on synaptic function and LTP recorded *in vivo* from the dentate gyrus (M. E. Gilbert and C. M. Mack, unpublished observations).

D. Aluminum

A link between aluminum toxicity and Alzheimer's disease was originally suggested by the histological similarity in plaques that develop following aluminum encephalopathy in animals and plaques that characterize postmortem hippocampal tissue of Alzheimer's patients (Farnell *et al.*, 1982; McLachlan and Massiah, 1992; Yokel *et al.*, 1988). Severe impairments in cognitive function also accompany Alzheimer's disease and aluminum neurotoxicity (McLachlan and Massiah,

1992). Although no *in vivo* electrophysiological testing of aluminum-exposed animals has been reported, a decrease in maximal population spike amplitude, an increase in threshold, and a reduction in incidence and maintenance of LTP have been demonstrated in CA1 recordings from hippocampal slices taken from rabbits treated with aluminum (Farnell *et al.*, 1982; 1985; Franceshetti *et al.*, 1990). A decrease in potassium-stimulated glutamate release (Provan and Yokel, 1992; Yokel *et al.*, 1988) and a disruption of second messenger signal transduction (Jope, 1988; Shafer *et al.*, 1993) have also been reported in hippocampal slices bathed in low concentrations of aluminum. Preliminary studies, however, have failed to find evidence of compromised synaptic function or impaired LTP in CA1 recordings from slices bathed in 100 μM of aluminum chloride. The reduction in glutamate release in slices bathed in aluminum-containing solution has not been replicated (Shafer and Gilbert, 1993). These discrepant findings suggest that the alterations in synaptic function following *in vivo* exposure may be related to subtle histopathological alterations that accompany aluminum-induced encephalopathy instead of a direct effect of aluminum on mechanisms supporting LTP.

IX. Solvents

A. Ethanol

Although the solvent literature in neurotoxicology is extensive, few examples of solvent-induced changes in hippocampal field potentials have been reported. There is, however, a history of using these techniques to study the effects of ethanol exposure on brain function. Acute administration of ethanol reduces excitatory synaptic transmission recorded from the CA1 region and dentate gyrus *in vivo* (Steffensen and Hendriksen, 1992; Steffensen *et al.*, 1992). An increase in early and late inhibition and a reduction in paired pulse facilitation accompany reduced excitability in the dentate gyrus (Weisner *et al.*, 1987; Steffensen and Hendriksen, 1992). A similar increase in paired pulse inhibition has been detected in response to acute treatment with methanol (C. M. Mack and M. E. Gilbert, unpublished observations). A decrease in the magnitude of LTP has also been reported in hippocampal slices bathed in low concentrations of ethanol (Blitzer *et al.*, 1990). An augmentation in GABA-mediated synaptic transmission and a reduction in NMDA-mediated synaptic transmission are two possible sites where ethanol may exert its acute effects on hippocampal function (see review by Little, 1991). In contrast, persistent effects of chronic ethanol exposure, including a depression in dentate granule cell excitability (Abraham *et al.*, 1984) and a reduction in the magnitude of LTP recorded from CA1 hippocampal slices (Durand and Carlen, 1984; Swartzwelder *et al.*, 1988), have been attributed to the documented cell loss following months of ethanol treatment.

X. Summary and Conclusions

The physiological basis of *in vivo* field potential recordings has been reviewed, and both the advantages and limitations of these measures in characterizing the summated influences of excitatory and inhibitory synaptic transmission in two regions of the hippocampus have been described. By necessity, an overly simplified view of the circuitry and cellular interactions that influence hippocampal activity has been presented. The original concept of a trisynaptic serial relay through each of the three principal cell populations has been challenged by evidence of the increasingly complex integration of information from higher and lower brain centers that occurs in each subregion. Hippocampal field potentials *in vivo* have contributed to our understanding of the significant, but frequently neglected, moderating influences of commissural and subcortical afferents on the excitability of a given cell population.

Application of these measures to neurotoxicological investigations has aided in identifying treatment effects exerted primarily through excitatory or inhibitory synaptic pathways and has provided a foundation for, and direction to, more mechanistic studies. Neurophysiological investigations in conscious animals provide a functional index in neurotoxicological assessments, affording a frame of reference for interpreting *in vitro* findings and the consequences of static, terminal observations of neuropathological studies. Field potential analyses, particularly in hippocampus, serve as a primary interface between functional brain changes and neurobehavioral endpoints. The importance of studying the functional output of an entire network is perhaps best exemplified by the advances made in our understanding of the cellular processes underlying the synaptic plasticity believed necessary for learning and memory. Progress in this area has recently provided the tools needed to begin identifying the more subtle impairments that frequently accompany low-level chronic exposure to neurotoxic chemicals.

Acknowledgments

The authors thank Drs. William K. Boyes, Robert M. Joy, and Stephen M. Lasley for their comments on an earlier draft of this manuscript. The contribution of Dr. Lasley to Fig. 5 is also gratefully acknowledged.

References

Abalis, I. M., Eldefrawi, M. E., and Eldefrawi, A. T. (1985). High-affinity stereospecific binding of cyclodiene insecticides and gamma-hexachlorocylohexane to gamma-aminobutyric acid receptors of rat brain. *Pestic. Biochem. Physiol.* **24,** 95–102.

Abalis, I. M., Eldefrawi, M. E., and Eldefrawi, A. T. (1986). Effects of insecticides on GABA-induced chloride influx into rat brain microsacs. *J. Toxicol. Environ. Health* **18,** 13–23.

Abraham, W. C., Rogers, C. J., and Hunter, B. E. (1984). Chronic ethanol-induced decreases in the response of dentate granule cells to perforant path input in the rat. *Exp. Brain Res.* **54,** 406–414.

Acsady, L., Halasy, K., and Freund, T. F. (1993). Calretinin is present in non-pyramidal cells of the rat hippocampus- III. Their inputs from the median raphé and medial septal nuclei. *Neuroscience* **52,** 829–841.

Akers, R. M., Lovinger, D. H., Colley, P. A. Linden, D. J., Routtenberg, A. (1986). Translocation of protein kinase C activity may mediate hippocampal long-term potentiation. *Science* **231,** 587–589.

Alger, B. E. (1991). Gating of GABAergic inhibition in hippocampal pyramidal cells. *Ann. N.Y. Acad. Sci.* **627,** 249–263.

Alger, B. E., and Nicoll, R. A. (1982a). Feed-forward dendritic inhibition in rat hippocampal pyramidal cells studied in vitro. *J. Physiol.* **328,** 105–123.

Alger, B. E., and Nicoll, R. A. (1982b). Pharmacological evidence for two kinds of GABA receptor on rat hippocampal pyramidal cells studied in vitro. *J. Physiol.* **328,** 125–141.

Allen, C. N., and Fonnum, R. (1984). Trimethyltin inhibits the activity of hippocampal neurons recorded in vitro. *Neurotoxicology* **5,** 23–30.

Altmann, L., Sveinsson, K., and Wiegand, H. (1991). Long-term potentiation in rat hippocampal slices is impaired following acute lead perfusion. *Neurosci. Lett.* **128,** 101–112.

Altmann, L., Weinsberg, F., Sveinsson, K., Lilienthal, H., Wiegand, H., and Winneke, G. (1993). Impairment of long-term potentiation and learning following chronic lead exposure. *Toxicol. Lett.* **66,** 105–112.

Amaral, D. G. (1978). A golgi study of cell types in the hilar region of the hippocampus in the rat. *J. Comp. Neurol.* **182,** 851–914.

Andersen, P., Bliss, T., and Skrede, K. K. (1971). Lamellar organization of hippocampal excitatory pathways. *Exp. Brain Res.* **13,** 222–238.

Andersen, P., Eccles, J. C., and Loyning, Y. (1964). Location of postsynaptic inhibitory synapses on hippocampal pyramids. *J. Neurophysiol.* **27,** 592–607.

Andersen, P., Gross, G. N., Lomo, T. and Sveen, O. (1969). Participation of inhibitory and excitatory interneurons in the control of hippocampal cortical output. In *The Interneuron* (M. A. B. Brazier, Ed.), pp. 415–465, University of California Press, Los Angeles.

Andersen, P., Holmqvist, B. Voorhoeve, P. E. (1966). Entorhinal activation of dentate granule cells. *Acta Physiol. Scand.* **66,** 448–460.

Andrade, R., and Chaput, Y. (1991). 5-hydroxytryptamine-like receptors mediate the slow excitatory response to serotonin in the rat hippocampus. *J. Pharmacol. Exp. Ther.* **257,** 930–937.

Andreasen, M., and Lambert, J. D. C. (1991). Noradrenaline receptors participate in the regulation of GABAergic inhibition in area CA1 of the rat hippocampus. *J. Physiol.* **439,** 649–669.

Arai, A., and Lynch, G. (1992). Factors regulating the magnitude of long-term potentiation induced by theta pattern stimulation. *Brain Res.* **598,** 173–184.

Armstrong, D. L., Read, H. L., Cork, A. E., Montemayor, F., and Wayner, M. J. (1987). Effects of trimethyltin on evoked potentials in mouse hippocampal slices. *Neurotoxicol. Teratol.* **9,** 359–362.

Austin, K. B., Beiswanger, C., Bronzino, J. D., Austin-LaFrance, R. J., Galler, J. R., and Morgane, P. J. (1992). Prenatal protein malnutrition alters behavioral state modulation of inhibition and facilitation in the dentate gyrus. *Brain Res. Bull.* **28**, 245–255.

Balaban, C. D., O'Callaghan, J. P., and Billingsley, M. L. (1988). Trimethyltin-induced neuronal damage in the rat brain: comparative studies using silver degeneration stains, immunocytochemistry and immunoassay for neuronotypic and gliotypic proteins. *Neuroscience* **26**, 337–361.

Bilkey, D. K., and Goddard, G. V. (1987). Septohippocampal and commissural pathways antagonistically control inhibitory interneurons in the dentate gyrus. *Brain Res.* **405**, 320–325.

Blanpied, T. A., and Berger, T. W. (1992). Characterization in vivo of the NMDA receptor-mediated component of dentate granule cell populations synaptic responses to perforant path input. *Hippocampus* **2**, 373–388.

Bliss, T. V. P., Douglas, R. M., Errington, M. L., and Lynch, M. A. (1986). Correlation between long-term potentiation and release of endogenous amino acids from dentate gyrus of anesthetized rats. *J. Physiol.* **377**, 391–408.

Bliss, T. V. P., and Collingridge, G. L. (1993). A synaptic model of memory: long-term potentiation in the hippocampus. *Nature* **361**, 31–39.

Bliss, T. V. P., and Lomo, T. (1973). Long-lasting potentiation of synaptic transmission in the dentate area of the anaesthetized rabbit following stimulation of the perforant path. *J. Physiol.* **232**, 331–356.

Blitzer, R. D., Gil, O., and Landau, E. M. (1990). Long-term potentiation in rat hippocampus is inhibited by low concentrations of ethanol. *Brain Res.* **537**, 203–208.

Boyes, W. K., and Moser, V. C. (1988). An alpha-2 adrenergic mode of action of chlordimeform on rat visual function. *Toxicol. Appl. Pharmacol.* **92**, 402–418.

Burdette, L. J., and Gilbert M. E. Stimulus parameters affecting paired pulse depression of dentate granule cell field potentials: I. Stimulus intensity. *Brain Res.* in press.

Burdette, L. J., Hart, G. J., and Masukawa, L. M. (1991). Recovery of paired pulse inhibition following kindled seizures. *Epilepsia* **32**, 32.

Buzsaki, G. (1984). Feed-forward inhibition in the hippocampal formation. *Prog. Neurobiol.* **22**, 131–153.

Buzsaki, G., and Czeh, G. (1981). Commissural and perforant path interactions in the rat hippocampus. *Exp. Brain Res.* **43**, 429–438.

Buzsaki, G., and Eidelberg, E. (1982). Convergence of association and commissural pathways on CA1 pyramidal cells of the rat hippocampus. *Brain Res.* **237**, 283–295.

Cao, F., and Leung, L. S. (1991). Behavior-dependent paried-pulse responses in the hippocampal CA1 region. *Exp. Brain Res.* **87**, 553–561.

Casida, J. E., Gammon, D. W., Glickman, A. H., and Lawrence, L. J. (1988). Mechanisms of selective action of pyrethroid insecticides. *Ann. Rev. Pharmacol. Toxicol.* **24**, 413–438.

Chang, L. W., and Annau, Z. (1984). Developmental neuropathology and behavioral teratology of methyl mercury. In *Neurobehavioral Teratology* (J. Yanai Ed.), Elsevier Science Publishers, Amsterdam.

Chang, L. W., and Dyer, R. S. (1985). Septotemporal gradients of trimethyltin-induced hippocampal lesions. *Neurobehav. Toxicol. Teratol.* **7**, 43–49.

Choi, D. (1988). Glutamate neurotoxicity and diseases of the nervous system. *Neuron* **1**, 623–634.

Choi, D. (1990). The role of glutamate neurotoxicity in hypoxic-ischemic neuronal death. *Ann. Rev. Neurosci.* **13**, 171–182.

Coan, E. J., and Collingridge, G. L. (1985). Magnesium ions block an N-methyl-D-aspartate receptor-mediated component of synaptic transmission in rat hippocampus. *Neurosci. Lett.* **53**, 21–26.

Cole, A. J., Saffen, D. W., Baraban, J. M., and Worley, P. F. (1989). Rapid increase in an immediate early gene messenger RNA in hippocampal neurons by synaptic NMDA receptor activation. *Nature* **340**, 474–476.

Colley, P. A., and Routtenberg, A. (1993). Long-term potentiation as synaptic dialogue. *Brain Res. Rev.* **18**, 115–122.

Collingridge, G. L. (1992). The mechanism of induction of NMDA receptor-dependent long-term potentiation in the hippocampus. *Exp. Physiol.* **77**, 771–797.

Collingridge, G. L., and Bliss, T. V. P. (1987). NMDA receptors—their role in long–term potentiation. *Trends Neuropharmacol.* **10**, 288–293.

Collingridge, G. L., and Lester, R. A. J. (1989). Excitatory amino acid receptors in the vertebrate central nervous system. *Pharmacol. Rev.* **40**, 143–210.

Conn, P. J., and Desai, M. A. (1991). Pharmacology and physiology of metabotropic glutamate receptors in mammalian central nervous system. *Drug Dev. Res.* **24**, 207–229.

Costa, L., and Murphy, S. D. (1988). Alpha-2-adrenoreceptors as a target for formamidine pesticides: in vitro and in vivo studies in mice. *Toxicol. Appl. Pharmacol.* **93**, 319–328.

Creager, R., Dunwiddie, T., and Lynch, G. (1980). Paired-pulse and frequency facilitation in the CA1 region of the in vitro rat hippocampus. *J. Physiol.* **299**, 409–424.

Crofton, K. M., Howard, J. L., Moser, V. C., Gill, M. W., Reiter, L. W., Tilson, H. A. and MacPhail, R. C. (1991). Interlaboratory comparison of motor activity experiments: implications for neurotoxicological assessments. *Neurotoxicol. Teratol.* **13**, 599–609.

Davies, C. H., Davies, S. N., and Collingridge, G. L. (1990). Paired-pulse depression of monosynaptic GABA-mediated inhibitory postsynaptic responses in rat hippocampus. *J. Physiol.* **424**, 513–521.

Deadwyler, S. A., West, J. R., Cotman, C. W., and Lynch, G. S. (1975). A neurophysiological analysis of commissural projections to dentate gyrus of the rat. *J. Neurophysiol.* **38**, 1667–1684.

deJonge, M., and Racine, R. J. (1986). The development and decay of kindling-induced increases in paired pulse depression in the dentate gyrus. *Brain Res.* **412**, 318–328.

Diamond, D. M., Dunwiddie, T. V., and Rose, G. M. (1988). Characteristics of hippocampal primed burst potentiation in vitro and in the awake rat. *J. Neurosci.* **8**, 4079–4088.

Dingledine, R. (1983). N-methyl-d-aspartate activities voltage-dependent calcium conductance in rat hippocampal pyramidal cells. *J. Physiol.* **343**, 385–405.

Dingledine, R. (1986). NMDA Receptors: what do they do? *Trends Neurosci.* **9**, 47–49.

Douglas, R. M., McNaughton, B. L., and Goddard, G. V. (1983). Commissural inhibition and facilitation of granule cell discharge in fascia dentata. *J. Comp. Neurol.* **219**, 285–294.

Dunwiddie, T. V., and Lynch, G. (1979). The relationship between extracellular calcium concentration and the induction of hippocampal long-term potentiation. *Brain Res.* **169**, 103–110.

Durand, D., and Carlen, P. L. (1984). Impairment of long-term potentiation in rat hippocampus following chronic ethanol treatment. *Brain Res.* **308**, 325–332.

Dutar, P., and Nicoll, R. A. (1988). A physiological role for GABA$_{BI}$ receptors in the central nervous system. *Nature* **332**, 156–158.

Dyer, R. S., and Boyes, W. K. (1984). Trimethyl tin reduces recurrent inhibition in rats. *Neurobehav. Toxicol. Teratol.* **6**, 369–371.

Dyer, R. S., Walsh, T. J., Wonderlin, W. F., and Bercegeay, M. (1982). The trimethyl tin syndrome in rats. *Neurotoxicol. Teratol.* **4**, 127–133.

Eccles, C. U., and Annau, Z. (1987). Prenatal exposure to methyl mercury. In *The Toxicity of Methyl Mercury* (C. U. Eccles and

Z. Annau Eds.), pp. 114–130, The Johns Hopkins University Press, Baltimore.

Eichenbaum, H., Otto, T., and Cohen, N. J. (1992). The hippocampus—What does it do? *Behav. Neural Biol.* **57,** 2–36.

Eichenbaum, H., and Otto, T. (1993). LTP and memory: can we enhance the connection? *Trends Neurosci.* **16,** 163–164.

Fagg, G. E., and Foster, A. C. (1983). Amino acid neurotransmitters and their pathways in the mammalian central nervous system. *Neuroscience* **9,** 701–719.

Fantie, B. D., and Goddard, G. V. (1982). Septal modulation of the population spike in the fascia dentata produced by perforant path stimulation in the rat. *Brain Res.* **252,** 227–237.

Farnell, B. J., Crapper-McLachlan, D. R., Baimbridge, K., deBoni, U., Wong, L. and Wood, P. L. (1985). Calcium metabolism in aluminum encephalopathy. *Exp. Neurol.* **88,** 68–83.

Farnell, B. J., DeBoni, U., and Crapper McLachlan, D. R. (1982). Aluminum neurotoxicity in the absence of neurofibrillary degeneration in CA1 hippocampal pyramdial neurons in vitro. *Exp. Neurol.* **78,** 241–258.

Fazeli, M. S., Corbet, J., Dunn, M. J., Dolphin, A. C., and Bliss, T. V. P. (1993). Changes in protein synthesis accompanying long-term potentiation in the dentate gyrus. *J. Neurosci.* **13,** 1346–1353.

Fifkova, E., Eason, H., and Schaner, P. (1992). Inhibitory contacts on dendritic spines of the dentate fascia. *Brain Res.* **577,** 331–336.

Fonnum, F. (1984). Glutamate: A neurotransmitter in mammalian brain. *J. Neurochem.* **42,** 1–11.

Franceschetti, S., Bugiani, O., Panzica, F., Tagliavini, F., and Avanzini, G. (1990). Changes in excitability of CA1 pyramidal neurons in slices prepared from AlCl₃-treated rabbits. *Epilepsy Res.* **6,** 39–48.

Freund, T. F., and Antal, M. (1988). GABA-containing neurons in the septum control interneurons in the hippocampus. *Nature* **336,** 170–173.

Freund, T. F., Gulyas, A. I., Acsady, L. Gorcs, T., and Toth, K. (1990). Serotonergic control of the hippocampus via local inhibitory interneurons. *Proc. Natl. Acad. Sci.* **87,** 8501–8505.

Frotscher, M. and Leranth, C. (1985). Cholinergic innervation of the rat hippocampus as revealed by choline acetyltransferase immunocytochemistry: a combined light and electron microscopic study. *J. Comp. Neurol.* **239,** 237–246.

Frotscher, M., and Zimmer, J. (1983). Commissural fibers terminate on non-pyramidal neurons in the guinea pig hippocampus—A combined Golgi/EM degeneration study. *Brain Res.* **265,** 289–293.

Gahwiler, B. H. (1983). The action of neuropeptides on the bioelectric activity of hippocampal neurons. In Neurobiology of the Hippocampus (W. Seifert, Ed.), pp. 157–173, Academic Press, New York.

Geinisman, Y., deToledo-Morrell, L., and Morrell, F. (1991). Induction of long-term potentiation is associated with an increase in the number of axospinous synapses with segmented postsynaptic densities. *Brain Res.* **566,** 77–88.

Germroth, P., Schwerdtfeger, W. K., and Buhl, E. H. (1989). GABAergic neurons in the entorhinal cortex project to the hippocampus. *Brain Res.* **494,** 187–192.

Gilbert, M. E. (1991). Potentiation of inhibition with perforant path kindling: An NMDA-dependent process. *Brain Res.* **564,** 109–116.

Gilbert, M. E. (1992a). Neurotoxicants and limbic kindling. In *The Vulnerable Brain and Environmental Risks. Malnutrition and Hazard Assessment.* Vol. 1. pp. 173–192. (R. L. Isaacson and K. F. Jensen, Eds.), Plenum Press, New York.

Gilbert, M. E. (1992b). Proconvulsant activity of endosulfan in amygdala kindling. *Neurotoxicol. Teratol.* **14,** 143–149.

Gilbert, M. E. (1992c). A characterization of chemical kindling with the pesticide endosulfan. *Neurotoxicol. Teratol.* **14,** 151–158.

Gilbert, M. E. (1992d). Chemical kindling with lindane. *The Toxicologist* **12,** 275.

Gilbert, M. E., and Dyer, R. S. (1988). Increased hippocampal excitability produced by amitraz. *Neurotoxicol. Teratol.* **10,** 229–235.

Gilbert, M. E., and Lasley, S. M. (1993). Developmental lead exposure increases LTP threshold in rat dentate gyrus in vivo. *Soc. Neurosci. Abstr.* **19,** 910.

Gilbert, M. E., and Mack, C. M. (1989). Enhanced susceptibility to kindling by chlordimeform may be mediated by a local anesthetic action. *Psychopharmacology* **99,** 163–167.

Gilbert, M. E., and Mack, C. M. (1990). The NMDA-receptor antagonist, MK-801, suppresses long-term potentiation, kindling, and kindling-induced potentiation in the perforant path of the unanesthetized rat. *Brain Res.* **519,** 89–96.

Gilbert, M. E., Mack, C. M., and Burdette, L. J. (1992). Stimulus intensity and pharmacologic effects on late inhibition in dentate granule cell field responses in vivo. *Epilepsia* **33,** 33.

Gilbert, M. E., Mack, C. M., and Crofton, K. M. (1989). Pyrethroids and enhanced inhibition in the hippocampus of the rat. *Brain Res.* **477,** 314–321.

Gordon, C. J., Mohler, F. S., Watkinson, W. P. and Rezvani, A. H. (1988). Temperature regulation in laboratory mammals following acute toxic insult. *Toxicology* **53,** 161–178.

Gottlieb, D. I., and Cowan, W. M. (1973). Autoradiographic studies of the commissural and ipsilateral association connections of the hippocampus and dentate gyrus of the rat. *J. Comp. Neurol.* **149,** 393–422.

Green, E. J., Barnes, C. A., and McNaughton, B. L. (1993). Behavioral state dependence of homo- and hetero-synaptic modulation of dentate gyrus excitability. Exp. *Brain Res.* **93,** 55–65.

Halasy, K., Miettinen, R., Szabat, E., and Freund, T. F. (1992). GABAergic interneurons are the major postsynaptic targets of median raphé afferents in the rat dentate gyrus. *Eur. J. Neurosci.* **4,** 144–153.

Hargreaves, E. L., Cain, D. P., and Vanderwolf, C. H. (1990). Learning and behavioral-long-term-potentiation: importance of controlling for motor activity. *J. Neurosci.* **10,** 1472–1478.

Harkins, A. B., and Armstrong, D. L. (1992). Trimethyltin alters membrane properties of CA1 hippocampal neurons. *Neurotoxicology* **13,** 569–582.

Harley, C. W., and Sara, S. J. (1992). Locus coeruleus bursts induced by glutamate trigger delayed perforant path spike amplitude potentiation in the dentate gyrus. *Exp. Brain Res.* **89,** 581–587.

Hassan, Z., Zimmer, L., and Woolley, D. (1984). Time course of the effects of trimethyltin on limbic evoked potentials and distribution of tin in blood and brain in the rat. *Neurotoxicology* **5,** 217–244.

Heinneman, U., Beck, H., Drier, J. P., Ficker, E., Stabel, J. and Zhang, C. L. (1992). The dentate gyrus as a regulated gate for the propagation of epileptiform activity. In *The Dentate Gyrus and Its Role in Seizures* (C. E. Ribak, C. M. Gall, and I. Mody, (Eds.), Epilepsy Res., Suppl. 7:273–280.

Hess, G., Kuhnt, U. and Voronin, L. L. (1987). Quantal analysis of paired pulse facilitation in guinea pig hippocampal slices. *Neurosci. Lett.* **77,** 187–192.

Hotson, J. R. and Prince, D. A. (1980). A calcium-activated hyperpolarization follows repetitive firing in hippocampal neurons. *J. Neurophysiol.* **43,** 409–419.

Jahr, C. E. and Stevens, C. F. (1990). A quantitative description of NMDA receptor-channel kinetic behavior. *J. Neurosci.* **10,** 1830–1837.

Jope, R. S. (1988). Modulation of phosphoinositide hydrolysis by NaFl and aluminum in rat cortical slices. *J. Neurochem.* **51,** 1731–1736.

Joy, R. M., Stark, L. G. and Albertson, T. E. (1982). Proconvulsant effects of lindane: Enhancement of amygdaloid kindling in the rat. *Neurobehav. Toxicol. Teratol.* **4,** 347–354.

Joy, R. M., and Albertson, T. E. (1985). Effects of lindane on excitation and inhibition evoked in dentate gyrus by perforant path stimulation. *Neurobehav. Toxicol. Teratol.* **7,** 1–8.

Joy, R. M., and Albertson, T. E. (1987). Factors responsible for increased excitability of dentate gyrus granule cells during exposure to lindane. *Neurotoxicology* **8,** 517–528.

Joy, R. M., and Albertson, T. E. (1991). Interactions of GABAa antagonists with deltamethrin, diazepam, pentobarbital, and SKF100330A in the rat dentate gyrus. *Toxicol. Appl. Pharmacol.* **109,** 251–262.

Joy, R. M. and Albertson, T. E. (1993). NMDA receptors have a dominant role in population spike paired pulse facilitation in the dentate gyrus of urethane-anesthetized rats. *Brain Res.* **604,** 273–282.

Joy, R. M., Albertson, T. E. and Ray D. E. (1989). Type I and Type II pyrethroids increase inhibition in the hippocampal dentate gyrus of the rat. *Toxicol. Appl. Pharmacol.* **98,** 398–412.

Joy, R. M., Stark, L. G., and Albertson, T. E. (1983). Proconvulsant actions of lindane: Effects on afterdischarge thresholds and durations during amygdaloid kindling in rats. *Neurotoxicology* **2,** 211–220.

Kandel, E. R., Spencer, W. A. and Brinley, F. J. (1961). Electrophysiology of hippocampal neurons. I. Sequential invasion and synaptic organization. *J. Neurophysiol.* **24,** 225–242.

Klančnik, J. M. and Phillips, A. G. (1991). Modulation of synaptic plasticity in the dentate gyrus of the rat by electrical stimulation of the median raphé nucleus. *Brain Res.* **557,** 236–240.

Konnerth, A., and Heinemann, U. (1983). Presynaptic involvement in frequency facilitation in the hippocampal slice. *Neurosci. Lett.* **42,** 255–260.

Korn, H., Oda, Y., and Faber, D. S. (1992). Long-term potentiation of inhibitory circuits and synapses in the central nervous system. *Proc. Natl. Acad. Sci.* **89,** 440–443.

Lambert, J. D. C. and Jones, R. S. G. (1990). A reevaluation of excitatory amino acid-mediated synaptic transmission in the rat dentate gyrus. *J. Neurophysiol.* **64,** 119–132.

Larson, J., Wong, D., and Lynch, G. (1986). Pattern stimulation at the theta frequency is optimal for the induction of hippocampal long-term potentiation. *Brain Res.* **358,** 347–350.

Lasley, S. M., Polan-Curtain, J., and Armstrong, D. L. (1993). Chronic exposure to environmental levels of lead impairs in vivo induction of long-term potentiation in rat hippocampal dentate. *Brain Res.* **614,** 347–351.

Lawrence, L. J., and Casida, J. E. (1984). Interaction of lindane, toxephine cyclodienes with brain specific t-butylbicyclophosphorothionate receptor. *Life Sci.* **35,** 171–178.

Leranth, C., Malcolm, A. J. and Frotscher, M. (1990). Afferent and efferent synaptic connections of somatostatin-immunoreactive neurons in the rat fascia dentata. *J. Comp. Neurol.* **295,** 111–122.

Leung, L. S., Shen, B., and Kaibara, T. (1992). Long-term potentiation induced by patterned stimulation of the commissural pathway to hippocampal CA1 region in freely moving rats. *Neuroscience* **48,** 63–74.

Little, H. J. (1991). Mechanisms that may underlie the behavioral effects of ethanol. *Prog. Neurobiol.* **36,** 171–194.

Lomo, T. (1971). Patterns of activation in a monosynaptic cortical pathway: the perforant path input to the dentate area of the hippocampal formation. *Exp. Brain Res.* **12,** 18–45.

Lupica, C. R., and Dunwiddie, T. V. (1991). Differential effects of mu- and delta-receptor selective opioid agonists on feedforward and feedback GABAergic inhibition in hippocampal brain slices. *Synapse* **8,** 237–248.

Lynch, M. A., Clements, M. P., Voss, K. L., Branham, C. R., and Bliss, T. V. P. (1991). Is arachidonic acid a retrograde messenger in long-term potentiation? *Lipid Turnover in the Central Nervous System Biochem. Soc. Trans.* **19,** 391–396.

Lynch, M. A., Kessler, M., Arai, A., and Larson, J. (1990). The nature and causes of hippocampal long-term potentiation. *Prog. Brain Res.* **83,** 233–250.

Malenka, R. C. (1992). The role of postsynaptic calcium in the induction of long-term potentiation. *Mol. Neurobiol.* **5,** 289–295.

Malenka, R. C., Kauer, J. A., Perkel, D. J., Mauk, M. D., Kelly, P. T., Nicoll, R. A. and Waxham, M. N. (1989). An essential role for postsynaptic calmodulin and protein kinase activity in long-term potentiation. *Nature* **340,** 554–557.

Malenka, R. C., Kauer, J. A., Perkel, D. J., and Nicoll, R. A. (1990). Long-term potentiation in the hippocampus. *Prog. Cell Res.* **1,** 263–277.

Malinow, R., Schulman, H., and Tsien, R. W. (1989). Inhibition of postsynaptic PKC or CaMKII blocks induction but not expression of LTP. *Science* **245,** 862–866.

Massicotte, G. and Baudry, M. (1991). Triggers and substrates of hippocampal synaptic plasticity. *Neurosci. Biobehav. Rev.* **15,** 415–423.

Mayer, M. L., and Westbrook, G. L. (1987). The physiology of excitatory amino acids in vertebrate CNS. *Prog. Neurogiol.* **28,** 197–296.

McLachlan, D. R. and Massiah, J. (1992). Aluminum ingestion: a risk factor for Alzheimer's disease? In *The Vulnerable Brain and Environmental Risks*, Vol. 2, *Toxins in Food*. (R. L. Isaacson and K. F. Jensen (Eds.), pp. 49–60. Plenum Press, New York.

McNaughton, B. I. (1983). Activity dependent modification of hippocampal synaptic efficacy. In *Neurobiology of the Hippocampus* (W. Seifert Ed.), pp. 223–252. Academic Press, New York.

McNaughton, B. I. (1982). Long-term synaptic enhancement and short-term potentiation in rat fascia dentata act through different mechanisms. *J. Physiol.* **324,** 249–262.

McNaughton, B. L. (1993). The mechanism of expression of long-term enhancement of hippocampal synapses: current issues and theoretical implications. *Ann. Rev. Physiol.* **55,** 375–396.

McNaughton, N., and Miller, J. J. (1984). Medial septal projections to the dentate gyrus of the rat: electrophysiological analysis of distribution and plasticity. *Exp. Brain Res.* **56,** 243–256.

Miettinen, R., and Freund, T. F. (1992). Neuropeptide Y-containing interneurons in the hippocampus receive synaptic input from median raphé and GABAergic septal afferents. *Neuropeptides* **22,** 185–193.

Milgram, N. W., Yearwood, T., Khurgel, M., Ivy, G. O., and Racine, R. J. (1991). Changes in inhibitory processes in the hippocampus following recurrent seizures induced by systemic administration of kainic acid. *Brain Res.* **551,** 236–246.

Morrisett, R. A., Mott, D. D., Lewis, D. V., Schwartzwelder, H. S. and Wilson, W. A. (1991). GABAb-receptor-mediated inhibition of the N-methyl-d-aspartate component of synaptic transmission in the rat hippocampus. *J. Neurosci.* **11,** 203–209.

Moser, E., Mathiesen, I., and Andersen, P. (1993). Association between brain temperature and dentate field potentials in exploring and swimming rats. *Science* **259,** 1324–1326.

Mott, D. D., and Lewis, D. V. (1991). Facilitation of the induction of long-term potentiation by GABAb receptors. *Science* **252,** 1718–1720.

Muller, D., Arai, A., and Lynch, G. (1992). Factors governing the potentiation of NMDA receptor-mediated responses in hippocampus. *Hippocampus* **2,** 29–38.

Muller, D., and Lynch, G. (1988). N-methyl-d-aspartate receptor-mediated component of synaptic responses to single-pulse stimulation in rat hippocampal slices. *Synapse* **2,** 666–668.

Muller, D., and Lynch, G. (1989). Synaptic modulation of N-methyl-d-aspartate in hippocampus. *Synapse* **5**, 94–103.

Narahashi, T. (1987). Neuronal target sites on insecticides. In *Sites of Action for Neurotoxic Pesticides* (R. M. Hollingworth, and M. B. Green, (Eds.), pp. 226–250. American Chemical Society, Washington, D.C.

Nathan, T., and Lambert, J. D. C. (1991). Depression of the fast IPSP underlies paired-pulse facilitation in area CA1 of the rat hippocampus. *J. Neurophysiol.* **66**, 1704–1715.

Nathan, T., Jensen, M. S., and Lambert, J. D. C. (1990). GABAb receptors play a major role in paired-pulse facilitation in area CA1 of the rat hippocampus. *Brain Res.* **531**, 55–65.

Newberry, N. R., and Nicoll, R. A. (1984). A bicuculline-resistent inhibitory postsynaptic potential in rat hippocampal pyramidal cells in vitro. *J. Neurophysiol.* **348**, 239–254.

Newberry, N. R., and Nicoll, R. A. (1985). Comparison of the action of baclofen with gamma-aminobutyric acid on rat hippocampal pyramidal cells in vitro. *J. Physiol.* **360**, 161–185.

Nowak, L., Bregestovski, P., Ascher, P., Herbert, A., and Prochiantz, A. (1989). Magnesium gates glutamate-activated channels in mouse central neurones. *Nature* **307**, 462–465.

O'Callaghan, J. P., Niedzweicki, D., Gilbert, M. E., Miller, L. P., and Ornstein, P. (1988). Trimethyltin-induced neurotoxicity may not be mediated through an excitotoxic mechanism. *Soc. Neurosci. Abstr.* **14**, 1082.

O'Dell, T. J., Hawkins, R. D., Kandel, E. R., and Araneio, O. (1991). Tests of the role of two diffusible substances in long-term potentiation: Evidence for nitric oxide as a possible early retrograde messenger. *Proc. Nat. Acad. Sci.* **88**, 1285.

Pacelli, G. J., Su, W., and Kelso, S. R. (1989). Activity-induced depression of synaptic inhibition during LTP-inducing patterned stimulation. *Brain Res.* **486**, 26–32.

Pacelli, G. J., Su, W., and Kelso, S. R. (1991). Activity-induced decrease in early and late inhibitory synaptic conductances in hippocampus. *Synapse* **7**, 1–13.

Provan, S. D., and Yokel, R. A. (1992). Aluminum inhibits glutamate release from transverse rat hippocampal slices: role of G proteins, Ca channels and protein kinase C. *Neurotoxicology* **13**, 413–420.

Qian, Z., Gilbert, M. E. and Kandel, E. R. (1994). Differential induction neuronal activity in rat brain of a protein tyrosine phosphatase containing a nuclear localization signal. *Learning and Memory* **1**, 180–188.

Qian, Z., Gilbert, M. E., Colicos, M., Kandel, E. R. and Kuhl, D. (1993). Tissue-plasminogen activator is induced as an immediate-early gene during seizure, kindling and long-term potentiation. *Nature* **361**, 453–457.

Racine, R. J., and Milgram, N. W. (1983). Short-term potentiation phenomena in the rat limbic forebrain. *Brain Res.* **260**, 201–216.

Racine, R. J., Milgram, N. W., and Hafner, S. (1983). Long-term potentiation phenomena in the rat limbic forebrain. *Brain Res.* **260**, 217–231.

Racine, R. J., Moore, K.-A., and Wicks, S. (1991). Activation of the NMDA-receptor: a correlate in the dentate gyrus field potential and its relationship to long-term potentiation and kindling. *Brain Res.* **556**, 226–239.

Rausche, G., Sarvey, J. H., and Heinemann, U. (1989). Slow synaptic inhibition in relation to frequency habituation in dentate granule cells of rat hippocampal slices. *Exp. Brain Res.* **78**, 233–242.

Ribak, C. E., and Seress, L. (1983). Five types of basket cell in the hippocampal dentate gyrus: a combined Golgi and microscopic study. *J. Neurocytol.* **12**, 577–597.

Rich-Bennett, E., Dahl, D., and LeCompte, B. B. (1992). Modulation of paried-pulse activation in the hippocampal dentate gyrus by cholecystokinin, baclofen, and bicuculline. *Neuropeptides* **24**, 263–270.

Riess, J. A., and Needleman, H. L. (1992). Cognitive, neural, behavioral effects of low-level lead exposure. In *The Vulnerable Brain and Environmental Risks*, Vol. 2, *Toxins in Food*. (R. L. Isaacson and K. F. Jensen, Eds.), Plenum Press, New York.

Robinson, G. B., and Racine, R. J. (1986). Interactions between septal and entorhinal inputs to the rat dentate gyrus: facilitation effects. *Brain Res.* **379**, 63–67.

Robinson, G. B., and Reed, G. D. (1992). Effect of MK-801 on the induction and subsequent decay of long-term potentiation in the unanesthetized rabbit hippocampal dentate gyrus. *Brain Res.* **569**, 78–85.

Robinson, G. B., Sclabassi, R. J., and Berger, T. W. (1991). Kindling-induced potentiation of excitatory and inhibitory inputs to hippocampal dentate granule cells. I. Effects on linear and nonlinear response characteristics. *Brain Res.* **562**, 17–25.

Ropert, N., and Guy, N. (1991). Serotonin facilitates GABAergic transmission in the CA1 regions of rat hippocampus in vitro. *J. Physiol.* **441**, 121–136.

Sara, S. J., and Bergis, O. (1991). Enhancement of excitability and inhibitory processes in hippocampal dentate gyrus by noradrenaline: a pharmacological study in awake, freely moving rats. *Neurosci. Lett.* **126**, 1–5.

Scharfman, H. E. (1991). Dentate hilar cells with dendrites in the molecular layer have lower thresholds for synaptic activation by perforant path than granule cells. *J. Neurosci.* **11**, 1660–1673.

Schwartzkroin, P. A., Scharfman, H. E., and Sloviter, R. S. (1990). Similarities in circuitry between Ammon's horn and dentate gyrus: local interactions and parallel processing. *Prog. Brain Res.* **83**, 269–286.

Seress, L., and Ribak, C. E. (1984). Direct commissural input to the basket cells of the hippocampal dentate gyrus: evidence for feed-forward inhibition. *J. Neurocytol.* **13**, 215–225.

Shafer, T. J., and Gilbert, M. E. (1994). Aluminum chloride fails to alter long-term potentiation or glutamate release in rat hippocampal slices. *The Toxicologist,* in press.

Shafer, T. J., Mundy, W. R., and Tilson, H. A. (1993). Aluminum decreases muscarinic, adrenergic, and metabotropic receptor-stimulated phosphoinositide hydrolysis in hippocampal and cortical slices from rat brain. *Brain Res.* **629**, 133–140.

Sloviter, R. S., Koebetitz, K., Walsh, T. J., and Dempster, D. W. (1986). On the role of seizure activity in the hippocampal damage produced by trimethyltin. *Brain Res.* **367**, 169–182.

Steffensen, S. C., and Henriksen, S. J. (1991). Effects of baclofen and bicuculline on inhibition in the fascia dentata and hippocampus regio superior. *Brain Res.* **538**, 46–53.

Steffensen, S. C., and Henriksen, S. J. (1992). Comparison of the effects of ethanol and chloridiazepoxide on electrophysiological activity in the fascia dentata and hippocampus regio superior. *Hippocampus* **2**, 201–211.

Steffensen, S. C., Yeckel, M. F., Miller, D. R., and Henriksen, S. J. (1992). Ethanol-induced suppression of hippocampal long-term potentiation is blocked by lesions of the septohippocampal nucleus. *Alcohol. Clin. Exp. Res.* **17**, 655–659.

Swartzwelder, H. S., Farr, K. L., Wilson, W. A., and Savage, D. D. (1988). Prenatal exposure to ethanol decreases physiological plasticity in the hippocampus of the adult rat. *Alcohol* **5**, 121–124.

Tang, C. M., Dichter, M., and Morad, M. (1989). Quisqualate activates a rapidly inactivating high conductance ionic channel in hippocampal neurons. *Science* **243**, 1474–1477.

Thalmann, R. H., and Ayala, G. F. (1982). A late increase in potassium conductance follows synaptic stimulation of granule neurons of the dentate gyrus. *Neurosci. Lett.* **29**, 243–248.

Teyler, T. J., Alger, B. E., Bergman, T., and Livingston, K. (1977). A comparison of long-term potentiation in the in vitro and in vivo hippocampal preparations. *Behav. Biol.* **19**, 24–34.

Tusell, J. M., Sunol, C., Gelpi, E., and Rodriguez-Farre, E. (1987). Relationship between lindane concentration in blood and brain and convulsant response in rats after oral or intraperitoneal administration. *Arch. Toxicol.* **60**, 432–437.

Tusell, J. M., Sunol, C., Gelpi, E., and Rodriguez-Farre, E. (1988). Effect of lindane at repeated low doses. *Toxicology* **49**, 375–379.

Wallace, C. S., Hawrylak, N., and Greenough, W. T. (1991). Studies of synaptic structural modifications after long-term potentiation and kindling: context for a molecular morphology. In *Long-term Potentiation: A Debate of Current Issues* (M. Baudry, and J. Davis Eds.), pp. 190–232. MIT Press, Cambridge.

Weisner, J. B., Henriksen, S. J., and Bloom, F. E. (1987). Ethanol enhances recurrent inhibition in the dentate gyrus of the hippocampus. *Neurosci. Lett.* **79**, 169–173.

White, W. F., Nadler, J. V., and Cotman, C. W. (1979). Analysis of short-term plasticity at the perforant path-granule cell synapse. *Brain Res.* **178**, 41–53.

Wieraszko, A. (1983). Glutamic and aspartic acid as putative neurotransmitters: Release and uptake studies on hippocampal slices. In *Neurobiology of the Hippocampus* (W. Seifert, Ed.), pp. 175–196. Academic Press, New York.

Wigstrom, H., and Gustafsson, B. (1983). Large long-lasting potentiation in the dentate gyrus in vitro during blockade of inhibition. *Brain Res.* **275**, 153–158.

Winson, J., and Azburg, C. (1978). Neuronal transmission through hippocampal pathways is dependent on behavior. *J. Neurophysiol.* **41**, 716–732.

Yeckel, M. F., and Berger, T. W. (1990). Feedforward excitation of the hippocampus by afferents from the entorhinal cortex: redefinition of the role of the trisynaptic pathway. *Proc. Natl. Acad. Sci.* **87**, 5832–5836.

Yokel, R. A., Provan, S. D., Meyer, J. J., and Campbell, S. R. (1988). Aluminum intoxication and the victim of Alzheimer's disease: similarities and differences. *Neurotoxicology* **9**, 429–442.

Yuan, Y., and Atchison, W. D. (1993). Disruption by methylmercury of membrane excitability and synaptic transmission of CA1 neurons in hippocampal slices of the rat. *Toxicol. Appl. Pharmacol.* **120**, 203–215.

Zalutsky, R. A., and Nicoll, R. A. (1990). Comparison of two forms of long-term potentiation in single hippocampal neurons. *Science* **248**, 1619–1624.

Zhou, M., Small, S. A., Kandel, E. R., and Hawkins, R. D. (1993). Nitric oxide and carbon monoxide produce activity-dependent long-term synaptic enhancement in hippocampus. *Science* **260**, 1946–1950.

Zimmer, L., Woolley, D., and Chang, L. (1985). Does phenobarbital protect against trimethyltin-induced neuropathology of limbic structures? *Life Sci.* **36**, 851–858.

Zipp, F., Nitsch, R., Soriano, E., and Frotscher, M. (1989). Entorhinal fibers form synaptic contacts on parvalbumin-immunoreactive neurons in the rat fascia dentata. *Brain. Res.* **495**, 161–166.

Zucker, R. S. (1989). Short-term synaptic plasticity. *Ann. Rev. Neurosci.* **12**, 13–31.

CHAPTER

9

Electrophysiological Analysis of Complex Brain Systems: Sensory-Evoked Potentials and Their Generators

DAVID W. HERR
Neurotoxicology Division
U.S. Environmental Protection Agency
Research Triangle Park, North Carolina 27711

WILLIAM K. BOYES
Neurotoxicology Division
U.S. Environmental Protection Agency
Research Triangle Park, North Carolina 27711

I. Introduction

In order to understand the functional consequences of exposure to neurotoxic compounds, it is advantageous to understand their actions on complex neural systems. The nervous system is an intricate communication and information processing network, in which the operation of the individual elements (neurons and their components) and their interactions on both a simple and system scale are critical. Understanding the operation of the elements helps to interpret the operation of the networks. Thus, the preceding chapters in this volume provide information critical to understanding the actions of neurotoxic compounds on the constituent elements of the nervous system. Studying the

actions of neurotoxicants at a multifarious network or systems scale is also important because the integrated outputs of neural systems are the likely determinants of behavior, and it is at this level that physiological measures are best likely to correlate with behavioral outcomes. In addition, neurophysiological studies of complex systems may provide a link to help bridge the gap between actions of compounds at the cellular or subcellular level, and those on the ultimate behavioral outcome. Finally, for neurotoxicologists interested in hazard identification, one approach has been to start investigations with apical measures which reflect the combined contribution of many possible sites of neurotoxicant action in order to have a greater possibility of detecting adverse effects of a test substance with many potential actions.

The study of sensory function using sensory-evoked potentials is an example of the use of electrophysiological procedures to examine the function of complex neural systems. The sensory systems are available for controlled stimulation through a variety of experimentally manipulated artificial and naturalistic stimuli, en-

This manuscript has been reviewed by the Health Effects Research Laboratory, U.S. Environmental Protection Agency, and approved for publication. Mention of trade names and commercial products does not constitute endorsement or recommendation for use.

205

abling investigation of the functional properties of these systems in both laboratory animals and humans. In addition, study of sensory function in its own right is important for neurotoxicologists because of the high prevalence of sensory dysfunction following exposure to neurotoxic compounds (Crofton and Sheets, 1989).

Sensory-evoked potentials represent a class of electrophysiological procedures which involve stimulation of sensory receptors or afferent nerves, and recording the evoked electrical activity from some portion or portions of the neural pathways. Typical stimuli might be a flash of light or a changing visual pattern for eliciting visual-evoked potentials, a click or tone of a specific frequency for auditory-evoked potentials, or an electrical shock to a peripheral nerve for somatosensory-evoked potentials. Somatosensory research has expanded recently to include stimuli which invoke the participation of sensory receptors in a more natural fashion, such as vibration (Hashimoto *et al.*, 1991; Hamano *et al.*, 1993; Snyder, 1992). The recording of sensory-evoked potentials must overcome signal-to-noise problems due to the small voltage of most of the evoked responses relative to other ongoing neural and nonneural electrical activity. This is typically accomplished through a combination of optimizing electrode placements, amplification and filtering of signals, and signal averaging or processing in phase with repetitive stimulation. The technical aspects of conducting sensory-evoked potential studies and the utility of these techniques in the detection, characterization, and localization of clinical disorders have been reviewed extensively by others (Boyes, 1993; Cracco and Bodis-Wollner, 1986; Regan, 1988).

Sensory-evoked potentials have been applied in numerous studies of neurotoxicity in both laboratory animals and humans, and the use of these procedures has also been frequently reviewed. Three initial reviews, one by Fox and colleagues (1982), which dealt with sensory-evoked potentials among other electrophysiological techniques, and two subsequent reviews dedicated exclusively to use of sensory-evoked potentials in toxicology by Rebert (1983) and Dyer (1985) firmly established the rationale for, and utility of, these procedures. Subsequently, reviewers have discussed the correspondence between electrophysiological and behavioral measures (Dyer, 1986; Boyes, 1993), the correspondence between electrophysiological and pathological assessments of neurotoxicity (Mattsson *et al.*, 1989, 1990), the application of evoked potentials in routine toxicity testing (Mattsson *et al.*, 1992), the use of evoked potentials in exposed human populations (Arezzo *et al.*, 1985; Otto, 1986; Seppalainen, 1988), and the relationships between measures in experimental animals and humans (Hudnell and Boyes, 1991;

Boyes, 1994). Rebert (1983) and Mattsson and Albee (1988) have stressed the need to evaluate multiple sensory systems in a comprehensive evaluation of neurotoxicity. Others have focused on testing particular sensory systems (Dyer, 1987b; Boyes, 1992) or particular classes of neurotoxic agents (Dyer, 1987a). Draft proposed testing guidelines have been developed for use by the Environmental Protection Agency for cases in which sensory toxicity is suspected (Boyes, 1990). The sensitivity of sensory-evoked potentials to neurotoxic insults has ranged from extremely sensitive to less sensitive than other procedures. One possibility for the range of sensitivity of sensory-evoked potentials is the extent to which neurotoxic alterations affect the neural elements which act as generators for the evoked response.

One topic that has not been covered extensively in the previous neurotoxicology literature is the neurological generators of sensory-evoked potentials. An understanding of the tissues contributing to evoked potential peaks is crucial for interpretation of neurotoxicant-induced changes. Among other things, sensory-evoked potential data may help localize sites of lesions or dysfunction, separate central from peripheral nervous system damage, or provide direction for more extensive neurophysiological, biochemical, or pathological studies. The ability to provide this guidance is predicated on understanding, as well as possible, the tissues that give rise to the distinct evoked potential components. In addition, decisions as to whether sensory-evoked potential changes are interpreted as ''adverse'' in a risk assessment situation might be aided by knowledge of the generators of the potentials. Because sensory-evoked potentials reflect measures of complex systems, the answers to generator localization questions are frequently not simple. On the other hand, in some cases relatively clear relationships exist between neural structures and evoked potential components. Given the importance of understanding the neurological generators of sensory-evoked potentials for interpreting the actions of neurotoxic compounds and the lack of such information in available review articles, this chapter focuses on compiling the available literature regarding the neural sources of selected sensory-evoked potentials.

II. Localization of Evoked Potential Generators

A generator of an evoked response can be defined as the source of neural electrical activity which produces the observed recording. The physiological basis for the electrical activity can range from generator potentials, excitatory postsynaptic potentials (EPSPs), inhibitory postsynaptic potentials (IPSPs), action potentials, and glial cell transmembrane currents. The amplitude of the recorded potential is inversely related to the square of the distance, and is directly proportional to the cosine of the angle, between the active electrode and the flow of neuronal electrical activity (Lopes da Silva and Van Rotterdam, 1982; Morris and Lüders, 1985). "Open field" generators are sources in which the predominantly parallel arrangement of dendrites results in a dipole layer when the neurons are activated. Such current dipoles can be recorded from electrodes placed at a distance from the generator site. In contrast, "closed field" generators are produced when the random or opposing arrangement of neural dendrites produces a canceling of signals generated by the individual EPSPs and no net voltage can be measured at sites distant to the generator. A closed field generator can only be recorded when the electrodes are placed within the closed field (Lopes da Silva and Van Rotterdam, 1982; Wood and Allison, 1981).

Several methods can be used to help localize the neural generators of the electrical responses. Surface mapping techniques record the distribution of the evoked response from an electrode array placed on the surface of the cortex or skull. Although the distribution of electrical activity can be uniquely defined if the source generator is known (forward prediction), there is no unique solution to equations which attempt to localize source generators based solely on surface potentials (inverse prediction) (Helmholz, 1853; Nunez, 1990). The recorded potential may be due to the supposition of the individual generator responses produced by multiple sites of simultaneous neural activity (Darcey *et al.,* 1980; Gulrajani *et al.,* 1984; Wood, 1982). Additionally, the effects of sequential and/or simultaneous activation of multiple generators over the sampling time of the evoked response must also be considered when inverse prediction methods are used (Snyder, 1991). To assist in localizing evoked potential generators, investigators often model current flow instead of the actual voltages by calculating the Laplacian derivative of the recorded potentials. This procedure assists in localizing the current flux by subtracting the average potential gradient from regions surrounding the center electrode (Hjorth, 1975; MacKay, 1984; Srebro, 1985). Thus, the results represent activity originating in the local area of the surface electrodes. Recently, spline Laplacians have been developed, which simultaneously use the data from all of the electrodes to model the distribution of the surface potential at different time points. The derived model is then used to calculate the surface Laplacian (Nunez, 1990; Perrin *et al.,* 1987a,b, 1989, 1990). This methodology has the advantage of

having a sharp current distribution, thus better localizing the generator sources. Although such surface mapping techniques can be used to determine the regions of the brain which show the largest amount of electrical activity, potentials recorded from the surface of the brain do not localize the cellular elements which physically generate the evoked responses.

Invasive techniques can also be used to help determine the neural generators of evoked potentials. One such method determines current source density (CSD) depth profiles for evoked potentials, thereby localizing the sources and sinks which produce surface-recorded potentials. This method relies on recording from closely spaced electrodes placed at various depths from the cortical surface (Barna *et al.,* 1981; Jellema and Weijnen, 1991). The second spatial derivative of the recorded voltage profiles is used to derive the CSD profile (Nicholson, 1973; Freeman and Nicholson, 1975; Mitzdorf, 1985). If the evoked potential is generated in the cortex, the polarity of the response will reverse when the electrode array transverses the dipole generator as the depth from the cortical surface increases. Simultaneous multiunit recordings can assist in determining if the sources and sinks of the CSD profile represent areas of neuronal excitation or inhibition. Initially, CSD analysis was believed to detect only EPSPs and their active sinks, along with the associated passive sources (Mitzdorf, 1985, 1986). It has been postulated recently that inhibitory postsynaptic potentials can result in active current sources (Schroeder *et al.,* 1990b) and that presynaptic elements and closed field generators can contribute to the CSD (Tenke *et al.,* 1993). When coupled with neuroanatomical information, the data gathered from CSD studies can be used to infer which populations of neurons contribute to potentials recorded from the surface of the brain (Kraut *et al.,* 1985; Schroeder *et al.,* 1991; Tenke *et al.,* 1993; Vaknin *et al.,* 1988). Thus, CSD analysis addresses a concern of neurophysiologists as to whether the responses that are recorded from the surface of the brain represent neural activity in cortical tissue or whether the potentials are volume conducted from subcortical sites.

Direct recordings from various regions of the neuraxis can help determine the generators of surface-recorded potentials (Achor and Starr, 1980a; Arezzo *et al.,* 1981; Desmedt, 1986; Møller and Jannetta, 1986; Møller *et al.,* 1990; Morioka *et al.,* 1991; Sen and Møller, 1991). The timing and polarity of the locally recorded response should match the predicted values based on surface recordings. Manipulation of stimulus parameters should produce changes in the locally recorded responses (latency, frequency, amplitude, duration), which should be interpretable with respect to the changes observed in surface recordings. Additional experimental manipulations may involve lesioning portions of the neuraxis to determine which portions of surface-recorded potentials change (Achor and Starr, 1980b; Allison *et al.*, 1991b; Chen and Chen, 1991; Dyer *et al.*, 1987b; Fullerton and Kiang, 1990; Glenn and Kelly, 1992; Wada and Starr, 1983a,b). If a discrete lesion is made, and selected portions of the evoked response are predictably altered, an argument for a localized generator site can be rationalized. Depth recordings and lesion studies are normally amendable only to animal experimental studies. Occasionally, human studies are performed as a part of necessary surgical procedures. The presence of a preexisting illness may limit the conclusions of some of the human studies. Thus, extrapolation of the generators from animal models to humans remains an important issue in neurophysiology.

III. Electroretinograms

Electroretinograms (ERG) are evoked responses recorded from the cornea in response to visual stimulation. The typical ERG consists of a negative a-wave, a positive b-wave, a late negativity, and a positive c-wave (Fig. 1; Table 1). If wick electrodes (Sieving *et al.*, 1978), a very bright stimulus, and great care in recording procedures are employed, an early receptor potential can be seen to precede the a-wave. Additionally, superimposed on the b-wave are a series of oscillatory potentials (Armington, 1980; Carr and Siegel, 1990; Potts, 1972). The early receptor potential is generated at the outer portions of the rods and cones (Brown *et al.*, 1965). The a-wave is believed to result from hyperpolarization of the proximal portions of stimulated rod and cone photoreceptors (Carr and Siegel, 1990). This conclusion is supported by CSD analysis which indicates a current source in the outer regions of the receptor layer (Heynen and van Norren, 1985). The b-wave arises from cells in the middle layers of the retina. It is thought to be generated by extracellular and transmembrane potassium ion currents involving Müller (glia) cells, which occur as a consequence of neuronal activity primarily in the retinal bipolar cells (Miller and Dowling, 1970; Dowling, 1987). This hypothesis corresponds to the CSD current sink found in the outer nuclear layer (Heynen and van Norren, 1985) and the source–sink pair reported in the receptor cell layer to the outer plexiform layer of the retina (Baker *et al.*, 1988). Similar depth profiles of activity have been reported for pattern ERGs (PERGs) (Riemslag and Heymen, 1984), although other authors feel that PERGs involve some different generators than flash

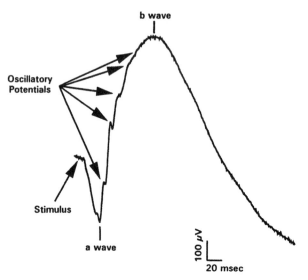

FIGURE 1 Corneal electroretinogram (ERG) recorded from an adult male Long Evans rat produced by a single flash stimulus. The sampling duration was not long enough to record the c wave in this example. The active electrode was a platinum-iridium wire loop (4 mm diameter) placed in contact with the cornea, referenced to a subdermal needle electrode placed in the lateral canthus. The stimulus was generated by a Grass PS22 Photostimulator, channeled through a fiber optic bundle, and focused in the plane of the dilated pupil (i.e., Maxwellian view), stimulating a circular region 30° in diameter on the retina. Flash intensity was 14.29 lux-sec, measured at the cornea. The ERG was amplified 1000×, analog filtered (0.1–1000 Hz, 6 dB/octave rolloff), and sampled at 5000 Hz. The waveform was graciously donated by Dr. Andrew Geller and Katherine Osborne.

ERGs (Maffei and Fiorentini, 1986). These authors (Maffei and Fiorentini, 1986) feel that retinal ganglion cells may contribute to the PERG. This hypothesis is supported by the source–sink pair found in the nerve fiber layer to the middle of the inner plexiform layer of the retina (Baker *et al.*, 1988; Sieving and Steinberg, 1987). The oscillatory potentials represent a series of events originating in the bipolar layer (Wachtmeister and Dowling, 1978) and appear to have a different origin than the b-wave (Wachtmeister, 1980, 1981a,b). How-

TABLE 1 Proported Generators of Electroretinograms

Component/stimulation	Origin
Early receptor potential/flash	Outer portions of rods and cones
a-wave/flash	Photoreceptor hyperpolarization
b-wave/flash	Middle retina: Bipolar and/or Müller cells
c-wave/flash	Pigmented epithelium
Oscillatory potentials/flash	Middle retina, cells unknown
ERG/pattern	Retinal ganglion cells and possible middle retinal cells

ever, the presence of oscillatory potentials may contribute to the ERG b-wave (Lachapelle, 1990). The ERG c-wave originates in the pigmented epithelium of the retina and is related to the change in extracellular potassium produced by receptor cell activity (Oakley and Green, 1976). One should note that the flash ERG does not include contributions from retinal ganglion cells (Maffei and Fiorentini, 1981). Thus, flash ERGs represent a methodology with which to assess the function of portions of the retina that do not involve ganglion cells, whereas PERGs may assess retinal responses inclusive of ganglion cells.

IV. Flash-Evoked Potentials

Flash-evoked potentials (FEPs; Fig. 2; Table 2) represent a cortical response to visual stimulation, which have been mapped using surface recordings to regions over the visual cortex and surrounding regions (Dyer *et al.*, 1987a; Montero, 1973; Onofrj *et al.*, 1985). These results have been supported by CSD analysis and depth recordings (Brankačk *et al.*, 1990; Ducati *et al.*, 1988; Kraut *et al.*, 1985; Mitzdorf, 1985; Schroeder *et al.*,

1991) and lesion studies (Dyer *et al.*, 1987b). Because the CSD work in Fig. 2 was performed in nonhuman primates, the peak nomenclature for the peaks commonly identified in this species will be used. An additional early latency positive peak (P_{21}) is often identified in FEPs. The exact generator of peak P_{21} has not been determined, although polarity reversal within the visual cortex (Brankačk *et al.*, 1990; Kraut *et al.*, 1985), reduced amplitudes with cortical lesions (Dyer *et al.*, 1987b), and a cortical distribution of sources and sinks (Siegel and Sisson, 1989) argue for a cortical origin. Peak N_{40} is believed to be produced by depolarization of cells in cortical lamina 4 (Brankačk *et al.*, 1990; Kraut *et al.*, 1985; Mitzdorf, 1985; Schroeder *et al.*, 1991) by thalamocortical afferents. Peak P_{65} may represent subsequent hyperpolarization of stellate cells in lamina 4 (Kraut *et al.*, 1985; Schroeder *et al.*, 1991). The middle portions of the FEP (peak N_{95}) may be related to stellate cell input to supragranular elements in lamina 3 (Kraut *et al.*, 1985). This portion of the FEP is unlikely to be due to cortical input from the superior colliculus (Barnes and Dyer, 1986), as was previously believed (Rose and Lindsley, 1965; Creel *et al.*, 1974). However, recent work in monkeys has

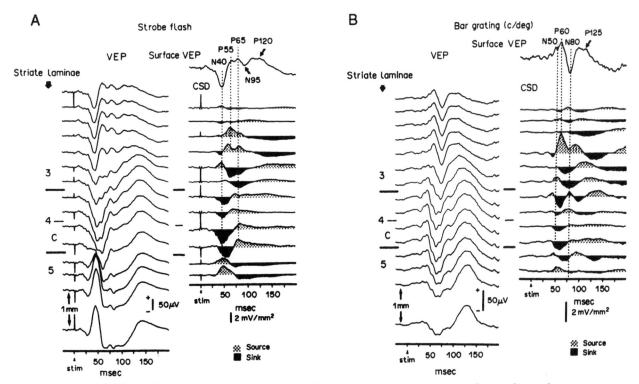

FIGURE 2 (A) Depth profile of FEP and derived CSD recorded from striate cortex of an awake monkey. The surface FEP is located above the CSD profile. (B) Depth profile of PEP and derived CSD recorded from striate cortex of an awake monkey. The surface PEP is located above the CSD profile. Note the reversal of FEP and PEP components over striate laminae and the associated sources and sinks of the CSD. (From Schroeder *et al.*, 1991, with permission from Elsevier Science Ltd., The Boulevard, Langford Lane, Kidlington OX5 1GB, UK.)

TABLE 2 Proported Generators of
Visual-Evoked Potentials

Component/stimulation	Origin
P_{21} (rat)/flash	Visual cortex
N_{40}/flash	Visual cortex lamina 4, Depolarization following thalamocortical input
P_{65}/flash	Visual cortex lamina 4, stellate cell hyperpolarization
N_{95}/flash	Stellate cell input to supragranular layers
Later portions of FEPs/flash	Lamina 4 activity resulting from thalamocortical loops (rats), and also extrastriate cortex (monkeys)
N_{50}/pattern	Visual cortex lamina 4C, stellate cell activation
P_{60}/pattern	Visual cortex lamina 3, pyramidal cell activation
N_{80}/pattern	Outside area 17
P_{125}/pattern	Inside and outside area 17

indicated that peak N_{95} may originate in cortical area V4. This peak was associated with current sinks and increased multiunit activity in the middle cortical laminae, reflecting an excitatory FEP component in the extrastriate cortex (Givre *et al.*, 1994). The later portions of the FEP in rats (peak N_{160} and beyond) are believed to be generated by depolarizations in lamina 4, like peak N_{40} (Brankačk *et al.*, 1990), which are produced by a reverberating thalamocortical circuit set in motion by the retinogeniculate afferent volley (for reviews see Bigler, 1977; Sumitomo and Klingberg, 1972). However, late FEP components (peak P_{120} and beyond) may be generated in the extrastriate cortex in monkeys (Kraut *et al.*, 1985; Givre *et al.*, 1994). Givre and co-workers (1994) have found a polarity inversion, a source/sink CSD configuration, and increased multiunit activity in the upper laminae of cortical area V4 associated with FEP peak P_{120} in monkeys. They concluded that peak P_{120} reflects the excitatory response of pyramidal cells in extrastriate area V4. Thus, although FEPs are generated at the cortical level, damage to the visual system in the periphery, or central portions of the visual pathway, may be reflected by changes in the cortical recordings.

V. Pattern-Evoked Potentials

Visual stimulation with pattern stimuli affords the opportunity to selectively activate different subelements of the visual system. As such, it is not surprising that different investigators, using different pattern stimuli, have reported divergent data for the localization and generators of pattern-evoked potentials (PEPs). Because of the wide variety of pattern stimuli that have been employed, only an overview of findings will be presented.

Pattern-evoked potentials, like FEPs, are believed to be cortically generated responses (Fig. 2; Table 2). Surface mapping experiments have found that PEPs occur in visual area 17 (Dyer *et al.*, 1987a; Onofrj *et al.*, 1985). However, the distributions of PEPs and unpatterned stimuli may differ (Dyer *et al.*, 1987a; Regan and Heron, 1969). Additionally, other investigators have reported that portions of PEPs can be recorded outside of area 17 (Dyer *et al.*, 1987a; Jeffreys and Axford, 1972; Manahilov *et al.*, 1992; Srebro and Purdy, 1990). Invasive experiments have generated data supporting the hypothesis of a cortical generation of PEPs. Surface lesions are able to alter PEPs (Dyer *et al.*, 1987b). Also, most components of PEPs have been found to invert over the depth of the cortex, indicating cortical dipoles are involved (Schroeder *et al.*, 1991). Similar polarity inversion for PEPs have been reported in humans, with the PEP generators located slightly more superficially than FEP generators (Ducati *et al.*, 1988). The sinks and sources for FEPs and PEPs generated by CSD analysis are very similar in monkeys and cats (Mitzdorf, 1986; Schroeder *et al.*, 1991). The peak nomenclature employed here is that of monkeys, from which the waveforms in Fig. 2 were recorded. PEP peak N_{50} is due to a sink at the base of lamina 4 and may represent activation of stellate cells in lamina 4C. Peak P_{60} inverts, has a source/sink configuration, and is associated with excitatory neuronal activity in lamina 3 (activation of pyramidal cells) (Schroeder *et al.*, 1991). Peak N_{80} did not invert in cortical area 17, but was temporally associated with neuronal activation (sinks) in several cortical layers. This PEP peak may be associated with activity in regions outside of area 17 (Schroeder *et al.*, 1990a, 1991). PEP peak P_{125} is also associated with several sources and sinks in area 17, does not undergo polarity inversion over the cortical lamina, and may also be influenced by neural activity in regions other than area 17 (Schroeder *et al.*, 1990a, 1991). Thus, the CSD data support the conclusion the PEPs are cortical potentials that have some generators in primary visual cortex, and additional neural activity in other cortical areas which may contribute to surface-recorded responses.

VI. Brain Stem Auditory-Evoked Responses

Brain stem-evoked auditory responses (BAERs) are often recorded from electrodes located dermally, sub-

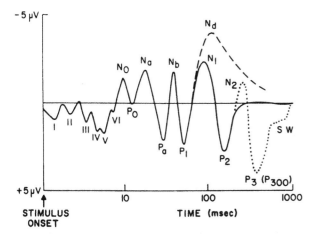

FIGURE 3 Schematic auditory-evoked potential showing BAER, MLR, and long latency auditory potentials. The solid lines represent "exogenous" components generated by the auditory stimulus. The dashed and dotted lines represent components that are generated by the cortex and are heavily influenced by the cognitive state of the individual. Note the logarithmic time scale. (From Hillyard *et al.*, 1984.)

dermally, or near the cortical surface, but are believed to reflect volume conducted electrical activity from brain stem generators (Fig. 3; Table 3). If appropriate recording conditions are used, a cochlear microphonic can also be recorded, which is generated in the cochlea (Møller and Jannetta, 1985; Starr and Zaaroor, 1990).

TABLE 3 Proported Generators of Auditory-Evoked Potentials

Component	Origin
BAER peak 1 (human)	Distal VIII nerve
BAER peak 2 (human)	Proximal VIII nerve
BAER peaks, 1a and 1b (animal)	VIII nerve
BAER peak 3 (human) BAER peak 2 (animal)	Cochlear nucleus
BAER peak 4 (human) BAER peak 3 (animal)	Superior olivary N./trapezoid/ cochlear N.
BAER peak 4 (animal)	Nucleus of lateral lemniscus
BAER peaks, 5, 6, 7 (human) BAER peak 5 (animal)	Inferior colliculus
MLR P_0	Dendrite depolarization in cortical lamina 2–3
MLR N_a	Dendrite depolarization in cortical lamina 4–5 or inhibition of pyramidal cells in layer 3
MLR P_a	Activation of pyramidal cells in layers 3 and 5
Long latency auditory potentials	Auditory cortex and associated brain regions

The cochlear microphonic represents electrical activity of cochlear hair cells (Cheatham and Dallos, 1982; Nuttall and Dolan, 1991). The BAER is composed of several peaks occurring within approximately 10 msec of stimulation, which are believed to represent sequential activation of auditory structures. However, some differences in nomenclature and/or generators have been suggested for humans and animals. Buchwald (1983) has postulated that BAERs represent summed postsynaptic potentials. Other investigators feel that action potentials are involved in generating BAERs (Fullerton and Kiang, 1990; Møller and Jannetta, 1985, 1986). Data indicate that the auditory nerve is the generator of peaks 1 (distal nerve) and 2 (proximal nerve) in humans, and peak 1 (including 1a and 1b) in animals (Buchwald, 1983; Chen and Chen, 1991; Jewett, 1970; Jewett and Williston, 1971; Meurice *et al.*, 1991; Møller and Jannetta, 1985, 1986; Møller *et al.*, 1988; Starr and Zaaroor, 1990). The inability to detect a second auditory nerve peak in some animal species has been attributed to a shorter length of nerve (Møller and Jannetta, 1986). The activity associated with BAER peak 1 (and 2 in humans) may either represent graded generator potentials in the cochlea (Buchwald, 1983) or action potentials in the auditory nerve (Møller and Jannetta, 1985, 1986). Peak 3 in humans (peak 2 in animals) is hypothesized to be generated either by action potentials in the cochlear nucleus or by postsynaptic potentials in neurons of the cochlear nucleus (Buchwald, 1883; Chen and Chen, 1991; Henry, 1979; Jewett, 1970; Møller and Jannetta, 1985, 1986). Peak 4 in humans (peak 3 in animals) is believed to be generated in the superior olivary/trapezoid body complex (Achor and Starr, 1980a,b; Buchwald, 1983; Chen and Chen, 1991; Henry, 1979; Jewett, 1970; Møller and Jannetta, 1985, 1986; Wada and Starr, 1983a,b). However, contributions from the cochlear nucleus and the lateral lemniscus should not be overlooked (Achor and Starr, 1980a,b; Møller and Jannetta, 1985, 1986). The nuclei of the lateral lemniscus may be involved in the generation of peak 4 in animals (Buchwald, 1983; Fullerton and Kiang, 1990; Henry, 1979). Peak 5 of the BAER may be generated in the inferior colliculus of animals (Buchwald, 1983; Jewett, 1970; Henry, 1979). However, it is postulated that activity in the inferior colliculus in humans is associated with the vertex negative potential (N_0/SN_{10}) and waves 5, 6, and 7 of the BAER (Møller and Jannetta, 1985). Several authors have warned that more than one structure can be involved in the generation of a single (or several) BAER peak, especially those representing transmission "upstream" from the cochlear nucleus (Achor and Starr, 1980a,b; Jewett, 1970; Møller and Jannetta, 1985, 1986; Wada and Starr, 1983a,b). The data clearly support the conclusion that the early BAER peaks are generated by the auditory nerve and the cochlear nucleus. Subsequent BAER peaks seem to be generated by activity in several structures simultaneously, ipsilaterally as well as contralaterally. Thus, while the generators of some BAER peaks are not uniquely defined, they do seem to represent higher level processing of the auditory signal at the level of the brain stem.

VII. Middle and Late Auditory Potentials

Auditory-evoked potentials that occur approximately 10–50 msec after stimulation are generally termed "middle latency responses" (MLRs; Fig. 3; Table 3). In animals, these potentials are recorded over the auditory cortex (Arezzo et al., 1975; Kaga et al., 1980; Barth and Di, 1990, 1991; Di and Barth, 1992). However, in humans a more superior temporal site is indicated (Cohen, 1982; Picton et al., 1974; Wood and Wolpaw, 1982). Different authors have used various nomenclatures to describe the peaks recorded in the MLR. We will attempt to compare the results using the nomenclature in Fig. 3 and provide the author's description when appropriate. The MLR typically consists of three to four components, labeled P_O (P_1, $P_{14/15}$), N_a (N_1, N_{19}), and P_a (P_2, P_{30-35}). In rodents and nonhuman primates, peak P_O has a latency of approximately 10–15 msec, peak N_a has a latency of approximately 15–25 msec, and peak P_a has a latency of approximately 23–50 msec (Arezzo et al., 1986; Barth and Di, 1990; Knight et al., 1985). In monkeys, an earlier cortical component (approximately 8 msec after stimulation; N_8) has been identified, which appears to reflect depolarization of thalamocortical afferents and activation of lamina 4 stellate cells (Arezzo et al., 1975; Steinschneider et al., 1992). Both cortical and subcortical generators have been proposed for MLRs (Barth and Di, 1990; Di and Barth, 1992; Hinman and Buchwald, 1983; Knight et al., 1985; McGee et al., 1992; Shaw, 1991). Current source density analysis of the MLR in rats has indicated polarity reversal for all peaks in the cortex. The data indicate that peak P_O is generated by depolarization of supragranular dendrites in cortical layers 2–3 (Barth and Di, 1990; Arezzo et al., 1986), and that peak N_a is generated either by depolarization of infragranular dendrites in cortical layers 4–5 (Barth and Di, 1990) or by inhibition of the lamina 3 pyramidal cells (Arezzo et al., 1986). The generators for peak P_a include a sink in supragranular layers (activation of lamina 3 pyramidal cells) and a sink in the infragranular layers (activation of lamina 5 pyramidal cells) (Barth and Di, 1990; Arezzo et al.,

1986). It is proposed that MLRs are produced by both sequential and parallel activation of pyramidal cells in auditory cortex. The afferent input that generated the cortical potentials has been postulated to originate in the thalamus (Di and Barth, 1992), although other authors have suggested a role of brain stem–thalamus–cortex projections (Buchwald et al., 1991; Dickerson and Buchwald, 1991; Kraus et al., 1992). Therefore, it appears that the MLR may represent continued processing of auditory input at the level of brain stem–thalamic–cortical levels, as recorded from the cortical surface.

Following the MLR are long latency auditory potentials that are believed to be cortically generated. These potentials occur over the range of approximately 50–300 msec following stimulation (Fig. 3). Arezzo and co-workers (1986) have defined a negative peak with a latency of approximately 60 msec poststimulation in awake monkeys. Their CSD analysis indicated that neuronal inhibition in lamina 3 (source), coupled with a passive sink in the superficial cortex, was responsible for peak N_{60} (possibly corresponding to peak N_b in Fig. 3). Barth and Di (1990) have performed CSD analysis on a negative peak (N2) occurring over 150–175 msec poststimulation in anesthetized rats. Their data indicated that this peak began with a distal sink and a proximal source in supragranular layers (lamina 3), followed by a distal sink and proximal source in infragranular layers (lamina 5). They interpreted the results to indicate that these responses were the result of active inhibition. Knight and co-workers (1985) defined the late auditory peaks of N_{50}, N_{80}, and P_{130} (possibly corresponding to peaks N_b, N_1, and P_2 in Fig. 3, respectively) in awake rats. These peaks were attenuated by stimulation rates in excess of 1 Hz and during slow wave sleep, and were maximally recorded in vertex and frontal brain areas. These data are in agreement with the optimal recording sites and effects of stimulation frequency observed in human late auditory potentials (Nelson and Lassman, 1973; Picton et al., 1974; Wolpaw and Wood, 1982). The late latency potentials are influenced by cognitive processes, one example of which is the P_{300} (P_3) potential produced in "oddball" paradigms involving stimuli from auditory or other sensory modalities (Ehlers et al., 1991; Polich, 1989; Wesensten et al., 1990; Yamaguchi et al., 1993). Although the generators of later auditory potentials have not been conclusively defined, they do appear to represent higher order processing of auditory signals and as such could serve as indices of toxicant-induced changes in higher sensory function. Presumably, such potentials reflect the activity of sensory and associative areas of cortex along with possible communication with other brain regions.

VIII. "Far-Field" Somatosensory Potentials

Somatosensory-evoked potentials (SEPs) are often elicited by the electrical stimulation of nerves, producing a large and synchronous afferent volley. When recorded from cranial electrodes, early latency components generated in the peripheral nerves, spinal cord, brain stem, and thalamus can be discriminated. These portions of SEPs are often referred to as far-field SEPs because the potentials are volume conducted from the generator site to the recording electrodes. Latencies associated with the various peaks vary depending on the site of stimulation (forelimb/arm, hindlimb/leg, tail) and species studied. An attempt will be made to identify the peaks based on their conventional polarity and latency in humans. Where possible, data from animal studies will be included, but the peaks will be identified using the best matching peak in human recordings (Fig. 4; Table 4).

The first SEP peak, known as P_9 (median nerve stimulation) or P_{17} (tibial nerve stimulation), is believed to represent the peripheral action potential approaching the spinal column (Allison and Hume, 1981; Cracco and Cracco, 1976; Desmedt and Cheron, 1981a; Kimura *et al.*, 1986; Pratt and Starr, 1981) and may not be observed in animal recordings (Arezzo *et al.*, 1979; Boyes and Cooper, 1981; Wiederholt and Iragui-Madoz, 1977). The second positive peak (P_{11} with median nerve stimulation or P_{24} with tibial nerve stimulation) is generated by the afferent volley propagating up the dorsal

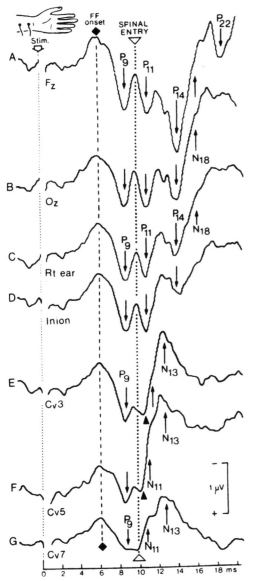

FIGURE 4 Early somatosensory-evoked potential (SEP) produced by electrical stimulation of median nerve at the wrist. Active electrodes were located on midline front (Fz; A), midline occipital (Oz; B), opposite earlobe (C), 2 cm above inion (D), and the spinous processes of spinal vertebrae Cv3 (E), Cv5 (F), and Cv7 (G). The reference electrode was located on the opposite dorsal hand. "FF onset" denotes the beginning of the P_9 potential. (From, Generator Sources of SEP in man. In *Evoked Potentials. Frontiers of Clinical Neuroscience* (R. Q. Cracco and I. Bodis-Wollner, Eds.), © 1986, John Wiley & Sons, Inc. Reprinted by permission of John Wiley & Sons, Inc.

TABLE 4 Proported Generators of Somatosensory-Evoked Potentials

Component/Stimulation	Origin
P_9 (human median nerve) P_{17} (human tibial nerve)	Peripheral nerve compound action potentials
P_{11} (human median nerve) P_{24} (human tibial nerve)	Dorsal columns of the spinal cord
N_{13}/P_{13} (human median nerve) N_{27}/P_{27} (human tibial nerve) (cervical recording)	Excitatory postsynaptic potentials in interneurons of layers 4–5 of dorsal horn of spinal cord
P_{13}/P_{14} (human median nerve) P_{31} (human tibial nerve) (cephalic recording)	Afferent volley in medial lemniscus from brain stem nuclei to thalamus (human) or brain stem dorsal column nuclei (rat?)
Several peaks, 15–20 msec (human median nerve)	Thalamocortical radiations
N_{18} (human median nerve)	Subcortical, involving brain stem dorsal column nuclei
N_{20}/P_{30} (P_{20}/N_{30}) (human median nerve)	Somatosensory cortex area 3b, depolarization following thalamocortical input
P_{25} and N_{35} (human median nerve)	Somatosensory cortex area 1
Later latency peaks	Somatosensory cortex areas 1 and 3b, secondary somatosensory cortex, and other cortical areas

columns of the spinal cord (Allison and Hume, 1981; Anziska and Cracco, 1981; Arezzo *et al.*, 1979; Desmedt, 1986; Desmedt and Cheron, 1980; Kimura *et al.*, 1986) and possibly by activity in the dorsal column nuclei of the brain stem (Iragui-Madoz and Wiederholt, 1977). This wave can be recorded as a negative peak when measured in a "near-field" configuration from the spinal cord (Allison and Hume, 1981; Desmedt, 1986). In recordings from the dorsal neck/spinal cord area, a N_{13} spinal potential (median nerve stimulation) can be observed (N_{27} with tibial nerve stimulation; Kimura *et al.*, 1986). This potential has a positive polarity (P_{13}/P_{27}) when recorded from the ventral (esophageal) surface. The N_{13}/P_{13} peak is believed to represent EPSPs on interneurons in layers 4 and 5 of the dorsal horn and is not related to a scalp-recorded P_{13} (Desmedt, 1986; Desmedt and Cheron, 1981b). This wave is sometimes called the cervical N_{13}/P_{13} to differentiate it from cortical peaks with approximately the same latency. The next peak in the far-field SEP (P_{13}/P_{14} with median nerve stimulation and P_{31} with tibial nerve stimulation) may be generated by the afferent volley in the medial lemniscus from the dorsal column brain stem nuclei toward the thalamus (Allison and Hume, 1981; Arezzo *et al.*, 1979; Desmedt, 1986; Desmedt and Cheron, 1980, 1981a; Kimura *et al.*, 1986). Alternatively, an origin for peak P_{13}/P_{14} in dorsal column nuclei has been proposed (Iragui-Madoz and Wiederholt, 1977; Møller *et al.*, 1986, 1989, 1990; Morioka *et al.*, 1991; Pratt and Starr, 1981) and possibly activity in the inferior cerebellar peduncle (Iragui-Madoz and Wiederholt, 1977).

Several investigators have reported multiple peaks in the 15- to 20-msec poststimulation range (following median nerve stimulation). These waves often require high-pass filtering for adequate isolation (Green *et al.*, 1986; Maccabee *et al.*, 1986). The proposed generator for these peaks is the afferent volley in the thalamocortical radiations propagating toward the somatosensory cortex (Arezzo *et al.*, 1979; Eisen *et al.*, 1984; Iragui-Madoz and Wiederholt, 1977) and possibly activity in the cerebellum (Iragui-Madoz and Wiederholt, 1977). A negative peak occurring approximately 18 msec after stimulation (N_{18}) can be recorded from many areas of the cranium. This wave is generated subcortically and is believed to represent either efferent activity from the brain stem dorsal column nuclei to other brain stem sites (Desmedt and Cheron, 1981a; Hashimoto, 1984; Tomberg *et al.*, 1991; Urasaki *et al.*, 1990) or depolarization of primary afferent dorsal column fibers by interneurons and collateral fibers from dorsal column nuclei efferent pathways (Sonoo *et al.*, 1992). Because this portion of the SEP is so widespread, it is best recorded using a noncephalic reference. Thus, genera-

tors for far-field somatosensory potentials have been identified that begin with peripheral action potentials and progress to components originating at nearly every step of the somatosensory pathway, from the spinal cord to the somatosensory cortex.

IX. Cortical Somatosensory Potentials

Somatosensory-evoked potentials that occur later than approximately 20 msec after median nerve stimulation are considered to be cortically generated (Figs. 5 and 6; Table 4). The major peaks of the cortical SEPs are N_{20}/P_{20}, P_{25}, P_{30}/N_{30}, and N_{35}. Anterior to the central sulcus, a P_{20}–N_{30} waveform is recorded, whereas posterior to the central sulcus the polarity is inverted (N_{20}–P_{30}) (Arezzo *et al.*, 1981, 1986; Desmedt and Cheron, 1981a; Desmedt *et al.*, 1987; Onofrj *et al.*, 1990). This complex is created by depolarization of cortical neurons by thalamocortical input. The alternate polarity configuration is believed to be created by a tangential dipole in the posterior bank of the sulcus in somatosensory cortical area 3b (Fig. 5) (Allison *et al.*, 1991a,b; Arezzo *et al.*, 1981; McCarthy *et al.*, 1991; Nagamine *et al.*, 1992; Wood *et al.*, 1985). The waveforms are thus "mirror images" of the dipole as measured on either side of the isopotential line. Inversion of the N_{20}–P_{30} complex has been observed in depth recordings in somatosensory cortex area 3b (Arezzo *et al.*, 1981; McCarthy *et al.*, 1991; Vanderzant *et al.*,

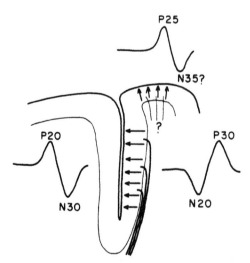

FIGURE 5 Tangential and somatosensory radial generators model for cortical somatosensory potentials in humans following median nerve stimulation. The N_{20}–P_{30}/P_{20}–N_{30} complex is believed to be generated by a tangential dipole in somatosensory area 3b. The P_{25}–N_{35} complex is hypothesized to be generated by a radial dipole in somatosensory area 1. Note that the two components overlap in time. (From Allison *et al.*, 1980, with permission of S. Karger AG, Basel.)

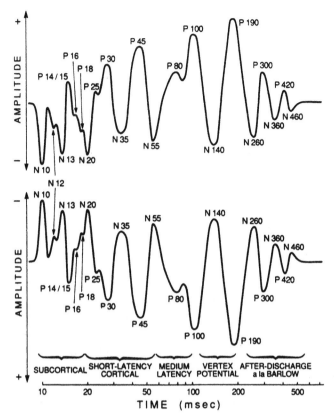

FIGURE 6 Somatosensory-evoked potential evoked by median nerve stimulation in humans illustrating both the early subcortical peaks and the later cortically generated peaks. Note the logarithmic time scale and the nonlinear amplitude axis, allowing all peaks to be displayed in the same graph. The waveform is a composite of recordings from Erb's point (N_{10}), neck (N_{12}, N_{13}), and parietal scalp. The waveform is illustrated with both positive voltage plotted upward (upper trace) and negativity plotted upward (lower trace). The schematic was drawn by T. Allison, based on data from Allison *et al.* (1983) and Goff *et al.* (1977). (From Regan, 1989, p. 289, with permission of Appleton & Lange, CT.)

1991), providing further evidence for a cortical origin. The P_{25}–N_{35} complex occurs simultaneously with the N_{20}–P_{30} waveform (Fig. 6), by activation of a separate generator source. The generator for the P_{25}–N_{35} complex is believed to be a radial dipole in somatosensory area 1 (Fig. 5) (Allison *et al.*, 1991a,b; Arezzo *et al.*, 1986; McCarthy *et al.*, 1991). Inversion of the P_{25}–N_{35} complex has been found in somatosensory cortex areas 1 and 2, supporting the cortical origin of this waveform (Arezzo *et al.*, 1981; McCarthy *et al.*, 1991; Vanderzant *et al.*, 1991). In contrast to the studies of Allison and co-workers (1991a,b; McCarthy *et al.*, 1991), other authors have suggested that activity in the motor cortex (area 4) can contribute to the SEP (peak P_{22}) (Arezzo *et al.*, 1981; Desmedt *et al.*, 1987). However, the contribution of motor area to cortical SEPs remains to be clarified.

Current source density analyses of SEPs in rats (lissencephalic brains, therefore no tangential generators produced by brain sulci or gyri) have reported a positive–negative cortical SEP. Evoked activity began with a source in cortical layers 1–2 and a sink in layers 3–5. This response was followed by a sink in layers 1–4 and a source in layers 5–6 (Di *et al.*, 1990). These authors conclude that the data support a sequential intralaminar depolarization process. The application of CSD analysis to animals with gyrencephalic brains will allow comparison of these results with those of the proported SEP generators in cortical areas 3b, 1, and 2.

Following peak N_{35} are the later components of the cortical SEP (Fig. 6). Less is known about the generators of these potentials, but several mapping studies have been performed. A positive peak at about 80–100 msec and a negative peak at about 150–200 msec, following stimulation of the medial nerve over the contralateral somatosensory cortex in humans, have been reported (Goff *et al.*, 1980). Allison and co-workers (1992) have identified several late somatosensory components in humans. They proposed that the scalp-recorded P_{45}, N_{60}, and P_{100} potentials were mainly generated in the contralateral (to stimulation) somatosensory cortical area 1. A N_{70}/P_{70} peak (precentrally/postcentrally recorded) appeared to be generated in the contralateral somatosensory cortical area 3b. A N_{120}/P_{120} peak (temporal/frontal and parietal recording) was proposed to be generated bilaterally in the upper wall of the secondary somatosensory cortex. Finally, the N_{140} and P_{190} portions of the SEP were proposed to be generated bilaterally on the frontal cortical regions. Arezzo and co-workers (1981, 1986) have reported several late SEP peaks in monkeys. Peak P_{40} was largest between the intraparietal and sylvian fissures and was associated with increases in neuronal activity in region 7b. Peak N_{45} was generated in the posterior bank of the central sulcus and the anterior portion of the postcentral crown, and may extend into the anterior bank of the intraparietal sulcus. The last peak, P_{110}, appeared to be generated over a wide cortical region, including areas 3, 1, 2, 5, and 7b. Thus, while some work has been performed to localize the late SEP components, the exact neuronal generators of these components remain in question.

X. Cerebellar Somatosensory Potentials

Very little is known regarding the generators of SEPs recorded from the cerebellum. It has been suggested that portions of the rat far-field SEP, possibly corresponding to the human SEP components in the 15- to 20-msec range, may originate in the inferior cerebellar

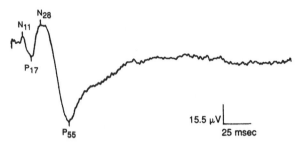

FIGURE 7 Somatosensory-evoked potential recorded from the cerebellum following electrical stimulation of ventral caudal tail nerves in adult male Long Evans rats ($n = 10$). An active electrode was located 3 mm posterior to lambda, on the midline. Reference electrode was located 7 mm anterior to bregma, and 2 mm laterally. The stimulus (delivered at 1.1 Hz) was a 50-μsec biphasic (cathodal–anodal) constant current 4 mA pulse, delivered from needle electrodes embedded (1–2 mm in the ventral tail. The cathode and anode were located 12 and 13 cm posterior to the tail hairline, respectively. The ground electrode was a needle located 8 cm posterior to the tail hairline. The EEG was analog filtered (0.3–10,000 Hz, 6 dB/octave rolloff), amplified 10,000×, A/D converted at 2048 Hz, and 100 trials were averaged. Stimulus was presented at the initiation of sampling. Waveform was donated by Dr. David Herr.

peduncles and cerebellum (Iragui-Madoz and Wiederholt, 1977; Wiederholt and Iragui-Madoz, 1977). The afferent activity would arrive via the spinocerebellar tracts, thus providing the opportunity to monitor somatosensory pathways not included in the traditional far-field or cortical SEPs (Fig. 7). Recordings from the cerebellar surface, or depth recordings from the cerebellum in monkeys and/or cats, have identified potentials generated in the cerebellum following stimulation of the spinal cord or peripheral nerves. The initial component appears to be generated by afferents in the spinocerebellar tracts (Carrea and Grundfest, 1954; Dow, 1939). The latter potentials appear to represent the responses of cerebellar neurons (Carrea and Grundfest, 1954). Following stimulation of spinocerebellar nerves in cats, the cerebellar SEP appears to be predominately distributed over the anterior cerebellar lobe (Dow, 1939). It is likely that portions of the cerebellar response would be amendable to CSD analysis; however, these experiments remain to be performed.

XI. Future Research Needs

Unambiguous identification of generators of sensory-evoked potentials is of high importance to the toxicologist. When an unknown compound is shown to alter an evoked response, knowledge of the generators of the evoked response could permit localization of the compound's site of action. This knowledge can then be applied to focused studies in these brain re-

gions. Application of methodologies such as CSD analysis and surface mapping techniques to evoked potentials whose generators are not defined is therefore a pressing need in toxicology. Once a generator for an evoked response is defined, correlation of changes in the response of the generator with changes in surface-recorded responses following toxicant exposure is essential. Changes in the generator should predict the observed alterations in the surface potentials. Methods such as CSD and surface mapping have been employed in neurobiological experiments and in some human disease conditions. However, neurotoxicologists have yet to exploit the application of these techniques to examine which evoked potential generator sites have been affected by exposure to xenobiotics. Toxicologists also need a better understanding of the correlation between pathological changes in neural generator tissues and modifications in surface-recorded potentials. Finally, a better understanding of the pharmacology of evoked responses is necessary. Little is known about which neurotransmitters and receptor types are involved in mediating the neurotransmission producing sensory-evoked potentials. Understanding which neurotransmitter systems have been altered, and the generator sites involved, will allow toxicologists to better define the toxic effects of a compound.

References

Achor, L. J., and Starr, A. (1980a). Auditory brain stem responses in the cat. I. Intracranial and extracranial recordings. *Electroencephalogr. Clin. Neurophysiol.* **48,** 154–173.

Achor, L. J., and Starr, A. (1980b). Auditory brain stem responses in the cat. II. Effects of lesions. *Electroencephalogr. Clin. Neurophysiol.* **48,** 174–190.

Allison, T., Goff, W. R., Williamson, P. D., and VanGilder, J. C. (1980). On the neural origin of early components of the human somatosensory evoked potential. In *Clinical Uses of Cerebral, Brainstem and Spinal Somatosensory Evoked Potentials. Progress in Clinical Neurophysiology* (J. E. Desmedt, Ed.), Vol. 7, pp. 51–68, Karger, Basel.

Allison, T., and Hume, A. L. (1981). A comparative analysis of short-latency somatosensory evoked potentials in man, monkey, cat, and rat. *Exp. Neurol.* **72,** 592–611.

Allison, T., McCarthy, G., and Wood, C. C. (1992). The relationship between human long-latency somatosensory evoked potentials recorded from the cortical surface and from the scalp. *Electroencephalogr. Clin. Neurophysiol.* **84,** 301–314.

Allison, T., McCarthy, G., Wood, C. C., and Jones, S. J. (1991a). Potentials evoked in human and monkey cerebral cortex by stimulation of the median nerve. A review of scalp and intracranial recordings. *Brain* **114**(6), 2465–2503.

Allison, T., Wood, C. C., and Goff, W. R. (1983). Brain stem auditory, pattern-reversal visual, and short-latency somatosensory evoked potentials: latencies in relation to age, sex, and brain and body size. *Electroencephalogr. Clin. Neurophysiol.* **55,** 619–636.

Allison, T., Wood, C. C., McCarthy, G., and Spencer, D. D. (1991b). Cortical somatosensory evoked potentials. II. Effects of excision of somatosensory or motor cortex in humans and monkeys. *J. Neurophysiol.* **66**(1), 64–82.

Anziska, B., and Cracco, R. Q. (1981). Short latency SEPs to median nerve stimulation: comparison of recording methods and origin of components. *Electroencephalogr. Clin. Neurophysiol.* **52**, 531–539.

Arezzo, J. C., Legatt, A. D., and Vaughan, H. G., Jr. (1979). Topography and intracranial sources of somatosensory evoked potentials in the monkey. I. Early components. *Electroencephalogr. Clin. Neurophysiol.* **46**, 155–172.

Arezzo, J. C., Pickoff, A., and Vaughan, H. G., Jr. (1975). The sources and intracerebral distribution of auditory evoked potentials in the alert rhesus monkey. *Brain Res.* **90**, 57–73.

Arezzo, J. C., Simson, R., and Brennan, N. E. (1985). Evoked potentials in the assessment of neurotoxicology in humans. *Neurobehav. Toxicol. Teratol.* **7**, 299–304.

Arezzo, J. C., Vaughan, H. G., Jr., Kraut, M. A., Steinschneider, M., and Legatt, A. D. (1986). Intracranial generators of event-related potentials in the monkey. In *Evoked Potentials. Frontiers of Clinical Neuroscience* (R. Q. Cracco and I. Bodis-Wollner, Eds.), Vol. 3, pp. 174–189, Wiley-Liss, New York.

Arezzo, J. C., Vaughan, H. G., Jr., Legatt, A. D. (1981). Topography and intracranial sources of somatosensory evoked potentials in the monkey. II. Cortical components. *Electroencephalogr. Clin. Neurophysiol.* **51**, 1–18.

Armington, J. C. (1980). Electroretinography. In *Electrodiagnosis in Clinical Neurology* (M. J. Aminoff, Ed.), pp. 305–347, Churchill Livingstone, New York.

Baker, C. L., Jr., Hess, R. R., Olsen, B. R., and Zrenner, E. (1988). Current source density analysis of linear and non-linear components of the primate electroretinogram. *J. Physiol.* **407**, 155–176.

Barna, J. S., Arezzo, J. C., and Vaughan, H. G., Jr. (1981). A new multielectrode array for the simultaneous recording of field potentials and unit activity. *Electroencephalogr. Clin. Neurophysiol.* **52**, 494–496.

Barnes, M. I., and Dyer, R. S. (1986). Superior colliculus lesions and flash evoked potentials from rat cortex. *Brain Res. Bull.* **16**(2), 225–230.

Barth, D. S., and Di, S. (1990). Three-dimensional analysis of auditory-evoked potentials in rat neocortex. *J. Neurophysiol.* **64**(5), 1527–1536.

Barth, D. S., and Di, S. (1991). The functional anatomy of middle latency auditory evoked potentials. *Brain Res.* **565**(1), 109–115.

Bigler, E. D. (1977). Neurophysiology, neuropharmacology and behavioral relationships of visual system evoked after-discharges: a review. *Behav. Rev.* **1**(2), 95–112.

Boyes, W. K. (1990). Proposed test guidelines for using sensory evoked potentials as measures of neurotoxicity. Office of Research and Development, United States Environmental Protection Agency. Deliverable 2555. EPA Document Number 600x-90166.

Boyes, W. K. (1992). Testing visual system toxicity using evoked potential technology. In *The Vulnerable Brain and Environmental Risks*, Vol. 1, *Malnutrition and Hazard Assessment* (R. L. Isaacson and K. F. Jensen, Eds.), pp. 193–222, Plenum Press, New York.

Boyes, W. K. (1993). Sensory-evoked potentials: measures of neurotoxicity. In *Assessing Neurotoxicity of Drugs of Abuse, NIDA Research Monograph 136* (L. Erinoff, Ed.), pp. 63–100. U.S. Department of Health and Human Services. Rockville, Maryland.

Boyes, W. K. (1994). Rat and human sensory evoked potentials and the predictability of human neurotoxicity from rat data. *Neurotoxicology,* **15**(3), 569–578.

Boyes, W. K., and Cooper, G. P. (1981). Acrylamide neurotoxicity. Effects on far-field somatosensory evoked potentials in rats. *Neurobehav. Toxicol. Teratol.* **3**(4), 487–490.

Brankačk, J., Schober, W., and Klingberg, F. (1990). Different laminar distribution of flash evoked potentials in cortical areas 17 and 18b of freely moving rats. *J. Hirnforsch.* **31**(4), 525–533.

Brown, K. T., Watanabe, K., and Murakami, M. (1965). The early and late receptor potentials of monkey cones and rods. *Cold Spring Harbor Symp. Quant. Biol.* **30**, 457–482.

Buchwald, J. S. (1983). Generators. In *Bases of Auditory Brain-Stem Evoked Responses* (E. J. Moore, Ed.), pp. 157–195, Grune & Stratton, New York.

Buchwald, J. S., Rubinstein, E. H., Schwafel, J., and Strandburg, R. J. (1991). Midlatency auditory evoked responses: differential effects of a cholinergic agonist and antagonist. *Electroencephalogr. Clin. Neurophysiol.* **80**(4), 303–309.

Carr, R. E., and Siegel, I. M. (1990). *Electrodiagnostic Testing of the Visual System: A Clinical Guide.* F. A. Davis, Philadelphia.

Carrea, R. M. E., and Grundfest, H. (1954). Electrophysiological studies of cerebellar inflow. I. Origin, conduction and termination of ventral spino-cerebellar tract in monkey and cat. *J. Neurophysiol.* **17**, 208–238.

Cheatham, M. A., and Dallos, P. (1982). Two-tone interactions in the cochlear microphonic. *Hear. Res.* **8**, 29–48.

Chen, T.-J., and Chen, S.-S. (1991). Generator study of brainstem auditory evoked potentials by radiofrequency lesion method in rats. *Exp. Brain Res.* **85**, 537–542.

Cohen, M. M. (1982). Coronal topography of the middle latency auditory evoked potentials (MLAEPs) in man. *Electroencephalogr. Clin. Neurophysiol.* **53**, 231–236.

Cracco, R. Q., and Bodis-Wollner, I. (1986). *Evoked Potentials. Frontiers of Clinical Neuroscience,* Vol. 3, Wiley-Liss, New York.

Cracco, R. Q., and Cracco, J. B. (1976). Somatosensory evoked potential in man: far field potentials. *Electroencephalogr. Clin. Neurophysiol.* **41**, 460–466.

Creel, D., Dustman, R. E., and Beck, E. C. (1974). Intensity of flash illumination and the visually evoked potential of rats, guinea pigs and cats. *Vision Res.* **14**, 725–729.

Crofton, K. M., and Sheets, L. P. (1989). Evaluation of sensory system function using reflect modification of the startle response. *J. Am. Coll. Toxicol.* **8**, 199–211.

Darcey, T. M., Ary, J. P., and Fender, D. H. (1980). Methods for the localization of electrical sources in the human brain. *Prog. Brain Res.* **54**, 128–134.

Desmedt, J. E. (1986). Generator sources of SEP in man. In *Evoked Potentials. Frontiers of Clinical Neuroscience* (R. Q. Cracco and I. Bodis-Wollner, Eds.), Vol. 3, pp. 235–245, Wiley-Liss, New York.

Desmedt, J. E., and Cheron, G. (1980). Central somatosensory conduction in man: neural generators and interpeak latencies of the far-field components recorded from neck and right or left scalp and earlobes. *Electronecephalogr. Clin. Neurophysiol.* **50**, 382–403.

Desmedt, J. E., and Cheron, G. (1981a). Non-cephalic reference recording of early somatosensory potentials to finger stimulation in adult or aging normal man: differentiation of widespread N18 and contralateral N20 from the prerolandic P22 and N30 components. *Electroencephalogr. Clin. Neurophysiol.* **52**, 553–570.

Desmedt, J. E., and Cheron, G. (1981b). Prevertebral (oesophageal) recording of subcortical somatosensory evoked potentials in man: The spinal P13 component and the dual nature of the spinal generators. *Electroencephalogr. Clin. Neurophysiol.* **52**, 257–275.

Desmedt, J. E., Nguyen, T. H., and Bourguet, M. (1987). Bit-mapped color imaging of human evoked potentials with reference to the N20, P22, P27 and N30 somatosensory responses. *Electroencephalogr. Clin. Neurophysiol.* **68**, 1–19.

Di, S., and Barth, D. S. (1992). The functional anatomy of middle-latency auditory evoked potentials: thalamocortical connections. *J. Neurophysiol.* **68**(2), 425–431.

Di, S., Baumgartner, C., and Barth, D. S. (1990). Laminar analysis of extracellular field potentials in rat vibrissa/barrel cortex. *J. Neurophysiol.* **63**(4), 832–840.

Dickerson, L. W., and Buchwald, J. S. (1991). Midlatency auditory-evoked responses: effect of scopolamine in the cat and implications for brain stem cholinertic mechanisms. *Exp. Neurol.* **122**(2), 229–239.

Dow, R. S. (1939). Cerebellar action potentials in response to stimulation of various afferent connections. *J. Neurophysiol.* **2**, 543–555.

Dowling, J. E. (1987). *The Retina: An Approachable Part Of The Brain.* The Belknap Press of Harvard University Press, Cambridge, Massachusetts.

Ducati, A., Fava, E., and Motti, E. D. F. (1988). Neuronal generators of the visual evoked potentials: intracerebral recording in awake humans. *Electroencephalogr. Clin. Neurophysiol.* **71**(2), 89–99.

Dyer, R. S. (1985). The use of sensory evoked potentials in toxicology. *Fundam. Appl. Toxicol.* **5**, 24–40.

Dyer, R. S. (1986). Interactions of behavior and neurophysiology. In *Neurobehavioral Toxicology* (Z. Annau, Ed.), pp. 193–213, Johns Hopkins University Press, Baltimore.

Dyer, R. S. (1987a). Macrophysiological assessment of organometal neurotoxicity. In *Neurotoxicants and Neurobiological Function: Effects of Organoheavy Metals*, (H. A. Tilson and S. B. Sparber, Eds.), pp. 137–184, John Wiley and Sons, New York.

Dyer, R. S. (1987b). Somatosensory evoked potentials. In *Electrophysiology in Neurotoxicology*, (H. E. Lowndes, Ed.), Vol II, pp. 1–69, CRC Press, Boca Raton, Florida.

Dyer, R. S., Clarke, C. C., and Boyes, W. K. (1987a). Surface distribution of flash-evoked and pattern reversal-evoked potentials in hooded rats. *Brain Res. Bull.* **18**(2), 227–234.

Dyer, R. S., Jensen, K. F., and Boyes, W. K. (1987b). Focal lesions of visual cortex—Effects on visual evoked potentials in rats. *Exp. Neurol.* **95**, 100–115.

Ehlers, C. L., Wall, T. L., and Chaplin, R. I. (1991). Long latency event-related potentials in rats: effects of dopaminergic and serotonergic depletions. *Pharmacol. Biochem. Behav.* **38**(4), 789–793.

Eisen, A., Roberts, K., Low, M., Hoirch, M., and Lawrence, P. (1984). Questions regarding the sequential neural generator theory of the somatosensory evoked potential raised by digital filtering. *Electroencephalogr. Clin. Neurophysiol.* **59**, 388–395.

Fox, D. A., Lowndes, H. E., and Bierkamper, G. G. (1982). Electrophysiological techniques in neurotoxicology. In *Nervous System Toxicology* (C. L. Mitchell, Ed.), pp. 299–335, Raven Press, New York.

Freeman, J. A., and Nicholson, C. (1975). Experimental optimization of current source density technique for anuran cerebellum. *J. Neurophysiol.* **38**, 369–382.

Fullerton, B. C., and Kiang, N. Y. S. (1990). The effect of brainstem lesions on brainstem auditory evoked potentials in the cat. *Hear. Res.* **49**(1-3), 363–390.

Givre, S. J., Schroeder, C. E., and Arezzo, J. C. (1994). Contribution of extrastriate area V4 to the surface-recorded flash VEP in the awake macaque. *Vision Res.* **34**(4), 415–438.

Glenn, S. L., and Kelly, J. B. (1992). Kainic acid lesions of the lateral lemniscus: effects on binaural evoked responses in rat auditory cortex. *J. Neurosci.* **12**(9), 3688–3699.

Goff, G. D., Matsumiya, Y., Allison, T., and Goff, W. R. (1977). The scalp topography of human somatosensory and auditory evoked potentials. *Electroencephalogr. Clin. Neurophysiol.* **42**, 57–76.

Goff, W. R., Williamson, P. D., VanGilder, J. C., Allison, T., and Fisher, T. C. (1980). Neural origins of long latency evoked poten-tials recorded from the depth and from the cortical surface of the brain in man. In *Clinical Uses of Cerebral, Brainstem and Spinal Somatosensory Evoked Potentials. Progress in Clinical Neurophysiology* (J. E. Desmedt, Ed.), Vol. 7, pp. 126–145, Karger, Basel.

Green, J. B., Nelson, A. V., and Michael, D. (1986). Digital zero-phase-shift filtering of short-latency somatosensory evoked potentials. *Electroencephalogr. Clin. Neurophysiol.* **63**, 384–388.

Gulrajani, R. M., Roberge, F. A., and Savard, P. (1984). Moving dipole inverse ECG and EEG solutions. *IEEE Trans. Biomed. Eng.* **BME-31**, 903–910.

Hamano, T., Kaji, R., Diaz, A. F., Kohara, N., Takamatsu, N., Uchiyama, T., Shibasaki, H., and Kimura, J. (1993). Vibration-evoked sensory nerve action potentials derived from Pacinian corpuscles. *Electroencephalogr. Clin. Neurophysiol.* **89**(4), 278–286.

Hashimoto, I. (1984). Somatosensory evoked potentials from the human brainstem: origins of short latency potentials. *Electroencephalogr. Clin. Neurophysiol.* **57**, 221–227.

Hashimoto, I., Gatayama, T., Yoshikawa, K., Sasaki, M., and Nomura, M. (1991). Compound activity in sensory nerve fibers is related to intensity of sensation evoked by air-puff stimulation of the index finger in man. *Electroencephalogr. Clin. Neurophysiol.* **81**, 176–185.

Helmholz, H. (1853). Über einige gesetze der verthheilung elektrischer stome inkörperlichen leitern, mit anwendung auf de thier-ischerlektrischen versuche. *Ann. Phys. Chem.* **29**, 211–233; 353–377.

Henry, K. R. (1979). Auditory brainstem volume-conducted responses: origins in the laboratory mouse. *J. Am. Audit. Soc.* **4**(5), 173–178.

Heynen, H. and van Norren, D. (1985). Origin of the electroretinogram in the intact macaque eye—II. Current source-density analysis. *Vision Res.* **25**(5), 709–715.

Hillyard, S. A., Simpson, G. V., Woods, D. L., Van Voorhis, S., and Münte, T. F. (1984). Event-related brain potentials and selective attention to different modalities. In *Cortical Integration, IBRO Series* (F. Reinoso-Suárez and C. Ajmone-Marsan, Eds.), Vol. II, pp. 395–414, Raven Press, New York.

Hinman, C. L. and Buchwald, J. S. (1983). Depth evoked potential and single unit correlates of vertex midlatency auditory evoked responses. *Brain Res.* **264**, 57–67.

Hjorth, B. (1975). An on-line transformation of EEG scalp potentials into orthogonal source derivations. *Electroencephalogr. Clin. Neurophysiol.* **39**, 526–530.

Hudnell, H. K., and Boyes, W. K. (1991). The comparability of rat and human visual-evoked potentials. *Neurosci. Biobehav. Rev.* **15**, 159–164.

Iragui-Madoz, V. J., and Wiederholt, W. C. (1977). Far-field somatosensory evoked potentials in the cat: Correlation with depth recording. *Ann. Neurol.* **1**, 569–574.

Jeffreys, D. A., and Axford, J. G. (1972). Source localization of pattern specific components of human visual evoked potentials. *Exp. Brain Res.* **16**, 1–21.

Jellema, T., and Weijnen, J. A. W. M. (1991). A slim needle-shaped multiwire microelectrode for intracerebral recording. *J. Neurosci. Methods* **40**(2-3), 203–209.

Jewett, D. L. (1970). Volume-conducted potentials in response to auditory stimuli as detected by averaging in the cat. *Electroencephalogr. Clin. Neurophysiol.* **28**, 609–618.

Jewett, D. L., and Williston, J. S. (1971). Auditory-evoked far fields averaged from the scalp of humans. *Brain* **94**, 681–696.

Kaga, K., Hink, R., Shinoda, Y., and Suzuki, J. (1980). Evidence for a primary cortical origin of a middle latency auditory evoked potential in cats. *Electroencephalogr. Clin. Neurophysiol.* **50**, 254–266.

Kimura, J., Kimura, A., Machida, M., Yamada, T., and Mitsudome, A. (1986). Model for far-field recordings of SEP. In *Evoked Potentials. Frontiers of Clinical Neuroscience* (R. Q. Cracco and I. Bodis-Wollner, Eds.), Vol. 3, pp. 246–261, Wiley-Liss, New York.

Knight, R. T., Brailowsky, S., Scabini, D., and Simpson, G. V. (1985). Surface auditory evoked potentials in the unrestrained rat: component definition. *Electroencephalogr. Clin. Neurophysiol.* **61**, 430–439.

Kraus, N., McGee, T., Littman, T. and Nicol, T. (1992). Reticular formation influences on primary and non-primary auditory pathways as reflected by the middle latency response. *Brain Res.* **587**(2), 186–194.

Kraut, M. A., Arezzo, J. C., and Vaughan, H. G. Jr. (1985). Intracortical generators of the flash VEP in monkeys. *Electroencephalogr. Clin. Neurophysiol.* **62**, 300–312.

Lachapelle, P. (1990). Oscillatory potentials as predictors to amplitude and peak time of the photopic b-wave of the human electroretinogram. *Docum. Ophthalmol.* **75**, 73–82.

Lopes da Silva, R., and Van Rotterdam, A. (1982). Biophysical aspects of EEG and MEG Generation. In *Electroencephalography, Basic Principles, Clinical Applications, and Related Fields* (E. Niedermeyer, and F. Lopes da Silva, Eds.), pp. 15–26, Urban and Schwarzenberg, Baltimore.

Maccabee, P. J., Hassan, N. F., Cracco, R. Q., and Schiff, J. A. (1986). Short latency somatosensory and spinal evoked potentials: Power spectra and comparison between high pass analog and digital filter. *Electroencephalogr. Clin. Neurophysiol.* **65**, 177–187.

MacKay, D. M. (1984). Source density mapping of human visual receptive fields using scalp potentials. *Exp. Brain Res.* **54**, 579–581.

Maffei, L., and Fiorentini, A. (1981). Electroretinographic response before and after section of the optic nerve. *Science* **211**, 953–954.

Maffei, L., and Fiorentini, A. (1986). Generator sources of the pattern ERG in man and animals. In *Evoked Potentials. Frontiers of Clinical Neuroscience* (R. Q. Cracco and I. Bodis-Wollner, Eds.), Vol. 3, pp. 101–116, Wiley-Liss, New York.

Manahilov, V., Riemslag, F. C. C., and Spekreijse, H. (1992). The laplacian analysis of the pattern onset response in man. *Electroencephalogr. Clin. Neurophysiol.* **82**(3), 220–224.

Mattsson, J. L., and Albee, R. R. (1988). Sensory evoked potentials in neurotoxicology. *Neurotoxicol. Teratol.* **10**, 435–443.

Mattsson, J. L., Albee, R. R., and Eisenbrandt, D. L. (1989). Neurological approach to neurotoxicological evaluation in laboratory animals. *J. Am. Coll. Toxicol.* **8**, 271–286.

Mattsson, J. L., Boyes, W. K., and Ross, J. F. (1992). Incorporating evoked potentials into neurotoxicity test schemes. In *Target Organ Toxicology Series: Neurotoxicology* (H. A. Tilson and C. L. Mitchell, Eds.), pp. 125–145, Raven Press, New York.

Mattsson, J. L., Eisenbrandt, D. L., and Albee, R. R. (1990). Screening for neurotoxicity: complementarity of functional and morphological techniques. *Toxicol. Pathol.* **18**, 115–127.

McCarthy, G., Wood, C. C., and Allison, T. (1991). Cortical somatosensory evoked potentials I. Recordings in the monkey *Macaca fascicularis. J. Neurophysiol.* **66**(1), 53–63.

McGee, T., Kraus, N., Littman, T., and Nicol, T. (1992). Contributions of medial geniculate body subdivisions to the middle latency response. *Hear. Res.* **61**(1–2), 147–154.

Meurice, J. C., Paquereau, J., and Marillaud, A. (1991). Same location of the source of P_1 of BAEPs and N_1 of CAP in guinea pig. *Hear. Res.* **53**(2), 209–216.

Miller, R. F., and Dowling, J. E. (1970). Intracellular responses of the Müller (glial) cells of mudpuppy retina: their relation to b-wave of the electroretinogram. *J. Neurophysiol.* **33**(3), 323–341.

Mitzdorf, U. (1985). Current source-density method and application in cat cerebral cortex: investigation of evoked potentials and EEG phenomena. *Physiol. Rev.* **65**(1), 37–100.

Mitzdorf, U. (1986). The physiological causes of VEP: current source density analysis of electrically and visually evoked potentials. In *Evoked Potentials. Frontiers of Clinical Neuroscience* (R. Q. Cracco and I. Bodis-Wollner, Eds.), Vol. 3, pp. 141–154, Wiley-Liss, New York.

Møller, A. R., and Jannetta, P. J. (1985). Neural generators of the auditory brainstem response. In *The Auditory Brainstem Response* (J. T. Jacobson, Ed.), pp. 13–31, College-Hill Press, San Diego.

Møller, A. R., and Jannetta, P. J. (1986). Simultaneous surface and direct brainstem recordings of brainstem auditory evoked potentials (BAEP) in man. In *Evoked Potentials. Frontiers of Clinical Neuroscience* (R. Q. Cracco and I. Bodis-Willner, Eds.), Vol. 3, pp. 227–234, Wiley-Liss, New York.

Møller, A. R., Jannetta, P. J., and Burgess, J. E. (1986). Neural generators of the somatosensory evoked potentials: recording from the cuneate nucleus in man and monkeys. *Electroencephalogr. Clin. Neurophysiol.* **65**, 241–248.

Møller, A. R., Jannetta, P. J., and Jho, H. D. (1990). Recordings from human dorsal column nuclei using stimulation of the lower limb. *Neurosurgery* **26**(2), 291–299.

Møller, A. R., Jannetta, P. J., and Sekhar, L. N. (1988). Contributions from the auditory nerve to the brain-stem auditory evoked potentials (BAEPs): results of intracranial recording in man. *Electroencephalogr. Clin. Neurophysiol.* **71**(3) 198–211.

Møller, A. R., Sekiya, T., and Sen, C. N. (1989). Responses from dorsal column nuclei (DCN) in the monkey to stimulation of upper and lower limbs and spinal cord. *Electroencephalogr. Clin. Neurophysiol.* **73**, 353–361.

Montero, V. M. (1973). Evoked responses in the rat's visual cortex to contralateral, ipsilateral and restricted photic stimulation. *Brain Res.* **53**, 192–196.

Morioka, T., Shima, F., Kato, M., and Fukui, M. (1991). Direct recording of somatosensory evoked potentials in the vicinity of the dorsal column nuclei in man: their generator mechanisms and contribution to the scalp far-field potentials. *Electroencephalogr Clin. Neurophysiol.* **80**(3), 215–220.

Morris, H. H. III, and Lüders, H. (1985). Electrodes. In *Long-term Monitoring in Epilepsy (EEG Suppl. No. 27)* (J. Gotman, J. R. Ives, P. Gloor, Eds.), pp. 3–26, Elsevier Science Publishing New York.

Nagamine, T., Kaji, R., Suwazono, S., Hamano, T., Shibasaki, H., and Kimura, J. (1992). Current source density mapping of somatosensory evoked responses following median and tibia nerve stimulation. *Electroencephalogr. Clin. Neurophysiol.* **84**(3), 248–256.

Nelson, D., and Lassman, F. (1973). Combined effects of recovery period and stimulus intensity on the human auditory evoked vertex response. *J. Speech Hear. Res.* **16**, 297–308.

Nicholson, C. (1973). Theoretical analysis of field potentials in anisotropic ensembles of neuronal elements. *IEEE Trans. Biomed. Eng.* **20**, 278–288.

Nunez, P. L. (1990). Localization of brain activity with electroencephalography. In *Advances in Neurology* Vol. 54, *Magnetoencephalography* (S. Susumu Ed.), pp. 39–65, Raven Press, New York.

Nuttall, A. L., and Dolan, D. F. (1991). Cochlear microphonic enhancement in two tone interactions. *Hear. Res.* **51**(2), 235–246.

Oakley, B., II. and Green, D. C. (1976). Correlation of light-induced changes in retinal extracellular potassium concentration with C-wave of the electroretinogram. *J. Neurophysiol.* **39**(5), 1117–1133.

Onofrj, M., Basciani, M., Fulgente, T., Bazzano, S., Malatesta, G., and Curatola, L. (1990). Maps of somatosensory evoked potentials (SEPs) to mechanical (tapping) stimuli: comparison with P14, N20, P22, N30 of electrically elicited SEPs. *Electrocencephalogr. Clin. Neurophysiol.* **77**(4), 314–319.

Onofrj, M., Harnois, C., and Bodis-Wollner, I. (1985). The hemispheric distribution of the transient rat VEP: a comparison of flash and pattern stimulation. *Exp. Brain Res.* **59**, 427–433.

Otto, D. (1986). The use of sensory evoked potentials in neurotoxicity testing of workers. *Semin. Occupat. Med.* **1**, 175–183.

Perrin, F., Bertrand, O., and Pernier, J. (1987a). Scalp current density mapping: Value and estimation from potential data. *IEEE Trans. Biomed. Eng.* **BME-34**(4), 283–288.

Perrin, F., Pernier, J., Bertrand, O., Giard, M. H., and Echallier, J. F. (1987b). Mapping of scalp potentials by surface spline interpolation. *Electroencephalogr. Clin. Neurophysiol.* **66**, 75–81.

Perrin, F., Pernier, J., Bertrand, O., and Echallier, J. F. (1989). Spherical splines for scalp potential and current density mapping. *Electroencephalogr. Clin. Neurophysiol.* **72**, 184–187.

Perrin, F., Pernier, J., Bertrand, O., and Echallier, J. F. (1990). Corrigenda. *Electroencephalogr. Clin. Neurophysiol.* **76**, 565.

Picton, T. W., Hillyard, S. A., Krausz, H. I., and Galambos, R. (1974). Human auditory evoked potentials. I. Evaluation of components. *Electroencephalogr. Clin. Neurophysiol.* **36**, 179–190.

Polich, J. (1989). Habituation of P300 from auditory stimuli. *Psychobiology* **17**(1), 19–28.

Potts, A. M. (1972). Electrophysiological measurements. In *The Assessment of Visual Function* (A. M. Potts, Ed.), pp. 187–206. C. V. Mosby, St. Louis.

Pratt, H. and Starr, A. (1981). Mechanically and electrically evoked somatosensory potentials in humans: scalp and neck distributions of short latency components. *Electroencephalogr. Clin. Neurophysiol.* **51**, 138–147.

Rebert, C. S. (1983). Multisensory evoked potentials in experimental and applied neurotoxicology. *Neurobehav. Toxicol. Teratol.* **5**, 659–671.

Regan, D. (1989). *Human Brain Electrophysiology. Evoked Potentials and Evoked Magnetic Fields in Science and Medicine.* Appleton & Lange, Norwalk, Connecticut.

Regan, D., and Heron, J. R. (1969). Clinical investigations of lesions of the visual pathway: a new objective technique. *J. Neurol. Neurosurg. Psychiatry* **32**, 479–483.

Riemslag, F. C. C., and Heymen, H. G. M. (1984). Depth profile of pattern LERG in macque. *Docum. Ophthalmol.* **40**, 143–148.

Rose, G. H., and Lindsley, D. B. (1965). Visually evoked electrocortical responses in kittens: development of specific and nonspecific systems. *Science* **148**, 1244–1246.

Schroeder, C. E., Tenke, C. E., and Givre, S. J. (1990a). Extrastriate contributions to surface VEP in the awake macaque. *Invest. Ophthalmol. Vis. Sci.* **30**, 258.

Schroeder, C. E., Tenke, C. E., Givre, S. J., Arezzo, J. C., and Vaughan, H. G., Jr. (1990b). Laminar analysis of bicuculline-induced epipleptiform activity in area 17 of the awake macaque. *Brain Res.* **515**, 326–330.

Schroeder, C. E., Tenke, C. E., Givre, S. J., Arezzo, J. C., and Vaughan, H. G., Jr. (1991). Striate cortical contribution to the surface-recorded pattern-reversal VEP in the alert monkey. *Vision Res.* **31**(7/8), 1143–1157.

Sen, C. N., and Møller, A. R. (1991). Comparison of somatosensory evoked potentials recorded from the scalp and dorsal column nuclei to upper and lower limb stimulation in the rat. *Electroencephalogr. Clin. Neurophysiol.* **80**(5), 378–383.

Seppalainen, A. M. (1988). Neurophysiological approaches to the detection of early neurotoxicity in humans. *CRC Crit. Rev. Toxic.* **18**, 245–298.

Shaw, N. A. (1991). A possible thalamic component of the auditory evoked potential in the rat. *Brain Res. Bull.* **27**(1), 133–136.

Siegel, J., and Sisson, D. F. (1989). Comparison of visually evoked potentials generated with light flash and optic nerve shock in rat. *Soc. Neurosci. Abstr.* **15**(1), 117.

Sieving, P. A., Fishman, G. A., and Maggiano, J. M. (1978). Corneal wick electrode for recording bright flash electroretinograms and early receptor potentials. *Arch. Ophthalmol.* **96**, 899–900.

Sieving, P. A., and Steinberg, R. H. (1987). Proximal retinal contribution to the intraretinal 8-Hz pattern ERG of cat. *J. Neurophysiol.* **57**(1), 104–120.

Snyder, A. Z. (1992). Steady-state vibration evoked potentials: Description of technique and characterization of responses. *Electroencephalogr. Clin. Neurophysiol.* **84**(3), 257–268.

Snyder, A. Z. (1991). Dipole source localization in the study of EP generators: a critique. *Electroencephalogr. Clin. Neurophysiol.* **80**(4), 321–325.

Sonoo, M., Genba, K., Zai, W., Iwata, M., Mannen, T., and Kanazawa, I. (1992). Origin if the widespread N18 in median nerve SEP. *Electroencephalogr. Clin. Neurophysiol.* **84**(5), 418–425.

Srebro, R. (1985). Localization of visually evoked cortical activity in humans. *J. Physiol.* **360**, 233–246.

Srebro, R., and Purdy, P. D. (1990). Localization of visually evoked cortical activity using magnetic resonance imaging and computerized tomography. *Vision Res.* **30**(3), 351–358.

Starr, A., and Zaaroor, M. (1990). Eighth nerve contributions to cat auditory brainstem responses (ABR). *Hear. Res.* **48**(1-2), 151–160.

Steinschneider, M., Tenke, C. E., Schroeder, C. E., Javitt, D. C., Simpson, G. V., Arezzo, J. C., and Vaughan, H. G., Jr. (1992). Cellular generators of the cortical auditory evoked potential initial component. *Electroencephalogr. Clin. Neurophysiol.* **84**, 196–200.

Sumitomo, L., and Klingberg, F. (1972). The role of lateral geniculate body in the generation of photically and electrically evoked after-discharges in freely moving rats. *Acta Biol. Med. Germ.* **29**, 43–54.

Tenke, C. E., Schroeder, C. E., Arezzo, J. C., Vaughan, H. G., Jr. (1993). Interpretation of high-resolution current source density profiles: simulation of sublaminar contributions to the visual evoked potential. *Exp. Brain Res.* **94**(2), 183–192.

Tomberg, C., Desmedt, J. E., Ozaki, I., and Noel, P. (1991). Nasopharyngeal recordings of somatosensory evoked potentials document the medullary origin of the N18 far-field. *Electrocephalogr. Clin. Neurophysiol.* **80**, 496–503.

Urasaki, E., Wada, S., Kadoya, C., Yokota, A., Matsuoka, S., and Shima, F. (1990). Origin of scalp far-field N18 of SEP's in response to median nerve stimulation. *Electroencephalogr. Clin. Neurophysiol.* **77**, 39–51.

Vaknin, G., DiScenna, P. G., and Teyler, T. J. (1988). A method for calculating current source density (CSD) analysis without resorting to recording sites outside the sampling volume. *J. Neurosci. Methods* **24**(2), 131–135.

Vanderzant, C. W., Beydoun, A. A., Domer, P. A., Hood, T. W., and Abou-Khalil, B. W. (1991). Polarity reversal of N20 and P23somatosensory evoked potentials between scalp and depth recordings. *Electroencephalogr. Clin. Neurophysiol.* **78**(3), 234–239.

Wachtmeister, L. (1980). Further studies of the chemical sensitivity of the oscillatory potentials of the electroretinogram (ERG). I. GABA- and glycine antagonists. *Acta Opthalmologica* **58**, 712–725.

Wachtmeister, L. (1981a). Further studies of the chemical sensitivity of the oscillatory potentials of the electroretinogram (ERG). II. Glutamate-aspartate- and dopamine antagonists. *Acta Opthalmologica* **59**, 247–258.

Wachtmeister, L. (1981b). Further studies of the chemical sensitivity of the oscillatory potentials of the electroretinogram (ERG). III. Some Ω amino acids and ethanol. *Acta Opthalomologica* **59**, 609–619.

Wachtmeister, L., and Dowling, J. E. (1978). The oscillatory potentials of the mudpuppy retina. *Invest. Ophthalmol. Vis. Sci.* **17**, 1176–1186.

Wada, S-I., and Starr, A. (1983a). Generation of auditory brain stem responses (ABRs). II. Effects of surgical section of the trapezoid body on the ABR in guinea pigs and cat. *Electroencephalogr. Clin. Neurophysiol.* **56**, 340–351.

Wada, S-I., and Starr, A. (1993b). Generation of auditory brain stem responses (ABRs). III. Effects of lesions of the superior olive, lateral lemniscus and inferior colliculus on the ABR in guinea pig. *Electroencephalogr. Clin. Neurophysiol.* **56**, 352–366.

Wesensten, N. J., Badia, P., and Harsh, J. (1990). Time of day, repeated testing, and interblock interval effects on P300 amplitude. *Physiol. Behav.* **47**(4), 653–658.

Wiederholt, W. C. and Iragui-Madoz, V. J. (1977). Far field somatosensory potentials in the rat. *Electroencephalogr. Clin. Neurophysiol.* **42**, 456–465.

Wolpaw, J. R., and Wood, C. C. (1982). Scalp distributions of human auditory evoked potentials II. *Electroencephalogr. Clin. Neurophysiol.* **39**, 609–620.

Wood, C. C. (1982). Application of dipole localization methods to source identification of human evoked potentials. *Ann. N.Y. Acad. Sci.* **388**, 139–155.

Wood, C. C. and Allison, T. (1981). Interpretation of evoked potentials: a neurophysiological perspective. *Can. J. Psychol.* **35**(2), 113–135.

Wood, C. C., Cohen, D., Cuffin, B. N., Yarita, M., and Allison, T. (1985). Electrical sources in human somatosensory cortex: identification by combined magnetic and potential recordings. *Science* **227**, 1051–1053.

Wood, C. C., and Wolpaw, J. R. (1982). Scalp distribution of human auditory evoked potentials. II. Evidence for overlapping sources and involvement of auditory cortex. *Electroencephalogr. Clin. Neurophysiol.* **54**, 25–38.

Yamaguchi, S., Globus, H., and Knight, R. T. (1993). P3-like potential in rats. *Electroencephalogr. Clin. Neurophysiol.* **88**(2), 151–154.

PART
III

Neurobehavioral Toxicology

INTRODUCTORY OVERVIEW

A. General Approaches

B. Biochemical Correlates for Neurobehavioral Toxicity

Neurobehavioral Toxicology
An Introductory Overview

DEBORAH A. CORY-SLECHTA
Department of Environmental Medicine
University of Rochester School of Medicine and Dentistry
Rochester, New York 14642

ROBERT C. MACPHAIL
Division of Neurotoxicology
Health Effects Research Laboratories
U.S. Environmental Protection Agency
Research Triangle Park, North Carolina 27711

I. Behavioral Toxicology: The Descriptive Stage

Research in behavioral toxicology is often thought to be relatively descriptive in nature in that many studies seek primarily to determine the dose and/or time dependence of the effects of a toxicant upon some particular behavior. Although such research, not specifically deemed mechanistic, is often accorded second-class citizenship, studies aimed at characterizing behavioral effects of toxicants are entirely appropriate for a young discipline such as behavioral toxicology. Although a historical perspective is not the intended purpose of these remarks, it is probably fair to say that behavioral toxicology as a formal discipline is less than 20 years old, in contrast to its counterpart, behavioral pharmacology, which has received extensive experimental attention over the past 40 or so years.

In fact, the necessity for and appropriateness of descriptive studies at this stage of the discipline relate specifically to the search for mechanisms of action.

Attempting to understand the mechanism(s) of a behavioral change is not a worthwhile pursuit unless a convincing demonstration can be provided that the toxicant induces a specific, well-defined reproducible behavioral effect and that the particular behavioral change represents a direct effect of the toxicant rather than an indirect manifestation of changes in some other behavioral functions. For example, changes in accuracy on a designated learning (acquisition) paradigm could result from a toxicant-induced change in sensory capabilities or in motor function, or in motivational status of the organism. In this case, determination of the mechanism of a presumed learning impairment would be both premature and misguided. Therefore, advances in the understanding of the behavioral effects of environmental chemicals are first and most importantly advanced through confirmation and a specific delineation of the nature of the behavioral impairment. Indeed, the more precisely defined the behavioral aberration, the greater the specificity with which studies of biochemical, cellular, molecular, and behavioral mechanisms of action can be directed.

II. Moving toward Mechanisms of Action

Despite the overriding perception that mechanisms of action in science are necessarily defined at a cellular or molecular level, mechanisms in fact can be defined at all levels of analysis of nervous system function, including the behavioral level. The rate, form, and pattern of operant behavior are determined by antecedent conditions (i.e., past experience) and by current stimulus conditions, including the consequences maintaining the behavior. An understanding of the behavioral mechanisms of the effects of the toxicant is sought through a determination of which of these and other controlling factors are modified by the compound. For example, has the toxicant altered the deprivation state (an antecedent variable), stimulus control (current stimulus conditions), or the perceived magnitude of reinforcement (consequence variable)? Assuming a single, necessarily molecular, mechanism of action for a behavioral effect may, instead, markedly oversimplify the nature of the behavioral change and result in the loss of significant information furthering our understanding of the relationship between behavior and the nervous system. Instead of capitalizing on the richly complex assortment of variables controling behavior, such a research tactic would inevitably freeze conditions and result in neurobiological accounts of behavior that lack generality.

Toxicant-induced changes in accuracy on a learning paradigm may again be a useful example in this context. As mentioned earlier, learning may be changed by altering any or all of the behavioral processes that contribute to its expression. In other words, learning impairments may have different or even multiple behavioral mechanisms of action, arising from changes in one or more sensory functions, i.e., deficits in environmental stimulus control. An inability to hear adequately can certainly retard learning in a classroom as well as in other situations. A similar problem arises with visual deficits. Alternatively, or additionally, changes in learning may be a result of altered attentional processes, quite a different type of impairment of appropriate control over behavior by environmental stimuli. For this reason, some learning and memory tasks include an obligatory response with which the subject initiates each trial to ensure it is attending to the relevant stimuli. Motor function deficits can likewise lead to apparent changes in learning and/or memory functions. Motor incoordination or weakness may increase the difficulty in executing the desired response. Increased motor activity may decrease the latency of responding, causing responses to occur before attending to relevant environmental stimuli. Motiva-

tional changes may also influence learning by changing the perceived quality or magnitude of the reinforcer. The ability to parcel out underlying behavioral processes necessarily means that an important principle in the selection of a test method should be flexibility in parametric variation of key parameters in order.

Schedule-controlled operant behavioral offers a standardized yet flexible paradigm for investigating behavior process and the behavioral mechanisms of actions of toxicants. Table 1 presents the essential elements of the paradigm along with the important controlling variables that have been identified (MacPhail, 1990). Also shown are the neurobiological processes that are accessible through manipulation of the controlling variables. In general terms, the paradigm involves presentation of a discriminative exteroceptive stimulus (S^D), in the presence of which responses are reinforced. Although put forth by Skinner 50 years ago, it still remains remarkable that this conceptually simple paradigm can offer a rich complexity of processes for experimental investigation. If a toxicant is suspected of altering sensory function, manipulations can be made in either the intensity or the modality of the discriminative stimulus. A disruption in operant behavior due to an auditory deficit, for example, may be lessened or magnified by appropriately changing the intensity of the auditory stimulus, and would not be expected to occur if a visual stimulus were used. Alternatively, if motivational changes were suspected, suitable manipulations of the magnitude and/or quality of the reinforcer should both mimic the effect of the toxicant and modulate the toxicant's effects on performance. Similar manipulations of any or all of the other controlling variables are possible and are limited only by the ingenuity

TABLE 1   Schedule-Controlled Operant Behavior
$(S^D \cdot R \rightarrow S^R)$

Controlling variables	Neurobiological function
1. S^D: Discriminative stimuli	
Qualitative	Sensory
Quantitative	
2. R: Response	
Topography	Motor
Force	
Ongoing rate	
3. S^R: Reinforcer	
Qualitative	Motivation
Quantitative	
4. $R \rightarrow S^R$: Reinforcement schedule	Learning and attention
Differing schedules	
Schedule parameter	
5. Historical variables	
Long-term	Memory
Short-term (context)	

and drive of the investigator to identify the processes by which toxicants affect behavior. In this regard, the approach is virtually identical to that used by the neurochemist in isolating, say, an enzyme under well-defined (controlled) experimental conditions for studies on the effects of a toxicant (cf. Dews, 1960).

A precise understanding of such behavioral mechanisms of action not only leads to a further elaboration of the experimental analysis of behavior, but also provides significant and critical guidance as to the direction of mechanistic studies at other levels of analysis of the nervous system. Knowing, for example, that changes in auditory function rather than in attentional processes are the source of learning impairments will render auditory system investigations a far more profitable direction for identifying neuroanatomical substrates than would investigations of hippocampal dysfunction. All too often, unfortunately, behavioral mechanisms are ignored in behavioral toxicological studies. Without such information, however, efforts aimed at delineating mechanisms of effects at a cellular or molecular level of analysis, including determination of brain region(s) involved in such effects, may well represent premature and even misguided efforts.

Achieving an understanding of behavioral mechanisms experimentally can be facilitated via efforts on at least two fronts. The first involves a more precise identification and elaboration of the specific behavioral processes contributing to the measure of performance in any particular behavioral paradigm. Several examples of such approaches can be mentioned. Passive avoidance procedures are often said to evaluate changes in memory function. The dependent variable is the time elapsing before an animal reenters a compartment of the apparatus in which it previously received foot shock. When toxicant treatment precedes behavioral testing, changes in pain sensitivity could conceivably alter the perceived magnitude of the shock using this paradigm. Similarly, increased motor activity levels may result in a more rapid entry into the compartment in which shock was delivered. An experimental assessment of the extent to which changes in time to reenter the shocked compartment reflects any or all of these behavioral processes must be undertaken. As another example, the rotarod is a rotating dowel on which mice or rats are tested. The time at which the animal falls off the rod is measured and presumed to reflect motor endurance (Cory-Slechta, 1989). Mice can be frequently seen, however, scrambling up the sides of the device, and even running backwards, causing them to fall off. These and other such behaviors are incompatible with staying on the rotarod and thus decrease the total time score, thereby contaminating the variable as a measure of endurance. Performance

in a water maze is also construed, depending upon experimental conditions, to represent a measure of learning or memory. In one widely used version of the water maze, the time and number of trials to find an escape platform hidden below an opaque body of water is measured. Aging-induced memory loss in rodents has been repeatedly reported using this paradigm, as reflected in a slower decay of the curve relating time to find the escape platform across trials. In a study investigating the possible mechanisms of this behavioral change, however, Lindner and Gribkoff (1991) found that this aging-related deficit arose from hypothermia due to the temperature of the maze water. In fact, when aged animals were warmed between trials in the water maze, there were no differences between young and aged rats. As such examples demonstrate, it should not be presumed a priori that an alteration obtained using a behavioral paradigm necessarily reflects specific changes in the behavioral function that paradigm purports to measure. Specificity must await experimental verification through the use of parametric modifications of the paradigm or, where possible, via microanalyses and/or the use of behavioral probe or challenge sessions, or alternatively through the use of other behavioral paradigms.

Microanalyses of operant behavior represent a second approach for facilitating the goal of precisely defining the nature of a behavioral deficit. Microanalysis refers to a finer-grained analysis of the behavior than is often provided in behavioral toxicology studies where only a single, and perhaps even a relatively global, measure of behavior is utilized. Analyzing the pattern of errors in a learning task provides one such example. Some authors have analyzed patterns of errors in a radial arm maze to attempt to further define the nature of memory impairments. Cohn and colleagues (1992) and Cohn and Cory-Slechta (1992), using detailed analyses of error patterns in the repeated learning paradigm, found that compounds that similarly decreased overall accuracy did so via substantially different patterns of errors, suggesting differences as well in the mechanisms by which overall accuracy decrements were achieved. These authors also used microanalyses of error patterns to show that changes in learning produced by low level lead exposure arise from perseverative behavior, much like those noted in response to the glutamate antagonist, MK-801 (Cohn et al., 1993).

An additional strategy for facilitating an elaboration of the specific nature of toxicant-induced behavioral changes can be achieved by superimposing behavioral probes on stable behavioral baselines. One example comes from the study by Mele and associates (1986) in which, after testing a series of fixed interval values in control and PCB-exposed monkeys, a fixed interval

60-sec schedule was arranged with reinforcement omitted after 25% of the intervals. This behavioral probe resulted in differential effects between PCB-treated and control monkeys, which were not apparent on the fixed interval schedule in the absence of reinforcement omission, thus suggesting potential PCB-induced alterations in consequence variables that control behavior. A more direct example is provided by Cohn and colleagues (1993) in an experiment that attempted to determine whether Pb-induced deficits in repeated learning were because of a loss of stimulus control. In that study, Pb-exposed rats exhibited decreased accuracy levels during the repeated learning but not during the performance components of a multiple repeated learning and performance schedule. The deficits noted during the repeated learning component included a perseveration or repetition of the response sequence that was appropriate during the performance component. This suggested that Pb-exposed rats did not attend to the changes in the light stimuli (on or off) that signaled the transition between the repeated learning and performance components of the schedule. To evaluate this possibility, a loud tone stimulus was added to the light stimuli signaling the component transitions during probe sessions, based on the contention that if deficits in stimulus control were the basis of learning impairments, then addition of the tone stimulus should improve performance of Pb-exposed rats to the same levels of accuracy exhibited by controls. There are, of course, numerous different possibilities for using various types of behavioral probes to address questions related to behavioral mechanisms of action (MacPhail *et al.,* 1983). Further efforts in this area are clearly warranted in behavioral toxicology research.

Complicating an understanding of mechanisms of action is the possibility that there may not be a single behavioral mechanism of action for a behavioral toxicologic effect; instead, there may be multiple behavioral mechanisms underlying a specific behavioral change. The same complexity may apply, however, to mechanisms at any other level of analysis of the nervous system, whether it be the physiological, cellular, or molecular level. In addition, it should be remembered that different behavioral effects produced by toxicant exposure may be the result of a single behavioral mechanism of action. Such potential complexities must necessarily be considered in defining mechanisms of effects.

III. Caveats for Behavioral Toxicology Research

One variable that has been found to, in some cases, rather dramatically modify both behavior and the effects of acute drug administration in behavioral pharmacological studies is prior history of reinforcement. Response-produced shock, for example, can evolve from behavior maintained by shock termination (e.g., Morse and Kelleher, 1977). Prior behavioral histories on various schedules of reinforcement are well known to influence response rates on subsequent reinforcement schedules (e.g., LeFrancois and Metzger, 1993). An accumulating literature also attests to the numerous conditions under which drug effects are modified by prior behavioral and/or pharmacological experience (e.g., Barrett, 1977; Nader *et al.,* 1992).

The potential for such confounds must be taken into consideration in both the design and the interpretation of behavioral toxicology studies that use experimental subjects repeatedly in different behavioral paradigms as well as in those cases in which between group designs are necessitated. Frequently, situations involving repeated use of the same subjects across behavioral paradigms involve nonhuman primates. The similarity of nonhuman primates as experimental models for humans is, of course, a decided advantage over the use of experimental species such as rodents. But the repeated use of the same subjects across behavioral paradigms, an approach dictated by economic and ethical considerations, raises the possibility of encountering problems related to behavioral history. This also occurs frequently where between groups designs must be employed for the assessment of some irreversible toxicants. In such studies, the behavior of groups of toxicant-treated and control animals may be evaluated, in a cohort fashion, on a series of behavioral baselines over time. If differences in performance on any of these behavioral baselines are noted, then these become differences in behavioral history that are carried over into the next behavioral baseline evaluated. At this point, and from there on, it becomes almost impossible to determine whether any group differences subsequently observed represent the effects of differences in behavioral history, an effect of the toxicant itself on behavior, or an interaction of the two.

A potential example of the importance of history has emerged in the literature describing the behavioral toxicity of lead. Levin and Bowman noted an impairment of delayed alternation in the first cohort of lead-exposed monkeys that were tested on this baseline (1986). A subsequent cohort, which had a different behavioral history, actually exhibited a facilitation of performance (Levin and Bowman, 1989). Similarly, Rice and Karpinski (1988) reported that Pb exposure adversely affected delayed alternation performance in a cohort of monkeys that likewise had a relatively extensive behavioral history. A subsequent cohort with a different behavioral history exhibited a marginal improvement in performance relative to control (Rice, 1992). A similar facilitation of delayed alternation performance was also noted in rodents exposed to lead that had no prior experience on any behavioral baseline (Cory-Slechta *et al.*, 1991). These discrepancies were not readily explained by differences in lead exposure levels and were attributed both by Rice (1992) and Cory-Slechta and colleagues (1991) to differences in behavioral history and, in the case of rodents with no prior history on another behavioral baseline, to the specifics of the behavioral methods used to train the rats for the delayed alternation baseline.

Obvious, but not particularly desirable, solutions to such a problem are the inclusion of additional experimentally naive subjects at the outset of testing on each new behavioral baseline, testing subjects in a counterbalanced order across these behavioral baselines, or to completely randomize behavioral history of reinforcement. The fact that performances on later behavioral baselines are equivalent between groups should not necessarily be taken as evidence that prior differences in behavioral history are no longer operative since behavioral pharmacological studies that have imposed different histories of schedule-controlled behavior have subsequently revealed differential drug effects even in the face of indistinguishable baseline performances (Nader *et al.*, 1992). It is clear that efforts are warranted to understand the role of behavioral history as a determinant of both performance and toxicant effects. Great care must be taken, in the meanwhile, in behavioral toxicology experiments to ascertain the extent to which prior history may play a role in the observations of toxicant-induced behavioral impairments.

A related problem can arise in between-group designs involving changes in parametric values of a behavioral paradigm when a fixed numbers of sessions are arranged prior to implementing the change. Consider an experiment in which groups of toxicant-treated and control animals are tested for a specified number of sessions on a discrimination task, after which a new discrimination task is introduced. Unless all animals in both the control and treated groups have achieved equivalent accuracy levels on the initial discriminaton, differences in behavioral history will be introduced as a confounding variable into the subsequent discrimination. It will be particularly problematic if the original discrepancy in accuracy values is treatment related, as it will be impossible to determine in any of the subsequent discriminations whether group differences reflect toxicant exposure or differences in behavioral history and their subsequent interaction with prevailing behavioral contingencies. Behavioral baselines that utilize

measures of accuracy as a dependent variable may be particularly vulnerable to this problem, but so will differences in response rate. Kreitzer (1980), for example, found that endrin-exposed quail made more errors than controls on a discrimination reversal task and that the differences increased exponentially with successive reversals.

While these caveats would seem to argue against the use of chronic, repeated or complex designs in behavioral toxicology experiments, in fact an enormous amount may be learned from such endeavors, given adequate experimental controls. These remarks are simply intended as cautionary advice in achieving the objectives of specifically and precisely defining the behavioral effects imposed by toxicant exposure. Furthermore, achieving these goals can only be facilitated by making use of advances in our understanding of behavior itself, such as in the experimental analysis of behavior, as well as in behavioral pharmacology, since toxicants and drugs have many commonalties in their actions.

The first series of chapters in this volume address many of the issues raised earlier as they relate to specific behavioral processes. The goal of these chapters is to provide useful guidance as to methodologies utilized in the analysis of many significant behavioral functions and to provide assistance in interpretation of such data. The second series of chapters extends to work aimed at furthering our knowledge of the relationships between behavior and its underlying neurochemical substrates, a goal which will facilitate our understanding of the interaction of toxicants with behavior.

References

Barrett, J. E. (1977). Behavioral history as a determinant of the effects of d-amphetamine on punished behavior. *Science* **198**, 67–69.

Cohn, J., and Cory-Slechta, D. A. (1992). Differential effects of MK-801, NMDA and scopolamine on rats learning in a four-member repeated acquisition paradigm. *Behav. Pharmacol.* **3**, 403–413.

Cohn, J., Ziriax, J., Cox, C., and Cory-Slechta, D. A. (1992). Comparison of the error patterns produced by scopolamine and MK-801 on a repeated acquisition and repeated transition baseline. *Psychopharmacology* **107**, 243–354.

Cohn, J., Cox, C., and Cory-Slechta, D. A. (1993). The effects of lead exposure on learning in a multiple repeated acquisition and performance schedule. *Neurotoxicology* **14**, 329–346.

Cory-Slechta, D. A. (1989). Behavioral measures of neurotoxicity. *Neurotoxicology* **10**, 271–296.

Cory-Slechta, D. A., Pokora, M. J., and Widzowski, D. V. (1991). Behavioral manifestations of prolonged lead exposure initiated at different stages of the life cycle. II: Delayed spatial alternation. *Neurotoxicology* **12**, 761–776.

Kreitzer, J. F. (1980). Effects of toxaphene and endrin at very low dietary concentrations on discrimination acquisition and reversal in bobwhite quail, Colinus Virginianus. *Environ. Pollution* **23**, 217–230.

LeFrancois, J. R., and Metzger, B. (1993). Low-response rate conditioning history and fixed-interval responding in rats. *J. Exp. Anal. Behav.* **59**, 543–549.

Levin, E. D., and Bowman, R. E. (1986). Long-term lead effects on the Hamilton Search Task and delayed alternation in monkeys. *Neurotoxicol. Teratol.* **8**, 219–224.

Levin, E. D., and Bowman, R. E. (1989). Long-term effects of chronic postnatal lead exposure on delayed spatial alternation in monkeys. *Neurotoxicol. Teratol.* **10**, 505–510.

Lindner, M. D., and Gribkoff, V. K. (1991). Relationship between performance in the Morris water task, visual acuity, and thermoregulatory function in aged F-344 rats. *Behav. Brain Res.* **45**, 45–55.

MacPhail, R. C. (1990). Environmental modulation of neurobehavioral toxicity. In *Behavioral Measures of Neurotoxicity.* (R. W. Russell, P. E. Flattau, and A. M. Pope Eds.), pp. 184–190. National Academy Press, Washington, D.C.

MacPhail, R. C., Crofton, K. M., and Reiter, L. W. (1983). Use of environmental challenges in behavioral toxicology. *Fed. Proc.* **42**, 3196–3200.

Mele, P. C., Bowman, R. E., and Levin, E. D. (1986). Behavioral evaluation of perinatal PCB exposure in rhesus monkeys: fixed-interval performance and reinforcement omission. *Neurotoxicol. Teratol.* **8**, 131–138.

Morse, W. H., and Kelleher, R. T. (1977). Determinants of reinforcement and punishment. In *Handbook of Operant Behavior* (W. K. Honig and J. E. R. Staddon, Eds.), pp. 174–200, Prentice-Hall, New Jersey.

Nader, M. A., Tatham, T. A., and Barrett, J. E. (1992). Behavioral and pharmacological determinants of drug abuse. In *The Neurobiology of Drug and Alcohol Addiction* (P. W. Kalivas and H. H. Samson, Eds.), *Ann. N.Y. Acad. Sci.* **654**, 368–385. The New York Academy of Sciences, New York.

Rice, D. C. (1992). Behavioral effects of lead in monkeys tested during infancy and adulthood. *Neurotoxicol. Teratol.* **14**, 235–246.

Rice, D. C., and Karpinski, K. F. (1988). Lifetime low-level lead exposure produces deficits in delayed alternation in adult monkeys. *Neurotoxicol. Teratol.* **10**, 207–214.

10

Behavioral Screening Tests: Past, Present, and Future

R. C. MACPHAIL
Neurotoxicology Division
Health Effects Research Laboratory
U.S. Environmental Protection Agency
Research Triangle Park, North Carolina 27711

H. A. TILSON
Neurotoxicology Division
Health Effects Research Laboratory
U.S. Environmental Protection Agency
Research Triangle Park, North Carolina 27711

All scientific work is incomplete—whether it be observational or experimental. All scientific work is liable to be upset or modified by advancing knowledge. That does not confer upon us a freedom to ignore the knowledge we already have, or to postpone the action that it appears to demand at a given time. (Sir A. B. Hill, 1965)

The development and application of behavioral screening tests has been an area of intense activity in recent years, and a considerable amount has been written on the topic (e.g., Alder *et al.,* 1986; Buelke-Sam *et al.,* 1985; Dews, 1978; Moser and MacPhail, 1990; Tilson and Moser, 1992). Screening generally refers to the development of preliminary information on the effect of a chemical on intact animals or some organ system (Zbinden *et al.,* 1984). Behavioral screening methods have long been applied in assessing the

effects of pharmaceutical and medicinal compounds. Behavioral screening methods have also been applied for some time to assess the effects of a number of environmental compounds. More recently, there has been a considerable increase in screening efforts in the field of neurotoxicology. This increase has been largely due to a growing recognition that the nervous system is vulnerable to a wide array of chemicals and that most existing chemicals have not been evaluated for neurotoxic potential (NAS, 1984).

The purpose of this chapter is to review behavioral screening methods and many of the issues surrounding their development and application. Some preliminary remarks regarding terminology and the purpose of screening are followed by some historical antecedents to contemporary screening research. Screens used today in neurotoxicology research are described next along with their general attributes and criteria for evaluating their utility. This is followed by coverage of topics that have not, in comparison, received much attention previously: quantitative analyses and interpretation, and incorporation of screening tests into a tiered-testing strategy.

The research described in this article has been reviewed by the Health Effects Research Laboratory, U.S. Environmental Protection Agency, and approved for publication. Approval does not signify that the contents necessarily reflect the views of the Agency nor does mention of trade names or commercial products constitute endorsement or recommendation for use.

Webster's Third New International Dictionary (1981) defines the verb screen "To examine usually methodically in order to make a separation into different groups," and screening as "The act of examining in order to make a separation into different groups." Since screening is undertaken to provide preliminary data on a chemical's effects, relatively simple observational techniques have so far played a predominant role in screening. It is instructive to note that observation is defined as "an act of recognizing and noting some fact or occurrence often involving the measurement of some magnitude with suitable instruments" (Webster's, 1981). Screening may therefore be construed as preliminary research to separate chemicals that have an effect of interest or concern from those that do not. Because of the preliminary nature of screening studies, some chemicals may be separated into a third category of having indeterminant or equivocal effects.

Behavioral screening tests are used to evaluate the functional status of an organism and the effects a chemical has on that functional status. Although functional assessments are now well established and widely used, their inclusion in toxicity testing evolved gradually. Early testing for toxicity was confined largely to a search for obvious effects: death, tumors, reproductive deficits, and birth defects. The prime characteristic of these early studies was postmortem evaluations of the structural integrity of organ systems. Structural abnormalities were considered unequivocal evidence of toxicity. It almost seemed that the in-life phase of chemical exposure represented a nuisance, albeit necessary, phase until organs could be collected and assessed. Any observations made during exposure were limited to obvious ones, for example, survival, body weight, and litter size. In 1951, Frazer commented on existing practices in toxicology and drew attention to the importance of more thorough, in-life evaluations:

Changes in functional activity which are equally as important [as histopathology] . . . are often not studied or are regarded as relatively unimportant. (p. 2)

Frazer (1951) continued:

Many investigators appear to take little notice of such functional changes as loss of appetite, nausea, vomiting, diarrhea, decreased activity, abnormal behaviour pattern, and other indications of changed function. (p. 5)

Barnes and Denz (1954) also commented on the importance of observations of the "behaviour of animals, including activity and functional pattern" (p. 195) in toxicity assessments. Ruffin (1963) further noted the general lack of appreciation of the importance of behavioral assessments in toxicology. In fact, since that time, many investigators have acknowledged the importance

of behavioral assessments in toxicity testing (e.g., Fox, 1977; Arnold *et al.*, 1977), although details of how these assessments should be made have generally been lacking.

In 1978, the Office of Pesticide Programs of the U.S. Environmental Protection Agency (EPA) provided a somewhat more detailed description of the clinical observations that should be made in evaluating the toxic potential of pesticides:

Cage-side observations should include, but not be limited to, changes in:
(A) The skin and fur;
(B) Eyes and mucous membranes;
(C) Respiratory system;
(D) Circulatory system;
(E) Autonomic and central nervous system;
(F) Somatomotor activity; and
(G) Behavior pattern.
(H) Particular attention should be directed to observation of tremors, convulsions, diarrhea, lethargy, sleep and coma.

This wording is remarkably similar to that found in Guideline No. 408 of the Organization of Economic Cooperation and Development (OECD) for evaluating the toxic effects of subacute exposures to chemicals. Similar but abbreviated wording is also found in van den Heuvel and colleagues' (1987) description of an interlaboratory study sponsored by the British Toxicological Society on the acute oral toxicity of several chemicals:

Cage-side observations should include changes in the skin and fur, eyes and mucous membranes, and also respiratory, circulatory, autonomic and central nervous systems, and somatomotor activity and behaviour pattern.

Collectively, then, these sources clearly state the need for determining the behavioral consequences of chemical exposures in laboratory studies. None of these sources, however, provide details on the types of observations or evaluations to make or how these should be carried out, analyzed, and reported (see also Sette and MacPhail, 1992).

Behavioral screening methods have long played a key role in pharmacology, particularly in assessing the effects of drugs acting on the central nervous system (CNS). Fowler and colleagues (1979), for example, identified a number of signs to monitor in drug studies, including ones that could be unobtrusively observed (e.g., urination, tremors), ones that could be scored following provocation (e.g., sensorimotor reflexes), and ones that could be monitored electronically (e.g., motor activity). Brimblecombe (1979) recommended a more thorough evaluation that progressed from observations of behavior and measurement of motor activity to assessment of conditioned behavior, modification of drug-induced behavior, and electrophysiological assessments. Brimblecombe (1979) noted that the progression involved increasing levels of technical complexity:

Some of these procedures are simple to carry out and involve the minimum of interference with the animal. Others are much more complex, involving specialized equipment and skills and often major surgical intervention. The latter are not appropriate for routine use in a toxicological laboratory but they undoubtedly have a role to play in detailed investigations of particular compounds. (p. 413)

Irwin (1968) described an extensive battery of tests for assessing the behavioral effects of drugs in mice, many of which involved interactive assessments (e.g., neuromuscular tone, sensorimotor reflexes). In many respects, the Irwin battery represented a formal elaboration of screening batteries that were extensively used in several CNS pharmaceutical laboratories and can be considered the basis for the screening batteries subsequently developed for assessing the effects of neurotoxic substances found in the environment.

Based on these early efforts, it is clear that the types of behavioral end points of concern in screening should include general appearance, autonomic signs, reactivity, coordination, muscle tone, and activity level. These early batteries varied, however, in terms of the specific tests as well as the extent of guidance provided for conducting the tests. As a consequence, although the screening batteries may have been highly standardized in individual laboratories, the overall degree of standardization and specification was generally low. Many reports submitted to the U.S. EPA, for example, provided no details of the clinical evaluations made in assessing an industrial chemical, and tersely concluded that no clinical signs were seen. A reader of such reports would be hard-pressed to state what sort of signs were evaluated, how they were evaluated, and whether the technical staff was skilled in detecting the signs if indeed they were present.

The relatively recent recognition of the importance of neurotoxicity testing has produced in many regards a fundamental change in how chemicals are assessed for potential health hazards. For example, the U.S. EPA promulgated in 1985 several guidelines for assessing the neurotoxic potential of industrial chemicals that included a functional observational battery (FOB) and an automated assessment of motor activity. Development of these testing guidelines was stimulated in part by recommendations made by a number expert committees for tests to screen chemicals for neurotoxic potential (see Sette, 1989; WHO, 1986). These recommendations have consistently identified FOBs and motor activity as useful tests for identifying the neurotoxic

potential of chemicals. In contrast to previous practices, the U.S. EPA guidelines provide a structured framework for carrying out behavioral evaluations, including appropriate tests and testing conditions and the use of explicit scoring criteria, as well as the use of positive control compounds in order to demonstrate testing proficiency.

The advantages of behavioral assessments in screening chemicals are enormous. First and foremost, behavioral assessments are noninvasive and do not require training, surgical intervention, or other preparation of the animals prior to assessment. Being noninvasive, behavioral tests allow repeated assessment of chemical-exposed animals over time. A much more complete description of the onset and progression of a disease state can be provided instead of a static account at the end of dosing. Behavioral assessments also permit determination of the recovery of function following exposure, as well as any compensation for compromised function that may occur during exposure (MacPhail, 1994). These features also permit integration of explicit behavioral assessments into a wide variety of ongoing studies, for example, in standard systemic toxicity studies and chronic bioassays. Indeed, there is perhaps no other end point in toxicology that is so versatile and accessible as the behavior of intact animals.

In many respects FOB and motor activity tests represent complementary approaches to behavioral assessment (Dourish, 1987). Tests comprising a FOB are subjective in that they depend on observation by technicians as well as direct interaction with the animal. Many of the FOB tests involve semiquantitative measurements, either noting the presence or absence of signs or the relative intensity of a sign. In addition, the FOB involves intensive data collection, with several diverse measures collected within a relatively brief period of time. Motor activity assessments, on the other hand, are automated and objective. As Maurissen and Mattsson (1989) noted,

Automation of motor activity tests has eliminated the potential problems associated with the interaction between the observer and the animal being tested. (p. 196)

Being automated, motor activity data involve ratio and interval scales of measurement, and therefore provide more sensitive and numerically rigorous descriptions of behavioral effects. Most motor activity assessments also entail relatively few measures collected over a longer period of time than with the FOB. The length of time motor activity is measured allows assessment of a chemical's effect on exploratory tendencies and an organism's habituation to the test environment.

Although the introduction of explicit behavioral tests into toxicological screening has been salutory, for the science as well as public health, many questions have arisen regarding their composition, analysis, and interpretation. A review of the literature indicates, for example, a diverse array of screening tests (see Tilson and Moser, 1992). A rudimentary screen might involve simple cage-side observations of undisturbed animals, noting whether they are behaving normally or not. This is the sort of screening battery that has dominated standard systemic toxicity studies. Some batteries provide more guidance on the types of end points to be evaluated. The battery described by Mattsson and co-workers (1989) includes about a dozen end points that are scored as increased, decreased, or unchanged in comparison to control animals. The batteries developed by Gad (1982), the U.S. EPA (1985), and Moser and colleagues (1988) involve approximately 30 end points and generally provide explicit scoring criteria. At the other extreme, the batteries described by Irwin (1968) and O'Donoghue (1989) involve in excess of 50 end points with explicit scoring criteria. Such a diversity inevitably raises fundamental questions regarding the comparability of the batteries and chemical effects.

Assessing the comparability of behavioral screening batteries can be accomplished by applying some basic criteria. Everyone would agree that a screening battery should be reliable, sensitive, and specific. In other words the battery should provide reproducible results within a laboratory over time as well as between laboratories. Until relatively recently (Buelke-Sam et al., 1985; Moser and MacPhail, 1992), interlaboratory comparisons of screening battery results had been almost nonexistent in neurotoxicology. Batteries should also be sensitive to detecting the toxicant effects of interest, and specific in allowing separation of compounds with the effect of interest from those without. Issues of sensitivity and specificity are best addressed through the use of model chemicals, including both positive and negative controls. Of course, the battery should also provide reproducible data on the effects of positive and negative control chemicals.

Additional criteria for evaluating behavioral screening batteries include efficiency and diagnostic power. To what extent, for example, can the results of behavioral screening be used to identify the neurobiological functions affected by a chemical and/or the neural substrates damaged? Increasing the number of end points evaluated in a screening battery may increase its diagnostic capability, but at the expense of efficiency. Given the large number of existing chemicals (NAS, 1984), as well as the number of new ones introduced each year, it seems that compromises will inevitably

have to be made in creating batteries that are both efficient and diagnostic.

Also important in evaluating the adequacy of behavioral screening batteries is the context in which the battery is to be applied. Mention was made earlier of the extensive use of FOBs in the CNS pharmaceutical industry. In this context, however, FOBs have been used to identify agents that may have an intended effect (e.g., analgesia) and to determine whether there may be unwanted side effects (e.g., respiratory difficulty). Synthetic chemists can provide researchers with an enormous number of new chemicals (and chemical series or classes) to screen for desirable and undesirable effects (Irwin, 1962; Boren, 1966). In the former case, accumulation of knowledge regarding chemicals having the intended effect may lead to refinement of the tests in a screening battery to more efficiently detect only the signs sought. In the latter case, the battery must provide a broad range of end points to ensure that the chemical does not produce undesirable side effects.

In contrast to the pharmaceutical industry, many chemicals found in the environment are developed for their usefulness in manufacturing or in the control of pests and not because of their possible effects on humans. Under these circumstances, behavioral screening batteries must remain broad in order to provide as much data as possible on the undesirable effects likely to be seen in exposed humans. Any modification of a test battery in toxicology will likely involve more tests instead of fewer tests. It is simply not possible at this time to adequately describe the range of possible signs, let alone patterns of signs, that may be produced by environmental chemicals. It is conceivable, however, that further experience in behavioral screening will eventually lead to the identification of many of the major patterns of effect produced by environmental chemicals. Concurrent accumulation of knowledge on structure–activity relationships could result in a more flexible approach to screening in which selected groups of tests are applied to evaluate particular chemicals or chemical classes.

Analysis of behavioral screening data poses unique opportunities and challenges. Behavioral screening tests often provide data that are quantal (i.e., a sign is either present or absent), ordinal (i.e., the sign is ranked for magnitude), or continuous (i.e., the magnitude of the sign is measured) in nature. In addition, data on many signs are collected in individual animals, often at several times relative to exposure. Comparatively little attention has been given so far to the analysis of behavioral screening data and, as a result, a variety of different approaches have been used. In considering these different approaches it is important to understand that screening studies differ from studies designed to more fully characterize the effects of a bona fide toxic substance. The preliminary nature of screening studies argues against the use of many established statistical practices. Take, for instance, the issue of correlated measures. With so many end points collected in the same organism, an isolated finding could easily be due to chance. Statistical reasoning might dictate correction of the α level for the number of tests conducted. It should be obvious, however, that this approach could lead to adverse consequences. Simply increasing the number of tests within a battery could result in a complete lack of significant effects, traditionally defined, whatsoever. An α of 0.05, divided by 50, is a very small number. Indeed, Gad (1989) has recommended loosening, rather than tightening, the α level in behavioral screening studies.

On the other hand, the sheer number of tests and test times increases the likelihood of finding effects that may not be directly related to chemical exposure. Further investigation of such isolated "false alarms" could be counterproductive and divert scarce resources from screening additional chemicals (Boren, 1966).

One approach to quantitative analysis, and a possible solution to the problem of false-positive results, has been suggested by Moser (1991). Screening tests are first grouped into plausible functional domains, for example, autonomic, neuromuscular, and sensorimotor. Data from individual animals for each of the tests are transformed to a four-point severity scale. Severity scores are assigned on the basis of data from control animals. A score of 1, for example, is assigned to a test result commonly observed in control rats, whereas a score of 4 is assigned to a test result rarely observed in control rats. Scores are then averaged for each rat's data from the tests comprising each domain and then averaged across rats within each treatment group. Statistical tests of significance are next applied to determine effective doses and times relative to dosing. Patterns of effect can then be described based on affected domains, with further statistical tests performed to determine the test results contributing to a significant domain effect. This type of analysis was applied successfully to the results of testing a relatively large number of chemicals (Moser *et al.*, 1995).

The potential advantages of this analytical approach lie in its synthesis of the large amounts of data ordinarily collected in behavioral screening. An isolated finding with a single test would not likely contribute to an effect at the domain level. For example, Moser and colleagues (1992) compared the effects of subchronic acrylamide and 2-hydroxyethyl acrylate. Acrylamide produced dose- and time-related ef-

fects on several of the tests comprising the neuromuscular domain. 2-Hydroxyethyl acrylate produced some effects on these tests, but no overall domain effect. Moser and colleagues (1992) concluded that 2-hydroxyethyl acrylate produced little evidence of neurotoxicity, a finding that was supported by the results of histopathological analysis.

Approaches to data synthesis and pattern analysis may offer considerable advantages in evaluating screening data. Isolated findings are not likely to contribute significantly to an overall domain effect, thereby reducing the likelihood of false-positive results. By the same token, small but consistent effects on many tests within a domain are likely to summate in a significant domain effect. It is possible, however, that a chemical may produce a specific behavioral effect that would be diluted in a domain analysis. For example, diethylaminopropionitrile produces a specific inhibition of urination (O'Donoghue, 1989). It remains to be determined whether such a specific effect would be robust enough to be revealed as an overall effect on autonomic function. Researchers should be encouraged to explore further a variety of ways to analyze behavioral screening data that could aid in decision making without unnecessary sacrifice of details on how chemicals affect behavior (e.g., Geyer, 1990; Koek *et al.*, 1987; Pryor *et al.*, 1983; Tamborini *et al.*, 1990).

Interpretation of behavioral screening data is a complex issue, especially in the context of neurotoxicology. No other issue can polarize the scientific community more quickly and reliably. Critics have argued that behavioral changes could occur in chemical-treated animals for a variety of reasons not necessarily related to an action on the nervous system. Proponents, on the other hand, have argued that the analysis of behavior has played a pivotal role for many years in evaluating the effects of centrally acting drugs as well as direct destruction of regions of the nervous system. Several strategies may prove useful in sharpening the interpretation of behavioral screening data.

Standardization of screening batteries results in a wealth of historical control data that grows with each new determination of the effects of a chemical. As a result, comparison of control data from an ongoing or recent experiment should give investigators insight into the comparability of current test conditions relative to those of the past and should aid in chemical data interpretation. For example, a significant test result within the range of historical control values, along with a spuriously low (or high) concurrent control value, should be viewed with suspicion. This is not to say the result should be dismissed, but a direct replication would certainly be appropriate.

Analyses of historic control data have so far received surprisingly little attention in neurotoxicology. Notable exceptions have been presented for motor activity by MacPhail and colleagues (1989) and Crofton and associates (1991). The analyses were undertaken largely in response to the concerns of critics over the reproducibility of motor activity assessments. Data were collected from several laboratories that routinely used motor activity assessments in research. Comparisons were made of the variability between rats in control groups as well as between control group means across experiments. Since most of the data had already been collected, no attempt was made to standardize many of the conditions (e.g., type of monitoring device, strain of rat) that could affect the outcome (Reiter and MacPhail, 1979; 1982). The results showed an impressive consistency in historical control values. Similar analyses of results from the literature showed an equally good, if not better, degree of consistency (Crofton and MacPhail, 1995).

Positive control data for reference substances also aid in interpreting behavioral screening data. Screening laboratories should be encouraged to test the effects of a number of positive control chemicals. The results of screening a new chemical could then be compared to determine whether it has effects like chemical X, Y, or Z. To the extent that considerably more is ordinarily known about the effects of a reference chemical, more accurate statements can be made about the toxic potential of the new chemical. Follow-up investigations could then be made to further establish the similarity of effects of the two chemicals, as well as delve into the mechanism(s) underlying the new chemical's effect on behavior.

Support for concluding a chemical's effect on behavior was due to an action on the nervous sytem would be enhanced by additional data on the structural, biochemical, and/or electrophysiological effects of exposure. Although multidisciplinary data of this sort are becoming increasingly important in neurotoxicology (e.g., Becking et al., 1993), either the absence of such data or a lack of correlation should in no way undermine conclusions regarding the potential toxicological significance of the behavioral test results. The multiplicity of levels of organization of the nervous system, along with our fragmentary knowledge of the mechanisms of action of most chemicals, does not auger well for the prospects of establishing these sorts of correlations, especially in a screening context. Screening organophosphates for their ability to produce delayed neurotoxicity in hens, for instance, is a well-established practice, but there has been little evidence of correlations between the clinical signs and structural damage in the peripheral nervous system or spinal cord produced by organophosphates.

Deciding whether a chemical's effect on behavior is due to an action on the nervous system should also involve evaluations of other organ systems (MacPhail, 1994). Establishing a correlation between the magnitude of a behavioral sign and the degree of damage to the liver or kidney would in all likelihood raise valid suspicions regarding the neurotoxic potential of the chemical. It is, however, ironic that this type of strategy has not been pursued so far, especially by critics of the use of behavioral tests in neurotoxicology. Evidence suggests, however, that there is little correlation among the systemic, developmental, and behavioral effects of a number of solvents, pesticides, and industrial chemicals (MacPhail et al., 1995).

In addition to the development and application of screening tests, several of the expert committees mentioned earlier also recommended that screening studies be carried out in a tiered approach to testing. Tiered testing appears to be logical, especially in light of the enormous number of chemicals and the economy of technical resources currently available for screening. Ideally, the results of screening studies would lead to more in-depth investigations of a chemical's effects and ultimately establish the chemical's mechanism(s) of action. Few attempts have been made, however, to establish explicit tier testing schemes. Although a wide range of behavioral tests is available for follow-up studies (e.g., Cory-Slechta, 1989; Peele and Vincent, 1989; Tilson, 1990, 1993), few attempts have been made in comparing the results of screening studies with those from in-depth investigations. These types of comparisons will, nevertheless, be indispensible in enhancing our understanding of the significance of behavioral screening data.

A related problem in tiered testing is the composition of tests in a screening battery. In-depth follow-up presupposes a sufficient number of nervous system functions have been adequately evaluated to guide further research. But what about functions not evaluated in a screening battery? Is that function assessed through follow-up on the chemicals shown to be positive in the screening battery or on those shown to be negative? Take, for instance, evaluations of acquired behavior (learning and performance). Current screens in neurotoxicology do not test for chemical effects on acquired behavior, even though acquired behavior reflects some of the most important functions of the nervous system of animals and humans. Anger (1990) reviewed a large number of studies of humans exposed to chemicals in the workplace and concluded that some of the most common deficits were found in tests of intelligence, memory, and spatial relations. These findings should raise concerns over the adequacy of current behavioral screening batteries for predicting the types of hazards likely to occur in humans. An especially pessimistic scenario involves a new chemical judged negative on the basis of screening data, only to find significant cognitive effects in humans following commercial introduction. No data are available to either refute or support this scenario. Systematic comparisons of acquired and naturally occurring behavior (e.g., Reiter *et al.*, 1981; Prior *et al.*, 1983; Peele, 1989; Moser and MacPhail, 1990) have, however, generally shown overlapping dose ranges for a number of chemicals. Nevertheless, the scenario cannot be discounted outright and should serve to stimulate efforts to incorporate tests of acquired behavior into screening batteries (NAS, 1984; see also Sette, 1989).

References

Alder, S., P. Candrian, J. E., and Zbinden, G. (1986). Neurobehavioral screening in rats: validation study. *Methods Find. Exp. Clin. Pharmacol.* **8**, 279–289.

Anger, W. K. (1990). Worksite behavioral research: results, sensitive methods, test batteries and the transition from laboratory data to human health. *Neurotoxicology* **11**, 629–720.

Arnold, D. L., Charbonneua, S. M., Zawidzka, Z. Z., and Grice, H. C. (1977). Monitoring animal health during chronic toxicity studies. *J. Environ. Pathol. Toxicol.* **1**, 227–239.

Barnes, J. M., and Denz, F. A. (1954). Experimental methods in determining chronic toxicity: a critical review. *Pharmacol. Rev.* **6**, 191–242.

Becking, G. C., Boyes, W. K., Damstra, T., and MacPhail, R. C. (1993). Assessing the neurotoxic potential of chemicals—A multidisciplinary approach. *Environ. Res.* **61**, 1–12.

Boren, J. J. (1966). The study of drugs with operant techniques. In *Operant Behavior: Areas of Research and Application* (W. K. Honig, Ed.), Appleton-Century-Crofts, New York.

Brimblecombe, R. W. (1979). Behavioral tests in acute and chronic studies. *Pharmacol. Ther.* **5**, 413–415.

Buelke-Sam, J., Kimmel, C. A., and Adams, J. (1985). Design considerations in screening for behavioral teratogens: results of the collaborative behavior teratology study. *Neurobehav. Toxicol. Teratol.* **7**, 537–673.

Cory-Slechta, D. A. (1989). Behavioral measures of toxicity. *Neurotoxicology* **10**, 271–296.

Crofton, K. M., Howard, J. L., Moser, V. C., Gill, M. W., Reiter, L. W., Tilson, H. A., and MacPhail, R. C. (1991). Interlaboratory comparison of motor activity experiments: implications for neurotoxicology assessments. *Neurotoxicol. Teratol.* **13**, 599–609.

Crofton, K. M., and MacPhail, R. C. (1995). Reliability of motor activity assessments. In *Motor Activity and Movement Disorders* (K. P. Ossenkopp, M. Kavaliers, and P. R. Sanberg Eds.), Elsevier Publishers, Amsterdam, in press.

Dews, P. B. (1978). Epistemology of screening for behavioral toxicity. *Environ. Health Perspect.* **26**, 37–42.

Dourish, C. T. (1987). Effects of drugs on spontaneous motor activity. In *Experimental Psychopharmacology* (A. J. Greenshaw and C. T. Dourish Eds.), Humana Press, Clifton, New Jersey.

Fowler, J. S. L., Brown, J. S., and Bell, H. A. (1979). The rat toxicity screen. *Pharmacol. Ther.* **5**, 461–466.

Fox, J. G. (1977). Clinical assessment of laboratory rodents on long term bioassay studies. *J. Environ. Pathol. Toxicol.* **1**, 199–226.

Frazer, A. C. (1951). Synthetic chemicals and the food industry. *J. Sci. Food Agric.* **2**, 1–7.

Gad, S. C. (1982). A neuromuscular screen for use in industrial toxicology. *J. Toxicol. Environ. Health* **9**, 691–704.

Gad, S. C. (1989). Principles of screening in toxicology with special emphasis on applications to neurotoxicology. *J. Amer. Coll. Toxicol.* **8**, 21–27.

Geyer, M. A. (1990). Approaches to the characterization of drug effects on locomotor activity in rodents. In *Modern Methods in Pharmacology*, Vol. 6, Wiley-Liss, New York.

Hill, A. B. (1965). The environment and disease: association or causation? *Proc. R. Soc. Med.* **58**, 295–300.

Irwin, S. (1962). Drug screening and evaluative procedures. *Science* **136**, 123–128.

Irwin, S. (1968). Comprehensive observational assessment: Ia. A systematic, quantitative procedure for assessing the behavioral and physiologic state of the mouse. *Psychopharmacology* **13**, 222–257.

Koek, W., Woods, J. H., and Ornstein, P. (1987). A simple and rapid method for assessing similarities among directly observable behavioral effects of drugs: PCP-like effects of 2-amino-5-phosphonovalerate in rats. *Psychopharmacology* **91**, 297–304.

MacPhail, R. C. (1992). Principles of identifying and characterizing neurotoxicity. *Toxicol. Lett.* **64/65**, 209–215.

MacPhail, R. C. (1994). Behavioral analysis in neurotoxicology. In *Neurobehavioral Toxicity: Analysis and Interpretation* (B. Weiss and J. O'Donoghue Eds.), Raven Press, New York.

MacPhail, R. C., Berman, E., Elder, J., Kavlock, R., Narotsky, M., Moser, V., Slicht, M., and Sumrell, B. (1995). An interdisciplinary approach to toxicological screening. IV. Comparison of results. *J. Toxicol. Environ. Health,* in press.

MacPhail, R. C., Peele, D. B., and Crofton, K. M. (1989). Motor activity and screening for neurotoxicity. *J. Amer. Coll. Toxicol.* **8**, 117–125.

Mattsson, J. L., Albee, R. R., and Eisenbrandt, D. L. (1989). Neurological approach to neurotoxicological evaluation in laboratory animals. *Toxicol. Indust. Health* **5**, 195–202.

Maurissen, J. P. J., and Mattsson, J. L. (1989). Critical assessment of motor activity as a screen for neurotoxicity. *Toxicol. Indust. Health* **5**, 195–201.

Moser, V. C. (1991). Applications of a neurobehavioral screening battery. *J. Amer. Coll. Toxicol.* **10**, 661–669.

Moser, V. C., Anthony, W. C., Sette, W. F., and MacPhail, R. C. (1992). Comparison of subchronic neurotoxicity of 2-hydroxyethyl acrylate and acrylamide in rats. *Fundam. Appl. Toxicol.* **18**, 343–352.

Moser, V. C., and MacPhail, R. C. (1990). Comparative sensitivity of neurobehavioral tests for chemical screening. *Neurotoxicology* **11**, 335–344.

Moser, V. C., and MacPhail, R. C. (1992). International validation of a neurobehavioral screening battery: The IPCS/WHO collaborative study. *Toxicol. Lett.* **64/65**, 217–223.

Moser, V. C., McCormick, J. P., Creason, J. P., and MacPhail, R. C. (1988). Comparison of chlordimefom and carbaryl using a functional observational battery. *Fundam. Appl. Toxicol.* **11**, 189–206.

Moser, V. C., Cheek, B. M., and MacPhail, R. C. (1995). A multidisciplinary approach to toxicological screening. III. Neurobehavioral evaluations. *J. Toxicol. Environ. Health,* in press.

National Academy of Sciences, (1984). *Toxicity testing: Strategies to determine needs and priorities.* National Academy Press, Washington, D.C.

O'Donoghue, J. L. (1989). Neurotoxicology. In *A Guide to General Toxicology* (J. K. Marquis, Ed.), 2nd ed., Karger, New York.

O'Donoghue, J. L. (1989). Screening for neurotoxicity using a neurologically based examination and neuropathology. *J. Amer. Coll. Toxicol.* **8**, 97–116.

Peele, D. B. (1989). Learning and memory: considerations for toxicology. *J. Amer. Coll. Toxicol.* **8**, 213–223.

Peele, D. B., and Vincent, A. (1989). Strategies for assessing learning and memory, 1978–1987: a comparison of behavioral toxicology, psychopharmacology and neurobiology. *Neurosci. Biobehav. Rev.* **13**, 317–322.

Pryor, G. T., Uyeno, E. T., Tilson, H. A., and Mitchell, C. L. (1983). Assessment of chemicals using a battery of neurobehavioral tests: a comparative study. *Neurobehav. Toxicol. Teratol.* **5**, 91–117.

Reiter, L. W., and MacPhail, R. C. (1982). Factors influencing motor activity in neurotoxicology. In *Nervous System Toxicology* (C. L. Mitchell, Eds.), Raven Press, New York.

Reiter, L. W., and MacPhail, R. C. (1979). Motor activity: a survey of methods with potential utility in toxicity testing. *Neurobehav. Toxicol. Teratol.* **1**, 53–66, [Suppl.]

Reiter, L. W., MacPhail, R. C., Ruppert, P. H., and Eckerman, D. A. (1981). Animal models of toxicity: some comparative data on the sensitivity of behavioral tests. In *Behavioral Consequences of Exposure to Occupational Environments,* Proc. 11th Conf. Environ. Toxicol. *11,* 11–23.

Ruffin, J. B. (1963). Functional testing for behavioral toxicity: A missing dimension in experimental environmental toxicology. *J. Occup. Med.* **5**, 117–121.

Sette, W. F. (1989). Adoption of new guidelines and data requirements for more extensive neurotoxicity testing under FIFRA. *Toxicol. Indust. Health* **5,** 181–194.

Sette, W. F., and MacPhail, R. C. (1992). Qualitative and quantitative issues in assessment of neurotoxic effects. In *Neurotoxicology* (H. Tilson and C. Mitchell Eds.), Raven Press, New York.

Tamborini, P., Sigg, H., and Zbinden, G. (1990). Acute toxicity testing in the nonlethal dose range: a new approach. *Reg. Toxicol. Pharmacol.* **12**, 69–87.

Tilson, H. A. (1990). Behavioral indices of neurotoxicity. *Toxicol. Pathol.* **18**, 96–104.

Tilson, H. A. (1993). Neurobehavioral methods used in neurotoxicological research. *Toxicol. Lett.* **68**, 231–240.

Tilson, H. A., and Moser, V. C. (1992). Comparison of screening approaches. *Neurotoxicology* **13**, 1–14.

U.S. Environmental Protection Agency, (1985). Toxic Substances Control Act Test Guidelines: Final Rule. Health Effects Testing Guidelines. Subpart G—Neurotoxicity. *Fed. Reg.* **50**, 39458–39470.

van den Heuvel, M. J., Dayan, A. D., and Shillaker, R. O. (1987). Evaluation of the BTS approach to the testing of substances and preparations for their acute toxicity. *Hum. Toxicol.* **6**, 279–291.

World Health Organization, (1986). Principles and methods for the assessment of neurotoxicity associated with exposure to chemicals. *Environmental Health Criteria Document 60.* World Health Organization, Geneva.

Zbinden, G., Elsner, J., and Boelsterli, U. A. (1984). Toxicological screening. *Reg. Toxicol. Pharmacol.* **4**, 275–286.

CHAPTER

11

Neurobehavioral Methods for the Evaluation of Sensory Functions

JACQUES P. J. MAURISSEN
The Dow Chemical Company
Midland, Michigan 48674

I. Introduction

Sensory systems are the necessary interfaces through which information is acquired. In neurotoxicology, they are important to consider for several reasons. First, they are end point measurements in themselves. Second, they are affected by neurotoxicants, sometimes in a preferential manner (compared to motor functions, for example). Finally, sensory integrity is usually a *sine qua non* condition for a meaningful interpretation of other higher order behavioral tests. In the presence of sensory deficits, the data interpretation of a great number of tests is more a matter of conjecture than of scientific statement.

The purpose of this chapter is to provide guidance in the development and application of methods for sensory evaluation in toxicology, to review the current state of research in sensory neurotoxicology, and to identify practices and principles that will advance this field of investigation.

This chapter will first give a brief history of psycho-physics, pointing out that most of the classical psychophysical methods in use today were designed more than 100 years ago. A wealth of practical and theoretical information is therefore available to whoever plans on testing sensory functions. The cardinal features of signal detection theory are then described, and its contributions to the field of psychophysics emphasized.

Because the occupational medicine literature on sensory testing is rife with confusion and inaccuracies, it was felt necessary first to summarize the psychophysical methods and to distinguish them clearly from the response paradigms used in the sensory evaluation process. The next section reviews some of the techniques available in animal psychophysics. Salient examples are then given of the use of sensory assessment in human and animal neurotoxicology.

Later, a number of technical and methodological flaws are reviewed and discussed. This exercise leads to an enunciation of general principles in the design and conduct of neurotoxicological studies. Finally, the chapter closes with some specific and general recommendations.

II. Historical Survey of Psychophysics

Since the halcyon times of the Greek hegemony, much has been written on sensory systems from a philosophical point of view (Boring, 1942). However, it was not before the 19th century that an experimental approach to the study of sensation and sensory organs evolved under the aegis of anatomists, physiologists, physicists, and philosophers. In 1822, François Magendie described the sensory function of the posterior spinal roots. In 1826, the German physiologist Johannes Müller gave a physiological meaning to the Aristotelian classification of the senses, and further developed and expanded the doctrine of the specific energies of nerves (first proposed by Charles Bell), i.e., a given nerve generates one and only one type of sensation.

German philosophy also affected the discourse on sensory phenomena. In 1809, Johann Friedrich Herbart was given Immanuel Kant's chair of philosophy in Köningsberg (Germany). He favored the application of mathematics to psychology, but denied the possibility of psychological experimentation. He developed the concept of limen (threshold) in the context of consciousness, which later became a fundamental concept (Herbart, 1824–1825).

Under the impetus of sensory physiology, Ernst Heinrich Weber (1795–1878), professor of anatomy and later of physiology at Leipzig (Germany), made critical contributions to the understanding and measurement of the tactual senses (Weber, 1834, 1846). In one of his most notable experiments, Weber showed that the smallest difference between two weights could be expressed as a ratio between the weights. In 1849, Hermann Helmholtz became professor of physiology at Köningsberg, Kant's old university, and performed a monumental series of experiments in the fields of vision and hearing.

By the middle of the 19th century, mental measurement was not a new idea. But it was Gustav Theodor Fechner's (1801–1887) originality to combine Herbart's concept of limen and use of mathematics in psychological measurement, with Weber's use of experimentation. Fechner's interests in physics led him to become lecturer, then full professor of physics at the University of Leipzig in 1833. Fechner also indulged in the study of philosophy. On October 22, 1850, while lying abed, he was reflecting upon the physical and mental worlds, and was struck by their basic harmony.

He was then laying the groundwork for what will be recognized later as his major achievement, the foundation of a new scientific discipline that united physical and psychological universes, and that he called psychophysics (from the Greek ψυχή, mind, and φυσικός, physics). In 1860, he published the *Elemente der Psychophysik* where he described the three fundamental psychophysical methods (method of limits, method of constant stimuli, and method of adjustment). In his book, he also stated that sensation is proportional to the logarithm of excitation: $S = k \log I$ (Fechner's law), where a sensation S is elicited by a stimulus of an intensity I and where k is a constant.

Fechner's law quickly attracted controversy. Hering, one of both Fechner's and Weber's students, pointed out that Fechner's logarithmic law could not be generalized to all sensory processes. Plateau, professor at the University of Gent (Belgium), conceived of an experiment where he had asked several subjects to paint cardboards with a gray exactly halfway between a white and black model. After photometric measurement, the brightness of the gray cardboard was approximately one-eighth of the white cardboard. Plateau provisionally concluded that, possibly, sensation is proportional to the cubic root of brightness. Plateau even gave his formula a more general expression: $S = k I^p$, where a sensation S is elicited by a stimulus I raised to a power $p < 1$.

Delboeuf, professor at the University of Liège (Belgium), also recognized some of the mathematical and physical difficulties of Fechner's logarithmic law. He pursued and expanded Plateau's work, but concluded that a logarithmic law (modified to prevent the occurrence of a negative sensation, among other things) instead of a power law better described the data (Delboeuf, 1873). Delboeuf (1883) did not revise his position following an experiment of Thompson Lowne who proposed that the sensation of brightness was proportional to the square root of the excitation (Thompson, 1877), another example of a power law.

Following the influence of the positivist philosophy of Auguste Comte and the development of objective psychology that John B. Watson called behaviorism (Watson, 1913), psychologists at Harvard started using the word "operationism" in the 1930s. Operationism is a technique that rejects the use of private experience and ethereal mentalism from the explanation process in psychology. It denied the use of metaphors in science, dealt only with public events, and defined scientific hypotheses in terms of concrete testable operations. Stevens became the leader of this new approach (Stevens, 1935). Another major achievement of Stevens' career was to revitalize the field of psychophysics and to expand it to magnitude estimation and cross-modality matching (Stevens, 1975). He revisited Fechner's logarithmic law and showed that equal stimulus ratios produced equal sensation ratios. The ratio invariance is the expression of a power function instead of a logarithmic function.

In the 1950s, psychophysics benefited from some new advances in signal detection theory (Green and Swets, 1974), which was developed by engineers who integrated statistical decision theory and electronic communications. More recently, under the influence of Duncan Luce, the concept of psychophysics has been examined in the framework of measurement theory and new mathematical models have been conceptualized (Baird and Noma, 1978; Falmagne, 1985).

III. Psychophysics and Signal Detection Theory

Classical Fechnerian psychophysics centered more around the probability that a "Yes, I hear the tone" response was made instead of around the probability that an observer truly detected a stimulus. Signal detection theory (SDT) broadened the scope of psychophysical measurement.

One of the most fundamental premises of SDT is that several factors can affect the observer's response. For example, whether or not one will be able to hear a phone ring cannot be answered only by knowing how "loud" the phone rings (Baird and Noma, 1978). How easily a phone ring (referred to as signal) can be detected partially depends on how loud the stereo is playing in the same room, i.e., it depends on the presence of other extraneous stimulations (referred to as noise). In SDT, "noise" refers to any background stimulation of any sensory modality other than the stimulus to be detected. Noise is always present, and the task of the observer is to decide whether a perceived event arose from noise alone or from a combination of noise and signal; in other words, detection depends on the signal-to-noise ratio.

Cognitive or decisional factors also affect the report of a detection:

1. expectation: an expected ring will have more chance to be detected than an unexpected one;
2. motivation: if missing a call can have disastrous effects, the probability of detecting it will be very high (payoff matrix).

Figure 1 summarizes the different outcomes in a SDT experiment. It shows the relations between the presence (+) or the absence (−) of a signal vs a response reporting detection (+) or no detection (−) of the signal. Conceivably, two observers can have the same

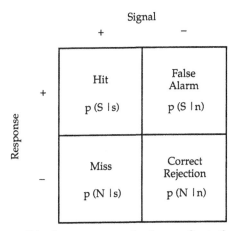

FIGURE 1 Stimulus response matrix where *s* refers to the presence of a signal added to the ambient noise, *n* to the noise alone (no signal), *S* to the response that the signal is believed to be present, and *N* to the response that the signal is believed to be absent.

sensitivity, but different response strategies. Suppose a conservative observer wants to be absolutely certain that a signal was presented before reporting the presence of the signal. In this situation, the number of misses and correct rejections will be relatively high. In contrast, under the same stimulus conditions, a liberal observer who wants to be sure not to miss a signal will most likely have a relatively high number of hits and false alarms. If such response strategies were to be ignored, the liberal observer would incorrectly appear to have a much lower threshold than a conservative one because of the higher number of hits.

The major contribution of SDT was to define strategies that can be used to separate an observer's true sensitivity from a host of nonsensory factors that are known to influence the observer's response. SDT refers to the pure measure of sensitivity as d' (i.e., an index of distinctiveness between "noise" and "noise + signal" distributions), and opposes it to β, a response criterion (i.e., an index of the observer's criterion or bias as affected by motivation, expectation, attitude, decision strategy, cost or benefit of a particular response, etc.). Psychophysical studies need to maximize strategies to evaluate observers' sensitivity in a manner relatively uncontaminated by response criteria.

IV. Psychophysical Evaluation

The way Fechner defined psychophysics (i.e., study of the relation between stimulus and sensation) was consonant with the widely held position that sensation could only be viewed as a subjective experience and could not be studied in animals (Bernard, 1856). A number of scientists categorically (but incorrectly) jettisoned the concept of sensory measurement when they wrote:

> The whole notion of measuring sensations numerically, remains in short, a mere mathematical speculation . . . (James, 1890, Vol. 1, p. 539)

> . . . there can be no question of an actual measurement of sensations. (Mach, 1906, p. 81)

> How much stronger or weaker one sensation is than another, we are never able to say. (Wundt quoted in James, 1890, Vol. 1, p. 534)

> We have no way of measuring sensation, in the sense that physical quantities are measured. . . . Consequently, there is no possibility of expressing sensation mathematically as a function of stimulus . . . (Moon, 1961, p. 537)

> Evidently, then, the idea of measuring sensation is utter naïveté. (Moon, 1961, p. 538)

Finally, and more recently, Stokinger (1974) also stated:

> Resort to animal experimentation instead of human eliminates one whole area of response, namely, the organoleptic or sensory response . . . (pp. 18–19)

The preceding statements are based on an introspective approach of sensations and clearly lead to an impasse. Sensation can only be objectively studied in terms of behavior or specific responses; psychophysics therefore is defined as the scientific study of the relationship between the physical dimensions of a stimulus and the behavioral response it elicits. This definition has the advantage of eliminating the introspectionistic, mental, and anthropomorphic approaches, which had hampered scientific endeavor.

Stevens (1958) best captured the essence of the psychophysics when he stated:

> What we can get at in the study of living things are the responses of the organisms, not some hyperphysical mental stuff, which, by definition, eludes objective test. Consequently, verifiable statements about sensation become statements about responses—about differential reactions of organisms. (. . .) I know nothing about your sensations except what your behavior tells me. (p. 386)

As a consequence, the realm of psychophysics has become quantitative in nature and therefore amenable to animal experimentation.

The psychophysical approach to the evaluation of sensory systems can be conceptually divided into three main domains:

1. Concept of threshold
2. Method for the presentation of stimuli in a prespecified order (psychophysical methods)

3. Codified response to the stimuli according to a specific format (response paradigms)

A. Concept of Threshold

At each stimulus level, there is a different probability that a stimulus will be detected, e.g., as stimulus intensity increases, so does the probability that the stimulus will be detected. The normal ogive (sigmoidal curve) serves as a model for the relationship between magnitude of stimulation and probability of detection. It is referred to as the psychometric function (also known as the poikilitic function), an example of which is given in Fig. 2. The psychometric function can be used to calculate the sensory threshold.

One of the objects of psychophysics is the determination of thresholds (or limens). The threshold refers to the dimension of a stimulus along a continuum that corresponds to a prespecified detection rate (usually the midpoint of the psychometric function). Two types of thresholds can be determined. When the subject has to report the presence or absence of a stimulus, the absolute threshold is usually defined as the point at which a stimulus has reached a magnitude such that it is detected on 50% of the trials (Fig. 2). When a subject is presented with two stimuli in a sequential or concurrent manner, the differential threshold is usually defined as the change in the stimulus magnitude that is detected on 75% of the trials. Because sensitivity of the organism to stimulation varies from moment to moment, the notion of threshold is best deemed statistical rather than physiological in nature.

The application of psychophysical methods is not restricted to these two measures, but these methods can also be applied to suprathreshold measurements to determine, for example, intramodal or cross-modal intensity matching (e.g., adjusting the intensity of a

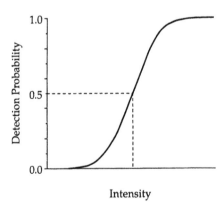

FIGURE 2 Psychometric function relating stimulus intensity with probability of a report of detection (usually represented by a cumulated Gaussian curve).

sound to make it subjectively equal to the intensity of a sound of a different frequency, or adjusting the intensity of a light to make it subjectively equal to a reference sound). Other quantitative methods have also been devised to measure sensation magnitude in humans (Corso, 1967; Gescheider, 1988) and in animals (Moody, 1970; Moody and Stebbins, 1982; Pfingst *et al.*, 1975).

B. *Psychophysical Methods*

Psychophysical methods describe the order in which stimuli of different intensities are presented to the observer (e.g., random, ascending, descending, or any other order). A number of psychophysical methods are available for the determination of absolute and differential thresholds.

1. Method of Limits

The method of limits (also called the method of just noticeable differences) consists in presenting a series of stimuli in sequence along a continuum, starting either above or below the threshold that is to be determined (Fig. 3). In a descending series, the first stimulus presented clearly exceeds the detection level; then the following stimuli are decreased along a continuum (e.g., intensity, frequency, etc.) until the observer reports lack of stimulus detection. In an ascending series, the first stimulus is chosen below the assumed detectability threshold and is increased along the same continuum until the observer reports detection of the stimulus. Ascending and descending series should be equal in number. The midpoint (or some other interpolated point) between the last two stimulus levels of each series is calculated, and the average of these calculated midpoints is reported as the threshold.

The method of limits is prone to several sources of error:

1. The observer may prematurely change the response from "no detection" to "detection" in view of the number of stimuli already presented in any particular stimulus series (error of anticipation).

2. The observer has a propensity to repeat the preceding response (error of habituation) and is, therefore, more likely to keep reporting "no detection" in an ascending series and "detection" in a descending series.

3. The observer may tend to count the number of trials needed to go from no detection to detection if the series always starts at the same stimulus level. This strategy could be used by the observer as an incidental cue to judgment and could artificially decrease threshold variability.

With the method of limits, it is therefore important to alternate ascending and descending series, and to change the starting point of each series of stimuli.

2. Method of Constant Stimuli

This method (also called method of right and wrong cases) does not incorporate some of the potentials errors inherent to the method of limits. On the basis of preliminary exploration, a number of stimulus values, equally spaced, are chosen so that they will likely encompass the expected threshold. Instead of being given in a serial order, all stimuli of different magnitudes are presented in a random order (without replacement) within each set of trials (Fig. 4).

A psychometric function can be derived from the percentage of correct detections at each stimulus level. The threshold is calculated as the stimulus magnitude corresponding to a prespecified correct detection rate (most often by interpolation between two levels). This method of constant stimuli has an important disadvantage, however, compared to the method of limits and some other methods: all predetermined stimulus levels must be tested many times before a sensory threshold can be determined with some accuracy, but the extreme levels (i.e., those that are always or never de-

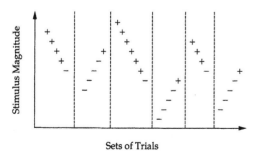

FIGURE 3 The method of limits where the stimuli are presented in a descending and ascending order and where the report of the presence or absence of a signal is indicated by "+" or "−", respectively.

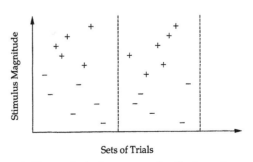

FIGURE 4 The method of constant stimuli where the report of the presence or absence of a signal is indicated by "+" or "−", respectively.

tected) increase the testing time and do not typically contribute to the threshold calculation.

3. Method of Adjustment

The method of adjustment (also called method of average error) is a variation on the method of limits, and, like this method, can be used to determine a threshold without the need for establishing a psychometric function. Typically, the observer, as opposed to the experimenter, takes active participation and has full control over one of the dimensions of the stimulus to be evaluated (Békésy, 1947).

The observer is requested to adjust one dimension of a stimulus to reach some criterion (e.g., to make the stimulus barely detectable or barely undetectable, or to make it equal to a comparison stimulus). The average magnitude of the settings determined by the observer represents the threshold. One of the advantages of the method of adjustment is that it provides a rapid way of measuring thresholds, and therefore best applies when dealing with a quickly changing threshold. However, the method of adjustment is extremely sensitive to bias, and the test results can be influenced more by nonsensory than by sensory determinants. One of the primary uses of this method is to establish a range of stimulus values to be used with other methods.

4. Adaptive Methods

The adaptive methods (also called tracking methods) postdate Fechner, but are variants of his method of limits. The stimulus value on any trial is determined by both the stimulus value and the observer's performance on the preceding trial. These techniques do not assume the existence of a psychometric function.

a. Simple Up–Down Method Historically, this method (also called staircase method) arose in tests of explosives. If the observer detects the presence of the stimulus, the stimulus magnitude is decreased at the next trial; if the observer does not detect the presence of the stimulus, its magnitude is increased at the next trial (Dixon and Mood, 1948). A sequence of stimulus changes in one direction is referred to as a run. The point at which the direction of a run changes is called a reversal. From trial to trial, the step size remains constant (on any chosen scale), and testing is stopped when a predetermined number of trials or runs is reached (Fig. 5). An even number of runs is recommended for the simple (as well as transformed; see the following section) up–down methods to minimize some of the biases listed earlier for the method of limits.

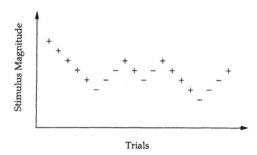

FIGURE 5 The *simple* up–down method where the report of the presence or absence of a signal is indicated by a "+" or a "−", respectively. Each correct response is followed by a signal decrease, and each incorrect response is followed by a signal increase.

Because this method automatically concentrates testing around threshold, it typically requires less time than the methods of limits or the method of constant stimuli. To annihilate expectations on the direction of a change in a stimulus dimension, a further refinement of the staircase method is the double staircase method where two staircases, one starting above and the other starting below threshold, are interweaving (Cornsweet, 1962).

With the simple (as well as transformed) up–down rule, the threshold is typically estimated either by averaging all stimulus levels from the second to the last run (Brownlee *et al.,* 1953) or by averaging peaks and valleys (or reversals) for the study (Wetherill and Levitt, 1965).

The up–down rule converges on some point such that the probability of increasing the stimulus value, P, equals the probability of decreasing it, 1-P. Therefore, the average value obtained for threshold happens to correspond to the stimulus level that is detected in 50% of the trials (for a demonstration, see part A of Table 1).

b. Transformed Up–Down Methods The transformed up–down methods are an extension of the simple up–down method. Observations are placed around the stimulus level such that points other than those corresponding to the 50% correct detection point can be derived in a manner similar to the threshold calculation used with a simple up–down method. For example, the following stimulus presentation strategy has been used in a number of studies: whenever an observer is incorrect, the stimulus value is increased, but it decreases only if two consecutive correct responses are given (Fig. 6). Although the step size may change initially (for logistics reasons), it has to remain the same during the trials used for threshold determination.

Table 1 (part B) shows that with this rule, the threshold corresponds to the stimulus level correctly detected

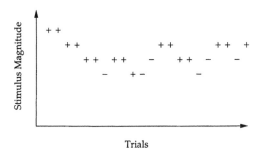

FIGURE 6 A *transformed* up–down method where the report of the presence or absence of a signal is indicated by a " + " or a "−", respectively. Two consecutive responses are followed by a signal decrease, and each incorrect response is followed by a signal increase.

in 70.7% of the cases. Example C illustrates a rule where three consecutive responses are required to decrease the stimulus level, whereas a single incorrect response is enough to increase it. Further details on these procedures can be found in Wetherill and Levitt (1965). More complex computerized strategies based on probabilities allow for a threshold estimation corresponding to any specific correct detection rate (e.g., 75%).

c. Parameter Estimation by Sequential Testing (PEST) The concept of the PEST methods originated in statistical sequential sampling analysis (Wald, 1947). PEST methods were developed by Taylor and Creelman (1967) and have been revisited on several occasions since (Findlay, 1978; Pentland, 1980; Shelton *et al.*, 1982; Watson *et al.*, 1983).

The PEST is a variation of the up–down method. However, it uses a variable step size (that either in-

creases or decreases as a function of the response to the previous stimulus) and a variable number of trials (or reversals) to determine the location of the threshold as efficiently as possible. The session ends when a selected criterion has been met, e.g., when some predetermined minimum step size has been obtained. In general, these methods require a smaller number of trials for threshold calculation.

C. Response Paradigms or Procedures

While the previous section dealt with methods for stimulus presentation, this section deals with another important element in sensory threshold determination, i.e., the type of response given by the observer to indicate detection or no detection of the stimulus. The value of psychophysical measurement strongly depends on the type of response selected.

1. Yes–No Paradigm

This procedure consists in asking the observer whether the stimulus of interest was detected. The subject answers "Yes" or "No," orally or through the aid of a manipulandum (this paradigm can be used in animals). No other response is usually accepted in a simple detection task. This method is called the "yes–no paradigm" (or "yes–no procedure"). Each test trial consists in the presentation of one observation interval, during which the stimulus under test is or is not presented. The observer gives a "Yes" or "No" answer, then receives feedback about the correctness of the answer. Schematic representation of a trial is presented in Fig. 7. The threshold is defined as the point on the stimulus scale that corresponds to the

---

TABLE 1 Up–Down Simple and Transformed Rules

Stimulus sequence	Symbolic representation	Probability of changing level[a]	Condition of convergence[b]	Probability of correct response at convergence
A. One correct R^c ⇒ decrease S^d	+ ⇓	P	$P = 1 - p$	$P = 0.5$
One incorrect R ⇒ increase S	− ⇑	$1 - P$		
B. Two consecutive correct R ⇒ decrease S	+ + ⇓	P^2	$P^2 = 1 - P + P (1 - P)$	$P = 0.707$
One incorrect R ⇒ increase S	− ⇑	$1 - P$	∴	
	+ − ⇑	$P (1 - P)$	$P^2 = 1 - P^2$	
C. Three consecutive correct R ⇒ decrease S	+ + + ⇓	P^3	$P^3 = 1 - P + P (1 - P)$	
	− ⇑	$1 - P$	$+ P^2 (1 - P)$	
One incorrect R ⇒ increase S	+ − ⇑	$P (1 - P)$	∴	$P = 0.794$
	+ − ⇑	$P^2 (1 - P)$	$P^3 = 1 - P^3$	

[a] P = probability of a correct response; $1 - P$ = probability of an incorrect response.
[b] The series converges or stabilizes at the level where the sum of the probability of a decrease equals the sum of the probabilities of an increase.
[c] R is response.
[d] S is stimulus level.

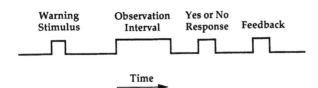

FIGURE 7 A trial of the yes–no paradigm.

midpoint on the psychometric function, i.e., to the 50% "detection" as measured by the number of corrected "yes" responses in the yes–no paradigm (Fig. 2).

The yes–no procedure elicits an introspective report. The phenomenological nature of the response makes this procedure highly susceptible to criterional shifts or biases, unconscious suggestion, and outright malingering. For example, nothing prevents a subject from reporting "no" even if the signal was felt ("miss") or reporting "yes" even if the signal was not felt ("false alarm").

However, the use of trials where no stimuli are presented (also called "catch" or "blank" trials) allows the experimenter to "correct" the threshold calculation for the "liberal" attitude of the subject who reports the presence of the signal in its absence. Unfortunately, the yes–no paradigm is often used without catch trials (or without enough of these) so that the subject's criterion cannot be evaluated and the data are influenced in an unknown fashion. To minimize the effects of this bias on threshold, there should be approximately the same number of trials with and without signal, and a correction procedure that adjusts the observed proportion of hits according to the observed proportion of false alarms should be used to obtain an estimate of the true proportion of hits (Green and Swets, 1974, p. 129). Table 2 illustrates how the actual threshold changes with observed proportions of false alarms of 0, 10, 20, or 30%.

When the yes–no paradigm is used in conjunction with the method of limits without "catch trials" and/or without correction, sensory effects cannot be dissected out from the observer's personal criterion or bias, and it is not possible to determine how much the "reported" sensory threshold is affected by extraneous nonsensory variables.

2. Forced-Choice Paradigm

In the forced-choice paradigm (contrary to the yes–no paradigm), more than one observation interval (also referred to as alternative) is presented, and the observer is forced to choose an alternative among the several alternatives presented. This method is referred to as the forced-choice paradigm or procedure. The stimulus is always presented in one of several alternative locations (spatial forced choice) or during one of

TABLE 2 Corrected Proportion of Hits (%) as a Function of False Alarms[a]

Stimulus levels	Observed proportion of hits (%)	Observed proportion of false alarms			
		0%	10%	20%	30%
10	100	100	100	100	100
9	100	100	100	100	100
8	90	90	89	88	86
7	70	70	67	63	57
6	60	60	56	50	43
5	40	40	33	25	14
4	35	35	28	19	7
3	30	30	22	13	0
2	10	10	0	—	—
1	0	0	—	—	—
Stimulus level threshold		5.5	5.7	6	6.5

[a] According to the formula $p^*(S \mid s) = [p(S \mid s) - p(S \mid n)]/[1 - p(S \mid n)]$, where $p^*(S \mid s)$ is the true (corrected) proportion of hits, $p(S \mid s)$ is the observed (uncorrected) proportion of hits, and $p(S \mid n)$ is the proportion of false alarms. The threshold is expressed in stimulus level units as a function of false alarms and is calculated by linear interpolation (when needed).

several time intervals (temporal forced choice). Each test trial typically consists in the presentation of a warning stimulus followed by two or more observation intervals. The subject is then forced to choose the alternative containing the appropriate signal and is given some feedback concerning the correctness of the response. The importance of feedback has been stressed by Lukaszewski and colleagues (1962). A schematic representation of a two-alternative forced-choice paradigm is given in Fig. 8.

Whereas the yes–no procedure leads to an introspective process, no phenomenologic report is sought in the forced-choice procedure. This forced-choice paradigm has been preferred by many because of its relative insensitivity to personal bias or criterion change. In contradistinction to the yes–no procedure, an objective reference point (i.e., a standard) is provided for the forced-choice procedure: while the comparison stimulus is always presented with the standard in the forced-choice procedure, the subjects have to use their own "internal" standards in the yes–no procedure. Forced-

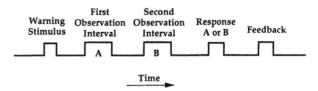

FIGURE 8 A trial of the forced-choice paradigm.

choice paradigms most effectively circumvent criterion problems.

Conceptually, the threshold in a two-alternative forced choice is calculated similarly to the yes–no procedure. It represents the stimulus dimension corresponding to the point of maximal uncertainty on the psychometric function. In a two-alternative forced choice, this function starts at 50% because this point represents the expected percentage of correct responses due to chance (Fig. 9). The threshold is defined as the point on the stimulus scale that corresponds to or is close to the midpoint on the psychometric function, i.e., to the 75% correct as measured by the number of correct responses. In other words, the cumulative distribution of the percentages of correct responses is normal about the 75% rate for the two-alternative forced choice (Fig. 9) and about the 50% rate for the yes–no procedure (Fig. 2).

3. Rating Paradigm

In the rating procedure, a signal or no signal is presented to the observer who is instructed to report the likelihood of its presence or absence on a several-point scale. The procedure of a rating scale experiment follows that of a yes–no paradigm. However, the answer is more than binary. This procedure has only been applied to humans. However, although it has a long history, it has seldom been used (if ever) in the context of toxicological assessment.

V. Techniques in Animal Sensory Evaluation

A number of techniques have been devised to evaluate sensory functions in animals, ranging from simple observations to rather sophisticated computerized conditioning methods. They can be divided into unconditioned and conditioned responses.

A. Unconditioned Responses

1. Observations

Rats have been evaluated for gross auditory deficits by watching their reactions to a sound in a quiet surrounding. A movement of the head or the body (acoustic startle), or the ears (Preyer's or pinna reflex) coincident with the sound has often been taken as evidence that the sound was detected. Mattsson and colleagues (1992), however, reported on a study where the startle response to a sharp noise did not reveal ototoxic effects of exposure to high levels of an aromatic hydrocarbon. However, the same rats displayed severely disrupted (or missing) auditory brainstem responses at middle and high frequencies. Similarly, the pinna reflex cannot be used as a test of hearing functions. Under some circumstances, the pinna reflex can be elicited even when there is severe hearing loss. Furthermore, the absence of this reflex does not necessarily mean that the animal has a pronounced hearing impairment (Anderson and Wedenberg, 1965). Use of the pupillary light reflex is an equally inappropriate test of visual function (Conquet *et al.*, 1979).

The main advantage of simple observational methods is that they require no training of the subject, therefore minimizing the time for evaluation. However, quantification of both stimuli and responses is usually rather crude. Determining the limits of sensory capacities is clearly outside the scope of such methods.

A number of more complex observational procedures have been devised to assess specific sensory deficits. The visual cliff apparatus, for example, was designed by Walk and Gibson (1961) to measure visual depth perception. This device uses the principle of a drop-off. The animal is placed on a center board or platform and can choose between a short drop-off on one side and a long drop-off on the other side. A sheet of glass is placed under the center board and is extended on both sides across the shallow and deep drop-offs to equalize visual (e.g., reflections) and tactual cues on both sides of the center board. Several models of visual cliffs have been proposed. Confounders, such as light reflection in the glass, vibrissae's tactile sensitivity, or echolocation cues, for example, have to be considered before making a statement about depth perception.

The optokinetic drum has been used to test some visual functions by taking advantage of some innate reflexes (Smith and Bojar, 1938). Typically, the animal is placed in the center of a vertical drum. Visual stimuli located inside the drum are presented as vertical grat-

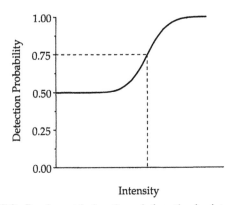

FIGURE 9 Psychometric function relating stimulus intensity with probability of a report of detection for a two-alternative forced-choice paradigm.

ings and move in one direction. The animal's response consists in moving the head or eyes in the direction of motion (optokinetic or optomotor nystagmus) when the stimulus is detected. This response is not usually easy to quantify in this context. A more complex apparatus has allowed for the evaluation of spectral sensitivity as well as spatial contrast sensitivity. In such a device, the stimulus of interest travels in one direction whereas a standard grating of adjustable contrast travels in the opposite direction. The amount of contrast of the standard grating necessary to match the stimulus of interest can be compared under different stimulus conditions. However, calibration and engineering difficulties cannot be underestimated (Wallman, 1975).

A variety of pain sensitivity assessment devices and procedures originated from the need in psychopharmacology to have a model for testing analgesic drugs. In 1954, Bianchi and Franceschini used the tail-clip method developed by Haffner (1929). They enclosed the arms of an artery clip in thin rubber and applied the clip to the base of the mouse's tail. While control mice tried to dislodge the clip, mice treated with analgesics did not react to it. A better quantitative system incorporating a mercury manometer was assembled by Green and colleagues (1951) to produce caudal compression. Woolfe and MacDonald (1944) invented the hot plate in which a small cage with a metal floor was electrically heated usually to 55°C. Time to paw licking was the dependent variable purported to reflect pain. Ben-Bassat and colleagues (1959) and Janssen and associates (1963) measured the time needed before mice or rats raised their tail out of a hot water bath. Later on, Pryor and co-workers (1983) and Hunskaar and colleagues (1986) also used a similar system to induce a tail withdrawal reflex in rats. Electrical stimulation of the rat's feet through a grid floor was used by Evans (1961) to evoke a flinch/jump response.

2. Reflex Modulation

Hoffman and Ison (1980) defined "reflex modulation" or "reflex modification" as ". . . *the phenomenon whereby the reflex elicited by one stimulus is modified by the prior presentation, withdrawal, or change of another (usually weaker) stimulus*" (p. 175). Reflex modulation is present the first time the modifying stimulus is applied; therefore, it is not the expression of any learning process. Figure 10 illustrates the principle of reflex modulation.

As early as 1863, Sechenov showed that reflexes of frogs to painful cutaneous stimuli could be inhibited by stimulation of the brain. One of the first applications of reflex modulation to the evaluation of auditory sensitivity was demonstrated by Yerkes (1905) who showed that ringing a bell before tapping on a frog's head modi-

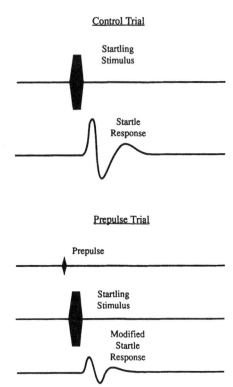

FIGURE 10 Representation of a startle reflex (top) and prepulse inhibition of the same reflex or reflex modulation (bottom).

fied the flexor jerk reflex. Yerkes therefore concluded that frogs could hear.

More refined reflex modulation techniques have been used in a number of laboratories. For example, Young and Fechter (1983) studied hearing in rats and guinea pigs with the method of constant stimuli. The reflex-eliciting stimulus was a 20-msec 115-dBA white noise burst, and the reflex modifying signal was a 100-msec pure tone of several intensities and frequencies. Control trials were mixed with prepulse (stimulus) trials. A control trial consisted in the presentation of the startle-eliciting stimulus alone. In a stimulus trial, presentation of a prepulse preceded presentation of the reflex-eliciting stimulus (Fig. 10). A psychometric function relating relative startle amplitude to prepulse intensity was generated for each frequency. These data were comparable to those collected with operant procedure in rats by Kelly and Masterton (1977) and in guinea pigs by Prosen and colleagues (1978).

The sensory threshold is usually defined as the lowest intensity of the reflex-modifying stimulus that produces a statistically significant reduction in the mean amplitude of response to the reflex-eliciting stimulus.

One of the advantages of reflex modulation is that it can be performed in a short period of time. A number of factors can alter the amplitude of the startle reflex:

1. parameters and nature of the stimulation (Blumenthal, 1988),
2. circadian rhythms (Horlington, 1970; Davis and Sollberger, 1971; Ison *et al.*, 1991),
3. food and water deprivation (Fechter and Ison, 1972),
4. duration of the lead interval between the startle-modifying and a startle-eliciting stimuli (Hoffman and Ison, 1980),
5. processes of habituation and sensitization (Moyer, 1963), and
6. background stimulation present at the time of or prior to startle elicitation (Davis, 1974; Stitt *et al.*, 1976).

Reflex modulation has been demonstrated in a number of animal species, such as humans (Krauter *et al.*, 1973), rabbits (Ison and Leonard, 1971), guinea pigs (Young and Fechter, 1983), rats (Hoffman and Searle, 1965; Pilz *et al.*, 1987), mice (Storm *et al.*, 1981), and pigeons (Stitt *et al.*, 1976). The method of constant stimuli (Hoffman and Searle, 1968) and the tracking method (Hoffman, 1984) have been used in conjunction with reflex modulation.

While reflex inhibition due to a prepulse unequivocally reflects detection of the prepulse, the absence of a reflex modification can be due to a series of events besides lack of detection, such as effects on the motor or central nervous systems. Methods are, however, available to distinguish between sensory vs motor effects. While absolute startle amplitude during control trials reflects an effect of the test compound on the motor system, sensory detection thresholds are quantified by relating relative startle inhibition to prestimulus intensity (Fechter and Young, 1983).

A more extensive coverage of reflex modulation can be found in Hoffman and Ison (1980), Ison and Hoffman (1983), Hoffman (1984), Fechter and Young (1986), Crofton (1990, 1992), and Crofton and Sheets (1989). A thorough review of the mammalian startle response can be found in Davis (1984) and in Davis and associates (1982).

B. Conditioned Responses

1. Classical Conditioning

Pavlov (1927) opened some new avenues for the assessment of sensory functions in animals. An unconditioned stimulus (i.e., capable of provoking a reaction by itself) is followed by an unconditioned response (or reflex). The principle governing conditioned reflexes is the following: when a neutral stimulus (i.e., that has no observable effects by itself) precedes an unconditioned stimulus and is repeatedly paired with it, the once neutral stimulus becomes capable of inducing alone the response previously associated with the unconditioned stimulus, and is referred to as a conditioned stimulus. This method of pairing is called Pavlovian or classical conditioning. Pavlovian conditioning has lent itself to a series of approaches that have enriched the field of sensory testing.

a. Method of Contrasts The specialization (as opposed to generalization) of a conditioned reflex is evidence of a discrimination process that Pavlov obtained with the method of contrasts (or method of successive presentation of stimuli). The principle underlying this method is the following: a conditioned stimulus, always associated with an unconditioned stimulus, is differentiated from a neighboring neutral stimulus never associated with the unconditioned stimulus. Differential reaction to the conditioned and neutral stimuli has been propounded as evidence of sensory discrimination. Some of Pavlov's studies involved the visual discrimination of a circle or a piece of white paper (the reinforced conditioned stimuli) from an ellipse or a piece of gray paper (the neutral never reinforced stimuli). After discrimination was acquired between the circle vs the ellipse, or between the white vs the gray paper, intermediate shapes or shades of gray were used to evaluate the limits of discrimination.

Even though most of the work from Pavlov's school (Razran and Warden, 1929) has been performed on dogs, the principles of classical conditioning have been used in other species (Razran, 1933). For example, Northmore and Muntz (1974) measured the sensitivity of fish to moving and stationary bars of light and to diffuse light of different wavelengths. The unconditioned stimulus was an electric shock, which produced an increase in the activity of the fish. Jamison (1951) measured auditory intensity thresholds in the rat by conditioning of an autonomic response. Ammonia was used as the unconditioned stimulus to decrease heart rate. Pavlovian conditioning established a conditioned heart rate decrement. This procedure was applied to determine hearing thresholds by the method of limits.

b. Autoshaping The autoshaping technique, a classical conditioning procedure, was adapted by Passe (1981) to study the absolute threshold of visual sensitivity in pigeons. The animals were trained in a Skinner box. During the training phase, key illumination (conditioned stimulus) was followed by food presentation (unconditioned stimulus) a number of times. The pigeons, then, pecked the key only when it was illuminated, although key pecking was not required to obtain food. The intensity of key illumination was varied, and the threshold of sensitivity was defined as the intensity at which the pigeon pecked the response key on 50% of the conditioned stimuli. More information on autoshaping can be found in Schwartz and Gamzu (1977).

c. Conditioned Emotional Response   In 1941, Estes and Skinner described the "conditioned emotional response" (CER). The CER (also referred to as "conditioned suppression") consists of the suppression of a steady rate of free-operant behavior in the presence of a warning (conditioned) stimulus that has been associated and terminates with a shock (unconditioned stimulus). Even though the CER takes place in an operant context, it can be classified as a Pavlovian conditioned response because of its basic associative properties. This procedure has been fully exploited to assess sensory systems. More details about parametric manipulation of the conditioned and unconditioned stimuli can be found in Kamin (1965).

The suppressing properties of the conditioned stimulus are measured by the suppression ratio (SR):

$$SR = \frac{\text{prestimulus responses} - \text{stimulus responses}}{\text{prestimulus responses}}.$$

Prestimulus responses refer to the number of responses during a fixed time period preceding the onset of the conditioned stimulus, and the stimulus responses are the number of responses during the duration of the conditioned stimulus (Hoffman *et al.*, 1963). This ratio reflects the probability of detection of each stimulus level (a suppression ratio of 0 indicates no detection, whereas a suppression ratio of 1 reflects detection). Figure 11 illustrates the conditioned suppression paradigm.

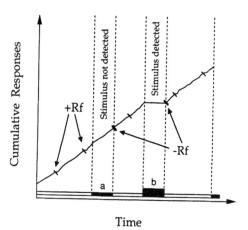

FIGURE 11   Representation of the CER during a variable interval schedule of reinforcement. Each occurrence of a response moves the response-marking pen one unit toward the top of the paper. Positive reinforcements (+Rf) are indicated by a thin oblique line and negative reinforcements (−Rf), typically a mild electrical shock, are indicated by a thick oblique line. The horizontal black lines indicate the presence of a low (a) vs a high (b) magnitude warning stimulus. Suppression ratios of 0 (a) and 1 (b) are illustrated.

When the suppression ratio rapidly changes as a function of a stimulus parameter, a threshold is usually and arbitrarily defined as the magnitude of the stimulus parameter under study corresponding to a suppression ratio of 0.5 (Hendricks, 1966; Kelly and Masterton, 1977; Masterton *et al.*, 1968; Smith, 1970). Others have defined detection in terms of an arbitrary conditioned suppression rate at each trial (e.g., 12.5%) and have calculated a threshold as the stimulus magnitude corresponding to approximately 50% detection rate (Sidman *et al.*, 1966).

Conceptually, a statistically significant difference between response rate before and during the warning stimulus can also be considered. Sidman and colleagues (1966), for example, studied hearing and vision in mice using this procedure. Mice were trained to lick a tube and they received a drop of milk after a variable number of responses. Such contingencies of reinforcement generated a very stable response rate (i.e., licking rate). Under such conditions, an extraneous stimulus preceding an electric shock interferes with the response rate when the animal detects the presence of the stimulus. Kelly and Masterton (1977) studied the auditory sensitivity of the albino rat with the same technique. A series of psychophysical methods have been used with the CER, such as the method of limits (Hendricks, 1966), the method of constant stimuli (Masterton *et al.*, 1969), and the tracking method (Sidman *et al.*, 1966). Reviews on the use of the conditioned emotional response in psychophysics can be found in Smith (1970), Blough and Blough (1977), and Blough and Young (1985).

2. Instrumental Conditioning

In 1898, Thorndike published a thesis on "Animal Intelligence" in which he used a number of puzzle boxes and enunciated what would be known later as the "law of effect." Simply stated, the law of effect refers to a modification of the relationship between a stimulus and a response as a result of the consequences of this response. Later, Skinner (1938) systematized Thorndike's approach and developed methods for the experimental analysis of behavior. Skinner made the rate of responding the major dependent variable, whereas Thorndike used the time needed by the subject to reach the solution. The term "operant conditioning" has been used specifically to refer to Skinner's conceptual approach. Placing a response under stimulus control (i.e., modifying the probability of a response in the presence of a discriminative stimulus) opened the way to sensory process evaluation by modifying the physical characteristics of the controlling stimulus. The instrumental procedures described next emphasize more the discrete trial approach described by Thorndike rather than the free operant approach favored by Skinner. A more complete review on operant conditioning techniques in animal psychophysics can be found in Blough and Blough (1977).

a. Early Developments One of the early prototypes of discrimination devices was Yerkes' apparatus (1907, 1912): the rat is placed in an observation chamber that opens to two alleys at the end of which two stimuli are presented. Close to the stimuli, a side door gives access to a compartment where food is delivered for responding to the appropriate stimulus. A shock was administered when the wrong alley was chosen.

Lashley (1930) later devised the jumping stand to study pattern vision in the rat. This apparatus is made of a platform facing two apertures which served both as covers as well as stimuli. The rat is placed on the jumping stand and is trained to jump through one of the apertures. If the rat chooses the correct stimulus covering the aperture, it is reinforced with food on the platform located behind the aperture. If the incorrect choice is made, the rat falls into a net and does not have access to food. The Lashley jumping stand was modified later by Sutherland (1964). More recently, an apparatus similar to the Sutherland box was computerized by Passig and co-workers (1992).

b. Go/No Go Procedures This procedure is the operant analog of the yes–no signal detection theory procedure when it is applied to animal studies. The animal is trained to respond in the presence of a standard stimulus configuration. It is also trained to withhold the same response in the presence of a different stimulus configuration.

Terman and Terman (1972) trained rats to tease apart sensitivity to pure-tone auditory stimuli from response bias. The animals were trained to press a lever in the presence of a light. This event was followed by the presentation of a standard or of a nonstandard (i.e., comparison) auditory stimulus and by a 3-sec choice interval, during which the light was off. A lever press during the choice interval that followed the presentation of the standard stimulus was positively reinforced with brain stimulation. This is equivalent to the "hit" of the yes–no procedure. The absence of a lever press during the choice interval that followed the presentation of the nonstandard stimulus was also positively reinforced. This corresponds to the "correct rejection" of the yes–no procedure. Failure to press the lever following presentation of the standard signal or pressing the lever following presentation of the nonstandard stimulus was followed by a time out. In terms of signal detection theory these last two events correspond to a "miss" and a "false alarm," respectively.

A shock avoidance conditioning paradigm has also been used with a single-lever technique in monkeys (Clack and Herman, 1963). In this experiment, the authors generated audiometric curves in monkeys using the tracking up–down procedure for presentation of the stimuli. The results are comparable to the data published by others with positive reinforcement techniques.

c. Forced-Choice Procedures The forced-choice paradigm described earlier also can be used in studies of sensory processes in animals. In contradistinction to the go/no go procedure, two or more response manipulanda (or alleys in a maze) are available with the forced-choice paradigm.

For example, tactile roughness discrimination in rats was evaluated with a T-maze by Finger and Frommer (1968). On the floor of the arms of the maze was a removable cardboard strip. First, the floor of one arm was covered with sandpaper, and the floor of the other with the smooth back-surface of the same material. Half of the rats were reinforced with food for choosing the coarser surface and the other half were reinforced for choosing the smoother surface. The position of the surfaces was randomly switched for each animal. Later, a more difficult roughness discrimination was established with two fine grades of aluminum oxide paper. The relatively easy discrimination was mastered in approximately 100 trials, whereas the more difficult task required a minimum of 250 trials.

Yager and Thorpe (1970) evaluated photopic spectral and saturation sensitivity in fish. They trained common goldfish (*Carassius auratus*) to press an "observing key" at one end of the tank. In the spectral sensitivity experiment, one of the two choice keys, located at the other end of the tank, was illuminated. If the fish pressed the illuminated choice key, a reinforcement occurred. If the other choice key was pressed, an incorrect response was recorded and there was no reinforcement. In the saturation discrimination experiment, the correct response key was illuminated with "white" plus monochromatic lights, whereas the incorrect key was illuminated by "white" light only. A modified method of constant stimuli was used in these studies. The thresholds were determined as the energy level corresponding to the interpolated 75% correct responding on the psychometric curves. A correction procedure prevented position bias.

VI. Behavioral Assessment of Sensory Functions in Toxicology

A number of chemicals affect sensory functions (Ison, 1984; Crofton, 1992), and numerous books and review articles have been written on the effects of chemicals on vision (Merigan, 1979; Merigan and Weiss, 1980; Evans, 1982), audition (Miller, 1985; Crofton, 1990, 1992), tactile sensitivity (Maurissen, 1979; 1985; Maurissen and Weiss, 1980), olfaction, and gustation (Grossman, 1968; Halpern, 1982; Amoore, 1986; Cometto-Muñiz and Cain, 1991). The following paragraphs give a few examples of the contributions of psychophysical approaches to toxicology.

A. Vision

Visual processes are multifaceted and encompass color vision, spatial and temporal contrast sensitivity (including visual acuity and flicker fusion frequency), size, depth, and directional movement perception. Central or peripheral vision can also be tested under photopic or scotopic conditions.

Methylmercury causes visual system changes. Hunter and colleagues (1940), for example, emphasized the importance of visual field constriction in the diagnosis of methylmercury intoxication. Merigan (1980) used a perimetry apparatus to measure visual fields in monkeys. A monkey that had blood mercury levels approximating 3 ppm during a period of 4 months showed a visual field constriction to about 25 to 40° of eccentricity bilaterally. In contrast, the control monkey had fields extending 55° nasally to 70° temporally. Visual constriction in the exposed monkey was accompanied by loss of sensitivity to high flicker frequencies at all luminance levels tested. Evans and Garman (1980) used a three-choice discriminative procedure and showed in the same monkeys that visual discrimination accuracy was decreased at scotopic levels, but was unaffected at photopic levels.

Heavy long-term exposure to carbon disulfide can cause a variety of signs and symptoms of both central and peripheral origins. Among these signs is impaired color recognition (Grant, 1975). Raitta and associates (1981) administered the Farnsworth–Munsell 100-hue test to 62 carbon disulfide exposed and 40 nonexposed workers. This test consists of 85 caps of saturated colors distributed in four boxes (Farnsworth, 1943), and the subject has to arrange the colored caps in chromatic order. The authors reported that color discrimination was impaired more often in the carbon disulfide-exposed group than in the control group. However, it should be emphasized that individual exposure levels were unknown and that the number of years of employment in the chemical departments did not correlate with the total error score in this color discrimination test.

Acrylamide is well known for its potential effects on the peripheral nervous system (Spencer and Schaumburg, 1974a,b). However, central visual functions can also be affected, as demonstrated by Merigan and co-workers (1982, 1985) in primates. Macaque monkeys faced two oscilloscopes with evenly illuminated displays. Two types of visual stimuli were used. In any test session, a fine grating or a flickering unpatterned stimulus was compared to a blank scope of the same average luminance. The monkey's task consisted of pressing the pushbutton corresponding to the stimulus display. Stimulus position changed randomly from trial to trial. Stimulus difficulty was adjusted according to a staircase procedure in a two-alternative forced-choice paradigm. Four monkeys were tested in both modalities. One monkey served as a time-matched control and the other three monkeys received an oral dose of 10 mg/kg/day of acrylamide, 5 days a week until the appearance of overt toxic signs. The time course of effects revealed a marked reduction in visual acuity and flicker fusion. Following cessation of dosing, flicker-fusion thresholds returned to normal, whereas visual acuity thresholds recovered only partially and remained stable below control values for more than 10 weeks. Morphological examination of a monkey euthanatized at the end of the dosing period revealed distal axonal swelling in the optic tract, the lateral geniculate nucleus, and the retino-geniculate pathway (Merigan and Eskin, 1986; Eskin et al., 1985).

B. Audition

Psychoacousticians have measured auditory sensations for a long time. Sound is pervasive, ubiquitous, and protean. Pure tones are mainly characterized by their pitch and loudness; however, other subjective attributes of sounds are their density, localization, pleasantness, annoyance, smoothness, and roughness. Hearing in vertebrates and invertebrates has been reviewed by Fay (1988) and Stebbins (1983).

The potential ototoxicity of aminoglycoside antibiotics and salicylates is well known. Prosen and colleagues (1978) studied the effects of kanamycin in guinea pigs. The animals were trained to press a response key with their nose (observing response). In the presence of a tone, they had to press another key. The method of constant stimuli was used. Catch trials were presented on 25% of all trials. Several frequencies and amplitudes were tested. The threshold was defined as the intensity to which the subject responded 50% of the time. Following establishment of a stable audiogram, the animals received a daily subcutaneous injection of kanamycin sulfate (200 mg/kg) after the test session for 2 to 3 weeks. By that time a significant threshold shift was recorded at high frequencies (above 32 kHz). Threshold testing continued for approximately 6 more weeks to ensure that thresholds were asymptotically stable. Then, a cytocochleogram was constructed and confirmed the results of the latest audiogram. No outer hair cells were present at the basal end of the cochlea, and thresholds were elevated by approximately 50 dB for frequencies above 16 kHz.

In 1983, Pryor and co-workers were the first to describe the ototoxic effects of toluene in experimental animals. They exposed rats initially to 1400 ppm (for 5 days) and subsequently to 1200 ppm of toluene for 14 hr a day for 5 weeks. The rats learned to pull or climb a pole suspended from the ceiling of the test chamber to avoid or escape a 1-mA electrical stimulation delivered to the floor. This stimulus was always preceded by a warning nonaversive stimulus (e.g., a tone). A response during the warning stimulus (evidence that this stimulus was perceived) terminated the trial and prevented delivery of the electrical shock (conditioned avoidance response). Five intensities of the tone at five frequencies (4 to 20 kHz) were evaluated. Although the ability of the rat to respond to nonauditory warning stimuli was unchanged, its ability to respond to tones was markedly impaired at high frequencies (above 8 kHz). No differences between control and exposed groups were observed at a low frequency of 4 kHz. This high-frequency hearing loss has also been confirmed by electrophysiological (Rebert *et al.,* 1983) and anatomical methods (Sullivan *et al.,* 1989).

Hearing impairment has been reported following chronic glue sniffing (Ehyai and Freemon, 1983), although it appears to be rare (Fornazzari *et al.,* 1983). The ototoxic potential of other solvents, such as mixed xylene and styrene, has also been demonstrated in experimental animals (Pryor *et al.,* 1987).

Trimethyltin can also cause alterations in the auditory system. Young and Fechter (1986) gave a single i.p. injection of 0, 2, 4, and 6 mg/kg of trimethyltin chloride to Long-Evans hooded rats, and tested their hearing with the reflex-modulation technique. They presented tones of 10 and 40 kHz at several intensities prior to presentation of a white noise burst. The lead time between the prestimulus (tone) and the eliciting stimulus (white noise burst) was 100 msec. Intertrial intervals ranged from 30 to 60 sec. Baseline trials, where only the eliciting stimulus was presented, were randomly interspersed. The method of constant stimuli was used for presentation of the auditory stimuli. From the psychometric functions, threshold inhibition was calculated as the minimum prestimulus intensity that reliably produced inhibition of the reflex response. Data were collected from all groups prior to treatment, and post-tests were conducted at various intervals following treatment. The authors showed that a single

trimethyltin injection produced a frequency-specific, dose-dependent auditory impairment (i.e., sensory effect), as well as a decrease in the amplitude of the baseline startle response (i.e., motor effect). Loss of auditory sensitivity at high frequencies was related to the loss of outer hair cells in the basal turn of the cochlea (Hoeffding and Fechter, 1991). A similar impairment of auditory functions was also demonstrated in albino rats (Eastman *et al.,* 1987).

A repertory of ototoxic agents has been assembled by Worthington and co-workers (1973) and Rybak (1992).

C. Cutaneous (Somesthetic) Sensitivity

Cutaneous sensitivity is a multisensory modality that encompasses sensitivity to pain, temperature (cold and warm), and contact (touch, pressure, and vibration). Several classes of primary afferents have been described on the basis of either their morphology (Iggo, 1982) or their physiology (Vallbo and Johansson, 1978).

Other more complex somatosensory functions deal with spatial and/or ballistic characteristics, such as two-point discrimination, roughness or texture discrimination, cutaneous pattern of activity, and three-dimensional shape recognition. Although gnosia functions depend on the integrity of exteroceptive and proprioceptive endorgans and of their primary parietal projection area, they also heavily depend on the integrity of the adjacent integrative sensory areas and on the motor precentral gyrus.

Advances in the neurobiology of somesthetic sensitivity have unfortunately not been paralleled by the use of these more integrative techniques in animal or human neurotoxicology. Somatosensory functions are, however, known to be impaired by industrial agents (e.g., acrylamide, *n*-hexane, some metals), pharmacological agents (e.g., vinca alkaloids, some antimicrobials), and natural agents [e.g., cassava flour (Obidoa and Obasi, 1991)].

In sufficient doses, acrylamide causes central–peripheral distal axonopathy. Because of the predominancy of sensory signs and symptoms, Maurissen and colleagues (1983) trained six monkeys to report detection of a vibratory stimulus using the up–down method for the presentation of stimuli. Each monkey served as its own control. After obtaining baseline data in all monkeys, the experimental group received oral doses of 10 mg/kg/day 5 days a week until overt signs of intoxication were apparent. The time-matched control group received the vehicle alone. Transient visuomotor impairment and body weight loss resolved after 5–8 weeks. Vibration sensitivity decreased during the same period, but this impairment outlasted all the other effects by many weeks. It provided unique information on the clinical status of the peripheral nervous system that could not have been inferred from the other observations made in this study.

Tilson and Burne (1981) studied the effects of triethyl tin on pain reactivity and neuromotor functions. The rats were tested with three different techniques.

1. a hot plate where the dependent measure was the time before the rat lifted or licked both forepaws;

2. a warm water beaker where the time before the rat removed its tail from the beaker was measured (tail flick response);

3. a Plexiglas restrainer tube (patterned after Dallemagne and Richelle, 1970) where the rats were trained on an operant shock titration schedule (Weiss and Laties, 1958). At one end of the tube, a hole allowed the animal to break a photobeam with its nose (response). At the other end of the tube, electrodes were attached snugly to the rat's tail. Initially, a low-intensity electrical stimulation was delivered to the tail and increased in the absence of a response. Each response decreased the shock intensity in discrete steps.

The rats were injected subcutaneously 5 days a week for 3 weeks with 0, 0.25, or 0.5 mg/kg of triethyl tin bromide. Although triethyl tin increased latencies in the hot plate and tail flick tests shortly after the high dose, it did not, however, affect the shock intensity the rats stabilized at, on the titration schedule. Because paw licking and tail flicking represented more involved motor capabilities than nose poking, the authors interpreted their results in terms of neuromotor rather than somatosensory effects. Even though the operant titration technique is sensitive to the pharmacological effects of analgesic agents (Burne and Tilson, 1980), the intensities the rats maintained most likely acted as warning signals and did not represent a true pain threshold. The titration schedule therefore seems to generate discriminative avoidance behavior rather than escape behavior (Dallemagne and Richelle, 1970).

D. Chemical Senses

Olfaction and gustation respond to direct contact with chemicals. These chemical senses normally work in concert with one another as a functional unit. Olfactory dysfunctions appear, however, to be more prevalent than gustatory disorders. Marked disturbances in the ability to smell and taste can be caused by a number of pathological conditions, either related or unrelated to chemical exposure.

1. Olfaction

The detection of chemical substances in the environment is conveyed by two anatomically distinct systems: the olfactory nerve for the perception of odors (olfaction) and the trigeminal nerve for the perception of pungency (common chemical sense or chemesthesis). A number of nosological conditions can lead to hyperosmia, hyposmia, anosmia, or dysosmia, unrelated to chemical exposure (e.g., upper respiratory infection, head trauma, nasal and paranasal sinus diseases, etc.). Even though Amoore (1986) compiled a list of approximately 200 chemicals that are purported to have an effect on the olfactory system, very little attention has been devoted to the quantitative assessment of olfactory functions in toxicological research.

Methyl bromide, an industrial fumigant and insecticide, is a well-known neurotoxicant. Severe selective damage to the olfactory epithelium was recorded by Hurtt and associates (1988) after exposure of rats to methyl bromide. Hastings and colleagues (1991) studied the effects of methyl bromide on the rat olfactory system. Prior to exposure, the rats were food-deprived and trained to retrieve pieces of food buried beneath the bedding in the cage. The experimental group was exposed to 200 ppm of methyl bromide for 4 hr per day, 4 days per week for 2 weeks, whereas the control group was exposed to filtered air only. After only 4 hr of exposure to methyl bromide, olfactory function (as measured by the latency of food retrieval) appeared severely impaired. However, latencies decreased during the first week of exposure even though the rats were still being exposed. After the first week of exposure and through several weeks of recovery, control and experimental groups did not significantly differ. Extensive (but not complete) morphological damage to the olfactory epithelium was present following 4 hr of exposure. Overall, morphological repair lagged far behind recovery of the ability to detect a suprathreshold odor. The explanation for such a phenomenon is not clear.

Adams and Crabtree (1961) studied alkaline battery workers, exposed to cadmium and nickel dust, and compared them against a control group. The workers were asked to evaluate subjectively their own sense of smell (good, diminished, and none). They were then presented with several concentrations of phenol solutions and were asked to report detection of the presence or absence of phenol. They were not requested to refrain from smoking, were not tested in an odorless room, and were not blind to the goal of the study. However, the symptomatology corresponded well with the olfactory testing results. A significant positive association was also recorded between olfactory measurement and proteinurea, suggesting that cadmium was the offending dust.

2. Gustation

A number of diseases (such as cancer, cancer therapy, degenerative diseases, etc.) can be responsible for gustation disorders, i.e., hypergeusia, hypogeusia, ageusia, and dysgeusia (or parageusia). Taste disorders linked to industrial chemicals appear to be rare. Studies reporting a change in the preference for a chemical flavor should not be considered *ipso facto* as evidence of an altered taste acuity.

Sodium lauryl sulfate (present in toothpaste) was studied because of its detergent and surfactant properties. DeSimone and colleagues (1980) presented human volunteers with aqueous solutions of sucrose, sodium chloride, citric acid, and quinine, representing primary taste modalities. Each session was divided into two parts. In the first part, each test solution was tasted (but not swallowed). In the second part, a 0.05% solution of sodium lauryl sulfate was held in the mouth for 1 min, followed by rinsing with deionized water and then by tasting the four test solutions. The subjects were asked to estimate the intensity of a taste (method of magnitude estimation). Sodium lauryl sulfate was shown to decrease the sweetness of sucrose, the saltiness of sodium chloride, the bitterness of quinine while adding a bitterness dimension to the sourness of citric acid.

Systemic chelating agents, such as penicillamine, have been linked to hypogeusia in patients (Keiser *et al.*, 1968). This reduced taste sensitivity, reported as a subjective taste impairment, was also measured with a three-drop forced-choice method (Henkin, 1963). There has been some speculations that copper and/or zinc depletion, caused by penicillamine treatment, was the mechanism producing the taste abnormality. In one patient with cystinuria treated with this drug, ageusia coincided with a decrease in circulating zinc concentration. Zinc therapy restored gustatory function. Copper excretion due to penicillamine has also been proposed as decreasing taste acuity (Russell *et al.*, 1983).

VII. Critical Appraisal of the Sensory Evaluation Approach

Even though psychophysics dates back to 1860, a good part of the sensory literature in human neurotoxicology and in occupational medicine has ignored the scientific advancements made during the last 130 years in psychophysics and has disregarded the more recent teachings from the signal detection theory. The following paragraphs exemplify some of the terminological, conceptual, methodological, and technological difficulties encountered in testing sensory functions.

A. Terminology

Table 3 summarizes the main psychophysical methods and response paradigms. A number of papers pub-

TABLE 3 Commonly Used Modes of Stimulus Presentation and Response Paradigms

Psychophysical methods[a]	Response paradigms
Method of limits	Forced-choice paradigm
Method of constant stimuli	Yes–no paradigm
Method of adjustment	Rating paradigm
Adaptive methods	

[a] Choice of a psychophysical method is conceptually independent of the choice of a response paradigm.

lished mainly in the fields of audiology and occupational medicine, as well as a number of technical manuals, have fostered confusion between these psychophysical methods and response paradigms (Arezzo *et al.*, 1986; Gerr *et al.*, 1990; Sosenko *et al.*, 1987). For example, the "method of limits" has been contrasted with the "forced-choice paradigm," although they do not pertain to the same conceptual domain. While the former describes the order in which the stimuli are presented, the latter refers to the response expected from the subject.

The forced-choice paradigm has not always been understood, even by some who were specifically writing a textbook chapter about threshold determinations (Wilber, 1979).

In the forced-choice method the examinee knows when the stimulus may have been presented, but does not necessarily know it was. In this procedure during a fixed time interval, which is usually defined by a light cue, the observer is required to indicate whether or not the stimulus was present. (p. 8)

In fact, this is the definition of the yes–no paradigm.

The same author also used the terms "constant stimuli" in the context of psychophysical methods and seemed to confuse the method of constant stimuli with the method of adjustment. She wrote:

Some methods use constant stimuli instead of discrete stimulus presentation. In these procedures the stimulus is always present . . . (p. 9)

Similar types of erroneous statements are illustrated in some operating manuals for commercially available test devices. One of these describes the forced-choice paradigm as allowing the choice between two intervals with the option to abstain from choosing. By definition, however, abstention is not an option in the forced-choice paradigm.

B. Units of Measurement

Many investigators testing sensory functions do not seem to be aware of the importance and the meaning of the measurement units. In the field of vision it is common to see authors refer to the flash intensity in terms of device settings on a photostimulator (Onofrj *et al.*, 1985) instead of using appropriate optical units. Others fail to give any information on the overall level of illumination when administering a color vision test such as the Farnsworth–Munsell 100-hue test (Fallas *et al.*, 1992).

To take a few examples in the field of vibration sensitivity, some amplitude thresholds have been reported in inappropriate or erroneous units such as:

1. no units (Elderson *et al.*, 1989);
2. arbitrary units (Sosenko *et al.*, 1987);
3. Biothesiometer units (LeQuesne and Fowler, 1986);
4. vibration units (Gerr and Letz, 1988);
5. volts (Anger *et al.*, 1986; Arezzo and Schaumburg, 1980; Arezzo *et al.*, 1983, 1985; Bleecker, 1983; Jensen *et al.*, 1991; Kardel and Nielsen, 1974; Lipton *et al.*, 1987; Svendsgaard *et al.*, 1987);
6. volts square (Nielsen, 1972);
7. Hertz (Bleecker and Moreland, 1984); and
8. negative displacement expressed in micrometers (Demers *et al.*, 1991).

Displacement (not negative, however) is the real dimension of interest when expressing an amplitude threshold for vibration sensitivity, but the units just mentioned are not measures of displacement. Given that voltage is a measure of the difference of potential between the two poles of the vibrator, a mechanical load applied to the vibrating probe will have a more or less pronounced effect on the total displacement depending on the viscoelastic properties of the skin, the flexure stiffness of the vibrator, and the vibrator transfer functions relating electrical and mechanical energies. Such a load will not affect, however, the potential difference applied to the vibrator. Usually, users do not address these topics in their reports; sometimes, they may not be aware of the concepts. Goldberg and Lindblom (1979) also endorse the notion that vibration sensitivity thresholds should be expressed in terms of amplitude of stimulator movement rather than in volts, as is commonly done.

Because of the aforementioned problems, it is often impossible either to replicate the results of another investigator or to compare results among published manuscripts. For example, Bleecker (1983) and Anger and colleagues (1986) studied vibration sensitivity and reported their results in volts. The subjects' age (31–40 vs 36 years on average) and the device were similar in both studies. The reported thresholds were 3.67 V vs. 8.8 V in the two control groups. Does the voltage dis-crepancy reflect a normal variation in actual displacement threshold in two similar groups of subjects? Or does it reflect a difference between two devices of the same make? An amplitude calibration could have answered the question.

C. Concept of Threshold

A number of investigators using commercially available technology for vibration or electrical sensitivity have used the two-alternative forced-choice procedure in conjunction with the simple up–down method. They have calculated a "threshold" defined as the average amplitude of a number of high and low scores (reversals). As shown earlier (Table 1A), the simple up–down method is designed to converge on the amplitude level that is detected on 50% of the trials. On the other hand, Fig. 9 shows that the 50% point does not correspond to a threshold, but to chance level. It is therefore erroneous to define the energy level corresponding to a 50% detection rate as a threshold in the present situation. This approach would be akin to defining the threshold in a yes–no paradigm by reference to 0% rather than to 50%.

A number of authors are clear about this point. Rose and associates (1970) wrote:

. . . it makes no sense to use the yes–no rule in a forced-choice situation since there is not just one point, but a continuum of stimulus values for which the forced-choice psychometric function has the value 0.5. (p. 200)

Kershaw (1985) commented:

In a few papers other UDTR [up–down transformed rule] rules are used with the 2IFC [2-interval forced-choice] procedure. (. . .) It is difficult to see what interpretation can be made of the threshold estimates that they obtain. If the Up-and-Down rule is used in 2IFC experiments, then levels will drift toward low stimulus levels for which there is little or no chance of detecting the stimulus. (p. 36)

Penner (1978) made the following statements about the forced-choice procedure:

Obviously, it is necessary to track stimulus levels for which a unique detection probability exists. Since infinitely many stimulus levels result in 100% and 50% detection, these levels are of no interest as a dependent variable. We cannot compare thresholds obtained in various conditions if these thresholds are not unique. We must restrict our measures to stimulus levels that have a unique detection probability associated with them. (p. 115)

Finally, Levine and Shefner (1981) succinctly commented about the forced-choice paradigm:

As 50% is the score expected for an invisible stimulus, it cannot be taken as the threshold. (p. 17)

Other transformed up–down methods have been invented and used without regard for the probability the new rule would converge on. Arezzo and colleagues (1986) and Sosenko and colleagues (1987) used the two-alternative forced-choice procedure and presented the stimuli in the following manner: if two correct responses (out of three consecutive trials) are recorded at a particular stimulus level, the intensity of the stimulus is decreased at the next trial. Conversely, if two incorrect responses (out of three consecutive trials) are recorded at a stimulus level, the intensity of the stimulus is increased at the next trial. If two consecutive errors or correct detections are recorded, a third trial is not necessary to effect a change in stimulus level. Table 4 expresses this rule. Binomial analysis shows that the probability of a correct response at convergence still is 50%, i.e., the threshold approximates the stimulus level corresponding to the 50% detection rate. Given that these studies used a two-alternative forced-choice paradigm, the calculated "threshold" corresponded to a stimulus level that *ipso facto* could not be detected.

D. Number of Trials

The total number of trials to be used in a test session should neither be left to chance nor be determined on the basis of convenience. Simulation studies have provided guidance on the minimum number of trials to be used in sensory evaluation to ensure threshold convergence on a stimulus value. It is clear that variability of threshold estimates decreases with an increasing number of test trials and a decreasing step size. Too small a step size will waste observations whereas too large a step size will not allow precise evaluation of the threshold. Brownlee and associates

(1953) and Dixon and Mood (1948) recommended that the step size approximate one standard deviation of the variate under investigation (e.g., the amplitude levels tested at all trials with the up–down rule).

As early as 1948, Dixon and Mood recommended a minimum of 40–50 trials for threshold determinations using the simple up–down method. Later, Shelton and colleagues (1982) showed that neither PEST nor the transformed up–down rule for the presentation of stimuli will converge on a steady level (i.e., the threshold) with only 10 trials. They recommended a minimum of 40–60 trials to determine a threshold with the transformed up–down rule to obtain consistent measurements. However, some investigators have concluded their tests after as few as five or six incorrect responses, and have calculated their "thresholds" of thermal or vibratory sensitivity as the average of the levels of four to six correct and incorrect responses (Arezzo and Schaumburg, 1980; Katims *et al.*, 1986, 1991; Sosenko *et al.*, 1987). Reliability of such measurements may be very poor (Dixon and Mood, 1948).

E. Other Miscellaneous Problems

Published reports often do not include sufficient methodological details necessary to understand how evaluation of sensory functions was performed. Some publications do not even describe either the psychophysical method or the response paradigm used in their studies (Lupolover *et al.*, 1984).

When the yes–no procedure is used, usually the investigators do not use an appropriate number of catch trials (to minimize bias, it is recommended to use a number of blank trials approximately equal to the number of signal trials). Rarer are the occurrences when false positives are taken into consideration and a cor-

TABLE 4 Transformed Up–Down Rule[a]

Stimulus sequence	Symbolic representation	Probability of changing level[b]	Condition of convergence[c]	Probability of correct response at convergence
Two out of three correct $R^d \Rightarrow$ decrease S^e	$+ + \Downarrow$	P^2		
	$- + + \Downarrow$	$P^2(1 - P)$		
	$+ - + \Downarrow$	$P^2(1 - P)$	$P^2 + 2P^2(1 - P) =$	
			$2P(1 - P)^2 + (1 - P)^2$	$P = 0.5$
Two out of three incorrect $R^d \Rightarrow$ increase S	$- - \Uparrow$	$(1 - P)^2$	$\therefore$	
	$+ - - \Uparrow$	$P(1 - P)^2$	$6P^2 - 4P^3 - 1 = 0$	
	$- + - \Uparrow$	$P(1 - P)^2$		

[a] Rule used by Arezzo *et al.* (1986), Gerr *et al.* (1990), and Sosenko *et al.* (1987).
[b] P = probability of a correct response; $1 - P$ = probability of an incorrect response.
[c] The series converges or stabilizes at the level where the sum of the probability of a decrease equals the sum of the probabilities of an increase.
[d] Response.
[e] Stimulus level.

rection procedure is used to provide a better estimate of sensory threshold.

VIII. General Principles in the Design and Conduct of Scientific Studies

Methodological fallacies encumber every step of scientific endeavor, including the field of sensory evaluation. In his *Introduction à la Médecine Expérimentale*, Claude Bernard (1856) reminds us that the fundamental principle in science is the "experimental doubt":

Le grande principe expérimental est donc le doute . . . (p. 73)

Bernard stated that it is not enough to prove a hypothesis, but it is essential to try to disprove it.

A few other general rules of scientific evidence, which hold true for other scientific domains, merit to be reemphasized in the context of the design and conduct of scientific studies:

1. Preexisting differences among experimental and control groups should be avoided or addressed before studying the potential effects of a substance on a toxicological end point.

2. Experimental and control groups should be treated exactly in the same manner, except for the independent variable of interest: it is not always recognized in human studies that the group exposed to a potential neurotoxicant may already have an expectation for the results of the study, whether through previous industrial hygiene briefings or through an informed consent form (Brownell and Stunkard, 1982; Myers *et al.*, 1987). This potential bias should be discussed in such studies.

3. Data should be incontrovertible, i.e., the collected data should precisely measure what they are purported to measure in the appropriate units. It ensues that equipment should be calibrated regularly and that the investigator should have at least a minimum amount of competence in metrology and technology.

4. Experimental data should result from explicit manipulations rather than stem out of some other unknown or even unsuspected factor(s), such as competing etiologies, experimenter or observer bias, malingering, or abnormal control values.

5. The report should inform the reader of *all* the analyses performed, whether significant or not, so that the true significance of the results can be appreciated (McNemar, 1960; Neher, 1967).

6. Data should not be analyzed in originally unintended ways; this kind of *a posteriori* analysis cannot

be used in any way to draw any conclusions from the data. Such an analysis can only generate hypotheses. A replication of the study *on another independent data set* is necessary to test the new hypotheses (Freedman, 1983).

7. The larger the number of variables examined, the greater the false positive rate: computer technology (unfortunately) makes it easy to collect and analyze data that may be of little interest. This practice can inflate enormously the type I error rate. Performing many tests at the 0.05 level will escalate the experimentwise error rate and increase the probability of generating fallacious positive results (Cohen, 1990; Muller *et al.*, 1984; Ryan, 1959). The same concept holds for multiple analyses of the same data sets.

8. The study should have enough power to unmask an effect of the size it is supposed to be able to detect: power analysis allows the investigator to adopt a sample size such that a significant difference can be found for a meaningful effect magnitude at a given α (Cohen, 1988; Muller *et al.*, 1984; Muller and Benignus, 1992). When the null hypothesis is accepted, the investigator has to provide evidence that, had an effect reached a meaningful size, it would have been detected.

9. Hypothesis-generating studies should be identified as such instead of being presented as hypothesis testing studies: a study is called an hypothesis testing (or confirmatory) study only if all the hypotheses and type I error rates have been specified before the data are collected. Furthermore, Muller and colleagues (1984) added that no transformations of the dependent or independent variables may be performed *a posteriori*. They specified that if any of these criteria were violated, the study should be considered as hypothesis generating or exploratory.

10. The statistical analysis should be appropriate for the experimental study design: simple designs should be analyzed with simple statistical methods, whereas complex designs should be analyzed with complex statistical methods.

Beyond these principles of scientific evidence, it should also be reminded that the research, whether performed in animal or in humans, shall be conducted in a manner consonant with all applicable international, federal, and/or local regulations and accepted codes of ethical conduct.

IX. Conclusions and Recommendations

Psychophysics has significantly contributed to our knowledge of sensory functions in health and disease. In psychophysics, like in other scientific domains, one crucial point is to assure that the subject's behavior is under the strict control of some specific dimension of the stimulus of interest and is not the result of other sensory or nonsensory controlling variables (e.g., bias, malingering, attention fluctuations, etc.). The experimenter has to eliminate or control all extraneous variables that can confound interpretation of the sensory data. When studying the effects of a test compound on sensory functions, the investigator further faces the problem of distinguishing specific sensory loss from other nonsensory toxic effects (e.g., motor or motivational changes).

To minimize the effects of extraneous variables in sensory testing, it is recommended that:

1. the subject be tested under the restricted conditions of a controlled environment (e.g., laboratory conditions, standard instructions, etc.);
2. the interaction between tester and testee be minimized by automation of stimulus delivery and response recording (Dyck *et al.*, 1978);
3. the forced-choice response paradigm be used (Dyck *et al.*, 1990; Haughton *et al.*, 1979; Sekuler *et al.*, 1973; Vaegan and Halliday, 1982);
4. the test method incorporates an effective algorithm to prevent malingering; and
5. the stimulus parameters be frequently calibrated and rigorously controlled during the study and appropriately defined in the publication.

From a more general point of view, the investigator should have adequate psychophysical, behavioral, toxicological, technological, and statistical competency before planning a sensory study. More specifically, it is also recommended that the investigator:

1. design the study to maximize the information content of the result by balancing both type I and type II errors;
2. be aggressive and inquisitive in pursuing alternate plausible hypotheses as an explanation for the data;
3. discuss the statistical power of negative studies;
4. report the results of all statistical analyses, whether significant or not;
5. address the effect of multiplicity of tests on the *P* values in publications; and
6. report the actual *P* values (e.g., $P = 0.014$) instead of inequalities (e.g., $P < 0.05$) in publications.

It is hoped that utilization of these principles and practices will benefit the field of neurotoxicology by improving research standards and by increasing the unassailability of data and the defensibility of their interpretation.

References

Adams, R. G., and Crabtree, N. (1961). Anosmia in alkaline battery workers. *Br. J. Indust. Med.* **18,** 216–221.

Amoore, J. E. (1986). Effects of chemical exposure on olfaction in humans. In *Toxicology of the Nasal Passages* (C. S. Barrow, Ed.), pp. 155–190, Hemisphere Corporation, Washington, D.C.

Anderson, H., and Wedenberg, E. (1965). A new method for hearing tests in the guinea pig. *Acta Otolaryngol.* **60,** 375–393.

Anger, W. K., Moody, L., Burg, J., Brightwell, W. S., Taylor, B. J., Russo, J. M., Dickerson, N., Setzer, J. V., Johnson, B. L., and Hicks, K. (1986). Neurobehavioral evaluation of soil and structural fumigators using methyl bromide and sulfuryl fluoride. *Neurotoxicology* **7,** 137–156.

Arezzo, J. C., and Schaumburg, H. H. (1980). The use of the Optacon as a screening device. A new technique for detecting sensory loss in individuals exposed to neurotoxins. *J. Occup. Med.* **22,** 461–464.

Arezzo, J. C., Schaumburg, H. H., and Laudadio, C. (1985). The Vibratron: a simple device for quantitative evaluation of tactile/vibratory sense. *Neurology* **35** (Suppl. 1), 169.

Arezzo, J. C., Schaumburg, H. H., and Laudadio, C. (1986). Thermal sensitivity tester. Device for quantitative assessment of thermal sense in diabetic neuropathy. *Diabetes* **35,** 590–592.

Arezzo, J. C., Schaumburg, H. H., and Petersen, C. A. (1983). Rapid screening for peripheral neuropathy: a field study with the Optacon. *Neurology* **33,** 626–629.

Baird, J. C., and Noma, E. (1978). *Fundamentals of Scaling and Psychophysics.* John Wiley and Sons, New York.

Békésy, G. von (1947). A new audiometer. *Acta Otolaryngol.* **35,** 411–422.

Ben-Bassat, J., Peretz, E., and Sulman, F. G. (1959). Analgesimetry and ranking of analgesic drugs by the receptacle method. *Arch. Int. Pharmacodyn.* **122,** 434–447.

Bernard, C. (1856). *Introduction à l'Etude de la Médecine Expérimentale*, p. 212. Editions Pierre Beltond, Paris (Reprinted in 1966).

Bianchi, C., and Franceschini, J. (1954). Experimental observations on Haffner's method for testing analgesic drugs. *Br. J. Pharmacol.* **9,** 280–284.

Bleecker, M. L. (1983). Optacon—a new screening device for peripheral neuropathies. In *Neurobehavioral Methods in Occupational Health* (R. Gilioli, M. G. Cassito, and V. Foà, Eds.), pp. 41–46, Pergamon Press, New York.

Bleecker, M. L., and Moreland, R. (1984). Carpal tunnel syndrome: comparisons of alterations in vibration threshold versus electrodiagnostic studies. In *Conference on Medical Screening and Biological Monitoring for the Effects of Exposure in the Workplace*, National Technical Information Service, No. PB86-242641, p. 27, Springfield, Virginia.

Blough, D., and Blough, P. (1977). Animal psychophysics. In *Handbook of Operant Behavior* (W. K. Honig and J. E. R. Staddon, Eds.), pp. 514–539, Prentice-Hall, Englewood Cliffs, New Jersey.

Blough, P. M., and Young, J. S. (1985). Psychophysical assessment of sensory dysfunction in nonhuman subjects. In *Toxicology of the Eye, Ear, and Other Special Senses* (A. W. Hayes, Ed.), pp. 79–90, Raven Press, New York.

Blumenthal, T. D. (1988). The startle response to acoustic stimuli near startle threshold: effects of stimulus rise and fall time, duration, and intensity. *Psychophysiology* **25,** 607–611.

Boring, E. G. (1942). *Sensation and Perception in the History of Experimental Psychology.* Irvington Publishers, New York.

Brownell, K. D., and Stunkard, A. J. (1982). The double-blind in danger: untoward consequences of informed consent. *Am. J. Psychiatry* **139,** 1487–1489.

Brownlee, K. A., Hodges, J. L., Jr., and Rosenblatt, M. (1953). The up-and-down method with small samples. *J. Am. Stat. Assoc.* **48**, 262–277.

Burne, T. A., and Tilson, H. A. (1980). Titration procedure with rats using a nose poke response and tail shock. *Pharmacol. Biochem. Behav.* **13**, 653–656.

Clack, T. D., and Herman, P. N. (1963). A single-lever psychophysical adjustment procedure for measuring auditory thresholds in the monkey. *J. Audit. Res.* **3**, 175–183.

Cohen, J. (1988). *Statistical Power Analysis for the Behavioral Sciences*. Lawrence Erlbaum Associates, Hillsdale, New Jersey.

Cohen, J. (1990). Things I have learned (so far). *Am. Psychol.* **45**, 1304–1312.

Cometto-Muñiz, J. E., and Cain, W. S. (1991). Influence of airborne contaminants on olfaction and the common chemical sense. In *Smell and Taste in Health and Disease* (T. V. Getchell, R. L. Doty, L. M. Bartoshuk, and J. B. Snow, Eds.), pp. 765–785, Raven Press, New York.

Conquet, P., Tardieu, M., and Durand, G. (1979). Evaluation of ocular reflexes during toxicity studies. *Pharmacol. Ther.* **5**, 585–591.

Cornsweet, T. N. (1962). The staircase-method in psychophysics. *Am. J. Psychol.* **75**, 485–491.

Corso, J. F. (1967). *The Experimental Psychology of Sensory Behavior*, Holt, Rinehart and Winston, New York.

Crofton, K. M. (1990). Reflex modification and the detection of toxicant-induced auditory dysfunction. *Neurotoxicol. Teratol.* **12**, 461–468.

Crofton, K. M. (1992). Reflex modification and the assessment of sensory dysfunction. In *Neurotoxicology* (H. A. Tilson and C. L. Mitchell, Eds.), pp. 181–211, Raven Press, New York.

Crofton, K. M., and Sheets, L. P. (1989). Evaluation of sensory system function using reflex modification of the startle response. *J. Amer. Coll. Toxicol.* **8**, 199–211.

Dallemagne, G., and Richelle, M. (1970). Titration schedule with rats in a restraining device. *J. Exp. Anal. Behav.* **13**, 339–348.

Davis, M. (1974). Sensitization of the rat startle response by noise. *J. Comp. Physiol. Psychol.* **87**, 571–581.

Davis, M. (1984). The mammalian startle response. In *Neural Mechanism of Startle Behavior* (R. C. Eaton, Ed.), pp. 287–351, Plenum Press, New York.

Davis, M., Gendelman, D. S., Tischler, M. D., and Gendelman, P. M. (1982). A primary acoustic startle circuit: lesion and stimulation studies. *J. Neurosci.* **2**, 791–805.

Davis, M., and Sollberger, A. (1971). Twenty-four-hour periodicity of the startle response in rats. *Psychon. Sci.* **25**, 37–39.

Delboeuf, J. (1873). *Etude psychophysique. Recherches Théoriques et Expérimentales sur la Mesure des Sensations et Spécialement des Sensations de Lumière et de Fatigue.* F. Hayez, Bruxelles.

Delboeuf, J. (1883). *Examen Critique de la Loi Psychophysique, sa Base et sa Signification.* Librairie Germer Baillière, Paris.

Demers, R. Y., Markell, B. L., and Wabeke, R. (1991). Peripheral vibratory sense deficits in solvent-exposed painters. *J. Occup. Med.* **33**, 1051–1054.

DeSimone, J. A., Heck, G. L., and Bartoshuk, L. M. (1980). Surface active taste modifiers: a comparison of the physical and psychophysical properties of gymnemic acid and sodium lauryl sulfate. *Chem. Senses* **5**, 317–330.

Dixon, W. J., and Mood, A. M. (1948). A method for obtaining and analyzing sensitivity data. *J. Am. Stat. Assoc.* **43**, 109–126.

Dyck, P. J., Karnes, J. L., Gillen, D. A., O'Brien, P. C., Zimmermam, I. R., and Johnson, D. M. (1990). Comparison of algorithms of testing for use in automated evaluation of sensation. *Neurology* **40**, 1607–1613.

Dyck, P. J., Zimmerman, I. R., O'Brien, P. C., Ness, A., Caskey, P. E., Karnes, J., and Bushek, W. (1978). Introduction of automated systems to evaluate touch-pressure, vibration, and thermal cutaneous sensation in man. *Ann. Neurol.* **4**, 502–510.

Eastman, C. L., Young, J. S., and Fechter, L. D. (1987). Trimethyltin ototoxicity in albino rats. *Neurotoxicol. Teratol.* **9**, 329–332.

Ehyai, A., and Freemon, F. R. (1983). Progressive optic neuropathy and sensorineural hearing loss due to chronic glue sniffing. *J. Neurol. Neurosurg. Psychiatry* **46**, 349–351.

Elderson, A., Gerritsen van der Hoop, R., Haanstra, W., Neijt, J. P., Gispen, W. H., and Jennekens, F. G. I. (1989). Vibration perception and thermoception as quantitative measurements in the monitoring of cisplatin induced neurotoxicity. *J. Neurol. Sci.* **93**, 167–174.

Eskin, T. A., Lapham, L. W., Maurissen, J. P. J., and Merigan, W. H. (1985). Acrylamide effects on the macaque visual system. II. Retinogeniculate morphology. *Invest. Ophthalmol. Vis. Sci.* **26**, 317–329.

Estes, W. K., and Skinner, B. F. (1941). Some quantitative properties of anxiety. *J. Exp. Psychol.* **29**, 390–400.

Evans, H. L. (1982). Assessment of vision in behavioral toxicology. In *Nervous System Toxicology* (C. L. Mitchell, Ed.), pp. 81–107, Raven Press, New York.

Evans, H. L., and Garman, R. H. (1980). Scotopic vision as an indicator of neurotoxicity. In *Neurotoxicity of the Visual System* (W. H. Merigan, and B. Weiss, Eds.), pp. 135–147, Raven Press, New York.

Evans, W. O. (1961). A new technique for the investigation of some analgesic drugs on a reflexive behavior in the rat. *Psychopharmacologia* **2**, 318–325.

Fallas, C., Fallas, J., Maslard, P., and Dally, S. (1992). Subclinical impairment of colour vision among workers exposed to styrene. *Br. J. Ind. Med.* **49**, 679–682.

Falmagne, J.-C. (1985). *Elements of Psychophysical Theory.* Oxford University Press, New York.

Farnsworth, D. (1943). The Farnsworth-Munsell 100 hue and the dichotomous tests for color vision. *J. Opt. Soc. Am.* **33**, 568–578.

Fay, R. R. (1988). *Hearing in Vertebrates.* Hill-Fay Associates, Winnetka, Illinois.

Fechner, G. T. (1860). *Elemente der Psychophysik,* Vol. 1 and 2, Breitkopf und Härtel, Leipzig.

Fechter, L. D., and Ison, J. R. (1972). The inhibition of the acoustic startle reaction in rats by food and water deprivation. *Learn. Motiv.* **3**, 109–124.

Fechter, L. D., and Young, J. S. (1983). Discrimination of auditory from nonauditory toxicity by reflex modulation audiometry: Effects of triethyltin. *Toxicol. Appl. Pharmacol.* **70**, 216–227.

Fechter, L. D., and Young, J. S. (1986). Reflexive measures. In *Neurobehavioral Toxicology* (Z. Annau, Ed.), pp. 23–42, Johns Hopkins University Press, Baltimore.

Findlay, J. M. (1978). Estimates on probability functions: a more virulent PEST. *Percept. Psychophys.* **23**, 181–185.

Finger, S., and Frommer, G. P. (1968). Effects of somatosensory thalamic and cortical lesions on roughness discrimination in the albino rat. *Physiol. Behav.* **3**, 83–89.

Fornazzari, L., Wilkinson, D. A., Kapur, B. M., and Carlen, P. M. (1983). Cerebellar, cortical, and functional impairment in toluene abusers. *Acta Neurol. Scand.* **67**, 319–329.

Freedman, D. A. (1983). A note on screening regression equations. *The Am. Stat.* **37**, 152–155.

Gerr, F., Hersham, D., and Letz, R. (1990). Vibrotactile threshold measurement for detecting neurotoxicity: reliability and determination of age- and height-standardized normative values. *Arch. Environ. Health* **45**, 148–154.

Gerr, F., and Letz, R. (1988). Reliability of a widely used test of peripheral cutaneous vibration sensitivity and a comparison of two testing protocols. *Br. J. Ind. Med.* **45**, 635–639.

Gescheider, G. A. (1988). Psychophysical scaling. *Ann. Rev. Psychol.* **39**, 169–200.

Goldberg, J. M., and Lindblom, U. (1979). Standardised method of determining vibratory perception thresholds for diagnosis and screening in neurological investigation. *J. Neurol. Neurosurg. Psychiatry* **42**, 793–803.

Grant, W. M. (1975). *Toxicology of the Eye.* Thomas, Springfield.

Green, A. F., Young, P. A., and Godfrey, E. I. (1951). A comparison of heat and pressure analgesiometric methods in rats. *Br. J. Pharmacol.* **6**, 572–587.

Green, D. M., and Swets, J. A. (1974). *Signal Detection Theory and Psychophysics.* Krieger Publishing Co., Huntington.

Grossman, S. P. (1968). Drug effects on taste, olfaction, and food intake. In *Drugs and Sensory Functions* (A. Herxheimer, Ed.), pp. 101–130, Churchill, London.

Haffner, F. (1929). Experimentelle Prüfung schmerzstillender Mittel. *Deutsche medizinische Wochenschrift* **55**, 731–733.

Halpern, B. P. (1982). Environmental factors affecting chemoreceptors: an overview. *Environ. Health Perspect.* **44**, 101–105.

Hastings, L., Miller, M. L., Minnema, D. J., Evans, J., and Radike, M. (1991). Effects of methyl bromide on the rat olfactory system. *Chem. Senses* **16**, 43–55.

Haughton, P. M., Lewley, A., Wilson, M., and Williams, R. G. (1979). A forced-choice procedure to detect feigned or exaggerated hearing loss. *Br. J. Audiol.* **13**, 135–138.

Hendricks, J. (1966). Flicker thresholds as determined by a modified conditioned suppression procedure. *J. Exp. Anal. Behav.* **9**, 501–506.

Henkin, R. I., Gill, J. R., and Bartter, F. C. (1963). Studies on taste thresholds in normal man and in patients with adrenal cortical insufficiency: the role of adenal cortical steroids and of serum sodium concentration. *J. Clin. Invest.* **42**, 727–735.

Herbart, J. F. (1824–1825). *Psychologie als Wissenschaft, neu gegründet auf Erfahrung, Metaphysik und Mathematik.* A. W. Unzer, Königsberg.

Hoeffding, V., and Fechter, L. D. (1991). Trimethyltin disrupts auditory function and cochlear morphology in pigmented rats. *Neurotoxicol. Teratol.* **13**, 135–145.

Hoffman, H. S. (1984). Methodological factors in the behavioral analysis of startle. The use of reflex modification procedures and the assessment of threshold. In *Neural Mechanisms of Startle Behavior* (R. C. Eaton, Ed.), pp. 267–285, Plenum, New York.

Hoffman, H. S., Fleshler, M., and Jensen, P. (1963). Stimulus aspects of aversive controls: the retention of conditioned suppression. *J. Exp. Anal. Behav.* **6**, 575–583.

Hoffman, H. S., and Ison, J. R. (1980). Reflex modification in the domain of startle. I. Some empirical findings and their implications for how the nervous system processes sensory input. *Psych. Rev.* **87**, 175–189.

Hoffman, H. S., and Searle, J. L. (1965). Acoustic variables in the modification of startle reaction in the rat. *J. Comp. Physiol. Psychol.* **60**, 53–58.

Hoffman, H. S., and Searle, J. L. (1968). Acoustic and temporal factors in the evocation of startle. *J. Acoust. Soc. Am.* **43**, 269–282.

Horlington, M. (1970). Startle response circadian rhythm in rats: lack of correlation with motor activity. *Physiol. Behav.* **5**, 49–53.

Hunskaar, S., Berge, O.-G., and Hole, K. (1986). A modified hotplate test sensitive to mild analgesics. *Behav. Brain Res.* **21**, 101–108.

Hunter, D., Bomford, R. R., and Russell, D. S. (1940). Poisoning by methylmercury compounds. *Q. J. Med.* **9**, 193–213.

Hurtt, M. E., Thomas, D. A., Working, P. K., Monticello, T. M., and Morgan, K. T. (1988). Degeneration and regeneration of the olfactory epithelium following inhalation exposure to methyl bromide: pathology, cell kinetics, and olfactory function. *Toxicol. Appl. Pharmacol.* **94**, 311–328.

Iggo, A., and Andres, K. H. (1982). Morphology of cutaneous receptors. *Ann. Rev. Neurosci.* **5**, 1–31.

Ison, J. R. (1984). Reflex modification as an objective test for sensory processing following toxicant exposure. *Neurobehav. Toxicol. Teratol.* **6**, 437–445.

Ison, J. R., and Hoffman, H. S. (1983). Reflex modification in the domain of startle. II. The anomalous history of a robust and ubiquitous phenomenon. *Psychol. Bull.* **94**, 3–17.

Ison, J. R., and Leonard, D. W. (1971). Effects of auditory stimuli on the amplitude of the nictitating membrane reflex of the rabbit (*Oryctolagus cuniculus*). *J. Comp. Physiol. Psychol.* **75**, 157–164.

Ison, J. R., Bowen, G. P., and Kellogg, C. (1991). Potentiation of acoustic startle in the rat (*Rattus norvegicus*) at the onset of darkness. *J. Comp. Psychol.* **105**, 3–9.

James, W. (1890). *The Principles of Psychology.* Henry Holt, New York.

Jamison, J. H. (1951). Measurement of auditory intensity thresholds in the rat by conditioning of an autonomic response. *J. Comp. Physiol. Psychol.* **44**, 118–125.

Janssen, P. A. J., Niemegeers, C. J. E., and Dony, J. G. H. (1963). The inhibitory effect of Fentanyl and other morphine-like analgesics on the warm water induced tail withdrawal reflex in rats. *Arzneimittel-Forsch.* **13**, 502–507.

Jensen, T. S., Bach, F. W., Kastrup, J., Dejgaard, and Brennum, J. (1991). Vibratory and thermal thresholds in diabetics with and without clinical neuropathy. *Acta Neurol. Scand.* **84**, 326–333.

Kamin, L. J. (1965). Temporal and intensity characteristics of the conditioned stimulus. In *Classical Conditioning* (W. F. Prokasy, Ed.), pp. 118–147, Appleton-Century-Crofts, New York.

Kardel, T., and Nielsen, V. K. (1974). Hepatic neuropathy. A clinical and electrophysiological study. *Acta Neurol. Scand.* **50**, 513–526.

Katims, J. J., Naviasky, E. H., Ng, L. K. Y., Rendell, M., and Bleecker, M. L. (1986). New screening device for assessment of peripheral neuropathy. *J. Occup. Med.* **28**, 1219–1221.

Katims, J. J., Patil, A. S., Rendell, M., Rouvelas, P., Sadler, B., Weseley, S. A., and Bleecker, M. L. (1991). Current perception threshold screening for carpal tunnel syndrome. *Arch. Environ. Health* **46**, 207–212.

Keiser, H. R., Henkin, R. I., Bartter, F. C., and Sjoerdsma, A. (1968). Loss of taste during therapy with penicillamine. *JAMA* **203**, 381–383.

Kelly, J. B., and Masterton, B. (1977). Auditory sensitivity of the albino rat. *J. Comp. Physiol. Psychol.* **91**, 930–936.

Kershaw, C. D. (1985). Statistical properties of staircase estimates from two interval forced choice experiments. *Br. J. Math. Stat. Psychol.* **38**, 35–43.

Krauter, E. E., Leonard, D. W., and Ison, J. R. (1973). Inhibition of the human eyeblink by a brief acoustic stimulus. *J. Comp. Physiol. Psychol.* **84**, 246–251.

Lashley, K. S. (1930). The mechanism of vision: I. A method for rapid analysis of pattern-vision in the rat. *J. Genet. Psychol.* **37**, 453–460.

LeQuesne, P. M., and Fowler, C. J. (1986). A study of pain threshold in diabetics with neuropathic foot lesions. *J. Neurol. Neurosurg. Psychiatry* **49**, 1191–1194.

Levine, M. W., and Shefner, J. M. (1981). *Fundamentals of Sensation and Perception.* Addison-Wesley Publishing Company, Reading, Massachusetts.

Lipton, R. B., Galer, B. S., Dutcher, J. P., Portenoy, R. K., Berger, A., Arezzo, J. C., Mizruchi, M., Wiernik, P. H., and Schaumburg, H. H. (1987). Quantitative sensory testing demonstrates that subclinical sensory neuropathy is prevalent in patients with cancer. *Arch. Neurol.* **44,** 944–946.

Lukaszewski, J. S., and Elliott, D. N. (1962). Auditory threshold as a function of forced-choice technique, feedback, and motivation. *J. Acoust. Soc. Am.* **34,** 223–228.

Lupolover, Y., Safran, A. B., Desangles, D., de Weisse, C., Meyer, J. J., Bousquet, A., and Assimacopoulos, A. (1984). Evaluation of visual function in healthy subjects after administration of Ro 15-1788. *Eur. J. Clin. Pharmacol.* **27,** 505–507.

Mach, E. (1906). *The Analysis of Sensations and the Relation of the Physical to the Psychical.* Republication of the English translation of the 5th edition. Dover, New York.

Magendie, F. (1822). Expériences sur les fonctions des racines des nerfs rachidiens. *J. Physiol. Exp. Pathol.* **2,** 276–279; 366–371.

Masterton, B., Heffner, H., and Ravizza, R. (1969). The evolution of human hearing. *J. Acoust. Soc. Am.* **45,** 966–985.

Mattsson, J. L., Boyes, W. K., and Ross, J. F. (1992). Incorporating evoked potentials into neurotoxicity test schemes. In *Neurotoxicology* (H. A. Tilson and C. L. Mitchell, Eds.), pp. 125–145, Raven Press, New York.

Maurissen, J. P. J. (1979). Effects of toxicants on the somatosensory system. *Neurobehav. Toxicol.* **1** (Suppl. 1), 23–31.

Maurissen, J. P. J. (1985). Psychophysical testing in human populations exposed to neurotoxicants. *Neurobehav. Toxicol. Teratol.* **7,** 309–317.

Maurissen, J. P. J., and Weiss, B. (1980). Vibration sensitivity as an index of somatosensory function. In *Experimental and Clinical Neurotoxicology* (P. S. Spencer and H. H. Schaumburg, Eds.), pp. 767–774, Williams and Wilkins, Baltimore.

Maurissen, J. P. J., Weiss, B., and Davis, H. T. (1983). Somatosensory thresholds in monkeys exposed to acrylamide. *Toxicol. Appl. Pharmacol.* **71,** 266–279.

McNemar, Q. (1960). At random: sense and non-sense. *Am. Psychol.* **15,** 295–300.

Merigan, W. H. (1979). Effects of toxicants on visual systems. *Neurobehav. Toxicol.* **1** (Suppl. 1), 15–22.

Merigan, W. H. (1980). Visual field and flicker fusion thresholds in methylmercury-poisoned monkeys. In *Neurotoxicity of the Visual System* (W. H. Merigan and B. Weiss, Eds.), pp. 149–163, Raven Press, New York.

Merigan, W. H., Barkdoll, E., and Maurissen, J. P. J. (1982). Acrylamide-induced visual impairment in primates. *Toxicol. Appl. Pharmacol.* **62,** 342–345.

Merigan, W. H., Barkdoll, E., Maurissen, J. P. J., Eskin, T. A., and Lapham, L. W. (1985). Acrylamide effects on the macaque visual system. I. Psychophysics and electrophysiology. *Invest. Ophthalmol. Visual Sci.* **26,** 309–316.

Merigan, W. H., and Eskin, T. A. (1986). Spatio-temporal vision of macaques with severe loss of P_β retinal ganglion cells. *Vision Res.* **26,** 1751–1761.

Merigan, W. H., and Weiss, B. (Eds.) (1980). *Neurotoxicity of the Visual System.* Raven Press, New York.

Miller, J. J. (1985). *Handbook of Ototoxicity.* CRC Press, Boca Raton, Florida.

Moody, D. B. (1970). Reaction time as an index of sensory function. In *Animal Psychophysics: The Design and Conduct of Sensory Experiments* (W. C. Stebbins, Ed.), pp. 277–302, Appleton-Century-Crofts, New York.

Moody, D. B., and Stebbins, W. C. (1982). Detection of the effects of toxic substances on the auditory system by behavioral methods. In *Nervous System Toxicology* (C. L. Mitchell, Ed.), pp. 109–131, Raven Press, New York.

Moon, P. (1961). *The Scientific Basis of Illuminating Engineering.* Dover Publications, New York.

Moyer, K. E. (1963). Startle response: habituation over trials and days, and sex and strain differences. *J. Comp. Physiol. Psychol.* **56,** 863–865.

Muller, K. E., Barton, C. E., and Benignus, V. A. (1984). Recommendations for appropriate statistical practice in toxicologic experiments. *Neurotoxicology* **5** (2), 113–126.

Muller, K. E., and Benignus, V. A. (1992). Increasing scientific power with statistical power. *Neurotoxicol. Teratol.* **14,** 211–219.

Müller, J. (1826). *Zur vergleichenden Physiologie des Gesichtssinnes des Menschen und der Thiere.* C. Cnobloch, Leipzig.

Myers, M. G., Cairns, J. A., and Singer, J. (1987). The consent form as a possible cause of side effects. *Clin. Pharmacol. Ther.* **42,** 250–253.

Neher, A. (1967). Probability pyramiding, research error and the need for independent replication. *Psychol. Rec.* **17,** 257–262.

Nielsen, V. K. (1972). The peripheral nerve function in chronic renal failure. IV. An analysis of the vibratory perception threshold. *Acta Med. Scand.* **191,** 287–296.

Northmore, D. P. M., and Muntz, W. R. A. (1974). Effects of stimulus size on spectral sensitivity in fish (*Scardinius erythrophthalmus*), measured with a classical conditioning paradigm. *Vision Res.* **14,** 503–514.

Obidoa, O., and Obasi, S. C. (1991). Coumarin compounds in cassava diets: 2 health implications of scopoletin in gari. *Plant Foods Hum. Nutr.* **41,** 283–289.

Onofrj, M., Harnois, C., and Bodis-Wollner, I. (1985). The hemispheric distribution of the transient rat VEP: a comparison of flash and pattern stimulation. *Exp. Brain Res.* **59,** 427–433.

Passe, D. H. (1981). Autoshaping as a psychophysical paradigm: absolute visual sensitivity in the pigeon. *J. Exp. Anal. Behav.* **36,** 133–139.

Passig, C., Pinto-Hamuy, T., Jenkins, W. M., Arraztoa, J. A., Guadagno, H., and Olivares, R. (1992). Development of a computerized system for simultaneous visual discrimination in rats. *Behav. Brain Res.* **47,** 199–201.

Pavlov, I. P. (1927). *Conditioned Reflexes: An Investigation of the Physiological Activity of the Cerebral Cortex* (translated by G. V. Anrep), 1960. Dover Publications, New York.

Penner, M. J. (1978). Psychophysical methods and the minicomputer. In *Minicomputers in Sensory and Information-Processing Research* (M. S. Mayzner and T. R. Dolan, Eds.), pp. 100, 115. Lawrence Erlbaum Associates, Hillsdale, New Jersey.

Pentland, A. (1980). Maximum likelihood estimation: the best PEST. *Percept. Psychophys.* **28,** 377–379.

Pfingst, B. E., Hienz, R., and Miller, J. (1975). Reaction-time procedure for measurement of hearing. II. Threshold functions. *J. Acoust. Soc. Am.* **57,** 431–436.

Pilz, P. K. D., Schnitzler, H.-U., and Menne, D. (1987). Acoustic startle threshold of the albino rat (*Rattus norvegicus*). *J. Comp. Psychol.* **101,** 67–72.

Prosen, C. A., Petersen, M. R., Moody, D. B., and Stebbins, W. C. (1978). Auditory thresholds and kanamycin-induced hearing loss in the guinea pig assessed by a positive reinforcement procedure. *J. Acoust. Soc. Am.* **63,** 559–566.

Pryor, G. T., Dickinson, J., Howd, R. A., and Rebert, C. S. (1983). Transient cognitive deficits and high-frequency hearing loss in weanling rats exposed to toluene. *Neurobehav. Toxicol. Teratol.* **5,** 53–57.

Pryor, G. T., Rebert, C. S., and Howd, R. A. (1987). Hearing loss in rats caused by inhalation of mixed xylenes and styrene. *J. Appl. Toxicol.* **7,** 55–61.

Pryor, G. T., Uyeno, E. T., Tilson, H. A., and Mitchell, C. L. (1983). Assessment of chemicals using a battery of neurobehavioral tests: a comparative study. *Neurobehav. Toxicol. Teratol.* **5,** 91–117.

Raitta, C., Teir, H., Tolonen, M., Nurminen, M., Helpiö, E., and Malmström, S. (1981). Impaired color discrimination among viscose rayon workers exposed to carbon disulfide. *J. Occup. Med.* **23**, 189–192.

Razran, G. H. S. (1933). Conditioned responses in animals other than dogs. *Psychol. Bull.* **30**, 261–326.

Razran, H. S., and Warden, C. J. (1929). The sensory capacities of the dog as studied by the conditioned reflex method. *Psychol. Bull.* **26**, 202–222.

Rebert, C. S., Sorenson, S. S., Howd, R. A., and Pryor, G. T. (1983). Toluene-induced hearing loss in rats evidenced by the brainstem auditory-evoked response. *Neurobehav. Toxicol. Teratol.* **5**, 59–62.

Rose, R. M., Teller, D. Y., and Rendleman, P. (1970). Statistical properties of staircase estimates. *Percept. Psychophys.* **8**, 199–204.

Russell, R. M., Cox, M. E., and Solomons, N. (1983). Zinc and the special senses. *Ann. Intern. Med.* **99**, 227–239.

Ryan, T. A. (1959). Multiple comparisons in psychological research. *Psychol. Bull.* **56**, 26–47.

Rybak, L. P. (1992). Hearing: the effects of chemicals. *Otolaryngol. Head Neck Surg.* **106**, 677–686.

Schwartz, B., and Gamzu, E. (1977). Pavlovian control of operant behavior. In *Handbook of Operant Behavior* (W. K. Honig and J. E. R. Staddon, Eds.), pp. 53–97, Prentice-Hall, Englewood Cliffs, New Jersey.

Sechenov, I. M. (1863). *Reflexes of the Brain* (translated by S. Belsky), 1965, M.I.T. Press, Cambridge.

Sekuler, R., Nash, D., and Armstrong, R. (1973). Sensitive, objective procedure for evaluating response to light touch. *Neurology* **23**, 1282–1291.

Shelton, B. R., Picardi, M. C., and Green, D. M. (1982). Comparison of three adaptive psychophysical procedures. *J. Acoust. Soc. Am.* **71**, 1527–1533.

Sidman, M., Ray, B. A., Sidman, R. L., and Klinger, J. M. (1966). Hearing and vision in neurological mutant mice: a method for their evaluation. *Exp. Neurol.* **16**, 377–402.

Skinner, B. F. (1938). *The Behavior of Organisms. An Experimental Analysis*, Appleton-Century-Crofts, New York.

Smith, J. (1970). Conditioned suppression as an animal psychophysical technique. In *Animal Psychophysics: The Design and Conduct of Sensory Experiments* (W. C. Stebbins, Ed.), pp. 125–159, Appleton-Century-Crofts, New York.

Smith, K. U., and Bojar, S. (1938). The nature of optokinetic reactions in mammals and their significance in the experimental analysis of the neural mechanisms of visual functions. *Psychol. Bull.* **35**, 193–219.

Sosenko, J. M., Gadia, M. T., Natori, N., Ayyar, D. R., Ramos, L. B., and Skyler, J. S. (1987). Neurofunctional testing for the detection of diabetic peripheral neuropathy. *Arch. Intern. Med.* **147**, 1741–1744.

Spencer, P. S., and Schaumburg, H. H. (1974a). A review of acrylamide neurotoxicity. Part I. Properties, uses and human exposure. *Can. J. Neurol. Sci.* **1**, 143–150.

Spencer, P. S., and Schaumburg, H. H. (1974b). A review of acrylamide neurotoxicity. Part II. Experimental animal neurotoxicity and pathologic mechanisms. *Can. J. Neurol. Sci.* **1**, 152–169.

Stebbins, W. C. (1983). *The Acoustic Sense of Animals.* Harvard University Press, Cambridge.

Stevens, S. S. (1935). The operational definition of psychological concepts. *Psychol. Rep.* **42**, 517–527.

Stevens, S. S. (1958). Measurement and man. *Science* **127**, 383–389.

Stevens, S. S. (1975). *Psychophysics. Introduction to its Perceptual, Neural, and Social Prospects.* John Wiley and Sons, New York.

Stitt, C. L., Hoffman, H. S., Marsh, R. R., and Schwartz, G. M. (1976). Modification of the pigeon's visual startle reaction by the sensory environment. *J. Comp. Physiol. Psychol.* **90**, 601–619.

Stokinger, H. E. (1974). Behavioral toxicology in the development of threshold limit values. In *Behavioral Toxicology. Early Detection of Occupational Hazards* (C. Xintaras, B. L. Johnson, and I. de Groot, Eds.), HEW Publication No. (NIOSH) 74-126, U.S. Government Printing Office, Washington, D.C.

Storm, J. E., Hulebak, K. L., and Fechter, L. D. (1981). Modification of acoustic startle reflex amplitude by background level and prestimuli in two strains of mice. *Soc. Neurosci. Abstr.* **7**, 658.

Sullivan, M. J., Rarey, K. E., and Conolly, R. B. (1989). Ototoxicity of toluene in rats. *Neurotoxicol. Teratol.* **10**, 525–530.

Sutherland, N. S. (1964). The learning of discrimination by animals. *Endeavor* **23**, 69–78.

Svendsgaard, D. J., Soliman, S. A., Soffar, A., and Otto, D. A. (1987). Screening pesticide plant workers for organophosphorus induced delayed neuropathy (OPIDN). *Toxicologist* **7**, 134.

Taylor, M. M., and Creelman, C. D. (1967). PEST: efficient estimates on probability functions. *J. Acoust. Soc. Am.* **41**, 782–787.

Terman, M., and Terman, J. S. (1972). Concurrent variation of response bias and sensitivity in an operant-psychophysical test. *Percept. Psychophys.* **11**, 428–432.

Thompson Lowne, B. (1877). On the quantitative relation of light to sensation. A contribution to the physiology of retina. *J. Anat. Physiol.* **11**, 707–719.

Thorndike, E. L. (1898). *Animal Intelligence: An Experimental Study of the Associative Processes in Animals.* Psychol. Rev. [Monograph Suppl., 2, N°8].

Tilson, H. A., and Burne, T. A. (1981). Effects of triethyl tin on pain reactivity and neuromotor function of rats. *J. Toxicol. Environ. Health* **8**, 317–324.

Vaegan, and Halliday, B. L. (1982). A forced-choice test improves clinical contrast sensitivity testing. *Br. J. Ophthalmol.* **66**, 477–491.

Vallbo, A. B., and Johansson, R. S. (1978). The tactile sensory innervation of the glabrous skin of the human hand. In *Active Touch. The Mechanism of Recognition of Objects by Manipulation* (G. Gordon, Ed.), pp. 29–54, Pergamon Press, New York.

Wald, A. (1947). *Sequential Analysis.* John Wiley and Sons, New York.

Walk, R. D., and Gibson, E. J. (1961). A comparative and analytical study of visual depth perception. *Psychol. Monog.* **75**, N° 15, 1–44.

Wallman, J. (1975). A simple technique using an optomotor response for visual psychophysical measurements in animals. *Vision Res.* **15**, 3–8.

Watson, A. B., and Pelli, I. G. (1983). QUEST: a Bayesian adaptive psychometric method. *Percept. Psychophys.* **33**, 113–120.

Watson, J. B. (1913). Psychology as the behaviorist sees it. *Psychol. Rev.* **20**, 158–177.

Weber, E. H. (1834). *De Pulsu, Resorptione, Auditu et Tactu. Annotationes Anatomicae et Physiologicae.* C. F. Koehler, Lipsiae.

Weber, E. H. (1846). Der Tastsinn und das Gemeingefühl. In *Handwörterbuch der Physiologie* (R. Wagner, Ed.), Vol. 3, pp. 481–588, Vieweg, Braunschweig, Germany.

Weiss, B., and Laties, V. G. (1958). Fractional escape and avoidance on a titration schedule. *Science* **128**, 1575–1576.

Wetherill, G. B., and Levitt, H. (1965). Sequential estimation of points on a psychometric function. *Br. J. Math. Stat. Psychol.* **18**, 1–10.

Wilber, L. A. (1979). Threshold measurement methods and special considerations. In *Hearing Assessment* (W. F. Rintelmann, Ed.), pp. 1–28, University Park Press, Baltimore.

Woolfe, G., and MacDonald, A. D. (1944). The evaluation of the analgesic action of pethidine hydrochloride (Demerol). *J. Pharmacol. Exp. Ther.* **80**, 300–307.

Worthington, E. L., Lunin, L. F., Heath, M., and Catlin, F. I. (1973). *Index-Handbook of Ototoxic Agents 1966–1973*. The Johns Hopkins University Press, Baltimore.

Yager, D., and Thorpe, S. (1970). Investigations of goldfish color vision. In *Animal Psychophysics: The Design and Conduct of Sensory Experiments*, (W. C. Stebbins, Ed.), pp. 259–275, Appleton-Century-Crofts, New York.

Yerkes, R. M. (1905). The sense of hearing in frogs. *J. Comp. Neurol. Psychol.* **15**, 279–304.

Yerkes, R. M. (1907). *The Dancing Mouse: A Study in Animal Behavior.* Macmillan, New York. Reprinted in Classics in Psychology: *The Dancing Mouse and the Mind of a Gorilla,* (1973), pp. 92, 119, and 153. Arno Press, New York.

Yerkes, R. M. (1912). The discrimination method. *J. Anim. Behav.* **2**, 142–144.

Young, J. S., and Fechter, L. D. (1983). Reflex inhibition procedures for animal audiometry: a technique for assessing ototoxicity. *J. Acoust. Soc. Am.* **73**, 1686–1693.

Young, J. S., and Fechter, L. D. (1986). Trimethyltin exposure produces an unusual form of toxic auditory damage in rats. *Toxicol. Appl. Pharmacol.* **82**, 87–93.

12

Motor Function and the Physical Properties of the Operant: Applications to Screening and Advanced Techniques

M. CHRISTOPHER NEWLAND
Department of Psychology
Auburn University
Auburn, Alabama 36849

All visible behavior comprises motor acts representing the terminal stages of a complex of neural events, so all behavior has a component of motor function embedded in it. This simple observation alone suggests that the characterization of motor function must include behavioral testing, but the justification for behavioral testing is deeper. The rationale for studying behavior is analogous to that provided for the study of evolution. Just as nothing makes sense in biology except in light of evolution, it might be said that nothing in the nervous system makes sense except in light of behavior. Although the nervous system mediates behavior, both the cause and effect of nervous system activity can be found in the contact that the nervous system makes with the environment, both inside and outside of the skin, through behavior. It is through behavior that the nervous system senses, is molded by, and changes the environment. Neurotoxicants that damage the nervous system become of general concern only when that damage is manifested in behavior; a lesion in the visual system is a problem only if vision is impaired, and basal ganglia lesions may be noticed only when it results in clumsiness.

Defining a phenomenon might be considered a first step in studying it, but defining behavior is more difficult than sometimes appears. Definitions that capture functional interactions with the nervous system such as "all visible activity of the nervous system" confuses behavior with other nervous system activity, such that measured by electrophysiologists. In a definitive source, Catania (1992) defines behavior as "anything an organism does." This definition captures the observation that those who study behavior usually are interested in the performance of an intact organism, but also it is one that excludes practically nothing. Such a definition provides little guidance to an investigator about how to study behavior, what dimensions are important, how behavior changes, and how to relate behavior to events in the nervous system or to events in the environment. More specific definitions arise from considerations of how behavior is acquired and maintained.

An adequate definition of *behavior* that is not overly inclusive may be impossible, but precise and useful definitions of *types* of behavior certainly can be made using physical or functional units. Physical units include the temporal and spatial extent of a response as well as the force or speed with which a response is executed. Functional units include the different ways that behavior contacts the environment. For example, the single most important distinction between operant and respondent conditioning is a functional one: sensitivity to consequences. Operants are sensitive to the consequences of prior responses whereas respondents are not.

There is a long list of books and chapters that relate specific behavioral deficits to lesions that act at particular sites, neurochemical systems, or metabolic processes in the nervous system (e.g., Bannister, 1992; Pincus and Tucker, 1978; Spencer and Schaumburg, 1980). Clinical neurologists form precise diagnoses on the basis of these signs, and experimental neuroscientists characterize the function of nervous systems by producing precisely aimed lesions. Conclusions are usually grounded on the careful observation of behavior. This chapter does not revisit that literature but, instead, describes how an appreciation of the behavior on which diagnoses and analyses are based might improve their precision and enhance our understanding of how brain and behavior interact at the level of motor output. Such an appreciation can guide neurotoxicologists attempting to design simple screens or synthesize new behavior, that is to mold novel behavior from different elements, for advanced testing. An analysis of the determinants of behavior, and of chemical influences over them, will enlarge our understanding of movement disorders, their treatment, and prevention by placing them in a behavioral context.

Applications of the principles of behavior to synthesize (form complex behavior from simpler units) and analyze (break up complex behavior into simpler units) behavior have resulted in a better understanding of functional deficits invoked by chemical exposure. The contributions include characterizations of deficits whose effects are diffuse, such as the interactions between chemical exposure and schedule-controlled behavior, as well as those that can be traced to specific neural mechanisms, as in the characterization of sensory function. A theme that often appears when interactions are sought between chemical exposure and behavior is that specificity can often be found in behavior, even when it is hard to identify in the nervous system (Branch, 1984, 1991; MacPhail, 1985).

Elements of the ongoing stream of behavior can be traced to specific neural mechanisms, and the identification of mechanisms supports comparisons and generalities across species. Psychophysical techniques, which can be applied to animal or human species, exemplify the ability to relate function, as expressed in behavior, to neural mechanisms (Blough and Blough,

1977; Harrison, 1991; Stebbins, 1970; Stebbins and Coombs, 1975) as well as an ability to isolate purely sensory phenomena from contextual influences (McCarthy and Davison, 1988; Maurissen, 1988; Nevin, 1981). These behavioral techniques include a thorough understanding of the three-term contingency defining operant behavior as well as the two-term, stimulus–stimulus associations defining respondent conditioning. The sensitivity of these behavioral techniques rivals and often exceeds that available with *in vitro* techniques. Even unconventional stimuli, such as the internal state produced by a drug or chemical, have been explored and have uncovered discriminative effects mediated by substrates that may be more like a dopamine receptor than like a conventional transducer (Schuster and Balster, 1977). The precision with which it is now possible to study conventional senses might have been predicted by a sensory physiologist of a generation or two ago. The ability to study private events such as how a drug feels would probably not even have been thought of.

The study of motor function could benefit from the general approach taken by psychophysicists. An informed application of behavioral techniques can contribute to an understanding of motor function in ways that parallel the contributions made by psychophysical techniques with animals and in ways that might not be fully appreciated yet. Behavioral techniques have already been used to train animals to perform tasks with great precision, but there are certainly more opportunities for characterization. Context, availability of alternative responses, effort, and prevailing reinforcement contingencies all could play a role in the expression of neurotoxicity and in the ability to detect it. In addition, motor function, like sensory function, can be measured in physically meaningful units amenable to mathematical treatment that, in turn, can inform an analysis of neural mechanisms (Fowler, 1987).

This chapter covers some of these possible influences of the expression of motor function. Instead of attempting to review all the ways in which motor function can be measured or disrupted, this chapter focuses on applications of behavioral principles to the general problem of detecting and characterizing motor deficits.

I. Behavioral Screens for Motor Deficits

Before discussing advanced applications, some discussion of the possibilities of successfully screening for motor deficits is in order. The problem posed by the presence of a huge number of untested chemicals suggests that some sort of screening should be conducted to identify neurotoxicants (Pryor *et al.*, 1983;

Tilson *et al.,* 1979). Since motor-related disorders constitute a large portion of complaints associated with potential neurotoxicants, tests for motor effects should be an essential component of screening strategies (Anger, 1984). The mechanics of conducting screens and relating screens to advanced testing can assume many forms, but common elements derive from two simple but sometimes conflicting dictates: screens should provide suggestions about how to proceed with testing, and advanced applications are too expensive and time consuming to conduct with every chemical. The conflicts lie in the information desired from a screen, its price, and at which point one screening stops and advanced testing begins.

An assumption underlying this discussion is that the results of screens are the beginning, not the end, of inquiry. This must be the case because screens are necessarily imperfect and are almost always relatively insensitive. One reason for the insensitivity of screens is the simple fact that they rarely take advantage of conditioning to reduce variability. Application of operant or respondent conditioning techniques permits an investigator to gain control over the target behavior and eliminate irrelevant sources of behavioral control, with their contributions to variability in the measures, so that a more refined end point is available for investigation. Thus, operant running in a wheel may well be more reliable and sensitive a measure than unconditioned running (e.g., Youssef *et al.,* 1993) and adding an electrified grid to a treadmill task may enhance sensitivity (Gibbons *et al.,* 1968). Training a monkey to execute an effortful rowing task identified manganese effects at exposure levels that were two or more orders of magnitude lower than previously reported using careful clinical observation. By using magnetic resonance imaging to visualize the central nervous system of the same monkeys on which behavioral testing was accomplished, it was determined that these effects corresponded to the accumulation of manganese in the globus pallidus (Newland and Weiss, 1992; Newland *et al.,* 1989).

A second assumption is that a battery of tests that can begin narrowing down the mechanism of toxicity and guide the design of advanced tests is worthwhile. Advanced applications, although more precise, must also be more focused and so the risk of missing an effect is greater.

One approach to screening can be exemplified as an attempt to discover the ''apical test,'' a test that can identify every neurotoxicant regardless of what portion of the nervous system is affected. The major benefit of the apical test is that it is relatively inexpensive to conduct. The disadvantage is that its lack of selectivity provides little guidance in the design of advanced tests.

The other approach entails a battery of tests with portions designed to identify different components of the nervous system. The advantage of the battery is that it provides better guidance for further testing, but these batteries are more time consuming and therefore more expensive in the short run. Determining the ''cost'' of testing is not a straightforward matter, however. Advanced testing is sometimes not performed and, consequently, regulatory decisions are based on the results of relatively broad and insensitive tests (Maurissen and Mattsson, 1989). The long-term costs of a high threshold limit value may be greater than the short-term costs of applying a detailed test battery.

A. Measurement Considerations

A neurobehavioral screen should be inexpensive, reliable, valid, and free of false negatives and false positives (Gad, 1989; MacPhail, 1987; Sette and Levine, 1986; Tilson *et al.,* 1979; Vorhees, 1987), and the analysis must respect the measurement scale on which the numbers describing the test are based. The first criterion, inexpensive, implies, among other things, that a technician should be able to conduct the test and score the results with relatively little time devoted to each individual test. Reliability is a fundamental measurement consideration since it pertains to the trustworthiness of the measurement. A test that is reliable will show good test–retest reliability (same score on different measurement occasions) and good interrater reliability (two raters arrive at the same conclusion), but this is not enough. Interrater reliability is often presented as a single score, such as a correlation or a percent agreement, but more revealing measures reflecting bias are necessary. The 2×2 arrangement of possibilities (effect present or absent, scorer rates it as present or absent) suggests that a simple score describing percent agreement or reliability omits important information. Even a 2×2 conceptualization makes it difficult to assess situations where there are more than two raters. For this, then, a κ statistic might be useful, coupled with explicit statements about biases apparent in the errors.

Measurement entails the assigning of numbers to an outcome. The numbers may merely substitute for a name (e.g., assigning a 1 for a female) but typically they have quantitative implications. Statistical tests are performed on these numbers to help quantify a decision about whether an effect is present or absent. Attention has been given in the toxicology literature to the proper application of these tests (e.g., Gad, 1989; Moser *et al.,* 1988). A collection of measurement scales has been characterized, based on axiomatic principles, to identify the kinds of arithmetic operations that can be supported by different scales, and these scales are called ordinal (Stevens, 1951; Stine, 1989). *Ordinal* scales apply where order is important. The magnitude of the number specifies some degree of magnitude, but distances between numbers are unimportant. Describing "amount of tremor" on a scale of 1 to 5 is an example of an ordinal scale. A value of 4 represents more tremor than 1 or 2, but the difference between 4 and 5 may not be the same as the difference between 1 and 2. Consequently, "mean" tremor in a group makes no sense because calculating a mean assumes that information between intervals is maintained; if a measure of central tendency is required, then a median is called for (Siegel and Castellan, 1988). To exemplify how means can be misleading if the scale is ordinal, note that assigning the values [1, 2, 3, 4, 5] and [1,2, 3, 5, 10] provides identical ordinal information. Medians are preserved in each scale, but the "means" calculated from each scale are different.

Interval scales (degrees Centigrade, for example) support calculations of means since distances on the scale reflect real differences in what is being measured. *Ratio* scales (degrees Kelvin, body weight, grip strength, tremor power) not only preserve differences but also ratios. A ratio scale of tremor based on a spectral analysis of acceleration or force (Elble and Koller, 1990; Newland, 1988) would be far more valuable, if more difficult to obtain, than an ordinal ranking of tremor because such a scale supports more precise comparisons about differences and meaningful ratios can be calculated.

Statistical inferences may be incorrect if scaling considerations are ignored. Statistical power is lowered when statistics based on rank, such as a median, are conducted on numbers that would support means. Sometimes medians are necessary as measures of central tendency, even if the data are interval or ratio scaled, as when the data are not normally distributed. In this case, techniques like randomization testing (Edgington, 1987) preserve the power of interval scales without requiring such parametric assumptions of normally distributed data. On the other hand, tests based on means, such as ANOVAs, calculated on ordinal data cannot be assessed at all as to whether they are prone to Type I and Type II errors. The scale simply does not support the computations performed so their accuracy or the direction of error cannot be ascertained.

Measurement considerations shape the development of advanced measures of motor end points . Although screens are usually distinguished from advanced applications according to the ease of testing, they might also be distinguished by the dominance of numbers that fall on nominal (naming) or ordinal scales.

One goal of an advanced application is the measurement of a phenomenon more precisely and with numbers that fall on interval or ratio scales. This is not a perfect distinction since body weight and grip strength are both ratio scales and are parts of screens, but it captures the intent of developing advanced techniques for measuring neurobehavioral toxicity.

Reasonable measurement properties are necessary for a test to have any utility, but they are not sufficient for a test to be informative. The screen, like any test, must be valid. Validity is the degree to which the test measures what it is supposed to measure. Validity can be determined by producing a deficit and then detecting it with the test. This is the rationale behind the use of positive controls or drawing from the literature to select screens. Specific lesions, chemical insults with known effects, or other interventions must be performed in order to confirm the validity of a test. As with reliability, the full assessment of validity requires a consideration of a 2×2 array of possibilities. There are two ways to be accurate (correctly detect an effect, correctly report that one is absent) and there are two ways to make an error (false positives and misses). Whenever errors are possible, one must determine explicitly which type of error is more tolerable. Criteria for asserting whether an effect is present can be designed to err on the side of missing present effects (which could allow a toxic chemical to pass the screen) or incorrectly identifying an effect as being present (which could impose unnecessary expenses on a manufacturer). The decision about what errors are tolerable must derive from societal considerations, hopefully with an appreciation of the consequences that different classes of error can have (Gad, 1989; Swets, 1988, 1992).

B. Apical Tests of Motor Function: Locomotive Activity and Tremor

Perhaps the most common approach to screening for neurotoxicity, including motor deficits, is some variant of locomotor activity, and its widespread use has provided a large literature to draw from when designing and interpreting the test. Locomotor activity has face validity: on the face of it, severe motor deficits should be revealed. Face validity is a weak, nearly irrelevant argument in favor of any test, but locomotor activity has other desirable measurement properties, including replicability, sensitivity, and measurability on a ratio scale (MacPhail *et al.,* 1989; Pryor *et al.,* 1983).

Locomotor activity is sensitive to a large number of nonchemical, environmental interventions (MacPhail, *et al.,* 1989; Rafales, 1986; Reiter and MacPhail, 1982), a broad sensitivity that implies a degree of nonspecificity. The test involves quantifying the activity of an animal, often by making a grid on the floor of an arena and counting the number of squares crossed during a time-limited session. Many agents can change the number of squares crossed during some specified amount of time, so that measure by itself reveals little more than a first approximation of what doses have nervous system activity. Adding a measure of the number of times that the animal rises on its hind legs (rears) could help identify compounds that cause peripheral neuropathies since such disorders would likely reduce the functionality of the hind legs enough that rearing is reduced (Reiter and MacPhail, 1982). For example, peripheral neuropathy that impairs hindlimb function or a cerebellar disorder that disrupts gait should be identifiable as reduced locomotor activity as well as rearing. In a sense, locomotor activity comes close to being an "apical" test, a characteristic that is both an advantage and a disadvantage. It is testimony to the nonspecific nature of locomotor activity that rearing has also been offered as an indicator of central nervous system excitability (Moser *et al.*, 1988). If the test is not followed up on, and there is little information available to reveal how follow-up should proceed, then interpretation is impossible (Maurissen and Mattsson, 1989).

It is sometimes argued that locomotor activity is unconditioned behavior; this is strictly true in most applications since investigators rarely condition or train responses. But locomotor activity is conditionable behavior since the movement can certainly be placed under stimulus control or the control of consequences, a characteristic revealed in the sensitivity of this measure to environmental or historical variables. Nevertheless, conditioning is rarely performed and, indeed, the test is selected in order to avoid the time required to accomplish such conditioning, so it is not always clear what the determinants of movement are (MacPhail, 1987).

Tremor might be close to an apical test for motor function. Lesions at many different components of motor systems often produce tremor so, at the very least, the presence of tremor is evidence that some motor pathway is involved in the effects of the toxicant under investigation. The converse does not hold, however. The absence of tremor does not imply that there is no motor effect. Tremor can assume different forms that can be associated with specific interventions. That is, a certain intervention may result in a specific pattern of tremor. This is not to say that neural mechanisms underlying tremor are well understood, they are not (Elble and Koller, 1990), but certain types of tremor are associated with specific interventions or disease

processes. Table 1 summarizes some of these relationships and shows that tremor can be defined according to magnitude, frequency, and the conditions under which tremor occurs (Newland, 1988; Elble and Koller, 1990).

The interpretation of Table 1 should be made cautiously, as with the interpretation of any test. The presence of a 3- to 5-Hz kinetic tremor following a cerebellar lesion does not mean that a cerebellar lesion is always the cause of that type of tremor. Neural interconnections are too complicated to support such a simple inference. The presence of that type of tremor can guide further investigation, and a reasonable start would be the cerebellum, as well as central and peripheral pathways that the cerebellum participates in, but the sign is not definitive.

Table 1 also indicates that reporting the mere presence of tremor is only a beginning. A full characterization is required and this includes the conditions under which tremor appears as well as amplitude and frequency. The first characterization might be obtained from a well-designed battery but the last two will be difficult. There are some reports of tremor obtained from activity chambers (e.g., Gerhart *et al.*, 1983) but one must be careful to eliminate the mechanics of the chamber and other components of the system when implementing this tactic (for discussion of techniques see, e.g., Newland, 1988; Elble and Koller, 1990). For example, some commercially available devices are designed to assess locomotor activity of a rat by resting

TABLE 1 Effects of Some Representative Interventions on Tremor[a]

Intervention	Type of tremor	Frequencies affected
Cerebellar lesions	Increased kinetic ("intention")	3–5 Hz
	Postural	3–5 Hz, maybe slower
Basal ganglia, Parkinson's disease	Increased resting, especially "pill-rolling"	3–5 Hz
Peripheral neuropathy	Increased physiological; lack of exacerbation with anxiogenics	8–12 Hz
Anxiogenics, ethanol withdrawal	Increased broad band; influenced by anxiolytics	8–12 and others
Acute ethanol	Decreased broad band	2–25
Cholinergic agonist (e.g., chlordecone, oxotremorine)	Increased high frequency	7–20 Hz

[a] From Elble and Koller (1990), Findley and Capildeo (1984), and Newland (1988).

a platform on load cells. Therefore, they also provide a continuous measure of the force exerted by the rat as it moves about on the platform. Unfortunately, the measurement system comprises not only the rat but also the mechanics of the platform, and the properties of the system are always part of the measures taken. Simply tapping the platform with a hammer causes it to oscillate for tens of seconds at frequencies that would be relevant to tremor (unpublished observations). If these natural oscillations are not accounted for then they could be misinterpreted as tremor or could mask real tremors that are occurring.

Some tactics for designing behavioral preparations for detecting increases and decreases in tremor are described in the section on advanced applications, below. Some compounds, acutely administered alcohol, for example (Newland and Weiss, 1991) decrease tremor and refined techniques may be essential for detecting this effect.

C. Specific Tests of Motor Function

Test batteries more specific than locomotor activity are required to guide the interpretation of test results and the design of advanced applications. Tremor, which is more directly linked to motor function, provides more information, but the links between types of tremor and specific lesions are still poorly understood. Tilson et al. (1979) suggested designing test batteries according to the effects of known neurotoxicants on human. A variant of this strategy, elaborated here with respect to motor systems, is to design batteries according to different functional components of mammalian nervous system (see also Mattsson et al., 1989). The primate is widely thought to be the most informative species for motor testing, especially for basal ganglia neurotoxicants. For economic reasons the most common laboratory subject is a rodent so tests applicable to rodents are used most frequently.

Motor systems comprise the cortex, especially the precentral gyrus, subcortical structures like the basal ganglia, cerebellum, peripheral nerve, spinal reflexes, and the neuromuscular junction. Because movement represents the integrated output of the nervous system, no single test can be uniquely associated with a specific functional neural component. Nevertheless, some relatively specific tests are possible and these can guide interpretation and more refined testing. Some tests with potential benefit are listed in Table 2. The structure visible in Table 2 is similar to that taken in the description of tremor and requires the same caution in interpretation. The tests have detected behavioral consequences of lesions or chemical intervention. The backward inference taken when saying that an effect reflects specific neural damage should always be undertaken with great caution.

1. Basal Ganglia

A challenge to the endeavor of identifying motor neurotoxicants with nonprimate species is embedded in the basal ganglia neurotoxicants. It is likely that this class of neurotoxicants would be missed by most existing neurological screens. This is the case now as it was when MPTP was tested as a *treatment* for Parkinson's disease in the early 1950s. After primitive toxicity testing in animals, that compound was given to patients, who probably got worse (Lewin, 1984). It is now known that MPTP toxicity, which closely resembles Parkinson's disease, is most fully expressed in primates and that identifying MPTP toxicity, although possible, is difficult in nonprimate species. Important differences in primate and rodent motor function are apparent in the functionality of basal ganglia structures, suggesting that reliance on screens alone may miss certain motor disorders, especially those involving the basal ganglia and if only rodents are used (King et al., 1988). Accordingly, several of the entries associated with basal ganglia disorders in Table 2 refer to studies using nonhuman primates because of their prevalence in basal ganglia research and because of concern about the degree to which it is possible to identify such disorders specifically and sensitively with rodents (Burns et al., 1983; King et al., 1988; O'Keeffe and Lifshitz, 1989).

The basal ganglia are thought to orchestrate the sequences of muscle movements required to carry out a motor act and may also be involved in the assembly of other complex behavior, including those designated as cognitive (Delong et al., 1984). The structures comprising the basal ganglia—substantia nigra, caudate, putamen, and globus pallidus—also use sensory information to update the direction of movements and postures as the act is carried out. Disruption of basal ganglia function can interfere with the sequencing of motor acts, and the specific characteristics of the disruption depend on where the damage occurs.

Among the signs of basal ganglia damage are changes in muscle tone, coordination, the speed and quantity of movement, and rate of blinking, although the particular manifestations depend on the location of damage. Some of these have been identified in human and nonhuman (mostly primate) species. Some potentially simple measures, like rate of eye blinking (Karson, 1983), have not been evaluated for their potential as a screen. Table 3, modified from Thach and Montgomery (1990), summarizes the clinical manifestations of lesions in different regions of the basal ganglia. The table is an overview of basal ganglia function and masks important details of the rich interconnections seen in these struc-

TABLE 2   Motor Effects of Lesions in Certain Parts of the Nervous System

Motor system	Clinical signs	Tests	Intervention	Reference
Cortical	Palsy. Specific losses of movement. Loss of integrated movement.	Licking (tongue reach, but not grooming). Forelimb reach. Subtle impairment of grooming. Postural change. Flexed limbs when held by tail. Trapped by corner.	Surgical lesions.	Whishaw (1990); Vanderwolf *et al.* (1978)
Basal ganglia	Low frequency resting or action tremor. Bradykinesia. Akinesia. Loss of postural reflexes. Oral-facial movement. Dystonia. Athetosis. Altered blinking. Stereotypies.	Turning. Catalepsy. Stereotypy. Tremor. Blink rate. Licking (tongue control). Forelimb reach. Dystonic postures.	Haloperidol. Manganese. Carbon disulfide. MPTP. Hypoxia.	Richter (1945); O'Keeffe and Lifshitz (1989); Pisa (1988); Creese and Iverson (1973, 1974); Karson (1983)
Cerebellum	Intention tremor at physiological frequency. Lowered tremor amplitude? Nystagmus. Loss of postural reflexes. Uncoordination. Disordered gait.	Gait analysis. Rotorod. Action tremor. Nystagmus. Running wheel. Rotorod. Landing foot spread.	Ethanol. 3-Acetyl-pyridine. Inorganic mercury.	Joliceur *et al.* (1979); Gilbert *et al.* (1982); Kulig *et al.* (1985); Youssef *et al.* (1993)
Peripheral nerve. Neuromuscular junction	Tremor, often low frequency or broad-band high frequency. Weakness. Loss of use of limbs. Peripheral neuropathy.	Grip strength. Rotorod. Gait analysis. Landing foot spread. Low-frequency tremor. Leg splay.	2,5-Hexanedione. Acrylamide.	Spencer and Schaumberg, (1980); Joliceur *et al.* (1979); Gilbert *et al.* (1982); Pryor *et al.* (1983)

tures, but it can help interpret the clinical signs of basal ganglia damage. The effects of an intervention depend on what part of the basal ganglia is affected and whether the effect is excitatory or inhibitory.

Globus pallidus lesions or damage resulting from carbon monoxide, carbon disulfide, or manganese exposure produces comparable (not identical) effects: dystonic postures, slow postural adjustments, and action tremor (Barbeau, 1984; Barbeau *et al.*, 1976; Jellinger, 1986; Richter, 1945; Wood, 1981). Parkinsonian signs, including increased muscle tone appearing as rigidity or spasticity, or bradykinesia appearing as slowed or absent movement and tremor, follow damage to the substantia nigra induced by acute neuroleptic

TABLE 3   Simplified Model of the Sequence of Connections with Basal Ganglia, and the Effect of a Lesion in a Particular Region.[a]

Substantia nigra/pars compacta	Inhibits ⇒	Cortex Excites ⇓ Caudate/putamen	Inhibits ⇒	Globus pallidus/internal segment	Inhibits ⇒	Thalamus/pons	Producing → →	Bradykinesia. Muscle tone
Lesion		↑ Lesion		↓ ↑ Lesion		↑ ↓ ↑		↑ ↓ ↑

[a] Adapted from Thach and Montgomery (1990).

exposure or MPTP administration. Nearly opposite effects result from damage to the striatum (comprising caudate and putamen) or upregulation of dopamine receptors in the substantia nigra due to chronic neuroleptic administration. These interventions may result in pallidal excitation and result in involuntary jerky or twitching movements (chorea), writhing (athetosis), or ballistic movements of limbs, especially of one side of the body (hemiballismus), as well as voluntary movements and postural adjustments that are irregular and rapid. The differences among these syndromes that have similar gross basal ganglia effects makes it clear that Table 3 is only a simplification and that subtle differences among these regions will require different tests to detect.

The bidirectionality in basal ganglia dysfunction can be seen in the different effects of dopamine agonists and antagonists. The acute administration of dopamine agonists at high doses elevates locomotor activity and produces stereotypies. The latter effect seems to be linked to the function of the striatum, a component of the basal ganglia, and has been characterized by an ordinal scale, describing the occurrence of peculiar, vigorous, almost mechanical repetitive emission of a small range of head, neck, mouth, or tongue movements (Creese and Iversen, 1973, 1974). There are species differences in the degree to which stereotypies and dyskinetic movements appear. Costall and colleagues (1975) report dyskinetic movements, with relatively little stereotypy, after administration of dopamine agonists into the striatum of guinea pigs. Chronic blockade of dopamine receptors in the basal ganglia results in loss of motor control and altered sensitivity to acute administration of dopamine agonists. The latter effect can be detected with the use of drug challenges and an appropriate end point (on the use of drug challenges see Walsh and Tilson, 1986).

Compounds such as haloperidol that inhibit dopamine activity in the substantia nigra reduce locomotor activity at low doses and produce sedation, complete immobility, and catalepsy at higher doses (Mason *et al.*, 1978; Mason, 1984). Catalepsy is reasonably specific to dopamine blockage by acute administration by a compound such as haloperidol. A rat is placed in an unusual posture such as positioning forelimbs on a ledge or a hindlimb raised onto a 3-cm-high platform. In a dose-related manner, haloperidol and other neuroleptics greatly prolong the duration with which a rat will hold this position.

Loss of motor control associated with chronic dopamine blockade often appears in movements of the tongue, jaw, and face of rodents (Bures and Bracha, 1990; Fowler and Mortell, 1992; Pisa, 1988; Waddington *et al.*, 1983). Chronic impairment of function

of these regions is also visible in humans with tardive dyskinesia, which results from chronic neuroleptic administration. Whishaw and Kolb (1983) devised a simple test of control over tongue movements. They place a sweet mash of chocolate cookies on a ruler and positioned it on one edge of the wire mesh side of a cage. The extent to which the tongue could be extended could be read directly from the ruler: the ruler was cleaned of the mash as far as the tongue could reach. A similar strategy involving operant reaching with a forelimb through a narrow opening might detect striatal lesions (Pisa, 1988, Whishaw *et al.*, 1986). Such reaching can be impaired even if the complex movements associated with grooming remain intact. Care must be taken in interpretation. A normal reach would include extension of the limb, wrapping the paws around the food object, and placing it in the mouth. It is quite possible that the extent of reach could be undamaged even if some components are undamaged. In this case, a reach may be possible, but the retraction would involve a raking movement instead of coordinated grabbing in a primate (Kuyper, 1981).

The difficulty in detecting basal ganglia disorders in nonprimate species may lie in the *tests* used with other species and not in the failure of these disorders to appear. The decline in limb movements central to the complex movement impaired when the basal ganglia are damaged is relatively easy to observe in nonhuman primates since so much of their movement critically depends on precise control of the extremities. Identifying such subtle movement impairments in rats could require advanced applications, but these can be performed in rodents and even pigeons. Fowler and colleagues have developed sophisticated ways of measuring the movement of forelimb and tongue movements in rats with the result that haloperidol and other compounds with basal ganglia activity can be detected (Fowler, 1977; Fowler and Mortell, 1992). Even pigeons have been trained to mimic a "catching" task by anticipating where a moving stimulus will appear (Rilling and LaClaire, 1989).

2. Cerebellum

The cerebellum, like the basal ganglia, coordinates movement. It composes motor acts before their execution and directs their onset, progress, and offset (Brooks 1986; Thatch and Montgomery, 1990). However, the particulars of cerebellar and basal ganglia participation differ, as evidenced by anatomical connections and different consequences of damage. It has been suggested that the basal ganglia are responsible for orchestrating contractions across different muscles whereas cerebellar damage is seen in the uncoordinated contraction of a single muscle (Brooks, 1986).

The cerebellum contains three functional units. The *vestibulocerebellum vestibular nuclei* (fastigial nuclei) receive afferents from the labyrinths of the vestibular apparatus which detect gravity and acceleration of the head. Efferents go to oculomotor muscles and muscles of the limbs and trunk responsible for posture. Damage to these regions results in disordered gaze, impairment in the degree to which the eye muscles compensate for head movements to maintain gaze (oculomotor reflex), and cyclical, involuntary movements of the eye (nystagmus). Damage also results in impairments in balance, walking, and postural adjustments required to stand in a gravitational field as seen in the clumsiness, imbalance, and poor reflexes associated with alcohol intoxication.

The *spinalcerebellum* (interposed nucleus) receives afferents from proprioceptors in joints and muscles and from the motor cortex. Efferents eventually influence α and γ motor neurons and initiate and modulate the progress of movements under operant control (voluntary movement). Damage to these regions appears to be responsible for the tremors associated with cerebellar damage. This region of the cerebellum dampens the natural tendency to tremor that is present in most limbs. Damage here slows or impedes the modulation of muscle movements that occurs in response to the position or velocity of that muscle and removesthe dampening of tremor. It also impairs the execution of repetitive flexion/extension movements.

The *cerebrocerebellum* (dentate nucleus) contains afferent from the cerebral cortex (and none from sensory fibers) and efferents return to the motor cortex. These regions are active immediately before the motor cortex, which in turn are active immediately before movement is executed. Damage to this region might be difficult to detect except with more advanced testing. Longer reaction times and impairment in well-learned movements are the results of damage to these regions.

Cerebellar activity can be influenced by cortical activity preceding a motor act, the muscular or subcutaneous consequences of that act, and by the activity of the vestibular system. Cerebellar damage is often accompanied by disturbances in antigravity reflexes important for balance, postural reflexes, and gait. The loss of these reflexes is a sign of ethanol intoxication, many of whose motor effects of which are mediated by the cerebellum. Characteristic cerebellar-related disturbances also appear in the initiation and execution of motor acts.

"Intention" tremor, a 3- to 5-Hz oscillation that appears immediately before or during the initial stages of motor acts like touching one's nose with a finger or tracing a line on a piece of paper, is often a consequence of cerebellar lesions. Tremor may also appear as the target of movement is approached. In both cases, the magnitude of tremor increases in amplitude during the course of movement. This distinguishes cerebellar tremor from others on simple screens or in the neurological examination. Intention tremor and other increases in tremor are readily detectable if the magnitude of the increase is sufficiently large. Decreases in tremor, which occur, for example, after exposure to relatively low doses of ethanol (Newland and Weiss, 1991; Sinclair *et al.*, 1982), are more difficult to detect and may require advanced applications.

Gait can be examined by inking the pads on a rat's back paws and providing an opportunity for the rat to walk through an alleyway. The footprints left can be described according to stride length, angle, and symmetry (deMedinaceli *et al.*, 1982; Jolicoeur *et al.*, 1979). The compound 3-acetylpyridine, which damages the medulla oblongata and climbing fibers of the cerebellum, produces a diverse group of effects, including increased landing foot spread, decreased length of stride, and increased width (which might be described as waddling). Pryor (1991) suggested that the acute effects of toluene on cerebellar function were responsible for that chemical's disruption of gait and landing foot splay in the absence of effects on a rotorod and with grip strength.

An effect of a toxicant may preclude acquiring some types of data and the analysis of gait provides an example of this. Rats with severe cortical damage will sometimes become trapped in corners or by walls if their vibrissae touch a wall. Hydrocephalic rats (exposed neonatally to cadmium) show a severe thinning of cortex (Newland *et al.*, 1986) and a waddling gait that is easily identifiable by a gait analysis (Newland *et al.*, 1983), but obtaining several successive strides required for gait analysis from the hydrocephalic rats proved difficult (unpublished observations). Automated measures, such as a running wheel with steps on it (Kulig *et al.*, 1988; Youssef *et al.*, 1993), might be effective in some cases.

Rotating rods on which a rat must maintain balance have been used as a screen for motor effects. Accelerating the rotorod can improve the ability of the rotorod to detect effects at low levels of exposure, although in either case performance is highly dependent on such details of testing as width of the rods, rate of speed, rate of acceleration, and amount of experience (Bogo *et al.*, 1981). Rotating rods can identify effects that might be cerebellar or peripheral in nature, including the ataxia and loss of coordination produced by ethanol or organic mercury (e.g., Gilbert *et al.*, 1982), but there is some question about its sensitivity for detecting cerebellar effects. In an experiment described by Gilbert and Maurissen (1982), effects on an accelerating rotating rod appeared at about the same time as changes in body weight, suggesting that the rotorod might not be as sensitive as body weight. This could be interpreted as indicating that the rotorod provides no additional information over body weight in identifying low-dose effects. However, the information could also contribute to decisions about further applications that might be more sensitive, if effects on, say, rotorod are seen in the absence of effects on other end points.

3. Cortex

The motor cortex interacts with cerebellum and basal ganglia and contains cell bodies of the axons that descend through the cortical-spinal tracts. A map of all striated muscles can be described on the cortex by stimulating individual neurons and determining what muscle twitches. Plasticity of function is even present here, as it is in the cerebellum (Thompson *et al.*, 1984). The activity of individual neurons can be and is susceptible to operant conditioning. The resulting condition spreads to nearby regions and associated muscles are found to contract, providing the neural basis for response induction (Fetz and Baker, 1973). Cortical damage can be closely linked to control over particular muscles, so neglect or lack of use of a collection of muscles could reflect specific cortical damage and could be detected by inspection.

Other motor characteristics of cortical damage can be identified. Simple signs visible in a rat include postural abnormalities or poor appearance due to inadequate grooming (Whishaw, 1990). Cortical or basal ganglia damage sufficiently impairs coordinated jaw movements so that a rat cannot trim its claws, a normal component of grooming, and consequently their claws become long (Whishaw and Kolb, 1983). Other components of grooming remain intact in decorticate rats, so cortical damage might be distinguished from basal ganglia damage according to the specifics of how grooming is impaired.

Severe cortical damage also produces subtle alterations in the pattern of grooming: the rat may look groomed, but the postural adjustments seen during grooming differ (Whishaw, 1990). Decorticate rats, when held briefly by the tail, show flexion in the hindlimbs. These effects are seen with damage so severe that eating and drinking are also impaired, so the appearance of such signs should not be used in identifying no-effect levels. Impairment of forelimb and tongue reaching has been reported following both basal ganglia and cortical type of damage, but decorticate rats, and not basal ganglia-damaged rats, can still groom (Wishaw *et al.*, 1986).

4. Descending Pathways and Neuromuscular Junction

For a motor act to be carried out, an action potential must arrive at the neuromuscular junction and cause muscles to contract. Therefore, the descending fibers must be intact and the neuromuscular junction must be functioning. The descending fibers, especially long ones, are sensitive to damage arising from a wide variety of sources, including nutritional deficits, metabolic disorders, immune disorders, genetic disorders, infections, ischemic damage, and exposure to chemotherapeutics, heavy metals, organophosphates, carbon disulfide, acrylamide, or 2,5-hexanedione (Spencer and Schaumberg, 1980). The mechanism of damage can take many forms. Demyelination slows or stops the conduction of action potentials down the fibers. Impairment in the transport of nutrients and wastes between the cell body and peripheral regions of the fiber can damage the terminal regions of the fiber and result in a dying back of the fiber. Constrictions in the fiber block the flow of materials from terminal regions to the cell body, leading to axonal swelling and the eventual destruction of the fiber. The neuromuscular junction uses acetylcholine as a neurotransmitter, so chemicals such as acetylcholinesterase inhibitors, nicotine, or botulinum toxin, as well as disease processes that target the receptor, can alter the activity of this junction. Also important for rapid reflexes and postural adjustments are various spinal reflexes, comprising two or three synapses originating in a muscle proprioceptor and terminating at the neuromuscular junction.

Many of the tests now used in conventional screening batteries are well suited to identify these peripheral effects, in part because toxicants that cause peripheral neuropathies are commonly used in validating test batteries. The strengths and weaknesses of some of these tests have already been described, including observation and measurement of gait (Jolicoeur *et al.*, 1979). Tests such as grip strength, locomotor activity, and landing foot spread are part of many recommended test batteries (Moser, 1989, 1991; Pryor *et al.*, 1983). As described earlier, the pattern of effects might help decide whether an effect is peripheral or is related to other causes.

Tremor can be particularly sensitive to damage to the descending motor pathways and the neuromuscular junction. Slow tremors could result from damage to these pathways and these can be relatively easy to detect. Decreases in broad-band, high frequency tremor may reflect damage to specific motor units, and these effects require advanced testing to detect (Freund, 1983; Freund *et al.*, 1984a,b; Hömberg *et al.*, 1987). Acute ethanol intoxication also decreases tremor, which also requires advanced testing (described later). Some related peripheral effects of ethanol, such as its disruption of spinal reflexes, may be more easily detected and could be responsible for its effects on gait and coordination (Kucera and Smith, 1971; Lathers and Smith, 1976; Marsden *et al.*, 1967).

II. Behavioral Principles in the Design of Advanced Applications

The preceding sections reviewed some approaches to screening for neurotoxicants using techniques that require a minimal amount of intervention because the behavior studied usually consisted of responses already in the subject's repertoire. Much information can be gained from such studies, but in order to isolate mechanisms (behavioral or neural), examine such dynamic effects as tolerance, or characterize the consequences of exposure to low doses of a toxicant, it is necessary to synthesize specific responses tailored to the toxicant and its hypothesized neural effects. The material in this and following sections will emphasize considerations in the design and interpretations of such advanced applications.

Much is known about behavior and behavior change, and a few powerful determinants of behavior change can be isolated. Behavior can change across generations through mechanisms of natural selection or within a generation through such mechanisms as respondent (Pavlovian) or operant (instrumental) conditioning. Nonassociative forms of behavior change, such as reflex modification, sensitization, and habituation are also powerful determinants of transient behavior change in specific settings that can be exploited in neurotoxicity assessments (Fechter and Young, 1986).

A. Operant Behavior

The plasticity provided by the probable evolutionary advantage of being susceptible to operant conditioning offers many advantages to an investigator. A few elementary principles of operant conditioning can be exploited to synthesize a response in order to study it or relate it to events in the nervous system. If a complete description of the conditions maintaining behavior is provided, then these same principles can help characterize the conditions under which disturbances in the physical characteristics of responding are likely to appear and shed understanding on the lines along which behavior fractures.

1. The Three Term Contingency

Operants are responses that can be arbitrarily defined and created. Susceptibility to operant conditioning renders an organism sensitive to the consequences of behavior and to discriminative stimuli whose relevance changes too rapidly for evolutionary processes to be effective. The resulting flexibility in behavior provides the substrate required to synthesize a response according to the needs of the investigator. If the response can be defined both physically and functionally then the three term contingency defining operant conditioning can be applied to differentiated a response.

Functional units of operant behavior are organized around the relationship between behavior and the environment (Schick, 1971). The lever press provides a simple example of how functional and physical definitions interact. There are many ways to press a lever, and in most experiments any of these ways will satisfy the schedule contingencies since the lever press is defined, functionally, according to whether or not it registers and contributes to fulfilling a reinforcement contingency. When one is concerned about some physical property of behavior, such as force differentiation in the forelimb, than a narrowly defined physical criterion becomes important.

The contingencies defining operant conditioning can be stated simply: in the presence of a discriminable stimulus and other antecedent conditions, a response occurs that has a consequence. If the future occurrence of the response is changed by the consequence, then that response is called an *operant, i.e.,* a response that *operates* on the environment (Catania, 1973, 1991). The operant itself is defined in terms of its plasticity. The antecedent conditions can include stimuli present concurrently or previously (as in remembering) as well as specific behavioral histories. The consequences are defined as reinforcing or punishing according to whether they increase or decrease the likelihood of the operant reoccurring. Since the operant, consequence, and discriminative stimuli that provide the setting are all defined functionally, tactics for manipulating any element are explicit in their definitions.

2. Schedules of Reinforcement

An operant might have only intermittent specified consequences and the way that the consequences are arranged has powerful effects on the type of responding that appears, as well as on how chemicals influence responding (Lattal, 1991; Zeiler, 1977). The arrangement of such intermittent consequences is called the schedule of reinforcement, which describes the antecedent conditions, the operant, and the consequence(s). Intermittency can be arranged in units of time, responses, or some combination of the two (Zeiler, 1977). For example, the delivery of a reinforcer might occur after a fixed number of responses has occurred. The resulting *fixed ratio* schedule of reinforcement characteristically maintains a high rate of vigorous responding, as do many response-based schedules. On many schedules, the delivery of a reinforcer might occur following the first response after a certain amount of time has passed. The resulting *fixed interval* schedule of reinforcement results in a characteristic pattern of responding described as a pause followed by a moderate rate of responding or a gradually increasing rate of responding as the interval elapses. The common fixed interval and fixed ratio schedules have the advantage of being somewhat unnatural reinforcement schedules, so when a pattern emerges in the laboratory the environmental determinants can be isolated with minimal contamination from other biological factors.

The important role that the schedule of reinforcement plays in the emergence of behavior and in mediating drug or toxicant effects has been reviewed in several sources (see, for example, Laties and Wood, 1986; MacPhail, 1985; Morse and Kelleher, 1977; Rice, 1988; Zeiler, 1977). Such important physical properties as peak force, effort, response durations, or sheer physical exertion are all influenced by the schedule (Fowler *et al.*, 1977; Fowler, 1987; Weiss and Laties, 1964; Notterman and Mintz, 1965; Newland, 1994). If the conditions maintaining responding are specified inadequately, then evaluating an experiment becomes difficult. If they are specified, then an investigator can tailor the response according to the needs of the question being examined. To produce responding that occurs as rapidly as is physically possible, a ratio schedule of reinforcement might be applied. If responding is to occur at a steady rate, with little other temporal pattern apparent, then a *variable* schedule, in which the delivery of the reinforcer is unpredictable, would be more suitable. A variable ratio schedule should be used if a constant, high rate is required and a variable interval schedule should be used if a constant, moderate rate is required. Fixed interval and fixed ratio schedules also maintain different levels of physical exertion (Newland and Weiss, 1990), as well as highly characteristic patterns of responding that show great species generality.

B. Direct Schedule Effects and the Synthesis of a Response

1. Measuring the Response

The operant studied most frequently in laboratory settings using mammals is a discrete event defined by a contact closure produced by the depression of a lever (a lever press). The convenience of such a simple, discrete event has led to such widespread use of levers that the term "operant" is sometimes used synonymously with the term "lever press" and the general term "operant" is confused with the specific class of operants comprising lever presses (e.g., Norton, 1989). Nothing in the definition of operants is so restrictive. An operant is a physical event that can be defined along any of several dimensions or combinations of dimensions. The functional definition specifies only that to be defined as an operant, a response must be sensitive to consequences. Whatever the response or its duration, complexity, or form, if it is sensitive to consequences, then an extensive behavioral literature can be brought to bear on manipulating the operant or interpreting changes in it.

The physical specification of a response has traditionally been of secondary interest in investigations of operant conditioning, but the physical specification of the response is of utmost importance when studying motor function. The selection of the transducer and the design of the response device directly affect the conclusions drawn. Table 4 shows some of the physical dimensions that can and have been examined in studies of motor function, the relevant dimensions, and some representative transducers that have been used to measure these dimensions (see also Weber and McLean, 1975, or similar texts). The choice of the transducer must be determined by the question under study, but the principles of behavior apply whatever choice is made.

2. Shaping, Differentiation, and Variability

Measurement is only one step in synthesizing a response, the second step is shaping or differentiating the response. Response differentiation and induction are processes analogous to the more familiar stimulus-control processes of discrimination and generalization. Discrimination refers to the degree to which one stimulus decreases (or increases) the likelihood of a particular response occurring, as compared to its likelihood in the presence of a training stimulus. Generalization refers to the degree to which stimuli resembling the training stimulus along some physical dimension occasion the same response. In studies of discrimination and generalization, the stimulus is defined precisely where the response is generally a lever press of some minimum force and unspecified duration.

Differentiation entails the shaping or specifying the form of a response by applying consequences according to precisely defined physical dimensions. Induction refers to the degree to which the response changes along such physical dimensions as force, duration, or speed. Responses have been differentiated along many physical dimensions, including force (Clark *et al.*, 1962; Elsner *et al.*, 1988; Fowler, 1987; Johanson *et al.*, 1979; Notterman and Mintz, 1965), duration (La-

TABLE 4 Some Physical Dimensions of Responding and Transducers for Measuring Them

	Quality	Dimensions	Transducer
Elementary units	Time	Seconds	Clock
	Linear displacement	Centimeters	Linear displacement differential transformer; linear resistor
	Angular displacement	Degrees, radians	Rotary variable differential transformer; potentiometer
	Mass	Grams	Strain gauge; load cell[a]
Combinations	Linear velocity	Centimeters/sec	Speedometer; derivative[b] of position
	Angular velocity	Degrees/sec	Tachometer; integral[b] of acceleration
		Revolutions/sec	
	Acceleration	cm/sec/sec	Accelerometer
		g or milli-**g**[c]	Derivative[b] of velocity
			Second derivative[b] of position.
	Force above a threshold	Responses	Switch closure
	Force	g *g	Strain gauge; load cell
		cm/sec/sec	
		Newtons	

[a] Technically, these measure force. If acceleration is constant (or if gravity does not change) then measures of force and mass are easily interchanged.

[b] Derivatives and intergrals are with respect to time.

[c] A boldface **g** represents the acceleration due to gravity. A milli-**g** is a thousandth of a **g**.

ties and Weiss, 1962; Weiss and Laties, 1964; Zeiler *et al.*, 1980), spatial location (Eckerman *et al.*, 1980; Hori and Watanabe, 1987; Iversen and Morgensen, 1988), linear position (Clark, *et al.*, 1962); velocity (Rilling and LeClaire, 1989; Zhuravin and Bures, 1986), and angular position (Newland and Weiss, 1991). Even variability in responding can be defined as an operant and targeted for shaping. Page and Neuringer (1985) delivered reinforcers according to variability in the spatial location of responses, and Weiss and Laties (1965) shaped responding according to variability in successive interresponse times.

The operant can be differentiated according to a lower bound (a floor), an upper bound (a ceiling), or a band. The lever press is a response defined according to a lower bound since minimum force and displacement are required to define the response but no upper limits apply. If a ceiling applies, then all responses must be less than some value, e g., if only those responses less than 30 g in force are eligible for reinforcement. The band is nicely illustrated by the force band experiments (Notterman and Mintz, 1965; Samson and Falk, 1974) or by temporal differentiation procedures, in which a subject is trained to execute a response whose duration is specified by a minimum and maximum (Zeiler *et al.*, 1980). In either case, a band is defined by a minimally and maximally acceptable requirement. The relationship between the physical requirement and the obtained response can be quite precise.

The emitted response is often related to the required one in a fashion that can be described by a power function with an exponent less than 1 (e.g., Notterman and Mintz, 1965; Platt *et al.*, 1973), indicating both some variability in behavior and that the response produced is likely to fall short of the requirement. Variability is important since shaping a response is possible only because there is always some variability in behavior. The degree of variability is related to the reinforcement schedule in place, even when variability is not directly specified in the schedule (Notterman, 1959; Eckerman and Lanson, 1969). Under certain circumstances, shaping and a seemingly trivial response requirement can drastically reduce variability. Figure 1 from Notterman (1959, reproduced in Notterman and Mintz, 1965) illustrates the role of response variability in differentiation. A rat was first placed in a chamber with a load cell present. The operant level of forces exerted, that is, those forces exerted before training commenced, was recorded. Then, a minimal force of 3 g was required to produce a reinforcer. Later no reinforcers were ever presented (extinction). Response forces are shown in Fig. 1 on a response-by-response basis. Initially, response force varied over a range of about 50 g. After the criterion of 3 g gained control over behavior, the median force decreased and variability was reduced to about 10 g, even though this minimum force requirement was well below the forces previously produced. During extinction, responding continued for a while, and variability in response forces increased considerably, returning approximately to pretraining levels.

Shaping and differentiation are possible because of the variability in response forces. If the rat in the exam-

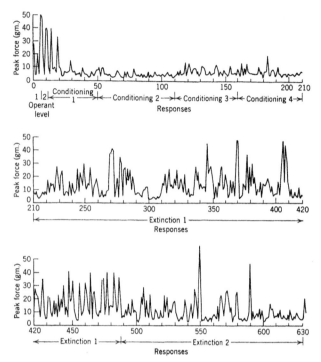

FIGURE 1   Peak force of successive responses for a single subject during operant level, conditioning, and extinction. (From Notterman, 1959. Reprinted by permission of John Wiley & Sons, Inc.)

ple had emitted the same force each time then shaping would have been impossible. Once 3 g was required for reinforcement, the emitted response force was much lower than the rats had been producing prior to this small contingency. Reductions in the variability of responding appear to be a natural consequence of reinforcement contingencies (Schwartz, 1980, 1982) and occur unless variability is explicitly reinforced (Page and Neuringer, 1985) or under conditions in which the reinforcement contingencies change (Fig. 1). Progressively finer differentiation of the target force is certainly possible by applying a force band. These principles are not restricted to force, however. The basic principles of shaping apply across the different response dimensions that can be differentiated.

Variability is an ever-present property of behavior that renders it sensitive to change. When a response has been shaped within a band, some responses usually fall outside of the band (unless the band is extremely broad). Often these are called errors but such a designation is misleading since they represent the extremes of variability. It would be revealing to examine variability by considering whether it promotes appropriate behavior change or whether it is nonfunctional.

3. Maintenance

Reinforcement contingencies differentiate responses precisely by shaping and then maintaining

them. All three terms of the contingency contribute to the maintenance of a response. Some characteristics of maintained behavior can be illustrated by examining responding on the response device illustrated in Fig. 2, which was designed to assess positioning and tremor in a squirrel monkey (Newland and Weiss, 1991). A squirrel monkey was trained to grip a bar attached to an angular transducer and to hold it within a prescribed defined band for at least 8 sec. When the angle requirement was fulfilled a tone sounded, and when the temporal requirement was fulfilled a different tone sounded (higher frequency) and the reinforcement cycle began. Some variability in performance always appeared in these experiments, even though the position band was fairly broad, the temporal requirement was set to a minimum, and there was ample stimulus support for the whole behavior.

Figure 3 contains histograms showing the amount of time spent holding the bar during a session. These distributions were weighted by the time spent in a response class and are called dwell-time distributions. Dwell times are constructed from the duration histogram and are the number of responses of duration t multiplied by t. The raw histogram that forms the basis for Fig. 3 would be dominated by the very short responses, even though these comprise only a small portion of the total session time. Weighting the histogram by the duration allows the amount of time spent in a response to have a visual influence that corresponds

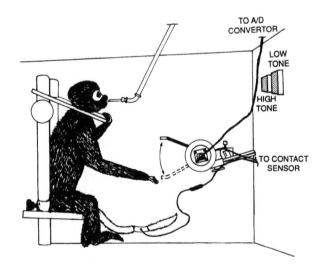

FIGURE 2   An illustration of a squirrel monkey and a response bar. The bar is insulated everywhere except the handle to restrict the response topography. A contact sensor attached to the monkey's tail and the bar registers when the bar is gripped. Bar position is digitally converted and stored for later analysis. When the bar is held in range, a low-frequency tone sounds. Every completed response results in a high-frequency tone burst and, on a random ratio 2 schedule of reinforcement, a squirt of juice. (From Newland and Weiss, 1992. Reprinted by permission of Rutgers University Press.)

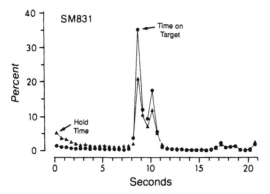

FIGURE 3 Typical dwell-time distributions illustrating the time during a session spend holding the bar and the time on target for one monkey. The distribution is in 0.5-sec-wide bins beginning at 0.25 sec. The dwell time in a particular class interval is obtained by multiplying the number of bar holds (or TOTs) falling in that interval by the duration represented by that interval. This procedure weights the distribution by the duration of the event and shows the amount (or percent) of time spent in each class interval. Normalizing was achieved by dividing the total time spent holding the bar. (From Newland and Weiss, 1992. Reprinted by permission of Rutgers University Press.)

to the amount of behavior that it represents (Iversen, 1991; Weiss, 1970).

Two separate classes of response durations were present, even after extensive training aimed at differentiating long durations within an angular band. One class, short-duration responses off the target, appeared as an elevation in dwell times shorter than about 3 sec in the "hold" distributions of Fig. 3. This response class was especially sensitive to ethanol. In a dose-related fashion, ethanol elevated the number of short-duration holds. This drug had relatively little effect on the class of response for which there was much stimulus support: the targeted response that resulted in the occurrence of a discriminative stimulus and reinforcer delivery (Newland and Weiss, 1991). The resistance of that class of responding to ethanol made it possible to evaluate tremor.

The presence of different response classes in temporally defined procedures, and of differential drug effects on these response classes, may be a general phenomenon. Response differentiation schedules all produce response classes that seem to be functionally similar, even when the topographics and species under investigation are different. Skjoldager and Fowler (1988), for example, reported sensitivity of a similarly defined class (sustained responses within a force band) of responses to pimozide. Platt and colleagues (1973) trained rats to hold a lever for a minimum requirement. The median duration was related to the required duration according to a power function relationship over a range of 0.8 to 6.4 sec, but the number of very short

response durations was greater than predicted by a power function, indicating that this response class is separable from those targeted for reinforcement. Bigelow (1971) reinforced long interresponse time (IRTs) under a ratio schedule and also reported two classes of responses: targeted IRTs and short IRTs. Only the rate of long IRTs was sensitive to other schedule contingencies imposed.

The class of short, unreinforced responses has been described as anticipatory by Skjoldager and Fowler (1988), as false starts by Newland and Weiss (1991), and as adventitiously reinforced durations by Platt and colleagues (1973). The frequency of this short class of responses is sensitive to reinforcement contingencies (suggested by Platt *et al.*, 1973) and duration requirement (Skjoldager and Fowler, 1988) so they are not reflexive in nature. They appear to be under operant control, even though they were not specified in the schedule. Their presence in such a variety of response topographies and their sensitivity to reinforcement contingencies suggest that this response class is related functionally to the behavior and not structurally to the anatomy supporting it.

Responding can be differentiated according to arbitrarily defined dimensions but evidently some general principles hold over many physical dimensions. Since differentiation in time is relatively simple to accomplish, these principles have been examined mostly with focus on this dimension. Power functions whose exponents are about 0.8 to 1.0 describe the relationship between median duration and required duration as well as the relationship between variability and required duration. To the extent that this has been studied with other response dimensions, such as force, similar power–function relationships and variability have been reported (e.g., Notterman and Mintz, 1965). This generally is important because it implies that influences over the physical properties of responding could apply to many different topographies. As long as one is studying a response that can be defined as an operant, then determinants of operant behavior should apply.

4. Synthesizing a Complex Response: Chain and Second-Order Schedules

Many studies of motor function entail the assembly of units of behavior, such as the protracted positioning response illustrated in Fig. 2, that are far more elaborate than a simple lever press. In that example, a monkey held a bar within an angular range for at least 8 sec. When the response criterion was met, a low frequency tone sounded, and when the 8 sec passed, a brief, high frequency tone, paired with reinforcer delivery, sounded. This response unit was then placed under an intermittent schedule of primary reinforcement such that only a random half of the successful responses resulted in fruit juice delivery. Such performance can be conceptualized as a second-order schedule and the literature investigating performance under these complex schedule arrangements influenced decisions about the shaping and maintenance of that response. Investigations of chain and second-order schedules of reinforcement reveal the necessary and sufficient conditions for assembling such complex behavior patterns.

A second-order schedule is a schedule within a schedule (Marr, 1979). For example, the conventional designation "FR 30" is used to describe a schedule in which 30 lever presses produce a food pellet (*i.e.*, there is a fixed ratio between responses and reinforcers). Under a second-order schedule, a FR 30 requirement might be applied to completing a single fixed interval 1-min schedule (FI 1 min). Each response unit is a FI 1' schedule and 30 such units are needed to secure food delivery. The consequence for completing the FI 1 min might be a brief stimulus change such a flash of light or a tone. The primary reinforcer, like a food pellet, is delivered only after the entire schedule has been satisfied. Such a second-order schedule is designated as FR 30 (FI 1 min : S^P). The FI 1 min is called a unit schedule or sometimes a within-component schedule. The FR is now called the overcomponent schedule. The symbol "S^P" indicates that completing the unit schedule results in a stimulus and that stimulus is paired with the primary reinforcer by presenting it at the end of the last unit schedule. Many schedule arrangements could be described, sometimes ones in which conflicting influences over behavior are specified. For example, FR 5 (DRL t'' : S^P) directly pits the high rate maintained by a FR schedule against the low rate maintained by the DRL t'' schedule, which specifies that responses must be separated by 5 sec (de-Lorge, 1971; Bigelow, 1971). When this occurs, response classes defined by a response separated by 5 sec occur at a high rate.

A coherent, unitary structure develops if a stimulus is presented at the end of a unit schedule, and especially if that stimulus is explicitly paired with a primary reinforcer (Byrd and Marr, 1969; DeLorge, 1971; Kelleher and Gollub, 1962; Marr, 1979; Findley and Brady, 1965). For example, under the FR 30 (FI 1 min : S^P) schedule, each component FI schedule contain a scallop and generally look, "FI like." The overall structure of the emission of FI unit schedules look FR like, in that a pause will precede the emission of a series of FI 1-min schedules. If that second-order schedule is inverted to a FI 20 min (FR 100 : S^P), then FR-like responding appears in the component schedule and the emission of FR-100 schedules assumes a FI-like pattern

with a gradually increasingly likelihood of occurrence as the interval elapses. The unit schedules and the overcomponent schedules are not always independent (for examples and greater discussion see Marr, 1979).

A chain schedule is a variant of second-order schedules, except that a series of responses must be completed, in lock step, before delivery of the consequence (Catania, 1991; Thomas, 1964). As the name implies, a break in any link invalidates the entire chain. Under a chain schedule, a different stimulus is associated with each link. If no stimulus is provided, then the schedule is called tandem. Complex sequences of dissimilar responses can be established by training the chain, gradually, beginning with the link that ultimately will be closest to primary reinforcement and working backward. The role that discriminative stimuli play in supporting responding can then be examined by adding or removing exteroceptive stimulus. Laties (1972) demonstrated that the presence of exteroceptive stimuli can shift the dose–effect curve for many drugs to the right. Stated differently, removing exteroceptive stimuli makes behavior more sensitive to the disruptive effects of drugs and toxicants. Interestingly, basal ganglia damage associated with Parkinson's disease can make a patient more dependent on such exteroceptive stimuli (described later),

The judicious use of stimuli permits the assembly of extended chains or complex patterns that otherwise would be difficult, or impossible, to produce. Injudicious use can prevent even simple target responses from ever occurring. As a simple example consider several ways of defining a fixed ratio schedule of reinforcement. In the nomenclature of second-order schedules, the conventional FR 100 schedule might be designated FR 100 (FR 1 : S^P), where the paired stimulus is the mechanical click of the lever as it is depressed. If the relay click is ignored, then the FR 100 could also be designated as a tandem FR 10 FR 10 FR 10 . . . (10 times). If a different stimulus is associated with each link then it would be a chain FR 10 FR 10 FR 10 . . . . If a brief tone is presented at the end of each FR 10, then this schedule would be designated as a FR10 (FR 10 : S^P). The response requirement is always the same, the differences among the different implementation of the 100-response requirement are in how these responses are chunked or in the stimulus environment supporting them.

Awareness of these properties of second-order schedules could help an investigator synthesize complex responses and avoid pitfalls. An investigator might think that a FR 100 schedule would be easier to perform if broken into smaller chunks, following the rule-of-thumb that large jobs should be broken into smaller ones. This could be a mistake since the chain version of this schedule results in ragged, poorly maintained performance as compared with the tandem version (which is a simple FR 100). The stimuli associated with the early portions of the chain are never associated with the reinforcer, but instead are consistently associated with extinction (the absence of reinforcement). Performance *can* be enhanced enormously by presenting a brief stimulus, *paired* with primary reinforcement, at the end of successive 10-response units [FR 10 (FR10: S^p)], and large ratios can be maintained with this arrangement (Thomas, 1964; Findley and Brady, 1962). In all cases the response requirement is the same, but the stimulus arrangements differ in subtle but behaviorally very important ways.

A complex performance can be assembled from collections of simpler units with the result that a parsimonious account of what appears to be complex behavior is possible and impressively complicated behavior can be understood by relatively well-understood and simple processes. Even "erroneous" responses can be viewed as a natural result of the schedule maintaining behavior. An experiment reported by Newland and Marr (1985) exemplifies this principle, even though it is not directly related to motor function. In this experiment, a complex performance called match-to-sample (in which pigeons peck a key whose color matched a sample key) was placed under an overall schedule of reinforcement, and thereby could be conceptualized as a unit of a second-order schedule. The control (no drug) performance, as well as that after administration of different doses of chlorpromazine or imipramine, was influenced by the schedule under which performance was maintained. "Errors" occurred with less frequency as the schedule requirement was nearly fulfilled, and hence, as the reinforcer was close, although many errors occurred at the beginning of the FR and FI schedules. Perfect performance was always possible during control conditions, but its occurrence depended on prevailing enviornmental contingencies. Drug-induced performance changes interacted with the particulars of the schedule in effect. Massey and colleagues (1993) reported an identical phenomenon with drug discrimination, not match-to-sample, as the unit response. When assembling complex response repertoires, the roles of discriminative stimuli, as well as the schedule under which the behavior is maintained, all determine how effectively the target response can be synthesized, the variability seen in behavior, the pattern of errors, and the effects of chemical exposure.

5. Linking Complex Behavior and Neural Function

Discriminative stimuli and consequences have powerful influences over chemical-induced perturbations of behavior. Conclusions about the ability of the subject to perform the task cannot be rendered in the absence of knowledge of how performance is maintained. As indicated in the discussion of the different ways of producing a FR100 schedule, an animal capable of behaving is a particular way may fail to do so simply because of the way that performance is maintained. As indicated in the schedule of match-to-sample, so-called errors probably reflect natural variability that is subject to environmental control.

Conditions influencing the integrity of unit schedules in the face of chemical exposure and the role of the brief stimulus could help both methodologically and conceptually in synthesizing a response and understanding how chemicals interact with motor function (Marr, 1979). Pharmacological intervention can influence within-component or overcomponent schedules differently, depending on the drug and the schedule. The preservation of schedule patterns in both the unit schedule and the over component schedule after drug challenge has been reviewed by Marr (1971, 1979). The influence of chlorpromazine on a fixed-interval unit schedule or on a fixed ratio over component schedule resembled the drug's effects when these schedules are presented in isolation. This observation suggests that knowledge of drug or toxicant effects on simple schedules will apply to complex assemblies of schedules.

The arrangement of discriminative and reinforcing stimuli can account for behavioral units sometimes called motor programs that arise in studies of "knowledge of results" and "closed-loop systems" (reviewed in Brooks, 1986). For example, motor programs are said to develop with repeated training, especially when the behavior is supported by discriminative and consequential stimuli (Brooks *et al.*, 1978); this emergence has even been called "understanding" the behavioral requirements (Brooks *et al.*, 1983). The attribution entailed by "understanding" raises the interesting possibility that the word refers in part to the emission of a structured response whose occurrence is more likely with adequate stimulus support. The emergence of coherent response units with repeated training is a consistent feature seen in vertebrates and is sensitive to the arrangement of discriminative and reinforcing stimuli, even when the sensory stimuli are proprioceptive ones like muscle spindles or stretch receptors. The private nature of proprioception raises difficult methodological problems (mostly involving ways of making the private stimuli public), but the principles of behavioral control still apply.

Motor programs may require an intact basil ganglia and seem to be susceptible to damage in these areas (see following discussion). An appreciation of the determinants of behavior under chain or second-order schedules could provide a link between a basic area of behavioral research and research on a region of the brain, whose damage results in such disorders as Parkinson's disease. Schneider and associates (1988) reported a distinction between classical Parkinsonian signs (cogwheeling, rigidity) and other behavioral manifestations of MPTP exposure. Before the classical signs appeared, several monkeys had great difficulty in executing a task that could be described as a chain: the monkeys held a lever down until one of four buttons lit, then the monkey had to press the lit button. Prior to exposure the response was executed rapidly but after exposure, and before other signs appeared, there were long pauses between responses, with occasional periods of freezing in mid-response. Responding could be reinstated by providing physical guidance. Motoric freezing or powerful interactions between stimulus support for behavior and Parkinson's disease (e.g., Stern *et al.*, 1983; Flowers, 1978; Bloxham *et al.*, 1984) as well as with drugs, such as chlorpromazine, that produce acute Parkinsonian effects (Laties, 1972). Flowers (1978) and Stern and colleagues (1983) both noted that difficulty in performing sequential tasks was exacerbated by the absence of stimulus support. Experimental analysts of behavior quickly recognize this as a distinction between chain and tandem schedules, and that the signs of Parkinson's disease were exaggerated under a tandem schedule.

The role of stimulus factors (''closed loop systems'') might be examined by comparing chain and tandem schedules or other schedules in which the presence of exteroceptive stimuli is manipulated (Laties, 1972; Thomas, 1964). This approach has been used in examination of stimulus control and it could also be useful in motor function. The identification of consistent behavioral determinants underlying these three ways of describing motor function might suggest a common link among them.

There are important advantages to conceptualizing the deficits described in this section as being related to distinctions among chain, tandem, and second-order schedules. If the conceptualizing is correct, then it could permit a straightforward application of what is known about these schedule arrangements to understanding the cognitive disruption seen in this and related disease processes. This could go a long way in organizing a variety of toxic effects under one or a small number of behavioral mechanisms. The linkage works in the other direction as well. Research on these higher-order schedules has identified highly orderly relationships in the absence of neural mechanisms. The possibility of relating order in behavior to neural mechanisms is always exciting.

C. Indirect Effects of Conventional Schedules

The procedures described earlier for differentiating a response or synthesizing response units into second-order schedules are examples of reinforcement schedules that shape a response topography directly. In these schedules, the form of the response is a *direct* variable specified in the schedule. Physically distinct operants also emerge from schedule-controlled behavior as an *indirect* variable (Zeiler, 1977). That is, they are not a target for shaping but instead emerge as a natural part of the reinforcement contingencies. Awareness of the degree to which reinforcement schedules can indirectly influence response topography can be particularly helpful in neurotoxicological assessments (Newland, 1994). If schedule-controlled operant behavior is part of a testing protocol, then awareness that the commonly used fixed ratio and fixed interval schedules indirectly produce different response topographies can be exploited to identify effects of a neurotoxicant on physical characteristics of responding.

1. Interresponse Time Distributions (IRTs)

Historically, molecular features of response topography have been examined by inspecting distributions of IRTs. Short IRTs correspond to high response rates and long IRTs to low rates. The histogram used to show IRT distributions provides detailed information about the topography and pattern of responding. Newland and Weiss (1990, 1992) reported on drug and toxicant effects on molar and molecular characteristics of effortful responding. The manipulandum employed supported a rowing-like movement in which the animal had to pull with its arms and push with its feet against

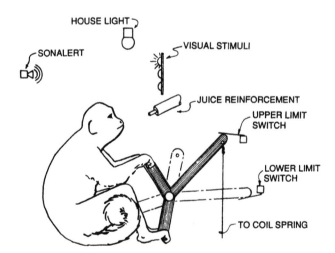

FIGURE 4 The ''rowing'' manipulandum. The monkey pulls the lever with its arms while simultaneously pushing with its legs. A response was defined in three steps: (1) closing the lower limit switch, (2) opening the upper limit switch, and (3) opening the lower limit switch. To accomplish this, the manipulandum had to be moved through an arc length of 10 cm against a 40 N (4 kg) spring. A brief Sonalert pulse followed each successful response. The visual stimuli indicated which schedule component was in effect. (From Newland and Weiss, 1990. Reprinted by permission of Elsevier Science Ltd.)

a spring that approximated its body weight (Fig. 4). Behavior was maintained by two different schedules: a fixed ratio and a fixed interval schedule. These schedules were presented in strict alternation so the two topographies were produced by an animal during a single experimental session.

The fixed ratio schedule maintained a vigorous pattern of responding: the monkey perched on the lever and quickly executed the response without pause until the response requirement was fulfilled. Such high-rate performance is characteristic of fixed ratio performance, and there are two reasons for this. First, under a fixed ratio schedule there is a direct relationship between response rate and reinforcement rate: the more responses, the more reinforcers. The second reason is more subtle but probably more fundamental. Under a fixed ratio schedule, a reinforcer is likely to be delivered during a burst of responses, which produces a string of short interresponse times. Short interresponse times are reinforced, are more likely to reoccur, and so dominate responding (Zeiler, 1977). Short interresponse times produce high rates.

The fixed interval schedule maintained a different topography. The monkey was observed sitting in a relaxed fashion, occasionally pulling the lever until the interval lapsed. The determinants of this response pattern may be the inverse of those seen with the high-rate FR performance. First, the rate of reinforcement is less directly linked to the rate of responding (as long as at least one response occurs every t seconds). Second, under the fixed interval schedule, long interresponse times (waiting) are more likely to terminate in a reinforcer whereas short interresponse times (bursts of high-rate responses) are unlikely to. Thus, moderate-to-long interrespone times are reinforced and reoccur. These produce lower rates.

Examples of IRT histograms from fixed ratio and fixed interval schedules are shown in Fig. 5 (from Newland and Weiss, 1990). Although the operant was quite different from that seen with conventional response devices such as levers and buttons, the pattern of IRTs resembles those reported with other response devices and other species (Weiss and Gott, 1972; Rice, 1988; Rice *et al.*, 1979). The histogram from the fixed ratio component was characterized by a sharp peak at 0.25 to 0.5 sec with very few IRTs at longer times. This type of responding can only be produced by unbroken bursts of rapid responses. Pauses in responding appear as long IRTs, and erratic IRT histograms, which appear when no single class dominates, are often seen in the fixed interval schedule. The distribution from the fixed interval schedule showed multiple peaks and a broader distributions. Few bursts of responses occurred, but, instead, responding was more uneven, longer IRTs ap-

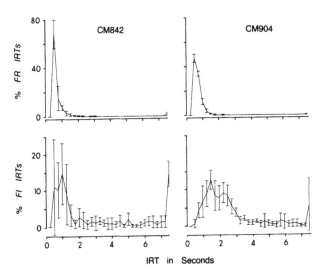

FIGURE 5 Interresponse time distributions in 0.25-sec intervals from five sequential drug-free sessions for two animals from the fixed ratio and fixed interval components. Error bars show ± two standard errors. All interresponse times longer than 7 sec are summed in the rightmost point. Interresponse times include both the time between responses and response durations. (From Newland and Weiss, 1990. Reprinted by permission of Elsevier Science Ltd.)

peared, and no IRT class dominated. Error bars indicate how much variability there is from session to session, and these, too, reflect differences in the two schedules. Greater session-to-session variability appeared in the fixed interval schedule than in the fixed ratio schedule.

2. Separating Interresponse Times from Other Measures of Topography

The interresponse time has often been used to describe the fine structure of behavior as well as response topography, but other measures, such as response duration, sometimes reveal other features of performance. Response duration is the elapsed time between the beginning and the end of a response. And usually defined, the interresponse time includes both duration and the time between responses; when interresponse times are defined this way it is impossible to isolate the effect of the toxicant on these two measures. Isolating these measures is informative if they relate to different properties of behavior and if chemicals differentially influence durations and interresponse times. A true interresponse time, then, represents the time between the offset of one response and the onset of the next, which truly is the time between responses.

Both interresponse times and durations are influenced by the schedule and by chemical exposure, but in different ways. Figure 6 shows separate histograms of durations and true interresponse times. This figure illustrates how the schedule of reinforcement, and man-

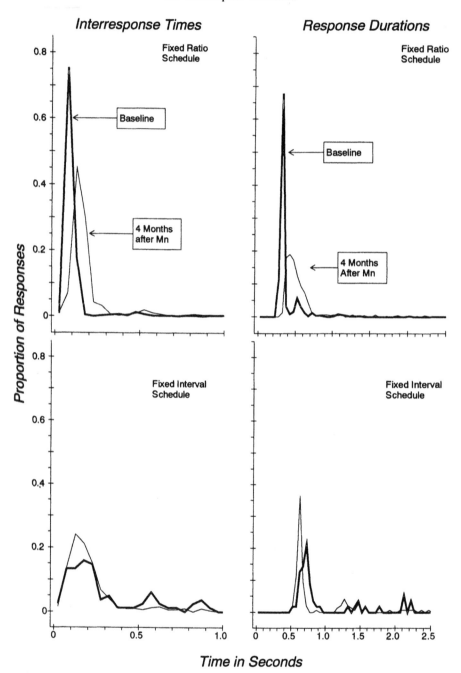

FIGURE 6 Interresponse time (left) and response duration (right) distributions from a single session for one monkey immediately prior to manganese and 4 months after manganese administration. Data from the fixed ratio component (top) and from the fixed interval component (bottom). Interresponse time is only the time between responses and does not include response durations. The response device is the one illustrated in Fig. 4.

ganese exposure, affects these two measures of performance. These data were taken from an experiment described in Newland and Weiss (1992), using the rowing device shown in Fig. 4. Under the fixed ratio schedule, both durations and interresponse times were very short and narrowly distributed before manganese expo-

sure. Two modes appeared in the distribution of durations: a large one at about 0.3 sec and another (a harmonic) at about 0.6 sec. The harmonic probably represents durations of responses that did not meet the displacement criterion and therefore comprised two nearly complete back-and-forth operations of the manipulandum. Under the fixed interval schedule, both duration and IRT histograms were similar. Both interresponse times and durations were longer and more broadly distributed than those seen under the fixed ratio schedule.

In the absence of data, one might speculate that if the response required considerable effort, then durations would be so sensitive to the physical requirement that schedule contingencies would have relatively little influence over them. However, even when responding was effortful, the schedule exerted such powerful control over topography that the modal duration under the fixed interval schedule was twice as long as under the fixed ratio schedule and some responses had durations greater than a second (see also Fowler, 1987; Fowler *et al.*, 1977).

Manganese affected durations and interresponse times differently. Manganese shifted the fixed ratio interresponse time distribution to the right, toward longer interresponse times. The peaked shape of the fixed ratio IRT distribution remained intact after manganese exposure, as did the distinction between the interresponse time and the duration distribution. The distribution of response durations was shifted to the right and the sharp peak seen in control conditions was broadened after manganese exposure. This smearing of the distribution of response durations indicates difficulty in the execution of the response. The appearance of a longer, broader distribution of durations corresponded to a large increase in the number of incomplete responses, responses that failed to meet the displacement criterion, and so probably is caused by the presence of several incomplete responses required to produce a single complete response. This effect would occur because the duration timer began with the activation of the first limit switch and continued until a response was registered (see the legend for Fig. 6 for the definition of a response). Response durations continued to be much shorter than seen in the FI schedule, even after manganese exposure, so schedule control remained intact. These effects correspond to the general conclusion that manganese affected motor function without affecting other properties of behavior. A distinction remained between the fixed ratio and fixed interval components, so both schedule control and stimulus control were unaffected. Since the animal continued to respond at a high rate, motivation was also unaffected. In this subject, and in others in the experiment (see Newland and Weiss, 1992), the effect of manganese was mostly visible in the vigorous behavior maintained by the fixed ratio schedule. Such separation between the physical executes a response and

the conditions maintaining a response have also been reported with MPTP (Schneider 1988).

D. *Isolating Motor from Other Effects*

Rate, force, duration, interresponse time, incomplete responses, or other measures describe different properties of behavior, but they are not completely independent of each other. They may be differentially affected by the conditions maintaining behavior or by toxicant exposure (Newland, 1992; Fowler, 1988). While keeping track of all these end points can present a challenge, doing so is necessary in order to identify specificity. Separations can be accomplished functionally when designing the experiments, statistically using multivariate analysis or, perhaps, theoretically by drawing on a basic literature on determinants of behavior.

1. Functional Separation

Functional separation of motor and other behavioral effects can be achieved after synthesizing an operant that has specific physical characteristics and then demonstrating that an intervention separately alters motor or other effects. For example, changes in the rate or pattern of responding in the absence of changes in response force would demonstrate that motor function is intact even when a manipulation alters the rate of responding. Conversely, changes in the execution of a response in the absence of perturbations of rate or pattern of responding under reinforcement schedules would indicate specific motor deficits in the absence of effects on, motivation, learned behavioral patterns, or adjustment to reinforcement contingencies.

When interresponse times are separated from response durations, IRTs may reflect the influence of the schedule whereas durations reflect motor abilities. Clear separation between motor and other behavioral effects can be quite difficult to obtain because of the interrelatedness of different measures of motor function and behavior, but separation is possible. For example, Fowler and colleagues (1991) used manipulanda that sustained either a pushing or a grasp-and-pull response against a force transducer, and recorded IRTs and durations separately. IRTs and durations were affected differently by the drugs decamethonium and haloperidol. Response durations and force were affected similarly to one another and differently from the way IRTs were affected. Since durations and IRTs provide separate information about toxicant and schedule effects, it can be useful, even necessary, to separate motor effects from schedule or even from motivational variables, and different approaches can be taken to separate these different measures of behavior.

Newland and Weiss (1992) reported a sharp separation of motor and other behavioral effects produced by relatively low levels of manganese exposure using the rowing device exemplified in Fig. 4. Cebus monkeys exposed to cumulative doses of 5 to 30 mg/kg of manganese, doses that produced no overt signs, nevertheless showed considerable changes in how the device was moved against a spring adjusted to approximate the body weight of the animal. A 10 to 100-fold elevation in the number of uncompleted responses appeared, even though the rate and pattern of responding were unchanged. As pointed out earlier, these incomplete responses were responsible for the broadening of the distribution of response durations in Fig. 6. In the face of increased difficulty executing the response, the monkeys continued to respond at high rates on the fixed ratio schedule. Since low rates continued to appear on the fixed interval schedules, it is possible to surmise that differential stimulus control by the stimuli associated with the schedules was unaffected. That is, since differential responding remained intact, stimulus control and the specifics of schedule performance were unaffected.

In another example of functional separation and drug interactions, Falk (1969) trained rats to maintain a force within a narrow band for a reinforcer. d-Amphetamine generally increased a measure called "work rate" (percentage of session time spent with paw on the lever) at doses that had little effect on other measures of motor function, suggesting a behavior effect of this drug in the absence of clearly defined motor effects. Chlorpromazine had roughly opposite effects, motor effects appeared at doses that did little to overall response rates. In addition, chlorpromazine, at doses that otherwise were behaviorally inactive, blocked the extremely disruptive effects of high doses of d-amphetamine.

Conclusions about the separation between motor and other behavior effects were possible in Falk's (1969) report in part because the investigators trained unusual responses. But another important feature is the collection of multiple measures of responding. Monitoring motor function is a data-intensive endeavor that may be necessary to demonstrate selectivity since demonstrating selectivity requires that one dependent measure varies while another does not (Fowler, 1987).

Sometimes separating motor from other effects can be accomplished by the deliberate production of a motor deficit. Laties and Evans (1980) identified changes in discriminative performance in pigeons induced by methyl mercury exposure. The disruptions were accompanied by response-rate changes and ataxia and the investigators wished to determine the extent to which ataxia accounted for the effect of methyl mer-

cury. Accordingly, they produced a reversible ataxia in two different ways: hobbling the bird by taping one foot or administering different doses of ethanol. In neither case did the changes in discriminative performance resemble the changes seen after methyl mercury, indicating that the effect of methyl mercury had different sources and was not an artifact of motor impairment.

2. Explicitly Removing Behavioral Influences: The Measurement of Tremor

Functional separation can also require the isolation of motor end points that have relatively little behavioral influence, such as tremor. Tremor is a property of the motor system that can reflect certain nervous system disorders (Findley and Capildeo, 1984). It is also relatively stable, within individuals and across different measurement settings, suggesting minimal influence from the sorts of conditioning factors that influence other measures of motor function (Marsden et al., 1969a,b). Measurement considerations, and artifact, play an important role in the evaluation of tremor and can influence the conclusions drawn.

In order to detect normal tremor it is necessary to have precise control over the position of the limb being measured. If not, then the isolation of behavioral influences can be difficult and "movement artifact" may appear in the measure (for example, see the earlier discussion on evaluating tremor with activity chambers). With linguistically competent humans, some behavioral control can be exerted through directions, but other species require greater sophistication in behavior control. If the animal can be trained to hold a limb still or to move it only at certain times, then tremor can be detected by careful application of the behavioral considerations reviewed earlier. Newland (1988) reviewed both measurement and behavioral considerations in tremor assessment with nonhuman animals.

The apparatus shown in Fig. 2 was used to evaluate tremor in squirrel monkeys. Figure 7 shows how a combination of measurement and behavioral considerations can be applied to the measurement of motor function. The behavior of maintaining angular position within a sufficiently restricted range was not difficult to establish. Once this was accomplished, angular position was filtered, digitized, and stored for spectral analysis.

Tremor might be described as fluctuations in position (top row of Fig. 7), but this description is relatively uninformative of the tremor spectrum since it emphasizes low frequencies (1 Hz and less). Close inspection of the position record after 1 mg/kg of oxotremorine reveals tremorous deviations, but the magnitude of the deviation is small compared with the slow changes in position. The expression of these same data in units

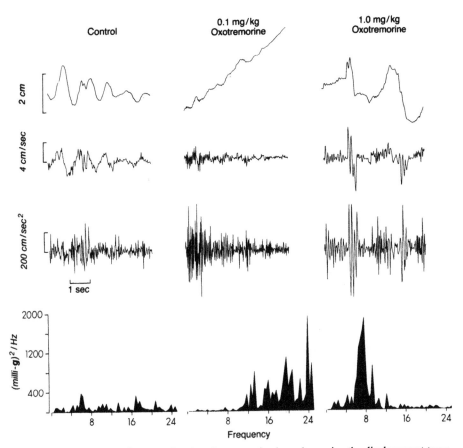

FIGURE 7   Representative records taken from a squirrel monkey using the displacement transducer illustrated in Fig. 2. Each column contains data from a different dosing condition. The first row shows displacement of the bar through the 5.12-sec sampling interval. The second row shows the first derivative with respect to time (velocity) of the records in row 1. The third row shows the second derivative in units of acceleration. The bottom row contains the power spectrum of the acceleration record in row 3. Oxotremorine (0.1 mg/kg) elevated power in the higher frequencies and reduced power at lower frequencies; 1.0 mg/kg resulted in a single, pronounced mode at about 6.0 Hz. (From Newland, 1988. Reprinted by permission of Elsevier Science.)

of acceleration, such as would be obtained from an accelerometer, highlights those frequencies of interest, those in the 4- to 25-Hz band. Since acceleration is the second derivative of positive, it can be obtained from a position signal by differentiating position twice with respect to time (third row in Fig. 6). Filtering and other smoothing techniques must be applied to handle the amplification of noise that accompanies such differentiating (Newland, 1988; Bendat and Piersol, 1971).

Figure 8 shows tremor spectra after different doses of oxotremorine and Fig. 9 shows spectra from different doses of ethanol. In each case, dose- and frequency-specific effects can be seen, effects that could reflect some of the well-known motor components of the effects of these two drugs. For example, the broad-band reduction in power at frequencies higher than 1 Hz seen after ethanol could account for motor slowing, retarded reflexes, and increased reaction time seen

with this drug since tremor appears to provide the nervous system with the ability to respond quickly (Elble and Koller, 1990; Marsden *et al.,* 1967; Newland and Weiss, 1991). Detecting a reduction in normal tremor, which is already small in amplitude, requires a precise measurement strategy but it can also reveal important drug effects, such as those seen with ethanol. The effects of oxotremorine probably reflect frequency-specific muscarinic influences over tremor (Fowler *et al.,* 1990; Gerhart *et al.,* 1982; Pinder, 1984).

The analysis of tremor has been used to isolate motor effects of other compounds, including those with basal ganglia effects in rodents. Fowler and colleagues have adapted a similar strategy with rats to that reported in this section. They differentiated the emission of forelimb force in a rat such that a specific force within a band was required for reinforcement. High-frequency deviations reflecting tremor were spectral analyzed and

tranquilizing and motor effects) from two sedatives (whose tranquilizing effects were accompanied by qualitatively different motor effects). Remarkably, the

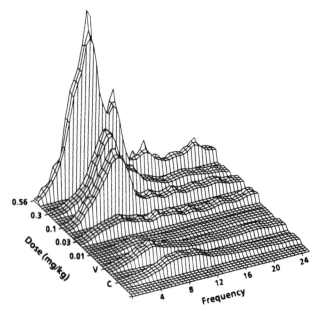

FIGURE 8 Power spectra of acceleration records over a range of frequencies up to 25 Hz (not including 0 Hz) for control, vehicle, and five doses of oxotremorine. Each ''wall'' representing the vehicle and drug sessions indicates an ensemble average from at least two sessions. The control record represents an ensamble average of many sessions.

were used to provide detailed separation of the effects of dopaminergic and cholinergic compounds (Fowler *et al.*, 1972, 1990).

3. Statistical Separation

One way of managing the large quantities of data acquired with investigations of motor function is statistical reduction through multivariate techniques as discriminant analysis or factor analysis. Walker and colleagues (1981) provide an example of this approach. They examined chlorpromazine, haloperidol, chlordiazepoxide, and pentobarbital on duration and rate of lever pressing maintained under a fixed ratio 10 schedule of reinforcement. Durations and response rates provided ''separate and nonredundant'' information about the effects of chlorpromazine, haloperidol, chlordiazepoxide, and pentobarbital on responding on conventional levers. One discriminant function produced 100% separation between neuroleptics (which had both

FIGURE 9 Spectral analysis of the second derivative of bar position obtained after control, vehicle, and those doses of ethanol (p.o.) that differ from control. Confidence intervals of 95% for the geometric mean of the control sessions are represented by the shaded areas. Symbols represent the geometric mean of the spectral estimate from ethanol or vehicle sessions. A symbol is filled if (1) the spectrum is different in shape from control conditions *and* (2) if the 95% confidence intervals do not overlap. (From Newland and Weiss, 1992. Reprinted by permission of Rutgers University Press.)

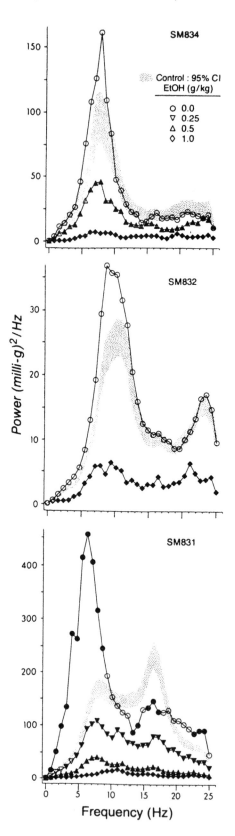

separation was accomplished using only two measures, rate and duration, and nonlinear combinations of them (squares of the measures, and interactions). A second function provided some separation between chlordiaz-epoxide and pentobarbital. Their analysis resulted in a reasonable reduction of a fairly large collection of measures to two dimensions. Multiple measures of the physical characteristics of responding, coupled with measures of rate and pattern, multiplied by different doses of different classes of chemicals can quickly result in a forest of data that can benefit from an empiri-cally based scheme for classification.

The value of a statistical approach in classifying mea-sures of behavior has been demonstrated by Gentry and associates (1983), who examined two classes of mea-sures of fixed interval performance and confirmed that a factor-analytic approach can produce the same conclu-sion that experimental approaches can. (This does not imply that statistical correlations substitute for experi-mental control!) It is well known that fixed interval schedule performance can be described in measures of overall rate and temporal pattern, and that these charac-teristics of behavior can be differentially influenced, de-pending on the intervention. That is, they can be sepa-rated functionally. Gentry and associates (1983) factor analyzed different measures of fixed interval perfor-mance and found two factors; one that can be described as a rate and a second as a pattern measure (e.g., an in-creasing probability of responding as the interval elapses). That the factor analytic approach was able to separate measures that in other experimental settings are separable supports the potential value of this strategy.

4. Theoretical Separation

A rich literature has developed around a class of procedures called concurrent schedules of reinforce-ment (de Villiers, 1977; Davison and McCarthy, 1988) in which, as the name implies, two or more schedules are presented simultaneously. Typically some variant of a variable interval or random interval schedule is used, in part because it maintains relatively steady rates of responding. For example, under a schedule described as a Conc VI 15″ VI 60″ schedule, responses on the left lever produce a reinforcer at unpredictable times but with an average interreinforcer interval of 15 sec. The schedule operating on the other lever operates similarly but with a 60-sec average interreinforcer inter-val. A response cost is almost always imposed on switching levers to enhance the discriminability be-tween the schedules available on the levers. The study of these schedules has spawned attempts to describe response choice quantitatively and to use the resulting

mathematical description of behavior to provide theo-retical separation of different influences over behavior.

Concurrent schedules have received little attention from behavioral toxicologists, and only in recent years have they been used in analyses of drug effects (Hey-man, 1983; Heyman and Seiden, 1985; Heyman and Monahgan, 1990; Shah *et al.*, 1990; Ziriax *et al.*, 1993). An approach pertinent to the current topic is one that has been taken by Heyman and Monaghan (1990). Un-der certain conditions, response rates on a particular lever can be described by the following equation:

$$B_{\text{left}} = \frac{B_{\max} R_{\text{left}}}{R_{\text{half}} + R_{\text{left}}},$$

where B_{left} is the response rate (behavior) on the left lever, R_{left} is the reinforcement rate for responding on the left lever, R_{half} is the fitted constant interpreted as the reinforcement rate corresponding to the half-maximal response rate, and $B_{\max}$ is the fitted constant interpreted as maximum response rate.

This equation arose from early research on concur-rent schedules and applies to multiple variable interval schedules, in which VI schedules with different rein-forcement densities are presented in sequence, as well as to concurrent VI schedules, in which they are pre-sented simultaneously, under certain conditions (Herrnstein, 1970; McCarthy and Davison, 1989; de Villiers, 1977). It has been suggested that this equation permits the separation of motor capabilities from moti-vational influences through the constants $B_{\max}$ and R_{half}. The positive asymptote, $B_{\max}$, should be influenced by physical characteristics of the lever or by impairment of coordination, strength, or the ability to execute the response, at least insofar as it represents the maximal rate at which the animal could respond. R_{half} was said to be influenced by the availability of other reinforcers and represents the efficacy of the reinforcer used (Herrnstein, 1974).

Heyman and Monaghan (1987) showed that $B_{\max}$ is sensitive to lever weight but not deprivation under some conditions, whereas R_{half} is sensitive to depriva-tion but not lever weight; the ability of these interven-tions to produce separate effects is supportive of the utility of this analysis. Neuroleptics decreased both $B_{\max}$ and R_{half} in a dose-dependent fashion whereas am-phetamine usually had opposite effects. Other drugs were investigated, but with less consistent results (for a summary see Heyman and Monaghan, 1990).

Peele and Crofton (1987) attempted to separate mo-toric and motivational effects of cypermethrin and per-methrin using the parameters $B_{\max}$ and R_{half}, and the results were generally consistent with previously estab-lished behavioral effects of the compounds. The disrup-tion of behavior of permethrin was unrelated to the schedule and appeared to represent motor deficits, especially fine tremor, produced by this drug. Accordingly, its effect on B_{max} was greater and appeared at lower doses than its effect on R_{half}. The effect of cypermethrin on R_{half} was greater than its effect on B_{max}, and the authors interpreted this as consistent with the tendency of cypermethrin to increase the rate of other activities and, accordingly, decrease the efficacy of the reinforcer under investigation with relatively little effects on motor function.

The application of concurrent schedules to an evaluation of chemical exposure has not been fully developed and no secure predictions can be offered about their ultimate value. The goal described in Heyman and Monaghan (1990) resembles that of others in attempting to separate motor effects from "hedonic" or other schedule effects. Although the theory from which the equation is drawn successfully accounts for behavior under a few reinforcement schedules, its extension to chemical perturbations is less certain.

Some difficulties in applying this equation have been pointed out by Heyman and Monaghan (1990). A practical consideration is that the experiments take a long time to conduct. Five or more different reinforcement rates need to be used to fit the curve described in the equation and, if multiple doses are used, that can take a very long time. Testing all five reinforcement rates in a single session overcomes some of these difficulties. Peele and Crofton (1987) achieved a stable baseline after 60 sessions, making it possible to develop a dose–effect curve much more quickly since the effect of one dose on all five reinforcement rates could be tested in a single session.

The separation of hedonic and motor effects depends on the independence and stability of B_{max} and R_{half}. Heyman and Monaghan (1989) provided evidence that these parameters are independent but others have reported interdependence (McDowell and Wood, 1985; Snyderman, 1984; McSweeney, 1975; see discussion in McCarthy and Davison, 1989). The magnitude of the parameters can drift somewhat from session to session and a stable baseline is essential in evaluating acute chemical effects or in evaluating perturbations due to a chronic effect. Drifting baselines present many problems.

Concurrent schedules are likely to be valuable additions to the behavioral toxicologist's repertoire (Newland et al., 1994). The class of procedures provides a technique for presenting a choice to a subject, for gaining control over and accounting for other reinforcers, and perhaps for isolating sensory, motor, and even memory from other behavioral influences. But the application of these procedures to behavioral pharmacology and toxicology is still in its nascent stages so it is

too early to say exactly where it will prove most useful. Associated with concurrent schedules is a theoretical framework, called the matching law. Whether that theory will be of value in separating motor effects from other chemical effects remains to be seen.

E. Dynamic Effects during Response Maintenance

1. Stimulus Support for the Response

The three term contingency described earlier makes it clear that operant behavior falls under the influence of stimuli correlated with the availability of a consequence. These stimuli can be exteroceptive, proprioceptive, or they can derive from the organism's own behavior. The degree to which exteroceptive stimuli support behavior, that is, the degree to which behavior is under exteroceptive stimulus control can modulate the effects of chemical exposure (Laties, 1972; Laties and Evans, 1980; Rees et al., 1985; Wood et al., 1983). This phenomenon has played an important role in characterization of chemical-induced impairment of stimulus control over behavior.

Rees and colleagues (1985) and Wood and associates (1983) examined stimulus support for responding with a fixed consecutive number 8 (FCN-8) schedule. In this arrangement, a rat was presented with two levers. After eight or more presses on the left lever a press on the right lever resulted in access to a reinforcer. The behavior of switching from the left to the right lever came under the control of the number of left lever presses, or some covariate like elapsed time. Thus, behavior was under the control of the animal's own prior behavior (or elapsed time). Under some conditions, a light was turned on after the eighth left lever press. This light supported the behavior of switching from the left to the right lever while greatly reducing the number of premature switches. In other words, the light exerted stimulus control over the behavior of switching and enhanced the precision of switching. In the presence of light, behavior was less disrupted by acute doses of toluene, amphetamine, and methyl mercury (Laties and Evans, 1980; Rees et al., 1985; Wood et al., 1983). That is, the presence of an exteroceptive stimulus shifted the dose–effect curves to the right. Moreover, the exteroceptive stimulus also facilitated the development of behavioral tolerance to the degrading effects of amphetamine.

Behavior under strong stimulus control is relatively resistant to chemically induced perturbations (Laties, 1975; Thompson, 1976, 1978; Thompson and Moerschbaecher, 1979). Stimulus support also facilitates the acquisition and maintenance of motor acts. Conceptually there is much overlap between this phenomenon and the study of "closed-loop" systems or sensory feedback in the motor-learning literature (e.g., Brooks, 1986). When proprioceptive control over muscle movement is removed by deafferentation, other sources of exteroceptive control develop (Taub *et al.*, 1975; Taub and Berman, 1968; Wylie and Tyner, 1981, 1989) and this underlies adjustment to the impairment produced by the lesion. Proprioceptive control over limb movements or positions is important in guiding limbs, but if such stimulus support is removed then other control from other stimulus modalities, such as vision, take its place and can facilitate adjustment to impairment, when increased effort or loss of reinforcement is involved.

Another similarity between proprioceptive and exteroceptive stimulus control is suggested by an experiment reported by Fowler and colleagues (1991). They trained rats to press against a force transducer with either a low (8 g) or high (32 g) force. The high force was more resistant to disruption by decamethonium and haloperidol than the low force condition, an observation that Fowler and colleagues (1991) attributed to enhanced proprioceptive control exerted by pressing with a high force. If so, then analysis of such effects can be achieved similarly to the analysis of how discriminative stimuli alter other behavioral effects of drugs.

Diseases that selectively degrade portions of the nervous system can reveal the role of those regions by showing how the nervous system performs in their absence. The importance of strong stimulus control over movement is particularly apparent when the basal ganglia have suffered damage, such as in Parkinson's disease, which damages basal ganglia, especially the substantia nigra. In these cases, exteroceptive stimulus control plays an important role in the control of movement. Clinical signs of Parkinson's disease include rigidity, difficulty initiating or maintaining a stride, and abrupt stops when executing a motor act. Special problems arise when the patient attempts to accomplish an act without visual guidance or when attempting to do two things at once (Flowers, 1976; Brooks, 1986). An example that is sometimes used is that a Parkinsonian patient can get out of a chair or can shake hands, but if rising from a chair to shake hands is attempted simultaneously then the patient freezes in mid-movement. He or she will describe the experience by saying that movements that once could be accomplished without thought now require conscious planning and attention.

Damage to the substantia nigra in Parkinson's disease or to another output region of the basal ganglia, the globus pallidus, impairs simple tracking and the impairment is greatly enhanced if visual guidance is not provided, indicating greater reliance on visual stimuli

(Bloxham *et al.*, 1984; Flowers, 1976; Stern *et al.*, 1983). Similar deficits have been noted when the globus pallidus is damaged (Hore *et al.*, 1977; Vaillet *et al.*, 1987). Hore *et al.*, (1977) investigated a well-trained step-tracking response in monkeys. Monkeys were trained to move the forearm with the elbows as the pivot between two locations. Successful execution of this response did not require exteroceptive stimuli when the nervous system was intact; the training history and proprioceptive afferent from muscle were sufficient to produce adequate performance prior to intervention. However, when the globus pallidus was abruptly, and acutely, made inoperative by cooling, visual support became critical to the successful execution of the task. Restoring the visual stimulus restored baseline performance. The behavioral impairment associated with profound damage to these output regions of the basal ganglia was overcome by providing a visual stimulus guiding movement, a sort of visual prosthesis.

Even the profound disruption in walking caused by Parkinson's disease can be overcome by placing parallel strips of tape, at about stride's length, across the patient's path so that the patient walks across them like one would walk across ties along a railroad track (Forssberg *et al.*, 1984). The effect is striking; at the beginning of such a path the patient walks effortlessly to the end of the strips, and then akinesia reappears.

The general conclusions that exteroceptive stimuli can act prosthetically to overcome impairment has been seen in a variety of settings and accordingly points to a generality that might unify some observations reviewed in this section. Although still very speculative, it seems that some of the roles of the basal ganglia and of exteroceptive stimuli overlap and that removing the exteroceptive stimulus or impairing the basal ganglia might produce similar behavioral manifestations.

2. Tolerance

Tolerance "might be viewed as a special form of adaptation in which continued exposure to a chemical agent that results in an increased resistance to the noxious consequences of the exposure" (Hammond and Beliles, 1980, pp. 414–415). Tolerance, then, would appear as a shift to the right in the dose–effect resulting from repeated exposure. The phenomenon of tolerance can be traced to any of a number of mechanisms, including physiologically based mechanisms such as induction of microsomal enzymes or stimulation of metallothionine, as well as behavioral mechanisms of tolerance resulting from the opportunity to behave while under the influence of a chemical.

Behavioral tolerance has been demonstrated with a wide range of drugs (Balster, 1985; Corfield-Summer and Stolerman 1978) and some toxicants (Wood *et al.*,

1983), and the phenomenon enjoys a solid empirical and conceptual foundation. Behavioral tolerance can be separated from physiological tolerance on the basis of whether the exposed organism has had an opportunity to perform while under the influence of a chemical. For example, repeated, acute exposure to a compound can be achieved by administering the compound either before or after an experimental session. If tolerance appears only when exposure is before the experimental session, i.e., if the subject has had an opportunity to behave while under the influence of a compound, then behavioral tolerance is responsible for the reduced sensitivity to the chemical.

Behavioral tolerance is most likely to appear if reinforcement loss is a result of the behavioral actions of a compound (Corfield-Summer and Stolerman, 1978). In an early demonstration of the phenomenon, Schuster and colleagues (1966) trained rats to respond to either a differential reinforcement of a low rate schedule (DRL 30″), in which reinforcers were delivered only if responses were spaced 30 sec apart, or a fixed interval 30″ schedule. Amphetamine increased the rate of responding on both schedules, but the rate increase resulted in reinforcement loss only under the DRL schedule. Accordingly, after five repeated dosings, tolerance to the rate-increasing effect of amphetamine appeared only on the DRL schedule.

Schuster and colleagues (1966) examined tolerance to reductions in response rates, but tolerance has also been reported on the motor effects of drugs. Johanson and associates (1979) reported tolerance to both motor and behavioral effects of amphetamine, and Tang and colleagues (1988) successfully distinguished behavioral tolerance to midazolam's disruption of discriminative motor control from tolerance to nonspecific rate-reducing effects. The latter study demonstrated not only that behavioral tolerance can be distinguished from other forms of tolerance, but also that specific forms of behavioral tolerance can be specified. In that study, rats were trained to press on a leaf spring attached to a load cell transducer with a force of between 0.147 and 0.265 Newtons (equivalent to 14.7–26.5g) and sustain that force band for 1.5 sec. Midazolam was administered before each session to one group of rats or after each session to another group. A measure called "work rate" (analogous to response rate) was constructed from the proportion of the session time spent with the paw on the lever. Other measures of motor function included the number of entrances into the force band and "tonic accuracy," or proportion of total response time that was within the criterion.

Complete tolerance to midazolam's decrease in work rate appeared in both the "before" and "after"

groups, indicating that the opportunity to respond was not necessary for the development of tolerance. However, partial tolerance appeared to midazolam's disruption of discriminative motor function, and it appeared only in the "before" group. The appearance of tolerance in the "before" group but not in the "after" group suggested that the opportunity to behave while under the influence of the drug was required for tolerance to the motor disruption to develop. This study is important because of its clear demonstration that tolerance to one effect of midazolam did not directly carry over to tolerance to another and that different manifestations of tolerance can be distinguished.

3. Adjustment to Impairment

Tolerance, behavioral or otherwise, results from chronic exposure to a compound and is expressed as reduced sensitivity to the effects of that compound. The mechanism of behavioral tolerance, overcoming reinforcement loss, can accommodate adjustments to chronic impairment, which can endure long after chemical exposure has ended. Adjustment to this impairment does not necessarily make the organism less sensitive to the chemical, but it does entail "resistance to noxious consequences of exposure," which forms part of some definitions of tolerance. A simple experiment by Taub and Berman (1968) illustrates the conditions under which adjustment to impairment develops. They deafferented the arm of a monkey by severing sensory fibers in the dorsal root ganglia. The limb fell into disuse and the other arm was thereafter used almost exclusively. If both limbs were deafferented, however, use of the limbs, in the absence of proprioceptive feedback, emerged over the course of months. When the animal was required to use the arm it did so, but only then did adjustments to the sensory impairment appear.

Fowler and colleagues (1993) presented unambiguous evidence for tolerance to the physical execution of a response even when no evidence appeared for tolerance to overall response-rate reductions. Rats pressed a conventional lever under a fixed ratio schedule of reinforcement. Triazolam administered before behavioral sessions nearly doubled the response durations in rats with limited opportunity to behave under the drug's influence. However, little increase was seen in rats administered triazolam chronically before each session. Although increases in response duration may have increased the density of reinforcement somewhat, the net effect would be small compared with decreases seen in overall response rates, which displayed no tolerance. The tolerance may have been related to the increase in the overall effort exerted as durations increased. Whatever the mechanism, this work is a clear demonstration that adjustment to impairment in the molecular characteristics of response execution that would have been overlooked if the measurement of response execution had been omitted. This work suggests that motor properties of that anxiolytic may be susceptible to tolerance even when sedation is not.

Newland and Weiss (1992) reported data suggesting an adjustment to impairment subsequent to manganese exposure. Cebus monkeys received acute injections of manganese in doses of 5 to 10 mg/kg/injection. After cumulative doses ranging from 5 to 20 mg/kg of manganese, an increased number of incomplete response was shown when the rowing device (described earlier) was operated under a fixed ratio schedule of reinforcement but there was no reinforcement loss under this schedule. The monkeys continued to collect as many reinforcers as before, although the increased response durations and elevated number of uncompleted responses suggest that they expended more effort for the same rate of reinforcement. Figure 10 illustrates the in-

creased response durations that accompanied the elevated number of uncompleted responses for one monkey. Since this monkey was pulling its body weight, the increased durations required additional force so the monkey was working harder for the same number of reinforcers. Several months later, response durations assumed a different pattern; they became shorter, less variable (note the smaller error bars), especially for the first responses in the ratio, and more like the pattern seen before manganese exposure. Little change, other than less variability, was seen in the mean interresponse time.

The data in Fig. 10 are consistent with an adjustment to impairment that took two forms. An adjustment associated with a reduced effort for each reinforcer could have produced the decreases in response durations seen after many months of exposure. No such adjustment appeared in the interresponse times, but these did reflect reduced variability and a reinstatement of schedule effects visible before manganese administration.

Adjustment to impairment can be mediated by mechanisms similar to behavioral tolerance. If chronic impairment can be likened to chronic drug exposure the extension is direct. Reinforcement loss or increased effort per reinforcer might be required for adjustment to occur. As Taub and Berman (1968) showed, if there is an alternate way to collect a reinforcer then adjustment is less likely to develop.

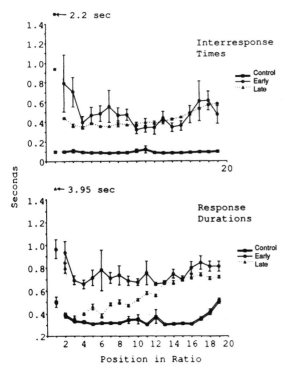

FIGURE 10 Interresponse time (IRT; top) and response durations (bottom) from the fixed ratio component of sessions on days 5 (control), 104 ("early"), and 173 ("late") for one monkey. The abscissa shows the location of the duration or IRT in the ratio. Postreinforcer pauses are not included. For example, the response duration over point 1 is the average duration of the first response in the ratio averaged across the session. The IRT over point 1 is the average time between responses 1 and 2 with the average taken across the session. Only 19 durations are shown because sometimes the subject perched on the bar to collect the reinforcer, thereby producing a very long duration. All IRTs are shown; 20 responses contain 19 IRTs. The error bars show the standard error across the mean taken from the session. (From Newland and Weiss, 1992).

III. Some General Conclusions

This chapter has attempted to describe some ways in which principles derived from the experimental analysis of behavior, and of chemical interactions with behavior, can contribute to the characterization and understanding of motor disfunction. Some contributions may be new to the study of motor effects. For example, the potential role of reinforcement loss or increased effort per reinforcer could bring a large body of research on behavioral tolerance to bear on the problem of adjustment to impairment. Behavioral principles could also unify phenomena in cases where similar effects have been studied under different names. An example is the examination of the role of exteroceptive stimulus control in mediating drug effects, which resemble discussion of closed-loop systems and cortical control by exteroceptive stimuli in overcoming deficits associated with basal ganglia disorders. In the latter case, either discipline can inform the other.

There is a good chance that initial identification of motor neurotoxicants will be accomplished by epidemiological studies or, preferably, by screens. Accordingly, a portion of this chapter was devoted to some issues that must be considered when designing screening strategies, including the selection of tests that tap each of the motor systems likely to be affected, proper attention to measurement and the inference that derives from the measures taken, and caution when drawing conclusions on the basis of screens alone. Even if a test used in a battery is selected because of evidence that it reveals the effect of a type of lesion, it should always be realized that single, simple tests tend to be insensitive to dose while being sensitive to a variety of neural and nonneural lesions (Gerber and O'Shaughnessy, 1986). The identification of neural and behavioral mechanisms and the full characterization of dose–effect and dose–response relationships require advanced testing.

Another goal of this chapter has been to present a case for investigating behavioral mechanisms of disruption and recovery. Motor function and dysfunction have traditionally been investigated according to principles of heterogeneous (classical) reduction (Marr, 1990). That is, behavioral phenomena have been traced to neurochemical or pathological changes in the nervous system, a strategy that has contributed greatly to our understanding of brain and behavior. However, behavioral mechanisms of action, such as enhanced reliance on exteroceptive stimulus control or reinforcement loss, can also mediate chemically induced decrements. Reducing a complex phenomenon to a parsimonious statement at the same level of analysis has been called homogeneous reduction and is exemplified by the laws of motion in classical mechanics, evolution by natural selection, or reducing volitional behavior to the three term contingency of reinforcement. When such a reduction can be accomplished then a large literature, and a variety of complex phenomena, can be accounted for by a small number of natural processes.

A final goal has been to identify some of the ways in which an understanding of behavior can contribute to the synthesis of a response for the study of dysfunction. To study disruption of motor acts, it is necessary to train a motor act in the first place. The technology to do so, and the conceptual framework supporting it, is in place and can be applied. It should be expected that investigation setting out to study a behavioral phenomenon will not feel it necessary to reinvent techniques or to create tasks according to a subjective understanding of the behavior under study. Enough is known about behavior to tailor preparations to the question and to apply concepts, such as the three term contingency, to understanding how behavior changes with chemical exposure. Careful application coupled with understanding can clarify mechanisms of change,

simplify preparations, and, when mechanisms are identified, bring a large literature to bear on a problem.

References

Anger, W. K. (1984). Neurobehavioral testing of chemicals: impact on recommended standards. *Neurotoxicol. Teratol.* **6**, 147–153.

Balster, R. L. (1985). Behavioral studies of tolerance and dependence. In *Behavioral Pharmacology: The Current Status* (L. S. Seiden and R. L. Balster, Eds.), pp. 403–418, Liss, New York.

Bannister, R. (1992). *Brain and Bannister's Clinical Neurology*. Oxford University Press, Oxford.

Barbeau, A. (1984). Manganese and Extrapyramidal Disorders. *Neurotoxicology* **1**, 13–36.

Barbeau, A., Inoue, N., and Cloutier, T. (1976). Role of manganese in dystonia. *Adv. Neurol.*, **14**, 339–352.

Bendat, J. S., and Piersol, A. G. (1971). *Random Data: Analysis and Measurement Procedures*. Wiley-Interscience, New York.

Bigelow, G. (1971). Fixed-ratio reinforcement of spaced responding. *J. Exp. Anal. Behav.* **16**, 23–30.

Blough, D., and Blough, P. (1977). Animal Psychophysics. In *Handbook of Operant Behavior* (W. K. Honig and J. E. R. Staddon, Eds.), pp. 514–539, Prentice-Hall, Englewood Cliffs, New Jersey.

Bloxham, C. A., Mindel, T. A., and Frith, C. D. (1984). Initiation and execution of predictable and unpredictable movements in Parkinsons disease. *Brain* **107**, 371–384.

Bogo, V., Hill, T. A., and Young, R. W. (1981). Comparison of accelerod and rotorod sensitivity in detecting ethanol- and acrylamide- induced performance decrement in rats: review of experimental considerations of rotating rod systems. *Neurotoxicology* **2**, 765–787.

Branch, M. N. (1984). Rate dependency, behavioral mechanisms, and behavioral pharmacology. *J. Exp. Anal. Behav.* **42**, 511–522.

Branch, M. N. (1991). Behavioral pharmacology. In *Techniques in the Behavioral and Neural Sciences: Experimental Analysis of Behavior*. Part 2 (I. H. Iversen and K. A. Lattal, Eds.), pp. 21–78, Elsevier, Amsterdam.

Brooks, V. B. (1986). *The Neural Basis of Motor Control*. Oxford University Press, New York.

Brooks, V. B., Kennedy, P. R., and Ross, H. G. (1983). Movement programming depends on understanding of behavioral requirements. *Physiol. Behav.* **31**, 561–563.

Brooks, V. B., Reed, D. J., and Eastman, M. J. (1978). Learning of pursuit visuo-motor tracking by monkeys. *Physiol. Behav.* **21**, 887–892.

Bures, J., and Bracha, V. (1990). The control of movements by the motor cortex. In *The Cerebral Cortex of the Rat* (B. Kolb and R. C. Tees, Eds.), pp. 213–238, MIT Press, Cambridge.

Burns, R. S., Chiueh, C. C., Markey, S. P., Ebert, M. H., Jacobowitz, D. M., and Kopin, I. J. (1983). A primate model of parsinsonism: selective destruction of dopaminergic neurons in the pars compacta of the substantia nigra by N-methyl-4-phenyl 1236-tetrahydoropyridine. *Proc. Natl. Acad. Sci.* **80**, 4546–4550.

Byrd, L. D., and Marr, M. J. (1969). Relations between patterns of responding and presentation of stimuli under second-order schedules. *J. Exp. Anal. Behav.* **12**, 713–722.

Catania, A. C. (1973). The nature of learning. In *The Study of Behavior: Learning, Motivation, Emotion, and Instinct* (J. A. Nevin and G. S. Reynolds, Eds.), pp. 31–70, Scott, Foresman, Glenview, Illinois.

Catania, A. C. (1991). Glossary. In *Experimental Analysis of Behavior: Part 2*. (I. H. Iverson and K. A. Lattal, Eds.), pp. G1–G44, Elsevier, Amsterdam.

Catania, A. C. (1992). *Learning*. Prentice-Hall, Englewood Cliffs, New Jersey.

Clark, R., Jackson, J. A., and Brady, J. V. (1962). Drug effects on lever positioning behavior. *Science* **135**, 1132–1133.

Corfield-Summer, P. K., and Stolerman, I. P. (1978). Behavioral tolerance. In *Contemporary Research in Behavioral Pharmacology* (D. E. Blackman and D. J. Sanger, Eds.), pp. 391–448, Plenum, New York.

Costall, B., Naylor, R. J., and Pinder, R. M. (1975). Dyskinetic phenomena caused by the intrastriatal injection of phenylethylamine, phenylpiperazine, tetrahydroisoquinoline, and tetrahydronapthalese derivatives in the guinea pig. *Eur. J. Pharmacol.* **31**, 94–99.

Creese, I., and Iversen, S. D. (1973). Blockage of amphetamine induced motor stimulation and stereotypy in the adult rat following neonatal treatment with 6-hydroxydopamine. *Brain Res.* **55**, 369–82.

Creese, I., and Iversen, S. D. (1974). The role of forebrain dopamine systems in amphetamine induced stereotyped behavior in the rat. *Psychopharmacology* **39**, 345–357.

Culberson, J. W., Tang, M., Lau, C. E., and Falk, J. L. (1990). Diazepam and discriminative motor control: acute chronic and withdrawal effects. *Pharmacol. Biochem. Behav.* **35**, 419–427.

Davison, M., and McCarthy, D. (1988). *The Matching Law*. L. Erlbaum, Hillsdale, New Jersey.

de Villiers, P. A. (1977). Choice in concurrent schedules and a quantitative formulation of the law of effect. In *Handbook of Operant Behavior* (W. K. Honig and J. E. R. Staddon, Eds.), pp. 233–287, Prentice-Hall, Englewood Cliffs, New Jersey.

Delong, M. R., and Crutcher, M. D., and Georgopoulos, A. (1985). Primate globus pallidus and subthalamic nucleus: functional organization. *J. Neurophysiol* **53**, 530–543.

Delong, M. R., Georgopoulos, A. P., Crutcher, M. D., Mitchell, R. T., Richardson, R. T., and Alexander, G. E. (1984). Functional organization of the basal ganglia: contributions of single-cell recording studies. In *Functions of the Basal Ganglia* (D. Evered and M. O'Connor, Eds.), pp. 74–77, Pitman, London.

deLorge, J. O. (1971). The effects of brief stimuli presented under a multiple schedule of second-order schedules. *J. Exp. Anal. Behav.* **15**, 19–25.

deMedinaceli, L., Freed, W. J., and Wyatt, R. J. (1982). An index of the functional condition of rat sciatic nerve based on measurements made from walking tracks. *Exp. Neurol.* **77**, 634–643.

Eckerman, D. A., Hienz, R. D., Stern, S., and Kowlowitz, V. (1980). Shaping the location of a pigeon's peck: effect of rate and size of shaping steps. *J. Exp. Anal. Behav.* **33**, 299–310.

Eckerman, D. A., and Lanson, R. N. (1969). Variability of response location for pigeons responding under continuous reinforcement, intermittent reinforcement, and extinction. *J. Exp. Anal. Behav.* **12**, 73–80.

Edgington, E. S. (1987). *Randomization Tests*. Dekker, New York.

Elble, R. J., and Koller, W. C. (1990) *Tremor*. Johns Hopkins University Press, Baltimore.

Elsner, J., Fellmann, C., and Zbinden, G. (1988). Response force titration for the assessment of the neuromuscular toxicity of 2,5-hexanedione in rats. *Neurotoxicol. Teratol.* **10**, 3–14.

Falk, J. L. (1969). Drug effects on discriminative motor control. *Physiol. Behav.* **4**, 421–427.

Fechter, L. D., and Young, J. S. (1986). Reflexive Measures. In *Neurobehavioral Toxicology* (Z. Annau, Ed.), pp. 23–42, Johns Hopkins University Press, Baltimore.

Fetz, E., and Baker, M. A. (1973). Operantly conditioned patterns of precentral unit activity and correlated responses in adjacent cells and contralateral muscles. *J. Neurophysiol* **36**, 179–204.

Findley, J. D., and Brady, J. V. (1962). Facilitation of large ratio performance by use of conditioned reinforcement. *J. Exp. Anal. Behav.* **5**, 113–166.

Findley, L. J., and Capildeo, R. (1984). *Movement Disorders: Tremor*. Oxford University Press, New York.

Flowers, K. (1978). Lack of prediction in the motor behavior of Parkinsonism. *Brain*, **101**, 35–52.

Flowers, K. A. (1976). visual closed-loop and open-loop characteristics of voluntary movement in patients with parkinsonism and intention tremor. *Brain* **99**, 269–310.

Forssberg, H., Johnels, B., and Steg, G. (1984). Is parkinsonian gait caused by a regression to an immature walking pattern? *Adv. Neurol.* **40**, 375–379.

Fowler, S. C. (1987). Force and duration of operant response ad dependent variables in behavioral pharmacology. In *Advances in Behavioral Pharmacology* (T. Thompson and P. B. Dews, Eds.), Vol. 6, Erlbaum, New York.

Fowler, S. C., Bowen, S. E. and Kallman, M. J. (1993). Practice-augmented tolerance to triazolam: evidence from an analysis of operant response durations and interresponse times. *Behav. Pharmacol.* **4**, 147–157.

Fowler, S. C., Filewich, R. J., and Leberer, M. R. (1977). Drug effects upon force and duration of response during fixed-ratio performance in rats. *Pharmacol. Biochem. Behav.* **6**, 421–426.

Fowler, S. C., Liao, R. M., and Skjoldager, P. (1990). A new rodent model for neuroleptic-induced pseudo-Parkinsonism: low doses of haloperidol increase forelimb tremor in the rat. *Behav. Neurosci.* **104**, 449–456.

Fowler, S. C., Liao, R. M., and Skjoldager, P. (1990). A new rodent model for neuroleptic-induced pseudo-Parkinsonism: low doses of haloperidol increase forelimb tremor in the rat. *Behav. Neurosci.* **104**, 449–456.

Fowler, S. C., Morgenstern, C., and Notterman, J. M. (1972) Spectral analysis of variations in force during a bar-pressing time discrimination. *Science* **176**, 1126–1127.

Fowler, S. C., and Mortell, C. (1992). Low doses of haloperidol interfere with rat tongue extensions during licking: a quantitative analysis. *Behav. Neurosci.* **106**, 386–395.

Fowler, S. C., Skjoldager, P. D., Liao, R. M., Chase, J. M. and Johnson, J. S. (1991). Distinguishing between haloperidol's and decamethonium's disruptive effects on operant behavior in rats: use of measurements that complement response rate. *J. Exp. Anal. Behav.* **56**, 239–260.

Freund, H. J. (1983). Motor unit and muscle activity in voluntary motor control. *Physiol. Rev.* **63**, 387–436.

Freund, H. J., Hefter, H., Homberg, V., and Reiners, K. (1984a). Determinants of tremor rate. In *Movement Disorders: Tremor* (L. J. Findley and R. Capildeo, Ed.), pp. 195–204, Oxford University Press, New York.

Freund, H. J., Hefter, H., Homberg, V., and Reiners, K. (1984b). Differential diagnosis of motor disorders by tremor analysis. In *Movement Disorders: Tremor* (L. J. Findley and R. Capildeo, Eds.), pp. 27–36, Oxford University Press, New York.

Gad, S. C. (1989). Screens in neurotoxicity: objectives, design, and analysis, with a functional observational battery as a case example. *J. Am. Coll. Toxicol.* **8**, 287–300.

Gentry, G. D., Weiss, B., and Laties, V. G. (1983). The microanalysis of fixed-interval responding. *J. Exp. Anal. Behav.* **39**, 327–343.

Gerber, G. J., and O'Shaughnessy, D. (1986). Comparison of the behavioral effects of neurotoxic and systemically toxic agents: how discriminatory are behavioral tests of neurotoxicity. *Neurobehav. Toxicol. Teratol.* **8**, 703–710.

Gerhart, J. M., Hong, J. S., and Tilson, H. A. (1983). Studies on the possible sites of chlordecone-induced tremor in rats. *Toxicol. Appl. Pharmacol.* **70**, 382–389.

Gibbons, R. J., Kalant, H., and LeBlanc, A. E. (1968). A technique for accurate measurement of moderate degrees of alcohol intoxication in small animals. *J. Pharmacol. Exp. Ther.* **159**, 236–242.

Gilbert, S. G., and Maurissen, J. P. J. (1982). Assessment of the effects of acrylamide, methylmercury, and 2,5-hexanedione on motor functions in mice. *J. Toxicol. Env. Health* **10**, 31–41.

Hammond, P. B. and Beliles, R. P. (1980). Metals. In *Casarett and Doull's Toxicology: The Basic Science of Poison* (J. Doull, C. D. Klaassen, and M. O. Amdur, Eds.), pp. 409–467, Macmillan, New York.

Harrison, J. M. (1991). Stimulus Control. In *Techniques in the Behavioral and Neural Sciences: Experimental Analysis of Behavior. Part 1* (I. H. Iversen and K. A. Lattal, Eds.), pp. 251–300, Elsevier, Amsterdam.

Herr, E. W., Hong, J., Chen, P., Tilson, H. A., and Harry, G. J. (1986). Pharmacological modification of DDT-induced tremor and hyperthermia in rats: distributional factors. *Psychopharmacology* **89**, 278–283.

Herrnstein, R. J. (1970). On the law of effect. *J. Exp. Anal. Behav.* **13**, 243–266.

Herrnstein, R. J. (1974). Formal properties of the matching law. *J. Exp. Anal. Behav.* **21**, 159–164.

Heyman, G. M. (1983). A parametric evaluation of the hedonic and motoric effects of drugs: pimozide and amphetamine. *J. Exp. Anal. Behav.* **40**, 113–122.

Heyman, G. M., and Monaghan, M. M. (1990). Contributions of the matching law to the analysis of the behavioral effects of drugs. In *Advances in Behavioral Pharmacology* (J. E. Barrett, T. Thompson, and P. B. Dews, Eds.), Vol. 7, pp. 39–78, Lawrence Erlbaum, Hillsdale, NJ

Heyman, G. M., and Seiden, L. S. (1985). A parametric description of amphetamine's effect on response rate: changes in reinforcement efficacy and response topography. *Psychopharmacology* **85**, 154–161.

Hömberg, V., Hefter, H., Reiners, K., and Freund, H. J. (1987). Differential effects of changes in mechanical limb properties on physiological and pathological tremor. *J. Neurol. Neurosurg. Psychiatry* **50**, 568–579.

Hore, J., Meyer-Lohmann, J., and Brooks, V. (1977). Basal ganglia colling disables learned arm movements of monkeys in the absence of visual guidance. *Science* **195**, 584–586.

Hori, K., and Watanabe, S. (1987). An application of the image processing system for detecting and controlling pigeon's peck location. *Behav. Brain Res.* **26**, 75–78.

Iversen, I. H. (1991). Methods of analyzing behavior patterns. In *Techniques in the Behavioral and Neural Sciences. Vol. 6. Experimental Analysis of Behavior. Vol. 2.,* (I. H. Iversen and K. A. Lattal, Eds.), pp. 193–242, Elsevier, Amsterdam.

Iversen, I. H., and Mogensen, J. (1988). A multipurpose vertical holeboard with automated recording of spatial and temporal visit patterns for rodents. *J. Neurosci. Methods* **25**, 251–263.

Jellinger, K. (1986). Exogeneous lesions of the pallidum. In *Handbook of Clinical Neurology: Vol. 5(49)* P. J. Vinken, G. W. Bruin, H. L. Klawans (Eds), pp. 465–489. Elsevier, New York.

Johanson, C. E., Aigner, T. G., Seiden, L. S., and Schuster, C. R. (1979). The effects of methamphetamine on fine motor control in rhesus monkeys. *Pharmacol. Biochem. Behav.* **11**, 273–278.

Jolicoeur, F. B., Rondeau, D. B., Hamel, E., Butterworth, R. F., and Barbeau, A. (1979). Measurement of ataxia and related neurological signs in the laboratory rat. *Le Journal Canadien Des Sciences Neurologiques,* **6**, 209–215.

Karson, C. N. (1983). Spontaneous eye-blink rates and dopaminergic systems. *Brain* **106**, 643–653.

Katoh, Z. (1988). Slowing effects of alcohol on voluntary eye movements. *Aviat. Space Environ. Med.* **59**, 606–610.

Kelleher, R. T., and Gollub, L. R. (1962). A review of positive conditioned reinforcement. *J. Exp. Anal. Behav.* **5**, 543–597.

King, F. A., Yarbrough, C. J., Anderson, D. C., Gordon, T. P., and Gould, K. G. (1988). Primates. *Science* **240**, 1475–1482.

Kucera, J., and Smith, C. M. (1971). Excitation by ethanol of rat muscle spindles. *J. Pharmacol. Exp. Ther.* **179**, 301–311.

Kulig, B. M., Vanwersch, R. P. A., and Wolthuis, O. L. (1985). The automated analysis of coordinated hindlimb movement in rats during acute and prolonged exposure to toxic agents. *Toxicol. Appl. Pharmacol.* **80**, 1–10.

Kuyper, H. G. J. M. (1981). Anatomy of the descending pathways. In *Motor Control, Sect. 1,* Vol. 2, *Handbook of Physiology* (V. B. Brooks, Ed.), pp. 597–566, American Physiological Society, Bethesda, Maryland.

Lathers, C. M., and Smith, C. M. (1976). Ethanol effects on muscle spindle afferent activity and spinal reflexes. *J. Pharmacol. Exp. Ther.* **197**, 126–134.

Laties, V. G. (1972). The modification of drug effects on behavior by external discriminative stimuli. *J. Pharmacol. Exp. Ther.* **183**, 1–13.

Laties, V. G. (1975). The role of discrimination stimuli in modulating drug action. *Fed. Proc.* **34**, 1880–1888.

Laties, V. G., and Evans, H. L. (1980). Methylmercury-induced changes in operant discrimination by the pigeon. *J. Pharmacol. Exp. Ther.* **214**, 620–628.

Laties, V. G., and Wood, R. W. (1986). Schedule-controlled behavior in behavioral toxicology. In *Neurobehavioral Toxicology* (Z. Annau, Ed.), pp. 69–93, Johns Hopkins University, Baltimore.

Lattal, A. (1991). Scheduling positive reinforcers. In *Techniques in the Behavioral and Neural Sciences: Experimental Analysis of Behavior* (I. H. Iverson and K. A. Lattal, Eds.), pp. 87–134. Elsevier, Amsterdam.

Lewin, R. (1984). Trail of ironies to Parkinson's disease. *Science* **224**, 1083–1085.

MacPhail, R. C. (1985). Effects of pesticides on schedule-controlled behavior. In *Behavioral Pharmacology: the Current Status* (L. S. Seiden and R. L. Balster, Eds.), pp. 519–535. Liss, New York.

MacPhail, R. C. (1987). Observational batteries and motor activity. *Zbl. Bakt. Hyg. B.* **185**, 21–27.

Marr, M. J. (1971). Effects of chlorpromazine in the pigeon under a second-order schedule of food presentation. *J. Exp. Anal. Behav.* **13**, 291–300.

Marr, M. J. (1979). Second-order schedules and the generation of unitary response sequences. In *Advances in the Analysis of Behavior.* Vol. 1, *Reinforcement and the Organization of Behavior* (M. D. Zeiler and P. Harzem, Eds.), pp. 223–260. Wiley, New York.

Marr, M. J. (1990). Behavioral pharmacology: issues of reductionism and causality. In *Advances in Behavioral Pharmacology* (J. E. Barrett, T. Thompson, and P. B. Dews, Eds.), Vol. 7, pp. 1–12, Erlbaum, Hillsdale, New Jersey.

Marsden, C. D., Foley, T. H., Owen, D. A. L. and McAllister, R. G. (1967). Peripheral beta-adrenergic receptors concerned with tremor. *Clin. Sci.* **33**, 53–65.

Marsden, C. D., Meadows, J. C., Lange, G. W., and Watson, R. S. (1969a). Variations in human physiological finger tremor, with particular reference to changes with age. *Electroenceph. Clin. Neurophysiol.* **27**, 169–178.

Marsden, C. D., Meadows, J. C., Lange, G. W., and Watson, R. S. (1969b). The relation between physiological tremor of the two hands in healthy subjects. *Electroenceph. Clin. Neurophysiol.* **27**, 179–185.

Marsden, C. D., Meadows, J. C., Lange, G. W., and Watson, R. S. (1969c). Variations in human physiological finger tremor with particular reference to changes with age. *Electroenceph. Clin. Neurophysiol.* **27**, 169–178.

Mason, S. T. (1984). *Catecholamines and Behavior*. Cambridge University Press, Cambridge.

Mason, S. T., Roberts, D. C. S., and Fibiger, H. C. (1978). Noradrenergic influences on catalepsy. *Psychopharmacology* **60**, 53–57.

Massey, B. W., McMillan, D. E. and Wessinger, W. D. (1992). Discriminative-stimulus control by morphine in the pigeon under a fixed-interval schedule of reinforcement. *Behav. Pharmacol.* **3**, 475–488.

Mattsson, J. L., Albee, R. R., and Eisenbrandt, D. L. (1989). Neurological approach to neurotoxicological evaluation in laboratory animals. *J. Am. Coll. Toxicol.* **8**, 271–286.

Maurissen, J. P. J. (1988). Quantitative sensory assessment in toxicology and occupational medicine: applications, theory, and critical appraisal. *Toxicol. Lett.* **43**, 321–343.

Maurissen, J. P. J., and Mattsson, J. L. (1989). Critical assessment of motor activity as a screen for neurotoxicity. *Toxicol. Indust. Health* **5**, 195–201.

McCarthy, D., and Davison, M. C. (1980). Independence of sensitivity to relative reinforcement rate and discriminability in signal detection. *J. Exp. Anal. Behav.* **34**, 273–284.

McDowell, J. J., and Wood, H. M. (1985). Confirmation of linear system theory prediction: Rate of change of herrnstein's k as a function of response-force requirement. *J. Exp. Anal. Behav.* **48**, 61–71.

McSweeney, F. K. (1975). Concurrent schedule responding as a function of body weight. *Anim. Learn. Behav.* **3**, 264–270.

Morse, W. H., and Kelleher, R. T. (1977). Determinants of reinforcement and punishment. In *Handbook of Operant Conditioning* (W. K. Honig and J. E. R. Staddon, Eds.), pp. 174–200, Prentice-Hall, Englewood Cliffs, New Jersey.

Moser, G. C., McCormick, J. P., Creason, J. P., and MacPhail, R. C. (1988). Comparison of chlordimeform and carbaryl using a functional observational battery. *Fund. Appl. Toxicol.* **11**, 189–206.

Moser, V. C. (1989). Screening approaches to neurotoxicity: a functional observational battery. *J. Am. Coll. Toxicol.* **8**, 85–93.

Moser, V. G. (1991). Investigations of amitraz neurotoxicity in rats: IV. Assessment of toxicity syndrome using a functional observational battery. *Fund. Appl. Toxicol.* **17**, 7–16.

Nevin, J. A. (1981). Psychophysics and reinforcement schedules: an integration. In *Quantitative Analyses of Behavior: Discriminative Properties of Reinforcement Schedules* (M. L. Commons and J. A. Nevin, Eds.), pp. 3–30. Ballinger, Cambridge.

Newland, M. C. (1988). Quantification of motor function in toxicology. *Toxicol. Lett.* **43**, 295–319.

Newland, M. C. (1994). Operant behavior and the measurement of motor dysfunction. In *Neurobehavioral Toxicity: Analysis and Interpretation*. (B. Weiss and J O'Donoghue, Eds.), Raven Press, New York, in press.

Newland, M. C., Ceckler, T. L., Kordower, J. H., and Weiss, B. (1989) Visualizing manganese in the primate basal ganglia with magnetic resonance imaging. *Exp. Neurol.* **106**, 251–258.

Newland, M. C., and Marr, M. J. (1985). The effects of chlorpromazine and imipramine on rate and stimulus control of matching to sample. *J. Exp. Anal. Behav.* **44**, 49–68.

Newland, M. C., Ng, W. W., Baggs, R. B., Gentry, G. D., Weiss, B., and Miller, R. K. (1986). Operant behavior in transition reflect neonatal exposure to cadmium. *Teratol.*, **34**, 231–241.

Newland, M. C., Ng, W. W., Baggs, R. B., Miller, R. K., Infurna, R. N., and Gentry, G. D. (1983). Acute behavioral toxicity of cadmium in neonatal rats. *Teratology* **27**, 65A–66A.

Newland, M. C., Sheng, Y., Logdberg, B., and Berlin, M. (1994). Prolonged behavioral effects of in utero exposure to lead or methylmercury: reduced sensitivity to changes in reinforcing stimuli during behavioral transitions and in steady state. *Toxicol. Appl. Pharmol.*, in press.

Newland, M. C., and Weiss, B. (1986). The effects of oxotremorine on tremor and operant behavior in squirrel monkeys. *The Toxicologist* **6**, 25.

Newland, M. C., and Weiss, B. (1990). Drug effects on an effortful operant: pentobarbital and amphetamine. *Pharmacol. Biochem. Behav.* **36**, 381–387.

Newland, M. C., and Weiss, B. (1991). Ethanol's effects on tremor and positioning in squirrel monkeys. *J. Stud. Alcohol* **52**, 492–499.

Newland, M. C., and Weiss, B. (1992). Persistent effects of manganese on effortful responding and their relationship to manganese accumulation in the primate globus pallidus. *Toxicol. Appl. Pharmol.* **113**, 87–97.

Norton, S. (1989). Methods for behavioral toxicology. In *Principles and Methods of Toxicology* (A. W. Hayes, Ed.), Raven Press, New York.

Notterman, J. M., and Mintz, D. E. (1965). *Dynamics of Response*. Wiley, New York.

O'Keeffe, R. T., and Lifshitz, K. (1989). Nonhuman primates in neurotoxicity: screening and neurobehavioral toxicity studies. *J. Am. Coll. Toxicol.* **8**, 127–140.

Page, S., and Neuringer, A. (1985). Variability is an operant. *J. Exp. Anal. Behav.* **11**, 429–452.

Peele, D. B., and Crofton, K. M. (1987). Pyrethroid effects on schedule-controlled behavior: time and dosage relationships. *Neurotoxicol. Teratol.* **9**, 387–394.

Pincus, J. H., and Tucker, G. J. (1978). *Behavioral Neurology*. Oxford University Press, New York.

Pinder, R. M. (1982). Drug-induced tremor. In *Movement Disorders: Tremor* (L. J. Findley and R. Capildeo, Eds.), pp. 445–462, Oxford University Press, New York.

Pisa, M. (1988). Motor functions of the striatum of the rat: critical role of the lateral region in tongue and forelimb reaching. *Neuroscience* **24**, 453–463.

Platt, J. R., Kuch, D. O., and Bitgood, S. C. (1973). Rats' leverpress durations as psychophysical judgements of time. *J. Exp. Anal. Behav.* **19**, 239–250.

Pryor, G. T. (1991). A toluene-induced motor syndrome in rats resembling that seen in some human solvent abusers. *Neurotoxicol. Teratol.* **13**, 387–400.

Pryor, G. T., Uyeno, E. T., Tilson, H. A., and Mitchell, C. L. (1983). Assessment of chemicals using a battery of neurobehavioral tests: a comparative study. *Neurobehav. Toxicol. Teratol.* **5**, 91–117.

Rafales, L. S. (1986). Assessment of locomotor activity. In *Neurobehav. Toxicol.* (Z. Annau, Ed.), pp. 54–68, Johns Hopkins, Baltimore.

Rees, D. C., and Balster, R. L. (1988). Attenuation of the discriminative stimulus properties of ethanol and oxazepam, but not of pentobarbital, by Ro 15-4513 in mice. *J. Pharmacol. Exp. Ther.* **224**, 592–598.

Rees, D. C., Wood, R. W., and Laties, V. G. (1985). The roles of stimulus control and reinforcement frequency in modulating the behavioral effects of d-amphetamine in the rat. *J. Exp. Anal. Behav.* **43**, 243–255.

Reiter, L. W., and MacPhail, R. C. (1982). Factors influencing motor activity measurements in neurotoxicology. In *Nervous System Toxicology* (C. L. Mitchell, Ed.), pp. 45–66. Raven Press, New York.

Rice, D. C. (1988). Quantification of operant behavior. *Toxicol. Lett.* **43**, 361–379.

Rice, D. C., Gilbert, S. G. and Willes, R. F. (1979). Neonatal low-level lead exposure in monkeys: locomotor activity, schedule-controlled behavior, and the effects of amphetamine. *Toxicol. Appl. Pharmol.* **51**, 503–513.

Richter, R. (1945). Degeneration of the basal ganglia in monkeys from chronic carbon disulfide poisoning. *J. Neuropath. Exp. Neurol.* **4**, 324–353.

Rilling, M. E., and LaClaire, T. L. (1989). Visually guided catching and tracking skills in pigeons: a preliminary analysis. *J. Exp. Anal. Behav.* **52**, 377–386.

Samson, H. H., and Falk, J. L. (1974). Ethanol and discriminative motor control: Effects on normal and dependent animals. *Pharmacol. Biochem. Behav.* **2**, 791–801.

Schick, K. (1971). Operants. *J. Exp. Anal. Behav.* **15**, 413–423.

Schneider, J. S. (1988). Deficits in operant behaviour in monkeys treated with N-Methyl-4-Phenyl-1236-Tetrahydropyridine (MPTP). *Brain* **111**, 1265–1285.

Schuster, C. R. and Balster, R. L. (1977). The discriminative stimulus properties of drugs. In *Advances in Behavioral Pharmacology* (T. Thompson and P. B. Dews, Eds.), pp. 86–139, Academic Press, New York.

Schuster, C. R., Dockens, W. S., and Woods, J. H. (1966). Behavioral variables affecting the development of amphetamine tolerance. *Psychopharmacology* **9**, 170–182.

Schwartz, B. (1980). Development of complex, stereotyped behavior in pigeons. *J. Exp. Anal. Behav.* **33**, 153–166.

Schwartz, B. (1982). Reinforcement-induced stereotypy: how not to teach people to discover rules. *J. Exp. Psychol. Gen.* **111**, 23–59.

Sette, W. F., and Levine, T. E. (1986). Behavior as a regulatory endpoint. In *Neurobehavioral Toxicology* (Z. Annau, Ed.), pp. 391–403, Johns Hopkins Press, Baltimore.

Shah, K., Bradshaw, C. M., and Szabadi, E. (1990). Interaction between antidepressants and d-amphetamine on variable-interval performance. *Psychopharmacology* **100**, 548–554.

Siegel, S., and Castellan, N. J. (1988). *Nonparametric Statistics for the Behavioral Sciences.* McGraw-Hill, New York.

Sinclair, J. G., Lo, G. F., and Harris, D. P. (1982). Ethanol effects on the olivocerebellar system. *Can. J. Physiol. Pharmacol.* **60**, 610–614.

Skjoldager, P., and Fowler, S. C. (1988). Effects of pimozide, across doses and within sessions, on discriminated lever release performance in rats. *Psychopharmacology* **96**, 21–28.

Snyderman, M. (1984). Body weight and response strength. *Behav. Anal. Lett.* **3**, 255–265.

Spencer, P. S., and Schaumburg, H. H. (1980). *Experimental and clinical neurotoxicology.* Williams and Wilkins, Baltimore.

Stebbins, W. C. (1970). Studies of hearing and hearing loss in the monkey. In *Animal Psychophysics.* (W. C. Stebbins, Ed.), Prentice-Hall, Englewood Cliffs, New Jersey.

Stebbins, W. C., and Coombs, S. (1975). Behavioral assessment of ototoxicity in nonhuman primates. In *Behavioral Toxicology* (B. Weiss and V. G. Laties, Eds.), Plenum Press, New York.

Stern, Y., Mayeux, R., Rosen, J., and Ilson, J. (1983). Perceptual motor dysfunction in Parkinson's Disease: a deficit in sequential and predictive voluntary movement. *Neurol. Neurosurg. Psychiatry* **46**, 145–151.

Stevens, S. S. (1951). *Handbook of Experimental Psychology.* Wiley, New York.

Stine, W. W. (1989). Meaningful inference: the role of measurement in statistics. *Psychol. Bull.* **105**, 147–155.

Swets, J. A. (1988). Measuring the accuracy of diagnostic systems. *Science* **240**, 1285–1293.

Swets, J. A. (1992). The science of choosing the right decision threshold in high-stake diagnostics. *Am. Psychol.* **47**, 522–532.

Tang, M., and Falk, J. L. (1979). Ethanol withdrawal and discriminative motor control: effect of chronic intake level. *Pharmacol. Biochem. Behav.* **11**, 581–584.

Tang, M., Lau, C. E., and Falk, J. L. (1988). Midazolam and discriminative motor control: chronic administration, withdrawal, and modulation by the antagonist RO 15-1788. *J. Pharmacol. Exp. Ther.* **246**, 1053–1060.

Taub, E., and Berman, A. J. (1968). Movement and learning in the absence of sensory feedback. In *The neuropsychology of spatially oriented behavior* (S. J. Freedman, Ed.), pp. 173–192, Dorsey, Homewood, Illinois.

Taub, E., Goldberg, I. A., and Taub, P. (1975). Deafferentation in monkeys: pointing at a target without visual feedback. *Exp. Neurol.* **46**, 178–186.

Thach, W. T., and Montgomery, E. B. (1990). Motor systems. In *Neurobiology of Disease* (A. L. Pearlman and R. C. Collins, Eds.), pp. 168–196, Oxford University, New York.

Thomas, J. R. (1964). Multiple baseline investigation of stimulus functions in an FR chained schedule. *J. Exp. Anal. Behav.* **7**, 241–245.

Thompson, D. M. (1976). Repeated acquisition of behavioral chains: effects of methylphenidate and imipramine. *Pharmacol. Biochem. Behav.* **4**, 671–677.

Thompson, D. M. (1978). Stimulus control and drug effects. In *Contemporary Research in Behavioral Pharmacology* (D. E. Blackman and D. J. Sanger, Eds.), pp. 159–207, Plenum, New York.

Thompson, D. M. and Moerschbaecher, J. M. (1979). An experimental analysis of the effects of d-amphetamine and cocaine on the acquisition and performance of response chains in monkeys. *J. Exp. Anal. Behav.* **32**, 433–444.

Thompson, R. F., Clark, G. A., Donegan, N. H., Lavond, D. G., Madden, I. V., Mamounas, L. A., Mauk, M. D., and McCormick, D. A. (1984). Neuronal substrates of basic associative learning. In *Neuropsychology of Memory* (L. Squire and N. Butters, Eds), pp. 49–85. Elsevier, New York.

Tilson, H. A., Mitchell, C. L., and Cabe, P. A. (1979). Screening for neurobehavioral toxicity: The need for and examples of validation of testing procedures. *Neurobehav. Toxicol.* **1**, 137–148.

Vaillet, F., Trouche, E., Beaubaton, D., and Legallet, E. (1987). The role of visual reafferents during a pointing movement: comparative study between open loop and closed-loop performances in monkeys before and after unilateral electrolytic lesion of the substantia nigra. *Exp. Brain Res.* **65**, 399–410.

Vanderwolf, C. H., Kolb, B., and Cooley, R. K. (1978). Behavior of the rat after removal of the neocortex and hippocampal formation. *J. Comp. Physiol. Psychol.* **92**, 156–175.

Vorhees, C. V. (1987). Reliability, sensitivity, and validity of behavioral indices of neurotoxicity. *Neurobehav. Toxicol. Teratol.* **9**, 445–464.

Waddington, J. L., Cross, A. J., Gamble, S. J., and Bourne, R. C. (1983). Chronic orofacial dyskinesia and dopaminergic function in rats after 6 months of neuroleptic treatment. *Science* **220**, 530–532.

Walker, C. H., Faustman, W. O., Fowler, S. C., and Kazar, D. B. (1981). A multivariate analysis of some operant variables used in behavioral pharmacology. *Psychopharmacology* **74**, 182–186.

Walsh, T. J., and Tilson, H. A. (1986). The use of pharmacological challenger. In *Neurobehavioral Toxicology* (Z. Annau, Ed.), pp. 244–267, Johns Hopkins University Press, Baltimore.

Weber, L. J., and Mclean, D. L. (1975). *Electrical Measurement Systems for Biological and Physical Scientists.* Addison-Wesley, Reading, Massachusetts.

Weiss, B. (1970). The fine structure of operant behavior during transition states. In *The Theory of Reinforcement Schedules* (W. N. Schoenfeld, Ed.), pp. 277–311, Appleton-Century-Crofts, New York.

Weiss, B., and Gott, C. T. (1972). A microanalysis of drug effects on fixed-ratio performance in pigeons. *J. Pharmacol. Exp. Ther.* **180,** 189–202.

Weiss, B., and Laties, V. G. (1964). Effects of amphetamine, chlorpromazine, pentobarbital, and ethanol on operant response duration. *J. Pharmacol. Exp. Ther.* **144,** 17–23.

Weiss, B., Santelli, S., and Lusink, G. (1977). Movement disorders induced in monkeys by chronic haloperidol treatment. *Psychopharmacology* **53,** 289–293.

Whishaw, I. Q. (1990). The decorticate rat. In *The cerebral cortex of the rat* (B. Kolb and R. C. Tees, Eds.), pp. 239–267, MIT Press, Cambridge.

Whishaw, I. Q., and Kolb, B. (1983). "Stick out your tongue": Tongue protrusion in neocortex and hypothalamic damaged rats. *Physiol. Behav.* **30,** 471–480.

Whishaw, I. Q., O'Connor, W. T., and Dunnett, S. T. (1986). The contributions of motor cortex, nigrostriatal dopamine, and caudate-putamen to skilled forelimb use in the rat. *Brain* **109,** 805–843.

Whishaw, I. Q., Schallert, T., and Kolb, B. (1981). An analysis of feeding and sensorimotor abilities of rats after decortication. *J. Comp. Physiol. Psychol.* **95,** 85–103.

Wood, R. W. (1981). Neurobehavioral toxicity of carbon disulfide. *Neurotoxicol. Teratol.* **3,** 397–405.

Wood, R. W., Rees, D. C., and Laties, V. G. (1983). Behavioral effects of toulene are modulated by stimulus control. *Toxicol. Appl. Pharmacol.* **68,** 462–472.

Wylie, R. M., and Tyner, C. F. (1981). Weight-lifting by normal and deafferented monkeys: evidence for compensatory changes in ongoing movements. *Brain Res.* **219,** 172–177.

Wylie, R. M., and Tyner, C. F. (1989). Performance of a weight-lifting task by normal and deafferented monkeys. *Behav. Neurosci.* **103,** 273–282.

Youssef, A. F., Weiss, B., and Cox, C. (1993). Neurobehavioral toxicity of methanol reflected by operant running. *Neurotoxicol. Teratol.* **15,** 223–227.

Zeiler, M. (1977). Schedules of reinforcement. In *Handbook of Operant Behavior* (W. K. Honig and J. E. R. Staddon, Eds.), Prentice-Hall, Englewood Cliffs, New Jersey.

Zeiler, M. D., Davis, E. R., and DeCasper, A. J. (1980). Psychophysics of key-peck duration in pigeons. *J. Exp. Anal. Behav.* **34,** 23–34.

Zhuravin, I. A., and Bures, J. (1986). Physiology and behavior. *Physiol. Behav.* **36,** 611–617.

Ziriax, J. M., Snyder, J. R., Newland, M. C., Weiss, B. (1993). Amphetamine modifies the microstructure of behavior. *Exp. Clin. Psychopharm.* **1,** 121–132.

CHAPTER

13

Approaches to Utilizing Aspects of Cognitive Function as Indicators of Neurotoxicity

MERLE G. PAULE
Behavioral Toxicology Laboratory
Division of Neurotoxicology
National Center for Toxicological Research
Jefferson, Arkansas 72079
and
Complex Brain Function Laboratory
Center for Applied Research and Evaluation
Arkansas Children's Hospital
Little Rock, Arkansas 72202

I. Introduction

A primary charge for toxicologists is to make risk assessments or predictions about the circumstances under which a particular compound will be toxic to humans. In the ideal case, the data needed to make such predictions are obtained from laboratory animal models in well-controlled experiments under known conditions of exposure *prior* to the occurrence of human exposures. The use of appropriate animal models is critical in the risk assessment process (McMillan and Owens, 1995). Unfortunately, many neurotoxicants (e.g., methyl mercury, lead) have been identified as such, not because of their observed effects in laboratory animals, but because of their adverse effects on human brain function as evidenced by observable changes in behavior. This situation arose because, historically, little effort was made to determine *a priori* the action of a given compound on nervous system function. This absence of effort likely derived from a variety of factors, including the lack of awareness

about the potential hazards associated with exposure, the lack of appropriate technology for making relevant observations, the lack of consensus on which measures were relevant, and the lack of regulatory demand. This chapter addresses the issue of relevant measures and outlines an approach for future research that may help develop a consensus concerning subsequent research in this area. Given the very large number of compounds in the chemical universe that are estimated to be neuroactive (Anger and Johnson, 1985) and thus potentially neurotoxic, it seems prudent that a focused effort be brought to bear on further development of approaches for determining their effects on important aspects of brain function.

A major difficulty for the toxicologist interested in assessing neurotoxic risk using cognitive end points obtained from laboratory animals may be illustrated by the following comparison. Scientists interested in predicting the carcinogenicity of a particular compound in humans often choose tumor formation as a relevant end point for use in their animal models since the end point of interest in humans is also a tumor: liver tumors in mice, rats, dogs, monkeys, and humans are very similar or nearly identical and the relevance of one to the other is obvious. For neurotoxicologists interested in modeling human cognitive function in animals, the task is much less straightforward. Many would argue that the state of the science is such that it is unclear which measures of "cognitive" function in animals are directly relevant to aspects of cognitive function in humans. Most animal researchers agree that studies of cognitive function in animals have direct relevance to the human situation, yet the direct comparisons that are needed to demonstrate the links between species are generally nonexistent. It is also apparent that some scientists who study human cognitive function generally ignore important research findings from animal studies. Additionally, researchers who study human cognitive function have, until recently, put little effort into adapting or developing appropriate human cognitive tests for use in nonhumans.

II. Complex Brain Function

Because the term "cognitive function" invokes different emotions and interpretations in different individuals depending on their history and bias (an issue addressed previously by Eckerman and Bushnell, 1992), the more generic term "complex brain function" will be used instead. Complex brain function as used here refers to those functions that include learning, short-term memory, and attention; those that are measured in humans using traditional intelligence tests (e.g., the Wechsler intelligence scale for children, Wechsler, 1974); and those referred to as "executive functions" as described by Welsh and Pennington (1988). Executive functions are those that do not correlate highly with traditional measures of intelligence (i.e., IQ scores) but include "the ability to maintain an appropriate problem solving set for attainment of a future goal" (Welsh and Pennington, 1988). Complex brain functions as discussed here will not include autonomic or reflex activities, even though they can be complex and they are clearly important and relevant measures of nervous system integrity.

Maintenance of normal complex brain function after treatment with an experimental compound would represent a situation analogous to the absence of a cancerous tumor after chemical exposure; decrements in complex brain function after exposure to a neurotoxicant would represent a situation analogous to the induction of tumor growth after treatment with a carcinogen. Aspects of complex brain function can thus be used as biomarkers of neurotoxicant effect. It is the identification and discussion of examples of relevant complex

brain functions that serve as the focus of this chapter. Examples of methods will be mentioned in discussing experimental approaches that may prove fruitful in addressing issues fundamental to the discipline of neurotoxicology as it relates to the assessment of the effects of chemicals on complex brain function. Details of specific methods used in assessing the effects of neurotoxicants on complex brain function will generally not be presented here.

III. Behavior as Biomarker: Behavioral Toxicology

Since complex brain function as defined here cannot be directly observed [IQ cannot be seen but is measured by scoring responses (behaviors) to specific situations (problems)], it must be inferred from the behavior of the organism being studied. Thus, the measurements highlighted here are those associated with behaviors thought to be dependent on specific brain functions or functional domains. A main goal of this chapter is to propose research approaches that utilize complex brain functions (modeled by specific behaviors) as biomarkers of neurotoxicity. The focus will be on efforts to increase the use of laboratory animals, instead of humans, as sources of experimental data. Principles underlying the discussion have their foundation in the disciplines of behavioral pharmacology and toxicology (Annau, 1986; Johnson, 1990; Dews, 1972; Weiss and Laties, 1972; Weiss and O'Donoghue, 1994), focusing specifically on those efforts that deal with complex brain function.

IV. Selection of Human Brain Functions That Can Be Modeled in Animals

The behavioral repertoire of humans includes complex facilities such as verbal, problem-solving, and social skills. The language skills of humans are not found in other animals, and although it may be possible to observe nonverbal behaviors that are highly correlated with verbal ability, identification of such behaviors and the subsequent development of appropriate animal models have not yet been accomplished. Thus, animal models of verbal ability are not likely to become available for use in determining the neurotoxic effects of chemicals. On the other hand, social abilities are most certainly observable in many laboratory animal species, but the measurement of such abilities is often tedious, time consuming, labor intensive, hard to automate, and, in some cases, may have little apparent relevance for species other than those in which they are observed. The study of the effects of toxicants on the social behaviors of some species is very important and relevant to humans (e.g., Burbacher *et al.,* 1990b; Laughlin *et al.,* 1991); however, the widespread use of such behaviors as end points in routine risk assessment for complex brain function does not appear to be imminent.

Nonverbal and nonsocial problem-solving abilities, however, are common to a variety of species and can be modeled relatively easily in laboratory animals using operant behavioral techniques (so-called because subjects must "operate" something in their environment, such as a lever or a button, to indicate a response). The great power of operant behavioral technology derives in part from the tools it provides for the training of experimental subjects to perform very specific tasks depending on the rules (contingencies) of reinforcement (i.e., a reinforcer or reward can be obtained by pressing only the lever under the red light after a minimum of 10 sec has passed since the last time it was pressed). Thus, by requiring subjects to make specific responses to specific stimuli in order to obtain a reinforcer (banana-flavored food pellet, peanut, candy, money, etc.), tasks can be devised that require subjects to use what are thought to be specific complex brain functions such as learning and visual discrimination. Operant behaviors are also very amenable to automation and eliminate or drastically reduce the need for potentially confounding examiner–subject interactions. In-depth discussions on the use of operant behavior in toxicology can be found elsewhere (Cory-Slechta 1992, 1994; Laties and Wood, 1986; Weiss and O'Donoghue, 1994).

The number of problems that can be modeled using operant behavioral techniques will likely be limited more by the imagination of the experimenter than by the capacity of animal subjects to solve them. Problem-solving abilities represent rich functional domains shared by both laboratory animals and humans and, by invoking the concept of face validity, their relevance to human complex brain function seems clear.

Problem-solving tasks can be modeled in animals in exactly the same way as they can be modeled in humans. For example, short-term memory problems may be presented to subjects such that at the start of a particular problem or trial, a specific form or sample stimulus, such as a white-on-black triangle, is illuminated for their observation. Once the subject has observed the sample form (the triangle), it is extinguished and a time delay of variable length is imposed until three different geometric symbols are illuminated for the subject to view, one of which matches the sample (triangle) stimulus. A choice of the triangle by the subject represents a correct solution to the memory problem, and determining choice accuracies over a variety of time delays provides one with metrics of processes associated with short-term memory. With this approach, the same problems can be presented to both humans and laboratory animals. The maintenance of task continuity across species allows for the quantitative determination of interspecies similarities and differences and assists in the extrapolation of data from laboratory animals to humans. Use of identical behavioral measures in both laboratory subjects and humans is also important because it provides laboratory animal test validity. Several interesting examples of animal task utilization in human subjects have been reported (Chavoix *et al.,* 1991; Irle *et al.,* 1987; Kessler *et al.,* 1986; Overman *et al.,* 1992; Peuster *et al.,* 1991) and some have been reportedly useful in detecting abnormal behavior in certain clinical populations such as those with Alzheimer's or Parkinson's disease (Sahakian *et al.,* 1988).

V. A Test Battery Approach: Interspecies Comparisons

In studies at the Food and Drug Administration's National Center for Toxicological Research (NCTR) and at the Center for Applied Research and Evaluation at Arkansas Children's Hospital, a battery of operant tasks is used in both laboratory animals and humans. The NCTR operant test battery (OTB) was originally developed for use with laboratory animals (rhesus monkeys responding for banana-flavored food pellets) and was devised to allow the assessment of several complex brain functions in single experimental sessions. The specific tasks contained in the NCTR OTB and the apparatus have been described in detail elsewhere (Paule *et al.,* 1988a; Schulze *et al.,* 1988). The tasks include those thought to allow assessment of aspects of motivation, color and position discrimination, time estimation, short-term memory and attention, and learning. The motivation task requires that more work be performed for each subsequent reinforcer earned. For example, the first reinforcer costs only two lever presses, but the next one costs four, the next six, the next eight, and so on. The color and position discrimination task requires that subjects discriminate a stimulus color (red, yellow, blue, or green) and then respond at an appropriate position (left or right, depending on the color presented). For the time estimation task, subjects must hold a response lever in the depressed position for some minimum amount of time (10 sec) but no longer than some maximum amount of time (14 sec). For the short-term memory task, subjects observe a sample stimulus (geometric form) and then choose a matching geometric form from several choice stimuli presented some time later (e.g., 2–30 sec). For the learning task, subjects are presented with four response levers and over the course of each test session must learn a specific predetermined sequence of lever presses.

Extensive utilization of the NCTR OTB in both acute (Frederick *et al.*, 1994; Ferguson and Paule, 1992, 1993; Buffalo *et al.*, 1994; Schulze and Paule, 1990) and chronic drug studies in monkeys (Paule *et al.*, 1992a, 1994) has demonstrated its utility in the animal laboratory. Additionally, evidence suggests that performance in each of the tasks in the NCTR OTB is relatively independent of performance in the other tasks (Paule, 1990, 1994), a desirable feature for any test battery. Such observations also support the hypothesis that each of the behaviors monitored in the NCTR OTB is representative of a different domain of complex brain function.

Studies that are conducted with children as subjects (Paule *et al.*, 1988b; Paule and Cranmer, 1990a; Paule, 1994) utilize identical behavioral apparatus and tasks but substitute money (nickels) for food pellets as reinforcers. Additionally, children are shown a videotape that demonstrates in a few minutes how to perform each of the tasks, whereas it takes several months to train monkeys how to perform the tasks correctly. Comparative data clearly demonstrate that the NCTR OTB performance of well-trained rhesus monkeys is generally indistinguishable from that of children (Paule *et al.*, 1990b). Thus, analogous complex behaviors in laboratory animals and humans can be measured using computerized instruments such as the NCTR OTB. The similarity of monkey and human performance has been observed by others (e.g., Evans 1982) and likely results from the anatomical similarities between the species. Several investigators have reported on similarities between human and monkey performance of other complex tasks (Burbacher *et al.*, 1990a; Hopkins *et al.*, 1990; Ringo *et al.*, 1986). Those such as object permanence (Burbacher *et al.*, 1986) and object retrieval behaviors (Diamond and Goldman-Rakic, 1985; Goldman-Rakic, 1987) have been used to examine the effects of chemical exposure and neuroanatomical lesions. In the object permanence task, subjects observe a preferred object being hidden in one of two identical wells. After a delay of a few seconds, the subjects are allowed to retrieve the object and after a correct retrieval the object is hidden in the other well. Early in development, both human and monkey infants will tend to search for the object in the well where it was found on the previous trial; later in development the correct choice will be made without error. In object

retrieval tasks, subjects are required to retrieve items from clear plexiglass boxes. This problem is solved by reaching around a barrier (a side of the box) to obtain the item which, though in clear view of the subject, cannot be obtained by reaching directly for it. Again, developmental progression of performance in this task is remarkably similar in both humans and monkeys (Diamond and Goldman-Rakic, 1985). These studies clearly demonstrate cross-species similarities in complex brain function, but the procedures used are generally labor intensive, often involve infant subjects, and require direct interaction of the examiner with the subject. Removal of the examiner from the testing situation, as is generally done in operant behavioral analyses, eliminates a potentially confounding variable from the environment in which the data are collected and thus may serve to decrease data variability. Additionally, it has been shown that operant behavioral procedures can be automated for use with infant monkeys (Rice, 1979; Rice and Gilbert, 1990). Utilizing a battery of operant tasks to monitor and quantify aspects of several complex brain functions within single test sessions is a demonstration of possibilities not yet fully explored with respect to interspecies comparisons. Future efforts in this area should focus on determining the utility of these OTB tasks in adult humans and developing additional tasks that will allow the assessment of other functional domains (i.e., the creation of other tasks whose performance is independent of the ones currently being measured) in both laboratory animals and humans.

VI. The Relevance Issue

Discussion of the NCTR OTB has focused on specific behavioral methods originally developed for use in the animal laboratory and then taken into the clinic for use in children. In developing these tasks for animals, however, it was intended that they would model brain functions thought to be relevant to humans. Once it became evident that the NCTR OTB tasks were appropriate for assessing complex behaviors in children, the opportunity of making a direct assessment of OTB task relevancy presented itself. Acknowledging that standardized IQ tests have been aptly criticized for cultural bias and other aspects of content (e.g., Garcia, 1975), it is still generally accepted that IQ scores are related to important complex brain functions in humans (Weinberg, 1989). Additionally, changes in IQs have been reported as a consequence of low-level lead exposure in humans (Bellinger *et al.*, 1992). Statistical analyses of data from about 100 children have demonstrated highly significant correlations between many NCTR

OTB measures and IQ (Paule *et al.*, 1992b; Paule, 1994), thus confirming their relevance to brain function. Others have also reported significant correlations between operant measures (reaction times) and intelligence (Sen *et al.*, 1983). OTB and IQ data obtained from children have been used in mathematical models (rough sets and modified rough sets, unpublished results); to estimate children IQs based on OTB data alone. Subsequently, the same models were used to estimate monkey IQs based on OTB data alone (Hashemi *et al.*, 1994). Such models should serve useful in providing metrics of brain function in animals that can then serve as the focus of important comparative (interspecies) studies of brain function. Additionally, they should prove useful for grouping animals according to behavioral ability to minimize bias in behavioral experiments.

With respect to the OTB IQ correlation analyses, it is important to note that the correlation coefficients between OTB measures and IQ scores were never above 0.6 and were generally much less. Some OTB measures had no correlation with IQ. These observations indicate that the OTB tasks are modeling brain functions not accounted for by IQ and perhaps represent types of executive functions referred to earlier. Future efforts should involve examination of the degree of correlation of OTB behaviors with measurements from other instruments often used to assess cognitive function in humans, e.g., standard neuropsychological test batteries such as the Halstead–Reitan battery, the Luria–Nebraska neuropsychological battery, and computerized instruments such as the neurobehavioral evaluation system [for an overview of these and other batteries and how they are used in the assessment of the effects of neurotoxicants in humans see Hartman (1988)]. Additionally, efforts should be made to capitalize on current technology to automate existing batteries developed for human use so that the potential for large-scale and routine data collection can be realized. Toward this end, efforts are underway (Anger and Sizemore, personal communication) to develop a computerized test battery for use in humans that will contain tasks that are routinely modeled in animals, as well as some from the more commonly used neuropsychological instruments for humans mentioned earlier. There is no reason that complex brain function assessments cannot become routine parts of human health examinations in much the same way cardiovascular, hematopoietic, and other physiologic functions are currently monitored. It should be emphasized that for the purposes of neurotoxic risk assessment, development of tasks to be used in assessing complex brain function should also be appropriate for use in animals since it is from animals that the predictive data will come.

Development of complex brain function tasks sensitive to known neurotoxicant exposures in humans may be valuable for assessing damage already done, for identifying the neurological substrates affected, for monitoring the effectiveness of intervention, and for providing means for monitoring the time course of neurotoxic insult. But unless they provide measures that can be modeled in the animal laboratory, they will fall short of enhancing the basic science tools that are needed to improve our ability to predict adverse outcome in humans using data from animal surrogates.

A compilation of normative OTB data for children of several ages is currently underway. These data will be important not only for comparison with and continued validation of laboratory animal models, but also for future behavioral studies in humans. If normative OTB data are thoroughly developed and prove to be adequately descriptive for a given age and sex, then they can be used as behavioral standards of specific complex brain functions. As such they should then be useful in identifying subjects with abnormal function.

VII. Multispecies Similarities

Much of the previous discussion has focused on the similarities between human and monkey performance of complex behavioral tasks and may leave the impression that important or analogous comparisons can only be made between humans and other primates. That this is clearly not the case may be illustrated by considering an example of a complex behavior that has been modeled in at least three different species including humans, monkeys, and rats. McMillan and Patton (1965) were the first to describe a multispecies comparison of precise timing behavior, clearly a behavior thought to require complex brain function. In their studies, subjects were trained to press and hold a bar for at least 1.00 sec but no longer than 1.27 sec. It was observed that acquisition curves for this behavior followed similar functions in all three species. Additionally, after training to stable patterns of responding, all three species pressed the bar for a duration only slightly longer than the minimum requirement. With continued testing, this aspect of behavior became more prominent in rats than in either monkeys or humans, but the species similarities in precise timing behavior remained clear. In recent studies of timing behavior completed at the NCTR and Arkansas Children's Hospital, detailed cross species comparisons were made among rats, monkeys, and children. In a temporal response differentiation or time estimation task, reinforcers were obtained for holding a response lever in the depressed position for at least 10 sec but no more than 14 sec. Children worked for nickels; rats and rhe-

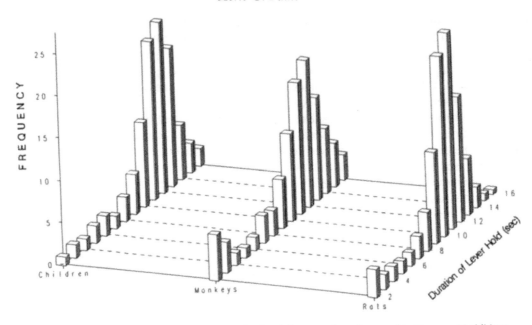

FIGURE 1 Comparison of time estimation behavior in adult rats and monkeys and in 11-year-old children. Data presented were generated during performance of the temporal response differentiation task used in the NCTR OTB. In this task, subjects must hold a response lever in the depressed position for at least 10 sec but no more than 14 sec in order to obtain a reinforcer (food pellets for rats and monkeys, nickels for children). Lever holds of less than 2 sec are not shown as they represent rapid bursts of lever presses and are not thought to be associated with timing behavior. Note the remarkable similarity in the pattern of lever holds. Frequency is defined as the percentage of total 2- to 16-sec responses.

sus monkeys worked for food pellets. Figure 1 shows data for all three species for the distribution of lever holds longer than 2 sec in duration (shorter holds are not indicated as they represent bursts of lever presses that are thought to be unrelated to processes associated with time estimation). Note the dramatic similarity in time estimation behavior across all three species. These observations should make it clear that exactly the same behavioral end points can be obtained in laboratory animals and humans and that analogous brain functions can be modeled across a variety of species. In using models of brain function, i.e., complex behaviors, as end points in neurotoxicological studies (those generating data for use in risk assessments) it will be important for the discipline of behavioral toxicology that, whenever possible, those end points be analogous to human behaviors. Such procedures will allow a more thorough understanding of species differences and similarities and will provide the data necessary for validating risk assessments in animals as a useful approach for determining adverse effects of chemicals on human brain function.

VIII. Examples Using Learning Paradigms

The ways in which specific brain functions can be modeled are many. By way of example, a discussion of

some of the behavioral tasks thought to model learning behavior might serve to illustrate. [Detailed discussions of these kinds of procedures can be found in earlier reviews (Miller and Eckerman, 1986; Eckerman and Bushnell, 1992).] Any behaviors that require training, for example, all of the tasks in the NCTR OTB, can be used to generate learning or acquisition curves during the time subjects are learning how to perform a particular task. Percent correct responding can be plotted versus test session, trial number, week of training, etc., and measures of learning can be obtained. Likewise, subjects can be trained to perform specific learning tasks such as that used in the NCTR OTB. This task is a type of typical repeated acquisition procedure (see Boren, 1963; Boren and Devine, 1968) in which subjects repeatedly acquire (learn) specific response sequences [e.g., lever presses (Thompson, 1970), maze navigation (Furuya et al., 1988) etc.] that vary from test session to test session. Learning curves can be generated every time a subject is tested, and learning behavior generated in these types of tasks stabilizes after a period of training, after the subject has learned the rules of reinforcement. There can be a variety of response sequences applicable to repeated learning procedures. A timely contribution to the field would involve a detailed comparison of data generated from repeated learning procedures that use disparate response sequences. For example, do learning curves generated from subjects making repeated alley selections in a maze correlate with those generated from animals making repeated lever selections in a Skinner box? Are aspects of task acquisition curves (the learning-how-to-learn curves) correlated with the daily learning curves generated by animals performing repeated acquisition tasks? Over the past three decades, repeated acquisition procedures that employ a variety of different behaviors have been developed. It might now prove more useful to explore in detail the similarities between these procedures than to continue to focus on the differences between them. In this way, it may be possible to determine to what degree task generalizations can be made.

Not only should comparisons of learning be made between tasks under the repeated acquisition rubric, but they should also be made with other measures of learning. Thus, learning how to avoid a signaled foot shock should be compared with learning how to respond in a color and position discrimination task, etc. Similar kinds of comparisons should also be made between tasks thought to share important attributes. For example, there are a variety of tasks that generate behaviors that are thought to be time based: fixed interval schedules under which responding is reinforced only after a specific amount of time has elapsed; differential reinforcement of low rate schedules in which responding is reinforced only if it occurs a certain minimum amount of time since the last response; and temporal response differentiation responding described earlier as requiring maintenance of a response for more than some minimum amount of time but less than some maximum amount of time. An examination of the behavioral correlations among these tasks might yield metrics common to all time-based behaviors. Additionally, a comparison of chemical effects on a variety of time-based behaviors may demonstrate similar effects in several, thus greatly facilitating the interpretation of intertest comparisons.

IX. Summary

The need to use animal models for predicting the effects of neuroactive agents on complex brain functions in humans is absolute and has led to the development of automated systems for administering identical behavioral tasks to both laboratory animals and humans. The maintenance of task continuity across species allows for the quantitative determination of interspecies similarities and differences in complex brain function and assists in the extrapolation of data from laboratory animals to humans. Efforts to develop be-

haviors for modeling additional complex brain functions in animals will increase the number of functional domains amenable to the risk assessment process.

References

Anger, W. K., and Johnson, B. L. (1985). Chemicals affecting behavior. In *Neurotoxicity of Industrial and Commercial Chemicals* (J. L. O'Donoghue, Ed.), pp. 52–148, CRC Press, Boca Raton, Florida.

Annau, Z. (Ed.) (1986). *Neurobehavioral Toxicology.* The Johns Hopkins University Press, Baltimore, Maryland.

Bellinger, D., Stiles, K. M., and Needleman, H. L. (1992). Low level lead exposure, intelligence, and academic achievement: a long term follow-up study. *Pediatrics* **90**(6), 855–861.

Boren, J. J. (1963). The repeated acquisition of new behavioral chains. *Am. Psychol.* **18**, 421–422.

Boren, J. J., and Devine, D. D. (1968). The repeated acquisition of behavioral chains. *J. Exp. Anal. Behav.* **11**, 651–660.

Buffalo, E. A., Gillam, M. P., Allen, R. R., and Paule, M. G. (1994). Acute behavioral effects of MK-801 in rhesus monkeys: Assessment using an operant test battery. *Pharmacol. Biochem. Behav.* **48**(4), 935–940.

Burbacher, T. M., Grant, K. S., and Mottet, N. K. (1986). Retarded object permanence development in methylmercury exposed Macaca fascicularis infants. *Dev. Psychol.* **22**(6), 771–776.

Burbacher, T. M., Rodier, P. M., and Weiss, B. (1990a). Methylmercury developmental neurotoxicity: a comparison of effects in humans and animals. *Neurotoxicol. Teratol.* **12**(3), 191–201.

Burbacher, T. M., Sackett, G. P., and Mottet, N. K. (1990b). Methylmercury effects on the social behavior of Macaca fascicularis infants. *Neurotoxicol. Teratol.* **12**, 65–71.

Chavoix, C., Hagger, C., Sirigu, A., Gravelle, M., and Aigner, T. (1991). An automated delayed nonmatching-to-sample task to assess visual recognition memory in humans. *Soc. Neurosci. Abstra.* **17**, 476.

Cory-Slechta, D. A. (1992). Schedule-controlled behavior in neurotoxicology. In *Neurotoxicology* (H. Tilson and C. Mitchell, Eds.), pp. 271–294, Raven Press, New York.

Cory-Slechta, D. A. (1994). Neurotoxicant-induced changes in schedule-controlled behavior. In *Principles of Neurotoxicology* L. W. Chang, Ed.), pp. 313–344, Dekker, New York.

Dews, P. B. (1972). An overview of behavioral toxicology. In *Behavioral Toxicology* (B. Weiss and V. G. Laties, Eds.), pp. 439–445, Plenum Press, New York.

Diamond, A., and Goldman-Rakic, P. S. (1985). Evidence for involvement of prefrontal cortex in cognitive changes during the first year of life: comparison of human infants and rhesus monkeys on a detour task with transparent barrier. *Neurosci. Abstr.* **11**(2), 832.

Eckerman, D. A., and Bushnell, P. J. (1992). The neurotoxicology of cognition: attention, learning, and memory. In *Neurotoxicology* (H. A. Tilson, and C. L. Mitchell, Eds.), pp. 213–270, Raven Press, New York.

Evans, H. L. (1982). Assessment of vision in toxicology. In *Nervous System Toxicology* (C. L. Mitchell, Ed.), pp. 81–107, Raven Press, New York.

Ferguson, S. A., and Paule, M. G. (1992). Acute effects of chlorpromazine in a monkey operant behavioral test battery. *Pharmacol. Biochem. Behav.* **42**(1), 333–341.

Ferguson, S. A., and Paule, M. G. (1993). Acute effects of pentobarbital in a monkey operant behavioral test battery. *Pharmacol. Biochem. Behav.* **45**, 107–116.

Frederick, D. L., Schulze, G. E., Gillam, M. P., and Paule, M. G. (1994). Acute effects of physostigmine on complex operant behavior in rhesus monkeys. *Pharmacol. Biochem. Behav.*, in press.

Furuya, Y., Yamamoto, T., Yatsugi, S., and Ueki, S. (1988). A new method for studying working memory by using the three-panel runway apparatus in rats. *Jpn. J. Pharmacol.* **46**, 183–188.

Garcia, J. (1975). The futility of comparative IQ research. *UCLA Forum Med. Sci.* **18**, 421–442.

Goldman-Rakic, P. S. (1987). Development of cortical circuitry and cognitive function. *Child Dev.* **58**, 601–622.

Hartman, D. E. (1988). *Neuropsychological Toxicology, Identification and Assessment of Human Neurotoxic Syndromes.* Pergamon Press, New York.

Hashemi, R. R., Pearce, B. A., Hinson, W. G., Young, J. F., and Paule, M. G. (1994). Estimation of monkey IQ based on human data using rough sets. *Proceed. 3rd Intl. Wrkshp. Rough Sets Soft Comp,* in press.

Hopkins, W. D., Washburn, D. A., and Rumbaugh, D. M. (1990). Processing of form stimuli presented unilaterally in humans, chimpanzees (Pan troglodytes), and monkeys (Macaca mulatta). *Behav. Neurosci.* **104**(4), 577–582.

Irle, E., Kessler, J., Markowitsch, H. J., and Hofmann, W. (1987). Primate learning tasks reveal strong impairments in patients with presenile or senile dementia of the Alzheimer type. *Brain Cogn.* **6**(4), 429–449.

Johnson, B. L. (ed.) (1990). *Advances in Neurobehavioral Toxicology: Applications in Environmental and Occupational Health,* Lewis Publishers, Chelsea, Michigan.

Kessler, J., Irle, E., and Markowitsch, H. J. (1986). Korsakoff and alcoholic subjects are severely impaired in animal tests of associative memory. *Neuropsychologia* **24**(5), 671–680.

Laties, V. G., and Wood, R. W. (1986). Schedule-controlled behavior in behavioral toxicology. In *Neurobehavioral Toxicology* (Z. Annau, Ed.), pp. 69–93, The Johns Hopkins University Press, Baltimore.

Laughlin, N. K., Bushnell, P. J., and Bowman, R. E. (1991). Lead exposure and diet: differential effects on social development in the rhesus monkey. *Neurotoxicol. Teratol.* **13**(4), 429–440.

McMillan, D. E., and Owens, S. M. (1995). Extrapolating scientific data from animals to humans in behavioral toxicology and behavioral pharmacology. In *Neurotoxicology: Approaches and Methods* (L. W. Chang, Ed.), Academic Press, Orlando, Florida, in press.

McMillan, D. E., and Patton, R. A. (1965). Differentiation of a precise timing response. *J. Exp. Anal. Behav.* **8**, 219–226.

Miller, D. B., and Eckerman, D. A. (1986). Learning and memory measures. In *Neurobehavioral Toxicology* (Z. Annau, Ed.), pp. 94–149, The Johns Hopkins University Press, Baltimore.

Overman, W. H., Carter, L., and Thompson, S. (1992). Development of place memory in children as measured in a dry morris maze. *Soc. Neurosci. Abstr.* **18**(1), 332.

Paule, M. G. (1990). Use of the NCTR operant test battery in nonhuman primates. *Neurotoxicol. Teratol.* **12**(5), 413–418.

Paule, M. G. (1994). Analysis of brain function using a battery of schedule-controlled operant behaviors. In *Neurobehavioral Toxicity: Analysis and Interpretation* (B. Weiss and J. O'Donoghue, Eds.), pp. 331–338, Raven Press, New York.

Paule, M. G., Schulze, G. E., and Slikker, W., Jr. (1988a). Complex brain function in monkeys as a baseline for studying the effects of exogenous compounds. *Neurotoxicology* **9**(3), 463–470.

Paule, M. G., Cranmer, J. M., Wilkins, J. D., Stern, H. P., and Hoffman, E. L. (1988b). Quantitation of complex brain function

in children: preliminary evaluation using a nonhuman primate behavioral test battery. *Neurotoxicology* **9**(3), 367–378.

Paule, M. G., Allen, R. R., Bailey, J. R., Scallet, A. C., Ali, S. F., Brown, R. M., and Slikker, W., Jr. (1992a). Chronic marijuana smoke exposure in the rhesus monkey II: Effects on progressive ratio and conditioned position responding. *J. Pharmacol. Exp. Ther.* **260**(1), 210–222.

Paule, M. G., Blake, D. J., Allen, R. R., and Casey, P. H. (1992b). NCTR operant test battery (OTB) performance in children: correlation with IQ. *Soc. Neurosci. Abstr.* **18**(1), 332.

Paule, M. G., and Cranmer, J. M. (1990a). Complex brain function in children as measured in the NCTR monkey operant test battery. In *Advances in Neurobehavioral Toxicology: Applications in Environmental and Occupational Health* (B. L. Johnson, ed.), pp. 433–447, Lewis Publishers, Chelsea, Michigan.

Paule, M. G., Forrester, T. M., Maher, M. A., Cranmer, J. M., and Allen, R. R. (1990b). Monkey versus human performance in the NCTR operant test battery. *Neurotoxicol. Teratol.* **12**(5), 503–507.

Paule, M. G., Allen, R. R., Binienda, Z., Rountree, R. L., and Slikker, W., Jr. (1994). Selective behavioral toxicity of chronic dideoxycytidine treatment in the monkey. *Neurotoxicol. Teratol.,* in press.

Peuster, A., Overman, A. H., Caulfield, S., and Hakan, R. (1991). Radial arm maze measures development of spatial memory in young children. *Soc. Neurosci. Abstr.* **17**(2), 1044.

Rice, D. C. (1979). Operant conditioning of infant monkeys (Macaca fascicularis) for toxicity testing. *Neurobehav. Toxicol.* **1**, (Suppl 1), 85–92.

Rice, D. C., and Gilbert, S. G. (1990). Automated behavioral procedures for infant monkeys. *Neurotoxicol. Teratol.* **12**, 429–439.

Ringo, J. L., Lewine, J. D., and Doty, R. W. (1986). Comparable performance by man and macaque on memory for pictures. *Neuropsychologia* **24**(5), 711–717.

Sahakian, B. J., Morris, R. G., Evenden, J. L., Heald, A., Levy, R., Philpot, M., and Robbins, T. (1988). A comparative study of visuospatial memory and learning in Alzheimer-type dementia and Parkinson's disease. *Brain* **111**, 695–718.

Schulze, G. E., McMillan, D. E., Bailey, J. R., Scallet, A. C., Ali, S. F., Slikker, W., Jr., and Paule, M. G. (1988). Acute effects of delta-9-tetrahydrocannabinol (THC) in rhesus monkeys as measured by performance in a battery of cognitive function tests. *J. Pharmacol. Exp. Ther.* **245**(1), 178–186.

Schulze, G. E., and Paule, M. G. (1990). Acute effects of d-amphetamine in a monkey operant behavioral test battery. *Pharmacol. Biochem. Behav.* **35**, 759–765.

Sen, A., Jensen, A. R., Sen, A. K., and Arora, I. (1983). Correlation between reaction time and intelligence in psychometrically similar groups in American and India. *Appl. Res. Ment. Retard.* **4**, 139–152.

Thompson, D. M. (1970). Repeated acquisition as a behavioral baseline. *Psychonom. Sci.* **21**(3), 156–157.

Weinberg, R. A. (1989). Intelligence and IQ. Landmark issues and great debates. *Am. Psychol.* **44**(2), 98–104.

Weiss, B., and Laties, V. G. (Eds.) (1972). *Behavioral Toxicology.* Plenum Press, New York.

Weiss, B., and O'Donoghue, J. L. (Eds.) (1994). *Neurobehavioral Toxicity, Analysis and Interpretation,* Raven Press, New York.

Welsh, M. C., and Pennington, B. F. (1988). Assessing frontal lobe functioning in children: views from developmental psychology. *Dev. Neuropsych.* **4**(3), 199–230.

Wechsler, D. (1974). Wechsler intelligence scale for children. Psychological Corporation, New York.

CHAPTER

14

Use of Drug Discrimination Learning in Behavioral Toxicology: Classification and Characterization of Toxins

ANTHONY L. RILEY
Psychopharmacology Laboratory
Department of Psychology
The American University
Washington, D.C. 20016

In a review of behavioral measures of neurotoxicity, Cory-Slechta (1989) noted several specific agendas of behavioral toxicology, specifically, the screening (or indexing) and characterization of toxicological agents. Although the specific agendas for the field can be clearly identified, it is less easy to identify a common or best methodology in such screening and characterization. As noted by Cory-Slechta (1989), the techniques are multifaceted and are typically chosen on the basis of the specific issue being addressed by individual investigators. For example, the use of rotarods, operant discrimination tasks, or the radial arm maze is in some sense predicated on one's interest in characterizing the motoric, sensory, or memorial effects of the toxin in question [for an illustration of various test methods used in the assessment of the effects of toxins on behavior and neuromotor function see Geller *et al.* (1979) and Tilson (1987)]. Such a plethora of methods and designs need not be seen as a lack of consensus as to the most appropriate design to be used in the behavioral assessment of toxicity, but instead should be viewed as a flexibility in research methodology in

the evaluation of the multiple and various effects produced by toxins (see Weiss and Laties, 1979).

Some designs widely used in the behavioral assessment of toxicity are adaptations of standard operant procedures developed by Skinner in his earlier demonstrations of the effects of schedules of reinforcement on behavior (Ferster and Skinner, 1957; Skinner, 1938; see also Laties, 1982; Tilson and Harry, 1982). For example, baselines on which animals have been reinforced under specific reinforcement contingencies (after every *n*th response, i.e., a fixed ratio schedule, or for the first response after the lapse of a specific duration, i.e., a fixed interval schedule) are typically stable and characteristic of the schedule under which they are generated. Perturbations of such baselines by various compounds have been used extensively to index their toxicity (Cory-Slechta and Weiss, 1985; MacPhail, 1985). Other operant procedures have also provided information regarding the nature of the toxic effects (see Geller *et al.*, 1979) as well as the biological mechanism underlying the effects (Cory-Slechta, 1992). For example, operant discrimination designs in

309

which animals are reinforced for responding in the presence of one stimulus and not in its absence (or in the presence of a second stimulus) have been utilized to assess the effects of toxins on sensory (Brady *et al.*, 1979; Maurissen, 1979) and learning and memory processes (Cabe and Eckerman, 1982; Eckerman and Bushnell, 1992).

I. Drug Discrimination Learning

As a group, the aforementioned operant designs have been fruitful both in the screening and characterization of toxins, providing a stable and quantifiable baseline for addressing the agendas noted earlier. Although behavioral toxicologists have used such behavioral methods in their evaluation of toxins, one operant design widely used in behavioral pharmacology has not been utilized in toxicology, i.e., the drug discrimination procedure (for reviews see Jarbe, 1987, 1989; Overton, 1984, 1987; Stolerman, 1993). This design, like the discrimination baseline described earlier, utilizes specific stimuli to signal the availability of reinforcement. Unlike traditional discrimination designs in which lights and tones may serve as discriminative stimuli, within this design drugs serve such functions. For example, in a drug discrimination procedure an animal is typically reinforced for making a specific response, e.g., a press on the right lever, following the administration of a specific drug. Responses on the other lever have no programmed consequences during this time. On other sessions, the animal is reinforced for making responses on the second lever, in this case the left lever, following administration of the drug vehicle. As described earlier, responses on the other lever have no programmed consequences during these sessions. Under these conditions, the animal acquires the drug discrimination and behavior comes under control of the drug, i.e., the animal makes drug- and vehicle-appropriate behavior following administration of the drug or its vehicle, respectively (see Colpaert and Balster, 1988; Colpaert and Rosecrans, 1978; Colpaert and Slangen, 1982; Glennon *et al.*, 1992; Ho *et al.*, 1978; Lal, 1978; Lal and Fielding, 1989; Thompson and Pickens, 1971; for a bibliography, see Samele *et al.*, 1992).

Following the initial demonstration of drug discrimination learning with alcohol (Conger, 1951; for historical reviews of drug discrimination learning see Overton, 1991, 1992), a wide range of compounds have been utilized by a number of different species within a variety of experimental designs and parameters (Jarbe, 1987, 1989; Overton, 1987; Stolerman, 1993). That animals can acquire discriminative control by a drug extends the range of stimuli which support such learning.

The interest in drug discrimination learning, however, goes considerably beyond this extension. Specifically, within behavioral pharmacology the design has been useful in both the classification and the characterization of drugs.

A. Classification

Once discriminative control has been established to a particular drug and animals are making drug-appropriate responses, other drugs can be administered and tested for their ability to substitute for the training drug, i.e., to determine if the stimulus control established to the training drug generalizes to the test drugs. Under such conditions, drugs from the same class as the training drug substitute and engender drug-appropriate responding (Barry, 1974). For example, Lal and colleagues (1978) reported that animals trained to discriminate morphine from its vehicle generalized this control to a range of other narcotics, including methadone, fentanyl, codeine, heroin, and butorphanol (see also Holtzman, 1985; Holtzman and Locke, 1988). These same animals, however, displayed vehicle-appropriate responding when administered one of a number of nonopiate compounds (e.g., amphetamine, atropine, cocaine, LSD, and pentobarbital). Similar within-class generalization has been reported for a variety of other drug classes, including the hypnotic/sedatives, stimulants, psychedelics, and anesthetics (Jarbe, 1989).

The fact that compounds from the same general class (or subclass) generalize to each other suggests that the compounds share some stimulus property, presumably the interoceptive and subjective effects produced by the generalized compounds. Such generalization, however, does not imply that the stimulus effects are identical. This, too, can be revealed by the drug discrimination procedure. For example, while animals trained to discriminate ethanol from its vehicle generalize this control to the barbiturate pentobarbital, the reverse is not true. That is, pentobarbital-trained animals do not generalize control to ethanol (Barry, 1974; Barry and Krimmer, 1977, 1978). Such asymmetry suggests that the two drugs share some stimulus property, but that there is not complete overlap in the subjective effects of the two compounds (Barry, 1992). The specific stimulus property establishing discriminative control would, therefore, depend on the specific training drug. Such within-class differences can also be revealed by explicit discrimination training between drugs belonging to the same class. For example, although lysergic acid diethylamide (LSD) and lisurgide hydrogen maleate (LHM), a structural analog of LSD, generalize to each other, when these two compounds serve different discriminative functions in a drug vs drug vs saline discrimination animals acquire a discrimination between the two compounds (Appel *et al.*, 1992), a discrimination possibly based on the differential effects they have on dopamine.

B. Characterization

Although the behavioral classification of drugs is an important application of the drug discrimination procedure, the design is not limited to this use. The design is also extensively used in drug characterization. For example, once discriminative control has been established to a particular drug, other compounds can be coadministered with the training drug to determine if and to what extent these compounds either antagonize or potentiate the discriminative control of the training drug. In relation to the antagonism of the training drug, the benzodiazepine antagonists RO 15-1788 and CGS 8216 selectively antagonize the stimulus properties of the benzodiazepines and not alcohol or the barbiturates (see Herling and Shannon, 1982; Young and Dewey, 1982). Although the assessment of drug interactions by the drug discrimination procedure has focused primarily on drug antagonism, assessments of potentiation have also been made. For example, Snoddy and Tessel (1985) have noted that the noradrenergic antagonist yohimbine potentiated the stimulus effects of amphetamine, a potentiation thought to be mediated by the antagonistic effects of yohimbine at α_2 presynaptic autoreceptors. Given that the amphetamine stimulus may be noradrenergically mediated, the antagonism of presynaptic autoreceptors should increase overall noradrenergic activity and potentiate its stimulus effects (see Goudie, 1992). The interacting effects of drugs within the drug discrimination design are not limited to these acute preparations. For example, France and Woods (1988) have documented changes in the sensitivity to the opiate antagonist naltrexone by a single prior administration of a low dose of morphine. That is, animals trained to discriminate naltrexone from water required smaller doses of the training drug to produce drug-appropriate responding following a single dose of morphine 24 hr prior to the test, a supersensitivity to naltrexone consistent with *in vitro* evidence noting receptor upregulation in opiate-exposed subjects.

The drug discrimination procedure has also been utilized extensively in attempts to determine the biochemical bases of drug activity (see Glennon *et al.*, 1992). One drug that has received considerable attention is cocaine. Animals trained to discriminate cocaine from its vehicle generalize stimulus control to a range of compounds which act either directly or indirectly on dopamine. Such generalization patterns implicate

dopamine as the mediator of the stimulus effects of cocaine, an implication consistent with the reported biochemical action of cocaine as a dopamine transport inhibitor. For example, Woolverton (1992) has reported that other dopamine transport inhibitors (such as nomifensine, buproprion, GBR 12909) completely substitute for cocaine in cocaine-trained subjects. Further, cocaine agonists (such as quinpirole and apomorphine) generally substitute fully for the cocaine cue, whereas dopamine antagonists (such as SCH 23,390 and haloperidol) attenuate the stimulus effects of cocaine. From such assessments, it has been suggested that the stimulus properties of cocaine are produced by inhibiting the reuptake of dopamine and thereby increasing the level of dopamine at D1 and D2 receptor subtypes of the dopamine receptor. Thus, through the use of the drug discrimination design, the specific biochemical effects of cocaine are being determined. Given the relationship between the stimulus and reinforcing effects of abused drugs (for a discussion see Brady *et al.*, 1990; Schuster and Johanson, 1988), such a determination may be useful in the introduction and evaluation of compounds that can selectively and completely block both the stimulus and reinforcing properties of cocaine (see Overton, 1987).

A more recent application of the drug discrimination design has been in establishing structure–activity relationships. In such studies, the ability of compounds that are structural variations and molecular modifications of the training drug to produce drug-appropriate responding (i.e., substitute for the training drug) is examined. In so doing, the relationship between the chemical structure of the compound and its ability to serve a specific discriminative function can be determined. Such information may be useful in identifying which structural aspects of a compound are necessary for producing its discriminative effects and in designing agents that could antagonize or potentiate these effects (Glennon *et al.*, 1987). Such an analysis has now been done for a number of compounds, including tryptamine, phenylisopropylamine, and benzodiazepine derivatives (see Glennon, 1992; Glennon *et al.*, 1987).

The drug discrimination procedure, thus, is a useful tool in the classification of compounds that share similar stimulus properties and the further pharmacological, biochemical, and chemical characterization of their stimulus effects.

II. Procedural Variations: The Conditioned Taste Aversion Baseline

Although the drug discrimination procedure is often described in general terms to illustrate its behavioral applications, it should be noted that such learning is influenced by a number of parametric variations, including among other things the dose of the training drug, presession interval between training and testing, route of administration of the training drug, the schedule of reinforcement (and the specific reinforcing event) maintaining the discrimination, the specific sequencing of the training sessions, and the species being evaluated (see Stolerman, 1993). In addition to these more parametric variations, the specific training design itself is an important factor in the acquisition and degree of discriminative control (Stolerman, 1993). For example, a range of procedural variations have been introduced and utilized in drug discrimination learning, including dose vs dose discriminations, drug vs drug discriminations, drug vs drug vs vehicle discriminations, and compound discriminations (in which an animal is trained to discriminate a compound of two drugs from its individual elements (Stolerman and Mariathasan, 1990).

A new procedure that utilizes a modification of the conditioned taste aversion design to assess the stimulus properties of drugs has been introduced (Lucki, 1988; Martin *et al.*, 1990; Mastropaolo *et al.*, 1989). In the typical taste aversion design (Garcia and Ervin, 1968; Revusky and Garcia, 1970; Riley and Tuck, 1985a; Rozin and Kalat, 1971; for a bibliography see Riley and Tuck, 1985b), an animal is given a novel solution to drink and is then injected with a sickness-inducing compound, e.g., LiCl. After only a few such pairings, the animal associates the taste with the sickness and acquires an aversion to the sickness-associated taste, avoiding its consumption on subsequent exposures. To apply this basic procedure to drug discrimination learning requires only a few modifications (see Table 1). Specifically, one now gives the taste–sickness pairing only when the pairing is preceded by a training drug. That is, the animal is administered a training drug immediately prior to being given a taste–sickness pairing. On other days, the animal is administered the drug vehicle prior to the same taste, but on these days the

animal is not made sick. Under these conditions, the training drug serves as a stimulus signaling the subsequent taste–sickness pairing, whereas the drug vehicle serves as a stimulus signaling a safe exposure to the same taste.

In one of the first demonstrations of drug discrimination learning within this design (Mastropaolo *et al.*, 1989), rats were injected every fourth day with 1.8 mg/kg phencyclidine (PCP) 15 min prior to 20-min access to a novel saccharin solution. Immediately following access to saccharin, subjects were injected with the emetic LiCl (experimental subjects) or the LiCl vehicle (control subjects). On intervening recovery days, all subjects were injected with distilled water (the PCP vehicle) 15 min prior to saccharin alone. After only four such conditioning/recovery cycles, experimental subjects significantly decreased saccharin consumption following the injection of PCP while consuming the same saccharin solution at high levels following distilled water (see Fig. 1). Control subjects drank saccharin following both PCP and the PCP vehicle, indicating that the suppressed consumption of saccharin by the experimental subjects following PCP was a function of its conditioning history and not a result of unconditioned suppression produced by PCP. That discriminative control of a relatively low dose of PCP (1.8 mg/kg) was rapidly established within the taste aversion baseline suggests that this procedure may be a sensitive index of stimulus control.

A wide range of compounds have now been shown to establish discriminative control within the aversion baseline, e.g., alcohol (Kautz *et al.*, 1989), alprazolam (Glowa *et al.*, 1991), amphetamine (Revusky *et al.*,

TABLE 1 General Procedure for Group L (Experimental) and Group W (Control) during the Acquisition of Drug Discrimination Learning within the Conditioned Taste Aversion Baseline

Day	Procedure	Group[a]	
		L	W
—	Habituation	W-S	W-S
1	Conditioning	D-S-L	D-S-W
2–4	Recovery	W-S	W-S

[a] D, drug; L, LiCl; W, water; and S, saccharin.

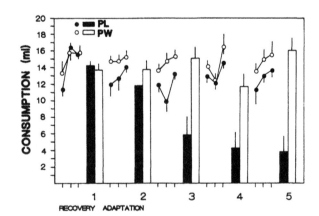

FIGURE 1 The mean amount of saccharin consumed (± SEM) for experimental subjects injected with PCP prior to a saccharin–LiCl pairing and the PCP vehicle prior to saccharin alone (Group PL) during adaptation and throughout the repeated conditioning/recovery cycles. Control subjects (group PW) were injected with PCP and the PCP vehicle prior to saccharin alone. (From Mastropaolo *et al.*, 1989 with permission of Elsevier Science Ltd.)

1982), buprenorphine (Pournaghash and Riley, 1993), chlordiazepoxide (van Hest *et al.*, 1992; Woudenberg and Hijzen, 1991), cholecystokinin (Melton *et al.*, 1993; Melton and Riley, 1993), cocaine (Geter and Riley, 1993), diprenorphine (Smurthwaite and Riley, 1992), 8-OH-DPAT (Lucki, 1988; Lucki and Marcoccia, 1991; van Hest *et al.*, 1992), estradiol (de Beun *et al.*, 1991), fentanyl (Jaeger and Mucha, 1990), morphine (Jaeger and van der Kooy, 1993; Martin *et al.*, 1990, 1991; Pournaghash and Riley, 1993; Skinner and Martin, 1992; Stevenson *et al.*, 1992), nalorphine (Smurthwaite and Riley, 1994a), naloxone (Kautz *et al.*, 1989; Smurthwaite *et al.*, 1992), pentobarbital (Jaeger and Mucha, 1990; Riley *et al.*, 1989), testosterone (de Beun *et al.*, 1992), and TFMPP (Lucki, 1988). Similar to that reported with PCP, in most of these cases control is established rapidly, typically within four to eight conditioning trials.

The sensitivity of the taste aversion baseline of drug discrimination learning is also indexed by the dose at which stimulus control can be established. This relative sensitivity has been noted with work on the acquisition of discriminative control with the opiate antagonist naloxone hydrochloride (Kautz *et al.*, 1989). Prior work with naloxone as a discriminative stimulus in operant baselines of drug discrimination learning has generally resulted in only marginal evidence of discriminative control. For example, Lal and colleagues (1978), as well as Overton and Batta (1979), have been unable to establish discriminative control with naloxone, prompting Overton (1987) to list naloxone as a compound "virtually undiscriminable." Although these reports have questioned the ability of naloxone to serve as a discriminative cue, Carter and Leander (1982) were able to train pigeons to discriminate naloxone from its vehicle. Even here, however, such control was evident only with extremely high doses of naloxone (30 mg/kg) and with extensive training.

Based on the apparent sensitivity of the averison baseline, Kautz and colleagues (1989) attempted to establish discriminative control with naloxone. In her design, rats were injected every fourth day with low doses of naloxone (1 mg/kg) 15 min prior to a saccharin–LiCl pairing and on intervening days with the naloxone vehicle prior to saccharin alone. Control subjects received the same sequence of injections, but were not injected with LiCl. After only four conditioning trials, the experimental subjects significantly decreased saccharin consumption following naloxone and consumed the same saccharin solution following the naloxone vehicle. Control subjects consumed saccharin following both injections. Thus, naloxone was able to serve as a discriminative stimulus at a low dose and with only limited training (see also Smurthwaite *et al.*,

1992). Similar rapid acquisition with low doses of the opiate antagonists diprenorphine (Smurthwaite and Riley, 1992) and nalorphine (Smurthwaite and Riley, 1994a) has also been reported. Thus, the aversion design seems sensitive not only in terms of the rate of acquisition of stimulus control, but also in terms of the dose sufficient to establish the control (see also Skinner and Martin, 1992).

A. Drug Classification and Characterization

As described earlier, within the traditional assessments of discrimination learning stimulus control by the training drug typically generalizes to drugs from the same class. Similar patterns have been reported within the aversion baseline. The procedure utilized in these assessments is similar to that described earlier for the training of the drug discrimination with the exception that on one of the recovery sessions following a conditioning trial, animals are administered a test drug to assess its ability to substitute for the training drug. On these days, no injections of LiCl are given following saccharin consumption (see Table 2). In one of the initial tests for generalization within the aversion baseline, Mastropaolo and associates (1989) reported that animals trained to discriminate PCP from its vehicle generalized this control to ketamine, a compound from the same pharmacological class as PCP. In this case, subjects avoided saccharin consumption when it was preceded by either the training drug, PCP, or the test compound, ketamine (see Fig. 2). On the other hand, amphetamine, a drug from a different class than PCP did not substitute for the PCP stimulus, i.e., subjects consumed saccharin at control levels when it was preceded by amphetamine (see Fig. 3). Similarly, Woudenberg and Hijzen (1991) have reported within the aversion design that animals trained to discriminate the anxiolytic chlordiazepoxide from its vehicle generalized this control to diazepam, but not to clonidine, haloperidol, or pentobarbital.

TABLE 2 General Procedure for Group L (Experimental) and Group W (Control) during Generalization Testing within the Conditioned Taste Aversion Baseline of Drug Discrimination Learning

Day	Procedure	Group[a]	
		L	W
1	Conditioning	D-S-L	D-S-W
2	Recovery	W-S	W-S
3	Probe	X-S	X-S
4	Recovery	W-S	W-S

[a] D, drug; L, LiCl; W, water; S, saccharin; and X, probe.

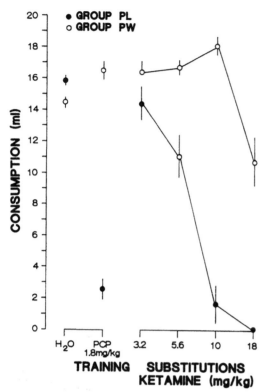

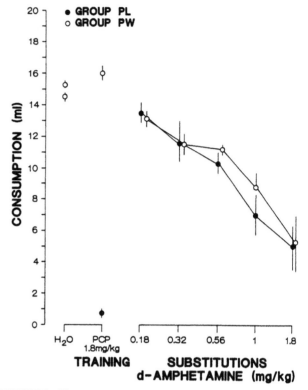

FIGURE 2   The mean amount of saccharin consumed (± SEM) for subjects injected with PCP prior to a saccharin–LiCl pairing and the PCP vehicle prior to saccharin alone (group PL) following the training dose of PCP and the PCP vehicle and following 1/4 log doses of ketamine during generalization testing. Control subjects (group PW) were injected with PCP and the PCP vehicle prior to saccharin alone. (From Mastropaolo et al., 1989 with permission of Elsevier Science Ltd.)

FIGURE 3   The mean amount of saccharin consumed (± SEM) for subjects injected with PCP prior to a saccharin–LiCl pairing and the PCP vehicle prior to saccharin alone (group PL) following the training dose of PCP and the PCP vehicle and following 1/4 log doses of amphetamine during generalization testing. Control subjects (group PW) were injected with PCP and the PCP vehicle prior to saccharin alone. (From Mastropaolo et al., 1989 with permission of Elsevier Science Ltd.)

Such generalization indicates that the training and the test drugs share some property that underlies the generalization. However, as noted earlier such classes of compounds can often be further differentiated into subclasses by the drug discrimination procedure, presumably by virtue of their differing biochemical and behavioral activity. Such compounds can be differentiated by the taste aversion baseline of drug discrimination as well. For example, in a modification of the aversion baseline, Kautz and Riley (1990) have reported that experimental animals administered the relatively selective μ opioid antagonist naloxone prior to a saccharin–LiCl pairing and the broad-based opiate antagonist diprenorphine prior to saccharin alone acquired the two-drug discrimination, avoiding saccharin when it was preceded by naloxone and consuming the same saccharin solution when it was preceded by diprenorphine (see Fig. 4; see also, Kautz and Riley, 1992, 1993; Kautz et al., 1992). This discrimination is interesting in that in generalization tests, these compounds typically generalize to each other. Thus, the

procedure can be modified to allow for a differentiation between compounds that may share sufficient properties to underlie generalization, but that nevertheless differ in their activity.

Although work within the aversion baseline has focused primarily on demonstrating such control and determining what drugs serve as discriminative stimuli, there is some evidence that this specific drug discrimination design may be useful in the characterization of the drug as well. Further, findings within this design are generally in agreement with those reported in more traditional assessments of discrimination learning. For example, Stevenson and colleagues (1992) reported that naloxone blocked the stimulus effects of morphine within the aversion baseline when administered 10 to 60 min prior to the injection of the training dose of morphine (see Fig. 5; see also Martin et al., 1991; Pournaghash and Riley, 1993). By 180 min, the antagonistic effects of naloxone were less evident. Thus, not only were the antagonistic actions of naloxone on morphine demonstrated within the aversion design, but the de-

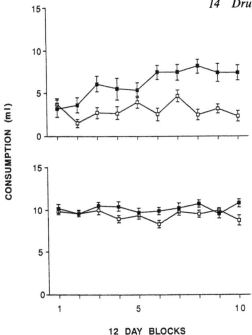

FIGURE 4 The mean amount of saccharin consumed for experimental subjects (group L) injected with naloxone (□) prior to saccharin–LiCl pairings and with diprenorphine (■) prior to saccharin alone (top). Control subjects (group W) were injected with naloxone (□) and diprenorphine (■) prior to saccharin alone (bottom).

sign also allowed for an assessment of the temporal parameters of this antagonism. Antagonism within the aversion design has been reported with other drugs, e.g., chlordiazepoxide (Woudenberg and Hijzen, 1991), cholecystokinin (Melton and Riley, 1993, 1994), and the opiate agonsit buprenorphine (Pournaghash and Riley, 1993).

Reports on potentiation within the taste aversion baseline are limited; however, such effects have been noted. For example, Melton and Riley (1993) reported that the opiate antagonist naloxone failed to substitute for the gut peptide cholecystokinin (CCK) in animals trained within the aversion baseline to discriminate CCK from its vehicle. Interestingly, when naloxone was given in combination with doses of CCK that were not sufficient to maintain stimulus control, animals responded as if they were performing under the training dose of CCK, i.e., naloxone potentiated the effects of CCK in the absence of having any effect on its own (see Fig. 6).

A related behavioral characterization concerns determining the specific biochemical mediation of a drug's stimulus properties. This, too, has been reported for the aversion baseline. For example, Melton and Riley (1994) have reported that only the CCK-A antagonist devazepide (and not the CCK-B antagonist L-365,260) blocked the stimulus properties of CCK, suggesting that its stimulus effects are mediated by the

CCK-A receptor. The biochemical mediation of buprenorphine (Pournaghash and Riley, 1993), cocaine (Geter and Riley, 1993), and naloxone (Smurthwaite *et al.*, 1992) has also been addressed using this baseline.

Although the specific procedure described earlier for the aversion design is the one that has been used most extensively in assessments of drug discrimination learning, several modifications of this procedure have been introduced to address specific issues or problems noted with the basic taste averison baseline of drug discrimination learning. For example, discriminative training with the drug signaling the taste alone and the drug vehicle signaling the taste-illness pairing has been utilized to control for possible interactions between the training drug and poison (Jaeger and Mucha, 1990; Mastropaolo *et al.*, 1989). Others have allowed access during generalization tests to both the poison-associated taste and water to circumvent the effects of unconditioned suppression by the test drug (Lucki, 1988; Lucki and Marcoccia, 1991). Further, cumulative dosing procedures have been used to allow for rapid dose–response assessments during generalization testing (Smurthwaite and Riley, 1994b). Although drug discrimination learning can apparently occur across several modifications of the aversion design, it should not be assumed that the acquisition of such discriminations is not affected by variations in the experimental procedure. Similar to discriminations acquired within the more traditional operant baseline (see earlier discussion), it is likely that a number of variables, e.g., dose of the training drug, duration of training, species tested, testing procedure, and route of administration, affect the rate of acquisition of the discrimination as well as subsequent generalization functions and drug interactions (see Jarbe, 1989; Overton, 1987; Riley *et al.*, 1991). To date, little is known as to the influence of such procedural variations on drug discriminations within the aversion design (though see Jaeger and Mucha, 1990; Riley *et al.*, 1993).

III. Applications to Behavioral Toxicology

This chapter has focused on a research design, i.e., drug discrimination learning, used almost exclusively in behavioral pharmacology. Within behavioral pharmacology, this general design has been used primarily in the assessment of the stimulus properties of recreational and abused drugs. This may be due in part to the belief that since recreational drugs are used because of their subjective effects and that drug discrimination learning relies on subjective effects of drugs, this specific research tool can be used to identify and characterize recreational compounds. Given that this chapter

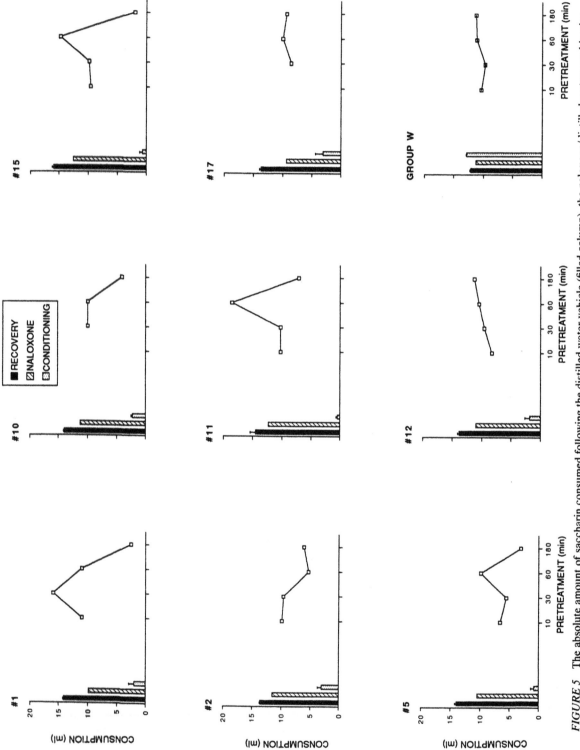

FIGURE 5 The absolute amount of saccharin consumed following the distilled water vehicle (filled column), the naloxone/distilled water combination (hatched column), morphine (stippled column), and the naloxone/morphine combinations (open squares) with naloxone pretreatment intervals of 10, 30, 60, and 180 min for individual experimental subjects trained to discriminate morphine from its vehicle. Bars represent SEM. The mean amount consumed following these same injections for control subjects is presented in the lower right panel. (From Stevenson *et al.*, 1992 with permission of Elsevier Science Ltd.)

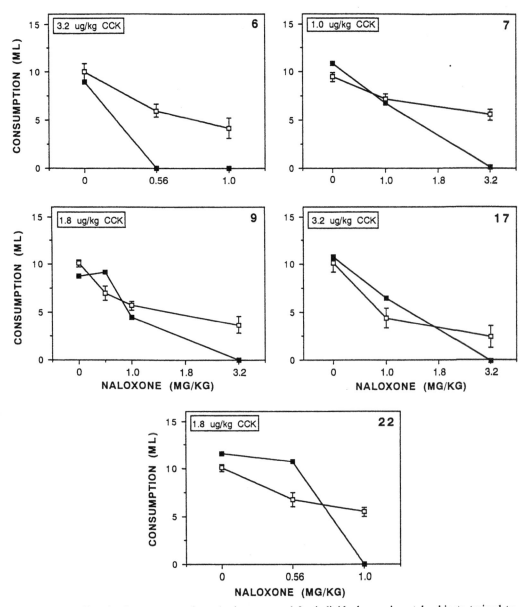

FIGURE 6 The absolute amount of saccharin consumed for individual experimental subjects trained to discriminate CCK from its vehicle (filled squares) following various doses of naloxone (0–3.2 mg/kg) administered in combination with a dose of CCK (noted in insert) ineffective in producing saccharin avoidance when administered alone. The mean amount (± SEM) of saccharin consumed for control subjects is represented by open squares. (From Melton and Riley, 1993 with permission of Elsevier Science Ltd.)

began with a promissory note that this procedure can also be used in behavioral toxicology, the questions of how and whether the design can be used in behavioral toxicology need to be addressed.

In relation to how drug discrimination learning can be applied to behavioral toxicology, the procedure should provide information regarding toxins similar to that provided in the classification and characterization of recreational and abused drugs. As noted, animals could be trained to discriminate known toxins and then

be assessed for the ability of other compounds to substitute for the training drug. The presence or absence of generalization of the training toxin to these other compounds could then be used in part to classify the test drugs. If the training drug is a known toxin, this procedure then provides a behavioral screen or index of toxicity, placing compounds into major groupings as well as specific subclasses. Further, the design could be used to assess the effects of potential blocking and potentiating agents on the discrimination of a known toxin. With special training procedures (see previous discussion), one could isolate the biochemical and biological systems on which the compounds are acting to produce their stimulus effects and establish structure–activity relationships.

To date, however, there is little evidence that toxins do in fact support such learning. As such, the actual utility of this design remains untested. Although classical toxins have not generally been examined within drug discrimination learning, considerable evidence exists that suggests that toxins could serve as discriminative stimuli. First, although the majority of the compounds examined within the drug discrimination procedure are recreational and abused drugs, compounds that are not typically considered recreational, e.g., imipramine, haloperidol, and ketamine, do in fact support drug discrimination learning (Overton, 1982, 1987). As described earlier, both the opiate antagonist naloxone (Kautz *et al.*, 1989) and the gut peptide CCK (Melton *et al.*, 1993) are effective in the drug discrimination procedure. More direct support for the possibility that toxins may be examined within the drug discrimination design comes from the limited work on aversive or toxic compounds as discriminative stimuli. For example, strychnine and tetraethylammonium (Overton, 1982) and LiCl (Winsauer *et al.*, 1993) are effective drug stimuli. Further, pentylenetetrazol as well as barbiturate and opiate withdrawal support drug discrimination learning (for a review see Emmett-Oglesby *et al.*, 1990). That nonrecreational drugs (as well as the state associated with the absence of dependence-producing drugs) can serve as discriminative stimuli argues against the position that the reinforcing subjective effects of a drug are a necessary condition for such learning (for a discussion see Brady *et al.*, 1990; Schuster and Johanson, 1988) and argues for the possibility that the drug discrimination design can be used for drugs outside of the general class of recreationally used and abused compounds.

IV. Conclusion

This chapter has attempted to present a general research design (and a specific modification of that design) that may serve as a behavioral baseline to classify and characterize toxins. The drug discrimination design, as described, is widely used as such a tool in behavioral pharmacology to classify recreational compounds and to characterize their stimulus effects. Although little is currently known regarding the ability of toxins to serve as discriminative stimuli, the application of this design to behavioral toxicology appears to be a natural extension of that used in behavioral

pharmacology and should provide an efficient method for their identification and classification.

Acknowledgments

The research on which this paper is based was supported by a grant from the Mellon foundation and a National Research Service Award from the National Institute on Drug Abuse (1 F31 DA04577-01). Requests for reprints should be sent to Anthony L. Riley, Psychopharmacology Laboratory, Department of Psychology, The American University, Washington, D.C. 20016.

References

Appel, J. B., Baker, L. E., Barrett, J. B., Michael, E. M., Riddle, E. E., and van Groll, B. J. (1992). Use of drug discrimination in drug abuse research. In *Drug discrimination: Applications to drug abuse research* (R. A. Glennon, T. U. C. Jarbe, and J. Frankenheim, Eds.), pp. 369–397, NIDA Monograph 116. U.S. Government Printing Office, Washington, D.C.

Barry, H. (1974). Classification of drugs according to their discriminable effects in rats. *Fed. Proc.* **33**, 1814–1824.

Barry, H. (1992). Distinctive discriminative effects of alcohol. *NIDA Res. Monogr.* **116**, 131–144.

Barry, H., and Krimmer, E. C. (1977). Discriminable stimuli produced by alcohol and other CNS depressants. In *Discriminative stimulus properties of drugs* (H. Lal, Ed.), pp. 73–92, Plenum, New York.

Barry, H., and Krimmer, E. C. (1978). Pharmacology of discriminative drug stimuli. In *Drug discrimination and state dependent learning* (B. T. Ho, D. W. III Richards, and D. L. Chute, Eds.), pp. 3–32. Academic Press, New York.

Brady, J. V., Bradford, L. D., and Hienz, R. D. (1979). Behavioral assessment of risk-taking and psychophysical functions in the baboon. *Neurobehav. Toxicol.* **1** (Suppl. 1), 73–84.

Brady, J. V., Hienz, R. D., and Ator, N. A. (1990). Stimulus functions of drugs and the assessment of abuse liability. *Drug Dev. Res.* **20**, 231–249.

Cabe, P. A., and Eckerman, D. A. (1982). Assessment of learning and memory dysfunction in agent-exposed animals. In *Nervous system toxicology* (C. L. Mitchell, Ed.), pp. 133–198, Raven Press, New York.

Carter, R. B., and Leander, J. D. (1982). Discriminative stimulus properties of naloxone. *Psychopharmacology* **77**, 305–308.

Colpaert, F. C., and Balster, R. L. (Eds.) (1988). *Transduction mechanisms of drug stimuli*. Springer-Verlag, Berlin.

Colpaert, F. C., and Rosecrans, J. A. (Eds.) (1978). *Stimulus properties of drugs: Ten years of progress*. Elsevier, Amsterdam.

Colpaert, F. C., and Slangen, J. L. (Eds.) (1982). *Drug discrimination: Applications in CNS pharmacology*. Elsevier, Amsterdam.

Conger, J. J. (1951). The effects of alcohol on conflict behavior in the albino rat. *Q. J. Stud. Alcohol* **12**, 1–29.

Cory-Slechta, D. A. (1989). Behavioral measures of neurotoxicity. *Neurotoxicology* **10**, 271–296.

Cory-Slechta, D. A. (1992). Schedule-controlled behavior in neurotoxicology. In *Neurotoxicology* (H. Tilson and C. Mitchell, Eds.), pp. 271–294, Raven Press, New York.

Cory-Slechta, D. A., and Weiss, B. (1985). Alterations in schedule-controlled behavior of rodents correlated with prolonged lead exposure. In *Behavioral pharmacology: The current status* (L. S. Seiden and R. L. Balster, Eds.), pp. 487–501, Liss, New York.

de Beun, R., Heinsbroek, R. P. W., Slangen, J. L., and van de Poll, N. E. (1991). Discriminative stimulus properties of estradiol in male and female rats revealed by a taste-aversion procedure. *Behav. Pharmacol.* **2,** 439–445.

de Beun, R., Jansen, E., Slangen, J. L., and van de Poll, N. E. (1992). Testosterone as appetitive and discriminative stimulus in rats: Sex- and dose-dependent effects. *Physiol. Behav.* **52,** 629–634.

Eckerman, D. A., and Bushnell, P. J. (1992). The neurotoxicology of cognition: Attention, learning, and memory. In *Neurotoxicology* (H. Tilson and C. Mitchell, Eds.), pp. 213–270. Raven Press, New York.

Emmett-Oglesby, M. W., Mathis, D. A., Moon, R. T. Y., and Lal, H. (1990). Animal models of drug withdrawal symptoms. *Psychopharmacology* **101,** 292–309.

Ferster, C. B., and Skinner, B. F. (1957). *Schedules of reinforcement.* Appleton-Century-Crofts, New York.

France, C. P., and Woods, J. H. (1988). Acute supersensitivity to the discriminative stimulus effects of naltrexone in pigeons. *J. Pharmacol. Exp. Ther.* **244,** 599–605.

Garcia, J., and Ervin, F. R. (1968). Gustatory-visceral and telereceptor-cutaneous conditioning: adaptations in internal and external milieus. *Commun. Behav. Biol.* **1,** 389–415.

Geller, I., Stebbins, W. C., and Wayner, M. J. (Eds.) (1979). Test methods for definition of effects of toxic substances on behavior and neuromotor function. *Neurobehav. Toxicol.* **1,** (Suppl. 1).

Geter, B., and Riley, A. L. (1993). The ability of a D1 and D2 antagonist combination to antagonize the discriminative stimulus properties of cocaine. In *Problems of drug dependence, 1992: Proceeding of the 54th annual scientific meeting* (L. Harris, Ed.), p. 95. U.S. Government Printing Office, Washington, D.C.

Glennon, R. A. (1992). Discriminative stimulus properties of hallucinogens and related designer drugs. In *Drug discrimination: Applications to drug abuse research.* NIDA Monograph 116 (R. A. Glennon, T. U. C. Jarbe, J. Frankenheim, Eds.), pp. 25–44. U.S. Government Printing Office, Washington, D.C.

Glennon, R. A., and Young, R. (1987). The study of structure-activity relationships using drug discrimination methodology. In *Methods of assessing the reinforcing properties of abused drugs* (M. A. Bozarth, Ed.), pp. 373–390. Springer-Verlag, Berlin.

Glennon, R. A., Jarbe, T. U. C., and Frankenheim, J., (Eds.) (1992). *Drug discrimination: Applications to drug abuse research.* NIDA Monograph 116. U.S. Government Printing Office, Washington, D.C.

Glowa, J. R., Jeffreys, R. D., and Riley, A. L. (1991). Drug discrimination using a conditioned taste-aversion paradigm in rhesus monkeys. *J. Exp. Anal. Behav.* **56,** 303–312.

Goudie, A. J. (1992). Discriminative stimulus properties of amphetamine, cathinone, and related compounds. In *Drug discrimination: Applications to drug abuse research.* NIDA Monograph 116 (R. A. Glennon, T. U. C. Jarbe, J. Frankenheim, Eds.), pp. 45–60. U.S. Government Printing Office, Washington, D.C.

Herling, S., and Shannon, H. E. (1982). Ro 15-1788 antagonizes the discriminative stimulus effects of diazepam in rats but not similar effects of pentobarbital. *Life Sci.* **31,** 2105–2112.

Ho, B. T., Richards, W., III and Chute, D. L. (Eds.) (1978) *Drug discrimination and state dependent learning.* Academic Press, New York.

Holtzman, S. G. (1985). Discriminative stimulus properties of opioids that interact with mu, kappa and PCP/sigma receptors. In *Behavioral pharmacology: The current status* (L. S. Seiden, R. L. Balster, Eds.), pp. 131–147. Liss, New York.

Holtzman, S. G., and Locke, K. W. (1988). Neural mechanisms of drug stimuli: Experimental approaches. In *Transduction mecha-*

nisms of drug stimuli (F. C. Colpaert, and R. L. Balster, Eds.), pp. 139–153. Springer-Verlag, Berlin.

Jaeger, T. V., and Mucha, R. F. (1990). A taste aversion model of drug discrimination learning: Training drug and condition influence rate of learning, sensitivity and drug specificity. *Psychopharmacology* **100,** 145–150.

Jaeger, T. V., and van der Kooy, D. (1993) Morphine acts in the parabrachial nucleus, a pontine viscerosensory relay, to produce discriminative stimulus effects. *Psychopharmacology* **110,** 76–84.

Jarbe, T. U. C. (1987). Drug discrimination learning: Cue properties of drugs. In *Experimental Psychopharmacology* (A. J. Greenshaw, and C. T. Dourish, Eds.), pp. 433–479, The Humana Press, Clifton, New Jersey.

Jarbe, T. U. C. (1989). Discrimination learning with drug stimuli: Methods and applications. In *Neuromethods,* Vol. 13, *Psychopharmacology* (A. A. Boulton, G. B. Baker, and A. J. Greenshaw, Eds.), pp. 513–563, The Humana Press, Clifton, New Jersey.

Jeffreys, R. D., Pournaghash, S., Glowa, J. R., and Riley, A. L. (1990). The effects of Ro 15-4513 on ethanol-induced taste aversions. *Pharmacol. Biochem. Behav.* **35,** 803–806.

Kautz, M. A., Geter, B., McBride, S. A., Mastropaolo, J. P., and Riley, A. L. (1989). Naloxone as a stimulus for drug discrimination learning. *Drug Dev. Res.* **16,** 317–326.

Kautz, M. A., Logan, J. P., Romero, A. E., Schwartz, M. D., and Riley, A. L. (1989). The effects of Ro 15-4513 on ethanol drug discrimination learning. *Soc. Neurosci. Abstr.* **15,** 633.

Kautz, M. A., and Riley, A. L. (1990). Drug-drug discrimination with the opiate antagonists naloxone and diprenorphine. Paper presented at the Society for Neuroscience, St. Louis.

Kautz, M. A., and Riley, A. L. (1993). Morphine/nalorphine discrimination learning within a conditional two-drug discrimination procedure. In *Problems of drug dependence, 1992: Proceeding of the 54th annual scientific meeting* (L. Harris, Ed.), p. 247, U.S. Government Printing Office, Washington, D.C.

Kautz, M. A., and Riley, A. L. (1993). Opiate generalization in rats trained to discriminate nalorphine vs. morphine within the conditioned taste aversion procedure. Paper presented at the College of the Problems of Drug Dependence, Toronto, Canada.

Kautz, M. A., Smurthwaite, S. T., and Riley, A. L. (1991). Drug/drug discrimination learning within the conditioned taste aversion procedure. In *Problems of drug dependence, 1991: Proceeding of the 53rd annual scientific meeting* (L. Harris, Ed.), p. 382, U.S. Government Printing Office, Washington, D.C.

Lal, H. (Ed.) (1978). *Discriminative stimulus properties of drugs.* Plenum Press, New York.

Lal, H., and Fielding, S. (Eds.) (1989). *Drugs as interoceptive stimuli.* Liss, New York.

Lal, H., Gianutsos, G., and Miksic, S. (1978). Discriminative stimuli produced by narcotic analgesics. In *Discriminative stimulus properties of drugs* (H. Lal, Ed.), pp. 23–45. Plenum, New York.

Lal, H., Miksic, S., and McCarten, M. (1978). A comparison of discriminative stimuli produced by naloxone, cyclazocine and morphine in the rat. In *Stimulus Properties of Drugs: Ten Years of Progress* (F. C. Colpaert and J. A. Rosecrans, Eds.), pp. 177–180, Elsevier/North Holland Press, Amsterdam.

Laties, V. G. (1982). Contributions of operant conditioning to behavioral toxicology. In *Nervous system toxicology* (C. L. Mitchell, Ed.), pp. 67–79. Raven Press, New York.

Lucki, I. (1988). Rapid discrimination of the stimulus properties of 5-hydroxytryptamine agonists using conditioned taste aversion. *J. Pharmacol. Exp. Therp.* **247,** 1120–1127.

Lucki, I., and Marcoccia, J. M. (1991). Discriminated taste aversion with a 5-HT1A agonist measured using saccharin preference. *Behav. Pharmacol.* **2,** 335–344.

MacPhail. R. S. (1985). Effects of pesticides on schedule-controlled behavior. In *Behavioral Pharmacology: The Current Status* (L. S. Seiden, and R. L. Balster, Eds.), pp. 519–535. Liss, New York.

Martin, G. M., Bechara, A., and van der Kooy, D. (1991). The perception of emotion: Parallel neural processing of the affective and discriminative properties of the opiates. *Psychobiology* **19**, 147–152.

Martin, G. M., Gans, M., and van der Kooy, D. (1990). Discriminative properties of morphine that modulate associations between taste and lithium chloride. *J. Exp. Psych: Anim. Behav. Pro.* **16**, 56–68.

Mastropaolo, J. P., Moskowitz, K. H., Dacanay, R. J., and Riley, A. L. (1989). Conditioned taste aversions as a behavioral baseline for drug discrimination learning: An assessment with phencyclidine. *Pharmacol. Biochem. Behav.* **32**, 1–8.

Maurissen, J. P. J. (1979). Effects of toxicants on the somatosensory system. *Neurobehav. Toxicol* **1**, (Suppl. 1), 23–31.

Melton, P. M., Kopman, J. A., and Riley, A. L. (1993). Cholecystokinin as a stimulus in drug discrimination learning. *Pharmacol. Biochem. Behav.* **44**, 249–252.

Melton, P. M., and Riley, A. L. (1993). An assessment of the interaction between cholecystokinin and the opiates within a drug discrimination procedure. *Pharmacol. Biochem. Behav.* **44**, 249–252.

Melton, P. M., and Riley, A. L. (1994). Receptor mediation of the stimulus properties of CCK within the conditioned taste aversion baseline of drug discrimination learning. *Pharmacol. Biochem. Behav.* **48**, 275–279.

Mitchell, C. L. (Ed.) (1982). *Nervous system toxicology.* Raven Press, New York.

Overton, D. A. (1982) Comparison of the degree of discriminability of various drugs using the T-maze drug discrimination paradigm. *Psychopharmacology* **76**, 385–395.

Overton, D. A. (1984). State dependent learning and drug discriminations. In *Handbook of Psychopharmacology* (L. L. Iversen, S. D. Iversen, and S. H. Snyder, Eds.), pp. 59–127. Plenum Press, New York.

Overton, D. A. (1987). Applications and limitations of the drug discrimination method for the study of drug abuse. In *Methods of assessing the reinforcing properties of drugs* (M. A. Bozarth, Ed.), pp. 291–340, Springer-Verlag, New York.

Overton, D. A. (1991). Historical context of state dependent learning and discriminative drug effects. *Behav. Pharmacol.* **2**, 253–264.

Overton, D. A. (1992). A historical perspective on drug discrimination. In *Drug discrimination: Applications to drug abuse research*. NIDA Monograph 116 (R. A. Glennon, T. U. C. Jarbe, and J. Frankenheim, Eds.), pp. 5–24, U.S. Government Printing Office, Washington, D.C.

Overton, D. A., and Batta, S. K. (1979). Investigation of narcotics and antitussives using drug discrimination techniques. *J. Pharmacol. Exp. Ther.* **211**, 401–408.

Pournaghash, S., and Riley, A. L. (1993). Buprenorphine as a stimulus in drug discrimination learning: An assessment of mu and kappa receptor activity. *Pharmacol. Biochem. Behav.* **46**, 593–604.

Revusky, S. Coombes, S., and Pohl, R. W. (1982). Drug states as discriminative stimuli in a flavor-aversion learning experiment. *J. Comp. Physiol. Psychol.* **96**, 200–211.

Revusky, S., and Garcia, J. (1970). Learned associations over long delays. In *Psychology of Learning and Motivation: Advances in Research and Theory*, Vol. 4 (G. Bower and J. Spence, Eds.), pp. 1–83, Academic Press, New York.

Riley, A. L., Jeffreys, R. D., Pournaghash, S., Titley, T. L., and Kufera, A. M. (1978). Conditioned taste aversions as a behavioral baseline for drug discrimination learning: Assessment with the dipsogenic compound pentobarbital. *Drug Dev. Res.* **16**, 229–236.

Riley, A. L., Kautz, M. A., Geter, B., Smurthwaite, S. T., Pournaghash, S., Melton, P. M., and Ferrari, C. M. (1991). A demonstration of the graded nature of the generalization function of drug discrimination learning within the conditioned taste aversion procedure. *Behav. Pharmacol.* **2**, 323–334.

Riley, A. L., and Tuck, D. L. (1985a). Conditioned taste aversions: A behavioral index of toxicity. *Ann. N. Y. Acad. Sci.* **443**, 272–292.

Riley, A. L., and Tuck, D. L. (1985b). Conditioned food aversions: A bibliography. *Ann. N.Y. Acad. Sci.* **443**, 381–437.

Riley, A. L., Wetherington, C. L., and Sobel, B. X. (1993). The temporal analysis of the generalization function within the conditioned taste aversion baseline of drug discrimination learning. Paper presented at the Society for Neuroscience, Washington, D.C.

Rozin, P., and Kalat, J. W. (1971). Specific hungers and poison avoidance as adaptive specializations of learning. *Psychol. Rev.* **78**, 459–486.

Samele, C., Shine, P. J., and Stolerman, I. P. (1992). Forty years of drug discrimination research: A bibliography for 1951-1991. NIDA Adminsitrative Document, National Clearinghouse for Alcohol and Drug Information.

Schuster, C. R., and Johanson, C. E. (1988). Relationship between the discriminative stimulus properties and subjective effects of drugs. In *Transduction Mechanisms of Drug Stimuli* (F. C. Colpaert and R. L. Balster, Eds.), pp. 161–175, Springer-Verlag, Berlin.

Skinner, B. F. (1938). *Behavior of organisms.* Appleton-Century-Crofts, New York.

Skinner, D. M., and Martin, G. M (1992). Conditioned taste aversions support drug discrimination learning at low dosages of morphine. *Behav. Neurol. Biol.* **58**, 236–241.

Smurthwaite, S. T., Kautz, M. A., Geter, B., and Riley, A. L. (1992). Naloxone as a stimulus in drug discrimination learning: Generalization to other opiate antagonists. *Pharmacol. Biochem. Behav.* **41**, 43–47.

Smurthwaite, S. T., and Riley, A. L. (1992). Diprenorphine as a stimulus in drug discrimination learning. *Pharmacol. Biochem. Behav.* **43**, 839–846.

Smurthwaite, S. T., and Riley, A. L. (1994a). Nalorphine as a stimulus in drug discrimination learning. *Pharmacol. Biochem. Behav.*, **48**, 635–642.

Smurthwaite, S. T., and Riley, A. L. (1994b). Animals trained to discriminate morphine from naloxone generalize morphine (but not naloxone) control to nalorphine. Paper presented at the College of the Problems of Drug Dependence, West Palm Beach, Florida.

Snoddy, A. M., and Tessel, R. E. (1985). Prazosin: Effect of psychomotor-stimulant cues and locomotor behavior in mice. *Eur. J. Pharmacol.* **116**, 221–228.

Stevenson, G. W., Pournaghash, S., and Riley, A. L. (1992). Antagonism of drug discrimination learning within the conditioned taste aversion procedure. *Pharmacol. Biochem. Behav.* **41**, 245–249.

Stolerman, I. P. (1993). Drug discrimination. In *Methods in Behavioral Pharmacology* (F. van Haaren, Ed.), pp. 217–243, Elsevier/North Holland Press, Amsterdam.

Stolerman, I. P., and Mariathasan, E. A. (1990). Discrimination of an amphetamine-pentobarbitone mixture by rats in an AND-OR paradigm. *Psychopharmacology* **102**, 557–560.

Thompson, T., and Pickens, R. (Eds.) (1971). *Stimulus Properties of Drugs*. Appleton-Century-Crofts, New York.

Tilson, H. A. (1987). Behavioral indices of neurotoxicity: What can be measured? *Neurotoxicol. Teratol.* **9,** 427–443.

Tilson, H. A., and Harry, C. J. (1982). Behavioral principles for use in behavioral toxicology and pharmacology. In *Nervous system toxicology* (C. L. Mitchell, Ed.), pp. 1–27, Raven Press, New York.

Tilson, H., and Mitchell, C. (Eds.) (1992). *Neurotoxicology*. Raven Press, New York

van Hest, A., Hijzen, T. H., Slangen, J. L., and Oliver, B. (1992). Assessment of the stimulus properties of anxiolytic drugs by means of the conditioned taste aversion procedure. *Pharmacol. Biochem. Behav.* **42,** 487–495.

Weiss, B., and Laties, V. G. (1979). Assays for behavioral toxicity: A strategy for the Environmental Protection Agency. *Neurobehav. Toxicol.* **1,** (Suppl. 1), 213–215.

Winsauer, P. J., Verrees, J. F., and Mele, P. C. (1993). Discriminative stimulus properties of lithium chloride. Paper presented at the Society for Neuroscience, Washington, D.C.

Woolverton, W. L. (1992). Discriminative stimulus properties of cocaine. In *Drug Discrimination: Applications to Drug Abuse Research*. NIDA Monograph 116 (R. A. Glennon, T. U. C. Jarbe, and J. Frankenheim, Eds.), pp. 61–74, U.S. Government Printing Office, Washington, D.C.

Woudenberg, F., and Hijzen, T. H. (1991). Discriminated taste aversion with chlordiazepoxide. *Pharmacol. Biochem. Behav.* **39,** 859–863.

Young, R., and Dewey, W. L. (1982). Differentiation of the behavioural responses produced by barbiturates and benzodiazepines by the benzodiazepine antagonist Ro 15-1788. In *Drug Discrimination: Applications in CNS Pharmacology* (F. C. Colpaert and J. L. Slangen, Eds.), pp. 235–240, Elsevier/North Holland Press, Amsterdam.

CHAPTER

15

Extrapolating Scientific Data from Animals to Humans in Behavioral Toxicology and Behavioral Pharmacology

DONALD E. MCMILLAN
S. MICHAEL OWENS
Department of Pharmacology and Toxicology
University of Arkansas for Medical Sciences
Little Rock, Arkansas 72205

I. Introduction

Whether an investigator is screening new chemicals for possible therapeutic activity as drugs or is testing chemicals found in the environment for their toxicological effects, the testing of the chemical almost always begins in a subhuman animal. Although sometimes the investigator is only interested in knowing the direct benefits or dangers that it produces in the animal species studied, more often the investigator is using the animal as a model of some aspect of human function that cannot safely be studied in humans. Based on the effects of the chemical in the animal model, predictions or "extrapolations" are made about possible effects of the chemical in humans.

The assumption that one can extrapolate effects from animals to man goes back to the notions of Darwin who in *The Origin of the Species* (1859) suggested a continuity between all animal species. That Darwin also believed this continuity included behavioral responses seems clear from his subsequent publication *Expression of the Emotions in Man and Animals* (1872).

Others, such as Galton and Romanes (Boring, 1950), developed these ideas further to form the conceptual foundation of comparative psychology. Although the term "comparative psychology" is not widely used today, investigators in behavioral toxicology are studying comparative behaviors whenever they attempt to extrapolate their animal findings to humans in the risk assessment process. Unfortunately, animal models can sometimes be less than perfect predictors of human responses as shown by the thalidomide tragedy. In this case rats exposed to the drug failed to predict the tragic human teratology (Manson, 1986).

The purpose of this chapter is to discuss some of the factors that can contribute to successful animal modeling, as well as some of the limitations of animal models in predicting human behavioral responses to chemical exposure. This chapter also suggests some areas into which behavioral toxicologists and behavioral pharmacologists might direct their research as the field of molecular biology continues to develop. From this point on the discussion will be simplified by using the term "chemical" to describe both drugs and xenobiotics.

II. Factors That Contribute to the Validity of Animal Models

Differences between the effects of chemicals that interact with behavior in different species can arise at a number of levels, including the way in which the animal processes the chemical (species differences in uptake, distribution, metabolism, excretion, etc.), the physiological and biochemical substrates with which the chemical interacts (species differences in neurotransmitters, receptors, second messengers, etc.), or species differences in behavior, which may be genetic, learned, or imposed by the investigator's procedures for behavioral measurement.

An example of how pharmacokinetic factors can contribute to species differences in the response to a chemical can be seen in the sleeping-time response of four species to hexobarbital (Alvares and Pratt, 1990). In these experiments the sleeping time of animals given a standard dose of hexobarbital varied from 12 min in the mouse to 315 min in the dog. These effects were directly proportional to the hexobarbital half-life and indirectly proportional to the clearance of the chemical, which in this case was largely determined by liver enzyme activity. Similar effects can be seen across strains within a species Festing (1990). Because animals within the same strain should have similar physiology, strain differences in metabolism would presumably be due to genetically controlled differences in isoenzyme expression. Although a range of sleeping-time differences can be observed in different strains of mice (Festing, 1990), the variance among strains is still much smaller than the variance across species.

In addition to metabolic differences across species and strains, animals may also differ in basic physiological and biochemical substrates. An example of how differences in physiological substrates can lead to species differences in responses to chemicals has been discussed by Dews (1976). In both birds and man, papillary constriction occurs with increasing light levels. In human subjects the pupil is constricted by smooth muscle that receives cholinergic innervation of muscarinic receptors, but in birds the constrictor muscle is striated with cholinergic innervation by somatic motor nerves of nicotinic receptors. Thus, atropine, a muscarinic blocking agent, dilates the human pupil, but is without effect on the nicotinic receptors controlling pupil dilation in birds.

Examples of how behavioral differences across species can limit the validity of animal-to-human extrapolations in behavioral toxicology can be trivially obvious. For example, flight is largely confined to birds among animals that are commonly used for toxicity testing. A chemical that affects flight in birds has no absolute parallel in humans, although clearly there are human motor behaviors that share common elements with flight. Thus the effects of chemicals on flight still might be used to model the motor effects of chemicals in humans. It is not necessary that a model show an absolute homology before it can be useful as a predictor for human responses to chemicals.

An important source of variability across animals can be differences in their behavioral history. It is quite obvious that animals of different species have very different experiences during their life times. Although it is not always as obvious, animals of the same species and strain may also have different behavioral histories. The elegant work of McKearney (1979) has emphasized the important role that behavioral history can play in the response of the animal to the effects of chemicals. For example, amphetamines usually do not increase punished responding; however, after a history of avoidance responding, amphetamines produce large increases in punished responding. In this case the behavioral history of the animal was manipulated systematically, but there are potentially many instances where unknown events in the behavioral history of the animal can be important determinants of interactions between chemicals and behavior.

With all of the potential metabolic, receptor, neurotransmitter, second messenger, behavioral, genetic, and behavioral history differences that are seen across species it is amazing that any degree of extrapolation is possible at all. Yet as Dews (1976) has pointed out, the effects of chemicals on behavior tend to ". . . run true across species, and major species differences are the exception rather than the rule." Other investigators have suggested (McMillan, 1987) that species differences tend to be quantitative rather than qualitative in most instances. If this were not true, our whole risk assessment process for neurobehavioral toxicology would be invalid. Fortunately, this is not the case. The remainder of this chapter discusses some of the factors that control the extrapolation of data from animals to humans and suggests some ways to deal with the information.

III. Species Scaling of Drug and Chemical Disposition

Although most biologists believe that experimental data collected in animals can be extrapolated to humans, they may not realize the biological reasons underlying the pharmacokinetic similarities among species. Indeed, most mammals show more commonalty in anatomical and physiological makeup than they show diversity (Dedrick, 1973; Dedrick and Bischoff, 1980). For instance, a schematic diagram of blood flow through the whole animal is virtually the same for all species, and the rate of blood flow to the organs varies in a systematic fashion that is related to animal body size. As other examples of animal commonalty, the liver is the major site for metabolic inactivation in all mammals, and the lung is an important intake and elimination site for volatile chemicals. Although seemingly minor physiological and anatomical differences can be important (e.g., rats do not have a gall bladder), the continuity in how animals absorb, distribute, metabolize, and eliminate xenobiotics is quite remarkable.

Because of the striking similarities in the anatomy and physiology among species, the disposition of chemicals in various animals can be derived with reasonable accuracy if the problem is approached from an animal engineering point of view. This means we need to understand the animal species scaling factors involved in such things as delivering blood to organs and the systematic increase in organ size as we go from laboratory animals to humans in our experiments.

In a now classic article, Adolf (1949) showed that anatomical and physiological processes correlate among species as an exponential function of animal body weight. He also found that anatomical parameters like body fluid volumes and organ sizes were nearly proportional to body weight. However, metabolic and physiological properties such as blood flow and creatinine clearance per unit of organ weight tend to be less than directly proportional to animal weight. This relatively simple finding started the important field of allometric analysis. In allometry, a series of empirical equations are used to show that the logarithms of anatomical or physiological variables correlate with the logarithms of animal body weight across species.

With a full understanding of the important implications of these simple allometric concepts, investigators have shown that the pharmacokinetics of most chemicals are also predictable across species (see reviews by Dedrick, 1973; Dedrick and Bischoff 1980; Boxenbaum, 1982; Mordenti, 1986). Organ volumes, fluid compartment volumes, and volumes of distribution of chemicals show a nearly proportional increase with body weight (an allometric exponent of about 1.0), whereas organ blood flows and consequently chemical clearances can be scaled to body weight with an exponent of 0.6–0.7. Because the terminal elimination $t_{1/2}$ of chemicals depends on the volume of distribution (V_d) and systemic clearance (Cl$_s$) of the chemical (i.e., $t_{1/2} = (0.693 \times V_d)/\text{Cl}_s$), $t_{1/2}$ scales to body weight with an exponent of 0.25–0.33. Figures 1–3 show representative examples of phencyclidine V_d, Cl$_s$, and $t_{1/2}$ correlated to animal body weight (Owens *et al.*, 1987).

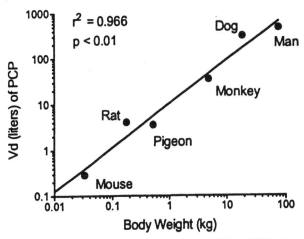

FIGURE 1   The relationship between phencyclidine (PCP) volume of distribution (V_d) and animal body weight across several species. The allometric relationship for the best-fit line through these data is $V_d = 10B^{0.96}$, where B is animal body weight and 0.96 is the so-called allometric exponent. Data are from Owens *et al.* (1987). Reprinted with permission of Williams and Wilkins.

These pharmacokinetic species scaling factors can be very useful for developing risk assessments of chemicals or for the design of experiments in animals. For instance, when the effects of a chemical in an animal species are studied for the first time, it is extremely helpful to know the volume of distribution and clearance of the chemical. The use of these values can allow us to make estimates of the dose to administer to each animal and the blood concentrations over time in each animal, before we conduct the experiments. Or, for example, if we can estimate the $t_{1/2}$ of a toxic chemical from animal data, it might

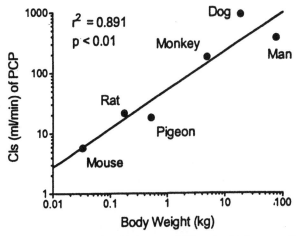

FIGURE 2   The relationship between PCP systemic clearance (Cl$_s$) and animal body weight across several species. The allometric relationship for the best-fit line through these data is $\text{Cl}_s = 50B^{0.64}$. Data are from Owens *et al.* (1987). Reprinted with permission of Williams and Wilkins.

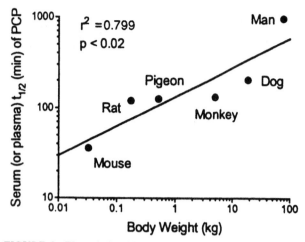

FIGURE 3 The relationship between PCP $t_{1/2}$ and animal body weight across several species. The allometric relationship for the best-fit line through these data is $t_{1/2} = 126B^{0.32}$. Data are from Owens *et al.* (1987). Reprinted with permission of Williams and Wilkins.

help us to decide how frequently factory workers can be safely exposed to a chemical.

IV. Physiologically Based Pharmacokinetic Models of Chemicals

Although correlations and species scaling based on body weight are important first steps, in the development of meaningful animal models for pharmacokinetics and risk assessment, additional techniques might be used to more accurately access the consequences of chemical exposure in animals. An important technique for these purposes is physiologically based pharmacokinetic (PB-PK) models. PB-PK models are designed to quantify the disposition of foreign compounds or chemicals throughout the organ systems of an animal model. These PB-PK models yield chemical flow diagrams in an animal for the calculation of various measures of chemical disposition (i.e., absorption, distribution, metabolism, and elimination).

To acquire data for a PB-PK model, the chemical concentrations at multiple tissue sites are extensively characterized in the body of one animal species over time. Rats are often used for such studies. Typically, the concentration of the chemical is measured in several tissues and body fluids such as liver, lung, kidney, brain, testes, blood, cerebrospinal fluid, urine, and feces. In addition, if there are known or suspected sites of toxicity (e.g., the spleen and lymph glands for immunotoxicants) these organs will also be studied.

As described by the name, both the physiological and pharmacokinetic aspects of the chemical must be simultaneously accounted for in the model. Therefore, a major

consideration in a PB-PK model is the way the chemical is delivered to the tissues. Consequently, PB-PK models are of two general types: blood flow-limited models or membrane rate-limited models (Fig. 4). These terms describe the rate-limiting effects of either transport of chemical to the organ (or tissue) or the ability of a chemical to transfer across a barrier such as the membrane of a cell. Most chemicals fall into the category of a blood flow-limited model; this simply means that the rate of delivery of the chemical to the organ by the bloodstream is the major rate-limiting step.

In addition to the pharmacokinetic data, the solubility of the chemical in blood and tissues and the *in vitro* metabolism kinetic constants (i.e., K_m and V_{max}) are determined. These various data sets (i.e., organ blood flows, chemical concentration versus time data, chemical partition coefficients) are then incorporated into a mathematical model that depicts the entire disposition of the xenobiotic over time in a large variety of tissues and sites of action (Fig. 5). This can become quite a complex undertaking since a large set of mass balance equations for the processes of uptake, distribution, metabolism, and elimination must be simultaneously solved.

When these data are then coupled with species scaling data from several species, they can provide an extremely powerful technique for predicting the risk of chemical exposure in larger mammals (such as humans) based on experimental data collected in smaller mammals (such as rodents, dogs, and monkeys). In this way, reasonably accurate concentration data at specific time points in different tissues and organs can be predicted in another species without sacrificing the animal. In addition, this type of model can help to identify target sites for toxicity or tissue accumulation sites for the chemical.

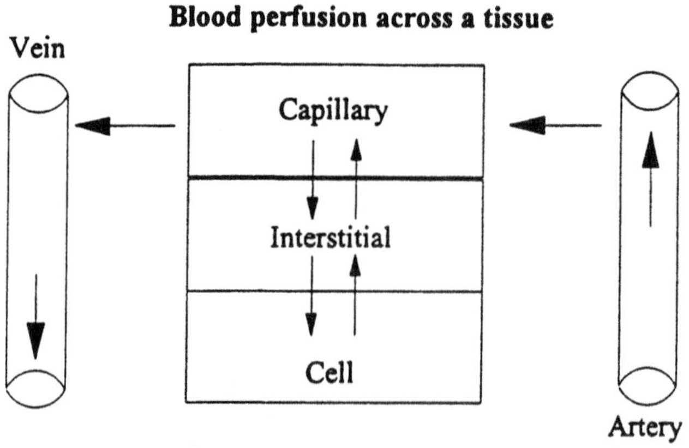

Blood perfusion across a tissue

FIGURE 4 Model for blood flow and perfusion of a local tissue area. The rate of chemical transport between the capillary, interstitial space, and cell will determine if the model is "blood-flow limited" or "membrane-rate limited." (Adapted from Gibaldi and Perrier, 1982.)

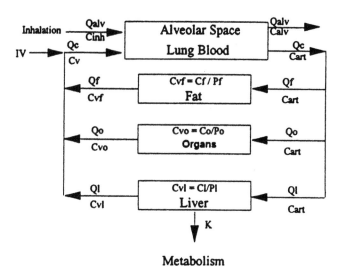

Metabolism

FIGURE 5 A representative PB-PK model consisting of lung and tissue compartments. In this case the chemical can enter through the inspired air or blood stream. C, concentration; Q, flow; art, arterial; v, vein; alv, alveolar; f, fat; o, organs; l, liver; P, tissue:blood partition coefficient; K, elimination rate constant.

These data can also greatly aid our ability to interpret the results of biomarkers that predict an underlying adverse effect. A biomarker might be as simple as an elevated concentration of a chemical or metabolite in human serum or a chemical covalent adduct found in the urine.

As a final note about the influence of absorption, distribution, and elimination on responses to chemicals, it should be realized that both age and sex or gender of the animal are major factors to be considered. For instance, the very young of all species do not have fully developed abilities to metabolize chemicals and the very old have a decreased ability to clear chemicals. In addition, age can affect the distribution volume of chemicals due to differences in the ratio of muscle to fat over time. Although these factors are not always a problem, age of the animal must be at least matched in the design or interpretation of animal experiments.

V. Differences in Response and Toxicity at the Sites of Action

The exact mechanism of action of most chemicals is not known, and behavioral effects are often produced by a combination of mechanisms. However, if the site of action is known, a major factor to consider is the affinity (K_d) and concentration of chemical receptors (B_{max}) among species. Species differences in receptor affinity and number have not been extensively examined like species scaling of pharmacokinetic data. How-

ever, some studies have been conducted. For example, 5-[^{3}H]hydroxytrytamine binding affinity (K_d) for the serotonin receptors is virtually the same value in frontal cortex and hippocampus in rat, rabbit, guinea pig, and cat (2.4–2.9 nM), and B_{max} only varies from 5.4 to 23.7 fmol/mg in the same animals (Schnellmann *et al.,* 1984). Phencyclidine (PCP)-binding characteristics to the PCP receptor (K_d and B_{max}) are nearly the same across a wide variety of mammals and lower invertebrates (Vu *et al.,* 1990).

In contrast, [^{3}H]mepyramine binding to H$_1$ receptors in membranes from the brains of human, rat, guinea pig, rabbit, and mouse indicate greater species differences. The guinea pig and human H$_1$ receptors have about three to six times greater affinity for [^{3}H]mepyramine than rat, mouse, and rabbit H$_1$ receptors (Chang *et al.,* 1979). These species differences in affinity for H$_1$ receptor sites occur for some antihistamines, but not for others, and the regional distribution of sites differs considerably in various species. Consequently, to study differences in behavioral response across species, investigators should attempt to administer doses of the chemicals that will lead to equivalent effects among species at the site of action or determine the chemical concentrations at the sites of action after equivalent behavioral or toxicological effects among species.

VI. Behavioral Factors Influencing Species Comparisons

In his extensive review of interspecies differences in response to chemicals, Dews (1976) noted that the control pattern of behavioral activity is a major determinant of the effects of a chemical on behavior. If the control patterns of responding differ across species, it becomes very difficult to make comparisons of chemical–behavior interactions across species. Therefore, it is exceedingly important to make the behavioral measurements as similar as possible across species when making comparisons or extrapolations in behavioral toxicology.

Unfortunately, this is not always easy. Using schedule-controlled behavior as an example, one might try to make comparisons across species under "identical" conditions; however, differences in the animals size, which we have already seen can be a confounding factor in the pharmacokinetic aspects of behavioral toxicology, can also be important in behavior. For example, if we wish to compare rats, pigeons, monkeys, and humans performing under the commonly used multiple fixed-interval fixed-ratio schedule, which has been used in a multitude of studies in behavioral pharmacol-

ogy and behavioral toxicology, some of the basic problems in extrapolating animal data to humans can be
illustrated.

First there is the choice of a reinforcer. Food is
commonly used in such studies, but there is a potential
problem in equating the value of a food reinforcer
across studies. Does a 4-sec access to grain for a pigeon
have the same reinforcing strength for the pigeon as
does an M&M candy for a human? Fortunately, the
literature strongly suggests that the scheduling of the
occurrence of reinforcers is a more powerful determinant of behavior than the specific nature of the reinforcer (Dews, 1958). Furthermore, there is a large
amount of literature available that suggests the effects
of chemicals on behavior are not powerfully influenced
by the particular reinforcer used to maintain behavior
(Kelleher and Morse, 1968). This indicates that extrapolation across species is not severely limited by the
use of different reinforcers in different studies.

Can we equate the level of food deprivation across
the species in such studies? It has been shown that the
level of food deprivation can be an important determinant of the effects of chemicals on behavior (Kelly and
Thompson, 1988). The contribution of the relationship
between food deprivation and reinforcer magnitude to
the interaction between chemicals or other chemicals
on behavior has not been systemically investigated.
At least in most instances in the study of chemical–
behavior interactions, animals have been food deprived
to about the same percentage of their free feeding
weight.

Another consideration is the response that will be
reinforced. It is customary in behavioral pharmacology
and behavioral toxicology for a response that closes a
lever switch or opens a response-key contact to be
reinforced and maintained under schedule control. Is
the key peck response usually employed to study pigeon behavior equivalent to the lever-press response
for the rat? Should the force requirements for responding for all species be determined by the animal
with the lowest absolute strength or should response
force be determined proportionally to weight? Again,
there is no data base to answer such questions, although under most schedules of reinforcement the
baseline performance does not seem to be heavily influenced by the force of the response or other topo-
graphic considerations (Dews, 1958). However, there
are a few exceptions to this rule (McMillan, 1990).

There are other behavioral factors that might influence chemical–behavior interactions. When establishing multiple schedule control in the pigeon, performance is probably best controlled by visual stimuli,
whereas a rat is often more easily controlled by auditory stimuli. Again the question might be asked as to
whether or not factors such as the sensory modality
by which stimulus control occurs is an important determinant of chemical effects. Fortunately, the answer to
this question appears to be a qualified no. Dews (1958)
has compared patterns of fixed-interval responding
across a range of fixed-interval durations in several
species, under different stimulus conditions using different manipulanda for the response, and different reinforcers to maintain responding. Despite all of these
differences, patterns of responding across species were
essentially similar. This is an encouraging validation
for those attempting to extrapolate behaviors across
species.

VII. Examples of Species and Strain Differences in Behavior in Response to Chemical Exposure

The data base comparing chemical effects across
species is extensive, although rarely have there been
formal attempts to make systematic comparisons.
More frequently it has become necessary to search the
literature to find the appropriate data which have been
collected in different laboratories under a wide variety
of conditions. Given these difficulties it is surprising
how consistent the effects of some chemicals have
been. The remainder of this section gives some examples of both similarities and differences in species and
strains for the effects of chemicals on behavior.

To illustrate species differences in the behavioral
response to chemical exposure we refer to a previous
analysis of the discriminative stimulus effects of the
animal anesthetic and widely abused chemical, PCP
(McMillan, 1990). The data are summarized in Table
1. In these studies, rats, squirrel monkeys, and pigeons
were trained to discriminate PCP from saline by reinforcing response on one manipulandum when PCP had
been given before the session and on a different manipulandum when saline had been given before the session. Subsequently, the animals given different doses
of PCP and a range of doses of the optical isomers of
N-allylnormetazocine and cyclazocine with both manipulanda available to determine the extent to which the
stimulus properties of these chemicals would generalize to those of the training dose of PCP. Only in the
pigeon were all five compounds active, so data in the
table show potencies for each chemical relative to the
potency for that chemical in the pigeon. PCP,
(+)NANM, and (+)cyclazocine were approximately
four to five times more potent in the squirrel monkey
than in the pigeon, but (−)NANM and (−)cyclazocine
were inactive in squirrel monkeys up to doses that
suppressed responding. The rat resembled the squirrel

TABLE 1 Dose Producing 50% Responding on the PCP Key and the Potency of Chemicals in the Rat and the Squirrel Monkey Relative to the Pigeon

| | Dose producing 50% chemical key responding[a] | | | Relative potency | |
Chemical	Rat	Squirrel monkey	Pigeon	Rat/pigeon	Squirrel monkey/pigeon
PCP[b]	0.9	0.09	0.5	0.56	5.56
PCP[b]	1.3	0.07	0.32	0.24	4.57
(+)NANM	0.56	0.33	1.7	3.04	5.15
(−)NANM	IA[c]	IA	4.1	—	—
(+)CYC	10.0	0.5	2.0	0.20	4.00
(−)CYC	1.2	IA	1.0	0.83	—

[a] All doses are in milligrams per kilogram.
[b] Data are the results from two separate analysis run on different occasions.
[c] Inactive at the doses tested.

monkey in its responses to NANM with the (+)isomer of NANM being three times more potent in rat than pigeon and the (−)isomer being inactive; however, PCP itself appeared to be less active in rat than pigeon. Furthermore, the (−)isomer of cyclazocine was slightly less potent in rat than pigeon. Thus in general these chemicals were more potent in squirrel monkeys trained to discriminate PCP from saline than they were in pigeons, except that the (−)isomers of NANM and cyclazocine appeared to be inactive. In contrast, these chemicals were less potent as discriminative stimuli in rats trained to discriminate PCP from saline than in pigeons with the exception of (+)NANM, where the rat most closely resembled the squirrel monkey.

How are these data to be interpreted given the previous discussion? The lack of an adequate data base makes it difficult to answer this question, although several possibilities can be considered. Among the possibilities are pharmacokinetic differences. In the original investigations upon which Table 1 is based, rats usually received the chemical intraperitoneally, whereas pigeons and monkeys were injected intramuscularly. These differences in the route of administration could contribute to chemical distribution differences, which in turn could result in differences in behavioral responses to the chemicals. Similarly, the pigeons were usually tested during cumulative dosing whereas rats and squirrel monkeys were not. Although half-lives for PCP have been determined for rats and pigeons (Owens et al., 1987), details of PCP metabolism are available only for rats. Thus it is possible, perhaps even likely, that the data in Table 1 are powerfully influenced by pharmacokinetic variables, although it seems less likely that the differences in squirrel monkey and pigeon responses to the optical isomers of cyclazocine and *n*-allylnormetazocine are primarily determined by pharmacokinetic differences.

Neuroanatomical differences across these three species are also profound. As pointed out previously (McMillan, 1990), pigeons have a comparatively large cerebellum and striatum compared to rats. The squirrel monkey has a considerably more convoluted cerebral cortex than the rat or pigeon. To what extent gross neuroanatomical differences might influence responses to the chemicals in Table 1 is difficult to determine. These data seem to suggest that squirrel monkeys are considerably more "stereospecific" than pigeons, but the basis for this difference is unknown. It should also be mentioned that these species show many other anatomical differences that may be important determinants of chemical responses, such as appendages, eye placement, and many others. As discussed earlier, comparisons of PCP receptors show mammals have very similar K_d and B_{max} values for the PCP receptor in the central nervous system (Vu et al., 1990), which would perhaps suggest that species differences in response to PCP-like chemicals are determined more by the dose delivered to the receptor than by differences in affinity of the chemical for the receptor or the number of receptors in different species.

There are also many other differences in these experiments that might limit comparisons across the three species. Some of these differences were imposed by the experimeter, but some were not. The response that is reinforced in pigeons is usually a key peck, whereas the reinforced response is usually a lever press in the rat and squirrel monkey. The rat and monkey were reinforced for responding at a particular position, whereas the pigeons were required to track the location of a key color. Different reinforcers also were used across these experiments. As stated earlier, with all of these differences in animals and procedures, it is remarkable that the data are as similar across species as they are. This continuity suggests that extrapolation across species can be valid even when there are many small differences in the experimental conditions under which the experiments were conducted.

Some of the same problems that potentially influence attempts to extrapolate across species must also be considered when comparisons are made across strains. For example, it has been reported (Shelnutt *et al.*, 1993) that both PCP and MK-801 produce much more pronounced motor activity increases in DBA mouse strain than in the C57/B1 mouse strain. Binding studies showed no differences across strains in B_{max} or K_d for binding of the high-affinity PCP receptor ligand, TCP. However, following a single dose of PCP, serum PCP levels were higher in DBA mice than in C57/BL mice. Although this pharmacokinetic difference across strains might be an important contributor to the behavioral difference in response to PCP, other factors must be involved since an inverted U-shaped curve in both strains described the dose–response relationship for PCP and it is difficult to see how differences in plasma levels of PCP could determine its efficacy across strains as opposed to its potency. One such factor might be the interaction of PCP with other neurotransmitter systems. For example, Carney and colleagues (1992) have observed differences between DBA/2J and C57BL/6ByJ mice in motor activity following PCP administration which were accompanied by an increased release of endogenous dopamine by PCP (Carney *et al.*, 1992).

Although the discussion of behavioral factors of species scaling has mostly focused on the sources of species differences in response to the behavioral effects of chemicals, it must be remembered that the overwhelming evidence from the literature is that when the effects of chemicals on behavior are studied in different species under similar conditions, the effects of chemicals on behavior also seem to be comparable. Some prime examples might include the use of the chemical self-administration model to predict abuse liability in humans and the use of responding suppressed by punishment to screen for chemicals with antianxiety activity in humans. For example, despite variations in schedule of reinforcement, unit dose, and other experimental differences, rats, squirrel monkeys, rhesus monkeys, pigtail macaques, baboons, cats, dogs, and humans all self administer cocaine, suggesting that similar molecular substrates are shared across species for the reinforcing efficacy of cocaine (Johanson and Fischman, 1989). Similarly, barbiturates and other antianxiety agents increase punished responding in rats, cats, pigeons, and monkeys under a variety of experimental conditions (McMillan, 1975), although with some of the newer antianxiety agents, such as buspirone, there may be some efficacy differences across species (Witkin and Perez, 1990). There are many other

such examples that could have been chosen to support the position that the effects of chemicals on behavior show a high degree of continuity across species despite minor procedural differences. Clearly, extrapolation of chemical effects on behavior across species is the rule and not the exception.

VIII. The Future

Although it can be difficult to make comparisons of the behavioral effects of chemicals across strains and species, sometimes the differences between strains and species can be valuable to the basic researcher. For example, a large number of inbred strains of rats and mice are now available for use in behavioral research and they offer some important advantages for behavioral toxicologists and behavioral pharmacologists (Festing, 1990), but most investigators have been slow to take advantage of this opportunity.

One possible advantage in the use of inbred strains is that they are likely to show a lower within-group variability than outbred strains (Festing, 1990), although this has yet to be convincingly demonstrated in behavioral toxicology and behavioral pharmacology. If a low within-group variability can be demonstrated, it follows that it would be likely that small effects of chemicals could be more easily demonstrated in an inbred group. Furthermore, the use of inbred strains of the same species removes some of the problems encountered in cross-species comparisons. For example, the experimeter can use the same species-appropriate test apparatus, schedule of reinforcement, and reinforcer, thereby eliminating some of the sources of variability that are frequently imposed in cross-species comparisons.

The use of inbred strains also provides an opportunity to compare an inbred strain for which a chemical effect has been established with an inbred strain of the same species in which the effect has not been observed. Among the most extensive studies of this type have been the studies of inbred strains of rodents that accept or do not accept alcohol, such as the P and NP rats strains developed by Li and colleagues (1993). The availability of inbred strains showing and not showing a particular response to chemicals opens the opportunity to study the biochemical mechanism underlying the differences in the behavioral effects as well as the genetic basis for that mechanism. By using recombinant inbred strains (fully inbred descendants of a second-generation cross between two inbred strains), a powerful tool can be applied that is useful in detecting and mapping major gene loci (Bailey, 1981). These opportunities need to be exploited by behavioral toxicologists and pharmacologists to enable us to begin to study the genetic contributions to interactions between chemicals and behavior.

The behaviorist need not be satisfied with the available inbred strains in these days of rapid developments in molecular biology. The development of transgenic animal models may soon give us the option of "building our own animals" for a given experiment. In the past, many of the interesting chemicals have been nonselective in that they produce a series of effects at a variety of receptor types such as adrenergic chemicals binding to α, β, and dopamine receptors or to narcotic analgesics binding to μ, κ, and δ receptors. The historical strategy for understanding the mechanism of action of these chemicals has been to synthesize a variety of congeners to develop agonists and antagonists with a specific affinity for one receptor subtype for comparison with the prototype chemical. For example, this strategy has been quite successful in helping to develop a number of new analgesics with decreased abuse liability and has led to the development of clinically useful new narcotic antagonists (Martin, 1983). With the availability of transgenic animals a new strategy appears to be possible. For example, a transgenic animal model with a decreased number of μ receptors might be developed to study the analgesic effects of morphine. This might allow us to determine which effects of morphine are and are not mediated by μ receptors. Although such transgenic animals are not yet generally available, the approach has already proven feasible as demonstrated by the development of many transgenic animals including transgenic pigs whose transgene expression results in an increase in growth hormone (Pursel *et al.,* 1989). Similar innovations are likely to prove invaluable for understanding both the toxicity and the pharmacology of chemicals by allowing us to build the animal that we need for a particular type of testing.

Until transgenic animals are more readily available, behavioral toxicologists and pharmacologists should utilize the interesting animals strains and species that are already available to continue to validate the extrapolation process. Somewhat different strategies may be needed for behavioral pharmacology and behavioral toxicology. In behavioral pharmacology it has been common for the activity of an effective chemical to be discovered in humans, sometimes by accident. The behavioral pharmacology of the chemical is then studied to determine its activity profile and then a series of chemical congeners are synthesized and tested in animals. On the basis of this profile, promising compounds are selected and then tested for their therapeutic usefulness in humans. Certain monoamine oxidase inhibitors, benzodiazepines, and a number of other therapeutic discoveries have more or less followed this pattern. Although some behavioral toxins have followed a similar pattern of accidental discovery of their activity in humans, followed by the study of their activity in animal models, the systematic return of these toxins and their congeners to humans is generally unethical. Furthermore, our general lack of knowledge about the neurobehavioral effects of many of the chemicals in our environment has led to the inclusion of neurobehavioral toxicity testing in the "risk assessment industry" whereby the study of behavioral toxicity is confined to the testing of the chemicals in animals models, sometimes with little hope of validating these observations in humans. Under such circumstances the emergence of a comparative neurobehavioral toxicity becomes a pressing need. With the demonstration of neurobehavioral toxicity (or lack thereof) in each species, our confidence in the extrapolation process to humans increases. The development of a systematic comparative neurobehavioral toxicology has a far greater chance of improving estimates of risk assessment and protecting public health than does the artificial application of arbitrary safety factors applied to data from a single strain or species.

Acknowledgments

This work was supported by grants DA02251 (D.E.M.), DA04136 (S.M.O.) and a Research Scientist Development Award (K02 DA00110 to S.M.O.) from the National Institute on Drug Abuse.

References

Adolf, E. F. (1949). Quantitative relationships in the physiological constitutions of mammals. *Science* **109,** 579–585.

Alveres, A. P., and Pratt, W. B. (1990). Pathways of drug metabolism. In *Principles of Drug Action* (W. B. Pratt and P. Taylor, Eds.), pp. 365–422, Churchill Livingstone, New York.

Bailey, D. W. (1981). Recombinant unbred strains and bilneal congenic strains. In *The mouse in Biomedical Research,* Vol. I, (H. L. Foster, J. D. Small, and J. G. Fox Eds.), pp. 225–229, Academic Press, New York.

Boring, E. G. (1950). A history of experimental psychology. Appleton Century Crofts, New York.

Boxenbaum, H. (1982). Interspecies scaling, allometry, physiological time and the ground plan for pharmacokinetics. *J. Pharmacokinet. Biopharm.* **10,** 210–227.

Carney, J. M., Seale, T. W., Bardo, M. J., and Dwoskin, L. P. (1992). Qualitative and quantitiative differences in the behavioral effects of phencyclidine in inbred mice. In *Multiple sigma and PCP receptor ligands: Mechanisms for neuromodulation and neuroprotection* (J-M. Kamenka and E. F. Domino, Eds.), pp. 607–618, NPP Books, Ann Arbor.

Chang, R. L., Tran, V. T., and Snyder, S. H. (1979). Heterogeneity of histamine H1-receptors: species variation in [^{3}H]mepyramine binding of brain membranes. *J. Neurochem.* **32,** 1653–1663.

Darwin, C. (1859). On the origin of species by means of natural selection. London.

Darwin, C. (1872). The expression of the emotions in man and animals. London.

Dedrick, R. L. (1973). Animal scale-up. *J. Pharmacokinet. Biopharm.* **1,** 435–461.

Dedrick, R. L., and Bischoff, K. B. (1980). Species similarities in pharmacokinetics. *Fed. Proc.* **39,** 54–59.

Dews, P. B. (1976). Interspecies differences in drugs effects: behavioral. In *Psychotheraputic drugs* (E. Usdin and I. S. Forrest Eds.), pp, 175–224, Dekker, New York.

Dews, P. B. (1958). Analysis of effects of psychopharmacological agents in behavioral terms. *Fed. Proc.* **17,** 1024–1030.

Festing, M. F. W. (1990). Use of genetically heterogeneous rats and mice in toxicological research: a personal perspective. *Toxicol. Appl. Pharmacol.* **102**(2), 197–204.

Gibaldi, M., and Perrier, D. (1982). *Pharmacokinetics.* Dekker, New York.

Johanson, C. E., and Fischman, M. W. (1989). The pharmacology of cocaine related to its abuse. *Pharmacol. Rev.* **4,** 3–52.

Kelleher, R. T., and Morse, W. H. (1968). Determinants of the specificity of behavioral effects of drugs. *Ergeb. Physiol. Biol. Chemie Exper. Pharmacol.* **60,** 1–56.

Kelly, T. H., and Thompson, T. (1988). Food deprivation and methadone effects on fixed-interval performance by pigeons. *Arch. Intern. Pharmacodyn. Ther.* **293,** 20–36.

Li. T-K., Lumeng, L., and Doolittle, D. P. (1993). Selective breeding for alcohol preference and associated responses. *Behav. Gene.* **12,** 163–169.

Manson, J. M. (1986). Teratogens. In *Casarett and Doull's Toxicology* (C. D. Klaassen, M. O. Amdu, and J. Doull Eds.), pp. 195–222, Macmillan Publishing, New York.

Martin, W. R. (1983). Pharmacology of Opioids. *Pharmacol. Rev.* **35,** 283–323.

McKearney, J. W. (1979). Interrelations among prior experience and current conditions in the determination of behavior and the effects of drugs. In *Advances in behavioral Pharmacology*, Vol. 2, (T. Thompson and P. B. Dews, Eds.), pp. 39–64, Academic Press, New York.

McMillan, D. E. (1975). Determinants of drug effects on punished responding. *Fed. Proc.* **34,** 1870–1879.

McMillan, D. E. (1990). The pigeon as a model for comparative behavioral pharmacology and toxicology. *Neurotoxicol. Teratol.* **12,** 523–529.

McMillan, D. E. (1987). Risk assessment for neurobehavioral toxicity. *Environ. Health Perspect.* **76,** 155–161.

Mordenti, J. (1986). Man versus beast: pharmacokinetic scaling in mammals. *J. Pharmacol. Sci.* **75,** 1028–1040.

Owens, S. M., Hardwick, W. C., and Blackall, D. (1987). Phencyclidine pharmacokinetic scaling among species. *J. Pharmacol. Exp. Ther.* **242,** 96–101.

Pursel, V. G., Pinkert, C. A., Miller, K. F., Bolts, D. J., Campbell, R. G., Palmitor, R. D., Brinster, R. L., and Hammer, R. E. (1989). Genetic engineering of livestock. *Science* **244,** 1281–1288.

Schnellmann, R. G., Waters, S. J., and Nelson, D. L. (1984). [^{3}H]5-Hydroxytrytamine binding sites: species and tissue variation. *J. Neurochem.* **42,** 65–70.

Shelnutt, S. R., Owens, S. M., Hardwick, W. L., Rogers, S. E., and McMillan, D. W. (1993). Strain differences in motor activity response to phencyclidine (PCP) in mice: relationship to PCP receptor binding and pharmacokinetics. *Coll. Prob. Drug Dep.* **141,** 247.

Vu, T. H., Weismann, A. D., and London, E. D. (1990). Pharmacological characteristics and distributions of σ- and phencyclidine receptors in the animal kingdom. *J. Neurochem.* **54,** 598–604.

Witken, J. W., and Perez, L. A. (1990). Comparison of effects of buspirone and gepirone with benzodiazepines and antagonists of dopamine and serotonin receptors on punished behavior of rats. *Behav. Pharmacol.* **1,** 247–254.

Role of Dopaminergic and Glutamatergic Neurotransmitter Systems in Lead-Induced Learning Impairments

DEBORAH A. CORY-SLECHTA
Department of Environmental Medicine
University of Rochester School of Medicine and Dentistry
Rochester, New York 14642

I. Introduction

The neurotoxic properties of lead have been known for hundreds of years. In fact, many of the neurotoxic symptoms arising from lead exposure were well described by the ancient Romans and Greeks. Despite this knowledge, however, the problems arising from environmental lead exposure remain very much with us today. It was, for example, only 25–30 years ago in the United States that episodes of acute lead encephalopathy in children resulting from exposures to relatively high levels of lead, levels producing blood lead concentrations well above 80 μg/dl, were still being frequently reported. Such children would present with signs of toxicity that included a swollen, edematous, and hemorrhagic brain. If these children were fortunate enough to survive these episodes of acute encephalopathy, they faced the possibility of sustaining permanent sequelae that could include marked mental retardation.

It was such repeated episodes of acute encephalopathy that partially prompted the eventual removal of lead from paint and subsequently from gasoline, the two major sources of human environmental lead exposure. These attempts to abate human environmental lead exposure were no doubt of assistance in decreasing the reported incidence of overt neurotoxicity in pediatric populations. Unfortunately, however, this improvement in a widespread public health problem was offset to a great extent by the increasing awareness, based on cross-sectional and subsequent prospective epidemiological studies, that lead-induced changes in cognitive functions, as least as manifest in altered IQ scores and other psychometric indices, occurred at far lower levels of lead exposure than had previously been thought (e.g., Needleman *et al.,* 1979; Bellinger *et al.,* 1984; Dietrich *et al.,* 1993). These studies, in fact, eventually prompted a lowering of the definition of blood lead levels of concern in children in the United States, first to 40 μg/dl in blood and, more recently, to 10–15 μg/dl (U.S. EPA, 1986).

The findings described in the prospective epidemiological studies were largely corroborated by results of experimental animal studies, both in rodents (Cory-Slechta *et al.,* 1985; Cohn *et al.,* 1993) and in nonhuman primates (e.g., Rice, 1985) describing comparable types of behavioral toxicity. Moreover, the reported changes in cognitive as well as other behavioral functions were observed in the experimental animal studies at blood lead levels (10–15 μg/dl) that directly paralleled those of concern as defined by human studies. It is important to recognize that these blood lead levels currently associated with lead-induced behavioral toxicity in the experimental animal studies do not represent thresholds for neurobehavioral toxicity, but simply the lowest levels of exposure at which such changes have yet been studied.

What remains unclear even today, however, is the neurobiological bases of these lead-induced changes in cognitive functions. Numerous efforts over the years to delineate a neuroanatomical bases of the behavioral manifestations resulting from lead exposure have proven largely unsuccessful. It is most likely this very lack of success that has prompted the shift in focus to thinking instead about biochemical or neurochemical bases for lead-induced behavioral toxicity.

In that regard, it should be pointed out that in experimental animal studies, lead exposure has been reported to impact a wide variety of neurotransmitter systems, including dopaminergic, opiate, cholinergic, GABAergic, norepinephrine systems, and serotonin systems (e.g., Shellenberger, 1984; Winder and Kitchen, 1984). Moreover, for any given neurotransmitter system, lead exposure has been reported to impact numerous of its functions, altering, for example, synthesis of neurotransmitter, release and reuptake mechanisms, and numbers of receptors.

While such indications of changes in neurotransmitter systems are of course provocative in terms of providing guidance to potential neurobiological mechanisms of cognitive deficits, they leave several unanswered questions with respect to the relationship of such changes, expressed at a biochemical/cellular level, to lead-induced behavioral toxicity. For example, which of these various neurotransmitter systems impacted by Pb exposure might serve as the basis of the described behavioral effects? Even if likely candidate neurotransmitter systems can be identified on the basis of the relevant neuroscience literature, the question of which of the numerous effects of lead on that particular neurotransmitter system's functioning is the candidate mechanism for behavioral toxicity still must be addressed. Is it, for example, the change in synthesis of neurotransmitter or changes at the reuptake site?

Another question that must be resolved is whether effects of lead that are reported at a biochemical/cellular or *in vitro* level of analysis are of sufficient biological magnitude and/or clinical relevance to even serve as neurobiological mechanisms of behavioral toxicity. Consider the example of Parkinson's disease, where it

has been reported that losses of dopamine on the order of 80–90% are required before signs of the disease are manifest. Are changes, then, on the order of 15, 20, or even 25% in levels of synthesis of a given neurotransmitter, or in numbers of receptors, or uptake sites, of sufficient biological magnitude and clinical relevance to serve as a mechanism of lead-induced behavioral toxicity?

The remainder of this chapter describes the highlights of studies carried out designed to address such questions (see also Cory-Slechta *et al.*, 1993). If focuses on the potential involvement of lead-induced changes in two particular neurotransmitter systems, dopaminergic and glutamatergic [specifically the *N*-methyl-D-aspartate (NMDA) receptor complex] systems, in one particular behavioral manifestation of lead toxicity, i.e., learning deficits. The possibility of dopaminergic system involvement in these cognitive deficits was considered because of the substantial number of reports of lead-induced changes in dopaminergic function (e.g., Govoni *et al.*, 1979; Missale *et al.*, 1984; Moresco *et al.*, 1988; Lasley, 1992) and because of the involvement of dopaminergic systems in various cognitive processes (e.g., LeMoal and Simon, 1991; Cador *et al.*, 1991; Beninger, 1989; Packard and White, 1989). The role of glutamatergic systems, specifically the NMDA receptor complex, was investigated because it has been extensively implicated in learning and memory processes (Cohn and Cory-Slechta, 1992; Cotman *et al.*, 1989; Upchurch and Wehner, 1990) and because of recent reports of lead impacting NMDA receptor complex function (Alkondon *et al.*, 1990; Guilarte and Miceli, 1992; Ujihara and Albuquerque, 1993).

The studies highlighted herein focus on lead exposure occurring only at one particular stage of development in the rodent, i.e., postweaning, beginning at 21 days of age. Parallel studies are underway examining the impact of Pb exposure administered during the postnatal period of development (e.g., Cory-Slechta *et al.*, 1992), i.e., during ontogeny of these neurotransmitter systems, but these efforts have not yet been investigated to the same extent as has been done with postweaning Pb exposures. These postweaning Pb exposures were chronic and occurred via drinking water. In these studies, the effects of two different exposure concentrations of Pb were compared to controls, with the lower Pb exposure concentration producing blood lead levels of 15–25 μg/dl, and the higher exposure concentration associated with blood lead levels of 40–60 μg/dl. The former represents a blood lead level just above that of current concern for human pediatric populations, whereas blood lead levels of

40–60 μg/dl are still consistent with occupational Pb exposure in this country.

II. Does Pb Exposure Selectively Impair Learning Processes?

Although as noted earlier, numerous studies, including both human prospective epidemiological studies and experimental animal studies, suggest Pb-induced changes in cognitive processes, the extent to which such changes represented direct effects of Pb exposure on learning processes per se or whether they were secondary to changes in motivational levels, or sensory or motor functions, had yet to be firmly established. Before proceeding to implement studies designed to determine the neurochemical bases of lead-induced learning impairments, it was first imperative to determine that Pb exposure indeed produced direct effects on learning processes per se that were not simply a secondary consequence of some other behavioral change. In the absence of direct effects of lead on learning processes per se, it would hardly be worth pursuing their neurochemical bases. Moreover, these efforts were undertaken to more precisely define the nature and behavioral mechanisms of Pb-induced learning impairments. This was based on the contention that the more precise the determination of the nature of a behaviorally toxic effect, the more accurate the hypotheses that can be generated with respect to its underlying neurobiological substrates.

To determine whether postweaning Pb-induced changes in learning were direct or indirect behavioral effects, a multiple repeated acquisition and performance schedule paradigm was utilized (described in detail in Cohn *et al.*, 1993). Daily 1-hr behavioral test sessions were conducted in operant test chambers that included three response manipulanda (levers) arranged horizontally from left to right on the front wall of the chamber. Each behavioral test session included two different parts or components: a learning (repeated acquisition) component and a performance (P) component. During the learning (repeated acquisition, RA) component of the session, rats were required to complete a sequence of three responses to produce food delivery. With each successive behavioral test session, the sequence of three responses defined as the correct sequence changed in an unpredictable way. For example, in one session the correct sequence might be to press the left (L) lever, then the right (R) lever, and then the center (C) lever. During the subsequent session, the correct sequence during the learning component would change, e.g., to CLR. These sequences were chosen from a list including LRC, RCL, RLC, CLR, and CRL.

As can be noted from the list, sequences requiring repetitive responses on a single lever, e.g., LLR, were excluded to preclude any reinforced history of perseverative (repetitive) responding. The learning component therefore required the rat to learn a new sequence of three responses during each successive behavioral test session, allowing the measuring of learning repeatedly across time. The performance component likewise required the correct completion of a sequence of three responses for each food delivery, but in this case the sequence remained constant (LCR) across sessions. In contrast to the learning component, then, the performance component only required rats to perform a sequence of three responses that had already been learned.

In this paradigm, both the learning and the performance components of the session required intact motivational levels as well as intact sensory and motor capabilities of the organism, but learning per se was only required in the learning component of the session where the correct sequence of three responses changed daily. The contention of these studies, therefore, was that if Pb exposure produced direct effects on learning processes per se, which were not secondary changes due to alterations in motivational, sensory, or motor effects, then Pb-induced changes in accuracy would only be noted during the learning components of the session, the only component in which learning was required. In contrast, if Pb exposure resulted in nonspecific changes in behavior, e.g., changes in motivational levels, or sensory or motor capabilities, accuracy changes would be manifest in both the learning and performance components since these behavioral functions were necessary for behavior in both components of the paradigm.

Typical performance during a 1-hr behavioral test session as maintained by this paradigm is depicted in Fig. 1. During each such test session, the learning and performance components were each presented twice, with each component presentation lasting either 15 min or until 25 reinforcers had been delivered, whichever occurred first. As Fig. 1 shows, behavior during the performance components of the session was characterized by a high rate of reinforcement (food) delivery, as indicated by the pips in the top tracing, and by the emission of relatively few errors (bottom tracing). The first presentation of the learning component clearly resulted in a decline in the rate at which reinforcement delivery occurred, and a corresponding increase in the frequency of errors, as the animal began to learn the correct sequence for this particular session. Progress in learning the sequence designated as correct for this session was evident during the second presentation of the learning component, where the rate of reinforce-

FIGURE 1 Cumulative record of representative responding on the multiple repeated acquisition and performance schedule over the course of a 1-hr behavioral session. The learning and performance components were presented twice each, with components alternating after 25 reinforcer deliveries or 15 min, whichever occurred first. Time is represented horizontally. Correct responses, depicted in the upper tracing, cumulate vertically, and the slope of the lines during each component presentation indicate rate of responding. Pips in the upper tracing indicate a food delivery for the correct completion of the sequence of three responses. The bottom tracing depicts errors. (From Cory-Slechta, unpublished data.)

ment delivery gradually increased, and error rates declined, resulting in a typical learning curve when considered across the two learning components.

Comparison of Pb-exposed to control animals on this schedule revealed that Pb exposure did indeed result in selective effects on learning processes that were not due to changes in motivational levels, or sensory or motor capabilities (detailed in Cohn *et al.,* 1993). The nature of this selective effect is illustrated in Fig. 2, which shows the comparative cumulative records of responding of a typical control rat (top tracings) as compared to a rat chronically exposed to 250 ppm Pb acetate (40–60 μg/dl blood Pb) that exhibited rather severe behavioral effects of exposure. Behavior of the Pb-exposed rat was quite normal during both performance components of the schedule, i.e., resulting in a high rate of food deliveries and a relatively low rate of errors. However, during the learning components of the schedule (labeled "A"), the Pb-exposed rat shown here earned virtually no food deliveries for

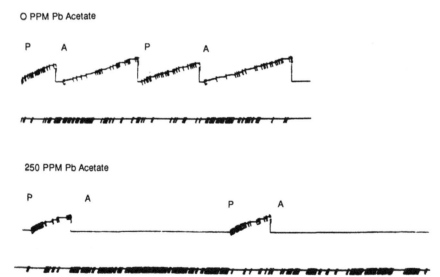

FIGURE 2 Cumulative record of representative performance of a typical control (top cumulative record) and Pb-exposed (250 ppm) rat (bottom cumulative record) on the multiple repeated acquisition and performance schedule. The learning (A) and performance (P) components were presented twice each, with components alternating after 25 reinforcer deliveries or 15 min, whichever occurred first. Time is represented horizontally. The top tracing of each cumulative record depicts correct responses, which cumulate vertically, with the slope of the lines during each component presentation indicative of response rate. Pips in the upper tracings show food delivery for a correct completion of the sequence of three responses. The bottom tracing of each cumulative record shows occurrences of errors. (From Cory-Slechta, unpublished data.)

correctly completing the new sequence of three responses. This was not due to a lack of responding on the part of the rat since, as the bottom tracing shows, the rat continued to emit a very high rate of errors throughout both presentations of the learning component.

Analyses of the patterns of errors contributing to the decline in overall accuracy during the learning component by Pb-exposed rats compared to controls revealed an increase in two types of perseverative errors, i.e., repetitive responding (Cohn *et al.,* 1993). Specifically, Pb-exposed rats continued to respond repeatedly on a single lever (e.g., LLL) despite the fact that sequences requiring any repetitive responses were purposely excluded precisely to preclude engendering a history of perseverative responding. In addition, Pb-exposed rats continued to emit the three-member sequence LCR, that was correct during the performance component of the schedule, even during the learning components of the session, i.e., they perseverated on the performance component sequence. This was a pattern of behavior that was obviously incompatible with learning the correct sequence for that session's learning component.

These experiments confirmed that Pb exposure does indeed produce impairments in learning processes that cannot be ascribed to other types of behavioral changes and also provided a validation to pursue the potential neurochemical bases for such effects.

III. Are Dopaminergic Systems Involved in Pb-Induced Learning Impairments?

Current thinking holds that there are two major classes of dopamine receptors, a D1- and a D2-type family, each of which contains multiple subtypes. Although D1 receptors appear to be entirely postsynaptic, D2-type dopamine receptors may be either presynaptic (autoreceptors) or postsynaptic. This distinction is important with respect to the differences that result from activation at these two different sites. Specifically, activation of presynaptic D2 receptors results in a decline in dopamine synthesis and release, and consequently a decline in postsynaptic dopamine receptor activation. In contrast, activation of D2 postsynaptic receptors results in a very different cascade of effects, including inhibition of adenylate cyclase activity (in some brain regions) and modulation of phosphoinositide metabolism. An additional distinction lies in sensitivity to dopamine and dopaminergic ligands. Presynaptic D2 receptors are more sensitive to such compounds than are postsynaptic D2 receptors, so that the presence of low concentrations of dopamine or dopaminergic ligands

results in preferential activation of presynaptic D2 receptors and, hence, diminished dopaminergic function, whereas in the presence of higher concentrations of dopamine or dopaminergic ligands, there is presumably a preferential activation of postsynaptic D2 receptors and enhanced dopaminergic function.

The finding that Pb exposure resulted in selective effects on learning processes provided the rationale for exploring the role that neurotransmitter systems might play in such effects. Dopaminergic neurotransmitter systems served as a potential candidate for involvement in lead-induced learning deficits for at least two reasons. The first was the numerous reports of changes in dopaminergic system function produced by Pb exposures. Changes have been described, for example, in the synthesis and turnover of dopamine (e.g., Memo *et al.,* 1980, 1981), in the regulation of DA synthesis (Lasley and Lane, 1988), in dopamine reuptake parameters (e.g., Misale *et al.,* 1984), in levels of DA and metabolites (e.g., Jason and Kellogg, 1981), and in the number of dopamine D1 and D2 receptors (e.g., Lucchi *et al.,* 1981; Rossouw *et al.,* 1987; Moresco *et al.,* 1988). In addition, dopaminergic systems do play a role in various cognitive functions as indicated both by the results of lesion studies, as well as behavioral pharmacological studies (e.g., LeMoal and Simon, 1991; Cador *et al.,* 1991; Beninger, 1989; Packard and White, 1989; Wolterink *et al.,* 1993).

While the reports of Pb-induced changes in dopaminergic function expressed at the biochemical/cellular level were, of course, provocative in providing potential bases for learning impairments, the first question that had to be addressed with respect to the involvement of dopaminergic systems in lead-induced learning impairments was whether such changes were of sufficient biological magnitude and clinical relevance to even serve as mechanisms of learning impairments. In order to address this issue, the contention was that if Pb-induced biochemical changes in dopaminergic systems were of sufficient biological magnitude and clinical relevance, then they should produce changes in dopaminergic sensitivity at the level of the whole animal, that is, changes in dopaminergic sensitivity that could be "described" by the whole animal.

For that purpose, drug discrimination procedures were utilized as a type of functional or behavioral neurochemical assay. This procedure has proven extremely useful for studying the functional properties of dopamine and other receptors in a highly selective fashion. Basically, the paradigm establishes a discrimination (in the simplest case) between a drug stimulus and a saline stimulus by reinforcing one behavioral response following the drug injection and reinforcing an alternative behavioral response when it is preceded by a saline injection. In these studies, rats were taught to report whether they had received a D2 receptor agonist (quinpirole) or saline, or, for other groups of rats, whether they had received an injection of a D1 receptor agonist (SKF 38393) or saline, i.e., they were trained to discriminate the stimulus properties of these specific dopaminergic agonists from saline. A schematic of the procedure by which this is accomplished is presented in Fig. 3. Put simply, in this paradigm, the rat is injected with the drug or saline before being placed in an operant chamber. The order of injections of drug or saline across sessions is random, such that the rat has no way to predict whether it will be receiving drug or saline on any given session. Following some interval of time postinjection that allows adequate uptake of the drug, the rat is placed in the operant chamber. At that point, a response on what is arbitrarily designated as the drug lever (response manipulandum) is rewarded with food delivery, i.e., the rat is rewarded for correctly reporting that it received drug after a drug injection. Likewise, if the rat had received saline, then responding on what is arbitrarily defined as the saline lever results in food delivery, i.e., the rat is rewarded for correctly reporting that it received saline after a saline injection. [The specific procedures are described in detail in Cory-Slechta and Widzowski (1991).] After some number of sessions (depending on the particular training drug and training dose), rats come to acquire the discrimination to a specified level of accuracy (77% accuracy in 8 out of 10 consecutive sessions), i.e., they learn to reliably report whether they have received the dopaminergic agonist or saline prior to the session.

Of obvious importance to such studies is the question as to whether responding on the drug lever has anything to do with D1 receptors in the case of SKF 38393, and with D2 receptors in the case of quinpirole, the respective D1 and D2 agonists used in these studies. Prior efforts have already clearly established the specificity of the stimulus properties of these compounds for their respective receptor subtypes (Cory-Slechta *et al.*, 1989; Weathersby and Appel, 1986; Kamien and Woolverton, 1985; Cunningham *et al.*, 1985). Specifically, other D2 agonists, such as low doses of apomorphine, completely substitute for quinpirole, whereas the D1 agonist SKF 38393 does not substitute and instead produces saline lever responding. Moreover, the stimulus properties of the D2 agonist quinpirole are blocked by the D2 antagonist haloperidol, but not by the D1 antagonist SCH 23390. Conversely, D2 agonists such as apomorphine at low doses and quinpirole do not substitute for the stimulus properties of the D1 agonist SKF 38393, and SKF 38393 responding is blocked by the D1 antagonist SCH 23390 but not by the D2 antagonist haloperidol.

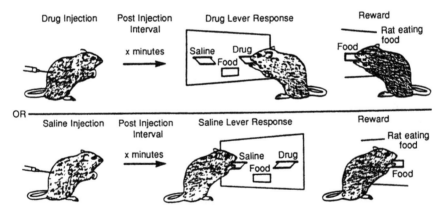

FIGURE 3 The drug discrimination procedure. (Top) Rat is rewarded for responding on the "drug" lever if it has received the drug injection prior to the start of the experimental session. (Bottom) Rat is rewarded for responding on the "saline" lever if it has received a saline injection prior to the session. The sequence of saline and drug injections is random across the daily (M–F) 10-min experimental sessions. Ten responses on the correct lever are required for each food delivery; percentage correct is determined only from responses occurring up to the time of the first food delivery. Criterion accuracy is defined as three or less incorrect responses before a completion of 10 responses on the correct lever (77%). Acquisition of the discrimination is defined as 8 of 10 consecutive sessions in which the 77% accuracy criterion is obtained. At that point, "test" sessions, i.e., sessions in which lower doses of the training drug or of other drugs, or of antagonist preadministration, are interspersed between training drug and saline sessions. The 3-min "test" session ends, after a short delay with reinforcer delivery regardless of the pattern of responding during the test. (From Cory-Slechta *et al.*, 1993.)

In the context of Pb exposure, the question of interest relates to Pb-induced changes in dopaminergic sensitivity. To evaluate sensitivity, lower doses of the training drug are substituted for the training dose after the discrimination has been learned. Typically, as the dose of the training drug declines, responding on the drug lever likewise declines, as it becomes increasingly difficult to discriminate drug from saline. This substitution of lower doses results in a sensitivity dose–effect curve relating drug lever responding to drug dose. Shifts of this dose–effect curve and its corresponding ED_{50} value can then be assessed in response to Pb exposure, with shifts of the curve to the right consistent with pharmacological subsensitivity and shifts of the dose–effect curve to the left indicative of supersensitivity.

As can be seen in Fig. 4, percent responding on the drug lever declined with decreasing doses of quinpirole (D2) and with SKF 38393 (D1) in these experiments, as expected. Postweaning Pb exposure did result in changes in dopaminergic sensitivity producing both a D2- and a D1-type supersensitivity. This was indicated by the significant left shift of the dose–effect curves for drug lever responding of both Pb-treated groups (50 and 250 ppm) relative to the control group (0 ppm), both in those groups that had been trained to discriminate the D2 agonist quinpirole from saline and those

that had been trained to discriminate the D1 agonist SKF 38393 from saline.

One of the questions raised by the D2/saline drug discrimination findings was whether the observed D2 supersensitivity produced by Pb exposure represented a presynaptic or a postsynaptic D2 supersensitivity, a difference of importance both with respect to the mechanisms of the effects of Pb, as well as for the consequences of the supersensitivity. To address that question, an additional group of normal, i.e., non-Pb-exposed rats, was trained to discriminate the same training dose of the D2 agonist quinpirole from saline (Widzowski and Cory-Slechta, 1993) and, subsequently, a series of pharmacological manipulations was undertaken to evaluate the hypothesis that if the stimulus properties of this dose of quinpirole were mediated by presynaptic D2 receptors (autoreceptors), then any pharmacological manipulations that resulted in a decline in DA release or synthesis and consequent decreased postsynaptic dopamine receptor activation (i.e., mimicked autoreceptor function) should result in quinpirole lever responding. In contrast, any pharmacological manipulations that resulted in postsynaptic receptor activation should result in saline lever responding. A postsynaptic mediation of this training dose of quinpirole would result, instead, in an opposite pattern of effects.

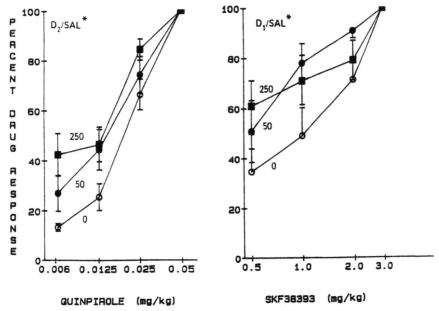

FIGURE 4 Sensitivity dose–effect curves for the control (open circles), 50 ppm Pb-exposed (filled circles), and 250 ppm Pb-exposed (filled squares) groups trained to discriminate 0.05 mg/kg of the D2 agonist quinpirole from saline (left) or 3.0 mg/kg of the D1 agonist SKF 38393 from saline (right). Each data point represents a group mean ± SE based on at least two determination per dose per rat, with 10 rats per Pb exposure concentration. (From Cory-Slechta and Widzowski, 1991.)

The results of the various pharmacological manipulations undertaken were consistent with a presynaptic basis for quinpirole mediation (Table 1), suggesting that one net effect of the biochemical/cellular changes produced by Pb exposure on dopaminergic systems is an autoreceptor supersensitivity. Presynaptic D2 supersensitivity in conjunction with D1 supersensitivity is indicative of a net effect of Pb consisting of diminished dopamine release. This would eventually deprive all dopamine receptors of their endogenous ligand and ultimately result in a supersensitivity. Thus Pb exposure per se, through some as yet unidentified cascade of effects, appears to mimic chronic autoreceptor agonism.

Another goal of these experiments was to attempt to relate changes in dopaminergic sensitivity produced by Pb exposure to other indices of dopaminergic system function. An aspect of dopaminergic function examined in this context was D1 and D2 receptor binding in various dopaminergic regions of brain. One particularly interesting relationship that emerged from these assessments, both with postweaning (Widzowski and Cory-Slechta, unpublished data) and postnatal Pb exposure (Widzowski et al., 1994), was an apparent linear relationship between increases in quinpirole (D2 autoreceptor or presynaptic receptor) sensitivity, as assessed using the drug discrimination paradigm, and increased D2 B_{max} values in one particular brain region, namely the nucleus accumbens, raising the possibility that the nucleus accumbens may be a preferential site for the actions of Pb. That such a preferential vulnerability might reflect regional central nervous system differences in Pb exposure was ruled out, at least in the case of postnatal exposure, by a study that revealed relative homogeneity of 12 different brain regions (Widzowski and Cory-Slechta, 1994).

This finding of a relationship between increased dopaminergic D2 presynaptic (quinpirole) sensitivity and increased D2 receptor binding in nucleus accumbens was a provocative finding for several reasons. First of all, the nucleus accumbens is a terminal dopaminergic projection area in which autoreceptor regulation of dopamine synthesis is known to be particularly important (Westfall et al., 1983; Demarest et al., 1983). Second, the importance of the nucleus accumbens to certain dopaminergic drug stimulus properties has recently been demonstrated. Specifically, in animals trained to discriminate systemic injections of either D-amphetamine (Nielsen and Scheel-Kruger, 1986) or cocaine (Wood and Emmett-Oglesby, 1989) from saline, direct injection of these compounds into the nucleus accumbens, but not into the striatum, engenders full substitution for the systemically administered training dose, i.e., substitutes fully, provoking high levels of drug lever responding. Finally, lesions of the nucleus accumbens have been reported to produce perseverative responding under a variety of behavioral conditions (Taghzouti et al., 1985a,b; Reading and Dunnett,

TABLE 1   Pharmacological Basis of Quinpirole Drug Stimulus Mediation[a]

D2 autoreceptor agonism		DA depletion	
NPA	QUIN	AMPT alone	QUIN
Low doses of APO	QUIN	AMPT with QUIN	QUIN
Low doses of QUIN	QUIN		

Postsynaptic DA receptor blockade		Postsynaptic DA receptor activation	
AMPT alone	QUIN	SKF 38393 (D1)	SAL
AMPT with QUIN	QUIN	High doses of APO (D1)	SAL
SCH23390 alone	QUIN	d-AMPH (D1 and D2)	SAL
High does of HAL	?	SCH23390 and AMPH (D2)	SAL
		QUIN and SKF38393 (D1)	SAL
		QUIN and AMPH (D1 and D2)	SAL

[a] Summary of the pharmacological challenges used to determine whether a training dose of 0.05 mg/kg of the D2 agonist quinpirole was mediated by D2 presynaptic or postsynaptic receptors. The left side of each row in each column defines the pharmacological challenge whereas the right side in each row defines the outcome as either predominantly quinpirole level responding (QUIN) or saline lever responding (SAL). Data from Widzowski and Cory-Slechta (1993).

1991), consistent with the perseverative basis of Pb-induced learning impairments described earlier (Cohn *et al.*, 1993) and which has also been described as a basis of accuracy impairments produced by Pb on other behavioral baselines (e.g., Cory-Slechta *et al.*, 1991; Rice, 1985; Bushnell and Bowman, 1979).

Having determined that Pb-induced changes in dopaminergic systems expressed at a biochemical/cellular level were certainly of sufficient biological magnitude and clinical relevance to provoke changes in dopaminergic sensitivity at the level of the whole animal, the next question that must be addressed is whether or not the resulting dopaminergic D1 and/or D2 supersensitivity plays any role in Pb-induced learning impairments. One indication that these dopaminergic sensitivity changes might be involved in Pb-induced learning impairments would be the finding of differential changes in accuracy in control and Pb-exposed rats on the learning baseline in response to administration of dopaminergic compounds. That is, acute administration of dopaminergic compounds might evoke differential changes in accuracy of learning in control vs Pb-treated rats.

To test this assertion, the effects of acute administration of the D2 agonist quinpirole, the D1 agonist SKF 38393, and the catecholamine depleter α-methyl-*p*-tyrosine were assessed in control and Pb-exposed rats on the learning and performance baseline described earlier (Cohn and Cory-Slechta, 1994b). As illustrated in Fig. 5 for the D1 agonist SKF 38393, all three of these compounds decreased accuracy in a generally dose-dependent fashion in the learning component of the learning and performance baseline. However, the impairments of accuracy noted in response to these dopaminergic compounds did not differ in magnitude between the control and Pb-exposed groups.

Taken together, these findings, as summarized in Table 2, clearly indicate that Pb-induced changes in the dopaminergic neurotransmitter system have a functional counterpart at the level of the whole animal, expressed as changes in D1 and D2 dopaminergic sensitivity, but that these alterations in dopaminergic sensitivity do not appear to play a role in Pb-induced learning impairments. At least two caveats with respect to such a conclusion must be pointed out, however. The first is that the approach utilized here to delineate the relationship between changes in dopaminergic sensitivity and learning impairments produced by Pb exposure should not be considered the only or the final resolution of this question. It is entirely conceivable that the pattern of dopaminergic changes induced by Pb exposure is far more complex than was tapped by assessment of the effects of acute administration of dopaminergic

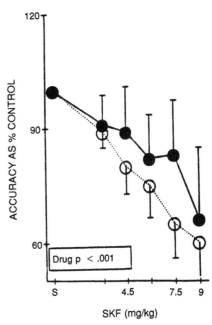

FIGURE 5 Overall accuracy ± SE during the second presentation of the learning component as a fucntion of dose (3.0, 4.5, 6.0, 7.5, 9.0 mg/kg) of SKF 38393 plotted as a percentage of saline values for the control (filled circles) and Pb-exposed group (open circles). Boxed inset indicates significant P values from the accompanying repeated measures analysis of variance. (From Cohn and Cory-Slechta, 1994b.)

compounds on the learning baseline. It is possible, for example, that only the administration of mixtures of dopaminergic compounds or subchronic or extended administration of dopaminergic compounds, i.e., dosing protocols sufficient to invoke changes in dopaminergic sensitivity, B_{max}, etc., would have resulted in differential changes in learning accuracy in control vs Pb-exposed rats.

A second caveat is that the lack of any differences between control and Pb-treated rats in the effects of acute administration of dopaminergic agonists on the learning baseline does not rule out the possibility that Pb-induced dopaminergic sensitivity changes are involved instead in other manifestations of Pb-induced behavioral toxicity. Problems such as distractibility, impulse control, and attentional deficits have been ascribed to Pb exposure in numerous studies (Byers and Lord, 1943; Bellinger *et al.*, 1984; Fergusson *et al.*, 1988; Hatzakis *et al.*, 1979; Needleman *et al.*, 1979; Silva *et al.*, 1988; Thomson *et al.*, 1989), and dopaminergic function also seems to be related to certain types of attentional deficits (LeMoal and Simon, 1991). Such possibilities must be explored in future experiments.

TABLE 2 Summary of Pb-Induced Changes in Dopaminergic and Glutamatergic Systems and their Relationship to Pb-Induced Changes in Learning

	Dopaminergic	Glutamatergic
Drug discrimination	↑ D1 sensitivity ↑ D2 sensitivity	↓ MK-801 sensitivity
Multiple repeated acquisition and performance	SKF effect on accuracy Quinpirole effect on accuracy	↓ MK-801 effect on accuracy ↑ NMDA effect on accuracy
FI schedule		
Receptor binding	↑ D1 binding in frontal cortex ↑ D2 binding in nucleus accumbens	↓ MK-801 binding in CA1, dentate

IV. Are Glutamatergic Systems Involved in Pb-Induced Learning Impairments?

The glutamatergic system, in particular the *N*-methyl-D-aspartate receptor complex, has been extensively implicated in learning and memory processes (e.g., Cotman *et al.,* 1989) and therefore it serves as a likely candidate to subserve Pb-induced learning impairments. Moreover, several previous studies have now reported changes in the NMDA receptor complex following Pb exposure. Changes in glutamate synthetase activity have been reported in guinea pigs at blood lead levels as low as 13 μg/dl (Sierra and Tiffany-Castiglioni, 1991; Sierra *et al.,* 1989). *In vitro* studies show that NMDA-evoked whole-neuron and single-channel currents are inhibited by Pb exposure in a concentration-dependent manner, with an IC_{50} of 10 μM Pb (Alkondon *et al.,* 1990). Studies by Ujihara and Albuquerque (1993) suggest that Pb may be acting as a noncompetitive antagonist of the glycine site. Lead exposure has also been reported to inhibit MK-801 binding under several different conditions (Johnson *et al.,* 1992; Guilarte and Miceli, 1992; Alkondon *et al.,* 1990).

Again, such changes, expressed at a biochemical/cellular level, are certainly provocative with respect to potential NMDA involvement in Pb-induced learning impairments, but it must, of course, first be determined that such effects are of sufficient biological magnitude or clinical relevance to serve in a mechanistic capacity for behavioral toxicity. The approach taken to address that issue was the same as that utilized in the aforementioned dopaminergic studies. Specifically, if NMDA receptor complex changes produced by Pb exposure were of sufficient import, they should result in changes in NMDA receptor complex sensitivity at the level of the whole animal, i.e., changes in NMDA sensitivity that could be described by the whole animal.

Given several reports of inhibited MK-801 binding following Pb exposure (Johnson *et al.,* 1992; Guilarte and Miceli, 1992; Alkondon *et al.,* 1990), initial efforts focused on the possibility of Pb-induced changes in MK-801 sensitivity at the level of the whole animal using drug discrimination procedures. Previous studies had already established that MK-801 could indeed serve as a discriminative stimulus, and, further, that in MK-801-trained animals, other noncompetitive NMDA antagonists, such as phencyclidine, known to act at the MK-801 binding site on the NMDA receptor complex, substitute completely for MK-801 in a drug discrimination paradigm, whereas competitive antagonists such as CPP, CGS 19755 and NPC 12626, which act at the glutamate binding site on the NMDA receptor complex, produce at most a partial substitution (Ferkany *et al.,* 1989; Jackson and Sanger, 1988; Koek *et al.,* 1990; Tricklebank *et al.,* 1989; Willets *et al.,* 1991), thus establishing the specificity of MK-801 drug lever responding for the MK-801 site of the NMDA receptor complex.

To assess the extent of any Pb-induced changes in MK-801 sensitivity, rats chronically exposed to 0, 50, or 150 ppm Pb acetate in drinking water from weaning were trained to discriminate the stimulus properties of a dose of 0.05 mg/kg MK-801 from saline using standard operant food-reinforced drug discrimination procedures (Johnson and Cory-Slechta, 1993; Cory-Slechta, in press). Following acquisition of the discrimination, lower doses of MK-801 were substituted for the training dose to generate a dose–effect curve for MK-801 sensitivity. As expected, responding on the MK-801 lever declined with a decreasing dose of MK-801 in all three groups.

ED$_{50}$ values for the resulting dose–effect curves for the Pb-exposed groups were notably larger than those of controls (0.013 ± 0.006, 0.0219 ± 0.0066, and 0.0257 ± 0.003 mg/kg for the 0, 50, and 150 ppm groups, respectively), indicative of MK-801 subsensitivity. This subsensitivity was most pronounced in the group exposed to the higher concentration of Pb, 150 ppm. Interestingly, this subsensitivity is a finding consistent with the Pb-induced inhibition of MK-801 binding described in the several studies mentioned earlier. One unexplained as yet finding was that while phencyclidine (PCP) substituted completely for MK-801, as expected, engendering dose-related increases in drug lever responding, there was no comparable subsensitivity to PCP in the Pb-exposed groups relative to control as would be predicted on the basis of MK-801 subsensitivity. Whether this absence of subsensitivity in Pb-exposed groups represents the additional effects of PCP in blocking DA reuptake (Doherty *et al.*, 1980; Garey and Heath, 1976; Hitzemann *et al.*, 1973) remains to be determined.

The finding of Pb-induced subsensitivity to the stimulus properties of MK-801 in the drug discrimination procedure clearly indicates that indeed the effects of Pb on the NMDA receptor complex, expressed at a biochemical/cellular level, are of sufficient biological magnitude and clinical relevance to impact sensitivity at the level of the whole animal. What must next be determined, however, is whether such changes play any role in Pb-induced learning impairments. To assess that possibility, it was again postulated that one possible indication of an involvement of MK-801 subsensitivity in Pb-induced learning impairments would be a differential effect of glutamatergic compounds on accuracy in the learning paradigm in control vs Pb-exposed rats.

For that purpose, the effects of acute administration of MK-801 on the learning and performance baseline described earlier were examined. Acute administration of MK-801 clearly resulted in dose-dependent decreases in accuracy during the second presentation of the learning component, as shown in Fig. 6. In line with the hypothesis described earlier, moreover, these effects also differed in control and Pb-exposed rats, with the impairment of accuracy produced by MK-801 being significantly attenuated in Pb-exposed rats relative to controls (Cohn and Cory-Slechta, 1993). These findings can be described as providing initial support for the possibility of an involvement of glutamatergic system changes in Pb-induced learning impairments. It should also be noted that this attenuation of the accuracy-impairing properties of MK-801 in Pb-

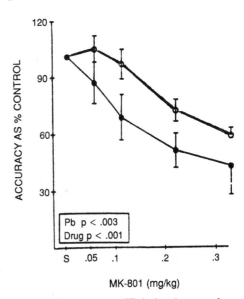

FIGURE 6 Overall accuracy ± SE during the second presentation of the learning component as a function of dose (0.05, 0.1, 0.2, 0.3 mg/kg) of MK-801 plotted as a percentage of saline values for the control (filled circles) and Pb-exposed group (open circles). Boxed inset indicates significant p values from the accompanying repeated measures analysis of variance. (From Cohn and Cory-Slechta, 1993).

exposed rats was consistent with the previous findings both in terms of the inhibition of MK-801 binding produced by Pb exposure and with the MK-801 subsensitivity observed in the drug discrimination paradigm, as described earlier.

Evidence from previous studies had already suggested that glutamatergic systems might be involved in Pb-induced learning impairments (Fig. 7). An evaluation of the acute effects of drugs on the learning vs performance baseline described earlier in normal non-Pb-exposed rats revealed that acute administration of MK-801 impaired accuracy. Moreover, it did so primarily by increasing the frequency of perseverative errors (Cohn *et al.*, 1992; Cohn and Cory-Slechta, 1992). This was in contrast to compounds such as scopolamine, a cholinergic antagonist, which also impaired accuracy in the learning component of the baseline, but did so by increasing the frequency of skipping errors, i.e., skipping forward through the three response sequence (Cohn *et al.*, 1992; Cohn and Cory-Slechta, 1992). Thus, acute MK-801 administration and Pb exposure may impair learning accuracy by similar means.

The finding that MK-801 results in a differential impairment of accuracy of learning in Pb-exposed vs control rats also raised the possibility that Pb exposure could be exerting effects at other sites on the NMDA receptor complex. For example, subsensitivity at the noncompetitive antagonist site and the consequent inhibition of complex activity might result in an upregula-

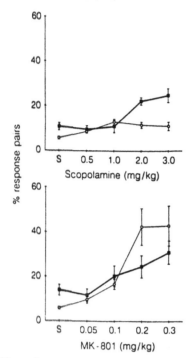

FIGURE 7 Type of error produced as a function of dose of scopolamine (top graph) and MK-801 (bottom graph) depicted as group mean ± SE. Percentages are the proportion of all pairs of responses consisting of a correct lever press followed by either a perseverative (repetitive) error (open circles) or by a skipping error (closed squares). (From Cohn *et al.*, 1992.)

occurring at a biochemical/cellular level have a functional counterpart in altered MK-801 sensitivity at the level of the whole animal, but, additionally, that these Pb-induced glutamatergic changes may indeed play a role in the learning impairments ascribed to Pb exposure. Again, however, it needs to be stressed that additional studies to evaluate the contribution of Pb-induced NMDA receptor complex changes to Pb-induced learning and other behavioral impairments are clearly warranted.

V. Future Research Agendas

These research approaches outlined here are intended to describe new strategies to enhance our understanding of the neurochemical bases of the various behavioral manifestations resulting from Pb exposure. At the same time that they provide new insights into the potential bases and brain regions involved in lead-induced learning impairments, they generate numerous additional questions to be addressed. For example, given the accumulating information regarding neurotransmitter system interactions, the question of whether Pb-induced changes in dopaminergic systems, or in glutamatergic systems, represent direct or indirect effects remains to be determined. Moreover, the extent to which Pb-induced D1 and D2 supersensitivity represent direct effects or occur as a function of known D1–D2 receptor interactions is as yet unclear.

As mentioned earlier, further attention must also be given to the involvement of neurotransmitter system changes in the various behavioral effects arising from exposure to Pb. The studies to date have focused on the involvement of dopaminergic and glutamatergic system changes in just one such effect, namely learning impairments. But other behavioral manifestations have been described as well in various human studies, including changes in attention, reaction time, and, more recently, perseveration. In addition, other approaches to directly evaluate the involvement of neurotransmitter sensitivity changes as a basis of behavioral impairments must be implemented as well before any firm conclusions about the involvement of specific neurotransmitter systems in any particular behavioral manifestation of Pb exposure can be concluded. One additional approach would be to attempt to induce similar changes in neurotransmitter sensitivity in non-Pb-exposed (i.e., control) animals and to assess the extent to which it results in behavioral deficits analogous to those produced by Pb exposure. Finally, it must be noted that disturbances of more than one neurotransmitter system may well be involved in a particular behavioral manifestation of Pb exposure. The findings presented here, then, represent

tion at the glutamate (NMDA)-binding site in an attempt to overcome this inhibition. While an assessment of NMDA sensitivity (NMDA vs saline drug discrimination) in Pb-exposed rats is currently in progress, an evaluation of the effects of NMDA on the learning vs performance baseline has been completed (Cohn and Cory-Slechta, 1994a).

In that study, acute administration of NMDA (10–50 mg/kg) resulted in significant decrements in accuracy on the second presentation of the learning component of the schedule. Like MK-801, these effects of NMDA again differed significantly in control vs Pb-exposed rats. In this case, however, acute administration of NMDA produced a significant potentiation of the impairment of accuracy in Pb-exposed rats compared to controls, decreasing accuracy to levels of about 45% of saline control values at a dose of 35 mg/kg, whereas similar effects in control rats emerged only after a dose of 50 mg/kg. These findings then are consistent with the possibility of an upregulation or supersensitivity at the glutamate-binding site of the NMDA receptor complex.

Thus, the findings to date with glutamatergic systems, summarized in Table 2, suggest not only that Pb-induced changes in the NMDA receptor complex only a beginning to unraveling the relationships between Pb-induced neurotransmitter systems changes and Pb-induced behavioral deficits.

Acknowledgments

The efforts of Mary Jane Pokora, Jeffrey Cohn, Daniel Widzowski, and Timothy Greenamyre to these studies are gratefully acknowledged.

References

Alkondon, M., Costa, A. C. S., Radharkrishnan, V., Aronstam, R. S., and Albuquerque, E. X. (1990). Selective blockade of NMDA-activated channel currents may be implicated in learning deficit caused by lead. *FEBS Lett.* **261,** 124–130.

Bellinger, D. C., Needleman, H. L., Leviton, A., Waternaux, C., Rabinowitz, M. B., and Nichols, M. L. (1984). Early sensory-motor development and prenatal exposure to lead. *Neurotoxicol. Teratol.* **6,** 387–402.

Beninger, R. J. (1989). Dissociating the effects of altered dopaminergic function of performance and learning. *Brain Res. Bull.* **23,** 365–371.

Bushnell, P. J., and Bowman, R. E. (1979). Reversal learning deficits in young monkeys exposed to lead. *Pharmacol. Biochem. Behav.* **10,** 733–742.

Byers R. K., and Lord, E. E. (1943). Later effects of lead poisoning on mental development. *Am. J. Dis. Child* **66,** 417–494.

Cador, M., Taylor, R. J., and Robbins, T. W. (1991). Potentiation of the effects of reward-related stimuli by dopaminergic-dependent mechanisms in the nucleus accumbens. *Psychopharmacology* **104,** 377–385.

Cohn, J., and Cory-Slechta, D. A. (1994a). Lead exposure potentiates the accuracy-impairing and response rate-altering effects of *N*-methyl-D-aspartate on a multiple schedule of repeated acquisition and performance. *Neurotoxicol. Teratol.* **16,** 455–465.

Cohn, J., and Cory-Slechta, D. A. (1994b). Assessment of the role of dopamine systems in lead-induced learning impairments using a repeated acquisition and learning baseline. *Neurotoxicology* **15,** 913–926.

Cohn, J., and Cory-Slechta, D. A. (1992). Differential effects of MK-801, NMDA and scopolamine on rats learning a four-member repeated acquisition paradigm. *Behav. Pharmacol.* **3,** 403–413.

Cohn, J., Ziriax, J. M., Cox, C., and Cory-Slechta, D. A. (1992). Comparison of error patterns produced by scopolamine and MK-801 on repeated acquisition and transition baselines. *Psychopharmacology* **107,** 243–254.

Cohn, J., Cox, C., and Cory-Slechta, D. A. (1993). The effects of lead exposure on learning in a multiple repeated acquisition and performance schedule. *Neurotoxicology* **14,** 329–346.

Cory-Slechta, D. A. Mk-801 subsensitivity following postweaning lead exposure. *Neurotoxicology* (in press).

Cory-Slechta, D. A., and Widzowski, D. V. (1991). Low level lead exposure increases sensitivity to the stimulus properties of dopamine D1 and D2 agonists. *Brain Res.* **553,** 65–74.

Cory-Slechta, D. A., Widzowski, D. V., and Newland, M. C. (1989). Behavioral differentiation of the stimulus properties of a dopaminergic D1 agonist from a D2 agonist. *J. Pharmacol. Exp. Ther.* **250,** 800–808.

Cory-Slechta, D. A., Weiss, B., and Cox, C. (1985). Performance and exposure indices of rats exposed to low concentrations of lead. *Toxicol. Appl. Pharmacol.* **78,** 291–299.

Cory-Slechta, D. A., Pokora, M. J., and Widzowski, D. V. (1991). Behavioral manifestations of prolonged lead exposure initiated at different stages of the life cycle: II. Delayed spatial alternation. *Neurotoxicology* **12,** 761–776.

Cory-Slechta, D. A., Pokora, M. J., and Widzowski, D. V. (1992). Postnatal lead exposure induces supersensitivity to the stimulus properties of a D2-D3 agonist. *Brain Res.* **598,** 162–172.

Cory-Slechta, D. A., Widzowski, D. V., and Pokora, M. J. (1993). Functional alterations in dopamine systems assessed using drug discrimination procedures. *Neurotoxicology* **14,** 105–114.

Cotman, C. W., Bridges, R. J., Taube, J. S., Clark, A. S., Geddes, J. W., and Monaghan, D. T. (1989). The role of the NMDA receptor in nervous system plasticity and pathology. *J. NIH Res.* **1,** 65–74.

Cunningham, K. A., Callahan, P. M., and Appel, J. B. (1985). Dopamine D1 receptor mediation of the discriminative stimulus properties of SKF38393. *Eur. J. Pharmacol.* **119,** 121–125.

Demarest, K. T., Lawson-Wendling, K. L., and Moore, K. E. (1983). d-Amphetamine and gamma-butyrolactone alteration of dopamine synthesis in the terminals of nigrostriatal and mesolimbic neurons. *Biochem. Pharmacol.* **32,** 691–697.

Dietrich, K. N., Berger, O. G., Succop, P. A., Hammond, P. B., and Bornschein, R. L. (1993). The developmental consequences of low to moderate prenatal and postnatal lead exposure: Intellectual attainment in the Cincinnati lead study cohort following school entry. *Neurotoxicol. Teratol.* **15,** 37–44.

Doherty, J. D., Simonivic, M., So., R., and Meltzer, H. Y. (1980). The effect of phencyclidine on dopamine synthesis and metabolism in rat striatum. *Eur. J. Pharmacol.* **65,** 139–149.

Fergusson, D. M., Fergusson, J. E., Horwood, L. J., and Kinzett, N. G. (1988). A longitudinal study of dentine lead levels, intelligence, school performance and behaviour. *J. Child Psychol. Psychiatry* **2,** 793–809.

Ferkany, J. W., Kyle, D. J., Willets, J., Rzeszotarski, W. J., Guzewska, M. E., Ellengerger, S. R., Jones, S. M., Sacaan, A. I., Snell, L. D., Borosky, S., Jones, B. E., Johnson, K. M., Balster, R. L., Burchett, K., Kawasaki, K., Hoch, D. B., and Dingledine, R. (1989). Pharmacological profile of NPC 12626, a novel, noncompetitive N-methyl-D-aspartate receptor antagonist. *J. Pharmacol. Exp. Ther.* **250,** 100–109.

Garey, R. E., and Heath, R. G. (1976). The effects of phencyclidine on the uptake of ³H-catecholamines by rat striatal and hypothalamic synaptosomes. *Life Sci.* **18,** 1105–1110.

Govoni, S., Memo, M., Spano, P. F., and Trabucchi, M. (1979). Chronic lead treatment differentially affects dopamine synthesis in various rat brain regions. *Toxicology* **12,** 343–349.

Guilarte, T. R., and Miceli, R. C. (1992). Age-dependent effects of lead on [³H]MK801 binding in NMDA receptor-gated ionophore: in vitro and in vivo studies. *Neurosci. Lett.* **148,** 27–30.

Hatzakis, A., Kokkevi, A., Maravellas, C., Katsouyanni, K., Salaminios, P., Kalandidi, A., Koutselinis, A., Stefanis, C., and Trichopoulos, D. (1989). Psychometric intelligence deficits in lead exposed children. In *Lead Exposure and Child Development. An International Assessment* (M. Smith, L. Grant, and A. I. Sors, Eds.), pp. 260–270, Kluwer, Dordtecht.

Hitzemann, R. J., Loh, H. H., and Domino, E. F. (1973). Effect of phencyclidine on the accumulation of ¹⁴C-catecholamines formed from ¹⁴C-tyrosine. *Arch. Int. Pharmacodyn.* **202,** 252–258.

Jackson, A., and Sanger, D. J. (1988). Is the discriminative stimulus produced by phencyclidine due to an interaction with N-methyl-D-aspartate receptors? *Psychopharmacology* **96,** 87–92.

Jason, K. M., and Kellogg, C. K. (1981). Neonatal lead exposure: effects on development of behavior and striatal dopamine neurons. *Pharmacol. Biochem. Behav.* **15,** 641–649.

Johnson, S. C., and Cory-Slechta, D. A. (1993). Postnatal (PN) and postweaning (PW) lead exposures differentially affect sensitivity to the noncompetitive NMDA antagonist MK801. *The Toxicologist* **13**, 166.

Johnson, S. C., Greenamyre, J. T., and Cory-Slechta, D. A. (1992). Effects of postweaning lead exposure on [^{3}H]dizocilpine (MK801) binding in rat brain. *Soc. Neurosc. Abstr.* **18**, 978.

Kamien, J. B., and Woolverton, W. L. (1985). The D1 dopamine agonist SKF38393 functions as a discriminative stimulus in rats. *Psychopharmacology* **87**, 368–379.

Koek, W., Woods, J. H., and Colpaert, F. C. (1990). N-methyl-D-aspartate antagonism and phencyclidine-like activity: a drug discrimination analysis. *J. Pharmacol. Exp. Ther.* **253**, 1017–1025.

Lasley, S. M. (1992). Regulation of dopaminergic activity, but not tyrosine hydroxylase, is diminished after chronic inorganic lead exposure. *Neurotoxicology* **13**, 625–636.

Lasley, S. M., and Lane, J. D. (1988). Diminished regulation of mesolimbic dopaminergic activity in rats after chronic inorganic lead exposure. *Toxicol. Appl. Pharmacol.* **95**, 474–483.

Lucchi, L., Memo, M., Airaghi, M. L., Spano, P. F., and Trabucchi, M. (1981). Chronic lead treatment induces in rat a specific and differential effect on dopamine receptors in different brain areas. *Brain Res.* **213**, 397–404.

LeMoal, M., and Simon, H. (1991). Mesocorticolimbic dopaminergic network: functional and regulatory roles. *Physiol. Rev.* **17**, 155–234.

Memo, M., Lucchi, L., Spano, P. F., and Trabucchi, M. (1980). Lack of correlation between the neurochemical and behavioural effects induced by d-amphetamine in chronically lead-treated rats. *Neuropharmacology* **19**, 795–799.

Memo, M., Lucchi, L., Spano, P. F., and Trabucchi, M. (1981). Dose-dependent and reversible effects of lead on rat dopaminergic system. *Life Sci.* **28**, 795–799.

Missale, C., Battaini, F., Govoni, S., Castelletti, L., Spano, P. F., and Trabucchi, M. (1984). Chronic lead exposure differentially affects dopamine transport in rat striatum and nucleus accumbens. *Toxicology* **33**, 81–90.

Moresco, R. M., Dall'olio, R., Gandolfi, O., Govoni, S., DiGiovine, S., and Trabucchi, M. (1988). Lead neurotoxicity: a role for dopamine receptors. *Toxicology* **53**, 315–322.

Needleman, H. L., Gunnoe, C., Leviton, A., Reed, R., Peresie, H., Maher, C., and Barrett, P. (1979). Deficits in psychologic and classroom performance of children with elevated dentine lead levels. *N. Engl. J. Med.* **300**, 689–695.

Nielsen, E. B., and Scheel-Kruger, J. (1986). Cueing effects of amphetamine and LSD: elicitation by direct microinjection of the drugs into the nucleus accumbens. *Eur. J. Pharmacol.* **125**, 85–92.

Packard, M. G., and White, N. M. (1989). Memory facilitation produced by dopamine agonists: role of receptor subtype and mnemonic requirements. *Pharmacol. Biochem. Behav.* **33**, 511–518.

Reading, P. J., and Dunnett, S. B. (1991). The effects of excitotoxic lesions of the nucleus accumbens on a matching to position task. *Behav. Brain Res.* **46**, 17–29.

Rice, D. C. (1985). Chronic low-lead exposure from birth produces deficits in discrimination reversal in monkeys. *Toxicol. Appl. Pharmacol.* **77**, 201–210.

Rossouw, J., Offermeier, J., and Van Rooyen, J. M. (1987). Apparent central neurotransmitter receptor changes induced by low-level lead exposure during different developmental phases in the rat. *Toxicol. Appl. Pharmacol.* **91**, 132–139.

Shellenberger, M. K. (1984). Effects of early lead exposure on neurotransmitter systems in the brain. *Neurotoxicology* **5**, 177–212.

Sierra, E. M., and Tiffany-Castiglioni, E. (1991). Reduction of glutamine synthetase activity in astroglia exposed in culture to low levels of inorganic lead. *Toxicology* **65**, 295–304.

Sierra, E. M., Rowles, T. K., Martin, J., Bratton, G. R., Womac, C., and Tiffany-Castiglioni, E. (1989). Low level lead neurotoxicity in a pregnant guinea pig model: neuroglial enzyme activities and brain trace metal concentrations. *Toxicology* **59**, 81–96.

Silva, P. A., Hughes, P., Williams, and Faed, J. M. (1988). Blood lead, intelligence, reading attainment and behaviour in eleven year old children in Dunedin, New Zealand. *J. Child Psychol. Psychiatry* **29**, 43–52.

Taghzouti, K., Louilot, A., Hermal, J. P., LeMoal, M., and Simon, H. (1985a). Alternation behavior, spatial discrimination and reversal disturbances following 6-hydroxydopamine lesions in the nucleus accumbens of the rat. *Behav. Neural Biol.* **44**, 354–363.

Taghzouti, K., Simon, H., Louilot, A., Herman, J. P., and LeMoal, M. (1985b). Behavioral study after local injection of 6-hydroxydopamine in the nucleus accumbens of the rat. *Brain Res.* **344**, 9–20.

Thomson, G. O., Raab, G. M., Hepburn, W. S., et al. (1989). Blood lead levels and children's behaviour: results from the Edinburgh lead study. *J. Child Psychol. Psychiatry* **30**, 515–528.

Tricklebank, M. D., Singh, L., Oles, R. J., Wong, E. H. F., and Iversen, S. D. (1989). A role for receptors of the N-methyl-D-aspartic acid in the discriminative stimulus properties of phencyclidine. *Eur. J. Pharmacol.* **141**, 497–501.

Ujihara, H., and Albuquerque, E. X. (1993). Developmental change of the inhibition by lead of NMDA-activated currents in cultured hippocampal neurons. *J. Pharmacol. Exp. Ther.* **263**, 868–875.

Upchurch, M., and Wehner, J. M. (1990). Effects of N-methyl-D-aspartate antagonism on spatial learning in mice. *Psychopharmacology* **100**, 209–214.

Weathersby, R. T., and Appel, J. B. (1986). Dopamine D2 receptor mediation of the discriminative stimulus properties of LY171555 (quinpirole). *Eur. J. Pharmacol.* **132**, 87–91.

Westfall, T. C., Naes, L., and Paul, C. (1983). Relative potency of dopamine agonists on autoreceptor function in various brain regions of the rat. *J. Pharmacol. Exp. Ther.* **224**, 199–205.

Widzowski, D. V., and Cory-Slechta, D. A. (1994). Homogeneity of regional brain lead concentrations. *Neurotoxicology* **15**, 295–308.

Widzowski, D. V., and Cory-Slechta, D. A. (1993). Apparent mediation of the stimulus properties of a low dose of quinpirole by dopaminergic autoreceptors. *J. Pharmacol. Exp. Ther.* **266**, 526–534.

Widzowski, D. V., Finkelstein, J. N., Pokora, M. J., Johnson, S. C., and Cory-Slechta, D. A. (1994). Time course of postnatal lead-induced changes in dopamine receptors and their relationship to changes in dopamine sensitivity. *Neurotoxicology*, **15**, 853–866.

Willets, J., Balster, R. L., and Leander, J. D. (1991). The behavioral pharmacology of NMDA receptor antagonists. *Trends Pharmacol. Sci.* **11**, 423–428.

Winder, C., and Kitchen, I. (1984). Lead neurotoxicity: a review of the biochemical, neurochemical and drug-induced behavioral evidence. *Prog. Neurobiol.* **22**, 59–87.

Wolterink, G., Phillips, G., Cador, M., Donselaar-Wolterink, I., Robbins, T. W., and Everitt, B. J. (1993). Relative roles of ventral striatal D1 and D2 dopamine receptors in responding with conditioned reinforcement. *Psychopharmacology* **110**, 355–364.

Wood, D. M., and Emmett-Oglesby, M. W. (1989). Mediation in the nucleus accumbens of the discriminative stimulus produced by cocaine. *Pharmacol. Biochem. Behav.* **33**, 453–457.

CHAPTER
17

Dopaminergic Bases of Polychlorinated Biphenyl-Induced Neurotoxicity

RICHARD F. SEEGAL
New York State Department of Health
Wadsworth Center for Laboratories and Research
Albany, New York 12201-0509
and
School of Public Health
State University of New York at Albany
Albany, New York 12203-3727

I. Introduction

Polychlorinated biphenyls (PCBs) are members of a large class of persistent environmental contaminants known as halogenated aromatic hydrocarbons (HAHs). This class not only includes the PCBs, but also structurally related compounds including polychlorinated dibenzo-p-dioxins (PCDDs) and polychlorinated dibenzofurans (PCDFs). Although HAHs have a worldwide distribution in biota, water, and sediment, the majority of the work describing the potential neurotoxic actions of the HAHs has focused on the PCBs because they are present at the highest concentrations in the environment (Hansen, 1987). Because of their widespread distribution, their accumulation in wildlife and man, and their potential toxicity, the manufacture of PCBs in the United States was banned in 1976 (National Research Council, 1979). However, between 170 and 230 million kilograms of PCBs have been released into the environment (Erickson, 1986; Hansen, 1987) and may be responsible for continuing birth defects and increased mortality in wildlife (Hoffman *et*

al., 1987) and neurobehavioral dysfunctions in humans (Jacobson *et al.,* 1990a).

PCBs consist of 209 theoretically possible congeners that differ in the number and position of the chlorines on the biphenyl ring structure (Mullin *et al.,* 1984). Based on their structure and the ability of the congeners to bind to the aryl hydrocarbon (Ah) receptor and induce hepatic enzymes and immunotoxic effects (Safe, 1990), PCBs may be divided into three major classes. The first class consists of the coplanar dioxin-like congeners that bind to the Ah receptor and induce aryl hydrocarbon hydroxylase and 7-ethoxyresorufin *O*-deethylase (EROD), the second class consists of the *ortho*-substituted congeners that neither bind to the receptor nor induce enzyme activity, and the third class consists of mono-*ortho* coplanar compounds that are intermediate between the two previous classes in their ability to induce enzyme activity. The *ortho*-substituted congeners, because of their inability to bind to the Ah receptor, were thought to be largely inactive (Parkinson and Safe, 1987). However, this chapter reviews work that demonstrates that the *ortho*-substituted congeners are neurologically active and may induce neurochemical changes in both the adult and developing central nervous system (CNS) by different mechanisms than the coplanar congeners.

Data suggesting that PCBs are neurotoxicants are based primarily on two major series of studies. The first are epidemiological and suggest an association between perinatal exposure of humans to PCBs and delays in motor reflex development and cognitive dysfunctions (Gladen *et al.*, 1988; Jacobson *et al.*, 1985). The second series describes behavioral and neurochemical changes in nonhuman primates and laboratory rodents exposed either perinatally or as adults to PCBs (Schantz *et al.*, 1991; Pantaleoni *et al.*, 1988; Seegal *et al.*, 1991a,b).

The first epidemiological studies examined the neurological consequences of exposure of Japanese to cooking oil contaminated with high concentrations of PCBs, their thermal degradation products, and polychlorinated naphthalenes. Consumption of contaminated cooking oil was associated with decreased peripheral nerve conduction velocities in the exposed adults and significant decreases in birth weight and IQ of perinatally exposed offspring (Kuroiwa *et al.*, 1969; Yoshimura and Ikeda, 1978).

A similar contamination of cooking oil occurred in Taiwan in 1978. The epidemiological studies were conducted by Rogan and Hsu (Rogan *et al.*, 1988). Perinatally exposed infants weighted less than unexposed controls, were more likely to have cola-colored skin pigmentation and nail-bed deformities, and, most importantly, performed more poorly on tests of cognitive function. Furthermore, infants born to mothers more than 7 years after the initial exposure to the contaminants were also adversely affected (Chen *et al.*, 1992). These results suggest that perinatal exposure to HAHs yields persistent effects on important CNS functions. As with the Japanese rice-oil exposure, the cooking oil was contaminated with high concentrations of PCBs as well as their thermal degradation products, polychlorinated dibenzofurans, quaterphenyls, and naphthalenes, making it difficult to unequivocally determine which of these agents were responsible for the observed somatic and cognitive dysfunctions. However, in a review of the somatic and behavioral consequences of exposure of Taiwanese children to contaminated cooking oil, Rogan and Gladen (1992) state that the children with the most obvious physical findings (e.g., those with somatic effects thought to be due to exposure to dibenzofurans and/or dioxin-like PCB congeners) were not those with the greatest neurobehavioral deficits, suggesting that the development cognitive dysfunctions may be induced by exposure to unidentified contaminants other than dioxin-like HAHs. This potential conclusion is supported by nonhuman primate studies that have demonstrated cognitive and

neurochemical changes following exposure to Aroclor 1016 (Schantz *et al.*, 1991; Seegal *et al.*, 1991a), which is composed primarily of lightly chlorinated *ortho*-substituted congeners (Erickson, 1986) and which were qualitatively different from the deficits seen following exposure to 2,3,7,8-TCDD (TCDD).

The second series of epidemiological studies examined the health effects of low-level exposure to PCBs and other contaminants and were conducted in Michigan by the Jacobsons (Schwartz *et al.*, 1983; Jacobson *et al.*, 1984a, 1985) and in North Carolina by Rogan and Gladen (Rogan *et al.*, 1986; Gladen and Rogan, 1988). The Michigan study examined infants born to mothers who consumed contaminated fish from the Great Lakes. These authors found a statistical relationship between cord serum PCB levels and decreases in birth weight, delays in motor reflex development, and cognitive dysfunctions (Jacobson *et al.*, 1990a,b; 1992) that persisted at least through 4 years of age (Jacobson *et al.*, 1984b; Fein *et al.*, 1984). The North Carolina study documented an association between prenatal PCB exposure and decreases in birth weight and delays in motor reflex development in infants through 24 months of age, but failed to detect significant changes in cognitive function (Gladen *et al.*, 1988; Gladen and Rogan, 1991). Both studies demonstrated a statistical relationship between cord serum PCB levels and developmental delays, although the role of other putative environmental contaminants cannot be ruled out. The discrepancies between these two studies may involve not only differences in methodology and age of the infants at testing, but also differences in the contaminants to which the infants were exposed.

These epidemiological studies suggest a relationship between perinatal exposure of humans to PCBs and related HAHs and alterations in CNS function. However, there are inherent limitations in the ability of epidemiological studies to experimentally determine whether perinatal exposure to PCBs and HAHs or other uncontrolled factors, including alterations in nutritional status, cigarette smoking, or the use of drugs of abuse, were responsible for the alterations in CNS development. Laboratory studies that provide experimenter control over important variables such as knowledge of which toxicants the animals were exposed to and the nutritional status of the mother and offspring thus become very important. The studies that most closely approximate the human exposure conditions and hence provide the most relevant data to support or refute these epidemiological studies are those that have perinatally exposed nonhuman primates to either commerical mixtures of PCBs or to TCDD.

Bowman and co-workers at the University of Wisconsin exposed female nonhuman primates (*Macaca mulatta*) prior to mating to commerical mixtures of PCBs [Aroclor 1016 : 0.25 ppm diet (estimated to be 7μg/(kg·day) or 1 ppm diet (estimated to be 30 μg/kg·day) for approximately 22 months; Aroclor 1248 : 2.5 ppm diet (estimated to be 90 μg/(kg·day) for approximately 18 months] and examined the behaviorial sequelae in their offspring. Changes in locomotor activity were assessed only following exposure to Aroclor 1248 and resulted in hyperactivity compared with age-matched controls (Bowman and Heironimus, 1981a; Bowman *et al.*, 1981b). In subsequent studies, the authors noted decreases in spatial discrimination reversal learning in offspring exposed to diets containing either 1 ppm Aroclor 1016 or 2.5 ppm Aroclor 1248 that persisted more than 4 years after exposure (Levin *et al.*, 1988). To relate the nonhuman primate exposure to human exposure, the 0.25-ppm Aroclor 1016 diet would be equivalent to a human consuming approximately 0.75 kg of freshwater fish per month that was contaminated with 2 ppm PCBs. The types of deficits in these tasks and the duration of change are similar to the cognitive dysfunctions seen by Jacobson in human infants exposed perinatally to contaminants found in Greak Lakes fish (e.g., deficits in visual recognition memory). The neurobehaviorial changes were qualitatively and quantitatively different from those induced by perinatal exposure to TCDD (Bowman *et al.*, 1990), suggesting that developmental exposure to PCBs may alter cognitive function via different mechanisms from those involving dioxins. These behavioral findings also provide experimental support for the conclusion reached by Rogan and Gladen (1992), who noted a dichotomy between the somatic effects of perinatal exposure to contaminated cooking oil (an Ah receptor-mediated response) and the cognitive dysfunctions that may be mediated by non-Ah receptor-mediated events. The total intake of PCBs by nonhuman primates in Bowman's highest dose group was half of the 1 g that was estimated to have been ingested by the exposed Taiwanese (Chen *et al.*, 1992).

Although these epidemiological and laboratory studies have documented changes in CNS function following exposure to complex mixtures of PCBs and related contaminants, they have not investigated the biochemical correlates of changes in motor activity and cognitive function.

A common factor underlying many of the neurobehaviorial changes seen following perinatal exposure to PCBs is alterations in concentrations and activity of the neurotransmitter dopamine (DA) (Seegal, 1994). Elevations in DA concentrations, induced by pharmacologic agents such as *d*-amphetamine, increase locomotor activity (Dominic and Moore, 1969) whereas iontophoretic administration of DA agonists and

antagonists in the dorsal prefrontal cortex of nonhuman primates results in decrements in peformance in object and spatial discrimination tasks (Brozoski *et al.*, 1979; Sawaguchi *et al.*, 1989). Most importantly, both perinatal and adult exposure of rodents and nonhumans primates to commerical mixtures of PCBs and individual congeners alters brain concentrations of DA (Agrawal *et al.*, 1981; Seegal, 1992; Seegal *et al.*, 1991a,b, 1994). Thus, changes in brain DA function may be one of the key biochemical factors responsible for the behavioral changes seen in humans and laboratory animals following exposure to PCBs. In order to better understand the potential link between PCB-induced changes in DA concentrations and behavioral dysfunction, the remainder of this chapter is devoted to a description of the neurochemical changes seen in rodents, nonhuman primates, and tissue culture preparations exposed to PCBs and to a discussion of the possible biochemical mechanisms responsible for the changes in dopaminergic function.

II. Neurochemical Effects of Exposure of Adult Rodents and Nonhuman Primates to PCBs

A number of studies documenting changes in CNS concentrations of biogenic amine neurotransmitters in the adult laboratory rodent have been conducted. The first studies (Seegal *et al.*, 1985, 1986a,b) were undertaken to determine (i) the toxicokinetics of PCBs following a single oral exposure and (ii) the direct neurochemical effects of PCBs, in the absence of potential changes in maternal behavior and development due to perinatal exposure to PCBs. Adult male rats were orally gavaged with Aroclor 1254, at doses of 500 or 1000 mg/kg, and were sacrificed on postgavage days 1, 3, 7, and 14. Regional brain concentrations of biogenic amines, their metabolites, and PCB concentrations were determined using high-performance liquid chromatography with electrochemical detection and glass-capillary, high-resolution gas chromatography, respectively, Brain DA, norepinephrine (NE), and serotonin (5-HT) concentrations were significantly depressed on postgavage days 1 and 3 and returned to control levels by day 14. Brain PCB concentrations, on the other hand, reached their maximum concentrations by day 3, and by day 14 were only slightly elevated compared to values obtained in control animals.

In a second study (Seegal *et al.*, 1991b), adult male rats were exposed to Aroclor 1254 via contaminated rat chow (containing either 500 or 1000 ppm of Aroclor 1254) 30 days. On the basis of an average daily comsumption of 20 g of chow per day the 500 ppm animals consumed approximately 10 mg of Aroclor 1254/day or a total of 300 mg over the 30-day experiment. Given the lower daily intake of PCBs and the ability of rodents to metabolize PCBs (Sundstrom *et al.*, 1976), particularly following chronic exposure to PCBs, the applied dose is considerably less than that used in the previous experiment. Unlike the previous experiment, where significant changes in concentrations of all biogenic amine neurotransmitters were noted, only changes in DA and its metabolites were detected. These results suggest that (i) PCBs are capable of altering an important neurotransmitter in the adult CNS and (ii) PCB-induced alterations in brain DA concentrations are more sensitive than alterations in other biogenic amine neurotransmitter systems.

Similar decreases in brain DA concentrations have also been in the adult nonhuman primate *Macaca nemestrina* (pig-tailed macaque) following exposure to commercial mixtures of PCBs. Seegal and colleagues (1991a) exposed nonhuman primates to either Aroclor 1016 or Aroclor 1260, at doses of 0.8, 1.6, or 3.2 mg/(kg·day) for 20 weeks. Average total consumption of PCBs for each dose group ranged from approximately 1.12 to 4.48 g of PCBs. Following exposure, the animals were sacrificed and concentrations of biogenic amine neurotransmitters, their metabolites, and individual PCB congeners were determined using previously described methods (Seegal *et al.*, 1991a). Subchronic exposure to either commercial mixture of PCBs resulted in significant decreases in only brain concentrations of DA (Fig. 1) that were most evident at the highest dose. Furthermore, only a limited number of PCB congeners accumulated in brain (Table 1) and these congeners were all ortho or mono-*ortho* chlorine-substituted, suggesting that these congeners may have been responsible for the observed decrease in brain DA concentrations

In an additional study (Seegal *et al.*, 1994), designed to determine whether the observed alterations in dopaminergic function persisted following removal of the animals from PCBs, nonhuman primates were exposed to 3.2 mg/(kg·day) of either Aroclor 1016 or Aroclor 1260 for 20 weeks and then followed for an additional 24 weeks before sacrifice. Although brain PCB concentrations in the postexposure animals were reduced by approximately 75% compared with values obtained in similarly dosed animals sacrificed immediately after exposure (Table 2), brain DA concentrations did not return to preexposure control values and were statistically indistinguishable from DA concentrations seen in animals sacrificed immediately following exposure (Fig. 1). These results suggest that subchronic exposure of the adult nonhuman primate to levels of PCBs that yield serum concentrations similar to those seen in

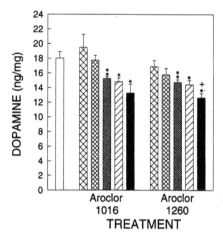

FIGURE 1 Effects of PCB exposure on dopamine concentrations in caudate nucleus from adult *Macaca nemestrina*. Animals were exposed to either Aroclor 1016 or Aroclor 1260 at 0.8 (⊠), 1.6 (⊘), or 3.2 (▨) mg/(kg·day) for 20 weeks before sacrifice. An additional group of animals was exposed to either Aroclor 1016 or Aroclor 1260 at 3.2 mg/(kg·day) for 20 weeks and sacrificed either 24 or 44 weeks postexposure (◪). A third group of animals was exposed to either Aroclor 1016 or Aroclor 1260 at 3.2 mg/(kg·day) for 66 weeks (■) before sacrifice. Control animals (□) were exposed to vehicle only. Each bar represents the mean of the data from two to five animals. Data were statistically analyzed using the nonparametric Kruskal–Wallis test. Significantly different from control: *($P \leq 0.05$), **($P \leq 0.01$); significantly different from animals exposed for 20 weeks to 3.2 mg/(kg·day) Aroclor 1260 +($P \leq 0.05$).

occupationally exposed workers (Lawton *et al.*, 1985) produces long-term, if not permanent, alterations in DA concentrations.

The effects of prolonged continuous exposure to PCBs on brain DA concentrations have also been examined to determine whether the previously observed decreases in brain DA concentrations were asymptotic. Nonhuman primates were exposed to 3.2 mg/(kg·day) of either Aroclor 1016 or Aroclor 1260 for 66 weeks—a

TABLE 1 PCB Congeners Constituting More Than 3% of the Total Concentration in Caudate Nucleus from Nonhuman Primate Brain Following a 20-Week Exposure to 3.2 mg/(kg·day) of either Aroclor 1016 or 1260

Aroclor 1260		Aroclor 1016	
Congener	% in Brain	Congener	% in Brain
2,3,4,2',4',5'	12	2,4,4'	76
2,3,4,5,2',4',5'	11	2,4,2',4'	14
2,4,5,2',4',5'	10	2,5,2',5'	7
2,3,4,5,6,2',6'	4		
2,3,4,5,2',3',5',6'	3		
2,3,4,5,2',3',4'	3		
2,3,4,5,6,3',4'	3		

TABLE 2 Brain PCB Concentrations in Caudate Nucleus of Nonhuman
Primates Following Exposure to either Aroclor 1016 or 1260

Exposure conditions		PCB concentration (ppm)	
Dose	Exposure period	Aroclor 1016	Aroclor 1260
0.8 mg/(kg·day)	20-week exposure	0.5 ± 0.04	3.8 ± 1.3
1.6 mg/(kg·day)	20-week exposure	1.3 ± 0.2	6.2 ± 1.6
3.2 mg/(kg·day)	20-week exposure	4.1 ± 0.4	19.4 ± 0.4
3.2 mg/(kg·day)	20-week exposure + 24-week postexposure	1.2 ± 0.5	3.8 ± 1.0

period of exposure three times longer than used previously. Following the 66-week exposure, brain DA concentrations in both the Aroclor 1016 and 1260-exposed animals were 24 and 29% lower, respectively, than concentrations seen in the control animals (Fig. 1). Despite this prolonged exposure, the animals gained weight at the same rate as controls and showed no evidence of chlorobiphenyl poisoning (e.g., chloracne, nail-bed deformities, or alopecia).

These studies suggest that (i) exposure of adult animals to mixtures of PCBs results in statistically significant reductions in brain concentrations of DA; (ii) the dopaminergic system may be more sensitive to PCBs than other biogenic amine neurotransmitter systems; (iii) because only *ortho*-substituted congeners are detected in brain following exposure to complex mixtures of PCBs, this class of congeners may be responsible for the observed decreases in brain DA concentrations; and (iv) the reductions in brain DA concentrations persist following removal of the animal from PCBs. However, these studies provide neither information on the biochemical bases for the reductions in neurotransmitter nor definitive evidence of which PCB congeners were responsible for the decreases in brain DA concentrations.

III. Neurochemical Effects of PCBs in *in Vitro* Preparations

To answer these questions, Seegal and Shain (1992) have carried out a series of studies using pheochromocytoma (PC12) cells. There are several advantages to using this *in vitro* test system. First, PC12 cells are a continuous cell line derived from a rat adrenal gland tumor that synthesizes, stores, releases, and metabolizes DA in a manner similar to that of the mammalian central nervous system (Greene and Rein, 1977; Kittner *et al.*, 1987). Second, PC12 cells consist of a single phenotype, are considerably less complex than

the mammalian brain, and thus allow determination of the neurochemical effects of a toxicant on a single cell type. Third, only microgram quantities of test compounds are required. Thus, expensive and potentially highly toxic compounds can be tested in a closed system. Finally, there is evidence (manuscript in preparation) that PC12 cells only poorly metabolize PCBs; thus we can determine whether the parent compound or its metabolites are the active agent(s) responsible for the observed alterations in cellular DA content.

In the first study (Seegal *et al.*, 1989), PC12 cells were exposed to Aroclor 1254 (0.1 to 100 ppm in RPMI growth medium) for 6 hr. This exposure resulted in a statistically significant, dose-dependent decrease in the cell content of DA and NE (Table 3). These changes in neurotransmitter content were not because of cell death or reduced cellular metabolism because incorporation of neither [³H]-leucine nor [¹⁴C]-deoxyglucose into PCB-treated cells was affected.

PC12 cells were then used to determine the potential neurotoxicity of the PCB congeners that had accumulated in nonhuman primate brain following a 20-week exposure to Aroclor 1016 (Seegal *et al.*, 1990). Cells were exposed to varying concentrations of the three congeners found in nonhuman primate brain (2,4,4'; 2,4,2',4'; or 2,5,2',5'), either singly or as a mixture of congeners in the same concentration ratio as found in

TABLE 3 Effects of Aroclor 1254 on
Cellular Biogenic Amine Concentrations
Determined in PC-12 Cells in Culture
(Expressed as a Percentage of DMSO
Vehicle Control)

Aroclor 1254 exposure (ppm)	Dopamine	Norepinephrine
1	73 ± 6.0	65 ± 5.6
10	60 ± 5.5	70 ± 6.3
100	27 ± 3.8	45 ± 6.2

brain. Significant reductions in cell DA content were noted with all congeners, although the greatest decreases were observed following exposure of the cells to the mixture of the three congeners. These results suggest that the congeners found in nonhuman primate brain may be responsible for the observed decreases in CNS concentrations of DA.

Shain *et al.* (1991) then determined the relationship between the structure of PCB congeners and their ability to alter cellular DA content. Dose-dependent decreases in cellular DA content were determined following exposure of PC12 cells to more than 55 individual PCB congeners. In support of the findings seen in the just-described nonhuman primate studies, the congeners that decreased cellular DA content were all *ortho*-chlorine substituted. The most active congener was the di-*ortho*-substituted congener 2,2', whereas the coplanar, dioxin-like congeners 3,4,3',4' and 3,4,5,3',4' were inactive at the highest concentrations that were tested. These results are summarized in Table 4. Because these congeners do not bind to the Ah receptor nor induce AHH or EROD (Sawyer and Safe, 1982), it has been suggested that the biochemical mechanisms by which *ortho*-substituted congeners alter *in vitro* DA function are distinct from those by which coplanar dioxin-like congeners alter CNS function. These *in vitro* results (i) aid in explaining the findings of Schantz *et al.* (1991), who noted qualitative differences in behavorial effects following exposure to Aroclor 1016 and TCDD, and (ii) support the dichotomy between cognitive dysfunctions and Ah receptor-mediated somatic effects noted by Gladen and Rogan (1992).

An additional series of experiments was conducted to begin to determine the biochemical mechanisms responsible for the observed decreases in cellular DA concentrations. Again the tissue culture system was the preparation of choice for these investigations because, unlike experiments carried out in whole animals, DA and its metabolities cannot move into another compartment (e.g., blood or cerebrospinal fluid). Therefore,

any changes in cellular or medium concentrations of DA and its metabolites can be accounted for. The PCB-induced reductions in cellular DA content can, therefore, only be due to one of three possible reasons: (i) enhanced release of DA into medium; (ii) enhanced catabolism of DA to its end products [homovanillic acid (HVA) or 3,4-dihydroxyphenylacetic acid (DOPAC), which would result in elevated medium concentrations of these metabolites]; or (iii) alterations in the synthesis of DA. Following exposure to Aroclor 1254, medium concentrations of both DA and its metabolites were significantly depressed, suggesting that the decreases in cellular DA concentrations were due to inhibition of DA synthesis instead of to either enhanced release of cellular DA into media or increased catabolism of DA.

If, as suspected, PCBs decrease cellular DA content by inhibiting the synthesis of DA, then PCBs can only induce these changes by decreasing either (i) the activity of the two enzymes responsible for the synthesis of DA [e.g., tyrosine hydroxylase (TH), the rate-limiting enzyme that converts tyrosine to L-dopa, and aromatic amino acid decarboxylase (AADC), which converts L-dopa to DA (McGeer *et al.*, 1978)] or (ii) the availability of the essential amino acid precursor, tyrosine. The effects of 2,2', the most active congener, on the congener, on the biosynthesis of DA have been investigated using two separate approaches.

In the first series of experiments (Seegal *et al.*, 1991c), neuroblastoma cells (N1E-N115) were exposed to 2,2'. These cells were used because they express high levels of TH producing large quantities of L-dopa, but only poorly express AADC, and therefore synthesize little or no biogenic amine neurotransmitters (e.g., DA) (Amano *et al.*, 1972). Thus, if PCBs decrease cellular DA concentrations by inhibiting the activity of TH, a decrease in medium concentrations of L-dopa would be expected. On the other hand, if PCBs affect AADC activity, no changes in the concentrations of L-dopa would be expected. When N1E-N115 cells were exposed to 2,2' there were significant reductions in

TABLE 4 Representative IC_{50} Values for PCB Congener-Mediated Decreases in Cellular Dopamine Content Determined *in Vitro*

<100 μM		100–200 μM		>200 μM		No effect
Congener	IC_{50} (μM)	Congener	IC_{50} (μM)	Congener	IC_{50} (μM)	Congener
2----,2----	64	2-4--,2-4--	115	2-4--,-34--	253	--4--,--4--
2-4-6,2----	71	2-4-6,----	150	-3-5-,--4--	310	-34--,-34--
2--5-,2----	88	2----,-----	182	--4--,----	335	-345-,-34--
2--5-,2--5-	86	2-4--,--4--	196	-3---,--4--	410	

medium concentrations of L-dopa, strongly suggesting that PCBs reduce cellular DA concentrations by interfering with the synthesis of DA at the level of TH.

In a second series of experiments (Seegal and Shain, 1992), the TH activity in disrupted PC12 cells supplied with excess tyrosine and the cofactor tetrahydrobiopterin required for the synthesis of DA was measured. Exposure to 2,2' significantly reduced the conversion of tyrosine to DA compared with similarly treated cells exposed to vehicle only, providing further evidence for the hypothesis that PCBs interfere with the synthesis of DA at the level of TH. Because an excess of both tyrosine and its cofactor (tetrahydrobiopterin) were provided, the observed decreases in DA production cannot be attributed to a decrease in the availability of tyrosine.

Thus, in both adult rodents and nonhuman primates, as well as in tissue culture preparations, exposure to PCBs results in decreases in brain or cellular concentrations of DA. *In vitro* findings strongly suggest that the decreases in cellular DA concentrations are due to a reduction in the ability of dopaminergic neurons to synthesize DA. It has been suggested that a similar mechanism may be responsible for the decreases in brain DA concentrations seen in adult rodents and nonhuman primates exposed to PCBs.

IV. Neurochemical Effects of Perinatal Exposure to PCBs

Epidemiological studies (Jacobson *et al.*, 1985; Gladen *et al.*, 1988) have demonstrated a statistical relationship between perinatal exposure to PCBs and deficits in cognitive function and motor reflex behavior. It is therefore important to determine the neurochemical effects of perinatal exposure to both commercial mixtures of PCBs and individual PCB congeners in a laboratory setting. It is hoped that information gathered from these experiments will aid in understanding the reported behavioral deficits. To this end, studies where pregnant dams have been exposed to powdered rat chow adulterated with Aroclor 1016 [30 and 100 mg/(kg·day)] during gestation (GD8–21), lactation (birth to weaning), or combined gestational and lactational exposure have been completed. Pregnant dams are also being exposed to individual PCB congeners from GD6 through weaning to examine the developmental effects of structurally different congeners. Because of the large number of congeners to be tested, it was decided to expose dams and pups for the entire gestational and lactational period to different doses of the congeners instead of determining the effects of exposure to a

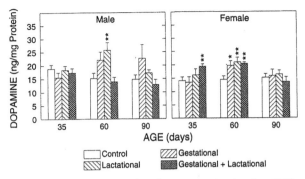

FIGURE 2 Effects of perinatal exposure to Aroclor 1016 on dopamine concentrations in the substantia nigra of male and female rats. Dams were exposed to 100 ppm adulterated chow during gestation (GD8–21), lactation (D0–21), or gestation + lactation, and offspring were sacrificed on days 35, 60, or 90 of age. $N = 6–10$ animals per group. Data were statistically analyzed using one-way analysis of variance with post hoc *t* tests. Significantly different from control: *($P \leq 0.05$) and **($P \leq 0.01$).

single concentration of PCB congeners during different periods of development.

Perinatal exposure to 30 mg/(kg·day) of Aroclor 1016 failed to significantly alter biogenic amine concentrations in the brains of the offspring (data not shown). However, exposure to 100 mg/(kg·day) of Aroclor 1016 significantly elevated brain biogenic amine concentrations, with the most consistent increases seen for DA. This dose of Aroclor 1016 did not significantly alter body weight or growth in the dams or offspring. The results of exposure to 100 mg/(kg·day) on DA concentrations in the substantia nigra (SN) of male and female offspring are shown in Fig. 2.

In a second series of experiments, rat dams were exposed to 3,4,3',4' [TCB; 0.1 or 1.0 mg/(kg·day)] from gestational day 6 through weaning. As shown in Fig. 3, TCB resulted in significant increases in DA concen-

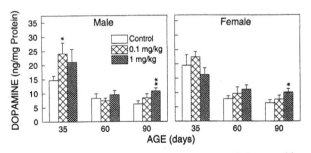

FIGURE 3 Effects of perinatal exposure to 3,4,3',4'-tetra-chloro-biphenyl (TCB) on dopamine concentrations in the substantia nigra of male and female rats. Dams were exposed to either 0.1 or 1 mg/kg of TCB from gestational day 6 through lactation, and offspring were sacrificed on days 35, 60, or 90 of age. $N = 7–8$ animals per group. Data were statistically analyzed using one-way analysis of variance with post hoc *t* tests. Significantly different from control: *($P \leq 0.05$) and **($P \leq 0.01$).

trations in the SN of both male and female offspring without adversely affecting pup body weight. Unlike the elevations in brain DA concentrations seen in the Aroclor 1016-exposed offspring, brain DA concentrations, compared with age-matched controls, remained elevated at least through 90 days of age. Although elevations, instead of decreases in brain DA concentrations seen by Agrawal and associates (1981) (dams and offspring were exposed to concentrations of TCB 30 to 300 times less than the doses used by Agrawal and co-workers), were noted, the persistence of the changes in brain DA concentrations, seen in both studies, suggest that TCB yields long-lasting changes in brain DA concentrations with the direction of change in neurotransmitter concentrations dependent on the dose of TCB. Future studies, using *ortho*-substituted and additional coplanar PCBs, administered perinatally, will help in determining whether the structure of the congener is as important a variable in altering DA function as is shown in the just described *in vitro* studies.

V. Summary

The finding that TCB is active when administered during development stands in sharp contrast to the lack of effects observed following administration to the adult rat. Tilson and colleagues (1979) were unable to produce any behavioral or neurochemical effects in the adult. Similarly, Chishti and Seegal (1992) found that TCB did not alter striatal DA concentrations following direct injection into the striata of adult rats. Thus, TCB, and perhaps other structurally similar PCB congeners, alter CNS function when administered during development, but are inactive when administered to the adult. However, as reported here, both classes of congeners elevate brain biogenic amine concentrations when administered perinatally. Indeed, Eriksson and Fredriksson (personal communication) noted increased locomotor activity in mice following early postnatal exposure to both 2,4,4' and 3,4,3',4', demonstrating that both *ortho* and coplanar PCB congeners have similar behavioral effects when administered during development. Furthermore, the increases in locomotor activity are in the predicted direction based on our observed increases in brain biogenic amine concentrations.

This age-dependent dichotomy in neurochemical effects [e.g., elevations in biogenic amine concentrations following perinatal exposure to both *ortho*-substituted (Aroclor 1016) and coplanar congeners (TCB), decreases in biogenic amine concentrations following adult exposure to *ortho*-substituted congeners, and no

effect seen following adult exposure to TCB] suggests that the biochemical mechanisms responsible for the elevations in brain biogenic amines and their metabolites following perinatal exposure differ from those mechanisms responsible for the reductions in brain concentrations of DA following adult exposure to PCBs. This age-dependent dichotomy may be due to the ability of PCBs to alter concentrations of circulating hormones or bind to steroid receptors (Gellert, 1978; Jansen *et al.*, 1993). In turn, alterations in hormones, including steroid and thyroid hormones, particularly during development, have profound and long-lasting effects on neuronal organization (Fox *et al.*, 1978; Puymirat, 1985; Kaylor *et al.*, 1984) and activity of CNS neurotransmitters, including DA (Di Paolo *et al.*, 1985; Van Hartesveldt and Joyce, 1986). Indeed, *ortho*-substituted congeners and polychlorinated hydroxybiphenyls are weakly estrogenic (Jansen *et al.*, 1993; Korach *et al.*, 1988) and decrease striatal DA concentrations by inhibiting the activity of TH (Foreman and Porter, 1980). Furthermore, TCB, other coplanar congeners, and TCDD are antiestrogenic, due primarily to their ability to enhance the metabolism of estrogens to their hydroxy metabolites (Gierthy *et al.*, 1988). Because alterations in concentrations of steroid hormones, estrogens in particular, influence the development and function of biogenic amine neurotransmitters (Kaylor *et al.*, 1984), PCB-induced changes in steroid function might have long-lasting effects following perinatal exposure, but yield either minimal or no effect following adult exposure. In addition, PCBs, administered during development, alter thyroid hormone function (van den Berg *et al.*, 1988; Ness *et al.*, 1993), which would also have profound consequences on neurochemical development (Puymirat, 1985; Vaccari *et al.*, 1990). However, given the paucity of information on the potential mechanisms responsible for alteration in CNS function following perinatal exposure to PCBs, it is not yet possible to determine which of the just-mentioned hypothesized mechanisms (or indeed other as yet unidentified mechanisms) may be responsible for alterations in neuronal development.

Thus, neurochemical data that have been gathered in adult nonhuman primates, rodents, and in tissue culture document the ability of *ortho*-substituted congeners to decrease brain and cellular content of biogenic amines and, in particular, DA. Furthermore, the *in vitro* data suggest that one mechanism responsible for these decreases may involve alterations in the ability of the neuron or cell to synthesize DA. On the other hand, neurochemical data gathered following perinatal exposure of the rodent to both Aroclor mixtures and individual congeners strongly suggest that additional biochemical mechanisms may be responsible for the unexpected increases in brain biogenic amine concentrations. Nevertheless, the epidemiological and laboratory-based studies in nonhuman primates and rodents provide strong evidence of a relationship between perinatal exposure to HAH, and in particular PCBs, and alterations in CNS development. Because these PCB-induced changes in motor and cognitive behavior are similar to changes due to manipulations of the dopaminergic system (Dominic and Moore, 1969; Brozoski *et al.*, 1979; Sawaguchi *et al.*, 1988, 1989), they provide support for the hypothesis that alterations in neurochemical and, in particular, biogenic amine function may be a primary mechanism responsible for these changes. Although PCB-induced changes in DA function appear to be sufficient to explain the observed neurobehavioral changes, the neurochemical and anatomical complexity of the brain make it likely that other neurotransmitters may also be altered (Eriksson *et al.*, 1991). A more complete description of the neurochemical, endocrine, and behavioral changes induced by perinatal exposure to PCBs will allow an understanding of the underlying physiological and biochemical processes responsible for the potentially important changes in CNS function induced by exposure to HAHs, as well as providing clues to therapeutic strategies that may be developed to ameliorate these dysfunctions (Levin, 1993).

Acknowledgments

Research support for this work was provided in part by NIH Grants ESO3884 and ESO4913 and EPA Contract R-813830.

References

Agrawal, A. K., Tilson, H. A., and Bondy, S. C. (1981). 3,4,3′,4′-Tetrachlorobiphenyl given to mice prenatally produces long term decreases in striatal dopamine and receptor binding sites in the caudate nucleus. *Toxicol. Letts.* **7**, 417–424.

Amano, T., Richelson, E., and Nirenberg, M. (1972). Neurotransmitter synthesis by neuroblastoma clones. *Proc. Natl. Acad. Sci. U.S.A.* **69**, 258–263.

Bowman, R. E., and Heironimus, M. P. (1981a). Hypoactivity in adolescent monkeys perinatally exposed to PCBs and hyperactive as juveniles. *Neurobehav. Toxicol. Teratol.* **3**, 15–18.

Bowman, R. E., Heironimus, M. P., and Barsotti, D. A. (1981b). Locomotor hyperactivity in PCB-exposed rhesus monkeys. *Neurotoxicology* **2**, 251–268.

Bowman, R. E., Schantz, S. L., Ferguson, S. A., Tong, H. Y., and Gross, M. L. (1990). Controlled exposure of female rhesus monkeys to 2,3,7,8-TCDD; Cognitive behavioral effects in their offspring. *Chemosphere* **20**, 1103–1108.

Brozoski, T. J., Brown, R. M., Rosvold, H. E., and Goldman, P. S. (1979). Cognitive deficit caused by regional depletion of dopamine in prefrontal cortex of rhesus monkey. *Science* **205**, 929–931.

Chen, Y.-C. J., Guo, Y.-L., Hsu, C.-C., and Rogan, W. J. (1992). Cognitive development of Yu-Cheng ('oil disease') children prenatally exposed to heat-degraded PCBs. *JAMA* **268**, 3213–3218.

Chishti, M. A., and Seegal, R. F. (1992). Intrastriatal injection of PCBs decreases striatal dopamine concentrations in rats. *Toxicologist* **12**, 320.

Di Paolo, T., Rouillard, C., and Bedard, P. (1985). 17β-Estradiol at a physiological dose acutely increases dopamine turnover in rat brain. *Eur. J. Pharmacol.* **117**, 197–203.

Dominic, J. A., and Moore, K. E. (1969). Acute effects of α-methyltyrosine on brain catecholamines and on spontaneous and amphetamine stimulated motor activity in mice. *Arch. Int. Pharmacodyn. Ther.* **178**, 166–176.

Erickson, M. D. (1986). *Analytical Chemistry of PCBs.* Butterworth, Boston.

Eriksson, P., Lundkvist, U., and Fredriksson, A. (1991). Neonatal exposure to 3,3′,4,4′-tetrachlorobiphenyl: changes in spontaneous behaviour and cholinergic muscarinic receptors in the adult mouse. *Toxicology* **69**, 27–34.

Fein, G. G., Jacobson, J. L., Jacobson, S. W., Schwartz, P. M., and Dowler, J. K. (1984). Prenatal exposure to polychlorinated biphenyls: effects on birth size and gestational age. *J. Pediatr.* **105**, 315–320.

Foreman, M. M., and Porter, J. C. (1980). Effects of catechol estrogens and catecholamines on hypothalamic and corpus striatal tyrosine hydroxylase activity. *J. Neurochem.* **34**, 1175–1183.

Fox, T. O., Vito, C. C., and Wieland, S. J. (1978). Estrogen and androgen receptor proteins in embryonic and neonatal brain: hypothesis for roles in sexual differentiation and behavior. *Am. Zool.* **18**, 525–537.

Gellert, R. J. (1978). Uterotrophic activity of polychlorinated biphenyls (PCB) and induction of precocious reproductive aging in neonatally treated female rats. *Environ. Res.* **16**, 123–130.

Gierthy, J. F., Lincoln, D. W., Kampcik, S. J., Dickerman, H. W., Bradlow, H. L., Niwa, T., and Swaneck, G. E. (1988). Enhancement of 2- and 16α-estradiol hydroxylation in MCF-7 human breast cancer cells by 2,3,7,8-tetrachlorodibenzo-p-dioxin. *Biochem. Biophys. Res. Commun.* **157**, 515–520.

Gladen, B., and Rogan, W. (1988). Decrements on six-month and one-year Bayley scores and prenatal polychlorinated biphenyls (PCB) exposure. *Am. J. Epidemiol.* **128**, 912.

Gladen, B. C., Rogan, W. J., Hardy, P., Thullen, J., Tingelstad, J., and Tully, M. (1988). Development after exposure to polychlorinated biphenyls and dichlorodiphenyl dichloroethene transplacentally and through human milk. *J. Pediatr.* **113**, 991–995.

Gladen, B. C., and Rogan, W. J. (1991). Effects of perinatal polychlorinated biphenyls and dichlorodiphenyl dichloroethene on later development. *J. Pediatr.* **119**, 58–63.

Greene, L. A., and Rein, G. (1977). Release, storage and uptake of catecholamines by a clonal cell line of nerve growth factor (NGF) responsive pheochromocytoma cells. *Brain Res.* **129**, 247–263.

Hansen, L. G. (1987). Environmental toxicology of polychlorinated biphenyls. In *Polychlorinated Biphenyls (PCBs): Mammalian and Environmental Toxicology* (S. Safe and O. Hutzinger, Eds.), pp. 15–48, Springer-Verlag, New York.

Hoffman, D. J., Rattner, B. A., Sileo, L., Docherty, D., and Kubiak, T. J. (1987). Embryotoxicity, teratogenicity, and aryl hydrocarbon hydroxylase activity in Forster's terns on Green Bay, Lake Michigan. *Environ. Res.* **42**, 176–184.

Jacobson, J. L., Fein, G. G., Jacobson, S. W., Schwartz, P. M., and Dowler, J. K. (1984a). The transfer of polychlorinated biphenyls (PCBs) and polybrominated biphenyls (PBBs) across the human placenta and into maternal milk. *Am. J. Public Health* **74**, 378–379.

Jacobson, J. L., Jacobson, S. W., Schwartz, P. M., Fein, G. G., and Dowler, J. K. (1984b). Prenatal exposure to an environmental toxin: A test of the multiple effects model. *Dev. Psychol.* **20**, 523–532.

Jacobson, J. L., Jacobson, S. W., and Humphrey, H. E. B. (1990a). Effects of in utero exposure to polychlorinated biphenyls and related contaminants on cognitive functioning in young children. *J. Pediatr.* **116**, 38–45.

Jacobson, J. L., Jacobson, S. W., and Humphrey, H. E. B. (1990b). Effects of exposure to PCBs and related compounds on growth and activity in children. *Neurotoxicol. Teratol.* **12**, 319–326.

Jacobson, J. L., Jacobson, S. W., Padgett, R. J., Brumitt, G. A., and Billings, R. L. (1992). Effects of prenatal PCB exposure on cognitive processing efficiency and sustained attention. *Dev. Psychol.* **28**, 297–306.

Jacobson, S. W., Fein, G. G., Jacobson, J. L., Schwartz, P. M., and Dowler, J. K. (1985). The effect of intrauterine PCB exposure on visual recognition memory. *Child Dev.* **56**, 853–860.

Jansen, H. T., Cooke, P. S., Porcelli, J., Liu, T.-C., and Hansen, L. G. (1993). Estrogenic and antiestrogenic actions of PCBs in the female rat: in vitro and in vivo studies. *Reprod. Toxicol.* **7**, 237–248.

Kaylor, W. M., Jr., Song, C. H., Copeland, S. J., Zuspan, F. P., and Kim, M. H. (1984). The effect of estrogen on monoamine systems in the fetal rat brain. *J. Reprod. Med.* **29**, 489–492.

Kittner, B., Brautigam, M., and Herken, H. (1987). PC12 cells: a model system for studying drug effects on dopamine synthesis and release. *Arch. Int. Pharmacodyn. Ther.* **286**, 181–194.

Korach, K. S., Sarver, P., Chae, K., McLachlan, J. A., and McKinney, J. D. (1988). Estrogen receptor-binding activity of polychlorinated hydroxybiphenyls: conformationally restricted structural probes. *Mol. Pharmacol.* **33**, 120–126.

Kuroiwa, Y., Murai, Y., and Santa, T. (1969). Neurological and nerve conduction velocity studies on 23 patients with chlorobiphenyls poisoning. *Fukuoka Igaku Zasshi* **60**, 446–462.

Lawton, R. W., Ross, M. R., Feingold, J., and Brown, J. F., Jr. (1985). Effects of PCB exposure on biochemical and hematological findings in capacitor workers. *Environ. Health Perspect.* **60**, 165–184.

Levin, E. D. (1993). Development of treatments for toxicant-induced cognitive deficits. *Neurotoxicol. Teratol.* **15**, 203–206.

Levin, E. D., Schantz, S. L., and Bowman, R. E. (1988). Delayed spatial alteration deficits resulting from perinatal PCB exposure in monkeys. *Arch. Toxicol.* **62**, 267–273.

McGeer, P. L., Eccles, J. C., and McGeer, E. G. (1978). Catecholamine neurons. In *Molecular Neurobiology of the Mammalian Brain*, pp. 233–293, Plenum Press, New York.

Mullin, M. D., Pochini, C. M., McCrindle, S., Romkes, M., Safe, S. H., and Safe, L. M. (1984). High-resolution PCB analysis: synthesis and chromatographic properties of all 209 PCB congeners. *Environ. Sci. Technol.* **18**, 468–476.

National Research Council (1979). *Polychlorinated Biphenyls.* National Academy of Sciences, Washington, D.C.

Ness, D. K., Schantz, S. L., Mostaghian, J., and Hansen, L. G. (1993). Effects of perinatal exposure to specific PCB congeners on thyroid hormone concentrations and thyroid histology in the rat. *Toxicol. Letts.* **68**, 311–323.

Pantaleoni, G., Fanini, D., Sponta, A. M., Palumbo, G., Giorgi, R., and Adams, P. M. (1988). Effects of maternal exposure to polychlorinated biphenyls (PCBs) on F1 generation behavior in the rat. *Fundam. Appl. Toxicol.* **11**, 440–449.

Parkinson, A., and Safe, S. (1987). Mammalian biologic and toxic effects of PCBs. In *Polychlorinated Biphenyls (PCBs): Mammalian and Environmental Toxicology* (S. Safe and O. Hutzinger, Eds.), pp. 49–75, Springer-Verlag, New York.

Puymirat, J. (1985). Effects of dysthyroidism on central catecholaminergic neurons. *Neurochem. Int.* **7**, 969–977.

Rogan, W. J., Gladen, B. C., McKinney, J. D., Carreras, N., Hardy, P., Thullen, J. D., Tinglestad, J., and Tully, M. (1986). Neonatal effects of transplacental exposure to PCBs and DDE. *J. Pediatr.* **109**, 335–341.

Rogan, W. J., Gladen, B. C., Hung, K. L., Koong, S. L., Shih, L. Y., Taylor, J. S., Wu, Y. C., Yang, D., Rogan, N. B., and Hsu, C. C. (1988). Congenital poisoning by polychlorinated biphenyls and their contaminants in Taiwan. *Science* **241**, 334–336.

Rogan, W. J., and Gladen, B. C. (1992). Neurotoxicology of PCBs and related compounds. *Neurotoxicology* **13**, 27–36.

Safe, S. (1990). Polychlorinated biphenyls (PCBs), dibenzo-*p*-dioxins (PCDDs), dibenzofurans (PCDFs), and related compounds: environmental and mechanistic considerations which support the development of toxic equivalency factors (TEFs). *CRC Crit. Rev. Toxicol.* **21**, 51–88.

Sawaguchi, T., Matsumura, M., and Kubota, K. (1988). Dopamine enhances the neuronal activity of spatial short-term memory task in the primate prefrontal cortex. *Neurosci. Res.* **5**, 465–473.

Sawaguchi, T., Matsumura, M., and Kubota, K. (1989). Delayed response deficits produced by local injection of bicuculline into the dorsolateral prefrontal cortex in Japanese macaque monkeys. *Exp. Brain Res.* **75**, 457–469.

Sawyer, T., and Safe, S. (1982). PCB isomers and congeners: Induction of aryl hydrocarbon hydroxylase and ethoxyresorufin *O*-deethylase enzyme activities in rat hepatoma cells. *Toxicol. Letts.* **13**, 87–94.

Schantz, S. L., Levin, E. D., and Bowman, R. E. (1991). Long-term neurobehavioral effects of perinatal polychlorinated biphenyl (PCB) exposure in monkeys. *Environ. Toxicol. Chem.* **10**, 747–756.

Schwartz, P. M., Jacobson, S. W., Fein, G. G., and Jacobson, J. L. (1983). Lake Michigan fish consumption as a source of polychlorinated biphenyls in human cord serum, maternal serum, and milk. *Am. J. Public Health* **73**, 293–296.

Seegal, R. F. (1992). Perinatal exposure to Aroclor 1016 elevates brain dopamine concentrations in the rat. *Toxicologist* **12**, 320.

Seegal, R. F. (1994). The neurochemical effects of PCB exposure are age-dependent. *Arch. Toxicol., Supplement 16: Use of Mechanistic Information on Risk Assessment* (H. M. Bolt, B. Hellman, L. Dencker, Eds.), pp. 128–137. Springer-Verlag, Berlin.

Seegal, R. F., Brosch, K., Bush, B., Ritz, M., and Shain, W. (1989). Effects of Aroclor 1254 on dopamine and norepinephrine concentrations in pheochromocytoma (PC-12) cells. *Neurotoxicology* **10**, 757–764.

Seegal, R. F., Brosch, K. O., and Bush, B. (1986a). Regional alterations in serotonin metabolism induced by oral exposure of rats to polychlorinated biphenyls. *Neurotoxicology* **7**, 155–166.

Seegal, R. F., Brosch, K. O., and Bush, B. (1986b). Polychlorinated biphenyls produce regional alterations of dopamine metabolism in rat brain. *Toxicol. Letts.* **30**, 197–202.

Seegal, R. F., Bush, B., and Brosch, K. O. (1985). Polychlorinated biphenyls induce regional changes in brain norepinephrine concentrations in adult rats. *Neurotoxicology* **6**, 13–24.

Seegal, R. F., Bush, B., and Shain, W. (1990). Lightly chlorinated ortho-substituted PCB congeners decrease dopamine in nonhuman primate brain and in tissue culture. *Toxicol. Appl. Pharmacol.* **106**, 136–144.

Seegal, R. F., Bush, B., and Brosch, K. O. (1991a). Comparison of effects of Aroclors 1016 and 1260 on nonhuman primate catecholamine function. *Toxicology* **66**, 145–163.

Seegal, R. F., Bush, B., and Brosch, K. O. (1991b). Sub-chronic exposure of the adult rat to Aroclor 1254 yields regionally-specific changes in central dopaminergic function. *Neurotoxicology* **12**, 55–66.

Seegal, R. F., Bush, B., and Shain, W. (1991c). Neurotoxicology of ortho-substituted polychlorinated biphenyls. *Chemosphere* **23**, 1941–1949.

Seegal, R. F., Bush, B., and Brosch, K. O. (1994). Decreases in dopamine concentrations in adult non-human primate brain persist following removal from polychlorinated biphenyls. *Toxicology* **86**, 71–87.

Seegal, R. F., and Shain, W. (1992). Neurotoxicity of polychlorinated biphenyls: the role of ortho-substituted congeners in altering neurochemical function. In *The Vulnerable Brain and Environmental Risks*, Vol. 2, *Toxins in Food* (R. L. Isaacson and K. F. Jensen, Eds.), pp. 169–191, Plenum Press, New York.

Shain, W., Bush, B., and Seegal, R. F. (1991). Neurotoxicity of polychlorinated biphenyls: structure-activity relationship of individual congeners. *Toxicol. Appl. Pharmacol.* **111**, 33–42.

Sundstrom, G., Hutzinger, O., and Safe, S. (1976). The metabolism of chlorobiphenyls - a review. *Chemosphere* **5**, 267–298.

Tilson, H. A., Davis, G. J., McLachlan, J. A., and Lucier, G. W. (1979). The effects of polychlorinated biphenyls given prenatally on the neurobehavioral development of mice. *Environ. Res.* **18**, 466–474.

Vaccari, A., Rossetti, Z. L., De Montis, G., Stefanini, E., Martino, E., and Gessa, G. L. (1990). Neonatal hypothyroidism induces striatal dopaminergic dysfunction. *Neuroscience* **35**, 699–706.

van den Berg, K. J., Zurcher, C., and Brouwer, A. (1988). Effects of 3,4,3′,4′-tetrachlorobiphenyl on thyroid function and histology in marmoset monkeys. *Toxicol. Letts.* **41**, 77–86.

Van Hartesveldt, C., and Joyce, J. N. (1986). Effects of estrogen on the basal ganglia. *Neurosci. Biobehav. Rev.* **10**, 1–14.

Yoshimura, T., and Ikeda, M. (1978). Growth of school children with polychlorinated biphenyl poisoning or yusho. *Environ. Res.* **17**, 416–425.

CHAPTER

18

Bases of Excitatory Amino Acid System-Related Neurotoxicity

HUGH A. TILSON
Neurotoxicology Division
Health Effects Research Laboratory
U.S. Environmental Protection Agency
Research Triangle Park, North Carolina 27711

WILLIAM R. MUNDY
Neurotoxicology Division
Health Effects Research Laboratory
U.S. Environmental Protection Agency
Research Triangle Park, North Carolina 27711

I. Introduction

L-Glutamate is a common amino acid, existing in millimolar concentrations in the cytoplasm of most vertebrate neurons (Mayer and Westbrook, 1987). A number of biological functions, including intermediary metabolism in neural tissues and detoxification of ammonia in the brain, involve glutamate. In addition, glutamate is a constituent in the synthesis of proteins and peptides and the precursor for γ-aminobutyric acid, an inhibitory neurotransmitter. There are several precursors for glutamate synthesis and multiple metabolic pools of glutamate in the central nervous system.

In the brain, glutamate is found in higher levels than any other amino acid. Generally, only about 20% of the total amount is localized in nerve terminals, although in areas with dense glutamatergic innervation, the trans-

mitter pool may be over 40% (Greenamyre, 1986). The releasable pool of L-glutamate is separate from the metabolically active pool. After L-glutamate is released, it is removed from the synapse by a high-affinity uptake system located on nerve terminals and astrocytes. Some L-glutamate may be processed to form glutamine in astrocytes, which is then released into the extracellular space where it is taken up by nerve terminals and converted to glutamate. Curtis and Watkins (1963) conclusively showed that L-glutamate has excitatory or stimulatory effects on neurons, and subsequent research has indicated that L-glutamate fulfills several criteria for a neurotransmitter, including Ca^{2+}-dependent release upon stimulation, high-affinity uptake into nerve terminals, presence of the amino acid and synthetic enzymes in nerve terminals, blockade of synaptic transmission by glutamate receptor antagonists, and cellular and molecular effects of receptor activation (Cotman *et al.,* 1987).

Excitatory amino acid (EAA) neurotransmitter systems are almost exclusively localized in the brain and spinal cord (Table 1). Many pathways in the brain are believed to use glutamate as the neurotransmitter, in-

This manuscript has been reviewed by the Health Effects Research Laboratory, U.S. Environmental Protection Agency, and approved for publication. Mention of trade names or commercial products does not constitute endorsement or recommendation for use.

359

TABLE 1   Examples of Pathways Utilizing Glutamate as
the Neurotransmitter

Allocortical–CA3 hippocampal to lateral septum
Auditory nerve
Cerebellum–granule cells, climbing fibers
Corticostriatal pathway–caudate, substantia nigra
Hippocampus–entorhinal cortex to hippocampus, Schaffer
  collaterals within hippocampus
Retina–photoreceptor cell, bipolar cells, rods
Thalamus–ventral lateral thalamic nuclei from cortex
Vagus nerve–afferents of arterial baroreceptors
Visual cortex–ipsilateral to lateral geniculate

---

cluding afferent, intrinsic, and efferent hippocampal pathways; cerebellar pathways; corticocortical association fibers and corticofugal fibers from the cortical pyramidal neurons to the basal ganglia and thalamus; spinal cord pathways, such the substantial gelatinosa of the dorsal horn; intrinsic pathways of the striatum; and visual, olfactory, and auditory sensory systems (Herrling, 1992; Watkins and Evans, 1981).

At least five subtypes of receptors are sensitive to EAA (Fig. 1) (cf. Farooqui and Horrocks, 1991; Herrling, 1992). Subtypes of EAA receptors have been categorized according to their effects on ionic balance (ionotropic) or second messenger systems (metabotropic), sensitivity to receptor agonists and antagonists, and/or location pre- or postsynaptically (Table 2).

The *N*-methyl-D-aspartate (NMDA) receptor is activated or gated by the transmitter in a voltage-

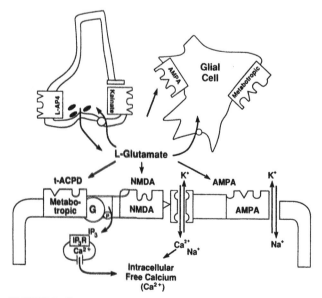

FIGURE 1   Type and putative location of various subtypes of excitatory amino acid receptors.

dependent manner (Farooqui and Horrocks, 1991). (Fig. 2). The NMDA receptor consists of several domains, including the transmitter recognition site, a cation-binding site located inside the channel where magnesium ions can bind and block ion fluxes, a phencyclidine (PCP)-binding site which requires agonist binding to the transmitter recognition site and interacts with the cation-binding site, and glycine-binding site which appears to allosterically modulate the interaction between the transmitter and the recognition and PCP-binding sites. Radioligand studies have also suggested the presence of a polyamine-binding site in the NMDA receptor complex, whereas Zn^{2+} appears to act as an inhibitory modulator of channel function at a separate site near the mouth of the ion channel. The NMDA receptor is ionotropic, permitting an intracellular influx of Ca^{2+}, as well as an exchange of Na^+ and K^+ across the cell membrane. 2-Amino-5-phosphonovalerate (D-AP5) competes directly with L-glutamate for the recognition-binding site (competitive inhibitor), whereas dibenzocyclohepteneimine (MK-801) binds to a channel site following activation of the recognition site with L-glutamate (noncompetitive inhibitor). Cycloleucine and kynurenate antagonize glycine at the glycine-binding site (Cotman *et al.*, 1989).

A second EAA receptor has been found to be selectively sensitive to α-2-amino-3-hydroxy-5-methylisoxazole-4-proprionate (AMPA), which appears to be involved in fast excitatory synaptic transmission (Young *et al.*, 1991). AMPA receptors are voltage independent and ionotrophic, permitting exchange of Na^+ and K^+ when activated. Competitive AMPA receptor antagonists are 6-cyano-7-nitroquinoxaline (CNQX) and 6,7-dinitroquinoxaline-2,3-dione (DNQX) (Farooqui and Horrocks, 1991). AMPA receptors have been found on glial cells, as well as on neurons.

The kainate receptor is also ionotrophic and, although similar to the AMPA receptor, is differentially sensitive to domoate. The kainate receptor is selectively inhibited by 2-amino-3[-(carboxymethoxy)-5-methylisoxazol-4-yl]proprionic acid (AMOA) and 2-amino-3-[2-(3-hydroxy-5-methylisoxazol-4-yl]-proprionic acid (AMNH).

The AP4 receptor is characterized pharmacologically by the inhibitory effect of 2-amino-4-phosphonobutyrate (L-AP4) on synaptic potentials evoked both in spinal cord and hippocampus, suggesting a presynaptic localization that regulates the release of EAAs (Herrling, 1992). There is no known AP4 receptor antagonist and the physiological consequences of receptor activation are not fully understood (cf. Farooqui and Harrooks, 1991).

The fifth EAA receptor type that has been identified thus far is the metabotropic receptor in which the

TABLE 2   Characteristics of EAA Receptor Subtypes

Type	Receptor Class	Agonist	Antagonist	Ion permeability
Ionotropic	NMDA	NMDA, L-Glu, L-Asp, polyamines	D-AP5 (competitive) MK801 and ketamine (noncompetitive) Kynurenate (glycine site)	Ca^{2+}, K^+, Na^+
Inotropic	AMPA	AMPA, L-Glu quisqulate	CNQX, DNQX	Na^+, K^+
Ionotropic	Kainate	KA, L-Glu domoate	AMNH, AMOA	Na^+, K^+
Ionotrophic	AP4	L-AP4, L-Glu	Unknown	Unknown
Second Messenger	Metabotropic	t-ACPD, L-Glu, L-BMAA, ibotenate quisqualate	L-AP3, L-AP4	None

agonist-binding site is not coupled to an ion channel, but rather to a second messenger system. Evidence from molecular biology indicates at least five subtypes of metabotropic glutamate receptors, some of which are coupled negatively to cAMP and some to stimulation of phosphoinositide metabolism (cf. Farooqui and Horrocks, 1991). Metabotropic receptor agonists include *trans*-1-aminocyclopentyl-1,3-dicarboxylate (t-ACPD), ibotenic acid, and β-*N*-methylamino-L-alanine (L-BMAA). L-2-Amino-3-phosphonopropanoic acid (L-AP3) is a known metabotropic receptor antagonist (Herrling, 1992). Like AMPA receptors, metabotropic receptors may be located on glial cells, as well as on neurons.

Autoradiographic-binding techniques with selective ligands for the various EAA receptor subtypes have permitted a quantitative determination of the relative distribution of EAA receptors in the brain (Young *et al.*, 1991). Table 3 indicates that the AMPA receptor is found in relatively high concentrations in the cerebral cortex, molecular layer of the cerebellum, CA1 and CA3 regions of the hippocampus, striatum, and nucleus accumbens. Kainate receptors are found in relatively high concentrations in the CA3 region of the hippocampus, whereas high concentrations of NMDA receptors are seen in the cerebral cortex and CA1 region of the

hippocampus. Most of the area CA3 and dentate gyrus have moderate levels, whereas the mossy fiber termination zone is low in NMDA receptors. The highest concentration of metabotropic receptors is seen in the molecular layer of the cerebellum. Little information is available concerning the distribution of the AP4 receptor.

Some EAA synapses appear to function via a coordinated action of NMDA and non-NMDA receptors. Non-NMDA receptors, such as the AMPA type, generate fast depolarizing responses that then facilitate voltage-dependent NMDA receptor-gated currents. NMDA-induced currents are slower and involve influx of Ca^{2+} (Cotman *et al.*, 1989). Two hypotheses have been made concerning the localization of EAA receptors at the synaptic level (Herrling, 1992). The first model suggests that the synaptic membrane contains both NMDA- and non-NMDA receptors. At low levels of synaptic activity, relatively low levels of the neuro-

TABLE 3   Relative Distribution of EAA Receptor Subtypes in the Brain[a]

Region	Relative binding concentration			
	AMPA	Kainate	NMDA	Metabotropic
Cerebral cortex	+ + +	+	+ +	+
Cerebellum				
Molecular	+ + +	+/−	+/−	+ +
Granule	+	+	+	+
Hippocampus				
CA1	+ + + +	+/−	+ +	+
CA3	+ + +	+ + +	+	+
Striatum	+ + +	+	+	+
Nucleus accumbens	+ + +	+	+	+
Globus palladus	+	+/−	+/−	+/−
Ventral pallidum	+	+	+	+
Substantia nigra	+	+/−	+/−	+/−
Subthalamus	+	+/−	+/−	+
Pons	+/−	+/−	+/−	+/−

[a] From Young *et al.* (1991).

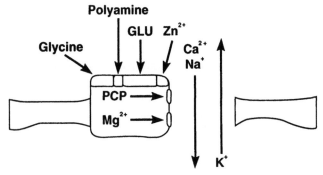

FIGURE 2   Principal working components of the NMDA receptor complex.

transmitter L-glutamate are released, which binds to both types of EAA receptors. It is assumed that magnesium ions may be present in sufficient quantities to block the consequences of low-level NMDA receptor activation, whereas activation of non-NMDA receptors causes ion channels to open and Na^+ influx to occur, resulting in depolarization of the membrane. At higher presynaptic activity, the membrane is depolarized enough to remove voltage-dependent magnesium block, causing Ca^{2+}, as well as Na^+, to flow into the cell.

Herrling (1992) also describes a second model, which places non-NMDA receptors in the subsynaptic membrane. These receptors are responsible for fast excitatory transmission. NMDA receptors are thought to be located either in a separate synapse or extrasynaptically. Activation of non-NMDA receptors causes fast excitatory postsynaptic potentials. If the same synaptic activity occurs when either synaptic or nonsynaptic NMDA receptors are activated, the resulting excitatory postsynaptic potentials will be potentiated. Additional research is required to determine if these are mutually exclusive models of the EAA receptors or whether these two models exist in some proportion in various parts of the nervous system.

Recent research involving molecular cloning has found that the vast majority of EAA receptors in the brain are NMDA, AMPA–kainate, and metabotropic receptors. It is likely that glutamatergic transmission in different synapses is mediated through distinct receptors and combinations of different EAA receptors. AMPA–kainate receptors appear to mediate fast, voltage-independent synaptic responses and eventually promote voltage-dependent NMDA receptor activation (Nakanishi, 1992).

II. Functional Significance of EAA Receptors

A. Neurobiological Functions

Research has demonstrated that EAA receptor systems play an important role in several critical neurobiological functions, including long-term potentiation (LTP), learning and memory, and developmental plasticity. LTP is a persistent enhancement of the postsynaptic response following a short epoch of high-frequency stimulation applied to the same group of fibers (Collingridge and Bliss, 1987). It has been hypothesized that LTP may underlie learning and memory. NMDA receptor antagonists applied prior to high-frequency stimulation prevent the development, but not maintenance, of LTP, suggesting that NMDA re-

ceptor activation is critical for the initiation of LTP (Cotman *et al.*, 1989). Whether LTP is presynaptic or postsynaptic is a matter of intense research and debate. It has been postulated that high-frequency stimulation depolarizes postsynaptic neurons, removing the magnesium block of NMDA receptors. This would lead to the activation of the NMDA receptor complex and an influx of Ca^{2+}. Increased intracellular Ca^{2+} activates a number of calcium-mediated processes, such as Ca^{2+}/ calmodulin kinase II or protein kinase C. Increased calcium-mediated processes may increase postsynaptic responsiveness associated with LTP. Some work suggests that the early development of LTP may be presynaptic (Davies *et al.*, 1989). Evidence also exists that LTP may not be NMDA mediated in all pathways, such as the mossy fiber to the CA3 pathway in the hippocampus (Nicoll *et al.*, 1988).

LTP has been proposed as a neurochemical/neurophysiological model of learning and memory (Collingridge and Bliss, 1987). Since NMDA receptors appear to play an important role in LTP, it has been postulated that NMDA receptor activation may play an important role in learning and memory (Bliss and Lynch, 1988). Areas of the brain that are involved in learning and memory, such as the hippocampus and cortex, also express LTP and have a high concentration of NMDA receptors (cf. Willner *et al.*, 1992). Research has found that NMDA receptor antagonist can block the acquisition of behavioral tasks. Morris and colleagues (1986) first reported that the intraventricular application of AP-5 interfered with the ability of rats to learn a spatial memory task in a water maze. The systemic administration of MK-801 can also selectively interfere with the acquisition of behavioral tasks (Robinson *et al.*, 1989). However, it is not clear that NMDA receptor activation underlies acquisition of all tasks since MK-801 was found to interfere with taste-potentiated odor aversions but had no effect on the acquisition of taste–illness or odor–shock aversions (Robinson *et al.*, 1989). More research is needed to determine the precise role of NMDA and non-NMDA EAA receptor activation in learning and memory, as well as different cognitive functions mediated in various regions of the central nervous system.

It has been suggested that the NMDA receptor complex may also play an important role in developmental plasticity. NMDA receptors appear to be crucial for determining the relative positions of synaptic inputs during early development (Cline *et al.*, 1987) and may be a prerequisite for the formation of ocular dominance columns in the visual cortex (Klienschmidt *et al.*, 1987). Other research has suggested that NMDA may have a trophic influence on developing neurons. NMDA has been reported to enhance neurite outgrowth and cell survival in tissue culture (Pearce *et al.*, 1987). NMDA receptor activation might also mediate programmed cell loss that is present during normal development (Cotman *et al.*, 1989).

B. Neurological and Degenerative Disorders

Other research has suggested that overactivation of EAA receptors may produce excitotoxicity, leading to cell death, and may be implicated in several neurological disorders, including epilepsy (Sloviter, 1983), Huntington's disease (Coyle and Schwarcz, 1976), olivopontocerebellar atrophy (Plaitakis *et al.*, 1982), sulfite oxidase deficiency (Olney *et al.*, 1975), stroke (Rothman, 1984), hypoglycemic encephalopathy (Wieloch, 1985), Alzheimer's disease (Advokat and Pellegrin, 1992), Guam-type amyotrophic lateral sclerosis/parkinsonism dementia (ALS/PD) (Spencer *et al.*, 1987), and lathyrism (Spencer and Schaumburg, 1983; Spencer *et al.*, 1986) (Table 4).

Epilepsy is characterized by abnormal synchronization and amplification of neuronal firing. Neurochemical studies have indicated that glutamate levels in the brain increase during epilepsy (Nadi *et al.*, 1987). Direct infusion of glutamate and its analogs can produce seizures and neurohistophathological changes similar to those in the brains of patients with epilepsy (Olney *et al.*, 1986). Development of epilepsy appears to be a disruption of the normal responsiveness of EAA receptor activation. Several studies suggest that EAA receptors may affect phospholipid metabolism in neural membranes and contribute to epilepsy (cf. Farooqui and Horrocks, 1991). Activation of kainate/AMPA subtype receptors is associated with initial synchronization of neuronal firing, whereas activation of NMDA receptors is associated with dendritic depolarization, triggering repetitive calcium spikes, and burst-firing of the soma. Disease may modify this sequence of events leading to synchronized bursts of neuronal firing or epilepsy. Cell death associated with epilepsy may be related to an overstimulation of glutamate receptors

TABLE 4 Neurological Disorders Possibly Related to Overstimulation of Excitatory Amino Acid Receptors

Alzheimer's disease
Epilepsy
Guam-type amyotrophic lateral sclerosis/parkinsonism dementia
Huntington's disease
Hypoglycemic encephalopathy
Olivopontocerebellar atrophy
Stroke
Sulfite oxidase deficiency

and an accumulation of intracellular calcium (Wyler *et al.*, 1987).

Huntington's disease is an autosomal dominant neurodegenerative disorder characterized by progressive neurological impairments consisting of chorea, psychiatric symptoms, and dementia (Pearson and Reynolds, 1992). Neurodegenerative changes in the brain have been observed in Huntington's disease, including selective neuronal loss in the basal ganglia, cortex, and limbic system. It is been proposed that a neurotoxic metabolite of tryptophan, quinolinic acid, may be produced in Huntington's disease (Schwarcz *et al.*, 1984). It has been suggested that quinolinic acid may act on NMDA receptors since it has been reported that NMDA receptors are affected significantly in the striatum of Huntington's disease patients (Young *et al.*, 1988). Recent work, however, has found that quinolinic acid concentrations are not significantly elevated in the striatum of individuals with Huntington's disease (Reynolds *et al.*, 1988). Other work suggests that a further intermediate in kynurenine metabolism, 3-hydroxykynurenine, is neurotoxic (Pearson and Reynolds, 1992) and may play a role in Huntington's disease.

Patients with olivopontocerebellar degeneration have been found to have an impairment in glutamate metabolism. Because of a deficiency of glutamic dehydrogenase, patients that ingest glutamate via their diet experience abnormally high levels of glutamate, which may cause neurotoxicity (Plaitakis *et al.*, 1982). Olney and colleagues (1975) reported that a sulfite oxidase deficiency may be related to a neurological disorder consisting of blindness, spastic quadriplegia, and death in early infancy. A deficiency in sulfite oxidase could lead to an accumulation of cysteine-*S*-sulfate, which has been shown to be excitotoxic in rats (Olney *et al.*, 1975).

Decreased availability of oxygen is associated with a number of neurochemical changes, including the release of free fatty acids from various phospholipids and the accumulation of lysophopholipids and diacylglycerols (Farooqui and Horrocks, 1991). Studies using microdialysis techniques have shown that extracellular levels of glutamate increase during ischemia (Globus *et al.*, 1988), which may lead to overexcitation of EAA receptors and neurotoxicity (Choi, 1990). Glutamate neurotoxicity appears to have three phases, including induction or overstimulation of EAA receptors and rapid cellular alterations, amplification or secondary changes promoted by the initial overstimulation of EAA receptors, and expression or destructive changes causally related to neuronal cell degeneration and death. Overstimulation of NMDA and non-NMDA receptors is implicated in the brain damage induced by focal brain ischemia, as well as hypoglycemia and mechanical trauma and spinal cord injury (Choi, 1988; Albers *et al.*, 1989) and dementia pugilistica (Olney, 1990).

With regard to Alzheimer's disease, Maragos and associates (1987) have proposed that glutamatergic overstimulation due to excessive synthesis or release of glutamate, decreased inhibition of glutamatergic neurons, or impaired reuptake or catabolism of glutamate can produce an excitotoxic effect leading to neuronal degeneration and the eventual formation of neurofibrillary tangles. Advokat and Pellegrin (1992) have also suggested that disease-associated degeneration in neurotransmitter systems may be associated with an increase in L-glutamate release which could be neurotoxic. On the other hand, others (Hardy and Cowburn, 1987) have suggested that Alzheimer's disease may be related to glutamatergic understimulation. The mechanism(s) associated with glutamatergic-induced neurotoxicity and the development of neurofibrillary tangles, a hallmark of Alzheimer's disease, is not completely understood.

Guam-type amyotrophic lateral sclerosis/parkinsonism dementia is a neurodegenerative disease of the motor neurons and corticospinal tract (Garruto and Yase, 1986) characterized by muscle weakness, progressive atrophy of the muscles, paralysis, and spasticity. Increased plasma glutamate levels have been reported in patients with ALS/PD (Plaitakis and Caroscio, 1987). Furthermore, the diet of patients with ALS/PD has been found to contain a nonprotein amino acid, L-BMAA (Spencer *et al.*, 1987). Feeding studies with nonhuman primates have demonstrated neurological dysfunction similar to that observed in humans with guam ALS/PD (Spencer *et al.*, 1987). *In vitro* studies using cultured mouse cortical neurons have found that the neurotoxicity produced by BMAA was substantially decreased by pretreatment with D-APV, a competitive NMDA receptor antagonist (Weiss *et al.*, 1989). The *in vivo* neurotoxicity of BMAA has also been reported to be blocked by AP7, a NMDA receptor antagonist (Ross and Spencer, 1987).

Spencer and colleagues (1986; Spencer and Schaumburg, 1983) have reported that consumption of large quantities of the chick pea (*Lathyrus sativus*) is associated with a paralytic disease known as lathyrism. β-N-Oxalylamino-L-alanine (BOAA) has been isolated from the seeds of *L. sativus* and appears to be an excitotoxicant; subsequent studies have found that BOAA-induced neurotoxicity in cortical neurons *in vitro* was not selectively blocked by D-APV (Weiss *et al.*, 1989). Kynurenate, which antagonizes non-NMDA receptors, attenuated BOAA-induced neurotoxicity, suggesting that BOAA differs from BMAA by acting on non-NMDA receptors. This conclusion is supported by the results of Ross and Spencer (1987) who reported that the neurotoxicity of BOAA is blocked by *cis*-2,3-piperadine dicarboxylic acid, an antagonist of quisqualate and kainate-preferring glutamate receptors.

III. Mechanism(s) of EAA-Induced Neurotoxicity

Glutamate-induced neurotoxicity in the mouse retina was first reported by Lucas and Newhouse (1957). Olney and colleagues (1971) were the first to propose that glutamate neurotoxicity is associated with overstimulation of EAA receptors. Choi and colleagues (Choi, 1987; Choi *et al.*, 1987) have used the dissociated murine cortical cell culture preparation to show the excitotoxic effects of L-glutamate and related EAA agonists and the ability of competitive and noncompetitive NMDA receptor antagonists to protect against the excitotoxicity. Several factors contribute to the expression of EAA-induced neurotoxicity, including the density of EAA receptors, the mix of EAA receptors present in the preparation, and the efficiency of high-affinity transport mechanisms to clear EAA from the synapse.

Research on the mechanism of EAA-induced neurotoxicity suggests that two separate processes may be involved (Farooqui and Horrocks, 1991; Cotman *et al.*, 1989). Excessive depolarization resulting from excess synaptic EAA results in an accumulation of high levels of Na^+ ions (Fig. 3). Associated with the influx of Na^+ is increased intracellular concentrations of Cl^- and water, which initiates osmotic damage. Generally, this

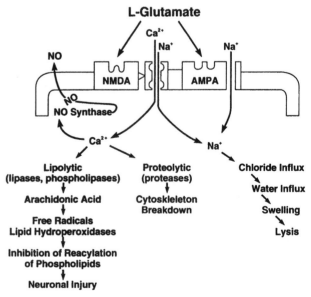

FIGURE 3 Putative mechanism(s) of excitotoxicity.

process is reversible if the EAA is removed from the system.

High synaptic levels of EAA also increase the activity of voltage-dependent NMDA receptors, which could lead to an increase in intracellular Ca^{2+} and eventual cell death (Nicotera *et al.*, 1992). Increased intracellular Ca^{2+} affects the activity of lipolytic and proteolytic enzymes. Activation of lipolytic enzymes results in the synthesis of prostaglandins, leukotrienes, and thromboxanes and promotes the formation of free radicals and lipid hydroperoxides, which inhibit reacylation of phospholipids in neuronal membranes and could lead to neuronal injury and death. Ca^{2+}-induced protease activity may cause the breakdown of the cytoskeleton, leading to cellular damage.

Recent evidence has suggested that NMDA receptor-mediated Ca^{2+} influx into neurons can result in the direct activation of nitric oxide (NO) synthase (Kiedrowski *et al.*, 1992). NO can diffuse across neuronal membranes and act as a second messenger, affecting the activity of enzymes such as guanylate cyclase to increase cyclic GMP content. It has been postulated that NO may serve as an intracellular messenger by regulating enzymes such as guanylate cyclase which may be important in processes such as synaptic plasticity and excitotoxicity. Dawson and associates (1991) reported that NO may mediate NMDA-induced neurotoxicity in primary cultures of rat cerebral cortical neurons. Data published by Pauwels and Leysen (1992) suggest that NO may be neurotoxic in an indirect way by activation of astrocytes. Certainly, more research is needed to identify and characterize the complex cellular and molecular changes resulting from overactivation of EAA receptors leading to cell damage and death.

IV. EAA Overstimulation: The Hippocampus and Cognitive Dysfunction

A. *Direct Central Application of NMDA*

Although cognitive dysfunction resulting from overstimulation of EAA receptors has been suspected in several neurodegenerative diseases in humans (Olney, 1990; Farooqui and Horrocks, 1991), it has only been recently that the effects of EAA overactivation were studied systematically in animal models (Schwarz *et al.*, 1984). This is particularly true with regard to overstimulation of NMDA receptors, which is thought to play a crucial role in the mechanism of EAA-induced excitotoxicity.

Animal studies of EAA-induced memory dysfunction have typically employed stereotaxic application of EAA agonists into regions of the brain known to play

a crucial role in cognitive function. One such area is the hippocampus, which is intimately involved in learning and memory (O'Keefe and Nadel, 1978). Characterization of EAA-containing pathways in the hippocampus has also been well-studied (Fonnum *et al.*, 1983; Young *et al.*, 1991). The main EAA input to the hippocampus arises from the entorhinal cortex via the perforant pathway which terminates on the dendrites of the granule cells in the area dentata and the molecular layer of the hippocampus. The Schaffer collaterals of the hippocampal pyramidal neurons utilize EAAs and terminate mainly in the stratum radiatum of the CA1 area of the hippocampus, whereas mossy fibers from dentate granule cells innervate pyramidal cells in the CA3 region of the hippocampus. NMDA receptors have been reported in high concentrations in the CA1 region of the hippocampus, whereas kainate receptors predominate in the CA3 region (Table 3). AMPA receptors occur in relatively high concentrations in both the CA1 and CA3 regions (Young *et al.*, 1991).

Impairment of memory following direct application of kainic or ibotenic acid into the hippocampus was reported by several investigators in the early 1980s (Handelman and Olton, 1981; Jarrard *et al.*, 1984; Sutherland *et al.*, 1981). Rogers and colleagues (1989a), however, were the first to report that bilateral infusion of NMDA into the dorsal hippocampus interfered with the ability of rats to acquire a spatial memory task in a water maze when tested 4 weeks after treatment. These behavioral effects were observed in animals having significant cell loss in the area near the site of the injection. Rogers and colleagues (1989a) also reported that application of NMDA to the surface of the cortex, which resulted in significant tissue destruction, had no effect on learning capability.

In another study, Rogers and Tilson (1989) reported that the neurobehavioral and histological effects produced by intrahippocampal infusion of NMDA could be blocked by pretreatment with systemic injections of MK-801, a noncompetitive NMDA receptor antagonist. In these studies, NMDA was applied stereotaxically into the dorsal hippocampus of rats; some rats were pretreated with various doses of MK-801 prior to NMDA application. Four weeks after NMDA treatment, rats were tested for the ability to learn a spatial task in a water maze. NMDA significantly affected acquisition in a water maze and resulted in neuronal cell loss near the site of the intrahippocampal injection. MK-801 produced a dose-related blockade of both the cognitive deficits and cell loss in the hippocampus.

Subsequent experiments (Rogers and Tilson, 1990a) found that coinfusion of 7-(D-2-aminophosphonoheptanoic acid (D-2-APH), a NMDA competitive receptor antagonist, with NMDA blocked the behavioral and histological effects of the excitotoxicant. These effects were found to be stereospecific as the less efficacious L-(+)-APH form afforded little protection against intrahippocampal infusion of NMDA. Other studies found that the noncompetitive NMDA receptor blocker MK-801 afforded much less protection against the behavioral and histological effects of intrahippocampal infusions of kainic acid (Rogers and Tilson, 1990b).

Rogers and colleagues (1989b) also investigated the effects of repeated stimulation of the perforant path on hippocampal neurotoxicity. This study was based on the theory that sustained electrical stimulation of the perforant path would cause an excess release of EAA, probably glutamate, in CA1 and CA3 regions of the hippocampus, resulting in cell death and subsequent cognitive deficits. This study found that rats undergoing perforant path stimulation were impaired in the ability to learn a spatial task in a water maze 3 weeks after stimulation. In addition, there was significant pyramidal cell loss in CA1 and CA3 regions of the hippocampus. Prior treatment with MK-801 attenuated the cell loss in CA1, whereas MK-801 was less protective against stimulation-induced CA3 pyramidal cell loss.

In summary, these studies support the hypothesis that excessive stimulation of NMDA receptors either by direct application of NMDA or by sustained stimulation of the perforant path results in cell destruction and behavioral changes. When the overstimulation of NMDA receptors occurs in a region of the brain such as the hippocampus, significant impairments of learning capabilities are observed. That the behavioral and histological effects of NMDA can be blocked by noncompetitive and competitive receptor antagonists specific for the NMDA receptor further supports the conclusion that overstimulation of EAA receptors as a consequence of disease or exposure to environmental agents can result in neuronal damage and cognitive dysfunction.

B. Tris(2-chloroethyl)phosphate (TRCP) as a Hippocampal Neurotoxicant

TRCP is a weak inhibitor of brain acetylcholinesterase, serum cholinesterase, and neurotoxic esterase (Eldefrawi *et al.*, 1977) and was used as a flame-retardant in plastics, polymeric foams, and synthetic fibers (Matthews *et al.*, 1990). Human exposure to TRCP is expected to occur in the workplace and as a result of trace exposure in the environment.

Experiments in animals have found that systemic administration of TRCP produces epileptiform convulsions (Clayton and Clayton, 1981), and recent toxicological studies on the effects of repeated exposure to TRCP in rats found a loss of pyramidal neurons in the

CA1 region of the hippocampus; other regions of the hippocampus were affected to a lesser extent (Matthew *et al.*, 1990). This pattern of cell loss in the hippocampus is consistent with transient forebrain ischemia in rats (Volpe *et al.*, 1984; Davis *et al.*, 1986).

In experiments designed to characterize the neurotoxicological effects of TRCP, Tilson *et al.* (1990) gave rats convulsive doses of TRCP and found consistent loss of CA1 hippocampal pyramidal cells. Less extensive effects were observed on pyramidal cells in the CA3 and CA4 regions and on granule cells in the dentate gyrus. The hippocampal damage produced by systemic administration of TRCP was associated with impairment in the performance of a working memory task in a water maze. Similar types of memory deficits have been reported in rats following transient ischemia and resultant CA1 hippocampal pyramidal cell damage (Davis *et al.*, 1986; Volpe *et al.*, 1984). Hippocampal CA1 pyramidal neurons are selectively vulnerable to short periods of global ischemia and contain a high number of NMDA receptors (Monaghan and Cotman, 1985).

The seizure-related and neuropathological effects of TRCP were significantly attenuated by pretreatment with atropine or chlordiazepoxide, suggesting that the hippocampal damage was related to the seizures. Subsequent work found that pretreatment with MK-801 to block NMDA receptors was only partially successful against TRCP-induced neurotoxicity. These results are similar to those reported in animal models of global ischemia where MK-801 pretreatment does not afford complete protection; MK-801 is generally believed to be more protective against focal ischemia (Choi, 1990). It has been suggested that overstimulation of NMDA receptors may be more involved in neurotoxicity produced by focal ischemia (Choi, 1990), whereas non-NMDA receptors may play a significant role in the pathogenesis of global ischemia. This conclusion is supported by data showing that 2,3-dihydroxy-6-nitro-7-sulfamoyl-benzo(F)quinoxaline (NBQX), a non-NMDA receptor antagonist, significantly decreases the neurotoxicity produced by global ischemia (Sheardown *et al.*, 1990).

In summary, the results are consistent with the possibility that TRCP produces a global ischemic insult associated with both NMDA and non-NMDA receptor activation. Additional research will be needed to elucidate more fully the mechanism of TRCP-induced neurotoxicity.

C. Hippocampal Neurotoxicity and Domoic Acid

During a 2-month period in 1987, over 100 Canadians reported gastrointestinal problems following ingestion of mussels grown in Prince Edward Island (Perl *et al.*, 1990). Some of the patients reported neurological signs and symptoms, including headache, confusion, disorientation, and memory loss. Neuropsychological testing of some patients months after acute exposure revealed severe anterograde memory loss (Teitelbaum *et al.*, 1990). There was also evidence of motor or sensorimotor neuronopathy and axonopathy, as well as decreased glucose metabolism in the medial temporal lobes. Teitelbaum and associates (1990) also found neuronal damage in the amygdala and hippocampus of brains from patients who died after ingestion of the mussels. It was suggested that the intoxication may be due to the presence of domoic acid, produced during an algal bloom and accumulated by the mussels (Perl *et al.*, 1990; Wright *et al.*, 1990). Domoic acid was isolated from *Chondria armata* as an anthelmintic in the 1950s (Daigo, 1958) and is an analog of L-glutamate.

Electrophysiological studies in cultured rat hippocampal neurons have shown that currents induced by domoic acid in cultured hippocampal neurons are essentially identical to those produced by kainic acid (Stewart *et al.*, 1990). Furthermore, domoic acid-induced currents are inhibited by the non-NMDA receptor antagonist CNQX, but are insensitive to the competitive NMDA antagonist D-AP5. Stewart and colleagues (1990) also reported that the pattern of degeneration in chick retina produced by domoic acid was identical to that of kainic acid and differed from the pattern produced by NMDA and quisqualic acid. Domoic acid was about seven times more potent than kainic acid in these studies. Furthermore, the noncompetitive receptor, antagonist MK801 did not block demoic acid-induced retinal degeneration, whereas CNQX prevented domoic acid neurotoxicity.

In vivo electrophysiological studies have also indicated that domoic acid is similar to kainic acid in its mechanism of action (Debonnel *et al.*, 1989). Unitary extracellular recordings obtained from CA1 and CA3 pyramidal cells in the rat hippocampus were used to compare domoic acid with other EAA agonists. In both the CA1 and CA3 regions, the activation of pyramidal cells by domoic acid was about 3 times more potent than that produced by kainic acid. Domoic acid and kainic acid were more than 20 times more potent in the CA3 than in the CA1 region. This difference in subregional sensitivity is consistent with the high density of kainate receptors in the CA3 region of the hippocampus. Terrian and colleagues (1991) have reported that domoic acid significantly increased K^+-evoked release of endogenous glutamate from superfused guinea pig mossy fiber synaptosomes, an effect antagonized by prior treatment with CNQX.

Stewart and colleagues (1990) have also reported that the seizure-brain damage syndrome induced by systemic administration of domoic acid is similar to that of kainic acid. Furthermore, domoic acid produces bilateral damage in numerous forebrain structures, while sparing the midbrain, hindbrain, and spinal cord. Stewart and colleagues (1990) also reported that domoic acid affected pyramidal neurons in CA1, while having less of an effect in CA3. Dentate gyrus neurons were relatively unaffected. With the exception of domoic acid-induced cytopathological changes in the neocortical areas, the pathological changes produced by domoic acid closely resemble those produced by kainic acid. Tryphonas and co-workers (1990) have also reported that the systemic administration of domoic acid developed selective encephalopathy characterized by neuronal degeneration in the limbic system. These authors also reported that domoic acid produced a retinopathy characterized by neuronal hydropic degeneration of the inner nuclear layer and vacuolation of the external plexiform layer.

Animal models of domoic acid-induced cognitive dysfunction have not been well-studied. Sutherland and colleagues (1990) reported that direct injection of domoic acid directly into the hippocampus of rats produced large decreases in pyramidal cells in CA3. Consistent with the distribution of kainate receptors in the hippocampus, intrahippocampal infusion of domoic acid had less of an effect on pyramidal cells in CA1. Sutherland and co-workers (1990) also reported that animals receiving intrahippocampal infusions of domoic acid were deficient in the acquisition of a spatial memory task in a water maze. Other investigators (Dakshinamurti *et al.*, 1991) have reported that direct application of domoic acid into the CA3 region of the hippocampus produces electrical seizure discharge activity which was reduced with local infusion of GABA. These data suggest that domoic acid might be related to decreased GABAergic inhibition.

In summary, considerable evidence exists to indicate that domoic acid is an environmentally relevant neurotoxicant. Humans ingesting contaminated mussels show significant memory loss and neuropathological changes in the hippocampus. Domoic acid has been isolated from the contaminated mussels and, in animal models, has been shown to be a kainic acid-like excitotoxicant.

V. Summary and Conclusions

Excitatory amino acids such as L-glutamate are involved in normal neurotransmission in the mammalian nervous system and are believed to be critical for many important neurobiological functions, including long-term potentiation, learning and memory, and developmental plasticity. Much is known about the distribution of the various types of EAA receptor subtypes, which include NMDA, L-AP4, t-ACPD or metabotropic, AMPA, and kainate receptors. The role that the various subtypes of receptors play in the normal functioning of the nervous system is currently being studied by several laboratories.

Evidence suggests that some neurodegenerative and neurological diseases may be related to abnormalities of EAA neurotransmission. The exact involvement of EAA in these disease processes is not known, although it has been postulated that excitotoxicity resulting from disease-related overstimulation of glutamate receptors may be a causative factor. Subsequent research has focused on the possible mechanism(s) by which glutamate receptor overstimulation causes excitotoxicity and eventual cell death. At the present time, it is generally believed that osmotic-related cell lysis and increases in intracellular calcium are critical events eventually leading to cell death. Research is currently underway to understand secondary and tertiary steps leading to excitotoxicity.

It has also been suggested that cognitive dysfunction associated with some neurodegenerative and neurological disorders is related to overstimulation of glutamate receptors in key areas of the brain. The involvement of the NMDA subtype of receptors seems likely given their critical role in long-term potentiation and learning and memory, as well as their relatively high distribution in regions of the brain known to be involved in cognitive function such as the hippocampus. *In vivo* studies have shown that direct application of NMDA into the hippocampus produces neuronal destruction and associated cognitive dysfunction. The neuroanatomical and neurobehavioral effects of NMDA are blocked by NMDA receptor blockers such as MK-801 and 2-(D)-APH. Cognitive dysfunction can also be produced by stimulation of a perforant path, a major glutamate-containing afferent pathway into the hippocampus.

Recent reports indicate that humans may be exposed to chemicals that are excitotoxicants, including agents such as tris(2-chloroethyl)phosphate, which is not chemically related to glutamate, and domoic acid, which is structurally similar to kainic acid. In animal models, these agents produce cognitive dysfunction and neuronal loss in the hippocampus. The extent to which humans are exposed to other chemicals that are excitotoxicants is, of course, not known. It is, however, noteworthy that recent research has found that animals exposed to lead during development are hypersensitive to systemic administration of NMDA (Petit *et al.*, 1992). Cohn and Cory-Slechta (1994) reported

that rats exposed to lead during development were differentially sensitive to the effects of systemically administered NMDA on a multiple schedule of repeated learning and performance. Cohn and Cory-Slechta (1993) also reported that rats exposed to lead during development were subsensitive to the behavioral effects of MK-801, a noncompetitive NMDA receptor antagonist. These results, using behavioral endpoints, appear to be consistent with reports indicating that lead exposure inhibits MK-801 binding (Alkondon *et al.*, 1990), NMDA receptor-ion channel function (Guilarte and Miceli, 1992), and long-term potentiation in hippocampal slices (Lasley *et al.*, 1992). Additional research on the effects of lead and other neurotoxicants on EAA receptor function and structure seems warranted.

References

Advokat, C., and Pellegrin, A. I. (1992). Excitatory amino acids and memory: evidence from research on Alzheimer's Disease and behavioral pharmacology. *Neurosci. Biobehav. Rev.* **16,** 13–24.

Albers, G. W., Goldberg, M. P., and Choi, D. W. (1989). N-methyl-D-aspartate antagonists: ready for clinical trial in brain ischemia? *Ann. Neurol.* **25,** 398–403.

Alkondon, M., Costa, A. C. S., Radharkrisnan, V., Aronstam, R. S., and Albuquerque, E. X. (1990). Selective blockade of NMDA-activated channel currents may be implicated in learning deficit caused by lead. *FEBS* **261,** 124–130.

Bliss, T. V. P., and Lynch, M. A. (1988). Long-term potentiation of synaptic transmission in the hippocampus: Properties and mechanisms. In *Long-Term Potentiation: From Biophysics to Behavior* (P. W. Landfield and S. A. Deadwyler, Eds.), pp. 3–72, Liss, New York.

Choi, D. W. (1987). Ionic dependence of glutamate neurotoxicity. *J. Neurosci.* **7,** 369–379.

Choi, D. W. (1988). Glutamate neurotoxicity and diseases of the nervous system. *Neuron* **1,** 623–634.

Choi, D. W. (1990). Cerebral hypoxia: some new approaches and unanswered questions. *J. Neurosci.* **10,** 2493–2501.

Choi, D. W., Maulucci-Gedde, M., and Kriegstein, A. R. (1987). Glutamate neurotoxicity in cortical cell culture. *J. Neurosci.* **7,** 357–368.

Clayton, G. D. and Clayton, F. E. (1981). *Patty's Industrial Hygiene and Toxicology.* 3rd revised ed., Vol 2A, p. 2393, Wiley, New York.

Cline, H., Debski, E., and Constantine-Paton, M. (1987). N-Methyl-D-aspartate receptor antagonist desegregates eye-specific stripes. *Proc. Natl. Acad. Sci.* **84,** 4342–4345.

Cohn, J., and Cory-Slechta (1994). Lead exposure potentiates the effects of NMDA on repeated learning. *Neurotoxicol. Teratol.* **16,** 455–465.

Cohn, J., and Cory-Slechta, D. A. (1993). Subsensitivity of lead-exposed rats to the accuracy-impairing and rate-altering effects of MK-801 on a multiple schedule of repeated learning and performance. *Brain Res.* **600,** 202–218.

Collingridge, G. L., and Bliss, T. V. (1987). NMDA receptors-their role in long-term potentiation. *Trends Neurosci.* **10,** 288–293.

Cotman, C. W., Monaghan, D. T., Ottersen, O. P., and Storm-Mathisen, J. (1987). Anatomical organization of excitatory amino acid receptors and their pathways. *Trends Neurosci.* **10,** 273–279.

Cotman, C. W., Bridges, R. J., Taube, J. S., Clark, A. S., Geddes, J. W., and Monaghan, D. T. (1989). The role of the NMDA receptor in central nervous system plasticity and pathology. *J. NIH Res.* **1**, 65–74.

Coyle, J. T., and Schwarcz, R. (1976). Lesion of striatal neurones with kainic acid provides a model for Huntington's chorea. *Nature* **263**, 244–246.

Curtis, D. R., and Watkins, J. C. (1963). Acidic amino acids with strong excitatory actions on mammalian neurons. *J. Physiol.* **166**, 1–14.

Daigo, K. (1958). Studies on the constituents of Chondria armata. II. Isolation of an anthelmintical constituent. *Yakugaku Zasshi* **79**, 353.

Dakshinamurti, K., Sharma, S. K., and Sundaram, M. (1991). Domoic acid induced seizure activity in rats. *Neurosci. Letts.* **127**, 193–197.

Davies, S. N., Lester, R. A. J., Reyman, K. G., and Collingridge, G. L. (1989). Temporally distinct pre- and postsynaptic mechanisms maintain long-term potentiation. *Nature* **338**, 500–503.

Davis, H. P., Tribuna, J., Pulsinelli, W. A., and Volpe, B. T. (1986). Reference and working memory of rats following hippocampal damage induced by transient forebrain ischemia. *Pharmacol. Biochem. Behav.* **37**, 387–392.

Dawson, V. L., Dawson, T. M., London, E. D., Bredt, D. S., and Snyder, S. H. (1991). Nitric oxide mediates glutamate neurotoxicity in primary cortical cultures. *Proc. Natl. Acad. Sci.* **88**, 6368–6371.

Debonnel, G., Beauchesne, L., and DeMontigny, C. (1989). Domoic acid, the alleged "mussel toxin", might produce its neurotoxic effect through kainate receptor activation: an electrophysiological study in the rat dorsal hippocampus. *Can. J. Physiol. Pharmacol.* **67**, 29–33.

Eldefrawi, A. T., Mansour, N. A., Brattsten, L. B., Ahrens, V. D., and Lisk, D. J. (1977). Further toxicologic studies with commercial and candidate flame retardant chemicals, Part II. *Bull. Environ. Contam. Toxicol.* **17**, 720–726.

Farooqui, A. A., and Horrocks, L. A. (1991). Excitatory amino acid receptors, neural membrane phospholipid metabolism and neurological disorders. *Brain Res. Rev.* **16**, 171–191.

Fonnum, F., Fosse, V. M., and Allen, C. N. (1983). Identification of excitatory amino acid pathways in the mammalian nervous system. In *Excitotoxins* (K. Fuxe, P. Roberts and R. Schwarcz, Eds.), pp. 3–18, Plenum Press, New York.

Garruto, R. M., and Yase, Y. (1986). Neurodegenerative disorders of the western Pacific: the search for mechanisms of pathogenesis. *Trends Neurosci.* **9**, 368–374.

Globus, M., Dietrich, W. D., Martinez, E., Valdes, I., and Ginsberg, M. D. (1988). Effect of ischemia on the in vivo release of striatal dopamine, glutamate and γ-aminobutyric acid studied by intracerebral microdialysis. *J. Neurochem.* **51**, 1455–1464.

Greenamyre, J. T. (1986). The role of glutamate in neurotransmission and in neurologic disease. *Arch. Neurol.* **43**, 1058–1063.

Guilarte, T. R., and Miceli, R. C. (1992). Lead inhibits N-methyl-D-aspartate (NMDA) receptor channel function: a potential for lead neurotoxicity. *Neurotoxicology* **13**, 493.

Handelmann, G. E., and Olton, D. S. (1981). Spatial memory following damage to hippocampal CA3 pyramidal cells with kainic acid: impairment and recovery with preoperative training. *Brain Res.* **217**, 41–58.

Hardy, J., and Cowburn, R. (1987). Glutamate neurotoxicity and Alzheimer's disease. *Trends Neurosci.* **10**, 406.

Herrling, P. L. (1992). Synaptic physiology of excitatory amino acids. *Arzneim.-Forsch./Drug Res.* **42**, 202–208.

Jarrard, L. E., Okaichi, H., Steward, O., and Goldschmidt, B. (1984). On the role of hippocampal connections in the performance of place and cue tasks: comparisons with damage to the hippocampus. *Behav. Neurosci.* **98**, 946–954.

Kiedrowski, L., Costa, E., and Wroblewski, J. T. (1992). Glutamate receptor agonists stimulate nitric oxide synthase in primary cultures of cerebellar granule cells. *J. Neurochem.* **58**, 335–341.

Kleinschmidt, A., Bear, M., and Singer, W. (1987). Blockade of "NMDA" receptors disrupts experience-dependent plasticity of kitten striate cortex. *Science* **238**, 355–358.

Lasley, S. M., Armstrong, D. L., and Polan-Curtain, J. (1992). In vivo induction of hippocampal long-term potentiation (LTP) in rats is impaired by chronic exposure to environmental levels of lead. *Neurotoxicology* **13**, 497.

Lucas, D. R., and Newhouse, J. P. (1957). The toxic effect of sodium L-glutamate on the inner layers of the retina. *Arch. Ophthalmol.* **58**, 193–201.

Maragos, W. F., Greenamyre, J. T., Penney, J. B., and Young, A. B. (1987). Glutamate dysfunction in Alzheimer's disease: an hypothesis. *Trends Neurosci.* **10**, 65–68.

Matthews, H. B., Dixon, D., Herr, D. W., and Tilson, H. A. (1990). Subchronic toxicity studies indicate that tris(2-chloroethyl)phosphate administration results in lesions in the rat hippocampus. *Toxicol. Indust. Health* **6**, 1–15.

Mayer, M. L., and Westbrook, G. L. (1987). Cellular mechanisms underlying excitotoxicity. *Trends Neurosci.* **10**, 59–61.

Monaghan, D. T., and Cotman, C. W. (1985). Distribution of N-methyl-D-aspartate sensitive L-[3H]glutamate ginding sites in rat brain. *J. Neurochem.* **5**, 2909–2919.

Morris, R. G. M., Anderson, E., Lynch, G. S., and Baudry, M. (1986). Selective impairment of learning and blockade of long-term potentiation by an N-methyl-D-asparate receptor antagonist. *Nature* **319**, 774–776.

Nadi, N. S., Wyler, A. R., and Porter, R. J. (1987). Amino acids and catecholamines in the epileptic focus from the human brain. *Neurology* **37** (Suppl.), 106.

Nakanishi, S. (1992). Molecular diversity of glutamate receptors and implications for brain function. *Science* **258**, 597–603.

Nicoll, R. A., Kauer, J. A., and Malenka, R. C. (1988). The current excitement in long-term potentiation. *Neuron* **1**, 97–103.

Nicotera, P., Bellomo, G., and Orrenius, S. (1992). Calcium-mediated mechanisms in chemically induced cell death. *Ann. Rev. Pharmacol. Toxicol.* **32**, 449–470.

Olney, J. W. (1990). Excitotoxic amino acids and neuropsychiatric disorders. *Ann. Rev. Pharmacol. Toxicol.* **30**, 47–71.

Olney, J. W., Ho, O. L., and Rhee, V. (1971). Cytotoxic effects of acidic and sulphur containing amino acids on the infant mouse central nervous system. *Exp. Brain Res.* **14**, 61–76.

Olney, J. W., Misra, C. H., and de Gubareff, T. (1975). Cysteine-S-sulfate: brain damaging metabolite in sulfite oxidase deficiency. *J. Neuropathol. Exp. Neurol.* **34**, 167–176.

Olney, J. W., Collins, R. C., and Sloviter, R. S. (1986). Excitotoxic mechanisms of epileptic brain damage. *Adv. Neurol.* **44**, 857–877.

O'Keefe, J., and Nadel, L. (1978). *The Hippocampus as a Cognitive Map*. Oxford University Press, Oxford.

Pauwels, P. J., and Leysen, J. E. (1992). Blockade of nitric oxide formation does not prevent glutamate-induced neurotoxicity in neuronal cultures from rat hippocampus. *Neurosci. Lett.* **143**, 27–30.

Pearce, I., Cambray-Deakin, M., and Burgoyne, R. (1987). Glutamate acting on NMDA receptors stimulates neurite outgrowth from cerebellar granule cells. *FEBS Lett.* **233**, 143–147.

Pearson, S. J., and Reynolds, G. P. (1992). Increased brain concentrations of a neurotoxin, 3-hydroxykynurenine, in Huntington's Disease. *Neurosci. Lett.* **144**, 199–201.

Perl, T. M., Bedard, L., Kosatsky, T., Hockin, J. C., Todd, E. C. D., and Remis, R. S. (1990). An outbread of toxic encephalopathy caused by eating mussels contaminated with domoic acid. *N. Engl. J. Med.* **322**, 1775–1780.

Petit, T. L., LeBoutillier, J. C., and Brooks, W. J. (1992). Altered sensitivity to NMDA following developmental lead exposure in rats. *Physiol. Behav.* **52**, 687–693.

Plaitakis, A., Berl, S., and Yahr, M. D. (1982). Abnormal glutamate metabolism in an adult-onset degenerative neurological disorder. *Science* **216**, 193–196.

Plaitakis, A., and Caroscio, J. T. (1987). Abnormal glutamate metabolism in amyotrophic lateral sclerosis. *Ann. Neurol.* **22**, 575–579.

Reynolds, G. P., Pearson, S. J., Halket, J., and Sandler, M. (1988). Brain quinolinic acid in Huntington's Disease. *J. Neurochem.* **50**, 1959–1960.

Robinson, G. S., Crooks, G. B., Shinkman, P. G., and Gallagher, M. (1989). Behavioral effects of MK-801 mimic deficits associated with hippocampal damage. *Psychobiology* **17**, 156–164.

Rogers, B. C., Mundy, W. R., Pediaditakis, P., and Tilson, H. A. (1989a). The neurobehavioral consequences of N-methyl-D-aspartate (N-MDA) administration in rats. *Neurotoxicology* **10**, 671–674.

Rogers, B. C., Barnes, M. I., Mitchell, C. L., and Tilson, H. A. (1989b). Functional deficits after sustained stimulation of the perforant path. *Brain Res.* **493**, 41–50.

Rogers, B. C., and Tilson, H. A. (1989). MK-801 prevents cognitive and behavioral deficits produced by NMDA receptor overstimulation in the hippocampus. *Toxicol. Appl. Pharmacol.* **99**, 445–453.

Rogers, B. C., and Tilson, H. A. (1990a). Stereospecificity of N-MDA-induced functional deficits. *Neurotoxicology* **11**, 13–22.

Rogers, B. C., and Tilson, H. A. (1990b). Kainate-induced functional deficits are not blocked by MK-801. *Neurosci. Lett.* **109**, 335–340.

Ross, S. M., and Spencer, P. S. (1987). Specific antagonism of behavioral action of "uncommon" amino acids linked to motor-system diseases. *Synapse* **1**, 248–253.

Rothman, S. (1984). Synaptic release of excitatory amino acid neurotransmitter mediates anoxic neuronal death. *J. Neurosci.* **4**, 1884–1891.

Schwarcz, R., Whetsell, W. L., and Mangano, R. M. (1983). Quinolinic acid: an endogenous metabolite that produces axon-sparing lesions in rat brain. *Science* **219**, 316–318.

Schwarcz, R., Foster, A. C., French, E. D., Whetsell, W. O., and Kohler, C. (1984). Excitotoxic models of neurodegenerative disorders. *Life Sci.* **35**, 19–32.

Sheardown, M. J., Nielson, E. O., Hansen, A. J., Jacobsen, P., and Honore, T. (1990). 2,3-Dihydroxy-6-nitro-7-sulfamoyl-benzo(F)quinoxaline: a neuroprotectant for cerebral ischemia. *Science* **247**, 571–574.

Sloviter, R. S. (1983). "Epileptic" brain damage in rats induced by sustained electrical stimulation of the perforant pathway. I. Acute electrophysiological and light microscopic studies. *Brain Res. Bull.* **10**, 675–697.

Spencer, P. S., and Schaumburg, H. H. (1983). Lathyrism: a neurotoxic disease. *Neurobehav. Toxicol. Teratol.* **5**, 625–629.

Spencer, P. S., Ludolph, A., Swivedi, M. P., Roy, D. N., Hugon, J., and Schaumburg, H. H. (1986). Lathyrism: evidence for role of the neuroexcitatory amino acid BOAA. *Lancet* **2**, 1066–1067.

Spencer, P. S., Nunn, P. B., Hugon, J., Ludolph, A. C., Ross, S. M., Roy, D. N., and Robertson, R. C. (1987). Guam amyotrophic lateral sclerosis-Parkinsonism-dementia linked to a plant excitant neurotoxin. *Science* **237**, 517–522.

Stewart, G. R., Zorumski, C. F., Price, M. T., and Olney, J. W. (1990). Domoic acid: a dementia-inducing excitotoxic food poison with kainic acid receptor specificity. *Exp. Neurol.* **110**, 127–138.

Sutherland, R. J., Whishaw, I. Q., and Kolb, B. (1981). A behavioral analysis of spatial localization following electrolytic, kainate-, or colchcine-induced damage. *Behav. Brain Res.* **217**, 41–58.

Sutherland, R. J., Hoesing, J. M., and Whishaw, I. Q. (1990). Domoic acid, an environmental toxin, produces hippocampal damage and severe memory impairment. *Neurosci. Lett.* **120**, 221–223.

Teitelbaum, J. S., Zatorre, R. J., Carpenter, S., Gendron, D., Evans, A. C., Gjedde, A., and Cashman, N. R. (1990). Neurologic sequelae of domoic acid intoxication due to the ingestion of contaminated mussels. *N. Engl. J. Med.* **322**, 1781–1787.

Terrian, D. M., Conner-Kerr, T. A., Privette, T. H., and Gannon, R. L. (1991). Domoic acid enhances the K + -evoked release of endogenous glutamate from guinea pig hippocampal mossy fiber synaptosomes. *Brain Res.* **551**, 303–307.

Tilson, H. A., Veronesi, B., McLamb, R. L., and Matthews, H. B. (1990). Acute exposure to tris(2-chloroehtyl)phosphate produces hippocampal neuronal loss and impairs learning in rats. *Toxicol. Appl. Pharmacol.* **106**, 254–269.

Tryphonas, L., Truelove, J., Nera, E., and Iverson, F. (1990). Acute neurotoxicity of domoic acid in the rat. *Toxicol. Pathol.* **18**, 1–9.

Volpe, B. T., Pulsinelli, W. A., Tribuna, J., and Davis, H. P. (1984). Behavioral performance of rats following transient forebrain ischemia. *Stroke* **15**, 558–562.

Watkins, J. C., and Evans, R. H. (1981). Excitatory amino acid transmitters. *Ann. Rev. Pharmacol. Toxicol.* **21**, 165–204.

Weiss, J. H., Koh, J. Y., and Choi, D. W. (1989). Neurotoxicity of B-N-methylamino-L-alanine (BMAA) and B-N-oxalylamino-L-alanine (BOAA) on cultured cortical neurons. *Brain Res.* **497**, 64–71.

Wieloch, T. (1985). Hypoglycemia-induced neuronal damage prevented by an N-methyl-D-aspartate antagonist. *Science* **230**, 681–683.

Willner, J., Gallagher, M., Graham, P. W., and Crooks, G. B. (1992). N-Methyl-D-aspartate antagonist D-APV selectively disrupts taste-potentiated odor aversion learning. *Behav. Neurosci.* **106**, 315–323.

Wright, J. L. C., Bird, C. J., DeFreitas, A. S. W., Hampson, D., McDonald, J., and Quilliam, M. A. (1990). Chemistry, biology and toxicology of domoic acid and its isomers. *Can. Dis. Wkly. Rep.* **16S1E**, 21–26.

Wyler, A. R., Nadi, N. S., and Porter, R. J. (1987). Acetylcholine, GABA, benzodiazepine, glutamate receptors in the temporal lobe of epileptic patients. *Neurology* **37**, 103.

Young, A. B., Greenamyre, J. T., Hollingsworth, Z., Albin, R., and D'Amato, C. (1988). NMDA receptor losses in putamen from patients with Huntington's disease. *Science* **241**, 981–983.

Young, A. B., Sakurai, S. Y., Albin, R. L., Makowiec, R., and Penney, J. B. (1991). Excitatory amino receptor distribution: Quantitative autoradiographic studies. In *Excitatory Amino Acids and Synaptic Transmission* (H. V. Wheal and A. M. Thompson, Eds.), pp. 19–31, Academic Press, New York.

Role of Serotonergic Systems in Behavioral Toxicity

W. SLIKKER, JR.
Division of Neurotoxicology
National Center for Toxicological
Research/Food and Drug
Administration
Jefferson, Arkansas 72079-9502

M. G. PAULE
Division of Neurotoxicology
National Center for Toxicological
Research/Food and Drug
Administration
Jefferson, Arkansas 72079-9502

H. W. BROENING
Division of Neurotoxicology
National Center for Toxicological
Research/Food and Drug
Administration
Jefferson, Arkansas 72079-9502

I. Introduction

The serotonergic or 5-hydroxytryptamine (5-HT)-containing cells of the nervous system were one of the first subsystems to be described and yet many of their functional aspects remain unresolved (Sjoerdsma and Palfreyman, 1990). Over the 4 decades since the 5-HT system was first discovered, evidence has accumulated that implicates this system in several important nervous system functions (Table 1). Although each of these functions impacts everyday life, it is the role of the serotonergic system on learning and memory and behavior in general that is the focus of this chapter. After the basic anatomy, electrophysiology, and neurochemistry of the serotonergic system are surveyed, the behavioral effects of selected chemicals or drugs that affect the system are examined. A special emphasis is placed on the acute and chronic effects of serotonergic-like agents and the neurochemical bases of any resultant behavioral toxicity.

II. The Serotonergic System

A. Anatomy

The anatomy of the serotonergic system of the mammalian brain has been described as a dual and highly redundant axonal system (Törk, 1990). One component, the fine varicose axon system, arises from the dorsal raphe nucleus (Fig. 1). The second component, the beaded or basket axon system, arises from the median raphe nucleus and both systems project to the forebrain. These two systems probably coexist in several forebrain regions but in differing ratios. The striatum appears to receive most of its serotonergic input from the fine varicose fiber system. The granule and polymorph layers of the hippocampal dentate gyrus receive high concentrations of the beaded axon system arising from the median raphe. In contrast, the cerebral cortex receives about equal innervation from both systems.

TABLE 1   Aspects of Nervous System
Functions Thought to Be Modulated by
Serotonergic Systems

Sleep/wakefulness
Nociception
Endocrine
Temperature regulation
Learning and memory

TABLE 2   Summary of the Distribution of the Different
Subtypes of 5-HT Receptors in the Mammalian Brain[a]

	5-HT$_{1A}$	5-HT$_{1B}$	5-H$_{1C}$	5-H$_{1D}$	5-HT$_2$	5-HT$_3$
Hippocampus	+ + +	+	+	+	+	+
Hypothalamus	+ +	+ +	+ +	+ +	+	
Striatum		+ +	+	+ +	+ +	
Neocortex	+ + +	+ +	+	+ +	+ + +	
Raphé	+ + +	+		+		

[a] From Palacios et al. (1990).

B. Neurochemistry

Radioligand binding and molecular biological techniques have identified three families of serotonergic receptors with at least six different subtypes (Palacios et al., 1990; Zifa and Fillion, 1992). These subtypes or receptor classes are not observed homogeneously throughout the brain but have a differential distribution (Table 2). The hippocampus has a high concentration of 5-HT$_{1A}$ receptors whereas the striatum has a preponderance of 5-HT$_{1B}$, 5-HT$_{1D}$, and 5-HT$_2$ receptors and

few 5-HT$_{1A}$ receptors. Because of this differential distribution of receptor subtypes and because the different types of receptors diverge in their drug binding profiles and signal transduction mechanisms, serotonin itself can have differential effects in different brain regions.

C. Electrophysiology

5-HT neurons discharge tonically in an almost clock-like fashion with an intrinsic frequency of one to five spikes per second. The ionic currents and channels and membrane properties that account for this regular activity have been described previously (Jacobs and Azmitia, 1992). The basic activity of 5-HT neurons is in general altered by a state of arousal with activities being high during waking and high activity periods and low or inactive during low activity or sleep periods (McGinty and Harper, 1976; Jacobs and Fornal, 1991).

At least three different transduction or coupling mechanisms have been described for the serotonergic receptor subtypes (Aghajanian et al., 1990). These coupling effects can be inhibitory of facilitory and are believed to be mediated by the opening and closing of K$^+$ channels via G-protein or phosphotidalinositol (PI)-mediated events, or by the direct interaction of 5-HT with the ion channels (Table 3).

III. Behavioral/Neurochemical Linkage

A. Behavioral Pharmacology

Because of the differential location and coupling of the various 5-HT receptor subtypes, the variety of drugs that interact with 5-HT systems and their effects on the serotonergic system is large (Table 4). Therapeutic agents may range in classification from anxiolytics to antidepressants or anorectics to memory enhancers. One serotonergic uptake inhibitor, zimelidine, has been reported to attenuate the disruptive effects of ethanol on memory and learning by stimulating serotonergic activity (Weingartner et al., 1983). Another uptake in-

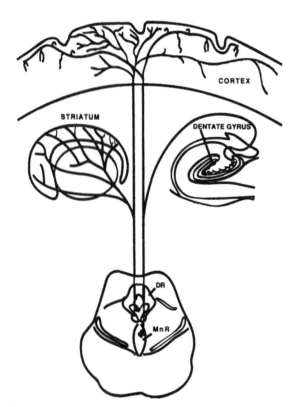

FIGURE 1   Summary diagram of the main features of the dual serotonergic system innervating the forebrain (from Törk, 1990). The fine varicose axons system (D-fibers) arises from the dorsal raphe nucleus (DR) with fibers that branch profusely in their target areas. The second system (also called basket axons), with large, round or oval varicosities (M-fibers), arises from the median raphe nucleus (MnR) with thick nonvaricose axons, giving rise to branches with the characteristics beaded varicose axons.

TABLE 3 Proposed Serotonergic Signal
Transduction Mechanisms[a]

5-HT$_1$ receptors	Inhibitory effects mediated by opening of K$^+$ channels via G-proteins and/or cyclic AMP.
5-HT$_2$ receptors	Facilitory effects mediated by closing K$^+$ channels via PI turnover to IP$_3$ and diacylglycerol and subsequent activation of protein kinase C.
5-HT$_3$ receptors	Fast excitations involving a direct interaction of 5-HT with an ion channel.

[a] From Aghajanian *et al.* (1990).

hibitor, fluoxetine or Prozac, is widely prescribed as an antidepressant (Fuller *et al.*, 1974; Kramer, 1993), while a third drug, fenfluramine, is prescribed as an anorectic agent (Pinder *et al.*, 1975).

A classic behavioral response to direct serotonergic agonists or serotonin-releasing agents is a specific locomotor-related response observed in rodents (Green and Backus, 1990). This response, known as the serotonin motor syndrome, consists of head weaving, hindlimb abduction, straub tail, piloerection, and reciprocal forepaw treading. It can be produced by serotonin-releasing agents such as 3,4-methylenedioxymethamphetamine (MDMA) (Slikker *et al.*, 1989) or fenfluramine (Green and Kelly, 1976) and by serotonergic agonists such as 5-methoxy-*n,n*-dimethyltryptamine (5-MeODMT) or quipazine (Grahame-Smith, 1971; Green *et al.*, 1976). Microdialysis experiments have confirmed the enhanced release of 5-HT into the extracellular space after MDMA administration in the rat (Gough *et al.*, 1991). The time course of enhanced 5-HT release (20–120 min) correlated well with the time course of the MDMA-induced serotonin motor syndrome (Slikker *et al.*, 1989). Subsequent decreases in 5-HT concentrations in the caudate nucleus can also be confirmed in the extracellular perfusate and in the caudate tissue homogenate at 3 hr after MDMA administration (Fig. 2).

Linkages between this chemically induced serotonin motor syndrome and the selective serotonin receptor

TABLE 4 Selected Therapeutic Drugs That Affect the Serotonergic System

Drug	Effect	Interaction
Buspirone	Anxiolytic	5-HT$_{1A}$ partial agonist
Ritanserin	Antidepressant/anxiolytic	5-HT$_2$ antagonist
Fenfluramine	Anorectic	Indirect 5-HT agonist
Zimelidine	Antipanic, memory enhancer	5-HT uptake inhibitor
Fluoxetine	Antidepressant	5-HT uptake inhibitor

interaction have been demonstrated. The selective 5-HT$_{1A}$ receptor agonist 8-hydroxy-2-(di-*n*-propylamino) tetralin (8-OH-DPAT) produces the serotonin motor syndrome (Arvidsson *et al.*, 1981). In addition, selective receptor interactions can also be shown in that ritanserin, a 5-HT$_2$ antagonist, does not block the 8-OH-DPAT-induced motor syndrome but instead causes a dose-related enhancement of the syndrome when administered before 8-OH-DPAT in the rat (Green and Backus, 1990).

B. Toxicological Effects

Several 5-HT-related agents have been reported to result in axonal or nerve terminal degeneration and/or to decreases in tryptophan hydroxylase activity (the rate-limiting enzyme in 5-HT synthesis) or 5-HT concentrations (Lorens, 1978; Slikker *et al.*, 1986, 1988, 1989; Scallet *et al.*, 1988; Gibb *et al.*, 1986; O'Hearn *et al.*, 1988; Fuller, 1985). Acute behavioral effects resulting from exposure to some of these same agents have also been described (Lorens, 1978; Slikker *et al.*, 1989). The route and duration of chemical exposure as well as the assessment methodology, however, are known to dramatically influence the behavioral effects of these compounds (Lorens, 1978).

The 5-HT-releasing agents such as methylenedioxyamphetamine (MDA), MDMA, *p*-chloroamphetamine (PCA), and fenfluramine have been investigated frequently since the mid-1980s when MDMA was reported to produce long-lasting 5-HT depletions and nerve terminal degeneration in selected brain regions of the rat (Ricaurte *et al.*, 1985, 1991; Gibb *et al.*, 1990). Although acute behavioral responses have been frequently described for animal models, long-term or residual behavioral effects after MDMA or PCA exposure have been difficult to identify (Slikker *et al.*, 1989; Nencini *et al.*, 1988; Ricaurte *et al.*, 1993; Kehne *et al.*, 1992). An attempt to clarify the relationship between the acute pharmacological and the long-term residual effects of these 5-HT-releasing agents was devised by Broening and colleagues (1994) using developmental age as an independent variable.

Tissue levels of 5-HT, its metabolite 5-hydroxyindoleacetic acid (5-HIAA), and 5-HT reuptake sites were measured in the brains of rats exposed to MDMA at selected stages of development. MDMA exposure at postnatal day (PND) 10 did not result in altered 5-HT or 5-HIAA levels 1 week after administration in any brain region examined. However, MDMA exposure at PND 40 and PND 70 resulted in dose-dependent reductions in 5-HT and 5-HIAA levels that were still evident 1 week after treatment. Time course studies revealed that at PND 10, MDMA reduced 5-HT levels

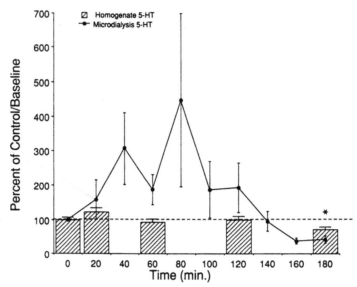

FIGURE 2 MDMA-induced changes in serotonin. Female Sprague–Dawley rats (250–300 g), previously implanted with microdialysis guide cannula, were dosed with MDMA (10 mg/kg, i.p.) at time zero. Line represents results of microdialysis; bars represent striatal (caudate/putamen) homogenate levels. Both are shown as percentage of baseline/control ± SEM. The asterisk indicates points that are significantly different from respective control/baseline. Control values for homogenate were 1.35 ng/mg; dialysis baseline content was 28 pgm/10 μl.

acutely (≤24 hr) but that levels later recovered to control levels (Fig. 3). MDMA also reduced 5-HT levels acutely when given at PND 40 or PND 70, but at these ages 5-HT levels were persistently depressed (≥168 hr). Analysis of 5-HT reuptake site concentrations revealed that after MDMA exposure at PND 10 or PND 40, there was little or no effect on these sites. However, if given at PND 70, MDMA significantly reduced 5-HT reuptake site concentrations within 24 hr after administration (Fig. 4). These experiments demonstrate that MDMA acutely reduces 5-HT levels when given as early as PND 10 in rats; sensitivity to MDMA-induced presistent reductions in 5-HT and 5-HIAA levels develops between PND 10 and PND 40; and sensitivity of 5-HT reuptake sites to persistent reductions by MDMA develops between PND 40 and PND 70. These results indicate that the acute pharmacological responses to MDMA (i.e., the serotonin motor syndrome and 5-HT release) are not inextricably linked to the subsequent residual depletions and neurohistological damage associated with MDMA exposure in the adult rat (Scallet *et al.,* 1988; Slikker *et al.,* 1988; O'Hearn *et al.,* 1988).

In an attempt to reconcile the apparent discrepancy between long-lasting neurochemical alterations produced by MDMA and the relatively acute behavioral effects of such exposure, a series of experiments were performed in the monkey. Behavior in an operant test battery (OTB) was used to model the complex brain functions of time perception, short-term memory, motivation to work for food reinforcers, learning, and color and position discrimination (for details of the procedure see Schulze *et al.,* 1988; Frederick *et al.,* 1994a). Testing occurred 30 min after intramuscular injections of a variety of doses of MDMA. The behavioral end points monitored for each of these tasks included percentage

FIGURE 3 Time course of MDMA-induced 5-HT reductions in frontal cortex (FRCX). MDMA (0 or 40 mg/kg) was administered orally at PND 10 (A), PND 40 (B), or PND 70 (C), and rats were then sacrificed at 3, 6, 12, 24, 72, 120, and 168 hr after MDMA exposure. Saline control rats are represented by the solid line with open squares and MDMA treated rats (40 mg/kg) are represented by the dashed line with filled squares. 5-HT levels were determined by HPLC-ED. Values are represented as means ± SE. *P < 0.05. †P < 0.005 versus saline control.

FIGURE 4 Time course of alterations of 5-HT reuptake site populations after MDMA administration. MDMA [0 (solid bars) or 40 (right hatched bars) mg/kg] was orally administered to rats on PND 10 (A), PND 40 (B), or PND 70 (C), and animals were sacrificed at 12, 24, 72, and 168 hr after MDMA exposure. 5-HT reuptake site populations were assayed using the methods of Marcusson *et al.* (1988). *P < 0.05. †P < 0.005 versus saline control.

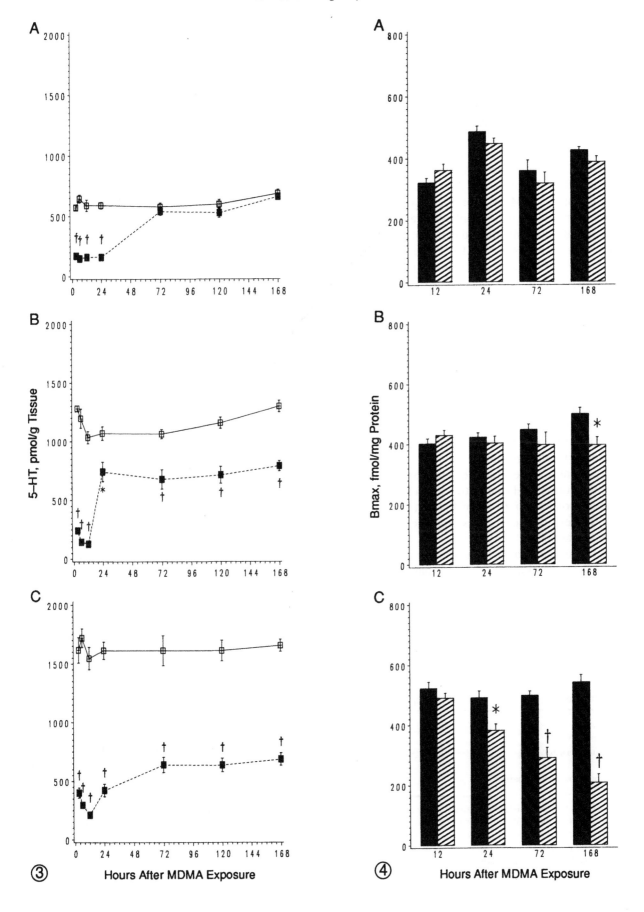

③   Hours After MDMA Exposure

④   Hours After MDMA Exposure

task completed, response rate or latency, and response accuracy. Table 5 is a summary of the effects of MDMA on the performance of these tasks. The results indicate that the functions of time estimation, motivation to work for food reinforcers, and learning are more sensitive to the acute effects of MDMA than are short-term memory and color and position discrimination (Frederick *et al.*, 1994a).

With this information concerning the acute effects of MDMA exposure as background, further studies were designed to examine the effects of chronic or repeated exposure to MDMA on these same operant behaviors. Three adult male monkeys, trained to perform the operant tasks mentioned previously, were exposed to repeated doses of MDMA (Paule *et al.*, 1991; Frederick *et al.*, 1994b). These studies were conducted in order to try to determine the effects of 5-HT depletion on the complex brain functions being modeled. Each dose of MDMA (up to and including 20 mg/kg) was given intramuscularly twice daily for 14 consecutive days in ascending order and the chronic exposure lasted for several months (see Fig. 5). Although there were clear acute effects of MDMA (generally doses at ≥1.0 mg/kg) to disrupt performance of all tasks, marked tolerance to these effects was demonstrated as exemplified by a learning task (Fig. 5). Operant behaviors generally returned to predrug values a few weeks after chronic treatment ended, when 5-HT systems would presumably be severely depleted (Slikker *et al.*, 1988). Thus, normal 5-HT neurotransmitter concentrations would not appear to be of fundamental importance for the performance for these specific rather complex behavioral tasks.

The behavioral tolerance to the repeated daily doses of MDMA in the monkey was expected based on earlier

studies in adult rodents (Slikker *et al.*, 1989). Serotonin motor syndrome scores observed on days 2 and 3 of MDMA or PCA treatment declined toward control levels and by day 4 were not significantly different from those of control rats, even though 5-HT concentrations in the frontal cortex and hippocampus of MDMA-treated subjects were reduced by 50% as determined 1 month after the end of MDMA or PCA treatment. Thus, behavioral tolerance was observed in the same animals that had demonstrated significant acute behavioral responses on day 1 of dosing and that demonstrated significant depletions of 5-HT and 5-HIAA in selected forebrain regions at sacrifice after behavioral assessment.

Although behavioral tolerance can be demonstrated after a pharmacological challenge of the serotonergic system, behavioral alterations in nonchallenged animals have not been routinely observed in 5-HT-depleted subjects. Several reasons may underlie this apparent discrepancy between the neurochemical and behavioral end points: (1) a larger than 50% reduction in 5-HT content may be required before behavioral alterations can be detected, (2) an adequately sensitive behavioral test of 5-HT function has yet to be used, (3) the brain areas in which the 5-HT depletions occur are not subserving the behaviors that were monitored, or (4) the loss of function in these areas is being compensated for by other brain areas.

The degree of 5-HT depletion produced by electrolytic lesions in the medial forebrain bundle has been correlated by others with behavioral indices such as jump thresholds in the rat. These investigators reported that an 18 to 35% decrease in whole brain 5-HT concentrations accompanied a 17 to 54% decrease in jump threshold (Lints and Harvey, 1969). Moreover, lesions lying outside the medial forebrain bundle did not decrease the jump threshold and had no effect on brain 5-HT concentrations (Lints and Harvey, 1969). Thus, a depletion of <50% in brain 5-HT content appears to be sufficient to alter behavior in this task (thought to be associated with nociception). Data from two different laboratories (Slikker *et al.*, 1989; Nencini *et al.*, 1988) fail to demonstrate any alterations in nociception 2–4 weeks after MDMA treatment, even though forebrain 5-HT was depleted 50% or more. Part of the explanation as to why 5-HT reductions reported in these studies did not result in behavioral effects associated with nociception may reside in the more refined anatomical selectivity of MDMA-induced toxicity (5-HT depletion) versus the perhaps less selective effects of electrolytic lesions on 5-HT systems (Lints and Harvey, 1969; Nencini *et al.*, 1988; Slikker *et al.*, 1989).

Thus, the behavioral and neurochemical observations with MDMA and PCA (Slikker *et al.*, 1989) may

TABLE 5 Acute Effects of MDMA on Complex Brain Function in Monkeys[a]

	Dose of MDMA		
Brain function	0.10 mg/kg	0.30 mg/kg	1.00 mg/kg
Time estimation	−[b]	−	*[c]
Short-term memory	−	−	−
Motivation to work for food	−	−	*
Learning	−	−	*
Color and position discrimination	−	−	−

Note. MDMA was administered intramuscularly to each of three adult male rhesus monkeys 30 min prior to behavioral assessment.
[a] Adapted from Frederick *et al.* (1994a).
[b] Denotes no effect.
[c] Denotes significant decrease compared to control (saline vehicle) performance.

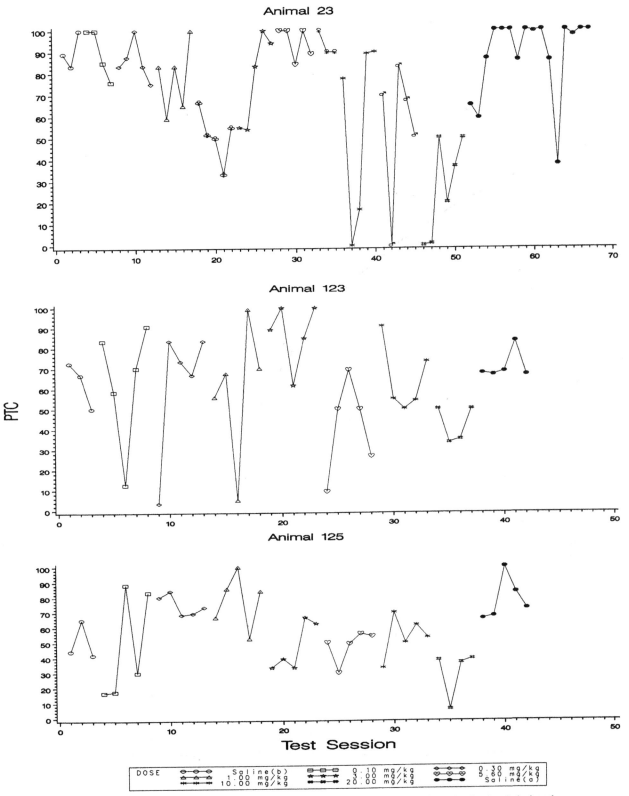

FIGURE 5 Effects of chronic treatment with MDMA on performance of a complex behavioral task designed to model learning in adult male rhesus monkeys. Each panel represents data from an individual subject. Each dose of MDMA was given intramuscularly twice per day, 7 days per week for 2 weeks. Data are represented in chronological order. Note the development of tolerance to the behavioral effects of treatment: in some instances repeated administrations of the same dose have less effect than earlier administration of the same dose whereas in other cases, higher doses have less effect than lower doses. Saline (b) is the vehicle treatment before the start of chronic MDMA administration. Saline (a) is the vehicle treatment after chronic MDMA administration.

be more appropriately compared with results from other studies employing chemical rather than electrolytic lesions (Lorens, 1978). Using p-chlorophenylalanine (an inhibitor of 5-HT synthesis), Tenen (1967) demonstrated depletion of whole brain 5-HT content and an increased sensitivity to electric shock as measured by the flinch-jump method. The whole brain 5-HT depletion was large (88%), however, compared to the decrease in jump threshold (32%). Messing and colleagues (1976), using PCA, demonstrated a decrease of 5-HT concentration in regional brain areas (23–49% depletion) and an increase in motor activity at 2 days after treatment. At 7 days after PCA treatment, however, brain 5-HT concentrations were less (but significantly) depleted (11–41%), whereas motor activity was not different from control values. Thus, the degree of 5-HT depletion as well as the anatomical selectivity of the depletion may provide an explanation for the apparent discrepancy between neurochemical and behavioral end points.

Although several different assessments of spontaneous (e.g., Slikker et al., 1989) and operant (e.g., Frederick et al., 1994a,b) behaviors have failed to identify residual effects for even 1–4 weeks after the last dose of MDMA or during repeated doses of MDMA, pharmacological challenge studies have demonstrated a shift in the MDMA dose–response curve at several months after an earlier chronic MDMA exposure. When the monkeys that had been repeatedly exposed to MDMA in the escalating, chronic dose paradigm described earlier (Paule et al., 1991; see Fig. 5) were challenged with MDMA several months after their last exposure, they were less sensitive to the acute effects of the drug than they were prior to chronic treatment (Fig. 6). These results suggest that although these complex behaviors had returned to prechronic "baseline" levels, the sensitivity of these behaviors to disruption by MDMA was "persistently" decreased. This residual "tolerance" to MDMA is consistent with the hypothesis that chronic MDMA treatment somehow altered the 5-HT system such that its acute activation by subsequent MDMA treatment was diminished. Therefore, pharmacological challenge in combination with behavioral testing before and after exposure to serotonergic neurotoxicants may be the most effective tool for the evaluation of behavioral alterations. Kirby and colleagues (1993) report that after reducing striatal 5-HT concentrations to less than 10% of control values in rats using 5,7-dihydroxytryptamine (5,7-DHT), serotonin release in the striatum, as measured by microdialysis, was not different from untreated rats. Only after challenge with fenfluramine were there significant alterations observed in serotonergic function in 5,7-DHT-treated animals. These neurochemical data may pro-

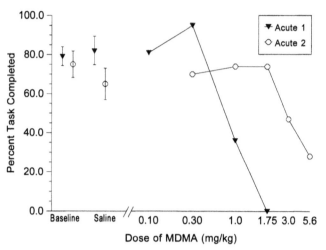

FIGURE 6   Residual or long-term tolerance to the acute effects of MDMA on performance of a complex behavioral task designed to model learning in adult male rhesus monkeys. MDMA was given intramuscularly 30 min prior to behavioral assessment. Dose-response curve number one was determined prior to chronic treatment with MDMA (as described in Fig. 5 legend). Dose–response curve number two was determined approximately 15 months after the end of chronic MDMA treatment. Points above baseline and saline are control data for noninjected and injected days, respectively. Note that there is a persistent tolerance to the acute effects of MDMA (for details see Frederick et al., 1994b).

vide some insight into why so few long-term behavioral alterations have been observed after exposure to putative serotonergic neurotoxicants such as MDMA or fenfluramine. In effect, the serotonergic neurotransmitter system may be able to compensate for the loss of a majority of its central innervation during nonstressful or normal conditions. However, when challenged, the remaining components of the serotonergic systems no longer have the capability of responding in a "normal" fashion.

As was mentioned previously, drugs acting through serotonergic mechanisms are used therapeutically for a variety of illnesses. However, it should be noted that the illnesses that these agents are used to ameliorate are primarily psychiatric in nature (Siever et al., 1992; Kramer, 1993) and include depression, obsessive-compulsive behaviors, anxiety, and schizophrenia. Several rat models exist which have been suggested to be relevant for the study of anxiety disorders (Broekkamp and Jenck, 1989; Broekkamp et al., 1989). These include conflict tests, elevated plus mazes, defensive burying, and social interaction models. These behaviors have been demonstrated to be sensitive to serotonergic compounds such as 8-OH-DPAT, busprione, m-chlorophenyl-piperazine (mCPP), fluoxetine, and sertraline. Several studies have also linked serotonergic function, or rather the lack thereof, with aggression in humans (Brown et al., 1982; Linnoila et al., 1983).

Animal models of aggression include mouse killing behavior in rats which is also sensitive to alterations in serotonergic function (Gibbons *et al.*, 1981, 1978). Future studies on the effects of compounds such as MDMA, fenfluramine, or PCA might utilize such behaviors to evaluate the functional effects of lasting reductions in 5-HT concentrations and/or serotonergic nerve terminal degeneration.

References

Aghajanian, G. K., Sprouse, J. S., Sheldon, P., and Rasmussen, K. (1990). Electrophysiology of the central serotonin system: receptor subtypes and transducer mechanisms. *N.Y. Acad. Sci.* **600**, 93–103.

Arvidsson, L. E., Hacksell, U., Nilsson, J. L. G., Hjorth, S., Carlsson, A., Lindberg, P., Sanchez, D., and Wilkstrom, H. (1981). 8-Hydroxy-2-(di-n-propylamino)tetralin, a new centrally acting 5-hydroxytryptamine receptor agonist. *J. Med. Chem.* **24**, 921–923.

Broekkamp, C. L., and Jenck, F. (1989). The relationship between various animal models of anxiety, fear-related psychiatric symptoms and response to serotonergic drugs. In *The Behavioral Pharmacology of 5-HT* (T. Archer, P. Bevan, and A. R. Cools, Eds.), pp. 321–335, Lawrence Erlbaum Associates, Hillsdale, New Jersey.

Broekkamp, C. L., Berendsen, H. H. G., Jenck, F., and Van Delft, A. M. L. (1989). Animal models for anxiety and response to serotonergic drugs. *Psychopathology* **22** (Suppl. 1), 2–12.

Broening, H. W., Bacon, L., and Slikker, W., Jr. (1994). Age modulates the long-term but not the acute effects of the serotonergic neurotoxicant 3,4-methylenedioxymethamphetamine. *J. Pharmacol. Exp. Ther.* **271**(1), 285–293.

Brown, G. L., Ebert, M. H., Goyer, P. F., Jimmerson, D. C., Klein, W. J., Bunney, W. E., and Goodwin, F. K. (1982). Aggression, suicide and serotonin: relationships to CSF amine metabolites. *Am. J. Psychiatry* **139**, 741–746.

Frederick, D. L., Gillam, M. P., Allen, R. R., and Paule, M. G. (1994a). Acute effects of methylenedioxymethamphetamine (MDMA) on several complex brain functions in monkeys. *Pharmacol. Biochem. Behav.*, in press.

Frederick, D. L., Ali, S. F., Slikker, W., Jr., Gillam, M. P., Allen, R. R., and Paule, M. G. (1994b). Effects of chronic MDMA administration on operant test battery performance in rhesus monkeys. Submitted for publication.

Fuller, R. W., Perry, K. W., Snoddy, H. D., and Molloy, B. B. (1974). Comparison of the specificity of 3-(p-trifluoromethylphenoxy)-N-methyl-3-phenylpropylamine and chlorimipramine as amine uptake inhibitors in mice. *Eur. J. Pharmacol.* **28**, 233–236.

Fuller, R. W. (1985). Persistent effects of amphetamine, p-chloroamphetamine, and related compounds on central dopamine and serotonin neurons in rodents. *Psychopharmacol. Bull* **21**(3), 528–532.

Gibb, J. W., Hanson, G. R., and Johnson, M. (1986). Effects of (+)-3,4-methylenedioxymethamphetamine [(+)MDMA] and (−)-3,4-methylenedioxymethamphetamine [(−)MDMA] on brain dopamine, serotonin and their biosynthetic enzymes. *Neuroscience* **12**, 608.

Gibb, J. W., Johnson, M., Stone, D., and Hanson, G. R. (1990). MDMA: Historical perspectives. *The N.Y. Acad. Sci.* **600**, 601–612.

Gibbons, J. L., Barr, G. A., and Bridger, W. H. (1978). Effects of parachlorophenylalanine and 5-hydroxytryptophan on mouse

killing behavior in killer rats. *Pharmacol. Biochem. Behav.* **8**, 91–98.

Gibbons, J. L., Barr, G. A., Bridger, W. H., and Leibowitz, S. F. (1981). L-Tryptophan's effects on mouse killing, feeding, drinking, locomotion and brain serotonin. *Pharmacol. Biochem. Behav.* **15**, 201–206.

Gough, B., Ali, S. F., Slikker, W., Jr., and Holson, R. R. (1991). Acute effects of 3,4-methylenedioxymethamphetamine (MDMA) on monoamines in rat caudate. *Pharmacol. Biochem. Behav.* **39**, 619–623.

Grahame-Smith, D. G. (1971). Inhibitory effects of chlorpromazine on the syndrome of hyperactivity produced by L-tryptophan or 5-methoxy-N,N-dimethyltryptamine in rats treated with a monoamine oxidase inhibitor. *Br. J. Pharmacol.* **43**, 856–864.

Green, A. R., Youdim, M. B. H., and Grahame-Smith, D. G. (1976). Quipazine: its effects on rat brain 5-hydroxytryptamine metabolism, monoamine oxidase activity and behavior. *Neuropharmacology* **15**, 173–179.

Green, A. R., and Kelly, P. H. (1976). Evidence concerning the involvement of 5-hydroxytryptamine in locomotor activity produced by amphetamine or tranylcypromine plus L-dopa. *Br. J. Pharmacol.* **57**, 141–147.

Green, A. R., and Backus, L. I. (1990). Animal models of serotonin behavior. *N.Y. Acad. Sci.* **600**, 237–249.

Jacobs, B. L., and Azmitia, E. C. (1992). Structure and function of the brain serotonin system. *Physiol. Rev.* **72**, 165–229.

Jacobs, B. L., and Fornal, C. A. (1991). Activity of brain serotonergic neurons in the behaving animal. *Pharmacol. Rev.* **43**, 563–578.

Kehne, J. H., McCloskey, T. C., Taylor, V. L., Black, C. K., Fadayel, G. M., and Schmidt, C. J. (1992). Effects of the serotonin releasers 3,4-methylenedioxymethamphetamine (MDMA), 4-chloroamphetamine (PCA) and fenfluramine on acoustic and tactile startle reflexes in rats. *J. Pharmacol. Exp. Ther.* **260**, 78–89.

Kirby, L. G., Kreiss, D. S., Singh, A., and Lucki, I. (1993). The effect of 5,7-dihydroxytryptamine lesions on basal and fenfluramine-induced serotonin release in striatum. *Soc. Neurosci. Abstr.* **19**, 126.7.

Kramer, P. D. (1993). *Listening to Prozac: a psychiatrist explores antidepressant drugs and the remaking of the self*. Penguin Books USA, New York.

Linnoila, M., Virkkunen, M., Scheinen, M., Nuutila, A., Rimon, R., and Goodwin, F. K. (1983). Low cerebrospinal fluid 5-hydroxyindoleacetic acid concentration differentiates impulsive from nonimpulsive violent behavior. *Life Sci.* **33**, 2609–2614.

Lints, C. E., and Harvey, J. A. (1969). Altered sensitivity to foot shock and decreased brain content of serotonin following brain lesions in the rat. *J. Comp. Physiol. Psychol.* **67**, 23–31.

Lorens, S. A. (1978). Some behavioral effects of serotonin depletion depend on method: a comparison of 5,7-dihydroxytryptamine, and electrolytic raphe lesions. *N.Y. Acad. Sci.* **305**, 532–553.

Marcusson, J. O., Bergström, M., Eriksson, K., and Ross, S. B. (1988). Characterization of [^{3}H]paroxetine binding in rat brain. *J. Neurochem.* **50**, 1783–1790.

McGinty, D. J., and Harper, R. M. (1976). Dorsal raphe neurons: Depression of firing during sleep in cats. *Brain Res.* **101**, 569–575.

Messing, R. B., Phebus, L., Fisher, L. A., and Lytle, L. D. (1976). Effects of p-chloroamphetamine on locomotor activity and brain 5-hydroxyindoles. *Neuropharmacology* **15**, 157–163.

Nencini, P., Woolverton, W. L., and Seiden, L. S. (1988). Enhancement of morphine-induced analgesia after repeated injections of methylenedioxymethamphetamine. *Brain Res.* **457**, 136–142.

O'Hearn, E., Battaglia, G., DeSouza, E. B., Kuhar, M. J., and Molliver, M. E. (1988). Methylenedioxyamphetamine (MDA) and methylenedioxymethamphetamine (MDMA) cause selective ablation of serotonergic axon terminals in forebrain: immunocytochemical evidence for neurotoxicity. *J. Neurosci.* **8**(8), 2788–2803.

Palacios, J. M., Waeber, C., Hoyer, D., and Mengod, G. (1990). Distribution of serotonin receptors. *N.Y. Acad. Sci.* **600**, 36–52.

Paule, M. G., Gillam, M. P., and Allen, R. R. (1991). Effects of chronic MDMA (Methylenedioxymethamphetamine) administration on rhesus monkey performance in an operant test battery (OTB). *Soc. Neurosci. Abstr.* **17**, 597.5.

Pinder, R. M., Brogden, R. N., Sawywe, P. R., Speight, T. M., and Avery, G. S. (1975). Fenfluramine: a review of its pharmacological properties and therapeutic efficacy in obesity. *Drugs* **10**, 241–323.

Ricaurte, G. A., Bryan, G., Strauss, L., Seiden, L., and Schuster, C. (1985). Hallucinogenic amphetamine selectively destroys brain serotonin nerve terminals. *Science* **229**, 986–988.

Ricaurte, G. A., Molliver, M. E., Martello, M. B., Katz, J. L., Wilson, M. A., and Martello, A. L. (1991). Dexfenfluramine neurotoxicity in brains of non-human primates. *Lancet* **388**, 1487–1497.

Ricaurte, G. A., Markowska, A. L., Wenk, G. L., Hatzidimitriou, G., Wlos, J., and Olton, D. S. (1993). 3,4-methylenedioxymethamphetamine, serotonin and memory. *J. Pharmacol. Exp. Ther.* **266**, 1097–1105.

Scallet, A. C., Lipe, G. W., Ali, S. F., Holson, R. R., Frith, C. H., and Slikker, W., Jr. (1988). Neuropathological evaluation by combined immunohistochemistry and degeneration-specific methods: application to methylenedioxymethamphetamine. *Neurotoxicology* **9**(3), 529–538.

Schulze, G. E., McMillan, D. E., Bailey, J. E., Scallet, A. C., Ali, S. F., Slikker, W., Jr., and Paule, M. G. (1988). Acute effects of Δ-9-tetrahydrocannabinol in rhesus monkeys as measured by performance in a battery of complex operant tests. *J. Pharmacol. Exp. Ther.* **245**, 178–186.

Siever, L. J., Kahn, R. S., Lawlor, B. A., Trestman, R. L., Lawrence, T. L., and Coccaro, E. F. (1992). Critical issues in defining the role of serotonin in psychiatric disorders. *Pharmacol. Rev.* **43**, 509–525.

Sjoerdsma, A., and Palfreyman, M. G. (1990). History of serotonin and serotonin disorders. *N.Y. Acad. Sci.* **600**, 1–8.

Slikker, W., Jr., Ali, S. F., Scallet, A. C., and Frith, C. H. (1986). Methylenedioxymethamphetamine (MDMA) produces long lasting alterations in the serotonergic system of rat brain. *Neuroscience* **12**, 363.

Slikker, W., Jr., Ali, S. F., Scallet, A. C., Frith, C. H., Newport, G. D., and Bailey, J. R. (1988). Neurochemical and neurohistological alterations in the rat and monkey produced by orally administered methylenedioxymethamphetamine (MDMA). *Toxicol. Appl. Pharmacol.* **94**, 448–457.

Slikker, W., Jr., Holson, R. R., Ali, S. F., Kolta, M. G., Paule, M. G., Scallet, A. C., McMillan, D. E., Bailey, J. R., Hong, J. S., and Scalzo, F. M. (1989). Behavioral and neurochemical effects of orally administered MDMA in the rodent and nonhuman primate. *Neurotoxicology* **10**, 529–542.

Tenen, S. S. (1967). The effects of p-chlorophenylalanine, a serotonin depleter, an avoidance acquisition, pain sensitivity and related behavior in the rat. *Psychopharmacologia* **10**, 204–219.

Törk, I. (1990). Anatomy of the serotonergic system. *N.Y. Acad. Sci.* **600**, 9–35.

Weingartner, H., Rudorfer, M. V., Buchsbaum, M. S., and Linnoila, M. (1983). Effects of serotonin on memory impairments produced by ethanol. *Science* **221**, 472–474.

Zifa, E., and Fillion, G. (1992). 5-hydroxytryptamine receptors. *Pharmacol. Rev.* **44**(3), 401–458.

PART
IV

Neurochemical and Biomolecular Approaches

Neurochemical and Biomolecular Approaches
An Introductory Overview

WILLIAM SLIKKER, JR.
Division of Neurotoxicology
National Center for Toxicological Research/Food and Drug Administration
Jefferson, Arkansas 72079-9502

The assessment of toxicity must address the diverse cellular and biochemical nature of the nervous system and thus requires a variety of approaches. Unlike organ systems that are the sum of repeating, often identical, building blocks, the nervous system is diverse in cellular structure and neurochemical activity. The genomic regulation responsible for this diverse cellular structure and function is also distinguishable from that of other somatic cells, most notably by the lack of neuronal division characterizing the adult central nervous system. The size of nervous system cells are an example of this diversity, since axonal length may be measured in microns or meters. In order to reveal alterations in the nervous system, therefore, a wide diversity of techniques is required.

The molecular, cellular, and integrated subunits of the nervous system also require a variety of techniques for comprehensive assessment. Some alterations may be observed at the level of mitochondrial DNA, as is the case proposed by Drs. Parker and Cheng to describe the effects of 2',3'-dideoxycytidine on nerve cells in culture, whereas other effects may be observed at the level of extracellular fluid concentrations of neu-

rotransmitters collected by microdialysis from active caudate cells of the waking rodent, as described by Dr. Bowyer. Isolation of homogeneous cell types by culture techniques as described by Drs. Aschner, Kimelberg, and Vitarella serve to reduce biological variation through eliminating hormonal, hemodynamic, and exposure differences. These various approaches each have their own technical and theoretical advantages that may be applied to assess the diverse nature of the nervous system.

The quantitative output of many of these traditional and sophisticated techniques is especially appropriate for the risk assessment process of neurotoxicants. Biomarkers of effect are described by Drs. Ali and Slikker for general approaches and by Dr. Johannessen for molecular and cellular indices. These chemical markers of neuropathology include quantification of induced proteins and degradative products, either of which may be incorporated into the risk assessment process at the hazard identification or the dose–response level.

Other approaches or techniques are useful in describing mechanisms of neurotoxicity. The influence of environmental chemicals and drugs on axonal transport and microtubules can be described by quantifying a variety of neurofilament building blocks and associated proteins and enzymes, as proposed by Dr. Sabri. Molecular techniques focusing on nucleic acid hybridization offer still another insight into the mechanisms of neurotoxicity. Examples presented by Dr. Billingsley concerning subtractive hybridization can be used to isolate rare mRNAs that relate specifically to neurotoxic insult. Another approach, presented by Dr. Lo-Pachin, uses electron probe X-ray microanalysis to reflect both the distribution and the quantification of elements and water associated with nerve and glial cell injury. These mechanistic data may be used to categorize potential neurotoxicants and to provide important information to enhance the risk assessment process for previously identified neurotoxicants.

In the following chapters, several important neurochemical and biomolecular approaches to the study of neurotoxicity are presented. These approaches permit systematic evaluation of the diverse nervous system characteristics and also provide biomarkers and mechanistic data for the risk assessment of neurotoxicants.

Basic Biochemical Approaches in Neurotoxicology: Assessment of Neurotransmitters and Neuroreceptors

SYED F. ALI

Division of Neurotoxicology
National Center for Toxicological Research/FDA
Jefferson, Arkansas 72079

WILLIAM SLIKKER, JR.

Division of Neurotoxicology
National Center for Toxicological Research/FDA
Jefferson, Arkansas 72079

I. Introduction

Since the discovery of neurotransmitter systems in the mid-1940s, many advances in the field of neurotoxicology have occurred. A great many of these advances have resulted from the full understanding of the effects of neurotoxicants on the various aspects of neurotransmitter systems. Basic biochemical and neurochemical methods for the assessments of the nervous system are numerous. Indeed, most of the general biochemical methods for cellular toxicity assessments (e.g., methods for the assessment of protein phosphorylation, calcium homeostasis, free radical formation, ion channels, mitochondrial function, nucleic acid metabolism, lipid synthesis, etc.) are all applicable for the evaluation of neural tissues. All these general approaches and methods are widely published and need not be introduced in this chapter. This chapter focuses on the assessment of some of the more "unique" aspects of the nervous system. One of the unique properties of nerve cells is their ability for intercellular communication via neurotransmitter and neuroreceptor systems. The general approaches needed to assess these "neural communication systems" are presented and discussed in this chapter.

A. Brain Dissection

For neurochemical analysis, brain tissue is dissected into specific brain areas. Regional dissection is generally necessary because brain regions may vary dramatically in their relative concentration of neurotransmitters (e.g., high concentrations of dopamine in the striatum/caudate nucleus as compared to other brain regions). After sacrifice of the animals by institutional animal care and use committee (IACUC) and NIH guidelines, brain can be removed and immersed in ice-cold saline or Tris buffer for 1–2 min. The chilled brain is then dissected into different regions either by free hand or by serial sectioning and then the different nuclei are punched out with the use of an anatomical atlas (Palkovits and Brownstein, 1990). Free hand brain dissection is used to provide the regional areas such as caudate nucleus, nucleus accumbens, prefrontal cortex, septum, hippocampus, hypothalamus, thalamus, substantial nigra, entorinal cortex, cerebellum, and brain stem. The punch technique is applied to isolate additional nuclei. For neurochemical analysis, all tissues are frozen on dry ice and stored at −70°C until processed.

Monkey brains are dissected either by free hand or by the slice and punch technique to produce the desired regions as described by Slikker and colleagues (1988) using the anatomical atlas of Szebenyl (1970). The cortex is removed as four large pieces: the prefrontal cortex, the premotor/precentral cortex, the preoptic/parietal cortex, and the occipital cortex. The basal ganglia, thalamus, hypothalamus, and brain stem are removed as described by Slikker and colleagues (1988). The basal ganglia can be further dissected into the caudate nucleus, putamen, nucleus accumbens, and the hippocampi can be rolled out from the surrounding temporal gyrus and frozen as two entire, separate pieces, one from each hemisphere. As each sample is removed, it is immediately frozen on dry ice and each dissection should be completed within 5 min of removal of the brain from the animal. The anatomical consistency of the dissection and the time to complete the dissection are critical for reproducibility of results. A schematic for rat and monkey brain dissection is presented in Fig. 1.

II. Neurotransmitter Analysis

A. Tissue Preparation

To measure the neurotransmitter concentrations or receptor-binding parameters, the tissue needs to be prepared in different ways. For neurotransmitter concentrations the tissue is homogenized in perchloric acid whereas membrane is used for neurotransmitter receptor binding. For uptake studies, crude synaptosomal or mitochondrial preparations are prepared. Therefore, the detailed description of tissue preparation for each specific assay is provided.

B. Monoamine Analysis

Concentrations of dopamine (DA), serotonin (5-HIAA), and their metabolites 3,4-dihydroxyphenylacetic acid (DOPAC), homovanillic acid (HVA), and 5-hydroxyindoleacetic acid (5-H1AA) are quantified by high-performance liquid chromatography (HPLC) combined with electrochemical detection (EC) as described by Ali *et al.* (1993a). Each region of the brain is weighed and diluted with a measured volume (10%,

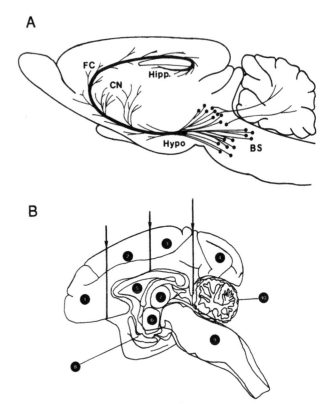

FIGURE 1 Diagram of a rat brain (A) (modified from Cooper *et al.*, 1984) and a monkey brain (B) (modified from *Atlas of Macaca mulatta*, Szevenyl, 1970). Rat: Hipp, hippocampus; FC, frontal cortex; CN, caudate nucleus; Hypo, hypothalamus; BS, brain stem. Monkey: (1) Frontal cortex, (2) premotor cortex and precentral gyrus, (3) parietal cortex and postcentral gyrus, (4) occipital cortex, (5) caudate nucleus, (6) hypothalamus, (7) thalamus, (8) hippocampus, (9) brain stem, and (10) cerebellum (adapted from Ali *et al.*, 1989).

w/v) of 0.2 *N* perchloric acid containing 100 ng/ml of the internal standard 3,4-dihydroxybenzylamine (DHBA). Brain tissue is disrupted by ultrasonication for 10 sec, centrifuged (15,000*g*; 7 min), and 150 μl of the supernatant is removed and filtered through a 0.2-μm Nylon-66 microfilter [MF-1 microcentrifuge filter, Bioanalytic System (BAS), W. Lafayette, IN]. Aliquots of 25 μl representing 2.5 mg of brain tissue are injected directly onto the HPLC/EC system for separation of the neurotransmitters DA and 5-HT and their metabolites DOPAC, HVA, and 5-HIAA.

The analytical system includes a Waters Associates 510A pump (Milford, MA), a Rheodyne 7125 injector (Rheyodyne, Inc., Cotati, CA), a Supelco Supelcosil LC-18, a 3-μm (7.5 cm × 4.6 mm) analytical column, a LC-4B amperometric detector, and a LC-17 oxidative flow cell (BAS) consisting of a glassy carbon electrode (TL-5) versus a Ag–AgCl₂ reference electrode maintained at a potential of 0.75 V. The mobile phase consists of 0.07 *M* potassium phosphate, pH 3.0, 8% methanol, and an ion pairing reagent of 1.02 m*M* 1-heptane sulfonic acid. Chromatograms are recorded and integrated on a Perkin-Elmer LCl-100 integrator (Perkin-Elmer Corp., Norwalk, CT). The concentration of DA and 5-HT and their metabolites DOPAC, HVA, and 5-HIAA is calculated using a standard curve. The standard curves are generated by determining in triplicate the ratio among three different known amounts of each amine or its metabolites and a constant amount of the internal standard. Figure 2A is a representative chromatogram of monoamine standards, Fig. 2B is a representative chromatogram of monoamines from a control tissue extract, and Fig. 2C is a representative chromatogram of monoamines from a MPTP-treated mouse striatum.

C. Amino Acids Analysis

Quantification of brain amino acids requires that they first be extracted and derivatized as described by Lipe and colleagues (1991). Each brain region is weighed and sonicated in 1 ml of 0.05 *M* HCl in 47.5% methanol after adding 20 μl of DL-homoserine (50 m*M*/ml) as an internal standard. The mixture is centrifuged at 31,000*g* at 0°C for 10 min, and 400 μl of 0.5 *M* KHCO₃ is added and centrifuged at 15,000*g* at 0–4°C for 10 min. The supernatant is filtered through a 0.45-μm pore size microfilter [MF-1 microcentrifuge filter (BAS)] by centrifuging at 4000*g* at 0–4°C for 5 min.

Precolumn derivatization with *O*-phthalaldehyde-*tert*-butylthiol (OPA) is used to quantitate the amino acid concentrations. The OPA reagent is prepared by dissolving 27 mg of *O*-phthalaldehyde in 2 ml methanol and adding 20 μl of *tert*-butylthiol followed by 4.5 ml of 100 m*M* borate buffer, pH 9.5 (Zielke, 1985). The OPA reagent should be kept in a tightly sealed vial at 4°C and used within 5 days.

1. Analytical System

The analytical system consists of a Waters Model 6000A solvent delivery system (Waters Associated), a Rheodyne 7125 injector (Rainin Instrument, Woburn, MA), a Biophase ODS, a 5-μm particle size (250 ×

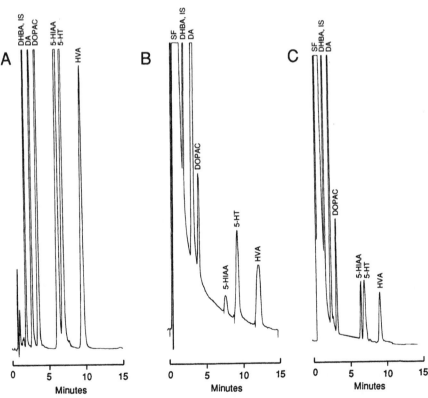

FIGURE 2 (A) Chromatogram of a standard mixture of monoamines and their metabolites. SF, solvent front; DHBA,IS, dihydroxybenzylamine, internal standard; DA, dopamine; DOPAC, dihydroxyphenylacetic acid; 5-HIAA, 5-hydroxyindoleacetic acid; 5-HT, 5-hydroxytryptamine; HVA, homovanillic acid. See text for details of chromatographic conditions. Chromatograms of control mouse striatum extracts (B), and MPTP-treated mouse striatum extracts (C).

4.6 mm) analytical column (BAS), and a BAS Model LC-4B electrochemical detector with a low volume flow cell containing a glassy carbon electrode with a potential of 700 mV relative to a Ag/AgCl$_2$ reference electrode.

The mobile phase consists of two solutions, with the flow being switched from one solution to the other at a predetermined time in a step gradient manner. Solution "A" is made up of 50 mM K$_2$HPO$_3$ and 1 mM Na$_2$–EDTA adjusted to pH 7.0, 2% tetrahydrofuran (THF), and 20% acetonitrile. Solution "B" differs from "A" only in that it contains 10% THF. The flow rate is 1.5 ml per min at ambient temperature. The switch from "A" to "B" is made 5 min after beginning flow of solution "A." After a chromatogram is complete, solution "A" is pumped through the column for at least 7 min before the next sample is injected. Chromatograms are integrated and recorded on a Perkin-Elmer LCI-100 Integrator (Perkin-Elmer Corp.).

2. Assay Procedure

A 20-μl aliquot of tissue extract, plasma sample, or standard amino acid is mixed in a 1.5-ml plastic centrifuge tube with 100 μl of the OPA reagent, allowed to react for 2 min, and then 10 μl is injected onto the column. The amino acid concentrations are calculated using a standard curve for each amino acid. Standard amino acid solutions are prepared in a water/methanol mixture (1/1) containing 1 mM Na$_2$–EDTA. The standard curves are generated by triplicate determination of the peak area ratio among three different amounts of each amino acid and a constant amount of the internal standard. Concentrations of aspartate, glutamate, glutamine, glycine, taurine, and GABA are determined. Figure 3A is a representative chromatogram of the standard mix of different amino acids, Fig. 3B is a representative chromatogram of amino acid neurotransmitters in a tissue extract, and Fig. 3C is a representative chromatogram of amino acids from a TMT-treated mouse hippocampus.

III. Neuroreceptors Analysis

A. Membrane Preparation

For the neurotransmitter receptor-binding assay, each region of the brain is homogenized in 20 vol (w/v) of 0.32 M sucrose followed by centrifugation (50,000g for 10 min). The pellet is rehomogenized in deionized distilled water (pH adjusted to 7.4) and centrifuged (50,000g for 10 min). The pellet is then resuspended in 50 mM Tris–HCl (pH 7.4) and centrifuged (50,000g for 10 min). This final pellet is sus-

pended in the incubation buffer (50 mM Tris–HCl containing 2.5 mM CaCl$_2$, 1 mM MgCl$_2$, 5 mM KCl, 120 mM NaCl, pH 7.4) at a concentration of 50 mg/ml wet weight of tissue as described by Ali *et al.* (1986).

B. Synaptosome Preparation

All synaptosomal isolation procedures are performed using ice-cold (2–4°C) solutions. "Purified" synaptosomes are rapidly prepared as described by Minnema and Michaelson (1985). Rat brain regions [e.g., striatum for [^{3}H]dopamine release; cortex for γ-[^{3}H]aminobutyric acid release; and hippocampus for [^{3}H]acetylcholine release, [^{3}H]deoxyglucose phosphate efflux, and ^{45}Ca efflux] are homogenized in 0.32 M sucrose. The homogenates are centrifuged at 1000g for 10 min and the resulting supernatant is then layered onto a two-step discontinuous sucrose gradient. Following centrifugation at 223,000g for 25 min, the isolated synaptosomes are harvested from the 0.8–1.2 M sucrose interphase and are kept at 2°C until used. It is recommended that the synaptosomal preparation be used immediately or within 2–4 hr of preparation because overnight freezing may cause the loss of the integrity of the synaptosomal preparation.

C. Mitochondria Preparation

Mitochondria are isolated from whole brain or specific brain regions by using a modification of the methods of Briggs and Cooper (1981) and Matlib and Schwartz (1983). This isolation technique has been used to obtain a viable mitochondrial preparation not contaminated by synaptosomes, a problem common to many mitochondria isolation procedures (Biesold, 1974). This technique eliminates both hyperosmotic gradients and lengthy preparation times that can compromise the functional integrity of mitochondrial preparations (Biesold, 1974). Rat brain is homogenized in 10 vol of ice-cold (2°C) medium containing 300 mM mannitol, 10 mM Hepes, 1 mM EGTA, and 1 mg BSA/ml, pH 7.4. The homogenate is centrifuged at 1000g for 10 min in a refrigerated centrifuge (2°C). After centrifugation, equal volume aliquots of the resulting supernatant are layered onto three precooled "metrizamide gradients" (0.8 ml) in 1/2 × 2 in. The "metrizamide gradient" is prepared by mixing 0.8 ml of 40% (w/v) metrizamide solution with 1.8 ml of the 300 mM mannitol medium described earlier. The tubes are centrifuged at 35,000 rpm for 15 min using an ultracentrifuge. This centrifugation should result in approximately 11,000g at the gradient interface. The supernatants are removed and the remaining pellets are gently rinsed with ice-cold (2°C) medium consisting

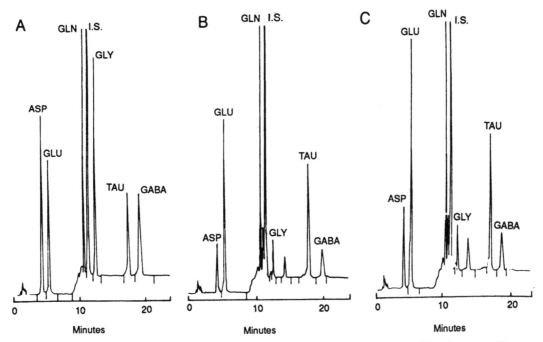

FIGURE 3 (A) Chromatogram of a standard amino acid mixture. Asp, aspartate; Glu, glutamate; Gln, glutamine; I.S., internal standard (DL—homoserine); Gly, glycine; Tau, taurine; GABA, γ-amino-butyric acid. See text for details of chromatographic conditions. Chromatograms of mouse hippocampus extracts: untreated mouse (B) and TMT-treated mouse (C) (adapted from Lipe *et al.,* 1991).

of 200 m*M* mannitol, 40 m*M* KCl, 0.5 mg BSA/ml, and 10 m*M* Hepes, pH 7.4. The pellets are pooled and gently resuspended in the 200 m*M* mannitol medium described earlier to approximately 1 ml/g original tissue weight. Mitochondria prepared in this fashion is examined for purity and morphological integrity by electron microscopy, using the tissue fixing, staining, and sectioning procedure described by Miller and colleagues (1986). Mitochondrial viability is tested by examining succinate oxidation (state 4 respiration) and the stimulation of oxygen utilization by ADP addition (state 3 respiration).

D. Receptor–Ligand-Binding Assay

The development of a series of binding assays that are used for both *in vivo* and *in vitro* evaluation of neurotoxicity requires, as a first step, the concise delineation of the binding phenomenon for each ligand to be used (Yamamura *et al.,* 1984; Bondy and Ali, 1992). These basic criteria include demonstration of:

1. The specificity of the binding reaction, as judged by competition with various pharmacological agents.
2. An appropriate regional distribution of binding sites within the brain.
3. The attainment of a reversible equilibrium of the receptor–ligand interaction.

4. The extent of binding is proportional to the amount of membrane in the incubation tube.
5. The stability of isotope throughout the incubation.
6. A relatively low dissociation constant of the binding reaction and saturable receptor sites.
7. If possible, receptor binding should be coupled to, and correlated with, another relevant biological event. This could include stimulation of adenyl cyclase (in the case of catecholamine agonist binding) or modulation of the rate of transmitter release from an isolated synaptosomal preparation.

A receptor–ligand screening system has been developed from a variety of binding reactions. These can be studied at close to optimal conditions in the same tissue preparation, using essentially parallel procedures. This allows the simultaneous assay of several receptors in an objective and nonbiased way. This approach is essential if one is to find a transmitter or modulator species that is selectively altered by a toxic agent. The underlying hypothesis is that some neural circuits are more sensitive than others to metabolic disturbances. This is in contrast to a specifically predictive approach for which the interaction of a toxicant with only a single transmitter species is studied.

Rat brain regions can be dissected by the previously described methods, homogenized in 20 vol of 0.32 *M* sucrose, and centrifuged (40,000*g* for 10 min). The precipitate is resuspended in the same volume of deionized water and recentrifuged (40,000g, 10 min). This final pellet is suspended in 50 mM Tris, pH 7.4, for immediate use or can be stored at −20°C. In the case of GABA and diazepam assays, the final pellet is washed twice more in Tris buffer (40 mM, pH 7.4) to assure the removal of endogenous inhibitory materials. For NMDA/glutamate receptor binding the endogenous glutamate can be removed by incubating the membrane with Triton X-100 and by washing several times with Tris–HCl buffer (Ali *et al.*, 1993b). This preparation consists of a crude membrane instead of purified synaptosomal membranes so that binding in tissues that may be in differing states can be compared. The degree of myelination of a tissue is likely to alter the proportion of synaptosomes recovered. Treated animals may have brains that are abnormal with respect to their water or myelin content, or may be developmentally delayed. In such cases, subcellular fractionation may make comparison with tissues from control animals unreliable.

Binding measurement is carried out in a 1-ml incubation mixture containing 40 mM Tris, pH 7.4, and appropriate labeled and unlabeled pharmacological agents. The amount of tissue per tube corresponds to 5 mg original wet tissue and contains around 200–400 μg protein. Incubation can vary from 15 to 60 min at 37°C. In all cases, equilibrium is reached in this period. At the end of incubation, samples are filtered on glass fiber filters (25 mm diameter, 0.3 μm pore size) and washed rapidly three times with 5 ml of Tris buffer using a Brandell cell harvester (Gaithesburg, MD). In the case of the strychnine assay, filters are only washed twice because the receptor–ligand complex formed has a very rapid rate of dissociation. The stability of the isotope solution is important, and fresh isotope dilutions should be made every few days. Filters are then dried and radioactivity is estimated in 5 ml of a scintillation mixture.

Table 1 shows the concentrations of tritiated ligands that are commonly used for the various assays. The incubations are carried out in the presence and absence of a large excess of a competing chemical whose concentrations are also presented in Table 1. The difference between these two values is taken as an index of specific binding. Specific binding constitutes over 65% of total binding in these assays. Preliminary criteria of binding have been established for many of the previously listed assays. Some of these have been illustrated using the binding of [^{3}H]spiroperidol to brain membranes (Bondy and Ali, 1992). This interaction is proportional to the amount of membrane in the incubation tube, is stereospecific with respect to competition with the isomers of butaclamol, and reaches a relatively rapid equilibrium that is reversible (Bondy and Ali, 1992). Binding is heterogeneous within brain regions, being highest in the striatum, a region rich in dopaminergic terminals (Bondy and Ali, 1992). As newer ligands with narrower specificity become available, the series of compounds listed must be continually updated.

The use of pharmacological agents instead of endogenous biological molecular species has several advantages (Bondy and Ali, 1992):

1. The synthetic chemicals often have a higher affinity for the receptor site than the endogenous substance. This frequently accounts for their pharmacological efficacy.
2. The synthetic chemicals are generally less likely to be degraded by catabolic enzymes, such as monoamine oxidase, and are also less prone to oxidation than are endogenous catecholamines and indolamines.
3. The specificity of pharmacological agents can be much greater than that of neurotransmitters. Thus norepinephrine will bind to α^1, α^2, β^1, and β^2 adrenergic sites, although these can be distinguished by use of drugs. Homogenization of a tissue destroys the normal anatomical separation of sites, but careful design of incubation conditions allows partial analysis of subclasses of receptors.

The specificity of binding reactions can be improved by selection of brain regions. Thus, [^{3}H]spiroperidol largely binds to dopamine D2 receptors in striatal membranes, whereas serotonin receptors are largely involved in the frontal cortex (Leysen *et al.*, 1978). Another means of enhancing specificity is by the utilization of appropriate unlabeled agents. For example, [^{3}H]spiroperidol in the presence of 1 μM ketanserin will largely bind to dopamine receptors and, conversely, in the presence of 10 μM (−)sulpiride will ensure a predominance of interaction of the labeled ligand with serotonin receptors (Altar *et al.*, 1985).

E. Scatchard Analysis

The Scatchard plot (Scatchard, 1949) is one of the most popular methods to represent the graphical analysis of radioligand receptor-binding studies. With the increasing number of neurotransmitter, as well as neurotransmitter receptor-binding studies in neuroscience, it is very important to use this method correctly and to analyze and interpret the data appropriately.

TABLE 1 Pharmacological Agents Used in Neurotransmitter Receptor-Binding Studies

Labeled ligand	Specific activity/concentration (Ci/mmol)(nM)	Unlabeled competitor	Concentration/incubation (μM/condition)	Neurotransmitter receptor species assayed
DL-Benzilic-[4,4'-³H]-quinuclidinyl benzilate	30.1/1.0	Atropine	1.0/60 min/37°C	Muscarinic Cholinergic
N-[methyl-³H]SCH-23390	74.5/1.0	(+)Butaclomol	1.0/20 min/37°C	Dopamine (D1)
[1-phenyl-4-³H]Spiroperidol	24.5/1.0	(+)Butaclamol	1.0/20 min/37°C	Dopamine (D2)
Methylene-[³H(N)]muscimol	20.6/2.0	GABA	2.0/30 min/4°C	GABA
[methyl-³H]Diazepam	83.0/0.75	Diazepam/flunitrazepam	1.0/45 min/4°C	Benzodiazepine
[1,2-³H(N)]Serotonin	26.3/3.0	Serotonin	1.0/15 min/37°C	Serotonin
[G-³H]Strychnine sulfate	15.0/4.0	Strychnine	1.0/20 min/4°C	Glycine
9-10-[0,10,³11-³H(N)]Dihydro-α-ergocryptine	23.0/1.3	Ergocryptine	1.0/30 min/25°C	α-Adrenergic
levo-[proply-1,2,3-³H]	47.5/0.7	Alprenolol	1.0/30 min/25°C	β-Adrenergic
[Piperidinyl-³H]CGS-19755 dihydroalprenolol	60.5/10.0	L-Glutamate	1000/15 min/4°C	NMDA/glutamate
[N-ally-2,3-³H]Naloxone	50.2/1.0	Levalorphan	1.0/45 min/25°C	Opiate
Opiate subtype				
Tyrosyl-[3,5-³H(N)]DAGO	40.1/1.0	Etorphine	1.0/45 min/25°C	μ
[9-³H(N)]Bremazocine	33.3/1.0	Etorphine	1.0/45 min/25°C	κ
Tyrosyl-[3,5-³H(N)]enkephalin	30.5/1.0	Etorphine	1.0/45 min/25°C	δ
Piperidyl-[3,4-³H(N)]TCP	42.2/5.0	Phencyclidine	5.0/45 min/25°C	σ (PCP)

IV. Effects of Selective Neurotoxicants on Neurotransmitter and Neuroreceptor System

A. MPTP

1-Methyl-4-phenyl-1,2,3,6-tetrahydropyridine (MPTP) causes neurotoxicity in rodents and nonhuman primates by affecting primarily the dopamine neurotransmitter system and selectively depleting dopamine in the striatum of rodents and nonhuman primates (Ali *et al.*, 1993a, 1994a; Johannessen *et al.*, 1985; Johannessen, 1991; Irwin *et al.*, 1992, 1993; Sonsalla *et al.*, 1989; Javitch, 1985; Heikkila *et al.*, 1984, 1985; Mitchell *et al.*, 1985; Jenner *et al.*, 1984; Burns *et al.*, 1983). It has been demonstrated in 1-year-old C57/B6N mice that multiple, i.p., injections of MPTP (4× 10 mg/kg) produce a significant reduction in striatal dopamine concentrations 1 hr postdosing that continues until 72 hr (Ali *et al.*, 1993a). The effects on dopamine metabolites are transient. Similar effects were observed in 7-month-old animals, where a significant decrease of dopamine and its metabolites, DOPAC and HVA, was found 24 hr after the administration of MPTP (Fig. 4).

The depletion of dopamine, however, was not observed in very young animals (postnatal day 23). The dopamine depletion results clearly demonstrate that MPTP-induced neurotoxicity is age related (Ali *et al.*, 1993a, 1994a). Similar effects have been reported by others and it has been demonstrated that MPTP-induced age-related neurotoxicity and the accumulation of MPP⁺, the neurotoxic metabolite of MPTP, can be correlated with the depletion of dopamine in striatum (Ricaurte *et al.*, 1987; Langston *et al.*, 1987; Irwin *et al.*, 1988, 1992, 1993). Although the mechanism of action of MPTP is still unclear, depletion of dopamine remains the best and most reliable indicator of neurotransmitter system disruption that has been reported in the literature.

It has been demonstrated that a single dose of 40 mg/kg,i.p. MPTP administered to mice produces an elevation of reactive oxygen species (ROS) in the striatum within 15 min and dopamine depletion at 60 min after the dose administration. Again, this effect

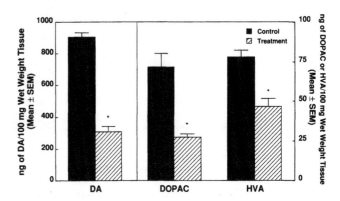

FIGURE 4 Effects of MPTP on 7-month-old mouse striatal dopamine and its metabolites (DOPAC and HVA) 24 hr postdosing. *$P < 0.001$, significantly different from control (adapted from Ali *et al.*, 1993a).

was observed in older (1 year) but not younger (1 month) mice (Ali *et al.*, 1994a). These data are consistent with the hypothesis that MPTP induces its neurotoxic effects by producing ROS, which in turn are responsible for producing the dopamine depletion. In any case, dopamine depletion remains the hallmark of MPTP-induced neurotoxicity in rodents, dogs, monkeys, and humans.

B. MDMA

3,4-Methylenedioxymethamphetamine (MDMA, Ecstasy) is a drug of abuse and is known to selectively effect the serotonergic system in the rodent and nonhuman primate. Data demonstrate that oral administration of MDMA, the most common route of administration in human, produces neurotoxicity in rat and monkey (Slikker *et al.*, 1988, 1989; Ali *et al.*, 1987, 1989, 1990a, 1993b). The effects of MDMA are quite marked (55 and 77% reduction in 5-HT concentration in frontal cortex in rat and monkey, respectively) as long as 4 weeks after treatment. These data indicate that MDMA effects on the serotonergic system are not transient but are long lasting. Studies confirm the long-lasting effects of MDMA in that a significant reduction of 5-HT concentration was observed in the hippocampus of female rats 4 months after eight successive oral doses of MDMA (40 mg/kg) (Ali *et al.*, 1989). In contrast, MDMA produced only a transient (2 weeks) alteration in the DA and norepinephrine systems that recovered to control levels by 4 weeks. These transient effects of MDMA on catecholeaminergic systems are consistent with the lack of effects of MDMA on tyrosine hydroxylase (Gibb *et al.*, 1986) and dopamine receptors (Ali *et al.*, 1990a).

To evaluate whether neurochemical changes can be correlated with behavioral alterations and neuropathological changes after exposure to this neurotoxic agent, a multidisciplinary approach was used that evaluated behavioral, neurochemical, and neuropathological end points (Fig. 5). It was found that oral administration of MDMA in rats produced a significant increase in a specific behavior, the serotonin motor syndrome (Slikker *et al.*, 1989; Ali *et al.*, 1991), a significant depletion of 5-HT in several brain regions (Gollamudi *et al.*, 1989), and also a significant increase in the abundance of abnormal nerve terminals (Scallet *et al.*, 1988). These studies suggest that neurochemical end points can complement other end points and may help in elucidating the mechanism of action of neurotoxic agents.

When data were compared among species, it was found that oral doses of MDMA (5 or 10 mg/kg) in the monkey resulted in a dramatic dose-related decrease (71–83% reduction) in 5-HT concentrations in the frontal cortex and hippocampus. As in the rat, the brain stem was least affected, exhibiting only a 25% reduction of 5-HT compared to control values. These lasting changes were observed 1 month after MDMA dosing and occurred in the absence of body weight alterations. The regional effects of MDMA on the serotonergic system

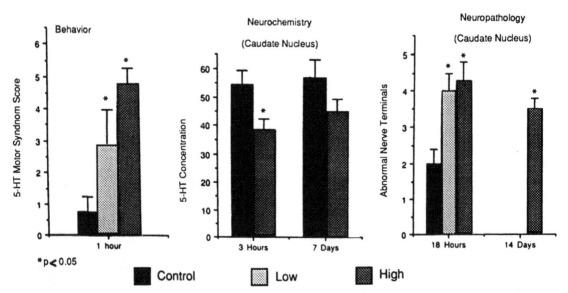

FIGURE 5 Effects of oral administration of MDMA on the serotonin motor syndrome score, concentration of serotonin in the caudate nucleus, and density of silver impregnation in the caudate nucleus of the rat. Note that there was a significant dose-related increase in the serotonin motor syndrome score, a depletion of serotonin concentration, and a marked increase in silver impregnation in the caudate nucleus. *$P < 0.01$, significantly different from saline control (adapted from Ali *et al.*, 1989).

are shown in Fig. 6 for both the rat and the monkey. It is interesting to note that the greatest reduction of 5-HT content occurred in those brain areas containing an abundance of serotonergic axons and nerve terminals (Cooper *et al.*, 1984). Those areas supporting serotonergic cell bodies (brain stem) were less severely affected by MDMA treatment. These neurochemical data suggest that the effects of MDMA on the serotonergic system are predominantly directed toward the terminals rather than the soma of the neuron. It was also demonstrated that an intracerebral injection of MDMA in rat did not produce a significant alteration in 5-HT and 5-HIAA concentrations (Ali *et al.*, 1990b), suggesting that a peripheral metabolite of MDMA may be responsible for the observed serotonergic neurotoxicity. Regardless of the mechanism, the quantification of 5-HT and 5-HIAA levels remains one of the most frequently used approaches to assess MDMA-induced effects.

C. Metals

1. Trimethyl Tin (TMT)

Organometals, such as TMT and triethyl tin (TET), have been used for a variety of industrial and agricul-

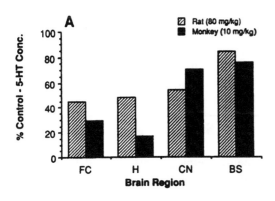

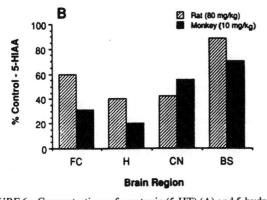

FIGURE 6 Concentrations of serotonin (5-HT) (A) and 5-hydroxy-indoleacetic acid (5-HIAA) (B) expressed as percentages of control in different regions of the rat and monkey brain after oral administration of MDMA at 80 and 10 mg/kg, respectively. FC, frontal cortex; H, hippocampus; CN, caudate nucleus, BS, brain stem (adapted from Ali *et al.*, 1990a).

tural purposes such as plastics stabilizers, chemosterilants, and also as biocides to control fungus, bacteria, and insects (Smith and Smith, 1975). These organotins, particularly TMT and TET, produce a variety of neurotoxic effects in both laboratory animals and humans. A single injection of TMT produces a significant alteration in muscarinic cholinergic receptor binding in different brain areas in mice (Ali *et al.*, 1986). Two days after a single dose of TMT, the affinity of muscarinic cholinergic receptor binding in frontal cortex increased. This increase, however, was transient because it returned to control values after 2 weeks (Fig. 7). This dose of TMT also produced a significant decrease in the rate of dopamine turnover in caudate nucleus of mice (Ali *et al.*, 1986). A single injection of TMT also produced a dose-dependent increase in aspartic acid, glutamine, and glycine in several brain areas of mice (Lipe *et al.*, 1991). Similar effects have also been reported in rat brain. According to these reports, a single injection of TMT produces a significant alteration in the concentration of several amino acids in different regions of rat brain (Hikal *et al.*, 1988; Wilson *et al.*, 1986).

2. Manganese (Mn)

Mn is an essential element. Both deficiency and excess of Mn can produce neurotoxicity. Chronic exposure to Mn has been shown to produce significant alterations in monoamine levels in adult mice (Gianutsos and Murray, 1982), rat (Autissier *et al.*, 1982; Bonilla and Dietz-Edward, 1974; Gianutsos and Murray, 1982), rabbit (Mustafa and Chandra, 1971), monkey (Neff *et al.*, 1969; Bird *et al.*, 1984), and human (Hornykiewicz, 1972). In rat brain, Mn also inhibits the synaptosomal uptake of dopamine, norepinephrine, serotonin, glutamate, and GABA (Lai *et al.*, 1984). Mn also produces a significant alteration in dopamine, serotonin, and muscarinic cholinergic receptor binding in rat brain (Seth *et al.*, 1981; Donaldson and LaBella, 1984). Intrastriatal injection of Mn also produces significant alterations in dopamine (Lista *et al.*, 1986; Sloot *et al.*, 1994). Therefore, the use of these basic techniques in assessing neurochemical alterations are well accepted and are frequently used by neurotoxicologists.

D. Methamphetamine (METH)

METH is a drug of abuse and causes neurotoxicity by affecting the dopamine system. METH has been reported to produce the long-term depletion of dopamine and its metabolites (Fuller and Hemrick-Luecke, 1980; Kogan *et al.*, 1976; Seiden *et al.*, 1975), to alter high-affinity dopamine uptake sites (Kovachich *et al.*, 1989; Ricaurte *et al.*, 1980; Steranka and Sanders-

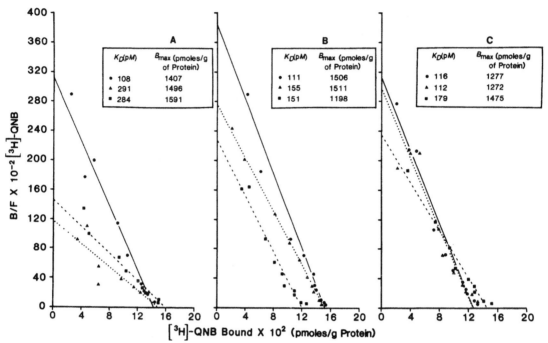

FIGURE 7 Scatchard plots of [³H]quinuclidinyl benzilate binding to frontal cortex membrane preparation from mouse brain 48 hr (A), 1 week (B), and 2 weeks (C) after acute administration of 0 (●), 1 (▲), or 3 (■) mg/kg TMT. Each point is the mean of duplicate determinations from a typical saturation analysis experiment of total and nonspecific binding. The lines were fitted to the data points by linear regression analysis with *r* values ranging from 0.829 to 0.998 (adapted from Ali *et al.*, 1986).

Bush, 1980; Wagner *et al.*, 1980), and to decrease the activity of tyrosine hydroxylase in striatum (Gibb and Kogan, 1979; Hotchkiss and Gibb, 1979; Fibiger and McGeer, 1980). It has been reported that repeated injections of METH produce a significant alteration in dopamine and serotonin concentrations in striatum of two strains of mice (Ali *et al.*, 1994a,b,c). O'Callaghan and Miller (1994) and Miller and O'Callaghan (1994) also reported the depletion of dopamine after multiple injections of METH. Similar effects have been reported in the rat (Bowyer *et al.*, 1992, 1993, 1994). The mechanism of METH-induced depletion of dopamine in the striatum is still unclear. Cadet and colleagues (1994, 1995) reported the attenuation of METH-induced neurotoxicity in Cu/Zn superoxide dismutase transgenic mice as compared to control mice. The authors used DA depletion as an index of METH-induced neurotoxicity. These data suggest that these METH-induced effects may be mediated through the generation of free radicals. The METH-induced depletion of dopamine can be blocked by pretreatment with NMDA receptor antagonist (+)MK-801 in mice and rat (Ali *et al.*, 1994a,b,c; Bowyer *et al.*, 1992; Weihmuller *et al.*, 1992; O'Callaghan and Miller, 1994) or by lowering the ambient temperature (Ali *et al.*, 1994b; Bowyer *et al.*, 1992, 1994). Figure 8 demonstrates the depletion of dopamine

and serotonin in striatum of mice after METH administration.

V. Summary and Conclusion

The neurochemical methods and their application described in this chapter represent a basic approach in the field of neurotoxicology. Although other more sophisticated techniques as described in other chapters are available, these basic biochemical techniques are powerful tools in describing the selective neurotoxicity of specific neurotoxicants. As part of an integrated, multidisciplinary approach for the assessment of neurotoxicants, these techniques can be used to describe biomarkers of effect and enhance our understanding of the mechanism(s) of neurotoxicity. As presented in this chapter, some neurotoxicants may produce relatively selective effects: MPTP effects on the dopamine system, MDMA effects on the serotonin system, TMT effects on muscarinic cholinergic receptor binding and amino acids concentrations, manganese effects on the dopamine system, and methamphetamine effects on both the dopamine and the serotonin systems. Therefore, these basic assessment tools of the neurotransmitter system can be used to build a foundation of under-

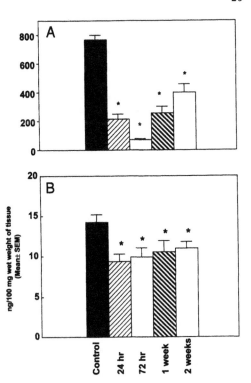

FIGURE 8 Time course of the effects of multiple injections of METH (4 × 10 mg/kg, i.p.) on striatal dopamine (A) and serotonin (B) concentrations in CD mice at room temperature (23°C). Each value is the mean ± SEM derived from six to eight animals/group. * Significantly different from control (modified from Ali *et al.*, 1994b).

standing on which to base further more sophisticated investigations.

References

Ali, S. F., Chang, L. W., and Slikker, W., Jr. (1991). Biogenic amino as biomarkers for neurotoxicity. *Biomed. Environ. Sci.* **4**, 207–216.

Ali, S. F., David, S. N., and Newport, G. D. (1993a). Age-related susceptibility to MPTP-induced neurotoxicity in mice. *Neurotoxicology* **14**, 29–34.

Ali, S. F., and David, S. N., Newport, G. D., Cadet, J. L., and Slikker, W., Jr. (1994a). MPTP-induced oxidative stress and neurotoxicity are age-dependent: evidence from measures of reactive oxygen species and striatal dopamine levels. *Synapse,* **18**, 27–34.

Ali, S. F., Holson, R. R., Newport, G. D., Slikker, W., Jr., and Bowyer, J. F. (1993b). Development of dopamine and N-methyl-D-aspartate systems in rat brain: the effect of prenatal phencyclidine exposure. *Brain Res.* **73**, 25–33.

Ali, S. F., Newport, G. D., Bailey, J. R., and Slikker, W., Jr. (1990a). Oral administration of MDMA produces selective serotonergic neurotoxicity in rodent and nonhuman primates. *SAAS Bull. Biochem.* **3**, 48–53.

Ali, S. F., Newport, G. D., Holson, R. R., Slikker, W. J., and Bowyer, J. F. (1994b). Low environmental temperatures of pharmacologic agents which produce hypothermia decreases methamphetamine neurotoxicity in mice. *Brain Res.,* **658**, 33–38.

Ali, S. F., Newport, G. D., O'Callaghan, J. P., and Miller, D. B. (1994c). Age-related susceptibility to methamphetamine-induced neurotoxicity in mice. *Toxicologist* **14**(1), 321.

Ali, S. F., Newport, G. D., Scallet, A. C., Binienda, Z., Ferguson, S. A., Bailey, J. R., Paule, M. G., and Slikker W., Jr. (1993c). Oral administration of 3,4-methylenedioxymethamphetamine (MDMA) produces selective serotonergic depletion in nonhuman primate. *Neurotoxicol. Teratol.* **15**, 91–95.

Ali, S. F., Scallet, A. C., Newport, G. D., Lipe, G. W., Holson, R. R., and Slikker, W. Jr. (1989). Persistent neurochemical and structural changes in rat brain after oral administration of MDMA. *Res. Comm. Subst. Abuse* **10**, 225–23.

Ali, S. F., Slikker, W., Jr., Newport, G. D., and Goad, P. T. (1986). Cholinergic and dopaminergic alterations in the mouse central nervous system following acute trimethyltin exposure. *Acta Pharmacol. Toxicol.* **59**, 179–188.

Ali, S. F., Slikker, W., Jr., Scallet, A. C., and Frith, C. H. (1987). Neurochemical and neuropathological changes produced by methylenedioxymethamphetamine (MDMA) in different regions of rat brain. *J. Neurochem.* **48**, S 112.

Ali, S. F., Tandon, P., Tilson, H. A., Lipe, G. W., Newport, G. D., and Slikker, W., Jr. (1990b). Intracerebral and oral administration of methylenedioxymethamphetamine (MDMA): distribution and neurochemical alterations in rat brain. *Eur. J. Pharmacol.* **183**, 450.

Altar, C. A., O'Neil, S., Walter, R. J., and Marshall, J. F. (1985). Brain dopamine and serotonin receptor sites revealed by digital subtraction autoradiography. *Science* **228**, 597–600.

Autissier, N., Rochette, L., Dumas, P., Beley, A., Loireau, A., and Bralet, J. (1982). Dopamine and norepinephrine turnover in various regions of the rat brain after chronic manganese chloride administration. *Toxicology* **24**, 175–182.

Biesold, D. (1974). Isolation of brain mitochondria. In *Research Methods in Neurochemistry* (N. Marks, and R. Rodnight, Eds.) pp. 39–52, Plenum, New York/London.

Bird, E., Anton, A. H., and Bullock, B. (1984). The effect of manganese inhalation on basal ganglia dopamine concentrations in Rhesus monkey. *Neurotoxicology* **5**, 59–68.

Bondy, S. C., and Ali, S. F. (1992). Neurotransmitter receptor. In *Neurotoxicology* (M. B. Abou-Donia, Ed.), pp. 121–154, CRC Press, Boca Raton, Florida.

Bonilla, E., and Dietz-Ewald, M. (1974). Effects of L-Dopa on brain concentrations of dopamine and homovanillic acid in rats after chronic manganese chloride administration. *J. Neurochem.* **22**, 297.

Bowyer, J. F., Davies, D. L., Schmued, L., Broening, H. W., Newport, G. D., Slikker, W. Jr., and Holson, R. R. (1994). Further studies of the role of hyperthermia in methamphetamine neurotoxicity. *J. Pharmacol. Exp. Ther.* **268**, 1571–1580.

Bowyer, J. F., Gough, B., Slikker, W., Jr., Lipe, G. W., Newport, G. D., and Holson, R. R. (1983). Effects of a cold environmental or age on methamphetamine-induced dopamine release in the caudate putamen of female rats. *Pharm. Biochem. Behav.* **44**, 87–98.

Bowyer, J. F., Tank, A. W., Newport, G. D., Slikker, W., Jr., Ali, S. F., and Holson, R. R. (1992). The influence of environmental temperature on the transient effects of methamphetamine on dopamine levels and dopamine release in rat striatum. *J. Pharmacol. Exp. Ther.* **260**, 817–824.

Briggs, C., and Cooper, J. (1981). A synaptosomal preparation from the guinea pig pleum myenteric plexus. *J. Neurochem.* **36**, 1097–1108.

Burns, R. S., Chiueh, C. C., Markey, S. P., Ebert, M. H., and Jacobowitz, D. M., and Kopin, I. J. (1983). A primate model of parkinsonism: selective destruction of dopaminergic neurons in pars compacta of the substantia nigra by 1-methyl-4-phenyl-1,2,3,6-tetrahydropyridine. *Proc. Natl. Acad. Sci. USA* **80**, 4546–4550.

Cadet, J. L., Ali, S., Rothman, R., and Epstein, C. J. (1995). Neurotoxicity, drugs of abuse and the Cu/Zn-superoxide dismutase transgenic mice. *Mol. Neurobiol.*, in press.

Cadet, J. L., Sheng, P., Ali, S. F., Rothman, R., Carlson, E., and Epstein, C. J. (1994). Attenuation of methamphetamine-induced neurotoxicity in copper/zinc superoxide dismutase transgenic mice. *J. Neurochem.* **62**, 380–383.

Cooper, J. R., Bloom, E. E., and Roth, R. H. (1984). *The Biochemical Basis of Neuropharmacology*, 4th ed., pp. 233–235. Oxford University Press, New York.

Donaldson, J., and LaBella, F. S. (1984). The effect of manganese on the cholinergic receptor in vivo and in vitro may be mediated through modulation of free radicals. *Neurotoxicology* **5**, 105–113.

Fibiger, H. C., and McGeer, E. G. (1980). Effect of acute and chronic methamphetamine treatment on tyrosine hydroxylase activity in brain and adrenal medulla. *Eur. J. Pharmacol.* **16**, 176–180.

Fuller, R. W., and Hemrick-Luecke, S. (1980). Long-lasting depletion of striatal dopamine by a single injection of amphetamine in iprindole-treated rats. *Science* **209**, 305–307.

Gianutsos, G., and Murray, M. T. (1982). Alterations in brain dopamine and GABA following inorganic or organic manganese administration. *Neurotoxicology* **3**, 75–84.

Gibb, J. W., Hanson, G. R., and Johnson, M. (1986). Effects of (+)-3,4-methylenedioxymethamphetamine [(+)MDMA] and (−)-3,4-methylenedioxymethamphetamine [(−)MDMA] on brain dopamine serotonin and their biosynthetic enzymes. *Neuroscience* **12**, 608–617.

Gibb, J. W., and Kogan, F. J. (1979). Influence of dopamine synthesis on methamphetamine-induced changes in striatal and adrenal hydroxylase activity. *NS Arch. Pharmacol.* **310**, 185–187.

Gollamudi, R., Ali, S. F., Lipe, G. W., Newport, G. D., Webb, P., Lopez, M., Leakey, J. E. A., Kolta, M. G., and Slikker, W., Jr. (1989). Influence of inducers and inhibitors on the metabolism in vitro and neurochemical effects in vivo of MDMA. *Neurotoxicology* **10**, 455–466.

Heikkila, R. E., Hess, A., and Duvoisin, R. D. (1984). Dopaminergic neurotoxicity of 1-methyl-4-phenyl-1,2,3,6-tetrahydropyridine in mice. *Science* **221**, 1451–1453.

Heikkila, R. E., Manzino, L., Cabbat, F. S., and Duvoisin, R. C. (1985). Studies on the oxidation of the dopaminergic neurotoxin 1-methyl-1,2,3,6-phenyltetrahydropyridine by monoamine oxidase-B. *J. Neurochem.* **45**, 1049–1054.

Hikal, A. H., Lipe, G. W., Slikker, W., Jr., Scallet, A. C., Ali, S. F., and Newport, G. D. (1988). Determination of amino acids in different regions of the rat brain. Application to the acute effects of tetrahydrocannabinol (THC) and trimethyltin (TMT). *Life Sci.* **42**, 2029–2035.

Hornykiewicz, O. (1972). Neurochemistry of Parkinsonism. In *Handbook of Chemistry* (S. Lajtha, ed.), Vol. 4, p. 465, Plenum Press, New York.

Hotchkiss, A. J., and Gibb, J. W. (1979). The long-term effects of multiple doses of methamphetamine on neostriatal tryptophan hydroxylase, tyrosine hydroxylase, choline acetyletransferase and glutamate decarboxylase activities. *J. Pharmacol. Exp. Ther.* **214**, 257–262.

Irwin, I., Delanney, L. E., and Langston, J. W. (1993). MPTP and aging: studies in the C57BL/6N mouse. *Adv. Neurol.* **60**, 197–206.

Irwin, I., Finnegan, K. T., Delanney, L. E., DiMonte, D., and Langston, J. W. (1992). The relationships between aging, monoamine oxidase, striatal dopamine and the effects of MPTP in C57BL/6 mice: a critical reassessment. *Brain Res.* **572**, 224–231.

Irwin, I., Ricaurte, G. A., Delanney, L. E., and Langston, J. W. (1988). The sensitivity of nigrostriatal dopamine neurons to MPP$^+$ does not increase with age. *Neurosci. Lett.* **87**, 51–56.

Jarvis, M. F., and Wagner, G. C. (1985). Age-dependent effect of 1-methyl-4-phenyl-1,2,3,6-tetrahydropyridine (MPTP). *Neuropharmacology* **24**, 581–583.

Javitch, J. A., D'Amato, R. J., Strittmatter, S. M., and Snyder, S. H. (1985). Parkinsonism-inducing neurotoxin, N-methyl-4-phenyl-1,2,3,6-tetrahydropyridine: uptake of the metabolite N-N-methyl-4-phenylpyridinium by dopamine neurons explains selective toxicity. *Proc. Natl. Acad. Sci. USA* **82**, 2173–2177.

Jenner, P., Rupniak, N. M. J., Rose, S., Kelley, E., Kilpatrick, G., Lees, A., and Marsden, C. D. (1984). 1-methyl-4-phenyl-1,2,3,6-tetrahydroypridine-induced parkinsonism in the common marmoset. *Neurosci. Lett.* **50**, 85–90.

Johannessen, J. N. (1991). A model of chronic neurotoxicity: long-term retention of the neurotoxin 1-methyl-4-phenylpyridium (MPP$^+$) within catecholaminergic neurons. *Neurotoxicology* **12**, 285–302.

Johannessen, J. N., Chiueh, C. C., Burns, R. S., and Markey, S. P. (1985). Differences in the metabolism of MPTP in the rodent and primate parallel differences in sensitivity to it neurotoxic effect. *Life Sci.* **36**, 219–224.

Kogan, F. J., Nichols, W. K., and Gibb, J. W. (1976). Influence of methamphetamine on nigral and striatal tyrosine hydroxylase activity and on striatal dopamine levels. *Eur. J. Pharmacol.* **36**, 363–371.

Kovachich, G. B., Aronson, C. E., and Brunswick, D. J. (1989). Effects of high-dose methamphetamine administration on serotonin uptake sites in rat brain measured using [^{3}H] cyanoimipramine autoradiography. *Brain Res.* **505**, 123–129.

Lai, J. C. K., Leung, T. K., and Lim, L. (1984). Differences in neurotoxic effects of manganese during development and aging: some observations on brain regional neurotransmitter and non-neurotransmitter metabolism in a developmental rat model of chronic manganese encephalopathy. *Neurotoxicology* **5**, 37–51.

Langston, J. W., Irwin, I., and Delanney, L. E. (1987). The biotransformation of MPTP and disposition of MPP$^+$: the effects of aging. *Life Sci.* **40**, 749–754.

Leysen, J. E., Niemegeers, C., Tollenaere, J. P., and Laduron, P. M. (1978). Serotonergic component of neuroleptic receptors. *Nature* **272**, 168–171.

Lipe, G. W., Ali, S. F., Newport, G. D., Scallet, A. C., and Slikker, W. Jr. (1991). Effect of trimethyltin on amino acid concentrations in different regions of the mouse brain. *Pharmacol. Toxicol.* **68**, 450–455.

Lista, A., Abarca, J., Ramos, C., and Daniels, A. J. (1986). Rat striatal dopamine and tetrahydrobiopterin content following an intrastriatal injection of manganese chloride. *Life Sci.* **38**, 2121–2127.

Matlib, M. A., and Schwartz, A. (1983). Selective effects of dilitiazem, a benzothiazepine calcium channel blocker, and diazepam, and other benzodiazepines on the Na$^+$/Ca$^+$ exchange carrier system of heart and brain mitochondria. *Life Sci.* **32**, 2837–2842.

Miller, M. L., Andringa, A., Adams, W., and Radlike, M. J. (1986). Intracisternal protein in the type II pneumocyte of the ferret, guinea pig, and mongrel dog. *J. Ultrastruct. Mol. Struct. Res.* **95**, 131–141.

Miller, D. B., and O'Callaghan, J. P. (1994). Environmental-, drug- and stress-induced alterations in body temperature affect the neurotoxicity of substituted amphetamine in the C57BL/6J mouse. *J. Pharmacol. Exp. Ther.*, **270**, 752–769.

Minnema, D., and Michaelson, I. A. (1985). A superfusion apparatus for the examination of neurotransmitter release from synaptosomes. *J. Neurosci. Methods* **14**, 193–206.

Mitchell, I. J., Cross, A. J., Sambrook, M. A., and Crossman, A. R. (1985). Sites of the neurotoxic action of 1-methyl-4-phenyl-

1,2,3,6-tetrahydropyridine in the macaque monkey includes the ventral tegmental area and the locus coeruleus. *Neurosci. Lett.* **61**, 195–200.

Mustafa, S. J., and Chandra, S. V. (1971). Levels of 5-hydroxytryptamine, dopamine and norepinephrine in whole brain of rabbits in chronic manganese toxicity. *J. Neurochem.* **18**, 931–939.

Neff, N. H., Barrett, R. E., and Costa, E. (1969). Selective depletion of caudate nucleus dopamine and serotonin during chronic manganese dioxide administration to squirrel monkeys. *Experientia* **25**, 1140–1146.

O'Callaghan, J. P., and Miller, D. B. (1994). Neurotoxicity profiles of substituted amphetamines in the C57BL/6J mouse. *J. Pharmacol. Exp. Ther.* **270**, 741–751.

Palkovits, M., and Brownstein, M. J. (1990). Maps and guide to microdissection of the rat brain. Academic Press, San Diego.

Ricaurte, G. A., Irwin, I., Forno, L. S., Delanney, L. E., and Langston, J. W. (1987). Aging and 1-methyl-4-phenyl-1,2,3,6-tetrahydropyridine-induced degeneration of dopaminergic neurons in the substantia nigra. *Brain Res.* **403**, 43–51.

Ricaurte, G. A., Schuster, C. R., and Seiden, L. S. (1980). Long-term effects of repeated methamphetamine administration on dopamine and serotonin neurons in the rat brain: a regional study. *Brain Res.* **193**, 153–163.

Scallet, A. C., Ali, S. F., Holson, R. R., Lipe, G. W., and Slikker, W. Jr. (1988). Neuropathological evaluation by combined immunohistochemistry and degeneration-specific methods: application to methylenedioxymethamphetamine. *Neurotoxicology* **9**, 529–538.

Scatchard, G. (1949). The attraction of proteins for small molecules and ions. *Ann. N.Y. Acad. Sci.* **51**, 660–672.

Seiden, L. S., Fischman, M. W., and Schuster, C. R. (1975). Long-term methamphetamine induced changes in brain catecholamines in tolerant rhesus monkeys. *Drug Alcohol Depend.* **1**, 215–219.

Seth, P. K., Hong, J. S., Kilts, C., and Bond, S. C. (1981). Alteration of cerebral neurotransmitter receptor function by exposure of rats to manganese. *Toxicol. Lett.* **9**, 247–254.

Slikker, W., Jr., Ali, S. F., Scallet, A. C., Frith, C. H., Newport, G. D., and Bailey, J. R. (1988). Neurochemical and neuropathological alteration in the rat and monkey produced by orally administered methylenedioxymethamphetamine (MDMA). *Toxicol. Appl. Pharmacol.* **94**, 448–457.

Slikker, W., Jr., Holson, R. R., Ali, S. F., Kolta, M. G., Paule, M. G., Scallet, A. C., McMillan, D. E., Bailey, J. R., Hong, J. S., and Scalzo, F. M. (1989). Behavioral and neurochemical effects of orally administered MDMA in the rodent and nonhuman primate. *Neurotoxicology* **10**, 529–542.

Sloot, W. N., van der Sluijs-Gelling, A. J., and Gramsbergen, J. B. P. (1994). Selective lesions by manganese and extensive damage by iron after injection into rat striatum or hippocampus. *J. Neurochem.* **62**, 205–216.

Smith, P., and Smith, L. (1975). Organotin compounds and applications. *Chem. Br.* **11**, 208–226.

Sonsalla, P. K., Nicklas, W. J., and Heikkila, R. E. (1989). Role of excitatory amino acids in methamphetamine induced nigrostriatal dopaminergic toxicity. *Science* **243**, 398–400.

Steranka, L. R., and Sanders-Bushy, E. (1980). Long-term effects of continuous exposure to amphetamine on brain dopamine concentrations and synaptosomal uptake in mice. *Eur. J. Pharmacol.* **65**, 439–443.

Szebenyl, E. S. (1970). *Atlas of Macaca mulatta.* University Press, Rutherford, New Jersey.

Wagner, G. C., Ricaurte, G. A., Seiden, L. S., Schuster, C. R., Miller, R. J., and Westley, J. (1980). Long-lasting depletions of striatal dopamine and loss of dopamine uptake sites following repeated administration of methamphetamine. *Brain Res.* **181**, 151–160.

Weihmuller, F. B., O'Dell, S. J., and Marshall, S. J. (1992). MK-801 protect against methamphetamine-induce striatal dopamine terminal injury is associated with attenuated dopamine overflow. *Synapse* **11**, 155–163.

Wilson, W. E., Hudson, P. M., Kamamatsu, T., Walsh, T. J., Tilson, H. A., Hong, J. S., Marenpot, R. R., and Thompson, M. (1986). Trimethyltin-induced alterations in brain amino acids, amines and amine metabolites: relationship to hyperammonemia. *Neurotoxicology* **7**, 63–74.

Yamamura, H. I., Enna, S. G., and Kuhar, M. J. (1984). *Neurotransmitter receptor binding.* Raven Press, New York.

Zielke, H. R. (1985). Determination of amino acids in the brain by high performance liquid chromatography with isocratic elution and electrochemical detection. *J. Chromatogr.* **347**, 320–324.

21

Biomolecular Approaches to Neurotoxic Hazard Assessment

JAN N. JOHANNESSEN
Neurobehavioral Toxicology Team
Division of Toxicological Research
Center for Food Safety and Applied Nutrition
U.S. Food and Drug Administration
Laurel, Maryland 20708

I. Introduction

The development of rapid, convenient, and sensitive methods which reliably illuminate the potential for adverse neurological actions of a compound is a high priority within the neurosciences. A recent Office of Technological Assessment report (U.S. Congress, 1990) indicates that little or no data on possible adverse neurological effects exist for a large number of chemicals that are in common use. Epidemiological studies indicate that the incidence of several neurological diseases such as Parkinson's disease has increased significantly over the last several decades (Lillianfeld *et al.*, 1990). The idiopathic nature of these diseases has led some to speculate that the rapid increase in the post-World War II industrial and agricultural use of potentially hazardous chemicals may play a role (Barbeau *et al.*, 1987; Koller *et al.*, 1990; Tanner *et al.*, 1989).

Examples of environmentally induced neurological disorders are well documented. Extrapyramidal disorders in manganese miners, mercury poisoning in consumers of fish tainted with methyl mercury, lathyrism from excessive consumption of the chick pea (U.S. Congress, 1990), and amnesic shellfish poisoning associated with domoic acid contamination (Perl *et al.*, 1990) are but a few examples. While there is no question that focal populations exposed to unusual environmental chemicals or to high levels of neurotoxicants in the workplace may manifest overt symptoms of neurological disease, the broader question of the long-term effects of sporadic or chronic low-dose exposure remains largely unanswered. Several epidemiological studies have sought to identify environmental conditions that may contribute to the incidence of Parkinson's disease (Barbeau *et al.*, 1987; Koller *et al.*, 1990; Tanner *et al.*, 1989). Although some statistically significant correlations have been noted, the risk factors are broad, including such conditions as rural versus urban living, source of drinking water, and local use of agricultural chemicals. Thus, the contributions of individual chemicals to the calculated risk and the relationship of increased risk to individual exposure rates are difficult to discern. Further, the increased risk from exposure to these conditions is often low—perhaps less than

1.5 times that seen in the general population—and the results of these studies often reach conflicting conclusions. Clearly, a more thorough understanding of mechanisms by which toxicants exert their effects and the development of reliable markers which can flag potential neurotoxicants is necessary.

In approaching the tasks of developing methods to detect potential adverse neurological effects of xenobiotics and studying their mechanisms of action, one must consider methods suited to assessing the integrity of the brain at varying levels of organizational complexity. Behavior would constitute a measure of the highest level of organizational complexity, requiring the proper functioning and integration of a number of neuronal subsystems. Thus, methods for detecting decrements in such things as learning, motor activity, appetitive behavior, and sensory discrimination are the most broadly applied methods to screen for potential neurotoxicity (Sobotka, 1991) and are covered elsewhere in this series. There are potential pitfalls in using behavioral methods exclusively. Because of the redundancy built into the nervous system, extensive damage to a particular neural system may be necessary before overt behavioral changes are detected. For example, it is estimated that there must be an 80% reduction in striatal dopamine levels before parkinsonian symptoms appear. To complicate matters, the severity of the behavioral perturbations induced by the loss of a given system may vary with species. Lesioning the primate nigrostriatal dopamine system with the neurotoxin 1-methyl-4-phenyl-1,2,3,6-tetrahydropyridine (MPTP) results in a severe and permanent parkinsonian syndrome, whereas the dog exhibits a more mild motor disturbance from which it rapidly recovers, despite a permanent and severe (>96%) loss of the nigrostriatal dopamine system (Burns et al., 1983; Johannessen et al., 1989). Using the nigrostriatal system as an example, the opposite situation—behavioral deficits in the absence of neural damage—can also occur. The functional output of this system can be blocked for long periods of time (weeks) by such pharmacologic agents as reserpine. Thus, behavioral testing may give clues as to which neuronal system is affected, but a functional assessment together with an examination of the biochemical changes induced within that system is required for a meaningful description of a compound's effects.

The second organizational level at which neurotoxicity can be assessed is that of the individual neurons and nuclei which form the basic functional units of the nervous system. The histologic examination of brain sections is a major adjunct to behavioral testing in the identification of potentially neurotoxic compounds. A number of factors suggest that conventional histopathologic screening, while clearly an essential part of the hazard identification process, also may have limitations when it comes to the identification of potential neurotoxins. It has been estimated that the brain has over 600 distinct neuronal types in addition to many nonneuronal cell types (Switzer, 1993). The diversity of energy requirements, specific uptake systems, dependence on specific trophic factors, and other attributes render individual neuronal types selectively vulnerable to different insults. Thus, MPTP selectively targets nigrostriatal dopaminergic neurons, certain amphetamines target either serotonergic or dopaminergic neurons, and excitatory amino acids preferentially affect subsets of cortical and hippocampal neurons. This recognition that neurotoxins may affect only particular neuronal types (selective vulnerability) makes the ability to detect the loss of small groups of cells an essential capability. In order to screen across all neuronal phenotypes, these classical staining methods require time-consuming examination of a tremendous number of slides per animal by a highly trained pathologist. Further, subtle neurotoxic effects on neurons, which may be a harbinger of more severe toxicity at higher doses (e.g., loss of terminal portions of the axon), require special staining techniques for detection (Switzer, 1993).

The measurement or detection of biochemical changes within tissue and cells of the nervous system represents a third level at which damage from potential neurotoxicants may be detected and characterized. The biochemical end points discussed in this chapter which may prove useful in the assessment of a potential neurotoxic hazard fall into several broad categories. Perhaps the most widely used neurochemical end points have been those molecules that comprise the machinery of chemical neurotransmission: neurotransmitters and neuromodulators along with their attendant synthetic and degradative enzymes, receptors, reuptake sites, and storage vesicles. Because specific transmitters and related molecules are associated with certain subsets of neurons, they will be referred to as *phenotypic markers,* although this term is not entirely accurate. In addition, the measurement of certain proteins which are common to all neurons, but are not found in nonneuronal cells (*neurotypic markers*), has also been evaluated for the assessment of neural damage.

A second category of potential markers for neurotoxicant-induced damage are those biochemical constituents of neural tissue that exhibit dramatic increases in concentration in response to neuronal stress or damage. Ideally one would like a marker that would detect neuronal injury regardless of neuronal type or the nature of the insult. In addition, the marker should give a high signal against a normally low background and be technically convenient. Such candidate markers considered here include prolonged changes in ion homeostasis or gene expression. The third type of biochemical end point considered here is one that reflects extensive breakdown of essential cellular macromolecules such as DNA and structural proteins.

II. Phenotypic and Neurotypic Markers

The measurement or histologic visualization of phenotypic specific neuronal markers has been one of the more productive and widely used biochemical approaches to identifying and characterizing the effects of potential neurotoxins. Measuring the loss of neurotransmitters and related molecules provides a sensitive and technically routine approach to detecting toxicity manifested in a biochemically defined set of neurons. The fact alluded to earlier that neurotoxins generally exhibit fairly narrow selectivities, i.e., by exploiting neuronal macromolecules for bioactivation and selective neuronal uptake, clearly places extreme limits on the usefulness of examining a narrow range of phenotypic markers as a general screen for neurotoxicity.

Once a specific, biochemically defined set of neurons has been identified as a target population for a particular neurotoxin, the phenotypic markers become extremely valuable in several aspects of characterizing the effects and mechanism of the toxin. Sensitive methods for the rapid quantitative analysis of neurotransmitters and related molecules, e.g., high-performance liquid chromatography (HPLC), radio enzymatic assays, radioimmunoassays and enzyme-linked immunosorbent assays (RIA and ELISA), and receptor-binding assays, have been extant for some time and access to these methods is widely available. Antibodies that specifically recognize a large number of neurotransmitters, related enzymes, and, more recently, receptors can be obtained commercially, enabling the immunohistochemical examination of a desired subset of neurons within a given brain region.

Use of the rapid, sensitive, and quantifiable end points attainable with the use of phenotypic markers is essential for the full characterization of a compound's potential adverse effects. Although the definition of what constitutes an adverse effect may vary, once defined, the quantitation of a given end point is necessary for determining the dose–response, the no-effect level (NOEL), and the time course of action. Measuring changes in the levels of phenotypic markers is also useful in revealing differences in sensitivity across

brain regions, in making interspecies comparisons, and for mechanistic studies.

The loss of molecules common to most neuronal types (neurotypic markers) has also been evaluated as a biochemical end point. An example is the marked hippocampal loss of synapsin I and p38, two proteins associated with neurotransmitter storage vesicles, which occurs following the pronounced hippocampal damage induced by treatment with trimethyltin (Brock and O'Callaghan, 1987). The loss of other neurotypic proteins, including neurofilament proteins and neuron specific enolase, has been used as an index of neuronal damage. Because neurotypic markers are normally occurring components of many neuronal types and are thus present in significant amounts in normal tissue, the loss of one cell type from a brain area containing an admixture of many neuronal types may result in a quantitatively insignificant decrease in the concentration of that neurotypic marker. Thus, a dramatic loss of neurons from a dissected region may be accompanied by permanent and quantifiable decreases in the levels of neurotypic markers, but minor damage may go undetected.

The need for new biochemical and molecular approaches, which would enable the detection of neural damage induced by neurotoxins without regard to the nature of the target neuron or the mechanism of the toxin's action, is clearly established. The remainder of this chapter is devoted to an examination of several approaches which may be deserving of consideration as potentially useful neurotoxicologic end points, including the localization and measurement of calcium uptake, altered neuronal (protooncogenes, stress genes) and glial (glial fibrillary acidic protein; laminin-binding protein) gene expression, and markers of protein and DNA breakdown. The list of approaches is by no means exhaustive and two caveats are in order. First, measuring neurochemical changes should not be confined to test tubes and gels. Because the extreme heterogeneity of the brain has such tremendous functional significance, many qualitative and quantitative methods have been adapted for use directly on tissue sections. The use of chemical neuroanatomy is an important adjunct to the use of standard methods of quantitative analysis and is an important part of any neurochemical analysis. Second, the usefulness of the end points described herein, with the notable exception of glial fibrillary acidic protein, has not been evaluated using a large number of neurotoxins. Because of the tremendous clinical emphasis on stroke, the potential markers of neural degeneration have been primarily correlated with neuronal damage secondary to ischemia, which can be viewed as a form of excitotoxicity, or to treatment with excitotoxins such as kainate or *N*-

methyl-D-aspartate (NMDA). It is hoped that bringing attention to these approaches will spark a broader evaluation of their potential as useful tools in neurotoxicology.

III. Uptake of Calcium by Injured Neuronal Tissue *in Vivo*

The theory that sustained elevated intracellular calcium concentrations secondary to massive calcium influx constitutes a final common pathway leading to cell death has received considerable attention (Cheung *et al.,* 1986; Schanne *et al.,* 1979). Small local fluctuations in intracellular calcium concentrations, mediated by calcium channels or release from intracellular stores, are an important intracellular trigger for biochemical processes which underlie essential neurophysiological functions such as neurotransmitter release. In addition to normal physiological signaling, calcium can also activate a number of lytic enzymes, including proteases (Siman and Noszek, 1988), endonucleases (McConkey *et al.,* 1988), and phospholipases (Farooqui *et al.,* 1990), which can, if overstimulated, damage a cell (Nicotera *et al.,* 1989; Siesjö and Bengtsson, 1989). Thus, the ability to detect focal areas of excessive calcium accumulation in the brain *in vivo* may prove a valuable approach to identifying neuronal populations that are vulnerable to a given neurotoxin. The development of calcium-sensitive fluorescent dyes, coupled with confocal microscopy, has sparked numerous investigations of the relationship of intracellular free calcium levels to physiological and degenerative events within cultured neurons. However, these techniques have not been adapted to the study of Ca^{2+} uptake in the intact brain.

A diversity of *in vivo* approaches has been used to assess focal accumulation of calcium in brain. These methods have, in many cases, yielded extremely concordant results, despite the widely varying complexity of the methodologies employed and the anatomical resolution of the approaches. The model of neuronal degeneration in which the role of calcium accumulation has been the most extensively studied, and for the purposes of this discussion therefore provides the most complete data set, is that of delayed degeneration of hippocampal CA1 pyramidal cells secondary to transient forebrain ischemia (Pulsinelli *et al.,* 1981). Although particulars of the paradigm vary between investigators, ischemia induced by occlusion of the blood supply to the forebrain for 5 to 30 min generally results in the delayed neuronal degeneration of hippocampal CA1 pyramidal cells. The postischemic loss of CA1 neurons appears largely due to the excessive release of

endogenous glutamate as evidenced by the protective effects of lesioning the glutamatergic inputs to CA1 or administrating MK-801, which blocks the actions of glutamate at the NMDA receptor (Jørgensen *et al.,* 1987; Wieloch *et al.,* 1985). Thus the ischemic model of CA1 degeneration can be viewed as a form of excitotoxicity.

Methods for measuring changes in postischemic intracellular calcium in CA1 neurons generally employ postmortem methodologies; however, real-time monitoring of calcium uptake has been indirectly measured by the use of calcium-sensitive microelectrodes inserted into the interstitial space (Benveniste *et al.,* 1988). Within 2 min of the induction of ischemia, the extracellular calcium concentration rapidly drops to roughly 10% of its resting level, remaining at this concentration for the duration of the ischemic episode, recovering rapidly to basal levels on reperfusion. This rapid loss of extracellular calcium is thought to reflect rapid and massive cellular influx via ligand-gated calcium channels since the drop does not occur in animals in which the glutamatergic inputs to CA1 have been severed or in animals that received infusions of the NMDA antagonist AVP (Benveniste *et al.,* 1988). Although this technique, which has also been used to monitor perturbations in extracellular calcium concentrations during spreading depression (Hansen and Zeuthen, 1981), allows for precise real-time measurements, it reflects Ca^{2+} uptake in an anatomically minute area and does not reveal the identity of the cell type(s) (e.g., neurons, glia) in which calcium influx is occurring. For this reason, the method is most useful for certain mechanistic studies, but is not generally applicable for identifying neurotoxic compounds.

A second approach to detecting increased uptake of Ca^{2+} in neural tissue following ischemia has been to measure total calcium content in hippocampus, either by dissection followed by digestion and subsequent determination of ion content by atomic absorption spectrometry or by particle-induced X-ray emission (PIXE) using a proton microprobe in tissue slices (Deshpande *et al.,* 1987; Martins *et al.,* 1988). Both of these approaches yield accurate measurements of tissue calcium, but do not yield information on the localization of calcium within specific cell types. Increased levels of tissue calcium after ischemia follow a similar time course observed using other methods. One to 2 days following the ischemic insult calcium levels begin to rise markedly. In addition to the coincidence of high calcium concentrations and areas of cellular degeneration, areas in which degenerating fibers, but not cell bodies, occurred also exhibited increased calcium levels. This approach has also been applied to the study of long-term changes in tissue calcium levels following focal injections of kainic acid (Korf and Postema, 1984). Increased tissue Ca^{2+} levels have been observed for at least 56 days after kainate treatment; thus they far outlast the period of neural degeneration. The magnitude of the increased calcium concentrations in kainate-treated brain areas can reach 10-fold that measured in control tissue.

Histochemical methods have also been used to evaluate postischemic neuronal calcium influx, both at the light and electron microscopic level (Dux *et al.*, 1987; Mies *et al.*, 1990; Nakamura *et al.*, 1993). By applying these techniques to postischemic brain tissue, the ultrastructural localization of calcium can be identified and correlated with the morphological criteria for cellular degeneration. These techniques have revealed that degenerating CA1 neurons exhibit two temporally distinct periods of calcium influx: one that begins during the ischemic episode and lasts less than 30 min, and a second, more prolonged period that begins roughly 24 hr after the ischemic episode and persists throughout the period of neuronal degeneration. In addition to neuronal accumulation, significant glial and extracellular deposits of calcium are also seen following ischemia. This observation may account for the persistence of high tissue calcium levels long after the vulnerable neurons have degenerated. The histochemical approaches provide a convenient method of simultaneously observing a chemical outcome (i.e., calcium influx) and neuronal damage at the cellular level. One disadvantage of this method is the inability to accurately quantify extracellular or intracellular calcium concentrations; however, the relative scale based on staining intensity has yielded data that are consistent with those obtained using chemical methods.

The method for visualizing the accumulation of calcium in areas that have sustained neural damage that best defines the spatial and temporal limits of calcium influx as a marker of degenerating neurons, is the uptake of $^{45}Ca^{2+}$ and other radionuclides, followed by autoradiography of brain slices (Fig. 1) (Dienel, 1984; Mies *et al.*, 1990; Nakamura *et al.*, 1993). Using an ischemic model, Dienel has carefully explored the effects of varying the time of $^{45}Ca^{2+}$ injection relative to the period of ischemia, and total $^{45}Ca^{2+}$ recirculation time, on accumulation by damaged tissue (Dienel, 1984). His studies have revealed a number of important characteristics of $^{45}Ca^{2+}$ uptake by damaged brain tissue. The accumulation of $^{45}Ca^{2+}$ in brain is very selective and time dependent. Regional and temporal differences in $^{45}Ca^{2+}$ uptake correlate with the extent of neuronal damage, which in turn depends on the extent of the ischemia and the sensitivity of neuronal subpopulations to ischemia. Uptake is not restricted to the soma

of injured neurons, but includes dentritic areas. A much larger difference between injured and uninjured tissue is seen when comparing $^{45}Ca^{2+}$ uptake than by comparing tissue calcium levels as measured by atomic absorption spectrometry. Thus the use of $^{45}Ca^{2+}$ may enhance the contrast between injured and uninjured neural tissue. In addition to localization of postischemic damage, increased uptake and long-term retention of $^{45}Ca^{2+}$ have been observed in brain areas undergoing degeneration following administration of the neurotoxins, kainate, 6-hydroxydopamine, and capsaisin (Gramsbergen *et al.*, 1988; Janscó *et al.*, 1984; Van Den Berg and Gramsbergen, 1993).

The use of radionuclides as probes for injured neural tissue has been extended in an attempt to adapt its use to clinically useful applications, such as positron emission tomography (PET). A number of candidate ions have been assessed using the ischemic rat model (Dienel and Pulsinelli, 1986; Gramsbergen *et al.*, 1988). Several, especially ^{63}Ni, were found to reveal injured neuronal tissue in a manner similar to $^{45}Ca^{2+}$ (Fig. 1). The accumulation of radioactive forms of cobalt has also been evaluated as a marker of neural damage. In experimental degeneration of the frontal cortex induced by microinjections of kainate in the cat, visualization of ^{55}Co with PET following systemic administration showed selective accumulation within damaged areas (Gramsbergen *et al.*, 1988). This approach to localizing neuronal damage has been applied to the detection of stroke damage in humans. Suspected damage, which did not show up on MRI scans, was revealed by the increased uptake of ^{55}Co in the affected areas as visualized with PET scanning (Jansen *et al.*, 1993).

Evaluating the uptake of calcium in discrete brain areas, either by measuring the tissue levels of calcium or by visualizing the uptake of $^{45}Ca^{2+}$, clearly has potential as a method for identifying neurotoxicant-induced neurodegeneration. Although the majority of this work has been done with ischemia, the utility of this approach in identifying degenerating neuronal populations has been demonstrated with a limited number of neurotoxins as well as following traumatic spinal injury. The observation of significant extracellular and nonneuronal calcium deposits in areas showing neuronal damage extends the usefulness of this approach beyond the original premise that increased calcium accumulation would reflect neuronal influx. Additional studies which seek to investigate the universal nature of increased calcium uptake in neural tissue damaged by a number of neurotoxins with diverse modes of action would help to address the general nature of this phenomenon.

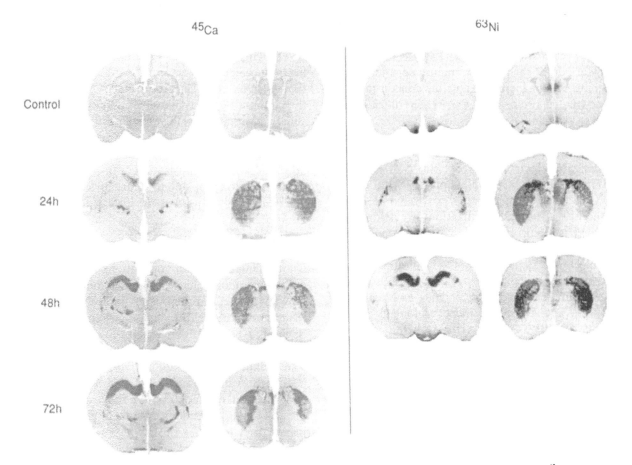

FIGURE 1 Effects of 30 min of ischemia (four-vessel occlusion) in the rat on regional accumulation of ^{45}Ca (left) or ^{63}Ni (right). The times at left are the hours of recirculation prior to sacrifice. Autoradiographs of sections at the level of the caudate and anterior hippocampus are shown. Radionuclides were injected 5 hr prior to sacrifice. Note that the early accumulation of these radionuclides in the striatum precedes the maximum accumulation in hippocampus. This sequence parallels the times at which significant pathology develops in these two areas. (Figure reproduced from Dienel, 1984; Dienel and Pulsinelli, 1986.)

IV. Altered Gene Expression

Toxic insults induce rather extreme perturbations in cellular homeostasis which can trigger equally extreme reactive changes. Toxins may induce rapid influxes of Ca^{2+}, generate free radicals, cause fluctuations in intracellular pH, or cause a rapid drop in energy stores. Either directly or through a number of transduction mechanisms, these types of perturbations can either induce or suppress the expression of neuronal genes and profoundly influence neuronal protein synthesis.

Overall, neuronal protein synthesis is greatly reduced following neuronal injury, as reflected in the greatly reduced Nissl staining (chromatolysis) seen in degenerating neurons. The expression of subsets of neuronal proteins, however, appears to be greatly enhanced following neuronal injury. Some of the neuronal proteins exhibiting enhanced synthesis following injury appear to have roles in protecting and repairing vital cellular machinery, whereas others may play a role in the degradative process. The profile of genes exhibiting increased expression following neuronal injury as well as the time course of increased expression may vary depending on the neuronal type involved and the nature of the insult. Despite these uncertainties, monitoring the appearance of certain neuronal proteins and their respective mRNAs may prove a useful tool in neurotoxicology.

The process of neural degeneration also triggers marked reactive changes in nonneuronal elements, such as glia and microglia, inducing changes in the expression of proteins specifically found in these cell types. Assuming that the neurotoxin being studied is not toxic to glial cells, monitoring changes in the expression of glial proteins is uncomplicated by the potentially confounding effects of the toxin on protein synthetic machinery.

A. Approaches to Monitoring Gene Expression

Estimating the overall protein synthesis rate can be accomplished by measuring the incorporation of radio-labeled amino acids into protein. This approach can be used *in vivo* to further define specific loci of depressed protein synthesis and has been successfully applied to the study of protein synthesis following ischemic damage (Fig. 2) (Bodsch *et al.,* 1985; Dienel *et al.,* 1980; Mies *et al.,* 1990). Decreased incorporation of radiolabeled amino acids into protein following ischemia, as measured by autoradiographic density, is confined to the areas of neural damage, is related to the length of ischemic insult, and is less pronounced in young animals, which are resistant to the neurodegenerative effects of ischemia in the hippocampus. The observation of decreased protein synthesis in areas of the thalamus following ischemia has led to the consideration that minor damage to these areas, previously unrecognized, may occur.

While observing anatomically defined changes in overall protein synthesis may prove useful in delimiting areas affected by a given neural insult, little is learned about the dynamics of individual gene expression. It is of interest to identify the very small population of genes that may undergo greatly enhanced expression. Two approaches to this question have been successful in identifying certain genes which exhibit enhanced expression following neural injury: one utilizes

analysis of protein gene products and the other, mRNA.

Analysis of protein synthesis can be studied by incorporating labeled amino acids into protein *in vivo,* then separating individual proteins by high resolution two-dimensional polyacrylamide gel electrophoresis (2-D PAGE) (Hochstrasser *et al.,* 1988). The method involves dissection of the tissue of interest, followed by solubilization of the proteins and their separation by isoelectric point on an isoelectric-focusing tube gel (first dimension). The extruded focusing gels are then laid on top of a sodium dodecyl sulfate (SDS) slab gel and separated electrophoretically by molecular weight. The result, visualized by exposing the gel to film or staining for protein using sensitive silver staining techniques, is a complex pattern of spots which may number over 2000, each of which represents a distinct protein or protein subunit (Fig. 3). Densitometric scans of each film or stained gel can then be digitized such that the position, size, and optical density of each spot is recorded in a separate computer file. Sophisticated pattern recognition software, specifically written for this purpose, can then make a digital "overlay" of two gels, identifying those proteins that show increased expression. This approach has been used to identify differences (both increases and decreases) in protein expression in disease states (Harrington *et al.,* 1986), ischemic damage (Dienel *et al.,* 1986; Nowak, 1985), and neurotoxic damage (Brock and O'Callaghan, 1987).

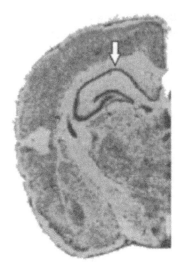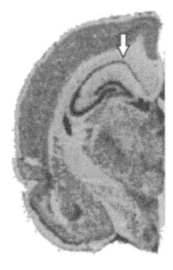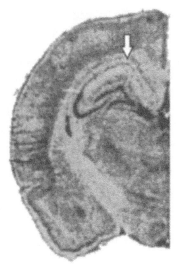

FIGURE 2 Graded decreases in protein synthesis in rat hippocampus 48 hr after increasing periods of ischemia induced by four-vessel occlusion. Rats were injected with [³H]valine at the indicated times and sacrificed 70 min later. Sections were cut and exposed to film, thus darker areas indicate greater incorporation of [³H]valine into protein. Arrows indicate the pyramidal cell layer of CA1. Progressive decreases in [³H]valine incorporation are seen with longer ischemic episodes. (Figure from Dienel *et al.,* 1980.)

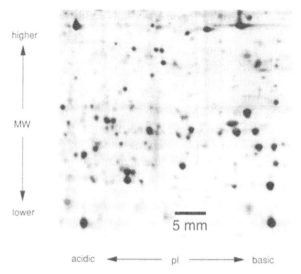

higher

↑

MW

↓

lower

5 mm

acidic ◄————— pI —————► basic

FIGURE 3   Scanned digital image from a small section (approximately 3 × 3 cm) of a silver-stained two-dimensional gel of proteins from monkey caudate. After solubilizing the proteins, they are first separated by isoelectric point on isoelectric-focusing tube gels. They are then extruded onto the top of a polyacrylamide slab gel, then electrophoresed to separate proteins by molecular weight. The result is a series of spots, each of which represents a single protein or protein subunit. In an entire silver-stained gel (in this case 16 × 16 cm), well over a thousand spots may be visible. Computer programs can make a digital overlay of two gels, noting differences in protein expression between samples. Significant post-translational changes in proteins may result in shifts in the position of a protein.

Since the addition or removal of phosphate groups or carbohydrate moieties can result in changes in molecular weight and/or isoelectric point, post-translation changes which may or may not be accompanied by changes in levels of protein expression can also be detected using this method.

A second approach to the identification of genes that may be preferentially activated by neurotoxic insults focuses on defining differences in the transcription products isolated from the neural tissue of treated and untreated animals. By isolating mRNA from injured brain tissue, constructing the complementary DNAs (cDNAs), then allowing these cDNAs to hybridize with control mRNA, the single-stranded cDNAs which remain unhybridized represent gene products uniquely associated with damaged tissue. This technique (subtractive hybridization), described in detail in the chapter by Billingsley, has revealed a number of unique gene products associated with neural damage, several of which await full characterization (Quach et al., 1993). The two methods just described are clearly not routine approaches easily applied to screening large numbers of potential neurotoxins or brain tissue samples. Both are labor intensive, require special skills and equipment, and are primarily intended as global discovery mechanisms.

Once potential protein markers are identified, more rapid assay methodologies that enable the processing of large numbers of samples are required for the experiments designed to validate the usefulness of a given marker. If the function of a given protein is known (e.g., specific enzyme), an assay based on functional activity can be developed. This approach has been used to follow the temporal changes in expression of a number of proteins following ischemia [e.g., the postischemic increase (30-fold) in ornithine decarboxylase activity (Dienel et al., 1985)]. A more common approach has been exploitation of the specific, high-affinity protein binding of antibodies to specific proteins. The obvious advantage of this approach over functional assays is that the methodologies for detection and quantitation can be adapted to all proteins, only requiring the availability of antibodies with the required specificity for the protein of interest.

Two major approaches to the quantification of individual proteins using antibody-binding techniques have been applied to the study of protein expression following neural injury. The first involves separation of solubilized proteins by molecular weight using SDS–PAGE, followed by Western transfer of the proteins to a nitrocellulose or PVDF membrane (Towbin et al., 1992). The membranes containing the immobilized proteins are then exposed to a specific antibody which binds the protein of interest. The desired antigens are visualized using appropriate enzyme-linked secondary antibodies followed by chromogens or enhanced chemiluminescence. Developed blots or films can then be scanned to determine the position and optical density of each immunoreactive protein band. Semiquantitative comparisons can be made by running tissue proteins from animals exposed to various treatment conditions side by side. The technique has the advantage that any cross-reactivity of the antibody with proteins other than the desired one can be monitored. In addition, proteolytic products of the protein of interest may be detected and quantified. The disadvantage is that the technique is time consuming and relatively few samples can be run at one time. In addition, the quantitation of proteins is relative. Variation in the optical density of a given protein band will depend on a number of variables, including efficiency of the transfer from gel to membrane, binding of the antibody to the protein of interest, and length of development time during the visualization process. For these reasons, within gel comparisons are preferred when using this technique.

Once the specificity of an antibody for the protein of interest has been established, a number of more rapid and accurate assay methods can be developed. The RIA is rapid, sensitive, and relatively easy to develop. Following solubilization of the sample, an aliquot is added to a tube containing a fixed dilution of the antibody and a fixed amount of the pure protein of interest, labeled with a tracer such as ^{125}I. The relative concentration of unlabeled protein in the solubilized sample and ^{125}I-labeled protein will determine how much ^{125}I-labeled protein will be bound to the antibody (Fig. 4). Following precipitation of the antibody-bound protein, the amount of ^{125}I-labeled protein bound to the antibody is determined in a gamma counter. A quantitative estimate of the amount of the target protein in the homogenate can then be determined by interpolation from a standard curve generated using the same fixed concentration of antibody and ^{125}I-labeled protein, and by varying the amounts of the pure, unlabeled protein. Similar antibody-binding assays, termed ELISA, utilize colorimetric detection methods. A number of variations of the ELISA method are routinely used, but labeled antibody rather than labeled ligand is the basis for quantitation (Fig. 5). The development of rapid, accurate assay methods must be a part of validation efforts which target potential protein markers of neural damage. It is important that these quantitative approaches be complemented with immunohistochemical studies to confirm that the potential marker protein exhibits temporal and anatomic patterns of staining which match those of the neural degeneration.

Even neuronal proteins that exhibit greatly enhanced expression during injury may not be expressed

during very severe forms of injury. The inducible form of the 72-kDa heat shock protein is one example (see Section IV,C,1). Following insults such as ischemia, the synthesis of mRNA continues, despite drastic reductions in protein synthesis in vulnerable areas (Bodsch and Takahashi, 1983). If the protein sequence is known, an oligonucleotide probe that will hybridize to a complementary region of the mRNA coding for that protein or a cDNA probe derived from the specific mRNA of interest can be synthesized using radiolabeled nucleotides. These labeled probes can then be hybridized to mRNA which has been isolated, separated by size on an agarose gel, and blotted onto an appropriate support (Northern blotting) or can be hybridized to mRNA directly in a tissue section (Emson, 1993), allowing autoradiographic visualization and quantitation. The relative increase in the mRNA levels transcribed from a given gene in response to neural damage is often far greater than the increase in the corresponding protein level, especially in situations where the metabolic integrity of the cell is severely compromised (Nowak, 1993).

A recent method of monitoring the expression of given genes is the use of transgenic animals in which copies of a "reporter" gene such as *lacZ*, the gene coding for β-galactosidase, is incorporated into the genome of a host animal in such a way that transcription factors which induce expression of the target gene induce coexpression of the reporter gene. The reporter gene product is usually an enzyme that can easily be identified by a simple colorimetric method, thus simplifying the histologic localization of cells expressing the protein of interest or allowing rapid quantification of gene expression by a simple test tube method. This approach has been successfully applied to the study of c-*fos* expression following excitotoxic damage (Smeyne *et al.*, 1993).

In choosing genes for which the monitoring of mRNA or protein levels has potential usefulness in neurotoxicology, one would ideally like those that are vigorously expressed following neural damage relative to their constitutive expression levels. In addition, since the increased expression of a particular protein may occur only in a subset of vulnerable neuronal phenotypes in a mixed population of cells, the background expression should be low. The marker would also ideally be expressed in response to a broad range of toxic insults and be able to detect damage to most neuronal phenotypes. The candidate proteins discussed later in no way comprise a comprehensive list; rather they represent examples of proteins which have potential usefulness in neurotoxicology and which have been examined as markers of neural injury. There is a tremendous amount of work yet to be done, both in identifying new

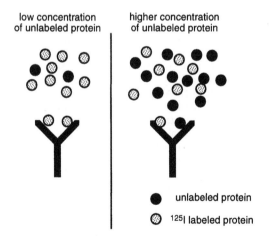

low concentration
of unlabeled protein

higher concentration
of unlabeled protein

● unlabeled protein

◉ ^{125}I labeled protein

FIGURE 4 Competitive protein binding to antibody is the basis of the radioimmunoassay. At left, low concentrations of target protein will displace little of the pure, radiolabeled protein. Increasing concentrations of the unlabeled protein in the sample, when mixed with the same concentration of labeled protein, will result in less labeled protein being bound to the antibody. Separation of the protein bound to antibody from protein in solution enables quantification of the amount of labeled protein bound to the antibody. From this information, the amount of target protein in the sample can be calculated by interpolation using a standard curve generated using known amounts of pure target protein as the unlabeled protein.

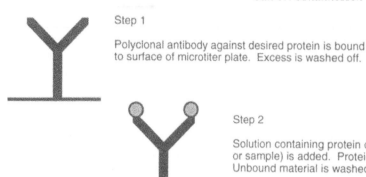

Step 1

Polyclonal antibody against desired protein is bound to surface of microtiter plate. Excess is washed off.

Step 2

Solution containing protein of interest (standard or sample) is added. Protein binds to antibody. Unbound material is washed off plate.

Step 3

A second antibody (e.g. mouse monoclonal) directed against a different epitope on the protein than the first, is added and allowed to bind to the protein. The second antibody is bound to an enzyme (ENZ), such as horseradish peroxidase. Excess is washed off.

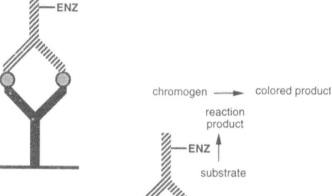

chromogen ⟶ colored product

reaction product

substrate

Step 4

The substrate for the enzyme is added along with a chromogen. A colored reaction product forms when the chromogen is exposed to the reaction product.

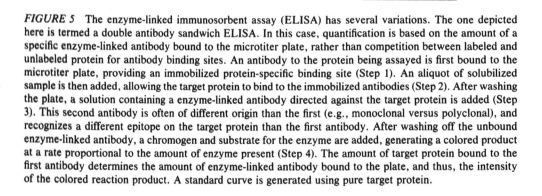

FIGURE 5 The enzyme-linked immunosorbent assay (ELISA) has several variations. The one depicted here is termed a double antibody sandwich ELISA. In this case, quantification is based on the amount of a specific enzyme-linked antibody bound to the microtiter plate, rather than competition between labeled and unlabeled protein for antibody binding sites. An antibody to the protein being assayed is first bound to the microtiter plate, providing an immobilized protein-specific binding site (Step 1). An aliquot of solubilized sample is then added, allowing the target protein to bind to the immobilized antibodies (Step 2). After washing the plate, a solution containing a enzyme-linked antibody directed against the target protein is added (Step 3). This second antibody is often of different origin than the first (e.g., monoclonal versus polyclonal), and recognizes a different epitope on the target protein than the first antibody. After washing off the unbound enzyme-linked antibody, a chromogen and substrate for the enzyme are added, generating a colored product at a rate proportional to the amount of enzyme present (Step 4). The amount of target protein bound to the first antibody determines the amount of enzyme-linked antibody bound to the plate, and thus, the intensity of the colored reaction product. A standard curve is generated using pure target protein.

candidate genes and in expanding our understanding of how these genes and their protein products respond to a broad range of neurotoxic insults.

B. Cellular Immediate Early Genes

The earliest apparent increases in gene expression following excitotoxic insults involve members of the protooncogene family. This group of genes, referred to as the cellular immediate early (cIE) genes, code for a number of proteins. For purposes of this discussion, two of these, c-*fos* and c-*jun* (the respective protein products are referred to as Fos and Jun), will be focused on. The protein products of the cIE genes (e.g., Fos and Jun) translocate to the nucleus where they subsequently combine in pairs to form heterodimers, which act as transcription factors, binding to specific transcription elements on DNA (Morgan and Curran,

1989). It is important to note that different members of the cIE gene family are expressed at various times after cellular activation. Because cIE gene products can combine in a number of ways to form heterodimers (Jun can combine with cIE gene products other than Fos), a potentially large number of unique transcription factors can result (Morgan and Curran, 1989). The expression of cIE genes seems to follow a significant change in the homeostatic status of the cell. Considering the temporal aspects of cIE gene activation and the number of possible heterodimeric combinations, it is clear that the cIE gene family is clearly capable of initiating a program of sequential transcriptional events from a single stimulus which might constitute a reactive change to the initial perturbation.

Expression of c-*fos* is regulated by two independent pathways (Fig. 6). One is activated by increased calcium influx via voltage or ligand-gated calcium channels (Curran and Morgan, 1986; Morgan and Curran, 1986, 1988). The calcium influx results in the activation of a transcription factor by a calmodulin-dependent mechanism. The activated transcription factor then translocates to the nucleus and binds to a region of DNA 60 bp upstream from the c-*fos* gene, initiating transcription (Sheng *et al.*, 1988). A second pathway,

which does not utilize calcium, is activated by peptide growth factors, such as FGF and NGF, and utilizes a binding element (serum response element) located about 300 bp upstream from the c-*fos* gene (Cochran *et al.*, 1984; Müller *et al.*, 1984; Sheng *et al.*, 1988).

As might be expected from the ubiquitous role of calcium as a signaling mechanism, rapid induction of the Fos protein has been demonstrated following increased neuronal activity induced by such diverse stimuli as stress (Deutch *et al.*, 1991) and electrical stimuli (Morgan *et al.*, 1987; Shin *et al.*, 1990) as well as with the excitotoxins NMDA and kainate (Li *et al.*, 1992; Sonnenberg *et al.*, 1989; Szekely *et al.*, 1987) and electrical stimulation of peripheral nerves (Birder *et al.*, 1991). Thus, rapid appearance of c-*fos* mRNA or Fos protein following treatment with an agent cannot be used to infer toxicity; it may simply indicate increased activity.

Studying the temporal relationship between c-*fos* expression and cell death has, however, provided a more useful approach to identifying dying cells using Fos (Smeyne *et al.*, 1993). In contrast to the transient (hours) appearance of Fos, which immediately follows cellular activation, cells that are undergoing programmed cell death during development exhibit contin-

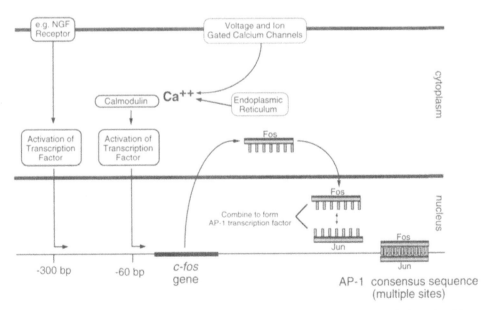

FIGURE 6 Dual transduction pathways activate the c-*fos* gene. Transcription can be initiated by a calcium–calmodulin-dependent pathway which can be activated by the opening of ligand or voltage-activated calcium channels. Alternatively, certain peptide trophic factors can induce c-*fos* expression by a pathway that does not require the influx of extracellular calcium. Once Fos protein has been synthesized, it combines with Jun to form a heterodimer that acts as a transcription factor, binding to segments of DNA that have the AP-1 consensus sequence. It is not clear at this time which of the c-*fos* activation pathways is responsible for the secondary prolonged expression of c-*fos* seen in injured neurons. However, it is interesting to note that the time course of the secondary c-*fos* expression observed in hippocampus following kainate treatment parallels the increased accumulation of calcium seen under similar conditions.

uous expression of c-*fos* prior to degenerating. By monitoring c-*fos* expression following peripheral nerve transection, continuous expression of c-*fos* is observed selectively in those neurons destined for degeneration. Following treatment with kainate, there is a transient appearance of Fos protein 2 hr after treatment, which disappears within 24 hr of treatment, but reappears in degenerating hippocampal pyramidal neurons from 4 to 10 days after treatment. This prolonged secondary expression of c-*fos*, which coincides with the appearance of degenerative changes, has been confirmed in other studies using ischemia and excitotoxicity (Jørgensen *et al.*, 1991; Schreiber *et al.*, 1993). Whether the increased c-*fos* expression is induced by the marked calcium uptake described earlier or is dependent on some other stimuli is unknown. The destruction of neuronal systems which tonically inhibit other neurons may lead to a sustained appearance of Fos in the denervated target cells (Jian *et al.*, 1993). Thus, in certain instances sustained c-*fos* expression may be an indirect indicator of neuronal damage.

Another cIE gene that may prove useful in identifying neurons undergoing delayed degeneration is c-*jun*. Using models of neuronal degeneration secondary to status eptilepticus and hypoxic ischemia, c-*jun* expression follows a different temporal pattern of expression than c-*fos* (Dragunow *et al.*, 1993). Similar to the rapid appearance of Fos, the Jun protein was also expressed strongly within 1 hr of hypoxic ischemia; however, the distribution of Jun was widespread and thus was not restricted to those areas destined to undergo subsequent degeneration. Twenty-four hours after either hypoxic ischemia or status epilepticus, although Fos immunoreactivity had virtually disappeared, Jun was still strongly expressed, but was restricted to those neuronal groups that would subsequently degenerate. An additional finding was that pharmacologic treatments which enhanced c-*fos* expression did not induce c-*jun* expression. These findings, coupled with the observations of increased Jun expression in axotomized central and peripheral neurons (Dragunow, 1992), suggest that monitoring changes in the levels of c-*jun* mRNA or Jun protein may provide a useful marker of damaged neurons.

C. Stress Proteins

The coordinated sequence of events that occurs within a cell following an externally induced perturbation in cellular homeostasis is often referred to as the stress response, and the set of proteins that show a selective increase in expression are called stress proteins. While this generic term may include a number of known and unknown proteins, this discussion focuses on members of two gene families: the heat shock proteins (hsp) and the ubiquitins. Since certain members of these protein families show greatly enhanced expression in response to stress, they appear to be ideal candidates as useful markers of neuronal injury.

1. The 72-kDa Heat Shock Protein

Heat shock proteins (hsp) are some of the most conserved proteins yet identified. They occur in virtually all organisms, from bacteria to humans, and are expressed in virtually all mammalian cells (Morimoto *et al.*, 1990). The term "heat shock" does not convey the range of physiological functions these proteins carry out nor does it indicate the multitude of stressors which can induce the expression of heat shock proteins. The heat shock family includes a number of subgroups ranging in size from 28 to 110 kDa (Welch, 1990). Some of these are constitutively expressed, whereas others are stress induced. Many members of the hsp family have protein- and DNA-binding properties. The physiological role for the constitutively active forms is thought to be that of molecular chaperone, aiding in the post-translational processing of proteins within the endoplasmic reticulum, shuttling proteins through membranes, aiding in the proper folding during translation, and aiding in proteolysis. The forms that are induced by stress seem to have similar functions: stabilizing and refolding partially denatured proteins and hastening the degradation of others.

Recent interest has been focused on the 70-kDa hsp family as markers of neuronal stress, particularly hsp72. hsp72 is normally expressed at very low levels, but a number of stressors and toxins, such as hyperthermia, hypoxia, excitotoxins, and heavy metals, can induce robust expression of this gene. In contrast to c-*fos*, calcium may not be the primary signal that initiates transcription of hsp72. The activation of a heat shock transcription factor appears responsible for the rapid translation. The heat shock transcription factor may normally be bound to other cytoplasmic proteins, preventing it from binding to DNA. When cellular injury leads to slightly denaturing intracellular conditions, the heat shock transcription factor complex may dissociate, allowing the transcription factor to bind to the heat shock element on DNA, initiating transcription.

It is not surprising that the expression of heat shock proteins can be induced by hyperthermia in mammals, given the nearly universal occurrence of this response. In brain, the expression of hsp72 has been demonstrated by visualizing changes in the levels and distribution of hsp72 mRNA or protein. Localization of the cells exhibiting increased expression following hyperthermia has revealed a discrete set of labeled cell types and locations. Increased diffuse glial expression of hsp72 is seen throughout fiber tracts in the brain and in endothelial cells (Brown, 1990; Marini *et al.*, 1990; Sprang and Brown, 1987). In contrast, neurons that exhibit elevated synthesis of hsp72 mRNA are highly localized (Blake *et al.*, 1990; Marini *et al.*, 1990). High levels of hsp72 expression, as visualized by *in situ* hybridization, coincide with the granule cell layer of the dentate gyrus, the paraventricular nucleus of the hypothalamus, and the median eminence. Expression of the hsp72 gene in these discrete neuronal populations following hyperthermia does not signal the subsequent degeneration of these cell groups, but may indicate their relative sensitivity to heat stress or extremely heightened activity as they participate in a coordinated neural response to this stressor. It has been pointed out that many neuroactive agents with neurotoxic potential may induce changes in body temperature, and thus increased expression of hsp72 following administration of these agents may be secondary to the hyperthermic effects induced by the test substance (Nowak, 1993).

Increased synthesis of several heat shock proteins after ischemia is found in damaged hippocampus by using 2-D PAGE, including hsp72 (Dienel *et al.*, 1986; Nowak, 1985). A predominantly neuronal distribution of hsp72 protein is seen with immunohistochemical staining following ischemia; however, variable results concerning the relationship between hsp72 staining and subsequent degeneration have been obtained. Hippocampal neurons in CA3 and the dentate gyrus, which are resistant to forebrain ischemia, exhibit strong postischemic hsp72 immunoreactivity (Chopp *et al.*, 1991; Tomioka *et al.*, 1993; Vass *et al.*, 1988). The vulnerable CA1 neurons may exhibit strong hsp72 immunoreactivity following ischemia prior to degeneration or may remain virtually devoid of immunoreactivity, depending on such experimental conditions as the animal model used and the length of ischemia. This apparent discrepancy is resolved when the severity of the neuronal injury is considered. If hsp72 mRNA is visualized by *in situ* hybridization in postischemic brains in which the vulnerable CA1 hippocampal neurons do not show hsp72 immunoreactivity, a very strong and persistent hsp72 mRNA signal is detected in these neurons (Fig. 7) (Nowak, 1991; Nowak *et al.*, 1990). Thus, in very severely injured neurons the translational capabilities can become so severely impaired that heat shock proteins cannot be synthesized, despite the presence of abundant quantities of message. This result emphasizes the utility of examining the effects of potential neuronal stressors on the transcriptional portion of protein expression.

Studies that have addressed the function of hsp72 expression following ischemia add support to the relationship of hsp72 expression and neuronal stress. Short

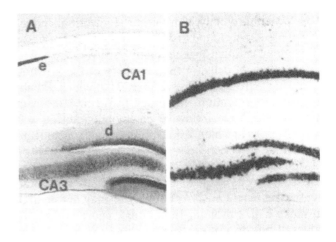

FIGURE 7 Comparison of the hippocampal distribution of hsp72 protein as visualized with immunocytochemistry (A) and hsp72 mRNA visualized with *in situ* hybridization (B) 24 hr after transient ischemia in the gerbil. Note that the vulnerable CA1 neurons are unable to synthesize significant quantities of hsp72 protein (A) despite very high levels of hsp72 mRNA (B). The cells in the dentate gyrus (d) and CA3, which survived the ischemic insult, show strong staining for hsp72 protein. Some ependymal (e) staining for hsp72 protein can also be seen in (A). (From Nowak, 1993.)

periods of ischemia, which are sufficient to induce strong hsp72 expression in CA1 neurons, but not severe enough to cause degeneration of these neurons, will protect CA1 neurons from a subsequent ischemic episode which would cause massive neurodegeneration in CA1 (ischemic tolerance) (Kitagawa *et al.*, 1991; Liu *et al.*, 1993; Nakagami *et al.*, 1993; Simon *et al.*, 1993). By varying the length of the conditioning ischemic episode, a tight correlation between hsp72 expression and subsequent neural protection has been established (Kitagawa *et al.*, 1991).

The relationship between hsp72 expression and neuronal degeneration following treatment with neurotoxins has also been examined. Systemic or brain microinjection of kainate induces increased hsp72 staining and hsp72 mRNA in areas susceptible to this neurotoxin (Gonzalez *et al.*, 1989; Longo *et al.*, 1993; Schreiber *et al.*, 1993; Uney *et al.*, 1988; Vass *et al.*, 1989; Wang *et al.*, 1993). As with ischemia, hsp72 staining is not always seen in those cells that subsequently degenerate. Microinjection of the excitotoxin NMDA into the rat entorhinal cortex induces neuronal degeneration in the surrounding tissue. Expression of hsp72, as estimated by immunocytochemical methods, was found to precede neuronal degeneration and was confined largely to those neurons that subsequently underwent degeneration (Yee *et al.*, 1993). In contrast to the rapid induction of Fos protein, which appeared and disappeared in neurons surrounding the injection site within 8 hr, hsp72 followed a more gradual increase in expression, peaking roughly 48 hr after the injection. Induction of hsp72 was found to be a generally good predictor of eventual cellular degeneration; however, there was a population of cells on the periphery of the affected area which strongly expressed hsp72 but which did not undergo degeneration. Increased hsp72 immunoreactivity is also seen following bicuculline-induced seizures in areas in which lesions subsequently occur (Shimosaka *et al.*, 1992).

Activation of the hsp72 gene is clearly related to stress. Neurons subjected to lethal insults appear to synthesize hsp72 mRNA, but may or may not be able to make hsp72 protein. In contrast, neurons that are stressed but do not degenerate show sharp increases in hsp72 mRNA and protein. Thus, evidence of hsp72 gene activation cannot be used as presumptive evidence of subsequent neuronal degeneration, but may act as a useful marker of neuronal stress. The usefulness of detecting neurons that may be stressed by a given agent is illustrated by the observation that following treatment with phencyclidine, ketamine, or MK-801, cingulate neurons which exhibit reversible damage characterized by the appearance of vacuoles (Olney *et al.*, 1991) are also hsp72-positive following treatment with these agents (Sharp *et al.*, 1993). In interpreting the significance of an observed increase in hsp72 expression following treatment with a potential neurotoxic agent, consideration should be given to the possibility that the increase may be secondary to hyperthermia induced by the test compound.

2. Ubiquitin

Expression of ubiquitin following exposure to heat and other stressors often parallels that of the heat shock proteins and thus it has been considered a heat shock/stress protein (Bond and Schlesinger, 1985, 1986). The levels of ubiquitin expression in brain increase following heat stress, though modestly compared to the degree of change seen in the periphery (Nowak *et al.*, 1990). As with many of the heat shock proteins, the ubiquitins are constitutively expressed and perform vital functions related to protein processing, movement, and catabolism. Ubiquitin, an 8-kDa polypeptide, is synthesized either as a monomer attached to a tail protein (ubiquitin fusion protein) or as a polymer, consisting of many ubiquitin repeats. Separate genes (and thus mRNAs) direct the synthesis of mono- and polyubiquitins. The polyubiquitin genes are induced following stress. Once expressed, the polymeric form of ubiquitin can be disassembled into ubiquitin monomers, providing for a rapid increase in the availability of ubiquitin monomers (Caday *et al.*, 1993).

The interaction of monomeric ubiquitin with protein is different from that of the heat shock proteins. Ubiquitin monomers attach covalently to protein via ε amino groups on lysine residues in an ATP-requiring enzymatic process. The ubiquitination process appears to aid in the degradation of damaged and short-lived proteins, but conversely appears to stabilize others, such as histones and cytoskeletal proteins. Thus, increased ubiquitin expression and conjugation during neuronal injury could conceivably contribute to the rapid degradation of damaged proteins and the stabilization of others. Because ubiquitin attaches covalently to proteins, separation of proteins by SDS–PAGE followed by immunoblotting with ubiquitin antibodies identifies which proteins are ubiquitinated (Bond *et al.*, 1988). Thus, potentially valuable information can be gained by studying changes in both polyubiquitin gene expression and the pattern of protein–ubiquitin conjugates. The apparent role of ubiquitination in the stress response, coupled with the association of ubiquitin immunoreactivity with abnormal features identified in several neurodegenerative diseases (Garofalo *et al.*, 1991; Kuzuhara *et al.*, 1988; Lowe *et al.*, 1988; Mori *et al.*, 1987), has prompted recent investigations into changes in ubiquitin expression and conjugation during experimentally induced neuronal damage.

Increased amounts of ubiquitin mRNA have been demonstrated in the hippocampus following ischemia (Caday *et al.*, 1993; Nowak *et al.*, 1990). Consistent with stress responses in other systems, it is the high molecular weight polyubiquitin transcripts that are increased, whereas monoubiquitin mRNA levels remain unchanged (Caday *et al.*, 1993). The increased polyubiquitin expression precedes the delayed neuronal death. Although the time course of increased ubiquitin gene expression follows a time course similar to that of hsp72, the magnitude of the increase is modest in comparison, possibly because the constitutive levels of ubiquitin gene expression are much higher than hsp72.

Synthesis of ubiquitin seems to follow the generalized decrease in protein synthesis induced by ischemia (Dienel *et al.*, 1980; Nakagomi *et al.*, 1993). Immunoreactive ubiquitin disappears rapidly (within 4 hr) from vulnerable neurons in the hippocampus following ischemia, returning only in those neurons that recover (Kato *et al.*, 1993; Yamashita *et al.*, 1991). While synthesis of protein may decrease, the pattern of ubiquitinated proteins is altered, with increases in high molecular weight protein–ubiquitin conjugates being observed (Hayashi *et al.*, 1991, 1992).

Treatment with neurotoxins may in some cases increase the synthesis of immunoreactive ubiquitin. Localized infusions of NMDA cause an increase in ubiquitin immunoreactivity which parallels that seen for hsp72, both temporally and spatially (Yee *et al.*, 1993). Following treatment with trimethyltin, increases in ubiquitin immunoreactivity are seen in the hippocampal pyramidal and cerebellar Purkinje cells, but an analysis of ubiquitin conjugates indicates a selective increase in ubiquitination of high molecular weight proteins in the hippocampus (Anderson *et al.*, 1992). This change is not accompanied by increased levels of polyubiquitin mRNA. Identifying specific ubiquitinated proteins which may indicate irreversible stress is an approach that has not yet been fully pursued.

D. Glial Proteins

Degenerating neurons create a distinct alteration in the biochemical milieu of the neuropil. Drastic changes in secreted proteins, calcium homeostasis, and metabolic by-products, along with fragments of cellular organelles and macromolecules liberated by degenerating neurons, must have marked influences on surrounding cells, including neuroglia. Indeed, classical histopathologic methods of assessing CNS damage have consistently considered gliosis and glial scarring as indicators of the neurodegenerative process. Gliosis is observed in neural damage secondary to a number of distinct etiologies, including neurodegenerative disease, physical trauma, and CNS pathogens (Lindsay, 1986). It is this response of glia to neural damage from a wide range of causes that makes glial proteins attractive candidate markers of neuronal injury from a number of distinct neurotoxins.

1. Glial Fibrillary Acidic Protein

Glial fibrillary acidic protein (GFAP) is the major intermediate filament protein of astrocytes, and has been isolated and characterized for some time (Eng, 1985). Antibodies raised against GFAP recognize astrocytes, and the greatly enhanced GFAP immunoreactivity characteristic of reactive astrocytes has led to the routine use of GFAP immunostaining as a method for the neuropathologic determination of gliosis. GFAP has become perhaps the most thoroughly characterized biochemical marker of neurotoxicant-induced neural damage, thanks largely to the efforts of O'Callaghan (1993). Recognizing that the use of immunostaining for GFAP is subject to fixation-induced variations in staining intensity and that the methods ultimately are time consuming and not well adapted to rapid quantitation, O'Callaghan (1991) has investigated the relationships among neuronal injury, gliosis, and quantitative levels of GFAP by developing a rapid, specific, and sensitive ELISA.

The GFAP signal induced by neural injury (i.e., relative increase in GFAP levels compared to background or control levels) is robust. Treatment with neurotoxins that produce overt neural damage as defined by histo-logic criteria, such as TMT, can result in 60-fold increases in GFAP concentrations within severely damaged areas (Brock and O'Callaghan, 1987). Of additional importance is the observation that GFAP concentrations increase in areas that do not display overt evidence of histologic damage, suggesting that increases in GFAP may detect areas with minimal neuronal damage not seen in routine Nissl-stained sections.

The effort to validate the use of GFAP levels as an indicator of neurotoxin-induced neural damage has been extensive. Findings by O'Callaghan (1993) support the use of GFAP as a general neurotoxin screen. O'Callaghan observed that increases in GFAP are induced by roughly 20 neurotoxins, which include a diversity of types: heavy metals, excitotoxins, nitriles, amphetamines, and selective catecholaminergic toxins. The quantitation of GFAP has also proved useful in the assessment of traumatic brain injury. In contrast, increases in GFAP are not induced by a wide variety of neuropharmacologic agents, including those such as reserpine, which induce long-term reductions in neurotransmitter content but no cell loss. Comparisons of the distribution of GFAP immunostaining and quantitative measurements of GFAP in microdissected tissue samples indicate a good correspondence between the two. As mentioned previously, the sensitivity of GFAP as a marker is such that degeneration of minor elements within the neuropil, which would not be detected by Nissl staining, results in increased levels of GFAP. This is an important point since selective neurotoxins may result in degeneration of terminals and axons, leaving the soma intact [e.g., MPTP in the mouse, 6-hydroxydopamine injected into the striatum, 3,4 methylene dioxymethamphetamine (MDMA)] or in degeneration of a subset of neurons which constitute a minor component of the total neuronal population. By exploiting the two-step process required for the MPTP toxicity toward dopaminergic neurons—bioactivation to the active toxin (MPP$^+$) within astrocytes followed by selective accumulation of MPP$^+$ by dopaminergic terminals via the dopamine reuptake site–O'Callaghan has been able to demonstrate that the increased striatal GFAP induced by MPTP is due solely to the degeneration of dopaminergic terminals and is not due to direct effects of the active toxin on astrocytes (O'Callaghan *et al.*, 1990).

A number of additional considerations regarding the use of GFAP as a general indicator of neuronal damage are also being addressed. The time course of the GFAP increase may vary depending on the toxin studied. Evidence suggests that compounds that have a prolonged course of action, such as TMT, result in a prolonged elevation in GFAP, whereas those that are cleared rapidly may induce a more transient rise in GFAP levels (O'Callaghan, 1993). There are exceptions, however, such as the prolonged (>6 month) increase in hippocampal GFAP induced by a single injection of kainate (Van Den Berg and Gramsbergen, 1993). The time course of elevated GFAP levels induced by a given toxin may also vary with the age of the animal (Barone, 1993). Thus, when using GFAP as a neurotoxic screen, a time course would provide the most complete information. The quantitative assay of GFAP has been applied to a variety of other species, including fish, birds, and primates, and the utility of this approach to detecting neuronal damage in those species has been demonstrated (Evans *et al.*, 1992, 1993).

2. Laminin and Laminin-Binding Protein

Laminin is a glycoprotein associated with basement membranes. In brain, immunoreactive laminin is most often seen surrounding blood vessels. Following neural damage induced by toxins, laminin immunoreactivity of blood vessels increases, and laminin also appears in reactive astrocytes which are also GFAP positive (Liesi *et al.*, 1984). Laminin immunoreactivity in reactive astrocytes following local injections of neurotoxin does not exhibit the persistence of GFAP staining, returning to basal levels within 1 month.

A protein which binds to laminin has recently been isolated and antibodies to it have been generated (Kleinman *et al.*, 1991). This laminin binding protein (LBP) has not been sequenced, so changes observed while using antibodies to LBP cannot at this time be linked to a specific functional change. In contrast to laminin, this LBP is not found in the basement membranes, but in normal rat brain is confined to a few neuronal elements, specifically neurons within layers II/III and V of the cerebral cortex and mossy fibers and terminals within the hippocampal CA3 region (Jucker *et al.*, 1991). Following injury, however, strong LBP immunoreactivity is seen within reactive astrocytes. LBP-positive astrocytes are confined to the region of injury and are seen following neuronal damage induced by mechanical injury or ischemia. The intense immunoreactivity of LBP which accompanies neural degeneration appears to hold potential as a

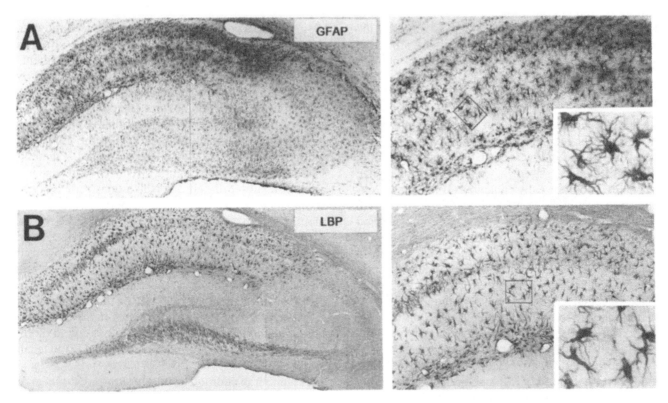

FIGURE 8   Comparison of gliosis in rat hippocampus 32 days after transient ischemia, as visualized immunocytochemically using antibodies directed against glial fibrillary acidic protein (GFAP; A) and laminin-binding protein (LBP; B). Higher magnifications are to the right. In these adjacent sections, note that both GFAP- and LBP-like immunoreactivity labels the reactive astrocytes within the damaged CA1 region. The GFAP antibodies also label glia in less affected areas, such as CA2 and CA3, whereas LBP-like immunoreactivity in glia is not prominent in these areas. (From Jucker *et al.*, 1993.)

marker of reactive gliosis. The staining in astrocytes from control brain is very low, suggesting a good signal can be obtained relative to background. The LBP immunoreactivity is also persistent, being evident in hippocampus for at least 1 month after the induction of ischemic damage (Jucker *et al.,* 1993). The applicability of LBP as a marker for neurotoxic damage has not been tested; however, the positive results from mechanical and ischemic injury suggest that the results may parallel those reported for GFAP. A comparison of GFAP and LBP immunoreactivity in reactive gliosis in the gerbil hippocampus 1 month after transient ischemia is shown in Fig. 8.

V. Markers of Macromolecular Catabolism

Previous sections have focused on biochemical changes associated with initial perturbations (e.g., Ca^{2+} influx), reactive changes (e.g., altered gene expression), and the loss of phenotypic markers (e.g., depletion of neurotransmitters and synthetic enzymes) induced by neurotoxic agents. Once a cell has been damaged beyond repair, a number of catabolic events occur. Destruction or disassembly and resorption of the major cellular macromolecules are required as dying cells are cleared from the brain. Monitoring the breakdown of selected proteins or DNA, the major cellular macromolecules, may prove useful in the detection and quantification of neuronal damage.

A. Proteolysis

Breakdown and removal of cellular protein may be accomplished by several routes, including phagocytosis of cellular debris by infiltrating macrophages, digestion of cellular components by neuronal lysosomes (autophagy), or activation of cytoplasmic (neutral) proteases. The latter mechanism has been studied in models of neuronal degeneration. The degradation of several of the major proteins associated with the neuronal cytoskeleton exhibits a time course and specificity for degenerating neurons that suggest potential usefulness as a method for monitoring neuronal degeneration.

As with so many of the biochemical and molecular events associated with the process of neurodegeneration, proteolysis of neuronal structural proteins is, in many instances, triggered by marked increases in intracellular calcium concentrations. There are several calcium-activated neutral proteases, often termed calpains, which appear to subserve diverse functions within neurons (Nixon, 1986; Nixon *et al.,* 1986; Schlaepfer *et al.,* 1985; Siman and Noszek, 1988; Siman *et al.,* 1989). Neuronal proteases which are active at

low (physiological) calcium concentrations play an apparent role in the constant remodeling of the cytoskeletal proteins required for such activities as neurite extension and axonal transport. Calpains which require calcium concentrations which exceed the norm have a broad substrate specificities, cleaving structural elements such as neurofilament proteins, spectrin, and microtubule-associated proteins.

Detection and quantification of neuronal proteins and their breakdown products are most often accomplished by separation of solubilized protein by SDS–PAGE, followed by Western transfer of the proteins to a membrane and the visualization of the proteins and breakdown products using antibody probes that recognize epitopes common to both the native protein and its proteolytic products (Fig. 9). For example, this approach has been used to characterize the proteolysis induced by calpain I in the hippocampus following administration of excitotoxins *in vivo* (Roberts-Lewis and Siman, 1993; Siman and Noszek, 1988). Administration of kainate or NMDA into the rat lateral ventricle results in the marked degeneration of hippocampal CA3 neurons. This degeneration was accompanied by loss of the intact structural protein spectrin and the concomitant appearance of spectrin immunoreactive proteolytic products.

Several observations support the relationship of the appearance of this proteolytic pattern to neural degeneration (Siman *et al.,* 1989). The spatial location of maximal proteolysis, as determined by microdissection followed by SDS–PAGE and immunoblot analysis, coincides with the area of maximal neuronal degeneration, as visualized histologically. Specific pharmaco-

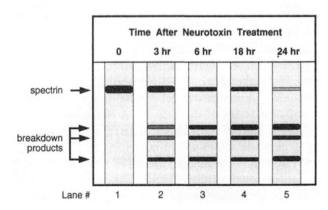

FIGURE 9 Illustration of the pattern of protein breakdown products which might be seen with Western blotting following SDS–PAGE of proteins from a damaged brain area, at increasing intervals after injury. The lane numbers are indicated at the bottom, and the interval between insult and sample collection at the top. The antibody used would be able to recognize both the intact protein (e.g., spectrin) and several breakdown products. (Adapted from Siman *et al.,* 1989.)

logic blockade of the NMDA receptor prevents both spectrin proteolysis and the appearance of histologic damage. Temporally, spectrin proteolysis precedes the appearance of degenerating neurons exhibiting positive silver staining (agyrophilia). While monitoring of spectrin proteolysis proved to be a very sensitive marker for degenerating neurons in this model of excitotoxic damage to hippocampal pyramidal cells, it is also apparent that the usefulness of this approach for monitoring all forms of neural degeneration may be limited. Although pyramidal cell damage secondary to ischemia is accompanied by considerable spectrin proteolysis, the extensive degeneration of granule cells within the dentate gyrus induced by lateral ventricular injections of colchicine is not (Siman *et al.*, 1989). Thus, the type of neurotoxic insult as well as the affected cell type may determine whether spectrin proteolysis occurs. It should be noted that the distribution of the calpains is not uniform throughout the brain (Siman *et al.*, 1985). Calpain I is mainly found in neurons but is also seen in a small number of specific glial populations. Subpopulations of neurons in the ventral horn of the spinal cord, cerebellum, reticular formation, substantia nigra pars compacta, subthalamus, and subiculum stain intensely with a monoclonal antibody directed against calpain I. Curiously, little immunoreactivity is seen in the pyramidal cells of the hippocampus, in contrast to the robust calpain I-mediated proteolysis seen following excitotoxin-induced degeneration.

B. DNA Fragmentation

During fetal development, a number of "excess" neurons die (programmed cell death) by a process termed apoptosis. The biochemical hallmark of apoptosis is the fragmentation of DNA into discrete sizes separated by approximately 185 bp by calcium-activated endonuclease activity (McConkey *et al.*, 1988; McConkey *et al.*, 1989). Separation of these DNA fragments on an agarose gel results in a pattern composed of a series of discrete DNA bands, known as laddering. In contrast, DNA isolated from cells undergoing necrosis shows a continuous smear of degraded DNA, indicative of random breaks (Kerr and Harmon, 1991). Neuronal cultures are well suited for examining DNA fragmentation patterns following toxic insults, and positive and negative evidence for neurotoxin-induced apoptosis has been obtained (Dessi *et al.*, 1993; Dipasquale *et al.*, 1995; Forloni *et al.*, 1993). However, characterizing DNA fragmentation patterns *in vivo* in discrete neuronal populations selectively damaged by a given insult may be difficult since the DNA of the cells of interest is diluted considerably with DNA from unaffected neurons, glia, and endothelial cells (Ignatowicz *et al.*, 1991).

Apoptosis as a form of induced neuronal degeneration can be identified using morphological criteria; however, ultrastructural examination is required (Harmon and Allan, 1988). A technique has recently been developed that allows the *in situ* labeling of DNA breaks (Gavrieli *et al.*, 1992). The method, which can be applied to slide mounted sections cut from conventionally fixed and embedded tissue, involves first enhancing access to the DNA by treating the sections with a proteolytic enzyme. Sections are then treated with terminal-deoxynucleotidyl transferase (TdT) which binds to the 3'-OH ends of the nicked DNA, initiating the synthesis of a polydeoxynucleotide extension. By supplying biotinylated deoxyuridine as the substrate, biotin is incorporated into the polymer. Visualization is then achieved by exposing the section to an avadin–HRP complex which binds to the biotin and can be subsequently visualized using conventional chromogens such as diaminobenzidine. The technique, TdT-mediated dUTP-biotin nick end labeling (TUNEL), stains the nuclei of cells which are undergoing programmed cell death during development, dying cells in tissues such as intestine which have a normal turnover, and cells in which apoptosis is chemically induced (Gavrieli *et al.*, 1992).

TUNEL has recently been applied to the nervous system, labeling cells undergoing programmed cell death in the developing spinal cord (Taginuma and Takashita, 1993). In addition, neuronal degeneration in the facial nucleus secondary to transection of the facial nerve is also accompanied by strong neuronal labeling of broken DNA with TUNEL (Wilcox *et al.*, 1993). Although the ability of TUNEL to detect cells undergoing apoptosis has been clearly demonstrated, TUNEL may also label cells containing random DNA breaks. Because the technique is relatively new, the full utility of this technique for labeling degenerating neurons has not yet been assessed.

VI. Future Directions

The mechanisms underlying cell death in the nervous system secondary to neurotoxic insults have recently undergone a fresh examination. Two major modes of cell death have been described: necrosis and apoptosis (Kerr and Harmon, 1991). Morphologically, necrosis is characterized by clumping of chromatin, gross swelling of the cell and cellular organelles, and a general breakdown of membranes. Apoptosis is characterized by shrinkage of the cell and nucleus, compaction and segregation of the chromatin, and development of convolutions in the cell membrane which eventually break off, forming apoptotic bodies which are absorbed by neighboring cells. Although necrosis is viewed as a passive process initiated by massive perturbations in cellular homeostasis which cannot be compensated for, apoptosis can be viewed as an active, gene-directed program which requires active protein synthesis. In neuronal cultures, cells may undergo apoptosis when deprived of trophic factors, a process which presumably plays a role *in vivo* in the pruning of excess neurons which have failed to make contact with their target cells, the source of a requisite trophic factor(s) (Johnson and Deckwerth, 1993). Since trophic factor deprivation-induced apoptosis is prevented by inhibitors of protein synthesis, it has been hypothesized that trophic factors actively suppress the expression of a "suicide" program (Martin *et al.*, 1988). Thus, neural degeneration following a neurotoxic insult could lead to necrosis if the initial insult is severe and causes a general breakdown in cellular membranes or energy production. Alternatively, if a compound independently initiated activation of this "suicide" program, or were to interfere with the production or release of a growth factor, the proper interaction of a trophic factor with its receptor, or the transduction process, cellular degeneration might follow the apoptosis pathway.

Recognition of apoptosis as a potential mode of neurotoxicant-induced neural degeneration forces a continued investigation of methods which detect this form of degeneration. The marked glial reaction seen following insults such as ischemia is not generally accompanied by the loss of large numbers of neurons during development. Loss of neurons by apoptosis, which can be induced by such insults as X-irradiation (Harmon and Allan, 1988), can be subtle. A small subset of neurons within a defined group may undergo apoptosis and the distribution may be scattered, such that dying cells are intermingled with a larger population of healthy cells. In addition, apoptosis appears to be a rapid process and the apoptotic bodies formed are removed from the brain quickly. Thus the detection of cells undergoing apoptosis is difficult, but it is important to know how extensively this form of cell loss is induced by neurotoxic agents. Because of the subtlety of apoptosis, the possibility exists that one neurotoxic end point may simply be a reduction in cell counts within a given neural population, an end point that is difficult to discern unless it is dramatic. Clearly, future studies should be directed at developing methods that can evaluate more subtle forms of neurotoxicity.

References

Anderson, V., Hajimohammadreza, I., Gallo, J., Anderton, B., Uney, J., Brown, A., Nolan, C., Cavanagh, J., and Leigh, P.

(1992). Ubiquitin, PGP 9.5 and dense body formation in trimethyltin intoxication: differential neuronal responses to chemically induced cell damage. *Neuropathol. Appl. Neurobiol.* **18**, 360–375.

Barbeau, A., Roy, M., Bernier, G., Campanella, G., and Paris, S. (1987). Ecogenetics of Parkinson's disease: prevalence and environmental aspects in rural areas. *Can. J. Neurol. Sci.* **14**, 36–41.

Barone, S., Jr. (1993). Developmental differences in neural damage following trimethyltin as demonstrated with GFAP immunohistochemistry. In *Markers of Neuronal Injury and Degeneration* (J. N. Johannessen, Ed.), pp. 306–316, The New York Academy of Sciences, New York.

Benveniste, H., Jorgensen, M. B., Diemer, N. H., and Hansen, A. J. (1988). Calcium accumulation by glutamate receptor activation is involved in hippocampal cell damage after ischemia. *Acta Neurol. Scand.* **78**, 529–536.

Birder, L. A., Roppolo, J. R., Iadorola, M. J., and de Groat, W. C. (1991). Electrical stimulation of visceral afferent pathways in the pelvic nerve increases c-fos in the lumbosacral spinal cord. *Neurosci. Lett.* **129**, 193–196.

Blake, M. J., Nowack, T. S., Jr., and Holbrook, N. J. (1990). In vivo hyperthermia induces expression of HSP70 mRNA in brain regions controlling the endocrine response to stress. *Mol. Brain Res.* **8**, 89–92.

Bodsch, W., and Takahashi, K. (1983). Selective neuronal vulnerability to cerebral protein- and RNA-synthesis in the hippocampus of the gerbil brain. In *Cerebral Ischemia* (A. Bes, P. Braquet, R. Paoletti, and B. K. Siesjö, Eds.), pp. 197–208, Elsevier, New York.

Bodsch, W., Takahashi, K., Barbier, A., Ophoff, B. G., and Hossmann, K.-A. (1985). Cerebral protein synthesis and ischemia. *Prog. Brain Res.* **63**, 197–210.

Bond, U., Agell, N., Haas, A. L., Redman, K., and Schlesinger, M. J. (1988). Ubiquitin in stressed chicken embryo fibroblasts. *J. Biol. Chem.* **263**, 2384–2388.

Bond, U., and Schlesinger, M. J. (1985). Ubiquitin is a heat shock protein in chicken embryo fibroblasts. *Mol. Cell. Biol.* **5**, 949–956.

Bond, U., and Schlesinger, M. J. (1986). The chicken ubiquitin gene contains a heat shock promoter and expresses an unstable mRNA in heat-shocked cells. *Mol. Cell Biol.* **6**, 4602–4610.

Brock, T. O., and O'Callaghan, J. P. (1987). Quantitative changes in the synaptic vesicle proteins synapsin I and p38 and the astrocyte-specific protein glial fibrillary acidic protein are associated with chemical-induced injury to the rat central nervous system. *J. Neurosci.* **4**, 931–942.

Brown, I. (1990). Induction of heat shock (stress) genes in the mammalian brain by hyperthermia and other traumatic events: a current perspective. *J. Neurosci. Res.* **27**, 247–255.

Burns, R. S., Chiueh, C. C., Markey, S. P., Ebert, M. H., Jacobowitz, D. M., and Kopin, I. J. (1983). A primate model of parkinsonism: selective destruction of dopaminergic neurons in the pars compacta of the substantia nigra by N-methyl-4-phenyl-1,2,3,6-tetrahydropyridine. *Proc. Natl. Acad. Sci. USA* **80**, 4546–4550.

Caday, C. G., Sklar, R. M., Berlove, D. J., Kemmou, A., R. H. Brown, J., and Finkelstein, S. P. (1993). Polyubiquitin gene expression following cerebral ischemia. In *Markers of Neuronal Injury and Degeneration* (J. N. Johannessen, Ed.), pp. 188–194, The New York Academy of Sciences, New York.

Cheung, J. Y., Bonventre, J. V., Malis, C. D., and Leaf, A. (1986). Calcium and ischemic injury. *N. Engl. J. Med.* **314**, 1670–1676.

Chopp, M., Li, Y., Dereske, M., Levine, S., Yoshida, Y., and Garcia, J. (1991). Neuronal injury and expression of 72-kDa heat-shock protein after forebrain ischemia in the rat. *Acta Neuropathol.* (*Berl.*) **83**, 66–71.

Cochran, B. H., Zullo, J., Verma, I. M., and Stiles, C. D. (1984). Expression of the c-fos gene and of an fos-related gene is stimulated by platelet-derived growth factor. *Science* **226,** 1080–1082.

Curran, T., and Morgan, J. I. (1986). Barium modulates c-fos expression and post-translational modification. *Proc. Natl. Acad. Sci. USA* **83,** 8521–8524.

Deshpande, J. K., Siesjo, B. K., and Wieloch, T. (1987). Calcium Accumulation and neuronal damage in the rat hippocampus following cerebral ischemia. *J. Cereb. Blood Flow Metab.* **7,** 89–93.

Dessi, F., Charriaut-Marlangue, C., Khrestchatisky, M., and Ben-Ari, Y. (1993). Glutamate-induced neuronal death is not a programmed cell death in cerebellar culture. *J. Neurochem.* **60,** 1953–1955.

Deutch, A. Y., Lee, M. C., Gillham, M. H., Cameron, D. A., Goldstein, M., and Iadorola, M. J. (1991). Stress selectively increases fos protein in dopamine neurons innervating the prefrontal cortex. *Cereb. Cortex.* **1,** 273–292.

Dienel, G. A. (1984). Regional accumulation of calcium in postischemic rat brain. *J. Neurochem.* **43,** 913–925.

Dienel, G. A., Curz, N. F., and Rosenfeld, S. J. (1985). Temporal profiles of proteins responsive to transient ischemia. *J. Neurochem.* **44,** 600–610.

Dienel, G. A., Kiessling, M., Jacewicz, M., and Pulsinelli, W. A. (1986). Synthesis of heat shock proteins in rat brain cortex after transient ischemia. *J. Cereb. Blood Flow Metab.* **6,** 505–510.

Dienel, G. A., and Pulsinelli, W. A. (1986). Uptake of radiolabeled ions in normal and ischemia-damaged brain. *Ann Neurol.* **19,** 465–472.

Dienel, G. A., and Pulsinelli, W. A., and Duffy, T. E. (1980). Regional protein synthesis in rat brain following acute hemispheric ischemia. *J. Neurochem.* **35,** 1216–1226.

Dipasquale, B., Marini, A. M., and Youle, R. J. (1995). Apoptosis and DNA degradation induced by 1-methyl-4-phenylpyridinium (MPP⁺) in neurons. Submitted for publication.

Dragunow, M. (1992). Axotomized medial septal-diagonal band neurons express Jun-like immunoreactivity. *Brain Res. Mol. Brain Res.* **15,** 141–144.

Dragunow, M., Young, D., Hughes, P., G., M., Lawlor, P., Singleton, K., Sirimanne, E., Beilharz, E., and Gluckman, P. (1993). Is c-Jun involved in nerve cell death following status epilepticus and hypoxic-ischaemic brain injury? *Mol. Brain Res.* **18,** 347–352.

Dux, E., Mies, G., Hossmann, K.-A., and Siklos, L. (1987). Calcium in the mitochondria following frief ischemia of gerbil brain. *Neurosci. Lett.* **78,** 295–300.

Emson, P. C. (1993). In-situ hybridization as a methodological tool for the neuroscientist. *Trends Neurosci.* **16,** 9–16.

Eng, L. F. (1985). Glial fibrillary acidic protein: the major protein of glial intermedite filaments in differentiated astrocytes. *J. Neuroimmunol.* **8,** 203–214.

Evans, H. L., Jortner, B. S., and El-Fawal, H. A. N. (1992). Glial response to trimethyl lead in the macaque monkey. *Toxicologist* **12,** 317.

Evans, H. L., Little, A. R., Gong, Z. L., Duffy, J. S., Wirgin, I., and El-Fawal, H. A. N. (1993). Glial fibrillary acidic protein indicates in vivo exposure to environmental contaminants: PCBs in the Atlantic Tomcod. In *Markers of Neuronal Injury and Degeneration* (J. N. Johannessen, Ed.) pp. 402–406, The New York Academy of Sciences, New York.

Farooqui, A. A., Liss, L., and Horrocks, L. A. (1990). Elevated activities of lipases and lysophospholipase in Alzheimer's disease. *Dementia* **1,** 208–214.

Forloni, G., Chiesa, R., Smiroldo, S., Verga, L., Salmona, M., Tagliavini, F., and Angeretti, N. (1993). Apoptosis mediated neurotoxicity induced by chronic application of b amyloid fragment 25–35. *Mol. Neurosci.* **4,** 523–526.

Garofalo, O., Kennedy, P. G. E., Swash, M., Martin, J. E., Luthert, P., Anderton, B. H., and Leigh, P. N. (1991). Ubiquitin and heat shock protein expression in amyotrophic lateral sclerosis. *Neuropathol. Appl. Neurobiol.* **17,** 39–45.

Gavrieli, Y., Sherman, Y., and Ben-Sasson, S. A. (1992). Identification of programmed cell death in situ via specific labeling of nuclear DNA fragmentation. *J. Cell Biol.* **119,** 493–501.

Gonzalez, M., Shiraishi, K., Hisanaga, K., Sagar, S., Mandabach, M., and Sharp, F. (1989). Heat shock proteins as markers of neural injury. *Brain Res. Mol. Brain Res.* **6,** 93–100.

Gramsbergen, J. B. P., Duin, L. V.-V. D., Loopuijt, L., Paans, A. M. J., Vaalburg, W., and Korf, J. (1988). Imaging of the degeneration of neurons and their processes in rat or cat brain by 45CaC12 autoradiography or 55CoC12 positron emission tomography. *J. Neurochem.* **50,** 1798–1807.

Hansen, A. J., and Zeuthen, T. (1981). Extracellular ion concentration during spreading depression and ischemia in the rat brain cortex. *Acta Physiol. Scand.* **113,** 437–445.

Harmon, B. V., and Allan, D. J. (1988). X-ray-induced cell death by apoptosis in the immature rat cerebellum. *Scanning Microscopy* **2,** 561–568.

Harrington, M. G., Merril, C. R., Asher, D. M., and Gajdusek, D. C. (1986). Abnormal proteins in the cerebrospinal fluid of patients with Creutzfeldt-Jakob disease. *N. Engl. J. Med.* **315,** 279–283.

Hayashi, T., Takada, K., and Matsuda, M. (1991). Changes in ubiquitin and ubiquitin-protein conjugates in the CA1 neurons after transient sublethal ischemia. *Mol. Chem. Neuropathol.* **15,** 75–82.

Hayashi, T., Takada, K., and Matsuda, M. (1992). Post-transient ischemia increase in ubiquitin conjugates in the early reperfusion. *Neuroreport* **3,** 519–520.

Hochstrasser, D. F., Patchornik, A., and Merril, C. R. (1988). Development of polyacrylamide gels that improve the separation of proteins and their detection by silver staining. *Anal. Biochem.* **173,** 412–423.

Ignatowicz, E. Vezzani, A.-M., Rizzi, M., and D'Incalci, M. (1991). Nerve cell death induced in vivo by kainic acid and quinolinic acid does not involve apoptosis. *Mol. Neurosci.* **2,** 651–654.

Janscó, G., Karcsú, S., Kiraly, E., Szebini, A., Tóth, L., Bácsy, E., Joó, F., and Párdicz, A. (1984). Neurotoxin induced nerve cell degeneration: possible involvement of calcium. *Brain Res.* **295,** 211–216.

Jansen, H. M. L., Pruim, J., Paans, A. M. J., Franssen, E. J., Jong, B. M. D., Haaxma, R., and Korf, J. (1993). Visualization of ischemic brain damage with ⁵⁵Co positron emission tomography in man. *J. Cereb. Blood Flow Metab.* **13,** (S707).

Jian, M., Staines, W. A., Iadarola, M. J., and Robertson, G. S. (1993). Destruction of the nigrostriatal pathway increases Fos-like immunoreactivity predominantly in striatopallidal neurons. *Brain Res. Mol. Brain Res.* **19,** 156–160.

Johannessen, J. N., Chiueh, C. C., Bacon, J. P., Garrick, N. A., Burns, R. S., Weise, V. K., Kopin, I. J., Parisi, J. E., and Markey, S. P. (1989). Effects of 1-methyl-4-phenyl-1,2,3,6-tetrahydropyridine in the dog: effect of pargyline pretreatment. *J. Neurochem.* **53,** 582–589.

Johnson, E. M., Jr., and Deckwerth, T. L. (1993). Molecular mechanisms of developmental neuronal death. *Annu. Rev. Neurosci.* **16,** 31–46.

Jørgensen, M., Johansen, F., Diemer, N., and Jørgensen, M. (1991). Post-ischemic and kainic acid-induced c-fos protein expression in the rat hippocampus. *Acta Neurol. Scand.* **84,** 352–356.

Jørgensen, M. B., Johansen, F. F., and Diemer, N. H. (1987). Removal of the entorhinal cortex protects hippocampal CA-1 neurons from ischemic damage. *Acta Neuropathol. (Berlin)* **73**, 189–194.

Jucker, M., Bialobok, P., Kleinman, H. K., Walker, L. C., Hagg, T., and Ingram, D. K. (1993). Lamini-like and lamini-binding protein-like immunoreactive astrocytes in rat hippocampus after transient ischemia: Antibody to lamini-binding protein is a sensitive marker of neural injury. In *Markers of Neuronal Injury and Degeneration* (J. N. Johannessen, Ed.), pp. 245–252, The New York Academy of Sciences, New York.

Jucker, M., Kleinman, H. K., Hohmann, C. F., Ordy, J. M., and Ingram, D. K. (1991). Distinct immunoreactivity to 110 kDa laminin-binding protein in adult and lesioned rat forebrain. *Brain Res.* **555**, 305–312.

Kato, H., Chen, T., Liu, X., Nakata, N., and Kogure, K. (1993). Immunohistochemical localization of ubiquitin in gerbil hippocampus with induced tolerance to ischemia. *Brain Res.* **619**, 339–343.

Kerr, J. F. R., and Harmon, B. V. (1991). Definition and incidence of apoptosis: an historical perspective. In *Apoptosis: The Molecular Basis of Cell Death* (L. D. Tomei and F. O. Cope, Eds.), pp. 5–29, Cold Spring Harbor Laboratory Press, Cold Spring Harbor.

Kitagawa, K., Matsumoto, M., Kuwabara, K., Tagaya, M., Ohtsuki, T., Hata, R., Ueda, H., Handa, N., Kimura, K., and Kamada, T. (1991). 'Ischemic tolerance' phenomenon detected in various brain regions. *Brain Res.* **561**, 203–211.

Kleinman, H. K., Weeks, B. S., Cannon, F. B., Sweeney, T. M., Sephel, G. C., Clement, B., Zain, M., Olson, M. O. J., Jucker, M., and Burrous, B. A. (1991). Identification of a 110 kD nonintegrin cell surface laminin-binding protein which recognizes an A chain neurite promoting peptide. *Arch. Biochem. Biophys.* **290**, 32–325.

Koller, W., Vetere-Overfield, B., Gray, C., Alexander, C., Chin, T., Dolezal, J., Hassanein, R., and Tanner, C. (1990). Environmental risk factors in Parkinson's disease. *Neurology* **40**, 1218–1221.

Korf, J., and Postema, F. (1984). Regional calcium accumulation and cation shifts in rat brain by kainate. *J. Neurochem.* **43**, 1052–1060.

Kuzuhara, S., Mori, H., Izumiyama, N., Yoshimura, M., and Ihara, Y. (1988). Lewey bodies are ubiquitinated. *Acta Neuropathol. (Berl.)* **75**, 345–353.

Li, X., Song, L., Kolasa, K., and Jope, R. S. (1992). Adrenalectomy potentiates immediate early gene expression in rat brain. *J. Neurochem.* **58**, 2330–2333.

Liesi, P., Kaakkola, S., Dahl, D., and Vaheri, A. (1984). Laminin is induced in astrocytes of adult brain by injury. *EMBO J.* **3**, 683–686.

Lillianfeld, D., Chan, E., Ehland, J., Godbold, J., Landrigan, P., Marsh, G., and Perl, D. (1990). Two decades of increasing mortality from Parkinson's disease among the US elderly. *Arch Neurol.* **47**, 731–734.

Lindsay, R. M. (1986). Reactive gliosis. In *Astrocytes: Cell Biology and Pathology of Astrocytes* (S. Federoff and A. Vernadakis, Eds.), pp. 231–262, Academic Press, Orlando, Florida.

Liu, Y., Kato, H., Nakata, N., and Kogure, K. (1993). Temporal profile of heat shock protein 70 synthesis in ischemic tolerance induced by preconditioning ischemia in rat hippocampus. *Neuroscience* **56**, 921–927.

Longo, F., Wang, S., Narasimhan, P., Zhang, J., Chen, J., Massa, S., and Sharp, F. (1993). cDNA cloning and expression of stress-inducible rat hsp 70 in normal and injured rat brain. *J. Neurosci. Res.* **36**, 325–335.

Lowe, J., Blanchard, A., Morrell, K., Lennox, G., Reynolds, L., Billet, M., Landon, M., and Mayer, R. J. (1988). Ubiquitin is a common factor in intermediate inclusion bodies of diverse type in man, including those of Parkinson's disease, Pick's disease,

and Alzheimer's disease, as well as Rosenthal fibers in cerebellar astrocytomas, cytoplasmic bodies in muscle, and Mallory bodies in alcoholic liver disease. *J. Pathol.* **155**, 9–15.

Marini, A. M., Kozuka, M., Lipsky, R. L., and Nowak, T. S. J. (1990). 70-Kilodalton heat shock protein induction in cerebellar astrocytes and granule cells in vitro: Comparison with immunocytochemical localization after hyperthermia in vivo. *J. Neurochem.* **54**, 1509–1516.

Martin, D. P., Schmidt, R. E., DiStefano, P. S., Lowry, O. H., Carter, J. G., and Johnson, E. M. (1988). Inhibitors of protein synthesis and RNA synthesis prevent neuronal death caused by nerve growth factor deprivation. *J. Cell Biol.* **106**, 829–844.

Martins, E., Inamura, K., Themner, K., Malmqvist, K. G., and Seisjö, B. K. (1988). Accumulation of calcium and loss of potassium in hippocampus following transient cerebral ischemia: a proton microprobe study. *J. Cereb. Blood Flow Metab.* **8**, 531–538.

McConkey, D. J., Hartzell, P., Duddy, S. K., Hakansson, H., and Orrenius, S. (1988). 2,3,7,8-Tetrachlorodibenzo-p-dioxin kills immature thymocytes by Ca^{2+}-mediated endonuclease activation. *Science* **242**, 256–259.

McConkey, D. J., Nicotera, P., Hartzell, P., Bellomo, G., Wyllie, A. H., and Orrenius, S. (1989). Glucocorticoids activate a suicide process in thymocytes through an elevation of cytosolic Ca^{2+} concentration. *Arch. Biochem. Biophys.* **269**, 365–370.

Mies, G., Kawai, K., Saito, N., Nagashima, G., Nowack, T. S., Jr., Ruetzler, C. A., and Klatzo, I. (1990). Cardiac arrest-induced complete cerebral ischemia in the rat: dynamics of post-ischemic in vivo calcium uptake and protein synthesis. *Neurol Res.* **15**, 253–263.

Morgan, J. I., Cohen, D. R., Hempstead, J. L., and Curran, T. (1987). Mapping patterns of c-fos expression in the central nervous system after seizure. *Science* **237**, 192–197.

Morgan, J. I., and Curran, T. (1986). Role of ion flux in the control of c-fos expression. *Nature* **322**, 552–555.

Morgan, J. I., and Curran, T. (1988). Calcium as a modulator of the immediate-early gene cascade in neurons. *Cell Calcium* **9**, 303–311.

Morgan, J. I., and Curran, T. (1989). Stimulus-transcription coupling in neurons: role of cellular immediate-early genes. *TINS* **12**, 459–462.

Mori, H., Kondo, J., and Ihara, Y. (1987). Ibiquitin is a component of paired helical filaments in Alzheimer's disease. *Science* **235**, 1641–1644.

Morimoto, R. I., Tissières, A., and Georgopoulis, C. (Eds.). (1990). *Stress Proteins in Biology and Medicine.* Cold Spring Harbor Laboratory Press, Cold Spring Harbor.

Müller, R., Bravo, R., Burckhardt, J., and Curran, T. (1984). Induction of c-fos gene and protein by growth factors precedes activation of c-myc. *Nature* **312**, 716–720.

Nakagomi, T., Kirino, T., Kanemitsu, H., Tsujita, Y., and Tamura, A. (1993). Early recovery of protein synthesis following ischemia in hippocampal neurons with induced tolerance in the gerbil. *Acta Neuropathol. (Berl.)* **86**, 10–15.

Nakamura, K., Hatakeyama, T., Furuta, S., and Sakaki, S. (1993). The role of early Ca^{2+} influx in the pathogenesis of delayed neuronal death after brief forebrain ischemia in gerbils. *Brain Res.* **613**, 181–192.

Nicotera, P., Conkey, D. J. M., Dypbukt, J. M., Jones, D. P., and Orrenius, S. (1989). Ca^{2+} Activated mechanisms in cell killing. *Drug Metab. Rev.* **20**, 193–201.

Nixon, R. A. (1986). Fodrin degradation by calcium-activated neutral proteinase (CANP) in retinal ganglion cell neurons and optic glia: Preferential localization of CANP activities in neurons. *J. Neurosci.* **6**, 1264–1271.

Nixon, R. A., Quackenbush, R., and Vitto, A. (1986). Multiple calcium-activated neutral proteinases (PANP) in mouse retinal ganglion cell neurons: Specificities for endogenous neuronal substrates and comparison to purified brain CANP. *J. Neurosci.* **6,** 1252–1263.

Nowak, T. S., Jr. (1985). Synthesis of a stress protein following transient ischemia in the gerbil. *J. Neurochem.* **45,** 1635–1641.

Nowak, T. S., Jr. (1991). Localization of 70 kDa stress protein mRNA induction in gerbil brain after ischemia. *J. Cereb. Blood Flow Metab.* **11,** 432–439.

Nowak, T. S., Jr. (1993). Synthesis of heat shock/stress proteins during cellular injury. In *Markers of Neuronal Injury and Degeneration* (J. N. Johannessen, Ed.), pp. 142–156, The New York Academy of Sciences, New York.

Nowak, T. S., Jr., Bond, U., and Schlesinger, M. J. (1990). Heat shock RNA levels in brain and other tissues after hyperthermia and transient ischemia. *J. Neurochem.* **54,** 451–458.

O'Callaghan, J. P. (1991). Quantification of glial fibrillary acidic protein: comparison of slot-immunobinding assays with a novel sandwich ELISA. *Neurotoxicol. Teratol.* **13,** 275–281.

O'Callaghan, J. P. (1993). Quantitative features of reactive gliosis following toxicant-induced damage of the CNS. In *Markers of Neuronal Injury and Degeneration* (J. N. Johannessen, Ed.), pp. 195–210, The New York Academy of Sciences, New York.

O'Callaghan, J. P., Miller, D. B., and Reinhard, J. F., Jr. (1990). Characterization of the origins of the astrocyte response to injury using the dopaminergic neurotoxicant MPTP. *Brain Res.* **521,** 73–80.

Olney, J. W., Labruyere, J., Wang, G., Wozniak, D. F., Price, M. T., and Sesma, M. A. (1991). NMDA antagonist neurotoxicity: Mechanism and prevention. *Science* **254,** 1515–1518.

Perl, T. M., Bedard, L., Kosatsky, T., Hockin, J. C., Todd, E. C. D., and Remis, R. S. (1990). An outbreak of toxic encephalopathy caused by eating mussels contaminated with domoic acid. *N. Engl. J. Med.* **322,** 1775–1780.

Pulsinelli, W. A., Brierley, J. B., and Plum, F. (1981). Temporal profile of neuronal damage in a model of transient forebrain ischemia. *Ann. Neurol.* **11,** 491–498.

Quach, T. T., Schrier, B. K., and Duchemin, A. M. (1993). Gene expression in brain injury: identification of a new cDNA structurally related to adhesive and trophic agents. In *Markers of Neuronal Injury and Degeneration* (J. N. Johannessen, Ed.), pp. 423–430, The New York Academy of Sciences, New York.

Roberts-Lewis, J. M., and Siman, R. (1993). Spectrin proteolysis in the hippocampus: a biochemical marker for neuronal injury and neuroprotection. In *Markers of Neuronal Injury and Degeneration* (J. N. Johannessen, Ed.), pp. 78–86, The New York Academy of Sciences, New York.

Schanne, F. A. X., Kane, A. B., Young, E. E., and Farber, J. L. (1979). Calcium dependence of toxic cell death: a final common pathway. *Science* **206,** 456–470.

Schlaepfer, W. W., Lee, C., Lee, V. M.-Y., and Zimmerman, U.-J. P. (1985). An immunoblot study of neurofilament degeneration in situ and during calcium-activated proteolysis. *J. Neurochem.* **44,** 502–509.

Schreiber, S., Najm, I., Tocco, G., and Baudry, M. (1993). Co-expression of HSP72 and c-fos in rat brain following kainic acid treatment. *Neuroreport* **5,** 269–272.

Sharp, F. R., Butman, M., Wang, S., Koistinaho, J., Graham, S. H., Sagar, S. M., Berger, P., and Longo, F. M. (1993). Heat shock proteins used to show that haloperidol prevents neural injury produced by ketamine, MK 801, and phencyclidine. In *Markers of Neuronal Injury and Degeneration* (J. N. Johannessen Ed.), pp. 288–290, The New York Academy of Sciences, New York.

Sheng, M., Dougan, S. T., McFadden, G., and Greenberg, M. E. (1988). Calcium and growth factor pathways of c-fos Transcriptional activation require distinct upstream regulatory sequences. *Mol. Cell. Biol.* **8,** 2787–2796.

Shimosaka, S., So, Y., and Simon, R. (1992). Distribution of HSP72 induction and neuronal death following limbic seizures. *Neurosci. Lett.* **138,** 202–206.

Shin, C., McNamara, J. O., Morgan, J. I., Curran, T., and Cohen, D. R. (1990). Induction of c-fos mRNA expression by afterdischarge in the hippocampus of naive and kindled rats. *J. Neurochem.* **55,** 1050–1055.

Siesjö, B. K., and Bengtsson, F. (1989). Calcium fluxes, calcium antagonists, and calcium-related pathology in brain ischemia, hypoglycemia, and spreading depression: a unifying hypothesis. *J. Cereb. Blood Flow Metab.* **9,** 127–140.

Siman, R., Gall, C., Perlmutter, L. S., Christian, C., Baudry, M., and Lynch, G. (1985). Distribution of calpain I, an enzyme associated with degenerative activity, in rat brain. *Brain Res.* **347,** 399–403.

Siman, R., and Noszek, J. C. (1988). Excitatory amino acids activate calpain I and induce structural protein breakdown in vivo. *Neuron* **1,** 279–287.

Siman, R., Noszek, J. C., and Kegerise, C. (1989). Calpain I activation is specifically related to excitatory amino acid induction of hippocampal damage. *J. Neurosci.* **9,** 1579–1590.

Simon, R., Niiro, M., and Gwinn, R. (1993). Prior ischemic stress protects against experimental stroke. *Neurosci. Lett.* **163,** 135–137.

Smeyne, R. J., Vendrell, M., Hayward, M., Baker, S. J., Miao, G. G., Schilling, K., Robertson, L. M., Curran, T., and Morgan, J. I. (1993). Continuous c-fos expression precedes programmed cell death in vivo. *Nature* **363,** 166–169.

Sobotka, T. J. (1991). Screening for neurotoxicity: application in the toxicological evaluation of regulated food chemicals. *J. Am. Coll. Toxicol.* **10,** 671–676.

Sonnenberg, J. L., Mitchelmore, E., Macgregor-Leon, P. F., Hempstead, J., Morgan, J. I., and Curran, T. (1989). Glutamate receptor agonists increase the expression of Fos, Fra, and AP-1 DNA binding activity in the mammalian brain. *J. Neurosci. Res.* **24,** 72–80.

Sprang, G. K., and Brown, I. R. (1987). Selective induction of a heat shock gene in fiber tracts and cerebellar neurons of the rabbit brain detected by in situ hybridization. *Mol. Brain Res.* **3,** 89–93.

Switzer, R. C., III (1993). Silver staining methods: their role in detecting neurotoxicity. In *Markers of Neuronal Injury and Degeneration* (J. N. Johannessen, Ed.), pp. 341–348, The New York Academy of Sciences, New York.

Szekely, A. M., Barbaccia, M. L., and Costa, E. (1987). Activation of specific glutamate receptor subtypes increases c-fos proto-oncogene expression in primary cultures of neonatal rat cerebellar granule cells. *Neuropharmacology* **26,** 1779–1782.

Taginuma, H., and Takashita, N. (1993). Neuronal death in the cervical cord of embryonic chick-quail chimeras. *Soc. Neurosci. Abstr.* **19,** 440.

Tanner, C. M., Chen, B., Wang, W., Peng, M., Liu, Z., Liang, X., Kao, L. C., Gilley, D. W., Goetz, C. G., and Schoenberg, B. S. (1989). Environmental factors and Parkinson's disease: a case-control study in China. *Neurology* **39,** 660–664.

Tomioka, C., Nishioka, K., and Kogure, K. (1993). A comparison of induced heat-shock protein in neurons destined to survive and those destined to die after transient ischemia in rats. *Brain Res.* **612,** 216–220.

Towbin, H., Staehelin, T., and Gordon, J. (1992). Electrophoretic transfer of proteins from polyacrylamide gels to nitrocellulose sheets: procedure and some applications. *Biotechnology* **24**, 145–149.

U.S. Congress, O. O. T. A. (1990). *Neurotoxicity: Identifying and Controlling Poisons of the Nervous System. OTA-BA-436.* U.S. Government Printing Office, Washington, D.C.

Uney, J., Leigh, P., Marsden, C., Lees, A., and Anderton, B. (1988). Stereotaxic injection of kainic acid into the striatum of rats induces synthesis of mRNA for heat shock protein 70. *FEBS Lett.* **235**, 215–218.

Van Den Berg, K. J., and Gramsbergen, J. B. P. (1993). Long-term changes in glial fibrillary acidic protein and calcium levels after a single systemic dose of kainic acid. In *Markers of Neuronal Injury and Degeneration* (J. N. Johannessen, Ed.), pp. 394–401, The New York Academy of Sciences, New York.

Vass, K., Berger, M., Nowak, T. J., Welch, W., and Lassmann, H. (1989). Induction of stress protein HSP70 in nerve cells after status epilepticus in the rat. *Neurosci Lett.* **100**, 259–264.

Vass, K., Welch, W. J., and Nowak, T. S., Jr. (1988). Localization of 70 kDa stress protein induction in gerbil brain after ischemia. *Acta Neuropathol.* **77**, 128–135.

Wang, S., Longo, F., Chen, J., Butman, M., Graham, S., Haglid, K., and Sharp, F. (1993). Induction of glucose regulated protein (grp 78) and inducible heat shock protein (hsp 70) mRNAs in rat brain after kainic acid seizures and focal ischemia. *Neurochem. Int.* **23**, 575–582.

Welch, W. J. (1990). The mammalian stress response: cell physiology and biochemistry of stress proteins. In *Stress Proteins in Biology and Medicine* (R. I. Morimoto, A., Tissieres and C. Geogopoulos, Eds.), pp. 223–278, Cold Spring Harbor Press, Cold Spring Harbor.

Wieloch, T., Lindvall, O., Blomqvist, P., and Gage, F. H. (1985). Evidence for amelioration of ischaemic neuronal damage in the hippocampal formation by lesions of the perforant path. *Neurol. Res.* **7**, 24–26.

Wilcox, B. J., Clatterbuck, R. E., Price, D. L., and Koliatsos, V. E. (1993). Evidence for programmed cell death in motor neurons in the rat CNS. *Soc. Neurosci. Abstr.* **19**, 441.

Yamashita, K., Eguchi, Y., Kajiwara, K., and Ito, H. (1991). Mild hypothermia ameliorates ubiquitin synthesis and prevents delayed neuronal death in the gerbil hippocampus. *Stroke* **22**, 1574–1581.

Yee, W., Frim, D., and Isacson, O. (1993). Relationship between stress protein induction and NMDA-mediated neuronal death in the entorhinal cortex. *Exp. Brain Res.* **94**, 193–202.

Nucleic Acid Hybridization Techniques and Neurotoxicity Assessment

MELVIN L. BILLINGSLEY
Department of Pharmacology
Macromolecular Core Facility
Milton S. Hershey Medical Center
Penn State College of Medicine
Hershey, Pennsylvania 17033

STEPHANIE M. TOGGAS
Department of Neuropharmacology
Division of Virology
The Scripps Research Institute
La Jolla, California 92037

I. Overview of Nucleic Acid Hybridization Techniques

Nucleic acid hybridization techniques allow for sensitive and specific detection, quantitation, and purification of DNA and RNA. Such techniques have been applied to measurement of changes in levels of nucleic acids resulting from exposures to neurotoxicants (Billingsley and O'Callaghan, 1994). The central premise governing nucleic acid hybridization is that single-stranded DNA and RNA molecules can form thermodynamically stable, double-stranded complexes based on their degree of sequence complementarity. The formation of double-stranded structures is a reversible process, dependent on biophysical parameters such as salt concentration, temperature, and presence of chemical denaturants such as formaldehyde and formamide. Denaturation of double-stranded hybrids involves the disruption of the hydrogen bonds maintaining Watson-Crick base pairing between A–T (or A and U in the case of RNA) and G–C. DNA : DNA hybrids, RNA : RNA

hybrids (homoduplexes), RNA : DNA hybrids (heteroduplexes), and intrastrand loops containing double-stranded regions can form. This has been exploited to measure RNA levels with DNA molecules, to isolate populations of specific molecules such as mRNAs via oligonucleotides (i.e., oligo(dT) to isolate poly(A)$^+$-enriched RNA), and to compare the relatedness of DNA and RNA populations. Hybridizations can occur in both liquid and solid phases; both are germane to neurotoxicology.

A. Solution-Phase Hybridizations

1. DNA–DNA and DNA–RNA Solution Hybridization

Conditions under which most biologically relevant nucleic acid hybrids form are those found in biological fluids, which contain ionized salts (primarily NaCl) at approximately 150 mM concentration. Temperatures in living cells are also maintained within a narrow range. Biologically relevant hybrids include double-stranded homodimers of DNA found in the nucleus, as well as molecules of tRNA, rRNA, and mRNA which, although single-stranded, display elaborate secondary structure, forming stem loop and cloverleaf configurations via intrastrand hydrogen bonds. Thus, nucleic acid hybrids exist physiologically in a solution phase, under tightly controlled conditions of temperature, salt, and chemical composition. Experimentally, solution-phase hybridizations can be easily adjusted to affect the extent of hybridization by varying ionic strength, temperature, and chemical conditions. The chemical composition of the DNA molecule affects the stability of a DNA : DNA hybrid; G–C base pairs form three hydrogen bonds, whereas A–T base pairs form only two hydrogen bonds.

In order to appreciate some of the empirical methods used to manipulate duplex formation, definition of some terminology and concepts is needed. *Denaturation* is the loss of duplex state between two nucleic acid chains or between two regions of the same nucleic acid chain caused by the sequential disruption of hydrogen bonds (Cantor and Schimmel, 1980). Hence, regions of DNA with a high content of C and G are more resistant to denaturation than regions rich in A and T. In addition, base pairs at the end of double-stranded DNA duplexes are less stable than those in the middle of the helix. High temperatures and chemical denaturants such as urea, formaldehyde, formamide, and guanidine hydrochloride disrupt hydrogen bonds and can be used in concert to regulate conditions of denaturation.

The transition from double helix to the single-stranded DNA coil is best described using the *melting temperature* (T_m) (Beltz *et al.*, 1983). This value is used to denote the temperature at which the denaturation of the complex is 50% complete, under defined conditions of salt concentration and chemical denaturants. One key determinant for T_m is the degree of complementarity between two strands of nucleic acid. For instance, two strands of DNA that are completely complementary will have a higher T_m than those that have 10% mismatch of nucleotides (90% homologous). Thus, increasing temperatures promote strand separation and denaturation, with strands containing the greatest complementarity requiring higher temperatures to separate.

Ionic strength also affects the T_m; concentrations of ionic constituents can either stabilize or destabilize the nucleic acid duplex (Seed, 1982). The negatively charged phosphate moieties in the nucleic acid chain are shielded by Na^+ ions in the solution, minimizing intra- and interstrand repulsion. Consequently, increasing concentrations of NaCl stabilize hybrids and effectively raise the T_m; low concentrations (<100 mM) lower the T_m. Thus, at a constant temperature,

NaCl concentrations can be manipulated experimentally to create conditions which either favor formation of hybrids having some degree of mismatch (high salt) or demand that hybrids have nearly complete complementarity (low salt).

Manipulations of temperature, ion concentration, and the presence of chemical denaturants are collectively used to determine the level of stringency. *Stringency* refers to the conditions that must be overcome in order to allow formation of a given hybrid molecule (Zoller and Smith, 1982). Low stringency conditions are those of low temperature (30–45°C), high salt (>150 mM NaCl), and an absence of denaturants. Under these conditions, mismatches between hybrid strands are more readily tolerated. Conditions of high stringency are those at higher temperatures (>50°C), low NaCl (<100 mM), and the presence of denaturants such as formamide. DNA hybrids must have a nearly complete degree of complimentarity to form stable double-stranded species under conditions of high stringency.

The final important concept is the process of *renaturation;* literally, this means that the original native structure of the nucleic acid duplex is reformed after removal of denaturing conditions (Freifelder, 1976). Renaturation can be exploited by the investigator such that conditions can be established which promote the association of denatured nucleic acids to form hydrogen bonds with denatured, radiolabeled nucleic acids which, although complementary, are added exogenously. Labeled nucleic acids added exogenously for purposes of detection or quantitation are often referred to as "*probes*" and the DNA or RNA of interest present in the tissue is termed "*target*" sequences. The formation of DNA : DNA hybrids and DNA : oligonucleotide hybrids are shown schematically in Fig. 1, under conditions of low and high stringency.

The advent of advanced cloning techniques and synthetic oligonucleotides has greatly facilitated the use of nucleic acid hybridization techniques. Several specific techniques will be discussed relative to neurotoxicological assessments in the following sections. For more technical detail, the reader is referred to several lab manuals which are widely used in molecular biology laboratories (Sambrook *et al.*, 1989; Ausubel *et al.*, 1993).

2. RNase Protection Assays

One specific application of solution-phase hybridization is the RNase protection assay, which can be used to simultaneously assay for multiple species of mRNA (Berk and Sharp, 1977). The principles behind this technique are illustrated in Fig. 2. Briefly, RNA is extracted from the tissue under conditions that prevent RNA degradation (Chornezynski and Sacchi, 1987). The to-

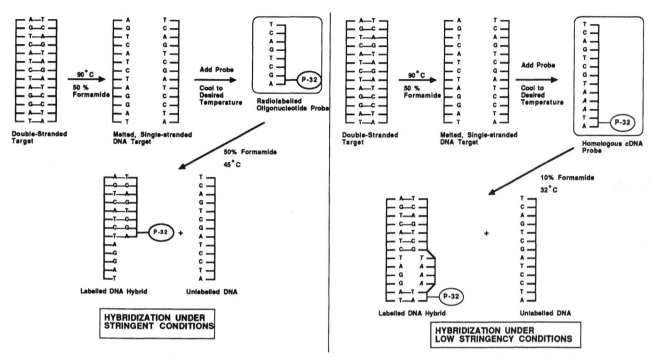

FIGURE 1 Hybridization occurs between complementary regions of nucleic acid and can be controlled by manipulating conditions of ionic strength, temperature, and the presence of denaturants to create conditions of high and low stringency.

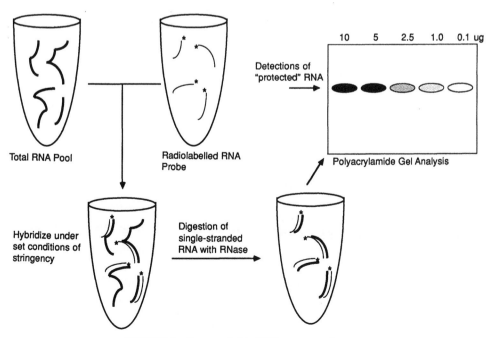

FIGURE 2 Strategy behind RNase protection assays.

tal RNA population, consisting of tRNA, rRNA, and mRNA (2–5% of total), is then incubated with a radiolabeled recombinant RNA probe (riboprobe) of defined length. The probe is referred to as an antisense probe because, in order to hybridize to the target (sense) mRNA, its sequence must be the complement of the sense mRNA and thus, the probe sequence must be in the antisense orientation. The riboprobe is derived from a plasmid vector which contains the cDNA sequence of interest downstream from a promotor which is recognized by procaryotic RNA polymerases (SP6, T3, T7) (Melton *et al.*, 1984). It is crucial to know both the size of the cDNA insert and its orientation relative to the promoter (antisense). Several commercially available plasmids have been engineered as vehicles for production of antisense and sense RNA from investigator-subcloned cDNA inserts.

An excess amount of the radiolabeled RNA antisense riboprobe is added to the solution of total RNA, heated to high temperature (90–100°C) to disrupt any secondary structure, and allowed to hybridize overnight under controlled conditions of temperature (42–60°C) and NaCl (>100 mM). During this time, the radiolabeled RNA probe binds to mRNA targets, forming a double-stranded complex (RNA : RNA) for the length of the probe : target hybrid. Following hybridization, purified RNase is added. This enzyme recognizes and digests only single-stranded RNA species and regions of RNA, leaving the double-stranded probe : target complex intact (hence the name "protection"). The RNAse is inactivated by digesting it with a protease such as proteinase K. The protected species are then purified from the mixture by protein extraction and ethanol precipitation, and electrophoresed on an acrylamide gel. The size-separated, radiolabeled RNA products are visualized using autoradiography and should correspond to the length of the riboprobe that is complimentary to the target sequence.

Since the size of a given riboprobe is determined by the size of the cDNA insert within the plasmid containing the RNA polymerase promotor, it is possible to assay for several different mRNA species at once. One application in neurotoxicity is for determination of alterations in mRNA levels following neurotoxicant exposure. By using RNase protection assays, one could determine such alterations by using a series of riboprobes specific for a variety of toxicant-specific and injury-response mRNAs, each detected with a probe of a different size.

The advantages of RNase protection assays are the ability to monitor multiple mRNAs simultaneously, the overall sensitivity relative to Northern blot analysis (one order of magnitude), and the tolerance of some RNA degradation in the samples. In addition, the sensitivity of the assay allows for the use of total RNA in most cases, obviating the time and expense of preparing poly(A)$^+$-enriched RNA. Care must be taken in the design of riboprobes, avoiding riboprobes that are rich in AU regions (which may cross-hybridize with 3′ ends of mRNAs in general) or have considerable homology to other target mRNAs. This latter problem is of some concern when measuring members of a multigene family. However, the specificity of the assay is such that probes with even a few areas of mismatch will form hybrids which will not be completely protected from RNase digestion and, thus, cleaved into smaller fragments. Therefore, careful choice of sequences used for probes will allow this method to be used to distinguish even very similar target sequences.

3. Gel-shift Assays: Transcription Factor Analysis

A more sophisticated use of "hybridization" has resulted in the development of so-called gel shift assays to determine alterations in transcription factors which bind to DNA in a sequence-specific manner (Fried and Crothers, 1981). These assays, called gel-shift assays, determine the binding of protein transcription factors to specific target sequences in DNA. The principle behind this assay is that DNA–protein complexes migrate more slowly in an electrophoretic field than either protein or target DNA. Several toxicants, such as kainate, induce the expression of immediate-early genes such as c-*fos* and c-*jun* (Pennypacker *et al.*, 1993). These proteins form heterodimers which then bind to specific sequences on the DNA molecule. Because of the sequence-specific nature of the binding, it is possible to synthesize oligonucleotides containing consensus-binding sites for specific transcription factors. The synthetic oligonucleotide is radiolabeled and then allowed to interact with cell extracts containing the putative transcription factors. If a transcription factor is present in extracts following toxicant administration, it will then bind to target DNA sequences, leading to a slower mobility of the complexes in an electric field. Using autoradiography, it is possible to locate both the free oligonucleotide and the protein–DNA complexes.

B. Solid-Phase Approaches

1. Southern Blotting: DNA Analysis

The use of solid-phase nucleic acid hybridization for measurement of specific target sequences was initiated by E.M. Southern in 1975 (Southern, 1975). This technique involves restriction enzyme digestion of DNA into specific fragments, followed by size separation of DNA fragments using agarose gel electrophoresis. Following incubation in an alkaline solution (i.e.,

0.5 *M* NaOH/1.0 *M* NaCl) to denature the DNA and rinsing in a neutralizing buffer, the gel is directly apposed to a membrane, consisting of either nylon or nitrocellulose. The DNA is transferred from the gel to the membrane via either capillary fluid movement or vacuum. The DNA interacts with the membrane through combinations of hydrophobic and electrostatic bonds. Exposure of the size-fractionated, membrane-bound DNA to ultraviolet light causes formation of covalent crosslinks between the membrane matrix and the target DNA, minimizing loss of the target from the matrix during hybridization incubations and posthybridization washes.

Once target DNA has been immobilized, nonspecific sites which could bind labeled probe DNA are blocked by prehybridization of the membrane in a solution consisting of excess nonspecific unlabeled DNA (i.e., salmon sperm DNA), proteins, and detergents. Specific target sequences of DNA can be detected by hybridizing the membrane with a solution containing a radiolabeled complimentary DNA probe (cDNA), a riboprobe, or an oligonucleotide probe. Conditions of hybridization are then adjusted based on the considerations mentioned earlier. Specific formulae have been derived for calculating the temperature of hybridization, based on the Na$^+$ concentration, the content of G and C in the probe, the length of the probe, and the formamide concentration. One such equation is shown here for cDNA-derived probes:

$$T_m = 81.5° \text{C} + 16.6 \times \log[\text{Na}^+] + 0.41 \times (\% \text{ G} + \text{C content}) - 0.61 (\% \text{ formamide}) - 500/\text{No. base pairs.}$$

Hybridization is carried out at temperatures 20–25° C less than the T_m.

After hybridization, stringency is experimentally controlled by washing the membrane at temperatures similar to the hybridization temperature, in a graded series of buffers, starting with buffers of low stringency and ending with the most stringent set of conditions required. The degree of stringency is increased by sequentially reducing the Na$^+$ content and sequentially increasing the temperature of the wash buffer. Low stringency washes are useful for revealing multiple targets of DNA which have some degree of homology with the probe. Indeed, low stringency hybridizations of DNA have been used to isolate family members in multigene families (Lefkowitz and Caron, 1988). High stringency washes are used to determine target sites most homologous to the probe DNA.

Target : probe interactions can be detected by a number of means. Most commonly, the probe is radiolabeled with either ^{35}S- or ^{32}P-deoxynucleotides; the hybrids are then visualized using autoradiography or

phosphorimaging and quantitated using densitometry. Numerous alternative methods of probe labeling are available including biotinylated DNA probes, digoxigenin DNA probes, and fluorescent probes (Langer *et al.*, 1981). Advantages of these approaches are the long stability of chemically modified DNA probes and the reduction of radioisotopic exposure and wastes. However, issues of sensitivity relative to radioisotopic detection can be problematic when levels of target DNA are low. Hence, the choice between modes of detection centers on needs for sensitivity, background vs signal, and probe stability.

The prime use of Southern blot analysis is to determine the presence or absence of a given DNA target. As a result, direct application of Southern blotting techniques in neurotoxicity is uncommon. However, Southern blots are essential in the evaluation of transgenic animals and mice bearing targeted gene disruptions. In addition, genetic strains of mice which lack particular neuronal proteins and display aberrant responses to neurotoxicants may reflect altered genomic DNA, and would thus be analyzed using Southern blots. Further, neurotoxicant-associated genes cloned from a particular species can be analyzed for potential conservation across a wide variety of species by this method.

2. Northern Blotting: mRNA Analysis

Immobilization of size-fractionated mRNA on membranes is termed Northern blotting; this technique is widely used to demonstrate levels as well as sizes of mRNA transcripts (Thomas, 1981). It is performed by subjecting RNA populations to denaturing agarose gel electrophoresis. Depending on the abundance of target RNA, the pool of RNA can be either total RNA (composed of tRNA, rRNA, and mRNA) or poly(A)$^+$-enriched RNA (mRNA = 2–5% of total). RNA is very labile to the effects of RNase from tissues, fingers, lab contamination, etc. This procedural point is perhaps the single largest source of error and frustration for novices. Purification methods employ baked glassware, solutions that are rendered RNase free by treatment with 0.1% diethylpyrocarbonate (DEPC) and subsequent autoclaving, and denaturants for extraction which minimize RNase activity.

Once RNA has been electrophoresed, it is transferred from the gel to a nitrocellulose or nylon membrane via capillary action using high salt buffers. RNA is immobilized on the membrane by baking *in vacuo* at 80°C. For nylon membranes, electrotransfer techniques using low salt buffers may be used, and RNA can be covalently linked to the membrane using ultraviolet crosslinking. Detection is accomplished using radiolabeled or colorimetric probes complementary to the target sequences. Blots are prehybridized to block nonspecific sites and are then hybridized with the labeled probe, in a fashion similar to that described for Southern blots. However, RNA:DNA hybrids are more thermodynamically stable than DNA:DNA hybrids; thus, more stringent conditions than those used for Southern blots may be needed to reduce background levels of hybridization in certain cases.

3. *In Situ* Hybridization

Localization of the target mRNA to specific cell populations in the central nervous system (CNS) is accomplished by the technique of *in situ* hybridization (Wilson and Higgins, 1989). In this case, mRNA is not extracted from a brain homogenate; instead, histologic sections of the brain itself serve as the source of the target mRNA (hence the term "*in situ*"). The mRNA is immobilized in the tissue by fixation with a crosslinking fixative, which can be accomplished by perfusion of the animal with paraformaldehyde. The tissue is then either cryoprotected in sucrose-containing solutions and frozen for cryostat sections or embedded in paraffin and sectioned. Probes can be cDNAs, oligonucleotides, or antisense riboprobes. Nonspecific cDNAs, oligonucleotides, and sense riboprobes are used as negative controls for hybridization. The probe is hybridized with the section overnight in solutions similar to those used for Southern and Northern hybridizations, using the same considerations for conditions of stringency and maintaining an RNase-free environment. If riboprobes are used, posthybridization washes with RNase digest excess probe and decrease background signal. Target mRNA is detected by radioisotopic or colorimetric methods, with ^{35}S commonly used. Signals are visualized using autoradiography and/or coating of the sections with photographic emulsion and subsequent development. *In situ* hybridization is a valuable technique for mapping changes in gene expression following toxicant administration and for identifying regions and cellular subsets most sensitive to the insult.

II. Applications of Hybridization Techniques to Issues in Neurotoxicology

A. *Changes in mRNA Expression Caused by Neurotoxicants*

Perhaps the most common use of hybridization technology in neurotoxicology has been the use of Northern blot analysis to measure changes in mRNA levels following toxicant administration. Such mRNAs can be of targets of the toxicant, of response genes such as c-*fos* and GFAP which increase as a result of damage,

or of genes which compensate for losses of neuronal inputs. Numerous examples in the literature document an increase of c-*fos* mRNA and protein following noxious and/or toxic insult to the CNS (Morgan and Curran, 1991). As such, specificity for given toxicants is lacking, but c-*fos* and other immediate-early genes can be used to outline which brain regions (Northern blot) or neuronal populations (*in situ* hybridization) are activated by the toxicant.

There are several problems associated with Northern blot analysis—some are technical and others interpretational. Technically, performing reliable Northern blots requires attention to maintaining an RNase-free environment and having sufficient amounts of tissue available for analysis. The first issue is easily surmounted by experience in this area. However, interpretation of the resulting data must be carefully viewed in context with the experiment. Using the case of a neuron-specific protein, loss of signal as a consequence of toxicant exposure can indicate the following: (a) general toxicity to neurons in the brain region; (b) selective loss of the target mRNA/protein in neurons, without consequence to other neuronal proteins; or (c) compensatory change resulting from losses of neuronal input to the specific cell. Several of these issues can be resolved by using a battery of probes for other neuron-specific and glial-specific mRNAs and by performing *in situ* hybridization studies in parallel. Thus, a well-designed experiment would analyze RNA from specific brain regions using Northern analysis following toxicant exposures. Several probes for glial- and neuronal-specific mRNAs would be used, such as GFAP and synapsin, as well as any specific neuronal probe of experimental interest. Parallel experiments using *in situ* hybridization would be used to determine the cellular patterns of hybridization which accompany neurotoxicant-induced damage.

For example, the selective neurotoxicant 1-methyl-4-phenyl-1,2,3,6-tetrahydropyridine (MPTP) damages dopaminergic nerve terminals in the striatum. The mechanism of action of MPTP is thought to involve selective uptake into dopaminergic neurons via the dopamine uptake pump (DP), followed by oxidative metabolism by monoamine oxidase-B (MAO-B) of MPTP to MPP$^+$ (Heikkila *et al.*, 1984). Tyrosine hydroxylase (TH) is the rate-limiting step in catecholamine biosynthesis and is a marker for dopamine and other catecholaminergic neurons (O'Callaghan, 1988). Glial cells respond to the chemical injury and express higher levels of GFAP (Brock and O'Callaghan, 1987). Thus, these four gene products are used to illustrate how hybridization techniques can be used to document CNS damage, to identify patterns of gene expression following neurotoxicity, and to isolate additional genes which are specific to the toxicant-induced damage. Table 1 shows a partial list of four categories of genes: neuron specific; brain region specific; injury activated; and functional (housekeeping) genes, found in all cells. Each can be used to assay for specific patterns of loss and reactivity.

After administration of a toxicant to an experimental animal, the choice of when to look for damage is a critical decision. In the early phase of MPTP toxicity (0–48 hr), numerous compensatory changes exist which occur in the levels of many proteins in sensitive regions. However, loss of specific neuronal populations takes several days following intoxication, and a robust glial response may be seen only after 7 days. Assuming that one had nucleotide probes to TH, DP, MAO-B, and GFAP, which itself requires that something be known *a priori* about the regions damaged by the toxicant, both Northern blot analysis and RNase protection assays could be used to document changes caused by neuronal loss and glial reactions. *In situ* hybridization would reveal anatomical areas of greatest change. By 7 days following MPTP intoxication, in the nigrostriatal regions, one would expect to see a loss of signal for TH and DP. Because MAO-B is widely distributed in cells other than those damaged, there would be little or no change in this gene product. Finally, there would be a marked increase in mRNA for GFAP. Both RNase protection assays, and Northern blots would show similar trends. If probes are chosen carefully, the RNase protection assay could be used to simultaneously analyze all four of the target mRNAs. However, these techniques do not allow assignment of causality and cannot be used to isolate novel gene products since sequence information is needed for successful analysis using these approaches.

B. Isolation of Gene Products Using Advanced Hybridization Techniques

Often it is desirable to isolate gene products that are either related to the mechanisms of toxicant action or may serve as biomarkers for cells sensitive to a given toxicant. Selective neurotoxicants serve as good examples to illustrate how two different methods can be used to isolate gene products related to toxicity.

1. Differential Hybridization

One of the first approaches that was used to isolate brain-specific genes was differential hybridization (Sutcliffe *et al.,* 1983). In this strategy, a differential signal between control and toxicant-treated populations of mRNA can be used to isolate genes that are up- or downregulated following toxicant administration. In this case, the probe is a total population of radiolabeled cDNAs, which is used to screen cDNA libraries and/or Northern blots. Implicit in this method is that levels of mRNAs of differentially expressed genes are altered following neuronal damage and that these can be detected by comparison with controls. By arranging cDNAs from a given library as a grid, colonies displaying differential patterns can be identified and isolated, and the cDNA of interest amplified and sequenced. This approach is useful for detection of mRNAs of moderate to high abundance (0.06% of total and greater); the key limitation is that rare mRNAs are not detected by such differential screening (Dworkin and Dawid, 1980). As a result, more recent studies have favored the use of alternative techniques which allow for sensitive detection and easy isolation of gene products that are differentially altered as a result of toxicant administration.

2. Subtractive Hybridization

Subtractive hybridization methods were pioneered by developmental biologists interested in isolating developmentally specific mRNAs which were expressed at low levels and at specific times (Sargent and Dawid, 1983). Although the specifics of subtractive hybridization cloning methods differ, all protocols involve cDNAs isolated based on the following principle: a

TABLE 1 Examples of Potentially Useful Gene Products for Assessment of Neurotoxicity

Class I (neuron specific)	Class II (region specific)	Class III (injury reactive)	Class IV (housekeeping)
Synapsins	SNAP-25	GFAP	Actin
Synaptophysins	Stannin	HSP-72	G-6PDH
Tau	Cerebellin	c-*fos*/*jun*	Tubulin
Neurofilaments	DARPP-32	Sulfated GP-2	Nuclear lamins

Note. This is a partial list of proteins that have been characterized and sequenced. Note that class II genes are not necessarily confined to a given brain structure, but are differentially expressed in some neurons at high levels relative to others and, hence, can be used to document changes in regions of enrichment.

single-stranded cDNA representative of mRNA from one cell population hybridized to an excess of mRNA from another cell population results in the formation of double-stranded hybrids between complementary strands of common sequences. cDNA species remaining single stranded should represent unique sequences of their respective cell population, which can then be separated from the hybrids in the mixture for use in cloning procedures. Subtractive hybridization has great utility in situations where the amount of sharing material is limited. Using a modification of the basic procedure developed by Travis and Sutcliffe (1988), it has been estimated that these techniques can effect isolation of clones representative of mRNAs of 0.001% abundance.

A variety of subtractive hybridization cloning protocols have been developed since the technique was first used. With the recent advances in molecular techniques and increased use of subtractive hybridization as a cloning tool, new methods of subtractive cloning are now replacing those first employed. Advantages of new methods include the ability to achieve greater degrees of enrichment of target sequences and decreased carryover of false positives.

Once unique, single-stranded cDNAs are purified from the double-stranded hybrids in a subtractive hybridization mixture, they can be used for cloning in several ways. The most straightforward application is to use the population of single-stranded cDNAs as a "subtracted probe" to screen cDNA libraries of interest. This approach was used by Bernal and colleagues (1990) to identify cDNAs representative of mRNAs expressed predominantly in the cerebral cortex of monkeys. However, a disadvantage of this method is that the radiolabeled probe, not present in a cloning vector, has a limited life.

To circumvent this, another method of subtractive cloning can be used in which subtracted cDNA libraries are established from the unique cDNAs isolated. This approach was used by Krady and co-workers (1990). In this study, the unique single-stranded cDNA pool was subjected to various cloning manipulations which resulted in the presence of the cDNAs in plasmid vectors, which were then used to generate significant quantities of probe cDNA for use in the isolation of larger cDNAs.

Another approach, exemplified by studies of Miller and co-workers (1987), combines subtractive hybridization and differential colony hybridization procedures that result in an increased enrichment for unique sequences. The goal of these experiments was to isolate clones expressed at high levels in rat brain at embryonic day 16 (E16). Criteria for selection of E16 cDNA clones were set as those cDNAs that were expressed at a 10-

fold or greater level than corresponding cDNAs in adult brain. Subtractive hybridization was performed, and the single-stranded cDNAs that were isolated were then used as a subtracted probe to screen an E16 rat brain cDNA library. In addition, the remaining double-stranded fraction (the common probe) was used to screen the same library. Clones that were detected by the subtracted probe but not by the common probe were further tested by differential colony hybridization to determine which were recognized by radiolabeled cDNA derived from E16 mRNA but not recognized by that derived from adult mRNA. Northern blot and *in situ* hybridization analysis of clones isolated enabled classification by spatial patterns of expression.

A similar approach was used by Oyler and colleagues (1990) in an attempt to isolate hippocampal-specific cDNAs (30). In these experiments, hippocampal cDNA libraries were constructed and subtractive hybridization was performed using mRNA from whole brain minus hippocampus. Following isolation of the unique cDNA fraction, hippocampal libraries were screened with this subtracted probe. One cDNA clone was found to encode a novel neuron-specific protein termed SNAP-25 (synaptosomal-associated protein of 25 kDa). Subsequent experiments have demonstrated that this protein is identical to the major protein undergoing fast axonal transport in neuronal subsets. In addition, this protein is a major "docking" protein involved in synaptic transmission and is a target for proteolysis by botulinum neurotoxins A and E (Binz *et al.*, 1994). This type of parallel convergence of separate areas of research underscores how aspects of neurobiology, namely molecular cloning of neuron-specific genes and the study of proteins involved in fast axonal transport, can increase the understanding of neuronal function and of molecular targets of neurotoxins.

A modification of subtractive cloning methods, described by Wieland and co-workers (1990), exploits the use of the polymerase chain reaction (PCR) to amplify single-stranded cDNAs yielded from subtractive hybridization. Enrichment of target sequences by this protocol was reported to be from 100- to 700-fold. Use of this technique should further facilitate cloning of rare mRNA species, avoiding the limitations of protocols using conventional hydroxyapatite separations.

Consequently, molecular biologic approaches represent an attractive alternative to the classical descriptive techniques used to study common features of cells destroyed by neurotoxicants. In addition, these sensitive molecular techniques afford a further advantage in that genes common only among sensitive cells can be isolated and cloned. Furthermore, since these gene products can be sequenced, information gained can be used to study possible mechanisms of toxicity. Such molecular approaches have been used (Krady *et al.*, 1990; Toggas *et al.*, 1992) to isolate putative genes which encode for products associated with trimethyl tin (TMT) toxicity.

C. Isolation and Characterization of Gene Products Related to Mechanisms of Neurotoxicant Action via Subtractive Hybridization

Our laboratory is interested in determining the mechanisms of organotin neurotoxicity. Given that past studies have shown that TMT is a selective neurotoxicant (Balaban *et al.*, 1988), it has been hypothesized that sensitive cells express a gene products(s) that predisposes them to the actions of TMT (Bernal *et al.*, 1990; Toggas *et al.*, 1992, 1993). Avidin/biotin-based subtractive hybridization was performed to enrich for cDNAs representative of mRNAs expressed in TMT-sensitive cell populations. The cDNAs obtained were used as probes for Northern blots containing poly(A)$^+$-enriched mRNA from both normal and TMT-treated brains to screen for cDNAs that recognized mRNAs in normal brains which were absent or decreased due to TMT exposure. These cDNAs were then used to isolate full-length clones from rat brain cDNA libraries (Toggas *et al.*, 1992). The full-length clones were sequenced and found to encode a small, 88 amino acid peptide, which was termed stannin. The deduced amino acid sequence was used to derive a synthetic peptide which was used to produce antipeptide antisera. This antisera was used to immunolocalize stannin in TMT-sensitive tissues. Stannin immunoreactivity is present in many TMT-sensitive neurons. This pattern of distribution is consistent with known patterns of TMT-induced damage, and it is provocative to note that stannin is also found in spleen and kidney, two peripheral tissues that are sensitive to TMT (Toggas *et al.*, 1993). However, ongoing experiments using transfection of plasmids encoding stannin into TMT-sensitive and -insensitive cells will ultimately determine whether stannin expression, *per se*, is necessary or sufficient for conferring TMT sensitivity.

III. Advantages and Limitations of Hybridization Techniques

One key limitation of hybridization analysis is determining cause versus effect when studying neurotoxicants. Often, changes in expression of a given gene product following toxicant administration are not directly associated with the cause of toxicity. For exam-

ple, the appearance of products of immediate-early genes following administration of a variety of toxicants is correlative with cellular "stress" and is not necessarily related to the proximate cause of damage. Careful interpretation of such data allows one to determine which neuronal pathways are activated, however.

It is important to match appropriate techniques with appropriate neurotoxicants; for instance, rapidly acting neurotoxicants such as cholinesterase inhibitors are best studied at the level of neurotransmission. Alternatively, agents that cause delayed neurotoxic effects can be studied using a range of hybridization techniques.

There are numerous advantages of approaches that generate DNA sequence information. For instance, if a mechanism or target for a toxicant is not known, subtractive hybridization techniques allow the development of reagents based on cDNA and predicted protein sequence information. The sequence information can then be used for rapid searching of DNA and protein databases for comparative analysis. For example, such computer-based comparisons between laboratories allowed SNAP-25 to be rapidly identified as a synaptic docking protein and a target for clostridial neurotoxins (Binz *et al.*, 1994; Oyler *et al.*, 1990). Thus, hybridization techniques are being employed by a wider range of neurotoxicologists and, when used appropriately, can afford considerable insight into the actions of and consequences of exposure to neurotoxicants.

References

Ausubel, F., Brent, R., Kingston, R., Moore, D., Seidman, J., Smith, J., and Struhl, K. (1989). *Current Protocols in Molecular Biology*, Vol. 1, John Wiley and Sons, New York.

Balaban, C. D., O'Callaghan, J. P., and Billingsley, M. L. (1988). Trimethyltin-induced neuronal damage in the rat: comparative studies using silver degeneration stains, imunocytochemistry and immunoassay for neuronotypic and gliotypic proteins. *Neuroscience* 26, 337–361.

Beltz, G. A., Jacobs, K. A., Eickbush, T. H., Cherbas, P. T., and Kafatos, F. C. (1983). In *Methods in Enzymology* (R. Wu, L. Grossman, and K. Moldave, Eds.) Academic Press, NY. 100: 266–285.

Berk, A. J., and Sharp, P. A. (1977). Sizing and mapping of early adenovirus mRNAs by gel electrophoresis of S1 endonuclease-digested hybrids. *Cell* 12, 721–732.

Bernal, J., Godbout, M., Hasel, K. W., Travis, G. H., and Sutcliffe, J. G. (1990). Patterns of cerebral cortex mRNA expression. *J. Neurosci. Res.* 27, 153–158.

Binz, T., Blasi, J., Yamasaki, S., Baumeister, A., Zink, E., Sudhof, T. C., Jahn, R., and Niemann, H. (1994). Proteolysis of SNAP-25 by types E and A Botulinal neurotoxins. *J. Biol. Chem.* 269, 1617–1620.

Billingsley, M. L., and O'Callaghan, J. P. (1994). Molecular Neurotoxicology. In *Principles of Neurotoxicology* (L. W. Chang, Ed.), pp. 551–562, Dekker, New York.

Brock, T. O., and O'Callaghan, J. P. (1987). Quantitative changes in the synaptic vesicle proteins synapsin I and p38 and the astrocyte-specific protein glial fibrillary acidic protein are associated with chemical-induced injury to the rat central nervous system. *J. Neurosci.* **7,** 931–942.

Cantor, C. R., and Schimmel, P. R. (1980). *Biophysical Chemistry Part I,* W. H. Freeman, New York.

Chomczynski, P., and Sacchi, N. (1987). Single-step method of RNA isolation by acid guanidinium thiocyanate-phenol-chloroform extraction. *Anal. Biochem.* **162,** 156–159.

Dworkin, M. B., and Dawid, I. B. (1980). Use of a cloned library for the study of abundant poly A$^+$ RNA during *Xenopus laevis* development. *Dev. Biol.* **76,** 449–464.

Freifelder, D. (1976). *Physical Biochemistry.* W. H. Freeman, New York.

Fried, M., and Crothers, D. M. (1981). Equilibria and kinetics of lac repressor-operator interactions by polyacrylamide gel electrophoresis. *Nucleic Acids Res.* **9,** 6505–6525.

Heikkila, R. E., Manzino, L., Cabbat, F. C., and Duvosin, R. C. (1984). Inhibition of monoamine oxidase produces protection against the dopaminergic toxicity of 1-methyl-4-phenyl-1,2,3,6,tetrahydropyridine. *Nature* **311,** 467–469.

Krady, J. K., Oyler, G. A., Balaban, C. D., and Billingsley, M. L. (1990). Use of avidin-biotin subtractive hybridization to characterize mRNA common to neurons destroyed by the selective neurotoxicant trimethyltin. *Mol. Brain Res.* **7,** 287–297.

Langer, P. R., Waldrop, A. A., and Ward, D. C. (1981). Enzymatic synthesis of biotinylated polynucleotides: novel nucleic acid affinity probes. *Proc. Natl. Acad. Sci. USA* **78,** 6633–6637.

Lefkowitz, J., and Caron, M. G. (1988). Adrenergic receptors. *J. Biol. Chem.* **267,** 4993–4996.

Melton, D. A., Krieg, P. A., Rebagliati, M. R., Maniatis, T., Zinn, K., and Green, M. R. (1984). Efficient in vitro synthesis of biologically active RNA and RNA hybridization probes from plasmids containing a bacteriophage SP6 promoter. *Nucleic Acids Res.* **12,** 7035–7056.

Miller, F. D., Naus, C. C. G., Higgins, G. A., Bloom, F. E., and Milner, R. J. (1987). Developmentally regulated rat brain mRNAs: molecular and anatomical characterization. *J. Neurosci.* **7,** 2433–2444.

Morgan, J. I., and Curran, T. (1991). Stimulus-transcription coupling in the nervous system: involvement of the inducible protooncogens fos and jun. *Annu. Rev. Neurosci.* **14,** 421–451.

O'Callaghan, J. P. (1988). Neurotypic and gliotypic proteins as biochemical markers of neurotoxicity. *Neurotoxicol. Teratol.* **10,** 445–452.

Oyler, G. A., Higgins, G. A., Hart, R. A., Battenberg, E., Bloom, F. E., Billingsley, M. L., and Wilson, M. C. (1990). The identification of a novel synaptosomal associated protein, SNAP-25, differentially expressed by neuronal subpopulations. *J. Cell Biol.* **109,** 3039–3052.

Pennypacker, K. R., Walczak, D., Thai, L., Fannin, R., Mason, E., Douglass, J., and Hong, J. S. (1993). Kainate-induced changes in opoid peptides and AP-1 protein expression in the rat hippocampus. *J. Neurochem.* **60,** 204–211.

Sambrook, J., Fritsch, E. F., and Maniatis, T. (1989). *Molecular Cloning: A Laboratory Manual.* Cold Springs Harbor Press, New York.

Sargent, T. D., and Dawid, I. B. (1983). Differential gene expression in the gastrula of Xenopus laevis. *Science* **222,** 135–139.

Seed, B. (1982). *Genetic Engineering: Principles and Methods.* (K. Setlow and A. Hollaender, Eds.), pp. 91–102, Plenum, New York.

Southern, E. M. (1975). Detection of specific sequences among DNA fragments separated by gel electrophoresis. *J. Mol. Biol.* **98,** 503–517.

Sutcliffe, J. G., Milner, R. J., Shinnick, T. M., and Bloom, F. E. (1983). Identifying the protein products of brain-specific genes using antibodies to chemically synthesized peptides. *Cell* **33,** 671–682.

Thomas, P. S. (1980). Hybridization of denatured RNA and small DNA fragments transferred to nitrocellulose. *Proc. Natl. Acad. Sci. USA* **77,** 5201–5205.

Toggas, S. M., Krady, J. K., and Billingsley, M. L. (1992). Molecular neurotoxicology of trimethyltin: identification of stannin, a novel protein expressed in trimethyltin-sensitive cells. *Mol. Pharmacol.* **42,** 44–56.

Toggas, S. M., Krady, J. K., Thompson, T. A., and Billingsley, M. L. (1993). Molecular approaches for determination of mechanisms of selective neurotoxicants. *Ann. N.Y. Acad. Sci.* **679,** 157–177.

Travis, G. H., and Sutcliffe, J. G. (1988). Phenol emulsion-enhanced DNA-driven subtractive cDNA cloning: isolation of low abundance monkey cortex-specific mRNAs. *Proc. Natl. Acad. Sci. USA* **85,** 1696–1700.

Wieland, I., Bolger, G., Asouline, G., and Wigler, M. (1990). A method for difference cloning: gene amplification following subtractive hybridization. *Proc. Natl. Acad. Sci. USA* **87,** 2720–2724.

CHAPTER

23

Application of the Technique of Brain Microdialysis in Determining the Mechanisms of Action of Neurotoxicants

JOHN F. BOWYER
Division of Neurotoxicology
National Center for Toxicological Research
Jefferson, Arkansas 72079-9502

I. Introduction

Microdialysis is the technique of passing an artificial physiological buffer through a semipermeable membrane embedded within tissue to enable the fractional recovery of substances within the tissue and/or the administration of substances to that tissue. This brief overview describes previous, present, and future applications of *in vivo* microdialysis for determining the mechanisms involved in the neurotoxicity produced by neurotoxicants. The use of microdialysis for investigating the mechanisms by which amphetamine (AMPH) and methamphetamine (METH) produce neurotoxicity will be emphasized; however, many of the microdialysis approaches and techniques cited herein have also been applied to other types of neurotoxicity produced by such insults as ischemia and excitotoxins (Benveniste, 1989). For a thorough review of how microdialysis can be applied to neuroscience see Robinson and Justice (1991).

II. Basic Methods and Techniques

Methods and equipment developed by Bioanalytical Systems Inc. (West Lafayette, IN) were used for the studies of *in vivo* microdialysis (Gough *et al.*, 1991, 1993; Bowyer *et al.*, 1993a,b). To implant the microdialysis guide cannula in the caudate/putamen (CPU), male rats ranging from 4 to 12 months of age were anesthetized with sodium pentobarbital (50 mg/kg, i.p.) and placed into a stereotaxic frame (Köpf, Topanga, CA). The level dorsal skull surface was exposed and a small hole was drilled to allow implantation of the intracerebral guide cannula (CMA/microdialysis; Stockholm, Sweden) in the caudate. Coordinates were AP 0.2 mm; Lat 3.0 mm; and DV 5.5 mm relative to Bregma for placement of the guide cannula. The guide cannula was fixed to the skull of the rat with dental acrylic and two anchor screws. Body temperature during surgery and recovery from anesthesia was maintained at 37°C with a small heating pad

433

under the animal. To avoid effects of anesthesia and to allow recovery from surgical trauma, the dialysis experiments were performed between 3 and 7 days after surgery.

On the morning of the test, the animal was hand-held and the dialysis probe was inserted through the guide cannula and into the underlying CPU. CMA/12 microdialysis probes were used (Bioanalytical Systems) and measured 14 mm in length and 0.65 mm in diameter. The membrane tip was 2.0×0.5 mm. The probes had an *in vitro* recovery efficiency of between 16 and 20% for either serotonin (5-HT) and dopamine (DA) or amino acids at a flow rate of 1.0 μl/min. A complete Krebs–Ringer bicarbonate buffer, pH 7.4 (140 mM Na$^+$; 1.25 mM Ca^{2+}; 1.5 mM Mg^{2+}; 4.5 mM K$^+$; 121 mM Cl$^-$; 25 mM CO$_3^-$; 10 mM D-glucose), was employed whenever possible. A less complete Ringer's buffers with a slightly lower pH (6.2) was also used, and it did not significantly affect microdialysis experiments involving neurotoxic doses of METH (Bowyer *et al.*, 1993a). Furthermore, aromatic monoamines and their metabolites were more stable in the pH 6.2 buffer. However, large changes in the [Ca^{2+}] in the microdialysis buffer away from the 1.25 mM physiological range can affect DA release (Moghaddam and Bunney, 1989; Westerink *et al.*, 1989). The microdialysate was perfused at a flow rate of 1.0 μl/min, using a CMA/100 microinjection pump (CMA/microdialysis, Stockholm, Sweden). Animals undergoing testing were housed in an awake animal system (20 in. diameter, 12 in. deep glass bowl with balancing arm) with wire tether and single channel swivel. If dialysates are collected at 20-min intervals both monoamines and amino acids levels can be determined. The methods of Muusze (1982) were used to determine DA, 5-HT, and their metabolites in 10-μl dialysate aliquots and methods similar to Zielke (1985) were used to detect glutamate and other amino acids in 10-μl microdialysate aliquots.

III. Applications to the Neurotoxicity of Amphetamines

Relatively stable baseline values for DA, 5-HT, and their metabolites recovered from CPU microdialysate can usually be established (less than 10 pg/10 μl microdialysate) within 2 to 3 hr after initiation of microdialysis after which time animals can be dosed with METH. Stable baselines for recovery of amino acids in microdialysate, including glutamate, glutamine, and taurine, can be reached within the same time frame in rats of 4 months or less of age (consult the microdialysis reviews cited herein). However, in rats older than 5 months, longer times may be necessary to reach stable

recovery rates and the basal rates may be higher (Bowyer *et al.*, 1993a; unpublished data). The changes in the levels of DA, 5-HT, and their metabolites, as well as the amino acid levels in the microdialysate, can be monitored simultaneously.

In order to produce METH neurotoxicity, four doses of 5 mg/kg METH (i.p., spaced 2 hr apart) were administered to either male or female rats 4 to 12 months of age. METH neurotoxicity rarely occurs using this paradigm if the environmental temperature is less than 20°C. At an environmental temperature of 24.5°C, 60% of male rats (4 to 6 months old) administered 4 $\times$ 5 mg/kg will develop significant hyperthermia by the third or fourth dose, subsequently showing signs of METH neurotoxicity (Bowyer *et al.*, 1994). If the degree of hyperthermia becomes extreme ($\geq$ 41.3°C), it is easily noticed because these rats normally collapse and no longer exhibit the METH-induced stereotype. In order to reduce the mortality of the rats suffering from extreme hyperthermia, without drastically interfering with the neurotoxic effects of METH, rats attaining a rectal temperature of greater than 41.3°C should be placed on crushed ice until their temperature is below 39.5°C (15 to 40 min).

It should be noted that in all neurotoxicity studies, including those involving microdialysis, care must be taken to ensure that the neurotoxicant administered actually produces neurotoxicity. Although histologic and biochemical techniques are necessary to assure that neurotoxicity has occurred, in the case of METH and AMPH neurotoxicity, the occurrence of pronounced hyperthermia during repetitive dosing is highly correlative with neurotoxicity (Bowyer *et al.*, 1994). In addition, the appearance of prolonged seizure activity during exposure to excitotoxins such as kainic acid correlates greatly with neurotoxic effects (Lothman and Collins, 1981; Ben-Ari, 1984).

Most of the rats used for microdialysis were sacrificed at 1, 3, or 14 days after either METH or saline administration. Animals were sacrificed by decapitation, and brains were removed and immediately immersed in 4°C normal saline. The right and left cerebral hemispheres were than separated, and the right cerebral hemisphere was fixed in formalin (10%)/normal saline solution for later histological verification of cannulae and probe location. The CPU and hippocampus of the left hemisphere were immediately dissected and stored at -70°C until the total monoamine content could be determined.

Initial microdialysis studies reported the effects of single doses of METH/AMPH and changes in the extracellular levels of aromatic monoamines. However, these studies were not sufficient to evaluate neurotoxicity because the dosing paradigms employed did not normally produce METH/AMPH neurotoxicity, which is characterized by large depletions in striatal dopamine coupled with reactive gliosis and nerve terminal degeneration. Furthermore, in these studies, the duration of microdialysis employed was not sufficient to monitor all the events producing neurotoxicity.

To consistently produce neurotoxicity in rats of 6 months of age or less, either METH or AMPH levels need to be elevated for more than 4 hr. Therefore, the neurotoxic events may either be the cumulative effects of METH/AMPH and/or neurotransmitters released by METH/AMPH over a 5- to 10-hr time period or be the result of specific events that occur between 4 and 10 hr of exposure. In extended duration microdialysis experiments designed to look at the neurotoxic effects of METH/AMPH, increases in the extracellular levels of both DA (O'Dell *et al.*, 1991; Weihmuler *et al.*, 1992) and glutamate (Nash and Yamamoto, 1992) that occurred 5 hr or more after initiation of METH or AMPH administration have been correlated with neurotoxicity. However, data have shown (Bowyer *et al.*, 1993a) that environmental temperature and age are important factors modulating neurotoxicity and that increases in extracellular DA and/or glutamate levels alone are not sufficient to predict the degree of neurotoxicity produced by METH. Figure 1 shows that the extracellular levels produced by METH in 12-month-old female rats were less than that produced in 6-month-old female rats, yet it was the 12-month-old rats that showed the greatest neurotoxicity. The extracellular levels of glutamate did increase more in the 12-month-old rats (Fig. 2), but taurine was not elevated by METH treatment. Furthermore, since the completion of these experiments, it has been observed that the extracellular glutamate rise does not always correlate with the degree of dopamine depletions produced by METH in 6-month-old male rats (unpublished data).

These initial studies have just begun to show some of the ways in which microdialysis can be used to resolve the mechanisms behind METH/AMPH neurotoxicity. To date, these studies have dealt primarily with evaluating the changes in striatal extracellular levels of the neurotransmitters DA, 5-HT, and amino acids such as glutamate and taurine. Changes in striatal extracellular levels of adenosine, ATP, and metabolites after neurotoxic doses of METH/AMPH still need to be evaluated. The technical ability to measure adenosine using *in vivo* microdialysis has been demonstrated (Chen *et al.*, 1993), and experiments evaluating changes in adenosine levels during METH/AMPH administration are underway in this and other laboratories.

Other important uses for microdialysis in research involving METH/AMPH neurotoxicity include: (1) de-

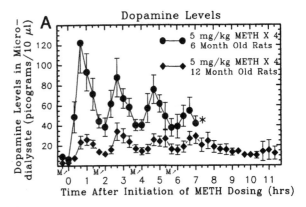

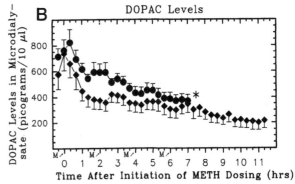

FIGURE 1 Differences between 12- and 6-month-old rats in the extracellular CPU levels of DA and DOPAC during METH administration. Changes in the extracellular levels of DA (A) and DOPAC (B) during METH administration (23°C ET) are shown as fluctuations of either DA or DOPAC in microdialysate collected at 20-min intervals. The M and the arrows on the *x* axis indicate the time of METH administration. The levels of DA for 12-month-old rats were significantly ($P < 0.0075$) less over the time course of METH administration. Mean DA levels for controls over 7 hr were 1.8 pg/10 μl from the older rats and 16.5 pg/10 μl for the younger animals. Although METH decreased DOPAC levels to the same extent in both age groups, the initial DOPAC levels were lower in the 12-month-old rats prior to METH ($P < 0.006$). The number of METH-exposed rats was nine for the 6-month-old rats and eight for the 12-month-old rats. (Reprinted from Bowyer *et al.*, 1993a with permission of Elsevier Science Ltd.)

termining the role of neurotoxic substances such as reactive oxidative species (ROS) or nitric oxide (NO) in the CPU, (2) directly exposing CPU to compounds that may either protect against or potentiate neurotoxicity, (3) determining the brain extracellular levels of AMPH during multiple administrations of AMPH, and (4) continuously exposing either the CPU or substantia nigra/ventral tegmentum to neurotrophic factors or antisense RNA to neurotrophic factors during METH/AMPH exposure to determine the role of different growth factors in the neurotoxicity. These will be discussed in turn.

Microdialysis offers the opportunity to measure the generation/extracellular levels of neurotoxic mediators such as ROS and NO. However, both ROS and NO are short-lived compounds and are not readily detect-

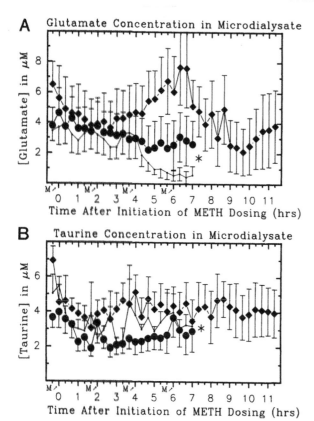

A Glutamate Concentration in Microdialysate

B Taurine Concentration in Microdialysate

FIGURE 2 Extracellular CPU levels of glutamate and taurine in 6- and 12-month-old rats administered METH. There were no significant increases in the extracellular levels of either [glutamate] or [taurine] during the course of METH administration, although [glutamate] did tend to increase over the course of METH exposure in the younger rats. The number of rats per groups was four for 6-month control (dots) and five for 6-month 4 × 5 mg/kg METH (filled circles) groups whereas four rats were in the 12-month 4 × 5 mg/kg METH (filled diamonds) group. (Reprinted from Bowyer *et al.,* 1993a with permission of Elsevier Science Ltd.)

able. In the case of NO it is also possible to measure the formation of NO_x^- from NO (Lou *et al.,* 1993). For ROS, their detection using microdialysis depends on including a compound within the microdialysate that can (1) diffuse into the extracellular space, (2) react with ROS or NO to form a stable compound, and (3) diffuse back across the microdialysis probe tip for subsequent collection. At present there are several other laboratories looking at ROS using microdialysis and compounds such as salicylate and cysteine which can react with ROS to form relatively stable compounds which can be quantitated. Although it has been possible to indirectly measure NO_x and cGMP levels in cerebellum to monitor NO generation using microdialysis techniques (Lou *et al.,* 1993, 1994; Vincent, 1994), it is more difficult applying these methods in the CPU because the number of cells producing NO and the levels of nitric oxide synthetase are considerably lower in the CPU (Vincent, 1994; Bredt *et al.,* 1990).

As an indirect means of evaluating the involvement of the neurotoxic mediator NO in METH/AMPH neurotoxicity, this laboratory has been evaluating the effects of including either the nitric oxide synthetase (NOS) inhibitors N^G-nitro-L-arginine and N^G-nitro-L-arginine methyl ester or the NO_x generators sodium nitroprusside and isosorbide dinitrate in the microdialysate buffer during METH/AMPH exposure (Gough *et al.,* 1993, 1994). The inclusion of inhibitors of NO in the microdialysis buffer allows the determination of how these inhibitors affect DA, glutamate, and adenosine release without producing the peripheral effects that occur through systemic NO inhibitor administration that might also affect neurotoxicity. These techniques have been used in the cerebellum by other investigators (Vincent, 1994). One added feature of including NO inhibitors or other putative neuroprotective or neurotoxic substances in the microdialysate is that the effects are localized to the probe vicinity and may be compared to the contralateral CPU. Furthermore, systemic administration of either NOS inhibitors or NO_x generators may affect systemic and brain vascular tone and perfusion (Moncada *et al.,* 1991), which could alter neurotoxicity without affecting NO generation by neurons and glia within brain. A drawback of this approach is that immunohistological techniques are necessary to determine whether the compounds have affected neurotoxicity (whole striatal DA levels will not accurately reflect regional changes in striatal DA levels around the microdialysis probe).

Microdialysis is well suited for determining the extracellular levels of either AMPH or METH achieved after systemic administration and it has been used to determine extracellular concentrations of drugs since its inception (Ungerstedt *et al.,* 1982). The use of microdialysis will enable the determination of whether increased age and body temperature potentiate AMPH/METH lethality and neurotoxicity primarily through their effects on the pharmacokinetics of either AMPH or METH. Detection of AMPH concentrations as low as 50 n*M* in 10 μl of microdialysate ($\simeq$ 0.5 pmol) is possible by modifying the technique of *o*-pthaldialdehyde/3-mercaptopropionic derivitization of amino acids for use in derivatization of AMPH and subsequent HPLC separation coupled with fluorescent detection (Bowyer *et al.,* 1994). Furthermore, the use of "mass transfer" techniques, which can be more accurate in determining extracellular levels of compounds (Jacobson *et al.,* 1985; Morrison *et al.,* 1991; Benveniste and Hansen, 1991), allows determination of AMPH levels without requiring as great a detection sensitivity. These techniques permit detection of striatal extracellular levels as low as 1.5 μM during AMPH administration and are sensitive enough to permit detection of extracellular AMPH levels after 1 mg/kg i.p. doses (Clausing *et al.*, 1994). At present only gas chromatography analysis or radioimmunoassays enable quantitation of the METH levels expected to be recovered in the microdialysate. For a complete description of how microdialysis can be used to determine extracellular drug levels see Stähle (1991). Unfortunately, METH cannot be suitably derivatized using the previously described methods.

Microdialysis may permit the exposure of either the CPU or midbrain to neurotrophic/growth factors (NTFs). Although NTFs can be directly injected into these brain regions, there is the possibility that direct injection may produce damage unless the injection rate is very slow. By modification of the microdialysis probe tip (removal of dialysis membrane and occlusion of the probes outflow), microdialysis can be used for slow infusion of NTF (≤0.1 µl/min). Furthermore, using a dialysis membrane with a high molecular weight cutoff (40 kDa) and a reduced microdialysis flow rate (0.1 µl/min) to enhance the extraction of NTF, microdialysis itself can be used for brain exposure to NTF. NTF candidates that appear to have the highest probability of playing a role in METH/AMPH neurotoxicity at present are related NTFs, glial-derived neurotrophic factor (Lin *et al.*, 1993), and several members of the transforming growth factor-β family (Burt, 1992). However, other NTFs may also play an important role in either METH neurotoxicity or protection/recovery from the neurotoxic effects. Along with directly infusing NTFs in the microdialysis buffer in an effort to ameliorate the neurotoxic effects of METH/AMPH, the inclusion of antisense oligodeoxynucleotides (ODNs) and phosphorothioate ODNs, which have been shown to block the synthesis of the proteins targeted (Woolf *et al.*, 1990; Coleman, 1990) to the previously mentioned NTFs, may aid in determining the roles of various NTFs in METH/AMPH neurotoxicity. That is, if the loss of a particular NTF plays a prominent role in METH/AMPH neurotoxicity, then exposure of the CPU or substantia nigra to the antisense ODN to the NTF (in the absence of METH) may produce reductions in striatal dopamine.

In conclusion, the technique of brain microdialysis has already provided insights into the roles of aromatic monoamine, glutamate, age, and body temperature in METH/AMPH neurotoxicity and have an even longer and more extensive use in investigating the mechanisms of excitotoxic compounds. The technique may be used to evaluate potential neurotoxicants to determine (1) the generation/brain levels and effects of ROS, NO, and other neurotoxic mediators; (2) the effects of age and body temperature on neurotoxicant levels and their effects on extracellular neurotransmitter levels;

(3) the interactions of neurotoxicants and neuroprotectants; and (4) the role of NTFs in neurotoxicity. The studies on METH/AMPH and excitotoxin neurotoxicity presented here, as well as microdialysis studies on the neurotoxicity of other compounds, demonstrate that microdialysis can play a prominent role in determining the mechanisms involved in neurotoxic compounds.

References

Ben-Ari, Y. (1984). The role of seizures in kainic acid-induced brain damage. In *Excitotoxins* (K. Fuxe, P. Roberts and R. Schwarcz, Eds.), pp. 184–198, Plenum, New York/London.

Benveniste, H. (1989). Brain microdialysis. *J. Neurochem.* **52,** 1667–1679.

Benveniste, H., and Hansen, A. J. (1991). Practical aspects of using microdialysis for determination of brain interstitial concentrations. In *Techniques in the Behavioral and Neural Sciences* (T. E. Robinson and J. B. Justice Jr., Eds.), Vol. 7, pp. 81–100, Elsevier, Amsterdam/New York.

Bowyer, J. F., and Clausing, P. (1994). Using Orthophthaldialdehyde (OPA) and 3-Mercaptopropionic Acid (MERA) Derivatization to Quantify D-Amphetamine (AMPH) Levels in Striatal Microdialysates by High Performance Liquid Chromatography (HPLC). *Soc. Neurosci. Abs.* **20,** 1027.

Bowyer, J. F., Davies, D. L., Schmued, L., Broening, H. W., Newport, G. D., Slikker, W., Jr., and Holson, R. R. (1994). Further studies of the role of hyperthermia in methamphetamine neurotoxicity. *J. Pharmacol. Exp. Ther.* **268,** 1571–1580.

Bowyer, J. F., Gough, B., Slikker, W., Jr., Lipe, G. W., Newport, G. D., and Holson, R. R. (1993a). Effects of a cold environment or age on methamphetamine-induced dopamine release in the caudate putamen of female rats. *Pharmacol. Biochem. Behav.* **44,** 87–98.

Bowyer, J. F., Gough, B., Broening, H. W., Newport, G. D., and Schmued, L. (1993b). Fluoro-Gold and pentamidine inhibit the *in vitro* and *in vivo* release of dopamine in the striatum of rat. *J. Pharmacol. Exp. Ther.* **266,** 1066–1074.

Bredt, D. S., and Snyder, S. H. (1990). Isolation of nitric oxide synthetase, a calmodulin-requiring enzyme. *Proc. Natl. Acad. Sci. USA* **87,** 682–685.

Burt, D. W. (1992). Evolutionary grouping of the transforming growth factor-β superfamily. *Biochem. Biophys. Res. Comm.* **184,** 590–595.

Chen, L-S, Fujitaki, J., and Dixon, R. (1993). An improved assay for adenosine in rat brain microdialysis using microbore high performance liquid chromatography. *J. Liq. Chromatogr.* **16,** 2791–2796.

Clausing, P., Holson, R. R., Slikker, W., Jr., Gough, B., and Bowyer, J. F. (1994). D-Amphetamine (AMPH) levels in microdialysate recovered from caudate-putamen (CPU) after doses that produce either behavioral or neurotoxic effects. *Soc. Neurosci. Abstr.* **20,** 1026.

Coleman, A. (1990). Antisense strategies in cell and developmental biology. *J. Cell Sci.* **97,** 399–409.

Gough, B., Ali, S. F., Slikker, W., Jr., and Holson, R. R. (1991). Acute effects of 3,4-methylenedioxymethamphetamine (MDMA) on monoamines in rat caudate. *Pharmacol. Biochem. Behav.* **39,** 619–623.

Gough, B., Holson, R. R., Slikker, W., Jr., and Bowyer, J. F. (1993). Effects of nitric oxide synthetase (NOS) inhibition and adenosine

(AD) receptor antagonism on extracellular dopamine (DA) levels during methamphetamine (METH) exposure. *Neurosci. Abstr.* **19,** 1679.

Jacobson, I., Sandberg, M., and Hamberger, A. (1985). Mass transfer in brain dialysis devices—a new method for estimation of extracellular amino acid concentration. *J. Neurosci. Methods* **15,** 263–268.

Lin, L.-F. H., Doherty, D. H., Lile, J. D., Bektesh, S., and Collins, F. (1993). GDNF: a glial cell line-derived neurotrophic factor for midbrain dopaminergic neurons. *Science* **260,** 1130–1132.

Lou, D., Knezevich, S., and Vincent, S. R. (1993). N-methyl-D-aspartate-induced nitric oxid release: an *in vitro* microdialysis. *Neuroscience* **57,** 897–900.

Lou, D., Leung, E., and Vincent, S. R. (1994). Nitric oxide-dependent efflux of cGMP in rat cerebellar cortex: an *in vivo* microdialysis study. *J. Neurosci.* **14,** 263–271.

Lothman, E. W., and Collins, R. C. (1981). Kainic acid-induced limbic seizures: Metabolic, behavioral, electroencephalographic, and neuropathological correlates. *Brain Res.* **218,** 299–318.

Moghaddam, B., and Bunney, B. S. (1989). Ionic composition of microdialysis perfusing solution alters the pharmacological responsiveness and basal outflow of striatal dopamine. *J. Neurochem.* **53,** 652–654.

Moncada, S., Palmer, R. M. J., and Higgs, E. A. (1991). Nitric oxide: Physiology, pathophysiology, and pharmacology. *Pharmacol. Rev.* **43,** 109–142.

Morrison, P. F., Bungay, P. M., Hsiao, J. K., Mefford, I. N., Dykstra, K. H., and Dedrick, R. L., (1991). Quantitative microdialysis. In *Techniques in the Behavioral and Neural Sciences* (T. E. Robinson and J. B. Justice Jr., Eds.), Vol. 7, pp. 47–80, Elsevier, Amsterdam/New York.

Muusze, R. G. (1982). The LC analysis of catecholamine metabolites in urine. *Chromatogr. Sci.* **20,** 257–278.

Nash, J. F., and Yamamoto, B. K. (1992). Methamphetamine neurotoxicity and striatal glutamate release: comparison to 3,4-methylenedioxymethamphetamine. *Brain Res.* **581,** 237–243.

O'Dell, S. J., Weihmuller, F. B., and Marshall, J. F. (1991). Multiple methamphetamine injections induce marked increases in extracellular striatal dopamine which correlate with subsequent neurotoxicity. *Brain Res.* **564,** 256–260.

Robinson, T. E., and Justice, J. B., Jr., Eds. (1991). Microdialysis in the neurosciences. In *Techniques in the Behavioral and Neural Sciences,* Vol. 7. Elsevier, Amsterdam/New York.

Stähle, L. (1991). The use of microdialysis in pharmacokinetics and pharmacodynamics. In *Microdialysis in the Neurosciences,* Vol. 7, *Techniques in the Behavioral and Neural Sciences* (T. E. Robinson and J. B. Justice Jr., Eds.), pp. 155–174, Elsevier, Amsterdam/New York.

Ungerstedt, U., Herrera-Marschitz, M., Jungnelius, U., Stahle, L., Tossman, U., and Zetterstrom, T. (1982). Dopamine synaptic mechanisms reflected in studies combining behavioral recordings and brain dialysis. *Adv. Biosci.* **37,** 219–231.

Vincent, S. R. (1994). Nitric oxide: a radical neurotransmitter in the central nervous system. *Progr. Neurobiol.* **42,** 129–160.

Weihmuller, F. B., O'Dell, S. J., and Marshall, J. F. (1992). MK-801 protection against methamphetamine-induced striatal dopamine terminal injury is associated with attenuated dopamine overflow. *Synapse* **11,** 155–163.

Westerink, B. H. C., Hofsteede, R. M., Tuntler, J., and de Vries, J. B. (1989). Use of calcium antagonism for the characterization of drug-evoked dopamine release from the brain of conscious rats determined by microdialysis. *J. Neurochem.* **52,** 722–729.

Woolf, T. D., Jennings, C. G. B., Rebagliati, M., and Melton, D. A. (1990). The stability, toxicity and effectiveness of unmodified and phosphorothioate antisense oligodeoxynucleotides in xenopus oocytes and embryos. *Nucleic Acids Res.* **18,** 1763–1769.

Zielke, R. H. (1985). Determination of amino acids in the brain by high-performance liquid chromatography with isocratic elution and electrochemical detection. *J. Chromatogr.* **347,** 320–324.

Combined Electrical Resistance Method for Cell Volume Measurement and Continuous Perfusion for the Measurement of the Release of Endogenous Substances: An in Vitro Assay for Cytotoxicity

M. ASCHNER
Department of Physiology
and Pharmacology
Bowman Gray School of Medicine
Winston-Salem, North Carolina 27106

H. K. KIMELBERG
Department of Pharmacology
and Toxicology, and
Division of Neurosurgery
Albany Medical College
Albany, New York 12208

D. VITARELLA
Department of Pharmacology
and Toxicology, and
Division of Neurosurgery
Albany Medical College
Albany, New York 12208

I. Introduction

Over the last two decades, toxicity testing has extensively relied on tissue culture methods to describe pathologic consequences of chemical exposure, to elucidate mechanisms of action, and to ascertain hazards to human health. The evaluation of chemically induced cytotoxicity, particularly in a heterogeneous system such as the brain, is often difficult in the intact animal because numerous factors (neural, hormonal, and hemodynamic) are not under experimental control. Hence, a simplified model, such as tissue culture, is indispensable as a tool for understanding basic physiology and pathology. Once the cellular purity, content, and degree of maturation have been established, tissue cultures afford a host of advantages over *in vivo* techniques. Cell mor-

phology, protein synthesis and release, energy metabolism, receptor interaction, and neurotransmitter uptake and release, as well as electrolyte and nonelectrolyte uptake and release, can be easily studied and manipulated. Circumventing metabolic degradation, known pharmacological concentrations of toxic compounds can be administered to cultures. Furthermore, toxic effects on juxtaposed cells or other organs which can interact with and buffer or exacerbate the toxic effect of the xenobiotic on the central nervous system (CNS) can be eliminated altogether or factored in sequentially. Toxic compounds can be easily added and withdrawn from the cultures and long-term effects may be studied.

Since the early 1980s, the functions of astrocytes have begun to be delineated. The prevailing view that astrocytes predominantly function as passive physical support for neurons has rapidly faded. It is now clear that to understand the normal and abnormal brain, one must also understand the roles assumed by astrocytes, for they function prominently not only in normal brain physiology and development, but also in the pathology of the CNS. The neuronal–astrocytic functional partnership proposed by Hydèn in 1961 has now been accepted. Work using primary monolayer cultures of astrocytes derived from neonatal rodent brains has led to major advantages in our understanding of astroglial physiology, and the understanding of astrocyte-associated mechanisms of neurotoxicity is no exception. This chapter is not intended to review glial functions or extensive methodologies for their culturing. These issues have already been discussed (Aschner and Vitarella, 1993; Aschner et al., 1993). Instead, a new version of an available technique (Mazzoni et al., 1989) developed by O'Connor and colleagues (1993) is presented which allows for dynamic measurement of changes in cell volume of substratum-attached monolayer cell cultures. When combined with release measurements of endogenous cell markers it affords a powerful tool for rapid measurements of cytotoxicity.

Since a primary focus of our laboratory has been studying the effects of heavy metals on astrocytic homeostasis, this chapter presents our version of combined volume change measurements with simultaneous on-line amino acid release.

II. METHODS

A. Cell Culture

Briefly, the cerebral hemispheres of newborn rats (Sprague–Dawley) are removed, the meninges are carefully dissected off, and the tissue is dissociated using Dispase II (Boehringer-Mannheim Biochemicals, neutral protease, Dispase Grade II). Cultures are prepared as described by Frangakis and Kimelberg (1984) and are grown on a plastic cell support film (Bellco Biotechnology, Vineland, NJ) (for efflux measurements) or on No. 1 coverslips (3.0 × 1.4 cm) (for volume combined with efflux measurements). The monolayers reach confluency after approximately 3 weeks. Immunocytochemically, ≥95% of the cells stain positively for the astrocytic marker, glial fibrillary acidic protein (GFAP), using a previously reported procedure (Frangakis and Kimelberg, 1984). Cell viability is also routinely assessed by the trypan blue exclusion method (20% vol/vol of 0.4% staining solution).

B. Combined Volume with D-[2,3-³H]Aspartate and Na₂⁵¹CrO₄ Efflux Measurements

Astrocytes grown on the slides are transferred to a 60-mm dish and are loaded overnight by adding 5 ml of warmed minimum essential medium containing 10% horse serum, 40 μCi of $Na_2^{51}CrO_4$ (radioactive concentration 1 μCi/ml; specific activity 50 mCi/mg Cr) and 20 μCi D-[2,3-³H]aspartic acid (radioactive concentration 1 μCi/ml; specific activity 300 mCi/mg D-aspartate). Radiolabeled D-aspartate is used as a marker for intraastrocytic glutamate and aspartate. All of these amino acids are likely to be transported on the same carrier protein and, if there is no major compartmentalization problems, the radioactive probe used at low concentrations but high specific activity should equilibrate with the entire pool of glutamate and aspartate, labeling it uniformly.

The cells are continuously perfused and the affluent is collected at 1-min intervals (approximately 1 ml/min). The fractions (1 ml each) are first counted in a Clini-Gamma LKB 1272 (Pharmacia, Gaithersburg, MD) to determine ^{51}Cr radioactivity. Subsequently, Ecoscint (National Diagnostics, Manville, NJ) is added to each fraction and the beta activity is determined by a Beckman LS 3801 liquid scintillation analyzer (Beckman Instruments, Irvine, CA). Results are expressed as the fractional release of the respective emitter contained in the astrocytes at each time point. This was calculated by summing the effluxed radioactivity to that point including the remaining radioactivity counted in the cell support film. A computer program (Microsoft Excel) was adopted for the calculations.

To assess the effect of mercuric chloride (MC) on aspartate release, astrocytes are preloaded overnight with both D-[³H]aspartate and $Na_2^{51}CrO_4$. An inherent difficulty with continuous perfusion methods is that loss or lysis of cells also contributes to radioactivity in the perfusate. To distinguish efflux from intact cells of preloaded D-[³H]aspartate from MC-induced cell lysis or sloughing, the astrocytic release of D-[³H]aspartate and $Na_2^{51}CrO_4$ was compared (Richt and Stitz, 1992). As shown in Fig. 1, the ^{51}Cr release from preloaded astrocytes exposed to MC is negligible (range 0.00–0.11) at 5 μM MC whereas a progressive release of D-[³H]aspartate is observed. In the absence of MC, the loss of radioactive aspartate from the astrocytes exceeds that of ^{51}Cr (range 0.22–0.30), attesting to a net efflux of aspartate from preloaded astrocytes. [Using the same experimental paradigm for D-aspartate release with methyl metanethiosulfonate (100 μM), the $Na_2^{51}CrO_4$ method was found to be a valid indicator of nonspecific efflux (results not shown).]

C. Measurements of Astrocyte Volume Changes

Astrocytic volume is measured by a novel dynamic method which measures electrical resistance (O'Con-

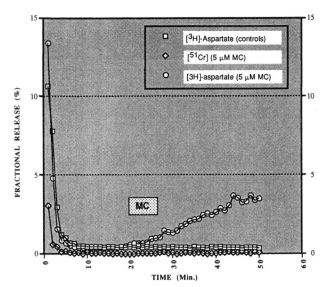

FRACTIONAL RELEASE (%)

□ [³H]-Aspartate (controls)
◇ [⁵¹Cr] (5 μM MC)
○ [3H]-aspartate (5 μM MC)

MC

TIME (Min.)

FIGURE 1 Effects of MC on the D-[2,3-³H]aspartate and Na₂⁵¹CrO₄ release from astrocytes. A 5-min (20–25 min) perfusion of astrocytes with MC (5 μM) enhanced the release of D-[2,3-³H]aspartate above control levels. The MC-induced release of D-[2,3-³H]aspartate was not accompanied by cell sloughing as indicated by efflux measurements of Na₂⁵¹CrO₄. Astrocytes were preloaded overnight with 20 μCi of D-[2,3-³H]aspartate and 40 μCi of Na₂⁵¹CrO₄ prior to the actual efflux measurements. Cultures were then washed and, at time zero, were perfused with HEPES–acid buffer. MC was added at 20–25 min. The perfusate was collected as described under Section II. The data are expressed as a percentage of the remaining label left in the cell at each time point (see Section II). Values shown are means of triplicate experiments (SEM <5%).

nor *et al.,* 1993). This, as well as other approaches to volume measurement highlighting their advantages and disadvantages, is reviewed extensively by Kimelberg and colleagues (1990). Briefly, astrocytes are grown on coverslips and placed in a perfusion chamber (Fig. 2) containing a channel (channel height is 100 μm) bridged between two silver wire electrodes (approximately 5 cm in length). Leads from the silver electrodes are made of insulated copper wire and are soldered to the silver with pure indium and the connections are covered with wax (a molten solution of 50% Apiezon M grease and 50% paraffin wax). The silver electrodes are connected through a large resistor (1 MΩ) to a lock-in amplifier (5301, EG & G Princeton Applied Research, Princeton, NJ) that supplies a 500-Hz, 5-V signal to the system. The lock-in amplifier is used because it is able to resolve small voltage changes with high noise rejection. At the 500-Hz frequency, the cell membranes are insulated and current will travel over and not through the cell monolayer. Apart from the solution in the channel, the two chambers are insulated from each other (Fig. 2). The control bathing medium for all experiments consists of 22 m*M* NaCl, 3.3 m*M* KCl,

0.4 m*M* MgSO₄, 1.3 m*M* CaCl₂, 1.2 m*M* KH₂PO₄, 10 m*M* D-(+)-glucose, 25 m*M* HEPES (*N*-2-hydroxyethylpiperazine-*N*′-2-ethanesulfonic acid), and 200 m*M* mannitol. HEPES-buffered solutions are maintained at pH 7.4 by adding 1 *N* NaOH. The osmolality of the solutions is approximately 300 mosmol as measured by a freezing point osmometer (Advanced Instruments, Inc., Needham Heights, MA). Hyposmotic solutions are made by removing the mannitol. The replacement of part of the NaCl with mannitol in isosmotic solutions has the important feature of maintaining the same electrolyte concentration in the respective iso- and hyposmotic solutions, assuring identical conductivity properties. However, a small correction has to be made for the small decrease in solution conductance due to mannitol, which is corrected for by adding a small amount of water. It is clearly critical to balance all the experimental solutions to the same resistivity so that the resistance (and voltage) differences measured when the solutions are changed can be accounted for solely due to changes in cell volume. Prior to experimentation, the resistivities of the solution are checked by measuring the solutions over a blank coverslip in the perfusion chamber. The resistivities of paired isosmotic and hyposmotic solutions (or any other experimental solution) are then balanced to within 0.5% by adding small amounts of water to the hypotonic buffer.

As cells swell their volume increases such that the volume of the solution within the channel available for current flow decreases proportionally, resulting in an increase in measured resistance in the channel above the cells. Since $V = IR$ (where V is voltage, I is current, and R is resistance) and I is constant (500 Hz), changes in V are directly proportional to changes in R. The resting height of the astrocytic monolayer is normally about 5 μ*M*, as determined by this method. The chamber is designed to have a height of approximately 100 μm above the cells. Since the percentage change is measured in voltage (and resistance), a 1% change in the measurement translates to approximately a 1-μm change in the average cell height of the monolayer. A recorded increase in the voltage (and thus resistance) means that the volume through the channel above the cells available for current flow has decreased by the same amount as the volume of the monolayer cell height has increased.

Each experiment is initiated by placing a No. 1 coverslip (2.5 × 1.0 cm) on which a confluent monolayer of astrocytes is growing in the channel and continuously perfusing it with isotonic solution at room temperature. If desired, the chamber may be placed in an incubator and experiments can be conducted at 37°C. The flow through the chamber is driven by hydrostatic pressure

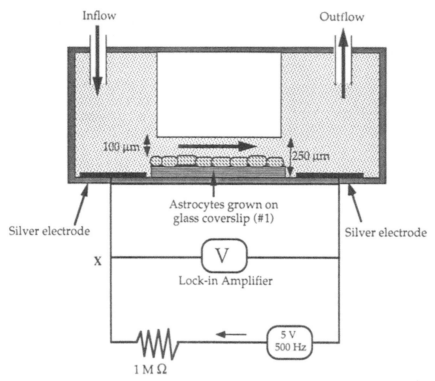

Inflow   Outflow

100 µm   250 µm

X   V
Lock-in Amplifier

Silver electrode   Astrocytes grown on
glass coverslip (#1)   Silver electrode

5 V
500 Hz

1 MΩ

FIGURE 2 The chamber used for the electrical resistance measurement of volume changes in substratum-attached astrocytes. The lock-in amplifier both supplies a reference voltage and measures changes in the signal voltage at point X, thus acting as a voltage divider. The system is set up with a large external resistance (1 MΩ) much greater than the chamber resistance so that the current *i* is constant. The height of the fluid-filled channel above the cells is approximately 100 μM. Dimensions of the chamber are not to scale (for further details on the method refer to the text).

(due to the height difference between the solutions and chamber). A pump may also be utilized.

A typical profile of D-[^{3}H]aspartate, ^{86}RbCl, and Na$_2$^{51}CrO$_4$ release from astrocytes upon exposure to hypotonic buffer is illustrated in Fig. 3. Release of taurine, glutamate, aspartate (Fig. 3), and other amino acids has been shown to occur during regulatory volume decrease (RVD) in a number of vertebrate and invertebrate cell types (Gilles *et al.*, 1991). Pasantes-Morales and Schousboe (1988) showed a hypotonic media-induced release of radiolabeled and endogenous taurine. Kimelberg and colleagues (1990) reported that exposure of primary astrocyte cultures to hypotonic media led to the release of label after cells had been allowed to accumulate L-[^{3}H]glutamate, D-[^{3}H]aspar-tate, or [^{3}H]taurine, as well as release of endogenous glutamate and taurine as measured by high-performance lipid chromatography (HPLC). Volume regulation under hypotonic conditions is due primarily to the cellular loss of KCl (for references see O'Connor *et al.*, 1993). Both electroneutral cotransporters and electroconductive channels are be-

lieved to be responsible for the volume-sensitive KCl fluxes (Hoffman and Simonsen, 1989). A role for K$^+$ in RVD is also supported by the finding that when the typical release profile of ^{86}Rb (a specific marker for K$^+$ release) (Fig. 3) is blocked by quinine, a known blocker of Ca^{2+}-activated K$^+$ channels, the swelling status of the cell is sustained, with no apparent RVD (results not shown; Vitarella *et al.*, 1994). As noted in Fig. 3, the swelling-induced release of D-[^{3}H]aspartate and ^{86}RbCl occurs in the absence of Na$_2$^{51}CrO$_4$ release, "authenticating" D-[^{3}H]aspartate and ^{86}RbCl release and distinguishing it from hypotonic-induced cell lysis or sloughing.

Since mercuric chloride is known to induce astrocytic release of preloaded D-aspartate (Mullaney *et al.*, 1993) and preloaded ^{86}Rb (unpublished observations), it was next examined whether MC-induced amino acid release is a result of astrocytic swelling. Utilizing the same direct electrical resistance method, the cell volume changes upon exposure to MC-containing buffer (5 μM), hypotonic buffer, and the latter two combined were measured. When exposed to

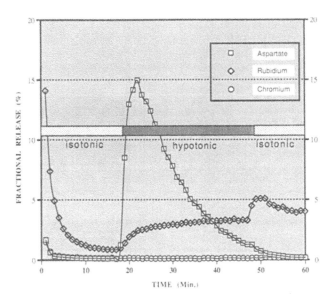

FIGURE 3 Time course of hypotonic-induced release of D-[2,3-³H]aspartate, $Na_2^{51}CrO_4$, and $^{86}RbCl$ (a marker for K^+ transport). Astrocytes grown on glass slides (No. 1) were preloaded overnight with 20 μCi of D-[2,3-³H]aspartate or 40 μCi of $^{86}RbCl$ and 40 μCi of $Na_2^{51}CrO_4$ prior to the actual efflux measurements. They were transferred to the electrical resistance chamber and perfused with the appropriate buffers at 1 ml/min. The perfusate was collected at 1-min intervals. The radioactivity of each sample was determined by a liquid scintillation counter (aspartate and rubidium) or a gamma counter (chromium). The data are expressed as a percentage of the remaining label left in the cell at each time point (see Section II). Exposure to hypotonic media (15–45 min) caused the release of D-[2,3-³H]aspartate and ^{86}Rb.

hypotonic buffer, swelling occurs rapidly, reaching a peak level at about 3 min. Thereafter, the cells begin to volume regulate (RVD). The 4% increase in voltage would thus indicate an approximate 4-μm change in average cell monolayer height, or an approximate doubling of the volume, if cell height is proportional to cell volume (O'Connor *et al.*, 1993). There is no evidence for MC (5 μM)-induced changes in cell volume. [A similar concentration caused a marked increase in preloaded D-aspartate release (Mullaney *et al.*, 1993).] However, the addition of MC (5 μM) to hypotonically swollen astrocytes prevents the typical regulatory volume decrease seen upon exposure to hypotonic media (Fig. 4A). Rather, the cells maintain their swollen state throughout the experimental period. [Peak swelling in these experiments (hypotonic vs hypotonic plus MC) differs because of variation in astrocyte cell culture densities (i.e., the fact that the change in voltage is lower in MC-treated cells does not necessarily imply less swelling compared to hypotonic-induced swelling).] As noted in Fig. 4B, although MC *per se* does not induce astrocytic swelling, its addition to hypotonic

solutions completely abolishes RVD when each time point is expressed as percent maximal swelling.

III. SUMMARY

The electrical resistance technique offers an elegant means to study cellular mechanisms of swelling and RVD. When combined with the continuous perfusion method, it offers the added feature of release measurements of preloaded radiolabeled ions and amino acids. This is a promising assay system with great potential as an *in vitro* tool for the study of cytotoxicity. The use of $Na_2^{51}CrO_4$ provides a convenient determination of cytotoxicity, replacing more elaborate protocols, such as release of lactate dehydrogenase. A modification of the existing chamber with suitable optics through the cell channel could also allow the simultaneous measurement of changes in intracellular ion concentrations using fluorescent probes. Thus, by combining volume measurements with intracellular ion measurements, one could not only determine the role of ions such as calcium or potassium in RVD, but also determine how interference with such processes can have an impact on volume regulation and the release of endogenous substances.

In the experiments described in this chapter, the swelling of confluent astrocytic monolayers upon exposure to hypotonic solutions and MC was examined. This methodology is applicable to studying volume regulation not only in astrocytes, but any substratum-attached cell types because most mammalian cell types, when exposed to hypotonic media, will swell and subsequently return to their normal volume. Furthermore, the technique can be utilized to screen chemicals which induce cell swelling and for the release of endogenous substances. With this method, it is not necessary to have a single confluent monolayer of cells to study swelling and release, as both multilayers of cells or less than confluent cell cultures can be studied, although the sensitivity of the system would change. Finally, from an economic point of view, the system is inexpensive compared to other cytotoxic-screening methods. A lock-in amplifier can be purchased for approximately $3000. The method is extremely reproducible and allows for many experiments to be run in a single day. Unlike many other methods which offer a "snap shot" in time, the electrical resistance technique combined with efflux measurements allows for continuous correlation of relative changes in cell volume and release of endogenous markers, allowing for sensitive measurements of cytotoxicity.

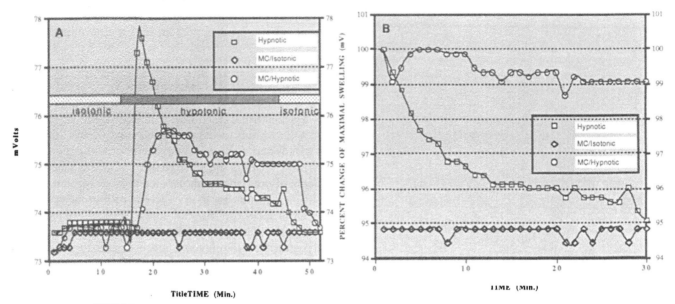

FIGURE 4 Astrocytic volume upon exposure to MC (5 μM, 20–25 min), hypotonic HEPES buffer (15–45 min), or a combination of MC and hypotonicity. As described in the text, the changes in voltage are related to changes in average volume and height of the astrocytic monolayer. No change in voltage was ascertained upon exposure to MC, whereas exposure to the hypotonic solution resulted in a peak volume change around 3 min with a return to normal volume within 30 min, i.e., RVD. Exposure to hypotonicity and MC resulted in rapid astrocytic swelling with lack of RVD. (A) The change in the recorded voltage in the three experimental paradigms. (B) The voltage at each time point relative to maximal swelling (for further details refer to the text).

Acknowledgments

This work was supported in part by PHS Grant NIEHS 05223 (awarded to MA), NS 23750 (awarded to HKK), and U.S. EPA R-819210 (awarded to MA).

References

Aschner, M., Aschner, J. L., and Kimelberg, H. K. (1993). The role of glia in CNS-induced injuries. In *Handbook of Neurotoxicology*, Vol. 1, *Basic Principles and Current Concepts* (L. W. Chang, ed.), Dekker, New York.

Aschner, M., and Vitarella, D. (1993). CNS glial cell cultures for neurotoxicological investigations: methodologies and approaches. In *Handbook of Neurotoxicology*, Vol. 3, *Approaches and Methodologies* (L. W. Chang, Ed.), Dekker, New York.

Frangakis, M., and Kimelberg, H. K. (1984). Dissociation of neonatal rat brain by dispase for preparation of primary astrocyte cultures. *Neurochem. Res.* **9,** 1689–1698.

Gilles, R., Hoffman, E. K., and Bolis, L. (Eds.) (1991). *Advances in Comparative and Environmental Physiology: Volume and Osmolality Control in Animal Cells.* Springer Verlag, Berlin.

Hydèn, H. (1961). Satellite cells in the nervous system. *Sci. Am.* **205,** 62–70.

Kimelberg, H. K., Goderie, S. K., Higman, S., Pang, S., and Waniewski, R. A. (1990). Swelling-induced release of glutamate,

aspartate, and taurine from astrocyte cultures. *J. Neurosci.* **10,** 1583–1591.

Mazzoni, M. C., Lundgren, E., Arfors, K. E., and Intaglietta, M. (1989). Volume changes of an endothelial cell monolayer on exposure to anisotonic media. *J. Cell Physiol.* **140,** 272–280.

Mullaney, K. J., Vitarella, D., Albrecht, J., Kimelberg, H. K., and Aschner, M. (1993). Stimulation of D-aspartate efflux by mercuric chloride from rat primary astrocyte cultures. *Dev. Brain Res.* **75,** 261–268.

O'Connor, E. R., Kimelberg, H. K., Keese, C. R., and Giaever, I. (1993). An electrical resistance method for measuring changes in monolayer cultures applied to astrocyte swelling. *Am. J. Physiol.* **264,** C471–C478.

O'Connor, E. R., and Kimelberg, H. K. (1993). Role of calcium in astrocyte volume regulation and in the release of ions and amino acids. *J. Neurosci.* **13,** 2638–2650.

Pasantes-Morales, H. P., and Schousboe, A. (1988). Volume regulation in astrocytes: a role for taurine as an osmoeffector. *J. Neurosci. Res.* **20,** 505–509.

Richt, J. A., and Stitz, L. (1992). Borna disease virus-infected astrocytes function *in vitro* as antigen-presenting and target cells for virus-specific CD4-bearing lymphocytes. *Arch. Virol.* **124,** 95–109.

Vitarella, D., DiRisio, D. J., Kimelberg, H. K., and Aschner, M. (1994). Potassium and taurine release are highly correlated with regulatory volume decrease in neonatal primary rat astrocyte cultures. *J. Neurochem.* **63,** 1143–1149.

CHAPTER

25

Electron Probe X-Ray Microanalysis as a Tool for Discerning Mechanisms of Nerve Injury

RICHARD M. LOPACHIN
Department of Anesthesiology
Montefiore Medical Center
Albert Einstein Medical School
Bronx, New York 10467

I. Introduction

This section discusses electron probe X-ray microanalysis (EPMA) and the relevance of this technique to neurotoxicology. Principles of microprobe analysis are briefly summarized, as well as methods of quantification, sample preparation, and potential technical problems. The section concludes with specific examples of how EPMA has been applied to neurotoxicology studies and the subsequent mechanistic insights gained.

EPMA is a quantitative electron microscope technique that simultaneously measures total concentrations of biologically relevant elements (Na, Mg, P, S, K, Cl, Ca) and water in different cellular compartments (e.g., cytoplasm, mitochondria). Application of EPMA can provide important information relevant to issues in nervous system toxicology and physiology. For example, it is understood that elements and their ionic species play a significant role in most aspects of nerve and glial cell physiology. Yet despite this apparent familiarity, very little is actually known concerning the precise subcellular distributions and concentrations of

elements and water in nervous tissue cells. EPMA studies have begun to define how elements distribute among cellular compartments of myelinated axons, glial cells, and nerve cell bodies of the central and peripheral nervous systems (Andrews *et al.*, 1987; LoPachin *et al.*, 1988, 1991; Saubermann and Scheid, 1985; Somlyo *et al.*, 1985; Wroblewski, 1989). Such information can help confirm and extend our current understanding of elements (ions) and their involvement in neurophysiological processes.

Substantial evidence indicates that the normal subcellular distribution of elements and water is disrupted following a variety of injurious processes and that the structural and functional consequences of injury are mediated by transmembrane shifts in Na, K, Ca, and other ions (see reviews by Dubinsky, 1993, LoPachin *et al.*, 1992c; Macknight, 1984; Trump *et al.*, 1979, 1989; Waxman *et al.*, 1991). However, as was the case with physiological processes, very little specific information is available concerning how translocated elements and associated water distribute among subcellular anatomical compartments following injury. EPMA quantification of the elemental changes associated with chemical-, mechanical-, and metabolic-induced injury

445

of nervous tissue is currently a burgeoning area of research (Garruto *et al.*, 1984; Hirsch *et al.*, 1991; Jancso *et al.*, 1984; Landsberg *et al.*, 1992; LoPachin *et al.*, 1990, 1992a,b, 1993a,b,c, 1994a,b; Wroblewski, 1989). The information generated by these studies has helped to shape our current understanding of certain neurotoxic and neuropathic conditions (see Section III).

The ability of EPMA to measure compartmental water content is particularly important for studies of neurotoxicity. Alterations in elements and ions are accompanied by osmoregulatory changes in water content (Macknight, 1984). The direction and magnitude of change in water content can indicate whether the system is undergoing deregulation, i.e., loss of osmoregulation and elemental composition, or homeostatic compensation, i.e., coordinated changes in elements and water (see, for example, LoPachin *et al.*, 1992a, 1994a). Moreover, since EPMA is an electron microscopy technique, different cell types that comprise nervous tissue (i.e., nerve and glial cells) can be optically discriminated on the basis of structural correlates. Thus, the individual *in situ* responses of glial cells and neurons to neurotoxicant exposure can be documented and, in doing so, the potential involvement of glia in neurotoxic expression (Aschner and LoPachin, 1993; LoPachin and Aschner, 1993) can be explored. This represents a substantial advantage over other techniques where such cell type discrimination is impossible. Finally, EPMA provides a unique opportunity to assess the status of cell regions (e.g., Schwann cell cytoplasm, nodes of Ranvier) and organelles (e.g., mitochondria, smooth endoplasmic reticulum) that might be involved in the etiology of cell injury and which are otherwise inaccessible to direct biochemical or electrophysiological evaluation. Since EPMA data reflect both distribution and quantitation, they represent detailed and comprehensive information necessary to define the role of elements and water in nerve and glial cell injury.

The next section describes the principles and methodologies of EPMA. Other microanalysis methods are available, e.g., laser microprobe analysis, particle-induced X-ray emission, and electron energy loss spectroscopy. However, discussion of these exciting techniques is beyond the scope of this chapter.

II. PRINCIPLES AND METHODS OF EPMA

EPMA was developed in the late 1940s by Guinier (1949) and was originally intended for use in the physical sciences, e.g., physics, geology, and metallurgy. However, in the 1960s, pioneering work by T. A. Hall and associates (see review by Hall, 1989) demonstrated the utility of this technique for elemental analysis of biological specimens. Continued technical refinements and the development of cryoultramicrotomy have expanded the applicability of this method into cell biology, pathology, and clinical medicine (LeFurgey *et al.*, 1988; Moreton, 1981).

A. Production and Detection of X-Rays

EPMA is based on the collection of X-rays that result when electrons of an electron microscope beam interact with elemental atoms comprising the tissue section (for comprehensive discussions see Chandler, 1977; Goldstein *et al.*, 1981; Ingram *et al.*, 1989; LeFurgey *et al.*, 1988; Somlyo *et al.*, 1985). Two types of X-rays are produced by electron–electron interactions: characteristic X-rays and continuum radiation. Characteristic X-rays are produced when high-energy electrons of the microscope beam strike and eject inner shell electrons of an atom. The resulting inner orbital ionization is a labile state that is stabilized by decay of an outer shell electron into the lower energy inner orbital. This transition requires release of energy in the form of an X-ray photon. The energy of this photon is equivalent to the energy difference between the two shells and is characteristic for that element since the atomic number of the element determines the discreet binding energies of each shell. Continuum radiation, also called bremsstrahlung or white radiation, occurs when incident beam electrons are inelastically scattered by the electromagnetic field of resident atomic nuclei. Because of deceleration, the incident electron loses an amount of energy which ranges from zero to the initial ionization energy of primary electrons. Continuum is a function of the total number of all atoms in the analyzed compartment and is therefore a measure of corresponding mass.

In biological studies, X-rays (characteristic and continuum) are most frequently detected using an energy-dispersive spectrometer that can perform simultaneous multielemental analyses ($Z > 10$) with low, nondamaging beam currents (e.g., 1 nA). An energy dispersive spectrometer or EDS system consists of a lithium-drifted silicon [Si(Li)] semiconductor that can collect X-rays in the form of charge pulses. Each charge pulse is converted to a voltage pulse that is proportional to the energy of the arriving X-ray. A multichannel analyzer converts the analog signal from the detector into a digital signal that can be sorted based on predetermined energy bands. The resulting spectrum relating the number of counts per channel (energy band) can be displayed on a video monitor with the characteristic peaks for each element of interest marked. Si(Li) detectors are not stable at room temperature and, therefore, must be cooled by a cold finger attached to a liquid nitrogen Dewar. The detector is isolated from the microscope chamber by a beryllium window which, unfortunately, absorbs low-energy X-rays from elements lighter than Na ($Z = 11$). This problem has been circumvented by the recent development of thin plastic windows and windowless detectors that permit measurements of elements as light as boron ($Z = 5$).

B. Instrumentation and Cryopreparative Methods

EDS systems can be interfaced with either transmission (TEM) or scanning electron microscopes (SEM). The use of EDS with TEM systems requires the addition of a scanning attachment that allows the electron beam to be rastered within chosen anatomical compartments. TEM provides superior spatial resolution (<1 nm), especially when ultrathin (<200 nm) sections are used. However, the use of such sections imposes quantitative and statistical problems due to the low generation rate of X-ray counts and correspondingly poor signal-to-noise ratios. It is also difficult to maintain unfixed ultrathin sections in a frozen-hydrated state (see Section II,C) since the high surface-to-volume ratios of these sections promote facile sublimation of water both during cyrosectioning and while in the vacuum of the microscope column (see Section II,D). Consequently, determinations of wet weight concentrations and water content are difficult since they require measurements of continuum in the hydrated state. In contrast to TEM systems, SEM have lower spatial resolution (10 nm) but the ability to use thicker sections (up to 2 μm) offers better peak-to-continuum ratios for quantitative analyses. Thick sections can also be maintained in a fully hydrated state because they are less susceptible to uncontrolled dehydration. For identification and analysis of ultrastructure, scanning microscopes can be fitted with transmitted electron detectors that provide scanning transmission electron microscopy (STEM).

Early EPMA studies used tissues prepared by wet chemical methods; however, recent work has shown that such fixation promotes artifactual translocation of elements (Moreton, 1981; Morgan, 1979; Roos and Barnard, 1985; Somlyo, 1985). Therefore, cryopreparative methods were developed to ensure rapid freezing of tissue samples with preservation of normal elemental distribution and morphological structures (Parsons *et al.*, 1985; Saubermann, 1980; Saubermann *et al.*, 1981a). Tissue can be rapidly frozen by a variety of methods, e.g., clamp freezing using liquid nitrogen-cooled polished metal surfaces (LoPachin *et al.*, 1988),

quench freezing with super-cooled liquids such as freon-12 (Saubermann and Scheid, 1985), or by slam freezing on helium-cooled polished metal (Escaig, 1982). The objective of these freezing techniques is to remove heat rapidly and thereby reduce ice crystal growth. Crystal size is a function of (1) volume of tissue frozen; (2) shape of tissue; (3) water content of tissue; and (4) rate of heat removal at the sample surface. Severe ice crystal formation in dehydrated cryosections is manifest as a lacy appearance of anatomical compartments and can make recognition and analysis of cell regions difficult.

C. Cryoultramicrotomy

Once frozen the tissue is cryosectioned on a cryomicrotome. Cryosectioning of frozen biological material is a complex and poorly understood process. Controversy surrounds the temperature at which tissues are sectioned (Frederik and Busing, 1981; Karp *et al.*, 1982; Saubermann *et al.*, 1977). Many research groups cut at very low temperatures (<−100°C) based on the theory that vitreous ice formed during the freezing procedure will recrystallize at warmer temperatures and, therefore, cause damage. Other groups cut at relatively warm temperatures (e.g., −45°C) where frozen material is more ductile and, consequently, easier to cut. Once cryosections are cut, they are transferred from the knife to a specimen grid (e.g., beryllium or copper). For TEM systems, the grid is part of a mobile cold stage unit (e.g., gatan cryotransfer system, Gatan, Inc.) that is inserted into the cryochamber of the microtome. When the grid is full, the cold stage is simply transported to the microscope and inserted into the vacuum column. For SEM systems, the grid is held by a transport system in the microtome and when full the system can be transferred using a Delrin evacuated device directly onto a cold stage located within the vacuum column of the microscope (Saubermann *et al.*, 1981a).

D. Standardization and Quantitation

This section discusses methods of quantification and standardization. Other methods are available and the reader is referred to Gupta and Hall (1978), Rick and colleagues (1982), and Shuman and colleagues (1976).

Following placement of hydrated cryosections within the vacuum column, the first step in the analysis routine is to determine the continuum for large areas (e.g., 60 × 60 μm) of the section. To do this, the beam is rastered within each area and the corresponding data (continuum) are stored on a computer disk for later retrieval. Elemental analyses of individual anatomical compartments are not performed because it is difficult to identify morphological structures in the hydrated state (Fig. 1A) and because hydrated samples are more sensitive to radiation damage than dried samples. In addition, elemental mass fractions in the hydrated state are lower than when dried, consequently the peak-to-background ratio of an element is less favorable for accurate detection and measurement (Saubermann *et al.*, 1977, 1981a; Saubermann, 1988b; Zierold, 1982). Once regional hydrated continuums have been acquired, the section is freeze dried in the vacuum column of the microscope by raising the temperature of the cold stage from -185 to $-60°C$. After dehydration, the temperature is returned to $-185°C$ for analysis.

Morphological compartments are easily recognized in dehydrated thin cryosections (Fig. 1B) and quantitation is continued by rastering the beam within chosen compartments (e.g., mitochondria, axoplasm). Characteristic X-ray counts (minus background) are then determined over specific energy ranges (e.g., Na = $0.96–1.12$ keV, K = $3.24–3.40$ keV, Ca = $3.60–3.76$ keV). Generally, for quantitation of dry weight concentrations in thin biological sections, the Hall method (1973) of continuum normalization is used. This method is based on the finding that the continuum is directly related to the mass of the specimen and, therefore, the ratio of characteristic counts to continuum counts is proportional to the mass fraction (mass of element/mass of specimen) of element in the analyzed volume. This is expressed by the following equation:

$$R_x = P_x - b_x / W_t - W_e,$$

where R_x is the mass fraction of element x, P_x is the characteristic X-ray counts for element x, b_x is the background counts under the characteristic peak, W_t is the continuum counts in a selected region, and W_e is the extraneous continuum counts. The absolute dry weight concentration of element x (mmol/kg dry weight) can be derived by standardization where an R_x value from an unknown concentration of x is compared to a standard curve of mass fraction ratios from known concentration of x. Typical standards contain varying concentrations of element dissolved in a matrix such as albumin, gelatin, or polyvinyl pyrrolidone which presumably mimics the dry weight mass of the speci-

men (for details see Chandler, 1977; Hagler and Buja, 1984; Saubermann *et al.*, 1981b). Water content (% H_2O) and wet weight concentrations (mmol element/kg wet weight) can be determined by the following algorithm:

$$\% \ H_2O = (1 - W_d / W_h) \times 100\%$$

absolute wet weight mass fraction:

$$Cx_w = Cx_d (1 - \% \ H_2O)$$

where W_d and W_h are the corrected continuum counts for dried and hydrated section areas, respectively, and Cx_w and Cx_d are the wet and dry weight concentrations for element x, respectively. These algorithms have proven to be reliable and have been applied to the study of elements and water in both nervous (e.g., LoPachin *et al.*, 1988, 1990, 1992a,b, 1993a,b; Saubermann and Scheid, 1985; Saubermann and Stockton, 1988) and nonnervous tissues (e.g., Bulger *et al.*, 1981; Saubermann *et al.*, 1986a,b).

E. Modes of Operation

EPMA data can be collected by two methods: static probe analysis and digital X-ray imaging. For static analysis, the operator manually places the beam over a morphological compartment and then initiates collection of X-rays by the EDS system. Table 1 shows the dry and wet weight elemental composition of medium diameter myelinated axons and glial areas in rat spinal cord white matter. Chemical maps of subcellular elemental distribution can be generated using digital X-ray imaging (Fiori *et al.*, 1988; Gorlen *et al.*, 1984; Ingram *et al.*, 1989; Saubermann and Heyman, 1987). Here the beam is computer controlled and is moved point by point across a specimen region in accordance with a preselected analysis matrix (e.g., 64×64 point matrix). The collected data are used to form digital images where each point or pixel is fully quantitative with respect to elemental concentrations and water content. Digital imaging has been used to study the distribution of elements and water in normal leech ganglion (Saubermann and Heyman, 1987; Saubermann and Stockton, 1988; Saubermann, 1988a, 1989), myelinated axons, cell bodies, and glia in the central nervous system (CNS) and peripheral nervous system of rat (PNS) (Lo-

FIGURE 1 Frozen, unfixed hydrated (A) and dehydrated (B) cryosections (500 nm nominal thickness) of normal rat tibial nerve are presented. Tibial nerve samples (approximately 0.5 cm in length) were excised and immediately frozen in melting freon. Frozen samples were then cryosectioned at $-55°C$ in a specially designed microtome cryochamber. Sections were placed on a nylon-coated beryllium grid and then transferred onto the cold stage ($-185°C$) in the vacuum column of a scanning electron microscope equipped with a STEM detector and an EDS system. After acquiring area continuums from the hydrated sample (A), the sample was dehydrated (B) by raising the temperature of the stage to $-60°C$ for 30 min. Following dehydration, elemental concentrations and water contents were determined in individual anatomical compartments.

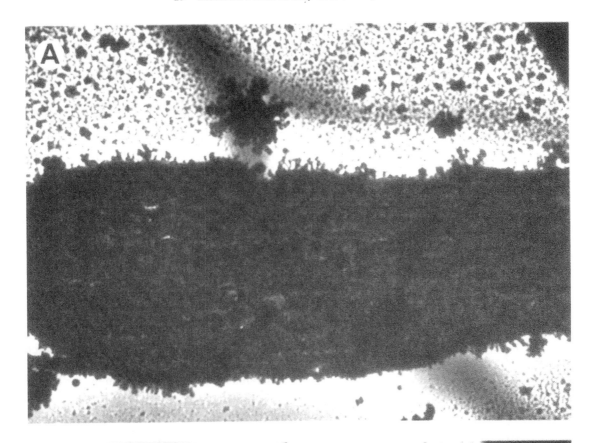

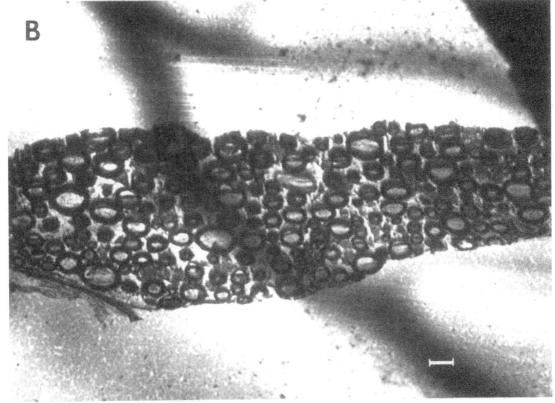

TABLE 1 Dry and Wet Weight Elemental
Composition of Several Anatomical
Compartments in Rat Cervical Spinal Cord[a]

Element	mmol/kg dry wt ± SEM	mmol/kg wet wt ± SEM
Medium diameter axoplasm (91 ± 1% water)		
Na	219 ± 17	20 ± 2
P	429 ± 14	39 ± 1
Cl	538 ± 22	48 ± 2
K	1947 ± 62	175 ± 6
Mg	35 ± 5	3 ± 0
Mitochondria (73 ± 2% water)		
Na	189 ± 21	50 ± 6
P	483 ± 26	130 ± 7
Cl	632 ± 47	171 ± 13
K	1937 ± 128	533 ± 35
Mg	29 ± 8	8 ± 2
Glial areas (73 ± 2% water)		
Na	78 ± 10	27 ± 6
P	580 ± 26	178 ± 23
Cl	90 ± 7	30 ± 6
K	422 ± 43	127 ± 18
Mg	17 ± 5	6 ± 2
Myelin (33 ± 3% water)		
Na	99 ± 9	69 ± 6
P	682 ± 20	470 ± 14
Cl	63 ± 4	43 ± 3
K	204 ± 15	140 ± 10
Mg	12 ± 2	8 ± 1

[a] From LoPachin *et al.* (1991).

Pachin et al., 1988, 1991) and cellular areas in mouse cerebellar cortex (Fiori *et al.*, 1988). X-ray maps are especially useful in studies of nerve cell injury since the quantitative and spatial information provided can show precisely how damage causes decompartmentalization of both water and elements (LoPachin *et al.*, 1990, 1992a; Saubermann and Stockton, 1988).

III. Application of EPMA to Neurotoxic Mechanisms

EPMA has been used to study a variety of neurotoxic and neuropathic conditions in the PNS and CNS (see reviews by LoPachin and Saubermann, 1990; LoPachin *et al.*, 1992). In particular, EPMA studies have provided insight into the mechanisms of experimental diabetic neuropathy and acrylamide distal axonopathy.

The peripheral neuropathy associated with diabetes has been suggested to be a result of impaired axolemmal Na,K-ATPase activity (Greene *et al.*, 1987, 1988, 1989). If this scenario is valid, microprobe analysis of peripheral nerves from experimental diabetic animals should detect a loss of axoplasmic K and gain in Na, Cl, and Ca in association with axonal swelling (Table 2; see also Macknight, 1984; Saubermann and Stockton, 1988). However, EPMA of tibial nerve cryosections from streptozocin-treated diabetic (10 and 20 weeks hyperglycemic) rats revealed a selective increase in axoplasmic K and Cl concentrations with no change in water content (Table 2; Lowery *et al.*, 1990; Lo-Pachin *et al.*, 1993c). This pattern of elemental alteration is not consistent with the inhibition of Na,K-ATPase activity (see Table 2). In the proximal sciatic nerve of diabetic rat, axoplasmic K and Cl levels were significantly decreased, although increases in Na and water content that might indicate Na pump inhibition were not observed (Table 2; Lowery *et al.*, 1990; Lo-Pachin *et al.*, 1993c). Thus, the reported changes in intraaxonal elemental composition do not support a functional inhibition of Na,K-ATPase activity in peripheral nerve myelinated axons of diabetic rat. Instead the data suggest new avenues of research involving mechanisms that might influence K regulation in myelinated axons (e.g., K^+ channels, ion exchangers, axonal transport).

Exposure of humans and laboratory animals to an acrylamide (ACR) monomer produces a distal axonopathy of peripheral and central myelinated axons (Spencer and Schaumburg, 1974). This type of nerve damage is characterized by paranodal axon swellings that are filled with excess neurofilaments and fragments of smooth endoplasmic reticulum (Prineas, 1969). As ACR exposure continues, axonal regions below the swelling degenerate. Microprobe analysis of swollen axon areas indicate marked elemental deregulation and a loss of osmoregulation (LoPachin *et al.*, 1992a). Most

TABLE 2 Ouabain Exposure vs Experimental Diabetes: Axonal Elemental Composition and Water Content

Element	Diabetic tibial[a]	Control[b]	Ouabain[c]	Control[d]
Na	225 ± 9	233 ± 8	1553 ± 123*	272 ± 14
Cl	415 ± 21*	356 ± 10	929 ± 76*	694 ± 46
K	1931 ± 78*	1595 ± 32	472 ± 53*	2096 ± 135
Ca	8 ± 2*	1.3 ± 1	13 ± 5	14 ± 3
H_2O	92 ± 1	91 ± 1	95 ± 1*	91 ± 1

[a] Medium diameter tibial nerve axons from experimentally diabetic rats (10 weeks post-streptozocin; LoPachin *et al.*, 1993c).

[b] Medium diameter tibial nerve axons from aged-matched control rats (LoPachin *et al.*, 1993c).

[c] *In vitro* ouabain-exposed (3 mM, 1-hr incubation) tibial nerve axons (Lehning *et al.*, unpublished).

[d] *In vitro* control (1-hr incubation) tibial nerve axons (Lehning *et al.*, unpublished).

* Significantly different from appropriate control ($P < 0.05$).

prominently, mitochondria and axoplasm from swollen regions exhibit substantial increases in Ca content. In contrast, nonswollen internodal axon areas display selective decreases in K and Cl which might be related to the development of axonal atrophy (Gold *et al.,* 1992). Parallel studies of the distal axonopathy produced by 2,5-hexanedione showed that corresponding axonal swellings were associated with only minimal changes in elements (LoPachin *et al.,* 1994a). This finding suggests that the elemental derangement associated with ACR-induced axonal swellings is a specific effect and not a general phenomenon associated with axoplasmic expansion. Based on this and evidence from additional EPMA and biochemical studies (LoPachin *et al.,* 1993a,b; Lehning *et al.,* 1994), a new hypothesis has been developed for distal axon degeneration caused by ACR intoxication (LoPachin and Lehning, 1994). It is proposed that ACR decreases the delivery of Na,K-ATPase to distal paranodal axon sites. The reduction in enzyme activity is associated with a loss of axoplasmic K and an influx of Na. This loss of transmembrane ion gradients promotes reversal of Na/Ca exchangers which promotes an influx of Ca and initiates cytoskeletal disruption and distal axon degeneration. It is hoped that the insight provided by EPMA studies will provide a foundation for the molecular and biochemical research necessary to delineate the mechanism of ACR neurotoxicity.

Acknowledgment

Some of the research presented in this chapter was supported by a grant from the National Institute of Environmental Health Sciences (RO1-ES03830-01-07).

References

Andrews, S. B., Leapman, R. D., Landis, D. M. D., and Reese, T. S. (1987). Distribution of calcium and potassium in presynaptic nerve terminals from cerebellar cortex. *Proc. Natl. Acad. Sci. USA* **84,** 1713–1717.

Aschner, M., and LoPachin, R. M. (1993). Astrocytes: targets and mediators of chemical-induced CNS injury. *J. Toxicol. Environ. Health* **38,** 329–342.

Bulger, R. E., Beeuwkes, R., and Saubermann, A. J. (1981). Application of scanning electron microscopy to X-ray analysis of frozen-hydrated sections. III. Elemental content of cells in the rat renal papillary tip. *J. Cell Biol.* **88,** 274–280.

Chandler, J. A. (1977). X-Ray Microanalysis in the Electron Microscope. In *Practical Methods in Electron Microscopy,* Vol. 5, (A. M. Glauert, Ed.), North-Holland, Amsterdam.

Dubinsky, J. M. (1993). Examination of the role of calcium in neuronal death. In *Markers of Neuronal Injury and Degeneration* (J. N. Johannessen, Ed.), pp. 34–42, The New York Academy of Sciences, New York.

Escaig, J. (1982). New instruments which facilitate rapid freezing at 83K and 6K. *J. Microsc.* **126,** 221–229.

Fiori, C. E., Leapman, R. D., Swyt, C. R., and Andrews, S. B. (1988). Quantitative X-ray mapping of biological cryosections. *Ultramicroscopy* **24,** 237–250.

Frederik, P. M., and Busing, W. M. (1981). Strong evidence against section thawing whilst cutting on the cryo-ultratome. *J. Microsc.* **122,** 217–220.

Garruto, R. M., Fukatsu, R., Yanagihara, R., Gajdusek, D. C., Hook, G., and Fiori, C. E. (1984). Imaging of calcium and aluminum in neurofibrillary tangle-bearing neurons in Parkinsonism-dementia of Guam. *Proc. Natl. Acad. Sci. USA* **81,** 1875–1879.

Gold, B. G., Griffin, J. W., and Price, D. L. (1992). Somatofugal axonal atrophy precedes development of axonal degeneration in acrylamide neuropathy. *Arch. Toxicol.* **66,** 57–66.

Goldstein, J. I., Newbury, D. E., Echlin, P., Joy, D. C., Fiori, C., and Lifshin, E. (1981). *Scanning Electron Microscopy and X-ray Microanalysis.* Plenum Press, New York.

Gorlen, K. E., Barden, L. K., Del Priore, J. S., Fiori, C. E., Gibson, C. C., and Leapman, R. D. (1984). Computerized analytical electron microscope for elemental imaging. *Rev. Sci. Instrum.* **55,** 912–921.

Greene, D. A., Lattimer, S. A., and Sima, A. A. F. (1987). Sorbitol, phosphoinositides and sodium-potassium-ATPase in the pathogenesis diabetic complications. *N. Engl. J. Med.* **316,** 599–606.

Greene, D. A., Lattimer, S. A., and Sima, A. A. F. (1988). Are disturbances of sorbitol, phosphoinositide, and Na-K-ATPase regulation involved in pathogenesis of diabetic neuropathy? *Diabetes* **37,** 688–693.

Greene, D. A., Lattimer-Greene, S. A., and Sima, A. A. F. (1989). Pathogenesis of diabetic neuropathy, role of altered phosphoinositide metabolism. *Crit. Rev. Neurobiol.* **5,** 143–219.

Gupta, B. L., and Hall, T. A. (1978). Electron microprobe X-ray analysis of calcium. In *Calcium Transport and Cell Function* (A. Scarpa, and E. Carafoli, Eds.), pp. 28–51, New York Academy of Sciences, New York.

Hagler, H. K., and Buja, L. M. (1984). New techniques for the preparation of thin freeze dried cryosections for X-ray microanalysis. In *Science of Biological Specimen Preparation,* pp. 161–166. SEM, Chicago.

Hall, T. A. (1989). The history of electorn probe microanalysis in biology. In *Electron Probe Microanalysis* (K. Zierold, and H. K. Hagler, Eds.), pp. 1–15, Springer-Verlag, New York.

Hall, T. A., Anderson, H. C., and Appleton, T. (1973). The use of thin specimens for X-ray microanalysis of biology. *J. Microsc.* **99,** 177–182.

Hirsch, E. C., Brandel, J.-P., Galle, P., Javoy-Agid, F. and Agid, Y. (1991). Iron and aluminum increase in the substantia nigra of patients with Parkinson's disease: an X-ray microanalysis. *J. Neurochem.* **56,** 446–451.

Ingram, P., Nassar, R., LeFurgey, A., Davilla, S., and Sommer, J. R. (1989). Quantitative X-ray elemental mapping of dynamic physiologic events in skeletal muscle. In *Electron Probe Microanalysis* (K. Zierold, and H. K. Hagler, Eds.), pp. 251–264, Springer-Verlag, New York.

Jancso, G., Karcsu, S., Kiraly, E., Szebeni, A., Toth, L., Bacsy, E., Joo, F., and Parducz, A. (1984). Neurotoxin induced nerve cell degeneration: possible involvement of calcium. *Brain Res.* **295,** 211–216.

Karp, R. D., Silcox, J. C., and Somlyo, A. V. (1982). Cryoultramicrotomy: evidence against melting and the use of a low temperature cement for specimen orientation. *J. Microsc.* **125,** 157–165.

Landsberg, J. P., McDonald, B., and Watt, F. (1992). Absence of aluminum in neuritic plaque cores in Alzheimer's disease. *Nature* **360,** 65–68.

LeFurgey, A., Bond, M., and Ingram, P. (1988). Frontiers in electron probe microanalysis: application to cell physiology. *Ultramicroscopy* **24,** 185–220.

Lehning, E. J., Mathew, J., Eichberg, J., and LoPachin, R. M. (1994). Changes in Na-K ATPase and protein kinase C activities in peripheral nerve of acrylamide-treated rats. *J. Toxicol. Environ. Health* **42**, 331–342.

LoPachin, R. M., Lowery, J., Eichberg, J., Kirkpatrick, J. B., Cartwright, J., and Saubermann, A. J. (1988). Distribution of elements in rat peripheral axons and nerve cell bodies determined by X-ray microprobe analysis. *J. Neurochem.* **51**, 764–775.

LoPachin, R. M., LoPachin, V. R. and Saubermann, A. J. (1990). Effects of axotomy on distribution and concentration of elements in rat sciatic nerve. *J. Neurochem.* **54**, 320–332.

LoPachin, R. M., and Saubermann, A. J. (1990). Disruption of cellular elements and water in neurotoxicity: studies using electron probe X-ray microanalysis. *Toxicol. Appl. Pharmacol.* **106**, 355–374.

LoPachin, R. M., Castiglia, C. M. and Saubermann, A. J. (1991). Elemental composition and water content of myelinated axons and glial cells in rat central nervous system. *Brain Res.* **549**, 253–259.

LoPachin, R. M., Castiglia, C. M., and Saubermann, A. J. (1992a). Acrylamide disrupts elemental composition and water content of rat tibial nerve. I. Myelinated axons. *Toxicol. Appl. Pharmacol.* **115**, 21–34

LoPachin, R. M., Castiglia, C. M., and Saubermann, A. J. (1992b). Acrylamide disrupts elemental composition and water content of rat tibial nerve. II. Schwann cells and myelin. *Toxicol. Appl. Pharmacol.* **115**, 35–43.

LoPachin, R. M., Castiglia, C. M. and Saubermann, A. J. (1992c). Perturbation of axonal elemental composition and water content: implication for neurotoxic mechanisms. *Neurotoxicology* **13**, 123–138.

LoPachin, R. M., Lehning, E. J., Castiglia, C. M., and Saubermann, A. J. (1993a). Acrylamide disrupts elemental composition and water content of rat tibial nerve. III. Recovery. *Toxicol. Appl. Pharmacol.* **122**, 54–60.

LoPachin, R. M., Castiglia, C. M., Lehning, E. J., and Saubermann, A. J. (1993b). Effects of acrylamide on subcellular distribution of elements in rat sciatic nerve myelinated axons and Schwann cells. *Brain Res.* **608**, 238–246.

LoPachin, R. M., Castiglia, C. M., Saubermann, A. J., and Eichberg, J. (1993c). Ganglioside treatment modifies abnormal elemental composition in peripheral nerve myelinated axons of experimentally diabetic rats. *J. Neurochem.* **60**, 477–486.

LoPachin, R. M., and Aschner, M. (1993). Glial-neuronal interactions: relevance to neurotoxic mechanisms. *Toxicol. Appl. Pharmacol.* **118**, 141–158.

LoPachin, R. M., Lehning, E. J., Stack, E. C., Hussein, S. and Saubermann, A. J. (1994a). 2,5-Hexanedione alters elemental composition and water content of rat peripheral nerve myelinated axons. *J. Neurochem.* **63**, 2266–2278.

LoPachin, R. M., Lehning, E. J., Stack, E. C., and Saubermann, A. J. (1994b). Disruption of Schwann cell elemental composition is not a primary neurotoxic effect of 2,5-Hexanedione. *Neurotoxicology* **15**, 927–934.

LoPachin, R. M., and Lehning, E. J. (1994). Acrylamide-induced distal axon degeneration: a proposed mechanism of action. *Neurotoxicology* **15**, 247–260.

Lowery, J. M., Eichberg, J., Saubermann, A. J., and LoPachin, R. M. (1990). Distribution of elements and water in peripheral nerve of streptozotocin-induced diabetic rats. *Diabetes* **39**, 1498–1503.

Macknight, A. D. C. (1984). Cellular response to injury. In *Edema* (N. C. Staub, and A. E. Taylor, Eds.), pp. 489–520, Raven Press, New York.

Moreton, R. B. (1981). Electron-probe X-ray microanalysis: techniques and recent applications in biology. *Biol. Rev.* **51**, 409–461.

Morgan, A. J. (1979). Non-freezing techniques of preparing biological specimens for electron microprobe X-ray microanalysis. *Scanning Elect. Microsc.* **II**, 635–648.

Parsons, K., Bellotto, D. J., Schulz, W. W., Buja, L. M., and Hagler, H. K. (1985). Towards routine cryoultramicrotomy. *EMSA Bull.* **14**, 49–60.

Prineas, J. (1969). The pathogenesis of dying-back polyneuropathies. II. An ultrastructural study of experimental acrylamide intoxication in the cat. *J. Neuropathol. Exp. Neurol.* **28**, 598–621.

Rick, R., Dorge, A., and Thurau, K. (1982). Quantitative analysis of electrolytes in frozen dried sections. *J. Microsc.* **125**, 239–247.

Roos, N., and Barnard, T. (1985). A comparison of subcellular element concentrations in frozen-dried, plastic-embedded, dry-cut sections and frozen-dried cryosections. *Ultramicroscopy* **17**, 335–344.

Saubermann, A. J., Riley, W., and Beeuwkes, R. (1977). Cutting work in thick section cryomicrotomy. *J. Microsc.* **111**, 39–49.

Saubermann, A. J. (1980). Application of cryosectioning to X-ray microanalysis of biological tissue. *Scanning Elect. Microsc.* **II**, 421–430.

Saubermann, A. J., Echlin, P., Peters, P. D., and Beeuwkes, R. (1981a). Application of scanning electron microscopy to X-ray analysis of frozen-hydrated sections I. Specimen handling techniques. *J. Cell Biol.* **88**, 257–267.

Saubermann, A. J., Beeuwkes, R., and Peters, P. D. (1981b). Application of scanning electron microscopy to X-ray analysis of frozen-hydrated sections II. Analysis of standard solutions and artificial electrolyte gradients. *J. Cell Biol.* **88**, 268–273.

Saubermann, A. J., and Scheid, V. L. (1985). Elemental composition and water content of neuron and glial cells in the central nervous system of the north american medicinal leech (*Macrobdella decora*). *J. Neurochem.* **44**, 825–834.

Saubermann, A. J., Dobyna, D. C., Scheid, V. L., and Bulger, R. E. (1986a). Rat renal papilla: comparison of two techniques for X-ray analysis. *Kidney Int.* **29**, 675–681.

Saubermann, A. J., Scheid, V. L., Dobyan, D. C., and Bulger, R. E. (1986b). Simultaneous comparison of techniques for X-ray analysis of proximal tubule cells. *Kidney Int.* **29**, 682–688.

Saubermann, A. J. and Heyman, R. V. (1987). Quantitative digital X-ray imaging using frozen hydrated and frozen dried tissue sections. *J. Microsc.* **146**, 169–182.

Saubermann, A. J. (1988a). X-Ray mapping of frozen hydrated and frozen dried cryosections using electron microprobe analysis. *Scanning* **10**, 239–244.

Saubermann, A. J. (1988b). Quantitative electron microprobe analysis of cryosections. *Scanning Microsc.* **2**, 2207–2218.

Saubermann, A. J. and Stockton, J. D. (1988). Effects of increased extrcellular K on the elemental composition and water content of neuron and glial cells in leech CNS. *J. Neurochem.* **51**, 1797–1807.

Saubermann, A. J. (1989). X-Ray microanalysis of cryosections using image analysis. In *Electron Probe Microanalysis* (K. Zierold, and H. K. Hagler, Eds.), pp. 73–85. Springer-Verlag, New York.

Shuman, H., Somlyo, A., and Somlyo, A. P. (1976). Quantitative electron probe microanalysis of biological thin sections: methods and validity. *Ultramicroscopy* **1**, 317–339.

Somlyo, A. P., Urbanics, R., Vadasz, G., Kovach, A. G. B., and Somlyo, A. V. (1985). Mitochondrial calcium and cellular electrolytes in brain cortex frozen in situ: electron probe analysis. *Biochem. Biophys. Res. Commun.* **132**, 1071–1078.

Somlyo, A. P. (1985). Cell calcium measurement with electron probe and electron energy loss analysis. *Cell Calcium* **6**, 197–212.

Spencer, P. S., and Schaumburg, H. H. (1974). A review of acrylamide neurotoxicity. Part II. Experimental animal neurotoxicity and pathologic mechanisms. *Can. J. Neurol. Sci.* **1**, 151–169.

Trump, B. F., Berezesky, I. K., Chang, S. H., Pendergrass, R. E., and Mergner, W. J. (1979). The role of ion shifts in cell injury. *Scanning Elect. Microsc.* **III**, 1–14.

Trump, B. F., Berezesky, I. K., Smith, M. W., Phelps, P. C. and Elliget, K. A. (1989). The relationship between cellular ion dereg-ulation and acute and chronic toxicity. *Toxicol. Appl. Pharmacol.* **97**, 6–22.

Waxman, S. G., Ransom, B. R., and Stys, P. K. (1991). Non-synaptic mechanisms of Ca^{2+}-mediated injury in CNS white matter. *TINS* **14**, 461–468.

Wroblewski, R. (1989). *In situ* elemental analysis and visualization in cryofixed nervous tissues. *J. Microsc.* **155**, 81–112.

Zierold, K. (1982). Cryopreparation of mammalian tissue for X-ray microanalysis in STEM. *J. Microsc.* **125**, 149–156.

CHAPTER

26

Glutamate Receptor-Mediated Neurotoxicity

JOHN W. OLNEY
Departments of Psychiatry and Pathology
Washington University School of Medicine,
St. Louis, Missouri 63110

I. Introduction

Glutamate (Glu), the major excitatory neurotransmitter in the mammalian central nervous system (CNS), has vitally important beneficial functions, but also harbors treacherous neurotoxic potential. Glu neurotoxicity can be expressed in several different ways, depending on which excitatory amino acid (EAA) receptor subtype is involved and whether the receptor is excessively activated or inhibited. Classical excitotoxicity, the most extensively studied type of neurotoxicity caused by Glu or related EAA, results from excessive activation of EAA ionotropic receptors. Another entirely different form of neurotoxicity is seen following excessive inhibition of a specific subtype of EAA ionotropic receptor, the *N*-methyl-D-aspartate (NMDA) receptor. Two additional mechanistically distinctive forms of EAA neurotoxicity have been described; one associated with excessive activation and the other with excessive inhibition of the EAA metabotropic receptor. The rich diversity of EAA receptor subtypes discovered by receptor cloning methods suggests the likelihood that EAA receptors may be associated with yet additional forms of neurotoxic expression that remain to be discovered. In evaluating the neurotoxic potential of therapeutic drugs and food additives or contaminants, it is essential to understand the mechanism(s) underlying each of these forms of neurotoxic expression and to devise appropriate testing methods that are tailored to address unique features of each form of neurotoxicity.

II. Historical Perspective

It has only been within the past decade that Glu has become widely accepted as a major neurotransmitter in the vertebrate CNS, and only over that same period that the neurotoxic properties of Glu have generated significant interest among neuroscientists. In the preceding decades, it was shown (Curtis and Watkins, 1963) that Glu and several of its structural analogs depolarize (excite) CNS neurons and that systemic administration of Glu to animals of various species causes acute degeneration of neurons in the retina (Lucas and Newhouse, 1957; Olney, 1969a) or several regions of brain that lack blood–brain barriers (Olney, 1969b, 1971). Only Glu analogs possessing the neuroexcitatory properties of Glu reproduced its neurotoxic effects (Olney *et al.*, 1971), and these analogs displayed a parallel order of potencies for their excitatory and toxic actions. Moreover, ultrastructural studies (Olney, 1969a, 1971) localized the apparent site of toxic action to postsynaptic dendrosomal membranes known to be responsive to the excitatory actions of Glu. From these and related observations the excitotoxic concept was proposed (Olney *et al.*, 1971; Olney, 1974) which holds that an excitatory mechanism and EAA synaptic receptors mediate Glu neurotoxicity.

Identification of EAA receptor subtypes differentially sensitive to specific agonists [NMDA, quisqualic acid (Quis), and kainic acid (KA)] and discovery of antagonists that block the excitatory actions of EAA agonists at such receptors (Watkins, 1978; Watkins and Evans, 1981) permitted the excitotoxic hypothesis to be rigorously tested. Shortly after the first EAA antagonists were identified electrophysiologically (e.g., α-amino adipate, D-2-amino-5-phosphonovalerate), it was shown that they protect neurons in the *in vivo* mouse hypothalamus against the neurotoxic actions of Glu or its more potent analog, NMDA (Olney *et al.*, 1979, 1981). By *in vitro* methods (Rothman, 1984, Olney *et al.*, 1986, 1987a,b; Choi *et al.*, 1988), many EAA antagonist candidates were subsequently screened systematically and were found to have anti-excitotoxic activities corresponding in potency and receptor specificity to their known anti-excitatory activities.

A. EAA Receptors, Pharmacologically and Electrophysiologically Defined

Of the three types of EAA receptor (NMDA, Quis, KA) that are capable of mediating excitotoxic events, the NMDA receptor has received the most attention and, as is discussed later may be of particular interest in relation to developmental neuropathological processes. Several features of the NMDA receptor distinguish it from other subtypes of EAA receptor. This receptor is linked to a cation channel which has a much higher Ca^{2+} conductance than ion channels associated with other EAA receptor subtypes (Dingledine 1983; MacDermott *et al.*, 1986; but see Hume *et al.*, 1991), and the NMDA ion channel is subject to a voltage-dependent Mg^{2+} blockade (Mayer *et al.*, 1984). The NMDA receptor is closely associated with a strychnine-insensitive glycine receptor (Johnson and Ascher, 1987) and with a polyamine receptor (Ransom and Stec, 1988; Reynolds and Miller, 1989; Romano *et al.*, 1991). When the former is activated by glycine and the latter by spermine or spermidine, this facilitates opening of the NMDA ion channel. Phencyclidine (PCP) receptors (Lodge *et al.*, 1982, 1987) are positioned within the NMDA ion channel permitting PCP agonists to perform an open channel block (MacDonald *et al.*, 1987). In addition, there is evidence that Zn^{2+}, acting at one or more separate sites in or near the NMDA ion channel, acts as an inhibitory modulator of channel function (Westbrook and Mayer, 1987). Thus, the NMDA receptor system is a remarkably complex entity, the normal function of which depends on a dynamic equilibrium among multiple facilitative and inhibitory factors. It follows that any pathological process abnormally increasing any of the facilitative forces or decreasing inhibitory forces might create an imbalance rendering the system prone to an expression of excitotoxicity. For example, such an imbalance might occur during development based on imperfect synchrony between the maturational schedules of excitatory versus inhibitory factors.

In addition to the NMDA receptor, two other EAA receptors linked to ion channels (ionotropic receptors) were initially described. These are the Quis and KA receptors, which are often referred to collectively as non-NMDA receptors. Initially the Quis receptor was characterized as an ionotrophic receptor but more recently it has been recognized that Quis activates two separate receptors; one is the previously mentioned non-NMDA ionotropic receptor and the other is a metabotropic receptor that is coupled by a G-protein to phosphoinositide hydrolysis (Sladeczek *et al.*, 1985; Sugiyama *et al.*, 1987). The Quis ionotropic receptor has been renamed after AMPA (amino-3-hydroxy-5-methylisoxazole-4-proprionic acid), an agonist which acts selectively at this receptor without activity at the Quis metabotropic receptor. This metabotropic receptor is discussed in greater detail later in relation to two distinctive forms of neurotoxicity: one induced by drugs that activate and one by drugs that inhibit this receptor subtype.

A Glu receptor that is differentially responsive to 2-amino-4-phosphonobutyrate (AP4) has been described in the retina (Miller and Slaughter, 1986) and dentate hippocampal gyrus (Koerner and Cotman, 1981). Relatively little is understood about this receptor and there is no evidence at present that it participates in neurotoxicological processes.

B. EAA Receptors, Genetically Defined

Molecular biology techniques have recently revolutionized our way of viewing and classifying EAA receptors. Recently, one NMDA receptor which appears to have most if not all biological properties of the NMDA receptor channel complex was cloned and sequenced

(Moriyoshi *et al.*, 1991), then four additional NMDA receptor subunits (Monyer *et al.*, 1992; Meguro *et al.*, 1992; Yamazaki *et al.*, 1992) were described which, when combined with the first subunit, modify its functional properties in important ways (Hollmann *et al.*, 1991; Hume *et al.*, 1991, Monyer *et al.*, 1992). Four receptor subunits that are differentially sensitive to AMPA (Boulter *et al.*, 1990; Keinanen *et al.*, 1990), five that are differentially sensitive to kainic acid (Bettler *et al.*, 1990; Egebjerg *et al.*, 1991; Werner *et al.*, 1991), and five glutamate metabotropic receptor subunits (Masu *et al.*, 1991; Houamed *et al.*, 1991; Watson and Abbot, 1991; Abe *et al.*, 1992) have been discovered. It remains to be determined how these various subunits are combined into assemblies that comprise functional receptors in the mammalian CNS and how many different assemblies are expressed and/or are functionally active in any given CNS region at any given time in ontogenesis. However, it is quite likely that the heteromeric composition of a given EAA receptor complex will prove to be an important determinant of the type(s) of neurotoxicity it expresses. For example, it is likely that receptors during development might function differently from more mature receptors by virtue of a developmentally unique heteromeric assembly pattern or by virtue of certain receptor subunits being present at one age but not another.

III. Excitotoxins and Neurodegenerative Diseases

Given the abundance of excitotoxins in the environment, the high concentration of these agents in the CNS, their intrinsic neurotoxic potential, and the several mechanisms by which such potential might be expressed, excitotoxins are logical candidates for complicity in neurodegenerative conditions. Over the past decade, evidence has begun to accumulate for the involvement of EAA and an excitotoxic mechanism in the pathophysiology of a wide variety of neurological disorders, including food poisoning (neuroendocrinopathies, motor neuron disorders, and/or dementias), sulfite oxidase deficiency, olivopontocerebellar degeneration, amyotrophic lateral sclerosis, epilepsy, hypoglycemia, hypoxia/ischemia, CNS trauma, dementia pugilistica, Huntington's disease, Alzheimer's disease, and parkinsonism (reviewed in Choi, 1988; Olney, 1989).

IV. Excitotoxin Antagonists

Evidence implicating excitotoxins in neurodegenerative disorders stimulated interest in the possibility that EAA antagonists might prove valuable as neuroprotective agents in clinical neurology. The first generation of EAA antagonists identified were competitive NMDA antagonists which compete with NMDA agonists for binding at NMDA receptors. The most powerful antiexcitotoxic drugs identified thus far are noncompetitive NMDA antagonists which act at phencyclidine receptors to block both the excitatory and toxic actions of NMDA. MK-801, a drug developed by Merck, Sharp, and Dohme, is the most potent known compound in this category (Olney *et al.*, 1987a). Certain currently marketed drugs, including dextromethorphan (Goldberg *et al.*, 1987) and several anti-parkinsonian agents (Olney *et al.*, 1987b), are moderately potent noncompetitive NMDA antagonists. Mixed antagonists, such as kynurenic acid and 7-chloro-kynurenic acid, block the excitotoxic effects of both NMDA and non-NMDA agonists, but block the former more effectively than the latter. These compounds are of particular interest for their ability to block the excitotoxic effects of NMDA noncompetitively by an action at the strychnine-insensitive glycine site (Kemp *et al.*, 1987). Ifenprodil is a noncompetitive NMDA antagonist that blocks NMDA receptor function by an action at the polyamine site (Carter *et al.*, 1988). The quinoxalinediones, CNQX, DNQX, and NBQX, have the important distinction of being the first agents found to block the excitatory (Honore *et al.*, 1988; Sheardown *et al.*, 1990) and excitotoxic (Mosinger *et al.*, 1991) actions of non-NMDA agonists more powerfully than they block those of NMDA. A new family of 2,3-benzodiazepine molecules, of which GYKI 52466 is prototypic, represents the first noncompetitive antagonists of non-NMDA receptors and the only known agents that selectively block KA and Quis receptors without any blocking activity at NMDA receptors (Olney *et al.*, 1992; Zorumski *et al.*, 1993).

V. Neurotoxicity Syndromes Associated with EAA Receptors

A. Classical Excitotoxicity

"Excitotoxicity" refers to the paradoxical property, shared by Glu and specific excitatory amino acid analogs, of causing acute neuronal degeneration by excessive stimulation of postsynaptic EAA ionotrophic receptors—receptors through which Glu functions physiologically as a transmitter. The precise mechanism(s) by which excessive EAA receptor activation leads to acute neuronal death is not well understood, although increased membrane permeability and abnormal Na^+, Cl^-, and Ca^{2+} influx are assumed to play important roles. Two decades ago, when the excito-toxic concept was first advanced, the EAA receptor was a taxonomically undefined and purely hypothetical entity, and Glu was not recognized, as it is today, as the natural transmitter released at the majority of excitatory synapses in the mammalian CNS. In the late 1970s, three subtypes of the EAA receptor (NMDA, Quis, and KA) were pharmacologically characterized, all of which are ionotropic receptors, i.e., receptors linked to ion channels, and all of which mediate classical excitotoxic reactions. It is believed that most of the neurons in the CNS possess one or more subtype of the EAA ionotropic receptor on their dendritic or somal surfaces, which makes them vulnerable to excitotoxic degeneration and which explains why the degenerative process conspicuously involves dendrosomal but not axonal portions of the neuron.

Classical excitotoxity is the cytopathological process by which subcutaneously administered Glu destroys neurons in certain brain regions that lack blood–brain barriers and by which various EAA agonists produce experimental ''axon-sparing'' lesions when injected directly into specific brain regions; it is also the mechanism by which Glu or related EAA are thought to participate in various neurodegenerative conditions. A major impetus in understanding how the excitotoxic potential of endogenous Glu may be unleashed to cause neuronal degeneration came with the discovery (Benveniste *et al.*, 1984; Rothman, 1984) that conditions such as hypoxia/ischemia and head trauma cause a marked outpouring of Glu and aspartate from the intracellular to the extracellular compartment of brain where they interact with EAA receptors to trigger an excitotoxic cascade (increased excitation begets increased excitotoxin release which begets more excitation, and so forth). Alternatively, recent evidence (Henneberry *et al.*, 1989; Beal *et al.*, 1993) suggests that energy deficiency, or related adverse conditions, can compromise an inhibitory mechanism (the voltage-dependent Mg^{2+} blockade) by which the NMDA receptor is held in check, thereby allowing even normal physiological concentrations of Glu to induce an excitotoxic reaction.

The discovery of numerous EAA receptor antagonists that block both the excitatory and the neurotoxic actions of EAA agonists has provided ample confirmation of the excitatory nature of Glu neurotoxicity and has prompted efforts to develop anti-excitotoxic drugs for neurotherapeutic purposes. Anti-excitotoxins have not been tested for their ability to prevent neuronal degeneration in human populations; however, in *in vivo* animal studies they have been shown to protect against brain damage associated with several conditions, including status epilepticus, hypoxia/ischemia, hypoglycemia, and CNS trauma (Choi, 1988; Olney, 1989).

B. New EAA-Related Experimental Neurotoxic Syndromes

1. NMDA Antagonist Neurotoxicity

It has been demonstrated that both competitive and noncompetitive NMDA antagonists have neurotoxic side effects consisting of acute neurodegenerative changes in pyramidal neurons of the posterior cingulate and retrosplenial cortices (Olney *et al.*, 1989). The toxic reaction is reversible at low doses, but neurons are killed at higher doses (Allen and Iversen, 1990; Fix *et al.*, 1993; Wozniak *et al.*, 1993). The reversible reaction is accompanied by the induction of a 72-kDa heat shock protein (hsp) localized to the same cingulate/retrosplenial neurons that manifest pathomorphological changes (Sharp, *et al.*, 1991; Olney *et al.*, 1991). Both the pathomorphological and hsp responses to NMDA antagonists can be prevented by either anticholinergic agents or GABAergic agents (benzodiazepines and certain barbiturates) (Olney *et al.*, 1991). To explain these findings, an indirect mechanism operating through a trisynaptic circuit has been proposed: Glu acting at a NMDA receptor on a GABAergic interneuron normally maintains tonic inhibitory control over the release of acetylcholine at a muscarinic receptor on the surface of the cingulate neuron; blockade of the NMDA receptor abolishes the inhibitory control over acetylcholine release and subjects the cingulate neuron to a state of persistent cholinergic hyperstimulation, which putatively is the proximate cause of the pathomorphological and hsp reactions in the cingulate neuron. Thus, NMDA antagonist neurotoxicity can be considered an excitotoxicity syndrome, but it is a new form of excitotoxicity that is triggered by the inhibition (rather than excitation) of Glu receptors and must be classified as cholinergic (rather than glutamatergic) excitotoxicity. An interesting implication is that either hyperstimulation or suppression of a specific ionotropic Glu receptor, the NMDA receptor, can result in injury and/or the destruction of CNS neurons. Another interesting implication is that both of the major excitatory transmitters in the mammalian CNS (Glu and acetylcholine) are capable of exciting neurons to death.

2. Metabotropic Receptor Neurotoxicity

Other evidence implicates the Glu metabotropic receptor in two types of neurotoxic reaction: one associated with suppression and the other with excessive activation of this receptor subtype.

a. Neurotoxicity Associated with Suppression of Metabotropic Receptor Function When the metabotropic receptor is activated physiologically by Glu or experimentally by quisqualate or the more selective agonist, *trans*-ACPD [*trans*-(±)-1-amino-1,3-cyclopentanedicarboxylic acid], it causes phosphoinositide hydrolysis (PiH) which triggers mobilization of intracellular Ca^{2+} stores and activation of protein kinase C, which mediates phosphorylation of proteins that may be vitally important for both the function and survival of CNS neurons. Schoepp and colleagues (1991) have shown that AP3 (2-amino-3-phosphonopropionate) blocks ACPD-stimulated PiH in hippocampal slices and that daily subcutaneous administration of AP3 to infant rats for several consecutive days causes rats to grow up with almost a complete absence of optic nerves (Tizzano *et al.*, 1990; Fix *et al.*, 1993b). The primary impact of AP3 on the developing visual system is on the retina which shows gradual evolution of cytopathological changes over a 2- to 3-day period and total degeneration of the developing retina after five to seven daily treatments (Tizzano *et al.*, 1991; Price *et al.*, 1992). In the brain, toxic changes were usually limited to regions that lack blood–brain barriers, but in some animals there was a more generalized pattern of damage affecting neurons scattered throughout much of the neuraxis. Tentatively, these findings are interpreted as: The immature retina lacks blood–retina barriers so the entire organ, following subcutaneous administration of AP3, is flooded with the compound, and this provides an *in vivo* testing ground that demonstrates what happens to CNS neurons if they are continuously exposed during critical periods of development to an agent that blocks the physiological action of Glu at its metabotropic receptor. As mentioned earlier, at this receptor, Glu stimulates a sequence of second messenger functions which may be vital for maturation and survival of the neuron.

b. Neurotoxicity Associated with Hyperactivation of the Metabotropic Receptor Other experiments (Price *et al.*, 1992) have revealed that the specific isomer of *trans*-ACPD (1S,3R) that acts as a pure agonist in driving the Glu metabotropic receptor and stimulating its G-protein-linked second messenger functions induces pathomorphological changes when injected directly into the lateral ventricle of either the immature or adult rat brain. These changes bear some resemblance to those associated with the blockade of NMDA receptors, and it appears that only a select population of neurons is susceptible, namely certain large neurons in the lateral septal nucleus.

It is noteworthy that the second messenger functions of the Glu (Quis) metabotropic receptor are identical to

those that have been described for the M1 cholinergic muscarinic receptor which has been postulated to be involved in the cingulate cortical neurotoxicity syndrome induced by NMDA antagonists. Thus, the findings are consistent with the hypothesis that excessive stimulation of G-protein linked second messenger systems may be involved in both ACPD and NMDA antagonist neurotoxicity, despite the fact that the reaction is mediated through a cholinergic receptor on cerebrocortical neurons in one case and a Glu receptor on lateral septal neurons in the other.

C. Implications for Drug Testing

It is both desirable and necessary to have standard screening approaches by which environmental agents can be screened for safety. However, inevitably the method of testing must be tailored to the task, i.e., it must be designed to be both sensitive and relevant to the specific types of neurotoxic effects to be expected from a given agent.

Concerning Glu itself, the testing protocol must address the following facts: that Glu freely enters circumventricular (CVO) brain regions that lack blood–brain barriers (Perez and Olney, 1972); that it only requires ingestion of Glu on a one-time basis for CVO neurons to be rapidly destroyed (Olney and Ho, 1970; Olney *et al.*, 1980); that blood Glu levels rise much higher in humans than in any other known species following a given oral load of Glu (Stegink *et al.*, 1979); that Glu blood levels rise much higher following intake of Glu in liquid preparations than when mixed in solid foods; that humans ingest Glu-laden liquid preparations together with other Glu-laden foods in a bolus fashion (the whole dose being ingested within a short-term interval) instead of in a rodent style by *ad lib* nibbling throughout the 24-hr period; that CNS neurons, including CVO neurons, possess NMDA receptors that render them hypersensitive to Glu neurotoxicity during development (Olney, 1976; Ikonomidou *et al.*, 1989; Wang *et al.*, 1990); and that subtoxic doses of Glu which do not produce brain damage can nevertheless influence growth and development by inducing erratic perturbations in blood levels of growth and gonadotrophic hormones (Olney *et al.*, 1976; Olney and Price, 1980; Plant *et al.*, 1989).

It follows from the just-mentioned discussion that agents suspected of having Glu-like properties should be tested by protocols aimed particularly at evaluating the potential of the agent for inducing adverse effects during development and that differences between the animal species used and humans must be carefully considered. For example, in order to clarify the potential toxic interaction with neuroendocrine regulatory systems, it is desirable to conduct the studies in subhuman primates because they are more humanoid in neuroendocrine regulatory mechanisms than are other species. However, it is essential to recognize that monkeys are not at all humanoid in gastrointestinal absorption and/or metabolism of Glu following oral intake. Since monkeys do not display high blood Glu elevations following even massive oral doses of Glu, whereas humans develop high blood levels from relatively low oral doses (Steginck *et al.*, 1979), the appropriate protocol would be one in which Glu is administered parenterally to monkeys in doses required to elicit the specific blood Glu levels that would be induced by various oral doses in humans, then measure the perturbations that such dosing causes in neuroendocrine systems at any given age during monkey ontogenesis. Interestingly, even though the toxic interactions between Glu and neuroendocrine systems were first described in the late 1960s, no such primate studies have ever been done. Yet Glu continues to be one of the most widely and heavily used food additives in the world.

Other important features of Glu neurotoxicity testing include the administration of antagonists specific for various subtypes of Glu receptor to clarify which receptor subtypes mediate each aspect of the Glu neurotoxic syndrome. However, it goes without saying that more sophisticated experiments of this kind would be futile if superimposed upon a basic protocol that does not take into consideration the fundamentals just addressed.

It is not safe to assume that all excitotoxins will have all of the same properties as Glu. For example, L-cysteine mimics Glu in being an excitotoxin that acts at NMDA receptors to destroy neurons, especially in the immature brain, but it differs markedly from Glu in that it freely penetrates not only gastrointestinal but placental and blood–brain barriers to induce widespread excitotoxic damage throughout the forebrain of infant or fetal animals (Olney *et al.*, 1972, 1990). Obviously, a drug testing protocol aimed only at evaluating Glu-like neurotoxic interactions with CVO neurons would be inappropriate for agents that share these unique properties of L-cysteine.

Evaluating the neurotoxic potential of agents that block Glu receptors promises to be an even more complicated challenge than evaluating the neurotoxic properties of Glu agonist analogs. For example, since agents that act at the PCP receptor within the NMDA receptor ion channel are the most potent NMDA antagonists known and are able to freely penetrate blood–brain barriers, they are of great potential interest as neuroprotective agents in acute brain injury syndromes or in more chronic conditions such as epilepsy. However, evaluating the neurotoxic potential of such agents must

first take into conderation the several important physiological functions that NMDA Glu receptors normally perform, including regulation of neuroendocrine parameters such as somatic growth, reproductive functions such as onset of puberty and ovarian cycling, and cognitive functions such as memory. The role of NMDA receptors in other cognitive functions is suggested by the fact that blockade of NMDA receptors by either competitive or noncompetitive NMDA antagonists induces psychotic symptoms in normal subjects. The situation becomes more complicated by the discovery that NMDA antagonists cause pathomorphological changes in pyramidal neurons of the cingulate and retrosplenial cortices (Olney *et al.*, 1989). The more this interesting neurotoxic syndrome has been studied the more complicated it has become. Interesting features observed so far that have bearing on the design of neurotoxicity testing protocols include an apparent hypersensitivity of adult nonpregnant females (Olney *et al.*, 1989; Wozniak *et al.*, 1993), a lack of susceptibility of immature rats until they are almost fully mature (Farber *et al.*, 1992), and the tendency of the neurotoxic reaction to be reversible following low doses but irreversible following high doses (Allen and Iversen, 1990; Fix *et al.*, 1993; Wozniak *et al.*, 1993). The latter point is particularly important because there is an interaction among dose, age, and sex such that adult female nonpregnant rats are so sensitive that they sustain considerable destruction of neurons following even low doses that are only slightly above threshold for inducing a reversible reaction in males. Other features of the neurotoxic syndrome are that the acute vacuole reaction can most readily be detected 4–6 hr following treatment in thin (1 μm) plastic histological sections, whereas the neuron-necrosing action is most readily detected 4–7 days following treatment in routine paraffin-embedded H&E-stained sections. Although these may seem like very elementary methodological points, they are important enough that failure to be guided by them will cause one to miss detecting either the acute vacuole reaction or the cell killing effect. Finally, it is important to recognize that the pathomorphological changes are accompanied by an abnormal expression of heat shock protein which, under some treatment circumstances, can be detected in neurons throughout several regions of the forebrain (Sharp *et al.*, 1993) in addition to the cingulate and retrosplenial cortices.

VI. Summary

Accumulating evidence suggests that either excessive activation or suppression of either ionotropic or metabotropic EAA receptors can have neurotoxic consequences and that a variety of different mechanisms may be involved. The distinctive features of each type of mechanism must be taken into consideration in devising protocols for toxicological evaluation of environmental agents suspected of interacting with EAA receptor systems.

References

Abe, T., Sugihara, H., Nawa, H., Shigemoto, R., Mizuno, N., and Nakanishi, S. (1992). Molecular characterization of a novel metabotropic glutamate receptor mGluR5 coupled to inositol phosphate/Ca^{2+} signal transduction. *J. Biol. Chem.* **267**, 13361–13368.

Allen, H. L., and Iversen, L. L. (1990). Phencyclidine, dizocilpine and cerebrocortical neurons. *Science* **247**, 221.

Beal, M. F., Hyman, B. T., and Koroshetz, W. (1993). Do defects in mitochondrial energy metabolism underlie the pathology of neurodegenerative disease? *Trends Neurosci.* **16**, 125–131.

Benveniste, H., Drejer, J., Schousboe, A., and Diemer, N. M. (1984). Elevation of the extracellular concentrations of glutamate and aspartate in rat hippocampus during transient cerebral ischemia monitored by intracerebral microdialysis. *J. Neurochem.* **43**, 1369–1374.

Bettler, B., Boulter, J., Hermans-Borgmeyer, I., O'Shea-Greenfield, A., Deneris, E. S., Moll, C., Borgmeyer, U., Hollmann, M., and Heinemann, S. (1990). Cloning of a novel glutamate receptor subunit, GluR5: expression in the nervous system during development. *Neuron* **5**, 583–595.

Bolden, C., Cusack, B., and Richelson, E. (1992). Antagonism by antimuscarinic and neuroleptic compounds at the five cloned human muscarinic cholinergic receptors expressed in Chinese hamster ovary cells. *J. Pharm. Exp. Ther.* **260**, 576–580.

Boulter, J., Hollmann, M., O'Shea-Greenfield, A., Hartley, M., Deneris, E., Maron, C., and Heinemann, S. (1990). Molecular cloning and functional expression of glutamate receptor subunit genes. *Science* **249**, 1033–1037.

Carter, C., Benavides, J., Legendre, P., Vincent, J. D., Noel, F., Thuret, F., Lloyd, K. G., Arbilla, S., Zivkovic, B., MacKenzie, E. T., Scatton, B., and Langer, S. Z. (1988). Ifenprodil and SL 82.0715 as cerebral anti-ischemic agents. II. Evidence for *N*-methyl-D-aspartate receptor antagonist properties. *J. Pharmacol. Exp. Ther.* **247**, 1222–1232.

Choi, D. W., Koh, J., and Peters, S. (1988). Pharmacology of glutamate neurotoxicity in cortical cell culture: attenuation by NMDA antagonists. *J. Neurosci.* **8**, 185–196.

Choi, D. W. (1988). Glutamate neurotoxicity and diseases of the nervous system. *Neuron* **1**, 623–634.

Curtis, D. R., and Watkins, J. C. (1963). Acidic amino acids with strong excitatory actions on mammalian neurons. *J. Physiol.* **166**, 1–14.

Dingledine, R. (1983). *N*-methylaspartate activates voltage-dependent calcium conductance in rat hippocampal pyramidal cells. *J. Physiol.* **343**, 385–405.

Egebjerg, J., Bettler, B., Hermans-Borgmeyer, I., and Heinemann, S. (1991). Cloning of a cDNA for a glutamate receptor subunit activated by kainate but not AMPA. *Nature* **351**, 745–748.

Farber, N. B., Price, M. T., Labruyere, J., Fuller, T. A., and Olney, J. W. (1992). Age dependency of NMDA antagonist neurotoxicity. *Neurosci. Abstr.* **18**, 1148.

Farber, N. B., Price, M. T., Labruyere, J., Nemnich, J., St. Peter, H., Wozniak, D. F., and Olney, J. W. (1993). Antipsychotic drugs block phencyclidine receptor-mediated neurotoxicity. *Biol. Psychiatry* **34**, 119–121.

Fix, A. S., Horn, J. W., Wightman, K. A., Johnson, C. A., Long, G. G., Storts, R. W., Farber, N., Wozniak, D. F., and Olney, J. W. (1993). Light and electronmicroscopic evaluation of neuronal vacuolization and necrosis induced by the non-competitive *N*-methyl-D-aspartate (NMDA) antagonist MK(+)801 (Dizocilpine maleate) in the rat retrosplenial cortex. *Exp. Neurol.* **123**, 204–215.

Goldberg, M. P., Viseskul, V., and Choi, D. W. (1988). Phencyclidine receptor ligands attenuate cortical neuronal injury after *N*-methyl-D-aspartate exposure or hypoxia. *J. Pharmacol. Exp. Ther.* **245**, 1081–1086.

Henneberry, R. C. (1992). Minireview: cloning of the genes for excitatory amino acid receptors. *Bioessays* **14**, 465–471.

Henneberry, R. L., Novelli, A., Cox, J. A., and Lysko, P. G. (1989). Neurotoxicity at the *N*-methyl-D-aspartate receptor in energy-compromised neurons. An hypothesis for cell death in aging and disease. *Ann. N.Y. Acad. Sci.* **568**, 225–233.

Hollmann, M., Hartley, M., and Heinemann, S. F. (1991). Ca^{2+} permeability of KA-AMPA-gated glutamate receptor channels depends on subunit composition. *Science* **252**, 851–853.

Honore, T., Davies, S. N., Drejer, J., Fletcher, E. J., Jacobson, P., Lodge, D., and Nielsen, F. E. (1988). Quinox-alinediones: potent competitive non-NMDA glutamate receptor antagonists. *Science* **241**, 701–703.

Houamed, K. M., Kuijper, J. L., Gilbert, T. L., Haldeman, B. A., O'Hara, P. J., Mulvihill, E. R., Almers, W., and Hagen, F. S. (1991). Cloning, expression, and gene structure of a G protein-coupled glutamate receptor from rat brain. *Science* **252**, 1318–1321.

Hume, R. I., Dingledine, R., and Heinemann, S. F. (1991). Identification of a site in glutamate receptor subunits that controls calcium permeability. *Science* **253**, 1028–1031.

Ikonomidou, C., Mosinger, J. L., Shahid Salles, K., Labruyere, J., and Olney, J. W. (1989). Sensitivity of the developing rat brain to hypobaric/ischemic damage parallels sensitivity to N-Methyl-Aspartate neurotoxicity. *J. Neurosci.* **9**, 2809–2818.

Johnson, J. W., and Ascher, P. (1987). Glycine potentiates the NMDA response in cultured mouse brain neurons. *Nature* **325**, 529–531.

Keinanen, K., Wisden, W., Sommer, B., Werner, P., Herb, A., Verdoorn, T. A., Sakmann, B., and Seeburg, P. H. (1990). A family of AMPA-selective glutamate receptors. *Science* **249**, 556–558.

Kemp, J. A., Foster, A. C., and Wong, E. H. F. (1987). Non-competitive antagonists of excitatory amino acid receptors. *TINS* **10**, 294–299.

Koerner, J. F., and Cotman, C. W. (1981). Micromolar L-2-amino-4-phosphonobutyric acid selectively inhibits perforant path synapses from lateral entorhinal cortex. *Brain Res.* **216**, 192–198.

Lodge, D., and Anis, N. A. (1982). Effects of phencyclidine on excitatory amino acid activation of spinal interneurons in the cat. *Eur. J. Pharmacol.* **77**, 203–204.

Lodge, D., Aram, J. A., Church, J., Davies, S. N., Martin, D., O'Shaughnessy, C. T., and Zeman, S. (1987). Excitatory amino acids and phencyclidine-like drugs. In *Excitatory Amino Acid Transmission* (T. P. Hicks, D. Lodge, and H. McLennan, Eds.), pp. 83–90, Liss, New York.

Lucas, D. R., and Newhouse, J. P. (1957). The toxic effect of sodium L-glutamate on the inner layers of the retina. *AMA Arch. Ophthalmol.* **58**, 193–201.

MacDermott, A. B., Mayer, M. L., Westbrook, G. L., Smith, S. J., and Barker, J. L. (1986). NMDA-receptor activation increases cytoplasmic calcium concentration in cultured spinal cord neurones. *Nature* **321**, 519–522.

MacDonald, J. F., Miljkovic, A., and Pennefather, P. (1987). Use-dependent block of excitatory amino acid current in cultured neurons by ketamine. *J. Neurophysiol.* **58**, 251–267.

MacDonald, R. L., and Barker, J. L. (1978). Different actions of anticonvulsant and anesthetic barbiturates revealed by use of cultured mammalian neurons. *Science* **200**, 775–777.

Magbagbeola, J. A. O., and Thomas, N. A. (1974). Effect of thiopentone on emergence reactions to ketamine anaesthesia. *Can. Anaesth. Soc. J.* **21**, 321–324.

Masu, M., Tanabe, Y., Tsuchida, K., Shigemoto, R., and Nakanishi, S. (1991). Sequence and expression of a metabotropic glutamate receptor. *Nature* **349**, 760–765.

Mayer, M. L., Westbrook, G. L., and Guthrie, P. B. (1984). Voltage-dependent block by Mg^{2+} of NMDA responses in spinal cord neurones. *Nature* **309**, 261–263.

Meguro, H., Mori, H., Araki, K., Kushiya, E., Kutsuwada, T., Yamazaki, M., Kumanishi, T., Arakawa, M., Sakimura, K., and Mishina, M. (1992). Functional characterization of a heteromeric NMDA receptor channel expressed from cloned cDNAs. *Nature* **357**, 70–74.

Miller, R. F., and Slaughter, M. M. (1986). Excitatory amino acid receptors of the retina: diversity of subtypes and conductance mechanisms. *Trends Neurosci.* **9**, 211–218.

Monyer, H., Sprengel, R., Schoepfer, R., Herb, A., Higuchi, M., Lomeli, H., Burnashev, N., Sakmann, B., and Seeburg, P. H. (1992). Heteromeric NMDA receptors: molecular and functional distinction of subtypes. *Science* **256**, 1217–1221.

Moriyoshi, K., Masu, M., Ishii, T., Shigemoto, R., and Mizuno, N. (1991). Molecular cloning and characterization of the rat NMDA receptor. *Nature* **354**, 31–37.

Mosinger, J. L., Price, M. T., Bai, H. Y., Xiao, H., Wozniak, D. F., and Olney, J. W. (1991). Blockade of both NMDA and non-NMDA receptors is required for optimal protection against ischemic neuronal degeneration in the in vivo adult mammalian retina. *Exp. Neurol.* **113**, 10–17.

Olney, J. W. (1969a). Glutamate-induced retinal degeneration in neonatal mice. Electron microscopy of the acutely evolving lesion. *J. Neuropathol. Exp. Neurol.* **28**, 455–474.

Olney, J. W. (1969b). Brain lesions, obesity and other disturbances in mice treated with monosodium glutamate. *Science* **164**, 719–721.

Olney, J. W. (1971). Glutamate-induced neuronal necrosis in the infant mouse hypothalamus: an electron microscopic study. *J. Neuropathol. Exp. Neurol.* **30**, 75–90.

Olney, J. W. (1974). Toxic effects of glutamate and related amino acids on the developing central nervous system. In *Heritable Disorders of Amino Acid Metabolism* (W. H. Nyhan, Ed.), pp. 501–512, John Wiley, New York.

Olney, J. W. (1976). Brain damage and oral intake of certain amino acids. In *Transport Phenomena in the Nervous System: Physiological and Pathological Aspects* (G. Levi, L. Battistin and A. Lajtha, Eds.), Plenum, New York, pp. 497–506.

Olney, J. W. (1980). Excitatory neurotoxins as food additives: an evaluation of risk. *Neurotoxicology* **2**, 163–192.

Olney, J. W. (1989). Excitatory amino acids and neuropsychiatric disorders. *Biol. Psychiatry* **26**, 505–525.

Olney, J. W., and Ho, O.L. (1970). Brain damage in infant mice following oral intake of glutamate, aspartate or cysteine. *Nature* **227**, 609–610.

Olney, J. W., and Price, M. T. (1980). Neuroendocrine interactions of excitatory and inhibitory amino acids. *Brain Res. Bull.* **5**, (Suppl 2), 361–368.

Olney, J. W., Ho, O. L., and Rhee, V. (1971). Cytotoxic effects of acidic and sulphur-containing amino acids on the infant mouse central nervous system. *Exp. Brain Res.* **14**, 61–76.

Olney, J. W., Ho, O. L., Rhee, V., and Schainker, B. (1972). Cysteine-induced brain damage in infant and fetal rodents. *Brain Res.* **45**, 309–313.

Olney, J. W., Cicero, T. J., Meyer, E. R., and deGubareff, T. (1976). Acute glutamate-induced elevations in serum testosterone and luteinizing hormone. *Brain Res.* **112**, 420–424.

Olney, J. W., deGubareff, T., and Labruyere, J. (1979). a-Aminoadipate blocks the neurotoxic action of N-methylaspartate. *Life Sci.* **25**, 537–540.

Olney, J. W., Labruyere, J., and deGubareff, T. (1980). Brain damage in mice from voluntary ingestion of glutamate and aspartate. *Neurobehav. Toxicol.* **2**, 125–129.

Olney, J. W., Labruyere, J., Collins, J. F., and Curry, K. (1981). D-Aminophosphonovalerate is 100-fold more powerful than D-alpha-aminoadipate in blocking N-methylaspartate neurotoxicity. *Brain Res.* **221**, 207–210.

Olney, J. W., Price, M. T., Fuller, T. A., Labruyere, J., Samson, L., Carpenter, M., and Mahan, K. (1986). The anti-excitotoxic effects of certain anesthetics, analgesics and sedative-hypnotics. *Neurosci. Lett.* **68**, 29–34.

Olney, J., Price, M., Shahid Salles, K., Labruyere, J., and Frierdich, G. (1987a). MK-801 powerfully protects against N-methyl aspartate neurotoxicity. *Eur. J. Pharmacol.* **141**, 357–361.

Olney, J. W., Price, M. T., Labruyere, J., Shahid Salles, K., Frierdich, G., Mueller, M., and Silverman, E. (1987b). Anti-parkinsonian agents are phencyclidine agonists and N-methyl aspartate antagonists. *Eur. J. Pharmacol.* **142**, 319–320.

Olney, J. W., Labruyere, J., and Price, M. T. (1989). Pathological changes induced in cerebrocortical neurons by phencyclidine and related drugs. *Science* **244**, 1360–1362.

Olney, J. W., Zorumski, C., Price, M. T., and Labruyere, J. (1990). L-Cysteine, a bicarbonate-sensitive endogenous excitotoxin. *Science* **248**, 596–599.

Olney, J. W., Labruyere, J., Wang, G., Sesma, M. A., Wozniak, D. F., and Price, M. T. (1991). NMDA antagonist neurotoxicity: mechanism and Protection. *Science* **254**, 1515–1518.

Olney, J. W., Sesma, M. A., and Wozniak, D. F. (1993). Glutamatergic GABAergic and cholinergic systems in posterior cingulate cortex: Interactions and possible mechanisms of limbic system disease. In *The Neurobiology of Cingulate Cortex and Limbic Thalamus* (B. A. Vogt and M. Gabriel, Eds.), Birkhäuser, Boston, pp. 557–580.

Perez, J. V., and Olney, J. W. (1972). Accumulation of glutamic acid in arcuate nucleus of infant mouse hypothalamus following subcutaneous administration of the amino acid. *J. Neurochem.* **19**, 1777–1781.

Plant, T. M., Gray, V. L., Marshall, G. R., and Arslan, M. (1989). Puberty in monkeys is triggered by chemical stimulation of the hypothalamus. *Proc. Natl. Acad. Sci.* **86**, 2506–2510.

Price, M. T., Ikonomidou, C., Labruyere, J., Izumi, Y., and Olney, J. W. (1992). Neurotoxicity linked to the glutamate metabotropic receptor. *Neurosci. Abstr.* **18**, 83.

Ransom, R. W., and Stec, N. L. (1988). Cooperative modulation of MK-801 binding to the N-methyl-D-aspartate receptor-ion channel complex by L-glutamate, glycine, and polyamines. *J. Neurochem.* **51**, 830–836.

Reynolds, I. J., and Miller, R. J. (1989). Ifenprodil is a novel type of N-methyl-D-aspartate receptor antagonist: interaction with polyamines. *Mol. Pharmacol.* **36**, 758–765.

Romano, C., Sesma, M. A., McDonald, C. T., Price, M. T., and Olney, J. W. (1993). Immunochemical studies of a metabotropic glutamate receptor (mGluR5) in rat brain. *Soc. Neurosci. Abstr.* **19**, 926.

Rothman, S. M. (1984). Synaptic release of excitatory amino acid neurotransmitter mediates anoxic neuronal death. *J. Neurosci.* **4**, 1884–1891.

Schoepp, D. D., and Johnson, B. G. (1991). In vivo 2-amino-3-phosphonopropionic acid administration to neonatal rats selectively inhibits metabotropic excitatory amino acid receptors ex vivo in brain slices. *Neurochem. Int.* **18**, 411–417.

Sharp, F. R., Jasper, P., Hall, J., Noble, L., and Sagar, S. M. (1991). MK-801 and ketamine induce heat shock protein HSP72 in injured neurons in posterior cingulate and retrosplenial cortex. *Ann. Neurol.* **30**, 801–809.

Sharp, F. R., Butman, M., Wang, S., Koistinaho, J., Graham, S. J., Noble, L., Sagar, S. M., Berger, P., and Longo, F. M. (1992). Haloperidol prevents induction of the HSP-70 heat shock gene in neurons injured by phencyclidine, MK-801, and ketamine. *J. Neurosci. Res.* **33**, 605–616.

Sharp, F. R., Butman, M., Aardalen, K., Nakki, R., Koistinaho, J., Massa, S., and Sagar, S. M. (1993). Phencyclidine induces the HSP70 heat shock gene in injured cortical neurons via multiple receptors: disinhibition as a mechanism of injury. *Soc. Neurosci. Abstr.* **19**, 1697.

Sheardown, M. J., Nielsen, E. O., Hansen, A. J., Jacobsen, P., and Honore, T. (1990). 2,3-Dihydroxy-6-nitro-7-sulfamoyl-benzo(-f) quinoxaline: a neuroprotectant for cerebral ischemia. *Science* **247**, 571–574.

Sladeczek, F., Pin, J. P., Recasens, M., Bockaert, J., and Weiss, S. (1985). Glutamate stimulates inositol phosphate formation in striatal neurons. *Nature* **317**, 717–719.

Stegink, L. D., Reynolds, W. A., Filer, L. J., Baker, G. L., Daabees, T. T., and Pitkin, R. (1979). Comparative metabolism of glutamate in the mouse, monkey and man. In *Glutamic Acid: Advances in Biochemistry and Physiology* (L. J. Filer, Jr., S. Garratini, M. R. Kare, W. A. Reynolds, and R. J. Wurtman, Eds.), pp. 85–102, Raven Press, New York.

Sugiyama, H., Ito, I., and Hirono, C. (1987). A new type of glutamate receptor linked to inositol phospholipid metabolism. *Nature* **312**, 531–533.

Tizzano, J. P., Bailey, R. L., Engelhardt, J. A., and Schoepp, D. D. (1990). Effect of postnatal D,L-2-amino-3-phosphonoproopionate (D,L-AP3) treatment on physical development, behavior and phosphoinositide hydrolysis in the rat. *Soc. Neurosci. Abstr.* **16**, 547.

Tizzano, J. P., Schoepp, D. D., Price, M. T., and Olney, J. W. (1991). Widespread degeneration induced in the developing rodent CNS by D,L-2-amino-3-phosphonopropionate (AP3). *Neurosci. Abstr.* **17**, 70.

Wang, G. J., Labruyere, J., Price, M. T., and Olney, J. W. (1990). Extreme sensitivity of infant animals to glutamate toxicity: role of NMDA receptors. *Neurosci. Abstr.* **16**, 198.

Watkins, J. C., and Evans, R. H. (1981). Excitatory amino acid transmitters. *Ann. Rev. Pharmacol. Toxicol.* **21**, 165–204.

Watkins, J. C. (1978). Excitatory amino acids. In Kainic Acid as a Tool in Neurobiology (E. McGreer, J. W. Olney, and P. McGreer, Eds.), pp. 37–69, Raven Press, New York.

Watson, S., and Abbot, A. (1990). TiPS receptor nomenclature supplement. *Trends Pharmacol. Sci.* (Suppl), 1–30.

Weissman, A. D., Casanova, M. F., Kleinman, J. E., London, E. D., and DeSouza, E. B. (1991). Selective loss of cerebral cortical sigma, but not PCP binding sites in schizophrenia. *Biol. Psychiatry* **29**, 41–54.

Westbrook, G. L., and Mayer, M. L. (1987). Micromolar concentrations of Zn^{++} antagonize NMDA and GABA responses of hippocampal neurones. *Nature* **328**, 640–643.

Werner, P., Voigt, M., Keinanen, K., Wisden, W., and Seeburg, P. H. (1991). Cloning of a putative high-affinity kainate receptor expressed predominantly in hippocampal CA3 cells. *Nature* **351**, 742–744.

Wozniak, D. F., McEwen, M., Sesma, M. A., Olney, J. W., and Fix, A. S. (1993). MK-801 induces extensive neuronal necrosis in posterior cingulate/retrosplenial cortices. *Neurosci. Abstr.* **19**, 1770.

Yamazaki, M., Mori, H., Araki, K., Mori, K. J., and Mishina, M. (1992). Cloning, expression and modulation of a mouse NMDA receptor subunit. *Fed. Eur. Biochem. Soc.* **300**, 39–45.

Zorumski, C. F., Yamada, K. A., Price, M. T., and Olney, J. W. (1993). A benzodiazepine recognition site associated with the non-NMDA glutamate receptor. *Neuron* **10**, 61–67.

Assessment of Neurotoxicity via Chemical Perturbation of Axonal Transport

MOHAMMAD I. SABRI
Center for Research on Occupational
and Environmental Toxicology,
and Department of Neurology,
Oregon Health Sciences University,
Portland, Oregon 97201

I. Introduction

Neurons, unlike other mammalian cells, have long processes which often extend considerable distances from the cell body. This extraordinary structural makeup poses special problems for intra- and intercommunication for these cells. Although the concept that materials are translocated from the cell body to the axon was initially advanced by the studies of Waller, Cajal, and others, investigations by Weiss and Hiscoe (1948) provided the first experimental evidence of intraaxonal transport by demonstrating the movement of axoplasm (axoplasmic transport) in peripheral nerves. These classical studies provided proof for intracellular movement of materials (axonal transport) from the cell body into the axon. Subsequent experiments by these authors allowed calculation of the rate of axonal transport in peripheral nerves.

Initially, it was thought that the entire axoplasm moves distally at a constant rate. Studies using radioactive precursors have provided evidence that materials in the axon are transported at distinct rates, ranging from a fraction of a millimeter to several hundred millimeters per day (Grafstein and Forman, 1980). Video-enhanced microscope studies of axonal transport have shown that organelles move along microtubules (Allen *et al.*, 1982; Brady *et al.*, 1982; Schnapp and Reese, 1986). It is now universally accepted that axonal transport is a necessary and ongoing function of normal neuronal cells.

Since axonal transport is vital for neuronal integrity, interference by neurotoxicants may lead to profound adverse effects on neuronal structure and function. A number of reviews and books have been published on axonal transport in relation to toxic chemicals and drugs (Grafstein and Forman, 1980; Ochs, 1982; Weiss, 1982; Elam and Cancalon, 1984; Iqbal, 1986; Smith and Bisby 1987; Ochs and Brimijoin, 1993). This chapter summarizes current knowledge of axonal transport and the adverse effect of selected neurotoxic agents.

II. Rates and Mechanisms of Axonal Transport

Axonal transport is now a well-established phenomenon vital for the integrity of the neuron. Materials required for the maintenance of nerve fibers are synthesized in neuronal cell bodies and transported along the axon (*via* anterograde transport) to the sites of utilization (for review *see* Sabri, 1986). Substances are also transported from the axon terminal toward the cell body (*via* retrograde transport) to maintain neuronal structure and function (Bisby, 1987). Although the anterograde transport system has five distinct transport groups (groups I–V) (Grafstein and Forman, 1980), a simple classification of fast anterograde, slow anterograde, and retrograde axonal transport remains useful.

A. Fast Anterograde Transport

Earlier studies using radioactive precursors and subcellular fractionation (see Sabri, 1986) and more recent investigations using Allen's video-enhanced contrast-differential interference contrast (AVEC-DIC) microscope techniques (Brady *et al.*, 1982, Fahim *et al.*, 1985; Allen, 1987) have collectively shown that fast-transported materials are predominantly membrane bound. Membrane-bound enzymes, neurotransmitters, glycoproteins, lipids, amino acids, and Ca^{2+} also undergo fast axonal transport. Intraaxonal transport of acetylcholine esterase and dopamine β-hydroxylase are used as endogenous markers of fast axonal transport (Ochs, 1982). Although fast transport occurs in both the peripheral (PNS) and central nervous systems (CNS), the velocities are somewhat different in each system. For example, mammalian peripheral nerves show a fast transport rate of 410 mm per day regardless of the size of the animal (Ochs, 1982). By comparison, fast transport velocity in the optic system might reach only 250 mm per day (Grafstein and Forman, 1980). In squid giant axons, fast-transported organelles, visualized by AVEC-DIC, move at rates comparable to that in mammalian nerve fibers (Allen, 1987; Allen *et al.*, 1982; Brady *et al.*, 1982; Reese, 1987). The reason for these differences in fast transport rates is unknown. Membrane-bound components with rates of 200–410 mm/day are classified as group I polypeptides (Grafstein and Forman, 1980). Fast-transported materials, including mitochondrial components which are transported at 34–68 mm/day, are termed group II polypeptides. A large number of chemicals and drugs block fast axonal transport resulting in distal axonal degeneration of the central and peripheral nerves known as central–peripheral–distal axonopathy (CPDA).

Axonal transport is an active process driven by ATP (Ochs, 1982; Adams, 1982; Brady, 1985). Ochs has proposed a transport filament model where microtubules, ATP, and Ca^{2+} play key roles in transport mechanisms (Ochs and Brimijoin, 1993). In this model (Fig. 1), ATP is hydrolyzed by a Ca^{2+}-Mg^{2+} ATPase associated with side arms of microtubules to supply energy for transport. Various axonally transported materials bind to carriers cyclically attached to microtubule side arms; ATP is utilized by the side arms to drive the carriers along microtubules (Ochs and Brimijoin, 1993). In this model, all kinds of materials that are bound to the carrier can be transported. The rate of transport is determined by the materials dropping off from the carriers. The presence of Ca^{2+} and calmodulin appears to be necessary since the transport can be blocked with trifluoperazine, an agent that reduces the action of calmodulin (Ekstrom *et al.*, 1987).

In addition to ATP and microtubules, other factors are also necessary for axonal transport. Kinesin, a brain protein with ATPase activity that promotes fast anterograde transport of organelles, is referred to as a "motor protein." 5'-Adenylimidodiphosphate, a nonhydrolyzable ATP analog, inhibits kinesin-ATPase activity and blocks fast axonal transport (Brady, 1985). Another motor protein, cytoplasmic dynein (MAP-1C), promotes retrograde transport. However, the mechanisms by which kinesin and dynein promote axonal transport in the anterograde and retrograde directions are speculative. It has been suggested that these proteins are bound to organelles and allow them to move along microtubules by a cyclic action of cross-bridges

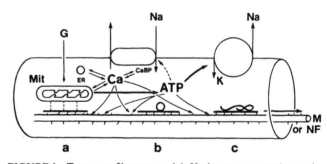

FIGURE 1 Transport filament model: Various transported materials, mitochrondria (a), organelles and vesicles (b), and soluble proteins and peptides (c), are bound to filament carries (thick bars) attached to microtubules with side arms. Glucose (G) is metabolized to generate ATP in the mitochondria (Mit). The ATP is hydrolyzed by ATPase in the side arms to generate chemical energy for axonal transport. The side arms, by cyclically attaching to filament carriers, utilize ATP to drive filament carriers with their cargo along the microtubules (M). ATP also provides energy to the sodium pump controlling the level of Na^+ and K^+. Calcium levels are regulated by mitochondria, endoplasmic reticulum (ER), and calcium-binding proteins (CaBP). Ca^{2+}, in association with calmodulin, activates ATPase. (From Ochs and Brimijoin, 1993.)

between them and microtubules (Vallee and Bloom, 1991). The attachment of motor proteins to organelles moving in anterograde and retrograde directions is illustrated in Fig. 2. In this model, kinesin and cytoplasmic dynein bound to different organelles move along microtubules in an anterograde and retrograde direction, respectively. The direction of movement is determined by the motor protein present. One hypothesis is that both of these motor proteins may be present on the same organelle. Inactivation of kinesin and activation of dynein at the terminal by some accessory factor(s) could regulate anterograde/retrograde transport (Sheetz *et al.*, 1989). However, studies have shown that both anterograde and retrograde transport are blocked when giant axons are injected with the anti-kinesin antibody (Brady *et al.*, 1990).

B. Fast Retrograde Transport

Materials transported down the axon *via* fast anterograde transport turn around at axon terminals or at a site of constriction and proceed back toward the perikaryon by retrograde transport. Retrograde transport plays an important role not only in recycling materials from the cell body to the axon terminal and back again, but also in transferring information about axonal status and terminal environment to the cell body (Bisby, 1987). Materials transported in a retrograde direction are mainly membrane-bound organelles resembling lysosomal structures and are much larger in size than those carried by anterograde transport. Approximately 50% of fast-transported proteins turn around and retrogradely move toward the perikaryon. Retrograde transport is thought to be a mechanism by which nerve growth factor, neurotoxins, and viruses enter the nervous system.

Although most materials in retrograde transport are carried by fast transport, slow retrograde transport of proteins has also been reported in sciatic nerves (Fink and Gainer, 1980). Studies have shown that neurofilaments also undergo retrograde transport following transection of mouse sciatic nerves (Glass and Griffin, 1991). A number of investigators have studied retrograde movement of particles by AVEC-DIC microscopy (Forman, 1983; Reese, 1987; Fahim *et al.*, 1985; Vallee *et al.*, 1989; Vallee and Bloom, 1991). Since the loss of retrogradely transported target signals may be an important factor for neuronal integrity, abnormalities in retrograde transport may play a key role in the pathogenesis of the CPDA. MAP-1C promotes retrograde transport of organelles toward negative (−) ends of microtubules (Vallee *et al.*, 1989). Dynein hydrolyzes ATP to generate energy for retrograde movement (Schnapp and Reese, 1989).

C. Slow Transport

Slow transport carries mainly metabolic enzymes and the proteins of the cytoskeleton (Sabri, 1986; Hammerschlag and Brady, 1989). Quantitatively, the amount of materials transported in slow transport is much larger than in fast axonal transport (Grafstein and Forman, 1980). Long after neuronal cell bodies are injected with protein precursors, a wave of slowly traveling radioactivity (0.2–4.0 mm/day) is observed in the axon (see Grafstein and Forman, 1980). Subcellular fractionation and sodium dodecyl sulfate–polyacrylamide gel electrophoresis (SDS–PAGE) studies show that as much as 50% of labeled proteins carried by slow transport are soluble as opposed to predominantly membrane-associated materials carried by fast transport. The slowest subcomponent of the slow transport, slow component a (SCa) or group V (Grafstein and Forman, 1980; Hoffman *et al.*, 1987), transports mainly neurofilament (NF) and microtubule (MT) protein subunits at a rate of 0.2–1.0 mm/day. While NF, the major 10-nm intermediate filaments found selectively in neurons, as composed of light (NF-L; 68 kDa), medium (NF-M), and heavy (NF-H; 200 kDa) protein subunits, MT are present in all cell types and are composed of α-tubulin (57 kDa); and β-tubulin (53 kDa). The faster subcomponent of slow transport, slow component b (SCb) or group IV, is complex in nature and carries more than 200 proteins and enzymes moving at 2–4 mm/day. A third subcomponent of slow transport, termed "intermediate" or group III, is transported at 4–8 mm/day (see Grafstein and Forman, 1980). Lasek and colleagues (1993) have reported a fast rate of neu-

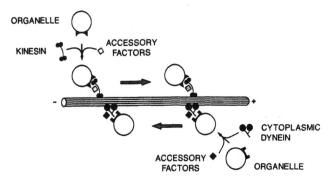

FIGURE 2 The role of kinesin, cytoplasmic dynein, and accessory factors in organelles moving along microtubules. Motor proteins (kinesin or cytoplasmic dynein) and an accessory factor(s) are both required for the transport of organelles. The attachment of kinesin or cytoplasmic dynein and an accessory factor(s) to microtubules activates ATPase activity and hydrolyzes ATP to generate the force necessary to move organelles toward the positive (+) or negative (−) ends of microtubules, respectively. (From Sheetz *et al.*, 1989.)

rofilament transport (72–144 mm/day) in mouse optic axons. This study shows that slow transport machinery is capable of transporting assembled NF about 100 times faster than previously reported (Hoffman and Lasek, 1975). Xu and colleagues (1993) have reported that increased expression of neurofilament subunit NF-L in the ventral horn region of the spinal cord of transgenic mice causes proximal axon swelling, axonal degeneration, and skeletal muscle atrophy similar to human motor neuron disease.

The mechanism of slow axonal transport are not as well understood as those of fast axonal transport. Dynamin, an enzyme with mechanochemical properties, has been suggested to play a role in slow axonal transport (Shpetner and Vallee, 1989). GTP, instead of ATP, may be the physiological substrate for dynamin (Shpetner and Vallee, 1989). The components of SCa (*i.e.*, tubulin and neurofilament triplet proteins) are synthesized in neuronal cell bodies and transported down the axon at a rate of approximately 1 mm/day (Lasek *et al.*, 1984). These investigators proposed a "structural hypothesis" and suggested that both NF and MT are translocated as an assembled cytoskeleton from the cell body to the axon (Lasek *et al.*, 1984) and that the entire MT–NF network is transported as a unit through the axon (Lasek *et al.*, 1992). Support for this has been obtained by the recent photoactivation method showing that fluorescent MT are translocated at rates of 0.9–2.8 mm/day in the axon of cultured *Xenopus* neurons (Reinsch *et al.*, 1991). These authors have visualized the movement of intact MT tagged with a photoactivatable fluorescent compound in growing *Xenopus* axons (Reinsch *et al.*, 1991). However, the concept that the entire cytoskeletal network is transported in the mammalian axon has been challenged. Nixon (1992) has suggested that while MT are mobile in *Xenopus* axons, these structures are stationary in mammalian sensory axons and exhibit exchange with moving tubulin subunits as has been shown by photobleaching experiments (Lim *et al.*, 1989, 1990; Okabe and Hirokawa, 1990). Photobleaching experiments in PC12 cells have shown that MT are indeed quite stationary and are not transported in neurites (Okabe & Hirokawa, 1990; Hollenbeck, 1990). A number of investigators have reported that NF in the axon are also stationary and that cytoskeletal elements are neither assembled in neuronal perikaryon nor transported as assembled structures in the axon (Nixon, 1992). Evidence for the transport of NF subunits required to replenish NF turnover in the axon has been presented (Nixon and Logvinenko, 1986; Nixon, 1991; Weisenburg *et al.*, 1987; Ochs, 1982; Ochs *et al.*, 1989; Ochs and Brimijoin, 1993). The results of these studies are controversial and are at variance with the structural

hypothesis of Lasek and colleagues (1984, 1992). Archer and colleagues (1994) data are consistent with the thesis that NF are largely stationary and that the transported cytoskeletal protein pool may exchange materials with the bulk of the cytoskeleton in the axon. Support of a stationary or near stationary cytoskeleton in peripheral axons has also been reported (Watson *et al.*, 1990, 1993).

III. Axonal Transport Blockers

Defects in axonal transport have been reported in a number of naturally occurring neurodegenerative diseases as well as in experimental animals exposed to a variety of toxic chemicals. Selected neurotoxic agents, which affect neuronal integrity by disrupting axonal transport, are discussed next.

A. Acrylamide

Acrylamide (Fig. 3) is one of the most extensively studied neurotoxic compounds (for reviews *see* (LeQuesne, 1980; Tilson, 1981; Sabri, 1983b; Miller and Spencer, 1985). The neurotoxicity of acrylamide was first recognized among workers engaged in the production of this chemical for the polymer industry. Early signs of acrylamide neuropathy are sensory loss in the fingers followed by hindlimb weakness and paralysis (Spencer and Schaumburg, 1980). Axons display abnormally large numbers of 10-nm neurofilaments and smooth vesicular profiles (Spencer and Schaumburg, 1980), and tubulovesicular structures and dense and multivesicular bodies (Chretien *et al.*, 1981). Neuronal perikaryon display some changes but their degeneration is rarely seen in acrylamide neuropathy (Sterman, 1982a,b, 1983).

Pharmacokinetic studies show that radiolabeled acrylamide is rapidly distributed throughout the body, metabolized to glycidamide (Calleman *et al.*, 1990; Bergmark *et al.*, 1991, 1993), and readily excreted in the urine (Miller *et al.*, 1982). Only small quantities of free acrylamide are detectable in peripheral nerves (Poole *et al.*, 1981) but a large amount of radioactivity is found in the blood (Miller *et al.*, 1982). Biochemical studies have focused on the action of acrylamide on glycolytic enzyme inhibition (Sabri and Spencer, 1980;

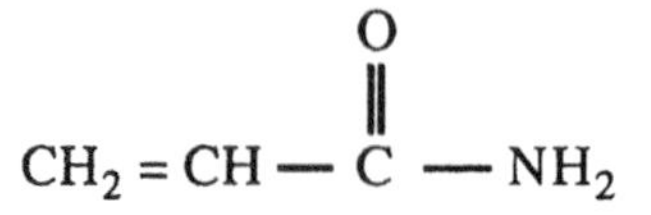

$$CH_2 = CH - \overset{\overset{\textstyle O}{\|}}{C} - NH_2$$

FIGURE 3 Chemical structure of acrylamide.

Howland, 1981; Sabri, 1983a) and sulfhydryl reactivity (Dixit *et al.*, 1981a,b; Srivastava *et al.*, 1986; Shivakumar and Ravindranath, 1992). Inhibition of many sulfhydryl enzymes, such as alcohol dehydrogenase (Dixit *et al.*, 1981a), phosphofructokinase (PFK), and GAPDH, occurs when crystalline or tissue enzymes are treated with acrylamide *in vitro* (Sabri, 1983a; Ravindranath and Pai, 1991). Inhibition of GAPDH and PFK can be prevented by prior addition of the sulfhydryl protectant dithiothreitol (Howland *et al.*, 1980a,b; Howland, 1981; Sabri, 1983a; Ravindranath and Pai, 1991; Shivakumar and Ravindranath, 1992).

Neurochemical studies have shown that GAPDH activity in sciatic nerves is selectively reduced while normal levels are present in brain and liver of animals with acrylamide neuropathy (Sabri, 1983a). Howland (1981) demonstrated that both GAPDH and neuron-specific enolase (NSE) are inhibited by acrylamide and, most significantly, that the inhibition of NSE is higher in areas more distal to the site of axonal degeneration. Although acrylamide inhibits GAPDH and NSE, loss of ATP is not seen (Brimijoin and Hammond, 1985). Acrylamide inhibits oxidative metabolism (Sickles and Goldstein, 1986) and reduces mitochondrial respiration in the cerebellum (Medrano and LoPachin, 1989). Animals are partially protected from the neurotoxic effects of acrylamide by dietary sodium pyruvate, an alternative substrate beyond the position of putative enzyme blockade for the generation of ATP (Sabri *et al.*, 1989). Acrylamide alters phosphorylation of sciatic nerve proteins (Berti-Mattera *et al.*, 1990), alters neurofilament proteins (Howland and Alli, 1986), inhibits glutathione-*S*-transferase, and affects dopamine receptors in rat brain (Srivastava *et al.*, 1986). It has also been demonstrated that acrylamide induces the expression of c-*fos* and c-*jun* (Endo *et al.*, 1993), alters neurofilament protein gene expression (Endo *et al.*, 1994), and depletes MAP-1 and MAP-2 immunoreactivity in rat brain (Chauhan *et al.*, 1993a,b).

Several investigators have examined the effect of acrylamide on axonal transport in laboratory animals. Pleasure and colleagues (1969) found the blockade of axonal transport in acrylamide neuropathy. A modest decrease was reported in slow transport of proteins following a single injection of acrylamide in rats (Gold *et al.*, 1985). Courand and colleagues (1982) reported a 60% decrease in the fast transport of acetylcholinesterase. Wier and colleagues (1978) observed a 68% inhibition of fast transport of proteins in cats exposed to acrylamide. Acrylamide-treated chicks revealed multifocal blockade of fast transport in swollen paranodes and internodes filled with 10-nm neurofilaments and other axoplasmic organelles (Chretien *et al.*, 1981; Souyri *et al.*, 1981).

Sahenk and Mendell (1981) observed that the rate of retrograde transport was reduced along the distal portion of the axon. Similar conclusions were drawn by Jacobsen and co-workers (Jacobsen and Sidenius, 1983; Jacobsen *et al.*, 1983). Miller and associates (1983; Miller and Spencer, 1984) showed that single injections of acrylamide produce a dose-dependent decrease in the retrograde transport of ^{125}I-labeled NGF and tetanus toxin. Moretto and Sabri (1988) observed an early inhibition of retrogradely transported ^{125}I-labeled tetanus toxin in sciatic nerves of hens treated with acrylamide; this inhibition is transient and returns to normal within 48 hr.

Although acrylamide causes no significant change in either fast or slow anterograde axonal transport of proteins in sciatic nerves (Sidenius and Jacobsen, 1983), both fast and slow axonal transport are inhibited by acrylamide in rat optic axons (Sabri and Spencer, 1990). Sickles (1989b, 1991) reported a reduction in both rate and amount of fast-transported proteins in animals treated with acrylamide. Acrylamide reduces phosphorylation of neurofilament proteins in dorsal root ganglion following sciatic nerve transection, presumably by blocking a retrogradely transported "trophic" factor(s) which triggers perikaryal response (Gold *et al.*, 1991, 1993). Acrylamide is able to inhibit fast axonal transport of glycoproteins in peripheral nerves (Harry *et al.*, 1989, 1992; Storm-Dickerson, 1992). However, Brat and Brimijoin (1993) have shown that acrylamide and its metabolite, glycidamide, inhibit neurite outgrowth without impairing bidirectional transport of neuroblastoma cells. These investigators have suggested that mechanisms other than direct inhibition of fast transport are responsible for acrylamide neurotoxicity. Martenson and colleagues (1993) have reported that acrylamide produces no adverse effect on the activity of either dynein or kinesin.

B. Aliphatic Hexacarbon Compounds

n-Hexane and its metabolites, methyl-*n*-butyl ketone (M*n*BK), 2,5-HD, and its 3,4-dimethyl derivative, DMHD (Fig. 4), produce a common pattern of axonal degeneration in the central and peripheral nervous system, referred to as γ-diketone neuropathy (for review, *see* Spencer *et al.*, 1980). The cardinal initial feature of hexacarbon neuropathy is the appearance of giant axonal swellings filled with 10-nm neurofilaments (Spencer *et al.*, 1980). The accumulation of neurofilaments presumably reflects an abnormality of slow axonal transport (SCa). As the disease progresses, axonal swellings appear in more proximal portions of the nerve and the distal region undergoes Wallerian-like degeneration (CPDA) (Fig. 5).

n-HEXANE $CH_3-CH_2-CH_2-CH_2-CH_2-CH_3$

Mn-BK $CH_3-\overset{\overset{O}{\|}}{C}-CH_2-CH_2-CH_2-CH_3$

2,5-HD $CH_3-\overset{\overset{O}{\|}}{C}-CH_2-CH_2-\overset{\overset{O}{\|}}{C}-CH_3$

3-M-2,5-HD $CH_3-\overset{\overset{O}{\|}}{C}-\overset{\overset{CH_3}{|}}{CH}-CH_2-\overset{\overset{O}{\|}}{C}-CH_3$

3,4-DM-2,5-HD $CH_3-\overset{\overset{O}{\|}}{C}-\overset{\overset{}{|}}{\underset{\underset{CH_3}{|}}{CH}}-\overset{\overset{}{|}}{\underset{\underset{CH_3}{|}}{CH}}-\overset{\overset{O}{\|}}{C}-CH_3$

FIGURE 4   n-Hexane and its neurotoxic metabolites.

Mendell and Sahenk (1980) reported a decrease in the velocity of fast transport in MnBK-exposed animals. Morphological studies revealed abnormal accumulation of 10-nm neurofilaments in axonal swellings. Sahenk and Mendell (1981) observed a decrease in bidirectional transport in neuropathic animals. Sickles (1989a) reported that 2,5-HD causes significant reductions in fast transport and that transport velocity returns to control levels after 24 hr (Sickles, 1991). Jacob-

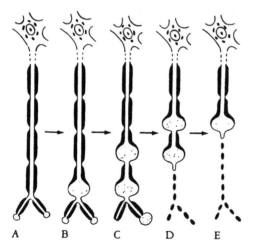

FIGURE 5   Distal (dying-back) axonopathy produced by neurotoxic hexacarbon compounds (i.e., MnBK and 2,5-HD). (A) Normal axon. (B) During chronic intoxication, axonally transported materials (stippling) accumulate, causing giant axonal swellings and myelin retraction on the proximal sides of nodes of Ranvier in distal portions of the fiber. (C) Paranodal swellings appear at more proximal loci with time, as transport fails at these sites. (D and E) Distal portions below axonal swellings undergo Wallerian-like degeneration. (From Spencer et al., 1979.)

sen and colleagues (1986) and Braendgaard and Sidenius (1986b) have observed that transport is impaired before the appearance of clinical symptoms and electrophysiological signs of neuropathy in 2,5-HD intoxication. Sabri (1992) found inhibition of retrograde transport in rodents treated with 2,5-HD and proposed that the inhibition of retrograde transport interferes with the signaling mechanism by which the perikaryon responds to axonal injury (Sabri et al., 1987). Support for the role of retrograde axonal transport in nerve cell responses to axonal injury is provided by experiments where sciatic nerve crush-induced ODC activity was attenuated by 2,5-HD treatment (Nennesmo and Kristensson, 1986; Myall et al., 1990). Braendgaard and Sidenius (1986a) observed that fast transport was normal in hexacarbon neuropathy but the velocity of SCa was increased in neuropathic animals. Watson and colleagues (1991) also reported an increased rate of slow transport with reduced numbers of neurofilaments in rat sensory axons. Increased velocity of neurofilament transport in the optic and peripheral nerves of 2,5-HD-treated animals was first reported by Monaco and colleagues (1985, 1989a). Although this result is unexpected, it may account for the massive increase of 10-nm neurofilaments in axonal swellings in the distal axon. DMHD, which has a greater neurotoxic potency than 2,5-HD, also accelerated the rate of neurofilaments transport in the optic system (Monaco et al., 1989b). These investigators do not support crosslinking of neurofilaments. Instead, they suggest that neurofilaments and microtubules are disconnected from each other during the course of axonal transport and may account for increased velocity in neuropathic animals (Monaco et al., 1989b). Decreased velocity of slow transport has been observed in sensory and motor axons of rats receiving intrathecal injections of 2,5-HD (Hammond-Tooke, 1992).

The biochemical mechanisms of hexacarbon neuropathy are not fully understood (Sayre, 1985). Toxic inhibition of selected enzymes in glycolysis has been reported in hexacarbon neuropathy (Sabri, 1984). An impairment of glucose utilization and mitochondrial respiration has been reported in rat brain following 2,5-HD intoxication (Couri and Milks, 1982; Planas and Cunningham, 1987; Medrano and LoPachin, 1989). Reduced ATP levels were also observed in rat sciatic nerves treated in vitro with 2,5-HD (Sabri, 1984; Sickles et al., 1990). A number of investigators have suggested that crosslinking of neurofilament protein with γ-diketones impairs slow axonal transport and causes axonal degeneration in hexacarbon neuropathy (Anthony et al., 1983b, Graham et al., 1982, 1984; DeCaprio and O'Neill, 1985). The pathogenesis of 2,5-HD toxicity is postulated to proceed as follows:

(i) 2,5-HD reacts with lysyl-amino groups on neurofilaments to form imines; (ii) the imines cyclize to form pyrroles; and (iii) the pyrroles oxidize, resulting in protein–protein crosslinking within or between neurofilaments (DeCaprio, 1986, Genter *et al.*, 1987). This causes the blockade of axonal neurofilaments and their accumulation in axonal swellings. A large number of studies have shown that neurofilaments are the target and that pyrrolation of neurofilaments is an absolute requirement in 2,5-HD neurotoxicity (Anthony *et al.*, 1983a; Genter *et al.*, 1987; DeCaprio *et al.*, 1982, 1988; Lapadula *et al.*, 1988; Abou-Donia *et al.*, 1988).

C. Organophosphorus Compounds

Numerous organophosphorus (OP) compounds cause neurodegeneration in the central and peripheral nervous systems, often referred to as OP-induced delayed polyneuropathy (OPIDP). The most well-known neurotoxic OP compounds are tri-*o*-cresyl phosphate (TOCP, *see* Fig. 6) and diisopropyl fluorophosphate (DFP). The complex biochemical, pharmacological, and toxicological effects of OP compounds have been reviewed elsewhere (Johnson, 1975b; Abou-Donia and Lapadula, 1990; Davis and Richardson, 1980). Although OP compounds are potent inhibitors of AChE, this inhibition is unrelated to the pathogenesis of OPIDP. AChE inhibition causes accumulation of acetylcholine and overstimulation of muscarinic and nicotinic receptors within hours of OP exposure. Cholinergic reactions, nor OPIDP, are the most common features of OP toxicity. OPIDP is predominantly a motor neuropathy, but sensory loss has been noted. Humans exposed to TOCP suffer nerve fiber degeneration in both the PNS and the CNS indicative of CPDA. OP neuropathy is readily produced in chickens, but rats are relatively insensitive to OP exposure. The reasons for differences in vulnerability between species are unknown. OPIDP is thought to be initiated by modification of neuropathy target esterase (NTE) 7–21 days after initial OP exposure. Johnson (1975a, 1982) proposed that the initial biochemical event underlying OPIDP involves the phosphorylation and inhibition of NTE, an enzyme distinct from AChE. Inhibition of 70–80% of brain NTE is a requisite for OPIDP, although inhibition alone is not enough for the development of neuropathy. Johnson (1975a) suggested that the development of OPIDP requires "aging" of inhibited (phosphorylated) NTE, leaving a charged monosubstituted phosphoric acid residue at the active side. This essential step apparently occurs in the axon and is key for the development of OPIDP. The mechanism by which "aged inhibited" NTE causes OPIDP is unknown (Fig. 7).

While early studies showed no adverse effects of OP compounds on axonal transport (Pleasure *et al.*, 1969), Carrington and colleagues (1989) showed OP-induced acceleration of anterograde transport in sciatic nerves. These investigators proposed the involvement of (i) increased activity of Ca^{2+}/calmodulin kinase II and (ii) enhanced phosphorylation and disassembly of cytoskeletal elements in the mechanisms of OPDIP (Abou-Donia and Lapadula, 1990; Abou-Donia *et al.*, 1988). Moretto and colleagues (1987) have shown a correlation between aged inhibited NTE and the inhibition of retrograde transport (RT) in sciatic nerves of hens treated with a neurotoxic dose of di-*n*-butyl-2,2-dichlorovinyl phosphate (DBDCVP). Treatment of animals with phenylmethylsulfonyl fluoride, an agent that prevents the development of OPIDP by inhibiting NTE without the "aging" reaction, had no adverse effect on RT, axonal integrity, or clinical status when given prior to a neurotoxic dose of DBDCVP. Lotti and associates (Bertolazzi *et al.*, 1993) have reevaluated Johnson's NTE hypothesis (1975b) and have suggested that aging of NTE may not always be necessary for the expression of OPIDP. These authors have proposed that all inhibitors capable of inhibiting NTE have the potential of causing OPIDP. Thus mipafox and methamidiphos and certain protective inhibitors (e.g., carbamates and sulfonyl fluoride) which inhibit NTE without aging produce OPIDP (Lotti *et al.*, 1993).

D. β-β'-Iminodipropionitrile (IDPN)

Single intraperitoneal injection or repeated administration of IDPN (Fig. 8) produces proximal axonal swellings and progressive atrophy of the distal axon (Griffin *et al.*, 1982). IDPN neuropathy is a prime example of a nervous system disease caused by defective axonal transport.

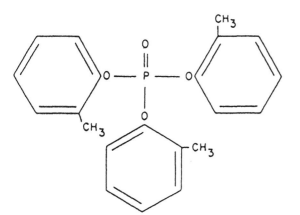

FIGURE 6 Chemical structure of tri-*o*-cresyl phosphate.

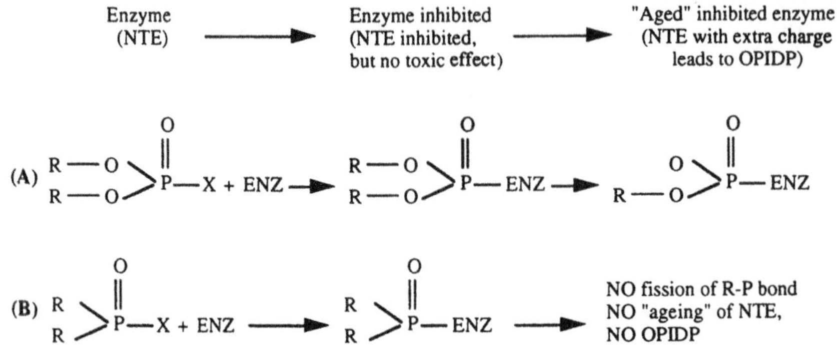

Enzyme ⟶ Enzyme inhibited ⟶ "Aged" inhibited enzyme
(NTE)   (NTE inhibited,   (NTE with extra charge
                         but no toxic effect)   leads to OPIDP)

FIGURE 7 Suggested consequences of inhibiting NTE with (A) phosphate (neurotoxic) and (B) phosphinate (nonneurotoxic) compounds. (Modified from Johnson, 1975b.)

Neurofilament accumulation in the proximal axon and reduced axonal caliber are hallmark features of IDPN neuropathy (Clark *et al.*, 1980; Papasozomenos *et al.*, 1981). IDPN causes a selective blockade of neurofilament triplet proteins carried in slow axonal transport (Yokoyama *et al.*, 1980). Inhibition of slow transport was observed whether IDPN was given before or after the slow component was labeled. Fast anterograde transport is unaffected in IDPN neuropathy (Griffin and Price, 1980; Griffin *et al.*, 1982). However, one study has shown a small but significant decrease in the rate of fast transport in IDPN-treated animals (Sickles, 1989a). Blockage of retrogradely transported proteins labeled with *N*-[³H]succinimidyl propionate was produced by IDPN (Fink *et al.*, 1986).

Griffin and co-workers (1982) observed that the defect in slow transport following IDPN intoxication is preceded by neurofilament and microtubule segregation in the axon. Whether given systemically or injected directly into rat sciatic nerve, IDPN causes a displacement of neurofilaments to the periphery and microtubules toward the center of the axon (Papasozomenos *et al.*, 1981; Griffin *et al.*, 1983). These studies suggest that IDPN may not impair transport mechanisms *per se* but rather alter the properties of neurofilaments or their subunits so that they are unable to be transported. The underlying biochemical mechanisms of IDPN action remains unknown. It has been suggested that cyanoacetaldehyde or another reactive metabolic intermediate may form a covalent bond with the ε amino group of lysine residues on neurofilament subunits (Sayre *et al.*, 1985) and cause their increased aggregation. Another mechanism may involve increased autophosphorylation of neurofilaments (Eyers *et al.*, 1989).

E. Zinc Pyridinethione

Zinc pyridinethione (ZPT) (Fig. 9) is a broad spectrum antifungal and antibacterial agent used in certain antibacterial and antidandruff hair shampoos (for review, *see* Sahenk and Mendell, 1980a). ZPT is the water-insoluble salt of 1-hydroxy-2(1*H*)-pyridinethione. In rodents, ZPT or its parent compound induces dying-back neuropathy and hindlimb paralysis. Although the mechanism of action is unknown, ZPT inactivates certain enzymes by chelating metal cofactors which presumably accounts for its biological activity (Sahenk and Mendell, 1980a,b). The neurotoxic property of ZPT

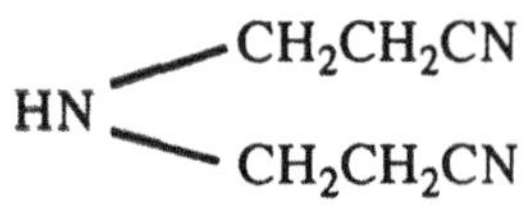

HN
⟨ —CH₂CH₂CN
⟨ —CH₂CH₂CN

FIGURE 8 Chemical structure of β-β'-iminodipropionitrile.

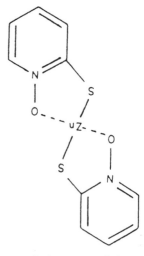

FIGURE 9 Chemical structure of zinc pyridinethione.

may be related to its action as an antimetabolite of nicotinic acid. The parent compound of ZPT inhibits glycolytic enzymes which may reduce intracellular ATP levels in fungi. ZPT causes distal axonopathy with accumulation of tubulovesicular membranes in motor nerve terminals. At an early stage, ZPT impairs fast axonal transport of proteins in sensory axons (Mendell and Sahenk, 1980; Sahenk and Mendell, 1980a). Some investigators have reported that the accumulation of membranous materials at axon terminals is related to an abnormality in the turnaround of fast-transported materials, thereby reducing retrograde transport (Sahenk and Mendell, 1980a,b). It has been proposed that the zinc present in ZPT binds to sulfhydryl groups and inactivates thiol proteases which are necessary for turnaround (Sahenk and Lasek, 1988). These investigators have shown that a direct application of protease inhibitors on axonal tips causes profound inhibition of turnaround of transported membranous materials and causes their accumulation, similar to that caused by ZPT. It is apparent that deficits in fast anterograde and retrograde transport appear to play an important role in the pathogenesis of ZPT neuropathy as well as in human neuroaxonal dystrophy where peripheral nerves and nerve terminals show similar abnormalities (Sahenk and Mendell, 1980b).

F. Doxorubicin

Doxorubicin (Fig. 10), also known as adriamycin, is a glycosidic anthracycline antibiotic (for review *see* Cho *et al.*, 1980a,b). Doxorubicin causes a dose-related cardiomyopathy and progressive congestive heart failure in humans. In experimental animals, a single intra-venous injection of doxorubicin causes ataxia and selective necrosis in dorsal root ganglia (Cho *et al.*, 1980b). Doxorubicin forms a complex with DNA and inhibits DNA-dependent DNA and RNA polymerases (Cho *et al.*, 1980a). A morphological study of doxorubicin-treated rats revealed Wallerian degeneration secondary to necrosis of dorsal root ganglia (Cho *et al.*, 1980b). Doxorubicin eventually causes the breakdown of the neuronal perikaryon, swelling in the proximal axon, and nerve fiber degeneration in a proximo-distal (dying-forward) manner. The effect of doxorubicin on fast axonal transport in sciatic nerves and on protein synthesis in dorsal root ganglia has been studied by Mendell and Sahenk (1980). These investigators found that doxorubicin delays the processing time of labeled materials in neuronal perikaryon without affecting the rate of fast axonal transport in peripheral nerves. Sidenius (1986) reported abnormalities in both fast and slow axonal transport in doxorubicin-treated animals. An increased processing time of [³H]leucine by dorsal root ganglia, without alteration in the rate of fast axonal transport, was found in doxorubicin-treated animals (Mendell and Sahenk, 1980)).

G. p-Bromophenylacetylurea (BPAU)

BPAU (Fig. 11) produces a model dying-back neuropathy with hindlimb paralysis in rats (Ochs and Brimijoin, 1993). Morphological studies 2 weeks after intoxication revealed widespread axonal degeneration in the central and peripheral nervous systems. The mechanisms of BPAU neurotoxicity have not been elucidated. The presence of tubulomembranous debris in the distal axon points toward a defect in fast axonal transport (Ochs and Brimijoin, 1993). Studies of Jacobsen and Brimijoin (1981) and Brimijoin and colleagues (1982) reported changes in fast axonal transport of materials in BPAU-treated animals. These investigators found that while fast anterograde transport was normal, retrograde transport was impaired in BPAU-treated animals (Ochs and Brimijoin, 1993). Axonal transport studies of Nagata and Brimijoin (1986a,b) and Nagata and colleagues (1987) have shown that the neurotoxicity of BPAU is linked to an abnormal onset of rapid axonal transport (Nagata and Brimijoin, 1986b). BPAU induces a dose-dependent reduction in retrograde

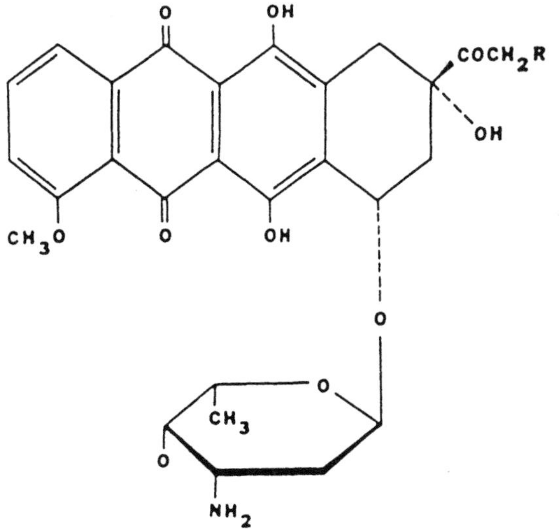

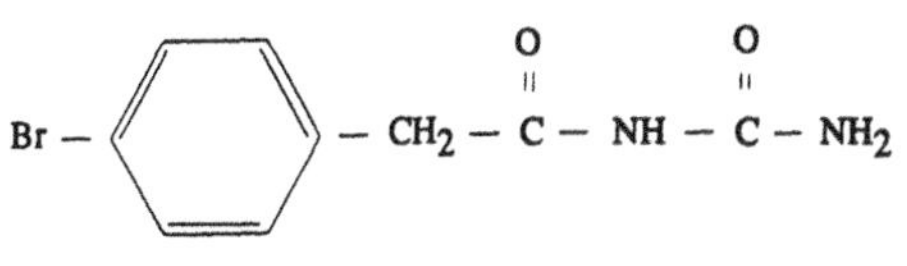

FIGURE 10 Chemical structure of doxorubicin (adriamycin).

FIGURE 11 Chemical structure of *p*-bromophenylacetylurea.

transport, while the rate of fast anterograde transport was unaffected (Nagata and Brimijoin, 1986a,b). Furthermore, BPAU-induced deficits in retrograde transport were due to reduced rather than delayed turnaround of the materials.

Electron microscope autoradiography studies indicated that the accumulation of tubulomembranous materials might be due to local stasis of fast-transported materials in BPAU neuropathy (Oka and Brimijoin, 1992). BPAU neuropathy appears to be initiated by abnormal organelles and the premature onset of fast axonal transport (Oka and Brimijoin, 1990).

H. Carbon Disulfide

Carbon disulfide (CS_2) is used mostly in the production of viscose rayon fibers for the textile industry and for cellophane film for packaging. Inhaled vapor of CS_2 is hazardous to occupational workers (for review, *see* Seppäläinen and Haltia, 1980; O'Donoghue, 1985). CS_2-induced symptoms of central nervous system damage include memory loss, insomnia, tremors, suicidal tendencies, and extreme irritability. CS_2-exposed workers show decreased nerve conduction velocity of motor fibers. Chronic CS_2 exposure produces axonal damage in both the central and peripheral nervous systems. The most prominent feature of CS_2 intoxication is a dying-back pattern of giant axonal swellings containing masses of neurofilaments and reduced numbers of microtubules. Swollen mitochondria and accumulated glycogen granules and tubules are seen in long spinal tracts and peripheral nerves of intoxicated animal (O'Donoghue, 1985).

The mechanisms of CS_2 neurotoxicity are unknown. However, its reactivity with a variety of nucleophilic substances suggests that several mechanisms may be involved in its neurotoxicity. The involvement of copper- and zinc-dependent enzyme systems in the expression of CS_2 neurotoxicity has been suggested. CS_2 inactivates dopamine-β-hydroxylase, a copper-dependent enzyme. CS_2 neurotoxicity may also be related to the depletion of pyridoxine and thiamine since supplementation with these vitamins delays the onset and reduces the severity of neurotoxicity. Inhibition of glycolytic enzymes by CS_2 may reduce ATP generation, impair energy-dependent fast axonal transport, and cause axonal degeneration. Tarkowski and colleagues (1980) observed an uncoupling of oxidative phosphorylation in brain mitochondria of rats intoxicated with CS_2. Ultrastructural studies revealed severe mitochondrial lesions in both central and peripheral axons. Systemically injected labeled CS_2 binds to axonal neurofilament proteins, suggesting that neurofilaments may be damaged directly by CS_2. Although the mechanisms are

unknown, CS_2 neurotoxicity may be associated with the blockade of retrograde axonal transport (Anzil *et al.*, 1982).

I. Erythro-9-[3-(2-hydroxynonyl)]adenine (EHNA)

EHNA (Fig. 12) is a structural analog of adenosine (Penningroth, 1986). EHNA inhibits microtubule-activated ATPase activity of brain MAP-1C (Shpetner *et al.*, 1988). This inhibitory effect is selective for dynein-like ATPases. EHNA blocks actin assembly and inhibits actin-dependent motility *in vivo*. Millimolar concentrations of EHNA appear to be cytotoxic to cells in tissue culture. The mechanism of cytotoxicity is unclear. Forman and colleagues (1983) observed a selective blockade of retrograde transport with little or no inhibition of anterograde transport of microscopically visible organelles in isolated lobster axons. EHNA inhibits retrograde transport velocity both in living and in permealized axons.

J. Microtubule Poisons

Microtubules (MT) are intimately involved in mediating all kinds of axonal transport, and MT poisons are potent inhibitors of transport. The mechanisms by which MT poisons act is poorly understood. Selected MT poisons are discussed next.

1. Spindle Inhibitors (Colchicine)

Colchicine (Fig. 13) is one of the most powerful and specific spindle inhibitors. Colchicine has been used in the treatment of *gout* for a long time, but its chemical structure has been established only recently (for details *see* Dustin, 1984). The mechanism of action of colchicine is not fully understood. It is also not yet certain

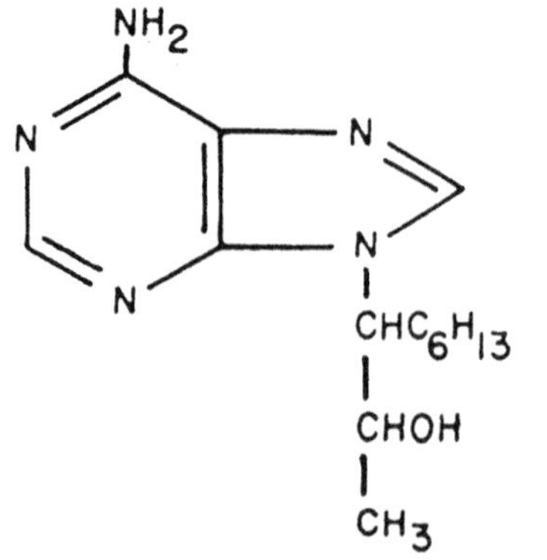

FIGURE 12 Chemical structure of erythro-9-[3-(2-hydroxynonyl)]adenine.

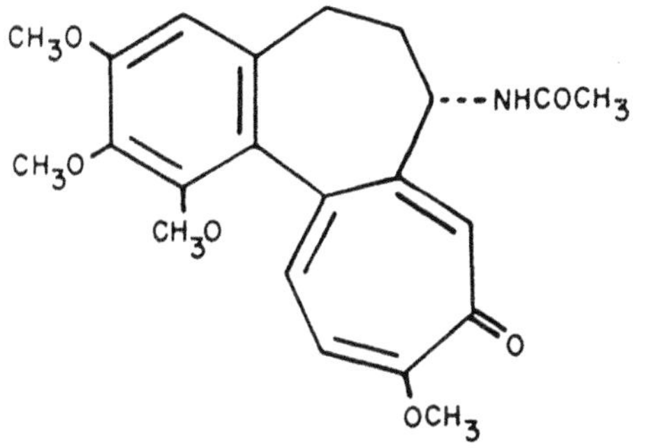

FIGURE 13 Chemical structure of colchicine.

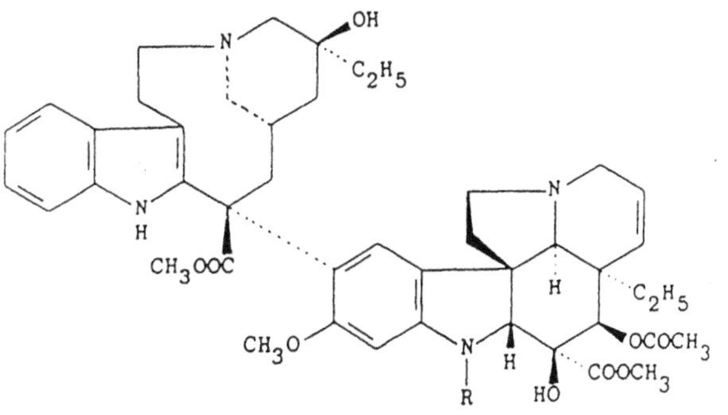

FIGURE 14 Chemical structures of vinblastine and vincristine.

Vinblastine   R=CH₃

Vincristine   R=CHO

if the beneficial effect of colchicine is mediated through its effect on MT. Colchicine binds to the MT subunits (i.e., tubulin) and inhibits MT assembly. *Gout* patients treated with colchicine often develop neurotoxicity. The main toxic effects are myopathy and peripheral neuropathy (Riggs *et al.*, 1986; Kuncl *et al.*, 1987). Axonal degeneration of large-diameter fibers in sural nerves has been reported following colchicine treatment (Kuncl *et al.*, 1987). Intraocular injection of colchicine in rabbits blocks fast axonal transport within 5 h (see Dustin, 1984). In the rat, only large doses (approximately 1 mg) of colchicine inhibit fast transport. Colchicine blocks slow axonal transport of proteins by inhibiting the growth of MT in the chick sciatic nerves (see Dustin, 1984). Blockade of fast transport and disruption of MT by millimolar concentratons of colchicine and VLB has been observed in sciatic nerve axons. Larger doses of colchicine inhibit fast transport of proteins and disrupt MT and neurofilaments (see Dustin, 1984). In salamander nerves, the rate of fast axonal transport was reduced by colchicine exposure. Topical application of colchicine on sympathetic nerves impairs fast axonal transport of catecholamine granules (Dahlström, 1971). Direct injection of colchicine in the hypogastric nerve alters the transport of noradrenaline without any adverse effect on MT (see Dustin, 1984). Colchicine given intracisternally also impairs the transport of secretory granules toward the pituitary, but changes in MT are not seen (see Dustin, 1984).

2. Vinca Alkaloids

Vinblastine (VLB) and vincristine (VCR) (Fig. 14) are potent and widely used vinca alkaloids in cancer chemotherapy (for review *see* Dustin, 1984). Vinca alkaloids bind to tubulin at a different site than colchicine and lead to the formation of loose spiral macrotubules. The properties of vinca alkaloids are quite similar to that of colchicine, although their chemical structures are quite different. Derivatives of VBL (i.e., vindesine

and desacetylvinblastine) are also potent antitumor agents (Dustin, 1984). Treatment with vinca alkaloids causes neurotoxocity which may be related to their action on neuronal MT and inhibition of axonal transport. VLB impairs fast axonal transport in the rat optic system and sciatic nerve *in vitro* (see Dustin, 1984). Intrathecal injections of VCR block the transport of neurosecretory granules in the hypothalamus, but disassembly of MT does not occur. Local application of VLB has been reported to block retrograde transport of rabies virus in mouse sciatic nerve following its injection in the foot pad (see Dustin, 1984). VCR was found to be more effective than VBL in blocking fast axonal transport of proteins and causing MT disassembly in cat sciatic nerve. The blockade of fast axonal transport and disassembly of MT by VBL in cat sciatic nerves shows an intimate relationship between MT and the fast axonal transport system (Ghetti *et al.*, 1982).

3. Taxol

Taxol (Fig. 15), an experimental antitumor agent, favors the formation of MT and increases its stability

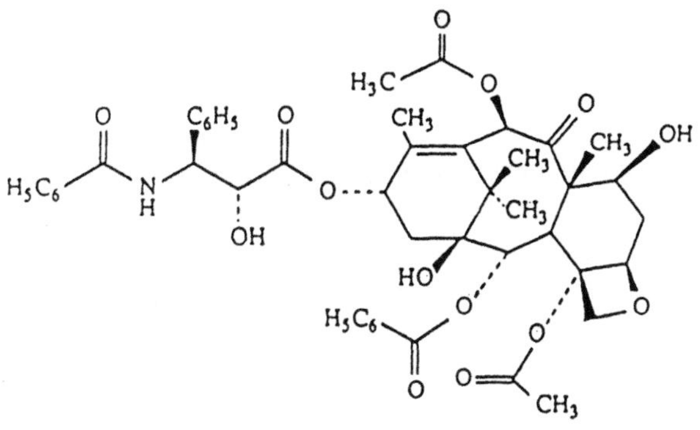

FIGURE 15 Chemical structure of taxol.

(Schiff and Horwitz, 1980). Taxol binds to tubulin and MT *in vitro* (see Dustin, 1984). MT formed by the treatment of taxol cannot be disassembled with Ca^{2+} or by a low temperature. Taxol does not interfere with the binding of colchicine and increases the number of MT in cells following treatment. Studies conducted by Roijitta and colleagues (1984) and Royatta and Raine (1986) have shown that local injection of taxol causes some aggregation of MT, but axonal degeneration was not seen up to 21 days. Horie and colleagues (1987) reported that pretreatment with taxol counteracts colchicine blockage of axonal transport in neurites of cultured dorsal root ganglia. The mechanism of protection appears related to a taxol-induced reduction in the concentration of free tubulin and the inability of colchicine to disrupt MT. Komiya and Tashiro (1988) found that the axonal transport of tubulin was completely blocked by a single subepineurial injection of taxol in rat sciatic nerve. A surprising finding was that fast axonal transport was essentially intact. In human, high doses of taxol produce a sensory neuropathy, presumably by producing MT aggregates in dorsal root ganglia, axons, and Schwann cells (Lipton *et al.*, 1989).

IV. Conclusions

This chapter reviews the progress of axonal transport that has been made since 1948. It is apparent that substantial progress has been made since then in (1) dividing the axonal transport system into various rates, (2) identifying types of materials transported, and (3) examining the consequences of its blockade on the integrity of neuronal perikaryon, axon, and myelinating cells. AVEC-DIC microscopy has revolutionized the study of axonal transport in that it is now possible to visualize particle movement in the axoplasm. AVEC-DIC has facilitated the discovery of motor proteins and has enhanced our knowledge of the mechanisms of axonal transport. This chapter also discusses the effect of selected neurotoxic chemicals on axonal transport. Neurons with its large processes are quite vulnerable to a variety of toxic chemicals. However, it is not yet clear if the chemical inhibition of axonal transport is a primary effect or occurs secondary to structural changes in nerve fibers. A number of studies have shown that interference with cellular energy (ATP) generation, Ca^{2+} concentration, and cytoskeletal integrity disrupts the axonal transport system (Ochs and Brimijoin, 1993). Chemicals that block retrogradely transported trophic signals, disrupt MAPS, and perturb oncogene expression, loading, and turnaround mechanisms have adverse effect on axonal transport. Some chemicals may impair axonal transport by inhibiting the function of specific "motor" proteins.

V. Future Research Directions

Recent technical advances have greatly influenced the study of axonal transport in several areas: (1) The introduction of AVEC-DIC microscopy has revolutionized research on axonal transport by allowing direct visualization of organelle transport in anterograde and retrograde directions. This capability should accelerate the discovery and development of specific agents which regulate the movement of organelles carrying specific neurotransmitter substances. (2) The discovery of motor proteins and associated factors provides important new data in clarifying the molecular mechanisms of axonal transport and its precise role in neuronal structure and function. (3) The use of molecular probes (*e.g.*, monoclonal antibodies, cDNA and mRNA probes) will help create animal models of various neurodegenerative diseases (see Studelska *et al.*, 1987; Xu *et al.*, 1993). (4) The application of fluorescent and microinjection techniques should resolve the question of cytoskeleton stability, synthesis, and transport in the near future. The molecular cytochemical technique using tubulin labeled with a fluorescent group to visualize MT *in vivo* holds considerable promise for yielding information to enhance our understanding of cytoskeletal assembly and turnover and their critical role in neuronal integrity (see Okabe and Hirokawa, 1990; Reinsch *et al.*, 1991). These research advancements should lead to a greater understanding of the pathogenesis of nervous system disorders of unknown etiology.

Acknowledgments

I am grateful to Dr. Peter S. Spencer for making useful suggestions in the manuscript; Ilse Schoffstoll for drawing the chemical structures and illustrations, and for typing the manuscript; and to Monica Fenton for valuable editorial assistance. This chapter was written with the support of NIH Grant NS 19611 and a grant from the Medical Research Foundaton of Oregon.

References

Abou-Donia, M. B., and Lapadula, M. (1990). Mechanisms of organophosphorus ester-induced delayed neurotoxicity. Type I and Type II. *Annu. Rev. Pharmacol. Toxicol.* **30,** 405–440.

Abou-Donia, M. B., Lapadula, D. M., and Suwita, E. (1988). Cytoskeletal proteins as targets for organophosphorus compounds and aliphatic hexacarbon-induced neurotoxicity. *Toxicology* **49,** 469–477.

Adams, R. J. (1982). Organelle movement in axons depends on ATP. *Nature* **297,** 327–329.

Allen, R. D. (1987). The microtubule as an intracellular engine. *Sci. Am.* **256,** 42–49.

Allen, R. D., Travis, J. L., Hayden, J. H., Allen, N. S., Breuer, A. C., and Lewis, L. J. (1982). Cytoplasmic transport: moving ultrastructural elements common to many cell types revealed by bideo-enhanced microscopy. *Cold Springs Harbor Symp. Quant. Biol.* **46,** 85–87.

Anthony, D. C., Boekelheide, K., Anderson, C. V., and Graham, D. G. (1983a). The effect of 3,4-dimethyl substitution on the neurotoxicity of 2,5-hexanedione. II. Dimethyl substitution accelerates pyrrole formation and protein cross-linking. *Toxicol. Appl. Pharmacol.* **71,** 372–382.

Anthony, D. C., Giangaspero, F., and Graham, D. G. (1983b). The spatio-temporal pattern of the axonopathy associated with the neurotoxicity of 3,4-dimethyl-2,5-hexanedione in the rat. *J. Neuropathol. Exp. Neurol.* **42,** 548–560.

Anzil, A. P., Isenburg, G., and Kreutzberg, G. W. (1982). Disulfiram-induced impairment of horseradish peroxidase retrograde transport in rat sciatic nerve. In *Axoplasmic Transport in Physiology and Pathology* (D. G. Weiss and A. Gorio, Eds.), pp. 119–122, Springer-Verlag, Basel.

Archer, D. R., Watson, D. F., and Griffin, J. W. (1994). Phosphorylation-dependent immunoreactivity of neurofilaments and the rate of slow axonal transport in the central and peripheral axons of the rat dorsal root ganglion. *J. Neurochem.* **62,** 1119–1125.

Bergmark, E., Calleman, C. J., and Costa, L. G. (1991). Formation of hemoglobin adducts of acrylamide and its epoxide metabolite glycidamide in the rat. *Toxicol. Appl. Pharmacol.* **111,** 352–363.

Bergmark, E., Calleman, C. J., He, F., and Costa, L. G. (1993). Determination of hemoglobin adducts in humans occupationally exposed to acrylamide. *Toxicol. Appl. Pharmacol.* **120,** 45–54.

Berti-Mattera, L. N., Eichberg, J., Schrama, L., and LoPachin, M. (1990). Acrylamide administration alters protein phosphorylation and phospholipid metabolism in rat sciatic nerve. *Toxicol. Appl. Pharmacol.* **103,** 502–511.

Bertolazzi, M., Moretto, A., and Lotti, M. (1993). Inhibition of neuropathy target esterase is not required to promote organophosphate polyneuropathy. *Toxicologist* **13,** 124 (Abstract).

Bisby, M. A. (1987). Does recycling have functions other than disposal. In *Axonal Transport* (R. V. Smith and M. A. Bisby, Eds.), pp. 365–384, Liss, New York.

Brady, S. T. (1985). A novel brain ATPase with properties expected for the fast axonal transport motor. *Nature* **317,** 73–75.

Brady, S. T., Lasek, R. J., and Allen, R. D. (1982). Fast axonal transport in extruded axoplasm from squid giant axon. *Science* **218,** 1129–1131.

Brady, S. T., Pfister, K. K., and Bloom, G. S. (1990). A monoclonal antibody to the heavy chain of kinesin inhibits anterograde and retrograde axonal transport in isolated squid axoplasm. *Proc. Natl. Acad. Sci. USA* **87,** 1061–1065.

Braendgaard, H., and Sidenius, P. (1986a). Anterograde components of axonal transport in motor and sensory nerves in experimental 2,5-hexanedione neuropathy. *J. Neurochem.* **47,** 31–37.

Braengaard, H., and Sidenius, P. (1986b). The retrograde fast component of axonal transport in motor and sensory nerves of the rat during administration of 2,5-hexanedione. *Brain Res.* **378,** 1–7.

Brat, D. J., and Brimijoin, S. (1993). Acrylamide and Glycidamide impair neurite outgrowth in differentiating N1E.115 Neuroblastoma without disturbing rapid bidirectional transport of organelles observed by video microscopy. *J. Neurochem.* **60,** 2145–2152.

Brimijoin, S., Dyck, P. J., Jacobsen, J., and Lambert, E. H. (1982). Axonal transport in human nerve disease and in the experimental neuropathy induced by P-bromophenyl acetyl urea. In *Axoplasmic Transport* (D. G. Weiss and A. Gorio, Eds.), pp. 124–130, Springer-Verlag, Berlin.

Brimijoin, W. S., and Hammond, P. I. (1985). Acrylamide neuropathy in the rat: effects on energy metabolism in sciatic nerve. *Mayo Clin. Proc.* **60,** 3–8.

Calleman, C. J., Bergmark, E., and Costa, L. G. (1990). Acrylamide is metabolized to glycidamide in the rat: evidence from hemoglobin adduct formation. *Chem. Res. Toxicol.* **3,** 406–412.

Carrington, C. D., Lapadula, D. M., and Abou-Donia, M. B. (1989). Acceleration of anterograde transport in cat sciatic nerve by diisopropylphosphofluoridate. *Brain Res.* **476,** 179–182.

Chauhan, N. B., Spencer, P. S., and Sabri, M. I. (1993a). Effect of acrylamide on the distribution of microtubule-associated proteins (MAP1 and MAP2) in selected regions of rat brain. *Mol. Chem. Neuropathol.* **18,** 225–245.

Chauhan, N. B., Spencer, P. S., and Sabri, M. I. (1993b). Acrylamide-induced depletion of microtubule-associated proteins (MAP1 and MAP2) in the rat extrapyramidal system. *Brain Res.* **602,** 111–118.

Cho, E.-S., Spencer, P. S., and Jortner, B. S. Doxorubicin. (1980a). In *Experimental and Clinical Neurotoxicology* (P. S. Spencer and H. H. Schaumburg, Eds.), pp. 430–439, Williams and Wilkins, Baltimore.

Cho, E. S., Spencer, P. S., Jortner, B. S., and Schaumberg, H. H. (1980b). A single intravenous injection of doxorubicin (Adriamycin®) induces sensory neuronopathy in rats. *Neurotoxicology* **1,** 583–591.

Chretien, M., Patey, G., Souyri, F., and Droz, B. (1981). Acrylamide-induced neuropathy and impairement of axonal transport of proteins. II. Abnormal accumulation of smooth endoplasmic reticulum at sites of focal retention of fast transported proteins. Electron microscope radioautographic study. *Brain Res.* **205,** 15–28.

Clark, A. W., Griffin, J. W., and Price, D. L. (1980). The axonal pathology in chronic IDPN intoxication. *J. Neuropathol. Exp. Neurol.* **39,** 42–55.

Courand, J. Y., Di Giambernardino, L., Chretien, M., Souri, F., and Fardeau, M. (1982). Acrylamide neuropathy and changes in the axonal transport and muscular content of the molecular forms of acetylcholinesterase. *Muscle and Nerve* **5,** 302–312.

Couri, D., and Milks, M. (1982). Toxicity and metabolism of the neurotoxic hexacarbons n-hexane, 2-hexanone and 2,5-hexanedione. *Annu. Rev. Pharmacol. Toxicol.* **22,** 145–166.

Dahlström, A. (1971). Axoplasmic transport (with particular reference to adrenergic neurons). *Philos. Trans. R. Soc. Lond. Biol. Sci.* **261,** 325–358.

Davis, C. S., and Richardson, R. J. (1980). Organophosphorus compounds. In *Experimental and Clinical Neurotoxicology* (P. S. Spencer and H. H. Schaumburg, Eds.), pp. 527–544, Williams and Wilkins, Baltimore.

DeCaprio, A. P. (1986). Mechanisms of in vitro pyrrole adduct autooxidation in 2,5-hexanedione-treated protein. *Mol. Pharmacol.* **30,** 452–458.

DeCaprio, A. P., Briggs, R. G., Jackowski, S. J., and Kim, J. C. S. (1988). Comparative neurotoxicity and pyrrole-forming potential of 2,5-hexanedione and perdeuterio-2,5-hexanedione in the rat. *Toxicol. Appl. Pharmacol.* **92,** 75–85.

DeCaprio, A. P., Olajos, E. S., and Weber, P. (1982). Covalent binding of neurotoxic n-hexane metabolite: Conversion of primary amines to substituted pyrrole adducts by 2,5-hexanedione. *Toxicol. Appl. Pharmacol.* **65,** 440–450.

DeCaprio, A. P., and O'Neill, E. A. (1985). Alterations in rat axonal cytoskeletal proteins induced by in vitro and in vivo 2,5-hexanedione exposure. *Toxicol. Appl. Pharmacol.* **78,** 235–247.

Dixit, R., Mukhtar, H., and Seth, P. K. (1981a). *In vitro* inhibition of alcohol dehydrogenase by acrylamide: interaction with enzyme-SH groups. *Toxicol. Lett.* **1,** 487–498.

Dixit, R., Mukhtar, H., Seth, P. K., and Krishnamerti, C. R. (1981b). Conjugation of acrylamide with glutathione catalyzed by glutathione-s-transferase of rat liver and brain. *Biochem. Pharmacol.* **30**, 1739–1748.

Dustin, P. P. (1984). *Microtubules*. Springer-Verlag, Berlin.

Ekstrom, P., Kanje, M., and McLean, W. G. (1987). The effect of trifluoperazine on fast and slow axonal transport in the rabbit vagus nerve. *J. Neurobiol.* **18**, 283–292.

Elam, J. S., and Cancalon, P. (1984). *Axonal Transport in Neuronal Growth and Regeneration*. Plenum Press, New York.

Endo, H., Sabri, M. I., Stephens, J. M., Pekala, P. H., and Kittur, S. (1993). Acrylamide induces immediate-early gene expression in rat brain. *Brain Res.* **609**, 231–236.

Endo, H., Kittur, S., and Sabri, M. I. (1994). Acrylamide alters neurofilament protein gene expression in rat brain. *Neurochem. Res.* **19**, 815–820.

Eyers, J., McLean, W. G., and Leterrier, J. F. (1989). Effect of a single dose of β,β'-iminodipropionitrile *in vivo* on the properties of neurofilaments *in vitro*: Comparison with the effect of iminodipropionitrile added directly on neurofilaments *in vitro*. *J. Neurochem.* **52**, 1759–1766.

Fahim, M. A., Lasek, F. J., Brady, S. T., and Hodge, A. J. (1985). AVEC-DIC and electron microscopic analysis of axonally-transported particles in cold-blocked squid giant axons. *J. Neurocytol.* **14**, 689–704.

Fink, D. J., Purkiss, D., and Mata, M. (1986). Beta,beta'-iminodipropionitrile impairs retrograde axonal transport. *J. Neurochem.* **47**, 1032–1038.

Fink, D. J., and Gainer, H. (1980). Retrograde axonal transport of endogenous proteins in sciatic nerves demonstrated by covalent labeling *in vivo*. *Science* **208**, 303–305.

Forman, D. S., Brown, K. J., and Promersberger, M. E. (1983). Selective inhibition of retrograde axonal transport by erythro-9-[3-(hydroxynonyl)] adenine. *Brain Res.* **272**, 194–197.

Forman, D. S. (1983). New approaches to the study of the mechanism of fast axonal transport. *Trends Neurosci.* **7**, 112–116.

Genter, M. B., Szakal-Quin, G., Anderson, C. W., Anthony, D. C., and Graham, D. G. (1987). Evidence that pyrrole formation is a pathogenetic step in γ-diketone neuropathy. *Toxicol. Appl. Pharmacol.* **87**, 351–362.

Ghetti, B., Alyea, C., Norton, J., and Ochs, S. (1982). Effects of vinblastine on microtubule density in relation to axoplasmic transport. In *Axoplasmic Transport* (D. G. Weiss, Ed.), pp. 322–327, Springer-Verlag, Berlin.

Glass, J. D., and Griffin, J. W. (1991). Neurofilament redistribution in transected nerves: evidence for bidirectional transport of neuroligaments. *J. Neuroscience* **11**, 3146–3154.

Gold, B. G., Griffin, J. W., and Price, D. L. (1985). Slow axonal transport in acrylamide neuropathy: different abnormalities produced by single-dose and continuous administration. *J. Neurosci.* **5**, 1755–1768.

Gold, B. G., Austin, D. R., and Griffin, J. W. (1991). Regulation of aberrant neurofilament phosphorylation in neuronal perikara. II. Correlation with continued axonal elongation following axotomy. *J. Neuropathol. Exp. Neurol.* **50**, 627–648.

Gold, B. G., and Spencer, P. S. (1993). Neurotrophic function in normal nerve and in peripheral neuropathies. In *Neuroregeneration* (A. Gorio, Ed.), pp. 101–122, Raven Press, New York.

Grafstein, B., and Forman, D. S. (1980). Intracellular transport in neurons. *Physiol. Rev.* **60**, 1167–1283.

Graham, D. G., Anthony, D. C., Boekelheide, K., Maschmann, N. A., Richards, R. G., Wolfram, J. W., and Shaw, B. R. (1982). Studies of the molecular pathogenesis of hexane neuropathy. II. Evidence the pyrrole derivatization of lysyl residues leads to protein crosslinking. *Toxicol. Appl. Pharmacol.* **64**, 415–422.

Graham, D. G., Szakal-Quin, G., Priest, J. W., and Anthony, D. C. (1984). *In vitro* evidence that covalent crosslinking of neurofilaments occurs in gamma-diketone neuropathy. *Proc. Natl. Acad. Sci. USA* **81**, 4979–4982.

Griffin, J. W., Fahnestock, K. E., Price, D. L., and Hoffman, P. N. (1983). Microtubule-neurofilament segregation produced by beta,beta'-iminodipropionitrile: evidence for the association of fast axonal transport with microtubules. *J. Neurosci.* **3**, 557–566.

Griffin, J. W., Hoffman, P. N., and Price, D. L. (1982). Axonal transport in β,β'-iminodipropionitrile neuropathy. In *Axoplasmic Transport in Physiology and Pathology* (D. G. Weiss and A. Gorio, Eds.), pp. 109–118, Springer-Verlag, Berlin.

Griffin, J. W., and Price, D. L. (1980). Proximal axonopathies induced by toxic chemicals. In *Experimental and Clinical Neurotoxicology* (P. S. Spencer and H. H. Schaumburg, Eds.), pp. 161–187, Williams and Wilkins, Baltimore.

Hammerschlag, R., and Brady, S. T. (1989). Axonal transport and neuronal cytoskeleton. In *Basic Neurochemistry*, 4th ed., (G. Silgel, B. Agranoff, R. W. Albers, and P. Molinoff, Eds.), pp. 457–478, Raven Press, New York.

Hammond-Tooke, G. D. (1992). Slow axonal transport is impaired by intrathecal 2,5-hexanedione. *Exp. Neurol.* **116**, 210–217.

Harry, G. J., Goodman, J. F. Bouldin, T. W., Toews, A. D., and Morell, P. (1989). Acrylamide-induced increases in deposition of axonally-transported glycoproteins in rat sciatic nerve. *J. Neurochem.* **52**, 1240–1247.

Harry, G. J., Morell, P., and Bouldin, T. W. (1992). Acrylamide exposure impairs axonal transport of glycoproteins in myelinated axons. *J. Neurosci. Res.* **31**, 554–560.

Hoffman, P. N., Cleveland, D. W., Griffin, J. W., Landes, P. W., Cowan, N. J., and Price, D. L. (1987). Neurofilament gene expression: a major determinant of axonal caliber. *Proc. Natl. Acad. Sci. USA* **84**, 3472–3476.

Hoffman, P. N., and Lasek, R. J. (1975). The slow component of axonal transport. Identification of major structural polypeptides of the axon and their generality among mammalian neurons. *J. Cell Biol.* **66**, 351–366.

Hollenbeck, P. J. (1990). Cytoskeleton on the move. *Nature* **343**, 408–409.

Horie, H., Takenaka, T., Ito, S., and Kim, S. U. (1987). Taxol counteracts colchicine blockade of axonal transport in neurites of cultured dorsal root ganglion cells. *Brain Res.* **420**, 144–146.

Howland, R. D. (1981). The etiology of acrylamide neuropathy: Enolase, phosphofructokinase, and glycereldehyde-3-phosphate dehydrogenase activity in peripheral nerve, spinal cord, brain and skeletal muscle of acrylamide-intoxicated cats. *Toxicol. Appl. Pharmacol.* **60**, 324–333.

Howland, R. D., and Alli, P. (1986). Altered phosphorylation of rat neuronal cytoskeletal proteins in acrylamide-induced neuropathy. *Brain Res.* **363**, 33–339.

Howland, R. D., Vyas, I. L., Lowndes, H. E., and Angentieri, T. M. (1980a). The etiology of toxic peripheral neuropathies: *in vitro* effects of acrylamide and 2,5-hexanedione on brain enolase and other enzymes. *Brain Res.* **202**, 131–142.

Howland, R. D., Vyas, I. L., and Lowndes, H. E. (1980b). The etiology of acrylamide neuropathy: possible involvement of neuron specific enolase. *Brain Res.* **190**, 529–535.

Iqbal, Z. (1986). *Axoplasmic Transport*. CRC Press, Boca Raton, Florida.

Jacobsen, J., and Brimijoin, S. (1981). Axonal transport of enzymes and labeled proteins in experimental axonopathy induced by p-bromophenylacetylurea. *Brain Res.* **229**, 103–122.

Jacobsen, J., Brimijoin, S., and Sidenius, P. (1983). Axonal transport in neuropathy. *Muscle Nerve* **6**, 164–166.

Jacobsen, J., and Sidenius, P. (1983). Early and dose-dependent decreases of retrograde axonal transport in acrylamide intoxicated rats. *J. Neurochem.* **10**, 447–454.

Jacobsen, J., Sidenius, P., and Brændgaard, H. (1986). A proposal for a classification of neuropathies according to their axonal transport abnormalities. *J. Neurol. Neurosurg. Psychiatry* **49**, 986–990.

Johnson, M. K. (1975a). Organophosphorus esters causing delayed neurotoxic effects: mechanism of action and structure/activity studies. *Arch. Toxicol.* **34**, 259–288.

Johnson, M. K. (1975b). The delayed neuropathy caused by some organophosphorus esters: mechanism and challenge. *CRC Crit. Rev. Toxicol.* **3**, 218–316.

Johnson, M. K. (1982). The target for initiation of delayed neurotoxicity by organophosphorus esters: biochemical studies and toxicological applications. *Rev. Biochem. Toxicol.* **4**, 141–212.

Komiya, Y., and Tashiro, T. (1988). Effects of taxol on slow and fast axonal transport. *Cell Motil. Cytoskeleton* **1**, 151–156.

Kuncl, R. W., Duncan, G., Watson, D., Alderson, K., Rogawski, M. A., and Peper, M. (1987). Colchicine myopathy and neuropathy. *N. Engl. J. Med.* **316**, 1562–1568.

Lapadula, D. M., Suwita, E., and Abou-Donia, M. B. (1988). Evidence for multiple mechanisms responsible for 2,5-hexanedione-induced neuropathy. *Brain Res.* **458**, 123–131.

Lasek, R. J., Garner, J. A., and Brady, S. T. (1984). Axonal transport of the cytoplasmic matrix. *J. Cell. Biol.* **99**, 212–221.

Lasek, R. J., Paggi, P., and Katz, M. (1992). Slow axonal transport mechanisms move neurofilaments relentlessly in mouse optic axons. *J. Cell. Biol.* **117**, 607–616.

Lasek, R. J., Paggi, P., and Katz, M. J. (1993). The maximum rate of neurofilament transport in axons: a view of molecular transport mechanisms continuously engaged. *Brain Res.* **616**, 58–64.

LeQuesne, P. M. (1980). Acrylamide. In *Experimental and Clinical Neurotoxicology* (P. S. Spencer and H. H. Schaumburg, Eds.), pp. 309–325, Williams and Wilkins, Baltimore.

Lim, S. S., Sammak, P. J., and Borisy, G. G. (1989). Progressive and spatially differentiated stability of microtubules in developing neuronal cells. *J. Cell Biol.* **109**, 253–263.

Lim, S. S., Edson, K. J., Letourneau, P. C., and Borisy, G. G. (1990). A test of microtubule translocation during neurite elongation. *J. Cell Biol.* **111**, 123–130.

Lipton, R. B., Apfel, S. C., Dutcher, J. P., Rosenberg, R., Kaplan, J., Berger, A., Einzig, A. I., Wiernick, P., and Schaumburg, H. H. (1989). Taxol produces a predominantly sensory neuropathy. *Neurology* **39**, 368–373.

Lotti, M., Moretoo, A., Capodicasa, E., Bertolazzi, M., Peraica, M., and Scapellato, M. L. (1993). Interaction between neuropathy target esterase and its inhibitors and the development of polyneuropathy. *Toxicol. Appl. Pharmacol.* **122**, 165–171.

Martenson, C. H., Sheetz, M. P., and Graham, D. G. (1993). Effect of acrylamide on the activity of purified dynein and kinesin. *Toxicologist* **13**, 124 (Abstract).

Medrano, C. J., and LoPachin, R. M. (1989). Effect of acrylamide and 2,5-hexanedione on brain mitochondrial respiration. *Neurotoxicology* **10**, 249–256.

Mendell, J. R., and Sahenk, Z. (1980). Interference of neuronal processing and axoplasmic transport by toxic chemicals. In *Experimental and Clinical Neurotoxicology* (P. S. Spencer and H. H. Schaumburg, Eds.), pp. 139–160, Williams and Wilkins, Baltimore.

Miller, J. J., Carter, D. E., and Spencer, I. G. (1982). Pharmacokinetics of acrylamide in Fisher-344 rats. *Toxicol. Appl. Pharmacol.* **63**, 36–46.

Miller, M. S., Miller, M. J., Burks, T. F., and Sipes, I. G. (1983). Altered retrograde axonal transport of nerve growth factor after single and repeated doses of acrylamide in the rat. *Toxicol. Appl. Pharmacol.* **69**, 96–101.

Miller, M. S., and Spencer, P. S. (1984). Single doses of acrylamide reduce retrograde transport velocity. *J. Neurochem.* **43**, 1401–1408.

Miller, M. S., and Spencer, P. S. (1985). The mechanisms of acrylamide axonopathy. *Annu. Rev. Pharmacol. Toxicol.* **25**, 643–666.

Monaco, S., Autilio-Gambetti, L., Lasek, R. J., Katz, M. J., and Gambetti, P. (1989a). Experimental increase of neurofilament transport rate: decreases in neurofilament number and in axon diameter. *J. Neuropathol. Exp. Neurol.* **48**, 23–32.

Monaco, S., Autilio-Gambetti, L., Zabel, D., and Gambetti, P. (1985). Giant axonal neuropathy: acceleration of neurofilament transport in optic axons. *Proc. Natl. Acad. Sci. USA* **82**, 920–924.

Monaco, S., Jacob, J., Jenich, H., Patton, A., Autilio-Gambetti, L., and Gambetti, P. (1989b). Axonal transport of neurofilament is accelerated in peripheral nerve during 2,5-hexanedione intoxication. *Brain Res.* **491**, 328–334.

Moretto, A., Lotti, M., Sabri, M. I., and Spencer, P. S. (1987). Progressive deficit of retrograde axonal transport is associated with the pathogenesis of di-*n*-butyl dichlorovos axonopathy. *J. Neurochem.* **49**, 1515–1522.

Moretto, A., and Sabri, M. I. (1988). Progressive deficits in retrograde axon transport precede degeneration of motor axons in acrylamide neuropathy. *Brain Res.* **440**, 18–24.

Myall, O. T., Allen, S. L., and McLean, W. G. (1990). The effect of 2,5-hexanedione on the induction of ornithine decarboxylase in the dorsal root ganglion of the rat. *Neurosci. Lett.* **114**, 305–308.

Nagata, H., and Brimijoin, S. (1986a). Axonal transport in the motor neurons of rats with neuropathy induced by p-bromophenyl acetyl urea. *Ann. Neurol.* **19**, 458.

Nagata, H., and Brimijoin, S. (1986b). Neurotoxicity of halogenated phenylacetyl ureas is linked to abnormal onset of rapid axonal transport. *Brain Res.* **385**, 136–142.

Nagata, H., Brimijoin, S., Low, P., and Schmelzer, J. D. (1987). Slow axonal transport in experimental hypoxia and in neuropathy induced by p-bromophenyl acetyl urea. *Brain Res.* **422**, 319–327.

Nennesmo, I., and Kristensson, K. (1986). Cytofluorometric quantification of somatopetal axonal transport: Effects of a conditioning lesion and 2,5-hexanedione. *Neuropathol Appl. Neurobiol.* **12**, 379–387.

Nixon, R. A. (1991). Axonal transport of cytoskeltal proteins. In *The Neuronal Cytoskeleton* (R. Burgoyne, Ed.), pp. 283–307, Wiley-Liss, New York.

Nixon, R. A. (1992). Slow axonal transport. *Curr. Opin. Cell Biol.* **4**, 8–14.

Nixon, R. A., and Logvinenko, K. B. (1986). Multiple fates of newly synthesized neurofilament proteins: evidence for a stationary neurofilament network distributed non-uniformly along axons of retinal ganglion cell neurons. *J. Cell Biol.* **102**, 647–658.

Ochs, S. (1982). *Axoplasmic Transport and its Relation to Other Nerve Functions.* John Wiley and Sons, New York.

Ochs, S., and Brimijoin, W. S. (1993). Axonal transport. In *Peripheral Neuropathy* (P. J. Dyck, P. K. Thomas, J. W. Griffin, P. A. Low, and J. F. Poduslo, Eds.), 3rd ed., Vol. 1, pp. 331–360, Saunders, Philadelphia.

Ochs, S., Jersild, A. Jr., and Li, J-M. (1989). Slow transport of freely movable cytoskeletal components shown by beading partition of nerve fibers in the cat. *Neuroscience* **33**, 421–430.

O'Donoghue, J. L. (1985). Carbon disulfide and organic sulfur-containing compounds. In *Neurotoxicity of Industrial and Commercial Chemicals* (J. L. O'Donoghue, Ed.), Vol. II, pp. 39–60, CRC Press, Boca Raton, Florida.

Oka, N., and Brimijoin, S. (1990). Premature onset of fast axonal transport in bromophenyl acetyl urea neuropathy: an electrophoretic analysis of proteins exported into motor nerves. *Brain Res.* **509**, 107–110.

Oka, N., and Brimijoin, S. (1992). Tubulomembranous lesions in BPAU neuropathy reflect local stasis of fast axonal transport: evidence from electron microscope autoradiography. *Mayo Clin. Proc.* **67**, 341–352.

Okabe, S., and Hirokawa, N. (1990). Turnover of fluorescently-labeled tubulin and actin in the axon. *Nature (London)* **343**, 479–482.

Papasozomenos, S. C., Autilio-Gambetti, L., and Gambetti, P. (1981). Reorganization of axoplasmic organelles following β,β'-iminodipropionitrile administration. *J. Cell. Biol.* **91**, 866–871.

Penningroth, S. M. (1986). Erythro-9-[3-(2-hydroxynonyl] adenine and vanadate as probes for microtubule-based cytoskeletal mechanochemistry. *Meth. Enzymol.* **134C**, 477–487.

Planas, A. M., and Cunningham, V. J. (1987). Uncoupling of cerebral glucose supply and utilization after hexane-2,5-dione intoxication in the rat. *J. Neurochem.* **48**, 816–823.

Pleasure, D. E., Mishler, K. C., and Engle, W. K. (1969). Axonal transport of proteins in experimental neuropathies. *Science* **66**, 524–525.

Poole, C. F., Sye, W.-F., Zlatkis, A., and Spencer, P. S. (1981). Determination of acrylamide in nerve tissue homogenates by electron-capture gas chromatography. *J. Chromatogr.* **217**, 239–245.

Price, D. L., and Griffin, J. W. (1980). Neurons and sheathing cells as targets of disease processes. In *Experimental and Clinical Neurotoxicology* (P. S. Spencer and H. H. Schaumburg, Eds.), pp. 2–23, Williams and Wilkins, Baltimore.

Ravindranath, V., and Pai, K. S. (1991). The use of rat brain slices as an *in vitro* model for mechanistic evaluation of neurotoxicity. *Neurotoxicology* **12**, 225–234.

Reese, T. S. (1987). The molecular basis of axonal transport in the squid giant axon. In *Molecular Neurobiology in Neurology and Psychiatry* (E. R. Kandel, Ed.), pp. 89–102, Raven Press, New York.

Reinsch, S. S., Mitchison, T. J., and Kirschner, M. (1991). Microtubule polymer assembly and transport during axonal elongation. *J. Cell Biol.* **115**, 365–379.

Riggs, J. E., Schochet, S. S., Gutmann, L., Crosby, T. W., and DiBartolomeo, A. G. (1986). Chronic human colchicine neuropathy and myopathy. *Arch. Neurol.* **43**, 521–523.

Roijitta, M., Horwitz, S. B., and Raine, C. S. (1984). Taxol-induced neuropathy: short-term effects of local injection. *J. Neurocytol.* **13**, 685–701.

Royatta, M., and Raine, C. S. (1986). Taxol-induced neuropathy: chronic effects of local injection. *J. Neurocytol.* **15**, 483–496.

Sabri, M. I. (1983a). *In vitro* and *in vivo* inhibition of glycolytic enzymes by acrylamide. *Neurochem. Pathol.* **1**, 179–191.

Sabri, M. I. (1983b). Mechanism of action of acrylamide on the nervous system. *Biol. Mem.* **8**, 16–27.

Sabri, M. I. (1984). *In vitro* effect of *n*-hexane and its metabolites on selected enzymes in glycolysis, pentose phosphate pathway and citric acid cycle. *Brain Res.* **297**, 145–150.

Sabri, M. I. (1986). Chemical neurotoxins and disruption of the axonal transport system. In *Axoplasmic Transport* (Z. Iqbal, Ed.), pp. 185–208, CRC Press, Boca Raton, Florida.

Sabri, M. I. (1992). Effects of 2,5-hexanedione and 3,4-dimethyl-2,5-hexanedione on retrograde axonal transport in sciatic nerves. *Neurochem. Res.* **17**, 835–839.

Sabri, M. I., Dairman, W., Fenton, M., Juhasz, L., Ng, T., and Spencer, P. S. (1989). Effect of exogenous pyruvate on acrylamide neuropathy in rats. *Brain Res.* **483**, 1–11.

Sabri, M. I., Soiefer, A. I., Moretto, A., Lotti, M., Miller, M. S., and Spencer, P. S. (1987). Early retrograde transport defects induced by primary axonal toxins. In *Axonal Transport* (R. S. Smith and M. A. Bisby, Eds.), pp. 459–472, Liss, New York.

Sabri, M. I., and Spencer, P. S. (1980). Toxic distal axonopathy: Biochemical studies and hypothetical mechanisms. In *Experimental and Clinical Neurotoxicology* (P. S. Spencer and H. H. Schaumburg, Eds.), pp. 206–219, Williams and Wilkins, Baltimore.

Sabri, M. I., and Spencer, P. S. (1990). Acrylamide impairs fast and slow axonal transport in rat optic system. *Neurochem. Res.* **15**, 603–608.

Sahenk, Z., and Lasek, R. J. (1988). Inhibition of proteolysis blocks anterograde-retrograde conversion of axonally-transported vesicles. *Brain Res.* **460**, 199–203.

Sahenk, Z., and Mendell, J. R. (1980a). Axoplasmic transport in zinc pyridinethione neuropathy: evidence for an abnormality in distal turnaround. *Brain Res.* **186**, 343–353.

Sahenk, Z., and Mendell, J. R. (1980b). Zinc pyridinethione. In *Experimental and Clinical Neurotoxicology* (P. S. Spencer and H. H. Schaumburg, Eds.), pp. 578–592, Williams and Wilkins, Baltimore.

Sahenk, Z., and Mendell, J. R. (1981). Acrylamide and 2,5-hexanedione neuropathies: abnormal bidirectional transport rate in distal axons. *Brain Res.* **219**, 397–405.

Sayre, L. M., Autilio-Gambetti, L., and Gambetti, P. L. (1985). Pathogenesis of experimental giant neurofilamentous axonopathies: a unified hypothesis based on chemical modification of neurofilaments. *Brain Res.* **10**, 69–83.

Schiff, P. B., and Horwitz, S. B. (1980). Taxol stabilizes microtubules in mouse fibroblast cells. *Proc. Natl. Acad. Sci. USA* **77**, 1561–1565.

Schnapp, B. J., and Reese, T. S. (1986). New developments in understanding rapid axonal transport. *Trends Neurosci.* **9**, 155–162.

Schnapp, B. J., and Reese, T. S. (1989). Dynein is the motor for retrograde axonal transport of organells. *Proc. Natl. Acad. Sci. USA* **86**, 1548–1552.

Seppäläinen, A. M., and Haltia, M. (1980). Carbon disulfide. In *Experimental and Clinical Neurotoxicology* (P. S. Spencer and H. H. Schaumburg, Eds.), pp. 356–373, Williams and Wilkins, Baltimore.

Sheetz, M. R., Stener, E. R., and Schroer, T. A. (1989). The mechanism and regulation of fast axonal transport. *Trends Neurosci.* **12**, 474.

Shivakumar, B. R., and Ravindranath, V. (1992). Selective modulation of glutathione in mouse brain regions and its effect on acrylamide-induced neurotoxicity. *Biochem. Pharmacol.* **43**, 263–269.

Shpetner, H. S., Paschal, B. M., and Vallee, R. B. (1988). Characterization of the microtubule-activated ATPase of brain cytoplasmic dynein (MAP 1C). *J. Cell. Biol.* **107**, 1001–1009.

Shpetner, H. S., and Vallee, R. B. (1989). Identification of dynamin, a novel mechanochemical enzyme that mediates interactions between microtubules. *Cell* **59**, 421–432.

Sickles, D. W. (1989a). Toxic neurofilamentous axonopathies and fast anterograde axonal transport. II. The effects of single doses of neurotoxic and non-neurotoxic diketones and β,β'-iminodipropionitrile (IDPN) on the rate and capacity of transport. *Neurotoxicology* **10**, 103–112.

Sickles, D. W. (1989b). Toxic neurofilamentous axonopathies and fast anterograde axonal transport. I. The effects of single doses of acrylamide on the rate and capacity of transport. *Neurotoxicology* **10**, 91–102.

Sickles, D. W. (1991). Toxic neurofilamentous axonopathies and fast anterograde axonal transport. III. Recovery from single injections and multiple dosing effects of acrylamide and 2,5-hexanedione. *Toxicol. Appl. Pharmacol.* **108**, 390–396.

Sickles, E. W., Fowler, S. R., and Testine, A. R. (1990). Affects of neurofilamentous axonopathy-producing neurotoxicants on *in vitro* production of ATP by brain mitochondria. *Brain Res.* **528**, 25–31.

Sickles, D. W., and Goldstein, B. D. (1986). Acrylamide produces a direct dose-dependent and specific inhibition of oxidative metabolism in motoneurons. *Neurotoxicology* **7**, 187–196.

Sidenius, P. (1986). The effect of doxorubicin on slow and fast components of the axonal transport system in rat. *Brain* **109**, 885–889.

Sidenius, P., and Jacobsen, J. (1983). Anterograde axonal transport in rats during intoxication with acrylamide. *J. Neurochem.* **40**, 697–704.

Smith, R. S., and Bisby, M. A. (1987). *Axonal Transport.* Liss, New York.

Souyri, F., Chretien, M., and Droz, B. (1981). Acrylamide-induced neuropathy and impairment of axonal transport of proteins. I. Multifocal retention of fast transported proteins of the peripheral axons as revealed by light microscope radioautography. *Brain Res.* **205**, 1–13.

Spencer, P. S., Sabri, M. I., Scaumburg, H. H., and Moore, C. L. (1979). Does a defect of energy metabolism in the nerve fiber underlie axonal degeneration in polyneuropathies? *Ann. Neurol.* **5**, 501–505.

Spencer, P. S., and Schaumburg, H. H. (1980). *Experimental and Clinical Neurotoxicology.* Williams and Wilkins, Baltimore.

Spencer, P. S., Schaumburg, H. H., Sabri, M. I., and Veronesi, B. (1980). The enlarging view of hexacarbon neurotoxicity. *CRC Crit. Rev. Toxicol.* **7**, 179–356.

Srivastava, S. P., Sabri, M. I., Agrawal, A. K., and Seth, P. K. (1986). Effect of single and repeated doses of acrylamide and bis-acrylamide on glutathione-s-transferase and dopamine receptors in rat brain. *Brain Res.* **371**, 319–323.

Sterman, A. B. (1982a). Acrylamide induces early morphologic reorganization of neuronal cell body. *Neurology* **32**, 1023–1026.

Sterman, A. B. (1982b). Cell body remodeling during dying-back axonopathy: DRG changes during advanced disease. *J. Neuropathol. Exp. Neurol.* **41**, 400–411.

Sterman, A. B. (1983). Altered sensory ganglia in acrylamide neuropathy. Quantitative evidence of neuronal reorganization. *Exp. Neurol.* **42**, 166–176.

Storm-Dickerson, T., Spencer, P. S., and Gold, B. G. (1992). Early and selective impairment in the fast antergrade transport (FAT)

rate of glycoproteins in acrylamide (AC) neuropathy. *Soc. Neurosci. Abstr.* **18**, 1084.

Studelska, D. R., Oakes, S. G., and Brimijoin, S. (1987). Monoclonal antibodies to rapidly transported, particle-associated antigens of rat sciatic nerves: production and characterization. In *Axonal Transport* (R. S. Smith and M. A. Bisby, Eds.), pp. 279–290, Liss, New York.

Tarkowski, S., Kolakowski, J., Gorny, R., and Opacka, J. (1980). Content of high energy phosphates and ultrastructure of mitochondria in the brain of rats exposed to carbon disulfide. *Toxicol. Lett.* **5**, 207–213.

Tilson, H. A. (1981). The neurotoxicity of acrylamide: an overview. *Neurobehav. Toxicol. Teratol.* **3**, 445–461.

Vallee, R. B., and Bloom, G. S. (1991). Mechanisms of fast and slow axonal transport. *Annu. Rev. Neurosci.* **14**, 59–92.

Vallee, R. B., Shpetner, H. S., and Paschal, B. M. (1989). The role of dynein in retrograde axonal transport. *Trends Neurosci.* **12**, 66–70.

Watson, D. F., Fittro, K. P., Hoffman, P. N., and Griffin, J. W. (1991). Phosphorylation-related immunoreactivity and the rate of transport of neurofilaments in chronic 2,5-hexanedione intoxication. *Brain Res.* **539**, 103–109.

Watson, D. F., Glass, J. D., and Griffin, J. W. (1993). Redistribution of cytoskeletal proteins in mammalian axons disconnected from their cell bodies. *J. Neurosci.* **13**, 4354–4360.

Watson, D. F., Hoffman, P. N., and Griffin, J. W. (1990). The cold stability of microtubules increases during axonal maturation. *J. Neurosci.* **10**, 3344–3352.

Wier, R. L., Glaubiger, G., and Chase, T. N. (1978). Inhibition of fast axoplasmic transport by acrylamide. *Environ. Res.* **17**, 251–255.

Weisenburg, R. C., Flynn, J., Gao, B., Awodi, S., Skee, F., Goodman, S. R., and Riederer, B. M. (1987). Microtubule gelatin-contraction: Essential components and relation to slow axonal transport. *Science* **23**, 1119–1121.

Weiss, D. G. (1982). *Axoplasmic Transport.* Springer-Verlag, Berlin.

Weiss, P., and Hiscoe, H. B. (1948). Experiments on the mechanisms of nerve growth. *J. Exp. Zool.* **107**, 315–395.

Xu, Z., Cork, L. C., Griffin, J. W., and Cleveland, D. W. (1993). Increased expression of neurofilament subunit NF-L produces morphological alterations that resemble the pathology of human motor neuron disease. *Cell* **73**, 23–33.

Yokoyama, K., Tsukita, S., Ishikawa, H., and Kurokawa, M. (1980). Early changes in the neuronal cytoskeleton caused by β,β'-iminodipropionitrile: selective impairment of neurofilament polypeptides. *Biomed. Res.* **1**, 537–547.

28

Disruption of Energy Metabolism and Mitochondrial Function

WILLIAM B. PARKER
Southern Research Institute
Birmingham, Alabama 35205

YUNG-CHI CHENG
Department of Pharmacology
Yale University School of Medicine
New Haven, Connecticut 06510

I. Biochemistry of Mitochondria and ATP Production in the Brain

Neurons have a very high metabolic rate relative to other cell types due to the need to maintain transmembrane ionic gradients and other vital functions. As in all cells, oxidative phosphorylation in mitochondria is primarily responsible for the generation of the ATP that is required as the energy source for most cellular functions. Therefore, neurons are very sensitive to the toxic actions of agents which disrupt mitochondrial production of ATP. For example, the classic inhibitors of oxidative phosphorylation, such as cyanide, sodium azide, and dinitrophenol, are all very toxic to neural tissue. These compounds kill animals by inhibiting oxidative phosphorylation in the central nervous system resulting in the disruption of nerve function and the cessation of breathing. Chronic toxicity with these agents results in optic atrophy, deafness, ataxia, seizures, myoclonus, basa ganglia degeneration, and movement disorders (Wallace, 1992a). In addition, there are a number of neurological diseases associated with defective mitochondria (Zeviani and Antozzi, 1992; Lestienne, 1992; Wallace, 1992b), which further supports the idea that neurons are particularly sensitive to the loss of mitochondrial function.

Mitochondria are the most abundant cellular organelle. There can be as many as 10,000 mitochondria in a single cell and they can account for 10 to 20% of all cellular protein. They are small cellular organelles that are composed of an outer and inner membrane. The inner membrane is extensively folded, giving it a very large surface area. Mitochondria were once free-living organisms and can perform many functions of a whole cell, such as RNA, DNA, and protein synthesis. Mitochondria replicate independently from cellular replication.

The mitochondria contain a genome consisting of a closed circular DNA molecule approximately 16 kb in length. Each mitochondrion can contain as many as 1000 separate genomes. The mitochondrial genome is known to code for 37 genes: 22 code for tRNA molecules, 2 code for 15S and 21S rRNAs, and 13 code for proteins (12 are components of the multimeric complexes of the inner membrane and function in the elec-

483

tron transport chain and 1 is a component of the mitochondrial ribosome). Interestingly, mitochondria have a slightly different genetic code for translating mRNA into protein than that of mammalian cells (Wallace, 1992c). All other mitochondrial proteins are coded on nuclear genes and are synthesized from the mRNA transcript with cytoplasmic ribosomes, which are then transported into the mitochondria. Approximately 90% of the proteins found in mitochondria are coded in the nucleus.

The primary function of mitochondria is the production of ATP that is used in various cellular processes. ATP is composed of two phosphodiester bonds, which when hydrolyzed release a large amount of energy. In most cells the energy required to produce ATP is obtained from the energy that is released in the oxidation of sugars, fats, and proteins. However, neural tissues are only able to utilize glucose as an energy source. In the cytoplasm of the cell, glucose is converted to pyruvic acid by a process referred to as glycolysis, which results in the production of two molecules of ATP. Under aerobic conditions, pyruvic acid enters the mitochondria and is converted to acetyl-CoA. Acetyl-CoA is oxidized to two molecules of CO_2 by the enzymes of the citric acid cycle within the inner compartment of the mitochondria, resulting in the reduction of three molecules of NAD^+ and one molecule of FAD. The electrons from NADH and $FADH_2$ are then fed into the electron transport chain in the inner mitochondrial membrane. The ultimate electron acceptor from the electron transport chain is oxygen, which is converted to water in the process. The electron transport chain is coupled to the pumping of protons across the inner mitochondrial membrane into the space between the inner and outer mitochondrial membranes, which generates an electrochemical gradient. The return of the protons to the inner portion of the mitochondria is coupled to the synthesis of ATP. Eleven molecules of ATP are produced for each molecule of acetyl-CoA that is oxidized to CO_2 and H_2O.

Under anaerobic conditions, pyruvic acid from glucose is converted to lactic acid instead of acetyl-CoA, and the sole source of cellular ATP is from glycolysis. Lactic acid is secreted from the cell and must be removed. Many more molecules of ATP are produced from oxidative phosphorylation than glycolysis, but when oxygen becomes limiting, a considerable amount of ATP can be generated from glycolysis only.

II. Inhibitors of Mitochondrial Function

There are two classes of neurotoxic compounds whose mechanism of toxicity has been extensively studied and the inhibition of mitochondrial function has been implicated as the cause of the neurotoxicity. The mechanism of action of these compounds will be discussed in detail. There are a number of other neurotoxins that are known to disrupt energy metabolism. However, in these cases there is not enough data to state with certainty that the mechanism of toxic action of these compounds is due to the disruption of energy metabolism. These compounds will only be briefly discussed. More detail about their mechanism of action can be obtained by consulting the quoted references.

A. 2',3'-Dideoxyinosine and 2',3'-Dideoxycytidine

Only three compounds are approved for the treatment of AIDS: 3'-azido-3'-deoxythymidine (AZT, zidovudine), 2',3'-dideoxyinosine (ddI, didanosine), and 2',3'-dideoxycytidine (ddC, zalcitabine). Treatment with two of these compounds, ddI and ddC, results in a painful peripheral neuropathy that limits their effectiveness. These compounds are analogs of 2'-deoxynucleosides that do not contain a 3'-hydroxyl (Fig. 1). They are phosphorylated by cellular enzymes to their respective 5'-triphosphates (ddATP and ddCTP), which are then used as substrates for DNA synthesis by the HIV reverse transcriptase, resulting in chain termination. HIV reverse transcriptase is responsible for copying the HIV RNA genome into the double-stranded DNA molecule that is integrated into the host genome, which is a vital process in HIV replication. Therefore, inhibition of this function by these agents results in the inhibition of virus replication. There are many other anti-HIV nucleoside analogs being considered for use as anti-AIDS compounds that inhibit HIV replication in essentially the same manner (DeClerq, 1992; Schinazi, 1992). Because the anti-HIV nucleoside analogs are potent inhibitors of HIV replication, the most important distinguishing factor among them is their differing toxicities.

The primary dose-limiting toxicity of ddI and ddC is a painful sensory-motor peripheral neuropathy. This toxicity is dose and time related, occurring 6 to 14 weeks after beginning of therapy and presents as a dysesthesia of the feet. Patients complain of numbness, tingling, burning, and pain in the feet. In the beginning the discomfort is intermittent, but becomes more constant with continued therapy. Some show a decrease in the vibratory sense. The condition is reversible upon cessation of treatment.

Few studies have been conducted in experimental animals due to the rapid clinical development of these agents and the lack of suitable animal models for peripheral neuropathy. Therefore, most of the mechanis-

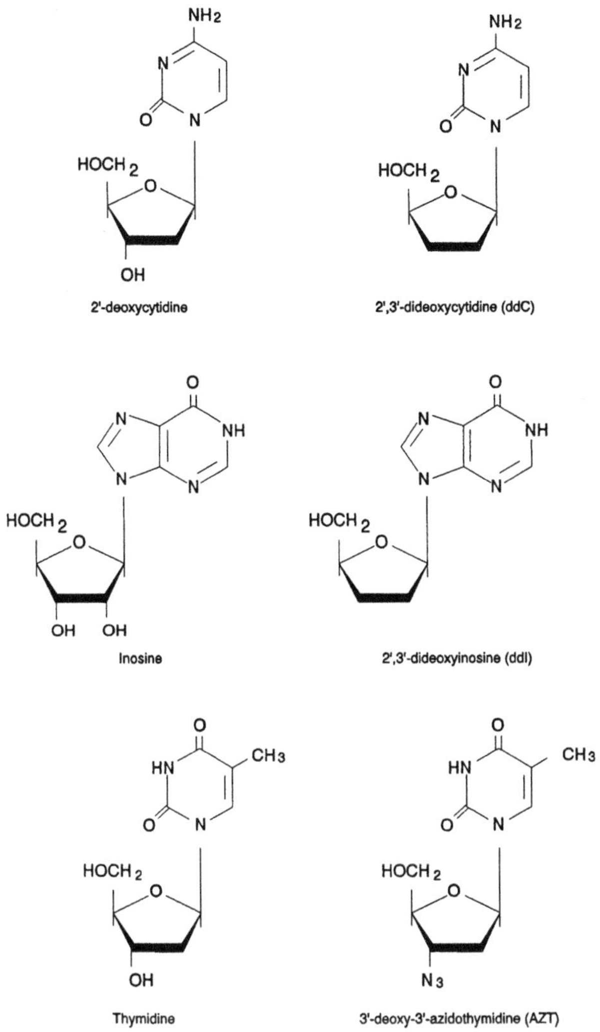

2'-deoxycytidine

2',3'-dideoxycytidine (ddC)

Inosine

2',3'-dideoxyinosine (ddI)

Thymidine

3'-deoxy-3'-azidothymidine (AZT)

FIGURE 1 Structures of anti-HIV nucleoside analogs.

tic information about the toxic effects of these agents has been learned from studies in isolated cell culture systems. ddI is phosphorylated in cells by the 5'nucleotidase to ddIMP (Johnson and Fridland, 1989), which is converted to ddATP by the purine synthetic enzymes that convert IMP to ATP. ddIMP is not converted to ddGTP. ddC is phosphorylated to ddCTP by sequential action of deoxycytidine kinase, cytosine nucleoside monophosphate kinase, and the nonspecific nucleoside diphosphate kinase (Fig. 2).

The formation of ddATP and ddCTP from ddI and ddC is necessary for their anti-HIV activity and is also responsible for their toxicities. These metabolites are analogs of the substrates that are used by the cellular DNA polymerases. Fortunately, ddATP and ddCTP are very poor inhibitors of the DNA polymerases responsible for the replication of chromosomal DNA (DNA polymerases α, δ, and ε). The K_i for the inhibition of these polymerases by ddCTP or ddATP is ap-

proximately 100 μM, and inhibition of these polymerases in intact cells by these agents is not a problem at concentrations required to inhibit the HIV reverse transcriptase. Only at concentrations of ddC greater than those used to treat patients is enough ddCTP produced to inhibit these enzymes. Agents that inhibit these enzymes are useful as anticancer agents and result in toxicity to rapidly proliferating cells such as those of the bone marrow. ddCTP and ddATP are more potent inhibitors of DNA polymerase β, the enzyme responsible for DNA repair, with K_i's of approximately 1 μM. Both ddATP and ddCTP are potent inhibitors of DNA polymerase γ, the enzyme responsible for the mitochondrial DNA synthesis (Starnes and Cheng, 1987). The K_i's for inhibition of DNA polymerase γ by ddCTP and ddATP (approximately 0.02 μM) are similar to their K_i's for the inhibition of the HIV reverse transcriptase.

There are great differences in the potency of these two drugs against both HIV replication and mitochondrial replication. These differences in potency are due to differences in the production of the active form of these compounds (Hao *et al.*, 1988). ddC is not degraded by any cellular enzymes to any appreciable degree and is readily phosphorylated in cells, whereas ddI is rapidly degraded by the ubiquitous enzyme, purine nucleoside phosphorylase, and is only poorly phosphorylated in cells to its active form.

ddI and ddC inhibit mitochondrial DNA replication in intact cells (Chen *et al.*, 1991), but ddC does not inhibit mitochondrial DNA synthesis in isolated preparations of mitochondria (Keilbaugh *et al.*, 1990). The lack of activity against mitochondrial DNA replication in isolated mitochondria is because of the inefficient synthesis of ddCTP by mitochondria. Mitochondria contain nucleoside and nucleotide kinases which can phosphorylate nucleosides, but the ddCTP that inhibits

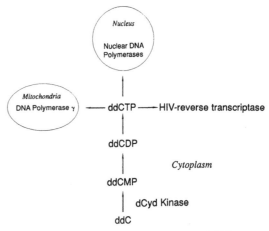

FIGURE 2 Cellular metabolism of ddC.

mitochondrial replication in intact cells is produced in the cytoplasm and is transported into the mitochondria (Chen and Cheng, 1992). Because the 5'-nucleotidase and the synthetic enzymes for ATP synthesis are also located in the cytoplasm, it is likely that ddATP found in the mitochondria is also produced in the cytoplasm and is transported into the mitochondria (Fig. 2). However, experimental evidence for this has not yet been shown.

Inhibition of mitochondrial DNA synthesis of proliferating cells does not immediately result in an effect on cell growth (Chen and Cheng, 1989). Treatment of Molt-4F cells with 0.2 μM ddC for 6 days decreased mitochondrial DNA content by more than 95%, having no effect on cell growth. During this time ATP levels remained constant and the production of lactic acid was increased, which indicated that the cells were producing the needed ATP by glycolysis. However, after treatment of cells for 6 days with 0.2 μM ddC, the growth rate of the cells began to decline, indicating that the loss of mitochondria fell below some critical level that could not be compensated for by increased ATP production through glycolysis.

Although there is much experimental evidence to suggest that the peripheral neuropathy caused by ddI and ddC is due to the inhibition of mitochondrial replication, this hypothesis is still not proven because of the lack of studies in an in vivo model for peripheral neuropathy and the difficulty of doing experiments in humans. However, there are a couple of factors that support this hypothesis: (1) Peripheral neuropathy is caused by a number of different anti-HIV nucleoside analogs composed of different bases, such as ddC, ddI, and D4T. The only common metabolic features of these compounds is their potent inhibition of DNA polymerase γ and their relative lack of activity against the DNA polymerases responsible for chromosomal DNA replication. (2) Mitochondrial deficiencies are known to result in peripheral neuropathy (Pezeshkpour et al., 1987; Peyronnard et al., 1980).

PC12 cells are a rat pheochromocytoma cell line that has been developed as an in vitro model for peripheral neuropathy (Chen et al., 1991; Keilbaugh et al., 1991). Upon addition of nerve growth factor these cells will stop growing and differentiate into neuron-like cells. The effect of drug treatment during this differentiation stage on neurite formation can be used to assess neurotoxicity. When anti-HIV nucleoside analogs were evaluated in this model, the inhibition of neurite formation correlated with the ability of the compound to cause peripheral neuropathy in patients (Keilbaugh et al., 1991). ddC and ddI inhibit mitochondrial DNA replication in differentiated PC12 cells (Chen et al., 1991). These results indicate that ddC and ddI also deplete

mitochondria in nondividing nerve cells in culture and suggest that this could also occur in vivo. A higher concentration of ddC was required to inhibit mitochondrial DNA replication in PC12 cells than in CEM cells, probably due to decreased phosphorylation by dCyd kinase. Few other biochemical studies have been done in this model to characterize the mechanism responsible for the inhibition of neuron differentiation.

Treatment of AIDS patients with AZT does not result in peripheral neuropathy. AZT can inhibit the replication of isolated mitochondria, but is only a marginal inhibitor of mitochondrial DNA synthesis when tested in intact cells. More importantly, AZT inhibits cell growth at concentrations that do not affect mitochondrial DNA synthesis, indicating that the inhibition of mitochondrial DNA synthesis by AZT is not a limiting toxicity. This result is consistent with the clinical toxicities associated with AZT. In patients, the rate-limiting toxicity for AZT is bone marrow suppression. These results are also consistent with the inhibition of DNA polymerase γ by AZT-TP; DNA polymerase γ is approximately 100-fold less sensitive to AZT-TP than it is to ddCTP (Vasquez-Padua et al., 1990). Treatment with AZT results in myopathy that is believed to be due to the inhibition by AZT-TP of DNA polymerase γ in muscle cells (Arnaudo et al., 1991; Weissman et al., 1992; Lewis et al., 1992). These results indicate that there could be differences in the metabolism of the anti-AIDS nucleoside analogs which may also explain the different toxicity profiles of these agents.

One animal model has been developed to study the peripheral neuropathy induced by ddC (Anderson et al., 1991; Feldman et al., 1991). Treatment of rabbits with ddC results in symptoms that are similar, but not identical, to the peripheral neuropathy seen in patients. The neuropathy in rabbits is characterized by neurological and electrophysiological deficits of peripheral nerves with myelin and axonal pathology. The most significant pathological findings were myelin splitting and intramyelinic edema, demyelination, and remyelination, with axonal loss. No effects of ddC were detected in the central nervous system. These studies did not evaluate the inhibition of mitochondrial replication in the development of the peripheral neuropathy. The authors speculate that if inhibition of mitochondrial replication was relevant to this pathology, then pathologic changes in neurons, axons, or Schwann cells should also occur. However, it is possible that different tissues have different metabolisms of anti-HIV nucleoside analogs or different requirements for mitochondria and that peripheral nerves are the most sensitive to the depression of mitochondrial numbers. Direct studies in this or other *in vivo* models are needed to assess the role of the inhibition of mitochondrial DNA replication in the development of peripheral neuropathy.

Inhibition of mitochondrial DNA replication should not affect mitochondria that are not in the process of replication. Therefore, mitochondria must turn over in nerve cells, if the peripheral neuropathy induced by ddC and ddI is due to the inhibition of mitochondrial replication. Very little has been done to study the turnover of mitochondria in animal tissues. One study has shown that mitochondria in the rat brain turn over with a half-life of approximately 30 days (Gross *et al.*, 1969). Miller and colleagues (1985) were able to detect considerable incorporation of [^{3}H]thymidine into the mitochondrial DNA in cat brains, which indicated that mitochondrial DNA synthesis does occur in nonproliferating neural cells. It is also possible that these compounds could inhibit the repair of mitochondrial DNA and that this could lead to mitochondrial destruction. Very little is known about the repair of mitochondrial DNA. Chromosomal DNA is a dynamic molecule that is in constant need of repair. It is likely that mitochondrial DNA is also continuously repaired. Because DNA polymerase γ is the only DNA polymerase in mitochondria, it is likely that this polymerase is responsible for the repair of mitochondria.

In summary, the available data support the following hypothesis to explain the mechanism of the peripheral neuropathy induced by many of the dideoxynucleosides utilized in the treatment of AIDS. Because these drugs do not cure AIDS, they must be administered continually to maintain levels of the active forms to permanently inhibit HIV reverse transcriptase and inhibit viral replication. However, the active forms of the drugs, which are responsible for the inhibition of the viral reverse transcriptase, also are very potent inhibitors of the mitochondrial DNA polymerase. Indeed, there is very little selectivity between the inhibition of purified DNA polymerase γ and the HIV reverse transcriptase by these compounds. Furthermore, there is a constant rate of mitochondrial turnover in nonproliferating tissues. Therefore, long-term treatment with these compounds not only inhibits viral DNA synthesis but also inhibits mitochondrial DNA synthesis. With the inhibition of mitochondrial DNA synthesis, mitochondrial numbers decline. At some pont the numbers of mitochondria fall below some threshold level which can no longer produce enough ATP to support vital cellular processes. It is also possible that the decline in mitochondria results in the loss of some other unknown vital function and that loss of this activity results in cell death. Although all tissues require mitochondria and decreases in mitochondrial numbers have been ob-

served in muscle cells of patients treated with some of these compounds, it is likely that there are differences between the different tissues which make the peripheral nerves particularly sensitive to the decline in mitochondrial numbers. These differences have not been identified, but they could relate to a number of characteristics. It is possible that (a) the threshold amount of mitochondria needed to maintain metabolic needs of nerve cells is greater than it is in other cells, (b) other cells are more able to compensate for the loss of mitochondria by generating ATP by glycolysis, (c) the mitochondrial turnover rate is greater in neurons than it is in other cell types, (d) mitochondria in the axons are more vulnerable to insult than are mitochondria in cell bodies, or (e) there could be differences in the metabolism of these agents in the different tissues which could account for their specificity. Although there is much information that supports this hypothesis, there are many questions concerning the effect of these compounds in intact animals on nerve cells that need to be answered before this hypothesis can be considered fact.

B. 1-Methyl-4-phenyl-1,2,3,6-tetrahydropyridine (MPTP)

MPTP was first recognized as a neurotoxin after the appearance of a parkinsonian syndrome in four heroin addicts in 1982 (Langston *et al.*, 1982). Within 1 week of the use of what was thought to be MPPP, an analog of meperidine, these people exhibited many of the symptoms of Parkinson's disease: jerking of limbs, stiffness, generalized slowing, and difficulty in moving. Within 2 to 6 weeks, three of the four patients were hospitalized with the following parkinsonian-like symptoms: near total immobility, increased tone, inability to speak intelligibly, fixed stare, diminution of blinking, facial seborrhea, constant drooling, positive glabellar tap test, and cogwheel rigidity in the upper extremities. One patient exhibited a "pill-rolling" tremor, and all patients had a flexed posture typical of fully developed Parkinson's disease. Although the fourth patient was not treated in a hospital, his symptoms were similar to the three hospitalized patients. Similar to patients with Parkinson's disease, all of these people responded to therapy with L-dopa, and the condition is irreversible. Analysis of the brain tissue of another person who had taken MPPP, who had developed this Parkinson's disease-like syndrome, and who had later died from a drug overdose revealed nerve cell damage in the substantia nigra (Davis *et al.*, 1979). The cell loss was comparable in severity to that seen in brains from patients that suffered from idiopathic Parkinson's disease. However, unlike the brains from patients with idiopathic Parkinson's disease, the locus coeruleus and dorsal motor vagus nucleus were not affected.

After much work the following story has been pieced together to explain the development of this particular toxicity, and this work has been reviewed (Kindt *et al.*, 1986; Singer *et al.*, 1987; Singer and Ramsay, 1990; Nicklas *et al.* 1992). The narcotic MPPP, 1-methyl-4-propion-oxypiperidine (Fig. 3), was produced in an unregulated "clandestine" laboratory to circumvent narcotic laws, which did not regulate its use. Unfortunately for the individuals who used this drug, it is a very unstable compound that decomposes to MPTP at high temperatures or low pH. MPTP is not only a narcotic, but is also the agent responsible for the parkinsonian syndrome described earlier. The MPPP used by the people in the original study of this toxocity was found to contain mostly MPTP and very little MPPP (Langston *et al.*, 1982).

MPTP readily crosses the blood–brain barrier and is oxidized to MPDP⁺ by the mitochondrial monoamine oxidase (MAO-B) throughout the brain. MPDP⁺ is then oxidized further to MPP⁺ both enzymatically and non-enzymatically, which is concentrated by the dopamine transport system in dopaminergic nerve endings of cells in the substantia nigra. MPP⁺ is then actively transported into the mitochondria by an energy-dependent, carrier-mediated transport system where it accumu-

lates to concentrations greater than 40-fold that in the cytoplasm. MPP⁺ reversibly inhibits the NADH dehydrogenase associated with complex I of the mitochondrial electron transport chain. The concentration of MPP⁺ required to inhibit the oxidation of pyruvate by 50% is between 200 and 500 μM. The inhibition of oxidative phosphorylation results in the decline of cellular ATP levels and ultimately in cell death. The selective accumulation of MPP⁺ in the dopaminergic neurons of the substantia nigra results in the selective killing of these cells. It is not known whether or not MPP⁺ is transported to the cell body or if it remains in the nerve terminal. In experimental animals and cell culture systems, the toxicity of MPTP can be prevented by treatment with inhibitors of either MAO-B or dopamine uptake.

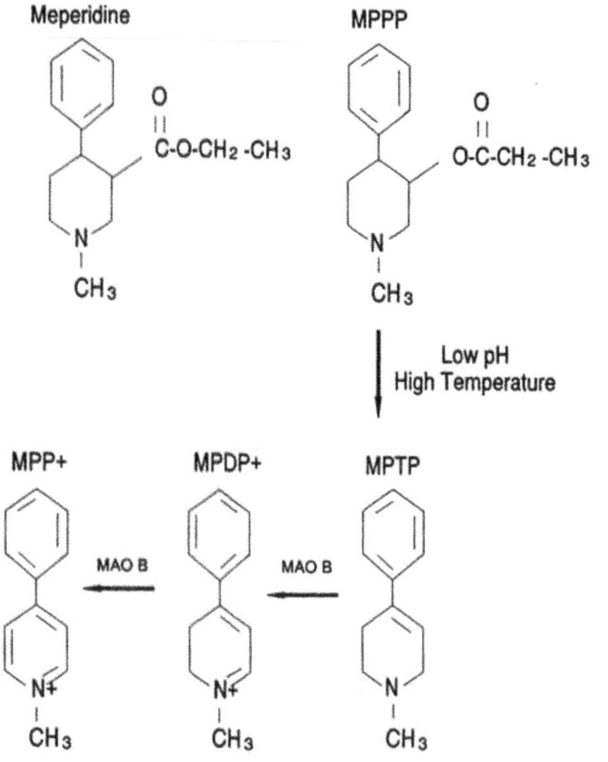

This unfortunate occurrence has had some positive effects for our understanding of Parkinson's disease. The ability to reproduce Parkinson's disease in animals has aided in the understanding of the disease and drug development. The toxicity to subhuman primates is very similar to that seen in humans. From this episode it has been suggested that idiopathic Parkinson's disease may result from environmental exposure to some unknown compounds. However, at this time there is no known environmental agent that has been linked to idiopathic Parkinson's disease, and it is possible that there are many causes of Parkinson's disease. If Parkinson's disease is caused by environmental agents, then it may be possible to prevent this disease by decreasing the exposure to these agents.

A unique approach to the treatment of Parkinson's disease is being tested in two people who suffer from MPTP-induced Parkinson's disease. Human fetal brain tissue has been transplanted into the brains of these people and has resulted in improved mobility and quality of life (Fahn, 1992; Widner *et al.*, 1992). If these studies are successful in reversing the damage, then it is possible that a similar approach may be useful in the treatment of people with idiopathic Parkinson's disease (Freed *et al.*, 1992; Spencer *et al.*, 1992).

C. Methyl Mercury

Methyl mercury is a well-known neurotoxin due to the environmental disaster that occurred in Minamata, Japan in the 1950s. Methyl mercury that was discharged into Minamata bay contaminated fish and shellfish and was consumed by the inhabitants which resulted in severe neurotoxicological and developmental effects. The biochemical actions of methyl mercury that are responsible for this toxicity are still not well understood. However, methyl mercury does inhibit mitochondrial production of ATP by inhibiting cyto-

FIGURE 3 Conversion of MPPP to MPP⁺.

chrome-*c* of the electron transport chain (Verity *et al.*, 1975), although it is still not certain whether this effect is the cause of the observed toxicity or if it is secondary to some other cellular effects (Komulainen, 1988).

D. Acetyl-ethyl-tetramethyl-tetralin, Triethyl Tin, Hexachlorophene, and Halogenated Salicylanilides

These compounds of dissimilar structure result in a similar neurotoxicity, bubbling, and vacuolation of the myelin sheath and ceroid inclusions, and are all known uncouplers of oxidative phosphorylation (Cammer, 1980). This correlation suggests that the uncoupling of oxidative phosphorylation is responsible for the observed toxicity of these compounds. However, these actions occur at concentrations that have minimal effect of ATP levels which indicates that this action of these compounds is not primarily responsible for the observed toxicity. This has been reviewed by Cammer (1980).

E. Chloramphenicol

Chloramphenicol is a wide spectrum antimicrobial agent, which selectively kills bacterial cells due to its preferential binding of the bacterial ribosomes and subsequent inhibition of protein synthesis (Smith and Weber, 1983). Because of the similarity between bacteria and mitochondrial protein synthesis, chloramphenicol also inhibits mitochondrial protein synthesis. This effect of chloramphenicol is believed to be responsible for much of the toxicity of this agent. Treatment with chloramphenicol can also result in some minor neurotoxicity, such as blurring of vision, digital paresthesia, and optic neuritis (Godel *et al.*, 1980), although these toxicities are not the major toxicity of this drug (Sande and Mandell, 1990).

References

Anderson, T. D., Davidovich, A., Arceo, R., Brosnan, C., Arezzo, J., and Schaumburg, H. (1991). Peripheral neuropathy induced by 2′,3′-dideoxycytidine. A rabbit model of 2′,3′-dideoxycytidine neurotoxicity. *Lab. Invest.* **66**, 63–74.

Arnaudo, E., Dalakas, M., Shanske, S., Moraes, C. T., Dimauro, S., and Schon, E. A. (1991). Depletion of muscle mitochondrial DNA in AIDS patients with zidovudine-induced myopathy. *Lancet* **337**, 508–510.

Cammer, W. (1980). Toxic demyelination: biochemical studies and hypothetical mechanisms. In *Experimental and Clinical Neurotoxicity* (P. S. Spencer and H. H. Schaumburg, Eds.), pp. 239–256, Williams and Wilkins Company, Baltimore.

Chen, C. H., and Cheng, Y.-C. (1989). Delayed cytotoxicity and selective loss of mitochondrial DNA in cells treated with the anti-human immunodeficiency virus compound 2′,3′-dideoxycytidine. *J. Biol. Chem.* **264**, 11934–11937.

Chen, C.-H., Vazquez-Padua, M., and Cheng, Y.-C. (1991). Effect of anti-human immunodeficiency virus nucleoside analogs on mitochondrial DNA and its implication for delayed toxicity. *Mol. Pharmacol.* **39**, 625–628.

Chen, C.-H., and Cheng, Y.-C. (1992). The role of cytoplasmic deoxycytidine kinase in the mitochondrial effects of the anti-human immunodeficiency virus compound, 2′,3′-dideoxycytidine. *J. Biol. Chem.* **267**, 2856–2859.

Davis, G. C., Williams, A. C., Markey, S. P., Ebert, M. H., Caine, E. D., Reichert, C. M., and Kopin, I. J. (1979). Chronic parkinsonism secondary to intravenous injection of meperidine analogues. *Psychiatry Res.* **1**, 249–254.

DeClercq, E. (1992). HIV inhibitors targeted at the reverse transcriptase. *AIDS Res. Hum. Retroviruses* **8**, 119–134.

Fahn, S. (1992). Fetal-tissue transplants in Parkinson's Disease. *N. Engl. J. Med.* **327**, 1589–1590.

Feldman, D., Brosnan, C., and Anderson, T. D. (1991). Ultrastructure of peripheral neuropathy induced in rabbits by 2′,3′-dideoxycytidine. *Lab. Invest.* **66**, 75–85.

Freed, C. R., Breeze, R. E., Rosenberg, N. L., Schneck, S. A., Kriek, E., Qi, J.-X., Lone, T., Zhang, Y.-B., Snyder, J. A., Wells, T. H., Ramig, L. O., Thompson, L., Mazziotta, J. C., Huang, S. C., Grafton, S. T., Brooks, D., Sawle, G., Schroter, G., and Ansari, A. A. (1992). Survival of implanted fetal dopamine cells and neurologic improvement 12 to 46 months after transplantation for Parkinson's Disease. *N. Engl. J. Med.* **327**, 1549–1555.

Godel, V., Nemet, P., and Lazar, M. (1980). Chloramphenicol optic neuropathy. *Arch. Ophthalmol.* **98**, 1417–1421.

Gross, N. J., Getz, G. S., and Rabinowitz, M. (1969). Apparent turnover of mitochondrial deoxyribonucleic acid and mitochondrial phospholipids in the tissues of the rat. *J. Biol. Chem.* **244**, 1552–1562.

Hao, Z., Cooney, D. A., Hartman, N. R., Perno, C. F., Fridland, A., DeVico, A. L., Sarnagadharan, M. G., Broder, S., and Johns, D. G. (1988). Factors determining the activity of 2′,3′-dideoxynucleosides in suppressing human immunodeficiency virus *in vitro*. *Mol. Pharmacol.* **34**, 431–435.

Hooper, C. (1992). Fetal Transplants improve MPTP Parkinson's Patients. *J. NIH Res.* **4**, 31–34.

Johnson, M. A., and Fridland, A. (1989). Phosphorylation of 2′,3′-dideoxyinosine by cytosolic 5′-nucleotidase of human lymphoid cells. *Mol. Pharmacol.* **36**, 291–295.

Keilbaugh, S. A., Prusoff, W. H., and Simpson, M. V. (1991). The PC12 cell as a model for studies of the mechanism of induction of peripheral neuropathy by anti-HIV-1 dideoxynucleoside analogs. *Biochem. Pharmacol.* **42**, R5–R8.

Keilbaugh, S. A., Moschella, J. A., Chin, C. D., Mladenovic, J., Lin, T.-S., Prusoff, W. H., Rowehl, R. A., and Simpson, M. V. (1990). Role of mtDNA replication in the toxicity of 35-azido-3′deoxythymidine (AZT) in AIDS therapy and studies on other anti-HIV-1 dideoxynucleosides. *Biochem. Pharmacol.* **38**, 1033–1036.

Kindt, M. V., Nicklas, W. J., Sonsalla, P. K., and Heikkila, R. E. (1986). Mitochondria and the neurotoxicity of MPTP. *TIPS* **7**, 473–475.

Komulainen, H., (1988). Neurotoxicity of methyl mercury: cellular and subcellular aspects. In *Metal Neurotoxicity* (S. C. Bondy and K. Prassad, Eds.), pp. 167–182, CRC Press, Boca Raton, Florida.

Langston, J. W., Ballard, P., Tetrud, J. W., and Irwin, I. (1982). Chronic parkinsonism in humans due to a product of Meperidine-analog synthesis. *Science* **219**, 979–980.

Lestienne, P. (1992). Mitochondrial DNA mutations in human diseases: a review. *Biochimie* **74**, 123–130.

Lewis, W., Gonzalez, B., Chomyn, A., and Papoian, T. (1992). Zidovudine induces molecular, biochemical, and ultrastructural changes in rat skeletal muscle mitochondria. *J. Clin. Invest.* **89**, 1354–1360.

Miller, C. T., Krewski, D., and Tryphonas, L. (1985). Methylmercury-induced mitochondrial DNA synthesis in neural tissue of cats. *Fundam. Appl. Toxicol.* **5**, 251–264.

Nicklas, W. J., Saporito, M., Basma, A., Geller, H. M., and Heikkila, R. E. (1992). Mitochondrial mechanisms of neurotoxicity. *Ann. N.Y. Acad. Sci.* **648**, 28–36.

Peyronnard, J.-M., Charron, L., Bellavance, A., and Marchand, L. (1980). Neuropathy and mitochondrial myopathy. *Ann. Neurol.* **7**, 262–268.

Pezeshkpour, G., Krarup, C., Buchtal, F., DiMauro, S., Bresolin, N., and McBurney, J. (1987). Peripheral neuropathy in mitochondrial disease. *J. Neurol. Sci.* **77**, 285–304.

Sande, M. A., and Mandell, G. L. (1990). Antimicrobial agents: tetracyclines, chloramphenicol, erythromycin, and miscellaneous antibacterial agents. In *The Pharmacological Basis of Therapeutics* (A. G. Goodman, T. W. Rall, A. S. Nies, and P. Taylor, Eds.), pp. 1125–1130, Pergamon Press, New York.

Schinazi, R. F., Mead, J. R., and Feorino, P. M. (1992). Insights into HIV chemotherapy. *AIDS Res. Hum. Retrovir.* **8**, 963–990.

Singer, T. P., and Ramsay, R. R. (1990). Mechanism of the neurotoxicity of MPTP: an update. *FEBS Lett.* **274**, 1–8.

Singer, T. P., Trevor, A. J., and Castagnoli, N., Jr. (1987). Biochemistry of the neurotoxic action of MPTP: or how a faulty batch of "designer drug" led to parkinsonism in drug abusers. *TIBS* **12**, 266–270.

Smith, A. L., and Weber, A. (1983). Pharmacology of chloramphenicol. *Pediatr. Clin. North Am.* **30**, 209–236.

Spencer, D. D., Tobbins, R. J., Naftolin, F., Marek, K. L., Vollmer, T., Leranth, C., Roth, R. H., Price, L. H., Gjedde, A., Bunney, B. S., Sass, K. J., Elsworth, J. D., Kier, E. L., Makuch, R., Hoffer, P. B., and Redmond, D. E. (1992). Unilateral transplantation of human fetal mesencephalic tissue into the caudate nucleus of patients with Parkinson's disease. *N. Engl. J. Med.* **327**, 1541–1548.

Starnes, M. C., and Cheng, Y.-C. (1987). Cellular metabolism of 2',3'-dideoxycytidine, a compound active against human immunodeficiency virus *in vitro*. *J. Biol. Chem.* **262**, 988–991.

Vazquez-Padua, M. A., Starnes, M. C., and Cheng, Y.-C. (1990). Incorporaton of 3'-azido-3'-deoxythymidine into cellular DNA and its removal in a human leukemic cell line. *Cancer Commun.* **2**, 55–62.

Verity, M. A., Brown, W. J., and Cheung, M. (1975). Organic mercurial encephalopathy: in vivo and in vitro effects of methyl mercury on synaptosomal respiration. *J. Neurochem.* **25**, 759–766.

Wallace, D. C. (1992a). Mitochondrial genetics: a paradigm for aging and degenerative diseases. *Science* **256**, 628–632.

Wallace, D. C. (1992b). Diseases of the mitochondrial DNA. *Annu. Rev. Biochem.* **61**, 1175–1212.

Wallace, D. C. (1992c). Diseases of the mitochondrial DNA. *Annu. Rev. Biochem.* **61**, 1175–1212.

Weissman, J. D., Constantinitis, I., Hudgins, P., and Wallace, D. C. (1992). ^{31}P magnetic resonance spectroscopy suggests impaired mitochondrial function in AZT-treated HIV-infected patients. *Neurology* **42**, 619–623.

Widner, H., Tetrud, J., Rehncrona, S., Snow, B., Brundin, P., Gustavii, B., Bjorklund, A., Lindvall, O., and Langston, J. W. (1992). Bilateral fetal mesencephalic grafting in two patient with Parkinsonism induced by 1-methyl-4-phenyl-1,2,3,6-tetrahydropyridine (MPTP). *N. Engl. J. Med.* **327**, 1556–1563.

Zeviani, M., and Antozzi, C. (1992). Defects of mitochondrial DNA. *Brain Pathol.* **2**, 121–132.

PART V

In Vitro Models

In Vitro Neurotoxicology
An Introductory Overview

ALAN M. GOLDBERG
Johns Hopkins School of Public Health
Baltimore, Maryland 21205

DANIEL ACOSTA, JR.
Department of Pharmacology and Toxicology
College of Pharmacy
The University of Texas
Austin, Texas 78712

In vitro models offer unique opportunities to elucidate basic understandings of neurotoxicity or neuronal cell injury. Complex biochemical phenomena can be satisfactorily studied with CNS neural tissue *in vitro*. Slices, organotypic cell cultures, cells in suspension, cell lines, and primary cell cultures as several techniques currently in use.

There is a clear and pressing need for rapid mechanistically based tests to evaluate the potential for xenobiotics to damage the nervous system. The chapters in this section provide toxicity mechanisms with basic methodological approaches to better understand the neurotoxicology of selected agents.

The chapter by Drs. Campbell and Abdulla provides a strategic overview of the area of toxic substances that affect the nervous system, the use of specific diease models, and the role of trophic factors. This chapter sets the stage for another chapter by the same authors on calcium homeostasis and calcium-induced cell damage. Here the authors document, in detail, the role

calcium can play in eliciting toxicity and the authors close with an approach to fully evaluate calcium-related toxicological mechanisms.

The chapter by Drs. Honegger and Schilter looks at the use of serum-free aggregating brain cell cultures as a tool for studying neurotoxicology. These authors demonstrate that aggregating brain cell cultures can form three-dimensional brain cell units that mimic aspects of normal brain development. They show how a variety of toxicants can be examined in developmental, as well as for long-term, chronic studies.

Drs. Fountain and Teyler provide great insight into the use of brain slice techniques in neurotoxicology. The reader is provided with an opportunity to understand the use of brain slices for many different areas and insight is provided into many neurotoxic agents and the effects of these agents on the brain slice system. Finally, a screening method is proposed for the evaluation of neurotoxicity.

The chapter by the late Dr. van den Bercken and

Dr. Vijverberg *et al.* in this section examines an *in vitro* electrophysiological approach to neurotoxicity. Using pyrethroids as model compounds, these authors elegantly and clearly demonstrate how classical electrophysiological approaches can be used in the elucidation of mechanisms of toxicity. In addition to examining pyrethroids, this chapter also provides details of lead neurotoxicity.

Dr. Verity provides an in-depth discussion on the value of neuronal cells in suspension as experimental models, whereas Dr. Ronnett reviews the use of human cell lines for neurotoxicity studies. The importance of the so-called nonneuronal cell types (glial cells and Schwann cells) in gaining a better understanding of neurotoxic mechanisms is described by Drs. Aschner and Vitarella (glial cells) and DeVries (Schwann cells). Dr. Bornstein provides a historical perspective on the use of organotypic cultures of nerve tissue for neurotoxicity studies.

In vitro sciences in general and *in vitro* neurotoxicolcogy in particular offer the unique opportunity to provide necessary information on potential agents that can affect the nervous system in a timely and meaningful way. This section of the series just highlights some possible routes of development and methodological approaches that allow us to address questions of current societal need. It should be clear to the reader that these are but the first beginnings of the use of these methods in safety assessment and hazard evaluation. As we more fully understand the basic underlying mechanisms of nervous system function, we will be able to design methods and approaches that will allow understanding the potential physiological implications of exposure to agents that affect the nervous system.

CHAPTER

29

Strategic Approaches to in Vitro Neurotoxicology

IAIN C. CAMPBELL
Institute of Psychiatry
London SE5 8AF
United Kingdom

ELIZABETH M. ABDULLA*
Wellcome Research Laboratories
Beckenham, Kent BR3 3BS
United Kingdom

I. Introduction

A. The Uniqueness of Neurons

Intact neurons have an extremely large surface-to-volume area because of their dendrites and axons. They are physiologically very responsive to electrical stimulation, i.e., they are excitable. In addition, the neuron is one of the most metabolically active cells in the body (Kaplan, 1982) with complex, highly organized anterograde and retrograde systems for the fast and slow transport of cellular proteins and transmitters. Any toxic insult to such a highly active cell with such a large area of membrane, which differs in susceptibility as it differs in function (cell soma, dendrite, synapse), which does not divide, is serious. The neuron is unlike any other cell in the body in the sense that it has an absolute requirement for glucose and cannot metabolize lipids. In addition, neurons share the distinction with the lens of the eye of not requiring insulin for transport of glucose into the cell. Neurons then do not divide and many undergo apoptosis in the normal course of development. Plasticity is another very im-portant specialized property whereby changes in stimulation (in the pattern of signals, transmitters, pressure damage, toxic damage) lead to permanent functional changes in the neuron which could perhaps include neurite outgrowth, selective expression of a particular group, subgroup, or type of receptor, and activation of genes for new specialized transmitters. "The age of plasticity" in children is the period when neurons are maximally capable of forming new connections and synapses, of learning, and memory.

B. In Vivo versus in Vitro Neurotoxicology

Neurons in culture differ from neurons *in vivo* for a number of reasons which must be kept in mind when interpreting data from *in vitro* systems. Primary neurons in culture are metabolically, physiologically, and morphologically stunted; they are not subject to normal excitatory and inhibitory inputs. The susceptibility to toxic insult of these relatively "resting" neurons may then be reduced. Artificial "electrical" stimulation could perhaps be partially achieved by depolarization using ~15 mM K$^+$ (slightly above the normal 5 mM),

495

but this can never really mimic the constant changing electrical events *in vivo*.

Two basic options for growing neurons in culture are available. The first option uses primary neurons. However, these do not divide and survive only for a limited time. In addition, it is difficult to obtain homogeneous, well-defined, and reproducible cultures. The advantage of primary neurons (particularly fetal and neonatal neurons) is that they extend neurites rapidly with minimal requirement for extracellular matrix (ECM) proteins, neural cell adhesion molecules (N-CAM, N-cadherin), or nerve growth factor (NGF), which may provide a useful system for measuring neurotoxicity *in vitro*. The alternative approach is the use of neuroblastomas, but these are transformed cells and are likely to differ substantially from normal neurons. Advantages in using cell lines include convenience and reproducibility and a reduction in the use of animals. Cell lines are also amenable to longer-term culture for chronicity studies. Cell lines, unlike primary neurons, need to be manipulated with, for example, NGF and ECMs (e.g., laminin) or with N-CAMs to encourage them to express differentiated characteristics. However, the *in vivo* importance of NGF and other neurotrophic signals is unclear and thus it is possible that their presence *in vitro* could increase or decrease neuronal susceptibility to toxic insult. The relative anoxia which exists in culture also leads to loss of the differentiated phenotype, although this problem can be lessened by "roller culture techniques" (in which cultures are exposed to the air nine times per hour) or "millipore filter culture" which allows medium to reach both sides of a monolayer. Coculture of neuronal cells with differentiated astroglial cells or coculture in mixed cell brain cultures (reaggregate, micromass, and explant) encourage expression of features of the neuron *in situ*. These features include adhesion to the substratum, neurite outgrowth, expression of NGF receptors, integrins, cessation of proliferation, and downregulation of c-*myc*.

The life history of a normal neuron *in vivo* is characterized by a temporal progression that may be viewed as a discrete series of neurogenic steps through which virtually all cells pass. These steps include induction, proliferation, migration, restriction and determination, differentiation (expression), the formation of axonal pathways and synaptic connections, and onset of physiological function (Oppenheim, 1991). In many parts of the peripheral and central nervous systems (CNS) roughly one-half of the neurons can be further characterized by their expression of an additional terminal step in the neurogenic process: a cascade of cellular and molecular events leading to regression and ultimate degeneration and cell death (apoptosis). Cell lines may provide the means to probe the steps in this cascade.

II. Effects and Use of Different Toxic Substances

The most obvious way to study neurotoxicity *in vitro* is to begin by examining the effects of known toxic agents. It is clear from even a cursory glance at such substances that they are from a chemically diverse group. This necessitates and allows a multipronged approach which may elucidate a few final discrete common pathways leading to cell death. Information on the postulated mode of action of some of these well-known neurotoxic compounds is described in the following sections. An *in vitro* test of neurotoxicity, which is simple to assay, yet physiologically complex and therefore composed of a series of discrete biochemically definable steps (such as neurite outgrowth), is needed to test for the neurotoxic potential of such a wide range of substances.

A. Free Radicals

Free radicals are chemical species that have a single unpaired electron in an outer orbital. In such a state the free radical is extremely reactive and unstable and enters into reactions with inorganic or organic chemicals, proteins, lipids, and carbohydrates, particularly with key molecules in membranes and nucleic acids. Moreover, free radicals initiate autocatalytic reactions whereby molecules with which they react are themselves converted into free radicals and thus propagate a chain of damage.

Sources of free radicals include hydrogen peroxide, hydroxyl radicals, superoxide, and oxygen-derived radicals. The main effect of these reactive species is on membrane lipids, sulfhydryl bonds of proteins, and nucleotides of DNA; they may cause peroxidation of lipids within cellular and organellar membranes and cause damage to endoplasmic reticulum, mitochondria, and other microsomal components.

There are both endogenous and exogenous antioxidants (scavengers of free radicals), including vitamin E, sulfhydryl-containing compounds such as cysteine, glutathione, and D-penicillamine, and serum proteins such as ceruloplasmin and transferrin. Transferrin acts by binding free iron which could catalyze free radical formation. Both superoxide dismutase (which converts superoxide to hydrogen peroxide) and catalase in peroxisomes (which decomposes hydrogen peroxide to oxygen and water) dispose of free radicals. There is now a postulated role for free radicals in the etiology of motor neuron disease where a mutation in superoxide dismutase type I is found in 10% of cases of amyotrophic lateral sclerosis (ALS) (Rosen *et al.*, 1993) (see Section III, B).

Carbon tetrachloride can cause free radical-mediated toxicity. Inflammation produces reactive oxygen from the leukocyte oxygen burst, a mechanism mediating microbial killing. Irradiation injury results from production of hydrogen and hydroxyl-free radicals which cause damage by altering nucleic acids. The "free radical theory of aging" states that free radicals are formed more frequently in the aging organism due to the loss of inactivating enzymes (Cotran *et al.*, 1989).

1-Methyl-4-phenyl-1,2,3,6-tetrahydropyridine (MPTP) is toxic when converted by brain monoamine oxidase-B (MAO-B) to MPP$^+$ (1-methyl-4-phenylpyridinium ion) which acts by a free radical-mediated mechanism to damage the mitochondrial complex 1 of dopaminergic cells of the substantia nigra: it inhibits NAD$^+$-linked substrates at the same site as rotenone (Ramsay *et al.*, 1986).

Kuroda and colleagues (1990) cultured tyrosine hydroxylase-positive dopaminergic neurons from fetal monkey substantia nigra. MPP$^+$ decreases neurons to 2–25% of controls. Hence, this *in vitro* system may be useful for detecting primate-specific neurotoxins and for studying primate-specific neural diseases at both the molecular and the cellular levels.

In another *in vitro* study, explant cultures from midbrain regions of 16-day rat embryos were grown for 8 days before MPP$^+$ or MPTP was added. Effects were monitored by visual assessment of neurite outgrowth and catecholamine fluorescence and by measurement of [^{3}H]dopamine (DA) uptake. MPP$^+$ (10 μM/7 days) had no clear morphological effects but abolished catecholamine fluorescence and reduced DA uptake; effects were apparent within 24 h. MPTP was less effective (Mytilineou *et al.*, 1985). Similar effects were found with dissociated cultures of embryonic rat midbrain (Friedman and Mytilineou, 1987).

Neurons were used in a dissociated culture prepared from the ventral tegmentum of the mesencephalon (which includes the substantia nigra) of 15- to 16-day-old rat embryos. MPP$^+$ and MPTP were added to 7-day-old cultures and effects were monitored by catecholamine fluorescence and tyrosine hydroxylase histochemistry. With MPP$^+$ and MPTP at 10 to 100 μM there was a decrease in fluorescence but no loss of cells whereas with 0.1–10 μM MPP$^+$ there was a selective loss of tyrosine hydroxylase-positive cells. Generalized toxic effects were seen at concentrations above 100 μM for MPTP and above 30 μM for MPP$^+$. These results were obtained with rat neurons which are rela-

tively resistant to MPTP *in vivo*, perhaps due to a subtle difference in their energy metabolism (Sanchez-Ramos, 1986). The same authors used the MPTP antagonist mazindol, a selective dopamine uptake blocker, and showed partial protection against MPP$^+$ whereas desipramine, a blocker of noradrenaline uptake, and deprenil, an inhibitor of MAO-B, gave no protection (Walum *et al.*, 1990). Studies have been done with cocultures of rat cerebellar granule cells and astrocytes; astrocyte MAO-B converts MPTP to MPP$^+$ (Martini *et al.*, 1989).

MPTP had no morphological effect on actively growing neuroblastoma N2AB-1 cells or C6 glioma cells nor did it affect cell numbers. However, a low dose of MPP$^+$ (33.7 μM) was cytotoxic to mitotic N2AB-1 cells inducing vacuole formation and cell lysis and inhibiting cell growth over a 3-day period. Protein synthesis was reduced by 50% after 5 hr by 33.7 μM MPP$^+$.

Differentiated, neurite-bearing N2AB-1 cells lose neurites and change in size and shape following exposure to MPP$^+$ (0.33–33.7 μM/24 hr) and some cells appear to be mitogenically stimulated, indicating that MPP$^+$ may act as a teratogen. C6 glioma cells, however, were resistant to MPP$^+$. Although mitotic N2AB-1 cells incubated with MPTP produced only trace amounts of MPP$^+$, C6 gliomia cells generated significant amounts of this metabolite (3.6 μM). Moreover, although the morphology and cell number of coculture did not change in the presence of MPTP, glioma–neuroblastoma cocultures produced 2.9 μM MPP$^+$, which decreased protein synthesis by 18%. The C6 glioma cells perhaps delay the regression of neurites. The early changes in protein synthesis and the loss of neurite in cells treated with MPP$^+$ may reflect the drop in catecholamine levels and the dying back of fibers seen after *in vivo* administration of MPTP (Notter *et al.*, 1988).

In rat pheochromcytoma cell lines (PC12), DA uptake blockade (by 50%) was seen at 2 μM MPTP, but choline and 2-deoxyglucose transport were unaffected. Incubation with MPTP (100 μM, 3 days) decreased stored DA, but 1.5 mM was needed for 7 days to kill 95% of the cells (Denton and Howard, 1984).

Further studies with PC12 revealed that some clones are relatively resistant to MPP$^+$, which led the authors to conclude that the *in vivo* neuronal selectivity of MPTP results from subtle differences in the energy metabolism of different classes of neurons (Denton and Howard, 1987).

Several conclusions can be drawn from these observations. Species, cell type, and the stage of development of the cell may be critically important to the development of an *in vitro* test of neurotoxicity. In addition, the "read out," e.g., catecholamine uptake, cellular morphology, and inhibition of neurite outgrowth, should be chosen carefully as all parameters may not be changed by the toxic insult.

B. Excitatory Amino Acids

Excitatory amino acids (EAA) cause neuropathological changes by acting on postsynaptic receptors which causes persistent depolarization (Dykens, 1987) and receptor-mediated increases in intracellular-free calcium ($[Ca^{2+}]_i$) (Seubert *et al.*, 1988). This results in toxicity, possibly by a hyperosmotic effect and/or by a calcium-mediated activation of catabolic enzymes (e.g., phospholipases, proteases, and endonucleases), and an increase in intracellular-free calcium which can lead to cell death. Kainic acid causes receptor-mediated increases in intracellular-free calcium that results in cellular toxicity (Berdkowsky *et al.*, 1983). Some food and food additives may contain potentially neurotoxic compounds, e.g., tartrazine and β-*N*-methylamino-L-alanine (L-BMAA) from the false sago palm *Cycas circanalis* in Guam and β-*N*-oxylamino-L-alanine (L-BOAA) from the chick pea *Lathyrus sativus*. These two nonprotein amino acids contribute to the pathology of Guam disease or ALS/parkinsonian/dementia (L-BMAA) and lathyrism (L-BOAA), respectively (Harvey, 1988).

Excitatory amino acids fall into five main subgroups (Meldrum and Garthwaite, 1990) and the susceptibility of neurons to excitatory amino acids can change markedly with age. Different neurons often exhibit special development profiles of vulnerability to each of the selective excitatory amino acid receptor agonists (Meldrum and Garthwaite, 1990), which should be considered when developing *in vitro* tests.

Intense glutamate exposure produces immediate neuronal swelling due to the toxic influx of extracellular Ca^{2+}, which can be prevented by removing extracellular Na^+ or Cl^- (Olney *et al.*, 1986). However, even without acute swelling, most neurons exposed to glutamate (even briefly) will die in a delayed fashion if extracellular Ca^{2+} is present (Choi, 1985). Removal of extracellular calcium substantially attenuates EAA-induced neuronal loss in cortical and hippocampal cultures (Choi, 1985) and in cerebellar slices (Garthwaite and Garthwaite, 1986). Furthermore, glutamate-induced $^{45}Ca^{2+}$ uptake by cortical neurons is highly correlated with resultant neuronal degeneration (Marcoux *et al.*, 1988). These findings provide a link between glutamate neurotoxicity and data suggesting that excess Ca^{2+} influx accompanies hypoxic–ischemic neuronal injury *in vivo* (Siesjo, 1988).

Glutamate-induced neurotoxicity may be blocked, in particular by *N*-methyl-D-aspartate (NMDA)-type glutamatergic receptor antagonists (Rothman, 1984). Glutamate activates several subtypes of receptor ionophore complexes, named for their preferred pharmacological agonists: NMDA, kainate, and quisqualate (Watkins and Olverman, 1987). Both NMDA and non-NMDA (kainate and quisqualate) receptors mediate the ability of glutamate to excite neurons or to produce excitotoxic neuronal swelling. Selective antagonism of NMDA receptors alone, however, although unable to prevent either neuroexcitation or acute neuronal swelling, is sufficient to block the late degeneration induced by brief glutamate exposure (Choi, 1988). The prominent role of the NMDA subtype of the glutamate channel in glutamate neurotoxicity is partially explained by the latter's unique ability to open a membrane channel that is highly permeable to calcium (McDermott *et al.*, 1986). The NMDA receptor-activated calcium channel may be the major route by which glutamate induces a toxic Ca^{2+} influx, although additional Ca^{2+} entry probably occurs through voltage-activated Ca^{2+} channels, the Na^+/Ca^{2+} exchanger, and nonspecific membrane leakage (Choi, 1988). Ca^{2+} may also be released from intracellular stores by the action of glutamate on metabotropic (nonchannel-linked) quisqualate receptors (Sladaczek *et al.*, 1985; Nicoletti *et al.*, 1986). Glutamate neurotoxicity can be attenuated by antagonists added after glutamate exposure (Choi, 1988). Toxic exposure to glutamate triggers further neuronal injury by promoting the excessive leakage of endogenous glutamate stores.

The addition of high concentrations of magnesium (which blocks transmitter release) to hippocampal neurons in culture (which have developed excitatory glutamatergic synapses) protects against the extensive neuronal degeneration seen in its absence: neurons without glutamatergic synapses are not sensitive to EAA toxicity (Rothman and Samaie, 1985).

Neuronal vulnerability to EAA-induced injury is not uniform. Cortical neurons containing NADPH-diaphorase, striatal neurons containing γ-aminobutyric acid (GABA), and striatal neurons containing acetylcholine esterase all possess some intrinsic resistance to injury by NMDA *in vitro* and a concomitant heightened vulnerability to kainate and quisqualate.

Another factor (in addition to glutamatergic receptor density) likely to influence neuronal susceptibility to EAA toxicity is the corelease of zinc, which can influence the receptor distribution activated by glutamate (Koh and Coi, 1988).

C. Metals and Metal Complexes

The role of metals in neurotoxicity is unclear, but they can catalyze free radical chain reactions (Brown *et al.*, 1979). The copper levels in the cerebrospinal fluid of patients with Parkinson's disease are very high. Trimethyl tin (TMT) causes neuronal necrosis in the hippocampus and in the amygdaloid and pyriform cortex (Brown *et al.*, 1979). In rat, TMT is bound to hemoglobin which renders it nontoxic, but in marmoset, hamster, and gerbil, TMT is toxic as it does not bind to their hemoglobin (Brown *et al.*, 1984).

Some exogenous substances, including heavy metals, have been demonstrated to modulate long-term potentiation (LTP) (Woolley *et al.*, 1984), producing a response facilitation in the dentate gyrus that is quite similar to LTP, although perhaps induced by the convulsant activity of these agents.

Aluminium, depending on its route of administration, induces accumulation of neurofilaments in proximal axons or in the perikaryon (Ramsay *et al.*, 1986). Aluminium toxicity has been implicated in Alzheimer's disease (AD) because (a) there is increased aluminium in the brain correlated with aging; (b) aluminium is present in neurofibrillary tangles of AD (Good *et al.*, 1992) and Guam disease brains (Perl, 1986; using laser microprobe mass analysis LAMMA); and (c) aluminium is seen in senile plaques of AD (Halliwell and Gutteridge, 1988).

The pathogenesis of AD has been linked to the use of aluminium utensils and antacids. However, there is a lack of AD in patients suffering from aluminium intoxication or in aluminium factory workers. Thus aluminium in the lesions associated with the β-amyloid protein may be a secondary phenomenon. However, concern over heavy metal contamination in the atmosphere and in drinking water is likely to result in continuing interest in this area.

D. Acrylamide

The acrylamide monomer causes central and peripheral neurological defects, characterized by nerve terminal deformation, distal axonal swelling, and retrograde axonal degeneration (Lequesne, 1980; Cavanagh, 1982). As acrylamide is water soluble, it is distributed evenly throughout the body and does not require metabolic conversion by the liver; *in vitro* data may therefore be very pertinent.

In acrylamide-treated neuroblastoma N1E115 cultures, growth cone activities cease and the cones thicken. This is followed by neurite swelling and a progressive distal degeneration of the processes (Walum and Peterson, 1984). Inhibition of nerve fiber outgrowth is also seen in chick dorsal root ganglion cultures in the presence of acrylamide (Sharma and Obersteiner, 1977), and Walum and Flint (1988) found an inhibition of morphological differentiation in the rat midbrain micromass culture at concentrations lower than those causing general cytotoxicity.

Despite the morphological and functional differences between growing and synapsing nerve endings, *in vivo* swelling of terminals both peripherally and centrally is an early sign of acrylamide intoxication (Lequesne, 1980). The mechanism by which acrylamide causes axonal degeneration is still a matter for conjecture. The compound is highly reactive and will bind to a number of sites in neurons, and one hypothesis is that neurofilaments are modified (Sayre, 1985). Acrylamide was found to increase the number and size of mitochondria *in vivo* by a mechanism thought to involve an impairment of calcium regulation (Jones and Cavanagh, 1984). Acrylamide also causes increased respiration in cultured neuroblastoma cells (Nyberg and Walum, 1984). In addition, $[Ca^{2+}]_i$ is increased in acrylamide-treated neuroblastoma cells following depolarization (Walum *et al.*, 1987) and thus the impaired sequestration of $[Ca^{2+}]_i$ may contribute to the neuropathy.

E. Organophosphates

Organophosphorus esters (OPs) cause phosphorylation of a hydroxyl group on serine in the active site of certain enzymes depending on their specific shape size and charge. Phenylmethylsulfonyl fluoride (PMSF) inhibits this reaction for some OPs. Perhaps the most famous OP is Lewisite or "nerve gas." OPs are divided into two classes based on various parameters, e.g., chemical structure, species selectivity, age sensitivity and length of latent period, clinical signs, morphology and distribution of lesions, protection with PMSF, inhibition of neurotoxic esterase, and effect on catecholamine secretion from adrenodullary chromaffin cells. Type I has a pentavalent phosphorus and Type II has a trivalent phosphorus.

Exposure to organophosphates (to tri-*o*-tolyl phosphate, TOCP, Type I) has occurred in several ways, including administration of creosote for tuberculosis (59 cases in 1989), ingestion of ginger extract (10,000–20,000 cases in the United States in 1930), and use of contaminated cooking oil in Morocco (10,000 cases in 1959) and in India (1000 cases in 1960). There have also been 12 cases of exposure to leptophos [O-(4-bromo-2,5-dichlorophenyl)O-methyl phenylphosphonothioate] when it was manufactured in the United States in 1974–1975. Other organophosphates have been used in suicide attempts. Some exposure has also occurred when organophosphates have been used as insecticides or in animal dips (Abou-Donia and Lapadula, 1990).

OPs induce direct or indirect inhibition of acetylcholine esterase and hence cause excessive activity in cholinergic neurons; in addition, Type I compounds phosphorylate calcium/calmodulin kinase II, which phosphorylates various cytoskeletal proteins such as microtubules, α- and β-tubulin, neurofilaments, and MAP 2, causing disassembly of these elements and damage to the axon. Enhanced phosphorylation of accumulated cytoskeletal proteins prevents their assembly into polymers; instead they aggregate into solid masses and may undergo Ca^{2+}-activated proteolysis. Dissociation of cytoskeletal proteins causes increased fast axonal transport in treated animals: this results in accumulation of mitochondria at distal portions of the axon. Calcium is released from damaged mitochondria and/or endoplasmic reticulum, leading to overloading and disruption of intracellular, extracellular ionic gradients and to entry of water into the axon. Focal internodal swelling and the Ca^{2+}-activated proteolysis that follows, result in focal axonal degeneration that spreads somatofugally to involve the entire axon (Abou-Donia and Lapadula, 1990).

F. Hexane

The toxicity of acrylamide and 2,5-hexanedione (the oxidation product of *n*-hexane) may be caused by damage to neurofilaments. In the case of 2,5-hexanedione, damage may be caused by formation of a pyrrole adduct with the protein γ-lysyl-amino group on the neurofilaments (Sayre *et al.*, 1985); subsequent oxidation of the pyrrole ring leads to a crosslinking reaction of the neurofilaments (Rosenberg *et al.*, 1987), resulting in their accumulation and to subsequent axonal swelling (Di-patre and Butcher, 1991). 2,5-Hexanedione causes well-defined peripheral neuropathy in rats and humans, as well as testicular dysfunction (Spencer *et al.*, 1980).

III. The Use of Disease Models

A. Parkinsonism

MPTP causes toxicity when it is converted by brain MAO-B to MPP^+ which acts by a free radical-mediated mechanism to damage the mitochondrial complex 1 of dopaminergic cells of the substantia nigra; it inhibits NAD^+-linked substrates at the same site as rotenone (Ramsay *et al.*, 1986). MPP^+ is the active component of "street crack." MPP^+ selectively kills dopaminergic neurons in the substantia nigra of humans and certain laboratory animals, resulting in a Parkinson's-like syndrome. MPP^+ is a substrate for the catecholamine uptake system. Inhibition of either MAO-B or dopamine uptake blocks the toxic activity of MPTP (Wilson *et al.*, 1991). In addition to the toxic action of MPTP and MPP^+, they are able to induce central and peripheral release of catecholamines (Wilson *et al.*, 1991) in a non-MAO-B-dependent way. In the rat atrium *in vitro* this positive ionotropic effect is blocked by propranolol and nomifensine. In a study with MPTP parkinsonian monkeys, *Macaca fascicularis*, the D2 agonist (+)-4-propyl-9-hydroxynapthoxasine [(+)-PHNO] was found to alleviate the condition and a concomitant 40–70% decrease in the number of D2 sites in the caudate and putamen was seen (Alexander *et al.*, 1991).

Phenothiazines such as chlorpromazine (or largactyl), used for treatment of schizophrenia, cause parkinsonian tremor. Haloperidol, a nonphenothiazine, is structurally related to MPTP and is also converted to a pyridinium metabolite, but by a non-MAO-B mechanism. Intrastriatal administration of the metabolite into rat shows it to be 10% as effective as MPP^+ in causing the irreversible depletion of striatal nerve terminal dopamine; this raises the possibility that some neurological disorders seen in experimental animals and in man during chronic treatment with haloperidol may be due to this metabolite (Sabramanyam *et al.*, 1990).

B. Amyotrophic Lateral Sclerosis and Lathyrism

Guam disease occurs in the Chamorros, living on the Pacific island of Guam, who have a high incidence of ALS, the most common of the motor neuron diseases, often associated with clinical signs of parkinsonism or senile dementia. The neuropathology is similar to that of sporadic ALS (axonal and myelin degeneration symmetrically affecting the motor tracts of the cortex, brain stem, and spinal chord) except for a high incidence of neurofibrillary tangles; senile plaques are absent. Guam disease is sometimes also referred to as ALS/parkinsonian/dementia. The cycad, *C. circinalis* (false sago palm), grows in the western pacific and its seeds are used to make the flour in Guam. This flour contains an amino acid, L-BMAA, that has been proposed to be the exogenous excitotoxin in Guam disease (Spencer, 1987). It has also been found that zinc was also ingested with the L-BMAA because of the way the flour was prepared on Guam so the possibility exists that zinc may have contributed to the etiology of Guam disease (Duncan *et al.*, 1992), perhaps by potentiating the excitotoxicity (Koh and Choi, 1988).

L-BMAA becomes toxic *in vitro* in the presence of bicarbonate, at 0.3–3 m*M* (Ross *et al.*, 1987; Weiss *et al.*, 1989), perhaps through the formation of the α-methyl carbamate. The excitotoxicity of L-BMAA *in vitro* can be blocked by 2-amino-(5 or 7)-phosphonoheptanoic acid, AP5 and AP7, indicating an action on NMDA receptors (Ross *et al.*, 1987; Weiss and Choi, 1988). At low concentrations L-BMAA shows a selective excitotoxic action on NADPH-diaphorase-positive neurons, suggesting an action on non-NMDA receptors (Weiss *et al.*, 1989).

When rats are given high doses of L-BMAA (i.p.) there is cerebellar pathology and selective degeneration of stellate basket and Purkinje cells (Meldrum and Garthwaite, 1990). When monkeys are given L-BMAA orally they suffer immobility, tremor, weakness, and possible chromatolysis of Betz cells (Spencer *et al.*, 1987).

Rosen and colleagues (1993) have reported a tight genetic linkage between familial ALS (FALS) and a genetic defect in the gene on chromosome 21q for the cytosolic Cu/Zn-binding superoxide dismutase (SOD1) which catalyzes the dismutation of the toxic superoxide anion $O_2^{\cdot-}$ to O_2 and H_2O_2. No data are present to show a consistent increase or decrease in the activity of this enzyme. A slight possibility exists that the mutations cause a measurable change in the enzyme activity and may contribute to FALS by (1) increasing neuronal levels of superoxide anion or (2) increasing cellular levels of H_2O_2. Another possibility is that the correlation has no causal effect on the development of the disease.

Lathyrism or neurolathyrism is a syndrome that includes symptoms of spastic paraplegia and can be of either acute or chronic onset. It is linked to dietary consumption of L-BOAA from *Lathyrus sativus* which is consumed in large quantities (e.g., 1 g/kg/day) during drought in East Africa and southern Asia, when the incidence of the disease suddenly and concomitantly increases. L-BOAA acts on glutamate receptors, predominantly those on spinal neurons. When given systemically to immature mice, L-BOAA causes periventricular pathology. When L-BOAA is given to monkeys intrathecally, it causes changes in anterior horn cells and when given orally it causes tremors, myoclonus, and spasticity (Spencer *et al.*, 1986).

C. Alzheimer's Disease

Several types of neurons, including cholinergic, somatostinergic, and noradrenergic, degenerate in Alzheimer's disease. Loss of basal forebrain cholinergic neurons that project to the cerebral cortex, hippocampus, and amygdala is a striking feature of the neuropathology of AD; it can be reproduced in experimental animals by injecting an excitotoxin into the basal forebrain region where these cells are located (Coyle, 1983). This has provided a useful model for studying the role of cholinergic neurons in the cognitive defects

of AD. The fact that excitotoxins are effective in destroying cholinergic neurons implies that these neurons have EAA receptors through which endogenous excitotoxins could act pathologically to destroy these neurons. Alternatively, topical application of an excitotoxin to the cerebral cortex causes retrograde degeneration of those basal forebrain cholinergic neurons that project to the cortex (Sofroniew and Pearson, 1985). Thus an excitotoxic process can cause neuronal degeneration by either a direct or an indirect mechanism; since EAA receptors are present on many types of CNS neurons, an excitotoxic, either direct or indirect, could explain the death of somatostatinergic and noradrenergic neurons as well as cholinergic neurons. Arendash and colleagues (1987) described late-occurring pathological changes resembling neuritic plaques and neurofibrillary tangles in various limbic and neocortical regions of rat brain 14 months after injection of an excitotoxin into the basal forebrain to destroy cholinergic neurons. If this observation can be corroborated, it becomes a tenable hypothesis that not only can the primary degeneration of various types of neurons in AD be explained by an excitotoxic process, but other aspects of AD neuropathology (e.g., plaques and tangles) may arise as a delayed manifestation of or secondary reaction to this primary degenerative process (Olney, 1990). Reinforcing this line of conjecture is the finding of Deboni and McLachlan (1985) that exposure of cultured human spinal neurons to abnormal concentrations of glutamate or aspartate causes these neurons to produce paired helical filaments of the type that make up neurofibrillary tangles in AD. In addition, Procter and colleagues (1988) have presented evidence for a possible metabolic defect affecting glutamatergic neurons.

D. Stroke/Epilepsy

Human brain requires continuous oxygen and glucose. Hence, irreversible damage occurs if blood flow is reduced below 10 ml/100 g tissue/min; if blood flow is completely interrupted, damage will occur in a few minutes (Choi and Rothman, 1990). In stroke the ischemia and hypoxia is localized to individual vascular territories. Certain neuronal subpopulations are highly sensitive to hypoxic–ischemic insult, particularly hippocampal CA1 cells and neocortical layers 3, 5, and 6. Some of this special vulnerability may be accounted for by the central neurotoxicity of the endogenous excitatory amino acid neurotransmitter, glutamate, released into extracellular space under hypoxic–ischemic conditions.

Under normal conditions, powerful neuronal and glial uptake systems remove synaptically released glutamate from the extracellular space before toxicity occurs (Schousboe, 1981). Glutamate at $100 \mu M$ for 5 min destroys large numbers of cultured cortical neurons (Choi, 1987). There is a protective effect of late agonist administration against excitatory amino acid neurotoxicity *in vivo* (Foster *et al.*, 1988) and it is possible that glutamate antagonists may be protective against ischemic–hypoxic damage *in vivo*. Cortical NADPH-diaphorase-containing neurons (non-NMDA receptor-bearing) appear to be selectively spared in a neonatal rat model of ischemic injury (Ferriero *et al.*, 1988).

Glutamate receptor involvement in initiation and propagation of seizure (Dingledine *et al.*, 1990) and in massive neuronal death during periods of ischemia and hypoglycemia (Choi *et al.*, 1988) is well established. The first clear demonstration that glutamate plays a role in hypoxic–ischemic brain damage came from studies with rats showing that intrahippocampal injection of the competitive NMDA antagonist 2-amino-7-phosphonoheptanoate (AP7) reduced the loss of Ca1 pyramidal neurons produced by transient carotid ligation (Simon *et al.*, 1984); AP7 also protected the rat caudate from injury induced by hypoglycemia.

Dysfunction of glutamatergic pathways has also been implicated in epilepsy (Weiss and Choi, 1988). The majority of cholinergic synapses in the CNS are muscarinic, and acetylcholine is thought to play a key role in neural mechanisms underlying memory, learning, arousal, control of movement, and possibly the generation of epileptic foci (Choi, 1988). Decreased numbers of muscarinic receptors in the CNS have been reported following induction of kindling, a model for epilepsy in which the repeated administration of initially subconvulsive electrical stimuli eventually induces seizure activity. This decrease in receptor number has been suggested to be an agonist-independent phenomenon secondary to the increased electrical activity that occurs during seizures (Nathanson, 1987). Decreased numbers of receptors in the mouse mutant, *totterer,* an animal model of epilepsy, have been observed with a developmental time course that suggests the changes in receptor number are also secondary to the increased electrical activity that occurs during seizures (Liles *et al.*, 1988). Direct electrical or pharmacological stimulation can decrease muscarinic receptor number in the CNS (Savage *et al.*, 1985).

IV. Role of Trophic Factors in Neuronal Well-Being

The neurotrophic theory (Abdul-Ghani *et al.*, 1981) describes the massive death of neurons that occurs in many parts of the nervous system during normal development. With the discovery of nerve growth factor, it was recognized that neurons compete for trophic agents (Oppenheim, 1989; Barde, 1989). Because there is now evidence for the derivation of trophic agents or trophic-like effects from afferents, glia, and the extracellular matrix (Walicke, 1989; Johnson *et al.*, 1988), it seems likely that the targets of neurons may not be the only source of neurotrophic support for neuronal survival.

Neurological diseases that occur in later stages of life, long after the cessation of normal cell loss, and that involve the death of discrete populations of neurons may be the result of defective processes similar or even identical to those that regulate neuronal survival during development. For instance, neuronal loss in diseases such as ALS, parkinsonism, and AD may be partly due to abnormal neurotrophic or related cellular or molecular mechanisms (Oppenheim, 1989). The loss of cholinergic neurones in AD has been suggested to involve NGF deprivation (Oppenheim, 1989; Snider and Johnson, 1989). In ALS and Huntington's disease, synaptically mediated neurotoxic mechanisms have been postulated to kill discrete populations of cells (Mochetti, 1989). Increases in NGF gene expression have been observed in hippocampus *in vivo*, following an experimental seizure in which it is presumed that EAA transmission overtakes inhibitory GABAergic transmission. Diazepam blocks the kainate-induced NGF mRNA increase, which suggests that the balance between excitatory and inhibitory transmission may be a critical aspect of NGF regulation. These observations support the hypothesis that NGF is synthesised and released in response to GABAergic and glutamatergic activity (Mochetti, 1991).

Evidence also exists that adrenal steroids may be the common pathway by which a variety of drugs lead to NGF (and brain-derived neurotrophic factor, BDNF) expression. Clenbuterol, a lipophilic β-adrenergic agonist able to cross the blood–brain barrier when administered to 21-day-old rats, causes a transient threefold increase in NGF mRNA in cerebral cortex 5 hr later. The effect is blocked by concomitant dosing with (-)propanolol. The cyclophilin mRNA (examined as a control to ensure that there is no general increase in mRNA) remained unchanged. Clenbuterol also increases NGF protein levels. These data suggest that noradrenaline acts on postsynaptic astroglial β-adrenergic receptors to induce NGF (Mochetti, 1989).

An apparent inverse relationship exists between $[Ca^{2+}]_i$ and the trophic factor dependence observed *in vivo* and *in vitro* as neurons mature (Johnson and Deckworth, 1993). Developmental "apoptosis" of neurons requires the cells to be exquisitely sensitive to

NGF but the mature nervous system cannot be so vulnerable.

There are two genes in *Caenorhabditis elegans*: *ced-3* and *ced-4*; a mutation in either of these genes causes survival of almost all cells that would normally die in the normal course of development. A third gene, *ced-9*, appears to repress the expression of *ced-3* and *ced-4* (Ellis and Horvitz, 1986). In mammalian cells, similar genes have not yet been found but a gene product of a putative "killer gene" called "thanatin" has been proposed. Sympathetic neurons deprived of NGF can be prevented from dying by general inhibitors of protein and RNA synthesis. However, there is a time when these inhibitors are unable to rescue the cells but NGF (and cAMP and depolarization with K^+) is still active as a saving agent (Edwards *et al.*, 1991). It is thought that NGF may cause a post-translational switch that inactivates thanatin at this late stage (Johnson and Deckworth, 1993). Whether apoptosis will emerge as a valid means of assessing neurotoxicity remains to be established; however, the interrelationship between growth factors and gene expression demands that the topic be further investigated.

References

Abdul-Ghani, A. S., *et al.* (1981). Effects of Tityus toxin and sensory stimulation on muscarinic cholinergic receptors in vivo. *Biochem. Pharmacol.* **30**, 2713–2714.

Abou-Donia, M. B., and Lapadula, D. M. (1990). Mechanisms of organophosphorous ester-induced delayed neurotoxicity Type I and Type II. *Annu. Rev. Pharmacol. Toxicol.* **30**, 405–440.

Alexander, G. M., *et al.* (1991). Dopamine receptor changes in untreated and (+)-PHNO-treated MPTP parkinsonian primates. *Brain Res.* **547**, 181–189.

Arendash, G. W., *et al.* (1987). Long-term neuropathological and neurochemical effects of nucleus basalis lesions in the rat. *Science* **238**, 952–956.

Barde, Y. A. (1989). Trophic factors and neuronal survival. *Neuron* **2**, 1525–1534.

Berdkowsky, E., *et al.* (1983). Kainate, *N*-methyl-D-aspartate, and other excitatory amino acids increase calcium influx into rat brain cortex cells in vitro. *Neurosci. Lett.* **36**, 75–80.

Brown, A. W., *et al.* (1979). The behavioural and neuropathological sequelae of intoxication by trimethyltin compounds in the rat. *Am. J. Pathol.* **97**, 59–82.

Brown, A. W., *et al.* (1984). The neurotoxicity of trimethyltin chloride in hamsters, gerbils and marmosets. *J. Appl. Toxicol.* **4**, 12–21.

Cavanagh, J. B. (1982). The pathokinetics of acrylamide intoxication: a reassessment of the problem. *Neuropathol. Appl. Neurobiol.* **8**, 315–336.

Chalmers, D. T., *et al.* (1990). Differential alteration in cortical glutaminergic binding sites in senile dementia of the Alzheimer's type. *Proc. Natl. Acad. Sci. U.S.A.* **87**, 1352–1356.

Choi, D. W. (1985). Glutamate neurotoxicity in cortical cell culture is calcium dependent. *Neurosci. Lett.* **58**, 293–297.

Choi, D. W. (1987). Ionic dependence of glutamate neurotoxicity in cortical cell culture. *J. Neurosci.* **7**, 369–379.

Choi, D. W. (1988a). Glutamate neurotoxicity and diseases of the nervous system. *Neuron* **1**, 623–634.

Choi, D. W. (1988b). Calcium-mediated neurotoxicity: Relationship to specific channel types and role in ischaemic damage. *Trends Neurosci.* **11**, 465–469.

Choi, D. W., and Rothman, S. M. (1990). The role of glutamate neurotoxicity in hypoxic-schemic neuronal death. *Annu. Rev. Neurosci.* **13**, 171–182.

Cotran, R. S., Kumar, V., and Robbins, R. S. (1989). Robbins Pathological Basis of Disease. Saunders, Philadelphia.

Coyle, J. T., Price, D. L., and De Long, M. A. (1983). Alzheimer's disease: a disorder of cortical cholinergic innervation. *Science* **219**, 1184–1190.

Deboni, U., and McLachlan, D. R. C. (1985). Controlled induction of paired filaments of the Alzheimer's type in cultured human neurons by glutamate and aspartate. *J. Neurol. Sci.* **68**, 105–118.

Denton, T., and Howard, B. D. (1984). Inhibition of dopamine uptake by *N*-methyl-4-phenyl-1,2,3,6-tetrahydropyridine, a cause of Parkinsonism. *Biochem. Biophys. Res. Commun.* **119**, 1186–1190.

Denton, T., and Howard, B. D. (1987). A dopaminergic cell line variant resistant to the neurotoxin 1-methyl-4-phenyl-1,2,3,6-tetrahydropyridine. *J. Neurochem.* **49**, 622–630.

Dingledine, R., McBain, C. J., and McNamara, J. O. (1990). Excitatory amino acid receptors in epilepsy. *Trends Pharmacol. Sci.* **11**, 334–338.

Di-patre, P. L., and Butcher, L. L. (1991). Cholinergic fiber perturbations and neurite outgrowth produced by intrafimbrial infusion of neurofilament-disrupting agent 2,5-hexanedione. *Brain Res.* **539**(1), 126–132.

Duncan, M. W., *et al.,* (1992). Zinc, a neurotoxin to cultured neurons, contaminates cycad flour prepared by traditional guamanian methods. *J. Neurosci.* **12**(4), 1523–1537.

Dykens, J. A., Stern, A., and Trenker, E. (1987). Mechanism of kainate toxicity to cerebellar neurons in-vitro is analogous to reperfusion tissue injury. *J. Neurochem.* **49**, 1222–1228.

Edwards, S. N., Buckmaster, A. E., and Tolkovsky, A. M. (1991). The death programme in cultured sympathetic neurons can be suppressed at the post translational level by nerve growth factor, cyclic AMP and depolarization. *J. Neurochem.* **57**, 2140–2143.

Ellis, H. M., and Horvitz, H. R. (1986). Genetic control of programmed cell death in the nematode *C elegans*. *Cell* **44**, 817–829.

Ferriero, D. M., *et al.* (1988). Selective sparing of NADPH-diaphorase neurons in neonatal hypoxia-ischaemia. *Ann. Neurol.* **24**, 670–676.

Foster, A., Gill, R., and Woodruff, G. N. (1988). Neuroprotective effects of MK-801 in-vivo: Selectivity and evidence for delayed degeneration mediated by NMDA receptor activation. *J. Neurosci.* **8**, 4745–4754.

Friedman, L., and Mytilineou, C. (1987). The toxicity of MPTP to dopamine neurons in culture is reduced at high concentrations. *Neurosci. Lett.* **79**, 65–72.

Garthwaite, G., and Garthwaite, J. (1986). Neurotoxicity of excitatory amino acid receptor agonists in rat cerebellar slices: Dependence on calcium concentration. *Neurosci. Lett.* **66**, 193–198.

Glenner, G. G. (1989). The pathobiology of Alzheimer's disease. *Annu. Rev. Med.* **40**, 45–51.

Good, P. F., Perl, D. P., Bierer, L. M., and Schmeidler, J. (1992). Selective accumulation of aluminum and iron in the neurofibrillary tangles of Alzheimer's disease: a laser microprobe (LAMMA) study. *Ann. Neurol.* **31**(3), 286–292.

Good, P. F., and Perl, D. P. (1993). Aluminium in Alzheimer's. *Nature* **362**, 418.

Halliwell, B., and Gutteridge, J. M. C. (1988). Free radicals and antioxidant protection: Mechanisms and significance in toxicology and disease. *Human Toxicol.* **7**, 7–13.

Harvey, A. L. (1988). Possible developments in neurotoxicity testing in-vitro. *Xenobiotica* **18**(6), 625–632.

Jones, H. B., and Cavanagh, J. B. (1984). The evolution of intracellular response to acrylamide in rat spinal ganglion neurons. *Neuropathol. Appl. Neurobiol.* **8**, 355–370.

Johnson E. M., Taniuchi, M., and Distefano, P. S. (1988). Expression and possible function of nerve growth factor receptors on Schwann cell. *Trends Neurosci.* **11**, 299–304.

Johnson, E. M., Jr., and Deckworth, T. L. (1993). Molecular mechanisms of developmental neuronal death. *Annu. Rev. Neurosci.* **16**, 31–46.

Kaplan, B. B. *et al.* (1982). Analysis of gene expression in brain: comparative and developmental studies. In *Basic and Clinical Aspects of Molecular Neurobiology* (A. M. Guiffrida-Stella, G. Gombos, G. Banzi, and H. S. Bachelard, Eds.) pp. 87–97, Fondazione Internationale Menarini, Milano.

Koh, J., and Choi, D. W. (1988). Zinc alters excitatory amino acid neurotoxicity on cortical neurons. *J. Neurosci.* **8**, 2164–2171.

Kuroda, Y., *et al.* (1990). Assay system for neurotoxicants causing parkinson's disease: 1-methyl-4-pyridinium ion (MPP+) inhibits survival of cultured neurons from substantia nigra of fetal monkey (Macaca fascicularis). *Adv. Behav. Biol.* **38A**(1), 305–308.

Lequesne, P. M. (1980). Acrylamide. In *Experimental and Clinical Neurotoxicology* (P. S. Spencer, and H. H. Schaumburg, Eds.) pp. 309–325, Williams and Wilkins, Baltimore.

Liles, W. C., *et al.* (1988). Decreased muscarinic acetylcholine receptor number in the central nervous system of tottering (tg/tg) mouse. *J. Neurochem.* **46**, 977–982.

Marcoux, F. W., *et al.* (1988). Ketamine prevents glutamate-induced calcium influx and ischaemic nerve cell injury. In *Sigma and Phenylcyclidine-like Compounds as Molecular Probes in Biology* (E. F. Domino, and J. M. Kamenka, Eds.) NPP Books, Ann Arbour.

MacDermott, A. B., *et al.* (1986). NMDA-receptor activation increases cytoplasmic calcium concentration in cultured spinal chord neurones. *Nature* **321**, 519–522.

Martini, A. M., Schwartz, J. P., and Kopin, I. J. (1989). The neurotoxicity of 1-methyl-4-phenylpyridinium in cultured cerebellar granule cells. *J. Neurosci.* **9**, 3665–3672.

Meldrum, B., and Garthwaite, J. (1990). Excitatory amino acid neurotoxicity and neurodegenerative disease. *Trends Pharmacol. Sci.* **11**, 379–387.

Mochetti, I. (1991). Theoretical basis for a pharmacology of nerve growth factor biosynthesis. *Annu. Rev. Pharmacol. Toxicol.* **32**, 303–328.

Mytilineou, C., Cohen, G., and Heikkila, R. E. (1985). 1-Methyl-4-phenylpyridine (MPP+) is toxic to mesencephalic dopamine neurons in culture. *Neurosci. Lett.* **57**, 19–24.

Nathanson, N. M. (1987). Molecular properties of the muscarinic acetylcholine receptor. *Annu. Rev. Neurosci.* **10**, 195–236.

Nicoletti, F., *et al.* (1986). The activation of inositol phospholipid metabolism as a signal-transducing system for excitatory amino acids in primary cultures of cerebellar granule cells. *J. Neurosci.* **6**, 1905–1911.

Notter, M. F. D., *et al.* (1988). Neurotoxicity of MPTP and MPP+ in vitro: characterization using specific cell lines. *Brain Res.* **456**, 254–262.

Nyberg, E., and Walum, E. (1984). On the application of neuroblastoma cells in neurotoxicological studies: implications of acrylamide induced neurite disintegrations. *ATLA* **11**, 194–203.

Olney, J. W., Price, M. T., and Samson, L. (1986). The role of specific ions in glutamate neurotoxicity. *Neurosci. Lett.* **65**, 65–71.

Olney, J. W. (1990). Excitotoxins and brain disorders. *Annu. Rev. Pharmacol. Toxicol.* **30**, 47–71.

Oppenheim, R. W. (1989). The neurotrophic theory and naturally occurring motoneuron death. *Trends Neurosci.* **12**, 252–255.

Oppenheim, R. W. (1991). Cell death during development of the nervous system. *Annu. Rev. Neurosci.* **14**, 453–501.

Perl, D., *et al.* (1986). Laser microprobe mass analysis of aluminium in Guam disease. *J. Neuropath. Exp. Neurol.* **45**, 379–383.

Procter, A. W., *et al.* (1988). Evidence of glutamatergic denervation and possible abnormal metabolism in Alzheimer's disease. *J. Neurochem.* **50**, 790–801.

Purvess, D. (1988). *Body and Brain, A Trophic Theory of Neural Connections.* Harvard, Cambridge, Massachusetts.

Ramsay, R. R., *et al.* (1986). Inhibition of mitochondrial NADH dehydrogenase by pyridine derivatives and its possible relation to experimental and idiopathic parkinsonism. *Biochem. Biophys. Res. Commun.* **135**, 269–275.

Rosen, D. R., *et al.* (1993). Mutations in Cu/Zn superoxide dismutase gene are associated with familial amyotrophic lateral sclerosis. *Nature* **362**, 59–62.

Rosenberg, C. K., *et al.* (1987). Hyperbaric oxygen accelerates the neurotoxicity of 2,5-hexanedione. *Toxicol. Appl. Pharmacol.* **87**, 374–379.

Ross, S. M., Seelig, M., and Spencer, P. S. (1987). Specific antagonism of excitotoxic action of "uncommon" amino acids assayed in organotypic cortical cultures. *Brain Res.* **425**, 120–127.

Rothman, S. M. (1984). Synaptic release of excitatory amino acid neurotransmitter mediates anoxic neuronal death. *J. Neurosci.* **4**, 1884–1891.

Rothman, S. M., and Samaie, M. (1985). The physiology of excitatory synaptic transmission in cultures of dissociated rat hippocampus. *J. Neurophysiol.* **54**, 701–713.

Sabramanyam, B., *et al.* (1990). Identification of a potentially neurotoxic pyridinium metabolite of haloperidol in rats. *Biochem. Biophys. Res. Commun.* **166**(1), 238–244.

Sanchez-Ramos, J., *et al.* (1986). 1-Methyl-4-pyridinium (MPP+) but not 1-methyl-4-phenyl-1,2,3,6-tetrahydropyridine (MPTP) selectively destroys dopaminergic neurons in cultures of dissociated rat mesencephalic neurons. *Neurosci. Lett.* **72**, 215–220.

Savage, D. D., Rigsbee, L. C., and McNamara, J. O. (1985). Knife cuts of entorhinal cortex: effects on development of kindling and seizure-induced decrease of muscarinic cholinergic receptor. *J. Neurosci.* **5**, 408–413.

Sayre, L. M., Autilio-Gambetti, L., and Gambetti, P. (1985). Pathogenesis of experimental giant neurofilamentous axonopathies: a unified hypothesis based on chemical modification of neurofilaments. *Brain Res. Rev.* **10**, 69–83.

Schousboe, A. (1981). Transport and metabolism of glutamate and GABA in neurons and glial cells. *Int. Rev. Neurobiol.* **22**, 1–45.

Seubert, P., *et al.* (1988). Stimulation of NMDA receptors induces proteolysis of spectrin in hippocampus. *Brain Res.* **460**, 189–194.

Sharma, R. P., and Obersteiner, E. J. (1977). Acrylamide cytotoxicity in chick ganglion cultures. *Toxicol. Appl. Pharmacol.* **42**, 149–156.

Siesjo, B. K. (1988). Historical overview: calcium, ischaemia, and death of brain cells. *Ann. N.Y. Acad. Sci.* **522**, 638–661.

Simon, R. P., *et al.* (1984). Blockade of N-methyl-D-aspartate receptors may protect against ischaemic damage in the brain. *Science* **226**, 850–852.

Sladaczek, F., *et al.* (1985). Glutamate stimulates inositol phosphate formation in striatal neurones. *Nature* **317**, 717–719.

Snider, W. D., and Johnson, E. M. (1989). Neurotrophic molecules. *Ann. Neurol.* **26**, 489–506.

Sofroniew, M. V., and Pearson, R. C. A. (1985). Degeneration of cholinergic neurones in the basal nucleus following kainic or

N-methyl-D-aspartic acid application to the cerebral cortex of the rat. *Brain Res.* **339,** 186–190.

Spencer, P. S., *et al.* (1980). The enlarging view of hexacarbon neurotoxicity. *CRC Crit. Rev. Toxicol.* **7,** 279–356.

Spencer, P. S., *et al.* (1986). Lathryism: evidence for role of the neuroexcitatory amino acid BOAA. *Lancet* Vol 2, 1066–1067.

Spencer, P. S., *et al.* (1987). Guam amyotrophic lateral sclerosis-parkinsonism-dementia linked to a plant excitant neurotoxin. *Science* **237,** 517–522.

Teyler, T. J., and Discenna, P. (1987). Long term potentiation. *Annu. Rev. Neurosci.* **10,** 131–161.

Walicke, P. A. (1989). Novel neurotrophic factors, receptors and oncogenes. *Annu. Rev. Neurosci.* **12,** 103–126.

Walum, E., *et al.* (1987). Cultured neuroblastoma cells as neurotoxicological models: acrylamide induced neurite disintegrations. In *Model Systems in Neurotoxicology: Alternative Approaches to Animal Testing* (S. Shahar, and A. M. Goldberg, Eds). pp. 212–236, Alan R. Liss, New York.

Walum, E., and Flint, O. P. (1988). Acrylamide, 2,5-hexanedione and beta-aminopropionitrile toxicity tested in rat embryo midbrain cell cultures. *ATLA* **15,** 238–244.

Walum, E., Hansson, E., and Harvey, A. L. (1990). In-vitro testing of neurotoxicity. *ATLA* **19,** 153–179.

Walum, E., and Peterson, A. (1984). Effects of subacute concentrations of acrylamide on the morphology of cultured mouse neuroblastoma N1E115 cells: a time-lapse cinematographic study. *ATLA* **12,** 33–66.

Watkins, J. C., and Olverman, H. J. (1987). Agonists and antagonists for excitatory amino acids. *Trends Neurosci.* **10,** 265–272.

Weiss, J. H., and Choi, D. W. (1988). β-N-Methylamino-L-alanine neurotoxicity: requirement for bicarbonate as a co-factor. *Science* **241,** 973–975.

Weiss, J. H., Koh, J-Y., and Choi, D. W. (1989). Neurotoxicity of β-N-methylamino-L-alanine (BOAA) on cultured cortical neurons. *Brain Res.* **497,** 64–71.

Wilson, J. S., *et al.* (1991). Mechanisms of the inotropic actions of MPTP and MPP+ on isolated atria of rat. *Toxicol. Appl. Pharmacol.* **111,** 49–57.

Woolley, D., *et al.* (1984). Do some insecticides and heavy metals produce long term potentiation in the limbic system? In *Cellular and Molecular Neurotoxicology* (T. Narahashi, Ed.) pp. 45–69, Raven, New York.

CHAPTER

30

The Use of Serum-Free Aggregating Brain Cell Cultures in Neurotoxicology

PAUL HONEGGER
Institute of Physiology
University of Lausanne
Switzerland

BENOÎT SCHILTER
Nestec Ltd.
Research Centre
Vers-chez-les-Blanc
Switzerland

I. Introduction

Rotation-mediated aggregating cell cultures (also called reaggregating or aggregate cell cultures) were introduced by Moscona (1960), after he realized that freshly dissociated cells of various fetal organs are able to reaggregate spontaneously and that the resulting three-dimensional cell structures are able to express organotypic features if kept in a rotation-mediated suspension culture for several weeks. Initially, such cultures were used mainly for morphological investigations of organ-specific pattern formation (Moscona, 1965; DeLong, 1970). Subsequently, Seeds and collaborators (Seeds, 1973; Seeds *et al.*, 1977) demonstrated the usefulness of aggregating brain cell cultures as a model for developmental studies using biochemical techniques of investigation. Several modifications of the original culture technique, notably the use of a mechanical (nonenzymatic) tissue dissociation procedure (Honegger and Richelson, 1976) and the design of a serum-free, chemically defined medium, greatly facilitated the culture method and at the same time

increased both the yield and the reproducibility of the cultures. The possibility to prepare and grow aggregating brain cell cultures in a chemically defined medium (Honegger *et al.*, 1979) gave new impetus to the study of soluble endogenous factors (e.g., hormones, growth factors) and their role in the brain, and the availability of large quantities of highly reproducible cultures was advantageous for developmental studies using multidisciplinary assay techniques.

Because of their three-dimensional structure permitting maximal cell-to-cell interactions, aggregating brain cell cultures are able to reconstruct tissue-specific patterns and to reproduce a series of developmental processes similar to those observed during normal brain ontogeny (e.g., cell proliferation and migration, synaptogenesis, myelination). During the first 2 weeks *in vitro* many aggregate-forming cells (i.e., the majority of astrocytes and oligodendrocytes and some neurons such as GABAergic neurons) are able to proliferate. Thereafter, the mitotic activity decreases rapidly, and most of the cells undergo extensive maturation (Honegger, 1985, 1987; Honegger and Matthieu, 1990). The neuronal and glial maturation continues for several weeks. At maximal differentiation (i.e., after 40 to 50 days *in vitro*), the aggregates resemble small spheres of mature brain tissue. The organotypic structural and functional maturation of aggregating brain cell cultures has been demonstrated using both morphological (Moscona, 1965; Seeds and Vatter, 1971; Seeds and Haffke, 1978; Trapp *et al.*, 1979; Lu *et al.*, 1980) and biochemical (Seeds, 1973; Seeds *et al.*, 1977; Honegger and Richelson, 1976; Honegger *et al.*, 1979; Honegger, 1985, 1987; Honegger and Matthieu, 1990) techniques of investigation.

Previously, aggregating brain cell cultures have been used almost exclusively for basic neurobiological investigations. However, several lines of work suggested that this culture system is also suitable for studies in toxicology (Trapp and Richelson, 1980; Wehner *et al.*, 1985; Jacobs *et al.*, 1986; Atterwill, 1987, 1989; Honegger and Werffeli, 1988; Honegger and Schilter, 1992). This chapter therefore focuses on the properties of this culture system that may be of interest for toxicological investigations.

II. General Methodology

A. Culture Preparation

Aggregating brain cell cultures are prepared usually from brains or brain parts of fetal rats or mice. The mechanical dissociation procedure requires embryonal or early fetal brain tissue. The use of relatively undifferentiated tissue also permits obtaining cultures with a high capacity for growth and differentiation. For routine work, cultures are prepared from brains or brain parts (e.g., from the telencephalon) of 15- to 16-day rat fetuses as previously described in detail (Honegger and Richelson, 1976; Honegger, 1985). The dissection and mechanical dissociation of the tissue is performed in ice-cold, sterile solution D (modified Puck's D solution; Wilson *et al.*, 1972), pH 7.4, 340 mOsm, containing NaCl (138 mM), KCl (5.4 mM), Na$_2$HPO$_4$ (0.17 mM), KH$_2$PO$_4$ (0.22 mM), glucose (5.5 mM), sucrose (58.4 mM), and gentamicin (20 μg/ml) as antibiotic. The excised and pooled brain tissue is dissociated by a first passage through a 200-μm pore nylon mesh layer by gently squeezing the nylon bag from the outside with a glass rod. The dispersed tissue is gently triturated with a 5-ml plastic pipette and then filtered by gravity flow through a second nylon mesh with 115-μm pores. The resulting single cell suspension is sedimented by centrifugation (15 min, 290 g, 4°C) and washed once or twice (depending on the amount of tissue processed) with ice-cold solution D by gentle trituration and subsequent centrifugation. The washed cells are finally re-

suspended in cold culture medium. Using 15-day fetal telencephalon cells, the final cell density is about 1.5 × 10^7 cells/ml. Aliquots of this cell suspension (usually 4 ml, but depending on the type of culture vessel used, this volume may need to be adjusted, see later) are transferred to 25-ml DeLong culture flasks. The flasks are then placed on a gyratory shaker (stroke radius 2.5 cm) within a CO$_2$ incubator, and kept under constant agitation at 37°C in an atmosphere of 10% CO$_2$ and 90% humidified air. Gas-permeable closures are used to ensure an adequate gas supply. The rotation speed initially set at 68 rpm is gradually increased to 80 rpm during the first week (70 rpm 4 hr after inoculation; 72 and 74 rpm the next day in the morning and the evening, respectively; and then daily increases by 2 rpm up to 80 rpm). The use of uniform culture vessels and correct gyratory agitation are the most important factors for the formation of highly reproducible cultures. To obtain optimal agitation for a given type of culture vessel it is recommended to modify the volume of culture medium per flask instead of the initial rotation speed.

Aggregating brain cell cultures prepared from fetal rat brain contain usually similar proportions of the different cell types as the original brain tissue. For specific problems it may be advantageous to study cultures enriched in certain cell types. Aggregate cultures highly enriched in either neurons or glial cells (astrocytes and oligodendrocytes) can be prepared by early treatment of the cultures with selective agents (Honegger *et al.*, 1986; Honegger and Werffeli, 1988; Corthésy-Theulaz *et al.*, 1990). Selective elimination of most of the glial cells is achieved by treating cultures twice (on days 2 and 4) with the antimitotic drug cytosine arabinoside (Ara-C; 0.4 μM final concentration). The effect of Ara-C is enhanced by a single treatment of the cultures on day 2 with epidermal growth factor (EGF: 5 ng/ml), which is able to partially synchronize glioblast proliferation (Guentert-Lauber and Honegger, 1985). Selective removal of the majority of the neurons is obtained by treating serum-free aggregate cultures of fetal rat telencephalon on day 7 for 24 hr with cholera toxin (10^{-7} M final concentration).

B. Culture Maintenance

Usually, the cultures are grown from the outset in serum-free, chemically defined medium (Honegger *et al.*, 1979; Honegger, 1985; Honegger and Matthieu, 1990) consisting of Dulbecco's modified Eagle's medium (DMEM) supplemented with transferrin (1 μg/ml), insulin (0.8 μM), L-triiodothryonine (30 nM), hydrocortisone-21-phosphate (20 nM), BME vitamins, vitamin B$_{12}$, lipoic acid (1 μM), trace amounts of retinol and α-tocopherol, L-carnitine (12 μM), and various quantities of trace elements (Honegger, 1985).

Immediately after incubation, the dissociated cells begin to reaggregate. Within 2 days, uniform cell clusters with a diameter of about 100 μm are formed. They continue to increase in size during the following 2 weeks to attain a maximal diameter of 300 to 400 μm. At day 2 the cultures are transferred to 50-ml DeLong flasks and 4 ml of fresh medium is added. Thereafter, the medium is replenished regularly by the exchange of 5 ml of medium per flask. With the progress of maturation, aggregate cultures greatly increase their metabolic activity, causing deficiencies of certain substrates (e.g., glucose, some amino acids). Therefore, the cultures are split at day 19 by distributing the aggregates from each flask into two or more separate cultures. Excessive fluctuations in glucose levels can be prevented by increasing the glucose concentration in the medium from 4.5 to 6.0 mg/ml. Furthermore, it was found that the addition of a mixture of lipids bound to albumin (Albu-Max, GIBCO-BRL) to long-term cultures increases the stability of the differentiated phenotype of neurons as well as glial cells (Zurich *et al.*, 1993).

III. Use of Aggregating Brain Cell Cultures in Neurotoxicology

Aggregating brain cell cultures possess several unique features that appear to be useful for neurotoxicological investigations:

1. The high yield and high reproducibility of the cultures. Typically, from the fetuses of 20 time-pregnant rats of gestational day 15 some 100 flasks of telencephalon cell aggregates can be obtained. The amount of tissue present in each flask (i.e., 1000 to 2000 aggregates containing a total of 4 to 8 mg of protein and 0.7 to 1.4 mg of DNA) is sufficient for multidisciplinary investigations combining, for example, assay techniques of molecular biology with biochemical and morphological analyses. The excellent reproducibility of the cultures (due to the mechanical tissue dissociation technique and the use of a chemically defined culture medium) is evidenced by highly reliable results using triplicate or quadruplicate cultures (flasks) for assays. In this context it is worth mentioning that the so-called micromass cultures (Flint, 1983) currently used in toxicological routine tests represent a simplified version of aggregating cell cultures. However, only the latter are suitable for mechanistic toxicological studies of both differentiating and mature brain cells. These two sys-

tems could therefore be used simultaneously for complementary investigations (Atterwill *et al.*, 1992).

2. The developmental potential of aggregating brain cell cultures. A great number of drugs are thought to affect distinct developmental processes in the brain. However, most of the *in vitro* systems currently used in teratology (including whole embryo culture and micromass cell culture) are severely limited with respect to the developmental stages that can be studied. Since aggregating brain cell cultures are able to form an organotypic cell architecture, thus reproducing a series of morphogenetic events occurring in the normal brain *in vivo*, a wide range of developmental processes can be studied in this system. Furthermore, since these cultures are able to express a highly differentiated phenotype, they may also be used for toxicological investigations of differentiated brain cells.

3. Long-term maintenance of aggregating brain cell cultures. Many environmental neurotoxicants may be present in the brain at relatively low concentrations and affect the brain only after continued chronic exposure. Aggregate cultures are able to maintain a highly differentiated state for many months *in vitro*, thus providing a unique model to study long-term effects of toxic or teratogenic drugs. Furthermore, they may serve as a model to study cellular and molecular processes involved in brain ageing.

A. Screening for Potentially Toxic or Teratogenic Substances

In a validation study, serum-free aggregating cell cultures prepared from fetal rat telencephalon have been used to test a series of potentially teratogenic chemicals. The test compounds were provided as coded solutions of unknown content and concentration for the evaluation and ranking of their potential toxicity/teratogenicity. This study has now been completed (Kucera *et al.*, 1993) and some general aspects of this work are presented here.

In order to obtain rapidly a first estimate of the dose-dependent toxicity with a minimum of cultures, aggregates were treated from culture day 5 to day 14 with aliquots of the coded solutions using four different dilutions per drug and duplicate flasks for each assay point. The dilutions were chosen to cover a wide concentration range (i.e., dilutions between 500- and 200,000-fold of the original stock solution were used). By following this scheme, up to 12 different compounds could be prescreened simultaneously with only one culture batch of 100 flasks. In his first approach only a minimal set of biochemical test criteria was used to evaluate toxicity: the activity of LDH in the culture medium serving as an indicator for cell damage or cell death, respectively, and the activities of two cell type-specific enzymes (i.e., the GABAergic marker GAD, and the glial marker GS) determined in the homogenates of the cultures at day 14. The results thus obtained for one of the test drugs, ketoconazole, are presented in Fig. 1. As shown in Fig. 1A, elevated levels of LDH were found in the media of cultures treated

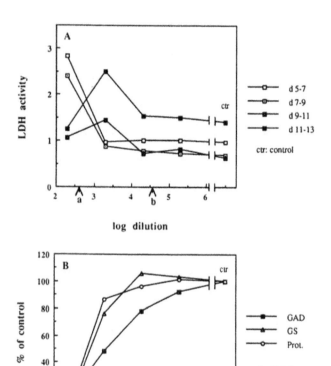

FIGURE 1 Preliminary evaluation of the toxicity of an unknown substance using serum-free aggregate cultures of fetal rat telencephalon as a test system. The treatment of the cultures with aliquots of the dissolved substance over a wide range of dilutions began on day 5 and continued until day 14. (A) Determination of the release of lactate dehydrogenase (LDH) into the culture medium as an indicator for cell damage. The levels of LDH activity were determined at days 7, 9, 11, and 13 before the medium was replenished. LDH activity is expressed as total optical density per minute per flask. (B) Assay of two cell type-specific enzyme activities and the total protein content in homogenates of the cell cultures prepared on day 14. Glutamic acid decarboxylase (GAD) activity was determined according to Wilson *et al.* (1972); glutamine synthetase (GS) was assayed by a modification (Patel *et al.*, 1982) of the method of Pishak and Phillips (1979); protein was determined by the Folin phenol method (Lowry *et al.*, 1951) using bovine serum albumin as the standard. Enzyme activities are expressed as percentages of the total activities per culture flask measured in the control cultures (ctr) which were treated with an equal volume of the solvent only. The concentration range a–b shown on the abscissa was chosen for the subsequent experiment (Fig. 2).

with relatively high concentrations of this compound (i.e., at 500-fold dilution already within the first 2 days of treatment and at 2000-fold dilution after 4 days). At these doses all three biochemical values (i.e., protein content, GAD, and GS activities) determined in the cultures on day 14 were greatly reduced (Fig. 1B), in accord with the findings of increased LDH release, suggesting a strong cytotoxic reaction. On the other hand, cultures receiving more diluted samples showed only a reduction of GAD activity (Fig. 1B), indicating a selective neurotoxic effect.

Based on these first results a new concentration range (shown as range a–b in Figs. 1A and 1B) could be defined for a more detailed analysis. In this second experiment the cultures were treated again between days 5 and 14 as before. For the biochemical analysis on day 14, more experimental points (triplicate cultures and seven different drug dilutions) were chosen. Furthermore, a larger set of biochemical criteria was analyzed (i.e., total protein and DNA content; GAD, ChAT, AChE, GS, and CNP). The results obtained for ketoconazole are shown in Fig. 2A. At low drug concentrations (i.e., at relatively high dilutions), selective neurotoxic effects could be observed (e.g., reduction of GAD and ChAT activities but no significant change for GS and CNP). In contrast, at higher doses all biochemical parameters were greatly reduced, indicating a general (unspecific) cytotoxicity. These results confirmed and extended the findings obtained in the prescreening experiment. After disclosure of the identity and concentration of the coded drug, its dose-dependent toxicity could finally be deduced (Fig. 2B).

The entire validation study comprised six pairs of compounds belonging to different classes of chemicals with known or suspected teratogenic activity (i.e., ketoconazole and metronidazole; 6-aminonicotinamide and 4-hydroxypyridine; methoxyacetic acid and methoxyethanol; sulfadiazine and sulfanilamide; theophylline and caffeine; and the retinoids Ro 13-6307 and Ro 1-5488). Based on the data of the biochemical analyses, a rank order for the potential toxicity/teratogenicity was established by determining the drug concentration where at least one of the parameters showed a reduction of 50%. The results thus obtained were in good agreement with those of two other groups participating in this validation study using whole embryo cultures of chicken and rats, respectively (Kucera *et al.*, 1993).

B. Developmental Studies

Screening tests as described in Section III,A allow a first rapid evaluation of a series of chemicals for their potential toxicity or teratogenicity. The cultures used

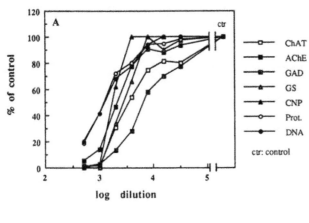

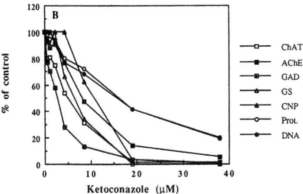

FIGURE 2 The effect of ketoconazole on cell type-specific enzyme activities, total protein, and DNA in serum-free aggregate cultures of fetal rat telencephalon. Cultures were treated from day 5 to day 14 with different doses of ketoconazole corresponding to the concentration range a–b shown in Fig. 1. (A) The values are expressed as a function of the dilution factor for a unknown substance. (B) The values are expressed as a function of the actual concentration of the test substance (ketoconazole). Enzyme activities were determined in the homogenates of the cultures harvested on day 14 and expressed as percentages of the total activities (per culture flask) measured in the control cultures (ctr) treated with an equal volume of solvent. The three neuron-specific enzymes, choline acetyltransferase (ChAT), acetylcholinesterase (AChE), and glutamic acid decarboxylase (GAD), were assayed according to Wilson *et al.* (1972); glutamine synthetase (GS) representing the astrocytes was determined by a modification (Patel *et al.*, 1982) of the method of Pishak and Phillips (1979); 2′,3′-cyclic nucleotide 3′-phosphohydrolase, used as the marker for oligodendrocytes, was assayed according to Sogin (1976); DNA was determined by a fluorimetric assay using Hoechst dye 33258, and protein by the Folin phenol method (Lowry *et al.*, 1951).

for this purpose contained relatively undifferentiated brain cells which were either proliferating or at an early stage of differentiation. However, the sensitivity of cells to a given drug may alter as a function of cell maturation. Furthermore, certain drugs may affect developmental processes that occur only at a more advanced stage of maturation (e.g., synaptogenesis, myelination). Therefore, further experiments are required in order to study drug effects on successive develop-

mental events and, ultimately, to elucidate the mechanisms underlying drug toxicity.

As a follow-up of the screening experiment with ketoconazole (Section III,A), developmentally more advanced aggregate cultures were used to examine the influence of cell differentiation on drug sensitivity. Thus, aggregate cultures were treated between days 25 and 34 with ketoconazole at the same concentrations as before and were assayed on day 34 for the various biochemical parameters. The results presented in Fig. 3 show that compared to immature cultures (Fig. 2B), the more differentiated cultures require higher doses of ketoconazole for comparable effects. In particular, the glial parameters (GS, CNP) remained practically unaffected up to 20 μM of ketoconazole, the concentration that caused an almost total loss of enzymatic activity in early cultures. On the other hand, the neuronal parameters were also greatly reduced in the differentiated cultures. These results suggest that the dividing brain cells (i.e., astrocytes, oligodendrocytes, and some of the GABAergic neurons) are highly sensitive to ketoconazole, whereas among the differentiated brain cells neurons are more sensitive to this drug than glial cells. The finding that in the differentiated cultures both the protein and DNA levels remained practically unchanged (Fig. 3) suggests that ketoconazole affected cell functions without causing immediate cell death. This view is supported by the finding that ketoconazole treatment caused relatively little LDH release (Fig. 4).

Analogous developmental studies using 6-aminonicotinamide as a test substance revealed a different pattern of reactivity (Fig. 5). In immature cultures (Fig.

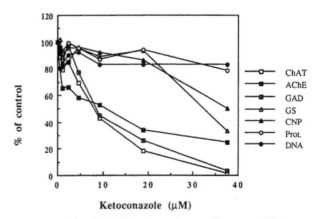

FIGURE 3 The effect of ketoconazole on cell type-specific enzyme activities, total protein, and DNA in serum-free aggregate cultures of fetal rat telencephalon assayed at a more advanced developmental stage. The treatment with different doses of ketoconazole began on day 25 and continued until day 34 (day of harvest). The values are expressed as percentages of the total activities per culure flask measured in control cultures treated with equal volumes of solvent. The assay methods used are given in the legend to Fig. 2.

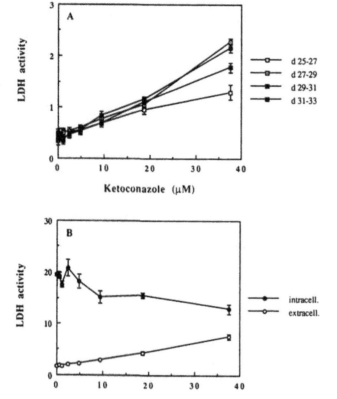

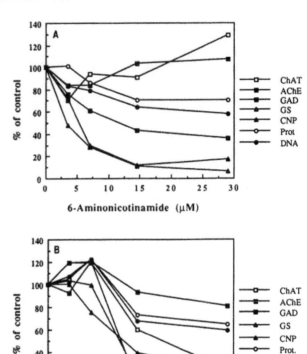

FIGURE 4 The effect of ketoconazole on the release of lactate dehydrogenase (LDH) in serum-free aggregate cultures of fetal rat telencephalon. Cultures were treated with different doses of ketoconazole from day 25 to day 34 as stated in the legend to Fig. 3. (A) Levels of LDH activity determined in the culture medium at days 28, 30, 32, and 34 before replenishment of the media. (B) Comparison between the total LDH activity released (i.e., the sum of the LDH activities measured in the media between days 25 and 34 at a particular ketoconazole concentration) and the LDH activity determined in the homogenates of the corresponding cultures on day 34 (termed intracellular LDH). LDH activity is expressed as total optical density per minute per flask (bars indicate SEM).

FIGURE 5 The effect of 6-aminonicotinamide on cell type-specific enzyme activities, total protein, and DNA in serum-free aggregate cultures of fetal rat telencephalon. (A) Cultures were treated with different doses of 6-aminonicotinamide from day 5 to day 14. (B) Cultures were treated with different doses of 6-aminonicotinamide from day 26 to day 35. The values are expressed as percentages of the total activities per culture flask measured in control cultures treated with an equal volume of solvent. More details concerning the methods used are given in the legend to Fig. 2.

5A), a significant decrease of the glial (GS, CNP) and GABAergic (GAD) marker enzymes was obtained with treatments at concentrations ≤7 μM, whereas the cholinergic enzyme activities (ChAT, AChE) remained unaffected up to the highest concentrations tested (30 μM). On the other hand, in differentiated cultures (Fig. 5B), most parameters were greatly affected at concentrations ≥15 μM 6-aminonicotinamide. This finding was corroborated by a massive release of LDH (Fig. 6) indicating increased cell death. These results taken together suggest that in early cultures 6-aminonicotinamide affected particularly the dividing cells, whereas in differentiated cultures the drug was highly cytotoxic for both glial cells and neurons.

Developmental studies with diazepam revealed another pattern of toxicity characteristic for this drug

(Fig. 7). Undifferentiated brain cells (Fig. 7A) were found to be relatively sensitive to diazepam, showing an almost total loss of neuronal (ChAT, GAD) and glial (CNP, GS) enzyme activities at 10 μg/ml (35 μM) of diazepam. In differentiated cultures (Fig. 7B), neurons and astrocytes appeared to be more sensitive to diazepam than oligodendrocytes. In a similar study (not shown) where aggregate cultures were treated between days 20 and 28 with different anticonvulsants [i.e., carbamazepine, diphenylhydantoin, losigamone (Stein *et al.*, 1991), phenobarbital, and valproate] it was found that these drugs (with the exception of valproate which up to 80 μg/ml did not produce any significant effect) reduced selectively the neuronal parameters, in particular the activity of the GABAergic marker enzyme GAD. The order of potency (on a weight per volume basis) thus established was: diphenylhydantoin > losigamone > carbamazepine > phenobarbital > valproate (Schilter *et al.*, 1995).

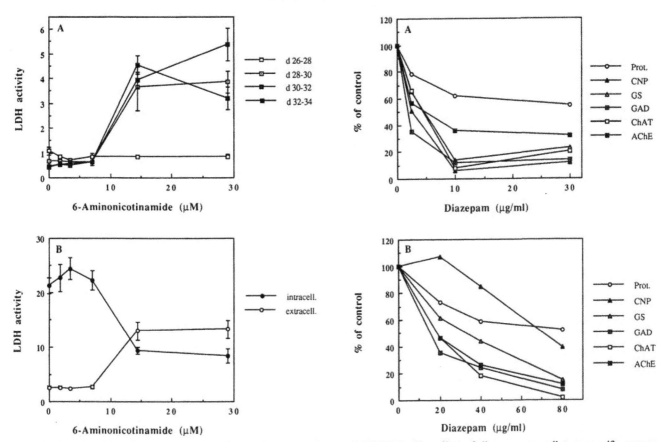

FIGURE 6 The effect of 6-aminonicotinamide on the release of lactate dehydrogenase (LDH) in serum-free aggregate cultures of fetal rat telencephalon. Cultures were treated with different doses of ketoconazole from day 26 to day 35. (A) Levels of LDH activity determined in the culture medium at days 28, 30, 32, and 34 before replenishment of the media. (B) Comparison between the total LDH activity released and the remaining LDH activity in the corresponding cultures on day 35 as described in the legend to Fig. 4. LDH activity is expressed as total optical density per minute per flask (bars indicate SEM).

FIGURE 7 The effect of diazepam on cell type-specific enzyme activities and total protein in serum-free aggregate cultures of fetal rat telencephalon. (A) Cultures were treated with different doses of diazepam from day 5 to day 14. (B) Cultures were treated with different doses of diazepam from day 20 to day 28. The values are expressed as percentages of the total activities per culture flask measured in control cultures treated with an equal volume of solvent. More details concerning the methods used are given in the legend to Fig. 2.

Until now some 30 different compounds have been tested in aggregating brain cell cultures by applying the protocol described here. In general, it could be observed that the patterns of drug effects are highly reproducible in repeated tests and that they exhibit characteristic features for individual drugs or drug classes. Cultures at advanced developmental stages were also found potentially useful in studying agents affecting myelin formation (Almazan *et al.*, 1985) and maintenance (Honegger and Matthieu, 1990; Kerlero de Rosbo *et al.*, 1990; Matthieu *et al.*, 1990) and the expression, modification, and stability of cytoskeletal proteins (Riederer *et al.*, 1992).

C. Long-Term Drug Effects

The potential risk of long-term exposure to low levels of xenobiotics is difficult to assess. In the living individual, the effects may not exceed the limits of normality and thus remain undetected. Furthermore, the mechanisms of action of chronic drug effects may differ from those of acute intoxication. *In vitro* models are therefore needed; however, most of the *in vitro* systems currently used in toxicology/teratology (e.g., monolayer cell cultures, micromass cultures, whole embryo cultures) are not suitable for long-term investigations. Actually, the two most promising *in vitro* systems for long-term studies in neurotoxicology appear to be slice cultures (Gähwiler *et al.*, 1992) and aggregating cell cultures. Aggregating brain cell cultures show progressive maturation for 6 to 7 weeks *in vitro*, and their final level of cellular organization and differentiation can be maintained in chemically defined medium for many more weeks. Therefore, work is in progress using this culture system as a model to study long-term effects of low levels of drugs and heavy metal compounds in the brain. The following example, using 6-aminonicotinamide as a model drug, may illustrate how the prolonged exposure of brain cells to low doses of chemicals may increase the selective toxic effects. Based on the previous developmental study with 6-aminonicotinamide (Fig. 5), a concentration range was chosen (i.e., between 1.0 and 4.0 μM) where no significant effects were apparent after treatment between days 26 and 35. Aggregate cultures were treated with 6-aminonicotinamide within this concentration range for either 9 days (i.e., between days 26 and 35) or 29 days (i.e., between days 26 and 55). The results show that 9 days of treatment (Fig. 8A) had little or no effect on the biochemical parameters measured, except for CNP activity which was lightly decreased (by 20%) at the highest dose (4 μM) applied. In contrast, after 29 days of treatment (Fig. 8B), CNP activity was slightly but significantly affected already at 1 μM of 6-aminonicotinamide, and all enzymatic activities mea-

sured were drastically reduced at 4 μM. These results show that the prolonged exposure to low doses of 6-aminonicotinamide increased the toxic action on brain cells. Furthermore, since both the total protein content and total DNA remained unaffected (Fig. 8B), it can be concluded that at low dose 6-aminonicotinamide affected the function but not the viability of differentiated brain cells.

IV. Conclusion

Aggregating brain cell cultures form three-dimensional cell structures in which the fetal cells are able to interact and differentiate in an organ-specific manner, thus reproducing a series of developmental processes occurring in normal brain ontogeny. This culture technique has been considerably simplified and improved so that it has become possible to prepare and grow large batches of highly reproducible cultures. Furthermore, the cultures are able to develop in a chemically defined medium, and they can be maintained for months at a highly differentiated stage. For toxicological investigations, aggregating brain cell cultures present several interesting features. The large quantity of replicate cultures that can be obtained permits the simultaneous analysis of over a dozen compounds using multidisciplinary test criteria. Aggregating brain cell cultures may therefore be used as principal or complementary systems for routine evaluations of potential neurotoxicants. The remarkable developmental potential of aggregating cell cultures makes them a unique model to study agents affecting specific developmental processes and, in particular, the consequences of toxic insults in the developing brain. In addition, long-term cultures of brain cell aggregates can be used to examine the influence of chronic exposure to low levels of xenobiotics.

Clearly, aggregating brain cell cultures offer a wide range of possibilities for application in neurotoxicology. To exploit these potentialities most profitably, care should be taken to define a set of appropriate test parameters. In the practical examples presented in this chapter, drug effects were monitored essentially by measuring changes in enzymatic activities. Although this approach proved to be valuable in assessing cytotoxic as well as cell type-specific effects, additional criteria have to be developed for more detailed analyses. For example, the expression, post-translational modification, and turnover of cytoskeletal proteins such as neurofilament proteins, tubulin, microtubule-associated proteins, spectrin, synapsin, and synaptophysin may be used as indicators for alterations in neuronal maturation, impairment of the neuronal con-

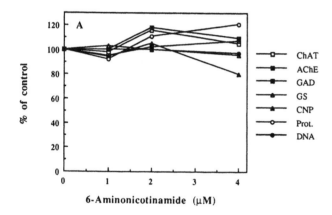
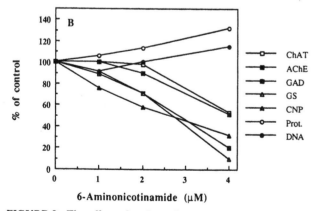

FIGURE 8 The effect of prolonged treatment with low doses of 6-aminonicotinamide on cell type-specific enzyme activities, total protein, and DNA in aggregate cultures of fetal rat telencephalon. (A) Cultures were treated from day 26 to day 35. (B) Cultures were treated from day 26 to day 55. The values are expressed as percentages of the total activities per culture flask measured in control cultures treated with an equal volume of solvent. More details concerning the methods used are given in the legend to Fig. 2.

nectivity, or structural and functional changes at synaptic junctions. In aggregating brain cell cultures, cytoskeletal proteins were found to be expressed in a development-dependent fashion resembling the situation *in vivo* (Riederer *et al.*, 1992), and some of these proteins (e.g., brain spectrin and high molecular weight neurofilament protein) appear to be highly sensitive to toxic insults (to be published). Functional and structural changes in glia cells may be indicated by the altered expression of glia cell-specific proteins (e.g., glial fibrillary protein, myelin-related proteins). For example, variations in myelin synthesis and stability can be monitored by measuring myelin basic protein levels (Almazan *et al.*, 1985; Kerlero de Rosbo *et al.*, 1990; Honegger and Matthieu, 1990; Matthieu *et al.*, 1990).

Notwithstanding the unique potentialities of aggregating brain cell cultures for predictive toxicological studies, this system may turn out to be most valuable for the elucidation of the mechanisms underlying the neurotoxic and teratogenic effects. Although cultures prepared from fetal rodent brains serve to study fundamental mechanistic aspects, aggregate cultures of fetal human brain (Pulliam *et al.*, 1988) may be required to address species-specific problems.

Acknowledgments

The authors thank Ms. M. Buvelot and Ms. D. Tavel for excellent technical assistance. This work was supported by the Swiss National Science Foundation (Grant 32-30214.90).

References

Almazan, G., Honegger, P., Matthieu, J. M., and Guentert-Lauber, B. (1985). Epidermal growth factor and bovine growth hormone stimulate differentiation and myelination of brain cell aggregates in culture. *Dev. Brain Res.* **21**, 257–264.

Atterwill, C. K. (1987). Brain reaggregate cultures in neurotoxicological investigations. In *In Vitro Methods in Toxicology* (C. K. Atterwill and C. E. Steele, Eds.), pp. 133–164, University Press, Cambridge.

Atterwill, C. K. (1989). Brain reaggregate cultures in neurotoxicological investigations: studies with cholinergic neurotoxins. *ATLA* **16**, 221–230.

Atterwill, C. K., Hillier, G., Johnston, H., and Thomas, S. M. (1992). A tiered system for in vitro neurotoxicity testing: a place for neural cell line and organotypic cultures? In *The Brain in Bits and Pieces: In Vitro Techniques in Neurobiology, Neuropharmacology and Neurotoxicology* (G. Zbinden, Ed.), pp. 81–113, MTC Verlag, Zollikon, Switzerland.

Corthésy-Theulaz, I., Mérillat, A. M., Honegger, P., and Rossier, B. C. (1990). Na$^+$-K$^+$-ATPase gene expression during in vitro development of rat fetal forebrain. *Am. J. Physiol.* **258**, C1062–C1069.

DeLong, G. R. (1970). Histogenesis of fetal mouse isocortex and hippocampus in reaggregating cell cultures. *Dev. Biol.* **22**, 563–583.

Downs, T. R., Wilfinger, W. W. (1983). Fluorometric quantification of DNA in cells and tissues. *Anal. Biochem.* **131**, 538–547.

Flint, O. P. (1983). A micromass culture method for rat embryonic neuronal cells. *J. Cell Sci.* **61**, 247–262.

Gähwiler, B. H., Knöpfel, T., Marbach, P., Müller, M., Rietschin, L., Scanziani, M., Staub, C., Vranesic, I., and Thompson, S. M. (1992). Use of organotypic slice cultures in neurobiological research. In *The Brain in Bits and Pieces. In vitro Techniques in Neurobiology, Neuropharmacology and Neurotoxicology* (G. Zbinden, Ed.), pp. 153–180, MTC Verlag, Zollikon, Switzerland.

Guentert-Lauber, B., and Honegger, P. (1985). Responsiveness of astrocytes in serum-free aggregate cultures to epidermal growth factor: dependence on the cell cycle and the epidermal growth factor concentration. *Dev. Neurosci.* **7**, 286–295.

Honegger, P. (1985). Biochemical differentiation in serum-free aggregating brain cell cultures. In *Cell Culture in the Neurosciences* (J. E. Bottenstein and G. Sato, Eds.), pp. 223–243, Plenum, New York.

Honegger, P. (1987). Oligodendrocyte development and myelination in serum-free aggregating brain cell cultures. In *A Multidisciplinary Approach to Myelin Disease* (G. Serlupi-Crescenzi, Ed.), NATO-ASI series. Plenum, New York.

Honegger, P., Du Pasquier, P., and Tenot, M. (1986). Cholinergic neurons of fetal rat telencephalon in aggregating cell culture respond to NGF as well as to protein kinase C-activating tumor promoters. *Dev. Brain Res.* **29**, 217–223.

Honegger, P., Lenoir, D., and Favrod, P. (1979). Growth and differentiation of aggregating fetal brain cells in a serum-free defined medium. *Nature* **282**, 305–308.

Honegger, P., and Matthieu, J. M. (1990). Aggregating brain cell cultures: a model to study myelination and demyelination. In *Cellular and Molecular Biology of Myelination* (G. Jeserich, H. H. Althaus, and T. V. Waehneldt, Eds.), pp. 155–170, Vol. 43, NATO-ASI series, Springer, Berlin.

Honegger, P., and Richelson, E. (1976). Biochemical differentiation of mechanically dissociated mammalian brain in aggregating cell culture. *Brain Res.* **109**, 335–354.

Honegger, P., and Schilter, B. (1992). Serum-free aggregate cultures of fetal rat brain and liver cells: methodology and some practical applications in neurotoxicology. In *The Brain in Bits and Pieces. In vitro Techniques in Neurobiology, Neuropharmacology and Neurotoxicology* (G. Zbinden, Ed.), pp. 51–79, MTC Verlag, Zollikon, Switzerland.

Honegger, P., and Werffeli, P. (1988). Use of aggregating cell cultures for toxicological studies. *Experientia* **44**, 817–823.

Jacobs, A. L., Maniscalco, W. M., and Finkelstein, J. N. (1986). Effects of methylmercury chloride, cycloheximide and colchicine on the reaggregation of dissociated mouse cerebellar cells. *Toxicol. Appl. Pharmacol.* **86**, 362–371.

Kerlero de Rosbo, N., Honegger, P., Lassmann, H., and Matthieu, J. M. (1990). Demyelination induced in aggregating brain cell cultures by a monoclonal antibody against myelin-ologodendroglial glycoprotein (MOG). *J. Neurochem.* **55**, 583–587.

Kucera, P., Cano, E., Honegger P., Schilter, B., Zijlstra, J. A., and Schmid, B. (1993). Validation of whole chick embryo cultures, whole rat embryo cultures and aggregating embryonic brain cell cultures using six pairs of coded compounds. *Toxicol. in Vitro* **7**, 785–798.

Lowry, O. H., Rosebrough, N. J., Farr, A. L., and Randall, R. J. (1951). Protein measurement with the Folin phenol reagent. *J. Biol. Chem.* **193**, 265–275.

Lu, E. J., Brown, W. J., Cole, R., and deVellis, J. (1980). Ultrastructural differentiation and synaptogenesis in aggregating rotation cultures of rat cerebral cells. *J. Neurosci. Res.* **5**, 447–463.

Matthieu, J. M., Roch, J. M., Torch, S., Tosic, M., Carpano, P., Insirello, L., Giuffrida Stella, A. M., and Honegger, P. (1990).

Triiodothyronine increases the stability of myelin basic protein mRNA in aggregating brain cell cultures. In *Regulation of Gene Expression in the Nervous System* (A. M. Guiffrida Stella, J. de Vellis, and J. R. Perez-Polo, Eds.), pp. 109–121, Liss, New York.

Moscona, A. A. (1960). Patterns and mechanisms of tissue reconstruction from dissociated cells. In *Developing Cell Systems and their Control* (D. Rudnick, Ed.), pp. 45–70, Ronald Press, New York.

Moscona, A. A. (1965). Recombination of dissociated cells and the development of cell aggregates. In *Cells and Tissues in Culture* (E. Willmer, Ed.), Vol. 1, pp. 489–529, Academic Press, New York.

Patel, A. J., Hunt, A., Gordon, R. D., and Balazs, R. (1982). The activities in different neural cell types of certain enzymes associated with the metabolic compartmentation glutamate. *Dev. Brain Res.* **4**, 3–11.

Pishak, M. R., and Phillips, A. T. (1979). A modified radioisotopic assay for measuring glutamine synthetase activity in tissue extracts. *Anal. Biochem.* **94**, 82–88.

Pulliam, L., Berens, M. E., and Rosenblum, M. L. (1988). A normal human brain cell aggregate model for neurobiological studies. *J. Neurosci. Res.* **21**, 521–530.

Riederer, B. M., Monnet-Tschudi, F., and Honegger, P. (1992). Development and maintenance of the neuronal cytoskeleton in aggregated cell cultures of fetal rat telencephalon and influence of elevated K⁺ concentrations. *J. Neurochem.* **58**, 649–658.

Schilter, B., Nöldner, M., Chatterjee, S. S., and Honegger, P. (1995). Anticonvulsant drug toxicity in rat brain cell aggregate cultures. *Toxicol. in Vitro*, in press.

Seeds, N. W. (1973). Differentiation of aggregating brain cell cultures. In *Tissue Culture of the Nervous System* (G. Sato, Ed.), pp. 35–53, Plenum Press, New York.

Seeds, N. W., Marks, M. J., and Ramirez, G. (1977). Aggregate cultures: A model for studies of brain development. In *Cell Culture and Its Application* (R. Acton and D. Lynn, Eds.), pp. 23–37, Academic Press, New York.

Seeds, N. W., and Haffke, S. C. (1978). Cell junction and ultrastructural development of reaggregated mouse brain cell cultures. *Dev. Neurosci.* **1**, 69–79.

Seeds, N. W., and Vatter, A. E. (1971). Synaptogenesis in reaggregating brain cell cultures. *Proc. Natl. Acad. Sci. U.S.A.* **68**, 3219–3222.

Sogin, D. C. (1976). 2′,3′-Cyclic NADP as a substrate for 2′,3′-cyclic nucleotide 3′-phosphohydrolase. *J. Neurochem.* **27**, 1333–1337.

Stein, U., Klessing, K., and Chatterjee, S. S. (1991). Losigamone. In *New Antiepileptic Drugs* (F. Pisani, E. Perucca, G. Avanzini, and A. Richens, Eds.), Epilepsy Res Suppl 3, pp. 129–133. Elsevier, Amsterdam.

Trapp, B. D., Honegger, P., Richelson, E., and Webster, H. deF. (1979). Morphological differentiation of mechanically dissociated fetal rat brain in aggregating cell cultures. *Brain Res.* **160**, 117–130.

Trapp, B. D., and Richelson, E. (1980). Usefulness for neurotoxicology of rotation-mediated aggregating cell cultures. In *Experimental and Clinical Neurotoxicology* (P. S. Spencer and H. H. Schaumberg, Eds.), pp. 803–819, Williams and Wilkins, Baltimore.

Walum, E., Wang, L., Jones, K., Nordin, M., Clemedson, C., and Varnbo, I. 1992). Cellular neuronal development in vitro—neurobiological and neurotoxicological studies in cultured model systems. In *The Brain in Bits and Pieces: In Vitro Techniques in Neurobiology, Neuropharmacology and Neurotoxicology* (G. Zbinden, Ed.), pp. 115–135, MTC Verlag, Zollikon, Switzerland.

Wehner, J. M., Smolen, A., Ness-Smolen, T., and Murphy, C. (1985). Recovery of acetylcholinesterase activity after acute organophosphate treatment of CNS reaggregate cultures. *Fundam. Appl. Toxicol.* **5**, 1104–1109.

Wilson, S. H., Schrier, B. K., Farber, J. L., Thompson, E. J., Rosenberg, R. N., Blume, A. J., and Nirenberg, M. W. (1972). Markers for gene expression in cultured cells from the nervous system. *J. Biol. Chem.* **247**, 3159–3169.

Zurich, M. G., Matthieu, J. M., and Honegger, P. (1993). Improvement of culture conditions for aggregating brain cells to study remyelination. *Schw. Arch. Neurol. Psychiatry*, **144**, 228–232.

Brain Slice Techniques in Neurotoxicology

STEPHEN B. FOUNTAIN
Department of Psychology
Kent State University
Kent, Ohio 44242

TIMOTHY J. TEYLER
Department of Neurobiology
Northeastern Ohio Universities
College of Medicine
Rootstown, Ohio 44272

I. Introduction

Since their development beginning in the 1960s, explant methods for maintaining central nervous system (CNS) tissue *in vitro* have become well-developed and are finding wide use in the fields of neurobiology, neuropharmacology, and, more recently, neurotoxicology. The first demonstration of a method for maintaining physiologically viable slices of mammalian CNS tissue was described by Yamamoto and McIlwain (1). In 1966, they demonstrated that thin sections of mammalian CNS tissue could be studied *in vitro* using electrophysiological methods comparable to those for studying the same tissue *in vivo*. Since then, these and related *in vitro* methods have been used for analyzing and characterizing the physiology of CNS tissue (2,3), for studying the mechanisms of synaptic plasticity thought to underlie learning and memory (4), and for analyzing the impact of drugs and toxic chemicals on neural circuits, single cells and synapses, and even specific receptors and ion channels (3,5–7).

The hippocampal formation, a limbic system structure, is particularly well-suited for use in brain slice methods because of its familiar lamellar architecture. When cut in thin sections in the appropriate plane, the classic trisynaptic circuit of the hippocampus is preserved for study in each slice. This circuit can be studied using standard electrophysiological methods, and the results from *in vitro* brain slice studies generally parallel the results from comparable *in vivo* studies (8). For these reasons, the hippocampal slice has become the most widely used brain slice preparation. In recent years, similar brain slice techniques have been employed to study diverse CNS structures, and *in vitro* methods have been refined and extended to deal with new and specialized problems.

Several investigators have suggested that a brain slice preparation could profitably be used as a screen for neurotoxicity (9–13) in addition to its more common use as an analytical tool (2,5). This chapter explores the various applications of brain slice methods in neurotoxicology. The next section describes the methodology employed for maintaining slices of CNS tissue *in vitro* using the hippocampal slice preparation as a representative example. Later sections introduce common electrophysiological approaches for studying the hippocampal slice and deal with the potential role of brain slice preparations as analytical and screening tools for neurotoxicology.

II. The Brain Slice Preparation: Hippocampus

The methods used here for preparing and studying slices of hippocampus *in vitro* are comparable to well-established methods that are widely used for studying explant tissue from hippocampus, cerebral and cerebellar cortex, and other areas of the CNS (14–17). In general, these methods involve (a) preparing slices of tissue that preserve many of the critical features of the *in vivo* CNS structure of interest, (b) providing an adequate environment for maintaining tissue viability for several hours, and (c) employing an appropriate battery of tests, commonly electrophysiological, of the viability and functioning of the tissue. To illustrate these ideas, a description of the hippocampal slice preparation follows.

A. Preparing the Hippocampal Slice

The hippocampal formation is surgically removed from an adult or juvenile rat brain, then it is chopped into slices 400–450 μm thick. Slices are taken transverse to the long (septo–temporal) axis of the hippocampus to preserve the classic trisynaptic circuit of the hippocampal formation (8,11,12,14). After brain slices are prepared, they are quickly transferred to a chamber designed to maintain tissue viability. Slices are then placed on nylon netting stretched over a pool of flowing artificial cerebrospinal fluid (aCSF) medium so that the slices lie at the interface of medium and chamber atmosphere. The resulting slice, shown in Figs. 1 and 2, preserves the two relatively distinct areas and the three major somatic fields that comprise the hippocampal formation: the CA1 and CA2 pyramidal cell fields of the hippocampus proper and the granule cell field of dentate gyrus. When a healthy slice is illuminated from below during study, all of these major structural features of the hippocampal formation, including somatic layers and fiber tracts, can be visualized using a dissecting microscope (8).

It should be mentioned that the hippocampal slice, like the intact hippocampus, contains systems for serving the major functions of the CNS: (a) transmission and modulation of information flow (via excitatory and inhibitory circuits) and (b) plasticity. In the intact CNS, these functions are subserved by a number of different neurotransmitter and neuromodulatory systems. The

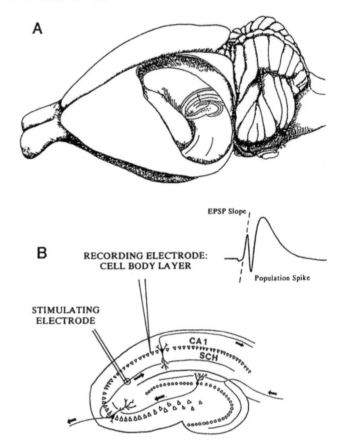

FIGURE 1 (A) Cutaway showing location of the hippocampal formation under the cortex of the rat brain. (B) Schematic representation of the classic trisynaptic circuit of the hippocampal formation. Stimulating and extracellular recording electrodes are shown positioned in the afferent Schaffer collaterals (SCH) and pyramidal cell body layer of area CA1, respectively. Also shown is a representative waveform recorded from the CA1 cell body layer showing a negative-going population spike superimposed on a positive-going population EPSP. Whereas spike height can be used as a direct measure of spike amplitude, often EPSP slope must be used as the measure of EPSP amplitude because the population spike often obscures the EPSP peak.

hippocampal slice, like the intact hippocampus, contains receptors for many, if not all, of the ligands involved in these systems. In addition to the first-messenger systems mentioned earlier, hippocampal CA1 neurons express a variety of second-messenger systems that control or modulate specific ionic conductances. Intracellular mechanisms in hippocampus that serve to link receptors to channels include G-proteins regulating protein kinase C, protein kinase A, and cAMP/cGMP-dependent alterations in protein phosphorylation and Ca^{2+}/calmodulin-dependent channel phosphorylation—all intracellular mediators of receptor activation expressed throughout the CNS. Thus, the *in vitro* hippocampal slice provides a good model of hippocampal physiology *in situ*. However, it should be remembered

FIGURE 2   A hippocampal slice on a nylon net in a slice chamber.

that (a) the slice is a sample of tissue recovering from trauma and hypoxia at preparation (18), (b) it lacks extrinsic influences found in the intact brain (e.g., afferents, efferents, hormones, and other compounds), (c) it may experience washout of important endogenous factors caused by the flowing medium, and (d) it may lack homeostatic regulatory mechanisms. All of these may influence the vulnerability of the slice to neurotoxicants.

B. The Slice Chamber and aCSF Medium

Slices may be maintained either at the interface of aCSF and an oxygen-enriched atmosphere or they may be completely submerged in aCSF. Interface slices are usually placed on netting stretched over glass or plastic. It has been found that nylon netting stretched over the concavity of a glass depression microscope slide works well (19). A ring of epoxy holds the netting in place. The depression slide rests on a stage that is located in a chamber (see Fig. 3) providing a humid atmosphere with an adequate oxygen content (95% O_2/ 5% CO_2). A small hole in the top of the chamber pro-

vides easy access to the tissue for positioning electrodes and other probes. Many variations on this design have been used successfully (8,14–17). Medium flows under and around the slice in a laminar manner at low rates (less than 1 ml/min); the medium is drawn off via a perfusion pump. In this procedure, this is accomplished quite simply by positioning input and output lines of the pump on opposite sides of the aCSF pool. Here again, a variety of methods have been used success-

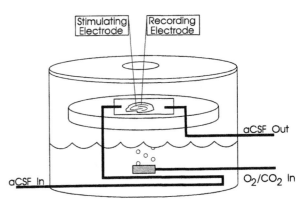

FIGURE 3   The chamber used for maintaining brain slices *in vitro*.

fully. Another method that has been used employs a static pool and a push–pull pump to exchange aCSF when required (8,11,20). Some laboratories use submerged slices (5); higher aCSF flow rates are required, presumably to provide an adequate supply of oxygen to the submerged tissue. Slices from different areas of brain appear to have different requirements in this regard; for example, cerebral cortical slices appear to require higher aCSF flow rates than hippocampal slices. For submerged slices, choosing an appropriate flow rate of the aCSF medium through the pool is important for optimal slice viability.

The aCSF medium we use is similar to that used by many laboratories, though aCSF recipes vary from lab to lab. The aCSF medium used here is composed of 124 mM NaCl, 3.3 mM KCl, 1.25 mM NaH$_2$PO$_4$, 1.2 mM MgSO$_4$, 2.4 mM CaCl$_2$, 25 mM NaHCO$_3$, and 10 mM glucose (19). A random survey of 22 publications using hippocampal slices yielded 18 different formulae for aCSF (21). One reason for this diversity is that the aCSF is often manipulated to obtain special conditions, for example, different ion or gas concentrations (22). A common manipulation is to eliminate Ca^{2+} from the aCSF to eliminate synaptic transmission. Others have manipulated aCSF temperature in order to manipulate tissue temperature (16).

Toxicants are often introduced to the aCSF line where they will affect the entire slice. However, determining exactly when the toxicant reaches the recording electrode is difficult. Better spatial and temporal control of toxicant application can be achieved using pressure injection or iontophoresis of toxicant via a micropipette. A purported advantage of submerged slices is in the more rapid access of the toxicant to the tissue (due to diffusion from both top and bottom of the slice).

C. Equilibration, Parameters for Electrophysiological Recording, and Other Concerns

Once the slices are placed in the chamber, they require a minimum of 30–60 min to equilibrate before consistent recordings can be obtained. An hour is typically allowed for equilibration. After slices have been allowed to equilibrate, electrodes may be positioned in the tissue using a dissecting microscope. To study properties of the CA1 area of hippocampus using field potential recordings (described later), a concentric microbipolar-stimulating electrode approximately 50–100 μm in diameter (23) is positioned in the Schaffer collaterals of stratum radiatum. A glass micropipette (5–20 μm tip diameter) filled with 2 M NaCl (2–4 MΩ resistance) is used to record extracellularly from the CA1

pyramidal cell body layer of the hippocampus (see Fig. 1 for approximate electrode placements). Stimulus pulses used to produce evoked potentials in CA1 range from 0 to 10 V with a duration of 0.1 msec. Monosynaptically driven field potentials are amplified (usually at 1K), filtered (1 Hz–3 kHz), then digitized (5- to 10-kHz sampling rate), analyzed, and displayed on a CRT using a microcomputer equipped with an analog-to-digital converter and a software system developed in-house. Waveforms are archived on magnetic media for more detailed analysis after the experiment is completed.

It appears that rates of success with this preparation vary from lab to lab and over time within the same laboratory (21). Many have speculated as to the causes of this variation with concerns voiced over factors ranging from the condition of animals prior to sacrifice to the humidity of the laboratory. Most researchers agree, however, that general cleanliness and consistent gentle handling of the tissue seem to be the best predictors of success (16,21). A number of investigators (8,14,16,17,21) have discussed in detail other factors that may be important for success with this method. Using the foregoing techniques, slices typically remain viable for 6–8 hr and remain alive with altered responses for 24 hr or more. Slices can then be removed for anatomical or chemical analyses.

III. The Hippocampal Slice as an Analytic Tool

The hippocampal slice is a versatile tool for the analysis of the cellular mechanism(s) of neurotoxicant action. While using the hippocampal slice to study toxicant action is not widespread, the preparation is a commonly employed tool for the study of many other aspects of neurobiology. The hippocampal slice has been used for the study of electrophysiological, biochemical, molecular biological, and anatomical phenomena.

A. Electrophysiology: Extracellular Methods

The extracellular approaches used for studying *in vitro* brain slice electrophysiology are essentially the same as those used for studying the same neural systems *in vivo*. Field potential recordings measure the population response of hundreds of neurons in the vicinity of the recording electrode (either a glass micropipette or a metal microelectrode). Field potential recordings are useful for sampling the activity of a large population of cells and for measuring synaptic responses (EPSPs and IPSPs) on the remote dendrites of hippocampal pyramidal cells–responses that are difficult to measure with intracellular methods. Intracellular recordings may severely attenuate these electrotonically distant responses, distorting their magnitude and duration due to the cable properties of large pyramidal cells. Field potential recordings are generally used to measure stimulus-induced activity and are commonly employed *in vivo* as well. They are most often used to study properties of and agent-induced changes in (a) excitatory systems, (b) inhibitory systems, and (c) neuronal plasticity within the hippocampal formation. Single or multiple unit activity measures are less useful because of the general lack of background spontaneous activity against which to measure the effects of added agents.

1. Excitatory Systems

When stimulating stratum radiatum of hippocampus and recording extracellular field potentials in the pyramidal cell body layer of area CA1 (see Fig. 1), low-intensity stimulation produces a slow, positive-going excitatory postsynaptic potential (EPSP). The population EPSP increases in amplitude and slope with increasing stimulus intensity until it triggers a fast, negative-going population spike. Evoked EPSP and spike amplitudes are stable over time when no pharmacological or toxicological agents are introduced, a fact that results in the ability to determine the input–output (I/O) function that relates stratum radiatum stimulus intensity and the amplitude of the response recorded in the CA1 pyramidal cell body layer. The stability of the I/O function can be tested by periodically collecting I/O data. Low levels of stimulation produce consistent population EPSPs, whereas higher levels of stimulation produce consistent population spikes. Thus, for example, the data necessary to characterize the I/O function (i.e., CA1 responses for several stimulation intensities) can be collected periodically. In addition, between I/O sampling, CA1 pyramidal cell responsivity can be monitored even more closely by stimulating stratum radiatum with a constant intensity stimulus (a "standard" stimulus) more frequently, perhaps ever 1–5 min. This sampling rate allows relatively close monitoring of CA1 excitability, but is not so frequent that it causes changes in baseline response (too frequent stimulation may induce synaptic plasticity). The relative magnitude of the field potential population spike and EPSP are the primary measures of hippocampal excitability. Secondary measures can include spike latency, spike width at half amplitude, and area under the spike and EPSP.

2. Inhibitory Systems

Neurotoxicant or drug effects on inhibitory systems are frequently found to be independent of effects on the amplitude of the population spike and population EPSP, the latter being taken as measures of the status of excitatory neuronal systems. Inhibition in hippocampal CA1 is produced by GABAergic systems, including local inhibitory basket cells (recurrent inhibition) and feedforward processes (24–26). The effects of activating recurrent and feedforward inhibition are not generally seen following a single stimulus pulse because inhibition reduces the magnitude of responses to stimulus events occurring only shortly after the initial stimulus. However, pairs of stimuli can be used to demonstrate inhibitory processes if the interval between pulses is of short duration. Paired pulses with a 25-msec interstimulus interval are typically employed (19). Thus, the status of local inhibitory systems can be assessed by periodically collected waveforms generated using the paired pulse method.

3. Neuronal Plasticity

Long-term potentiation (LTP) is a stable, relatively long-lasting increase in synaptic efficacy at monosynaptic junctions, occurring as the result of brief afferent fiber tetanization or behavioral learning (4,27). Initially observed in the hippocampus, a region long implicated in learning and memory, LTP has now been documented in a variety of brain structures and in a variety of species. The existence of a neural phenomenon that persists for considerable lengths of time and that occurs both as a result of electrical stimulation of afferents and in conjunction with behavioral learning has led many to consider the hypothesis that LTP underlies memory storage in the brain (4,27–29). This allows an investigation of the effect of toxicants on synaptic processes thought to be central to cognitive processes.

In hippocampal slices, LTP can be induced in area CA1 by a brief tetanus (100 Hz for 1 sec, applied twice) to the afferent fibers (the Schaffer collaterals). LTP develops within 10 min and is maintained for several hours with minimal decay. Thus, the status of mechanisms underlying LTP can be assessed by observing the expression and time course of decay of LTP by stimulating regularly (every 5 min, for example) over an hour post-tetanus and recording the evoked waveforms.

4. Current Source Density Analysis

Field potentials can be recorded from linear array electrodes that span the dendritic trees of the pyramidal cell. Potentials recorded from these multisite electrodes can be subjected to a current source density (CSD) analysis (30). The CSD analysis identifies the location and magnitude of current leaving and entering cells and could be useful for determining the effect of toxicants on synaptic currents and channel activity at specific cellular loci. CSD analyses can be performed *in vivo* and *in vitro,* although the electrode configurations differ markedly.

5. Control Procedures

Since brain slices have a limited and variable life span, methods have been developed to monitor slice viability over time. One method involves maintaining a separate stimulating electrode to monitor a baseline response over time. The use of this approach requires that the toxicant under study be excluded from this region of the slice, generally by applying it via microinjection to the experimental site some distance away. A varient of this approach utilizes separate sets of experimental and control stimulating and recording electrodes. This requires four electrodes and four microelectrode manipulators, the maximum that can be conveniently used in a slice experiment without crowding. A different approach is to monitor the constancy of the afferent fiber volley before and after the experimental treatment in low calcium medium (for toxicants that do not affect axonal conduction). It is also possible to use the latter method to distinguish changes in synaptic function and changes in general tissue viability. This test should be used with caution, however, as the fiber volley can be maintained for periods of time when synaptic transmission is clearly altered.

B. Electrophysiology: Intracellular and Patch Clamp Methods

Intracellular recordings from slice preparations (5) using small penetrating sharp micropipettes provide detailed information about transmembrane potential, current, and resistance. They can be used to clamp the membrane at fixed potential or currents, thus facilitating the identification of particular response components. For example, clamping the membrane at a series of different potentials can be used to determine the reversal potential for the response under study and thus the identity of the charge-carrying ions. Membrane-impermeant compounds can be delivered intracellularly by adding them to the electrolyte inside the pipette. The compounds then diffuse into the cell where they have their effect. The calcium chelator BAPTA, for example, has been used to block calcium-dependent processes in studies of LTP (31–33). Intracellular recordings are more difficult *in vivo* because of the mechanical instability associated with respiratory and cardiovascular activity.

Whole cell patch recordings (31) use larger nonpenetrating micropipettes that seal to the surface of the cell and allow for greater electrical clamping of membrane potential and current. Whole cell patch electrodes are also a good way to diffuse substances into cells, for example, drugs that block particular protein kinases. Unfortunately, the same forces that facilitate the diffusion of material into cells also facilitate the diffusion of material from the cell to the pipette. Thus, whole cell recordings are of short duration unless a perforated-patch electrode is used. Whole cell recordings are generally employed only *in vitro.*

C. Biochemistry and Molecular Biology

Hippocampal slices have been used for a variety of biochemical measures from protein synthesis to neurochemical measurement to receptor-binding assays. A practical advantage of brain slices for many biochemical measurements is that the amount of chemical needed is small. For example, protein synthesis is usually measured by introducing a radiolabeled amino acid into the aCSF and measuring its subsequent incorporation into protein on electrophoretic gels. The quantity of an expensive radioisotope (or rare compound) required is minimal due to the small amount of tissue (a hippocampal slice weighs under 5 mg). The effect of a toxicant on neurotransmitter release can be determined by high-performance liquid chromatography by either analyzing the aCSF perfusate for compounds of interest or utilizing a push–pull or dialysis probe inserted into specific areas of the slice. The small size of the hippocampal slice, however, limits the effectiveness of both methods as the amount of product to be detected is correspondingly small. Many biochemical studies are directed at uncovering the role of second messengers and/or transducing elements (G-proteins) in normal and abnormal situations. These studies often employ blockers or activators of specific kinase activity and can be used to advantage in determining the effects of a toxicant on intracellular biochemical processes. Brain slices also appear to be well-suited for the analysis of molecular genetic mechanisms (34–36), though this approach has received little attention in neurotoxicology.

D. Anatomy

Most toxicant effects at the anatomical level develop too slowly to be revealed in brain slices unless the toxicant is applied some time prior to preparing slices. If this is done, then one can measure physiological or biochemical changes in the slice prior to processing it for anatomy. Virtually all of the anatomical techniques used in whole brain have been successfully employed with brain slices. However, the small size of the tissue poses difficulties with respect to sectioning and surface artifact. Some silver impregnations, the rapid Golgi, for example, will produce so much surface artifact on the tiny slice that no usable tissue remains. The solution is to make a "sandwich" by placing the slice between larger pieces of brain and tying the bundle together. The bundle is processed like a normal brain and the artifact-free slice is removed later. Brain slices can be readily sectioned in a cryostat by placing slices directly onto a frozen base where they will instantly freeze. Light and electron microscopy techniques have been used to assess the effects of lead acetate on hippocampal slice morphology (37).

IV. The Brain Slice: Other Tissues

Although the hippocampus is, by far, the most commonly used brain slice preparation, many other tissues have also been used in slice preparations. Historically, the hippocampal slice has been especially appealing because it is among the easiest to study in electrophysiological experiments due to its anatomy; the hipopcampal slice preserves an intact trisynaptic circuit with clearly defined cell fields and fiber systems. The ability to preserve a multisynaptic circuit in a "planar" slice is also possible in slices of neocortex and spinal cord.

A. Neocortical Slices

Neocortical slices (38–40) are prepared by blocking the region of cortex and obtaining slices cut to preserve the columnar anatomy of this tissue. Such slices from rat visual cortex, for example, include cortical layers I–VI and white matter from visual areas 17 and 18. When the slice is obtained from a correctly oriented block, afferents from subcortical structures are preserved. A stimulating electrode placed in the white matter at the base of the column will activate the afferents producing multisynaptic activity in the cortical lamina. However, this stimulating electrode placement will also antidromically activate cortical efferents. Thus the resultant cortical-evoked response will contain both antidromic and orthodromic activation of cortical neurons—a potentially confounding problem. This can be addressed in two ways. Calcium can be removed from the aCSF to eliminate synaptic responses, leaving only fiber volleys and antidromic cell firing. Thus the contribution of antidromic activation to the response can be measured. Cortical efferents send collaterals back into cortex, however, and this procedure does not control for the synaptic activation elicited by collaterals of the fibers antidromically stimulated. In a slice cut to preserve cortical afferents, however, the stimulating electrode can be moved laterally (off-line) so that cells in the column being recorded

from are minimally antidromically activated. Several differences are present in preparing slices of neocortex from adult animals. The most important is that slice viability is enhanced by first reducing tissue metabolism to prevent damage from the ischemia inherent in preparing slices. This can be done using anesthesia (methyoxyfluorine inhalation), hypothermia (reducing core temperature to about 18°C), or both. This extra step is not required for immature cortex, which has anaerobic metabolism, or hippocampus.

B. Spinal Cord Slices

It is also possible to preserve afferents synapsing on the cells of interest in transverse slices of the spinal cord (41,42). Two difficulties are present in preparing cord slices. The first relates to the laminectomy that must be performed to expose the cord and the necessity of prolonged exposure to anesthesia during laminectomy. The second is the necessity of carefully removing the meninges that surround the cord that compromise the ability to obtain cleanly cut transverse sections.

C. Brain Slices of Other Tissues

Other tissues, including cerebellum (43–46), lateral septum (47), hypothalamus (48), locus coeruleus (49), basal ganglia (50), and central vestibular neurons (51), among others, have all been used successfully in slice experiments. For many of these tissues, the intrinsic anatomical organization makes the placement of stimulating electrodes on isolated afferents difficult.

V. Advantages and Disadvantages of Brain Slice Preparations

As with all experimental procedures, brain slice preparations have pros and cons. The major advantage is in the degree of control that the experimenter has over the chemical environment of the slice. The major disadvantage is that the tissue is no longer in its normal environment, resulting in altered physiology and viability. In determining whether a brain slice preparation is an appropriate approach for a given problem, one should consider the advantages and disadvantages of this approach.

A. Advantages of Brain Slice Preparations for Neurotoxicology Research

1. The composition of the extracellular fluid can be altered at will in known ways. The aCSF can be tailored

to the requirements of the experiment and changed whenever desired. This can often facilitate the identification of toxicant effects. For example, removing calcium from the aCSF eliminates synaptic responses, allowing the determination of toxicant effects on axonal conduction.

2. Many brain slice preparations are well-characterized, and the hippocampal slice is a particularly well-characterized tissue. Its physiology, chemistry, and anatomy are among the best understood in the brain. The hippocampal slice displays a range of physiological responses from fiber volleys to paired pulse inhibition to associative LTP (where a weak synaptic input is strengthened following coactivation with a strong input). Thus it can be used to test the effects of a compound on a variety of physiological responses ranging from basic membrane physiology to cognitive responses such as associative LTP.

3. For toxicant studies a major advantage of brain slices is that known concentrations of toxicant can be applied either to the entire slice by adding the toxicant to the aCSF or to limited domains of the slice by micropressure ejection (distal dendrites of hippocampal pyramidal cells, for instance). The effects of the toxicant can be measured by electrophysiological, biochemical, molecular biological, or anatomical methods (see Sections III and VI,G).

4. The mechanical stability of the slice offers long duration electrophysiological recordings using intracellular, whole cell patch, field potential, or CSD recordings. The placement of electrodes is under visual control, eliminating the inaccuracies of stereotaxic placements.

5. The anatomy of the slice can be manipulated surgically prior to study. For example, hippocampal slices can be prepared that selectively preserve certain afferents, allowing the experimenter to stimulate these fibers singly or in combination with other fibers.

6. No anesthetics are present. Animals are usually sacrificed by rapid cervical dislocation, a fast and painless method that avoids the potentially confounding effects of anesthesia.

B. Disadvantages of Brain Slice Preparations

1. Brain slices are devoid of extrinsic afferents and efferents resulting in altered physiological activity. For example, in the hippocampal slice this problem is most obvious in the reduction of spontaneous activity. Because spontaneous activity is low in hippocampal slices, the analysis of the effects of toxicant on single- or multiple-unit firing rates is often impractical.

2. As slices are in an artificial environment and are exposed to constantly flowing aCSF (from 0.2 to 2.0 ml/min in various preparations), they may experience washout of important endogenous compounds and factors. The absence of these factors may contribute to the reduced viability of brain slices or, potentially, to erroneous conclusions regarding toxicant action.

3. Brain slices are short lived (typically 8–12 hr). It is possible that long-latency toxicant effects may not be seen (unless the toxicant is administered long before slice preparation). For example, a toxicant that exerts its effect by blocking protein synthesis may require several hours before its action is detectable. Sequelae requiring days or longer to develop could not be followed in the same preparation, though serial sampling of slices from exposed animals might provide a means of overcoming this limitation (18).

4. Because there is no circulatory system, compounds transported by blood, such as hormones, are no longer able to modulate brain slice activity. Similarly, the blood–brain barrier is absent, a characteristic of the slice preparation that may be viewed as either an advantage or a disadvantage (12).

5. Although electrophysiological responses may appear normal, the tissue may differ biochemically in important ways from the same tissue in the intact nervous system (18). In addition, because cells in slices receive oxygen diffused from the surface of the tissue, neurons at various depths within the slice may experience environments that range from hyperoxic to hypoxic (18,52,53).

VI. Brain Slices as Tools in Neurotoxicology

A. Heavy Metals: Tin, Lead, and Mercury

Hippocampal slices have been used to study the effects of organometals on CNS tissue. Trimethyl tin (TMT) is a neurotoxic compound that produces impairments of learning and memory in both humans and animals (54–60). Neuronal damage resulting from TMT exposure in rats is most extensive in the hippocampus, though neuropathology is also reported in other areas. In the hippocampal slice, TMT suppresses population spike amplitude recorded from CA1 pyramidal cells and in dentate gyrus at concentrations as low as 0.6 μM (13,58,60,61) and increases spontaneous activity in dentate gyrus cells through calcium-dependent mechanisms (62). Work with slices taken from TMT-treated rats indicates that TMT produces long-term changes in the cholinergic septo-hippocampal system that are correlated with behavioral dysfunction (60).

Triethyl tin (TET) is a neurotoxic organometal that reliably produces myelinotoxicity in adult laboratory rats, mice, and rabbits (63–65). *In vivo* TET intoxication also leads to decreased levels of norepinephrine, dopamine, and serotonin in several brain regions within 2 hr postexposure (66,67). Changes in CNS catecholamine concentrations are observed at time points preceding the appearance of myelin vacuoles (68). *In vitro* assays of TET neurotoxicity have shown that TET disrupts cellular energy levels by inhibiting mitochondrial and neuronal ATPase activity, thereby interfering with processes of oxidative phosphorylation (67,69–73). However, TET disruption of cellular energy systems (namely, cellular cAMP levels) probably does not account for changes in brain catecholamine levels following TET exposure *in vivo* (74). Therefore, the basis for the dissociation in time course of the neurochemical and anatomical alterations following TET exposure *in vivo* are not yet understood. Work with hippocampal slices exposed to TET demonstrated a rapid and direct effect of TET on neurotransmission which occurred without apparent perturbation of presynaptic afferent fibers (20,75). Extracellular recordings showed that 10 μM TET suppressed evoked hippocampal CA1 EPSPs elicited by Schaffer collateral stimulation whereas evoked afferent fiber volley potentials remained stable (20). TMT also suppressed local inhibitory systems assessed using a paired pulse test (20). Further studies used the analytical aspects of the hippocampal slice to study the calcium dependency of this phenomenon (75). Elucidation of the mechanism underlying TET-induced suppression of evoked synaptic potentials may provide a better understanding of the lethargy and neurochemical changes observed shortly after TET exposure *in vivo*.

Lead (Pb) is a known neurotoxicant, but its mechanism of action is still under investigation. Hippocampal slices exposed to lead acetate for 30 min show suppression of the CA1 pyramidal cell population EPSP and spike in a dose-dependent manner (76). However, the presynaptic fiber volley and the anidromically evoked population spike of CA1 pyramidal cells were unaffected by lead at the highest concentration used (53 μM) (76). In another study, light microscopic and electron microscopic examination of hippocampal slices exposed to 40 μM lead acetate for 4 hr revealed that electrophysiological changes caused by short-term lead exposure are not related to specific observable neuronal damage (37). Generally, then, the results are consistent with the idea that lead interferes with calcium-mediated neurotransmitter release in the hippocampal slice as it does in the neuromuscular junction (76).

In another study, a "cumulative dose–response function" (cDRF) procedure was used. Hippocampal slices were exposed to 0.1, 1, 10, and 100 μM lead acetate or triethyl lead (TEL) in successive 30-min periods (19). Lead acetate produced a slight increase in excitability until the 100 μM exposure began, then an abrupt suppression of excitability to approximately 70% of baseline. In contrast, TEL appeared considerably more toxic than lead acetate; initial indications of suppression of excitability appeared with concentrations as low as 10 μM and profound suppression was observed at the 100 μM level (19). The results were consistent with the known *in vivo* neurotoxicity of lead acetate and TEL, where TEL is considered to be much more neurotoxic than lead acetate.

Methyl mercury (MeHg) continues to receive attention because there are still a number of open questions regarding its mechanism of action in the CNS. Evidence from studies of neuromuscular junction and cerebrocortical synaptosomes suggests that MeHg interferes with Ca^{2+}-mediated synaptic mechanisms via effects on Ca^{2+} stores within the nerve terminal ($[Ca^{2+}]_i$) (77–82). Other studies suggest that MeHg interferes with cellular aerobic respiration (83–85). Hippocampal slices were exposed to 0.1, 1, 10, and 100 μM concentratons of MeHg in successive 30-min epochs in a cDRF procedure. MeHg exposure produced enhanced tissue excitability at lower concentrations followed by suppression of excitability at the 100 μM concentration (19). In another experiment, the same method was used to expose slices to 10, 25, 50, 75, and 100 μM MeHg in successive epochs. MeHg enhanced excitability during 10 and 25 μM exposures, then produced profound suppression beginning with the 50 μM concentration. Thus, over both a broader and narrower range of concentrations, MeHg first enhanced then suppressed tissue excitability. Comparable effects on excitatory neurotransmission in hippocampal area CA1 without recovery following a washout procedure have been reported using a between-slice dose–response determination method (86).

Both experiments also provided intriguing evidence that inhibitory neuronal systems in the hippocampal slice were more susceptible to suppression by MeHg than excitatory neuronal systems. In the first experiment, paired pulse inhibition was suppressed after 5 min of exposure to 100 μM MeHg, whereas excitatory systems produced responses comparable to preexposure baseline at the same time point. By 5 min later, both excitatory and inhibitory neuronal systems were suppressed by MeHg (19). In the second experiment, inhibitory systems were suppressed throughout the 25 μM concentration, whereas excitatory systems were facilitated, not suppressed, during the same period. Thus, using the narrower range of concentrations (10–100 μM MeHg) allowed the dissociation of effects on excitatory and inhibitory systems to be observed for approximately 30 min. The results are consistent with the idea that MeHg suppressed local inhibitory systems prior to suppressing excitatory systems in hippocampal area CA1.

Finally, MeHg has been shown to suppress both the presynaptic (FV) and the postsynaptic (EPSP) responses with the same time course, suggesting that the effects of MeHg in the hippocampal slice were not primarily synaptic. These results are more consistent with the view that MeHg affected processes of general cellular viability such as cellular respiration than with the idea that MeHg has effects specific to synaptic function. A related hypothesis is that inhibitory systems could be more sensitive than excitatory systems to a disruption of cellular respiration caused by MeHg, resulting in suppression of inhibitory before excitatory systems. On the other hand, facilitation of excitability observed at lower concentrations of MeHg may reflect disruption of Ca^{2+}-mediated synaptic mechanisms.

B. Acrylamide

Recent work with acrylamide (ACR) sought to obtain initial information concerning dose–response relationships in the hippocampal brain slice preparation. Using standard procedures, hippocampal slices were exposed to 0, 1, 3, 5, 7, or 15 m*M* ACR in the aCSF medium for 2 hr. The results indicated that 1, 3, and 5 m*M* ACR produced little change in excitatory systems, whereas 7 and 15 m*M* ACR suppressed excitability, with the suppression appearing earlier for 15 m*M* compared to 7 m*M* ACR. EPSP data paralleled the population spike results. No effects on inhibitory systems were observed at any concentration (19).

In two additional studies, the effects of *N,N'*-methylenebisacrylamide (MBA) and methacrylamide (MTA) were compared to the effects of ACR (19). Hippocampal slices were exposed to 15 m*M* ACR, MBA, or MTA for 2 hr. ACR suppressed excitability in hippocampal area CA1, MBA produced a slight increase in excitability, and MTA produced a larger increase in excitability (19). In a later experiment, hippocampal slices were exposed for 2 hr as before, but the concentration was increased every 30 min. In this case, slices were exposed to 1, 3, 7, and 15 m*M* concentrations of ACR, MBA, or MTA. The results were generally consistent with those from the between-slice experiment. ACR produced an increase in excitability at lower concentrations (1 and 3 m*M*) and suppression at the 7 m*M* concentration, MBA produced little change from baseline, and MTA produced a step-wise increase in excitability corresponding to increasing concentrations (19).

If one were to speculate on the neurotoxic potential of these chemicals based on these data, ACR would be judged to be more neurotoxic than MBA. The facili-

tatory effects of MTA are reminiscent of those of lower concentrations of ACR, and thus it might be concluded that MTA should be viewed as having more neurotoxic potential than MBA but less than ACR. This ordering of the agents' neurotoxicity, MBA < MTA < ACR, parallels the known neurotoxicity of these agents *in vivo*. This is particularly surprising (and encouraging) given that other *in vitro* screening methods (using protein content, LDH activity, and cumulative glucose consumption) have found MBA to be more neurotoxic than both ACR and MTA (87) despite its low *in vivo* neurotoxicity. Although the authors of the latter study urge caution in relying on *in vitro* methods for neurotoxicity assessment because of the unexpected neurotoxicity of MBA in their *in vitro* studies, acute exposure of these agents in the hippocampal slice produced results that parallel those obtained *in vivo*.

C. Anticholinesterases

Hippocampal slices have been used to investigate the effects of cholinesterase inhibitors (anti-ChEs), including the organophosphates, diisopropylfluorophosphate (DFP) and pinacolyl methylphosphonofluoridate (soman), and the carbamate, eserine. DFP and eserine caused the development of a second population spike which was reversible for eserine, but not for DFP (88). Nicotinic and muscarinic antagonists did not block this effect at all concentrations of the anti-ChEs, suggesting that the observed effects were not the result of cholinergic mechanisms (88). Other studies have shown that DFP and soman induce epileptiform discharges in hippocampal slices and that some of these effects were reversible (89). The pattern of effects observed with DFP and soman is not similar to that observed with drugs that interefere with inhibitory systems, but is more similar to that produced by potassium channel blockers (89).

D. Anoxia, Hypoxia, and Carbon Monoxide

Hippocampal slices have been used to study the effects of anoxia, hypoxia, and carbon monoxide on CNS tissue. Studies of anoxia are easily accompliished by using different gas mixtures to produce anoxic conditions. A typical approach involves maintaining two reservoirs of aCSF, one equilibrated with a gas mixture of 95% O_2/5% CO_2 for control conditions and the other with 95% N_2/5% CO_2 for anoxic conditions (90). Similarly, different levels of hypoxia can be achieved by gassing the aCSF with gas mixtures with intermediate proportions of oxygen and nitrogen, such as a mixture of 75% N_2/20% O_2/5% CO_2, keeping carbon dioxide levels constant for buffering (91). When using submerged slices, simply switching medium lines at the appropriate times allows control of the duration of the anoxic or hypoxic episode. When using slices at the interface of aCSF and atmosphere, an additional switch is needed to change the gas source for the chamber. Several slices can also be maintained together in containers such as capped beakers under positive pressure with the gas mixture to prevent access to atmospheric oxygen. This approach is suitable for incubating larger numbers of slices before, during, and after an anoxic or hypoxic episode in biochemical studies, for example (90).

Hippocampal slices exposed to anoxia for 10 min show irreversible loss of the population spike recorded from pyramidal cells in area CA1 of hippocampus and from granule cells recorded in dentate gyrus, whereas slices exposed for shorter periods show differential effects on pyramidal cells and granule cells (90,92). As observed *in vivo*, pyramidal cells appear more sensitive to anoxia than dentate granule cells. With moderate hypoxia (75% N_2/20% O_2/5% CO_2 for 20 min), CA1 pyramidal cells show a posthypoxic increase in excitability (91). This effect is not the result of loss of inhibition; inhibitory systems recover with the same time course as excitatory systems (91).

More severe hypoxia results in spreading depression (SD) (91), and the hippocampal slice appears to provide an ideal model system for studying effects related to this phemomenon. Prolonged hypoxia or anoxia can produce irreversible neuronal damage. Minimizing Ca^{2+} influx by using Ca^{2+}-free aCSF protects slices against anoxic damage (90). Similarly, other studies have explored mechanisms of protection against hypoxia (53,93–96) with sometimes conflicting results. To explain variations in results obtained by different labs, it has been proposed that differences in methodology produce different anoxic or hypoxic conditons for slices. It is argued that sudden, severe anoxia triggers SD and is followed by relatively rapid tissue damage, whereas gradual, less severe hypoxia does not trigger SD and produces tissue damage only after prolonged periods of hypoxia (96).

Traditionally, the toxicity of carbon monoxide (CO) has been attributed primarily to tissue hypoxia resulting from the decreased oxygen-carrying capacity of the blood following CO exposure. This decrease in oxygen-carrying capacity is due to the high affinity of carboxyhemoglobin for CO, resulting in relatively little free hemoglobin for oxygen transport, and also to the fact that in the presence of carboxyhemoglobin there is an apparent increase in the binding strength for oxygen in the remaining hemoglobin. The result of the latter effect is a shift of the dissociation curve to the left and an increase in the severity of tissue hypoxia

beyond that expected from the loss of oxygen-carrying capacity attributable to the CO itself. However, it has also been proposed that CO toxicity could be the result of CO dissolved in blood plasma interfering with cellular respiration by binding to cytochrome-a_3 rather than by the presence of carboxyhemoglobin and resulting tissue hypoxia. The hippocampal slice provided a straightforward means of testing this idea. When hippocampal slices were exposed to various mixtures of CO or N_2, O_2, and CO_2 (the latter held constant at 5%), aCSF prepared with equivalent concentrations of CO and N_2 produced equivalent effects (22). It was concluded that short-term CO exposure was not toxic to hippocampal tissue.

E. Ethanol

A variety of studies of both acute and chronic alcohol exposure have demonstrated that alcohol has direct effects on hippocampal slice morphology and electrophysiology. For example, electrophysiological analysis of hippocampal slices from rats receiving chronic ethanol exposure (20 weeks) demonstrated shrinkage of stratum radiatum commissural input to CA1 pyramidal cells (97), reduction of recurrent inhibition in area CA1 (98,99), and tolerance to the alcohol-induced suppression of the evoked population spike in area CA1 (100). Chronic exposure to alcohol (9 months) in rats suppressed neuronal plasticity normally observed as LTP of pyramidal cell responses following high-frequency stimulation (101). Although the suppression of LTP was found to be reversible when rats were allowed a 2-month alcohol-free period to recover from 7 months of alcohol exposure (101), no study has systematically manipulated the alcohol exposure interval or the alcohol-free interval to characterize the time course of the development of suppression or the recovery of synaptic plasticity.

Studies of acute alcohol exposure in hippocampal slices indicate that alcohol suppresses the evoked population spike in hippocampal area CA1 (102), probably through potentiation of tonic GABA mechanisms (102). Exposing hippocampal slices to acute doses of alcohol in the medium potentiates GABA-mediated neuronal processes such as recurrent inhibition (102). This conclusion is consistent with the finding that acute intoxication in rats results in increased brain GABA levels.

Little attention has been directed toward characterizing the recovery of these excitatory and inhibitory processes following withdrawal of alcohol, and no studies of the effect of acute alcohol exposure and withdrawal on LTP have appeared. Two experiments (102), however, support the contention that such studies are necessary. In one test, slices were exposed to 20 m*M* alcohol in the incubating medium for 30 min and the status of inhibitory systems was assessed by the paired pulse method. Asymptotic potentiation of inhibition was observed by 20 min following the onset of exposure. Reversal to baseline levels of inhibition was not observed because the experiment was terminated after 20 min following removal of alcohol, but some recovery to near-baseline levels was observed by the end of testing (20-min postwithdrawal). In a second study reported in the same paper, slices were exposed to 100 m*M* alcohol for 30 min and the amplitude of the orthodromic population spike was assessed at regular intervals during exposure and for 10 min following termination of exposure. The population spike was suppressed to asymptotic levels by approximately 20 min following the onset of exposure, some recovery (tolerance) was observed by the end of alcohol exposure, and recovery to baseline levels of excitability was observed almost immediately following termination of exposure (102). Although the effect of ethanol on the population spike is correlated with the initial level of inhibition observed in the slice prior to alcohol exposure (102), the time course of recovery is clearly quite different for excitatory versus inhibitory effects of alcohol. This effect may be obtained because alcohol appears to have a direct effect on pyramidal cells, in addition to its effects on inhibitory systems, by altering intracellular Ca^{2+} mechanisms (103). In addition, the time course of initial effects of alcohol exposure may differ for excitatory and inhibitory effects, and the relationship of tolerance observed at high doses (102) to recovery following termination of alcohol exposure has not been explored.

F. Other Agents

Hippocampal slices have been used to study potential effects of the artificial sweetener aspartame (APM) and methylpyridines (104), among other agents. In these procedures, slices were exposed to the agents from the onset of exposure until the experiment was terminated (i.e., there was no "washout" procedure to remove the agent). Exposure to 0.01–10 m*M* APM increased the amplitude of the evoked population EPSP and population spike following exposure (105). In contrast to these effects, 2-, 3-, and 4-methylpyridine produced an entirely different pattern of effects on excitatory systems within the hippocampal slice. Exposure to 100 μM exposures of methylpyridines produced approximately 30–40% suppression following a 3-hr exposure (104).

1-Methyl-4-phenyl-1,2,3,6-tetrahydropyridine (MP-TP) is an example of a potent neurotoxicant that causes degeneration of nigrostriatal dopamine neurons, resulting in Parkinson's disease-like symptoms in hu-

mans (106,107). Two studies showed that MPTP and its metabolites, MPP^+ and $MPDP^+$, blocked synaptic transmission in mouse neostriatal slices and in guinea pig hippocampal slices (108–110). The effects of MPP^+ were reversible by washing out the agent, whereas the effects of MPTP and $MPDP^+$ were not (108,109). Other brain slices studies have focused on mechanisms of protection against irreversible effects of exposure to MPTP and it metabolites (111,112). Finally, MPTP, $MPDP^+$, and MPP^+ cause decreases in dopamine content in mouse brain slices independent of mechanisms involved in irreversible changes in synaptic transmission (113).

G. A Brain Slice Screen for Neurotoxicity?

In recent years, several investigators have proposed that a brain slice preparation might be employed for screening chemicals for neurotoxic potential (9–12,19). In particular, it has been proposed that the hippocampal slice might serve this function. The strategy of the proposed hippocampal slice screen for neurotoxicity is to use a battery of tests to assess agent-induced electrophysiological changes in the status of a model neural system (namely, the hippocampal slice preparation) following exposure to chemical agents. It has been argued in the past that this model system has as its primary conceptual advantage the fact that the *in vitro* hippocampal slice reflects the complexity of the *in vivo* nervous system (12). The hippocampal slice screen concept, then, is founded on the idea that chemical neurotoxicity, whatever its relative specificity within the CNS, should be expressed as some measurable perturbation of function within a system as complex as the hippocampal slice. Whether this assumption will ultimately be found to be proper is a matter of empirical test.

One approach to assessment is to distinguish between a screening phase and an analytical phase of assessment (12). In the screening phase, one would use relatively few tests that are broadly sensitive to a potentially large number of causative factors. The screening phase would be designed to determine that some agents require closer scrutiny because of their potential neurotoxicity. In the analytical phase, determining the mechanism of action of the agent's effects in the screening phase would be the paramount goal. This is the general approach that has met with considerable success in behavioral screening, and this is the approach that has been adopted for the proposed *in vitro* screening method. One prerequisite for this strategy to succeed, however, is to assemble a battery of tests extensive enough to detect many different neurotoxic mechanisms of action. Such a test battery will be somewhat imprecise with regard to specifying the mechanism of action of agents, but it allows much more rapid screening than would be possible with a more analytical approach. The logic is to develop a quick, but sensitive, screen for neurotoxicants, without regard to identifying the underlying mechanism of action. However, a practical and conceptual advantage of the hippocampal slice preparation is that it can be used during both phases of assessment, both as the primary method of screening and, following screening, as an analytical tool for more accurately determining the mechanism of action of neurotoxicants. One additional advantage of this arrangement is that results of screening with the hippocampal slice preparation can provide initial hypotheses to guide work in the analytical stage of research.

1. A Proposed Test Battery for *in Vitro* Screening

The proposed test battery would screen for agent-induced changes in (1) excitatory systems, (2) inhibitory systems, and (3) neuronal plasticity. The measures of pyramidal cell excitability in hippocampal area CA1 (postexposure change in population EPSP and spike) reflect a multitude of processes which subserve normal excitability and neurotransmission in neurons. For example, changes in cellular energetics or metabolism, membrane properties, and axonal transport, among others, would be reflected in changes in excitability as assessed by the screen. Mechanisms essential for neurotransmission alone include those underlying (1) axonal transport; (2) alteration of specific ionic conductances; (3) synthesis, storage, and release of neurotransmitters; and (4) neurotransmitter reuptake, to name but a few. Assessing the status of excitatory systems by observing changes in population EPSP and spike provides a simple means of assessing the status of all of these systems without attempting to pinpoint the particular mechanism targeted by a potentially neurotoxic agent. The results of such a test provide information concerning the potential neurotoxicity of an agent while presumably suggesting hypotheses concerning the mechanism of action. Actually determining the mechanism of action of an agent would be relegated to a later analytical phase of investigation.

A common approach to assessing the status of hippocampal inhibitory systems is to stimulate the same pathway twice using a stimulating electrode located in the Schaffer collaterals of stratum radiatum. This method is suitable for assessing postexposure changes in inhibitory systems as part of the proposed screen. One potential objection to this paired pulse method is that such stimulation activates both local recurrent and feedforward inhibitory systems. This should be viewed as a problem for characterizing agent effects on particu-

lar neuronal systems in an analytical phase of experimentation. However, the fact that both systems may be assessed simultaneously using this simple method is, in fact, advantageous for screening purposes. Agent-induced changes in one or both systems should be detected using this test, and initial evidence indicating agent effects on inhibitory systems can be used to prompt more rigorous analysis using well-established methods (2).

Toxicological data show that hippocampal damage results in behavioral learning and memory impairments. A number of neurotoxic agents (heavy metals, for example) have their most prominent effects on the hippocampus and related limbic structures and, consequently, manifest dysfunctions of learning, memory, and cognitive function in those exposed. For these reasons, evaluation of the status of the mechanisms underlying LTP is the final component of the proposed screen to assess the effects of suspected neurotoxicants on neuronal plasticity. A comparison between neurotoxicant-exposed and control slices would allow a determination of any effects of the agent on LTP induction, expression, and maintenance (each reflecting a different aspect of cellular plasticity). Thus, the status of mechanisms underlying LTP can be assessed by observing the time course of the decay of post-tetanic potentiation by sampling regularly (every 5 min, for example) over an hour post-tetanus.

Hippocampal LTP is dependent on excitatory amino acid neurotransmission for its induction. Additionally, LTP induction and expression are modified by a variety of modulatory agents (4). Both aspects of LTP mechanisms provide multiple opportunities for neurotoxicant action and, thus, neurotoxicant detection. The modulatory actions of a number of endogenous systems have been reviewed (4). Agents that influence these endogenous modulatory systems (principally the catecholamine, cholinergic, and opiate systems; adrenal and gonadal steroid hormones; brain-specific proteins/peptides; and gangliosides) also have the potential of influencing LTP and thus being detected by this means. For example, some agents may have specific effects on LTP, yet display little or no effects on other aspects of the synaptic response. The exogenous agent Δ-9-tetrahydrocannabinol has no effects on other response characteristics, but affects the decay constant of LTP dramatically (114).

The inclusion of a neuronal plasticity measure such as LTP in the proposed screen allows for the detection of agents whose primary or only effects are on the higher cognitive functions of learning and memory. Such an ability may prove advantageous in detecting the subtle effects of neurotoxic agents—effects that might not be detected by less sensitive biological assays.

2. Cumulative Dose–Response Functions

Any practical *in vitro* assay must efficiently assess the neurotoxicity of agents whose biological properties are unknown. An important step in screening is to characterize the dose–response function quickly and to determine the appropriate dose range for more analytical studies. The utility of the cDRF procedure is being evaluated, a method in which the dose of an agent is incremented periodically within the same slice preparation. This procedure can potentially provide initial dose–response information quickly using a minimum number of hippocampal slice samples.

The cDRF procedure monitors brain slice excitability beginning with a low dose of an agent. Periodically, the tissue is exposed to successively higher doses of the agent under study. The resulting cDRF may not necessarily produce the dose–response function that would be obtained from independent samples (due to potential cumulative effects of the weaker doses on responses observed for higher doses). However, the dose–response function obtained in this way may suffice to characterize an approximation of between-samples effects.

To determine the feasibility of this approach, the first neurotoxin to be tested in this way should have a known dose–response function determined with separate slices. cDRF data for acrylamide were collected after collecting dose–response data using the standard between-slice protocol. The results showed that the cDRF produced results comparable to the between-slice protocol over a range of doses from those producing no effects to those producing asymptotic effects (19). In other experiments, the cDRF procedure was used to determine whether the hippocampal slice preparation would detect neurotoxic effects of lead acetate, triethyl lead, and methyl mercury (19) and whether cDRF results could provide a rationale for choosing an appropriate range of doses for further analytical study. Although the limitations of this method are obvious, the results were consistent with the idea that this method could be used for rapidly determining the general characteristics of the dose–response function to guide more analytical between-sample studies (19).

3. Suitability to the Task

The proposed hippocampal slice screen can be used to assess neurotoxic effects of chemicals on a well-studied CNS circuit containing excitatory, inhibitory, and plastic properties. In principle, any deleterious effect of a chemical on this *in vitro* neural network will be reflected in changes in the electrophysiological properties of the tissue assessed by this measure. In addition, tests to assess the plasticity of the system (12) are thought to provide information regarding the status of systems underlying learning and memory processes in the normal brain (4,115). Such a test battery, though consisting of only a few tests, should be broadly sensitive to a multitude of neurotoxic mechanisms by assessing the status of the primary functions of neural tissue: excitatory transmission of information, inhibition, and plastic change. Results to date from hippocampal slice studies of the electrophysiological effects of exposing slices to various known or suspected neurotoxicants have illustrated the utility of the proposed hippocampal slice screen (11,19,20,105).

Potential practical advantages of a hippocampal slice screen include the fact that multiple tests of a variety of neuronal properties can be administered in a single slice, many comparable slice "samples" may be obtained from each animal donor, and testing should be easily automated. In addition, the hippocampal slice screen may have important practical and conceptual advantages over other assay methods because of the putative importance of the hippocampus to a variety of behavioral processes. For example, in addition to possessing excitatory and inhibitory neuronal systems comparable to those found throughout the CNS, the hippocampus is involved in the modulation and experience of emotions and is essential for specific kinds of memory processing. Thus the *in vitro* hippocampal slice preparation as a screen for neurotoxicity offers advantages of *in vitro* methods while allowing guarded, but relatively direct, extrapolation to dysfunction of learning, memory, and other behavioral processes.

There are clearly a number of questions yet to be answered before the hippocampal slice screen concept could be adopted for general neurotoxicity screening. First, how well will the hippocampal slice screen predict the neurotoxic potential of a broad range of agents representing many general classes of neurotoxicants and mechanism of action? Second, how will the problems common to many *in vitro* screening methods that must also be solved for the hippocampal slice preparation be managed? For example, the problem of biotransformation of agents by organs or tissues outside the CNS must be addressed (12) (e.g., biotransformation by the liver). Third, will the hippocampal slice screen be useful for toxicants with delayed or cumulative action? Will early electrophysiological effects permit the detection of more delayed effects typified by these agents? Finally, can the hippocampal slice preparation detect the neurotoxicity of chemicals that have specific effects on neuronal plasticity? The method for assessing neurotoxic effects on plasticity likewise should be tested with a broad, representative set of neurotoxic chemicals and nonneurotoxic controls.

VII. Organotypic Slice Cultures

The foregoing discussion has focused on the preparation, maintenance, and study of "fresh" slices of CNS tissue. An alternative method that has much promise for toxicology is the organotypic brain slice culture (116). Whereas fresh slices of tissue may be maintained for up to 24 hr, organotypic cultures using roller-tube techniques (117–122) can maintain viable CNS tissues for several weeks. The method for preparing these cultures differs significantly from that used for fresh slice preparations. Organotypic slices for cultures are best prepared from infant animals (119), whereas fresh slices may be prepared from animals of nearly any age. Organotypic slices must be prepared under aseptic conditions (119), embedded in a substrate such as chicken plasma on coverslips, then maintained in an appropriate culture medium that is much more complete than the salt/sucrose aCSF commonly used with fresh slices. Organotypic slices are maintained in a roller drum to provide alternating exposure to medium and aeration approximately 10 times an hour (117). Slices prepared in this way will, over the first 2–3 weeks in culture, thin to a monolayer culture with viable cells retaining much of their *in vivo* organization. For example, organotypic cultures of hippocampus retain differentiation of granule cells of dentate gyrus and pyramidal cell fields and appropriate connectivity between these fields (117). Because the tissue is nearly monolayer, it is possible to visualize living cells in the culture using phase-contrast microscopy (116,117). These methods have been used to study explanted tissue from a variety of brain areas, including spinal cord, hippocampus, cerebellum, hypothalamus, medulla, locus coeruleus, neurohypophysis, and cortex (116–123). A unique feature of this method is the possibility of establishing cocultures that develop innervation in culture. This approach has been used successfully with cocultures of septum and hippocampus, locus coeruleus and hippocampus, cerebellum and inferior olive, and substantia nigra and striatum, among others (116,117). These monolayer cocultures can be studied using electrophysiological, biochemical, and anatomical techniques.

Organotypic slice cultures have received little attention in toxicology to date, but initial work with agents such as bismuth (Bi) suggests that organotypic slice cultures may provide advantages over other approaches for some purposes. In one study, for example, insoluble Bi salts in concentrations up to 100 μM ap-

plied acutely to hippocampal cultures had no effect, but chronic exposure (19 days) to 10 μM Bi in culture produced neuronal degeneration (116). Thus, organotypic cultures, which retain characteristics of the architecture of the tissue *in vivo*, may provide an *in vitro* method for studying agents whose effects require days to weeks to develop.

VIII. Conclusions

Brain slice methods are excellent tools for studying neurotoxicant effects from a variety of perspectives and at a variety of levels of analysis. Brain slices, and the effects of potential toxicants on slices, can be studied using a variety of methods from electrophysiological and biochemical to anatomical. Slice methods are appropriate both for rapidly screening for potential toxic effects and for exploring mechanisms of action. The analytical power of the slice allows the same hippocampal preparation to be used to investigate the mechanisms of action in considerable depth and detail. The ability to use the same preparation for both screening and analysis can potentially increase the power and efficiency of the search for toxicant action. In addition, organotypic brain slice cultures offer unique opportunities for studying chronic effects of toxicants in an *in vitro* preparation that retains many of the features of *in vivo* and "fresh" slice neuronal networks. All in all, brain slice methods represent a formidable battery of tools for the neurotoxicologist.

Acknowledgments

Preparation of this chapter was supported by a grant from the Johns Hopkins Center for Alternatives to Animal Testing. Additional support was provided by the U.S. Environmental Protection Agency. Although preparation of this article has been funded in part by the Health Effects Research Laboratory, U.S. Environmental Protection Agency through cooperative agreement No. 813394 to Northeastern Ohio Universities College of Medicine, it has not been subjected to the Agency's peer and policy review and therefore does not necessarily reflect the views of the Agency and no official endorsement should be inferred.

References

1. Yamamoto, C., and McIlwain H. (1966). Electrical activities in thin sections from the mammalian brain maintained in chemically defined media *in vitro*. *J. Neurochem.* **13,** 1333–1343.
2. Dunwiddie, T. V. (1986). The use of *in vitro* brain slices in neuropharmacology. In *Electrophysiological Techniques in Pharmacology* (H. M. Geller, Ed.), pp. 65–90, Alan R. Liss, New York.
3. Schwartzkroin, P. A., and Altschuler, R. L. (1977). Development of kitten hippocampal neurons. *Brain Res.* **134,** 429–444.

4. Teyler, T. J., and DiScenna, P. (1987). Long-term potentiation. *Annu. Rev. Neurosci.* **10**, 131–161.
5. Kelly, J. S. (1982). Intracellular recording from neurons in brain slices *in vitro.* In *New Techniques in Psychopharmacology* (L. L. Iverson, S. D. Iverson, and S. H. Snyder, Eds.), pp. 95–183, Plenum, New York.
6. Gray, R., Fisher, R., Spruston, N., and Johnston, D. (1990). Acutely exposed hippocampal neurons: a preparation for patch clamping neurons from adult hippocampal slices. In *Preparations of Vertebrate Central Nervous System in vitro* (H. Jahnsen, Ed.), pp. 3–24, Wiley, New York.
7. Simmonds, M. A. (1990). Use of slices for quantitative pharmacology. In *Preparations of Vertebrate Central Nervous System in vitro* (H. Jahsen, Ed.), pp. 49–75, Wiley, New York.
8. Teyler, T. J. (1980). Brain slice preparation: Hippocampus. *Brain Res. Bull.* **5**, 391–403.
9. Kuroda, Y. (1980). Brain slices, assay systems for the neurotoxicity of environmental pollutants and drugs on mammalian central nervous system. In *Mechanisms of Toxicity and Hazard Evaluation* (B. Holmstedt, R. Lauwerys, M. Mercier, and M. Roberfroid, Eds.), pp. 59–62, Elsevier, Amsterdam.
10. Rowan, M. J. (1985). Central nervous system toxicity evaluation *in vitro:* neurophysiological approach. In *Neurotoxicology* (K. Blum, and L. Manzo, Eds.), pp. 585–612, Dekker, New York.
11. Fountain, S. B., and Teyler, T. J. (1987). Characterizing neurotoxicity using the in vitro hippocampal brain slice preparation: heavy metals. *Prog. Clin. Biol. Res.* **253**, 19–31.
12. Fountain, S. B., Ting, Y. L. T., and Teyler, T. J. (1992). The in vitro hippocampal slice preparation as a screen for neurotoxicity. *Toxic in Vitro* **6**, 77–87.
13. Armstrong, D. L. (1991). The hippocampal tissue slice in animal models of CNS disorders. *Neurosci. Biobehav. Rev.* **15**, 79–83.
14. Langmoen, I. A., and Andersen, P. (1981). The hippocampal slice *in vitro.* A description of the technique and some examples of the opportunities it offers. In *Electrophysiology of Isolated Mammalian CNS Preparations* (G. A. Kerkut, and H. V. Wheal, Eds.), pp. 51–105, Academic Press, London.
15. Richards, C. D. (1981). The preparation of brain tissue slices for electrophysiological studies. In *Electrophysiology of Isolated Mammalian CNS Preparations* (G. A. Kerkut, and H. V. Wheal, Eds.), pp. 107–132, Academic Press, London.
16. Schwartzkroin, P. A. (1981). To slice or not to slice. In *Electrophysiology of Isolated Mammalian CNS Preparations* (G. A. Kerkut, and H. V. Wheal, Eds.), pp. 15–50, Academic Press, London.
17. Dingledine, R. (Ed.) (1984). *Brain slices.* Plenum, New York.
18. Somjen, G. G., Aitken, P. G., Balestrino, M., and Schiff, S. J. (1986). Uses and abuses of in vitro systems in the study of the pathophysiology of the central nervous system. In *Brain Slices: Fundamentals, Applications and Implications* (A. Schurr, T. J. Teyler, M. T. Tseng, Eds.), pp. 89–104, Karger, Basel.
19. Fountain, S. B., and Rowan, J. D. (1993). Development of an *in vitro* hippocampal brain slice screen for neurotoxicity. In *In vitro toxicology: Tenth anniversary symposium of CAAT* (A. M. Goldberg, Ed.), pp. 27–40, Liebert, New York.
20. Fountain, S. B., Ting, Y. L. T., Hennes, S. K., and Teyler, T. J. (1988). Triethyltin exposure suppresses synaptic transmission in area CA1 of the rat hippocampal slice. *Neurotoxicol. Teratol.* **10**, 539–548.
21. DiScenna, P., (1986). Method and myth in maintaining brain slices. In *Brain slices: Fundamentals, Applications, and Implications* (A. Schurr, T. J. Teyler, and M. T. Tseng, Eds.), pp. 10–21, Karger, Basel.
22. Doolette, D. J., and Kerr, D. I. B. (1992). Comparison of carbon monoxide and nitrogen induced effects on synaptic transmission in the rat hippocampal slices. *Neurosci. Lett.* **138**, 9–13.
23. Chiaia, N., and Teyler, T. J. (1983). A simple method for fashioning small diameter concentric bipolar electrodes for stimulation of nervous tissues. *J. Neurosci. Methods* **7**, 269–273.
24. Andersen, P., Eccles, J. C., and Loyning, Y. (1964). Pathway of postsynaptic inhibition in the hippocampus. *J. Neurophysiol.* **27**, 608–619.
25. Buzsaki, G. (1984). Feedforward inhibition in the hippocampal formation. *Prog. Neurobiol.* **22**, 131–153.
26. Hesse, G. W., Shashoua, V. E., and Jacob, J. N. (1985). Inhibitory effect of cholesteryl *gamma*-aminobutyrate on evoked activity in rat hippocampal slices. *Neuropharmacology* **24**, 139–146.
27. Teyler, T. J., and DiScenna, P. (1984). Long-term potentiation as a candidate mnemonic device. *Brain Res.* **319**, 15–28.
28. Eccles, J. C. (1983). Calcium in long-term potentiation as a model for memory. *Neuroscience* **4**, 1071–1081.
29. Lynch, G., and Baudry, M. (1984). The biochemistry of memory: a new and specific hypothesis. *Science* **224**, 1057–1063.
30. Vaknin, G., DiScenna, P. G., and Teyler, T. J. (1988). A method for calculating current source density (CSD) analysis without resorting to recording sites outside the sampling volume. *J. Neurosci. Methods* **24**, 131–135.
31. Katsuki, H., Kaneko, S., Tajima, A., and Satoh, M. (1991). Separate mechanisms of long-term potentiation in two input systems to CA3 pyramidal neurons of rat hipopcampal slices as revealed by the whole-cell patch-clamp technique. *Neurosci. Res.* **12**, 393–402.
32. Ito, M., and Karachot, L. (1990). Messengers mediating long-term desensitization in cerebellar Purkinje cells. *Neuroreport* **1**, 129–132.
33. Komatsu, Y., and Iwakiri, M. (1992). Low-threshold Ca2+ channels mediate induction of long-term potentiation in kitten visual cortex. *J. Neurophysiol.* **67**, 401–410.
34. Wray, S., Zoeller, R. T., and Gainer, H. (1989). Differential effects of estrogen on luteinizing hormone-releasing hormone gene expression in slice explant cultures prepared from specific rat forebrain regions. *Mol. Endocrinol.* **3**, 1197–1206.
35. Wray, S., Kusano, K., and Gainer, H. (1991). Maintenance of LHRH and oxytocin neurons in slice explants cultured in serum-free media: effects of tetrodotoxin in gene expression. *Neuroendocrinology* **54**, 327–339.
36. Baudry, M., Shahi, K., and Gall, C. (1988). Induction of ornithine decarboxylase in adult rat hippocampal slices. *Brain Res.* **464**, 313–318.
37. Brinck, U., and Wechsler, W. (1989). Microscopic examination of hippocampal slices after short-term lead exposure in vitro. *Neurotoxicol. Teratol.* **11**, 539–543.
38. Whittingham, T. S., Lust, W. D., and Passonneau, J. V. (1984). An in vitro model of ischemia: metabolic and electrical alterations in the hippocampal slice. *J. Neurosci.* **4**, 793–802.
39. Berry, R. L., Teyler, T. J., and Han, T. Z. (1989). Induction of LTP in rat primary visual cortex: tetanus parameters. *Brain Res.* **481**, 221–227.
40. Berry, R. L., Nowicky, A., and Teyler, T. J. (1990). A slice preparation preserving the callosal projection to contralateral visual cortex. *J. Neurosci. Methods* **33**, 171–178.
41. Miletic, V., and Randic, M. (1981). Neonatal rat spinal cord slice preparation: postsynaptic effects of neuropeptides on dorsal horn neurons. *Brain Res.* **254**, 432–438.
42. Magnuson, D. S., Johnson, R., Peet, M. J., Curry, K., and McLennan, H. (1987). A novel spinal cord slice preparation from the rat. *J. Neuroscis. Methods* **19**, 141–145.

43. Garthwaite, G., and Garthwaite, J. (1984). Differential sensitivity of rat cerebellar cells in vitro to the neurotoxic effects of excitatory amino acid analogues. *Neurosci. Lett.* **48**, 361–367.

44. Kimura, H., Okamoto, K., and Sakai, Y. (1985). Climbing and parallel fiber responses recorded intracellularly from Purkinje cell dendrites in guinea pig cerebellar slices. *Brain Res.* **348**, 213–219.

45. Jahnsen, H. (1986). Electrophysiological characteristics of neurones in the guinea-pig deep cerebellar nuclei in vitro. *J. Physiol.* (*Lond*) **372**, 129–147.

46. Konnerth, A., Obaid, A. L., and Salzberg, B. M. (1987). Optical recording of electrical activity from parallel fibres and other cell types in skate cerebellar slices in vitro. *J. Physiol.* (*Lond*) **393**, 681–702.

47. Urban, I. J. (1987). Rat lateral septum in slice preparation with viable transmission. *Exp. Neurol.* **95**, 1–12.

48. Hatton, G. I., Doran, A. D., Salm, A. K., and Sweedle, C. D. (1980). Brain slice preparation: Hypothalamus. *Brain Res. Bull.* **5**, 405–414.

49. Henderson, G., Pepper, C. M., and Shefner, S. A. (1982). Electrophysiological properties of neurons contained in the locus coeruleus and mesencephalic nucleus of the trigeminal nerve in vitro. *Exp. Brain Res.* **45**, 29–37.

50. Griffith, W. H. (1988). Membrane properties of cell types within guinea pig basal forebrain nuclei in vitro. *J. Neurophysiol.* **59**, 1590–1612.

51. Lewis, M. R., Gallagher, J. P., and Shinnick Gallagher, P. (1987). An in vitro brain slice preparation to study the pharmacology of central vestibular neurons. *J. Pharmacol. Methods* **18**, 267–273.

52. Bingmann, D., and Kolde, G. (1982). PO2-profiles in hippocampal slices of the guinea pig. *Exp. Brain Res.* **48**, 89–96.

53. Reid, K. H., Schurr, A., and West, C. A. (1986). Effects of duration of hypoxia, temperature and ACSF potassium concentration on probability of recovery of CA1 synaptic function in the in vitro rat hippocampal slices. In *Brain slices: Fundamentals, Applications and Implications* (A. Schurr, T. J. Teyler, and M. T. Tseng, Eds.), pp. 143–146, Karger, Basel.

54. Fortemps, E., Amand, G., Bomboir, A., Lauwerys, R., and Laterre, E. C. (1978). Trimethyltin poisoning: report of two cases. *Int. Arch Occup. Environ. Health* **41**, 1–6.

55. Walsh, T. J., Gallagher, M., Bostock, E., and Dyer, R. S. (1982). Trimethyltin impairs retention of a passive avoidance task. *Neurobehav. Toxicol. Teratol.* **4**, 163–167.

56. Walsh, T. J., Miller, D. B., and Dyer, R. S., (1982). Trimethyltin, a selective limbic system neurotoxicant, impairs radial-arm maze performance. *Neurobehav. Toxicol. Teratol.* **4**, 177–183.

57. Fountain, S. B., Schenk, D. E., and Annau, Z., (1985). Serial-pattern-learning processes dissociated by trimethyltin exposure in rats. *Physiol. Psychol.* **13**, 53–62.

58. Cohen, C. A., Messing, R. B., and Sparber, S. B. (1987). Selective learning impairment of delayed reinforcement autoshaped behavior caused by low doses of trimethyltin. *Psychopharmacology Berlin* **93**, 301–307.

59. Hagan, J. J., Jansen, J. H., and Broekkamp, C. L. (1988). Selective behavioural impairment after acute intoxication with trimethyltin (TMT) in rats. *Neurotoxicology* **9**, 53–74.

60. Segal, M. (1988). Behavioral and physiological effects of trimethyltin in the rat hippocampus. *Neurotoxicology* **9**, 481–489.

61. Allen, C. N., and Fonnum, F. (1984). Trimethyltin inhibits the activity of hippocampal neurons recorded in vitro. *Neurotoxicology* **5**, 23–30.

62. Armstrong, D. L., Read, H. L., Cork, A. E., Montemayor, F., and Wayner, M. J. (1986). Effects of iontophoretic application of trimethyltin on spontaneous neuronal activity in mouse hippocampal slices. *Neurobehav. Toxicol. Teratol.* **8**, 637–641.

63. Lee, J. C., and Bakay, L. (1965). Ultrastructural changes in the edematous central nervous system. *Arch. Neurol.* **33**, 48–57.

64. Magee, P. N., Stoner, H. B., and Barnes, J. M. (1957). The experimental production of oedema in the central nervous system of the rat by triethyltin compounds. *J. Pathol. Bacteriol.* **73**, 107–124.

65. Torack, R. M., Terry, R. D., and Zimmerman, H. M. (1960). The fine structure of cerebral fluid accumulation. II. Swelling produced by triethyltin poisoning and its comparison with that in the human brain. *Am. J. Pathol.* **36**, 273–287.

66. Bentue Ferrer, D., Reymann, J. M., van den Driessche, J., Allain, H., and Bagot, H. (1985). Effect of triethyltin chloride on central aminergic neurotransmitters and their metabolites: relationship with pathophysiology of aging. *Dev. Aging Res.* **11**, 137–141.

67. Moore, K. E., and Brody, T. M. (1961). The effect of triethyltin on oxidative phosphorylation and mitochondrial adenosine triphosphate activation. *Biochem. Pharmacol.* **6**, 125–133.

68. Torack, R., Gordon, J., and Prokop, J. (1970). Pathobiology of acute triethyltin intoxication. *Int. Rev. Neurobiol.* **12**, 45–86.

69. Aldridge, W. N., and Cremer, J. E. (1955). The biochemistry of organotin compounds—diethyltin dichloride and triethyltin sulphate. *Biochemistry* **61**, 406–418.

70. Cremer, J. E. (1957). The metabolism *in vitro* of tissue slices from rats given triethyltin compounds. *Biochem. J.* **67**, 87–96.

71. Selwyn, M. J., Stockdale, M., and Dawson, A. P. (1970). Multiple effects of trialkyltin compounds on mitochondria. *Biochem. J.* **116**, 15–22.

72. Selwyn, M. J., Dawson, A. P., Stockdale, M., and Gains, N. (1970). Chloride-hydroxide exchange cross mitochondria, erythrocyte and artificial lipid membranes mediated by trialkyl- and triphenyltin compounds. *Eur. J. Biochem.* **14**, 120–126.

73. Wassenaar, J. S., and Kroon, A. M. (1973). Effects of triethyltin on different ATPases, 5'-nucleotidase and phosphodiesterases in grey and white matter of rabbit brain and their relation with brain edema. *Eur. Neurol.* **10**, 349–370.

74. Leow, A. C., Anderson, R. M., Little, R. A., and Leaver, D. D. (1979). A sequential study of changes in the brain and cerebrospinal fluid of the rat following triethyltin poisoning. *Acta. Neuropathol. Berl.* **47**, 117–121.

75. Ting, Y. L. T., Fountain, S. B., and Teyler, T. J. (1992). Extracellular Ca2+ modulation of triethyltin neurotoxicity in area CA1 of the rat hippocampal slice. *Toxic in Vitro* **6**, 159–164.

76. Altmann, L., Lohmann, H., and Wiegand, H. (1988). Acute lead exposure transiently inhibits hippocampal neuronal activities in vitro. *Brain Res.* **455**, 254–261.

77. McKay, S. J., Reynolds, J. N., and Racz, W. J. (1986). Effects of mercury compounds on the spontaneous and potassium-evoked release of dopamine from mouse striatal slices. *Can. J. Physiol. Pharmacol.* **64**, 1507–1514.

78. Levesque, P. C., and Atchison, W. D. (1988). Effect of alteration of nerve terminal Ca2+ regulation on increased spontaneous quantal release of acetylcholine by methyl mercury. *Toxicol. Appl. Pharmacol.* **94**, 55–65.

79. Atchison, W. D. (1988). Effects of neurotoxicants on synaptic transmission: lessons learned from electrophysiological studies. *Neurotoxicol. Teratol.* **10**, 393–416.

80. Kauppinen, R. A., Komulainen, H., and Taipale, H. (1989). Cellular mechanisms underlying the increase in cytosolic free calcium concentration induced by methylmercury in cerebrocortical synaptosomes from guinea pig. *J. Pharmacol. Exp. Ther.* **248**, 1248–1254.

81. Komulainen, H., and Bondy, S. C. (1987). Increased free intra-synaptosomal Ca^{2+} by neurotoxic organometals: distinctive mechanisms. *Toxicol. Appl. Pharmacol.* **88**, 77–86.

82. Levesque, P. C., Hare, M. F., and Atchison, W. D. (1992). Inhibition of mitochondrial Ca^{2+} release diminishes the effectiveness of methyl mercury to release acetylcholine from synaptosomes. *Toxicol. Appl. Pharmacol.* **115**, 11–20.

83. Ahammadsahib, K. I., Ramamurthi, R., and Dusaiah, D. (1987). Mechanism of inhibition of rat brain (Na^+-K^+)-stimulated adenosine triphosphatase reaction by cadmium and methyl mercury. *J. Biochem. Toxicol.* **2**, 169–180.

84. Ally, A., Phipps, J., and Miller, D. R. (1984). Interaction of methylmercury chloride with cellular energetics and related processes. *Toxicol. Appl. Pharmacol.* **76**, 207–218.

85. Yee, S., and Choi, B. H. (1992). Effects of methylmercury on respiratory activity of immature cerebral cortical neurons, cerebellar granule cells, purkinje cells, oligodendrocytes and astrocytes of prenatal and neonatal rat in-vitro. *Soc. Neurosci. Abstr.* **18**, 1609.

86. Yuan, Y., and Atchison, W. D. (1992). Methylmercury-induced disruption of rat hippocampal synaptic transmission and long-term potentiation. *Soc. Neurosci. Abstr.* **18**, 1609.

87. Hayashi, M., Tanii, H., Horiguchi, M., and Hashimoto, K. (1989). Cytotoxic effects of acrylamide and its related compounds assessed by protein content, LDH activity and cumulative glucose consumption of neuron-rich cultures in a chemically defined medium. *Arch Toxicol.* **63**, 308–313.

88. Williamson, A. M., and Sarvey, J. M. (1985). Effects of cholinesterase inhibitors on evoked responses in field CA1 of the rat hippocampus. *J. Pharmacol. Exp. Ther.* **235**, 448–455.

89. Lebeda, F. J., and Rutecki, P. A. (1985). Characterization of spontaneous epileptiform discharges induced by organophosphorous anticholinesterases in the *in vitro* rat hippocampus. *Proc. West. Pharmacol. Soc.* **28**, 187–190.

90. Kass, I. S. (1986). The hippocampal slice: an in vitro system for studying irreversible anoxic brain damage. In *Brain Slices: Fundamentals, Applications and Implications* (A. Schurr, T. J. Teyler, and M. T. Tseng, Eds.), pp. 105–117, Karger, Basel.

91. Reid, K. H. (1986). Ion changes associated with transient hypoxia in the hippocampal slice preparation. In *Brain Slices: Fundamentals, Applications and Implications* (A. Schurr, T. J. Teyler, and M. T. Tseng, Eds.), pp. 118–128, Karger, Basel.

92. Aitken, P. G., and Schiff, S. J. (1986). Selective vulnerability to hypoxia *in vitro*. *Neurosci. Lett.* **67**, 92–96.

93. Schurr, A., and Rigor, B. M. (1986). The hippocampal slice preparation in the study of brain protection against hypoxia. In *Brain Slices: Fundamentals, Applications and Implications* (A. Schurr, T. J. Teyler, and M. T. Tseng, Eds.), pp. 129–142, Karger, Basel.

94. Clark, G. D., and Rothman, S. M. (1987). Blockade of excitatory amino acid receptors protects anoxic hippocampal slices. *Neuroscience* **21**, 665–671.

95. Rothman, S. M., Thurston, J. H., Hauhart, R. E., Clark, G. D., and Solomon, J. S. (1987). Ketamine protects hippocampal neurons from anoxia in vitro. *Neuroscience* **21**, 673–678.

96. Aitken, P. G., Balestrino, M., and Somjen, G. G. (1988). NMDA antagonists: lack of protective effect against hypoxic damage in CA1 region of hippocampal slices. *Neurosci. Lett.* **89**, 187–192.

97. Abraham, W. C., Manis, P. B., Hunter, B. E., Zornester, S. F. and Walker, D. W. (1982). Electrophysiological analysis of synaptic distribution in CA1 of rat hippocampus after chronic ethanol exposure. *Brain Res.* **237**, 91–105.

98. Abraham, W. C., Hunter, B. E., Zornester, S. F., and Walker, D. W. (1981). Augmentation of short-term plasticity in CA1 of rat hippocampus after chronic ethanol treatment. *Brain Res.* **221**, 271–287.

99. Durand, D., and Carlen, P. L. (1984). Decreased neuronal inhibition *in vitro* after long-term administration of ethanol. *Science* **224**, 1359–1361.

100. Carlen, P. L., and Corrigall, W. A. (1980). Ethanol tolerance measured electrophysiologically in hippocampal slices and not in neuromuscular junctions from chronically ethanol-fed rats. *Neurosci. Lett.* **17**, 95–100.

101. Durand, D., and Carlen, P. L. (1984). Impairment of long-term potentiation in rat hippocampus following chronic ethanol treatment. *Brain Res.* **308**, 325–332.

102. Durand, D., Corrigall, W. A., Kujtan, P., and Carlen, P. L. (1981). Effect of low concentrations of ethanol on CA1 hippocampal neurons in vitro. *Can. J. Physiol. Pharmacol.* **59**, 979–984.

103. Carlen, P. L., Gurevich, N., and Durand, D. (1982). Ethanol in low doses augments calcium-mediated mechanisms measured intracellularly in hippocampal neurons. *Science* **215**, 306–309.

104. Fountain, S. B., and Teyler, T. J. (1991). Suppression of hippocampal slice excitability by 2-, 3-, and 4-methylpyridine. *Soc. Neurosci. Abstr.* **17**, 1460.

105. Fountain, S. B., Hennes, S. K., and Teyler, T. J. (1988). Aspartame exposure and *in vitro* hippocampal slice excitability and plasticity. *Fundam. Appl. Toxicol.* **11**, 221–228.

106. Langston, J. W., Irwin, I., Langston, E. B., and Forno, L. S. (1984). 1-Methyl-4-phenyl-pyridinium ion (MPP^+): identification of a metabolite of MPTP, a toxin selective to the substantia nigra. *Neurosci. Lett.* **48**, 87–92.

107. Markey, S. P., Johannessen, J. N., Chiueh, C. C., Burns, R. S., and Herkenham, M. A. (1984). Intraneuronal generation of a pyridinium metabolite may cause drug-induced parkinsonism. *Nature* **311**, 464–467.

108. Wilson, J. A., Wilson, J. S., and Weight, F. F. (1986). MPTP causes a non-reversible depression of synaptic transmission in mouse neostriatal brain slice. *Brain Res.* **368**, 357–360.

109. Galvan, M., Kupsch, A., and ten Bruggencate, G. (1987). Actions of MPTP and MPP^+ on synaptic transmission in guinea pig hippocampal slices. *Exp. Neurol.* **96**, 289–298.

110. Wilson, J. A., Wilson, J. S., and Weight, F. F. (1987). $MPDP^+$ causes a non-reversible decrease in neostriatal synaptic transmission in mouse brain slice. *Brain Res.* **425**, 376–379.

111. Pai, K. S., and Ravindranath, V. (1991). Protection and potentiation of MPTP-induced toxicity by cytochrome P-450 inhibitors and inducer: in vitro studies with brain slices. *Brain Res.* **555**, 239–244.

112. Wilson, J. A., Lau, Y. S., Gleeson, J. G., and Wilson, J. S. (1991). The action of MPTP on synaptic transmission is affected by changes in Ca^{2+} concentrations. *Brain Res.* **541**, 342–346.

113. Wilson, J. A., Doyle, T. J., and Lau, Y. S. (1990). MPTP, $MPDP^+$ and MPP^+ cause decreases in dopamine content in mouse brain slices. *Neurosci. Lett.* **108**, 213–218.

114. Nowicky, A. V., Teyler, T. J., and Vardaris, R. M. (1987). The modulation of long-term potentiation by Δ-9-tetrahydrocannibinol in the rat hippocampus, in vitro. *Brain Res. Bull.* **19**, 663–672.

115. Teyler, T. J., and Fountain, S. B. (1987). Neuronal plasticity in the mammalian brain: relevance to behavioral learning and memory. *Child Dev.* **58**, 698–712.

116. Gähwiler, B. H., Knopfel, T., and Marbach, P., et al. (1992). Use of organotypic slice cultures in neurobiological research. In *The Brain in Bits and Pieces* (G. Zbinden, Ed.), pp. 153–180, Verlag, Zollikon, Switzerland.

117. Gähwiler, B. H. (1981). Organotypic monolayer cultures of nervous tissue. *J. Neurosci. Methods* **4**, 329–342.

118. Gähwiler, B. H. (1984). Slice cultures of cerebellar, hippocampal and hypothalamic tissue. *Experientia* **40**, 235–243.

119. Robertson, R. T., Zimmer, J. and Gähwiler. B. H. (1989). Dissection procedures for preparation of slice cultures. In *A Dissection and Tissue Culture Manual of the Nervous System* (A. Shahar, J. de Vellis, A. Varnadakis, and B. Haber, Eds.), pp. 1–15, Alan R. Liss, New York.

120. Gähwiler, B. H. (1989). Slice cultures of nervous tissue. In *A Dissection and Tissue Culture Manual of the Nervous System* (A. Shahar, J. de Vellis, A. Vernadakis, and B. Haber, Eds.), pp. 65–67. Alan R. Liss, New York.

121. De Boni, U., and Chong, A. A. (1989). Organotypic culture of neonate rabbit hippocampus in roller tubes. In *A Dissection and Tissue Culture Manual of the Nervous System* (A. Shahar, J. de Vellis, A. Vernadakis, and B. Haber, Eds.). pp. 68–71, Alan R. Liss, New York.

122. Gähwiler, B. H., and Knöpfel, T. (1990). Cultures of brain slices. In *Preparations of Vertebrate Central Nervous System In Vitro* (H. Jahnsen, Ed.), pp. 77–100, Wiley, New York.

123. Hendelman, W. J. (1989). Dissection guide for cerebellum-locus ceruleus organotypic cultures. In *A Dissection and Tissue Culture Manual of the Nervous System* (A. Shahar, J. de Vellis, A. Vernadakis, and B. Haber, Eds.), pp. 16–22, Alan R. Liss, New York.

CHAPTER

32

Cell Suspension Techniques in Neurotoxicology

M. ANTHONY VERITY
Department of Pathology (Neuropathology)
and Brain Research Institute
University of California Los Angeles Medical Center
Los Angeles, California 90024

I. Introduction

A shift in neurotoxicologic studies has occurred whereby the discipline has moved from a descriptive to a more mechanistic science. Early studies relied heavily on animal protocols, with morbidity and lethality as measures of toxicity which have been replaced by methods capable of assessing neurotoxic damage at molecular and cellular levels. This chapter does not address the concerns surrounding the use of laboratory animals in biomedical research or as models of neurotoxic injury. Instead, emphasis has been directed to identifying systems which have provided an increased understanding of cellular and molecular toxicology.

A variety of *in vitro* cellular systems have been devised, many of which are applicable to subsequent culturing procedures. While the use of cell culture for neurotoxicologic investigation and more broadly in the analysis of cell injury has been well described (Nelson, 1978; Goldberg, 1980; Veronesi, 1992a,b), the use of cell suspensions for the elucidation of physiological or neurotoxicological end points is less used. It is not

the purpose of this chapter to extol the advantages or disadvantages of *in vitro* toxicity testing or to define the various primary cell cultures used in neurotoxicological research or compare the advantages and disadvantages of *in vivo* versus *in vitro* studies. The primary goals of this chapter are to identify those neural-derived cell suspensions which have provided meaningful physiological and neurotoxicological data and to provide examples of typical studies proven to be of value in elucidating neurotoxic injury.

II. Cell Suspension Models

Table 1 lists established cell suspension systems for use in neurobiological and neurotoxicological investigations. Such cell suspensions may be obtained as primary cell dissociates, for instance, directly derived from cerebrum, cerebellum, dorsal root ganglia, or even brain slice preparations in contrast to the resuspension of cells secondarily derived from a tissue culture system. Cell suspensions obtained from certain cell cultures allow for a variety of special preparations,

TABLE 1 Cell Suspension Systems Used in Neurotoxicological Studies

Neuronal
    Cortical isolates
    Cerebellar granule neurons
    PC12
Glia
    Glial isolates
    Cultures
        $1°$
        C_6 glioma
Cerebral microvessels
Reaggregate suspensions

not easily available from the primary dissociate due to yield or cell contamination.

What value attaches to the use of cell suspensions compared to tissue culture? Table 2 summarizes potential advantages and disadvantages of cell suspensions compared to tissue culture studies. In the case of cerebellar granule cell suspension studies (vide infra) the rapidity and ease of preparation in under 2 hr contrasts to the need for dissociation, establishment in culture, addition of anti-mitotic agents, and stabilization which may take 2 days prior to effective, reproducible use of the immature culture system. A further 7–10 days is needed for appropriate maturation, early synaptogenesis, and minimal glial contamination to allow for reproducible studies of this specific neuronal culture. Both systems provide excellent fixation for immunocytochemistry and ultrastructure. In each case, the rapidity and ease of immediate fixation contrasts sharply with the need for perfusion to provide equivalent optimum ultrastructural observations *in vivo*. Acute or subacute observations up to 6 hr in the case of granule cell suspension studies are permitted in contrast to the subacute/chronic studies available in culture, which also permits developmental paradigms (Verity and Verity, 1991). Both systems allow for appropriate biochemical and biophysical methods but the ease and greater cell

yield associated with bulk separation and suspension provide for broader biochemical investigations.

III. The PC12 Pheochromocytoma Cell

Neurons and chromaffin cells are related cell types (Greene and Tischler, 1982). The PC12 line of rat pheochromocytoma cells has proven useful in studying noradrenergic properties of chromaffin cells and neurons. In particular, the line can be modulated between a replicating or a nondividing state analogous to a fully differentiated neuron. The chromaffin cell-like properties, including the ability to synthesize, store, and secrete catecholamines, have provided an excellent *in vitro* system for studying such mechanisms leading to an understanding of such processes in catecholaminergic neuronal populations. A most striking property of the PC12 line is its capacity to respond to nerve growth factor (NGF), a protein profoundly influencing the growth, maturation, and induced neuronal characteristics (Greene and Tischler, 1982). The majority of observations have been made using PC12 culture often during NGF-induced neuronal differentiation. Certainly, the use of the PC12 cell in neurotoxicology has been established (Veronesi, 1992a). Conversely, suspensions of PC12 cells have received little attention but recent observations have attested to their value. For instance, Ransom and colleagues (1991) used a flow cytometric analysis of PC12 cell suspensions to examine single cell Ca^{2+} mobilization. The response of the PC12 population to bradykinin was very heterogeneous and allowed for the isolation of subclones which revealed robust Ca^{2+} transients as a function of bradykinin receptor stimulation. This procedure provides a method to establish stable subclones from a heterogeneous population based on a functional response. Subclones of cells may be isolated that have optimum responses to virtually any signaling pathway that is coupled to Ca^{2+} flux. Subsequent genetic analysis of such subclones will allow for close study of the genetic

TABLE 2 Comparative Features of Suspension or Culture Studies of Cerebellar Granule Cells

Suspension	Culture
Rapid preparation, 2 hr	Dissociation, establishment in culture, $\approx$ 48 hr.
Ease of procedure. Low cost.	Special techniques, equipment, maintenance.
High yield.	Low yield, variable survivorship.
Excellent morphology, ultrastructure.	Excellent morphology, immunocytochemistry.
Acute, subacute toxicity studies.	Acute-chronic, low-dose toxicity. Developmental toxicology.
Broad range biochemical studies.	Biochemistry limited by yield. Biophysical methods, e.g., patch clamp.

molecular pathways relevant to individual cell signal transduction processes. In an analogous study, Appel and Barefoot (1989) used estimates of intracellular $[Ca^{2+}]$ in PC12 cell suspensions to demonstrate that Ca^{2+} mobilization was receptor mediated and coupled to neurotransmitter release. Systems such as these warrant attention in the neurotoxicological literature due to their extreme sensitivity and value in interrelating neurotoxicant action with signal transduction pathways leading to defined neurotransmitter events.

IV. Cerebral Microvessels

Recent techniques for the isolation of cerebral microvessels (Brendel *et al.*, 1974; Goldstein *et al.*, 1975) have provided a metabolically active preparation from which endothelial cell culture has been established. Both metabolically active isolated microvessels in suspension and cultures of endothelial cells are suitable for studies of cellular and metabolic properties (DeBault, 1982). A close relationship has been found between the function of microvessels or endothelial cells in culture and known activities of the blood–brain barrier, and such data has supported the hypothesis that the cerebral microvessel is the site responsible for blood–brain barrier activity. Techniques for the isolation of cerebral microvessels (Goldstein *et al.*, 1975; Williams *et al.*, 1980) led naturally to the establishment of pure endothelial cell cultures (DeBault, 1982; Spatz *et al.*, 1980). Glucose uptake into isolated cerebral microvessels in suspension is stereospecific and is mediated by a Na^+-independent, carrier-mediated process (Goldstein *et al.*, 1975, 1977; Kolber *et al.*, 1979). Such uptake was inhibited by cytochalasin B and phlorizin, properties established in characterization of glucose transport at the blood–brain barrier *in vivo*. Betz and Goldstein (1978, 1980) demonstrated the existence of both L- and A-systems for amino acid uptake in isolated cerebral microvessels and retinal-derived microvessels. The L-system is characterized as a Na^+-independent bidirectional transmembrane transport whereas it is likely that the A-system only occurs from the antiluminal side, supported by the presence of Na^+-K^+ ATPase in the basement membrane side of the microvessel (Eisenberg *et al.*, 1980). It is noteworthy that endothelial cell culture, devoid of γ-GTP activity, also possesses L- and A-systems for amino acid uptake (Cancilla and DeBault, 1980). Although γ-GTP activity has been detected in microvessels and endothelial cell culture, its enrichment in culture appears dependent on the presence of glial cells (DeBault and Cancilla, 1980). This observation highlights the possible role of cell–cell communication, in this case glial–endothelial in the preservation of blood–brain barrier function or at least in glial factor induction of γ-GTP. Goldstein

(1979) and Eisenberg and colleagues (1980) showed that brain microvessels were able to take up K^+ in a temperature, oxygen-dependent and ouabain-sensitive manner. Fatty acids instead of glucose served as a primary source of energy for ^{86}Rb uptake, a radioactive analog of K^+. The capillaries also contained a Ca^{2+}-ATPase sensitive to cyclic AMP inhibition (Joo, 1979). Isolated cerebral microvessels are able to take up monoamines in a temperature, ionic, and ouabain-sensitive, carrier-mediated process (Hardebo and Owman, 1980).

Such neurobiological observations on isolated microvessel suspensions provide the essential framework upon which mechanistic neurotoxicological studies may be initiated. The value of such studies in the elucidation of toxicant interaction in blood–brain barrier dysfunction cannot be too strongly emphasized. For instance, Goldstein and coworkers (1977), using the isolated brain capillary preparation, examined the effect of lead. In subsequent studies Goldstein (1992) developed methods to isolate microvessels from immature brain, propagate endothelial cells in tissue culture, and coculture endothelial cells and astrocytes in order to model angiogenesis *in vitro*. Such studies have implicated defects in second-messenger functioning in the pathogenesis of the microvascular dysfunction occurring in the immature brain with lead poisoning.

V. Brain-Derived Reaggregating Cell Suspensions

Reaggregating cultures derived from single cell suspensions of fetal brain have proven useful in neurobiological and neurotoxicological study (Trapp and Richelson, 1980; Atterwil, 1987). In common with other cell suspension or tissue culture methods, both advantages and disadvantages may be identified. It is useful to stress at the outset that the model provides data at a histogenic level as compared to cellular and gives opportunities for developmental neurotoxicology. Aggregates formed from different brain regions exhibit region-specific patterns of organization and provide a cell suspension for the analysis of cell interaction and mechanics involved in brain histogenesis (Levitt *et al.*, 1976). Advantages of reaggregate suspensions include the development of a three-dimensional cytoarchitecture more closely allied to the *in vivo* phenotype; a close *in vivo* association between different cell species; biochemical parameters more closely mimic the *in vivo* state but do not reflect individual cell activity; maturational studies more likely reflect the *in vivo* state; myelination may occur under controlled circumstances; and direct accessibility of compounds including neurotoxins with appropriate ease of administration, withdrawal, and concentration are easily controlled. Disadvantages include the possibility of diffusional gradients between periphery and center of the aggregates including both toxin exposure and metabolite accessibility; the mixture of cell types may complicate apparent single cell marker activities; and individual aggregates may differ significantly due to the nature of original cells participating in aggregate formation, especially when aggregates are derived from small brain regions, e.g., mesencephalon, not allowing for reasonable cell representation. A further obvious limitation is the inability to perform electrophysiological examination as the tissue aggregates in suspension.

Experimental protocols indicate that the rotation-mediated aggregating cell culture is a useful model system for developmental toxicological studies (Honegger and Richelson, 1977a,b; Wehner et al., 1985). Cellular differentiation is linked to maturation of the aggregate. The maturation of neurons or different glial species may be followed closely (Trapp et al., 1979) and may reflect specific brain areas, e.g., hippocampus (DeLong, 1970). Immunocytochemical procedures suggest that the biochemical differentiation of neuronal and glial cells resembles that found in vivo and also reflects changing phenotypic expression of cell markers as a function of maturation. Myelination may be followed morphologically and biochemically and appropriate biochemical analyses may also be used (Sheppard et al., 1978; Matthieu et al., 1979). The presence of a variety of synapses has been demonstrated and semiquantitative analysis correlated an increase of synapse formation with in vitro age of culture (Fig. 1). Even complex synapses characteristic of the cerebellum have been found in cerebellar cell reaggregation suspensions (Stefanelli et al., 1977).

This suspension system represents a powerful tool for the investigation of long-term synaptogenic control and molecular biology. In particular, the analysis of disturbed synaptogenesis as a function of neurotoxic insult cannot be overemphasized given the specific advantages of the system as discussed. Limited neurotoxicologic or environmental "modifications" have been assessed using reaggregate suspensions. K^+- or veratridine-induced depolarization demonstrated [^{3}H]norepinephrine release (Majocha et al., 1981); thyroid hor-

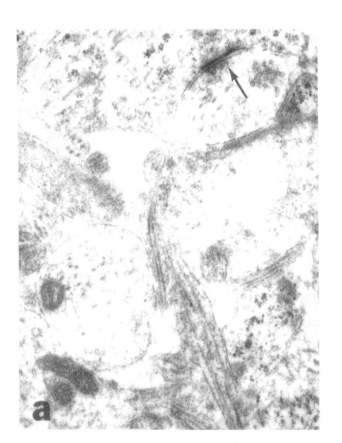 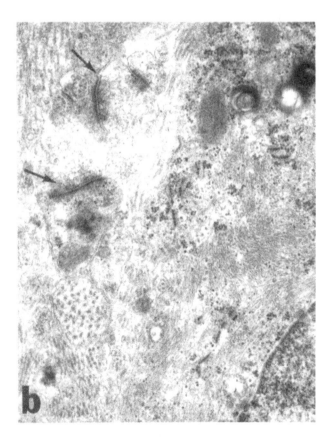

FIGURE 1 (a) Immature, rat cortical 5-day aggregate revealing watery pre- and postsynaptic complexes, longitudinal neurotubules in dendritic processes, small clusters of presynaptic vesicles, and wide interstitial space (×20,000). (b) Rat 28-day cortical aggregate revealing mature synaptic complex formation, numerous longitudinally disposed masses of neurotubules, and polysome-filled astrocyte cytoplasm containing masses of intermediate filaments (×20,000).

mone (T3) enhanced the activity of $Na^+ + K^+$-ATPase (Atterwil *et al.*, 1985) or choline acetyltransferase (Atterwil *et al.*, 1984); and kainic acid revealed a greater reduction in cholinergic and GABA-related enzyme activities in aggregates derived from cholinergic neuron-enriched telencephalon (Honneger and Richelson, 1977b), suggesting a relatively greater neurotoxicity on cholinergic and GABAergic cells in the more differentiated culture. Wehner and colleagues (1985) demonstrated a 95% reduction in acetylcholinesterase activity in the presence of organophosphorus compounds which rapidly returned to normal levels after 7 days, a recovery requiring active protein synthesis. Dimberg and colleagues (1992) examined morphological and biochemical parameters of differentiation in aggregating mouse brain suspensions after low dose (0.5 Gy) radiation.

VI. Bulk Separation of Neuron Perikarya or Glial Cells for Suspension Studies

Primary cultures of neural tissue begin with an enzymatic or mechanical dissociation step to allow the production of single cell preparations prior to culture. Beginning with the studies of Rose (1969) followed by the enzymatic or mechanical dissociation procedures introduced by Norton and Poduslo (1970) or Sellinger and colleagues (1971), methods for the separation of neuronal cell bodies or glial cells derived from cortex have proven satisfactory for application in cellular neurochemistry. It is not the purpose of this chapter to describe in detail the methodology underlying such bulk separation procedures [the reader is referred to methods as described in detail by Sellinger and Azcurra (1974)]. Although the advantages are evident, especially in allowing for subsequent primary culture, some limitations are inherent in the procedure. For instance, neuronal perikarya are obtained instead of intact nerve cells; varied perturbations in endogenous metabolic pools will occur; unknown yields referrable to the *in situ* composition will be obtained; and metabolic parameters may change as a function of the dissociation procedure whether enzymatic or mechanical. However, the availability of purified preparations of neurons and/or glial cells provides the neurotoxicologist with a valuable resource, especially for subsequent application to primary tissue culture.

Studies utilizing dissociated neuronal preparations have included centrifugal fractionation of isolated perikarya, metabolic, and enzyme investigations. Differential centrifugation of homogenates of neuronal perikarya has yielded nuclear, mitochondrial, microsomal, and soluble cell fractions (Johnson and Sellinger, 1971; Kohl and Sellinger, 1972), and neurochemical investi-

gations of the neuronal lysosome resulted from the development of such isolation procedures.

Specific organelle isolation from neuronal perikarya may prove of value in neurotoxicological studies. Bonnefoi (1992) used primary neural cell culture for the determination of both cytosolic and mitochondrial glutathione content following exposure to the monohalomethane, methyl iodide. The method is based on controlled digitonin disruption of the plasma membrane (due to the higher cholesterol content of the plasma membrane relative to that of the mitochondrial membrane) thereby leaving the mitochondrial inner membrane intact. Unfortunately, the cerebrocortical culture used was a combined neuronal–glial culture, thereby precluding definitive interpretations on cell specificity. However, an analogous use in neuronal suspension, e.g., cerebellar granule cell suspension, PC12 cells, may be more advantageous due to a higher degree of cell specificity, greater yield of cells in suspension, and use of neonate compared to fetal animals at 15–16 days gestation.

Johnson and Sellinger (1971, 1973) examined the effect of age on the protein synthetic activity of neurons and glial cells in suspension and on the nature of the protein synthesized by neurons. Using a protein-synthesizing system derived from cortical neurons, Kohl and Sellinger (1972) observed that [^{3}H]proline was incorporated into neuronal proteins more efficiently than either [^{3}H]phenylalanine or [^{3}H]leucine; all three amino acids in turn being more effectively incorporated in the 18-day-old compared to 8-day old cortical neurons. These results are contrary to those of Gilbert and Johnson (1972) and our own (see later) who found that neurons from newborn mice incorporated [^{14}C]arginine at a higher rate than for brain cells obtained from 15-day-old animals. Gordon and Balazs (1983) separated cell types from the developing rat cerebellum and compared the transport of glutamate into preparations enriched in Purkinje cells, granule neurons, or astrocytes. Cerebella from 8-day old rats were dissociated using mild trypsinization (Cohen *et al.*, 1978), and the separation of cell types was obtained by unit gravity sedimentation. These studies showed that the predominant site of [^{3}H]glutamate uptake in cerebellar *cultures* enriched in interneurons were the astrocytes and the rank order for V_{max} was astrocytes > Purkinje cells > replicating granule cells > granule cells with an astrocyte : granule cell ratio > 25 : 1. It is of interest to note that the rate of [^{3}H]glutamate transport [V_{max}] was similar in the astrocyte suspension and astrocyte culture but the apparent affinity was one order of magnitude lower in the cultured cells (Table 3), similar to the apparent difference for [^{3}H]GABA uptake in cultured compared to separated astrocytes (Cohen *et al.*, 1979; Gordon, 1982). These ob-

TABLE 3 Kinetic Parameters of L-[³H]Glutamate Uptake into Suspension of Astrocyte Perikarya or Astrocyte Cultures[a]

	Astrocytes	
	Suspension	Cultured
V_{max} (nmol/min per mg protein)	18 ± 2.8	20 ± 6.9
K_m (μM)	2.9 ± 0.1	27.5 ± 5

[a] Data rerived from Gordon and Balazs (1983). Values are mean ±SEM.

servations suggest that the monolayer culture with plastic substratum may alter cell membrane–cytoskeletal relationships influencing kinetics of the transport systems. In all instances, however, astrocytes in suspension or culture have an outstanding Na^+-dependent high capacity for [³H]glutamate uptake reflecting on the integrity of the isolated glial suspension preparation.

A more usual suspension of astrocytes may be derived from primary or clonal astrocyte culture. Select examples of such studies may be presented. Erecinska and Silver (1986) examined transmembrane potentials, ion gradients, and energy parameters in C_6 glial cells harvested for suspension experiments from culture and incubated under Dubnoff metabolic conditions for up to 1 hr. Numerous studies were performed, including measurement of ATP, ADP, creatine phosphate, cytochrome c reduction, oxygen uptake, intracellular water and ion content, and microelectrode measurements on randomly selected single cells adhering to the floor of plastic petri dishes. Such data gathered in suspension may be compared with similar data derived from astrocyte culture (Table 4). Kauppinen and colleagues (1988) removed cells from culture flasks following incu-

TABLE 4 Energy Parameters and Ion Gradients in Astrocyte Suspension or Culture

	Astrocytes	
	Suspension	Cultured
Respiration rate (μmol min⁻ g⁻¹)	1.3[a]	2.2[b]
ATP (mM)	5–6.5[a]	5[c]
[K^+] (mM)	150–160[a]	145[d,f,g]
Membrane potential (mV)	−90[a]	−60 to −90[d,e,f,g]

[a] Ericinska and Silver (1986).
[b] Hertz (1977).
[c] Passonneau et al., (1977).
[d] Kukes et al., (1976).
[e] Sugaya et al., (1979).
[f] Waltz et al., (1984).
[g] Kimmelberg et al., (1979).

bation in 0.2% EDTA in DBS for 30 min at 37°C with gentle scraping. Following this treatment, greater than 90% of the cells excluded trypan blue. Cells maintained at 4°C in a modified experimental medium retained viability for 6 hr. Apart from measurement of respiration, glycolysis, adenylates, and L-[³H]glutamate gradients, such astrocyte suspensions may be used to monitor plasma membrane potential using the potential-sensitive cationic cyanine dye, 3,3′-diethylthiadicarbocyanine iodide. Glucose deprivation depolarized the plasma membrane of astrocytes in suspension and collapsed the transmembrane K^+ and glutamate gradients.

The author has recently had the opportunity to study a human-derived astrocytoma cell line. The cells may be grown in large yield culture flasks and are easily dislodged into suspension by trypsinization and, following this procedure, astrocytes are rounded, exclude trypan blue, may be easily washed, and resuspended for numerous studies, including intracellular [Ca^{2+}], and membrane potential studies using specific potential-sensitive dyes. The procedure is valuable in comparing the suspension metabolic properties of human astrocyte populations obtained from differing brain regions in a variety of neuropathological states.

VII. Cerebellar Granule Cells in Suspension

This laboratory has a long-term interest in the use of cerebellar granule cells in suspension or culture for investigations of putative mechanisms underlying the neurotoxicity of methyl mercury (MeHg). Granule cells in suspension have allowed for correlations of cellular macromolecular synthesis and energy metabolism with interpretive observations previously gained in slice or cell-free systems. The *in vitro* suspension system provides a highly enriched neuronal cell type known to be targeted by MeHg *in vivo;* allowed for correlations of mercurial-induced metabolic perturbation with the onset and magnitude of cytotoxicity and more recently provided the starting cell suspension for cerebellar granule cell culture.

Bulk isolated neonatal cerebellar perikarya may be prepared using a mechanical (Sellinger and Azcurra, 1974; Sarafian and Verity 1986) or enzymatic (Messer, 1977; Sarafian and Verity, 1986) technique. The author has used both procedures but currently favors the enzymatic digestion procedure because of the greater bulk yield of cerebellar granule cells and the slightly extended neonatal window for isolation (4–10 days versus 4–8 days for mechanical disruption). A comparative study of mechanical versus enzymatic preparative techniques on [³H]phenylalanine incorporation into protein or [³H]uridine incorporation into RNA is pre-

TABLE 5 Comparison of Enzymatic or Mechanical Dissociation of Cerebellar Granule Cells on Protein or RNA Synthesis *In Vitro*[a]

	Suspension preparation	
	Mechanical	Enzymatic
[^{3}H]Phenylalanine incorp. cpm 10^{-4}/ min per mg protein	1.5 ± 0.3	4.0 ± 0.6
[^{3}H]uridine incorp. cpm 10^{-3}/min per mg protein	3.5 ± 0.4	2.0 ± 0.5

[a] Data represent mean ±SEM collected from Sarafian *et al.*, (1984) and Sarafian and Verity (1985, 1986).

sented (Table 5). The specific activity of [^{3}H]phenylalanine incorporation, normalized per milligram of protein, is slightly better in the enzymatic system whereas no significant difference was observed with [^{3}H]uridine incorporation into RNA. [^{3}H]Phenylalanine incorporation is linear up to 12 min in either mechanical or enzymatic systems whereas [^{3}H]uridine incorporation is sigmoidal with relatively low rates for the first 5 min, presumably reflecting the rate of uridine phosphorylation. Approximately 50% inhibition of [^{3}H]phenylalanine incorporation was observed at 12 μM MeHg whereas IC$_{50}$ values using the enzymatic digestion procedure were approximately 8 μM for both [^{3}H]phenylalanine and [^{3}H]uridine incorporation (Sarafian and Verity, 1985). Syversen (1977, 1982) used isolated cerebral and cerebellar neurons to study changes in macromolecular synthesis following MeHg injection in 25-day-old rats. These data suggest that for each cell type the sensitivity of RNA synthesis to MeHg was equal to or greater than that of protein synthesis, an observation confirmed in our bulk suspension cerebellar granule cell experiments and in subsequent studies of transcription inhibition (Sarafian and Verity, 1986). Protein syn-

thesis in synaptosome preparations (Verity *et al.*, 1977; Cheung and Verity, 1981) is strictly dependent upon internal/external ion gradients, and MeHg inhibition of *in vitro* translation may reflect a primary disturbance of such ion gradients. Isolated cerebellar granule cells in suspension provided an excellent intact cellular system to investigate this hypothesis. Determinations of intracellular [Na$^+$] and [K$^+$] were measured in granule cell suspensions as a function of [MeHg]. At 20 μM MeHg, a significant elevation in [Na$^+$] and reduction of [K$^+$] was observed accompanied by a major inhibition of protein synthesis. However, incubation in 200 μM ouabain revealed similar changes in Na$^+$/K$^+$ content without inhibition of protein synthesis, thereby dissociating the mercurial effect on ion flux from protein synthesis inhibition (Sarafian *et al.*, 1984). Similarly, the ability to determine intracellular adenine nucleotide pools allowed for the simultaneous assessment of ATP, ADP, cellular energy charge, and protein synthesis as a function of [MeHg] and suggested a close correlation between ATP decline and protein synthesis inhibition but prior to a major decrease in cellular energy charge. However, these experiments did invoke the possibility that minor changes in energy charge may have a major controlling influence on *in situ* translation reflecting modulation of synthetic rate (Hucul *et al.*, 1985; Mendelsohn *et al.*, 1977).

In a subsequent investigation, the granule cell suspension was used in correlate observations of mercurial-induced impaired cellular energetics, macromolecular synthesis, and the evolution of cytotoxicity assessed by trypan blue exclusion or LDH release (Sarafian *et al.*, 1989). Freshly isolated cells exclude trypan blue >93% and the evolution of cytotoxicity, measured by trypan blue exclusion, may be assessed over 3–4 hr. We have found a moderate degree of aggregation occurring over this time likely due to DNA which can

TABLE 6 Ca^{2+}-Ionophore A23187 Inhibits Methyl Mercury-Induced Lipoperoxidation in Cerebellar Granule Cell Suspensions

	GSH (μg/mg protein)	Lipoperoxidation (nmol malonaldehyde/mg protein)	Viability (%)
Control	3.29 ± 0.6 (5)	2.24 ± 0.3 (6)	9.8 ± 2 (5)
Methylmercury (10 μM)	1.85 ± 0.5 (5)*	3.38 ± 0.6 (6)*	48.4 ± 9 (5)*
A23187 (0.25 μM)	2.78 ± 0.7 (5)	1.58 ± 0.4 (4)*	83.0 ± 4 (4)*
Methyl mercury ± A23167	1.32 ± 0.7 (4)**	0.88 ± 0.3 (4)**	50.6 ± 7 (4)*

Note. Values are means ±SEM (no experiments) performed at 3 hr, 37°C in Krebs–Ringer medium, Ca^{2+}-free. GSH content at zero time was 4.78 ± 0.7 (5) μg/mg protein. Malonaldehyde production is expressed as an increase from zero time determination.

*P < 0.05 compared to control; **P < 0.01 (Student's *t* test)

TABLE 7 Comparison of Endogenous GSH and ATP in Granule Cell Suspension or Culture

	Suspension	Culture
GSH (nmol/mg)	5.3 ± 0.3^a	23.6 ± 5.6^a
	4.1 ± 0.2^a	18.8 ± 1.3^b
	7.0 ± 0.7^c	
ATP (nmol/mg)	35 ± 6^d	39.1 ± 9.3^e
		20.1 ± 3.9^f
		22.2 ± 3^g
		(2-hr incubation in CSS)

[a] Sarafian and Verity (1991).
[b] Verity and Sarafian (1991).
[c] Verity (1992).
[d] Sarafian *et al.*, (1989).
[e] Verity *et al.*, (1991).
[f] Sarafian and Verity (1990).
[g] Verity *et al.*, (1990).

Note. GSH content of C_6 cells: 40.8 ± 2.4 (Cho and Bannai, 1990).

be minimized by the addition of deoxyribonuclease maintaining the cells in suspension. The results suggest that mercurial cytotoxicity cannot be strictly related to equivalent changes in the inhibition of protein or RNA synthesis or ATP reduction in the short term and suggest that additional cytotoxic mechanisms must exist. In this respect, several reports had indicated that MeHg was capable of producing membrane lipoperoxidation, an event known to be associated with cytotoxicity. However, we were able to dissociate the onset of lipoperoxidation from the mercurial-induced cytotoxicity (Verity and Sarafian, 1991) and an example of such a study utilizing the granule cell suspension system is summarized in Table 6. This study, using low cytotoxic concentrations of MeHg ($10 \mu M$), revealed accelerated lipoperoxidation after a 3-hr incubation with a reduction in reduced glutathione, GSH. Of note, however, was the apparent inhibition of lipoperoxidation induced by the Ca^{2+} ionophore A23187, without significant cytoprotection. Although this experiment confirms the

ability of MeHg to induce a lipoperoxidative signal, inhibition of such a signal, e.g., by the ionophore A23187, fails to protect, thereby dissociating a strict causal relationship between mercurial-induced lipoperoxidation and cell injury. The mechanism of A23187 inhibition of lipoperoxidation is unknown.

What differences may exist in the endogenous levels of GSH or adenine nucleotides between cerebellar granule cells in suspension compared to culture? Data derived in this laboratory provide for comparison due to analogous analytical techniques. While the ATP content in freshly isolated suspension is similar to that in a 10- to 14-day culture (Table 7), the activity of GSH in suspension is considerably less than that in culture. These observations suggest that minimal loss of ATP occurs during suspension preparation which also precludes loss of GSH. How then to account for the high activity of GSH in culture? It is likely that the GSH content in culture reflects either a significant induction in neurons or the presence of a high GSH concentration in a contaminant cell population, e.g., glia. Evidence to support this latter suggestion has been obtained (Weir *et al.*, 1990) and the GSH content of astrocytes in culture is two-fold that of granule cells (Cho and Bannai, 1990).

VIII. Morphologic Characteristics of Isolated Cerebellar Granule Cells

As indicated previously, mechanical dissociation usually provided less yield with more variable cell viability (30–80% trypan blue exclusion). Enzymatic dissociation provided higher yield with less membrane debris and free nuclei. Only rare Purkinje cells were seen, the remaining cells by light and electron microscopy demonstrating a uniformly small size, prominent nucleus, and a thin cytoplasmic envelope surrounding the nucleus (Fig. 2). Excellent integrity was obtained as evaluated by trypan blue exclusion; exogenously

FIGURE 2 Transmission electron micrograph of enzymatically dissociated 7-day-old cerebellar granule neurons following glutaraldehyde fixation suspension. (a) Control neurons reveal moderately electron-dense cytoplasmic matrix, plasmalemmal pseudopodia, and homogeneously dispersed nuclear euchromatin ($\times 8000$). (b) Cerebellar neurons in suspension following a 2-hr incubation at 37°C with $10 \mu M$ MeHg. Note. Cellular swelling, vesiculation of the endoplasmic reticulum with polysome dissociation, microfocal ruptures of plasmalemma, absence of filopodia, and early disruption of nuclear chromatin ($\times 8000$). (c) Detail of control granule cell cytoplasm containing numerous elongate, nonswollen mitochondria and evenly dispersed polysomal ribonucleic acid. The nuclear membrane is intact and marginated by regular euchromatin ($\times 28,000$). (d) Granule cell cytoplasm following $10 \mu M$ MeHg for 2 hr at 37°C. Prominent mitochondrial enlargement with focal swelling of cristae and numerous intramitochondrial-dense bodies are evident. Endoplasmic reticulum vesiculation especially close to the plasmalemmal membrane is seen. Nuclear chromatin is now clumped at the nuclear membrane suggesting early oligonucleosome formation ($\times 28,000$).

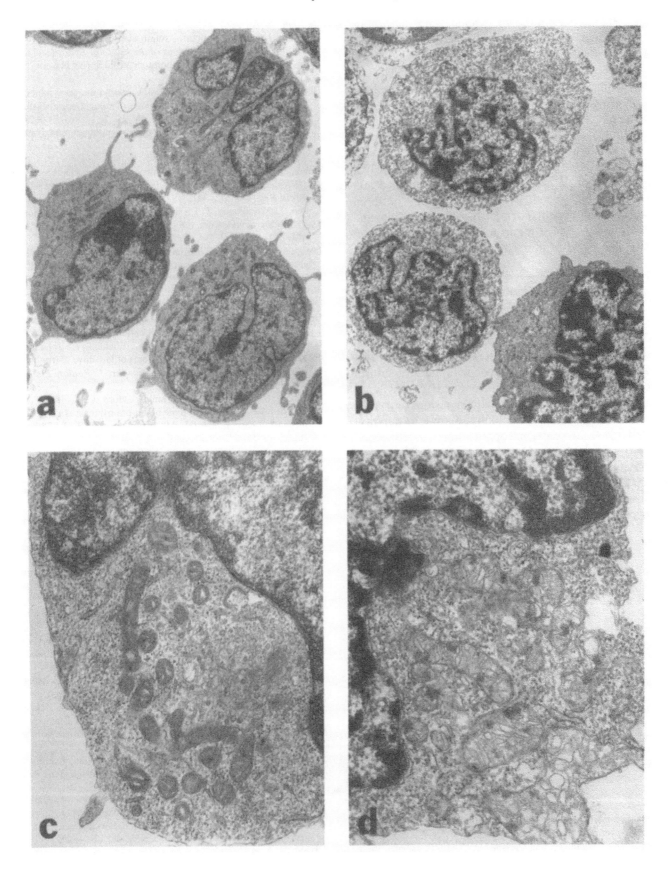

added ribonuclease or desoxyribonuclease did not affect the rate of protein synthesis while brief homogenization eliminated greater than 95% of protein synthesis measured by [³H]phenylalanine incorporation. The cell suspension provides excellent material for morphological neurotoxicology. Light microscopy may reveal plasmalemmal vesiculation and blebbing while ultrastructural examination is easily performed following dropwise addition of aliquots of cerebellar suspension into cold buffered 3% glutaraldehyde. After 1 hr at 4°C, the suspension is centrifuged at 4500 rpm for 5 min and the pellet is fixed overnight in freshly buffered glutaraldehyde, postfixed in 1% osmium tetroxide, and dehydrated in graded methanol prior to embedding in resin. The effect of MeHg is documented in Fig. 2.

IX. Conclusion

With the advent of characterized primary and clonal cultures, the use of dissociated neural cell systems has fallen into disfavor. Although it is unlikely that a resurgence of this technology will occur in the near future, the ease of preparation of some defined systems, e.g., cerebral and cerebellar neuronal suspensions or trypsin dislocated astrocytes from culture, allows for rapidity, relative ease, low cost, and application to biochemical, biophysical, and morphological methods. The use of such suspensions for short-term mechanistic studies is self-evident but the method also provides for the assessment of neurotoxicity as a primary screening procedure.

Acknowledgment

The author gratefully acknowledges the administrative and secretarial assistance of William Schram in preparation of this manuscript.

References

Appel, K., and Barefoot, D. S. (1989). Neurotransmitter release from bradykinin-stimulated PC-12 cells: Stimulation of cytosolic Ca²⁺ and neurotransmitter release. *Biochem. J.* **263,** 11–18.

Atterwil, C. K. (1987). Brain reaggregate cultures in neurotoxicological investigations (1987).

Atterwil, C. K., Atkinson, D. J., Bermudez, I., and Balazs, R. (1985). Effect of thyroid hormone and serum on the development of Na⁺, K⁺-ATPase and associated ion-fluxes in cultures from rat brain. *Neuroscience* **14,** 361–373.

Betz, A. L., and Goldstein, G. W. (1978). Polarity of the blood-brain barrier: Neutral amino acid transport into isolated brain capillaries. *Science* **202,** 225–226.

Betz, A. L., and Goldstein, G. W. (1980). Transport of hexoses, potassium and neutral amino acids into capillaries isolated from bovine retina. *Exp. Res.* **30,** 593–605.

Brendel, K., Megan, E., and Carlson, E. C. (1974). Isolated brain microvessels: a purified, metabolically active preparation from bovine cerebral cortex. *Science* **185,** 953–955.

Cancilla, P. A., and DeBault, L. E. (1980). Demonstration of an A and L system for neutral amino acid transport in a cerebral endothelial cell line. In *Biology of the Vascular Endothelial Cell,* p. 30, Cold Spring Harbor Laboratory, Cold Spring Harbor, New York.

Cheung, M., and Verity, M. A. (1981). Methyl mercury inhibition of synaptosome protein synthesis. Role of mitochondrial dysfunction. *Environ. Res.* **24,** 286–298.

Cho, Y., and Bannai, S., (1990). Uptake of glutamate and cysteine in C-6 glioma cells and in cultured astrocytes. *J. Neurochem.* **55,** 2091–2097.

Cohen, J., Balazs, R., Hajos, F., Currie, D. N., and Dutton, G. R. (1978). Separation of cell types from the developing cerebellum. *Brain Res.* **148,** 313–331.

Cohen, J., Woodhams, P. L., and Balazs, R. (1979). Preparation of viable astrocytes from the developing cerebellum. *Brain Res.* **161,** 503–514.

DeBault, L. E., (1982). Isolation and characterization of the cells of the cerebral microvessels. *Adv. Cell Neurobiol.* **3,** 339–371.

DeBault, L. E., and Cancilla, P. A. (1980). γ-Glutamyl transpeptidase in isolated brain endothelial cells: induction by glial cells *in vitro. Science* **207,** 653–655.

DeLong, G. R. (1970). Histogenesis of fetal mouse isocortex and hippocampus in reaggregating cell cultures. *Dev. Biol.* **22,** 536.

Dimberg, Y., Totmar, O., Aspberg, A., Ebendal, T., Johansson, K.-J., and Walinder, G. (1992). Effects of low-dose X-irradiation on mouse-brain aggregation cultures. *Int. J. Radiat. Biol.* **61,** 355–363.

Eisenberg, H. M., Suddith, R. L., and Crawford, J. S. (1980). Transport of Na+ and K+ across the blood-brain barrier. In *The Cerebral Microvasculature* (H. M. Eisenberg and R. L. Suddith, Eds.), pp. 57–67, Academic Press, New York.

Erecinska, M., and Silver, I. A. (1986). The role of glial cells in regulation of neurotransmitter amino acids in the external environment. 1. Transmembrane electrical and ion gradients and energy parameters in cultured glial-derived cell lines. *Brain Res.* **369,** 193–202.

Gilbert, B. E., and Johnson, T. C. (1972). Protein turnover during maturation of mouse brain tissue. *J. Cell Biol.* **53,** 143–147.

Goldberg, A. M. (1980). Mechanisms of neurotoxicity as studies in tissue culture systems. *Toxicology* **17,** 201–208.

Goldstein, G. W. (1979). Relation of K⁺ transport to oxidative metabolism in isolated brain capillaries. *J. Physiol. (Lond.)* **286,** 185–195.

Goldstein, G. W. (1992). Blood-brain barrier as a target for Pb toxicity. *Neurotoxicology* **13,** 484.

Goldstein, G. W., Csejtey, J., and Diamond, I. (1977). Carrier-mediated glucose transport in capillaries isolated from rat brain. *J. Neurochem.* **28,** 725–728.

Goldstein, G. W., Wolinsky, J. S., and Csejtes, J. (1977). Isolated brain capillaries: A model for the study of Pb encephalopathy. *Anal. Neurol.* **1,** 235–239.

Goldstein, G. W., Wolinsky, J. S., Csejtes, J., and Diamond, I. (1975). Isolation of metabolically active capillaries from rat brain. *J. Neurochem.* **25,** 715–717.

Gordon, R. D., (1982). Certain biochemical properties of cerebellar cells related to aminoacid-mediated neurotransmission. Ph.D. Thesis, London University.

Gordon, R. D., and Balazs, R. (1983). Characterization of separated cell types from developing rat cerebellum: Transport of glutamate and aspartate by preparations enriched in Purkinje cells, granule neurons and astrocytes. *J. Neurochem.* **40,** 1090–1099.

Greene, L. A., and Tischler, A. S. (1982). PC-12 pheochromocytoma cultures in neurobiological research. *Adv. Cell Neurobiol.* **3,** 373–414.

Hardebo, J. E., and Owman, C. (1980). Characterization of the *in vitro* uptake of monoamines into brain microvessels. *Acta Physiol. Scand.* **108**, 223–229.

Hertz, L. (1977). Energy metabolism of glial cells. In *Dynamic properties of glial cells* (E. Schoffeniels, G. Franck, L. Hertz, and D. B. Tower, Eds.), pp. 121–132, Pergamon Press, Oxford.

Honegger, P., and Richelson, E. (1977a). Biochemical differentiation of aggregating cell cultures of different fetal rat brain regions. *Brain Res.* **133**, 329.

Honegger, P., and Richelson, E. (1977b). Kainic acid alters neurochemical development in fetal rat brain aggregating cell cultures. *Brain Res.* **138**, 580.

Hucul, J. A., Henshaw, E. C., and Young, D. A. (1985). Nucleoside diphosphate regulation of overall rates of protein biosynthesis acting at the level of initiation. *J. Biol. Chem.* **260**, 15585–15591.

Johnson, D. E., and Sellinger, O. Z. (1971). Protein synthesis in neurons and glial cells of the developing rat brain: an *in vivo* study. *J. Neurochem.* **18**, 1445–1460.

Johnson, D. E., and Sellinger, O. Z. (1973). Synthesis of soluble neuronal proteins *in vivo*, age-dependent differences in the incorporation of leucine and phenylalanine. *Brain Res.* **54**, 129–142.

Joo, F. (1979). The role of adenosine triphosphatase in the maintenance of molecular organization of the basal lamina in the brain capillaries. *Front. Matrix Biol.* **7**, 166–182.

Kauppinen, R. A., Enkvist, K., Holopainen, I., and Akerman, K. E. O. (1988). Glucose deprivation depolarizes plasma membrane of cultured astrocytes and collapses transmembrane K+ and glutamate gradients. *Neuroscience* **26**, 283–289.

Kimelberg, H. K., Bowman, C., Biddlecome, S., and Bourke, R. S. (1979). Cation transport and membrane potential properties of primary astroglial cultures from neonatal rat brain. *Brain Res.* **177**, 533–550.

Kohl, H. H., and Sellinger, O. Z. (1972). Protein synthesis in neuronal perikarya isolated from cerebral cortex of the immature rat. *J. Neurochem.* **19**, 699–711.

Kolber, A. R., Bagnell, C. R., Krigman, M. R., Hayward, J., and Morell, P. (1979). Transport of sugars into microvessels isolated from rat brain: a model for the blood-brain barrier. *J. Neurochem.* **33**, 419–432.

Kukes, G., Elul, R., and DeVellis, J. (1976). The ionic basis of the membrane potential in a rat glial cell line. *Brain Res.* **104**, 71–92.

Levitt, P., Moore, R. Y., and Garber, B. B. (1976). Selective cell association of catecholamine neurons in brain aggregates *in vitro*. *Brain Res.* **111**, 311–320.

Majocha, R. E., Pearse, R. N., Baldessariani, R. J., DeLong, G. R., and Walton, K. G. (1981). The noradrenergic system in cultured aggregates of fetal rat brain cells: morphology of the aggregates and pharmacological indices of noradrenergic neurons. *Brain Res.* **230**, 235–252.

Matthieu, J. M., Honegger, P., Favrod, P., Gaultier, E., and Dolivo, M. (1979). Biochemical characterization of a myelin fraction isolated from rat brain aggregating cell cultures. *J. Neurochem.* **32**, 869.

Mendelsohn, S. L., Nordeen, S. K., and Young, D. A. (1977). Rapid changes in initiation-limited rates of protein synthesis in rat thymic lymphocytes correlate with energy change. *Biochem. Biophys. Res. Commun.* **79**, 53–60.

Messer, A., (1977). The maintenance and identification of mouse cerebellar granule cells in monolayer culture. *Brain Res.* **130**, 1–12.

Nelson, P. G., (1978). Neuronal cell cultures as toxicological test systems. *Environ. Health Pespect.* **26**, 125–133.

Norton, W. T., and Poduslo, S. E., (1970). Neuronal soma and whole neuroglia of rat brain: a new isolation technique. *Science* **167**, 1144–1145.

Passonneau, J. V., Schwartz, J. P., and Lust, W. D. (1977). Some aspects of intermediary metabolism in glioma cells in culture. In *Dynamic properties of glial cells.* (E. Schoffeniels, G. Franck, L. Hertz, and D. B. Tower, Eds.), pp. 121–132, Pergammon Press, Oxford.

Ransom, J. T., Cherwinski, H. M., Dunne, J. F., and Sharif, N. A. (1991). Flow cytometric analysis of internal Ca^{2+} mobilization via a B_2-bradykinin receptor on a subclone of PC-12 cells. *J. Neurochem.* **56**, 983–989.

Rose, S. P. R. (1969). Neurons and glia: Separation techniques and biochemical interrelationships. In *Handbook of Neurochemistry* (A. Lajtha, Ed.), Vol. 2, pp. 183–193, Plenum Press, New York.

Sarafian, T., Cheung, M. K., and Verity, M. A. (1984). *In vitro* methyl mercury inhibition of protein synthesis in neonatal cerebellar perikarya. *Neuropathol. Appl. Neurobiol.* **10**, 85–100.

Sarafian, T., Hagler, J., Vartavarian, L., and Verity, M. A. (1989). Rapid cell death induced by methyl mercury in suspension of cerebellar granule neurons. *J. Neuropathol. Exp. Neurol.* **48**, 1–10.

Sarafian, T., and Verity, M. A. (1985). Inhibition of RNA and protein synthesis in isolated cerebellar cells by *in vitro* and *in vivo* methyl mercury. *Neurochem. Pathol.* **3**, 27–39.

Sarafian, T., and Verity, M. A. (1986). Mechanism of apparent transcription inhibition by methyl mercury in cerebellar neurons. *J. Neurochem.* **47**, 625–631.

Schousboe, A., Booher, J., and Hertz, L. (1970). Content of ATP in cultivated neurons and astrocytes exposed to balanced and potassium-rich media. *J. Neurochem.* **17**, 1501–1504.

Sellinger, O. Z., and Azcurra, J. M. (1974). Bulk separation of neuronal cell bodies and glial cells in the absence of added digestive enzymes. In *Research Methods in Neurochemistry* (N. Marks and R. Rodnight, Eds.), Vol. 1, pp. 3–38, Plenum Press, New York.

Sellinger, O. Z., Ohlsson, W. G., Frankel, A. J., Azcurra, J. M., and Petiet, P. D. (1971). A study of the nascent polypeptides synthesized on the free polyribosomes of rat brain *in vivo*. *J. Neurochem.* **18**, 1243–1260.

Sheppard, J. R., Brus, D., and Wehner, J. M. (1978). Brain reaggregate cultures: biochemical evidence for myelin membrane synthesis. *J. Neurobiol.* **9**, 309.

Spatz, M., Bembry, J., Dodson, R. F., Hervonen, H., and Murray, M. R. (1980). Endothelial cells cultures derived from isolated cerebral microvessels. *Brain Res.* **191**, 577–582.

Stefanelli, A., Cataldi, E., and Ieradi, L. A., (1977). Specific synaptic systems in reaggregated spherules from dissociated chick cerebellum cultivated *in vitro*. *Cell Tissue Res.* **182**, 311.

Sugaya, E., Sekiya, Y., Kobori, T., and Noda, Y. (1979). Glial membrane potential and extracellular potassium concentration in cultured glial cells. *Exp. Neurol.* **66**, 403–408.

Syversen, T. L. M. (1977). Effects of methyl mercury on *in vivo* protein synthesis in isolated cerebral and cerebellar neurons. *Neuropathol. Appl. Neurobiol.* **3**, 225–236.

Syverson, T. L. M. (1982). Changes in protein and RNA synthesis in rat brain neurons after a single dose of methyl mercury. *Toxicol. Lett.* **10**, 31–34.

Trapp, B. D., Honegger, P., Richelson, E., and Webster, H. D. E. F. (1979). Morphological differentiation of mechanically dissociated fetal rat brain in aggregating cell cultures. *Brain Res.* **160**, 117.

Trapp, B. D., and Richelson, E. (1980). Usefulness for rotation-mediated aggregating cell cultures. In *Experimental and Clinical*

Neurotoxicology (P. S. Spencer and H. H. Schaumburg, Eds.), pp. 803–820, Williams and Wilkins Press, Baltimore.

Verity, M. A., Brown, W. J., Cheung, M., and Czer, G. (1977). Methyl mercury inhibition of synaptosome and brain slice protein synthesis: *In vivo* and *in vitro* studies. *J. Neurochem.* **29,** 673–679.

Verity, M. A., and Sarafian, T. (1991). Role of oxidative injury in the pathogenesis of methyl mercury neurotoxicity. In *Advances in mercury toxicology* (T. Suzuki, N. Imura, and T. W. Clarkson, Eds.), pp. 209–222, Plenum Press, New York.

Verity, M. A., and Verity, A. N. (1991). Methyl mercury modulation of cerebellar granule cell survival and neuritogenesis *in vitro* is Ca^{2+}-dependent. *Neurotoxicology* **12,** 799–800.

Veronesi, B., (1992a). The use of cell culture for evaluating neurotoxicity. In *Neurotoxicology* (H. Tilson, and C. Mitchell, Eds.). pp. 21–50, Raven Press, New York.

Veronesi, B. (1992b). In vitro screening batteries for neurotoxicants. In *Current Issues in Neurotoxicology* (A. Mutti, L. Manzo, L. G. Costa, and J. M. Cranmer, Eds.), pp. 185–195, Intox Press, Little Rock, Arkansas.

Walz, W., Wuttke, W., and Hertz, L. (1984). Astrocytes in primary cultures: Membrane potential characteristics reveal exclusive potassium conductance and potassium accumulator properties. *Brain Res.* **292,** 367–374.

Wehner, J. M., Smolen, A., Ness-Smolen, T., and Murphy, C. (1985). Recovery of acetylcholinesterase activity after acute organophosphate treatment of CNS reaggregate cultures. *Fundam. Appl. Toxicol.* **5,** 1104–1109.

Weir, K., Sarafian, T., and Verity, M. A. (1990). Methyl mercury induces paradoxical increases in reduced glutathione in cerebellar granule cell culture. *Toxicology* **10,** 137.

Williams, S. K., Gillis, J. F., Matthews, M. A., Wagoner, R. C., and Bitensky, M. W. (1980). Isolation and characterization of brain endothelial cells: morphology and enzyme activity. *J. Neurochem.* **35,** 374–381.

CHAPTER

33

Central Nervous System Glial Cell Cultures for Neurotoxicological Investigations

MICHAEL ASCHNER
Department of Physiology and Pharmacology
Bowman Gray School of Medicine
Winston-Salem, North Carolina 27157

DOMENICO VITARELLA
Department of Pharmacology and Toxicology
Albany Medical College
Albany, New York 12208

I. Introduction

Historically, it is customary to credit Rudolf Virchow (1846) with the discovery of neuroglia (for review see Somjen, 1988). As a practicing pathologist he was familiar with inflammatory processes that can afflict the ventricular cavities of the brain, compelling him to oppose the opinions of his contemporaries that the brain lacks connective tissue. Accordingly, Virchow hypothesized that underneath the single cell layer of the ependyma, the ventricles must be lined by connective tissue cells, and coined the name "Nervenkitt" (putty). The argument that the brain contains nonneuronal elements was later also advanced by Camillio Golgi (1885–1886). Utilizing his novel metallic impregnation technique, Golgi demonstrated that the brain contains regions of distinct nonneuronal tissue. Ramón y Cajal (1913) carried Golgi's initial findings further and developed a system for classifying glial cells. In this system, the overwhelming majority of the cells in the neuroglia are classified as macroglia, and a small minority,

thought to be derived from nonnervous tissue, are called microglia. The macroglia, in turn, are subdivided into astrocytes and oligodendrocytes. Astrocytes come in two distinct forms: protoplasmic astrocytes are found in the gray matter and fibrous astrocytes in the white matter. Functional differences between these two types are not well established, although they are known to differ anatomically.

Astrocytes appear to function significantly in the maintenance of normal brain physiology, during both development and adulthood. For example, it is postulated that cell–cell contact between astrocytes and neurons in the prenatal brain controls the number of different cell types and determines their correct proportions (reviewed by Rakic, 1991). Such contact also appears to regulate appropriate regional deployment of astrocytes, which presumably demarcates the boundaries for cellular compartmentalization and plays an important role in neuronal migration (Schmechel and Rakic, 1979; Levitt and Rakic, 1980). Astroglial–neuronal contacts have been implicated in the induction of nerve cell phenotypic expression (Patterson and Chun, 1974; Lauder and McCarthy, 1986), ensheathment of neuronal dendritic spines, and growth cone guidance (reviewed by Vernadakis, 1988). Moreover, the recently recognized phagocytic capacity of astrocytic processes might be involved in the rearrangement and elimination of synaptic junctions. This process is an important step in competitive synaptic elimination and neuronal plasticity (reviewed by Rakic, 1991).

Astrocytes ensheath blood–brain barrier capillaries, influencing endothelial cell differentiation and modulating endothelial transport processes (e.g., essential amino acid transport, glucose, electrolytes). Accordingly, they are viewed as metabolic intermediaries between endothelial cells and neurons (reviewed by Kimelberg and Norenberg, 1989). Astrocytes maintain a high concentration of intracellular K^+ with negative membrane potential close or equivalent to the K^+ equilibrium potential, indicating high K^+ conductance. Accordingly, they are presumed to function in K^+ spatial buffering (Orkand *et al.*, 1966). This is achieved by removal of K^+ from the extracellular fluid in regions of neuronal activity by its diffusion and transport across the astrocytic syncitium. In addition, astrocytes function prominently in the exchange of nutrients, second messengers, and growth factors modifying the neuronal–glial microenvironment, thereby playing a critical role in maintaining brain homeostasis (reviewed by Vernadakis, 1988). Finally, recent studies have implicated astrocytes as the antigen-presenting cells of the central nervous system (CNS), supporting intracerebral T cell activation processes and CNS-mediated immune responses (Fierz *et al.*, 1985; Fontana and Fierz 1985). New evidence favoring astrocytic participation in various CNS disorders such as epilepsy, parkinsonism, Huntington's disease, immune disorders, and heavy metal neurotoxicity is rapidly accumulating. [For the role of astrocytes in CNS disorders please refer to Aschner *et al.*, (1994).]

Oligodendrocytes synthesize the CNS myelin which envelopes axons, giving rise to the white matter (reviewed by Lopez-Cardozo *et al.*, 1989). Like its peripheral counterpart, the Schwann cell, oligodendrocytes are responsible for both the production and the maintenance of myelin. However, whereas individual Schwann cells form only one sheath of myelin around peripheral axons, a single oligodendrocyte can spiral around a number of central axons, forming up to 40 compact, multilamellar myelin sheaths around multiple axonal segments. Accordingly, oligodendrocytes can be envisioned as cytoskeletal support elements of the CNS, tying groups of axons together. It has been suggested that like its astrocytic counterpart, oligodendrocytes may also function in the control of K^+ concentra-

tions in the extracellular fluid of the white matter (Barres, 1988).

It is now clear that both astrocytes and oligodendrocytes represent more than mere passive cytoskeletal support elements of the CNS. However, for several decades, the prominent cellular heterogeneity within various regions of the brain severely impeded our understanding of the roles of neuroglia in both normal and pathological brain states. Today, much of our increased understanding can be attributed to the many neurobiological tools that our predecessors lacked. One of these tools is the availability of tissue culture techniques which offer major advantages in probing cell function and homeostasis.

A wide range of systems such as brain slices, isolated cells, perfused brains, and single cells are currently available for neurobiological studies. However, these systems allow only short-lived studies because of their limited life span. This poses a significant obstacle for neurotoxicological studies, where long-term studies at low concentrations of the toxic agent are physiologically most relevant. Cell culture systems provide the possibility of performing studies over prolonged periods of time.

This chapter is not meant to cover all research approaches and methodologies related to neuroglia. Rather, it is intended to highlight significant and recent concepts concerning the identity and nature of the CNS glia, which are relevant to current issues in neurobiology and neurotoxicology. This chapter focuses on methods for astroglial and oligodentrytic culturing, and their properties in culture, emphasizing both the advantages and disadvantages afforded by these cultures.

II. Cell Cultures: Definitions

Cells harvested directly from the organism and maintained *in vitro* for periods exceeding 24 hr are defined as primary cell cultures (Fedoroff, 1977). These cultures may be derived from tissue and dissociated into single cells before seeding into the culture vessel and hence are certified as cell cultures because the tissue organization has been disrupted. If tissue fragments are maintained *in vitro* they are referred to as explant cultures. Cell cultures can be monotypic (one cell type), they can consist of a number of cell types (mixed cultures), or they can contain reaggregates of several cell types. For an excellent review on glial cells and tissue culture definitions the reader is referred to Hertz and colleagues (1985).

Cell lines, by definition, are cultures that have been serially transplanted or subcultured from one culture vessel to another for a number of generations. These can be propagated indefinitely and therefore cost less to maintain than primary cultures (Hertz *et al.*, 1985). The lingering question, however, about the phenotypic identity and differentiation state of these cells after a number of serial propagation still remains. Consideration of this important issue, as well as a description of the various cell lines available, has been eloquently reviewed by Hertz and colleagues (1985) and will not be considered here. This chapter focuses on the preparation of primary astrocyte cultures, oligodendrocyte cultures, and aggregate cultures which, in addition to neurons, contain both astrocytes and oligodendrocytes.

III. Methods for Studying and Identifying Neuroglial Cells in Culture

Glial aggregates and individual neurons, even from small and well-defined brain regions, can be obtained by microdissection (Lowry, 1953; Hydén, 1959). Although the harvested cells are damaged by dissection, they appear to be metabolically active for several hours (Hydén and Rönnback, 1975). Studies of different glial cell subtypes without physical separation of the cells can be achieved by autoradiographic and histological methods (Altman, 1963; Droz and Leblond, 1963; Diamonds *et al.*, 1966).

Greater cellular yields of glia for immediate experimentation may be obtained by other bulk methods (Rose, 1965; Blomstrand and Hamberger, 1969; Hamberger and Sellström, 1975; Farooq and Norton, 1978). These methods allow for isolation of glial cells from specifically defined brain areas (Palmer, 1973; Henn *et al.*, 1977). The functional properties of these cellular isolates and the degree of cellular differentiation are ensured by previous interactions with other CNS cells. Albeit useful, the preceding methods are plagued by several limitations: low yield, contamination with other cell types, and cell damage during isolation (Hamberger and Sellström, 1975). Additional methods for bulk isolation of specific cell types may be achieved by affinity chromatography, automated cytofluorography in cell sorting, and magnetic microsphere cell separation. All employ the use of specific antibodies to identify specific markers on the surface of the cell of interest.

Alternative sources for large numbers of glial cells are cultured cell lines (Benda *et al.*, 1968; Silberstein *et al.*, 1972; Westermark *et al.*, 1973; Edström *et al.*, 1974; Murphy *et al.*, 1976; Pfeiffer *et al.*, 1978). Primary cultures may also be passaged. These primary cultures may be region specific and offer the possibility of

choosing cell type, tissue region, and desired species, as well as developmental age.

A problem central to all culture systems is cellular impurities and culture heterogeneity. It is therefore essential that both the characteristic morphology and antigenic phenotype of the cultured cells be characterized to permit the unique identification of these cell types. These may include either immunocytochemical or enzymatic markers which are associated with the plethora of CNS cells. Table 1 provides a list of various markers that can be used for distinguishing and identifying culture populations and for asessing their homogeneity.

A. Astrocytes

The functions of astrocytes have begun to be delineated only within the last 2 decades (for reviews see Fedoroff and Vernadakis, 1986a,b,c; Kimelberg and Norenberg, 1989; Abbott, 1991). The prevailing view that they exclusively function as passive physical support for neurons has rapidly faded. It is now clear that in order to understand the physiology and pathology of both the developing and the mature brain, the roles assumed by astrocytes must also be understood for they function prominently not only in normal brain physiology and development, but also in the pathology of the nervous system. Indeed, much of what is known about astrocyte function and the neuronal–astrocytic interrelationship comes from work with cell cultures. The following discussion focuses on the methodology for obtaining primary astroglial-enriched cultures from neonatal rat cortex and elaborates on their functional specialization.

1. Preparation of Primary Astrocyte Cultures

Several methods now exist for preparing primary astrocyte cultures, and these can be easily applied to different brain regions. There is some variability in culture composition reported from different laboratories and minor variations in procedure, including use of mouse vs rat. The consequences of these differences are not always clear. Another issue central to astrocytic culturing is the enzymatic vs mechanical cell dissociation. The neutral bacterial protease Dispase method (Frangakis and Kimelberg, 1984) is commonly used for the initial dissociation of neonatal 1-day-old Sprague–Dawley rat pup brains, which produces very uniform cultures with good yield and high viability (95%). This method uses successive removal of the sedimenting undissociated cells and therefore avoids filtration through nylon meshes. The growth of the cells is rapid after the initial plating and the cultures can be started from plating densities as low

TABLE 1 Proteins and Lipids Characteristic of CNS Cells

Cells	Markers	References
Astrocytes	Glial fibrillary acidic protein	Bignami and Dahl (1974a,b); Bock *et al.* (1977); Stein *et al.* (1980); Hansson *et al.* (1982); Hallermayer *et al.* (1981); Frangakis and Kimelberg (1984)
	S-100	Hydén and McEwen (1966); Benda *et al.* (1968); Haglid *et al.* (1976)
	α,α-Enolase	Langley and Ghandour (1981)
	Glutamine synthetase	Hallermayer *et al.* (1981); Hertz *et al.* (1978); Martinez-Hernandez *et al.* (1977); Warringa *et al.* (1988)
	A2B5	Bartlett *et al.* (1981)
	G_{D3}	Goldman *et al.* (1986)
	G_{Qib}	Freedman *et al.* (1984); Raff *et al.* (1983)
	Glutathione-*S*-transferase	Abramowitz *et al.* (1988)
	Pyruvate carboxylase	Shank *et al.* (1985)
	Vimentin	Fedoroff *et al.* (1983)
	Laminin	Liesi *et al.* (1983)
Oligodendrocytes	Myelin basic protein	Hansson *et al.* (1982); Trapp *et al.* (1982)
	Myelin-associated glycoprotein	Trapp *et al.* (1982)
	Glycerol-3-phosphate dehydrogenase	Leveille *et al.* (1980)
	GM_4 (sialogalactosylceramide)	Cochran *et al.* (1982)
	Galactosylglycerides	Schachner (1982); Ranscht *et al.* (1982)
	Sulfogalactosylceramides	Norton (1983, 1984)
	Sulfogalactosylcerides	Norton (1983, 1984)
	2',3'-Cyclic-nucleotide 3'-phosphohydrolase	Bansal and Pfeiffer (1985); Kurihara and Tsukada (1967, 1968); McCarthy and de Vellis (1980); Sprinkle *et al.* (1987)
	Neurite growth inhibitors 35 and 250	Schwab (1984)
	Wolfgram protein	Roussel *et al.* (1981); Roussel and Nussbaum (1981)
	Proteolipid proteins	Laursen *et al.* (1984)
	Transferrin	Connor and Fine (1986)
	Biotin	LeVine and Macklin (1988)
	Cholesterolester hydrolase	Bhat and Pfieffer (1985)
	Carbonic anhydrase II	Ghandour *et al.* (1980)
Endothelial cells	Factor VIII	Hoyer *et al.* (1973)
	Alkaline phosphatase	Bannister and Romanul (1963); Kanje *et al.* (1978); Hansson *et al.* (1980)
Fibroblasts	Fibronectin	Raff *et al.* (1979); Stieg *et al.*, (1980); Hallermayer and Hamprecht (1984)
Ependymal cells	Beating cilia	Hansson *et al.* (1980)
	A2B5	Bartlett *et al.*, (1981)
Neurons	Neuron-specific enolase	Bock and Dissing (1975); Grasso *et al.* (1977); Persson *et al.* (1978)
	Tetanus toxin-binding gangliosides	Raff *et al.* (1979)

as 1×10^4 cells/cm². The cells are grown in a 5% CO_2/95% air-humidified atmosphere at 37°C in Eagle's minimum essential medium (MEM) with 10% horse serum (GIBCO), plus penicillin (10,000 units/ml), streptomycin (100 μg/ml), and a BME vitamin solution. About 2–3% of the seeded cells attach to the culture dishes and attain a saturation density of 2–3 × 10^4 viable cells/cm² by 3–4 weeks. Cells are routinely stained for glial fibrillary acidic protein (GFAP). Phase microscopy should confirm that 90–100% of the cells stain positively for this protein; one component of the intermediate glial filaments is found only in the cytoplasm of astrocytes. Pure cultures of astrocytes are hence verified as such by the presence of this unique marker. Confluency is routinely achieved within 3–4 weeks of culturing.

The search for a physical and biochemical environment that optimizes the physiological functioning of the cells is a problem crucial to all cell cultures. It is crucial that the cell type in question predominates the culture and grows rapidly, albeit, with restraint. In addition, the degree of differentiation of the cells should coincide with *in vivo* conditions.

The growth of astrocytes with increasing time in culture can be determined by washing the culture dishes in serum-free media, trypsinizing the monolayer with 0.25% trypsin solution for 10 min at 37°C, and counting the total cells with a hemocytometer. After the initial 3 days in culture only 1.6 to 2.6% of the total cells seeded are recovered (Frangakis and Kimelberg, 1984). However, beyond this time point, the cells rapidly enter the long growth phase showing a doubling time of approximately 30 hr, with the cells seeded at the higher density entering this phase more quickly. It is further noted that for all seeding densities a similar saturation density is reached at $2.2–2.4 \times 10^5$ cells/dish or $1.1–1.2 \times 10^4$ cell/cm^2 by day 14.

2. Functional Characteristics of Cultured Astrocytes

Once immunocytochemical and other phenotypic characteristics of the cultures are identified, electrophysiological, morphological, and functional properties can be tested and compared to *in situ* astrocytic properties. In fact, many of the functional properties of *in situ* astrocytes are evident in cell culture. The following discussion reviews recent evidence to promote the usefulness of the culure model in identifying astrocytic functions, such as uptake and interaction with neurotransmitters, ion transport, and communication with neurons via neurotransmitter receptors.

Cultured astrocytes express receptors for various physiologically active agents and possess membrane channels for sodium, potassium, and chloride (Bevan *et al.*, 1985). Astrocytes possess a bumetanide and furosemide-sensitive Na$^+$, K$^+$/2Cl$^-$ cotransport system that is responsible for active transport of Cl$^-$ (Fig. 1). They also possess both 4-acetamido-4'-isothiocyanatostilbene-2,2'-disulfonic acid and furosemide-sensitive anion-exchange systems which mediate the exchange of intracellular for extracellular Cl$^-$ and HCO$_3^-$, equilibrating the intra- to extracellular ratios of these ions in the absence of other competing fluxes (reviewed by Kimelberg, 1991). *In vitro* studies have also shown the existence of norepinephrine receptors (Kimelberg and Pelton, 1983) and α-adrenergic-mediated opening of Cl$^-$ conductance channels (reviewed by Kimelberg and Walz, 1988), as well as inositol phosphate accumulation (Pearce *et al.*, 1985, 1986; Ritchie *et al.*, 1987). Other properties of cultured astrocytes include large negative, predominantly K$^+$-dependent membrane potential, averaging −70 mV, and norepinephrine-induced depolarization of the membrane potential, primarily due to increased conductance of Cl$^-$ (Fig. 1). β-receptors have also been directly localized

to GFAP-positive primary astrocytes in culture. These receptors appear to be primarily β_1 or may include both β_1 and β_2 receptor subtypes (Ebersolt *et al.*, 1981; Harden and McCarthy, 1982). Activation of these receptors results in large increases in intracellular cAMP (Kimelberg and Frangakis, 1983; Fig. 1).

Astrocytes in culture express a variety of classic neurotransmitter receptor types (and subtypes), e.g., GABAergic, dopaminergic, and cholinergic. Astrocytes also possess binding sites for various neuropeptides, e.g., ACTH, enkephalins, and glucagon. Many of these receptor types are linked to functional intracellular responses, such as changes in cAMP levels, and increases in polyphosphoinositide turnover and Ca^{2+} concentrations.

Astroglial cells can synthesize and, in some cases, release neurotransmitters and neuromodulators (Dennis and Miledi, 1974; Tucek *et al.*, 1978; see also Barres, 1991). For example, astrocytes synthesize and release GABA, glutamate, and aspartate, as well as the neuropeptides, enkephalin, somatostatin, and substance P. Astrocytes also release classic paracrine and endocrine substances, such as the eicosanoids (e.g., PGE$_2$) and steroids (e.g., pregnenolone). However, the mechanism of gliotransmitter release (e.g., vesicular, membrane transport reversal, tension-controlled release) has not been established. Furthermore, since much of the evidence supporting the putative neuromodulatory role of these chemicals is based on *in vitro* research, confirmation of this role necessarily awaits demonstration of *in vivo* relevance.

Avid uptake of glutamate, aspartate, taurine, and GABA by Na$^+$-dependent mechanisms occurs in cultured astrocytes (McLennan, 1976; Schousboe *et al.*, 1977, 1986; Fonnum *et al.*, 1980; Martin and Shain, 1979). Astrocytic uptake of both glutamate and GABA occupies a critical role in their metabolism and is an important mechanism for their removal from the synaptic cleft, limiting their extracellular concentration and the possibility of *en masse* excitation of *N*-methyl-D-aspartate receptors.

Glial preparations isolated from fresh brain tissue and a variety of astrocytic cultures show high affinity uptake of serotonin (5-HT) and catecholamines (reviewed by Kimelberg, 1986). The first report of uptake of catecholamines by glial cells in culture (Pfister and Goworek, 1977) showed, by histofluorescence, norepinephrine and dopamine uptake in explant cultures from neonatal rat cerebral cortices. Hoffman and Vernadakis (1979) also reported glial uptake of norepinephrine in whole brain cultures from 8-day chick embryos. Rat primary astrocyte cultures also take up norepinephrine and dopamine (reviewed by Kimelberg, 1986). High affinity uptake is greatest in the presence

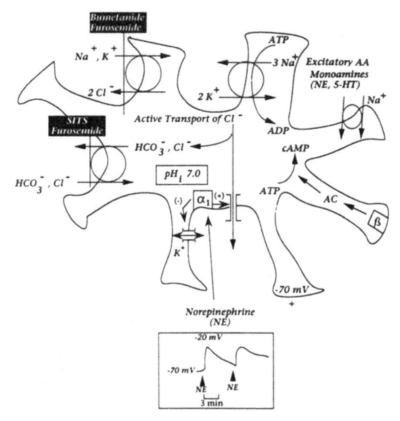

FIGURE 1 Transport systems (partial list) believed to be operative in primary astrocyte cultures. For details please refer to the text. NE, norepinephrine; AA, amino acids; SITS, 4-acetamido-4'-isothiocyanatostilbene-2,2'-disulfonic acid; 5-HT, serotonin.

of both pargyline and tropolone, inhibitors of the catecholamine metabolizing enzymes monoamine oxidase and catechol O-methyltransferase, respectively. A significant upake of 5-HT was also noted in rat primary astrocyte cultures with a high affinity for the Na^+-sensitive component of 5-HT uptake with a K_m of 0.40 μM (reviewed by Kimelberg, 1986).

The receptors, ion channels, and active transport systems mentioned earlier may not all be present on the same astrocytes. Variations due to regional astrocyte specialization have been reported. For example, both the number of β-adrenoreceptors per astroglial cell and their sensitivity seem to vary from one brain region to another. Similarly, regional astrocytes exhibit great variation in their ability to accumulate various amino acid neurotransmitters, supporting the concept of functional specialization of astrocytes and the maintenance of this specialization in cell culture.

B. Oligodendrocytes

Procedural improvements for isolating intact oligodendrocytes from CNS tissues (Poduslo and Norton,

1972) and for maintaining them in long-term culture (Gebicke-Härter *et al.*, 1981; Lisak *et al.*, 1981; Norton *et al.*, 1983; for reviews see Hertz *et al.*, 1985; Pfeiffer, 1984) have been instrumental in probing for oligodendrocytic functions. When attached to a substratum, oligodendrocytes synthesize myelin components (Yim *et al.*, 1986) and assemble membranes with the ultrastructural and biochemical characteristics of myelin.

Oligodendrocytes generate the CNS myelin from projections of their plasma membranes, which ensheathe axons to form the multilamellar configuration that is the hallmark of myelin (Peters *et al.*, 1976; Raine, 1984). Whereas myelin is one of the best characterized of all membranes (Morell, 1984; Norton, 1981), little is known about the oligodendrocyte from which myelin originates. Table 1 lists some of the lipids and proteins that are characteristic of oligodendrocytes and their product, myelin (Wolfgram, basic, and proteolipid proteins and the galactosylglycerides and sulfogalactoceramides and -glycerides). Interestingly, cultured oligodendrocytes elaborate all of these products even in the absence of neurons and axons. Although astrocytes and neurons accelerate the rate of production of myelin-like membranes by oligodendrocytes, their production by oligodendrocytes in the absence of the latter two cell types would suggest an intrinsically inherent property of oligodendrocytes to elaborate myelin-like membranes (Lopez-Cardozo *et al.*, 1989).

It is now possible to grow oligodendrocytes in monolayer cell cultures and to identify them with immunocytochemical methods (see Table 1). The enhanced accessibility afforded by the cell culture environment (Hertz *et al.*, 1985; Kettenmann *et al.*, 1983; Barres *et al.*, 1988) has allowed their electrophysiological properties in culture to be studied. The results to date strongly suggest that the basic physiological properties of oligodendrocytes are largely preserved *in vitro* (reviewed by Barres *et al.*, 1988).

1. Preparaton of Oligodendrocytes in Cultures

Morphological studies as well as immunocytochemical studies of Wolfgram protein localization demonstrate that recognizable oligodendrocytes do not appear within the rodent neopallium until postnatal day 8, with peak oligodendrogliogenesis occurring in the corpus callosum between 1 and 4 weeks of postnatal life and between 1 and 5 months in the neocortex. It is possible to establish primary monotypic oligodendroglial cultures by vigorously dissociating the neopallium of postnatal mouse or rat pups. As with astrocytes, the establishment of these cultures is based on the developmental characteristics of the CNS. By carefully selecting a spatial and temporal region of the CNS when the major proliferative population consists of oligodendrytic precursor cells and neurogenesis is completed, a vigorous cell dissociation procedure will destroy the majority of mature cells, and culture conditions can be chosen which allow the preferential attachment and proliferation of oligodendrocytic cells.

Early attempts to culture oligodendrocytes consisted of dissecting myelin-rich white matter from whole brains, followed by tissue trypsinization, disrupting the slender processes connecting the cell bodies with the myelin. However, the viability of these cultures was limited (Szuchet and Stefansson, 1980) and early events in oligodendroglogenesis were not amenable to experimentation. Accordingly, alternative methods have been developed, in which cells derived from late gestational or newborn brains are used (Booher and Sensenbrenner, 1972). If plated at low density in a culture medium containing 10–20% of fetal calf serum, the cells proliferate rapidly, reaching confluence within 1–2 weeks. If newborn calf serum is used the cultured cells reach confluency at 1 week in culture and proceed to differentiate into phase-dark cells which grow on the top of an astrocytic monolayer (Labourdette *et al.*, 1979). The two cell types can be separated,

and the top layer reveals the morphological and immunocytochemical characteristics of oligodendrocytes.

Several methods are available for the culturing of oligodendrocytes. The method of Szuchet and colleagues (1980a,b), as modified by Szuchet and Yim (1984), is described here. Briefly, the white matter of 6-month-old lamb (12 in total) brains are removed, minced, and softened by 0.1% trypsin for 1 h at 37°C. The tissue is further dissociated by passage through a series of nylon and stainless-steel screens from 350 μm down to 30 μm pore size. The final suspension is adjusted to 0.9 M sucrose and centrifuged. During this step, myelin, which constitutes approximately 50% of the total mass, floats to the top of the tube, while the cells form a pellet. The myelin is discarded and the cell pellet is resuspended in 0.9 M sucrose and resolved on a linear sucrose gradient (from 1.0 to 1.15 M) into three bands (one to three from top to bottom) and a pellet. Under these conditions, oligodendrocytes from band 3 remain as floating clusters and are better than 98% pure as assessed by immunocytochemical methods (Szuchet and Yim, 1984). The isolated cells are suspended in Dulbecco's modified Eagles' medium supplemented with 20% horse serum and 2 mM L-glutamine and are seeded on tissue culture plates at a concentration of 2×10^6 cells/ml. The cultures are maintained at 37°C in a 5% CO_2/95% air-humidified atmosphere.

Since the development of oligodendrocytes is impaired in the presence of high concentrations of serum (Raff *et al.*, 1983; Sykes and Lopez-Cardozo, 1988), the introduction of serum-free media, especially when cultures are poorly enriched in oligodendrocytes, optimizes the culturing conditions (Lopez-Cardozo *et al.*, 1989). Omission of serum also slows down astrocytic proliferation. The latter can be completely inhibited by arabinosylcytosine. Insulin (or insulin-like growth factor I), however, is essential for culturing oligodendrocytes in serum-free media. For a comprehensive review on the influence of chemically defined media on oligodendrocytic growth, the reader is referred to Lopez-Cardozo and colleagues (1989).

2. Oligodendrocyte Properties in Culture

Tissue culture techniques are well adapted to the study of metabolic events in oligodendrocytes. Lipid and glycolipid metabolism was studied by Szuchet and colleagues (1983), who followed the incorporation of radiolabeled acetate, glycerol, galactose, and sulfate into oligodendrocytes as a function of time for up to 35 days. The specific activity of 2',3'-cyclic-nucleotide 3'-phosphodiesterase rises rapidly from days 2 to 5 in culture (serum-free media; Lopez-Cardozo *et al.*, 1989), leveling thereafter. Induction of glycerol-3-phosphate dehydrogenase is somewhat slower, appearing after a 5-day lag. Upon their attachment to the substratum, oligodendrocytes initially synthesize membranes resembling their own cell membrane (Polak and Szuchet, 1988; Szuchet *et al.*, 1988). Later in the culturing period, the membranes synthesized by oligodendrocytes resemble myelin. Thus, it appears that cultured oligodendrocytes may reenact the early events that are associated with myelination.

Oligodendrocytes in culture express high densities of inwardly rectifying potassium currents and outwardly rectifying chloride channels. Because potassium channels preferentially conduct potassium from outside to inside of the cell and because they are open at the resting potential of the cell, Barres and colleagues (1988) postulated that they may serve in removing potassium from the extracellular space and regulating its extracellular concentrations within the white matter of the brain. However, since oligodendrocytes, unlike astrocytes (Kimelberg and Kettenmann, 1990), are only weakly electrically connected, Kettenmann and Ransom (1988) have alternatively postulated that instead of mediating potassium redistribution, they may provide for a slow intracellular exchange of ions and small soluble molecules mediated by diffusion.

C. Rotation-Mediated Aggregating Cell Cultures

The rotation-mediated aggregating cell culture system, first developed by Moscona (1961), consists of enzymatically or metabolically dissociating whole brain or specific brain regions into single cells. These cells, once introduced into flasks and continuously rotated, will reaggregate into clusters due to the vortex of rotation.

1. Preparation of Rotation-Mediated Aggregate Cultures

For optimal culturing conditions the tissue of choice should be derived from fetal brain. When rat brains are studied, cultures derived from 15- to 16-day-old gestation fetuses provide the greatest yield of aggregates capable of developing a high degree of biochemical and morphological differentiation. There is some variability in culture composition reported from different laboratories and minor variations in procedure, including use of mouse vs. rat and enzymatic vs mechanical dissociation. The consequences of these differences are not always clear.

The methodology associated with aggregating cell cultures is relatively straightforward. In brief, whole brains of rat embryos (15–16 days in gestation) are mechanically dissociated into single cells by either enzymatic or mechanical sieving. Brain cells are successively filtered through a 210- and 130-μm pore diameter filters (Trapp and Richelson, 1980). The resulting cell suspension is washed three times with Puck's D_1 solution by centrifugation and resuspended in Dulbecco's modification of Eagle's medium (Trapp and Richelson, 1980) with 10% fetal calf serum, 600 mg% glucose, and 50 U/ml penicillin. Suspensions containing approximately 4–5 $\times$ 10^7 cell/ml are inoculated into 25-ml De Long flasks and incubated at a 90% O_2/10% CO_2 humidified atmosphere (37°C). Under constant rotation of 80 rpm the cells reaggregate, and by 4 days *in vitro* form spherical aggregates of undifferentiated cells. By day 15 in culture, the cells in the aggregates differentiate morphologically to both immunocytochemically positive neurons and neuroglia. By 25 days, ultrastructurally, 90% of the cells in the rotating aggregates are morphologically identified as mature neurons, astrocytes, and oligodendrocytes (Trapp *et al.*, 1982).

2. Functional Properties of Brain Reaggregate Cultures

Cells within rotation-mediated aggregates will change over time from a population of undifferentiated neuroepithelial cells to an integrated population of neurons, astrocytes, and oligodendrocytes. Their differentiation resembles that found *in vivo*. Dendritic and axonal growth and synaptogenesis peak at 21–30 days *in vitro*, myelination occurs, and cell division is restricted (Atterwill, 1987). Astrocytes within the aggregates reveal intense GFAP-positive immunofluorescence. By 25 days in culture, all cells that are identified as neurons are intensely stained by neuron-specific enolase. Myelin basic protein, myelin associated glycoprotein, and 2',3'-cyclic-nucleotide 3'-phosphodiesterase are also localized immunocytochemically in oligodendrocytes. The protein composition of myelin purified from 30-day-old aggregates resembles that of myelin purified from immature rat brain (Trapp *et al.*, 1982).

IV. Advantages and Disadvantages of Astrocytes and Oligodendrocytes in Culture Systems; Implications for Neurotoxicology

The evaluation of chemically induced cytotoxicity, particularly in an heterogeneous system such as the brain, is often difficult in the intact animal because numerous factors (neural, hormonal, and hemodynamic) are not under experimental control. Hence, a simplified model, such as tissue culture, is indispensable as a tool for understanding basic physiology and pathology. The major advantages and disadvantages of glial tissue culture techniques over *in vivo* studies are summarized in Table 2.

Once the cellular purity, content, and degree of maturation have been established, astrocytic and oligodendrytic cultures afford a host of advantages over *in vivo* techniques. Cell morphology, protein synthesis and release (myelin), energy metabolism, receptor interaction, and neurotransmitter uptake and release, as well as electrolyte and nonelectrolyte uptake and release, can be easily studied. Dispersion of cells in culture permits access to clean membrane surfaces for electrophysiological studies utilizing patch clamping and allows for rapid and reliable exchange of solutions. These are but a few of the parameters that can be probed over time. Direct effects of chemicals on a relatively homogeneous population allow for study of specific aspects of the growth and differentiation of cells, as well as the kinetics of uptake and metabolism of the parent compound. The incubation period can be chosen to correspond to the period of mammalian CNS growth spurt during which glial replication is at its peak and cellular organization and differentiation are proceeding at a rapid rate or can be chosen to span over a period of enzymatic maturation. Hence, timing of the culture relative to developmental age can be followed with respect to the vulnerability of gliogenesis to the xenobi-

otic. Teratological mechanisms relative to the maturation of the culture can be probed. Mature glial cultures allow for the discovery of potential mechanisms associated with neurotoxicity (e.g., electrolyte and nonelectrolyte transport, altered metabolism, etc.).

The culture model also makes it possible to study regional specialization and can be extended to study astrocytic–neuronal interactions by coculturing astroglial and neuronal cells as two separate monolayers in the same culture dish (at a distance from each other). The system may provide information on how astrocytes respond to the neuronal environment, and vice versa, and how astrocytic homeostasis affects neuronal development and function.

Tissue cultures allow direct evaluation of the effects of toxic agents on the CNS. Circumventing metabolic degradation, known pharmacological concentrations of toxins can be administered to cultures. Furthermore, toxic effects on juxtaposed cells or other organs which can interact with and buffer or exacerbate the toxic effect of the xenobiotic on the CNS can be altogether eliminated or factored in sequentially. Toxins can be easily added and withdrawn from the cultures, and long-term effects may be studied.

As previously discussed, cultured glia express a variety of receptors. However, whether the receptor phenotype observed *in vitro* accurately reflects the *in vivo* situation remains to be determined. The observed phenotype might be a function of culture conditions or inherent astrocyte heterogeneity. Perhaps more important, glial membrane receptor composition appears to depend on neuron-induced differentiation (Barres, 1991; Bevan, 1990). The issue of differentiation is particularly important for *in vitro* studies of neurotoxicity. Since glial cells and neurons promote mutual functional differentiation of each other, cell types resulting from purified cultures may result in undifferentiated cells or cells with altered differentiaton, making results difficult to interpret. The sensitivity of undifferentiated cells to neurotoxicants has not been assessed.

Although the use of cultured astrocytes and oligodendrocytes in toxicity testing has emerged as a powerful tool to evaluate the responses of target cells at the cellular and molecular level, one must bear in mind some intrinsic pitfalls of culture systems. For example, cells can undergo varying degrees of differentiation. From the toxicologic viewpoint, the extent of cellular differentiation must be carefully defined since multiple phenotypic states may exhibit different toxicologic responsiveness and the phenotypic expression of cells in culture may itself be the target of toxic insult.

While chemicals can be easily added and withdrawn from the cultures, and their effects directly probed in culture systems, caution should be used when correlat-

TABLE 2 Primary Neuroglial Cultures: Advantages and Disadvantages in Neurotoxic Research

Property/Parameter	Advantages*	Disadvantages
Defined system (incubation media)	x	x
Selection for surviving cells	x	x
Partly differentiated cells	x	x
Cellular homogeneity in culture	x	x
Hypersensitivity and relative immaturity of cultures	x	x
Effects at the cellular level during proliferation or at confluence	x	x
Temporal assessment Long term vs. acute exposure	x	
Biochemical assessment Correlation between toxicity and parent compound or metabolite	x	
Endogenous metabolism	x	
Extracellular media composition	x	
Physiological contact with neighboring cells	x	x

* Many of the properties/parameters listed above can be both an advantage and disadvantage. For example, culture homogeneity, while allowing the study of a unique cell type, may mask protective roles afforded by other cells which are missing from the culture. For further details, please refer to the text.

ing effects occurring *in vitro* to those that are observed in the intact animal, where additive interactions are likely to occur. One must remember several concepts: (1) a number of different, sometimes competing, processes influence the ability of a toxin to attack and destroy specific cells. Metabolism of the administered agent by a nontarget cell or tissue may be responsible for bioactivation and/or detoxification of the compound or its metabolite, affecting the vulnerability of the cells to the neurotoxin. (2) A cell culture is manyfold more homogeneous and simpler than any tissues, particularly the CNS. Removal of many cell types and barriers can facilitate diffusion or even active transport of the compound or its metabolite, limiting or enhancing toxicity by determining at which sites the toxin can reach sufficiently high concentrations to interfere with vital cellular processes. (3) The capacity of the cell to repair or replace damaged organelles or enzymes can also be critical in determining cell survival after toxic insult and may obviously depend on neighboring cells and physical barriers, which may be absent in the culture altogether. Accordingly, characteristics that are described as advantageous in particular circumstances may be disadvantageous in others.

Aggregating cell cultures offer a distinct advantage for multidisciplinary investigation of the CNS *in vitro*. Because of the large yield of the technique, the biochemical characterization of aggregating cell cultures exceeds that of any other primary CNS culture system. The morphological structure of the reaggregate resembles that of the developing brain, allowing for cell–cell contact, migration, and architectural reorganization of the cells into patterns closely resembling the *in vivo* brain. Accordingly, many of the limitations outlined in Table II for *in vitro* methods are eliminated in this system.

V. Summary

Since primary oligodendrocyte and astrocyte cultures are more likely to closely resemble their corresponding cell type in the brain than do passaged cell lines, it is not too surprising that these culture techniques have become an indispensable tool in neurobiological studies. Fetal, newborn, or adult astrocytes and oligodendrocytes from a wide variety of animal species can be successfully established in primary culture. Although, ironically, many of the questions that were raised about glial cell function in the CNS by Cajal and his contemporaries almost a century ago are still echoed today, one can identify major advances in our understanding of the physiology of these cells. Tissue culture investigations have uncovered many new and

unexpected roles for neuroglial cells and, as a result, their aesthetically pleasing shape can now be coupled to function. The culture model makes it possible to study mechanisms of action and it can be extended to address astrocytic– and oligodentrytic–neuronal interactions. The potential is enormous and efforts employing tissue culture techniques to the role of neuroglia in neurotoxicity are likely to be rewarded.

Acknowledgment

Preparation of this manuscript was supported in part by NIEHS (ES05223) and USEPA (R819210) awarded to MA.

References

Abbott, N. J., (Ed.) (1991). *Glial-Neuronal Interactions. Ann. N.Y. Acad. Sci.* Vol. 633, New York.

Abramowitz, M., Homma, H., Ishigaki, S., Tansey, F., Cammer, W., and Listowsky, I. (1988). Characterization and localization of glutathione-S-transferases in rat brain and binding of hormones, neurotransmitters, and drugs. *J. Neurochem.* **50**, 50–57.

Altman, J. J. (1963). Regional utilization of leucine-³H by normal rat brain: microdensitometric evaluation of autoradiograms. *J. Histochem. Cytochem.* **11**, 741–750.

Aschner, M., Aschner, J. L., and Kimelberg, H. K. (1994). The role of glia in CNS-induced injuries. In *Handbook of Neurotoxicology*, Vol. 1, *Basic Principles and Current Concepts* (L. W. Chang, Ed.), Dekker, New York. pp. 93–109.

Atterwill, C. K. (1987). Brain reaggregate cultures in neurotoxicological investigations: adaptational and neuroregenerative processes following lesions. *Mol. Toxicol.* **1**, 489–502.

Bannister, R. G., and Romanul, F. C. A. (1963). The localization of alkaline phosphatase activity in cerebral blood vessels. *J. Neurol. Neurosurg. Psychiatry* **26**, 333–340.

Bansal, R., and Pfeiffer, S. W. (1985). Developmental expression of 2′,3′-cyclic-nucleotide 3′-phosphohydrolase in dissociated fetal rat brain cultures and rat brain. *J. Neurosci. Res.* **14**, 21–34.

Bartlett, P. F., Noble, M. D., Pruss, R. M., Raffm, M. C., Rattray, S., and Williams, C. A. (1981). Rat neural antigen-2 (RAN-2): a cell surface antigen on astrocytes, ependymal cells, Muller cells and lepto-meninges defined by a monoclonal antibody. *Brain Res.* **204**, 339–351.

Barres, B. A., Chun, L. L. Y., and Corey, D. P. (1988). Ion channel expression by white matter glia. I. Type 2 astrocytes and oligodendrocytes. *Glia* **1**, 10–30.

Barres, B. A. (1991). New roles for glia. *J. Neurosci.* **11**, 3685–3694.

Benda, P., Lightbody, J., Sato, G., Levine, L., and Sweet, W. (1968). Differentiated rat glial cell strain in tissue culture. *Science* **161**, 370–372.

Bevan, S. (1990). Ion channels and neurotransmitter receptors in glia. *Semin. Neurosci.* **2**, 467–481.

Bhat, S., and Pfeiffer, S. E. (1985). Subcellular distribution and developmental expression of cholesterol ester hydrolase in fetal rat brain cultures. *J. Neurochem.* **45**, 1356–1362.

Bevan, S., Chiu, S. Y., Gray, P. T. A., and Ritchie, J. M. (1985). The presence of voltage-gated sodium, potassium and chloride channels in rat cultured astrocytes. *Proc. R. Soc. Lond. Ser. B.* **225**, 299–313.

Bignami, A., and Dahl, D. (1974a). Astrocyte-specific protein and radial glia in the cerebral cortex of newborn rat. *Nature* **252**, 55–56.

Bignami, A., and Dahl, D. (1974b). Astrocyte-specific protein and neuroglial differentiation. An immunofluorescence study with antibodies to glial fibrillary acidic protein. *J. Comp. Neurol.* **153**, 27–37.

Blomstrand, C., and Hamberger, A. (1969). Protein turnover in cell-enriched fractions from rabbit brain. *J. Neurochem.* **16**, 1401–1407.

Bock, E., and Dissing, J. (1975). Demonstration of enolase activity connected to the brain-specific protein 14-3-2. *Scand. J. Immunol.* **4**, 31–36.

Bock, E., Møller, M., Nissen, C., and Sensenbrenner, M. (1977). Glial fibrillary acidic protein in primary astroglial cultures derived from newborn rat brain. *FEBS. Lett.* **83**, 207–211.

Booher, J., and Sensenbrenner, M. (1972). Growth and cultivation of dissociated neurons and glial cells from embryonic chick, rat and human brain in flask cultures. *Neurobiology* **2**, 97–105.

Cochran, F. B., Yu, R. K., and Ledeen, R. W. (1982). Myelin gangliosides in vertebrates. *J. Neurochem.* **39**, 773–779.

Connor, J. R., and Fine, R. E. (1986). The distribution of transferrin immunoreactivity in the rat central nervous system. *Brain Res.* **368**, 319–328.

Dennis, M. J., and Miledi, R. (1974). Electrically induced release of acetylcholine from denervated Schwann cells. *J. Physiol.* **237**, 431–452.

Diamonds, M. C., Law, F., Rhodes, H., Lindner, B., Rosenzweig, M. R., Krech, D., and Bennett, E. L. (1966). Increases in cortical depth and glia numbers in rats subjected to enriched environment. *J. Comp. Neurol.* **128**, 117–125.

Droz, B., and Leblond, C. P. (1963). Axonal migration of proteins in the central nervous system and peripheral nerves as shown by autoradiography. *J. Comp. Neurol.* **121**, 325–337.

Edström, A., Kanje, M., and Walum, E. (1974). Effects of dibutyryl cyclic AMP and prostaglandin E₁ on cultured human glioma cells. *Exp. Cell Res.* **85**, 217–223.

Ebersolt, C., Perez, M., Vassent, G., and Bockaert, J. (1981). Characteristics of the B₁-and B₂-adrenergic-sensitive adenylate cyclase in glial cell primary cultures and their comparison with β₂-adrenergic-sensitive adenylate cyclase of meningeal cells. *Brain Res.* **213**, 151–161.

Farooq, M., and Norton, W. T. (1978). A modified procedure for isolation of astrocyte- and neuron-enriched fractions from rat brain. *J. Neurochem.* **31**, 887–894.

Fedoroff, S., (1977). Primary cultures, cell lines, and cell strains: terminology and characteristics. In *Cell, Tissue and Organ Cultures in Neurobiology* (S. Fedoroff and L. Hertz, Eds.), pp. 265–286, Academic Press, New York.

Fedoroff, S., and Vernadakis, A. (Eds.) (1986a). Astrocytes: Development, Morphology, and Regional Specialization. Vol. 1, Academic Press, New York.

Fedoroff, S., and Vernadakis, A. (Eds.) (1986b). Astrocytes: Biochemistry, Physiology and Pharmacology of Astrocytes. Vol. 2, Academic Press, New York.

Fedoroff, S., and Vernadakis, A. (Eds.) (1986c). Astrocytes: Cell Biology and Pathology of Astrocytes. Vol. 3, Academic Press, New York.

Fedoroff, S., White, R., Neal, J., Subramanyan, L., and Kalnins, I. (1983). Astrocyte cell lineage.II. Mouse fibrous astrocytes and reactive astrocytes in culture have vimentin- and GFP-containing intermediate filaments. *Dev. Brain Res.* **7**, 303–315.

Fierz, W., Endler, B., Reske, K., Wekerle, H., and Fontana, A. (1985). Astrocytes as antigen presenting cells. I. Induction of Ia antigen expression on astrocytes by T cells via immune interferon and its effect on antigen presentation. *J. Immunol.* **134**, 3785–3793.

Fonnum, F., Karlsen, F. L., Malthe-Sorenssen, D., Sterri, S., and Walaas, I. (1980). High affinity transport systems and their role in transmitter action. In *The Cell Surface and Neuronal Function* (C. W. Cotman, G. Poste, and G. L. Nicolson, *et al.*, Eds.), pp. 455–504, North-Holland, Amsterdam.

Fontana, A., and Fierz, W. (1985). The endothelium-astrocyte immune control system of the brain. *Springer Semin. Immunopathol.* **8**, 57–70.

Frangakis, M. V., and Kimelberg, H. K. (1984). Dissociation of neonatal rat brain by dispase for preparation of primary astrocyte cultures. *Neurochem. Res.* **9**, 1689–1698.

Freedman, P., Magnani, J. L., Nirenberg, M., and Ginsburg, V. (1984). Monoclonal antibody A2B5 reacts with many gangliosides in neuronal tissue. *Arch. Biochem. Biophys.* **233**, 661–666.

Gebricke-Härter, P. J., Althaus, H. H., Schwartz, P., and Neuhoff, V. (1981). Oligodendrocytes from postnatal cat brain in cell culture. I. Regeneration and maintenance. *Dev. Brain Res.* **1**, 497–518.

Ghandour, M. S., Langley, O. K., Vincendon, G., Gombos, G., Filippi, D., Limozin, N., Dalmasso, C., and Laurent, G. (1980). Immunochemical and immunohistochemical study of carbonic anhydrase II in adult rat cerebellum: A marker for oligodendrocytes. *Neuroscience* **5**, 559–571.

Goldman, J. E., Geier, S. S., and Hirano, M. (1986). Differentiation of astrocytes and oligodendrocytes from germinal matrix cells in primary culture. *J. Neurosci.* **6**, 52–60.

Golgi, C., (1885–1886). Sulla fina anatomia della sistema nervosa. Riv Sper Freniatr **8**, 9. Also appeared in 1903 as Vol. 16 of Opera Omnia, U Hoepfli, Milano, and in German translation in 1984, Fischer, Jena. Partial English translation by J. Workman in 1883 in Alienist and Neurologist **4**, 236–269.

Grasso, A., Roda, G., Hogue-Angeletti, R. A., Moore, B. W., and Perez, V. J. (1977). Preparation and properties of the brain specific protein 14-3-2. *Brain Res.* **124**, 497–507.

Haglid, K. G., Hamberger, A., Hansson, H.-A., Hydén, H., Persson, L., and Rönbäck, L. (1976). Cellular and subcellular distribution of the S-100 protein in rabbit and rat central nervous system. *J. Neurosci.* **2**, 175–191.

Hallermayer, K., Harmening, C., and Hamprecht, B. (1981). Cellular localization and regulation of glutamine synthetase in primary cultures of brain cells from newborn mice. *J. Neurochem.* **37**, 43–52.

Hallermayer, K., and Hamprecht, B. (1984). Cellular heterogeneity in primary cultures of brain cells revealed by immunocytochemical localization of glutamine synthetase. *Brain Res.* **295**, 1–11.

Hamberger, A., and Sellström, Å. (1975). Techniques for separation of neurons and glia and their application to metabolic studies. In *Metabolic Compartmentation and Neurotransmission* (S. Berl, D. D. Clarke, and D. Schneider, Eds.) pp. 145–146, Plenum Press, New York.

Hansson, E., Sellström, A., Persson, L. I., and Rönbäck, L. (1980). Brain primary culture—a characterization. *Brain Res.* **188**, 233–246.

Hansson, E., Rönbäck, L., Lowenthal, A., Noppe, M., Alling, C., and Karlsson, B., and Sellström, A. (1982). Brain primary culture—A characterization (part II). *Brain Res.* **231**, 173–183.

Harden, T. K., and McCarthy, K. D. (1982). Identification of the beta adrenergic receptor subtype on astroglia purified from rat brain. *J. Pharmacol. Exp. Ther.* **222**, 600–605.

Henn, F. A., Anderson, D. J., and Sellström, Å. (1977). Possible relationship between glial cells, dopamine, and the effects of antipsychotic drugs. *Nature* **266**, 637–638.

Hertz, L., Schousboe, A., Boechler, N., Mukerji, S., and Fedoroff, S. (1978). Kinetic characteristics of the glutamate uptake into normal astrocytes in culture. *Neurochem. Res.* **3**, 1–14.

Hertz, L., Juurlink, B. H. J. and Szuchet, S. (1985). Cell cultures. In *Handbook of Neurochemistry* (A. Lajtha, Ed.), pp. 603–661, Plenum, New York.

Hertz, L., Juurlink, B. H. J., Szuchet, S., and Walz, W. (1985). Cell and tissue cultures. In *Neuromethods 1. General Neurochemical Techniques* (A. A. Boulton and G. B. Baker, Eds.), pp. 117–167, Humana Press, Clifton, New Jersey.

Hoffman, D. W., and Vernadakis, A. (1979). Biochemical characterization of [³H]norepinephrine uptake in dissociated brain cell cultures from chick embryos. *Neurochem. Res.* **4,** 731–746.

Hoyer, L. W., de los Santos, R. P., and Hoyer, J. R. (1973). Antihemophilic factor antigen. Localization in endothelial cells by immunofluorescent microscopy. *J. Clin. Invest.* **52,** 2737–2744.

Hydén, H. (1959). Quantitative assay of compounds in isolated fresh nerve cells and glial cells from control and stimulated animals. *Nature* **184,** 433–435.

Hydén, H., and McEwen, B. S. (1966). A glial protein specific for the nervous system. *Proc. Natl. Acad. Sci. U.S.A.* **55,** 354–358.

Hydén, H., and Rönbäck, L. (1975). Membrane-bound S-100 protein on nerve cells and its distribution. *Brain Res.* **100,** 615–628.

Kanje, M., Joo, F., and Edström, A. (1978). Alkaline phosphatase activity in cultured glia and glioma cells. *Proc. Eur. Soc. Neurochem.* **1,** 517.

Kettenmann, H., Sonhof, U., and Schachner, M. (1983). Exclusive potassium dependence of the membrane potential in cultured mouse oligodendrocytes. *J. Neurosci.* **3,** 500–505.

Kettenmann, H., and Ransom, B. R. (1988). Electrical coupling between astrocytes and between oligodendrocytes studied in mammalian cell cultures. *Glia* **1,** 64–73.

Kimelberg, H. K., and Pelton, E. W. (1983). High-affinity uptake of [³H]norepinephrine by primary astrocyte cultures and its inhibition by tricyclic antidepressants. *J. Neurochem.* **40,** 1265–1270.

Kimelberg, H. K., and Frangakis, M. V. (1983). Effect of acute and chronic treatment with antidepressants and norepinephrine on the β-receptor mediated cAMP response of primary astrocyte cultures from rat brain. *J. Neurochem.* **41,** S42.

Kimelberg, H. K., and Frangakis, M. V. (1985). Furosemide- and bumetanide-sensitive ion transport and volume control in primary astrocyte cultures from rat brain. *Brain Res.* **361,** 125–134.

Kimelberg, H. K. (1986). Occurrence and functional significance of serotonin and catecholamine uptake by astrocytes. *Biochem. Pharmacol.* **35,** 2273–2281.

Kimelberg, H. K., and Norenberg, M. D. (1989). Astrocytes. Sci. Am. **260,** 66–76.

Kimelberg, H. K., and Walz, W. (1988). Ion transport and volume measurements in cell cultures. In *Neuromethods: The Neuronal Microenvironment* (A. A. Boulton, G. B. Baker, and W. Walz, Eds.), pp. 441–492, The Humana Press, Clifton, New Jersey.

Kimelberg, H. K., and Kettenmann, H. (1990). Swelling-induced changes in electrophysiological properties of cultured astrocytes and oligodendrocytes. I. Effects on membrane potentials, input impedance and cell-cell coupling. *Brain Res.* **529,** 255–261.

Kimelberg, H. K. (1991). Swelling and volume control in brain astroglial cells. In *Advances in Comparative and Environmental Physiology* (R. Gilles, et al., Eds.), Vol. 9, pp. 81–117, Springer-Verlag, Berlin Heidelberg.

Kurihara, T., and Tsukada, Y. (1967). The regional and subcellular distribution of 2',3'-cyclic nucleotide 3'-phosphohydrolase in the central nervous system. *J. Neurochem.* **14,** 1167–1174.

Kurihara, T., and Tsukada, Y. (1968). 2',3'-cyclic nucleotide 3'-phosphohydrolase in the developing chick brain and spinal cord. *J. Neurochem.* **15,** 827–832.

Labourdette, G., Roussel, G., Ghandour, M. S., and Nussbaum, J. L. (1979). Cultures from rat brain hemispheres enriched in oligodendrocyte-like cells. *Brain Res.* **179,** 199–203.

Langley, O. K., and Ghandour, M. S. (1981). An immunocytochemical investigation of non-neuronal enolase in the cerebellum: a new astrocyte marker. *Histochem. J.* **13,** 137–148.

Lauder, J., and McCarthy, K. (1986). In *Astrocytes: Biochemistry, Physiology and Pharmacology of Astrocytes* (S. Fedoroff and A. Vernadakis, Eds.), Vol. 2, pp. 295–314, Academic Press, New York.

Laursen, R. A., Samiullah, M., and Lees, M. B. (1984). The structure of bovine brain myelin proteolipid and its organization in myelin. *Proc. Natl. Acad. Sci. U.S.A.* **81,** 2912–2916.

Leveille, P. J., McGinnis, J. F., Maxwell, D. S., and de Vellis, J. (1980). Immunocytochemical localization of glycerol-3-phosphate dehydrogenase in rat oligodendrocytes. *Brain Res.* **196,** 287–305.

LeVine, S. M., and Macklin, W. B. (1988). Biotin enrichment in oligodendrocytes in the rat brain. *Brain Res.* **444,** 199–203.

Levitt, P., and Rakic, P. (1980). Immunoperoxidase localization of glial fibrillary acidic protein in radial glial cells and astrocytes of the developing rhesus monkey brain. *J. Comp. Neurol.* **193,** 815–840.

Liesi, P., Dahl, H., and Vaheri, A. (1983). Laminine is produced by early rat astrocytes in primary culture. *J. Cell Biol.* **96,** 920–924.

Lisak, R. P., Pleasure, D. W., Silberberg, D. H., Manning, M. C., and Saida, T. (1981). Long term culture of bovine oligodendroglia isolated with percoll gradient. *Brain Res.* **223,** 107–122.

Lopez-Cardozo, M., Sykes, J. E. C., Van der Pal, R. H. M., and Van Golde, L. M. G. (1989). Development of oligodendrocytes. Studies of rat glial cells cultured in chemically-defined medium. *J. Dev. Physiol.* **12,** 117–127.

Lowry, O. H. (1953). The quantitative histochemistry of the brain. *J. Histochem. Cytochem.* **1,** 420–428.

Martin, D. L., and Shain, W. (1976). High affinity transport of taurine and β-alanine and low affinity transport of γ-aminobutyric acid by a single transport system in cultured glioma cells. *J. Biol. Chem.* **254,** 7076–7084.

Martinez-Hernandez, A., Bell, K. P., and Norenberg, M. D. (1977). Glutamine synthetase: glial localization in the brain. *Science* **195,** 1356–1358.

McCarthy, K. D., and De Vellis, J. (1980). Preparation of separate astroglial and oligodendroglial cell cultures from rat cerebral tissue. *J. Cell Biol.* **85,** 890–902.

McLennan, H. (1976). The autoradiographic localization of L-[³H]glutamate in rat brain tissue. *Brain Res.* **115,** 139–144.

Morell, P. (1984). Myelin. Plenum Press, New York.

Moscona, A. A. (1961). Rotation-mediated histogenetic aggregation of dissociated cells: a quantifiable approach to cell interactions in vitro. *Exp. Cell Res.* **22,** 455.

Murphy, D. L., Donnelly, C. H., and Richelson, E. (1976). Substrate and inhibitor-related characteristics of monoamine oxidase in C6 rat glial cells. *J. Neurochem.* **26,** 1231–1235.

Norton, W. T. (1981). Biochemistry of myelin. *Adv. Neurol.* **31,** 93–121.

Norton, W. T. (1984). Recent advances in the neurobiology of oligodendroglia. *Adv. Cell Neurobiol.* **4,** 3–55.

Norton, W. T. (1984). Oligodendroglia. In *Advances in Neurochemistry,* Vol. 5, Plenum Press, New York.

Norton, W. T., Farooq, M., Fields, K. L., and Raine, C. S. (1983). The long term culture of bulk-isolated bovine oligodendroglia from adult brain. *Brain Res.* **270,** 295–310.

Orkland, R. K., Nicholls, J. G., and Kuffler, S. W. (1966). Effect of nerve impulses on the membrane potential of glial cells in the central nervous system of amphibia. *J. Neurophysiol.* **29,** 788–806.

Palmer, G. C. (1973). Adenyl cyclase in neuronal and glial-enriched fractions from rat and rabbit brain. *Res Commun. Chem. Pathol. Pharmacol.* **5,** 603–613.

Patterson, P. H., and Chun, L. L. (1974). The influence of nonneuronal cells on catecholamine and acetylcholine synthesis and accumulation in cultures of dissociated sympathetic neurons. *Proc. Natl. Acad. Sci. U.S.A.* **71,** 3607–3610.

Pearce, B., Cambray-Deakin, M., Morrow, C., Grimble, J., and Murphy, S. (1985). Activation of muscarinic and of α_1-adrenergic receptors on astrocytes results in the accumulation of inositol phosphates. *J. Neurochem.* **45,** 1534–1540.

Pearce, B., Morrow, C., and Murphy, S. (1986). Receptor mediated inositol phospholipid hydrolysis in astrocytes. *Eur. J. Pharmacol.* **121,** 231–243.

Persson, L. I., Rönbäck, L., Grasso, A., Haglid, K. G., Hansson, H.-A. Dolonius, L., Molin, S. O., and Nygren, H. (1978). 14-3-2 protein in rat brain. *J. Neurol. Sci.* **35,** 381–390.

Peters, A., Palay, S. L., and Webster, H. D. E. F. (1976). In *The Fine Structure of the Nervous System: The Neurons and Supporting Cells.* Saunders, Philadelphia.

Pfeiffer, S. E. (1984). Oligodendrocyte development in culture systems. In *Oligodendroglia* (W. T. Norton, Ed.), pp. 233–298, Plenum, New York.

Pfeiffer, S. E., Betschart, B., Cook, J., Mancini, P., and Morris, R. (1978). Glial cell lines. In *Cell, Tissue and Organ Cultures in Neurobiology* (S. Fedoroff and L. Hertz, Eds.), pp. 287–346, Academic Press, New York.

Pfister, C., and Goworek, K. (1977). Fluoreszenzhistokemische Untersuchungen zur Aufnahme von exogenem Noradrenalin und Dopamin durch in vitro kultivierte Cerebrocortexexplantate der Ratte. *Z. Mikros. Anat. Foprsch.* **91,** 521–535.

Poduslo, S. E., and Norton, W. T. (1972). Isolation and some chemical properties of oligodendroglia from calf brain. *J. Neurochem.* **19,** 727–736.

Polak, P. E., and Szuchet, S. (1988). Plasma membrane of cultured oligodendrocytes: I. Isolation, purification and initial characterization. *Glia* **1,** 29–53.

Raff, M. C., Fields, K. L., Hakomori, S.-I., Mirsky, R., Prruss, R. M., and Winter, J. (1979). Cell-type specific markers for distinguishing and studying neurons and the major classes of glial cells in culture. *Brain Res.* **174,** 283–308.

Raff, M. C., Miller, R. H., and Noble, M. (1983). A glial progenitor cell that develops in vitro into an astrocyte or an oligodendrocyte depending on culture medium. *Nature* **303,** 390–396.

Raine, C. S. (1984). Morphology of myelin and myelination. In *Myelin* (P. Morell, Ed.), pp. 1–41, Plenum, New York.

Rakic, P. (1991). Glial cells in development. In *Glial Neuronal Interactions* (N. J. Abbott, Ed.), *Ann. N.Y. Acad. Sci.* **633,** 96–99.

Ramón y Cajal, S. (1913). Contribución al conocimiento de la neuroglia del cerebro humano. *Trab. Lab. Invest. Biol. Univ. Madrid* **11,** 255–315.

Ranscht, B., Clapshaw, P. A., Price, J., Noble, M., and Seifert, W. (1982). Development of oligodendrocytes and Schwann cells studied with a monoclonal antibody against galactocerebroside. *Proc. Natl. Acad. Sci. U.S.A.* **79,** 2709–2713.

Ritchie, T., Cole, R., Kim, H., deVellis, J., and Noble, E. P. (1987). Inositol phospholipid hydrolysis in cultured astrocytes and oligodendrocytes. *Life Sci.* **41,** 31–39.

Rose, S. P. R. (1965). Preparation of enriched fractions from cerebral cortex containing isolated metabolically active neuronal cells. *Nature* **206,** 621–622.

Roussel, G., Labourdette, G., and Nussbaum, J. L. (1981). Characterization of oligodendrocytes in primary cultures from brain hemispheres of newborn rats. *Dev. Biol.* **81,** 372–378.

Roussel, G., and Nussbaum, J. L. (1981). Comparative localization of Wolfgram W1 and myelin basic protein in the rat brain during ontogenesis. *Histochem. J.* **13,** 1029–1047.

Schmechel, D. E., and Rakic, P. (1979). A Golgi study of radial glial cells in developing monkey telencephalon: morphogenesis and transformation into astrocytes. *Anat. Embryol.* **156,** 115—152.

Schousboe, A., Fosmark, H., and Svenneby, G. (1976). Taurine uptake in astrocytes cultured from dissociated mouse brain hemispheres. *Brain Res.* **116,** 158–164.

Schousboe, A., Svenneby, G., and Hertz, L. (1977). Uptake and metabolism of glutamate in astrocytes cultured from dissociated mouse brain hemispheres. *J. Neurochem.* **29,** 999–1005.

Shank, R. P., Bennet, G. S., Freytag, S. O., and Campbell, G. L. (1985). Pyruvate carboxylase: an astrocyte-specific enzyme implicated in the replenishment of amino acid neurotransmitter pools. *Brain Res.* **329,** 364–367.

Silberstein, S. D., Shein, H. M., and Berv, K. R. (1972). Catechol-O-methyl-transferase and monoamine oxidase activity in cultured rodent astrocytoma cells. *Brain Res.* **41,** 245–248.

Somjen, G. G. (1988). Nervenkitt: notes on the history of the concept of neuroglia. *Glia* **1,** 2–9.

Sprinkle, T. J., Agee, J. F., Tippins, R. B., Chamberlain, C. R., Faguet, G. B., and De Vries, G. H. (1987). Monoclonal antibody production to human and bovine 2′:3′-cyclic nucleotide 3′-phosphodiesterase (CNPase): high specificity recognition in whole brain acetone powders and conservation of sequence between CNP1 and CNP2. *Brain Res.* **426,** 349–357.

Stieg, P. E., Kimelberg, H. K., Mazurkiewicz, J. E., and Banker, G. A. (1980). Distribution of glial fibrillary acidic protein and fibronectin in primary astroglial cultures from rat brain. *Brain Res.* **199,** 493–500.

Sykes, J. E. C., and Lopez-Cardozo, M. (1988). The effect of serum on lipid synthesis and on the expression of oligodendrocyte marker-enzymes in oligodendrocyte-enriched glial cells in culture. *Neurochem. Int.* **12,** 467–474.

Szuchet, S., and Stefansson, K. (1980). In vitro behaviour of isolated oligodendrocytes. *Adv. Cell Neurobiol.* **1,** 313–346.

Szuchet, S., Arnason, B. G. W., and Polak, P. E. (1980a). Separation of ovine oligodendrocytes into two distinct bands on a linear sucrose gradient. *J. Neurosci. Methods* **3,** 7–19.

Szuchet, S., Stefansson, K., Wollman, R. L., Dawson, G., and Arnason, B. G. W. (1980b). Maintenance of isolated oligodendrocytes in long-term culture. *Brain Res.* **200,** 151–164.

Szuchet, S., and Yim, S. H. (1984). Characterization of a subset of oligodendrocytes separated on the basis of selective adherence properties. *J. Neurosci. Res.* **11,** 131–144.

Szuchet, S., Polak, P. E., Yim, S. H., and Lange, Y. (1988). Plasma membrane of cultured oligodendrocytes. II. Possible structural and functional domains. *Glia* **1,** 54–63.

Trapp, B. D., and Richelson, E. (1980). Usefulness for neurotoxicology of rotation-mediated aggregating cell cultures. In *Experimental and Clinical Neurotoxicology* (P. S. Spencer, and H. H. Schaumburg, Eds.), pp. 803–819, Williams and Wilkins, Baltimore.

Trapp, B. D., Webster, H. D. E. F., Johnson, D., Quarles, R. H., Cohen, S. R., and Murray, M. R. (1982). Myelin formation in rotation-mediated aggregating cell cultures: immunocytochemical, electron microscopic, and biochemical observations. *J. Neurosci.* **2,** 986–993.

Tucek, S., Zelena, J., Ge, I., and Vyskocil, F. (1978). Choline acetyltransferase in transected nerves, denervated muscles and

Schwann cells of the frog: correlation of biochemical, electron microscopical and electrophysiological observations. *Neuroscience* **3,** 709–724.

Vernadakis, A. (1988). Neuron-glia interrelations. *Int. Rev. Neurobiol.* **30,** 149–225.

Virchow, R. (1846). Ueber das granulierte Ansehsn der Wandungen der Gehirnventrikel. *Allg. Z. Psychiatrie* **3,** 242–250.

Warringa, R. A. J., Van Berlo, M. F., Klein, W., and Lopez-Cardozo, M. (1988). Cellular localization of glutamine synthetase and lactate dehydrogenase in oligodendrocyte-enriched cultures from rat brain. *J. Neurochem.* **50,** 1461–1468.

Westermark, B., Pontèn, J., and Hugosson, R. (1973). Determinants for the establishment of permanent tissue culture lines from human gliomas. *Acta Pathol. Scand. Sec. A* **81,** 791–805.

Yim, S. H., Szuchet, S., and Polak, P. E. (1986). Cultured oligodendrocytes: a role for cell-substratum interaction in phenotypic expression. *J. Biol. Chem.* **261,** 11808–11815.

Neurotoxicology Studies Utilizing Schwann Cell–Neuronal Interactions in Vitro

GEORGE H. DEVRIES

Department of Biochemistry and Molecular Biophysics
Medical College of Virginia
Richmond, Virginia 23298

The peripheral nervous system (PNS) is a frequent target of environmental toxins. In addition, the PNS is often the target of toxic side effects of therapeutic drugs, including drugs used to treat the acquired immunodeficiency syndrome and chemotherapeutic drugs. The neuropathy associated with these toxic insults is due to the effect of these toxins on either PNS neurons or the Schwann cells which ensheathe the PNS axons (Rutkowski and Tennekoon, 1991). Schwann cells and axons are intimately related to each other, both structurally and metabolically. Therefore, in addition to direct targeting of toxins to either of these cell types, there often are indirect secondary effects on each cell type. For example, a number of toxins, such as *n*-hexane, acrylamide, carbon disulfide, arsenic, and mercury, are primarily targeted to neurons. However, this primary insult subsequently leads to a loss of Schwann cell functions, resulting in demyelination. The converse may also be true in that Schwann cells may be targeted by toxins such as isoniazid, hexachlorophene, and triethyl tin. Subsequent secondary neuronal effects lead to reduced nerve conduction velocity. Therefore, morphological examination of the PNS after toxic in-

sult does not allow one to unequivocally determine which cell type was the primary target and which cell was secondarily affected by the toxin. In addition, there is the possibility that both cell types could be simultaneously affected.

Schwann cell culture models are extremely useful in determining both the primary and secondary targets of PNS toxic insults (Mason *et al.*, 1991). *In vitro* studies can be carried out studying toxins on Schwann cells and PNS neurons individually as well as the combined Schwann cell–neuron interactions. The *in vitro* systems described in this chapter include an overview of the Schwann cell lines which are available and primary Schwann cells which can be isolated from a number of different species. The *in vitro* culture of PNS neurons and their associated neurites is also described. Since Schwann cell function is closely related to axonal contact (Neuberger and DeVries, 1992), *in vitro* systems useful for determining the ability of Schwann cells to respond to axonal contact, before and after a neurotoxic insult, are also described. Finally, *in vitro* systems are described in which Schwann cells can be cultured with two sources of the neuronal axonal stimulus:

living neurites associated with cultured dorsal root ganglia neurons and an isolated axolemma-enriched fraction (AEF).

As shown in Fig. 1, Schwann cells respond to axonal contact in two ways. The initial contact of Schwann cells with the axon surface results in a wave of proliferation which allows the initial limited population of Schwann cells to divide and populate the growing axons. This proliferative response is also evident after demyelination in order to provide adequate numbers of Schwann cells for secure remyelination. If Schwann cell proliferation is inhibited after demyelination, the Schwann cells are unable to redifferentiate and reform a myelin sheath ((Hall and Gregson, 1974).

Subsequent to the proliferative phase, Schwann cells become mitotically quiescent and ensheathe the axon. The Schwann cells are then stimulated by axonal contact to begin the remarkable process of myelinogenesis. The initial phases of this differentiation response are marked by increased myelin gene expression in Schwann cells, which ultimately leads to synthesis of a mature myelin sheath.

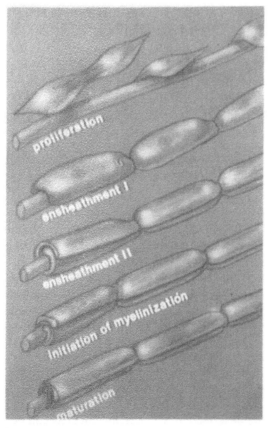

FIGURE 1 Response of Schwann cells to axonal contact. Initial contact results in a proliferative response followed by ensheathment and an axonal-stimulated myelination response ultimately resulting in a mature myelin sheath.

In vitro systems have been devised to study the proliferative and differentiation phases of the Schwann cell response to axonal contact. Schwann cells can be cocultured with neurites of dorsal root ganglion cells, followed by manipulation of the *in vitro* conditions to allow proliferation and differentiation to take place. In addition, a quiescent population of cultured Schwann cells can be stimulated to proliferate and differentiate via stimulation with AEF. It has been demonstrated that both the proliferative (Yoshino *et al.*, 1984) and differentiating (Knight *et al.*, 1993) effects of axonal contact can be mimicked by the addition of AEF to cultured Schwann cells. Comparison of the toxic effects of Schwann cells alone, neurons alone, and Schwann cells stimulated with either neuritic contact or AEF allows one to determine the *in vitro* cellular targets of a particular neurotoxin.

The most utilized source for the culture of Schwann cells is neonatal sciatic rat nerve, using the methodology initially described by Brockes and co-workers (1979). Neonatal sciatic nerves are dissociated enzymatically and placed in Dulbeccos minimal essential medium containing fetal calf serum. Under these conditions Schwann cells do not divide, whereas the other major cell type in the culture dish, the fibroblast, divides vigorously. Thus, with the judicious use of several cycles of antimitotic treatment, the dividing population of fibroblasts can be mostly eliminated from the mitotically quiescent Schwann cells. The small amount of remaining fibroblast contamination may be eliminated by complement-mediated lysis using the fibroblast-specific antibody anti-Thy1.1. However, it has been found that by several cycles of brief withdrawal of the antimitotic, effective fibroblast elimination can be achieved to the point of not requiring the complete mediated lysis step (Neuberger *et al.*, 1994). The second major method for the preparation of Schwann cells has been described by Wood (1976) and utilizes the dorsal root ganglion (DRG) as a Schwann cell "mother." In this method, embryonic DRG are cultured in the presence of an amount of antimitotic which is sufficient to eliminate fibroblasts, but not Schwann cells. The Schwann cells survive and proliferate vigorously on the neurites. After the Schwann cells have proliferated to occupy all of the neuritic surface, the DRG cell bodies are surgically excised from their associated neurites. Subsequently the neurites die, leaving a surviving population of Schwann cells. Although this method is effective, it is somewhat more time consuming than the method of Brockes and colleagues (1979) and is currently not utilized as extensively as other methods. Variations on these two major methods exist and have been summarized (Mason *et al.*, 1992).

Schwann cells are more difficult to prepare from the adult peripheral nerves. This difficulty of preparation is associated with the extensive tough surrounding connective tissue that must be removed by dissection prior to enzymatic dissociation of the adult peripheral nerve. However, a number of laboratories have described the successful methodology by which to successfully isolate Schwann cells from the adult peripheral nerve of rat (Scarpini *et al.*, 1988) or human (Morrisey *et al.*, 1991). The method described by Morrisey and co-workers (1991) requires preincubation of the desheathed nerve segments in tissue culture media for a period of time in order to "soften" the tissue and make it more amenable to subsequent dissociation. A persistent problem in the isolation of adult Schwann cells is the extent of fibroblast contamination which is substantially greater due to the increased connective tissue content of the starting tissue. However, the cloning of a new class of mitogens for Schwann cells, the neuregulins (Mudge, 1993; Peles and Yarden, 1993), has allowed for selective expansion of adult human Schwann cells while simultaneously not allowing fibroblast contamination to overrun the culture. With the use of this mitogen, adult human Schwann cell cultures with a purity greater than 95% can routinely be prepared (Bunge, 1994).

The metabolic properties of cultured Schwann cells are remarkably dependent on the species from which the Schwann cells are isolated. For example, mouse Schwann cells do not appear to be as regulated by axonal contact as other rodent Schwann cells (Manthorpe and Varon, 1989). Schwann cells derived from mouse peripheral nerve continuously proliferate and remain differentiated even in the absence of axonal contact. At the other end of the spectrum, human adult Schwann cells are quite unresponsive to a number of factors that are potent mitogens for cultured rodent cells (Morrisey *et al.*, 1991). Thus, it is important to utilize Schwann cell preparations derived from the species of interest for neurotoxicological studies instead of extrapolating data from one species to another. In this regard, Schwann cell lines from either human or rodent species are potentially very useful.

The strategy in creating Schwann cell lines is to immortalize the normally mitotically quiescent Schwann cells with a stably integrated gene whose expression will constitutively drive cell proliferation. For example, a rat Schwann cell line that has been transformed with a plasmid which codes for the large T antigen has been developed (Tennekoon *et al.*, 1987). Although this Schwann cell line constitutively divides in serum, surprisingly, the cells are able to ensheathe and segregate cultured neurites. In addition, a number of Schwann cell lines have been derived from human

dermal fibromas (Fletcher *et al.*, 1991). These Schwann cell lines variably reflect the intrinsic properties of the nontransformed human Schwann cells from which they have ultimately been derived. In contrast to the rat Schwann cell line, some of these cells cannot relate to cultured neurites (Klein and DeVries, unpublished observation). Therefore, since these cell lines cannot respond to a neuronal stimulus, they may be of limited usefulness in neurotoxicological studies.

A persistent problem in using Schwann cell culture methods is the availability of a sufficient amount of starting biological material from which to isolate Schwann cells in quantities sufficient to carry out neurotoxicological studies. A method has been developed whereby the tissue source (neonatal sciatic nerve) can be frozen, allowing for collection of sufficient biological material to do a large-scale preparation (Mason *et al.*, 1992). The frozen nerves are rapidly thawed and treated according to the standard paradigms for preparation of neonatal Schwann cells. It has been demonstrated that the yield and biological properties of frozen Schwann cells exactly parallel Schwann cells isolated from fresh sciatic nerve. This methodology then can be useful in neurotoxicological studies in which nerves of treated animals must be pooled to obtain sufficient material to study a particular parameter.

Cultured Schwann cells prepared from any species should be morphologically and antigenically characterized prior to being utilized in any studies. The morphological features of cultured Schwann cells can be easily assessed even with phase microscopy because of the distinctive bipolar nature of the Schwann cells which will align themselves end to end in fasicles in culture. In contrast, the fibroblasts are pleomorphic and are much more irregularly shaped. It has been shown that the nuclei of fibroblasts are also significantly larger and more rounded than the ovoid nuclei of Schwann cells (Baichwal *et al.*, 1987). Thus even a cursory examination of a final culture by phase microscopy gives some indication as to the purity of the cultures to be utilized in studies.

The morphological features of a rat Schwann cell preparation are shown in Fig. 2. This figure shows the predominantly ovoid shape of the Schwann cell bodies and the bipolar spindle-shaped appearance of the cells. Clearly the cells are aligned in parallel arrays of fasicles. No pleomorphic fibroblasts are evident. However, in order to completely characterize the properties of the preparation, antigenic markers should be utilized. Antigens commonly used to identify Schwann cells include the S100 antigen, CNPase, nerve growth factor, the low-affinity nerve growth factor receptor, and laminin. These antigens are present in appreciable quantities in cultured Schwann cells in the absence of

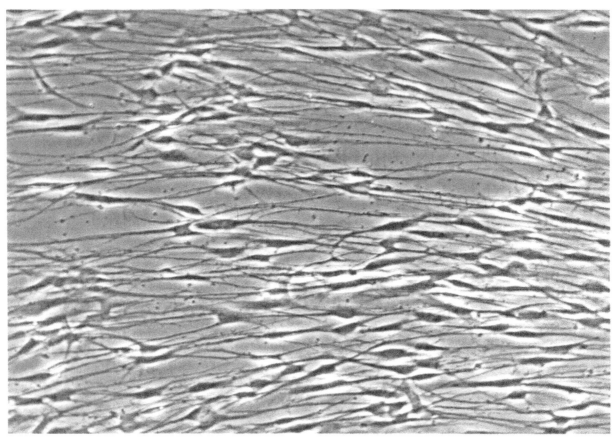

FIGURE 2 Phase microscopy of a Schwann cell preparation. The cells are spindle shaped and bipolar with ovoid cell bodies. Note the absence of any fibroblasts and the parallel alignment of the Schwann cells.

axonal contact. In addition, myelin-related proteins, such as the major myelin glycoprotein P_0 and myelin basic protein, can also be detected. However, it should be appreciated that these antigens are not present at as intense a level as other markers due to their down-regulation in the absence of axonal contact. In order to visualize the potential contamination with the fibroblasts, Thy1.1 staining is required. This antigen is present in the fibroblast surface membrane and is absent from the Schwann cell plasma membrane. A growing number of Schwann cell-related molecules are either upregulated or downregulated by axonal contact (Neuberger and DeVries, 1992). The ability of Schwann cells to upregulate such molecules in the presence or absence of potential neurotoxins is another useful marker by which neurotoxicity can be evaluated.

Although cultured Schwann cells, by themselves, can be used as targets for neurotoxins, it is useful to coculture them with a neuronal stimulus in order to more fully evaluate the functional capabilities of Schwann cells. In this regard, two preparations which provide a neuronal stimulus (neurites and AEF) will be described.

Two major types of neurites are available for coculture with Schwann cells. The laboratory of Bunge has pioneered in the preparation of cultured neurites derived from dorsal root ganglia, which are often referred to as "myelin competent" neurites (Wood, 1976). On the other hand, superior cervical ganglia provide a source of unmyelinated neurites, which will allow for only proliferation and not myelination to take place (Johnson and Argiro, 1983). Neurites from either source are prepared from embryonic material. In general, the dissected dissociated neuronal soma require antimitotic treatment to eliminate glia that are hidden in the neuronal clusters. These glial cells will grow out along the neurites and proliferate if they are not eliminated by antimitotic treatment. The antimitotic of choice in this regard is fluorodeoxyuridine, not β-cytosine arabinoside, which is toxic for cultured DRG neurons (Martin *et al.*, 1990).

The DRG neurite cultures require type 1 collagen substratum for successful culture (Wood, 1976). The ganglia are obtained from rat embryos (usually E-15) "plucked" clean of connective tissue and are dissociated with trypsin. The cells are then plated on the collagen substratum in a small bead of media. This small volume allows the clustered neurons to settle down as an aggregate unit so that after an overnight period they adhere to the collagen of substratum neurons as a group. The cultures are then flooded with media and cycled repeatedly through nonantimitotic-containing media (referred to as C-feed) or fluorodeoxyuridine-containing feed (referred to as F-feed). In all cases, nerve growth factor is added to the media to stimulate neurite outgrowth. The clustered neuronal cell bodies will then elaborate a "halo" of neurites along their edge. If the antimitotic treatment is successful, the edges of the neurites will be smooth and glial free. Lesser concentrations of antimitotic allow for survival of Schwann cells which will subsequently vigorously proliferate due to neuritic contact and overrun the culture. Neurites will continue to elongate with increased time in culture. Such neuritic preparations can be maintained almost indefinitely in culture with a good sterile technique. The phase-contrast microscopy characteristics are shown in Fig. 3, including the cultured neuronal soma and abundant neurites. The neurites are cell free as noted by their smooth edges. The neurites tend to "cluster" together and form fasicles.

Both the neuronal soma and the neurites derived from such soma are not a homogeneous preparation *in vitro*. The neuronal soma are clearly heterogeneous in size as they are also heterogeneous in metabolic properties. For example, studies of fibroblast growth factor immunoreactivity in cultured DRG neurons have demonstrated that certain soma contain more fibroblast growth factor than others (Neuberger and DeVries, 1993a). In addition, it is interesting to note that the surface of neurites can be dramatically altered with time and culture. For example, studies done on surface immunoreactivity to fibroblast growth factor have demonstrated that no bFGF immunoreactivity is noted after 3 days *in vitro* whereas marked surface bFGF immunoreactivity can be seen after 30 days *in vitro* (Neuberger and DeVries, 1993a). It is important to appreciate this heterogeneity with respect to time and culture and type of neuronal soma from which neurites are derived when considering effects of toxins on this heterogeneous neuronal source.

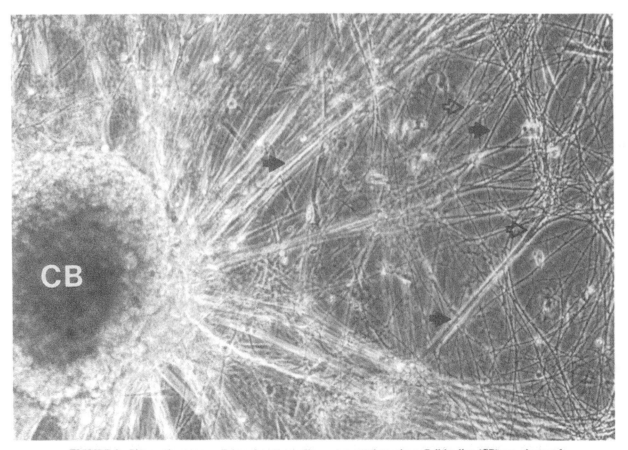

FIGURE 3   Phase microscopy of dorsal root ganglia neurons and neurites. Cell bodies (CB) are clustered together and support a dense outgrowth of neurites with smooth cell-free edges (filled arrows). The individual neurites often interact with each other to form fascicles (open arrows).

An alternative source of neuronal signals is an axolemma-enriched fraction, which can be prepared from a variety of sources (DeVries, 1981, 1984). A homogeneously well-myelinated CNS or PNS tissue source will allow for the highest purity preparation of axolemma (Detsky *et al.*, 1988; Yoshino and DeVries, 1993). The principle of this method is the homogenization of white matter in a buffered sucrose solution which maintains the integrity of the myelin sheath around the axon. Subsequent centrifugation of this homogenate allows flotation of the lipid-rich myelinated axons to the surface of the tube. This preparation, which is enriched in myelinated axons, is then disrupted osmotically and mechanically, resulting in a "shocked myelinated axon" preparation which contains a mixture of myelin, axolemmal vesicles, and axoplasm. These cellular components differ widely in their lipid content, allowing for their successful separation on a 10 to 40% sucrose density gradient. The axolemma-enriched fractions are characterized by an increased specific activity of acetylcholinesterase which is variably present in the axolemma-enriched fractions. The principles for the preparation of axolemma from peripheral nerve are somewhat different. Because of the connective tissue present in the nerve, flotation of intact myelinated axons is not compatible with successful initial disruption of the nerve. Therefore, in the procedure devised to isolate PNS axolemma, the nerves are vigorously and completely homogenized initially and the major cellular elements are separated by differential centrifugation. A crude microsomal pellet is then fractionated on a 10 to 40% sucrose gradient to obtain a PNS axolemma-enriched fraction (Yoshino and DeVries, 1993).

To date, CNS axolemma preparations have been isolated from a number of different species of mammalian brain. In addition, PNS axolemma-enriched fractions have been isolated from unmyelinated axons (splenic nerve) (DeVries *et al.*, 1994), homogeneously well-myelinated axons (spinal accessory nerve) (Yoshino and DeVries, 1993), or peripheral nerve which has variably myelinated axons (intradural dorsal root) (Yoshino and DeVries, 1993).

As noted previously, the addition of axolemma-enriched fractions to cultured glial cells results in both a proliferative and differentiation response. It is interesting to note that the response is neither species specific nor restricted to homologous axolemma-enriched fractions from either CNS or PNS (DeVries *et al.*, 1983). One interesting characteristic of AEF preparations is the ability of the cultured glial cell to act as a "biological affinity column." In a time-dependent process (Sobue and Pleasure, 1985), axolemma-enriched fractions specifically bind to Schwann cells

so that any irrelevant biologically active material in the preparation is not a problem. As noted in Fig. 4, the addition of neuritic vesicles to cultured Schwann cells results in a "coating" of this membrane specifically associated with the Schwann cells. This membrane is not associated with the noncellular elements of the culture. This "coated" appearance of Schwann cells is exactly the same in AEF-treated Schwann cell cultures (G. H. DeVries, unpublished observations). Examination of Schwann cell cultures after 72 hr of AEF treatment reveals increased proliferation and increased differentiation. This *in vitro* system can serve as a useful model for the *in vivo* response of Schwann cells to axonal contact.

The most intact system by which to study Schwann cell axon interactions is to coculture Schwann cells with neurites *in vitro*. This highly interactive system has been extensively studied in the laboratory of Bunge and detailed methods have been published to describe this preparation (Wood, 1976). The addition of Schwann cells to a non-glial cell-containing neuritic preparation allows the Schwann cells to rapidly and completely associate with neurites. As shown in Fig. 5 after a short time in culture, Schwann cells anywhere in the vicinity of neurites always rapidly associate with the neurites. Figure 5 shows the complete absence of any Schwann cells which have not associated with neurites.

One advantage of this neurite–Schwann cell interactive system is that the culture conditions can be manipulated to allow only proliferation or differentiation of Schwann cells. In serum-free media, Schwann cells will associate with neurites and only proliferate but not differentiate (Moya *et al.*, 1980). Further studies from the laboratory of Bunge describe the conditions required to allow Schwann cells to specifically differentiate (Eldridge *et al.*, 1987). These conditions include the addition of ascorbic acid to the media to allow the synthesis of hydroxyproline required for connective tissue synthesis and the addition of serum containing necessary and undefined factors. The ultimate differentiation response of Schwann cells results in the production of a myelin sheath in the neurite–Schwann cell cocultures. The myelin segments can be visualized with sudan black and quantitated morphologically as the number of myelinated segments observed within a given area of the field (Eldridge *et al.*, 1987).

Schwann cell culture methods, in conjunction with the coculture of Schwann cells with a neuronal stimulus, raise a number of possibilities for neurotoxicological studies. Some of these possible combinations are described in Table 1. Note that exposure to neurotoxin can be either *in vivo* or *in vitro* followed by a challenge with either neurites or axolemma. In a converse manner, the neurite may be exposed to the neurotoxin *in*

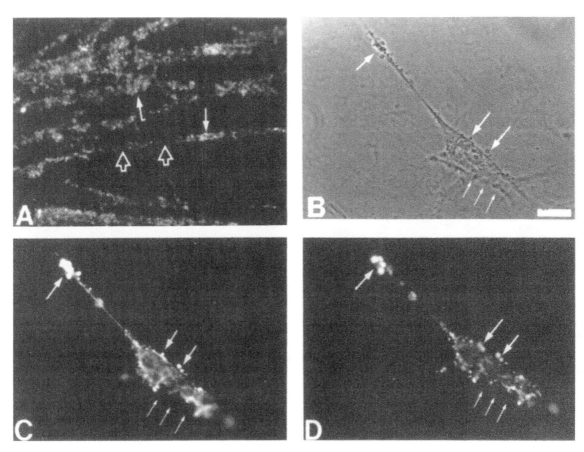

FIGURE 4 Interaction of neuritic vesicles with Schwann cell surface membrane. (A) Dorsal root ganglia neurites immunostained with anti-bFGF without prior membrane permeabilization. Aggregates of degenerating neuritic vesicles (arrows) are observed along the length of bFGF negative neuritic vesicles (open arrows) 3 days post-injury. These vesicles then associate with Schwann cells (large arrows, phase micrograph, panel B) while some of the Schwann cell plasma membrane is devoid of vesicles (small arrows). Panels C and D are of the same Schwann cell, immunostained with anti-bFGF without permeabilization and focused on different focal planes to provide perspective in three-dimensions. Note the association of bFGF immunoreactive vesicles with the Schwann cell surface membrane. Scale bar in panel B = μm. (Adaped from Neuberger and De Vries, 1993b).

vivo or *in vitro* followed by coculture with Schwann cells to evaluate appropriate metabolic activities related to proliferation and differentiation. One may also isolate axolemma-enriched fractions after exposure of the tissue source to a neurotoxic insult. The subsequent metabolic preparation of the isolated axolemma-enriched fraction will be compared to AEF isolated from neurotoxin-free tissue. Finally, the combined system, either Schwann cells plus intact neurites or Schwann cells plus axolemma, can be exposed *in vitro* followed by the evaluation of the requisite activity. Proper application to these different combinations will allow one to determine selectively which cellular element is being directly and indirectly affected by the neurotoxin.

The potential of these *in vitro* Schwann cell culture models has been used only in a very limited sense.

Mithen and colleagues (1990) have reported the effects of ethanol on rat Schwann cell proliferation and myelination using a neurite–Schwann cell coculture system which has been described earlier in this chapter. In these experiments the dorsal root ganglia were treated with mild exposure to antimitotics allowing enough Schwann cells to survive to repopulate the cultures when they are returned to serum-containing media. Exposure of such cultures for 1 month to ethanol concentrations that ranged from 43 to 172 mM resulted in compromised Schwann cell proliferation and myelin formation. This study points out not only the utility of the coculture systems, but problems that may be encountered relative to interpretation. The investigators were unable to distinguish between ethanol effects which inhibited neuronal production of neuronal mitogens for Schwann cells or ethanol effects on neuronal

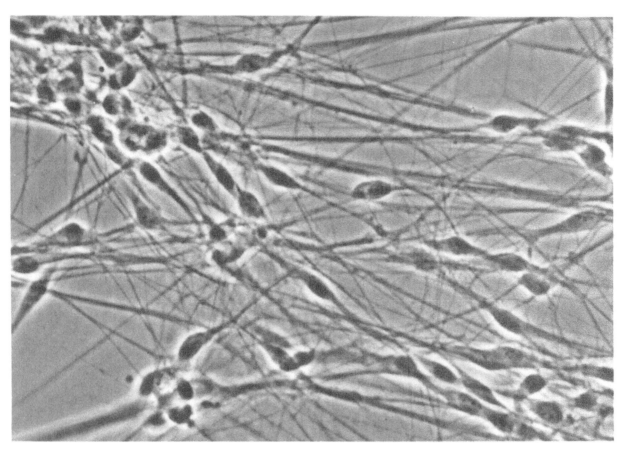

FIGURE 5 Phase microscopy of Schwann cell–dorsal root ganglia neurite cocultures. Note the frequent oval-shaped Schwann cell bodies that are entirely associated with neurites and the exact alignment of the Schwann cell processes along the neurites.

signals for myelination which impeded the ability of Schwann cells to respond to neuronal signals. Such information could have been obtained by individual studies of neurite and Schwann cell cultures alone. In an extension of this work, the laboratory of Johnson (Johnson and Covey, 1993) has studied the effect of ethanol on rat sympathetic neurons. These workers

TABLE 1 *In Vitro* Schwann Cell–Axon Combinations Useful for Neutoxicological Investigations

In vitro cell type	Exposure to neurotoxin	*In vitro* evaluation
SC[a]	*In vivo/in vitro*	SC + neurite/SC + AEF
Neurite	*In vivo/in vitro*	SC + neurite
AEF[b]	*In vivo*	SC + AEF
SC + neurite	*In vitro*	SC + neurite
SC + AEF	*In vitro*	SC + AEF

[a] Schwann cells.
[b] Axolemma-enriched fraction.

made the interesting observation that as the ethanol level was increased from 300 to 1000 mg% in these cultured neurons, axonal elongation was increased. However, ethanol concentrations greater than 300 mg% decreased the numbers of Schwann cells at the tips of axons whereas Schwann cells more toward the cell soma (500 μm) from the edge of the explant were not affected by ethanol treatment up to 300 mg%. The authors suggest that reduced Schwann cell numbers could result from direct effects of ethanol on Schwann cell mitosis, alteration of Schwann cell adhesion to axons, or an increase in Schwann cell death.

Obviously, another parameter that could be investigated in this Schwann cell culture model would be varying the time of ethanol treatment. For example, the cultures could be treated for shorter periods of time during which Schwann cells were actively dividing, then removed to see the subsequent effect on myelin initiation. Alternatively, the ethanol could have been added after extensive myelination has occurred to study its effect on demyelination.

It is also possible to pinpoint the exact metabolic target of neurotoxins using whole tissue preparations, provided that the molecules being evaluated are selectively found within the Schwann cells. For example, a study evaluating the effect of dideoxycytidine (ddC) on Schwann cell function in an *in vivo* rabbit model ddC intoxication has been recently reported (Anderson *et al.*, 1994). This study demonstrated that a Schwann cell-related mRNA was markedly reduced in animals that had been chronically treated with this anti-AIDS drug which often results in peripheral neuropathy. In turn, a decrease in this message, which coded for the major peripheral glycoprotein of myelin, led to a reduced capability of these cells to maintain an intact myelin sheath. Indeed, the splitting of the myelin sheath observed in this animal model was at the interperiod line, which is the exact localization of the P_0 protein. It is interesting to note that this neurotoxin selectively affected Schwann cells, no other cell type or tissue in the rabbit was affected. What is the basis of this selective vulnerability of Schwann cells? It is possible that these cells can selectively and rapidly metabolize ddC to the triphosphate metabolite which is required for entrance into the mitochondria. The mitochondria of these treated animals were greatly enlarged and contained paracrystaline arrays. Thus, these studies indicated that Schwann cells are selectively vulnerable, even in the absence of an *in vitro* study of cultured Schwann cells.

In summary, it should be evident that Schwann cells retain most, if not all, of their functional capabilities *in vitro*. Schwann cells can be functionally evaluated after stimulation with either intact axons (neurites) or axonal plasma membrane preparations (AEF) to proliferate and differentiate. Culture conditions can be manipulated to selectively increase or maintain cells in either a proliferative or differentiating state, allowing for the study of the effect of neurotoxins on either of these biological properties of Schwann cells. As outlined in Table 1, there are abundant possibilities for neurotoxicological investigations using appropriate combinations of Schwann cells, neurites, and axolemma-enriched fractions. Neurotoxin exposure can be either *in vitro* (to an individual cell type or combinations of cells) or *in vivo* followed by isolation of individual cells types and functional evaluations. It is anticipated that utilization of these *in vitro* interactive Schwann cell–neuronal systems will be useful in elucidating the *in vivo* targets of neurotoxins.

Acknowledgments

The studies carried out in the investigator's laboratory were supported by U.S. Public Service Health grants (NS10821, NS15408) and Neurotech, Richmond, Virginia. The author thanks Ms. Cathleen Finnerty for her expert secretarial assistance and Barbara Attema and Dr. Timothy Neuberger for the preparation of cultured cells depicted in Figs. 2–5.

References

Anderson, T. D., Davidovich, A., Feldman, D., Sprinkle, T. J., Arezzo, J., Brosan, C., Calderon, R. O., Fossom, L. H., DeVries, J. T., and DeVries, G. H. (1994). Mitochondrial schwannopathy and peripheral myelinopathy in a rabbit model of dideoxycytidine (ddC) neurotoxicity. *Lab. Invest.* **70**, 724–739.

Baichwal, R., Yan, L., Bosler, A., and DeVries, G. H. (1987). An automated method for radioautography of cultured Schwann cells. *J. Neurosci. Methods* **20**, 295–305.

Brockes, J. D., Fields, K. L., and Raff, M. J. (1979). Studies on cultured rat Schwann cells. I. Establishment of purified populations from cultures of peripheral nerve. *Brain Res.* **165**, 105–118.

Bunge, R. P. (1994). The multiple roles of Schwann cells: past, present and future. *Trans. Amer. Soc. Neurochem.* **25**(1), 330(S1).

Detsky, P., Bigbee, J. W., and DeVries, G. H. (1988). Isolation and characterization of axolemma-enriched fractions from discrete regions of bovine brain. *Neurochem. Res.* **13**, 449–454.

DeVries, G. H. (1981). Isolation of axolemma-enriched fractions from mammalian CNS. In *Research Methods in Neurochemistry* (N. Marks and R. Rodnight, Eds.), Vol. 5, pp. 3–38, Plenum Press, New York.

DeVries, G. H. (1984). The axonal plasma membrane. In *Handbook of Neurochemistry* (A. Lajtha, Ed.), pp. 341–360, Plenum, New York.

DeVries, G. H., Attema, B., Campbell, B., and Saunders, R. (1994). Isolation and partial characterization of an axolemma-enriched fraction from bovine splenic nerve. Submitted for publication.

DeVries, G., Minier, L., and Lewis, B. (1983). Further studies on the mitogenic response of cultured Schwann cells to rat CNS axolemma-enriched fractions. *Dev. Brain Res.* **9**, 87–93.

Eldridge, C. F., Bunge, M. B., Bunge, R. P., and Wood, P. M. (1987). Differentiation of axon-related Schwann cells in vitro: ascorbic acid regulates basal lamina assembly and myelin formation. *J. Cell Biol.* **105**, 1023–1034.

Fletcher, J. A., Kozakewich, H. P., Hoffer, F. A., Lage, J. M., Weidner, N., Tepper, R., Pinkus, G. S., Morton, C. C., and Corson, J. M. (1991). Diagnostic relevance of clonal cytogenetic abberations in malignant soft-tissue tumors. *N. Engl. J. Med.* **324**(7), 436–442.

Hall, S. M., and Gregson, N. A. (1974). The effects of mitomycin C on remyelination in the peripheral nervous system. *Nature* **252**, 303–305.

Johnson, I. M., and Argiro, V. (1993). Techniques in the tissue culture of rat sympathetic neurons. *Methods Enzymol.* **103**, 334–344.

Johnson, M., and Covey, R. (1993). Ethanol stimulates axonal growth but reduces Schwann cell numbers in cultures of rat sympathetic neurons. *Alcohol. Clin. Exp. Res.* **17**, 484.

Knight, R., Fossom, L., Neuberger, T., Attema, B., Tennekoon, G., Bharucha, V., and DeVries, G. H. (1993). Increased P_0 glycoprotein gene expression in primary and transformed rat Schwann cells after treatment with axolemma-enriched fractions. *J. Neurosci. Res.* **35**, 38–45.

Manthorpe, M., and Varon, S. (1989). Purification of neonatal mouse dorsal root ganglion Schwann cells. In *A Dissection and Tissue Culture Manual of the Nervous System* (A. Shahar, J. de Vellis,

A. Vernadakis, and B. Haber, Eds.), pp. 243–246, Alan R. Liss, New York.

Martin, D. P., Wallace, T. L., and Johnson, E. M. (1990). Cytosine arabinoside kills postmitotic neurons in a fashion resembling trophic factor deprivation. *J. Neurosci.* **10**, 184–194.

Mason, P., Attema, B., and DeVries, G. H. (1991). *In vitro* use of Schwann cells to elucidate neurotoxic injury. *Neurotoxicology* **12**, 459–472.

Mason, P., Attema, B., and DeVries, G. H. (1992). Isolation and characterization of neonatal Schwann cells from cryopreserved rat sciatic nerve. *J. Neurosci. Res.* **31**, 731–744.

Mithen, F., Reiker, M., and Birchem, R. (1990). Effects of ethanol on Schwann cell proliferation and myelination in culture. *In Vitro. Cell Dev. Biol.* **26**, 129–139.

Morrisey, T. K., Kleitman, N., and Bunge, R. P. (1991). Isolation and functional characterization of Schwann cells derived from adult peripheral nerve. *J. Neurosci.* **11**, 2433–2442.

Moya, F., Bunge, M. B., and Bunge, R. (1980). Schwann cells proliferate but fail to differentiate in defined medium. *Proc. Natl. Acad. Sci. U.S.A.* **77**, 6902–6906.

Mudge, A. W. (1993). New ligands for Neu? *Curr. Biol.* **3**(6), 361–364.

Neuberger, T. J., Kalini, O., Regelson, W., Kalimi, M., and DeVries, G. H. (1994). Glucocorticoids enhance the potency of Schwann cell mitogens. *J. Neurosci. Res.* **38**, 300–313.

Neuberger, T. J., and DeVries, G. H. (1992). Axonal contact as a determinant of oligodendrocyte and Schwann cell function. In *Myelin: Chemistry and Biology* (R. Martenson, Ed.), pp. 173–193, CRC Press, Boca Raton, Florida.

Neuberger, T., and DeVries, G. H. (1993a). Developmental regulation of fibroblast growth factor expression in cultured dorsal root ganglion neurons. I. *J. Neurocytol* **22**, 436–448.

Neuberger, T., and DeVries, G. H. (1993b). Distribution of fibroblast growth factor in cultured dorsal root ganglion neurons and Schwann cells. II. *J. Neurocytol.* **22**, 449–460.

Peles, E., and Yarden, Y. (1993). Neu and its ligands: from an oncogene to neural factors. *Bioessays* **15**(12), 815–824.

Rutkowski, J. L., and Tennekoon, G. (1991). An *in vitro* model for human peripheral nerve demyelination. In *Alternative Methods in Toxicology*, pp. 41–47, Liebert, New York.

Scarpini, E., Kreider, B., Lisak, R., and Pleasure, D. E. (1988). Establishment of Schwann cell cultures from adult rat peripheral nerves. *Exp. Neurol.* **102**(2), 167–176.

Sobue, G., and Pleasure, D. (1985). Adhesion of axolemma to Schwann cells: a signal and target specific process closely linked to axolemmal induction of Schwann cell mitosis. *J. Neurosci.* **5**, 379–387.

Tennekoon, G., Yoshino, J., Bigbee, J. W., Peden, K., Kishimoto, Y., Barbosa, E., Cornblough, D., DeVries, G. H., and McKhann, G. (1987). Transfection of neonatal rat Schwann cells with SV40 large T oncogene control of the metallothionein promoter. *J. Cell Biol.* **105**, 2315–2325.

Wood, P. M. (1976). Separation of functional Schwann cells and neurons from normal peripheral nerve tissue. *Brain Res.* **115**, 361–375.

Yoshino, J., Dinneen, M. P., Lewis, B. L., Meador-Woodruff, J. H., and DeVries, G. H. (1984). Differential proliferative responses of cultured Schwann cells to axolemma and myelin-enriched fractions. I. Biochemical studies. *J. Cell Biol.* **99**, 2303–2313.

Yoshino, J., and De Vries, G. H. (1993). Isolation and partial characterization of axolemma-enriched fractions from bovine and rabbit PNS. *Neurochem. Res.* **18**, 297–303.

Organotypic Cultures of Mammalian Nerve Tissues: A Model System for Neurotoxicological Investigations

MURRAY B. BORNSTEIN
Department of Neurology and Neuroscience
Albert Einstein College of Medicine
Bronx, New York 10461

This chapter is dedicated to Edith R. Peterson whose pioneer experiments first demonstrated that the specific intercellular relationships between axons and Schwann cells required to produce myelin can be reproduced *in vitro* by chick embryo dorsal root ganglia maintained in the Maximow slide assembly (1). Since that original demonstration, the technique has been extended to the mammalian central nervous system (CNS), first by Hild (2) in roller tube cultures of fragments of newborn kitten cerebellum and later by Bornstein in fragments of kitten and rodent cerebellum in the Maximow slide assembly (3). By now, fragments of virtually all levels of the neuraxis from the olfactory bulb to the neuromuscular junction have been maintained for periods of time ranging to a year or more. During their *in vitro* existence, they reproduce a relatively faithful model of the tissues as observed *in vivo*, demonstrating myelinogenesis and synaptogenesis as determined by structural, ultrastructural, functional (electrophysiological), and neurochemical characteristics. Cocultures such as dorsal root ganglia–spinal cord–striate muscle or of dorsal root ganglia–spinal cord–brain stem have demonstrated the capacity of the tissues to form specific synapses at a distance. A major advantage of the Maximow slide technique is the availability of the living tissue for constant, sequential observations and photographic recording of active or reactive events at magnifications up to 600 diameters. The entire population of cells—neurons, oligodendroglia, astroglia, and microglia—is presented. The availability of various levels of the neuraxis also offers the potential to evaluate regional vulnerability to experimental (neurotoxicological) manipulations. Finally, predetermined stages of development, reaction, or recovery can be selected by direct observation for examinations by other disciplines such as staining procedures, electron microscopy, electrophysiology, and neurochemistry.

Two previous publications have described this method (4,5). However, in view of the demanding nature of the method required to achieve differentiation, maturation, and long-term maintenance of the mature state of organization, the techniques will be presented in detail.

I. Tissue Culture Technique

The current method of maintaining central and peripheral nerve tissues is an adaptation of the one developed and originally used for dorsal root ganglia by Peterson and Murray (1) in the laboratory of cell physiology at Columbia's College of Physicians and Surgeons, New York City. The description of these techniques may appear exceedingly complex. Actually, they are quite simple when seen. Anyone interested in observing the methods is welcome to visit our laboratory.

A. Preparation and Sterilization of Glassware

Only Goldseal coverslips (Clay-Adams, a division of Becton-Dickinson Corp., Lincon Park, NJ; Cat. No. 3347) are used for the intimate glassware, i.e., 22-mm round coverslips of No. 1 thickness and 40-mm square coverslips of No. 3 thickness. The Maximow depression slide is made of plate glass, 75 × 45 mm 7–8 mm thick (Marsh Biomedical, Rochester, NY; Cat. No. S-1537). All other glassware is made of borosilicate (Pyrex, Corning, code 7740: Kimax, Kimble brand KG-33). When new, coverslips are cleaned by soaking for 24 hr in reagent-grade nitric acid at a 69–71% concentration. The Chen staining rack (Thomas Scientific; Cat. No. 8542-E40) is convenient for this procedure. Round coverslips are used only once. Square coverslips are reused until they break. To recycle the latter, wax and Vaseline are first removed by boiling repeatedly (twice is usually adequate) for 10 min in an Ivory soap solution and rinsing in running hot tap water for 10 min. They are then soaked for 24 hr in nitric acid, rinsed in cold running deionized water for 0.5 hr, and soaked in glass-distilled water overnight. Deionized water is supplied by a regenerative type, mixed-bed demineralizer (The Barnstead Co., Boston, MA; Model 3670) and a Model M3654 carbon prefilter. Glass-distilled water is obtained by passing the deionized water through a Corning AG-10 still. Only Tygon (U.S. Stoneware Co., Akron, OH) tubing and stainless-steel pipes are used to deliver deionized or distilled water.

When reused, Maximow slides are boiled in an Ivory soap solution and rinsed with hot tap water to remove Vaseline and wax before entering the usual wash procedure. Except for coverslips, which are cleaned in nitric acid as stated earlier, all glassware is cleaned by boiling in sodium metasilicate (Fisher Scientific Co., Pittsburgh, PA; Cat. No. S-408). A 20% stock solution of sodium metasilicate is prepared in deionized water and 40 ml of this is added to each 2 liter of deionized water for the boiling solution. The glassware is then rinsed once in cold deionized water and soaked overnight in 1% HCl in deionized water. It is then repeatedly rinsed with hot deionized water in a Turbomatic (Better-Built Machinery Corp., Saddle Brook, NJ; Model 3000) washing machine. After an overnight soak in glass-distilled water, it is dried by exposure to 250°C for 2 hr. Drying at this temperature also serves to burn any organic, potentially toxic, material which may have come into contact with and remained on the glass during the cleaning procedures. The material is then packaged for sterilizing. Small items are put into culture dish holders (A. H. Thomas, Cat. No. 4352). Larger pieces are wrapped in aluminum foil (1235-0-dry, fully annealed oil-free, dead-soft finish, 0.001-in. thick, Alu-Foil Products, Hauppauge, NY) and a heat-resistant paper (Steriroll, West Carolton Paper Inc., West Carolton, OH). For dry sterilization, the material is heated at 176°C for 2 hours. For wet sterilization of stoppers, solutions, etc., an American Sterilizer Co. (Erie, PA) autoclave (Model 57CR) is used; this employs the steam supplied by the institution to heat the outside chamber and to convert glass-distilled water into steam which is delivered to the interior of the chamber. The steam supplied by the institution may carry some of the chemicals used to pretreat the water as a deterrent to scale formation in the boilders. This special autoclave prevents the potentially toxic chemicals from contaminating solutions or glassware. Material is autoclaved at 120°C at 16 pounds pressure for times varying from 15 to 40 min, depending on the volume of solutions. Both dry and wet methods are controlled by Thermotubes sensitive to 166 and 121°C, respectively (Paper Thermometer Co., Greenfield, NH). Sterile supplies are stored in cabinets and drawers until used.

B. Preparation of Solutions and Medial

1. Sterility

All reagents and media are tested for sterility in thioglycollate medium and trypticase soy broth. These sterility checks are incubated, observed, and maintained for a minimum of 2 weeks and a maximum of 4 weeks to rule out the possiblity of bacterial or fungal contamination.

2. Hanks' Balanced Salt Solution

This solution (BSS) is prepared by dissolving two packages of Hanks' BSS dried powder (JRH Biosciences Inc, Lenexa, KS; Cat. No. 56-024-101) in 1990 ml of fresh glass-distilled water. It is stirred until dissolved while warmed to 37°C. A sodium bicarbonate solution (9.4 ml) of 7.5% (Grand Island Biological Co., Grand Island, NY; Cat. No. 670-5080AG) is added and the solution is gassed with CO_2 for about 4 min, using a large-mouth 10-ml pipette, until the solution turns distinctly yellow. The solution is then passed by vacuum through a Millipore glass filter (Millipore Corp., Bedford, MA; Cat. No. XX15 0447 00) fitted with a GS, 0.22-mm mean pore size filter, into a specially modified Bellco 1000-ml funnel (Bellco Glass, Inc., Vineland, NJ) which is designed to accept the Millipore filter at its top and to deliver the sterilized solution through a Teflon stopcock at the bottom. A sterility check is performed by adding 3–4 drops of filtered BSS to a tube of thioglycolate medium and tryptic soy broth (Baltimore Biological Laboratories, Baltimore, MD; Cat. No. 11716 and 11718, respectively). After sterilization, 100-ml aliquots of the BSS are dispensed into 125-ml Erlenmeyer flasks, closed with a No. 5 1/2 floating crepe rubber stopper (Thomas Scientific Supply Co., New York, NY; Cat. No. 8740-F-59) and refrigerated until use.

3. Serum

Unfiltered, sterile fetal calf serum can be purchased from various suppliers. Human placental cord serum is prepared by collecting blood from the umbilical vein at the base of the clamped umbilical cord of a placenta brought from the delivery room in a sterile bowl. The blood is slowly drawn into a 50-ml sterile, disposable syringe through a large needle (No. 18). When no more blood can be withdrawn easily (one may obtain from 10 to 60 ml from a placenta), the needle is removed and the blood is gently expressed down the wall of a 125-ml Erlenmeyer flask. Rapidly withdrawing and squirting the blood into the flask produces hemolysis. Large amounts of hemoglobin appear to be toxic to nerve tissue. The stoppered flask is refrigerated overnight. The red blood cells are removed by pipetting the serum into 15-ml centrifuge tubes and spinning at 2000 rpm in a refrigerated centrifuge (Dupont Sorvall, Inc., Newtown, CT; Model RC-2B; G2 head) for 0.5 hr. A few drops are removed for a sterility check. The serum is then withdrawn by pipette, placed in sterile 0.5 ml plastic centrifuge tubes (polypropylene) (Falcon, No. 2097), and refrigerated until use.

Eagle's minimum essential medium (MEM) with glutamine is purchased from commerical suppliers.

4. Preparation of Embryo Extract (EE 50)

Embryo extract can be bought from GIBCO BRL Lifte Technologies, Inc.

However, we prefer to make our own as it is less expensive and, we think, better for the tissues. We make a year's supply, distribute it in small amounts, and store it frozen at −90°C until use.

The method is somewhat tedious, but takes less than a working day to prepare a year's supply.

Nine-day-old embryonated PPLO-free pathogen-free eggs are delivered by SPAFAS, Inc. (Norwich, CT) on the morning of the preparation. Ten dozen eggs are ordered at a time.

The instrument kit contains flat-toothed forceps (Adson-Graefe tissue forceps 12.5 cm, Fine Science Tools (FST) Cat. No. 11030-12); large forceps to grip and remove the embryos; large curved forceps (FST Cat. No. 11001-16 16 cm) or straight (FST Cat. No. 11000-16 16 cm) for removing the shell; smaller curved forceps (FST Cat. NO. 11099-15 or 11052-10) for removing inner membranes; and a standard scalpel handle (FST Cat. No. 10003-12) for cracking the shell. These instruments are available through any major surgical instrument supplier or from Fine Science Tools, Inc. (Foster City, CA).

The procedure is as follows:

1. Eggs are placed into plastic trays, blunt side down, covered with gauze, and liberally sprinkled with 70% ethanol in distilled water.

2. Ten milliters of BSS is pipetted into dishes (crystallizing dishes, Kimble Cat. No. 23000-60 x 35 mm and 50.35 one over the other).

3. The egg is held blunt side up and wiped off with gauze wet with 70% alcohol and tapped with the scalpel handle or large forceps to crack the shell.

4. Large, curved forceps are used to remove pieces of the cracked shell over the air space, revealing the membrane. If shell debris has fallen onto the membrane, the egg is tipped so that the debris falls out before pealing off the membrane with the fine forceps.

5. Flat-toothed forceps are used to grab the embryo by the neck, removing the entire embryo, including eyes and head, and placed into a crystallizing dish containing 10 ml BSS. When the dish is filled with embryos, withdraw BSS and rinse with two changes of BSS.

6. Transfer embryos one by one to a sterile homogenizer jar (Virtis 45 Macro Cat. No. 6303-002, Gardiner, NY). Add 10 ml BSS, insert blades, and attach to homogenizer. Homogenize at slow speed, to avoid foaming, until homogenate is fairly smooth (5–10 min).

7. Measure the volume of embryo pulp with a pipette and transfer to a 150-ml Corex centrifuge flask with a Neoprene stopper. Add an equal volume of BSS minus the 20 ml to correct for the volume in step 6, yielding a 50% EE mixture.

8. Allow to sit at room temperature for 1 hr. Centrifuge at 6000 rpm for 30 min.

9. Draw off supernate and dispense 4 ml into small

test tubes. Test sterility. Store at −90°C. Thaw and centrifuge at times of preparing nutrient medium.

5. Nutrient Solutions

The nutrient solutions consist essentially of serum–saline mixtures with various additives. For the most part, they have been empirically determined and vary from tissue to tissue and, sometimes, from investigator to investigator. Antibiotics are not used. Those currently employed in our laboratory are:

1. For spinal cord, brain stem, muscle, cord-ganglion, and cord muscle combinations: 25% fetal bovine serum heat inactivated at 56°C for 0.5 hr; 10% 9-day chick embryo extract (EE-50); 50% Eagle's MEM plus L-glutamine; 7% BSS; the balance is made up with 600 mg% glucose; approximately 1.3 μg//ml achromycin (Lederle Laboratories, Pearl River, NY,) NDC 0005-4771-96; sterile tetracycline: 500 mg HCl, 1250 mg ascorbic acid).

Cultured dorsal root ganglia (DRG) serve to examine the effects of various exogenous and endogenous materials suspected of being involved in peripheral neuropathies. One may choose to explant DRG–spinal cord combinations. The inclusion of nerve growth factor (NGF) into the nutrient solution, at the time of explantation, will significantly increase the number of DRG neurons which survive for the life of the culture. The concentration of NGF ranges from 10 to 3000 unit/ml. Currently, the latter amount is routinely used. The normal nutrient solution, as noted earlier, is used for all future feedings.

2. For cerebellum: 40% fetal calf serum heat inactivated at 56°C for 0.5 hr; 25% Eagle's MEM with glutamine; 35% BSS; 600 mg% glucose; 0.1 units/ml of low zinc, glucagon-free insulin, supplied through the courtesy of the Squibb Institute for Medical Research (New Brunswick, NJ), approximately 1.3 μg/ml achromycin.

3. For cerebrum: The nutrient medium is the same as that for cerebellum except that human placental cord serum is used in place of fetal calf serum.

Recently, HEPES [4-(2-hydroxyethyl)-1-piperazineethane sulfonic acid, Boehringer-Mannheim Biochemicals, Indianapolis, IN: Cat. No. 223778] has been added to all nutrient solutions at a concentration of $10^{-2}M$ for its buffering ability.

6. Collagen-Coated Coverslips

The collagen solution is prepared from tendons obtained from the tails of 400- to 500-g Sprague–Dawley male rats. The tail is wrapped in gauze and soaked with 70% ethyl alcohol for 15 min. By successive fracture from the tip to the root, the tendons are pulled out, cut loose, and collected in a small amount of sterile distilled water in a petri dish. They are then transferred with forceps to a 150-ml Corex (Corning; Cat. No. 1265) bottle containing 1 : 1000 acetic acid solution. The volume of acetic acid solution is adjusted to 75–100 ml, depending on the total mass of tendon. The bottle is stoppered and stored at 4°C for 24 hr. It is then centrifuged at 6000–7000 rpm at 4°C for 3 hr. If the tendons have not separated from the supernate, more acetic acid solution is added and the contents are shaken. The mixture is spun again and the supernate is removed. The supernate is stored at refrigerator temperature (4°C) until dialysis. A dialysis setup is prepared by tying about 6 in. of 3.3-cm width dialyzer tubing (Fisher Scientific; Cat. No. 8-667-D) on to a glass tube which is then held in the center hole of a stopper of a 1-liter flask containing 700 ml of fresh glass-distilled water. The top is wrapped in steriroll paper and the unit is autoclaved for 30 min. After it has cooled, about 7 ml of the collagen–acetic acid solution is placed into the uncovered tube and allowed to run down into the bag of dialyzer tubing. The solution is dialyzed for 24 hr at 4°C and then removed by tearing the top of the bag and pipetting the thickened solution into tubes. If it has not reached an adequate viscosity, it is dialyzed again. The dialyzed collagen can again be stored at 4°C. To prepare collagen-coated coverslips, 1–2 drops of the dialyzed collagen solution are placed on a round coverslip and spread with a glass rod. It is then gelled by exposure to ammonia fumes for 2 min. The coverslips are repeatedly washed in sterile distilled water until free of ammonia, as indicated by phenol red in the water. They are then stored, seven coverslips to a Columbia staining jar (Thomas Scientific; Cat. No. 8542-C12), containing about 7 ml of BSS with glucose added to a final concentration of 600 mg% and 3 drops of fetal calf serum, which promotes the wettability of the collagen gel surface. They are stored at room temperature until use.

7. Sterile Room

Dissection and maintenance of tissue are performed in sterile rooms in which personnel wear cap, mask, and gown. The room is supplied with temperature-controlled air that has passed through an "absolute filter" calculated to be 96% effective in removing particulate matter, followed by a Precipitron electrostatic filter. The air is delivered into the rooms under positive pressure relative to the surrounding laboratory and at a rate sufficient to effect six exchanges an hour. The sterile room walls are coated with epoxy paint. Each morning the rooms are stocked with sterile supplies and all surfaces are wiped down with 80% ethyl alcohol. When not in use, the room is exposed to UV irradiation

(Model No. ST 2830, Hanovia Lamp Division of Englehard Industries, Inc., Newark NJ).

8. Preparation of Cultures

Most levels of the mammalian neuraxis, from the cerebral neocortex to the neuromuscular junction, may be cultured. There is an optimal time, empirically determined, for explantation of the various fragments.

The following list indicates the ages of tissue as usually employed for explantation in this laboratory:

Mouse cerebral cortex: 18-day embryo to 5 day postnatal
Mouse and rat cerebellum: newborn
Mouse and rat brain stem: 13- to 15-day embryo
Mouse and rat spinal cord: 13- to 15-day embryo
Mouse and rat dorsal root ganglion: 15- to 18-day embryo
Mouse and rat cord–ganglion combinations: 13- to 14-day embryo
Mouse and rat cord–muscle: 13- to 14-day embryo

Dissections are performed under sterile conditions. Newborn and older animals are etherized and soaked in 80% alcohol for 10 to 15 min. The desired tissue, e.g., cortex, cerebellum, is then removed *en bloc* and transferred to a petri dish containing BSS. To obtain embryos, the pregnant animal is anesthetized and soaked in 80% ethyl alcohol for 15 min. Sterile instruments are used to reflect the skin, open the peritoneum, and remove the uterus with its contained embryos, which is placed into a sterile petri dish. All further dissections are performed under microscopic control. The usual fine forceps, scissors, etc., are used for the gross dissections. For the finer dissections, jeweler's forceps (Dumont No. 5 stainless steel), iridectomy scissors, and new No. 11 scalpel blades are used. Care is taken to avoid traumatizing the tissue fragments. In preparing fragments, it is important to remember that nutrients and oxygen are supplied to the culture by diffusion. Therefore, one dimension of the fragment should not exceed 0.5 mm. The prepared tissue fragments are usually kept in a small volume (1–2 ml) of their nutrient solution in a Stender dish (VWR Scientific Corp.; Cat. No. 25467-02137 mm diameter, height 25 mm) until the dissection has been completed. To set up the Maximow slide assembly, 40-mm square coverslips are picked up with coverglass forceps (Fine Science Tools; Cat. No. 11073-10, or stainless steel, if available) and put down on a background piece of black filter paper (Thomas Scientific; Cat. No. 8613-1100, 11-cm diameter) in a petri dish (150 × 20 mm). The collagen-coated round coverslip is removed from its storage fluid by means of the same forceps, drained of excess fluid, and centered face up on the square

coverslip. With a Pasteur pipette (Bellco Glass, Inc.; Cat. No. 1271-S0021), a fragment or two of the prepared tissue is placed on the coverslip with a single drop of nutrient medium. The Maximow slide is touched with sterile (autoclaved) Vaseline at the four corners surrounding the depression and pressed onto the square coverslip which adheres and covers the depression. The assembly is lifted out and sealed with a paraffin–Vaseline mixture. Before sealing, newly explanted cultures are briefly, up to 3–5 min, exposed to an atmosphere of 5% CO_2–95% air. The sealed slides are placed in the lying-drop position into a rack that holds eight slides. The cultures are maintained in the lying-drop position and incubated at 34–35°C. (The usual 37°C incubation temperature is not well tolerated by nerve tissue.)

9. Maintenance of Cultures

Since the cultures are sealed into the Maximow slide assembly, there is no need to gas or humidify the incubator. The cultures may be removed from the incubator daily for observation with the light microscope. These periods at room temperature may be extended to at least an hour with no apparent harmful effect on the development of the tissue. Microscope objectives with long working distances are required to penetrate the two coverslips, the collagen gel, and the depth of the tissue without having to change any of the routine procedures.

Twice a week, the cultures are fed a fresh drop of nutrient medium. This involves removing the wax seal, which is performed outside the sterile room. In the sterile room, the square coverslip is lifted from the Maximow depression slide, turned over, and laid on to a small platform and the round coverslip is lifted off the old square coverslip by means of a flame-sterilized needle and a pair of coverglass forceps (Fine Scientific Tools; Cat. No. 11073-10). (Since these are also flame-sterilized during the feeding, it would be better to have stainless steel rather than chrome-plated forceps, but these are difficult to find in the United States. Occasionally, one can find suppliers in Europe.) The cultures are either drained of their drop of old nutrient or washed by placing them for 5–15 min in a Columbia staining jar containing BSS at room temperature (Thomas Scientific; Cat. No. 8542-C12), drained, and placed on a clean, sterile square coverslip, fed a drop of fresh medium, and, finally, incorporated and sealed into the Maximow slide assembly.

The cultured fragments have been maintained in this manner for periods ranging up to a year or more. At any time, the environment of the culture can be manipulated by either adding or withdrawing substances from the nutrient medium.

II. Neurotoxicological Studies

A number of neurotoxicological studies in organotypic cultures were peviously described by Yonezawa and colleagues (5). These included antimetabolities, heavy metals, thallium, β-bungarotoxin, alioquinel, chloraquine, and immunological factors related to multiple sclerosis and its laboratory animal counterpart, experimental allergic encephalomyelitis. Spencer and Veronesi and their co-workers (6.7) have continued to contribute important milestones to the program of *in vitro* neurotoxicological studies as examined in organotypic cultures.

Other investigators have also taken up this particular line of investigation. Whetsell and Seil and their collaborators (8–11) have followed many aspects of neurotoxicological effects in organotypic cultures that warrant study but are too numerous to detail here. Their accomplishments include kainic acid and quinolinic acid (8–11). Seil and his colleagues have also devoted considerable attention to ethanol, kainic acid (12–16), vincristine, and vinblastine (12–16).

Veronesi (17) has examined and presented an extensive review of cell culture models as experimental to address environmental neurotoxicity.

The interplay between *in vitro* and *in vivo* techniques has been mutually rewarding. The further extension of the interdisciplinary approaches into clinically oriented areas demonstrates the broadening influence of the organotypic model.

As previously mentioned, the clinical extensions of the laboratory data can be found in studies of the effects of immunological factors obtained from laboratory animals responding to challenge with whole CNS tissue in Freund's adjuvant, called experimental allergic encephalomyelitis (EAE), and from patients diagnosed as having multiple sclerosis (MS). For example, the cultures first demonstrated the capacity of mammalian CNS tissues to remyelinate following a total demyelinating experience induced by immunological factors (18). Since that first observation, CNS remyelination has been reported in many EAE-affected animals and MS patients. The similar patterns of demyelination produced by these factors as observed in organotypic cultures of rodent CNS tissues (18) led to Phase I and II clinical trials of a synthetic polypeptide in MS patients (19,20) and to the current extension to a Phase III trial.

The potential value of data derived from organotypic cultures is exemplified by a recent extension of some observation of Crain and Peterson. She and Masurovsky first observed that Taxol produced a specific toxic effect on cultures of rodent dorsal root and spinal cord (21). Peterson and Crain (22,23) then reported that nerve growth factor attenuated the neurotoxic effects of Taxol on the cultures tissues, Apfel and his co-workers (24) reported that nerve growth factor prevents the Taxol and toxic neuropathy in mice. In view of this attention being given to Taxol and other chemotherapeutic agents, such as vincristine, for their antineoplastic effects and the limitations placed on these clinical applications by the resulting toxic neuropathies, considerations are now being given to extend the protective effects of various growth factors in human clinical trials. In addition, the increasing availability of and interest in a wide variety of growth factors are stimulating investigations which use organtoypic culture model systems.

Finally, another long line of investigations utilizing this culture technique is yielding data that may be of signal importance in understanding some types of opiate addiction.

In 1956, Crain (25) first published a report on membrane resting and action potentials of chick embryo DRG neurons in long-term tissue culture. The close collaboration between Crain's electrophysiological laboratory and our tissue culture facilities has continued since then (26), involving studies of both central and peripheral neurons in long-term culture. Crain and his co-workers (27) have examined the excitatory effects of chronic exposure of DRG neurons to opioids. After chronic opioid exposure of DRG–spinal cord explants, the DRG neurons not only become tolerant to the inhibitory effects of high (μM) concentrations of opioids, but also show paradoxical excitatory responses when tested by further elevation of the opioid concentration (27). Subsequent studies on naive DRG neurons revealed that much lower (nM) concentrations of opioid agonists could elicit excitatory effects (e.g., prolongation of the action potential duration) which are generally masked by opioid inhibitory effects at higher concentrations (28). These excitatory effects are enhanced by a brief exposure of DRG neurons to GM1 ganglioside (29) as well as by chronic treatment with opioids (30), so that the cells become responsive to remarkably low opioid concentrations (fM–pM). In contrast, the cholera toxin-B subunit blocks the excitatory effects of opioids on the sensory neuron action potentials (31) by binding selectively to endogenous GM1 ganglioside in the DRG cell membrane. These studies are being pursued and may provide new methods for preventing some types of opiate addiction by treatments that block the upregulation of GM1 ganglioside that appears to occur in chronic opioid-treated neurons (30,32).

References

1. Peterson, E. R., and Murray, M. R. (1955). Myelin sheath formation in cultures of avian spinal ganglia. *Am. J. Anat* **96**, 319–356.

2. Hild, W. (1957). Myelogenesis in cultures of mammalian central nervous tissue. *Z. Zellforsch,* **46,** 71.

3. Bornstein, M. B., and Murray, M. R. (1958). Serial observations on patterns of growth, myelin formation, maintenance and degeneration in cultures of new-born rat and kitten cerebellum. *J. Biophys. Biochem. Cytol.* **4,** 499–504.

4. Bornstein, M. B. (1973). Organotypic mammalian central and peripheral nerve tissue. In *Tissue Culture: Methods and Applications* (I. F. Kruse, Jr. and M. K. Patterson, Jr. Eds.), pp. 86–97, Academic Press, New York and London.

5. Yonezawa, T., Bornstein, M. B., and Peterson, E. R. (1980). Organotypic cultures of nerve tissue as a model system for neurotoxicity investigation and screening. In *Experimental and Clinical Neurotoxicology* (P. S. Spencer and H. H. Schaumburg, pp. 788–802, Williams & Wilkins Press, Baltimore.

6. Ross, S. M., Seelig, M., and Spencer, P. S. (1987). Specific antagonisms of excitotoxic action of "uncommon" amino acids assayed in organotypic mouse cortical cultures. *Brain Res.* **425,** 120–127.

7. Veronesi, B., Lington, A. W., and Spencer, P. S. (1984). A tissue culture model of methyl ethyl ketone's potentiation of n-hexane neurotoxicity. *Neurotoxicology* **5,** 43–52.

8. Whetsell, W. O., Jr., Ecob-Johnston, M. S., and Nicklas, W. J. (1979). Studies of kainate-induced caudate lesions in organotypic tissue culture. *Adv. Neurol.* **23,** 645–654.

9. Whetsell, W. O., Jr., and Schwarcz, R. (1982). Mechanisms of excitotoxins examined in organotypic cultures of rat central nervous system. In *Wenner-Gren International Series.* (Fluxe, Roberts and Schwarcz Eds.), Vol. 39, pp. 207–219. MacMillan Press, New York.

10. Whetsell, W. O., Jr., and Schwarcz, R. (1983). The organotypic tissue culture model of corticostriatal system used for examining amino acid neurotoxicity and its antagonism: studies on kainic acid, quinolinic acid and (−)2-amino-7-phosphonoheptanoic acid. *J. Neural Trans. Suppl.* **19,** 53–63.

11. Whetsel, W. O., Jr., Kohler, C., and Schwarcz, R. (1988). Quinolinic acid: A glia-derived excitotoxin in the mammalian central nervous system. In *Biochem. Pathology of Astrocytes,* pp. 191–202.

12. Seil, F. J., Leiman, A. L., Herman, M. M., and Fisk, R. A. (1977). Direct effects of ethanol on central nervous system cultures: an electrophysiological and morphological study. *Exp. Neuro.* **55** (2), 390–404.

13. Seil, F. J., Blank, N. K., and Leiman, A. L. (1979). Toxic effects of kainic acid on mouse cerebellum in tissue culture. *Brain Res.* **161,** 253–265.

14. Seil, F. J., and Woodward, W. R. (1980). Kainic acid neurotoxicity in granuloprival cerebellar cultures. *Brain Res.* **197,** 285–289.

15. Seil, F. J., and Lampert, P. W. (1968). Neurofibrillary tangles induced by vincristine and vinglastine sulfate in central and peripheral neurons *in vitro. Exp. Neurol.* **21** (2), 219–230.

16 Seil, F. J., Lampert, P. W., and Klatzo, I. (1969). Neurofibrillary spheroids induced by aluminum phosphate in dorsal root ganglia neurons. *J. Neuropathol. Exp. Neurol.* **28,** 74–85.

17. Veronesi, B. (1992). Cell culture models as experimental tools to address environmental neurotoxicity. In *Handbook on in Vitro Toxicity.* (C. Nadolney, Ed.), CRC Press, Boca Raton, Florida, in press.

18. Bornstein, M. B. (1963). A tissue culture approach to demyelinative disorders. *NCI Monograph* **11,** 197–214.

19. Bornstein, M. B., Miller, A., Slagle, S., *et al.* (1987). A pilot trial of Cop 1 in exacerbating-remitting multiple sclerosis. *N. Engl. Jo. Med.* **317,** 408–414.

20. Bornstein, M. B., Miller, A., Slagle, S., *et al.* (1991). A placebo-controlled double-blind, randomized, two-center, pilot trial of Cop 1 in chronic progressive multiple sclerosis. *Neurology* **41,** 533–539.

21. Masurovsky, E. B., Peterson, E. R., Crain, S. M., and Horowitz, S. B. (1981). Microtubule arrays in taxol-treated mouse dorsal root ganglion-spinal cord cultures. *Brain Res.* **217,** 392–398.

22. Peterson, E. R., and Crain, S. M. (1982). Nerve growth factor attenuates neurotoxic effects of taxol on spinal cord-ganglion explants from fetal mice. *Science* **217,** 377–379.

23. Crain, S. M., and Peterson, E. R. (1984). Enhanced dependence of fetal mouse neurons on trophic factors after taxol exposure in organotypic cultures. In *Cellular and Molecular Biology of Neuronal Development* (I. B. Black Ed.), pp. 177–200.

24. Apfel, S. C., Lipton, R. B., Arezzo, J. C., and Kessler, J. A. (1991). Nerve growth factor prevents toxic neuropathy in mice. *Ann. Neurol.* **29,** 87–90.

25. Crain, S. M. (1956). Resting and action potentials of cultured chick embryo spinal ganglion cells. *J. Comp. Neurol.* **104,** 283–330.

26. Crain, S. M. (1976). *Neurophysiologic Studies in Tissue Culture.* Raven Press, New York.

27. Crain, S. M., Shen, K.-F., and Chalazonitis A. (1988). Opioids excite rather than inhibit sensory neurons after chronic opioid exposure of spinal cord-ganglion cultures. *Brain Res.* **455,** 99–109.

28. Crain, S. M., and Shen, K.-F (1990). Opioids can evoke direct receptor-mediated excitatory effects on sensory neurons. *Trends Pharmacol. Sci.* **11,** 77–81.

29. Shen, K.-F., Crain, S. M., and Ledeen, R. W. (1991). Brief treatment of sensory ganglion neurons with GM1 ganglioside enhances the efficacy of opioid excitatory effects on the action potential. *Brain Res.* **559,** 130–138.

30. Crain, S. M., and Shen, K.-F. (1992). After chronic exposure sensory neurons become supersensitive to the excitatory effects of opioid agonists and antagonists as occurs after acute elevation of GM1 ganglioside. *Brain Res.* **575,** 13–24.

31. Shen, K.-F., and Crain, S. M. (1990). Cholera toxin-B subunit blocks excitatory effects of opioids on sensory neuron action potentials indicating that GM1 ganglioside may regulate G_s-linked opioid receptor functions. *Brain Res.* **531,** 1–7.

32. Shen, K.-F. and Crain, S. M. (1992). Chronic selective activation of excitatory opioid receptor functions in sensory neurons results in opioid "dependence" without tolerance. *Brain Res.* in press.

Human Neuronal Cell Lines as in Vitro Models

GABRIELE V. RONNETT
Departments of Neuroscience and Neurology
Johns Hopkins University
School of Medicine
Baltimore, Maryland 21205

I. Introduction

The complexity of the central nervous system (CNS), and the heterogeneity of the cell types it contains have posed major limitations to the development of in vitro assay systems for the molecular and biochemical characterization of CNS processes. Perhaps a major drawback in the development of neuronal primary cultures or continuous cell lines has been the extreme difficulty in obtaining dividing cells in culture. Once transplanted into a culture dish, most cell types undergo a period of rapid log-phase cell division before mitosis slows and cells acquire a differentiated phenotype. Even when neurons are obtained from embryos during periods of active neurogenesis, it is difficult to maintain adequate populations of dividing cells in culture from which to establish cell lines, let alone for continuous culture. Neuronal cell lines expressing mature, differentiated phenotypes would be tremendously useful to delineate the cellular mechanisms involved in normal neuronal function. Additionally, cell cultures would be important resources for studying the effects of various agents and presumed neurotoxins on cellular function. Such neuronal cultures may obviate the need for more complicated studies involving animal testing.

Over the years, several strategies have emerged for the development of in vitro neuronal cell lines (Barlett et al., 1988; Cepko, 1988, 1989; Hammond et al., 1986; Hokfelt et al., 1980; Schubert et al., 1969, 1971; Augusti-Tocco and Sato, 1969b). Advances in molecular biological techniques have provided new approaches. Overall, methodologies can be grouped into one of two categories, either primary culture or continuous cell lines. Primary cultures are generated from cells directly taken from an organism, whereas continuous cell lines originate from transformed cells which can be passaged indefinitely. Depending on the tissue of origin, cells divide a limited number of times and may be passaged or propagated in culture. This has not been the experience with neurons or neuronal progenitors, as they do not divide in culture in high numbers, much less withstand repeated passaging.

For primary culture, cells are usually dispersed through enzymatic and/or mechanical means. As an alternative, cellular cytoarchitecture may be preserved in organotypic or explant culture by thinly slicing a particular area of the brain and placing it in culture (Walicke and Patterson, 1981b; Lumsden, 1968; Choi, 1985; Banker and Goslin, 1988, 1991; Moonen et al., 1982). Organotypic cultures have been useful for electrophysiologic studies in which a number of synaptic connections are preserved (Hendelman and Marshall, 1980; Hendelman et al., 1984; Hockberger et al., 1989; Huettner and Baughman, 1986; Wolf and Dubois-Dalcq, 1970). It is also possible to study the effects of various chemicals on neurite outgrowth and to coculture slices from different brain regions in a single dish using organotypic cultures. Although the preservation of heterogeneity may be useful for these kinds of study, biochemical analysis remains a problem in organotypic cultures. Additionally, reproducibility of protein content and cell number from slice to slice is hard to achieve.

Primary cultures of neurons are usually grown as individual dissociated cells by combining mechanical and enzymatic methods to disrupt the tissue to achieve suspensions of individual cells which may be growing in monolayer culture (Patterson, 1978; Martinou et al., 1989; O'Brien and Fischbach, 1986a,b; Huettner and Baughman, 1986; Hockberger et al., 1989). Neuronal processes are usually sheared off, but within several hours of plating neurons extend neurites, especially when neonatal tissues are used. Functional synapses may be established and cells often acquire the phenotypic characteristics of mature neurons. Coculture with nonneuronal cells may also be performed in order to study the effects of coculture on neuronal function. Primary cultures of the CNS are still heterogeneous, as the neurons in any region themselves are heterogeneous and are cultured along with other cells present in any brain such as glial and supporting cells. However, populations dissected from a specific brain region may be quite enriched in a specific class of neuron. This method has been particularly useful in generating sympathetic neurons (Higgins et al., 1991; Mains and Patterson, 1973; Walicke and Patterson, 1981a,b), which can be obtained in a fairly homogeneous form by manipulation of dissected sympathetic ganglia. The cellular heterogeneity of primary cultures may limit their usefulness. In addition, there is often a preparation-to-preparation variability in the quality and consistency of the cells obtained. However, these cell cultures have been useful for immunocytochemical and physiological studies because individual cells can be identified and studied.

The alternative to primary culture is the use of continuous cell lines. Upon repeated passage of primary cells, such as fibroblasts, a population of cells was sometimes selected that could be passaged indefinitely through spontaneous transformation (Christensen et al., 1984; Green et al., 1979; Quintanilla et al., 1986; Todaro and Green, 1963). Cell lines are usually clonal, meaning that the entire population originated from a single cell. These cell lines often have different characteristics from their normal counterparts. They tend to divide rapidly and often lose their normal chromosomal composition. The tissues that have been used to generate most cell lines since the mid-1970s originated from tumors. A priori this tissue was the best suited for development of cell lines, as these cells already possessed abnormal growth potentials. Tumors commonly used to generate neuronal cell lines were pheochromocytomas, from which originated the PC12 cell line, an extremely successful neuronal model system (Greene and Tischler, 1976, 1982; Guroff, 1985; Pfeiffer et al., 1977; Schubert, 1984; Schubert et al., 1969, 1971). These cells can be induced to stop dividing and differentiate upon exposure to nerve growth factor (NGF) (Anderson and Axel, 1986; Aloe and Levi-Montalcini, 1979; Doupe et al., 1985). Similarly, neuroblastoma cells have been used to generate cell lines (Augusti-Tocco and Sato, 1969b; Schubert et al., 1969, 1971). The most successful species used for neuroblastoma cell lines have been rat and human. Although cell lines produced in this manner have been extraordinarily useful, these cell lines are often not capable of expressing aspects of a mature or normal phenotype. Additionally, as their developmental program has been altered, the differentiation in these cells may not recapitulate the normal series of events.

More recently, several new techniques have emerged which may be used to generate cell lines. A particularly successful strategy has utilized oncogene-containing retroviruses (Barlett et al., 1988; Cepko, 1988, 1989). This requires a relatively high density of dividing cells. Viral oncogenes, which are able to transform cells and thereby immortalize them, are infected into recipient cells and during the process of DNA replication become inserted into the genome and are therefore stably transmitted to progeny. A number of cell lines have been developed, although it is often difficult to achieve a mature phenotype once cells have been transformed (Bartlett et al., 1988; Frederiksen et al., 1988; Lendahl and McKay, 1990; Mellon et al., 1990).

Another method that has been used to generate neuronal cell lines is somatic cell hybrids (Greene et al., 1975; Hammond et al., 1986; Lee et al., 1990a,b; Platika et al., 1985). A continuous cell line may be fused with a desired cell, such as a specific neuronal cell, and this may generate somatic cell fusions which are capable of continuous propagation. A problem with some somatic cell hybrids is that the chromosome content is unstable and cells may subsequently shed chromosomes, altering their phenotype.

Although tumors have been successfully used to generate cell lines, a number of other neurologic abnormalities exist which may possess low-level growth abnormalities, making them candidates for generating cell lines (Dambska *et al.*, 1984; Bignami *et al.*, 1968). Described for some time, there has been a recent resurgence in hemispherectomy as a treatment for intractable seizures (Goodman, 1986). As a result, brain tissue is obtained which may be used to establish human neuronal cultures (Ronnett *et al.*, 1990). Human cerebral cortical tissue has been utilized from a patient with unilateral megalencephaly, a low-grade proliferation, and migration disorder of neurons, as well as tissue from a patient with Rasmussen's encephalitis, to establish human cortical neuronal cell lines. In addition, similar methodologies have been utilized to establish several continuous cell lines from a human esthesioneuroblastoma, a tumor of olfactory receptor neuronal origin (L. D. Hester and G. V. Ronnett, manuscript in preparation). This chapter describes the general development of a culture strategy and the various parameters which may be tried in order to establish optimal cell growth and differentiation.

II. Selection of Tissue

Human cerebral cortical tissue has thus far been obtained from two patients undergoing hemispherectomy for intractable seizures. In the first cell line characterized, the patient had unilateral megalencephaly, a low-grade proliferation and migration disorder of neurons. Histologically, one sees microcalcifications, gliosis and heterotopia, ectopic foci of neurons that lack normal cytoarchitecture and/or are located in white matter (Ronnett *et al.*, 1990). Giant abnormal neurons are also seen. Despite this abnormal cytoarchitecture and low-grade atypical changes in the neurons, these cells still stained appropriately for neuronal markers such as neurofilament (NF) and neuron-specific enolase (NSE) and stained negatively for non-neuronal markers such as glial fibrillary acidic protein (GFAP) and S-100 protein (Ronnett *et al.*, 1990). A second patient diagnosed with Rasmussen's encephalitis presented at age 7 years for evaluation for intractable seizures and had similar surgery performed. In this case, histological examination revealed disorganization of the gray matter with microcalcifications and areas of heterotopia (G. V. Ronnett, L. D. Hester, J. Nye, and S. H. Snyder, manuscript in preparation).

The common feature in both these cases was the presence of heterotopia, which represented abnormally located collections of neurons. It may be that these neurons were not normally matured and therefore were able to be propagated in culture. Similar results may be obtained with other human cortical tissue. Tissue was also obtained and processed in a similar manner from two patients with esthesioneuroblastoma (L. D. Hester and G. V. Ronnett, manuscript in progress). Although malignant in nature, and perhaps not resulting in phenotypically as normal a cell line, it was of interest to try these methods for establishing cell lines.

Historically, cultures of neuronal or glial cells are prepared from embryonic or early postnatal animals. The exact embryonic or postnatal age used is determined by the phenotype of the cells desired in culture. Each type of neuron or glial cell undergoes neurogenesis and differentiation at a specific embryonic or postnatal time. Thus, cultures enriched in a certain type of cell can be obtained by selecting animals at that age. It has been the general experience that the cells of young animals are easiest to maintain in culture and have the highest capacity to sprout neurites. These factors are less crucial when starting with neoplastic tissues. In any case, the variables that can be controlled when attempting to culture human tissue are considerably less. In the two human cortical neuronal lines thus far established, the patients were 11 months and 7 years of age. Although hemispherectomy for intractable seizures is often performed at much later ages, cultures were attempted from younger patients for the aforementioned reasons.

III. Preparation of Cells for Plating

Establishing cell lines from human material may impose certain restrictions, depending on the individual institution. Guidelines are in the process of changing for the handling of human materials. In the case of material removed at surgery, informed consent may need to be obtained from the patient or family. In some cases, tissue is used for culture that is left over from surgical pathological analysis and may be considered "discarded." In this case, consent may not need to be obtained. Because of the potential risk of infection, handling human tissue usually requires a containment facility with a P-2 level rating or higher. This provides a negative-pressure room which minimizes any potential spread of infectious or harmful agents. It is always advisable to wear sterile disposable gloves when handling human tissue. Arrangements should be made ahead of time to be present at the time of biopsy or surgery so that the tissue can be immediately received after it has been processed and dissected. Tissue can often be stored for several days, refrigerated, and still used for cell culture. Given the difficulty in culturing neurons, it is advisable to take tissue as soon as possible and place it in a sterile balanced salt solution (BSS) with controlled pH for transport to a culture facility. All cultures of human neuronal cells thus far obtained have been handled in this manner.

Once the fragments or blocks of tissue have been received, material must be placed in sterile plastic culture dishes containing an osmotically balanced salt solution buffered to a physiologic pH. BSS are utilized for the short-term (hours or less) maintenance of cells. A buffering system containing bicarbonate, which requires equilibration with an atmosphere of 5% carbon monoxide to maintain a physiologic pH, is most often used. This is not optimal initially, when cellular material is exposed to ambient atmosphere. The pH rapidly changes and becomes quite basic under these conditions. To maintain physiologic pH, the buffering system of choice is 4-(2-hydroxyethyl)-1-piperazineethanesulfonic acid (HEPES). Minimal essential medium (MEM) is made, containing high glucose (4500 g/liter) and 4.8 g/liter HEPES, and the pH is adjusted to 7.3. This medium is designated MEM-AIR.

Single cells may be obtained in culture by placing tissue fragments onto a substrate and allowing individual cells to migrate out or by disaggregating the tissue physically or enzymatically to yield a cell suspension which can be directly plated onto substrate. Tissue fragments of 1 mm or less can be attached to collagen or gelatin. Over days to weeks, cells will migrate out and begin to propagate. Alternatively, a mixture of a number of enzymes can be used to disaggregate cells, and usually includes dispase, collagenase, hyaluronidase, pronase, or elastase. DNase is often useful, as any cells which have already fragmented and dispersed their nuclear contents will release DNA, which can damage cells. Often enzymes are mixed in a medium and kept cold to slow enzyme activity. Cellular disaggregation is also favored in the presence of low concentrations of divalent cations because many cell-to-cell contacts require the presence of calcium and magnesium.

A major problem in working with neuronal cells is their relatively fragile nature. For this reason, no enzymatic digestion procedures are used when culturing from human cortical cells. Discarded tissue is grossly dissected into gray and white matter, and the gray matter is placed in MEM containing D-valine (MDV, GIBCO, NY) and 15% (w/v) dialyzed fetal calf serum (dFCS, GIBCO). MDV, which contains the amino acid D-valine instead of L-valine, is especially useful for primary neuronal culture, as only cells of ectodermal

origin contain amino acid isomerase, which allows the D isomers of amino acids to be converted to the L isomers which are suitable for incorporation into proteins (Gilbert and Migeon, 1975, 1980; Oster-Granite and Herndon, 1978). Thus, fibroblasts, which are the principal contaminants in primary culture and which rapidly overgrow other cells in culture, are essentially eliminated as they cannot survive in MDV. Tissue is finely minced and pushed through a 150-μm mesh wire screen. These cell suspensions are then distributed among a number of 35-mm culture wells (Nunc, Naperville, IL) at a density of approximately 1×10^6 cells per cm^2 in 3 ml of culture medium and placed in a 7% CO$_2$ humidified incubator at 37°C. The medium in this case is slightly acidic, pH 7.2–7.3, as this favors neuronal growth. The medium used is MDV containing 15% dFCS. Antibiotics are not used at this stage for two reasons. Both penicillin and streptomycin may be neurotoxic at the concentrations used in culture, and if there is contamination present, it is desirable to be aware of it as soon as possible so that contaminated cultures may be removed before infection spreads to other wells. The choice of culture medium is discussed more thoroughly in the subsequent section.

In each well only a small percentage of cells may attach. Of the 3 ml of medium initially plated into each well, 1 ml is aspirated and replaced with fresh MDV containing 15% dFCS every third day. The timing of the next step depends on the ability of any of the cells to adapt to culture. For the human cortical neuronal cultures, nearly 3 weeks were required before cells that had survived in culture began to divide. After 3 weeks, nearly all the cells had died except for four small foci of growth (Ronnett *et al.*, 1990). In the case of the esthesioneuroblastomas, cells were also allowed to establish themselves in culture over 3 weeks. As this tissue initially demonstrated a more malignant phenotype, by 3 weeks large clones were noted which had already begun to propagate *in vitro*. Altogether, 46 clones of esthesioneuroblastoma were isolated. In general, at least 3 weeks should be allowed for cells to establish themselves in culture. At the end of this time, any clusters of cells can be isolated by cloning rings, replated, and allowed to propagate in MDV containing 15% dFCS. Cells are detached from substrate and passaged using 0.05% (w/v) trypsin in Hank's balanced salt solution (GIBCO) and split at a ratio of no greater than 1:4. It is extremely crucial not to stress any cells by low density plating. Initially, cells are plated at 25% or greater of confluence. At this point, subcloning may be performed by isolating individual cells or clones and placing them in a 96-well cluster dish in MDV containing 15% dFCS. These cell lines are frozen in dFCS containing 10% (w/v) dimethyl sulfoxide

(DMSO) and are reconstituted by thawing and plating in MDV containing 15% dFCS.

IV. Culture Media

Cells maintained in monolayer culture require a growth medium supplemented with serum, vitamins, hormones, and additional growth factors. In some cases, cells can be cultured in serum-free medium containing more vigorous supplementation and a more characterized protein such as bovine serum albumin (BSA). Standard formulations are available from most commercial suppliers. Defining the correct medium formulation for a certain cell type is often labor intensive, requiring a commitment of resources to sequentially analyze the growth requirements of that cell type. Conditions that permit serum-free culture may be even more tedious to determine, but serum-free culture is clearly optimal as this constitutes the most well-defined culture situation, permitting clear analysis of the effect of each reagent on cell viability and differentiation. On the other hand, each lot of serum varies considerably in its content, making culture with serum less reproducible.

A large number of defined media are commercially available, which vary in complexity from simple formulation such as Eagle's MEM containing essential amino acids, salts, and vitamins (Eagler, 1959) to rich media such as F12 (Ham, 1965), which contain larger numbers of amino acids, vitamins, minerals, and additional metabolites such as nucleosides. Serum-free culture conditions often combine to defined media, such as the mixture used for culture of sympathetic neurons, which combined Dulbecco's modified Eagle's MEM (D-MEM) (Barnes and Sato, 1980; Dulbecco and Freeman, 1959). An initial consideration is pH. Typically, cell culture is done at pH 7.4. As neurons are relatively fragile cells and perhaps prone to oxidative stress, culture from human cortical tissue has been performed, maintaining cultures at a slightly more acidic pH, generally 7.2–7.3. This is accomplished by preparing powdered medium with 2.2. g/liter of sodium bicarbonate and maintaining cells in a 7% CO_2-containing environment. The main energy source provided by defined medium is glucose. Media are generally provided containing high glucose (4500 g/liter) or low glucose (1000 g/liter). Neuronal culture usually utilizes basal media containing high glucose.

The amino acid and vitamin composition of basal media vary significantly. Usually, only essential amino acids are required, although many cell types require high glutamine concentrations. A specific problem encountered when culturing neuronal cells is that certain amino acids, namely glutamate, may act as excitotoxins, resulting in decreased viability of cultured cells. It may therefore be prudent to leave glutamate out of the medium. The basal media chosen for the culture of human cortical neuronal cell lines is a modification of D-MEM, which in addition to containing the D isomer of valine instead of the L isomer, contains no glutamate. This may explain some of the success in culturing neurons in this medium. Most medium preparations contain B vitamins, although other vitamins may need to be added, especially when serum-free culture or culture in reduced serum is done.

The source and concentration of serum used to supplement the basal medium varies significantly from cell line to cell line. The final concentration ranges from 5 to 20%, although neuronal cell cultures usually prefer a concentration on the high end. The most common sources of serum are horse and fetal calf serum. Serum requirements are probably quite different during the establishment and initial propagation of a cell line when compared to passage of an established cell line. The extreme lot-to-lot variability in serum often necessitates that different lots of serum be tested for their abilities to support growth of a specific cell line. In any case, it is advisable to heat inactivate serum prior to use, which involves heating serum in a water bath to a temperature of 56°C for 30 min, to destroy serum complement which may damage cells in culture. When MDV is used, it is necessary to supplement this with dFCS. The dialysis is performed using a membrane that allows passage of proteins of molecular weight 10,000–12,000. This effectively removes any endogenous L-valine, which may interfere with the action of MDV.

The use of MDV plus dFCS appears crucial for the establishment of cell lines, as mesenchymal elements are essentially eliminated from the culture using this medium. Although supplementation was not done for the human cortical neuronal lines thus far maintained, it may be necessary in other cases to supplement the media with amino acids or vitamins. A number of the components have been combined and are described for neuronal (Augusti-Tocco and Sato, 1969a; Liebermann and Sachs, 1978; Littauer *et al.,* 1979) or glial culture (Bornstein and Murray, 1958; Freshney, 1980; Ponten and Macintyre, 1968). Components of serum-free media developed by Sato and collaborators (Bottenstein, 1985; Bottenstein and Sato, 1979, 1985; Bottenstein *et al.,* 1980) may be added to MDV, such as insulin, transferrin, triiodthyroxine, 7-hydroxyprogesterone, corticosteroid, 17-β-estradiol, putrescine, and sodium selenite. After cultures are established, different media may be tried, depending on experimental requirements. A standard media used for HCN-1 and HCN-2 is D-

MEM (GIBCO) with high glucose containing 10% heat-inactivated FBS (Hyclone, Logan, Utah). Powdered medium is made with 2.2 g/liter of sodium bicarbonate, pH 7.3, and sterile filtered through a 0.2-μm filter. Media can be stored refrigerated, without serum added for up to a month. Cells are maintained in a 7% CO_2 incubator.

Whatever the initial culture conditions are for establishing neuronal cultures, it is desirable to determine what, if any, serum-free culture conditions may be utilized for a specific cell type. Cellular requirements for a number of nutrients and hormones were initially determined to support growth of neuroblastoma cell lines, although these media have proven useful for a wide variety of neuronal cells (Bottenstein, 1985; Bottenstein and Sato, 1979, 1985; Bottenstein et al., 1980; Honegger, 1985; Honegger and Lenoir, 1980; Honegger et al., 1979; Snyder and Kim, 1979). The basal medium is a mixture of D-MEM and F12. This mixture is supplemented with insulin, transferrin, selenium, progesterone, putrescine (a precursor of polyamines), thyroid hormone, and trace metals. In addition, protein is required and is often provided in the form of BSA. It is often necessary to use fatty acid-free BSA, as free fatty acids may be toxic to cells. Other agents that are often added to metabolize peroxides and superoxides, which can accumulate in culture medium, are catalase and superoxide dismutase, as well as vitamins C and E (O'Donell-Tormey et al., 1987; Saez et al., 1987; Walicke et al., 1986).

V. Maintenance of Cell Cultures

Once clones or groups of cells have been identified in the initial cultures, secondary cultures can be started. Initial clusters of cells are isolated by cloning rings, as previously described. These cells can be directly plated into small wells in MDV containing 15% dFCS. Once secondary cultures grow to confluence or 85% of confluence, more routine methods for maintenance of cultures can be employed.

To passage cultures, media are removed and cells are rinsed with 5 ml of 0.05% trypsin containing 0.53 mM EDTA (GIBCO). Several milliliters of fresh trypsin–EDTA solution are added to monolayers and placed in an incubator at 37°C for several minutes or until cells detach from the growing surface upon minimal agitation. Cells should be split at a ratio of no greater than 1:3. If the subcultivation ratio is greater, cells may spontaneously differentiate. Cell growth is enhanced when the medium is replaced five times per week. If cell growth does slow down and cell morphol-

ogy begins to change, it may be possible to recover cultures by passaging them down in container size.

To freeze cultures, cells are trypsinized as previously mentioned. The cell trypsin suspension is layered on an equal volume of cold serum and centrifuged. The pellet is resuspended in cold freezing medium containing 90% FBS and 10% DMSO and aliquoted quickly into sterile cryotubes and placed in a −5°C freezer overnight, then transferred to a −20°C freezer for 24 hr and finally placed in the vapor phase of liquid nitrogen storage.

To thaw cells, vials are quickly thawed in warm water and cells are layered over FBS, centrifuged, the supernatant discarded, and the cells are resuspended in the desired medium. Viability is enhanced if cells are placed in a small 35-mm dish when first thawed. Plating at too low a density will cause cells to stop dividing. It is possible that cells may be confluent within a day after thawing; if so, they are passaged into a 100-mm dish and expanded again when confluent. Medium must be changed the day after thawing to remove residual DMSO.

VI. Culture Substrates

Most cells require an artificial substrate when grown in vitro. Some transformed lines grow in suspension culture and are therefore anchorage independent. However, most cells require a substrate that is specially treated for attachment and propagation. Glass surfaces and coverslips are often used, as they can be removed from a culture well for electrophysiologic or microscopic analysis. Plastic substrate is currently the preferred support, and a great number of configurations of culture dishes and vessels exist for cultivation of cells (Barnes et al., 1984). In general, polystyrene is designated tissue culture grade, meaning that it has been chemically treated to minimize hydrophobicity. Neuronal cells may require more complicated modifications of tissue culture grade plastic. Although neurons may attach to plastic, it is not often adequate for neurite extension (Yavin and Yavin, 1980; Barde et al., 1978; Ebendal, 1979; Ebendal and Jacobson, 1977; Lindsay, 1979). A number of different treatments of modifications may be performed and tested. Some chemicals commonly used are polyornithine (PORN) or polylysine (PL), whereas more complicated modifications use extracellular matrix components such as laminin or fibronectin. The kind of substrate used may need to be modified, depending on the species of serum, the percentage of serum, and other growth factors used in culture. In the case of the HCN lines, HCN-1 grows best on laminin, whereas HCN-2 prefers a combination of collagen IV and laminin. Coating of slides with PL was performed in the same manner as treatment with PORN. For heparan sulfate, 1 ml of D-MEM containing 10 μg/ml heparan sulfate was added to slides, incubated overnight at 37°C, and rinsed once with sterile deionized distilled water prior to use.

To determine the best substrate for growth and morphology of a particular cell line, a number of different substrates and substrate combinations may be tested. Testing may be performed in a 24-well culture dish or on two-chamber slides (Nunc, Niles, IL). The following methods describe treatment conditions for successfully coating glass two-chamber slides.

Several standard modifications of plastic or glass culture vessels involve coating the surfaces with polymers of basic amino acids such as PL or PORN. The D isoforms are often used, as these are harder for cells to metabolize and remove from the surface. These two substrates have been successfully used for the culture of neurons (Letourneau, 1975; Yavin and Yavin, 1974, 1980). It should be noted that all the solutions mentioned herein will coat glass or plastic. However, whereas incubation times are on the order of hours to adequately coat plastic, overnight incubations are required for glass. For PORN (Sigma Chemicals, St. Louis, MO), a solution at a concentration of 1 mg/ml PORN in sterile deionized distilled water was added to slides, incubated for 1 hr at 37°C, rinsed twice with sterile deionized distilled water, and rinsed once with D-MEM prior to use.

It may be necessary to use more complicated modifications of the glass or plastic substrate. The first protein modification used relied on the principal that cells may attach well in a more natural setting, utilizing components of extracellular matrix (ECM). The first ECM component used was type I collagen (Ebendal, 1976; Bornstein, 1958; Elsdale and Bard, 1972; Masurovsky and Peterson, 1973). More recently, other forms of collagen have become commercially available and may be more suitable for neurite outgrowth and neuronal culture, as they are naturally associated with neuronal structures. In particular, collagen type IV has been useful (Turner *et al.*, 1987). Quite a number of procedures are available for the use of collagen or collagen IV. Examples are given here that utilize commercially available preparations. Collagen (GIBCO) or collagen IV (Collaborative Research) was prepared as a 50-μg/ml solution in sterile deionized distilled water, and 1 ml of this solution was used per well. In the case of collagen, the solution was allowed to air dry. Collagen IV (CIV) was maintained overnight at 37°C, aspirated, and air dried prior to use.

Other components of ECM have been successfully used to cultivate a variety of neurons (Carbonetto and Cochard, 1987; Carbonetto *et al.*, 1983; Chiu *et al.*, 1986; Condic and Bentley, 1989; Manthorpe *et al.*, 1983). Two of the most useful have been fibronectin and laminin. For laminin coating, 1 ml of D-MEM containing 25 μg/ml laminin was plated onto Labtek glass two-chamber tissue culture slides (Nunc Niles, IL) overnight at 37°C. Before use, the slides were rinsed with D-MEM. For fibronectin, slides were treated with a solution containing 1 ml of D-MEM containing 20 μg/ml fibronectin (Collaborative Research, Bedford, MA), incubated overnight at 37°C, and rinsed with D-MEM prior to use. Laminin or fibronectin may be used on top of glass or plastic previously coated with PORN or PL. This may be especially useful in the case of glass, where it has been found that laminin may not adhere well directly. When laminin is used directly with collagen IV or heparan suflate, collagen IV and heparan suflate are prepared as described and laminin is subsequently added in the aforementioned manner.

Evaluations of the results of substrate testing are somewhat objective by necessity. Cells are grown on two-chamber slides so that the upper well, which usually holds the media, may be removed to that the cells can be microscopically examined. For the HCN cell lines, criteria for optimal growth included the number of cells which attached as well as their morphology. On some substrates such as PORN, HCN cells adhered poorly and did not extend processes. Substrates were also judged on their ability to support differentiation of cells, as during this process cell bodies round up and are prone to detach from substrate.

VII. Differentiation of Neuronal Cell Lines

During the development of an organism, undifferentiated pluripotent cells differentiate into mature cells possessing specialized cellular machinery necessary for its function. As differentiation proceeds, these cells lose the ability to divide through a process of commitment, and they then undergo terminal differentiation which in general implies that cells can no longer revert to a less committed phenotype. For most cell types, a population of uncommitted stem cells or basal cells is maintained, which can be induced to differentiate upon the appropriate extracellular signals to repopulate mature cells as they are lost throughout adult life. These processes are not clearly understood, although we know they are complicated and rely upon precise signaling at each step in differentiation. By nature, a cell line that retains the ability to divide in culture cannot express most of the attributes of a terminally differentiated phenotype. However, it is possible to manipulate the culture environment and induce a number of cell types, including neurons to differentiate in culture. This makes them more useful for the study of neuronal processes. A number of parameters control cellular differentiation (Schleicher et al., 1993). The most difficult to duplicate in culture are cell-to-cell interactions which occur normally in the organism. Important roles may be played by heterologous and homologous cell interactions. As previously discussed, interactions between cells and the ECM are of critical importance in neurite outgrowth.

An explosive growth in the identification and characterization of agents which promote cellular differentiation and growth factors for neuronal and nonneuronal cells has occurred (Levi-Montalcini, 1964, 1979; McLean et al., 1986; Stockdale and Toppe, 1966; Wu and Wu, 1986). Growth factors such as NGF have been useful in differentiating cells such as PC12 cells in vitro, a process during which cells often show a more mature phenotype (Levi-Montalcini, 1982). HCN-1 and HCN-2 cells demonstrate the presence of neuronal markers, but the morphology is relatively bland, without significant processes. The use of a combination of agents, such as NGF, isobutylmethylxanthine, and cyclic AMP or phorbol ester, causes these cells to stop cell division and more dramatically express neuronal markers (Ronnett et al., 1990). To test the effect of growth factors and differentiating agents on a cell line, cells are seeded at a lower density, approximately 5×10^4 cells per cm^2 in plating medium, and after at least 24 hr in culture are changed to medium containing each reagent to be tested. Cell monolayers need to be feed with medium containing each individual agent at least every 3 days, as growth factors and chemicals may be degraded over this time. In the case of ascorbic acid, medium with fresh ascorbic acid must be added daily. After initial screening, agents may be combined to achieve a maximal effect. A list of these reagents and their methods of preparation and storage are provided in Table 1.

Often within 24 hr of initial exposure, neuronal cell cultures which differentiate become refractile and extend long branched processes. Although cells undergo several rounds of division at first, doubling time should slow and is often halted. Characterization of expression for neuronal markers may then be performed.

VIII. Characterization of Neuronal Cell Lines

After isolation and subcultivation of cells that have been propagated in vitro, one must undertake the time-consuming task of characterizing these cell lines. Of primary interest is the determination of neuronal cells versus nonneuronal cells. The principal nonneuronal cells might be expected to be glia and are more likely to be derivatives of astrocytes. On the other hand, neuronal precursors may be isolated from malignant tissues, or rather pathologic conditions.

The first criterion is morphological examination of the cells. This is done by growing cells on two-chamber culture slides on the desired substrate and in the desired medium. At no greater than 85% confluence, cell monolayers should be quickly rinsed three times in 37°C phosphate-buffered saline (PBS). Slides are then fixed and any of a variety of fixatives, i.e., for simplicity, 4% paraformaldehyde (PFA) in PBS, pH 7.3, may be utilized. Cell monolayers are allowed to incubate in PFA for 20 min, after which time slides may be stored in a 10% sucrose solution for subsequent microscopic examination. It is desirable to fix cells at a subconfluent stage so that the extent of their processes may be observed. If differentiation has been attempted, it is useful to process duplicate slides that have been subjected to a differentiation protocol in a similar manner to allow direct comparisons. Monolayers are then examined by phase-contrast or Nomarski optics.

Determination of a neuronal versus glial phenotype may be difficult by mere microscopic examination. Glial cells possess processes that are often highly branched and often quite long. Thus, they may appear similar to neurons. It is often more reliable to look at nuclear morphology to tell neurons from glia, although this is difficult at a light microscopic level. For these reasons, it is imperative to characterize the phenotype of cell lines by the use of any of a number of neuronal and nonneuronal markers. Markers can be classified as any of a number of proteins that are expressed specifically by neuronal or nonneuronal cells, and include intermediate filament proteins, enzymes necessary for function of a specific cell type, receptors, or ion channels (Fields, 1979, 1985; Loffner et al., 1986; Raff et al., 1978, 1979). A wide variety of intermediate filament proteins, microtubule proteins, and proteins associated with the cytoskeleton have been characterized and many are expressed exclusively in neurons. For example, mature neuronal cells express NF, and the microtubule-associated protein designated MAP2 and Tau are in general restricted to neuronal expression (Binder et al., 1985; Burgoyne, 1991; Burgoyne and Cumming, 1984; Calvert and Anderton, 1985; Couchie et al., 1988; Nunez, 1986, 1988; Olmsted, 1986; Riederer et al., 1986; Riederer and Matus, 1985). Vimentin is usually expressed by cells of mesenchymal origin, although many cells in culture express vimentin as well as other, more differentiated intermediate filaments. Therefore, the presence of vimentin is not generally viewed as diagnostic of tissue of origin. Glial cells usu-

TABLE 1   Preparation of Commonly Used Cell Culture Reagents

Reagent	Stock solution	Stability	Working dilution (fold dilution from stock)
Insulin	1 mg/ml in DW[a], pH 2.5	3 months sterile at 4°C	10 μg/ml (1 : 100)
Nerve growth factor	10 μg/ml in medium with serum	2 weeks at -20°C	25 ng/ml (1 : 400)
Dexamethasone	0.39 mg/ml in ethanol	Indefinite at 4°C	1 μM (1 : 1000)
Isobutylmethylxanthine	11.5 mg/ml in DW, add 1 N NaOH dropwise until solution is clear	Make fresh	0.5 mM (1 : 100)
Retimoic acid	3.2 mg/ml in DMSO[b]	2 months at -20°C, light sensitive	$1 \times 10^{-5}M$ (1 : 1000)
Dibutyryl cAMP	30 mg/ml in DW	Make fresh	0.5 mM (1 : 100)
TPA[c]	1.2 mg/ml in DMSO	2 months at -20°C, light sensitive	$2 \times 10^{-8}M$ (1 : 10,000)
Ascorbic acid	3.3 mg/10 ml in DW	Make fresh, keep on ice	30 μM (1 : 50)

[a] Glass-distilled water.
[b] Dimethyl sulfoxide.
[c] Phorbol 12-myristate 13-acetate.

ally express GFAP or S-100 protein. With passage in culture, the amounts of these proteins expressed may diminish, but may be reinduced with differentiation.

Neuronal cells express a variety of enzymes necessary for their specialized functions. The most ubiquitous of these is NSE (Marangos and Schmechel, 1987). Many specialized enzymes are required for neurotransmitter synthesis and may be used to determine the specific type of neuron present. In a similar manner, cells may be stained for the presence of neurotransmitters.

Rapid screening of multiple cell lines may be performed by immunocytochemistry using commercially available antibodies to neuronal and nonneuronal markers. A large number of these can be purchased from a number of vendors; some of their working dilutions are listed here.

Immunocytochemistry may be performed as previously described (Ronnett *et al.*, 1991) with modifications. Cells are plated at a density of 1×10^4 cells per cm^2 on two-chamber Lab-Tek tissue culture slides treated with MDV containing 25 μg/ml laminin. Cells are then treated according to individual protocol. Cells may be fixed in several ways, depending on the primary antibody used. All slides are quickly rinsed three times with PBS at 37°C. Neuronal and nonneuronal markers to be used include neurofilament (NF, Labsystems, Helsinki, Finland), vimentin (Boehringer-Mannheim), SM133 (Sternberger-Meyer, Inc., Jarrettsville, MD), neuron-specific enolase (Incstar, Stillwater, MN), tubulin (available from several vendors), neuron-specific tubulin (Sigma Chemicals), S-100 protein (Incstar), myelin basic protein (MBP, Incstar), and glial fibrillary acidic protein (Incstar). Fixation for staining with these antibodies is done by a 20-min incubation in PBS, pH 7.3, containing 4% (w/v) paraformaldehyde at 37°C.

Slides stained for phenylethylamine-*N*-methyl transferase (PNMT, Incstar), tyrosine hydroxylase (TH, a gift of Dr. Reinhard Grzanna, The Johns Hopkins University School of Medicine), dopamine-β-hydroxylase (a gift from Dr. Reinhard Grzanna, The Johns Hopkins University School of Medicine), serotonin (Incstar), vasoactive intestinal polypeptide (VIP, Instar), somatostatin (Incstar), and cholecystokinin-8 (CCK-8, Incstar) are fixed by incubation in PBS, pH 6, containing 4% PFA for 7 min, followed by PBS, pH 11, containing 4% PFA for 7 min at 37°C. Slides stained with antibody to γ-aminobutyric acid (GABA, Incstar) are fixed with PBS, pH 7.3, containing 4% PFA and 0.1% (w/v) glutaraldehyde for 15 min at 37°C. Fixation for staining for glutamate is done by a 5-min incubation with PBS containing 5% (w/v) carbodiimide (Sigma) at 37°C followed by a 1-hr incubation in 5% (w/v) glutaraldehyde at 37°C.

Antiserum may be used at the following dilutions: monoclonal anti-NF antibodies, 1 : 75; SMI 32, 1 : 1000; monoclonal anti-neurofilament SMI 33, 1 : 2000; monoclonal anti-vimentin antibody, 1 : 3; polyclonal anti-NSE antibody, 1 : 4; polyclonal anti-S-100 antibody, 1 : 4; polyclonal anti-tubulin antibody, 1 : 250; polyclonal anti-MBP, 1 : 800; and nonimmune serum controls (Vector Labs), 1 : 200.

Neurotransmitter antibodies are used at the following dilutions; anti-serotonin, 1 : 250; anti-CCK-8, 1 : 500; anti-GABA, 1 : 500; anti-VIP, 1 : 500; anti-somatostatin, 1 : 500; and anti-glutamate, 1 : 250. The specificity of staining was determined by preadsorbing each antiserum for 24 hr with 50 μg per ml of the appropriate neurotransmitter. In addition, the anti-PNMT antibody was used at 1 : 1000 dilution; anti-cholineacetyl transferase at 1 : 200; anti-tyrosine hydroxylase and anti-dopamine-β-hydroxylase at 1 : 1000 dilutions; polyclonal anti-methionine enkephalin antibody (Incstar) at 1 : 75 dilution; polyclonal anti-leucine enkephalin antibody (Incstar) at 1 : 800 dilution; and nonimmune serum as control at 1 : 200 dilution. Other antisera were used at the following dilutions: monoclonal anti-tau antibody (Sigma), 1 : 100; monoclonal anti-MAP2 antibody (Sigma), 1 : 50; and monoclonal anti-MAP5 antibody (Sigma), 1 : 250.

All slides are then rinsed three times for 5 min each in PBS, permeabilized by incubation in 0.1% (w/v) Triton X-100 for 5 min, and rinsed again three times in PBS. Endogenous peroxidase activity is quenched by incubation in PBS containing 2% (v/v) hydrogen peroxide, followed by three rinses in PBS. Nonspecific staining is blocked by incubation for 1 hr with nonimmune serum, appropriate for the secondary antibody, at a dilution of 1 : 100 in PBS containing 1% (w/v) BSA. Slides are then incubated in PBS containing 1% BSA and primary antiserum overnight at 4°C. The next day, slides are rinsed three times with PBS, blocked with PBS containing 1% BSA, and incubated for 2 hr at 25°C with the appropriate biotinylated secondary antibody using Vectastain kits (Vector Labs, Burlingame, CA). Slides are washed in PBS, blocked, incubated for 1 hr in PBS containing 1% BSA and avidin–biotin–horseradish peroxidase complex (Vector Labs), rinsed, and incubated for 5 min with chromogen, 3-amino-9-ethylcarbazole (Biomeda Corp., Foster City, CA).

Several sets of controls are used to evaluate the specificity of staining. Adult rat brain cryostat sections are fixed and permeabilized as previously described and stained with each of the aforementioned antiserum. In all cases, staining was specific for the appropriate neuronal (NF, SMI 33, NSE) or nonneuronal (MBP, S-100, GFAP, vimentin) regions. In addition, C6 glioma cells, primary cultures of rat neuronal cells, and NIH 3T3 cells can be plated, fixed, and stained as previously described (Ronnett *et al.*, 1991) to serve as immunocytochemical controls.

Once cell lines are identified, which demonstrate appropriate neuronal markers, it is advisable to perform karyotype analysis to determine chromosome number. The degree of variation from the normal chromosome number of 46 can be determined. Several standard methodologies are available for karyotype analysis in cultured cells (Freshey, 1987). It should be noted that for slowly dividing cells, a large number of flasks may be required to obtain enough mitotic cells for chromosomal analysis.

References

Aloe, L., and Levi-Montalcini, R. (1979). Nerve growth factor-induced transformation of immature chromaffin cells in vivo into sympathetic neurons, effect of antiserum to nerve growth factor. *Proc. Natl. Acad. Sci. U.S.A.* **76**, 1246–1250.

Anderson, D. J., and Axel, R. (1986). A bipotential neuroendocrine precursor whose choice of fate is determined by NGF and glucocorticoids. *Cell* **47**, 1079–1090.

Augusti-Tocco, G., and Sato, G. (1969a). Establishment of functional clonal lines of neurons form mouse neuroblastoma. *Proc. Natl. Acad. Sci. U.S.A.* **64**, 311–315.

Augusti-Tocco, G., and Sato, G. (1969b). Establishment of functional clonal lines of neurons from mouse neuroblatoma. *Proc. Natl. Acad. Sci. U.S.A.* **64**, 311–315.

Banker, G., and Goslin, K. (1988). Developments in neuronal cell culture. *Nature* **336**, 185–186.

Banker, G., and Goslin, K. (1991). *Culturing Nerve Cells*. MIT Press, Cambridge.

Barde, Y. A., Lindsay, R. M., Monard, D., and Thoenen, H. (1978). New factor released by cultured cells supporting survival and growth of sensory neurons. *Nature* **274**, 818.

Barlett, P. F., Reid, H. H., Bailey, K. A., and Bernard, O. (1988). Immortalization of mouse neural precursor cells by the c-*myc* oncogene. *Proc. Natl. Acad. Sci. U.S.A.* **85**, 3255–3259.

Barnes, D., and Sato, G. (1980). Methods for growth of cultured cells in serum-free medium. *Anal. Biochem.* **102**, 255–270.

Barnes, W. D., Sirbasku, D. A., and Sato, G. H. (1984). Methods for preparation of medic, supplements, and substrata for serum-free animal cell culture. In *Cell Culture Methods for Molecular and Cell Biology*, Vol. 1, (W. D. Barnes, D. A. Sirbasku, and G. H. Sato, Eds.), Alan R. Liss, New York.

Barlett, P. F., Reid, H. H., Bailey, K. A., and Bernard, O. (1988). Immortalization of mouse neural precursor cells by the c-*myc* oncogene. *Proc. Natl. Acad. Sci. U.S.A* **85**, 3255–3259.

Bignami, A., Palladini, G., and Zappella, M. (1968). Unilateral megalencephaly with nerve cell hypertrophy. *Brain Res.* **9**, 103–114.

Binder, L. I., Frankfurter, A., and Rebhun, L. I. (1985). The distribution of tau in the mammalian central nervous system. *J. Cell Biol.* **101**, 1371–1378.

Bornstein, M. (1958). Reconstituted rat tail collagen as a substrate for tissue cultures on conventional coverslips in Maximow slides and roller tubes. *Lab. Invest.* **7**, 134–140.

Bornstein, M. B., and Murray, M. R. (1958). Serial observations on patterns of growth, myelin formation, maintenance and degeneration in cultures of newborn rat and kitten cerebellum. *J. Biophys. Biochem. Cytol.* **4**, 499.

Bottenstein, J. (1985). Growth and differentiation of neural cells in defined media. In *Cell Culture in the Neurosciences* (J. E. Bottenstein and G. Sato, Eds.), pp. 3–44, Plenum, New York.

Bottenstein, J. E., Skaper, S. D., Varon, S. S., and Sato, G. H. (1980). Selective survival of neurons from chick embryo sensory ganglionic dissociates utilizing serum-free supplemented medium. *Exp. Cell. Res.* **125**, 183–190.

Bottenstein, J. E., and Sato, G. (1985). *Cell Culture in the Neuroscience*. Plenum, New York.

Bottenstein, J. E., and Sato, G. H. (1979). Growth of a rat neuroblastoma cell line in serum-free supplemented medium. *Proc. Natl. Acad. Sci. U.S.A.* **76**, 514–517.

Burgoyne, R. D. (1991). High molecular weight microtubule-associated proteins of brain. In *The Neuronal Cytoskeleton* (R. D. Burgoyne, Ed.), pp. 75–91, Wiley-Liss, New York.

Burgoyne, R. D., and Cumming, R. (1984). Ontogeny of microtubule-associated protein 2 in rat cerebellum:differential expression of the doublet polypeptides. *Neuroscience* **11**, 157–167.

Calvert, R., and Anderton, B. H. (1985). A microtubule-associated protein (MAP1) which is expressed at elevated levels during development of the rat cerebellum. *EMBO J.* **4**, 1171–1176.

Carbonetto, S., Gruver, M. M. S., and Turner, D. C. (1983). Nerve fiber growth in culture on fibronectin, collagen, and glycosaminoglycan substrates. *J. Neurosci.* **3**, 2324–2335.

Carbonetto, S., and Cochard, P. (1987). In vitro studies on the control of nerve fiber growth by the extracellular matrix of the nervous system. *J. Physiol. (Paris)* **82**, 258–270.

Cepko, C. (1988). Immortalization of neural cells via oncogene transduction. *Trends Neurosci.* **11**, 6–8.

Cepko, C. L. (1989). Immortalization of neural cells via retrovirus-mediated oncogene transduction. *Annu. Rev. Neurosci.* **12**, 47–65.

Chiu, A. Y., Matthew, W. D., and Patterson, P. H. (1986). A monoclonal antibody that blocks the activity of a neurite regeneration-promoting factor: studies on the binding site and its localization in vivo. *J. Cell Biol.* **103**, 1383–1398.

Choi, D. W. (1985). Glutamate neurotoxicity in cortical cell culture is calcium dependent. *Neurosci. Lett.* **58**, 293–297.

Christensen, B., Kieler, J., Villien, M., Don, P., Wang, C. Y., and Wolf, H. (1984). A classification of human urothelial cells propagated in vitro. *Anticancer Res.* **4**, 319–338.

Condic, M. L., and Bentley, D. (1989). Removal of the basal lamina in vivo reveals growth conebasal lamina adhesive interactions and axonal tension in grasshopper embryos. *J. Neurosci.* **9**, 2678–2686.

Couchi, D., Charriere-Bertrand, C., and Nunez, J. (1988). Expression of the mRNA for T-proteins during brain development and in cultured neurons and astroglial cells. *J. Neurochem.* **50**, 1894–1899.

Dambska, M., Wisniewski, K., and Sher, J. H. (1984). An autopsy care of hemimegalencephaly. *Brain Dev.* **6**, 60–64.

Doupe, A. J., Landis, S. C., and Patterson, P. H. (1985) Environmental influences in the development of neural crest derivatives: glucocorticoids, growth factors, and chromaffin cell plasticity. *J. Neurosci.* **5**, 2119–2142.

Dulbecco, R., and Freeman, G. (1959). Plague formation by the polyoma virus. *Virology* **8**, 396–397.

Eagler, H. (1959). Amino acid metabolism in mammalian cell cultures. *Science* **130**, 432.

Ebendal, T. (1976). The relative roles of contact inhibition and contact guidance in orientation of axons extending on aligned collagen fibrils in vitro. *Exp. Cell. Res.* **98**, 159–169.

Ebendal, T. (1979). Stage-dependent stimulation of neurite outgrowth exerted by nerve growth factor and chick heart in cultured embryonic ganglia. *Dev. Biol* **72**, 276.

Ebendal, T., and Jacobson, C. O. (1977). Tissue explants affecting extension and orientation of axons in cultured chick embryo ganglia. *Exp. Cell. Res.* **105**, 379–387.

Elsdale, T., and Bard, J. (1972). Collagen substrata for studies on cell behavior. *J. Cell Biol.* **54**, 626–637.

Fields, K. L. (1979). Cell type-specific antigens of cells of the central and peripheral nervous system. *Curr. Top. Dev. Biol.* **13**, 237–257.

Fields, K. L. (1985). Neuronal and glial surface antigens on cells in culture. In *Cell Culture in the Neurosciences.* (J. E. Bottenstein and G. Sato, Eds.), pp. 45–93, Plenum, New York.

Frederiksen, K., Jat, P. S., Valtz, N., Levy, D., and McKay, R. (1988). Immortalization of precursor cells from the mammalian CNS. *Neuron* **1**, 439–448.

Freshey, R. I. (1987). Characterization. In *Culture of Animal Cells: A Manual of Basic Technqiue,* Vol. 2, pp. 175–176, Alan R. Liss, New York.

Freshney, R. I. (1980). Culture of glioma of the brain. In *Brain Tumours, Scientific Basic, Clinical Investigation and Current Therapy* (D. G. T. Thomas and D. I. Graham, Eds.), pp. 21–50, Butterworths, London.

Gilbert, S. F., and Migeon, B. R. (1975). D-valine as a selective agent for normal human and rodent epithelial cells in culture. *Cell* **5**, 11–17.

Gilbert, S. F., and Migeon, B. R. (1980). Renal enzymes in kidney cells selected by D-Valine medium. *J. Cell. Physiol.* **92**, 161–168.

Goodman, A. R. (1986). Hemispherectomy and its alternatives in the treatment of intractable epilepsy in patients with infantile hemiplegia. *Dev. Med. Child Neurol.* **28**, 251–258.

Green, H., Kehinde, O., and Thomas, J. (1979). Growth of cultured human epidermal cells into multiple epithelia suitable for grafting. *Proc. Natl. Acad. Sci. U.S.A.* **76**, 5665–5668.

Greene, L. A., Shain, W., Chalazonitis, A., Breakfield, X., Minna, J., Coon, H. G., and Nirenberg M. (1975). Neuronal properties of hybrid neuroblastoma X sympathetic ganglion cells. *Proc. Natl. Acad. Sci. U.S.A.* **72**, 4923–4927.

Greene, L. A., and Tischler, A. S. (1976). Establishment of a nonadrenergic clonal line of rat adrenal pheochromocytoma cells which respond to nerve growth factor. *Proc. Natl. Acad. Sci. U.S.A.* **73**, 2424–2428.

Greene, L. A., and Tischler, A. S. (1982). PC12 pheochromocytoma cells in neurobiological research. *Adv. Cell. Neurobiol.* **3**, 373–414.

Guroff, G. (1985). PC12 cells as a model of neuronal differentiation. In *Cell Culture in the Neurosciences* (J. E. Bottenstein and G. Sato, Eds.), pp. 245–272, Plenum, New York.

Ham, R. G. (1965). Glonal growth of mammalian cells in a chemically defined synthetic medium. *Proc. Nalt. Acad. Sci. U.S.A.* **53**, 288.

Hammond, D. N., Wainer, B. H., Tonsgard, J. H., and Heller, A. (1986). Neuronal properties of clonal hybrid cell lines derived from central cholinergic neurons. *Science* **234**, 1237–1240.

Hendelman, W. J., Jande, S. S., and Lawson, D. E. (1984). Calcium-binding protein immunocytochemistry in organotypic cultures of cerebellum. *Brain Res. Bull.* **13**, 181–184.

Hendelman, W. J., and Marshall, K. C. (1980). Axonal projection patterns visualized with horseradish peroxidase in organized cultures of cerebellum. *Neuroscience* **5**, 1833–1846.

Higgins, D., Lein, P. J., Osterhout, D. J., and Johnson, M. I. (1991). Tissue culture of mammalian autonomic neurons. In *Culturing Nerve Cells* (G. Banker and K. Goslin, eds.), pp. 177–205, MIT Press, Cambridge.

Hockberger, P. E., Tseng, H.-Y., and Connor, J. A. (1989). Development of rat cerebellar Purkinje cells: electrophysiological properties following acute isolation and in long-term culture. *J. Neurosci.* **9**, 2258–2271.

Hokfelt, T., Johansson, O., Ljungdahl, A., Lundberg, J. M., and Schultzberg M. (1980). Peptidergic neurons. *Nature* **284**, 515–521.

Honegger, P., Lenoir, D., and Favrod P. (1979). Growth and differentiation of aggregating fetal brain cells in a serum-free defined medium. *Nature* **282**, 305–308.

Honegger, P. (1985). Biochemical differentiation in serum-free aggregating brain cell cultures. In *Cell Cutlure in the Neurosciences* (J. E. Bottenstein and G. Sato, eds.), pp. 223–243, Plenum, New York.

Honegger, P., and Lenoir, D. (1980). Triiodothyronine enhancement of neuronal differentiation in aggregating fetal rat brain cells cultured in a chemically defined medium. *Brain Res.* **199**, 425–434.

Huettner, J. E., and Baughman, R. W. (1986). Primary culture of identified neurons from the visual cortex of postnatal rats. *J. Neurosci.* **6**, 3044–3060.

Lee, H. J., Hammond, D. N., Large, T. H., Roback, J. D., Sim, J. A., Brown, D. A., Otten, U. H., and Wainer, B. H. (1990a). Neuronal properties and trophic activities of immortalized hippocampal cells from embryonic and young adult mice. *J. Neurosci.* **10,** 1779–1787.

Lee, H. J., Hammond, D. N., Large, T. H., and Wainer B. H. (1990b). Immortalized young adult neurons from the septal region: generation and characterization. *Dev. Brain Res.* **52,** 219–228.

Lendahl, U., and McKay, R. D. G. (1990). The use of cell lines in neurobiology. *Trends Neurosci.* **13,** 132–137.

Letourneau, P. C. (1975). Possible roles for cell-to-substratum adhesion in neuronal morphogenesis. *Dev. Biol.* **44,** 77–91.

Levi-Montalcini, R. (1964). Growth control of nerve cells by a protein factor and its antiserum. *Science* **143,** 105–110.

Levi-Montalcini, R. C. P. (1979). The nerve-growth factor. *Sci. Am.* **240,** 68.

Levi-Montalcini, R. (1982). Developmental neurobiology and the natural history of nerve growth factor. *Annu. Rev. Neurosci.* **5,** 341–361.

Liebermann, D., and Sachs, L. (1978). Nuclear control of neurite induction in neuroblastoma cells. *Exp. Cell. Res.* **113,** 383–390.

Lindsay, R. M. (1979). Adult rat brain astrocytes support survival of both NGF-dependent and NGF-insensitive neurons. *Nature* **282,** 80.

Littauer, U. Z., Giovanni, M. Y., and Glick, M C. (1979). Differentiation of human neuroblastoma cells in culture. *Biochem. Biophys. Res. Commun.* **88,** 933–939.

Loffner, F., Lohmann, S. M., Walckhoff, B., Walter, U., and Hamprecht, B. (1986). Immunocytochemical characterization of neuron-rich primary cultures of embryonic rat brain cells by established neuronal and glial markers and by monospecific antisera against cyclic nucleotide-dependent protein kinases and the synaptic vesicle protein synapsin I. *Brain Res.* **363,** 205–221.

Lumsden, C. E. (1968). The structure and function of nervous tissue. In *Nervous Tissue in Culture* (G. H. Bourne, ed.), pp. 67–140, Academic Press, New York.

Mains, R. E., and Patterson, P. H. (1973). Primary cultures of dissociated sympathetic neurons. *J. Cell Biol.* **59,** 329–345.

Manthorpe, M. E., Engvall, E., Rouslahti, E., Longo, F. M., Davis, G. E., and Varon, S. (1983). Laminin promotes neuritic regeneration from cultured peripheral and central neurons. *J. Cell Biol.* **97,** 1882–1890.

Marangos, P. J., and Schmechel, D. E. (1987). Neuron specific enolase, a clinically useful marker for neurons and neuroendocrine cells. *Annu. Rev. Neurosci.* **10,** 269–295.

Martinou, J. C., Le Van Thai, A., Cassar, G., Roubinet, F., and Weber, M. J. (1989). Characterization of two factors enhancing choline acetyltransferase activity in cultures of purified rat motoneurons. *J. Neurosci.* **9,** 3645–3656.

Masurovsky, E. B., and Peterson, E. R. (1973). Photoreconstituted collagen gel for tissue culture substrates. *Exp. Cell. Res.* **76,** 447–448.

McLean, J. S., Frame, M. C., Freshney, R. I., Vaughan, P. F. T., and Mackie, A. E. (1986). Phenotypic modification of human glioma and non-small cell lung carcinoma by glucocorticoids and other agents. *Anticancer Res.* **6,** 1101–1106.

Mellon, P. L., Windle, J. J., Goldsmith, P. C., Padula, C. A., Roberts, J. L., and Weiner, R. I. (1990). Immortalization of hypothalamic GnRH neurons by genetically targeted tumorigenesis. *Neuron* **5,** 1–10.

Moonen, G., Neale, E. A., MacDonald, R. L., Gibbs, W., and Nelson, P. G. (1982). Cerebellar macroneurons in microexplant cell culture. *Dev. Brain Res.* **5,** 59–73.

Nunez, J. (1986). Differential expression of microtubule components during brain development. *Dev. Neurosci.* **8,** 125–141.

Nunez, J. (1988). Immature and mature variants of MPA2 and tau proteins and neuronal plasticity. *Trends Neurosci.* **11,** 477–479.

O'Brien, R. J., and Fischbach, G. D. (1986a). Isolation of embryonic chick motoneurons and their survival in vitro. *J. Neurosci.* **6,** 3265–3274.

O'Brien, R. J., and Fischbach, G. D. (1986b). Excitatory transmission between interneurons and motoneurons in chick spinal cord cell cultures. *J. Neurosci.* **6,** 3284–3289.

O'Donell-Tormey, J., Nathan, C. F., Lanks, K., DeBoer, C. J., and De La Harpe, J. (1987). Secretion of pyruvate—an antioxidant defense of mammalian cells. *J. Exp. Med.* **165,** 500–514.

Olmsted, J. B. (1986). Microtubule-associated proteins. *Annu. Rev. Cell Biol.* **2,** 421–457.

Oster-Granite, M. L., and Herndon, R. M. (1978). Studies of cultured human and simian fetal brain cells. *Neuropath. Appl. Neurobiol.* **4,** 429–442.

Patterson, P. H. (1978). Environmental determination of autonomic neurotransmitter function. *Annu. Rev. Neurosci.* **1,** 1–17.

Pfeiffer, S. E., Betschart, B., Cook, J., Mancini, P., and Morris, R. (1977). Glial cell lines. In *Cell, Tissue, and Organ Cultures in Neurobiology* (S. Federoff and L. Hertz, eds.), pp. 287–346, Academic Press, New York.

Platika, D., Boulos, M. H., Baizer, L., and Fishman, M. C. (1985). Neuronal traits of clonal cell lines derived by fusion of dorsal root ganglia neurons with neuroblastoma cells. *Proc. Natl. Acad. Sci. U.S.A.* **82,** 3499–3503.

Ponten, J., and Macintyre, E. (1968). Interaction between normal and transformed bovine fibroblasts in culture. II. Cells transformed by polyoma virus. *J. Cell Sci.* **3,** 603–668.

Quintanilla, M., Brown, K., Ramsden, M., and Balmain, A. (1986). Carcinogen specific mutation and amplification of Ha-ras during mouse skin carcinogenesis. *Nature* **322,** 78–79.

Raff, M. C., Mirsky, R., Fields, K. L., Lisak, R. P., Dorfman, S. H., Silberberg, D. H., Gregson, N. A., Liebowitz, S., and Kennedy, M. C. (1978). Galactocerebroside is a specific cell-surface antigenic marker for oligodendrocytes in culture. *Nature* **274,** 813–816.

Raff, M. C., Fields, K. L., Hakomori, S. J., Mirsky, R., Pruss, R. M., and Winter, J. (1979). Cell-type-specific markers for distinguishing and studying neurons and the major classes of glial cells in culture. *Brain Res.* **174,** 283–308.

Riederer, B., Cohen, R., and Matus, A. (1986). MAP5: a novel microtubule-associated protein under strong developmental regulation. *J. Neurocytol.* **15,** 763–775.

Riederer, B., and Matus, A. (1985). Differential expression of distinct microtubule-associated proteins during brain development. *Proc. Natl. Acad. Sci. U.S.A.* **82,** 6006–6009.

Ronnett, G. V., Hester, L. D., Nye, J. S., Connors, K., and Snyder, S. H. (1990). Human cortical neuronal cell line: establishment from a patient with unilateral megalencephaly. *Science* **248,** 603–605.

Ronnett, G. V., Hester, L. D., and Snyder, S. H. (1991). Primary culture of neonatal rat olfactory neurons. *J. Neurosci.* **11,** 1243–1255.

Saez, J. C., Kessler, J. A., Bennett, V. L., and Spray, D. C. (1987). Superoxide dismutase protects cultured neurons against death by starvation. *Proc. Natl. Acad. Sci. U.S.A.* **84,** 3056–3059.

Schleicher, S., Boekoff, I., Arriza, J., Lefkowitz, R. J., and Breer, H. (1993). A β-adrenergic receptor kinase-like enzyme is involved in olfactory signal termination. *Pro. Natl. Acad. Sci. U.S.A.* **90,** 1420–1424.

Schubert, D., Humphreys, S., Baroni, C., and Cohn, M. (1969). In-vitro differentiation of a mouse neuroblastoma. *Proc. Natl. Acad. Sci. U.S.A.* **64,** 316–323.

Schubert, D., Humphreys, S., DeVitry, F., and Jacob, F. (1971). Induced differentiation of a neuroblastoma. *Dev. Biol.* **25,** 514–546.

Schubert, D. (1984). *Developmental Biology of Cultured Nerve, Muscle, and Glia.* Wiley, New York.

Snyder, E. Y., and Kim, S. U. (1979). Hormonal requirements for neuronal survival in culture. *Neurosci. Lett.* **13,** 225–230.

Stockdale, F. E., and Topper, Y. J. (1966). The role of DNA synthesis and mitosis in hormone dependent differentiation. *Proc. Natl. Acad. Sci. U.S.A.* **56,** 1283–1289.

Todaro, G. J., and Green, H. (1963). Quantitative studies of the growth of mouse embryo cells in culture and their development into established lines. *J. Cell Biol.* **17,** 299–313.

Turner, D. C., Carbonetto, S., and Flier, L. A. (1987). Magnesium-dependent attachment and neurite outgrowth by PC12 cells on collagen and laminin substrata. *Dev. Biol.* **93,** 285–300.

Walicke, P., Cowan, W. M., Ueno, N., Baird, A., and Guillemin, R. (1986). Fibroblast growth factor promotes survival of dissociated hippocampal neurons and enhances neurite extension. *Proc. Natl. Acad. Sci. U.S.A.* **83,** 3012–3016.

Walicke, P. A., and Patterson, P. H. (1981a). On the role of Ca^{2+} in the transmitter choice made by cultured sympathetic neurons. *J. Neurosci.* **1,** 343–350.

Walicke, P. A., and Patterson P. H. (1981b). The role of cyclic nucleotides in the transmitter choice made by cultured sympathetic neurons. *J. Neurosci.* **1,** 333–342.

Wolf, M. K., and Dubois-Dalcq, M. (1970). Anatomy of cultured mouse cerebellum. *J. Comp. Neurol.* **140,** 261–280.

Wu, R., and Wu, M. M. J. (1986). Effects of retinoids on human bronchial epithelial cells: Differential regulation of hyaluronate synthesis and keratin protein synthesis. *J. Cell. Physiol.* **127,** 73–82.

Yavin, E., and Yavin, Z. (1974). Attachment and culture of dissociated cells from rat embryo cerebral hemispheres on polylysine-coated surface. *J. Cell Biol.* **62,** 540–546.

Yavin, Z., and Yavin, E. (1980). Survival and maturation of cerebral neurons on poly-L-lysine surfaces in the absence of serum. *Dev. Biol.* **75,** 454–459.

CHAPTER

37

In Vitro Systems for the Investigation of Calcium Homeostasis and Calcium-Induced Cell Damage

IAIN C. CAMPBELL
Institute of Psychiatry
London SE5 8AF
United Kingdom

ELIZABETH M. ABDULLA
Wellcome Research Laboratories
Beckenham, Kent BR3 3BS
United Kingdom

I. Importance of Calcium in Cell Function

Intracellular-free calcium ($[Ca^{2+}]_i$) is vital for muscle contraction, gland secretion, neurotransmitter release, growth and differentiation, and the control of neuronal excitability (Smith and Augustine, 1988; Somlyo and Himpens, 1989; Tsien, 1988; Williamson and Monck, 1989). These diverse processes require a low, modulated, resting cytosolic calcium and mechanisms to regulate it both temporally and spatially (Rasmussen, 1989). The concentration of calcium in extracellular fluid is 1–5 mM whereas the concentration of free calcium in neurons (for example) is in the nanomolar range; a large increase in intracellular-free calcium can result in neuronal cell damage. This chapter describes the various processes which maintain the low levels of free calcium in neurons and some of the important reactions that are catalyzed by calcium. These reactions should be seen within the context of normal neu-

ronal function but also as systems which could produce cell damage if they were overstimulated as a consequence of abnormally elevated calcium.

II. Calcium Distribution

Neurons and other cells have a resting $[Ca^{2+}]_i$ of 10^{-8}–10^{-7} M, it is approximately 10^{-3} M in the extracellular fluid (Tsien, 1988). However, as the total Ca^{2+} in cells is estimated to be in the nanomolar range (Gibson and Peterson, 1987), most is either bound to membranes and cytosolic components or stored in intracellular organelles. Ca^2 is also bound externally on plasma membranes; at these sites, which may be phospholipids, proteins, sialic acids, gangliosides, and glycolipids, it generally exchanges rapidly and this may regulate membrane structure, potential, fluidity, and ability to communicate (Storch and Schachter, 1985). Neurons contain multiple Ca^{2+} pools, characterized by different kinetic properties. Most cytosolic Ca^{2+} is bound to proteins and this constitutes a considerable fraction (200 nM) of the total Ca^{2+}-buffering capacity (Rasmussen and Means, 1989). Nuclear Ca^{2+} in neurons is as high as the average cell Ca^{2+} content (Blaustein, 1988) and, like the cytosol, most of this is bound and may be involved in functions such as gene regulation (Meldolesi *et al.*, 1988). Ca^{2+} in mitochondria varies in brain regions but is relatively stable under resting conditions. Ca^{2+} in neuronal dense-cored vesicles and granules is largely complexed with other components and exchanges slowly with the cytosolic pool (Blaustein, 1988). Small synaptic vesicles can accumulate Ca^{2+} actively; nevertheless, the Ca^{2+} content in vesicles is not higher than in nerve terminals (Verhage *et al.*, 1989) and thus it appears that vesicular Ca^{2+} plays only a local role, possibly related to transmitter storage.

Endoplasmic reticulum (ER) and microsomes are probably the most important Ca^{2+} pools in terms of rapid exchange with the cytosol. For example, smooth ER in cerebellar cortex has a high Ca^{2+} content, which can increase fivefold after prolonged depolarization (Andrews *et al.*, 1987). ER organelles are heterogeneous in composition (and also in function) (Meldolesi *et al.*, 1988). In the ER, calciosomes exist in a number of cells. They are exclusively in a class of smooth vacuoles and are apparently distinct from, although often adjacent to, other ER organelles. Calciosomes are rich in Ca^{2+}-binding proteins such as calsequestrin and contain a Ca^{2+}-ATPase that is immunologically distinct. Because of their specific molecular components, calciosomes appear to be well suited to act as rapidly exchangeable, membrane-segregated Ca^{2+} pools (Meldolesi *et al.*, 1988).

It should be noted that the information just described is applicable to cells that are not stimulated by external signals. Information is now accumulating which indicates that when neurons in culture are subjected to electrical field stimulation (e.g., 1 Hz for 15 sec) or to depolarizing concentrations of K^+ there is a three- to fivefold increase in $[Ca^{2+}]_i$ but that this is unevenly distributed throughout the cell. The greatest increases apparently occur in the nucleus of the cell and at the terminal regions. The rise in nuclear $[Ca^{2+}]_i$ may be related to changes in gene transcription whereas the changes at the nerve endings are probably associated with neurotransmitter release or possibly with synaptic plasticity.

III. Calcium Movement

Cells regulate $[Ca^{2+}]_i$ mainly by controlling Ca^{2+} movement across the plasma membrane and/or the membranes of ER, mitochondria, or calciosomes. $[Ca^{2+}]_i$ increases are normally due to entry from the extracellular fluid via channels as a result of depolarization or by release of Ca^{2+} from intracellular stores (Nahorski, 1988).

The increased $[Ca^{2+}]_i$ initially interacts with its effectors and then multiple mechanisms operate to buffer it in the cytosol, sequester it in, and/or mobilize it from intracellular stores and extrude it across the plasma membrane. Eventually, all of this Ca^{2+} must be extruded to maintain Ca^{2+} homeostasis. Although Ca^{2+} entry by exchange and diffusion does not significantly contribute to influx under normal conditions, influx by these pathways in states such as ischemia is greatly enhanced and may contribute to irreversible cell damage (Blaustein, 1988).

IV. Ca^{2+} Buffering by Cytosolic Proteins

Cytosolic Ca^{2+}-buffering capacity has high- and low-affinity components. Various molecules, e.g., citrate, nucleotides, and inositol phosphates, account for the low-affinity binding of Ca^{2+} (<100 nM/mg protein) and hence are of dubious importance. The high affinity is due primarily to proteins, most of which share the typical EF-hand structure of their binding sites, e.g., calmodulin, parvalbumin, and vitamin D-dependent Ca^{2+}-binding protein; the distribution and concentration of these proteins vary in different neurons. Calmodulin, which is highly concentrated in brain (30–50 μM) and widely distributed among neurons, may account for a Ca^{2+}-buffering capacity of 120 nM (Pereschini *et al.*, 1989; Rasmussen and Means, 1989). Although Ca^{2+}-binding proteins have a relatively higher buffering capacity for Ca^{2+} in neurons than in smooth muscle and liver (Meldolesi *et al.*, 1988), they are only able to buffer Ca^{2+} that enters the neuron during the first few action potentials (about 5 pmol/mg protein/msec if Ca^{2+} is evenly distributed in the cytosol). In neurons firing at high frequency, and especially in those in which a relatively large fraction of the inward current during the rising phase of a action potential is carried by Ca^{2+}, these cytosolic Ca^{2+} buffers may saturate (Blaustein *et al.*, 1988).

V. Ca^{2+} Sequestration and Mobilization

Cytosolic proteins buffer transient rises in $[Ca^{2+}]_i$. Intracellular pools are then required to sequester it until it can be extruded. Because of the low affinity of Ca^{2+} for the mitochondrial Ca^{2+} transporter, there is only appreciable accumulation when $[Ca^{2+}]_i$ is increased by neuronal stimulation. When $[Ca^{2+}]_i$ rises as a result of neuronal activity, mitochondria respond by a net (ATP-dependent) uptake of Ca^{2+}; in the matrix it increases to approximately 1 μM (Nachshen, 1985). When $[Ca^{2+}]_i$ exceeds 5 μM (i.e., under pathological conditions), mitochondria sequester substantial amounts (Gibson and Peterson, 1987).

Synaptic vesicles sequester Ca^{2+} by an ATP-driven mechanism, but they have a low affinity and therefore probably contribute little to $[Ca^{2+}]_i$ regulation (McBurney and Neering, 1987).

ER and microsomes are the major nonmitochondrial Ca^{2+} stores and probably play the most important role in sequestering Ca^{2+} following neuronal activity; they accumulate it in an ATP-dependent fashion at a rate of 1–2 pmol/mg protein/msec (Meldolesi *et al.*, 1988), which is sufficient to remove Ca^{2+} from the cytosol following its entry (Blaustein, 1988). In neurons, calciosomes have not been investigated in detail; however, calciosome-like proteins are expressed, in cultured neurons at least, and appear to be localized in cell bodies and neurites (Meldolesi *et al.*, 1988). Depolarization, hormones, and ionomycin can induce increases in $[Ca^{2+}]_i$, even in the absence of $[Ca^{2+}]_o$ (Burgoyne *et al.*, 1990). These increases are believed to result from Ca^{2+} mobilization from the ER and mitochondria. The release of Ca^{2+} from the ER is quantal (Muallem *et al.*, 1989) and more is released as the ER fills. Indeed, the ER may need to be loaded in this way to contribute significantly to increased $[Ca^{2+}]_i$ during neuronal activity. Evidence suggests that during excitation, release of Ca^{2+} from the ER, in particular those rapidly exchangeable pools, is modulated by inositol 1,4,5-triphosphate (IP3) (Nahorski, 1988); decreasing $[Na^+]_o$ triggers IP3 production and Ca^{2+} mobilization (Smith and Augustine, 1988). However, whether Ca^{2+} release from calciosomes is triggered by IP3 remains to be established. Mitochondria are not involved in IP3-mediated Ca^{2+} mobilization, but release it by a Na^+/Ca^{2+} exchange mechanism (Nachshen, 1985).

VI. Calcium Extrusion

The concentration gradient, $[Ca^{2+}]_o >> [Ca^{2+}]_i$ (10^4-fold), and the electrical driving force (50–100 mV negative membrane potential) promote the net gain of Ca^{2+} at rest and during neuronal activity (Barritt, 1981). To maintain Ca^{2+} homeostasis, neurons have two parallel, independent mechanisms in their plasma membranes for extruding Ca^{2+}: Ca^{2+}/Mg^{2+} ATPase and a Na^+/Ca^{2+} exchanger.

Two general classes of Ca^{2+} pumps (Ca^{2+}/Mg^{2+} ATPase) have been identified: those in the ER, which participate in intracellular Ca^{2+} sequestration

(Burgoyne, 1990), and those in the plasma membrane, which extrude it from cells (Williamson and Monck, 1989) (Table 1). The ER systems are calmodulin-insensitive, Ca^{2+}-dependent ATPases with a Ca^{2+} : ATP coupling ratio of 2 : 1 and a molecular mass of about 105 kDa. The plasma membrane systems are calmodulin-modulated, Ca^{2+}-dependent ATPases with a Ca^{2+} : ATP coupling ratio of 1 : 1 and a molecular mass of about 140 kDa; they have a high affinity for Ca^{2+} ($K_{ca} = 0.2$–0.3 μM), but a low transport capacity and are proposed to be primarily responsible for Ca^{2+} efflux in the resting state (Exton, 1988).

An electrogenic Na^+/Ca^{2+} exchanger extrudes one Ca^{2+} in exchange for three Na^+ by utilizing the energy stored within the Na^+ gradient across the membrane, which is in turn restored by the Na^+/K^+ ATPase (Nachshen, 1986). The direction of Na^+/Ca^{2+} exchange can be either inward or outward depending on the Na^+ gradient. Initially, Ca^{2+} flux through this antiporter would be inward after depolarization because of the increase in intracellular Na^+. However, after repolarization promoted by Na^+/K^+ ATPase, there would primarily be an outward Ca^{2+} current. The rate of transport mediated by the Na^+/Ca^{2+} exchanger is controlled by factors such as the difference between the membrane potential and the reversal potential for the exchanger. It is also influenced by kinetic factors based on the fractional occupancy of the carriers by transported ions as well as by activating (nontransported) internal Ca^{2+}. The system has a low affinity for Ca^{2+} ($K_{ca} = 0.5$–1 μM) but a large capacity (Exton, 1988). Thus, when $[Ca^{2+}]_i$ is low, e.g., in resting neurons (-100 nM), the turnover of the exchanger is very low because the internal Ca^{2+} sites that participate in Ca^{2+} extrusion ($K_{ca} = 700$ nM), as well as those that activate exchanger-mediated Ca^{2+} entry ($K_{ca} = 600$ nM), are largely unoccupied (Blaustein, 1988). This means that the exchanger operates primarily to restore $[Ca^{2+}]_i$ during neuronal depolarization when $[Ca^{2+}]_i$ is raised.

TABLE 1   Properties of Ca^{2+} Transporters (in Squid Axons)[a]

Properties	Na^+/Ca^{2+} exchanger	Ca^{2+}/Mg^{2+} ATPase
K_{Ca}	0.5–1 μM	1 μM
ATP	Not linked	Obligatory
Mg^{2+}	Not linked	Essential
Stoichiometry	$Na^+ : Ca^{2+}$	ATP : Ca^{2+}
	3–4 : 1	1 : 1
Depolarization	Inhibition	No effect
Calmodulin	Not related	Related
Inhibition by La^{3+}	1 mM	1–100 μM
Inhibition by VO_4^{3-}	No effect	1–10 μM

[a] From Barritt (1981).

VII. Ca²⁺ Oscillations

Changes in $[Ca^{2+}]_i$ following agonist addition to neurons have shown that the responses may be oscillatory (Hallam and Rink, 1989). Mechanisms involved in the generation of Ca^{2+} oscillations appear to be of two types: oscillations that occur secondary to spontaneous action potentials, as in some secretory and retina cells, e.g., as in the salamander (Lamb et al., 1986), and oscillations induced by Ca^{2+}-mobilizing substances, e.g., IP3. The latter effect can be initiated in the absence of extracellular Ca^{2+}, and thus intracellular Ca^{2+} release is the primary source of Ca^{2+}, although extracellular Ca^{2+} is required for the maintenance of Ca^{2+} oscillations (Ambler et al., 1988). A minimum oscillating system requires a feedback loop, and for sustained oscillations, some delay step is necessary to generate the periodicity. A possible mechanism for negative feedback is provided by the inhibitory effect of Ca^{2+} on IP3 binding to its receptor, i.e., the time required for reaccumulation of Ca^{2+} into the IP3-sensitive pool could provide the delay which would allow the generation of the characteristic periodicity (Williamson and Monck, 1989).

VIII. Calcium-Linked Effector Mechanisms

An increase in $[Ca^{2+}]_i$ triggers the exocytotic release of neurotransmitters (Knight et al., 1989). Several ion channels that help to shape the frequency and duration of electrical responses in neurons are regulated directly by Ca^{2+} (Marty, 1989) or indirectly by Ca^{2+}-dependent protein kinases or phosphatases (Hemmings et al., 1989). Synapse formation is linked to $[Ca^{2+}]_i$. Actively elongating neurites and motile growth cones have a narrow range of $[Ca^{2+}]_i$ of 100–300 nM; a decrease below or an increase above this range produced by neurotransmitters or electrical activity is associated with the arrest of neurite elongation and cessation of growth cone movement (Ellis et al., 1990; Kater et al., 1988). Associative long-term potentiation (LTP), an activity-dependent, persistent increase in synaptic strength, is initiated by a local increase in $[Ca^{2+}]_i$ (Malenka et al., 1989). Finally, transient changes in $[Ca^{2+}]_i$ can trigger the transcription of "immediate-early genes" such as c-fos and c-jun in neuronal nuclei which encode DNA-binding proteins that alter the expression of other genes (Morgan and Curran, 1989).

IX. Immediate Targets of Calcium in Neurons

In neurons, there are two major classes of Ca^{2+} targets in the plasma membrane and three in the cytosol. In the plasma membrane, the first target group consists of channel proteins, i.e., certain K⁺ channels (Latorre et al., 1989), cation-selective channels, and Cl⁻ channels that are directly regulated by Ca^{2+} (Marty, 1989); these are distinct from the well-known receptor-operated calcium channels (N, L, P, and T). The second target group in the plasma membrane has two important families of membrane phospholipases: phospholipase C (PLC), which hydrolyzes phosphatidylinositol phosphates, and phospholipase A₂ (PLA2), which cleaves fatty acids from the glycerolipid backbone (Moskowitz et al., 1986). The products of PLC are well-established second messengers, e.g., diacylglycerol (DAG) and IP3. DAG activates the various isoforms of protein kinase C and IP3 mobilizes calcium from intracellular stores (Nahorski et al., 1988). PLA2 activation leads to the formation of the detergent lysolecithin and to various fatty acids including arachidonate. Arachidonic acid production is the rate-limiting step in eicosanoids production and is possibly involved in the generation of synaptic plasticity. Thus, prolonged activation of PLA2 is likely to result in uncontrolled membrane deacylction and the generation of excess second messengers (Moskowitz et al., 1986).

The three major cytosolic targets for calcium are protein kinase C (PKC), calpain (a Ca^{2+}-dependent protease), and calmodulin (CaM). PKC exists as a family of 83-kDa serine/threonine protein kinases, e.g., PKC1, PKC11, and PKC111, that are activated synergistically by Ca^{2+} and DAG (Huang, 1989). They are also activated by phorbol esters. In their activated state, cytosolic PKC isoforms (of which there are at least seven) move to the plasma membrane where they phosphorylate and regulate membrane proteins. In neurons, PKC regulates electrical excitability by phosphorylating certain ion channels (Berridge, 1987). In addition, they can regulate synaptic efficacy and appear to have a role in the initiation and/or maintenance of LTP (Nishizuka, 1986). PKC has been implicated in the intracellular changes that result from toxic exposure to methyl mercury. For example, in cell culture, the abnormal protein phosphorylation of neuronal (but not glial) proteins which follows treatment with methyl mercury is similar to that seen following addition of phorbol esters and is blocked by the addition of staurosporine (Sarafian and Verity, 1993).

Calpains are a group of neutral cysteine proteases that are activated directly by Ca^{2+} (Melloni and Pontremoli, 1989). They have a widespread distribution, are present in both membrane and cytosolic forms, and are implicated in the regulation of membrane proteins (e.g., receptors) and the cytoskeleton (e.g., tubulin and fodrin are substrates) in a number of cells, including neurons. Obviously, they are likely to be involved in the catabolism for a variety of proteins, including other enzymes, and thus the potential for causing cell damage is considerable; in fact there are some reports of increased calpain activity in age-related pathologies. In the hippocampus, excitotoxicity due to administration of NMDA results in loss of CA1 neurons (similar to that seen following global ischemia); these pathological changes are associated with calpain-mediated proteolysis of the major cytoskeletal spectrin (such changes are clearly seen using Western blots and image analysis) (Roberts-Lewis and Siman, 1993).

Calpains have an inactive form, procalpain, which is converted to the active enzyme by Ca-dependent proteolysis. This step is irreversible. The active form has increased sensitivity to Ca and under these conditions, Ca promotes dissociation of the natural inhibitor, calpastatin. (There are also reports of the existence of a natural promoter.)

CaM is a 15-kDa, ubiquitously distributed Ca^{2+}-binding regulatory protein of the "EF-hand" family (Pereschini *et al.*, 1989; Rasmussen, 1989). It is present in neuronal cytosol at a concentration of 30–50 μM. Each molecule has four Ca^{2+}-binding sites with dissociation constants (K_d) in the low micromolar range. At resting $[Ca^{2+}]_i$ (10^{-8} M), very little is bound to CaM. As the concentration rises to the micromolar level, the four binding sites are successively occupied and CaM becomes a multifunctional activator (Pereschini *et al.*, 1989).

X. Neural Calmodulin-Regulated Proteins

In its Ca^{2+}-bound form, CaM binds with different affinities to specific proteins and alters their functions. Among these proteins are a family of Ca^{2+}/CaM-dependent protein kinases. CaM kinase II is the predominant Ca^{2+}-dependent protein kinase in neurons of the mammalian cortex and hippocampus. CaM kinase II has a relatively broad substrate specificity, it can phosphorylate several neuronal proteins, and it is presumed to be a target for the postsynaptic Ca^{2+} current produced by activation of *N*-methyl-D-aspartate (NMDA) receptors. Furthermore, its rapid on/off regulation by autophosphorylation suggests that it could play a role in the initiation of LTP (Malenka *et al.*, 1989). In the presynaptic terminal, CaM kinase II may mediate synaptic transmission and increase glutamate and NA release from synaptosomes (Nichols *et al.*, 1990). One of its important presynaptic substrates is synapsin I, a protein that associates with synaptic vesicles and binds to the cytoskeleton. Phosphorylation by CaM kinase reduces the affinity of synapsin I for vesicles: this causes dissociation of vesicles from the cytoskeleton, making them available for fusion and leading to increased neurotransmitter release/action potential (Burgoyne *et al.*, 1990).

A second neural protein activated by Ca^{2+}/CaM is protein phosphatase-2B (calcineurin). Calcineurin is abundant in brain, but has a rather narrow substrate specificity. It was recently implicated in the Ca^{2+}-dependent inactivation of L-type Ca^{2+} channels in neurons (Armstrong, 1989). Phosphorylation of L-type Ca^{2+} channels by cAMP-dependent protein kinase enhances their activation by depolarization. Conversely, dephosphorylation by calcineurin desensitizes the channels. Calcineurin also dephosphorylates and inactivates DARRP-43, a protein inhibitor of the broad specificity brain protein phosphatase, phosphatase-I. Thus, activation of calcineurin may initiate a cascade of protein dephosphorylation.

Two additional targets of Ca^{2+}/CaM are isozymes of adenylate cyclase and cyclic nucleotide phosphodiesterase. Thus a rise in $[Ca^{2+}]_i$ may promote either production or degradation of cAMP depending on the nature of the local cyclases and phosphodiesterases. In addition, Ca^{2+}/CaM participates in the feedback control of $[Ca^{2+}]_i$ by activating membrane Ca^{2+}/Mg^{2+} ATPase that functions as a Ca^{2+} pump (Rasmussen, 1989).

XI. Growth Factors and β-Amyloid Effects on Calcium Homeostasis

There are reports that several growth factors can protect CNS neurons against a variety of insults. For example, fibroblast growth factor (FGF) protects hippocampal neurons against glutamate neurotoxicity and FGF, nerve growth factor (NGF), and insulin-like growth factors (IGFs) protect human cortical and rat hippocampal and septal neurons against hypoglycemic damage. In each case, the growth factor prevents the sustained elevation in intracellular calcium levels that normally mediate the cell damage (Mattson *et al.*, 1993). The processes that are involved in these protective effects are unclear but, for example, may involve enhanced calcium buffering or alterations in the expression of the NMDA-type glutamatergic receptor.

In contrast to the peptide-mediated effects just described, evidence is accumulating that β-amyloid protein (which accummulates in Alzheimer's disease) causes increased neuronal vulnerability to excitotoxicity and neurofibrillary degeneration; calcium responses to glutamate, depolarization, and calcium ionophores are markedly increased in amyloid-treated neurons (Mattson *et al.*, 1992).

Thus, the interaction of peptides, i.e., growth factors and β-amyloid, with factors known to alter intracellular free calcium is likely to become a major area of research in the immediate future.

XII. Relationship between Neurotoxicity and Increases in [Ca²⁺]ᵢ

In the preceding sections, the putative sites of the cytotoxic effects of calcium have been described. However, there are some important unresolved issues. For example, although excitotoxicity proceeds via a NMDA-type glutamatergic receptor-mediated increase in calcium influx, it is clear that high external K⁺ concentrations or cyanide exposure also increase [Ca²⁺]ᵢ but do not necessarily result in cell death. Thus, it is likely that excitotoxic damage is due to changes in [Ca²⁺]ᵢ in spatially or temporally restricted domains.

The importance of temporal factors is apparent from studies of hippocampal neurons exposed to toxic levels of glutamate, e.g., there is a period in which [Ca²⁺]ᵢ levels are apparently normal, even though irreversible injury may have occurred (Dubinsky, 1993).

The significance of spatial aspects of calcium distribution are strongly suggested by optical measurements which have shown that increases in [Ca²⁺]ᵢ are often confined to specific parts of a neuron (Nichols *et al.*, 1990). Therefore, responses to a rise in [Ca²⁺]ᵢ depend on the spatial organization of Ca²⁺ target proteins, their relative affinities for Ca²⁺ or for Ca²⁺/CaM, and the arrangement of more distal proteins in the response pathway. Factors that influence a local response include clustering of Ca²⁺ target proteins within the membrane and their association with the cytoskeleton. In this context, it should be noted that the affinities for Ca²⁺/CaM vary widely among its target proteins. Calcineurin and Ca²⁺/Mg²⁺ ATPase have relatively high affinities for Ca²⁺/CaM ($K_d = 5$ nM), whereas CaM kinase II has a considerably lower affinity ($K_d = 50$ nM) (Malenka *et al.*, 1989). Thus, calcineurin binds a larger proportion of available Ca²⁺/CaM than the CaM kinase II when the two enzymes are present at the same concentration. However, in general, information on the subcellular distribution and concentration of most neuronal Ca²⁺ target proteins is still inadequate to permit quantitative predictions of local cellular Ca²⁺ responses.

These various findings just described indicate that in most cases the mechanisms involved in calcium-mediated cell damage are still ill-defined in terms of local intracellular events. Finally, it should be kept in mind that elevation of [Ca²⁺]ᵢ could be a terminal consequence of cell damage, e.g., as a result of metabolic compromise, and thus the increase may not have any causal significance.

XIII. Techniques for Studying Calcium

Calcium is readily bound by proteins and phospholipids and hence the study of its functions is complex. Measurement of total tissue calcium is achieved by acid extraction and atomic absorption spectrometry (Gitelman, 1987). ⁴⁵Ca can be used to examine Ca²⁺ movement (Brass and Belmonte, 1969). Measuring [Ca²⁺]ᵢ is more difficult, but progress has been made as Ca²⁺-sensitive indicators have developed (Tsien, 1989). These indicators include the metallochromic dyes murexide and arsenazo III and the Ca²⁺-binding proteins aequorin and obelin; these emit light proportional to the square of [Ca²⁺]ᵢ. Ca²⁺-sensitive electrodes are also used; these contain membranes prepared with ion-selective ligands and a neutral carrier, e.g., polyvinyl chloride. The EMF is proportional to log[Ca²⁺]ᵢ. Various techniques have also been used to study the role distribution of calcium in neurons: (a) electron probe X-ray microanalysis, (b) subcellular fractionation, (c) cell permeabilization, and (d) immunocytochemistry of Ca²⁺-binding proteins.

Electrophysiological techniques ("voltage clamp" and "patch clamp") have been used for measuring Ca²⁺ currents in cultured neurons and in solubilized Ca²⁺ channels after reconstitution into artificial bilayers (Penner and Neher, 1989). The whole cell configuration allows intracellular application of drugs. Single channel "inside-out" patching permits access to the cytosolic face of the plasma membrane and experiments in the absence of cytosol (Rosenthal *et al.*, 1988).

Fluorescent indicators (e.g., Fura-2, Indo-1, and Fluo-3) Grynkiewicz *et al.*, 1985) contain Ca²⁺-selective-binding sites modeled on EGTA. Their stereochemistry enhances their quantum efficiency and photochemical stability. Compared to their predecessor, "Quin-2," the newer dyes offer up to 30-fold brighter fluorescence and improved selectivity for Ca²⁺ ($10^5 : 1$). With Fura-2, Ca²⁺ binding shifts the excitation spectrum from 380 to 340 nm, with little change in the 510-nm peak of the emission spectrum. These shifts permit [Ca²⁺]ᵢ to be deduced from the shape of the spectrum. The ratio of differences of excitation wavelengths cancels out variations in dye loading and local optical path length and compensates for changes in absolute illumination intensity and in detector sensitivity. These indicators can be loaded into the cytoplasm of cells without disrupting the plasma membrane because their hydrophobic carboxylate groups can be masked with labile esters (e.g., Fura-2 acetoxymethyl ester, Fura-2 AM), which are lipophilic and membrane permanent. Inside the cell, esterases release the free acid form of the dye (Tsien, 1988).

XIV. Conclusions

The concentration of extracellular calcium is in the millimolar range whereas free intracellular calcium is in the nanomolar range (measurement of free calcium in cells has become possible largely because of the development of calcium-binding fluorescent dyes). The massive external : internal concentration gradient is maintained by a large number of factors and this can probably be seen as a reflection of the importance of calcium homeostasis in cells. Homeostatic failure can obviously arise as a result of a large variety of toxic insults, e.g., it could result from metabolic changes or could be due to a specific effect on (for example) glutamatergic (NMDA) receptors. Therefore, in the former case, increases in intracellular free calcium are likely to be a result rather than a cause of neuronal damage.

Data show that increases in $[Ca^{2+}]_i$ resulting from toxic insult are not uniform throughout the cell but that the cytoxicity of calcium has both a spatial and temporal component.

There is now some evidence that β-amyloid protein potentiates the toxicity of glutamate which is produced by increases in neuronal-free calcium. This may be important in the etiology of Alzheimer's disease. There is also some evidence that growth factors such as FGF protect cells from glutamate-induced damage.

The consequence of increased levels of free calcium in cells is that a large number of enzymes become activated (e.g., proteases) and hence cell damage can occur. Elucidation of the specific effects of many of these enzymes in the damage process remains to be accomplished.

References

Ambler, S. K., Poenie, M., Tsien, R. Y., and Taylor, P. (1988). Agonist-stimulated oscillations and cycling of intracellular free calcium in individual cultured muscle cells. *J. Biol. Chem.* **263**, 1952–1959.

Andrews, S. B., Leapman, R. D., Landis, D. M. D., and Rees, T. S. (1987). The distribution of calcium and potassium in presynaptic nerve terminals from cerebellar cortex. *Proc. Natl. Acad. Sci. U.S.A.* **84**, 1713–1717.

Armstrong, D. L. (1989). Calcium channel regulation by calcineurin, a Ca^{2+}-activated phosphatase in mammalian brain. *Trends Neurosci.* **12**, 117–122.

Barritt, G. J. (1981). Calcium transport across cell membranes: progress toward molecular mechanisms. *Trends in Biol. Sci.* **2**, 322–325.

Berridge, M. J. (1987). Inositol triphosphate and diacylglycerol: two interacting second messengers. *Annu. Rev. Biochem.* **56**, 159–193.

Blaustein, M. P. (1988). Calcium transport and buffering in neurons. *Trends Neurosci.* **11**, 438–443.

Brass, L. F., and Belmonte, E. (1989). Calcium-45 exchange techniques to study calcium transport in intact platelets. In *Methods in Enzymology* (J. Hawiger, Ed.), Vol. 169, pp. 371–385. Academic Press, London.

Burgoyne, R. D. (1990). Secretory vesicle-associated proteins and their role in exocytosis. *Annu. Rev. Physiol.* **52**, 647–659.

Dubinsky, J. M. (1993). Increases in intracellular calcium and neurotoxicity: how are they related? In *Markers of Neuronal Injury and Degeneration. Ann. N.Y. Acad. Sci.* **679**, 34–42.

Ellis, C., Noran, M., McCormick, F., and Pawson, T. (1990). Phosphorylation of GAP and GAP-associated proteins by transforming and mitogenic tyrosine kinases. *Nature* **343**, 377–381.

Exton, J. H. (1988). Mechanisms of action of calcium-mobilizing agonists: some variations on a young theme. *FASEB J.* **2**, 2670–2676.

Gibson, G. E., and Peterson, C. (1987). Calcium and the aging nervous system. *Neurobiol. Aging* **8**, 329–343.

Gitelman, H. J. (1987). An improved automated procedure for the determination of calcium in biological specimens. *Anal. Biochem.* **18**, 521–531.

Grynkiewicz, G., Poenie, M., and Tsien, R. Y. (1985). A new generation of Ca^{2+} indicators with greatly improved fluorescence properties. *J. Biol. Chem.* **260**, 3440–3445.

Hallam, T. J., and Rink, T. J. (1989). Receptor-mediated Ca^{2+} entry; diversity of function and mechanism. *Trends Pharmacol. Sci.* **10**, 8–10.

Hemmings, J. C., Nairn, A. C., McGuinness, T. L., Huganir, R. L., and Greengard, P. (1989). Role of protein phosphorylation in neuronal signal transduction. *FASEB J.* **3**, 1583–1592.

Huang, K.-P. (1989). The mechanism of protein kinase C activation. *Trends Neurosci.* **12**, 425–432.

Kater, S. B., Mattson, M. P., Cohan, C., and Connor, J. (1988). Calcium regulation of the neuronal growth cone. *Trends Neurosci.* **11**, 315–320.

Knight, D. E., von Grafenstein, H., and Athayde, C. M. (1989). Calcium-dependent and calcium-independent exocytosis. *Trends Neurosci.* **12**, 451–461.

Lamb, T. D., Matthews, H. R., and Torre, V. (1986). Incorporation of calcium buffers into salamander retinal rods: a rejection of the calcium hypothesis of phototransduction. *J. Physiol.* **372**, 315–349.

Latorre, R., Oberhauser, A., Labarca, P., and Alvarez, O. (1989). Varieties of calcium-activated potassium channels. *Annu. Rev. Physiol.* **51**, 385–399.

Malenka, R. C., Kauer, J. A., Perkel, D. J., and Nicoll, R. A. (1989). The impact of postsynaptic calcium on synaptic transmission—its role in long-term potentiation. *Trends Neurosci.* **12**, 444–450.

Marty, A. (1989). The physiological role of calcium-dependent channels. *Trends Neurosci.* **12**, 420–424.

Mattson, M. P., Cheng, B., Davis, D., Bryant, K., Lieberburg, I., and Russell, E. R. (1992). Beta-amyloid peptides destabilize calcium homeostasis and render human cortical neurons vulnerable to excitotoxicity. *J. Neurosci.* **12**(2), 376–389.

Mattson, M. P., Smith-Swintosky, V., Cheng, B., Lieberburg, I., and Rydel, R. E. (1993). Neuronal calcium homeostasis: stabilization by growth factors and destabilization by beta-amyloid. In *Markers of Neuronal Injury and Degeneration. Ann. N.Y. Acad. Sci.* **679**, 1–21.

McBurney, R. N., and Neering, I. R. (1987). Neuronal calcium homeostasis. *Trends Neurosci.* **10**, 164–169.

Meldolesi, J., Volpe, P., and Pozzan, T. (1988). The intracellular distribution of calcium. *Trends Neurosci.* **11**, 449–452.

Melloni, E., and Pontremoli, S. (1989). The calpains. *Trends Neurosci.* **12**, 438–444.

Morgan, J. I., and Curran, T. (1989). Stimulus-transcription coupling in neurons: role of cellular immediate-early genes. *Trends Neurosci.* **12**, 459–462.

Moskowitz, N., Schook, W., and Puszkin, S. (1984). Regulation of endogenous calcium-dependent synaptic membrane phospholipase A_2. *Brain Res.* **290**, 273–280.

Muallem, S., Pandol, S. J., and Beeker, T. G. (1989). Hormone-evoked calcium release from intracellular stores is a quantal process. *J. Biol. Chem.* **264**, 206–212.

Nachshen, D. A. (1985). Regulation of cytosolic calcium concentration in presynaptic nerve endings isolated from rat brain. *J. Physiol.* **363**, 87–101.

Nachshen, D. A., Sanchez-Armass, S., and Weinstein, A. M. (1986). The regulation of cytosolic calcium in rat brain synaptosomes by sodium-dependent calcium efflux. *J. Physiol.* **381**, 17–28.

Nahorski, S. R. (1988). Inositol polyphosphates and neuronal calcium homeostasis. *Trends Neurosci.* **11**, 444–448.

Nichols, R. A., Sihra, T. S., Czernik, A. J., Nairn, A. C., and Greengard, P. (1990). Calcium/calmodulin-dependent protein kinase II increases glutamate and noradrenaline release from synaptosomes. *Nature* **343**, 647–651.

Nishizuka, Y. (1986). Studies and perspectives of protein kinase C. *Science* **233**, 305–312.

Penner, R., and Neher, E. (1989). The patch-clamp technique in the study of secretion. *Trends Neurosci.* **12**, 159–163.

Persechini, A., Moncrief, N. D., and Kretsinger, R. H. (1989). The EF-hand family of calcium-modulated proteins. *Trends Neurosci.* **12**, 462–467.

Rasmussen, C. D., and Means, A. R. (1989). Calmodulin, cell growth and gene expression. *Trends Neurosci.* **12**, 433–438.

Rasmussen, H. (1989). The cycling of calcium as an intracellular messenger. *Sci. Am.* **261**, 44–51.

Roberts-Lewis, J. M., and Siman, R. (1993). Spectrin proteolysis in the hippocampus: a biochemical marker for neuronal injury and neuroprotection. In *Markers of Neuronal Injury and Degeneration. Ann. N.Y. Acad. Sci.* **679**, 78–86.

Rosenthal, W., Hescheler, J., Trautwein, W., and Schultz, G. (1988). Control of voltage-dependent Ca^{2+} channels. *FASEB J.* **2**, 2784–2790.

Sarafian, T., and Verity, M. A. (1993). Changes in protein phosphorylation in cultured neurones after exposure to methyl mercury. In *Markers of Neuronal Injury and Degeneration. Ann. N.Y. Acad. Sci.* **679**, 65–77.

Smith, S. J., and Augustine, G. J. (1988). Calcium ions, active zones and synaptic transmitter release. *Trends Neurosci.* **11**, 458–464.

Somlyo, A. P., and Himpens, B. (1989). Cell calcium and its regulation in smooth muscle. *FASEB J.* **3**, 2266–2276.

Storch, J., and Schachter, D. (1985). Calcium alters the acyl chain composition and lipid fluidity of rat hepatocyte plasma membranes in vitro. *Biochim. Biophys. Acta* **812**, 473–484.

Tsien, R. Y. (1988). Fluorescent measurement and photochemical manipulation of cytosolic free calcium. *Trends Neurosci.* **11**, 419–424.

Tsien, R. Y. (1989). Fluorescent probes of cell signalling. *Annu. Rev. Neurosci.* **12**, 227–253.

Verhage, M., Besselsen, E., Lopes da Silva, F. H., and Ghijspen, W. E. J. M. (1989). Ca^{2+}-dependent regulation of presynaptic stimulus-secretion coupling. *J. Neurochem.* **53**, 1188–1194.

Williamson, J. R., and Monck, J. R. (1989). Hormone effects on cellular Ca^{2+} fluxes. *Annu. Rev. Physiol.* **51**, 107–124.

In Vitro Electrophysiological Studies in Neurotoxicology[1]

JOEP VAN DEN BERCKEN
Research Institute of Toxicology
University of Utrecht
NL 3508 TD Utrecht, The Netherlands

MARGA OORTGIESEN
Research Institute of Toxicology
University of Utrecht
NL 3508 TD Utrecht, The Netherlands

TRESE LEINDERS-ZUFALL[2]
Research Institute of Toxicology
University of Utrecht
NL 3508 TD Utrecht, The Netherlands

HENK P. M. VIJVERBERG
Research Institute of Toxicology
University of Utrecht
NL 3508 TD Utrecht, The Netherlands

I. Introduction

For many years excised nerve and muscle preparations of experimental animals have been the only choice for vertebrate electrophysiology and hence for electrophysiological studies in the field of *in vitro* neurotoxicology. In particular, preparations of the frog, i.e., excised myelinated nerves and single myelinated nerve fibers, ganglia, nerve-muscle preparations, the isolated spinal cord, and even the brain, have been used to study the effects of neurotoxic compounds. Although the toxicological gap between frog and mammals, including man,

is almost insurmountable from a more general viewpoint, electrophysiological studies in frog nerve preparations have contributed greatly to our present understanding of the functioning of the nervous system and of the mode of action of many neurotoxicants. In some cases, results obtained in frog preparations *in vitro* allowed the interpretation of effects of neurotoxicants on intact animals and even to extrapolate this interpretation to the human situation.

The utility of *in vitro* results may be illustrated by the example of pyrethroid insecticides. At an early stage of pyrethroid development it was shown that the cutaneous touch receptor in an isolated piece of frog skin is rather sensitive to the action of these insecticides (Akkermans *et al.*, 1975). After exposure of the outside of the skin to the pyrethroid allethrin, a small mechanical stimulus, which normally produces only one nerve impulse, caused a train of impulses in the afferent nerve fiber (Fig. 1). Further investigations showed that repetitive firing in sensory nerves was the most important effect of low concentrations of pyrethroids on various parts of the peripheral nervous system. The pyrethroid-induced repetitive nerve activity was studied in detail in

The manuscript was unfinished at the sudden death of Joep van den Bercken on February 20, 1992, and has been completed by the coauthors.

[1] This chapter has been revised and reprinted by permission from "Electrophysiological Approaches to *in vitro* Neurotoxicology" in *Alternative Methods in Toxicology,* Volume 9, edited by J. van den Bercken, *et al.* Copyright © 1990 by Mary Ann Liebert Inc.

[2] Present address: Yale University School of Medicine, Department of Neurobiology, New Haven, Connecticut 06510.

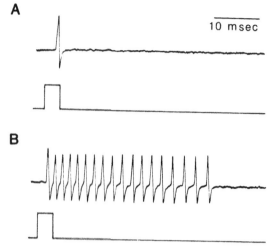

FIGURE 1 Repetitive nerve impulses induced by a pyrethroid insecticide in the peripheral sensory nervous system of the frog *Xenopus laevis*. (A) Nerve impulse evoked in a single sensory nerve fiber by a brief mechanical stimulus (lower trace) in the control situation. (B) Instead of a single nerve impulse a train of repetitive nerve impulses is evoked by the same stimulus after 30 min of *in vitro* exposure of the skin to 10 μM of the pyrethroid allethrin. (After Akkermans *et al.*, 1975.)

the lateral-line sense organ of the clawed frog, providing valuable information on the structure–activity relationship of these insecticides (Vijverberg *et al.*, 1982). Cyano pyrethroids, like cypermethrin, deltamethrin, and fenpropathrin, produced much more intense repetitive firing than non-cyano pyrethroids like allethrin, bioresmethrin, cismethrin, and permethrin. In the meantime, widespread practical application of these insecticides revealed that occupational exposure, in particular to cyano pyrethroids, frequently caused transient local burning or tingling sensations and also itching and numbness (paresthesia) mainly of the facial skin (He *et al.*, 1988; Kolmodin-Hedman *et al.*, 1982). It is now generally accepted that these skin sensations are caused by repetitive firing of sensory nerve endings in the skin, as was first observed in the frog (Lequesne *et al.*, 1980; Vijverberg and van den Bercken, 1990).

Since the 1980s improvements in the field of (neuronal) cell culture methods and revolutionary changes in electrophysiological techniques have advanced research into the mode of action of neurotoxicants to the molecular level. The effects of neurotoxicants on voltage-dependent and receptor-operated processes involved in the excitation of mammalian neurones can now be investigated in great detail under strictly controlled experimental conditons.

Results of recent electrophysiological investigations of effects of pyrethroid insecticides and lead, which are described next, illustrate that the *in vitro* approach contributes significantly to the understanding of the mode of

action of selective and nonselective neurotoxic compounds.

II. Neuronal Cultures: Neuroblastoma N1E-115 Cells

Techniques have been developed to dissociate neuronal tissues mechanically and enzymatically to obtain neuronal cultures, which can be used for neurotoxicological investigations. Dissociated cells, primary cultures, and tissue explant cultures may be obtained from animal species ranging from insects to mammals. These cultures generally contain mixed cell populations, including diverse types of neurones and various glial cells. The finite lifetime and changes of cell properties in the course of culture are disadvantages of these *in vitro* systems. The procedures to isolate neurones need to be repeated over and over again and require experimental animals.

For a number of reasons clonal cell lines are well-suited for electrophysiological studies. With the proper management of stock cultures in liquid nitrogen, clonal cell lines can be maintained almost indefinitely with fairly consistent phenotypes and can be obtained in any quantity. Differentiation can be induced readily by changes in culture conditions. Most importantly, the cultures contain a homogeneous cell population. Cell density can be kept low and culture conditions can be adapted to restrict the outgrowth of neurites without loss of functional characteristics. These latter features are important for electrical control of the membrane potential in voltage clamp experiments. In this connection it is noteworthy that, at least in N1E-115 cells, there is little correlation between morphological differentiation, i.e., cell enlargement and outgrowth of neurites, and electrophysiological differentiation, i.e., electrical and chemical excitability (Kimhi, 1981). Small and apparently undifferentiated neuroblastoma cells without neurites may show action potentials similar to those observed in mature neurones and may even be spontaneously active.

The neuroblastoma clone N1E-115, which is derived from the C-1300 neuroblastoma tumor of the mouse, has been used in electrophysiological investigations from a very early stage of cell culture. The C-1300 tumor originates from a part of the neural crest, which normally develops into sympathetic ganglia. Therefore, N1E-115 cells can be expected to have sympathetic properties. A detailed account of the characteristics of the C-1300 mouse neuroblastoma cells and their electrophysiology up to 1980 can be found elsewhere (Amano *et al.*, 1972; Kimhi, 1981; Spector, 1981). Four distinct types of ion currents were initially identified in these cells: two inward currents carried by voltage-dependent Na$^+$ and Ca^{2+} channels and two outward currents carried by voltage-dependent and by Ca^{2+}-activated K$^+$ channels (Moolenaar and Spector, 1978, 1979a,b). Physiological and pharmacological properties of specific ion channels involved have been characterized in detail by electrophysiological as well as by ion flux and radioligand-binding studies and are essentially similar to those of other mammalian nerve cells. Since the 1980s N1E-115 cells, as well as a variety of other neuronal cell lines, are being used in laboratories all over the world. N1E-115 cells contain voltage-dependent Na$^+$ channels and multiple subtypes of voltage-dependent K$^+$ and Ca^{2+} channels, Ca^{2+}-activated K$^+$ channels, and cation channels. In addition, they contain a variety of receptors for neurotransmitters, some of which are coupled to receptor-operated ion channels. These include nicotinic and muscarinic acetylcholine (ACh) receptor-operated ion channels and serotonin 5-HT$_3$ receptor-operated ion channels. A range of receptors for other neuromodulatory substances has been demonstrated by radioligand binding and second messenger studies.

The availability of N1E-115 cells and other clonal cell lines, together with the development of sophisticated electrophysiological techniques, i.e., the patch clamp technique, to study the functional properties of ion channels down to the level of the individual channel molecule, have started a new era in the field of *in vitro* neurotoxicology. On the one hand the action of neurotoxicants can be studied in great detail, while on the other hand their effects on a variety of ion channels can be studied in the same cell type under the same experimental conditions.

III. Patch Clamp Technique

The voltage clamp technique is a most powerful method to study the electrical properties of cell membranes and to investigate effects of neurotoxicants on excitable membranes. With this technique it is possible to control the voltage across the cell membrane and to simultaneously measure the ionic currents through the membrane. With the advancement of neuronal cell culture the voltage clamp technique has been adapted to record membrane currents in small isolated neurons and extended to record the miniature signals generated by the opening and closing of individual ion channels by means of the single channel patch clamp technique (Fig. 2) (Hamill *et al.*, 1981).

In the whole cell configuration the cell is gently sucked to the opening of a glass pipette with a (sub)micrometer tip opening. The pipette tightly seals to the cell and the remaining leakage between glass

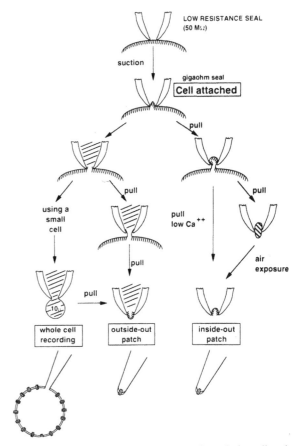

FIGURE 2 Patch pipette configurations for whole cell voltage clamp and for single channel patch clamp of outside-out and inside-out membrane patches. (Modified after Hamill *et al.*, 1981.)

and membrane, which has an electrical resistance in the order of GΩ, becomes negligible. Access to the cell interior is obtained by rupturing the patch of membrane inside the pipette by additional suction or by applying a current pulse, resulting not only in a reduction of the access resistance to the cell required for voltage clamp, but also in a rapid exchange of diffusible components between the pipette solution and the cytoplasm. Under these conditions the ionic composition of the intra- and extracellular solutions can be manipulated and drugs and poisons can be applied to either side of the membrane. By replacing permeable ions with impermeable ones and by adding selective ion channel blockers to the experimental solutions, currents through a specific type of ion channel can be measured without interference of other signals. For single channel recording the pipette is pulled away from the cell, resulting in the isolation of a patch of membrane that is attached to the pipette tip and contains few or only a single ion channel. Depending on experimental manipulations, "inside-out" or "outside-out" membrane patches can be obtained (See Fig. 2).

IV. Action of Pyrethroids

The pyrethroids constitute a major class of synthetic and highly active insecticides with a fairly selective neurotoxic mechanism of action. Pronounced excitatory effects are produced in insects as well as mammals, including man, whenever pyrethroids reach the nervous system in sufficient concentration. As mentioned earlier, the principal effect of pyrethroids is to induce repetitive activity in sense organs and in other parts of both the peripheral and the central nervous system. In addition, they may cause excessive neurotransmitter release due to depolarization of presynaptic nerve endings and, eventually, nerve conduction block (Vijverberg and van den Bercken, 1990).

A. Voltage-Dependent Na^+ Channels

In recent years it has become evident that all insecticidally active pyrethroids share the same mechanism of action on voltage-dependent Na^+ channels in the nerve membrane. In frog myelinated nerve fibers, pyrethroids induce a prolonged inward Na^+ tail current that follows the Na^+ current evoked by a step depolarization of the membrane (Vijverberg et al., 1982). A similar prolonged Na^+ tail current is observed in squid giant axons and other invertebrate nerve fibers after treatment with pyrethroids (Narahashi, 1986). The prolongation of the Na^+ current varies greatly with pyrethroid structure. Cyano-substituted pyrethroids prolong the Na^+ current to a much greater extent than non-cyano pyrethroids. The time course of decay of the Na^+ tail current in frog myelinated nerve fibers correlates well with the intensity of repetitive activity in the lateral-line sense organ of *Xenopus*. Depending on its size and time course the prolonged Na^+ current causes a depolarizing afterpotential, all or not associated with repetitive activity, or more permanent membrane depolarization, which may lead to excessive neurotransmitter release and to a complete block of membrane excitability. In N1E-115 cells exposed to pyrethroids the Na^+ current is prolonged both during depolarization and after repolarization (Fig. 3). In addition, pyrethroids also induce an increase in the peak amplitude of the inward Na^+ current in N1E-115 cells (Ruigt et al., 1987).

On the basis of macroscopic whole cell current recordings it was hypothesized that pyrethroids selectively affect the kinetic properties of the Na^+ channels and that they keep these channels open for a much longer time than is normal. Single channel patch clamp experiments in "outside-out" patches of N1E-115 cells have in fact confirmed that the open time of individual Na^+ channels is greatly prolonged by pyrethroids (Fig. 4) (Chinn and Narahashi, 1986; de Weille, 1986; Yamamoto et al., 1983). Similar results were obtained in dissociated frog spinal ganglion cells (de Weille and Leinders, 1989).

B. Other Ion Channels

It has been reported that high concentrations of pyrethroids have a suppressive effect on Na^+ as well as K^+ channels in invertebrate nerve preparations (Kiss, 1981; Nishimura et al., 1989; Omatsu et al., 1988; Wang et al., 1972). However, lower concentrations of pyrethroids, which significantly affect Na^+ channels, do not affect voltage-dependent K^+ channels (Salgado et al., 1989). It has also been reported that the pyrethroid tetramethrin partially blocks one type of Ca^{2+} channel in N1E-115 cells (Narahashi, 1986). This result has not been confirmed with other pyrethroids (Ruigt, 1984).

Several studies have suggested that pyrethroids, in addition to affecting Na^+ channels, also affect ion channels that are coupled to neurotransmitter receptors. High concentrations of cyano pyrethroids cause a stereoselective partial inhibition of radioligand binding to the GABA receptor–ion channel complex (Lawrence and Casida, 1983). From effects of pyrethroids on the properties of the ACh receptor–ion channel complex it has been suggested that these compounds delay the closing or desensitization of the ion channels coupled to the nicotinic ACh receptor (Sherby et al., 1988). Recent studies in voltage-clamped N1E-115 cells have shown that relatively high concentration of pyrethroids reduce the amplitude of the ACh response. Similar effects were observed on the 5-HT-induced response, which is mediated by an independent population of serotonin 5-HT₃ receptor-operated channels. These effects appeared to be nonspecific, as they were also produced by insecticidally inactive isomers (Oortgiesen et al., 1989). An inhibition of nicotinic ACh receptor- and serotonin 5-HT₃ receptor-mediated responses would cause inhibitory effects instead of the excitatory symptoms observed with pyrethroids.

Although the molecular aspects of pyrethroid action are not yet understood, it can be concluded that these insecticides primarily interact with voltage-dependent Na^+ channels in the nerve membrane in a highly selective manner, keeping these channels open for a much longer time than is normal.

V. Effects of Lead

Although many investigations have demonstrated effects of lead on the nervous system, the mechanism

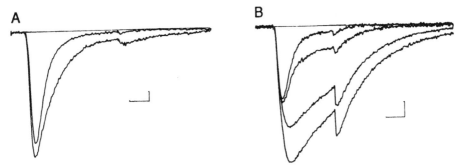

FIGURE 3 Effect of pyrethroid insecticides on Na$^+$ current evoked by a membrane depolarization to −5 mV in internally perfused, voltage-clamped N1E-115 cells. (A) Na$^+$ currents evoked by 10-msec step depolarizations of the whole cell membrane before and 20 min after external application of 100 μM phenothrin (holding potential −95 mV; calibrations: vertical 2 nA, horizontal 2 msec). (B) Four superimposed Na$^+$ current traces evoked by 15-msec step depolarizations to −5 mV before and 1.5, 4.5, and 30 min after external application of 1 μM fenfluthrin (calibrations: vertical 2 nA, horizontal 5 msec). The pyrethroids prolong Na$^+$ currents, enhance peak amplitudes, and induce prolonged Na$^+$ tail currents after termination of membrane depolarization. (After Ruigt *et al.*, 1987.)

of lead neurotoxicity at the cellular and molecular level is still poorly understood (Silbergeld, 1985). It appears that lead acts rather nonselectively, producing dysfunction at various sites within the nervous system to which it gains access in sufficient amounts.

Particular attention has been paid to potential effects of inorganic lead (Pb^{2+}) on synaptic transmission, both pre- and postsynaptically. Electrophysiological experi-

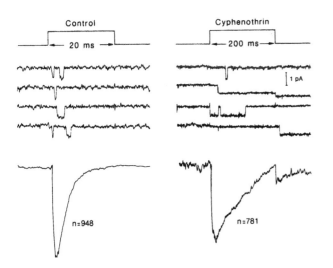

ments on frog and rat nerve-muscle preparations have shown that Pb^{2+} first blocks evoked neurotransmitter release and subsequently enhances spontaneous release (Atchison and Narahashi, 1984; Manalis *et al.*, 1984; Pickett and Bornstein, 1984). Similar results have been obtained in rat cortical synaptosomes (Minnema *et al.*, 1988; Suszkiw *et al.*, 1984). To explain this biphasic effect, it has been proposed that Pb^{2+} first blocks inward Ca^{2+} current in the presynaptic terminal, which is a prerequisite for neurotransmitter release, and subsequently penetrates into the cell and interferes with intracellular Ca^{2+} homeostasis by competing with Ca^{2+} for intracellular binding sites (Manalis *et al.*, 1984). In addition, it has been reported that high concentrations of Pb^{2+} reduce ACh-induced depolarizations in frog muscle (Manalis *et al.*, 1984; Shao and Suszkiw, 1991), whereas 1 μM Pb^{2+} in the mouse hemidiaphragm causes a transient decrease of the ACh response (Oortgiesen *et al.*, 1990a). Comparative effects of Pb^{2+} on the various types of voltage-dependent and receptor-operated ion channels have been investigated in N1E-115 cells.

A. Voltage-Dependent Ion Channels

containing 50 mM Ba^{2+}. In addition Ba^{2+} entry through open Ca^{2+} channels does not lead to the activation of Ca^{2+}-dependent K$^+$ current. Under these conditions membrane step depolarizations evoke fast transient and noninactivating inward Ba^{2+} current components that are carried by two distinct types of voltage-dependent Ca^{2+} channels. During superfusion with Pb^{2+} the amplitude of both Ba^{2+} current components is reversibly blocked in a concentration-dependent manner with an IC$_{50}$ of 4.8 ± 0.8 μM Pb^{2+} and complete block of the Ba^{2+} current is achieved with 100 μM Pb^{2+} (Fig. 5). Effects of Pb^{2+} on whole cell Ca^{2+} currents have been confirmed by a detailed investigation using N1E-115 cells (Audesirk and Audesirk, 1991) and similar results have been obtained with human neuroblastoma cells (Reuveney and Narahashi, 1991) and with rat dorsal root ganglion cells (Evans et al., 1991). The IC$_{50}$ obtained from N1E-115 cells is in the same order of magnitude as the value of 1 μM reported for Ca^{2+} influx inhibition by Pb^{2+} in rat synaptosomes (Minnema et al., 1988; Suszkiw et al., 1984) and as the

value of 1 μM of the dissociation constant of Pb^{2+} from presynaptic Ca^{2+}-binding sites in frog end plate (Maricq et al., 1991). Thus the present experiments corroborate the hypothesis that presynaptic Ca^{2+} channels are the target site for the block of neurotransmitter release by micromolar Pb^{2+} concentrations.

B. Ca^{2+}-Activated K$^+$ Channels

N1E-115 cells contain two types of Ca^{2+}-activated K$^+$ channels (CaK channels). Low conductance (SK) channels with a single channel conductance of 5.4 pS in a physiological K$^+$ gradient are blocked by nanomolar concentrations of the bee venom peptide apamin whereas large conductance (BK) channels with a single channel conductance of 98 pS are insensitive to apamin (Leinders and Vijverberg, 1992; Quandt, 1988). BK channels play a role in the repolarization phase of the action potential (Adams et al., 1982) whereas SK channels are involved in the afterhyperpolarization following the action potential (Hugues et al., 1982).

Single channel patch clamp experiments on "inside-out" patches of N1E-115 cells have revealed that Pb^{2+}, applied to the inside of the membrane, is able to activate CaK channels in the absence of Ca^{2+} (Fig. 6) (Leinders et al., 1992). Pb^{2+} is more potent than Ca^{2+} in activating both types of CaK channel. At a concen-

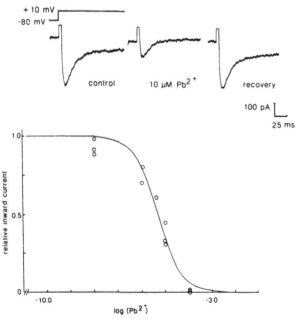

FIGURE 5 Pb^{2+} blocks voltage-dependent Ca^{2+} channels in N1E-115 cells. (Top) Ba^{2+} currents evoked by step depolarizations to +10 mV in control external solution, after 5 min of superfusion with 10 μM Pb^{2+}, and recovery after 5 min of washing with control external solution. The peak amplitude of the transient component of inward Ba^{2+} current was reduced to 34% of the control value by 10 μM Pb^{2+}. Holding potential is −80 mV. (Bottom) The concentration dependence of the blocking effect of Pb^{2+}. Ordinate represents the peak amplitude of the transient inward current normalized to control value. The IC$_{50}$ value and the slope factor estimated from the concentration–effect curve of block of the transient component of the Ba^{2+} current by Pb^{2+} are 4.8 ± 0.8 and -0.88 ± 0.14 μM, respectively. (After Oortgiesen et al., 1990b.)

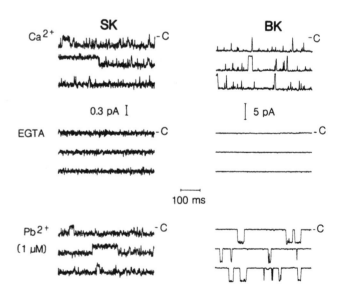

FIGURE 6 Effects of Pb^{2+} on Ca^{2+}-activated K$^+$ channels. Maximum open probability of single SK (left) and BK (right) channels in two inside-out excised patches by superfusion with 14.4 and 115.2 μM buffered free Ca^{2+}, respectively. Subsequent superfusion with Ca^{2+}-free EGTA-containing solution abolished single channel activity. In the same membrane patches superfusion with 1 μM buffered free Pb^{2+} induced full activation of the SK and partial (10% of maximum open probability) activation of the BK channel. Membrane potential was held at 0 mV. (After Leinders et al., 1992.)

tration of 1 μM Pb^{2+} the open probability of the SK channel is maximally enhanced, whereas the BK channel open probability in the presence of 1 μM Pb^{2+} is only 10% of the maximum attainable with Ca^{2+}.

C. Receptor-Operated Channels

Superfusion of voltage-clamped N1E-115 cells with 5-HT or ACh results in transient inward currents. The 5-HT-induced inward current is mediated by 5-HT$_3$ receptors. Pharmacological and physiological properties of the 5-HT$_3$ receptor-operated ion currents in N1E-115 cells have been described in detail (Neijt *et al.*, 1988a, 1989). It has been demonstrated that the 5-HT3 receptor-operated ion channel protein is a member of the class of directly ligand-gated ion channels and resembles the α-subunit of nicotinic receptors from *Torpedo californica* (Maricq *et al.*, 1991). The ACh-induced inward current in N1E-115 differs from that in the end plate by its insensitivity to block by α-bungarotoxin (α-BuTX) and its sensitivity to block by κ-BuTX (Oortgiesen and Vijverberg, 1989). The sensitivity to κ-BuTX is a pharmacological property of neuronal type nicotinic ACh receptors. The availability of agonists and antagonists selective to ACh and 5-HT receptors in N1E-115 cells and the heterologous desensitization of ACh and 5-HT receptor-mediated inward currents demonstrate the presence of two distinct and independent populations of ACh and 5-HT receptor-operated ion channels. The resemblance between the two may be exemplified by the fact that both ACh and 5-HT$_3$ receptor-operated ion currents in N1E-115 are blocked by the arrow poison *d*-tubocurarine, with IC$_{50}$ values of 0.5 μM and 0.8 nM, respectively (Oortgiesen and Vijverberg, 1989; Peters *et al.*, 1990). It is remarkable that the potency of *d*-tubocurarine to block 5-HT-induced ion current is much higher than that to block ACh-induced ion current.

Application of Pb^{2+} at concentrations ranging from 10 nM to 100 μM causes a reduction of the 5-HT-induced inward current, without affecting its time course. This blocking effect of Pb^{2+} is almost completely reversed by washing with external solution for 4–8 min. The estimated value of the IC$_{50}$ of the concentration–effect curve of block of Pb^{2+} is 49 $\pm$ 18 μM (Oortgiesen *et al.*, 1990b).

The peak amplitude of the ACh-induced inward current is 32% reduced by superfusion with 10 nM Pb^{2+} within 7 min, whereas the time course of the ACh response remains unaffected at this very low concentration of Pb^{2+}. At 3 μM Pb^{2+} the amplitude of the ACh-induced inward current reaches only 10% of the control value. However, after higher concentrations of Pb^{2+} the blocking effect is reversed and after superfu-

sion with 100 μM Pb^{2+} the peak amplitude of the ACh response amounts to almost 70% of the control value. In addition, the decay of the remaining ACh-induced inward current is markedly delayed by high concentrations of Pb^{2+}. Figure 7 shows the concentration dependence of the effect of Pb^{2+} on the inward current induced by 1 mM ACh in N1E-115 cells. The data can be fitted by the sum of a descending and an ascending concentration–effect curve with an IC$_{50}$ value of 19 $\pm$ 6.3 nM and an EC$_{50}$ of 21 $\pm$ 5.5 μM, respectively. The reversal of block of the ACh-induced current suggests a dual effect of Pb^{2+} on neuronal nicotinic. ACh receptor-activated channels (Oortgiesen *et al.*, 1990b).

In frog end plate, 10 mM Ni^{2+} also causes a dual effect on ACh receptor-operated ion channels; a reduction of single channel conductance and a simultaneous prolongation of channel open time (Magleby and Weinstock, 1980). Ca^{2+} has been shown to reduce single

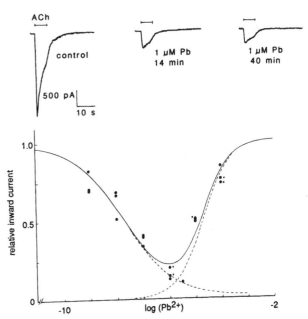

FIGURE 7 Effects of Pb^{2+} on nicotinic ACh-induced ion current. (Top left) An inward current induced by 1 mM ACh in control external solution. During superfusion with 1 μM Pb^{2+} the peak amplitude of the ACh-induced inward current was reduced to a steady level which amounts to 18% of the control value. Note that 1 μM Pb^{2+} also delays the time course of decay of the ACh-induced inward current. Membrane potential was held at -80 mV. Superfusion periods are indicated by bars. (Bottom) The concentration dependence of Pb^{2+} effects on the nicotinic ACh-induced ion current. Ordinate represents the inward current peak amplitude normalized to control value. The data were fitted by the sum of two sigmoidal concentration–effect curves. The estimated parameters of the fitted curve (solid line) are: IC$_{50}$ = 19 $\pm$ 6.3 nM; EC$_{50}$ = 21 $\pm$ 5.5 μM, and the slope factors are -0.45 ± 0.08 and 0.84 ± 0.15, respectively. The dashed lines represent the concentration–effect curve for the blocking effect of Pb^{2+} and for the reversal of block, according to the fitted parameters. (After Oortgiesen *et al.*, 1990b.)

channel conductance and to increase opening frequency of ACh receptor-operated ion channels in rat central neurones (Mulle *et al.*, 1992). Effects of Pb^{2+} on single cholinergic ion channels in N1E-115 cells remain to be investigated.

In addition to the effects described earlier, superfusion of N1E-115 cells with high concentrations of Pb^{2+} (>10 μM) induce a slow, noninactivating inward current, which is not mediated by a known type of receptor-operated or voltage-dependent ion channel. A similar current is induced by superfusion with Cd^{2+} and Al^{3+}. These currents appear mediated by a novel type of metal ion-activated channel, which has no known physiological function at present (Oortgiesen *et al.*, 1990c).

The results, which are summarized in Table 1, show that nueronal type nicotinic ACh receptors are the most sensitive to Pb^{2+} and are selectively blocked by nanomolar concentrations of Pb^{2+}. The same low concentrations of Pb^{2+} do not affect 5-HT$_3$ receptor-activated channels, voltage-dependent Na^+ and Ca^{2+} channels, or big Ca^{2+}-activated K^+ channels. Small Ca^{2+}-activated K^+ channels may constitute another sensitive target of Pb^{2+} after entering the cytoplasmic compartment of neurones. Table 1 also demonstrates that at various fixed concentrations Pb^{2+} selectively modifies subtypes of channels of the various classes investigated. In the micromolar range, external Pb^{2+} blocks voltage-dependent Ca^{2+} channels selectively as compared to Na^+ channels whereas internal Pb^{2+} selectively activates Ca^{2+}-dependent SK channels as compared to BK channels. The differential effects on neuronal type nicotinic ACh receptor-operated ion channels and 5-HT$_3$ receptor-operated channels also suggest selective and distinct interactions of Pb^{2+} with subtypes of directly ligand-gated ion channels. Although picomolar concentrations of intracellular Pb^{2+}

have been reported to activate protein kinase C (Markovac and Goldstein, 1988), neuronal type ACh receptors are among the most sensitive targets of Pb^{2+} known. Therefore, these effects should be considered in the interpretation of cognitive deficits resulting from low-level lead exposure (Needleman and Bellinger, 1991).

VI. Concluding Remarks

N1E-115 neuroblastoma cells are highly suited for detailed electrophysiological investigations into the mechanisms of action of neurotoxicants on various types of nerve membrane ion channels. These cells contain a variety of voltage-dependent, Ca^{2+}-activated and receptor-operated channels. The investigation of structure-related effects of selective agents at the target site provides information on their intrinsic potencies and such knowledge may assist in the interpretation of effects in intact animals. The diversity of ion channels in these cells can be exploited to investigate effects of nonselective neurotoxicants over a wide concentration range and on a variety of potential targets. The example of Pb^{2+} demonstrates that this is not only confirmatory but also leads to surprising results as the finding that postsynaptic neuronal type nicotinic ACh receptors constitute a highly sensitive target for Pb^{2+} and the discovery of ion channels that are yet to be supplied with a counterpart *in vivo*. Since it is highly unlikely that a single cell type can ever be regarded as a universal model of the mammalian neuron, additional models with complementary properties, including human cell lines, are required to extend *in vitro* neurotoxicological research. Investigation of basic physiological, biochemical, and pharmacological properties of distinct cell types is time consuming, but seems to be essential to adequately define and select complementary models.

Acknowledgments

The research has been financially supported by the University of Utrecht, the Netherlands Organization for Scientific Research, and by Shell Internationale Research Mij. B. V.

TABLE 1 Effects of Pb^{2+} on Subtypes of Voltage-Dependent, Receptor-Operated, and Ca^{2+}-Activated Ion Channels in N1E-115 Cells

Channel type and effect	EC$_{50}$ of Pb^{2+}
Voltage-dependent	
Ca^{2+} channel block	5μM
Na^+ channel block	>100 μM
Ca^{2+}-activated	
SK channel opening	<1 μM
BK channel opening	>1 <90 μM
Receptor-operated	
Nicotinic ACh channel block	20 nM
- reversal of block	20 μM

References

Adams, P. R., Constanti, A., Brown, D. A., and Clark, R. B. (1982). Intracellular Ca^{2+} activates a fast voltage sensitive K^+ current in vertebrate sympathetic neurones. *Nature* **296**, 746–749.

Akkermans, L. M. A., van den Bercken, J., and Versluijs-Helder, M. (1975). Comparative effects of DDT, allethrin, dieldrin and aldrin-transdiol on sense organs of *Xenopus laevis*. *Pestic. Biochem. Physiol.* **5**, 451–457.

Amano, T., Richelson, E., and Nirenberg, P. G. (1972). Neurotrans-

Atchison, W. D., and Narahashi, T. (1984). Mechanism of action of lead on neuromuscular junctions. *Neurotoxicology* **5**, 267–282.

Audesirk, G., and Audesirk, T. (1991). Effects of inorganic lead on voltage-sensitive calcium channels in N1E-115 neuroblastoma cells. *Neurotoxicology* **12**, 519–528.

Chinn, K., and Narahashi, T. (1986). Stabilization of sodium channel states by deltamethrin in mouse neuroblastoma cells. *J. Physiol.* **380**, 191–207.

de Weille, J. R. (1986). *The Modification of Nerve Membrane Sodium Channels by Pyrethroids.* Ph. D. Thesis, University of Utrecht.

de Weille, J. R., and Leinders, T. (1989). The action of pyrethroids on sodium channels in myelinated nerve fibres and spinal ganglion cells of the frog. *Brain Res.* **482**, 324–332.

Evans, M. L., Büsselberg, D., and Carpenter, D. O. (1990). Pb²⁺ blocks calcium currents of cultured dorsal root ganglion cells. *Neurosci. Lett.* **129**, 103–106.

Hamill, O. P., Marty, A., Neher, E., Sakmann, B., and Sigworth, F. J. (1981). Improved patch-clamp techniques for high-resolution current recording from cells and cell-free membrane patches. *Pflügers Arch.* **391**, 85–100.

He, F., Sun, H., Han, K., Wu, Y., Yao, P., Wang, S., and Liu, L. (1988). Effects of pyrethroid insecticides on subjects engaged in packaging pyrethroids. *Br. J. Ind. Med.* **45**, 548–551.

Hugues, M., Romey, G., Duval, D., Vincent, J. P., and Lazdunski, M. (1982). Apamin as a selective blocker of the calcium-dependent potassium channel in neuroblastoma cells: voltage-clamp and biochemical characterization of the toxin receptor. *Proc. Natl. Acad. Sci. U.S.A.* **79**, 1308–1312.

Kimhi, Y. (1981). Nerve cells in clonal systems. In *Excitable Cells in Tissue Culture* (P. G. Nelson and M. Lieberman, Eds.), pp. 173–245, Plenum, New York.

Kiss, T. (1988). Effect of deltamethrin on transient outward currents in identified snail neurones. *Comp. Biochem. Physiol. C* **91**, 337–341.

Kolmodin-Hedman, B., Swensson, A., and Åkerblom, M. (1982). Occupational exposure to some synthetic pyrethroids (permethrin and fenvalerate). *Arch. Toxicol.* **50**, 27–33.

Lawrence, L. J., and Casida, J. E. (1983). Stereospecific action of pyrethroid insecticides of the γ-aminobutyric acid receptor-ionophore complex. *Science* **221**, 1399–1401.

Leinders, T., and Vijverberg, H. P. M. (1992). Ca²⁺ dependence of small Ca²⁺-activated K⁺ channels in cultured N1E-115 mouse neuroblastoma cells. *Pflügers Arch.* **422**, 223–232.

Leinders, T., van Kleef, R. G. D. M., and Vijverberg, H. P. M. (1992). Divalent cations activate SK and BK channels in mouse neuroblastoma cells: selective activation of SK channels by cadmium. *Pflügers Arch.* **422**, 217–222.

Lequesne, P. M., Maxwell, I. C., and Butterworth, S. T. G. (1980). Transient facial sensory symptoms following exposure to synthetic pyrethroids: a clinical and electrophysiological assessment. *Neurotoxicology* **2**, 1–11.

Magleby, K. L., and Weinstock, M. M. (1980). Nickel and calcium ions modify the characteristics of the acetylcholine receptor-channel complex at the frog neuromuscular junction. *J. Physiol.* **299**, 203–218.

Manalis, R. S., Cooper, G. P., and Pomeroy, S. L. (1984). Effects of lead on neuromuscular transmission in the frog. *Brain Res.* **294**, 95–109.

Maricq, A. V., Peterson, A. S., Brake, A. J., Myers, R. M., and Julius, D. (1991). Primary structure and functional expression of the 5-HT₃ receptor, a serotonin-gated ion channel. *Science* **254**, 432–437.

Markovac, J., and Goldstein, G. W. (1988). Picomolar concentrations of lead stimulate brain protein kinase. *C. Nature* **334**, 71–73.

Minnema, D. J., Michaelson, I. A., and Cooper, G. P. (1988). Calcium efflux and neurotransmitter release from rat hippocampal synaptosomes exposed to lead. *Toxicol. Appl. Pharmacol.* **92**, 351–357.

Moolenaar, W. H., and Spector, I. (1978). Ionic currents in cultured mouse neuroblastoma cells under voltage-clamp conditions. *J. Physiol.* **278**, 265–286.

Moolenaar, W. H., and Spector, I. (1979a). The calcium action potential and a prolonged calcium dependent afterhyperpolarization in mouse neuroblastoma cells. *J. Physiol.* **292**, 297–306.

Moolenaar, W. H., and Spector, I. (1979b). The calcium current and the activation of a slow potassium conductance in voltage-clamped mouse neuroblastoma cells. *J. Physiol.* **292**, 307–323.

Mulle, C., Léna, C., and Changeux, J. P. (1992). Potentiation of nicotinic receptor response by external calcium in rat central neurones. *Neuron* **8**, 937–945.

Narahashi, T. (1986). Mechanisms of actions of pyrethroids on sodium and calcium channel gating. In *Neuropharmacology and Pesticide Action* (M. G. Ford, G. G. Lunt, R. C. Reay, and P. N. R. Usherwood, Eds.), pp. 36–60, Ellis Horwood, Chichester, England.

Needleman, H. L., and Bellinger, D. (1991). The health effects of low level exposure to lead. *Annu. Rev. Publ. Health* **12**, 111–140.

Neijt, H. C., te Duits, I. J., and Vijverberg, H. P. M. (1988a). Pharmacological characterization of serotonin 5-HT₃ receptor-mediated electrical response in cultured mouse neuroblastoma cells. *Neuropharmacology* **27**, 301–307.

Neijt, H. C., Karpf, A., Schoeffter, P., Engel, G., and Hoyer, D. (1988b). Characterization of 5-HT₃ recognition sites in membranes of NG108-15 neuroblastoma-glioma cells by radioligand binding. *Naunyn-Schmiedeb. Arch. Pharmacol.* **337**, 493–499.

Neijt, H. C., Plomp, J. J., and Vijverberg, H. P. M. (1989). Kinetics of the membrane current mediated by serotonin 5-HT₃ receptors in cultured mouse neuroblastoma cells. *J. Physiol.* **411**, 257–269.

Nishimura, K., Omatsu, M., Murayama, K., Kitasato, H., and Fujita, T. (1989). Neurophysiological effects of the pyrethroid insecticides bioresmethrin and kadethrin on crayfish giant axons. *Comp. Biochem. Physiol. C* **93**, 149–154.

Omatsu, M., Murayama, K., Kitasato, H., Nishimura, K., and Fujita, T. (1988). Effect of substituted benzyl chrysanthemates on sodium and potassium currents in the crayfish giant axon. *Pestic. Biochem. Physiol.* **30**, 125–135.

Oortgiesen, M., Lewis, B. K., Bierkamper, G. G., and Vijverberg, H. P. M. (1990a). Are postsynaptic nicotinic end-plate receptors involved in lead toxicity? *Neurotoxicology* **11**, 87–92.

Oortgiesen, M., van Kleef, R. G. D. M., Bajnath, R. B., and Vijverberg, H. P. M. (1990b). Nanomolar concentrations of lead selectivity block neuronal nicotinic acetylcholine responses in mouse neuroblastoma cells. *Toxicol. Appl. Pharmacol.* **103**, 165–174.

Oortgiesen, M., van Kleef, R. G. D. M., and Vijverberg, H. P. M. (1989). Effects of pyrethroids on neurotransmitter-operated ion channels in cultured mouse neuroblastoma cells. *Pestic. Biochem. Physiol.* **34**, 164–173.

Oortgiesen, M., van Kleef, R. G. D. M., and Vijverberg, H. P. M. (1990c). Novel type of ion channel activated by Pb²⁺, Cd²⁺ and Al³⁺ in cultured mouse neuroblastoma cells. *J. Membr. Biol.* **113**, 261–268.

Oortgiesen, M., and Vijverberg, H. P. M. (1989). Properties of neuronal type acetylcholine receptors in voltage clamped mouse neuroblastoma cells. *Neuroscience* **31**, 169–179.

Peters, J. A., Malone, H. M., and Lambert, J. J. (1990). Antagonism of 5-HT₃ receptor mediated currents in murine N1E-115 neuroblastoma cells by (+) tubocurarine. *Neurosci. Lett.* **110**, 107–112.

Pickett, J. B., and Bornstein, J. C. (1984). Some effects of lead at mammalian neuromuscular junction. *Am. J. Physiol.* **246**, C271–C276.

Quandt, F. N. (1988). Three kinetically distinct potassium channels in mouse neuroblastoma cells. *J. Physiol.* **395**, 401–408.

Reuveny, E., and Narahashi, T. (1991). Potent blocking action of lead on voltage-activated calcium channels in human neuroblastoma cells SH-SY5Y. *Brain Res.* **545**, 312–314.

Ruigt, G. S. F. (1984). *An Electrophysiological Investigation into the Mode of Action of Pyrethroid Insecticides.* Ph.D. Thesis, University of Utrecht.

Ruigt, G. S. F., Neijt, H. C., van der Zalm, J. M., and van den Bercken, J. (1987). Increase of sodium current after pyrethroid insecticides in mouse neuroblastoma cells. *Brain Res.* **437**, 309–322.

Salgado, V. L., Herman, M. D., and Narahashi, T. (1989). Interaction of the pyrethroid fenvalerate with nerve membrane sodium channels: temperature dependence and mechanism of depolarization. *Neurotoxicology* **10**, 1–14.

Shao, Z., and Suszkiw, J. B. (1991). Ca^{2+}-surrogate action of Pb^{2+} on acetylcholine release from rat brain synaptosomes. *J. Neurochem.* **56**, 568–574.

Sherby, S. M., Eldefrawi, A. T., Deshpande, S. S., Albuquerque, E. X., and Eldefrawi, M. E. (1988). Effects of pyrethroids on nicotinic acetylcholine receptor binding and function. *Pestic. Biochem. Physiol.* **26**, 107–115.

Silbergeld, E. K. (1985). Neurotoxicology of lead. In *Drug and Chemical Toxicology*, Vol. 3, *Neurotoxicology* (K. Blum and L. Manzo, Eds.), pp. 299–322, Dekker, New York.

Spector, I. (1981). Electrophysiology of clonal nerve cell lines. In *Excitable Cells in Tissue Culture* (P. G. Nelson and M. Lieberman, Eds.), pp. 247–277, Plenum, New York.

Suszkiw, J., Toth, G., Murawsky, K., and Cooper, G. P. (1984). Effects of Pb^{2+} and Cd^{2+} on acetylcholine release and Ca^{2+} movements in synaptosomes and subcellular fractions from rat brain and *Torpedo* electric organ. *Brain Res.* **323**, 31–46.

Vijverberg, H. P. M., and van den Bercken, J. (1990). Neurotoxicological effects and the mode of action of pyrethroid insecticides. *CRC Crit. Res. Toxicol.* **21**, 105–126.

Vijverberg, H. P. M., Ruigt, G. S. F., and van den Bercken, J. (1982). Structure-related effects of pyrethroid insecticides on the lateral-line sense organ and on peripheral nerves of the clawed frog, *Xenopus laevis*. *Pestic. Biochem. Physiol.* **18**, 315–324.

Vijverberg, H. P. M., van der Zalm, J. M., and van der Bercken, J. (1982). Similar mode of action of pyrethroids and DDT on sodium channel gating in myelinated nerves. *Nature* **295**, 601–603.

Wang, C. M., Narahashi, T., and Scuka, M. (1972). Mechanism of negative temperature coefficient of nerve blocking action of allerthrin. *J. Pharmacol. Exp. Ther.* **182**, 442–453.

Yamamoto, D., Quandt, F. N., and Narahashi, T. (1983). Modification of single sodium channels by the insecticide tetramethrin. *Brain Res.* **274**, 344–349.

PART
VI

Clinical Neurotoxicology

Principles and Issues
An Introductory Overview

NEIL L. ROSENBERG

Department of Medicine (Clinical Pharmacology and Medical Toxicology),
University of Colorado School of Medicine,
Denver, Colorado 80110

Neurotoxicology has emerged in recent years as a science of considerable breadth and as a discipline of great practical importance, both to society and to the practice of medicine. On the one hand, chemicals with neurotoxic potential represent defined experimental tools with which to probe the function of the nervous system and to model a variety of neurological disorders. On the other hand, the neurotoxic potential of occupational and environmental agents and many therapeutic and abused drugs constitute a growing problem on a global scale.

In 1990, the Office of Technology Assessment, the analytical arm of the U.S. Congress, published a manuscript regarding the problem of neurotoxins and how they can be identified and controlled. Previously, in 1983, the National Institute of Occupational Safety and Health listed neurotoxicity as 1 of the top 10 causes of work-related disorders in the United States. From these reports, it is estimated that 3–5% of the over 65,000 industrial chemicals, excluding pesticides, display neurotoxic potential. Of these, 28% are industrial chemicals for which occupational exposure standards are being developed and produce neurotoxic effects at some dose. Also, a large percentage of the 600 active pesticide ingredients registered with the U.S. Environmental Protection Agency have neurotoxic potential. In addition, a substantial number of therapeutic and abused drugs have significant neurotoxic effects.

Despite the tremendous impact that neurotoxins have had in science and medicine, neurotoxicology is arguably one of the most neglected areas of neurology. Interest, however, is growing as legislators and the public become educated in the subject.

Clinical neurotoxicology, as a distinct field of study in the neurosciences, is beginning to come of age. Although scientists have used neurotoxins in the laboratory for decades, until relatively recently, few have considered the threat that neurotoxins may play in our environment as worthy of serious investigation.

The effects of neurotoxins on health have been known since antiquity, with the earliest descriptions of heavy metal intoxication. Today, concerns about exposure sources range from occupational and environmental exposures to organic solvents and pesticides to the effects that naturally occurring compounds in foods may have in the development of neurodegenerative disorders. New, presumably toxic disorders, such as "sick building syndrome" and "multiple chemical sensitivity," have arisen seemingly out of nowhere to wreak havoc on the health care and legal systems. With these rapidly growing concerns there is an ever-increasing knowledge gap on how environmental chemicals may effect health and, specifically, the nervous system. We know the mechanism of nervous system injury for very few neurotoxins and have almost as little scientific evidence on which chemicals have neurotoxic potential.

Because of this problem of dealing with the clinical effects of neurotoxic chemicals, this section is designed to cover major areas of study, but to also cover basic principles of neurotoxicology.

Dr. Neil L. Rosenberg begins with an overview of basic principles of neurotoxicology. Many of these principles, such as dose–response relationships, are well-known among toxicologists, but are often neglected or ignored by clinicians when evaluating patients with possible neurotoxic injuries. Certain principles, such as neurotoxins not producing focal syndromes, are more specific to the field of neurotoxicology. One of the important messages from this chapter is that the focused history is still the most essential part of the evaluation of the patient with possible neurotoxic disease.

Dr. Ken Kulig writes on industrial and agricultural chemicals and their resultant neurotoxicity. The chemicals discussed are from four main groups and include solvents, pesticides, metals, and gases. The solvent syndrome, a controversial topic in neurotoxicology, is discussed in some detail, as well as the different organophosphate neurotoxic syndromes. These include acute OGP toxicity, the delayed neurotoxic syndromes, and the more recently described "intermediate syndrome." Overviews of metal poisoning, still a relatively common problem, and gases, particularly carbon monoxide poisoning, are also dealt with.

The chapter by Dr. Kevin T. Finnegan discusses neurotoxicologic concerns of drug abuse, an increasingly recognized problem. Both licit (ethanol) and illicit drugs of abuse have potential neurotoxicity. The ef-

fects of alcohol on the nervous system have been known for thousands of years, but it continues to be a major economic and public health problem throughout the world. Illicit drugs, primarily stimulants and opiates, particularly controlled substance analogs produced in clandestine drug-manufacturing laboratories, have been an important source of neurotoxicity in recent years and will, unfortunately, likely continue.

Dr. Terrence L. Cascino writes about the neurotoxic effects of antineoplastic agents, one of the more common causes of iatrogenic neurotoxic injuries. Agents with well-described neurotoxicity as well as newer agents are discussed. With an increasing number of antineoplastic agents as well as new combinations of standard agents, often in massive doses, potential neurotoxic injuries may be seen with increasing frequency in cancer patients.

Dr. Albert C. Ludolph's chapter is about a rapidly changing area of neurotoxicology: the association of neurotoxins and neurodegenerative disorders. Although both industrial and biological toxins have been used in the past to study pathogenetic mechanisms of naturally occurring neurodegenerative disorders, they are being increasingly sought as possible environmental agents in the cause of certain of these disorders. Most of the work to date has been in the study of Parkinson's disease, but studies are also being done in other sporadic neurodegenerative disorders such as motor neuron disease, hyperkinetic movement disorders, and Alzheimer's disease.

The chapters in this section represent the state of the art in these respective areas and are areas of ongoing active research. Continuing efforts are necessary to expand our understanding of possible pathogenetic mechanisms of neurotoxic injuries and ultimately the prevention and treatment of these disorders in man.

References

U.S. Congress, Office of Technology Assessment (1990). *Neurotoxicity: Identifying and Controlling Poisons of the Nervous System.* OTA-BA-436 U.S. Government Printing Office, Washington, D.C.

Centers for Disease Control (1983). Leading work-related diseases and injuries-United States. **MMWR 32,** 24–26.

Basic Principles of Clinical Neurotoxicology

NEIL L. ROSENBERG

Department of Medicine (Clinical Pharmacology and Medical Toxicology),
University of Colorado School of Medicine,
Denver, Colorado 80110

I. Introduction

Although recognition of neurotoxic injuries dates to antiquity with descriptions of heavy metal toxicity, it has only been since the 1970s that clinical neurotoxicology has grown as an increasingly important specialty. This has occurred as a consequence of the dramatic growth of industrial and agricultural chemicals as well as pharmaceuticals and drugs of abuse. With the increase of these groups of compounds has grown the number of situations that neurotoxic effects have been recognized.

Other chapters in this volume describe the basic science of neurotoxicology, areas that our knowledge is greatly advanced compared to our understanding of the adverse effects of toxic compounds on the human nervous system. This section discusses descriptions of the neurotoxicology of some of the major categories of drugs and chemicals encountered in clinical practice, including drugs of abuse (an alarming and growing area of concern), antineoplastic agents (some of the more commonly encountered iatrogenic neurotoxic injuries),

and industrial and agricultural chemicals. Environmental agents and their possible link to "idiopathic" neurodegenerative disorders, such as Parkinson's disease, Alzheimer's disease, and amyotrophic lateral sclerosis, are also discussed. This chapter will attempt to define the general field of clinical neurotoxicology, including some general principles. Case examples will be used to help illustrate these principles.

II. What Is Clinical Neurotoxicology?

For the purposes of this chapter, clinical neurotoxicology is defined as the study of the adverse effects of any chemical, pharmaceutical, or physical agent on the nervous system of a living organism.

In this chapter, the living organism will be human unless otherwise indicated. In addition, physical agents, such as radiation, electromagnetic fields, noise, and vibration, will not be discussed both because of their markedly different adverse effects on living organisms and because they are not generally considered "toxins."

Despite the lack of a controlled laboratory setting, a great deal of our understanding of the neurotoxic effects of a variety of agents has been gained by their often devastating consequences in humans. Often, the first suggestion that a substance has neurotoxic effects at all is because of an observation by a clinician. Following these observations, efforts are then directed by the basic scientist in the laboratory to try to better understand these effects, which will hopefully lead to more effective treatment or prevention.

Our understanding of clinical neurotoxicology comes from exposures occurring as the result of individual accidents (such as industrial settings), larger scale accidents affecting many individuals simultaneously (such as major industrial accidents affecting workers and/or surrounding communities), intentional infliction upon another individual (as in homicide attempts), unintentional infliction on another individual (as in iatrogenic neurotoxicity from prescribed drugs), intentional self-infliction (as in suicide attempts), and unintentional self-infliction (as in drug abuse).

Two additional definitions merit mention here as well. One is an overview definition of toxicology, which is very applicable to neurotoxicology (1):

Toxicology is a study of the interaction between chemicals and biologic systems in order to quantitatively determine the potential for the chemical(s) studied to produce injury which results in adverse effects in living organisms, and to investigate their nature, incidence, mechanism of production, factors influencing, and reversibility of such adverse effects.

Another definition, specific to neurotoxicology, has been used both by the Office of Technology Assessment and the Environmental Protection Agency (2):

Neurotoxicity or a neurotoxic effect is "an adverse change in the structure or function of the nervous system following exposure to a chemical agent."

All three definitions are slightly different; however, they all contain either the phrase "adverse effects" or "adverse change." There is disagreement as to what is meant by "adverse." Some would only interpret an adverse effect to be related to an irreversible, pathologic change in the nervous system, whereas others would consider any effect (including those that are completely reversible) to be adverse. For the purposes of this chapter, neurotoxic effects are those that cause irreversible, pathologic change of the nervous system, whereas adverse effects are any effect on neurologic function, even those that are transient and completely reversible. Therefore, all neurotoxic effects are adverse, but not all adverse effects are neurotoxic. Neurotoxic and adverse effects are further defined later in this chapter in relation to acute and chronic exposures.

III. Classification of Neurotoxic Disease

Several classification schemes for neurotoxic disease have been previously offered and include classification based on clinical effects (Table 1), classification by type of chemical (Table 2), or classification based on the cellular and subcellular target (Table 3). Each of these different classification schemes has both supporters and detractors and each offers a different approach to try to distill the vast array of neurotoxic substances into a logical, scientific classification. The most widely accepted classification scheme, though still imperfect, is the one that groups agents according to their primary cellular and subcellular neural targets because it incorporates the clinical effects and chemical type. However, one area that is not addressed well in any of these classification schemes is the area of neurodegenerative disorders, primarily because it is not yet known which toxins are capable of producing these "naturally occurring" diseases, with only a few well-recognized exceptions. These disorders are discussed in detail in Chapter 43.

IV. Basic Principles of Clinical Neurotoxicology

Generally speaking, up to this point, this chapter has only discussed basic principles of neurotoxicology; however, this section presents a detailed analysis of those principles that are most important in evaluating for the presence of neurotoxic disease. The basic principles of toxicology have been reviewed (3,4), but the following discussions list some basic principles of toxicology that are of particular relevance to the understanding of clinical neurotoxicology, specifically those that help in the recognition of neurotoxic illnesses.

How best to incorporate what is known about basic mechanisms of neurotoxicity into a rational approach to an individual with possible neurotoxic disease is less than perfect since these mechanisms are known for so few neurotoxins. In addition, the primary cellular and subcellular targets are also known for relatively few neurotoxins (Table 3).

Several important principles are listed as they apply to evaluating individuals with possible neurotoxic injuries. When appropriate, the case studies in the Appendix will be referred to as examples of how these principles are applied in the clinical setting.

A. Acute and Chronic Neurotoxicity

Many would define acute and chronic neurotoxicity as the two principle types of neurotoxicity. Understand-

TABLE 1 Major Neurologic Clinical Syndromes Produced by Various Categories of Neurotoxins with Specific Examples (Limited Listing)

Seizures
  Pharmaceuticals
    Psychiatric (antidepressants)
    Antineoplastic agents (alkylating agents)
    Neurologic drugs (anticholinergic agents)
  Drugs of abuse
    Narcotics (morphine)
    Stimulants/hallucinogens (amphetamine)
    Alcohol
  Industrial chemicals
    Solvents (methanol)
    Agricultural chemicals (organophosphates, organochlorines, carbamates)
    Metals (lead)
    Gases (carbon monoxide)
  Biological toxins (bacterial, venoms, plants, vitamin A)
Encephalopathy
  Pharmaceuticals
    Psychiatric (antidepressants)
    Antineoplastic agents (alkylating agents)
    Neurologic drugs (anti-epileptic drugs)
  Drugs of abuse
    Narcotics (all)
    Stimulants/hallucinogens (all)
    Alcohol
  Industrial chemicals
    Solvents (all)
    Agricultural chemicals (organophosphates, organochlorines, carbamates)
    Metals (aluminum)
    Gases (hydrogen sulfide)
  Biological toxins (bacterial, venoms, plant, vitamins)
Cerebellar dysfunction
  Pharmaceuticals
    Psychiatric (lithium)
    Antineoplastic agents (methotrexate)
    Neurologic drugs (anti-epileptic drugs)
  Drugs of Abuse
    Narcotics (all)
    Stimulants/hallucinogens (all)
    Alcohol
  Industrial Chemicals
    Solvents (all)
    Agricultural chemicals (organophosphates, organochlorines)
    Metals (mercury)
    Gases (methyl chloride)
  Biological toxins (bacterial, venoms, plant, vitamins)
Peripheral neuropathy
  Pharmaceuticals
    Antineoplastic agents (*cis*-platinum)
    Neurologic drugs (phenytoin)
  Drugs of abuse
    Alcohol
  Industrial chemicals
    Solvents (*n*-hexane)
    Agricultural chemicals (organophosphates, organochlorines)
    Metals (arsenic)
    Gases (nitrous oxide)

continued

TABLE 1 Continued

  Biological toxins (bacterial: diphtheria, vitamin B_6)
Parkinsonism and other movement disorders
  Pharmaceuticals
    Psychiatric (neuroleptics)
  Drugs of abuse
    Narcotics (MPTP)
  Industrial chemicals
    Solvents (methanol, carbon tetrachloride)
    Agricultural chemicals (organophosphates)
    Metals (manganese)
    Gases (carbon monoxide)

ing the difference between these two forms of neurotoxicity is of even more importance in neurotoxicology than in other areas of toxicology because the issue of reversibility versus permanent/irreversibility also occurs.

Some neurotoxic effects are reversible whereas others are irreversible. When a chemical produces pathologic injury to the central nervous system (CNS), the lack of ability of the CNS to regenerate indicates that most injuries are permanent. However, the peripheral nervous system (PNS), which retains the ability to regenerate, can suffer significant pathologic injury and still have most or all function recover.

Acute neurotoxic effects, commonly due to *physiologic* changes in the nervous system, and do not involve *pathologic* changes or degeneration of neuronal elements, are rapidly reversible. This situation typically develops after a single, large exposure and refers to the "adverse effects" defined earlier. Examples would be those effects due to pharmacologic modification of excitable neuronal membranes or of a neurotransmitter system.

Most chronic neurotoxic effects are associated with *pathologic* changes of the neuronal elements and are typically incompletely reversible or not reversible. This situation occurs after repeated moderate or large exposure and refers to the "neurotoxic effects" defined earlier.

These descriptions of acute (typically reversible) and chronic (typically irreversible) neurotoxicity are only generalizations, and many exceptions are known.

Although a descriptive animal toxicity test will usually define acute lethality (the first toxicity test performed on a new chemical), subacute (14 days), subchronic (90 days), and chronic exposures (6 months to 2 years) in terms of a defined time (3), it may not correlate well with human exposure. Because these are controlled laboratory situations, and no matter how much effort is taken to correlate the time exposure to humans, the results often fall far short of adequate modeling, particularly in the chronic, irreversible setting.

TABLE 2 Classification of Neurotoxins Based on Chemical Type (Limited Listing)

I. Pharmaceuticals
  A. Psychiatric
    1. Neuroleptics
    2. Antidepressants
    3. Lithium
    4. Benzodiazepines
    5. MAO inhibitors
  B. Antineoplastic agents
    1. Alkylating agents
    2. Methotrexate
    3. Vinca alkaloids
    4. *cis*-Platinum
    5. Cytosine arabinoside
    6. Procarbazine
    7. 5-Fluorouracil
  C. Antibiotics
  D. Neurologic drugs
    1. Anti-convulsants
    2. Anti-parkinsonian agents
    3. Centrally active cholinesterase inhibitors
  E. Other
II. Drugs of abuse
  A. Narcotics
    1. Opiates
    2. Barbiturates
    3. Controlled substance analogs: MPTP
  B. Stimulants/hallucinogens
    1. Amphetamine/methamphetamine
    2. Cocaine
    3. Controlled substance analogs: Ecstacy
    4. Phencyclidine (PCP)
    5. Marijuana
  C. Alcohol
III. Industrial chemicals
  A. Solvents
  B. Agricultural chemicals
    1. Pesticides: organophosphates, organochlorine, carbamates
    2. Herbicides
    3. Rodenticides
  C. Metals
  D. Gases
IV. Biological toxins
  A. Bacterial: diphtheria, tetanus, botulism
  B. Animal: venoms (scorpion, snake, and other reptile, bees, ticks, fish)
  C. Plant: mushrooms, cyanide, ergots, rapeseed oil, *Lathyrus sativus* and *Lathyrus cicera*, other plants
  D. Vitamin: A, D, and B_6
V. Miscellaneous

TABLE 3 Classification of Neurotoxins Based on Basic Mechanisms and Cellular and Subcellular Targets

I. Primary metabolic effects on neurons
  A. Neuronal membrane
    1. Membrane channel blockers
    2. Membrane channel openers
  B. Neurotransmitters
    1. Cholinergic system
      a. Cholinergic agonist
      b. Anticholinesterases
      c. Anticholinergic
      d. Presynaptic and postsynaptic cholinergic blockade
    2. Catecholaminergic system
      a. Cholinergic/adrenergic effects
      b. Catecholaminergic
      c. Depletion of catecholamines
    3. Serotonergic pathways
    4. GABAergic toxicity
    5. Glutaminergic pathways
  C. Structural integrity of neuron
    1. Neuronal Degeneration
      a. CNS
      b. PNS
    2. Axonal degeneration
      a. Central–peripheral distal axonopathy
      b. Central distal axonopathy
II. Primary metabolic effects on myelin
III. Primary metabolic effects on blood vessels
IV. Primary metabolic effects on neuromuscular junction
V. Primary metabolic effects on muscle
VI. Secondary effects due to hypoxia/ischemia
  A. Anoxic anoxia
  B. Ischemia anoxia, Secondary to cardiac arrest
  C. Cytotoxic anoxia

See cases A and B in the Appendix for a further discussion on this principle.

B. Strong Dose–Response Relationship

The dose–response relationship is a fundamental concept in toxicology, and simply states that within certain limits, under controlled conditions, there is a positive relationship between the amount (e.g., "dose") of a chemical that an individual or group of individuals is exposed to and the toxic effect (e.g., "response"). This concept also assumes that there will always be a certain amount of variability of the response since there is a great degree of variability with respect to biochemical, cellular, tissue, and organ system characteristics among individuals, especially in humans. This is also true, although to a lesser degree, among even inbred strains of animals used in toxicologic studies designed to evaluate the dose–response relationship.

In regard to the dose–response relationship for neurotoxic effects, it is generally felt that for most neurotoxins the variability of response is less than for other toxic effects (5). The reasons for this are not entirely clear, but it is a consistent observation in clinical practice. Since humans are typically exposed at much lower levels than those used in the laboratory in toxicity studies, it is necessary to extrapolate these dose–response relationships and to rely on those few clinical situations where a dose–response can be estimated. In addition to dosage extrapolations from laboratory to human exposures, one also needs to extrapolate for conditions of exposure, size, life span, metabolism, brain maturation rate, and absorption.

Much of the dose–response effects are concerned with carcinogenic effects instead of noncarcinogenic effects such as neurotoxicity. Neurotoxic effects are assumed to occur only when a certain level of exposure, threshold, has been exceeded; it is assumed that carcinogenic substances pose some risk at any level of exposure. Although some would argue that a similar risk exists for neurotoxic substances, clinical evidence for this assumption is lacking in most instances (6). Therefore, it appears that there is very little variability in the effects of most neurotoxins on different individuals. Although extremes of age or health (especially severe liver disease) may be related to different effects from the same level of exposure, this does not generally occur in the occupational setting (where most higher level exposures occur), where there is a narrower range of ages and most individuals are in good health.

In addition to the little variability of effects, neurotoxins are not associated with *allergic* (including hypersensitivity reactions) or *idiosyncratic* reactions. Chemical allergy results from previous sensitization to a chemical or one that is structurally similar. These allergic effects are not dose related, with even subthreshold levels producing a toxic response. With rare exceptions, neurotoxins are not associated with such allergic responses.

Chemical idiosyncrasy is a genetically determined abnormal reaction to a chemical that also rarely occurs with neurotoxins. A well-known example of an idiosyncratic neurotoxic effect occurs in individuals with a genetically determined atypical pseudocholinesterase. Inheriting this abnormal enzyme causes an individual to have a very prolonged, potentially fatal, effect to the use of succinylcholine, used to induce skeletal muscle relaxation during surgery. Chemical idiosyncrasy, like chemical allergy, does not follow normal dose–response effects and is uncommon with neurotoxic exposures.

The dose–response evaluation for neurotoxic substances is derived from the concept of a threshold effect (Fig. 1). This threshold is referred to as no observed effect level (NOEL) or no observed adverse effect level (NOAEL) in exposed experimental animals or humans. NOEL refers to that dose at or below which no biological effect of any kind is observed, whereas NOAEL refers to that dose at or below which no adverse effect is observed. NOAEL is currently the most commonly used threshold effect in neurotoxic substance evaluations.

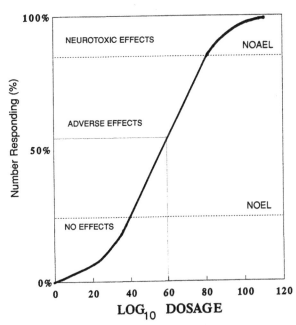

FIGURE 1 Diagram of a quantal dose–response relationship with the abscissa as a log dosage of the neurotoxin and the ordinate as the percent responding to the neurotoxin. The percent responding may be mortality, where an LD_{50} (50% mortality) is determined (where light line intersects curve at 50% responding) or some other monitored parameter. In the area below the no observed effect level (NOEL), no effect of any kind is seen. In the area between the NOEL and the no observed adverse effect level (NOAEL), adverse effects are seen, but are reversible. In the area above the NOAEL, potentially irreversible neurotoxic effects are seen (see text for definitions of adverse and neurotoxic effects).

See cases A and B in the Appendix for a further discussion on this principle.

C. *Proximity to Exposure and Improvement with Removal of Toxin and Delayed Neurotoxicity*

This refers to the observation that maximum symptoms generally occur with maximum exposure, with little delay in the onset of symptoms, and improvement rapidly occurs with either removal of the toxic substance or removal of the individual from the place of exposure.

Although this principle has been one of the cornerstones in our understanding of clinical neurotoxicology, it may need redefining, particularly in the area of neurodegenerative disorders such as Parkinson's disease and Alzheimer's disease. Based on findings of individuals exposed to 1-methyl-4-phenyl-1,2,3,6-tetrahydropyridine (MPTP) and the search for environmental toxins as possible etiologic agents for these disorders, remote exposures (i.e., those exposures which may have been symptomatic or even asymptomatic many years before the development of symptoms of a "neurodegenerative" disorder) may one day be linked to the much later development of progressive neurologic disorders. This concept also emphasizes the fear that the extremely low levels of industrial pollutants in the environment, levels that are insufficient to produce neurologic symptoms, may actually predispose susceptible individuals to the later development of a progressive neurodegenerative disorder.

Figure 2 demonstrates several possible mechanisms that remote exposures may lead to delayed neurotoxicity and the development of a progressive neurodegenerative disorder.

See cases A and B in the Appendix for a further discussion on this principle.

D. Selective Neurotoxicity

Selectivity refers to the notion that a chemical produces injury to one type of tissue or organism without harming some other tissue or organism, even though the two may exist in intimate contact. Selectivity can also refer to a toxin producing injury to a specific cell type or even a particular part of a cell without damaging another part of the cell. The cell type or part of the cell affected will determine the type of clinical effect that a neurotoxin may produce or whether there will be recovery of function or irreversible damage. Table 3 lists several neurotoxins and their sites of action.

See case A in the Appendix for a further discussion on this principle.

E. Multiple Syndromes from a Single Toxin

Depending on the level and duration of exposure (i.e., the dose–response relationship), one may develop widely different neurologic syndromes. For example, an acute high level exposure to *n*-hexane produces a transient nonspecific encephalopathy, characterized by euphoria, disorientation, and a cerebellar ataxia. Chronic, lower level exposure is associated with a peripheral neuropathy and a conspicuous lack

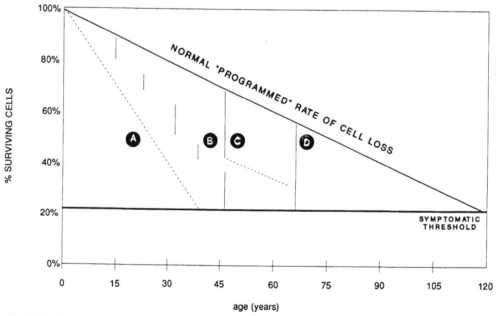

FIGURE 2 Possible mechanisms for toxic-"induced" neurodegenerative disorders. A well-accepted dogma in neuroscience is that any population of neurons has a normal rate of cell loss over the life of an organism. The slope of cell loss in this illustration (normal "programmed" rate of cell loss) crosses the symptomatic threshold (defined as 80% loss of initial cell population) at approximately 120 years, implying that at 120 years of age the organism would develop symptoms related to that cellular loss. An example of this would be loss of 80% of the substantia nigra dopaminergic neurons, which would result in Parkinson's disease. Possible mechanisms whereby a toxin(s) could cause an earlier expression of a "progressive degenerative" neurologic disorder, such as Parkinson's disease, include the following examples. (A) Constant environmental exposure to a substance toxic to a neuronal population that accelerates the normal rate of neuronal cell loss. (B) Multiple exposures to a substance toxic to a neuronal population that causes a demonstrable loss to some neurons, but still above the symptomatic threshold. Between exposures, the normal rate of cell loss continues. (C) A single, large exposure, still above the symptomatic threshold. After the exposure, the normal rate of cell loss continues. (D) A single, large exposure, which causes enough cell loss that the symptomatic threshold is crossed.

of CNS dysfunction. This principle emphasizes the need to not only know what an individual is exposed to, but also the general level of exposure.

F. A Chemical Formula May Not Predict Toxicity

Chemicals with similar structures may not have similar toxicities. Toluene (methyl benzene) produces a dramatic clinical syndrome with chronic exposure characterized by dementia, anosmia, and cerebellar ataxia. Removing the methyl group from toluene produces a chemical with no clear chronic neurotoxicity, benzene, but which is felt to be carcinogenic.

G. Enhancement of Neurotoxicity by "Innocent Bystanders"

This principle, although important to recognize, adds to the confusion in evaluating the clinical situation. Typically, individuals are exposed to multiple chemicals and/or chemical mixtures (more common with organic solvents than other chemicals). Although neurotoxicity may be known for an individual chemical, how it interacts with others in a mixture and how that may contribute to or enhance neurotoxicity is almost never known. Mixtures of chemicals are generally felt to be more neurotoxic than individual chemicals, but evidence of this is lacking. There are few examples of enhancement of neurotoxicity by "innocent bystanders," but one should keep in mind the enhancement of *n*-hexane neurotoxicity by methyl ethyl ketone to understand the concern of this occurring in other situations.

This concern can be minimized if one realizes that just because there is exposure to multiple chemicals or chemical mixtures, it does not mean that enhancement of neurotoxicity invariably occurs. The following possible effects should be kept in mind when mixtures of chemicals occur.

1. Potentiation

This occurs when one material of low or no toxicity enhances the toxic effect of another material, resulting in a more severe injury. An example of this is methyl ethyl ketone (no neurotoxic potential) potentiating the toxic neuropathy caused by *n*-hexane or methyl butyl ketone.

2. Antagonism

Antagonism occurs when one material interferes or reduces the toxic effect of another material, resulting in less or no neurotoxic injury. An example of this is

the use of ethyl alcohol to treat individuals with methyl alcohol toxicity.

3. Independent Effects

This is the situation with many neurotoxins, where materials exert their toxic effects independent of each other, particularly when they target different sites of the nervous system. Examples of this effect include nitrous oxide-induced myelopathy and aminoglycoside-induced ototoxicity.

4. Additive Effects

This effect occurs when materials with similar nervous system targets produce a toxic effect that is equal to the sum of the effects of the different materials. An important aspect to remember about additive effects is that each individual material has to be above the NOAEL in order for its toxic effect to be added to the total toxic effect. For example, if an individual is exposed to 10 different substances, but only 2 are above NOAEL, only those 2 are added to arrive at a final toxic effect. An example of this is peripheral neuropathy caused by chronic exposure to *n*-hexane in an individual taking toxic doses of vitamin B_6, which also produces a peripheral neuropathy.

5. Synergism

Synergism occurs when two or more materials with neurotoxic potential produce a greater toxic effect than either material would produce individually. A synergistic effect is always greater than an additive effect.

Determining the overall neurotoxicity from exposures to mixtures is complex in the clinical situation where laboratory control is not possible. The mixture itself may be complex and the parent compounds may contain degradation products and contaminants. Exposure routes may be multiple and include inhalation, ingestion, percutaneous absorption, and/or ocular absorption. A toxic effect to an individual may also be altered by prior exposure, overall health, and age. Although the complexity in trying to evaluate neurotoxicity related to a chemical mixture is formidable, referring back to these basic definitions and principles when confronted with an exposed individual will allow rational decision making in most instances.

H. Limited Clinical Laboratory Testing

In individuals with neurotoxic injuries, the clinical laboratory is generally not very useful. Clinical laboratory tests to assess industrial exposures are useful in acute exposures to assess the degree of exposure, but usually the history will identify the substance to which the individual was exposed. Acute exposures are usually completely reversible as far as neurotoxic injuries are concerned, and these individuals do not come to the attention of the neurologist.

Chronic low-level exposure, often associated with neurotoxicity, may occur at such low levels that clinical laboratory tests may fail to establish a body burden for that particular chemical. Establishing a link between exposure and clinical dysfunction relies on a focused history of exposure and a neurologic examination, and not on the clinical laboratory.

I. Nonfocal Syndromes

This principle is one of the most useful when evaluating a patient with presumed neurotoxic injury. Neurotoxins generally affect the central and peripheral nervous systems in a diffuse or bilaterally symmetric fashion. Examples of diffuse injury include the distal axonopathy from acrylamide (7) or the diffuse CNS white matter changes due to chronic toluene abuse (8). Bilaterally symmetric injuries are primarily related to those toxins that have specific cell targets, such as the effect of MPTP on the dopaminergic neurons of the substantia nigra (9).

See cases A and B in the Appendix for a further discussion on this principle.

J. Asymptomatic Neurotoxic Disease

Asymptomatic neurotoxic disease certainly occurs in nonoccupational settings, and therefore would be expected to occur in the occupational setting as well (5). Workers in paint manufacturing facilities have been found to have subclinical neuropsychological deficits (10), and further studies may reveal that asymptomatic disease may be an extremely common phenomenon.

See cases A and B in the Appendix for a further discussion on this principle.

V. Conclusions

Understanding some of the basic concepts of toxicology, those principles that are relatively specific to neurotoxic injuries, and how they apply in the clinical setting is necessary when trying to recognize neurotoxic disease. Defining the nature of the exposure and properly evaluating the clinical features (described in detail in Chapter 44) are the first critical steps in being able to recognize a neurotoxic injury from a naturally occurring disease.

APPENDIX

Case A

Chronic High Level Toluene Exposure in an Inhalant Abuser

A 28-year-old man, an inhalant abuser, was evaluated neurologically for effects related to chronic inhalation of toluene-containing products. The principle product abused was a clear spray lacquer containing 61% toluene by weight, and the vapors from an average of three 12-oz. cans of this product were inhaled on a daily basis for over 15 years. When he was first evaluated, he was acutely intoxicated and his neurologic examination at that time revealed a severe dementia (he was oriented to person but not place or time). He had such a severe cerebellar ataxia (both limb and

truncal) that he was unable to stand, sit, or even roll over in bed. His eye movements were extremely abnormal with nystagmus in all directions of gaze and opsoclonus was also present. Hearing (from a sensorineural hearing loss) and vision (from bilateral optic neuropathy) were severely impaired, and there was marked spasticity in all four extremities and bilateral Babinski signs were present.

Brain stem auditory-evoked potentials (BAEP) revealed marked delays in latencies bilaterally and were symmetric (Fig. 3). Magnetic resonance imaging (MRI) of the brain revealed diffuse cerebral atrophy and prominent changes in the white matter (Fig. 4).

The patient was evaluated repeatedly over a 6-year period of sustained abstinence from all organic solvents, other inhalants, and other drugs of abuse. His

LEFT EAR

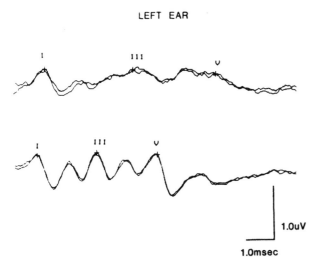

FIGURE 3 BAEP recordings from a chronic toluene abuser (upper tracing) and a normal individual. Note markedly prolonged latencies of waves III and V, and III–V and I–V interpeak latencies. This slowing was bilaterally symmetric. (From Rosenberg *et al.*, 1988.)

neurologic abnormalities improved gradually over that time, although most of the improvement occurred in the first 3 months of abstinence. His last neurologic examination (approximately 6 years after being initially evaluated) revealed cognitive impairment on detailed neuropsychological testing, but he was able to function sufficiently to maintain full-time employment as a custodian. He had some difficulty with mild cerebellar ataxia, but this was no longer functionally disabling. Reflexes were still hyperactive (indicative of residual spasticity) and Babinski signs were present, but these findings were also not associated with functional disability. Vision was improved, he had improved hearing (although he was using hearing aids), and eye movements were normal.

1. Acute and Chronic Neurotoxicity

This case illustrates both acute and chronic neurotoxicity. The patient was first evaluated acutely intoxicated, which resolved rapidly in a week's time. This rapid reversal of acute solvent intoxication has been previously reported (11–13) and is typical of most neu-

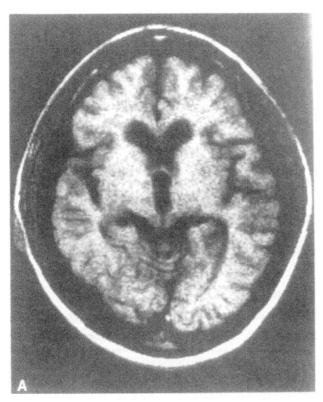
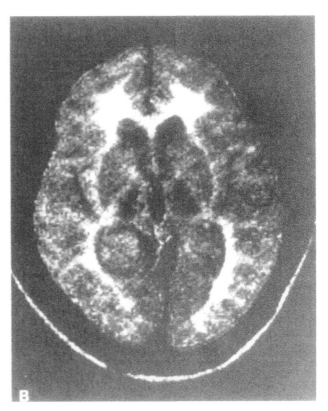

FIGURE 4 (A) A T1-weighted spin echo scan (TR = 30 msec) shows abnormally prominent gray/white differentiation throughout. (B) A T2-weighted spin echo scan (TR = 2000 msec, TE = 60 msec) showing severe, extensive hyperintense white matter. Note also loss of normal gray/white differentiation in areas that are not hyperintense. (From Rosenberg *et al.*, 1988.)

rologic effects resulting from acute intoxications. However, even after 6 years, despite abstinence, some irreversible effects still exist (i.e., neurotoxic effect). These irreversible changes have been widely reported in the literature (8,11–16).

2. Strong Dose–Response Relationship

In chronic solvent abuse, many neurotoxic effects have been demonstrated (8,11–16). In most cases with abnormalities noted on MRI or BAEP, the individual abuser inhaled highly concentrated amounts of solvent vapor several hours per day for many years. A dose–response relationship between neuropsychological abnormalities and MRI changes has also been reported (16). Individuals with less intense abuse (i.e., not daily) generally do not have the neurotoxic effects demonstrated on MRI or BAEP.

3. Proximity to Exposure and Improvement with Removal of Toxin and Delayed Neurotoxicity

Despite the irreversible neurotoxic effects observed in this case, after abstaining, much clinical improvement occurred. Once the improvement begins, providing no repeat exposure, it is sustained. Therefore, no delayed effects or late deterioration is seen.

4. Selective Neurotoxicity

Although no apparent specific cell population is affected by toluene, based on neuropathologic and MRI studies (8,14–16), the white matter appears to be selectively attacked.

5. Nonfocal Syndromes

Clinically, this patient had "multifocal" nervous system involvement, but had never demonstrated a "focal" neurologic deficit, such as hemiparesis, or other unilateral findings. The MRI revealed "diffuse" white matter changes and the BAEP demonstrated bilaterally symmetric slowing, both supportive of nonfocality.

6. Asymptomatic Neurotoxic Disease

Symptoms represent complaints by an individual. Even when this patient demonstrated marked abnormalities on his neurologic examination, he had no complaints of any kind. This is an extreme example of why asymptomatic neurotoxic disease may be extremely common.

Case B

Chronic Low-Level Exposure to Solvent Mixtures in an Occupational Setting

A 32-year-old autobody painter was evaluated for symptoms felt to be related to organic solvent exposure in his workplace. He was employed by the same company for 7 years and worked 10 hr per day, 5 days per week, spray painting autobody parts in a spray booth. The area was ventilated by several fans and the patient had always worn an appropriate respirator.

During the course of his employment he had never experienced any untoward effects except for occasional headaches. These headaches would appear during the course of the workday and gradually lessen after leaving work, resolving before going to sleep. The headaches occurred on average three times per month, but never occurred on weekends or holidays.

One month prior to this evaluation, however, he experienced an episode where he was painting without a respirator and was overcome by the solvent vapors. He did not lose consciousness but needed to be taken outside the building for fresh air. He rapidly improved and was able to continue working and complete the workday. Over the next several weeks he began to complain increasingly about headaches, fatigue, sleeping difficulties, memory impairment, and depression. On examination, he appeared depressed. His responses on memory testing were slow but accurate. The remainder of his neurologic examination was normal.

Because of persistent complaints the following tests were performed and were normal: MRI scan of the brain, brain stem auditory-evoked responses, and neuropsychological tests. The only abnormality on neuropsychological tests was evidence of depression.

The patient was started on antidepressants and underwent appropriate psychotherapy. After 3 months, he was functioning normally and working full-time.

1. Acute and Chronic Neurotoxicity

This case represents low-level chronic exposure to organic solvents, below the NOAEL (Fig. 1). Therefore, although this patient complained of acute symptoms while working, they were all rapidly reversible. No chronic neurotoxic effects were demonstrated on neurologic examination, neuropsychological testing, MRI, or BAEPs.

2. Strong Dose–Response Relationship

Whether exposure to organic solvents in the workplace produces any chronic neurotoxic effect is a hotly debated topic (10,17–25). The degree of exposure in the workplace is thousands of time lower than in the abuse setting (see case A). Therefore, given the strong dose–response relationship with neurotoxins, it is difficult to see how this low-level exposure can produce any neurotoxic effect, even though acute low-level reversible effects, as seen in this case, are common. Despite the fact that workplace exposures fall far short of the doses required to produce the effects seen in inhalant abusers, this will continue to be a controversial area.

3. Proximity to Exposure and Improvement with Removal of Toxin and Delayed Neurotoxicity

The occupational history of this patient demonstrates how symptoms first begin while at work, worsen during the workday, and gradually subside after leaving work. Symptoms did not occur when not working or on weekends and holidays. Once corrective changes were made at work and therapy was begun, the patient gradually improved, with no delayed effects or late deterioration.

4. Nonfocal Syndromes

In this case, nonfocality can only be demonstrated in his symptoms since his neurologic examination and all tests were normal.

5. Asymptomatic Neurotoxic Disease

Although overt neurotoxicity has not been convincingly demonstrated from low-level chronic organic solvent exposure in the workplace, others have shown evidence of "subclinical" effects (10).

References

1. Balantyne, B. (1989). Toxicology. In *Encyclopedia of Polymer Science and Engineering,* Vol. 16, pp. 879–930, John Wiley, New York.
2. *Neurotoxicology: Identifying and Controlling Poisons of the Nervous System* (1990). U.S. Congress, Office of Technology Assessment, OTA-BA-436. p. 44. U.S. Government Printing Office, Washington, D.C.
3. Klaassen, C. D. (1986). Principles of toxicology. In *Cassarett and Doull's Toxicology: The Basic Science of Poisons* (C. D. Klaassen, M. O. Amdur, and J. Doull, Eds.), 3rd ed., pp. 11–32. Macmillan, New York.
4. Ballantyne, B., and Sullivan, J. B. (1992). Basic principles of toxicology. In *Hazardous Materials Toxicology* (J. B. Sullivan, Jr., and G. R. Krieger, Eds.), pp. 9–23, Williams and Wilkins, Baltimore.
5. Schaumburg, H. H., and Spencer, P. S. (1987). Recognizing neurotoxic disease. *Neurology* **37,** 276–278.
6. Silbergeld, E. K. (1988). Developing formal risk assessment methods for neurotoxins. In *Neurobehavioral Methods in Occupational and Environmental Health.* Pan American Health Organization, Washington, D.C.
7. Spencer, P. S., and Schaumburg, H. H. (1974). Review of acrylamide neurotoxicity. II. Experimental animal neurotoxicity and pathologic mechanisms. *Can. J. Neurol. Sci.* **1,** 152–169.
8. Rosenberg, N. L., Kleinschmidt-DeMasters, B. K., Davis, K. A., Dreisbach, J. N., Hormes, J. T., and Filley, C. M. (1988). Toluene abuse causes diffuse central nervous system white matter changes. *Ann. Neurol.* **23,** 611–614.
9. Burns, R. S., Chiueh, C. C., Markey, S. P., Ebert, M. H., Jacobowitz, D. M., and Kopin, I. J. (1980). A primate model of parkinsonism: selective destruction of dopaminergic neurons in the pars compacta of the substantia nigra by *n*-methyl-4-phenyl-1,2,3,6-tetrahydropyridine. *Proc. Natl. Acad. Sci. U.S.A.* **80,** 4546–4550.
10. Bolla, K. I., Schwartz, B. S., Agnew, J., Ford, P. D., and Bleecker, M. L. (1990). Subclinical neuropsychiatric effects of chronic low-level solvent exposure in U.S. paint manufacturers. *J. Occup. Med.* **32,** 671–677.
11. Lazar, R. B., Ho, S. U., Melen, O., and Daghestani, A. N. (1983). Multifocal central nervous system damage caused by toluene abuse. *Neurology* **33,** 1337–1340.
12. Hormes, J. T., Filley, C. M., and Rosenberg, N. L. (1986). Neurologic sequelae of chronic solvent vapor abuse. *Neurology* **36,** 698–702.
13. Fornazzari, L., Wilkinson, D. A., Kapur, B. M., and Carlen, P. L. (1983). Cerebellar, cortical and functional impairment in toluene abusers. *Acta Neurol. Scand.* **67,** 319–329.
14. Rosenberg, N. L., Spitz, M. C., Filley, C. M., Davis, K. A., and Schaumburg, H. H. (1988). Central nervous system effects of chronic toluene abuse: clinical, brainstem evoked response and magnetic resonance imaging studies. *Neurotoxicol. Teratol.* **10,** 489–495.
15. Filley, C. M., Franklin, G. M., Heaton, R. K., and Rosenberg, N. L. (1989). White matter dementia: clinical disorders and implications. *Neuropsychiat. Neuropsychol. Behav. Neurol.* **1,** 239–254.
16. Filley, C. M., Heaton, R. K., and Rosenberg, N. L. (1990). White matter dementia in chronic toluene abuse. *Neurology* **40**(3), 532–534.
17. Grasso, P., Sharratt, M., Davies, D. M., and Irvine, D. (1984). Neurophysiological and psychological disorders and occupational exposure to organic solvents. *Fundam. Chem. Toxic.* **10,** 819–852.
18. Flodin, U., Edling, C., and Axelson, O. (1984). Clinical studies of psychoorganic syndromes among workers with exposure to solvents. *Am. J. Ind. Med.* **5,** 287–295.
19. Juntunen, J., Hupli, V., Hernberg, S., and Luisto, M. (1980). Neurological picture of organic solvent poisoning in industry: a retrospective clinical study of 37 patients. *Int. Arch. Occup. Environ. Health* **46,** 219–231.
20. Gregersen, P., Klausen, H., and Elsnab, C. V. (1987). Chronic toxic encephalopathy in solvent-exposed painters in Denmark 1976–1980: clinical cases and social consequences after a 5-year follow-up. *Am. J. Ind. Med.* **11,** 399–417.
21. Baker, E. L., Letz, R. E., Eisen, E. A., Pothier, L. J., Plantamura, D. L., Larson, M., and Wolford, R. (1988). Neurobehavioral effects of solvents in construction painters. *J. Occup. Med.* **30,** 116–123.
22. Fidler, A. T., Baker, E. L., and Letz, R. E. (1987). The neurobehavioral effects of occupational exposure to organic solvents among construction painters. *Br. J. Ind. Med.* **44,** 292–308.
23. Errebo-Knudsen, E. O., and Olsen, F. (1986). Organic solvents and presenile dementia (the painters' syndrome): a critical review of the Danish literature. *Sci. Total Environ.* **48,** 45–67.
24. Gade, A., Mortensen, E. L., and Bruhn, P. (1988). "Chronic painter's syndrome": a reanalysis of psychological test data in a group of diagnosed cases, based on comparisons with matched controls. *Acta Neurol. Scand.* **77,** 293–306.
25. Errebo-Knudsen, E. O., and Olsen, F. (1987). Solvents and the brain: explanation of the discrepancy between the number of toxic encephalopathy reported (and compensated) in Denmark and other countries. *Br. J. Ind. Med.* **44,** 71–72.

CHAPTER

40

Clinical Neurotoxicology of Industrial and Agricultural Chemicals

KEN KULIG
Division of Emergency Medicine and Trauma
University of Colorado Health Sciences Center
Morrison, Colorado 80465

Neurotoxic chemicals are commonly found in many workplaces, the environment, and the home. Thirty percent of the workplace chemicals for which the American Conference of Governmental Industrial Hygienists (ACGIH) have recommended maximum exposure concentrations have been so listed in part because of their neurotoxic potential. Many other common chemicals may have neurotoxic properties that have not yet been recognized.

The incidence of human chemical neurotoxicity is difficult to determine with accuracy for many reasons. In many cases symptoms and signs of neurotoxicity are insidious and not recognized. Concomitant ethanol, tobacco, street drug, prescription drug, family history of naturally occurring neurologic disease, normal aging, and other environmental factors may be offered as alternative explanations for a neurotoxic syndrome. A genetic predisposition to neurotoxicity may explain why one worker develops a neuropathy when co-workers similarly exposed do not. In the absence of a single massive exposure, determining which agents a given worker has been exposed to is sometimes difficult. Under right-to-know legislation this problem has

been alleviated somewhat but not eliminated. Finally, a long latent period from exposure to symptoms which is typical of many neurotoxic exposures may lead a physician to "rule out" neurotoxicity or to never even consider the possibility.

The litigious nature of our society makes this an extremely important topic at this time. Even minor exposures to known or potential neurotoxins can result in worker's compensation claims or civil litigation. Performing a battery of subjective tests on a worker who is convinced that his health is ruined, and who stands to win millions of dollars if he can convince a jury of this, often confuses rather than clarifies the scientific issues surrounding exposure to neurotoxins.

How then can a clinician perform a proper evaluation of a patient with a possible neurotoxic syndrome? This chapter provides an outline for such an evaluation and then focuses on several prototypical neurotoxins, describing the classic neurotoxic syndrome associated with each and possible diagnostic strategies the clinician can take. This discussion will emphasize the initial history and physical examination as well as initial laboratory studies which will assist in confirming the diagnosis. Detailed neurological evaluation is discussed in other parts of this book.

The chief complaint may vary from subtle generalized symptoms such as irritability or mood changes, to more alarming and emergent problems such as loss of coordination, peripheral neuropathy, or loss of consciousness. Whenever a physician determines that a patient has a neurologic syndrome without obvious cause, it is imperative that a detailed occupational and hobby history be obtained. The latter is commonly forgotten, but can be extremely important if the patient uses toxic chemicals in the home, such as lead glazing in stained glass or ceramic making. Types of questions to ask include the following:

1. What type of work do you do now, and for how long have you performed this exact job? What did you do before? Have you done other jobs in the same company in the past several months where you were exposed to chemicals?
2. What exactly do you do at work? What chemicals do you come in contact with? Can you provide material safety data sheets (MSDSs) for these? Do you use any type of protection at work (i.e., a respirator)? Do you wear it all the time? Who fitted it for you? Do you wear gloves, apron, boots, etc.?
3. Who is your immediate supervisor and may I speak to that person?
4. Has anybody else in the workplace complained of similar problems? Has anybody in your home complained of similar problems?
5. When are your symptoms the worst? When are they not so bad? What happens to your symptoms when you go on vacation? Is there anything you can do to make your symptoms better or worse?
6. Describe your use of the following: alcohol, tobacco, street drugs. Do you currently take any medications? What medications have you been on before?
7. Do you have any family history of any neurologic diseases?
8. What are your hobbies? Do you use or store any chemicals in your home or garage? Do you use pesticides or herbicides at home?
9. Have you ever seen a physician for these same symptoms? Was a work injury report filed?
10. Describe the ventilation in your workplace.
11. Did you have a preemployment physical? What did it show?
12. Is there any health surveillance done in your workplace? If so, what is tested and what have the results been in your case before?
13. Has an industrial hygienist visited your workplace? What were the results?
14. Have you ever filed a worker's compensation claim? What were your symptoms then? Has your labor union been notified of similar problems?
15. Do you live near a Superfund site or other hazardous waste sites?
16. What is the source of your drinking water? If well water, has it been tested?
17. Is your home in an area where air pollution is a particular problem? If so, do you eat locally or home-grown vegetables?

If suspicion is high for the existence of chemical neurotoxicity, it is then recommended that all exposure to the suspected agent(s) cease. This may require removing the worker from the workplace entirely until the diagnostic workup is complete and a definitive diagnosis is made. Continued exposure, even if better protection is used, is not safe in this setting.

Table 1 is a partial list of known neurotoxic chemicals divided arbitrarily into four classes: solvents (commonly used in industry for this purpose; common monomers are also listed here); pesticides (listed by general group and not by specific name); metals and metalloids; and gases (most common state of the named substance resulting in toxic exposure). Pharmaceuticals are not included. Prototypical chemical or chemical classes that are discussed in detail in the text are listed first. This discussion focuses on agents directly causing neurotoxicity that is not secondary to hypoxia from respiratory depression or from hypotension, dysrhythmias, or other systemic toxicity. The table does not distinguish between agents primarily toxic to the central nervous system (CNS) versus those primarily toxic to the peripheral nervous system (PNS). Many agents are toxic to both.

I. Solvents

The prototypical group discussed here is the chlorinated hydrocarbon (aliphatic) solvents; n-hexane is also reviewed. The aliphatic (straight-chain) solvents are grouped together, although there are differences among the group members. The most commonly encountered members of this group in both industry and the environment include methylene chloride (CH_2Cl_2), chloroform ($CHCl_3$), carbon tetrachloride (CCl_3), 1,1,1-trichloroethane (Cl_3CCH_3), trichloroethylene ($ClCH{=}CCl_2$), and perchloroethylene ($Cl_2C{=}CCl_2$).

All of these agents are potent anesthetics and can cause CNS damage via respiratory depression and re-

TABLE 1 Partial List of Known or Suspected Human Neurotoxins from Four Chemical Groups[a]

Solvents (related and common monomers)
Aliphatic-chlorinated hydrocarbon solvents

n-Hexane	Acrylamide	Aniline
Styrene	Methanol	β-Chloroprene
Benzene	Diethylaminoproprionitrile	Cyclohexanol
Toluene	Acetone	Cyclohexanone
Xylene	Acetonitrile	Dicyclopentadiene
Methyl *n*-butyl ketone	Acetates	Formaldehyde
Methyl ethyl ketone	Butanol	

Pesticides (group classifications)

Organophosphates	Methyl bromide
Chlorophenoxy herbicides	Organochlorines
Pyrethroids	TOCP
Carbamates	

Metals and metalloids

Lead	Thallium	Tin
Arsenic	Barium	Gold
Cadmium	Manganese	Platinum
Mercury	Chromium	Bismuth
Antimony	Selenium	Tellurium
Tungsten	Zinc	Aluminum

Gases

Carbon monoxide	Ammonia
Ethylene oxide	Carbon disulfide
Hydrogen cyanide	Nitrous oxide
Hydrogen sulfide	

[a] Refer to text for explanation.

sultant hypoxia. Aside from this mechanism, these agents have been implicated in a variety of neurotoxic syndromes primarily involving the central nervous system and higher cognitive functions. Causation issues are difficult because toxicity usually becomes evident only after many years and most affected patients have been exposed to numerous chemicals over that time period. Alcohol, tobacco, street drug, and naturally occurring disease can be complicating factors in this population. Most workers exposed to the chlorinated aliphatic hydrocarbon solvents are likely to have had significant exposure to other more potent CNS toxins such as toluene or styrene. Much of the epidemiologic work, particularly in the Scandinavian countries, has therefore referred to exposure to any of these agents as a "mixed solvent syndrome," which is used synonymously with "solvent syndrome."

The neurotoxic syndromes claimed to result from long-term exposure to these agents are often subjective complaints of headache, dizziness, fatigue, malaise, weakness, memory impairment, and anxiety. Difficulties in concentration and problem solving, general intellectual slowing, emotional lability, irritability, and sexual dysfunction are also common complaints.

Several attempts to standardize the nomenclature of the solvent syndromes have been made. In 1985 the World Health Organization (WHO) devised the following scheme to classify impairments (1):

Severity	WHO classification	Common use names
Minimal	Organic affective syndrome	Neurasthenic syndrome
Moderate	Mild chronic encephalopathy	Psycoorganic syndrome
Severe	Severe chronic encephalopathy	Solvent dementia

In general, patients were classified as being more severely affected based on the problems demonstrated on objective testing. There is a large amount of overlap among these classifications and a large degree of subjective interpretation on the part of the examiner.

A research conference held in the same year in North Carolina developed a different classification (2):

Type 1: fatigue, irritability, depression, anxiety, not accompanied by deficits on neuropsychological testing. Reversible.
Type 2a: disturbances of mood and personality, fatigue, impulse control, and motivation. May be reversible.
Type 2b: intellectual disturbances, memory loss, difficulty in learning and concentration, psychomotor disturbances. May be irreversible.
Type 3: progressive and global neuropsychological, intellectual, and emotional decline. Most problems irreversible.

The classifications just noted were intended to standardize communication and thereby enhance research capabilities in this difficult area. However, because the classifications themselves are subjective, and the objective tests used to determine the classification of a given patient are often not standardized, difficulties remain. These issues become of great importance in worker's compensation and civil litigation cases.

Solvent toxicity can be acute or chronic. The acute effects are generally those of intoxication followed by CNS and respiratory depression. Some of the chlorinated hydrocarbon solvents have been used therapeutically as anesthetic agents (i.e., trichloroethylene, methylene chloride). Their use in this capacity has ceased because safer agents have replaced them. A useful model for worker-related acute solvent intoxication is the deliberate solvent abuser who presents to medical attention for a variety of reasons. Regarding specifically the chlorinated aliphatic hydrocarbons, the most common presentation would be for coma or acute delerium. Typewriter correction fluid containing trichloroethane, trichloroethylene, or perchloroethylene, when deliberately abused by inhalation, has caused euphoria, confusion, and incoordination followed by coma and seizures (3,4). Accidental exposure to 1,1,1-trichloroethane in a 4-year-old child caused coma and possibly seizures (5); the child recovered with supportive therapy.

Multiple deaths have been reported from deliberate solvent abuse, including most of the aliphatic and aromatic solvents commonly used in industry (6). Some of the deaths appear to result from sudden cardiac dysrhythmias, as on occasion witnesses have seen the decedent suddenly fall after exertion. Most of these types of incidents involve abuse of freons and other fluorinated hydrocarbons (7,8). Nonfluorinated aliphatics may not have the same dysrhythmogenic potential and most of the acute emergencies or deaths appear to be from primary CNS and respiratory depression.

Industrial accidents and overexposures to these agents commonly result in a similar syndrome (9). Lesser exposures acutely not surprisingly result in typical intoxication syndromes involving higher cortical functions such as memory and problem solving (10).

Trichloroethylene (also known as trichlorothene) is well known for causing peripheral neuropathy, particularly of the trigeminal nerve (11,12). This effect has resulted in the chemical being used therapeutically for the treatment of tic douloureux in the past. Trigeminal somatosensory-evoked potentials have been proposed as a method for early detection of toxicity before clinical symptomatogy develops (11). The sural nerve conduction velocity and refractory period were abnormal in asymptomatic workers exposed to trichloroethylene when motor and autonomic nerve functions were normal (12).

As previously mentioned, it is unusual for a worker to be exposed over a long duration to only one solvent or solvent class. Most solvent-exposed workers over the course of their employment usually come in contact with a combination of aliphatics, aromatics, ketones, glycols, freons, and more complex hydrocarbons (13). It may be impossible to determine the cause of chronic toxicity even when industrial hygiene evaluation is carried out. Various methods of worker screening have been proposed. Detailed neuropsychiatric evaluation may reveal new disorders (14,15) which can be reversible when exposure is terminated (15). A target worker population may be those who have had recent accidents, as these may be more common among workers subjected to an exposure increase over the preceding several weeks (16).

Auditory-, visual-, and somatosensory-evoked potentials have been proposed as a screening tool for early nervous system dysfunction (17), as has detection of peripheral vibratory sensory deficits (18,19). The chlorinated hydrocarbon solvents as a class are particularly likely to cause hepatocellular damage, and therefore periodic measurement of hepatic enzymes may also be useful as a technique to detect definite overexposure (20,21). In the case of accidental or deliberate ingestion of the chlorinated hydrocarbons, it may be useful to note the fact that these materials are densely radiopaque when in the gastrointestinal tract (22).

n-Hexane is useful to discuss as the prototypical industrial solvent causing peripheral neuropathy as the major manifestation of toxicity. It is used primarily as a glue and adhesive solvent, although small amounts may be found in gasoline. Workers at risk include any that work with adhesives and include shoemakers, furniture makers, and adhesive manufacturers.

n-Hexane is unusual in several ways. Although it is an extremely potent peripheral neurotoxin, compounds quite similar in chemical structure such as *n*-pentane, *n*-heptane, and branched hexanes do not have significant peripheral neurotoxic properties. *n*-Hexane neuropathy has been described after both accidental occupational exposure and deliberate abuse. In addition, a common metabolite shared by *n*-hexane and methyl *n*-butyl ketone, 2,5-hexanedione, appears to be the cause of the neurotoxic properties shared by each (23).

Peripheral neuropathy is invariably initially recognized as the insidious onset of numbness of the toes and fingers. Continued exposure results in proximal spread of sensory loss, followed by loss of Achilles and patellar reflexes. Muscle weakness and atrophy develop after the sensory findings have become pronounced (24,25). Vibratory and position senses are frequently spared. Symptoms may progress even after removal of the patient from continued exposure, a phenomenon called "coasting." Other described symptoms from *n*-hexane have included malaise, muscle pain, headache, dizziness, anorexia, and emotional instability (26).

Nerve conduction studies of both motor and sensory nerves may be abnormal (26), and evoked potentials of the central nervous system may also be abnormal (25). Measurement of 2,5-hexanedione in the urine in concentrations above 5 mg/liter is strong evidence of significant exposure (27). The urine should be collected soon after exposure is terminated, however, as the urinary clearance of this metabolite is rapid.

The pathology of the peripheral nerves caused by *n*-hexane includes paranodal axonal swelling and retraction of the myelin sheaths with focal demyelination (28,29). Complete recovery is common after removal from exposure (26), although residual spinal cord damage has been reported (30).

II. Pesticides

Only the organophosphates (OGP) will be discussed here as the prototypical pesticides causing acute and delayed, peripheral and central neurotoxicity. Organophosphates have been recognized since they were first synthesized in the 1850s in Germany to be potent neurotoxins; their usefulness as insecticides depends on their ability to cause neurotoxicity in insects. An enormous amount of animal and clinical data have been published documenting their neurotoxic properties. However, some issues, particularly regarding treatment of acute poisoning and classification of syndromes (discussed later), remain controversial.

The organophosphates are toxic to both the CNS and the PNS after either acute or significant chronic exposure. The mechanism of both is generally felt to be by inhibition of the enzyme acetylcholinesterase which allows excess acetylcholine to accumulate at CNS, muscarinic, and nicotinic receptors. The acetylcholine receptors themselves may be subsequently altered, although this remains unproven.

The circulating acetylcholinesterases are commonly used as markers for the degree of OGP exposure, but these are not the enzymes whose inhibition results in the clinical symptoms. Red blood cell cholinesterase ("true cholinesterase") and serum cholinesterase ("pseudocholinesterase") are the two circulating and easily measurable enzymes. Neurotoxic esterase is an enzyme found in circulating lymphocytes that is more difficult to measure but appears to correlate better with subsequent development of neurotoxicity. Susceptibility to the toxic effects of OGP may in part be due to the wide interindividual variations in the activity of the esterases (31). In addition, because the range of "normal" is so wide, patients whose cholinesterase levels fall to the low normal range and then rise again as exposure to OGP is stopped may have demonstrated significant toxicity while maintaining their measurable cholinesterase levels in the normal range (32).

Symptoms of OGP toxicity depend on the sites of cholinesterase inhibition, which in turn are dependent on the route of exposure, the type of OGP, and as mentioned perhaps on the baseline activity of cholinesterase. The sites of cholinesterase inhibition and resultant toxicity include: CNS (confusion, dizziness, fatigue, ataxia, anxiety, respiratory depression, coma, and seizures), nicotinic (muscle weakness and fasciculations, hypertension, and tachycardia), and muscarinic (salivation, urination, diarrhea, gastrointestinal pain and cramping, emesis, bradycardia, bronchorrhea, miosis, sweating, and wheezing).

The specific neurotoxic syndrome seen is highly dependent on the amount and timing of exposure, as well as the OGP involved. It is useful to generally divide the syndromes into four categories, although significant overlap does occur.

A. Acute Exposure/Immediate Effect

Acute, massive exposure usually occurs after industrial or agricultural accidents, or after deliberate ingestion in a suicide attempt. Accidental exposure such as might occur from home pesticide spraying or wearing of clothing contaminated with organophosphates might cause similar but less severe effects. Initial CNS symptoms frequently include dizziness, ataxia, and confusion, which can rapidly progress to coma and seizures. Extreme muscle weakness and fasciculations may cause hypoventilation. The muscarinic findings are what may make the clinician suspect the diagnosis;

patients are frequently very "wet" with extreme diaphoresis, bronchial secretions, emesis, etc. Children may be more susceptible to the CNS toxicity of OGP than adults, and may present primarily with CNS manifestations and not the telltale muscarinic findings (33,34).

B. Acute Exposure/Delayed Neuropathy and Neuropsychiatric Toxicity

Organophosphate-induced delayed neuropathy (OPIDN) was perhaps first recognized in 1899 but was published in the medical literature in 1930 (35). Thousands of victims of ingestion of tri-o-cresyl phosphate in the United States and elsewhere developed weakness of the calf and wrist muscles 10–21 days after drinking contaminated alcoholic beverages. Severe cases developed irreversible paralysis of all extremities (36). Since this discovery, many, but not all, OGP have been found to cause OPIDN in man, perhaps the most sensitive species to this syndrome (37). Clinically, the syndrome begins 8–14 days following exposure with numbness and fatigue of the distal musculature. This may progress to the proximal muscles over the next several weeks; the condition does not cause a painful sensory neuropathy although numbness can occur. The cholinergic findings of acute poisoning may have been initially absent and therefore the diagnosis may be elusive. Improvement in neurologic function is typically slow and often incomplete (38). There may be residual atrophy of the interosseous muscles of the hands and the anterior tibial groups of muscles in the legs. The upper extremities may recover completely, yet the lower extremities may remain so weak as to make walking difficult (39). Residual motor conduction velocities and abnormal sensory-evoked potential amplitudes may be permanent (40).

Pathologically, the axon as well as the myelin sheaths are fragmented, although it appears from animal studies that the earliest functional disturbances occur in the most distal, nonmyelinated portion of the nerves (39). Demyelination occurs in the more proximal portions of the nerves, largely in a Wallerian pattern. These ultrastructural changes cause a "dying-back" phenomenon whereby the most distal portions of the longest nerve fibers appear to become nonfunctional. This is accompanied by a reparative process in the cell body called chromatolysis (39). The nucleus increases in size and the synthesis of nuclear RNA increases, thought to represent the cells attempt to repair its damaged axon.

The peripheral neuropathy discussed is easy to document on physical examination and by objective studies of peripheral nerve function. More difficult to document conclusively, but perhaps equally important, psychiatric and higher cerebral function abnormalities have been described after acute OGP exposure. Reported symptoms include depression, confusion, agitation, insomnia, irritability, memory dysfunction, and schizoid reactions (41–43). A matched-pair study of OGP-exposed versus unexposed individuals demonstrated abnormalities of memory, abstraction, and mood in the exposed population as well as greater distress and complaints of disability (44). These manifestations of residual toxicity can occur without symptoms or signs of OPIDN.

C. Acute Exposure Resulting in the "Intermediate Syndrome"

In between the acute cholinergic crisis of OGP poisoning and the OPIDN which occurs weeks later reputedly is the "intermediate syndrome." Initially described by Wadia in 1974 (45), it has recently resurfaced as an allegedly distinct entity by several case series (46,47) and reports (48). What distinguishes it from the OPIDN is the time of onset (1–4 days vs 2–3 weeks), it effects primarily the proximal musculature and cranial nerves, the electromyelogram shows tetanic fade versus denervation, the recovery time is 4–18 days versus 6–12 months, and the organophosphates involved are more likely to be fenthion, dimethoate, and monocrotophos (46).

It has been argued, however, that this syndrome represents the natural course of severe OGP poisoning when therapy with the cholinesterase regenerater pralidoxime (protopam, 2-PAM) has been absent or inadequate (49,50). In one case of possible intermediate syndrome, test doses of pralidoxime were effective in reversing the muscle weakness and cranial nerve palsies (50).

In summary, the organophosphate pesticides are clearly potent neurotoxins affecting both the brain and the PNS. Although a large amount of clinical and basic science research has been published on their effects and treatment, there is still much uncertainty, particularly regarding chronic low-level exposure.

III. Metals

The mechanisms by which metals and metalloids cause toxicity to specific nerve tissue are multifactorial, highly individualistic (implying a genetic susceptibility), and not predictably reversible. The two prototypical metals discussed here are lead and arsenic. Each has its own fascinating history, environmental controversies, and classically described nervous system target organs/tissues.

Lead is particularly controversial because of its neurotoxicity to the developing brain, and current public awareness of the dangers of lead is probably higher than at any other time in this country's history. Since 1991 new lead standards have been established by the Centers for Disease Control, a new lead-chelating agent has been approved by the Food and Drug Administration (Chemet), and whole new industries have been developed to remove leaded paint from residences. Lead is also somewhat unique in that it has no known biological function in man (i.e., it is not a trace element), yet can be found in every human in some concentration. It is therefore a political public health policy issue as to what concentration of this known neurotoxin we will permit in our citizens, particularly our children.

The sources of lead are innumerable and include the following, divided by the site of most common exposure:

The home

Leaded paint	Firearms	Soldering
Ceramics	Herbal	Soil
Leaded gasoline	remedies	House dust
Water (well or	Moonshine	Sinkers, curtain
tap)	alcohol	weights, etc.
	Contaminated	
	street drugs	

The workplace

Battery making	Firearm	Printing
and recycling	manufacture	Chemical plants
Brass/bronze	Metal grinding	Glass
foundries	Pipe cutting	manufacture
Jewelery	Welding	Cable
Smelters	Painting	manufacture
Pigment		Miscellaneous
manufacturing		

The environment

Contaminated soil, water, air, and dust from any of the just mentioned operations. Environmental contamination of areas downwind from smelters is probably the most common source.

Lead neurotoxicity is most likely to be from chronic exposure, although acute syndromes have been described. Assuming chronic low-level exposure, lead neurotoxicity in adults affects primarily the peripheral nerves of the upper extremities, whereas the brain appears to be the primary target organ in children. In adults, the classic description of lead neuropathy is of an asymmetrical weakness of the wrists (wrist drop),

with no loss of sensation (51). The extensors of the fingers and thumb may be affected before the wrist (52). Proximal muscle weakness of the upper extremities and foot drop may occur in severe adult cases, although this is more commonly seen in severely lead toxic children (53). In humans, Wallerian-type axonal degeneration is the most common pathologic finding, although segmental demyelination can be seen in laboratory animals (51). Peripheral nerve conduction velocities are usually slowed (54), even in workers who have not yet become symptomatic from lead intoxication (55–57). Likewise, subtle neurophysiological, psychological, and behavioral abnormalities can occur in lead-exposed workers before symptoms appear, and have been suggested as appropriate screening tests (58).

Adults may develop encephalopathy from exposure to organic lead, usually in the context of deliberate inhalation of leaded gasoline (59,60). Behavioral changes, including hallucinations, and extrapyramidal effects, including myoclonus, ataxia, tremor, and chorea, are prominent. There is usually a notable absence of peripheral neuropathy.

Chronic lead exposure in children is more likely to result in subtle cognitive dysfunction (61–63), including a lessening of measurable intelligence (64,65). Lead exposure *in utero* may have a similar effect (62). These findings have caused the Centers for Disease Control (in October of 1991) to lower the community intervention blood lead level to 10 μg/dl from its previous level of 25 μg/dl. This is the blood lead level which the CDC recommends should result in community-wide childhood lead poisoning prevention activities in an attempt to prevent CNS neurotoxicity in asymptomatic children.

The mechanism by which lead causes this deleterious effect on the developing brain is poorly understood. In severely poisoned children, mitochondrial disruption of brain cells occurs, resulting in cerebral edema, coma, and seizures (66). At lower blood lead concentrations, lead affects neurotransmitter activity, adenyl cyclase activity, and dendritic complexity (67–69). Which if any of these affects is responsible for the dose-dependent cortical dysfunction in children is unclear.

This is an important issue because it is currently unknown if these CNS changes are reversible and if their reversibility is affected by lead chelation therapy. If the mechanism of central neurotoxicity in children were known, then measuring the effect of chelation on this mechanism would be important. Clearly the elimination of lead from the body is enhanced by chelation therapy by both the classic drugs EDTA, BAL, and *d*-penicillamine(66) and the newer agent DMSA (dimercaptosuccinic acid, succimer, Chemet) (70). Whether or not this change in the body burden of lead results in improvement in brain function in lead-exposed children remains an important area of clinical research.

Arsenic, like lead, is notable for causing both CNS toxicity and peripheral neuropathy. However, the capacity of the human body to tolerate and rapidly excrete arsenic is much greater than for lead. Arsenic is a common element and can be found in association with lead (and cadmium) in many commercially mined ores. It therefore is of common concern in areas contaminated by smelter smoke and dust.

The neurotoxicity of arsenic is, however, much more likely to be secondary to a very large exposure, such as might be seen in a suicide or homicide attempt, and not to low-level environmental exposure. In addition, other organ systems are likely to be severely affected and to make the clinical picture more distinct. In a case of aresenic poisoning severe enought to cause neurotoxicity, for example, a patient is also likely to have severe vomiting and diarrhea, bone marrow suppression affecting all cell lines, Mee's lines of the fingernails, and keratoses of the palms and soles. Arsenic is also a known carcinogen.

Arsenic encephalopathy is an acute and long-lasting or permanent delerium which has been described after acute massive exposure (71), prolonged exposure from arsenic-contaminated alcoholic beverages (72), and from low-level occupational exposure (73). Clinical findings are nonspecific and may include confusion, disorientation, agitation, and hallucinations. Other CNS manifestations of acute arsenic intoxication include headache, drowsiness, memory loss, diplopia, optic neuritis, and seizures (74).

Far more common, predictable, and classic in its presentation is the peripheral neuropathy of arsenic. Symptoms usually begin with numbness, tingling, and burning paresthesias of the feet and later the hands in a stocking and glove distribution (74,75). The paresthesias are typically quite painful and are exacerbated by a pressure stimulus, such that walking might cause severe burning on the bottom of the feet (76). Vibration, position, and temperature sensation are also decreased. Sensory abnormalities spread proximally in a symmetrical pattern.

Within a few days of the onset of paresthesias, distal muscle weakness begins and spreads proximally. Foot drop and wrist drop are prominent early in the clinical course, and impressive wasting of the distal muscles occurs in proportion to the degree of other signs (74). Muscle tenderness, cramps, and fasciculations occasionally occur. Deep tendon reflexes are decreased and then disappear, with the ankle and patellar reflexes affected first (76).

Sensory nerve action potentials and motor nerve conductions are severely affected and may remain so for years (77). Nerve biopsies have revealed axonal degeneration with no segmental demyelination and a total reduction of both large and small diameter fibers (77). Unfortunately, even after only a single acute exposure, and even after rapid diagnosis and treatment, the peripheral neuropathy of arsenic poisoning may be permanent.

IV. Gases

The prototypical neurotoxic gases discussed in this section are carbon monoxide (CO) and ethylene oxide (EO).

Carbon monoxide annually causes more deaths than any other toxin or drug overdose. Exposure to excessive concentrations occurs in the home, the workplace, and the environment (in the form of automobile exhaust and cigarette smoke). Because carbon monoxide poisoning in humans is so common, large amounts of data have been published documenting its neurotoxicity. Many patients seem to recover from the acute effects of CO only to later develop significant neurotoxicity, commonly termed "delayed neurologic sequellae." The pathogenesis and prevention of this particular syndrome are matters of current controversy and hold great importance for many specialties of medicine.

The initial symptoms of headache, nausea, malaise, and lethargy from CO exposure may progress rapidly to loss of consciousness. This is commonly associated with high carboxyhemoglobin levels (>40% of the total hemoglobin), although high levels have been measured in patients without any signs of toxicity (78,79). The toxic effect on the brain during this acute stage seems to be only secondarily related to the functional anemia and hypoxia from CO exposure. The subsequent brain injury is also seemingly unrelated, as patients who have never been significantly hypoxic may still develop delayed sequellae. The most plausible current explanation is that CO poisoning results in a reperfusion injury which may be related to brain lipid peroxidation (80,81). These findings have important applications for prevention and treatment of delayed neurologic sequellae.

Clinically, this syndrome commonly develops 3 days to 3 weeks after recovery from the initial exposure has appeared to be complete. In one case series of 2360 CO poisonings in Korea resulting in a hospital visit (82), 65 patients were felt to have delayed neurologic sequellae, which included: mental deterioration, urinary/fecal incontinence, gait disturbance, mutism, tremor, weakness, speech disturbance, masked face, glabella sign, grasp reflex, increased muscle tone, short-step gait, retropulsion, intention tremor, increased DTRs, flaccid paresis, and ankle clonus. Seventy-five percent of the 36 patients from this group who could be followed for 2 years recovered.

Out of a series of 74 CO poisoned patients from the United Kingdom, 21 were described as having a deterioration of personality and 27 had a subsequent impairment of memory after initial recovery. The level of consciousness upon arrival to the hospital in this series correlated with the risk of development of sequellae (83).

An American case series which attempted to determine if hyperbaric oxygen (HBO) could prevent such sequellae described 213 CO-poisoned patients in which 131 were treated with HBO. None of the HBO-treated patients developed delayed neurologic symptoms. Of the patients not treated with HBO but receiving 100% oxygen and supportive therapy, 10/82 (12.1%) subsequently developed headaches, irritability, personality changes, confusion, and memory loss 1 to 21 days after admission (84). These symptoms resolved with late HBO treatment.

The same authors reported recovery from persistent coma from CO poisoning in six patients receiving late HBO therapy (85). Delayed sequellae apparently did not develop in these patients despite their being critically ill initially.

Other sequellae from CO poisoning include hearing loss (86), retinal hemorrhages (87,88), and basal ganglion infarcts (89,90). Children may be more susceptible to the initial CNS toxicity, with syncope occurring at levels lower than those commonly seen in adults (91). Delayed neurologic sequellae in children include chronic headaches, memory difficulties, and decline in school performance (91).

Ethylene oxide is the simplest of the reactive epoxides (chemical formula: CH_2—O—CH_2). It is one of the 25 most commonly manufactured chemicals in the United States, with annual production exceeding 6 billion pounds (92). Although interest in the toxicity of EO has been high because of its use in the health care field as an instrument sterilizer, the vast majority of exposed workers are employed in other industries. It is a strong irritant that can cause severe skin, mucous membrane, and pulmonary burns. In addition, it is a known mutagen, reproductive toxin, and carcinogen (92–95).

Like many of the other neurotoxins discussed in this chapter, the problems associated with ethylene oxide should be viewed as being caused by acute versus chronic exposure, and resulting in central or peripheral nervous system toxicity (or both). Acute, massive exposure commonly results in irritation, nausea, head-

ache, and then obtundation and seizures (96,97). Sequellae from this type of exposure may not occur if adequate supportive therapy is rendered (96). More chronic and moderate exposure concentrations have been reported to cause significant, long-lasting cognitive dysfunction (97,98) and peripheral neuropathy (97–101). The CNS toxicity, however, is particularly controversial. One patient, for example, seen by two different neurology, neurotoxicolgy, and neuropsychiatric groups and undergoing an extremely extensive workup, was concluded to have primarily a functional disorder by one group (102) and true CNS neurotoxicity by another (98). The central symptoms and signs in that case include headache, fatigue, dizziness, memory loss, difficulty concentrating, diminished intellectual function, depression, and irritability (102).

A blinded comparison of eight ethylene oxide-exposed hospital workers with nine unexposed workers did not show cognitive impairments in the exposed group, although the sample size was small and some of the differences did approach statistical significance (99). A neurobehavioral evaluation system using a computerized battery of tests did reveal more anxiety, tension, depression, fatigue, and confusion in the exposed group. There was not, however, a greater degree of anxiety demonstrated on the MMPI.

The peripheral neurotoxicity of ethylene oxide has been more conclusively demonstrated. It generally begins as limb weakness and fatigability, which may or may not be associated with central nervous system effects (97). On examination, distal limb weakness is obvious, and vibratory and position sense is decreased (97,100). Ataxia, wide-based gait, clumsy alternating hand movements, and decreased deep tendon reflexes are also common (97,100). Nerve conduction studies demonstrate abnormal slowing, and electromyelograms show scattered positive sharp waves and fibrillation potentials with denervation potentials. These findings may improve over time, but improvement does not appear to be predictable (97,100).

In an animal model, rats exposed to 250 ppm EO five times a week for 9 months did not demonstrate gross neurologic findings, but a histologic examination of nerves disclosed preferential distal axonal degeneration of myelinated fibers, although the extent of the distribution and the severity of the degenerative findings were variable (101).

V. Summary

This chapter has attempted to present a brief overview of common neurotoxic substances, with prototypical chemicals causing classic neurototoxic syndromes being briefly described. Toxicity to other organ systems and general patient management recommendations have not been discussed. In many of the situations described, significant controversy exists, and this chapter was not designed to settle these or to present an opinion on them.

References

1. World Health Organization (WHO). (1986). Principles and methods for the assessment of neurotoxicity associated with exposure to chemicals. Environmental Health Criteria Document 60. World Health Organization, Geneva.
2. Baker, E. L, and Seppalainen, A. M. (1986). Human aspects of solvent neurobehavioral effects. In *Proceedings of the Workshop on Neurobehavioral Effects of Solvents* (J. Cranmer and L. Golberg, Eds.), *Neurotoxicology* 7, 43–56.
3. Pointer, J. (1982). Typewriter correction fluid inhalation: a new substance of abuse. *J. Toxicol. Clin. Toxicol.* 19, 493–499.
4. Ranson, D. L., and Berry, P. J. (1986). Death associated with the abuse of typewriter correction fluid. *Med. Sci. Law* 26, 308–310.
5. Gerace, R. V. (1981). Near fatal intoxication by 1,1,1,-trichloroethane. *Ann. Emerg. Med.* 10, 533–539.
6. Garriott, J., and Patty, C. S. (1980). Death from inhalant abuse: Toxicological and pathological evaluation of 34 cases. *Clin. Toxicol.* 16, 305–315.
7. Bass, M. (1970). Sudden sniffing death. *JAMA* 212, 2075–2079.
8. Aviado, D. M. (1975). Toxicity of aerosols. *J. Clin. Pharmacol.* 15, 86–104.
9. McCarthy, T. B., and Jones, R. D. (1983). Industrial gassing poisoning due trichloroethane, perchloroethylene, and 1,1,1,-trichloroethane, 1961–80. *Br. J. Ind. Med.* 40, 450–455.
10. Anger, W. K., and Johnson, B. L. (1985). Chemicals affecting behavior. *Neurotoxic. Indust. Comm. Chem.* 1, 51–148.
11. Barrett, L., Garrel, S., Danol. V. *et al.* (1988). Chronic trichloroethylene intoxication: a new approach by trigeminal-evoked potentials. *Arch. Environ. Health* 42, 297–302.
12. Ruiyten, M. W., Verberk, M. M., and Sallé, H. J. (1991). Nerve function in workers with long-term exposure to trichloroethane. *Br. J. Ind. Med.* 48, 87–92.
13. Flodin, U., Edling, C., and Axelson, O. (1984). Clinical studies of psychoorganic syndromes among workers with exposure to solvents. *Am. J. Ind. Med.* 5, 287–295.
14. Axelson, O., Hane, M., and Hogstedt, C. (1976), A case-referent study on neuropsychiatric disorders among workers exposed to solvents. *Scand. J. Work. Environ. Health* 2, 14–20.
15. Ekberg, K., Barregard, L., Hagberg, S., *et al.,* (1986). Chronic and acute effects of solvents on central nervous system functions in floorlayers. *Br. J. Ind. Med.* 43, 101–106.
16. Hunting, K. L., Matanowski, G. M., *et al.* (1991). Solvent exposure and the risk of slips, trips, and falls among painters. *Am. J. Ind. Med.* 20, 353–370.
17. Massious, F. E., Lille, F., *et al.* (1990). Sensory and cognitive event related potentials in workers chronically exposed to solvents. *Clin. Toxicol.* 28(2), 203–219.
18. Demers, R. Y., Markell, B. L. and Wabake, R. (1991). Peripheral vibratory sense deficits in solvent-exposed painters. *J. Occup. Med.* 33(10), 1051.
19. Bleecker, M. L., Bolla, K. I. Agnew, J., *et al.,* (1991). Dose-Related subclinical neurobehavioral effects of chronic exposure to low levels of organic solvents. *Am. J. Ind. Med.* 19, 715–728.

20. Hodgson, M. J., Heyl, A. E., and Van Thiel, D. H. (1989). Liver disease associated with exposure to 1,1,1,-trichloroethane. *Arch. Intern. Med.* 149, 1793–1798.
21. Nakayama, H., Kobayashi, M., *et al.* (1988). Generalized eruption with severe liver dysfunction associated with occupational exposure to trichloroethylene. *Contact Dermatitis* 19, 48–51.
22. Dally, S. Garnier, R., and Bismuth, C. (1987). Diagnosis of chlorinated hydrocarbon poisoning by x-ray examination. *Br. J. Med.* 44, 424–425.
23. Couri, D., and Milks, M. (1982). Toxicity and metabolism of the neurotoxic hexacarbons n-hexane, 2-hexanone, and 2,5-hexanedione. *Annu. Rev. Pharmacol. Toxicol.* 22, 145–146.
24. Ruff, R. L., Petito, C. K., and Acheson, L. S. (1981). Neuropathy associated with chronic low level exposure to n-hexane. *Clin. Toxicol.* 18, 515–519.
25. Chang, Y. C. (1987). Neurotoxic effects of n-hexane on the human central nervous system: evoked potential abnormalities in n-hexane polyneuropathy. *J. Neurol. Neurosurg. Psychiatry* 50, 269–274.
26. Huang, C. Shih, T., and Cheng, S. (1991). *n*-Hexane polyneuropathy in a ball-manufacturing company. *J. Occup. Med.* 33(2), 129–139.
27. Governa, M., Calisti, R., Coppa, G. et al. (1987). Urinary excretion of 2,5-hexanedione and peripheral polyneuropathies in workers exposed to hexane. *J. Toxicol. Environ. Health* 20, 219–228.
28. Rizzuto, W., Terzian, H., and Galiazzo-Rizzuto, S. (1977) Toxic polyneuropathies in italy due to leather cement poisoning in shoe industries: a light and electron microscopic study. *J. Neurol. Sci.* 31, 343.
29. Korobkin, R., Asbury, A. L., Summer, A. J., *et al.* (1975). Glue-sniffing neuropathy. *Arch. Neurol.* 32, 158.
30. Oryshkevich, R. S., Wilcox, R., and Jhee, W. H. (1986). Polyneuropathy due to glue exposure: case report and 16-year follow up. *Arch. Phys. Med. Rehabil.* 67, 827–828.
31. Mutch, E., Blain, P. G., and Williams, F. M. (1992). Interindividual variations in enzymes controlling organophosphate toxicity in man. *Hum. Exp. Toxicol.* 11, 109–116.
32. Coye, M. J., Barnett, P. G., Midtling, J. E., *et al.* (1986). Clinical confirmation of organophosphate poisoning of agricultural workers. *Am. J. Ind. Med.* 10, 399–409.
33. Zweiner, R. J., and Ginsburg, C. M. (1988). Organophosphate and carbamate poisoning in infants and children. *Pediatrics* 81, 121–126.
34. Sofer, S., Tal, A., and Shahak, E. (1989). Carbamate and organophosphate poisoning in early childhood. *Pediatr. Emerg. Care* 5, 222–225.
35. Smith, M. I., Elvove, E., and Frazier, W. H. (1930). The pharmacological action of certain phenol esters with specific reference to the etiology of so-called ginger paralysis. *US Public Health Rep.* 45, 2509–2524.
36. Metcalf, R. L. (1982). Historical perspective of organophosphorus ester-induced delayed neurotoxicity. *Neurotoxicology* 3(4), 269–284.
37. Abou-Donia, M. B. (1981). Organophosphorus ester-induced delayed neurotoxicity. *Annu. Rev. Pharmacol. Toxicol.* 21, 511–548.
38. Barrett, D. S., and Oehme, F. W. (1985). A review of organophosphorus ester-induced delayed neurotoxicity. *Vet. Hum. Toxicol.* 27, 22–37.
39. Hayes, W. J. (1982). *Pesticides Studied in Man,* pp. 319–321, Williams & Wilkins, Baltimore.
40. Vasilescu, C., and Florence, A. (1980). Clinical and electrophysiological study of neuropathy after organophosphorus compounds poisoning. *Arch. Toxicol.* 43, 305–315.

41. Bowers, M. B., Goodman, E., and Sim, V. M. (1964). Some behavioral changes in man following anticholinesterase administration. *J. Nerv. Ment. Dis.* **138**, 383–389.

42. Clark, G. (1871). Organophosphate pesticide and behavior: a review. *Aerospace Med.* **42**, 735–740.

43. Hirshberg, A., and Lerman, Y. (1984). Clinical problems in organophosphorus insecticide poisoning: the use of a computerized information system. *Fundam. Appl. Toxicol.* **4**, S209–S214.

44. Savage, E. P., Keefe, T. J., and Mounce, L. M. (1988). Chronic neurological sequelae of acute organophosphate pesticide poisoning. *Arch. Environ. Health* **43**, 38–45.

45. Wadia, R. S., Sadagopalan, C., Amin, R. B., et al. (1974). Neurological manifestations of organophosphorus insecticide poisoning. *J. Neurol. Neurosurg. Psychiatry* **37**, 841–847.

46. Senanayake, N., and Karalliede, L. (1987). Neurotoxic effects of organophosphorus insecticide: an intermediate syndrome. *N. Engl. J. Med.* **316**, 761–763.

47. DeBlecker, J., Williams, J., VanDen Neucker, K., et al. (1992). Prolonged toxicity with intermediate syndrome after combined parathion and methyl parathion poisoning. *Clin. Toxicol.* **30**, 333–345.

48. DeBlecker, J., VanDen Neucker, K., and Williams, J. (1992). The intermediate syndrome in organophosphate poisoning: presentation of a case and review of the literature. *Clin. Toxicol.* **30**, 321–329.

49. Benson, B., Tolo, D., and McIntire, M. (1992). Is the intermediate syndrome in organophosphate poisoning the results of insufficient oxime therapy? *Clin. Toxicol.* **30**, 347–349.

50. Aaron, C. K., and Smilkstein, M. J. (1988). Organophosphate poisoning: intermediate syndrome or inadequate therapy? *Vet. Hum. Toxicol.* **30**, 370.

51. Taylor, J. R. (1984). Neurotoxicity of certain environmental substances. *Clin. Lab. Med.* **4**, 489–497.

52. Woltman, H. W. (1962). Neuritis and other diseases of peripheral nerves. In *Practice of Medicine*, (F. Tice, and S. Lightheaded, Eds.), **9**, 289.

53. Seto, D. S. Y., and Freeman, J. M. (1964). Lead neuropathy in childhood. *Am. J. Dis. Child.* **107**, 337.

54. Lille, F., Hazeman, P., Garnier R., et al. (1988). Effects of lead and mercury intoxications on evoked potentials. *Clin. Toxicol.* **26**, 103–116.

55. Seppalainen, A. M., Tola, S., Hernberg, S., et al. (1975). Subclinical neuropathy at "safe" levels of lead exposure. *Arch. Environ. Health* **30**, 180–183.

56. Buchtal, F., and Behse, F. (1979). Electrophysiology and nerve biopsy in men exposed to lead. *Br. J. Ind. Med.* **36**, 135–147.

57. Schwartz, J., Landrigan, P. J., Feldman, R. G., et al. (1988). Threshold effect in lead-induced peripheral neuropathy. *J Pediatri.* **112**(1), 12–17.

58. Pasternak, G., Becker, C. E., Lash, A. et al. (1989). Cross-sectional neurotoxicology study of lead-exposed cohort. *Clin. Toxicol.* **27**, 37–51.

59. Goldings, A. S., and Stewart, M. (1982). Organic lead encephalopathy: behavioral change and movement disorder following gasoline inhalation. *J. Clin. Psychiatry* **43**(2), 70–72.

60. Hansen, K. S., and Sharp, F. R. (1978). Gasoline sniffing, lead poisoning, and myoclonus. *JAMA* **240**(13), 1375–1376.

61. Perino, J., and Ernhart, S. B. (1974). The relation of subclinical lead exposure to cognitive and sensorimotor impairment in black preschoolers. *J. Learn. Disabil.* **7**, 616–620.

62. Bellinger, D., Leviton, A., Waternaux, C., et al. (1981). Longitudinal analyses of prenatal and postnatal lead exposure and early cognitive development. *N. Engl. J. Med.* **316**(17), 1037–1043.

63. Rosen, J. F. (1992). Health effects of lead at low exposure levels: expert consensus and rationale for lowering the definition of childhood lead poisoning. *Am. J. Dis. Child.* **146**, 1278–1281.

64. Needleman, H. L., and Gatsonis, C. A. (1990). Low-level lead exposure and the IQ of children. *JAMA* **263**(5), 673–678.

65. Baghurst, P. A., McMichael, A. J., Wigg, N. R., et al. (1992). Environmental exposure to lead and children's intelligence at the age of seven years. *N. Engl. J. Med.* **327**(18), 1279–1284.

66. Chisolm, J. J. (1968). The use of chelating agents in the treatment of acute and chronic lead intoxication in childhood. *J Pediatr.* **73**(1), 1–38.

67. Hrdina, P. D., Hanin, I., and Dubas, T. C. (1980). Neurochemical correlates of lead toxicity. In *Lead Toxicity*. (R. L. Singhal and J. A. Thomas, Eds.), pp. 273–300, Urban & Schwarzenberg, Baltimore.

68. Nathanson, J. A. (1977). Lead-inhibited adenylate cyclase, a model for the evaluation of chelating agents in the treatment of lead toxicity. *J. Pharm. Pharmacol.* **29**, 511–513.

69. Averill, D., and Needleman, H. L. (1988). Neonatal lead exposure retards cortical synaptogenesis in the rat. In *Low Level Lead Exposure: The Clinical Implications of Current Research*. (H. Needleman, Ed.), Raven Press, New York.

70. Mann, K. V., and Travers, J. D. (1991). Succimer, an oral lead chelator. *Clin. Pharmacol.* **10**, 914–922.

71. Fincher, R. E., and Koerker, R. M. (1987). Long-term survival in acute arsenic encephalopathy. *Am. J. Med.* **82**, 549–552.

72. Freeman, J. W., and Couch, J. R., (1978). Prolonged encephalopathy with arsenic poisoning. *Neurology* **28**, 853–855.

73. Beckett, W. S., Moore, J. L., Keogh, J. P., et al., (1986). Acute encephalopathy due to occupational exposure to arsenic. *Br. J. Ind. Med.* **43**, 66–67.

74. Heyman, A., Pfeiffer, J. B., Willett, R. W., et al. (1956). Peripheral neuropathy caused by arsenical intoxication: a study of 41 cases with observations on the effects of BAL (2,3 dimercaptopropanolol). *N. Engl. J. Med.* **254**(9), 401–409.

75. Hessl, S. M., and Berman, E. (1982). Severe peripheral neuropathy after exposure to monosodium methyl arsonate. *J. Toxicol. Clin. Toxicol.* **19**(3), 281–287.

76. Jenkins, R. B. (1966). Inorganic arsenic and the nervous system. *Brain* **89**, 479–498.

77. Le Quesne, P. M., and McLeod, J. G. (1977). Peripheral neuropathy following a single exposure to arsenic. *J. Neurol. Sci.* **32**, 437–451.

78. Benignus, V. A., Kafer, E. R., Muller, K. E., et al. (1987). Absence of symptoms with carboxyhemoglobin levels of 16–23%. *Neurotoxicol. Terato.* **9**, 345–348.

79. Davis, S. M., and Levy, R. C. (1984). High carboxyhemoglobin level without acute or chronic findings. *J. Emerg. Med.* **1**, 539–542.

80. Thom, S. R. (1990). Carbon monoxide-mediated brain lipid peroxidation in the rat. *J. Appl. Physiol.* **68**(3), 997–1003.

81. Thom, S. R. (1990). Antagonism of carbon monoxide-mediated brain lipid peroxidation by hyperbaric oxygen. *Toxicol. Appl. Pharmacol.* **105**, 340–344.

82. Choi, I. S. (1983). Delayed neurologic sequelae in carbon monoxide intoxication. *Arch. Neurol.* **40**, 433–435.

83. Smith, J. S., and Brandon, S. (1973). Morbidity from acute carbon monoxide poisoning at three-year follow up. *Br. Med. J.* **1**, 318–321.

84. Myers, R. A., Snyder, S. K., Emhoff, T. A., et al. (1985). Subacute sequelae of carbon monoxide poisoning. *Ann. Emerg. Med.* **4**(12), 1163–1167.

85. Myers, R. A., Snyder, S. K., Linberg S., et al. (1981). Value of hyperbaric oxygen in suspected carbon monoxide poisoning. *JAMA* **246**(21), 2478–2480.

86. Makishima, K., Keane, W. M., and Vernose, G. V. (1977). Hearing loss of a central type secondary to carbon monoxide poisoning. *Tr. Am. Acad. Ophth. Otol.* **84,** 452–457.

87. Kelley, J. S., and Sophocleus, G. J. (1978). Retinal hemorrhages in subacute carbon monoxide poisoning. *JAMA* **239,** 1515–1517.

88. Dempsey, L. C., O'Donnell, J. J., and Hoff, J. T. (1976). Carbon monoxide retinopathy. *Am. J. Ophthalmol.* **82,** 692–693.

89. MacMillan, V. (1977). Regional cerebral energy metabolism in acute carbon monoxide intoxication. *Can. J. Physiol. Pharmacol.* **55,** 11–116.

90. Jefferson, J. W. (1976). Subtle neuropsychiatric sequelae of carbon monoxide intoxication: two case reports. *Am. J. Psychiatry* **133,** 961–964.

91. Crocker, P. J., and Walker, J. S. (1985). Pediatric carbon monoxide toxicity. *J. Emerg. Med.* **3,** 433–448.

92. Landrigan, P. J., Meinhardt, T. J., Gordon, J., *et al.* (1984). Ethylene oxide: an overview of toxicologic and epidemiologic research. *Am. J. Ind. Med.* **6,** 103–115.

93. Sheikh, K. (1984). Adverse health effects of ethylene oxide and occupational exposure limits. *Am. J. Ind. Med.* **6,** 117–127.

94. Hertz-Picciotto, I., Neutra, R. R., and Collins, J. F. (1987). Ethylene oxide and leukemia. *JAMA* **257,** 2290.

95. Hogstedt, C., Malmquist, N., and Wadman, B. (1979). Leukemia in workers exposed to ethylene oxide. *JAMA* **241,** 1132–1133.

96. Salinas, E., Sasich, L., Hall, D. H., *et al.* (1981). Acute ethylene oxide intoxication. *Drug Intel. Clin. Pharmacol.* **15,** 384–386.

97. Gross, J. A., Haas, M. L., and Swift, T. R. (1979). Ethylene oxide neurotoxicity: report of four cases and review of the literature. *Neurology* **29,** 978–983.

98. Grober, E., Crystal, H., Lipton, R. B., and Schaumberg, H. (1992). EtO is associated with cognitive dysfunction. *J. Occup. Med.* **34,** 1114–1116.

99. Estrin, W. J., Bowler, R. M., Lash, A., *et al.* (1990). Neurotoxicogical evaluation of hospital sterilizer workers exposed to ethylene oxide. *Clin. Toxicol.* **28,** 1–20.

100. Finelli, P. F., Morgan, T. F., Yaar, I., *et al.* (1983). Ethylene oxide-induced polyneuropathy: a clinical and electrophysiologic study. *Arch. Neurol.* **40,** 419–421.

101. Ohnishi, A., Inoue, N., and Yamamoto, T. (1986). Ethylene oxide neuropathy in rats. *J. Neurol. Sci.* **74,** 215–221.

102. Dretchen, K. L., Balter, N. J., and Schwartz, S. L. (1992). Cognitive dysfunction in a patient with long-term occupational exposure to ethylene oxide. *J. Occup. Med.* **34,** 1105–1112.

CHAPTER
41

Clinical Neurotoxicological Concerns on Drugs of Abuse

KEVIN T. FINNEGAN
Departments of Psychiatry, Pharmacology, and Toxicology
University of Utah School of Medicine
Salt Lake City, Utah 84148

Dealing with the medical and neurologic problems related to the abuse of drugs continues to be one of the most challenging problems confronting the physician. Symptoms, signs, and complications vary tremendously depending on which drug is used, how much is used, and even the pattern of use. Relevant history is often not volunteered and may even be concealed. Further complicating this already bewildering clinical picture has been the proliferation of newly synthesized and untested street drugs, known as controlled substance analogs or "designer drugs." In this chapter, both traditional drugs of abuse as well as these synthetic analogs are discussed, with a special emphasis on their neurological complications and known neurotoxic effects.

I. Ethyl Alcohol

Ethanol is one of the two most frequently encountered drugs of abuse, the other being tobacco. Ethanol abuse constitutes a major economic and public health problem throughout much of the world, with recent estimates placing the costs of lost productivity and health care related to alcoholism in the United States alone in excess of 100 billion dollars per year (Department of Health and Human Services, 1987). The consumption of small amounts of ethanol produces a state of intoxication characterized by altered mood, mild incoordination, and impaired mentation. Chronic consumption of increasing amounts is associated with a number of adverse health effects, including the presence of a well-defined withdrawal syndrome that occurs with drinking cessation. The adverse effects of chronic ethanol on nonneural organ systems and the syndromes of ethanol intoxication and withdrawal have been extensively reviewed elsewhere (Thomas, 1986; Charness *et al.*, 1989; Lieber, 1991) and are not discussed here.

Although excessive ethanol consumption is capable of producing deleterious effects on virtually any organ, the nervous system appears particularly susceptible. The neural axis may be involved at any level, or combinations of levels, giving rise to clinical presentations characterized by a diverse array of signs and symptoms. Why the central nervous system (CNS) manifestations of excessive ethanol are so varied or why a given neurological complications develops in some alcoholics and not in others remains poorly understood, but a complex interplay among nutritional deficiencies, a direct toxic effect of ethanol, and genetic factors is presumed to be responsible. As well as being a psychoactive drug, for example, ethanol is also a food, and its consumption in large amounts causes primary malnutrition by displacing essential nutrients in the diet. Gastrointestinal complications associated with alcoholism can result in the malabsorption of certain vitamins, further depleting the nutrient pool. Vitamins are essential for the activity of many critical enzymes, and a decline in their availability is associated with several well-recognized diseases. Wernicke's encephalopathy and Korsakoff's syndrome, two important neurological complications associated with chronic alcoholism, are thought chiefly due to a deficiency in thiamine.

More recent evidence indicates that ethanol or its metabolites may also be directly capable of producing toxic effects on the nervous system. Chronic ethanol administration to animals and human volunteers fed nutritionally enriched diets, for example, is reported to produce alterations in brain morphology, diffuse reductions in cerebral blood flow, and learning impairment (Walker *et al.*, 1980; Hata *et al.*, 1987; McMullen *et al.*, 1987; Arendt *et al.*, 1988). Genetic factors may also influence the susceptibility of certain alcoholics to neurological complications. The activity of several enzyme systems, including transketolase, pyruvate dehydrogenase, and α-ketoglutarate dehydrogenase, depends on the availability of the cofactor thiamine. Variants of transketolase that differ in their binding affinity for thiamine have been described (Blass and Gibson, 1977). These findings raise the possibility that genetically determined alterations in thiamine-dependent enzymes may explain why, for example, only a relatively small proportion of thiamine-deficient alcoholics develop Wernicke's syndrome. It is to be emphasized, however, that at the cellular level the distinction among nutritional, toxic, and genetic effects becomes obscured. Nutritional deficiencies profoundly alter the metabolism and detoxification of ethanol; conversely, the induction of enzyme detoxification systems by ethanol enhances the degradation of certain vitamins, contributing to their depletion. Alterations in the biodisposition of ethanol in the genetically predisposed individual may result in either quantitative or qualitative changes in ethanol metabolites. Clearly, the interactions among these nutritional, genetic, and toxic components are exceedingly complex; further research will be required to delineate their precise role in the pathogenesis of the ethanol-related complications discussed next.

A. Wernicke's Encephalopathy and Korsakoff's Psychosis

Recognized since the 1880s, Wernicke's encephalopathy is a preventable disorder associated with a deficiency in thiamine. Thiamine deficiency is thought primarily responsible because animals fed a low thiamine diet display neuropathologic lesions similar to those seen in Wernicke's patients at autopsy and because thiamine replacement promptly alleviates clinical symptoms of the disorder (Mesulam *et al.*, 1977; Witt and Goldman-Rakic, 1983a,b). The syndrome is most commonly observed against a backdrop of chronic alcoholism, but may be associated with any condition resulting in a reduced vitamin intake (starvation, cancer, gastric plication, etc.). Alcoholics appear at greatest risk, perhaps because ethanol ingestion also hinders the absorption and biodisposition of thiamine, and may possibly interfere directly with critical thiamine-dependent enzymes (Abe and Itokawa, 1977). Deficits in energy metabolism arising from reductions in the activity of these thiamine-dependent pathways are hypothesized to result in neuronal damage, but the exact mechanism remains unclear. Antagonists of *N*-methyl-D-aspartate receptors, a glutamatergic receptor subtype involved in the neuronal-damaging effects of cerebral ischemia and hypoglycemia, have been reported to reduce the severity of the lesions produced by a thiamine-deficient diet in animals (Langlais *et al.*, 1988). These preliminary findings suggest that the mechanism by which thiamine deficiency produces neuropathology may not simply be related to reductions in energy metabolism, but that the participation of other neurotransmitter systems, such as glutamate, may also be required. Genetic factors are probably important as well since a significant number of thiamine-deficit alcoholics do not show evidence of Wernicke's encephalopathy. Other investigators emphasize the contribution of a direct toxic effect of ethanol to the development of Wernicke's. Supporting this view, the neuropathological lesions found in mice treated with a combination of ethanol and a thiamine-poor diet were significantly more extensive compared to animals fed the thiamine-poor diet alone (Phillips, 1987).

Classically, the disease is characterized by the clinical triad of oculomotor palsies (nystagmus, conjugate gaze, and lateral rectus palsies), gait disturbances (principally ataxia), and an encephalopathic picture typified by confusion, disorientation, and obtundation. Symptoms usually occur abruptly. Hypothermia and hypotension may be present. These signs and symptoms can occur together or in various combinations; two-thirds of patients present with less than the full clinical triad. A mortality rate of 17% was observed in one series of patients, chiefly due to concomitant infection (pneumonia, septicemia) or severe hepatic disease (Victor *et al.*, 1971). Neuropathologically, the syndrome is characterized by bilateral lesions in the diencephalon and brain stem that are distributed periventricularly. The affected areas include the mammillary bodies and structures surrounding the third ventricle, the aqueduct, and fourth ventricle; these may appear atrophied, demyelinated, or demonstrate capillary infiltration (Victor *et al.*, 1971). Petechial hemorrhages may occur. Loss of myelinated axons is more pronounced than loss of nerve bodies.

Importantly, autopsy studies indicate that significant pathology may occur long before the disease is manifested clinically. Harper and colleagues (1983) reported that only 26 of 131 patients showing pathological evidence of the syndrome at autopsy were correctly diagnosed during life. Similarly, a retrospective study revealed that only 16% of patients shown at autopsy to have pathological evidence of Wernicke's syndrome presented with the classic clinical triad; 19% showed none of the classic elements (Harper *et al.*, 1986). These findings emphasize that the physician must maintain a low threshold for the administration of thiamine to suspect alcoholics. In this regard, the reports indicating that several of the characteristics pathological changes of Wernicke's disease can be visualized in the living patient with the aid of computerized tomography or nuclear magnetic resonance scanning are encouraging, as this new technology may provide for early diagnosis and treatment (McDowell and LeBlanc, 1984; Charness and DeLaPaz, 1987).

Treatment with thiamine produces a rapid improvement in gaze palsies and ataxia; the confusional state clears more slowly. Ocular manifestations disappear almost completely in the vast majority of patients; however, one-half of patients may be left with a persistent fine horizontal nystagmus. Residual ataxia, characterized by a slow, shuffling, wide-based gait, is common. Lethargy, inattentiveness, and global confusion recover gradually over a period of 1 to 2 weeks; as these symptoms recede, however, a prominent deficit in memory, termed Korsakoff's psychosis, may become apparent.

Korsakoff's psychosis is a common but not an invariant sequel to Wernicke's encephalopathy. Anterograde amnesia (impairment of learning) is most obvious, but retrograde amnesia (disturbance of past memory) is affected as well. Treatment involves thiamine replacement, but once established the memory deficit recovers in only a small proportion of patients. The relationship between Wernicke's disease and Korsakoff's psychosis remains unsettled. Consistent with the failure of most patients to improve significantly, many investigators regard Korsakoff's psychosis as a manifestation of accumulated subclinical damage that occurs with repeated episodes of thiamine deficiency. Indeed, the classic studies carried out by Victor and colleagues (1971) show that the neuropathological changes observed in patients with either Wernicke's encephalopathy or Korsakoff's psychosis are very similar. As well as memory loss, however, Korsakoff patients also show apathy, inattentiveness, and reduced insight, signs customarily associated with frontal lobe dysfunction. In keeping with this, Korsakoff patients perform poorly on tests sensitive to frontal lobe function (verbal fluency, sorting tasks) compared to non-Korsakoff alcoholics (Jacobson, 1989). Korsakoff patients also demonstrate evidence of severe cortical atrophy (increased ventricular volume, enlarged sulci) on a computerized tomography scan. Since ethanol-induced cortical atrophy is considered by some to represent a direct toxic effect of ethanol (see later), the pathogenesis of Korsakoff's syndrome may be multifactorial, involving both a deficiency in thiamine and a toxic effect of ethanol.

B. Brain Atrophy

The use of sensitive imaging technology (computerized tomography, nuclear magnetic resonance scanning) has confirmed the results of earlier pneumoencephalographic studies suggesting that chronic ethanol abuse may induce structural alterations in the brain (Cala *et al.*, 1978; Carlen *et al.*, 1978). Observed changes include ventricular enlargement, widening of the hemispheric sulci, and an increase in the pericerebral space; these are found in one-half to two-thirds of alcoholic patients. The changes are partially reversible with abstinence; younger patients with the shortest drinking history respond most dramatically. The reversibility of the lesion has been offered as one explanation for the improvement in cognitive performance observed after drinking cessation. These findings, in particular the persistence of atrophic features in some ex-alcoholics, has led to the suggestion that ethanol itself may exert direct toxic effects on the brain.

Subsequent investigations offer some support for this concept. Quantitative morphometric imaging studies suggest that the hemispheric changes are due to a disproportionate loss of subcortical white matter compared to gray matter (Harper and Kril, 1985), an observation consistent with other reports demonstrating focal decreases in blood flow in the frontal and parietal brain regions of alcoholics (Risberg and Berglund, 1987). More directly, cortical neuron counts carried out on autopsy specimens show a loss of neurons from the superior temporal cortex; the motor, middle temporal, or anterior cingulate gyri appear spared, although neurons in these latter areas appear to be reduced in size (Kril and Harper, 1989).

Morphological alterations in the dendritic arborization of cerebellar Purkinje cells and neurons in the superior frontal and motor cortices of alcoholic patients have also been reported (Harper and Corbett, 1990), and somewhat similar changes are observed in well-nourished animals chronically administered ethanol (McMullen *et al.*, 1987; Arendt *et al.*, 1988).

Other findings, however, are less in keeping with the concept that ethanol acts directly as a neurotoxin. The severity of the cerebral atrophic changes, as assessed radiographically, might be expected to correlate with brain weight at autopsy, but this is an inconsistent finding (Torvik *et al.*, 1982; Harper *et al.*, 1988a). Atrophy changes have also been observed in heavy social drinkers (Cala *et al.*, 1983), but these alterations likewise are unrelated to brain weight or the size of the pericerebral or ventricular space at autopsy (Harper *et al.*, 1988b). Moreover, the most prominent manifestations of cerebral atrophy are noted in alcoholics with coexisting liver disease or Wernicke's encephalopathy, raising the possibility that factors associated with chronic alcoholism, such as head trauma, nutritional deficiency, or hepatic dysfunction, may be more responsible than a direct toxic effect of ethanol itself. Similarly, a relationship between the atrophic changes visualized by computerized tomography and the deficits in cognitive performance found in alcoholics is not firmly established. Corrected for age, the extent of the radiographic findings does not correlate very well with the psychometric deficits, or the length and severity of the drinking history (Lishman, 1990). Conversely, some alcoholics demonstrate psychometric deficits but have normal scans. Hence, the issue remains unresolved; a number of investigators regard the evidence linking cortical atrophy with a direct toxic action of ethanol as less than compelling.

C. Cerebellar Degeneration

A cerebellar syndrome, not accompanied by ophthalmoplegia or memory difficulty, has been described in association with chronic alcoholism (Victor *et al.*, 1959). The syndrome is characterized by ataxia of the legs and the upper extremities, and by dysarthria. Individuals with alcoholic histories in excess of 10 years appear most at risk. Clinical signs develop gradually over a period of weeks to months; however, the disorder may also appear abruptly. The lesion involves the degeneration of Purkinje cells within the cerebellum, and in almost all cases is restricted to the anterior and superior aspects of the vermis. Its anatomical distribution and histological appearance are identical to that observed in the cerebellum of patients with Wernicke's encephalopathy, suggesting that ethanol-induced cerebellar degeneration should probably be classified as a variant of Wernicke's instead of as a separate disorder. The pathogenesis is presumed to be related to thiamine deficiency, and the mainstays of treatment are thiamine replacement and abstinence from ethanol. These interventions will stabilize, and in some cases improve, the neurological deficit (Diener *et al.*, 1984).

D. Central Pontine Myelinolysis

Central pontine myelinolysis is a relatively uncommon disorder characterized by a symmetrical demyelination of the central portion of the pontine tegmentum (Adams *et al.*, 1959). Demyelinating foci in the putamen, internal capsule, and thalamus sometimes accompany the pontine lesion. Afflicted patients are generally alcoholic, but the syndrome has also been described in conjunction with hepatic disease, anorexia, cancer, burns, and severe hyponatremia. The majority of patients with autopsy-proven central pontine myelinolysis present with few clinical signs of the disorder during life, presumably because the size of the lesion is generally small, on the order of 2 to 3 mm in diameter. More extensive pontine involvement produces obvious clinical signs and symptoms, and the prognosis may become grave (Messert *et al.*, 1979). Damage to corticospinal tracts produces para- or quadraparesis, at times resulting in the complete paralysis of an otherwise alert and oriented patient ("locked-in" syndrome). Extension of the demyelinating lesion into the corticobulbar tracts produces dysphagia and dysarthria; parkinsonian features may also develop. The syndrome may be obscured by the presence of other ethanol-related complications, such as the withdrawal syndrome or Wernicke's encephalopathy. Computerized tomography or nuclear magnetic resonance scanning may visualize the pontine and extrapontine lesions in some patients; brain stem auditory-evoked potentials are usually abnormal.

Central pontine myelinolysis may not be so much ethanol related as it is associated with any medical illness that produces a serious hyponatremia. Burcar and colleagues (1977) note that the disorder is associated with the rapid or "over" correction of hyponatremia in the majority of patients, and point out that the first reports of central pontine myelinolysis appeared shortly after intravenous therapy was introduced for the correction of electrolyte abnormalities. Evidence from animals studies generally supports the concept that hyponatremia plays an important etiological role.

The rapid correction of experimentally induced hyponatremia in dogs, for example, is associated with the development of myelinolytic lesions similar to those found in humans (Laureno, 1983). Intracellular volume depletion, produced by rapid increases in the extracellular concentration of sodium, may be responsible. Other mechanisms evidently contribute, however, because the syndrome has also been reported in patients with only mild hyponatremia or whose electrolyte abnormalities were corrected slowly.

Clinicians disagree as to what therapeutic measures should be employed for the treatment of central pontine myelinolysis. Based on the concept that rapid shifts in electrolytes and water are responsible, most recommend that symptomatic hyponatremia should be treated by increasing the serum sodium concentration to not more than 120 to 130 mmol per liter (Narins, 1986). On the other hand, clinicians who favor the notion that it is the symptomatic hyponatremia itself that is responsible recommend the administration of sufficient hypertonic saline with diuretics to increase the serum sodium concentration by a rate of 2 mmol per liter per hr, with the cumulative increase not to exceed 25 mmol per liter in 48 hr (Ayus *et al.*, 1987). Still others advocate slower rates of correction (Laureno and Karp, 1988). The rarity of the disorder has made it difficult to systematically evaluate the effectiveness of one treatment modality over any other.

E. Ethanol-Induced Neuropathy and Myelopathy

Abnormalities in both peripheral nerve and skeletal muscle are associated with chronic alcoholism (Haller and Knochel, 1984). The peripheral nerve disorder generally presents as a symmetrical, distal, polyneuropathy of gradual onset. The legs are usually affected earlier and more severely than the upper extremities. Symptoms reflect an involvement of sensory, motor, and, occasionally, autonomic nerves. Numbness, paresthesia, pain, cramps, weakness, ataxia, and a burning sensation are frequent complaints. The neurological examination commonly shows the loss of deep tendon reflexes, reductions in touch and vibratory sensation, and muscle weakness. An optic neuropathy, characterized by a gradual impairment of vision and the development of scotomata, has also been described. The disorder is believed to result from a dietary deficiency in thiamine and perhaps other B vitamins. Symptoms usually remit with adequate vitamin supplementation and sobriety.

Ethanol-induced myopathy chiefly affects the proximal skeleton muscle system; the distribution may be asymmetrical. Alcoholic myopathy may present as an acute, necrotizing disorder that develops over the course of several hours or days, often following an episode of binge drinking. Pain, tenderness, weakness, and edema are the most common presentation. Ethanol-induced cardiomyopathy, with consequent congestive heart failure, has been reported. Elevated serum creatine kinase, myoglobinuria, and electromyogram abnormalities reflect the necrosis of muscle fibers observed on biopsy. A direct, toxic effect of ethanol on the muscle fiber is thought to be responsible (Song and Rubin, 1972). Treatment is supportive and directed at correcting renal failure caused by rhabdomyolysis, cardiac failure, and electrolyte abnormalities. The prognosis is that of a gradual recovery, provided that ethanol intake ceases.

A more common presentation of ethanol-induced myopathy develops gradually, over a period of weeks to months (Haller and Knochel, 1984). Pain is less prominent and the increase in serum creatine kinase levels tends to be smaller than in the acute form of the disorder. Myoglobinuria is not typical. Muscle weakness and atrophy, principally affecting the hip and shoulder girdles, are found on physical examinations. As mentioned previously, sobriety is associated with a gradual improvement in weakness, although recovery may not be complete.

F. Movement Disorders

Chronic severe intoxication with ethanol, or ethanol withdrawal, has been reported to produce transient parkinsonism, choreiform dyskinesias, or generalized choreiform movements (Mullin *et al.*, 1970; Carlen *et al.*, 1981; Fornazzari and Carlen, 1982). These observations suggest that ethanol is capable of affecting basal ganglia function. However, the rare reports of such complications suggest that the expression of these movement disorders requires the operation of other, probably preexisting, factors.

G. Marchiafava–Bignami Disease

This apparently rare complication of chronic alcoholism was originally described in Italian men addicted to red wine, but the disorder has also been described in association with other ethanol-based spirits and in other ethnic groups (Brion, 1976). Occurring in malnourished alcoholics, the disease is characterized by the degeneration of the corpus callosum and adjacent subcortical white matter. Symptoms include dementia, dysarthria, spasticity, and gait ataxia. The cause is presumed to be related to dietary deficiency. Coma and death may ensue in patients presenting with an acute onset of the disease; others may survive for many years in a demented state in the more chronic forms.

H. Ethanol and Vascular Disease

Recent epidemiological investigations have linked heavy ethanol consumption to the occurrence of cerebral infarction and subarachnoid hemorrhage (Hillbom and Kaste, 1981, 1983; Taylor, 1982). Adolescents and young adults appear at greatest risk. An effect of ethanol on hemocoagulation or on cerebral blood vessels may be involved, but the cause remains unknown.

I. Fetal Alcohol Syndrome

The fetal alcohol syndrome refers to a distinct cluster of congenital abnormalities observed in the offspring of alcoholic mothers (Rosett and Weiner, 1984). Development during the first trimester is thought to be the most vulnerable, a time when pregnancy may not yet be recognized. Commonly observed features include intrauterine growth retardation, microcephaly, and characteristic craniofacial anomalies consisting of short palpebral fissures, ptosis, mid-facial hypoplasia, microophthalmia, epicanthal folds, and small head circumference. Cutaneous, joint, cardiac, and genitourinary abnormalities have also been described. Neonatal mortality may be high; survivors manifest intellectual impairment, hyperactivity, attention deficits, and learning disabilities. The full expression of fetal alcohol syndrome is estimated to occur in about 6% of infants born to severely alcoholic mothers; however, a larger pool of such infants express one or more of the features of fetal alcohol syndrome in a less severe form (fetal alcohol effects). Why heavy intrauterine exposure to ethanol is not uniformly teratogenic is unknown, but parity, genetic factors, racial background, and a previous history of alcohol-related problems in the mother have been identified as risk factors. The physical and behavioral abnormalities of the syndrome have been reproduced in well-nourished animals administered ethanol, suggesting that a direct toxic effect of ethanol is responsible (Streissguth *et al.*, 1980).

More worrisome has been the recent reports suggesting that lesser amounts of ethanol may be associated with a variety of neuropsychological and behavioral effects. Deficits in intellectual performance, attention span, and reaction time were found in children born of nonalcoholic, but heavy drinking, mothers (Streissguth *et al.*, 1986, 1989). Intrauterine exposure to more than three drinks per day was reported to triple the risk for low IQ (<85) at age four. Unfortunately, the dose–effect relationship for the effects of ethanol on fetal development remain poorly described; spe-

cifically, whether there exists a threshold below which ethanol does not produce teratogenic effects is unknown. Therefore, it seems prudent for women to avoid all alcoholic beverages during pregnancy until the matter is clarified.

II. Methanol

Methanol (methyl alcohol) is a colorless liquid that is easily absorbed from the gastrointestinal tract. Methanol is a component of many household and workplace products, and because of its readily availability, it is often used as a substitute for ethanol by alcoholics when the latter is unavailable. Not uncommonly, it is also deliberately ingested in suicide attempts. Methanol is metabolized by hepatic alcohol dehydrogenase, forming formaldehyde and formate, and it is the accumulation of these metabolites that is responsible for its toxic effects (Suit and Estes, 1990). Formic acid accumulation produces a marked metabolic acidosis, as signaled by a precipitous drop in the serum bicarbonate concentration and the development of an increased anion gap. Formate also directly inhibits aerobic respiration, and the consequent increase in lactic acid production further exacerbates the acidosis.

Symptoms of methanol poisoning typically develop 12 to 24 hr after ingestion, although this may vary considerably among patients (Bennett *et al.*, 1953). The latency between ingestion and the appearance of symptoms reflects the relatively slow metabolism of methanol by hepatic alcohol dehydrogenase as compared with ethanol. Symptoms include malaise, nausea, weakness, and headache, progressing to confusion, stupor, and coma in severe cases of intoxication. Epigastric pain, similar to that described in acute pancreatitis, is common. Characteristic is the involvement of the visual system, with complaints of blurry, misty, or double vision being most frequent. Signs of ophthalmologic involvement include the presence of dilated, sluggish, or nonreactive pupils, hyperemia and edema of the optic disc, and peripapillary retinal edema. Optic atrophy may develop in cases of severe toxicity. The minimal lethal dose is about 80 g; visual symptoms can be expected with the ingestion of half this dose. Diagnosis is aided by the presence of an anion gap on laboratory examination, and most conclusively by a toxicological screen positive for methanol. Toxicological screening, however, may not always be available, and the clinician must maintain a high index of suspicion when confronted with a patient who complains of visual disturbances, is confused or obtunded, and has an unexplained metabolic acidosis.

Treatment is aimed at reversing the acidosis by using intravenous solutions of bicarbonate, removing methanol and formate with hemodialysis, and preventing the further oxidation of methanol to formate. The last is achieved by administering intravenous ethanol. Since ethanol is the preferred substrate for alcohol dehydrogenase, its administration prevents any further metabolism of methanol to formate. Currently under study as an alternative to ethanol therapy is 4-methylpyrazole, a potent inhibitor of alcohol dehydrogenase. The treatment of methanol poisoning has been reviewed in detail (Suit and Estes, 1990).

Methanol causes a variety of pathological changes in the CNS. A diffuse cerebral edema is common (Menne, 1938). The optic nerve and basal ganglia, in particular, are affected. The optic lesion is characterized by a selective demyelination of the retrolaminar segment of the optic nerve, with sparing of the axons, although the latter often appear edematous and swollen (Sharpe *et al.*, 1982). These pathological changes are thought to be related to the inhibitory actions of formate on mitochondrial energy production, with oligodendroglia being particularly vulnerable. Damage to oligodendroglia is hypothesized to produce demyelination, which in turn causes a loss of saltatory conduction and, ultimately, permanent visual damage. The selective vulnerability of the retrolaminar portion of the optic nerve is not well understood, but may be due to the selective accumulation of formate in this region.

A peculiar finding that serves to distinguish methanol poisoning from other intoxicants is the appearance of a bilateral putaminal hemorrhagic necrosis (Aquilonius *et al.*, 1980). The extent of the necrosis varies with the degree of poisoning; in severe cases the entire putamen and the surrounding white matter are destroyed. Extensive white matter necrosis without putaminal involvement, similar to that observed after carbon monoxide poisoning or hypotensive episodes, has also been described. The mechanism is presumed to involve a direct toxic effect of formate, but this remains speculative. Curiously, individuals with putaminal necrosis documented by computerized tomography have few clinical symptoms attributable to the lesion, although a parkinson-like syndrome has been reported in rare cases (Guggenheim *et al.*, 1971; Ley and Gali, 1983).

III. Illicit Drugs of Abuse

The recreational use of chemical substances that produce alterations in mood, thought, and feeling extends far back into recorded history. For much of this time abused substances were limited to those obtained from preparations of natural vegetative materials, which were generally of low potency and bioavailability. Drug abuse entered a new era early in the 19th century when advances in chemistry permitted the isolation and identification of individual alkaloids, such as morphine, codeine, and cocaine, from plant extracts. With this emerging technology, alkaloid extracts of relatively high purity became easily obtainable, and by the middle of the 19th century the abuse of pure alkaloids rather than crude preparations began to spread throughout Europe and the Far East.

More recently, legitimate efforts on the part of the pharmaceutical industry to develop new medicinal agents have led to an enormous growth in the number of psychoactive compounds with high abuse liability. The availability of these newer agents to the general public has been controlled by legislation designed to limit their production and distribution. Lured by the enormous profits involved, however, the last decade has seen an impressive growth in the illicit synthesis of controlled substances by domestic clandestine laboratories.

Indeed, illicitly synthesized controlled substances of every major class of psychoactive drugs are now found with alarming frequency in street samples. More worrisome has been the appearance of untested structural analogs of controlled substances, the abuse of which may expose the user to an altogether novel spectrum of medical hazards. This section reviews the clinical features and neurological complications associated with some of the more commonly abused illicit substances, with a particular focus on those drugs being produced by domestic clandestine laboratories.

The Clandestine Laboratory

Clandestine drug-manufacturing laboratories range from crude makeshift operations to highly sophisticated and technologically advanced facilities, some of which are mobile. They can be set up virtually anywhere and are often found in private residences, garages, motel rooms, apartments, house trailers, and the like. The drugs themselves can often be manufactured by relatively simple chemical additions or condensations of readily available precursors, using equipment comparable to that found in a high school chemistry laboratory. Because the synthetic schemes are so straightforward little specialized training is required. Not uncommonly, the "cooker" possesses no formal background in chemistry. "Cookbook" recipes for the manufacture of drugs such as methamphetamine, amphetamine, and phencyclidine provide the necessary instruction and are easily obtained from a variety of legal (underground newspapers and magazines) and illegal sources.

The *de novo* synthesis of complex alkaloids such as morphine or cocaine is technically difficult, and domestic clandestine laboratories have generally focused their efforts on the synthesis of simpler compounds such as methamphetamine and phencyclidine. Drug abuse has entered a new era, however, with the emergence of "designer drugs," structural analogs of controlled substances. Controlled substance analogs are agents that differ slightly in chemical structure from their regulated parent compounds, but which possess similar pharmacologic properties. The emergence of this phenomena marks, for the first time, the production of original narcotic drugs by illicit laboratories and not illicit copies of legitimate pharmaceutical products. The impetus for this industry grew out of the recognition that, at least until very recently, one could evade the penalties levied against those illegally trafficking a controlled substance by synthesizing or "designing" an unregulated structural analog. Since the 1980s analogs of every major class of abused drugs (opiates, stimulants, hallucinogens) have surfaced on the illicit drug market and now constitute a significant portion of the drug trade.

The ingestion of clandestinely manufactured drugs by the unsuspecting user carries the potential for serious toxicity. Clandestine laboratories make little effort to check the purity or even the identity of the final product; these can vary enormously from batch to batch. Given the level of expertise involved it is not uncommon that completely novel analogs are produced, usually as unintentional by-products of the reaction. Illicitly synthesized drugs may also be contaminated to a significant degree by the organic solvents (benzene, toluene, acetone) required for their synthesis and extraction. In addition to the uncertain identity of the drug itself, the powders are commonly diluted prior to sale ("cut," "stepped on") with crude adulterants, which by themselves may give rise to serious organ toxicity. Quinine, strychnine, or talc are popular additives. Although the potential adverse health consequences of most controlled substance analogs remain largely unknown, in at least one case they have proven serious: the ingestion of the meperidine analog 1-methyl-4-phenyl-1,2,3,6-tetrahydropyridine (MPTP) by a group of unsuspecting heroin addicts in 1982 resulted in the development of a severe, apparently permanent, parkinsonian syndrome (Langston *et al.*, 1983).

IV. Stimulants

Despite regulation restricting their manufacture and possession, abuse of stimulants continues to be a seri-

ous problem, forming about 10% of all drug-related visits to the emergency department.

A. Amphetamines

First introduced in the 1930s as a nasal decongestant, amphetamines were quickly appreciated for their ability to boost energy, decrease appetite, and build confidence. Because amphetamines are relatively inexpensive (the "poor man's cocaine"), their recreational use has proliferated. Known on the street as "speed," "uppers," "bennies," "dex," or "crystal," most of the drug marketed under these names is actually methamphetamine, the *N*-methyl derivative. Methamphetamine is easy to synthesize and over 350 methamphetamine laboratories were seized in California alone in 1988 (Joint Federal Task Force Report, 1990). Several schemes are used for their synthesis, depending on the availability of starting materials (Glennon, 1989). Most commonly employed is the amination of phenyl-2-propanone (a.k.a., phenylacetone or methyl benzyl ketone), using methylamine or hydroxylamine as the nitrogen donor. Methamphetamine is also frequently synthesized directly from the precursor ephedrine by catalytic hydrogenolysis. Enforcement efforts are directed at regulating the availability of these precursor compounds. This is not as easy as it might sound, however, as many of these precursors are also required for the synthesis of a large number of legitimate pharmaceuticals. The "analog game" is played here as well: recent efforts to reduce the illicit production of methamphetamine by regulating the availability of the precursor, ephedrine, are being circumvented by the use of *N*-methylephedrine, currently an unregulated ephedrine analog. This substitution has resulted in the recent appearance of the novel designer drug, *N,N*-dimethylamphetamine, on the illegal market. Despite being emergently classified as a Schedule I compound by the DEA in 1988, the illicit production of this compound continues unabated; seizure of 20 clandestine laboratories specializing in the manufacture of *N,N*-dimethylamphetamine in 1989 alone resulted in the confiscation of over 57 kg of drug (Drug Enforcement Administration, personal communication).

B. "ICE"

A new form of methamphetamine, appearing as large, crystal clear "rocks," has surfaced recently in the Far East, the Pacific Islands, and the West Coast. Called "ice," "glass," "Shabu," or "Hiropong," this form of the drug is prepared by slow evaporation of a saturated solution of the salt in water (or possibly isopropanol). The appearance of the resulting product differs dramatically from the traditional crystalline white powder, prepared by bubbling an etheric solution of the free base with hydrochloric acid. "Ice" is typically sold on the underground market in small "rocks," hence the additional street name "crack meth." Similar to "crack cocaine" the typical mode of ingestion is via smoking. The inhalational route provides an immediate and intense "high," and due to the pharmacokinetics of the drug, one that lasts for several hours. Because of these factors, "ice" carries the potential for serious addiction, and its use is expected to spread rapidly.

C. Analogs of Amphetamine

The phenethylamine structure permits a great number of substitutions, with about 2000 derivatives having been synthesized and characterized. Approximately 30 of these analogs have been tested for their therapeutic efficacy in clinical trials (Shulgin, 1970). Several of these analogs have found their way on to the illicit drug market, where they have become known as MDA (3,4-methylenedioxy-amphetamine), MDMA (3,4,-methylenedioxymethamphetamine; "ecstasy"), PMA (*p*-methoxyamphetamine), MMDA (3-methoxy-4,5-methylenedioxyamphetamine), DOM (4-methyl-2,5-dimethoxyamphetamine), and DOB (4-bromo-2,5-dimethoxyamphetamine). The Knoevenagel reaction, involving the condensation of an appropriately substituted benzaldehyde with either nitromethane or nitroethane, is the most popular route for the synthesis of these agents (Glennon, 1989). Interestingly, aromatic ring substitution of phenethylamines such as amphetamine converts these potent CNS stimulants into drugs possessing marked psychedelic and hallucinogenic properties (Shulgin, 1978). Although drugs such as DOB and DOM continue to enjoy some popularity, the recreational use of the methylenedioxy-substituted analogs (MDA, MDMA) is currently most widespread.

MDA became popular as a drug of abuse during the 1960s. With the classification of MDA as a Schedule I compound in 1970, its *N*-methyl derivative (MDMA) surfaced on the illicit drug market during the 1970s as an unscheduled substitute. MDMA currently is said to be the drug of choice among college students and young professionals (Adler *et al.*, 1985). A third analog in this series, 3,4-methylenedioxy ethyl amphetamine (MDE, "Eve"), has also been identified in street samples of confiscated material (Drug Enforcement Administration, personal communication).

D. "U4Euh"

Although not structurally related to the amphetamines, 4-methylaminorex ("U4Euh," or on the East Coast, "ICE") is a potent CNS stimulant drug. The parent compound, aminorex, was marketed in Europe as an anorectic agent before being withdrawn because of serious toxicological effects. Both drugs have surfaced recently in confiscated samples of street drugs. Appearing as a white or off-white powder, "U4Euh" is typically ingested via the oral or intranasal route; intravenous administration is uncommon. The drug is trivially synthesized by the condensation of phenylpropanolamine and cyanogen bromide. Because of its straightforward synthetic scheme and the ready availability of the requisite precursor chemicals, the illicit production and abuse of "U4Euh" are anticipated to increase.

E. Cocaine

Once considered a drug of the wealthy, cocaine ("coke," "snow," "blow," or "toot") now pervades all strata of society. In contrast to amphetamine and its analogs, which are purely synthetic compounds, cocaine is a structurally complex plant alkaloid extracted from the South American shrub *Erythroxylon coca*. The *de novo* synthesis of cocaine is technically difficult and the vast majority of the drug is prepared overseas and smuggled into this country. The alkaloid is typically extracted with a combination of water, dilute sulfuric acid, and kerosine as the organic solvent (Cooper, 1989). The aqueous and organic phases are separated, and ammonia, lime, or sodium carbonate is added to the aqueous phase to cause precipitation of the alkaloids; the residue is captured by filtration. At least 50% cocaine base, the Indians call this product "sulfato" or coca paste (also termed "bazooka"). Residual amounts of the solvent used to extract the alkaloid (acetone, benzene, toluene, ethyl ether) are almost always present as contaminates. The adverse health effects of these contaminates are unknown.

The hydrochloride salt is prepared by dissolving the coca paste in diethyl ether and bubbling hydrochloric gas through the solution, causing the precipitation of cocaine (called "flake"). Recently, dealers have begun selling a more potent free base form known as "rock" because of its hardened state. A portion is then "cracked" into smaller pieces prior to distribution. The preferred route of administration is smoking, producing an intense euphoria of rapid onset.

F. Clinical Features and Neurological Complications of Stimulant Abuse

The amphetamines and cocaine are potent central and peripheral sympathomimetic agents. The two main target organs are the brain and the heart. Differing primarily in their duration of action, amphetamine and cocaine produce similar psychological effects in the laboratory: users familiar with cocaine cannot distinguish between 8–10 mg of cocaine and 10 mg of amphetamine when both are given intravenously (Fischman *et al.*, 1976). Physical signs of cocaine or amphetamine abuse are also similar and represent a dose-related extension of their pharmacological actions. At low doses, the drugs produce euphoria and a sense of increased alertness and energy. Users may initially perceive themselves as unusually clever and having increased powers of insight, but with increasing dose and duration of abuse, grandiosity, irritability, and emotional lability are observed. Long-term abuse induces psychosis, characterized by paranoia, and visual, tactile, and auditory hallucinations, a syndrome that can be clinically indistinguishable from paranoid schizophrenia. Overdose with amphetamine or cocaine produces restlessness, insomnia, anxiety, and delirium. Physical findings include flushing or pallor, tremor, tachycardia, and mydriasis; eventually hyperreflexia, delirium, and seizures may be observed. Extreme anxiety and disorientation and an impending sense of doom commonly precede the seizures, but seizures may also occur suddenly and unexpectedly, leading to cardiopulmonary arrest. Palpitations progressing to ventricular arrhythmia have been reported. Hyperthermia is a life-threatening complication of amphetamine overdose, requiring aggressive treatment with ice and cooling blankets.

A CNS vasculitis after amphetamine use has been described (Salanova and Taubner, 1984). Frequently associated with this necrotizing vasculidity is the development of a subarachnoid or intracranial hemorrhage. Intracranial hemorrhage after amphetamine has also been reported without evidence of vasculitis (Harrington *et al.*, 1983). These authors advise that the presence of intracranial hemorrhage in an otherwise healthy young adult should raise the suspicion of amphetamine abuse. The vasculidity in some patients may also present as a mononeuropathy or mononeuropathy-ultiplex type of syndrome (Stafford *et al.*, 1975).

Chorea and athetosis are uncommon but well-documented complications of prolonged use of amphetamines (Caplan *et al.*, 1982). The etiology is unknown but may be related to an excessive stimulation of dopamine receptors. The disorder generally remits within a few days after cessation of intake but has been reported to persist for much longer periods. Rhabdomyolysis and disseminated intravascular coagulation, occasionally accompanied by evidence of noncardiogenic pulmonary edema, have been noted as an uncommon complication after amphetamine overdose (Kendrick *et al.*, 1977). Of uncertain etiology, the syndrome can carry a high risk of mortality unless treated aggressively.

Despite the fact that ring-substituted hallucinogenic analogs of amphetamine have gained a reputation for providing a tranquil psychedelic experience, a number of deaths after overdose with *p*-methoxyamphetamine and MDA have been reported (Simpson and Rumack, 1981). In all cases, the manifestations and complications of overdose were similar to those described for the parent compound amphetamine. Clinical experience with other clandestinely synthesized analogs is, as yet, scant, although the analog MDMA has been implicated as a cause of death in five individuals (Dowling *et al.*, 1987).

More worrisome is the numerous reports indicating that amphetamine and several of its analogs are capable of producing a selective degeneration of CNS dopaminergic and serotonergic neurons in experimental animals, including nonhuman primates (Seiden and Ricaurte, 1987; Finnegan and Schuster, 1989). The popular drug of abuse, MDMA, appears to be particularly potent: monkeys treated with MDMA at doses only slightly higher than those used recreationally by humans show large, apparently permanent, decrements in the levels of serotonin in the frontal cortex, hippocampus, and striatum. Immunocytochemical studies indicate that the reductions in transmitter levels are due to a selective degeneration of serotonergic axons and nerve terminals. Despite the compelling evidence obtained in animals, however, there is little published data to suggest that humans repeatedly exposed to MDMA suffer any persistent neurological or psychiatric sequelae. Differences in the metabolism and biodisposition of MDMA between the two species might account for the apparent lack of toxicity in humans. Alternatively, the role of serotonin in behavior and cognition is not well understood, and the available methods and instruments used for neurological and psychiatric assessment may simply lack the required sensitivity. These possibilities are currently the object of intense investigation.

V. Opiates and Opioid-Like Substances

The medicinal and psychological effects of opium have been known since at least the third century B.C. Crude opium is made from drying the milky exudate of the poppy plant *Papaver somniferum* and actually consists of over 20 different alkaloids, including morphine, codeine, thebaine, and papaverine. Many semisynthetic derivatives are made by relatively simple chemical modifications of the structurally complex botanical alkaloids, such as diacetylmorphine (heroin) from morphine or the potent analgestic agent oxycodone from thebaine. Although opium has been used as a recreational drug of abuse since the 18th century, it was with the isolation of morphine and the synthesis of its more potent derivative heroin in 1898 that opiate abuse assumed its most menacing dimensions.

A. Heroin

Known on the streets as "smack," "H," "Mexican brown," or "tar," heroin remains by far the most commonly abused opiate today. The two most favored routes of administration are subcutaneous ("skin popping") and intravenous (producing a rapid onset of action known as a "rush").

B. Clinical Features and Neurological Complications

The administration of small to moderate amounts of heroin induces drowsiness, lethargy, or stupor ("nodding out"). Some degree of respiratory depression and miosis are almost always present. Overdose produces impaired respiration, profound miosis, increasing coma, hypothermia, and occasionally bradycardia. Death occurs secondarily to respiratory arrest in virtually all cases. The clinical triad of coma, depressed respiration, and pinpoint pupils strongly suggest opiate poisoning. This clinical presentation may be obscured, however, by the coingestion of other drugs, most commonly methamphetamine or cocaine (the "speed ball"). The management of opiate overdose has been reviewed (Kulberg, 1986).

The intravenous use of any agent exposes the addict to a variety of infectious complications, including septic thrombophlebitis and arthritis, endocarditis, serum hepatitis, and, more recently, the acquired immune deficiency syndrome (AIDS). Recurrent bouts of lymphangitis, thrombophlebitis, and cellulitis may lead to obstruction of the venous and lymphatic drainage, resulting in an edematous and fibrotic appearance of the distal arms and hands. Other cutaneous stigmata include the formation of abscesses and coin-sized ulcers after the intradermal administration ("skin popping") of opiates.

Many noninfectious complications involve the nervous system. The anecdotal reports of cerebral or pituitary gland infarction, bilateral deafness, or encephalopathy represent likely sequelae of prolonged episodes of opiate-induced hypotension or hypoxemia instead of a direct toxic effect of the drug (Louria *et al.*, 1967; Pearson and Richter, 1979). Transverse myelopathy (sometimes a Brown–Sequard syndrome), as well as several types of peripheral neuropathy (principally the radial nerve and brachial plexus), has been described. Proposed explanations include an impaired circulation secondary to prolonged coma, the presence of toxic adulterants, or occasionally the direct injection of the drug itself into a nerve. One rather puzzling complication is a stroke-like syndrome evidently not related to coma, infection, or embolism (Caplan *et al.*, 1982). Noncardiogenic pulmonary edema may occur as an acute complication of parental opioid use (Steinberg and Karliner, 1968). Of unknown etiology, this pulmonary complication carries a significant risk of mortality and requires aggressive treatment. As mentioned previously, illicit drugs are commonly diluted with crude adulterants, which by themselves may be toxic. It is frequently unclear whether the toxicity is due to the drug itself, to the adulterant, or to an interaction between the drug and the adulterant.

VI. Opiate Analogs

Since the addictive potential of opiate drugs guarantees the clandestine chemist a captive audience for his synthetic wares, it is perhaps not surprising that several narcotic analgesics have served as templates for the design of new opiate analogs. The number of laboratories involved in the illicit manufacture of opiate analogs, however, is considerably less than that of amphetamine or phencyclidine (PCP), an observation perhaps related to the more complex chemistry of the opiates and the fact that heroin continues to be readily available. Nonetheless, analogs of the narcotic analgesics fentanyl and meperidine have been found with increasing frequency in samples of confiscated street material.

A. Fentanyl

Seven different analogs of fentanyl have so far surfaced on the illicit drug market, where they appear to be increasingly used as substitutes for heroin (Henderson, 1989). Known as "Persian white," "china white," or "synthetic heroin," approximately 10–20% of opiate addicts in California are currently thought to be using fentanyl analogs. The chief danger associated with the use of these agents is their remarkable potency. The synthetic derivative (+)3-methyl-fentanyl, for example, is approximately 7000 times as potent as morphine (Van Bever *et al.*, 1974); lethal doses of this agent are estimated to be on the order of a few micrograms. Minuscule amounts of very potent drugs can, therefore, result in serious overdose when used by the unsuspecting addict. Over 110 deaths have been attributed to fentanyl analog since 1979; in many cases the addicts were found with the needle still in the vein

(Henderson, 1989). The actual number of deaths may be higher, as few clinical laboratories are equipped for the analysis of such minute amounts of drug.

B. *Meperidine*

Three analogs of the opiate analgesic meperidine (Demerol) have been identified to date in confiscated street samples: 1-methyl-4-phenyl-4-propionoxypiperidine (MPPP), 1-phenethyl-4-phenyl-4-acetoxypiperidine (PEPAP), and perhaps the most notorious of all designer drugs MPTP. The accidental synthesis of MPTP illustrates the dangers associated with the clandestine manufacture of illicit drugs. Advertised as a new "synthetic heroin," this compound surfaced on the illegal drug market during the summer of 1982. Users rapidly developed a severe, apparently permanent, parkinsonian syndrome that was clinically indistinguishable from the idiopathic form of the illness (Langston *et al.*, 1983). Subsequent studies in animals has demonstrated that MPTP, in addition to its opioid activity, is a potent and highly selective nigrostriatal dopaminergic neurotoxin (Langston and Finnegan, 1987). While proving disastrous for the individuals afflicted, the discovery of MPTP has lead to a resurgence of interest in the etiology and treatment of Parkinson's disease. Based in part on the observation that the MAO-B inhibitor deprenyl prevents the parkinsonogenic effects of MPTP in animals, two clinical studies have examined the efficacy of this agent in early stage Parkinson's disease (Parkinson's Study Group, 1989; Tetrud and Langston, 1989). The results of these investigations suggest that deprenyl slows the progress of this neurodegenerative disorder, raising the possibility that drug therapy aimed at halting the disease process may soon be at hand.

VII. Hallucinogens

The use of psychedelic compounds, agents that alter perception and sensory experience, is not a new phenomenon. The trigger for their widespread use since the 1970s was the development of easily manufactured compounds, many of which emerged from earnest efforts of the pharmaceutical industry.

A. *Phencyclidine*

Known on the streets as "PCP," "the *PeaCe Pill*," "angel dust," "KJ," "crystal," "hog," and a variety of other names, PCP is a nonnarcotic, nonbarbiturate, dissociative anesthetic agent with hallucinogenic properties. First developed and marketed as a surgical anes-

thetic in the 1950s, the drug was quickly withdrawn when its psychotomimetic effects become evident. It reappeared in the late 1960s as a veterinary anesthetic, and shortly thereafter became popular as a recreational drug. Phencyclidine possesses a complex neurochemical profile and interacts with many neurotransmitter systems. Because the drug produces alterations in mental status that are highly reminiscent of schizophrenia, researchers have long been interested in its neurochemical and behavioral actions.

Chemically, PCP is 1-phenylcyclohexylpiperidine and is relatively easy to synthesize. Using similar synthetic methods, but employing slightly different precursors, clandestine chemists have generated over 20 different analogs of PCP since the 1970s (Scaplehorn, 1989). Possessing virtually the same psychopharmacological profile as the parent compound, these include cyclohexamine and phenylcyclohexylpyrrolidine, in which the piperidine ring has been replaced with an ethylamino group or a five-membered pyrrolidine ring, respectively. Other illicit PCP derivatives include phenylcyclopentylpiperidine and thienylcyclohexylpiperidine, in which the cyclohexyl or phenyl rings have been substituted with either a cyclopentyl or a thienyl group. Further, each of the rings may bear different substituents depending on the exact formula of the starting material. Clearly, controlling the availability of such an impressive number of structural analogs presents a formidable task for regulatory agencies.

B. *Clinical Features and Neurological Complications*

PCP is self-administered orally, intranasally, or by smoking. The latter route of administration is the most popular and affords the user some ability to titrate his level of intoxication. Phencyclidine quickly gained a reputation for creating a frightening emergency department experience for the uninitiated health professional. As a disassociative anesthetic, PCP reduces the perception of pain and radically alters the manner in which a stimulus is integrated and perceived. Thus, self-injurious behavior and trauma are common causes of death in PCP intoxication. Individuals display a marked variation in their response to the drug. Symptoms range from catatonic rigidity to unprovoked violent behavior associated with herculean displays of strength. Disorientation, incoordination, and confusion or fluctuating levels of consciousness during the course of intoxication are common. The PCP behavioral syndrome and its treatment have been reviewed (Rappolt *et al.*, 1979).

High doses of PCP cause profound coma, which may persist for days. Seizures may occur. Motor abnormalities include ataxia, myoclonus, dystonia, and opisthotonic or decerebrate posturing. Psychiatric manifestations include hallucinations, delusions, misperceptions, confusion, and delirium; these can persist for days to weeks after the initial ingestion. Some patients display classic signs of paranoid schizophrenia with persecutory delusions; scattered case reports suggest that with continued abuse the psychiatric syndrome may persist even in the absence of drug. Whether the chronic abuse of PCP predisposes to the development of schizophrenia in susceptible individuals or whether these individuals were destined to develop the illness anyway is uncertain. Neuropathological changes, other than those occurring secondarily to trauma or the medical complications associated with intoxication, have not been reported.

C. Lysergic Acid Diethylamide (LSD)

Drugs that induce psychedelic effects, principally the peyote cactus (containing mescaline) and mushrooms (containing psilocin), have been known for centuries. The discovery of the remarkably potent psychedelic effects of LSD in 1947 by Hoffman stimulated a great deal of interest among scientists in the late 1950s who hoped that its study might aid in the understanding of mental illness. By the 1960s, its recreational use had spread to students and artists, and reached a zenith during the summer of 1967. Despite being scheduled as a controlled substance in 1970, the drug, which is relatively easy to manufacture, remains available and is still used. In addition, chemists have synthesized dozens of congeners of LSD that cause similar subjective effects. Evidence suggests that a new generation of young people have discovered the remarkable psychedelic effects of LSD, producing a resurgence of abuse. Much of what is sold on the streets today as LSD or mescaline, however, is really PCP or one of its substituted congeners.

D. Clinical Features and Neurological Complications

CNS signs and symptoms of intoxication with LSD, mescaline, and related drugs are primarily psychiatric in nature. These agents produce a profound distortion in all sensory modalities (Kulberg, 1986). Synesthesias ("seeing" smells or "hearing" colors) are commonly reported. Loss of logical and sequential thinking, distortion of time perception, a sense of depersonalization, and loss of ego boundaries may combine to produce anxiety and panic (the "bad trip"). Mood may be labile, shifting rapidly from gaiety to depression, or elation to fear. The somatic signs of LSD ingestion are

largely sympathomimetic in nature: mydriasis, tachycardia, elevated blood pressure, hyperreflexia, nausea, tremor, and increased body temperature. Severe toxicity may result in coma, convulsions and respiratory arrest, but these are apparently rare. Most cases of LSD intoxication resolve without incident. Reduction of external stimuli by placing the patient in a quiet room, combined with positive reassurance ("talking down") can be helpful. Persistent neurologic complications have not been reported, although some investigators have argued that chronic use of LSD might predispose to the development of a chronic psychosis.

References

Abe, T., and Itokawa, Y. (1977). Effect of ethanol administration on thiamine metabolism and transketolase activity in rats. *Int. J. Nutr. Res.* **47**, 307–314.

Abramson, J., Abramson, S., Katz, S., and Hager, M. (1985). Getting high on ecstacy. *Newsweek* April 15, 96.

Adams, R. D., Victor, M., and Mancall, E. L. (1959). Central pontine myelinolysis: a hitherto undescribed disease occurring in alcoholic and malnourished patients. *Arch. Neurol. Psychiatry* **81**, 154–172.

Aquilonius, S. M., Bergstrom, K., and Enoksson, P. (1980). Cerebral computed tomography in methanol intoxication. *J. Comput. Assist. Tomogr.* **4**, 425–428.

Arendt, T., Henning, D., and Marchbanks, R. (1988). Loss of neurons in the rat basal forebrain cholinergic projection system after prolonged intake of ethanol. *Brain Res. Bull.* **21**, 563–569.

Ayus, J. C., Krothapalli, R. K., and Arieff, A. I. (1987). Treatment of symptomatic hyponatremia and its relation to brain damage: a prospective study. *N. Engl. J. Med.* **317**, 1190–1195.

Bennett, I. L., Cary, F. H., Mitchell, G. L., and Cooper, M. N. (1953). Acute methyl alcohol poisoning: a review based on experiences in an outbreak of 323 cases. *Medicine* **32**, 431–463.

Blass, J. P., and Gibson, G. E. (1977). Abnormality of a thiamine-requiring enzyme in patients with Wernicke-Korsakoff syndrome. *N. Engl. J. Med.* **297**, 1367–1370.

Brion, S. (1976). Marchiafava-Bignami syndrome. In *Handbook of Clinical Neurology* Vol. 28, 317–329.

Burcar, P. J., Norenberg, M. D., and Yarnell, P. R. (1977). Hyponatremia and central pontine myelinolysis. *Neurology* **27**, 223–226.

Cala, L. A., Jones, B., Mastaglia, F. L., and Wiley, B. (1978). Brain atrophy and intellectual impairment in heavy drinkers: a clinical, psychometric and computerized tomography study. *Aust. N.Z. J. Med.* **8**, 147–153.

Cala, L. A., Jones, B., Burns, P., Davis, R. E., Stenhouse, N., and Mastaglia, F. L. (1983). Results of computerized tomography, psychometric testing, and dietary studies in social drinkers, with emphasis on reversibility after abstinence. *Med. J. Aust.* **2**, 264–269.

Caplan, L. R., Hier, D. B., and Banks, G. (1982). Current concepts of cerebrovascular disease stroke: stroke and drug abuse. *Stroke* **13**, 869–872.

Carlen, P. L., Wortzman, G., Holgate, R. C., Wilkinson, D. A., and Rankin, J. G. (1978). Reversible cerebral atrophy in recently abstinent chronic alcoholics measured by computed tomographic scans. *Science* **200**, 1076–1078.

Carlen, P. L., Lee, M. A., Jacob, M., and Livshits, O. (1981). Parkinsonism provoked by alcoholism. *Ann. Neurol.* **9**, 84–86.

Charness, M. E., and DeLaPaz, R. L. (1987). Mammillary body atrophy in Wernicke's encephalopathy: antemortem identification using magnetic resonance imaging. *Ann. Neurol.* **22,** 595–600.

Charness, M. E., Simon, R. P., and Greenberg, D. A. (1989). Ethanol and the nervous system. *N. Engl. J. Med.* **321,** 442–454.

Cooper, D. (1989). Clandestine production processes for cocaine and heroin. In *Clandestinely Produced Drugs, Analogues and Precursors* (M. Klein, F. Sapienza, and H. McClain, Eds.), pp. 95–116, United States Department of Justice, Washington, D.C.

Diener, H. C., Dichgans, J., Bacher, M., and Guschlbauer, B. (1984). Improvement of ataxia in alcoholic cerebellar atrophy through alcohol abstinence. *J. Neurol.* **231,** 258–262.

Dowling, G. P., McDonough, E. T., and Bost, R. O. (1987). "Eve" and "Ecstasy": a report of five deaths associated with the use of MDEA and MDMA. *JAMA* **257,** 1615–1617.

Finnegan, K. T., and Schuster, C. R. (1989). The methylenedioxy amphetamines: pharmacological, behavioral and neurotoxic effects. In *Clandestinely Produced Drugs, Analogues and Precursors* (M. Klein, F. Sapienza, and H. McClain, Eds.), pp. 195–206, United States Department of Justice, Washington, D.C.

Fischman, M. W., Schuster, C., Rosnekov, L., Shick, J. F. E., and Krasnegor, N. A. (1976). Cardio-vascular and subjective effects of intravenous cocaine administration in humans. *Arch. Gen. Psychiatry* **33,** 983–989.

Fornazzari, L., and Carlen, P. L. (1982). Transient choreiform dyskinesias during alcohol withdrawal. *Can. J. Neurol. Sci.* **9,** 89–90.

Glennon, R. A. (1989). Synthesis and evaluation of amphetamine analogs. In *Clandestinely Produced Drugs, Analogues and Precursors* (M. Klein, F. Sapienza, I. McClain, and I. Khan, Eds.), pp. 39–66, United States Department of Justice, Washington, D.C.

Guggenheim, M. A., Couch, J. R., and Weinberg, W. (1971). Motor dysfunction as a permanent complication of methanol ingestion: presentation of a case with a beneficial response to levadopa treatment. Arch. Neurol. **24,** 550–554.

Haller, R. G., and Knochel, J. P. (1984). Skeletal muscle disease in alcoholism. *Med. Clin. North Am.* **68,** 91–103.

Harper, C. (1983). The incidence of Wernicke's encephalopathy in Australia—a neuropathological study of 131 cases. *J. Neurol. Neurosurg. Psychiatry* **46,** 593–598.

Harper, C. G., and Corbett, D. (1990). Changes in basal dendrites of cortical pyramidal cells from alcoholic patients—a quantitative Golgi study. *J. Neurol. Neurosurg. Psychiatry* **53,** 856–861.

Harper, C. G., and Kril, J. J. (1985). Brain atrophy in chronic alcoholic patients: a quantitative pathological study. *J. Neurol. Neurosurg. Psychiatry* **48,** 211–217.

Harper, C. G., Giles, M., and Finlay-Jones, R. (1986). Clinical signs in the Wernicke-Korsakoff complex: a retrospective analysis of 131 cases diagnosed at necropsy. *J. Neurosurg. Psychiatry* **49,** 341–345.

Harper, C. G., Kril, J. J., and Daly, J. M. (1988a). Brain shrinkage in alcoholics is not caused by changes in hydration: a pathological study. *J. Neurol. Neurosurg. Psychiatry* **51,** 124–127.

Harper, C. G., Kril, J. J., and Daly, J. M. (1988b). Does a "moderate" alcohol intake damage the brain? *J. Neurol. Neurosurg. Psychiatry* **51,** 909–913.

Harrington, H., Heller, H. A., Dawson, D., Caplan, L., and Rambaugh, C. (1983). Intracerebral hemorrhage and oral amphetamines. *Arch. Neurol.* **40,** 503–507.

Hata, T., Meyer, J. S., and Tanahashi, N. (1987). Three dimensional mapping of local cerebral perfusion in alcoholic encephalopathy with and without Wernicke-Korsakoff syndrome. *J. Cereb. Blood Flow Metab.* **7,** 35–44.

Henderson, G. L. (1989). Designer drugs: the california experience. In *Clandestinely Produced Drugs, Analogues and Precursors* (M. Klein, F. Sapienza, H. McClain, and I. Kahn, Eds.), pp. 7–20, United States Department of Justice, Washington, D.C.

Hillbom, M., and Kaste, M. (1981). Ethanol intoxication: a risk factor for ischemic brain infarction in adolescents and young adults. *Stroke* **12,** 422–425.

Hillbom, M., and Kaste, M. (1983). Alcohol intoxication: a risk factor for primary subarachnoid haemorrhage. *Neurology* **32,** 706–711.

Jacobson, R. (1989). Alcoholism, Korsakoff's syndrome and the frontal lobes. *Behav. Neurol.* **2,** 25–38.

Joint Federal Task Force Report (1990). *Guidelines for the Cleanup of Clandestine Drug Laboratories,* U.S. Department of Justice, Drug Enforcement Administration, Washington, D.C.

Kendrick, W. C., Hull, A. R., and Knochel, J. P. (1977). Rhabdomyolysis and shock after intravenous amphetamine administration. *Ann. Intern. Med.* **86,** 381–387.

Kril, J. J., and Harper, C. G. (1989). Neuronal counts from four cortical regions in alcoholic brains. *Acta Neuropathol. (Berl.)* **79,** 200–204.

Kulberg, A. (1986). Substance abuse: clinical identification and management. *Pediatr. Toxicol.* **33,** 325–360.

Langlais, P. J., Mair, R. G., and McEntee, W. J. (1988). Acute thiamine deficiency in the rat: brain lesions, amino acid changes and MK-801 pretreatment. *Soc. Neurosci. Abstr.* **14,** 744.

Langston, J. W., and Finnegan, K. T. (1987). The neurotoxicity of MPTP and its relationship to Parkinsonism. *ISI Atlas Sci. Pharmacol.* **1,** 147–150.

Langston, J. W., Ballard, P. A., Tetrud, J. W., and Irwin, I. (1983). Chronic parkinsonism in humans due to a product of meperidine-analog synthesis. *Science* **219,** 979–980.

Laureno, R. (1983). Central pontine myelinolysis following rapid correction of hyponatremia. *Ann. Neurol.* **13,** 232–242.

Laureno, R., and Karp, B. I. (1988). Pontine and extrapontine myelinolysis following rapid correction of hyponatremia. *Lancet* **1,** 1439–1441.

Ley, C. O., and Gali, F. G. (1983). Parkinsonian syndrome after methanol intoxication. *Eur. Neurol.* **22,** 405–409.

Lieber, C. S. (1991). Hepatic, metabolic and toxic effects of ethanol: 1991 update. *Alcohol. Clin. Exp. Res.* **15,** 573–592.

Lishman, W. A. (1990). Alcohol and the brain. *Br. J. Psychiatry* **156,** 635–644.

Louria, D., Hensle, T., and Rose, J. (1967). Major medical complications of heroin abuse. *Ann. Intern. Med.* **67,** 1–22.

McDowell, J. R., and LeBlanc, H. J. (1984). Computed tomographic findings in Wernicke-Korsakoff syndrome. *Arch. Neurol.* **341,** 453–454.

McMullen, P. A., Saint-Cyr, J. A., and Carlen, P. L. (1987). Morphological alterations in rat CA1 hippocampal pyramidal cell dendrites resulting from chronic ethanol consumption and withdrawal. *J. Comp. Neurol.* **225,** 111–118.

Menne, F. R. (1938). Acute methyl alcohol poisoning: a report of 22 instances with postmortem examinations. *Arch. Pathol.* **26,** 77–92.

Messert, B., Orrison, W. W., and Hawkins, M. J. (1979). Central pontine myelinolysis: considerations on etiology, diagnosis, and treatment. *Neurology* **29,** 1257–1260.

Mesulam, M. M., Hoesen, G. W., and Butters, N. (1977). Clinical manifestations of chronic thiamine deficiency in the rhesus monkey. *Neurology* **27,** 239–245.

Mullin, P. J., Kereshaw, P. W., and Bot, J. M. (1970). Choreoathetotic movement disorder in alcoholism. *Br. J. Med.* **4,** 279–281.

Narins, R. G. (1986). Therapy of hyponatremia: does haste make waste? *N. Engl. J. Med.* **314,** 1573–1575.

Parkinson's Study Group (1989). Effect of deprenyl on the progression of disability in early Parkinson's disease. *N. Engl. J. Med.* **321**, 1364.

Pearson, J., and Richter, R. (1979). Addiction of opiates: neurological aspects. In *Intoxication of the Nervous System* Part II, *Handbook of Clinical Neurology*. North Holland Publishing, Amsterdam.

Phillips, S. C. (1987). Neurotoxic interaction in alcohol-treated thiamine-deficient mice. *Acta Neuropathol. (Berl.)* **73**, 171–176.

Rappolt, R. T., Gay, G. R., and Farris, R. D. (1979). Emergency management of acute phencyclidine intoxication. *J. Am. Coll. Emerg. Phys.* **8**, 68–76.

Risberg, J., and Berglund, M. (1987). Cerebral blood flow and metabolism in alcoholics. In *Neuropsychology of Alcoholism* (O. A. Parsons, N. Butters, and P. E. Nathan, Eds.), pp. 64–75, Guilford Publications, New York.

Rosett, H. L., and Weiner, L. (1984). *Alcohol and the Fetus: A Clinical Perspective.* Oxford University Press, New York.

Salanova, V., and Taubner, R. (1984). Intracerebral hemorrhage and vasculitis secondary of amphetamine use. *Postgrad. Med. J.* **60**, 429–430.

Scaplehorn, A. W. (1989). Synthesis of phencyclidine (PCP), lysergic acid diethylamide (LSD), methaqualone and their analogues. In *Clandestinely Produced Drugs, Analogues and Precursors* (M. Klein, F. Sapienza, H. McClain, and I. Khan, Eds.), pp. 91–94, United States Department of Justice, Washington, D.C.

Seiden, L. S., and Ricaurte, G. A. (1987). Neurotoxicity of methamphetamine and related drugs. In *Psychopharmacology, A Third Generation of Progress* (H. Y. Meltzer, Ed.), pp. 359–399, Raven Press, New York.

Sharpe, J. A., Hostovsky, M., Bilbao, J. M., and Rewcastle, N. B. (1982). Methanol optic neuropathy: a histopathologic study. *Neurology* **32**, 1093–1100.

Shulgin, A. T. (1970). Psychotomimetic agents related to the catecholamines. *J. Psychoactive Drugs* **2**, 17–25.

Shulgin, A. T. (1978). Psychotomimetic drugs: structure-activity relationships. In *Handbook of Psychopharmacology* (L. L. Iversen, S. D. Iversen, and S. H. Snyder, Eds.), pp. 243–333, Plenum, New York.

Simpson, D. L., and Rumack, B. H. (1981). Methylenedioxy amphetamine: clinical description of overdose, death, and review of pharmacology. *Arch. Intern. Med.* **141**, 1507–1508.

Sixth Special Report to Congress on Alcohol and Health. (1987). DHHS Publication No. (ADM) 87-1519 Edn., Department of Health and Human Services, National Institute on Alcohol Abuse and Alcoholism, Rockville, MD.

Song, S. K., and Rubin, E. (1972). Ethanol produces muscle damage in human volunteers. *Science* **175**, 327–328.

Stafford, C. R., Bogdanoff, E. M., Green, L., and Spector, H. B. (1975). Mononeuropathy multiplex as a complication of amphetamine angitis. *Neurology* **25**, 570–572.

Steinberg, A. D., and Karliner, J. S. (1968). Clinical spectrum of heroin pulmonary edema. *Arch. Intern. Med.* **122**, 122–131.

Streissguth, A. P., Landesman-Dwyer, S., Martin, J. C., and Smith, D. W. (1980). Teratogenic effects of alcohol in humans and laboratory animals. *Science* **209**, 353–361.

Streissguth, A. P., Barr, H. M., Sampson, P. D., Parrish-Johnson, J. C., Kirchner, G. L., and Martin, D. C. (1986). Attention, distraction and reaction time at age 7 years and prenatal alcohol exposure. *Neurobehav. Toxicol. Teratol.* **8**, 717–725.

Streissguth, A. P., Barr, H. M., Sampson, P. D., Darby, B. L., and Martin, D. C. (1989). IQ at age 4 in relation to maternal alcohol use and smoking during pregnancy. *Dev. Psychol.* **25**, 3–11.

Suit, P. F., and Estes, M. L. (1990). Methanol intoxication: clinical features and differential diagnosis. *Cleve. Clin. J. Med.* **57**, 464–470.

Taylor, J. R. (1982). Alcohol and strokes. *N. Engl. J. Med.* **306**, 1111.

Tetrud, J. W., and Langston, J. W. (1989). The effect of deprenyl (selegiline) on the natural history of Parkinson's disease. *Science* **245**, 519.

Thomas, P. K. (1986). Alcohol and disease: central nervous system. *Acta Med. Scand.* **703**(Suppl), 251–264.

Torvik, A., Lindboe, C. F., and Rogde, S. (1982). Brain lesions in alcoholics: a neuropathological study with clinical correlations. *J. Neurol. Sci.* **56**, 233–248.

Van Bever, W. F. M., Neimegeers, C. J. E., and Janssen, P. A. J. (1974). Synthetic analgesics, synthesis and pharmacology of the diastereoisomers of *N*-[3]-methyl-1-(2-phenylethyl)-4-piperidyl]-*N*-phenylpropanamide. *J. Med. Chem.* **17**, 1047–1051.

Victor, M., Adams, R. D., and Mancall, E. L. (1959). A restricted form of cerebellar cortical degeneration occurring in alcoholic patients. *Arch. Neurol.* **71**, 579–688.

Victor, M., Adams, R. D., and Collins, G. H. (1971). The Wernicke-Korsakoff syndrome: a clinical and pathological study of 245 patients, 82 with post-mortem examinations. *Contemp. Neurol. Ser.* **7**, 1–206.

Walker, D. W., Barner, D. E., and Zornetzer, S. F. (1980). Neuronal loss in hippocampus induced by prolonged ethanol consumption in rats. *Science* **209**, 711–713.

Witt, E. D., and Goldman-Rakic, P. S. (1983a). Intermittent thiamine deficiency in the rhesus monkey. I. Progression of neurological signs and neuroanatomical lesions. *Ann. Neurol.* **13**, 376–395.

Witt, E. D., and Goldman-Rakic, P. S. (1983b). Intermittent thiamine deficiency in the rhesus monkey. II. Evidence for memory loss. *Ann. Neurol.* **13**, 396–401.

CHAPTER

42

Clinical Neurotoxic Concerns on Antineoplastic Agents

TERRENCE CASCINO
Department of Neurology
The Mayo Clinic
Rochester, Minnesota 55905

This chapter explores the wide spectrum of neurological complications of chemotherapy. Certain chemotherapeutic agents are commonly associated with particular neurological syndromes (Young, 1982; Sawicka, 1977; Weiss *et al.*, 1974; Young and Posner, 1980). The individual agents will be discussed reviewing specific complications.

Neurotoxicity of chemotherapy can produce disability and, at times, death. Neurotoxicity of these agents is common and occasionally is the dose-limiting toxicity. Occasionally, the toxicity is far worse than the original cancer.

Chemotherapeutic agents may cause toxicity which is indistinguishable from that of metastatic disease. It is vital that the clinician be aware of the potential for neurotoxicity and spare the patient unnecessary tests and treatments. Furthermore, neurotoxicity may aid in the understanding of how the nervous system operates.

Prior to discussing the effects of chemotherapy, it is important to note that the most common causes of neurotoxicity in cancer patients may be related to radiation side effects. Radiation necrosis of the brain, spinal cord, or plexus is not uncommon, occurring

acutely, subacutely within weeks or months, or delayed following a latency of months or years. (Rottenberg, 1977; Wilson, 1972; Reagan, 1968; Cascino, 1983).

I. Chemotherapy

Numerous chemotherapeutic agents are potential neurotoxins. The toxicity may clinically be apparent as a central nervous system (CNS) or peripheral nervous system (PNS) disease (Table 1). Fortunately, the CNS is often protected by the presence of the blood/brain barrier.

It is clear that certain cancers are now curable by chemotherapy. This has been possible through the development of more potent drugs and combinations of drugs, as well as a multimodality approach using both antineoplastic agents and radiation. It is because of these new approaches that it is more common to see neurotoxicity than in previous years. Patients are now exposed to higher concentrations of the drugs in the nervous system. As survival of patients increases, pa-

TABLE 1 Neurological Complications
of Chemotherapy

Central nervous system
  Encephalopathy
    Intrathecal methotrexate
    Ifosfamide
    BCNU
    L-Asparaginase
  Acute cerebellar syndrome
    5-Fluorouracil
    High-dose Ara-C
Peripheral nervous system
  Peripheral neuropathy
    Cisplatin
    Vincristine
    Taxol
    Suramin
  Myelopathy with paraparesis
    Intrathecal Ara-C
    Intrathecal methotrexate

tients are at a higher risk for developing long-term, irreversible toxicity.

Finally, drugs that do not normally have any neurotoxicity may show toxicity under special circumstances. High-dose chemotherapy with bone marrow rescue exposes the nervous system to very high doses of chemotherapy which may result in devastating complications, whereas low-dose chemotherapy may have no neurological complications.

II. Agents Causing Primarily CNS Toxicity

A. Methotrexate

Methotrexate is a folic acid analog that inhibits dihydrofolate reductase resulting in the depletion of intracellular pools of reduced folate which are necessary for thymidylate and purine metabolism which are necessary for the synthesis of DNA (Chabner and Meyers, 1982; Jolivet *et al.*, 1983).

Leucovorin is a reduced folate that prevents methotrexate toxicity to the bone marrow and gastrointestinal tract. Leucovorin does not prevent central nervous system toxicity. Methotrexate can be given orally, intrathecally, or parenterally and may be given in low doses or at high doses by continuous infusion. Low-dose oral or parenteral methotrexate results in little or no neurotoxicity.

Neurotoxicity may occur after intrathecal injection and high-dose (Rosen, 1977) intravenous methotrexate administration. The toxicity can be categorized by acute and subacute toxicity, stroke-like syndromes, and delayed neurotoxicity.

Up to 30% of patients receiving intrathecal methotrexate may develop neurotoxicity. The toxicity occurs regardless of whether or not the patient is receiving leucovorin.

1. Acute Toxicity

Aseptic meningitis occurs within hours of intrathecal administration of the drug and lasts 12 hr to 4 days (Bleyer *et al.*, 1981; Geiser *et al.*, 1975). The syndrome results from a chemical meningitis. This may occur regardless of the route of administration, either through a lumbar tap or through a ommaya reservoir (Abelson, 1978; Bleyer *et al.*, 1973).

The syndrome is characterized by headache, vomiting, lethargy, meningismus, and fever. Cerebrospinal fluid shows a sterile pleocytosis. The protein may be elevated.

The differential diagnosis is often that of infectious meningitis. The aseptic nature of the spinal fluid, as well as the short period of time following lumbar tap, mitigates against this diagnosis.

It has been estimated that the syndrome occurs between 10 and 60% of patients receiving intrathecal methotrexate. It often does not recur with subsequent injections in the same patient.

The syndrome is self-limited, resolving without specific treatment (Sullivan *et al.*, 1977). Pretreatment with intrathecal hydrocortisone has been advocated, but there is no clear evidence that this is necessary. The ideal treatment is prevention. Decreasing the concentration of methotrexate in the spinal fluid decreases the likelihood of aseptic meningitis. Dilution with large amounts of spinal fluid or a substance such as Ringer's lactate may be beneficial. Preservative-free methotrexate should be used, although there is no clear evidence that the preservatives cause the toxicity.

Less common complications of intrathecal methotrexate that may occur acutely include sudden death (Ten Hoeve and Twinjnstra, 1988), radiculopathy, seizures, encephalopathy, and pulmonary edema (Bernstein *et al.*, 1982). An intrathecal overdose is usually fatal.

2. Subacute Methotrexate Neurotoxicity

Subacute toxicity is rare. Two syndromes have been reported.

a. Transverse Myelopathy Transverse myelopathy occurring during the course of several injections may appear over a period of weeks (Skullerud and Halvorsen, 1985; Gagbano and Costani, 1976; Saiki *et al.*, 1972). This is a rare complication of methotrexate and may be reversible. Irreversible cases have been reported and may result in total paralysis. The myelopathy presents with pain, weakness, sensory level, and autonomic dysfunction of bowel and bladder. Occasionally, a meningoencephalopathy may accompany the myelopathy.

If this syndrome is suspected, methotrexate should be discontinued. The same syndrome has been seen in patients receiving β-cytosine arabinoside, a commonly used intrathecal agent (Wolff *et al.*, 1979). The cause of the syndrome is not clear, although it is felt to be idiosyncratic.

Patients developing myelopathy during the course of intrathecal methotrexate clearly need to be investigated for evidence of metastatic disease in the epidural space, meninges, or within the substance of the spinal cord. If no causative metastasis is uncovered, methotrexate should be discontinued. It may be hazardous to switch to cytosine arabinoside in such patients. Unfortunately, there is no treatment for the myelopathy.

There is some evidence that myelopathy is more common in patients with known meningeal disease or in patients receiving radiotherapy (Bleyer *et al.*, 1973).

b. Stroke-like Syndromes

Patients receiving high-dose parenteral methotrexate with leucovorin rescue may develop subacute, stroke-like syndromes. This is seen most often in patients being treated for osteogenic sarcoma as a primary form of treatment (Allen and Rosen, 1978).

The syndrome follows weekly administration of the high-dose methotrexate. A series in 1986 of 22 patients (Walker *et al.*, 1986) showed that most of the patients developed the syndrome 2 to 3 weeks following high doses of methotrexate. Patients developed hemiparesis and encephalopathy lasting minutes to 3 days. Patients normally recover without residual. Neuroimaging, cerebrospinal fluid examination, and electroencephalogram are normal or nonspecific. There is no clear evidence that these events are related to cerebrovascular disease or seizure.

The syndrome rapidly resolves with no other treatment needed. Patients who do develop this syndrome have been treated successfully later in the course of the disease without any recurrence.

The cause of the syndrome remains unclear. It has been postulated by Phillips and colleagues (1987) that there may be reduced cerebral glucose metabolism in these patients.

3. Delayed Neurotoxicity

The chronic leukencephalopathy associated with methotrexate is often associated with severe disability or even death (Glass *et al.*, 1986; Shapiro *et al.*, 1973; Kay *et al.*, 1972; Lucien *et al.*, 1975). The syndrome is dose related and often occurs in the setting of patients

heavily treated with doses of intrathecal methotrexate or high-dose parenteral methotrexate (Allen *et al.*, 1980). Most patients reported in the literature have underlying meningeal carcinomatosis or meningeal leukemia.

The incidence of delayed leukencephalopathy is rare in patients receiving less than 50 mg total dose of intrathecal methotrexate. Clearly, though, the incidence increases when patients receive concomitant cranioirradiation or parenteral methotrexate. The highest incidence reported of leukencephalopathy occurs in patients receiving greater than 2000 cGy cranioirradiation, intravenous methotrexate, and more than 50 mg total dose of intrathecal methotrexate.

Pathologically, the disease consists of diffuse white matter changes. A multifocal necrosis occurs in the white matter with microcalcifications and angionecrosis (Jacobs *et al.*, 1991; Shapiro *et al.*, 1973).

The clinical syndrome is similar to diffuse white matter diseases. The disease begins insidiously but may occur abruptly. The onset is usually many months or even years after the initiation of treatment. The initial complaint is that of loss of memory or change in personality. As the disease progresses, seizures, quadriparesis, lethargy, coma, and even death may occur.

Radiographic changes on the computerized tomography (CT) scan occur in the early stages of the disease. These changes include subependymal hypodensity around the frontal horns. As the disease progresses, diffuse white matter hypodensity, parenchymal calcification, and diffuse cerebral atrophy may occur. Contrast enhancment may be present (Di Chiro *et al.*, 1979; Shalen *et al.*, 1981).

A magnetic resonance imaging (MRI) scan would seem to be more sensitive. There are less reports of MRI scan abnormalities in the literature than CT scans. Chronic white matter changes, calcification, and brain atrophy may be seen (Ebner *et al.*, 1989).

Asymptomatic patients treated with intrathecal methotrexate may result in the just-mentioned radiographic changes (Duffner *et al.*, 1984; Peylan-Ramu *et al.*, 1978).

There is no cure or treatment for leukencephalopathy related to methotrexate. Discontinuation of the drug usually results in mild improvement. The key to prevention is early detection as well as limiting the amount of methotrexate given intrathecally. Phillips and co-workers (1989) have used high-dose leucovorin in an animal model. No clear evidence exists that high-dose leucovorin is beneficial in patients with methotrexate leukencephalopathy.

The etiology of the delayed complication of methotrexate is unclear. It has been postulated that this is related to a central nervous system folate deficiency

(Abelson, 1978). Other studies have suggested that methotrexate is a neuronal toxin directly (Gilbert *et al.*, 1989).

No evidence is available that the delayed complication is related to patients who have had acute or subacute neurotoxicity related to methotrexate.

It is clear that limiting the amount of drug used, avoiding the combination of simultaneous intrathecal methotrexate and cranioirradiation, and avoiding high-dose methotrexate are key factors. Unfortunately, it is not often possible to avoid all of these factors in patients with meningeal cancer. It is important to avoid high concentrations of methotrexate by administering drugs to areas of good spinal fluid drainage. Patients who have hydrocephalus are more likely to develop a leukencephalopathy when they receive intraventricular methotrexate. Similarly, if there is evidence of spinal fluid blockage, methotrexate given via the lumbar root may cause local damage.

B. Fludarabine Phosphate

Fludarabine phosphate is a purine nucleoside antineoplastic agent. It has been used in the past for hematological malignancies as well as solid tumors.

Central nervous system toxicity is subacute and presents as a diffuse white matter disease (Chun *et al.*, 1986; Merkel *et al.*, 1986). Evidence of quadriparesis, cortical blindness, and dementia occurs subacutely. If the drug is not discontinued, coma and death may ensue. Pathologically, there is evidence of demyelination. The cause is unclear.

C. 5-Fluorouracil

5-Fluorouracil is an antiprimidine, which is a fluorinated primidine. The main mechanism of action is to inhibit DNA synthesis by inhibiting thymidylate synthetase. 5-Fluorouracil crosses the blood/brain barrier with high concentrations in the cerebellum (Bourke *et al.*, 1973; Chadwick and Rogers, 1972). The drug is used to treat breast cancer, colorectal cancer, and other solid tumors.

The primary toxicity of 5-fluorouracil is to oral mucosa, intestine, and bone marrow. Stomatitis and diarrhea are common systemic toxicities and may be dose limiting. Bone marrow suppression is also frequent.

Neurological toxicity has been recognized since the early 1960s (Boileau *et al.*, 1971; Riehl and Brown, 1964; Moertel *et al.*, 1964). Toxicity to the cerebellum occurs acutely with ataxia, nystagmus, and dysarthria. This has been estimated to occur in about 5% of patients treated. The syndrome occurs a few weeks to many months following the start of therapy and is re-

versible when the drug is discontinued. The toxicity is related to the plasma level of the drug.

The differential diagnosis of acute cerebellar ataxia in patients receiving chemotherapy from cancer obviously includes a posterior fossa mass or paraneoplastic cerebellar degeneration. The treatment for 5-fluorouracil toxicity is discontinuation of the drug, which is usually followed from 1 to 6 weeks later by improvement.

The pathophysiology of this syndrome is not clear. No specific findings have been found at autopsy. Other central nervous system toxicities include extraocular palsies (Bixenman *et al.*, 1977), encephalopathy (Lynch *et al.*, 1981), extrapyramidal signs (Bergevin *et al.*, 1975), and optic neuropathy (Adams *et al.*, 1984). These also are dose dependent and reversible when the drug is discontinued.

A multifocal cerebral demyelinating disorder has been described in patients with colon cancer receiving 5-fluorouracil and levamisole (Forsyth *et al.*, 1991). Patients presented with encephalopathy, ataxia, dysarthria, and hemiparesis. There was improvement in these few patients with discontinuation of chemotherapy.

Neuroimaging studies are nonspecific but abnormal. Magnetic resonance imaging scans of the head demonstrate multiple white matter lesions which enhance after the administration of gadolinium in a periventricular area. Cerebral biopsies have been done on several patients showing demyelination and perivascular infiltration.

The etiology of this syndrome is not clear. It is unknown whether this represents 5-fluorouracil toxicity, levamisole toxicity, or a combination of both.

It is important that in patients receiving 5-fluorouracil and levamisole who develop multiple intracranial lesions that toxicity is considered. The differential diagnosis includes multiple metastasis, infection, or other types of demyelinating disease, such as multiple sclerosis.

Other drugs given with 5-fluorouracil have been reported to increase toxicity, including allopurinol (Campbell *et al.*, 1982) and PALA (Muggia *et al.*, 1987). Neurotoxicity associated with these drugs include encephalopathy as well as cerebellar disease.

Neurotoxicity also appears to be a feature of other fluorinated antiprimidine drugs. Patients receiving 5-fluorouracil derivatives such as carmofur or tegafur have been reported to develop reversible leukencephalopathy. Reports from Japan describe three cases with a subacute progressive leukencephalopathy in patients receiving carmofur. These patients appear clinically and pathologically similar to patients observed with toxicity secondary to 5-fluorouracil and levamisole.

D. ʟ-Asparaginase

The enzyme ʟ-asparaginase catalyzes the hydrolysis of the amino acid ʟ-asparagine to aspartic acid plus ammonia. This drug is useful in treating patients with acute lymphoblastic leukemia. Normal cells have the ability to synthesize their own asparagine and are spared. Toxicity is frequently hepatic or pancreatic in nature.

Encephalopathy was described as a common complication in the early 1970s Land *et al.*, 1972; Moure *et al.*, 1970; Oettgen *et al.*, 1970). The complication appears to be dose dependent. Since that time, lower cumulative doses have been used and neurotoxicity is less frequent. The encephalopathy occurs soon after the administration of the drug and is reversible. The clinical spectrum ranges from lethargy to delerium and hallucinations. Focal signs and seizures rarely occur.

A delayed encephalopathy has been reported 1 to 2 weeks after the drug is given. The cause of this is also unclear.

The etiology of ʟ-asparaginase-induced encephalopathy remains unclear. ʟ-Asparagine does not cross the blood/brain barrier well.

A more frequent clinical syndrome is that of thrombosis of cerebral veins. Intracranial venous thrombosis may lead to cerebral hemorrhage or infarct (Cairo *et al.*, 1980; Ott *et al.*, 1988; Feinberg and Swenson, 1988). Clinically the patients present with headache, seizures, focal weakness, or lethargy.

Venous thrombosis may occur during the course of therapy or soon after completion.

Diagnosis is based on clinical recognition. CT scanning may demonstrate hemorrhage or infarction. Patients with cerebral venous thrombosis may rarely have a "delta sign" which represents a clot in the venous sinus (Zilkha and Daiz, 1980). A MRI scan of the head (Moots *et al.*, 1987), especially with MRI angiography, seems to be the test of choice.

Coagulopathy related to a depletion of plasma proteins involved in coagulation and fibrinolysis appears to be the mechanism of this complication. The treatment for venous thrombosis in this setting is unclear. A trial of empiric steroids has been given on a number of occasions.

E. Procarbazine

Procarbazine is a methylhydrozine useful in treating hematological malignancies as well as brain tumor. The mechanism of antineoplastic action is unknown. The drug is a monomine oxidase inhibitor and crosses the blood/brain barrier quite easily (Oliverio *et al.*, 1964).

Encephalopathy occurs with the spectrum from lethargy to stupor and coma (Stolinsky *et al.*, 1970; Chabner *et al.*, 1973). The symptoms are reversible on discontinuation of the drug.

Of note, procarbazine has been known to potentiate the sedative effects of certain drugs, including phenothiazines and narcotics (Lee and Lucier, 1976).

The frequency of encephalopathy related to procarbazine remains unclear. The toxicity does appear to be dose related. There is no treatment.

Procarbazine has also been reported to cause peripheral neuropathy which is mild (Weiss *et al.*, 1974).

F. Ifosfamide

Ifosfamide is an oxazaphosphrine that is currently being used in a number of solid tumors. Encephalopathy is commonly seen in patients receiving the drug (Pratt *et al.*, 1986; Watkin *et al.*, 1989; Danesh *et al.*, 1989). The spectrum ranges from mild drowsiness to coma. The toxicity occurs during administration of the drug within a few hours and resolves when the drug is discontinued. There are rare reports of persistent encephalopathy. Toxicity is said to occur from 1 to 30% of patients treated. No treatment is necessary as it is reversible without treatment in most patients.

The cause of ifosfamide toxicity is unclear. It has been speculated that it may be related to toxic levels of chloracetaldehyde, which is an oxidation product of ifosfamide.

Patients with low serum albumin, prior administration of cisplatin, or renal dysfunction are more likely to develop encephalopathy. Patients who have had previous neurotoxicity with encephalopathy during ifosfamide infusion are also at risk for developing encephalopathy at the next treatment (Meanwell *et al.*, 1986).

Ifosfamide has been reported to cause cranial nerve dysfunction, seizures, and ataxia.

G. Etoposide

Etoposide (VP-16) is a semisynthetic podophyllotoxin derivative but is used for a wide range of malignancies. Neurotoxicity is unusual (Postmus *et al.*, 1984) but has been reported and includes dystonia, encephalopathy, and peripheral neuropathy (Filley *et al.*, 1982).

Acute neurological dysfunction has been seen when etoposide is used in high doses in patients with brain tumor (Leff *et al.*, 1988).

The cause of the neurotoxicity is unknown, but it has been speculated that binding microtubular protein leads to inhibition of axoplasmic flow.

H. Cytosine Arabinoside

Cytosine arabinoside is a primidine nucleoside analog that competitively inhibits DNA prelimerase and replicating cells. It is used in patients with leukemia.

Patients treated at low doses systemically rarely have any neurotoxicity. Intrathecal treatment is a common alternative to methotrexate treatment. Toxicity is similar to that seen with methotrexate.

I. Nitrosoureas

Nitrosoureas, especially N,N-bis(2-chloroethyl)-N-nitrosourea (BCNU), have been used for many years. These are cell cycle-specific agents that cross the blood/brain barrier well due to lipid solubility. These drugs given at normal doses intravenously rarely have any neurotoxicity.

1. High-Dose Chemotherapy

Chemotherapeutic agents are used in high doses in an effort to increase the concentration of the drug. Increasing the dose also increases the likelihood of toxicity. Neurotoxicity is present in a number of agents that are essentially nontoxic given at ordinary doses.

a. Cytosine Arabinoside Cytosine arabinoside is a primidine analog that inhibits DNA synthesis. As mentioned previously, given at low doses systemically, there is little toxicity. Spinal fluid levels are approximately 20% of serum concentrations (Slevin *et al.*, 1983).

High-dose administration (3 g/m^2 two times a day for 12 doses) is highly active when used to induce remission in patients with leukemia. Central nervous system toxicity consisting of acute cerebellar dysfunction has been reported to be associated with these high-dose treatments (Damon *et al.*, 1989; Lazarus *et al.*, 1981; Salinsky *et al.*, 1983; Boesen *et al.*, 1988).

The syndrome occurs within several days of drug administration in approximately 10 to 20% of patients treated. Risk factors may include hepatic (Nand *et al.*, 1986) or renal dysfunction (Herzig *et al.*, 1987; Damon *et al.*, 1989).

Leukencephalopathy and extraocular muscle palsies have been reported with high doses of cytosine arabinoside. Again, improvement may be incomplete after stopping the drug.

Autopsy studies reveal a decrease in the Purkinje cell population in the cerebellum (Winkelman and Hines, 1983). Degenerative and necrotic changes in Purkinje cells may be seen (Salinsky *et al.*, 1983).

b. BCNU The nitrosourea BCNU is widely used as a treatment for malignant brain tumor. High doses of BCNU have been employed. Encephalopathy similar to that seen with radiation necrosis and methotrexate encephalopathy has been reported (Burger *et al.*, 1981). The syndrome is uncommon. Once the syndrome is recognized, the drug should be discontinued. Often, severe disability has occurred.

Pathological studies show either foci of myelin vaculization with swollen axon cylinders or fibrinoid necrosis in gray and white matter. Optic neuropathy has also been reported, especially with cranioirradiation, including the anterior visual system.

2. Regional Chemotherapy

Intraarterial chemotherapy has been employed for 30 years. These injections are used to increase the concentration of the drug to the target organ (Hochberg *et al.*, 1985). Intraarterial injection via the carotid or vertebral arteries has been used as a form of therapy for brain tumor. The best results have been with nitrosoureas such as BCNU. Etoposide and cisplatin have also been administered.

Intraarterial BCNU produces severe neurotoxicity in approximately 30 to 50% of patients. The syndrome consists of an encephalopathy developing acutely or several weeks after therapy (Kleinschmidt-DeMasters, 1986; Shapiro *et al.*, 1987). Progressive neurological deficits of dementia, hemiparesis, or seizures are reported. In many patients, there is subclinical necrosis but no clinical change.

The pathology shows vasculopathy similar to radiation necrosis (Rosenblum *et al.*, 1989). The pathological changes are confined to the perfusion territory of the drug. It is often difficult to distinguish tumor recurrence, radiation necrosis, and BCNU toxicity. CT scan abnormalities show hypodensity of the white matter with occasional calcification or enhancement.

A recent brain tumor cooperative group study (Shapiro *et al.*, 1987) suggested that intraarterial BCNU had no distinct advantage over intravenous BCNU for high-grade glioma. Toxicity was increased in patients receiving intraarterial BCNU.

An additional toxicity seen with intraarterial BCNU is that of retinal toxicity. This is a vasculopathy resulting in blindness in about 20% of patients.

III. Peripheral Nervous System Toxicity

A. Cisplatin

cis-Diamminedichloroplatinum (cisplatin) is a highly effective agent against a variety of solid tumors. The primary systemic toxicities are renal suppression, myelosuppression, nausea, and vomiting. Ototoxicity is frequent, but vestibular toxicity is rare. Peripheral neuropathy may occur.

1. Peripheral Neuropathy

The most common neurological toxicity is peripheral neuropathy (Walsh *et al.*, 1982; Hemphill *et al.*, 1980; Thompson *et al.*, 1984; Kedar *et al.*, 1978; Hadley and Herr, 1979). The syndrome consists of a subacute sensory neuropathy with numbness or paresthesias beginning distally. The neuropathy is more likely to occur after a cumulative dose of 400 mg/m^2 or greater.

Neurologic examination reveals vibratory loss at the ankles with reduced deep tendon reflexes. Pain and temperature sensation may be reduced. Motor weakness is rare.

The course of the neuropathy is subacute. Occasionally, the neuropathy may progress or begin after cisplatin is discontinued (Mollman *et al.*, 1988a; Siegal and Haim, 1990; Grunberg *et al.*, 1989). The toxicity is dose dependent and incompletely reversible.

Neurophysiological studies suggest large myelinated sensory fiber involvement with demyelination and axonal loss (Hemphill *et al.*, 1980; Ongerboer and Visser, 1985). Electronmicroscopy shows axonal degeneration with secondary myelin breakdown. Platinum levels that are 10 to 20 times lower on the spinal cord and brain than in peripheral nerve have been found (Thompson *et al.*, 1984).

The differential diagnosis is usually not problematic. Patients receiving doses of cisplatin who are developing subacute sensory neuropathy are usually not a diagnostic challenge. Occasionally, the neuropathy could be confused with paraneoplastic neuropathy or meningeal cancer. Meningeal cancer is a polyradiculopathy and only rarely will simulate a peripheral neuropathy such as is seen in cisplatin. Paraneoplastic neuropathy affects all sensory fibers and does not improve on discontinuation of the drug.

Several studies have explored the use of agents to treat platinum neuropathy. WR-2721 (ethiofos) has been shown to reduce the incidence and severity of neuropathy in patients treated with platinum. ORG-2766, an ACTH analog, has been used to reduce or prevent platinum neuropathy in patients treated for ovarian cancer (Van der Hoop *et al.*, 1990). Given in high doses, fewer neurological signs and symptoms of neuropathy were seen as compared to patients receiving placebo. The response of the chemotherapy was similar in the two groups. If a cisplatin neuropathy could be circumvented, higher doses and more prolonged treatment of chemotherapy may be possible.

2. Ototoxicity

Ototoxicity affects primarily high tones. Frequently, patients will have a decrease in audiograms. It is less frequent that actual clinical hearing loss occurs. The hearing loss is related to loss of hair cells in the organ of corti (Stadnick *et al.*, 1975). It has been found that 75% of patients had audiometric hearing loss, but only 16% were symptomatic (Melamed *et al.*, 1985). The hearing loss is usually symmetric.

Hearing loss may be irreversible and is dose dependent. High doses of cisplatin have produced acute deafness (Guthrie and Gynther, 1985), including patients receiving intraarterial platinum.

Vestibular neuropathy is rare. Patients with vertigo or unsteadiness have been reported, but this appears to be an uncommon complication (Black *et al.*, 1982).

3. Visual Toxicity

Visual toxicity has rarely been reported as a complication of cisplatin. Papilledema with and without increased intracranial pressure has been seen (Walsh *et al.*, 1982). A single case of retrobulbar neuritis was reported (Ostrow *et al.*, 1978). Retinopathy in patients receiving high-dose (200 mg/m^2) cisplatin was reported in 1985 (Wilding *et al.*, 1985). Eight of 13 patients had blurred vision. These visual symptoms develop at a cumulative dose of 600 mg/m^2.

4. Lhermitte's Sign

Lhermitte's sign occurs secondary to demyelination of the posterior columns (Walther *et al.*, 1987; Eeles *et al.*, 1986). This complication may occur with or without peripheral neuropathy and has been reported in 21 to 40% of patients treated. The syndrome resolves over a period of weeks to months.

The differential diagnosis includes epidural spinal cord compression. The lack of pain or other neurological symptoms and signs, i.e., peripheral neuropathy, make epidural cancer an unlikely diagnosis. No treatment is necessary for this complication of cisplatin.

Cisplatin has been reported to cause central nervous system dysfunction. Focal brain syndromes or cortical blindness can occur (Pippitt *et al.*, 1981; Diamond *et al.*, 1982; Berman and Mann, 1980). A stroke-like onset has been reported. The syndromes may be transient and may be associated with seizures. The pathogenesis of CNS dysfunction is unclear.

Cerebral herniation has been reported in patients with intracranial mass lesions receiving cisplatin (Walker *et al.*, 1988).

B. Carboplatin

Carboplatin is an analog of cisplatin that is effective against a variety of tumors. There is little neurotoxicity related to carboplatin. Patients with thrombotic microangiopathy have been reported resulting in multiple cerebral infarcts (Delattre and Posner, 1989). Intracarotid carboplatin produces ocular toxicity and cerebral stroke-like spells (Stewart *et al.*, 1982).

C. Vinca Alkaloids

Vinca alkaloids such as vincristine, vindesine, and vinblastine are plant alkaloids derived from the periwinkle plant. These drugs inhibit microtubule formation, arrest cell division, and interfere with axonal transport, as well as secretory function of the neuron (Iqbal and Ochs, 1980). The cause of the neurotoxicity is unclear but is probably related to the main site of antineoplastic action (Sahenk *et al.*, 1987).

Vincristine is the most commonly used agent. These drugs are essentially bone marrow sparing. The major dose-limiting toxicities are usually neurological. Vinca alkaloids are used for a variety of human tumors.

1. Peripheral Neuropathy

A mixed motor sensory peripheral neuropathy related to vinca alkaloids is very common (Sandler *et al.*, 1969; Bradley *et al.*, 1970; Rosenthal and Kaufman, 1974; Casey *et al.*, 1973). These are dose related and are usually not very severe. In fact, a majority of patients who receive several courses of vincristine develop paresthesias and a decrease in ankle jerks. The neuropathy, especially if mild, is largely reversible when the drug is discontinued.

As the neuropathy progresses, there is a loss of sensory modalities distally with muscle weakness. As with cisplatin, the neuropathy may progress or even begin after the drug has been discontinued. For the most part, improvement will take place, if the drug is discontinued.

The neurologic examination, especially in patients with more severe diseases, reveals sensory loss to all modalities, distal weakness, and depression of the deep tendon reflexes.

Patients who have preexisting neuropathies such as hereditary sensory motor neuropathies have been reported to become severely worse with the administration of vincristine, leading to severe neurological involvement and death (Hogan-Dann *et al.*, 1984).

Vincristine neuropathy is axonal in type with no abnormalities being found in the dorsal root ganglion.

2. Autonomic Neuropathy

Autonomic dysfunction is very common in patients with vincristine toxicity (Holland *et al.*, 1973; Casey *et al.*, 1973). Constipation with resulting abdominal pain secondary to paralytic ileus occurs in almost half of the patients treated. Patients dying from paralytic ileus secondary to vincristine have been reported. The ileus usually resolves within days of administration. Less common autonomic manifestations include orthostatic hypotension, urinary incontinence, and impotence.

3. Other Toxicities

Rare reports of vincristine causing other toxicities are scattered throughout the literature. The syndrome of inappropriate antidiuretic hormone secretion following vincristine resulting in hyponatremia has been reported (Cutting, 1971). The syndrome can occur without other evidence of neurotoxicity.

Hoarseness following vincristine and vinblastine is uncommon but has been reported (Bohannon *et al.*, 1963). The cause of this is unclear but it may be secondary to lesion of the recurrent laryngeal nerve. Vocal cord paresis may be present.

Optic neuropathy, ocular muscle palsy, ptosis, fifth nerve palsy, and seventh nerve palsy have been reported. These complications are unusual and require extensive testing to be certain no other metastatic lesions are present.

Jaw pain, an uncommon symptom, may occur acutely with treatment. The cause of this remains unclear. It has been postulated that a myopathy occurs, but this is not clear.

Finally, intrathecal administration of vincristine is usually fatal. A single nonfatal case was treated with aggressive lavage of the cerebrospinal fluid by a 24-hr continuous infusion of lactated Ringer's solution through a catheter in the lateral ventricle (Dyke, 1989).

D. Taxol

Taxol is a new agent undergoing study. The dose-limiting toxicity is often peripheral neuropathy. Sensory neuropathy, including small and large fiber sensory loss, or painful dysesthesias occur within a day or a few weeks of drug administration. Improvement occurs with discontinuation of treatment. Less commonly, motor weakness or autonomic dysfunction may occur.

The mechanism of action of neuropathy is unclear but may be through action on microtubules at the dorsal root ganglion level.

E. Suramin

Suramin is a useful drug for prostate cancer. Peripheral neuropathy occurs frequently, including motor and sensory loss. Patients resembling Guillain–Barre syndrome have been reported.

IV. Summary

In summary, neurotoxicity secondary to antineoplastic agents is common and may be devastating. Multiagent protocols and multimodality treatments put patients at increasing risk of toxicity. It is vital that the neurologist and oncologist be familiar with these toxicities in patients being treated for systemic cancer.

References

Abelson, H. T. (1978). Methotrexate and central nervous system toxicity. *Cancer Treat. Rep.* **62**, 1999–2001.

Adams, J. R., *et al.* (1984). Recurrent acute toxic optic neuropathy secondary to 5-FU. *Cancer Treat. Rep.* **68**, 565.

Aguilar-Markulis, N. V., *et al.* (1981). Auditory toxicity effects of long-term cis-dichlorodiammineplatinum II therapy in genitourinary cancer patients. *J. Surg. Oncol.* **16**, 111–123.

Aisner, J., *et al.* (1974). Orthostatic hypotension during combination chemotherapy with vincristine (NSC-67574). *Cancer Chemother. Rep.* **58**, 927–930.

Albert, D. M., *et al.* (1967). Ocular complications of vincristine therapy. *Arch. Ophthal.* **78**, 709–713.

Allen, J. C., and Rosen, G. (1978). Transient cerebral dysfunction following chemotherapy for osteogenic sarcoma. *Ann. Neurol.* **3**, 441–447.

Allen, J. C., *et al.* (1980). Leukoencephalopathy following high-dose IV methotrexate chemotherapy with leucovorin rescue. *Cancer Treat. Rep.* **64**, 1261–1273.

Apfel, S. C., *et al.* (1991). Nerve growth factor prevents toxic neuropathy in mice. *Ann. Neurol.* **29**, 87–90.

Apfel, S. C., *et al.* (1992). Nerve growth factor prevents experimental cisplatin neuropathy. *Ann. Neurol.* **31**, 76–80.

Awidi, A. S. (1980). Blindness and vincristine. *Ann. Intern. Med.* **93**, 781.

Back, E. M. (1969). Death after intrathecal methotrexate. *Lancet* **2**, 1005.

Bagley, C. M. (1975). Single I.V. doses of 5-fluorouracil: a phase 1 study. *Proc. Am. Assoc. Cancer Res.* **16**, 12.

Bagshawe, K. D., *et al.* (1969). Intrathecal methotrexate. *Lancet* **2**, 1258.

Band, P. R., *et al.* (1973). Treatment of central nervous system leukemia with intrathecal cytosine arabinoside. *Cancer* **32**, 744–748.

Bates, S. E., *et al.* (1985). Ascending myelopathy after chemotherapy for central nervous system acute lymphoblastic leukemia: correlation with cerebrospinal fluid myelin basic protein. *Med. Pediatr. Oncol.* **13**, 4–8.

Benger, A., *et al.* (1985). Clinical evidence of a cummulative effect of high-dose cytarabine on the cerebellum in patients with acute leukemia: a leukemia intergroup report. *Cancer Treat. Rep.* **69**, 240–241.

Bergevin, P. R., *et al.* (1975). Neurotoxicity of 5-fluorouracil. *Lancet* Vol. **1**, 410.

Berman, I. J., and Mann, M. P. (1980). Seizures and transient cortical blindness associated with cis-platinum (II) diamminedichloride (PPD) therapy in a thirty-year old man. *Cancer* **45**, 746–766.

Bernstein, M. L., *et al.* (1982). Noncardiogenic pulmonary edema following injection of methotrexate into the cerebrospinal fluid. *Cancer* **50**, 866–868.

Bixenman, W. W., *et al.* (1977). Oculomotor disturbances associated with 5-fluorouracil chemotherapy. *Am. J. Ophthalmol.* **83**, 789–793.

Black, F. O., *et al.* (1982). Cisplatin vestibular ototoxicity: preliminary report. *Laryngoscope* **92**, 1363–1368.

Bleyer, W. A. (1977). Current status of intrathecal chemotherapy in human meningeal neoplasms. *Natl. Cancer Inst. Monogr.* **46**, 171–178.

Bleyer, W. A., *et al.* (1973). Neurotoxicity and elevated CSF MTX concentration in meningeal leukemia. *N. Engl. J. Med.* **289**, 770–773.

Bleyer, W. A., *et al.* (1981). Neurologic sequelae of methotrexate and ionizing radiation: a new classification. *Cancer Treat. Rep.* **65**, 89–98.

Blisard, K. S., and Harrington, D. A. (1989). Cisplatin-induced neurotoxicity with seizures in frogs. *Ann. Neurol.* **26**, 336–341.

Boesen, P., *et al.* (1988). Severe persistent cerebellar dysfunction complicating cytosine arabinoside therapy. *Acta Med. Scand.* **224**(2), 189–191.

Bohannon, R. A., *et al.* (1963). Vincristine in the treatment of lymphomas and leukemias. *Cancer Res.* **23**, 613–621.

Boileau, G., *et al.* (1971). Cerebellar ataxia during 5-fluorouracil (NSC-19893) therapy. *Cancer Chemother. Rep.* **55**, 595–598.

Borgeat, A., *et al.* (1986). Peripheral neuropathy associated with high-dose Ara-C therapy. *Cancer* **58**, 852–854.

Bourke, R. S., *et al.* (1973). Kinetics of entry and distribution of 5-fluorouracil in cerebrospinal fluid and brain following intravenous injection in a primate. *Cancer Res.* **33**, 1735–1746.

Brade, W. P., *et al.* (1985). Ifosamide-pharmacology, safety and therapeutic potential. *Cancer Treat. Rep.* **12**, 1–47.

Bradley, W. G., *et al.* (1970). Neuropathy of vincristine in man. *J. Neurol. Sci.* **10**, 107–131.

Breuer, A. C., *et al.* (1977). Paraparesis following intrathecal cytosine arabinoside. *Cancer* **40**, 2817–2822.

Breuer, A. C., *et al.* (1978). Multifocal pontine lesions in cancer patients treated with chemotherapy and CNS radiotherapy. *Cancer* **41**, 2112–2120.

Brook, J., and Schreiber, W. (1971). Vocal cord paralysis: a toxic reaction to vinblastine (NSC-49842) therapy. *Cancer Chemother. Rep.* **55**, 591–593.

Bunt, A. A., and Lund, R. D. (1974). Vinblastine-induced blockage of orthograde and retrograde axonal transport of protein in retinal ganglion cells. *Exp. Neurol.* **45**, 288–297.

Burger, P. D., *et al.* (1981). Encephalomyelopathy following high-dose BCNU therapy. *Cancer* **48**, 1319–1327.

Byfield, J. E. Ionizing radiation and vincristine: possible neurotoxic synergism. *Radiol. Clin. Biol.* **41**, 129–138.

Byrd, R. L., *et al.* (1981). Transient cortical blindness secondary to vincristine therapy in childhood malignancies. *Cancer* **47**, 37–40.

Cairo, M. S., *et al.* (1980). Intracranial hemorrhage and focal seizures secondary to use of L-asparaginase during induction therapy of acute lymphocytic leukemia. *J. Pediatr.* **97**, 829–833.

Campbell, T. N., *et al.* (1982). High-dose allopurinol modulation of 5-FU toxicity: phase 1 trial of an outpatient dose schedule. *Cancer Treat. Rep.* **66**, 1723–1727.

Campbell, T. N., *et al.* (1983). Clinical pharmacokinetics of intra-arterial cisplatin in humans. *J. Clin. Oncol.* **12**, 755–762.

Carmichael, S. M., *et al.* (1970). Orthostatic hypotension during vincristine therapy. *Arch. Intern. Med.* **126**, 290–293.

Carney, M. W. P., *et al.* (1982). Manic psychosis associated with procarbazine. *Br. Med. J.* **284**, 82–83.

Carpentieri, U., and Lockhart, L. H. (1978). Ataxia and athetosis as side effects of chemotherapy with vincristine in non-hodgkins lymphoma. *Cancer Treat. Rep.* **62**, 561–562.

Cascino, T. L., *et al.* (1983). CT of the brachial plexus in patients with cancer. *Neurology* **33**, 1553.

Casey, E. G., *et al.* (1973). Vincristine neuropathy: clinical and electrophysiological observations. *Brain* **96**, 69–86.

Cassady, J. R., *et al.* (1980). Augmentation of vincristine neurotoxicity by irradiation of peripheral nerves. *Cancer Treat. Rep.* **64**, 964–965.

Castellanos, A. M., *et al.* (1987). Regional nerve injury after intra-arterial chemotherapy. *Neurology* **37**, 834–837.

Chabner, B. A., and Myers, C. E. (1982). Clinical pharmacology of cancer chemotherapy. In *Cancer: Principles and Practice of Oncology* (V. T. Jr., DeVita, S. Hellman, and S. A. Rosenberg, Eds.), Lippincott, Philadelphia.

Chabner, B. A., *et al.* (1973). High-dose intermittent intravenous infusion of procarbazine (NSC-77213). *Cancer Chemother. Rep.* **57**, 361–363.

Chadwick, M., and Rogers, W. I. (1972). The physiologic disposition of 5-fluorouracil in mice bearing solid L1210 lymphocytic leukemia. *Cancer Res.* **32**, 1045–1056.

Chan, S. Y., *et al.* (1980). Block of axoplasmic transport in vitro by vinca alkaloids. *J. Neurobiol.* **11**, 251–264.

Chlien, L. T., *et al.* (1981). Progression of methotrexate-induced leukoencephalopathy in children with leukemia. *Med. Pediatr. Oncol.* **9**, 133–141.

Chun, H. G., *et al.* (1986). Central nervous system toxicity of fludarabine phosphate. *Cancer Treat. Rep.* **70**, 1225–1228.

Clark, A. W., *et al.* (1982). Paraplegia following intrathecal chemotherapy. *Cancer* **50**, 42–47.

Cohen, R. J., *et al.* (1983). Transient left homonymous hemianopsia and encephalopathy following treatment of testicular carcinoma with cisplatinum, vinblastine, and bleomycin. *J. Clin. Oncol.* **1**, 392–393.

Cruz-Sanchez, F. F., *et al.* (1991). Brain lesions following combined treatment with methotrexate and craniospinal irradiation. *J. Neurooncol.* **10**, 165–171.

Cutting, H. O. (1971). Inappropriate secretion of antidiuretic hormone secondary to vincristine therapy. *Am. J. Med.* **51**, 269–271.

Damon, L. E., *et al.* (1989). The association between high-dose cytarabine neurotoxicity and renal insufficiency. *J. Clin. Oncol.* **7**, 1563–1568.

Danesh, M. M., *et al.* (1989). Ifosfamide encephalopathy. *J. Toxicol. Clin. Toxicol.* **27**(4–5):293–298.

DeAngelis, L. M., *et al.* (1991). Evolution of neuropathy and myopathy during intensive vincristine/corticosteroid chemotherapy for non-hodgkin's lymphoma. *Cancer* **67**, 2241–2246.

Delattre, J., and Posner, J. B. (1989). Neurological complications of chemotherapy and radiation therapy. In *Neurology and General Medicine* (M. J. Aminoff, Ed.), Vol. 365, p. 387, Churchill Livingstone, New York.

Di Chiro, G., *et al.* (1979). Computed tomography profiles of periventricular hypodensity in hydrocephalus and leukoencephalopathy. *Radiology* **130**, 661–665.

Di Gregorio, F., *et al.* (1990). Efficacy of ganglioside treatment in reducing functional: alterations induced by vincristine in rabbit peripheral nerves. *Cancer Chemother. Pharmacol.* **26**, 31–36.

Diamond, S. B., *et al.* (1982). Cerebral blindness in association with cis-platinum chemotherapy for advanced carcinoma of the fallopian tube. *J. Obstet. Gynecol.* **59**, 84S–86S.

DiBella, N. H. (1980). Vincristine-induced orthostatic hypotension: a prospective clinical study. *Cancer Treat. Rep.* **64**, 359–360.

Djerassi, I., *et al.* (1977). High-dose methotrexate-citrovorin factor rescue in the management of brain tumors. *Cancer Treat. Rep.* **6**, 691–694.

Duffner, P. K., *et al.* (1984). CT abnormalities and altered methotrexate clearance in children with CNS leukemia. *Neurology* **34**, 229–233.

Dunton, S. F., *et al.* (1986). Progressive ascending paralysis following administration of intrathecal and intravenous cytosine arabinoside: a pediatric oncology group study. *Cancer* **57**, 1083–1088.

Dyke, R. W. (1989). Treatment of inadvertent intrathecal injection of vincristine. *N. Engl. J. Med.* **321**, 1270–1271.

Ebner, F., *et al.* (1989). MR findings in methotrexate-induced CNS abnormalities. *Am. J. Neuroradiol.* **10**, 959–964.

Eden, O. B., *et al.* (1978). Seizures following intrathecal cytosine arabinoside in young children with acute lymphoblastic leukemia. *Cancer* **42**, 53–58.

Eeles, R., *et al.* (1986). Lhermitte's sign as a complication of cisplatin-containing chemotherapy for testicular cancer. *Cancer Treat. Rep.* **70**, 905–907.

Feinberg, W. M., and Swenson, M. R. (1988). Cerebrovascular complications of L-asparaginase therapy. *Neurology* **38**, 127–133.

Filley, C. M., *et al.* (1982). Neurologic manifestations of podophyllin toxicity. *Neurology* **32**, 308–311.

Forsyth, P. A., *et al.* (1991). Multifocal inflammatory leukoencephalopathy in patients receiving adjuvant therapy with 5-fluorouracil and levamisole for adenocarcinoma of the colon. *Ann. Neurol.* **30**, 273.

Gagbano, R., and Costani, J. (1976). Paraplegia following intrathecal methotrexate: report of a case and review of the literature. *Cancer* **37**, 1663–1668.

Gagliano, R., and Costanzi, J. J. (1976). Paraplegia following intrathecal methotrexate. *Cancer* **37**, 1663–1668.

Gaidys, W. G., *et al.* (1983). Intrathecal vincristine: report of a fatal case despite CNS washout. *Cancer* **52**, 799–801.

Gangji, D., *et al.* (1980). Leukoencephalopathy and elevated levels of myelin basic protein in the cerebrospinal fluid of patients with acute lymphoblastic leukemia. *N. Engl. J. Med.* **303**, 19–21.

Geiser, C. F., *et al.* (1975). Adverse effects of intrathecal methotrexate in children with acute leukemia in remission. *Blood* **45**, 189–195.

Gerritsen Van Der Hoop, *et al.* (1990). Prevention of cisplatin neurotoxicity with an ACTH(4-9) analogue in patients with ovarian cancer. *N. Engl. J. Med.* **322**, 89–94.

Gerzun, K., *et al.* (1979). Polarity of vincristine, vindesine and vinblastine in relation to neurologic effects, abstracted. *Proc. Am. Assoc. Cancer Res.* **20**, 46.

Gilbert, M. R., *et al.* (1989). Methotrexate neurotoxicity: in vitro studies using cerebellar explants from rats. *Cancer Res.* **49**(9), 2502–2505.

Ginsgerg, S. J., *et al.* (1977). Vinblastine and inappropriate ADH secretion. *N. Engl. J. Med.* **294**, 941.

Glass, J. P., *et al.* (1986). Treatment-related leukoencephalopathy: a study of three cases and literature review. *Medicine* **65**, 154–162.

Gormley, P., *et al.* (1981). Pharmacokinetic study of cerebrospinal fluid penetration of cis-diamminedichloroplatinum (2). *Cancer Chemother. Pharmacol.* **5**, 247–260.

Gottlieb, D., *et al.* (1987). The neurotoxicity of high-dose cytosine arabinoside is age-related. *Cancer* **60**, 1439–1441.

Gottlieb, R. J., and Cuttner, J. (1971). Vincristine-induced bladder atony. *Cancer* **28**, 674–675.

Greenwald, E. S. (1976). Organic mental changes with fluorouracil therapy. *JAMA* **235**, 248–249.

Griffin, T. W., *et al.* (1977). The effect of photon irradiation on blood-brain barrier permeability to methotrexate in mice. *Cancer* **40**, 1109–1111.

Griffiths, J. D., *et al.* (1986). Vincristine neurotoxicity enhanced in combination chemotherapy including both teniposide and vincristine. *Cancer Treat. Rep.* **70**, 519–521.

Grossman, L., *et al.* (1983). Central nervous system toxicity of high-dose cytosine arabinoside. *Med. Pediatr. Oncol.* **11**, 246–250.

Grunberg, S. M., *et al.* (1989). Progressive paresthesias after cessation of therapy with very high-dose cisplatin. *Cancer Chemother. Pharmacol.* **25**, 62–64.

Guiheneuc, P., *et al.* (1980). Early phase of vincristine neuropathy in man. *J. Neurol. Sci.* **45**, 355–366.

Guthrie, T. H., and Gynther, L. (1985). Acute deafness: a complication of high-dose cisplatin. *Arch. Otolaryngol.* **111**, 344–345.

Gutin, P. H., *et al.* (1977). Treatment of malignant meningeal disease with intrathecal thio-TEPA: a phase 2 study. *Cancer Treat. Rep.* **61**, 885–887.

Hadley, D., and Herr, H. W. (1979). Peripheral neuropathy associated with cisdichlorodiammineplatinum (II) treatment. *Cancer* **44**, 2026–2028.

Hayes, F. A., *et al.* (1979). Tetany: a complication of cis-dichlorodiammineplatinum II therapy. *Cancer Treat. Rep.* **63**, 547–548.

Hemphill, M., *et al.* (1980). Sensory neuropathy in cisplatinum chemotherapy. *Neurology* **30**, 429.

Herzig, R. H., *et al.* (1987). Cerebellar toxicity with high-dose cytosine arabinoside. *J. Clin. Oncol.* **5**, 927–932.

Hirvonen, H. E., *et al.* (1989). Vincristine treatment of acute lymphoblastic leukemia induces transient autonomic cardioneuropathy. *Cancer* **64**, 801–805.

Hochberg, F. H. J., *et al.* (1985). The rationale and methodology for intra-arterial chemotherapy with BCNU as treatment for glioblastoma. *J. Neurosurg.* **63**, 876–880.

Hogan-Dann, C. M., *et al.* (1984). Polyneuropathy following vincristine therapy in two patients with Charcot-Marie-Tooth syndrome. *JAMA* **252**, 2862–2863.

Holland, J. F., *et al.* (1973). Vincristine treatment of advanced cancer: a cooperative study of 392 cases. *Cancer Res.* **33**, 1258–1264.

Horton, J., *et al.* (1970). 5-Fluorouracil in cancer: an improved regimen. *Ann. Intern. Med.* **73**, 897–900.

Howell, S. B., *et al.* (1983). Effect of allopurinol on the toxicity of high-dose 5-fluorouracil administered by intermittent bolus injection. *Cancer* **51**, 220–225.

Hwang, T., *et al.* (1985). Central nervous system toxicity with high-dose Ara-C. *Neurology* **35**, 1475–1479.

Iqbal, Z., and Ochs, S. (1980). Uptake of vinca alkaloids into mammalian nerve and its subcellular components. *J. Neurochem.* **34**, 59–68.

Jackson, D. V., *et al.* (1986a). Clinical trial of folinic acid to reduce vincristine neurotoxicity. *Cancer Chemother. Pharmacol.* **17**, 281–284.

Jackson, D. V., *et al.* (1986b). Clinical trial of pyridoxine to reduce vincristine neurotoxicity. *J. Neurooncol.* **4**, 37–41.

Jackson, D. V., *et al.* (1988). Amelioration of vincristine neurotoxicity by glutamic acid. *Am. J. Med.* **84**, 1016–1022.

Jacobs, P., *et al.* (1991). Methotrexate encephalopathy. *Eur. J. Cancer* **27**, 1061–1062.

Jaffe, N. (1972). Recent advances in the chemotherapy of metastatic osteogenic sarcoma. *Cancer* **30**, 1627–1631.

Jolivet, J., *et al.* (1983). The pharmacology and clinical use of methotrexate. *N. Engl. J. Med.* **309**, 1094–1104.

Jones, A. (1964). Transient radiation myelopathy. *Br. J. Radiol.* **37**, 727.

Kay, H. E. M., *et al.* (1972). Encephalopathy in acute leukemia associated with methotrexate therapy. *Arch. Dis. Child.* **47**, 344.

Kedar, A., *et al.* (1978). Peripheral neuropathy as a complication of cis-dichloriodiammineplatinum II treatment: a case report. *Cancer Treat. Rep.* **62**, 819–821.

Kleinschmidt-DeMasters, B. K. (1986). Intracarotid BCNU leukoencephalopathy. *Cancer* **57**, 1276–1280.

Koenig, H., and Patel, A. (1970). A biochemical basis for 5-Fluorouracil neurotoxicity. *Arch. Neurol.* **23**, 155–160.

Kori, S. H., *et al.* (1981). Brachial plexus lesions in patients with cancer: 100 cases. *Neurology* **31**, 45.

Kris, M. G., *et al.* (1985). Improved control of cisplatin-induced emesis with high-dose metoclopramide and with complications of metoclopramide, dexamethasone, and diphenhydramine. *Cancer* **55**, 527–534.

La Rocca, R. V., *et al.* (1990). Suramin-induced polyneuropathy. *Neurology* **40**, 954–960.

Land, V. J., *et al.* (1972). Toxicity of L-asparaginase in children with advanced leukemia. *Cancer* **30**, 339–347.

Lazarus, H. M., *et al.* (1981). Central nervous system toxicity of high-dose systemic cytosine arabinoside. *Cancer* **48**, 2577–2582.

Lee, I. P., and Lucier, G. W. (1976). The potentiation of barbiturate-induced narcosis by procarbazine. *J. Pharmacol. Exp. Ther.* **196**, 586–593.

Leonard, J. V., and Kay, J. D. S. (1986). Acute encephalopathy and hyperammonaemia complicating treatment of acute lymphoblastic leukemia. *Lancet* **1**, 162.

Leff, R. S., *et al.* (1988). Acute neurologic dysfunction after high-dose etoposide therapy for malignant glioma. *Cancer* **62(1)**, 32–35.

Lessner, H. E., *et al.* (1980). Phase 2 study of L-asparaginase in the treatment of pancreatic carcinoma. *Cancer Treat. Rep.* **64**, 1359–1361.

Levitt, L. P., and Prager, D. (1975). Mononeuropathy due to vincristine toxicity. *Neurology* **25**, 894–895.

Lipton, R. B., *et al.* (1989). Taxol produces a predominantly sensory neuropathy. *Neurology* **39**, 368–373.

Liu, H. M., *et al.* (1978). Methotrexate encephalopathy. *Hum. Pathol.* **9**, 635–648.

Lucien, J., *et al.* (1975). Disseminated necrotizing leukoencephalopathy: a complication of treated central nervous system leukemia and lymphoma. *Cancer* **35**, 219–305.

Luddy, R. E., and Gilman, P. A. (1973). Paraplegia following intrathecal methotrexate. *J. Pediatr.* **83**, 988–992.

Luque, F. A., *et al.* (1987). Parkinsonism induced by high-dose cytosine arabinoside. *Mov. Disord.* **2**, 219–222.

Lynch, H. T., *et al.* (1981). Organic brain syndrome secondary to 5-fluorouracil. *Dis. Colon Rectum* **24**, 130–131.

Macchi, P. J., *et al.* (1986). High field MR imaging of cerebral venous thrombosis. *J. Comp. Assist Tomogr.* **10**, 10–15.

Mahaley, M. S., *et al.* (1986). Central neurotoxicity following intracarotid BCNU chemotherapy for malignant gliomas. *J. Neuro-Oncol.* **3**, 297–314.

Majahan, S. L., *et al.* (1981). Acute acoustic nerve palsy associated with vincristine therapy. *Cancer* **47**, 2404–2406.

Martino, R. L., *et al.* (1984). Transient dysfunction following moderate-dose methotrexate for undifferentiated lymphoma. *Cancer* **54**, 2003–2005.

McHaney, V. A., *et al.* (1983). Hearing loss in children receiving cisplatin chemotherapy. *J. Pediatr.* **102**, 314–317.

McIntosh, S., and Aspnes, G. T. (1973). Encephalopathy following CNS prophylaxis in childhood lymphoblastic leukemia. *Pediatrics* **52**, 612–615.

McLeod, J. G., and Penny, R. Vincristine neuropathy: an electrophysiological and histological study. *J. Neurol. Neurosurg. Psychiatry* **32**, 297–304.

Meanwell, C. A., *et al.* (1986). Prediction of ifosfamide/mesna associated encephalopathy. *Eur. J. Cancer Clin. Oncol.* **22**, 815–819.

Melamed, L. B., *et al.* (1985). Cisplatin ototoxicity in gynecologic cancer patients. *Cancer* **55**, 41–43.

Mena, H., *et al.* (1981). Central and peripheral myelinopathy associated with systemic neoplasia and chemotherapy. *Cancer* **48**, 1724–1737.

Merkel, D. E., *et al.* (1986). Central nervous system toxicity with fludarabine. *Cancer Treat. Rep.* **70**, 1449–1450.

Miller, B. R. (1985). Neurotoxicity and vincristine. *JAMA* **253**, 2045.

Moertel, C. G., *et al.* (1964). Cerebellar ataxia associated with fluorinated pyrimidine therapy. *Cancer Chemother. Rep.* **41**, 15–18.

Mollman, J. E., *et al.* (1988a). Cisplatin neuropathy: risk factors, prognosis, and protection by WR-2721. *Cancer* **61**, 2192–2195.

Mollman, J. E., *et al.* (1988b). Unusual presentation of cis-platinum neuropathy. *Neurology* **38**(3), 488–490.

Moots, P. L., *et al.* (1987). Diagnosis of dural venous sinus thrombosis by magnetic resonance imaging. *Ann. Neurol.* **22**, 431–432.

Moress, G. R., *et al.* (1967). Neuropathy in lymphoblastic leukemia treated with vincristine. *Arch. Neurol.* **16**, 377–384.

Mott, M. G., *et al.* (1972). Methotrexate meningitis. *Lancet* **2**, 656.

Moure, J. M., *et al.* (1970). Electroencephalogram changes secondary to asparaginase. *Arch. Neurol.* **23**, 365–368.

Mubashir, B. A., and Bart, J. B. (1972). Vincristine neurotoxicity. *N. Engl. J. Med.* **287**, 517.

Muggia, F. M., *et al.* (1987). Weekly 5-Fluorouracil combined with PALA: toxic and therapeutic effects in colorectal cancer. *Cancer Treat. Rep.* **71**, 253–256.

Nand, S., *et al.* (1986). Neurotoxicity associated with systemic high-dose cytosine arabinoside. *J. Clin. Oncol.* **4**, 571–575.

Nelson, R. L., *et al.* (1980). Comparative pharmacokinetics of vindesine, vincristine, and vinblastine in patients with cancer. *Cancer Treat. Rep.* **7**(Suppl), 17–24.

Norton, S. W., and Stockman, J. A. (1979). Unilateral optic neuropathy following vincristine chemotherapy. *J. Pediatr. Ophthalmol. Strabismus* **16**, 190–193.

O'Callaghan, M. J., and Ekert, H. (1976). Vincristine toxicity unrelated to dose. *Arch. Dis. Child.* **51**, 289–292.

O'Dwyer, P. J., *et al.* (1985). Etoposide (VP-16-213): current status of an active anticancer drug. *N. Engl. J. Med.* **312**, 692–700.

Ochs, S., and Worth, R. (1975). Comparison of the block of fast axoplasmic transport in mammalian nerve by vincristine, vinblastine and desacetyl vinblastine amide sulfate, abstracted. *Proc. Am. Assoc. Cancer Res.* **16**, 70.

Oettgen, H. F., *et al.* (1970). Toxicity of *E. coli* L-asparaginase in man. *Cancer* **2**, 253–278.

Ojeda, V. J. (1982). Necrotizing leukoencephalopathy associated with intrathecal/intraventricular methotrexate therapy. *Med. J. Aust.* **2**, 289–293.

Oliverio, V. T., *et al.* (1964). Some pharmacologic properties of a new antitumor agent, *N*-isopropyl-α-(2-methylhydrazino)-*p*-toluamide, hydrochloride (NSC-77213). *Cancer Chemother. Rep.* **42**, 1–7.

Ongerboer de Visser, B. W., and Tiessens, G. (1985). Polyneuropathy induced by cisplatin. *Prog. Exp. Tumor Res.* **29**, 190–196.

Ostrow, S., *et al.* (1978). Ophthalmologic toxicity after cis-dichlorodiammineplatinum II therapy. *Cancer Treat. Rep.* **62**, 1591–1594.

Ott, N., *et al.* (1988). Sequelae of thrombotic or hemorrhagic complications following L-asparaginase therapy for childhood lymphoblastic leukemia. *Am. J. Pediatr. Hematol. Oncol.* **10**(3), 591–595.

Ozols, R. F., *et al.* (1985). High-dose cisplatin in hypertonic saline in refractory ovarian cancer. *J. Clin. Oncol.* **3**, 1246–1250.

Peylan-Ramu, N., *et al.* (1978). Abnormal CT scans of the brain in asymptomatic children with acute lymphocytic leukemia after prophylactic treatment of the central nervous system with radiation and intrathecal chemotherapy. *N. Engl. J. Med.* **298**, 815–818.

Phanthumchinda, K., *et al.* (1991). Stroke-like syndrome, mineralizing microangiopathy, and neuroaxonal dystrophy following intrathecal methotrexate therapy. *Neurology* **41**, 1847–1848.

Phillips, P. C., *et al.* (1986). Intensive 1,3-bis(2-chloroethyl)-1-nitrosourea (BCNU) monotherapy and autologous bone marrow transplantation for malignant gliomas. *J. Clin. Oncol.* **4**, 639–645.

Phillips, P. C., *et al.* (1987). Reduced cerebral glucose metabolism and increased brain capillary permeability following high-dose methotrexate chemotherapy: a positron emission tomographic study. *Ann. Neurol.* **21**, 59–63.

Phillips, P. C., *et al.* (1989). High-dose leucovorin reverses acute high-dose methotrexate neurotoxicity in the rat. *Ann. Neurol.* **25**(4), 365–372.

Pippit, C. H., *et al.* (1981). Cisplatin-associated cortical blindness. *Gynecol. Oncol.* **12**, 253–255.

Pizzo, P. A., *et al.* (1979). Neurotoxicities of current leukemia therapy. *Am. J. Pediatr. Hematol. Oncol.* **1**, 127–140.

Pochadly, C. (1979). Prophylactic CNS therapy in childhood acute leukemia: review of methods used. *Am. J. Ped. Hematol. Oncol.* **1**, 119–126.

Pomes, A., *et al.* (1986). Local neurotoxicity of cisplatin after intra-arterial chemotherapy. *Acta Neurol. Scand.* **73**, 302–303.

Postmus, P. E., *et al.* (1984). High-dose etoposide for refractory malignancies: a phase I study. *Cancer Treat. Rep.* **68**, 1471–1474.

Powell, B. L., *et al.* (1986). Peripheral neuropathy after high-dose cytosine arabinoside, daunorubicin, and asparaginase consolidation for acute nonlymphocytic leukemia. *J. Clin. Oncol.* **4**, 95–97.

Pratt, C. B., *et al.* (1986). Central nervous system toxicity following the treatment of pediatric patients with ifosfamide/mesna. *J. Clin. Oncol.* **4**, 1253–1261.

Pratt, C. B., *et al.* (1990). Ifosfamide neurotoxicity is related to previous cisplatin treatment for pediatric solid tumors. *J. Clin. Oncol.* **8**, 1399–1401.

Price, R. A., and Birdwell, D. A. (1978). The central nervous system in childhood leukemia. *Cancer* **42**, 717–728.

Price, R. A., and Jamieson, P. A. (1975). The central nervous system in childhood leukemia. II. Subacute leukoencephalopathy. *Cancer* **35**, 306–318.

Priest, J. R., *et al.* (1980). Thrombotic and hemorrhagic strokes complicating early therapy for childhood acute lymphoblastic leukemia. *Cancer* **46**, 1548–1554.

Raphaelson, M. I., *et al.* (1983). Vincristine neuropathy with bowel and bladder atony, mimicking spinal cord compression. *Cancer Treat. Rep.* **67**, 604–605.

Reagan, T. J., *et al.* (1968). Chronic progressive radiation myelopathy: its clinical aspects and differential diagnosis. *JAMA* **203**, 128.

Riehl, J. L., and Brown, W. J. (1964). Acute cerebellar syndrome secondary to 5-fluorouracil therapy. *Neurology* **14**, 961–967.

Ritch, P. S. (1988). Cis-dichlorodiammineplatinum II-induced syndrome of inappropriate secretion of antidiuretic hormone. *Cancer* **61**, 448–450.

Ritch, P. S., *et al.* (1983). Ocular toxicity from high-dose cytosine arabinoside. *Cancer* **51**, 430–432.

Robertson, G. L., *et al.* (1973). Vincristine neurotoxicity and abnormal secretion of antidiuretic hormone. *Arch. Intern. Med.* **132**, 717–720.

Rosen, G., *et al.* (1977). High-dose methotrexate with citrovorum factor rescue for the treatment of central nervous system tumors in children. *Cancer Treat. Rep.* **61**, 681–689.

Rosenblum, M. K., *et al.* (1989). Fatal necrotizing encephalopathy complicating treatment of malignant gliomas with intra-arterial BCNU and irradiation: a pathological study. *J. Neurooncol.* **7**(3), 269–281.

Rosenthal, S., and Kaufman, S. (1974). Vincristine neurotoxicity. *Ann. Intern. Med.* **80**, 733–734.

Rottenberg, D. A., *et al.* (1977). Cerebral necrosis following radiotherapy of extracranial neoplasms. *Ann. Neurol.* **1**, 339–357.

Rowinsky, E. K., *et al.* (1990). Taxol: a novel investigational antimicrotubule agent. *J. Natl. Cancer Inst.* **82**, 1247–1259.

Rubinstein, L. J., *et al.* (1975). Disseminated necrotizing leukoencephalopathy: a complication of treated central nervous system leukemia and lymphoma. *Cancer* **35**, 291–305.

Rudnick, S. A., *et al.* (1979). High dose cytosine arabinoside (HDARAC) in refractory acute leukemia. *Cancer* **44**, 1189–1193.

Russell, J. A., and Powles, R. L. (1974). Neuropathy due to cytosine arabinoside. *Br. Med. J.* **14**, 652–653.

Sahenk, Z., *et al.* (1987). Studies on the pathogenesis of vincristine-induced neuropathy. *Muscle Nerve* **10**, 80–84.

Saiki, J. H., *et al.* (1972). Paraplegia following intrathecal chemotherapy. *Cancer* **29**, 370–374.

Salinsky, M. C., *et al.* (1983). Acute cerebellar dysfunction with high-dose ARA-C Therapy. *Cancer* **51**, 426–429.

Sanderson, P. A., *et al.* (1976). Optic neuropathy presumably caused by vincristine therapy. *Am. J. Ophthal.* **81**, 146–150.

Sandler, S. G., *et al.* (1969). Vincristine-induced neuropathy: a clinical study of fifty leukemic patients. *Neurology* **90**, 367–374.

Sawicka, J., *et al.* (1977). Neurologic aspects of the treatment of cancer. In *Current Neurology* (H. R. Tyler, Ed.), Vol. 1, 301–305. Houghton Mifflin, Boston.

Schaeffer, S. D., *et al.* (1981). Cis-platinum vestibular toxicity. *Cancer* **47**, 857–859.

Scherokman, B., *et al.* (1985). Brachial plexus neuropathy following high-dose cytarabine in acute monoblastic leukemia. *Cancer Treat. Rep.* **69**, 1005–1006.

Schochet, S. S., *et al.* (1968). Neuronal changes induced by intrathecal vincristine sulfate. *J. Neuropath. Exp. Neurol.* **27**, 645–658.

Shalen, P. R., *et al.* (1981). Enhancement of the white matter following prophylactic therapy of the central nervous system for leukemia. *Radiology* **140**, 409–412.

Shapiro, W. R., and Green, S. B. (1987). Reevaluating the efficacy of intra-arterial BCNU. *J. Neurosurg.* **66**, 313–315. [Letter]

Shapiro, W. R., *et al.* (1973). Necrotizing encephalopathy following intraventricular instillation of methotrexate. *Arch. Neurol.* **28**, 96–102.

Shelanski, M. L., and Wisniewski, H. (1969). Neurofibrillary degeneration induced by vincristine therapy. *Arch. Neurol.* **20**, 199–206.

Shepherd, D. A., *et al.* (1978). Accidental intrathecal administration of vincristine. *Med. Pediatr. Oncol.* **5**, 85–88.

Shibutani, M., *et al.* (1989). Methotrexate-related multifocal axonopathy. *Acta. Neuropathol.* **79**, 333–335.

Siegal, T., and Haim, N. (1990). Cisplatin-induced peripheral neuropathy. *Cancer* **66**, 1117–1123.

Skullerud, K., and Halvorsen, K. (1985). Encephalomyelopathy following intrathecal methotrexate treatment in a child with acute leukemia. *Cancer* **42**, 1211–1215.

Slater, L. M., *et al.* (1969). Vincristine neurotoxicity with hyponatremia. *Cancer* **23**, 122–125.

Slevin, M. L., *et al.* (1983). Effect of dose and schedule on pharmacokinetics of high-dose cytosine arabinoside in plasma and cerebrospinal fluid. *J. Clin. Oncol.* **1**, 546–551.

Slyter, H., and Mason, R. (1980). Fatal myeloencephalopathy caused by intrathecal vincristine. *Neurology* **30**, 867–871.

Spiegel, R. J., *et al.* (1984). Treatment of massive intrathecal overdose by ventriculolumbar perfusion. *N. Engl. J. Med.* **311**, 386–388.

Stadnicki, S. W., *et al.* (1975). Cis-dichlorodiammineplatinum (II) (NSC-119875): hearing loss and other toxic effects in rhesus monkeys. *Cancer Chemother. Rep.* **59**, 467–480.

Stewart, D. J. (1991). Intra-arterial chemotherapy of primary and metastatic brain tumors. In *Neurological Complications of Cancer Treatment* (D. A. Rottenberg, Ed.), pp. 143–170, Butterworth-Heinemann, Boston.

Stewart, D. J., *et al.* (1982). Human central nervous system distribution of cis-diamminedichloroplatinum and use as a radiosensitizer in malignant brain tumors. *Cancer Res.* **42**, 2474–2479.

Stolinsky, D. C., *et al.* (1970). Clinical experience with procarbazine in Hodgkin's disease, reticulum cell sarcoma, and lymphosarcoma. *Cancer* **26**, 984–990.

Storm, A. J., *et al.* (1985). Effect of X-irradiation on the pharmacokinetics of methotrexate in rats: lateration of the blood-brain barrier. *Eur. J. Cancer Clin. Oncol.* **21**, 759–764.

Stuart, M. J., *et al.* (1975). Syndrome of recurrent increased secretion of antidiuretic hormone following multiple doses of vincristine. *Blood* **45**, 315–320.

Sullivan, M. P., *et al.* (1977). Combination intrathecal therapy for meningeal leukemia: two versus three drugs. *Blood* **50**, 471–479.

Suskind, R. M., *et al.* (1972). Syndrome of inappropriate secretion of antidiuretic hormone produced by vincristine toxicity (with bioassay of ADH level). *J. Pediatr.* **81**, 90–92.

Ten Hoeve, R. F., and Twinjnstra, A. (1988). A lethal neurotoxic reaction after intraventricular methotrexate administration. *Cancer* **62**, 2111–2113.

Thant, M., *et al.* (1982). Possible enhancement of vincristine neuropathy by VP-16. *Cancer* **49**, 859–864.

Thompson, S. W., *et al.* (1984). Cisplatin neuropathy: clinical, electrophysiologic, morphologic, and toxicologic studies. *Cancer* **54**, 1269–1275.

Tobin, W., and Sandler, S. G. (1968). Neurophysiologic alterations induced by vincristine (NSC-67574). *Cancer Chemother. Rep.* **52**, 519–526.

Van der Hoop, R. G., *et al.* (1990). Prevention of cisplatin neurotoxicity with an ACTH (4-9) analogue in patients with ovarian cancer. *N. Engl. J. Med.* **322**, 89–94.

Ventafridda, V., *et al.* (1994). On the significance of Lhermitte's sign in oncology.

Walker, R. W., *et al.* (1986). Transient cerebral dysfunction secondary to high-dose methotrexate. *J. Clin. Oncol.* **4**, 1845–1850.

Walker, R. W., *et al.* (1988). Cerebral herniation in patients receiving cisplatin. *J. Neurooncol.* **6**, 61–65.

Walker, R. W., *et al.* (1988). Complications of intra-arterial (IA) BCNU in the treatment of malignant gliomas. *Proceed ASCO* **7**, 84.

Walsh, T. J., *et al.* (1982). Neurotoxic effects of cisplatin therapy. *Arch. Neurol.* **39**, 719–720.

Walsh, T. J., *et al.* (1982). Neurotoxic effects of cisplatin therapy. *Arch. Neurol.* **39**, 719–720.

Walther, P. J., *et al.* (1987). The development of Lhermitte's sign during cisplatin neurotoxicity: possible drug-induced toxicity causing spinal cord demyelination. *Cancer* **60**, 2170–2172.

Wasserstrom, W. R., *et al.* (1982). Diagnosis and treatment of leptomeningeal metastases from solid tumors: experience with 90 patients. *Cancer* **49**, 759–772.

Watkin, S. W., *et al.* (1989). Ifosfamide encephalopathy: a reappraisal. *Eur. J. Cancer Clin. Oncol.* **25**(9), 1303–1310.

Watkins, S. M., and Griffin, J. P. (1978). High incidence of vincristine-induced neuropathy in lymphomas. *Br. Med. J.* **1**, 610–612.

Watson, P. R., *et al.* (1985). Severe central nervous system toxicity from high-dose cytarabine: expressive aphasia occurring after the second day of treatment. *Cancer Treat. Rep.* **69**, 313–314.

Weiden, P. L., and Wright, S. E. (1972). Vincristine neurotoxicity. *N. Engl. J. Med.* **286**, 1369–1370.

Weiss, H. D., *et al.* (1974). Neurotoxicity of commonly used antineoplastic agents. *N. Engl. J. Med.* **291**, 75–81, 127–133.

Weiss, R. B. (1991). Ifosfamide vs cyclophosphamide in cancer therapy. *Oncology* **5**, 67–76.

Whitelaw, D. M., *et al.* (1963). Clinical experience with vincristine. *Cancer Chemother. Rep.* **30**, 13–20.

Whittaker, J. A., and Griffith, I. P. (1977). Recurrent laryngeal nerve paralysis in patients receiving vincristine and vinblastine. *Br. Med. J.* **1**, 1251–1252.

Whittaker, J. A., *et al.* (1973). Coma associated with vincristine therapy. *Br. Med. J.* **4**, 335–337.

Wilding, G., *et al.* (1985). Retinal toxicity after high-dose cisplatin therapy. *J. Clin. Oncol.* **3**, 1683–1689.

Williams, M. E., *et al.* (1983). Ascending myeloencephalopathy due to intrathecal vincristine sulfate. *Cancer* **51**, 2041–2047.

Wilson, G. H., *et al.* (1972). Atrophy following radiation therapy for central nervous system neoplasms. *Acta Radiol.* **11**(Suppl), 361.

Wilson, W. B., *et al.* (1987). Sudden onset of blindness in patients treated with oral BCNU and low-dose cranial irradiation. *Cancer* **59**, 901–907.

Winkelman, M. D., and Hines, J. D. (1983). Cerebellar degeneration caused by high-dose cytosine arabinoside: a clinical pathological study. *Ann. Neurol.* **14**, 520–527.

Wolff, L., *et al.* (1979). Paraplegia following intrathecal cytosine arabinoside. *Cancer* **43**, 83–85.

Wolff, S. N., *et al.* (1983). High-dose VP-16-213 and autologous bone marrow transplantation for refractory malignancies: a phase 1 study. *J. Clin. Oncol.* **1**, 701–705.

Yim, Y. S., *et al.* (1991). Hemiparesis and ischemic changes of the white matter after intrathecal therapy for children with acute lymphocytic leukemia. *Cancer* **67**, 2058–2061.

Young, D. F. (1982). Neurological complications of cancer chemotherapy. In *Neurological Complications of Therapy* (A. Silverstein, Ed.), Mount Kisco: Futura Publishing, Mount Kisco, New York.

Young, D. F., and Posner, J. B. (1980). Nervous system toxicity of chemotherapeutic agents. In *Handbook of Clinical Neurology* (P. J. Vinken, and G. W. Bruyn, Eds.), Vol. 39, pp. 91–129, North Holland, New York.

Zahanko, D. S., and Dedirck, R. L. (1977). Antifolates: in vivo kinetic considerations. *Cancer Treat. Rep.* **61**, 513–518.

Zilkha, A., and Daiz, A. S. (1980). Computed tomography in the diagnosis of superior sagittal sinus thrombosis. *J. Comp. Assist. Tomogr.* **4**, 124–126.

Zimm, S., *et al.* (1984). Cytosine arabinoside cerebrospinal fluid kinetics. *J. Clin. Pharmacol. Ther.* **35**, 826–830.

CHAPTER

43

Neurotoxins and Neurodegenerative Diseases

A. C. LUDOLPH
Department of Neurology
Universitätsklinikum Charité
Medizinische Fakultät der
Humboldt-Universität zu Berlin
Neurologische Klinik und Poliklinik
10098 Berlin
Germany

I. Introduction

The use of industrial and naturally occurring neurotoxins has become a significant tool in studying pathogenetic mechanisms of neurodegeneration in the central nervous system (CNS). This chapter tries to show how clinical observations led to the study of toxic compounds in model systems. These studies have given clues to the understanding of endogenous mechanisms of neurodegeneration in some sporadic and inherited neurological diseases. This chapter also demonstrates that there is no present evidence which supports the assumption that exogenous chemicals play an *etiological* role in *progressive sporadic* neurodegenerative diseases.

Since toxic mechanisms of neurodegeneration are presently a matter of major interest, this chapter can only cover the subject partially. Therefore, it will be necessary to focus on compounds, mechanisms, and diseases which seem to be of major relevance for our understanding of human neurological diseases. The first part of this chapter discusses the possible impact of neurotoxic model systems for sporadic neurodegenerative diseases such as parkinsonism, hyperkinetic extrapyramidal movement disorders, and motor neuron diseases. In the second part of the chapter, some inherited metabolic diseases are briefly discussed which are presumably related to an endogenous neurotoxic mechanism as part of their pathogenesis.

II. Exogenous Toxins

A. MPTP and Parkinsonism

In the early 1980s it was reported that the meperidine analog 1-methyl-4-phenyl-1,2,3,6-tetrahydropyridine (MPTP) induces a nonreversible clinical picture in humans which closely resembles parkinsonism (Davis *et al.*, 1979; Langston *et al.*, 1983; Ballard *et al.*, 1985). Although first produced much earlier as an impurity and by-product of 1-methyl-4-phenyl-4-propionoxypiperidine synthesis (Langston *et al.*, 1989), only its self-administration as a heroin analog by drug addicts revealed the highly selective neurotoxic effect. Classical features of idiopathic Parkinson's disease such as rigidity, tremor, bradykinesia, and postural instability are present in patients intoxicated with MPTP (Ballard *et al.*, 1985). The apparent lower incidence of true resting tremor in MPTP parkinsonism remains a disparity, although it may be partially explained by age differences between patients who develop Parkinson's disease and MPTP-induced parkinsonism, respectively (Tetrud *et al.*, 1989). MPTP-induced parkinsonism clearly responds to L-Dopa treatment and the complications of therapy mirror those encountered in Parkinson's disease (Bloem *et al.*, 1990).

Neuropathological studies of the first human case of likely MPTP exposure, a 23-year-old patient, showed selective degeneration of dopaminergic cells of the pars compacta of the substantia nigra. In contrast to the neuropathology of idiopathic Parkinson's disease, other parts of the brain were largely spared (Davis *et al.*, 1979). Lewy bodies—the diagnostic hallmark of the neuropathology of idiopathic Parkinson's disease—have not yet been observed in humans exposed to MPTP. The rate of progression of the disease does not seem to be directly comparable to idiopathic Parkinson's disease (Tetrud *et al.*, 1989), although definite results of a long-term study are not available yet.

Major parts of the mechanisms underlying MPTP-induced neurotoxicity can be elucidated. The liphophilic compound readily crosses the blood–brain barrier. In rat brain preparations, Chiba *et al.* (1984) showed that oxidation of MPTP to 1-methyl-4-phenyl-2,3-dihydropyridinium ($MPDP^+$) and 1-methyl-4-phenylpyridinium (MPP^+) by monamine oxidase-B (MAO-B) is required for the development of nigral toxicity. Although monamine oxidase-A (MAO-A) also oxidizes MPTP (Chiba *et al.*, 1984; Salach *et al.*, 1984), only the B form plays a significant role in its bioactivation (Salach *et al.*, 1984; Singer *et al.*, 1985,1986) and only selective blockers of MAO-B prevent the neurotoxicity of MPTP (Heikkila *et al.*, 1984; Langston *et al.*, 1984; Markey *et al.*, 1984). The finding that MPP^+ accumulates in the primate brain supports the conclusion that MPP^+ is the metabolite mainly responsible for the neurotoxicity of MPTP (Markey *et al.*, 1984; Johannessen *et al.*, 1985; Irwin and Langston, 1984). In cultured cells, MPP^+ is toxic to dopaminergic neurons in mesencephalic explants (Mytilineou and Cohen, 1984; Mytilineou *et al.*, 1985; Sanchez-Ramos *et al.*, 1986), PC12 phaechromocytoma cells (Marongiu *et al.*, 1988), and hepatocytes (Di Monte, 1986a,b, 1987).

Since dopaminergic neurons lack or have little MAO-B activity (Westlund *et al.*, 1985; Konradi *et al.*, 1987), oxidation of the protoxin must take place at sites outside the target cells. Local astrocytes seem to be responsible for MPP^+ formation (Ransom *et al.*, 1987; Brooks *et al.*, 1989; Takada *et al.*, 1990) and a contribution of serotonergic neurons is possible (Brooks *et al.*, 1988). After biotransformation, the lipid-soluble MPP^+ reaches the extracellular space and is taken up by nerve terminals of the nigrostriatal neurons (Javitch and Synder, 1984). This uptake is selectively inhibited by the dopamine uptake blocker mazindol (Javitch *et al.*, 1985). The dopamine reuptake blocker nomifensine prevents the neurotoxic effect of MPTP in primates (Schultz *et al.*, 1986). Taken together, these data indicate that MPP^+ is actively transported into dopaminergic neurons. This effect may be partly responsible for the selective vulnerability of nigrostriatal neurons; however, MPP^+ shows an equally high affinity for other catecholamine uptake systems (Brooks *et al.*, 1988; Snyder and D'Amato, 1986). Finally, MPP^+ is accumulated by mitochondria as a result of the electrochemical gradient across their membrane (Ramsay *et al.*, 1986a,b,c). This uptake is potentiated by malate and glutamate and is reduced by respiratory inhibitors (Ramsay, 1986c). In mitochondria, MPP^+ inhibits NAD(H)-linked electron transport at the level of complex I at millimolar concentrations *in vitro* (Nicklas *et al.*, 1985; Mizuno *et al.*, 1987a), resulting in depletion of cellular ATP (Mizuno *et al.*, 1987b; Scotcher *et al.*, 1990). Studies show that cellular energy deficits are associated with an increased vulnerability to glutamate and its analog ("excitotoxicity") (Henneberry *et al.*, 1989) and therefore an excitotoxic mechanism might explain the histological features of the acute neurotoxic lesion (Turski *et al.*, 1991; Storey *et al.*, 1992). There is some speculation that free radicals might also be a major part of the pathogenetic mechanism induced by MPP^+. However, in contrast to experimental treatment with inhibitors of MAO-B, dopamine uptake blockers, and antagonists to glutamate receptors, therapeutic attempts with antioxidants, scavengers of free radicals, or manipulation of glutathionine levels did not have a convincing neuroprotective effect (Baldessarini *et al.*, 1986; Martinovits *et al.*, 1986; Perry *et al.*, 1986,1987) or even exacerbated MPTP neurotoxicity (Corsini *et al.*, 1985). Since the neurotoxic effect of MPTP in various species shows only a weak relation to the concentration of neuromelanin in the substantia nigra, early suggestions of a major role of melanin in MPTP toxicity are becoming unlikely.

Since the biochemical mechanism of MPTP neurotoxicity promised to yield insight into the etiopathogenesis of Parkinson's disease, a number of animal models were studied. A detailed discussion is beyond the scope of this chapter (see Zigmond and Stricker, 1989). The neurotoxic effect of MPTP shows a major variation between species, but also susceptibility differs between individual animals. Nonhuman primates are more susceptable to toxic effects than rodents. Neurotoxicity increases with age of the animal, an observation which is of obvious interest for the human disease. Explanations for this phenomenon still remain hypothetical (Langston *et al.*, 1989), but some authors favor the idea that the toxification process of MPTP to MPP$^+$ by MAO-B may be a major factor since MAO-B activity also increases with age (Strolin-Benedetti and Dostert 1989; Melamed *et al.*, 1990). After treatment with MPTP, nonhuman primates exhibit many features of human parkinsonism, particularly akinesia and rigidity. These symptoms are most similar to the consequences of the intoxication in humans, confirming the impression that resting tremor is only a minor feature of MPTP neurotoxicity. The most significant difference between MPTP neurotoxicity in humans and in experimental animals is the observation that the latter is always transient. The distribution of histopathological lesions induced in experimental animals also differs from the neuropathological findings in the only human case. In this *young* patient, lesions were confined to the substantia nigra, but in nonhuman primates, in particular in old animals, neuronal degeneration of the locus coeruleus and the ventral tegmental area (Mitchell *et al.*, 1985; Forno *et al.*, 1986) is part of the picture. Also, eosinophil intraneuronal inclusion bodies were described (Forno *et al.*, 1986). Neurochemically, MPTP induces severe reductions of biogenic amines in the striatum, but also in some extrastriatal regions like the nucleus accumbens and the ventral tegmental area (Ueki *et al.*, 1989; Pifl *et al.*, 1990; Russ *et al.*, 1991).

Despite some shortcomings, the MPTP model is the best available current model for Parkinson's disease (Zigmond and Stricker, 1989). Although progression similar to patients suffering from Parkinson's disease has not been claimed for patients suffering from MPTP intoxication, this model has given interesting insights into possible mechanisms of its prevention and symptomatic treatment. From the clinical viewpoint the most important contributions might have been the discovery that the activity of complex I is also decreased in patients with Parkinson's disease (Mizuno *et al.*, 1989). Of major interest is the development of experimental preventional strategies such as treatment with MAO-B inhibitors and glutamate antagonists, although their potential impact on the progression of sporadic parkinsonism remains hypothetical at the present time. The intensive search for compounds structurally related to MPTP which might play a role in the etiology of idiopathic Parkinson's disease remains unsuccessful. Until now therefore the observations on the exogenous neurotoxin MPTP did not increase in likelihood that similar exogenous factors contribute to the etiology of the idiopathic form of Parkinson's disease; however, the interest in this model and speculation on the relation of exogenous factors to this chronic progressive disease remain.

B. Mitochondrial Toxins and Basal Ganglia Necrosis

The selectivity of inhibition of mitochondrial metabolism induced by MPP$^+$ in substantia nigra cells seems to be dependent from its selective uptake by dopaminergic terminals. In contrast, some other mitochondrial toxins seem to inhibit chemical energy production of the brain nonselectively. It is a common effect of these neurotoxins that they induce symmetric lesions of the basal ganglia similar to those observed in some inherited metabolic diseases such as methylmalonic acidemia or glutaric aciduria type I (Aicardi *et al.*, 1985) or in hypoglycemic encephalopathies (Jellinger, 1986). Compounds of clinical interest are the industrial toxins cyanide and methanol, the plant and mycotoxin 3-nitropropionic acid, and, in an apparently more complex interaction, carbon monoxide. Cyanide is a direct inhibitor of complex IV of the mitochondrial chain whereas methanol produces its toxic effect after biotransformation to formic acid by alcohol dehydrogenase. Formic acid inhibits complex IV of the mitochondrial chain (Nicholls, 1975,1976). 3-Nitropropionic acid inhibits succinate dehydrogenase (a part of complex II) and carbon monoxide inhibits complex IV, although other toxic mechanisms complicate the picture. In a similar way, disulfiram intoxication induces a pallidoputaminal lesion in humans; a syndrome possibly due to the toxicity of its metabolite carbon disulfide (Rainey, 1977; Krauss *et al.*, 1991).

Experimental striatal necrosis in rodents and primates was obtained after metabolic inhibition by carbon disulfide (Richter, 1945), cyanide (Hurst, 1942; Hicks, 1950a), malonitrile (Hicks, 1950b), methanol (Potts *et al.*, 1955), sodium azide (Környvey, 1963; Miyoshi, 1967; Mettler, 1972), and 3-nitropropionic acid (Hamilton and Gould, 1987a,b). Chronic intoxication of monkeys with sodium azide, a potent inhibitor of complex IV, resulted in putaminal lesions which later also involved parts of the pallidum and thalamus (Mettler, 1972). Hyperkinesias accompanied these structural deficits (Mettler, 1972). In addition to effects on the retina, experimental administration of methanol to rhesus monkeys produced extensor rigidity and tremor associated with striatal lesions (Potts *et al.*, 1955); in these experiments treatment of systemic acidosis reportedly did not prevent basal ganglia damage. Putaminal necrosis induced by 3-nitropropionic acid was associated with hindlimb rigidity in rats and was not accompanied by systemic hypotension but a severe metabolic acidosis (Hamilton and Gould, 1987a,b; A. C. Ludolph et al., unpublished).

In humans, the toxic effects of mitochondrial poisons usually result in a pattern of symmetric basal ganglia damage. Two major clinical–anatomical syndromes can be distinguished. A dystonic picture, which may be accompanied by other features of a hyperkinetic syndrome, is ascribed by most authors to selective lesions of the putamen (Hawker and Lang, 1970). The pathogenetic relation of the lesion to the motor deficit is unexplained. Energy-depleting toxins can also induce anatomic lesions of the globus pallidus. This morphological alteration is often accompanied by an akinetic-rigid syndrome commonly explained by a lesion of the cortico–striato–pallido–thalamic–cortical loop which results in thalamic dysinhibition. The resulting parkinsonian syndrome poorly responds to L-Dopa, indicating a minor contribution of dopaminergic deficits. In some patients, lesions of putamen and globus pallidus may overlap and the pattern of clinical deficits may not consistently show the clinical–anatomical relation commonly seen.

Extrapyramidal deficits associated with bilateral putaminal lesions in the computerized tomography (CT) or magnetic resonance imagery scans are a possible consequence of methanol poisoning if the intoxication is prolonged. In the majority of patients, rigidity and akinesia, sometimes associated with dystonia, were described (Orthner, 1950; LeWitt et al., 1988). Spasticity and anterior horn cell necrosis are present in single cases (McClean et al., 1980; Guggenheim et al., 1971). Morphologically, 41 out of a large series of 124 patients showed symmetric necrosis predominantly of the lateral, basal, and caudal parts of the putamen (Orthner, 1950).

Cyanide intoxication also induces a, usually delayed, extrapyramidal deficit (Finelli, 1981; Uitti et al., 1985; Messing and Storch, 1988; Rosenberg et al., 1989; Grandas et al., 1989) which in some cases can be dominated by a Dopa-sensitive dystonic syndrome (Valenzuela et al., 1992). In a 6-fluorodopa positron emission tomography (PET)-scan study, Rosenberg and colleagues (1989) showed that the pattern of deficits was not only related to morphological changes in the putamen, but was also similar to that found in idiopathic Parkinson's disease. In contrast, the morphological integrity of the pigmented neurons of the zona compacta was preserved (Uitti et al., 1985). In an unknown proportion of 884 identified patients who were intoxicated by the mycotoxin 3-nitropropionic acid, a permanent dystonia with athetoid movements developed 2 weeks after patients had recovered from the initial acute encephalopathy. On CT scans these patients had bilateral

putaminal necrosis (He et al., 1990; Ludolph et al., 1991a).

Since the selective pattern of damage induced by metabolic toxins does not convincingly relate to vascular territories, metabolic and physiochemical characteristics of the target cell themselves ("pathoclisis") may explain the pattern of vulnerability (Vogt and Vogt, 1920,1922). Regional biochemical factors of possible impact seem to be local energy requirements, reserve capacities of energy-producing enzymes, metabolic acidosis, and the distribution of receptors which mediate cell injury. Some of these factors are age dependent. Although activation of excitatory amino acid receptors is part of the pathogenesis of lesions induced by cellular energy depletion associated with the application of mitochondrial poisons (Henneberry et al., 1989; Beal, 1992), the distribution of these receptors in the basal ganglia does not mirror the pattern of vulnerability induced by mitochondrial toxins (Hawker and Lang, 1990). But since energy-depleted cells can be partially protected by antagonists to glutamate (Turski et al., 1991; Storey et al., 1992; Ludolph et al., 1992), this approach may improve future therapeutic intervention. Further elucidation of the pathogenesis of basal ganglia necrosis after alterations of cerebral energy metabolism requires more sophisticated in vivo techniques than previously employed. This includes the development of model systems which monitor the impact of factors such as brain edema, hypotension, acidosis, and seizures. These factors may contribute to the characteristic pattern of vulnerability to a varying degree dependent from the individual toxin and its dosage.

C. Neurolathyrism and Neurocassavaism

Neurolathyrism is a neurotoxic disease that is still a significant cause of spastic paraparesis in parts of the Third World (Spencer and Schaumburg, 1983). When men are forced to rely on staple foods during times of drought and flood, heavy continuous consumption of Lathyrus spp. leads to the development of a stereotyped nonprogressive deficit of the pyramidal tract with prominent features of spasticity. Lathyrism has historically been reported in various regions of southern Eurasia and Africa, although recent outbreaks were confined to parts of Bangladesh, India, and Ethiopia (Ludolph et al., 1987; Haimanot et al., 1990). In Ethiopia, prevalence may reach 0.6–2.9% whereas in Bangladesh and India it may reach 0.3–2.5% (Ludolph et al., 1987; Haimanot et al., 1990).

The disease was already known to Hippocrates (1846) and the name lathyrism was first used by Cantani in the 19th century. In modern times the detailed natural course of neurolathyrism has been documented by the Israelian physician Kessler (1947). His work is based on the experience with an outbreak of neurolathyrism among 1400 prisoners of war in a forced labor camp of the German occupant troops in Wapniarka (Ukraine) during the World War II in 1942. Kessler himself was a victim of the intoxication, and 40 years later his clinical description is still complementary to modern understanding of human neurolathyrism. Three months after a continuous consumption of a daily ration of 400 g of *Lathyrus sativus* and 200 g of bread, 800 of 1400 patients developed the disease. Mostly the malnourished were affected but those who were taken from their homes in good health also experienced the typical sequence of symptoms: They initially developed global tonic cramps which predominantly affected the calf muscles. These cramps are characteristically provoked by coldness, physical exhaustion, and infections. Other acute and reversible signs and symptoms are urgency and frequency of urination, weakness, and pain and paresthesias of the extremities. Less commonly observed are myoclonic-like movements and short-term memory deficits (Kessler, 1947; Mertens, 1947; Ludolph *et al.*, 1987). Typically, then the permanent spastic paraparesis develops within days but may also appear insidiously over months. The clinical picture reportedly more often develops in males and consists of leg weakness, spontaneous cloni and bilaterally increased tendon jerks, a positive Babinski, and an increase of extensor tone to stretch of adductor and gastrocnemius muscles. This abnormal muscle tone forces the patients to walk with a scissoring gait on the balls of their feet with the knees slightly flexed. The most severely affected may also show abnormalities of the motor system supplying arm and bulbar muscles. Although described in individual patients, long-term deficits of the sensory and cerebellar system and sphincter dysfunction are seldom seen. Some patients were also studied 30–50 years after the acute intoxication; although some progression was seen in the peripheral motor system (Cohn *et al.*, 1977; Cohn and Streifler, 1981a,b) or individual patients complained about an increase of increasing spasticity in old age (S. Gimenez-Roldan *et al.*, unpublished manuscript), the resulting clinical picture was clearly distinct from sporadic amyotrophic lateral sclerosis (ALS).

In a long-standing case of lathyrism, Streifler and colleagues (1977) had neuropathological evidence of subclinical involvement of anterior horn cells. Studies of the CNS neuropathology of lathyrism are extremely rare; in 1926, Filimonoff emphasized that depletion of Betz cells in the motor cortex and degeneration of corticospinal and spinocerebellar tracts and dorsal columns are features of the pattern of vulnerability. The severe affection of corticospinal tracts in neurolathyrism has been documented by an electrophysiological study using magnetic cortex stimulation (Hugon *et al.*, 1990). In contrast, studies of somatosensory-evoked potentials after tibial nerve stimulation revealed only comparatively minor deficits in the central portion of the afferent pathway which were related to the severity of the disease (Hugon *et al.*, 1990).

The neurotoxicology of human neurolathyrism was previously confused by the discussion of compounds which cannot be detected in *L. sativus* and induce a pattern of vulnerability distinct from the human disease (Spencer and Schaumburg, 1983). The most prominent examples are β-aminopropionitrile, which induces osteolathyrism and only secondary changes of the spinal cord (Spencer *et al.*, 1984; Roy and Spencer, 1989), and β, β-iminodipropionitrile (Selye, 1957; Roy and Spencer, 1989), which is not present in *L. sativus* and experimentally induces a pattern of vulnerability which is distinct from neurolathyrism (Hugon *et al.*, 1988). An attempt has been made repeatedly to produce an animal model to improve the understanding of the pathogenesis of the disease (Spencer and Schaumburg 1983). The feeding of six *Macaca fascicularis* with *L. sativus* and a mineral- and vitamin-supplemented diet over 18 months produced a clinical picture which consisted of stimulus-sensitive myoclonic jerks, increased muscle tone, and weakness of the hindlimbs (Spencer *et al.*, 1986; Hugon *et al.*, 1988). Interindividual differences of susceptibility to the neurotoxic effect were observed and some of the signs of intoxication were reversible. Neuropathological examination of the tissue did not reveal major structural abnormalities, but clinical neurophysiological studies showed an impairment of corticospinal tract function. It was concluded that the early changes of human lathyrism were reproduced in this model system. Similar changes were produced after oral administration of the excitatory amino acid and the glutamate analog β-oxalylamino-L-alanine (BOAA) over weeks. The clinical picture, in particular the increase of muscle tone and the extensor posturing of the lower limbs, was comparable to the features of the intoxication induced by *L. sativus*, but clinical electrophysiological changes were present at a comparatively smaller degree. Morphological changes were also absent.

It seems that BOAA is the major candidate toxin associated with the etiology of neurolathyrism. This convulsant is the only neurotoxic chemical identified in the neurotoxic species. *L. sativus, cicera,* and *clymenum* reach concentrations approaching 1%. BOAA has biochemical and electrophysiological properties of an excitant excitatory amino acid and glutamate analog, selectively activates the α-amino-3-hydroxy-5-

methyl-4-isoxazole propionic acid (AMPA) subtype of ionotropic glutamate receptors in hippocampal neurons, and induces excitotoxic damage in mouse cortical neurons *in vitro* (Nunn *et al.*, 1987; Ross *et al.*, 1987). The neurotoxic effect is blocked by antagonists to the non-*N*-methyl-D-aspartate (non-NMDA) type of excitatory amino acid receptors (Ross and Spencer, 1987; Bridges *et al.*, 1989). Binding studies revealed that the mouse spinal cord contains only a few AMPA-binding sites and that lower concentrations of BOAA are required to replace AMPA-binding sites in the cortex than in the striatum and cerebellum and spinal cord (Ross *et al.*, 1989). Facilitation of presynaptic glutamate release (Gannon and Terrian, 1989), toxic effects on glial cells (Bridges *et al.*, 1991), and blocking of sodium-dependent glutamate transport (Ross *et al.*, 1985) may also contribute to the excitotoxic effect of BOAA.

Although BOAA appears to be the principal factor, other greatly unexplored risk factors may increase individual susceptibility. The male preponderance remains unexplained and it is unknown why young men in their twenties and thirties are the most vulnerable subjects. Since lathyrism occurs in patients suffering from food shortages, the role of malnutrition must also be defined. Further work is needed to explain the sudden onset of the disease which is sometimes associated with fever or hard physical work after months of toxic exposure. This presentation may simply be an expression of the cumulative effect of the toxin, but other factors may also have a significant role. Also, the appearance of neurolathyrism in a cluster is not necessarily an expression of common habits but may be indicative of genetically based differences in the metabolism and excretion of BOAA.

The results in experimental primates are similar to those obtained by others after feeding *L. sativus* under presumably less controlled conditions (Spencer and Schaumburg, 1983; Roy and Spencer, 1989). Also, previously performed animal experiments revealed major species differences (Spencer *et al.*, 1984), and it must be concluded that a satisfactory animal model for the chronic disease which shows clear-cut neuropathological changes under strictly controlled conditions (which includes definite absence of malnutrition) is not present.

It is another long-term goal of the investigations of human neurolathyrism to develop a toxin-free strain of *L. sativus*. However, before such a safe strain could reduce malnutrition in parts of the world traditionally reliant on *L. sativus* as a subsistence and emergency food, major open questions must be resolved concerning the biosynthesis and metabolic fate of BOAA in the plant. Possibly, elimination of BOAA from the plant

alone may not be sufficient for prevention of the disease. A better understanding of neurolathyrism and the mechanism leading to the clinical disease would have an impact on our understanding of the pathogenesis of clinically related diseases such as hereditary spastic paraplegia.

A clinical picture remarkably similar to neurolathyrism is associated with prolonged consumption of the bitter cassava, *Manihot esculenta*. The disease Konzo (Howlett *et al.*, 1989) is prevalent among rural populations in sub-Saharan Africa where the cassava root is used as staple food during times of drought. Outbreaks were reported from Mozambique (Ministry of Health, 1984), Tansania (Howlett *et al.*, 1989; Mlingi *et al.*, 1991), and Zaire (Trolli, 1938; Lucasse, 1952; Carton *et al.*, 1986; Tylleskär *et al.*, 1991) where local prevalence may reach 30 per 1000 (Rosling, 1990). The clinical picture consists of a sudden onset, symmetrical and permanent, but nonprogressive spastic paraplegia. As in lathyrism, males are predominantly affected. In contrast, hearing loss, visual impairment, and dysarthria are sometimes described as a feature of cassavaism but were not seen in neurolathyrism (Howlett *et al.*, 1989).

The roots and leaves of bitter varieties of cassava contain high cyanide levels which are bound in the form of the glycoside linamarin (Balagopalan *et al.*, 1988). Detoxification of these bitter varieties is necessary for consumption. Intensive soaking of roots removes cyanogenic glycosides, but in times of food shortage this procedure is often shortened and significant cyanogen concentrations remain in the flour (Tylleskär *et al.*, 1992). In humans, cyanide is converted to the less toxic compound thiocyanate. Affected patients have increased urinary thiocyanate and high blood cyanide levels (Tylleskär *et al.*, 1992). Since excessive intake of bitter cassava is often accompanied by low intake of sulfur amino acids because of a low protein diet (Lancaster *et al.*, 1982), sulfur deficiency presumably leads to poor detoxification of cyanide via thiocyanate formation (Cliff *et al.*, 1985) and may be a contributing or even necessary cofactor for motor neuron toxicity. Previous studies of the etiopathogenesis of cassavaism have been hampered by the lack of an adequate animal model. A priori, a disease that is associated with an increased intake of a compound which induces energy dysmetabolism by inhibiting complex IV would be expected to induce basal ganglia damage instead of being associated with outbreaks of spastic paraplegia. Possible explanations exist (Howlett *et al.*, 1989; Spencer *et al.*, 1992) but they require experimental confirmation in model systems.

The intriguing clinical similarity of the cassava-associated disease Konzo and neurolathyrism suggests that the gaps in our knowledge about their pathogenesis may have complementary aspects. Therefore, future research may have an advantage from comparative studies. Since neurolathyrism and Konzo permit the study of interesting model systems for upper motor neuron degeneration associated with prominent features of spasticity, previous and future results of clinical and experimental studies may have an impact on our understanding of the pathogenesis and future treatment of other human motor neuron diseases (Choi, 1992).

D. Domoic Acid

In November 1987, at least 145 patients in Canada developed an intoxication after consumption of cultured blue mussels (*Mytilus edilis*) containing high concentrations of the potent excitatory tricarboxylic amino acid domoic acid (Perl *et al.*, 1990). Acute neurological consequences of domoic acid intoxication included gastrointestinal distress, altered states of arousal, including coma, ophtalmoplegia, limbic seizures, and myoclonus (Teitelbaum *et al.*, 1990). The chronic neurological picture was characterized by selective impairment of memory and distal atrophy and by motor weakness usually not accompanied by sensory deficits (Teitelbaum *et al.*, 1990). Neurological complications were more frequently seen in elderly compared with younger patients. Two to 3 months after the intoxication PET studies in 4 subjects revealed a decrease of glucose metabolism in the structures of the mesial temporal lobe which were correlated with the memory scores of these patients. Electrophysiological studies clearly showed the presence of a motor neuronopathy or axonopathy. Neuropathological studies in 4 patients demonstrated a marked loss of neurons in the hippocampus and amygdaloid nucleus. In addition, lesions were present in the claustrum, secondary olfactory areas, the septal area, the nucleus accumbems septi, the dorsal medial nucleus of the thalamus, and the insular and subfrontal cortex. The pattern of hippocampal damage seemed to be similar to the abnormalities induced by the excitotoxin kainate in the same anatomical structures of experimental rodents (Teitelbaum *et al.*, 1990).

Electrophysiological (Biscoe *et al.*, 1975; Debonnel *et al.*, 1989; Stewart *et al.*, 1990) and receptor-binding studies (Slevin *et al.*, 1983) revealed that domoic acid is a potent agonist at the kainate glutamate receptor subtype. The neurotoxicity of domoic acid may be caused by a direct effect on kainate receptors but an excessive release of endogenous glutamate through the activation of a presynaptic kainate receptor may also play a role (Stewart *et al.*, 1990). A subcutaneous administration of domoic acid to adult rats resulted in a motor-behavioral syndrome similar to that induced by kainic acid (Stewart *et al.*, 1990). Neuropathological studies indicated the presence of degenerative changes in several amygdaloid nuclei, the hippocampus and subiculum, the olfactory bulb, the piriform and entorhinal cortices, the lateral septum, and the arcuate and middle dorsal thalamic nuclei (Tryphonas *et al.*, 1990a). In monkeys, excitatory damage was predominantly observed in the hypothalamus, hippocampus, and area postrema (Tryphonas *et al.*, 1990b). A kainate pattern of neuronal degeneration was observed in retinal explants treated with domoic acid (Stewart *et al.*, 1990).

In summary, domoic acid is a naturally occurring excitatory amino acid that seems to induce a pattern of acute neurobehavioral changes and permanent neurological deficits which resemble changes induced experimentally by kainic acid. Nothing is known about possible chronic neurotoxicity of poisoning by compounds similar to domoic acid and factors possibly potentiating or influencing their effect.

E. ALS/PD in the Western Pacific

Based on epidemiological criteria, the amyotrophic lateral sclerosis/parkinsonism–dementia complex (ALS/PD) in the Western Pacific is considered a classic example for environmentally induced neurodegeneration (Kurland, 1988). The disease occurred among three geographically and culturally separated population groups (Garruto and Yase, 1986): the Chamorro people of the Mariana islands of Guam and Rota (Elizan *et al.*, 1966), the Auyu and Jaqai of the southern lowlands of Irian Jaya (Gajdusek, 1963), and the Japanese residents of the Kii peninsula of the Honshu island (Yase, 1970). In all three foci, the incidence of ALS/PD is declining or the disease has disappeared among young individuals (Yase, 1972; Garruto *et al.*, 1985; Rodgers-Johnson *et al.*, 1986; Ludolph *et al.*, 1991). Migration studies on Gaum show that after immigration adult Filipinos who adopted the Chamorro life-style developed the clinical disease after 17 to 26 years of residency (Garruto *et al.*, 1981). After migration to the continental United States, Chamorros developed the disease up to 34 years after leaving the Marianas (Garruto *et al.*, 1980). A similar observation has been made in Irian Jaya (Gajdusek and Salazar, 1982). Shiraki and Yase (1975) reported that Japanese and Guamanian ALS patients showed ectopic and multinucleated cerebellar and other neurons indicating the presence of a pathogenetic factor during development of the central nervous system. Taken together, the data suggest that there is evidence that ALS/PD is disappearing and that this progressive disease develops after a latency.

The major clinical features of ALS/PD consist of motor neuron dysfunction, parkinsonism, dementia, and supranuclear palsy whereas the pathological picture is dominated by Alzheimer-like neurofibrillary degeneration (Hirano *et al.*, 1961). ALS/PD does not seem to be inherited (Kurland *et al.*, 1982), and a viral infection is virtually excluded by transmission studies (Gibbs and Gajdusek, 1982). Therefore, there is agreement that a nontransmissible enviornmental factor, which disappears with the modernization of the three foci, causes the disease. Earlier suggestions that changes in mineral metabolism, in particular aluminum, play a significant etiologic role have become unlikely (Steele *et al.*, 1990). In contrast, epidemiological studies and field observations (Reed *et al.*, 1987; Spencer *et al.*, 1990; Zhang *et al.*, 1990) demonstrate an association between use of the neurotoxic seed or seed kernel of *Cycas* spp. for food or medicine and the clinical appearance of ALS/PD.

The cycad hypothesis was originally proposed by Whiting (1963) and Kurland (1972) but was later abandoned since epidemiological evidence for an association between cycad consumption and ALS/PD was lacking in Japan and Irian Jaya at that time. Also, the majority of attempts to induce neurodegeneration by feeding studies in experimental animals were unsuccessful. However, a few exceptions existed: Dastur reported in 1964 that a monkey fed cycas flour developed severe weakness and wasting of one arm with neuropathological evidence of upper and lower motor neuron degeneration. In addition, after consumption of cycad leaves or seed, animals (cows, sheep, goats) developed ataxia and later hindlimb weakness. Although the pathology of cortical and spinal motor neurons was not described, neuropathology revealed degenerative changes of corticospinal, spinocerebellar tracts, and dorsal columns (Spencer and Dastur, 1989; Spencer *et al.*, 1990).

Studies show that cycad use not only plays a role in the Japanese and New Guinea foci (Spencer *et al.*, 1990), but that there is also experimental evidence that cycad chemicals induce changes *in vivo* and *in vitro* which are consistent with a possible etiopathogenetic role of these compounds. At present, two chemicals with neurotoxic potential are identified: (1) the amino acid β-*N*-methylamino-L-alanine (BMAA), a compound present in low concentrations in the seed (0.01% ww) (Polsky *et al.*, 1972; Spencer *et al.*, 1987; Duncan, 1992; Kisby *et al.*, 1992a); and (2) the glycoside cycasin and its aglycon methylazoxymethanol (MAM). Cycasin is present in cycad seed in concentration up to 2 to 4% (Matsumoto and Strong, 1963; Kobayashi and Matsumoto, 1965; Laqueur, 1968; Kisby *et al.*, 1992a).

BMAA has properties of a weak glutamate-like excitotoxin (Spencer *et al.*, 1987; Ross *et al.*, 1987; Ross and Spencer 1987; Weiss *et al.*, 1989) and its potency can be increased by physiological concentrations of bicarbonate (Weiss and Choi, 1988; Nunn and O'Brien, 1989). In rats, BMAA is a cerebellar toxin (Seawright *et al.*, 1990) and, if administered in high doses to primates, the compound induces clinical and morphological deficits of the pyramidal and extrapyramidal systems (Spencer *et al.*, 1987). Criticism on a causal role of BMAA centered on the high dosage required to induce experimental lesions in primates (Spencer *et al.*, 1987), which seems to be difficult to achieve in humans by the concentrations found in cycad flour (Duncan, 1992). Also, it is difficult to explain how a glutamate receptor-mediated toxic effect induces a progressive neurotoxic disease after a long latency (Spencer *et al.*, 1991b). However, studies show that BMAA not only has an effect on extracellular receptors, but is also transported into synaptosomes and accumulates in brain tissue *in vitro* (Kisby *et al.*, 1992b). The results of these studies indicate that the neurotoxic potential of BMAA is not confined to its extracellular effect and that further studies of its intracellular effect are necessary to define the toxic mechanism. Although the role of BMAA in the etiopathogensis of neurocycadism is still unclear, studies on its intracellular effect have the potential to define neurotoxic effects of a low dosage.

A second candidate toxin from the cycad plant with a possible intracellular effect is the glycoside and experimental carcinogen cycasin. Shimizu and colleagues (1986) reported that goats experimentally fed cycasin develop a motor disorder with spinal tract degeneration similar to that described in bovine cycadism (Hall *et al.*, 1968; Spencer and Dastur, 1989). It has been shown that cycasin is neurotoxic in mouse cortical explants and that the intact glycoside may enter neurons via the glucose transporter (Kisby *et al.*, 1992b). Then, in the cell, the aglycon of cycasin, methylazoxymethanol, appears. MAM alkylates nucleic acids (Spatz 1969; Morgan and Hoffman, 1983; Matsumoto, 1985) and has teratogenic, mutagenic, and carcinogenic properties. In mitotic cells, the alkylating properties of MAM result in the formation of 7-methylguanine and O^6-methylguanine adducts (Matsumoto and Higa, 1966; Nagata and Matsumoto, 1969; Fiala *et al.*, 1987). The effect of alkylation of slowly repaired neuronal macromolecules on protein synthesis in the postmitotic cell has been studied (Spencer *et al.*, 1991a).

In summary, based on the published epidemiological studies on ALS/PD, experimental investigations of molecular mechanisms of cycad neurotoxicity are of interest. Studies have been performed on parts of the neurotoxic mechanisms of two chemicals present in the cycad seeds. There are still no sufficient explanations how a long-latency effect of a toxic chemical could result in a rapidly progressive neurotoxic disease such as ALS/PD. An explanation may come from studies of the intracellular effects of cycad chemicals such as cycasin/MAM and possibly BMAA. Demonstration of such a chronic, possibly cumulative and irreversible, effect will have impact on future studies of the etiology and pathogenesis of chemically induced neurodegeneration.

III. Hereditary Diseases

Since the 1980s, the study of neurotoxic mechanisms became the testing ground for the elucidation of the pathogenesis of some inherited diseases. Huntington's disease is the most prominent example but other low-incidence diseases could also be associated with neurotoxic mechanisms of neurodegeneration.

A. Huntington's Disease

Huntington's disease is an autosomal dominant neurodegenerative disorder that is caused by a genetic defect localized on chromosome 4 (4p16.3). Clinical alterations usually begin in adulthood and initially consist of emotional disturbances, a reduction of memory and attentiveness, and slight abnormalities of movement. Later, severe choreoathetosis and dementia develop. The most striking neuropathological alterations are found in the striatum, but cortical structures are also affected in later stages. In the striatum, those neurons with prominent dendritic spines ("spiny neurons") are particularly vulnerable to the neurodegenerative process. Spiny neurons receive glutamatergic input from the cerebral cortex. In contrast, local aspiny interneurons containing neuropeptide Y and somatostatin and stained with the histochemical marker NADPH-diaphorase are relatively spared in Huntington's disease.

In 1976, two groups (Coyle and Schwarcz, 1976; McGeer and McGeer, 1976) showed that the pattern of vulnerability induced by intrastriatal injections of the non-NMDA glutamate analog kainic acid resembled the neurochemical and neuropathological changes seen in Huntington's chorea. These studies were the first to suggest that an "excitotoxic" mechanism of neurodegeneration (Olney, 1980) may be part of the pathogenesis. This model system was improved by intrastriatal injections of the endogenous tryptophan metabolite quinolinic acid, a potent excitotoxin acting via the NMDA glutamate receptor subtype (Beal *et al.*, 1986). Injection resulted in acute neuronal destruction

which, in contrast to the kainate-induced lesions, spares the NADPH-diaphorase-positive neurons. Also, similar to the neuropathological findings in Huntington's disease, aspiny neurons which stain with parvalbumin and γ-aminobutyric acid are relatively spared after injection of quinolinic acid or other NMDA agonists (Waldvogel *et al.*, 1991). Chronic experiments (6 or 12 months) revealed similar results (Beal *et al.*, 1991a). Administration of selective antagonists to the NMDA receptor prevents acute lesions (Beal *et al.*, 1988). These experimental results are supported by findings that kainate and NMDA receptors are depleted in Huntington's disease putamen (Young *et al.*, 1988). In an asymptomatic at-risk patient, the concentration of NMDA receptors was reduced by 50% (Albin *et al.*, 1990), indicating that loss of this receptor occurs early in the disease.

The pattern of vulnerability induced by injections of quinolinic acid suggested that an alteration of tryptophan metabolism might be the primary cause in Huntington's disease. However, detailed neurochemical studies of the kynurenine pathway in patients did not reveal major abnormalities. In particular, the concentration of quinolinic acid or of the tryptophan metabolite and glutamate antagonist kynurenic acid did not show major consistent alterations in Huntington's brain (Reynolds *et al.*, 1988; Beal *et al.*, 1990; Heyes *et al.*, 1991; Okuno *et al.*, 1991). Therefore, although some of the results of these studies indicate that an excitotoxic mechanism may be involved in the pathogenesis of the selective neuronal lesions, there is still no convincing evidence that an aberration of the kynurenine pathway of tryptophan metabolism is the cause of the disease.

Evidence has accumulated that the inhibition of mitochondrial energy production induces lesions which are mediated by receptors for excitatory amino acids (Henneberry *et al.*, 1989). Therefore, it may not be a surprise that such compounds also induce striatal lesions highly characteristic of Huntington's disease. Of current interest is aminooxyacetic acid which is an inhibitor of aspartate transaminase, an essential component of the malate–aspartate shuttle across the mitochondrial membrane (Beal *et al.*, 1991b). In the absence of a direct depolarizing effect on neurons (Riepe *et al.*, 1992), similar lesions are produced by the exogenous plant and mycotoxin 3-nitropropionic acid, an inhibitor of succinate dehydrogenase (Ludolph *et al.*, 1991a, 1992; Brouillet *et al.*, 1993).

Therefore, inhibition of chemical energy production which results in an excitotoxic pattern of injury may be a mechanism associated with selective neuronal death as observed in Huntington's disease (Beal, 1992). These hypotheses are based on the experience with neurotoxic models of neurodegeneration; further improvement of these models, which should also consider the latency of disease manifestation, may be the key for novel and potentially successful therapeutic approaches.

B. Other Selected Inherited Diseases

The pathogenesis of neurodegeneration in most inherited metabolic diseases is far from being clarified. Neurotoxicity is a likely factor in the pathogenesis of brain damage of some of these diseases.

Non-ketotic hyperglycemia (NKH) is a fatal neurological disorder of the newborn, clinically characterized by myoclonic seizures, muscle hypotonia, and respiratory distress. Survivors of the early postnatal period suffer from intractable seizures and severe mental retardation. Impaired catabolism of glycine results in a more than 10-fold increase of glycine in the cerebrospinal fluid of patients. Previous treatment attempts with strychnine were initiated under the assumption of a pathogenetic role of the inhibitory strychnine-sensitive glycine receptor but were unsuccessful. In contrast, independent groups have reported that seizures can be effectively treated by the noncompetitive glutamate antagonists and blockers of the NMDA-associated ion channel dextramorphane (Schmitt et al., 1992) and ketamine (Ohta et al., 1990). These reports indicate that the strychnine-insensitive glycine recognition site at the NMDA receptor complex plays a significant role in the pathogenesis of seizures in NKH. For opening of the ion channel associated with the NMDA receptor, the binding of synaptically released glutamate and two molecules of glycine is necessary. Elevated synaptic concentrations of glycine may potentiate NMDA responses, which results in seizures, intracellular calcium accumulation, and subsequent neuronal damage. However, systematic therapeutic trials are warranted to show whether seizure control is also associated with a preventive therapeutic effect on permanent neurological deficits such as mental retardation which would indicate that glycine neurotoxicity at the NMDA receptor contributes to the neurodegenerative process.

Organic acidemias such as glutaric aciduria type I are clinically characterized by acute metabolic crises associated with acidosis and/or hypoglycemia. Bilateral necrosis of basal ganglia and a permanent extrapyramidal movement disorder often result. It has been emphasized that the observed morphological alterations are consistent with an excitotoxic mechanism of damage (Soffer et al., 1992). There are well-defined neurotoxic effects of metabolites which definitely or potentially accumulate in glutaric aciduria type I; how-

ever, their impact in vivo is unknown. In concentrations found in the disease, the major metabolite glutaric acid inhibits glutamate decarboxylase (Stokke et al., 1976) and inhibits glutamate uptake in the synaptosomal fraction (Balcar and Johnston, 1972; Bennett et al., 1973). Both mechanisms might increase synaptic glutamate concentrations. Although the quinolinic acid concentration in the cerebrospinal fluid was described as normal in a single patient (Land et al., 1992), the characteristics of the metabolic block in glutaric aciduria type I make an accumulation of the neurotoxic compound quinolinic acid possible (Heyes, 1987). The potential neurotoxicity of this endogenous compound requires further studies, particularly during metabolic crises. Whether an effect of these potentially excitotoxic compounds is exacerbated in vivo during the characteristic acute metabolic alterations such as hypoglycemia remains to be determined.

IV. Summary and Conclusions

It is generally accepted that the study of selective neurotoxins plays a significant role in the improvement of our understanding of the pathogenesis of neurodegeneration in various diseases of the nervous system. The present and other future model systems will presumably reveal further insights into the pathogenesis of sporadically occurring neurodegenerative diseases. Also, an increasing number of metabolites with neurotoxic potential are identified which play a role in the destruction of neurons in hereditary degenerative diseases. However, the difficulty often remains to draw conclusions from an acute experimental testing system for a chronic metabolic alteration or disease. Also, the question remains unresolved whether exogenous chemicals such as MPTP play a role in the etiology of sporadic neurodegenerative diseases. At present, there is neither epidemiological evidence nor a mechanism in sight which could justify a positive answer. In particular, the question by which mechanism an exogenous chemical could cause a progressive neurodegenerative disease has not been seriously addressed. The answer to this question will show that neurotoxicology is not only a testing ground for the pathogenesis of neuronal degeneration, but can also partially explain its etiology.

References

Aicardi, J., Gordon, N., and Hagberg, B. (1985). Holes in the Brain. Dev. Med. Child. Neurol. 27, 249–252.

Albin, R., Young, A. B., Penney, J. B., Handelin, B., Balfour, R., Anderson, K. D., Markel, D. S., Tourtellotte, W. W., and Reiner, A. (1990). Abnormalities of striatal projection neurons and N-methyl-D-aspartate receptors in presymptomatic Huntington's disease. *N. Engl. J. Med.* **322**, 1293–1298.

Balagapolan, C., Padmaja, G., Nanda, S. K., and Moorthy, S. N. (1988). *Cassava in Food, Feed and Industry.* CRC Press, Boca Raton, Florida.

Balcar, V. J., and Johnston, G. A. R. (1972). The structural specificity of the high affinity uptake of L-glutamate and L-aspartate by rat brain slices. *J. Neurochem.* **19**, 2657–2666.

Baldessarini, R. J., Kula, N. S., Francoeur, D., and Finklestein, S. P. (1986). Antioxidants fail to inhibit depletion of striatal dopamine by MPTP. *Neurology* **36**, 735.

Ballard, P. A., Tetrud, J. W., and Langston, J. W. (1985). Permanent human parkinsonism due to 1-methyl-4-phenyl-1,2,3,6-tetrahydropyridine (MPTP): seven cases. *Neurology* **35**, 949–956.

Beal, M. F., Kowall, N. W., and Ellison, D. W. (1986). Replication of the neurochemical characteristics of Huntington's disease by quinolinic acid. *Nature* **321**, 168–171.

Beal, M. F., Kowall, N. W., Swartz, K. J., Ferrante, R. J., and Martin, J. B. (1988). Systemic approaches to modifying quinolinic acid striatal lesions in rats. *J. Neurosci.* **8**, 3901–3908.

Beal, M. F., Matson, W. R., Swartz, K. J., Gamache, P. H., and Bird, E. D. (1990). Kynurenine pathway measurements in Huntington's disease: evidence for reduced formation of kynurenic acid. *J. Neurochem.* **55**, 1327–1339.

Beal, M. F., Ferrante, R. J., Swartz, K. J., and Kowall, N. W. (1991a). Chronic quinolinic acid lesions in rats closely mimic Huntington's disease. *J. Neurosci.* **11**, 1649–1659.

Beal, M. F., Swartz, K. J., Hyman, B. T., Storey, E., Finn, S. F., and Koroshetz, W. (1991b). Aminooxyacetic acid results in excitotoxin lesions by a novel indirect mechanism. *J. Neurochem.* **57**, 1068–1073.

Beal, M. F. (1992). Does impairment of energy metabolism result in excitotoxic neuronal death in neurodegenerative illnesses? *Ann. Neurol.* **31**, 119–130.

Bennett, J. P., Logan, W. J., and Snyder, S. H. (1973). Amino acids as central nervous transmitters: the influence of ions, amino acid analogues, and ontogeny on transport systems for L-glutamic and aspartic acid and glycine into central nervous synaptosomes of the rat. *J. Neurochem.* **21**, 1533–1550.

Biscoe, T. J., Evans, R. H., Headley, P. M., Martin, M. R., and Watkins, J. C. (1975). Domoic acid and quisqualic acid as potent amino acid excitants of frog and rat spinal neurons. *Nature* **255**, 166–167.

Bloem, B. R., Irwin, I., Buruma, O. J. S., Haan, J., Roos, R. A. C., Tetrud, J. W., and Langston, J. W. (1990). The MPTP model: versatile contributions to the treatment of idiopathic Parkinson's disease. *J. Neurol. Sci.* **97**, 273–293.

Bridges, R. J., Stevens, D. R., Kahle, J. S., Nunn, P. B., Kadri, M., and Cotman, C. W. (1989). Structure-function studies on N-oxalyl-diamino-dicarboxylic acids and excitatory amino acid receptors: evidence that β-L-ODAP is a selective non-NMDA agonist. *J. Neurosci.* **9**, 2073–2079.

Bridges, R. J., Hatalski, C., Shim, S. N., and Nunn, P. B. (1991). Gliotoxic properties of the *Lathyrus* excitotoxin N-oxalyl-L-a,β-diaminopropionic acid (-L-ODAP). *Brain Res.* **561**, 262–268.

Brooks, W. J., Jarvis, M. F., and Wagner, G. C. (1988). Attenuation of MPTP-induced dopaminergic neurotoxicity by a serotonin uptake blocker. *J. Neural. Transm.* **71**, 85–90.

Brooks, W. J., Jarvis, M. F., and Wagner, G. C. (1989). Astrocytes as a primary locus for the conversion MPTP into MPP⁺. *J. Neural. Transm.* **76**, 1–12.

Brouilllet, E., Jenkins, B. G., Hyman, B. T., Ferrante, R. J., Kowall, N. W., Srivastava, R., Roy, D. S., Rosen, B. R., and Beal, M. F. (1993). Age-dependent vulnerability of the striatum to the mitochondrial toxin 3-nitropropionic acid. *J. Neurochem.* **60**, 356–359.

Cantani, A. (1873). Latirismo illustrata de tre casi clinici. *Il Morgagni* **XV**, 745.

Carton, H., Kayembe, K., Kabeya, O., Billiau, A., and Maertens, K. (1986). Epidemic spastic paraparesis in Bandundu (Zaire). *J. Neurol. Neurosurg. Psychiatry* **49**, 620–627.

Chiba, K., Trevor, A., and Castagnoli, N. Jr. (1984). Metabolism of the neurotoxic tertiary amine, MPTP, by brain monoamine oxidase. *Biochem. Biophys. Res. Commun.* **120**, 574–578.

Choi, D. W. (1992). Amyotrophic lateral sclerosis and glutamate—too much of a good thing? *N. Engl. J. Med.* **326**, 1493–1494.

Cliff, J., Lundqvist, P., Martensson, J., Rosling, H., and Sörbo, B. (1985). Association of high cyanide and low sulphur intake in cassava-induced spastic paraparesis. *Lancet* **ii**, 1211–1213.

Cohn, D. F., Streifler, M., and Schujman, E. (1977). Das motorische Neuron im chronischen Lathyrismus. *Nervenarzt* **48**, 127–129.

Cohn, D. F., and Streifler, M. (1981a). Human neurolathyrism, a follow-up study of 200 patients. 1. Clinical investigation. *Arch. Suisses Neurol. Neuroch. Psychiatry,* **128**, 151–156.

Cohn, D. F., and Streifler, M. (1981b). Human lathyrism, a follow-up study. 2. *Arch. Suisses Neurol. Neuroch. Psychiatry* **128**, 157–163.

Corsini, G. U., Pintus, S., Chiueh, C. C., Weiss, J. F., and Kopin, I. J. (1985). 1-Methyl-4-phenyl-1,2,3,6-tetrahydropyridine (MPTP) neurotoxicity in mice is enhanced by pretreatment with diethyldithiocarbamate. *Eur. J. Pharmacol.* **119**, 127–128.

Coyle, J. T., and Schwarcz, R. (1976). Lesions of striatal neurons with kainic acid provide a model for Huntington's chorea. *Nature* **263**, 244–246.

Dastur, D. K. (1964). Cycad toxicity in monkeys: clinical, pathological, and biochemical aspects. *Fed. Proc.* **23**, 1368–1369.

Davis, G. C., Williams, A. C., Markey, S. P., Ebert, M. H., Caine, E. D., Beichert, C. M., and Kopin, I. J. (1979). Chronic parkinsonism secondary to intravenous injection of meperidine analogues. *Psychiatry Res.* **1**, 249–254.

Debonnel, G., Beauchesne, L., and De Montigny, C. (1989). Domoic acid, the alleged "mussel toxin" might produce its neurotoxic effect through kainate receptor activation: an electrophysiological study in the rat dorsal hippocampus. *Can. J. Physiol. Pharmacol.* **67**, 29–33.

Di Monte, D., Jewell, S. A., Ekström, G., Sandy, M. S., and Smith, M. T. (1986a). 1-Methyl-4-phenyl-1,2,3,6-tetrahydropyridine (MPTP) and 1-methyl-4-phenylpyridine (MPP⁺) cause rapid ATP depletion in isolated hepatocytes. *Biochem. Biophys. Res. Commun.* **137**, 310–315.

Di Monte, D., Sandy, M. S., Ekström, G., and Smith, M. T. (1986b). Comparative studies on the mechanisms of paraquat and 1-methyl-4-phenylpyridine (MPP⁺) cytotoxicity. *Biochem. Biophys. Res. Commun.* **137**, 303–309.

Di Monte, D., Ekström, G., Shinka, T., Smith, M. T., Trevor, A. J., and Castagnoli, N. Jr. (1987). Role of 1-methyl-4-phenylpyridine (MPP⁺) ion formation and accumulation in 1-methyl-4-phenyl-1,2,3,6-tetrahydropyridine toxicity to isolated hepatocytes. *Chem. Biol. Interact.* **62**, 105–116.

Duncan, M. W. (1992). β-Methylamino-L-alanine (BMAA) and amyotrophic lateral sclerosis-parkinsonism dementia of the western Pacific. *Ann. N.Y. Acad. Sci.* **648**, 161–168.

Elizan, T. S., Hirano, A., and Abrams, B. M. (1966). Amyotrophic lateral sclerosis and parkinsonism-dementia complex on Guam: neurological reevaluation. *Arch. Neurol.* **14**, 356–368.

Fiala, E. S., Sohn, O. S., Puz, C., and Czerniak, R. (1987). Differential effects of 4-iodopyrazole and 3-methylpyrazole on the metabolic ectivation of methylazoxymethanol to a DNA methylating species by rat liver and rat colon mucosa *in vivo*. *J. Cancer Res. Clin. Oncol.* **113**, 145–150.

Filimonoff, I. N. (1926). Zur pathologisch-anatomischen Charakteristik des Lathyrismus. *Zbl. Ges. Neurol. Psychiatry* **105**, 76–92.

Finelli, P. (1981). Changes in the basal ganglia following cyanide poisoning. *J. Comput. Assist. Tomogr.* **5**, 755–756.

Forno, L. S., Langston, J. W., Delanney, L. E., Irwin, I., and Ricaurte, H. (1986). Locus coeruleus lesions and eosinophilic inclusions in MPTP-treated monkeys. *Ann. Neurol.* **20**, 449–455.

Gajdusek, D. C. (1963). Motor-neuron disease in natives of New Guinea. *N. Engl. J. Med.* **268**, 474–476.

Gajdusek, D. C., and Salazar, A. M. (1982). Amyotrophic lateral sclerosis and parkinsonism dementia in high incidence among the Auyu and Jakai people of West New Guinea. *Neurology* **32**, 107–126.

Gannon, R. L., and Terrian, D. M. (1989). BOAA selectively enhances L-glutamate release from guinea pig hippocampal mossy fiber synaptosomes. *Neurosci. Lett.* **107**, 289–294.

Garruto, R., Gajdusek, D. C., and Chen, K.-M. (1980). Amyotrophic lateral sclerosis among Chamorro migrants from Guam. *Ann. Neurol.* **8**, 612–619.

Garruto, R., Gajdusek, D. C., and Chen, K.-M. (1981). Amyotrophic lateral sclerosis and parkinsonism-dementia among Filipino migrants to Guam. *Ann. Neurol.* **10**, 341–350.

Garruto, R., Yanagihara, R., and Gajdusek, D. C. (1985). Disappearance of high-incidence amyotrophic lateral sclerosis and parkinsonism-dementia on Guam. *Neurology* **35**, 193–198.

Garruto, R. M., and Yase, Y. (1986). Neurodegenerative disorders of the western Pacific: the search for mechanisms of pathogenesis. *Trends Neurosci.* **9**, 368–374.

Gibbs, C. J. Jr., and Gajdusek, D. C. (1982). An update on long-term *in vivo* and *in vitro* studies to identify a virus as the cause of amyotrophic lateral sclerosis, parkinsonism dementia, and Parkinson's disease. In *Human Motor Neuron Diseases* (L. P. Rowland, Ed.), pp. 343–353, Raven Press, New York.

Grandas, F., Artieda, J., and Obeso, J. (1989). Clinical and CT-scan findings in a case of cyanide intoxication. *Mov. Disord.* **4**, 188–193.

Guggenheim, M. A., Couch, J. R., and Weinberg, W. (1971). Motor dysfunction as a permanent complication of methanol ingestion. *Arch. Neurol.* **24**, 550–554.

Haimanot, R. T., Kidane, Y., Wuhib, E., Kalissa, A., Alemo, T., Zein, Z. A., and Spencer, P. S. (1990). Lathyrism in rural northwestern Ethiopia: A highly prevalent neurotoxic disorder. *Int. J. Epidemiol.* **19**, 664–672.

Hall, W. T. K., and McGavin, M. D. (1968). Clinical and neuropathological findings in cattle eating the leaves of *Macrozamia lucida* or *Bowenia serrulata* (family Zamiaceae). *Path. Vet.* **5**, 26–34.

Hamilton, B. F., and Gould, D. H. (1987a). Nature and distribution of brain lesions in rats intoxicated with 3-nitropropionic acid: a type of hypoxic (energy deficient) brain damage. *Acta Neuropathol. (Berl)* **72**, 286–297.

Hamilton, B. F., and Gould, D. H. (1987b). Correlation of morphologic brain lesions with physiologic alterations and blood-brain barrier impairment in 3-nitropropionic acid toxicity in rats. *Acta Neuropathol. (Berl)*. **74**, 67–74.

Hawker, K., and Lang, A. E. (1990). Hypoxic-ischemic damage of the basal ganglia: case reports and a review of the literature. *Movement Disord.* **5**, 219–224.

He, F., Zhang, S., Zhang, C., Qian, F., Liu, X., and Lo, X. (1990). Mycotoxin induced encephalopathy and dystonia in children. In *Basic Science in Toxicology* (G. N. Volans, J. Sims, F. M. Sullivan, and P. Turner, Eds.), pp. 596–604, Taylor and Francis, London.

Heikkila, R. E., Manzino, L., Cabbat, F. S., and Duvoisin, R. C. (1984). Protection against the dopaminergic neurotoxicity of 1-methyl-4-phenyl-1,2,3,6-tetrahydropyridine (MPTP) by monoamine oxidase inhibitors. *Nature* **311**, 467–469.

Henneberry, R. C., Novelli, A., Cox, J. A., and Lysko, P. G. (1989). Neurotoxicity at the N-methyl-D-aspartate receptor in energy-compromised neurons: an hypothesis for cell death in aging and disease. *Ann. N.Y. Acad. Sci.* **568**, 225–233.

Heyes, M. P. (1987). Hypothesis: a role for quinolinic acid in the neuropathology of glutaric aciduria type I. *Can. J. Neurol. Sci.* **14**, 441–443.

Heyes, M. P., Swartz, K. J., Markey, S. P., and Beal, M. F. (1991). Regional brain and cerebrospinal fluid quinolinic acid concentrations in Huntington's disease. *Neurosci. Lett.* **122**, 265–269.

Hicks, S. P. (1950a). Brain metabolism *in vivo*, part I. *Arch. Pathol.* **49**, 111–137.

Hicks, S. P. (1950b). Brain metabolism *in vivo*, part II. *Arch. Pathol.* **50**, 545–561.

Hippocrates (1846). *Epidemarium*, 3rd ed., Vol. 5, book II, Sect IV, p. 126, Paris, Littre.

Hirano, A., Malamud, N., and Kurland, L. T. (1961). Parkinsonism-dementia complex, an endemic disease on the island of Guam. II. Pathological features. *Brain* **84**, 662–679.

Howlett, W. P., Brubaker, G. R., Mlingi, N., and Rosling, H. (1990). *Konzo*, an epidemic upper motor neuron disease studied in Tanzania. *Brain* **113**, 223–235.

Hugon, J., Ludolph, A. C., Roy, D. N., Schaumburg, H. H., and Spencer, P. S. (1988). Studies on the etiology and pathogenesis of motor neuron disease. II. Clinical and electrophysiologic features of pyramidal dysfunction in macaques fed *Lathyrus sativus* and IDPN. *Neurology* **38**, 435–442.

Hugon, J., Ludolph, A. C., Gimenez-roldan, S., Hague, A., and Spencer, P. S. (1990). Electrophysiological evaluation of human lathyrism-results in Bangladesh and Spain. In *New Advances in Toxicology and Epidemiology of ALS* (F. Clifford-Rose and F. Norris, Eds.), pp. 49–56, Smith-Gordan, London.

Hurst, E. W. (1942). Experimental demyelination of the nervous system. 3. Poisoning with potassium cyanide, sodium azide, hydroxylamine, narcotics, carbon monoxide, etc., with some consideration of bilateral necrosis occurring in the basal nuclei. *Aust. J. Exp. Biol. Med. Sci.* **20**, 297–312.

Irwin, I., and Langston, J. W. (1985). Selective accumulation of MPP$^+$ in the substantia nigra: a key to neurotoxicity? *Life Sci.* **36**, 207–212.

Javitch, J. A., and Snyder, S. H. (1984). Uptake of MPP$^+$ by dopamine neurons explains selectivity of parkinsonism-inducing neurotoxin, MPTP. *Eur. J. Pharmacol.* **106**, 455–456.

Javitch, J. A., D'amato, R. J., Strittmatter, S. M., and Snyder, S. H. (1985). Parkinsonism inducing neurotoxin, N-methyl-4-phenyl-1,2,3,6-tetrahydropyridine: uptake of the metabolite N-methyl-4-phenylpyridine by dopamine neurons explains selective toxicity. *Proc. Natl. Acad. Sci. U.S.A.* **82**, 2173–2177.

Jellinger, K. (1986). (Exogenous) striatal necrosis. In *Handbook of Clinical Neurology*, Vol. 5, *Extrapyramidal Disorders*. (P. J. Vinken, G. W. Bruyn, and H. L. Klawans, Eds.), pp. 499–518, Elsevier Science Publishers, Amsterdam.

Johannessen, J. N., Chiueh, C. C., Burns, R. S., and Markey, S. P. (1985). Differences in the metabolism of MPTP in the rodent and primate parallel differences in sensitivity in its neurotoxic effects. *Life Sci.* **36**, 219–224.

Kessler, A. (1947). Lathyrismus. *Monatsschr. Psychiatry Neurol.* **113**, 76–92.

Kisby, G., Ellison, M., and Spencer, P. S. (1992a). Content of the neurotoxins cycasin (methylazoxymethanol β-D-glucoside) and

BMAA (β-N-methylamino-L-alanine) in cycad flour prepared by Guam Chamorros. *Neurology* **42**, 1336–1340.

Kisby, G. E., Ross, S. M., Spencer, P. S., Gold, B. G., Nunn, P. B., and Roy, D. N. (1992b). Cycasin and BMAA: candidate neurotoxins for western pacific amyotrophic lateral sclerosis/ parkinsonism-dementia complex. *Neurodegeneration* **1**, 73–82.

Kobayashi, A., and Matsumoto, H. (1965). Studies on methylazoxy-methanol, the aglycone of cycasin. Isolation, biochemical, and chemical properties. *Arch. Biochem. Biophys.* **37**, 373–380.

Környvey, S. (1963). Patterns of CNS vulnerability in CO, cyanide and other poisonings. In *Selective Vulnerability of the Brain in Hypoxaemia* (J. P. Schade and W. H. McMenemery, Eds.), pp. 165–181, Blackwell, Oxford.

Konradi, C., Riederer, P., Jellinger, K., and Denney, R. (1987). Cellular actions of MAO inhibitors. *J. Neural Transm. (Suppl.)* **25**, 15–25.

Krauss, J. K., Mohadjier, M., Wakhloo, A. K., and Mundinger, F. (1991). Dystonia and akinesia due to pallidoputaminal lesions after disulfiram intoxication. *Mov. Disord.* **6**, 166–170.

Kurland, L. T. (1972). An appraisal of the neurotoxicity of cycad and the etiology of amyotrophic lateral sclerosis on Guam. *Fed. Proc.* **31**, 1540–1542.

Kurland, L. T., and Molgaard, C. A. (1982). Guamanian ALS: Hereditary or acquired? In *Human Motor Neuron Diseases* (L. P. Rowland, Ed.), pp. 165–171, Raven Press, New York.

Kurland, L. T. (1988). Amyotrophic lateral sclerosis and Parkinson's disease complex on Guam linked to an environmental neurotoxin. *Trends Neurosci.* **11**, 51–54.

Lancaster, P. A., Ingram, J. S., Lim, M. Y., and Coursey, D. G. (1982). Traditional cassava-based foods: survey of processing techniques. *Econ. Bot.* **36**, 12–45.

Land, J. M., Goulder, P., Johnson, A. W., and Hockaday, J. (1992). Quinolinate and glutaric aciduria type I. In *Abstracts of the 30th Annual Symposium of the Society for the Study of Inborn Errors of Metabolism*, p. 88.

Langston, J. W., Ballard, P., Tetrud, J. W., and Irwin, I. (1983). Chronic parkinsonism in humans due to a product of meperidine-analog synthesis. *Science* **219**, 979–980.

Langston, J. W., Irwin, I., Langston, E. B., and Forno, L. S. (1984). Pargyline prevents MPTP-induced parkinsonism in primates. *Science* **225**, 1480–1482.

Langston, J. A., Irwin, I., and Finnegan, K. T. (1989). Using neurotoxicants to study aging and Parkinson's disease. In *Parkinsonism and Aging* (D. B. Calne, G. Comi, D. Crippa, R. Horowski, and M. Trabucchi, Eds.), pp. 145–153, Raven Press, New York.

Laqueur, G. L. (1968). Toxicology of cycasin. *Cancer Res.* **28**, 2262–2267.

Lewitt, P. A., and Martin, S. D. (1988). Dystonia and hypokinesis with putaminal necrosis after methanol intoxication. *Clin. Neuropharmacol.* **11**, 161–167.

Lucasse, C. (1952). Le "Kitondji" (synonyme: Le "Konzo"): une paralysie spastique. *Ann. Soc. Belge Med. Trop.* **32**, 391–400.

Ludolph, A. C., Hugon, J., Dwivedi, M. O., Schaumburg, H. H., and Spencer, P. S. (1987). Studies on the aetiology and pathogenesis of motor neurone disease. 1. Lathyrism: clinical findings in established cases. *Brain* **110**, 149–165.

Ludolph, A. C., He, F., Spencer, P. S., Hammerstad, J., and Sabri, M. I. (1991a). 3-Nitropropionic acid: exogenous animal neurotoxin and possible human striatal toxin. *Can. J. Neurol. Sci.* **18**, 492–498.

Ludolph, A. C., Siddik, M., and Spencer, P. S. (1991b). Decline of ALS in the Irian Jaya epicenter. *Neurology (Suppl. 1)* **41**, 259.

Ludolph, A. C., Seelig, M., Ludolph, A. G., Novitt, P., Allen, C. A., Spencer, P. S., and Sabri, M. I. (1992). 3-Nitropropionic

acid decreases cellular energy levels and causes neuronal degneration in cortical explants. *Neurodegeneration* **1**, 155–161.

Markey, S. P., Johannessen, J. N. Chiueh. C. C., Burns, R. S., and Herkenham, M. A. (1984). Intraneuronal generation of a pyridinium metabolite may cause drug-induced parkinsonism. *Nature* **311**, 464–467.

Marongiu, M. E., Piccardi, M. P., Bernardi, F., Corsini, G. U., and Del-zompo, M. (1988). Evaluation of the toxicity of dopaminergic neurotoxins MPTP and MPP⁺ in PC12 phenochromocytoma cells: binding and biological studies. *Neurosci. Lett.* **94**, 349–354.

Matsumoto, H. (1985). Cycasin. In *CRC Handbook of Naturally Occurring Food Toxicants* (M. Reckcigl, Ed.), pp. 43–61, CRC Press, Boca Raton, Florida.

Matsumoto, H., and Strong, F. M. (1963). The occurrence of methyl-azoxymethanol in *Cycas ciricinalis, L. Arch. Biochem. Biophys.* **101**, 299–310.

Matsumoto, H., and Higa, H. H. (1966). Studies on methylazoxy-methanol, the aglycone of cycasin: methylation of nucleic acids *in vitro. Biochem. J.* **98**, 20C–22C.

Martinovits, G., Melamed, E., Cohen, O., Rosenthal, J., and Uzzan, A. (1986). Systemic administration of antioxidants does not protect mice against the dopaminergic neurotoxicity of 1-mthyl-4-phenyl-1,2,3,6-tetrahydropyridine (MPTP). *Neurosci. Lett.* **69**, 192–197.

McGeer, E. G., and McGeer, P. L. (1976). Duplication of biochemical changes of Huntington's chorea by intrastriatal injections of glutamic and kainic acids. *Nature* **263**, 517–519.

McLean, D. R., Jacobs, H., and Mielke, B. W. (1980). Methanol poisoning: a clinical and pathological study. *Ann. Neurol.* **8**, 161–167.

Melamed, E., Rosenthal, J., and Youdim, M. B. H. (1990). Immunity of fetal mice to prenatal administration of the dopaminergic neurotoxin 1-methyl-4-phenyl-1,2,3,6-tetrahydropyridine. *J. Neurochem.* **55**, 1427–1431.

Mertens, H. G. (1947). Zur Klinik des Lathyrismus. *Nervenarzt* **18**, 493–499.

Messing, B., and Storch, B. (1988). Computer tomography and magnetic resonance imaging in cyanide poisoning. *Eur. Arch. Psychiatry Neurol. Sci.* **237**, 139–143.

Mettler, F. A. (1972). Choreoathetosis and striopallidonigral necrosis due to sodium azide. *Exp. Neurol.* **34**, 291–308.

Ministry of Health, Mozambique (1984a). Mantakassa: an epidemic of spastic paraparesis associated with chronic cyanide intoxication in a cassava-staple area of Mozambique. 1. Epidemiology and clinical and laboratory findings in patients. *Bull. WHO* **62**, 477–484.

Mitchell, I. J., Cross, A. J., Sambrook, M. A., and Crossman, A. R. (1985). Sites of the neurotoxic action of 1-methyl-4-phenyl-1,2,3,6-tetrahydropyridine in the macaque monkey include the ventral tegmental area and the locus ceruleus. *Neurosci. Lett.* **61**, 195–200.

Miyoshi, K. (1967). Experimental striatal necrosis induced by sodium azide: a contribution to the problem of selective vulnerability and histochemical studies of enzyme activity. *Acta Neuropathol. (Berl.)* **9**, 199–216.

Mizuno, Y., Saitoh, T., and Sone, N. (1987a). Inhibition of mitochondrial NADH-ubiquinone oxidoreductase activity by 1-methyl-4-phenylpyridinium ion. *Biochem. Biophys. Res. Commun.* **143**, 294–299.

Mizuno, Y., Saitoh, T., and Sone, N. (1987b). Inhibition of mitochondrial α-ketoglutarate dehydrogenase by 1-methyl-4-phenylphyridinium ion. *Biochem. Biophys. Res. Commun.* **143**, 971–976.

Mizuno, Y., Ohta, S., Tanaka, M., Takamiya, S., Suzuki, K., Sato, T., Oya, H., Ozawa, T., and Kagawa, Y. (1989). Deficiencies in complex I subunits of the respiratory chain in Parkinson's disease. *Biochem. Biophys. Res. Commun.* **163**, 1450–1455.

Mlingi, N., Kimatta, S., and Rosling, H. (1991). *Konzo*, a paralytic disease observed in southern Tanzania. *Trop. Doct.* **21**, 24–25.

Morgan, R. W., and Hoffmann, G. R. (1983). Cycasin and its mutagenic metabolities. *Mutation Res.* **114**, 19–58.

Mytilineou, C., and Cohen, G. (1984). 1-Methyl-4-phenyl-1,2,3,6-tetrahydropyridine destroys dopamine neurons in explants of rat embryo mesencephalon. *Science* **225**, 529–531.

Mytilineou, C., Cohen, G., and Heikkila, R. E. (1985). 1-Methyl-4-phenylpyridine (MPP⁺) is toxic to mesencephalic dopamine neurons in culture. *Neurosci. Lett.* **57**, 19–24.

Nagata, Y., and Matsumoto, H. (1969). Studies on methylazoxymethanol, the aglycone of cycasin: methylation of nucleic acids in the rat fetal brain. *Proc. Soc. Exp. Biol.* **132**, 383–385.

Nicholls, P. (1975). Formate as an inhibitor of cytochrome c oxidase. *Biochem. Biophys. Res. Commun.* **67**, 610–616.

Nicholls, P. (1976). The effect of formate on cytochrome aa3 and on electron transpost in the intact respiratory chain. *Biochim. Biophys. Acta* **430**, 13–29.

Nicklas, W. J., Vyas, I., and Heikkila, R. E. (1985). Inhibition of NADH-linked oxidation in brain mitochondria by 1-methyl-4-phenylpyridine, a metabolite of the neurotoxin 1-methyl-4-phenyl-1,2,3,6-tetrahydropyridine. *Life Sci.* **36**, 2503–2508.

Nunn, P. B., Seelig, M., and Spencer, P. S. (1987). Stereospecific acute neuronotoxicity of "uncommon" plant amino acids linked to human motor-system disease. *Brain Res.* **410**, 375–379.

Nunn, P. B., and O'Brien, P. (1989). The interaction of β-N-methylamino-L-alanine with bicarbonate: a ¹H-NMR study. *FEBS Lett.* **251**, 31–35.

Ohta, Y., Ochi, N., Mizutani, N., Hayakawa, C., and Watanabe, K. (1990). Nonketotic hyperglycinemia: therapeutic attempt with NMDA receptor antagonist. In *Abstracts of the 5th International Congress on Inherited Metabolic Diseases*, p 79.

Okuno, E., Nakamura, M., and Schwarcz, R. (1991). Two kynurenine aminotransferases in human brain. *Brain Res.* **54**, 307–312.

Olney, J. W. (1980). Excitotoxic mechanisms of neurotoxicity. In *Clinical and Experimental Neurotoxicology* (P. S. Spencer and H. H. Schaumburg, Eds.), pp. 272–294, Williams and Wilkins, Baltimore.

Orthner, H. (1950). *Die Methylalkohol-Vergiftung mit besonderer Berücksichtigung neuartiger Hirnbefunde.* Springer, Berlin, Göttingen, Heidelberg.

Perl, T. M., Bedard, L., Kosatsky, T., Hockin, J. C. Todd, E. C. D., and Remis, R. S. (1990). An outbreak of toxic encephalopathy caused by eating mussels contaminated with domoic acid. *N. Engl. J. Med.* **322**, 1775–1780.

Perry, T. L., Young, V. W., Jones, K., and Wright, J. M. (1986). Manipulation of glutathionine contents fails to alter dopaminergic nigrostriatal neurotoxicity of N-methyl-4-phenyl-1,2,3,6-tetrahydropyridine (MPTP) in the mouse. *Neurosci. Lett.* **70**, 261–265.

Perry, T. L., Yong, V. W., Hansen, S., Jones, K., Bergeron, C., Foulks, J. G., and Wright, J. M. (1987). Alpha-tocopherol and beta-carotene do not protect marmosets against the dopaminergic neurotoxicity of N-methyl-4-phenyl-1,2,3,6-tetrahydropyridine. *J. Neurol. Sci.* **81**, 321–331.

Pifl, C., Bertel, O., Schingnitz, G., and Hornykiewicz, O. (1990). Extrastriatal dopamine in symptomatic and asymptomatic rhesus monkeys treated with 1-methyl-4-phenyl-1,2,3,6-tetrahydropyridine (MPTP). *Neurochem. Int.* **17**, 263–266.

Polsky, F. I., Nunn, P. B., and Bell, E. A. (1972). Distribution and toxicity of α-amino-β-methylaminopropionic acid. *Fed. Proc.* **31**, 1473–1475.

Potts, A. M., Praglin, J., Farkas, J., Orbison, L., and Chickering, D. (1955). Studies on the visual toxicity of methanol. VIII. Additional observations on methanol poisoning in the primate test object. *Am. J. Ophthal.* **40**, 76–83.

Rainey, J. M. (1977). Disulfiram toxicity and carbon disulfide poisoning. *Ann. J. Psychiatry,* **134**, 371–378.

Ramsay, R. R., Dadgar, J., Trevor, A., and Singer, T. P. (1986a). Energy-driven uptake of N-methyl-4-phenylpyridine by brain mitochondria mediates the neurotoxicity of MPTP. *Life Sci.* **39**, 581–588.

Ramsay, R. R., Salach, J. I., Dadgar, J., and Singer, T. P. (1986b). Inhibition of mitochondrial NADH dehydrogenase by pyridine derivatives and its possible relation to experimental and idiopathic parkinsonism. *Biochem. Biophys. Res. Commun.* **135**, 269–275.

Ramsay, R. R., Salach, J. I., and Singer, T. P. (1986c). Uptake of the neurotoxin 1-methyl-4-phenylpyridine (MPP⁺) by mitochondria and its relation to the inhibiton of the mitochondrial oxidation of NAD⁺-linked substrates by MPP⁺. *Biochem. Biophys. Res. Commun.* **134**, 734–738.

Ransom, B. R., Kunis, D. M., Irwin, I., and Langston, J. W. (1987). Astrocytes convert the parkinsonism inducing neurotoxin, MPTP, to its metabolite, MPP⁺. *Neurosci. Lett.* **75**, 323–328.

Reed, D., Labarthe, D., Chen, K.-M., and Stallones, R. (1987). A cohort study of amyotrophic lateral sclerosis and parkinsonism-dementia on Guam and Rota. *Am. J. Epidemiol.* **125**, 92–100.

Reynolds, G. P., Pearson, S. J., Halker, J., and Sandler, M. (1988). Brain quinolinic acid in Huntington's disease. *J. Neurochem.* **50**, 1959–1960.

Richter, R. (1945). Degeneration of basal ganglia in monkeys from chronic carbon disulfide poisoning. *J. Neuropathol. Exp. Neurol.* **4**, 324–353.

Riepe, M., Hori, N., Ludolph, A. C., Carpenter, D. P., and Spencer, P. S. (1992). Inhibition of energy metabolism by 3-nitropropionic acid activates ATP-sensitive potassium channels. *Brain Res.* **586**, 61–66.

Rodgers-Johnson, P., Garruto, R. M., Yanagihara, R., Chen, K.-M., Gajdusek, D. C., and Gibbs, C. J. Jr. (1986). Amyotrophic lateral sclerosis and parkinsonism-dementia on Guam: a 30-year evaluation of clinical and neuropathologic trends. *Neurology* **36**, 7–13.

Rosenberg, F., Myers, J., and Martin, W. (1989). Cyanide-induced parkinsonism: clinical, MRI, and 6-fluorodopa PET studies. *Neurology* **39**, 142–144.

Rosling, H. (1990). Cassava associated neurotoxicity in Africa. In *Basic Science in Toxicology* (G. N. Volans, J. Sims, F. M. Sullivan, and P. Turner, Eds.), pp. 605–614, Taylor and Francis, London.

Ross, S. M., Roy, D. N., and Spencer, P. S. (1985). β-N-Oxalylamino-L-alanine: action on high-affinity transport of neurotransmitters in rat brain and spinal cord synaptosomes. *J. Neurochem.* **44**, 886–892.

Ross, S. M., and Spencer, P. S. (1987). Specific antagonism of excitotoxic action of "uncommon" amino acids linked to motor system disease. *Synapse* **1**, 248–253.

Ross, S. M., Seelig, M., and Spencer, P. S. (1987). Specific antagonism of excitotoxic action of "uncommon" amino acids assayed by organotypic mouse cortical cultures. *Brain Res.* **425**, 120–127.

Ross, S. M., Roy, D. N., and Spencer, P. S. (1989). β-N-Oxalylamino-L-alanine action on glutamate receptors. *J. Neurochem.* **53**, 710–715.

Roy, D. N., and Spencer, P. S. (1989). Lathyrogens. In *Toxicants of Plant Origin* (P. S. Cheeke, Ed.), Vol. III, pp. 169–201, CRC Press, Boca Raton, Florida.

Russ, H., Mihatsch, W., Gerlach, M., Riederer, P., and Przuntek, H. (1991). Neurochemical and behavioral features induced by chronic low dose treatment with 1-methyl-4-phenyl-1,2,3,6-tetrahydropyridine (MPTP) in the common marmoset: implications for Parkinson's disease? *Neurosci. Lett.* **123,** 115–118.

Salach, J. I, Singer, T. P., Castagnoli, N. Jr., and Trevor, A. (1984). Oxidation of the neurotoxic amine 1-methyl-4-phenyl-1,2,3,6-tetrahydropyridine (MPTP) by monoamine oxidases A and B and suicide inactivation of the enzymes by MPTP. *Biochem. Biophys. Res. Commun.* **125,** 831–835.

Sanchez-Ramos, J., Barrett, J. N., Goldstein, M., Weiner, W. J., and Hefti, F. (1986). 1-Methyl-4-phenylpyridinium (MPP$^+$) but not 1-methyl-4-phenyl-1,2,3,6-tetrahydropyridine (MPTP) selectively destroys dopaminergic neurons in cultures of dissociated rat mesencephalic neurons. *Neurosci. Lett.* **72,** 215–220.

Schmitt, B., Steinmann, B., Gitzelmann, R., Thunhohenstein, L., and Dumermuth, G. (1992). Dextromethorphan, a N-methyl-D-asparatate antagonist, in the treatment of non-ketotic hyperglycinemia. In *Abstracts of the 30th Annual Symposium of the Society for the Study of Inborn Errors of Metabolism.* O 5.

Schultz, W., Scarnati, E., Sundstrom, E., Tsutsumi, T., and Jonsson, G. T. I. (1986). The catecholamine uptake blocker nomifensine protects against MPTP-induced parkinsonism in monkeys. *Exp. Brain Res.* **63,** 216–220.

Scotcher, K. P., Irwin, I., Delanney, L. E., Langston, J. W., and Di Monte, D. (1990). Effects of 1-methyl-4-phenyl-1,2,3,6-tetrahydropyridine and 1-methyl-4-phenylpyridinium ion on ATP levels of mouse brain synaptosomes. *J. Neurochem.* **54,** 1295–1301.

Seawright, A. A., Brown, A. W., Nolan, C. C., and Cavanagh, J. B. (1990). Selective degeneration of cerebellar cortical neurons caused by cycad neurotoxin, L-β-methylaminoalanine (L-BMAA), in rats. *Neuropathol. Appl. Neurobiol.* **16,** 153–169.

Selye, H. (1957). Lathyrism. *Rev. Can. Biol.* **16,** 1–82.

Shimizu, T., Yasuda, N., Kono, J., Yagi, F., Tadera, K., and Kobayashi A. (1986). Hepatic and spinal lesions in goats chronically intoxicated with cycasin. *Jpn. J. Vet. Sci.* **48,** 1291–1295.

Shiraki, H., and Yase, Y. (1975). ALS in Japan. In *Handbook of Clinical Neurology,* Vol. 22, *System Disorders and Atrophy,* Part 2 (P. J. Vinken, and G. W. Bruyn, Eds.), pp. 353–419, Elsevier Science Publishers, Amsterdam.

Singer, T. P., Salach, J. I., and Crabtree, D. (1985). Reversible inhibition and mechanism-based irreversible inactivation of monoamine oxidases by 1-methyl-4-phenyl-1,2,3,6-tetrahydropyridine (MPTP). *Biochem. Biophys. Res. Commun.* **127,** 341–346.

Singer, T. P., Salach, J. I., Castagnoli, N. Jr., and Trevor, A. J. (1986). Interactions of the neurotoxic amine 1-methyl-4-phenyl-1,2,3,6-tetrahydropyridine with monoamine oxidases. *Biochem. J.* **235,** 785–789.

Slevin, J. T., Collins, J. F., and Coyle, J. T. (1983). Analogue interactions with the brain receptor labeled by (^{3}H) kainic acid. *Brain Res.* **265,** 169–172.

Snyder, S. H., and D'amato, R. J. (1986). MPTP. a neurotoxin relevant to the etiology of Parkinson's disease. *Neurology* **36,** 250–258.

Soffer, D., Amir, N., Elpeleg, O. N., Gomori, J. M., Shalev, R. S., and Gottschalk-sabag, S. (1992). Striatal degeneration and spongy myelinopathy in glutaric acidemia. *J. Neurol. Sci.* **107,** 199–204.

Spatz, M. (1969). Toxic and carcinogenic alkylating agents from cycads. *Ann. N.Y. Acad Sci.* **163,** 848–859.

Spencer, P. S., and Schaumburg, H. H. (1983). Lathyrism: a neurotoxic disease. *Neurobehav. Toxicol. Teratol.* **5,** 625–629.

Spencer, P. S., Schaumburg, H. H., and Cohn, D. F., and Seth, P. K. (1984). Lathyrism: a useful model of primary lateral sclerosis. In *Research Progress in Motor Neurone Disease* (F. C. Rose, Ed.), pp. 312–327, Pitman, London.

Spencer, P. S., Roy, D. N., Ludolph, A., Hugon, J., Dwivedi, M. P., and Schaumburg, H. H. (1986). Lathyrism: evidence for role of the neuroexcitatory amino acid BOAA. *Lancet* **II,** 1066–1067.

Spencer, P. S., Nunn, P. B., Hugon, J., Ludolph, A. C., Ross, S. M., Roy, D. N., and Robertson, R. C. (1987). Guam amyotrophic lateral sclerosis-parkinsonism-dementia linked to a plant excitant neurotoxin. *Science* **237,** 517–522.

Spencer, P. S., and Dastur, D. K. (1989). Neurolathyrism and Neurocycadism. In *Neurological Sciences: An Overview of Current Problems* (D. K. Dastur, M. Shahani, and E. P. Bharucha, Eds.), Section VI, pp. 309–318, Interprint, New Delhi.

Spencer, P. S., Ross, S. M., Kisby, G., and Roy, D. N. (1990). Western Pacific amyotrophic lateral sclerosis: putative role of cycad toxins. In *Amyotrophic Lateral Sclerosis: Concepts in Pathogenesis and Etiology* (A. J. Hudson, Ed.), pp. 263–295, University of Toronto Press, Toronto.

Spencer, P. S., Kisby, G. E., and Ludolph, A. C. (1991a). Slow toxins, biologic markers, and long-latency neurodegenerative disease in the western Pacific region. *Neurology (Suppl. 2)* **41,** 62–66.

Spencer, P. S., Allen, C. A., Kisby, G., Ludolph, A. C., Ross, S. M., and Roy, D. N. (1991b). Lathyrism and western pacific amyotrophic lateral sclerosis: etiology of short and long latency motor system disorders. In *Advances in Neurology,* Vol. 56, *Amyotrophic Lateral Sclerosis and Other Motor Neuron Disease* (L. P. Rowland, Ed.), pp. 287–299, Raven Press, New York.

Spencer, P. S., Ludolph, A. C., and Kisby, G. E. (1992). Are human neurodegenerative disorders linked to environmental chemicals with excitotoxic properties? *Ann. N.Y. Acad. Sci.* **648,** 154–160.

Steele, J. C., Guzman, T. Q., Driver, M. G., Zolan, W., Heitz, L. F., Kilmer, F. H., Parker, C. M., Standal, B. R., Pobutsky, A. M., and Crapper McLachlan, D. R. (1990). Nutritional factors in amyotrophic lateral sclerosis on Guam: observations from Umatac. In *Amyotrophic Lateral Sclerosis: Concepts in Pathogenesis and Etiology* (A. J. Hudson, Ed.), pp. 193–223, Toronto University Press, Toronto.

Stewart, G. R., Zorumski, C. F., Price, M. T., and Olney, J. W. (1990). Domoic acid: a dementia-inducing excitotoxic food poison with kainic acid receptor specificity. *Exp. Neurol.* **110,** 127–138.

Stokke, O., Goodman, S. O., and Moe, P. G. (1976). Inhibition of brain glutamate decarboxylase by glutarate, glutaconate, and β-hydroxyglutarate: explanation of the symptoms in glutaric aciduria. *Clin. Chim. Acta* **66,** 411–415.

Storey, E., Hyman, B. T., Jenkins, B., Brouillet, E., Miller, J. M., Rosen, B. R., and Beal, M. F. (1992). 1-Methyl-4-phenylpyridinium produces excitotoxic lesions as a result of impairment of oxidative metabolism. *J. Neurochem.* **58,** 1975–1978.

Streifler, M., Coh, D. F., Hirano, A., and Schujman, E. (1977). The central nervous system in a case of neurolathyrism. *Neurology* **27,** 1176–1178.

Strolin-Benedetti, M., and Dostert, P. (1989). Monoamine oxidase, brain ageing and degenerative diseases. *Biochem. Pharmacol.* **38,** 555–561.

Takada, M., Li, Z. K., and Hattori, T. (1990). Astroglial ablation prevents MPTP-induced nigrostriatal neuronal death. *Brain Res.* **509,** 55–61.

Teitelbaum, J. S., Zatorre, R. J., Carpenter, S., Gendron, D., Evans, A. C., Gjedde, A., and Cashman, N. R. (1990). Neurologic sequelae of domoic acid intoxication due to the ingestion of contaminated mussels. *N. Engl. J. Med.* **322,** 1781–1787.

Tetrud, J. W., Langston, J. W., Garbe, P. L., and Ruttenber, A. J. (1989). Mild parkinsonism in persons exposed to 1-methyl-

4-phenyl-1,2,3,6-tetrahydropyridine (MPTP). *Neurology* **38,** 1483–1487.

Trolli, G. (1938). Paraplégie spastique épidémique, "Konzo" des indigènes du Kwabgo. In *Résumé des observations réunies, au Kwango, au sujet de deur affections d'origine indéterminée* (G. Trolle, Ed.), pp. 3–36, Fonds Reine Elisabeth, Brussels.

Tryphonas, L., Truelove, J., Nera, E., and Iverson, F. (1990a). Acute neurotoxicity of domoic acid in the rat. *Toxicol. Pathol.* **18,** 1–9.

Tryphonas, L., and Iverson, F. (1990b). Neuropathology of excitatory neurotoxins: the domoic acid model. *Toxicol. Pathol.* **18,** 165–169.

Turski, L., Bressler, K., Rettig, K.-J., Löschmann, P. A., and Wachtel, H. (1991). Protection of substantin nigra from MPP$^+$ neurotoxicity by N-methyl-D-aspartate antagonists. *Nature* **349,** 414–418.

Tylleskär, T., Banea, M., Bikangi, N. Fresco, L., Persson, L. A., and Rosling, H. (1991). Epidemiological evidence from Zaire for a dietary aetiology of *konzo,* an upper motor neuron disease. *Bull. WHO* **69,** 581–589.

Tyulleskär, T., Banea, M., Bikangi, N., Cooke, R. D., Poulter, N. H., and Rosling, H. (1992). Cassava cyanogens and *Konzo,* an upper motoneuron disease found in Africa. *Lancet* **i,** 208–211.

Ueki, A., Chong, P. N., Albanese, A., Rose, S., Chivers, J. K., Jenner, P., and Marsden, C. D. (1989). Further treatment with MPTP does not produce parkinsonism in marmosets showing behavioral recovery from motor deficits induced by an earlier exposure to the toxin. *Neuropharmacology* **28,** 1089–1097.

Uitti, R., Rajput, A., Ashenhurst, E., and Rozdilsky, B. (1985). Cyanide induced parkinsonism: a clinico-pathologic report. *Neurology* **35,** 921–925.

Valenzuela, R., Court, J., and Godoy, J. (1992). Delayed cyanide induced dystonia. *J. Neurol. Neurosurg. Psychiatry* **55,** 198–199.

Vogt, C., and Vogt, O. (1920). Zur Lehre der Erkrankungen des striären Systems. *J. Psychiatry Neurol.* **3,** 627–846.

Vogt, C., and Vogt, O. (1922). Erkrankungen der Großhirnrinde im Lichte der Topistik, Pathoklise und Pathoarchitektonik. *J. Psychol. Neurol.* **28,** 1–171.

Waldvogel, H. J., Faull, R. L. M., Williams, M. N., and Dragunow, M. (1991). Differential sensitivity of calbindin and parvalbumin immunoreactive cells in the striatum to excitotoxins. *Brain Res.* **546,** 329–335.

Weiss, J. H., and Choi, D. W. (1988). β-N-Methylamino-L-alanine neurotoxicity: requirement for bicarbonate as a cofactor. *Science* **241,** 973–975.

Weiss, J. H., Christine, C. W., and Choi, D. W. (1989). Bicarbonate dependence of glutamate receptor activation by β-N-methylamino-L-alanine: channel recording and study with related compounds. *Neuron* **3,** 321–326.

Westlund, K. N., Denney, R. M., Kochersperger, L. M., Rose, P. M., and Abell, C. W. (1985). Distinct monoamine oxidase A and B populations in primate brain. *Science* **230,** 181–183.

Whiting, M. G. (1963). Toxicity of cycads. *Econ. Bot.* **17,** 271–302.

Yase, Y. (1970). Neurologic disease in the Western Pacific islands, with a report on the focus of amyotrophic lateral sclerosis found in the Kii peninsula, Japan. *Am. J. Trop. Med.* **19,** 155–166.

Yase, Y. (1972). The pathogenesis of amyotrophic lateral sclerosis. *Lancet* **ii,** 292–296.

Young, A. B., Greenamyre, J. T., Hollingsworth, Z., Albin, R., D'amato, C., Shoulson, I., and Penney, J. B. (1988). NMDA receptor losses in putamen from patients with Huntington's disease. *Science* **241,** 981–983.

Zhang, Z. X., Anderson, D. W., and Mantel, N. (1990). Geographic patterns of parkinsonism-dementia complex on Guam: 1956 through 1985. *Arch. Neurol.* **47,** 1069–1074.

Zigmond, M. J., and Stricker, E. M. (1989). Animal models of parkinsonism using selective neurotoxins: clinical and basic implications. In *International Review of Neurobiology* (R. J. Bradley, Ed.), Vol. 31, pp. 1–79, Academic Press, New York.

Clinical Approaches
to Neurotoxicology
An Introductory Overview

ROBERT G. FELDMAN
Department of Neurology
Boston University School of Medicine
Boston, Massachusetts 02118

The manifestations of neurotoxicity vary according to the characteristics of the particular neurotoxic substance. The intensity and duration of exposure and the individual susceptibility of the exposed person influence the clinical effect. Intrinsic abilities to metabolize and excrete a toxin, and the ability of vulnerable structures and mechanisms to recover, determine the degree and persistence of impairment.

A diagnosis of a neurotoxic syndrome must be differentiated from a neurological disease of nonneurotoxic etiology. To do this requires an understanding of the pathogenesis of the neurological symptoms and the observed signs; an awareness that particular substances are capable of affecting certain nervous tissues; documenting the exposure in the environment and in tissues of an affected individual; and careful delineation of a time course relationship between exposure and the appearance of symptoms. A high index of suspicion helps in recognizing cases of neurotoxicologic injury. The clinician can focus on probable specific available substances which may have a greater predilection for one part or another of the nervous system when they know what is present in the patient's environment. Constellations of symptoms form syndromes associated with specific neurotoxic exposures. Prompt identification of an affected person is important in order to remove that individual from further exposure and additional injury, and to treat any existing symptoms. In addition, when the earliest presenting case is recognized among a group of individuals exposed to a common neurotoxin, preventative measures can be taken to protect the others from further exposure and/or to treat those who have mild or subclinical symptoms.

In this section, the contributors review parameters of neurologic function assessed by standardized neuropsychological tests and quantified neurophysiological measures. Along with supporting historical information and identifying biological markers of exposure, the clinical findings of a given case or group of individuals exposed to neurotoxins are placed into the context of these data to reach a conclusion and to formulate a proper diagnosis.

CHAPTER
44

The Recognition and Differentiation of Neurotoxic and Nonneurotoxic Syndromes

ROBERT G. FELDMAN
Department of Neurology
Boston University School of Medicine
Boston, Massachusetts 02118

I. Clinical Manifestations

Vulnerability of the various target neural structures, as well as the dose and type of neurotoxin involved, determines the clinical manifestations (Table 1). A neurotoxic effect may be direct or indirect, reversible or irreversible, depending on the specific substance and the nature and duration of exposure. Certain individuals in a commonly exposed group may be affected whereas others are not because of greater susceptibility and specific cellular biology. Often, natural protective mechanisms expected to prevent the effects of exposure to a neurotoxin by detoxifying and eliminating a nontoxic by-product may fail.

An alteration in protein or lipid content of cell membranes, disturbances in oxidative processes, ionic imbalances, defective neurotransmitter activity, or damage to capillary endothelium are some of the mechanisms that lead to the swelling of nerve cells or astrocytes and increased interstitial fluid which interfere with the transmission of nerve impulses and performance of the nervous system, resulting in dysfunction.

Irreversible damage to particular systems causes permanent clinical effects which interfere with function so much that the person is considered disabled.

Neurotoxic effects follow a time course from onset after an episode(s) of exposure to amelioration or recovery after cessation of exposure. An idiopathic (nontoxic) neurologic disease usually follows a progressive course with continued development of symptoms in the absence of any identified exposure to neurotoxins. The prompt appearance of neurologic effects after acute exposure to neurotoxins is more obvious than the emergence of neurologic symptoms following a latent period. In many cases, observations after a longer period of follow-up are necessary in order to determine persistence and/or progression of symptoms, improvement with resolution, or evidence of residual impairments. Locations of residence and employment, work tasks, hobbies, and all possible sources of exposure and injury, previous medical encounters, and accidents must be related to a time line. A neurotoxicological exposure questionnaire is useful in chronicling the onset of neurologic impairments in relation to circumstances of the probable presence (or absence) of a suspected neurotoxic sub-

TABLE 1 Chemical Exposures and Associated Neurotoxic Syndromes[a]

Neurotoxin	Sources of exposure	Clinical diagnosis	Locus of pathology
Metals			
Arsenic	Pesticides	Acute: Encephalopathy	Unknown (a)
	Pigments	Chronic: Peripheral	Axon (c)
	Antifouling paint	neuropathy	
	Electroplating industry		
	Seafood		
	Smelters		
	Semiconductors		
Lead	Solder	Acute: Encephalopathy	Blood vessels (a)
	Lead shot	Chronic: Encephalopathy and	Axon (c)
	Illicit whiskey	peripheral neuropathy	
	Insecticides		
	Auto body shop		
	Storage battery manufacturing		
	Foundries, smelters		
	Lead-based paint		
	Lead pipes		
Manganese	Iron, steel industry	Acute: Encephalopathy	Unknown (a)
	Welding operations	Chronic: Parkinsonism	Basal ganglia neurons (c)
	Metal-finishing operations		
	Fertilizers		
	Manufacturers of fireworks, matches		
	Manufacturers of dry cell batteries		
Mercury	Scientific instruments	Acute: Headache, nausea,	Unknown (a)
	Electrical equipment	onset of tremor	
	Amalgams	Chronic: Ataxia, peripheral	Axon (c)
	Electroplating industry	neuropathy, encephalopathy	Unknown (c)
	Photography		
	Felt making		
Tin	Canning industry	Acute: Memory defects,	Neurons of the limbic system
	Solder	seizures, disorientation	(a and c)
	Electronic components	Chronic:	Myelin (c)
	Polyvinyl plastics	Encephalomyelopathy	
	Fungicides		
Solvents			
Carbon disulfide	Manufacturers of viscose rayon	Acute: Encephalopathy	Unknown (a)
	Preservatives	Chronic: Peripheral	Axon (c)
	Textiles	neuropathy, parkinsonism	Unknown
	Rubber cement		
	Varnishes		
	Electroplating industry		
n-Hexane, methyl butyl ketone	Paints	Acute: Narcosis	Unknown (a)
	Lacquers	Chronic: Peripheral	Axon (c)
	Varnishes	neuropathy	
	Metal-cleaning compounds		
	Quick-drying inks		
	Paint removers		
	Glues, adhesives		
Perchloroethylene	Paint removers	Acute: Narcosis	Unknown (a)
	Degreasers	Chronic: Peripheral	Axon (c)
	Extraction agents	neuropathy, encephalopathy	Unknown
	Dry cleaning industry		
	Textile industry		
Toluene	Rubber solvents	Acute: Narcosis	Unknown (a)
	Cleaning agents	Chronic: Ataxia,	Cerebellum (c)
	Glues	encephalopathy	Unknown
	Manufacturers of benzene		
	Gasoline, aviation fuels		
	Paints, paint thinners		
	Lacquers		

(continued)

TABLE 1 Continued

Neurotoxin	Sources of exposure	Clinical diagnosis	Locus of pathology
Trichloroethylene	Degreasers Painting industry Varnishes Spot removers Process of decaffeination Dry cleaning industry Rubber solvents	Acute: Narcosis Chronic: Encephalopathy, cranial neuropathy	Unknown (a) Unknown (c) Axon (c)
Insecticides Organophosphates	Agricultural industry: Manufacturing and application	Acute: Cholinergic poisoning Chronic: Ataxia, paralysis, peripheral neuropathy	Acetylcholinesterase (a) Long tracts of spinal cord (c) Axon (c)
Carbamates	Agricultural industry: Manufacturing and application Flea powders	Acute: Cholinergic poisoning Chronic: Tremor, peripheral neuropathy	Acetylcholinesterase (a) Dopaminergic system (c)

a From Feldman (1990, 1994).

stance and the emergence of any medical complaint. Confirmation of the probable cause of the neurologic impairment or disability depends on the availability of reliable epidemiologic and environmental data concerning the neurotoxic substances. A clinical approach to the evaluation of neurotoxic effects begins with a careful history followed by a detailed neurological examination and the use of selected confirmatory tests (Feldman and Travers, 1983; Anger, 1990).

II. Neurological Examination

A neurological examination systematically assesses a person's ability to communicate, to comprehend, and to follow directions; to solve problems; to perform coordinated motor functions; and to perceive and identify sensations of various modalities. Involuntary reflex functions are tested to determine the intactness of afferent and efferent nerve pathways. Deviations from expected performance levels for an exposed individual are considered abnormal neurologic signs. Motor weakness, sensory loss, or altered mental status results from damage to specialized cells of the brain, spinal cord, and peripheral nerves. Similar neurologic signs and symptoms arise from damaged neural structures, regardless of the specific cause of the given malady. Neurotoxic symptoms may resemble those found in primary neurologic disease.

Careful consideration of clinical similarities must be given when making a differential diagnosis. For example, headache whether due to simple muscle strain, migraine, a tumor, or an unruptured aneurysm must be differentiated from symptomatic headache associated with toxic exposure to carbon monoxide, carbon diox-

ide, lead, zinc, or nitrates. Cognitive deficits associated with Alzheimer's disease, arteriosclerosis, or pseudomentia of depression must be differentiated from the neurobehavioral effects of toxic exposure to organic solvents, metals, or insecticides. Transient disturbances of awareness or epileptic seizures, with or without associated motor involvement, must be identified as a primary diagnosis distinct from the similarly appearing disturbances of consciousness or convulsions related to neurotoxic effects. Parkinson's disease is characterized by rigidity, slowness of movement, and tremor. However, a similar syndrome occurs in certain individuals exposed to phenothiazine medications, byproducts of meperidine production (1-methyl-4-phenyl-1,2,3,6-tetrahydropyridine), carbon disulfide, carbon monoxide, or manganese. Multiple sclerosis has protean clinical features due to the disseminated nature of white matter (myelin) lesions in the brain and spinal cord; clinical manifestations similar to those of multiple sclerosis may be misdiagnosed in an individual who develops demyelination of the spinal cord pathways after exposure to tri-*o*-ocresyl phosphate or mercury. Peripheral neuropathy developing in patients with diabetes shows a clinical picture similar to the neuropathy following methyl *n*-butyl ketone, *n*-hexane, or arsenic poisoning. Cases of motor neuropathy and/or anterior horn cell disease have been linked to exposure to lead and mercury.

A carefully performed neurological examination is followed by the appropriate electrodiagnostic, neuropsychologic, and imaging techniques. Results of general medical and neurologic examinations are assessed in relation to the significance of biological markers and exposure data in order to ascertain that a neurotoxic syndrome does, indeed, exist.

III. Neurotoxicological Syndromes

A. Central Nervous System

Severe central nervous system effects of acute exposure include headache, seizures, or behavioral changes. More chronic manifestations are less obvious than acute responses. In the initial stages of behavioral toxicity, mood changes are the earliest symptoms. Depression may also be observed along with disturbances in sleep, apathy, fatigue, ability, and diminished mental efficiency. This group of symptoms is referred to as organic affective syndrome, which is transient and disappears when exposure ends. A persistent mood disorder associated with impaired cognitive functioning indicates a mild chronic toxic encephalopathy. Manifestations begin insidiously and progress during exposure, subsiding gradually after cessation of exposure to neurotoxins. Serial neuropsychological testing is necessary to document impairment in these patients. Changes in routine activities of daily living, social interaction, and work performance reflect the effects of mild chronic toxic encephalopathy.

Continuous and/or intermittent exposure to neurotoxins results in the appearance of a severe chronic toxic encephalopathy, similar to other diffuse brain disorders. Disturbances in recent memory and learning affect the ability to adapt to new situations, thus interfering with many social and problem-solving skills. Although long-term memory is often preserved, abstract thinking, judgement, and personality changes are typically affected. Depressive symptoms are observed and may be a response to severe cognitive impairment experienced by the patient, similar to a post-traumatic stress effect. Severe chronic toxic encephalopathy develops slowly and insidiously while exposure takes place. However, it is not relentlessly progressive. In this way, it is distinct from other types of progressive dementia, such as Alzheimer's disease and multi-infarct dementia.

To be certain that the neurobehavioral findings are not due to structural abnormalities such as cerebrovascular disease or brain tumor, imaging techniques are needed. Computerized axial tomogram and/or magnetic resonance image will reveal the presence of such lesions, as well as the evidence of old brain trauma.

B. Peripheral Nervous System

Reduced or absent tendon reflexes and patterns of sensory loss greater in the lower extremities than upper are features of peripheral neuropathy. A gradual onset of intermittent tingling and numbness progressing to a lack of sensation and dysesthesia, muscle weakness, and atrophy results from damage to the motor and sensory nerve fibers. Involvement of nerve fibers in toxic polyneuropathies may be subdivided into those that affect primarily the axon (axonopathy), which present as distal sensory motor loss (most evident in the lower extremities where the axons are the longest); myelin sheath disorders (myelinopathies); those that result in spotty segmental demyelination; and possibly secondary axonopathy.

In the study of toxic neuropathies, an electrophysiologic survey of peripheral nerve function must include measurements of motor and sensory conduction in the upper and lower extremities. Special attention should be given to the primarily sensory-conducting characteristics of the sural nerve in the leg. This is of great value when the sural nerve is subsequently used for a nerve biopsy and the correlation between teased nerve fiber histology and the conduction characteristics. Differential diagnosis of proximal segments versus distal segments of nerve is used in identifying a distal toxic axonopathy or in identifying a focal neuropathic block of conduction. Understanding the pathophysiology in a suspected neurotoxic polyneuropathy has value because certain substances are recognized by their specific neurotoxic effect on peripheral nerve (Table 1).

IV. Formulation of Neurotoxicity Diagnosis

A clinical diagnosis is arrived at by the process which integrates subjective observations of the physical responses produced in the subject by the effects of a neurotoxin. Sets of physiological, anatomical, and behavioral concepts and principles accumulated from experience and based on a data base derived from previously published literature serve as the basis for formulating a diagnosis. The process of arriving at an opinion about causation of a diagnosis of neurotoxic disease is similar to the process of clinical diagnosis in the everyday practice of medicine, where professional judgement can determine the connections in a particular instance without a standard template routinely applied in every case.

The Boston University Environmental Neurology Assessment (Feldman *et al.*, 1993) offers a strategy to reach a diagnosis and to answer certain critical questions. Following this strategy (Table 2) has allowed for the recognition and ascertainment of neurotoxic cases.

TABLE 2 Boston University Environmental Neurology Assessment (BUENA)

I. Are the data sufficient to identify any or all complaints as being caused by a neurotoxin?
  A. List complaints and relate them on a time line to all possible exposures to sources of chemicals (work, home, hobby).
    1. Identify symptoms and functional changes expressed, experienced, and observed by others; list evidences of mood, anxiety, sleep disturbances, and effect on quality of life.
    2. Cite time of onset, duration, and intensity of complaints. Indicate when symptoms worsen or remit in relation to exposure (e.g., work week, weekend, time of shift, on vacation).
    3. Evaluate subject's family/genetic health, special sensitivities, and possible congenital factors.
  B. List all substances and how they are used (at workplace, home, hobbies).
    1. Obtain chemical names (not trade label names), material safety data sheets, and other identifying data concerning each substance.
    2. Review workplace information available, e.g., OSHA mandated material safety data sheets and employer training program materials; employer's medical records and exposure records which, if kept by employer, must be made available under OSHA rules. Review, if available, the following: employer's TSCA 8c and 8e reports to EPA, employer's community right to know reports to local officials regarding hazardous materials made, used, or sorted on site.
  C. Obtain environmental and industrial hygiene air measures or drinking water samplings to prove the presence of alleged chemicals in the alleged source. Current levels are important, but levels taken in relationship to occurrence of complaints are essential.
  D. Obtain urine and/or blood samples from the alleged affected individuals and from known unexposed control subjects of similar age and occupation, especially at time of complaints, to establish body burden of chemical.
  E. For suspect chemicals, develop information on dose–response relationships, animal studies, and toxicological and epidemiological studies.
II. Are the complaints substantiated by clinical neurological physical examination, standardized neuropsychological and neurophysiological tests, and appropriate blood and urine analyses? Also, are the complaints corroborated by epidemiological, toxicological, and animal studies; by NIOSH, OSHA, or EPA studies of the workforce or community; and by employer studies and reports to EPA or OSHA (e.g., TSCA 8c and 8e reports)?
III. Are the findings due to a primary neurological disease or other medical condition?
IV. Are the findings on examination explained by any other causal factors in past medical history, previous, and/or current unrelated exposures to substances from sources other than the one under consideration?
  A. Time line of past jobs, residences.
  B. Time line of past medical history.
V. Analyze individual cases for confirmatory studies; group data for cluster analysis and/or population statistical study.
VI. Identify and critically review previously published and/or reported case reports, case control studies, population studies, and animal studies concerning the alleged neurotoxins and relate documentation to case data.
VII. Estimate the damage consequences for the subject: disease, anxiety, loss of consort, functional impairments, need for special education, counseling, or medical surveillance, need for medical therapeutic measures, job disability, loss of earnings, etc.
VIII. Reevaluate after reasonable absence from all neurotoxic exposure to assess course of progression, recovery, or persistent impairment and/or disability.

References

Anger, K. W. (1990). Worksite behavioral research: results, sensitive methods, test batteries and the transition from laboratory data to human Health. *Neurotoxicology* **11,** 629–720.

Feldman, R. G., and Travers, P. H. (1984). Environmental and occupational neurology. In *Neurology: The Physicians Guide* (R. G. Feldman, Ed.), Thieme-Stratton, New York.

Feldman, R. G. (1990). Effects of toxins and physical agents on the nervous system. In *Neurology in Clinical Practice* (W. G. Bradley, R. B. Daroff, G. M. Fenichel, and C. D. Marsden, Eds.), pp. 1185–1209, Butterworth Publishers, Stoneham, Massachusetts.

Feldman, R. G., White, R. F., Eriator, I. I., Jabre, J. F., Feldman, E. S., and Niles, C. A. (1994). Neurotoxic effects of trichloroethylene in drinking water: approach to diagnosis. In *The Vulnerable Brain and Environment Risks,* Vol. 3, *Special Hazards from Air and Water,* (R. L. Isaacson and K. F. Jensen Eds.).

Feldman, R. G. (1994). Clinical syndromes associated with neurotoxicity. In *The 4th Edition of Encyclopedia of Occupational Health and Safety* (J. M. Stellman, Ed.), Geneva, Switzerland, in press.

CHAPTER
45

Exposure Assessment in Clinical Neurotoxicology: Environmental Monitoring and Biologic Markers

CHANG-MING CHERN
Neurology Institute
Veterans General Hospital
Taipai, Taiwan, Republic of China

SUSAN P. PROCTOR
Environmental and Occupational
Neurology Program,
Department of Neurology
Boston University School
of Medicine
Boston, Massachusetts 02118-5350

ROBERT G. FELDMAN
Department of Neurology
Boston University School
of Medicine
Boston, Massachusetts 02118

I. Introduction

In order for a neurotoxic syndrome to be attributed to exposure to a specific hazardous agent in a clinical setting, several requirements must be met: (1) the toxic substance is found in the environment to which the patient was exposed and/or measured by biologic markers indicative of exposure; (2) the detected neurologic deficits are compatible with the documented exposure; (3) the temporal sequence is correct such that exposure proceded the onset of clinical deficits; and (4) other explanations for similar neurologic manifestations have been satisfactorily eliminated. Thus, documentation of exposure to hazardous agents is a crucial step for diagnosis and differential diagnosis of a neurotoxic syndrome.

The traditional approach for assessing occupational or environmental exposure to neurotoxic substances has been based on measurement of the toxins in the appropriate environment, e.g., ambient air, soil, water, and foods, and may include numerous techniques to determine or estimate the contaminant, its source, the environmental media of exposure, routes of entry to the body, and the intensity, frequency, and duration of contact (NRC, 1991a). There is a growing interest in the use of biologic markers to supplement environment monitoring data in evaluating toxic exposures. In the clinical setting, the specific features of each individual case and the suspected hazardous agent(s) of interest determine the most appropriate and feasible exposure assessment method. Persons may report to a clinic with neurological symptoms experienced as a result of a specific acute exposure episode (e.g., chemical accident), some may have had a chronic exposure experience (e.g., exposure over many years in the workplace or through environmental sources such as drinking water), or perhaps the source of toxic exposure is undetermined. Exposure may be recent or it may have occurred sometime in the past.

The techniques that provide the most information about individual exposures are personal measurements. However, often it is not possible to obtain personal exposure or dose measures and more indirect, surrogate measures must be utilized in order to estimate exposures, including area samples at the work site or in a residence, estimates of drinking water exposure based on estimated intake levels, distance of the residence from the hazardous waste site, length of residence in a certain area, or duration of employment (NRC, 1991a). Quantifiable personal measurements can include personal air sampling (or personal monitoring, e.g., following an individual over time) and/or the use of biologic markers (e.g., biochemical, histological, or physiological indicators of neurotoxicant exposure or effect). Biologic markers serve as potential indicators of total uptake, reflecting intake from multiple routes and fluctuating exposure (Wilcosky and Griffith, 1990). Measurement of biologic markers in some cases has distinct advantages over environmental monitoring. Concentrations of a chemical in the air, measured by environmental monitoring method, often do not closely correlate with actual amounts absorbed because they do not account for certain physical and biological factors (e.g., variation in the concentration of chemicals at different locations and points in time, solubility, and particle size characteristics, use of personal protection devices, workload demand effects on respiratory volume, individual nutritional status, and body composition) which affect the uptake of chemical in the body (Aitio et al., 1988). However, biologic monitoring is of little value for assessing exposure to substances that are poorly absorbed or that exhibit toxic effects at the sites of first contact (Lauwerys, 1991) whereas environmental monitoring is more advantageous under such conditions. Therefore, environmental and biologic monitoring should be viewed as complementary methods (Ashford et al., 1990; Lauwerys, 1991).

In order to evaluate an association between a particular exposure and an identifiable outcome, it is necessary to define as accurately as possible the dose experienced by the patient. A distinction should be made between estimation of the exposure and of the dose to the individual. Exposure refers to the quantity or concentration of the agent present in the environment which is assumed to be absorbed by the individual and is assessed in reference to both the intensity encoun-

tered as well as the duration. Dose refers to the amount actually absorbed and is determined by both exposure features and characteristics of the individual (Checkoway et al., 1989; Johnson, 1992). Depending on the patient's exposure scenario, one may not be able to determine the actual dose.

In Section II, important points regarding the assessment of neurotoxicant exposure by environmental measures are outlined as they pertain to the clinical setting. Neurotoxic agents can be placed in three distinct categories, chemical, physical, and biological; however, the primary focus in this chapter is assessment of exposure to chemical agents. Sources of information to assist in the estimation of past exposures are described, along with industrial hygiene methods and environmental sampling strategies to identify the potential current neurotoxic exposures in the workplace or in the environment. The use of biologic markers in the assessment of exposure for the individual patient is examined in Section III. In Section IV, a case of arsenic neurotoxicity is presented to demonstrate the use of environmental and biologic measures in exposure assessment in the clinical setting and to discuss the strengths and weaknesses of these data.

II. Assessment of Exposure by Measuring the Neurotoxicant Levels in the Environment

The assessment of exposure by measurement of neurotoxicant levels in the environment may involve reconstructing exposure histories from data collected in the past or, when possible, may involve a determination of current exposure levels. In any case, the first step is to determine what is the chemical exposure(s) of concern. This may be straightforward; for example, a worker knows that he has been exposed to lead. However, in some cases, the neurotoxic agent is suspected but is unconfirmed or the source is undetermined. A review of the literature concerning the clinical signs and symptoms to possible known neurotoxic agents can offer a starting point. However, it may not be helpful since the full extent of neurologic effects due to exposure to many potentially toxic chemicals is still unknown (OTA, 1990). Few of the 70,000 chemicals in commercial use have been tested for neurotoxicity (NRC, 1992a) by any standardized research protocol. Also, in evaluations of hazardous waste sites, many potential contaminants have not been carefully studied. There are compounds that cannot be identified by standardized analytic methods, unidentified substances that result from in situ transformation processes, or are by-products of treatment processes (NRC, 1991a). Some identified sources of exposure to chemicals with the potential to adversely affect the nervous system are presented in Table 1.

Often it is necessary to combine the literature review with an extensive questionnaire of the patient's environmental, occupational, and medical history in order to identify potential hazardous substances thought to be capable of inducing the observed or reported effects. A well-designed and validated questionnaire which obtains information about the patient's occupational and environmental history can be extremely helpful in characterizing the circumstances of the suspected or reported exposures. The person should be asked about his/her current and past work history, use of personal protection equipment (e.g., masks, gloves, special work clothes), and residential history in order to ascertain if the individual experienced significant sources of exposure and, if so, to get an qualitative estimate of exposure dose.

A significant problem with questionnaire information is the fact that it is self-reported; hence, attempts to check the accuracy of the information reported should be undertaken if possible, e.g., review of environmental and occupational records, of the description of work practices, and of past medical records. Careful probing with a questionnaire and then a follow-up interview and record review by a trained environmental specialist can assist in identifying possible confounding exposures that need to be considered in addition to the suspected contact with neurotoxicants.

It is important to recognize the role of possible home and hobby exposures to accurately define exposure and to rule out other confounding exposures. For example, a person may work all day in a lead battery plant and also have a hobby of making stained glass. As this hobby involves working with lead solder, this person will have a higher dose of lead than his co-worker. In addition, most persons may not be aware of a chronic low level exposure to a chemical or chemicals everyday. Voluntary exposures to many legal and illegal materials can cause neurotoxic effects (NRC, 1992b). Substances with psychoactive effects such as alcohol, cocaine, toluene in glue, and amphetamines are just a few of the examples. Chronic alcohol abuse may cause symptoms indistinguishable from chronic organic solvent intoxication (Juntunen, 1982). In other circumstances, prescribed medication can have neurologic side effects as well (Sterman and Schaumburg, 1980). Also, some natural components in foods or food additives have been associated with specific neurotoxic syndromes (Abou-Donia, 1992).

A. Sources of Environmental Exposure Information Regarding the Neurotoxic Agent(s) of Concern

1. Past Exposure Levels

For occupational exposure determination, information on a subject's past job history at a particular work site can be obtained by review of company records, such as work history and production records. This type of information can also be used to validate the self-reported questionnaire responses and provide quantitative data not usually available from the subject's recollection, e.g., hours worked, number of sick days, or time spent in a specific job category. Within larger companies, the health and safety office may have exposure documentation from compliance testing or prior site visits. This is often the only information available about exact past exposure levels. Some work sites may have an environmental monitoring program in place to meet certain regulations for specific chemical exposures to detect actual or possible overexposure in the workplace and whether there is a related increased health risk (Ashford *et al.*, 1990). In addition to records of past exposure levels, other sources of information can be reviewed, e.g., industrial process records and information from chemical suppliers (material safety data sheets).

Procedures to measure workplace exposures depend largely on the main purpose of the monitoring program. There are two kinds of air sampling methods that may be in place: area sampling and personal sampling (Cohen, 1992; WHO, 1986). If the objective is to determine compliance with regulations or to locate a pollution source, workroom area samples would be taken from appropriate locations (WHO, 1986). In contrast, extensive and continuous data may be required for ongoing exposures and within the time frame of a daily work schedule can be gathered by personal sampling. Personal sampling estimates the individual's dose by sampling the air near the breathing zone (front part of a 2-feet-diameter sphere around the head), throughout the working hours (Cohen, 1992; WHO, 1986). The important factors to consider in critically reviewing records of environmental monitoring include (1) appropriate sampling and a valid analytic method, (2) type of samples (area versus personal sampling), (3) timing of samples, (4) use of personal protective equipment, and (5) consideration of all possible routes of exposure. It should not be assumed that exposure occurred only to the selected monitored substances (Melius *et al.*, 1988). Also, it is important to note that exposure to a combination of neurotoxicants may be of greater risk than exposure to each individually, e.g., hexane with

TABLE 1 Neurotoxins and Major Sources of Exposures[a]

Neurotoxin	Major uses/sources of exposure	
Metals		
Arsenic	Pesticides	Pigments
	Antifouling paint	Electroplating industry
	Seafood	Smelters
	Semiconductors	
Lead	Lead shot	Auto body shops
	Illicit whiskey	Storage battery manufacturing plants
	Solder	Foundries
	Lead-based paint	Smelters
	Lead-stained glass	Lead pipes
	Insecticides	
Manganese	Fertilizers/pesticides	Iron, steel industry
	Welding operations	Manufacturing fireworks, matches
	Metal finishing operations of high	Manufacturing using oxidation catalysts
	manganese steel	Manufacturing dry cell batteries
Mercury	Amalgams	Electroplating industry
	Scientific instruments/thermometers	Photography
	Textiles	Electrical equipment
	Pigments	Taxidermy
	Felt making	
Nickel	Paints	Electroplating industry
	Inks	Nickel-cadmium batteries
	Surgical and dental instruments	Alloys
	Coinage	
Tin	Silverware	Canning industry
	Solder	Electronic components
	Polyvinyl plastics	Fungicides
	Coated wire	
Tellurium	Rubber vulcanization	Coloring agent in glazes, glass
	Thermoelectric devices	Electronic industry
	Semiconductors	Foundries
Thallium	Rodenticides	Mercury and silver alloys
	Fungicides	Manufacturing special lenses
	Photoelectric cells	Infrared optical instruments
Solvents		
Carbon disulfide	Paints	Manufacturing vicose rayon
	Preservatives	Electroplating industry
	Rubber cement	Textiles
	Varnishes	
Methyl butyl ketone	Paints	Metal cleaning compounds
	Lacquers	Paint removers
	Varnishes	Quick-drying inks
Hexane	Lacquers	Pharmaceutical industry
	Rubber cement	Glues
	Printing inks	Stains
Perchloroethylene	Paint removers	Dry cleaning industry
	Degreasing solvents	Extraction agent for vegetables and mineral oils
	Textile industry	
Toluene	Rubber solvents	Cleaning agents
	Glues	Manufacturing benzene
	Gasoline	Paints
	Automobile, aviation fuels	Lacquers
	Paint thinners	
Trichloroethylene	Degreasing solvents	Painting industry
	Paints	Process of extraction of caffeine from coffee
	Rubber solvents	Dry cleaning industry
	Lacquers	Adhesive in shoe and boot industry
	Varnishes	

(continued)

TABLE 1 Continued

Neurotoxin	Major uses/sources of exposure	
Gases		
Carbon monoxide	Enclosed areas: mines, tunnels	Acetylene welding
		Exhaust fumes of internal combustion engines
Waste anesthetics	Operating rooms	Dental offices
Monomers		
Acrylamide	Paper, pulp industry	Photography
	Dyes	Water, waste treatment facilities
	Grouting material	
Pesticides		
Organophosphates	Agricultural industry	Field runoff
Chlorinated hydrocarbons (DDT, kepone, chlordane)	Agricultural industry	Dumping
Carbamates	Agricultural industry	Field runoff
	Spraying	

a From White *et al.* (1992).

methyl ethyl ketone vs hexane alone (Andrews and Snyder, 1986).

For determination of environmental exposures, records of drinking water quality or air quality required by law have become more available in recent years and include data for more chemicals. For example, the Agency for Toxic Substance and Disease Registry (ATSDR) has recently established a database (HAZDAT) of all the environmental contamination data collected by the U.S. Environmental Protection Agency (USEPA) for the hazardous waste sites on the national priority list. Also, the USEPA has a database system containing air monitoring data (Johnson, 1992). Environmental media evaluation guides (EMEGs), which are media-specific comparison values suggested by ATSDR, can help identify contaminants of concern at hazardous waste sites (ATSDR, 1992a).

2. Current Exposure Data

A comprehensive industrial hygiene (IH) investigation will help determine the range of possible current chemical exposures to the particular worker on the job and the level of exposure dose. In such an investigation, an IH team goes into the plant, observes the job(s) in action, and measures air sample concentrations for relevant chemicals used in the subject's work space and/or the actual amount he/she breathes in during the job. It requires a trained IH supervisor and staff as well as proper monitoring equipment. In the survey, the work processes, work conditions, and the raw materials used and any intermediate products should be described to provide information on where potential exposure is possible (WHO, 1986; Cohen, 1992). Job activities are reviewed with plant personnel to learn

where the worker is located when doing his/her jobs and when on break and exactly how the job assignments are carried out (Cohen, 1992). Observations of all plant operations and discussions with both plant personnel and workers who know the potential problems and are aware of complaints or symptoms among the workers are made to assist in establishing the source of the chemical contamination (WHO, 1986).

In environmental settings involving exposure to hazardous wastes in the water, soil, or air, often the only direct method to determine current exposures is a monitoring study of the environment through either personal or area sampling. As with an industrial hygiene investigation, this involves a trained team of engineers, hydrologists, and geologists to accurately perform and interpret measurements (ATSDR, 1992b).

B. Quantification of Exposure

Since many neurotoxic syndromes manifest after prolonged exposure, quantification of cumulative exposure is necessary for assessing the dose–response relationship in such cases. Borrowing from epidemiologic methods (Checkoway *et al.,* 1989; Smith, 1987) and depending on the availability of past environmental monitoring records, modeling using "historical" exposure records may be performed to estimate a patient's cumulative exposure dose. With the levels of the toxic agent at different times for different job assignments in the workplace and a list of job tasks for the patient as an employee in that company, a job matrix can be constructed and an estimate of the cumulative exposure can be determined. This represents a time-integrated measure which is the summation of concentration over time. For example, for a particular job during a specific period of time, the exposure can be quantified by multiplying the actual working time at that job (vacations and leaves not included), rate of intake of that agent (e.g., inhalation rate for airborne contaminants), and the level of the agent during that period. Then, by adding up each exposure at each job, the cumulative exposure to that specific agent for this patient can be estimated.

Past monitoring data may not be available for various reasons such as sampling and/or analytic limitations or workplace policies. Then, exposure status can only be assessed by measuring the current level of the toxic agent in the environment and making assumptions to estimate one's past exposure. Nevertheless, a semiquantitative cumulative exposure can be obtained in certain occasions.

C. Interpretation of the Exposure Data

In addition to reviewing the toxicologic, clinical, and epidemiologic literature, which describe previously reported examples of neurotoxic effects of known chemicals and often document exposure data, reference to occupational and environmental standards is helpful in interpreting the findings in suspected or recognized cases of neurotoxic syndromes. Current "reference levels" for many toxic agents have been based on data obtained from past environmental monitoring. The most readily available reference data are regulatory or advisory guidelines provided by national or international health and environmental agencies. These include the Occupational Safety and Health Administration's permissible exposure levels set for many toxic agents in the workplace and the National Institute for Occupational Safety and Health's (NIOSH) recommended exposure levels (NIOSH, 1990), the American Conference of Government Industrial Hygienists' (ACGIH) threshold limit values (ACGIH, 1992–1993), and the USEPA regulatory statements about limits of contamination for individual contaminants in various environmental media [e.g., pesticides in vegetables and lead levels in soil and drinking water (Johnson, 1992)]. Interested readers may also refer to the chapter on permissible exposure limits and recommended exposure limits in the book *Preventing Occupational Disease and Injury* (Weeks *et al.*, 1991). Environmental criteria can be obtained from the local U.S. EPA region office or by referring to related EPA reports, such as air quality criteria for lead (EPA, 1986).

However, there are some important factors to consider when interpreting regulatory standard or permissible exposure limits. First, the working environment that a reference level is assumed to represent may be different from the environment being assessed for the particular patient under evaluation. Under different working conditions, workers may be vulnerable to a level different than the permissible level, e.g., peak fluctuations in work processes. Also, persons may be working overtime or a second job so that comparison to the reference level, which is usually based on a 8-hr 5 days a week schedule, would not be comparable. Second, a daily 8-hr exposure to levels below these exposure limits may be erroneously regarded as harmless to the workers' health. In fact, permissible exposure levels are not based solely on current toxicologic and epidemiologic knowledge of adverse health effects but also on technical and economical limits. Finally, differences in individual variation and susceptibility must be considered.

III. Assessment of Exposure by Biologic Markers

Many biochemical profiles, such as liver function assays, have long been used in clinical practice as biologic markers of exposure and response after ethanol exposure. Recent technologies involving detection of changes at the cellular and molecular level, e.g., DNA adducts, have opened up a new field of research involving biologic markers of exposure to chemical carcinogens (Committee on Biologic Markers of the National Research Council, 1987). With these new markers there is the potential for greater analytical sensitivity that can lead to the identification of events that occur all along the continuum between exposure and clinical disease (NRC, 1991b). However, in clinical neurotoxicology there are relatively few biologic markers of exposure that have been studied, that give direct information about target organ dose, and that can be related to eventual clinical disease. More research funding at the "interface" between basic neurosciences and clinical epidemiology is needed (Silbergeld and Takeuchi, 1993) before potential markers can be validated in research studies for subsequent use in a clinical setting.

The following sections describe general characteristics of biologic markers and point out areas where insufficient research limits the use of these markers in clinical neurotoxicology.

A. Definition and Classification of Biologic Markers

Biologic markers are indicators of a change or variation in cellular or biochemical components or processes, structures, or functions measured in a biologic sample (NRC, 1992c). In clinical neurotoxicology, biologic markers can lead to earlier detection and treatment of adverse neurological effects or disease (Committee on Biologic Markers of the NRC, 1987; NRC, 1992c) as they can represent different stages along the progression from exposure to clinical disease (Fig. 1). Biologic markers of exposure are measures of the internal dose (ID), the biologically effective dose (BED), or of the product of an interaction between a substance and some target molecules or other nervous system receptor. Biologic markers of effect are measurable biochemical or physiological alterations within the nervous system that can be associated with an established or potential health impairment or disease. They include markers of early biologic effects, markers of altered structure and function changes more closely related to the development of disease, and markers of clinical disease. Biomarkers of susceptibility are indicators of increased (or decreased) risk and can affect the measurement and interpretation of other biomarkers along the continuum from exposure to disease (NRC, 1991b).

Whether a marker is an indicator of exposure, disease, or susceptibility may depend on the current state of knowledge (NRC, 1991b). Thus, the assignment of markers to one of these categories may overlap. For instance, blood acetylcholinesterase activity has been used as a biological marker of exposure to organophosphate insecticides, and inhibition greater than 30% is generally regarded as an indication of a significant acute effect (Baselt, 1988). Some, but not all, organophosphate pesticides can produce a delayed neurotoxicity. Thus, measurement of blood acetylcholinesterase activity may be an indicator of exposure to organophosphates but not necessarily serve as a marker of delayed neurotoxic effect. Studies have demonstrated that the peripheral neuropathic mechanism is not due to the ability of the chemical to inhibit acetylcholinesterase, but instead is due to phosphorylation and "aging" of the protein neuropathic target esterase (NTE) in nervous system tissue (Lotti *et al.*, 1984; Johnson, 1982). Investigations into the measurement of lymphocyte NTE as an independent marker of exposure and of delayed neuropathic effect have been initiated (Lotti *et al.*, 1983, 1986), but the relationship has not been clearly demonstrated (Kinebuchi, 1993).

1. Biologic Markers of Exposure

Basic pharmacokinetic principles, such as absorption, distribution, excretion, and metabolism, as well as individual variations due to gender and age, and in physiological functions involving blood flow, membrane permeability, respiratory rate, and nervous system accessibility all can affect markers of exposure such as ID and/or BED (NRC, 1992c).

When the target sites of toxicity are not known, or unavailable for sampling as are markers involving the nervous system, often a surrogate can be measured as an indicator of the BED. An example is the neurofilament crosslinking that follows exposure to *r*-diketones which is accompanied by dimerization of erythrocyte spectrin. The diketone structure is the site of neurotoxic action identified in 2,5-hexanedione, the major metabolite of *n*-hexane and methyl *n*-butyl ketone (Andrews and Snyder, 1986). Thus, testing of blood samples for spectrin will indicate the existence of a biologically effective dose to diketones since these compounds penetrate nervous system sites in proportion to the concentration of erythrocyte spectrin found in blood samples (NRC, 1992c). However, the significance of the spectrin concentration in terms of eventual clinical dysfunction has not been adequately studied.

2. Biologic Markers of Effect

A biologic marker of effect can be any qualitative or quantitative change that predicts nervous system impairment or damage resulting from exposure to a

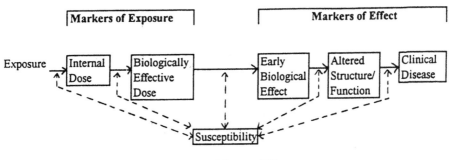

Markers of Susceptibility

FIGURE 1 Classifications of biologic markers. The solid lines indicate the progression from exposure to clinical disease and the dashed lines indicate that individual susceptibility can influence the rate of progression. (From NRC Committee on Biologic Markers, 1987. Taken from NRC, 1992, p. 46.)

substance (NRC, 1992c). In the clinical setting, electro-physiologic studies, including electroencephalogram, nerve conduction velocity (NCV), or evoked potentials serve as important biologic markers of effect to the nervous system. These specific diagnostic tests are discussed further elsewhere in the book.

3. Biologic Markers of Susceptibility

Various factors reflecting susceptibility, usually independent of exposure, can affect the relationship among the markers that represent events in the progression from exposure to disease. For instance, some individuals are more prone to intoxication with drugs such as isoniazid because they are slow acetylators due to a genetic difference in *N*-acetyl transferase activity. In other individuals, a preexisting disease state may affect the biotransformation of certain neurotoxicants, thus increasing the internal dose or the biologically effective dose. These markers of susceptibility may include inborn errors or differences in metabolism, variations in immunoglobulin concentrations, low organ reserve capacity, or other identifiable genetically, nutritionally, or environmentally induced factors that can influence the effect of the neurotoxic substances (NRC, 1992c).

B. Validation of Biologic Markers

As in the epidemiologic setting, the use of a particular biologic measurement as a marker of neurotoxicant exposure or neurotoxic effect requires validation to be used in a clinical setting. the validity of a marker is evaluated in terms of its sensitivity and specificity (Hennekens and Buring, 1987). Sensitivity is the ability of a marker to identify correctly those who have been exposed or have a disease or condition of concern. This is determined by calculating the proportion of the exposed or diseased persons who demonstrate the marker of exposure or effect. Specificity is the ability to identify correctly those who have not been exposed or do not have the disease or condition. It is determined by calculating the proportion of the unexposed or non-diseased persons who do not demonstrate the marker of exposure or effect. (Hennekens and Buring, 1987; NRC, 1992c). Animal studies are used to examine the sensitivity and specificity of a marker first before biological markers are tested in humans, and laboratory and population studies are conducted before a marker is used clinically (NRC, 1991b; Shulte, 1991; Wilcosky and Griffith, 1990). the validation process is time consuming and often difficult.

Deciding what the "critical effect" or threshold level in the progression from exposure to disease is crucial to the validation procedure (NRC, 1991b). For exam-ple, the decision about what should be the critical blood lead level that indicates increased exposure or preclinical effects will influence how this marker is interpreted. Lowering the cutoff point will mean more people who have actually been exposed or who have the neurotoxic effect will test positive (increased sensitivity), but so will the number of people who have not actually been exposed or who do not have the disease (lower specificity) (Hennekens and Buring, 1987). Because exposure to neurotoxic substances can result in a range of biologic effects, there is a need to have general agreement on which of these effects are critical, i.e., indicative of increased exposure that could lead to toxic effects.

C. Interpretation of Biomarker Data

There is only a limited number of neurotoxic compounds for which exposure can be tested by biologic sampling in the clinical setting. Most biomarker research studies in neurotoxicology have focused on the relationship between internal dose and external exposure instead of between internal dose and adverse effects. Therefore, the clinical relevance of these markers should be interpreted cautiously. It is necessary to understand the biochemical processes involved and to recognize that certain biomarkers may result from exposure to several related chemicals. For example, exposure to either trichloroethylene or perchloroethylene can result in urinary excretion of the same metabolites (e.g., trichloroethanol and trichloroacetic acid).

For those biomarkers whose clinical correlation have been well studied, relevant reference values have been established (Lauwerys, 1991). For example, some biological exposure indexes in the workplace have been periodically recommended and updated by ACGIH (1990).

D. Practical Issues in Measuring Biologic Markers

1. Appropriate Specimen

Three types of measurement of biologic media are usually selected for collection: the concentrations of the unchanged, suspected toxic agents; the concentrations of their metabolites; or the determination of biologic changes resulting from exposure. Identification of biologic markers from the nervous tissue that reflects health impairment or overt disease is the most representative and specific to the effects of neurotoxicants. Selected markers of neurotoxicity include general neuronal measures such as cell number; neuronal structure; general glial measures such as glial fibrillary acidic protein (GFAP) and an oligodendrocyte probe; and measures of neurotransmitter levels (NRC, 1992). Measurement of GFAP as a marker of nervous system damage resulting from chemical exposures has been tested in rats and mice (Stone, 1993). However, it is often not feasible to obtain nervous tissue specimens because detection of these markers is beyond the capability of most clinical laboratories and none have been validated for use in the clinical setting. Consequently, they are not routine clinical practices, and markers from specimens other than nervous tissue are more frequently selected.

Urine has the advantage of ease of collection. Chemicals or their metabolites are often present in urine in amounts proportional to the recently absorbed dose (Baselt, 1988). Breath samples can be easily obtained without discomfort to the patient, but their use is limited to the analysis of substances that are sufficiently volatile to appear in breath in measurable amounts. However, these substances may not be detectable in the breath soon after the end of exposure due to rapid elimination from the body in the exhaled air, thus collection of breath specimens should only be performed during or immediately after exposure. Blood levels usually demonstrate the best correlation with the atmospheric concentration of a substance, the amount absorbed (integrating all routes of exposure), and the severity of health effect and often are collected in the clinical setting. Other specimens that have been investigated include biopsy fat, saliva, breast milk, cerebrospinal fluid, hair, nails, teeth, and feces, but not all have been validated. Concentrations of polychlorinated biphenyls or some organic solvents in fat tissue are being assessed as biomarkers of exposure (Committee on National Monitoring of Human Tissues, 1991). An increased dentine lead level in deciduous teeth is associated with an impaired intelligence scale and neuropsychological performances in epidemiologic studies (Needleman, 1979), and can indicate previous exposure in children even if their blood lead levels may become normal. The level of heavy metals in segments of hair or nails can also be measured to indicate past exposure (Goldstein *et al.*, 1984; Wilcosky and Griffith, 1990). X-ray fluorescence has emerged as a promising tool to measure lead content in bones (Hu *et al.*, 1991). Each of these specimens offers certain advantage for its specific objective, but suffers its own difficulties in collection, storage, or analysis. In monitoring heavy metal intoxication, contamination of hair by external sources can invalidate the test result (Choucair and Ajax, 1988), and urine levels of mercury and arsenic increase shortly after ingestion of seafood (Lauwerys, 1983a). These difficulties must be recognized and controlled in the interpretation of exposure assessment in the clinical setting. Biomarkers for some representative neurotoxicants, appropriate specimen, and analysis methods are shown in Table 2.

2. Proper Timing for Markers

Time of occurrence and duration of marker presence must be carefully considered in biomarker analysis. The level of a marker can vary substantially over time, and different markers may persist for different periods of time after exposures end. Some markers may appear immediately in biologic materials, whereas others may take longer. Therefore, different markers may be selected at different stages along the clinical course. For instance, the blood lead level is the marker of choice in acute lead exposure, whereas erythrocyte or zinc protoporphyrin is a better index in subacute and chronic lead exposure, although it is also sensitive to iron deficiency (American Academy of Pediatrics, 1987). In many situations, failure to detect significant levels of biomarkers may signify ignorance about correct sampling time rather than absence of a neurotoxicant. For example, if exposure to trichloroethylene has ceased for about a week, urinary trichloroacetic acid is a better marker than is trichloroethanol (Lauwerys, 1983b). Also, fingernails, instead of urine, are the more appropriate specimen to test for arsenic months after the reported exposure period. Understanding the pharmacokinetics of each toxic substance is essential to make the correct judgment regarding when to collect an appropriate specimen for analysis.

3. Correct Device for Sample Collection

The appropriate sampling device is also important. In breath analysis, for example, breath samples should be stored at a constant temperature (preferably equal to body temperature) to avoid condensation of the moisture and loss of chemical solvent to be measured (Pasquini, 1978). For collection of blood, the containers, collection devices, and types of anticoagulants must be carefully selected for a specific application. Whole blood is needed for erythrocyte cholinesterase analysis (Baselt, 1988; WHO, 1986), but an appropriate anticoagulant that does not interfere with the analysis is crucial, e.g., fluoride cannot be used as an anticoagulant because it inhibits cholinesterase. Furthermore, most commercial vacuum blood tubes are unsuitable for trace metal analyses due to contamination (Baselt, 1988).

E. Ethical and Legal Issues

The use of biologic markers in clinical research or practice raises ethical and legal considerations. Bio-

TABLE 2 Measurement of Biomarkers for Selected Neurotoxicants: Appropriate Specimens, Markers, and Methods of Analysis[a]

Chemical class	Representative agents	Specimen	Markers	Method
Solvents	Acetaldehyde	Breath Blood Urine	Acetaldehyde	GC[b]
	Carbon disulfide (CS$_2$)	Urine	CS$_2$ metabolites	Iodine-azide test
		Blood	CS$_2$	GC
		Urine	2-Thiothiazolidine-4-carboxylic acid	LC[c]
	Carbon tetrachloride (CCl$_4$)	Breath Blood	CCl$_4$	Flame-ionization GC
	Ethylene glycol monoethyl ether	Urine	Methoxyacetic acid	Flame-ionization GC
	n-Hexane	Blood Urine	Hexane	Flame-ionizing GC
		Urine	2,5-Hexanedione	Flame-ionizing GC
	Trichloroethylene (TCE)	Breath Blood	TCE	GC
		Urine	Trichloroacetic acid and trichloroethanol	Colorimetry
		Urine	Trichloroacetic acid	Electron-capture GC
			Trichloroethanol	Electron-capture GC
Insecticides	Aldrine	Blood/plasma	Aldrin Dieldrine	Electron-capture GC
	Chlordane	Blood	Total chlordane metabolites	Electron-capture GC
	Carbaryl	Blood	Cholinesterase	Colorimetry
		Urine	1-Naphthol	Colorimetry
	Malathion	Blood	Cholinesterase	Colorimetry
		Urine	Organic phosphate	Colorimetry
Metals	Arsenic	Urine Hair Nail	Arsenic	Colorimetry or AAS[d]
	Cadmium	Urine Blood	Cadmium	AAS
	Lead	Blood	Lead	AAS
		Urine	Lead	EDTA chelation test
		Blood	Zinc protoporphorin	HPLC[e]
		Urine	*d*-ALA	Colorimetry
		Teeth	Lead	Anodic stripping voltammetry
		Bone	Lead	X-ray fluorescence
	Manganese	Urine Blood Hair	Manganese	AAS
	Mercury	Urine Blood	Mercury	AAS
		Blood	Methyl mercury	Electron-capture GC
Gases	Carbon monoxide	Blood	Carboxyhemoglobin	Spectrophotometry
	Ethylene oxide	Blood Breath	Ethylene oxide	GC
		Blood	Cytogenetic change	Direct microscopic exam
		Blood	Hemoglobin adducts	GC
Antibacterial agents	Hexachlorophene	Blood/plasma	Hexachlorophene	Electron-capture GC
Miscellaneous	Polychlorinated biphenyl (PB)	Blood Fat	PCB	Electron-capture GC

[a] From Baselt (1988).
[b] Gas chromatography.
[c] Liquid chromatography.
[d] Atomic absorption spectrometry.
[e] High performance liquid chromatography.

logic monitoring data could lead to possible discriminatory practices in the workplace or could be used in litigation over alleged health effects that result from exposure to hazardous chemicals. Therefore, it is crucial that the physician use validated markers as tools in assessing an individual's exposure and not feel pressured by legal or corporate influences to rely on markers before they are completely validated (Schulte, 1991).

IV. An Example of Exposure Assessment in the Clinical Setting: A Case of Arsenic Neurotoxicity

The following case report of exposure to arsenic illustrates the use of environmental monitoring and biological markers in clinical neurotoxicology.

A. Basis for Suspecting Arsenic Neurotoxicity: Literature Review

The most common neurologic deficit after arsenic exposure and intoxication is peripheral neuropathy (Feldman *et al.,* 1979; Schoolmeester and White, 1980), presenting as numbness, tingling, formications, and pain in the extremities 7–14 days after acute exposure (Jenkins, 1966) or as late as 8 weeks after a single exposure (Windebank, 1987). Onset is insidious in chronic exposure and onset of neuropathy after 2 years of repeated exposure has been reported (Goldstein *et al.,* 1984). Individual susceptibility to arsenic intoxication has been known to vary considerably and seems to depend largely on the amount and duration of previous exposure (Heyman *et al.,* 1956). Major pathologic changes occur in the spinal cord and peripheral nerves. Anterior horn cells decrease in number (Goldstein *et al.,* 1984), and axonal degeneration with secondary demyelination, especially in large myelinated fibers, is the major change in the peripheral nerves. Arsenic was found to exist in the sural nerve biopsied 2 months after the onset of arsenic polyneuropathy (Goebel *et al.,* 1990). In the 37 cases with neuropathy reported by Jenkins (1966), 32 (86%) had sensorimotor disturbances, 4 (11%) had only sensory, while only 1 (3%) had pure motor deficits. Sensory symptoms, which are common and more dominant, usually occur first. They may manifest as burning pain in the hands or soles of the feet. Thirty-nine of the 40 cases reported by Chhuttani and colleagues (1967) had objective sensory impairment to pin prick and touch. Such sensory disturbance is often followed by progressive weakness of the feet and hands (Klaassen, 1990).

In mild or early cases, impairment of vibration sense in the feet may be the only evidence of sensory disturbances (Heyman *et al.,* 1956). This occurred in 33 of the 37 cases (94%) reported by Jenkins (1966). In cases of severe damage, loss of vibration and position sense is common and may lead to ataxia and incoordination. Deep tendon reflexes disappear early. A predominantly distal symmetrical and severe muscular weakness of the extremities is also common. Wrist drop occurs in 10–40% of patients (Chhuttani and Chopra, 1979). Over 50% of the patients reported by Heyman and colleagues (1956) had evidence of foot drop. Cranial nerves are usually spared, but may be involved (Hotta, 1989; Diamant, 1958; Schenk and Stolk, 1967; Chhuttani *et al.,* 1967; Zolaga, 1985; Yamashita *et al.,* 1972; Bencko and Syman, 1977).

Both short-term and long-term exposure to arsenic can cause chronic encephalopathy (Klaassen, 1990) and it may be the presenting symptom in some cases. It may manifest with emotional distress and cognitive impairment, especially in attention and short-term memory. Some patients with arsenic encephalopathy may have organic psychosis with visual and auditory hallucinations, paranoid delusions, confusion, drowsiness, or delirium. In the past, chronic encephalopathy was more commonly seen after exposure to medicinal organic arsenic and was rarely seen with inorganic arsenic exposure (Goldstein *et al.,* 1984). However, more cases with subtle cognitive and/or affective disturbances are recognized with the availability of neuropsychological assessments. Chronic encephalopathy due to exposure to inorganic arsenic may be more common than the literature would suggest (Morton and Caron, 1989). The clinical course of arsenic encephalopathy usually shows improvement after discontinuation of exposure. However, permanent organic psychosis has been reported (Schenk and Stolk, 1967 as reported in Freeman and Couch, 1978).

Chromated copper arsenate (CCA) is one of the commonly used chemicals for wood treatment and preservation. It has been recognized as an occupational health hazard for workers in the wood preservation treatment industry (NIOSH, 1983). Two workers have been reported to develop acute arsenic intoxication after power sawing of CCA-treated wood (Peters *et al.,* 1986). Both showed systemic manifestations of acute poisoning, and one had neurologic involvement with blackout and memory impairment several days after exposure. The exposure continued for 3 weeks and the symptoms gradually ceased over the succeeding month. Hair and nail arsenic levels were markedly increased (about 150 times greater in the hair samples and over a 1000 times higher in the nails) compared to the reported laboratory normal range. Urine arsenic levels were normal. One of the workers resumed the job 1 year later and the symptoms recurred immediately after exposure. Neuropsychologic tests 6 months later revealed short-term memory impairment, which gradually improved. Sensory nerve conduction velocity studies 10 months after exposure still showed decreased sensory action potentials, which was compatible with her clinical complaints of increased tingling of the fingers.

Chronic nonoccupational exposure was reported in a family of eight who had burned CCA-treated wood in the stove to heat their house for 4 years. In addition to an array of systemic manifestations of arsenic intoxication, their common neurologic complaints were headache, muscle cramps, and sensory hyperesthesia. Blackouts and/or seizures were noted in four family members. Their symptoms were worse in the winter time. Arsenic levels were increased in their nails and hair, but not in the urine, with the highest reported in the parents. Arsenic levels ranged from 12 to 87 ppm in hair samples taken from family members (normal being less than or equal to 0.65 ppm or 65 mcg/100g) and ranged from 100 to 5000 times greater than normal in the nail samples. In samples taken of the dust and ash around their stove, arsenic was measured at 100–600 ppm. Ashes from the stove were greater than 1000 ppm. Arsenic measured in an air sample taken in the home was almost eight times the background sample. No objective evidence of peripheral neuropathy was documented despite subjective complaints of tingling (Peters *et al.*, 1984).

B. Case Report

A 62-year-old male (MD) visited the Environmental and Occupational Neurology Program at the Neurological Referral Center of Boston University Medical Center in September 1990. He stated that he had been suffering from numbness of hands and feet for 2 years and impaired memory for 1 year.

The probable exposure to arsenic began in May 1987 when MD used arsenic-impregnated lumber to build a new roof deck. He was not aware at that time that this wood had been treated with arsenate and was not informed of any precaution when he purchased the wood. He did all the cutting, sawing, and some nailing in the basement of his house. He purchased another batch of wood from the same company and again did similar work in his house in December 1987. He continued to be exposed in his house for another year until the contamination was suspected and cleaned out in January 1989.

The patient stated that his problems developed insidiously. In September 1988, he was found to have im-

paired vibration sensation over his toes and absent ankle jerk during a routine physical checkup. Two weeks later, he began to have paresthesia in his feet and then in his hands. In November 1988, he learned that the wood he had used in 1987 had been treated with chromated copper arsenate. A follow-up visit at that time revealed a poor performance in tandem gait, presence of the Romberg test, as well as "stocking and glove" impairment of all sensory modalities with absent ankle jerks and decreased knee jerks. Polyneuropathy was diagnosed, and arsenic intoxication was highly suspected after review of his medical history and exposure history.

Arsenic levels were measured in a 24-hr urine on repeat occasions and once in scalp hair and nail samples, both before and after his home was cleaned (Table 3). An above normal arsenic level was detected in a sample of hair in November 1988. Lead and mercury levels in 24-hr urine measurements in December 1988, before the home was cleaned, were within normal limits. The arsenic level in his dog's hair was 25 mcg/100 g. Also, bulk samples of wood shavings and dusts from his basement workshop floor were 2500 and 460 ppm, respectively. The levels in a second floor beam and furnace filter were 180 and 75 ppm, respectively. Other samples revealed arsenic levels ranging from 19 to 29 ppm.

A neurological consultation elsewhere in December 1988 confirmed the diagnosis of sensorimotor polyneuropathy. A sensory NCV study showed normal distal latencies which markedly decreased to almost absent sensory action potentials; a motor NCV study showed normal NCVs and distal latencies with borderline compound motor action potentials (CMAP) over the hands and markedly decreased CMAP over the feet.

MD arranged to have the arsenic dusts removed from his house in January 1989. Another cleanup was carried out in May 1989. Environmental analysis in his house showed marked improvement with the arsenic concentration of 43 ppm in the furnace filter and a range from 0.2 to 4.9 ppm at various other locations in the

TABLE 3 Biological Monitoring Measures in a Case of Arsenic Neurotoxicity

Arsenic levels (normal range)[a]	1988			1989	1990
	11/28	12/2	12/13	2/16	9/21
24-hr urine (0–100 mcg)		67	58		13
Hair (0–65 mcg/100 g)	67				
Nails (0–1.6 mcg/g)				0.66	

[a] Samples were tested at Bioran Laboratory (Boston, MA). Normal ranges given are from that laboratory.

home. However, his neurologic deficits failed to improve in the following years. In the fall of 1990, a neurologic examination was performed and a questionnaire asking about his occupational and residential history and his medical history was reviewed. Neurologic examination revealed persistence of sensorimotor polyneuropathy. Neuropsychological tests were done in December 1990. They revealed an overall superior intellectual ability; however, relative difficulties were seen on tasks requiring sustained attention, complex visuospatial organization, fine motor control, and memory. He demonstrated short-term memory problems that included impaired acquisition, sensitivity to interference, and uneven retrieval. These findings suggested a diffuse effect of deep cortical and/or subcortical dysfunction with frontal lobe involvement, which is compatible with arsenic encephalopathy.

Despite a course of chelating therapy with penicillamine in April 1991, his urinary arsenic output did not increase and his clinical conditions failed to improve.

C. Discussion

Documentation of exposure to a potential neurotoxic agent or agents is crucial for proper diagnosis of a neurotoxic syndrome. In this case, the potential for arsenic-treated wood as a source of this patient's symptoms was recognized so that an environmental exposure assessment could be performed at his home. Testing demonstrated increased levels of arsenic in dust and wood shaving samples, and they were subsequently controlled to lower levels after several thorough cleanups. Unfortunately, however, air samples for arsenic were not taken (either personal or area samples) as the primary route of exposure for arsenic in this scenario is through inhalation. Nevertheless, documentation of arsenic exposure in the home environment was obtained.

Biologic monitoring data are often difficult to interpret in the context of available relevant reference values because of possible confounding factors, including failure to follow standardized sampling techniques and existing differences between laboratories, as well as in exposure circumstances and in individual variation. "Normal data" for arsenic in various samples—less than 1 mcg/g in hair or nail and 10–30 mcg/liter in urine—is cited by Choucair and Ajax (1988). Other studies have also reported different reference levels (Bolla-Wilson and Bleecker, 1987; Goyer, 1986). In the clinical setting, repeat measures taken under the same sampling techniques in order to compare the individual's levels over time is strongly recommended, particularly if during that time the patient's potential exposure changes. It is important to ascertain information about

recent ingestion of seafood since organoarsenic levels may be significantly increased (half-life about 18 hr), and that can be reflected in urine sample analysis (Lauwerys, 1983). In this case, the questionnaire information is the only control for possible seafood intake. The method used in clipping the nails and hair for sampling (proximal vs distal segment) and possible environmental contamination to hair (Choucair and Ajax, 1988) should also be taken into consideration. In this case, no mention was made by the laboratory about the potential for contamination during collection or to proper sampling techniques.

Given the results of the environmental monitoring, arsenic was found in sufficient amounts in the environment to which the patient was exposed adequately concluded that increased arsenic intake probably occurred. The biological monitoring results do not confirm significantly increased levels of arsenic; although due to a lack of control for possible confounding factors as discussed earlier, they should not be relied upon exclusively as indicators of arsenic exposure in this patient. The case did reveal that indoor carpentry with treated wood can be the cause of toxic arsenic exposure. His neurologic deficits are compatible with arsenic intoxication and the onset of symptoms occurred after a year of flucuating, but chronic, exposure. Other suspected causes for his neuropathy and central nervous system deficits are unlikely on review of his clinical history and have been ruled out by tests for diabetes and questionnaire information regarding alcohol use and other solvent exposure.

Thus, regarding the four criteria presented at the outset of this chapter, the clinical conclusion is that this patient's peripheral and central nervous system findings are attributed to chronic arsenic exposure. This conclusion is based on the responses given in a self-reported environmental and occupational questionnaire, by the neurological findings on examination, by the presence of arsenic in home workshop environmental sampling and in biologic markers measured in the patient and his dog, and in finding these data consistent with similar cases of arsenic poisoning in the literature.

References

Abou-Donia, M. M. (1992). Nutrition and neurotoxicology. In *Neurotoxicology* (M. B. Abou-Donia, Ed.), pp. 319–335, CRC Press, Florida.

Aitio, A., Jarvisalo, J., Riihimaki, V., and Hernberg, S. (1988). Biologic monitoring. In *Occupational Medicine* (C. Zene, Ed.), 2nd ed., pp. 178–197, Year-Book Medical Publishers, Chicago.

American Academey of Pediatrics, Committees on Environmental Hazards and Accident and Poison Prevention. (1987). Statement on childhood lead poisoning. *Pediatrics* **79**(3), 457–465.

American Conference of Government Industrial Hygienists (AC-GIH) (1992–1993). Threshold limit values for chemical substances and physical agents and biological exposure indices. 5th ed. ACGIH Inc.

Andrews, L. S., and Snyder, R. (1986). Toxic effects of solvents and vapors. In *Casarett and Doull's Toxicology: The basic science of poisons* (C. Klaassen, M. Admur, and J. Doull, Eds.), pp. 636–688, Macmillan Publishing, New York.

Ashford, N. A., Spadafor, C. J., Hattis, D. B., and Caldart, C. C. (1990). *Monitoring the Worker for Exposure and Disease: overview and definitions.* The Johns Hopkins University Press, Baltimore.

Agency for Toxic Substances and Disease Registry (ATSDR) 1992a. Determining contaminants of concern. Public Health Assessment Guidance Manual. Chapter 5, Atlanta, Georgia.

Agency for Toxic Substances and Disease Registry (ATSDR) 1992b. Estimation of exposure dose. Public Health Assessment Guidance Manual. Appendix D, Atlanta, Georgia.

Baselt, R. C. (1988). *Biological Monitoring Methods for Industrial Chemicals.* PSG Publishers, Littleton, Massachusetts.

Bencko, V., and Symon K. (1977). Test of environmental exposure to arsenic and hearing changes in exposed children. *Environ. Health Perspect.* **19,** 95–101.

Bolla-Wilson, K., and Bleecker, M. L. (1987). Neuropsychological impairment following inorganic arsenic exposure. *J. Occup. Med.* **29**(6), 500–503.

Checkoway, H., Pearce, N., and Crawford-Brown, D. J. (1989). Characterizing the workplace environment. In *Research Methods in Occupational Epidemiology* (H. Checkoway, N. Pearce, and D. J. Crawford-Brown, Eds.), pp. 18–45. Oxford University Press.

Chhuttani, P., Chawla, L., and Sharma, T. (1967). Arsenical neuropathy. *Neurology* **17,** 269–274

Chhuttani, P., and Chopra, J. (1979). Arsenic poisoning. In *Handbook of Clinical Neurology, Vol. 36. Intoxication of the Nervous System.* (P. J. Vinken, and G. W. Bruyn, Eds.), pp. 199–216, Elsevier, North-Holland Biomedical Press, Amsterdam.

Choucair, A. K., and Ajax, E. T. (1988). Hair and nails in arsenical neuropathy. *Ann. Neurol.* **23**(6), 628–629.

Cohen, B. S. (1992). Industrial hygiene measurement and control. In *Environmental and Occupational Medicine* (W. Rom, Ed.), 2nd ed., pp. 1389–1404, Little, Brown and Company, Boston.

Committee on Biologic Markers of the National Research Council. (1987). Biological markers in environmental health research. *Environ. Health Perspect.* **74,** 3–9.

Committee on National Monitoring on Human Tissues. (1991). *Monitoring Human Tissues for Toxic Substances.* National Academy Press, Washington, D.C., pp. 29–45.

Diamant, H. (1958). The toxic action of some compounds on the inner and its vestibular connections. *Arch. Otolaryngol.* **67,** 546–552.

Environmental Protection Agency (EPA). (1986). Air quality criteria for lead. EPA Report no. EPA-600/8-83/028aF-DF, Environmental Criteria and Assessment Office, USEPA, Research Triangle Park, North Carolina.

Feldman, R. G., Niles, C. A., Kelly-Hayes, M., Dax, D. S., Dixon, W. J., Thompson, D. J., and Landau, E. (1979). Peripheral neuropathy in arsenic smelter workers. *Neurology* **29,** 939–944.

Freeman, J. W., and Couch, J. R. (1978). Prolonged encephalopathy with arsenic poisoning. *Neurology* **28,** 853–855.

Goebel, H. H., Schmidt, P. F., Bohl, J., Tettenborn, B., Kramer, G., and Gutman, L. (1990). Polyneuropathy due to acute arsenic intoxication: Biopsy studies. *J. Neuropathol. Exp. Neurol.* **49**(2), 137–149.

Goldstein, N. P., McCall, J. T., and Dyck, P. J. (1984). Metal neuropathy. In *Peripheral Neuropathy* (P. J. Dyck, P. K. Thomas, E. H. Lambert, and R. Bunge, Eds.), pp. 1227–1240. Saunders, Philadelphia.

Goyer, R. A. (1986). Toxic effects of metals. In *Casarett and Doull's Toxicology* (C. Klaassen, M. Admur, and J. Doull, Eds.), pp. 582–635, Macmillan Publishing, New York.

Hennekens, C. H., and Buring, J. E. (1987). In *Epidemiology in Medicine* (S. L. Mayrent, Ed.), pp. 327–345, Little, Brown and Company, Boston.

Heyman, A., Pfeiffer, J. B., Willett, R. W., and Taylor, H. (1956). Peripheral neuropathy caused by arsenical intoxication: A study of 41 cases with observations on the effects of BAL (2,3 Dimercapto-Propanol). *N. Engl. J. Med.* **254**(9), 401–409.

Hotta, N. (1989). Clinical aspects of chronic arsenic poisoning due to environmental and occupational pollution in and around a small refining spot. *Jpn. J. Constitutional Med.* **53,** 49–69.

Hu, H., Pepper, L., and Goldman, R. (1991). Effect of repeated occupational exposure to lead, cessation of exposure, and chelation on levels of lead in bone. *Am. J. Ind. Med.* **20,** 723–735.

Jenkins, R. B. (1966). Inorganic arsenic and the nervous system. *Brain* **89,** 479–498.

Johnson, B. L. (1992). A precis on exposure assessment. *J. Environ. Health* **55**(1), 6–9.

Johnson, M. K. (1982). The target of initiation of delayed neurotoxicity by organophosphate esters- biochemical studies and toxicological applications. *Rev. Biochem. Toxicol.* **4,** 141–212.

Juntunen, J. (1982). Alcoholism in occupational neurology: Diagnostic difficulties with special reference to the neurological syndromes caused by exposure to organic solvents. *Acta Neurol Scand.* **92**(Suppl), 89–108.

Kinebuchi, H. (1993). Summary of workshop: Pesticides. From the Fourth International Symposium of Neurobehavioral Methods and Effects in Occupational and Environmental Health. *Environ. Res.* **60,** 72–73.

Klaassen, C. (1990). Heavy metals and heavy metal antagonists. In *Goodman and Gilman's: The Pharmacological Basis of Therapeutics* (A. Gilman, T. Rall, A. Nies, and P. Taylor, Eds.), pp. 1602–1605, Pergamon Press, New York.

Lauwerys, R. (1983a). Arsenic. In *Industrial Chemical Exposure: Guidelines for biological monitoring.* Biomedical Publishers, Davis, California, pp. 12–15.

Lauwerys, R. (1983b). Trichloroethylene. In *Industrial Chemical Exposure: Guidelines for biological monitoring.* Biomedical Publishers, Davis, California, pp. 87–90.

Lauwerys, R. (1991). Occupational toxicology. In *Casarett and Doull's Toxicology* (C. Klaassen, M. Admur, and J. Doull, Eds.), pp. 947–969, Macmillan Publishing, New York.

Lotti, M., Becker, C. E., Aminiff, M. J., Woodrow, J. E., Seiber, J. N., Talcott, R. E., and Richarson, R. J. (1983). Occupational exposure to cotton defoliants DEF and Merphos- a rational approach to monitoring OPIDN. *J. Occup. Med.* **25**(7), 517–522.

Lotti, M., Becker, C. E., and Aminiff, M. J. (1984). Organophosphate polyneuropathy-pathogenesis and prevention. *Neurology* **34,** 658–662.

Lotti, M. *et al.* (1986). Inhibition of lymphocytic NTE predicts the development of OPIDN. *Arch. Toxicol.* **59,** 176–179.

Melius, J. M., Wallingford, K. M., and McCunney, R. J. (1988). The health hazard evaluation: Investigating occupational health problems. In *Handbook of Occupational Medicine* (R. J. McCunney, Ed.), pp. 362–373, Little, Brown and Company, Boston.

Morton, W. E., and Caron, G. A. (1989). Encephalopthathy: An uncommon manifestation of workplace arsenic posioning? *Am. J. Ind. Med.* **15,** 1–5.

National Institute for Occupational Health and Safety (NIOSH). (1983). Industrial hygiene surveys of occupational exposure to wood preservative chemicals.

National Institute for Occupational Health and Safety (NIOSH). (1990). Pocket Guide to Chemical Hazards. Superintendent of Documents, U. S. Government Printing Office, Washington, D. C.

National Research Council (NRC). (1987). Biologic markers in environmental health research. *Environ. Health Perspect.* **74**, 3–9.

National Research Council (NRC). (1991). Dimensions of the problem: Exposure assessment. In *Environmental Epidemiology* Vol. 1, Public health and hazardous waste. National Academy Press, Washington D. C.; pp. 103–153[1991a] and 219–255[1991b].

National Research Council (NRC). (1992). *Environmental Neurotoxicology*. National Academy Press, Washington D. C.; pp. 1–6[1992a], 9–19[1992b], and 43–51[1992c].

Needleman, H. L. (1979). Deficits in psychologic and classroom performance of children with elevated dentine lead levels. *N. Engl. J. Med.* **300**(13), 690–695.

Needleman, H. L. (1987). Introduction: Biomarkers in neurodevelopmental toxicology. *Environ. Health Perspect.* **74**, 149–151.

Office of Technology Assessment (OTA) U.S. Congress. (1990). *Neurotoxicity: Identifying and Controlling Poisons of the Nervous System*. OTA-BA-436, U.S. Government Printing Office, Washington, D.C.

Pasquini, D. A. (1978). Evaluation of glass sampling tubes for industrial breath analysis. *Am. Ind. Hyg. Asso. J.* **39**, 55–62.

Peters, H. A., Croft, W. A., Woolson, E. A., Darcey, B., and Olson, M. (1984). Seasonal arsenic exposure from burning chromium-copper-arsenate-treated wood. *JAMA* **251**(18), 2393–2396.

Peters, H. A., Croft, E. A., Woolson, E. A., Darcey, B., and Olson, M. (1986). Hematological, dermal, and neuropsychological disease from burning and power sawing chromium-copper-arsenic (CCA)-treated wood. *Acta Pharmacol. Toxicol.* **59**(7), 39–43.

Rinsky, R. A., and Young, R. J. *et al.* (1981). Leukemia in benzene workers. *Am. J. Ind. Med.* **2**, 217–245.

Schenk, V., and Stolk, P. (1967). Psychosis following arsenic (possibly thallium) poisoning: A clinical-pathological report. *Psych. Neurol. Neurosurg.* **70**, 31–37.

Schoolmeester, W. L., and White, D. R. (1980). Arsenic poisoning. *Southern Med. J.* **73**(2), 198–208.

Schulte, P. (1991). Contribution of biological markers to occupational health. *Am. J. Ind. Med.* **20**, 435–446.

Silbergeld, E. K., and Takeuchi, Y. (1993). Summary of Workshop: Biochemical methods in neurobehavioral toxicology. From the Fourth International Symposium of Neurobehavioral Methods and Effects in Occupational and Environmental Health. *Environ. Res.* **60**, 74–75.

Smith, T. (1987). Exposure assessment for occupational epidemiology. *Am. J. Ind. Med.* **12**, 249–268.

Sterman, A. B., and Schaumburg, H. H. (1980). Neurotoxicity of selected drugs. In *Experimental and Clinical Neurotoxicology* (P. S. Spencer, and H. H. Schaumburg, Eds.), pp. 593–612, Williams and Wilkins, Baltimore.

Stone, R. (1993). New marker for nerve damage. *Science* **259**(12), 1541.

Thomas, R. D. (1989). Epidemiology and toxicology of volatile organic chemical contaminants in water absorbed through the skin. *Reg. Toxicol. Pharmacol.* **8**(5), 779–795.

Weeks, J. L., Levy, B. S., and Wagner, G. R. (Eds.), (1991). Permissible exposure limits and recommended exposure limits. In *Preventing Occupational Disease and Injury*. American Public Health Association, pp. 653–696.

White, R. F., Feldman, R. G., and Proctor, S. P. (1992). Neurobehavioral effects of toxic exposure. In *Clinical Syndromes in Adult Neuropsychology* (R. F. White, Ed.), Elsevier, Amsterdam.

Wilcosky, T. C., and Griffith, J. D. (1990). Applications of biological markers. In *Biological Markers in Epidemiology* B. S. Hulka, T. C. Wilcosky, and J. D. Griffith, Eds.), pp. 16–27, Oxford University Press, New York/Oxford.

Windebank, A. (1987). Arsenic. In *Peripheral Neuropathy Due to Toxins. Handbook of Clinical Neurology* (P. J. Vinken, G. W. Bruyn, H. L. Klawans, and W. B. Matthews, Eds.), Vol. 51, pp. 263–292. Elsevier Science Publishers, B. V., Amsterdam.

World Health Organization (WHO). (1986). *Early Detection of Occupational Diseases*. WHO, Geneva, pp. 243–251.

Wrensch, M., Swan, S., Murphy, P., Lipscomb, J., Claxton, K., Epstein, D., and Neutra, R. (1990). Hydrogeologic assessment of exposure to solvent-contaminated drinking water: pregnancy outcomes in relation to exposure. *Arch. Environ. Health* **45**, 210–216.

Yamashita, N., Hojo, H., and Tanaka, M. (1972). Recent observations of Kyoto children poisoned by arsenic tinted morinage dry milk. *Jpn. J. Hyg.* **27**, 364–399.

Zaloga, G. P., Deal, J., Spurling, T., Richter, J., and Chernow, B. (1985). Case report: unusual manifestations of arsenic intoxication. *Am. J. Med. Sci.* **289**, 210–214.

46

Clinico–Neuropsychological Assessment Methods in Behavioral Neurotoxicology

ROBERTA F. WHITE
Department of Neurology
Boston University School of Medicine
Boston, Massachusetts 02118

SUSAN P. PROCTOR
Department of Neurology
Boston University School of Medicine
Boston, Massachusetts 02118

I. Reasons for Neuropsychological Assessment

Although neuropsychological assessment is generally associated with the administration of formal tests, the clinical assessment process is more complex. It includes a thorough review of medical, academic, occupational, and personal history through interviews with the patient and/or significant others and through review of medical and school records when appropriate. Often, neuropsychological assessment results are combined with information from history and physical examination and data from imaging and neurophysiological tests to determine the ultimate diagnosis.

In general, neurotoxicant-exposed patients are referred for neuropsychological testing in order to answer one or more of the following questions.

1. Does the patient demonstrate deficits on behavioral tests which would suggest the existence of encephalopathy?
2. If such deficits are found, how severe are they? Will they affect daily life functioning (ability to work, financial competency, safety, judgment)?
3. In patients with documented encephalopathy, does follow-up testing reveal any change in function (recovery of function in the absence of exposure, deterioration if exposure has continued)?
4. Does the patient have any behavioral abnormalities suggesting the existence of other disorders (neurological, psychiatric, motivational, developmental, medical)?
5. What are appropriate treatment approaches?

All of these questions imply the need for documentation of behavioral abnormalities and both diagnosis of any existing toxic related behavioral disorder and differential diagnosis of toxic-induced behavioral disorders versus other etiologies.

II. Cognitive Domains

The encephalopathy that occurs following occupational or environmental exposure to neurotoxicants in adults is usually characterized by deficits in one or more of the following functional areas: attention, executive function, fluency (verbal or visual), motor abilities, visuospatial skills, learning and short-term memory, and mood and adjustment. Table 1 presents a description of neuropsychological tests by functional domain. There is some variability between individuals in the extent of deficits which occurs following exposure to the same toxic agent. This variability most likely reflects a combination of factors, including exposure variables (duration of exposure, dosage, combinations of exposures to multiple neurotoxicants), developmental variables (age at exposure, cognitive maturity at exposure), and individual variables (differences in fragile cognitive skills, preexisting cognitive deficits or neuropsychiatric disorder, educational history, socioeconomic status). For example, we have seen many patients clinically with chronic exposure to trichloroethylene (TCE) in drinking water. The functional deficit patterns in patients with chronic residual TCE encephalopathy following exposure only during adulthood include a relatively mild form of encephalopathy characterized primarily by deficits in attention and executive function; a more severe form characterized by deficits in visuospatial abilities, motor function, fluency, and/or learning and memory; and a severe form in which there are multiple cognitive deficits, usually including difficulty in retaining newly learned information over delays (Baker *et al.*, 1985a).

Developmental exposure to neurotoxicants can produce a variety of deficits in any individual child, extending to most functional cognitive domains and affecting academic achievement and performance. Our clinical experience and the literature on toxicants such as lead suggest that the earlier the exposure the more devastating the effects are likely to be. Some children are, of course, exposed to toxicants from conception on. However, exposure does not have to occur at a very early age in order to affect a wide variety of cognitive functions. It appears that the developing brain is sensitive to neurotoxicants for an extended period of time.

Because the literature covering deficits on specific neuropsychological tests in toxicant-exposed subjects is too large to cover in a single chapter, this chapter summarizes research on a few common neurotoxic substances. For a more complete review of this literature the reader is referred elsewhere (Johnson, 1987; White *et al.*, 1990b). A highly informative review which covers the relationship between exposure to specific neurotoxicants and performance on specific behavioral tests is provided by Anger (1990).

A. General Intelligence

1. Research Findings

It has long been known that lead exposure in children affects general intelligence (Needleman *et al.*, 1979)

and that such effects are long lasting (Needleman *et al.*, 1988; White *et al.*, 1993), producing deficits in academic performance (Byers and Lord, 1943). In adults, exposure to the solvent carbon disulfide has been shown to be associated with deficits in performance IQ (Matthews *et al.*, 1990) and toluene exposure has been associated with verbal IQ/performance IQ discrepancies (Fornazzari *et al.*, 1983). Specific WAIS-R subtests, discussed under the relevant functional area, are sensitive to a number of neurotoxicants, including carbon disulfide (Peters *et al.*, 1982) and lead (Baker *et al.*, 1984, 1985b; Yokoyama *et al.*, 1988), as is the IQ score from the Raven's progressive matrices test (White *et al.*, 1993).

2. Clinical Experience

Depending on the severity of exposure and specific toxicant to which the patient was exposed, exposed patients can show a range of effects on general intelligence from none to a decrease in full-scale IQ. Patients classifiable as having a moderate to severe chronic toxic encephalopathy may show a mild decline in verbal IQ due to decrements in performance on reasoning tasks, diminished digit spans, and/or concreteness in vocabulary definitions whereas performance IQ may be more significantly affected due to impaired performance on any of the tasks (but especially block design, digit symbol, picture arrangement). In adults, IQ measures such as the Raven's may be affected by toxic exposure, but effects on a highly verbally mediated test such as the Peabody picture vocabulary test would be unusual unless the exposure produced a stroke or an anoxia.

B. Attention and Executive Function

1. Research Findings

Attention and executive functioning appear to be extremely susceptible to the effects of toxic exposure. The Trails task has been used in a number of behavioral neurotoxicology studies and has been shown to be sensitive to naphtha (White *et al.*, 1994), carbon disulfide (Peters *et al.*, 1982), and to the mixed solvent exposure experienced by silk screeners (White, NIOSH report, 1990a), shipyard painters (Valciukas *et al.*, 1985), and paint sniffers (Tsushima and Towne, 1977). Trails (Boey and Jeyaratnam, 1988), and other executive system tasks such as digit spans (Hanninen, 1978a; Baker *et al.*, 1984) and the Wechsler memory scale mental control task (Baker *et al.*, 1984) are also sensitive to lead exposure.

2. Clinical Experience

In clinical patients, deficits in simple attention, cognitive tracking, and cognitive flexibility are commonly

TABLE 1 Tests Commonly Used in Clinical Assessment of Possible Encephalopathy Secondary to Exposure to Toxic Chemicals: Organized by Functional Domain (Boston Extended Neurotoxicologic Battery; Clinical)

Domain	Description	Implication
General intellect		
Wechsler IQ tests (WAIS-R,[1,2,3] WISC, WPPSI)	Omnibus IQ measures	Overall level of cognitive function compared with population norms
Peabody picture vocabulary test[4]	Single word comprehension	Robust measure of verbal intelligence in adults, can be sensitive to exposure in children
Stanford-Binet[5]	Omnibus IQ measure	Similar to Wechsler tests
Wide range achievement test[6]	Academic skills in arithmetic, spelling, reading	Robust estimate of premorbid ability patterns in adults, can be sensitive to exposure in children
Attention and executive functioning		(Attention and executive functioning tasks are sensitive to many types of exposure)
Digit span (WAIS-R)[1]	Digits forward and backward	Measures simple attention and cognitive tracking
Arithmetic (Wechsler tests)[1]	Oral calculations	Assesses attention, tracking, and calculation
Trail making test[7]	Connect-a-dot task requiring sequencing and alternating sequences	Measures attention, sequencing, visual scanning, speed of processing
Continuous performance test[8]	Acknowledgment of occurrence of critical stimuli in a series of orally or visually presented stimuli	Assesses attention
Paced auditory serial addition[9]	Serial calculation test	Sensitive measure of attention and tracking speed
Wisconsin card sorting test[10]	Requires subject to infer decision-making rules	Tests ability to think flexibly
Verbal and language		(Language test are sometimes sensitive to exposure in children but are usually robust in adult exposure, except as noted)
Information (Wechsler tests)[1]	Information usually learned in school	Robust estimate of native abilities in adults
Vocabulary (Wechsler tests)[1]	Verbal vocabulary definitions	Fairly robust estimate of verbal intelligence although sensitive to concreteness associated with brain damage (including toxic encephalopathy)
Comprehension (Wechsler tests)[1]	Proverb definitions, social judgement, problem solving	Sensitive to reasoning skills; can be impaired after exposure to neurotoxicants
Similarities (Wechsler tests)[1]	Inference of similarities between nominative words	Sensitive to reasoning skills; can be impaired after exposure to neurotoxicants
Controlled oral word association[11]	Word list generation within alphabetical or semantic categories	Assesses flexibility, planning, arousal, processing speed, ability to generate strategies, somewhat sensitive to exposure
Boston naming test[12]	Naming of objects depicted in line drawings	Sensitive to aphasia, also sensitive to native verbal processing deficits or those acquired through childhood exposure
Reading comprehension (Boston[12] diagnostic aphasia exam)	A direct screening test of simple reading comprehension	Sensitive to moderate-to-severe dyslexia, usually insensitive to toxic exposure in adults
Writing sample[13]	Patient writes to dictation or describes a picture	Assesses graphomotor skills, spelling
Visuospatial and visuomotor		(Visuospatial and visuomotor tasks are frequently sensitive to exposure in adults and children)
Picture completion (Wechsler tests)[1]	Identification of missing details in line drawing	Measures perceptual analysis
Digit symbol (Wechsler tests)[1]	Coding task requiring matching symbols to digits	Complex task assessing motor speed, visual scanning, working memory

(continued)

TABLE 1 Continued

Domain	Description	Implication
Picture arrangement (Wechsler tests)[1,2,3]	Sequencing of cartoon frames to represent meaningful stories	Measures visual sequencing, ability to infer relationships from visuospatial/social stimuli
Block design (Wechsler tests)[1,2,3]	Assembly of 3-D blocks to replicate 2-D representations of designs	Assesses abstract visual construction ability and planning
Object assembly (Wechsler tests)[1,2,3]	Assembly of puzzles	Measure of concrete visual construction skills, Gestalt recognition
Boston visuospatial quantitative battery[13]	Drawings of common objects spontaneously and to copy	Measures constructional abilities, motor functioning
Hooper visual organization test[14]	Identification of correct outline of drawings of cut up objects	Sensitive to Gestalt integration processing
Rey-Osterreith complex figure (copy[15] condition)	Drawing of a complicated abstract visual design	Sensitive to deficits in visuospatial planning and construction
Santa Ana formboard test[16]	Knobs in a formboard are turned 180° with each hand individually and both hands together	Measures motor speed and coordination
Finger tapping[17]	Speed of tapping with each index finger	Sensitive to lateralized manual motor speed
Memory		(New learning and retention are often sensitive to toxicants; retrograde memory of prior events is more complexly related to exposure)
Logical memories–immediate and delayed recall (IR, DR) (Wechsler memory scales)[18,19]	Recall of paragraph information read orally on an immediate and 20-min delayed recall	Sensitive to new learning and retention of newly learned information
Verbal impaired associate learning[18,19] DR (Wechsler memory scales)	Two paired words are presented in a list of pairs; subject must recall second word; test is presented on immediate and delayed recall	Measures abstract verbal list IR, learning, retention
Figural memory[19] (Wechsler memory scales)	Multiple choice recognition of using recognition (not recall) performance measures	Assesses visual recognition and memory
Visual paired associate learning, IR, DR[19] (Wechsler memory scales)	Six visual designs are paired with six colors; recognition memory is tested immediately after the six are presented on learning trials and at delayed recall	Test of abstract visual learning using recognition (not recall) performance measures
Visual reproductions, IR, DR[18,19] (Wechsler memory scales)	Visual designs are drawn immediately after presentation and on delayed recall	Measures visual learning and retention
Delayed recognition span test[20]	Based on delayed nonmatching to sample paradigm, discs are moved about on a board to assess recognition memory for words, color, spatial locations	Assesses new learning
Peterson task[21]	Words or consonants are presented and must be recalled after a period of distraction	Measures sensitivity to interference in new learning
California verbal learning test[22]	Subject is presented with list of 16 words (which can be semantically related) over multiple learning trials and an interference list	Provides multiple measures of new learning, recall, recognition memory, use of strategies, and sensitivity to interference
Rey-Osterreith (IR, DR)[15]	Complex design is drawn from IR immediately after it has been copied and at a 20-min delayed recall	Assesses memory for visual information which is difficult to encode verbally
Personality and mood		
Profile of mood states[23]	Sixty-five single word descriptors of affective symptoms are endorsed by degree of severity on six scales	Sensitive to clinical mood disturbance and to affective changes secondary to toxicant exposure
Minnesota multiphasic personality inventory (R)[24]	True–false responses provided on personality inventory summarized on multiple clinical dimensions	Provides description of current personality function, some scales sensitive to exposure

(*continued*)

TABLE 1   *Continued*

¹ Wechsler, D., (1981). *Wechsler Adult Intelligence Scale-Rev.* Harcourt Brace Jovanovich, New York.

² Wechsler, D. (1991). *WISC-R Manual: Wechsler Intelligence Scale for Children-III.* Psychological Corporation, New York.

³ Wechsler, D. (1989). *Wechsler Preschool and Primary Scale of Intelligence-R.* Psychological Corporation, New York

⁴ Dunn, L. M., and Dunn, L. M. (1981). *Peabody Picture Vocabulary Test-Revised.* American Guidance Service, Circle Pines, Minnesota.

⁵ Terman, L. M., and Merrill, M. A. (1973). *Stanford-Binet Intelligence Scale Manual for the Third Revision Form L-M.* Houghton-Mifflin, Boston.

⁶ Jastak, J., and Jastak, S. (1993). *The Wide Range Achievement Test-3.* Jastak Association, Wilmington, Delaware.

⁷ Halstead, W. C. (1947). *Brain and Intelligence.* University of Chicago Press, Chicago.

⁸ Rosvold, H., Mirsky, A., Sarason, I., *et al.* (1956). A continuous performance test of brain damage. *J. Consult. Clin. Psychol.* **20,** 343–350.

⁹ Gronwell, D. M. A. (1977). Paced auditory serial-addition task: a measure of recovery from concussion. *Percept. Mot. Skills* **44,** 367–373.

¹⁰ Grant, D., and Berg, E. A. (1948). *The Wisconsin Card Sorting Test.* Department of Psychology, University of Wisconsin.

¹¹ Benton, A. L., and des Hamsher, S. (1976). *Multilingual Aphasia Examination.* University of Iowa, Iowa City, Iowa, (Manual, revised 1978).

¹² Kaplan, E., Goodglass, H., and Weintraub, S. (1983). *The Boston Naming Test.* Lea and Febiger, Philadelphia.

¹³ Goodglass, H., and Kaplan, E. (1983). *The Boston Diagnostic Aphasia Examination.* Lea and Febiger, Philadelphia.

¹⁴ Hooper, H. E. (1958). *The Hooper Visual Organization Test Manual.* Western Psychological Services, Los Angeles.

¹⁵ Osterrieth, P. A. (1944). Le test de copie d'une figure complexe. *Arch. Psychol.* **30,** 206–356.

¹⁶ Fleishman, E. A. (1954). Dimensional analysis of psychomotor abilities. *J. Exp. Psychol.* **48,** 437–454.

¹⁷ Reitan, R. M., and Davidson, L. A. (1974). *Clinical Neuropsychology: Current Status and Applications.* Hemisphere, New York.

¹⁸ Wechsler, D., and Stone, C. (1945). *Wechsler Memory Scale.* The Psychological Corporation, New York.

¹⁹ Wechsler, D. (1987). *Wechsler Memory Scale-Rev.* The Psychological Corporation, New York.

²⁰ Moss, M. B., Albert, M. S., Butters, N., and Payne, M. (1986). Differential patterns of memory loss among patients with Alzheimer's disease, Huntington's disease and alcoholic Korsakoff's syndrome. *Arch. Neurol.* **43,** 239–246.

²¹ Peterson, L. B., and Peterson, M. J. (1959). Short-term retention of individual verbal items. *J. Exp. Psychol.* **58,** 193–198.

²² Dellis, D., Kramer, J. H., Kaplan, E., and Ober, B. A. (1987). *California Verbal Learning Test Manual.* The Psychological Corporation.

²³ McNair, D. M., Lorr, M., and Droppleman, L. F. (1971). *Profile of Mood States.* Educational and Industrial Testing Services, San Diego, California.

²⁴ Hathaway, S. R., and McKinley, J. C. (1989). *The Minnesota Multiphasic Personality Inventory Manual-Rev.* The Psychological Corporation, New York.

seen in patients with all kinds of toxic exposures (including lead, carbaryl, chlordane, trichloroethylene, perchlorethylene, toluene, mixed solvents, mercury). In patients with a mild acute reversible encephalopathy or mild chronic residual encephalopathy, these deficits may represent the primary or only manifestations of the encephalopathy.

Cognitive tracking is often a significant problem for toxicant-exposed patients. Trails A or B is frequently sensitive to the tracking problems observed in these patients, although digit span (backwards), mental control, recurrent series writing (MN), and the WAIS-R arithmetic subtest are also sometimes sensitive. In the few patients tested with paced auditory serial addition (PASAT), PASAT has also been an effective measure of toxicant-induced tracking deficits. The Stroop test has been of variable utility with this kind of patient: in our experience there is great variability in patients' abilities to carry out the task demands and some patients who show problems on other executive tasks perform well on the Stroop.

Cognitive flexibility is also often affected in toxic encephalopathy. Although some patients show effects both on tracking and flexibility, others show a dissociation between these two types of dysfunction, exhibiting one but not the other. We have been unable to establish any methods of predicting which type of deficit will be seen; it does not appear to be related to type of exposure or even severity. Difficulties in cognitive flexibility may be seen on the Wisconsin card sorting test (WCST), Categories test, Trails B, and on the WAIS-R subtests requiring abstract reasoning.

Simple attention can be impaired in some cases. The continuous performance test is quite reliable in this assessment, although digit spans or visual spans are also helpful.

C. Motor Skills

1. Research Findings

Deficits in fine manual motor control have been reported following a wide variety of exposures, probably reflecting a special vulnerability of motor system structures such as the basal ganglia and cerebellum to a variety of toxins. Deficits in motor performance in mercury-exposed workers (Langolf *et al.*, 1978) have been reported. Impairments in finger tapping, grooved pegboard performance, and WAIS-R digit symbol were identified in agricultural workers with carbon disulfide exposure (Matthews *et al.*, 1990) and on the Santa Ana formboard task in rayon workers exposed to carbon disulfide (Hanninen, 1971); this chemical is known to affect the basal ganglia, producing a parkinsonian-like syndrome (Peters *et al.*, 1988). Likewise, lead exposure has been associated with relative impairment in Santa Ana performance (Hanninen, 1978a; Jeyaratnam *et al.*, 1986) and paint exposure has been associated with deficits in grooved pegboard performance (Tsushima and Towne, 1977).

2. Clinical Experience

Patients with identifiable toxic encephalopathy sometimes show no motor deficits at all. At the mildest end of the continuum of motor impairment are patients who show slowed performance only on relatively complex tasks such as formboard tests or digit symbol but not on simpler tasks such as finger tapping. Some patients have such impaired motor function that it is obvious on all tasks with a manual motor component, including Trails and constructional tasks. This is especially true of patients with parkinsonian-like symptoms secondary to exposure to organic solvents, carbon disulfide, or carbaryl and of patients with cerebellar dysfunction secondary to exposure to mercury or toluene. The finger tapping test is sometimes genuinely impaired in toxin-exposed patients, although this task is sensitive to symptomatic exaggeration on examination, e.g., we have seen several patients who do other tasks without motor deficits but who score a mean of 5–20 taps (normative mean is around 50) (Lezak, 1983) with each hand over five trials on the finger tapping test. Usually these patients cannot be prompted to increase their speed and have other instances of intertask performance variability in the testing protocol.

D. Language/Verbal Skills

1. Research Findings

Linguistic abilities have been rarely studied or reported in the behavioral neurotoxicology literature. Test batteries designed for studies in behavioral neurotoxicology have almost never included formal language testing such as naming or aphasia tests. More commonly, test batteries include vocabulary tests or verbal concept formation.

Most reports on the effects of neurotoxicants in adults conclude that basic language abilities are intact following such exposures (White *et al.*, 1990b). Thus, Peters and colleagues (1982), describing neuropsychological deficits in seven grain workers with extrapyramidal syndromes secondary to carbon disulfide exposure, reported that vocabulary performance was normal in contrast to impairments on psychomotor and visuospatial tasks. Adult patients with significant tri-

chloroethylene encephalopathy have repeatedly shown normal naming and other language functions (Feldman *et al.*, 1985; White *et al.*, 1990b). An exception to this general conclusion has been the report that subjects with carbon disulfide exposure showed "retarded language" (Hanninen, 1971) and the finding that the number of words generated on the controlled word association test (FAS) was inversely related to the severity of exposure to mixed solvents among silk screen workers (White *et al.*, 1990a). It seems likely, given the susceptibility of the basal ganglia, cerebellum, white matter, and frontal lobes to many types of intoxication, that verbal fluency and motor aspects of speech would be sensitive to toxic exposure. Naming, reading, linguistic aspects of writing, and verbal comprehension are not, however, sensitive to toxicants.

Two exceptions to the finding that vocabulary functions are generally intact have been the report of Tsushima and Towne (1977) that Peabody picture vocabulary test performance was impaired in paint sniffers relative to controls and the finding that WAIS vocabulary test performance was sensitive to acute exposure to lead (Baker *et al.*, 1984). Tsushima and Towne (1977) did not attribute their finding to exposure. The lowered scores on vocabulary items in the lead study were examined qualitatively; it was found that the subjects with higher lead levels tended to produce concrete, one-point responses but were aware of the usage and general definition of the words. Thus, the problem seemed to be more one of verbal flexibility than a loss in linguistic processing capability. Studies done since then suggest that vocabulary, when measured with multiple choice formats, is unaffected by exposure (Echeverria *et al.*, 1990; White *et al.*, 1993).

Tests of verbal concept formation such as WAIS similarities have been reported to be sensitive to carbon disulfide (Hanninen, 1971) and lead (Baker *et al.*, 1984).

It should be noted that developmental exposure to neurotoxicants has repeatedly been associated with deficits in basic language-processing capacity and acquisition of vocabulary (Needleman *et al.*, 1979; Shaheen, 1984).

2. Clinical Experience

Adult patients with toxic encephalopathy secondary to occupational or environmental exposure often show problems in verbal fluency or word retrieval, especially if they have developed white matter lesions secondary to exposure (White *et al.*, 1989). This may manifest itself in reduced scores on the FAS task. Problems with retrieval on naming tasks may lead to increased latencies to respond, but primary dysnomia or paraphasic errors attributable to toxic exposure have not been seen except in rare cases in which exposure has led to a stroke. Hesitancies may be noted in free speech, and patients with exposure affecting basal ganglia or cerebellum may be dysarthric and/or hypophonic. Motor aspects of writing may be affected, although linguistic aspects of writing and reading remain intact. In children, exposure to lead and other neurotoxicants such as trichloroethylene and mercury may affect language capabilities. It has not yet been ascertained the upper age at which linguistic deficits may appear; certainly they seem to occur even at exposures occurring at age 10 or later. A patient who was exposed to mercury at home from the ages 6 to 16 was evaluated; the mercury was brought home on the clothes of his mother from work. This patient underwent psychometric testing beginning with pre-kindergarten screening for school at age 4. Although his initial testing was normal, he began after 2 years of exposure to show problems with attention and language. After several more years, visuospatial and personality changes appeared. He performed normally on arithmetic until about fifth grade but then began to have problems with all academic subjects.

Verbally mediated reasoning tests such as WAIS-R similarities and comprehension are also frequently sensitive to neurotoxic effects. Although it is our impression that similarities is a more reliable measure of such reasoning deficits and although some patients show deficits on both tasks, there is interindividual variability in sensitivity of the tasks, with some patients showing effects on only one of them.

E. Visuospatial Abilities

1. Research Findings

Visuospatial functioning, particularly the ability to organize complex or abstract visual designs such as block designs, has long been established as a functional area that is particularly vulnerable to the effects of neurotoxic substances. Thus, an association between lead exposure and lowered scores on picture completion (Yokoyama *et al.*, 1988) has been reported. Carbon disulfide exposure has been associated with diminished scores on WAIS-R object assembly (Peters *et al.*, 1982) and block design (Hanninen *et al.*, 1978b). Relative deficits in performance on visuospatial tasks have been described following exposure to mercury (Vroom and Greer, 1972); this has been reported for block designs, object assembly, and on matching unfamiliar faces (Milner facial recognition test) (White *et al.*, 1990b). Likewise, trichloroethylene exposure (Feldman *et al.*, 1985) and perchloroethylene exposure (Echeverria and White, 1990) have been associated with deficits in visuospatial functioning.

2. Clinical Experience

Deficits in visuospatial functioning are seen following exposure to many types of toxicant, including

heavy metals, pesticides, organic solvents, and gases such as carbon monoxide. Such effects have been seen following exposure to lead, mercury, carbaryl, chlordane, arsenic, urea formaldehyde, trichloroethylene, perchloroethylene, toluene, methylene chloride, xylene, and many other substances. This is probably due to the vulnerability of subcortical and cerebellar structures to the effects of neurotoxicants. It has been repeatedly found that WAIS-R visuospatial subtests, especially block designs and object assembly, are sensitive to toxic exposure, producing a mild-to-moderate encephalopathy. With some kinds of exposure, especially mercury, deficits may be seen even in facial matching. The Rey-Osterreith construction, the Hooper visual organization test, degraded stimuli, and embedded figures tasks are also at times useful in evaluating patients with toxic exposures. Toxicant-induced changes in visual acuity, neuroophthalmologic function, manual motor dexterity, tactile sensitivity, cognitive flexibility and psychomotor speed may all contribute to impaired scores observed on many traditional neuropsychological visuospatial task, making them heuristically valuable in detecting central nervous system (CNS) effects of toxic exposure. However, when evaluating individual patients it becomes clear that primary problems in visuospatial analysis and organization occur in some of these patients.

F. Memory

1. Short-Term Memory

a. Research Findings Although memory function has been less extensively studied than other functional domains in behavioral neurotoxicology research, it is clear that anterograde memory function is vulnerable to the effects of numerous neurotoxic substances, especially heavy metals such as lead (Baker *et al.*, 1984; Hanninen *et al.*, 1978a) and mercury (Vroom and Greer, 1972), and carbon disulfide (Aaserud *et al.*, 1988) and mixed solvents (Ryan *et al.*, 1988). This is probably not a surprising finding given the extensive literature linking exposure to the self-administered neurotoxicant ethanol to changes in memory function (Brandt and Butters, 1986). In addition, it has been established through animal and human autopsy studies that toxicants such as tin (Besser *et al.*, 1987) are readily absorbed by the hippocampus.

b. Clinical Experience Exposure to neurotoxicants clearly can affect memory function at several levels, including the efficient encoding of new information, the ability to inhibit interference during learning, the retrieval of encoded information, and the retention of encoded information. Many patients with toxic exposures show difficulty in learning new information due to problems in attending to tasks and perhaps due to apathy or diminished motivation. Such patients show ineffective or retarded learning curves. Often there is exquisite sensitivity to interference which can be well documented with the Peterson task paradigm (in which information is retrieved from memory following a distractor task). Because retrieval can be uneven, recognition memory paradigms should be used to adequately assess the ability to retain new information. Even using recognition memory tasks, some patients with toxic exposures show significant forgetting of newly learned information over time, suggesting that the exposure has affected the limbic system. Other patients show uneven encoding and retrieval consistent with white matter or basal ganglia dysfunction. Because of the divergent patterns of memory impairment possible following toxic exposure, a rather extensive memory battery is used in assessing these patients. This may be particularly necessary if there is concurrent alchohol abuse and occupational or environmental exposure to neurotoxicants.

It is our impression after testing many patients that visual memory tasks and paired associate learning are especially sensitive to neurotoxicants. It is also clear that memory function may improve in the absence of exposure; a number of patients have shown a gradual improvement in anterograde memory functioning occurring over several years.

2. Retrograde Memory

a. Research Findings We have located almost nothing in the literature regarding formal assessment of retrograde memory in exposed patients. Studies that have used tasks such as the WAIS-R information subtest report no decrements in recall of information that was overlearned in childhood (e.g., Matthews *et al.*, 1990). However, the only paper we are aware of summarizing retrograde memory assessed with a systematic temporal presentation of stimuli is the case report linking retrograde memory loss as identified by the Albert's famous faces test to time of heaviest exposure to solvents (White, 1987).

b. Clinical Experience Because alcohol can cause retrograde memory loss, in the Korsakoff syndrome, with a temporal gradient for information previously learned (Brandt and Butters, 1986), it seems likely that exposure to some other neurotoxicants under the correct circumstances could also produce retrograde memory loss with a temporal gradient. Certainly, it is our experience that patients with exposure to solvents and metals sometimes show temporal gradients on formal

retrograde memory tasks such as Albert's famous faces and on interview. However, it remains unclear whether this represents forgetting of known information (true retrograde memory loss) or simply the effect of anterograde memory deficits at the time of exposure which resulted in lack of encoding of the information.

G. Affect and Personality

1. Research Findings

Changes in affect and behavior have been reported for centuries following neurotoxic exposure. These are often the most obvious effects of exposure to caretakers and can be extremely damaging. At high exposure levels, psychotic symptoms, including delusions and paranoia, have been reported in mercury intoxications; this was seen in hatters and led to the expression "mad as a hatter." Visual and auditory hallucinations and compulsive, repetitive, uncontrollable actions have been reported in manganese poisoning (Mena *et al.*, 1967), and paranoid delusions have been described in patients with arsenic exposure (Schenk and Stolk, 1967). At less than psychotic levels of severity, changes in mood and energy level have been among the symptoms reported in the "psychoneurasthenic" syndrome ascribed to exposure to carbon disulfide (Lilis, 1974) and to mixed solvents. A dose–effect relationship was reported between blood level and mood reports on the day lead levels were determined (Baker *et al.*, 1984). Mood symptoms such as irritability, anxiety, depression, and fatigue accompanied by listlessness have been reported in mercurialism (Hanninen, 1982); lassitude, fatigability, and anxiety have been reported in chronic arsenic exposure (Zettel, 1943; McCutchen and Utterback, 1966); and fatigue, irritability, and anorexia have been ascribed to perchloroethylene exposure (White *et al.*, 1990b). Dose–effect relationships between exposure and behavioral symptoms have been reported in toluene exposure (toluene is the active ingredient in glues often abused by sniffing), with low exposure producing fatigue and higher exposure causing euphoria, exhilaration, and excitement (Bor and Hurtig, 1977).

2. Clinical Experience

It has been our experience that personality/affective changes are almost always seen following toxic exposures of sufficient severity to produce an encephalopathy identifiable through neuropsychological assessment. These symptoms can be so dramatic and can appear to be so psychiatric to health care providers that patients can be misdiagnosed with a psychiatric disease instead of with an appropriate organic disorder secondary to exposure. It has not been our experience that exposures produce affective or psychotic symptoms in the absence of cognitive changes; changes in mood, personality, and reality testing appear to occur in conjunction with identifiable intellectual impairments. The two sets of symptoms are also usually localizable to the same brain structures or neurochemical processes. For example, cognitive changes localizable to the basal ganglia (e.g., slowing, diminished initiation, motor deficits, visuospatial impairments) may accompany similarly localizable affective and behavioral changes (e.g., depression, diminished arousal, visual hallucinations).

Probably the most common psychological symptom reported by patients with toxicant exposure is mood change. Even at relatively low levels of exposure, many toxicants, including metals and solvents, can affect mood. The most commonly reported symptoms include depression, irritability, fatigue, and confusion. Depressive symptoms reported often do not constitute a full-blown syndrome of clinical depression (with changes in appetite and sleeping), but are frequently confined to feelings of being low, sad, or overwhelmed. However, a subgroup of patients report anorexia, insomnia, or, more frequently, hypersomnia. Some patients report feeling tense or restless and many patients describe being apathetic or unmotivated. In some patients suicidal ideation or actual suicide attempts are seen; these often have a flavor of diminished impulse control in the face of aggressive/hostile feelings. Aggressive impulses are also at times expressed in homicidal or violent acts against others; we have seen rare patients with well-defined toxic exposures and clear encephalopathy who are both suicidal and homicidal. Euphoria is rarely seen as a residual symptom of toxic encephalopathy in the absence of exposure, though it is sometimes seen as an acute symptom during exposure. In rare cases, patients with toxic encephalopathy show manic/depressive cycles.

In patients with significant exposure, symptoms of organic psychosis may be apparent. We have seen patients with auditory, visual, and olfactory hallucinations secondary to exposure, although visual hallucinations are the most common. Patients at times describe odd tactile sensations which are usually symptomatic of peripheral neuropathy. Delusions, especially paranoid delusions, are sometimes seen in patients with toxic exposures. Sometimes it can be difficult to differentiate paranoid ideation from fact in patients with exposures whose paranoid ideas center on the exposure or events surrounding the exposure.

Patients with toxic encephalopathy at times develop unusual cognitive styles because of their intellectual impairments. Most commonly one sees circumstantial/tangential speech when the patient has experienced damage to the temporal or frontal lobes. Impulsive, superficial cognition can also be seen. Some patients appear to develop obsessive/ruminative/detail-oriented modes of thinking and speaking. In some cases, the cognitive style is clearly etiologically related to exposure, while in others it may represent an exacerbation of premorbid style or no change in style whatsoever. In patients with temporal lobe seizures secondary to exposure, development of intellectual/personality changes such as those seen in interictal temporal lobe personality disorder may occur.

Emotional reactions to the experience of being exposed to toxicants may also be observed in these patients. These may include preoccupation with the exposure and events surrounding it, preoccupation with physical and/or emotional symptoms thought to be related to exposure, and anger toward perceived perpetrators of the exposure, social agencies, the judicial system, health care providers, or insurance companies. In addition, post-traumatic stress syndromes can be seen postexposure in some cases, though in our experience this is somewhat overdiagnosed. It should be noted that any of these kinds of symptoms can occur in conjunction with or independent from organic affective or personality changes secondary to exposure. The occurence of these kinds of symptoms does not in itself rule out a concurrent encephalopathy, a point that is sometimes overlooked by clinicians.

Given the variety of organic and functional symptoms that occur following toxic exposures, it is not surprising that personality test results are often "abnormal." In the case of toxic exposures, as in some other neurologic disorders (e.g., multiple sclerosis), it can be dangerous to use Minnesota Multiphasic Personality Inventory (MMPI) cookbooks to interpret findings. Scales 1 (hypocondriasis) and 3 (hysteria) are often elevated in patients with toxic encephalopathy as are scales 2 (depression), 7 (anxiety), 8 (schizophrenia), and even sometimes scale 6 (paranoia). The F scale may be high, but it is unusual for scales L or K to be elevated as a direct consequence of exposure. Obviously, some patients will show patterns of long-standing personality disorder on this instrument whether they have a toxic encephalopathy or not. While the MMPI should not be used as a single diagnostic device in assessing toxic encephalopathy, it can be a helpful adjunct in defining secondary or concurrent behavioral issues. The Profile of Mood States is also frequently sensitive to the effects of toxic exposure. Its items are germane to the types of physiologic and emotional symptoms exposed patients experience. The scales measuring fatigue and confusion are especially sensitive to toxic encephalopathy as are depression, anger (irritability), and tension in some patients. The vigor scale is a difficult one to interpret because so many of the items are essentially measures of social desirability (SD); therefore, some patients who score high on fatigue also score high on vigor secondary to endorsing SD items whereas others score low due to low energy responses.

III. Methods of Assessment

A. History

When evaluating a patient with toxic exposures, information from medical records can be invaluable in reconstructing the patient's history, particularly if the patient experienced an acute exposure with confusion or alterations in consciousness or if the patient experienced an acute episode of symptoms or confusion following a chronic exposure. The medical records from treatment or hospitalization following such exposures may provide information on blood or urine levels of the chemical or metabolites; measurements of oxygen or carbon dioxide in the blood, which might suggest an hypoxic state; alterations in mental state or level of consciousness when admitted or examined with acute symptoms; electroencephalogram results; magnetic resonance imaging or computerized tomography results; neurologic examination findings; and other related medical findings (e.g., respiratory conditions, cardiac abnormalities, peripheral nerve disorders, metabolic problems, neuroophthalmologic abnormalities secondary to exposure). The medical record may also be helpful in providing information on coexisting neurologic or medical diseases such as hypertension, thyroid dysfunction, immune disorders, or infection, which might affect the testing, and can provide accurate information on medications.

The referral source may be extremely helpful in providing information on medical history, exposure history, and behavioral anomalies noted in the patient as well as providing a history on the course of development of symptoms thought to be related to the exposure in question.

When evaluating exposure cases it is important to know something about the patient's exposure: which chemicals were involved, was the exposure chronic or acute, what were the routes of entry (dermal, respiratory, oral), and likely dosages of exposure. Such information may be available from the referral source. However, at other times it is necessary to have the patient obtain material safety data sheets from the employer. These sheets, which employers must now have on hand for employees because of the right-to-know law, provide information on the chemical constituents of products with which the employee works and on the known hazardous effects of the chemicals. In some instances, because of OSHA requirements that factories and plants obtain information on exposure levels of employees to hazardous chemicals, it is possible to obtain information from the patient's place of employment on industrial hygiene studies and air monitoring of chemicals which have been done at the worksite.

Because many toxicants produce significant but nonetheless subtle effects on CNS functioning, it is important to estimate the patient's intellectual functioning preexposure as carefully as possible. School records and prior testing have been used to accomplish this. Interviews with family members can also be helpful in this regard, particularly if they are focused on information regarding developmental and school history, academic and occupational achievement, and notable behavioral and cognitive changes observed by the family postexposure. It is especially helpful to get a sense of the course of the patient's condition since exposure: declines in performance following periods of nonexposure are unusual in toxic cases and occur under only certain kinds of circumstances.

B. Interview

The patient can often provide substantial information on interview regarding the exposure being evaluated and any other exposures that the patient may have experienced. Many patients who are exposed to a neurotoxicant environmentally or occupationally that was thought to produce symptoms also have other sources of toxic exposure. Therefore, it is important to inquire about possible sources of exposure at home, through work, and in hobbies. The clinician needs to be aware of the many sources of exposure (e.g., working on one's house, car repairs, soldering) that may occur quite commonly and the names of frequently used chemicals in order to handle this inquiry accurately. We use a questionnaire which lists chemical and product names in order to help jog patients' memories regarding possible past exposures.

In addition to environmental and occupational exposures to chemicals, many patients experience exposures to self-administered neurotoxicants such as alcohol and street drugs. A careful inquiry into substance abuse is critical to accurately determining the contribution of an occupational or environmental exposure as the etiology of cognitive impairments observed neuropsychologically. It is important to find out when such substances were last used relative to testing. In addition, reports on the patient's reaction to using drugs such as alcohol can be important: some workers appear to use alcohol to treat CNS symptoms which they develop following toxic exposure at work (Juntunen, 1982) and some patients report greater sensitivity to alcohol following solvent exposures. Sometimes the distinction between self-administered drugs and occupational exposures becomes obscured. For example, some glues contain toluene, a potent neurotoxicant. Some patients use toluene recreationally, producing an encephalopathy, whereas others may be exposed occupationally or may prolong occupational exposure because they enjoy the high which may follow such exposure. We have also seen patients who have abused other chemicals with which they work, such as trichloroethylene.

It is essential to inquire about the patient's self-perceived changes in intellectual and mood functioning attributable to exposure. We ask a series of questions about cognitive changes in many intellectual spheres, some of which are often affected by exposures and

and answering interview questions or marked verbal or emotional disinhibition. Some patients are extremely angry about their exposure and the response of employers, health department authorities, or physicians to their concerns. These patients may feel compelled to relate every detail of these concerns and prior insensitivity to their problems, which must be handled delicately and with patience while still managing to elicit a careful history. When irritability or anger is combined with disinhibition, it may be necessary to calm the patient as much as possible to obtain a reasonable interview and assessment. Some patients are depressed due to a post-traumatic reaction to exposure, a secondary organic depression, or some other life event and may be difficult to interview because of apathy or unresponsiveness.

The issue of secondary gain is a critical one in toxic exposure work as the patient may be suing an employer and allowing for evaluation of inter- and intratask variability. Because memory tasks may be especially vulnerable to attempts at faking (Hawkins, 1990) and because memory tests are also particularly sensitive to neurotoxic exposure, we include a full memory battery which allows thorough examination of performance variability within and between tests as well as careful characterization of any memory processing deficits observed. Some patients show toxicant-induced amnestic disorders. For these patients, inclusion of retrograde memory testing may provide clues concerning the time of onset of anterograde memory deficits.

When testing patients with occupational exposures, demographic variables may affect test selection. Exposures often occur among factory workers or janitorial personnel who may have less education or less facility with verbal information processing than some other groups of patients referred for neuropsychological consultation. In addition, a significant proportion of exposed patients are immigrants whose language and educational history must be considered when designing and interpreting test batteries. Finally, many such patients have subcultural group membership which will affect testing. Given all of these considerations, it is helpful to include motor, memory, and visuospatial tasks that do not require verbal processing such as facial and design memory tasks, visual constructive tasks, motor coordination measures, and object memory tests in the test battery.

Because changes in mood state, perceptual experiences (i.e., hallucinations), energy level, and personality may be seen with toxic exposures as symptomatic of toxic encephalopathy, it is important to include an evaluation of mood and personality in the assessment procedures.

When designing the test battery for a patient, it is essential to know something about the effects of the specific neurotoxicant(s) to which the patient has been exposed. Some neurotoxicants produce specific effects on the hippocampus whereas others affect white matter, basal ganglia, cerebellum, frontal lobe, occipital cortex; obviously, each site of neuropathologic damage will be associated with different patterns of neuropsychological deficit which must be fully explored.

IV. Outcome

A. Referrals

Referral sources for neuropsychological assessment of possible toxic encephalopathy vary widely. We have received them from occupational health physicians, general internists, geriatricians, neurologists, psychia-

trists, industrial hygienists, occupational nurses, insurance claim personnel, rehabilitation counselors, and attorneys. Referral for further evaluation following the neuropsychological assessment depends on the differential diagnosis based on the assessment and on the extent of evaluation that has already been completed. However, following any convincing exposure, assessment of exposure is warranted through industrial hygiene studies and through blood or urine tests to assess current levels of the toxicant(s) in question or specific metabolites of said toxicant(s). In some cases exposure may be too remote in time to be accurately assessed by current levels. At such times, hair or teeth may be examined, a fat biopsy may be warranted, or X-ray fluoroscopy to detect bone levels of a toxin such as lead may be useful (Hu *et al.*, 1989).

Patients who evidence primary brain dysfunction (whether or not it is thought to be related to exposure) may require neurologic evaluation. Depending on the symptoms and history, we may also recommend assessment of vestibular function (if the patient reports dizziness or has a job involving climbing or working above ground); eye blink reflexes (especially following solvent exposures); electroencephalogram (to quantify seizure discharges and to detect focal and diffuse dysfunction); magnetic resonance imaging (to detect changes in basal ganglia, cerebellum, or white matter; to detect atrophy); proton emission tomography (to detect hypometabolism); nerve conduction assessment of latencies, amplitudes, and velocities (to rule out peripheral neuropathy); or neuroophthalmologic evaluation (to assess ocular or central perceptual processing). General medical examination with laboratory tests of metabolic function, cardiac status, or immunologic function may be warranted. Evaluation by an occupational and environmental health physician is frequently recommended if not already completed in order to assess the full range of possible exposure-related diseases.

Serial neuropsychological testing is often recommended after a specified interval in order to determine if there is recovery of function in the absence of exposure or deterioration in function under conditions of continuing exposure.

B. Treatment Referrals

Patients determined to have a neurologic or psychiatric disorder not related to exposure are obviously referred to the appropriate treatment source. In patients with toxic encephalopathy, few medications have been found to be helpful in the treatment of psychiatric or neurologic symptoms. Sedative drugs may contribute to memory problems and are generally not helpful. Antidepressant medications may be effective in some patients whose depression adds to functional impairment. Some patients benefit from supportive psychotherapy in accepting symptoms of depression or anxiety and/or losses in cognitive or vocational function which may accompany their encephalopathy. We have referred some patients for cognitive retraining in an attempt to remediate cognitive loss, but we have found such treatment to be largely unsuccessful. Strategies which sometimes prove useful in enhancing cognitive capacity in persons with a closed head injury are sometimes useful.

Social/Vocational Recommendations

Patients with toxic encephalopathies are often in need of concrete, specific referrals and recommendations in order to maintain vocational and financial security. If a patient is significantly impaired and is either permanently or temporarily unable to work, it may be necessary to suggest that the patient leave work and to encourage the patient to apply for social security disability payments and/or workman's compensation. In cases in which the damage is primarily related to the central nervous system, it is sometimes difficult to obtain these benefits, and the patient may need the help of a lawyer specializing in workman's compensation claims or toxic tort litigation. Sometimes the patient's union can help in such efforts, but at other times the patient may require assistance in obtaining reputable legal representation. In some cases, civil suit may be brought to acquire compensation for a patient's loss. In these legal situations the neuropsychological test report or testimony from the neuropsychologist may be essential in verifying the patient's disability, including severity of loss, prognosis, and etiology of loss (diagnosis). In milder cases, it may be appropriate to suggest that the patient's cognitive and health status be monitored while continuing to work or that the patient remain employed but that toxic exposure be avoided completely. For some patients, vocational counseling can be useful in assisting the patient to find work which he/she can successfully accomplish given the intellectual loss occurring with a toxic encephalopathy and/or which will allow the patient to work without experiencing hazardous exposure. For some patients with highly skilled vocations which garner them excellent pay (e.g., artists or artisans), a vocational change is within the patient's intellectual capabilities but may produce a significant loss in wage earning capacity unless the patient undergoes retraining in a lucrative vocational field.

C. Follow-Up Evaluation

For patients with an identifiable encephalopathy, it may be important to document recovery (or lack of recovery) from dysfunction over time. It has been our repeated finding that, in the absence of additional exposure, many patients recover slowly over a period of several years after an exposure. In other cases, mild or equivocal findings on occupational health assessment may produce the recommendation that the patient continue working with careful follow-up; in such cases it may be necessary to document whether deterioration in cognitive or affective function has occurred with continued exposure. At other times diagnosis may be uncertain at the time of initial testing and may become clarified by seeing the patient in follow-up after a period of nonexposure or continued exposure. Whatever the reason for follow-up testing, selection of follow-up test batteries can be problematic. After 15 years of experience with longitudinal testing of such patients, we generally have the patient repeat all or most of the battery at the first follow-up (an exception might be something like the WRAT, which can be omitted if performed normally at initial evaluation); this allows observation of a general improvement across all tests (even hold tests) sometimes seen due to motivational factors. At subsequent follow-ups, tasks which continue to show impairments are administered, unless there has been a reexposure or some other intervening medical or psychiatric event. Although practice effects are sometimes a problem, this is usually only the case for tasks with which the patient is having no difficulty. In our experience the most problematic test regarding practice effects has been the Wisconsin card sorting test: in some patients who have performed within normal limits on the WCST at a second or third follow-up, we have administered the Categories test and found significant impairment, suggesting that the improved performance was due to learning the WCST task demands and not due to improved reasoning or cognitive flexibility.

In order to carry out accurate differential diagnosis among patients who show deficits on testing, it is essential to include tasks that assess basic academic skills and intellectual capacity, tests that are known to be insensitive to the toxicant(s) at issue, and assessment of underlying or long-standing characterologic or psychiatric symptoms or disorders.

Obviously, we are discussing development of a battery to assess an individual patient clinically. Such batteries must be appropriate to the situation: use of epidemiologic study batteries designed for brief research testings or of computerized batteries which have been developed for epidemiologic studies and which have not been validated on brain-damaged populations is inappropriate and will not produce accurate clinical diagnosis or differential diagnosis (Proctor and White, 1990).

V. Differential Diagnosis

Issues in Differential Diagnosis of Toxic Encephalopathy

The most common problems encountered in this endeavor can be categorized as (1) characterizing premorbid cognitive status, (2) dissociating the effects of coexisting psychiatric disorders, (3) dissociating the effects of coexisting medical and neurologic disorders, and (4) differentiating the specific effects of exposure to different toxic agents in cases of multiple exposure (multitoxicant environmental or occupational exposure, substance abuse, prescription drug use). These issues have been covered extensively elsewhere in regard to solvents (White, 1993). The same principles apply to neurotoxicants in general, although the pattern of deficits associated with exposure may vary among toxicants.

References

Anger, W. K. (1990). Worksite behavioral research: results, sensitive methods, test batteries, and the transition from laboratory data to human health. *Neurotoxicology* II, 629–719.

Baker, E. L., Feldman, R. G., White, R. F., Harley, J. P., Niles, C., Dinse, G., and Berkey, K. (1984). Occupational lead neurotoxicity, a behavioral and electrophysiologic evaluation. I. Study design and year one results. *Br. J. Ind. Med.* 41, 352–361.

Baker, E. L., White, R. F., and Murawski, B. (1985a). Clinical evaluation of neurobehavioral effects of occupational exposure to organic solvents and lead. *Int. J. Mental Health* 14, 135–158.

Baker, E. L., White, R. F., Pothier, L. J., Berkey, C. S., Dinse, G. E., Travers, P. H., Harley, J. P., and Feldman, R. G. (1985b). Occupational lead neurotoxicity. II. Improvement in behavioral effects following exposure reduction. *Br. J. Ind. Med.* 42, 507–516.

Benton, A. L., and deS Hamsher, S. (1976). *Multilingual Aphasia Examination.* University of Iowa, Iowa City, Iowa, (Manual, revised 1978).

Besser, R., Kramer, G., Thumler, R., Bohl, J., Gutmann, L., and Hopf, H. C. (1987). Acute trimethyltin limbic cerebellar syndrome. *Neurology* 37, 945–950.

Binder, L. (1990). Malingering following minor head trauma. *Clin Neuropsychol.* 4, 25–36.

Boey, K. W., and Jeyaratnam, J. (1988). A discriminant function analysis of neuropsychological effort of low lead exposure. *Toxicology* 49, 309–314.

Bor, J. W., and Hurtig, H. I. (1977). Persistent cerebellar ataxia after exposure to toluene. *Ann. Neurol.* 2, 440–442.

Brandt, J., and Butters, N. (1986). The alcoholic Wernicke-Korsakoff syndrome and its relationship to long-term alcohol use. In *Neuropsychological Assessment of Neuropsychiatric Disorders* (I. Grant and K. Adams, Eds., pp. 441–477, Oxford University Press, New York.

Byers, R. K., and Lord, E. E. (1943). Late effects of lead poisoning on mental development. *Am. J. Dis. Child.* 66, 471–494.

Chong, J. P., Turpie, I., Haines, T., Muir, G., Farnworth, H., Cruttendon, K., Julian, J., Verma, D., and Hillers, T. (1989). Concordance of occupational and environmental exposure information

elicited from patients with Alzheimer's disease and surrogate respondents. *Am. J. Ind. Med.* 15, 73–89.

Cullen, M. R. (1987). The worker with multiple chemical sensitivities: an overview. In *Workers with Multiple Chemical Sensitivities* (M. Cullen, Ed.), Hanley and Belfus, Philadelphia.

Dellis, D., Kramer, J. H., Kaplan, E., and Ober, B. A. (1987). *California Verbal Learning Test Manual.* Psychological Corporation, New York.

Echeverria, D., White, R. F., and Sampao, C. (1990). A neurobehavioral evaluation of PCE exposure in patients and dry cleaners: a possible relationship between clinical and preclinical effects. Paper presented at Eighth International Neurotoxicology Conference, Little Rock, Arkansas.

Feldman, R. G., White, R. F., Currie, J. N., Travers, P. H., and Lessell, S. (1985). Long-term follow-up after single exposure to trichloroethylene. *Am. J. Ind. Med.* 8, 119–126.

Fleishman, E. A. (1954). Dimensional analysis of psychomotor abilities. *J. Exp. Psychol.* 48, 437–454.

Fornazzari, L., Wilkinson, D. A., Kapur, B. M., and Carlen, P. L. (1983). Cerebellar and functional impairment in toluene abusers. *Acta Neurol. Scand.* 67, 319–329.

Goodglass, H., and Kaplan, E. (1983). *The Boston Diagnostic Aphasia Examination.* Lea & Febiger, Philadelphia.

Grant, D., and Berg, E. A. (1948). *The Wisconsin Card Sorting Test.* Department of Psychology, University of Wisconsin.

Gronwell, D. M. A. (1977). Paced auditory serial-addition task: a measure of recovery from concussion. *Percept. Mot. Skills* 44, 367–373.

Halstead, W. C. (1947). *Brain and Intelligence.* University of Chicago Press, Chicago.

Hanninen, H. (1971). Psychological picture of manifest and latent carbon disulfide poisoning. *Br. J. Ind. Med.* 28, 374–381.

Hanninen, H. (1982). Behavioral effects of occupational exposure to mercury and lead. *Acta Neurol. Scand.* 66 (Suppl. 92), 167–175.

Hanninen, H., Hernberg, S., Mantere, P., Vesanto, R., and Jalkanen, M. (1978a). Psychological performance of subjects with low exposure to lead. *J. Occup. Med.* 20, 240–255.

Hanninen, H., Nurminen, M., Tolonen, M., and Martelin, T. (1978b). Psychological tests as indicators of excessive exposure to carbon disulfide. *Scand. J. Psychol.* 19, 163–174.

Hathaway, S. R., and McKinley, J. C. (1989). *The Minnesota Multiphasic Personality Inventory Manual-Rev.* The Psychological Corporation, New York.

Hawkins, K. A. (1990). Occupational neurotoxicology: some neuropsychological issues and challenges. *J. Clin. Exp. Neuropsychol.* 12, 664–680.

Hooper, H. E. (1958). *The Hooper Visual Organization Test Manual.* Western Psychological Services, Los Angeles.

Hu, H., Milder, F. L., and Burger, D. E. (1989). X-ray fluorescence: issues surrounding the application of a new tool for the burden of lead. *Environ. Res.* 49, 295–317.

Jastak, J., and Jastak, S. (1983). *The Wide Range Achievement Test-Rev.* Jastak Association, Wilmington, Delaware.

Jeyaratnam, J., Boey, K. W., Ong, C. N., and Phoon, W. O. (1986). Neuropsychological studies on lead workers in Singapore. *Br. J. Ind. Med.* 43, 626–629.

Johnson, B. L. (Ed.) (1987). *Prevention of Neurotoxic Illness in Working Populations.* John Wiley, New York, pp. 3–104.

Juntunen, J. (1982). Occupational neurology. *Acta Neurol. Scand.* 66, 89–102.

Kaplan, E., Goodglass, H., and Weintraub, S. (1983). *The Boston Naming Test.* Lea and Febiger, Philadelphia.

Koss, E. (1988). Occupational exposure to neurotoxins in Alzheimer's disease: metabolic and behavioral correlates. Paper presented at International Neuropsychological Society, New Orleans, Louisiana.

Langolf, G. D., Chaffin, D. B., Henderson, R., and Whittle, H. P. (1978). Evaluation of workers exposed to mercury using quantitative tests of tremor and neuromuscular functions. *Am. Ind. Hygiene Assoc. J.* **39,** 725–733.

Lezak, M. D. (1983). *Neuropsychological Assessment.* Oxford, New York.

Lilis, R. (1974). Behavioral effects of occupational carbon disulfide exposure. In *Behavioral Toxicology: Early Detection of Occupational Hazards* (C. Xintaras, B. L. Johnson, and I. de Groot, Eds.), NIOSH publication No. 74-126. U.S. Government Printing Office, Washington, D.C.

Matthews, C. G., Chapman, L. J., and Woodard, A. R. (1990). Differential neuropsychologic profiles in idiopathic versus pesticide-induced Parkinsonism. In *Advances in Neurobehavioral Toxicology* (B. L. Johnson, Ed.), pp. 323–330. Lewis Press, Chelsea, Massachusetts.

McCutchen, J. J., and Utterback, R. J. (1966). Chronic arsenic poisoning resembling muscular dystrophy. *Southern Med. J.* **59,** 1139–1145.

McNair, D. M., Lorr, M., and Droppleman, L. F. (1971). Profile of Mood States. Educational and Industrial Testing Service, San Diego, California.

Mena, I., Marin, O., Fuenzalida, S., and Cotzias, G. C. (1967). Chronic manganese poisoning: clinical picture and manganese turnover. *Neurology* **17,** 128–136.

Moss, M. B., Albert, M. S., Butters, N., and Payne, M. (1986). Differential patterns of memory loss among patients with Alzheimer's disease, Huntington's disease and alcoholic Korsakoff's syndrome. *Arch. Neurol.* **43,** 239–246.

Needleman, H. L., Gunnoe, C., Leviton, L. A., Reed, R., Peresie, H., Maher, C., and Barett, P. (1979). Deficits in psychologic and classroom performance of children with elevated dentin lead levels. *N. Engl. J. Med.* **300,** 689–695.

Needleman, H. L., Schell, A., Bellinger, D., Leviton, A., and Allred, E. N. (1988). The long-term effects of exposure to low doses of lead in childhood: an 11-year follow-up report. *N. Engl. J. Med.* **322,** 83–88.

Osterrieth, P. A. (1944). Le test de copie d'une figure complexe. *Arch. d. Psychol.* **30,** 206–356.

Peters, H. A., Levine, R. L., Matthews, C. G., Sauter, S. L., and Rankin, J. H. (1982). Carbon disulfide induced neuropsychiatric changes in grain storage workers. *Am. J. Ind. Med.* **3,** 373–391.

Peters, H. A., Levine, R. L., Matthews, C. G., and Chapman, L. J. (1988). Extrapyramidal and other neurologic manifestations associated with carbon disulfide fumigant exposure. *Arch. Neurol.* **45,** 537–540.

Peterson, L. B., and Peterson, M. J. (1959). Short-term retention of individual verbal items. *J. Exp. Psychol.* **58,** 193–198.

Proctor, S. P., and White, R. F. (1990). Psychoneurological criteria for the development of neurobehavioral test batteries. In *Advances in Behavioral Neurotoxicology* (B. L. Johnson, Ed.), pp. 273–281, Lewis Publishers, Chelsea, Massachusetts.

Reitan, R. M., and Davidson, L. A. (1974). *Clinical Neuropsychology: Current Status and Applications.* Hemisphere, New York.

Rosenberg, N. L., Kleinschmidt-DeMasters, B. K., Davis, K. A., Dreisbach, J. N., Hormes, J. T., and Filley, C. M. (1988a). Toluene abuse causes diffuse central nervous system white matter changes. *Ann. Neurol.* **23,** 611–614.

Rosenberg, N. L., Spitz, M. C., Filley, C. M., Davis, K. A., and Schaumberg, H. H. (1988b). Central nervous system effects of chronic toluene abuse—clinical, brain stem evoked response and magnetic resonance imaging studies. *Neurotoxicol. Teratol.* **10,** 489–495.

Rosvold, H., Mirsky, A., Sarason, I., et al. (1956). A continuous performance test of brain damage. *J. Consult. Psychol.* **20,** 343–350.

Rourke, B. P. (1985). *Neuropsychology of Learning Disabilities.* Guilford Press, New York.

Ryan, C. M., Morrow, L. A., and Hodgson, M. (1988). Cacosmia and neurobehavioral dysfunction associated with occupational exposures to mixtures of solvents. *Am. J. Psychiatry* **145,** 1442–1445.

Schenk, V. W., and Stolk, P. J. (1967). Psychosis following arsenic (possibly thallium) poisoning. *Psychol. Neurol. Neurochir.* II, **70,** 31–37.

Schottenfeld, R. S., and Cullen, M. (1985). Occupation-induced post traumatic stress disorders. *Am. J. Psychiatry* **142,** 198–202.

Shaheen, S. (1984). Neuromaturation and behavior development: the case of childhood lead poisoning. *Dev. Psychol.* **20,** 542–550.

Terman, L. M., and Merrill, M. A. (1973). *Stanford-Binet Intelligence Scale: Manual for the Third Revision.* Form L-M. Houghton-Mifflin, Boston.

Tsushima, W. T., and Towne, W. S. (1977). Effects of paint sniffing on neuropsychological test performance. *J. Abnorm. Psychol.* **86,** 402–407.

Valciukas, J. A., Lilkis, R., Singer, R. M., Glickman, L., and Nicholson, W. J. (1985). Neurobehavioral changes among shipyard painters exposed to solvents. *Arch. Environ. Health* **40,** 47–52.

Vroom, F. G., and Greer, M. (1972). Mercury vapor intoxication. *Brain* **95,** 305–318.

Wechler, D. and Stone, C. (1945). *Wechsler Memory Scale.* The Psychological Corporation New York.

Wechsler, D. (1967). *Wechsler Preschool and Primary Scale of Intelligence.* Psychological Corporation, New York.

Wechsler, D. (1974). WISC-R manual. *Wechsler Intelligence Scale for Children-Rev.* Psychological Corporation, New York.

Wechsler, D. (1981). *Wechsler Adult Intelligence Scale-Rev.* Harcourt Brace, Jovanovich, New York.

Wechsler, D. (1987). *Wechsler Memory Scale-Rev.* The Psychological Corporation, New York.

White, R. F. (1993). Clinical neuropsychological investigation of solvent neurotoxicity. In *Organic solvents and industrial chemicals.* (W. K. Anger, Ed.), Vol. II, *Effects and Mechanisms. Handbook of Neurotoxicology.* (L. W. Chang Ed.), Dekker, in press.

White, R. F. (1987). Differential diagnosis of probable Alzheimer's disease and solvent encephalopathy in older workers. *The Clinical Neuropsychologist* **1,** 153–160.

White, R. F., Moss, M. B., Proctor, S. P., and Feldman, R. G. (1989). MRI-detected white matter pathology in two cases of toxic encephalopathy. Presented at American Psychological Association, New Orleand, Alabama.

White, R. F., Feldman, R. G., Echeverria, D. E., and Schweikert, J. (1990a). Neuropsychological effects of chronic solvent exposure. Report to NIOSH on Grant #5K01 OH0028-03.

White, R. F., Feldman, R. G., and Travers, P. H. (1990b). Neurobehavioral effects of toxicity due to metals, solvents and insecticides. *Clin. Neuropharmacol.* **13,** 392–412.

White, R. F., Feldman, R. G., and Proctor, S. P. (1992a). Neurobehavioral effects of toxic exposures. In *Clinical Syndromes in Adult Neuropsychology: The Practitioner's Handbook* (R. F. White, Ed.), Elsevier, New York.

White, R. F., Feldman, R. G., Moss, M. B., Proctor, S. P. (1993). Magnetic resonance imaging (MRI), neurobehavioral testing and toxic encephalopathy: two cases. *Environ. Res.* **61,** 117–123.

White, R. F., Diamond, R., Proctor, S. P., Morey, C., and Hu, H. (1993). Residual cognitive deficits 50 years after childhood lead poisoning. *Br. J. Ind. Med.* **50**(7), 613–622.

White, R. F., Robins, T. G., Proctor, S. P., Echeverria, D., and Rocskay, A. Z. (1993). Neurobehavioral effects of naphtha exposure among automotive workers. *Occup. Environ. Med.* **51**, 102–112.

Yokoyama, K., Araki, S., and Aono, H. (1988). Reversibility of psychological performance in subclinical lead absorption. *Neurotoxicology* **9**, 405–410.

Zettel, H. (1943). The effects of chronic arsenic damage on heart and blood vessels. *Z. Klin. Med.* **142**, 689.

47

Behavioral Neurophysiology: Quantitative Measures of Sensory Toxicity

DONNA MERGLER
Centre pour
l'étude des interactions biologiques
entre la santé et l'environnement
CINBIOSE
Université du Québec à Montréal
Montréal, Canada HC3 3P8

Damage to visual, somatosensory, and olfactory systems has always been an important part of the clinical examination of neurotoxicity. Sensory systems include a receptor organ, afferent pathways, and central integrative networks, and neurotoxic substances may affect one or all of the three. Receptor organs are often directly in contact with neurotoxic substances. Olfactory receptors are the most evident since airborne pollutants pass directly over the receptor cells. Somatosensory receptors may be affected by percutaneous chemicals, which cross the protective barrier of the skin and seep into the underlying tissues. Even the retina may be directly affected by substances absorbed through the mucous membranes of the eye. Clinical reports, coupled with results from animal studies, have provided extensive evidence of the vulnerability of sensory receptors and fibers to neurotoxic assault.

Symptoms of sensory disorders, including blurred vision, tingling sensations, dizziness, and hyper- or hypoosmia, are often surveyed when assessing possible neurotoxic effects in exposed populations. However, it is only recently that quantitative measures of sensory functions have been used in a systematic fash-

ion to study early nervous system dysfunction associated with workplace and environmental exposure to toxic substances. These studies indicate that continued low to medium exposure can produce changes in sensory systems, which, in the early stages, are not necessarily accompanied by functional disorders. Progression appears to depend on the particular neurotoxin and exposure dose, as well as factors of individual susceptibility. Initially, alterations may be reversible or compensated through adaptive processes. However, as the impairment progresses, symptoms and signs, often nonspecific in nature, become apparent. Finally, damage may become so severe that a clear clinical syndrome, generally irreversible, is apparent. Identification of early nervous system dysfunction in exposed populations can lead to preventive actions in the environment and better public health strategies.

Efforts have been made to develop and validate quantitative behavioral neurophysiological tests of sensory functions that can be used under standard conditions in field studies of neurotoxin-exposed populations. These have focused primarily on vision, vibration sensitivity, and olfaction; however, there is evidence that deficits in hearing, temperature sensitivity, and vestibular functions may also be important indicators of early neurotoxicity.

I. Visual System

The visual system is composed of a complex network of interconnecting neurons, whose receptor organ, the retina, is one of the most intricate sensory organs; changes in retinal transmission, or in the optic pathways, can affect visual information processing. Subtle alterations in visual functions may prove particularly useful for the surveillance of early neurotoxic impairment and in determining etiology. In population studies of workers exposed to neurotoxic substances, two types of visual deficits, which can occur in the absence of loss of visual acuity, have been associated with exposure: acquired dyschromatopsia and reduced contrast sensitivity threshold at intermediate spatial frequencies.

A. Acquired Dyschromatopsia

Chromatic discrimination loss can result from changes in ocular structures, such as lens opacification or from altered functioning of the intricate neurooptic pathways (Pokorny et al., 1979). The chromal focus of the deficit can provide clues concerning the site of damage. Verriest (1963) classified acquired dyschromatopsia into three categories: Type I, loss in the red–green range; Type II, loss in the red–green and blue–yellow range; and Type III, loss in the blue–yellow range. Toxic retinopathies commonly produce Type III acquired dyschromatopsia, whereas toxic optic neuropathies tend to produce an acquired Type II or Type I deficit; however, in many diseases, Type III loss can develop over time into Type II, suggesting that the latter is a more advanced stage (Hart, 1987). This may be the case for certain types of neurotoxin-induced dyschromatopsia.

Chromatic discrimination appears to be particularly vulnerable to a wide range of organic solvents and may constitute an important early indicator of neurotoxic damage. Many studies of working populations in different industries have related acquired color vision loss to organic solvent exposures (Table 1). Color arrangement tests, such as the Lanthony D-15 desaturated hue panel (LD-15-D) (Lanthony, 1978) and the Farnsworth–Munsell 100 hue panel (FM-100) (Farnsworth, 1957), have been shown to be sensitive to neurotoxin-induced color vision loss. The former is a simple test, which takes approximately 5 min. The subjects are required to arrange 15 randomly ordered caps in order of chromatic similarity. Standardized lighting can be provided by a 1150 lux "daylight" placed 30 cm above the caps. The Farnsworth–Munsell 100 hue is a more extensive test, with 80 caps in four separate boxes. Again, the subjects are required to place the caps in order of color similarity. Both tests provide both quantitative (error scores) and qualitative (type of dyschromatopsia) information. Comparison of these tests in a neurotoxin-exposed population revealed that the results are highly correlated (Mergler et al., 1987). These tests identify persons with congenital color deficits, who can then be excluded from analysis in studies of neurotoxicity. It should be noted that tests commonly used to detect congenital color vision loss, such as the Ishihara pseudo-isochromatic plates, are ineffective in detecting acquired loss, particularly in the blue–yellow range (Pokorny et al., 1979).

Initial workplace studies on chromatic discrimination were done by Raitta and her collaborators (1978), who observed color vision loss, mainly in the blue–yellow range in 12 of 15 workers exposed to n-hexane. These workers had normal visual acuity and visual fields, but presented evidence of maculopathy (Raitta et al., 1978) and altered electroretinograms and visual-evoked potentials (Seppäläinen, 1979), although there was no correlation between the various deficits. Further studies on a large number of organic solvents suggest that the loss is a dose-dependent relationship, with initial damage to the distal retinal layers (Type III deficit), progressing to the more proximal layers and the optic nerve (Type II deficit) (Mergler, 1990). This pattern has also been observed with increasing alcohol consumption (Mergler et al., 1988).

Few studies have examined the reversibility of acquired color vision loss following short-term cessation of exposure. Gobba and co-workers, (1991), who had observed dose-related acquired dyschromatopsia among styrene-exposed workers, reported no change in a small subgroup of 20 workers who were tested following 1-month holidays. On the other hand, Mergler and Bélanger (1993), who examined color vision in solvent-exposed laboratory personnel over a 2-year interval, during which the ventilation system and work practices were improved, reported better color vision discrimination at the second testing period. These results suggest that color vision loss may be reversible in some circumstances, if exposure is reduced; however, more studies are needed to determine the factors that contribute to progression or regression, and the time course of reversibility.

There have been some workplace studies in which no relation was observed between color vision loss and exposure. Ruitjen and collaborators (1990), who examined color vision among viscose rayon workers,

TABLE 1 Studies of Acquired Dyschromatopsia in Working Populations

Industry	Exposure	Exposed (referents)	Test	Results	Reference
Vegetable oil extraction and adhesive plant	n-Hexane	15	FM-100[a]	12/15 manifested color vision loss	Raitta et al. (1978a)
Viscose rayon plant	Carbon disulfide	62 (40)	FM-100	Higher prevalence among exposed; no clear axes of loss	Raitta et al. (1981)
Household conditioners manufacture	Styrene, tetrachloroethylene	687	Rabkin polychromatic plates	Red/green loss: severity function of length of service	Alieva et al. (1985)
Explosives manufacture	Diethyl ether	89 (115)	LD-15-D[b]	Greater loss in more highly exposed workers	Blain et al. (1986)
Paint manufacture	Acetone, MEK, toluene, xylene, styrene	23	LD-15-D and FM-100	Type II and Type III loss: severity function of current exposure	Mergler et al. (1987)
Printshop	Alcohols, methylene chloride, xylene, Stoddard, toluene, perchloroethylene,	30 (30)	LD-15-D	Significant difference between exposed and referents: severity function of job type	Mergler et al. (1988)
Ink manufacture	Ethanol, methanol, MEK, naptha, toluene, xylene, isopropanol	26 (26)	LD-15-D	Significant difference between exposed and referents: severity function of current exposure	Legault-Bélanger et al. (1988)
Visocose rayon plant	Carbon disulfide	45 (37)	LD-15-D	No difference between groups	Ruitjen et al. (1990)
Disabled workers from Occupational Health Clinic	Different exposures	17	LD-15-D	14 dyschromatopsia, 10 with Type II loss	Mergler et al. (1991)
Aircraft manufacture (retirees)	Methylene chloride	25 (21)	LD-15-D	No significant differences between exposed and referents	Lash et al. (1991)
Microelectronics assembly (former workers)	Mixed organic solvent exposures	57 (57)	LD-15-D	No differences; prevalence high in both groups	Mergler et al. (1992)
		25			Hudnell et al. (1995)
		(25)			
Several plants	Toluene, tetrachloroethylene	261 (120)	New Lanthony	No difference between groups	Nakatsuka et al. (1992)
Reinforced plastics plants	Styrene	75 (60)	LD-15-D	Color vision loss increases with exposure	Gobba et al. (1991)
		128	LD-15-D		Mergler et al. (1992)
		60 (60)	FM-100		Fallas et al. (1992)

[a] Fransworth–Munsell 100 hue panel.
[b] Lanthony D-15 desaturated hue panel.

suggested that the levels of exposure in this particular plant were insufficient to produce the type of loss that had previously been observed by Raitta and colleagues (1980) among workers with much higher exposures. In an investigation of acquired dyschromatopsia in retirees with a history of exposure to methylene chloride and those with no exposure, Lash and colleagues (1991) did not observe differences between the two groups. Although the authors postulate that these levels of methylene chloride may not affect color vision, two other hypotheses should be examined. First, both groups, whose mean ages were in the upper sixties, presented high prevalences of color vision loss; age-related loss may mask solvent-related changes. Second, since these workers had not been exposed for a certain number of years, there may be improvement over time. In two studies of microelectronics workers (Mergler *et al.*, 1992; Hudnell *et al.*, 1995), the prevalence of acquired dyschromatopsia was high not only in the exposed group, but also in the reference populations. The authors suggest that this may be due to the geographical location, which was similar in both studies: a high desert plateau, with intense light and high levels of ultraviolet radiation (Mergler *et al.*, 1991).

In a major investigation of color vision among workers from several plants exposed to toluene and tetrachloroethylene, the authors report no differences between the exposed and referent populations (Nakatsuka *et al.*, 1992). In this study, the New Lanthony Panel was used. The negative results may indicate that exposure was insufficient to produce an effect, that these particular solvent mixtures do not affect color vision, or that this instrument is not sensitive to the type of color vision loss resulting from exposure. The New Lanthony Test, which requires subjects to discriminate between shades of gray, may not be sampling neuro-optic functions that are sensitive to organic solvent exposure. It is interesting to note that there was no evidence of age-related color vision loss in the study by Nakatsuka and colleagues (1992), which would support the idea that this test measures a different aspect of visual processing.

Although acquired dyschromatopsia has not been systematically assessed among workers with lead exposure, restricted color fields, especially for green, and macular lesions have been associated with lead exposure (Vints, 1975). In addition, a highly significant reduction in the visual field index, under mesopic adaptation conditions, was reported among workers exposed to lead stearate in the absence of changes in classical tests of visual field measurement, ophthalmoscopy, critical fusion frequency, or the blind spot area (Cavalleri *et al.*, 1982). No differences in chromatic discrimination were observed between manganese exposed workers and a matched referent population (Mergler *et al.*, 1994).

Unlike congenital color vision loss, acquired dyschromatopsia is variable and complex; it is age dependent and can be monocular or localized unequally in the two eyes (Pokorny *et al.*, 1979). In population studies on the possible relationship between chromatic discrimination loss and exposure, it is important to take into account these factors as well as the possible contribution of alcohol consumption and other environmental or workplace exposures, which may also affect color vision, such as high levels of ultraviolet radiation and working with certain colored screens of video display terminals (Mergler, 1990). Certain illnesses, such as diabetic retinopathy, ophthalmotoxic medications (Pokorny *et al.*, 1979), and Parkinson's disease (Price *et al.*, 1992), as well as congenital colorblindness and cataracts, can also affect chromatic discrimination, and study participants should be screened for these factors.

B. Contrast Sensitivity Loss

Measurements of near visual contrast sensitivity may provide another means of detecting subclinical changes in visual function. A loss of visual acuity, caused by optical phenomena, is generally associated with changes in high spatial frequencies of the contrast sensitivity pattern (Owsley *et al.*, 1983), whereas a loss in intermediate and/or low spatial frequencies reflects neural rather than optical efficiency (Greeves *et al.*, 1988). In patients with diabetic optic neuropathy or Parkinson's disease, diminished contrast sensitivity has been observed in the absence of loss of visual acuity (Regan and Neima, 1984; Bodis-Wollner *et al.*, 1987). A "notch" in intermediate spatial frequencies is one of the characteristic patterns of spatial frequency loss associated with multiple sclerosis (pre- and cortical patch of demyelination) (Regan, 1988). A more general spatial frequency loss, observed among patients with compressive lesions of the anterior visual pathways, seems to reflect a more diffuse disturbance of the optic nerve (Kupersmith *et al.*, 1982).

In a matched pair study of former microelectronics workers with a history of high exposure to mixed organic solvents, Mergler and colleagues (1991) observed contrast sensitivity loss over all spatial frequency bands, but it was the greatest in the intermediate frequencies between 3 and 12 cycles/degree. In a further analysis of a subgroup of these workers and referents with normal visual acuity, differences were only present in the intermediate frequency range, suggesting primarily neural rather than ocular damage (Frenette *et al.*, 1991). A study of another population of microelectronics workers reported similar findings (Hudnell *et al.*, 1995).

Tests for assessing contrast sensitivity threshold include targets of sinusoidal gratings at different spatial frequencies, in order of decreasing contrast. The simplest is a nonautomated version on cards. Subjects are asked to indicate the direction of the gratings; three cards, with different presentations, are used to ascertain the threshold level. More sophisticated automated and semiautomated versions are likewise available.

II. Vibration Sensitivity

Loss of somatosensory functions is an important sign of peripheral neuropathy and has been used for a long time in clinical diagnosis. The "glove and sock" pattern of paresthesia is a common finding following severe acute or chronic intoxication. Axonal degeneration, beginning in the distal ends of long, myelinated sensory axons, is a common characteristic of toxic neuropathy (Spencer *et al.*, 1985).

In the search for quantitative indicators of early nervous system alterations associated with exposure, a certain number of instruments have been developed to measure vibrotactile perception threshold. The tuning fork was the first quantitative measure for assessing vibrotactile perception, by noting the time taken for the subject to indicate that the stimulus had disappeared (Pearson, 1928). Since then, a wide variety of devices have been proposed (Table 2). The basic design includes a vibrating shaft, with varying stimulus amplitude. Most provide vibrations at a fixed frequency,

although stimulus frequency can be varied in more sophisticated devices (Era *et al.*, 1986; Lundborg *et al.*, 1987). There are many variations to both the physical structure (size of the shaft, pressure control, etc.) and to the method of administration (forced choice or method of limits); some can only be applied to fingertips, whereas others can be used on fingers and toes or on a wide variety of sites. The list in Table 2 is far from exhaustive and serves mainly to illustrate the varying parameters.

Testing protocols fall into two major categories: forced choice and the limits method. In the former the subject is presented with two stimuli and indicates which of the two he or she can feel. In the latter, the stimulus amplitude is increased and/or decreased and the subject indicates when he or she can feel it or no longer feel it. Gerr and Letz (1988) compared the two methods, using the same instrument, and found that the method of limits was more reliable and time efficient.

Many of these instruments have proved useful in population-based studies of workers with exposure to neurotoxic substances. Table 3 contains a list of some of the studies that have examined vibrotactile perception threshold in active working populations. Many of these studies reported a relationship between elevated vibrotactile perception thresholds and exposure, particularly for lower extremities. However, there is a clear need for the standardization of methods if interstudy comparisons are to be valid.

Vibrotactile perception loss can provide an important indication of peripheral neurotoxic damage.

TABLE 2 Specifications of Some of the Instruments Proposed for Assessing Vibrotactile Perception Threshold

Instrument	Tip diameter	Stimulus frequency (Hz)	Pressure	Method	Reference
Electromagnetic biothesiometer	6	100	Not controlled	Limits	Eloffson *et al.* (1980)
Biothesiometer	13	120	Not controlled	Limits	Blooms *et al.* (1984)
Modified biothesiometer	13	120	Constant	Limits	Goldberg and Lindblom (1979)
Optacon	144 vibrating pins	230	Not controlled	Limits	Arezzo and Schaumberg (1980)
Vibrameter	12	120	Not controlled	Limits	Verberk *et al.* (1985)
Vibratron	12	120	Not controlled	Forced choice	Arezzo and Schaumberg (1985)
Vibrator	13	50, 100, 250	Constant	Limits	Era *et al.* (1986)
Multirod	112 vibrating pins	230	Not controlled	Limits and forced choice	Muijser *et al.* (1987)
Vibrometer	5	8–500	Constant	Limits	Lundborg *et al.* (1987)
Vibratron II	15	120	Not controlled	Limits and forced choice	Gerr and Letz; (1988)
Vibratometer	13	120	Constant	Limits	Frenette *et al.* (1990)

TABLE 3 Studies of Vibration Sensitivity in Working Populations

Industry	Exposure	Exposed (referents)	Test	Results	Reference
Swedish Air Force	Jet fuel	242	Modified biothesiometer	Correlation between exposure to jet fuel and decreased VPT[a]	Lindblom and Goldberg (1980)
Car painters	Mixed organic solvents	102 (102)	Tuning fork	64% exposed vs 25% nonexposed with pathological VPT in lower limbs	Husman and Karli (1980)
Car painters	Mixed organic solvents	80 (80)	Electromagnetic biothesiometer	Significant differences in mean VPT for both upper and lower limbs	Elofsson et al. (1980)
Chloralki plant	Elemental mercury	138	Tuning fork	Eighteen subjects with mild sensory polyneurotpathy, including lower VPT	Albers et al. (1982)
Various workplaces	Organic solvents	65 (33)	Biothesiometer	VPT higher in exposed, but not statistically significant (5/6 sites)	Gregersen et al. (1984)
Shipyard	Xylene, toluene, ethyl glycol MBK	90 (202)	Vibrameter	Elevated VPT of lower extremities in 8.8% of exposed; no correlation between VPT and exposure duration	Halonen et al. (1986)
Furniture manufacture, automotive parts, printshops	Mixed solvents	124 (116)	Tuning fork	VPT significantly higher in lower extremities of exposed workers	Maizlish et al. (1987)
Painters	Mixed solvents	112 (105)	Biothesiometer, Vibratron	VPT significantly higher in painters; VPT related to cumulative exposure indices	Bove et al. (1989)
Painter	Mineral spirits, toluene, ketones, xylene, naptha	28 (20)	Vibrometer	Painters have significantly higher VPT	Demers et al. (1991)
Chemical factory	Acrylamide	82	Vibratron II	No differences in VPT with exposure	Bachmann et al. (1992)
Agriculture	Organophosphate pesticides	36	Vibratron II	Previously intoxicated had higher VPT, particularly in lower extremities	McConnell et al. (1994)
Manganese alloy production	Manganese	74 (74)	Vibratometer	No difference between exposed and referents	Mergler et al. (1994)

[a] Vibrotactile perception threshold.

Like the other senses, vibration sensitivity decreases with age (Pearson, 1928; Era *et al.*, 1986; Frenette *et al.*, 1990; Gerr *et al.*, 1990). It varies with height (Gerr *et al.*, 1990) and with gender (Frenette *et al.*, 1990), although the latter study did not account for intergender height differences. Vibration threshold is likewise affected by other factors, such as certain illnesses, including diabetes (Hokaday *et al.*, 1981; Bertelsmann *et al.*, 1986), as well as the use of vibrating tools (Verberk *et al.*, 1985; Brammer *et al.*, 1987; Lundborg *et al.*, 1987) and repetitive movement (Bleecker, 1986; Jetzer, 1991). These factors should be taken into account in the study design and/or statistical analyses when examining the relationship between vibrotactile perception loss and exposure to neurotoxic substances.

III. Olfactory Functions

Attention has been focused on smell perception, not only as an indicator of neural damage, but also because smell is used to identify the presence of hazardous substances; in the workplace, mask cartridge breakthrough is identified by odor. The particular vulnerability of the olfactory system to airborne toxins results from the anatomical location of receptor cells and possibly from the highly active transneuronal transport mechanisms within the olfactory nerve cells (Peri and Good, 1987; Hasting and Evans, 1991). Chemically induced olfactory loss has been classified into three categories (Amoore, 1986): respiratory, essential, and central. The first category groups hyposmias secondary to upper respiratory tract blockage, the second includes those that result from neurotoxic injury to olfactory mucosa and neural receptors, and the third refers to olfactory loss subsequent to central lesions.

Although as many as 120 substances have been reported to produce some form of temporary or permanent olfactory dysfunction (Amoore, 1986), it is relatively recently that valid and reproducible psychophysical tests, designed to assess odor identification and olfactory perception threshold, are being used more systematically in workplace studies. Odor identification includes elements of cognitive processing, requiring the subject to recognize and name different substances. Olfactory perception threshold provides a quantitative means of determining the lowest concentration at which the person can perceive an odor.

Olfactory functions have been systematically assessed in several work situations (Table 4). Some of the studies examined workers' capacity to identify odors, whereas others assessed odor perception threshold. The most commonly used odor identification test is the University of Pennsylvania smell identification kit (UPSIT), which consists of four booklets, each containing 10 microencapsulated odorants (Doty *et al.,* 1984a). The test is self-administered by scratching a pencil tip over the odorant label strip on the bottom of each page, thereby releasing the odor; the person is then required to choose one of four answers even if no apparent smell is perceived. The test has been shown to be reliable and compares well with more sophisticated test procedures (Doty *et al.,* 1985). Olfactory perception threshold is evaluated using serial dilutions of a particular substance. The forced choice method is used whereby the subject is required to choose between two bottles, one containing the odorant, the other a blank. Each bottle is placed below the subject's nostrils, who inhales when the bottle is squeezed. Both the test administrator and the subject are blinded as to which bottle contains the odorant, and the subject is blinded as to the concentration. Three consecutive correct answers are used to establish threshold levels. Commercial kits (Olfactolabs, El Cerrito) are available and provide reproducible responses (Fortier *et al.,* 1991).

Both odor identification and olfactory threshold perception appear to be affected by exposure to a large number of neurotoxic substances. Most neurotoxins are associated with loss of smell; however, hyperosmia has been observed with long-term exposure to manganese (Mergler *et al.,* 1994); the authors suggest that this might represent an excitatory phase of early manganese intoxication. Only one study examined both odor threshold and odor identification. Rose and colleagues (1992) reported a higher olfactory threshold to phenol in cadmium-exposed workers as compared to referents, although no differences were observed in odor identification. The authors suggest that cadmium affects peripheral receptor cell function and not the more central processing required for odor identification. Future studies relating smell and exposure should try to incorporate these two methods, which provide different and complementary information on the neural processing of olfactory information.

It is interesting that in the study by Ahlstrom and colleagues (1986) of petroleum workers, higher thresholds were observed for *n*-butanol and oil vapor, but not for pyridine or dimethyl disulfide; when thresholds were analyzed with respect to number of days since exposure, an improvement over time was noted. Both *n*-butanol and oil vapor were probably present in the environment in this plant, whereas the other two substances were not. This fact, coupled with the transient nature of the response, suggests that there was a temporary threshold shift, but not necessarily a permanent olfactory deficit. Temporary threshold shifts to the inhaled substance have been observed in 7-hr chamber inhalation studies in the absence of threshold changes to a structurally different substance (Mergler and Beauvais, 1992; Gagnon and Mergler, 1994). The question of odorant specificity and the relation of temporary threshold shift to permanent impairment are important to our understanding of the mechanisms that underlie olfactory deficits induced by chemical exposures.

Olfactory acuity is age related (Doty *et al.,* 1984b; Cain and Murphy, 1987; Fortier *et al.,* 1992). Controversy exists over the relation between smoking and olfaction, some studies have observed a positive association (Doty, 1979; Schwartz *et al.,* 1989; Fortier *et al.,* 1992) whereas others have not (Fordyce, 1961; Venstrom and Amoore, 1968; Dravnieks *et al.,* 1986). It is, however, an important variable to consider when assessing olfactory functions in working populations since there may be an interaction between environmental and smoking pollutants.

IV. Sensory Test Battery

Most studies on nervous system dysfunction among exposed workers have relied on neurobehavioral test batteries (Johnson, 1987), which have been effectively used to examine the association between workplace exposures and early alterations. The addition of sensory tests to such batteries provides an important complement for the detection and understanding of nervous system alterations. In a document put out by the U.S. Agency for Toxic Substances and Disease Registry on neurobehavioral test batteries for use in environmental field studies (Hutchinson *et al.,* 1992), the authors propose that tests of visual function (color vision and contrast sensitivity) and vibrotactile perception be included in the Level 1 "core" battery designed to screen for early neurotoxic effects of environmental pollutants. In the Level 2 battery, they suggest that other psychophysical measures of sensory function be considered: audiometry, olfactory functions (identification and threshold), visual critical fusion frequency and dark-adapted absolute luminance thresholds, postural sway, and thermal thresholds. The criteria of selection that has been proposed for the choice of neurobehavioral tests (Johnson, 1987) should likewise be applied to psychophysical measures: The tests should measure

TABLE 4　Studies of Olfactory Functions and Toxic Exposure in Working Populations

Industry	Exposure	Exposed (referents)	Test	Results	Reference
Alkaline battery	Cadmium and nickel	85 (75)	OPT to phenol	Significantly higher OPT among exposed	Adams and Crabtree (1961)
Not indicated	Lead	28	Olfactometer	Increased difficulty in odor differentiation with length of exposure	Popivanova and Kehavaiov (1980)
Cleaning oil tanks	Petroleum products	20 (40)	OPT to *n*-butanol, oil vapor, pyridine, dimethyl disulfide	Higher OPT among exposed to *n*-butanol and oil vapor, not pyridine or dimethyl disulfide	Ahlstrom *et al.* (1986)
Chemical plant	Acrylate and methylacrylate vapors	731	UPSIT[b]	Odor identification decreased with exposure level and duration	Schwartz *et al.* (1989)
Painters	Mixed solvents	54 (42)	UPSIT	Painters had lower score, but not significant	Sandmark *et al.* (1989)
Painters	Cumulative, 2–40% TLV	187	UPSIT	Cumulative lifetime dose related to odor identification loss	Schwartz *et al.* (1990)
Various workplaces	Hydrocarbons	264	Reported smell disturbances	Prevalence higher with higher exposure	Hotz *et al.* (1992)
Brazing operation	Cadmium	55 (16)	OPT to butanol and odor identification	Higher OPT among exposed; no difference in odor identification	Rose *et al.* (1992)
Printshop	Alcohols, methylene chloride, xylene, perchloroethylene, Stoddard, toluene	18 (18)	OPT to pm-carbinol (Olfactolabs kit No. 11)	Significantly higher OPT among exposed	Fortier *et al.*, unpublished data
Manganese alloy production workers	Manganese	74 (74)	OPT to pm-carbinol (Olfactolabs kit No. 11)	Significantly *lower* OPT among exposed	Mergler *et al.* (1994)

[a] Olfactory perception threshold.

[b] University of Pennsylvania odor identification test.

functions that are affected by several neurotoxic agents. They should have preferably yielded positive results in earlier studies, they must be reliable, with established construct validity, be advantageous in terms of cost (time, expertise and equipment) with respect to benefit (information provided), they should be relatively independent of subjects' cultural and educational background, and be reasonably motivating. Comprehensive batteries, including both neurobehavioral and psychophysical measures, would be useful not only for identifying early deficits, but also for improving our understanding of the mechanisms that underlie these outcomes.

References

Adams, R. G., and Crabtree, N. (1961). Anosmia in alkaline battery workers. *Br. J. Ind. Med.* **216,** 217–221.

Ahlström, R., Berglund, B., Berglund, U., Lindvall, T., and Wennberg, A. (1986). Impaired odor perception in tank cleaners. *Scand. J. Work Environ. Health* **12,** 574–581.

Albers, J. W., Caveder, D., Levine, S. P., and Langolf, G. D. (1982). Asympomatic sensorimotor polyneuropathy in workers exposed to elemental mercury. *Neurology* **32,** 1168–1174.

Amoore, J. E. (1986). Effects of chemical exposure on olfaction in humans. In *Toxicology of the Nasal Passages* C. S. Barrow (Ed.), pp. 155–190, Hemisphere Publishing Corp., Washington.

Arezzo, J. C., and Schaumberg, H. H. (1980). The use of the Optacon as a screening device. *J. Occup. Med.* **22,** 461–464.

Arezzo, J. C., and Schaumberg, H. H. (1985). The vibratron: a simple device for quantitative evaluation of tactile-vibratory sense. *Neurology* **35,** 169.

Bachmann, M., Myers, J. E., and Bezuidenhout, B. N. (1992). Acrylamide monomer and peripheral neuropathy in chemical workers. *Am. J. Ind. Med.* **21,** 217–222.

Bertelsmann, F. W., Heimans, J. J., Van Rooy, J. C. G. M., Heine, R. J., and Van Der Deer, E. A. (1986). Reproducibility of vibratory perception in patients with diabetic neuropathy. *Diabet. Res.* **3,** 463–466.

Bleecker, M. L. (1986). Vibration perception thresholds in entrapment and toxic neuropathies. *J. Occup. Med.* **28,** 991–994.

Blooms, S., Till, S., Sonksen, P., and Smith, S. (1984). Use of a biothesiovariation in 519 non-diabetic subjects. *Br. J. Ind. Med.* **288,** 1798–1795.

Bodis-Wollner, I., Marx, M., and Mitra, S. (1987). Visual dysfunction in Parkinson's disease. *Brain* **110,** 1675–1698.

Bove, F. J., Richard, L., and Baker, E. L. (1989). Sensory thresholds among construction trade painters: a cross-sectional study using new methods for measuring temperature and vibration sensitivity. *J. Occup. Med.* **31,** 320–325.

Brammer, A. J., Piercy, J. E., Auger, P. L., and Nohara, S. (1987). Tactile perception in hands occupationally exposed to vibration. *J. Hand Surg. Am.* **12,** 870–875.

Cain, W. S., and Murphy, C. L. (1987). Influence of aging on recognition memory for odors and graphic stimulation in "Olfaction and Taste IX." *Annu. N.Y. Acad. Sci.* **510,** 212–215.

Cavalleri, A., Trimarchi, F., Gelmi, C., Baruffini, A., Minola, C., Biscaldi, G., and Gallo, G. (1982). Effects of lead on the visual system of occupationally exposed subjects. *Scand. J. Work Environ. Health* **199**(Suppl 1), 148–151.

Demers, R. Y., Markell, B. L., Wabeke, G. L., and Wabeke, R. (1991). Peripheral vibratory sense deficits in solvent-exposed painters. *J. Occup. Med.* **33,** 1051–1054.

Doty, R. L., Newhouse, M. G., and Azzalina, J. D. (1985). Internal consistency and short-term test reliability of the University of Pennsylvania smell identification test. *Chemical Senses* **10,** 297–300.

Doty, R. L., Shaman, P., and Dann, M. (1984a). Development of the University of Pennsylvania smell identification test: a standardized microencapsulated test of olfactory function. *Physiol. Behav.* **32,** 489–502.

Doty, R. L., Shaman, P., Applebaum, S. L., Giberson, R., Sikorsji, L., and Rosenberg, L. (1984b). Smell identification ability: changes with age. *Science* **226,** 1441–1443.

Dravnieks, A. (1974). A building block model for the characterization of odorant molecules and their odors. *Annals N.Y. Acad. Sci.* **237,** 144–149.

Elofsson, S. A., Gamberale, F., Hindmarsh, T., Iregren, A., Isaksson, A., Johnsson, I., Knave, B., and Lydahl, E., Mindus, P., Persson, H., Philipson, B., Steby, M., Struwe, G., Soderman, E., Wennberg, A., and Widen, L. (1981). Exposure to organic solvents: a cross-sectional epidemiologic investigation on occupationally exposed car and industrial spray painters with special reference to the nervous system. *Scand. Work Environ. Health* **6,** 239–273.

Era, P., Jokela, J., Suominen, H., and Heikkinen, E. (1986). Correlates of vibrotactile thresholds in men of different ages. *Acta. Neuro. Scand.* **74,** 210–217.

Farnsworth, D. (1957). *The Farnsworth-Munsell 100 Hue Test Manual* (rev. ed.) Munsell Color Company, Baltimore, chapter 5.

Fordyce, I. D. (1961). Olfaction tests. *Br. J. Ind. Med.* **18,** 213–215.

Fortier, I., Ferraris, J., and Mergler, D. (1991). Measurement precision of an olfactory perception threshold test for use in field studies. *Am. J. Ind. Med.* **20,** 495–504.

Frenette, B., Mergler, D., and Bowler, R. (1991). Contrast sensitivity loss in a group of former microelectronics workers with normal visual acuity. *Optom. Vis. Sci.* **68,** 556–560.

Frenette, B., Mergler, D., and Ferraris, J. (1990). Measurement precision of a portable instrument to assess vibrotactile perception threshold. *Eur. J. Appl. Physiol.* **61,** 386–391.

Gagnon, P., Mergler, D., and Lagacé, S. (1994). Olfactory adaptation, threshold shift and recovery at low levels of exposure to methyl isobutyl kelone (MIBK). *Neurotoxicology* **15,** 632–642.

Gagner, and Mergler, D. (1994). Submitted for publication.

Gerr, F. E., and Letz, R. (1988). Reliability of a widely used test of peripheral cutaneous vibration sensitivity and a comparison of two testing protocols. *Br. J. Ind. Med.* **45,** 635–639.

Gerr, F. E., Hershman, D., and Letz, R. (1990). Vibrotactile threshold measurement for detecting neurotoxicity: reliability and determination of age and height-standardized normative values. *Arch. Environ. Health* **45,** 148–154.

Gobba, F., Galassi, C., Imbraini, M., Ghittori, S., Candela, S., and Cavalleri, A. (1991). Acquired dyschromatopsia among styrene-exposed workers. *J. Occup. Med.* **33,** 761–765.

Goldberg, J. M., and Lindblom, U. (1979). Standardized method of determining vibratory perception thresholds for diagnosis and screening in neurological investigation. *J. Neurol. Psychiatry* **42,** 793–803.

Greeves, A. L., Cole, B. L., and Jacobs, R. J. (1988). Assessment of contrast sensitivity of patients with macular disease using reduced contrast near visual acuity charts. *Ophthalmol. Physiol. Opt.* **8,** 371–377.

Gregerson, P., Angelsø, B., Nielsen, T. E., Nørgaard, B., and Uldall, C. (1984). Neurotoxic effects of organic solvents in exposed workers: an occupational, neuropsychological, and neurological investigation. *Am. J. Ind. Med.* **5,** 201–225.

Halonen, P., Halonen, J. P., Lang, H. A., and Karskela, V. (1986). Vibratory thresholds in shipyard workers exposed to solvents. *Acta Neurol. Scand.* **73,** 561–566.

Hart, W. M. (1987). Acquired dyschromatopsias. *Surv. Ophthalmol.* **32,** 10–31.

Hastings, L., and Evans, J. (1991). Olfactory primary neurons as a route of entry for toxic agents into the CNS. *Neurotoxicology* **12,** 707–714.

Hokaday, T. D. R., Holman, R. R., Hillson, R. M., Pim, B., and Smith, B. (1981). Comparison of vibration sense by biothesiometer and fasting glucose values during 3 years of diabetes. *Diabetologica* **20,** 670.

Hotz, P., Tschopp, A., Söderström, D., Holtz, J., Boillat, M. A., and Gutzwiller, F. (1992). Smell or taste disturbances, neurological symptoms, and hydrocarbon exposure. *Int. Arch. Occup. Health* **63,** 525–530.

Husman, K., and Karli, P. (1980). Clinical neurological findings among car painters exposed to a mixture of organic solvents. *Scand. J. Work Environ. Health* **6,** 33–39.

Hutchinson, L. J., Amler, R. W., Lybarger, J. A., and Chappell, W. (1992). Neurobehavioral test batteries for use in environmental health field study. U.S. Department of Health and Human Services, Agency for Toxic Substances and Disease Registry, Atlanta, Georgia.

Jetzer, T. C. (1991). Use of vibration testing in the early evaluation of workers with carpal tunnel syndrome. *J. Occup. Med.* **33,** 117–120.

Johnson, B. (1987). *Prevention of Neurotoxic Illness in Working Populations.* Wiley, Chichester.

Kupersmith, M. J., Siegel, M. I., and Ronald, E. C. (1982). Subtle disturbances of vision with compressive lesions of the anterior visual pathway measured by contrast sensitivity. *Am. Acad. Opthal.* **89,** 68–72.

Lanthony, P. (1978). The desaturated panel D-15. Doc. *Ophthal.* **46**, 185–189.

Lash, A. A., Becker, C. E., So, Y., and Shore, M. (1991). Neurotoxic effects of methylene chloride: are they long lasting in humans? *Br. J. Ind. Med.* **48**, 418–426.

Legault-bélanger, S., Bachand, M., Bédard, S., Brabant, C., de Grosbois, S., and Mergler, D. (1988) Perte de discrimination chromatique chez des travailleurs soumis à une exposition complexe et variable aux solvants organiques. *Arch. Mal. Prof.* **49**, 475–482.

Lindblom, U., and Goldberg, J. M. (1980). Screening for neurological symptoms and signs after exposure to jet fuel. *Acta Neurol. Scand.* **64**, 73–74.

Lundborg, G., Sollerman, C., Stromberg, T., Pyykko, J., and Rosen, B. (1987). A new principle for assessing vibrotactile sense in vibration-induced neuropathy. *Scand. J. Work Environ. Health* **13**, 375–379.

Maizlish, N. A., Fine, L. J., Albers, J. W., Whitehead, L., and Langolf, G. D. (1987). A neurological evaluation of workers exposed to a mixture of organic solvents. *Br. J. Ind. Med.* **44**, 14–25.

McDonnell, R., Keifer, M., and Rosenstock, L. (1994). Elevated quantitative vibrotactile threshold among workers previously poisoned with methamidophos and other organophosphate pesticides. *Am. J. Ind. Med.* **25**, 325–334.

Mergler, D., and Bélanger, S. (1993). Dépistage précoce et mesures prventives, dans le cadre de la surveillance des effets de l'exposition professionnelle. Forum Européen Sciences et Sécurité 1992, France, pp. 120–122.

Mergler, D., Blain, L., and Lagacé, J. P. (1987). Solvent related colour vision loss: an indicator of neural damage? *Int. Arch. Occup. Environ. Health* **59**, 313–321.

Mergler, D., Bélanger, S., de Grosbois, S., and Vachon, N. (1988). Chromal focus of acquired colour discrimination loss and solvent exposure among printshop workers. *Toxicology* **49**, 341–348.

Mergler, D. (1990). Color vision loss: a sensitive indicator of the severity of optic neuropathy. In *Advances in Neurobehavioral Toxicology: Applications in Environmental and Occupational Health* (B. L. Johnson, Ed.), pp. 175–182, Lewis Publishers, Chelsea, Mi.

Mergler, D., and Beauvais, B. (1992). Olfactory threshold shift following controlled 7-hour exposure to toluene and/or xylene. *Neurotoxicology* **13**, 211–216.

Mergler, D., Huel, G., Bowler, R., Iregren, A., Belanger, S., Baldwin, M., Tardif, R., Smargiassi, A., and Martin, L. (1994). Nervous system dysfunction among workers with long-term exposure to manganese. *Environ. Res.* **64**, 151–180.

Muijser, H., Hooisma, J., Hoogendijk, E. M. G., and Twisk, D. A. M. (1986). Vibration sensitivity as a parameter for detecting peripheral neuropathy I. Results in healthy workers. *Int. Arch. Environ. Occup. Health* **58**, 287–299.

Nakatsuka, H., Watanabe, T., Takeuchi, Y., Hisanaga, N., Shibata, E., Suzuki, H., Huange, Y. M., Chen, Z., Qu, Q. S., and Ikeda, M. (1992). Absence of blue-yellow color vision loss among workers exposed to toluene or tetrachloroethylene, mostly at low levels below occupational exposure limits. *Int. Arch. Occup. Environ. Health* **64**, 113–117.

Owsley, C., Sekular, R., and Siemens, D. (1983). Contrast sensitivity through adulthood. *Vis. Res.* **23**, 689–699.

Pearson, G. H. I. (1928). Effect of age on vibratory sensitivity. *Arch. Neurol. Psychiatry* **20**, 482–496.

Peri, D. P., and Good, P. F. (1987). Uptake of aluminum into central nervous system along nasal-olfactory pathways. *Lancet* **2**, 1028.

Popivaniova, P., and Kehaiov, A. N. (1980). Modifications de la sensitivité gustative et olfactive chez les ouvriers exposés à l'influence d'agents chimiques. *Acta Otorhinolaryngol. Belg.* **34**, 557–561.

Pokorny, J., Smith, V. C., Verrist, G., and Pinckers A. J. L. G. (Eds.) (1979). *Congental and Acquired Color Vision Defects.* New York, Grune and Stratton.

Price, M. J., Feldman, R. G., Adelberg, D., and Kayne, H. (1992). Abnormalities in color vision and contrast sensitivity in Parkinson's disease. *Neurology* **42**, 887–890.

Raitta, C., Teir, E., Tolonen, E., Helpio, E., and Malmstrom, S. (1981). Impaired color or vision discrimination among viscose rayon workers exposed to carbon bisulfide. *J. Occup. Med.* **23**, 189–192.

Raitta, C., Seppäläinen, A. M., and Huuskonen, M. S. (1978). *n*-Hexane maculopathy in industrial workers. *Graefes Archiv. Klin. Exp. Ophthalmol.* **209**, 99–110.

Regan, D., and Neima, D. (1984). Low-contrast letter charts in early diabetic retinopathy, ocular hypertension, glaucoma and Parkinson's disease. *Br. J. Ophth.* **68**, 885–889.

Regan, D. (1988). Low-contrast letter charts and sinewave gratings tests in ophthalmological and neurological disorders. *Clin. Vis. Sci.* **2**, 235–250.

Rose, C. S., Heywood, P. G., and Costanzo, R. M. (1992). Olfactoric impairment after chronic occupational cadmium exposure. *J. Occup. Med.* **34**, 600–605.

Ruijten, M. W. M. M., Sallé, H. J. A., Verberk, M. M., and Muijser, H. (1990). Special nerve functions and colour discrimination in workers with long term low level exposure to carbon disulphide. *Br. J. Ind. Med.* **47**, 589–595.

Sandmark, B., Broms, I., Lofgren, L., and Ohlson, C. G. (1989). Olfactory function in painters exposed to organic solvents. *Scand. J. Environ. Health* **15**, 60–63.

Schwartz, B. S., Doty, R., Monore, C., Frye, R., and Baker, S. (1989). Olfactory function in chemical workers exposed to acrylate and methacrylate vapors. *Am. J. Public Health* **79**, 613–618.

Schwartz, B. S., Ford, P. D., Bolla, K. I., Agnew, J., Rothman, N., and Bleecker, M. L. (1990). Solvent-associated decrements in olfactory function in paint manufacturing workers. *Am. J. Ind. Med.* **18**, 697–706.

Seppäläinen, A. M., Raitta, C., and Huuskonen, M. S. (1979). *n*-Hexane-induced changes in visual evoked potentials and electroretinograms of industrial workers. *Electroencephalogr. Clin. Neurophysiol.* **47**, 492–498.

Spencer, P. S., Path, M. R. C., and Schaumburg, H. (1985). Organic solvent neurotoxicity: facts and research needs. *Scand. J. Work Environ. Health* **11**(Suppl 1), 53–60.

Venstrom, D., and Amoore, J. E. (1968). Olfactory threshold in relation to age, sex or smoking. *J. Food Sci.* **33**, 264–265.

Verberk, M. M., Sallé, H. J. A., and Kempers, O. (1985). Vibratory and tactile sense of the fingers after working with sanders. *Int. Arch. Occup. Environ. Health* **56**, 217–223.

Verriest, G. (1963). Further studies on acquired deficiency of color discrimination. *J. Opt. Soc. Am.* **53**, 185–195.

Vints, L. A. (1975). The effect of lead on the visual organ. *Vestn Ofthmol.* **1**, 74–75.

48

Electrophysiological Investigations of Toxic Neuropathies

JOSEPH F. JABRE
Department of Neurology
Boston University School of Medicine
Boston, Massachusetts 02118

I. Introduction

The approach to the patient with neurotoxic exposure should take into account the diverse electrodiagnostic studies suited to document the toxic effect on the peripheral nervous system (PNS). Combined, these studies give an appraisal of the physiological status and offer insight into the nature and extent of PNS impairment.

Although these studies have achieved a great level of sophistication, there is no overall agreement on the techniques and methods used, let alone the interpretation of findings. Despite their widespread use in evaluating PNS lesions, a standard battery of tests is not applied uniformly. Thus, the reporting of findings from different laboratories cannot be compared without recognition of the limitations.

The usefulness of these techniques is limited by several factors, including the variability inherent to the data obtained, the interpretation of findings, and the mere definition of what is normal and how normal values are collected. The problem is made more complex when data obtained from different laboratories are compared. Differences in techniques, instrumentation, and variabilities among populations studied account for a great deal of inconsistencies among the data collected and its interpretation.

This chapter reviews routine nerve conduction studies (NCS), special nerve conduction analysis techniques, quantitative needle electromyography (EMG) including single fiber and macro EMG.

II. Routine Electrodiagnostic Studies

A. Nerve Conductions

The basic concept of nerve conductions is that when a nerve is electrically stimulated, a reaction occurs somewhere along its path, and, with appropriate recording electrodes, this reaction can be recorded and the time relationship between the stimulus and the response can be identified.

The motor response can be obtained by stimulating the nerve and recording from the muscle it innervates.

This response can be characterized by its latency, amplitude, duration, area, and wave form. To a large extent, the amplitude depends on the number and size of muscle fibers being stimulated. A normal motor response is a sign of a fairly synchronous discharge of the motor units. If there is desynchronization, the response is dispersed and its amplitude will be lowered.

Sensory nerve action potentials (NAP) are obtained by stimulating and recording from a nerve or one of its branches. The NAP is also characterized by its amplitude, duration, and wave form. The distal latency of the NAP is the interval measured from the stimulation of the distal-most accessible site on the nerve. Its conduction velocity is calculated by dividing the distance by the distal latency.

Nerves can be studied with routine NCS techniques used in the everyday practice of the EMG laboratory. The following studies were performed in the workup of toxic neuropathies (Jabre and Hackett, 1983).

1. The Median Nerve

The recording electrode is placed over the motor point of the abductor pollicis brevis muscle located in the middle third of the thenar eminence, close to the first metacarpal. The reference is placed distally over the first metacarpophalangeal joint. Distal stimulation of the median nerve is performed on the volar aspect of the wrist between the flexor carpi radialis and palmaris longus tendons. Proximally the nerve is stimulated above the elbow, just medial to the biceps tendon and brachial pulse. Optional stimulation points are at the axilla and Erb's point. Routinely, the conduction velocity is calculated between the elbow and the wrist.

The median sensory response is recorded antidromically from the digital nerve branches of the index finger. The active ring electrode is placed over the proximal phalanx and the reference ring electrode is placed over the intermediate phalanx of the index finger. The median nerve is stimulated 13 cm proximal to the active electrode.

2. The Ulnar Nerve

The active electrode is placed over the motor point of the abductor digiti minimi. The reference electrode is placed distally over the fifth digit. The nerve is stimulated at the wrist, lateral to the flexor carpi ulnaris tendon, below the elbow just distal to the medial epicondyle and above the elbow, at a point 12 cm proximal to the below-elbow point. The distance for the proximal segment is measured with the elbow held at 90° of flexion. Optional stimulation points are at the axilla and Erb's point. The conduction velocity is calculated between the wrist and below-elbow points and the below-elbow to above-elbow points (short segment).

The sensory response from the ulnar nerve is antidromically recorded from the digital branches to the fifth digit. The active ring electrode is positioned over the proximal phalanx of the fifth finger and the reference ring electrode is positioned over the intermediate phalanx. The ulnar nerve is stimulated at the wrist, lateral to the flexor carpi ulnaris tendon at a point 11 cm proximal to the active electrode.

The dorsal ulnar sensory response is recorded using a bar electrode placed over the dorsum of the hand between the fourth and fifth metacarpals. The nerve is stimulated at a point 8 cm proximal to the active recording electrode as it courses between the ulna and the flexor carpi ulnaris tendon.

3. The Radial Nerve

The sensory response from the radial nerve is recorded from a bar electrode placed in the web space on the dorsolateral aspect of the hand. The radial nerve is stimulated 10 cm proximal to the recording cathode as it traverses the radial shaft.

4. The Peroneal Nerve

The recording (disc) electrode is placed over the muscle belly of the extensor digitorum brevis on the lateral aspect of the dorsum of the foot. The reference electrode is placed over the base of the little toe. The nerve is stimulated distally at the ankle, lateral to the tendon of the tibialis anterior. Proximal stimulation sites are at the lateral aspect of the fibular head and in the popliteal fossa (knee), 10 cm above the fibular head point. Conduction velocities are calculated for the ankle to fibular head segment and for the fibular head to popliteal fossa segment.

5. The Superficial Peroneal Nerve

To record the superficial Peroneal sensory response, the active recording (bar) electrode is placed at the ankle about 4 cm anterior to the lateral malleolus. The stimulating electrode is placed on the anterolateral aspect of the leg against the fibula at a point 12 cm proximal to the active recording electrode.

6. The Posterior Tibial Nerve

The recording (disc) electrode is placed over the muscle belly of the abductor hallucis muscle. This site is located at a point 2 cm away from the navicular bone toward the heel. The reference electrode is placed over the base of the big toe. The nerve is stimulated distally at the ankle posterior to the medial malleolus. The proximal stimulation point is in the popliteal fossa about 2 cm lateral to the midline.

7. The Sural Nerve

To record the sural nerve sensory response, the surface recording (bar) electrode is placed at a point about

2 cm posteroinferior to the lateral malleolus and the reference electrode is placed more distally at the lateral aspect of the foot about 2.5 cm from the base of the little toe. Stimulation is performed at the posterolateral aspect of the leg, 14 cm proximal to the active recording electrode.

8. The F-Wave

The F-wave is a long latency muscle action potential obtained following supramaximal stimulation of motor axons. It is generally accepted that the F-wave is elicited by antidromic stimulation of the anterior horn cell at the axonal hillock. The F-wave is therefore always preceded by a M-wave. F-waves are routinely performed at the same setup for the motor nerve conduction study, using a slower sweep speed and a higher gain and stimulating the distal stimulation site. Ten impulses are delivered. The shortest latency potential is identified and selected as the F-wave for latency measurement.

9. The H-Reflex

The H-reflex is the electrical equivalent of a monosynaptic stretch reflex. It is obtained by selectively stimulating the Ia fibers which recruit the anterior horn cell or cells and generate a late response in the muscle, usually obtained before the direct motor response or M-wave. The H-reflex can be obtained by stimulating the posterior tibial nerve at the popliteal fossa. The response is recorded from the soleus, between the two heads of the gastrocnemii muscles. Minimal and maximal amplitude responses are obtained and the latency is measured to takeoff. H-reflexes can also be obtained in the forearm muscles, most notably, the flexor carpi radialis muscle.

10. The Blink Reflex

The blink reflex is elicited by stimulation of the supraorbital branch of the Vth nerve as it enters through the supraorbital foramen. On the ipsilateral side, both direct and indirect responses are seen. The direct (R1) has a latency of about 10.5 msec and is mono- or biphasic in configuration. The indirect (R2 ipsilateral) has a variable latency of about 30.5 msec and is polyphasic. On the contralateral side, only an indirect, long latency (R2 contralateral) polyphasic response is seen with a latency of about 30.5 msec.

These studies help determine the type of lesions affecting the peripheral nervous system, be it demyelination, axonal loss, or a mixture thereof.

Lesions, involving demyelination are characterized by prolonged distal latencies, slowed conduction velocities, and delayed late responses. Typically, sensory and motor action potentials are spread in time (sensory before motor) and the action potentials are desynchronized.

In lesions involving a focal demyelination, these findings are present when stimulating above the lesion but not below it. Similarly, conduction blocks will result in a lowered amplitude response when stimulating above the block but a normal amplitude when stimulating below it.

In diffuse demyelination, slowing (and amplitude changes) is present throughout, even though it is worse at pressure and entrapment sites.

In axonal loss lesions, sensory and motor action potential amplitudes are decreased but conduction velocities are preserved. Sensory amplitudes are reduced first, followed by a decrease in motor amplitude. These findings are present diffusely.

When axonal loss is severe, conduction velocities are slowed due to the involvement of the fastest conducting fibers.

B. Needle Examination

The effects of nerve lesions on the muscle can be studied with needle EMG techniques. These document the presence or absence of axonal injury by the study of denervation and the compensatory mechanisms available to the nervous system by the study of reinnervation. Needle EMG studies can document the presence or absence of axonal loss lesions as well as the age of the lesion and the regenerative capacity of the peripheral nervous system.

1. Fibrillations and Positive Waves

When a muscle fiber is denervated, the acetylcholine receptors spread all across the muscle fiber to attract new innervation to the denervated muscle fiber from adjacent nerves. The muscle fiber thus becomes more sensitive to the free acetylcholine and is depolarized and repolarized spontaneously as these molecules reach it. Each single depolarization is electrically detected as a single muscle fiber action potential recorded as a fibrillation or a positive wave. These discharge in a very rhythmic manner and usually start and stop abruptly.

As the muscle is reinnervated, both fibrillations and positive waves decrease in numbers and eventually disappear when the reinnervation is successfully completed.

2. Motor Unit Action Potentials

Following denervation, reinnervation is usually accomplished by collateral sprouting with the denervated muscle fibers seeking new nerve sprouts from adjacent nerves.

This reinnervation alters the motor unit in two ways: on the one hand, the motor unit now contains more muscle fibers; on the other hand, the newly acquired muscle fibers are asynchronous with those of the host unit and indeed among themselves. The newly formed end plates may not be stable in the beginning and many of them never reach maturity. Their respective muscle fibers either die or attract innervation from another source. This process of acquiring new muscle fibers and forming new endplates begins in the first 2 months after nerve injury and results in a prolongation of the motor unit potential duration and an increase in the number of its phases. The duration is prolonged simply because there are more fibers to depolarize, and the increase in the number of phases is due to the lack of synchronization between the host fibers and the newly acquired fibers.

III. Quantitative Techniques

A. Nerve Conductions

Recently described techniques enhance the diagnostic ability of nerve conduction studies by detecting abnormalities at an earlier stage or when routine conduction studies are normal. They do so by statistical analysis or by studying the distribution of conduction velocities of the different size axons within a nerve. Of these, the mean related values (MRV) (Jabre and Sato, 1990) and the distribution of conduction velocities (DCV) (Yokoyama *et al.*, 1990) techniques will be reviewed.

1. Mean Related Values Technique

The MRV technique statistically analyzes normal values, determines how they are collected, and decides whether or not they are Gaussian distributed before using them for the interpretation of results.

To this end, the frequency distribution of each electrophysiological variable studied is evaluated to determine if the variable has a Gaussian or non-Gaussian distribution. The purpose of this evaluation is to assure that the proper descriptive and comparative statistics are used during analysis.

If a variable has a gaussian frequency distribution, 68% of the individuals would fall within 1 SD around the mean (34% above and 34% below) whereas 95% of the individuals would fall within 2 SD of the mean. This leaves the remaining 5% distributed beyond the 2 SD above and below the mean. In addition, values ±3 SD from the mean are calculated for each parameter. If their predicted value is negative or falls outside the physiological range, their distribution is considered nongaussian.

The coefficient of variance for each variable (standard deviation/mean) is calculated and is considered significant if the value is greater than 0.15.

Variables are considered appropriate for transformation if they either had a non-Gaussian distribution or a coefficient of variance greater than 15% and/or if their diagnostic sensitivity was enhanced if they were related to another variable, a given distance, or another conduction variable.

This approach has many inherent benefits. (1) The enhancement of diagnostic sensitivity of NCS since the patient's values are now compared to "smoothed out" reference values which makes for a stricter definition of what is normal and abnormal. (2) The ability to express all data using a single unit (the standard deviation) instead of expressing them in mV, μV, ms, or m/s. Different parameters, or indeed nerves, can now be plotted on a single graph instead of many, giving the possibility of generating "profiles" of nerve conduction abnormalities in given pathologies. (3) The ability to assess degrees of severity. Nerve conductions can now be "graded"; for instance, results that fall between −2.5 and −2.9 MRV (clinically equivalent to mild) can be referred to as Grade I, those between −3 and −3.4 MRV (clinically equivalent to moderate) as Grade II, those between −3.5 and −3.9 MRV (clinically equivalent to severe) as Grade III, and an absent response could be called −4 MRV (absent) or Grade IV.

The drawbacks are mainly that it requires the availability of (and familiarity with) a personal computer and appropriate software to perform statistical analysis and compare data.

2. Distribution of Conduction Velocities Techniques

The techniques just described measure the conduction properties of the fastest sensory and motor nerve fibers given that the latency is measured to the takeoff of the action potential.

A more sophisticated, computer intensive technique known as distribution of conduction velocities (Yokoyama *et al.*, 1990) allows the estimation of conduction velocities in the medium and smaller as well. DCV attempts to quantify the conduction properties of the fiber population of a nerve resulting in a histogram of these conduction velocities. The techniques used involve mathematical modeling of the electric signal derived from a single nerve fiber. By in large, a nominal amplitude and waveform of such an action potential is assumed and given an appropriate weight. This allows the decomposition (a preferred term is "resolution") of the compound action potential of the nerve and thus the generation of a histogram of the distribution of conduction velocities of the nerve's fiber population.

This technique and near nerve recording techniques (Buchtal and Rosenfalck, 1966) are more sensitive in detecting abnormalities, but are more time consuming and computer intensive than the routine conduction studies described earlier.

B. Needle Examination

1. Single Fiber Jitter and Fiber Density Studies

Single fiber (Stålberg and Trontelj, 1979) and macro EMG (Stålberg, 1980; Stålberg and Fawcett, 1982) have added a great deal to our understanding of the basic microphysiology of the motor unit.

Single fiber EMG is performed with a special needle electrode. The electrode contains a 25-μm side port which allows for the recording of single muscle fibers. Using trigger and delay techniques, one can ensure the recording of single muscle fibers belonging to one motor unit. By measuring the variability between time-locked potentials, the jitter, one can measure the variability in neuromuscular transmission.

In early reinnervation, the jitter is increased because of the lack of maturity of the end plate and irregularity of conduction in the young nerve sprouts. When the end plate has matured and firmly established itself and when the axonal sprout has fully myelinated, the jitter becomes normal. Thus studies of the jitter allow the electromyographer to study the age and effectiveness of the reinnervation process.

A technique known as fiber density (Stålberg and Trontelj, 1979), also performed with a single fiber electrode, allows the measure of the number of muscle fibers belonging to the same motor unit in a 300 μm radius. Fiber density increases early in reinnervation because the number of muscle fibers belonging to the same motor unit in a given area of muscle is increased as a result of the collateral sprouting.

Fiber density and jitter studies are one of the more sensitive means of diagnosis and follow-up of the reinnervation process.

2. Concentric Macro EMG

Concentric macro EMG (Jabre, 1991; Guiheneuc *et al.*, 1988; Bauermeister and Jabre, 1992) is a technique that allows the study of the motor unit both in local and global terms. Using a Teflon-coated concentric needle, it is possible to record from a few (10–15) and a majority of the muscle fibers of the motor unit simultaneously. The study of these two areas allows one to evaluate the relationship between the small sample studied by concentric EMG and the large one studied by the macro recording surface (15 mm of cannula).

The question that can be implicitly addressed under those circumstances is how representative are the 10–15 fibers sampled by the concentric electrode of the motor unit as a whole. One may also ask what parameters are more or less representative and how are they affected in the disease of nerve and muscle.

The differences between the macro and concentric surface recording characteristics are quite revealing when there is local grouping or fractionation of the motor unit causing data obtained from the concentric action potential to be insufficient or even misleading. Macro EMG overcomes this limitation by showing the changes in the motor unit as a whole, whether distributed uniformly or not.

In neurogenic lesions (Gan and Jabre, 1992; Hilton-Brown and Stålberg, 1983; Stålberg and Sanders, 1983), macro EMG shows a great increase in the motor unit potential as compared to what could be gleaned from the concentric electrode alone.

In myopathic lesions (Gan and Jabre, 1992), macro EMG demonstrates a great reduction in the area and, more importantly, amplitude of the motor unit potential.

The combination of fiber density, jitter studies, and concentric macro EMG techniques can enhance the diagnostic capability of routine EMG techniques.

IV. Review of the Literature

The literature is replete with population and individual case studies of toxic exposure. The various electrophysiological manifestations are summarized in Table 1. The following is a summarization of these reports.

A. Metals

1. Arsenic

Feldman and colleagues (1979) conducted a double-blind controlled study of arsenic trioxide-exposed factory workers. In subjects falling within the subclinical and clinical groups, there was reduced nerve conduction velocities and amplitude measurements. Bansal and colleagues (1991) described a 35-year-old man with an arsenic neuropathy who had a severe polyneuropathy as well as a phrenic neuropathy which improved with *d*-penicillamine therapy. Oh (1991) described the electrophysiological abnormalities in 13 patients with arsenic neuropathy and found markedly abnormal sensory and mixed NCS along with moderately abnormal motor conductions. The findings, confirmed with Sural nerve biopsies, were typical of axonal loss lesions. The patients, who had ingested arsenic either in an

TABLE 1   Summary of Peripheral and Central Nervous System Involvement in Toxic Neuropathies of Various Etiologies

	Peripheral nervous system	Central nervous system
Arsenic	Sensory and mixed conductions; absent NAP or low Amp, slowed CV motor conductions; low Amp, slowed CV needle exam; Fibs and positive waves/polyphasic MUAPs sural nerve biopsy; axonal loss	
Lead	Sensory and mixed conductions; absent NAP or low Amp, slowed CV motor conductions; slowed, occasional conduction block (median and radial) needle exam; Fibs and positive waves/polyphasic MUAPs autonomic; significant decrease in R–R interval	VEP, SSEP, BAEP, and P300 latencies; increased encephalopathy, dementia, extrapyramidal synd Spasticity
TCE		Blink reflex; increased R1 and R2 latencies; trigeminal SEP; increased latencies
n-Hexane	Sensory conductions; low Amp, slowed CV motor conductions; low Amp, slowed CV, increased distal latency; DCV; V10 to V90 CV decreased. All nerve fibers are affected	VEP, BAEP, and SSEP; increased latencies
Acrylamide	Sensory conductions; low Amp, slowed CV motor conductions; low Amp, slowed CV baroreceptor and sympathetic sudomotor dysfunction	

Note. NAP, nerve action potential; Amp, amplitude; CV, conduction velocity; Fibs, fibrillations; MUAP, motor unit action potential; VEP, visual-evoked potentials; SSEP, somatosensory-evoked potentials; BAEP, brain stem auditory-evoked potentials; DCV, distribution of conduction velocities; V10 to 90, velocities from the 10th to 90th percentiles on the velocity distribution histogram.

attempted homicide or from poisoning from pesticide, developed subacute symmetrical sensory motor poly-neuropathy.

2. Lead

Feldman and colleagues (1977) demonstrated, that increased absorption of lead produces both central and peripheral nervous system changes. Their cases included neuropathy and mild encephalopathy, dementia and spastic paraparesis, isolated neuropathy, and extrapyramidal syndrome and lower motor neuron disease as well as personality changes.

Yagninas and colleagues (1992) studied the neuropathologic effects of low-level exposure to triethyl lead on young male rats. They noted randomly distributed light microscopic changes in the spinal cord as well as Wallerian degeneration but no lesions in the brain. Lumbosacral nerves were affected, showing reduced neurofilaments and neurotubules.

Murata and colleagues (1993) studied the effects of lead, zinc, and copper absorption on metal workers. They measured maximal motor and sensory conduction velocities in the radial and median nerves in 20 exposed by asymptomatic workers. They showed a slowing of the motor and sensory conductions as well as an inverse correlation between urinary δ-aminolevulinic and coproporphyrin levels and radial sensory conduction.

Matsumoto and colleagues (1993) studied the "tapping ability" of lead workers in Japan as a test (the tapping test) to investigate its persistence. They noted a decrease in the tapping ability in these exposed workers that was coincidental with higher blood levels. The recovery from fatigue was also worse in those with higher blood levels.

Kajiyama and colleagues (1993) described the case of a 25-year-old man with a 2-year history of exposure to lead stearate who had atrophy of the small hand muscles. His nerve conduction studies showed a block at the elbow, possibly due to a cubital tunnel syndrome. His findings improved after CaEDTA therapy.

Schwartz and colleagues (1988) demonstrated that maximal motor nerve conduction velocities may not be a very sensitive indicator for low-level lead toxicity where only regression studies showed the abnormality.

Araki and colleagues (1993) studied the interactive effects of lead, zinc, and copper on the peripheral nervous system and showed that zinc and copper antagonize the subclinical neurologic effects of lead.

Lead also affects the autonomic nervous system. Teruya and colleagues (1991) studied 172 lead-exposed workers and showed a significant decrease in the R–R interval in those whose lead blood levels exceeded 30 μg/dl.

Murata and colleagues (1987, 1991, 1993) studied the effects of lead on the central, peripheral, and autonomic nervous systems using a whole battery of tests which included visual, short-latency somatosensory, and brain stem auditory EPs as well as event related potential (P300), distribution of nerve conduction velocities, the R–R interval, and median and radial nerve conduction studies. They found that latencies of the visual-evoked potential, SSEP, and P300 were all increased in workers with blood lead levels below 65 μg/dl. It was also shown that these latencies, along with the BAEP latencies, were significantly correlated with the indicators for lead absorption. The R–R interval was significantly depressed, and conduction velocities were significantly slowed, indicating that lead not only affects the peripheral but also the central and autonomic nervous functions.

B. Solvents

1. Trichlorethylene (TCE)

Feldman and colleagues (1988) described a highly significant difference between the means of the R1 and R2 latencies in a TCE-exposed population and normal controls and suggested a subclinical alteration of the Vth cranial nerve function due to chronic environmental exposure to TCE.

Barret and colleagues (1991, 1992) showed that Vth nerve impairment was one of the main features of TCE and its breakdown product dichloroacetylene toxicity. This was demonstrated in Sprague–Dawley rats intoxicated by direct gastric administration of TCE that showed a significant decrease in the internode length and mean fiber diameter of nerve fibers. Barret and co-workers (1987) also reported abnormal trigeminal somatosensory-evoked potentials in 40 out of 104 occupationally exposed subjects to TCE.

In an attempt to explain the mechanism of cranial neuropathy associated with TCE exposure, Cavanaugh and Buxton (1989) raised the possibility that it may be due to the chemical causing the reactivation of a latent herpes simplex infection.

2. n-Hexane

Chang and colleagues (1993) described *n*-hexane-exposed workers of an offset printing factory with symptoms of peripheral neuropathy. The initial changes in the nerve conductions demonstrated reduced sensory amplitude followed by decreased motor amplitudes, slowed conduction velocities, and increased distal latencies. These authors attributed these changes to axonal degeneration and secondary demyelination.

Yokoyama and colleagues (1990) studied the distribution of conduction velocities (DCV) of sensory fibers in three patients with *n*-hexane poisoning at various time intervals after exposure. They observed that not only were faster fibers affected by *n*-hexane, as had been reported previously, but that all nerve fibers were also affected, probably due to a higher *n*-hexane exposure level. In their nerve biopsy, they confirmed the finding of axonal swelling, myelin retraction, and focal demyelination as well as loss of myelinated fibers.

Pryor (1991) showed that toluene causes a motor syndrome akin to "cerebellar ataxia" in rats. Pryor and colleagues (1992) studied the interactive effects of toluene and hexane on behavior and neurophysiologic responses in Fisher rats. They demonstrated that toluene greatly reduced the neuropathy caused by hexane by inhibiting the metabolism of hexane. In contrast, hearing loss and motor dysfunction induced by toluene were not reversed by hexane.

Chang (1990) followed up 11 patients with *n*-hexane-induced peripheral neuropathy that was moderate to severe in degree. After cessation of exposure, he noted that sensory functions improved earlier than motor functions and that all the patients regained their full motor capabilities within 1 to 4 years, although some retained muscle atrophy of the small hand and foot muscles. He noted that even the central nervous system involvement manifested by spasticity had improved and was therefore reversible.

Chang (1987) studied the neurotoxic effects of *n*-hexane on the central nervous system using multimodality evoked potentials in 34 exposed workers of printing factories. He noted that the absolute and interpeak latencies of patterned visual EPs in the polyneuropathy and subclinical groups were prolonged, whereas the interpeak amplitude was decreased in the subjects with polyneuropathy. Both central conduction time and the absolute latencies of the somatosenory EPs were also increased in the polyneuropathy and subclinical groups.

C. Polymer

1. Acrylamide

Experimental acrylamide neuropathy has been used as a model of "dying back" neuropathy since the early 1960s. A generalized sensorimotor peripheral neuropathy of the axonal degeneration type was described by Donaghy and colleagues (1991) in a dental technician who had been exposed to dental acrylic for over 30 years. The exposure consisted of handling as well as inhalation of fumes of methacrylate. Evidence of loss of unmyelinated axons was seen on Sural nerve biopsy.

The relationship between exposure to the acrylamide monomer and neurological presentation was investigated in 82 workers of a chemical industry by Bachmann and colleagues (1992). Symptoms included numbness and tingling, weakness, pain, peeling hand skin, and sweating.

Satchell (1990) found evidence of baroreceptor dysfunction by recording carotid sinus and depressor nerve activity in rabbits who had a mild axonal neuropathy induced by acrylamide.

By studying acrylamide-intoxicated rats, Gold and colleagues (1992) confirmed the somatofugal (proximal) nature of acrylamide intoxication in the presence of intact distal fibers. They suggested that acrylamide produces somatofugal atrophy of the axon by interfering with the delivery of a trophic which is transported retrogradely to the neuronal perikaryon. Sabri and Spencer (1990) showed that a deficit in fast and slow axonal transport was an important factor in the pathogenesis of axonal degeneration in acrylamide neuropathy.

Navarro and colleagues (1993) showed that the sympathetic sudomotor function was affected in experimental acrylamide neuropathy. They attributed the decrease in sweat gland response to the damage to the postganglionic sudomotor nerve fibers in acrylamide neuropathy.

V. Conclusion

Sensitive and adequately controlled neurophysiological investigations are needed to document the existence of peripheral nerve involvement in toxic and occupational neuropathies. These techniques will allow the early detection of abnormalities, but more importantly will document the extent, define the type, and shed some light on the prognosis of the peripheral nervous system lesions.

References

Araki, S., Murata, K., Uchida, E., Aono, H., and Ozawa, H. (1993). Radial and median nerve conduction velocities in workers exposed to lead, copper, and zinc: a follow-up study for 2 years. *Environ. Res.* **61**, 308–316.

Bachmann, M., Myers, J. E., and Bezuidenhout, B. N. (1992). Acylamide monomer and peripheral neuropathy in chemical workers. *Am. J. Ind. Med.* **21**, 217–222.

Bansal, S. K., Haldar, N., Dhand, U. K., and Chopra, J. S. (1991). Phrenic neuropathy in arsenic poisoning. *Chest* **100**(3), 878–880.

Barret, L., Garrel, S., Danel, V., and Debru, J. L. (1987). Chronic trichloroethylene intoxication: a new approach by trigeminal-evoked potentials? *Arch. Environ. Health* **42**(5), 297–302.

Barret, L., Torch, S., Usson, Y., Gonthier, B., and Saxod, R. (1991). A morphometric evaluation of the effects of trichloroethylene and dichloroacetylene on the rat mental nerve: preliminary results. *Neurosci. Lett.* **131**, 141–144.

Barret, L., Torch, S., Leray, Cl., Sarlieve, L., and Saxod, R. (1992). Morphometric and biochemical studies in trigeminal nerve of rat after trichloroethylene or dichloroacetylene oral administration. *Neurotoxicology* **13**, 601–614.

Bauermeister, W., and Jabre, J. F. (1992). The spectrum of concentric Macro EMG correlations. I. Normal subjects. *Muscle Nerve* **15**(10), 1081–1084.

Buchtal F., and Rosenfalck A. (1966). Evoked action potentials and conduction velocity in human sensory nerves. *Brain Res.* **3**, 1–122.

Cavanaugh, J. B., and Buxton, P. H. (1989). Trichloroethylene cranial neuropathy: is it really a toxic neuropathy or does it activate latent herpes virus? *J. Neurol. Neurosurg. Psychiatry* **52**, 297–303.

Chang, C. M., Yu, C. W., Fong, K. Y., Leung, S. Y., Tsin, T. W., Yu, Y. L., Cheung, T. F., and Chan, S. Y. (1993). *n*-Hexane neuropathy in offset printers. *J. Neurol. Neurosurg. Psychiatry* **56**, 538–542.

Chang, Y.-C. (1987). Neurotoxic effects of *n*-hexane on the human central nervous system: evoked potential abnormalities in *n*-hexane polyneuropathy. *J. Neurol. Neurosurg. Psychiatry* **50**, 269–274.

Chang, Y. C. (1990). Patients with *n*-hexane induced polyneuropathy: a clinical follow up. *Br. J. Ind. Med.* **47**, 485–489.

Donaghy, M., Rushworth, G., and Jacobs, J. M. (1991). Generalized peripheral neuropathy in a dental technician exposed to methyl methacrylate monomer. *Neurology* **41**, 1112–1116.

Feldman, R. G., Kelly-Hayes, M., Younes, R., and Aldrich, F. D. (1977). Lead neuropathy in adults and children. *Arch. Neurol.* **34**, 481–488.

Feldman R. G., Niles, C. A., Kelly-Hayes, M., Sax, D. S., Dixon, W. J., Thompson, D. J., and Landau, E. (1979). Peripheral neuropathy in arsenic smelter workers. *Neurology* **29**, 939–944.

Feldman, R. G., Chirico-Post, J., and Proctor, S. P. (1988). Blink reflex latency after exposure to trichloroethylene in well water. *Arch. Environ. Health* **43**(2), 143–148.

Gan, R., and Jabre, J. F. (1992). The spectrum of concentric macro EMG correlations. II. Patients with disease of muscle and nerve. *Muscle Nerve* **15**(10), 1085–1088.

Gold, B. G., Griffin, J. W., and Price, D. L. (1992). Somatofugal axonal atrophy precedes development of axonal degeneration in acrylamide neuropathy. *Arch. Toxicol.* **66**, 57–66.

Guiheneuc, P., Le Bastard, C., and Doncarli, C. (1988). Several unit macro (SUM) EMG: une nouvelle technique. In *Résumé des communications* Vlèmes Journées Françaises D' EMG, Lille, France. pp. 3–1. [Abstract]

Hilton-Brown, P., and Stålberg, E. (1983). Motor unit size in muscular dystrophy, a macro EMG and scanning EMG study. *J. Neurol. Neurosurg. Psychiatry* **46**, 996.

Jabre, J. F., and Hackett, E. R. (1983). *EMG Manual*. Charles C. Thomas, Springfield, Illinois. pp. 120.

Jabre, J. F., and Sato, L. (1990). The expression of electrophysiologic data as mean related values. *Muscle Nerve* **13**(9), 861–862.

Jabre, J. F. (1991). Concentric macro electromyography. *Muscle Nerve* **14**(9), 820–825.

Kajiyama, K., Doi, R., Sawada, J., Hashimoto, K., Hazama, T., Nakata, S., Hirata, M., Yoshida, T., and Miyajima, K. (1993). Significance of subclinical entrapment of nerves in lead neuropathy. *Environ. Res.* **60**, 248–253.

Matsumoto, T., Fukaya, Y., Yoshitomi, S., Arafuka, M., Kubo, N., and Ohno, Y. (1993). Relations between lead exposure and peripheral neuromuscular functions of lead-exposed workers—results of tapping test. *Environ. Res.* **61**, 299–307.

Murata, K., Araki, S., and Aono, H. (1987). Effects of lead, zinc and copper absorption on peripheral nerve conduction in metal workers. *Int. Arch. Occup. Environ. Health* **59**, 11–20.

Murata, K., and Araki, S. (1991). Autonomic nervous system dysfunction in workers exposed to lead, zinc, and copper in relation to peripheral nerve conduction: a study of R-R interval variability. *Am. J. Ind. Med.* **20,** 663–671.

Murata, K., Araki, S., Yokoyama, K., Uchida, E., and Fujimura, Y. (1993). Assessment of central, peripheral, and autonomic nervous system functions in lead workers: nueroelectrophysiological studies. *Environ. Res.* **61,** 323–336.

Navarro, X., Verdu, E., Guerrero, J., Buti, M., and Gonalons, E. (1993). Abnormalities of sympathetic sudomoter function in experimental acrylamide neuropathy. *J. Neurol. Sci.* **114**(1), 56–61.

Oh, S. J. (1991). Electrophysiological profile in arsenic neuropathy. *J. Neurol. Neurosurg. Psychiatry.* **54,** 1103–1105.

Pryor, G. T. (1991). A toluene-induced motor syndrome in rats resembling that seen in some human solvent abusers. *Neurotoxicol. Teratol.* **13,** 387–400.

Pryor, G. T., and Rebert, C. S. (1992). Interactive effects of toulene and hexane on behavior and neurophysiologic responses in Fischer-344 rats. *Neurotoxicology* **13,** 225–234.

Sabri, M. I., and Spencer, P. S. (1990). Acrylamide impairs fast and slow axonal transport in rat optic system. *Neurochem. Res.* **15**(6), 603–608.

Satchell, P. M. (1990). Baroreceptor dysfunction in acrylamide axonal neuropathy. *Brain* **113,** 167–176.

Schwartz, J., Landrigan, P. J., Feldman, R. G., Sibergeld, E. K., Baker, E. L., and von Lindern, I. H. (1988). Threshold effect in lead-induced peripheral neuropathy. *J. Pediatr.* **112,** 12–17.

Stålberg, E., and Fawcett, P. R. W. (1982). Macro EMG in healthy subjects of different ages. *J. Neurol. Neurosurg. Psychiatry.* **45,** 870–878.

Stålberg, E., and Sanders, D. B. (1983). The motor unit in ALS studied with different neurophysiological techniques. In *Progress in Motor Neurone Disease* (F. C. Rose, Ed.), pp. 103–122, Pitman Books, London.

Stålberg, E., and Trontelj, V. (1979). *Single Fiber Electromyography.* The Mirvalle Press, Old Working, Surrey, United Kingdom.

Stålberg, E. (1980). Macro EMG, a new recording technique. *J. Neurol. Neurosurg. Psychiatry* **43,** 475–483.

Teruya, Koji, Sakurai, Haruhiko, Omae, Kazuyuki, Higashi, Toshiaki, Muto, Takashi, Kaneko, and Yoko. (1991). Effect of lead on cardiac parasympathetic function. *Int. Arch. Occup. Environ. Health* **62,** 549–553.

Yokoyama, K., Feldman, R. G., Sax, D. S., Salzsider, B. T., and Kucera, J. (1990). Relation of distribution of conduction velecities to nerve biopsy findings in *n*-hexane poisoning. *Muscle Nerve* **13,** 314–320.

CHAPTER
49

Evoked Potential Testing in Clinical Neurotoxicology

JAMES A. D. OTIS
EEG and EP Laboratory, Boston University School of Medicine
Boston, Massachusetts 02118

JOSEPH S. HANDLER
Department of Neurology, Boston University School of Medicine
Boston, Massachusetts 02118

Evoked potentials (EP) are electrical cortical responses produced by specific stimulation of specific sensory pathways or by complex endogenous events. Sensory-evoked potentials (SEPs) refer to those responses produced by peripheral sensory stimuli. Event-related potentials (ERPs) are produced by the processing of complex data or by motor planning. Although ERPs provide interesting information about cognitive processing, they have little clinical utility. In contrast, SEPs are commonly used to assess the function of sensory pathways. Clinically, the most frequently used of these tests are the visual-evoked potentials (VEPs), brain stem- or auditory-evoked responses (BAERs), and the somatosensory-evoked potentials (SSEPs). These fundamental tests are used in practice to screen for multifocal diseases and to uncover clinically silent lesions. Chemosensory-evoked potentials (CSEPs) have been used to test additional sensory pathways in patients exposed to inhaled toxins.

The basic principle underlying evoked potentials is that peripheral sensory stimuli produce cortical responses with consistent, reproducible latencies. This is a function of large, heavily myelinated pathways. As the stimuli travel to the cortex, they activate local neuronal populations that produce local potentials. Abnormalities in these potentials allow localization of focal pathway dysfunction. Because these local potentials are comparatively small, amplification and computerized averaging are used to extract them from background electrical activity. Unlike EEG activity, EPs are time-locked to the stimulus and are reproducible. They can therefore provide a means of following the effect of a toxin over time.

This chapter reviews the use of EPs in neurotoxicology, discusses the anatomical basis of the tests, and indicates the effect of various known toxins.

I. Visual-Evoked Potentials

Visual-evoked potentials examine the integrity of the optic pathway. This consists of the optic nerve, the chiasm, and the optic tract and its projections to the geniculate nuclei and calcarine cortex. Because the optic pathways cross at the chiasm, the test is most reliable in assessing the optic nerve, anterior to the chiasm.

There are two types of VEPs: flash VEPs and pattern shift VEPs (PSVEPs). Flash EPs are crude tests that indicate only the presence of light perception. Pattern shift EPs provide more information and are the preferred test. Nevertheless, flash VEPs are useful in testing uncooperative subjects and children.

In both forms of VEPs, a visual stimulus is used to produce a cortical response that has a latency of approximately 100 msec. This is recorded over the occipital region and is termed the P100 peak. Other less consistent waveforms are also obtained but they have no clinical utility. In pattern shift EP testing, a checkerboard pattern that flickers on and off at a rate of 2 Hz is used as the stimulus. Each eye is tested separately. Flash EP testing is performed with a strobe. Both have been used to study the effects of neurotoxins (Otto and Hudnell, 1993).

PSVEPs have been found to be affected by several organic solvents, particularly n-hexane. The most common abnormality is delay of the P100 response. This indicates dysfunction of the optic nerve. It can often be seen as an early manifestation of toxicity even in asymptomatic individuals (Chang, 1987; Altman et al., 1990). P100 amplitude abnormalities have also been described but these are a less consistent abnormality (Urban, 1990).

Heavy metals also affect the P100 response in PSVEPs. Lead and mercury produce a delay in the P100 waveform in some studies (Sborgia et al., 1983), but not in others (Araki et al., 1987; Murata et al.,

1987). Children exposed to lead also show variable P100 responses (Lillienthal et al., 1990). The inconsistencies in these studies may be a result of duration of exposure and cumulative amount. It appears that workers exposed to a larger amount of lead or mercury are more likely to show anomalies, usually when they are symptomatic.

Flash EPs have been used as screening tests for neurotoxins in animals (Boyes, 1991). Small studies have looked at the effect of methyl mercury (Iwata, 1980), n-hexane (Seppalainen et al., 1979), carbon monoxide (Hosko, 1970), and carbon disulfide. In each case the number of individuals tested was small and the abnormalities were not consistent. These reports point to the potential efficacy of the use of VEPs in detecting toxicity. Further studies using PSVEPs are needed before any conclusions can be made regarding the clinical use of these tests for these toxins (Fig. 1).

II. Brain Stem Auditory-Evoked Potentials

BAERs are used to study the auditory pathways and their central projections. A clicking sound is used as the stimulus in one ear while the other ear is masked with white noise. Recording electrodes are placed over the earlobes and the vertex of the head. The eighth cranial nerve is activated followed by several brain stem structures. Activation occurs over 10 msec. As each brain stem level is activated, a wave is generated corresponding to a specific site (Chiappa, 1992). The most reproducible responses are the first five waves. Wave I occurs when the acoustic nerve is activated.

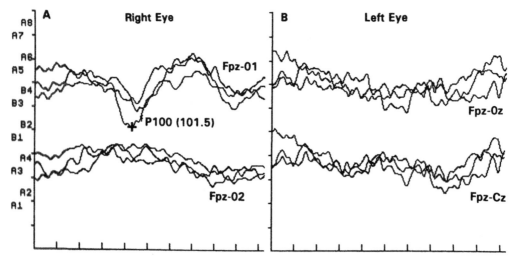

FIGURE 1   VEP in patient with trichloroethylene exposure. (A) Normal visual (right) EP with P100 wave at 101.5 msec. (B) Absent P100 wave in left eye.

Waves II and III reflect the activation of structures in the pontomedullary region. Waves IV and V are less clearly defined but seem to be functions of the upper pons and low midbrain, respectively. The absolute latencies of each wave are recorded but the interpeak latencies of waves I–III, III–V, and I–V are more consistent and reproducible.

BAERs are sensitive to processes that damage the acoustic nerve and the myelinated pathways in the brain stem. They are thus helpful in detecting insults caused by ototoxic substances as well as organic solvents that are likely to damage myelin.

Several studies have shown the ototoxic effects of lead. BAERs were found to be abnormal in children (Otto *et al.*, 1985; Holdstein *et al.*, 1986) and in industrial workers exposed to lead (Murata, 1987; Discalzi *et al.*, 1992). Interpeak latencies were prolonged as well as the absolute latency of wave I. This suggests a central and also a peripheral effect. Other studies have found no clear abnormalities (Lille *et al.*, 1988). Studies in monkeys suggest that BAERs become abnormal when serum levels of lead are high (mean of 55 µg/dl) (Lillienthal *et al.*, 1990). This may account for the lack of abnormalities in the latter studies, which studied only acute lead intoxication or low-level ingestion.

BAERs are also affected by organic solvent exposure. Toluene produces irreversible hearing loss due to destruction of hair cells leading to the prolongation or absence of wave I in animal studies (Rebert, 1986). This occurs after prolonged exposure to the solvent. In humans, BAERs seem to be unaffected by low levels (Massioui *et al.*, 1990). Higher levels of toluene and *n*-hexane produce auditory impairment (Rosenberg *et al.*, 1988; Chang, 1987). Other individual neurotoxins also produce abnormalities in BAERs. Xylene, styrene, and trichloroethylene produce high frequency hearing loss. Carbon disulfide produces isolated slowing of the conduction between waves III and V. This suggests damage to the myelinated tracts between the superior olivary complex and the lower midbrain (Rebert *et al.*, 1986).

It is clear that BAERs are helpful in detecting ototoxicity in cases of organic solvent exposure and in certain isolated toxins. It is unclear if they can detect more central damage consistently. More studies need to be done to see if BAERs have a higher yield than audiograms (Fig. 2).

III. Somatosensory-Evoked Potentials

SSEPs are obtained by recording a scalp potential after activation of a peripheral sensory or mixed nerve.

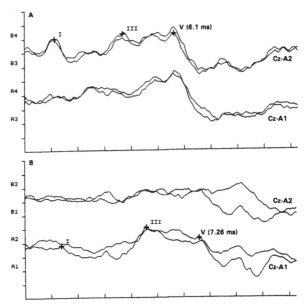

FIGURE 2 BAERs in patient with multiple solvent exposure. (A) Normal BAER. (B) Abnormal BAER.

Stimulation is achieved by application of a cutaneous electrical stimulus. The stimulus is conveyed proximally by the Ia fibers to the dorsal columns producing a local potential and then to the contralateral cortex. The absolute latencies of the local and cortical potentials can be influenced by limb length and peripheral neuropathy. The interpeak latencies are more consistent and can help localize pathology along the path from the peripheral nerve to the spinal cord, thalamus, and brain stem. It is usual to test both upper and lower extremities. This produces a higher yield because of the length of the lower extremity pathway.

The nomenclature of SSEPs is not as standardized as VEPs and BAERs. A full discussion of the technical aspects of SSEPs is beyond the scope of this chapter but is available in standard texts (Chiappa, 1992; Liveson, 1993).

The efficacy of SSEPs in detecting the effects of neurotoxins is unclear. Since many neurotoxins affect peripheral nerves it is not surprising that SSEPs would be altered. Organic solvents such as toluene, *n*-hexane, and trichloroethylene can produce slowing of the evoked response. This is not an early finding and may be more a function of peripheral axonopathy (Stetkarova *et al.*, 1993; Massioui *et al.*, 1990). At this time, SSEPs offer little advantage over standard nerve conduction velocities and the other forms of EPs. Nevertheless, they are abnormal in a variety of settings, including lead exposure, carbon monoxide poisoning, and in exposure to solvents (Valciukas, 1991) (Fig. 3).

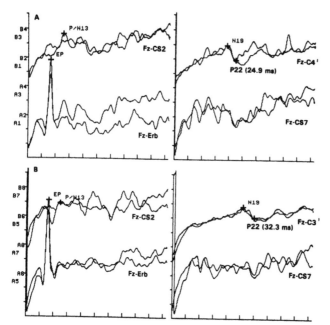

FIGURE 3   Upper limb SSEPs in patient with organic solvent exposure. (A) Normal N19/P22 at 24.9 msec on left. (B) Delayed N19/P22 at 32.3 msec on right.

IV. Chemosensory-Evoked Potentials

Many environmental toxins are irritating to the nasal mucosa and the olfactory nerve. Strong olfactory and trigeminal stimulation produced by inhaled substances produce cortical potentials that can be recorded.

Stimulation of the olfactory and trigeminal nerve often occurs simultaneously since many substances are both olfactants and irritants. The evoked responses obtained are different, however. Trigeminal irritation produces EPs which have their greatest amplitude contralateral to the stimulated side. The maximal amplitude is best recorded over the vertex. In contrast, olfactory stimulation produces EPs which have equal amplitude bilaterally and are of maximal amplitude over the parasagittal area (Hummel and Kobal, 1992). The technical aspects of these tests are discussed in detail in several reviews (Kobal and Hummel, 1988, 1991).

CSEPs have been used to provide objective evidence of trigeminal dysfunction in patients suffering from sick building syndrome (SBS) and from multiple chemical sensitivity (MCS). In both SBS and MCS, exposure to volatile organic compounds is presumed to be an important factor. It is clear that many patients with these disease states have a sensitivity to low levels of inhaled chemicals. It is difficult, however, to show that this always produces physiological changes. CSEPs have shown that changes do occur, at least in

some cases. Trichloroethylene produced changes in CSEP waveforms in 40/104 subjects. This correlated with subjective complaints of altered facial sensation (Barret *et al.*, 1987). Further studies are needed to assess the effectiveness of CSEPs as screening techniques. Blink reflexes, which provide an indirect means of testing trigeminal function, may be less cumbersome to use and require less specialized equipment.

V. Conclusions

Sensory-evoked potentials offer a reproducible, objective, and noninvasive means of assessing sensory pathways. Unlike nerve conduction velocities, they can detect central nervous system dysfunction. The area of damage can at times be localized and is often detected even when the patient is asymptomatic.

VEPs are the most reliable and consistent of the EPs. They are abnormal in cases of organic solvent exposure, particularly *n*-hexane, and in cases of high-level lead intoxication. BAERs are helpful in assessing brain stem pathways in cases of solvent exposure and carbon disulfide poisoning. Standard audiometry may be a more cost effective screening technique since most neurotoxins are also ototoxic. The utility of SSEPs is unproven in neurotoxicology. CSEPs are relatively untested and require specialized equipment which is difficult to standardize. They may prove to be useful in screening for MCS and SBS.

A reasonable approach would be to incorporate EPs as a part of a battery of neurophysiological testing. VEPs, audiometry, and nerve conduction velocities would provide rapid means of screening. If patients show signs of brain stem dysfunction on the examination or by history, BAERs would be added to the battery to obtain objective signs of pathology. Signs of history of spinal cord or posterior column problems should prompt SSEPs. This approach should lower the rate of unnecessary testing and allow for more effective and comfortable patient care.

References

Araki, S., Murata, K., and Aono, H. (1987). Central and peripheral nervous system dysfunction in workers exposed to lead, zinc and copper: a followup study of visual and somatosensory evoked potentials. *Int. Arch. Ocup. Environ. Health* **59**, 177–187.

Altman, L., Bottger, A., and Weigand, H. (1990). Neurophysiological and psychophysical measurements reveal effects of low-level organic solvent exposure in humans. *Int. Arch. Occup. Health* **62**, 493–499.

Barret, L., Garrel, S., *et al.* (1987). Chronic trichloroethylene intoxication: a new approach by trigeminal-evoked potentials? *Arch. Environ. Health* **42**, 297–302.

Boyes, W. (1991). Testing visual system toxicity using visual evoked potentials. In *The Vulnerable Brain* Vol. I, (R. Isaacson and K. Jensen, Eds.), Plenum, New York.

Chang, Y. C. (1987). Neurotoxic effects of *n*-hexane on the human nervous system: evoked potential abnormalities in *n*-hexane polyneuropathy. *J. Neurol. Neurosurg. Psychiatry* **50**, 269–274.

Chiappa, K. (Ed.) (1992). *Evoked Potentials in Clinical Medicine.* 2nd ed., Raven, New York.

Discalzi, G., *et al.* (1992). Auditory brainstem evoked potentials in lead exposed workers. *Neurotoxicology* **13**, 207–210.

Holdstein, Y., Pratt, H., *et al.* (1986). Auditory brainstem evoked potentials in asymptomatic lead-exposed subjects. *J. Laryngol. Otol.* **100**, 1031–1036.

Hosko, M. (1970). The effect of carbon monoxide on the visual evoked potential and the spontaneous electroencephalogram. *Arch. Environ. Health* **21**, 174–180.

Hummel, T., and Kobal, G. (1992). Differences in human evoked potentials related to olfactory or trigeminal chemosensory activation. *Electroencephalogr. Clin. Neurophysiol.* **77**, 190–198.

Iwata, K. (1980). Neuro-opthalmological indices of Minimata disease in Niigata. In *Neurotoxicity of the Visual System* (W. Merigan and B. Weiss, Eds.), Raven, New York.

Kobal, G., and Hummel, T. (1988). Cerebral chemosensory evoked potential elicited by stimulation of the human olfactory and respiratory mucosa. *Electroencephalogr. Clin. Neurophysiol.* **71**, 241–250.

Kobal, G., and Hummel, T. (1991). Olfactory evoked potentials in humans. In *Smell and Taste in Health and Disease* (T. Getchell, R. Duty, *et al.*, Eds.), pp. 255–275, Raven, New York.

Lille, F., Hazeman, P., *et al.* (1988). Effects of lead and mercury intoxications on evoked potentials. *Clin. Toxicol.* **26**, 103–116.

Lillienthal, H., Winneke, G., and Ewert, T. (1990). Effects of lead on neurophysiological and performance measures. *Environ. Health Perspect.* **89**, 21–25.

Liveson, J., and Ma, D. (1993). *Laboratory Reference for Clinical Neurophysiology.* F. A. Davis, Philadelphia.

Massioui, F., *et al.* (1990). Sensory and cognitive event-related potentials in workers chronically exposed to solvents. *Clin. Toxicol.* **28**, 203–219.

Murata, K., Araki, S., and Aono, H. (1987). Visual and brainstem auditory evoked potentials in lead-exposed workers. *Jpn. J. EEG EMG* **15**, 16–21.

Otto, D. A., Baumann, S., *et al.* (1985). Five year follow-up study of children with low-to-moderate lead absorption: electrophysiological evaluation. *Environ. Res.* **38**, 168–186.

Otto, D. A., and Hudnell, H. K. (1993). The use of visual and chemosensory evoked potentials in environmental and occupational health. *Environ. Res.* **62**, 159–171.

Rebert, C. S., and Beeker, E. (1986). Effects of inhaled carbon disulfide on sensory evoked potentials of Long-Evans rats. *Neurobehav. Toxicol. Teratol.* **8**, 533–541.

Rosenberg, M., Spitz, M., *et al.* (1988). Central nervous system effects of chronic toluene abuse: clinical, brainstem evoked response and magnetic resonance imaging studies. *Neurotoxicol. Teratol.* **10**, 489–495.

Sborgia, G., Assennato, G., *et al.* (1983). Comprehensive neurophysiological evaluation of lead-exposed workers. In *Neurobehavioral Methods in Occupational Health* (M. Gilioli, M. Cassito, and M. Foa, Eds.), pp. 283–294, Pergamon, London.

Seppalainen, A. M., *et al.* (1981). Changes induced by xylene and alcohol in human evoked potentials. *Electroencephalogr. Clin. Neurophysiol.* **51**, 148–155.

Stetkarova, I., Urban, P., *et al.* (1993). Somatosensory evoked potentials in workers exposed to toluene and styene. *Br. J. Ind. Med.* **50**, 520–527.

Urban, P., and Lucas, E. (1990). Visual evoked responses in photogravure printers exposed to toluene. *Br. J. Ind. Med.* **47**, 819–823.

Valciukas, J. (1991). *Foundations of Environmental and Occupational Neurotoxicology.* Van Norstrand, NY.

CHAPTER

50

Neuroimaging in Neurotoxicology

LEON D. PROCKOP
Department of Neurology
University of South Florida School of Medicine
Tampa, Florida 33612

I. Introduction

As discussed elsewhere in this book, the world abounds in potential neurotoxins. Neurotoxins can cause damage to the peripheral nervous system and/or the central nervous system (CNS). The damage may be focal, multifocal, or diffuse. Sites of focal and multifocal damage include peripheral nerve, spinal cord, brain stem, cerebellum, and cerebrum. In the cerebrum, damage may occur in the superficial gray matter cortex, the white matter, and/or the deep gray matter nuclei.

The adverse effects of neurotoxins can be completely reversible within hours or days after exposure. On the other hand, effects can be partially reversible or completely irreversible so that mild to moderate to severe nervous system damage can be permanent. Effects vary according to the potential neurotoxin, as the nervous system is more vulnerable to some substances than to others. Likewise, effects vary between chronic low-dose exposure and acute high-dose exposure. With the latter, associated or secondary damage

from hypoxia is more likely and complicates the sequelae documented by neuroimaging.

Because symptoms and signs of potential neurotoxicological damage usually are nonspecific to the disease process, clinicians often seek the assistance of laboratory diagnostic modalities to establish the diagnosis of neurotoxicity and/or to quantitate its extent. For example, a patient may complain of weakness and numbness after exposure to *n*-hexane, a potential neurotoxin known to affect peripheral nerves. Although the patient's symptoms may suggest a peripheral neuropathy, findings on neurological examination might not be diagnostic or might be equivocal. In this case, the clinician would order electromyography (EMG) and nerve conduction velocity (NCV) studies. These studies might confirm or deny the presence of neuropathy. Furthermore, EMG and NCV can differentiate a demyelinating from an axonal neuropathy. When neuropathy is produced, volatile hydrocarbons produce primarily axonal damage. Therefore, the electrophysiological studies may support a clinical suspicion that the patient is suffering neurotoxicity from *n*-hexane. However, since the patient might also be suffering from another disease causing axonal neuropathy, e.g., diabetes mellitus, the EMG findings do not confirm the diagnosis.

Because the symptoms and signs of CNS neurotoxicity are also nonspecific, clinicians have sought the assistance of neuroimaging procedures in their attempts to establish a diagnosis. For example, workers exposed to paint and paint thinners may complain of personality changes, headache, impaired memory, and other problems which suggest the presence of an encephalopathy. Historical data regarding the exact composition of the paint products and the degree of exposure may be lacking. Ongoing litigation may complicate assessment in that subjective complaints may be caused and/or magnified by a patient's desire for monetary gain. Physical findings might be within normal limits or might be equivocal whereas Neuropsychological assessment might be equivocal or nonspecific. The clinician yearns for "objective findings" and hopes that the brain-computed tomography (CT) scan, brain magnetic resonance imaging (MRI), or other tests will solve a diagnostic dilemma.

Although such tests have made quantum leaps in the definitive diagnosis of many neurological disorders, e.g., multiple sclerosis, they have been of comparatively little benefit in neurotoxicological diagnosis. Nonetheless, it is appropriate to review the positive perspectives they offer as well as the limitations of neuroimaging in neurotoxicology.

II. Neuroimaging in Neurotoxicology

A. Modalities

In 1891 Quinke performed the first lumbar puncture, providing physicians with the first test which, by its associated cerebrospinal fluid (CSF) analysis, provides specific diagnostic information related to CNS disease. For example, the clinical diagnosis of meningitis is confirmed by such tests. Since then, a plethora of tests has been introduced, including EEG, EMG, and many others.

Table 1 outlines testing modalities which produce images instead of the electrophysiological tracings as does the EMG or the numerical values as does CSF analysis. The images are captured on X-ray films, e.g., CT and MRI, or on other hard copy, e.g., transcranial Doppler (TCD). Table 1 also demonstrates that, since 1974, an increasing array of such tests has become available, taxing the clinician's ability to be aware of them, no less the cognizance of the potential value in a particular patient's diagnostic problem.

It has been said by many that an 80 to 90% probability of correct neurological diagnosis can be reached by

TABLE 1 Image-Producing Technologies

Technology	Year available
X-rays	1891
Pneumoencephalography	1896
Cerebral arteriography	1932
Echoencephalography	1954
Carotid Doppler	1962
Brain-computed tomography	1972
Carotid real-time ultrasound	1980
Magnetic resonance (MR) imaging	1984
Single photon emission-computed tomography	1984
Transcranial Doppler	1986
EEG/MEG	1986
Positron emission tomography	1987
Duplex color flow	1988
MR angiography	1990
Transcranial color flow	1992
MR three-dimensional volume reconstruction	1993
MR perfusion	1994

assessment of a thorough patient history and the data derived from neurological examination. Therefore, expensive and time-consuming tests add a 10 to 20% probability of diagnostic accuracy. As discussed later, some tests provide one type of data and some another type. Some tests provide information relative to neurotoxicological assessment whereas others are not particularly valuable or may provide no information relative to this diagnostic category. Judicious selection of neuroimaging modalities to be ordered in a specific patient is necessary.

It should be stated that the tests available since 1974 are not invasive, with little or no patient risk or discomfort accrued. This situation is in contrast to the now obsolete pneumoencephalography, (PEG), which caused patient distress, and the relatively little used angiography, which can cause distress and damage to the patient.

B. Uses and Limitations

Table 2 defines broadly the types of data provided by the various tests and indicates those which may

TABLE 2 Neuroimaging Data Most Applicable to Neurotoxicological Assessment

Data provided	Technology
Brain morphology	CT
	MRI
Vessel morphology and/or flow velocity	MRA
	Ultrasound
Brain metabolism	SPECT
	PET

supply information most applicable to the assessment of neurotoxicological disease.

The types of tests can be broadly categorized into those which provide primarily information concerning brain and/or spinal cord and spinal root morphology, i.e., CT and MRI. They can be categorized into those which provide information with respect to blood vessel lumen size and/or blood flow velocity, i.e., the neurosonology tests. Those include carotid Doppler, carotid real-time ultrasound, transcranial Doppler, duplex color flow imaging, and transcranial color flow. Among the other tests, single photon emission-computed tomography (SPECT) and positron emission tomography (PET) produce information on brain function. As mentioned, cerebral arteriography has little application to neurotoxicological diagnosis. Echoencephalography (ECHO) is obsolete. Electroencephalography/magnetoencephalography (EEG/MEG) has little present use but may have future application. It is too early to say whether techniques such as MR three-dimensional volume reconstruction will help define neurotoxicological damage. Likewise MR perfusion and spectroscopy tests may have future application but as yet are not applicable to neurotoxicology.

Three groups of neuroimaging procedures are discussed next: neurosonology, SPECT/PET, and CT/MRI.

1. Neurosonology

In general, neurotoxicological diseases do not cause alteration of cerebral blood flow (CBF) as could be detected by tests which are designed to evaluate the blood vessel lumen size and/or blood flow velocity. Therefore, this group of tests should be ordered in neurotoxicological evaluation only to "rule out" other conditions. For example, an elderly person with risk factors for cerebrovascular insufficiency might suffer from hemiparesis after exposure to a potential neurotoxin. Carotid ultrasound might find a complete or partial carotid occlusion on the appropriate side. Therefore, the physician might conclude that the development of hemiparesis was merely coincidental to the exposure. On the other hand, in the case of volatile hydrocarbon exposure or carbon monoxide, the concentration of the potential neurotoxin led to hypoxia and subsequently to stroke in a patient already suffering from cerebrovascular insufficiency, i.e., carotid stenosis.

2. SPECT/PET

SPECT and PET isotopic studies provide information about CBF and metabolism (1). In SPECT, administration of the isotopes xenon-133 and technetium 99m-HMPAO provides information about cerebral blood flow and, indirectly, cerebral metabolism. In PET, the use of 2-[^{18}F]fluoro-2-deoxy-D-glucose provides a measure of cerebral metabolic rate for glucose. Investigations of receptor ligands labeled with iodine-123 for use in PET are ongoing. Compounds under study include ligand binding within serotonergic systems, benzodiazepine receptors, γ-aminobutyric acid receptors, and muscarine cholinergic receptors. Thus, PET has the potential to become a definitive tool in the diagnosis of neurotoxicology.

Current neurotoxicological evaluation using SPECT is limited to severe cases of encephalopathy in which a pool of cerebral neurons have been damaged because of a neurotoxin with resulting decreased CBF. Although a normal SPECT study does not exclude the possibility of encephalopathy, its findings probably exclude severe encephalopathy. For example, the author has anecdotal experience with a man who was exposed occupationally to a mixture of volatile hydrocarbons in a poorly ventilated place. He provided data that he "nearly passed out." Subsequently he complained of persisting memory deficits. Neurological examination was normal and neuropsychological testing was equivocal. Although intellectual functioning was only in the low normal range, the definite premorbid data were not available. The man had received Cs and Ds in school before dropping out in the 10th grade. Brain MRI was normal as was the SPECT scan. By combining morphological and physiological studies, e.g., MRI and SPECT, the allegation of severe encephalopathy was stated to have no basis.

Overall, there is little substantive database reported in the literature concerning the results of SPECT analysis in neurotoxicological exposure.

PET scanning is more promising as a tool to quantitate brain damage which may occur as a result of human exposure to neurotoxins. Again, very little database has been reported.

3. CT/MRI

Described in 1961 (2) and introduced in 1973, X-ray computed tomography initiated a revolution in clinical medicine by imaging human tissues noninvasively. The impact on neurology was profound. With or without the use of contrast agents, lesions of the brain could be defined in a fashion previously impossible. Specific markers of some diseases, e.g., stroke, tumor, and trauma, could be defined. Unfortunately, CT has little sensitivity to detect neurotoxicological damage nor is it specificity apparent.

Magnetic resonance imaging uses the principles of nuclear magnetic resonance to produce excellent images of normal morphology as well as deviations from normal produced by disease. Its sensitivity is better than that of CT (1). In some diseases, specificity is exquisite and is markedly better than CT. For example, in multiple sclerosis, MRI markers of the disease are almost pathognomonic. Such specificity is lacking for neurotoxic disease.

III. Examples of Beneficial Uses of Neuroimaging in Neurotoxicology

The sensitivity of neuroimaging methodology in the detection of CNS correlates of neurotoxicological damage is limited. Often no changes from normal are seen despite patients' complaints of dysfunction. Other times, imaging abnormalities are seen in instances whereby it is difficult to determine whether the abnormalities seen are related to the neurotoxicological exposure or are preexisting to the exposure. Specificity of neuroimaging abnormalities in neurotoxicological evaluation is even more limited, i.e., it is difficult or sometimes impossible to ascribe an abnormality seen as being causally related to a toxic exposure. Nonetheless, in some circumstances imaging offers insight into focal or generalized abnormalities which can be correlated to toxicological exposure.

A. Carbon Monoxide

Carbon monoxide intoxication is often fatal. In survivors, CNS deficits are common. For example, Min (3) reported the effects of carbon monoxide intoxication in 86 patients. All had amnesia and 65% had apraxia. Physical lesions and neuropsychological symptoms have not been correlated, although some autopsy, CT, and MRI data may be meaningful.

Horowitz and co-workers (4) reported MRI findings for two patients. Both had high signal intensity areas in the globus pallidus. Four patients with carbon monoxide intoxication were studied using MRI and CT, and were described by Murata *et al.* (5). Three patients were unconscious initially and had notable long-term deficits. One patient did not lose consciousness but had some memory loss. In the three patients with continuing deficits, lesions were seen on T2-weighted MRI but not on T1-weighted MRI or CT (Fig. 1). The T1-weighted MRI of the fourth patient showed hyperintensity in bilateral global pallidal regions (Fig. 2).

On CT, cortical lesions are less commonly seen than white matter or globus pallidus lesions. Such CT findings are seen in patients without neuropsychological symptoms but who have had other neurological deficits. Bilateral lesions in the posterior watershed areas were reported in a patient with parkinsonian syndrome

following carbon monoxide intoxication (6). Such data suggest an important aspect of neuroimaging and neurotoxicology. Although the toxins may produce few or no specific signs of damage, CT or MRI may show more significant signs of damage resulting from secondary effects such as oxygen deprivation.

B. Organic Solvents

The relationship between excessive exposure to organic solvents to the subsequent development of chronic encephalopathy has been recognized since the 1890s. An abundant literature on the topic exists (7). Because of their affinities for lipid-rich brain tissue, organic solvents have been implicated in a variety of CNS disturbances. Unfortunately, the correlation of specific solvents with neuroimaging changes has been, as yet, poorly defined. Likewise, a correlation between duration and intensity of exposure, e.g., parts per million, and neurological deficits has not been defined.

For example, attention has been paid to the high relevance of neurasthenic symptoms in painters. These symptoms most frequently include memory loss and personality changes. Although neurologic and laboratory screening examinations have frequently showed no consistent abnormalities, psychological tests have documented short-term memory problems and an array of neuropsychological deficits. However, a consistent relationship of the severity of the deficits and degree of exposure have been lacking.

Some have questioned the validity of the neuropsychological findings such that the existence of a solvent-induced encephalopathy remains controversial. Dose–response relationships are lacking in behavioral dysfunction. Therefore, morphological evidence of brain damage by CT or MRI has been sought. A 1980 paper reported PEG changes, suggesting brain atrophy in 22 of 37 patients with suspected solvent toxicity (8). However, this now obsolete test has no specificity in defining loss of tissue bulk.

Triebig and Lang (9) reported on the use of brain imaging techniques in workers chronically exposed to solvents. Results were inconclusive when techniques such as cerebral blood flow, SPECT, CT, and MRI were used in patients with toxic encephalopathy. The authors carefully examined CT parameters in nonpainters and painters (using spray or nonspray techniques). On CT examination, chronic solvent exposure was not associated with brain atrophy in excess of that explained by aging.

As previously stated, it may be that chronic low-dose exposure produces physiological disruption of brain function as may be measured by a neuropsycho-

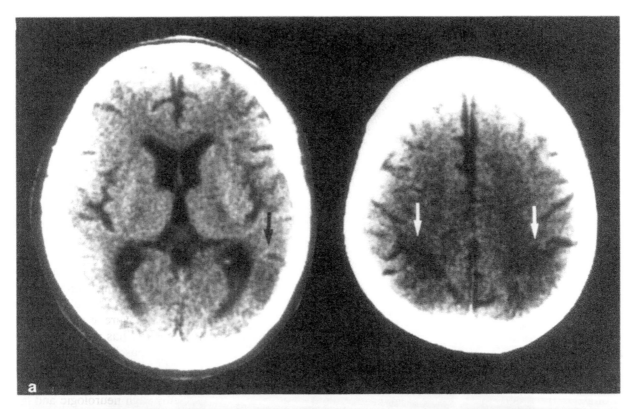

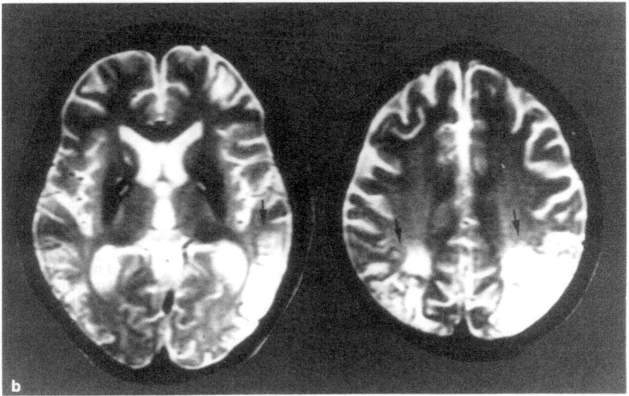

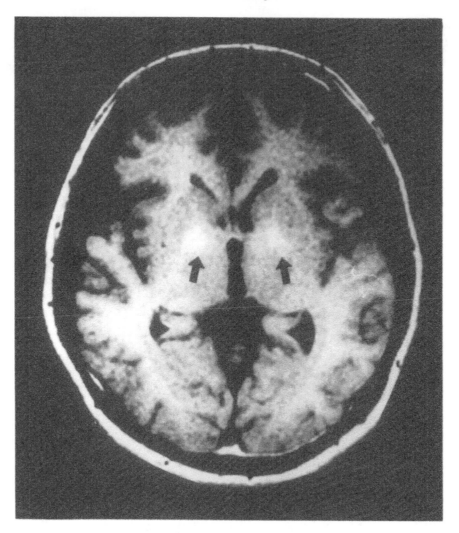

FIGURE 2 T1-weighted MRI: Arrows indicate high signal regions in both global pallidal regions. [From Murata *et al.* (5) with permission of Little, Brown and Company.]

logical test but no scientifically measurable degree of loss of brain structures can be defined by CT.

On the other hand, some investigators are of the opinion that other tests such as SPECT and PET do have a specific diagnostic value in neurotoxicological evaluation. For example, Morrow and colleagues (10) reported on a 31-year-old man who was exposed at work to tetrabromoethane, a halogenated aliphatic hydrocarbon. Subsequently, he did not return to work and reported such problems as headaches, memory loss, and depression. Neurobehavioral assessment revealed significant deficits such as learning ability, memory, and psychomotor speed. CT, EEG, and EMG were normal. However, a PET scan with ^{18}F-2-deoxyglucose showed that uptake was significantly decreased in multiple brain regions.

Subsequently, Callender and colleagues (11) reported 33 workers who had toxic encephalopathy after

acute exposure to pesticides, solvent mixtures, and other toxins. SPECT scans were abnormal in 31 (94%). Abnormalities were found most frequently in the temporal lobes, frontal lobes, basal ganglia, thalamus, and parietal lobes. The authors concluded that EEG, CT, and MRI did not have good sensitivity in this application and recommended other testing, such as metabolic imaging studies, to help evaluate other aspects of toxic effects such as dose, time, and mechanism.

Other workers report the value of MRI in neurotoxicological evaluations. Filley and co-workers (12) reported 14 chronic toluene abusers who were evaluated using neuropsychological assessments and brain MRI. Three patients were reported to be normal, three were in a borderline range, and eight were impaired. MRI analysis of white matter changes showed a strong correlation between abnormalities and neuropsychological impairment. In three of these cases dementia and toluene abuse appeared to be related to the severity of cerebral white matter involvement. These data indicate a need for more critical evaluation by MRI techniques of other patients exposed to volatile hydrocarbons.

In this regard, the report of White and co-workers (13) is of interest. MRI of a patient with toxic encephalopathy following exposure to 2,6-dimethyl-4-heptanone showed changes in the cerebral white matter and in the pons. The patient had a number of neuropsychological deficits. The MRI study showed subcortical rather than cortical sites of pathology, indicating, in this case, that the neurotoxin probably affects subcortical and temporal areas rather than the cortical gray matter.

Both of these reports suggest that MRI may be sensitive by visualizing abnormalities in individuals exposed to potential toxins. Whether the findings are specific to a particular toxin is unclear. A dose–effect relationship is not clear. Likewise, the effects of potential predisposing factors, e.g., age, alcohol consumption, and other disease states, have not yet been defined.

Furthermore, the differential effect between acute high-dose exposure and chronic low-dose exposure in the production of neuroimaging abnormalities has not been defined. This differential effect is illustrated by a 19-year-old man who was exposed to a mixed solvent occupationally while spray painting at an altitude greater than 5000 feet (14). In the poorly ventilated enclosed space, hydrocarbons would be expected to displace the already relatively impoverished oxygen content at that high altitude. Subsequent to 2 days of such exposure, he lapsed into coma. Brain CT scan performed within 24 hr was consistent with cerebral edema. Subsequently a CT scan demonstrated focal lesions of the cerebellum and basal ganglia (Fig. 3). He survived in a state of intellectual impairment, mutism, and spastic quadriplegia. CT and MRI scans taken 2 years after exposure demonstrated diffuse cortical atrophy as well as persisting basal ganglia and cerebellum lesions.

Although these imaging techniques demonstrated dramatic pathology, specificity between the toxins within the mixture to which he was exposed and the resulting neurological deficits and CT and MRI correlates could not be defined.

C. Heavy Metals

A relationship between MRI findings and heavy metal exposure has been reported. London and co-workers (15) performed *in vivo* studies of anesthetized rats treated with a single dose of manganese chloride. Manganese levels were high in the ventricular cerebrospinal fluid. The manganese apparently crossed the choroid plexus barrier, was absorbed in the ventricles, and then was transported to the subarachnoid space.

In a case study (16), an arc welder exposed to manganese-containing rods for many years developed confusion, poor memory, impaired cognition, and paranoid thoughts. Neurological findings included a right hemisensory deficit, right hemiparesis, and hyperreflexion of the legs. His blood and urine manganese concentrations were elevated even 10 months after stopping work. A variety of other studies, including brain CT scan, were normal. T1-weighted MRI showed hyperintense signals in the basal ganglia with normal T2 signals, followed by almost complete resolution 6 months later. A SPECT scan done at the time of the first MRI demonstrated decreased uptake diffusely in the cerebral cortex with local decreases in the basal ganglia.

White and colleagues (13) reported MRI findings of central and cortical atrophy and neuropsychological problems in a man exposed to inorganic mercury in his work at a thermometer factory.

Another example of the potential use of MRI in elucidation of neurotoxicity is the possible role of iron in the pathophysiology of Parkinson's disease. Evidence is accumulating to indicate that iron deposition occurs in some forms of the parkinsonian syndrome. Hauser and Olanow (17) reviewed this topic and have provided illustrative MRI images.

D. Substance Abuse

It is well known that the chronic abuse of ethyl alcohol may lead to profound neurological dysfunction, including peripheral neuropathy. A variety of focal and generalized deficits occur in the CNS; CT correlates of cortical atrophy and cerebellar degeneration are well recognized (18). Likewise, PET scan abnormalities indicate hypometabolism in the cerebellar vermes in patients with clinical and CT scan signs of alcoholic cerebellar degeneration (19).

The evaluation of other forms of substance abuse, including cannabis inhalation, opiates, psychedelics, amphetamines, and cocaine, is complicated by the fact that such drug abuse is almost always combined with heavy alcohol consumption. In one study (20), 23 drug abusers and 17 controls had clinical neurological examinations and MRIs of the brain. The drug abusers had significantly smaller cerebellar vermes than controls. White matter changes were also more common in the drug abusers. It was not clear whether the changes seen were related to alcohol or other abuse substances.

The role of SPECT and PET in other forms of substance abuse is under investigation and holds promise. For example, Weber and colleagues (21) performed brain SPECT in 21 crack abusers and 21 control sub-

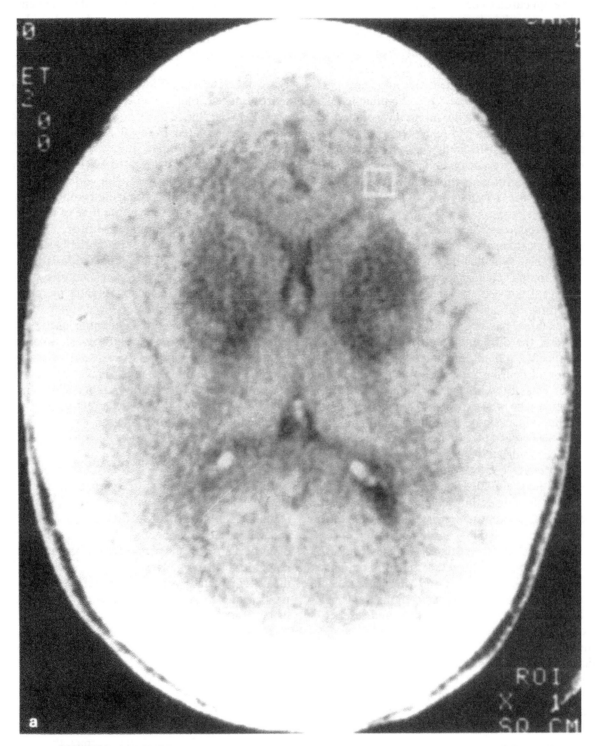

FIGURE 3 (a and b) CT without contrast administration taken 7 days after toxic exposure. Areas of abnormal hypodensity can be seen in the bilateral basal ganglia (a) and in the cerebellum (b). (c) Contralateral CT taken 14 days after toxic exposure. A more prominent hypodensity with associated contrast enhancement just anterior to those lesions is demonstrated. [From Trehan and Prockop (14) with permission of Little, Brown and Company.]

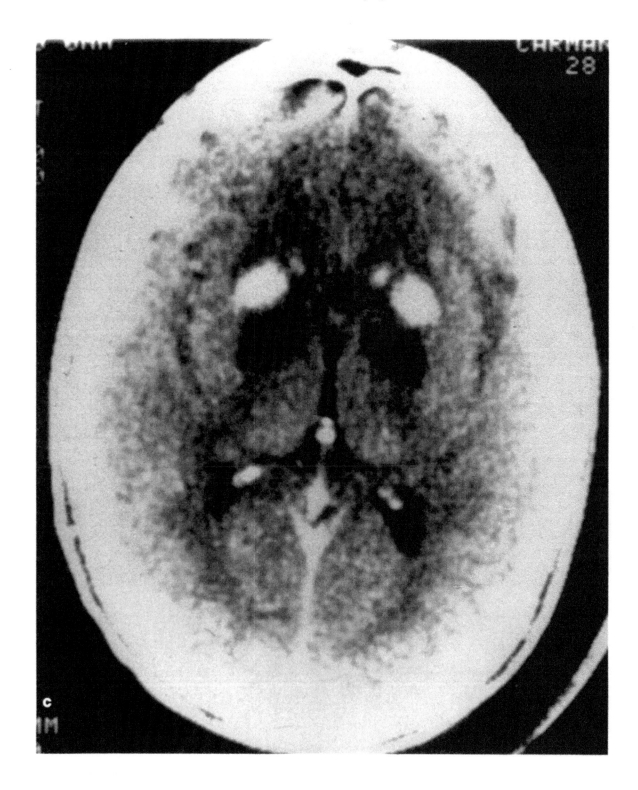

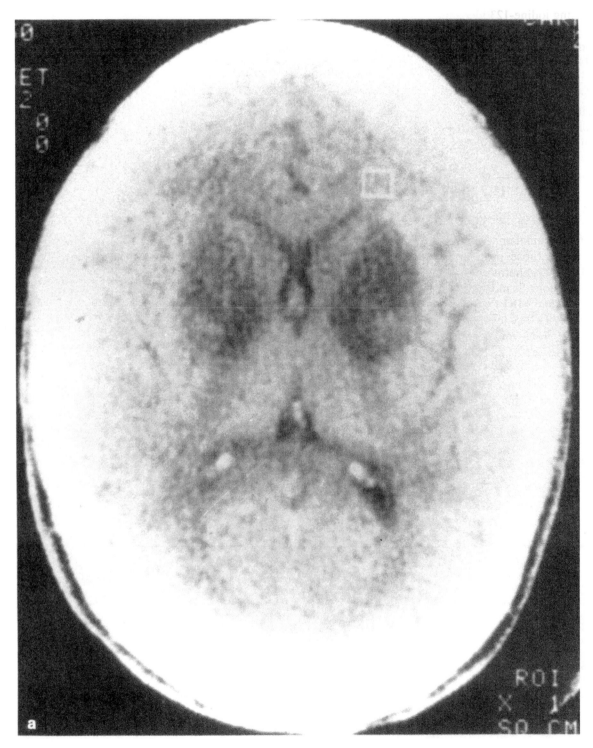

FIGURE 3   Continued

jects using iodine-123 ideoamphetamine (^{123}I-IMP) uptake and localization for their criteria. They concluded that quantitative measurements obtained from these SPECT scans were objective, sensitive, and useful for measuring cerebral blood flow abnormalities in crack abusers. For example, abnormally reduced ^{123}I-IMP activity occurred mainly in the frontal and parietooccipital cortex in 16 of 21 crack abusers, suggesting the disruption of regional cerebral blood flow.

IV. Summary

Neurotoxic chemicals sometimes cause focal and/or diffuse brain damage. Patient's symptoms and signs may mimic those of metabolic, degenerative, nutritional, and degenerative diseases. Neuroimaging, e.g., neurosonology, is most useful in ruling out other conditions. CT and MRI appear to have poor sensitivity to detection of CNS lesions related to chronic low-dose exposure. However, focal changes may occur in acute high-dose exposure. It is not yet clear whether the changes noted are specific to a particular neurotoxin or neurotoxic mixture. Evaluation of metabolic changes occurring with neurotoxicological exposure as measured by SPECT and PET is promising. Additional neuroimaging correlates including structural and metabolic studies are needed to explore dose, time, and specific toxic effects as well as the mechanism of toxicity.

References

1. Mazziotta, J. C., and Gilman, S. (eds.) (1992). *Clinical Brain Imaging: principles and applications.* F. A. Davis, Philadelphia.
2. Oldendorf, W. H. (1961). Isolated flying spot detection of radiodensity discontinuities displaying internal structural pattern of a complex object. *IRE Trans. Biomed. Electronics* BME. **8**, 68.
3. Min, S. K. (1986). A brain syndrome associated with delayed neuropsychiatric sequelae following acute carbon monoxide intoxication. *Acta Psychiatry Scand.* **73**, 80–86.
4. Horowitz, A. L., Kaplan, R., and Sarpel, G. (1987). Carbon monoxide toxicity: MR imaging in the brain. *Radiology* **162**, 787–788.
5. Murata, S., Asaba, H., Hiraishi, K., *et al.* (1993). Magnetic resonance imaging findings in carbon monoxide intoxication. *J. Neuroimag.* **3**, 128–131.
6. Klawans, H. L., Stein, R. W., Tanner, C. M., *et al.* (1982). A pure parkinsonian syndrome following acute carbon monoxide intoxication. *Arch. Neurol.* **39**, 302–304.
7. Prockop, L. (1979). Neurotoxic volatile substances. *Neurology* **29**, 862–865.
8. Juntunen, J., Hupli, V., Hernberg, S., *et al.* (1980). Neurological picture of organic solvent poisoning in industry. *Int. Arch. Occup. Environ. Health* **46**, 219–231.
9. Triebig, G., and Lang, C. (1993). Brain imaging techniques applied to chronically solvent-exposed workers: current results and clinical evaluation. *Environ. Res.* **61**, 239–250.
10. Morrow, L. A., Callender, T., Lottenberg, S., *et al.* (1990). PET and neurobehavioral evidence of tetrabromoethane encephalopathy. *J. Neuropsychiatry Clin. Neurosci.* **2**, 431–435.
11. Callender, T. J., Morrow, L., Subramanian, K., *et al.* (1993). Three-dimensional brain metabolic imaging in patients with toxic encephalopathy. *Environ. Res.* **60**, 295–319.
12. Filley, C. M., Heaton, R. K., and Rosenberg, N. L. (1990). White matter dementia in chronic toluene abuse. *Neurology* **40**, 532–534.
13. White, R. F., Feldman, R. G., Moss, M. B., *et al.* (1993). Magnetic resonance imaging (MRI), neurobehavioral testing, and toxic encephalopathy: two cases. *Environ. Res.* **61**, 117–123.
14. Trehan, R. R., and Prockop, L. D. (1994). Multifocal nervous system damage after exposure to volatile hydrocarbons. 17th annual meeting of American Society of Neuroimaging. [Abstract]
15. London, R. E., Toney, G., Gabel, S. A., *et al.* (1989). Magnetic resonance imaging studies of the brains of anesthetized rats treated with manganese chloride. *Brain Res. Bull.* **23**, 229–235.
16. Nelson, K., Golnick, J., Korn, T., *et al.*, (1993). Manganese encephalopathy: utility of early magnetic resonance imaging. *Br. J. Ind. Med.* **50**, 510–513.
17. Hauser, R. A., and Olanow, C. W. (1994). Magnetic resonance imaging of neurodegenerative diseases. *J Neuroimag.*, **4**, 146–158.
18. Bianco, F., Bozzao, L., Colonnese, C., and Fantozzi, L. (1983). The value of computerized tomography in the diagnosis of cerebellar atrophy. *Ital. J. Neurol. Sci.* **1**, 65–68.
19. Gilman, S., Adams, K., Koeppe, R. A., *et al.* (1990). Cerebellar and frontal hypometabolism in alcoholic cerebellar degeneration studies with positron tomography. *Ann. Neurol.* **28**, 775–785.
20. Aasly, J., Storsaeter, O., Nilsen, G., *et al.* (1993). Minor structural brain changes in young drug abusers: a magnetic resonance study. *Acta Neurol. Scand.* **87**, 210–214.
21. Weber, D. A., Franceschi, D., Ivanovic, M., *et al.* (1993). SPECT and planar brain imaging in crack abuse: iodine-123-iodoamphetamine uptake and localization. *J. Nucl. Med.* **34**, 899–907.

PART
VII

Risk Assessment for Neurotoxicity

Risk Assessment for Neurotoxicity

An Introductory Overview

HUGH A. TILSON

Neurotoxicology Division
Health Effects Research Laboratory
U.S. Environmental Protection Agency
Research Triangle Park, North Carolina 27711

I. Introduction

Because the public is greatly concerned about the possible adverse health effects that could result from exposure to chemicals in the environment, standards for regulating the manufacture, use, and release of chemicals have been promulgated by several regulatory agencies. Inherent in the regulation of chemicals is the issue of how to estimate the risk associated with exposure or ingestion of these agents. As an issue, risk has assumed legalistic as well as scientific connotations (Barnard, 1990). In the United States, regulatory agencies such as the U.S. Environmental Protection Agency (EPA), Food and Drug Administration (FDA), Occupational Safety and Health Administration (OSHA), and Consumer Product Safety Commission (CPSC) are mandated to estimate risk associated with

various products or chemicals that fall within their regulatory domain. The CPSC has published risk assessment guidelines for labeling art materials and other products (CPSC, 1991) and a draft version of neurotoxicology risk assessment guidelines is undergoing internal review at the U.S. EPA.

II. Risk Assessment and Risk Management

Risk assessment has been defined by the National Research Council (NRC, 1983) as the characterization of the potential adverse health effects of human exposures to environmental hazards. This concept is different from risk management, which describes the process of evaluating alternative regulatory actions and selecting among them. The NRC (1983) defined risk management as a process to determine whether the assessed risk should be reduced and, if so, to what extent. Risk management decisions are based on a balance of political, social, economic, and engineering considerations and data in the context of the regulatory legislative mandate specific for each regulatory agency.

This paper has been reviewed by the Health Effects Research Laboratory, U.S. Environmental Protection Agency, and approved for publication. Mention of trade names or commercial products does not constitute endorsement or recommendation for use.

767

III. Risk Assessment as a Four-Part Process

Risk assessment is typically viewed as an empirical process to determine the probability that an adverse health effect will be associated with exposure to a chemical. The NRC (1983) divided risk assessment process into four major steps, including hazard identification, dose–response assessment, exposure, assessment, and risk characterization.

As it pertains to neurotoxicology, hazard identification is the process of determining whether or not exposure to a chemical produces an adverse effect and whether that effect can be defined as neurotoxicity. Neurotoxic effects can be defined as adverse effects on neurobehavioral, neurophysiological, or neurochemical functioning and/or the structural integrity of the nervous system (Tilson, 1990). Data from human epidemiological studies and case studies can provide direct identification of hazards in humans. Case studies are sometimes limited, however, by inadequate characterization of demonstrable neurotoxic effects and/or exposure data. Although epidemiological studies lead to convincing evidence of neurotoxic hazard in human populations, precise exposure data are not often available. Controlled laboratory studies in humans could potentially provide accurate exposure and affect characterization information, but there are obvious ethical and practical limitations on such studies.

In general, hazard identification decisions are based on data obtained from animal studies. A number of uncertainties exist in animal-to-human extrapolation, including the problem of relative sensitivity across species, as well as between sexes of the same species. In some cases, such as cognitive function or verbal behavior, there may be qualitative differences between animal species and humans. Additional uncertainties exist with regard to data collected from adult animals to determine the presence of neurotoxicity. It is known that very young, aged, or chronically ill populations may be more sensitive to some classes of neurotoxicants than mature, healthy populations.

For purposes of hazard identification, chemical-induced changes in the function or structure of the nervous system that are irreversible or that result in a permanent change are of greatest concern. Neurotoxic agents can also produce slowly reversible or rapidly reversible changes in the structure or function of the nervous system. Neurotoxicity hazard identification must also take into consideration that nerve cells, once damaged, particularly in the central nervous system, may have a limited capacity for regeneration. Reversibility of effects following cell death or from the destruction of cellular processes is due to repair mechanisms or activation of homeostatic processes. Under such circumstances, future attempts to compensate to neurotoxic insult may be diminished. Neurotoxicity hazard identification must also consider that effects can get worse after exposure has ceased, i.e., progressive effects, and that neurotoxic effects may be observed at a time remote from the last contact with the chemical agent, i.e., latent effects. Some neurotoxic effects may be residual, capable of being observed only after environmental challenge or aging. Since it is unlikely that the mechanism of action of a neurotoxic agent will be known at the hazard identification state, there is an equal level of concern for chemicals that act directly or indirectly to produce neurotoxicity.

Neurotoxicology hazard identification may require a weight of evidence approach. More weight is given to studies that are appropriately designed and well executed and where the data are analyzed correctly. Other important factors include the species tested, the number of subjects used per group, testing of both genders, and assessment of possible susceptible populations. The route and duration of exposure are important factors, as well as the frequency of evaluation before, during, and after dosing or exposure. Decisions at the hazard identification level are frequently based on results from tests or batteries of tests that assess a number of functional and structural end points.

The second step of the risk assessment process is dose–response assessment in which dose–response relationships are used to estimate the association between the dose and effect for various conditions of exposure. This aspect of risk assessment as it relates to neurotoxicity is complicated by the multitude of possible effects that can be observed in nervous system structure and function following neurotoxic exposure. Thus, it is important to consider more than one dose–response relationship for neurotoxic agents.

Like other noncancer end points, the most frequently used approach for dose–response assessment is the quantification of risk through the use of uncertainty or safety factors (Barnes and Dourson, 1988). This approach results in an estimated exposure level or reference dose (RfD) believed to be unlikely to cause any neurotoxic effect in humans. A RfD is calculated by dividing a no observed adverse effect level (NOAEL) by uncertainty factors that account for inter- and intraspecies differences in sensitivities, animal-to-human extrapolation, and less than lifetime exposures. If a NOAEL cannot be determined, then a lowest observed adverse effect level (LOAEL) is used in the RfD calculation and an additional uncertainty factor of 10 is frequently utilized.

The uncertainty approach to establish reference doses is predicated on the likelihood that there is a threshold dose below which there is little or no adversity. Thresholds are largely dependent on the limit of sensitivity of the procedures used to measure neurotoxic effects, so thresholds may change as a function of methodological advancement. The determination of thresholds is also confounded by the statistical limit of detection, which depends on experimental design issues such as the number of subjects used in each treatment group. It has been argued that for some neurotoxicants such as developmental exposure to lead, no threshold dose may be evident (Bondy, 1985). Sette and MacPhail (1992), on the other hand, have argued that the concept of no threshold for effects is experimentally untestable and, therefore, not useful in neurotoxicology risk assessment.

Different approaches in quantitative neurotoxicology risk assessment have been proposed. Dews and colleagues (Dews, 1986; Glowa *et al.*, 1983) and Crump (1984) have proposed modeling the dose–response curve to estimate exposure-related changes (i.e., a 5 or 10% change) and using this benchmark dose to calculate a dose to estimate risk. Gaylor and Slikker (1990) have proposed estimating the probability of an adverse effect as a function of dose based on evaluation of normal population statistics and the mathematical relationship between a given biological change and the dose of a chemical.

In addition to identifying the NOAEL, LOAEL, or benchmark dose, dose–response assessments define the range of doses that are effective in producing neurotoxic effects for a given chemical, species, route of administration, and duration of exposure. Although dose–response functions in neurotoxicology are generally linear, U-shaped or inverted U-shaped curves have been reported (Davis and Svendsgaard, 1990). Such curves may reflect multiple mechanisms of action, the presence of homeostatic processes, or compensation to neurotoxic insult. Neurotoxicity risk assessments need to determine the uncertainty that inverted U-shaped functions might contribute to the estimate of the NOAEL or LOAEL and benchmark dose and the subsequent calculation of the RfD.

The third step in the risk assessment process is exposure assessment, which determines the source, route, dose, and duration of human exposure to an agent. This estimate of human exposure is combined with the dose–response assessment to generate a quantitative estimate of risk. If the probability of exposure is very low, then the estimate of risk is low. Guidelines for exposure assessment have been published by the U.S. EPA (1986, 1992). The importance of exposure has been underscored in a report by the National Research Council (1993) on pesticides in the diet of infants and children. The report indicated that infants and children differ quantitatively and qualitatively from adults in exposure to pesticide residues and foods and that such differences may account for some differences in pesticide-related health risks that exist between children and adults.

The last of the four steps in the risk assessment process is risk characterization, which integrates the dose–response and dose–effect information with the estimates of the nature, duration, and intensity of exposures, as well as the characteristics of the exposed populations (Sette and MacPhail, 1992). At the most basic level, risk characterization attempts to determine whether projected human exposure would exceed the reference dose or daily allowable intake. The risk characterization process involves a summary of the relative strengths and weaknesses in each step of the risk assessment process and includes a discussion of the major assumptions, scientific judgments, and qualitative/quantitative uncertainty estimates.

IV. Current Issues in Neurotoxicology Risk Assessment

Application of risk assessment methodologies to neurotoxicological end points is a relatively new area and a number of issues remain to be resolved. A draft report on principles of neurotoxicity risk assessment has been published to address some of these issues (U.S. EPA, 1993). The chapters contained in this section explore a number of important problems and advancements in neurotoxicology risk assessment. Drs. John R. Glowa and Robert C. MacPhail address neurotoxic risk by evaluating dose–effect data and determining the probability of observing a significant effect in a small proportion of the population. Drs. Dale Hattis and Kevin M. Crofton discuss how dynamic modeling can be used to extract risk assessment information from repeated dosing studies using multiple doses of prototypic neurotoxicants. Drs. David W. Gaylor and William Slikker discuss their approach to neurotoxicity risk assessment in which risk is defined as the proportion of a population whose levels of measurable neurotoxicity is equal to or exceeds a statistically derived abnormal level determined during the study. This approach provides estimates of risk as a function of dose of a neurotoxic agent. Dr. Hugh A. Tilson discusses some of the special problems of developmental neurotoxicology, whereas Dr. Bernard Weiss discusses some unique aspects of neurotoxicology risk assessment.

References

Barnard, R. C. (1990). Some regulatory definitions of risk: interaction of scientific and legal principles. *Reg. Toxicol. Pharmacol.* **11**, 201–211.

Barnes, D. G., and Dourson, D. G. (1988). Reference dose (RfD): description and use in human health risk assessments. *Reg. Toxicol. Pharmacol.* **8**, 471–486.

Bondy, S. C. (1985). Special considerations for neurotoxicological research. *CRC Crit. Rev. Toxicol.* **14**, 381–402.

Consumer Product Safety Commission (CPSC) (1991). Labeling requirements for art materials and other products: proposed rules. *Fed. Reg.* **56**, 15672–15710.

Crump, K. S. (1984). A new method for determining allowable daily intakes. *Fundam. Appl. Toxicol.* **4**, 854–871.

Davis, J. M., and Svendsgaard, D. J. (1990). U-shaped dose-response curves: their occurrence and implication for risk assessment. *J. Toxicol. Environ. Health* **30**, 71–83.

Dews, P. B. (1986). On the assessment of risk. In *Developmental behavioral Pharmacology* (N. Krasnegor, J. Gray, and T. Thompson Eds.), pp. 53–65. Lawrence Erlbaum, Hillsdale, New Jersey.

Gaylor, D. W., and Slikker, W. (1990). Risk assessment for neurotoxic effects. *Neurotoxicology* **11**, 211–218.

Glowa, J., DeWeese, J., Natale, M. E., and Holland, J. J. (1983). Behavioral toxicology of volatile organic solvents. I. Methods: acute effects. *J. Am. Col. Toxicol.* **2**, 175–185.

National Research Council (1983). *Risk Assessment in the Federal Government: Managing the Process.* National Academy Press, Washington, D.C.

Pesticides in the diets of infants and children. National Academy Press, Washington, D.C.

Sette, W. F., and MacPhail, R. C. (1992). Qualitative and quantitative issues in assessment of neurotoxic effects. In *Neurotoxicology* (H. Tilson and C. Mitchell Eds.), pp. 345–361, Raven Press, New York.

Tilson, H. A. (1990). Neurotoxicology in the 1990s. *Neurotoxicol. Teratol.* **12**, 293–300.

U.S. EPA (1986). Guidelines for estimating exposures. *Fed. Reg.* **51**, 34042–34054.

U.S. EPA (1992). Guidelines for exposure assessment. *Fed. Reg.* **57**, 22888–22938.

U.S. EPA (1993). Draft Report: principles of neurotoxicity risk assessment. *Fed. Reg.* **58**, 41556–41599.

CHAPTER
51

Concepts on Quantitative Risk Assessment of Neurotoxicants

WILLIAM SLIKKER, JR.
Division of Neurotoxicology
National Center for Toxicological Research/Food and Drug Administration
Jefferson, Arkansas 72079

DAVID W. GAYLOR
Division of Biometry and Risk Assessment
National Center for Toxicological Research/Food and Drug Administration
Jefferson, Arkansas 72079

I. Introduction

The need for quantitative risk assessment procedures for noncancer end points such as neurotoxicity has been discussed in reports written by the U.S. Congress (OTA, 1990), National Research Council (NRC, 1992), and a federal agency coordinating council (Reiter *et al.*, 1993). Although quantitative risk assessment procedures have been the focus of workshops and have been published in the open literature (Gaylor and Slikker, 1990; NRC, 1992), policy to implement these procedures has not been forthcoming due, in part, to the lack of validation and comparison to existing risk assessment procedures. In this chapter, a quantitative risk assessment procedure originally illustrated with a serotonergic neurotoxicant, MDMA (Slikker and Gaylor, 1990), is applied to another neurotoxicant, 1-methyl-4-phenyl-1,2,3,6-tetrahydropyridine (MPTP), that primarily affects the dopaminergic neurotransmitter system (Tipton and Singer, 1993). MPTP was selected as the agent for assessment because its neurotoxicity is well described in the literature and is comprised of behavioral, neurohistological, and neurochemical effects. The results of this assessment are systematically compared to those obtained with the currently used safety factor approach with and without the use of a benchmark dose (BMD) adjustment.

The neurotoxicant MPTP causes selective destruction of nigrostriatal dopaminergic neurons in rodents, nonhuman primates, and humans. The discovery that MPTP, a contaminant of a synthetic meperidine analog sold as a street drug, produced symptoms resembling Parkinson's disease stimulated a large number of studies on the mechanisms involved with the neurotoxic effects of MPTP and their possible relationship to idiopathic Parkinson's disease (Tipton and Singer, 1993).

The results from a host of studies reported since the 1980s have been used to formulate a mechanism of action for MPTP (McCrodden *et al.*, 1990; Johannessen, 1991; Tipton and Singer, 1993). In brief, MPTP is a lipophilic molecule that enters the brain rapidly after administration. A two-step reaction requires the enzyme monoamine oxidase (MAO)-B to convert MPTP to the pyridinium metabolite 1-methyl-4-phenylpyridinium (MPP$^+$). After this conversion, which occurs primarily in astrocytes, MPP$^+$ is sequestered within monoaminergic neurons by energy-dependent and selective monoaminergic transporters. MPP$^+$ is a potent inhibitor of mitochondrial oxidation of NAD$^+$-linked substrates. The blockade of respiration leads to a depletion of ATP and consequently the loss of membrane potential. The mechanisms underlying the subsequent destruction of nigrostriatal nerve cells may involve altered calcium homeostasis and perhaps free radical formation (Adams *et al.*, 1992).

Specific differences in susceptibility to MPTP (monkey > dog > mouse > rat) may be explained, at least in part, by the length of time the MAO-B-generated MPP$^+$ is retained in nigrostriatal tissue. Primates retain MPP$^+$ in the striatum with a half-life of about 10 days whereas MPP$^+$ concentrations in rat brain are negligible within hours after MPTP administration (Johannessen, 1991).

The mechanism of action of MPTP provides important information concerning the development of a biologically based, dose–response risk assessment model for this neurotoxicant, as was described previously for MDMA (Slikker and Gaylor, 1990). Several steps in the pathway for MPTP to produce neurotoxicity require either energy-dependent uptake of MPP$^+$ by monoaminergic transporters and/or stereospecificity (MAO-B metabolism). These proposed mechanisms or alternatives (e.g., oxidative stress) rely on pathways that are specific and saturable. Therefore, a risk assessment model that incorporates saturation kinetics appears appropriate.

II. Traditional and Benchmark Dose Estimations

Currently, the regulation of noncancer-causing toxicants is based on a reference dose (RfD) or, equivalently, the acceptable daily intake (ADI) as described by Barnes and Dourson (1988). The RfD is determined by dividing the no observed adverse effect level (NOAEL) as determined by animal studies by uncertainty or safety factors (usually values of 10) to account for possible greater average sensitivity in humans, sensitive human subpopulations, and inadequacies of experimental data. This NOAEL approach does not fully utilize the data of the complete dose–response curve and is limited by the statistical power of the experimental techniques utilized to detect differences in the means of the biological effects between treated and control animals. Unfortunately, there are many shortcomings with this approach, both theoretical (assumption of a threshold of effect) and practical (the less powerful the experiment the higher the RfD). These

limitations have been fully discussed by Slikker and Gaylor (1990) and Reiter and colleagues (1993).

In an attempt to reduce some of the shortcomings of the RfD approach, Crump (1984) proposed the use of a benchmark dose to replace the NOAEL. The BMD is defined as a lower confidence limit on a dose corresponding to a low level of risk, e.g., 1 to 10%. The advantage of this approach is that the confidence limit becomes tighter with better data, thereby resulting in appropriately larger values for the RfD. The use of a BMD and methods for calculating confidence limits to be used for BMDs have been discussed (Gaylor, 1992; Chen and Gaylor, 1992; Kodell and West, 1993; West and Kodell, 1993). Even with the improvement in risk assessment afforded with the use of the BMD, safety factors are still applied to obtain a "safe dose" and therefore the approach remains qualitative.

III. Quantitative Risk Estimation

In order to evaluate our quantitative biologically based, dose–response model, MPTP data sets from two mouse studies and one monkey study are utilized as previously reported for MDMA (Slikker and Gaylor, 1990). Alterations in neurotransmitter concentrations are used as end points of toxicity. It must be emphasized that behavioral (decreased motor activity) and neurohistological damage (increased degenerating nigrostriatal cells) also accompany MPTP administration (Ricaurte *et al.*, 1987; Johannessen *et al.*, 1989). Dopamine concentrations from several studies (Ricaurte *et al.*, 1987; Sonsalla *et al.*, 1987; W. Slikker *et al.*, unpublished observation) were used to develop the risk analysis (Table 1).

The quantitative risk estimation of a neurotoxicant requires four steps: (1) determination of a dose–response model, (2) determination of the distribution of measurements (variability) about the model, (3) determination of an adverse or abnormal level, and (4) estimation of the probability that a measure is beyond the abnormal level as a function of dose (Gaylor and Slikker, 1992). This approach is appropriate for nonquantal or continuous data (e.g., body weight, behavioral measures, or neurotransmitter concentrations). For continuous data a distinct value that separates normal levels from those associated with toxicity is seldom obvious. In the present example with MPTP, low levels of the neurotransmitter dopamine are associated with impaired motor function, but there is not an identifiable value below which brain/motor function is always impaired and above which brain/motor function is always normal. In addition, even compromised neurotransmitter levels may be sufficient for subjects to perform most

TABLE 1 Experimental Data

Total MPTP dose (mg/kg)	Average dopamine concentration (mg/g wet tissue)	Dopamine concentration as a proportion of control
Ricaurte *et al.* (1987)		
Male mice: Striatum		
0	10.4	1.00
10	9.7	0.93
20	7.7	0.74
40	5.2	0.50
80	3.9	0.38
Sonsalla *et al.* (1987)		
Male mice: Neostriatum		
0	11.2	1.00
40	8.1	0.72
60	4.1	0.37
80	1.4	0.12
Slikker *et al.* (1994)		
Monkey: Caudate nucleus		
0	11.2[a]	1.00
2.4	2.0[a]	0.18
3.6	0.8[a]	0.07

[a] Based on geometric means.

behavioral tasks but if they are "challenged," the same neurotransmitter levels may be too low to allow adaptation and normal functioning. Instead of a distinct threshold situation, a continuum exists with the probability of impaired brain/motor function increasing with decreasing levels of dopamine. In the absence of a dopamine level that is always indicative of impaired brain/motor function, it is nevertheless possible to statistically define an abnormal level of dopamine based on the distribution characteristics in a study population. Dopamine concentrations can be measured in the striatal areas of the brain in control (nontreated) animals and a normal range of dopamine concentrations can be established. Concentrations outside this normal range would be considered abnormal. Then, for animals administered a potentially toxic substance like MPTP, the proportion (risk of animals with abnormal levels of dopamine) can be observed.

IV. Abnormal Range

If measurements of a particular neurotransmitter level or other biological end points are available for a large number of untreated control animals, the percentiles of the distribution can be observed and an abnormal range can be selected. For a group of 20 animals, values below the smallest and above the largest measurements can be considered abnormal and this would

estimate the approximate 5th and 95th percentiles of the distribution of values. If the effects are described by a Gaussin (normal) distribution, then it is possible to calculate or look up the value corresponding to any percentile from the estimates of the mean $(\bar{x})$ and standard deviation (s): $\bar{x} - 2.33s$ and $\bar{x} + 2.33s$ estimate the 1st and 99th percentiles for a normal distribution, respectively.

As with many biological end points, neurotransmitter concentrations have only positive values that are relatively large. Such data tend to be described by a log-normal distribution (i.e., the logarithms of the values are normally distributed). Thus, converting the data to logarithms and then calculating the mean and standard deviation is an approach that can be used to estimate percentiles.

For such data (assuming a normal distribution) the mean $(\bar{x})$ and standard deviation (s) provide estimates of the percentiles. Each experimental observation (x) can be standardized to a normal deviate, $Z = (x - \bar{x})/s$. The probability that Z exceeds a certain value (risk) can be calculated or looked up in normal tables (Beyer, 1968). For example, the probability that Z exceeds 1.28, 1.64, and 2.33 is 0.10, 0.05, and 0.01, respectively. In this case, abnormal values for striatal dopamine concentrations have been defined as any value below the 1st percentile of those in the control animals. By definition, then, 1% of the control animals are regarded as abnormal. The mean value for the control is $\bar{x}_0$ and the abnormal level is $(\bar{x}_0 - 2.33s)$.

V. Dose–Response Model

It is usually necessary to estimate the risk for obtaining an abnormal value for doses other than those that were selected for the bioassay (e.g., lower doses). This requires the fitting of a dose–response curve to the experimental data in order to estimate the average response, $\bar{x}$, as a function of dose, $f(d)$. Because the mechanism of action of MPTP is at least partially understood, it is possible to select a particular mathematical form, $f(d)$, and estimate the unknown parameters. Because the biological effect of MPTP is reported to require an enzyme-mediated activation (MAO-B conversion of MPTP to MPP$^+$) and because MPP$^+$ is selectively taken up into dopaminergic cells via an energy-dependent transporter, saturation of the effects MPTP is anticipated. Therefore, a Michaelis–Menten type equation $\bar{x} = (K + V_d)/(K + d)$ appears appropriate. The minimum value, V, and K are estimated from the experimental data. The dose–response models for dopamine concentration versus total dose of MPTP are shown in Table 2.

TABLE 2 Dose–Response Models for Dopamine Concentration versus Total Dose of MPTP

Description	Ricaurte *et al.* (1987) (mouse)	Sonsalla *et al.* (1987) (mouse)	Slikker *et al.* (1994) (monkey)
Model[a]	$\dfrac{1}{1 + 0.020d}$	$\dfrac{1}{1 + 0.025d}$	$\dfrac{1}{1 + 2.3d}$
Average DA concentration of controls	10.4	11.2	11.2[b]
Standard deviation of controls	0.9	1.1	6.4
Abnormal level[c]	8.5	8.9	3.0
Abnormal/ average concentration	0.82	0.79	0.27

[a] DA = $(1 + ad)/(1 + bd)$. The plateau-type dose–response curve, where DA is dopamine concentration expressed as a proportion of the average concentration of control animals, d is total mg/kg of MPTP, and a and b are parameters estimated from bioassay data ($a = 0$ in these cases).

[b] Geometric mean.

[c] Estimate of first percentile of log-normal distribution of DA concentration for control animals.

An appropriate χ^2 goodness of fit test indicated an adequate fit of the model ($P < 0.19$) for the Ricaurte and colleagues (1987) mouse data. A significant ($P < 0.01$) lack of fit was found for the mouse data from Sonsalla *et al.* (1987). The similarity of the results for the two sets of mouse data provides some support for the choice of the saturation model. For the monkey data (Slikker *et al.*, 1990) there were too few dose levels to conduct a formal goodness-of-fit test. However, the estimated values were within 1 SD of the observed values, indicating an adequate fit of the saturation dose–response curve to the monkey data.

VI. Comparison of the Three Risk Assessment Approaches

The same three data sets concerning MPTP exposure were evaluated by the current RfD approach, the BMD approach, and the quantitative, biological model approach. These results are compared in Table 3. The traditional approach generated RfD levels of 0.1 and 0.04 mg/kg of MPTP from the two mouse studies and 0.008 mg/kg from the monkey data set. The safety factors used for the Ricaurte mouse study were $10 \times 10 = 100$ (10 for extrapolation from animal to human and 10 for heterogeneity among humans). The Sonsalla mouse study and the monkey study require an additional factor of 10 because the LOAEL

had to be used instead of the NOAEL. The monkey study used a combined safety factor of 300 ($10 \times 10 \times 3 = 300$) because of the generic assumption that phylogenetic similarity of the species allows for a safety factor of 3 instead of 10 for extrapolation from nonhuman primates to humans.

Using a quantitative model, the benchmark dose was set at the lower 95% confidence limit on the dose estimated to have a 5% risk of abnormal animals (LED_{05}). For the BMD approach, the estimated "safe" dose of MPTP is four to eight times lower than that obtained with the RfD approach at 0.018 and 0.005 mg/kg for the mouse studies and 0.002 mg/kg for the monkey study.

The quantitative approach was then used to provide an estimate of the dose that would result in a risk equal to or less than 1 in 10,000 subjects (i.e., only 1 in 10,000 subjects would suffer abnormally low levels of dopamine). This dose is two times lower than the BMD for the mouse studies and approximately seven times lower than the BMD for the monkey study.

VII. Discussion

This study of the neurotoxicant MPTP provides an example of the quantitative risk assessment process and the opportunity to compare the results obtained using this method to those obtained from the currently used safety factor method with and without a benchmark adjustment. By using three data sets from two different species, the degree of interstudy and interspecies variability may also be assessed. Beginning with the dose–response data (Table 2), it is obvious that the average dopamine concentration in control animals is very consistent across studies (10.4–11.2 mg/g wet tissue). The monkey dopamine values varied considerably and the geometric mean for these values is used. The monkey values probably vary most because they are derived from animals over a large age range that were not heavily inbred as were the mice. The mean abnormal concentrations of dopamine are again very similar in the two mouse studies (about 80% of normal control average). The average abnormal dopamine concentration for the monkey is lower because of the greater variability of the data from control monkeys. Because the experimental monkeys exhibited similar variability as compared to the control monkeys and greater sensitivity as compared to the mice, the overall risk estimation for MPTP in the monkey was more conservative.

The dose–response curve plateau or saturation model was selected because of the selective and saturable processes associated with the expression of the

TABLE 3 Acceptable Total Dose (mg/kg)

TABLE 3 Acceptable Total Dose (mg/kg)

Procedure	Ricaurte *et al.* (1987) (mouse)	Sonsalla *et al.* (1987) (mouse)	Slikker *et al.* (1994) (monkey)
Current: NOAEL/100 or LOAEL/1000[a]	$\frac{10}{100} = 0.100$ mg/kg	$\frac{40}{1000} = 0.040$ mg/kg	$\frac{2.4}{300^{b}} = 0.008$ mg/kg
Benchmark dose[c]	$\frac{1.8}{100} = 0.018$ mg/kg	$\frac{0.5}{100} = 0.005$ mg/kg	$\frac{0.066}{30^{b}} = 0.002$ mg/kg
Quantitative approach: Estimate of dose with risk ≤ 0.0001	0.009 mg/kg	0.002 mg/kg	0.0003 mg/kg

[a] Safety [uncertainty factor of 10 for extrapolation from animal to human, 10 for heterogeneity among humans, and 10 if no observed effect level (NOAEL) not obtained and lowest observed effect (LOAEL) is used].

[b] For primates a safety factor of 3 instead of 10 was used for extrapolation from animals to humans.

[c] Lower 95% confidence limit on the dose estimated to have a 5% risk of abnormal animals (LED$_{05}$).

neurotoxicity of MPTP, i.e., MAO-B enzymatic activation and MPP^{+} energy-dependent transporter uptake. Although statistical significance was not achieved for all data sets, the saturation model appeared to fit the Ricaurte and colleagues (1987) mouse data very well. In general, more than three data points are needed to verify the goodness of fit of data to the saturation model.

The MPTP dose determined acceptable or "safe" by the currently used qualitative risk assessment (RfD) procedure is considerably higher than the dose determined to produce a risk of 1 in 10,000 as defined by the quantitative method. The quantitative method applied to the mouse data resulted in a dose 10–20 times more conservative and the monkey comparison resulted in a dose 25 times more conservative than the currently used method. The benchmark adjustment of the current method resulted in intermediate values. Although several clinical cases of MPTP toxicity have been reported and behavioral, histological, and neurochemical adverse effects noted, no dose–response data exist for known exposures to MPTP in humans, and these cannot be compared to the animal model data (Kopin, 1986).

The relative effect of the overall variability of the experimental control data on the establishment of a benchmark dose for a particular end point was considered. The standard deviation of an end point measurement consists of two components: The true biological variance among animals that is constant for a specific experiment and the experimental measurement error that depends on a laboratory technique. Usually, the latter can be estimated from multiple measurements on the same animal. Even if a typical experimental measurement error (e.g., 5%) was doubled because of a poor experimental technique, and if the component

of variance due to biological variability among animals has a standard deviation of 15%, the benchmark dose would only be decreased by approximately 12% compared to the one based on tighter experimental control data.

For each of the three data sets, the proportion of animals with abnormally low dopamine concentrations was quite high, even at the lowest test dose of MPTP. For the study of Ricaurte and colleagues (1987), at the lowest dose of 10 mg/kg of MPTP, the estimated mean concentration was $1/(1 + 0.2) = 0.83$ of the control level. The abnormal level corresponding to the first percentile of the control animals was calculated to be 0.82. Hence, 46% of the animals were estimated to be abnormal at the NOAEL. From Sonsalla and colleagues (1987), the mean dopamine concentration at the lowest dose was estimated to be 0.5 of the control level. Since the abnormal level was calculated to be 0.79 of the control level, over 99.9% of the animals were estimated to be abnormal at the LOAEL. For the study of Slikker and colleagues (1994), the dopamine mean concentration at the lowest dose was estimated to be 0.15 of the control level. The abnormal level was estimated to be 0.25 of the control level. Here 83% of the animals were estimated to be abnormal at the LOAEL. Hence, in all three cases, the traditional approach of using the NOAEL or LOAEL for the calculation of a reference dose would have been based on MPTP doses resulting in large proportions of animals with abnormally low dopamine concentrations.

Shortcomings of the currently used qualitative or safety-factor risk assessment method include (1) the use of only one data point (NOAEL or LOAEL), (2) the need for the assumption that a threshold of effect exists, (3) the need for an assumption that factors of 10 are appropriate to account for interspecies and intraspecies variability, and (4) the inability to adjust the assessment to achieve a desired risk (e.g., 1 in 10,000). The advantages of a quantitative approach demonstrated here over the currently used RfD risk assessment procedures include the ability to (1) utilize continuous data, (2) utilize the whole of the dose–response curve data, (3) incorporate biological information into the dose–response model, and (4) provide an actual risk of exposure to a given dose.

In general, the described quantitative approach attempts to replace some of the assumptions of the currently used risk assessment process with biological and/or mechanistic information and utilize the dose–response data. Further studies to compare this quantitative approach with current procedures and to validate this novel approach are needed. To some extent this can be accomplished with sensitivity analyses conducted by computer simulations approximating experimental results.

References

Adams, J. D., Klaidman, L. K., and Cadenas, E. (1992). MPP+ redox cycling: a new mechanism involving hydride transfer. *Ann. N.Y. Acad. Sci.* **648**, 239–240.

Barnes, D. G., and Dourson, M. L. (1988). Reference dose (RfD): description and use in health risk assessments. *Regul. Toxicol. Pharmacol.* **8**, 471–486.

Beyer, W. H. (1968). *Handbook of Tables for Probability and Statistics,* 2nd ed., Chemical Rubber Co., Cleveland, Ohio.

Chen, J. J., and Gaylor, D. W. (1992). Dose-response modeling of quantitative response data for risk assessment. *Communications Statistics—Theory Methods* **21**, 2367–2381.

Crump, K. S. (1984). A new method for determining allowable daily intakes. *Fund. Appl. Toxicol.* **4**, 854–871.

Gaylor, D. W. (1992). Incidence of developmental defects at the no observed effect level (NOAEL). *Regul. Toxicol. Pharmacol.* **15**, 151–160.

Gaylor, D. W., and Slikker, W., Jr. (1990). Risk assessments for neurotoxic effects. *Neurotoxicology* **11**, 211–218.

Gaylor, D. W., and Slikker, W., Jr. (1992). Risk assessment for neurotoxicants. In *Neurotoxicology* (H. Tilson and C. Mitchell, Eds.), pp. 331–343, Raven Press, New York.

Johannessen, J. N., Chiueh, C. C., Bacon, J. P., Garrick, N. A., Burns, R. S., Weise, V. K., Kopin, I. J., Parisi, J. E., and Markey, S. P. (1989). Effects of 1-methyl-4-phenyl-1,2,3,6-tetrahydropyridine in the dog: effect of pargyline pretreatment. *Neurochemistry* **53**, 582–589.

Johannessen, J. N. (1991). A model of chronic neurotoxicity: long-term retention of the neurotoxin 1-methyl-4-phenylpyridinium (MPP+) within catecholaminergic neurons. *Neurotoxicology* **12**, 285–302.

Kodell, R. L., and West, R. W. (1993). Upper confidence limits on excess risk for quantitative responses. *Risk Anal.* **13**, 177–182.

Kopin, I. J. (1986). Toxins and Parkinson's disease: MPTP parkinsonism in humans and animals. *Adv. Neurol.* **45**, 137–144.

McCrodden, J. M., Tipton, K. F., and Sullivan, J. P. (1990). The neurotoxicity of MPTP and the relevance to Parkinson's disease. *Pharmacol. Toxicol.* **67**, 8–13.

National Research Council (NRC) (1992). *Environmental Neurotoxicology.* National Academy Press, Washington, D.C.

Office of Technology Assessment (OTA) (1990). Neurotoxicity: identifying and controlling poisons of the nervous system. U.S. Congress Office of Assessment (OTA-BA-436), U.S. Government Printing Office, Washington, D.C.

Reiter, L. W., Tilson, H. A., Dougherty, J., Harry, G. J., Jones, C. J., McMaster, S., Slikker, W., Jr. and Sobotka, T. J. (1993). Draft Report: principles of neurotoxicity risk assessment. *Fed. Reg.* **58**(148), 41556–51599.

Ricaurte, G. A., Irwin, I., Forno, L. S., Delanney, L. E., Langston, E., and Langston, J. W. (1987). Aging and 1-methyl-4-phenyl-1,2,3,6-tetrahydropyridine-induced degeneration of dopaminergic neurons in the substantia nigra. *Brain Res.* **403**, 43–51.

Slikker, W., Jr. and Gaylor, D. W. (1990). Biologically based dose-response model for neurotoxicity risk assessment. *Korean J. Toxicol.* **6**, 204–213.

Sonsalla, P. K., Youngster, S. K., Kindt, M. V., and Heikkila, R. E. (1987). Characteristics of 1-methyl-4-(2'-methylphenyl)-1,2,3,6-tetrahydropyridine-induced neurotoxicity in the mouse. *J. Pharmacol. Exp. Ther.* **242**, 850–857.

Tipton, K. F., and Singer, T. P. (1993). Advances in our understanding of the mechanisms of the neurotoxicity of MPTP and related compounds. *J. Neurochem.* **61**, 1191–1206.

West, W. H., and Rodell, R. L. (1993). Statistical methods of risk assessment for continuous variables. *Communications Statistics—Theory Methods* **22**, 3363–3376.

CHAPTER

52

Quantitative Approaches to Risk Assessment in Neurotoxicology

JOHN R. GLOWA

Behavioral Pharmacology Unit,
Laboratory of Medicinal Chemistry,
National Institute of Diabetes and Digestive and Kidney Diseases,
National Institutes of Health,
Bethesda, Maryland 20892

ROBERT C. MACPHAIL

Neurobehavioral Toxicology Branch,
Neurotoxicology Division,
Health Effects Research Laboratory,
U.S. Environmental Protection Agency,
Research Triangle Park, North Carolina 27711

I. Introduction

Recent advances in technological development have resulted in a high rate of the production of chemicals, with little or no knowledge of the potential health effects or ecological impact of the vast majority of these agents (NRC, 1984). Although the identification of the adverse effects of chemicals has been the subject matter of toxicology, the enterprise of attempting to predict whether those effects will have an impact on humans has become known as risk assessment. The particulars of how risk assessments are accomplished are a major concern for scientists, manufacturers, and regulatory officials alike. Ultraconservative approaches, resulting in the outright ban of an agent that produces any toxic effects at any dose, are unacceptable. All agents produce toxic effects at very high levels, and the rejection of any agent without prudent consideration of the dose levels that produce toxicity could result in the loss of chemicals useful to society. Thus, a fundamental principle in risk assessment is the characterization of dose–effect functions. However, the process of determining dose levels that

are likely to produce toxic effects in humans requires further consideration of these dose-effect data.

The first major consideration in risk assessment involves how one can predict toxic effects in humans. In many instances, experience has shown that we often first learn of the toxic effects of chemicals through accidental exposures to people (NRC, 1992). As a consequence, toxicologists have developed data necessary to predict effects in humans by first studying the effects of chemicals in animals. Animal-based predictions are necessary because it is often either impractical or unethical to test agents in humans, and quantifiable epidemiological data are ordinarily available only after extensive exposures have already occurred. The utility of animal data for predicting risks in humans, although heuristically practical, depends on many factors that are beyond the scope of this chapter. For further discussion of cross-species extrapolation the reader is referred to recent reviews (e.g., Rees and Glowa, 1994). However, it is clear that such comparisons depend on the ability to compare dose–effect data across species, again illustrating the fundamental importance of the dose–effect function in risk assessment.

The second consideration in risk assessment is the extent to which effects at high doses, often the only ones that produce observable effects, can be used to predict effects at the lower end of the dose–effect function. Prediction of the effects of low levels of agents has had a long tradition in toxicology, but not necessarily with much success. Low levels are of interest, of course, primarily because humans are exposed to low levels of almost countless agents. Humans typically avoid exposure to levels of chemicals that are clearly known to produce toxic effects, but it is difficult to know when a chance encounter to levels with barely detectable effects occurred. Little is known of the toxicity of chemicals at these levels, and there will always be concern that they are producing effects that could be measured should a more sensitive instrument be used. Long exposures are also of interest, primarily because humans may live a long time in the presence of low levels before symptoms appear. Several methods have evolved to characterize the chance of obtaining an adverse effect of low doses in humans using data from animals exposed to higher doses, but these have not been fully evaluated.

Another concern in risk assessment is that different types of toxicity may emerge as more and more chemicals are produced. One adverse health effect of toxicant exposure that has received considerable attention is neurotoxicity (NRC, 1992; OTA, 1990; Weiss, 1988). The possibility that neurotoxicants can result in functional impairment has led several investigators to attempt to apply traditional methods of risk assessment to assess behavioral toxicity, with mixed results (McMillan, 1987; Raffaele and Rees, 1990). At the same time, however, several newer approaches to risk assessment for behavioral toxicity have been developed (e.g., Dews, 1986a; Glowa et al., 1986; Gaylor and Slikker, 1990; Bogdan et al., 1995). This chapter assembles data on the neurotoxic effects of a pesticide using a measure of neurobehavioral function in animals and compares some of these methods for their utility to predict risks in humans.

II. Traditional Methods of Risk Assessment

A. Low-Dose Extrapolation

Some of the earliest methods to assess risk were developed to predict the carcinogenic and mutagenic effects of ionizing radiation. The clear and irreversible nature of these effects promoted assessments of the high end of the dose–response function to identify and characterize carcinogens. Increased public concerns over cancer, however, shifted the emphasis from the detection of clear effects to predicting ones expected to occur a long time after exposure or after long-term exposure to very low doses (i.e., to the low end of the curve). Unfortunately, cancers may spontaneously occur without identifiable cause. This makes the detection of a "threshold" dose (one that is just sufficient to increase the incidence of cancer) impossible. The inability to determine experimentally whether low doses of carcinogens actually produced cancer resulted in the notion that either no threshold existed for cancer or that the threshold could not be determined. This meant that any dose, no matter how small, had some probability of producing an effect. Compounding this problem was the issue of how low levels of risk could be predicted from data collected at the upper end of the function. Several families of low-dose extrapolation models, including (1) dichotomous response models; (2) linear, no-threshold models; (3) tolerance distribution models; (4) logistic models; (5) "hit" models; and (6) time-to-occurrence models, were developed to predict the effects of low levels of agents (NAS, 1980). Although almost any function can be made to fit the data well at the high end of the curve, estimates of effects at the low end of the curve varied enormously. For example, Cotruvo (1988) presented risk figures for trichlorethylene (TCE) based on four different models. Depending on the model, the risk estimates for a fixed concentration of TCE in drinking water varied by as much as five orders of magnitude (see also Krewski et al., 1984; Krewski and Van Ryzin, 1981). While it is possible to estimate the extreme low end of the curve, it has also become apparent that realistic attempts will be fruitless because (1) there are unavoidable experimental errors that can greatly influence the results, (2) the choice of the function is often arbitrary, and (3) a tremendous amount of data would be required to even approximate the attempt. For example, one experiment used over 24,000 mice in an attempt to characterize the carcinogenic effects of extremely low doses of 2-AAF (Farmer et al., 1980), only to find different dose–response functions for two types of cancer. These considerations have resulted in a general lack of consensus in the selection of low-dose extrapolation models for assessing cancer risks (Cranmer, 1981).

B. "Safety Factor" Approach

In contrast to low-dose extrapolation, the safety factor approach to risk assessment has relied on the use of an experimentally determined dose that produced no measurable effect, which was then divided by a safety factor(s) to calculate what could be considered an "acceptable" level of exposure to humans. Lehman and Fitzhugh (1954) first advocated a safety factor approach for establishing safe levels of chemical food additives. Lehman and Fitzhugh (1954) recommended the use of a safety factor of 100, which was the product of two factors of 10. Humans were considered to be 10 times more sensitive than laboratory species to toxicants, whereas the second factor of 10 was incorporated to account for individual differences in sensitivity (cf. Lewis, 1990).

Gaddum (1956) described a more quantitative approach to establishing safe dose levels in which a safe dose was calculated to be 6 SD below an easily observable toxic response (e.g., an LD_{50}). Gaddum's approach appears to be the first in which the attempt to establish safe doses was based on a more complete assessment of dose–effect data, instead of attempting to establish a no-effect level as recommended by Lehman and Fitzhugh (1954). Nevertheless, the LD_{50} is a point estimate and it provides no indication of the variability in the effect.

It seems fair to say that the approach of Lehman and Fitzhugh (1954) has provided the backbone for the current approach (the reference dose (RfD) approach) for establishing appropriate levels of toxicant exposures for humans. Although considerably more formalized than the approach taken by Lehman and Fitzhugh (1954), the RfD approach makes a fundamental assumption that a threshold exists for many toxic effects of concern. These toxic effects have generally been referred to as systemic (or noncancer) effects. In contrast to the "single hit" assumptions of cancer risk assessment, organ systems have a reserve or repair capacity that makes it inconceivable that a single molecular event could induce an effect. Since systemic toxicants rarely produce an observable effect at low doses (i.e., there is a threshold), it is possible that low exposures to some toxicants could be tolerated over a lifetime without deleterious effects.

C. Reference Dose Approach

Risk assessments are currently being assessed for systemic toxicity using the RfD approach. This approach first establishes a dose level that is without a statistical effect [i.e., no-observable-adverse-effect level (NOAEL)] or one with a minimal effect [i.e., lowest-observable-adverse-effect level (LOAEL)] in animals, and then divides this number by a series of uncertainty factors to calculate a RfD for humans (see Barnes and Dourson, 1988). A similar approach is used to assess the risks of hazardous air pollutants and is referred to as the reference concentration (RfC) approach. For example, the subchronic effects of a range of doses of a systemic toxicant could be assessed on a relevant end point (e.g., decreases in body weight gain in mice) and the results could be used to estimate a (threshold) dose without detectable effect (NOAEL). This dose would then be divided by a series of uncertainty factors (typically factors of 10) to compensate for less than lifetime exposure, species differences, and variability in the human population. The resulting RfD could be as much as several orders of magnitude smaller than the original threshold estimate (for further discussion see Barnes and Dourson, 1988).

There have been several criticisms of the RfD approach (Crump, 1984; Dews, 1986a,b; Dourson, 1986). (1) Primarily, the NOAEL or LOAEL is indeterminate. Over the portion of the dose–effect function that is nearly (but not completely) at asymptote, the ability to determine a dose-related effect becomes increasingly more difficult. (2) The NOAEL or LOAEL is directly dependent on the choice of doses actually used in an experiment. (3) The dependence of the NOAEL or LOAEL on sample size (1, above), makes it more likely the NOAEL and LOAEL will decrease with increases in "*n*." This, in effect, rewards those seeking less conservative risk figures for using smaller sample sizes. (4) The slope of the dose–response function, the metric relating the rate of change in effect to the rate of change in dose, is not taken into account. Agents producing dose–effect functions with steep slopes may warrant more concern, all other factors being equal. NOAELs or LOAELs from functions with shallow slopes may also be difficult to determine. (5) There is no firm biological justification for many of the uncertainty factors. For some agents, chronic low-level exposure may be less hazardous than acute exposure to higher levels. For some agents, humans may be less sensitive than a laboratory species. Although toxicokinetic and toxicodynamic factors can clearly differ between types of exposure and species, these differences should be measured directly rather than assumed. (6) It is not clear if the procedure can be applied to all types of effects. For example, it is not used for cancer risk assessment, as cancer is assumed to occur without a threshold. Likewise, for end points with considerable background variability, statistically significant effects occur only at relatively high doses. (7) The RfD is not a risk figure but rather a point estimate of an exposure at which risk is considered negligible. Given all these limitations, it is perhaps surprising that some RfDs have been shown to be within an order of magnitude of estimates produced under other established guidelines, such as the threshold limit value (TLV) (McMillan, 1987). Although the RfD approach produces a number that may allow risk managers to make decisions regarding the danger presented by exposure to an agent, for valid estimates of risk it is limited.

III. Quantitative Methods in Risk Assessment

The shortcomings of the RfD approach have led to the development of different approaches to apply dose–effect data to risk assessments. These approaches circumvent many of the problems associated with the RfD and low-dose extrapolation approaches, and provide quantitative estimates of risk. At least two basic approaches have been proposed, based on the different sources of variability in the dose–effect function. Dose tolerance models focus on differences in population sensitivity to an effect, as assessed by exposure to a range of toxicant doses, whereas effect tolerance models focus on the variability in the effect of a fixed dose and/or in comparison with background control variability. At least two variants of each basic approach exist, providing risk assessors with a range of options to the more traditional approaches. The basic features of each are presented next, and an example of the type of results each approach can produce is illustrated using a data set generated for these comparisons.

A. Dose Tolerance Models

The dose tolerance model was originally described by Dews (1980). In more definitive reports, Dews (1986a,b) subsequently described the bases for this approach, including its reliance on methods originally used to assess the replicability of dose–effect functions (see Dews and Berkson, 1954). Data from single experiments are first described by a mathematical function, and the dose resulting in a small but measurable effect is estimated from each function. Through independent replications of the experiment, a distribution of these point estimates is obtained that approximates the normal curve. The variability in these estimates is then used to predict the proportion of the population that could be expected to be affected at lower levels. Figure 1 illustrates this approach with data from four hypothetical animals exposed to a range of doses (ln dose 1–7). As the dose increases, the effect becomes greater in all animals. However, there are small differences. Differences in the dose producing a 10% decrement are described by the distribution at the 90% of control level. By calculating a Z score based on the variability of the sample distribution, an estimate of the dose likely to exhibit this effect in 1/100 individuals ($P = 0.01$) can be determined at the point of contact on the abscissa (arrow). This method has been used to assess the risks associated with acute exposure to several solvents found in the work place (Glowa *et al.*, 1986, 1987, 1992).

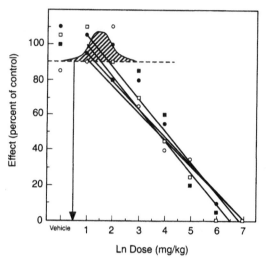

FIGURE 1 Representative dose–effect data illustrating some of the particulars of the risk assessment approach advocated by Dews. Effects (expressed as percentage of control) of a range of doses (ln converted) are assessed for four individual subjects. For each subject, a dose–effect function is calculated using linear regression (Dews advocated restricting the data used to calculate this function to the linear portion of the dose–effect curve, i.e., from 80 to 20%, to avoid ceiling and floor influences). From each curve a small, but measureable, effect (i.e., 10%) is determined, and the distribution of those effects (shaded curve) is used to estimate the dose (arrow) to produce effects in successively smaller proportions (i.e., $P = 0.1 \approx 1/10$, $P = 0.01 \approx 1/100$, $P = 0.001 \approx 1/1000$) of the population.

In those studies the concentration estimated to produce a 10% decrement in neurobehavioral functioning was determined from dose–effect data obtained in experiments using relatively small sample sizes ($n = 10$–12). In general these studies produced risk estimates that were in good agreement to established TLVs, although clear exceptions (e.g., ethyl acetate) were also found. A subsequent report (Glowa, 1991) compared estimates produced by this approach and with those of others (see later), using small ($n = 10$) and large ($n = 40$) sample sizes, to further validate the Dews/Glowa approach.

Another dose tolerance approach is currently being developed (Bogdan *et al.*, 1995) that extends the Dews/Glowa approach. Instead of requiring individual dose–effect data, it was developed to accommodate groups design data since they are more common in toxicological experiments. Dose–effect information is obtained by dosing separate animals with single doses over a range of doses. An iterative computer program randomly assigns each data point at a dose with every possible combination of one data point for each of the other doses to create a family of dose–effect functions. A 10% effect is calculated from each function and the variability of this large set of point estimates is used to determine dose tolerances. The method is illustrated,

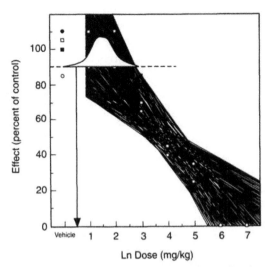

FIGURE 2 Representative dose–effect data illustrating the use of the iterative line-fitting program. Effects of a range of doses (ln converted) are assessed using four animals and seven doses plus vehicle. The effect in each animal of each dose is expressed as the percentage of vehicle. The program combines unique combinations of the effect at each dose with effects at the other doses to produce a set of all possible combinations of points. From each curve a small, but measureable, effect (i.e., 10%) is determined, and the distribution of those effects (shaded curve) is used to estimate the dose (arrow) to produce effects in successively smaller proportions (see Fig. 1) of the population.

again with modeled data, in Fig. 2. Here seven doses were assessed using four animals at each dose, generating 16,384 functions and corresponding point estimates (the figure does not show all the lines, for the purpose of illustration). The distribution of these point estimates is treated in a manner identical to the Dews/Glowa approach, producing estimates of the dose that would affect 1/10, 1/100, or 1/1000 individuals. As with the Dews/Glowa approach, sensitive outliers can significantly impact the risk estimates.

B. Effect Tolerance Models

Effect tolerance models are modifications of standard statistical or mathematical approachs to predict risks, which are based on variability in the effect of a fixed dose. Despite precedents (e.g., Imamura and Ikeda, 1973), the first attempt to systematically apply this method to risk assessment was described by Crump (1984). This approach uses maximum likelihood estimates (MLEs) to generate confidence intervals about a dose–effect function. In a program-driven (commercially available) form, a choice of various multistage polynomial functions can be used to fit the dose–effect data. A threshold parameter can be included or not, providing a number of options regarding the nature of the assumed underlying distribution. As

with the Dews/Glowa approach, a small but measurable effect (e.g., a 10% decrement) is typically chosen as a biologically relevant measure, and the intersection of the lower-bound confidence limit with this effect level establishes a "benchmark" dose. This method is illustrated in Fig. 3, where the intersection of the lower confidence interval (since we are interested in the lowest dose to possibly produce an effect) with the predetermined level of adverse effect (10%) defines the benchmark dose (arrow). Since calculation of the dose representing a 10% effect remains within the experimentally determined range of data, many functions can be used to estimate the point. This suggests that the levels of risk determined from this point will not be highly dependent on the type of function (i.e., linear, logistic, etc.) chosen. Crump (1984) also advocated the additional use of uncertainty factors for animal-to-human extrapolation.

Another effect tolerance approach has been described by Gaylor and Slikker (1990). This approach is also based on variability in the effect domain, but emphasizes the use of control variability to establish a boundary line for a critical, or "abnormal" effect. For example, in the original description, effects that exceeded 3 SD of the mean of the control values were considered abnormal. Various estimates of the probability of an effect at doses less than those producing this effect, or in ones to which a human population

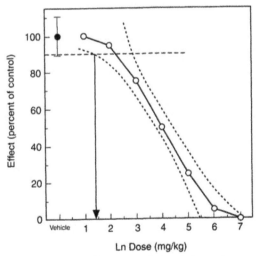

FIGURE 3 Representative dose–effect data illustrating particulars of the risk assessment approach advocated by Crump. Effects of a range of doses are assessed on an effect (expressed here as a percentage of control, but not in the text) in a traditional groups-design manner. The data are then used to fit a function (several options) and the confidence intervals about that function are then determined. From the function, a small but theoretically measurable effect (i.e., 10%, the "benchmark") is determined, and the intersect of the lower bound of the confidence interval is used to estimate the benchmark dose.

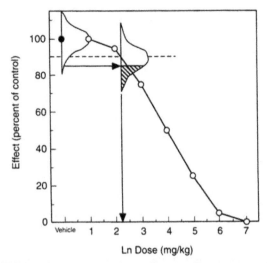

FIGURE 4 Representative dose–effect data illustrating particulars of the risk assessment approach advocated by Gaylor and Slikker (1990). Effects of a range of doses are assessed on some end point. A dose–effect function is calculated using a best-fitting curve. Particular attention is paid to the variability in control (vehicle) data, and a level of adverse effect (originally suggested to be 3 SD, horizontal arrow) is specified. A dose of interest is determined (here we have chosen the dose that produces a 10% effect), and the proportion of the population exhibiting an adverse effect at that dose is determined.

may be exposed, are gleaned using the distribution of the control data as a surrogate. Applying these estimates to data from exposed animals leads to conclusions about the proportion of the population affected at the lower doses or allows the calculation of doses that would affect a specified proportion of the population. This approach is illustrated in Fig. 4, again with the same generic dose–effect data used for the other approaches. An abnormal level of effect (3 SD) can be directly determined from the variability in the vehicle-treated control group (horizontal arrow). Taking the 10% effect level on the dose–effect function as an arbitrary point of interest, the figure suggests that if the same variability existed for treated animals at that dose as for control animals, some 40% (hatched area on the effect distribution) of the sample would be abnormally affected. One could also conceivably find the point on the curve where a small increase in risk would be expected (e.g., 1%) by sliding the effect distribution to the left toward the control distribution until the overlap between the area of the two distributions was 99%.

C. An Example with Carbaryl

Each of these quantitative approaches to risk assessment just described starts with a measurable effect on the dose–effect function, thereby avoiding extrapolation to the extreme low end of the curve. Each also

makes various assumptions about distributions in the different dimensions of the dose–effect function, although these are rarely tested (see Glowa, 1991). Most methods can also provide a true risk figure (i.e., proportion of the population expected to be affected at particular doses), and to varying degrees incorporate most of the data used to define the dose–effect relationship. Comparisons of how each approach behaves over a range of conditions would greatly aid in evaluating their relative merits, as well as their applicability for a variety of different end points (see also Pease *et al.*, 1991).

The goal of this chapter is to assess the risks of behavioral impairment following exposure to a selected neurotoxicant. In order to accomplish this, the effects of the carbamate insecticide carbaryl were assessed on motor activity. This agent was selected because it produces clear neurotoxic effects (Carpenter *et al.*, 1961). Motor activity was selected because of its widespread use in the assessment of neurotoxicant effects (e.g., Reiter and MacPhail, 1979; Crofton *et al.*, 1991; Crofton and MacPhail, 1994). Data were obtained for acute exposures and then used to estimate risks by each method in order that they could be compared directly.

Adult male Long–Evans rats (Charles River, Raleigh, NC) weighing 200–300 g were used in these experiments. Rats were placed in a motor activity monitor (see Crofton *et al.*, 1991) for 30-min daily sessions conducted 5 days/week. The general design of these experiments was a modified groups design. Sessions occurred for 1 week prior to dosing in order to establish a baseline of activity. Animals were then divided into four groups of nine that received either vehicle (5% Emulphor, 5% ethanol in saline) or one of three dosages of carbaryl (3, 10, or 30 mg/kg) given i.p. in a volume of 1 ml/kg body weight, 20 min before a session. The order of dosing was counterbalanced so that each rat received all treatments (one/week) over 4 weeks of testing. This design allowed full use of the current methods in a variety of different ways. For example, since traditional acute toxicological studies involve single exposures in a groups design format, the initial exposures of these animals (each group was initially exposed to a different dose) allowed comparisons of the effects of initial exposures to carbaryl with the effects of repeated intermittent exposures to the same doses. Since an earlier comparison of risk approaches indicated that some of these approaches may be sensitive to sample size (Glowa, 1991), this design also allowed comparison of results based on a small (initial exposure, *n* = 9) and a large (repeated intermittent exposure, *n* = 36) sample size. Finally, since the Dews/Glowa approach was designed to assess individual differences, results from the 36 independent replications were compared to those derived from the mean data from each of the four replications of the dose–effect function using different dosing orders (see Fig. 5).

Table 1 shows the raw data in terms of absolute counts. In the presence of vehicle, grand mean activity counts were just under 3500/session, with a SD of a little over 600. This yielded a coefficient of variation of approximately 18%. On the whole, carbaryl decreased activity over the full range of doses tested, with variability increasing slightly at the low and intermediate doses, and decreasing at the highest dose. These data are interesting because they challenge the assumption that variability in the dose data is similar to that in the control data, especially when slight increases in behavior above control (note some individuals had increased activity at 3 mg/kg) can add to low-to-intermediate dose variability and floor effects can decrease variability at high doses.

Although the data were collected as the absolute number of photocell counts/session, two transformations were used as illustrated in Figs. 1–4. First, the effects were transformed to a percentage of the vehicle-control value. Second, doses were transformed to natural logarithms (ln). Gaddum (1945) had previously advocated the use of log-normal distributions in transforming doses. Figure 5 illustrates these transformations and clearly shows that the effect of the lowest dose (3 mg/kg) would not be considered significant by conventional means. A linear function would be sufficient to describe the mean dose–effect data ($y = -37.77x + 133.354$) as it exhibited a reasonable goodness of fit ($r^2 = 0.736$). Based on these types of data alone, one could speculate that the NOAEL for carbaryl would be 3 mg/kg.

TABLE 1

| | | | Dose | | |
Cohort	N	Vehicle	3	10	30
Group 1					
Animal No. 1	—	4481	4101	2251	339
5		3803	2638	471	449
9		3219	3203	496	533
13		2771	3290	481	289
17		1921	2749	727	611
21		2571	2727	1112	504
25		3719	3710	588	311
29		2995	3791	2118	436
33		2206	957	1337	381
Mean	9	3076	3018	1065	428
SD		820	929	703	108
Group 2					
Animal No. 2		4166	3182	769	336
6		3085	2136	751	173
10		3861	3662	1328	389
14		4305	3951	1021	252
18		3021	3134	1193	359
22		3594	2810	933	869
26		3208	3432	1150	367
30		3594	2882	2518	229
34		4289	3646	4763	395
Mean	9	3680	3204	1603	374
SD		507	550	1298	201
Group 3					
Animal No. 3		3372	3915	1027	474
7		3577	3540	1650	148
11		3255	3333	455	98
15		3828	3809	362	440
19		3472	2336	211	271
23		3763	3242	921	416
27		3378	2888	392	219
31		3712	2890	546	349
35		2770	3190	2765	330
Mean	9	3459	3238	925	305
SD		323	493	821	131
Group 4					
Animal No. 4		4530	4162	1312	298
8		3500	4009	2160	90
12		3350	3579	863	206
16		4141	4391	988	634
20		3975	4894	262	739
24		2508	3699	507	477
28		3392	4156	753	236
32		3926	2623	833	880
36		3226	3773	893	434
Mean	9	3616	3921	952	444
SD		597	629	540	265
Grand mean	36	3458	3345	1136	388
SD		612	730	892	187

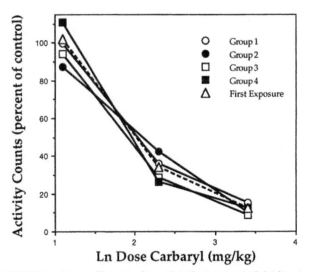

FIGURE 5 Dose–effect data for carbaryl on motor activity in rats. The mean effects of a range of doses (3–30 mg/kg, ln transformed) are assessed on a neurobehavioral end point (activity counts/session, expressed as a percentage of control) for four groups of nine rats. Each rat received vehicle and each dose in a counterbalanced manner. The data are plotted in two ways. Data from Groups 1–4 plot the effects of exposure to each dose, regardless of the order of exposure (circles and squares, within-subject design). The triangles represent the effects of the first exposure to carbaryl for each group (between-subjects design).

D. Dose Tolerance Approaches

When the effects of carbaryl were assessed by the method of Dews, a linear dose–effect function was fitted relating percentage of vehicle control to ln dose for each animal. These data could also be reasonably described by a linear function, in that the mean r^2 across animals was 0.841. For each function, a 90% of vehicle control level was determined, and the mean and standard deviation of these point estimates was determined. Mean parameters for the function were:

	a	b	90% (ln)	90% (mg/kg)
Mean	133.3	−37.84	0.969	2.63
SD			0.802	

Using "t" distribution tables, estimates of the dose of carbaryl producing a 10% effect in successively smaller proportions of the sample population were determined for $P = 0.1$ (1 out of 10), $P = 0.01$, and $P = 0.001$. The results were:

Probability	0.1	0.01	0.001
ln dose	−0.079	−0.908	−1.715
mg/kg dose	0.92	0.40	0.18

For purposes of comparison with the just-mentioned data ($n = 36$), the Dews/Glowa approach was applied to the same data in the modified groups-design format ($n = 4$). The mean of the vehicle effect was compared to the mean effect at each dose in each of the four groups. Thus, four mean dose–effect functions (% effect × ln dose) were obtained. The mean curve for calculating the data was well fit by a linear function ($r^2 = 0.840$) and similar to that of the larger sample (listed above).

	a	b	90% (ln)	90% (mg/kg)
Mean	131.75	−37.39	1.107	3.03
SD			0.146	

Using "t" distribution tables, estimates of the dose of carbaryl producing a 10% effect in successively smaller proportions of the sample population were determined for $P = 0.1$ (1 out of 10), $P = 0.01$, and $P = 0.001$. The results were:

Probability	0.1	0.01	0.001
ln dose	0.868	0.444	−0.385
mg/kg dose	2.38	1.56	0.68

The method of Dews was sensitive to the manipulation of collapsing individual subject data into group data, producing estimates from the groups-design data that were approximately three times larger than those of single-subject data (with a clear trend toward larger ratios being obtained for smaller, i.e., 1/1000, risk estimates). This is partly due to the nature of the change

in the "t" statistic value as "n" decreases. Whereas the variability of the 10% effect decreased by collapsing the individuals into four groups, the "t" statistic increased.

When the data were assessed using the iterative line-fitting program, four data points (arbitrarily chosen as the last animal in each group) had to be deleted because the program was originally designed to accept up to 32 animals per dose. Under these conditions the program generated all 32,768 of the possible combinations of a 32×3 array of points. Risk estimates were calculated based on a 10% decrease in motor activity. The mean dose to produce this effect was 2.75 mg/kg (ln = 1.012, ln SD = 0.532). The doses expected to produce this effect in successively smaller proportions of the population (i.e., 1/10 or $P = 0.1$, 1/100 or $P = 0.01$, and 1/1000 or $P = 0.001$) were:

Probability	0.1	0.01	0.001
mg/kg dose	1.40	0.79	0.51

When the data points from the smaller, initial exposure sample ($n = 9 \times 3$) were used, the program generated 729 functions, from which risk estimates were obtained. The doses estimated to produce a 10% effect in successively smaller proportions of the population were:

Probability	0.1	0.01	0.001
mg/kg dose	1.16	0.67	0.42

Recognizing that these were not identical data sets, it is tempting to conclude that the line-fitting program appeared relatively insensitive to manipulating the number of the subjects tested at each dose. Estimates from the small sample were only slightly lower than those of the large sample (with a just perceivable trend toward a larger ratio being obtained for lower, i.e., 1/1000, risk estimates). At these numbers of point estimates, there was considerably less influence by differences in the numbers derived from distribution tables compared to the Dews/Glowa approach.

E. Effect Tolerance Approaches

When the effects of carbaryl on motor activity were assessed by the method of Crump (1984), the THC program (for continuous data, Clement International Corp., Ruston, LA) was applied to the entire data set ($n = 36 \times 4$) and to the smaller, first-exposure data set ($n = 9 \times 4$). Raw data for vehicle effects and the effect of each dose were used to fit a single degree polynomial mean response model with a threshold. These data were used to calculate the benchmark dose at a 10% effect level, applying a 95% confidence interval (CI) derived from the MLE, as opposed to a least-squares approach. The parameters of the function for the entire data set were such that the MLE dose was 3.38 mg/kg for the large data set and 6.14 mg/kg for the smaller, first-exposure data set. The lower-bound confidence interval on the MLE for each data set was:

CI	Benchmark dose	Benchmark dose
95	($n = 36$)	($n = 9$)
	3.32 mg/kg	3.32 mg/kg

An interesting feature of the Crump procedure was that regardless of the difference in MLEs derived from the large and small data sets, the risk estimates were identical. No attempt was made to transform the dose parameter because the program requires a zero dose (control) and accomplishes transformations within the program. Nor was the use of percent control data attempted because the program provides relative response $\{[F(d) - F(0)]/F(0)\}$.

When the method of Gaylor and Slikker (1990) was applied, the effect of interest was arbitrarily set at a 10% decrease to allow comparison with the other approaches. Using the linear function initially describing the data, it was determined that a 10% decrement would occur at approximately 1.17 mg/kg. Imposing 18% variability (the SD from vehicle data) at this dose would result in a 17.4% downward shift in the distribution to align it with the 90% of vehicle mean. Using the criteria established by Gaylor and Slikker (1990) for an abnormal effect, the increase in risk at the 90% of vehicle effect level would be determined in the following manner. Under control conditions, a small percentage (0.13%) of the subjects would be expected to display abnormal levels of motor activity (3 SD below the vehicle-control mean). The corresponding value based on displacement of the control distribution to the 90% of vehicle level would be 0.54%. This would represent approximately a fourfold increase in the risk of having abnormally low activity after a dose of 1.17 mg/kg carbaryl.

IV. Conclusions

Quantitative models represent an exciting new trend in risk assessment for systemic toxicants. Unlike the traditional RfD approach, quantitative models generally describe dose–effect functions mathematically and specifically incorporate experimental variability into risk estimates. Perhaps the two most important advantages over the RfD approach are a shift in emphasis away from point estimates (NOAELs or LOAELs) toward using the entire data set in the context of the experimental design. As a consequence, with most of the approaches a premium is placed on well-controlled

experimental studies. Four approaches have been identified that can be collectively referred to as quantitative dose–effect approaches. Although there are many similarities, these approaches can be divided into those that focus on variability in an effect at a fixed exposure level (effect tolerance models) and those that focus on variability in exposure level producing a fixed effect (dose tolerance models). The effect tolerance models of Crump (1984) and Gaylor and Slikker (1990) can be differentiated by the source of variability that serves as the basis for risk estimates. The Crump (1984) approach characterizes variability in the dose–effect function to estimate a dose producing a small but measurable effect, whereas the Gaylor and Slikker (1990) approach uses the variability obtained under control conditions to define an extreme effect (i.e., greater than 3 SD removed from the control mean) and its associated dose. The more compact the variability about the dose–effect curve, the larger will be the resulting benchmark dose according to the Crump approach. On the other hand, the more compact the variability in control values, the *smaller* will be the dose estimated to produce an abnormal effect by the Gaylor–Slikker approach. As such, it appears that the Gaylor–Slikker approach differs from the Crump approach in penalizing instead of rewarding tight experimental control, in much the same way as does the RfD approach.

The dose tolerance model of Dews and Glowa has many features in common with the effect tolerance model of Crump (1984). The Dews/Glowa model mathematically describes a dose–effect function to estimate a dose of a toxicant that produces a small but measurable adverse effect (e.g., a 10% decrement in behavior). The main difference between the Dews/Glowa and the Crump model, however, is that the former focuses on variability in dose estimates to predict levels of exposure producing an adverse effect in successively smaller proportions of the population (e.g., a risk of 1 in 10, 1 in 100, 1 in 1000, etc.). Variability in experimental data using the Dews/Glowa approach is likely to affect dose estimates in much the same way as it does using the Crump model. That is, the more variability in the data, the lower will be the exposure levels estimated to produce a small measurable adverse effect.

The Dews/Glowa approach uses variability in the dose–effect functions between individual animals to arrive at risk estimates. This baseline (or within-subjects) design may often be impractical or impossible to apply in toxicology. For example, exposure may be perinatal, occur over long periods of time (months, years), or produce irreversible effects. Under these circumstances a groups (or between-subjects) design is appropriate. A modification of the Dews/Glowa approach, based on randomization analysis (Bogdan *et al.*, 1994), was applied to the first dose–effect determination for carbaryl in which separate groups of rats received different doses. When compared to the dose estimates calculated according to the original Dews/Glowa approach, the results showed a relatively good correspondence. For example, using the randomization approach the dose estimated to produce a 10% decrement in 1 out of 10 was only 26% higher than the estimate based on the Dews/Glowa approach (1.16 vs 0.92 mg/kg, respectively). The dose estimates calculated according to the two approaches systematically decreased with decreasing risk level, although the rate of decrease was greater for the Dews/Glowa (approximately fivefold) than for the randomization (approximately threefold) approach. Figure 6 presents the data obtained from these dose tolerance approaches in a somewhat different way. The dose estimated to produce an effect in 1/10 individuals is inversely related to coefficient of variation. This again illustrates the point that well-done experiments (those with little experimental error) reward the experimenter with higher dose estimates. The choice of an appropriate dose tolerance model must be made on the basis of the feasibility of establishing dose–effect functions in individual subjects, as well as the economics of between-subjects designs over within-subject designs, and the resulting greater volume of chemicals that could be tested using a between-subjects design.

It is beyond the scope of this chapter to consider how quantitative approaches could be incorporated into, or replace, current RfD/RfC calculations. It is, however, instructive to consider the nature of risk estimates derived from the various quantitative approaches. The dose tolerance models of Dews/Glowa and Bogdan *et al.* (1994) make risk estimates in the form of levels of exposure associated with a measurable effect (e.g., 10% change) in successively smaller proportions of a population. Dose tolerance approaches can also provide quantitative estimates of the range of sensitivity to a toxicant in a population. For example, the ratio of the highest to lowest dose (mean ± 3 SD) of carbaryl expected to produce a 10% decrement according to the Dews/Glowa approach was calculated to be 123. Of the effect tolerance approaches, the Gaylor/Slikker approach can also be used to estimate quantitatively the risk of an adverse effect at a particular exposure level. The approaches of Dews/Glowa and Gaylor/Slikker could both be used further to create surfaces for quantitatively predicting risk under varying exposure conditions. The main difference between the two approaches lies in the nature of the variability in the data used to predict risks. The Dews/Glowa approach directly establishes the variability in sensitivity to doses producing a fixed effect, whereas the Gaylor/Slikker approach uses the variability of control values as a surrogate for the variability in sensitivity to a fixed dose. The Crump approach, on the other hand, calculates a point estimate of exposure based on effect variability in the dose–effect function. It is unclear at this time how the benchmark dose estimate of Crump could be used to quantitatively describe the risk of an adverse effect at either fixed or varying exposure levels.

Acknowledgments

We gratefully appreciate the comments of Drs. P. B. Dews, J. Vandenberg, and R. Wyzga on an earlier version of this manuscript and thank Clement International Corporation (KS Crump Division) for providing data analysis for the Crump approach.

References

Barnes, D. G., and Dourson, M. (1988). Reference dose (RfD): description and use in health risk assessments. *Reg. Toxicol. Pharmacol.* **8**, 471–486.

Bogdan, M. A., MacPhail, R. C., and Glowa, J. R. (1995). A randomization test-based method of risk assessment for neurotoxicology, in preparation.

Carpenter, C. P., Weil, C. W., Polin, P. E., et al., (1961). Mammalian toxicity of 1-napthyl N-methylcarbamate (Sevin insecticide) *J. Agric. Food Chem.* **9**, 30–39.

Cotruvo, J. A. (1988). Drinking water standards and risk assessment. *Reg. Toxicol. Pharmacol.* **8**, 288–299.

Carnmer, M. F. (1981). Extrapolation from long term low dose animal studies. In *Measurement of Risks* (C. G. Berg and H. D. Maille, Eds.), pp. 415–441, Plenum Press, New York.

Crofton, K. M., and MacPhail, R. C. (1994). Reliability of motor activity assessments. In *Motor Activity and Movement Disorders*

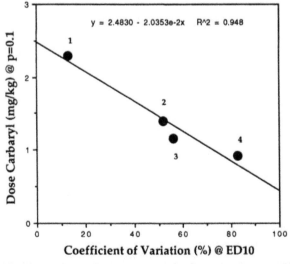

FIGURE 6   Changes in the estimate of the dose of carbaryl to affect 1/10 individuals using each of the approaches as a function of the coefficient of variation of the ED_{10} used in that approach (function is least-squares fit). Methods used were (1) Dews/Glowa ($n=4$); (2) Bogdan *et al.* ($n=32\times3$); (3) Bogdan *et al.* ($n=9\times3$); and (4) Dews/Glowa ($n=36$).

(K. P. Ossenkopp, M. Kavaliers, and P. R. Sanberg, Eds.), Elsevier Publishers, Amsterdam, in press.

Crofton, K. M., Howard, J. L., Moser, V. C., Gill, M. W., Reiter, L. W., Tilson, H. A., and MacPhail, R. C. (1991). Interlaboratory comparison of motor activity experiments: implications for neurotoxicological assessments. *Neurotoxicol. Teratol.* **13**, 599–609.

Crump, K. S. (1984). A new method for determining allowable daily intakes. *Fundam. Appl. Toxicol.* **4**, 854–871.

Crump, K. S., Allen, B. C., and Faustman, E. M. (1992). *The Use of the Benchmark Dose Approach in Health Risk Assessment.* Final Report prepared for USEPA Risk Assessment Forum.

Dews, P. B. (1980). Estimation of low risks. *The Pharmacologist* **22**, 159.

Dews, P. B. (1986a). On the assessment of risk. In *Developmental Behavioral Pharmacology* (N. Krasnegor, J. Gray, and T. Thompson, Eds.), pp. 53–65, Lawrence Erlbaum Associates, Hillsdale, New Jersey.

Dews, P. B. (1986b). Some general problems in behavioral toxicology. In *Neurobehavioral Toxicology* (Z. Annau, Ed.), pp. 424–434, The Johns Hopkins University Press, Baltimore, Maryland.

Dews, P. B., and Berkson, J. J. (1954). On the error of bio-assay with quantal response. In *Statistics and Mathematics in Biology* (O. Kempthorne, T. A. Bancroft, J. W. Gowen, and J. L. Lush, Eds.), pp. 361–370, Iowa State College Press, Ames, Iowa.

Dourson, M. L. (1986). New approaches in the derivation of acceptable daily intake (ADI). *Comments Toxicol.* **1**, 35–48.

Farmer, J. H., Kodell, R. L., Greenman, D. L. and Shaw, G. W. (1980). Dose and time response models for the incidence of bladder and lever neoplasms in mice fed 2-acetylaminofluorene continuously. *J. Environ. Pathol. Toxicol.* **3**, 55–68.

Gaddum, J. H. (1945). Log-normal distributions. *Nature* **156**, 463–466.

Gaddum, J. H. (1956). The estimation of the safe dose. *Br. J. Pharmacol.* **11**, 156–160.

Gaylor, D. W., and Slikker, W. (1990). Risk assessment for neurotoxic effects. *Neurotoxicology* **11**, 211–218.

Glowa, J. R. (1984). Behavioral toxicity of n-octane. In *Advances in Modern Environmental Toxicology*, Vol. 6, (M. A. Mehlman, Ed.), pp. 245–253, Princeton Scientific Publishers, Princeton, New Jersey.

Glowa, J. R., De Weese, J., Natale, M. E., Holland, J. J., and Dews, P. B. (1986). Behavioral toxicology of volatile organic solvents. I. Methods: acute effects of toluene. *J. Environ. Pathol. Toxicol. Oncol.* **6**, 153–168.

Glowa, J. R., and Dews, P. B. (1987). Behavioral toxicology of volatile organic solvents. IV. Comparisons of the behavioral effects of acetone, methyl ethyl ketone, ethyl acetate, carbon disulfide, and toluene the responding of mice. *J. Am. Coll. Toxicol.* **6**, 461–469.

Glowa, J. R. (1991). Dose-effect approaches to risk assessment. *Neurosci. Biobehav. Rev.* **15**, 153–158.

Glowa, J. R. (1992). Behavioral toxicology of volatile organic solvents. V. Comparisons of the behavioral and neuroendocrine effects among n-alkanes. *J. Am. Coll. Toxicol.* **10**, 639–646.

Imamura, T., and Ikeda, M. (1973). Lower fiducial limit of urinary metabolite level as an index of excessive exposure to industrial chemicals. *Br. J. Ind. Med.* **30**, 289–292.

Lehman, A. J., and Fitzhugh, O. G. (1954). 100-Fold margin of safety. *U.S. Quart. Bull.* **18**, 33–35.

Lewis, S. C., Lynch, J. R., and Nikiforov, A. I. (1990). A new approach to deriving community exposure guidelines from "no-observed-adverse-effect levels." *Reg. Toxicol. Pharmacol.* **11**, 314–330.

Krewski, D., Brown, C., and Murdoch, D. (1984). Determining "safe" levels of exposure: safety factors or mathematical models? *Fundam. Appl. Toxicol.* **4**, S383–S394.

Krewski, D., and Van Ryzin, J. (1981). Dose response models for quantal response toxicity data. In *Statistical and Related Topics* (M. Csorgo, D. Dawson, N. K. Rao, and A. K. Saleh, Eds.), North Holland, Amsterdam.

McMillan, D. E. (1987). Risk assessment for neurobehavioral toxicity. *Environ. Health Perspect.* **76**, 155–161.

National Academy of Science, Subcommittee on Risk Assessment of the Safe Drinking Water Committee. (1980). Problems of risk estimation. In *Drinking Water and Health*, Vol. 3, pp. 25–65, National Academy Press, Washington, D.C.

National Research Council. (1984). *Toxicity Testing: Strategies to Determine Needs and Priorities.* National Academy Press, Washington, D.C.

National Research Council. (1992). *Environmental Neurotoxicology.* National Academy Press, Washington, D.C.

Office of Technology Assessment. (1990). *Neurotoxicity: Identifying and Controlling Poisons of the Nervous System.* U.S. Government Printing Office, Washington, D.C.

Pease, W., Vandenberg, J., and Hooper, K. (1991). Comparing alternative approaches to establishing regulatory levels for reproductive toxicants: DBCP as a case study. *Environ. Health Perspect.* **91**, 141–155.

Raffaele, K. C., and Rees, D. C. (1990). Neurotoxicology dose/response assessment for several cholinesterase inhibitors: use of uncertainty factors. *Neurotoxicology* **11**, 237–256.

Rees, D. C., and Glowa, J. R. (1994). Extrapolations to humans for neurotoxicants. In *The Vulnerable Brain and Environmental Risks*, Vol. 3, *Toxins in Air and Water* (R. Isaacson and K. Jensen, Eds.), pp. 207–230, Plenum Press, New York.

Reiter, L. W., and MacPhail, R. C. (1979). Motor activity: a survey of methods with potential use in toxicity testing. *Neurobehav. Toxicol.* **1**(Suppl. 1), 53–66.

Weiss, B. (1988). Neurobehavioral toxicity as a basis for risk assessment. *Trends Pharmacol. Sci.* **91**, 59–62.

53

Use of Biological Markers in the Quantitative Assessment of Neurotoxic Risk

DALE HATTIS
Center for Toxicology, Environment, and Development
Hazard Assessment Group
Clark University
Worcester, Massachusetts 01610

KEVIN M. CROFTON
Division of Neurotoxicology
Health Effects Research Laboratory
U.S. Environmental Protection Agency
Research Triangle Park, North Carolina 27711

I. Concepts, Definitions, and Philosophy of Science Issues

"Biomarkers" (biological markers) are broadly defined as "indicators of events in biological systems or samples" (National Research Council, 1989a,b). Conventionally, the "events" that are "indicated" are conceived of as exposure, effects, or susceptibility to the effects of possible future exposure to biologically active substances, although it is recognized that there is a continuous gradation of events that occur between the uptake of a toxic substance into the body and the ultimate manifestation of impairments to health.

A. Potential Benefits of Biomarkers

There are four basic reasons why it is desirable to use biomarkers to open up the "black box" between exposure and effect:

1. It can lead to a more complete scientific understanding, incorporating more relevant information about causal mechanisms, than a simple input–output analysis based on external dose and the incidence of end effects.
2. It offers the eventual prospect of a better mechanism-based projection of risk beyond the range of possible direct observations in terms of dose rate, duration, and species.
3. It offers the possibility of greater sensitivity of detection and quantification of adverse effects in some cases—going from the whole organism to the cellular or subcellular level for the units in which damage is quantified and analyzed.
4. In cases where effects are detected in the form of long-term changes in higher nervous system functions that may be subject to a number of psychosocial confounding influences which are difficult to control for some exposed groups, physical or physiological biomarkers offer the prospect of confirming exposure-related effects and narrowing the search for causative agents.

Some more specific potential applications of biomarkers to accomplish these purposes are illustrated in the following sections.

789

B. Conceptual Hang-Ups and Practical Difficulties

As the word "indicators" in the biomarker definition implies, biomarkers tend to be windows on systems that are not completely understood and not directly amenable to complete and definitive observation [e.g., the precise relationship between hemoglobin adducts for acrylamide and acrylamide-induced neurotoxic damage is not at present known (Bailey *et al.*, 1986; Bergmark *et al.*, 1991)]. As such they tend to be imperfectly satisfying to both experimental scientists and risk analysts. Realizing the potential of biomarkers for both research and risk assessment requires overcoming some basic philosophical assumptions that are common in the scientific disciplines that must contribute the tools for both measuring biomarkers and studying health effects in human populations.

One difficulty for the basic experimental scientist is that full development of biomarkers of exposure and early effects generally requires the development of dynamic models:

1. of the appearance and disappearance/repair of the marker in relation to exposure or toxicant uptake, and
2. of the production of ultimate effects of concern in relation to the presence or quantitative level of the marker.

Some experimental scientists are uncomfortable with these types of dynamic model building exercises and have a Baconian philosophical reluctance (Kuhn, 1977) to incorporate modeling into their research program.[1]

---

[1] Near the beginning of the experimental scientific tradition in the 17th century, Francis Bacon fought against the classical and scholastic traditions of the middle ages, in which scientific truth was primarily sought by the quasideductive derivation of often elaborate "systems" and extensive rhetorical argument among the proponents of competing systems. As part of this, he tended to reject all kinds of theory building, particularly where the theorizing was in mathematical form. According to Kuhn (1977)

"Bacon himself was distrustful, not only of mathematics, but of the entire quasideductive structure of classical science. Those critics who ridicule him for failing to recognize the best science of his day have missed the point. He did not reject Copernicanism because he preferred the Ptolemic system. Rather, he rejected both because he thought that no system so complex, abstract, and mathematical could contribute to either the understanding or the control of nature. His followers in the experimental tradition, though they accepted Copernican cosmology, seldom even attempted to acquire the mathematical skill and sophistication required to understand or pursue the classical sciences."

The good Baconian carefully records and publishes his or her observations, designs experimental manipulations for "twisting the Lion's tail"—exhibiting nature under conditions it could never attain without forceful human intervention—but generally reserves judgment about propositions that cannot be subjected to relatively direct scrutiny.

As is illustrated in the acrylamide case study described next, however, some limited dynamic model building has the potential to shed new light on prior observations, provide clues for identifying key causal processes for toxicity, and pose new kinds of interesting questions for experimental research. Moreover, microcomputer-based modeling tools (e.g., Stella, by High Performance Systems for dynamic modeling; Crystal Ball by Decisioneering; or @Risk by Palisades Software for Monte Carlo simulation) have the promise to make the analysis of dynamic data much more manageable than has previously been the case, even for those without extensive experience in programming.

On the other hand, more mathematical/statistical workers, who have largely been in control of risk assessment procedures up to this point, often do not have the detailed familiarity with causal mechanisms to feel comfortable building realistic mechanism-based representations of complex biological processes. In any event, doing so would complicate the use of their usual "black box" curve-fitting approaches to analysis—introducing more variables than can be directly estimated from any single data set and therefore requiring relatively innovative (from a statistical standpoint) procedures to incorporate diverse information from different sources. Again, the great advances in computing hardware and software in recent years make it technically feasible, although not necessarily statistically straightforward, to make data/analytical systems that are much better reflections of defensible mechanistic theories of neurotoxic processes.

In addition to these discipline-based difficulties for biomarkers research, biomarkers projects must generally overcome serious challenges that go under the general heading of "validation." Schulte and Mazzuckelli (1991) and Schulte and Perera (1993) have explored many of the relevant technical issues in the validation of biomarkers.

Aside from the technical difficulties, however, there are obstacles that arise from the organization and funding of scientific research. In order to prove that a biomarker really is a good indicator of a pathological process leading to disease or impairment, there will generally be a need for cross-disciplinary projects that

1. transcend different levels of organization of the nervous system and fundamentally different kinds of end points (e.g., biochemical, morphological, and functional),
2. transcend differences in the methodologies that are available for study of humans vs other species *in vivo*,
3. transcend differences in the observations that can be made *in vitro* vs *in vivo*, or

4. transcend differences in the time scale over which different events occur (for example, a putative biomarker measuring the ongoing loss of neurons of the substantia nigra might only be capable of ultimate validation as a predictor of parkinsonism with the aid of a long-term prospective epidemiological study).

Organizing research teams, obtaining grants, and publication of results all present unusual hurdles for investigators wishing to pursue these integrative types of efforts. Academic departments may tend to include either epidemiologists or experimental neuroscientists, but not both. Thus organizational difficulties will be created by the need to organize research projects across existing academic units. Moreover, individual study sections conducting peer review for funding agencies and editorial boards reviewing papers for publication may often be dominated by researchers who are familiar and supportive of either the functional or the biochemical side of a biomarker research project: either the *in vitro* or the *in vivo* side or either the animal or human side, but not both. Adaptations in scientific institutional arrangements may need to be made to foster the kinds of cross-cutting research that can validate relationships between biomarkers and expectations for human risk.

II. How Can Biomarkers Be Used in Both Basic Scientific Research and Risk Assessment?

This section gives some further illustrations of the kinds of research questions that can be pursued with the aid of biomarkers and that address vital questions for risk assessment. Because of space limitations, only a few types of uses of biomarkers are focused on.

A. Elucidating Dose–Time–Response Relationships and Appropriate Dose Metrics for Predicting Response

As with carcinogenesis, many neurotoxic effects are most conveniently studied with experiments conducted at relatively high dose rates where detoxifying or activating enzymes, and other processes along the causal chain, may exhibit saturation phenomena. It is becoming increasingly recognized in the risk assessment of carcinogens that before it is possible to obtain information on the shape of the ultimate dose–response relationship arising from the multiple mutation mechanism of carcinogenesis, it is necessary to separate out the nonlinearities that are due to pharmacokinetic satura-

tion processes (Hoel *et al.*, 1983; Hattis, 1988). This is done with the aid of measurements of internal dosimeters [such as DNA or hemoglobin adducts (Perera and Weinstein, 1982; Perera *et al.*, 1986; Osterman-Golkar and Ehrenberg, 1983; Bailey *et al.*, 1986)] and/or physiologically based pharmacokinetic models (Fiserova-Bergerova, 1983; Ramsey and Andersen, 1984; Hattis, 1991). Similarly, it can be expected that measures of dosage of active forms of toxicants that are actually delivered to relevant sites in the nervous system will be important in sorting out what the dose–response relationships are likely to be for those portions of the causal chain of neurotoxicity that actually happen in nervous tissue.

In addition to the need to sort out nonlinearities in response with respect to dose rate, at least three other dynamic factors can be critical for understanding neurotoxic risks:

1. The dynamics of absorption, activation, inactivation, and excretion of the neurotoxic chemical,
2. The dynamics of repair/reversal processes for those types of damage that are reversible at low dosage (Hattis and Shapiro, 1990; also see Section II,A,1).
3. Differences in sensitivity of the organism at different stages of life (e.g., during development; NRC, 1989b).

1. Analyzing Dose–Time–Response Relationships for Classic Chronic Toxic Effects: The Case of Acrylamide

Risk assessments for chemicals producing adverse effects by "classic chronic" toxic damage processes [defined as those that are fundamentally reversible, at least in preclinical stages, but which take a relatively long time (weeks or months) for reversal/repair to occur (Hattis, 1986); this applies to some, but not all (Merigan *et al.*, 1985; Cavanagh and Nolan, 1982; De-Grandhap *et al.*, 1990), of the neurotoxic effects of acrylamide] need to address a number of significant issues:

1. What are the relationships between external dose and the generation of the internal damage/ toxicant accumulation?
2. What are the nature and dynamics of reversal of the "slow step" in the process that makes the process "chronic?"
3. What are the differences among species in both the generation of damage/toxin accumulation and the repair/reversal process?
4. How much interindividual variation can be expected among exposed people in both damage-

producing and repair processes (and therefore susceptibility to toxicity)?

The acrylamide case is interesting in that it indicates the potential helpfulness of an entirely theoretical modeling exercise in basic toxicological research. For this neurotoxicant, the exact physical form of the incipient damage that accumulates over weeks or months to ultimately lead to the grosser manifestations of peripheral neuropathy is not known. Three decades of experimental observations have yielded an extensive characterization of neuropathic effects on both morphological (Spencer and Schaumberg, 1974) and functional levels (Hopkins and Gilliat, 1971). At the key biochemical/molecular level, however, there is an almost embarrassing richness of candidates for causal intermediate processes in the generation of neurological damage. Among the most prominent of these are inhibition of retrograde transport systems (which convey material from the axons back to the cell body) (Gold *et al.*, 1985; Miller and Spencer, 1984) and anterograde transport systems (which convey material from the cell body to the axons) (Sickles, 1991), although a number of other mechanisms have also received serious study.

For initial modeling work published in 1990 (Hattis and Shapiro, 1990), we elected to return to some of the most classical studies of acrylamide neurotoxicity (Hopkins, 1970; Fullerton and Barnes, 1966; Kaplan and Murphy, 1972) and apply a simple dynamic analysis model to them. The data sets analyzed were those that have provided information on some specific manifestation of toxicity produced by different combinations of acrylamide dose rate and duration of exposure (e.g., see Table 1; similar data were available for some other effects and some other species). It was found that the pattern of increase in the time required to achieve a particular effect could provide us with two important pieces of information relevant to the assessment of risk:

1. The dynamics of repair of the incipient damage, i.e., how much of the past accumulated damage is

repaired per day? How does this calculated repair rate appear to change (a) across species, (b) for different adverse effect end points, and (c) with different amounts of calculated accumulated damage?
2. The dose of acrylamide that would be just barely able to produce each effect in each species, if the experiment were conducted over the entire life span of the animal.

Information of the first type may also be helpful in neurotoxicology research. Specific biomarkers for the main process causing a particular response should be repaired in different locations and in different species with the dynamics that are consistent with the repair rates calculated from the dose vs time of effect data.

Our initial model for analyzing these data (Fig. 1) was built around three assumptions:

1. A particular adverse effect occurs whenever a specific amount of damage is accumulated in the relevant portions of the nervous system. (That is, there is no appreciable delay between the production of damage and the manifestation of the resulting effects.)
2. Damage is produced at a rate that is approximately linear with the milligram per kilogram dose administered to the animals.
3. Repair of the accumulated damage occurs at a rate that depends directly on the amount of accumulated damage that needs to be repaired.

The first assumption provided us with our primary tool for quantitatively analyzing the data. Basically, by trial and error, for each data set, we determined the repair rate that made the amount of accumulated damage approximately equal for each of the dose and time

TABLE 1   Data of Fullerton and Barnes (1966) on Dose/Time Response for Development of Hindlimb Weakness in Rats

Mean dose (mg/kg-day)	Mean days to response	Mean cumulative dose (mg/kg)
7.5 (6–9)[a]	280	2100
12 (10–14)	84	1008
16.5 (15–18)	28	462
25 (20–30)	21	525

[a] Dose range.

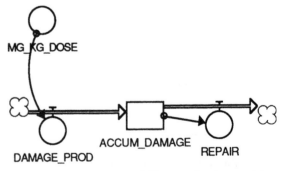

FIGURE 1   Where MG_KG_DOSE = mg/kg-day administered. DAMAGE_PROD = MG_KG_DOSE {INITIALLY ASSUME DAMAGE IS LINEAR WITH MG/KG DOSE}. ACCUM_DAMAGE = ACCUM_DAMAGE + *dt** (DAMAGE_PROD-REPAIR) where *dt* is the change in time. REPAIR = *K**ACCUM_DAMAGE where *K* is the rate of repair per day.

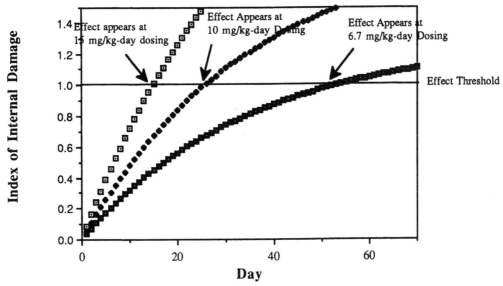

FIGURE 2 Theoretical dose–time internal damage functions for acrylamide, assuming linear repair of damage (2.9% repaired/day). □, 15 mg/kg-day; ◆, 10 mg/kg-day; ■, 6.7 mg/kg-day.

combinations that were observed to produce a particular response. Figure 2 illustrates the buildup of internal damage to levels producing an effect under this model, with a 2.9%/day rate of repair. (Some variations on the second and third assumptions were explored during the course of model development.) Table 2 gives the calculated repair rates and resulting estimates of long-term doses required to produce different toxic manifestations in different species.

Recently, new data have been collected (K. Crofton) on the dose rate and time responses of some functional end points for acrylamide neurotoxicity in rats. In this case, what was measured is not a simple presence or absence of enough "hindlimb weakness" to be noticeable, but a continuous parameter related to motor function. Rats were given acrylamide intraperitoneally for varying durations between 10 and 90 days, and verticle motor activity was monitored using an automated de-

vice. This activity was measured both during and after exposure in order to determine a time course of onset of acrylamide neurotoxicity, as well as the extent and dynamics of recovery following the termination of dosing.

The basic observations for the development of functional impairment are shown in Fig. 3. To provide a test of the original model for the accumulation of internal damage against the action of a hypothesized linear repair process, the modeler (DH) was initially only given the time course data on the development of the functional impairment (i.e., Fig. 3). He was then called upon to make predictions of the dynamics of recovery of function.

To avoid possible biases arising from an arbitrary choice of a criterion for defining a neurological "effect," these data were fit using a graded series of "effect" criteria—defined at every 5% decrement of ob-

TABLE 2 Repair Rates and Estimated "Accumulated Damage" Needed to Produce Effects for Different End Points and Different Species

	Best-fitting repair rate (day^{-1})	"Accumulated damage" for effect at lowest dose rate (mg/kg)	Predicted lowest lifetime dose rate yielding effect (mg/kg-day)
Baboon (Hopkins, 1970)			
Hindlimb weakness	0.030	278	8.34
Forelimb weakness	0.014	484	6.78
Rat (Kaplan and Murphy, 1972)			
Rotarod performance	0.045	295	13.3
Rat (Fullerton and Barnes, 1966)			
Hindlimb weakness	0.0175	425	7.44

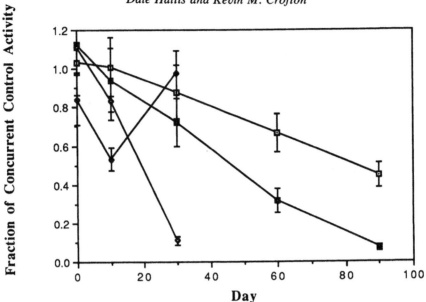

FIGURE 3   Decline in function during exposure to acrylamide (ranges represent ± 1 SE). □, 6.7 mg/kg-day; ◆, 10 mg/kg-day (30-day study); ■, 10 mg/kg-day (90-day study); ◆, 15 mg/kg-day.

served function between 70 and 45% of the concurrent control activity (see illustration in Fig. 4). The days of dosing required to achieve each of these effect levels were calculated by linear interpolation from all the data points for loss of function during the dosing period, as illustrated in Figs. 5–7. (The odd shape in the plot of the 10 mg/kg-day graph in Fig. 6 results from combining somewhat disparate results for this dose rate observed in separate 30- and 90-day dosing studies.) The fitting

was accomplished in a Microsoft Excel spreadsheet by choosing the repair rate that equalized the model-predicted amount of damage as much as possible at the time each effect criterion was passed for each dose rate. For each effect criterion level, the fitting function minimized the square of the logarithms of the ratio of the damage rate predicted for each dose rate to the geometric mean damage for all dose rates. For an example, see Table 3 for the 55% effect criterion.

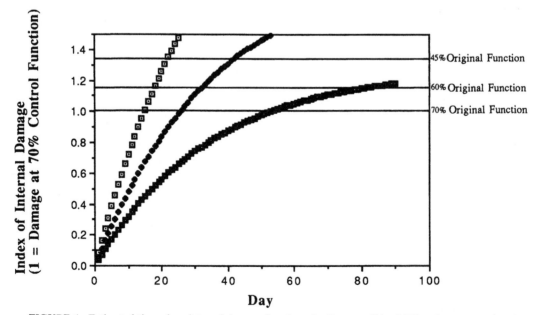

FIGURE 4   Estimated time–dose internal damage functions for the overall best-fitting damage repair rate (0.0288/day). □, 15 mg/kg-day; ◆, 10 mg/kg-day; ■, 6.7 mg/kg-day.

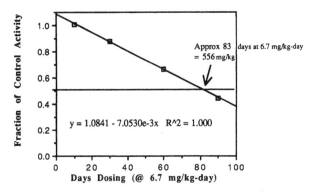

FIGURE 5 Interpolation of 6.7 mg/kg-day points to calculate the time required to reduce function to 50% of control activity.

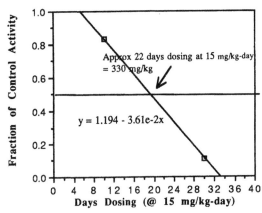

FIGURE 7 Interpolation of 15 mg/kg-day dosing points to calculate the time to reduce function to 50% of control activity.

The lower limit of 70% control function was chosen as the smallest amount of impairment that could be reliably distinguished from the control. (The amount of internal damage corresponding to this level of impairment was defined as one neurotoxic damage unit.) An "uptake coefficient" (relating the amount of internal damage produced per day per milligram per kilogram of acrylamide dose) was found to ensure that the metric of internal damage was 1 (as a geometric mean) at the times each dose group achieved the 70% level of control function.

Given this, Figs. 8 and 9 illustrate the predictions made for the recovery of function following the cessation of dosing for the 70% level of control function. For comparison with these predictions, "observed" values for the days on which different levels of function were recovered were derived by interpolating observed data points seen following the cessation of 30- and 90-day periods of dosing (e.g., see Fig. 10 for the recovery phase following 30-day dosing with 10 mg/kg-day.)

Table 4 shows the resulting comparison of model predictions with the observations for the times to recover different functional levels after the end of either 30- or 90-day dosing periods. A procedure has not been developed for formal statistical hypothesis testing for the results in Table 3. Nevertheless, it is clear that our original model is decisively rejected. In all 17 cases where a numerical comparison is possible between predicted and observed times for the recovery of specific levels of function, the observed time to recovery is greater than predicted, often by a considerable margin.

Figures 11 and 12 show the full-time pattern of loss and recovery of function. It can be seen that, contrary to our model expectations, in three out of four curves there appears to be some appreciable lag time before recovery of function begins to be observable. The effect is most noticeable for the two highest dose rates and the most profound levels of functional impairment.

In further exploration of this, it was hypothesized that the repair process might be saturable instead of strictly linear. That is, repair might be governed by Michaelis–Menten kinetics:

$$\text{Rate of repair of damage} = \frac{(\text{internal damage})(V_{\max})}{(K_m + \text{internal damage})},$$

where K_m is the amount of damage that produces half of the maximal rate of damage repair ($V_{\max}$). Using this formula, we were able to align predicted and observed times for recovery for the 90-day dosing experiment by assuming a $K_m = 1.6$ neurotoxic damage units. However, when this revised model was tested with the results of the 30-day exposures, the saturable model still greatly underestimated the observed times for recovery (Table 5).

The lack of adequate fit for these simple hypotheses leads to some biologically interesting hypotheses for possible exploration in further work:

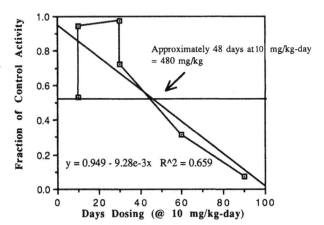

FIGURE 6 Interpolation of 10 mg/kg-day dosing points to calculate the time to reduce function to 50% of control activity.

TABLE 3

Dose rate (mg/kg-day)	Days dosing to reduce function to 55% control at given dose rate	Current model results (internal damage units—1 = damage producing 70% control function)	[log(ratio observed damage/geometric mean damage)]²
15	18	1.165	0.00033619
10	43	1.358	0.0023366
6.7	76	1.134	0.00090017
		Geometric mean = 1.215	Sum = 0.00357296

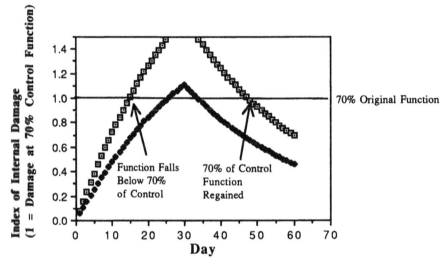

FIGURE 8   Prediction of damage and repair for 30 days of dosing with 10 (◆) or 15 (□) mg/kg-day.

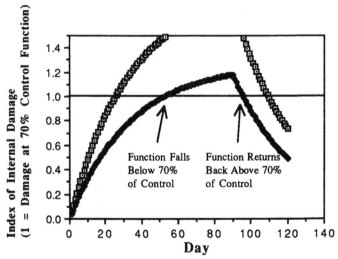

FIGURE 9   Prediction of recovery following 90 days of dosing at 6.7 (◆) or 10 (□) mg/kg-day.

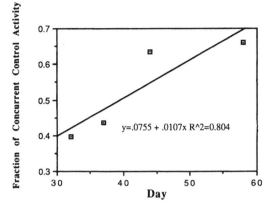

FIGURE 10 Recovery of function after 30 days of dosing with 10 mg/kg-day.

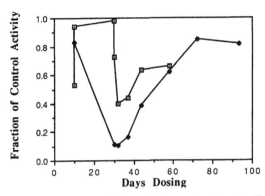

FIGURE 11 Thirty-day dosing: Full results for 10 (□) and 15 (◆) mg/kg-day.

TABLE 4 Comparison of Observed with Predicted Times for Recovery of Various Degrees of Original Function: Original Model Fit to the Data on the Observed Times for Development of Functional Loss (Fig. 3)

Dose rate (mg/kg-day) and duration	Model predicted day for recovery of function	Observed day for recovery of function	Neurotoxic damage units[a]
Effect criterion			
45% of control function			1.324
6.7 for 90 days	NA[b]	NA	
10 for 90 days	100	112	
10 for 30 days	NA	36	
15 for 30 days	38	50	
50% of control function			1.263
6.7 for 90 days	NA	NA	
10 for 90 days	102	114	
10 for 30 days	NA	38	
15 for 30 days	40	53	
55% of control function			1.215
6.7 for 90 days	NA	91	
10 for 90 days	103	116	
10 for 30 days	NA	41	
15 for 30 days	41	55	
60% of control function			1.140
6.7 for 90 days	92	94	
10 for 90 days	105	118	
10 for 30 days	NA	43	
15 for 30 days	43	58	
65% of control function			1.080
6.7 for 90 days	94	97	
10 for 90 days	107	121	
10 for 30 days	31	46	
15 for 30 days	45	60	
70% of control function			1.000
6.7 for 90 days	96	99	
10 for 90 days	110	123	
10 for 30 days	34	48	
15 for 30 days	48	63	

[a] 1 equals the amount of damage needed to reduce function to 70% of the function in concurrent control animals.

[b] Not applicable. The best-fitting model did not predict that sufficient damage would be done to reach this criterion of functional loss during the dosing period.

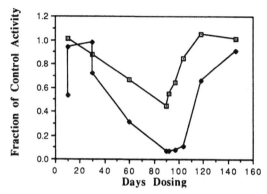

FIGURE 12   Ninety-day dosing and recovery: Full results for 6.7 (□) and 10 (◆) mg/kg-day.

1. There might be functional compensation or adaptation by the organism to overcome partial damage to a component of the nervous system.

2. There might be a two-stage damage process, with the second stage of damage representing less easily reversible (or even irreversible) damage. For such a two-stage process, the need for acrylamide to act on the same neuron twice might be expected to produce a greater proportion of stage 2 damage for the higher dose rates.

3. There might be a more complex relationship than we have assumed between the underlying pathological events and the resulting functional changes. Recent data suggest that recovery of function is not coincident with observable repair of physical damage

TABLE 5   Comparison of Observed with Predicted Times for Recovery of Various Degrees of Original Function: Saturable Repair Model (K_m = 1.6) Fit to Both the Observed Times for Development of Functional Loss and the Recovery Times Following 90-Day Exposures

Dose rate (mg/kg-day) and duration	Model predicted day for recovery of function	Observed day for recovery of function	Neurotoxic damage units[a]
Effect criterion			
45% of control function			1.413
6.7 for 90 days	NA[b]	NA	
10 for 90 days	112	112	
10 for 30 days	NA	36	
15 for 30 days	41	50	
50% of control function			1.327
6.7 for 90 days	NA	NA	
10 for 90 days	114	114	
10 for 30 days	NA	38	
15 for 30 days	43	53	
55% of control function			1.260
6.7 for 90 days	91	91	
10 for 90 days	116	116	
10 for 30 days	NA	41	
15 for 30 days	45	55	
60% of control function			1.170
6.7 for 90 days	93	94	
10 for 90 days	119	118	
10 for 30 days	NA	43	
15 for 30 days	48	58	
65% of control function			1.095
6.7 for 90 days	96	97	
10 for 90 days	121	121	
10 for 30 days	31	46	
15 for 30 days	50	60	
70% of control function			1.000
6.7 for 90 days	99	99	
10 for 90 days	125	123	
10 for 30 days	35	48	
15 for 30 days	54	63	

[a] 1 equals the amount of damage needed to reduce function to 70% of the function in concurrent control animals.

[b] Not applicable. The best-fitting model did not predict that sufficient damage would be done to reach this criterion of functional loss during the dosing period.

(K. M. Crofton *et al.*, manuscript in preparation). At a time point where animals display almost complete recovery of function from a 90-day exposure to acrylamide, histological studies yield little evidence of tissue repair in the peripheral nervous system. It is therefore possible that the functional recovery is due to an anatomical compensatory mechanism, such as sprouting at the neuromuscular junction by fibers that still work, resulting in fewer fibers innervating more muscular tissue (cf. DeGrandchamp and Lowndes, 1990; Madrid *et al.*, 1993).

These possibilities may be explored in further experimental and dynamic modeling work.

2. Analyzing the Apparent Special Sensitivity of the Developing Nervous System to Neurotoxic Risks of Methyl Mercury

It has long been recognized that the sensitivity of the developing organism to adverse effects can differ dramatically over short periods of time, especially during gestation. Developmental signals may be present over limited periods of time at the strength required to induce axons to find their targets and for other key events to occur. Understanding the physical and physiological bases of these differences will be a key to determining over what times periods exposures should be aggregated for best experimental and epidemiological quantification of risks. For example, Marsh and colleagues (1987) have published detailed information on the incidence of a variety of fetal methyl mercury effects in relation to the maximal levels of mercury found in the hair of the mothers during gestation. (The observations come from an Iraqi mass poisoning incident which resulted from the distribution of methyl mercury-treated Green Revolution seed grain.) Maximum mercury concentrations were assessed by a series of sequential measurements along the hair shafts during fetal development. Log probit dose–response fits to these data (e.g., Figs. 13 and 14) indicate very large amounts of interindividual variability in response[2] and could suggest appreciable risks at the much lower dosages that are present in the diets of people who consume relatively large amounts of fish with relatively large methyl mercury concentrations. An analysis of

---

[2] A log probit slope as low as 1 (as suggested by the plot in Fig. 14) would imply that 95% of the population would have thresholds for effect spread out over a span of about 10,000 fold in dosage, from 100-fold lower to 100-fold higher than the dose that would cause the effect in people of median susceptibility in an exposed population. A probit slope of 2 would suggest less, but still appreciable variability, with the thresholds of 95% of the population spread over a 100-fold range in dosage from 10-fold lower to 10-fold higher than the threshold for the median person.

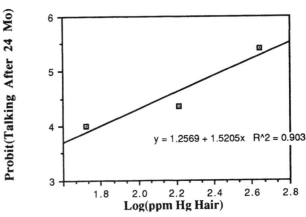

FIGURE 13 Log-probit dose–response relationship for talking after 24 months in relation to maternal hair mercury (Marsh *et al.*, 1987).

dose–response relationships for methyl mercury responses based on the Iraqi data (Institute of Medicine, 1991) indicates that

1. The traditional log probit model—long a staple for the analysis of animal acute lethality data—is quite compatible with the available human information on the response to methyl mercury (Table 6 and Fig. 15).

2. There is a pronounced tendency in these data for the apparent probit slopes of the fetal responses (Table 6A) to be less than the corresponding probit slopes for the adult effects (Table 6B). This indicates considerable potential for low dose fetal effects in the absence of apparent observable responses in adults. Such a conclusion, however, depends on whether the biomarker of exposure used in this case—the maximum hair mercury found at any time during gestation—is the most appropriate direct causal predictor of response that can be developed. Other possibilities might well include the concentration of mercury at a specific sensitive time during gestation or a weighted sum of con-

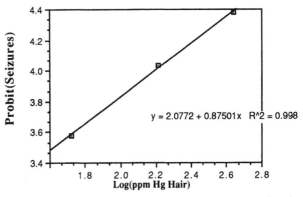

FIGURE 14 Log probit plot of seizures in children as a function of maternal hair mercury during pregnancy: Data of Marsh *et al.* (1987).

Dale Hattis and Kevin M. Crofton

TABLE 6 Maximum Likelihood Fits[a]

A. Marsh (1987) fetal effects data[b]

Effect	Background % Response[c]	Probit slope	Slope SE	Intercept	ED$_{50}$ (ppb blood)	ED$_{50}$ geometric SE	χ^2	Degrees freedom[d]	P[e]
Late walking	0	1.21	0.30	2.19	205	1.49	6.093	3	0.11
Late talking	7.3	1.76	0.71	0.81	244	1.38	0.689	1	0.41
Mental symptoms	2.4	0.99	0.76	1.88	1429	4.75	0.351	1	0.55
Seizures	0	1.10	0.53	1.54	1399	2.95	0.356	3	0.95
Neurol score >4	0	0.85	0.27	2.42	1047	2.54	0.874	3	0.83
	Average	1.18					Sum 8.363	11	0.68

B. Iraqi adult methyl mercury effects data[b]

Effect	Background % Response[c]	Probit slope	Slope SE	Intercept	ED$_{50}$ (ppb blood)	ED$_{50}$ geometric SE	χ^2	Degrees freedom[d]	P[e]
Paresthesias	7.5	2.17	0.63	−1.64	1145	1.24	1.155	2	0.76
Ataxias	2.5	3.92	0.76	−7.67	1687	1.11	4.95	3	0.18
Visual changes	0	2.19	0.43	−2.24	2006	1.16	3.695	3	0.3
Disarthria	5	4.67	1.36	−11.2	2952	1.10	5.608	3	0.13
Hearing defects	1.3	6.42	2.17	−18.05	3877	1.10	0.209	1	0.65
Deaths	0	7.58	3.19	−23.05	5007	1.18	0.83	1	0.36
					Sum: Adult effects		16.447	13	0.22
					Sum: Fetal and adult effects		24.81	24	0.42

[a] From the Institute of Medicine (1991). The equation fit is probit of excess risk over background = intercept + (slope) * log$_{10}$(blood Hg in ppb).

[b] Using the method of Finney (1971).

[c] Estimated from data in the lowest one to three dose groups.

[d] Number of dose groups available for analysis, less 2 for the number of parameters estimated from the data (the intercept and probit slope).

[e] The probability that a deviation as large as that observed between the log probit model and the data would have been expected by chance, even if the log probit model was a perfect description of the underlying dose–response function.

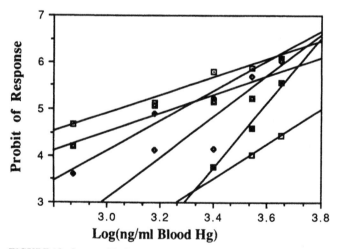

FIGURE 15 Log probit dose–response relationships for all effects in adults: Iraqi data. ⊡, probit XS paresthesia; ◆, probit XS ataxia; ■, probit XS visual changes; ◆, probit XS disarthria; ■, probit XS hearing defects; □, probit of percentage deaths.

centrations × duration over a specific set of sensitive periods. Accurate assessment of the degree of interindividual variability in susceptibility in humans, and consequent low dose risks, may depend on quantitative measurement and modeling of the dynamic causal processes responsible for the damage.

B. Elucidating Appropriate Interspecies Projection Rules

One of the most serious challenges for the use of animal models in neurotoxicology is the fact that many of the tasks that are performed by the human central nervous system are quite different than the tasks that can be assessed directly in animal systems. Moreover, even for tasks that are similar, it is not impossible that humans differ from specific experimental animals in the functional reserve capability and therefore the degree of impairment of function at some basic level that will be required to manifest itself as a particular degree of functional deficit on a specific test.

Use of appropriate biomarkers for changes in neurological functions that are similar between animals and humans may help break this kind of problem into more manageable subunits. A good analogy here is with the use of reductions in sperm counts as a biomarker for predicting possible changes in male fertility. Different species differ in the number of sperm they produce and in the degree of excess of sperm numbers over the amount required to achieve reliable fertilization (Meistrich, 1984; Amann, 1981; Galbraith *et al.*, 1983). Because of this, risk assessments for possible human male fertility effects from animal data have chosen to use information on sperm changes in preference to data on changes in animal male fertility itself (Meistrich and Brown, 1983). Sperm count changes are also a superior parameter for assessing possible human effects in epidemiological studies because they represent a continuous variable that can be studied in many individuals in an exposed population (Hattis *et al.*, 1988). Human male fertility performance, on the other hand, is only observable in the small subset of people who intend to reproduce at a specific time, and then only with considerable difficulty and statistical insensitivity because of the +/− nature of the data. Changes in sperm count distributions are also likely to be more sensitive in the sense that changes insufficient to cause appreciable reductions in fertility may be relatively readily detectable in cross-sectional comparative studies or serial studies of exposed groups.

C. Uncovering, Quantifying, and Tracing the Causes of Early Damage in Chronic Cumulative Neurotoxic Processes

It has often been stressed that neurons do not regenerate in adult life and that some important neurological conditions (including Alzheimer's disease, Parkinson's disease, and ALS) are caused by the chronic cumulative loss of specific types of neurons in older people. Because in many cases the brain can lose considerable numbers of the relevant cells before gross clinical signs of impairment are obvious, exposure to agents causing this category of conditions can proceed apparently innocuously for decades without being noticed. Even after symptoms appear, picking out the potentially relevant exposures from an unmeasured lifetime of different pursuits presents a practically insuperable challenge to clinicians and epidemiologists. More timely epidemiological research would be greatly facilitated if either or both of the following types of biomarkers were developed:

1. Measures of the past accumulation of relevant damage in people who have not yet developed clinical illness. For example, can our new tools for imaging the brain be applied to take a census of how many of the relevant cells are left in people of different ages with different past exposures? (This would be analogous to the use of chest X-rays, FEV1, and FVC to assess the accumulation of past respiratory damage of chronic cumulative types.)
2. Measures of the current rate of loss of the relevant cells. When the relevant cells die, do they release measurable amounts of a distinctive form of an enzyme or some other component that could be used as a biomarker? Do they perhaps in their death throes emit a distinctive type of electrical signal that could be picked up, identified, and measured?

Many years ago Weiss and Spyker (1974) suggested that methyl mercury might accelerate the loss of neurons in adult life and contribute to a "chronic cumulative" process that would fall within this category. An apparently very sound case/control epidemiological study among people in Singapore has found a strong association between blood levels of mercury and risk of Parkinson's disease (Ngim and Devathasan, 1989). This finding, using a biomarker of exposure, needs to be followed up in confirmatory studies and (hopefully) with the aid of new biomarkers of the types suggested earlier.

III. Summary/Conclusions

The potential contributions of biomarkers to the enrichment of our basic scientific understanding of the nervous system is best summarized with an analogy. Many writers of basic accounts of contemporary findings in neuroscience and neurotoxicology use very few numbers and not a single equation. The same could be said of most technical publications in neuroscience professional journals. The primary objective of most neuroscience research has been seen as to generate qualitative information on physical characteristics and causal connections—questions of "whether" and "how" (by what causal pathway) specific phenomena are related. Such qualitative facts, like "the shinbone is connected to the kneebone," certainly provide an essential framework for understanding phenomena like the kicking of field goals in football. However, much more is required for a full understanding of just why field goals of 40 yards are relatively common, field goals of 50 yards are relatively rare, and field goals over 60 yards are practically unknown, even for professionals who have trained all their lives for the task. To understand field goal performance quantitatively, one needs to know quantitative facts about the number of muscle fibers there are in the upper leg, the number that can be induced to contract at about the same time, the rate at which the fibers can shorten themselves, the accuracy with which the foot can be oriented to strike the football at the best place, and how all this translates into kinetic energy available to be delivered to the football.

The intent of this chapter has been to suggest that contemporary neuroscience and neurotoxicity can be enriched by directing more research attention to quantitative questions of "how much," "how fast," and "according to what quantitative functional relationship" among the series of causal intermediate processes involved. The role of "biomarkers" is to measure the causal intermediate processes at as many steps as is feasible along the causal chain.

References

Amann, R. P. (1980). Sperm production rates. In *The Testis* (A. Johnson, W. Gomes, and N. VanDenmark Eds.), Vol. I, pp. 433–482, Academic Press, New York.

Bailey, E., Farmer, P. B., Bird, I., Lamb, J. H., and Peal J. A. (1986). Monitoring exposure to acrylamide by the determination of S-(2-carboxyethyl)cysteine in hydrolyzed hemoglobin by gas chromatography-mass spectrometry. *Anal. Biochem.* **157**, 241–248.

Bergmark, E., Calleman, C. J., and Costa, L. G. (1991). Formation of hemoglobin adducts of acrylamide and its epoxide metabolite glycidamide in the rat. *Toxicol. Appl. Pharmacol.* **111**, 352–363.

Cavanagh, J. B., and Nolan, L. C. (1982). Selective loss of Purkinje cells from the rat cerebellum caused by acrylamide and the responses of beta-glucuronidase and beta-galactosidase. *Acta Neuropathol.* (Berl.) **58**, 210–214.

DeGrandchamp, R. L., and Lowndes, H. E. (1990). Early degeneration and sprouting at the rat neuromuscular junction following acrylamide exposure. *Neuropathol. Appl. Neurobiol.* **16**, 239–254.

Fiserova-Bergerova, V. (1983). *Modeling of Inhalation Exposure to Vapors: Uptake, Distribution, and Elimination* Vol. 1 and 2, CRC Press, Boca Raton, Florida.

Fullerton, P. M., and Barnes, J. M. (1966). Peripheral neuropathy in rats produced by acrylamide. *Br. J. Ind. Med.* **23**, 210–221.

Galbraith, W., Voytek, P., and Ryon, M. G. (1983). Assessment of risks to human reproduction and to the development of the human conceptus from exposure to environmental substances. In *Assessment of Reproductive and Teratogenic Hazards* (M. S. Christian, W. M. Galbraith, P. Voytek, and M. A. Mehlman, Eds.), Section II, Princeton Scientific Publishers, Princeton, New Jersey.

Gold, B. G., Griffin, J. W., and Price, D. L. (1985). Slow axonal transport in acrylamide neuropathy: different abnormalities produced by single-dose and continuous administration. *J. Neurosci.* **5**, 1755–1768.

Hattis, D. (1986). The promise of molecular epidemiology for quantitative risk assessment. *Risk Anal.* **6**, 181–193.

Hattis, D. (1988). *The Use of Biological Markers in Risk Assessment: Statistical Science*, Vol. 3, pp. 358–366.

Hattis, D. (1991). Use of biological markers and pharmacokinetics in human health risk assessment. *Environ. Health Perspect.* **89**, 230–238.

Hattis, D., and Shapiro, K. (1990). Analysis of dose/time/response relationships for chronic toxic effects—the case of acrylamide. *Neurotoxicology*, **11**, 219–236.

Hattis, D., Welch, L. S., and Schrader, S. M. (1988). Male fertility effects of glycol ethers—a quantitative analysis. CTPID 88-3, M.I.T. Center for Technology, Policy, and Industrial Development, Massachusetts.

Hoel, D. G., Kaplan, N. L., and Anderson, M. W. (1983). Implications of non-linear kinetics on risk estimation in carcinogenesis. *Science* **219**, 1032–1037.

Hopkins, A. (1970). The effect of acrylamide on the peripheral nervous system of the baboon. *J. Neurol. Neurosurg. Psychiatry* **33**, 805–816.

Hopkins, A. P., and Gilliatt, R. W. (1971). Motor and sensory nerve conduction velocity in the baboon: normal values and changes during acrylamide neuropathy. *J. Neurol. Neurosurg. Psychiatry* **34**, 415–426.

Institute of Medicine (1991). Seafood safety. In *Committee on the Evaluation of the Safety of Fishery Products, Food and Nutrition Board*, (F. E. Ahmed, Ed.), Institute of Medicine. National Academy Press, Washington, D.C.

Kaplan, M. L., and Murphy, S. D. (1972). Effect of acrylamide on rotarod performance and sciatic nerve β-glucuronidase activity of rats. *Toxicol. Appl. Pharmacol.* **22**, 259–268.

Kuhn, T. S. (1977). Mathematical versus experimental traditions in the development of physical science. In *The Essential Tension—Selected Studies in Scientific Traditions and Change* University of Chicago Press, Chicago, Illinois.

Madrid, R. G., Ohnishi, A., Hachisuka, K., and Murai, Y. (1993). Axonal sprouting of motor nerve in acrylamide-intoxicated rats with progressive weakness. *Environ. Res.* **60**, 233–241.

Marsh, D. O., Clarkson, T. W., Cox, C., Myers, G. J., Amin-Zaki, L., and Al-Tikriti, S. (1987). Fetal methylmercury poisoning: relationship between concentration in single strands of maternal hair and child effects. *Arch. Neurol.* **44**, 1017–1022.

Meistrich, M. L. (1984). Human reproductive risk assessment from results of animal studies, OPTS-62030 Glycol Ether Record, Exhibit 4-156, pp. 1–18, U.S. Environmental Protection Agency, Office of Pesticides and Toxic Substances, Washington, D.C.

Meistrich, M. L., and Brown, C. C. (1983). Estimation of the increased risk of human infertility from alterations in semen characteristics. *Fertil. Steril.* **40**, 220–230.

Merigan, W. H., Barkdoll, E., Maurissen, J. P. J., Eskin, T. A., and Lapham, L. W. (1985). Acrylamide effects on the macaque visual system. I. Psychophysics and electrophysiology. *Invest. Opthamol. Vis. Sci.* **26,** 309–316.

Miller, M. S., and Spencer, P. S. (1984). Single doses of acrylamide reduce retrograde transport velocity. *J. Neurochem.* **43,** 1401–1407.

National Research Council (1989a). *Biologic Markers in Pulmonary Toxicology.* National Academy Press, Washington, D.C.

National Research Council (1989b). *Biologic Markers in Reproductive Toxicology.* National Academy Press, Washington, D.C.

Ngim, C. H., and Devathasan, G. (1989). Epidemiologic study on the association between body burden mercury level and idiopathic Parkinson's disease. *Neuroepidemiology* **8,** 128–141.

Osterman-Golkar, S., and Ehrenberg, L. (1983). Dosimetry of electrophilic compounds by means of hemoglobin alkylation. *Annu. Rev. Public Health* **4,** 397–402.

Perera, F. P., and Weinstein, I. .B. (1982). Molecular epidemiology and carcinogen-DNA adduct detection: new approaches to studies of human cancer causation. *J. Chronic Dis.* **35,** 581–600.

Perera, F. R., Santella, R., and Poirier, M. (1986). Biomonitoring of workers exposed to carcinogens: immunoassays to benzo(a)-pyrene-DNA adducts as a prototype. *J. Occup. Med.* **28,** 1117–1123.

Ramsey, J. C., and Andersen, M. E. (1984). A physiologically based description of the inhalation pharmacokinetics of styrene in rats and humans. *Toxicol. Appl. Pharmacol.* **73,** 159–175.

Schulte, P. A., and Mazzuckelli, L. (1991). Validation of biological markers for quantitative risk assessment. *Environ. Health Perspect.* **90,** 239–246.

Schulte, P. A., and Perera, F. P. (1993). Validation. In *Molecular Epidemiology: Principles and Practices* (P. Schulte and R. Perera, Eds.) Academic Press, pp. 79–107.

Sickles, D. W. (1991). Toxic neurofilamentous axonopathies and fast anterograde axonal transport. III. Recovery from single injections and multiple dosing effects of acrylamide and 2,5-hexanedione. *Toxicol. Appl. Pharmacol.* **108**(3), 390–396.

Spencer, P. S., and Schaumburg, H. H. (1974). A review of acrylamide neurotoxicity. II. Experimental animal neurotoxicity and pathological mechanisms. *Can. J. Neurol. Sci.* **1,** 152–164.

Weiss, B., and Spyker, J. M. (1974). The susceptibility of the fetus and child to chemical pollutants. Behavioral implications of prenatal and early postnatal exposure to chemical pollutants. *Pediatrics* **53,** 851–859.

Developmental Neurotoxicology Risk Assessment

HUGH A. TILSON
Neurotoxicology Division
Health Effects Research Laboratory
U.S. Environmental Protection Agency
Research Triangle Park, North Carolina 27711

I. Introduction

It has been estimated that about one-half of all conceptions result in spontaneous abortion (Hertig, 1967), whereas 35% of postimplantation pregnancies end in embryonic or fetal loss (Wilcox *et al.*, 1985). Of the number of live births, approximately 3% have one or more congenital malformations at birth, whereas an additional 13% have serious developmental defects by the end of the first year of life (Shepard, 1986). It is now generally accepted that developmental toxicity can be detected at any time during the life span of the organism as a result of the perturbation(s) during gestation or after birth up to sexual maturity. Developmental toxicity includes death, growth retardation, structural abnormalities, or functional deficits (Schardein and Keller, 1989) (Table 1). Functional changes include severe and mild retardation, cerebral palsy, psychoses, epilepsy and abnormal neurological development, learning and memory deficits, sensory dysfunction, changes in motor activity, and disrupted maturational milestones.

Wilson (1977) estimated that about 20% of the developmental defects can be related to genetic causes, whereas 10% can be linked to specific external factors; the remaining 70% of developmental defects have no obvious cause. The contribution that environmental agents, genetic factors, nutritional deficiencies, drug abuse, exposure to tobacco and/or therapeutic agents, or combinations of all of these factors play in producing developmental defects is unknown.

It has long been thought that exposure to chemicals during development can adversely affect the structure and function of the nervous system, i.e., produce developmental neurotoxicity. Table 2 provides a partial list of chemicals generally regarded as developmental neurotoxicants. Lead is one of the most widely recognized developmental neurotoxicants, which has been shown to produce cognitive, sensory, motivational, and motoric deficits in all species evaluated (Riley and

This paper has been reviewed by the Health Effects Research Laboratory, U.S. Environmental Protection Agency, and approved for publication. Mention of trade names or commercial products does not constitute endorsement of recommendation for use.

TABLE 1   Examples of Developmental Toxic Effects

Death: Intrauterine or postnatal
Altered growth: Alteration in offspring organ or body weight at
  any time during development
Structural: Permanent structural change adversely affecting
  survival, development, or function or divergence beyond the
  normal range of structural development
Functional: Alterations or delays in physiological, biochemical, or
  behavioral competence

Vorhees, 1986). Generally recognized human developmental neurotoxicants include drugs of abuse (ethanol, cocaine, heroin, methadone), therapeutic agents (diphenylhydantoin), environmental agents (methyl mercury, polychlorinated biphenyls), and physical factors (X-radiation) (Riley and Vorhees, 1986; Rees *et al.,* 1990). A workshop on the comparability of human and animal developmental neurotoxicology found that the degree of qualitative comparability was good for the agents discussed, i.e., methyl mercury, polychlorinated biphenyls, ethanol, diphenylhydantoin, lead, and ionizing radiation (Stanton and Spear, 1990). Many chemicals having effects on sensory, motor, or cognitive function in humans have similar effects in animals. The degree of quantitative comparability, however, is not as good, possibly because of limited dose–response comparisons, incomplete or missing information concerning the actual internal dose relative to the administered dose, and insufficient information concerning the most sensitive end points (Francis *et al.,* 1990).

II. Regulation of Developmental Neurotoxicants

The probability or risk of developmental toxicity resulting from exposure to a chemical is one of several noncancer end points of concern to regulatory agen-

TABLE 2   Partial List of Chemicals Believed To Be Developmental Neurotoxicants

Alcohols	Methanol, ethanol
Antimitotics	Azacytidine
Insecticides	*p,p'*-DDT, chlordecone
Metals	Lead, methyl mercury
Polyhalogenated hydrocarbons	Polychlorinated biphenyls, polybrominated biphenyls
Psychoactive drugs	Cocaine, Methadone
Therapeutic agents	Diphenylhydantoin
Solvents	Carbon Disulfide, Toluene
Vitamins	Vitamin A

cies. In the United States, developmental toxicity is often considered to be a component of reproductive toxicity and typically involves exposure during major organogenesis, assessment of maternal toxicity during pregnancy, and evaluation of mother and offspring prior to term (U.S. EPA, 1982, 1985; FDA, 1966, 1970; OECD, 1981). Developmental toxicity may also be determined in studies involving exposure of both parents, prenatal and postnatal periods of development, or over several generations.

In recent years, the potential effects of chemicals on the development of the nervous system has received increased attention. It has been argued that the nervous system is especially vulnerable to chemical-induced perturbation (Grant, 1976; Spyker, 1975). Rodier (1976, 1979) has shown that there are specific times during maturation of the central nervous system when chemical-induced perturbation can cause developmental neurotoxicity. Dews (1986) has pointed out that the developing nervous system is generally resilient and may recover from some insults that would affect the mature nervous system permanently. That recovery can occur following developmental exposure does not, however, always mean that the risk has been decreased. Research has found that early developmental exposure to triethyl tin results in cognitive deficits that are exacerbated by the aging process (Barone *et al.,* 1993). In this study, rats were dosed with triethyl tin and showed marginal deficits at 3 months of age, no deficits at 12 months of age, and significant cognitive impairment relative to age-matched controls at 24 months of age. Delayed onset neurotoxicity has also been reported in monkeys exposed developmentally to methyl mercury (Rice, 1989).

Because of the perceived risk to human health, specific testing regulations and guidelines have evolved for developmental neurotoxicology studies. Behavioral testing of animals in developmental studies concerning new drugs has been required by Great Britain and Japan since 1975 (Kimmel, 1988). The World Health Organization (1984) published proposed testing guidelines for developmental neurotoxic effects of drugs and other agents, whereas the European Economic Community adopted testing protocols similar to those required by Great Britain and Japan (Kimmel, 1988). The U.S. Environmental Protection Agency has required developmental neurotoxicity data for several glycol ethers (Kimmel, 1988) and has published testing guidelines for developmental neurotoxicity studies (U.S. EPA, 1991b). In a workshop on the qualitative and quantitative comparability of human and animal developmental neurotoxicity, agents that are teratogenic to the central nervous system, neuropathic and neuroactive compounds, chemicals with hormonal activity and developmental toxicants were considered likely candidates for testing (Levine and Butcher, 1990).

The testing guidelines of the EPA provide direction for experimental design and dosing, as well as information concerning the types of assessments that should be performed. EPA's testing guidelines require measurements of maternal toxicity, growth and physical development of the offspring, developmental landmarks, motor activity, acoustic startle reactivity, learning and memory, and neuropathology (Table 3). Although the testing guidelines of the EPA provide direction as to the design and execution of developmental neurotoxicology studies, they do not specify how data should be interpreted or how risk assessments are to be performed on developmental neurotoxicological data. The U.S. EPA has published guidelines for developmental toxicity risk assessment which includes guidance for the interpretation of functional deficits following developmental exposure to chemicals (U.S. EPA, 1991a). Draft neurotoxicology risk assessment guidelines by the U.S. EPA also provide direction for the interpretation of structural and functional changes following developmental exposure to chemicals.

III. Developmental Neurotoxicology Risk Assessment

Risk assessment is a scientific process used to estimate the probability or possibility that exposure to an agent will be associated with an adverse effect. Risk assessment for developmental neurotoxicity follows the four steps described by the National Research Council (1983), including hazard identification, dose–response assessment, exposure assessment, and risk characterization.

TABLE 3 Measurements Included in U.S. EPA's Developmental Neurotoxicology Testing Guidelines

Period of dosing	Gestational day (GD) 6–postnatal day (PND)10
Route of exposure	Usually orally
Maternal measures	Observations once daily during dosing, body weights during dosing, at birth, and PND 11 and 21
Developmental parameters	Observations once daily, body weights at birth, PND 4, 11, 17, and 21, and biweekly thereafter Developmental milestones
Motor activity	PND 13, 17, 21, and 60
Acoustic startle	PND 22 and 60
Learning and memory	PND 21–24, 60
Neuropathology	PND 11 and end of study

Hazard identification deals with the qualitative assessment of experimental data to determine whether or not a hazard exists. Dose–response evaluations are concerned with the qualitative relationship between the magnitude of response as a function of dose and establishing reference doses. As pointed out by Kimmel and colleagues (1990), hazard identification for developmental neurotoxicity is usually done in the context of dose, route, and duration of exposure, all of which are important in extrapolating to potential hazards for human exposures. One advantage for considering hazard identification and dose–response assessment together is that the presence of an adverse effect may depend on whether a dose–response relationship exists. Hazard identification and dose–response assessments examine available experimental animal and human data to ascertain if developmental neurotoxicity is present. Also considered are the dose and exposure parameters (route, time, and duration of exposure) that produce such effects. All results from the hazard identification and dose–response assessment steps are evaluated to determine if sufficient evidence exists to judge whether or not a human developmental neurotoxicological hazard might exist. Positive data from epidemiologic studies would be sufficient to judge whether or not a human hazard exists. Judging that there are sufficient data to indicate a potential developmental hazard does not necessarily mean that the chemical would be a hazard under all exposure conditions. The real hazard will vary significantly depending on route and time of exposure, for example.

Data indicative of developmental neurotoxicity from animal studies are important in that a single, well-conducted study in a single experimental animal species might be sufficient to conclude that a human hazard might exist. Insufficient evidence exists if there is less than the minimum sufficient evidence necessary for determining potential developmental neurotoxicity. The quantitative aspect of dose–response assessment involves the derivation of a reference dose based on a no observed adverse effect level (NOAEL) or a lowest observed adverse effect level (LOAEL) corrected for uncertainties of intrapopulation variability, a less than lifetime exposure, and animal-to-human extrapolation (Barnes and Dourson, 1988).

Exposure assessment describes the magnitude, duration, frequency, and route(s) of exposure based on available monitoring data or modeling in order to estimate human exposure. There are some exposure parameters that are specific to developmental neurotoxicology studies. Developing organisms, for example, can be exposed to chemicals secondarily through placental transfer or breast milk, and measurements of a chemical in breast milk or cord blood may give a more precise estimate of developmental exposure. Another issue for exposure assessment is the duration and period of exposure as it relates to the stage of development or maturation. In effect, a single exposure during any one of many critical periods may produce developmental neurotoxicity.

Risk characterization integrates and summarizes all available information to determine the potential human health risk for any given exposure scenario. This step outlines limitations of the other steps of the risk assessment process, including the nature and human relevance of defined hazards, the conditions under which they are observed, the limitations of the exposure data or estimates, the comparison of the estimated exposures to the hazard evaluation studies, and the basis of the description of the potentially exposed populations. Quantitative estimates of risk are described in terms of the number of people whose estimated exposures exceed the reference dose. The risk characterization is communicated to the risk manager who uses the information to make public health decisions.

IV. Study Design Issues for Developmental Neurotoxicology

Conducting risk assessment for developmental neurotoxicology is heavily dependent upon a number of experimental or study design issues important for judging the sufficiency of data from each study. For human studies, a number of factors have been discussed (Bloom, 1981; Kimmel *et al.*, 1986) and include the power of the study; control of potential bias in data collection; collection of data on other risk factors, including age, smoking, alcohol consumption, drug use, and past reproductive history; and statistical factors related to sample selection and number of pregnancies per woman included in the study.

There are also a number of issues from developmental neurotoxicological studies in animals that should be considered when judging the adequacy of the data (Table 4). One of the most important issues concerns the health of the mother. Changes in maternal health could have a significant impact on the viability and development of the offspring. A change in maternal body weight is generally regarded as a good indicator of maternal toxicity in most species (U.S. EPA, 1991a). Changes in maternal body weight adjusted for gravid uterine weight may indicate whether the effect is maternal or intrauterine. Other measures of maternal toxicity include changes in organ weights, food and water consumption, clinical assessments, clinical chemistries, and histopathological changes. The potential for a chemical to produce maternal toxicity cannot be pre-

TABLE 4 Study Design Issues Important for Animal Developmental Neurotoxicology Studies

Presence of maternal toxicity
Developmental toxicity that affects nervous system structure and/ or function
Testing for developmental neurotoxic effects should assess full range of possible effects
Inherent variability of measures
Experimental design issues, including dose selection, species, age, weight, and health status
Appropriate statistical analysis
Control for genetic influences
Time and duration of exposure
"Unmasking" latent effects

dicted easily from other toxicity studies using adult animals, particularly male adult animals.

The interpretation of data from developmental toxicology studies can be confounded by maternal toxicity (Khera, 1984). Structural and functional alterations in offspring of exposed mothers are sometimes observed at doses producing some degree of maternal toxicity. In such cases, the dose–response curve for the mother should be compared to that of the offspring to determine if the offspring is differentially sensitive. If maternal and offspring toxicity occurs at similar doses, the qualitative nature of the toxicity in the offspring may be different from the maternal effects. The effects in the offspring may also be long lasting, whereas the maternal effects are transient (U.S. EPA, 1991a). Adverse effects in the offspring observed only at doses that are obviously toxic to the mother are difficult to interpret and may require additional research to determine if developmental toxicity in the offspring is secondary to the maternal toxicity.

Particular concern has been raised about interpreting developmental neurotoxicology data in the presence of postnatal maternal toxicity (Francis *et al.*, 1990). It is possible that exposure could influence maternal behavior toward the offspring and/or milk production or letdown and affect the development of the nervous system indirectly. Although alterations in maternal behavior and function could be affected by exposure to an agent, the information obtained concerning potential developmental neurotoxicity from such studies is very important in the hazard identification context (Francis *et al.*, 1990). In the case where significant maternal toxicity may appear to be a critical factor, other experimental designs (i.e., cross-fostering, weight-matched controls) should be considered (Tilson, 1992).

A second issue important for judging the sufficiency of experimental data is the presence of developmental toxicity that could confound functional and/or structural measurements of nervous system development. Developmental toxic effects in the offspring include altered viability and/or growth, morphological alterations, and functional deficits (Schardein and Keller, 1989; U.S. EPA, 1991a). Depending on the study design, there are several measures of viability and/or growth, including data concerning success or failure of implantation and information concerning the offspring (i.e., number, gender, and physical condition of offspring). Structural defects that may adversely affect survival, development, or function, as well as variations that indicate a divergence beyond the usual range of structural makeup of the live offspring, should be considered. Developmental toxic effects can also be expressed as alterations or delays in sensorimotor development, sexual dimorphism, and functioning of the central nervous system (i.e., learning and memory deficits), as well as other organ systems (e.g., kidneys).

It has also been argued that a battery of measurements, including functional and structural, is needed to evaluate the full range of possible neurotoxic effects (Tilson, 1992). Developmental neurotoxicity can be expressed as alterations in the structure, chemistry, or function of the peripheral and/or central nervous system. Structural changes include accumulation, proliferation, or rearrangement of neural elements, breakdown of cells, and gross changes in morphology. Of particular importance for developmental neurotoxicology studies is a change in the maturational appearance or rate or extent of growth of a particular brain region. Examples of possible neurochemical indicators of developmental neurotoxicity include changes in the maturational rate of the synthesis, release, uptake, or degradation of neurotransmitters, particularly if they are region-specific and/or are persistent. Neurophysiological end points of developmental neurotoxicity include altered maturation or persistent disruption of the action potential and/or sensory-evoked potential. Behavioral changes in the offspring are often taken as functional indicators of the net integrative output of the nervous system. Developmental neurotoxicants may affect the maturational appearance of specific sensorimotor reflexes or produce persistent alterations in higher cognitive functions, such as learning and memory.

Another important consideration in judging the sufficiency of data is that the power of the end point used to identify developmental toxic effects can vary significantly. Weight measurements, for example, are continuous data and may be able to detect a relatively small change (i.e., 5–10%) relative to a control group of nominal size (e.g., 20 animals), whereas tests such as presence or absence of a specific structural defect may require a greater change in treated animals if statis-

tical significance from control is to be obtained. The use of tests with a moderate degree of inherent variability will be more likely to detect the presence of neurotoxicity than tests with high background variability. Tests with very low variability, however, may be difficult to alter except at high doses (Butcher *et al.*, 1980).

The experimental and statistical design of the study is also important for judging the sufficiency of developmental neurotoxicity data. Selection of animal species can be an important variable in developmental studies. Mice, for example, are more sensitive to the developmental effects of methanol than rats (Andrews *et al.,* 1993). In addition, there is usually no a priori reason to predict that one gender would be more sensitive to developmental neurotoxicants than another, indicating that both should be evaluated for purposes of risk assessment. Assignment of animals to dose groups based on body weight reduces bias and facilitates statistical comparison. At least three doses of the chemical are selected, with the highest dose chosen to produce some indication of toxicity. As indicated in the preceding paragraph, developmental neurotoxicology studies frequently use a battery of tests to assess a wide range of possible neurobiological functions. Therefore, some consideration must be given to the possibility that significant effects might be obtained on the basis of chance alone. For example, some experiments may involve over 100 tests for significance of the treatment effect or time × treatment interactions. If a nominal P level of 0.05 is used, then there could be at least five or more significant effects based on chance alone. Some protection can be afforded against such false positives by evaluating the data for overall statistical significance using an analysis of variance and proceeding with post hoc tests after a significant interaction between treatment and some other main effect is obtained. Corrections for multiple tests may also be needed. A replicate study design can add confidence in the sufficiency of the experimental data.

Developmental neurotoxicology studies require that maternal and genetic influences are controlled. This is usually accomplished by using the litter as the statistical unit (Kimmel, 1988). The decision to use the litter as the statistical unit can influence the experimental design and the number of different measurements that can be performed on the offspring (Tilson, 1992).

The interpretation of data from developmental studies is influenced by the time during which exposure occurs. The nervous system undergoes a relatively precise and timed sequence of cell division, differentiation, and migration during the pre- and postnatal phase of development. It has been clearly shown (Rodier, 1976; Rodier *et al.,* 1979) that different regions of the brain develop at different times and that the presence or absence or the qualitative nature of a neurotoxic effect can depend on the specific time of exposure.

One of the problem areas in developmental neurotoxicology is that subtle or latent deviations in neural processes might not be observed with commonly used measurements, particularly some behavioral procedures. The nervous system does compensate to experimentally induced neurodegeneration and repeated exposure to chemical agents (Walsh and Tilson, 1986). If such compensation is suspected, a pharmacological or physiological challenge may be used to evaluate function and "unmask" effects that are not detectable. Hughes and Sparber (1978), for example, found that rats exposed to methyl mercury *in utero* learned a bar press task as rapidly as controls. When mercury-exposed rats were challenged with *d*-amphetamine, their operant performance was affected less than controls. Pharmacological challenges have been used frequently in developmental neurotoxicology studies to evaluate differences in sensitivity between treated and control animals (Walsh and Tilson, 1986).

Although interpretation of developmental neurotoxicology data may be limited in some cases, it is clear that structural and functional alterations occur following developmental exposure to chemicals. Such changes must be evaluated in light of other toxicity data, including maternal toxicity and other forms of developmental toxicity. Judging that an effect is sufficient evidence of developmental neurotoxicity is dependent on a number of study design issues and appropriate statistical analysis of the data. The level of confidence in such judgments may be increased by replicability of the effect, either in another study of the same type or by convergence of data from tests that are assumed to measure similar functions. A dose–response relationship is considered an important measure of chemical effect. Both monotonic and biphasic dose–response curves are possible, however, depending on the structural or functional measure evaluated.

V. Quantitative Developmental Neurotoxicology Risk Assessment

The ultimate purpose of the risk assessment process is to set acceptable levels for risk. Determination of an acceptable level for risk typically begins with the definition of a critical adverse effect based on evaluation of all the available human and/or animal experimental data. Such a process must take into consideration a number of issues, including extrapolation (animal to human, high to low dose, route to route) and dosing (acute versus chronic) parameters.

Based on the approach described by Barnes and Dourson (1988) for noncancer end points, the LOAEL or lowest dose at which there is a statistically or biologically significant increase in incidence of an adverse effect is determined. If possible, a NOAEL, which is defined as the highest dose at which there is no statistically significant increase in the presence of an adverse effect, is also determined.

The NOAEL or LOAEL is then used to calculate the reference dose (RfD), which is an estimate of a daily exposure to the human population that is likely to be without appreciable risk over a lifetime (Barnes and Dourson, 1988). Calculation of the RfD has historically been based on the use of safety factors, which are intended to estimate uncertainty in the data set. Derivation of a RfD through application of uncertainty factors is based on the assumption that there is a threshold for toxic effects (Sette and MacPhail, 1992). The NOAEL derived from a given study may lie above the actual threshold dose. Therefore, 10-fold safety factors have been utilized to account for uncertainties in the experimental data. These uncertainty factors include intraspecies (human) sensitivities, animal-to-human extrapolation, less than lifetime exposures, and extrapolating from a LOAEL in the absence of a NOAEL (Table 5). The mathematical model for calculating the exposure level considered to be without significant risk or the RfD is

$$RfD = NOAEL/UF,$$

A modifying factor ranging from 1 to 10 may also be used in the denominator to reflect the completeness of the database used to establish the critical effect (Sette and MacPhail, 1992). The value of an uncertainty factor may be decreased from 10 to another number (1,3, or 7) in certain cases. For example, data obtained from a nonhuman primate for a compound in which there are known similarities between humans and the test species might reduce the uncertainty factor for animal-to-human extrapolation.

TABLE 5 Uncertainty Factors Used in Calculating Reference Doses

Default factor	Uncertainty
10	Variation in sensitivity among members of human population
10	Animal-to-human extrapolation
10	Less than lifetime exposure
10	Extrapolating from LOAEL to NOAEL
MF	Modifying factor from <1–10 to account for scientific uncertainties such as quality of the data

In developmental studies, a developmental toxicology reference dose (RfDdt) is generally based on a short-term exposure instead of a chronic or lifetime exposure (U.S. EPA, 1991a) since it is well known that a single exposure at a critical period of development can produce significant developmental toxicity (Rodier, 1976; Rodier *et al.*, 1979). Therefore, the uncertainty factor of 10 for less than lifetime exposures is not used in developmental studies unless there is evidence for accumulation of a chemical. The terms of RfDdt or RfCdt are used to distinguish them from the oral and dermal RfD and inhalation reference dose (RfC) associated with chronic exposures (U.S. EPA, 1991a). A National Academy of Science panel recommended that because specific periods of vulnerability during development exists, that an uncertainty factor up to the 10-fold factor should be applied when there is evidence of postnatal developmental toxicity and when data from toxicity testing relative to children are incomplete (NRC, 1993).

The use of a NOAEL or LOAEL to calculate an estimate of exposure risk has several drawbacks (Kimmel, 1990). For example, the number and spacing of doses in a study might influence the dose selected for the NOAEL. The NOAEL, by definition, must be one of the doses in the experimental data set and the approach essentially ignores other data that have been collected, including the slope of the dose–response curve or the variability of the data set.

To circumvent some of these limitations in developmental studies, Kimmel and Gaylor (1988) suggested using the benchmark dose method described by Crump (1984). This approach establishes a dose–response curve and uses the shape of the dose–response curve to establish the dose required to increase the incidence of an effect at a particular level, i.e., a 10% change in response or ED_{10}. The benchmark dose is the lower confidence interval for the extrapolated effect (LED_{10}). Uncertainty factors may then be applied to the benchmark dose to calculate a RfD. The size of the uncertainty factor to be applied to the benchmark dose depends on the acceptable level of risk. The LED_{10} may be similar to a LOAEL and an uncertainty factor of 1000 might be applied. If a LED_{01} were calculated, this value might be similar to the NOAEL and a factor of 100 would be applied (Kimmel, 1990). Comparisons of the estimated risk derived from the use of the NOAEL and benchmark dose approaches have been made for developmental data sets (see Kimmel, 1990).

VI. Summary and Conclusions

The area of developmental neurotoxicology has evolved rapidly over the last several years. From initial experiments in animal models showing that developmental exposure can result in long-lasting behavioral changes (Butcher, 1970; Butcher *et al.*, 1972a,b), a number of studies performed in many different laboratories have demonstrated that exposure to a wide range of chemicals during development can have adverse effects on the structure and/or function of the nervous system (Riley and Vorhees, 1986). A number of generally recognized human developmental neurotoxicants now exist, including ethanol, methyl mercury, lead, heroin, methadone, cocaine, diphenylhydantoin, and polychlorinated biphenyls (Rees *et al.*, 1990).

Because of the potential human hazard, several countries require chemicals be tested for potential developmental neurotoxicity for premarket approval. In the United States, the EPA has promulgated testing guidelines for developmental neurotoxicology studies for the premarket approval of chemicals regulated under the Federal Insecticide, Fungicide, and Rodenticide Act (U.S. EPA, 1991b). U.S. EPA (1991a) has published guidelines for developmental toxicity risk assessment that cover developmental toxic effects, including structural and functional alteration in the nervous system. A draft of the neurotoxicity risk assessment guidelines is being reviewed by U.S. EPA which will supplement the developmental toxicity risk assessment guidelines.

Risk assessment in this area is hampered by the lack of a clear understanding of the biological mechanisms underlying developmental toxicity, intra/interspecies differences in the types of developmental alterations, appropriate pharmacodynamic information, and the influence of a number of confounding variables on the dose–response curve. Such information is needed if statistical models are to predict risk at levels of exposure.

One major assumption for risk assessment on noncancer end points such as developmental neurotoxicity is that there is a biological threshold. The presence of a threshold for a population may or may not be demonstrable because of unknown or undiscovered factors that increase the sensitivity of some individuals within the population. Research is needed to address the concern that exposure to a chemical may increase the risk for a population, but not necessarily for all members of the population. Additional research is needed to develop more appropriate animal models, determine the most appropriate study designs, better estimate target organ levels, and develop physiologically based dose–response models for developmental neurotoxicity.

References

Andrews, J. E., Ebron-McCoy, M., Logsdon, T. R., Kavlock, R. J., and Rogers J. M. (1993). Developmental toxicity of methanol in whole embryo culture: A comparative study with mouse and rat embryos. *Toxicology,* **81,** 205–215.

Barnes, D. G., and Dourson, D. G. (1988). Reference Dose (RfD): description and use in human health risk assessments. *Reg. Toxicol. Pharmacol.* **8,** 471–486.

Barone, S., Stanton, M. E., and Mundy, W. R. (1993). Latent neurotoxic effects of neonatal triethyl tin (TET) exposure are expressed with aging. *The Toxicologist* **13,** 300.

Bloom, A. D. (1981). Guidelines for reproductive studies in exposed human populations: report of Panel II. In *Guidelines for Studies of Human Populations Exposed to Mutagenic and Reproductive Hazards.* pp. 37–110, March of Dimes Birth Defects Foundation, White Plains, New York.

Butcher, R. E. (1970). Learning impairment associated with maternal phenylketonuria in rats. *Nature* **226,** 555–556.

Butcher, R. E., Brunner, R. L., Roth, T., and Kimmel, C. A. (1972a). A learning impairment associated with maternal hypervitaminosis-A in rats. *Life Sci.* **11,** 141–145.

Butcher, R. E., Vorhees, C. V., and Kimmel, C. A. (1972b). Learning impairment from maternal salicylate treatment in rats. *Nature* **236,** 211–212.

Butcher, R. E., Wootten, V., and Vorhees, C. V. (1980). Standards in behavioral teratology testing: test variability and sensitivity. *Teratogenesis Carcinog. Mutagen.* **1,** 49–61.

Crump, K. S. (1984). A new method for determining allowable daily intakes. *Fundam. Appl. Toxicol.* **4,** 854–871.

Dews, P. B. (1986). On the assessment of risk. In *Developmental Behavioral Pharmacology* (N. Krasnegor, J. Gray, and T. Thompson, Eds.), pp. 53–65, Lawrence Erlbaum, Hillsdale, New Jersey.

Food and Drug Administration (FDA) (1966). *Guidelines for Reproduction and Studies for Safety Evaluation of Drugs for Human Use.* Bureau of Drugs, Rockville, Maryland.

Food and Drug Administration (1970). Advisory Committee on Protocols for Safety evaluations. Panel on reproduction. Report on reproduction studies in the safety evaluation of food additives and pesticide residues. *Toxicol. Appl. Pharmacol.* **16,** 264–296.

Francis, E. Z., Kimmel, C. A., and Rees, D. C. (1990). Workshop on the qualitative and quantitative comparability of human and animal developmental neurotoxicity: summary and implications. *Neurotoxicol. Teratol.* **12,** 285–292.

Grant, L. D. (1976). Research strategies for behavioral teratology studies. *Environ. Health Perspect.* **18,** 85–94.

Hertig, A. T. (1967). The overall problem in man. In *Comparative Aspects of Reproductive Failure* (K. Benirschke, Ed.), pp. 11–41, Springer-Verlag, New York.

Hughes, J. A., and Sparber, S. B. (1978). d-Amphetamine unmasks postnatal consequences of exposure to methylmercury in utero: methods for studying behavioral teratogenesis. *Pharmacol. Biochem. Behav.* **8,** 365–375.

Khera, K. S. (1984). Maternal toxicity: a possible factor in fetal malformations in mice. *Teratology* **29,** 411–416.

Kimmel, C. A., Kimmel, G. L., and Frankos, V. (Eds.) (1986). Interagency Regulatory Liaison Group workshop on reproductive toxicity risk assessment. *Environ. Health Perspect.* **66,** 193–221.

Kimmel, C. A. (1988). Current status of behavioral teratology: science and regulation. *CRC Rev. Toxicol.* **19,** 1–10.

Kimmel, C. A., and Gaylor, D. W. (1988). Issues in qualitative and quantitative risk analysis for developmental toxicology. *Risk Anal.* **8,** 15–20.

Kimmel, C. A., Kimmel, G. L., Francis, E. Z., and Chitlik, L. D. (1990). An overview of the US EPA's proposed amendments to the guidelines for the health assessment of suspect developmental toxicants. *J. Am. Coll. Toxicol.* **9,** 39–47.

Kimmel, C. A. (1990). Quantitative approaches to human risk assessment for noncancer health effects. *Neurotoxicology* **11,** 189–198.

Levine, T. E., and Butcher, R. E. (1990). Workshop on the qualitative and quantitative comparability in human and animal developmental neurotoxicity. Work Group IV Report: triggers for developmental neurotoxicity testing. *Neurotoxicol. Teratol.* **12,** 281–284.

National Research Council (NRC) (1983). *Risk Assessment in the Federal Government: Managing the Process.* National Academy Press, Washington, D.C.

National Research Council (NRC) (1993). *Pesticides in the Diets of Infants and Children.* National Academy Press, Washington, D.C.

Organization for Economic Cooperation and Development (OECD) (1981). Guideline for testing chemicals' teratogenicity.

Rees, D. C., Francis, E. Z., and Kimmel, C. A. (1990). Scientific and regulatory issues relevant to assessing risk for developmental neurotoxicity overview. *Neurotoxicol. Teratol.* **12,** 175–181.

Rice, D. C. (1989). Delayed neurotoxicity in monkeys exposed developmentally to methylmercury. *Neurotoxicology* **10,** 645–650.

Riley, E. P., and Vorhees, C. V. (1986). *Handbook of Behavioral Teratology.* Plenum Press, New York.

Rodier, P. M. (1976). Critical periods for behavioral anomalies in mice. *Environ. Health Perspect.* **18,** 79–83.

Rodier, P. M., Reynolds, S. S., and Roberts, W. N. (1979). Behavioral consequences of interference with CNS development in the early fetal period. *Teratology* **19,** 327–365.

Schardein, J. L., and Keller, K. A. (1989). Potential human developmental toxicants and the role of animal testing in their identification and characterization. *CRC Rev. Toxicol.* **19,** 251–339.

Sette, W. F., and MacPhail, R. C. (1992). Qualitative and quantitative issues in assessment of neurotoxic effects. In *Neurotoxicology* (H. Tilson and C. Mitchell, Eds.), pp. 345–361, Raven Press, New York.

Shepard, T. H. (1986). Human teratogenicity. *Adv. Pediatr.* **33,** 225–268.

Spyker, J. A. (1975). Assessing the impact of low level chemicals on development: behavioral and latent effects. *Fed. Proc.* **34,** 1835–1844.

Stanton, M. E., and Spear, L. P. (1990). Workshop on the qualitative and quantitative comparability of human and animal developmental neurotoxicity, Work Group I report: comparability of measures of developmental neurotoxicity in humans and laboratory animals. *Neurotoxicol. Teratol.* **12,** 261–267.

Tilson, H. A. (1992). Study design considerations in developmental neurotoxicology. *Neurotoxicol. Teratol.* **14,** 199–203.

U.S. Environmental Protection Agency (EPA) (1982). Pesticide assessment guidelines. subdivision F. Hazard evaluation: human and domestic animals. Office of Pesticides and Toxic Substances, Washington, D.C. National Technical Information Service, Springfield, Virginia.

U.S. Environmental Protection Agency (EPA) (1985). Toxic Substances Control Act test guidelines: final rules. Fed. Reg. **50,** 39426–39428, 39433–39434.

U.S. Environmental Protection Agency (EPA) (1991a). Guidelines for developmental toxicology risk assessment. *Notice* **56,** 63798–63824.

U.S. Environmental Protection Agency (1991b). *Revised Neurotoxicity Testing Guidelines for Pesticides.* National Technical Information Service, Springfield, Virginia.

Walsh, T. J., and Tilson, H. A. (1986). The use of pharmacological challenges. In *Neurobehavioral Toxicology* (Z. Annau, Ed.), pp. 244–267, The Johns Hopkins University Press, Baltimore.

Wilcox, A. J., Weinberg, C. R., Wehmann, R. E., Armstrong, E. G., Canfield, R. E., and Nisula, B. C. (1985). Measuring early pregnancy loss: laboratory and field methods. *Fertil. Steril.* **44,** 366–374.

Wilson, J. G. (1977). Embryotoxicity of drugs in man. In *Handbook of Teratology* (J. G. Wilson and F. C. Fraser, Eds.), pp. 309–355, Plenum Press, New York.

CHAPTER

55

Unique Dimensions of Neurotoxic Risk Assessment

BERNARD WEISS

Department of Environmental Medicine
University of Rochester
School of Medicine and Dentistry
Rochester, New York 14642

Risk assessment won the mantle of a rigorous discipline with publication of the *Red Book,* the celebrated National Research Council (1983) report. But respectability exacted a price. According to the authors, they strove to codify a loose collection of tenets into a workable but flexible system that could be applied to regulation and policy. Many of their successors and interpreters reached for a less modest and more fallible end: pronouncing edicts. Instead of the supple tool its originators had claimed as their goal, some adherents of risk assessment turned it into a stereotyped exercise defined by a procession of boxes in flow charts. For neurotoxicology, it became as puzzling as trying to navigate in an unfamiliar city traversed only by one-way streets.

One major source of the incompatibility is the original hub of the report. It adopted cancer as its prototype. Cancer offered several virtues as a prototype besides its position in the vanguard of public anxieties about environmental poisons. Cancer is preceived as fundamentally a unitary process despite its appearance in many different guises. It is also perceived, despite its range of shadings, as a dichotomous phenomenon: it either is present or it is not. Also, numbers can be assigned to the public's unease in the form of probability statements. Recall how often journalists seriously report hypothetical outcomes as, for example, "three extra cases in a million." Many who apply the model find it easy to cast aside its complexities in the service of simplicity. Too many qualifications and uncertainties dim the impact of policy debates; better to subsidize ease of communication at the expense of conveying the true uncertainties.

Neurotoxicology often seemed to have become a willing co-conspirator. Even now, some of its practitioners seem content to adopt the circumscribed focus of the conventional model. Its limitations for neurotoxicology are overlooked. Neurotoxic risk assessment sometimes tends to be jammed into attire that is too tight for its unique and unruly bulges. If risk assessment is to attain more than its current peripheral status in neurotoxicology, it will have to adopt models more suitable for the peculiar contours of neurotoxicology.

A properly fashioned model can secure other advantages, too. One is a structure for framing research issues. Much current research assumes forms that are unlikely to yield information applicable to quantitative risk assessment. Some of it amounts primarily to demonstrations of toxicity based on the crudest of end points. Such tactics yield few surprises. Some of it, defended by the argument that mechanistic information is indispensable to rational risk assessment, offers acute experiments based on doses so massive that any reasonable mechanism would be engulfed. Although a comprehensive model able to encompass all the facets of neurotoxicity is impractical at this stage, at least we can survey the varieties of information that would be needed for such a model. One way to do so is to use the basic structure of the conventional model to determine how it would need to be modified to accommodate the special features of neurotoxicology. Especially, how should it be structured to achieve the ultimate goal of specifying exposure standards?

I. Hazard Identification and Detection

Hazard Identification is considered to be the first stage of the conventional model. It is basically a qualitative exercise that asks whether a particular agent is capable of inducing certain types of adverse health effects. To make such a judgment, several types of data are surveyed: clinical reports, epidemiologic studies, animal bioassays, and, in some instances, molecular structure. If a chemical is identified as a carcinogen, the process may come to an abrupt halt because of the Delaney Amendment or similar rulings.

Such a qualitative evaluation is less direct than the usual concise diagrams make it out to be. The *Red Book* provides detailed notes on how the different kinds of data should be weighed. Each element in the hazard identification appraisal elicits a number of probing questions because, most often, the data are packed with ambiguities. For example, in the interpretation of cancer data, how should the occurrence of rare tumors be treated in cases in which their incidence did not achieve statistical significance? Should benign and malignant tumors be counted equally? These are not trivial issues, and become even more complicated with neurotoxicants.

One current example offers such a challenge. Methyl *ter*-butyl ether (MTBE) was introduced as a gasoline additive to reduce carbon monoxide emissions. It underwent extensive preclinical toxicity testing, including an appraisal of neurotoxicity. Simple behavioral end points and neuropathology indicated adverse effects only at high exposure levels exceeding, by orders of magnitude, concentrations that populations would confront. On the basis of such data, the U.S. EPA (Gift, 1993) set a reference concentration (RfC) of

0.5 mg/m^3. With its introduction in Anchorage, Alaska, new questions erupted. Citizens of Anchorage began to complain of a gamut of symptoms, such as headaches and disorientation, resembling those provoked by recognized neurotoxicants. Are these valid? Was the preclinical evaluation adequate? Did it include behavioral assays that might prove sensitive to such effects? A comparison with ozone is instructive.

At high levels, ozone is a potent lung toxicant and severely damages lung tissue. Are similar, but less blatant, effects seen at low ambient levels? Lung damage, or even reduced pulmonary function, is ambiguous at the Clean Air Act level of 0.12 ppm. Behavioral effects in rats, however, can be seen even at levels of 0.08 ppm. But these effects consist of a reluctance to engage in physical activity, such as running in a wheel, which would increase ventilation rate and increase ozone dose (Weiss, 1989). The reduction in running, then, can be interpreted not as a direct toxic effect, but as the avoidance of aversive stimuli resulting from increased ventilation. Such an experimental situation might have been more germane to the MTBE risk assessment effort and human responses than the conventional preclinical assays on which the RfC was based.

II. Hazard Confirmation and Characterization

The complications introduced by neurotoxicity expand hazard identification to such an extent that another stage, labeled hazard confirmation and characterization, needs to be introduced. Consider how the questions posed by the usual sources of hazard information burgeon when end points are included that range over vastly more territory than the occurrence of cancer.

A. Epidemiology

All of the questions posed about epidemiology in the *Red Book* pertain to neurotoxicity as well. Issues such as conflicting positive and negative findings, appropriate control groups, prospective versus case control studies, and others are common to epidemiologic research. Neurotoxicology contends with an even wider range of puzzles. The pattern of toxicity is one such puzzle. Cancer is assumed to exhibit a prolonged delay in appearance because the emergence of a tumor is assumed to be the outcome of a multistage process. It may be due to chronic exposure, the typical bioassay approach, or it may represent the culmination of a sequence originated by a single event such as an acute exposure to ionizing radiation.

Neurotoxicity may also display a prolonged latency. Chronic exposure to acrylamide, for example, inflicts damage to susceptible nerve cells that eventually is severe enough to impair function, especially somatosensory acuity (although, oddly enough, the bulk of the animal literature relies on motor function as the relevant end point). Because of repair mechanisms (Hattis and Shapiro, 1990), the cumulative dose required to induce damage is a function of dose rate; low dose rates require a longer exposure period and higher total dose to exert toxic effects. How should this phenomenon be incorporated into exposure standards? Delayed effects may be seen with prenatal exposures (Weiss and Reuhl, 1994); such effects may seem to remain dormant until a later developmental stage. Methyl mercury is one example of such an agent. Perhaps the core problem is that the late manifestations can be difficult to specify. Consider the challenges posed by trying to link the incidence of cancer in a population with exposures that took place two decades before. Compare them with trying to link the incidence of learning disabilities in high school to lead exposure indices assayed in the first grade (Needleman *et al.*, 1990).

Two features of neurotoxic expression not seen with cancer are acute, immediate functional disorders and reversibility. Acute exposures of this kind must play a role in evaluating the health hazards posed by a chemical. Volatile organic solvents, for example, can act as anesthetics. For this reason, the American Conference of Governmental Industrial Hygienists (ACGIH), in addition to the chronic standards it defines as threshold limit values, also calculates a short-term exposure limit defined as, "The maximum concentration to which workers can be exposed for a period up to 15 min continuously without suffering from . . . narcosis of sufficient degree to increase accident proneness, impair self-rescue, or materially reduce work efficiency . . . ." Note that these are all behavioral criteria. Note, too, that they emphasize functional impairment of a nature that presents an immediate health threat. These effects, however, are different in character than those claimed to result from chronic exposure. Such qualitative discrepancies between acute and chronic and delayed effects do not fit easily into the conventional risk assessment scheme.

The other discrepancy with the cancer-based model is reversibility. The most prevalent statistical models of carcinogenesis exclude the issue even though it is observed in both epidemiological and laboratory studies. For example, the risk of lung cancer diminishes once a smoker becomes abstinent. Reversibility is a common feature of neurotoxicity. Recovery from a single episode of acute intoxication by an organic solvent

is usually rapid. Latent deficits from such episodes have not been examined, however; episodes of acute organophosphorous insecticide poisoning leave a legacy of persisting residual deficits (Savage *et al.*, 1988). But even some effects of chronic exposure may fade with time. Mercury vapor is a venerable neurotoxicant. Its cardinal neurological reflection is tremor. Exposed workers removed from the source of exposure show a gradual diminution of tremor amplitude (e.g., Wood *et al.*, 1973). Even victims of acrylamide poisoning exhibit at least partial recovery of somatosensory function, although, in the same monkeys, permanent visual damage may result (Merigan *et al.*, 1985).

Perhaps the greatest discrepancy with the cancer model is the nature of the assay. Cancer in humans is typically detected by the techniques of clinical medicine; biomarkers of preclinical cancer development have not yet evolved to a stage at which they are fully trusted. Neurobehavioral disorders, however, except in their most flagrant form, tend to lie beyond the reach of clinicians. Their detection usually requires the application of techniques from neuropsychology and experimental psychology. Correspondingly, the design of proper epidemiological studies presents much greater challenges. Cancer researchers are aware of, and control for, variables such as tobacco smoking. Neurotoxicologic evaluations may have to control for variables ranging from the diligence of housekeeping in the residences of children in lead studies to workers' years of schooling to, in some prospective studies, predispositions to depression. Most of the contentiousness over the relationship between lead and IQ arises from disagreement over the adequacy of statistical compensation for variables such as family socioeconomic status, parental IQs, maternal health habits, and other factors known to influence scores on IQ tests. Worker performance on psychological tests is clearly correlated with intelligence test scores before exposure on the job began. It is why Finnish investigators (Hanninen *et al.*, 1976) relied, for a baseline, on test scores obtained when Finnish men began their military service.

B. Animal Bioassays

The perplexities accompanying epidemiological data elevate laboratory animal observations to an even more critical role in neurobehavioral toxicology than they occupy in cancer risk assessment. Tumors may be histologically dissimilar in humans and rodents, they may be classified differently, or they may appear in different sites. Nonetheless, the appearance of tumors in experimental animals is usually assumed to reflect carcinogenic potential in humans. To undertake a preclinical assessment of neurotoxicity means making a choice among a plethora of approaches. Even the crudest approach requires as much attention to behavioral as to exposure history. Because neurotoxicity may emerge in a variety of ways, the typical strategy is to apply a battery of simple behavioral tests augmented by neuropathology. Most of the behavioral components are simple and direct. A functional observation battery and spontaneous motor activity, because of current regulatory requirements, are becoming universal. Proposals from EPA for the inclusion of schedule-controlled operant behavior are still pending. Its inclusion was sought specifically for the evaluation of volatile organic solvents (Environmental Protection Agency, 1991) because the agency maintains that a full accounting of potential neurotoxicity requires end points that reflect the capacity for complex performances. Some elements of industry, in particular, oppose its inclusion. They claim, for example, that it is too easily influenced by unrelated variables such as lack of appetite or other forms of illness arising from high doses producing generalized toxic effects. The core objection is that any chemical at a sufficiently high dose can impair behavior. One counter argument is that indices other than response rate, such as the distribution of responses within fixed intervals, tend to be resistant to variations in the reinforcing value of food (Weiss and Moore, 1956).

This dispute frames the conundrums of neurobehavioral toxicity. Which kinds of data are appropriate for prescribing exposure standards? Is it possible to find a chemical for which evidence of neurotoxicity, on the basis of the simple behavioral criteria used for conventional hazard identification, is at most ambiguous but which, with sensitive behavioral tests, is definitely neurotoxic? Lead is a useful model for such a debate. Only at extremely high doses does it yield evidence of morphological damage (Krigman *et al.*, 1980). By conventional clinical standards, effects on the hematopoietic system are more sensitive indicators than neuropathology. Behavioral toxicity, except for that resulting from encephalopathy produced by monumental doses, was not firmly established in animals until the late 1970s. By the middle of the 1980s, both animal and human investigations pointed to blood levels in the vicinity of 10 μg/dl as marking an elevated risk of adverse behavioral effects (Rice, 1990; Cory-Slechta, 1990). The clearest behavioral indicators came from schedule-controlled operant behavior.

Everyone will agree that the appraisal of hazard depends on the choice of the end point. At the same time, regulatory authorities are reluctant to press for the use of the more advanced, complex, and possibly more sensitive assay methods because of skepticism on the part of industry and others that their benefits exceed their costs. Do the data support that claim? Lead is an example that almost inverts the arguments against advanced criteria such as schedule-controlled operant behavior. One objection that could be leveled against identifying lead as primarily a neurotoxicant is that it impairs hematopoiesis, so that any defects in behavioral function can be ascribed to anemia. Now, take the next step: demonstrate that levels low enough to preclude interference with hematopoiesis produce behavioral effects. Accepting that conclusion suggests another approach to neurotoxic assessment. Begin with complex behavioral functions rather than assigning them to a second or later stage following identification of a chemical as neurotoxic. The feasibility of such a strategy was debated at a 1992 workshop on the interpretation of neurobehavioral toxicity (Weiss and O'Donoghue, 1994).

III. Exposure Assessment

Route, duration, and magnitude comprise the fundamental metrics of exposure assessment. Cancer modelers assume a monotonic relationship between cumulative exposure and probability of tumor. Exposure pattern is much more critical in neurotoxicology. A single exposure to a high dose of neurotoxicant can be disabling, in contrast to the same total dose distributed over a prolonged period. A single high dose of 1-methyl-4-phenyl-1,2,3,6-tetrahydropyridine (MPTP) induces a parkinsonian syndrome in monkeys and damages cells in the substantia nigra. An equivalent quantity administered in small doses over an extended period shows much less evidence of overt damage (Schneider and Kovelowski, 1990) but substantial evidence of behavioral impairment. Similarly, it might be expected that brief, intermittent episodes of high concentrations of a solvent such as carbon disulfide might inflict more damage than the equivalent total maintained at a lower, constant value. Some of the disagreements over the validity of the organic solvent syndrome (toxic encephalopathy and Painter's syndrome are virtual synonyms) might be attributable to such a factor. Exposure controls in the workplace in the 1950s were probably more susceptible to wayward elevations than those now in use. Did worker populations studied in the 1970s experience such episodic excursions in ambient concentrations? Did they result in small increments of cumulative damage?

Sufficient evidence even exists to implicate exposure pattern as an important variable in carcinogenesis.

Low-level, chronic exposure to ionizing radiation, for example, has been claimed to extend life span ("hormesis;" cf. Boxenbaum *et al.,* 1988). Furthermore, biological detoxification mechanisms can be overwhelmed by sufficiently high doses—one of the objections to the prevailing bioassay protocols for cancer. Pattern as an exposure variable tends to be ignored in neurotoxic risk assessment. However, models such as that of Hattis and Shapiro (1990), which incorporate repair mechanisms, should make it a more prominent issue.

The relationships in Table 1 show how different patterns may evoke different effects. At high concentrations, organic solvents can produce narcosis. At lower levels, chronic exposure may lead to persisting deficits revealed by psychological tests. MPTP has already been mentioned. Organophosphorous compounds can also evoke persistent deficits. The famous "Ginger Jake" episode of 1930 (Morgan, 1982), due to contamination of a soft drink flavoring by tri-*o*-cresyl phosphate, claimed perhaps 50,000 victims, many of whom were permanently disabled. As noted earlier, investigators such as Savage and colleagues (1988) discovered that farm workers who apparently had recovered from acute poisoning due to organophosphate insecticides demonstrated deficits on psychological tests even a year later. The effects of acrylamide on vibration sensitivity seem reversible (Maurissen *et al.,* 1983,1990), but its effects on the visual system are permanent (Merigan *et al.,* 1985). The category termed latent irreversible is exemplified by two agents. Cycasin is currently the leading candidate for the source of the Western Pacific syndrome labeled amyotrophic lateral sclerosis–parkinsonism dementia (Spencer, 1987). Even decades after leaving Guam, where most of the cases have been identified and where the offending agent presumably was consumed, individuals may first begin to evince the symptoms (Zhang *et al.,* 1990). Prenatal brain damage caused by methyl mercury may also not become apparent until much later in life (Spyker, 1975; Marsh *et al.,* 1987).

IV. Dose–Response Modeling

No other toxic end point has been the recipient of as intense mathematical modeling as cancer. Regulatory exposure standards, in fact, are based on the results of extrapolation modeling because no experiment is capable of yielding an estimate equivalent to a lifetime risk of 10^{-6}, the presumed regulatory goal. The data comprising the basis for modeling typically come from animal bioassays that rely on doses that are orders

TABLE 1 Exposure/Risk Classifications

Category	Example
Acute reversible	Solvents (anesthetic effects)
Acute irreversible	MPTP and parkinsonism; lingering effects of OP poisoning
Chronic reversible	Mercury vapor, clinical tremor; acrylamide, vibration sensitivity
Chronic irreversible	Acrylamide, visual function; carbon disulfide, psychomotor performance
Latent irreversible	Cycasin (ALS/PD); prenatal methyl mercury

of magnitude greater than those prevailing in the environment. Modeling efforts are aimed at risk estimates in the low dose range where observable data do not exist.

Quantitative modeling for risk was not applied to neurotoxicants until recently. Instead, calculations of exposure standards were consigned to the traditional formula by which a no observed adverse effect level (NOAEL), say, is buffered with a safety or uncertainty factor. Because of the arbitrary nature of this formula, toxicologists and statisticians have sought to replace it by approaches such as benchmark doses and a variety of other techniques (cf. Rees and Hattis, 1994). These share the common theme of greater reliance on the total dose–response function and, as much as feasible, confining the range of the model to the range of the observable data.

These advances in quantification, however, tend to conceal the even more fundamental question of what to model. Neurotoxicity is expressed in countless ways. In humans, subjective complaints and subjective indices assayed by psychological tests may be the first, and sometimes the only, manifestation of a toxic effect. The earliest indication of chlordecone neurotoxicity (Cannon *et al.,* 1978) consisted of complaints of nervousness. Higher scores or an inventory of anxiety differentiated farm workers exposed to organophosphate insecticides from control workers (Levin *et al.,* 1976). At high levels of exposure, many more effects would be visible. One of the most perplexing issues to face modelers is the change in the profile of neurotoxicity with dose. Examples come from many sources.

Clinically manifest manganese neurotoxicity in humans is characterized by dystonias reflecting some elements in common with Parkinson's disease. Sometimes preceding the emergence of neurological signs and sometimes accompanying them is a peculiar collection of psychological signs such as emotionally intense weeping and laughing. At lower levels of exposure, the indicators of neurotoxicity are only remotely predictive of the clinical syndrome, including fatigue, tinnitus, trembling of fingers, and irritability (Roels *et al.*, 1987). A parallel disconnection between overt and covert neurotoxicity appears in the laboratory. Monkeys administered high levels of manganese exhibit dystonias similar to those observed in humans (e.g., Suzuki *et al.*, 1975). Monkeys given doses of manganese orders of magnitude lower show changes in schedule-controlled operant behavior when the response consists of rowing against a resistance equivalent to the monkey's body weight (Newland and Weiss, 1992).

The manganese example of different effects at different doses can be multiplied many times. The challenge to extrapolation modeling, which is indispensable for associating risk with exposure, is how to choose the proper end points to model. Dose–response modeling inevitably has to make a U-turn, so to speak, to the stage of hazard confirmation and characterization. All of the developing apparatus of quantitative modeling will founder without the appropriate choice of an end point.

How to get to the appropriate choice of end point, not which statistical model to choose, remains the central issue for neurotoxic risk assessment. Perhaps one way to sketch a path is to consider how a sequence of decisions might be made in the selection of end points. A decision tree was sketched by Weiss and Spyker (1974) to try to accommodate a sequential strategy. Figure 1 shows that it begins with simple observational techniques, much like the functional observation battery. It then proceeds through various successive stages at each of which a decision is made about whether to deepen the question of neurotoxicity or to reject the agent under test. The objective of such a sequence was to provide increasingly assured estimates of risk.

The scheme still seems superficially reasonable and even consistent with current policies and outlooks, but it contains a serious flaw. Why, one could ask, invest so much time and effort to proceed along such a trail when it likely will culminate in a complex functional assay? What are the chances, if effects are found at low doses with the complex assays, that other effects will be found at higher doses with the simpler assays? Assume, hypothetically, that most of the time the progression is intransitive, so that the more complex, apical tests prove to be more sensitive. Do simple assays then play a really significant role in quantitative risk assessment beyond helping to identify an agent as possessing nervous system activity, which is not such a profound challenge? If modeling is to advance beyond what amounts to an elegant form of curve fitting, these issues will have to be confronted.

V. Risk Characterization

The outcome of the conventional risk assessment process is risk characterization, the step in which population exposure assessment is combined with dose–response assessment to yield a risk estimate. It is a logical goal for cancer risk assessment if all its assumptions are accepted. But how are neurotoxic outcomes to be addressed? A number of possibilities exist.

A. An Elevated Incidence or Prevalence of an Adverse Effect

The adverse effect can range from a symptom complex mostly composed of subjective complaints to serious and permanent disturbances such as mental retardation. This question is confounded with the question of susceptible subpopulations. One objective of the search by cancer researchers for biomarkers is to determine whether certain populations face an enhanced risk. Neurotoxicology cannot put off recognizing the importance of this issue for risk assessment.

Figure 2 sketches some of the implications of dealing with susceptible subpopulations. Figure 2A depicts a distribution of scores on some arbitrary test. Without a toxic challenge, a subpopulation equivalent to 30% of the total sample (lower curve) shows the same mean score. Figure 2B separates the responders and nonresponders after a toxic challenge that shifts the distribution of the responders by 1 SD. The experimenter, however, finds a distribution, after the challenge, like that of Fig. 2C. The distribution looks almost identical to that of Fig. 2A except for a small shift in the mean. Figure 2D reveals how a sample of 15 drawn randomly from this mixed population would respond. The effect would not be convincing to any experimenter.

To demonstrate a statistically significant ($P = 0.01$) effect 90% of the time, as noted in Fig. 3, a sample size of 265 would have to be tested. But there is another experimental strategy, applicable to humans, that can be used in some instances and that is similar to the approach adopted by Glowa (1991), Wood and Cox (1986), and others. Adopt what in essence is a single-subject design and test each subject repeatedly. Such a strategy proved useful in demonstrating that some children consistently exhibited behavioral disturbances

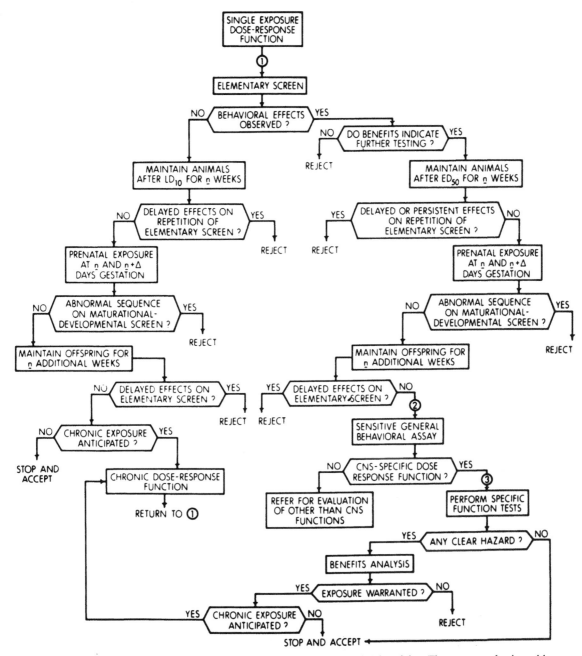

FIGURE 1 Proposed testing sequence for assessing neurobehavioral toxicity. The sequence begins with an elementary screen such as a functional observation battery and motor activity measurement. If behavioral effects are observed, and the advantages of the chemical warrant further testing, additional observations are undertaken. These include assays for delayed and developmental toxicity, and increasingly specific behavioral tests aimed at particular functions (Weiss and Spyker, 1974).

when challenged by a blend of artificial food dyes (Weiss *et al.*, 1980).

B. A Shift in the Population Distribution of Some Quality

For example, lead exposure levels are correlated with reduced IQ scores. A small shift in the IQ distribu-

tion exerts its most profound effects at the ends of the distribution, markedly reducing the proportion of superior scores and elevating the proportion of inferior scores (cf. Weiss, 1988). This would be one way to model a dose–response relationship. Another would be to model the relationship between the exposure index and the proportion of IQ score variance accounted for by exposure.

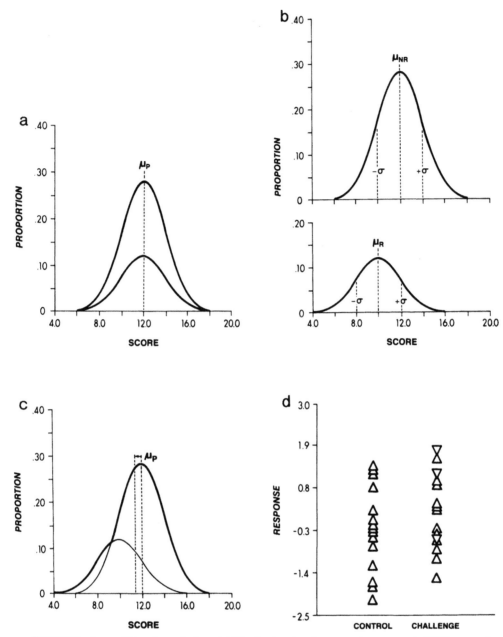

FIGURE 2 Susceptible subpopulations. (a) Distribution showing a sample composed of 30% responders (lower curve) and 70% nonresponders. (b) After challenge, the scores of the responders are shifted by 1 SD. (c) Because responders are not identified, the total distribution after challenge displays a negligible shift in the mean. (d) A randomly chosen sample of 15 individuals from the total population after challenge, showing the overlap despite the presence of a substantial incidence of responders. △, nonresponders; ▽, responders.

C. Impaired Development

This is an issue discussed extensively elsewhere in this volume. Again, impaired development encompasses effects as diverse as structural deformities and delayed language acquisition. Is the progression transitive or is it likely that an agent responsible for delayed language development would engender teratologies only at higher doses? If so, should risk estimates rely primarily on the less easily evaluated but more sensitive language functions or on the less sensitive but more concrete structural deformities?

RESPONSE AMPLITUDE ($\frac{d}{\sigma}$)

	1.0	2.0	3.0
10	2102	~550	265
20	~550	~150	65
30	265	65	30
40	~150	~37	16
50	90	23	12
100	23	7	4

PERCENTAGE OF RESPONDERS

FIGURE 3 Chart showing the sample size required to show a statistically significant effect ($P = 0.01$) 90% of the time for different magnitudes of response and different proportions contributed by sensitive subpopulations.

D. Accelerated Aging

This topic is discussed elsewhere (Weiss and Reuhl, 1994). Suppose an agent hastens the process of brain aging. How is the risk to be expressed? Rate of decline in some measure? Would not such an outcome of immense societal importance require longitudinal studies? Are there any substitutes for such studies?

E. Reduced Compensatory Capacity and Silent Damage

Damage may not be apparent until the organism is challenged, perhaps by another neurotoxicant, perhaps by aging. What role should this possibility play in risk assessment and experimental design?

F. Rate of Recovery from a Reversible Effect

Aging itself retards the speed with which organisms compensate for some adverse consequence. Old rats, for example, take much longer to recover from a dose of an acetylcholinesterase inhibitor (Michalek *et al.,* 1989), and old mice administered MPTP are less responsive to treatment with growth factors (Date *et al.,* 1990). This outcome is closely related to reduced compensatory capacity, but senescent animals are ignored in neurotoxic risk assessment.

VI. Implications for Investigational Strategies

Neurotoxicology has progressed beyond the stage at which it seemed sufficient simply to demonstrate that a particular agent could foster adverse effects. If neurotoxicology research is to exert any bearing on health risk assessment, meaning a quantitative characterization of risk, it will have to extend its horizon beyond the short-term, high-dose research that is so convenient for experimenters. Translated into experimental protocols it implies: (1) a retreat from massive, acute doses toward the low-level exposures most characteristic of environmental chemicals; and (2) more efforts to undertake longitudinal studies, simply because long latencies are so characteristic of neurological diseases. The future of neurotoxicology is certain to be entwined with current speculations about the relationship between neurotoxicants and neurodegenerative diseases. Demographic projections will compel even the most tenuous connections to be explored; neurotoxic risk assessment will jeopardize its stature if it ignores them.

Might a total reversal of strategy be the most effective way to deal with these labyrinthine issues? The conventional approach seeks first to determine, by fairly simple procedures, whether a particular agent is neurotoxic. More complex procedures are allocated to what are called second-tier or even third-tier assessments. Call this a top-down approach. Are there virtues to a bottom-up strategy? Could an assessment begin with apical criteria, then ascend to simpler, perhaps more specific and more mechanistic assays? Calculations of the costs and benefits of such an inversion could prove instructive.

Acknowledgments

This chapter was supported in part by Grants ES01247 and ES05433 from the National Institute of Environmental Health Sciences and Grant DA07737 from the National Institute on Drug Abuse.

References

Boxenbaum, H., Neafsey, P. J., and Fournier, D. J. (1988). Hormesis, Compertz functions, and risk assessment. *Drug Metab. Rev.* **19,** 195–229.

Cannon, S. B., Veazey, J. M., Jackson, R. S., Burse, V. W., Hayes, C., Straub, V. E., Landrigan, P. J., and Liddle, J. A. (1978). Epidemic kepone poisoning in chemical workers. *Am. J. Epidemiol.* **107,** 529–537.

Cory-Slechta, D. A. (1990). Bridging experimental animal and human behavioral toxicology studies. In *Behavioral Measures of Neurotoxicity* (R. W. Russell, P. E. Flatteau, and A. M. Pope, Eds.), pp. 137–158, National Academy Press, Washington, D.C.

Date, I., Notter, M. F. D., Felten, S. Y., and Felten, D. L. (1990). MPTP-treated young mice but not aging mice show partial recovery of the nigrostriatal dopaminergic system by stereotaxic injection of acidic fibroblast growth factor (aFGF). *Brain Res.* **526,** 156–160.

Environmental Protection Agency. (1991). Multi-substance rule for the testing of neurotoxicity. *Fed. Reg.* **56,** 9105–9119.

Gift, J. S. (1993). U.S. EPA reference concentration (RfC) for the gasoline additive methyl t-butyl ether (MTBE). *Toxicologist* **13,** 276.

Glowa, J. R. (1991). Dose-effect approaches to risk assessment. *Neurosci. Biobehav. Rev.* **15,** 153–158.

Hanninen, H., Eskelinen, M. A., Husman, K., and Nurminen, M. (1976). Behavioral effects of long-term exposure to a mixture of organic solvents. *Scand. J. Work Environ. Health* **2,** 240–255.

Hattis, D., and Shapiro, K. (1990). Analysis of dose/time/response relationships for chronic toxic effects—the case of acrylamide. *Neurotoxicology* **11,** 219–236.

Krigman, M. R., Bouldin, T. W., and Mushak, P. (1980). Lead. *Experimental and Clinical Neurotoxicology* (P. S. Spencer and H. H. Schaumburg, Eds.), pp. 490–507, Williams and Wilkins, Baltimore.

Levin, H. S., Rodnitzky, R. L., and Mick, D. L. (1976). Anxiety associated with exposure to organophosphate compounds. *Arch. Gen. Psychiatry* **33,** 225–228.

Marsh, D. O., Clarkson, T. W., Cox, C., Myers, G. J., Amin-Zaki, L., and Al-Tikriti, S. (1987). Fetal methylmercury poisoning: relationship between concentration in single strands of maternal hair and child effects. *Arch. Neurol.* **44,** 1017–1022.

Maurissen, J. P. J., Weiss, B., and Davis, H. T. (1983). Somatosensory thresholds in monkeys exposed to acrylamide. *Toxicol. Appl. Pharmacol.* **71,** 266–279.

Maurissen, J. P. J., Weiss, B., and Cox, C. (1990). Vibration sensitivity recovery after a second course of acrylamide intoxication. *Fundam. Appl. Toxicol.* **15,** 93–98.

Merigan, W. H., Barkdoll, E., Maurissen, J. P. J., Eskin, T. A., and Lapham, L. W. (1985). Acrylamide effects on the macaque visual system. I. Psychophysics and electrophysiology. *Invest. Opthal. Vis. Sci.* **26,** 309–316.

Michalek, H., Fortuna, S., and Pintor, A. (1989). Age-related differences in brain choline acetyltransferase, cholinesterases and muscarinic receptor sites in two strains of rat. *Neurobiol. Aging* **10,** 143–148.

Morgan, J. P. (1982). The Jamaica ginger paralysis. *JAMA* **248,** 1864–1867.

National Research Council (1983). *Risk Assessment in the Federal Government: Managing the Process.* National Academy Press, Washington, D.C.

Needleman, H. L., Schell, A., Bellinger, D., Leviton, A., and Allred, E. N. (1990). The long-term effects of exposure to low doses of lead in childhood: an 11-year follow-up report. *N. Engl. J. Med.* **322,** 83–88.

Newland, M. C., and Weiss, B. (1992). Persistent effects of manganese on effortful responding and their relationship to manganese accumulation in the primate globus pallidus. *Toxicol. Appl. Pharmacol.* **113,** 87–97.

Rees, D. C., and Hattis, D. (1994). Developing quantitative strategies for animal to human extrapolation. In *Principles and Methods of Toxicology* (A. W. Hayes, Ed.), pp. 1091–1155, 3rd ed., Raven Press, New York.

Rice, D. C. (1990). The health effects of environmental lead exposure. *Behavioral Measures of Neurotoxicity* (R. W. Russell, P. E. Flattau, and A. M. Pope, Eds.), pp. 243–267, National Academy Press, Washington, D.C.

Roels, H., Lauwerys, R., Buchet, J. P., Genet, P., Sarhan, M. J., Hanotiau, I., deFays, M., Bernard, A., and Stanescu, D. (1987). Epidemiological survey among workers exposed to manganese: effects on lung, central nervous system, and some biological indices. *Am. J. Ind. Med.* **11,** 307–327.

Savage, E. P., Keefe, T. J., Mounce, L. M., Heaton, R. K., Lewis, J. A., and Burcar, P. J. (1988). Chronic neurological sequelae of acute organophosphate poisoning. *Arch. Environ. Health* **43,** 38–45.

Schneider, J. S., and Kovelowski, C. J. (1990). Chronic exposure to low doses of MPTP. I. Cognitive deficits in motor asymptomatic monkeys. *Brain Res.* **519,** 122–128.

Spencer, P. S. (1987). Guam ALS/parkinsonism-dementia: a long-latency disorder caused by "slow toxin(s)" in food? *Can. J. Neurol. Sci.* **14,** 413–419.

Spyker, J. M. (1975). Behavioral teratology and toxicology. *Behavioral Toxicology* (B. Weiss and V. G. Laties, Eds.), pp. 311–344, Plenum Press, New York.

Suzuki, Y., Mouri, T., Suzuki, Y., Nishiyama, K., Fujii, N., and Yano, H. (1975). Study of subacute toxicity of manganese chloride in monkeys. *Tokushima J. Exp. Med.* **22,** 5–10.

Weiss, B. (1988). Neurobehavioral toxicity as a basis for risk assessment. *Trends Pharmacol. Sci.* **9,** 59–62.

Weiss, B., and Reuhl, K. (1994). Delayed neurotoxicity: a silent toxicity. *Handbook of Neurotoxicology* (L. Chang, Ed.), pp. 765–784, Dekker, New York.

Weiss, B., and Moore, E. W. (1956). Drive level as a factor in distribution of responses in fixed-interval reinforcement. *J. Exp. Psychol.* **52,** 82–84.

Weiss, B., and O'Donoghue, J. (Eds.) (1994). *Neurobehavioral Toxicity: Analysis and Interpretation.* Raven Press, New York.

Weiss, B. (1989). Behavior as an endpoint for inhaled toxicants. In *Concepts in Inhalation Toxicology* (R. O. McClellan and R. F. Henderson, Eds.), Hemisphere, New York.

Weiss, B., Williams, J. H., Margen, S., Abrams, B., Caan, B., Citron, L. J., Cox, C., McKibben, J., Ogar, D., and Schultz, S. (1980). Behavioral responses to artificial food colors. *Science* **207,** 1126–1128.

Weiss, B., and Spyker, J. M. (1974). Behavioral implications of prenatal and early postnatal exposure to chemical pollutants. *Pediatrics* **53,** 851–856.

Wood, R. W., Weiss, A. B., and Weiss, B. (1973). An analysis of hand tremor induced by industrial exposure to inorganic mercury. *Arch. Environ. Health* **26,** 249–252.

Wood, R. W., and Cox, C. C. (1986). A repeated-measures approach to the detection of the minimal acute effects of toluene. *Toxicologist* **6,** 221.

Zhang, Z. X., Anderson, D. W., Lavine, L., and Mantel, N. (1990). Patterns of acquiring parkinsonism-dementia on Guam 1944 through 1985. *Arch. Neurol.* **47,** 1019–1024.

INDEX

Printed and bound by CPI Group (UK) Ltd, Croydon, CR0 4YY

03/10/2024

01040313-0019